(4) 1-¼" blocks

(1) 3" block

TECHNICAL DRAWING

Third Edition

TECHNICAL DRAWING

Third Edition

David L. Goetsch

John A. Nelson

William S. Chalk

 Delmar Publishers Inc.

NOTICE TO THE READER

Publisher does not warrant or guarantee any of the products described herein or perform any independent analysis in connection with any of the product information contained herein. Publisher does not assume, and expressly disclaims, any obligation to obtain and include information other than that provided to it by the manufacturer.

The reader is expressly warned to consider and adopt all safety precautions that might be indicated by the activities described herein and to avoid all potential hazards. By following the instructions contained herein, the reader willingly assumes all risks in connection with such instructions.

The publisher makes no representations or warranties of any kind, including but not limited to, the warranties of fitness for particular purpose or merchantability, nor are any such representations implied with respect to the material set forth herein, and the publisher takes no responsibility with respect to such material. The publisher shall not be liable for any special, consequential or exemplary damages resulting, in whole or in part, from the readers' use of, or reliance upon, this material.

DEDICATION

From David L. Goetsch ....

.... To Savannah Day, Toby, Dustin, and Clifford Jay

From John A. Nelson ....

.... To my wife, Joyce

From William S. Chalk ....

.... To my wife, Joan

Delmar Staff
    Publisher: Michael McDermott
    Developmental Editors: Lisa A. Reale and Sheila Davitt
    Project Editors: Mary Beth Ray and Christopher Chien
    Production Supervisor: Wendy Troeger
    Senior Production Supervisor: Larry Main
    Art Supervisor: Judi Orozco
    Design Coordinator: Lisa Bower

For information, address Delmar Publishers Inc.
3 Columbia Circle Drive
Albany, New York 12212-5015

Printed in the United States of America
Published simultaneously in Canada
by Nelson Canada, a division of International Thomson Limited

    3 4 5 6 7 8 9 10 XXX 99 98 97 96 95

Library of Congress Cataloging-in-Publication Data
Goetsch, David L.
    Technical drawing / David L. Goetsch, John A. Nelson, William S. Chalk. — 3rd ed.
        p.   cm.
    Includes index.
    ISBN 0-8273-6285-4
    1. Mechanical drawing.   I. Nelson, John A., 1935-   .   II. Chalk, William.   III. Title.
T353.G63   1993   93-2026
604.2'4—dc20   CIP

Brief Contents

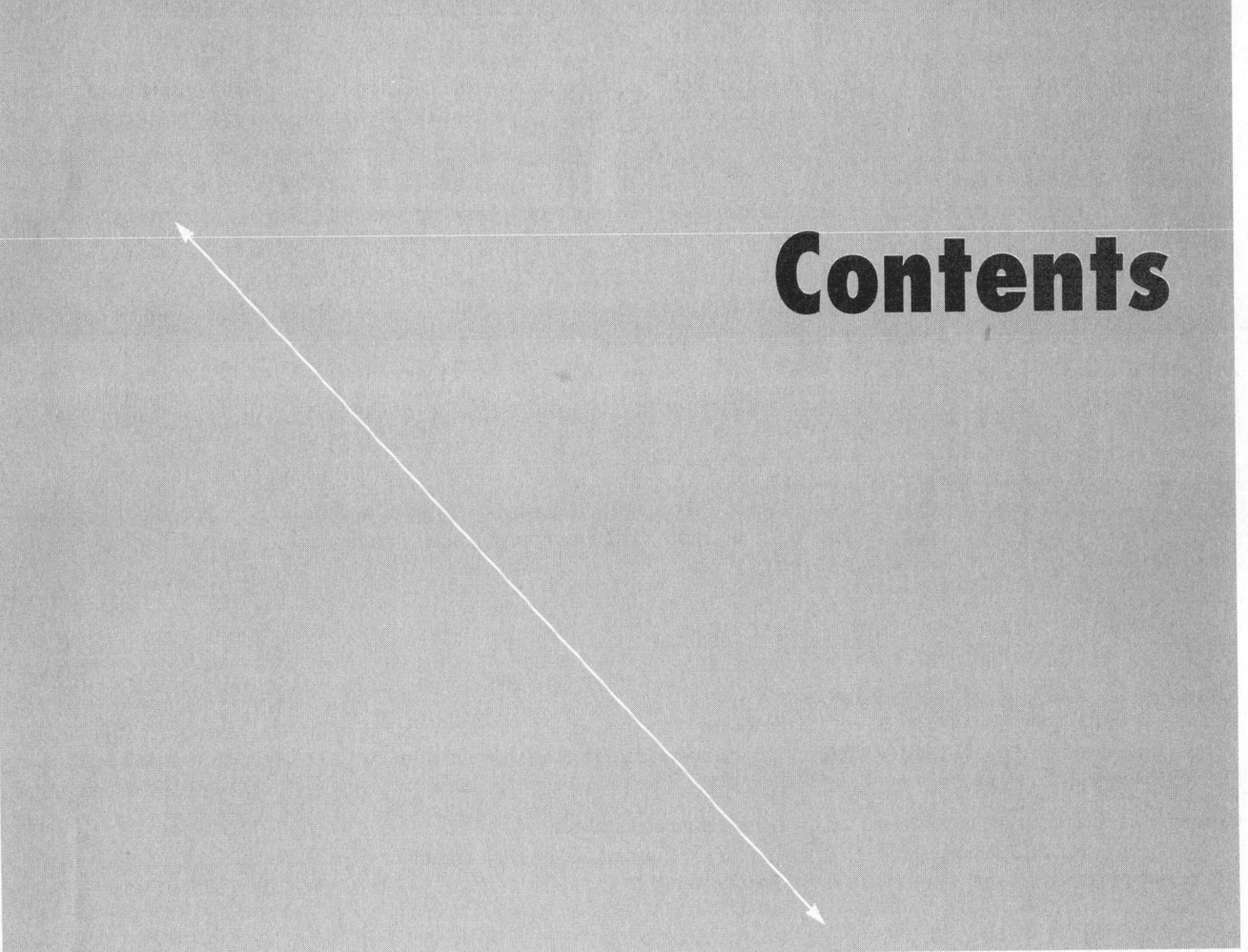

Contents

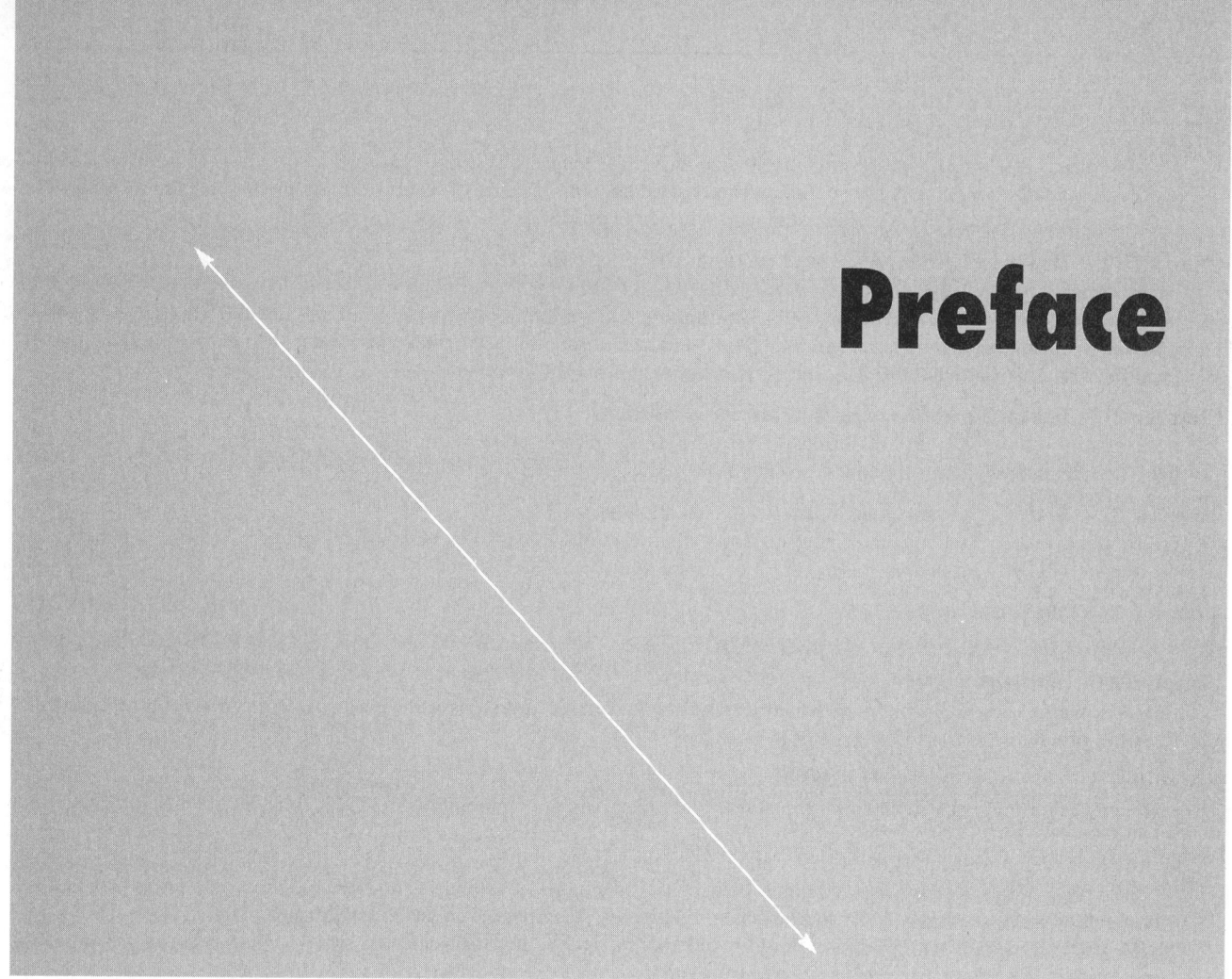

Preface

Purposes

Technical Drawing is intended for use in such courses as basic and advanced technical drawing, basic and advanced drafting, engineering graphics, descriptive geometry, mechanical drafting, machine drafting, tool and die design and drafting, and manufacturing drafting. It is appropriate for those courses offered in comprehensive high schools, area vocational schools, technical schools, community colleges, trade and technical schools, and at the freshman and sophomore levels in universities.

Prerequisites

There are no prerequisites. The text begins at the most basic level and moves step-by-step to the advanced levels. It is as well suited for students who have had no previous experience with technical drawing as it is for students with a great deal of prior experience.

Innovations

An advantage of the text is that it has evolved during a time when the world of technical drawing and design is undergoing a period of major transition from manual to automated techniques. Computer-aided drafting (CAD) is slowly but steadily gaining a foothold. This transitional period will last at least until the turn of the century, with CAD gaining greater acceptance every year.

This transition has created a need for a major text that deals with both traditional knowledge and skills and CAD-related knowledge and skills. *Technical Drawing* fills this need. Even when the world of technical drawing and design has become fully automated, drafters and designers will still need to know the traditional basics and technical drawing fundamentals. These basic factors will not change. Therefore, the traditional fundamentals are treated in depth in this text.

What is changing, and will continue to change, is the way that drafters and designers prepare technical drawings. For this reason, CAD is also treated in depth, and many of the drawings and illustrations were prepared on various CAD systems. Along with this treatment, *Technical Drawing* offers students and teachers a special blend of the manual and automated knowledge and techniques that are needed now through the turn of the century, and even beyond.

Another advantage of the text is that is was written after the latest update of the most frequently used drafting standard—ANSI series. Consequently, all dimensioning and tolerancing material in *Technical Drawing* is based on this most recent edition of the standard.

New Features in the 3rd Edition

- New chapter on visualization for computer modeling
- Expanded and updated design chapter with new design problems
- Design integrated throughout in all main drawing chapters with new design problems
- Type reset and art rescreened and registered
- New drawing problems added
- Structural drafting applications added
- Architectural drafting applications added
- Updated standards
- Advanced problems are identified with a special icon ❏.

Tested and Proven Features

- Step-by-step explanations of drawing procedures and techniques.
- Written in language your students will understand; technical terms are defined as they are used.
- Unique black and red color format depicts isometric views more clearly than "flat" black-and-white drawings; shaded effect is an excellent "depth" projector.
- Text and illustrations are located in *direct* relationship to each other.
- CAD techniques are introduced in a dedicated chapter.
- "Real-world" techniques and drawings are highlighted throughout the text.
- Although the emphasis is on mechanical drafting, other pertinent drafting subjects are included for a comprehensive, well-rounded approach to technical drawing.

The Package

1. Comprehensive *Textbook*.
2. Comprehensive, up-to-date manual *Workbook*.
3. *Instructor's Resource Guide* now includes:
   - Solutions to all text and workbook problems.
   - Answers to review questions.
   - Drawing tests.

Acknowledgments

The authors would like to acknowledge the efforts of many people without whose assistance this project would not have been completed. Special acknowledgment is made to Jerry L. Roiter, author of Chapter 11, "Engineering Visualization: Computational Design and Analysis"; Edward G. Hoffman, author of Chapter 20, "Shop Processes"; and Robert D. Smith, author of Appendix A, "Mechanical Drafting Mathematics." We thank Ray Adams, Dana Welch, Susan Wilkinson, and Ron Ryals for their assistance with illustrations. We thank Deborah M. Goetsch for her assistance with photography and typing.

The following individuals reviewed the manuscript and made valuable suggestions to authors. We and the publisher greatly appreciate their contributions to this textbook.

Larry Gilderhus	Ted Faye
Steve Huycke	William Denardo
Daniel Abbott	Ira Watson
Ken Deibert	Randy Blaschko
Edward Lavis	Ted Petry
Larry Lamont	Aaron Parsons
Barbara Williams	Pat Courington

About the Authors

David L. Goetsch is Dean of Technical Education and Professor of Computer-Aided Design and Drafting at Okaloosa-Walton Community College in Niceville, Florida. His drafting and design program has won national acclaim for its pioneering efforts in the area of computer-aided drafting (CAD). In 1984, his school was selected as one of only ten schools in the country to earn the distinguished Secretary's Award for an outstanding Vocational Program. Goetsch is a widely acclaimed teacher, author, and lecturer on the subject of drafting and design. He won Outstanding Teacher of the Year honors in 1976, 1981, 1982, 1983, and 1984. In 1986 he won the Florida Vocational Association's "Rex Gaugh Award" for outstanding contributions to technical education in Florida. He entered education full time after a successful career in design and drafting in the private sector where he spent more than eight years as a Senior Drafter and Designer for a subsidiary of Westinghouse Corporation. This is his 27th book.

David L. Goetsch

John A. Nelson has a strong background in industry and the classroom. Before entering education full time, Nelson spent 11 years in drafting and design in the private sector, beginning as a detailer and working up to a designer. He currently continues to freelance in this field. He has more than 24 years of teaching experience in the classroom at both the high school and college levels, and was New Hampshire's Vocational Teacher of the Year in 1982. He holds the associate in arts, bachelor of science, and master of education degrees. This is his 21st book.

John A. Nelson

Professor Emeritus William S. Chalk taught 31 years in the College of Engineering at the University of Washington in the departments of General, Mechanical, and Nuclear Engineering, and was Director of Nuclear Engineering Laboratories from 1976 to 1983. Prior to 1957 Professor Chalk had seven years of experience in industry with the Link-Belt Company and has worked summers with Boeing Airplane Company, The Puget Sound Naval Shipyard, and The Sandia Corporation.

Always interested in the practical applications of education, Professor Chalk has conducted a number of nation-wide design workshops for industry. He is a recipient of Western Electric's "Outstanding Instructor of the Year" award for 1968. In addition, Bill was associate editor of *Design Graphics Journal* in 1967.

Professor Chalk holds a master's degree in mechanical engineering from the University of Washington and has participated in design training workshops at Dartmouth College and Stanford University. He is the co-author of an engineering graphics text and workbook and is the author of an ASEE paper titled "A Realistic Approach to Teaching Engineering Design" (ASEE Nov. 1984).

Professor Chalk continues to be active in engineering by lecturing to engineers who plan to take engineering license examinations, and by lecturing to minority students in summer programs.

William S. Chalk

Jerry L. Roiter, author of Chapter 11, "Engineering Visualization: Computational Design and Analysis", has been teaching at the university level since 1972. He is currently a faculty member in the Department of Industrial Sciences at Colorado State University. Prior to beginning teaching he worked in industry as a designer and continues to work in industry on a regular basis as well as serving as a consultant. He was formerly Assistant Director of Visualization at the Cornell National Supercomputer Facility. He holds the bachelor of science, master of arts, and Doctor of Philosophy degrees.

Jerry L. Roiter

TECHNICAL DRAWING

Third Edition

Basics

Career Profile: Danny Varnum

Danny Varnum is a truly outstanding federal employee who just happens to be confined to a wheelchair. As a junior in high school, Danny was a leading basketball player in the county. A diving accident that year made him a quadriplegic with very limited use of his arms. But Danny persevered! Through his characteristic determination and hard work—and with the support and encouragement of his wife, Susie—Danny completed high school at age 19. Danny is working towards his applied associate of science degree from Okaloosa-Walton Community College. He then plans to continue his education toward a bachelor of science degree in computer science. Danny's determination is a guarantee he will make that goal!

That same determination made Danny an outstanding employee of the Air Force's Aeronautical Systems Center. For the last three years, Danny has been a key integrated product team member of the Joint Tactical Systems Program Office. As a packaging engineering drafter, Danny uses a state-of-the-art computer-aided design (CAD) workstation to design special containers for munitions and other vital Air Force systems. This year, our container design workload increased an unprecedented 500 percent as Danny's team tackled their most challenging job — designing containers for the top priority (1-1) MILSTAR program. The MILSTAR prime contractor was unable to produce critical shipping containers within the tight program schedule. The MILSTAR program director turned to Danny's team, spurred on by their lead drafter's (Danny's) dedication, the team worked to design and build the containers ahead of schedule and under budget. The team was awarded the Air Force In-House Value Engineering Award for delivering a quality product and saving the MILSTAR program over $3 million.

Danny not only led the team's drafter function, but defined the unique placement and sequencing of antenna containers onto the MILSTAR van. He worked hand in hand with the engineers to continually improve container designs to resolve space and size issues. Danny's work in container placement and sequencing required his TDY travel to Sandia National Laboratory to interface with the eventual users of the MILSTAR system. During the Sandia meeting, Danny met his MILSTAR users and sold them on his team's solution. His knowledge also helped him propose an innovative solution to still another MILSTAR user—a container between the two users. For his total performance and specific contributions to the MILSTAR program, Danny received a well-earned "superior" on his recent performance appraisal.

Although his professional skills are exceptional, Danny's most outstanding assets are his character, courage, and faith. Danny spends many hours supporting his son Scott's school activities. Scott is an excellent student and (following in his father's footsteps) a leading scorer on the basketball team. Although restricted in mobility, Danny is also an avid hunter and fisherman who encourages other disabled individuals in the workplace and community to join him.

Danny's unfailing good humor and positive outlook on life are absolutely contagious! A true inspiration to all who meet him, he is a role model for any employee. Danny is proof that the right attitude and determination can reduce any disability to an inconvenience that need not hinder performance—even superior performance—in the workplace.

Introduction: Graphic Communication and Technical Drawing

Chapter Outline

Graphic Communication

Graphic communication involves using visual material to relate ideas. Drawings, photographs, slides, transparencies, and sketches are all forms of graphic communication. Most children are able to draw before they are able to write. This is graphic communication. When one person sketches a rough map in giving another directions, this is graphic communication. Any medium that uses a graphic image to aid in conveying a message, instructions, or an idea is involved in graphic communication. One of the most widely used forms of graphic communication is the drawing.

Drawings Described

A *drawing* is a graphic representation of an idea, a concept or an entity which actually or potentially exists in life. The drawing itself is 1) a way of communicating all necessary information about an abstraction, such as an idea or a concept; or 2) a graphic representation of some real entity, such as a machine part, a house, or a tool, for example.

Drawing is one of the oldest forms of communication, dating back even farther than verbal communication. Cave dwellers painted drawings on the walls of their caves thousands of years before paper was invented. These crude drawings served as a means of communicating long before verbal communications had developed beyond the grunting stage. In later years, Egyptian hieroglyphics were a more advanced form of communicating through drawings.

The old adage "one picture is worth a thousand words" is still the basis of the need for technical drawings.

Types of Drawings

There are two basic types of drawings: artistic and technical. Some experts believe there are actually three types: the two mentioned and another type which combines these two. The third type is usually referred to as an illustration or rendering.

Artistic Drawings

Artistic *drawings* range in scope from the most simple line drawings to the most famous paintings. Regardless of their complexity or status, artistic drawings are used to express the feelings, beliefs, philosophies, or abstract ideas of the artist. This is why the lay person often finds it difficult to understand what is being communicated by a work of art.

In order to understand an artistic drawing, it is sometimes necessary to first understand the artist. Artists often take a subtle or abstract approach in communicating through their drawings. This gives rise to the various interpretations often associated with artistic drawings.

Technical Drawings

The technical drawing, on the other hand, is not subtle or abstract. It does not require an understanding of its creator, only an understanding of technical drawings. A *technical drawing* is a means of clearly and concisely communicating all of the information necessary to transform an idea or a concept into reality. Therefore, a technical drawing often contains more than just a graphic representation of its subject. It also contains dimensions, notes, and specifications.

The mark of a good technical drawing is that it contains all of the information needed by individuals for converting the idea or concept into reality. The conversion process may involve manufacturing, assembly, construction, or fabrication. Regardless of the process involved, a good technical drawing allows the conversion process to proceed without having to ask designers or drafters for additional information or clarification.

Figures I-1 and *I-2* contain samples of technical mechanical drawings which are used as guides by the

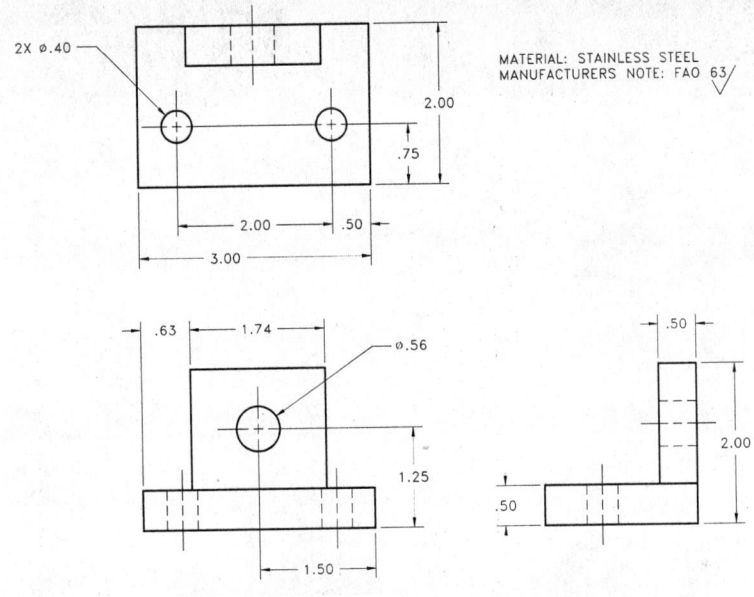

Figure I-1   Technical drawing (mechanical)

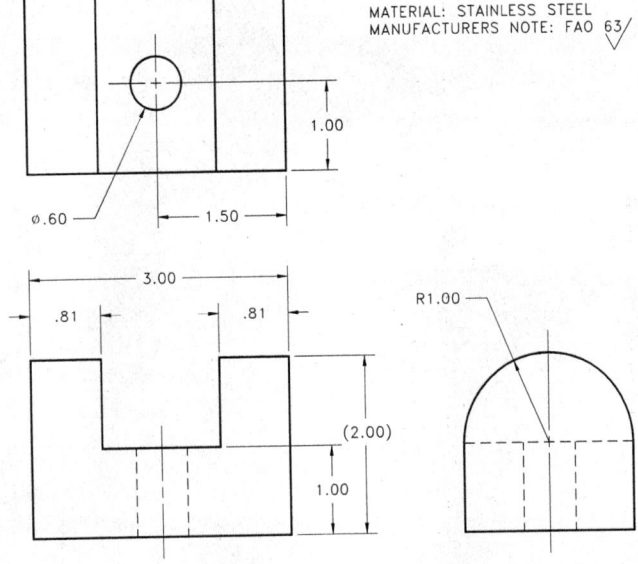

Figure I-2   Technical drawing (mechanical)

people involved in various phases of manufacturing the represented parts. Notice that the drawings contain a graphic representation of the part, dimensions, material specifications, and notes.

Illustrations or Renderings

Illustrations or renderings are sometimes referred to as a third type of drawing because they are neither complete-ly technical, nor completely artistic; they combine elements of both, as shown in *Figures I-3, I-4, I-5,* and *I-6*. They are technical in that they are drawn with mechanical instruments or on a computer-aided drafting system, and they contain some degree of technical information. However, they are also artistic in that they attempt to convey a mood, an attitude, a status, or other abstract, nontechnical feelings.

Figure I-3   Rendering

Figure I-4 Rendering

Figure I-4   Rendering

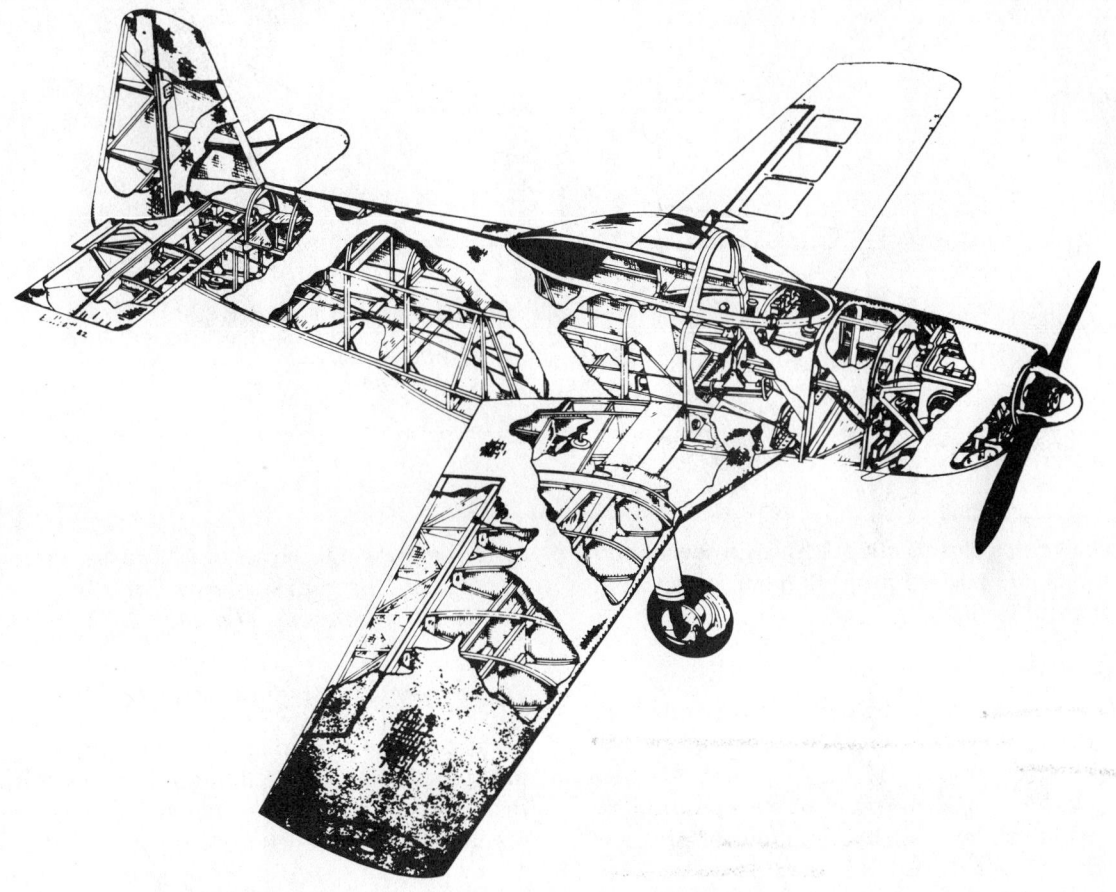

Figure I-5   Mechanical illustration *(Courtesy Ken Elliott)*

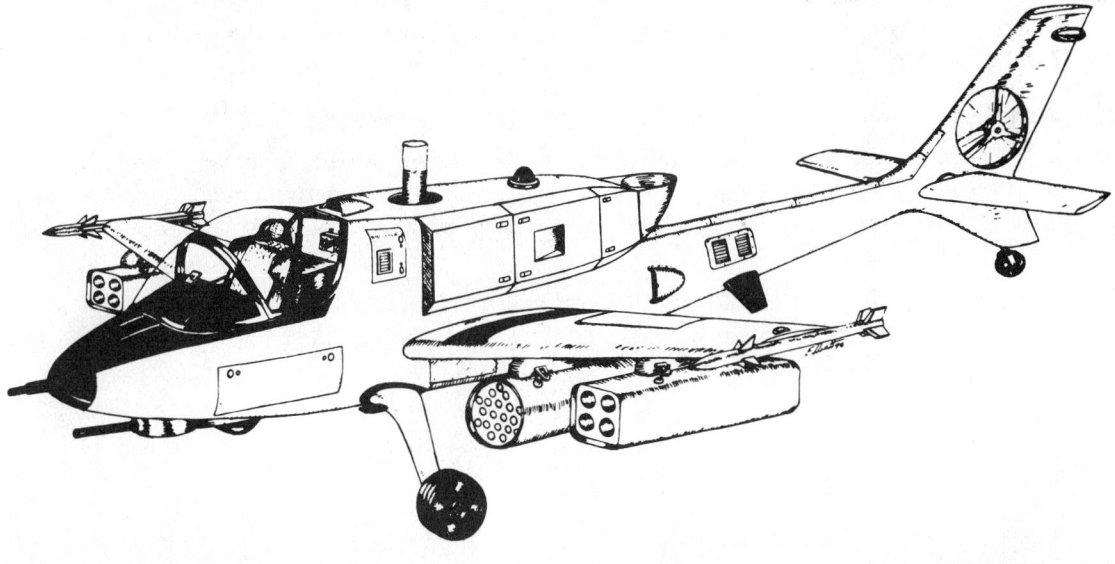

Figure I-6   Mechanical illustration *(Courtesy Ken Elliott)*

Types of Technical Drawings

Technical drawings are based on the fundamental principles of projection. A *projection* is a drawing or representation of an entity on an imaginary plane or planes. This
projection plane serves the same purpose in technical
drawing as is served by the movie screen in a theater.

As can be seen in ***Figure I-7***, a projection involves
four components: 1) the actual object that the drawing
or projection represents, 2) the eye of the viewer looking at the object, 3) the imaginary projection plane (the
viewer's drawing paper or the graphics display in a
computer-aided drafting system), and 4) imaginary lines
of sight called *projectors*.

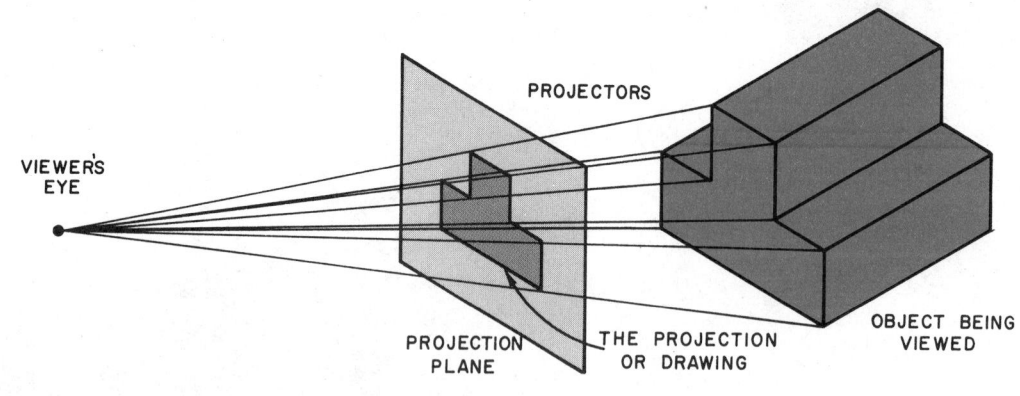

Figure I-7 The projection plane

Two broad types of projection, both with several sub-classifications, are parallel projection and perspective (converging) projection.

Parallel Projection

Parallel projection is subdivided into the following three categories: orthographic, oblique, and axonometric projections.

Orthographic projections are drawn as multiview drawings which show flat representations of principal views of the subject, *Figure I-8*. *Oblique projections* actually show the full size of one view and are of two varieties: *cabinet* (half-scale) or *cavalier* (full-scale) projections, *Figures I-9* and *1-10*. *Axonometric projections* are three-dimensional drawings, and are of three different varieties: isometric, dimetric, and trimetric, *Figures I-11, I-12,* and *I-13.*

Perspective Projection

Perspective projections are drawings which attempt to replicate what the human eye actually sees when it views an object. That is why the projectors in a per-

spective drawing converge. There are three types of perspective projections: one-point, two-point, and three-point projections, *Figures I-14, I-15,* and *I-16.*

Purpose of Technical Drawings

To appreciate the need for technical drawings, one must understand the design process. The design process is an orderly, systematic procedure used in accomplishing a needed design.

Any product that is to be manufactured, fabricated, assembled, constructed, built, or subjected to any other type of conversion process must first be designed. For example, a house must be designed before it can be built. An automobile must be designed before it can be manufactured. A printed circuit board must be designed before it can be fabricated.

The Design Process

The *design process* is an organized, step-by-step procedure in which mathematical and scientific principles, coupled with experience, are brought to bear in order

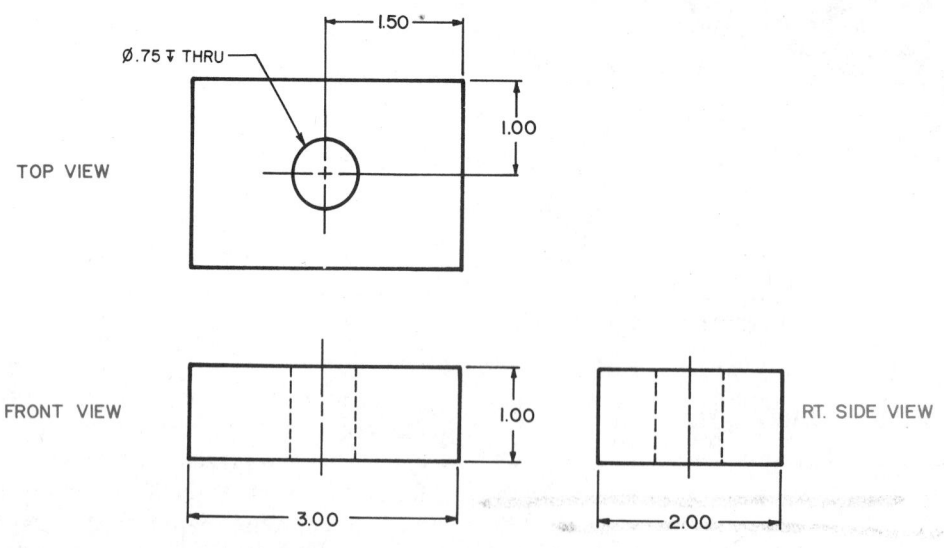

Figure I-8 Orthographic multiview drawing

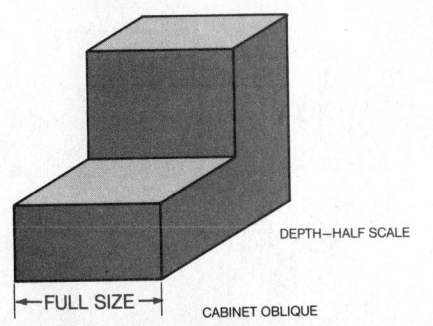

Figure I-9   Oblique projection (cabinet)

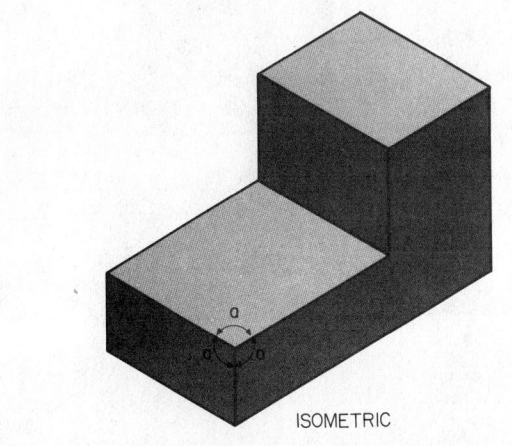

Figure I-10   Axonometric projection (isometric)

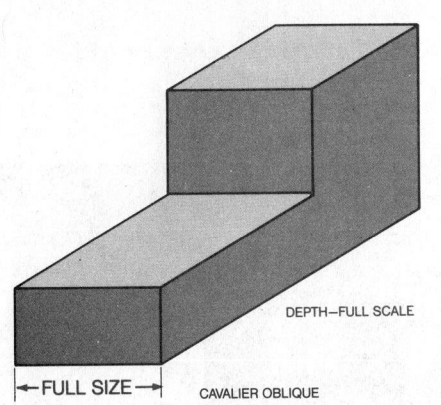

Figure I-11   Oblique projection (cavalier)

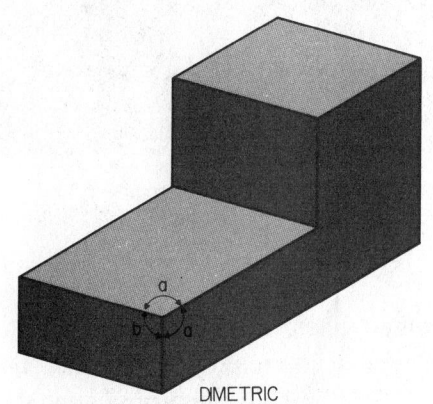

Figure I-12   Axonometric projection (dimetric)

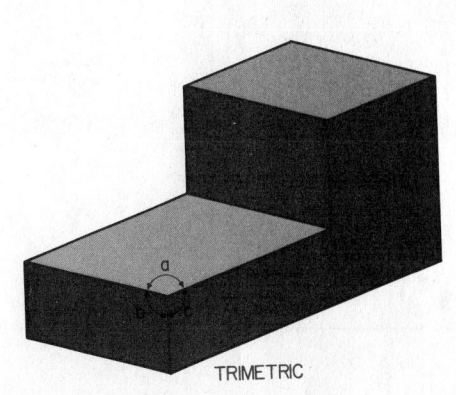

Figure I-13   Axonometric projection (trimetric)

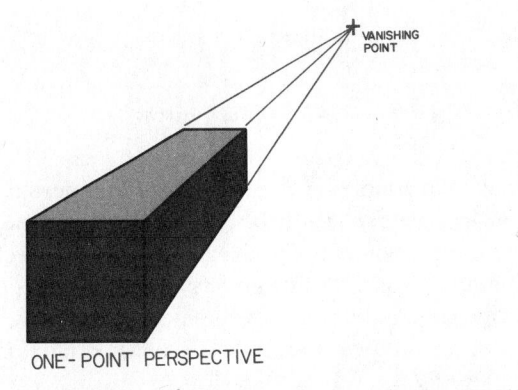

Figure I-14   One-point perspective projection

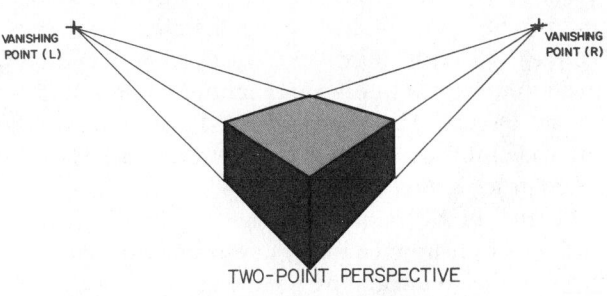

Figure I-15   Two-point perspective projection

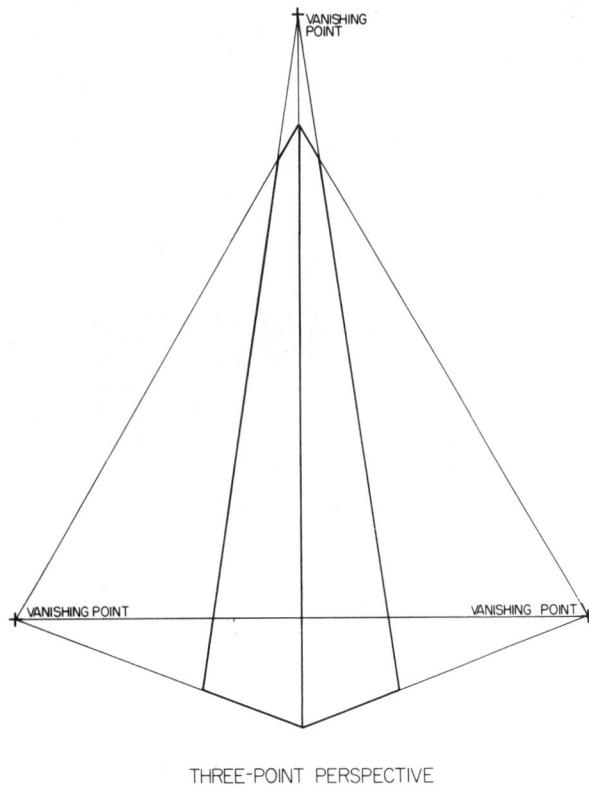

THREE-POINT PERSPECTIVE

Figure I-16 Three-point perspective projection

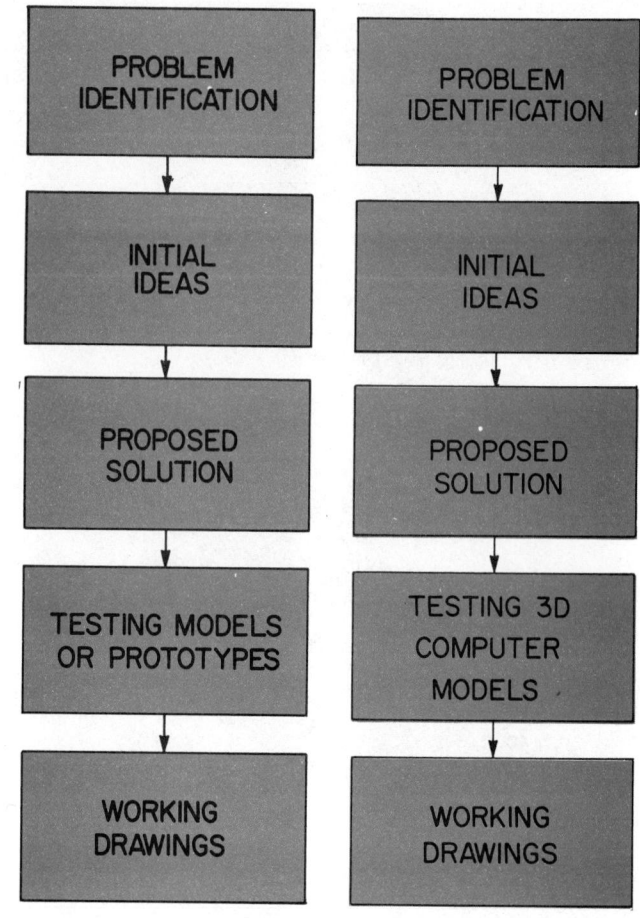

Figure I-17 The design process (manual)   ***Figure I-18*** The design process (CAD)

to solve a problem or meet a need. The design process has five steps. Traditionally, these steps have been 1) identification of the problem or a need, 2) development of initial ideas for solving the problem, 3) selection of a proposed solution, 4) development and testing of models or prototypes, and 5) developing working drawings, ***Figure I-17.***

The age of computers has altered the design process slightly for those companies which have converted to computer-aided design and drafting. For these companies, the expensive, time-consuming fourth step in the design process — the making and testing of actual models or prototypes — has been substantially altered, ***Figure I-18.*** This fourth step has been replaced with three-dimensional computer models which can be quickly and easily produced on a CAD system using the data base built up during the first three phases of the design process, ***Figure I-19.***

Whether in the traditional design process or the more modern computer version, in either case, working drawings are an integral part of the design process from start to finish.

The purpose of technical drawings is to document the design process. Creating technical drawings to support the design process is called *drafting*. People who do drafting are known as *drafters* or *drafting technicians*. The words "draftsman" or "draughtsman" are no longer used.

In the first step of the design process, technical drawings are used to help clarify the problem or the need.

Figure I-19 Three-dimensional computer model (*Courtesy Terak Corporation*)

The drawings may be old ones on file or they may be new ones created for the purpose of clarification. In the second step, technical drawings — often in the form of sketches or preliminary drawings — are used to document the various ideas and concepts formed. In the

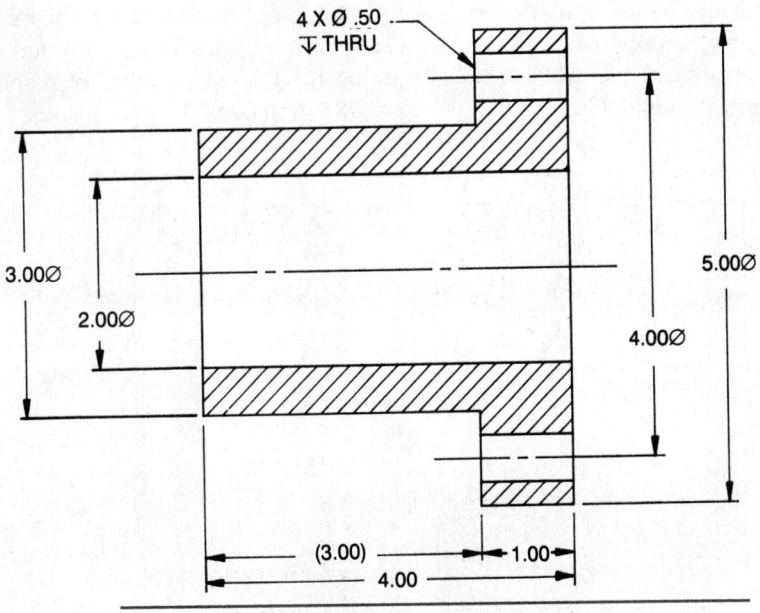

Figure I-20   Simple mechanical drawing (manual)

third step, technical drawings — again, usually preliminary drawings — are used to communicate the purposed solution.

If the traditional fourth step in the design process is being used, preliminary drawings and sketches from the first three steps will be used as guides in constructing models or prototypes for testing. If the more modern fourth step is being used, the data base built up during documentation of the first three steps can be used in developing three-dimensional computer models. In both cases, the final step is the development of complete working drawings for guiding individuals involved in the conversion process. *Figure I-20* is a working

drawing documenting the design of a simple mechanical part. The drawing was produced manually. *Figure I-21* is the same drawing produced on a CAD system.

Applications of Technical Drawings

Technical drawings are used in many different applications. They are needed in any setting which involves design, and in any subsequent form of conversion process. The most common applications of technical drawings can be found in the fields of manufacturing, engineering, architecture and construction, and all of their various related fields.

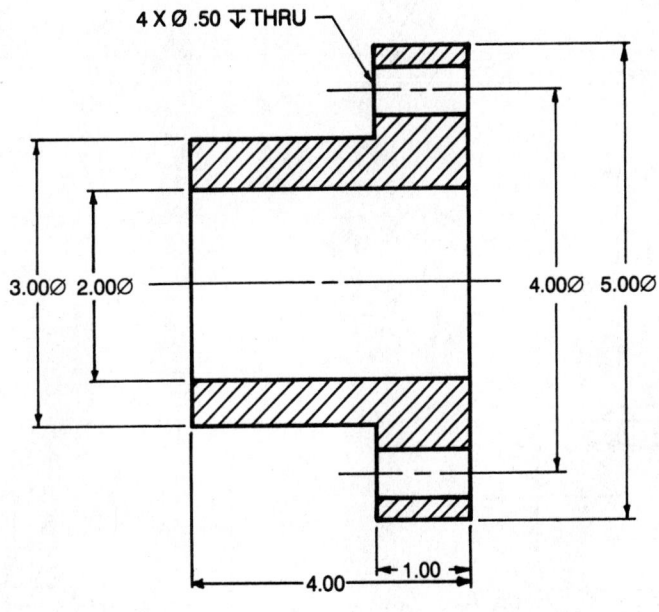

Figure I-21   Simple mechanical drawing (CAD)

Architects use technical drawings to document their designs of residential, commercial, and industrial buildings, *Figures I-22* and *I-23*. Structural, electrical, and mechanical [heating, ventilating, air conditioning (HVAC) and plumbing] engineers who work with architects also use technical drawings to document those aspects of the design for which they are responsible, *Figures I-24*, *I-25*, and *I-26*.

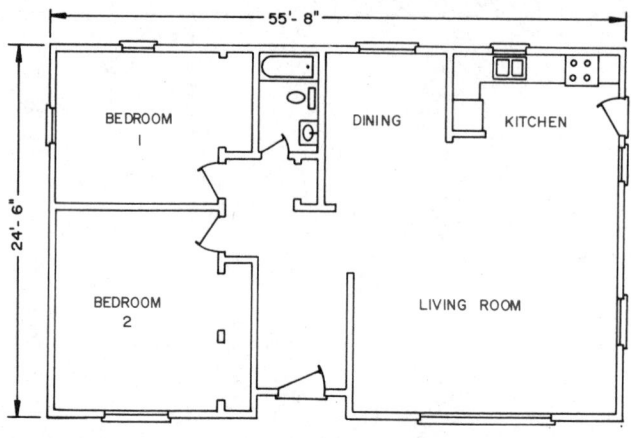

Figure I-22   Technical drawing (architectural)

Figure I-23   Technical drawing (architectural)

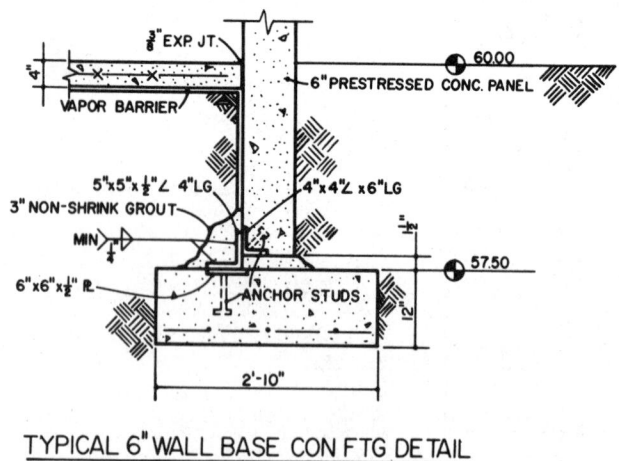

TYPICAL 6" WALL BASE CON FTG DETAIL

Figure I-24   Technical drawing (structural)

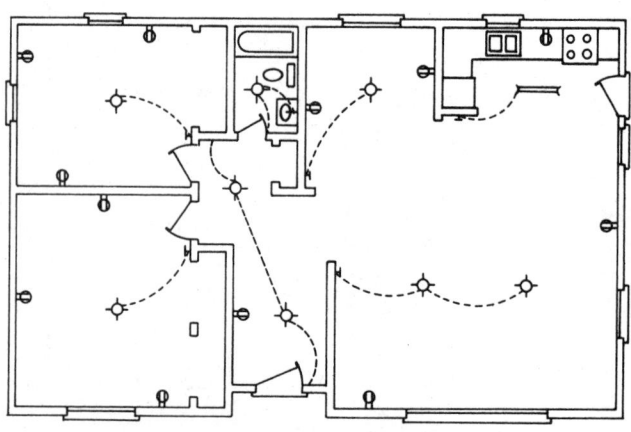

Figure I-25   Technical drawing (electrical)

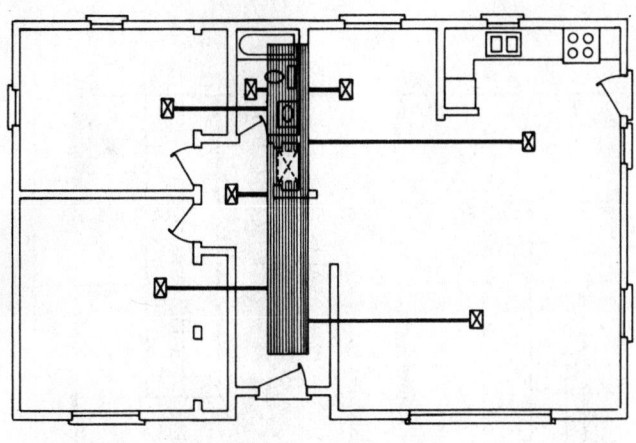

Figure I-26 Technical drawing (HVAC)

Surveyors and civil engineers use technical drawings to document such work as the layout of a new subdivision, or the marking-off of the boundaries for a piece of property, ***Figure I-27***. Contractors and construction personnel use technical drawings as their blueprints in converting architectural and engineering designs into reality, ***Figures I-28*** and ***I-29***.

Technical drawings are equally important to engineers, designers, and various other individuals working in the manufacturing industry. Manufacturing engineers use technical drawings to document their designs. Technical drawings guide the collective efforts of individuals who are concerned with the same common goal, ***Figures I-30*** and ***I-31***.

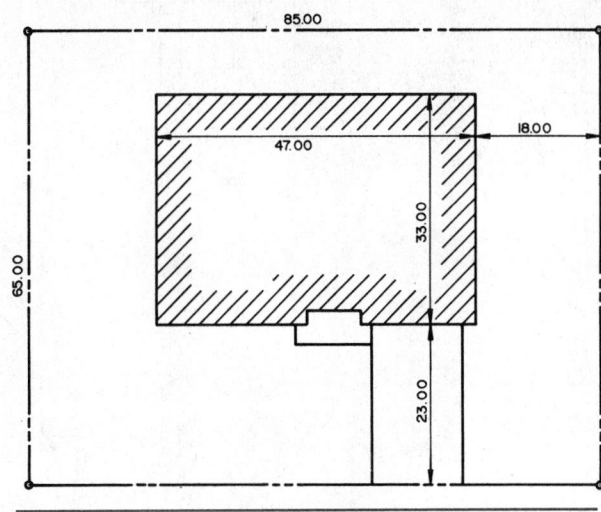

Figure I-27 Technical drawing (civil)

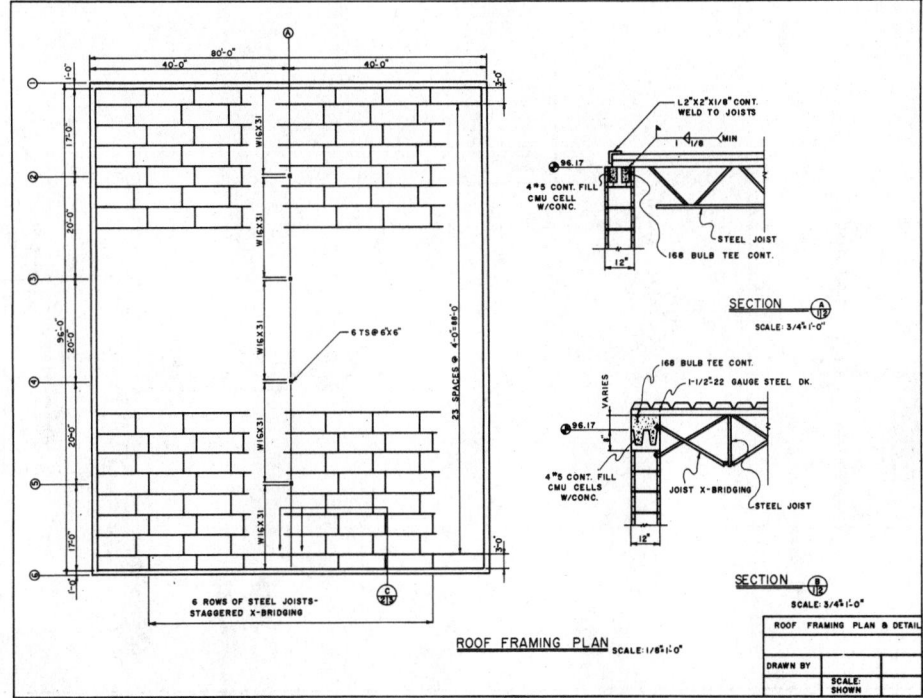

Figure I-28 Architectural/Engineering drawing

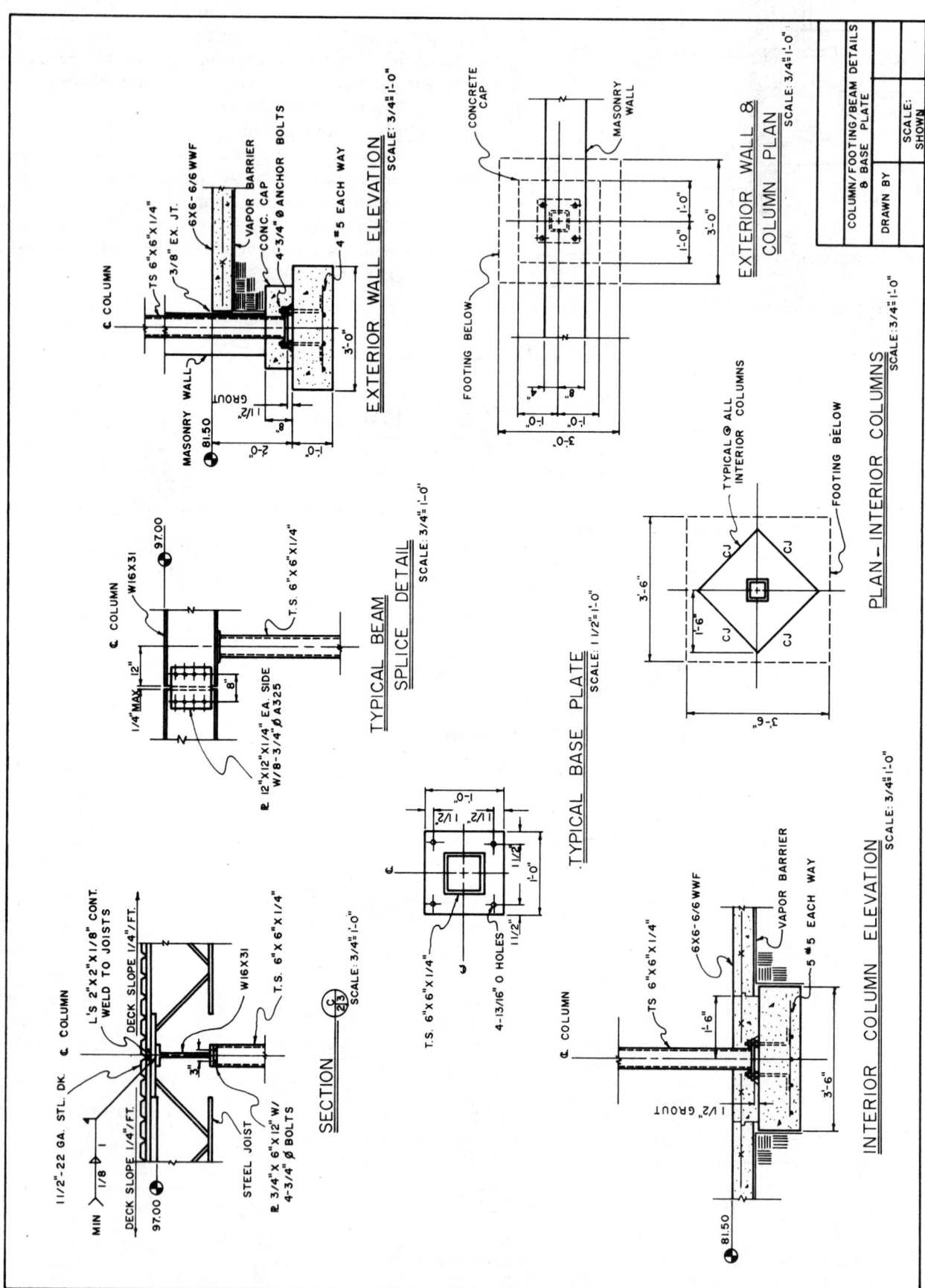

Figure I-29 Architectural/Engineering drawing

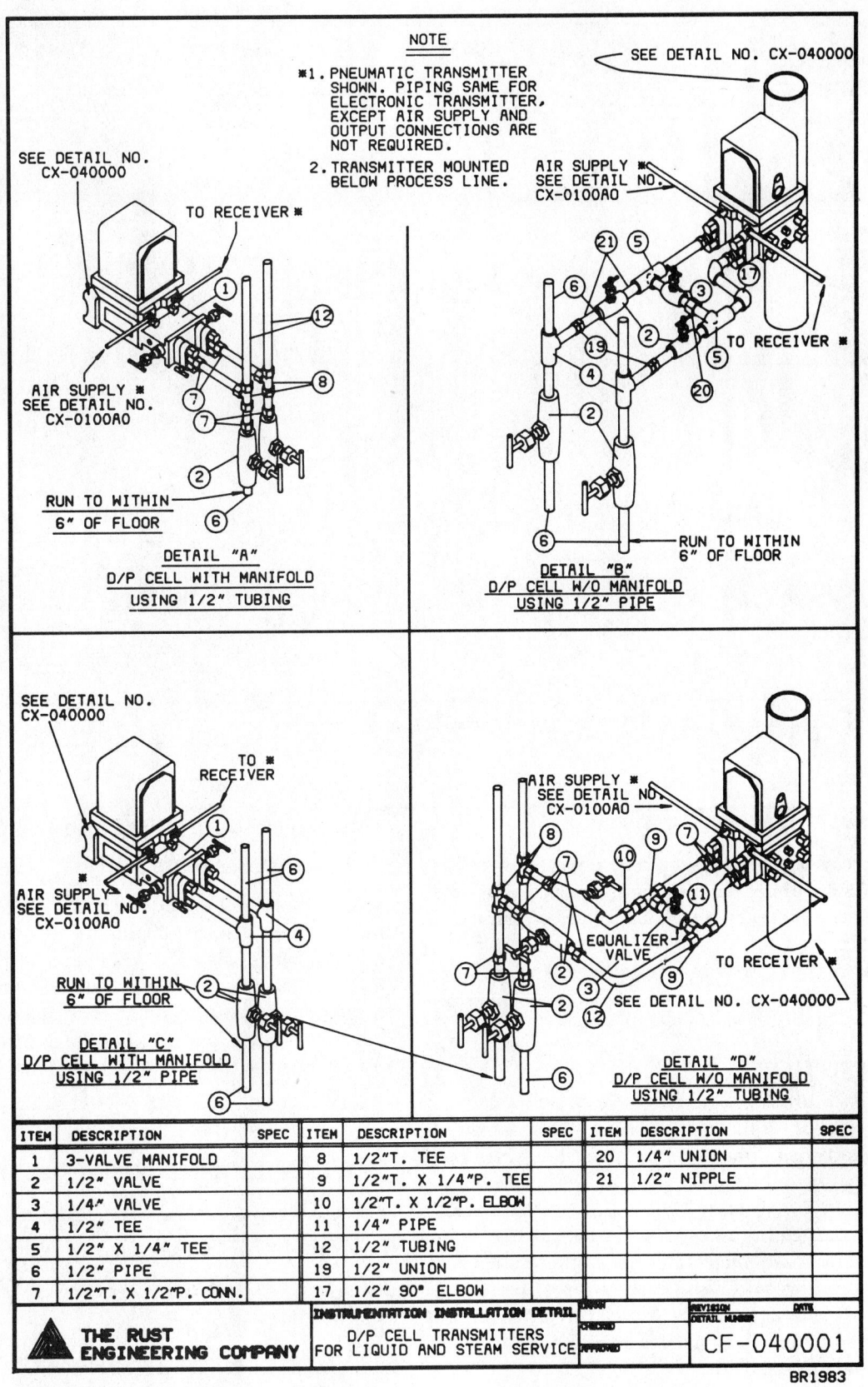

ITEM	DESCRIPTION	SPEC	ITEM	DESCRIPTION	SPEC	ITEM	DESCRIPTION	SPEC
1	3-VALVE MANIFOLD		8	1/2"T. TEE		20	1/4" UNION	
2	1/2" VALVE		9	1/2"T. X 1/4"P. TEE		21	1/2" NIPPLE	
3	1/4" VALVE		10	1/2"T. X 1/2"P. ELBOW				
4	1/2" TEE		11	1/4" PIPE				
5	1/2" X 1/4" TEE		12	1/2" TUBING				
6	1/2" PIPE		19	1/2" UNION				
7	1/2"T. X 1/2"P. CONN.		17	1/2" 90° ELBOW				

THE RUST ENGINEERING COMPANY

INSTRUMENTATION INSTALLATION DETAIL
D/P CELL TRANSMITTERS
FOR LIQUID AND STEAM SERVICE

CF-040001

BR1983

Figure I-30   Isometric mechanical drawing *(Courtesy The Rust Engineering Company)*

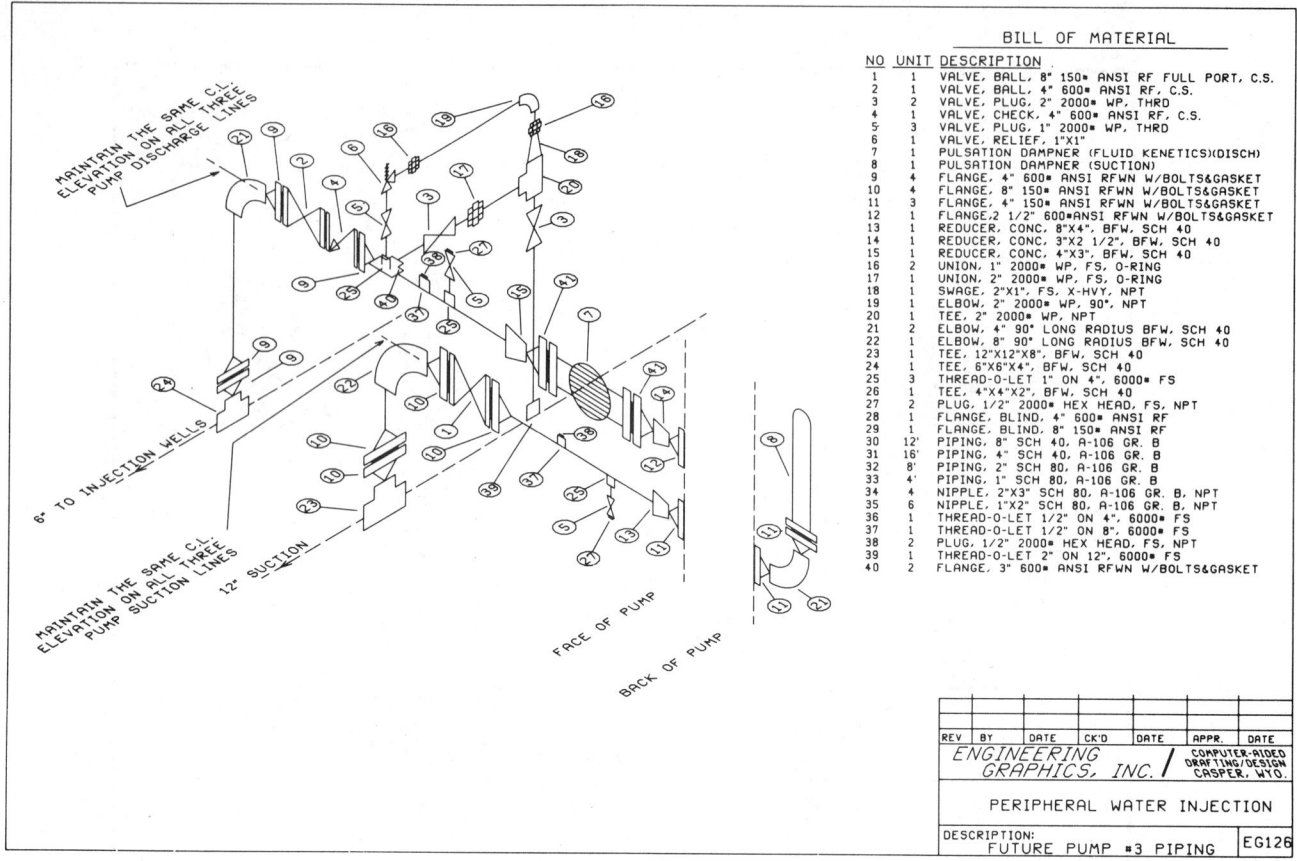

BILL OF MATERIAL		
NO	UNIT	DESCRIPTION
1	1	VALVE, BALL, 8" 150# ANSI RF FULL PORT, C.S.
2	1	VALVE, BALL, 4" 600# ANSI RF, C.S.
3	4	VALVE, PLUG, 2" 2000# WP, THRD
4	1	VALVE, CHECK, 4" 600# ANSI RF, C.S.
5	3	VALVE, PLUG, 1" 2000# WP, THRD
6	1	VALVE, RELIEF, 1"X1"
7	1	PULSATION DAMPNER (FLUID KENETICS)(DISCH)
8	1	PULSATION DAMPNER (SUCTION)
9	4	FLANGE, 4" 600# ANSI RFWN W/BOLTS&GASKET
10	4	FLANGE, 8" 150# ANSI RFWN W/BOLTS&GASKET
11	3	FLANGE, 4" 150# ANSI RFWN W/BOLTS&GASKET
12	1	FLANGE,2 1/2" 600#ANSI RFWN W/BOLTS&GASKET
13	1	REDUCER, CONC, 8"X4", BFW, SCH 40
14	1	REDUCER, CONC, 3"X2 1/2", BFW, SCH 40
15	1	REDUCER, CONC, 4"X3", BFW, SCH 40
16	2	UNION, 1" 2000# WP, FS, O-RING
17	1	UNION, 2" 2000# WP, FS, O-RING
18	1	SWAGE, 2"X1", FS, X-HVY, NPT
19	1	ELBOW, 2" 2000# WP, 90#, NPT
20	1	TEE, 2" 2000# WP, NPT
21	2	ELBOW, 4" 90° LONG RADIUS BFW, SCH 40
22	1	ELBOW, 8" 90° LONG RADIUS BFW, SCH 40
23	1	TEE, 12"X12"X8", BFW, SCH 40
24	1	TEE, 6"X6"X4", BFW, SCH 40
25	3	THREAD-O-LET 1" ON 4", 6000# FS
26	1	TEE, 4"X4"X2", BFW, SCH 40
27	2	PLUG, 1/2" 2000# HEX HEAD, FS, NPT
28	1	FLANGE, BLIND, 4" 600# ANSI RF
29	1	FLANGE, BLIND, 8" 150# ANSI RF
30	12'	PIPING, 8" SCH 40, A-106 GR. B
31	16'	PIPING, 4" SCH 40, A-106 GR. B
32	8'	PIPING, 2" SCH 80, A-106 GR. B
33	4'	PIPING, 1" SCH 80, A-106 GR. B
34	4	NIPPLE, 2"X3" SCH 80, A-106 GR. B, NPT
35	6	NIPPLE, 1"X2" SCH 80, A-106 GR. B, NPT
36	1	THREAD-O-LET 1/2" ON 4", 6000# FS
37	1	THREAD-O-LET 1/2" ON 8", 6000# FS
38	2	PLUG, 1/2" 2000# HEX HEAD, FS, NPT
39	1	THREAD-O-LET 2" ON 12", 6000# FS
40	2	FLANGE, 3" 600# ANSI RFWN W/BOLTS&GASKET

ENGINEERING GRAPHICS, INC.
COMPUTER-AIDED DRAFTING/DESIGN CASPER, WYO.
PERIPHERAL WATER INJECTION
DESCRIPTION: FUTURE PUMP #3 PIPING   EG126

Figure I-31   Isometric piping schematic *(Courtesy Engineering Graphics Inc.)*

Regulation of Technical Drawings

Technical drawing practices must be regulated because of the diversity of their applications. Just as the English language must have certain standard rules of grammar, the graphic language must have certain rules of practice. This is the only way to ensure that all people attempting to communicate using the graphic language are speaking the same language.

Standards of Practice

A number of different agencies have developed standards of practice for technical drawing. The most widely used standards of practice for technical drawing and drafting are those of the U.S. Department of Defense (DOD), the U.S. Military (MIL), and the American National Standards Institute (ANSI).

The American National Standards Institute does not limit its activities to the standardization of technical drawing and drafting practices. In fact, this is just one of the many fields for which ANSI maintains a continuously updated set of standards.

Standards of interest to drafters, designers, checkers, engineers, and architects are contained in the "Y" series of ANSI standards. ***Figure I-32*** contains a list of ANSI standards frequently used in technical drawing and drafting specifications.

What Students of Technical Drawing and Drafting Should Learn

Many people in the world of work use technical drawings in various forms. Engineers, designers, checkers, drafters, and a long list of related occupations use technical drawings as an integral part of their jobs. Some of these people must be able to actually make drawings; others are only required to be able to read and interpret drawings; some must be able to do both.

SIZE AND FORMAT————————— Y14.1

LINE CONVENTIONS AND LETTERING— Y14.2

PROJECTIONS—————————————— Y14.3

PICTORIAL DRAWING————————— Y14.4

DIMENSIONING AND TOLERANCING——— Y14.5M

SCREW THREADS——————————— Y14.6

GEARS, SPLINES AND SERRATIONS——— Y14.7

GEAR DRAWING STANDARDS—————— Y14.7.1

MECHANICAL ASSEMBLIES——————— Y14.14

Figure I-32   Sample list of drafting standards

What students of technical drawing and drafting should learn depends on how they will use technical drawings in their jobs. Will they make them? Will they read and interpret them? This textbook is written for students in the fields of engineering, design, drafting, and architecture, among others, who must be able to make, read, and interpret technical drawings. These students should develop a wide range of knowledge and skills, *Figure I-33*.

The learning required of technical drawing students can be divided into three categories: fundamental knowledge and skills, related knowledge, and advanced knowledge and skills.

In the "fundamentals" category, students of technical drawing and drafting should develop knowledge and skills in the areas of drafting equipment; such fundamental drafting techniques as line work, lettering, scale use, and sketching; geometric construction; multiview drawing; sectional views; descriptive geometry; auxiliary views; general dimensioning; and notation.

In the "related knowledge" category, students of technical drawing and drafting should develop a broad knowledge base in the areas of related math, welding, shop processes, and media and reproduction.

In the "advanced" category, students of technical drawing and drafting should develop knowledge and skills in the areas of development, geometric dimensioning and tolerancing, threads and fasteners, springs, cams, gears, machine design drawing, pictorial drawing,

LEARNING CHECKLIST FOR STUDENTS OF TECHNICAL DRAWING		
FUNDAMENTAL KNOWLEDGE AND SKILLS	**RELATED KNOWLEDGE**	**ADVANCED KNOWLEDGE AND SKILLS**
DRAFTING EQUIPMENT	MATH	DEVELOPMENT AND INTERSECTIONS
FUNDAMENTAL DRAFTING TECHNIQUES	WELDING	GEOMETRIC DIMENSIONING AND TOLERANCING
SKETCHING	MEDIA AND REPRODUCTION	
GEOMETRIC CONSTRUCTION	MANUFACTURING	THREADS AND FASTENERS
VISUALIZATION	MATERIALS AND PROCESSES	SPRINGS
MULTIVIEW DRAWING		CAMS
SECTION VIEWS		GEARS
GRAPHICAL DESCRIPTIVE GEOMETRY		MACHINE DESIGN DRAWING
AUXILIARY VIEWS		PICTORIAL DRAWING
GENERAL DIMENSIONING		DRAFTING SHORT-CUTS
NOTATION		CAD TECHNOLOGY

Figure I-33 Checklist for students of technical drawing

drafting shortcuts, and CAD/CAM technology and operations. The latter area represents a significant change in techniques used to create, maintain, update, and store technical drawings, *Figures I-34* and *I-35*.

Figures I-36 through *I-40* contain examples of several different kinds of technical drawings taken from the "real world" of drafting.

Figure I-34 Modern CAD technology *(Courtesy CADKEY)*

Figure I-35 Modern CAD system *(Courtesy Autodesk Inc.)*

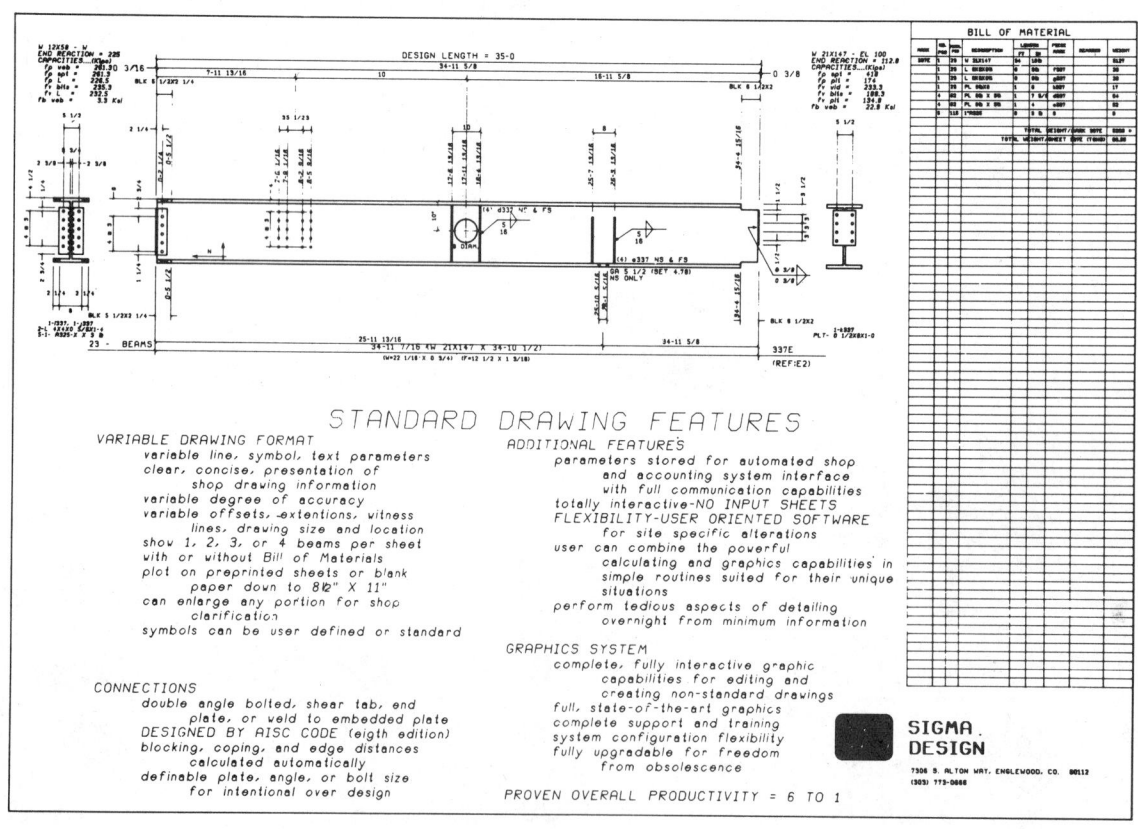

Figure I-36 Structural steel drawing *(Courtesy Sigma Design)*

Technical Drawing and Quality/Competitiveness

Companies that design and manufacture products or design and build structures operate in a competitive environment. Surviving and succeeding in the intensely competitive modern marketplace requires companies to focus more attention than ever before on quality. Such concepts as flexible manufacturing, computer-integrated manufacturing (CIM), just-in-time manufacturing (JIT), and statistical process control (SPC) are all attempts at improving quality and competitiveness.

As effective as these concepts may be at improving

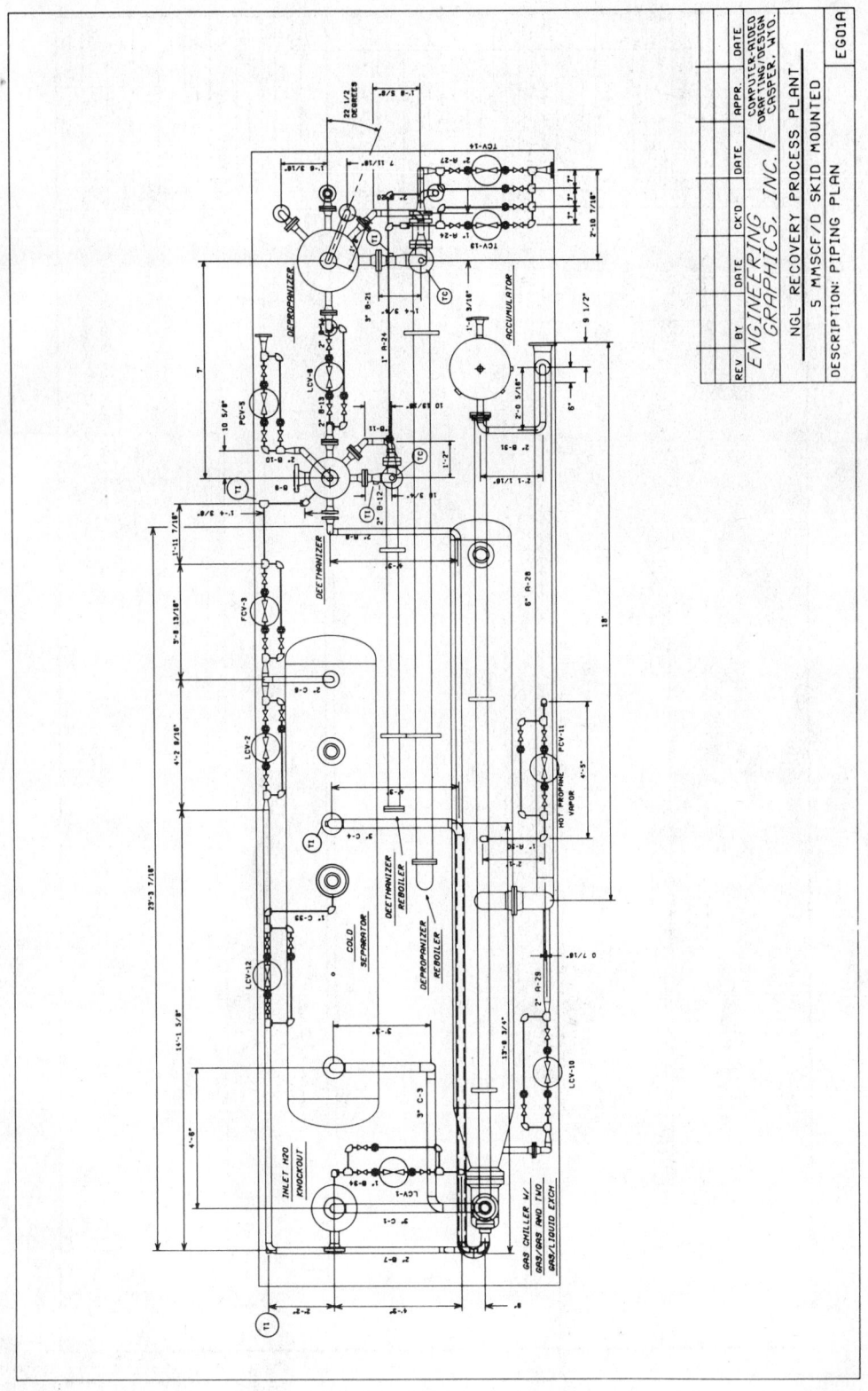

Figure I-37   Piping drawing *(Courtesy Engineering Graphics Inc.)*

the quality of what takes place during manufacturing processes, they cannot overcome a poor design or misinterpretations that result from poorly prepared technical drawings. The best way to ensure quality and the competitive edge it can give a company is to do the following: 1) design in quality from the outset; and 2) pre-

pare technical drawings that accurately and properly communicate the design to those who will manufacture or build it.

Quality can be defined as the extent to which a product conforms to the customer's expectations. In design and manufacturing, customer expectations are typically

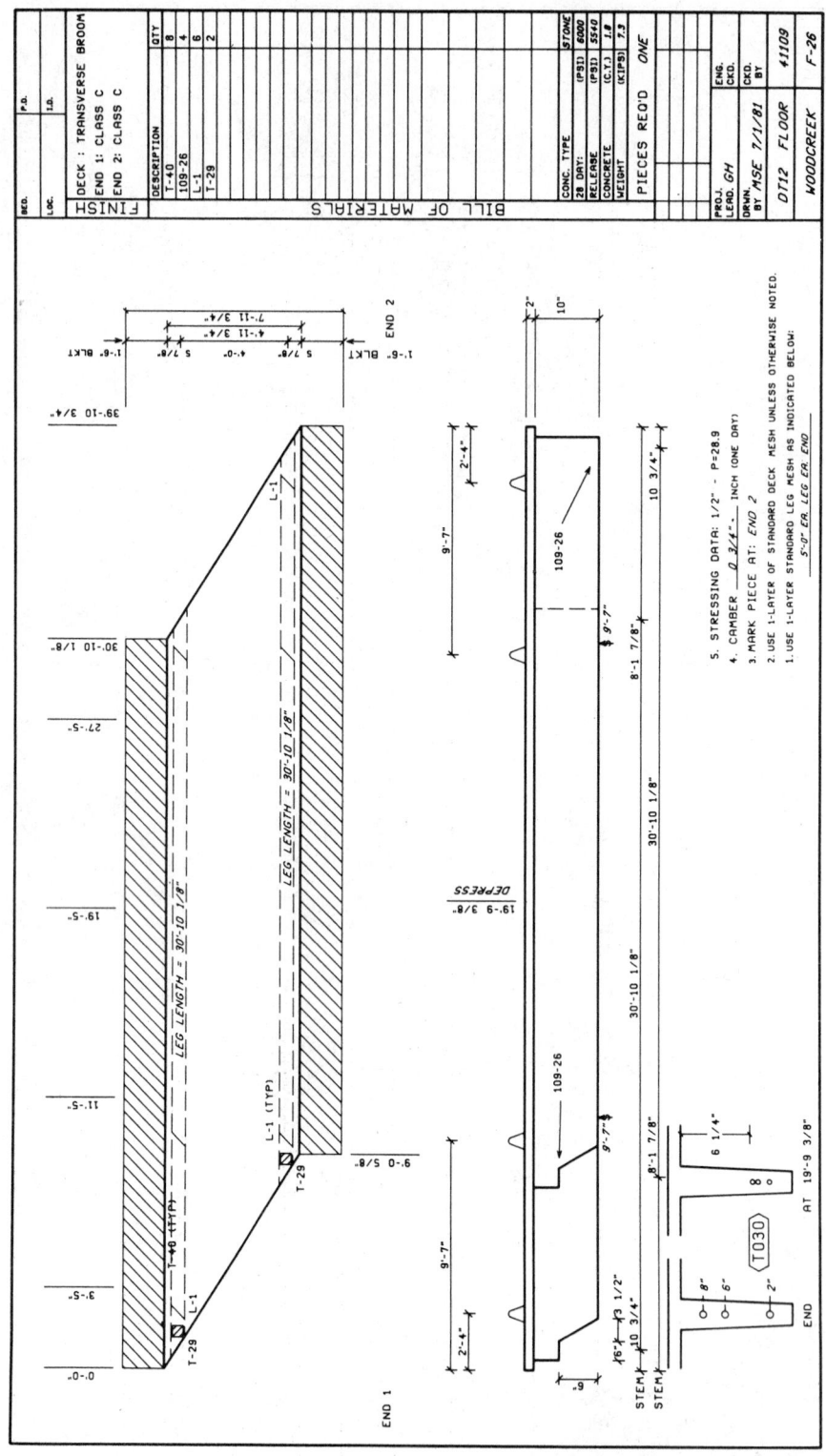

Figure I-38 **Prestressed concrete drawing** *(Courtesy Sigma Design)*

communicated in the form of specifications. Consequently, in this field, quality is often viewed as the extent to which specifications are met or exceeded. Companies that consistently meet or exceed customer expectations are more competitive.

In addition to designing in quality, it is important to develop a design that is as easy as possible to manufac-

ture. Often there will be several different design options that will meet or exceed customer expectations. When this is the case, the design chosen should be the one that is easiest to manufacture. This concept is known as *Design for Manufacturability* (DFM) and is covered in chapter 17.

Technical drawing plays a key role in determining the

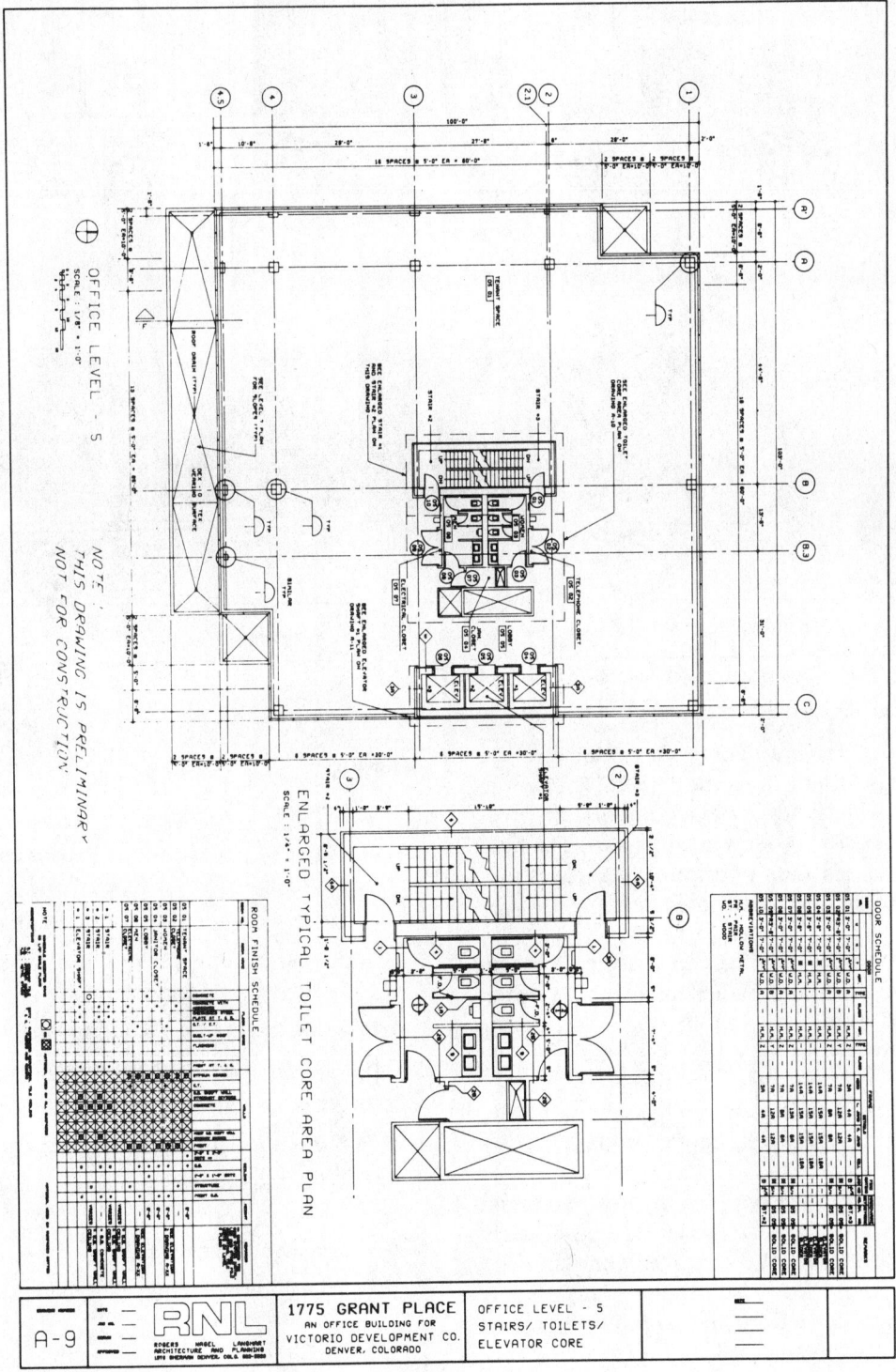

Figure I-39 *Architectural drawing (Courtesy Sigma Design)*

ultimate quality of a product. A good technical drawing is one that properly and conveniently communicates all of the information needed to transform a design into a product that meets or exceeds customer expectations. Anything less can detract from the quality of the product. The following examples illustrate this connection:

• An incorrect dimension on a technical drawing can cause the part to be produced at the wrong size. Each successive copy produced becomes waste.

• A dimension that is left off a technical drawing or just improperly placed can cause manufacturing

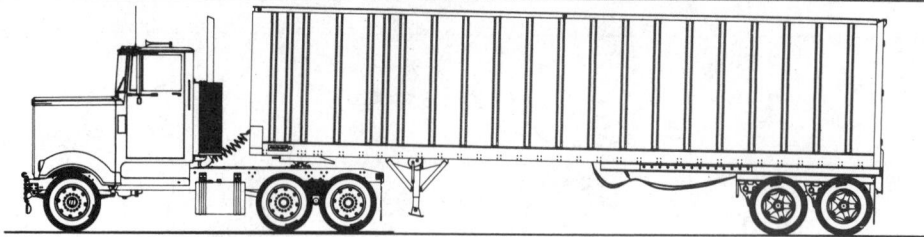

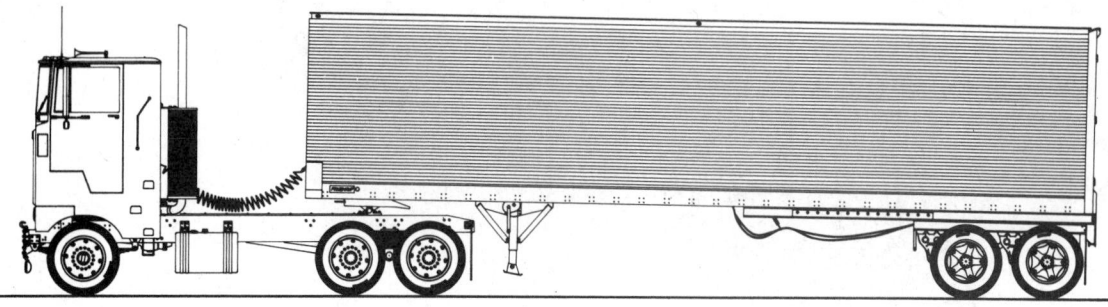

Figure I-40   Mechanical pictorial drawing *(Courtesy Fruehauf Corporation)*

personnel to interrupt their work to calculate the dimension or dimensions they need. The worst-case result is that they make a computation error and, consequently, produce a faulty part. The best-case result is that the interruption decreases productivity and, in turn, competitiveness.

• An entry left off a parts list can cause purchasing personnel to order an insufficient amount of material and, as a result, interrupt the flow of the manufacturing process.

• An improperly drawn view can cause a part to be manufactured in the wrong configuration.

These are only a few examples of how technical drawings can affect the ultimate quality of a product. By doing so they also affect the competitiveness of the company that manufactures the product. This book is designed to help students learn how to produce designs and technical drawings that make a positive contribution to quality and competitiveness.

Review

1. What is a drawing?
2. What old adage explains the basis of the need for technical drawings?
3. What are the two basic types of drawings?
4. Explain the major difference between the two basic types of drawings.
5. What are the four components of a projection?
6. Name the three subdivisions of parallel projection.
7. Name the three types of axonometric projection.
8. Name the three types of perspective projection.
9. What are the five steps in the design process?
10. Name four fields in which technical drawings are used extensively.
11. Explain how technical drawings differ from artistic drawings.
12. Name three organizations that regulate technical drawing practices.

Instruments, Hardware, and Systems

The advent of computer-aided drafting or computer-aided design (CAD) has radically changed the instruments used to produce technical drawings. In fact, one rarely hears the term instruments used in a modern technical drawing setting. Instead one hears such terms as hardware and systems. The tools used in a modern drafting department have changed continually over the years. Such changes are often associated with CAD, but in reality the tools used to produce technical drawings have always been in a state of continual evaluation and always will be.

Engineers, designers, and drafters who are old enough can remember when technical drawings were prepared in ink on linen using T-squares. Technological changes eventually saw T-squares superseded by parallel bars that were, in turn, superseded by drafting machines. Linen was replaced by vellum. Templates replaced various mechanical instruments. Overlay and scissors drafting techniques eliminated a certain amount of duplication.

These non-automated developments continued (and to a much lesser extent still continue) until the giant leaps in technical drawing technology brought about by CAD. With the advent of CAD, conventional drafting instruments began to be used less and less. However, they have not yet completely faded away and it will be years before they do.

Therefore, this chapter covers first the conventional instruments and then the hardware and systems associated with CAD. In this way students who wish to learn about the evolution of technical drawing instruments, techniques, and media from the past to the present may do so. This approach is recommended to help today's technical drawing personnel gain a perspective on where they fit into the evolutionary process with regard to their profession.

Conventional Instruments

In drafting, no lines are made freehand. Each and every line is drawn using some kind of a drafting tool. It is up to the drafter to own a complete set of standard drafting tools in order to be fully functional.

When purchasing conventional drafting equipment, care must be taken to obtain quality equipment from a reliable dealer. It is advisable to consult with an experienced drafter, a drafting instructor, or a reputable dealer. The following is a list of the minimum required drafting equipment. Special templates and special equipment must be added to this list, depending upon the field of drafting and, in some cases, the actual product manufactured by the company. Each of the following pieces of equipment is illustrated in this chapter.

- Drawing board — 24" x 36" (60 cm x 90 cm) minimum size
- T-square (parallel straightedge or drafting machine) to suit board
- 45° triangle — 8" (20 cm) size
- 30°–60° triangle — 8" (20 cm) size
- Triangular scale (depending upon the field of drafting)
- Center wheel bow compass — 6" (115 cm) with extension bar
- Drop bow compass (recommended)
- Irregular curves (two or three different configurations)
- Dividers — 6" (15 cm)
- Drafting brush
- Mechanical drafting pencils with lead
- Protractor or adjustable triangle
- Erasing shield
- Eraser
- Lead pointer with steel cutting wheel (pencil)
- Lead sandpaper or flat file (for compass lead)
- Circle template
- Ellipse template
- Drafting tape or drafting dots
- Calculator
- Dry cleaning pad (optional)

The following is a list of the required inking supplies.

- Technical pen #2½ (for lettering scriber)
- Technical pen #1 (for lettering scriber)
- Technical pen #0 (for lettering scriber)
- India ink — black
- Pen cleaning solvent
- Lettering scriber
- Lettering template #100
- Lettering template #175
- Lettering template #290
- Compass adaptor for pen

Note that some specialized or expensive equipment is often furnished by the company.

Conventional Drafting Requisites

Drawing Table

Drawing tables are available in a variety of styles. Most are adjustable up and down, and can tilt to almost any angle from vertical 90° to horizontal, *Figures 1-1* and *1-2*.

Drawing Surface

The drawing surface, whether it is a drawing tabletop or drawing board, must be flat, smooth and large

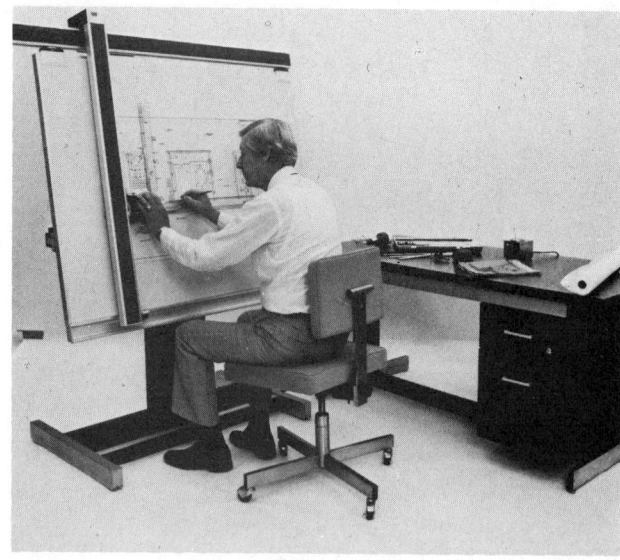

Figure 1-1   Drawing table and reference desk *(Courtesy Keuffel & Esser Co.)*

Figure 1-2   Small drawing table *(Courtesy Stacor Corp.)*

enough to accommodate the drawing and some drafting equipment. If a T-square is used on a drawing board, at least one edge of the board must be absolutely true. Most quality drafting boards have a metal edge to ensure against warpage and against which to hold the T-square securely.

Standard drafting boards range in size from small, 12" x 17" (30 cm x 43 cm), to large, 31" x 42" (78 cm x 105 cm). Standard drafting tabletops range in size from 31" x 42" (78 cm x 105 cm) to 37½" x 60" (94 cm x 150 cm).

Lighting.   It is important that the drawing surface be fully lighted without any shadows, *Figure 1-3*.

Top Cover.   The drafting board should have a top cover which protects the board surface and provides a perfect drawing surface. A good top cover actually seals over holes made by compass points, and it is easily cleaned, *Figure 1-4*.

Figure 1-3 Fluorescent lamp for drawing surface *(Courtesy Waldmann Lighting Co.)*

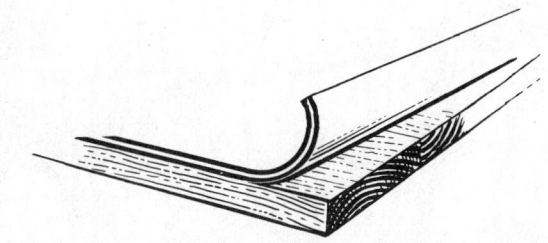

Figure 1-4 Drafting board top cover *(Courtesy Modern School Supplies Inc.)*

Figure 1-5 Equipment organizer *(Courtesy Stacor Corp.)*

Efficiency

To be fully efficient at drafting, all equipment must be clean, correctly adjusted and/or sharpened, and stored in a convenient location, ready for use at all times. It is good drafting practice to store each piece of equipment in a specific location and return it to its location after use. An organizer, such as the one shown in *Figure 1-5*, aids in keeping all equipment in its place.

Left-handed Drafters

Most drafting equipment is designed for the right-handed drafter, although left-handed types of drafting machines can be purchased. The T-square is simply placed on the right side of the drafting board by left-handed drafters, but everything else is right handed. The left-handed drafter has to adapt. The lettering scriber is especially difficult to manipulate.

T-Square

While T-squares are not used today in industry, they do provide a parallel straightedge for the beginning drafter. The T-square is used to draw horizontal lines. Draw lines only against the upper edge of the blade. Make sure the head is held securely against the left edge of the drawing board to guarantee parallel lines, *Figure 1-6*.

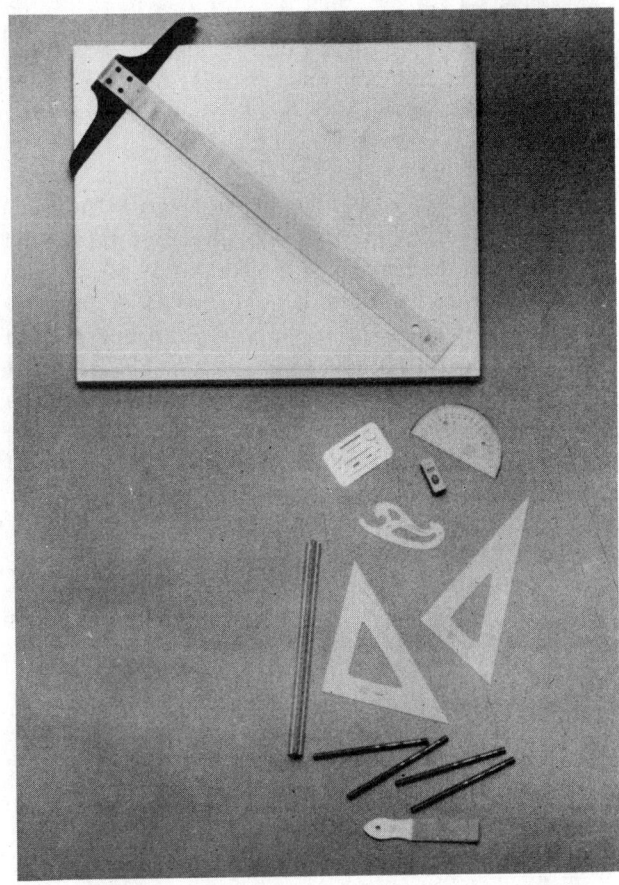

Figure 1-6 T-square in use *(Courtesy Teledyne Post)*

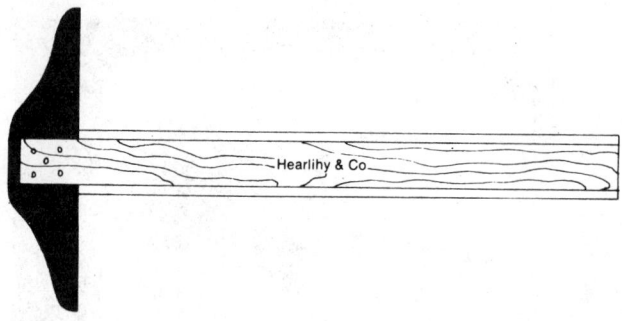

Figure 1-7 Parts of the T-square *(Courtesy Hearlihy and Co.)*

The T-square is composed of two parts: the head and the blade, ***Figure 1-7***. The two parts are fastened together at an exact right angle. The blade must be straight and free of any nicks or imperfections. A transparent acrylic edge is recommended since this allows the drafter to see the drawing underneath the edge. T-squares can be purchased with adjustable heads for drawing specific angles.

Parallel Straightedge

The parallel straightedge is always parallel, regardless of where it is placed upon the drawing surface. Parallel control is accomplished by a system of cords and pulleys, ***Figure 1-8***. The parallel straightedge replaced the T-square in industry, and is still used somewhat today. Most straightedges come with a transparent acrylic edge, and some have rollers for a smooth gliding action. Some have a locking brake that permits the straightedge to be locked in any position.

Drafting Machines

A drafting machine is a device that attaches to the drafting table and replaces both the T-square and the parallel straightedge. There are two basic kinds of drafting machines. One is the arm type, ***Figure 1-9***, and the other, the newer, is the track type, ***Figure 1-10***. On both types, a round head holds two straightedges at right angles to each other. The head can be rotated to set the straightedges at any angle. Most machines are available with interchangeable straightedges marked with different scales along their edges.

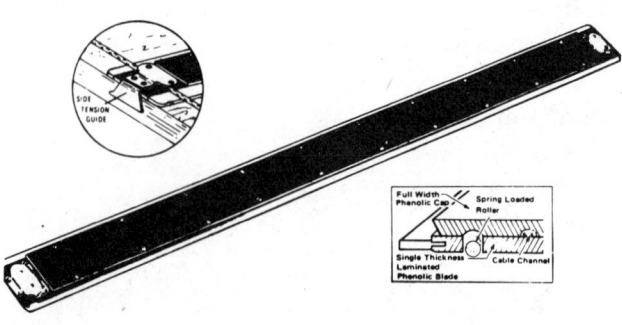

Figure 1-8 Parallel straightedge *(Courtesy Modern School Supplies Inc.)*

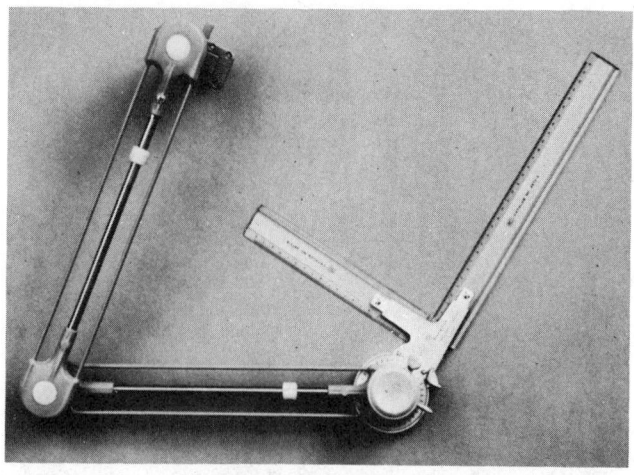

Figure 1-9 Arm-type drafting machine *(Courtesy Teledyne Post)*

1-10 Track-type drafting machine *(Courtesy Vemco Inc.)*

Drafting machines replace straightedges, scales, triangles, and protractors. They increase accuracy and greatly reduce drafting time. A drafting machine is one of the few tools that can be purchased either as a right-handed or a left-handed instrument.

Most drafting machines have a protractor and a vernier, which permit readings to 5 minutes of an arc, ***Figure 1-11***. Notice that zero on the protractor is in line with the zero on the vernier. The vernier is graduated in 5-minute increments from zero to 60 minutes. To read the vernier, first read the protractor, ***Figure 1-12***. In this example, the zero on the vernier points between 18° and 19°. On the vernier, notice that the only line that lines up with a line on the protractor is the 45;

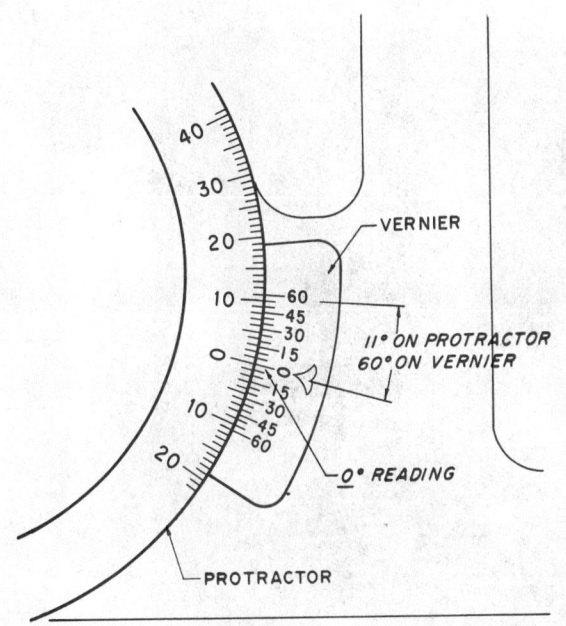

Figure 1-11   Protractor with vernier

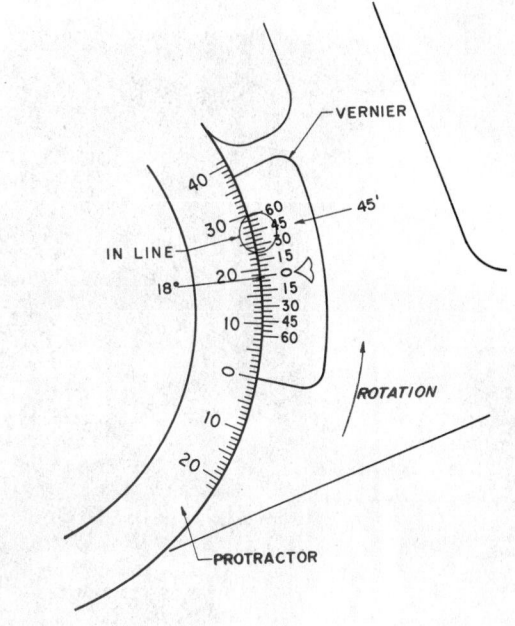

Figure 1-12   Reading the vernier

thus, this is read as 18°45′. Some drafting machine heads simplify this process by adding a digital readout, see ***Figure 1-13***.

Drafting machine straightedges come in sizes of 9″ (23 cm), 12″ (30 cm) and 18″ (45 cm), graduated or ungraduated, in both transparent plastic and aluminum scale, ***Figure 1-14***.

A drafting machine, although a precision instrument, should be checked for accuracy at least once a week. The instructions for checking and adjusting a drafting machine are included with the manufacturer's information.

Drawing Sets

Typical drawing sets include compasses, dividers, and a ruling pen, ***Figure 1-15***. Many sets include a variety of tools not normally used by the drafter. It is recommended that only those tools actually needed be purchased.

Figure 1-13   Drafting machine head with digital readout (*Courtesy Consul & Mutoh, Ltd.*)

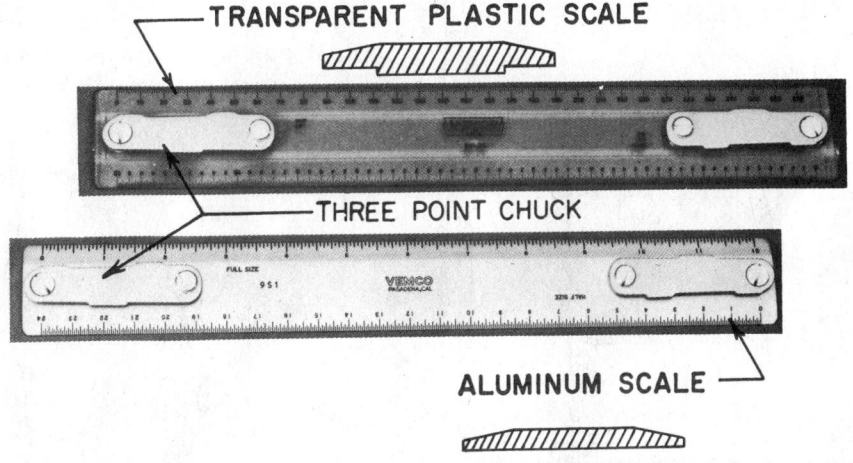

Figure 1-14   Transparent and metal drafting machine straightedges with scale (*Courtesy Vemco Inc.*)

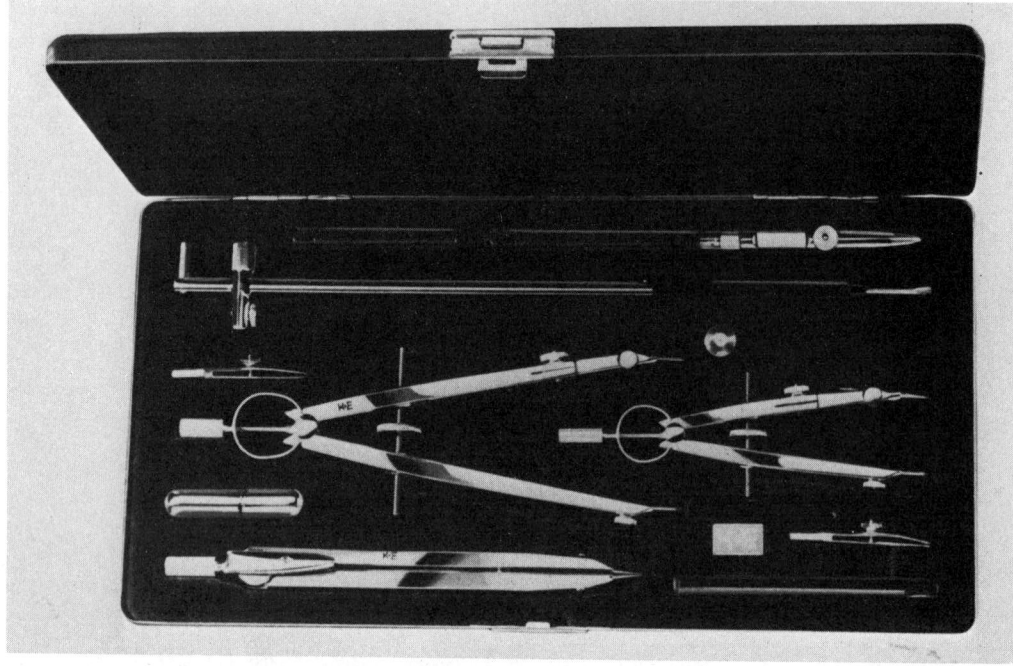

Figure 1-15 Complete drafting set *(Courtesy Keuffel & Esser Co.)*

Compasses

There are two main types of compasses: the friction-joint type, and the spring-bow type. The friction-joint type is still widely used for lightly laying out pencil drawings which will be inked. The disadvantage of this type of compass is that the setting may slip when strong pressure is applied to the lead.

The spring-bow type of compass, *Figure 1-16*, is best for pencil drawings and tracings as it retains its setting even when strong pressure is applied to obtain dark lines. The spring, located at the top of the compass, holds the legs securely against the adjusting screw. The adjusting screw is used to make fine adjustments.

Compass leads should extend approximately ⅜" (0.9 cm). The metal point of the compass extends slightly more than the lead to compensate for the distance the point penetrates the paper, *Figure 1-17*. The lead is sharpened with a sandpaper paddle to produce clean, sharp lines. The flat side of the lead faces *outward* in order to produce circles of very small diameter, *Figure 1-18*. Sharpen with paddle in direction of arrow, as shown, in order to keep the lead sharp longer.

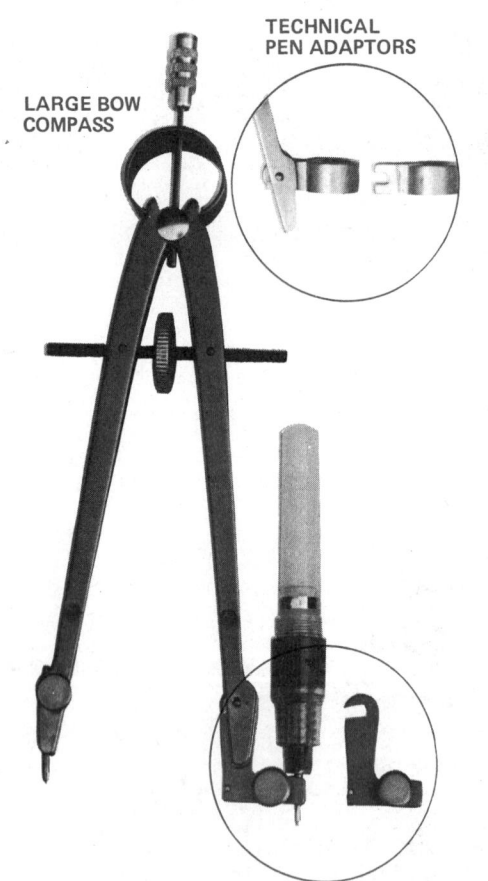

LARGE BOW COMPASS

TECHNICAL PEN ADAPTORS

Figure 1-16 Spring bow compass with pen adaptor
(Courtesy Vemco Inc.)

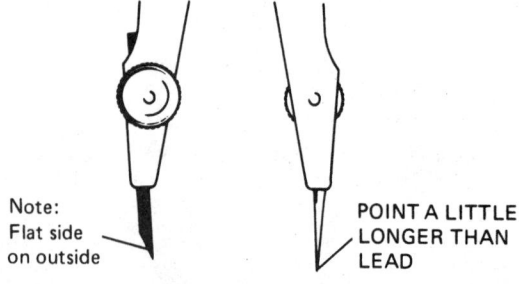

Note:
Flat side on outside

POINT A LITTLE LONGER THAN LEAD

Figure 1-17 Bow compass lead and metal points

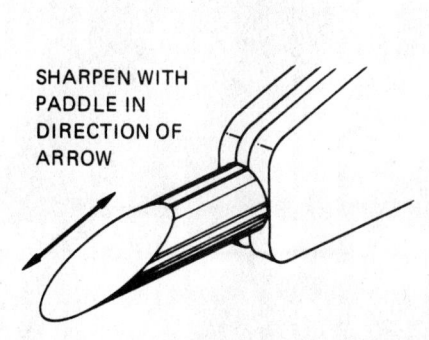

SHARPEN WITH
PADDLE IN
DIRECTION OF
ARROW

Figure 1-18 Bow compass lead point shape
(Courtesy Drafting for Trades and
Industry, Basic Skills, Nelson,
Delmar Publishers Inc.)

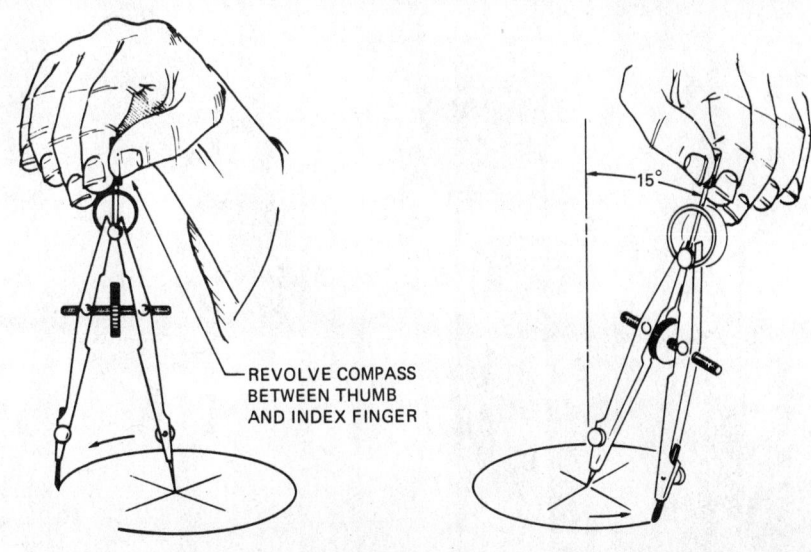

REVOLVE COMPASS
BETWEEN THUMB
AND INDEX FINGER

15°

Figure 1-19 Drawing a circle with a bow compass

To draw a circle with the bow compass, the compass is revolved between the thumb and the index finger, **Figure 1-19**. Pressure is applied downward on the metal point to prevent the compass from jumping out of the center hole.

Drop Bow Compass. The drop bow compass, **Figure 1-20**, is used for circles of .03" (0.08 cm) to .50" (1.3 cm) diameter. The compass is adjusted to the required radius. The center point is located on the circle or arc swing point and held in place with the index finger. Rotate the knurled head of the compass between the thumb and second finger.

Bow Compass with Lead Clutch. In order to eliminate the process of sharpening compass leads, some drafters use a compass with a lead clutch of 0.5-mm lead, **Figure 1-21**. This compass saves time, and ends messy lead sharpening. Special compasses are designed only for inking, **Figures 1-22** and **1-23**. An adaptor to attach to a standard compass to draw ink lines is illustrated in **Figure 1-24**.

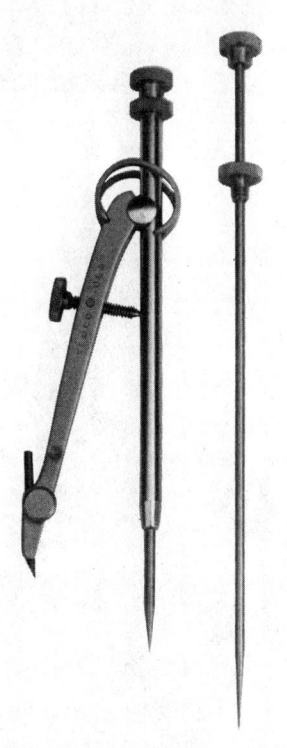

Figure 1-20 Drop bow compass
(Courtesy Vemco Inc.)

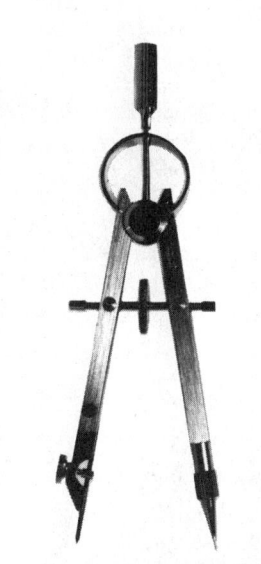

Figure 1-21 Bow compass with special
lead clutch (Courtesy B.

Figure 1-22 Inking bow compass
(Courtesy Koh-I-Noor

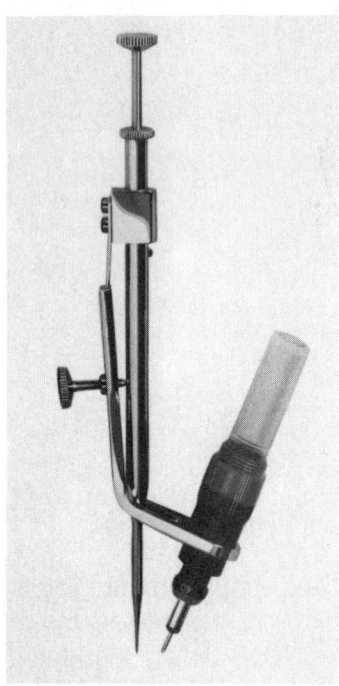

Figure 1-23 Inking drop bow compass
(*Courtesy Koh-I-Noor Rapidograph*)

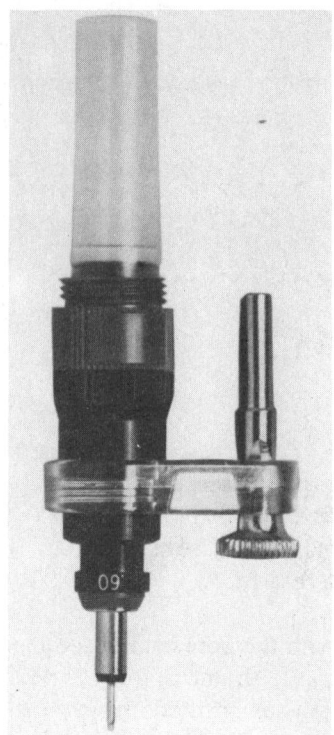

Figure 1-24 Standard compass ink adaptor
(*Courtesy Koh-I-Noor Rapidograph*)

Beam Compass.   A beam compass, *Figure 1-25*, is used to draw large circles or arcs. Fine line adjustments can be obtained and locked in place. Beam compasses come in sizes from 13" (33 cm) bars and upwards.

Adjustable Curve.   An adjustable curve, *Figure 1-26*, has a locking knob, and is used to draw any radius from 6.75" to 200" (17 cm to 500 cm). This tool takes over where the ordinary compass leaves off, and eliminates the beam compass.

Divider

A divider is similar to a compass except that it has a metal point on each leg. It is used to lay off distances and to transfer measurements, *Figure 1-27*.

Proportional Dividers.   Proportional dividers are used to enlarge or reduce an object in scale. This tool has a sliding, adjustable pivot which varies the proportions of the tips of each leg, *Figure 1-28*.

Triangle

Two standard triangles are used by drafters. One is a 30–60-degree triangle, usually written as 30°–60° triangle. The other is a 45-degree triangle and the protractor, written as 45° triangle. The 45° triangle consists of two 45-degree angles, and one 90-degree angle, *Figure 1-29A*. The 30°–60° triangle contains a 30-degree angle, a 60-degree angle, and a 90-degree angle, *Figure 1-29B*.

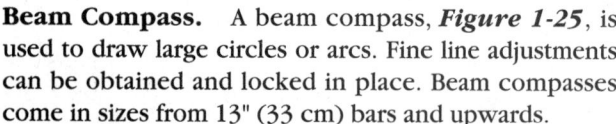

Figure 1-25   Beam compass (*Courtesy Vemco Inc.*)

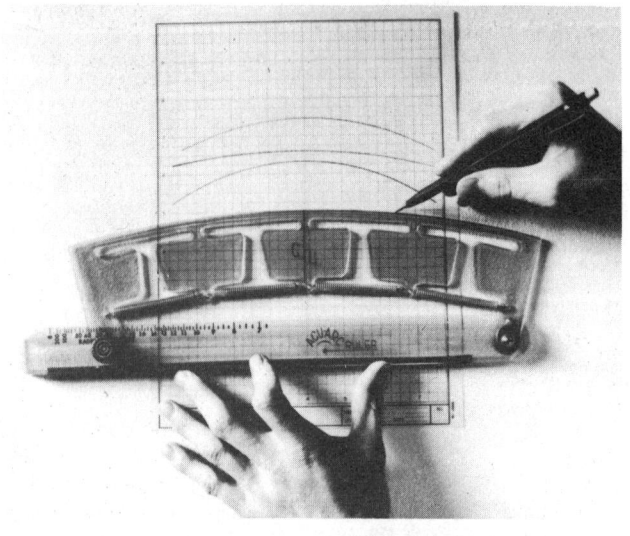

Figure 1-26   Adjustable curve (*Courtesy Hoyle Products Inc.*)

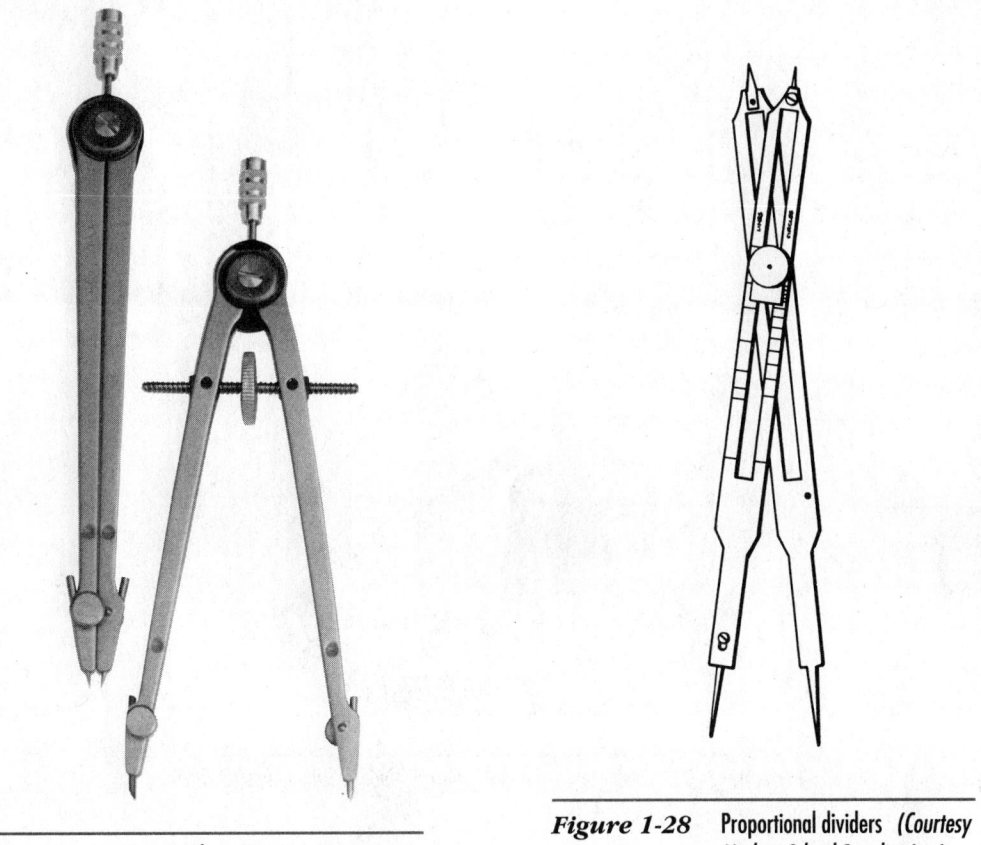

Figure 1-27   Divider (*Courtesy Vemco Inc.*)

Figure 1-28   Proportional dividers (*Courtesy Modern School Supplies Inc.*)

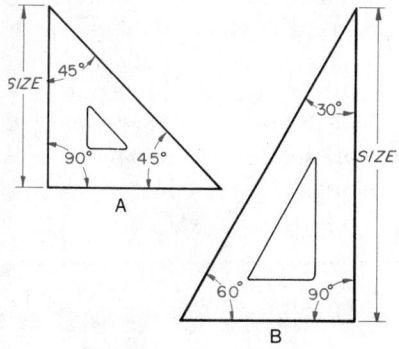

Figure 1-29   (A) 45° angle, and (B) 30°-60° triangle

Triangles are made of plastic and come in a variety of sizes other than those mentioned. When laying out lines, triangles are placed firmly against the upper edge of the straightedge. Pencils are placed against the left edge of the triangle and lines drawn upwards, away from the straightedge. Parallel angular lines are made by moving the triangle to the right after each new line has been drawn, ***Figure 1-30***.

Adjustable Triangle.   An adjustable triangle may take the place of both the 30°-60° and 45° triangles, ***Figure 1-31***. It is recommended, however, that this tool be used only for drawing angles that cannot be made with the two standard triangles. The adjustable triangle is set by eye and thus is not as accurate as the solid triangle.

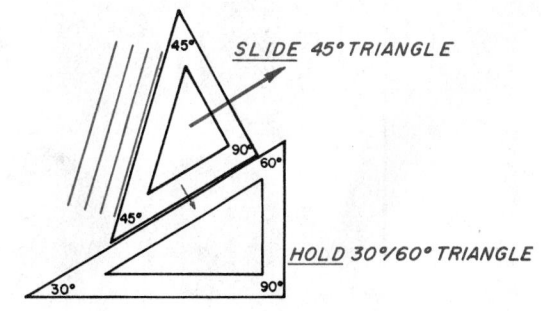

PARALLEL LINES

Figure 1-30   Drawing parallel angular lines

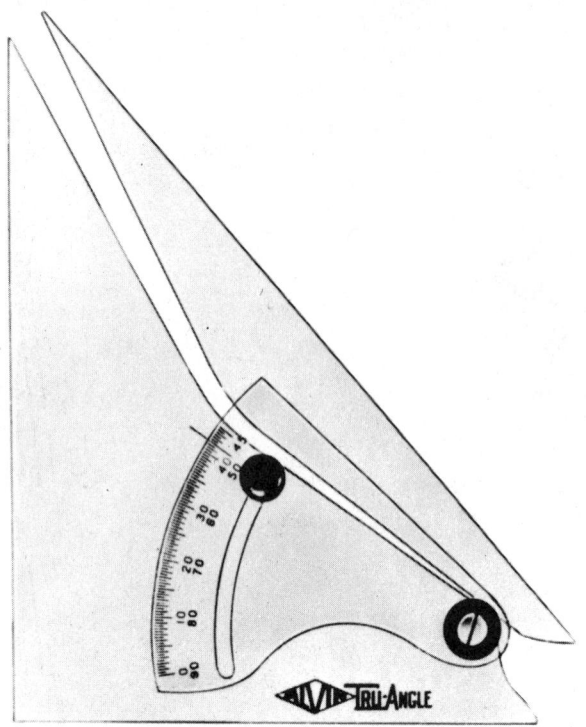

Figure 1-31   Adjustable triangle *(Courtesy Modern School Supplies Inc.)*

Template

A template is a thin, flat piece of plastic containing various cutout shapes, ***Figures 1-32***, ***1-33***, and ***1-34***. It is designed to speed the work of the drafter and to make the finished drawing more accurate. Templates are available for drawing circles, ellipses, plumbing fixtures, bolts and nuts, screw threads, electronic symbols, springs, gears, and structural metals, to name just a few uses.

Templates come in many sizes to fit the scale being used on the drawing. A template should be used wherever possible to increase accuracy and speed. It is preferable to purchase templates that are stamped and not molded, as molded templates become brittle in time and break.

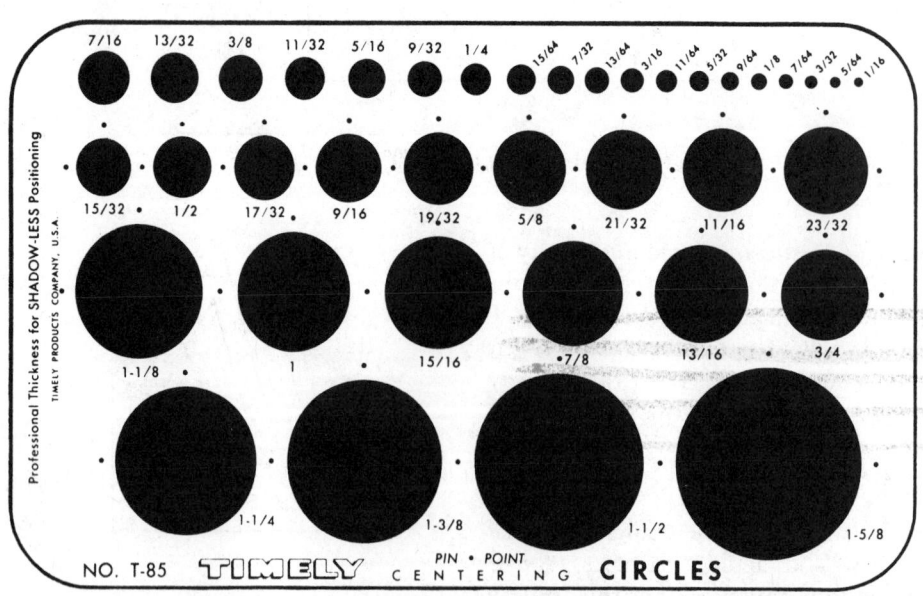

Figure 1-32   Circle template *(Courtesy Timely Products Co.)*

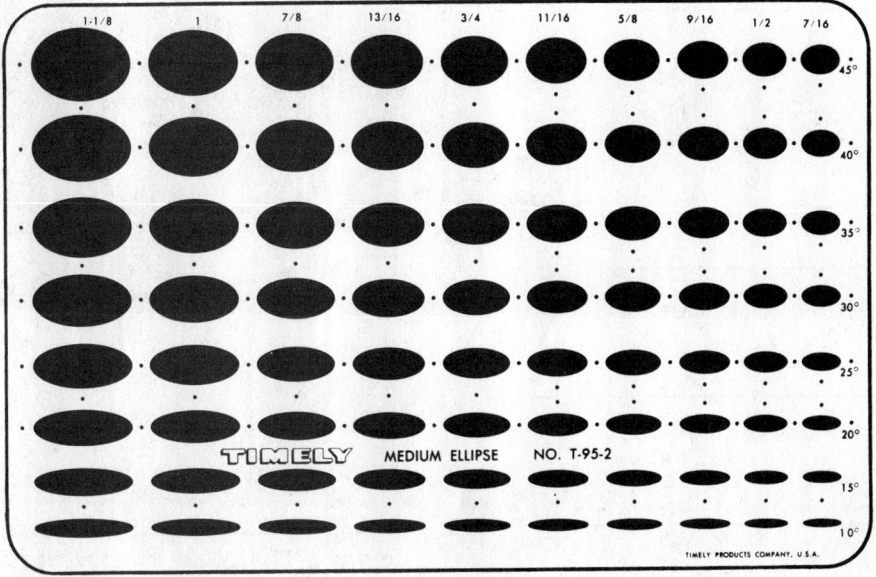

Figure 1-33 Ellipse template *(Courtesy Timely Products Co.)*

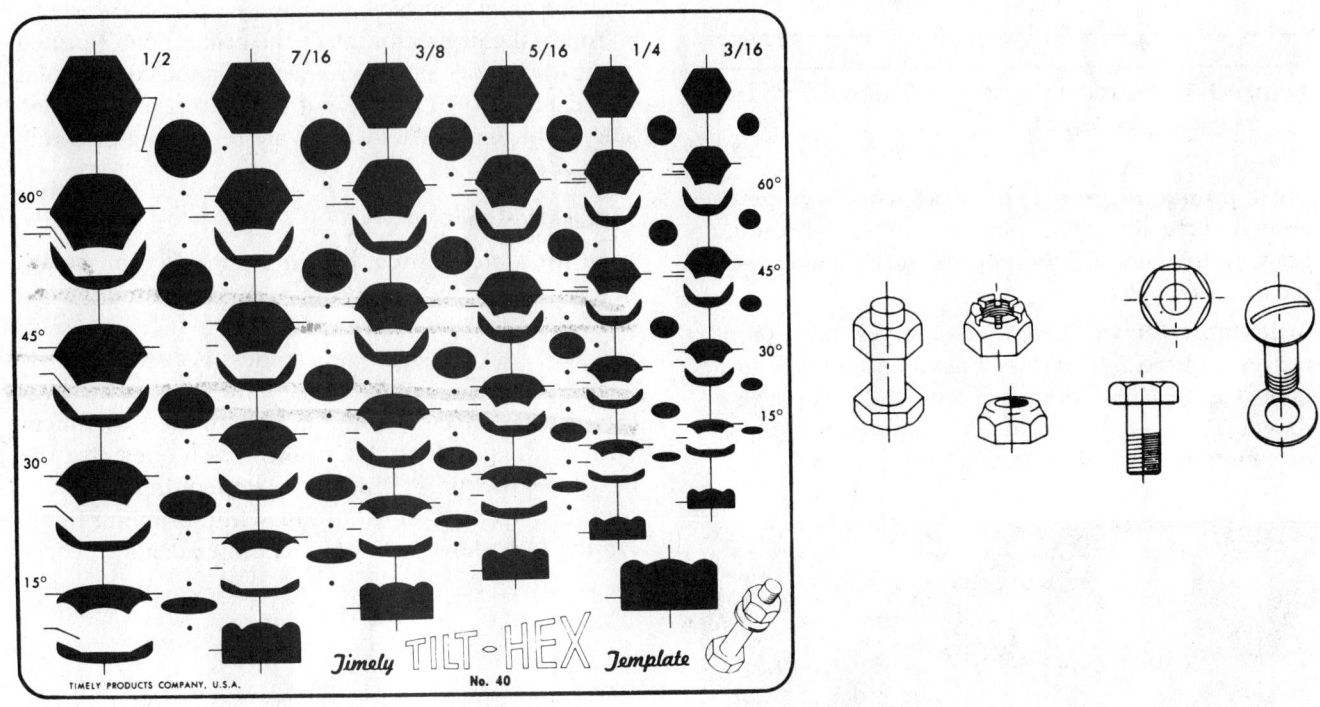

Figure 1-34 Bolt and nut template *(Courtesy Timely Products Co.)*

French Curves

French curves are thin, plastic tools that come in an assortment of curved surfaces, *Figure 1-35*. They are used to produce curved lines that cannot be made with a compass. Such lines are referred to as *irregular curves*. Most good French curves are actually segments of such geometric curves as ellipses, parabolas, hyperbolas, and the like.

Using a French Curve. To use a French curve, the irregular curve must be defined by a series of dots. *Lightly* connect straight lines to get a general idea of where the curved line is going. If the line makes an abrupt turn, a line *lightly* sketched in place of the straight lines may be more useful. Starting from one side or the other, line up the French curve along as many points as possible and draw a dark line connecting

Figure 1-35 Assortment of French curves *(Courtesy Modern School Supplies Inc.)*

these points, *Figure 1-36*. Readjust and align the French curve along all additional points, and continue drawing the curved line. Proceed in this manner until the line is completed.

Adjustable Curve. Adjustable curves form smooth curves. *Figure 1-37* shows a flexible steel measuring tape that measures the perimeter of the curve to be drawn. The curve is held by friction between many layers of interlocking channels.

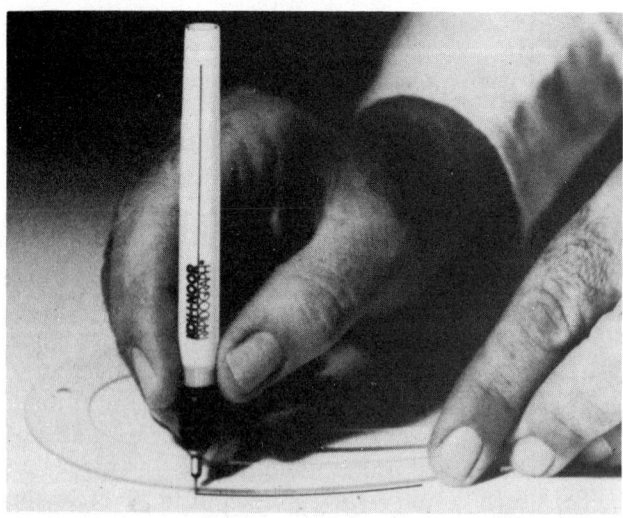

Figure 1-36 Drawing an irregular curve *(Courtesy Koh-I-Noor Rapidograph)*

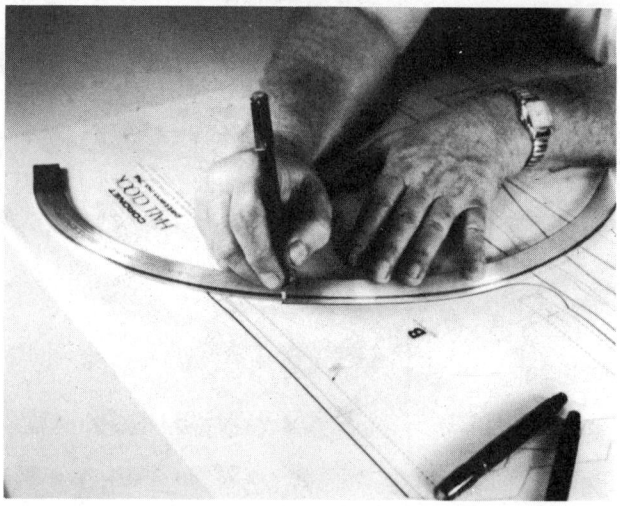

Figure 1-37 Using an adjustable curve *(Courtesy Hoyle Products Inc.)*

Protractor

A protractor is used to measure and lay out angles, *Figure 1-38*. It can be used in place of a drafting machine or an adjustable triangle.

To use the protractor, place the center point (located at the lower edge of the protractor) on the corner point of the angle. Align the base of the protractor along one side of the angle. The degrees are read along the semi-circular edge.

Pencils and Leads

Lead for a mechanical drafting pencil comes in 18 degrees of hardness, ranging from 9H, which is very hard, to 7B, which is very soft, *Figure 1-39*. For drafting purposes, the scale of hardness is as follows: 4H lead is recommended for layout work, and 2H, H, or HB leads are recommended for all other lines. Experiment with various leads to determine which ones give the best line thickness. This varies depending upon the pressure applied to the point while drawing lines. *Figure 1-40* shows lead for a mechanical drafting pencil.

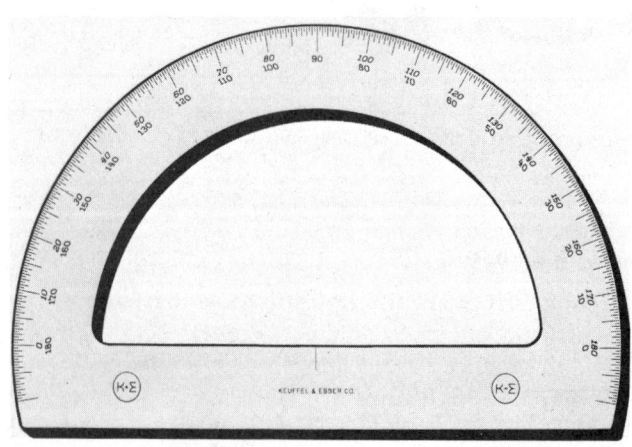

Figure 1-38 Protractor *(Courtesy Keuffel & Esser Co.)*

TASK	LEAD
CONSTRUCTION LINES	3H, 2H
GUIDE LINES	3H, 2H
LETTERING	H, F, HB
DIMENSION LINES	2H, H
LEADERLINES	2H, H
HIDDEN LINES	2H, H
CROSSHATCHING LINES	2H, H
CENTERLINES	2H, H
PHANTOM LINES	2H, H
STITCH LINES	2H, H
LONG BREAK LINES	2H, H
VISIBLE LINES	H, F, HB
CUTTING PLANE LINES	H, F, HB
EXTENSION LINES	2H, H
FREEHAND BREAK LINES	H, F, HB

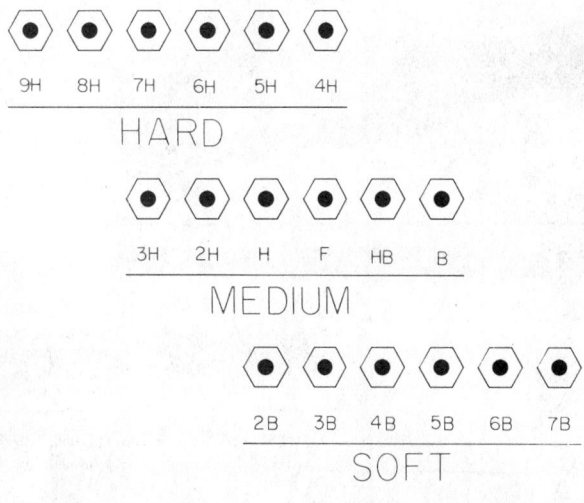

HARD — 9H 8H 7H 6H 5H 4H

MEDIUM — 3H 2H H F HB B

SOFT — 2B 3B 4B 5B 6B 7B

Figure 1-39   Grades of lead (left) and lead-lines chart (right)

Figure 1-40   Drafting lead *(Courtesy Staedtler Mars)*

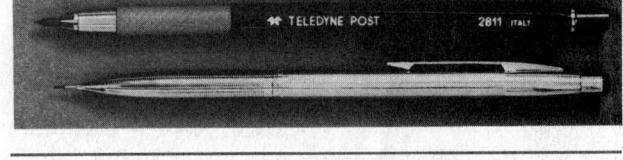

Figure 1-41   Lead holder *(Courtesy Teledyne Post)*

Regular pencils are sharpened with a pencil sharpener. It is important that enough wood is removed to ensure that the lead, not the wood, of the pencil comes in contact with the straightedge or triangle edges.

Lead Holders and Leads

Lead holders hold sticks of lead, ***Figure 1-41***. The leads designed for lead holders come in the same range of hardness as those for regular mechanical pencils, and are used for the same purposes. The main advantage is that they are more convenient to use. Leads are usually sharpened in a lead pointer. Electric lead pointers are fully automatic. A slight downward pressure of the lead starts the motor action, ***Figure 1-42***. This machine produces a perfectly tapered point, and eliminates all loose clinging graphite.

Sandpaper Paddle.   A sandpaper paddle consists of several layers of sandpaper attached to a small wooden holder, ***Figure 1-43***. The sandpaper is used to sharpen compass leads only. Do not sharpen leads over a drawing as the graphite will smear the drawing surface.

Erasers

Various kinds of erasers are available to a drafter. One of the most commonly used is a soft, white block-type eraser, ***Figure 1-44***. ***Figure 1-45*** shows a pencil-type eraser with an adjustable clutch. By developing good drawing habits, erasing can be kept to a minimum.

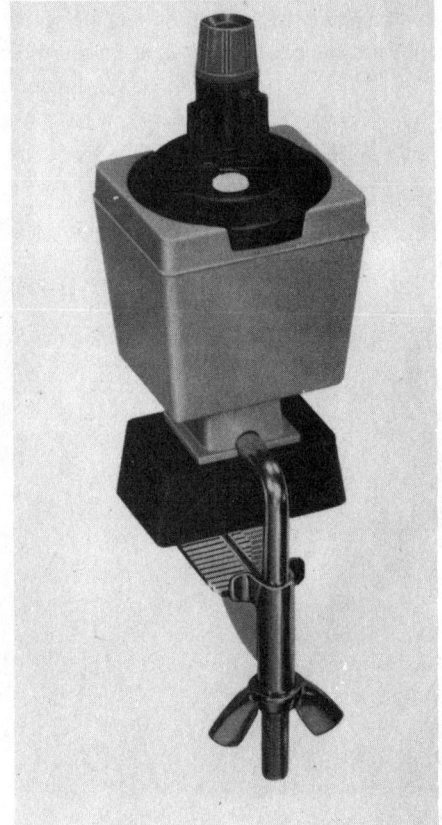

Figure 1-42   Lead pointer *(Courtesy Koh-I-Noor Rapidograph Inc.)*

Figure 1-43 Sandpaper paddle *(Courtesy Keuffel & Esser Co.)*

Figure 1-44 Block-type eraser *(Courtesy Staedtler Mars)*

Figure 1-45 Pencil-type eraser with clutch *(Courtesy Staedtler Mars)*

Electric Eraser.   An electric eraser speeds up corrections. Some models take a 7" (17.5 cm) long eraser strip. The model illustrated in *Figure 1-46* has a slip clutch to hold the eraser strip in place.

A cordless erasing machine can be used with or without the standard electric cord, and uses rechargeable NiCad batteries, *Figure 1-47*. As the eraser is placed in the stand, the batteries are recharged.

Electric erasers do save time, but care must be taken not to burn through the drawing paper. This can be avoided by using an erasing shield and placing a thick sheet of paper beneath the drawing to cushion it.

Erasing Shield.   An erasing shield restricts the erasing area so that correctly drawn lines will not be disturbed during the erasing procedure. It is made from a thin, flat piece of metal with variously sized cutouts, *Figure 1-48*. The shield is used by placing it over the line to be erased and erasing through the cutout.

Drafting Brush

The drafting brush is used to remove loose graphite and eraser crumbs from the drawing surface, *Figure 1-49*. Do not brush off a drawing surface by hand as this tends to smudge the drawing. Drafting brushes come in various sizes from 10½" to 14" (26 cm to 35 cm). The bristles can be either horsehair or nylon, and they can be cleaned with warm, soapy water.

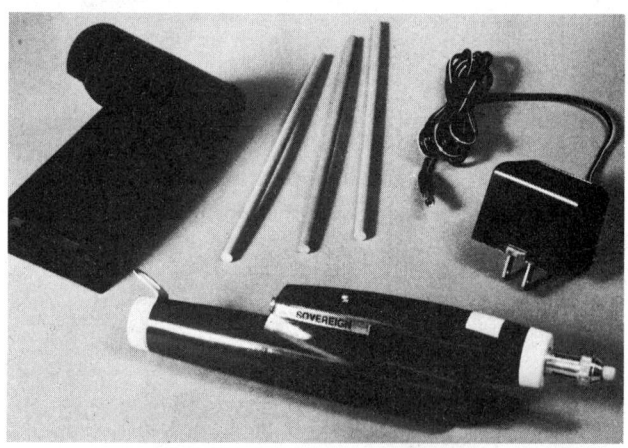

Figure 1-47 Rechargeable electric eraser *(Courtesy Rotex Co.)*

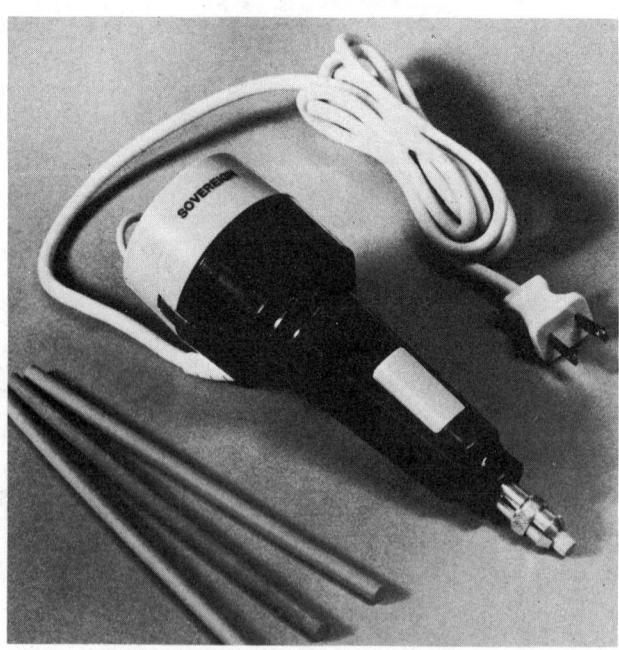

Figure 1-46 Electric eraser with slip clutch *(Courtesy Rotex Co.)*

Figure 1-48 Erasing shield *(Courtesy Staedtler Mars)*

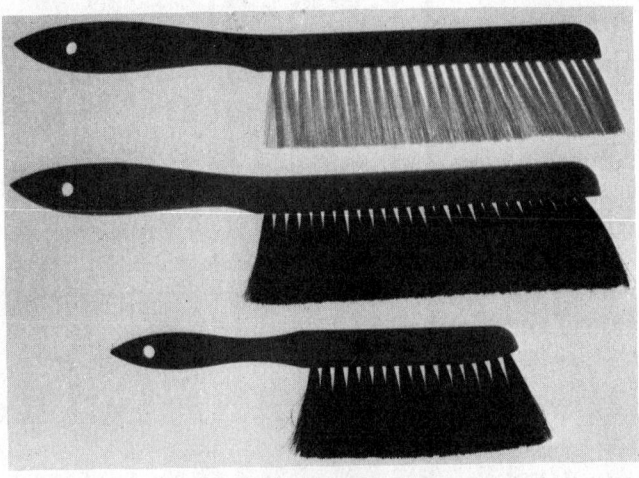

Figure 1-49 Drafting brush *(Courtesy Herlihy & Co.)*

Dry Cleaning Powder

Cleaning powder is used to help keep drawings clean, to avoid smearing, and to speed up the drafting process. Cleaning powder comes in a can or as pads, *Figure 1-50*, and is sprinkled over the drawing before starting. It is imperative that all cleaning powder is removed before placing the original drawing into a whiteprinter as the powder tends to stick to the roller. If good drafting habits are followed, the dry cleaning powder is not necessary.

Scales

Various kinds of scales are used by drafters, *Figure 1-51*. A number of different scales are included on each instrument. Scales save the drafter the work of computing new measurements every time a drawing is made larger or smaller than the original.

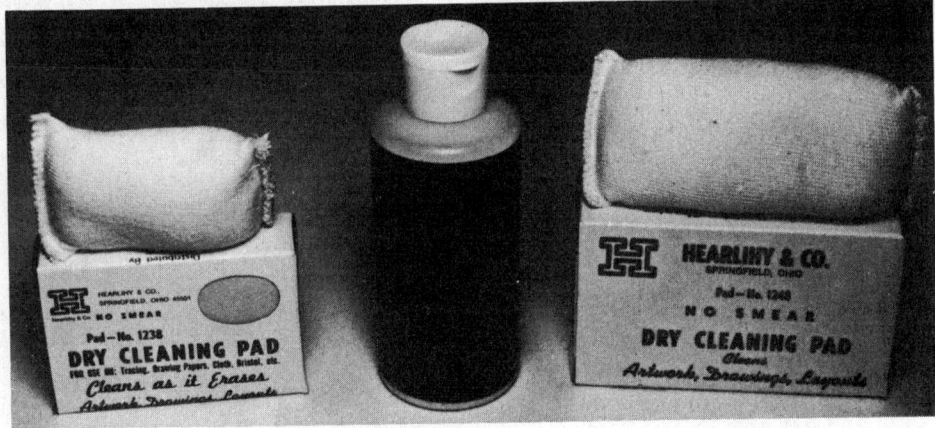

Figure 1-50 Dry cleaning powder *(Courtesy Herlihy & Co.)*

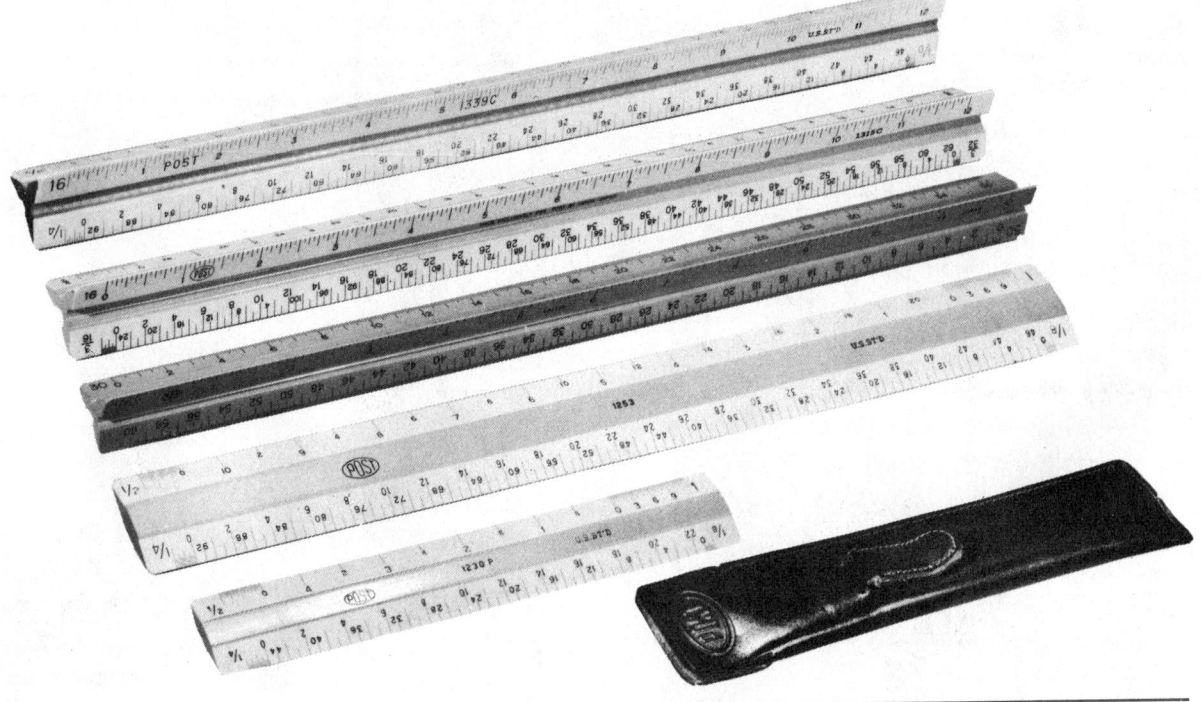

Figure 1-51 A variety of drafting scales *(Courtesy Teledyne Post)*

Scales come open divided and full divided. A *full-divided scale* is one in which the units of measurement are subdivided throughout the length of the scale. An *open-divided scale* has its first unit of measurement subdivided, but the remaining units are open or free from subdivision.

Mechanical Engineer's Scale

Mechanical engineer's scales are divided into inches and parts to the inch. To lay out a full-size measurement, use the scale marked 16. This scale has each inch divided into 16 equal parts or divisions of 1/16 inch. It is used by placing the 0 on the point where measurement begins, and stepping off the desired length, *Figure 1-52*.

To reduce a drawing 50 percent, use the scale marked 1/2. The large 0 at the end of the first subdivided measurement lines up with the other unit measurements that are part of the same scale. The large numbers crossed out in *Figure 1-53* go with the 1/4 scale starting at the other end. These numbers are ignored while using the 1/2 scale. To lay out 1¾ inches at the 1/2 scale, read full inches to the right of 0 and fractions to the left of 0.

The 1/4 scale is used in the same manner as the 1/2 scale. However, measurements of full inches are made to the left of 0 and fractions to the right, because the 1/4 scale is located at the opposite end of the 1/2 scale, *Figure 1-54*.

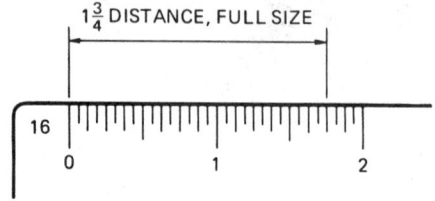

Figure 1-52 Regular full-size scale

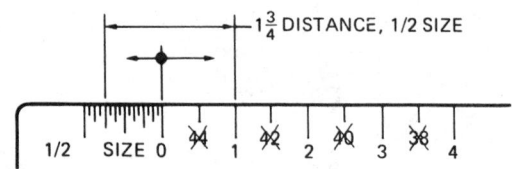

Figure 1-53 Half-size scale

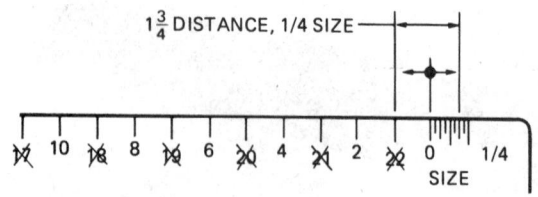

Figure 1-54 Quarter-size scale

Architect's Scale

The architect's scale is used primarily for drawing large buildings and structures. The full-size scale is used frequently for drawing smaller objects. Because of this, the architect's scale is generally used for all types of measurements. It is designed to measure in feet, inches, and fractions of an inch. Measure full feet to the right of 0; inches and fractions of an inch to the left of 0. The numbers crossed out in *Figure 1-55* correspond to the 1/2 scale. They can be used, however, as 6 inches as each falls halfway between full-foot divisions. Measurements from 0 are made in the opposite direction of the full scale, because the 1/2 scale is located at the opposite end of the scale, *Figure 1-56*.

With the 10 overlapping scales a drawing may be made in various sizes from full size to 1/28 size. Since the architect's scale is showing divisions reciprocating feet and inches, a scale marked $1''=1'0''$, makes a drawing with this scale 1/12 size.

Examples

¾"=1'0"	would be 1/16 size	(3/4 x 1/12)
¼"=1'0"	would be 1/48 size	(1/4 x 1/12)
½"=1'0"	would be 1/24 size	(1/2 x 1/12)
1"=1'0"	would be 1/12 size	(1 x 1/12)

Full size is measured on the full-size side of the architect's scale where each division is 1/16.

Half size is measured using the full-size side and dividing each measurement by 2.

Quarter size is measured by using the $3'' = 1'0''$ scale. $3'' = 1'0''$ would be 1/4 size (3 x 1/12 = 3/12 =1/4)

Scale does not = size

incorrect	1/4 scale ≠ 1/4 size
correct	1/4 scale = 1/48 size

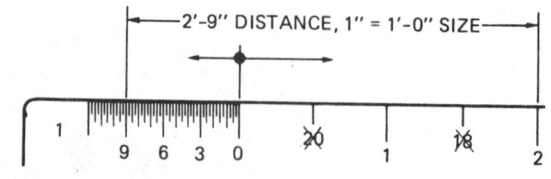

Figure 1-55 Architectural scale (full size)

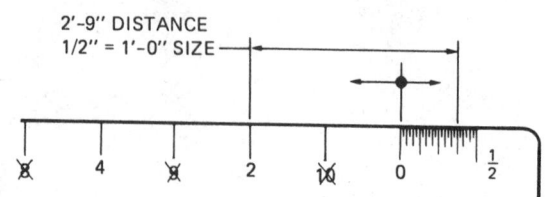

Figure 1-56 Architectural scale (half size)

Civil Engineer's Scale

A civil engineer's scale is also called a *decimal-inch scale*. The number 10, located in the corner of the scale in *Figure 1-57*, indicates that each graduation is equal to 1/10 of an inch or .1". Measurements are read directly from the scale. The number 20, located in the corner of the scale shown in *Figure 1-58*, indicates that it is 1/20 of an inch.

Using the same scale for civil drafting, one inch equals two hundred feet, *Figure 1-59*, and one inch equals one hundred feet, *Figure 1-60.*

A metric scale is used if the millimetre is the unit of linear measurement. It is read the same as the decimal-inch scale except that it is in millimetres, *Figure 1-61*.

Pocket Steel Ruler

The drafter should make use of a pocket steel ruler. The pocket steel ruler is the easiest of all measuring tools to use. The inch scale, *Figure 1-62*, is six inches long, and is graduated in 10ths and 100ths of an inch on one side and 32nds and 64ths on the other side.

The metric scale is 150 millimetres long (approximately six inches) and is graduated in millimetres and half millimetres on one side, *Figure 1-63*. Sometimes metric pocket steel rulers are graduated in 64ths of an inch on the other side.

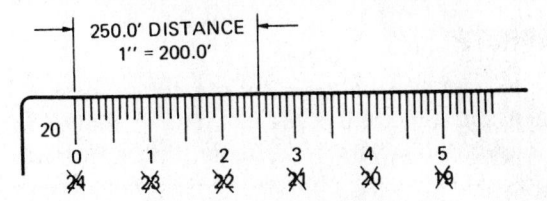

Figure 1-58   Civil engineer's scale (half scale)

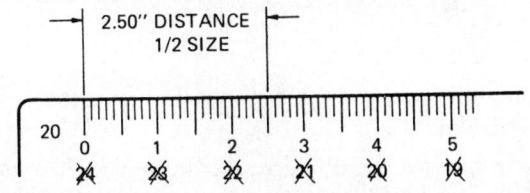

Figure 1-59   Mechanical scale (half size)

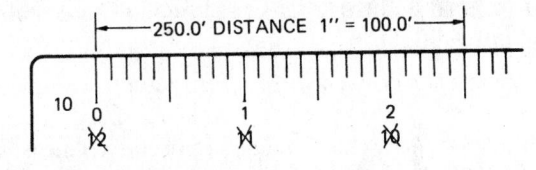

Figure 1-60   Civil scale

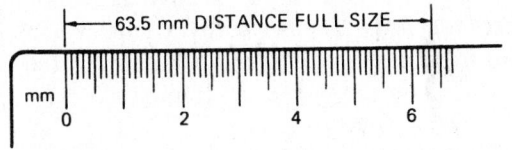

Figure 1-61   Metric scale

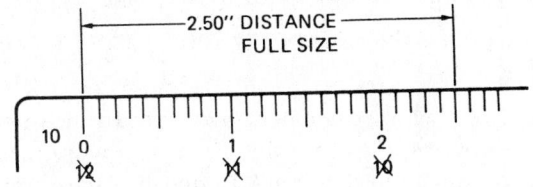

Figure 1-57   Civil engineer's scale

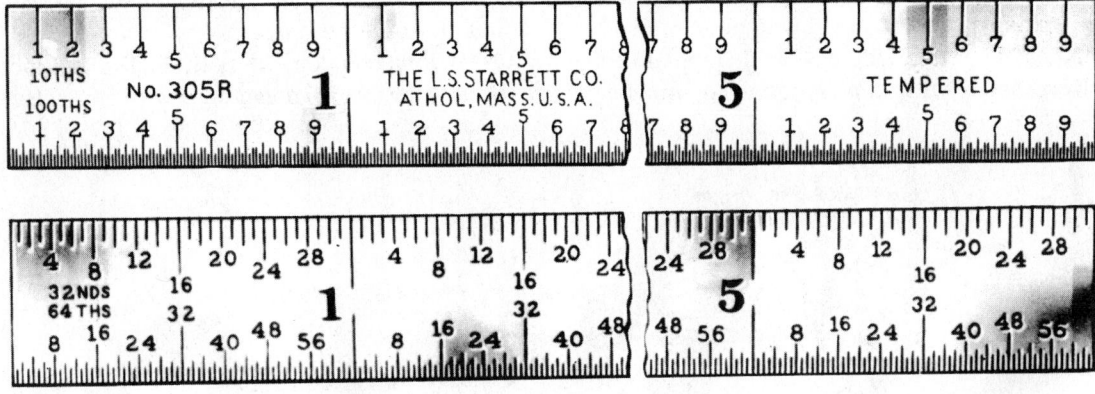

Figure 1-62   Steel scale (inch) *(Courtesy L. S. Starrett Co.)*

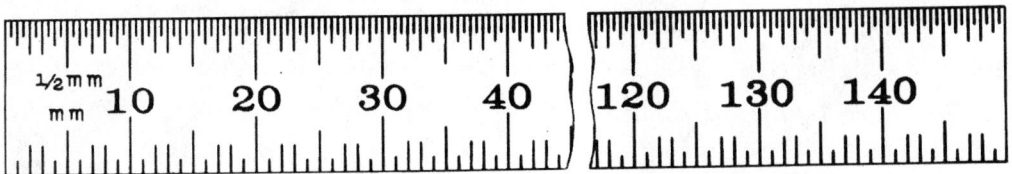

Figure 1-63   Steel scale (metric) *(Courtesy L. S. Starrett Co.)*

Measuring

The metric system uses the metre (m) as its basic dimension. A metre is 3.281 feet long or about 3⅜ inches longer than a yardstick. Its multiples, or parts, are expressed by adding prefixes. These prefixes represent equal steps of 1000 parts. The prefix for a thousand (1000) is *kilo*; the prefix for a thousandth (1/1000) is *milli*. One thousand metres (1000 m), therefore, equals one kilometre (1.0 km). One thousandth of a metre (1/1000 m) equals one millimetre (1.0 mm). Comparing metric to English then:

One millimetre (1.0 mm) = 0.001 metre (0.01 m) = .03937 inch

One thousand millimetres (1000 mm) = 1.0 metre (1.0 m) = 3.281 feet

One thousand metres (1000 m) = 1.0 kilometre (1.0 km) = 3281.0 feet

How To Read a Micrometer Graduated in Thousandths of an Inch (.001")

A micrometer consists of a highly accurate ground screw or spindle which is rotated in a fixed nut, thus opening or closing the distance between two measuring faces on the ends of the anvil and spindle, *Figure 1-64*. A piece of work is measured by placing it between the anvil and spindle faces, and rotating the spindle by means of the thimble until the anvil and spindle both contact the work. The desired work dimension is then found from the micrometer reading indicated by the graduations on the sleeve and thimble, as described in the following paragraphs.

Since the pitch of the screw thread on the spindle is 1/40" or 40 threads per inch in micrometers that are graduated to measure in inches, one complete revolution of the thimble advances the spindle face toward or away from the anvil face precisely 1/40 or .025 inch.

The reading line on the sleeve is divided into 40 equal parts by vertical lines that correspond to the number of threads on the spindle. Therefore, each vertical line designates 1/40 or .025 inch and every fourth line, which is longer than the others, designates hundreds of thousandths. For example: the line marked "1" represents .100", the line marked "2" represents .200", and the line marked "3" represents .300", and so forth.

The beveled edge of the thimble is divided into 25 equal parts, with each line representing .001" and every line numbered consecutively. Rotating the thimble from one of these lines to the next moves the spindle longitudinally 1/25 of .025" or .001 inch; rotating two divisions represents .002", and so forth. Twenty-five divisions indicate a complete revolution: .025 or 1/40 of an inch.

To read the micrometer in thousandths, multiply the number of vertical divisions visible on the sleeve by .025", and to this add the number of thousandths indicated by the line on the thimble which coincides with the reading line on the sleeve.

Example (See *Figure 1-65*):

The 1" line on the sleeve is visible, representing .100"

Three additional lines are visible, each representing .025". Thus, 3 x .025" = .075"

The third line on the thimble coincides with the reading line on the sleeve, each line representing .001". Thus, 3 x .001" = .003"

The micrometer reading is 100" + .075" + .003" = .178"

An easy way to remember how to read a micrometer is to think of the various units as if you were making change from a ten dollar bill. Count the figures on the sleeve as dollars, the vertical lines on the sleeve as quarters, and the divisions on the thimble as cents. Add up your change and put a decimal point instead of a dollar sign in front of the figures.

Micrometers come in both English and metric graduations. They are manufactured with an English size range

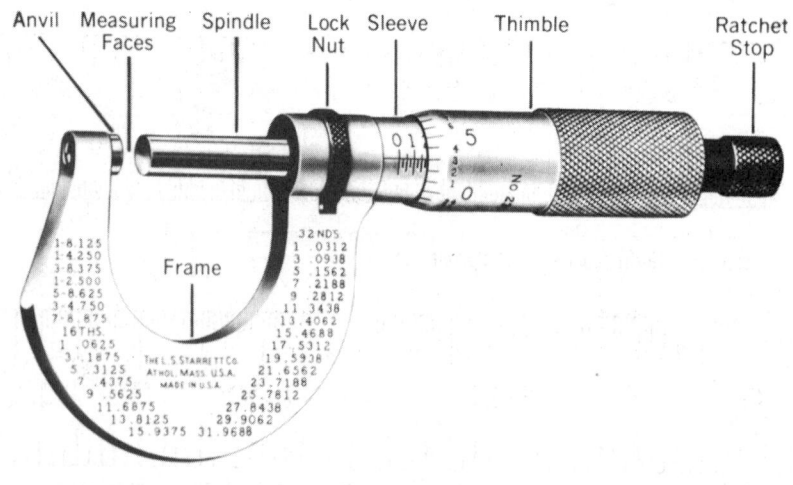

Figure 1-64   Micrometer *(Courtesy L. S. Starrett Co.)*

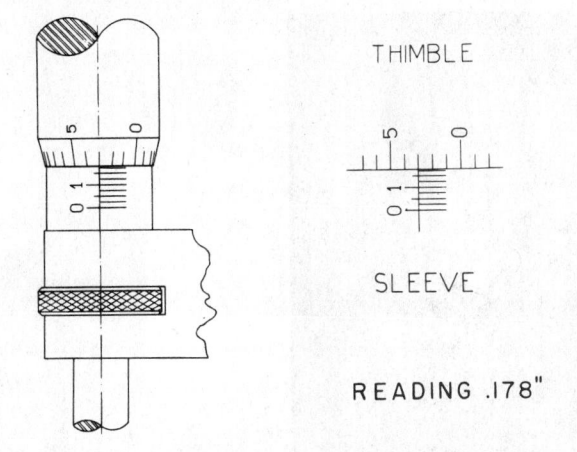

THIMBLE

SLEEVE

READING .178"

Figure 1-65   Reading a micrometer *(Courtesy L. S. Starrett Co.)*

of 1 inch through 60 inches, and a metric size range of 25 millimetres to 1500 millimetres. The micrometer is a very sensitive device and must be treated with extreme care.

Vernier Caliper

Vernier calipers have the ability of measuring both the outside and the inside measurements of an object, *Figures 1-66* and *1-67*. Use the bottom scale when measuring an outside size. Use the top scale when measuring an inside size.

Microfinish Comparator

The microfinish comparator is a handy tool for the drafter to approximate surface irregularities. Various kinds of microfinish comparators are available. *Figure 1-68* illustrates a comparator for cast surfaces.

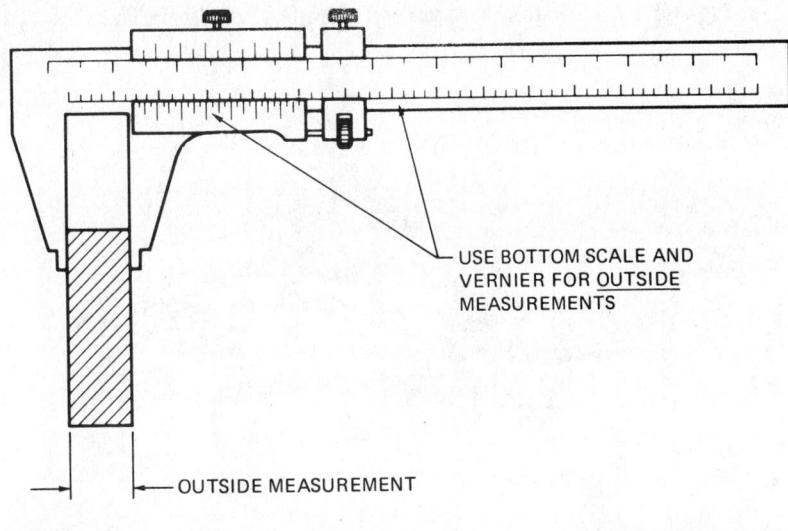

USE BOTTOM SCALE AND VERNIER FOR <u>OUTSIDE</u> MEASUREMENTS

OUTSIDE MEASUREMENT

Figure 1-66   Caliper—outside measurement

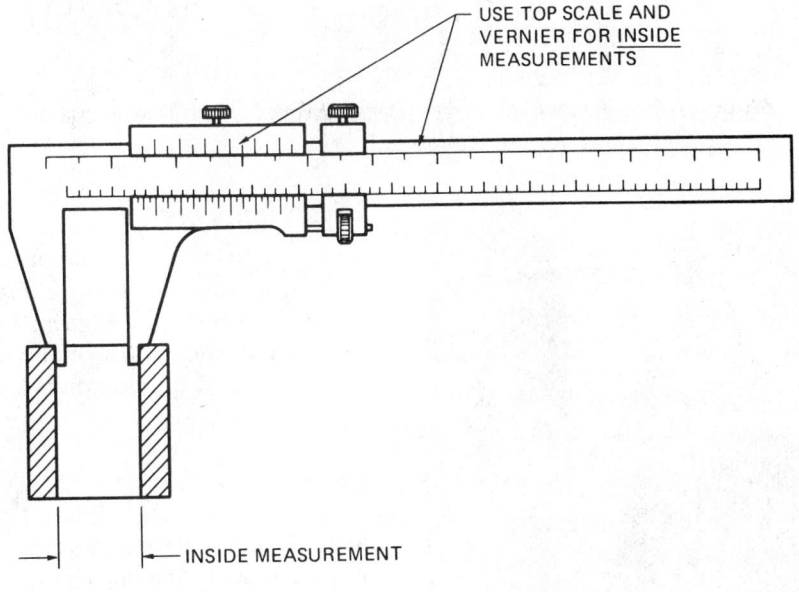

USE TOP SCALE AND VERNIER FOR <u>INSIDE</u> MEASUREMENTS

INSIDE MEASUREMENT

Figure 1-67   Caliper—inside measurement

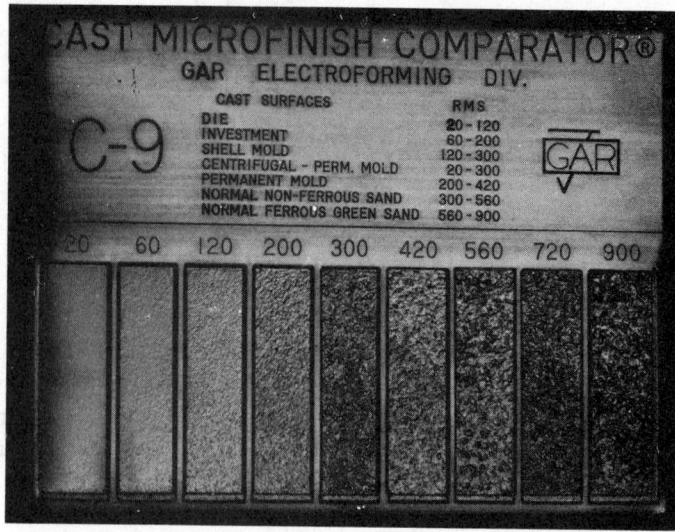

Figure 1-68 Microfinish comparator *(Courtesy GAR Electroforming Div., Mite Corp.)*

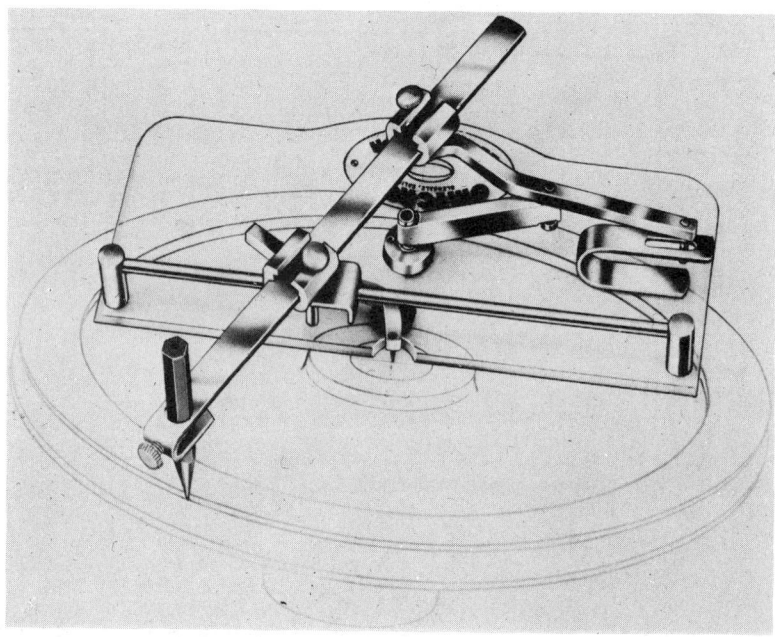

Figure 1-69 (A–above) Ellipsograph *(Courtesy Omicron Co.)* and (B–below) Oval compass *(Courtesy Oval Compass)*

Ellipses Instrument

Two unique instruments are used to draw large ellipses. An ellipsograph is shown in **Figure 1-69A**. The Oval Compass is shown in **Figure 1-69B**. With these tools, the height and width of the ellipse are measured, locked-in, and quickly drawn. A template is used to draw small ellipses.

Ink Tools

Some fields of drafting, such as civil (map) drafting, require that all drawings be done in ink. Some companies ink their drawings so that they can be reduced and filed on film. All artwork that is to be reproduced by

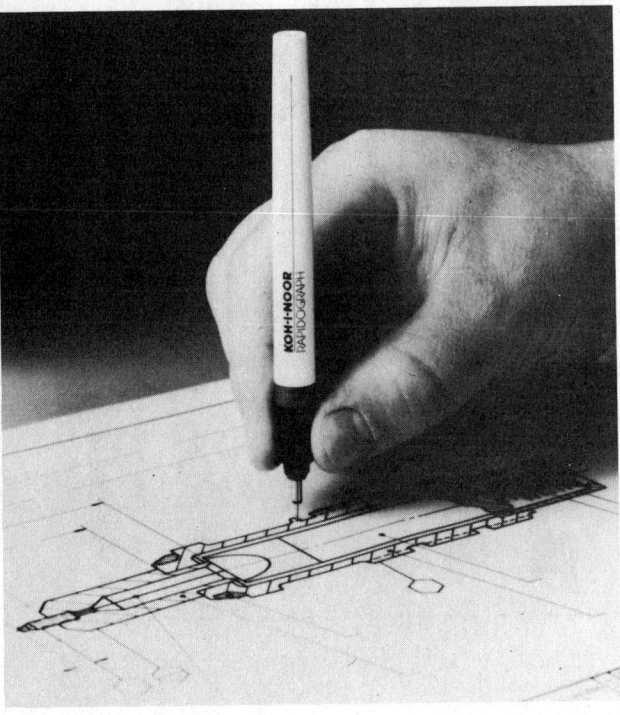

Figure 1-70   Technical inking pen *(Courtesy Koh-I-Noor Rapidograph)*

camera, such as in the field of technical illustration, is done in ink. Ink drawing is no more difficult than pencil drawing, *Figure 1-70*.

Technical Pens

The key to successful inking is a good technical pen, *Figure 1-71*. Technical pens are produced in two styles. Notice the ends of the two pens in the figure; one has a tapered end, the other a straight end. The tapered pen is used primarily for artwork; the straight end is used for drafting and mechanical lettering. Pens are available in various sizes and styles of pen-holder sets, *Figures 1-72* and *1-73*.

Technical pen points are manufactured of stainless steel, tungsten or jewels. The stainless steel point is chromium plated for use on tracing paper or vellum. Tungsten points are long wearing for use on abrasive, coated plotting film or triacetate. Jewel points are used on a plotter that has a controlled pen force.

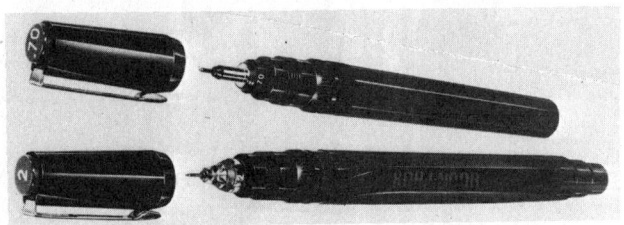

Figure 1-71   Drafting and art technical pens *(Courtesy Koh-I-Noor Rapidograph)*

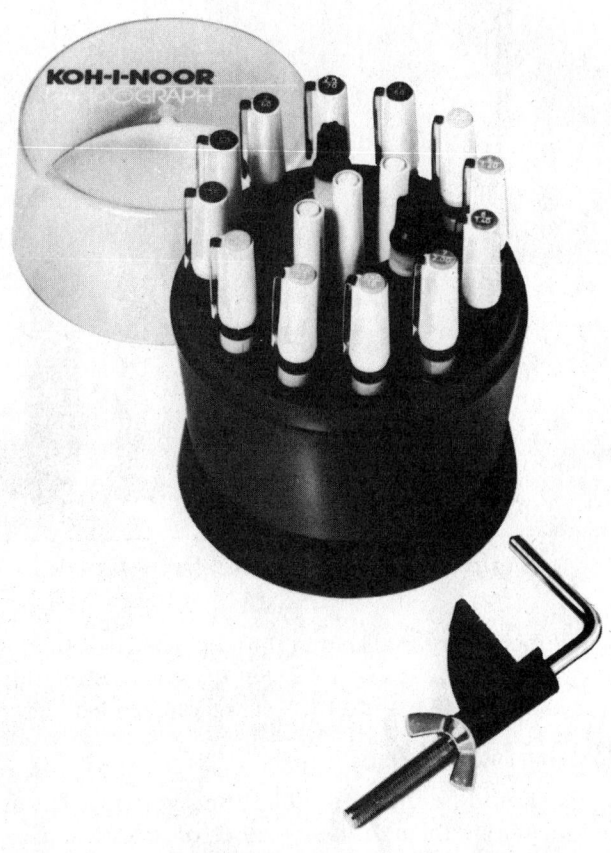

Figure 1-72   Revolving pen holder *(Courtesy Koh-I-Noor Rapidograph)*

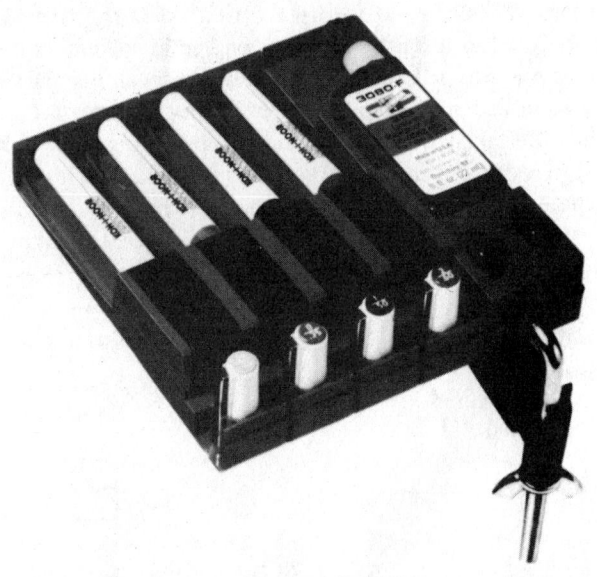

Figure 1-73   Flat pack pen holder *(Courtesy Koh-I-Noor Rapidograph)*

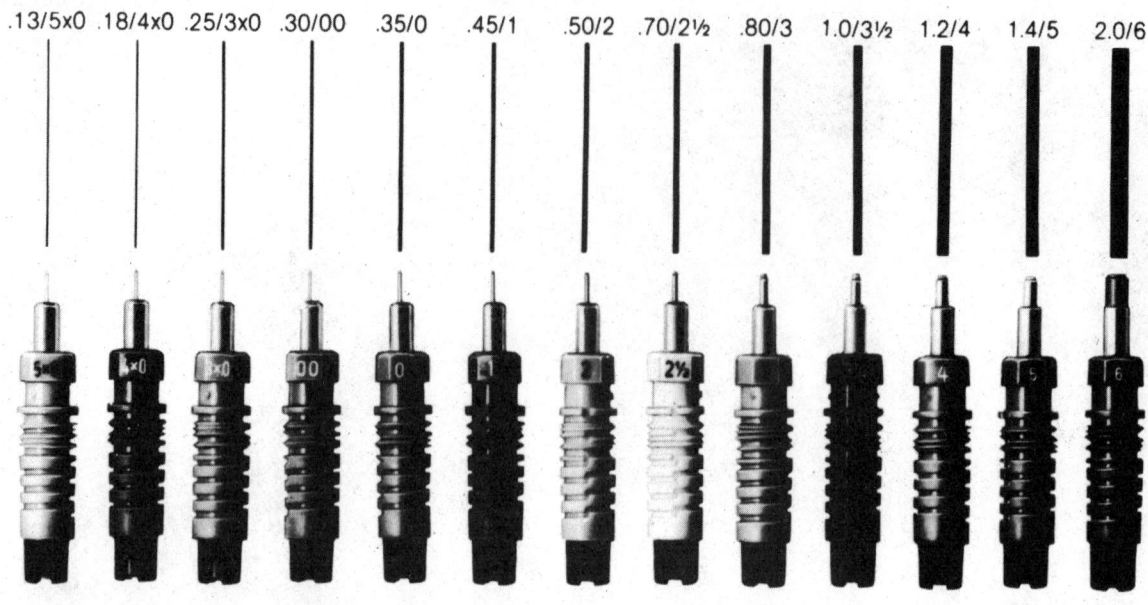

Figure 1-74 Pen sizes *(Courtesy Staedtler Mars)*

Pen points are available in thirteen standard sizes of varying widths, *Figure 1-74*. For general drafting inking, numbers .45/1 and .70/2½ are recommended.

Cleaning Technical Pens

Pens should be cleaned when they get sluggish or before storing them for long periods of time. The parts of most technical inking pens are similar to those shown in *Figure 1-75*. When not in use, technical pens should be kept in a storage clamp or else capped to prevent ink from drying in the point. If a pen does get clogged, remove the point and hold it under warm tap water. This normally softens the ink. If the ink has dried, use an ultrasonic cleaner or a mild solvent. If the pens will not be used for a week or more, all ink should be removed and the pens stored empty and clean. Care must be taken when removing and replacing the cleaning needle. An ultrasonic cleaner is used quickly and efficiently to clean technical pens, *Figure 1-76*.

When pens are to be cleaned by hand, use the following recommended steps:

Cleaning. Pens can be ruined by improper cleaning. Study Steps 1 through 5 and follow them closely when cleaning pens. (Refer again to Figure 1-75.)

Step 1   Remove the cap and the ink container.

Step 2   Soak the body of the pen in hot water. The ink container should also be soaked if ink has dried in it.

Step 3   After soaking, remove the pen body from the water. Hold the knurled part of the body with the top downward. Unscrew and remove the point section. Remove the end of the cleaning needle weight. Do not bend the cleaning needle or it will break.

Step 4   Immerse all body parts in a good pen-cleaning fluid or hot water mixed half with ammonia.

Step 5   Dry and clean.

Filling.   To fill the pen, follow Steps 1 through 5:

Step 1   Unscrew and remove the knurled lock ring.

Step 2   Remove the ink container. Leave the spacer ring in place.

Step 3   Fill the ink container with lettering ink. Do not fill it more than ¾ inch from top.

Step 4   Hold the filled container upright and insert the pen body into the container.

Step 5   Replace the knurled lock ring.

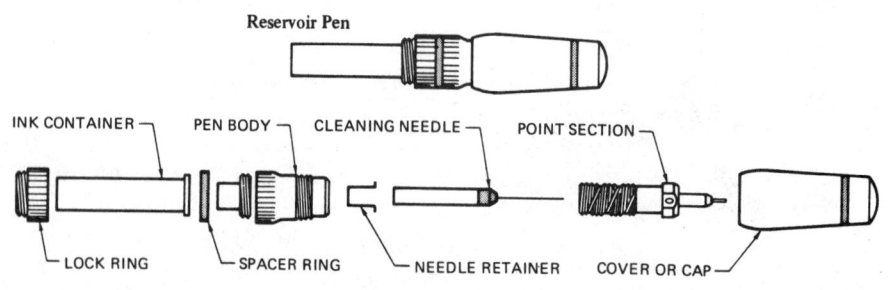

Figure 1-75   Internal parts of a technical inking pen

Figure 1-76 Technical pen ultrasonic cleaner *(Courtesy Keuffel & Esser Co.)*

Ink

A high-quality, fast-drying ink must be used in technical pens for the best results. The ink must be black and erasable, and it must not crack, chip or peel, ***Figure 1-77***. Keep inks out of extremely warm or cold temperatures. The bottles or jars should be kept airtight, and the excess ink should be cleaned from the neck of the container to keep it from drying in the cap. Inks in large containers should be transferred to smaller bottles or directly into pens, *away from working areas to avoid the possibility of spillage.*

Mechanical Lettering Sets

Lettering sets come in a variety of sizes and templates, ***Figure 1-78***. All sets contain a scriber, and various pen sizes and templates.

Scriber Templates

Scriber templates consist of laminated strips with engraved grooves which are used to form letters. A tracer pin moving in the grooves guides the scriber pen (or pencil) in forming the letters, ***Figure 1-79***.

Guides for different sizes and kinds of letters are available for any of the lettering devices. Different point sizes are made for special pens so that fine lines can be used for small letters and wide lines for large letters. Scribers may be adjusted to form vertical or slanted letters of several sizes from a single guide by

Figure 1-77 Drawing ink

simply unlocking the screw underneath the scriber and extending the arms, ***Figure 1-80***.

One of the principal advantages of lettering guides is that they maintain uniform lettering. This is especially useful where many drafters are involved. Another important use is for the lettering of titles, and note headings and numbers on drawings and reports.

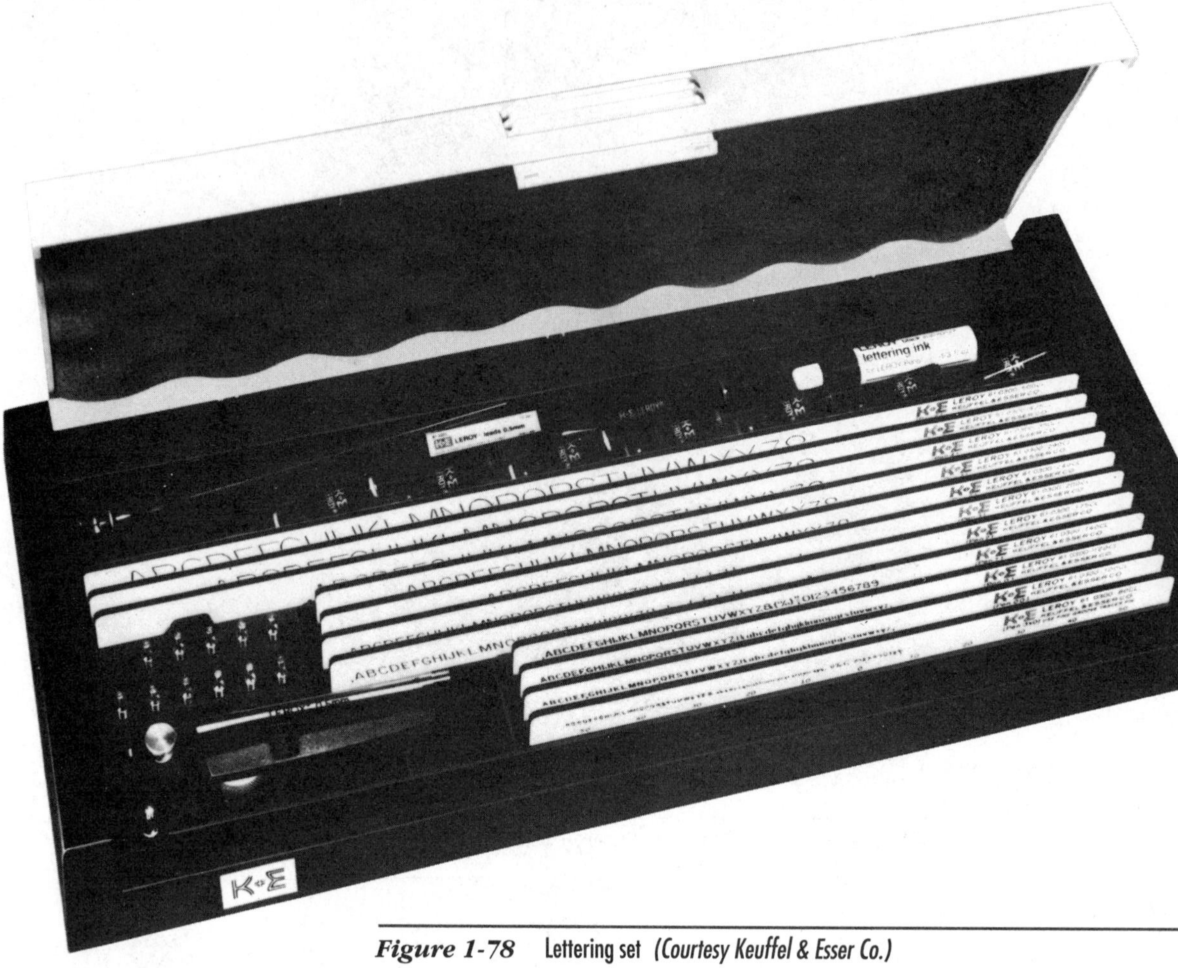

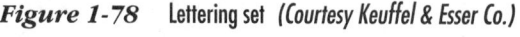

Figure 1-78 Lettering set *(Courtesy Keuffel & Esser Co.)*

Figure 1-79 Forming letters with a scriber *(Courtesy Koh-I-Noor Rapidograph)*

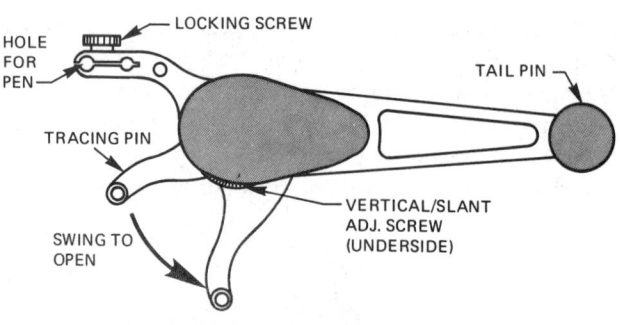

Figure 1-80 Scribers are adjustable

Letters used to identify templates are:

U = UPPERCASE
L = LOWERCASE
N = NUMBERS

Thus, a template identified as 8-ULN means it is 8/16 inch high (½"), and has uppercase letters and lowercase letters, and numbers.

Tracing Pin. Better, more expensive scribers use a double tracing pin, *Figure 1-81*. The blunt end is used

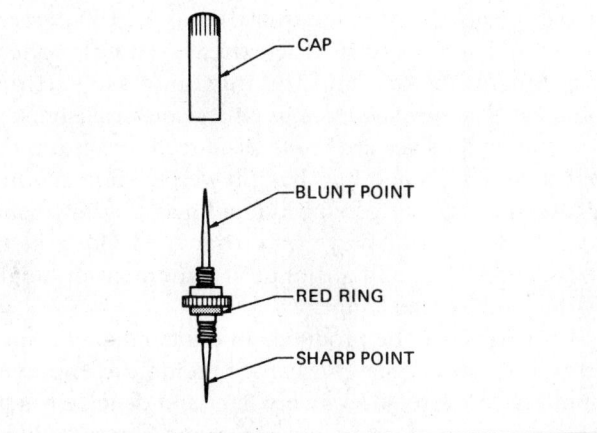

Figure 1-81   Double tracing pen

for single-stroke lettering templates or very large templates that have wide grooves. The sharp end is used for very small lettering templates, double-stroke letters, or script-type lettering using a fine groove. Most tracing pins have a sharp point, but some do not. Always screw the cap back on the unused end after turning the tracing pin to the desired tip. Be careful with the points as they will break if dropped and can cause a painful injury if mishandled.

Standard Template

Learning to form mechanical letters requires a great deal of practice. *Figure 1-82* shows a template having three sets of uppercase and lowercase letters. Practice forming each size letter and number until they can be made rapidly and neatly. Use a very light, delicate touch so as not to damage the template, scriber, or pen.

Size of Letters.   The size or height of the lettering on a template is called out by the number used to identify each set. Sizes are in thousandths of an inch. A #100 is .100 inch high, or slightly less than an eighth of an inch; a #240 is .240 inch high, or slightly less than a quarter of an inch.

Another system to determine template size uses simple numbers. These numbers are placed above the number 16 to indicate the fraction height of the letter. For instance, the number 3 placed above the number 16 would read as 3/16 inch in height.

Pens

There are two types of pens: the regular pen and the reservoir pen, *Figure 1-83*. The regular pen must be cleaned after each use. The reservoir pen should be cleaned when it gets sluggish or before being stored for long periods of time. This procedure is the same as it is for the cleaning of technical pens as described previously.

Butterfly Scriber

Basic Parts

The butterfly-type scriber shown in *Figure 1-84* is a delicate, precision tool that does its job without requir-

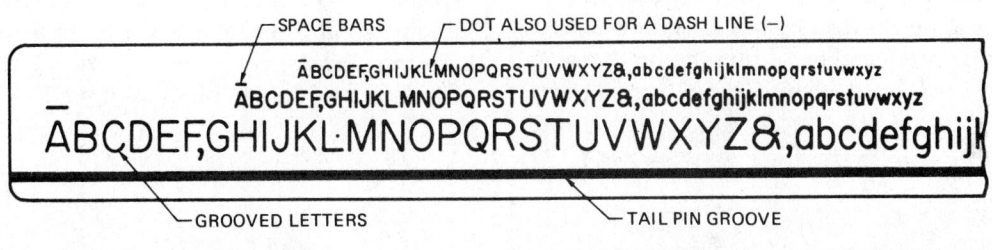

Figure 1-82   Lettering template

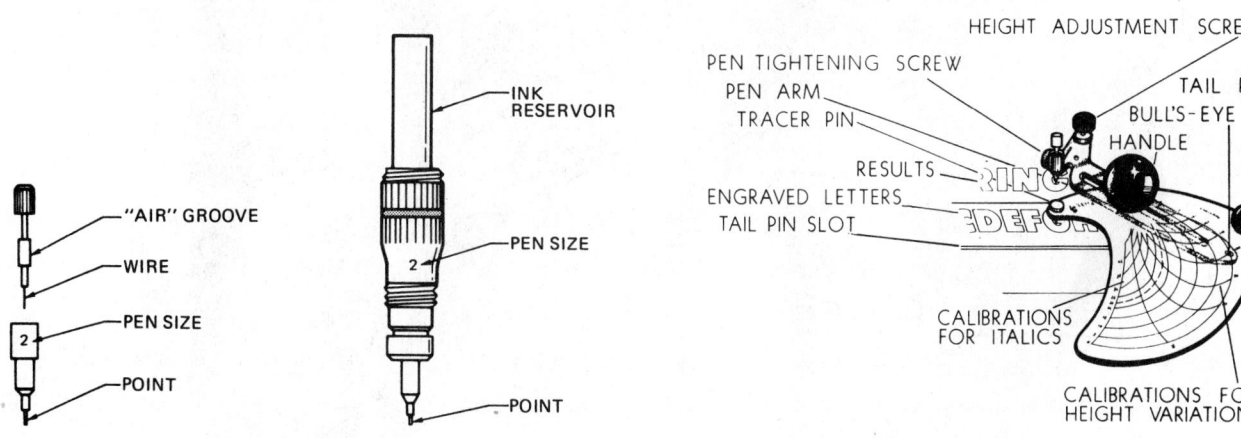

Figure 1-83   Ink pens used for lettering

Figure 1-84   Butterfly scriber *(Courtesy Letterguide Inc.)*

ing any adjustments, repairs, or maintenance. The clear plastic base of the scriber bears the setting chart used in adjusting the pen arm for enlargements, reductions, verticals, and slants to be produced by tracing the engraved letters of a letter guide template.

The pen arm of the scriber holds the pen accessories for the various jobs to be performed. The pen and the arm have a thumb-tightening screw device for securing the pen being used, and an adjustable pressure post screw with locking nut for controlling the amount of pressure at which the pen is set. The pressure post rides on the surface of the work when in use, and is used only in conjunction with the swivel knife. The bull's-eye setting marker at the opposite end of the pen arm offers a concise, accurate means of setting the scriber for the various percentages and angles desired.

The tracing pin is the hardened tool steel point used in tracing the template letter. The tail pin serves as the pivot point for the triangular action of the scriber. This pin travels in the center groove of the template.

Operation of the Butterfly Scriber

The butterfly scriber, a precision lettering tool, is the key to producing clean, sharp, controlled lettering. The setting chart, using the bull's-eye at the end of the pen arm for a marker, begins at the outer edge with a starting line marked "vertical." In this position, the scriber produces a vertical letter of normal size with the template being used. To enlarge this letter, set the bull's-eye at a position above the 100 percent intersection. At 120 percent, the scriber produces a letter 20 percent greater in height than it does at 100 percent. A reduction can be produced by setting the bull's-eye at a position below the 100 percent intersection.

Variations in height range from 100 up to 140 percent and down to 60 percent. The extreme settings produce condensed letters, and the intermediate settings produce headings, subheadings or large or small letters.

Slants in all sizes are easily produced by setting the bull's-eye on a line other than the vertical line. Normal slants or italics are produced in all height adjustments by setting the bull's-eye on either the 15-degree or 22½-degree line, and at the desired percent of height of the letter on the template.

Variations may be produced in slants ranging from 0 degree to 50 degrees forward. Tracing the engraved template letter requires a very light and delicate touch. This results in more accurately traced letters and less wear on the equipment. Each lettering application requires its own specific pen, and will place at the fingertips of the drafter the very best in standard typeface and hand-lettered alphabets for fast, easy rendering.

Special Effects. By using one's imagination many special effects can be achieved, *Figures 1-85*, *1-86*, and *1-87*.

Airbrush

Airbrush guns are used for such purposes as production designing, pictorial rendering, portrait figure rendering, architectural rendering, and technical illustration. There are two kinds of airbrushes: the single-action type and the double-action type. In the single-action airbrush, the trigger controls the flow of air only. The fluid control is adjusted in front by the nozzle. In the double-action airbrush, the trigger controls both the flow of air and the amount of fluid to be sprayed, *Figure 1-88*.

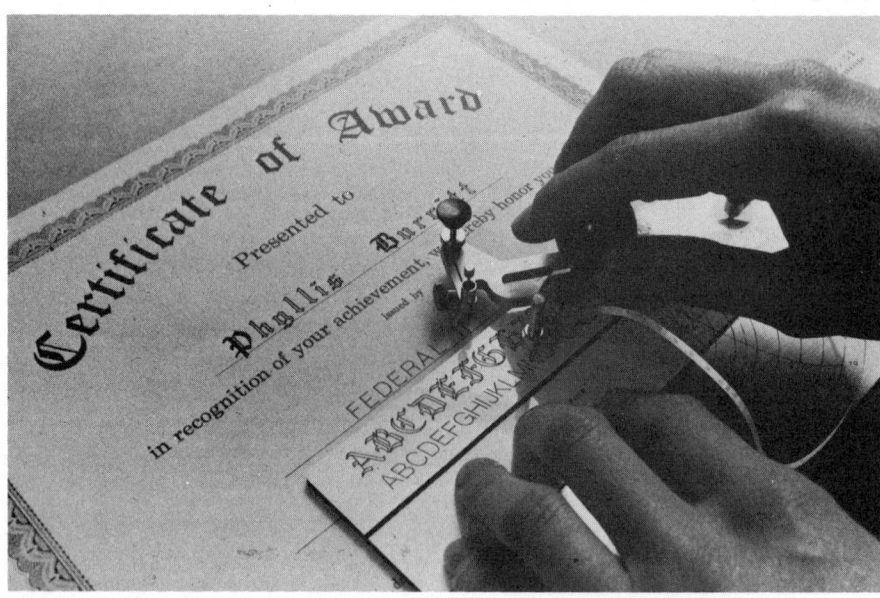

Figure 1-85 Adjustable scriber creates special effects *(Courtesy Letterguide Inc.)*

Figure 1-86   Sample lettering styles *(Courtesy Letterguide Inc.)*

Figure 1-87   Additional special effects *(Courtesy Letterguide Inc.)*

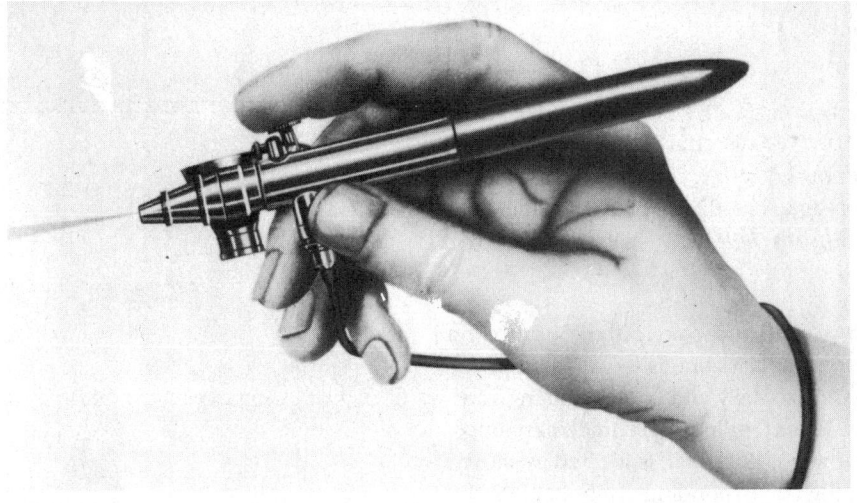

Figure 1-88   Airbrush *(Courtesy Badger Airbrush Co.)*

Paper Sizes

Two basic standard paper sizes are 8½ x 11 inches and 9 x 12 inches. The basic standard metric size, A-4, is 210 x 297 millimetres. See *Figure 1-89*. Examples of paper folded to A-size are shown in *Figure 1-90*.

	INCHES		MILLIMETRES	
SIZE	DIMENSIONS		SIZE	DIMENSIONS
A	8 1/2 x 11	9 x 12	A-4	210 x 297
B	11 x 17	12 x 18	A-3	297 x 420
C	17 x 22	18 x 24	A-2	420 x 594
D	22 x 34	24 x 36	A-1	594 x 841
E	34 x 44	36 x 48	A-0	841 x 1189

Figure 1-89   Paper sizes

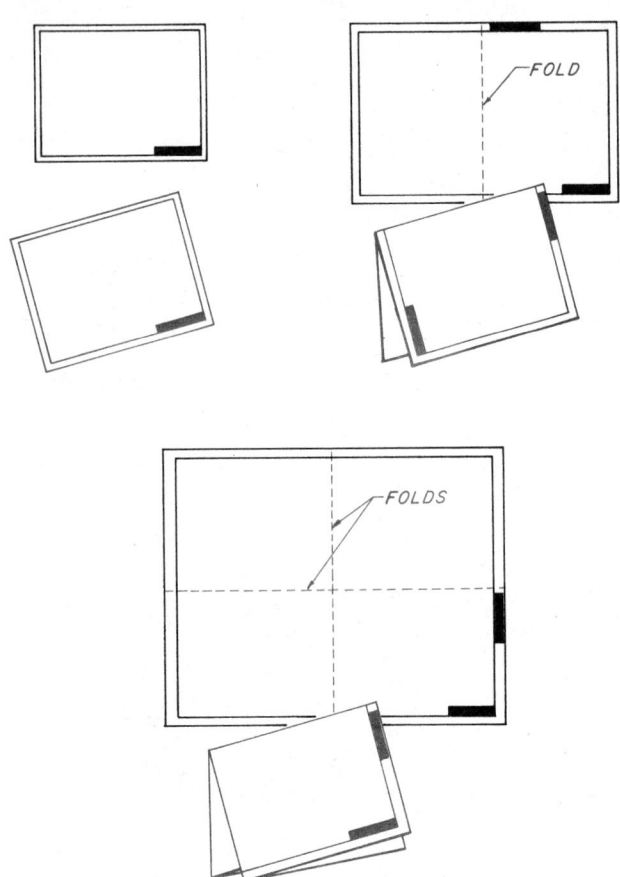

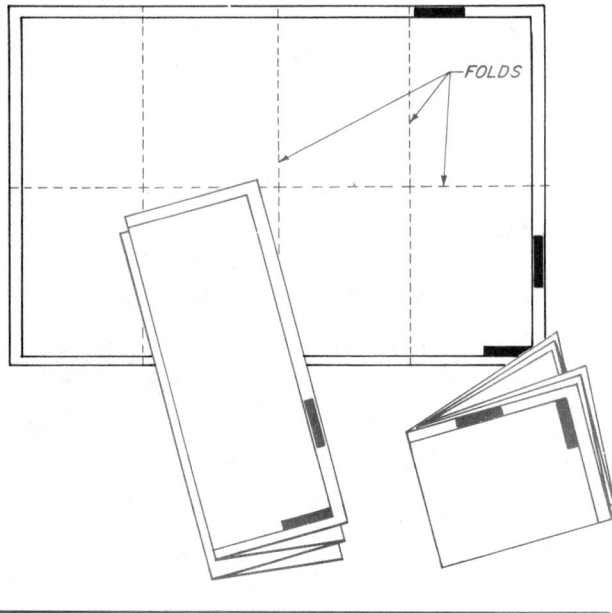

Figure 1-90 Paper folded to A-size

Borders

The location of the borders varies with each size sheet of paper, *Figure 1-91A*. This chart indicates the various standard borders used today. A standard horizontal border is shown in *Figure 1-91B*. A standard vertical border is shown in *Figure 1-91C*.

Zoning

Zoning is used to pinpoint a particular detail on a drawing. The exact rectangular zone is located by the use of numbers running horizontally and letters running vertically in the margins. By extending these imaginary lines, the exact rectangular zone, Zone 7-C, is located as shown in Figure 1-91A. See the corre-sponding symbol below the chart. The number at the left (1) indicates the page number, the number at the top right (7) indicates the corresponding number on the horizontal margin. The letter at the lower right (C) indicates the corresponding letter in the vertical margin.

Whiteprinter

Many types of whitepapers are available for use in drafting rooms. A *whiteprinter*, *Figure 1-92*, reproduces a drawing through a chemical process. Most of these machines work on the same basic principle, *Figure 1-93*. A bright light passes through the translucent original drawing and onto a coated whiteprint paper. The

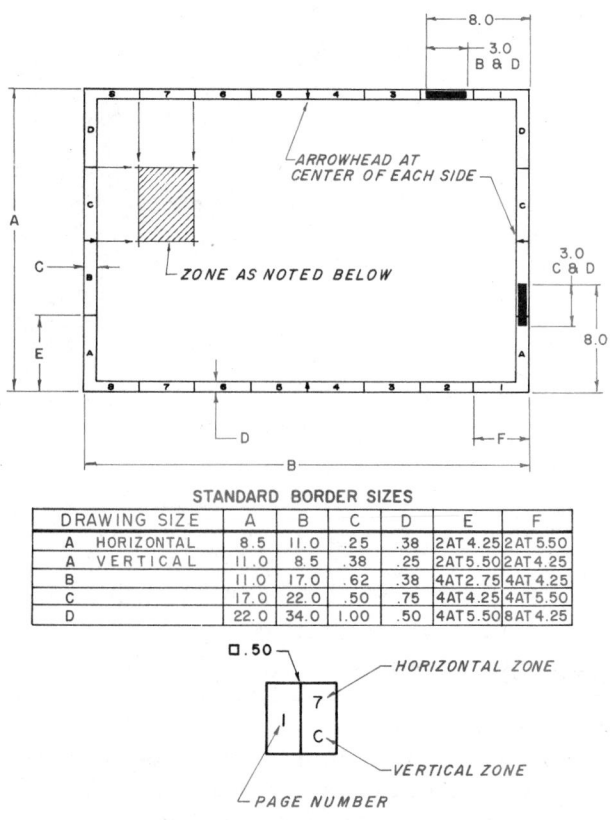

DRAWING SIZE		A	B	C	D	E	F
A	HORIZONTAL	8.5	11.0	.25	.38	2AT4.25	2AT5.50
A	VERTICAL	11.0	8.5	.38	.25	2AT5.50	2AT4.25
B		11.0	17.0	.62	.38	4AT2.75	4AT4.25
C		17.0	22.0	.50	.75	4AT4.25	4AT5.50
D		22.0	34.0	1.00	.50	4AT5.50	8AT4.25

Figure 1-91 (A) Standard border sizes *(Courtesy Bishop Graphics Co.)*

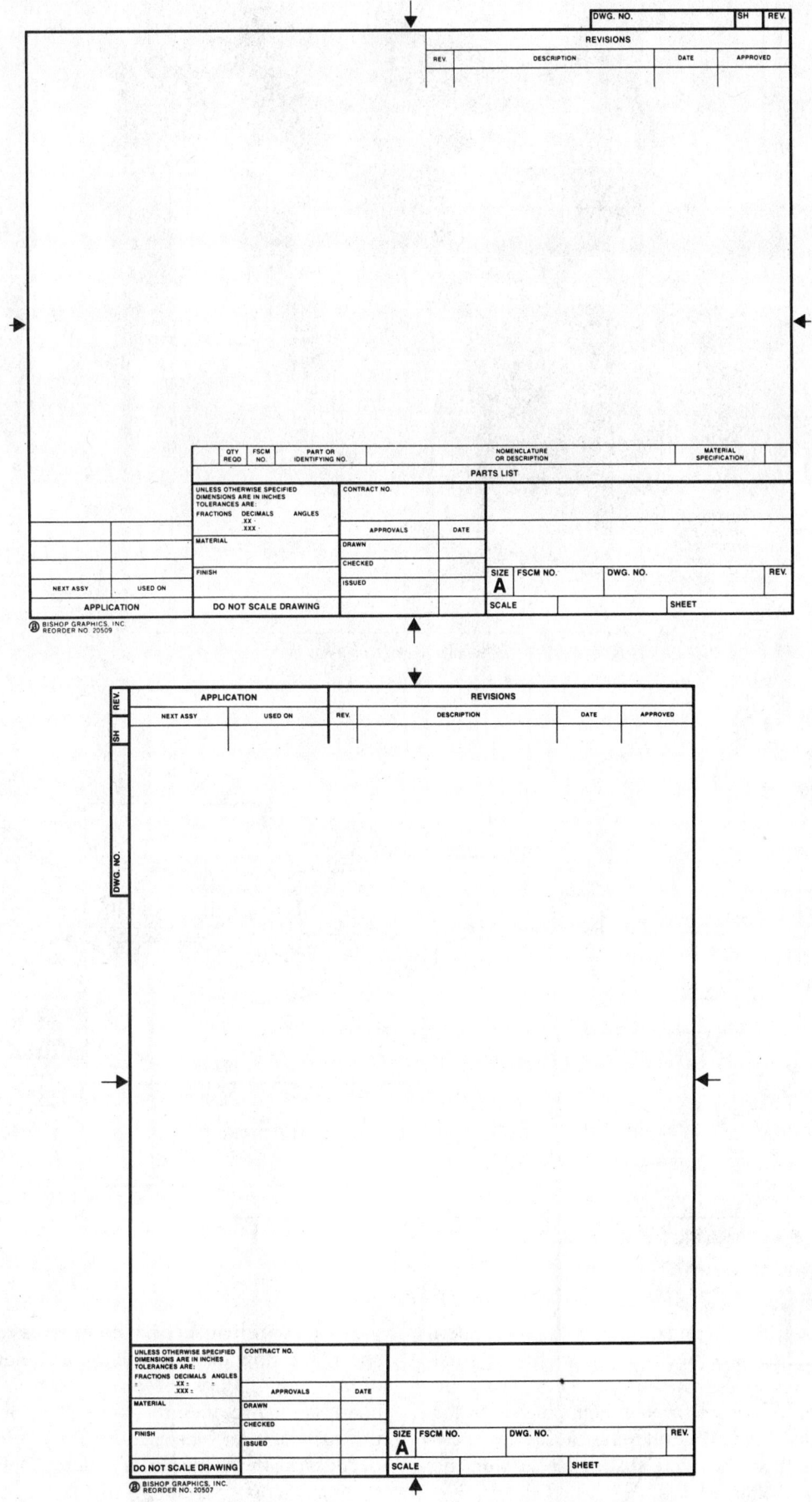

Figure 1-91   (B) Standard horizontal border and (C) Standard vertical border *(Courtesy Bishop Graphics Co.)*

Figure 1-92   Whiteprinter *(Courtesy Blu-Ray Inc.)*

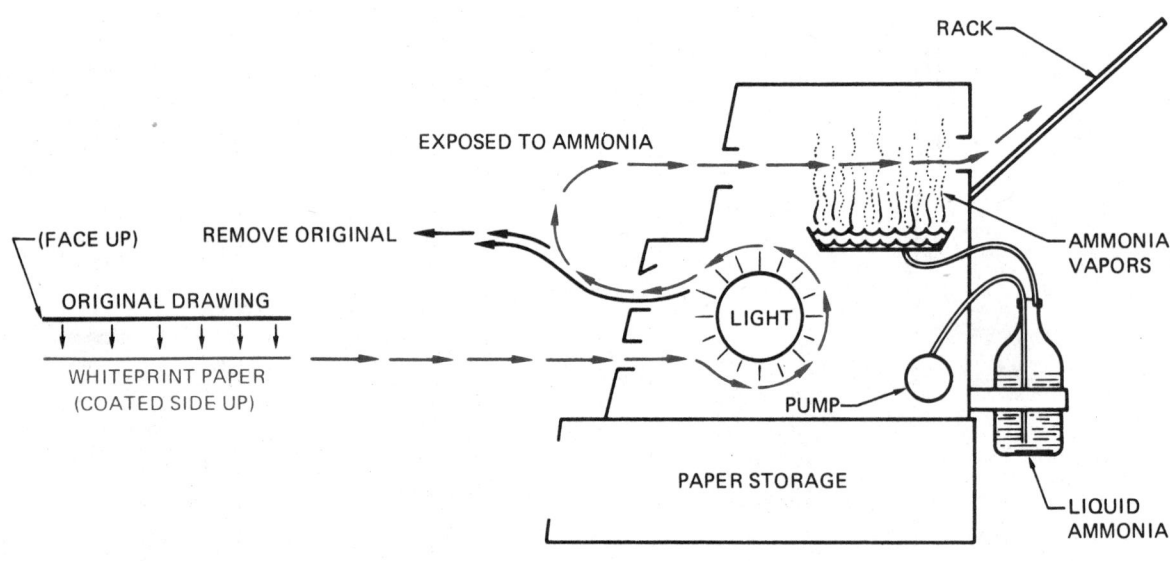

Figure 1-93   Whiteprinter process

light breaks down the coating on the whiteprint paper, but wherever lines have been drawn on the original drawing, no light strikes the coated sheet. Then the whiteprint paper is passed through ammonia vapor for developing. This chemical developing causes the unexposed areas — those that were shaded by lines on the original — to turn blue or black.

Most whiteprinters have controls to regulate the speed and flow of the developing chemical. Each type of machine requires different settings and has different controls. Before operating any whiteprinter, read all of the manufacturer's instructions.

Today, with the advent of new technology, copies are often made on a printer, *Figure 1-94*.

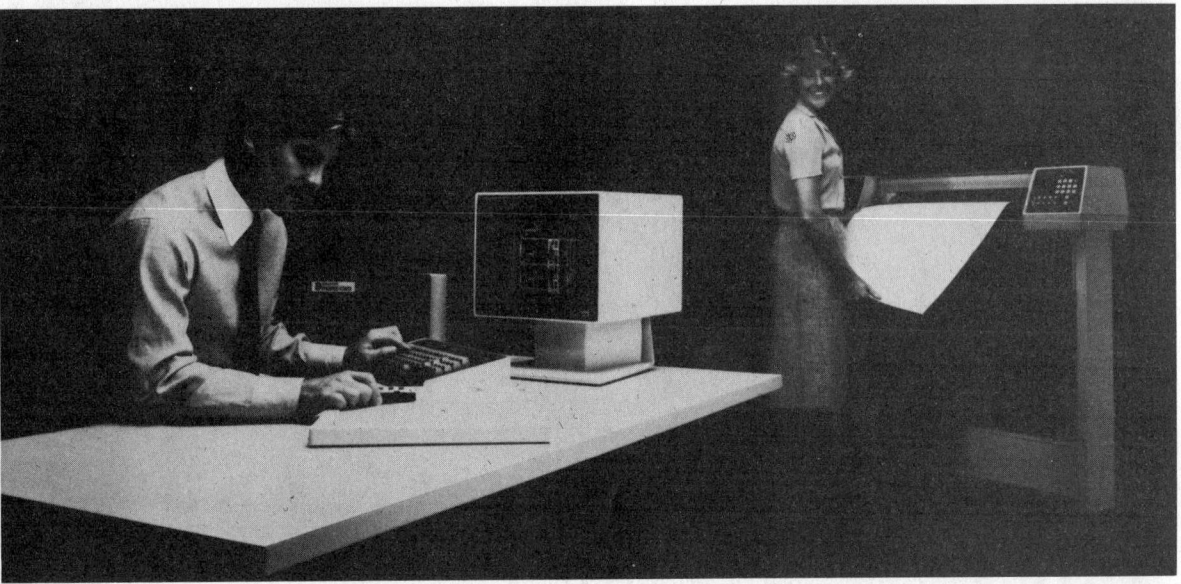

Figure 1-94   Copier *(Courtesy J. S. Staedtler Inc.)*

Files

A finished drawing represents a great deal of valuable drafting time and is, therefore, a costly investment. Drawings must be stored flat in a clean storage area, *Figure 1-95*. Vertical drawing storage is provided by hangers, *Figure 1-96*. Most engineering firms keep their files in fireproof and theftproof vaults.

Open-End Typewriter

A word processor-equipped, open-end typewriter is used to speed up the lettering process on a large drawing, *Figure 1-97*.

Figure 1-95   Drawing file system *(Courtesy Safco Products Inc.)*

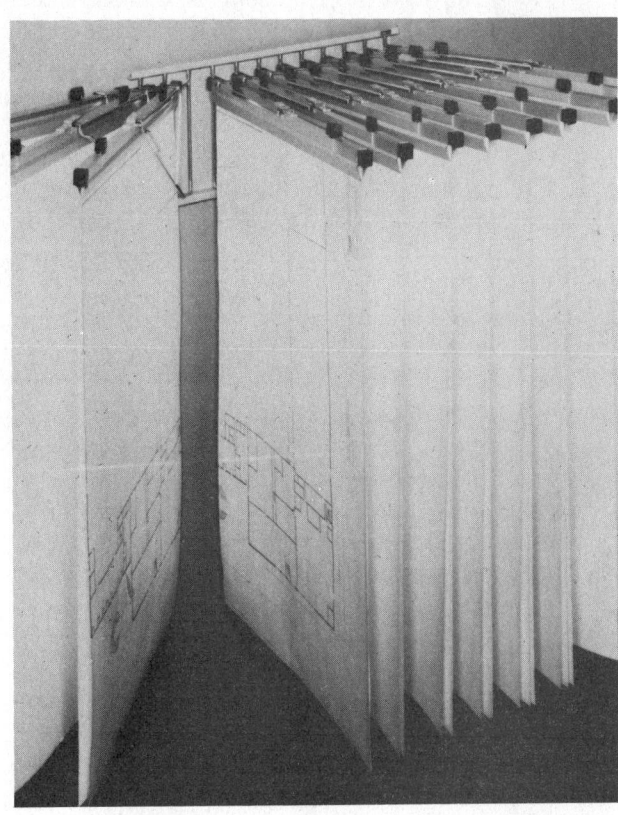

Figure 1-96   Vertical drawing file *(Courtesy Safco Products Inc.)*

Figure 1-97   Open-end typewriter *(Courtesy Diagram Corp.)*

Care of Drafting Equipment

Drafting tools are precision instruments, and the proper care will ensure that they last a lifetime.

Plastic Tools

Plastic drafting tools, such as T-squares, parallel straight-edges, templates and triangles, should be wiped immediately after use with a damp cloth to remove ink or graphite that may stain the tools or be carried to the next drawing. Once a plastic instrument is stained, a mild soap or ammonia solution will dissolve many water- and oil-based inks. Be careful *not* to use a solvent such as paint thinner, lacquer thinner or alcohol.

Plastic drafting tools should be kept out of direct sunlight and away from warm surfaces to prevent them from becoming brittle, cracked, and warped. They should be stored in a flat position with cloth or paper between them to reduce scratching the surface.

A great number of plastics are used in drafting instruments. Most are made from either styrene or acrylic plastic. Styrene is a more flexible and softer plastic than acrylic. Although acrylic instruments are harder, they are more prone to chipping. Because both types of plastic are relatively soft, plastic drafting instruments should never be used for a cutting edge.

Compasses

Almost all compasses are made of brass that is chrome- or nickel-plated. To clean these instruments, use a mild solution of soap and water to remove residue and dirt.

Compasses should not normally need oiling, unless they are kept in a damp area which could cause rust. If a compass is oiled unnecessarily, there is a risk of soiling the next drawing on which it is used.

Tables and Chairs

Wooden drafting furniture is cared for in the same manner as any other wooden furniture. It may be polished or waxed with ordinary products. Do not polish the insides of drawers or cabinets. These areas may retain the wax, which can then be transferred to drawings.

Steel furniture can be cleaned with soap and water, and then waxed.

The gears and joints on adjustable drafting tables are lubricated at the factory, and generally do not require further oiling. Additional oiling increases the risk of getting oil or grease on a drawing.

The tops of most drafting tables are coated with a vinyl film such as melamine or a phenol-laminate material. A glass cleaner or mild ammonia solution is used to clean these surfaces.

Use of Appliques

The word *applique* is a generic term used to describe a variety of shortcut products used in drafting. These products include such items as tapes, pads, and various other ready-made appliques for creating printed circuit board artwork, *Figure 1-98*. These same materials may also be used for a variety of tasks in other drafting fields, *Figure 1-99*. For example, architects use tapes for making lines and walls on floor plans.

Transfer cards are used primarily as substitutes for mechanical lettering, but any type of symbol or frequently used piece of graphic data can be placed on a transfer card. Transfer cards are especially designed to fit against a parallel bar, drafting rule or other straight edge for ease of alignment. Symbols are transferred from the card by rubbing them with a blunt point.

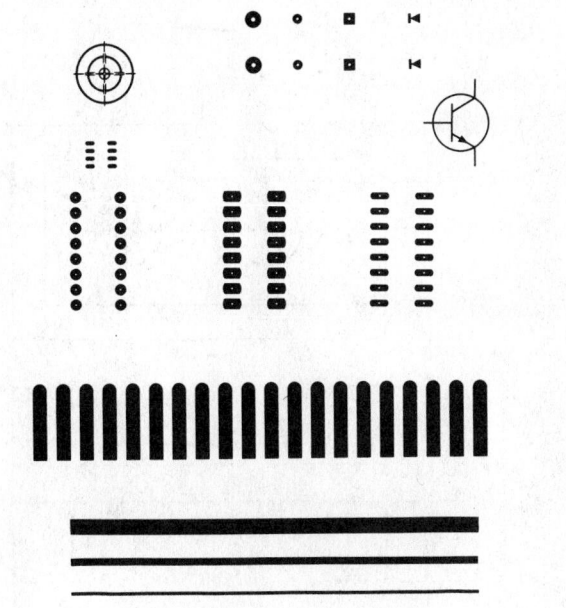

Figure 1-98 Tapes and pads for printed circuit board drafting

Dry transfer sheets are designed according to the same principles as transfer cards. The major differences are that transfer sheets are just that, sheets — not cards. Dry transfer sheets are used a great deal in architectural drafting and technical illustration. The transfer is made by rubbing the symbols on the sheet with a blunt, rounded point or a special burnisher.

Dry transfer materials do have some drawbacks. The heat of ammonia-developing print machines tends to lift dry transfer material from the sheet. In addition, the material may dry out and crack with age.

Use of Burnishing Plates

Burnishing is another shortcut for creating graphic symbology fast and easily. *Burnishing* involves placing an especially textured plate under the drafting medium (usually paper or vellum) and rubbing the drafting surface with a pencil. The pencil may be soft and dark or light and hard, depending on the amount of emphasis desired. Two of the most commonly used symbols on

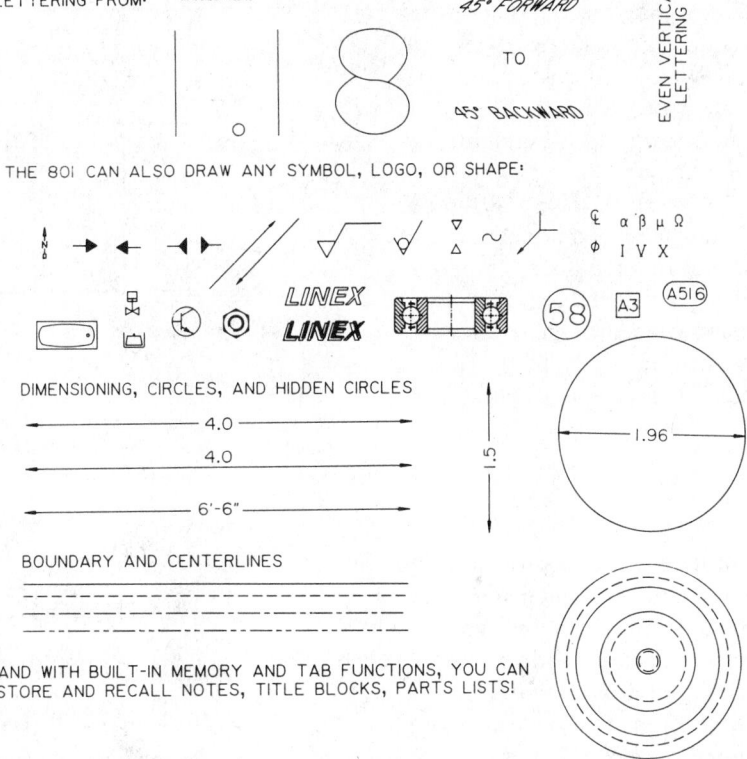

Figure 1-99 Sample applique *(Courtesy A. D. S./Linex Inc.)*

burnishing plates are bricks and stone, but any symbol could be made into a plate.

One weakness of burnishing plates is that the symbols they produce do not reproduce well.

Typewritten Text

Text on drawings and other types of documentation consists of dimensions, notes, and callouts. Creating text, or lettering, is one of the slowest, least productive manually performed tasks in drafting. Typewritten text is a shortcut for improving on lettering in manual drafting situations.

A number of different methods for typing text are used in drafting. The most frequently used are the open-carriage typewriter and the lettering machine. The *open-carriage typewriter* is any brand of typewriter that has been especially designed to hold larger-than-normal media, such as drawings, bills of material, and parts lists. Once a drawing is completed and ready for annotation, the drafter or engineering secretary rolls it into an open-carriage typewriter and types the required text. Typewritten text is fast, neat, and consistent, **Figures 1-100** and **1-101**.

Lettering machines provide another means for accomplishing typewritten text. Lettering machines are used primarily for titles on drawings, but may be used in any situation where Leroy lettering is used.

Lettering machines, such as the Kroy machine, output the letters on a clear tape that is pressed onto the drawing surface, **Figure 1-102**. Lettering machines are capable of producing text in a number of different sizes and styles. There are some drawbacks to this shortcut. The machine and its type fonts are expensive.

Another method used more and more frequently in modern drafting rooms for accomplishing typewritten text is the *computerized lettering machine*, **Figures 1-103** and **1-104**. This machine allows drafters to enter data through a keyboard and receive a simultaneous readout on a small display screen. If the text is correct as typed, an ENTER command will cause the lettering to be accomplished in ink automatically. If there are errors, they can be corrected before giving the ENTER command.

Overlay Drafting

Overlay drafting is a complete drafting process that uses advanced reproduction techniques and materials to reduce the amount of time spent in preparing drafting documentation. Actually more than a shortcut technique, the underlying principle of overlay drafting is nonrepetition. *Nonrepetition* means that once any type of graphic data or symbology has been drawn the first time, it should never have to be drawn again.

The special tools and materials of overlay drafting include pin bars, registration tabs, prepunched polyester film or a punch to punch holes in standard film, a

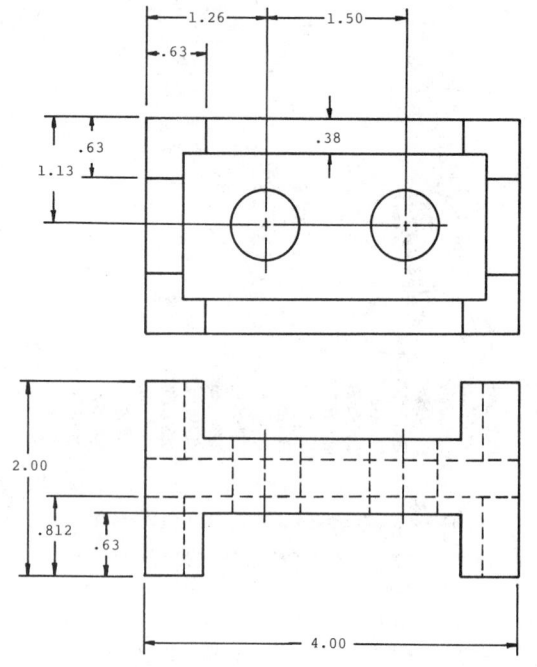

Figure 1-100   Typewritten text on drawings

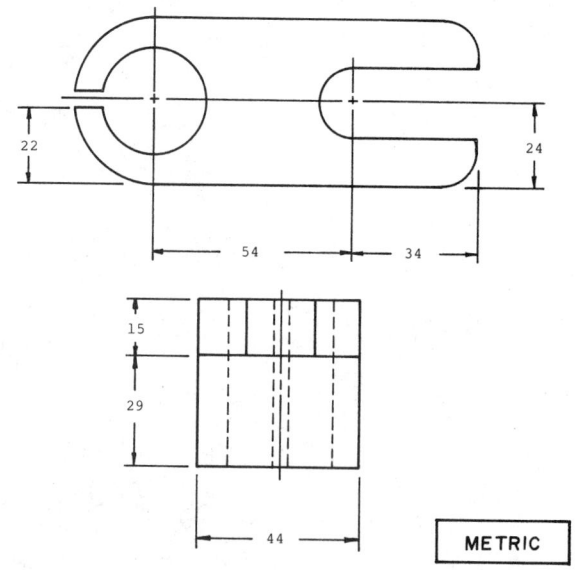

Figure 1-101   Typewritten text on drawings

Lettering Sample
Lettering Sample
Lettering Sample
Lettering Sample
LETTERING SAMPLE
LETTERING SAMPLE

Figure 1-102   Kroy lettering samples

Figure 1-103 Computerized lettering machine *(Courtesy Ozalid Corp.)*

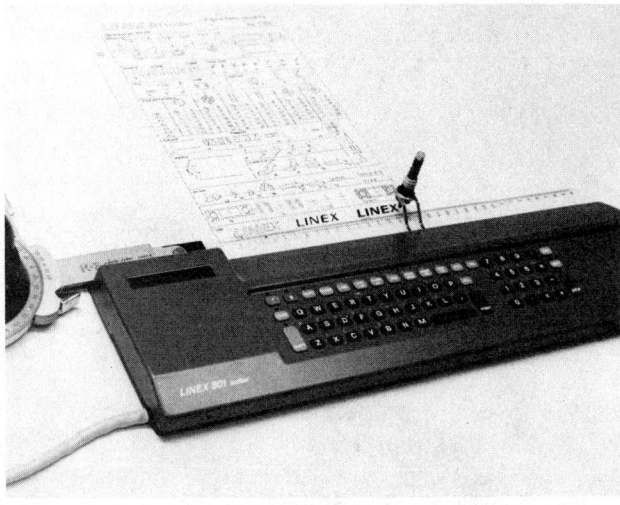

Figure 1-104 Computerized lettering machine *(Courtesy A. D. S./ Linex Inc.)*

flatbed vacuum frame printer, and a diazo print machine.

Overlay drafting involves placing a punched base sheet on a drafting table over a pin bar and placing various overlay sheets on top of it. Each successive overlay sheet is automatically lined up (registered) by the pin bar. The simplest example of how overlay drafting works is in preparing a set of commercial architectural plans.

In manual drafting, this preparation requires drawing the floor plan four separate times: once for the floor plan, once for the electrical plan, once for the plumbing plan, and once for the HVAC plan. Overlay drafting eliminates this time-consuming repetition. In overlay drafting, the floor plan is drawn once. Then, before it is dimensioned, three additional originals are created from it, using a special sensitized polyester film, a flatbed vacuum frame printer, and a diazo print

machine. In general terms, this is how the process works:

Step 1 A sheet of punched polyester film is placed on the drafting board. The holes across the top of the film fit over the pin bar which is permanently attached to the top edge of the drafting board. The floor plan is drawn on this base sheet, but, for the moment, the dimensions and all other annotation are left off. These will be added later, after the base sheet is used to reproduce several other originals that do not require dimensions or annotation.

Step 2 The base sheet is taken from the drafting board. A sheet of punched sensitized polyester film is placed on top of it. The two sheets are fastened together with plastic registration pins. Together, they are placed in the flatbed vacuum frame exposure unit. The lights in the unit expose the sensitized film, "burning away" all of the special light-sensitive emulsion, except where the light is blocked out by lines from the base sheet. The exposed polyester film is then run through the ammonia developing section of a print machine, producing what is called a slick. A *slick* is a polyester reproduction of the base sheet original. It is not drawn on; rather, it is used as a base sheet over which electrical, plumbing, and HVAC plans may be overlaid. The original base sheet of the floor plan may now be completed by adding dimensions and other annotation.

Step 3 The new slick may now be placed over the pin bar on a drafting board, and a clear sheet of polyester film placed on top of it. The electrical symbols for the electrical plan are added on this new sheet. Only information for the electrical plan is entered on this overlay sheet. To get a print of the electrical plan superimposed on the floor plan, the slick base sheet containing the floor plan and the electrical plan overlay are placed in the flatbed vacuum frame printer, along with a sheet of ammonia developing print paper. All three sheets are secured together with plastic registration pins. When the exposure step is completed, the sheets are separated and the print paper is run through the ammonia developing section of a diazo print machine, thus producing a print of the electrical plan. The same process is repeated for the plumbing plan and the electrical plan. To save even more time, three slicks of the floor plan could have been made and given to three different drafters; one of whom would create the electrical plan overlay, one the plumbing plan overlay, and the other the HVAC plan overlay. This process is illustrated in *Figure 1-105*.

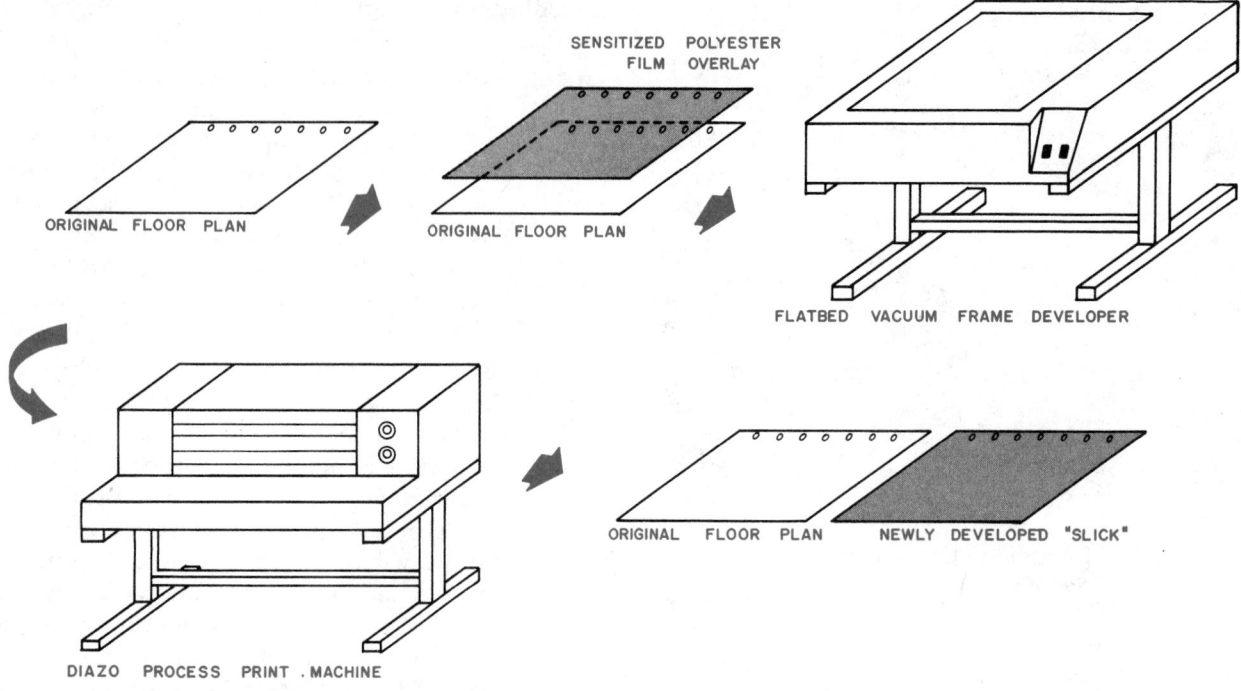

Figure 1-105　Overlay drafting process

Scissors Drafting Techniques

Scissors drafting is an extension of overlay drafting. The two combined can bring substantial productivity benefits in manual drafting situations. In *scissors drafting*, if any part of a set of drawings has ever been drawn before, say a typical detail or sectional view, it need not be redrawn. Rather, the techniques outlined previously to create a slick are used, and the details and other data needed from the slick are cut out and taped to a carrier sheet, *Figure 1-106*. A new slick is created using the carrier sheet as the original.

The scissors drafting technique is best illustrated by example. A mechanical drafter must construct a sheet of typical details and sectional views for a documentation package. All of the needed details have been drawn before, but in different jobs. Some of the details needed are in one job, some in another, and so on. Rather than redrawing, the drafter decides to use scissors drafting techniques.

First, the drafter locates all of the details needed and pulls the required sheets from the filing drawers. A slick of each sheet is then made. After cutting out the details from the slicks, the drafter assembles them on a carrier sheet. The first slicks allow the original drawings to be kept in case they are ever needed again.

Using the carrier sheet as an original, the drafter creates a slick containing all of the details. The slick may then be copied onto polyester film that accepts plastic lead or ink, and this new medium is used as any other original. The entire process takes about 20 minutes, whereas completely redrawing each detail would take hours.

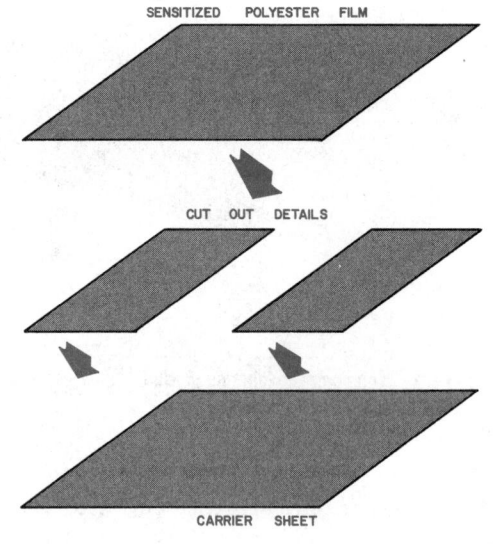

Figure 1-106　Scissors drafting

CAD Systems

The term *CAD system* is used a great deal in a modern technical drawing setting. A CAD system has three major components: 1) hardware, 2) software, and 3) users. The software component consists of computer programs (operating system and application packages), technical manuals, and user guides. *Users* are the technicians who operate a system and use it to prepare technical drawings and other forms of design documentation.

The *hardware* component consists of the tools available to users for input, processing, and output. These tools are introduced in the next section of this chapter. CAD and its various related concepts are covered in greater depth in Chapter 10.

CAD Hardware

The tools used in CAD to process and manipulate data and output, technical drawings, and other forms of design documentation are known collectively as hardware. Commonly used hardware components follow.

Processor

This is the actual computer. It stores the operating system, application software, and data that are currently being processed. It also processes data that are input and sends data that have been processed. It also processes data that are input and sends data that have been processed either to storage or to input devices. The processor includes both the hard drive and disk drive.

Input/Manipulation Devices

These are the devices used to interact with the processor and to manipulate data.

Keyboard.   Keyboards are used for entering commands, text, annotation, and dimensions.

Mouse.   A mouse is used to manipulate a cursor that appears on the graphics display. It can be used to select menu options, enter commands, position data, and identify data that are to be manipulated.

Digitizer.   Digitizers can be used for converting graphic data into X-Y coordinates that can be entered into the processor. When used in conjunction with a menu and/or a mouse, digitizers can also be used for issuing commands, selecting menu options, and identifying data that are to be manipulated.

Output Devices

These are the devices that convert digital data into graphics and text. Output can be hard or soft. Hard output is that which comes in paper form.

Graphics Display.   This is the display on which graphic data, text, menus, and various other types of information are displayed.

Plotter.   This is the device that actually makes a drawing by converting digital data into text and graphics. Lines are created by a pen that moves over the paper that can be stationary as with flatbed plotters or that moves as with drum plotters.

Printer.   This is the same type of device that is used with any other computer system. It converts digital data into alpha-numeric and/or graphic form. Printers may be the impact or laser type.

CAD systems as well as the hardware and software associated with them are covered in greater detail in Chapter 10. *Figure 1-107* is an example of a modern CAD hardware configuration.

Figure 1-107   A modern CAD hardware configuration *(Courtesy Autodesk Inc.)*

Review

1. Among the pencil lead grades from hard to soft, which two are recommended for drafting?
2. Describe the whiteprinter process used in today's drafting rooms.
3. List three kinds of drafting triangles.
4. Explain how to read a micrometer.
5. Why are the drawing surface and top cover so important?
6. List the various kinds of drafting tools used to draw horizontal lines, and explain which is the best and why.
7. List the standard paper sizes used in industry today.
8. What is the difference between an open-divided scale and a full-divided scale?
9. Which kind of compass is recommended and why?
10. For what is dry cleaning powder used, and why must it be removed from the drawing?
11. Why is a dusting brush used in drafting?
12. Explain how to read a vernier protractor. Why is it used?
13. Define the term *applique*.
14. What is the difference between a transfer card and a dry transfer sheet?
15. What is a Kroy lettering machine?
16. What tools and materials are needed to do overlay drafting?
17. What is scissors drafting?
18. What are the three basic components of a CAD system?
19. List the major pieces of hardware that might be included in a modern CAD configuration.

Chapter One Problems

Problems 1-1 through 1-6

Carefully measure each line in inches, or in millimeters if metric is indicated. Neatly enter your answers on a sheet of paper. For extra practice, measure each line full size as given, half size as given, quarter size as given, or ten-times scale as assigned by the instructor.

FULL SIZE

1)
2)
3)
4)
5)
6)
7)
8)
9)
10)

Problem 1-1

HALF SIZE

1)
2)
3)
4)
5)
6)
7)
8)
9)
10)

Problem 1-2

3/4 SCALE

1)
2)
3)
4)
5)
6)
7)
8)
9)
10)

Problem 1-3

METRIC (FULL SIZE) MILLIMETERS

1)
2)
3)
4)
5)
6)
7)
8)
9)
10)

Problem 1-4

METRIC (HALF SIZE) MILLIMETERS

1)
2)
3)
4)
5)
6)
7)
8)
9)
10)

Problem 1-5

METRIC (QUARTER SIZE) MILLIMETERS

1)
2)
3)
4)
5)
6)
7)
8)
9)
10)

Problem 1-6

Problems 1-7 through 1-22

Construct each object using the given dimensions. Use consistent thick, black object lines throughout. Keep all corners tight and sharp. Where indicated, draw thin, black center lines.

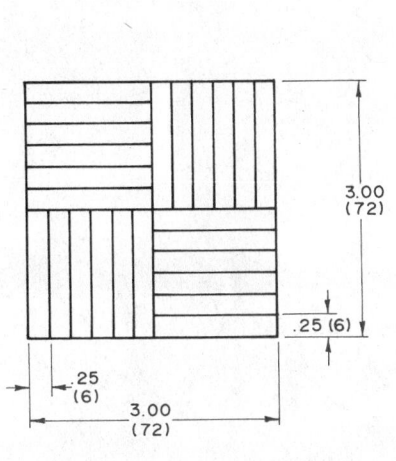

Problem 1-7

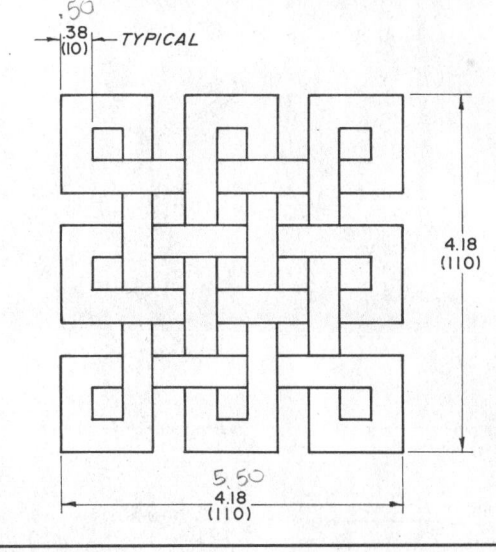

Problem 1-10

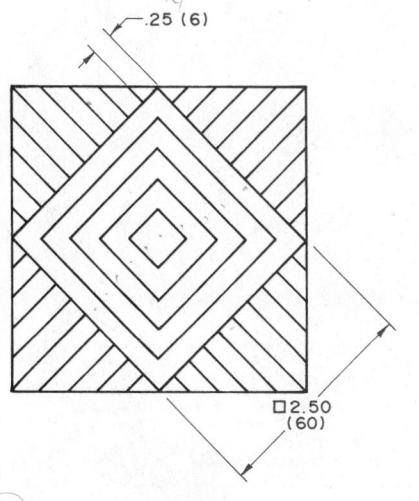

Problem 1-8

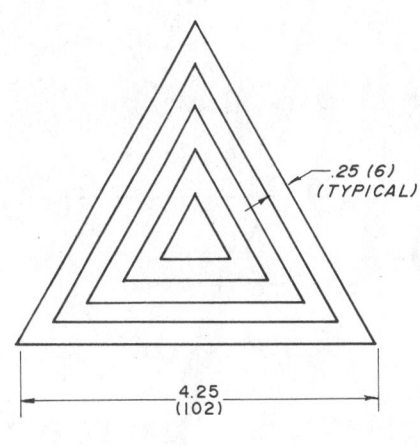

Problem 1-11

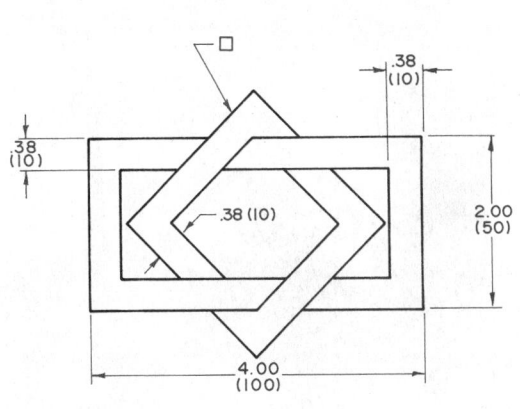

Problem 1-9

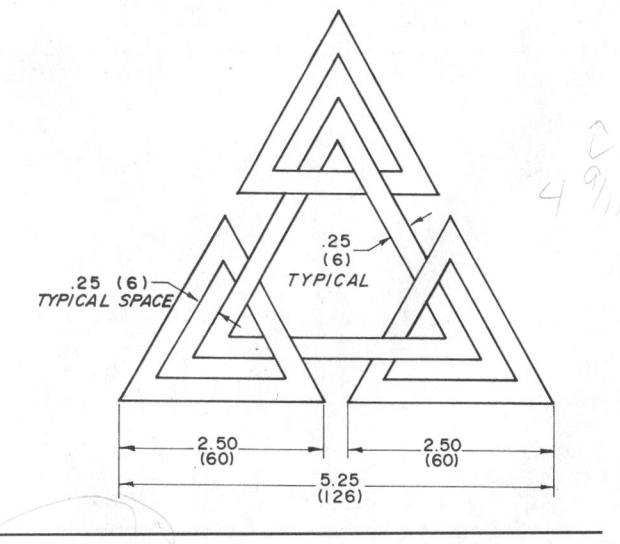

Problem 1-12

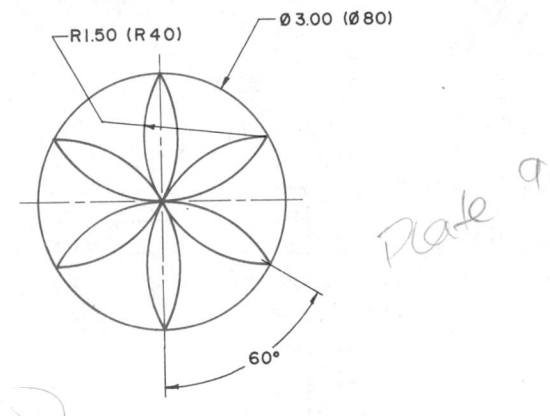

R1.50 (R40) Ø3.00 (Ø80)

60°

Problem 1-13

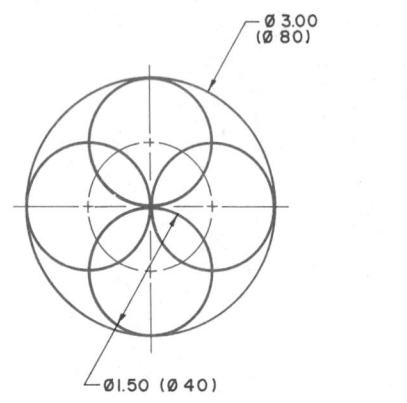

Ø 3.00 (Ø 80)

Ø1.50 (Ø 40)

Problem 1-14

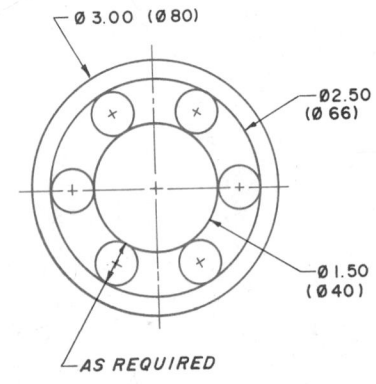

Ø 3.00 (Ø80)

Ø2.50 (Ø 66)

Ø 1.50 (Ø 40)

AS REQUIRED

Problem 1-15

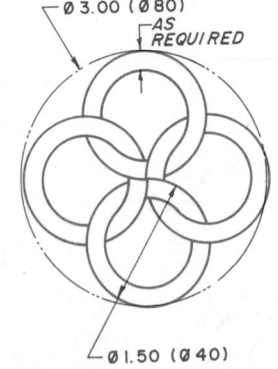

Ø 3.00 (Ø80)
AS REQUIRED

Ø1.50 (Ø 40)

Problem 1-16

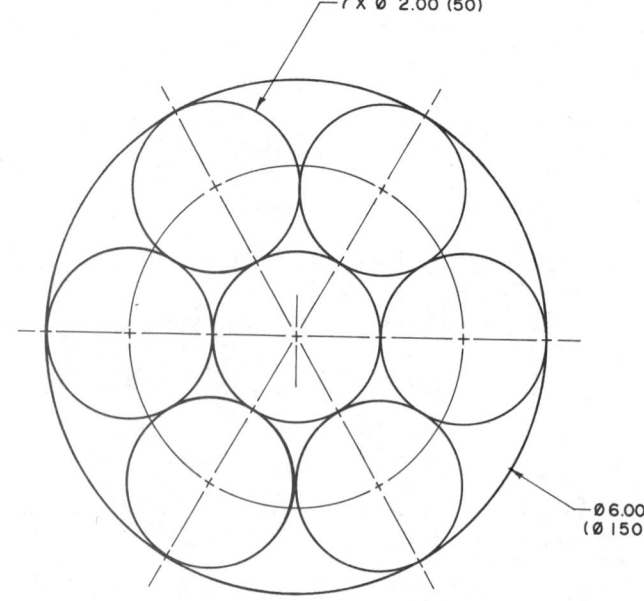

7 X Ø 2.00 (50)

Ø 6.00 (Ø 150)

Problem 1-17

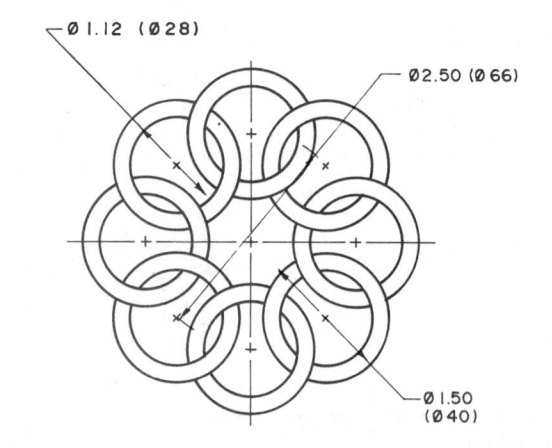

Ø 1.12 (Ø28)

Ø2.50 (Ø66)

Ø 1.50 (Ø40)

Problem 1-18

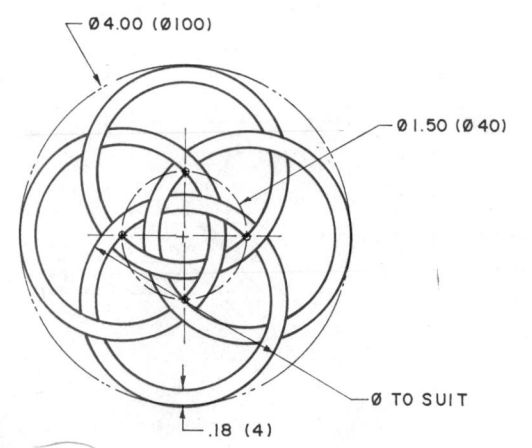

Ø 4.00 (Ø100)

Ø 1.50 (Ø 40)

Ø TO SUIT

.18 (4)

Problem 1-19

Ø 5.50 (Ø 132)

ALL SPACES = .25 (6)

Problem 1-20

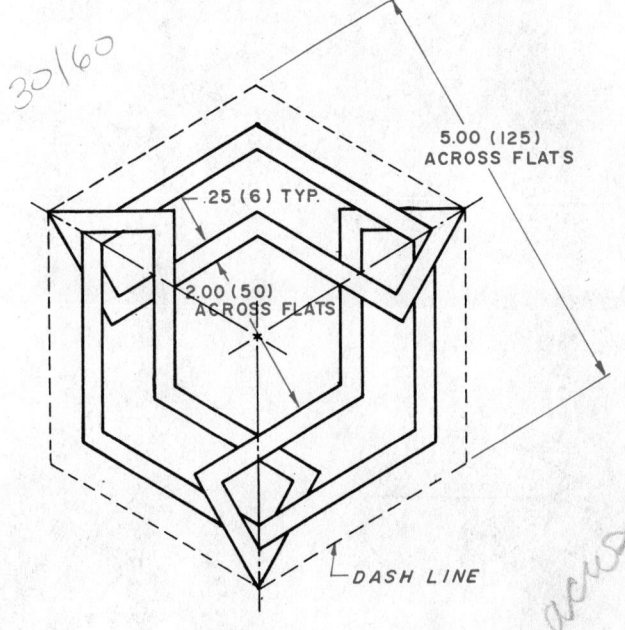

30/60

5.00 (125)
ACROSS FLATS

.25 (6) TYP.

2.00 (50)
ACROSS FLATS

DASH LINE

Problem 1-21

ctr. 30/60 measure marks

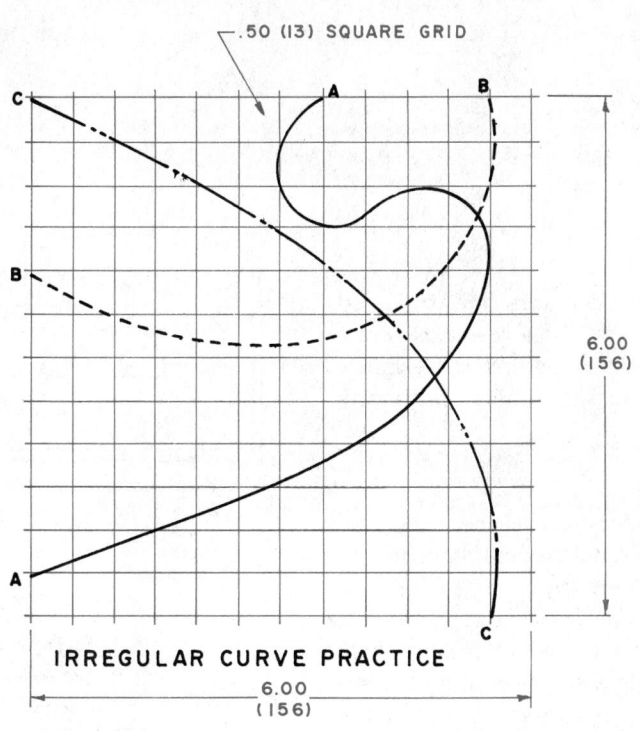

.50 (13) SQUARE GRID

C   A   B

B

6.00
(156)

A   C

IRREGULAR CURVE PRACTICE

6.00
(156)

Problem 1-22

Lettering, Sketching, and Line Techniques

Chapter Outline

Talking Sketching

Two types of sketching, which are often neglected in technical drawing texts but nevertheless complement the skills of a successful drafter, designer, or engineer, are conversational sketching and boardroom sketching.

Conversational Sketching

Conversational sketching occurs, usually between two or more individuals huddled around a drafting board, or a cafeteria table. The one doing the talking combines several types of sketches in one drawing as the idea is talked through. The first part of the sketch is most likely an orthographic view, the easiest projection to draw rapidly. Then, as the talking progresses, a pictorial is used to give the observers a feeling of depth. For example, a designer communicating an idea for an AM/FM receiver would begin with the orthographic view, **Figure 2-1**, and finish with a pictorial, **Figure 2-2**. A drafter, designer, or engineer sketches rapidly while talking, so doesn't have time to develop a single specific kind of sketching, such as orthographic views, an isometric, an oblique, or a two-point perspective.

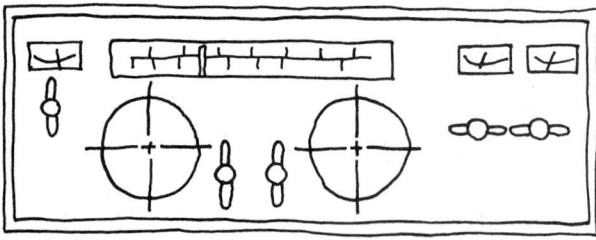

Figure 2-1   Talking sketch (orthographic view)

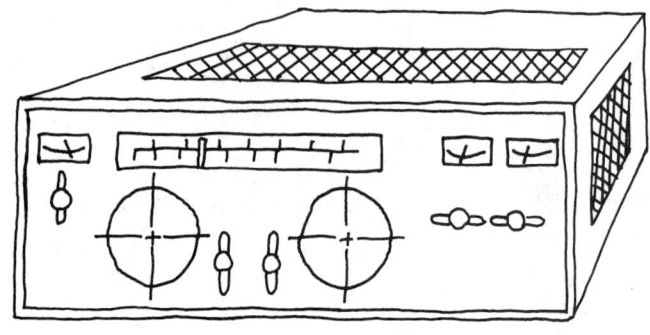

Figure 2-2   Talking sketch completed with pictorial

Boardroom Sketching

Communication with more individuals than a drafting board can accommodate requires a more visible surface for the larger group. The drawing surface in a boardroom sometimes is a chalkboard, but more often is one of three types: a pastel-colored porcelainized enamel, a glass panel, or a dry-erase surface. All of these surfaces require a special marker, and can be erased with a cloth.

Quality talking sketches, like quality sketching discussed earlier in this chapter, require practice to develop visible lines and especially, to develop lettering large enough to be read by those sitting at the back of the room.

Note that sketching and lettering on overhead projectors is another medium for talking sketching. Practice to make sure that all those present can read all the lettering and see the lines. Additional comments on the use of overhead projectors are in Chapter 24.

Freehand Lettering

Text is an important part of a technical drawing. Not all information required on technical drawings can be communicated graphically; the most obvious data being dimensions. *Text* on technical drawings consists of dimensions, notes, legends, and other data that are best conveyed using alphanumeric characters, *Figure 2-3*.

Several different ways are used to create text on technical drawings. The traditional method is by freehand lettering. Other methods include such mechanical lettering techniques as scriber templates, typewritten notation, and typed lettering generated by computer-aided drafting systems. This chapter focuses on freehand lettering. Other methods are described elsewhere in this text.

Lettering Styles

There are numerous different lettering styles or fonts, *Figure 2-4*. The standard style for freehand lettering on technical drawings, as established in American National Standards document Y14.2-1973, is single-stroke Gothic lettering. Vertical, single-stroke Gothic letters are the most universally used of the various styles available to drafters, *Figure 2-5*.

Some modifications of the standard Gothic configuration of letters are often made, without actually changing from the Gothic style of lettering. One way is through the use of uppercase and lowercase letters, *Figure 2-6*, but

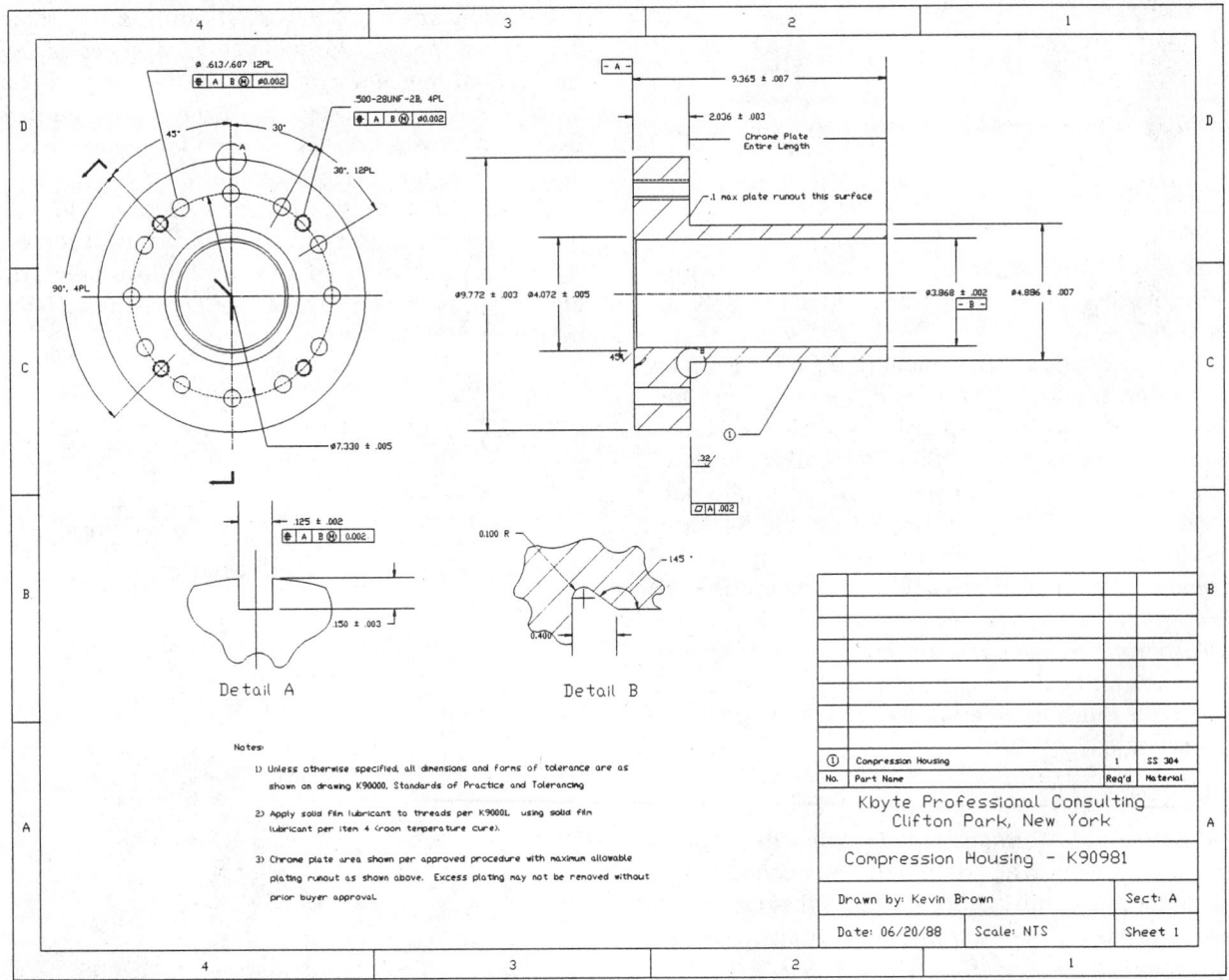

Figure 2-3   Examples of text on a technical drawing

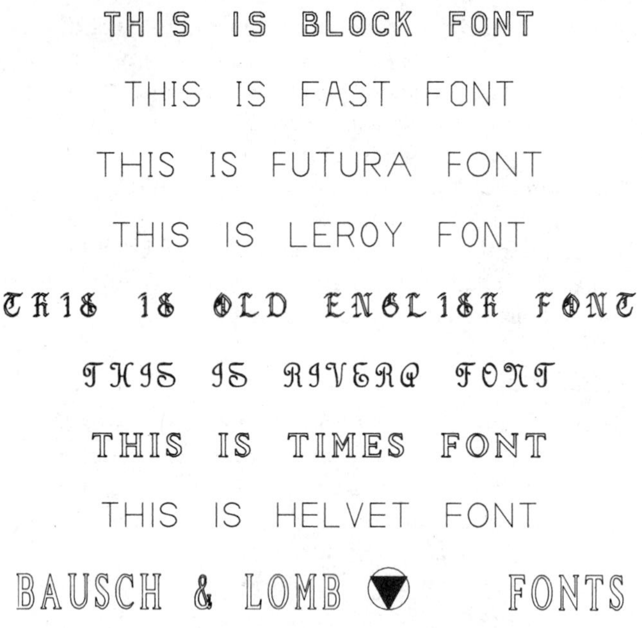

THIS IS BLOCK FONT

THIS IS FAST FONT

THIS IS FUTURA FONT

THIS IS LEROY FONT

THIS IS OLD ENGLISH FONT

THIS IS RIVERA FONT

THIS IS TIMES FONT

THIS IS HELVET FONT

BAUSCH & LOMB ▼ FONTS

Figure 2-4   Sample lettering fonts *(Courtesy Bausch & Lomb, Inc.®)*

SINGLE-STROKE GOTHIC LETTERING SAMPLE

Figure 2-5   Single-stroke Gothic lettering

this is seldom acceptable on technical drawings. Another method is to condense or extend the letters, *Figure 2-7*.

The most common way of modifying Gothic letters is by inclining them slightly to the right, *Figure 2-8*. Inclined lettering is easier to make as it lends itself to a natural direction of wrist action. The correct angle of the inclined elements is a two-unit incline to the right for each five units of letter height. Errors are not as detectable with inclined letters as they are with vertical elements. As the inclined elements are longer, they are easier to read. However, inclined lettering is not universally accepted, and caution must be exercised to not conflict with customary drafting styles. A backhanded or left-leaning inclination is never an acceptable modification.

Characteristics of Good Lettering

Good freehand lettering, regardless of whether it is uppercase or lowercase, condensed or extended or vertical or inclined, must have certain characteristics. These requisites include neatness, uniformity, stability, proper spacing, and speed.

Neat lettering is important so that the information being conveyed can be easily read. Few things detract

UPPERCASE GOTHIC
lowercase Gothic

Figure 2-6   Uppercase and lowercase Gothic lettering

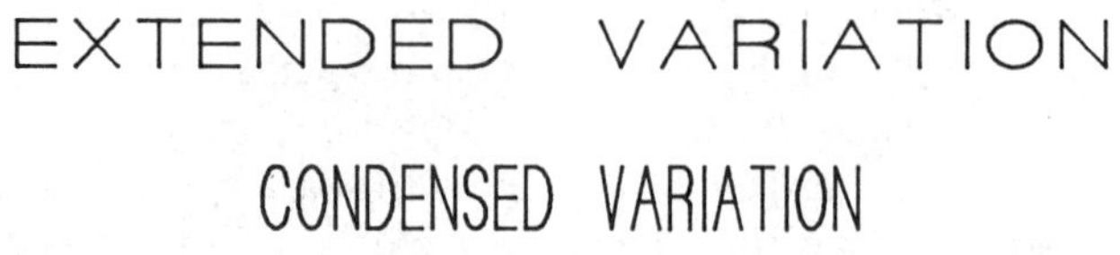

EXTENDED VARIATION

CONDENSED VARIATION

Figure 2-7   Extended and condensed variations of Gothic lettering

SAMPLE OF INCLINED LETTERING

Figure 2-8   Inclined Gothic lettering

from the appearance and quality of a technical drawing more than sloppy lettering, *Figure 2-9*.

For uniformity, all letters should be the same in height, proportion, and inclination. A necessary tactic for maintaining uniformity is the use of guidelines, *Figure 2-10*. The customary heights of characters in technical drawing are 1/8" (6 mm) for regular text, and 3/16" (9 mm) for headings and titles.

The proper stability or balance of letters is an important characteristic in freehand lettering. Each letter should appear balanced and firmly positioned to the human eye. Top-heavy letters are not balanced because they appear about to topple over, *Figure 2-11*.

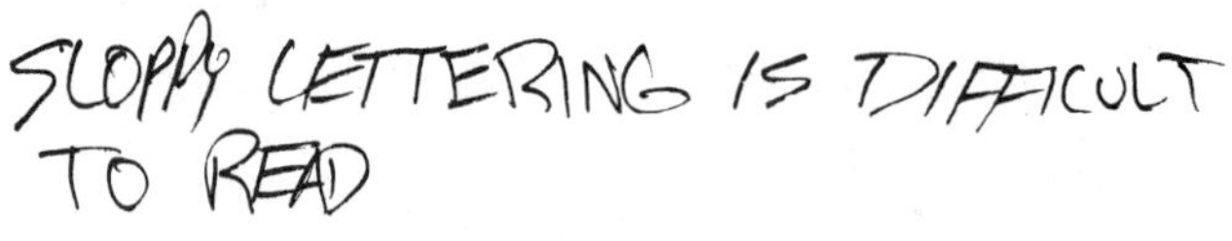

SLOPPY LETTERING IS DIFFICULT TO READ

Figure 2-9   Sloppy lettering is difficult to read

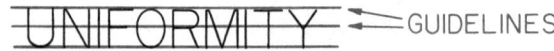

UNIFORMITY ⟵GUIDELINES

Figure 2-10   Guidelines improve uniformity

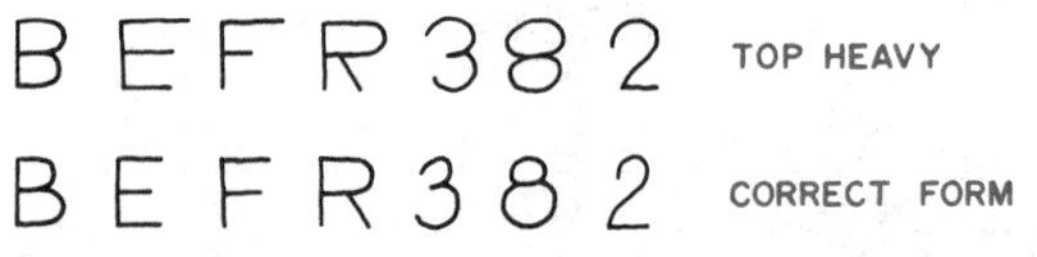

BEFR382   TOP HEAVY

BEFR382   CORRECT FORM

Figure 2-11   Top-heavy letters are not balanced

The proper spacing of letters and words is important, and it takes a lot of practice to accomplish. A good rule of thumb to follow in terms of spacing is to use close spacing within words, and far spacing between words, *Figure 2-12*. The proper positions of letters relative to one another in words is accomplished by spacing the letters in the word equally in the area, not by trying to equalize the spacing between letters. This becomes automatic if the drafter concentrates on the word being lettered, not on each letter. Another rule of thumb for spacing is to allow the width of one round letter, such as O, C, Q or G, between words. *Figure 2-13* illustrates how this type of spacing can make the lettering much easier to read.

In the modern drafting room, because time is money, speed in freehand lettering is critical. Typically, freehand lettering is one of the slowest, most time-consuming tasks drafters must perform. It takes many hours of practice to develop freehand lettering that is neat, uniform, balanced, properly spaced, and fast. Some drafters never reach this goal. Those who do, reach it through constant practice, coupled with continual efforts to improve.

Freehand Lettering Techniques

Freehand lettering techniques are learned by knowing what grades of lead to use, how to make the basic lettering strokes, and how to use guidelines; and by constantly practicing and trying to improve.

Lettering in ink has been greatly simplified in recent years. Old-fashioned tools, such as adjustable nib ruling pens and speedball pens, have been replaced by the less cumbersome, easier-to-use technical pen, *Figure 2-14*. When lettering in ink, drafters still use light guidelines made with pencil lead. The actual lettering is done with the desired pen point size. Commonly used pen points for lettering in ink are sizes 0, 1, and 2, which are standard American sizes. In metrics, these point sizes represent line widths of 0.35 mm, 0.50 mm, and 0.60 mm.

All letters and numbers are created using six basic strokes, *Figure 2-15*. The first stroke is a single stroke made downward and to the right at approximately 45 degrees. The second stroke is made downward and to the left at approximately 45 degrees. The third stroke is vertical, and is made from top to bottom. Stroke number four is horizontal, and is made from left to right. The fifth stroke is a half-circular stroke to the left, made from top to bottom. The sixth stroke is a half-circular stroke to the right, made from top to bottom. All alphanumeric characters can be created using combinations of these six strokes. *Figure 2-16* shows how these strokes are used for making selected characters.

Lettering Guidelines

Guidelines are a critical part of freehand lettering. Uniformity, neatness, and stability cannot be achieved without using guidelines.

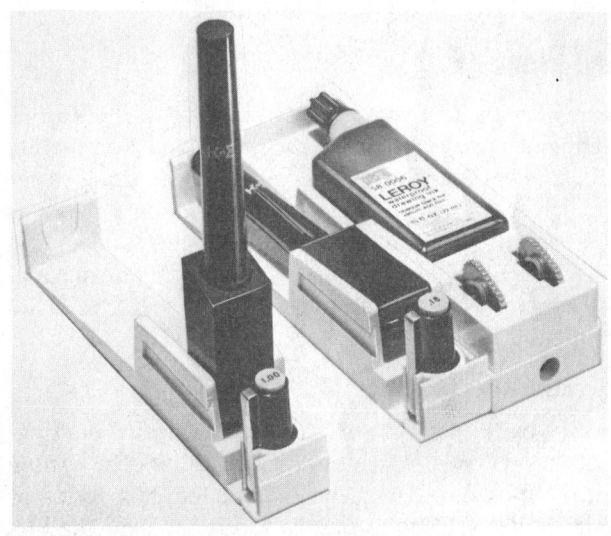

Figure 2-14   Modern technical pens *(Courtesy Keuffel & Esser Co.®)*

SPACING   WITHIN   WORDS
SHOULD   BE   CLOSE.

SPACING   BETWEEN   WORDS
SHOULD   BE   FAR.

Figure 2-12   The proper spacing of letters and words

THIS   IS   A   PROPERLY   SPACED   SENTENCE.

THISISANIMPROPERLYSPACEDSENTENCE.

Figure 2-13   Spacing between words is important

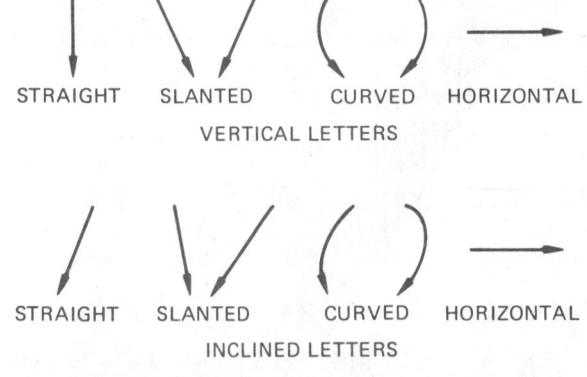

Figure 2-15   Basic strokes used for lettering

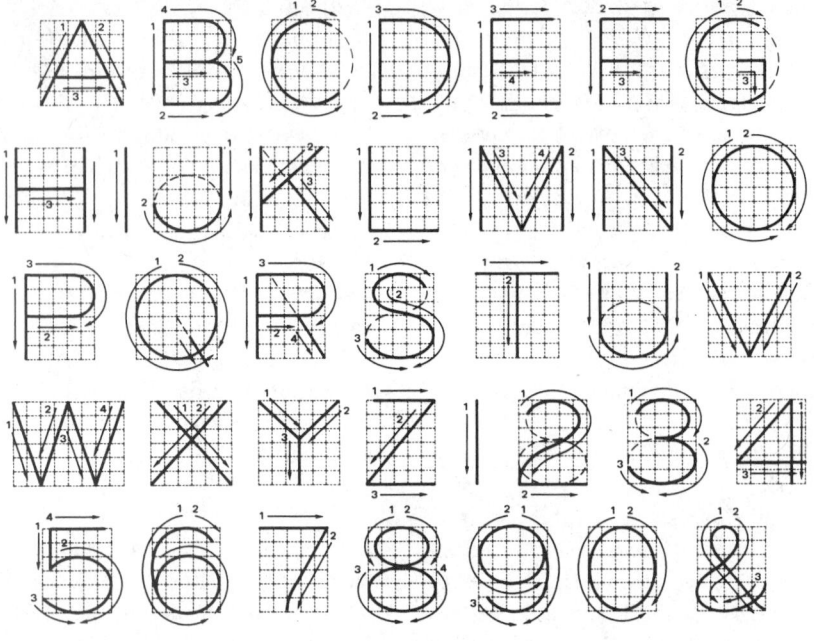

Figure 2-16 Forming uppercase Gothic letters and numerals—vertical style

Line Work

Line work is the generic term given to all of the various techniques used in creating the graphic data on technical drawings. Mechanical line work is made using either mechanical pencils or technical pens. Such devices as parallel bars, drafting machines, triangles, scales, and numerous other tools are used to guide the line-making. Since inking is dealt with in the first chapter, this chapter focuses on pencil line work.

Characteristics of Lines

Twelve basic types of lines are used in manual drafting. Each has its own individual characteristics. The *visible* line is thick and dark. The *hidden* line is a series of short dashes separated by even shorter breaks. The hidden line is thinner than the visible line. *Dimension* lines and *extension* lines are solid, thin lines of approximately the same width as hidden lines. *Dimension* lines should be broken for dimensions, and have arrowheads for terminations.

The *center* line is broken with one short dash in its center. It is the same width as the hidden, dimension, and extension lines. The *phantom* line is just like the center line except that it has two dashes. The dashes are repeated approximately every two inches. The *cutting plane* line is thick like the visible line and consists of a series of long, equally spaced dashes. All lines used on technical drawings should closely match those in *Figure 2-17*.

Horizontal and Vertical Lines

Horizontal lines are formed by pressing the straightedge (T-square, parallel bar, drafting machine scale, and so forth) against the worksheet with one hand and moving

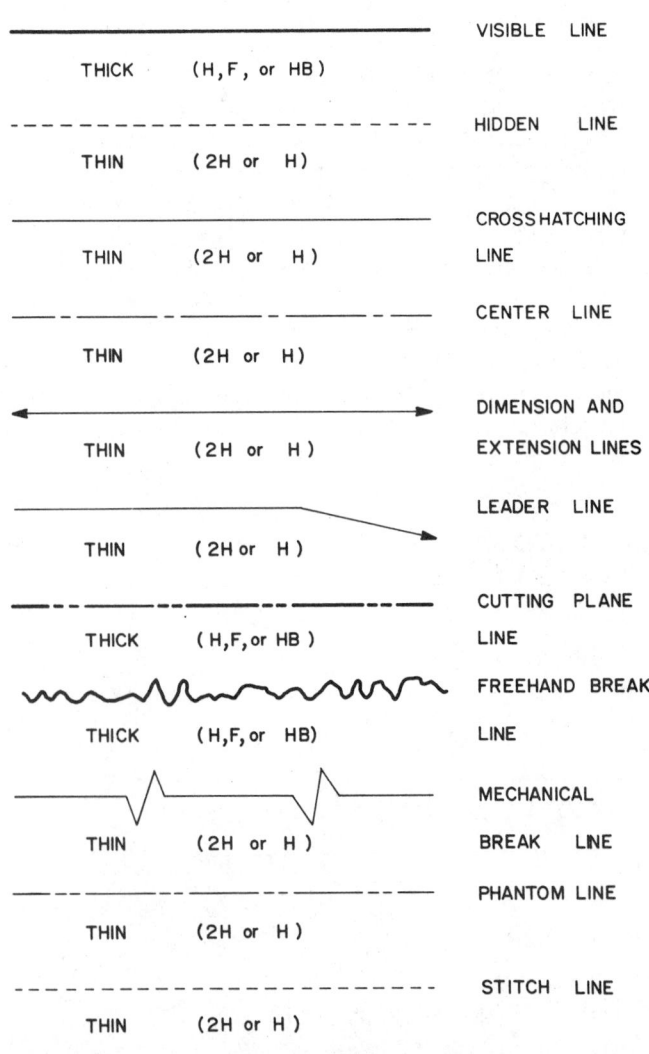

THICK　　(H, F, or HB)	VISIBLE　LINE
THIN　　(2H or　H)	HIDDEN　LINE
THIN　　(2H or　H)	CROSSHATCHING LINE
THIN　　(2H or　H)	CENTER　LINE
THIN　　(2H or　H)	DIMENSION AND EXTENSION LINES
THIN　　(2H or　H)	LEADER　LINE
THICK　　(H, F, or HB)	CUTTING　PLANE LINE
THICK　　(H, F, or HB)	FREEHAND BREAK LINE
THIN　　(2H or　H)	MECHANICAL BREAK　LINE
THIN　　(2H or　H)	PHANTOM LINE
THIN　　(2H or　H)	STITCH　LINE

Figure 2-17 Line types used on technical drawings

the pencil with the other. Uniformity of line widths and weights can be achieved by holding the pencil at approximately 60 degrees from the drawing surface, maintaining an even pressure downward, and slowly revolving the pencil axially as it moves across the drawing surface, *Figure 2-18*. This keeps the lead tip symmetrical.

Vertical lines are created according to the same principles, except that the drafter's hand moves upward rather than from left to right. The angle of inclination, the amount of pressure, and the rotating motion are the same as they are for horizontal lines.

Angular Lines

Many modern devices are available to assist drafters in making angular lines. These include protractors, adjustable triangles, and adjustable arms on drafting machines. However, most angular lines can be created simply by using the standard 30°-60° and 45° triangles alone, and in various combinations, *Figures 2-19, 2-20, 2-21, and 2-22.* These standard tools create angles of 15°, 30°, 45°, 60°, and 75°.

Parallel Lines

Parallel lines can be created in a number of different ways. Vertical (and horizontal) parallel lines are made by simply moving the straightedge the required distance and making each successive line, *Figure 2-23*.

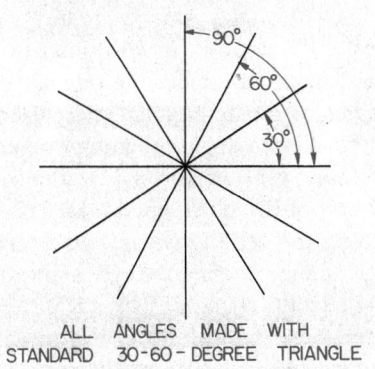

ALL ANGLES MADE WITH STANDARD 30-60-DEGREE TRIANGLE

Figure 2-20   Making angular lines with the 30°60° triangle

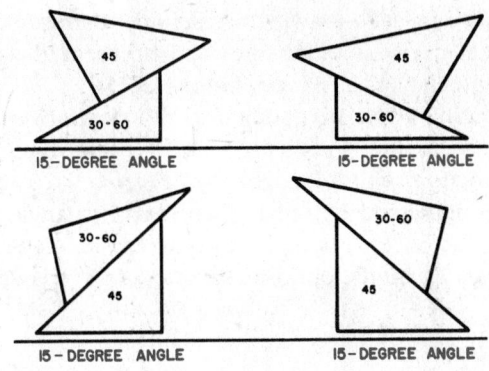

Figure 2-21   Making 15° angles

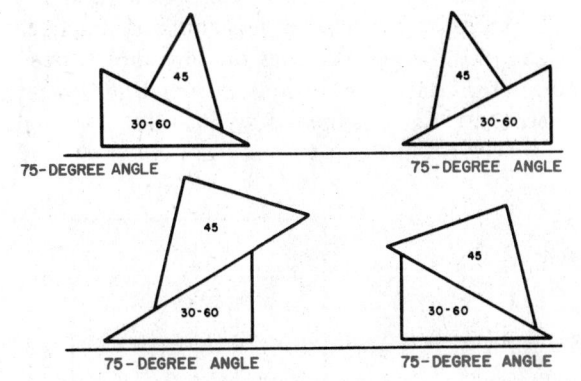

Figure 2-22   Making 75° angles

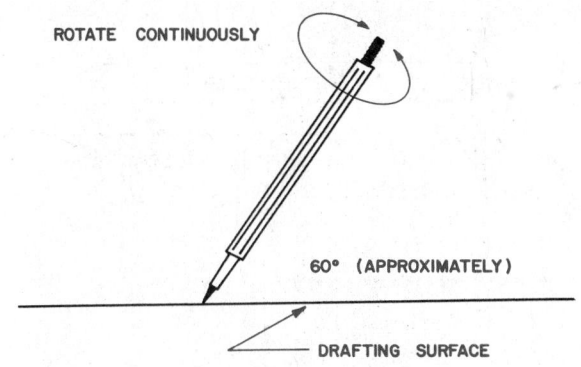

Figure 2-18   Maintaining uniformity of lines

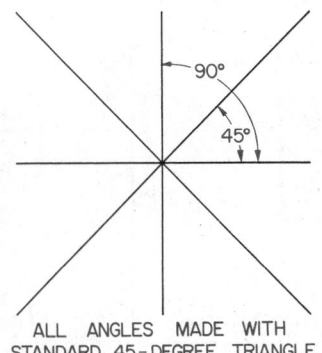

ALL ANGLES MADE WITH STANDARD 45-DEGREE TRIANGLE

Figure 2-19   Making angular lines with the 45° triangle

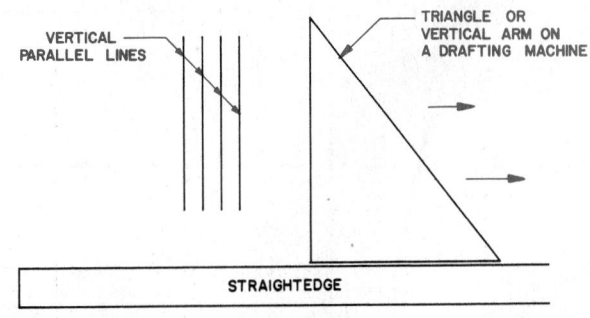

Figure 2-23   Making vertical parallel lines

Parallel lines at angles can be created by using the 30°-60° and 45° triangles in combination much the same as they are used for making angular lines. When using triangles to create angular lines, the first line is created at the desired angle. Aligning one edge of a triangle to the line, register any side of the second triangle against one of the nonaligned edges of the first triangle. Holding the second triangle to prevent it from moving, and sliding the first triangle along the engaged edge of the second triangle, will reposition the originally aligned edge to any desired parallel position. Successive parallel lines are created in the same way, *Figure 2-24*.

Perpendicular Lines

Drawing perpendicular lines can be accomplished in a manner similar to drawing parallel lines. Horizontal and vertical perpendicular lines can be created using a straightedge and a triangle, *Figure 2-25*.

Creating a line perpendicular to a nonhorizontal or nonvertical line is accomplished by using triangles in conjunction with a straightedge, *Figure 2-26*. Line 1 in this figure is drawn first. Then the 45° triangle is slid along the 30°-60° triangle, and the perpendicular line is created with the opposite side of the 45° triangle.

Sketching

Even in the world of high technology and computers, sketching is still one of the most important skills for drafters and designers. Sketching is one of the first steps in communicating ideas for a design, and it is used in every step thereafter. It is common practice for designers to prepare sketches that are turned over to drafters for conversion to finished working drawings. *Figure 2-27* is an example of a typical design sketch.

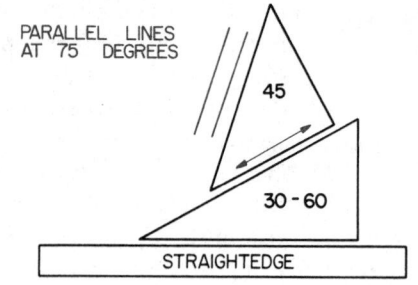

Figure 2-24   **Creating successive parallel lines**

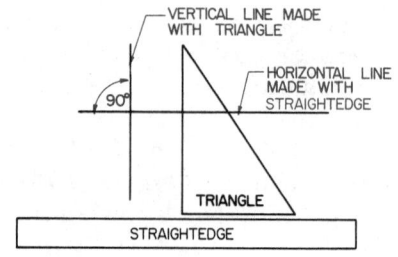

Figure 2-25   **Making perpendicular lines**

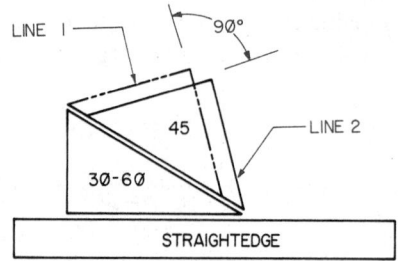

Figure 2-26   **Creating a line perpendicular to a nonvertical or nonhorizontal line**

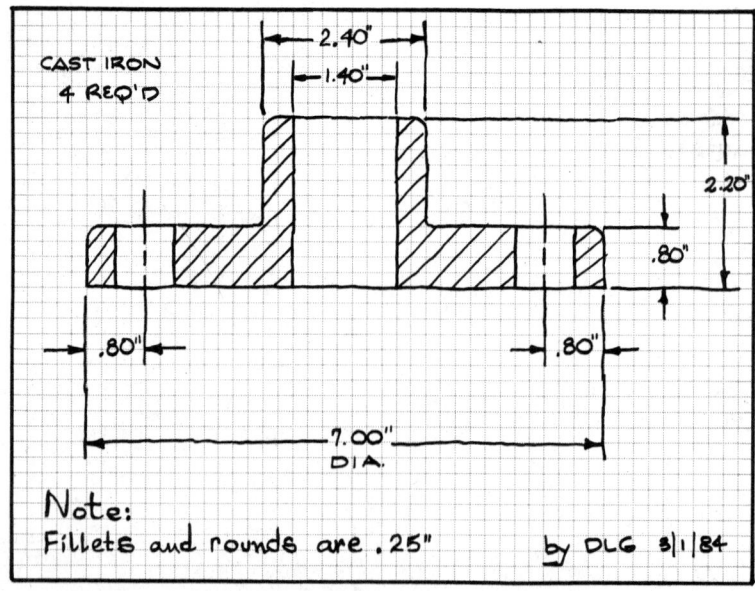

Figure 2-27   **Typical design sketch**

Sketching Lines

The lines used in creating sketches closely correspond to those used in creating technical drawings except, of course, that they are not as sharp and crisp. *Figure 2-28* illustrates the various types of lines used in making sketches.

The basic line types are: visible line, hidden line, center line, dimension line, sectioning line, extension line, and cutting plane line. These lines represent the various lines available for creating sketches. The character of each line, as illustrated in *Figure 2-29*, should be closely adhered to when making sketches.

Types of Sketches

The types of sketches correspond to the types of technical drawings. There are four types of sketching: orthographic, axonometric, oblique, and perspective, *Figures 2-30, 2-31, 2-32,* and *2-33.*

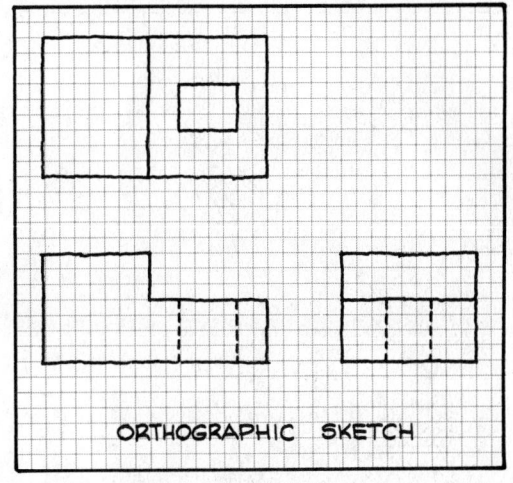

Figure 2-30   Orthographic sketch

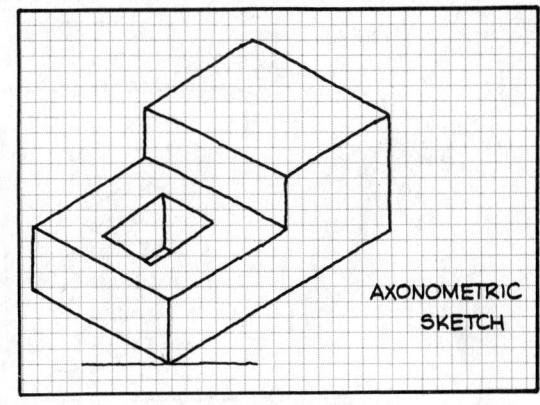

Figure 2-31   Axonometric sketch

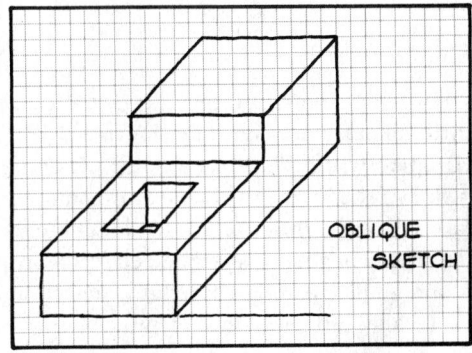

Figure 2-32   Oblique sketch

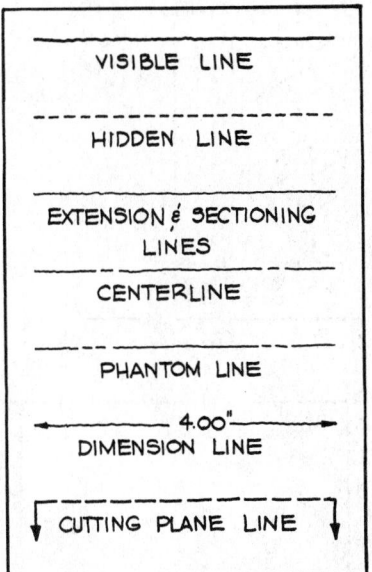

Figure 2-28   Lines used in sketching

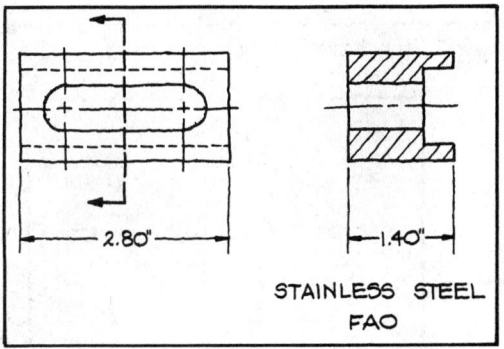

Figure 2-29   Sample design sketch

Orthographic sketching relates to flat, graphic facsimiles of a subject showing no depth. Six principal views of a subject may be incorporated in an orthographic sketch: top, front, bottom, back, right side, and left side, *Figure 2-34.* The views selected for use in a sketch depend on the nature of the subject and the judgment of the sketcher.

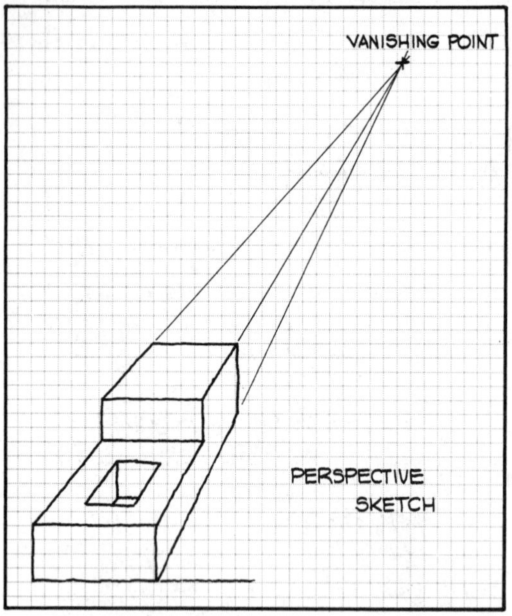

Figure 2-33 Perspective sketch

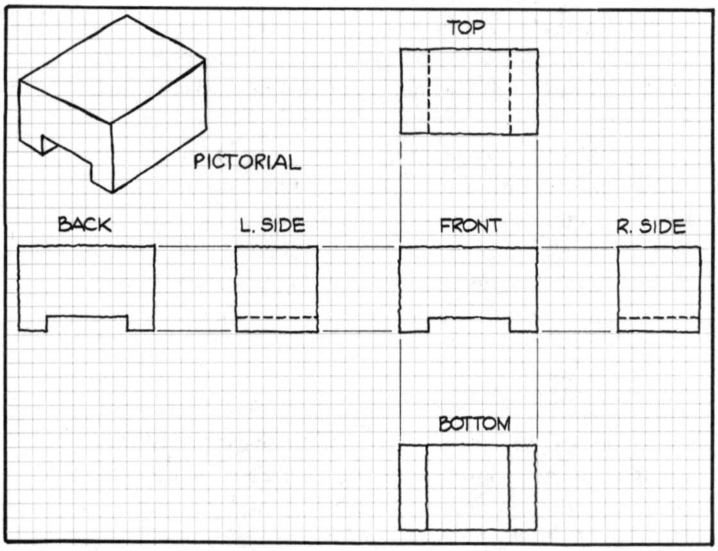

Figure 2-34 Six principal views in orthographic sketching

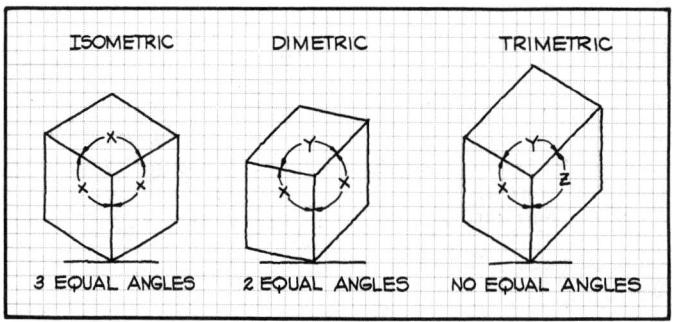

Figure 2-35 Three types of axonometric sketches

Axonometric sketching may be one of three types: isometric, dimetric, or trimetric, ***Figure 2-35.*** The type most frequently used is *isometric*, in which length and width lines recede at 30° to the horizontal and height lines are vertical, ***Figure 2-36.*** In sketching, the use of these terms is academic as they relate to proportional scales and angle positions of length, width, and depth, which are only estimated in sketching.

Oblique sketching involves a combination of a flat, orthographic front surface with depth lines receding at a selected angle (usually 45°), ***Figure 2-37.***

Perspective sketching involves creating a graphic facsimile of the subject. Consequently, depth lines must recede to a hypothetical vanishing point (or points), ***Figures 2-38*** and ***2-39.*** In fact, all pictorial sketches naturally tend to assume characteristics of perspective sketches as a result of how the eye views the apparent relative proportions of objects. This is not necessarily undesirable.

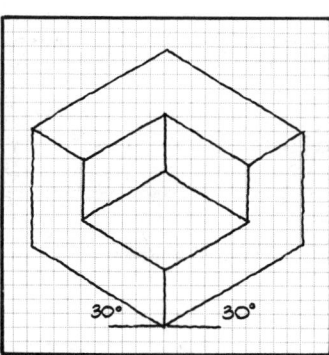

Figure 2-36 Isometric sketch

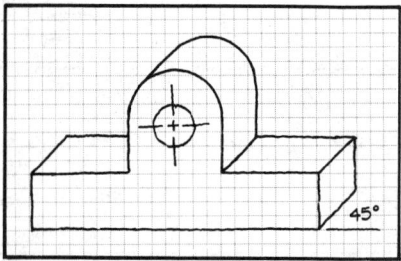

Figure 2-37 Oblique sketch

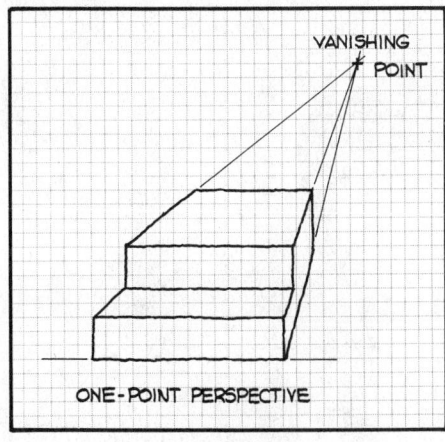

Figure 2-38   One-point perspective sketch

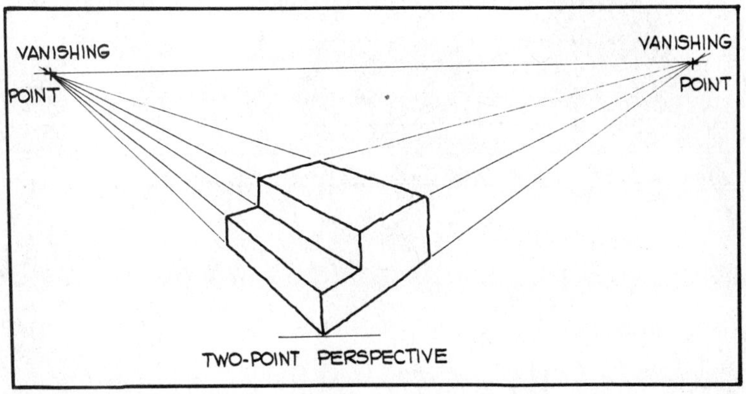

Figure 2-39   Two-point perspective sketch

Sketching Materials

An advantage of sketching is that it requires very few material aids. Whereas drafters must have a complete collection of tools, equipment, and materials in order to do working drawings, sketching requires nothing more than a pencil and a piece of paper. It is not uncommon for a sketch to be drawn on a paper napkin during a hurried luncheon meeting.

Sketching done in an office environment requires three basic materials: pencil, media (paper or graph paper), and an eraser. Graph paper simplifies the sketching process considerably, especially for students just learning, and should be used freely.

Sketching Techniques

Sketches, as with drawings, consist of straight and curved lines. With practice, drafters can become skilled in creating neat, sharp, clear examples — straight or curved — of all the various line types introduced previously. When sketching, the following general rules apply.

1. Hold the pencil firmly, but not so tightly as to create tension or hand fatigue.
2. Grip the pencil approximately one inch to one and one-half inches up from the point.
3. Maintain a comfortable angle between the pencil and the sketching strokes.
4. Draw horizontal lines from left to right using short, slightly overlapping strokes.
5. Draw vertical lines from top to bottom using short, slightly overlapping strokes.
6. Draw curved lines using short, slightly overlapping strokes.

In addition to these general rules, some specific techniques are used in making the various line types for sketching.

Sketching Straight Lines and Curves

Making straight lines on graph paper is a simple process of guiding the pencil using the existing lines. If graph paper is not available, pencil dots can be positioned to plot the path of the line, *Figure 2-40*. In this figure, the sketcher enters a series of pencil dots on the paper which provide a basic outline as to the shape of the object. Then, using a series of short, slightly overlapping strokes, the pencil dots are connected, *Figures 2-41, 2-42,* and *2-43*. This technique is also used for curved lines, *Figure 2-44*.

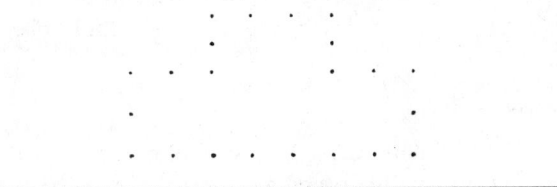

Figure 2-40   Dots used as guides in sketching

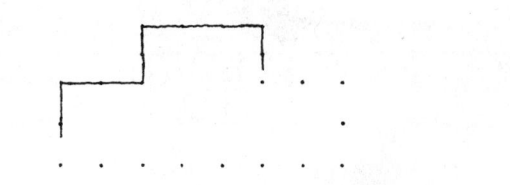

Figure 2-41   Using dots in sketching lines

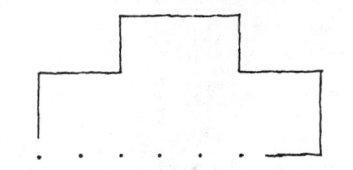

Figure 2-42   Continuing the sketch by connecting the dots

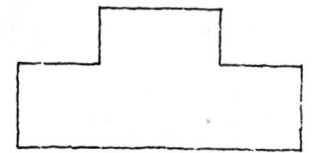

Figure 2-43   Completing the sketch made by using dots as guides

Figure 2-44   Using dots in sketching curved lines

Sketching Circles

Figure 2-45 illustrates a series of six steps that can be used for sketching a circle. Vertical and horizontal center lines are sketched, which positions the center of the circle (Step 1), and the radial distances of the desired circle size are marked on each of these lines, equidistant from the center (Step 2). A square is drawn symmetrically around the center, with the sides located at the radial line marks (Step 3). On the diagonals of the square (Step 4), the radial distances are again marked off from the center (Step 5). This provides four positions for the circumference to pass through at the sides of the square, and four more positions on the diagonals of the square. The right half of the circle is sketched in from top to bottom using short, slightly overlapping strokes, and then the left half is sketched in the same manner (Step 6).

Sketching Ellipses

A similar technique is used for sketching ellipses, except that the square becomes a rectangle, ***Figure 2-46***. Ellipses are oriented on the object being sketched as shown in the diagram in ***Figure 2-47***.

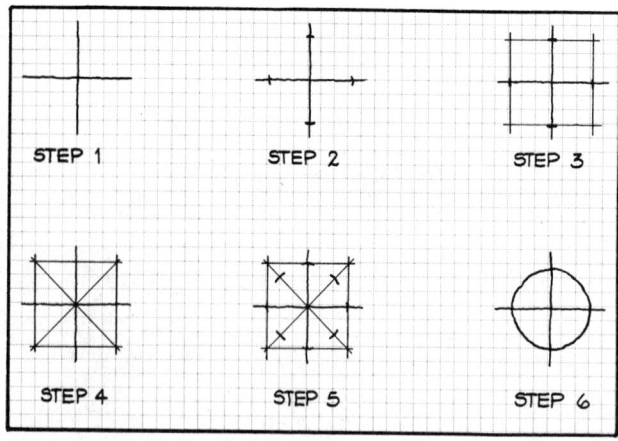

Figure 2-45   Sketching a circle

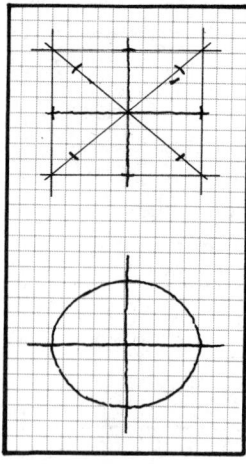

Figure 2-46   Sketching an ellipse

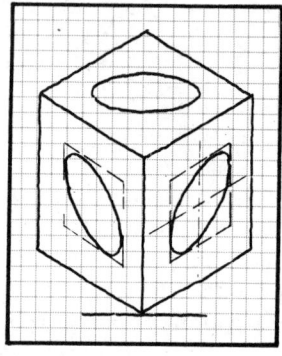

Figure 2-47   Orienting an ellipse on an object

Proportion in Sketching

Sketches are not done to scale, but it is important that they be made proportionately accurate. All of the various components of a sketch should be kept in proportion to those of the actual object. This technique takes a great deal of practice to master.

Some methods for achieving proportion recommend using a pencil or a strip of paper as a simulated scale. These techniques are not only unrealistic in terms of the real world, they defeat the very purpose of sketching. A skilled sketcher must learn to maintain proportion without the use of tools and aids. The best device for accomplishing proportion in sketching is the human eye. With practice, the drafter can become proficient in maintaining proportion without the use of extraneous, time-consuming devices. The following general rules relating to proportion will also help.

Step 1   In sketching, use graph paper whenever possible.

Step 2   Examine the object to be sketched and mentally break it into its component parts.

Step 3   Beginning with the largest components (length and width), estimate the proportion, such as the length is 4/3 times the width or 5/2 times the width, and so on.

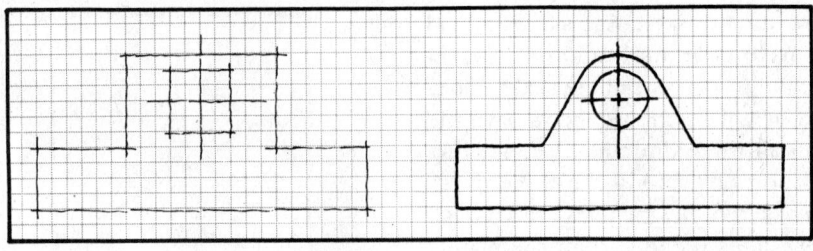

Figure 2-48   **Blocking in components**

Step 4   Lay out the largest component according to the proportions decided upon in Step 3. Use construction line squares and rectangles to block in irregularly shaped components, *Figure 2-48.*
Repeat Steps 3 and 4 until the entire object is finished.

Orthographic Sketching

Orthographic sketching may involve sketching any combination of the six principal views of the subject. The top, front, and right-side views are normally selected for representing an object in an orthographic sketch. However, these views are not always appropriate. The sketcher must learn to choose the most appropriate views. These are the views that show the most detail and the fewest hidden lines. A good rule of thumb to use in selecting views is to select the views which would give you all of the information you would need if you had to make the object yourself.

Once the views have been selected, the orthographic sketch may be laid out using the techniques set forth earlier in this chapter. To ensure that the sketched views align, the entire sketch should be blocked in before adding details, *Figure 2-49.* Once the layout is blocked in, the details can be added one view at a time, *Figures 2-50, 2-51,* and *2-52.*

Axonometric Sketching

As was mentioned earlier, there are three types of axonometric projection: isometric, dimetric, and trimetric.

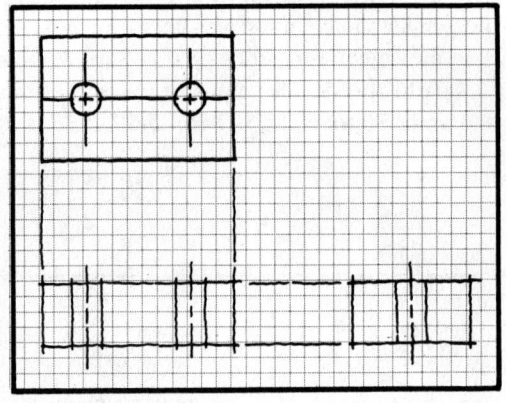

Figure 2-50   **Completing the top view**

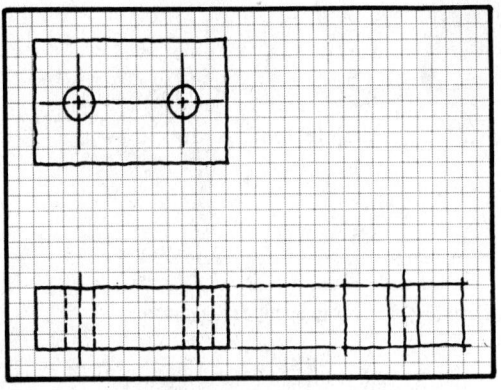

Figure 2-51   **Completing the front view**

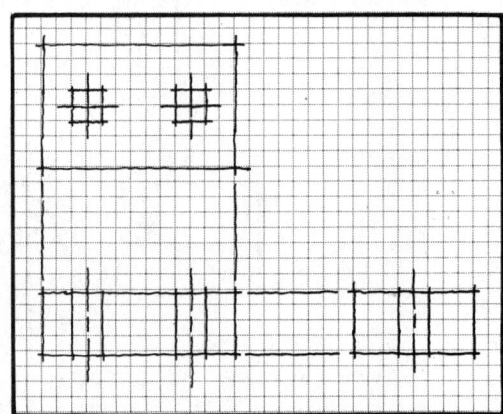

Figure 2-49   **Blocking in an orthographic sketch**

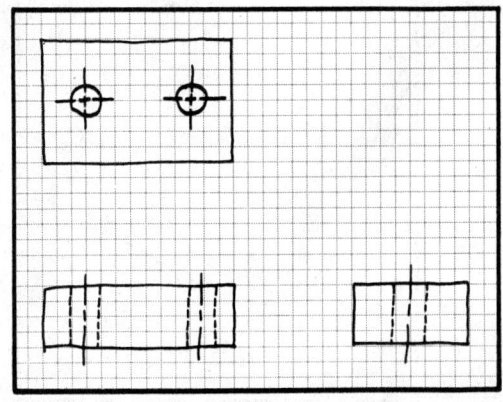

Figure 2-52   **Completing the sketch**

Isometric projection is used in sketching. Dimetric and trimetric projection have little application in sketching, due to the difficulty in proportioning scale values of length, width, and height. Isometric views have the same scaling value in all three directions, eliminating the need to vary proportions among the three directions. In an isometric sketch, height lines are vertical, and width and length lines recede at approximately 30° and 150° (180°-30°) from the horizontal.

The first step in creating an isometric sketch is to lay out the isometric axis, *Figure 2-53*. All normal lines will be parallel to one of the axis lines. The next step is to block in the object using construction lines, *Figure 2-54*. Five steps in the development of an isometric sketch are shown in *Figures 2-55, 2-56, 2-57, 2-58*, and *2-59*.

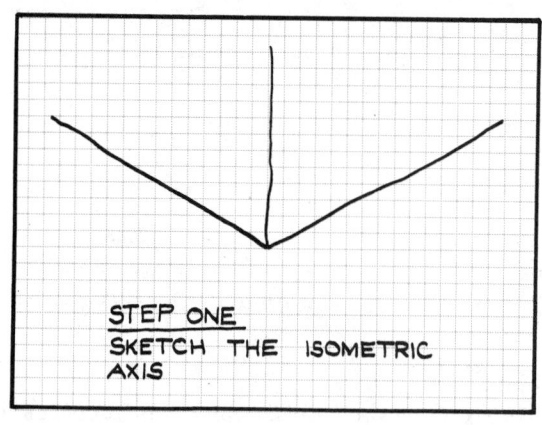

Figure 2-55   Step one in making an isometric sketch

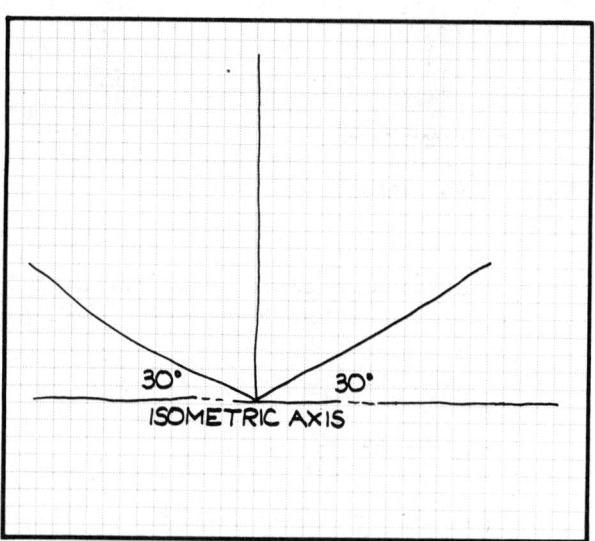

Figure 2-53   The isometric axis in sketching

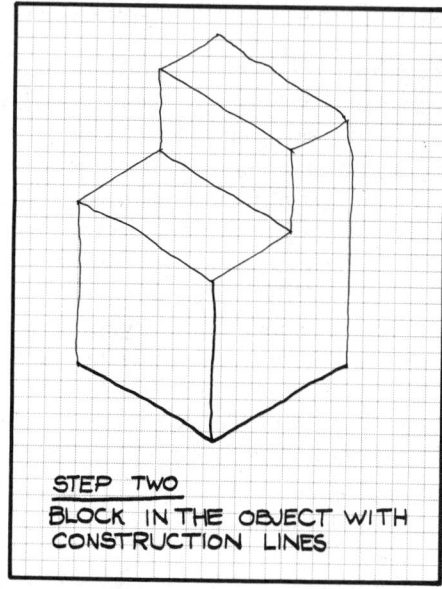

Figure 2-56   Step two in making an isometric sketch

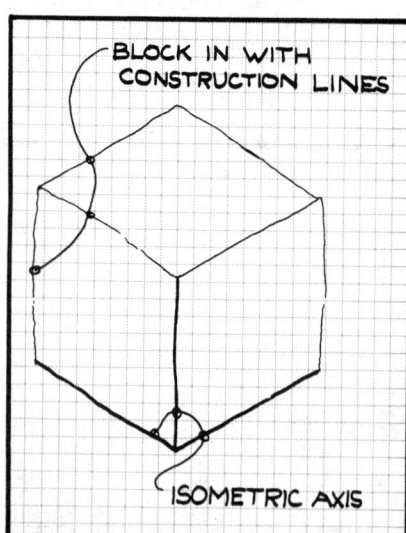

Figure 2-54   Blocking in the object

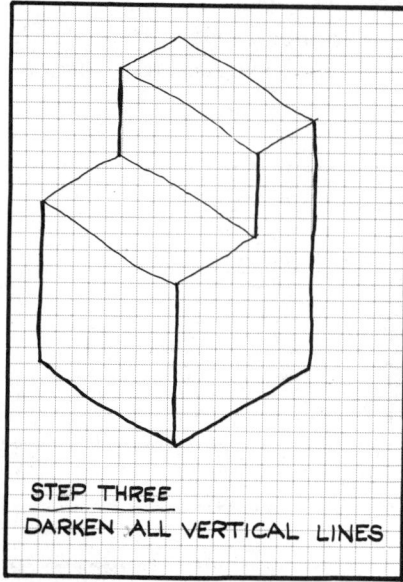

Figure 2-57   Step three in making an isometric sketch

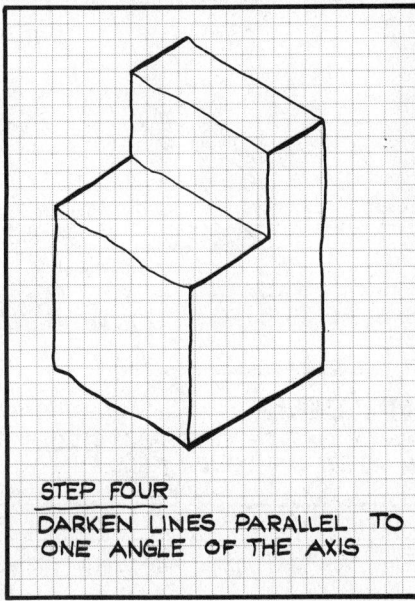

Figure 2-58   Step four in making an isometric sketch

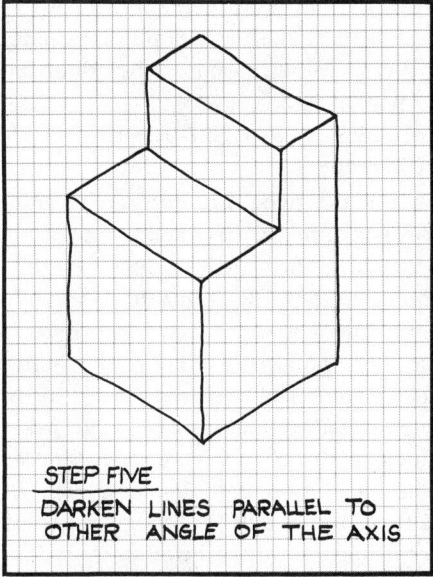

Figure 2-59   Step five in making an isometric sketch

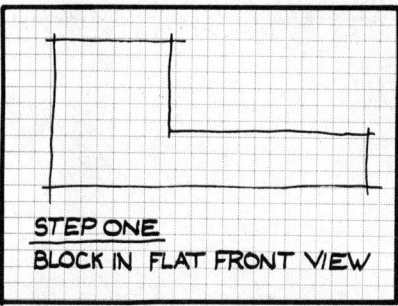

Figure 2-60   Step one in making an oblique sketch

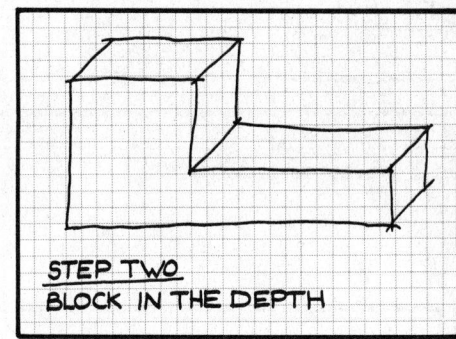

Figure 2-61   Step two in making an oblique sketch

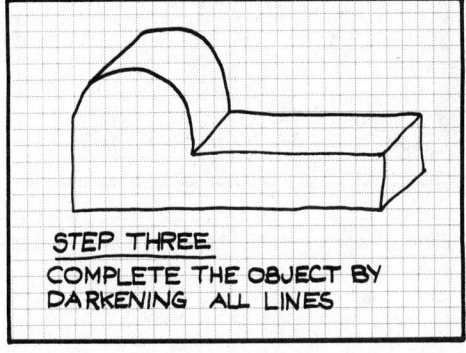

Figure 2-62   Step three in making an oblique sketch

Oblique Sketching

Oblique sketching involves laying out the front view of an object, and showing the depth lines receding at an angle (usually 45°) from the horizontal. Oblique sketching is particularly useful for dealing with an object having round components. Oblique sketching allows round components to be drawn round, rather than elliptical.

Using the blocking–in method, the flat front surface of the object is laid out. The depth is then blocked in using parallel lines, and the sketch is completed by outlining the exposed profile of the rear surface. *Figures 2-60, 2-61,* and *2-62* illustrate three steps in creating an oblique sketch.

Perspective Sketching

Perspective sketching closely approximates how the human eye actually sees an object. Two common types of perspective sketches are one-point and two-point perspectives.

A one-point perspective sketch is similar to an oblique sketch, except that depth lines recede to a vanishing point instead of receding parallel to one another, *Figure 2-63*. In constructing a one-point perspective, the following procedures apply.

Step 1   Lay out the flat front surface of the object using the blocking in method, *Figure 2-64*.

Step 2   Select and mark a single vanishing point. Project all points on the front surface back to the vanishing point, *Figure 2-65*.

Step 3   Estimate the depth of the object and mark it off on all line projectors, *Figure 2-66*.

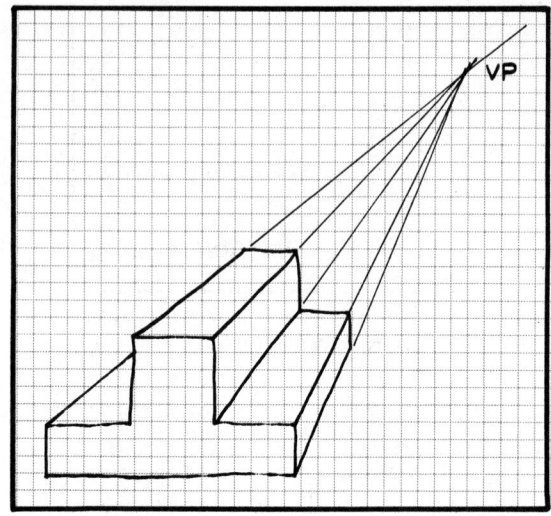

Figure 2-63   One-point perspective sketch

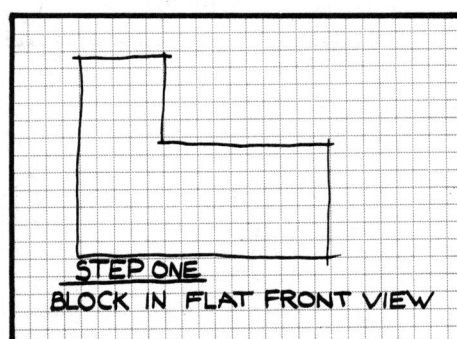

Figure 2-64   Step one in making a one-point perspective sketch

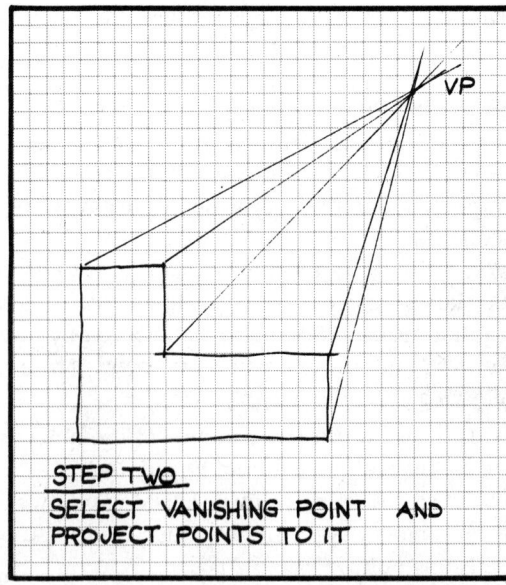

Figure 2-65   Step two in making a one-point perspective sketch

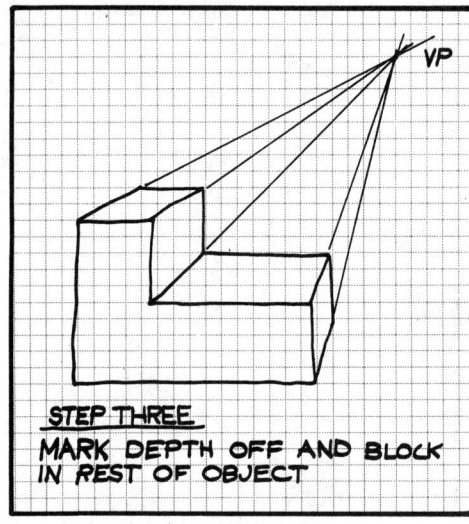

Figure 2-66   Step three in making a one-point perspective sketch

Step 4   Complete the sketch by outlining the exposed profile of the rear surface, *Figure 2-67*.

A two-point perspective resembles an isometric sketch, except that width and depth lines recede to the left and right vanishing points rather than receding in parallel, *Figure 2-68*.

In constructing a two-point perspective, the following procedures apply.

Step 1   Lay out the two-point perspective frame which consists of the vertical height line, the vanishing point left, the vanishing point right, and the receding lines (all estimated locations), *Figure 2-69*. The horizon line when positioned below the view provides a view of the bottom of the object, and when positioned above shows the top. The vanishing points must be on the horizon.

Step 2   Block in the object, estimating the length and width for proportion, *Figure 2-70*.

Step 3   Lay out the details, lightly giving special attention to proportion, *Figure 2-71*.

Step 4   Complete the two-point perspective sketch, *Figure 2-72*.

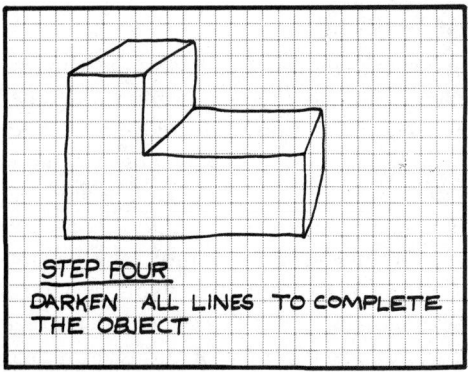

Figure 2-67   Step four in making a one-point perspective sketch

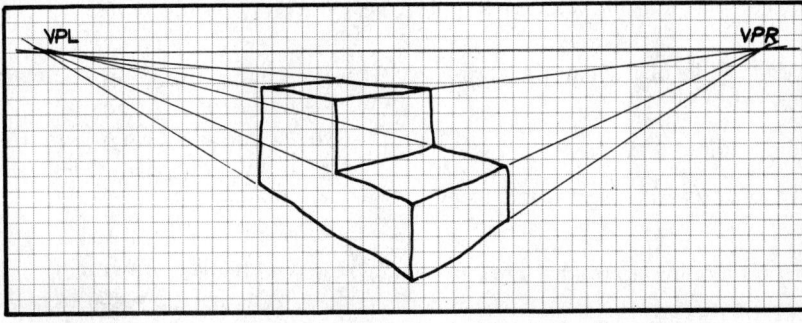

Figure 2-68   Two-point perspective sketch

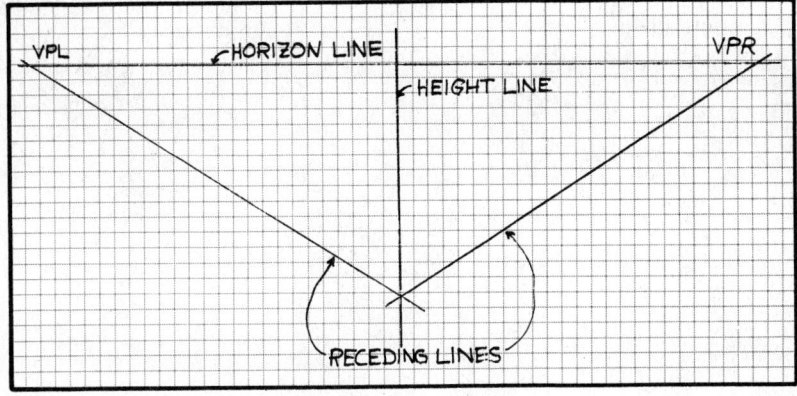

Figure 2-69   Laying out the two-point perspective frame

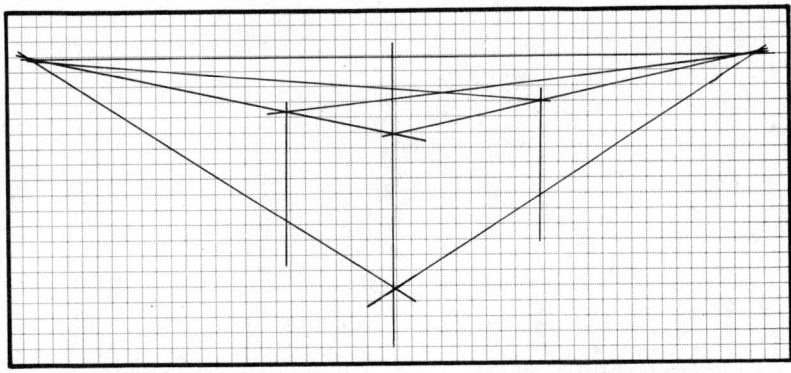

Figure 2-70   Blocking in the object

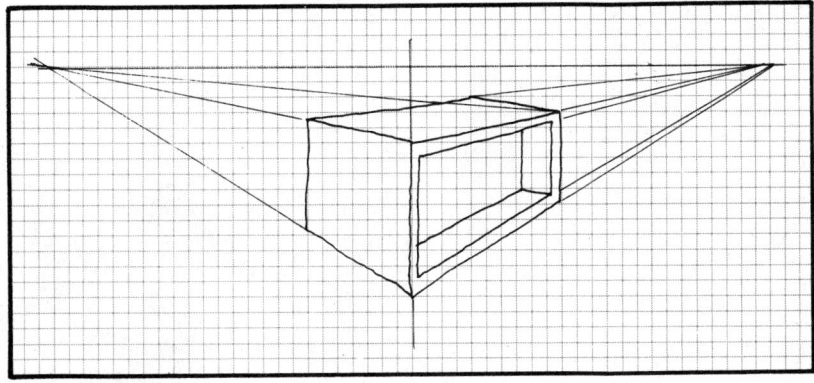

Figure 2-71   Laying out the details

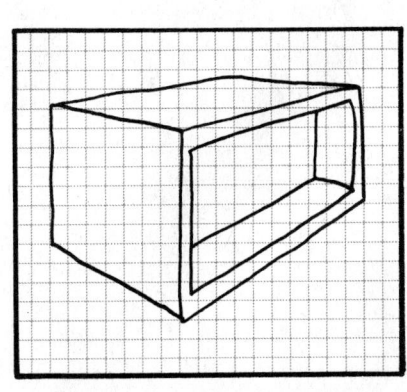

Figure 2-72   Completing the two-point perspective sketch

Review

1. What is the standard style of freehand lettering on technical drawings?

2. What is the slope of slanted lettering?

3. What are the five characteristics of good freehand lettering?

4. How many strokes are required to make a letter B?

5. How high should fractions be?

6. How much space should be left between words?

7. What are guidelines?

8. Why are guidelines important?

9. A setting of 8 on a lettering guide will produce letters of what height?

10. Define the term *line work*.

11. Name ten basic lines types used on technical drawings.

12. What are two advantages of sketching over mechanical drawing?

13. What are the six principal views of orthographic projection?

14. Which axonometric projection is preferred for sketching?

15. What kind of projection exhibits a circle as an elliptic shape?

Chapter Two Problems

Problems 2-1 through 2-4

The following problems are intended to give the beginning drafter practice in making neat uppercase, vertical, and Gothic-style letters and numbers. On an A-size sheet of vellum, carefully lay out light .12 spaces as shown. Use .50 borders all around and *be sure to skip a space* between lines of letters or numbers. Practice letters, numbers, words, sentences, and paragraphs, as given.

.12
(3)

YOUR FIRST NAME

YOUR LAST NAME

.09
(2)

DATE (EXAMPLE, 8 AUG 88)

YOUR SCHOOL – COLLEGE NAME

YOUR TOWN

YOUR STATE

YOUR MAIL ZIP CODE

YOUR TELEPHONE NUMBER

Problem 2-2

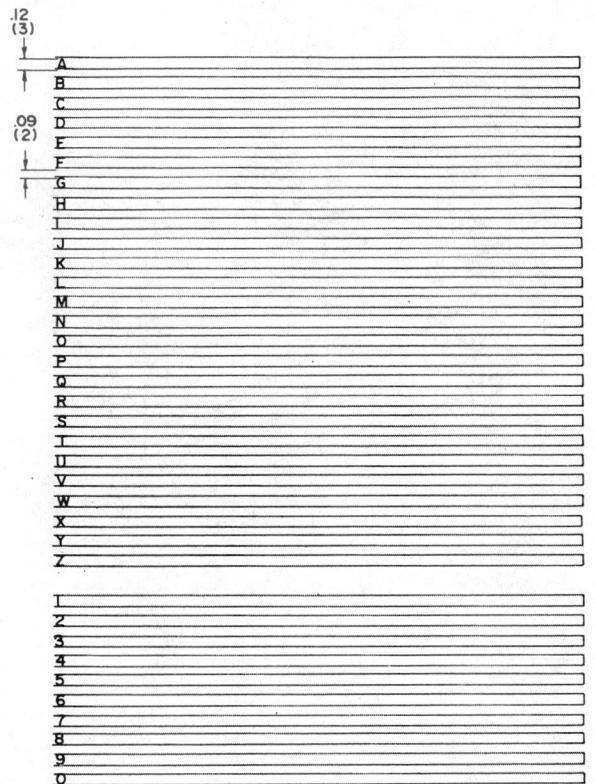

.12
(3)

A
B
C
.09
(2)
D
E
F
G
H
I
J
K
L
M
N
O
P
Q
R
S
T
U
V
W
X
Y
Z

1
2
3
4
5
6
7
8
9
0

Problem 2-1

.12
(3)

GOOD LETTERING TAKES PRACTICE

.09
(2)

IMPROVEMENT IN LETTERING REQUIRES EFFORT

PATIENCE AND HARD WORK EQUALS GOOD LETTERING

SINGLE – STROKE, UPPER CASE, VERTICAL GOTHIC
LETTERING IS USED IN THE MECHANICAL DRAFTING
FIELD

Problem 2-3

LETTERING IS EXTREMELY IMPORTANT. IT MUST BE NEA
LEGIBLE, CLEAR, IN PROPORTION AND MUST BE DONE
IN A REASONABLE AMOUNT OF TIME

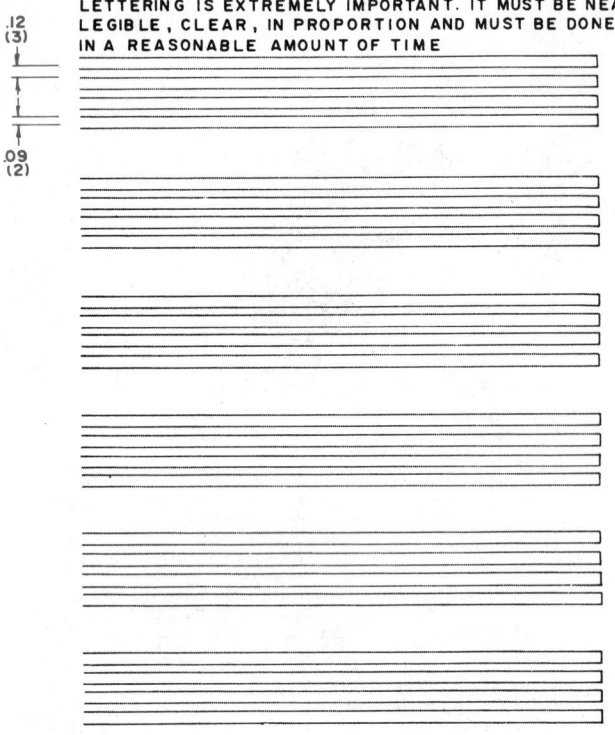

.12
(3)

.09
(2)

Problem 2-4

Problems 2-5 through 2-24

These problems are intended to give the beginning drafter practice in sketching. On an A-size sheet of vellum, neatly sketch the top, front, and right-side view of each pictorial figure shown. The grid is provided for proportion only. Try to keep the object in scale as you sketch; keep all parallel lines parallel, and sketch as neatly as possible. Do not use any drawing instruments. Add all lines, including hidden lines and center lines.

Next, sketch each object from your three-view drawings into isometric pictorial views.

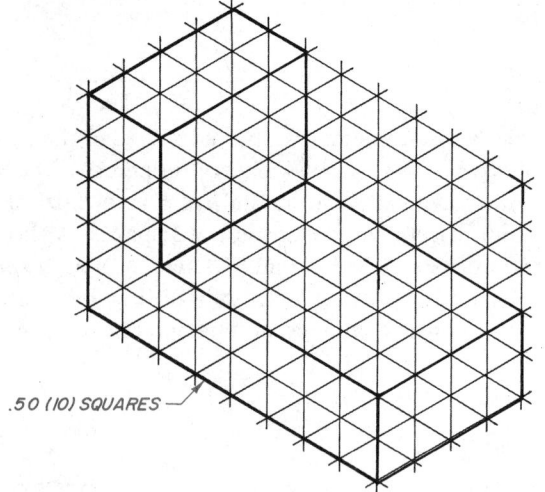

.50 (10) SQUARES

Problem 2-6

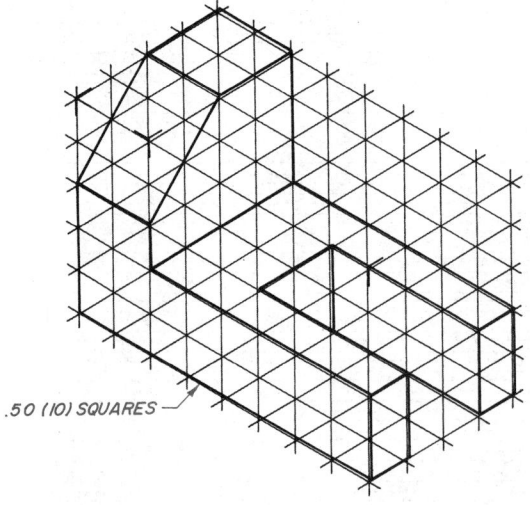

.50 (10) SQUARES

Problem 2-7

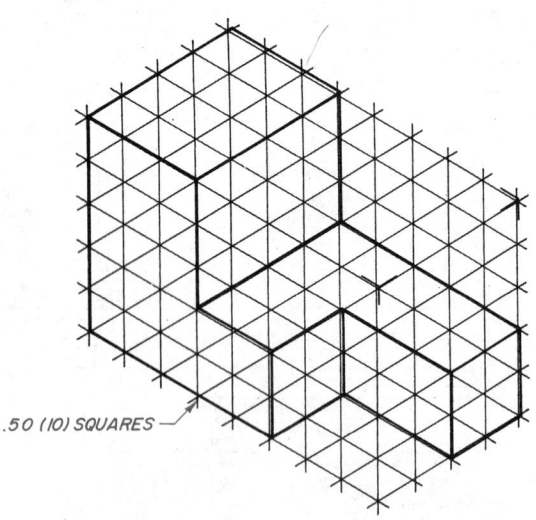

.50 (10) SQUARES

Problem 2-5

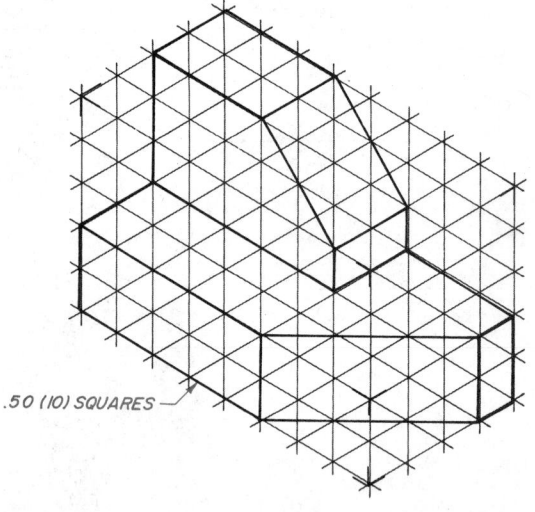

.50 (10) SQUARES

Problem 2-8

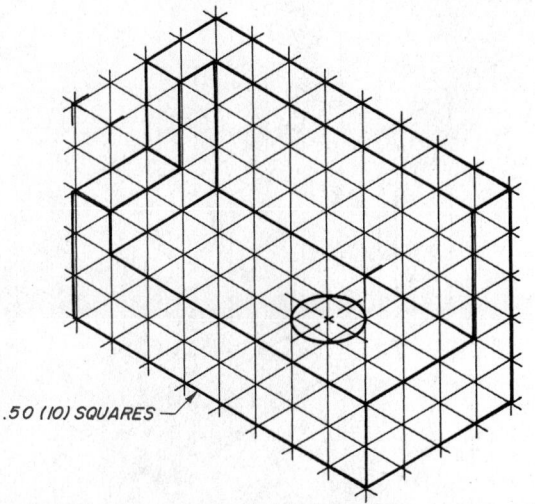

.50 (10) SQUARES

Problem 2-9

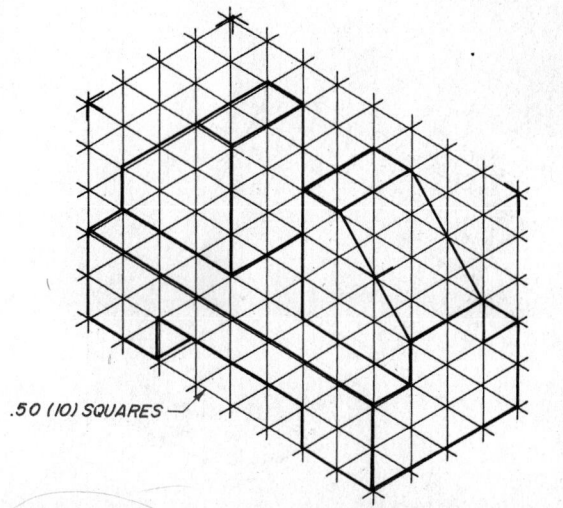

.50 (10) SQUARES

Problem 2-12

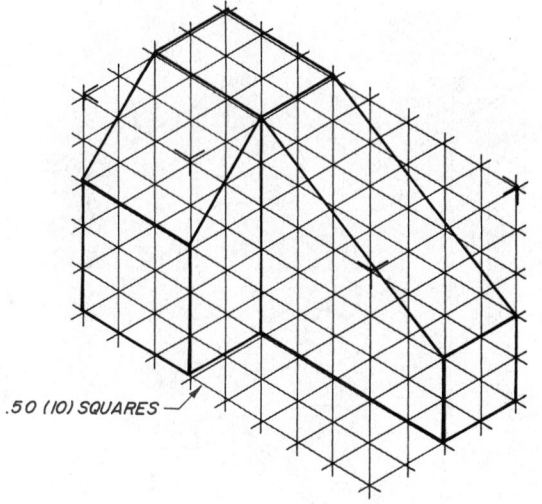

.50 (10) SQUARES

Problem 2-10

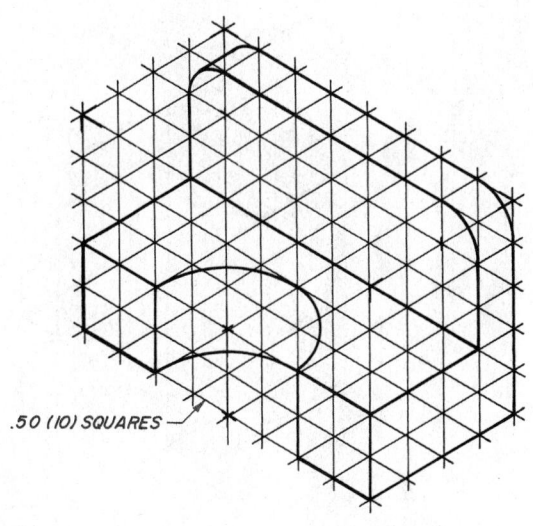

.50 (10) SQUARES

Problem 2-13

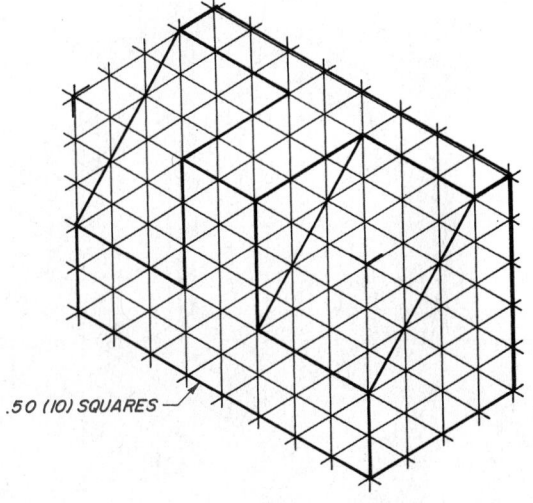

.50 (10) SQUARES

Problem 2-11

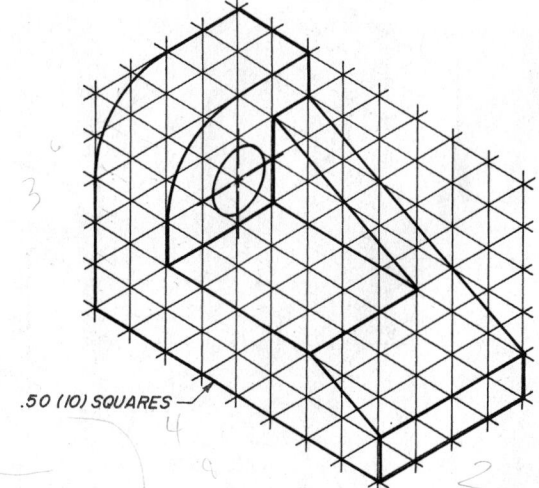

.50 (10) SQUARES

Problem 2-14

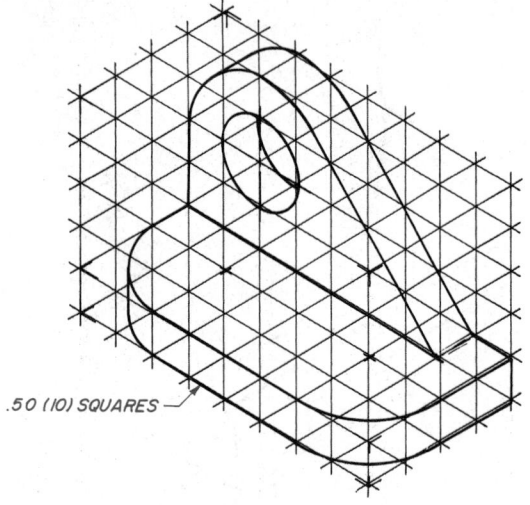

.50 (10) SQUARES

Problem 2-15

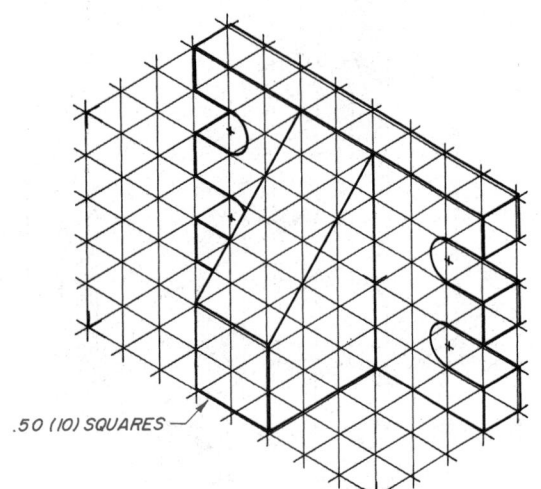

.50 (10) SQUARES

Problem 2-16

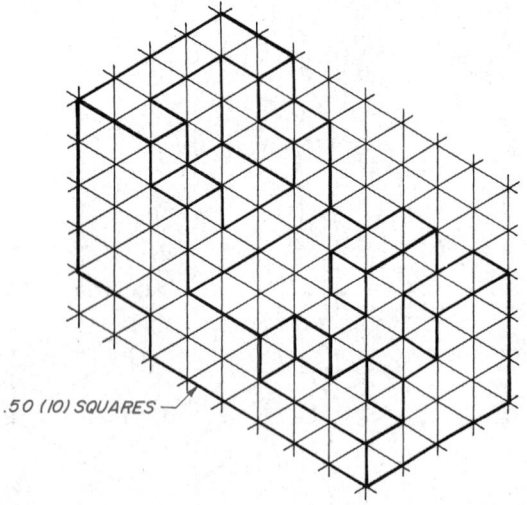

.50 (10) SQUARES

Problem 2-17

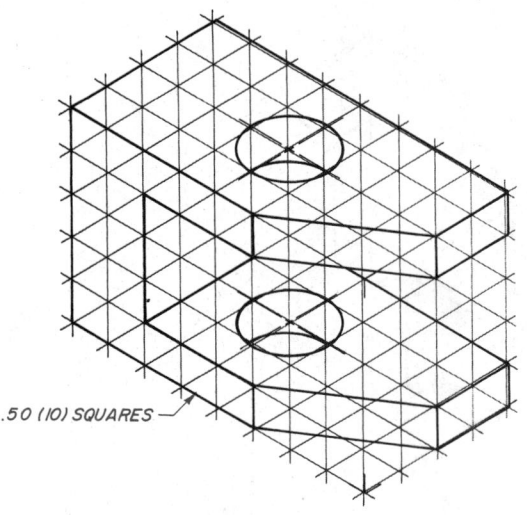

.50 (10) SQUARES

Problem 2-18

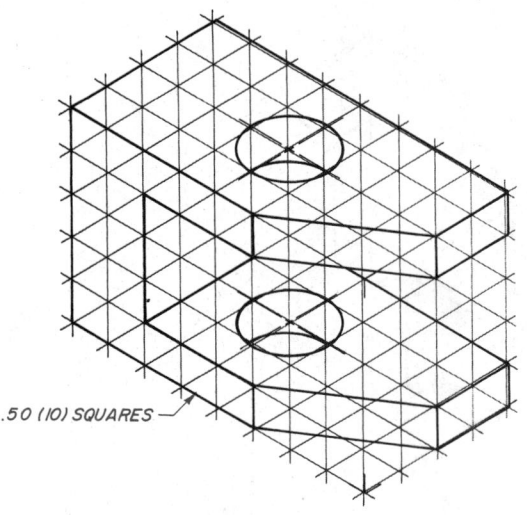

.50 (10) SQUARES

Problem 2-19

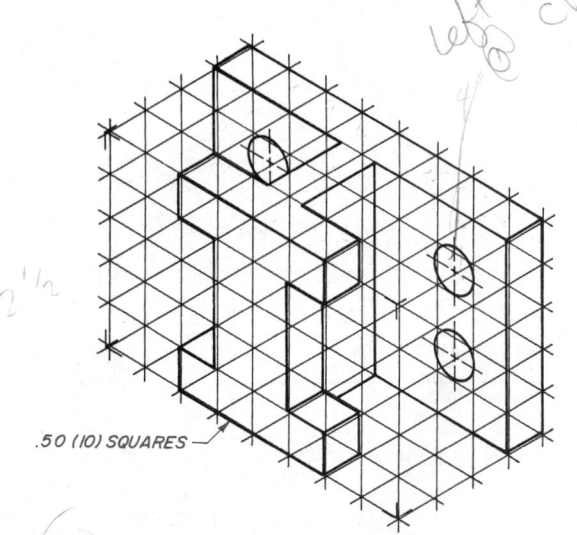

.50 (10) SQUARES

Problem 2-20

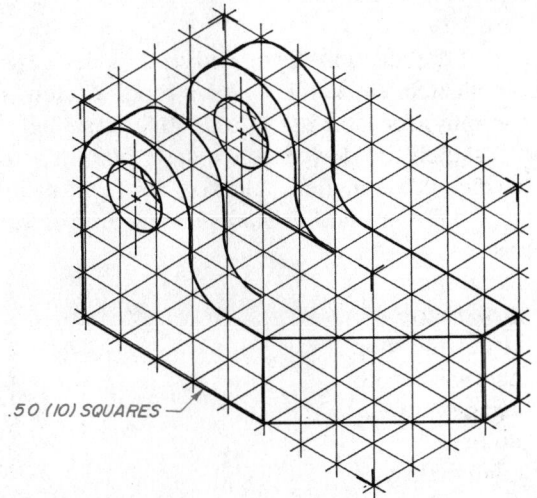

Problem 2-21

.50 (10) SQUARES

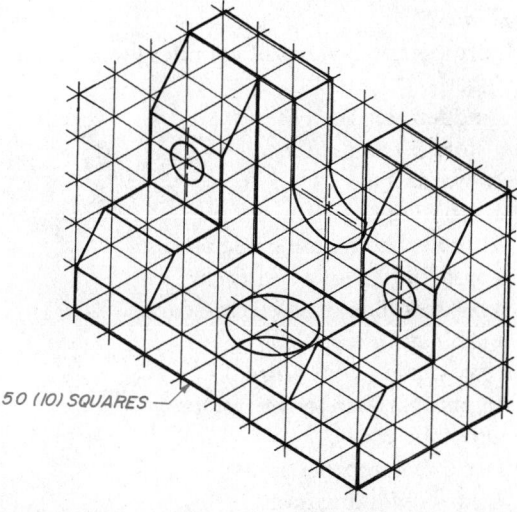

Problem 2-24

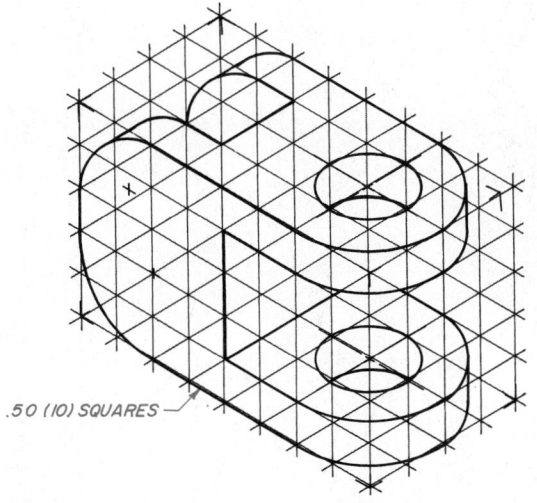

.50 (10) SQUARES

Problem 2-22

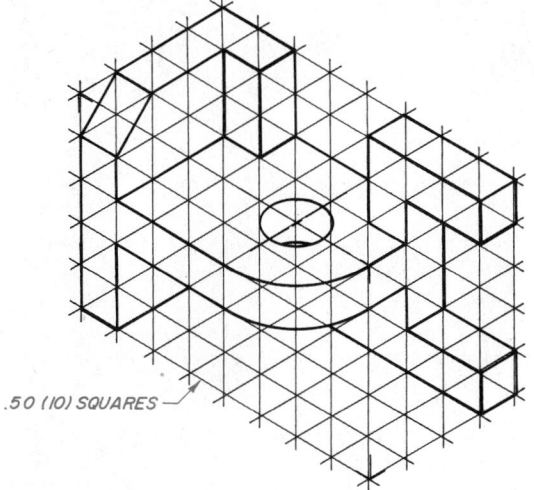

.50 (10) SQUARES

Problem 2-23

❏ Problems 2-25 through 2-28

These few problems are intended to provide future drafters, designers, and engineers experiences in developing an appreciation for sketching from memory, for recording preliminary design ideas, and for communicating orally, technical ideas while sketching.

❏ Problem 2-25

Visit a facility and mentally (no notes or rough sketches) study an object, such as a machine component, an object or a tool in a shop, a part of an automobile, a portion of sports equipment, etc., with a goal in mind that upon returning to class you are to sketch the object(s) from memory. Then, prepare orthographic or isometric type sketches.

The skill emphasized here is that if you observe and study an object as though you were required to sketch it later, you will *see more* than if you just casually look at it. Develop this skill collecting technical ideas for designing.

❏ Problem 2-26

Sketch orthographic views and an isometric of a representative object to be used as your nameplate for a desk, to match your career interests. For example; a bridge could represent a career in construction or civil engineering; a pressurized tank, for welding; an airplane, for flying; a computer, for computer science; and so on.

❏ Problem 2-27

Sketch orthographic views of your design of one of the following.

    a. a modern chair
    b. a storage cabinet for
        1. wardrobe
        2. drafting tools
        3. typical home use garden tools
    c. a basic outline of a sports car
    d. a basic outline of an electric car
    e. a cabinet for trophies
    f. a two-person skateboard
    g. an outdoor logo for your school
    h. a toolbox for
        1. car maintenance tools
        2. woodworking tools
        3. hobbies (fishing, models, etc.)

❏ Problem 2-28

Visit one of the locations listed and select a device, not too complicated, to sketch. Make a rough sketch (for your use only) for later use in a drafting class. Be prepared to sketch the device, while describing its attributes (shape, size, function, material, aesthetic properties, etc.) to either another student or at the board, to the class.

    a. metal shop
    b. wood shop
    c. electronic shop
    d. hardware store
    e. garage
    f. kitchen
    g. computer store

Geometric Construction

To be truly proficient in the layout of both simple and complex drawings, the drafter must know and fully understand the many geometric construction methods used. These methods are illustrated in this chapter, and are basically simple principles of pure geometry. These simple principles are used to actually develop a drawing with complete accuracy, and in the fastest time possible, without wasted motion or any guesswork. Applying these geometric construction principles give drawings a finished, professional appearance.

In laying out the various geometric constructions, it is important to use a very sharp, 4-H lead and to be extremely accurate at all times. Always draw light construction lines that can hardly be seen when held at arm's length. These light construction lines should not be erased as this takes up valuable drawing time and they can also be reused to check layout work if necessary.

Geometric Nomenclature

Points in Space

A *point* is an exact location in space or on a drawing surface, *Figure 3-1*. A point is actually represented on

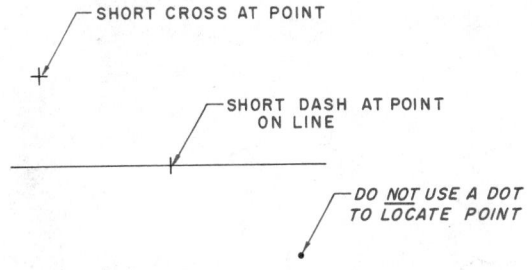

Figure 3-1   Points in space or on a surface

the drawing by a crisscross at its exact location. The exact point in (drawing) space is where the two lines of the crisscross intersect. When a point is located on an existing line, a light, short dashed line or crossbar is placed on the line at the location of the exact point. Never represent a point on a drawing by a dot, except for sketching locations. This is not accurate enough, and is considered to be poor drafting practice.

Line

A straight line is the shortest distance between two points, *Figure 3-2*. It can be drawn in any direction. A

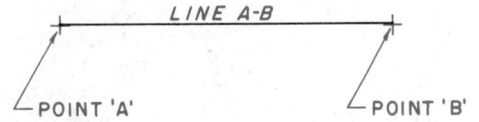

Figure 3-2   A straight line is the shortest distance between two points

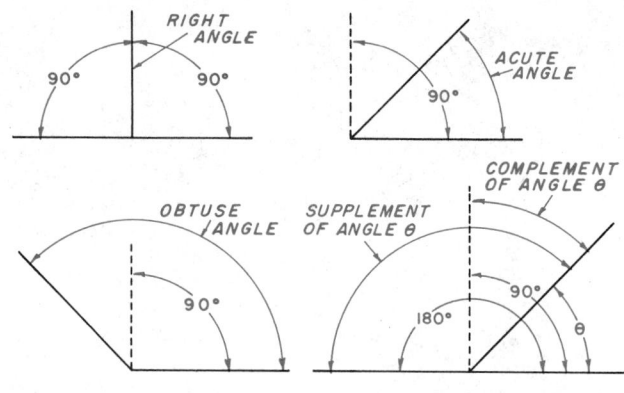

Figure 3-5   Kinds of lines

line can be straight, an arc, a circle or a free curve, as illustrated in *Figure 3-3*. If a line is indefinite, and the ends are not fixed in length, the actual length is a matter of convenience. If the end points of a line are important, they must be marked by means of small, mechanically drawn crossbars, as described by a point in space.

Straight lines and curved lines are considered to be parallel if the shortest distance between them remains constant. The symbol used for parallel lines is //. Lines that are tangent and at 90° are considered perpendicular. The symbol for perpendicular lines is ⊥ (singular), *Figure 3-4*, and ⊥'s (plural). The symbol for an angle is ∠ (singular) and ∠'s (plural). To draw an angle, use the drafting machine, a triangle, or a protractor. For extra accuracy, use the vernier on the drafting machine or a vernier protractor.

Angle

An *angle* is formed by the intersection of two lines. There are three major kinds of angles: right angles, acute angles, and obtuse angles, *Figure 3-5*. The *right*

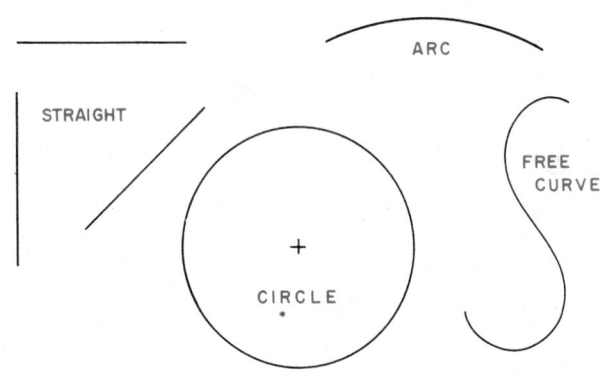

Figure 3-3   Kinds of lines

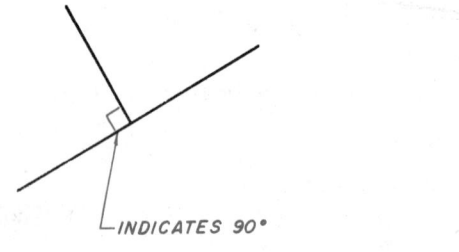

Figure 3-4   Perpendicular lines

angle is an angle of 90°. An *acute angle* is an angle at less than 90°, and an obtuse angle is an angle at more than 90°. Note that a straight line is 180°. Figure 3-5 also illustrates the complementary and supplementary angles of a given angle.

There are 360 degrees (360°) in a full circle. Each degree is divided into 60 minutes (60'). Each minute is divided into 60 seconds (60"). Example: 48°28'38" is read as 48 degrees, 28 minutes, and 38 seconds.

To convert minutes and seconds to decimal degrees, divide minutes by 60 and seconds by 3600.

Example:

$21°18'27" = 21° + (^{18}\!/_{60})° + (^{27}\!/_{3600})° = 21.3075°$

To convert decimal degrees to degrees, minutes, and seconds, multiply the degree decimal by 60 to obtain minutes, and the minute decimal by 60 to obtain seconds.

Example:

$77.365° = 77° + (.365 \times 60)' = 77°21'.9'$
$77°21'.9' = 77°21' + (.9 \times 60)" = 77°21'54"$

A vernier may be used to measure and read off minutes and seconds of a degree. The vernier scale is discussed in Chapter 1.

Triangle

A *triangle* is a closed plane figure with three straight sides and three interior angles. The sum of the three internal angles is always exactly 180°, half of the 360° in a full circle. *Figure 3-6* shows the various kinds of triangles: a right triangle, an equilateral triangle, an isosceles triangle, a scalene triangle, and an obtuse angle triangle.

A *right triangle* is a triangle having a right angle or an angle of 90°. The two sides forming the right angle are called legs, and the third side (the longest) is the hypotenuse. Any triangle inscribed in a semicircle is a right triangle, *Figure 3-7*.

An *equilateral triangle*, as its name implies, is a triangle with all sides of equal length. All of its interior angles are also equal. An *isosceles triangle* has two sides of equal length and two equal interior angles. A *scalene triangle* has no equal sides or angles. An *obtuse angle*

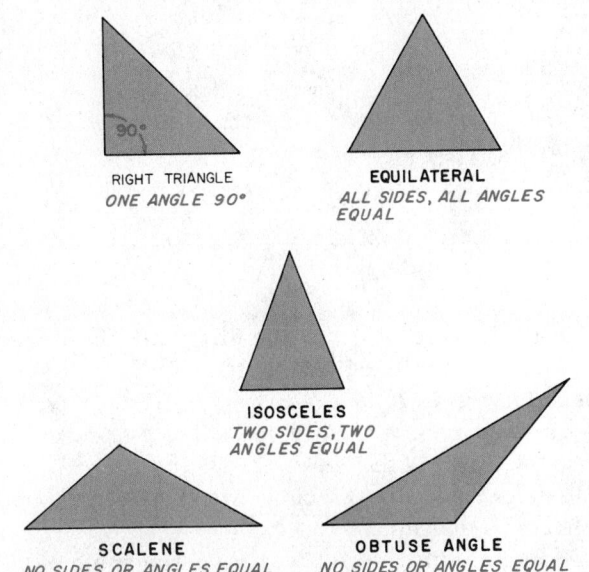

RIGHT TRIANGLE
ONE ANGLE 90°

EQUILATERAL
ALL SIDES, ALL ANGLES EQUAL

ISOSCELES
TWO SIDES, TWO ANGLES EQUAL

SCALENE
NO SIDES OR ANGLES EQUAL

OBTUSE ANGLE
NO SIDES OR ANGLES EQUAL

***Figure* 3-6** Kings of triangles

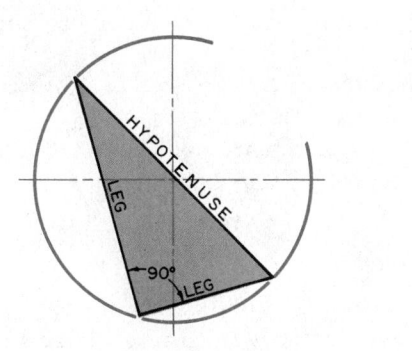

***Figure* 3-7** Any triangle inscribed within a semicircle is a right triangle

triangle is a triangle having an obtuse angle greater than 90°, with no equal sides.

Polygon

A *polygon* is a closed plane figure with three or more straight sides, ***Figure 3-8***. More specifically, shown in the figure are regular polygons, meaning that all sides are equal in each of these examples. The most important of these polygons as they relate to drafting are probably the triangle with three sides, the square with four sides, the hexagon with six sides, and the octagon with eight sides.

Quadrilateral

A *quadrilateral* is a plane figure bounded by four straight sides. When opposite sides are parallel, the quadrilateral is also considered to be a *parallelogram*, ***Figure 3-9***.

Circle

A *circle* is a closed curve with all points on the circle at the same distance from the center point. The major components of a circle are the diameter, the radius, and the circumference, ***Figure 3-10***.

The *diameter* of a circle is the straight distance from one outside curved surface through the center point to the opposite outside curved surface. The diameter of a circle is twice the size of the radius.

The *radius* of a circle is the distance from the center point to the outside curved surface. The radius is half the diameter, and is used to set the compass when drawing a diameter.

The *circumference* of a circle is the distance around the outer surface of the circle. To calculate the circumference of a circle, multiply the value of π (use the approximation 3.1416) by the diameter. A chart similar to the one found in the appendix of this text may be used.

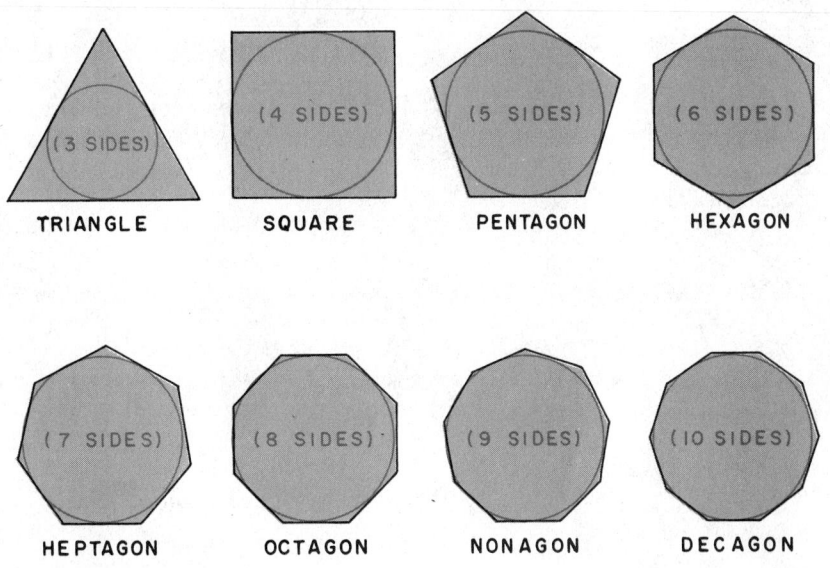

TRIANGLE (3 SIDES)

SQUARE (4 SIDES)

PENTAGON (5 SIDES)

HEXAGON (6 SIDES)

HEPTAGON (7 SIDES)

OCTAGON (8 SIDES)

NONAGON (9 SIDES)

DECAGON (10 SIDES)

***Figure* 3-8** Examples of polygons

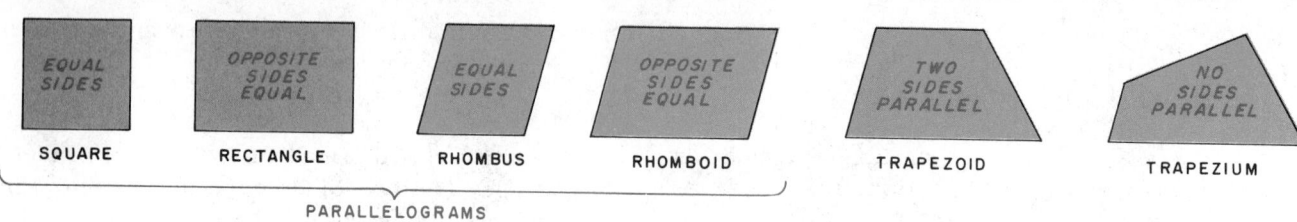

Figure 3-9 Quadrilaterals

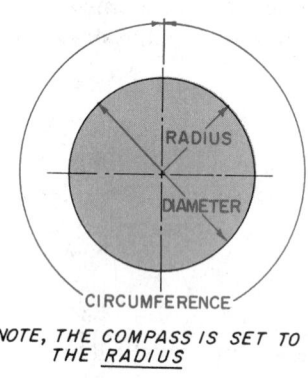

NOTE, THE COMPASS IS SET TO THE <u>RADIUS</u>

Figure 3-10 The major components of a circle

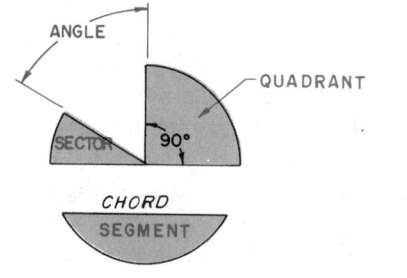

Figure 3-11 Other parts of a circle

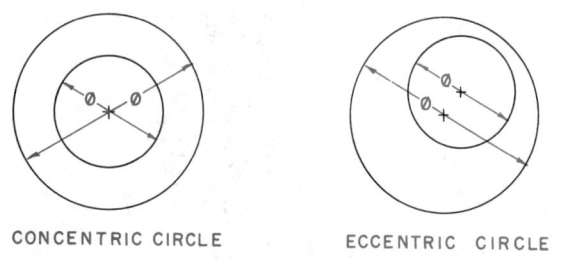

Figure 3-12 Concentric and eccentric circles

Other important parts of a circle are the central angle, the sector, the quadrant, the chord, and the segment, *Figure 3-11*.

A *central angle* is an angle formed by two radial lines from the center of the circle.

A *sector* is the area of a circle lying between two radial lines and the circumference.

A *quadrant* is a sector with a central angle of 90°, and usually with one of the radial lines oriented horizontally.

A *chord* is any straight line whose opposite ends terminate on the circumference of the circle. (A diameter is a chord passing through the center of the circle.)

A *segment* is the smaller portion of a circle separated by a chord.

Concentric circles are two or more circles with a common center point, *Figure 3-12*.

Eccentric circles are two or more circles without a common center point, Figure 3-12.

A *semicircle* is half of a circle.

Polyhedron

A *polyhedron* is a solid object bounded by plane surfaces. Each surface is called a face. If the faces are equal, regular polygons, the solid figure is a regular polyhedron, *Figure 3-13*.

Prism

A *prism* is a solid having ends that are parallel matched polygons, and sides that are parallelograms, *Figure 3-14*. This definition also applies to round or circular objects, such as a cylinder. When the polygon on one end of a prism is not parallel to the other end, it is said to be *truncated*. The *altitude* of a prism is the perpendicular distance between its end polygons (or bases).

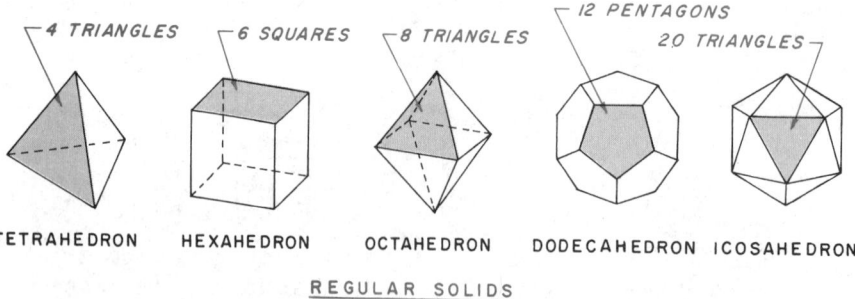

REGULAR SOLIDS

Figure 3-13 Polyhedrons (solids)

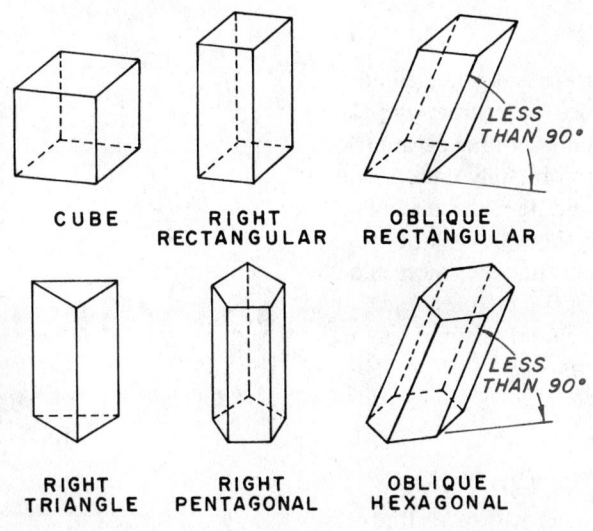

Figure 3-14   Prism

Pyramid and Cone

A *pyramid* is a polyhedron having a polygon as its base. Three or more triangles form its lateral sides which meet at a common vertex, ***Figure 3-15***. A *cone* is a pyramid with a central axis, and an infinite number of sides which form a continuous curved lateral surface. When the vertex of a pyramid or cone has been removed by a plane that intersects all the lateral sides (which forms a new polygon), the pyramid or cone is said to be *truncated*, ***Figure 3-16***.

Sphere

A *sphere* is a closed surface, every point on which is equidistant from a common point or center, ***Figure 3-17***. If a sphere is cut into two equal parts, the parts are called *hemispheres*. *Poles* are two reference measuring positions on the surface of the sphere on opposite sides of its center.

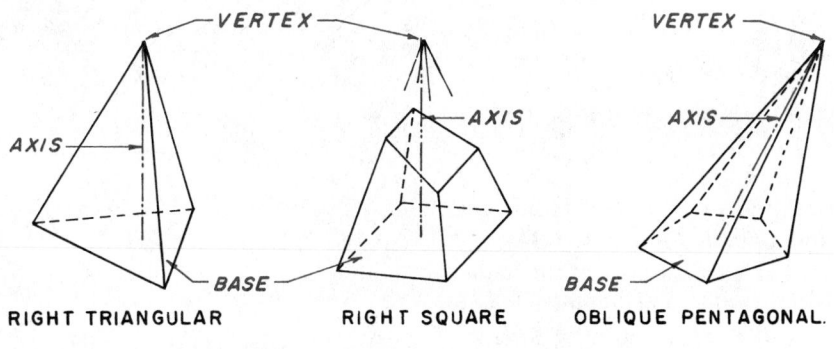

Figure 3-15   Pyramid

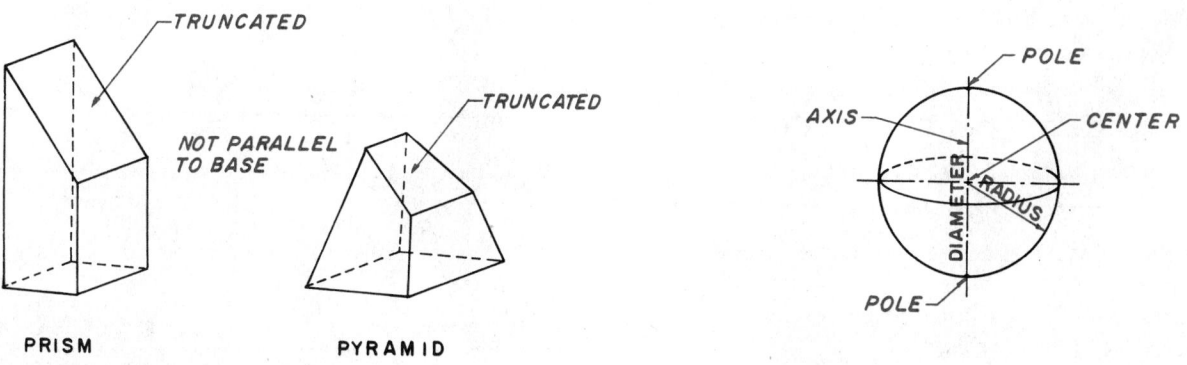

Figure 3-16   Truncated pyramid

Figure 3-17   Sphere

Elemental Construction Principles

The remaining portion of this chapter is devoted to illustrating step-by-step the many geometric construction principles used by the drafter to develop various geometric forms. It is important that each step be fully understood and followed. As the beginning drafter uses these geometric construction principles, various shortcuts will become evident, thus reducing the drawing time and increasing accuracy even more. At the end of the chapter, each of these techniques is incorporated or used in some way to complete the various given problems.

How To Bisect a Line

To *bisect* a line means to divide it in half or to find its center point. In the given process, a line will also be constructed at the exact center point at exactly 90°.

Given: Line A-B, *Figure 3-18A*.

Step 1 Set the compass approximately two-thirds of the length of line A-B and swing an arc from point A, *Figure 3-18B*.

Step 2 Using the exact same compass setting, swing an arc from point B, *Figure 3-18C*.

Step 3 At the two intersections of these arcs, locate points X and Y, *Figure 3-18D*.

Step 4 Draw a straight line connecting point X with point Y. Where this line intersects line A-B is the bisect of line A-B, *Figure 3-18E*. Line X-Y is also perpendicular to line A-B at the exact center point.

How To Bisect an Angle

To *bisect* an angle means to divide it in half or to cut it into two equal angles.

Given: Angle ABC, *Figure 3-19A*.

Step 1 Set the compass at any convenient radius and swing an arc from point B, *Figure 3-19B*.

Step 2 Locate points X and Y on the legs of the angle, and swing two arcs of the same identical length from points X and Y, respectively, *Figure 3-19C*.

Step 3 Where these arcs intersect, locate point Z. Draw a straight line from B to Z. This line will bisect angle ABC and establish two equal angles: ABZ and ZBC, *Figure 3-19D*.

How To Draw an Arc or Circle (Radius) through Three Given Points

Given: Three points in space at random: A, B, and C, *Figure 3-20A*.

Step 1 With straight lines, lightly connect points A to B, and B to C, *Figure 3-20B*.

Step 2 Using the method outlined for bisecting a line, bisect lines A-B and B-C, *Figure 3-20C*.

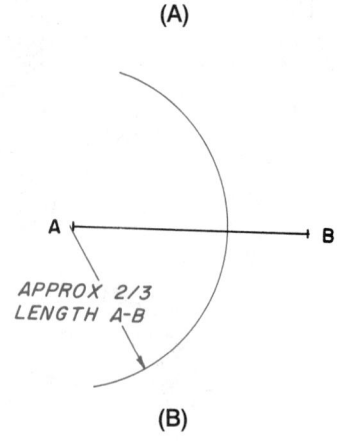
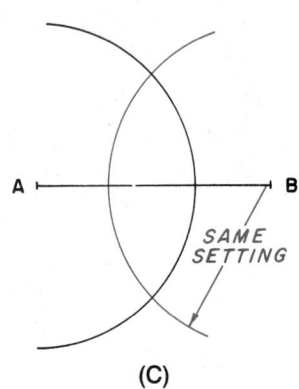
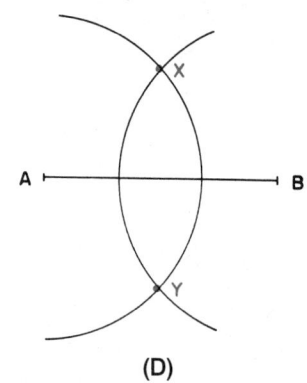
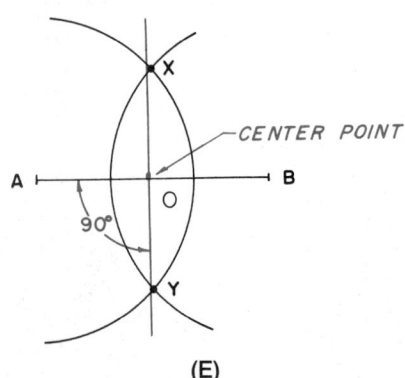

Figure 3-18 (A) How to bisect a line; (B) Step 1; (C) Step 2; (D) Step 3; (E) Step 4

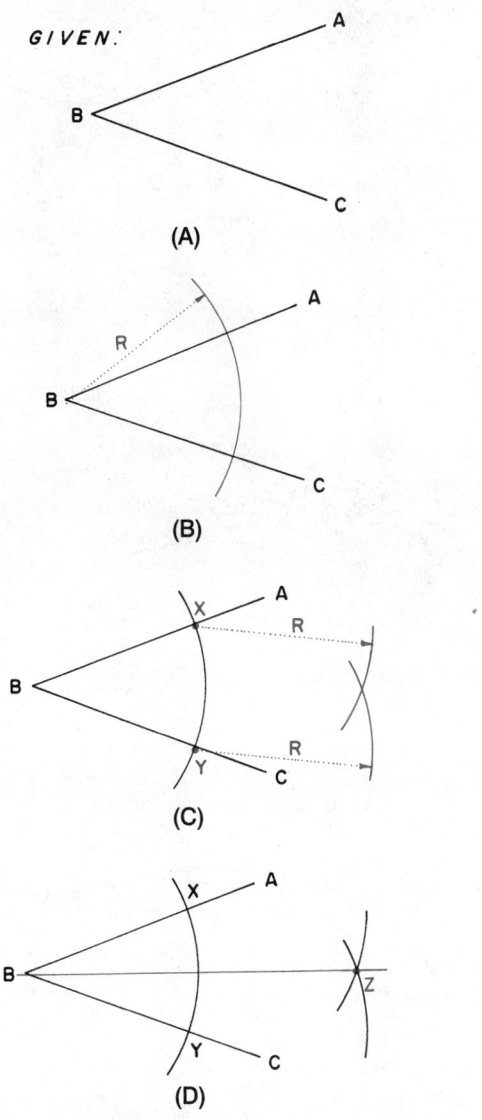

Figure 3-19   (A) How to bisect an angle;   (B) Step 1;   (C) Step 2;
(D) Step 3

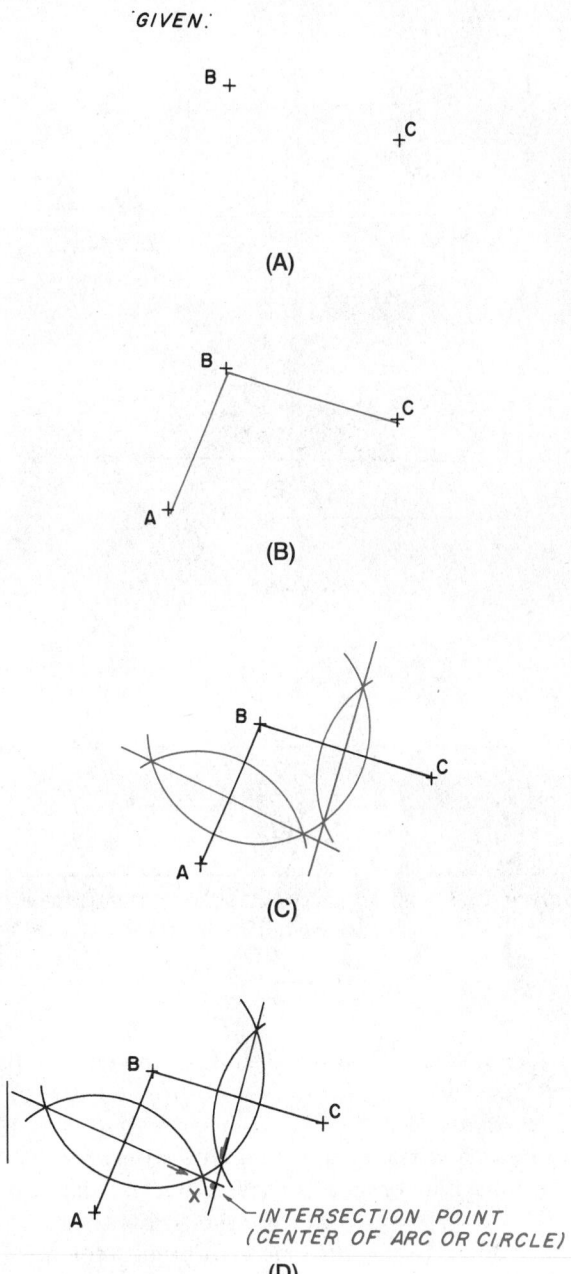

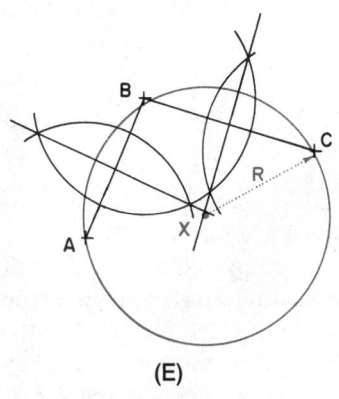

Figure 3-20   (A) How to draw an arc or circle (radius) through three
given points;   (B) Step 1;   (C) Step 2;   (D) Step 3;
(E) Step 4

Step 3   Locate point X where the two extended bisectors meet. Point X is the exact center of the arc or circle, **Figure 3-20D**.

Step 4   Place the point of the compass on point X and adjust the lead to any of the points A, B, or C (they are the same distance), and swing the circle. If all work is done correctly, the arc or circle should pass through each point, as shown in **Figure 3-20E**.

How To Draw a Line Parallel to a Straight Line at a Given Distance

Given: Line A-B, and a required distance to the parallel line, **Figure 3-21A**.

Step 1   Set the compass at the required distance to the parallel line. Place the point of the compass at

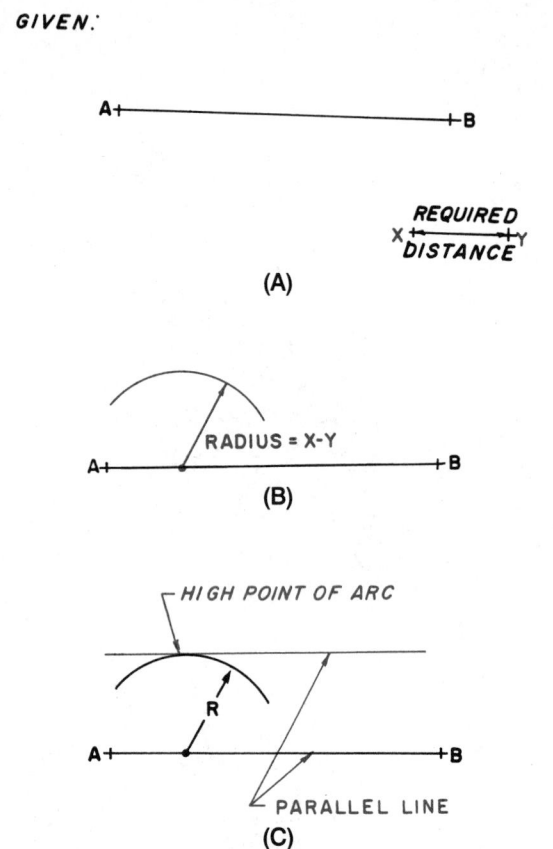

Figure 3-21 (A) How to draw a line parallel to a straight line at a given distance; (B) Step 1; (C) Step 2

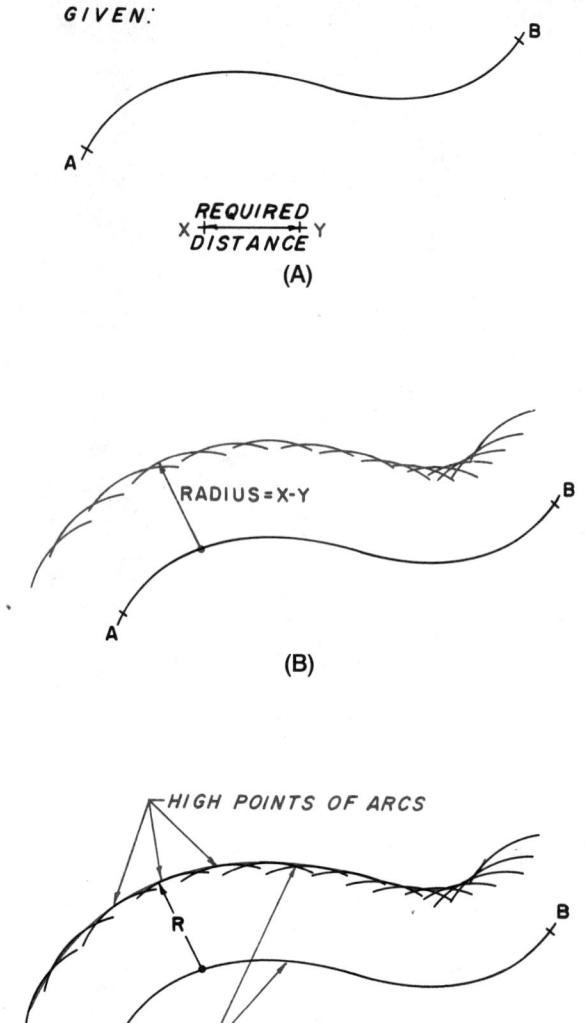

Figure 3-22 (A) How to draw a line parallel to a curved line at a given distance; (B) Step 1; (C) Step 2

any location on the given line, and swing a light arc whose radius is the required distance, *Figure 3-21B*.

Step 2 Adjust the straightedge of either a drafting machine or an adjustable triangle so that it lines up with line A-B, slide the straightedge up or down to the extreme high point, which is the tangent point, of the arc, then draw the parallel line, *Figure 3-21C*.

Note: The distance between parallel lines is measured on any line that is perpendicular to both.

How To Draw a Line Parallel to a Curved Line at a Given Distance

Given: Curved line A-B, and a required distance to the parallel line, *Figure 3-22A.*

Step 1   Set the compass at the required distance to the parallel line. Starting from either end of the curved line, place the point of the compass on the given line, and swing a series of light arcs along the given line, *Figure 3-22B*.

Step 2   Using an irregular curve, draw a line along the extreme high points of the arcs, *Figure 3-22C*.

How to Draw a Perpendicular to a Line at a Point (Method 1)

Given: Line A-B with point P on the same line, *Figure 3-23A*.

Step 1   Using P as the center, make two arcs of equal radius or one continuous arc (R1) to intercept line A-B on either side of point P, at points S and T, *Figure 3-23B*.

Step 2   Swing larger but equal arcs (R2) from each of points S and T to cross each other at point U, *Figure 3-23C*.

Step 3   A line from U to P is perpendicular to line A-B at point P, *Figure 3-23D*.

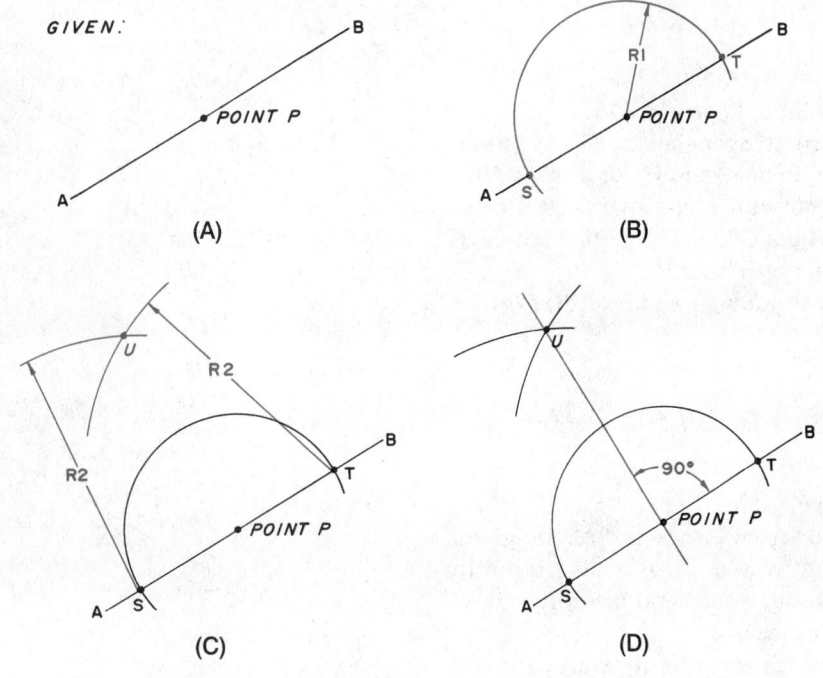

Figure 3-23   (A) How to draw a perpendicular to a line at a point (method 1);   (B) Step 1;   (C) Step 2;   (D) Step 3

How To Draw a Perpendicular to a Line at a Point (Method 2)

Given: Line A-B with point P on the line, ***Figure 3-24A***.

Step 1 Swing an arc of any convenient radius whose center O is at any convenient location NOT on line A-B, but positioned to make the arc cross line A-B at points P and Q, ***Figure 3-24B***.

Step 2 A line from point Q through center O intercepts the opposite side of the arc at point R, ***Figure 3-24C***.

Step 3 Line R-P is perpendicular to line A-B. (A right triangle has been inscribed in a semicircle.) See ***Figure 3-24D***.

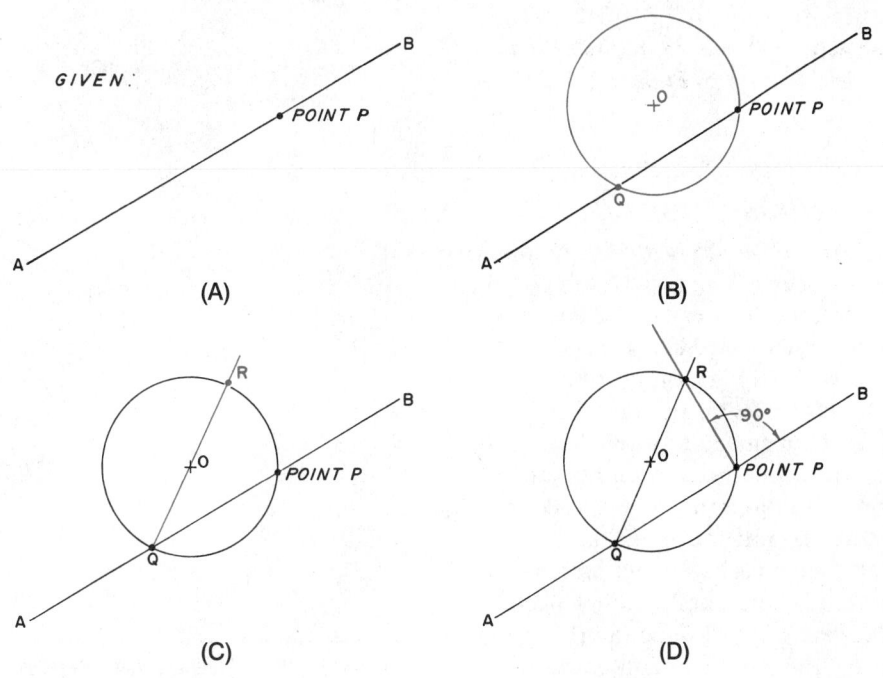

Figure 3-24   (A) How to draw a perpendicular to a line at a point (method 2);   (B) Step 1;   (C) Step 2;   (D) Step 3

How To Draw a Perpendicular to a Line from a Point Not on the Line

Given: Line A-B and point P, *Figure 3-25A*.

Step 1 Using P as a center, swing an arc (R1) to intercept line A-B at points G and H, *Figure 3-25B*.

Step 2 Swing larger, but equal length arcs (R2) from each of the points G and H to intercept each other at point J, *Figure 3-25C*.

Step 3 Line P-J is perpendicular to line A-B, *Figure 3-25D*.

How To Divide a Line into Equal Parts

Given: Line A-B, *Figure 3-26A*.

Step 1 Draw a line 90° from either end of the given line. A 90° line from point B is illustrated in *Figure 3-26B*, but either end or either direction will work as well.

Step 2 Place a scale with its zero at point A of the given line. If three equal parts are required, pivot the scale until the three-inch measurement (or any length representing three equal units of measure: 30, 60 and 90 mm, for example) is on the perpendicular line drawn in Step 1. Place a short dash at these points. The example in *Figure 3-26C* shows these dashes at the 1-inch and 2-inch marks.

Step 3 Project lines downward from these points and add short hash marks where these projected lines cross the original given line A-B. This divides the line A-B into three exact equal parts. Check all work, comparing your final step with *Figure 3-26D*. An example of equal spacing, where equally spaced holes are required with a given length, is illustrated in *Figure 3-26E*.

How To Divide a Line into Proportional Parts

Given: Line A-B, *Figure 3-27A*. *Problem:* Locate point X at 2/3 of the distance from point A to point B.

Step 1 Draw a line 90° from either end of the line. A 90° line from point B is illustrated in *Figure 3-27B*.

Step 2 Place a scale with its zero on point A of the given line. Because a 2/3 proportion is required, pivot the scale until any multiple of three units of measure intersects the perpendicular line drawn in Step 1. In this example, 6 is used, representing three 2-unit increments. The 2/3 position of this length is two 2-unit increments, or the 4 position, where a hash mark is made, as shown in *Figure 3-27C*. Projecting this point downward to line A-B, it becomes point X, which is 2/3 the overall distance from point A.

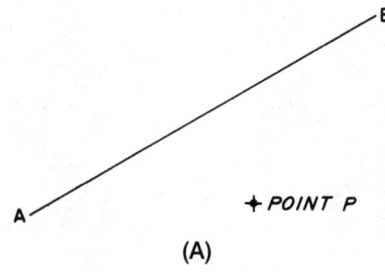

(A)

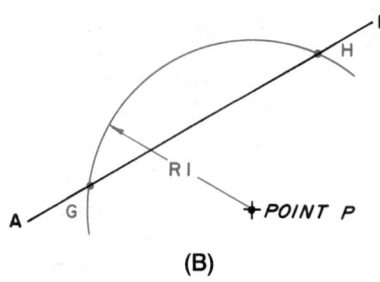

(B)

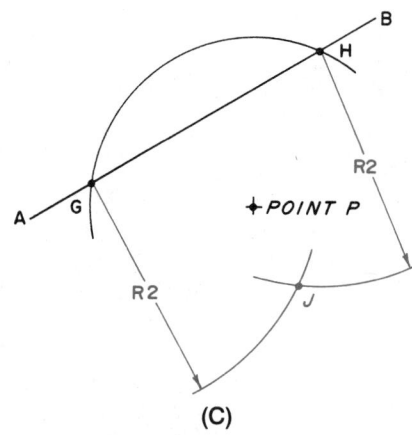

(C)

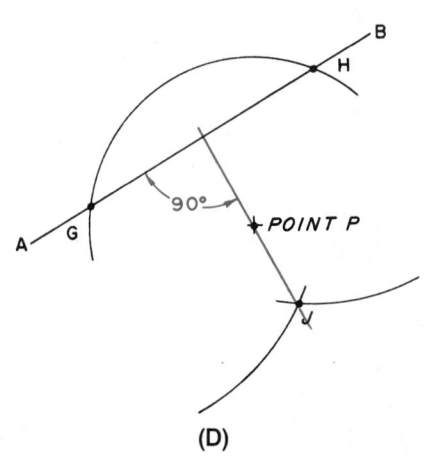

(D)

Figure 3-25 (A) How to draw a perpendicular to a line from a point not on the line; (B) Step 1; (C) Step 2; (D) Step 3

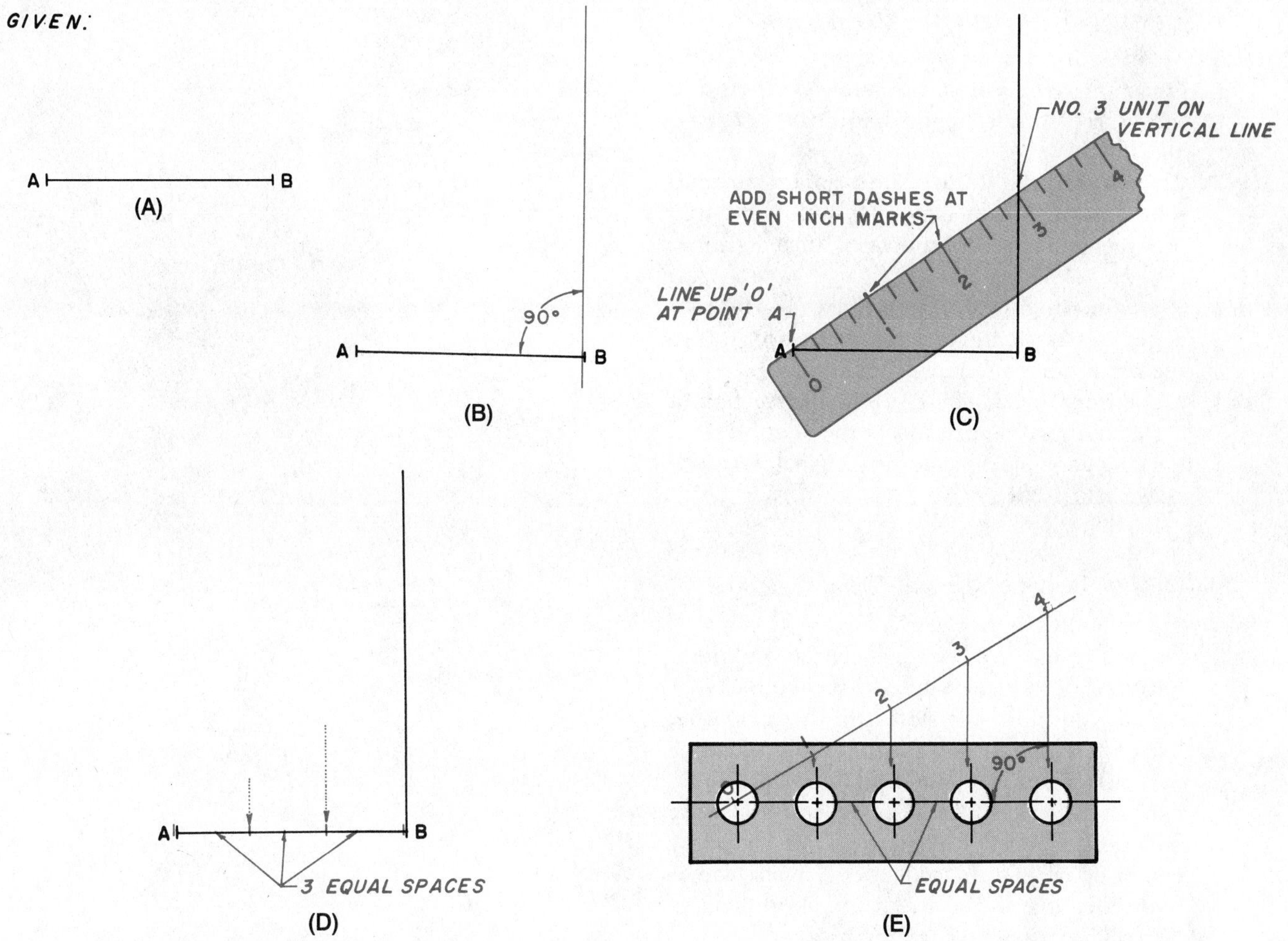

GIVEN:

A |————————————| B

(A)

90°

A |————————————| B

(B)

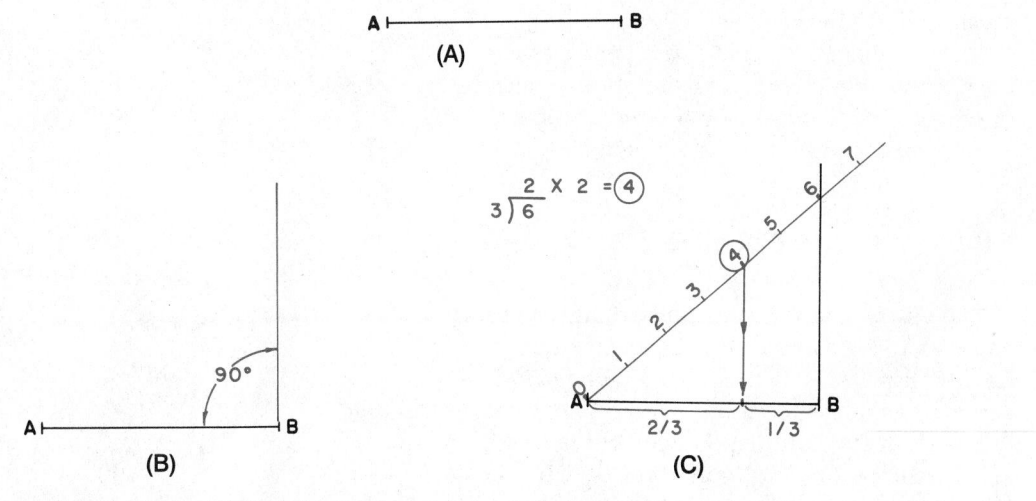

NO. 3 UNIT ON VERTICAL LINE

ADD SHORT DASHES AT EVEN INCH MARKS

LINE UP 'O' AT POINT A

(C)

A B

3 EQUAL SPACES

(D)

90°

EQUAL SPACES

(E)

Figure 3-26   (A) How to divide a line into equal parts;   (B) Step 1;   (C) Step 2;   (D) Step 3;   (E) An example of equal spacing within a given length

GIVEN:

A |————————————| B

(A)

90°

A |————————————| B

(B)

$\frac{2}{3 \overline{)6}} \times 2 = \textcircled{4}$

2/3   1/3

(C)

Figure 3-27   (A) How to divide a line into proportional parts;   (B) Step 1;   (C) Step 2

How To Transfer an Angle

Given: An angle formed by two straight lines, 0A-0B, *Figure 3-28A*, and one location of where the transferred angle begins (point 0'), *Figure 3-28B*.

Step 1   Refer to *Figure 3-28C*. Draw an arc through both legs of a given angle (R1) and then duplicate this radius at the transferred angle location, *Figure 3-28D*.

Step 2   Transfer the chord length between the two angle legs at the intersection of the arc (R2) to the arc at the transfer angle location.

Step 3   A line from the arc center to the intersection of arc and chord length forms the second line, forming an angle equal to the original, *Figures 3-28E* and *3-28F*.

How To Transfer an Odd Shape

Given: Triangle ABC, *Figure 3-29A*.

Step 1   Letter or number the various corners and point locations of the odd shape in counterclockwise order around its perimeter. In this example, place the compass point at point A of the original shape and extend the lead to point B. Refer back to *Figure 3-29A*. Swing a light arc at the new desired location, *Figure 3-29B*. Letter the center point as A' and add letter B' at any convenient location on the arc. It is a good habit to lightly letter each point as you proceed.

Step 2   Place the compass point at letter B of the original shape, *Figure 3-29C*, and extend the compass lead to letter C of the original shape.

Step 3   Transfer this distance, B-C, to the layout, *Figure 3-29D*.

Steps 4 and 5   Going back to the original object, place the compass point at letter A, *Figure 3-29E*, and extend the compass lead to letter C. Transfer the distance A-C as illustrated in *Figure 3-29F*. Locate and letter each point.

Step 6   Connect points A', B', and C' with light, straight lines. This completes the transfer of the object, *Figure 3-29G*. Recheck all work and, if correct, darken lines to the correct line weight.

How To Transfer Complex Shapes

A complex shape can be transferred in exactly the same way by reducing the shape into simple triangles and transferring each triangle using the foregoing method.

Given: An odd shape, A, B, C, D, E, F, G, *Figure 3-30A*. Letter or number the various corners and point locations of the odd shape in clockwise order around the perimeter, *Figure 3-30A*. Use the longest line or any convenient line as a starting point. Line A-B is chosen here as the example.

Step 1   Lightly divide the shape into triangle divisions, using the baseline if possible. Transfer each tri-

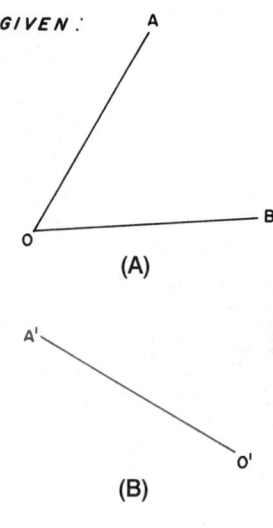
(A)

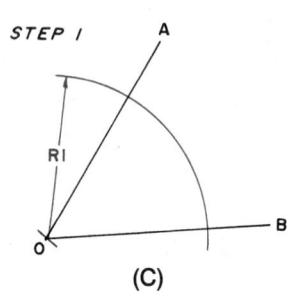
(B)

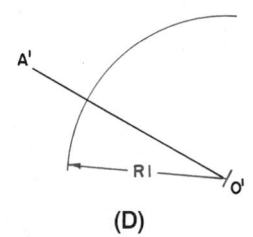
(C)

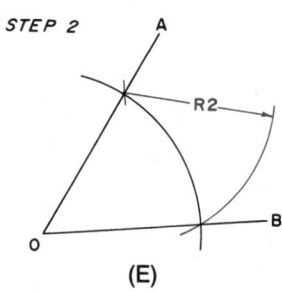
(D)

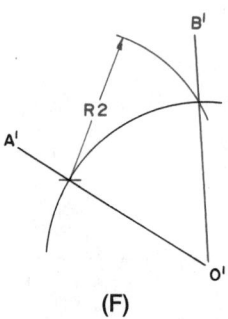
(F)

Figure 3-28   (A) How to transfer an angle;   (B) Given: Point 0';   (C) Step 1;   (D) Transferred angle;   (E) Step 2;   (F) Step 3

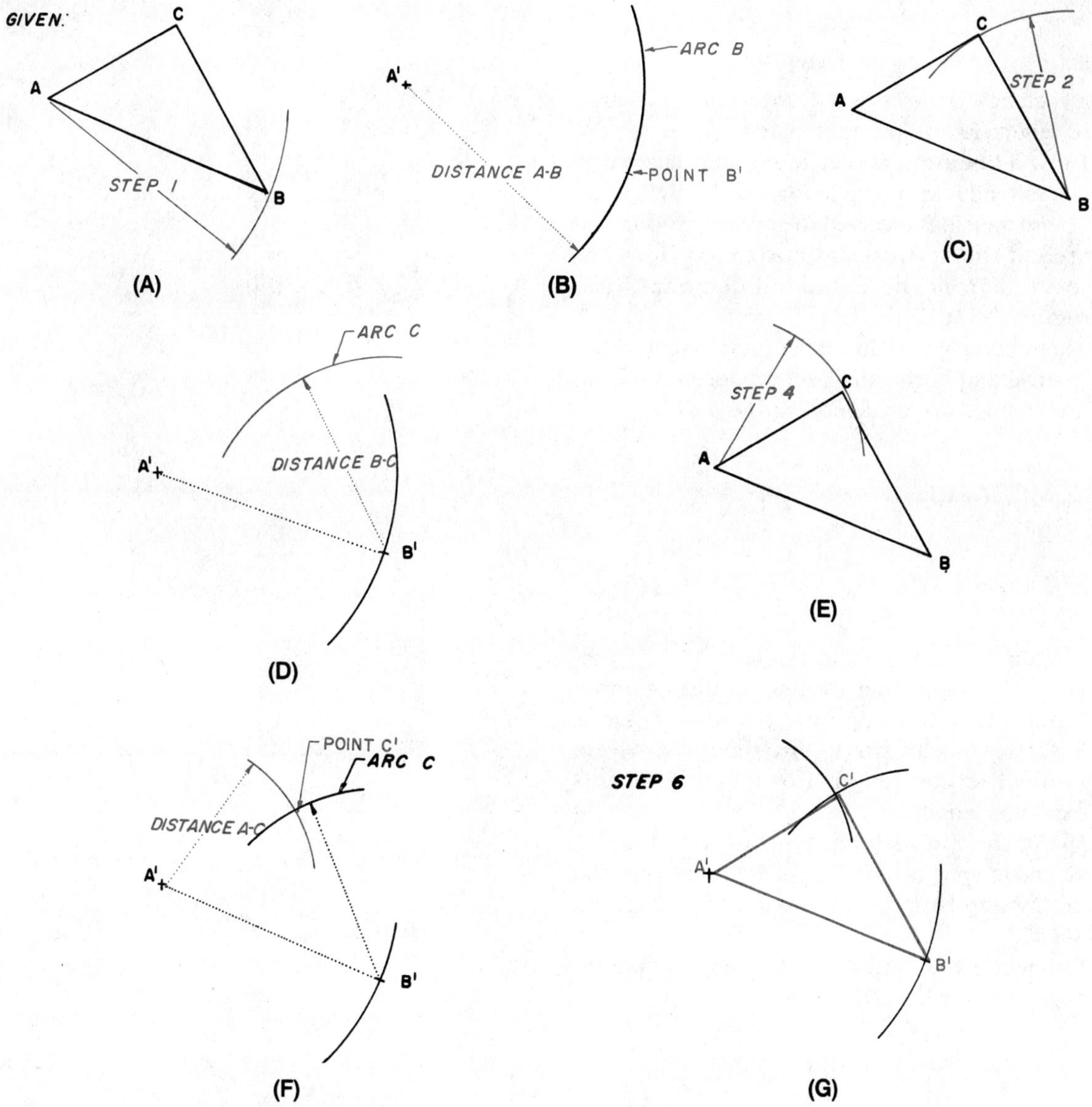

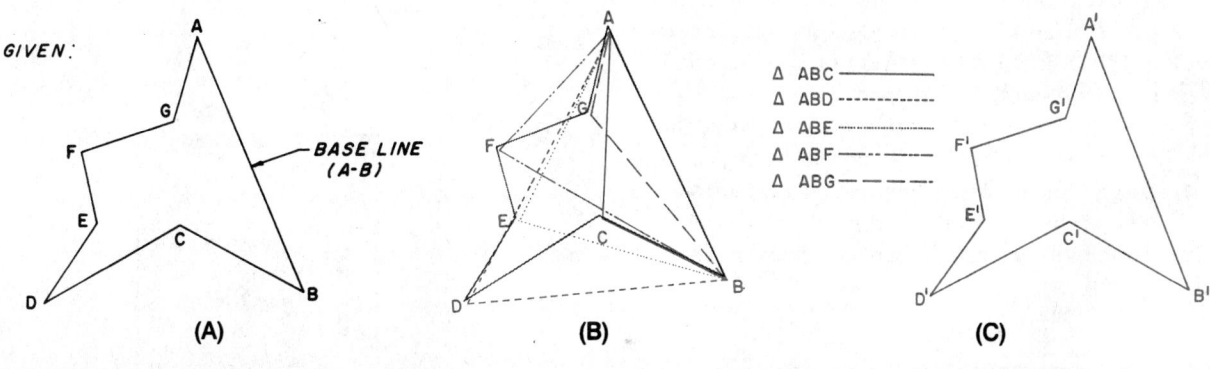

Figure 3-29 (A) How to transfer an odd shape; (B) Step 1; (C) Step 2; (D) Step 3; (E) Step 4; (F) Step 5; (G) Step 6

Figure 3-30 (A) How to transfer complex shapes; (B) Step 1; (C) Step 2

angle in the manner described in Figures 3-29A through 3-29G. Suggested triangles to be used in example *Figure 3-30A* are ABC, ABD, ABE, ABF, and ABG, *Figure 3-30B*.

Step 2 This completes the transfer. Check all work and, if correct, darken in lines to correct line thickness. See *Figure 3-30C*.

How To Proportionately Enlarge or Reduce a Shape

Given: A rectangle, *Figure 3-31A*. *Problem:* To enlarge or reduce its size proportionately.

Step 1 Draw a line from corner to corner diagonally, and extend it as shown in *Figure 3-31B*.

Step 2 The rectangle is enlarged to any size proportionately if the vertical and horizontal sides are located from the extended diagonal line, *Figure 3-31C*.

Step 3 The rectangle is reduced proportionately if the vertical and horizontal lines are located on the unextended diagonal line, *Figure 3-31D*.

Polygon Construction

How To Draw a Triangle with Known Lengths of Sides

Given: Lengths 1, 2, and 3, *Figure 3-32A*.

Step 1 Draw the longest length line, in this example length 3, with endpoints A and B, *Figure 3-32B*. Swing an arc (R1) from point A whose radius is either length 1 or length 2; in this example, length 1.

Step 2 Using the radius length *not* used in Step 1, swing an arc (R2) from point B to intercept the arc swung from point A at point C, *Figure 3-32C*.

Step 3 Connect A to C and B to C to complete the triangle, *Figure 3-32D*.

How To Draw an Equilateral Triangle

Given: A baseline and the given length of each side, *Figure 3-33A*.

Step 1 Either adjust the drafting machine angle to 60° or use a 30°–60° triangle and project from points A and B at 60° using light lines, *Figure 3-33B*. Allow light construction lines to cross as shown; do not erase them.

Step 2 Check to see that there are three equal sides and, if so, darken in the actual triangle using correct line thickness, *Figure 3-33C*. Care should be exercised in constructing the sharp corners. Again, do *not* erase the light construction lines.

How To Draw a Square

Given: The location of the center and the required distance across the "flats" of a square, *Figure 3-34A*.

Step 1 Lightly draw a circle with a diameter equal to the distance around the "flats" of the square. Set the compass at half the required diameter, *Figure 3-34B*.

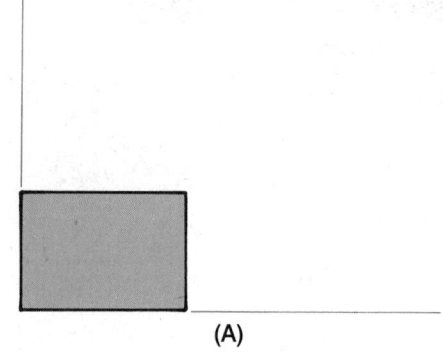
(A)

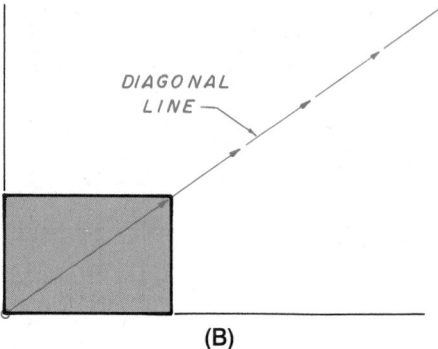
DIAGONAL LINE
(B)

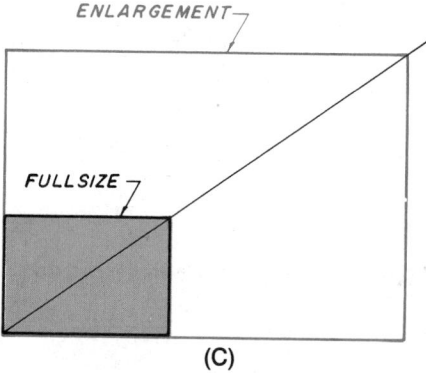
ENLARGEMENT
FULLSIZE
(C)

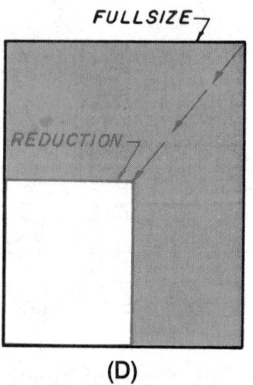
FULLSIZE
REDUCTION
(D)

Figure 3-31 (A) How to proportionately enlarge or reduce a shape; (B) Step 1; (C) Step 2; (D) Step 3

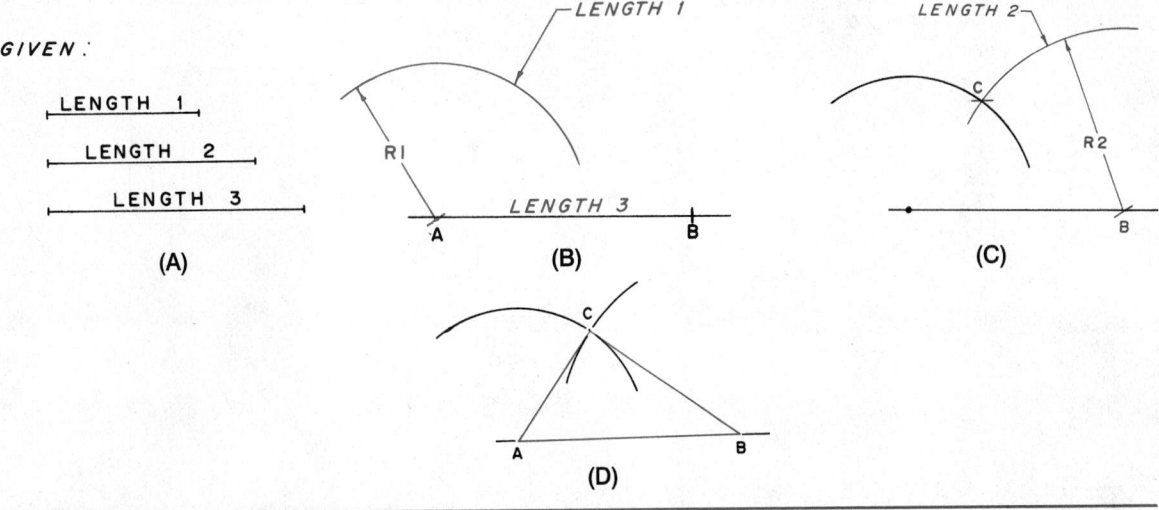

Figure 3-32   (A) How to draw a triangle with known lengths of sides;   (B) Step 1;   (C) Step 2;   (D) Step 3

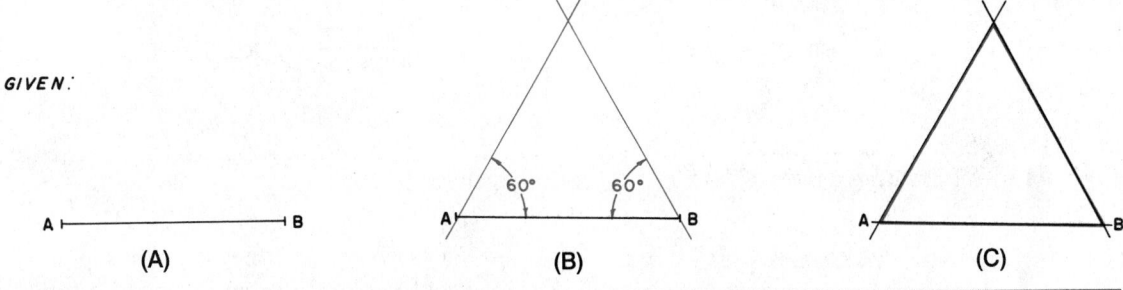

Figure 3-33   (A) How to draw an equilateral triangle;   (B) Step 1;   (C) Step 2

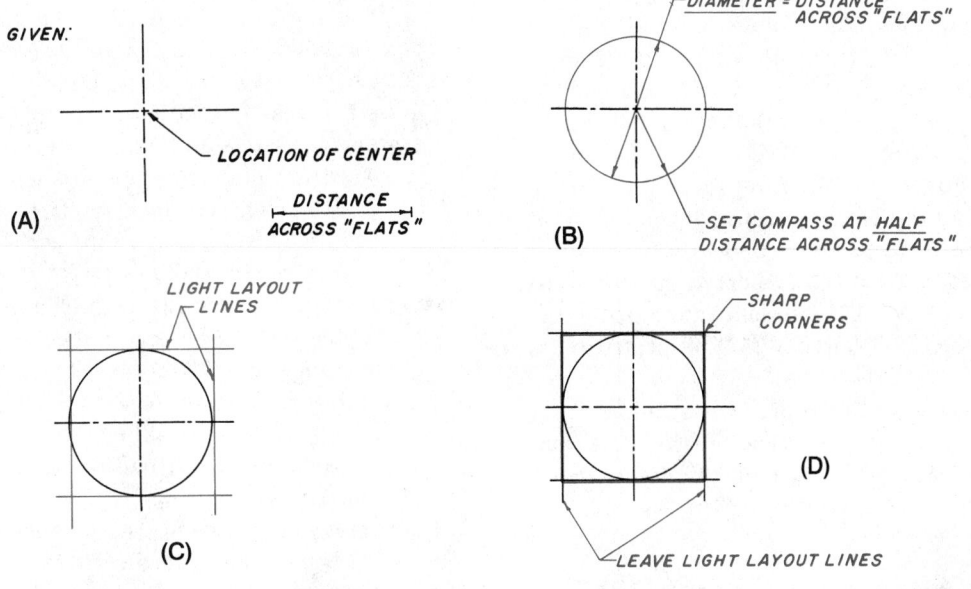

Figure 3-34   (A) How to draw a square;   (B) Step 1;   (C) Step 2;   (D) Step 3

Step 2 Using a triangle, lightly complete the square by constructing tangent lines to the circle. Allow the light construction lines to project from the square, as shown, without erasing them, *Figure 3-34C*.

Step 3 Check to see that there are four equal sides and, if so, darken in the actual square using the correct line thickness, *Figure 3-34D*. Care should be exercised in constructing the sharp corners. Again, do *not* erase the light construction lines.

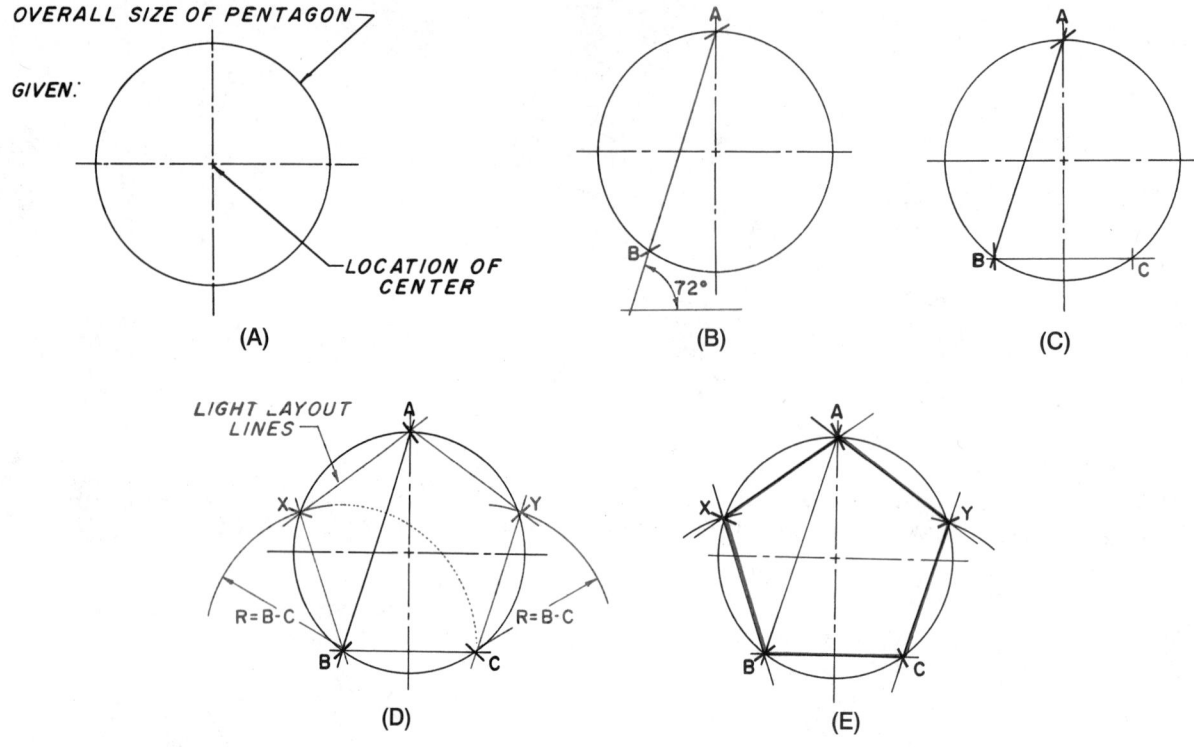

Figure 3-35 (A) How to draw a pentagon; (B) Step 1; (C) Step 2; (D) Step 3; (E) Step 4

How To Draw a Pentagon (5 Sides)

Given: The location of the pentagon center, and the diameter that will circumscribe the pentagon, *Figure 3-35A*.

Step 1 Locate point A at the top-center of the circle and, using a drafting machine, position an angle of 72° (360/5) from the horizontal (or 18° from the vertical) through point A to locate point B where the angle crosses the circumference of the circle, *Figure 3-35B*.

Step 2 Draw a horizontal line from point B to locate point C on the circumference of the circle on the opposite side, *Figure 3-35C*.

Step 3 Set the compass at the distance from point B to point C, and swing this distance from the points as illustrated in *Figure 3-35D* to locate points X and Y.

Step 4 Lightly connect the points. Check to see that there are five equal sides and, if correct, darken in the actual pentagon taking care to construct five sharp corners, *Figure 3-35E*.

How To Draw a Hexagon (6 Sides)

Given: The location of the required center and the required distance across the "flats" of a hexagon, *Figure 3-36A*.

Step 1 Lightly draw a circle with a diameter equal to the distance across the "flats" of the hexagon. Set the compass at half the required diameter, *Figure 3-36B*.

Step 2 Draw two horizontal lines tangent to the curve,

or two vertical lines if the hexagon is to be oriented at 90° to the illustrated position, *Figure 3-36C*.

Step 3 Using a 30°-60° triangle, lightly complete the hexagon by constructing tangent lines to the circle, *Figure 3-36D*. Allow the light construction lines to extend as shown; do not erase them.

Step 4 Check to see that there are six equal sides and, if so, darken in the actual hexagon using correct line thickness and taking care to construct six sharp corners, *Figure 3-36E*. Again, do *not* erase the light construction lines.

How To Draw an Octagon (8 Sides)

Given: The location of the required center and the required distance across the "flats" of an octagon, *Figure 3-37A*.

Step 1 Lightly draw a circle with a diameter equal to the distance across the "flats" of the octagon. Set the compass at half the required diameter, *Figure 3-37B*.

Step 2 Lightly draw two horizontal lines and two vertical lines tangent to the circle, as illustrated in *Figure 3-37C*.

Step 3 Using a 45° triangle, lightly complete the octagon by constructing tangent lines to the circle, *Figure 3-37D*. Allow the light lines to extend.

Step 4 Check that there are eight equal sides and, if so, darken in the actual octagon using correct line thickness and taking care to construct eight sharp corners, *Figure 3-37E*. Again, do not erase the light construction lines.

GIVEN:
LOCATION OF CENTER
DISTANCE ACROSS "FLATS"
(A)
DIAMETER = DISTANCE ACROSS "FLATS"
SET COMPASS AT HALF DISTANCE ACROSS "FLATS"
(B)
(C)
30°
LIGHT LAYOUT LINES
60°
90°
(D)
SHARP CORNERS
LEAVE LIGHT LAYOUT LINES
(E)

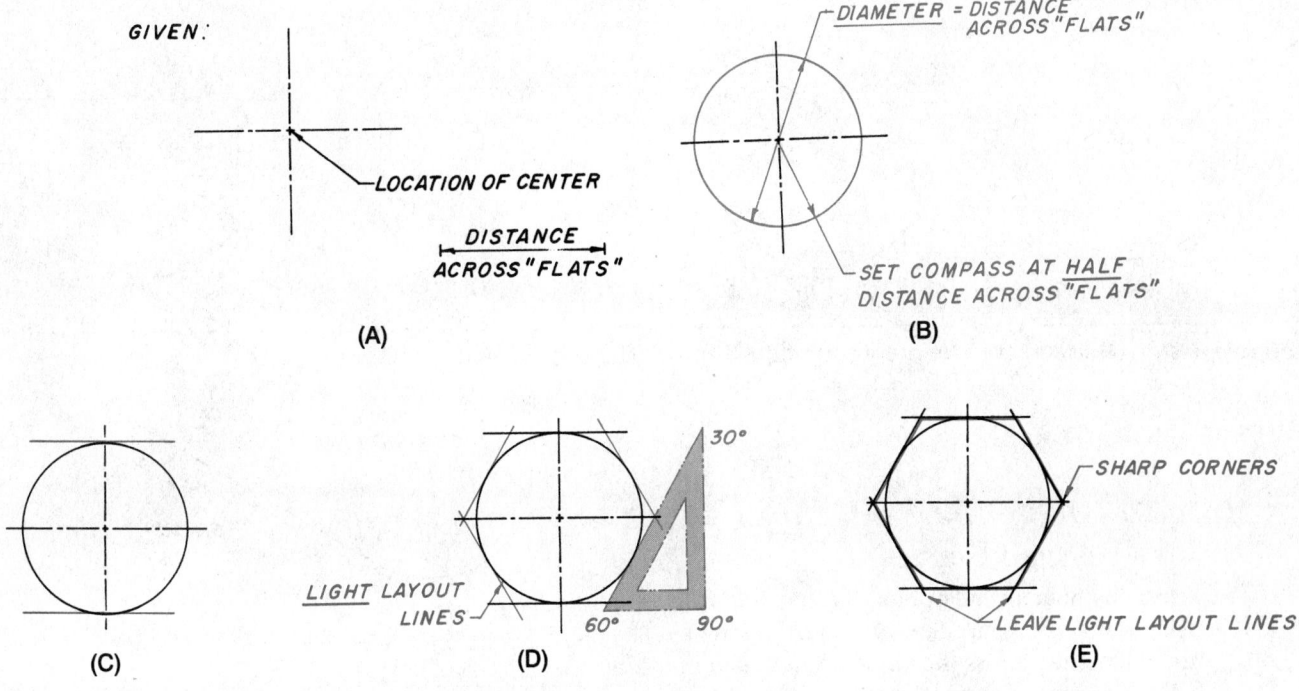

Figure 3-36 (A) How to draw a hexagon; (B) Step 1; (C) Step 2; (D) Step 3; (E) Step 4

GIVEN:
LOCATION OF CENTER
DISTANCE ACROSS "FLATS".
(A)
DIAMETER = DISTANCE ACROSS "FLATS"
SET COMPASS AT HALF DISTANCE ACROSS "FLATS"
(B)
(C)
45°
LIGHT LAYOUT LINES
45°
90°
(D)
SHARP CORNERS
LEAVE LIGHT LAYOUT LINES
(E)

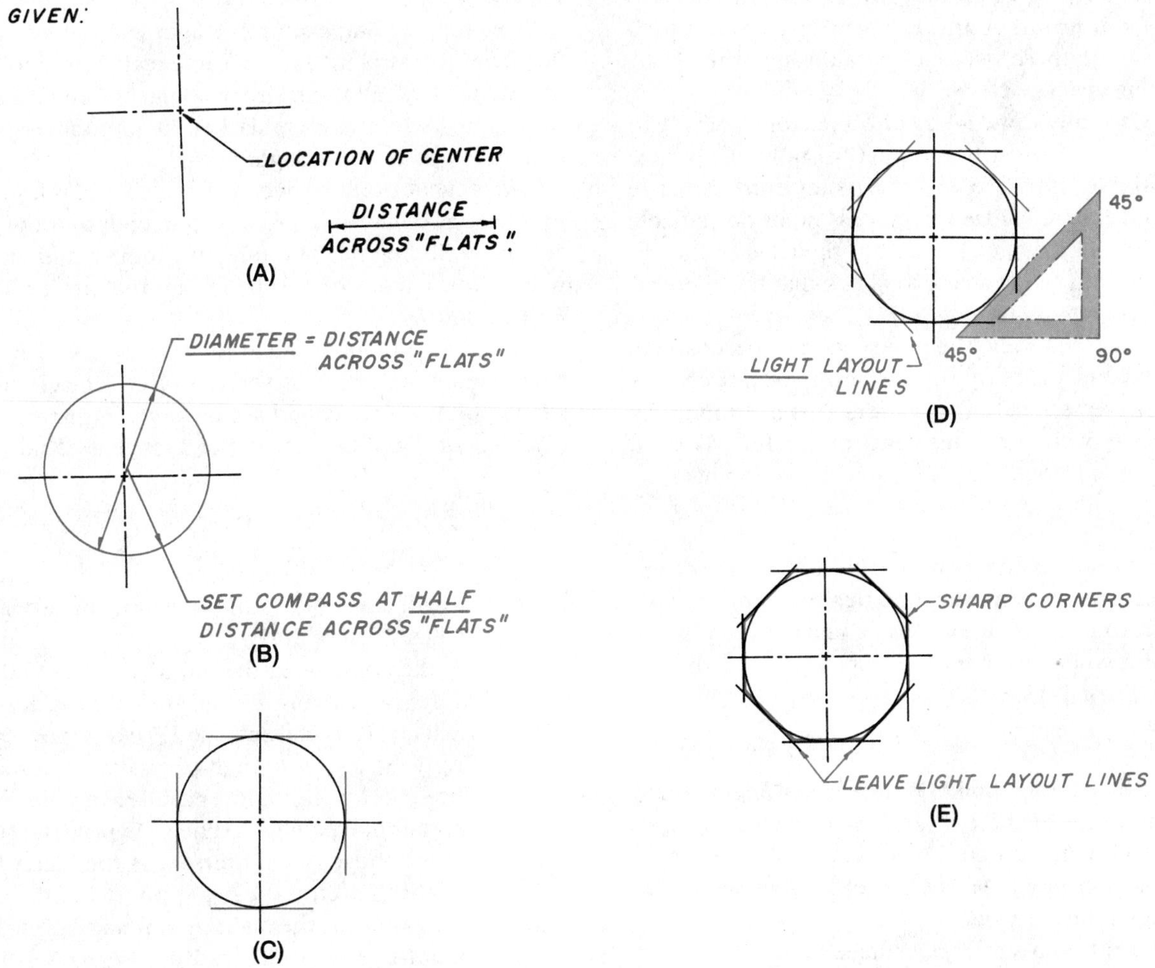

Figure 3-37 (A) How to draw an octagon; (B) Step 1; (C) Step 2; (D) Step 3; (E) Step 4

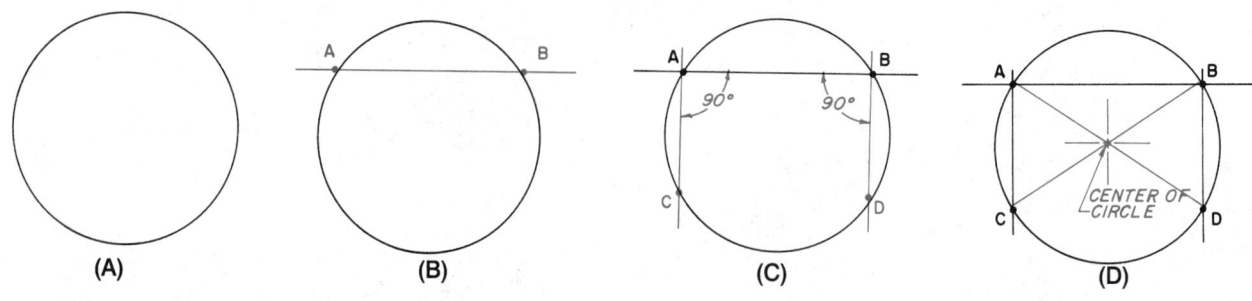

GIVEN:

(A) (B) (C) (D)

Figure 3-38   (A) How to locate the center of a given circle;   (B) Step 1;   (C) Step 2;   (D) Step 3

Circular Construction

How To Locate the Center of a Given Circle

Given: A circle without a center point, *Figure 3-38A*.

Step 1:   Using the drafting machine or T-square, draw a horizontal line across the circle approximately halfway between the *estimated* center of the given circle and the uppermost point on the circumference. Label the end points of the chord thus formed as A and B, *Figure 3-38B*.

Step 2:   Draw perpendicular lines (90°) downward from points A and B. Locate points C and D where these two lines pass through the circle, *Figure 3-38C*.

Step 3:   Carefully draw a straight line from point A to point D and from point C to point B. Where these lines cross is the exact center of the given circle. Place a compass point on the center point; adjust the lead to the edge of the circle and swing an arc to check that the center is accurate, *Figure 3-38D*.

Alternatively, the intersection of perpendicular bisectors of any two nonparallel chords serve to locate the center. This is a modification of the previous construction for passing a circle through any three nonaligned points.

Tangent Points

A *tangent point* is the exact location or point where one line stops and another line begins. Tangent also means to "touch." As an example, a tangent point is the exact point where a curved line stops and a straight line begins, *Figure 3-39*.

How To Locate Tangent Points

The tangent point is a point 90° from the straight line to the swing point of the curved line. Place a light hash mark at each tangent point. It is a good habit to always find *all* tangent points on the object in all views before darkening in the drawing.

The tangent points for a right angle bend are illustrated in *Figure 3-40A*. The tangent points for an acute angle bend are illustrated in *Figure 3-40B*. The tangent

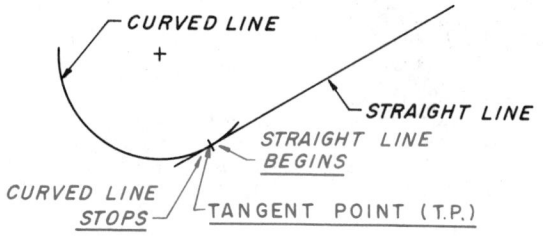

Figure 3-39 Tangent points

points for an obtuse angle bend are illustrated in *Figure 3-40C*.

In each preceding example, a light line is constructed 90° from the straight line to the exact swing point of the radius. Find all the tangent points before darkening in the final work. Always darken in all compass work first, followed by the straight lines.

The tangent point between two arcs or circles is the exact point where one arc or circle ends and the next arc or circle begins. The tangent point could also be where one arc or circle touches another arc or circle, *Figure 3-40D*.

The tangent point is found by drawing a straight line from the swing point of the first arc or circle to the swing point of the second arc or circle, *Figure 3-40E*. Place a short light dash at the exact tangent point.

How To Construct an Arc Tangent to a Right Angle (90°)

Given: A right angle (90°), lines A and B and a required radius, *Figure 3-41A*.

Step 1:   Set the compass at the required radius and, out of the way, swing a radius from line A and one from line B, as illustrated in *Figure 3-41B*.

Step 2:   From the extreme high points of each radius, construct a light line parallel to line A and another line parallel to line B, *Figure 3-41C*.

Step 3:   Where these lines intersect is the exact location of the required swing point. Set the compass point on the swing point and lightly construct the required radius, *Figure 3-41D*. Allow the radius swing to extend past the required area, as shown in the figure. It is

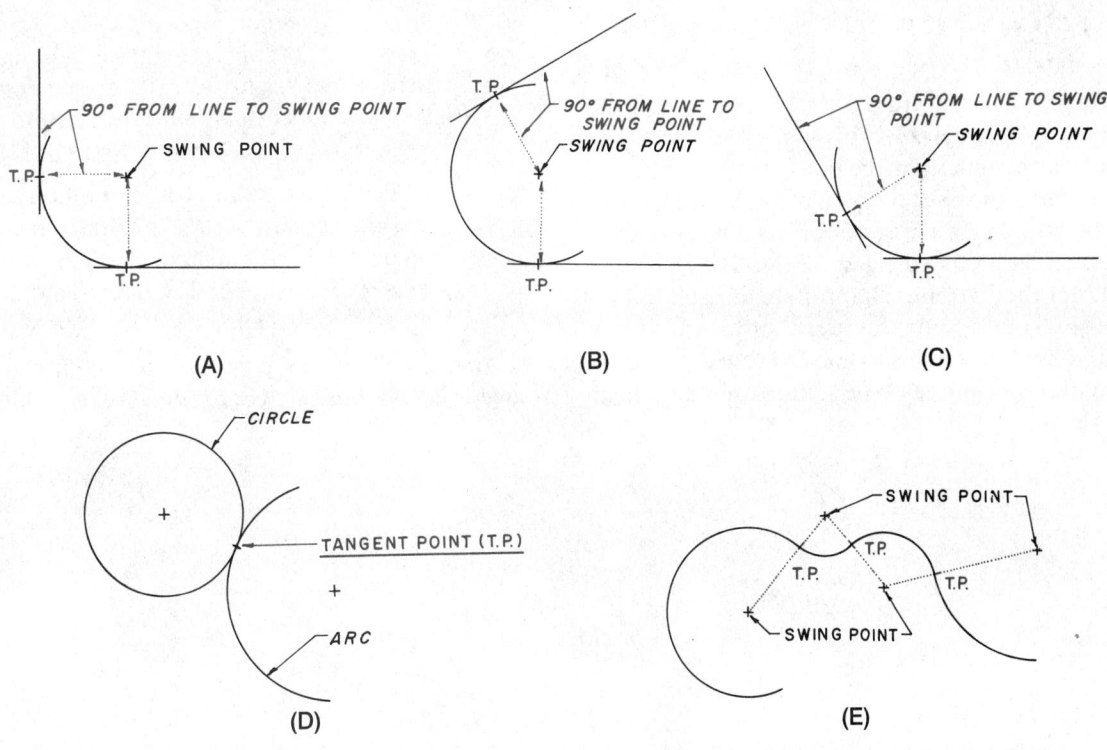

Figure 3-40   How to locate tangent points   (A) on a right angle;   (B) On an acute angle;   (C) On an obtuse angle;   (D) Between arcs or circles;   (E) Tangent points between arcs or circles

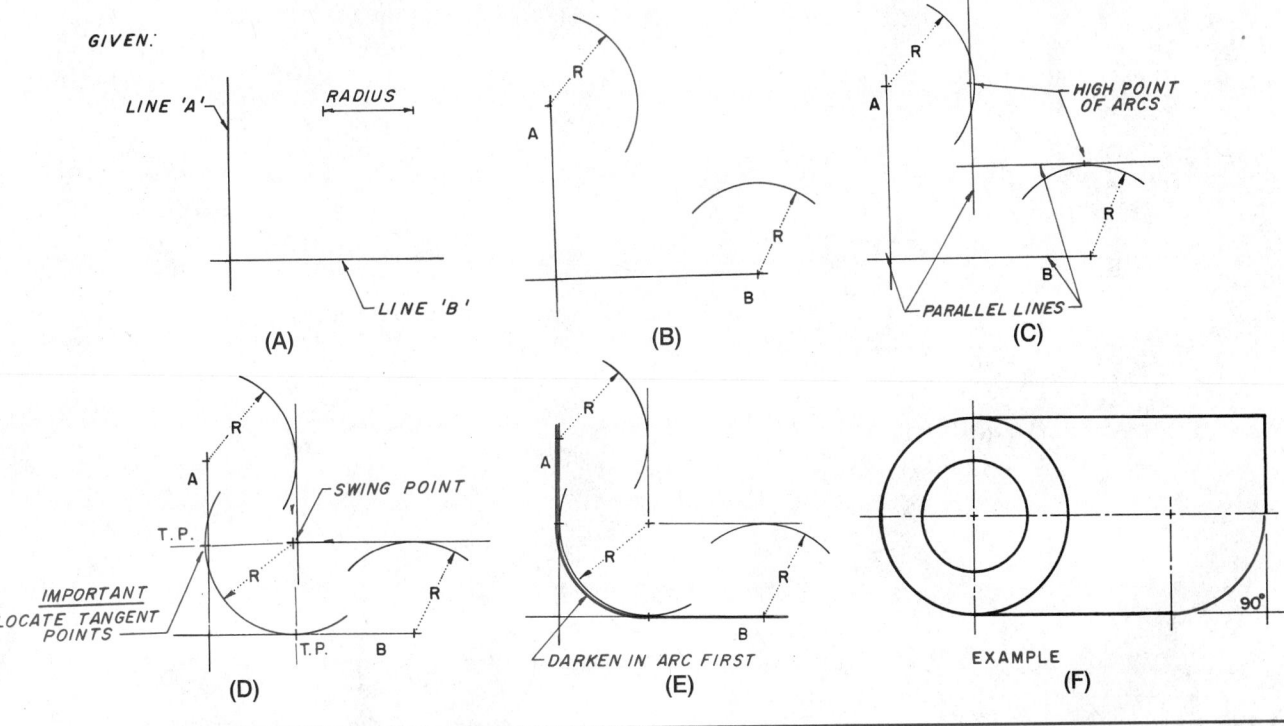

Figure 3-41   (A) How to construct an arc tangent to a right angle;   (B) Step 1;   (C) Step 2;   (D) Step 3;   (E) Step 4;   (F) Example of an arc tangent to two lines at a right angle

important to locate all tangent points (T.P.) before darkening in.

Step 4:   Check all work and darken in the radius using the correct line thickness. Darken in connecting straight lines as required. Always construct compass work first, followed by straight lines. Leave all light construction lines. See **Figure 3-41E**.

An example of an arc tangent to two lines at a right angle (90°) is illustrated in **Figure 3-41F**.

How To Construct an Arc Tangent to an Acute Angle (Less Than 90°)

Given: An acute angle, lines A and B, and a required radius, *Figure 3-42A*. Follow the exact same procedure as outlined in the preceding example.

Step 1: Set the compass at the required radius and, out of the way, swing a radius from line A and one from line B, as illustrated in *Figure 3-42B*.

Step 2: From the extreme high points of each radius, construct a light line parallel to line A and another line parallel to line B, *Figure 3-42C*.

Step 3: Where these lines intersect is the exact loca-

tion of the required swing point. Set the compass point on the swing point and lightly construct the required radius, *Figure 3-42D*. Allow the radius swing to extend past the required area, as shown in Figure 3-42D.

Step 4: Check all work, and darken in the radius using the correct line thickness. Darken in connecting straight lines as required. Always construct compass work first, followed by straight lines. Leave all light construction lines. See *Figure 3-42E*.

An example of an arc tangent to two lines at an acute angle (22°) is illustrated in *Figure 3-42F*.

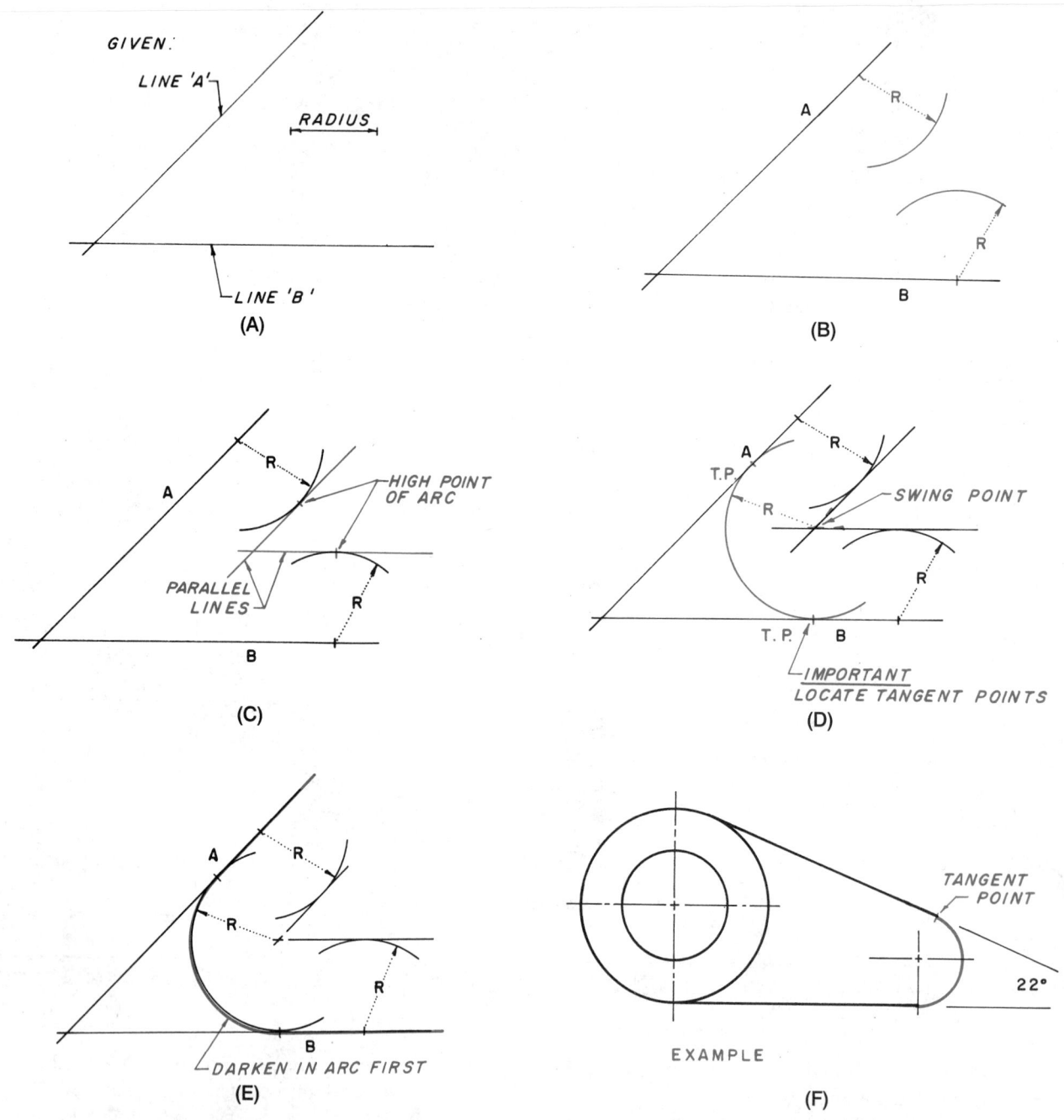

Figure 3-42 (A) How to construct an arc tangent to an acute angle; (B) Step 1; (C) Step 2; (D) Step 3; (E) Step 4; (F) Example of an arc tangent to two lines at an acute angle

How To Construct an Arc Tangent to an Obtuse Angle (More Than 90°)

Given: An obtuse angle between lines A and B, and a required radius, *Figure 3-43A*. Follow the exact same procedure as outlined in the two preceding examples.

Step 1: Set the compass at the required radius and, out of the way, swing a radius from line A and one from line B, as illustrated in *Figure 3-43B*.

Step 2: From the extreme high points of each radius, construct a light line parallel to line A and one parallel to line B, *Figure 3-43C*.

Step 3: Where these lines intersect is the exact location of the required swing point. Set the compass point on the swing point and lightly construct the required radius. Allow the radius swing to extend past the required area, as shown in *Figure 3-43D*.

Step 4: Check all work, darken in the radius using the correct line thickness. Darken in connecting straight lines as required. Always construct compass work first, followed by straight lines. Leave all light construction lines. See *Figure 3-43E*.

An example of an arc tangent to two lines at an obtuse angle (120°) is illustrated in *Figure 3-43F*.

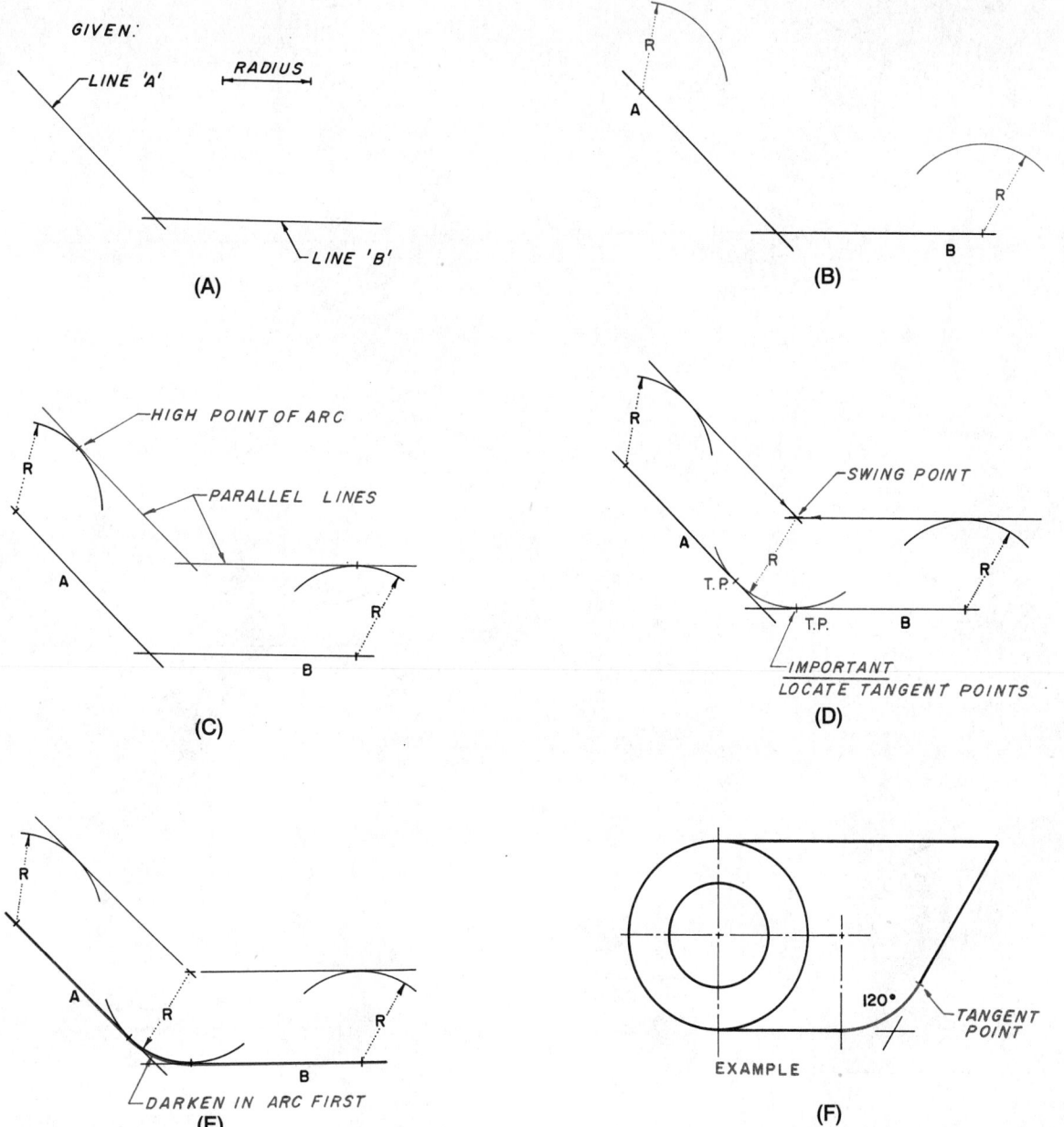

Figure 3-43 (A) How to construct an arc tangent to an obtuse angle; (B) Step 1; (C) Step 2; (D) Step 3; (E) Step 4; (F) Example of an arc tangent to two lines at an obtuse angle

How To Construct an Arc Tangent to a Straight Line and a Curve

Given: Straight line A, an arc B with a center point, and a required radius, *Figure 3-44A*.

Step 1: Set the compass at the required radius and, out of the way, swing a radius from the given arc B and one from the given straight line A, *Figure 3-44B*.

Step 2: From the extreme high points of each radius, construct a light straight line parallel to line A, and construct a radius outside the given arc B equal to the required radius, as illustrated in *Figure 3-44C*.

Step 3: Where these lines intersect is the exact location of the required swing point. Set the compass point on the swing point and lightly construct the required radius, *Figure 3-44D*. Allow the radius swing to extend past the required area as shown. Locate all tangent points (T.P.) before darkening in.

Step 4: Check all work, darken in the radius using the correct line thickness. Darken in the arcs first and the straight line last, *Figure 3-44E*.

An example of an arc tangent to a straight line and a curve is illustrated in *Figure 3-44F*.

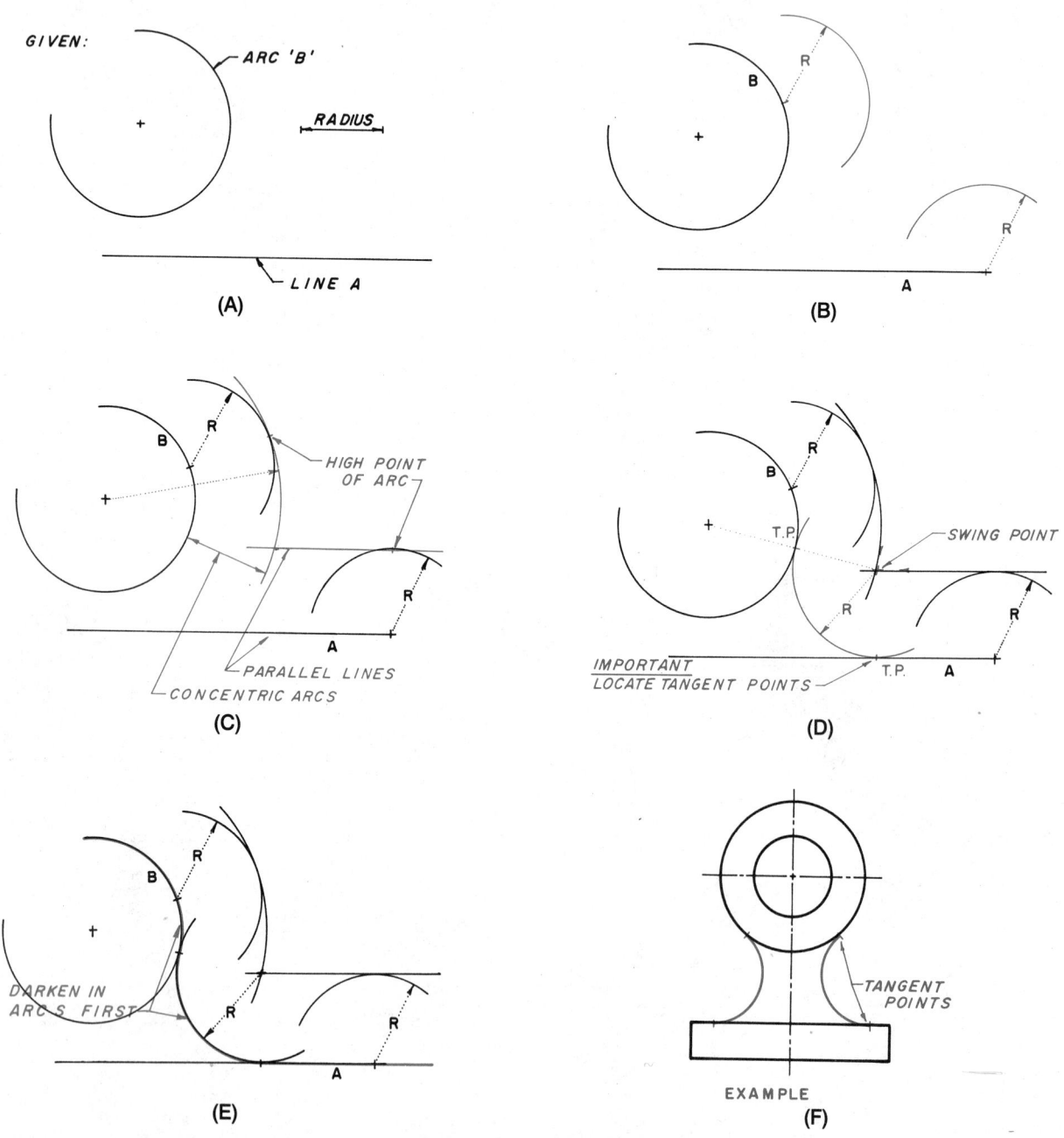

Figure 3-44   (A) How to construct an arc tangent to a straight line and a curve; (B) Step 1; (C) Step 2; (D) Step 3; (E) Step 4; (F) Example of an arc tangent to a straight line and a curve

How To Construct an Arc Tangent to Two Radii or Diameters

Given: Diameter A and arc B with center points located, and the required radius, *Figure 3-45A*.

Step 1: Set the compass at the required radius and, out of the way, swing a radius of the required length from a point on the circumference of given diameter A. Out of the way, swing a required radius from a point on the circumference of a given arc B, *Figure 3-45B*.

Step 2: From the extreme high points of each radius, construct a light radius outside of the given radii A and B, as illustrated in *Figure 3-45C*.

Step 3: Where these arcs intersect is the exact location of the required swing point. Set the compass point on the swing point and lightly construct the required radius, *Figure 3-45D*. Allow the radius swing to extend past the required area as shown. Before darkening in, it is important to locate all tangent points (T.P.), *Figure 3-45D*.

Step 4: Check all work, darken in the radii using the correct line thickness. Darken in the arcs or radii in consecutive order from left to right or from right to left, thus constructing a smooth connecting line having no apparent change in direction, *Figure 3-45E*.

An example of an arc tangent to two radii is illustrated in *Figure 3-45F*.

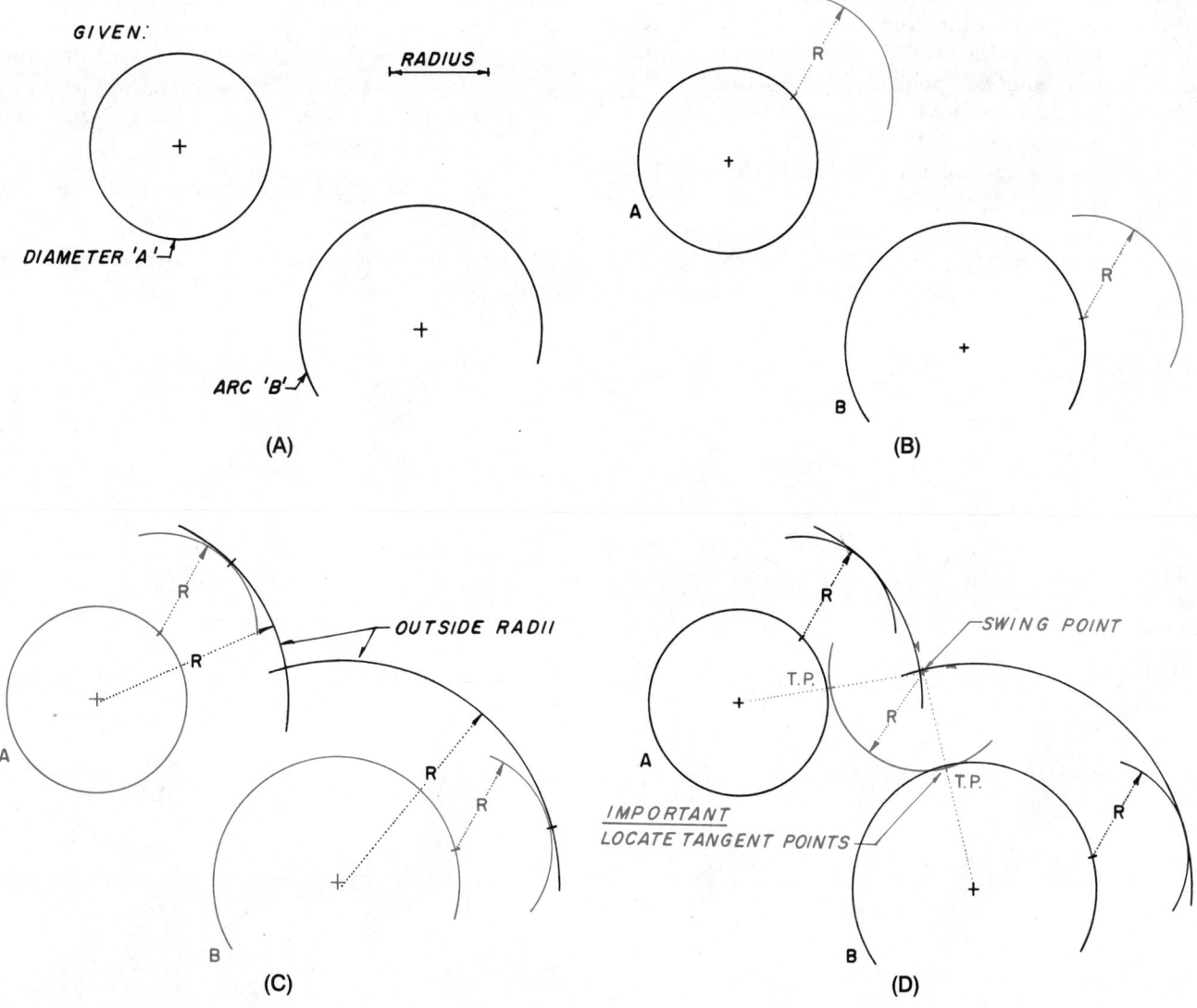

Figure 3-45   (A) How to construct an arc tangent to two radii or diameters; (B) Step 1; (C) Step 2; (D) Step 3

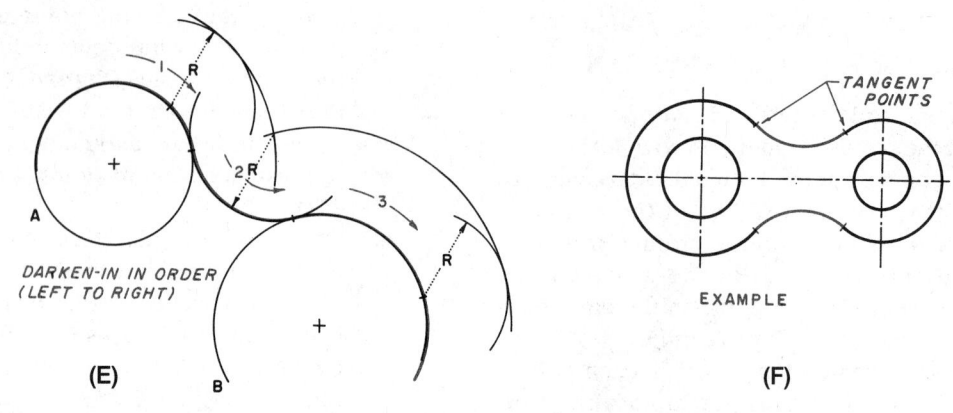

Figure 3-45 (Continued) (E) Step 4; (F) Example of an arc tangent to to two radii

How To Draw an Ogee Curve

An *ogee curve* is used to join two parallel lines. It forms a gentle curve that reverses itself in a neat symmetrical geometric form.

Given: Parallel lines A-B and C-D, *Figure 3-46A*.

Step 1: Draw a straight line connecting the space between the parallel lines. In this example, from point B to point C, *Figure 3-46B*.

Step 2: Make a perpendicular bisector to line B-C to establish point X, *Figure 3-46C*.

Step 3: Make perpendicular bisectors to the lines B-X and X-C, *Figure 3-46D*.

Step 4: Draw a perpendicular from line A-B at point B to intersect the perpendicular bisector of B-X, which locates the first required swing center.

Draw a perpendicular from line C-D at point C to intersect the perpendicular bisector of C-X, which locates the second required swing center, *Figure 3-46E*.

Step 5: Place the compass point on the first swing point and adjust the compass lead to point B, and swing an arc from B to X. Place the compass point on the second swing point and swing an arc from X to C, *Figure 3-46F*. This completes the ogee curve.

Note: point X is the tangent point between arcs. Check and, if correct, darken in all work.

An example of an ogee curve is illustrated in *Figure 3-46G*.

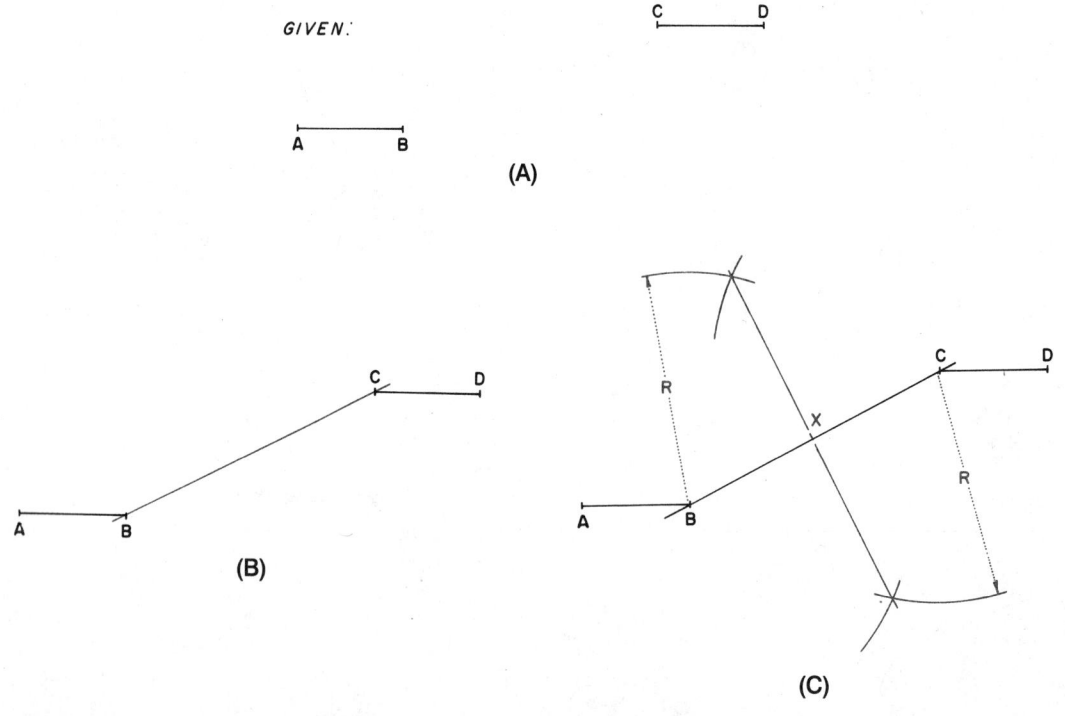

Figure 3-46 (A) How to draw an ogee curve; (B) Step 1; (C) Step 2

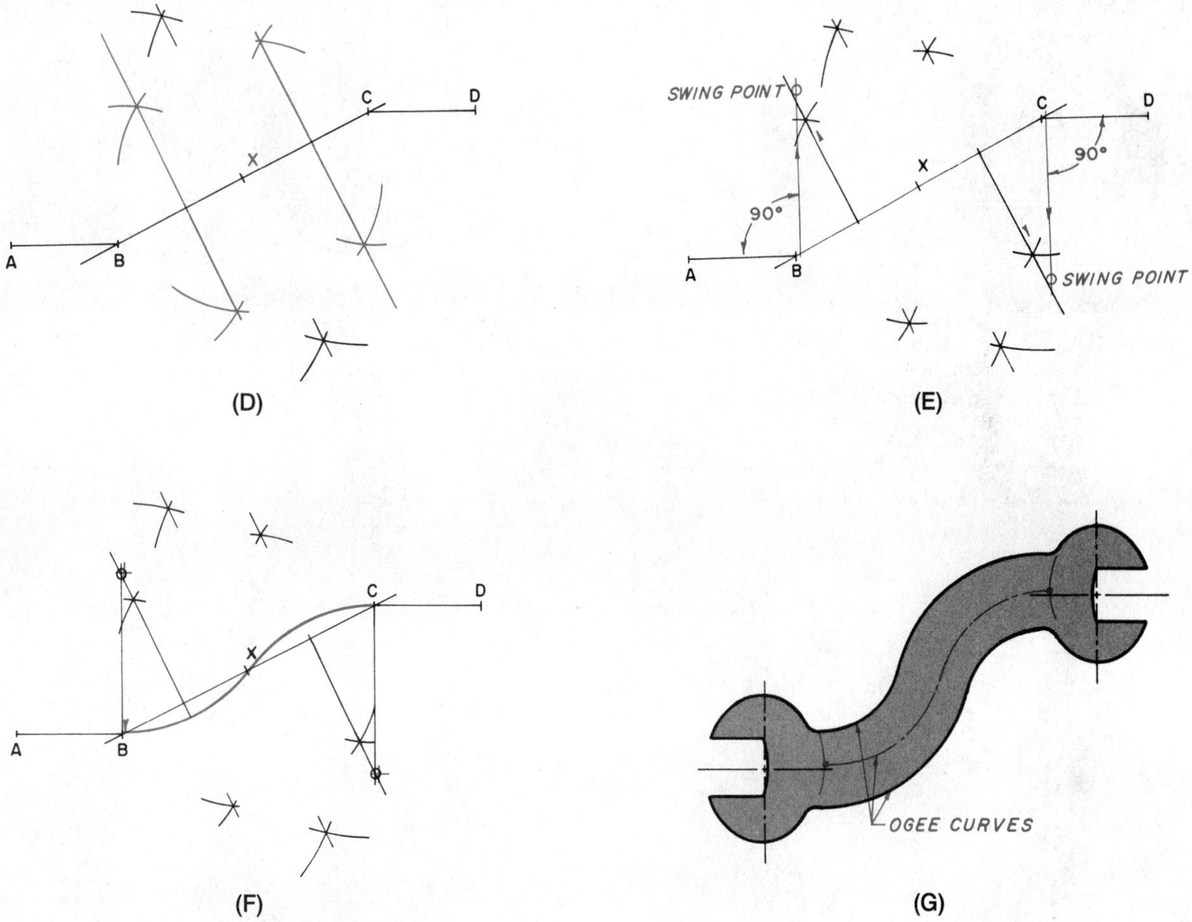

(D)

(E)

SWING POINT

90°

90°

SWING POINT

(F)

(G)

OGEE CURVES

Figure 3-46 (Continued) (D) Step 3; (E) Step 4; (F) Step 5; (G) Step 6

Conic Sections

A *conic section* is a section cut by a plane passing through a cone. These sections are bounded by various kinds of shapes. Depending upon where the section is cut, the various shapes can be a triangle, a circle, an ellipse, a parabola or a hyperbola, Figures 3-47A through 3-47E.

Triangle. *Figure 3-47A*. This shape results when a plane passes through the apex of the cone.

Circle. *Figure 3-47B*. This shape results when a plane passes through the cone, parallel with the base and perpendicular to the axis.

Ellipse. *Figure 3-47C*. This shape results when a plane passes through the cone inclined to the axis.

Parabola. *Figure 3-47D*. This shape results when a plane passes through the cone parallel to one element.

Hyperbola. *Figure 3-47E*. This shape results when a plane passes through the cone parallel with the axis of the cone.

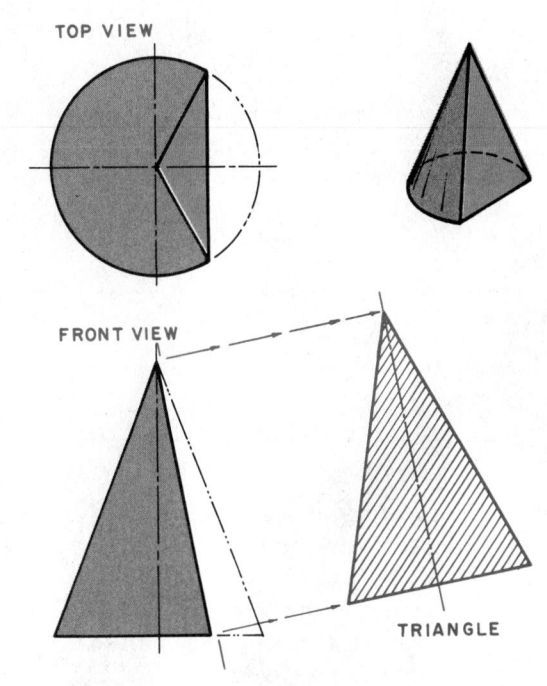

TOP VIEW

FRONT VIEW

TRIANGLE

Figure 3-47 (A) Triangle section

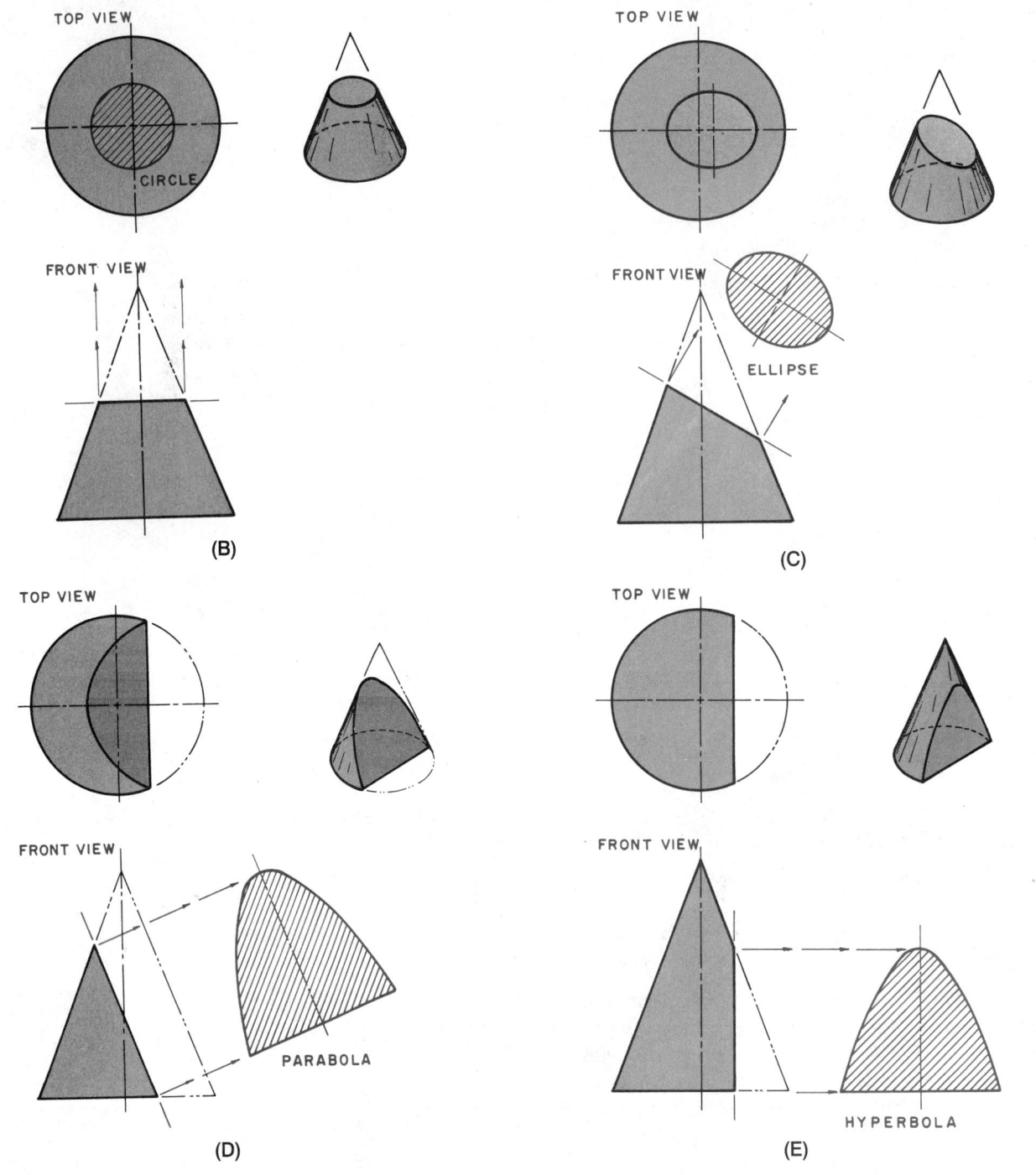

Figure 3-47 (Continued)   (B) Circle section;   (C) Ellipse section;   (D) Parabola section;   (E) Hyperbola section

How To Draw an Ellipse, Concentric Circle Method

Given: The location of the center point, the major diameter, and the minor diameter, ***Figure 3-48A***.

Step 1: Lightly draw one circle equal to the major diameter and another circle equal to the minor diameter, ***Figure 3-48B***.

Step 2: Divide both circles into 12 equal divisions by passing lines through the center at every 30

degrees. Number points 1 through 5 in clockwise consecutive order on the major diameter, positioning the 3 location on the right horizontal axis. Number points 7 through 11 in clockwise consecutive order, positioning the 9 location on the left horizontal axis. Point 6 is on the lower vertical axis at the minor diameter, and point 12 is on the upper vertical axis at the minor diameter, ***Figure 3-48C***. If the ellipse were to be positioned at 90° from this example, the point locations would be rotated accordingly.

Step 3: Project down from point 1 on the major diameter, and to the right from point 1 on the minor diameter. Where these lines intersect is the exact location where point 1 will be on the ellipse, *Figure 3-48D*.

Step 4: Project down from point 2 on the major diameter, and to the right from point 2 on the minor diameter. Where these lines intersect is the exact location where point 2 will be on the ellipse, *Figure 3-48E*.

Step 5: This completes the first quadrant of the ellipse. Continue around the circle to locate points 4, 5, 6, 7, 8, 9, 10, and 11 in the same manner except in reverse, *Figure 3-48F*.

Step 6: Lightly draw straight lines connecting points 1 through 12 in order to get a general idea of the ellipse outline, *Figure 3-48G*.

Step 7: Darken in the ellipse using an irregular curve. Carefully connect all the points with a smooth continuous line of the correct thickness, *Figure 3-48H*. It is sometimes helpful to divide into 10° or 15° spaces the two ends where the ellipse curves the fastest in order to have more points around the extreme ends. In this example, this is done between points 2-4 and 8-10.

Examples of ellipses are illustrated in *Figure 3-48I*, showing how a rotated circle would look at 0°, 15°, 45°, 60°, and 90°.

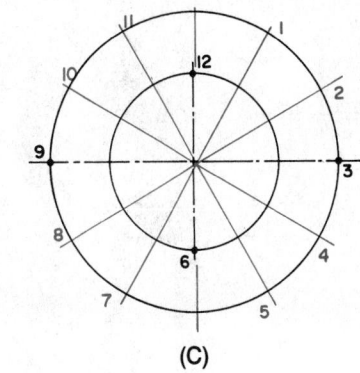

(C)

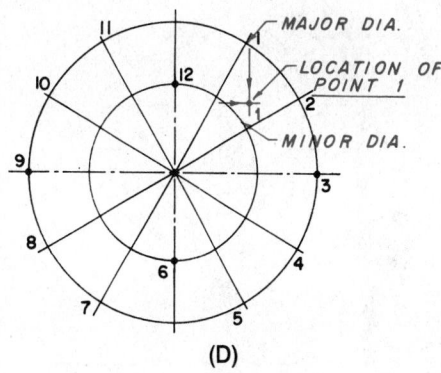

(D)

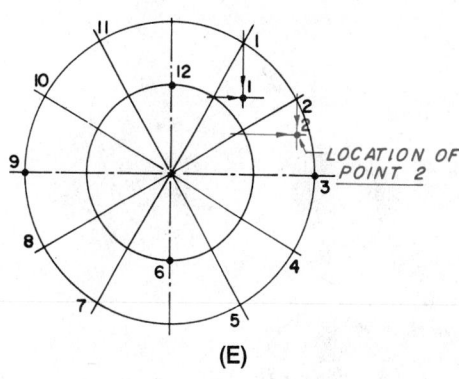

(E)

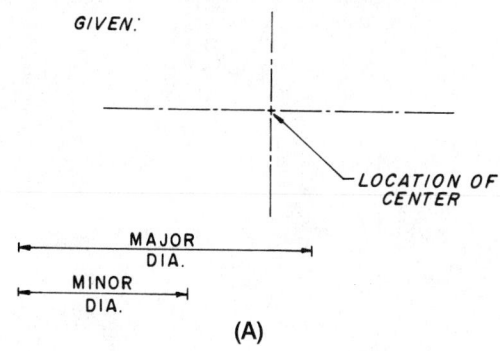

(A)

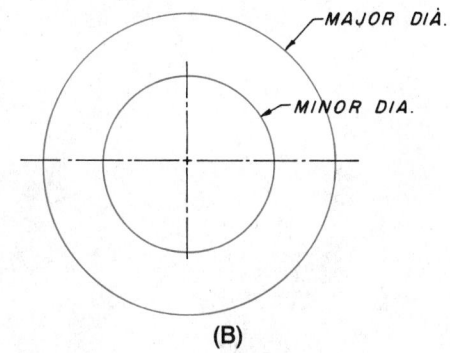

(B)

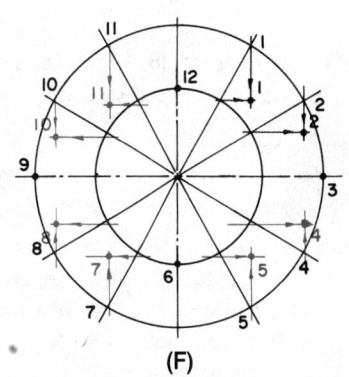

(F)

Figure 3-48   (A) How to draw an ellipse (concentric method); (B) Step 1; (C) Step 2; (D) Step 3; (E) Step 4; (F) Step 5

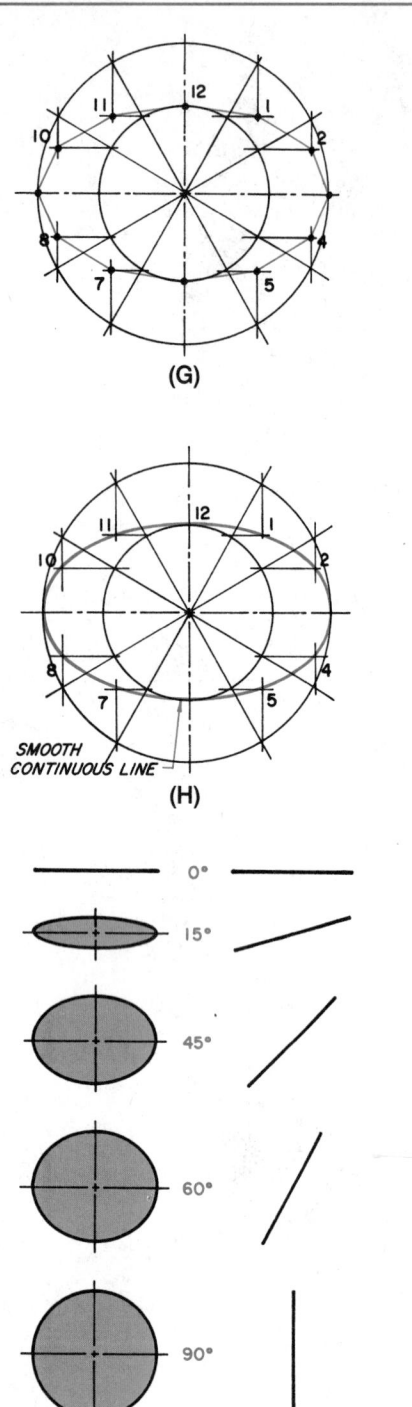

Figure 3-48 (Continued) (G) Step 6; (H) Step 7;
(I) Example of ellipses at various degrees

How To Draw a Trammel Ellipse

Given: A major or minor axis of a required ellipse, and any located point through which the ellipse must pass, **Figure 3-49A**. If a major axis is given, mark the end points A and B. If the minor axis is given, as is shown in the figure, mark the end points C and D. Mark as P the known location of where the ellipse must pass.

Step 1: Draw a perpendicular bisector of the given axis to locate the unknown axis position, crossing the known axis at a point that will become 0, **Figure 3-49B**.

Step 2: Mark an end corner of a separate strip of paper as point 0, and mark along the edge of the strip one-half of the distance A-B or C-D, whichever is known. Mark this point on the strip as A' or C' to correspond to the known axis, **Figure 3-49C**. Position the strip so that point 0 lies on the known point through which the ellipse must pass. Position the other strip mark on the axis that is not represented by the half-length on the strip, at the position (of two possible positions) that is nearest to the ellipse center.

Step 3: With the strip positioned as specified in Step 2, mark the strip at the location of where it crosses the other axis. The length of end point 0 to this new position represents one-half of the unknown axis length, and should correspondingly be labeled as A' or C', **Figure 3-49D**.

Step 4: Using the two marked strip locations, reposition the strip so that the mark representing one-half of the major axis always lies on the minor axis, and the mark representing one-half of the minor axis always lies on the major axis. Repeat as needed to establish enough points for constructing the ellipse, **Figure 3-49E**.

Step 5: Using a French curve, pass a smooth line through each of these established points, **Figure 3-49F**.

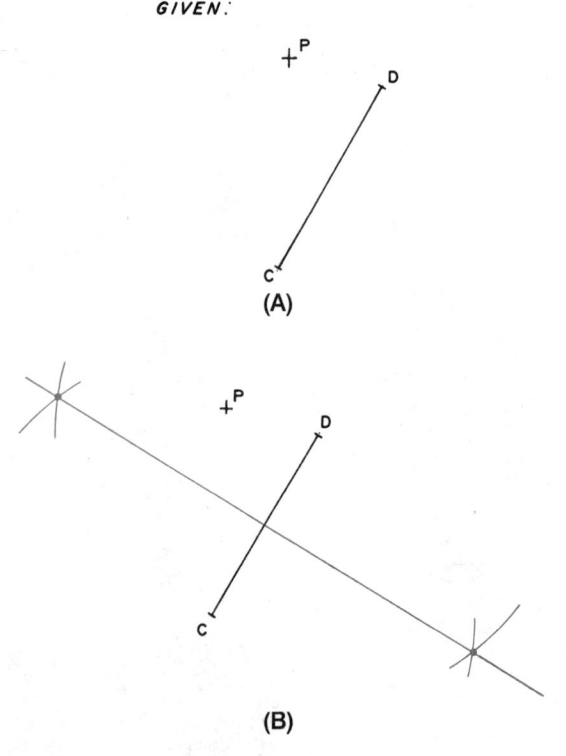

Figure 3-49 (A) Trammel ellipse; (B) Step 1

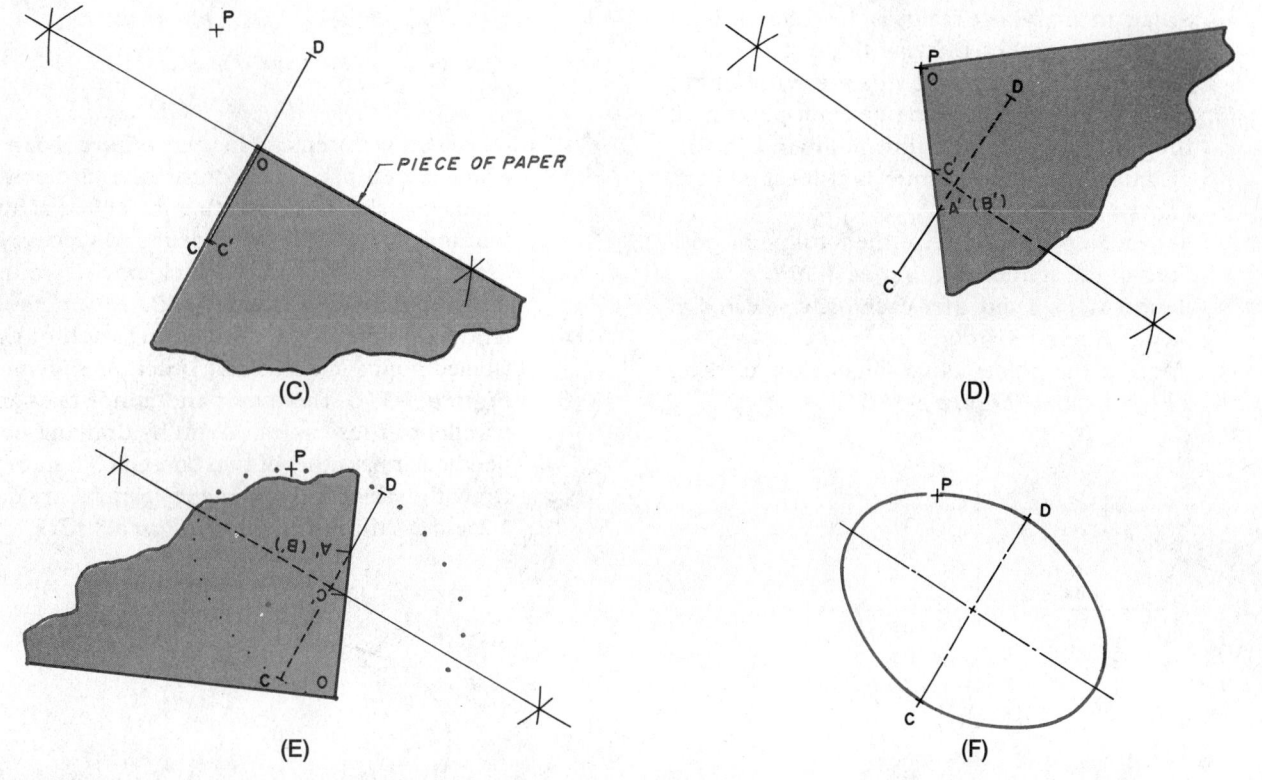

Figure 3-49 (Continued) (C) Step 2; (D) Step 3; (E) Step 4; (F) Step 5

How To Draw a Foci Ellipse

Given: Major and minor axes of a required ellipse. The end points of the major axis are labeled A and B, the end points of the minor axis are labeled C and D, and the ellipse center is labeled 0, *Figure 3-50A*.

Step 1: Swing a radius of one-half of the major axis (0A) from the end point of the minor axis (C or D) to locate the focal points E and F at the points where the arc crosses the major axis, *Figure 3-50B*.

Step 2: Select a random array of points on the major axis between one focal point and the ellipse center (space those at the ends more closely). Number these in consecutive order, *Figure 3-50C*.

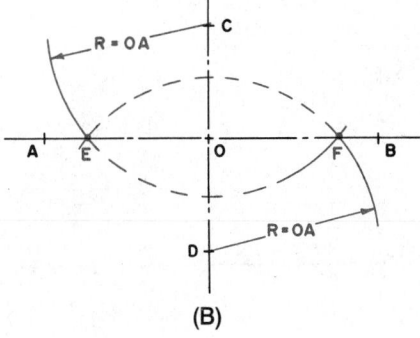

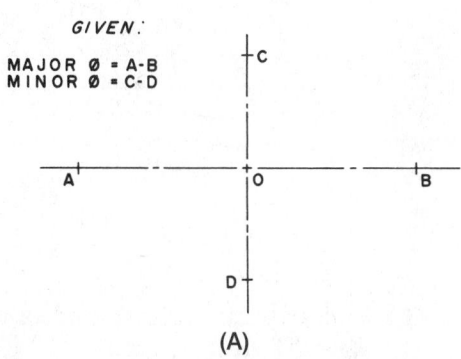

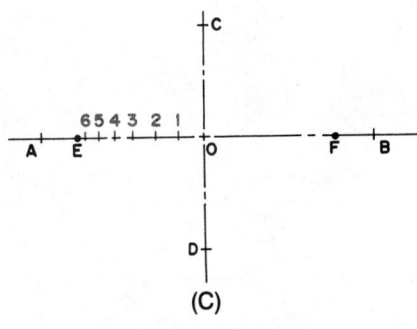

Figure 3-50 (A) How to draw a foci ellipse; (B) Step 1; (C) Step 2

Step 3: Swing an arc whose radius is the distance from point B to number 1. Swing this distance from focal point F. Intersect this arc with an arc whose radius is the distance from point A to the same numbered point, number 1 in this example, but whose center is at focal point E, *Figure 3-50D*.

Step 4: Repeat Step 3, but reverse the focal point positions of the arc centers, *Figure 3-50E*.

Step 5: Repeat Steps 3 and 4 for each point selected in Step 2, *Figure 3-50F*.

Step 6: Connect the points into a smooth curve, using a French curve, *Figure 3-50G*.

How To Locate the Major and Minor Axes of a Given Ellipse with a Located Center

Given: An ellipse, with center 0 located, *Figure 3-51A*.

Step 1: From the ellipse center 0, draw a circle at a radius that allows intersecting the ellipse at any four locations. Label these points in successive order 1, 2, 3, and 4, moving clockwise around the ellipse center, *Figure 3-51B*.

Step 2: Draw a rectangle by connecting each of the labeled points in successive order. As shown in *Figure 3-51C*, the major and minor axes are parallel to these sides, found by drawing perpendicular bisectors of two consecutive sides.

Step 3: Draw the major and minor axes parallel to sides 1-2 and 2-3 through point 0, *Figure 3-51D*.

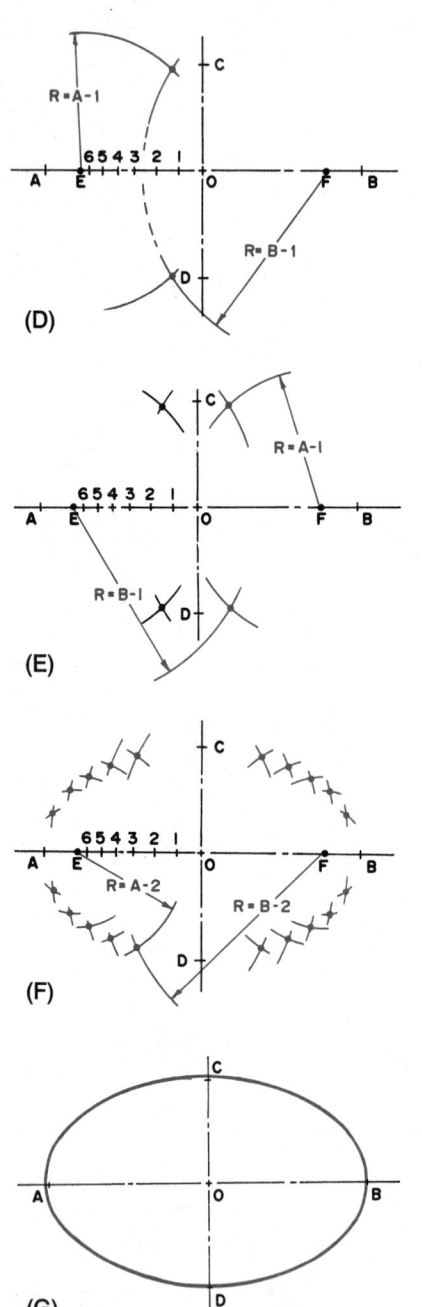

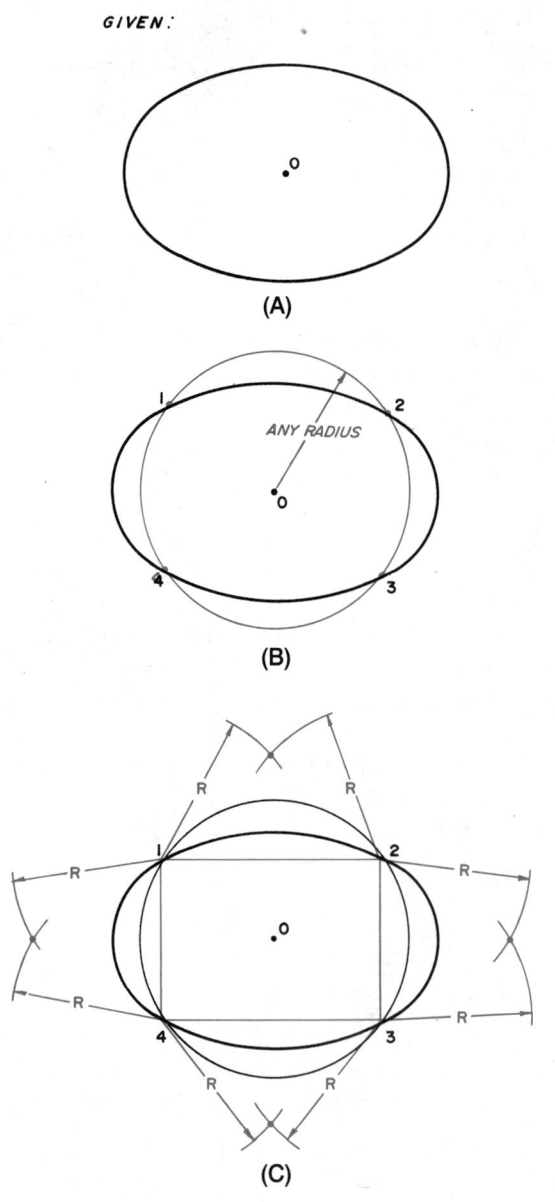

Figure 3-50 (Continued) (D) Step 3; (E) Step 4; (F) Step 5; (G) Step 6

Figure 3-51 (A) How to locate the major and minor axes of a given ellipse; (B) Step 1; (C) Step 2

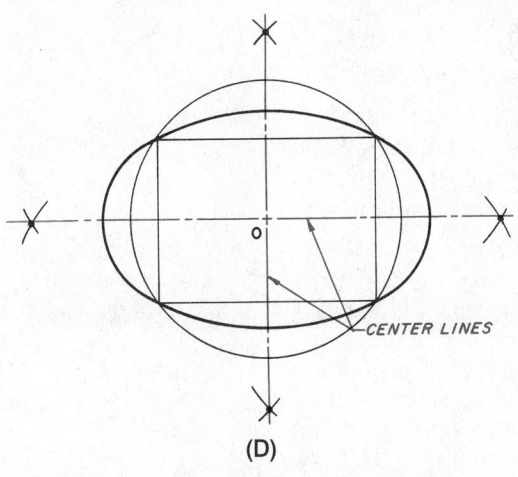

(D)

Figure 3-51 (Continued) (D) Step 3

How To Draw a Tangent to an Ellipse at a Point on the Ellipse

Given: An ellipse, and a point P on its perimeter, *Figure 3-52A*. If not provided, locate the major axis A-B and minor axis C-D, and the focal points E and F.

Step 1: Draw E-P and F-P, *Figure 3-52B*.

Step 2: Bisect the angle EPF, *Figure 3-52C*.

Step 3: Draw a perpendicular to the angle bisector of angle EPF at point P. This is the required tangent, *Figure 3-52D*.

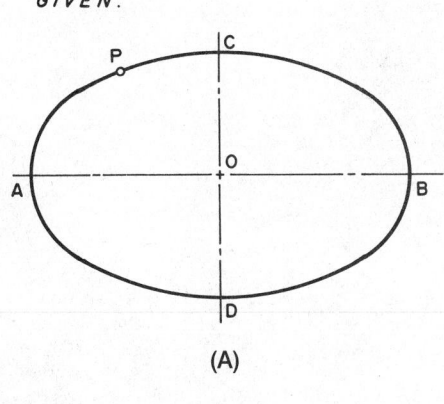

(A)

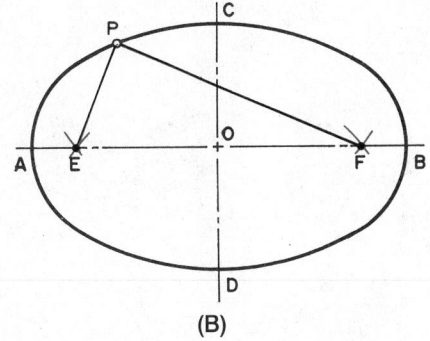

(B)

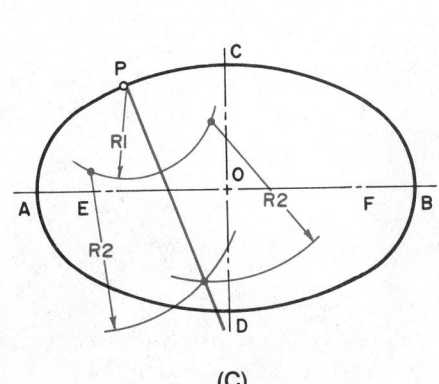

(C)

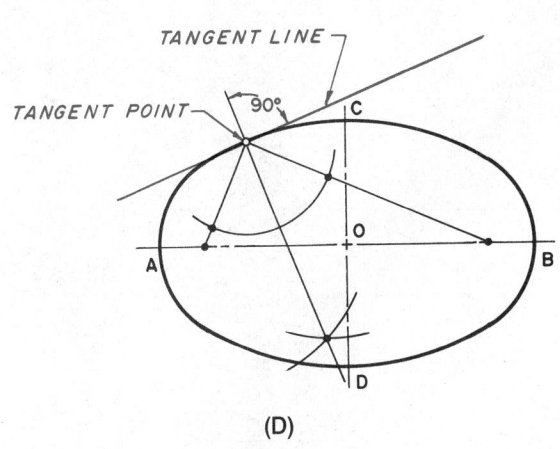

(D)

Figure 3-52   (A) How to draw a tangent to an ellipse at a point on the ellipse;   (B) Step 1;   (C) Step 2;   (D) Step 3

How To Draw a Tangent to an Ellipse from a Distant Point

Given: An ellipse and a distant point P, *Figure 3-53A*.

Step 1: If not provided, locate the major axis A-B and focal points E and F, *Figure 3-53B*.

Step 2: Swing an arc R1 from point P to pass through either focal point. F is selected for this example, *Figure 3-53C*.

Step 3: Swing an arc whose radius is equal to the major axis, from the other focal point (E in this example). Intersect the arc from Step 2 at points R and S, *Figure 3-53D*.

Step 4: Extend the lines from E to each of points R and S, to cross the ellipse at U and V, *Figure 3-53E*.

Step 5: The two possible points of tangency of the two possible lines from point P are U and V. Select and draw either, or both, *Figure 3-53F*.

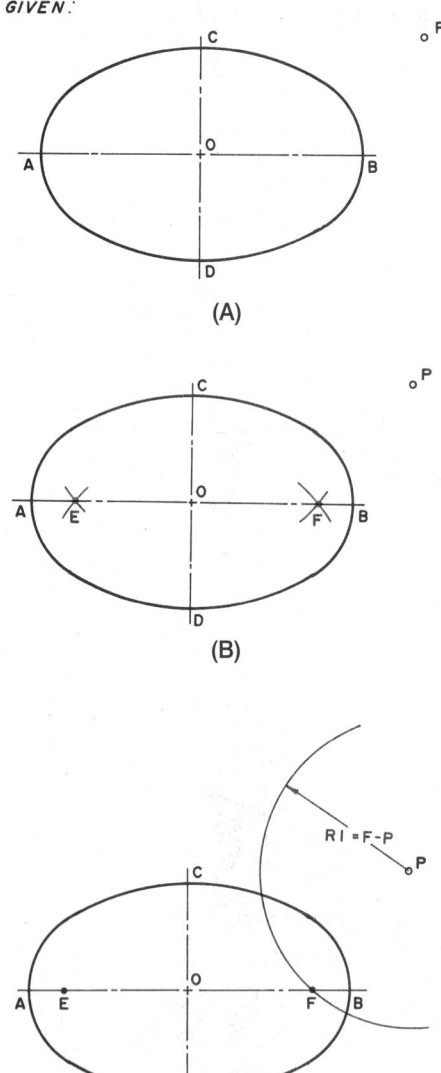

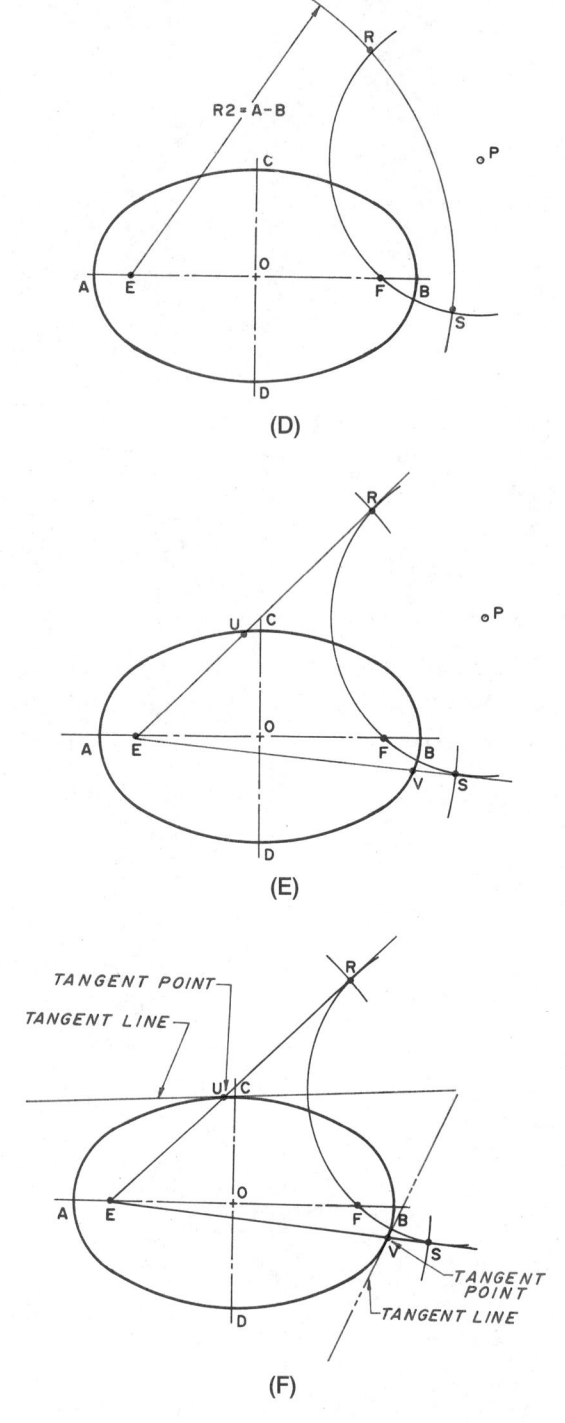

Figure 3-53   (A) How to draw a tangent to an ellipse from a distant point; (B) Step 1; (C) Step 2; (D) Step 3; (E) Step 4; (F) Step 5

How To Draw a Parabola (Method 1)

Given: The front view and plan view of cone ABC with cutting plane X-Y, *Figure 3-54A*.

Step 1: Locate points Y' and Y" in the plan view. In the front view, place the point of the compass on point Y and position the lead on point X. Swing point X to the extended baseline and down into the plan view, as illustrated in *Figure 3-54B*.

Step 2: Locate a point along line X-Y (in this example, point D) and, using point Y as a swing point, swing point D to the extended baseline and down into the plan view. In the front view, reset the compass to a distance equal to the horizontal distance from the cone axis to the outer edge (identified as length 1), illustrated in

Figure 3-54C. Transfer this distance to the plan view. Project point D down into the plan view and, where point D intersects the arc, locate points D' and D". Project these points to the right, intersecting with point D from above at the actual intersection points D' and D".

Step 3: Choose other various points along line X-Y in the front view, and project them over and down as described in Step 2, *Figure 3-54D*.

Step 4: Using an irregular curve, connect all points. This completes the parabola, *Figure 3-54E*.

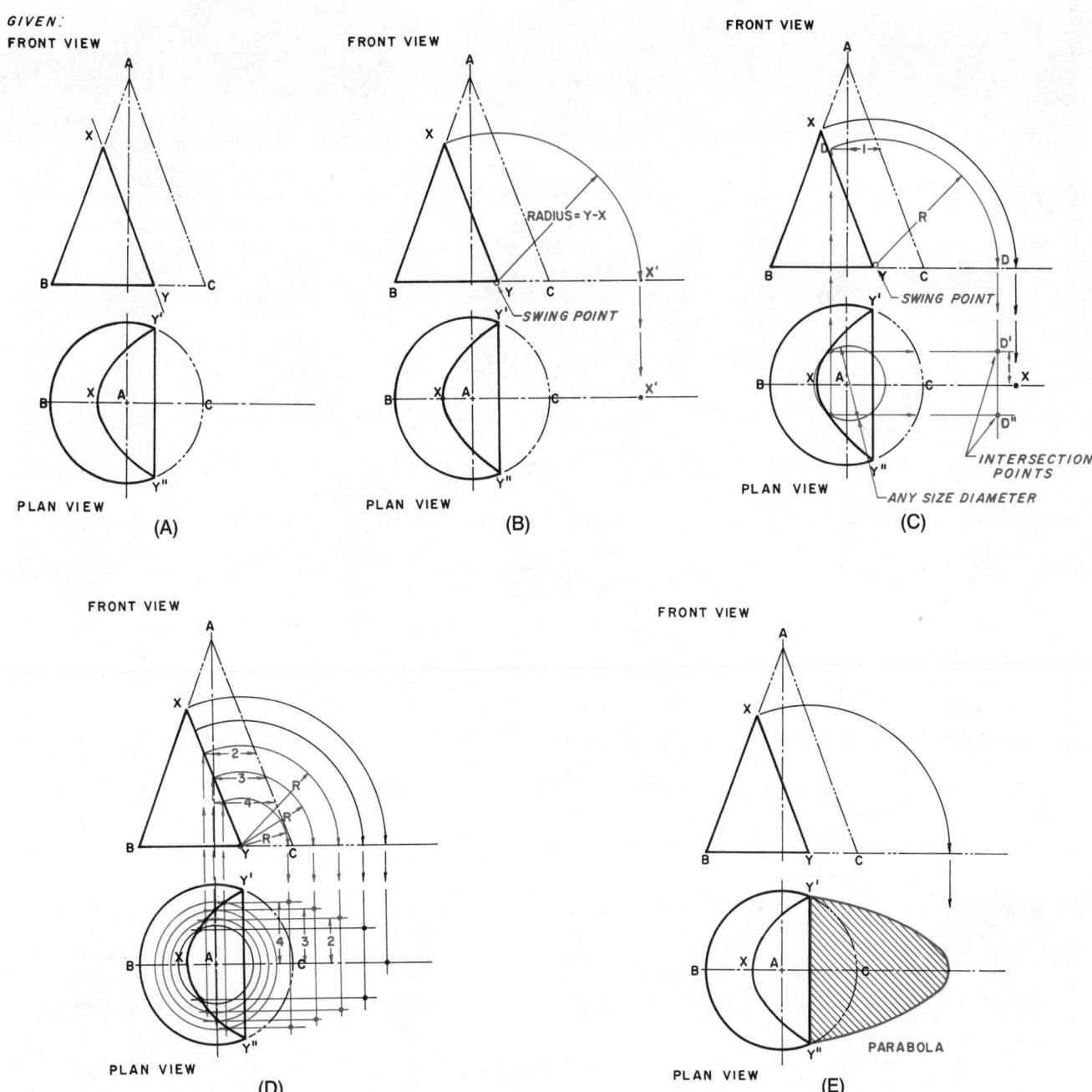

Figure 3-54 (A) How to draw a parabola (method 1); (B) Step 1; (C) Step 2; (D) Step 3; (E) Step 4

How To Draw a Parabola (Method 2)

Given: Required rise A-B, and required span A-C. *Problem:* Construct a parabola within the required rise and span, *Figure 3-55A*.

Step 1: Divide half the span distance A-0 into any number of equal parts. (In this example, 4 equal parts are used.) See *Figure 3-55B*.

Divide half the rise A-B into equal parts amounting to the square of the equal parts in Step 1 (in this example, $4^2 = 16$ equal parts).

Step 2: From line A-0, each point on the parabola is offset by a number of units equal to the square of the numbers of units from point 0, see *Figure 3-55C*. For example, point 1 projects 1 unit; point 2 projects 4 units; point 3 projects 9 units, and so forth.

Step 3: Using an irregular curve, connect the points to form a smooth parabolic curve, *Figure 3-55D*.

How To Find the Focus Point of a Parabola

Given: A parabolic curve with points A, 0, and B, *Figure 3-56A*. *Problem:* Find the focus point of the parabola A0B.

Step 1: Draw a line from point A to point B. Continue the center line to a point equal to length X, distance Y, to find point C, *Figure 3-56B*.

Step 2: Draw a line from point C to point A and bisect it. The intersection of the bisect line and the axis is the focus of the parabola, *Figure 3-56C*.

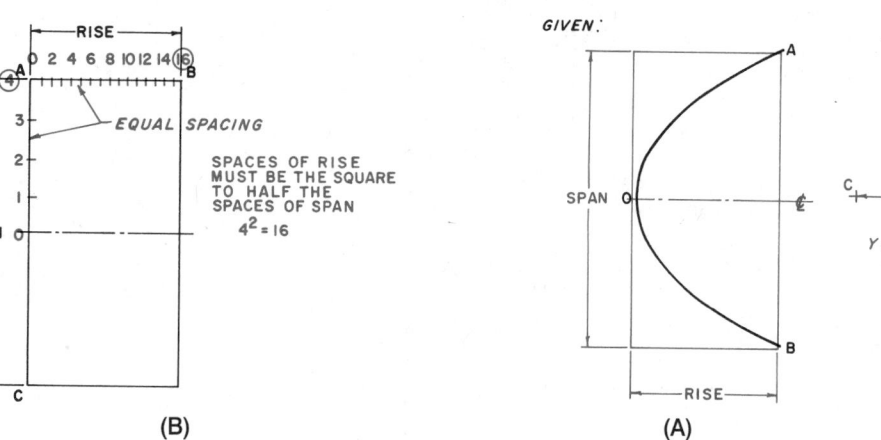

(A)

(B)

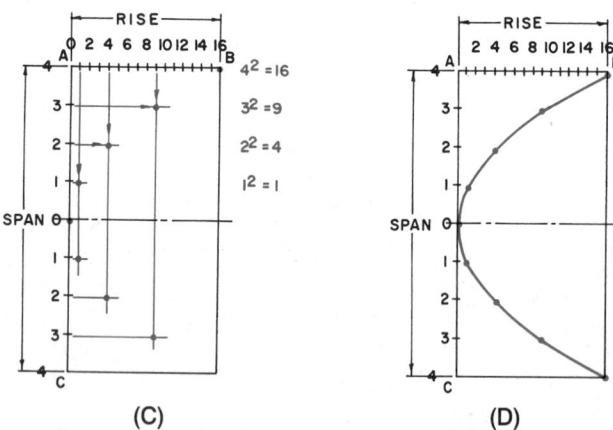

(C)   (D)

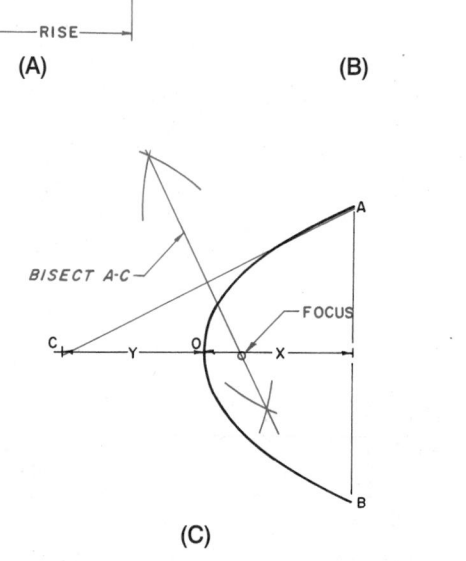

(C)

Figure 3-55   (A) How to draw a parabola (method 2);   (B) Step 1;   (C) Step 2;   (D) Step 3

Figure 3-56   (A) How to find a focus point of a parabola;   (B) Step 1;   (C) Step 2

How To Draw a Hyperbola (Method 1)

Given: The front view and plan view of cone ABC with cutting plane X-Y, *Figure 3-57A*.

Step 1: Locate points Y' and Y" in the plan view. In the front view, place the point of the compass on point Y and set the lead to point X. Swing point X to the extended baseline and down to the plan view, as illustrated in *Figure 3-57B*.

Step 2: Locate a point along line X-Y (in this example, point D). Using Y as a swing point, swing point D to the extended baseline and down into the plan view. In the front view, reset the compass to a distance equal to the horizontal distance from the cone axis to the outer edge (length 1), as illustrated in *Figure 3-57C*. Transfer the arc to the plan view; where the arc intersects with the cutting plane line X-Y, project to the right, and where these points intersect with point D from above is the actual location of points D' and D".

Step 3: Choose other points along line X-Y in the front view and project them over and down as described in Step 2, *Figure 3-57D*.

Step 4: Using an irregular curve, connect all points. This completes the parabola, *Figure 3-57E*.

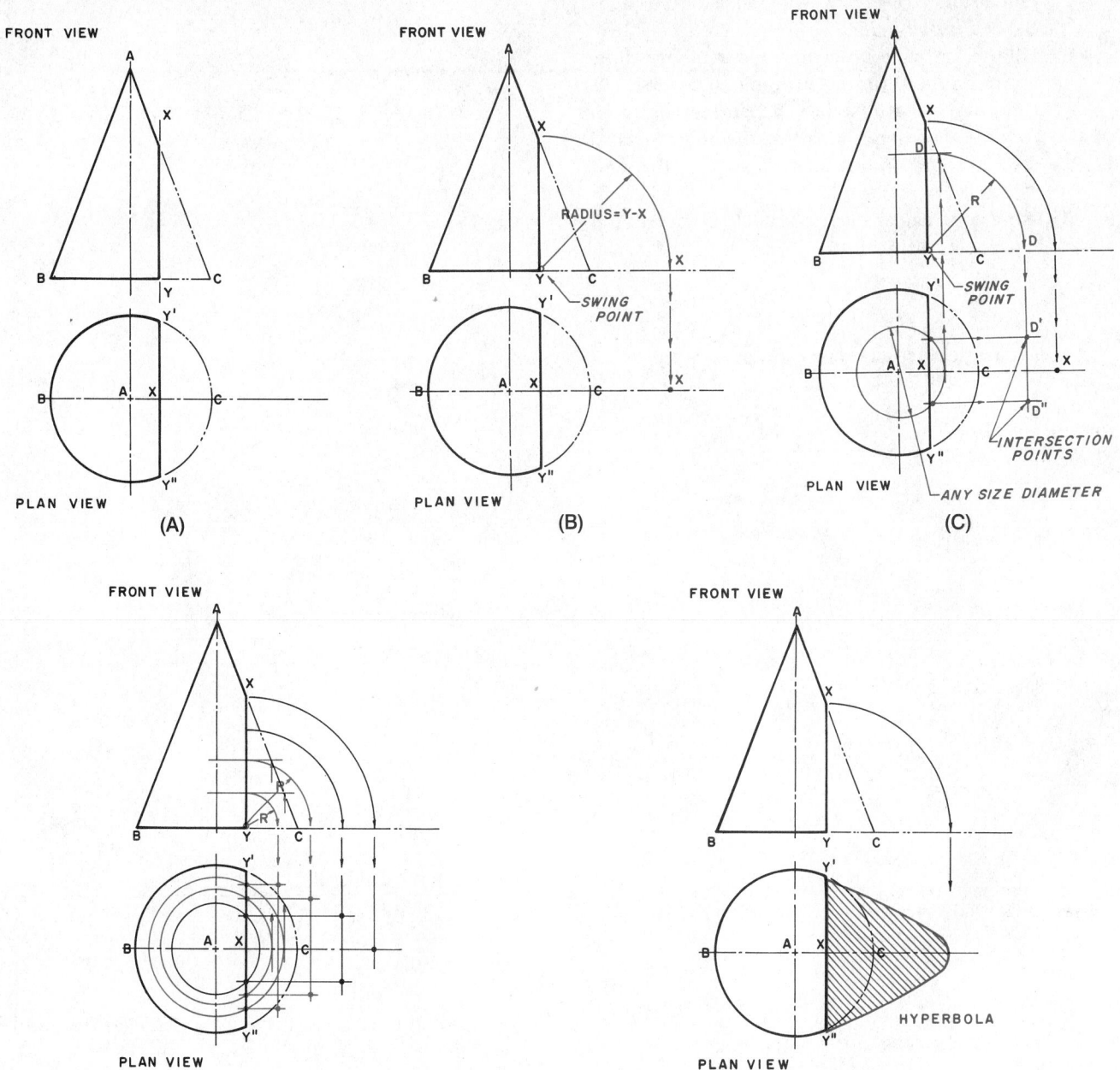

Figure 3-57 (A) How to draw a hyperbola (method 1); (B) Step 1; (C) Step 2; (D) Step 3; (E) Step 4

How To Draw a Hyperbola (Method 2)

Given: Coordinates, a square whose sides equal the transverse axis, and points A and B, *Figure 3-58A*.

Step 1: With the center at the intersection of the diagonal lines from opposite corners of the square, place the compass lead at the corner of the given square and swing arcs to intersect the horizontal line through the center. This locates focus A and focus B. Progressing outward along the horizontal line, mark off equal spaces of arbitrary length from the focus points, *Figure 3-58B*.

Step 2: Set the compass at dimension X (A-1) and swing an arc from focus A. Set the compass at dimension Y (B-1) and swing an arc from focus B. See *Figure 3-58C*.

Step 3: Follow the same procedure for as many points as are required (in this example, 6 points). Set the compass at distance A-2 and swing from focus A. Set the compass at distance B-2 and swing from focus B, and so on, *Figure 3-58D*.

Step 4: Using an irregular curve, carefully complete the hyperbola curve, *Figure 3-58E*.

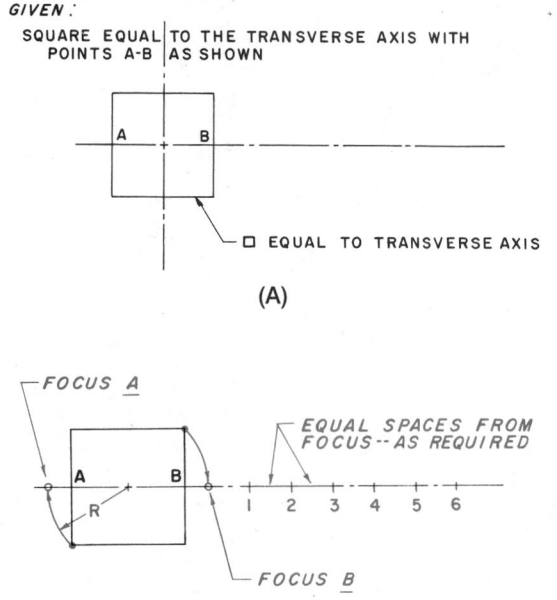

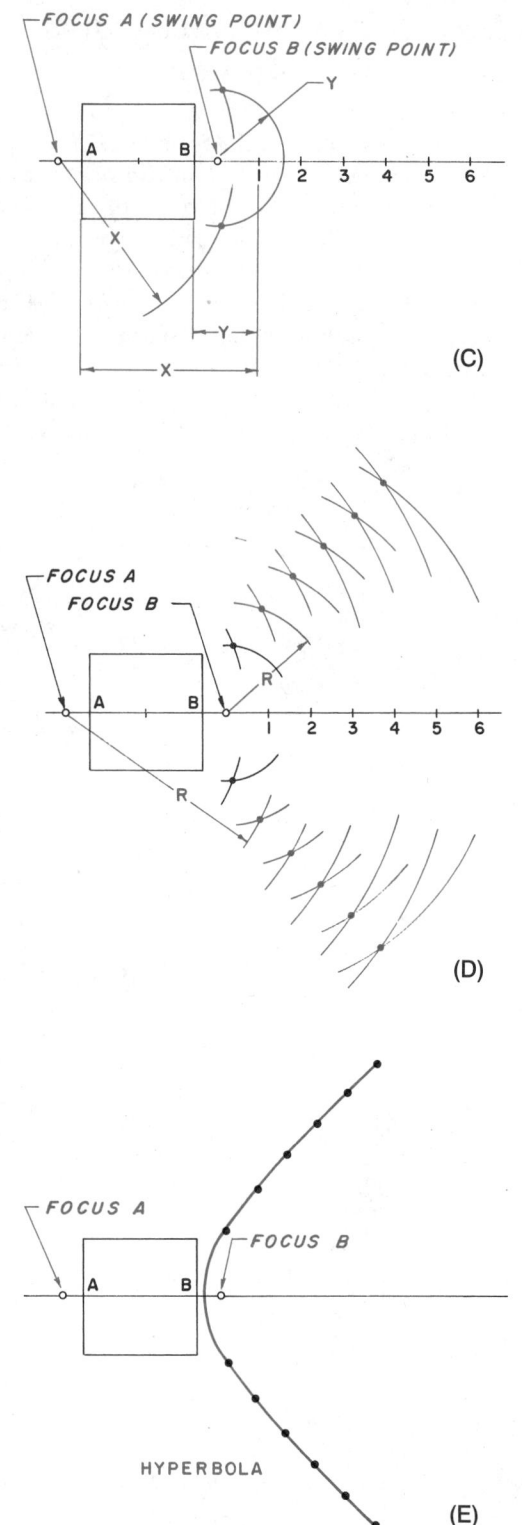

Figure 3-58 (A) How to draw a hyperbola (method 2); (B) Step 1; (C) Step 2; (D) Step 3; (E) Step 4

How To Join Two Points by a Parabolic Curve

Given: Points X and Y with 0 an assumed point of tangency, *Figures 3-59A*, *3-59B*, and *3-59C*.

Step 1: Divide line X-0 into an equal number of parts; divide 0-Y into the exact same number of equal parts. See *Figures 3-59D*, *3-59E*, and *3-59F*.

Step 2: Connect corresponding points, *Figures 3-59G*, *3-59H*, and *3-59I*. Using an irregular curve, draw the parabolic curve as a smooth flowing curve.

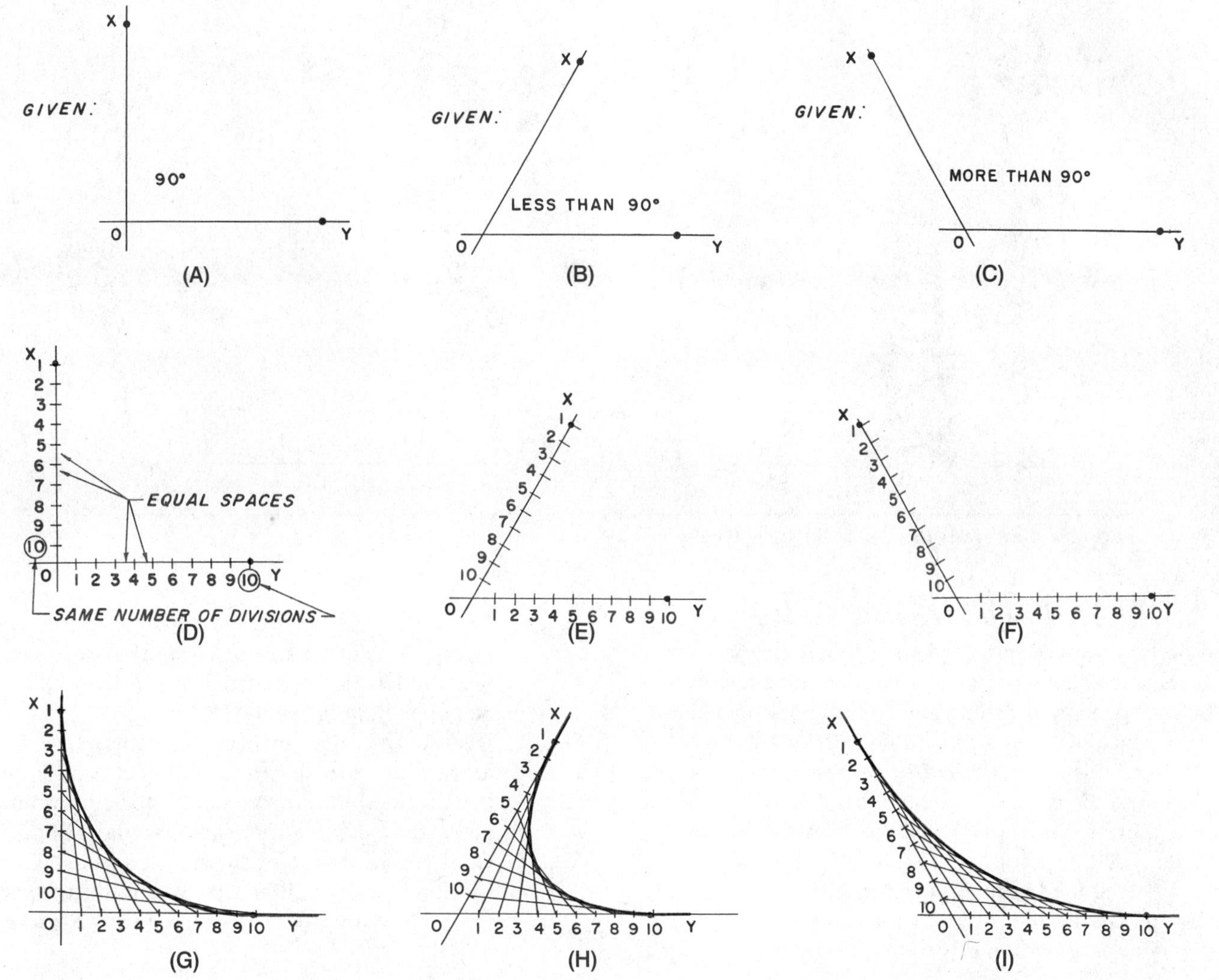

Figure 3-59 (A) How to join two points by a parabolic curve (given: 90° angle); (B) Given: Less than 90° angle; (C) Given: More than 90° angle; (D) Step 1; (E) Step 1; (F) Step 1; (G) Step 2; (H) Step 2; (I) Step 2

Supplementary Construction

How To Draw a Spiral

Given: Crossed axes, *Figure 3-60A*.

Step 1: Divide the circle into equal angles (in this example, 30°). Set the compass at a required radius and draw various diameters to suit, evenly spaced, as illustrated in *Figure 3-60B*.

Step 2: Starting any place along the angles, step over one angle and up one diameter until the required spiral is completed, *Figure 3-60C*.

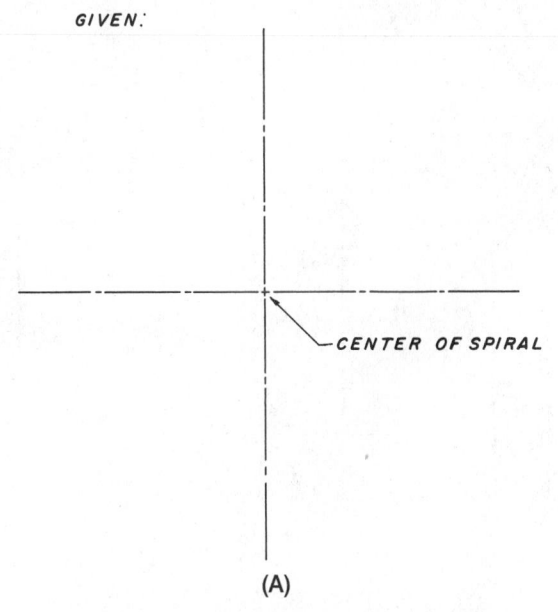

Figure 3-60 (A) How to draw a spiral

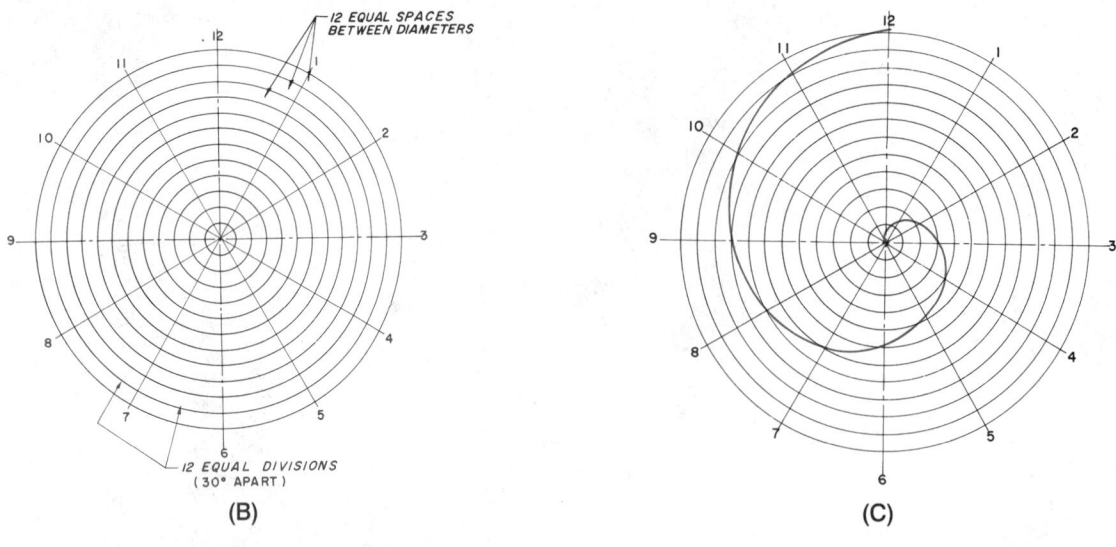

Figure 3-60 (Continued) (B) Step 1 (C) Step 2

How To Draw a Helix

A *helix*, a form of spiral, is used in screw threads, worm gears, and spiral stairways, to mention but a few uses. A helix is generated by moving a point around and along the surface of a cylinder with uniform angular velocity about its axis. A *cylindrical helix* is simply known as a helix, and the distance measured parallel to the axis traversed by a point in one revolution is called the *lead*.

Given: The top and front views of a cylinder with a required lead, *Figure 3-61A*.

Step 1: Divide the top view into an equal number of spaces (12 in this example). Draw a line equal to one revolution and/or circumference, project the lead distance from the end, and label all points per *Figure 3-61B*.

Step 2: Divide the circumference into the same amount of equal spaces, and project each from the inclined line to the corresponding point projected from the front view, *Figure 3-61C*.

Step 3: Connect all points of the helix, and draw as hidden lines the lines that would disappear from view on an actual helix form, *Figure 3-61D*.

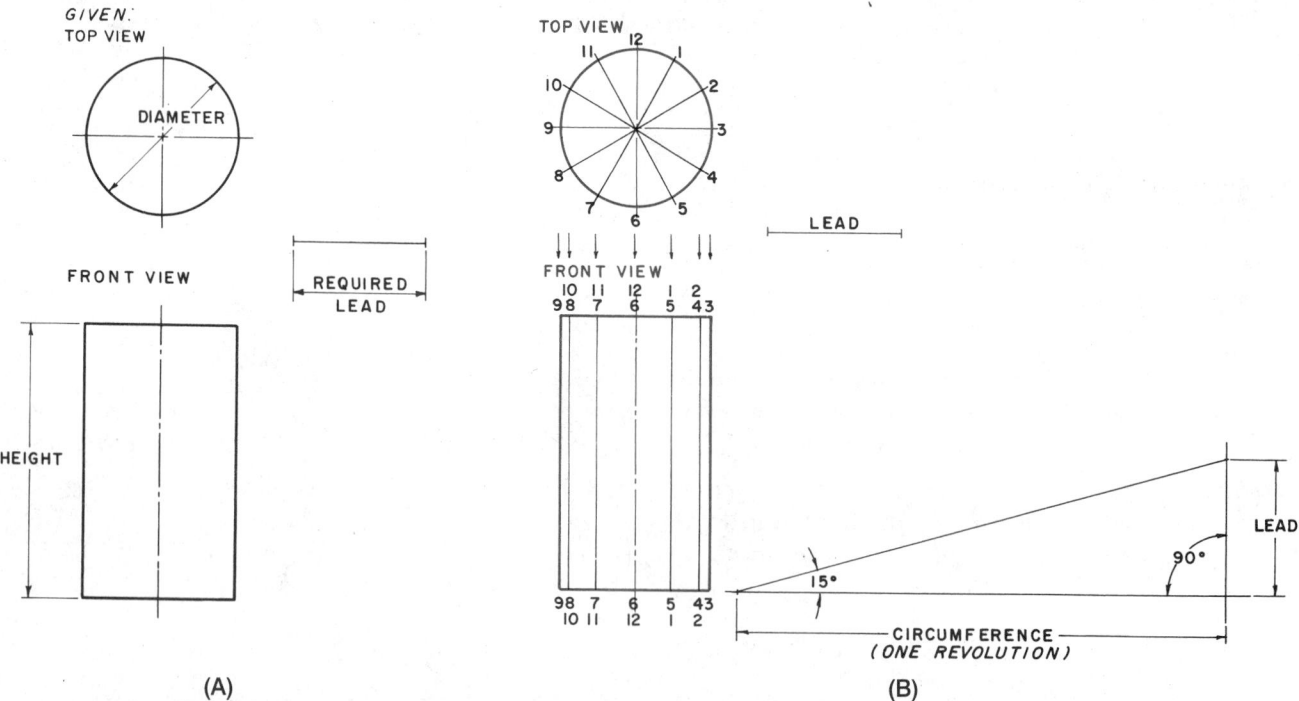

Figure 3-61 (A) How to draw a helix; (B) Given views

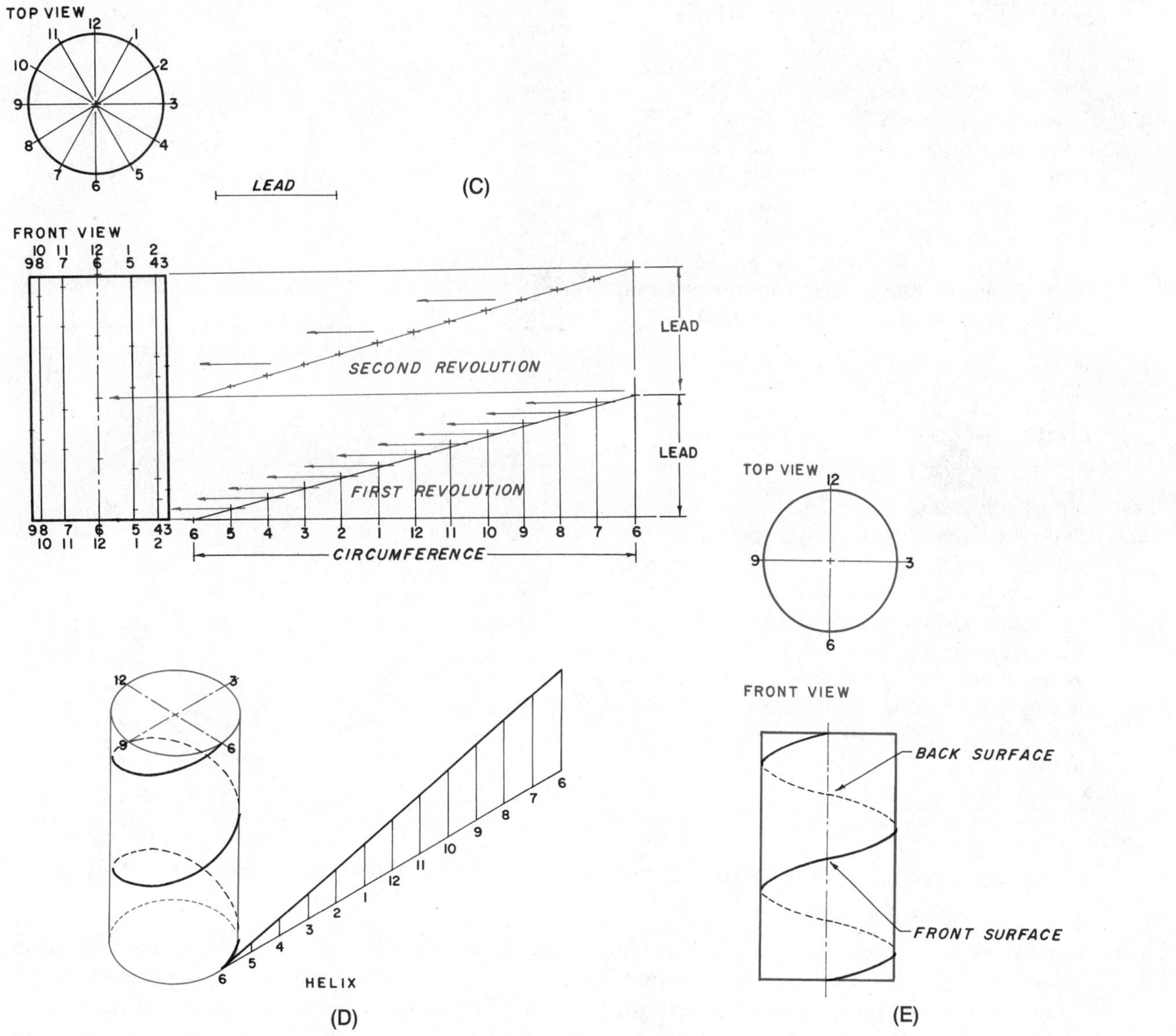

TOP VIEW
12
1
2
3
4
5
6
7
8
9
10
11
LEAD
(C)
FRONT VIEW
LEAD
LEAD
SECOND REVOLUTION
FIRST REVOLUTION
CIRCUMFERENCE
TOP VIEW
12
3
6
9
FRONT VIEW
BACK SURFACE
FRONT SURFACE
HELIX
(D)
(E)

Figure 3-61 (Continued) (C) Steps 1 and 2; (D) and (E)

Notice that a right-hand helix is illustrated. To construct a left-hand helix, simply project all the points in the opposite direction, ***Figure 3-61E***.

How To Draw an Involute of a Line

The curved path of a point on a string as it unwinds from a line, a triangle, a square or a circle is an *involute*. The involute is used to construct involute gears. The involute forms the face and part of the flank of the teeth of the gear.

Given: Line A-B, ***Figure 3-62A***.

Step 1: Extend line A-B, as illustrated in ***Figure 3-62B***. Set the compass at the length A-B, and, using point B as the swing point, swing semicircle A-C. With the compass set at length A-C, and, using point A as the swing point, swing semicircle A-D. Continue in this manner, alternating between points A and B until the required involute is completed, as shown in the figure.

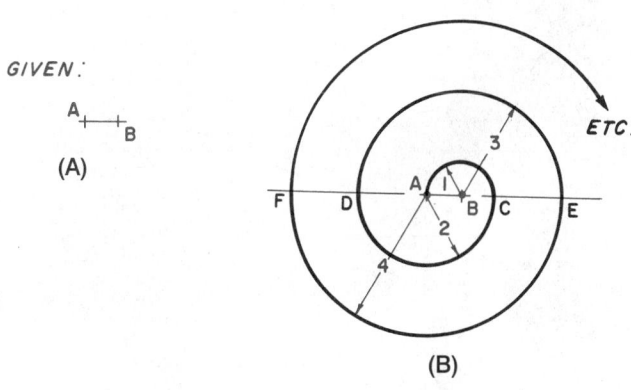

GIVEN:
A
B
(A)
ETC.
A
B
C
D
E
F
1
2
3
4
(B)

Figure 3-62 (A) How to draw an involute of a line; (B) Step 1

How To Draw an Involute of a Triangle

Given: Triangle ABC, *Figure 3-63A*.

Step 1: Extend straight lines from the triangle, as illustrated in *Figure 3-63B*. Set the compass at length A-C, and, using point A as the swing point, swing semicircle A-C. With the compass set at length B-D, and, using point B as the swing point, swing semicircle B-E. Continue similarly around the triangle until the required involute is completed, as shown in the figure.

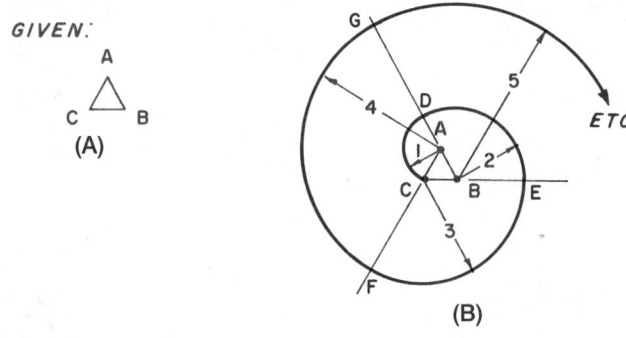

Figure 3-63 (A) How to draw an involute of a triangle; (B) Step 1

How To Draw an Involute of a Square

Given: Square ABCD, *Figure 3-64A*.

Step 1: Extend straight lines from the square, as illustrated in *Figure 3-64B*. Set the compass at length A-B, and using point A as the swing point, swing semicircle A-E. With the compass set at length B-E, and, using point B as the swing point, swing semicircle B-F. Continue in the same manner around the square until the required involute is completed, as shown in the figure.

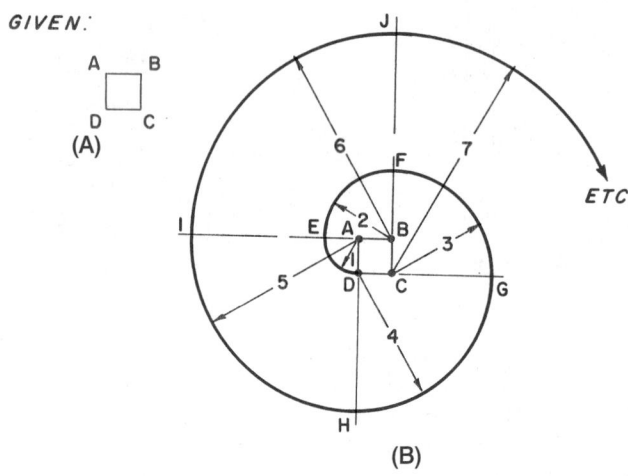

Figure 3-64 (A) How to draw an involute of a square; (B) Step 1

How To Draw an Involute of a Circle

Given: Circle A, *Figure 3-65A*.

Step 1: Divide the circle into a number of equal parts and number each point clockwise around the circle (in this example, 12 equal parts). Project a line perpendicular to the radius at each point in a clockwise direction, *Figure 3-65B*.

Step 2: Set the compass at length 1-2, and, using point 2 as the swing point, swing semicircle 1-2. With the compass set at length 1-2 plus 2-3, and, using point 3 as the swing point, swing semicircle 1-3. Continue in the same manner around the circle until the required involute is completed, as shown in *Figure 3-65C*.

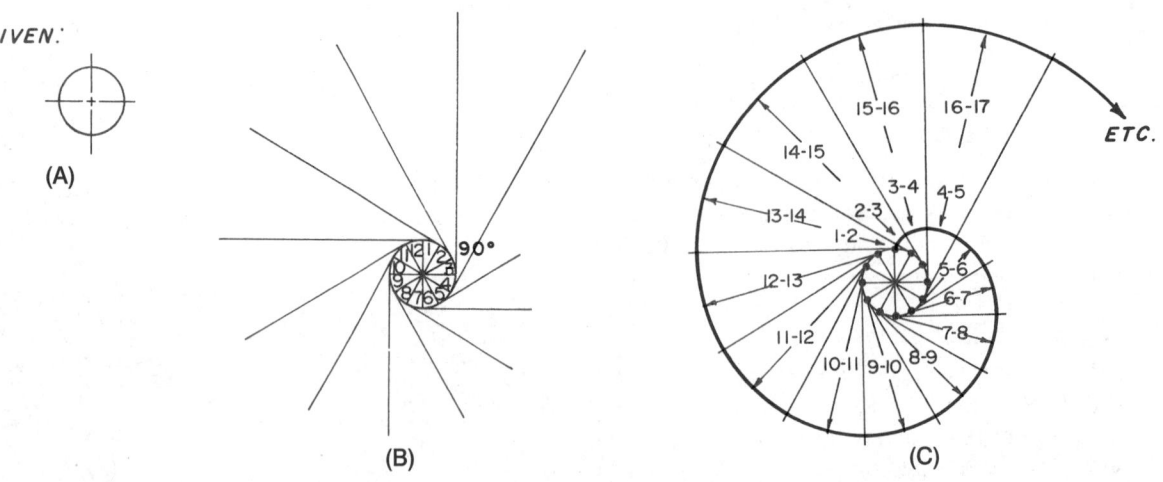

Figure 3-65 (A) How to draw an involute of a circle; (B) Step 1; (C) Step 2

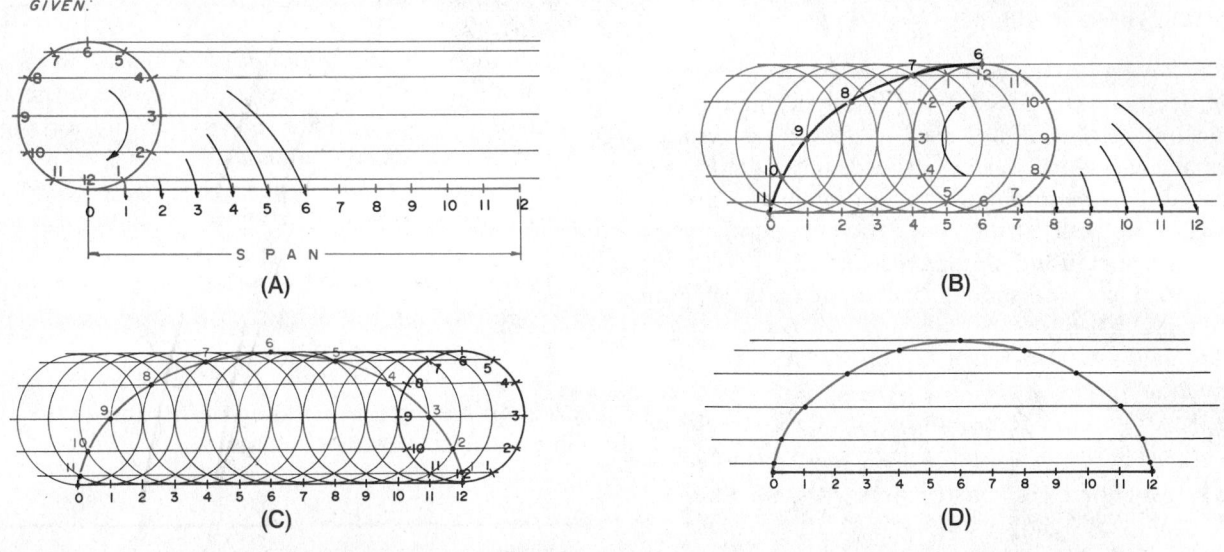

Figure 3-66 (A) How to draw a cycloidal curve (Step 1); (B) Step 2; (C) Step 3; (D) Step 4

How To Draw a Cycloidal Curve

Given: A required span of a cycloid.

Step 1: As the span represents the rolling distance of one revolution of a diameter, divide the span length by pi to find the required diameter. Divide the span into an equal number of divisions. Twelve is a convenient number, **Figure 3-66A**.

Step 2: Draw the rolling diameter tangent to the given span at point 0/12. Divide this diameter into the same number of equal spaces as done in the Step 1 span division. Consecutively number the radial end points of these divisions to make them meet their corresponding number of the span divisions as the diameter rolls along the span, **Figure 3-66B**.

Step 3: At each span division draw the rolling diameter tangent to the span in the rolled position, with the division numbers of span and diameter matching at their contact point. Locate the point 0/12 position on the diameter at each of these locations, marked with a small cross, **Figure 3-66C**.

Step 4: Connect all the point 1 diameter division positions with a smooth curve to complete the cycloidal curve, **Figure 3-66D**.

Note: If the given span is a concave arc, a *hypocycloid* will be drawn, **Figure 3-67**. If the span is convex, an *epicycloid* will be drawn, **Figure 3-68**.

Review

1. Explain the following terms:
   right angle, acute angle, obtuse angle
2. List six polygons.
3. Why are geometric construction procedures so important to the drafter?

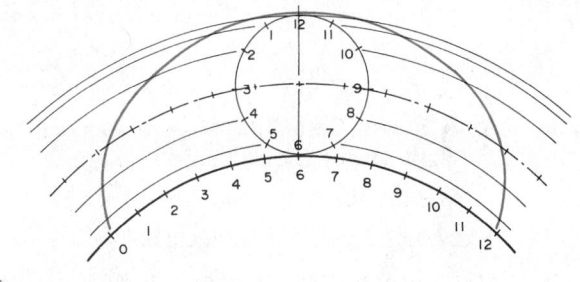

Figure 3-67 A hypocycloid

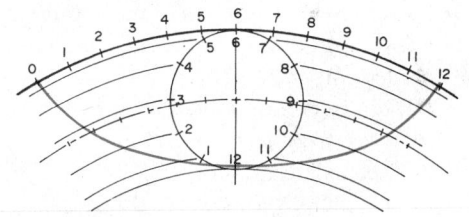

Figure 3-68 An epicycloid

4. What is meant by the term *tangent*, and why is it important to find all tangent points before darkening in a drawing?
5. Explain the following parts of a circle:
   central angle, sector, quadrant, segment
6. Explain the difference between a concentric circle and an eccentric circle.
7. How is an actual point in space represented on a drawing?
8. Completely describe an angle.
9. List five kinds of triangles.
10. What is meant by the circumference of a circle, and how is it calculated?
11. What is the sum of the three internal angles of a triangle?
12. Define a *line*.

Chapter Three Problems

The following problems are intended to give the beginning drafter practice in using the many geometric construction techniques used to develop drawings. Accuracy, line work, neatness, speed, and centering are stressed. It is recommended that dimensions not be added to problems at this time. The beginning drafter should practice using drafting instruments and correct line thicknesses, and should concentrate on developing good drafting habits.

The steps to follow in laying out all drawings throughout this book are:

Step 1 All geometric construction work should be made very accurately using a sharp 4-H lead.

Step 2 All geometric construction work must be laid out very lightly.

Step 3 Do not erase construction lines. If constructed lightly, they will not be seen on the whiteprint copy.

Step 4 Make a rough sketch of each problem before beginning to calculate the overall shape.

Step 5 Lightly draw each problem completely, first.

Step 6 Locate *all* tangent points as you proceed. Make short light dashes at each tangent point.

Step 7 Try to center the problem in the work area.

Step 8 Check each dimension for accuracy.

Step 9 Darken in the drawing using the correct line thickness and the following steps:
   - Locate and draw all center lines with a thin black line.
   - Darken in all diameters using a compass.
   - Darken in all radii using either a compass or a circle template.
   - Darken in all horizontal lines, either from right to left or from left to right.
   - Recheck all dimensions.
   - Check all lines for correct thickness.
   - Fill out the title block using light guidelines and neat lettering.

Problem 3-1

Divide the work area into four equal spaces. In the upper left-hand space, draw an inclined line 3.25 long. Bisect this line. Show all construction lines lightly.

In the upper right-hand space, draw an acute angle with intersecting lines, each approximately 2.75 long. Bisect this angle. Show all light construction lines.

In the lower left-hand space, draw an inclined line 2.88 long. Divide this line into 5 spaces. Show all light construction lines.

In the lower right-hand space, draw an inclined line 3.125 long. Divide this line into three proportional parts 2, 3, and 4. Show all light construction lines.

Problem 3-2

Divide the work area into four equal spaces. In the center of the upper left-hand space, draw an equilateral triangle having sides of 2.0. Bisect the interior angles. Show all light construction lines.

In the center of the upper right-hand space, draw a square having sides of 2.0. Show all light construction lines.

In the center of the lower left-hand space, draw a hexagon having the distance of 2.0 across the flats. Show all light construction lines.

In the center of the lower right-hand space, draw an octagon having the distance of 2.0 across the flats. Show all light construction lines.

Problem 3-3

Divide the work area into four equal spaces. In the upper left-hand space, draw two lines intersecting at 60° and draw an arc with a .88 radius tangent to the two lines. Show all light construction lines.

In the upper right-hand space, draw two lines intersecting at 120° and draw an arc with a .625 radius tangent to the two lines. Show all light construction lines.

In the lower left-hand space, draw two lines intersecting at 90° and draw an arc with 1.25 radius tangent to the two lines. Show all light construction lines.

In the lower right-hand space, draw an ellipse with a major diameter of 3.00 and a minor diameter of 1.75.

Problem 3-4

Divide the work area into four equal spaces. In the upper left-hand space, center a line 10 mm long and label it line A-B. Construct a straight line involute with five arcs.

In the upper right-hand space, center a triangle with sides equal to .25 and label the triangle ABC. Construct a triangular involute with five arcs.

In the lower left-hand space, center a square with sides equal to 10 mm and label the square ABCD. Construct a square involute with five arcs.

In the lower right-hand space, center a circle with a diameter of 1.00. Construct a circle involute with as many arcs as space will allow.

Problem 3-5

Divide the work area into four equal spaces. In the center of the upper left-hand space draw a spiral with 30° spaces, at 4–mm increments, for one 360° rotation.

In the center of the upper right-hand space, draw two lines at right angles intersecting a point 0. One line is to be 3.5 long, and the other 2.5 long. Draw a parabolic curve between these two lines.

In the center of the lower left-hand space, draw a rectangle 2.0 x 4.50 in size. Label the 2.0 side as the rise, and the 4.50 as the span. Construct a parabola (Method 2), and locate the focus point.

In the center of the lower right-hand space, draw a line 80 mm long, and label its end points as A and B. Locate a point on line A-B at 5/8 of the line's length from point A. Label the point X.

Problems 3-6 through 3-19

Using the art for problems 3-6 through 3-19, center each object within the work area using correct line thickness.

Do not erase all light construction lines.

Locate and draw a light short dash at all tangent points.

Do not dimension objects.

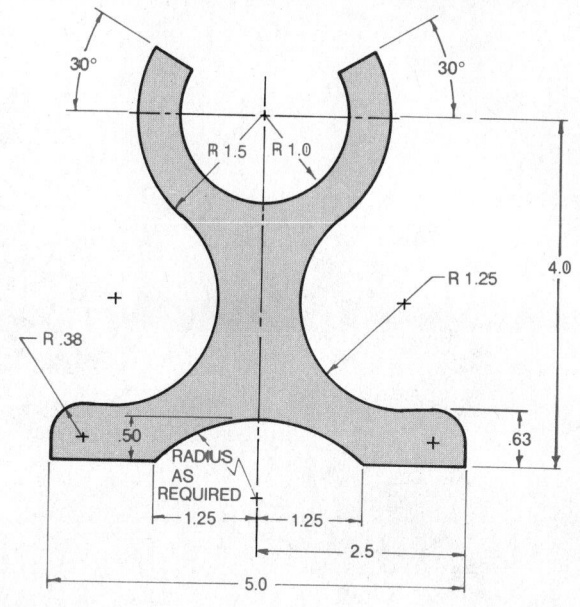

Problem 3-7

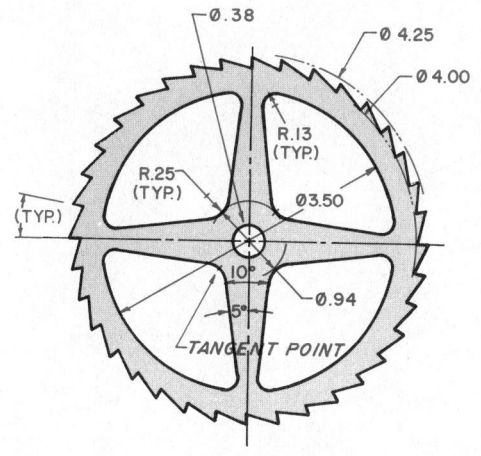

Problem 3-6

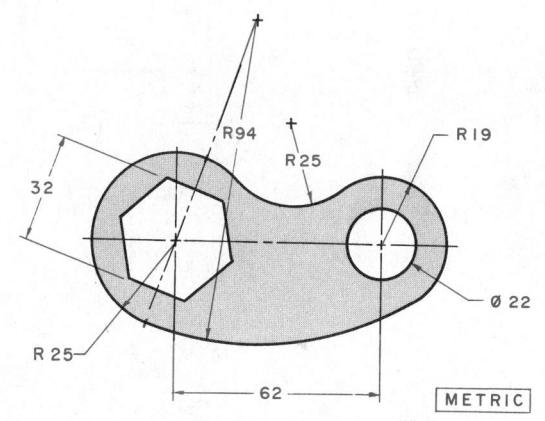

Problem 3-8

Problem 3-9

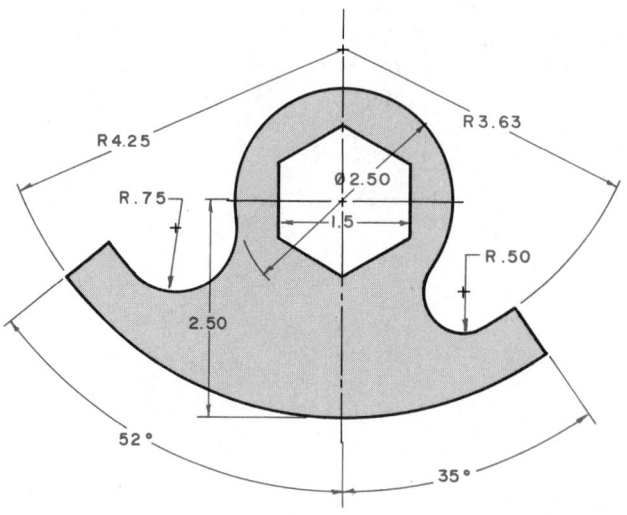

Problem 3-10

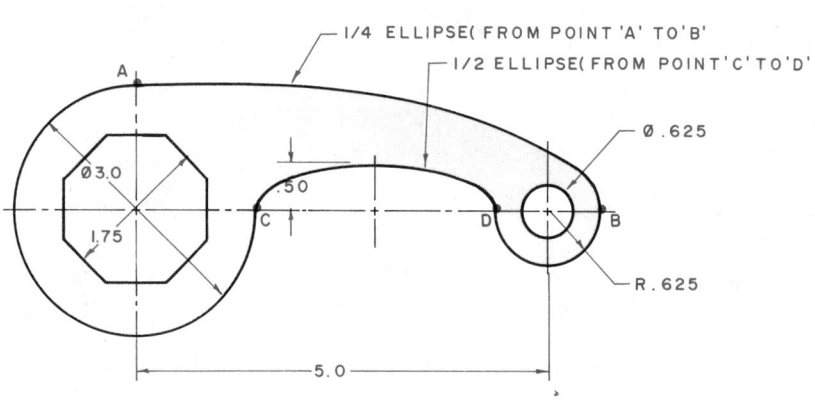

Problem 3-11

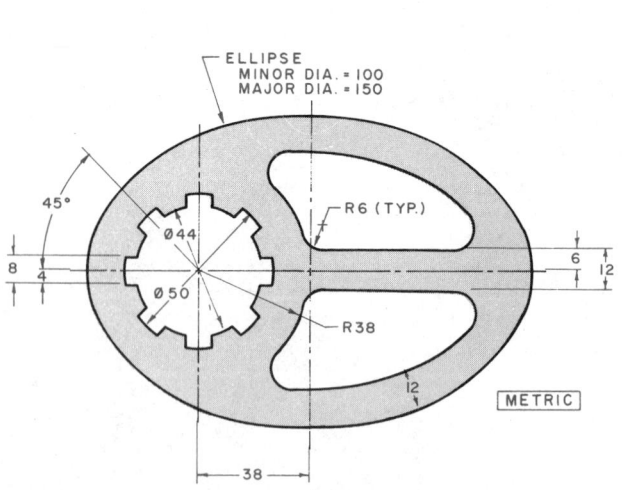

Problem 3-12

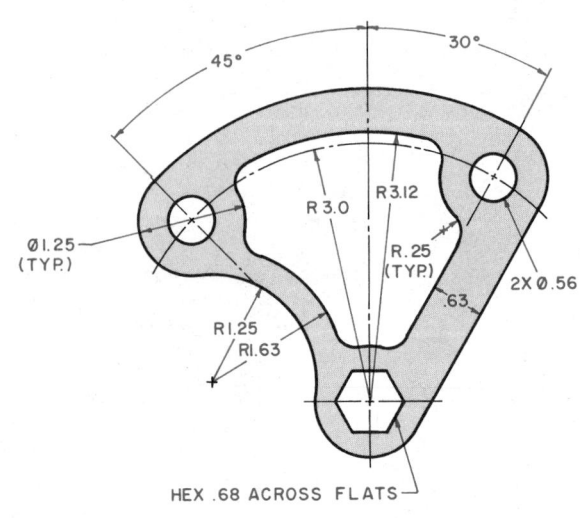

Problem 3-13

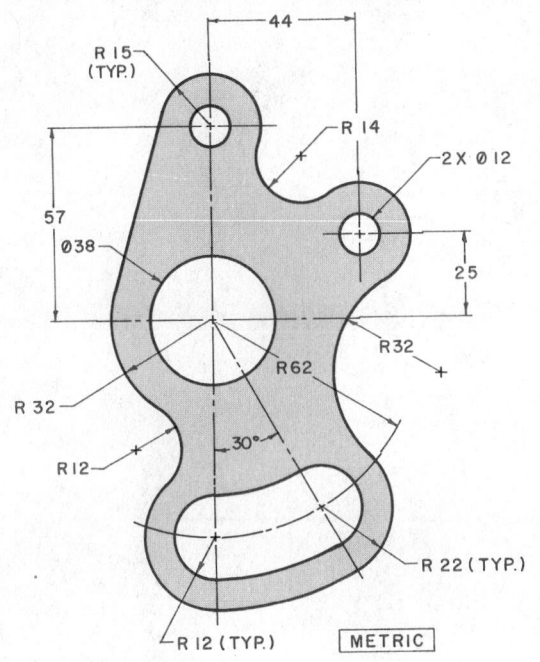

Problem 3-14

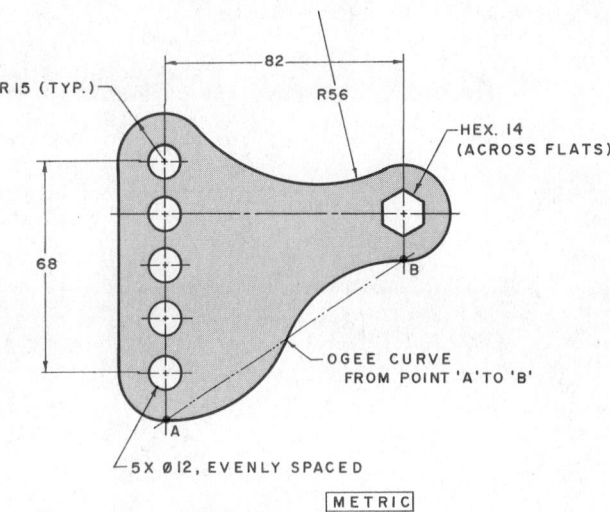

Problem 3-17

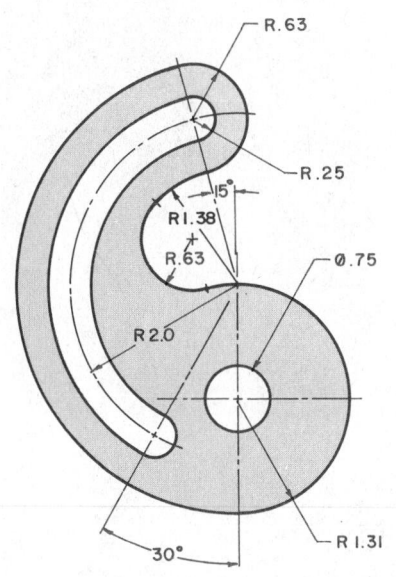

Problem 3-15

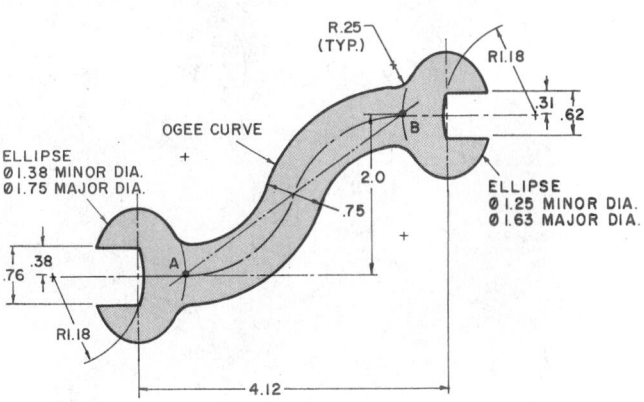

Problem 3-16

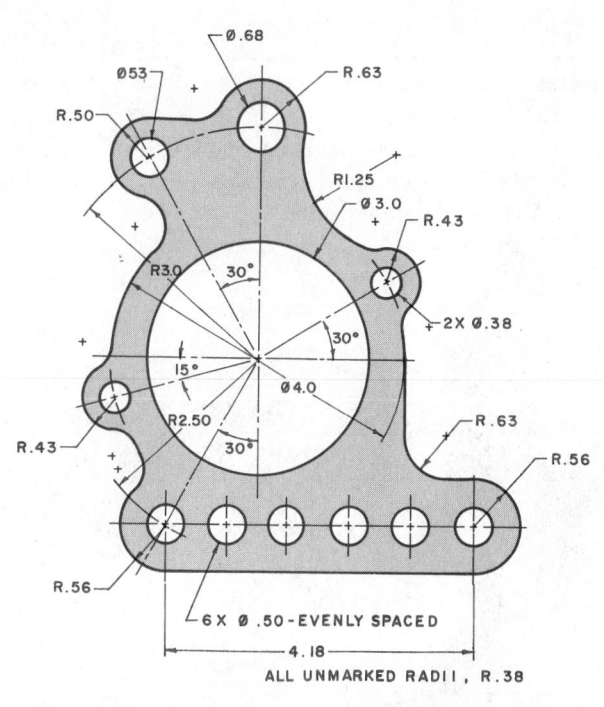

Problem 3-18

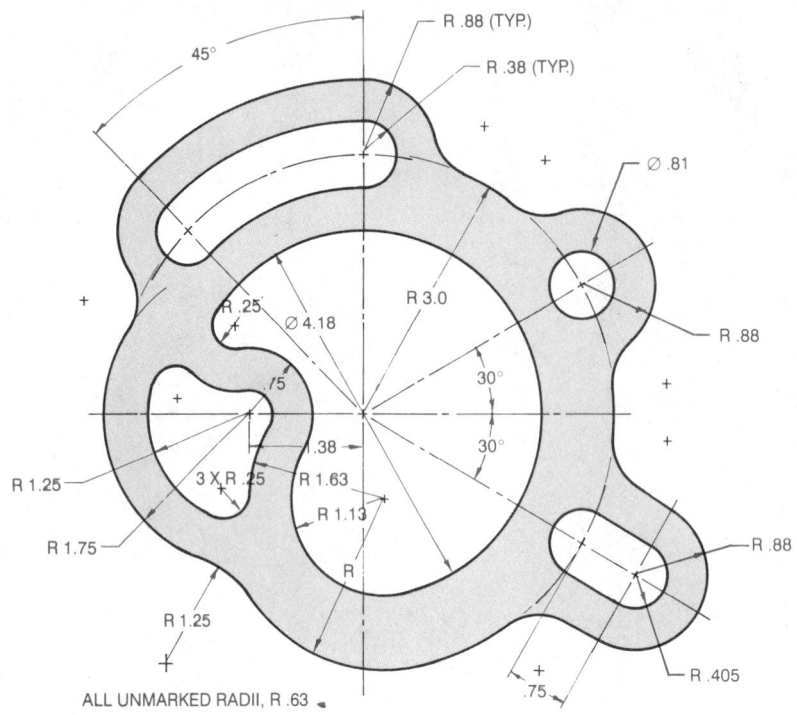

Problem 3-19

Problems 3-20 through 3-25

On an A-size sheet of vellum, carefully construct each of the assigned figures. Use thin, black center lines and thick, black object lines. Leave all light construction work. Do not add dimensions.

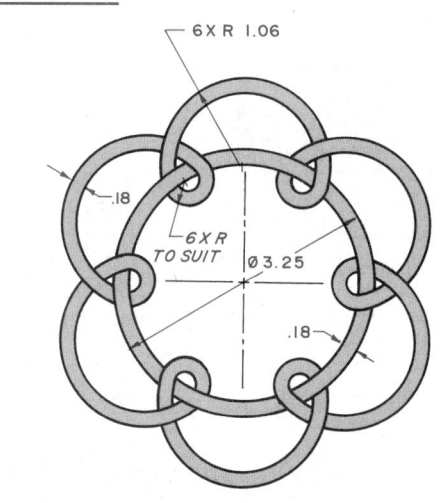

Problem 3-21

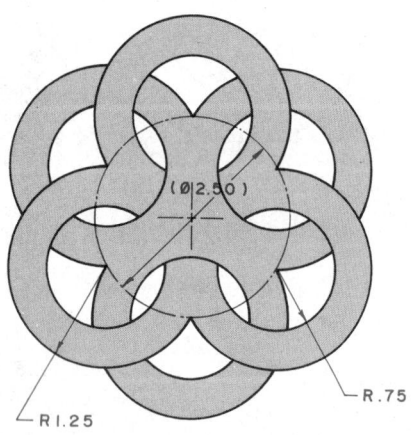

Problem 3-20

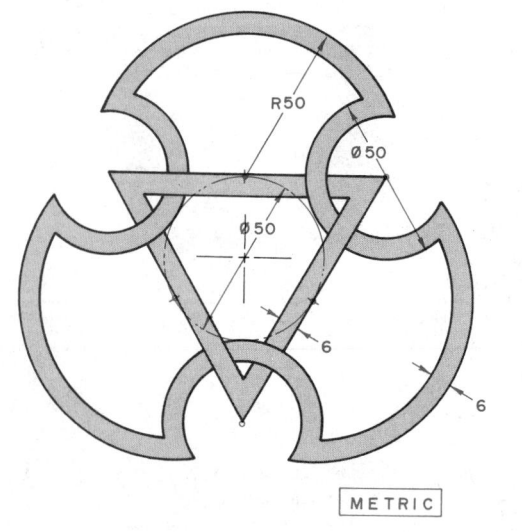

METRIC

Problem 3-22

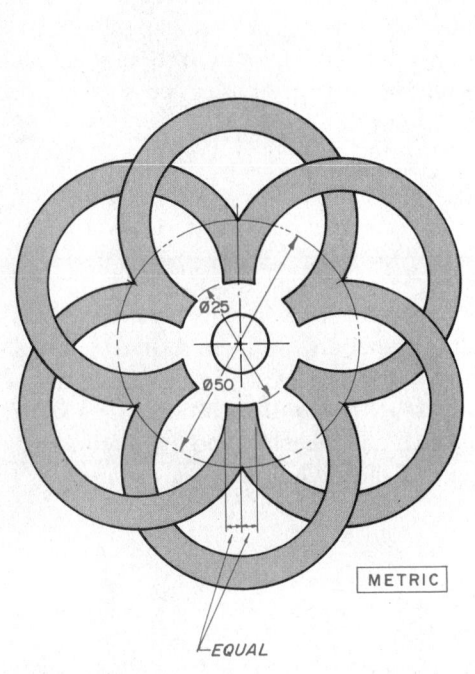

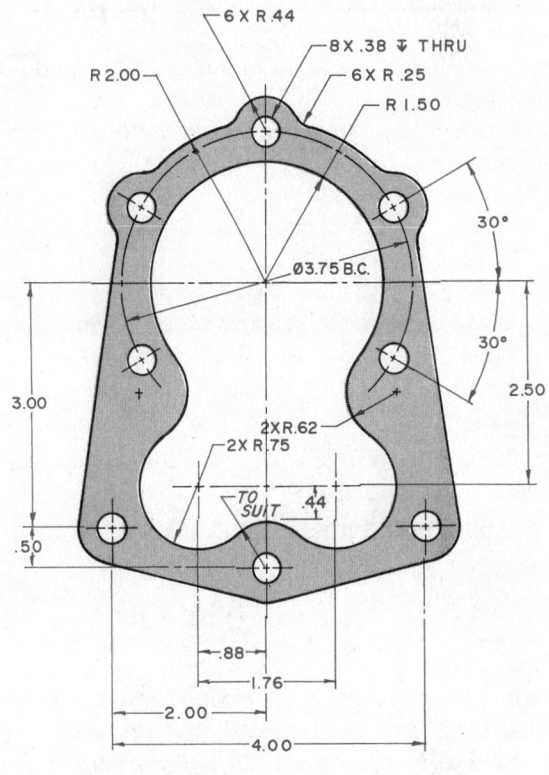

Problem 3-23

Problem 3-24

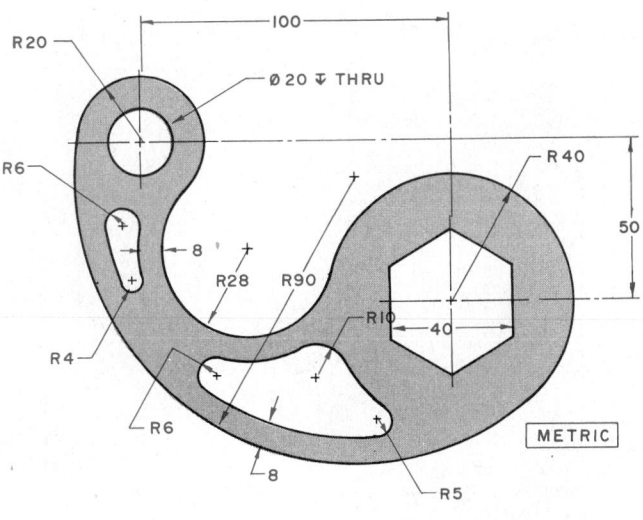

Problem 3-25

☐ **Problem 3-26**

Design and layout to scale, a clock face to match your interest(s) in technology.

☐ **Problem 3-27**

Find a logo for one of the following products, company, or organization. Prepare a drawing of it.

   a. an automobile

   b. a computer company

   c. an airline

   d. a product of your choice

   e. an organization

Problem 3-28

Refer to the figures indicated in this chapter and do the following:

a. Figure 3-22—Draw a curve parallel to a curve similar to the lower edge of the object for Problem 3-9.

b. Figure 3-30—Reproduce the drawing for Figure 3-30 and transfer it.

c. Figure 3-46—Change the spacing of lines a-b and c-d and draw an OGEE curve to join them.

d. Figure 3-61—Draw the helix shown.

e. Figure 3-65—Follow the steps and draw an involute curve.

Note: The involute curve is the basis for the profiles of spur gears and helical gears.

f. Figure 3-68—Construct the epicycloid shown.

Problem 3-29

The golden section shown is attributed to the Greeks who used it as a guide to create pleasing proportions for architecture, sculpture, and pottery. Three nested golden sections can be formed by following the construction pattern shown. Note that the ratio of the sides of each golden rectangle is 1.618 to 1.

$$(2.618)/(1.618) = 1.618; \text{ and } (4.236)/(2.618) = 1.618$$

(For interest check the ratios of the sides of credit cards and some food packaging.) Refer to the figure and reconstruct the three golden rectangles. Then, lay out some shelves for a living room for books and hi-fi equipment. Include vertical dividers that incorporate the golden section proportions.

Problem 3-30

The freehand sketch given shows a format for a typical classroom engineering problem statement and solution. On a sheet of 8.5 x 11 (A size) paper redo the statement (GIVEN and FIND) and the solution. Use a straight-edge for the sketch and diagrams and good lettering style for the statement and solutions.

Problem 3-31

You have been asked to help an individual layout a tentative plan for a service garage with parking facilities. The commercially zoned lot measures 25 m x 35 m. Sketch no. 1 shows a preferred arrangement of the service-pits building with parking. Layout the lot using the given information. Use a scale to make the best use of an A-size drawing.

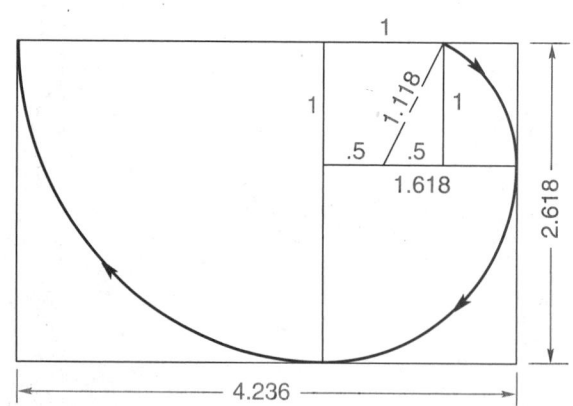

Problem 3-29

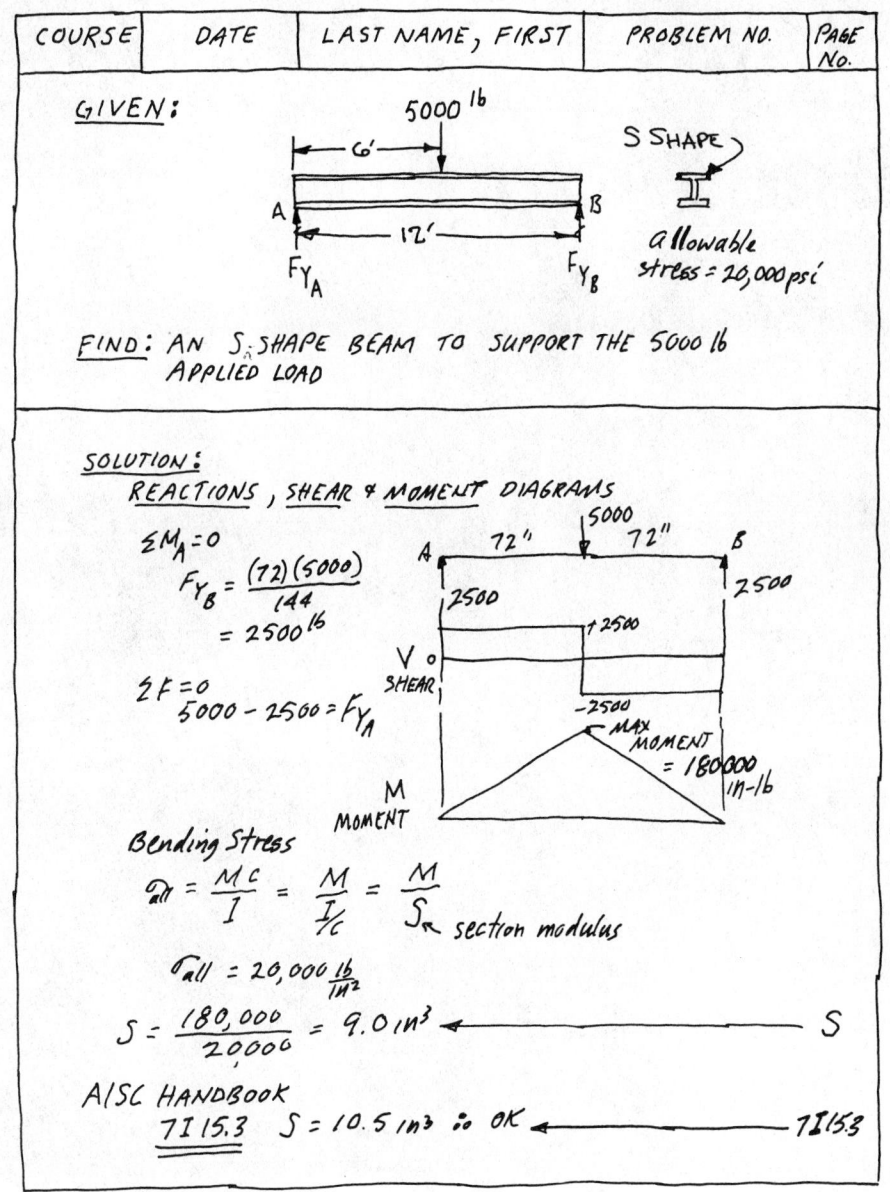

| COURSE | DATE | LAST NAME, FIRST | PROBLEM NO. | PAGE No. |

GIVEN:

5000^{lb}

S SHAPE

allowable stress = 20,000 psi

FIND: AN S-SHAPE BEAM TO SUPPORT THE 5000 lb APPLIED LOAD

SOLUTION:

REACTIONS, SHEAR & MOMENT DIAGRAMS

$\Sigma M_A = 0$

$F_{Y_B} = \dfrac{(72)(5000)}{144}$

$= 2500^{lb}$

$\Sigma F = 0$

$5000 - 2500 = F_{Y_A}$

Bending Stress

$\sigma_{all} = \dfrac{Mc}{I} = \dfrac{M}{\frac{I}{c}} = \dfrac{M}{S} \leftarrow$ section modulus

$\sigma_{all} = 20,000 \dfrac{lb}{in^2}$

$S = \dfrac{180,000}{20,000} = 9.0 \, in^3 \longleftarrow \quad S$

AISC HANDBOOK

$7 \text{I} 15.3 \quad S = 10.5 \, in^3 \, \therefore \, OK \longleftarrow \quad 7\text{I}15.3$

Problem 3-30

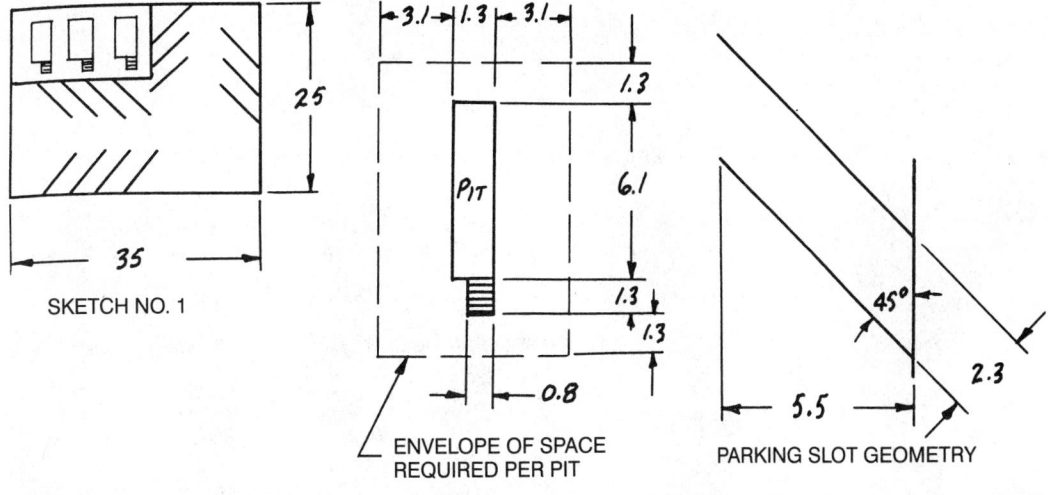

25

35

SKETCH NO. 1

3.1 — 1.3 — 3.1

1.3

PIT

6.1

1.3

1.3

0.8

ENVELOPE OF SPACE REQUIRED PER PIT

45°

2.3

5.5

PARKING SLOT GEOMETRY

Problem 3-31

Technical Drawing Fundamentals

Career Profile: Lisa Correa

While working towards her B.S. at San Jose University, Lisa Correa planned to utilize her industrial engineering degree with Disneyland or UPS. But as luck would have it, she had worked for Pacific Gas & Electric as a summer student and has been there ever since she graduated in 1990.

Currently, Lisa works as a field engineer with gas transmission and distribution engineering and construction. Most field engineers have a mechanical, civil, or construction management background. This is not a typical job for an industrial engineer. One of Lisa's first jobs was to convert conventional maps and blueprints of

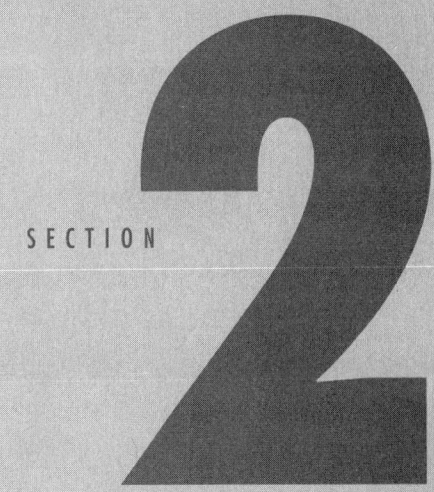
the gas line infrastructures into the computer network.

As time progressed, so did Lisa's responsibilities; but even now no particular day can be considered typical. When she is out in the field, she works with crews on various jobs. The gas reconstruction crews replace low-pressure steel lines with more efficient plastic lines. Transmission work involves high-pressure gas lines, regulators, and valve stations. It is on these types of jobs that Lisa earns her overtime. Each line, valve, and fitting must be pressure tested with water or nitrogen at anywhere from 600 to 1600 psig before any natural gas is allowed to flow. When up to two miles of pipe is to be tested, it can make for a very long day. Recently, Lisa has been supervising an electrical underground crew, which has been converting overhead power lines into an underground system. When this effects over one hundred businesses and over three miles of traffic, things can get hectic. Although dealing with frustrated clients is also part of Lisa's daily schedule that she did not plan on in college, dealing with the public is a part of her job that she does enjoy.

As each job has its own exclusive details, they all have basic requirements that Lisa must plan. Before a job starts, she has to acquire material that is not always in stock or takes a certain amount of time to expedite. While this is taking place, she reviews a job packet for estimates, permits, clearances, and release orders. She must also break down the accounting and be sure that enough money has been allotted to complete the job. The city, county, businesses, or residents to be effected must be notified before construction begins. After a review with the area and crew foremen for a construction schedule, the job is finally ready to begin.

Once the job has begun, Lisa must track its progress to ensure that design requirements are met, material is ordered and delivered on time, bills to subcontractors are paid, and the schedule is being followed. She visits each job site to discuss any questions or concerns that are important and must be dealt with.

Upon closing a job, Lisa reviews the blueprints, logs any changes, and redraws any as-builts or change orders. The cost of a job is reviewed to see where the most amount of money was spent, the reasons for the cost, and how it can be done more efficiently next time. Averaging six to ten jobs at a time is challenging, and it is always a new learning opportunity.

When confronted with a technical problem, she is able to glean information from her immediate supervisors, co-workers, various foremen in the field, vendors, and even the crews in the trenches who have pointed out things that would have otherwise been overlooked and unlearned. Though perhaps no one day can be called typical, Lisa can be assured of learning something new every day.

Multiview Drawings

D rafters, designers, and engineers follow a design process to solve technical product problems and use technical drawings to document and develop their ideas. The steps in the process leading to the technical drawings, primarily working drawings, were noted in the general comments in the Introduction. In the chapter following the Introduction, the drawing tools used to prepare the drawings were discussed, followed by another chapter that emphasized the first format drafters, designers, and engineers use to record their ideas, sketching techniques. This chapter focuses on the ever-present challenge to a design person—to prepare multiview drawings, to scale, of technical, 3-D concepts on a 2-D surface so that the users of the drawings will clearly understand, in 3-D, what the design person had in mind, and have sufficient information to accurately produce the concept.

Providing accurate shape descriptions of objects by drawing methods requires that three-dimensional object information be presented in a flat two-dimensional drawing space. This is usually done by drawing images of the object from multiple directions. Commonly shared methods and interpretations are essential for all who make or use such drawings. Technical drawings seldom use more than lines to outline an object's features. Visual qualities, such as color and texture, are more accurately specified by written requirements.

Viewing an object by eye creates a depth distortion. This phenomenon is recreated in a field of drawing called *perspective projection*, **Figure 4-1**. This distortion occurs because the visual rays used to create the image in the viewer's eye are not parallel. This distortion, while useful in providing the illusion of distance, results in a loss of image accuracy and fails to provide the information needed for most detailed technical communication.

Orthographic Projection

Orthographic projection is the most accurate description of shape description wherein an undistorted image of the object appears in a flat, transparent, but imaginary projection plane. A projection plane may be thought of as an imaginary pane of glass, with an object underneath it or behind it, **Figure 4-2A**. Viewed from the side, the pane of glass appears as a line, **Figure**

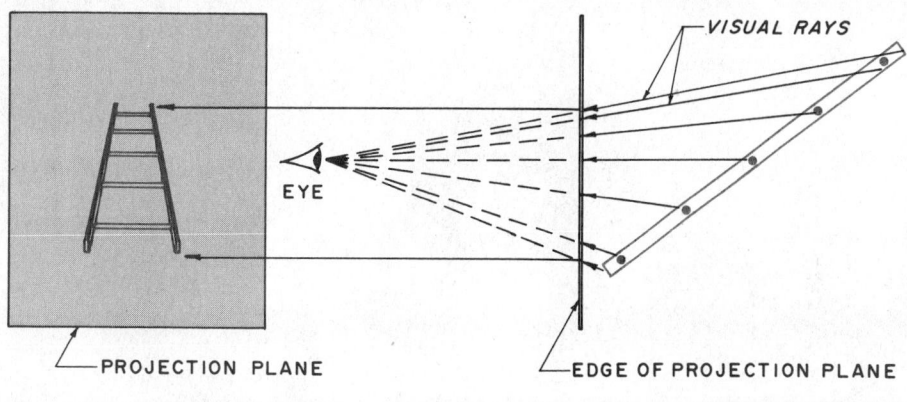

Figure 4-1   Perspective projection

4-2B. Assume that light beams emitting from the surfaces of the object are projected to the projection plane, and that each light beam from each exposed surface is directed toward the viewing plane. (See ***Figures 4-2C*** and ***D***.) The image shown in Figure 4-2A traces the path of each light beam as it intersects the viewing plane. This image is called a projection, and illustrates the features of the exposed surfaces of the object.

Normal Surfaces

The surface images of the figures just described do not represent the size and shape of their corresponding real surfaces. Rectangular surfaces appear as parallelograms. The lengths of edges are shorter in the image than their actual lengths (a phenomenon called *foreshortening*). However, if the viewing plane were positioned parallel

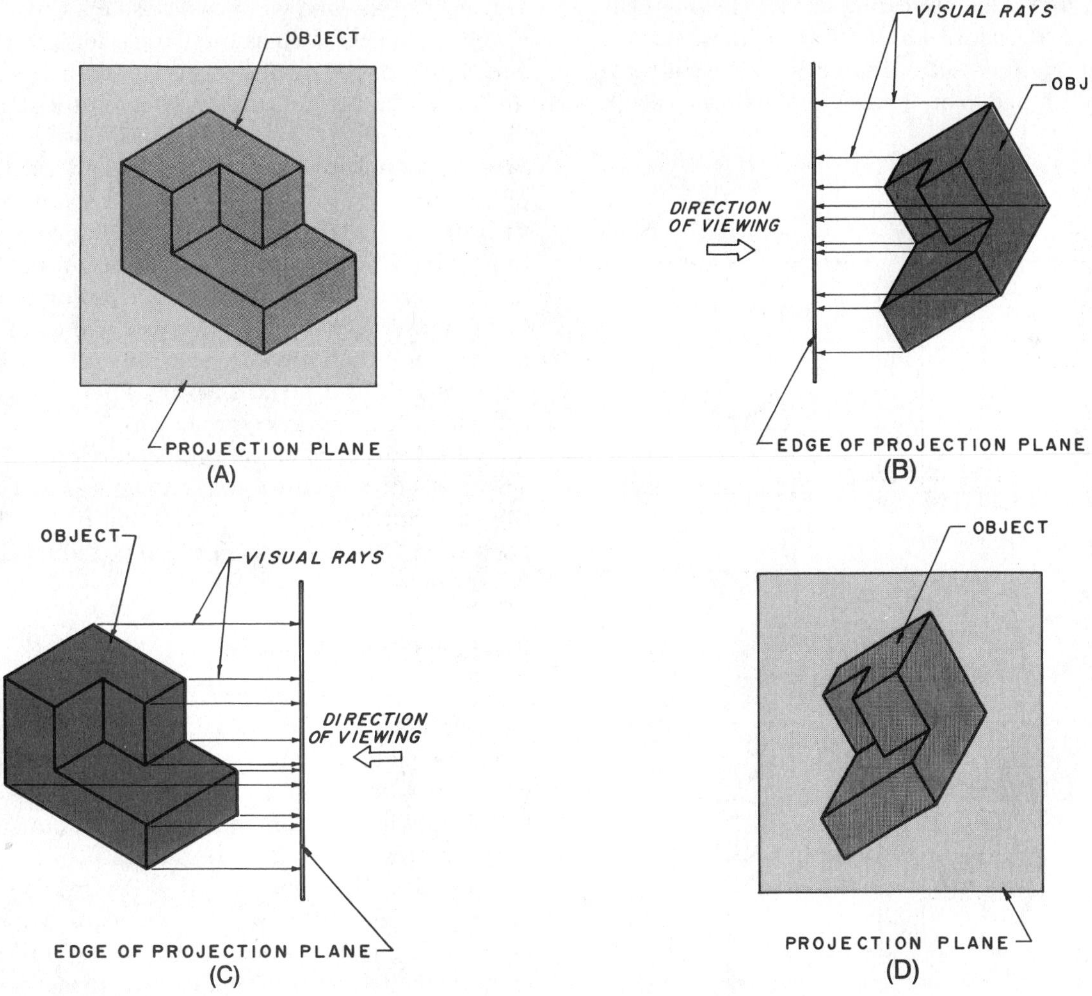

Figure 4-2   (A) Imaginary windowpane in front of object; (B) Side view of windowpane; (C) and (D) Reversed projection

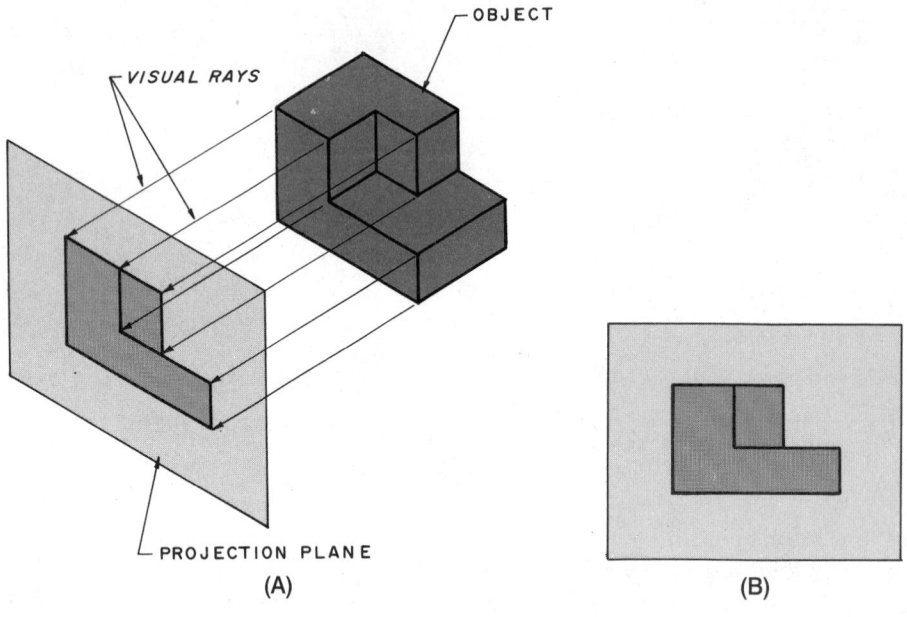

Figure 4-3 (A) Viewing plane parallel to surfaces; (B) Viewing plane provides actual size and shape of surfaces

to several of the object's surfaces, as shown in *Figure 4-3A*, then the image appearing in that viewing plane would provide the actual size and shape of those surfaces, as shown in *Figure 4-3B*. A surface that is parallel to a viewing plane is said to be "normal" to the viewing plane.

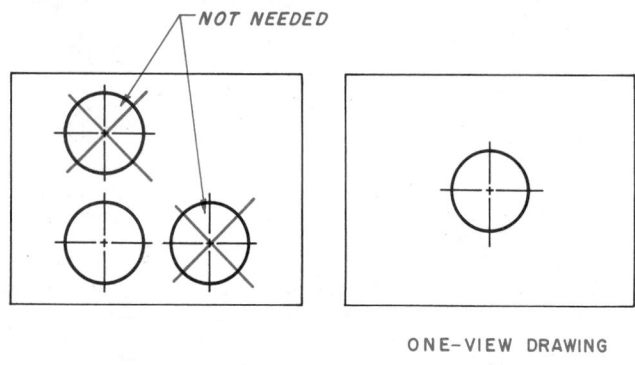

Figure 4-4 Simple one-view drawing

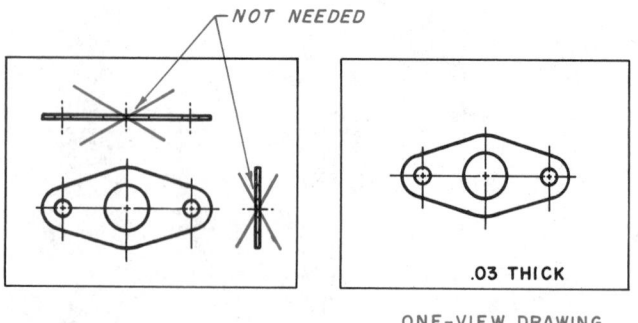

Figure 4-5 Simple one-view drawing

A simple object, such as the cannonball in *Figure 4-4*, needs only one view to describe its true size and shape. An object such as the flat gasket in *Figure 4-5* needs only one view and a callout stating its required thickness. The callout can be noted on the drawing, as illustrated, or listed in the title block under "material." *Figure 4-6A* shows an object with a viewing plane placed in a normal position relative to some surfaces of the object, resulting in the orthographic views shown in *Figure 4-6B*. The sizes and shapes of the object's normal surfaces are shown in perfect outline in the orthographic view, but the viewer still cannot determine other features of the object. For example, *Figures 4-6C*, *4-6D*, and *4-6E* are different objects that provide exactly the same projected view. Furthermore, not all the surfaces of the projected views of Figures 4-6C and 4-6E are shown in their real size and shape as they are not normal to the viewing plane. There is no way to fully distinguish these without additional information.

Two Orthographic Views

In order to present the images of each viewing plane in a flat drawing area, one of the viewing planes must be repositioned to lie in the same plane of the drawing as the other. The procedure used to do this is to create a *fold line* where the imaginary perpendicular-orthographic viewing planes meet each other along the straight edge of intersection, as shown in *Figure 4-7A*. Figure 4-7A shows the object in Figure 4-6A with a second viewing plane in position. The fold line acts as a hinge as the two intersecting viewing planes are swung into the same plane, *Figure 4-7B*. The resulting orthographic views are shown in *Figure 4-7C*. Each view

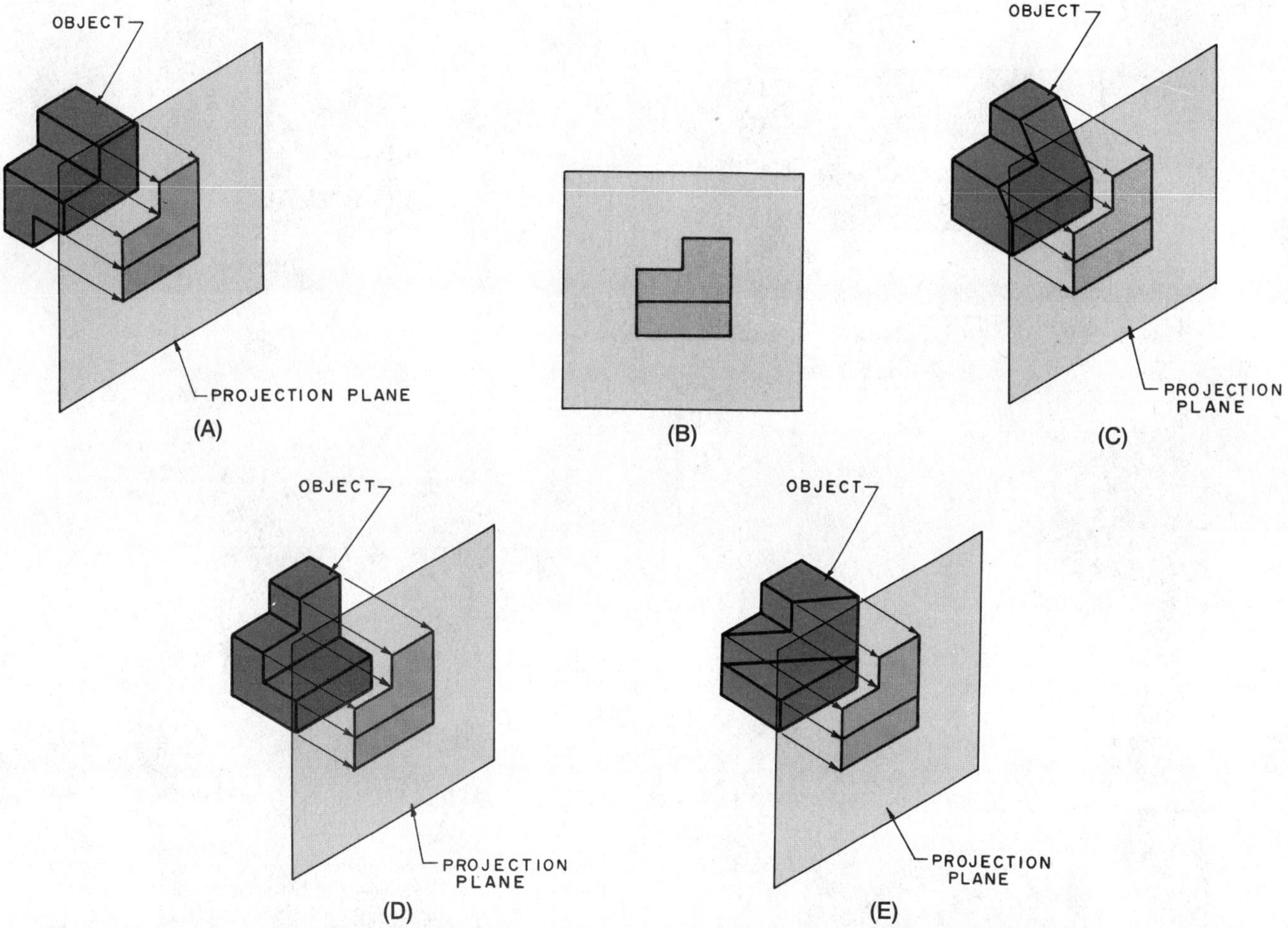

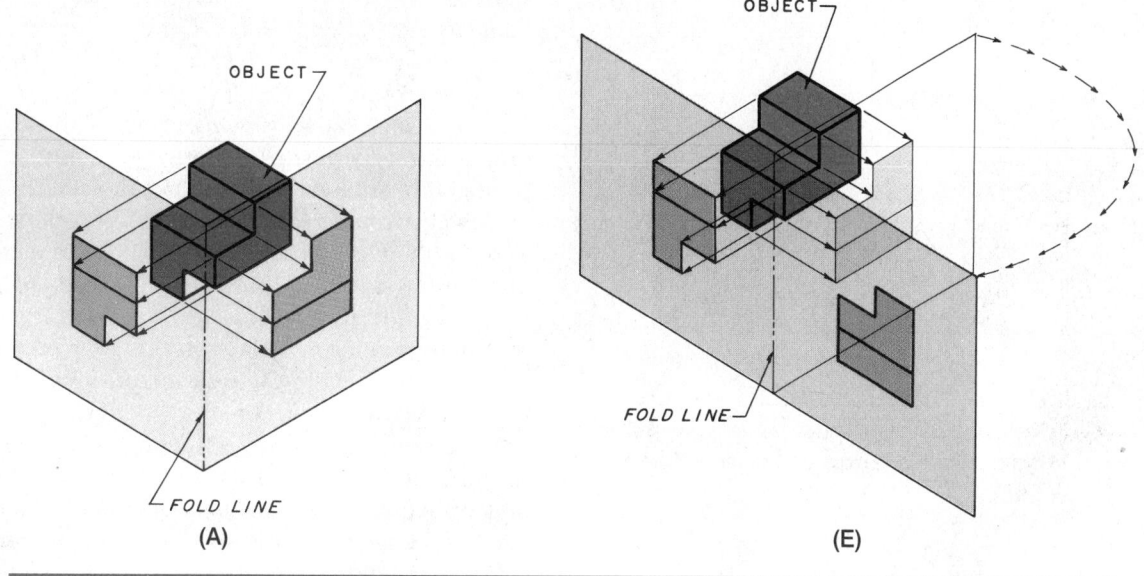

Figure 4-6   (A) Viewing plane in a normal position; (B) Orthographic view of the object; (C) Different object with the same orthographic view; (D) Another object with the same orthographic view; (E) Another object with the same orthographic view

Figure 4-7   (A) Fold line; (B) Fold line acts as a hinge

contains the image of a 90° rotated view or projection of the other. In this procedure, the features and surfaces are always aligned to each other and thus can be located.

Conversely, the same two orthographic views are shown in *Figures 4-7D*, *4-7E*, and *4-7F* in haphazard relationship to each other, putting them each in error and making it impossible to determine surface positions or the real shape of the object.

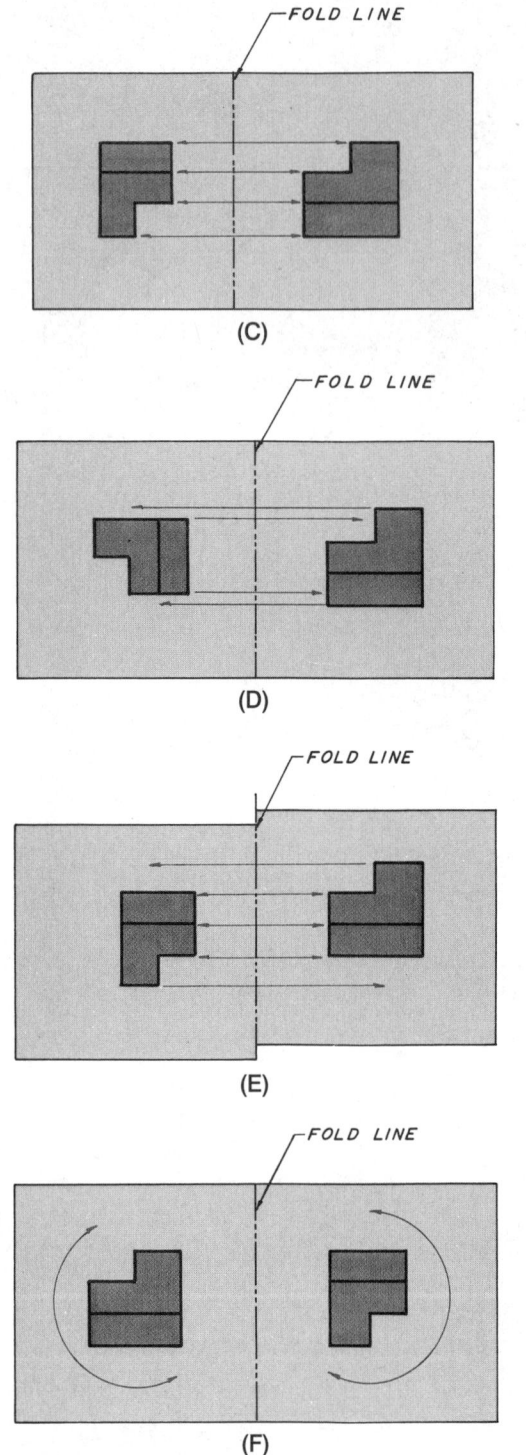

Figure 4-7 *(Continued)* (C) Projecting features from one view to the next view; (D), (E), and (F) Incorrect positioning of views

Labeling Two Views

Figures 4-8A and *B* shows the positioning of two adjoining orthographic viewing planes to describe the object shown in Figure 4-6C. The orthographic view that presents the most characteristic shape of the object is usually selected and identified. Once this view is

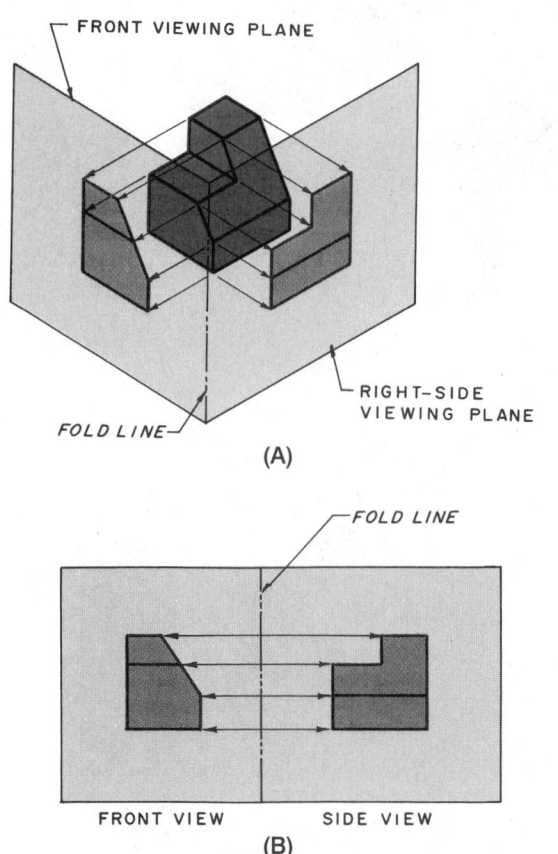

Figure 4-8 (A) Positioning of two adjoining orthographic viewing planes; (B) Two adjoining orthographic viewing planes flattened out

selected, the orthographic view that is projected to the right of the front view is identified as the *right-side view*. The term *front view* is often referred to as the front elevation and the *right-side view* is referred to as the *right profile*.

Figure 4-9A shows viewing planes with view images that accurately describe Figure 4-6D. But, finished orthographic drawings should not include the borders of viewing planes. For example, *Figure 4-9B* shows the views without the borders. Viewing planes were earlier identified as imaginary; therefore, the fold line is imaginary also. Often, fold lines are needed for view constructions, as illustrated later in this chapter and future chapters (especially Chapter 7), but in general, fold lines are removed on technical drawings.

The distance of the image from the fold line identifies the object's location from the viewing plane. The drafter decides what this distance will be in order to make the best use of the space on the drawing paper and to provide room for dimensions and notes. (See *Figure 4-10*.)

Multiple Orthographic Views

Figure 4-11A shows two orthographic views that do not completely describe the object. Multiple interpretations of the objects are possible, causing the viewer to

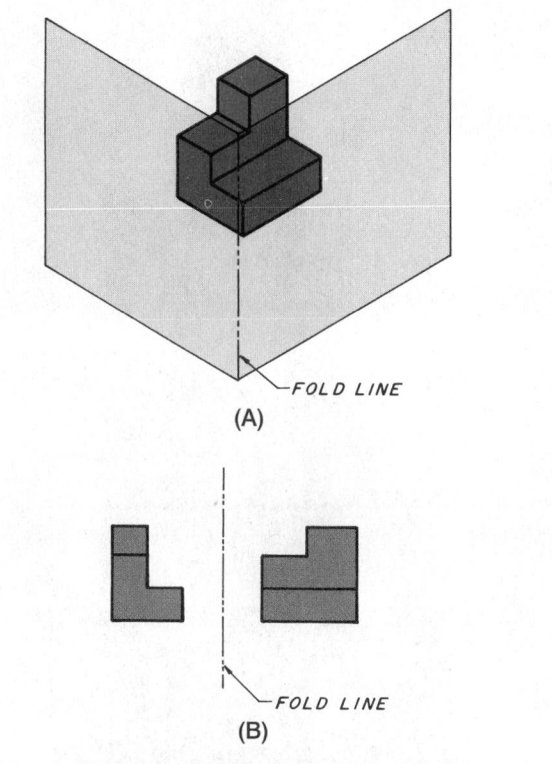

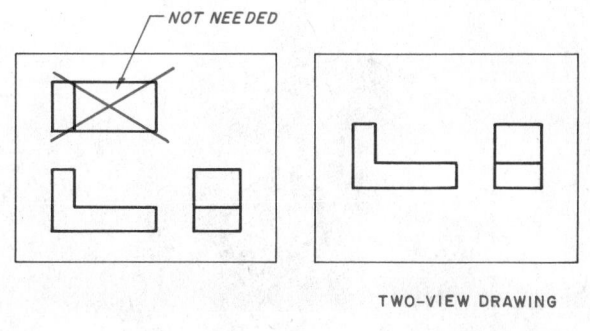

NOT NEEDED

TWO-VIEW DRAWING

Figure 4-10 Only two views are required

be misled by inadequate information. Correctly interpreted, the object appears as shown in *Figure 4-11B*. Incorrect interpretations are shown in *Figures 4-11C*, *4-11D*, and *4-11E*.

A third viewing plane is commonly used to ensure adequate definition of the represented object. The third viewing plane is usually above the object and perpendicular to the other two. It is called the *top view*. The top view is sometimes called the plan view in construction-related drawings. This third view is shown in *Figure 4-12*.

Select the views to best present the object's shape and make the most efficient use of the drawing area,

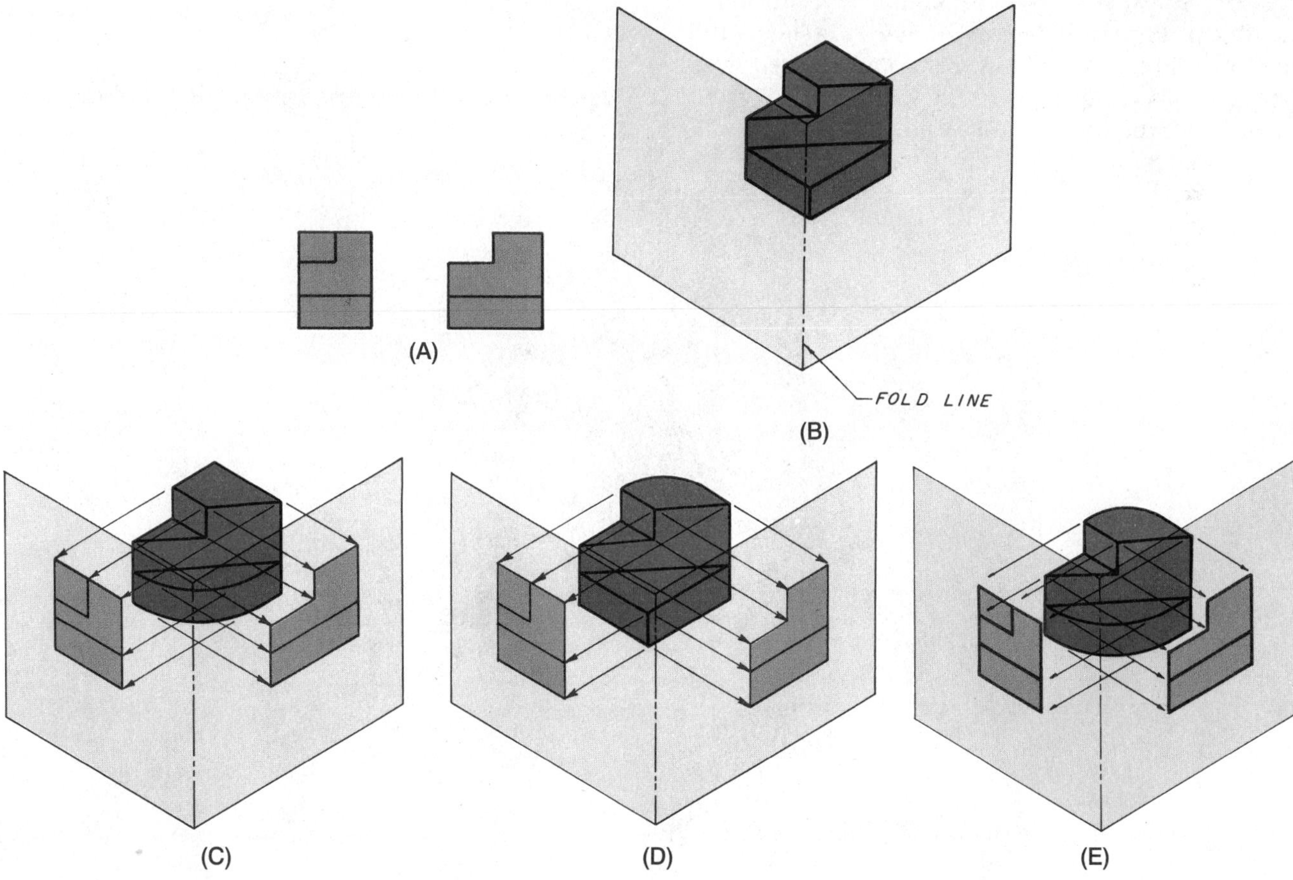

FOLD LINE
(A)

FOLD LINE
(B)

Figure 4-9 (A) Viewing planes and image; (B) Orthographic drawing omitting viewing planes

(A)

FOLD LINE
(B)

(C)　　　(D)　　　(E)

Figure 4-11 (A) Orthographic views of an object; (B) Correct interpretation; (C), (D), and (E) Incorrect interpretation

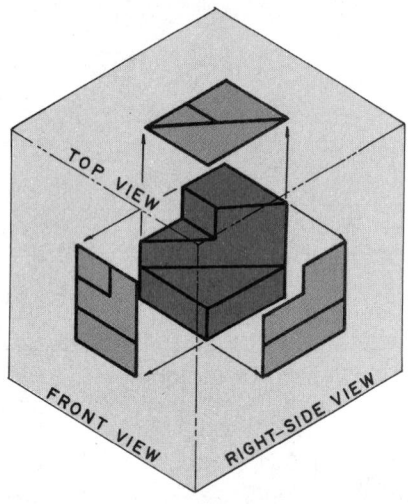

Figure 4-12   Third viewing plane added

Figure 4-13A, or to maintain the orientation of the object that is most easily understood by the reader, as illustrated in *Figure 4-13B*. Recall that the view selected should be an attempt to show the most characteristic shape of an object. For example, seeing a building drawn sideways makes reading the drawing very awkward.

The three viewing planes in Figure 4-12 are called *principal viewing planes*, as they are all perpendicular to one another. Orient the front view first. Three other principal planes complete the "box"; they make up the sides of a transparent box that completely surrounds the object, *Figure 4-14A*. *Figure 4-14B* shows the possible fold line selections that could be used with orthographic views on the principal planes. *Figure 4-14C* shows the six principal views.

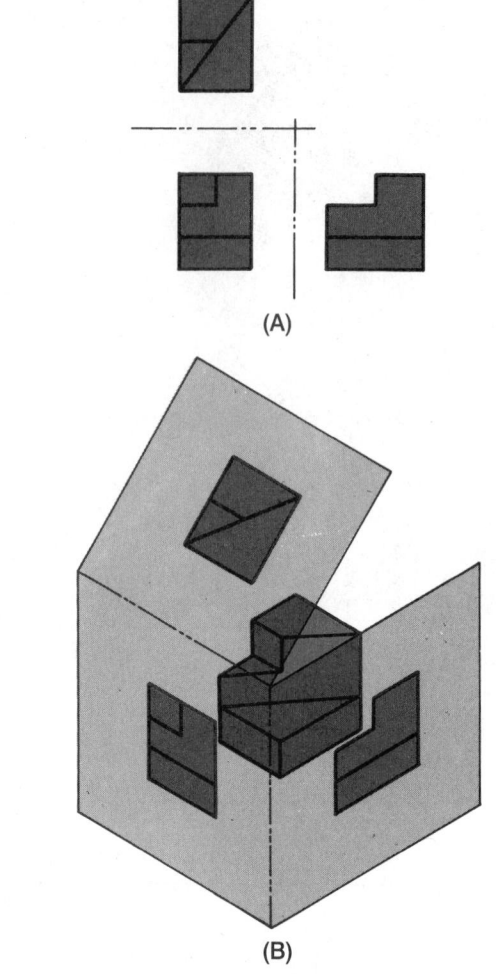

Figure 4-13   (A) Orthographic view of an object;   (B) Three principal planes

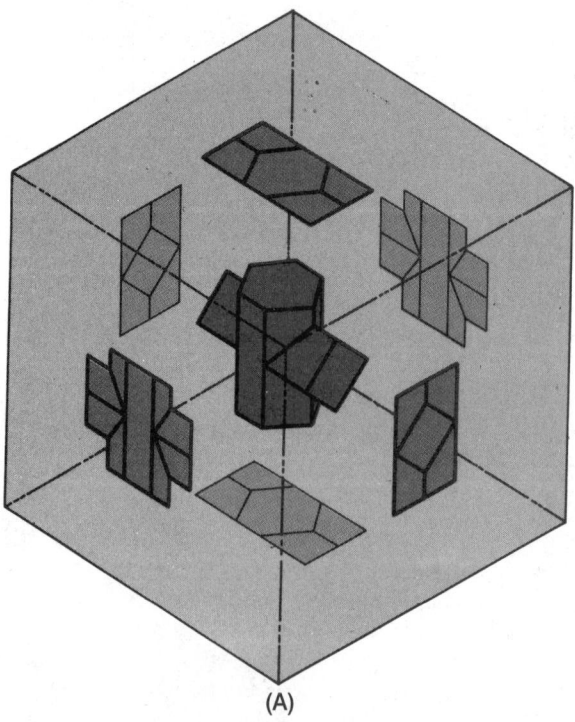

Figure 4-14   (A) Transparent box

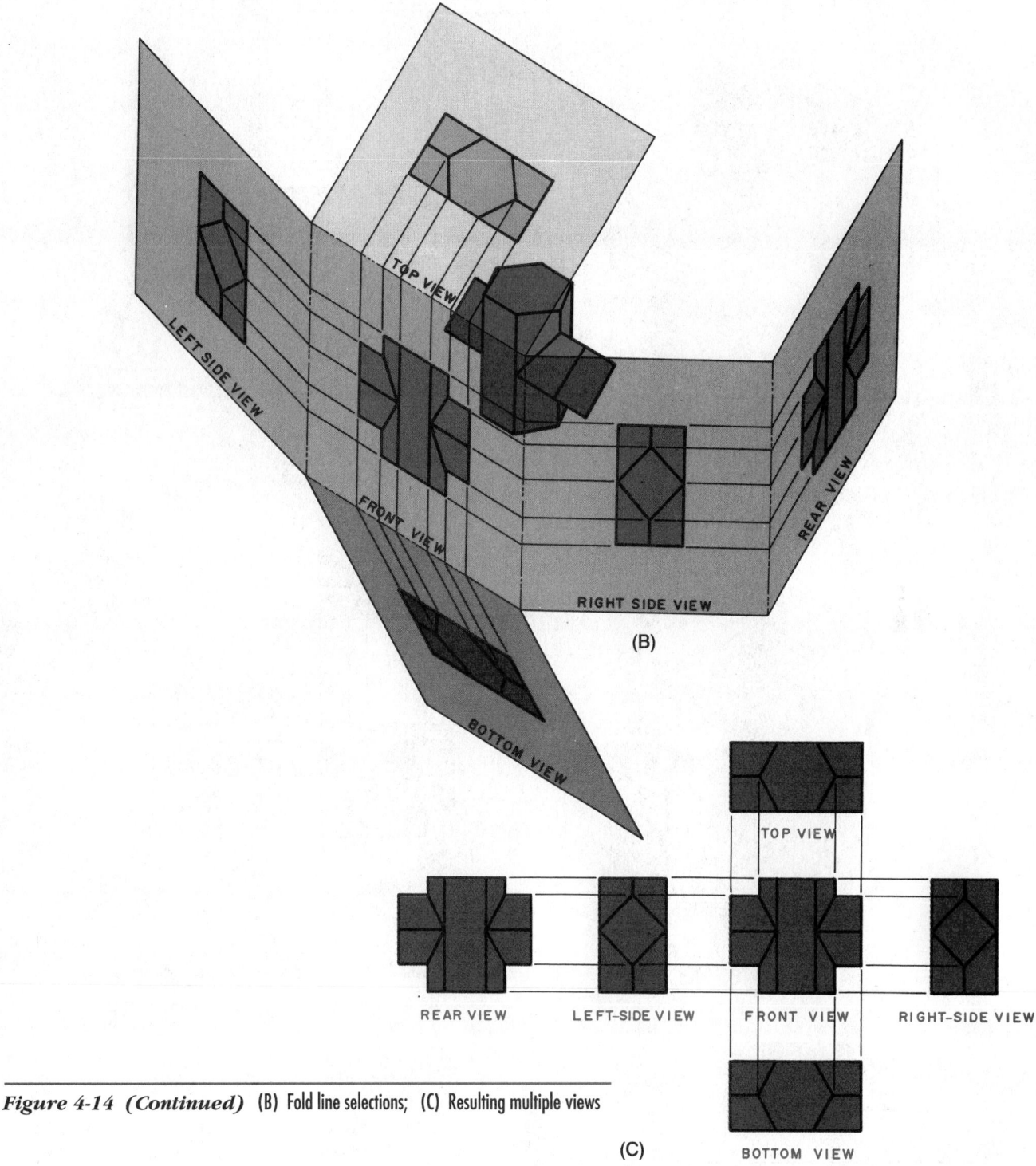

Figure 4-14 (Continued) (B) Fold line selections; (C) Resulting multiple views

(C)

Object Description Requirements

All six views of Figure 4-14C are not necessary to provide sufficient information; three are sufficient for this object. The common arrangement of top, front, and right-side views are shown in *Figure 4-15*. The top view aligns directly above the front view, and the right-side view aligns directly to the right of the front view. Both the top view and the right-side view images are the same distance behind the front view projection plane. See dimension b in *Figure 4-16*.

Dimension Transfer Methods

Figure 4-15 shows a method of transferring dimensions between the top and right-side views. This method, used only with principal views, employs a line at 45° to the fold lines of the principal planes.

Use a compass for a different method shown in Figure 4-16.

Figure 4-17 shows the transfer method, which is usually the most accurate. It is the method used in descriptive geometry discussed in Chapter 7. Draw the

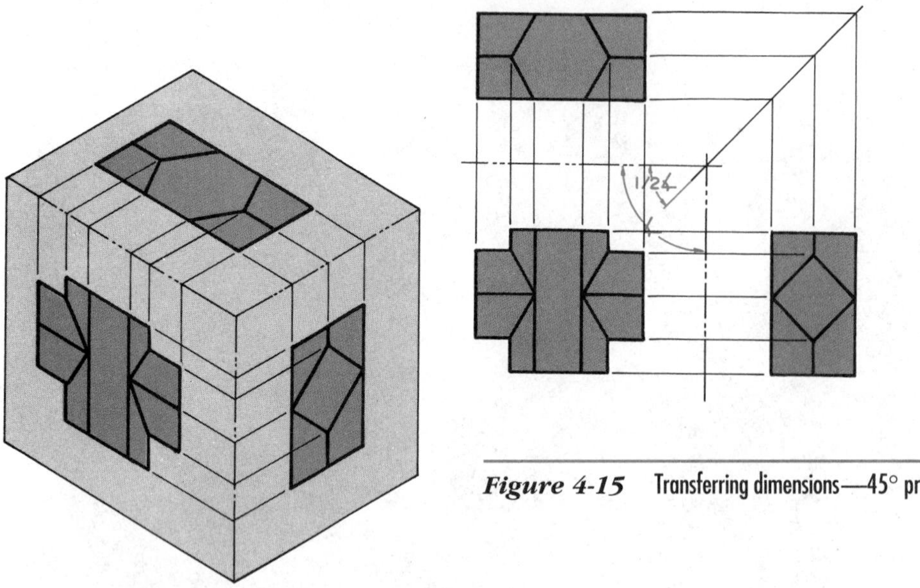

Figure 4-15   Transferring dimensions—45° projection method

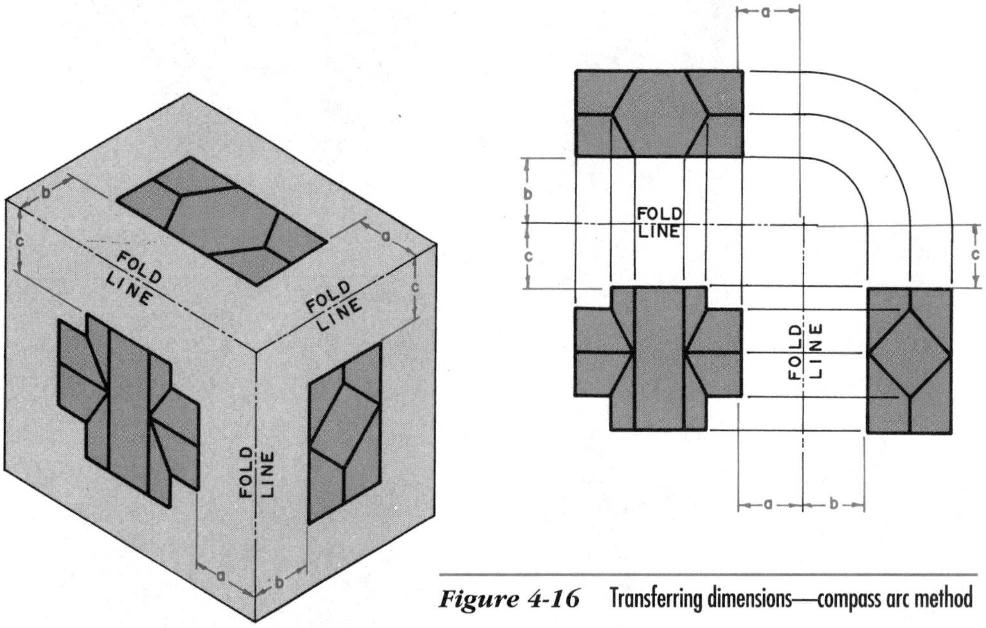

Figure 4-16   Transferring dimensions—compass arc method

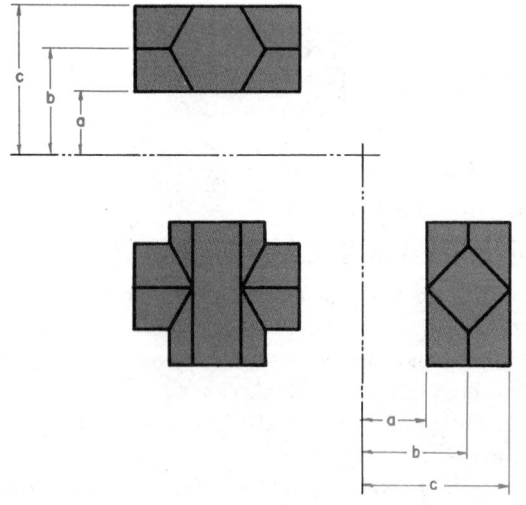

Figure 4-17   Transferring dimensions—transfer method

front and top views first and arbitrarily position a fold line between them. Recall that the fold line must be perpendicular to the projection lines between views. Construct a second fold line perpendicular to the projection lines that will go to the right to be used for the right-side view. This second fold line can be at any distance from the front view. Once the fold line is drawn, the distances from it to the right-side view must agree with distances to the top view, dimensions a, b, and c. Transfer these dimensions with a scale or dividers. If the front and right-side views are drawn first, then construct a fold line for the top view in a similar manner.

Hidden Lines

Figure 4-18 is an isometric view of an object. The orthographic top, front, and right-side views are illustrated in *Figure 4-19*. *Hidden lines* represent feature

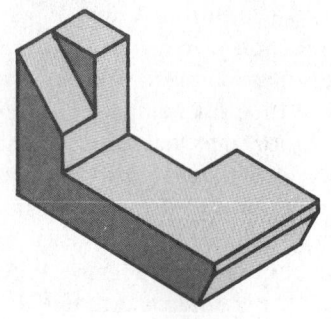

Figure 4-18   Isometric view of an object

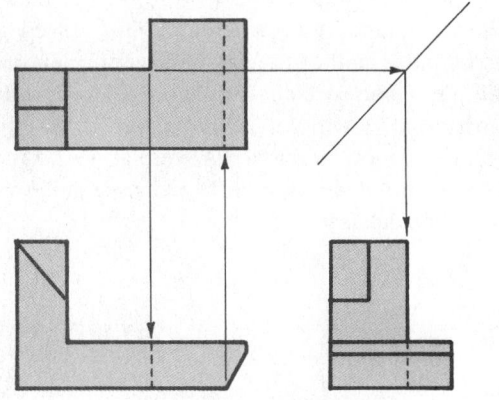

Figure 4-19   Orthographic views of the same object

outlines, for example, intersections of plane surfaces, real but invisible, and are used in each viewing plane where appropriate.

A hidden line may be covered by a visible (solid) object line in a particular view. This concept is shown in ***Figures 4-20A*** and ***4-20B***. Figure 4-20B shows surface B aligned behind the solid line of intersection of plane A with the inclined plane. As expected in orthographic projection, there are many hidden lines covered by solid lines. Mentally perceiving where the hidden lines lie in orthographic views is a process called visualization.

Figure 4-21 shows typical hidden line length and spacing, approximately ⅛" to ¼" long spaced approximately 1/32" to 1/16" (3.2 mm to 6.4 mm and .8 mm to 1.6 mm, respectively). The lengths and spaces can vary depending on the feature and the space available on the drawing, but consistency is encouraged. Hidden lines are narrower than solid lines. Use the accepted drafting practices of hidden line applications shown in ***Figure 4-22***.

Center Lines

Center lines are used primarily as locations of circular, cylindrical, or spherical features. They can also be used to specify locations of other principal symmetries as well, such as the center line of a ship or a rocket. Crossed perpendicular center lines indicate a two-coordinate location of where an axis exists. A single center line indicates the path of an axis. Identify all axes, on all

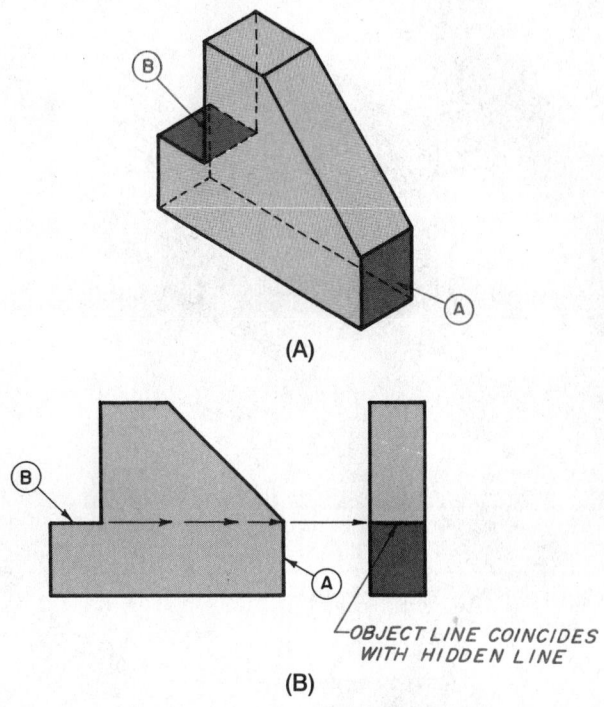

Figure 4-20   (A) Isometric view of an object;   (B) Object line takes precedence over hidden line

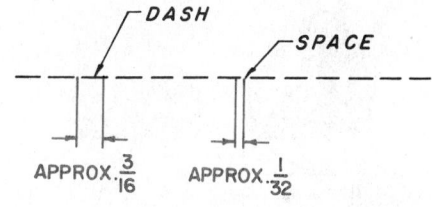

Figure 4-21   Average hidden line construction

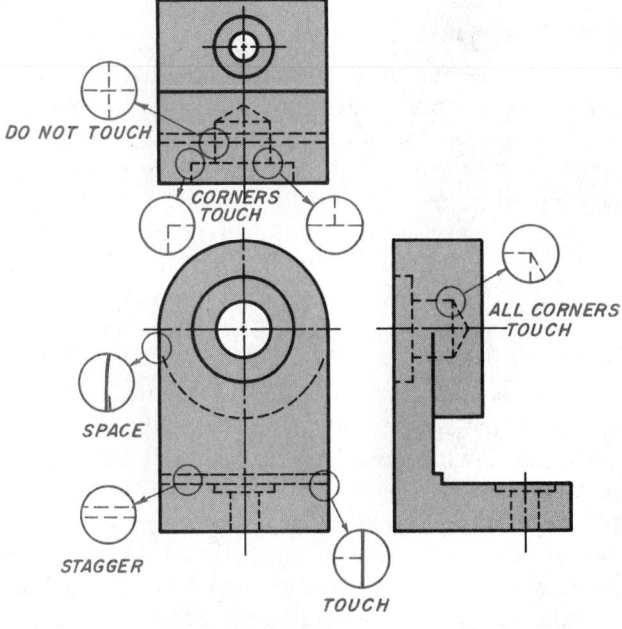

Figure 4-22   Hidden line drafting practices

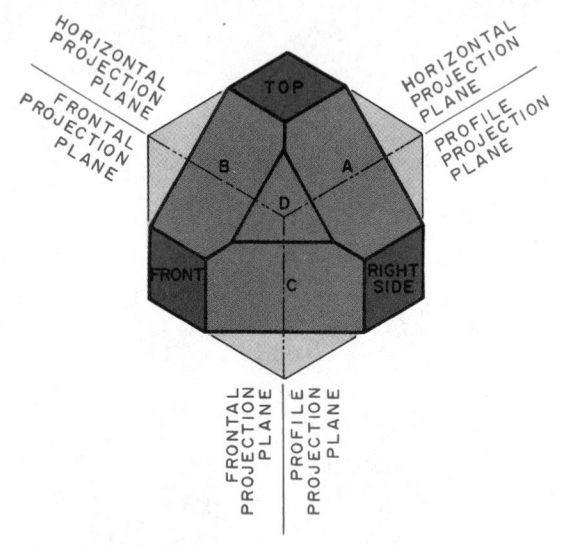

Figure 4-23 Surface categories

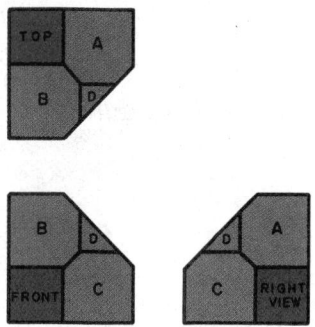

Figure 4-24 Orthographic views

views, of circular features except for noncritical radii, Figure 4-22.

Surface Categories

Recall that normal surfaces are parallel to one of the principal projection planes of the imaginary transparent box surrounding the object. In ***Figure 4-23***, these surfaces are identified as the top, front, and right side. The viewing planes containing these images are identified as horizontal, frontal and profile projection planes, respectively. A normal surface's true size and shape appears in one of the principal views, ***Figure 4-24***, but the surface appears as an edge view, a line, in the other principal views.

Surfaces A, B, and C, Figures 4-23 and 4-24, are inclined surfaces. They are not parallel to any one of the principal viewing planes, but each inclined surface is perpendicular to one of them. For example, inclined plane A is perpendicular to the frontal projection plane. Moreover, a surface that is perpendicular to a viewing plane will appear as an edge or line when projected on that plane. Surface A appears smaller than its true size in both the top and right-side views, Figure 4-24, because of foreshortening. Inclined surfaces B and C are

configured similarly to surface A, with B perpendicular to the profile projection plane, and C is perpendicular to the horizontal projection plane. Surface D is not perpendicular to any of the principal projection planes, and is therefore, categorized as an oblique plane. It appears foreshortened in all of the principal projection planes.

The transparent box of principal projection planes that surround cylindrical surfaces uses the cylindrical axis as the line of orientation, ***Figure 4-25***. If the cylindrical axis is made perpendicular to one of the principal viewing planes, then the cylindrical surface appears as a circle or an arc. Figure 4-25 shows a cylinder whose axis is perpendicular to the top horizontal principal projection plane, and the circular image that results in the top view is shown in ***Figure 4-26***. The curved surface of the cylinder cannot be classified as normal, inclined, or oblique. Plane surface A in Figure 4-26 is inclined to the horizontal viewing plane and shows as an ellipse in the right-side view. If surface A were at 45° to the horizontal viewing plane, a circle would appear as surface A in the right-side view.

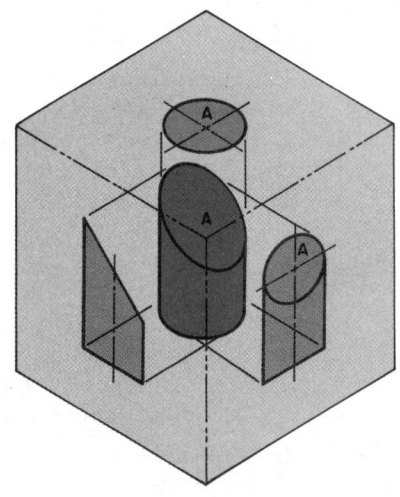

Figure 4-25 Object inside transparent box

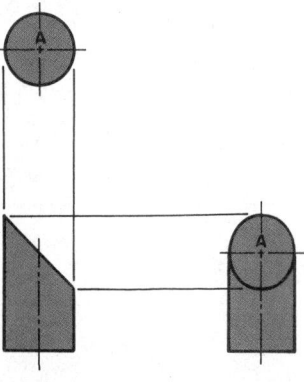

Figure 4-26 Circular image results in this top view

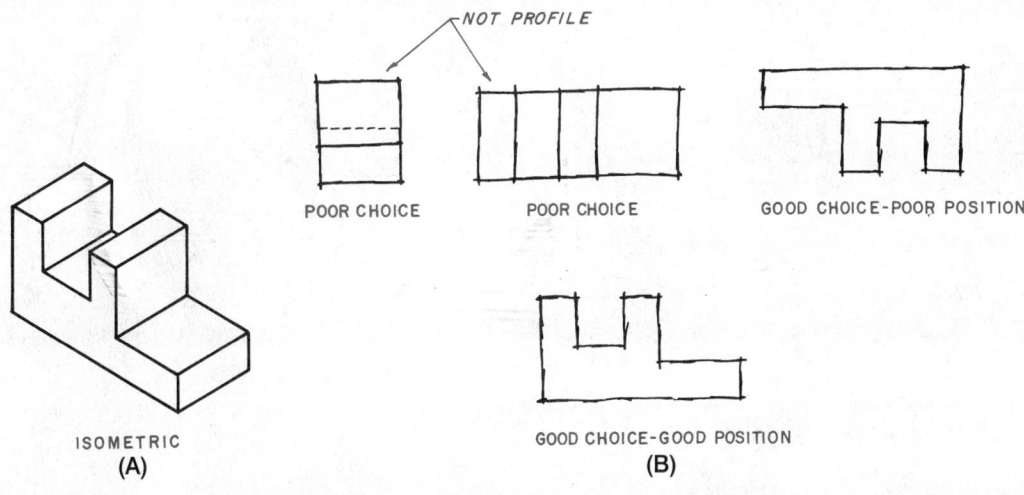

Figure 4-27   (A) Isometric view of an object;   (B) Sketches of object to determine front view

Planning the Drawing

Before beginning a three-view drawing of any object, make preliminary sketches of the object. Sketches of the object in *Figure 4-27A* are shown in *Figure 4-27B*. The sketches were used as an aid in selecting the best front view and its best position. Sketches are helpful in selecting the required views and their positions.

How To Prepare a Three-view Drawing

Step 1   Visualize the object to sure you have a good mental picture of it.

Step 2   Decide which view to use as the front view by sketching the object in several positions, Figure 4-27B. Keep in mind the following:
- the front view is the most important view; viewers see this one first.
- the front view should show the most basic profile; for example, use a side view of a vehicle with the wheels on the ground.
- the front view should appear stable; place the heavy part at the bottom.
- the front view should be positioned to minimize the number of hidden lines in all views; this enhances visualization.

Step 3   Decide how many views are needed to completely illustrate the object.

Step 4   Decide, keeping in mind the guidelines in step 2, which position to place the front view. Study *Figures 4-28*, *4-29*, *4-30*, and *4-31* to understand why 4-31 is the best position for the object from Figure 4-27A.

Step 5   Decide on the best placement of the views in the work area, usually centered. Do neat work.

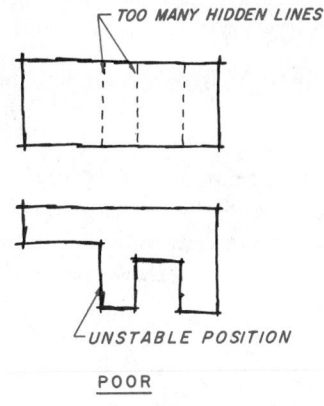

Figure 4-29   Poor front view—unstable, too many hidden lines

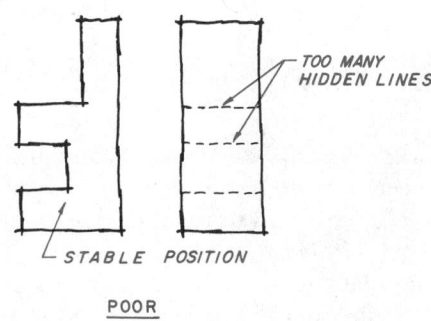

Figure 4-28   Poor front view—too many hidden lines

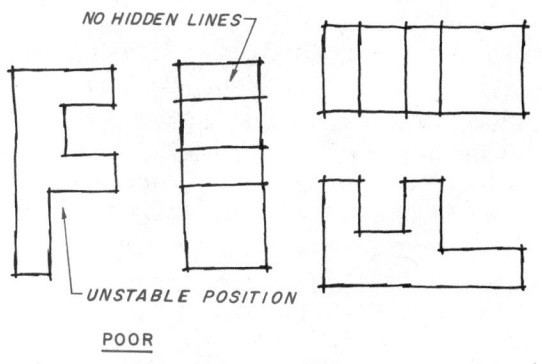

Figure 4-30 Poor front view—unstable

Figure 4-31 Best position—front view

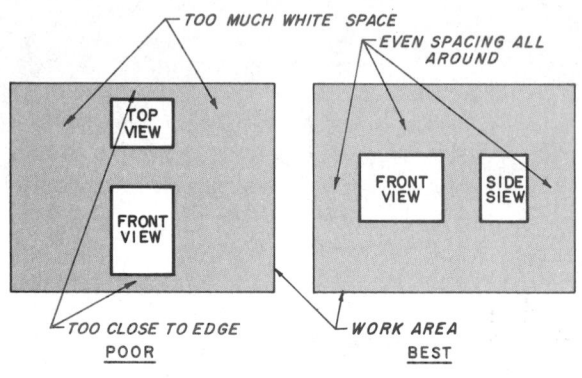

Figure 4-32   Positioning the views within the work area

Positioning the Views within the Work Area

The example shown on the left of *Figure 4-32* is centered, but poor judgment was exercised in using the space on the drawing. Space is wasted on both sides, and little room is available at the top and bottom for dimensions and notes. An inch (25.4 mm) of space near the borders is desirable. The example shown on the right in Figure 4-32 is balanced and well centered. Try to keep open spaces evenly distributed around views. A systematic method for centering views is discussed next, after some comments on a general sketching procedure for centering views.

Sketching Procedure

Sketch efficiently by doing it freehand, quickly and only to approximate scale on a sheet of practice drawing paper. First, select the front view, using the guidelines previously discussed, and sketch a light outline of the object in approximately the best position. Next, sketch outlines of the top and right side views. Locate center lines of important features. Finally, heavy in the outline and centerlines.

Centering the Drawing

Figures 4-33 and *4-34* show a simple object and a procedure used to center a sketch of the object within a specified work area. The simple object in Figure 4-33 is shown in a dimensioned isometric drawing. The views of the object in both figures are to be sketched freehand. Use an 8½ x 11 practice sheet, and follow the procedure given. There is no need to use a scale, but a drawing sheet with fade-out lines would be helpful.

Calculate the total horizontal distance needed for the front view plus the right-side view by adding the front view width of 4.0, plus 1.0 for the space between views, plus 2.0 for the depth of the right-side view, all for a total of 7.0 inches. To center these views, subtract 7.0 from 11.0, to get 4.0. Divide this in half for the spacing at each end, D = 2 inches. (See Figure 4-33.)

Calculate the total vertical distance needed for the front view plus the top view by adding the front view height of 2.5, plus 1.0 for the space between the views,

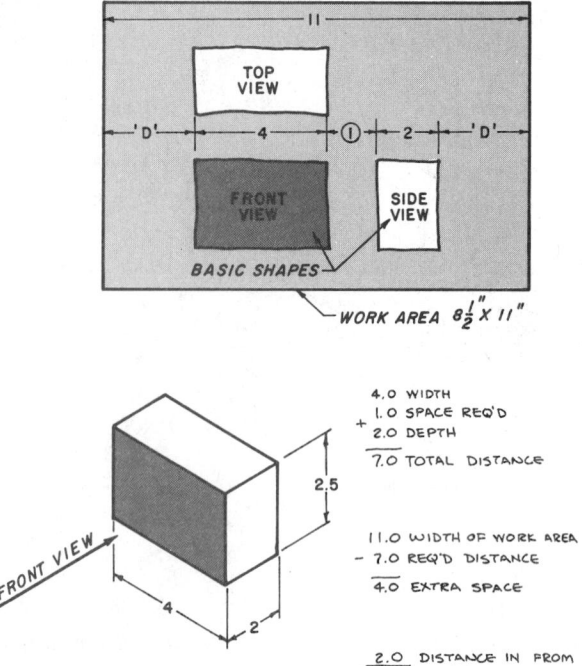

Figure 4-33   Centering views horizontally

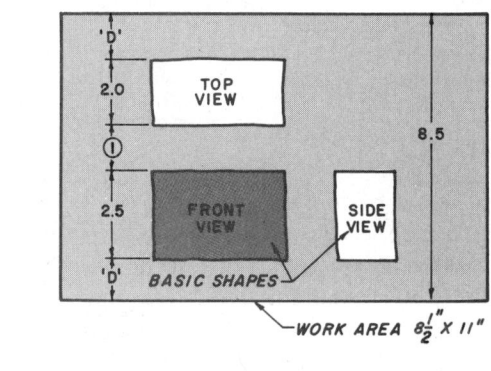

Figure 4-34   Centering views vertically

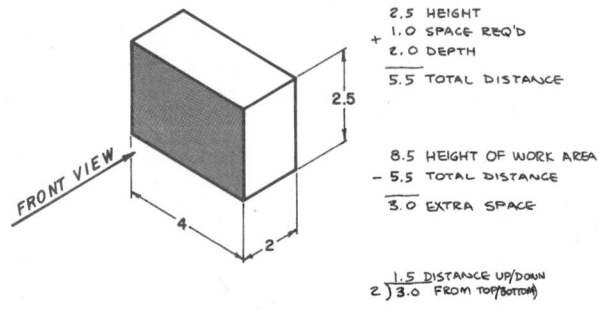

plus 2.0 for the depth of the object, all for a total of 5.5 inches. To center these views, subtract from 8.5 to get 3.0, and divide this in half for the upper and lower spacing, D = 1.5 inches. (See Figure 4-34.)

Try an alternate approach for centering a drawing. Lay out to scale on a practice sheet, with drawing instruments, outlines of the front view and the right-side view with a 1.0-inch space between the two views.

Next, layout the top view with a 1.0-inch space between the front and top view. Next, place a good drawing sheet over the practice sheet and move it around until the drawing is basically centered. Finally, trace the drawing onto the good sheet.

Numbering for Making Drawings

Sometimes a multiview drawing of an object, to be prepared from an isometric drawing, is so complicated that the person making the drawing is not quite sure of how some part of a view will look. These more difficult drawings are easier to visualize if points of the various features are numbered, then systematically analyzed. Think of the corners of the features as lines connecting points. For instance, think of each line as connecting two points in space, *Figure 4-35*. If the ends of a line can be found, all that needs to be done to complete the line is to connect the ends. Once the ends have been found and numbered in one view, the same can be done in other views by simple projection. *Figure 4-36* is an isometric view of a simple object, simple for discussion purposes. Each end of each line has been assigned a number.

Use the following steps for practice in numbering and analyzing drawings.

Step 1   Lightly draw the basic shape outline of the required views and the 45° projection line, *Figure 4-37*.

Step 2   Assign the numbers from the isometric sketch to each corner of the view.

Step 3   Project these points up to the top view, and number them as shown in *Figure 4-38*. Start with the number seen first, followed by a comma or slash, and then the number hidden.

Step 4   Project these points from the front view to the right-side view, and project the same numbered

points from the top view via the 45° line. The exact position of each point is located where the projection lines cross; number them, *Figure 4-39*. Projecting and locating points can be done with alternative methods as shown in Figures 4-16 and 4-17. *Note:* if a point can be located in any two of the three views, it can be projected into the third view.

Step 5   Connect points together, for example, point 1 to point 2, 3 to 4, and so on, until all the line connections are made. Check the planes made by the lines. Refer to the isometric view, Figure 4-36, as needed.

Comment: These steps describe how to make a complicated drawing. Read and understand a complicated drawing already made by first, locating and numbering points on an object, in at least three views. Connect the numbers and systematically study the object, line by line, plane by plane.

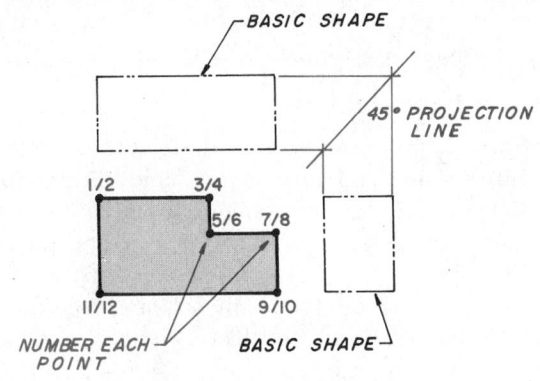

Figure 4-37   **Number one view**

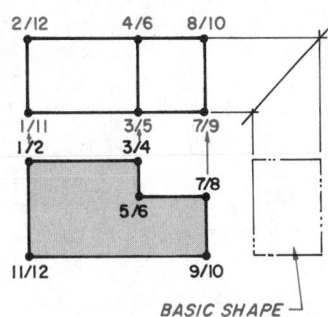

Figure 4-38   **Project numbers into the top view**

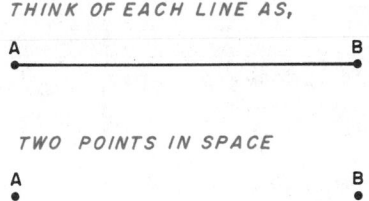

Figure 4-35   **A line is made up of two points in space**

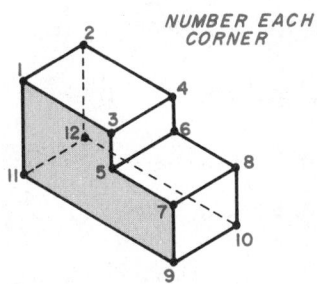

Figure 4-36   **Given: An isometric view of an object**

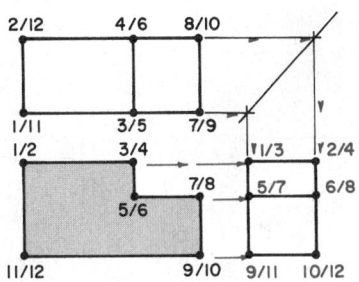

Figure 4-39   **Project numbers into the right-side view**

Technical Drawing Conventions

Assume now that preliminary sketching has been done to locate the best positions for views of an object for a multiview drawing, and that the views are going to be drawn, to scale, with the drawing tools discussed in chapter 1. The following topics are important for multiview, technical drawings.

- Rounds and Fillets
- Runouts
- Intersecting Surfaces
- Curve Plotting
- Cylindrical Intersections
- Incomplete Views
- Aligned Features
- How to Represent Holes
- Conventional Breaks

The following discussions on recommended techniques for preparing multiview technical drawings all emphasize the same message: effective communication of three-dimensional concepts, that is, to enhance the visualization process for the use of the drawings.

Rounds and Fillets

Metal casting processes require an object's material to flow into cavities and corners in a mold. (See chapter 20.) Sharp corners are not only undesirable on a casting, but are difficult to cast, so *rounds* are specified for outside corners, and *fillets*, for inside corners. Cast objects tend to crack due to the strain placed on the metal during cooling. Fillets distribute the strain and alleviate the tendency to crack. Rounds and fillets improve an object's ability to withstand cyclic fatigue loading, and also enhance the appearance of an object, *Figures 4-40* and *4-41*.

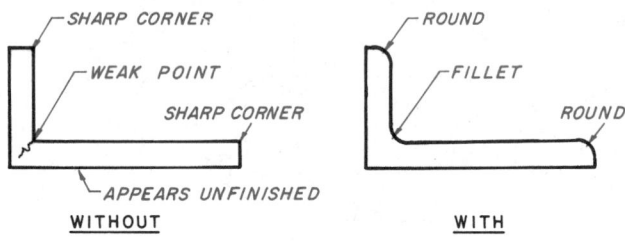

Figure 4-40   Rounds and fillets

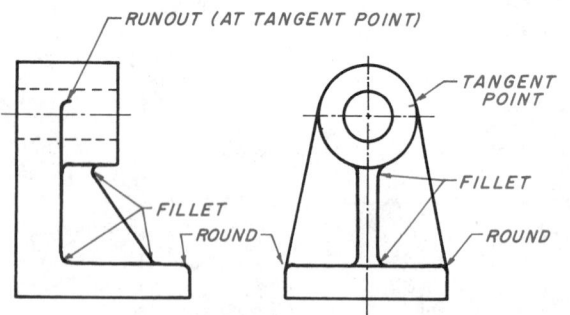

Figure 4-41   Runouts

Runouts

Runouts are curved surfaces formed where a flat and a curved surface meet, Figure 4-41. To find the exact intersection where a runout occurs, locate the tangent points of the curved surfaces, *Figure 4-42*. Project these tangent points into the next appropriate view and add the runouts as shown in *Figure 4-43*. Note the direction of each runout.

Treatment of Intersecting Surfaces

Figures 4-44 through *4-47* show how to illustrate various intersecting surfaces. In these examples, rounds, fillets, and runouts have been omitted to simplify the drawings. In actual practice, rounds, fillets, and runouts would have been added. In Figure 4-44, the flat surface meets the cylinder sharply (see top view), making a visible edge necessary in the front view. The meeting point in the top view is the exact location of the object line in the front view and right-side view.

In Figure 4-45, the flat surface blends into the cylinder, stopping at the point of tangency. In the top view, the tangent point is located at the center line and blends in at that location in the front view.

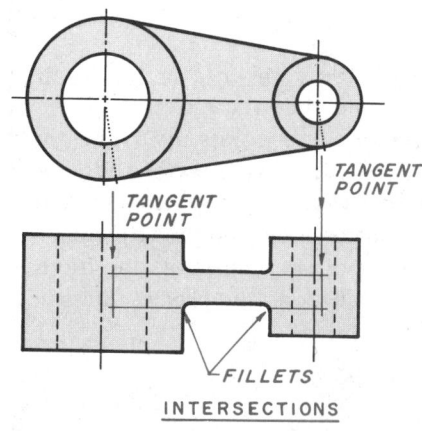

Figure 4-42   Locate the tangent points

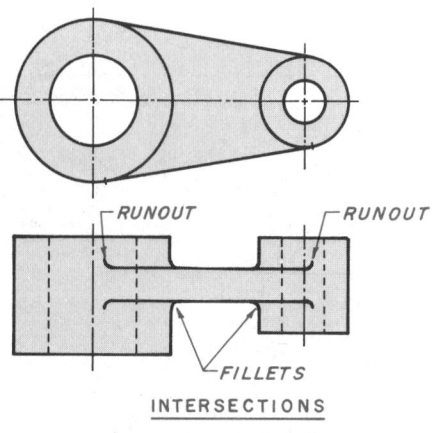

Figure 4-43   Complete runouts at tangent points

In Figure 4-46, the surface blends into the cylinder stopping at the point of tangency, similar to Figure 4-45 except on an angle. The tangent point is projected into the front view and right-side view.

In Figure 4-47, the radius blends into the base, stopping at the point of tangency. In the top view a line is drawn extending to the tangent point of the arc in the front view.

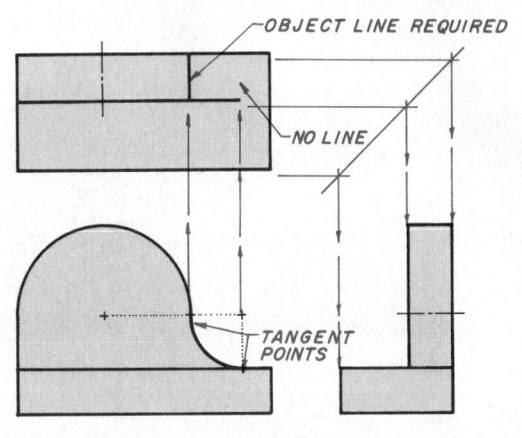

Figure 4-47   Intersecting surfaces

Curve Plotting

Curved surfaces in multiview drawings need to be projected into other views. Follow the steps listed to plot a curved surface.

Step 1   Lightly complete the basic outline of each view, **Figure 4-48**. Locate the center of an arc, point x.

Step 2   Draw the arc and divide it into equal spaces, **Figure 4-49**. Use 30° spacing for smaller arcs, but 10° spacing for larger ones. If accuracy is not needed, use larger spacing to save time.

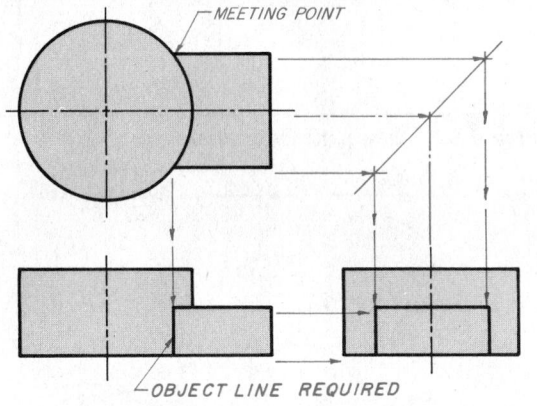

Figure 4-44   Intersecting surfaces

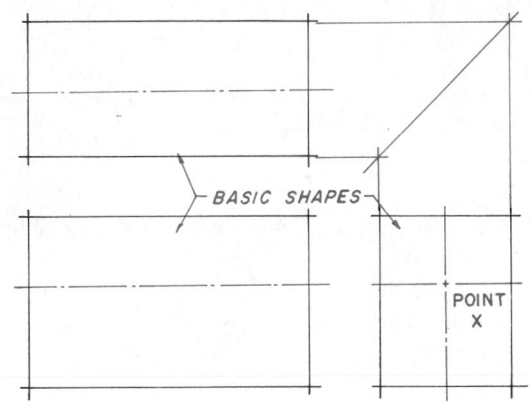

Figure 4-48   Curve plotting (Step 1)

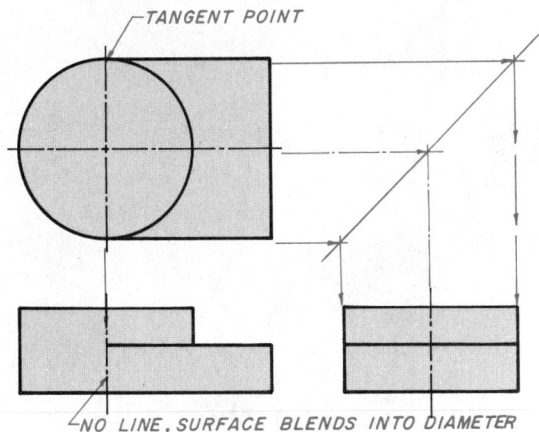

Figure 4-45   Intersecting surfaces

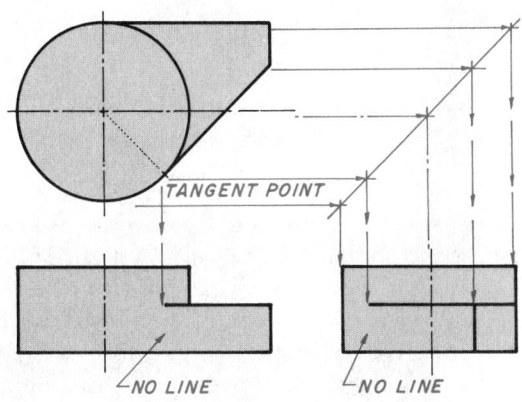

Figure 4-46   Intersecting surfaces

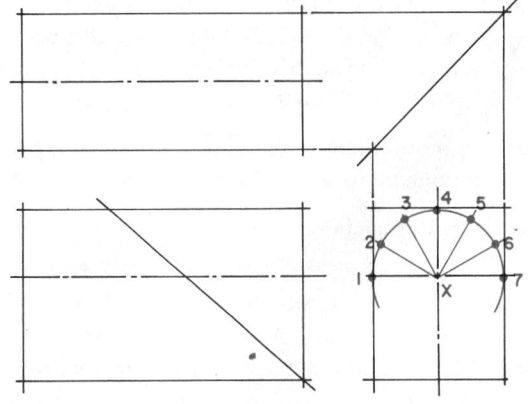

Figure 4-49   Curve plotting (Step 2)

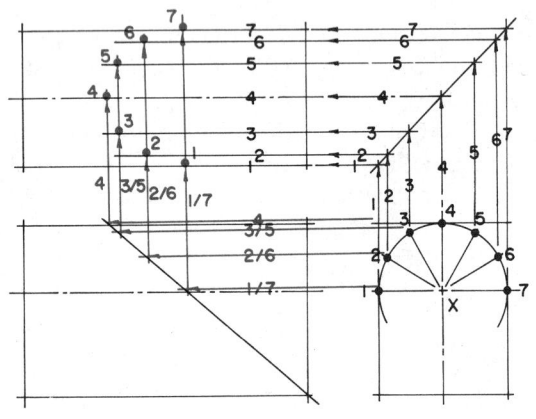

Figure 4-50 Curve plotting (Step 3)

Figure 4-51 Curve plotting (Step 4)

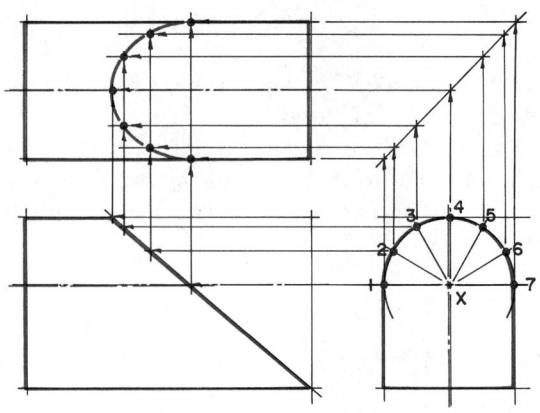

Figure 4-52 Curve plotting (completed)

Step 3 Number the points in order clockwise, **Figure 4-50**.

Step 4 Project these points up to the 45° axis line, and to the left to the top view, **Figure 4-51**.

Step 5 Project from the right-side view to the slanted surface in the front view and locate each point, Figure 4-51.

Step 6 Project these points up from the front view to the top view.

Step 7 Number the points 1 through 7 in the top view, where the numbered lines from the right-side view, intersect the numbered lines from the front view, **Figure 4-52**.

Note: The method shown in Figure 4-17 could be used for projecting and locating curves.

Cylindrical Intersections

If there is a considerable difference in the diameters of each of the intersecting cylinders, then use a straight line for the intersection, **Figures 4-53A** and **4-53B**. If there is little difference between the intersecting cylindrical diameters, then use a simple arc, constructed (see Figure 3-20.) or approximated through three points (**Figures 4-54A** and **4-54B**).

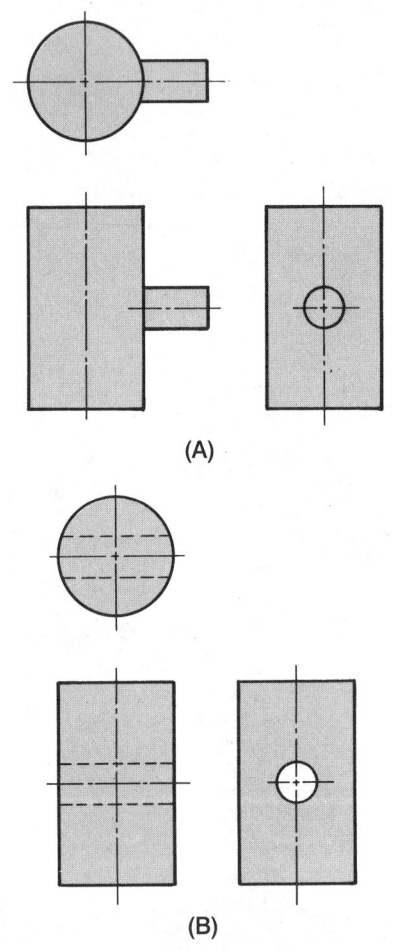

Figure 4-53 (A) Cylindrical intersections (straight line); (B) Cylindrical intersections (straight line)

In sheet metal construction, locate the actual edge of intersection as shown in **Figures 4-55** and **4-56**. Figure 4-55 shows line elements on each of two intersecting cylindrical surfaces. The line elements are parallel to their respective cylindrical axes and to a common (frontal) viewing plane. For example, line elements number 1 on each cylinder are the same distance from the frontal viewing plane. Also, the line elements num-

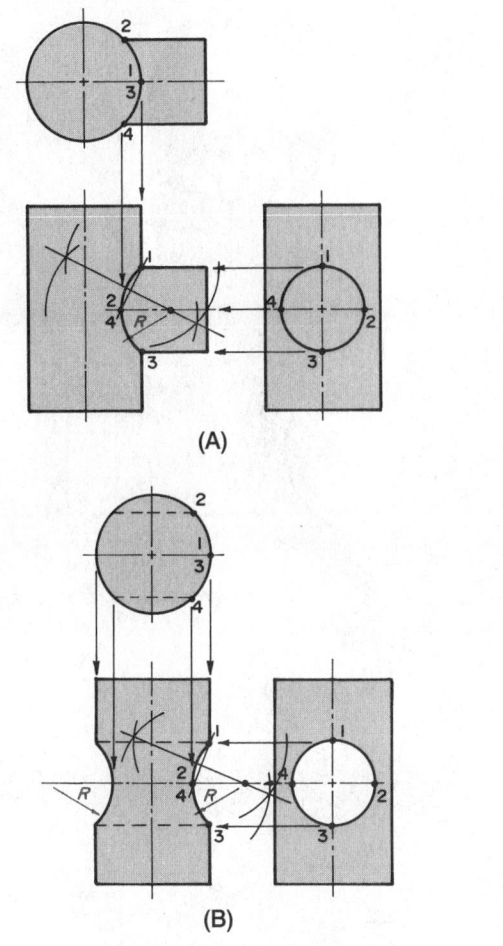

(A)

(B)

Figure 4-54   (A) Cylindrical intersections (arc);
(B) Cylindrical intersections (arc)

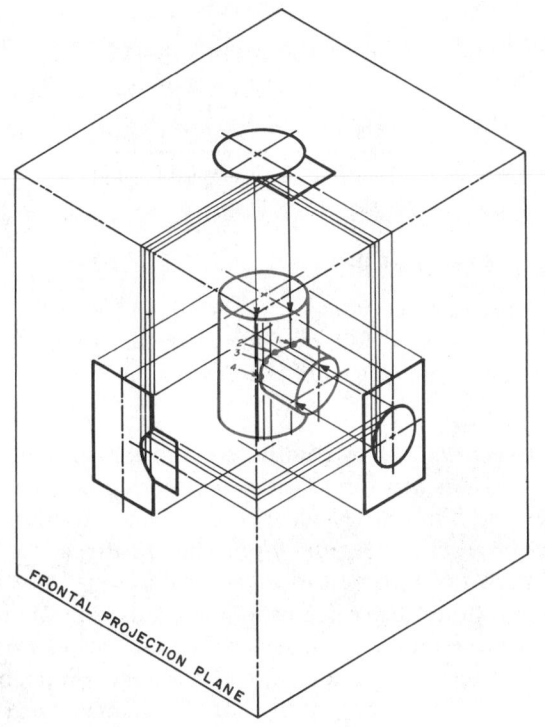

Figure 4-55   Two intersecting cylindrical surfaces

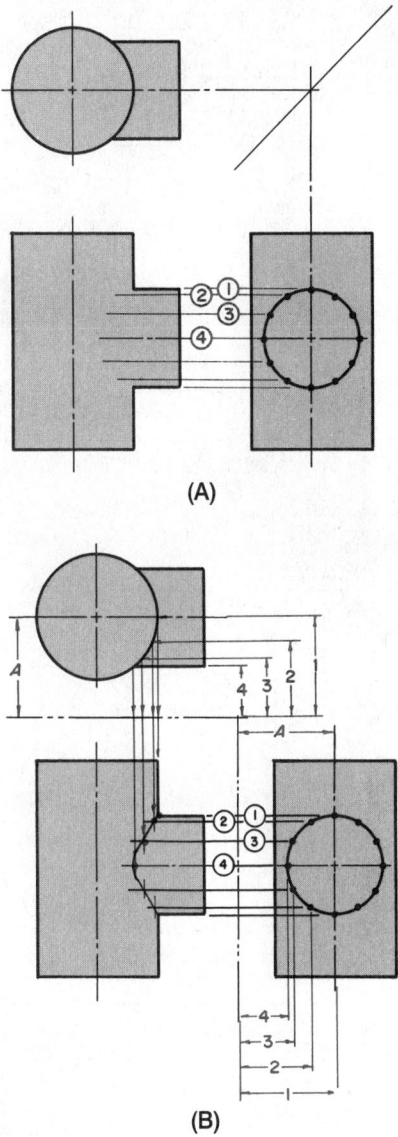

(A)

(B)

Figure 4-56   (A) Step 1; (B) Step 2

bers 2,3, and 4 are each the same distance from the frontal plane, respectively. Use this relationship of the elements to construct a curved line of intersection between the large cylinder and the smaller cylinder shown in Figure 4-55.

Step 1   Refer to ***Figure 4-56A*** and locate line elements 1, 2, 3, and 4 on the smaller cylinder in the right-side view. Project them into the front view.

Step 2   Locate a fold line between the front and right-side views, ***Figure 4-56B***, at a distance A from the center line of the large cylinder. Locate a fold line at the distance A from the top view. Transfer the distances labeled 1 through 4 in the right-side view to the top view as shown. Next, because each corresponding line element from each cylinder is parallel to the front view-ing plane and the same distance from it, they must intersect. Fit a curve through the points of intersection, Figure 4-56B. In a like manner complete the lower portion of the intersection.

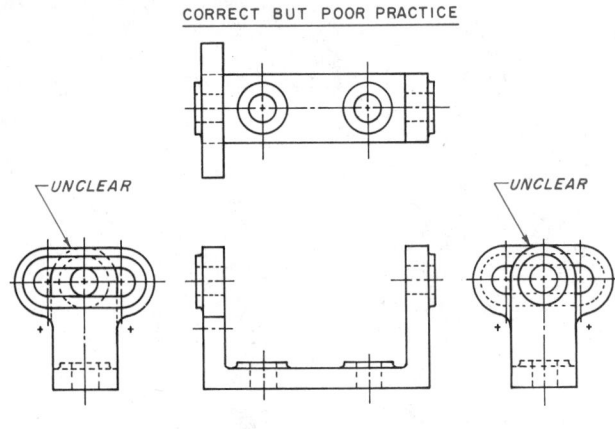

Figure 4-57   Overlapping views—poor practice

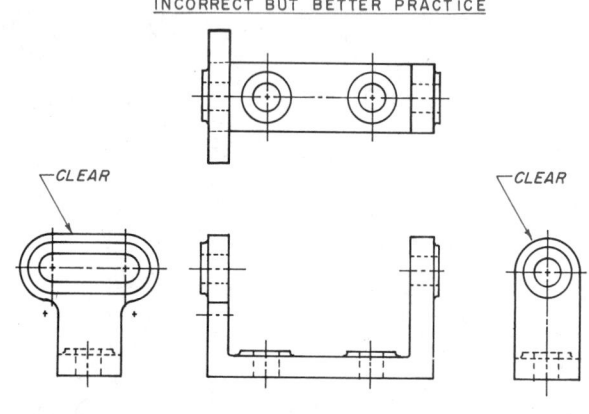

Figure 4-58   Omitting some details—better practice

Incomplete Views

A drawing can be absolutely correct, but confusing as in **Figure 4-57**. The left- and right-side views are correct, but difficult to read due to the overlapping of the details on both ends of the bracket. Omit some of the features in order to make a view clearer and easier to understand, **Figure 4-58**. For example, the right-side view shows only the right side details. Omitting some of the details in a view for clarity is a standard practice used in technical drawing. Sometimes a note is added such as, "only right side shown."

Aligned Features

Show features at their actual distance from an axis or center line, rather than in the projected or foreshortened position. **Figure 4-59A** is an isometric view of a symmetrical part. **Figure 4-59B** indicates how this symmetrical part appears confusing in conventional projection (top illustration), but gives a better indication of the part's symmetry when projected from the aligned position (lower illustration).

Chapter 5 has further information on aligning views for clarity.

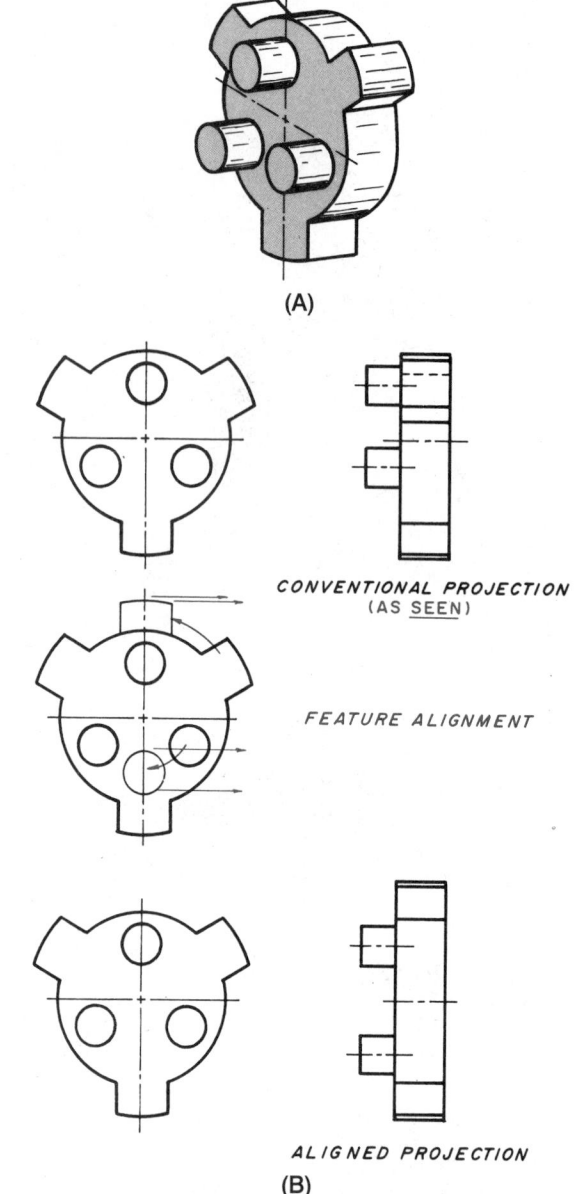

Figure 4-59   (A) Isometric view of a symmetrical part;
(B) Aligned projection

How to Represent Holes

Since products contain various kinds of holes, be able to draw each kind and, therefore, be able to identify different kinds of holes when reading drawings.

Plain Holes

A *through hole* goes completely through an object; a *blind hole* is machined to a specific depth. Through holes and blind holes constitute the majority of plain holes in products, **Figure 4-60**. The drawing at the left in Figure 4-60 shows a top view and a front view of a through hole. Holes are usually drilled. (See Chapter 20.) A drilled hole is not as accurate as a reamed hole. For example, the tolerance for a 1" diameter drilled hole might be ±.002", but ±.0002" for a 1" diameter reamed hole. A hole, specified in a callout to be reamed, is first drilled to a size approximately .015" smaller in diameter

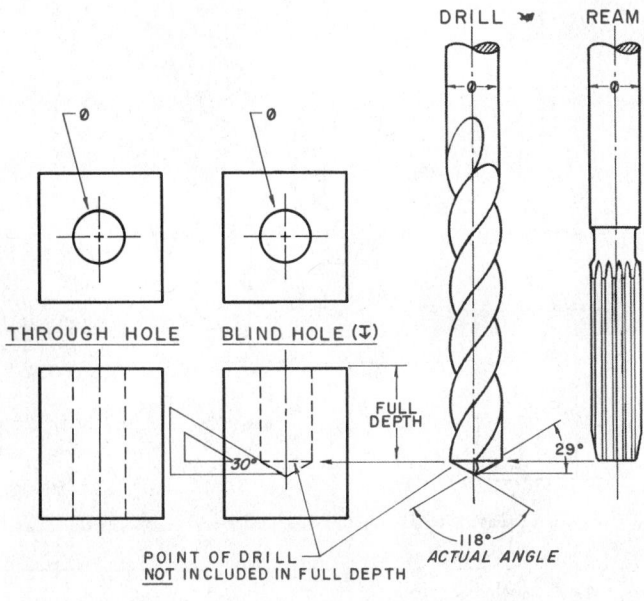

Figure 4-60 Through hole and blind hole

than the size specified, and then reamed to the callout specifications. The drafter, designer, or engineer specifies the reamed hole required; the machinist selects the proper drill.

A blind hole is drilled to a specific depth, Figure 4-60. Twist drills as shown in the figure have a conical point of approximately 118° for general-purpose drilling. The full depth of the hole is called out for the cylindrical portion of the hole, only.

Tapered Holes
A *tapered hole*, **Figure 4-61**, must be drilled first .015" smaller in diameter than the smallest diameter of the tapered hole, and then reamed.

Countersunk Holes
A *countersunk hole* is usually a plain through hole with an upper portion enlarged conically, **Figure 4-62**, most often to 82° to accommodate standard, flat head machine screws. A machinist makes the countersink to a depth that produces a specified diameter slightly larger than the head of the flat head screw.

Counterbored Holes
A *counterbored hole* is usually a plain through hole with an upper portion enlarged cylindrically, **Figure 4-63**, to a specified diameter and depth. A callout would specify the through hole to be slightly larger than the fastener, and the counterbore to be slightly

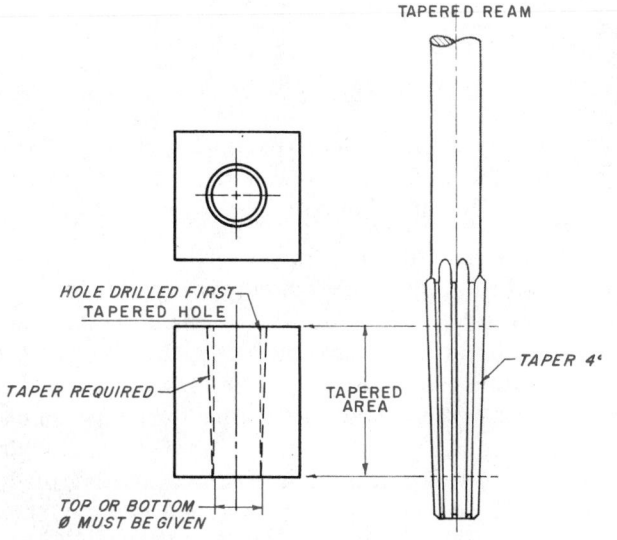

Figure 4-61 Tapered hole

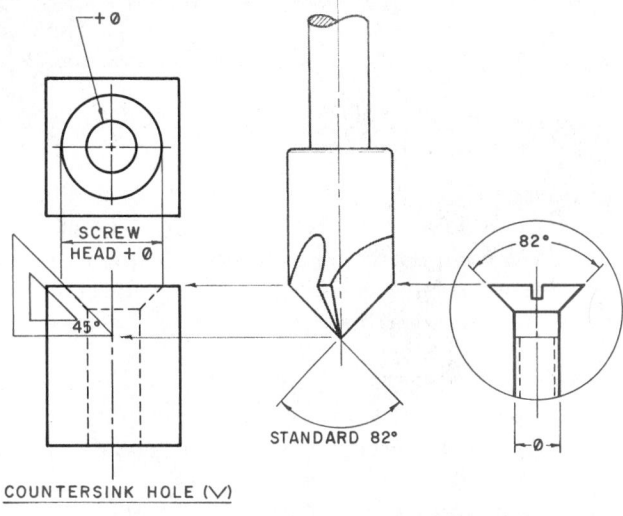

Figure 4-62 Countersunk hole

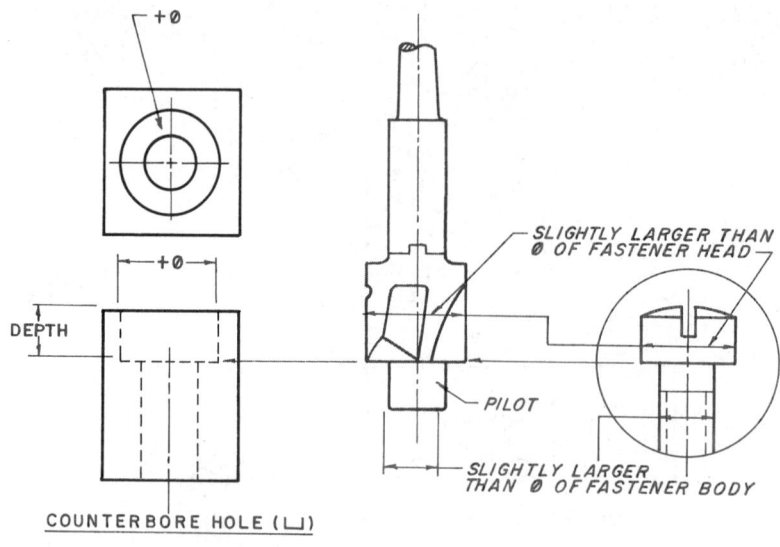

Figure 4-63   Counterbored hole

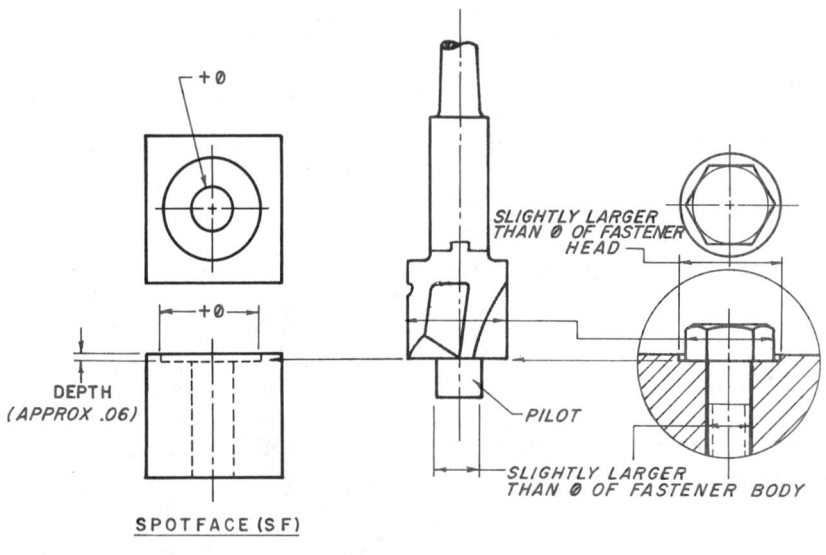

Figure 4-64   Spotfaced hole

larger than the head of the fastener and slightly deeper than the height of the head.

Spotface

A *spotfaced hole* is essentially a plain through hole with a shallow counterbore, *Figure 4-64*, deep enough to provide a flat surface for a head of a bolt, machine screw, or a nut for a threaded member.

Conventional Breaks

Use a *conventional break* for an effective way to draw long objects of constant cross section, such as a pipe, tubing, structural members, or shafting, on a relatively small sheet of drawing paper. A long pipe, drawn to 1/4 size, is shown in *Figure 4-65A*, but redrawn to full size, with a conventional break, in *Figure 4-65B*.

This presents a much clearer understanding of the pipe, or other member, especially if there are holes to be located and dimensioned at either end.

Kinds of Conventional Breaks

Three kinds of breaks are commonly used in technical drawings; the S, the Z and the freehand. Follow the steps in *Figure 4-66* for constructing an S break for a shaft or tube. The S break may be drawn with instruments or done freehand. Use *Figure 4-67A* as a guide for constructing a Z break, usually used for long, wide parts, of constant cross section. The Z part is usually done freehand. The freehand break, *Figure 4-67B*, is only one example of the use of the freehand break line. For example, broken-out sections will be discussed in Chapter 5.

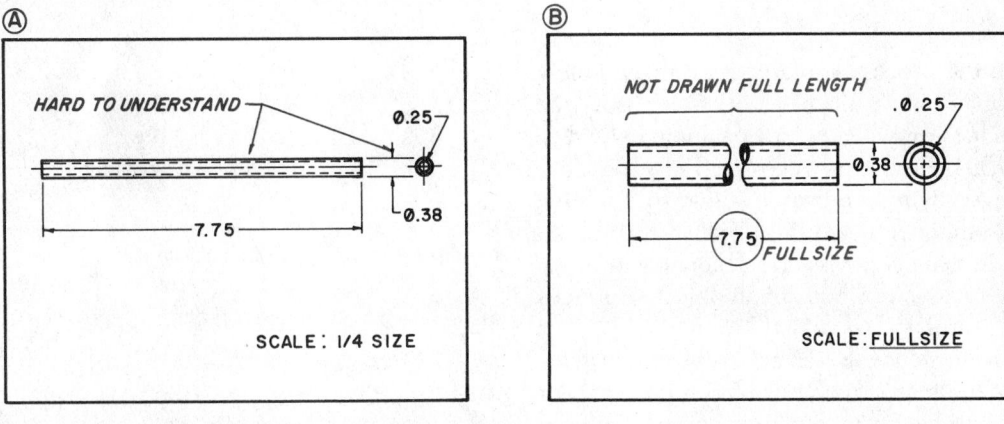

Figure 4-65   Conventional breaks

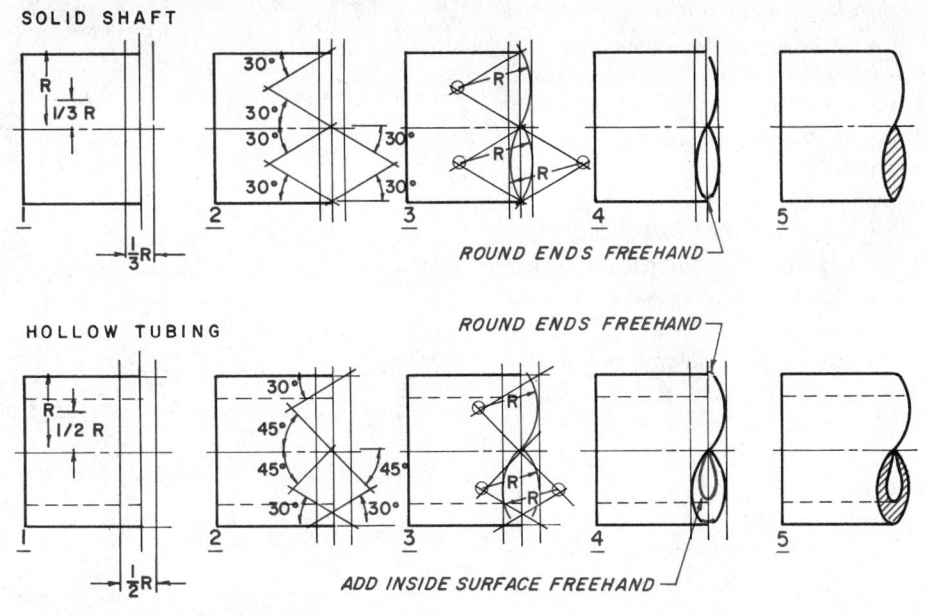

Figure 4-66   "S" break

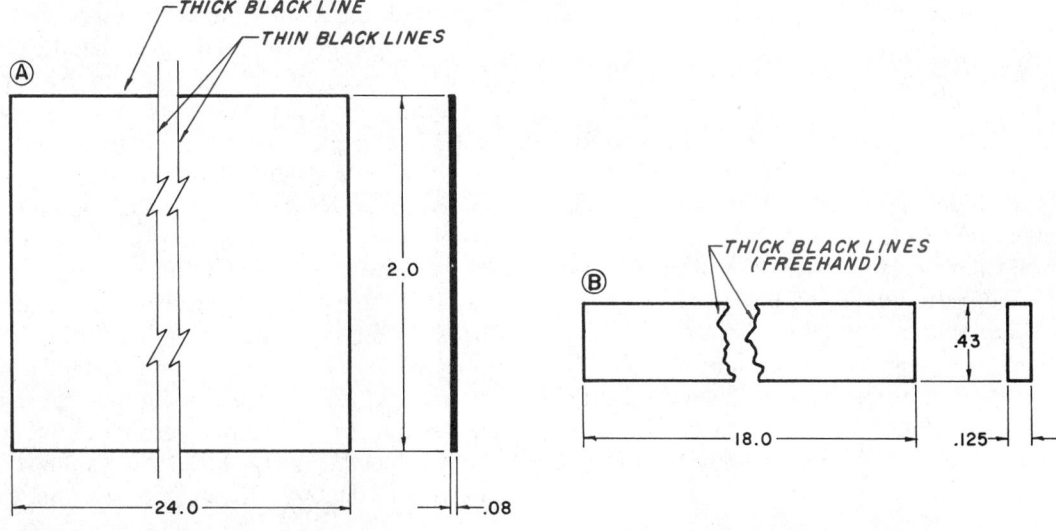

Figure 4-67   "Z" break and freehand break

Visualization

Visualization is the process of recreating a three-dimensional image of an object in a person's mind, using the evidence and clues provided by orthographic drawings. This process has been the focus of much investigation, often used for studying a person's ability to perceive (such as in intelligence tests). The goal of reading an orthographic drawing is to visualize accurately information about the relative positions of an object's surfaces and geometric features.

There is some evidence that an active what-if imagination is a key ingredient in the visualization process. For example, a solid line circle in a view causes a what-if visualizer to seek further evidence. If an adjoining view has a solid line rectangle, bisected by a center line (axis) that aligns with the circle's center, a solid cylinder is indicated, *Figure 4-68*. If the aligned rectangle has hidden lines, a hole is indicated, *Figure 4-69*. A single orthographic view seldom is capable of providing three-dimensional evidence, and the reader must consider two or more views simultaneously in seeking an understanding of feature outlines.

Engineering drawings all follow the same procedures, and none is any more or less difficult to read than another. However, the quantity of geometrical information varies considerably among drawings, and the time required to interpret the information varies accordingly. Visualization practice using simple drawings is required by most individuals to gain experience in mental three-dimensional image construction. Even experienced designers often use modeling clay to form a three-dimensional object from orthographic views. Some build models out of cardboard and glue. Some do quasi-isometric drawings of the overall object, based on the orthographic views, and then systematically try to draw in lines, planes, and solids until all parts of the puzzle fit into a three-dimensional concept. Chapter 18 has excellent guidelines for creating pictorial drawings from orthographic views.

First-Angle Projection

While the United States, Britain, and Canada use the third-angle projection multiview drawing system, most of the rest of the industrial world uses first-angle projection. The main difference between the two systems is the position of the projection plane. Recall the following sentence from the beginning of this chapter; "A projection plane may be thought of as an imaginary pane of glass, with an object underneath it or behind it, Figure 4-2A." That is, it is observer first, then projection plane, and last, the object. For first-angle projection, it is observer first, then the object, and last, the projection plane.

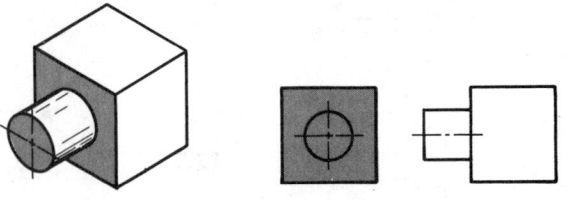

Figure 4-68   Visualization of an object

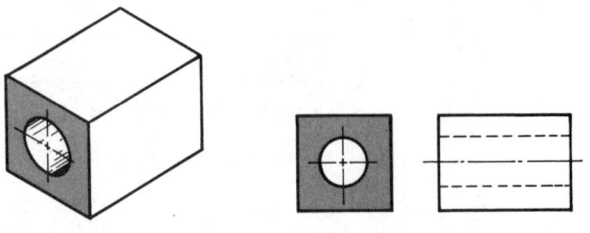

Figure 4-69   Visualization of an object

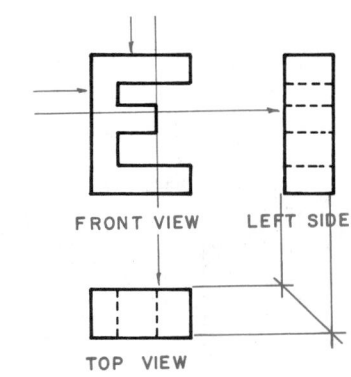

Figure 4-70   First-angle projection

First-angle projection, just as third-angle projection, starts with the front view as the most important view. Refer to *Figure 4-70* and imagine that the observer is at the left of the front view and looking toward the object, the letter "E." What the observer sees, looking at the object, is on the projection plane behind the object, i.e., observer-object-plane. Next, imagine that the observer is above the front view and looking toward the object. What the observer sees is the top view, which is placed below the object on the drawing.

Due to the increase in international exchange of parts and drawings, the projection method used should be indicated on the drawing, with the standard symbol, or the name of the country issuing the drawing. The standard symbol is usually placed on a drawing in or near the title block. *Figure 4-71A* shows the standard symbol used to denote third-angle projection. (See *Figure*

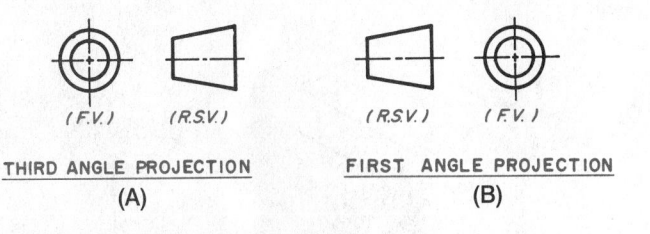

(F.V.)   (R.S.V.)   (R.S.V.)   (F.V.)

THIRD ANGLE PROJECTION   FIRST ANGLE PROJECTION
(A)   (B)

Figure 4-71   (A) Third-angle projection symbol;
(B) First-angle projection symbol

4-71B, first-angle projection.) Note that in Figure 4-71A the observer is assumed to be at the right of the front view (F.V.) and what the observer sees is shown in the right-side view (R.S.V.). In Figure 4-71B the observer is also assumed to be at the right of the object and what the observer sees is shown in a right-side view, but the projection plane is behind the object.

Review

1. Why is a sketch so important in planning a drawing?
2. Explain the following terms (use illustrations if necessary): runout, rounds, and fillets.
3. Explain the difference between a perspective view and a multiview of an object.
4. What is considered the work area on a drawing?
5. Where is the third-angle projection multiview system used? the first-angle?
6. What is the usual depth calloff for a counterbored hole?
7. For what is the lightly constructed 45° angle projection line used?
8. What is used in dimensioning a blind hole?
9. List three important considerations that must be used in positioning the front view.
10. How many views are used to describe an object?
11. To illustrate a countersunk hole, what angle is used to draw the countersunk portion of the hole?
12. When and why should a drawing be lightly numbered?
13. List three major kinds of conventional breaks and note why each is used.
14. When is it permissible to use an incomplete view?
15. Explain in full the mathematical procedure used to center a typical three-view drawing with a 1" (25 mm) space between views within the work area. Use a sketch.
16. What are the more frequent uses of center lines?
17. What is the difference between an inclined plane and an oblique plane?
18. Describe two ways of drawing cylindrical intersections.

Chapter Four Problems

The following problems are intended to give the beginning drafter practice in visualizing the multiview system, practice in choosing and sketching the required view, practice is using drafting instruments and using correct line thickness. As these are beginning problems, no dimensions will be used at this time.

The steps to follow in laying out all drawings throughout this book are:

Step 1 Study the problem carefully.

Step 2 Choose the view with the most detail as the front view.

Step 3 Position the front view so that there will be the least number of hidden lines in the other views.

Step 4 Make a sketch of all required views.

Step 5 Center the required views within the work area with a 1-inch (25 mm) space between each view.

Step 6 Use light projection lines. Do not erase them.

Step 7 Lightly complete all views.

Step 8 Check to see that all views are centered within the work area.

Step 9 Check to see that there is a 1-inch (25 mm) space between all views.

Step 10 Carefully check all dimensions in all views.

Step 11 Darken in all views using correct line thickness.

Step 12 Recheck all work, and, if correct, neatly fill out the title block using light guidelines and neat lettering.

Problems 4-1 through 4-11

Construct a 3-view drawing of each object, using the listed steps.

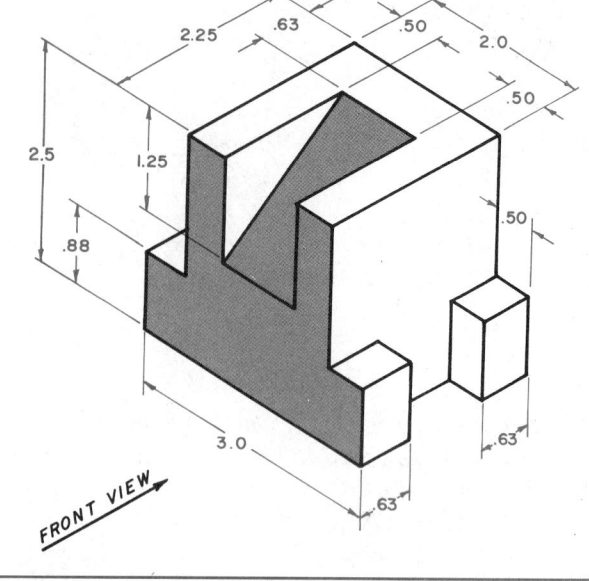

Problem 4-2

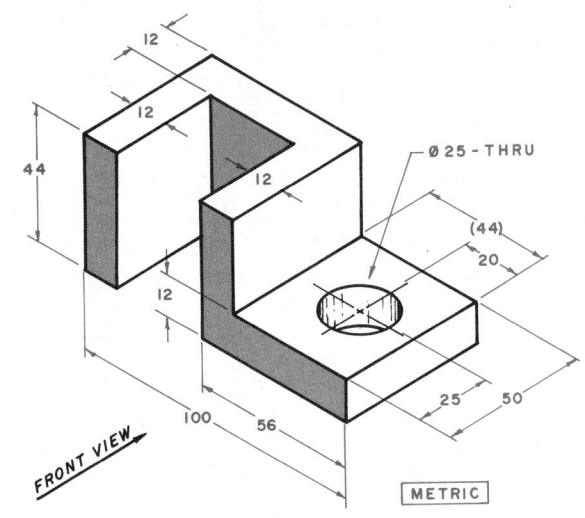

Problem 4-3

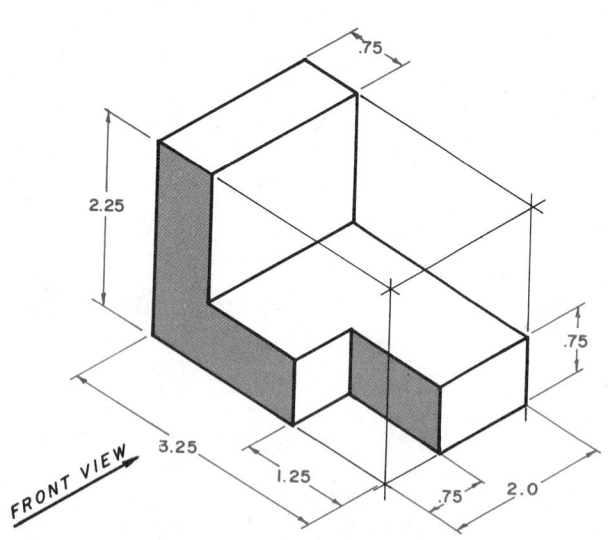

Problem 4-1

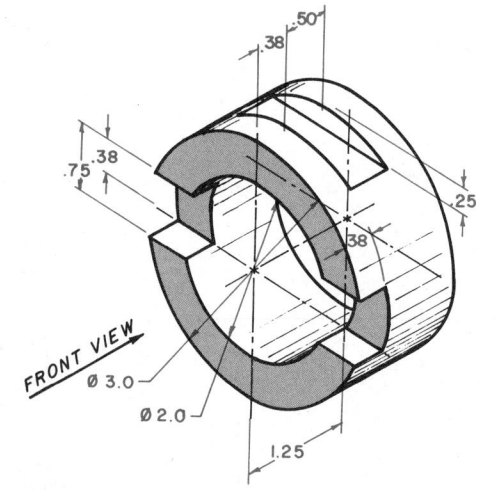

Problem 4-4

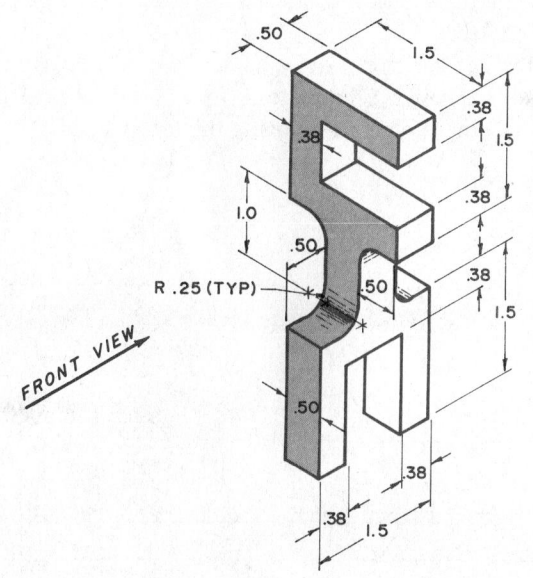

.50
1.5
.38
.38
1.5
1.0
.38
.50
.50
R .25 (TYP)
.38
1.5
.50
FRONT VIEW
.50
.38
.38
1.5

Problem 4-5

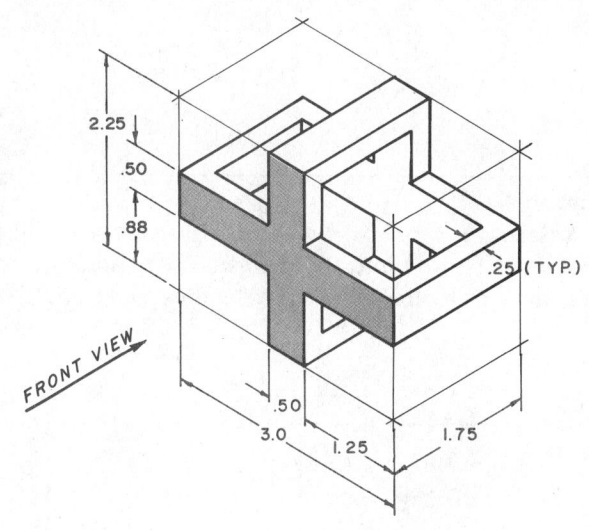

.62
2.0
2.5
.25
1.0
1.5
Ø .25 ⊤ THRU
⊔ Ø .38 ⊤ .31
.31
.50
.06
.12
Ø .88 THRU
.25
1.5
.31
FRONT VIEW
3.0
2 X Ø .25 THRU

Problem 4-8

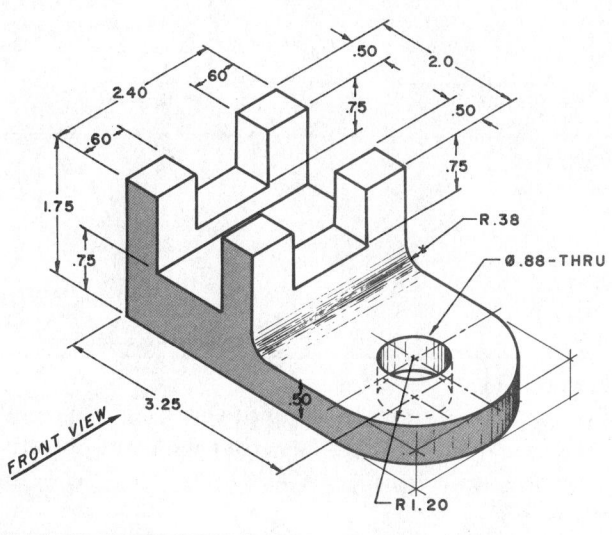

.50
2.0
2.40
.60
.75
.50
.60
.75
R .38
1.75
Ø .88–THRU
.75
3.25
.50
FRONT VIEW
R 1.20

Problem 4-6

2.25
.50
.88
.25 (TYP.)
FRONT VIEW
.50
3.0
1.25
1.75

Problem 4-9

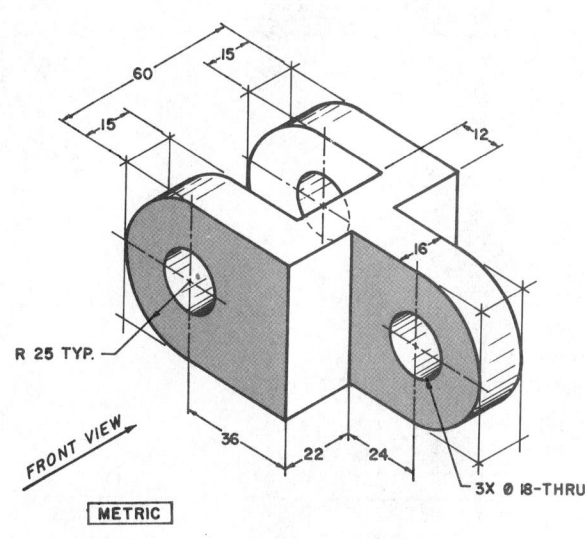

60
15
15
12
16
R 25 TYP.
FRONT VIEW
36
22
24
3X Ø 18-THRU
METRIC

Problem 4-7

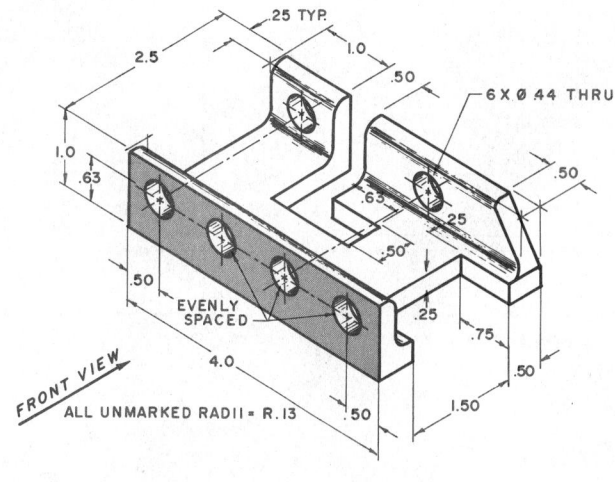

25 TYP.
1.0
2.5
.50
6 X Ø 44 THRU
1.0
.63
.50
.63
.25
.50
.50
.25
EVENLY
SPACED
.75
4.0
.50
.50
1.50
FRONT VIEW
ALL UNMARKED RADII = R .13

Problem 4-10

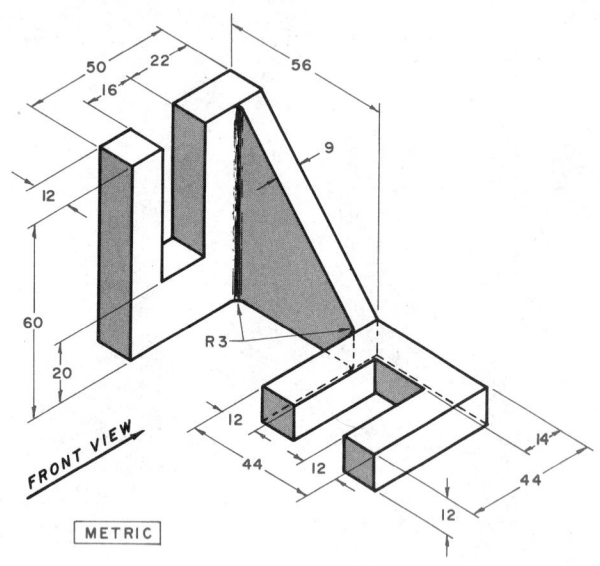

Problem 4-11

Problem 4-12

Construct a 3-view drawing of this object. Project points 1, 2 and 3 into the right-side view and into the top view. Complete all views using the listed steps.

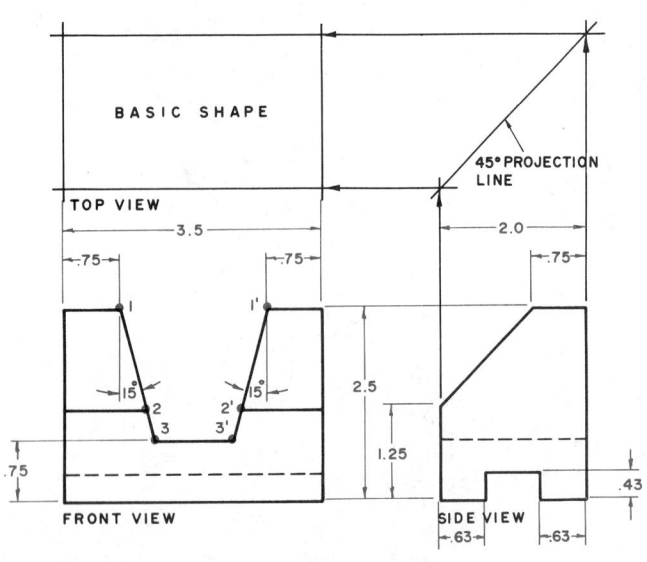

Problem 4-12

Problem 4-13

Construct a 3-view drawing of this object. Project points 1 through 8 into the top view and into the front view. Complete all views using the listed steps.

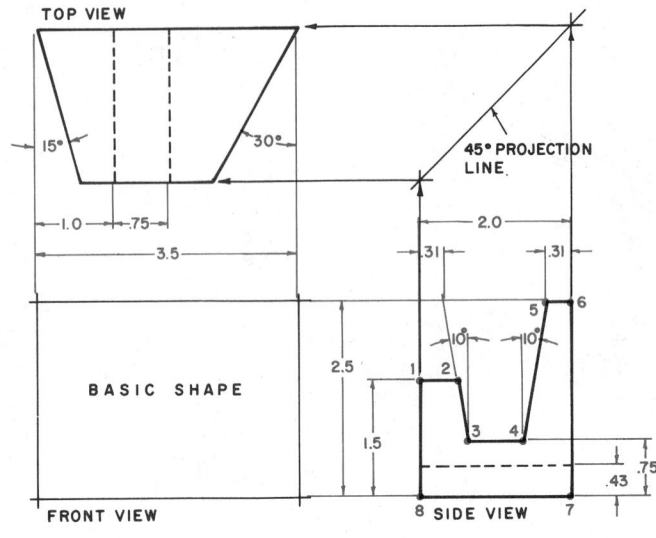

Problem 4-13

Problem 4-14

Construct a 3-view drawing of this object. Project points 1 through 6 into the front view and into the right-side view. Complete all views using the listed steps.

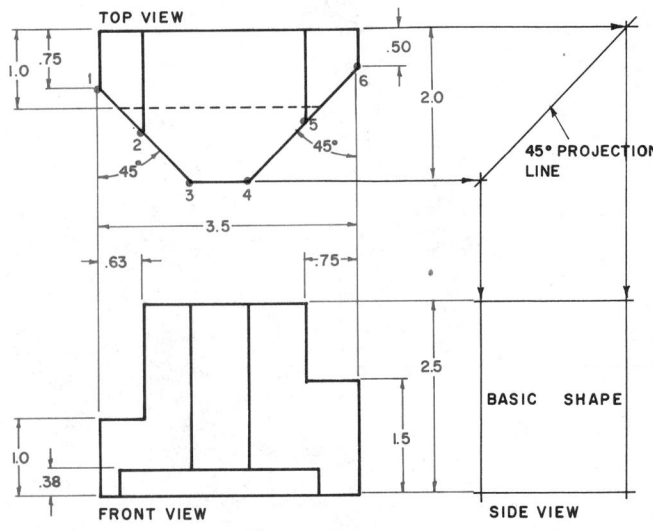

Problem 4-14

Problem 4-15

Construct a 3-view drawing of this object. Project points 1 through 7 into the right-side view and into the front view. Complete all views using the listed steps.

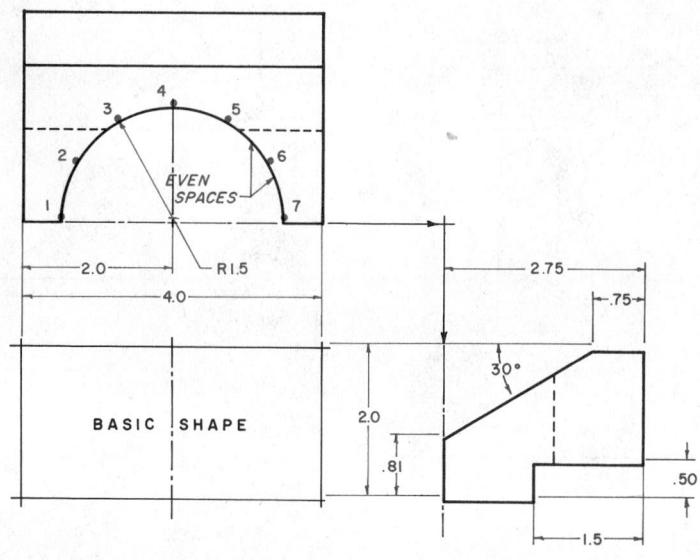

Problem 4-15

Problem 4-16

Construct a 3-view drawing of this object. Project points 1 through 7 into the right-side view and into the top view. Complete all views using the listed steps.

Problem 4-17

Construct a 3-view drawing of this object. Locate and number 12 points around the 1.5 diameter hole. Project points 1 through 12 into the front view and into the right-side view. Complete all views using the listed steps.

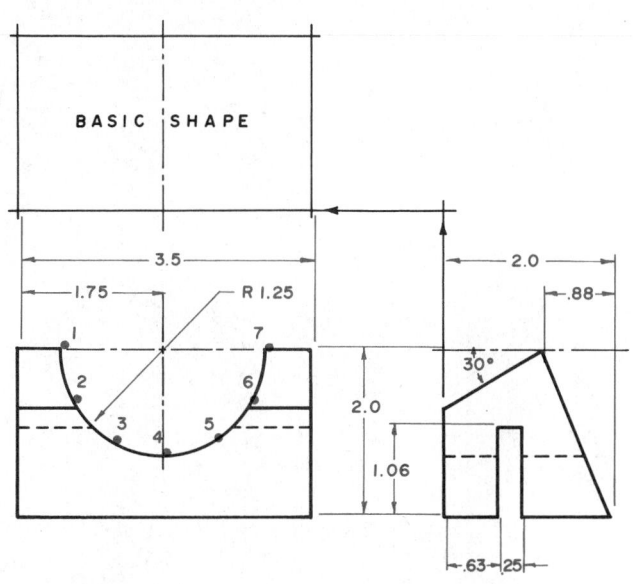

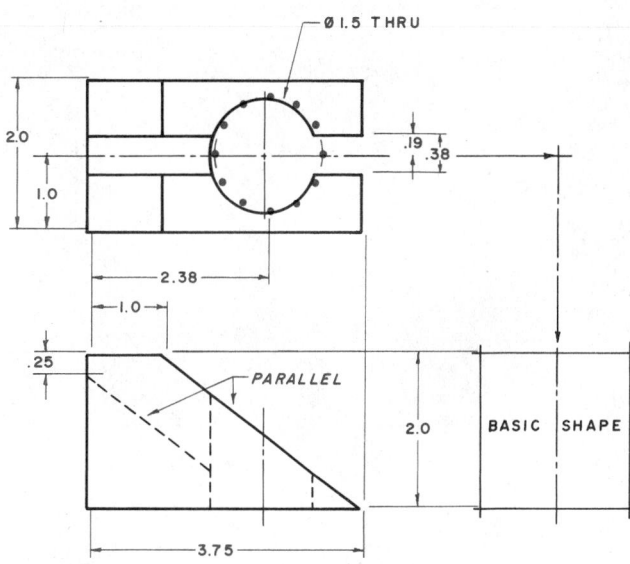

Problem 4-16

Problem 4-17

Problems 4-18 through 4-25
Construct a 3-view drawing of each object, using the listed steps.

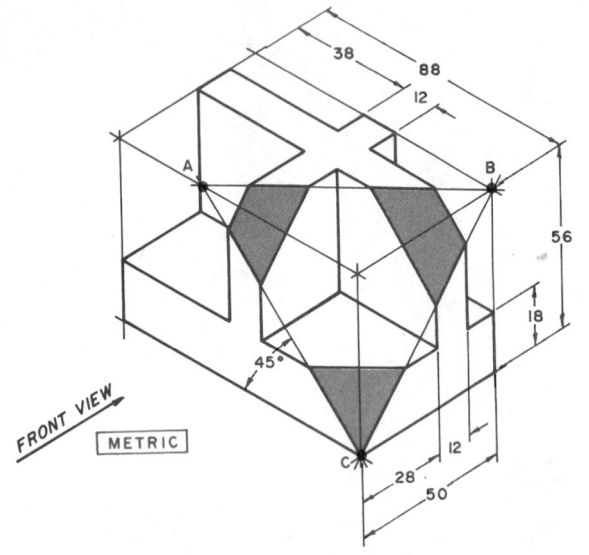

Problem 4-18

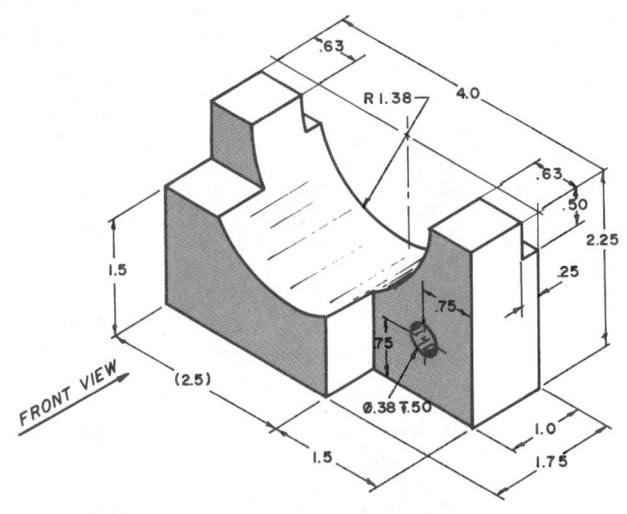

Problem 4-20

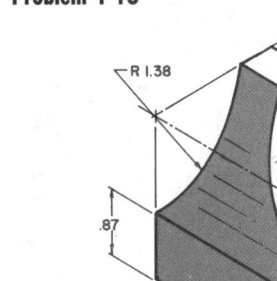

Problem 4-19

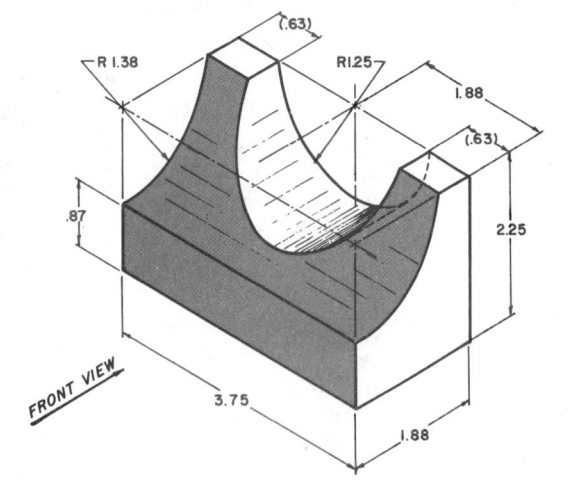

Problem 4-21

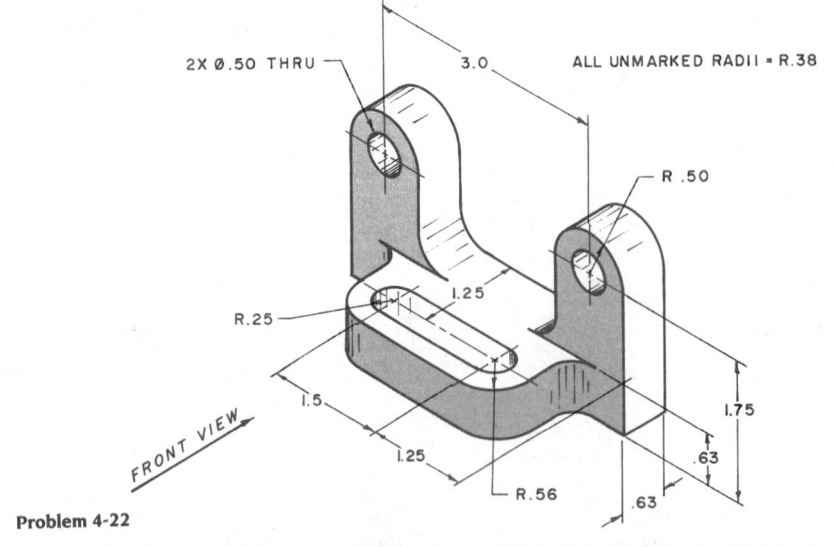

Problem 4-22

Problem 4-22

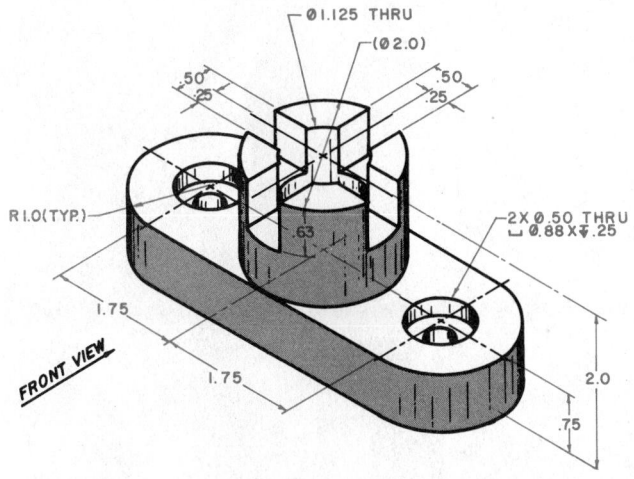

Problem 4-23

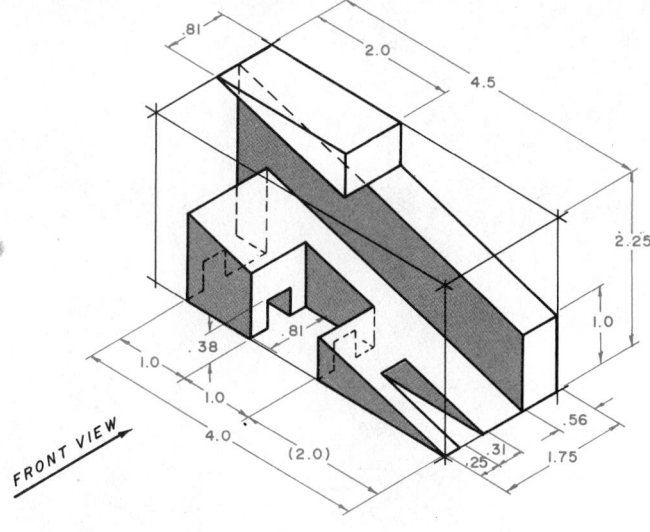

Problem 4-24

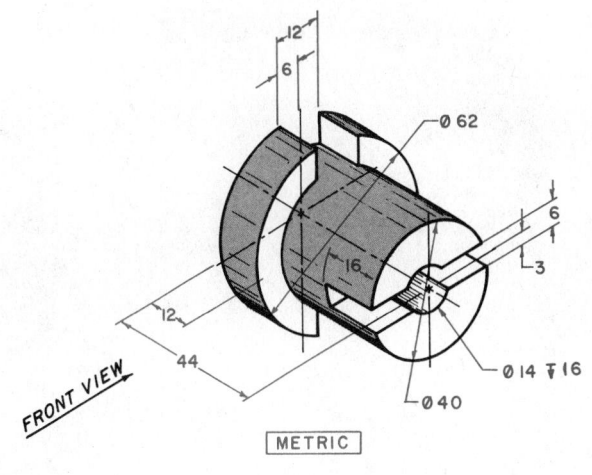

Problem 4-25

Problem 4-26
Construct a 4-view drawing of this object (front, top, right, and left-side views). Use the listed steps.

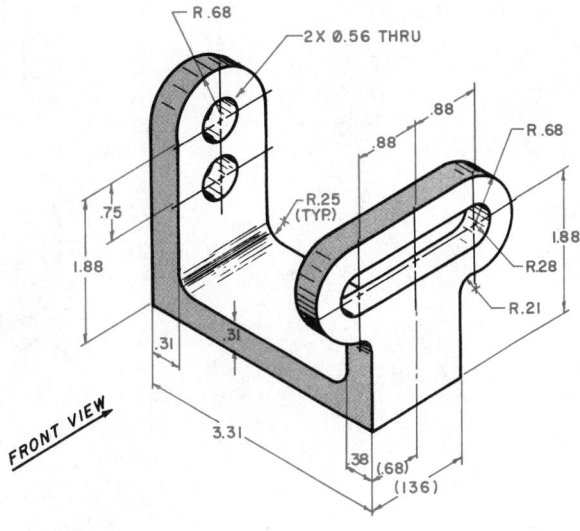

Problem 4-26

Problems 4-27 through 4-36
Construct a 3-view drawing of each object, using the listed steps.

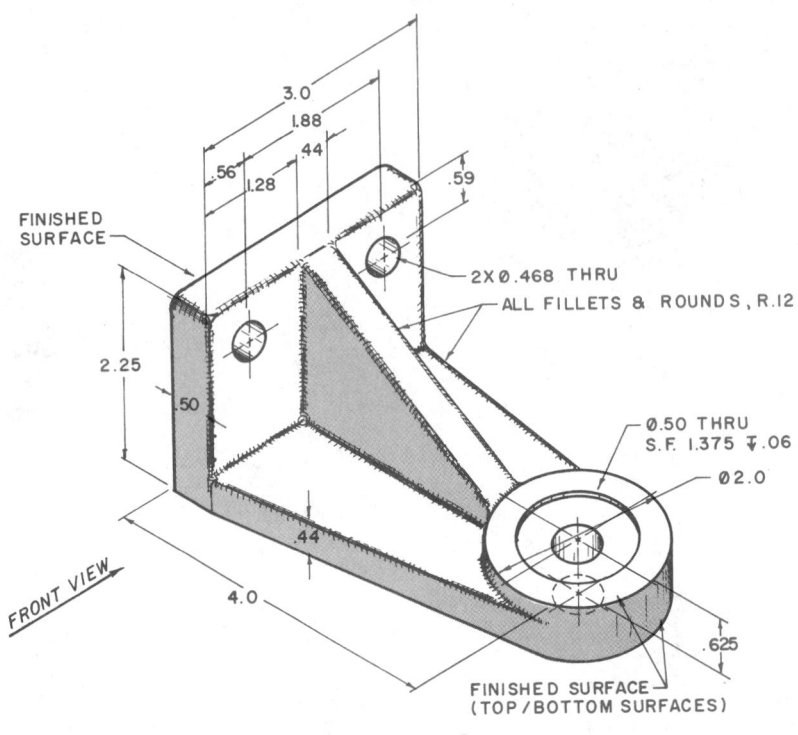

FINISHED SURFACE

3.0
1.88
.44
.56
1.28
.59

2.25

.50

2X Ø.468 THRU

ALL FILLETS & ROUNDS, R.12

Ø.50 THRU
S.F. 1.375 ▼.06

Ø2.0

FRONT VIEW

.44

4.0

.625

FINISHED SURFACE
(TOP/BOTTOM SURFACES)

Problem 4-27

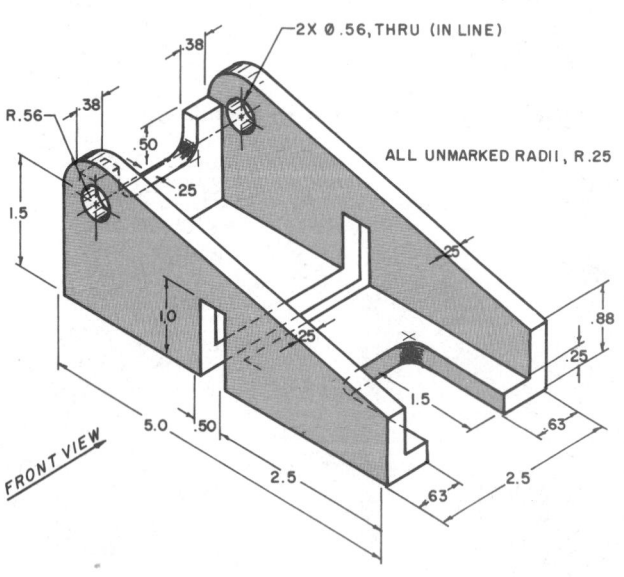

R.56

.38
.38
.50
.25

1.5

1.0

5.0
2X Ø.56, THRU (IN LINE)

ALL UNMARKED RADII, R.25

.25

.25

.88
.25

.50

1.5

FRONT VIEW

.63

.63

2.5

2.5

Problem 4-28

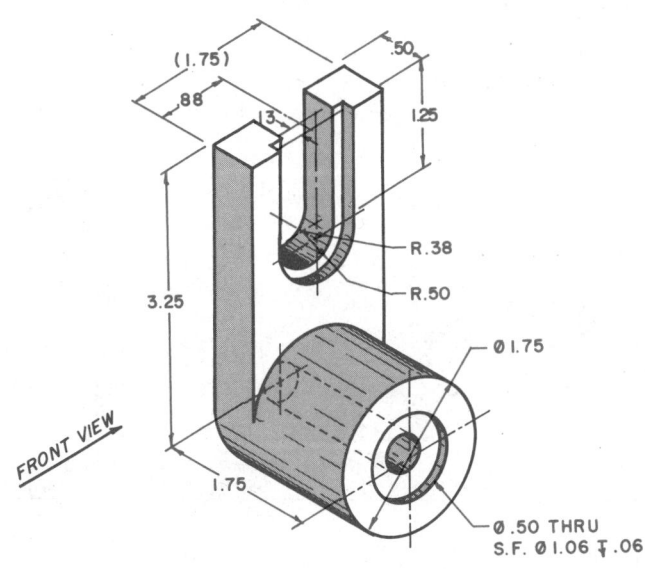

(1.75)

.88
.13

.50

1.25

3.25

R.38

R.50

Ø1.75

FRONT VIEW

1.75

Ø.50 THRU
S.F. Ø1.06 ▼.06

Problem 4-29

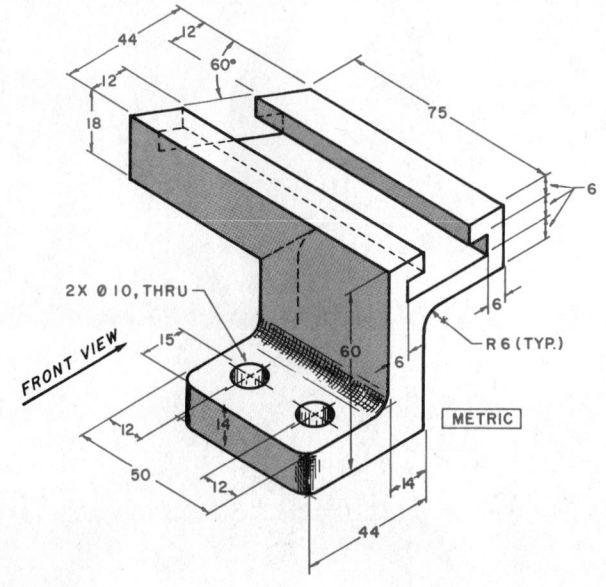

Problem 4-30

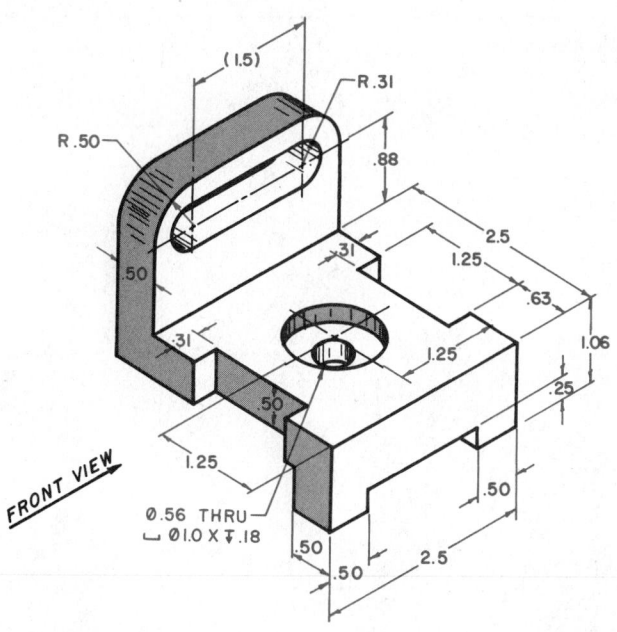

Problem 4-31

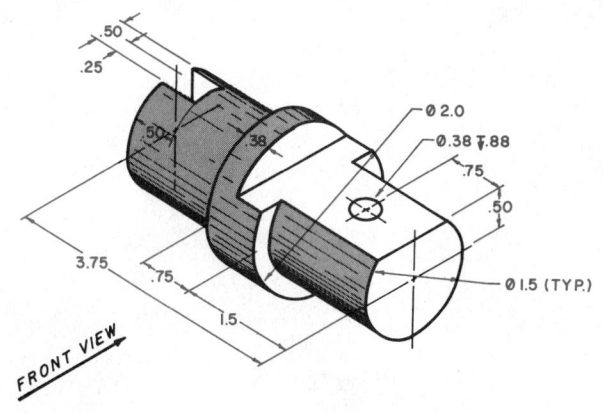

Problem 4-32

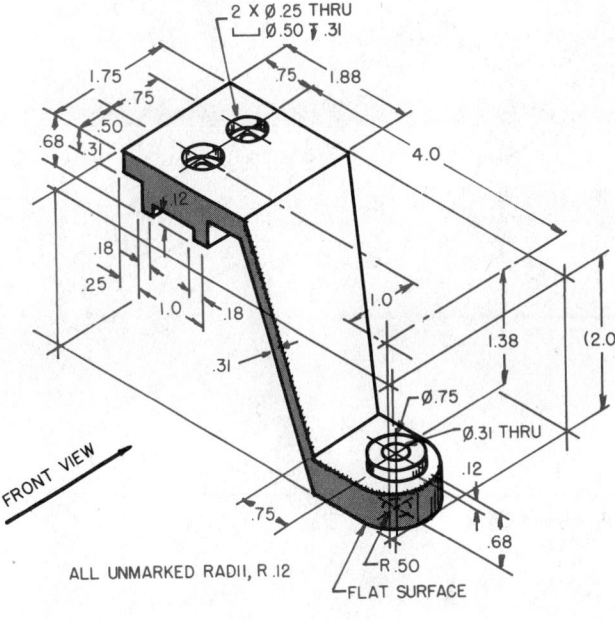

Problem 4-33

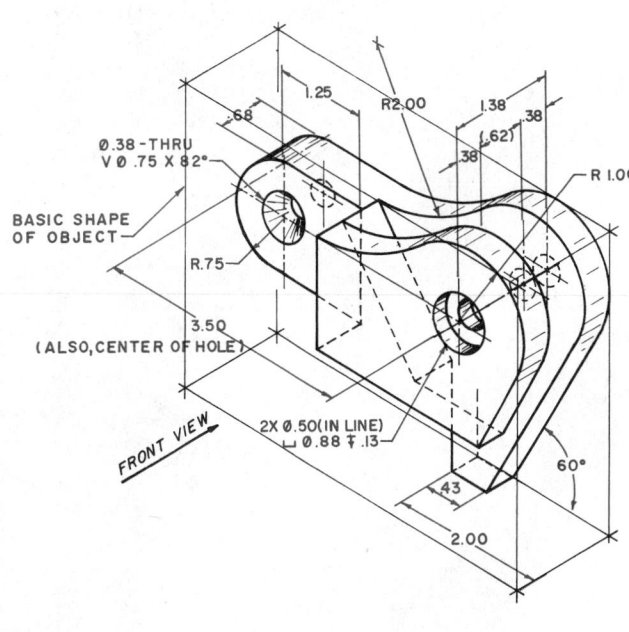

Problem 4-34

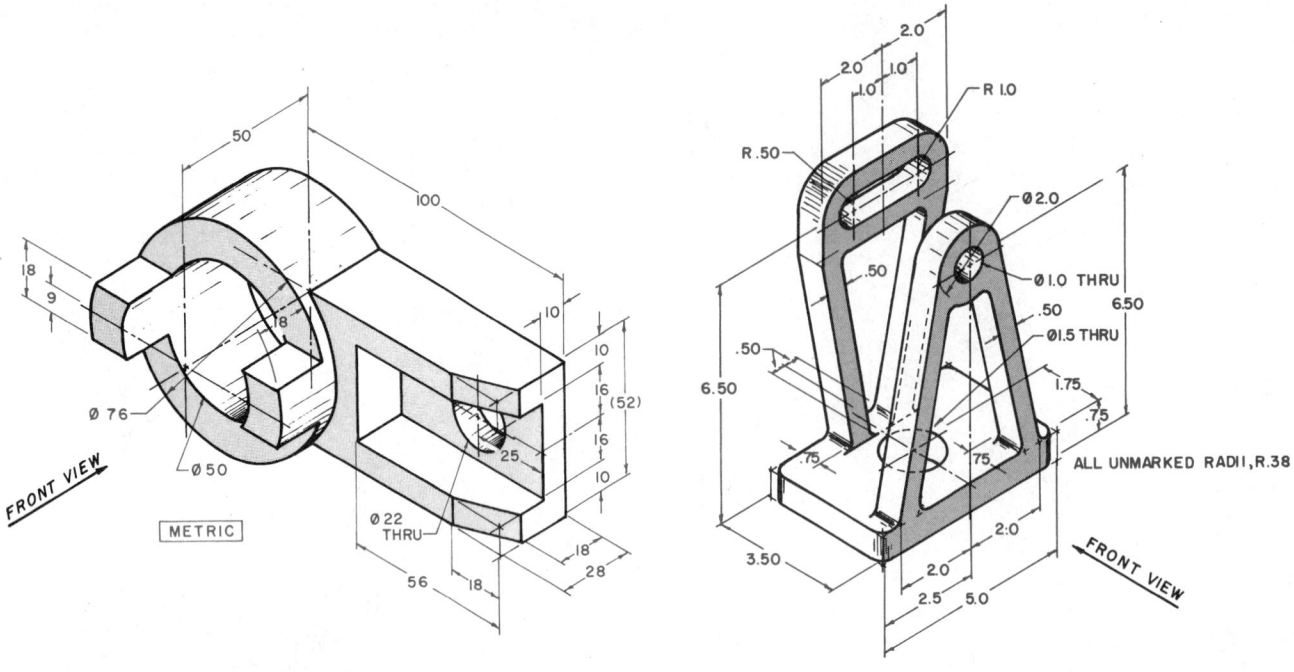

Problem 4-35

Problem 4-36

Problems 4-37 through 4-50

Choose the front view and construct a 3-view drawing of this object, using the listed steps.

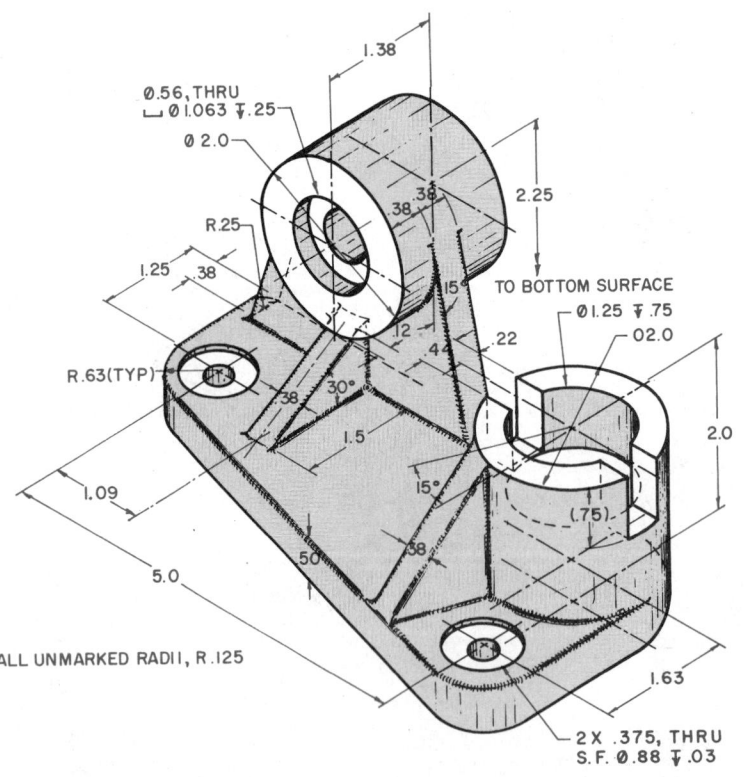

Problem 4-37

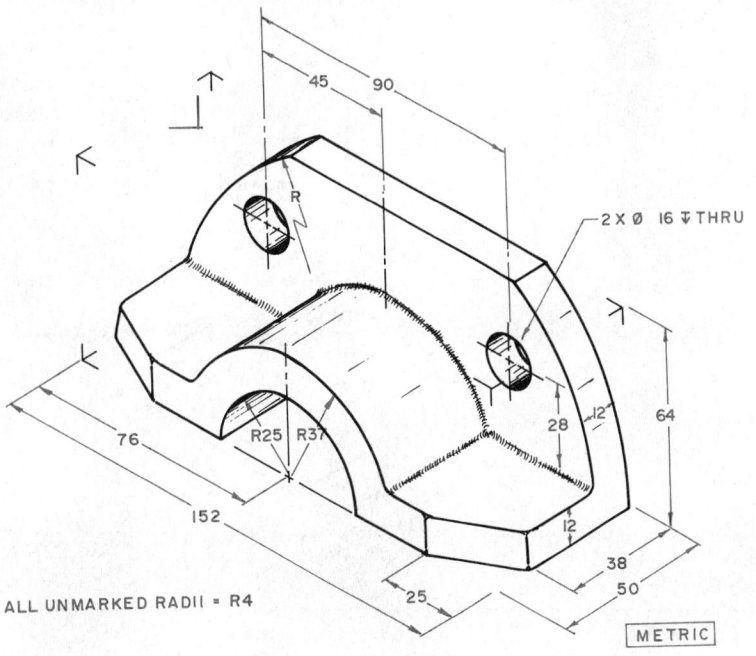

Problem 4-38

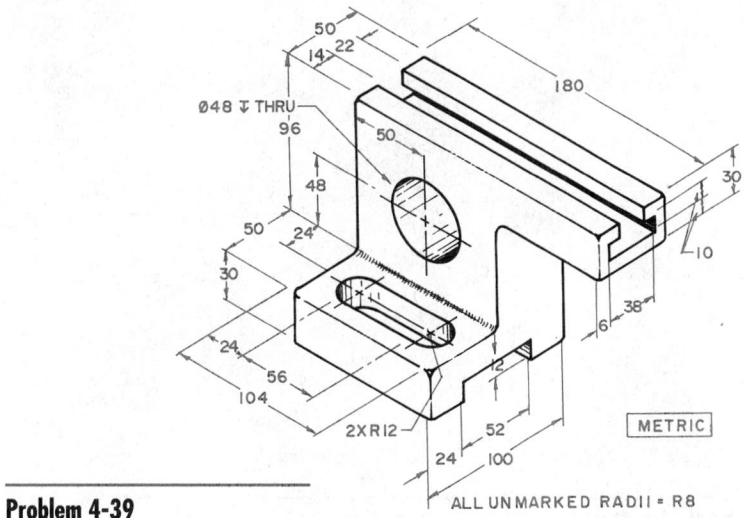

Problem 4-39

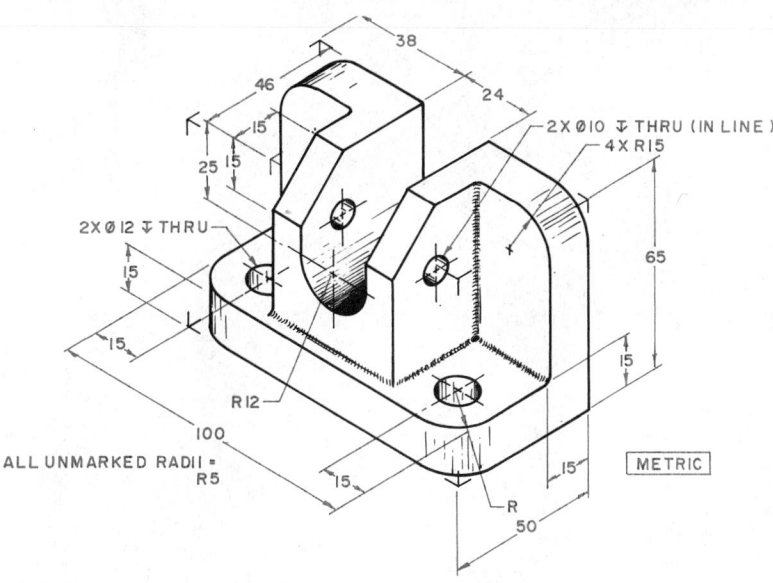

Problem 4-40

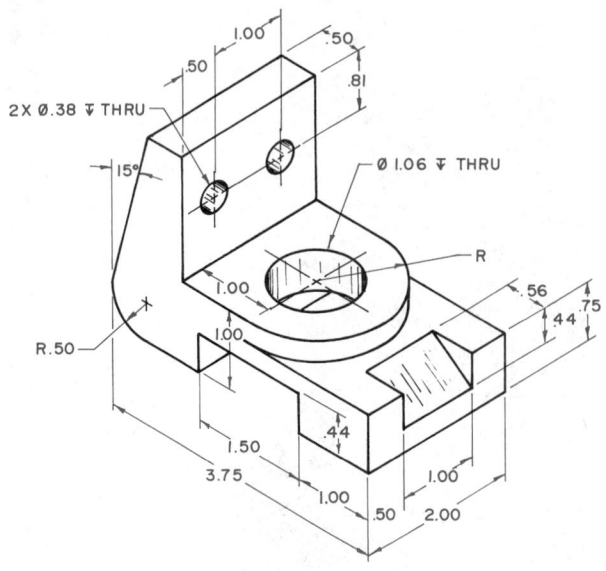

Problem 4-41

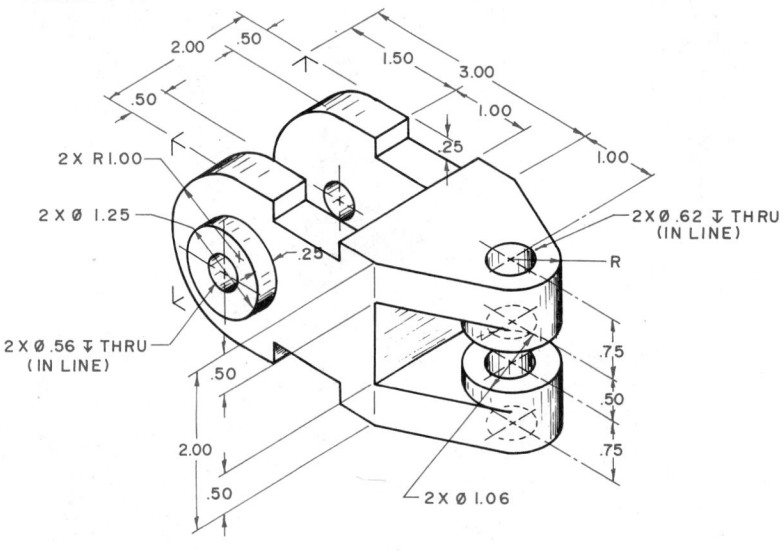

Problem 4-42

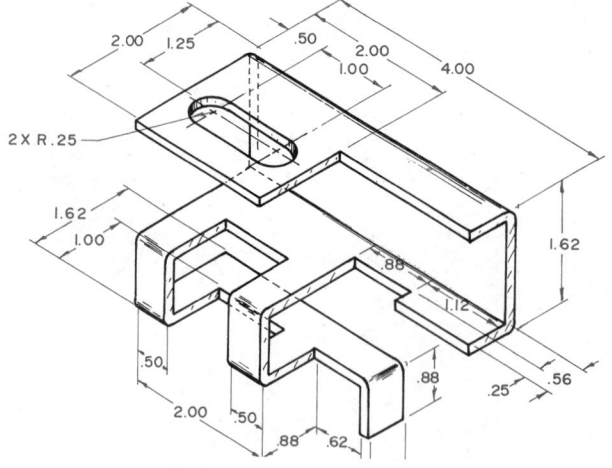

Material Thickness .0125
Internal Radii .0125

Problem 4-43

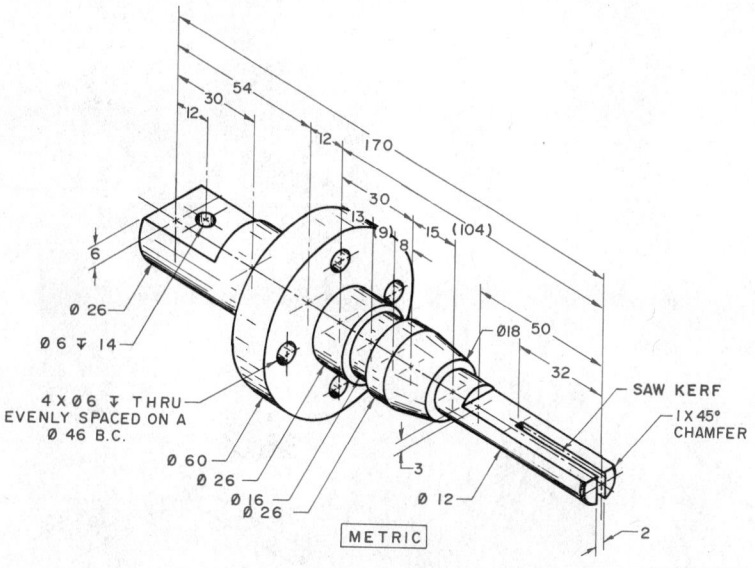

Problem 4-44

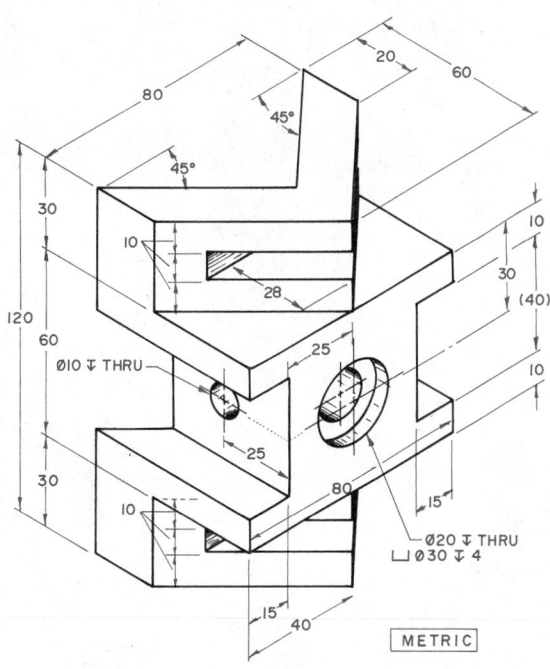

Problem 4-45

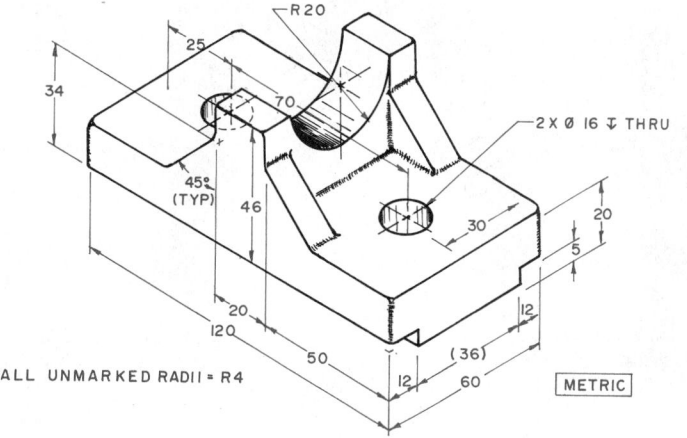

Problem 4-46

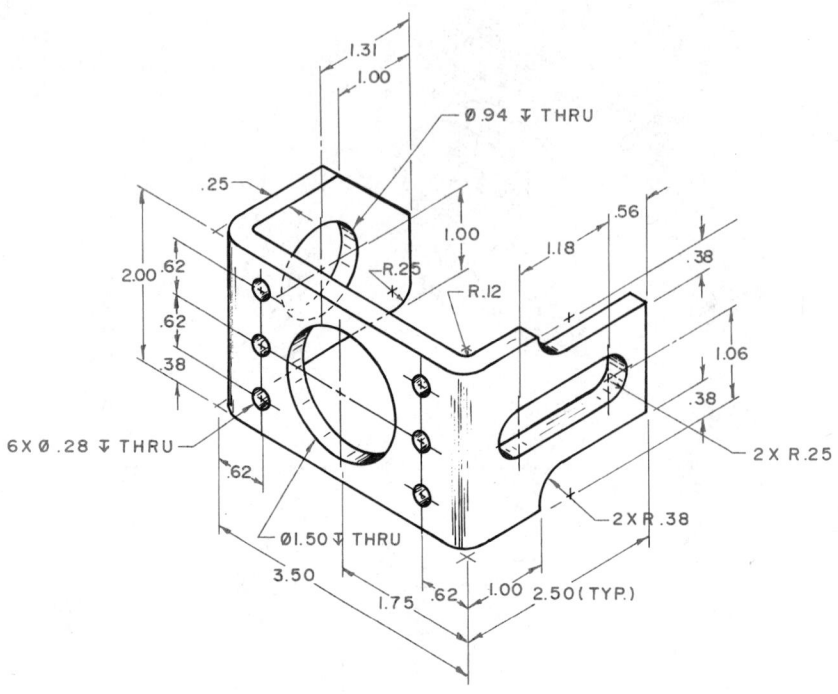

Problem 4-47

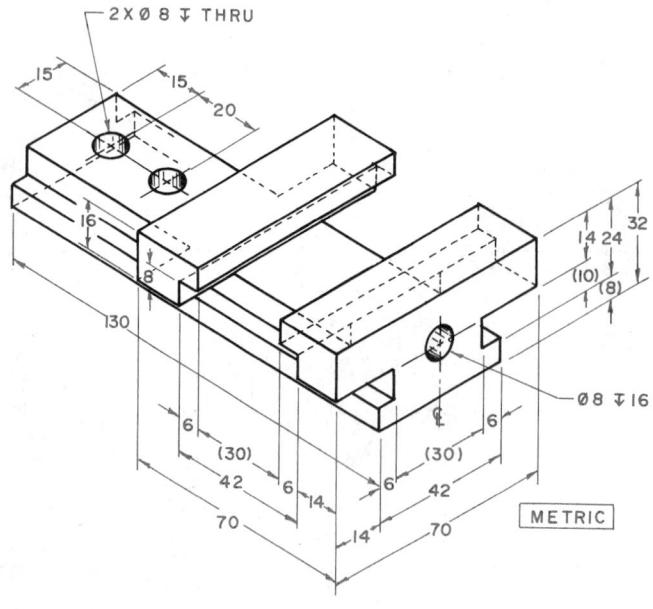

Problem 4-48

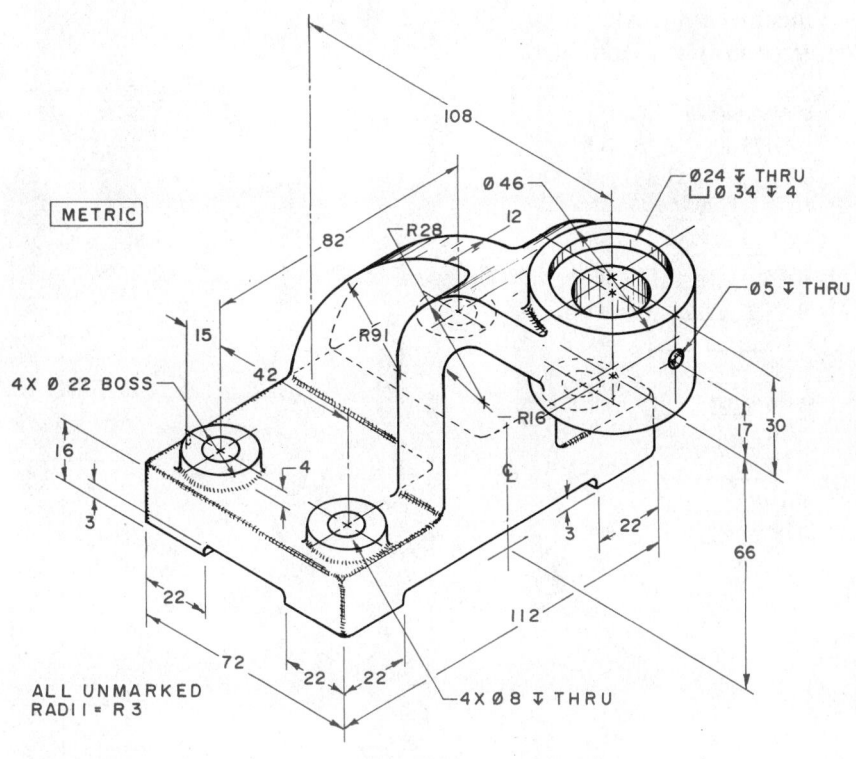

METRIC

108

Ø 46

Ø24 ⊥ THRU
⊔ Ø 34 ⊥ 4

82

R28

12

Ø5 ⊥ THRU

15

R91

42

4X Ø 22 BOSS

R16

16

4

C

17 30

3

3   22

22

66

22

112

72   22   22

ALL UNMARKED
RADII = R 3

4X Ø8 ⊥ THRU

Problem 4-49

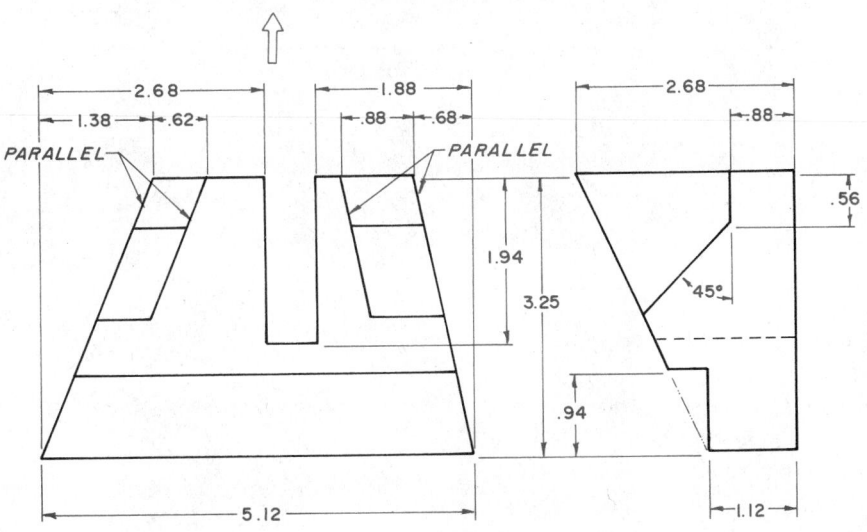

2.68

1.88

2.68

1.38   .62

.88   .68

.88

PARALLEL

PARALLEL

.56

1.94

3.25

45°

.94

5.12

1.12

Problem 4-50

❏ **Problem 4-51**

For practice in reading multiview drawings that have already been drawn, study the pattern of the sample responses related to the numbered multiview drawing, and then complete the surface analyses as indicated.

Sample Responses

1. Surface	25-26	in R is 5-6-10-9	in T
2. Surface	16-17	in F is 29-30	in R
3. Surface	16-17	in F is 2-3-6-10-11-7	in T

Complete the Responses

1. Surface	13-14-15-16	in F is _____	in T
2. Surface	24-25-26-27-28	in R is _____	in T
3. Surface	29-25-26-27-30	in R is _____	in T
4. Surface	8-7-12-11	in T is _____	in F
5. Surface	13-14	in F is _____	in T
6. Surface	29-25-24	in R is _____	in F
7. Surface	11-7-2	in T is _____	in F
8. Surface	11-7-2	in T is _____	in R
9. Surface	32-30-27-28	in R is _____	in T
10. Surface	6-5-9-10	in T is _____	in F

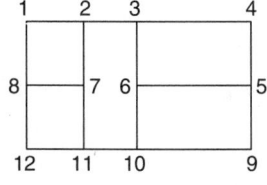

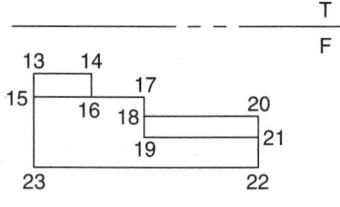 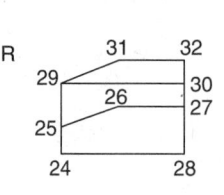

Problem 4-51

❏ **Problem 4-53**

The steel end-cover shown has eight through holes 9/16" diameter for 1/2" diameter fillister head cap screws. (See Appendix.) Change the callout for the holes, to be counterbored. (Refer to Chapter 9 for callouts.) The heads of the cap screws will not protrude above the surface of the cover. (This change is for safety reasons; some counterbored holes are specified for aesthetic reasons.) Prepare a two-view drawing of the cover, with the new callout and the dimensions. Show one hole in cross section, in a separate view.

❏ **Problem 4-52**

Six possible T views, all correctly drawn, are shown for the one F view. Draw an A view for each one of the T views.

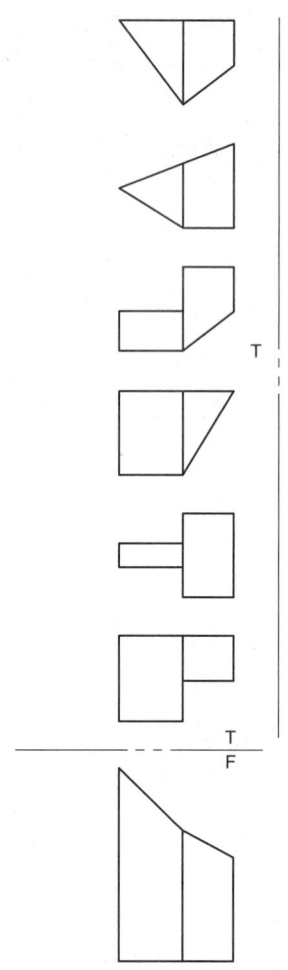

Problem 4-52

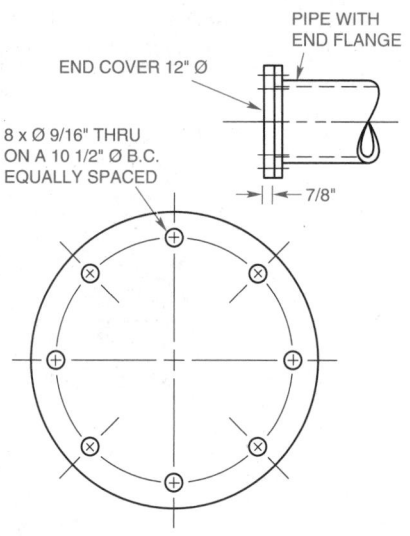

PIPE WITH
END FLANGE

END COVER 12" Ø

8 x Ø 9/16" THRU
ON A 10 1/2" Ø B.C.
EQUALLY SPACED

7/8"

Problem 4-53

❏ Problem 4-54

Visit a lab or shop listed, and do a field sketch (rough sketch) of a piece of apparatus (not too complicated). Later, use the sketch to prepare a multiview drawing of the apparatus. Show approximate dimensions and note general materials.

a. Chemistry lab
b. Physics lab
c. Metal shop
d. Wood shop
e. Electrical/Electronics shop
f. Computer lab

❏ Problem 4-55

Visit a track and field area and do a field sketch (rough sketch) of one of the following. Later, use the sketch to prepare a multiview drawing of the item sketched. Give approximate dimensions and note general materials.

a. Hurdle
b. Starting block
c. Javelin
d. High-jump post
e. Horse
f. Springboard (used with a horse)
g. Parallel bars
h. Other

❏ Problem 4-56

Visit a weight room and do a field sketch (rough sketch) of a portion of a complicated assembly or all of a relatively simple assembly. Later, use the sketch to prepare a multiview drawing of the item sketched. Give approximate dimensions and note general materials.

❏ Problem 4-57

Do a conceptual design of a contemporary desk lamp for a student's desk. That is, do the general outline of the lamp only, not the insides. Prepare a multiview drawing of the lamp and indicate approximate sizes.

❏ Problem 4-58

Design a ladder for use around a typical home. Use wood and fasten the rungs to the side rails with glue, unless otherwise directed. Prepare a multiview drawing of the ladder, to scale, with approximate dimensions. (A Z break may be useful here.)

❏ Problem 4-59

Utilize the following anthropometric data and design one of the objects listed. Use the dimensions given or use your own body dimensions as a guide. Prepare a multiview drawing of the design, with approximate dimensions and general materials.

A door (standard height 6'8" to 7'0"; door knob or handle, 38" above floor.)
A working counter for the center of a kitchen (standard height 36" to 40")
A bar stool (standard height 27" to 30")
A simple work bench (standard height 36")
A computer workplace (25" to 26" knee clearance)

❏ Problem 4-60

Design a simple, metal grill for use over an open fire, collapsible and small enough to fit in a backpack. Prepare a multiview drawing of the grill. Include approximate dimensions.

❏ Problem 4-61

Design, for a junior high school science class, an inexpensive, easy to assemble, portable, demonstration apparatus for illustrating the change of supporting force, F, (in pounds) in the diagonal member BC as the angle θ is changed, while the weight W is constant. Refer to the frame outline shown. Prepare a multiview drawing of the design, with approximate dimensions and note the materials used. (A scale may need to be designed, for example, use a spring with a pointer attached, calibrated for different forces pulling on the spring.)

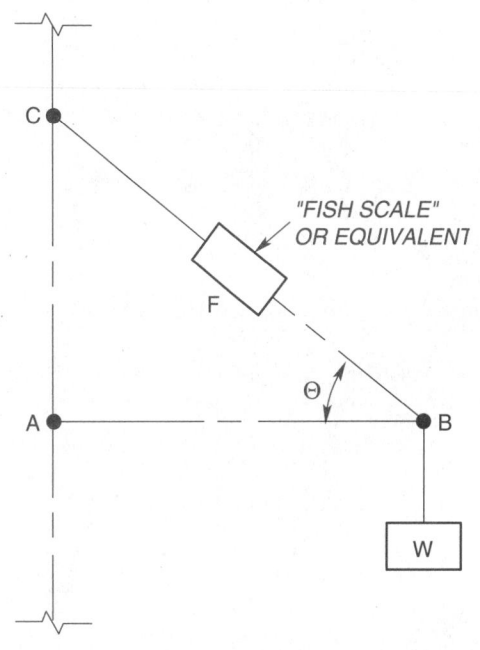

Problem 4-61

❏ Problem 4-62

Design an easy-to-assemble apparatus for a junior high school science class. It should be portable and inexpensive to demonstrate two facets of a pendulum. First, demonstrate how the period (a complete cycle back and forth) of a pendulum changes with a change in length. Second, demonstrate how the period doesn't change with a change in weight for any one length. Prepare a multiview drawing of the design, indicate approximate dimensions, and note the materials in general. The equation for the period T of a pendulum is T = 2π x square root of (length/g). The g in the equation is the gravity constant. If the length is in meters and the gravity constant equals 9.81 meters/second squared, then the period will be in seconds. For example, a swing (a type of pendulum) with a length of 12 meters and your best friend in the swing, would have a period of approximately 6.9 seconds. (This would be about a 40′ long swing.)

❏ Problem 4-63

The drawing shown is for a special aluminum, circular sealing plate, used on an experimental apparatus. The plate has a groove for an O-ring seal and threaded holes for a pressure transducer. A similar plate is needed for another project, but the outside diameter is to be changed to 200 mm, and the threaded holes on the 41-mm bolt circle are to be drilled and tapped for a metric fastener M8 x 1.25. The depth of the threaded portion of holes varies from 1 times the diameter of the fastener if the material is steel, to 2 times the fastener diameter for aluminum. Prepare a similar drawing for the circular plate and use your own judgment for the changes in the dimensions. Note that the title block has been omitted for convenience; however, the third-angle projection symbol is shown. Note: Problem 4-63 is a "real-world" drawing. It does not conform to standard drafting conventions in all cases.

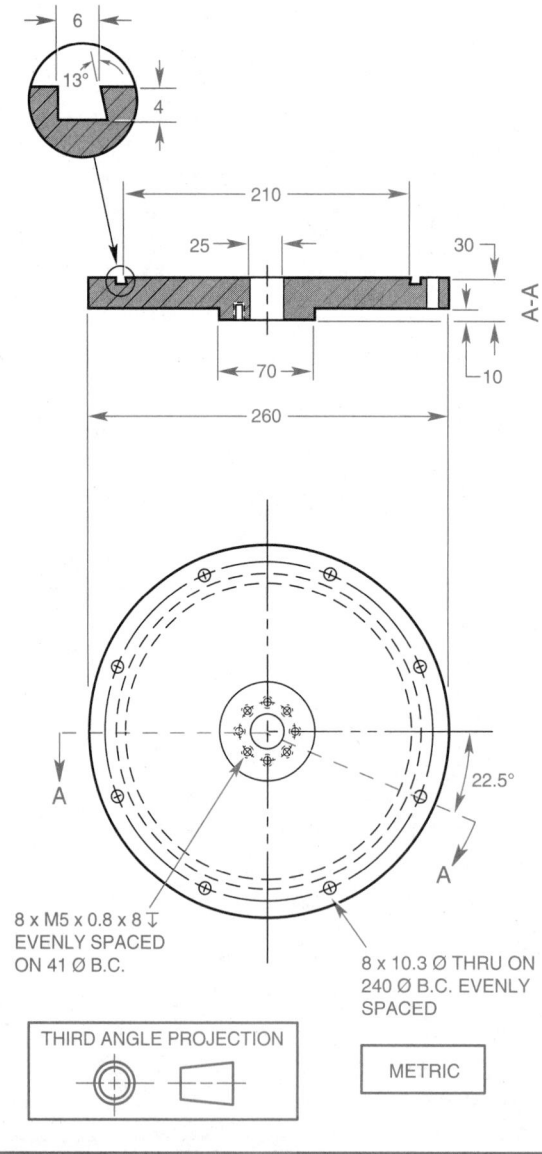

Problem 4-63

Sectional Views

Chapter Outline

Sectional Views

A *sectional view* communicates more information about objects with complicated internal features than the conventional multiview drawing method can with hidden lines. The multiview drawing method, discussed in Chapter 4, is an excellent way to illustrate various external features of an object; however, complicated internal features are difficult to illustrate with hidden lines only. Sectional views solve the problem.

A sectional view is a view of the surfaces that would be seen if an object could be cut open. The view is commonly referred to as a cross section, or simply section, *Figure 5-1*.

Hidden lines are omitted in sectional views, unless it is absolutely necessary to illustrate some unique feature.

Cutting-Plane Line

The *cutting-plane line* indicates the path that an imaginary cutting plane follows to slice through an object. Think of the cutting-plane line as a saw blade that is used to cut through the object. the cutting-plane line is represented by a thick black dashed line, as shown in *Figure 5-2*. The line on top is the newer cutting-plane line; sizes are approximated and spaced by eye. If the cutting-plane line passes through the center of the object, and it is very obvious where the cutting plane is located, it can be omitted. This is the drafter's decision.

Direction of Sight

The drafter must indicate the direction in which the object is to be viewed after it is sliced or cut through. This is accomplished by adding a short leader and arrowhead to the ends of the cutting-plane line, *Figure 5-3*. These arrowheads indicate the direction of sight. Letters are usually added to the ends of the cutting-plane line to indicate exactly what cutting plane is used. The cutting plane extends past the object by approximately ½″, as shown.

Section Lining

Section lining shows where the object is sliced or cut by the cutting plane line. Section lining is represented

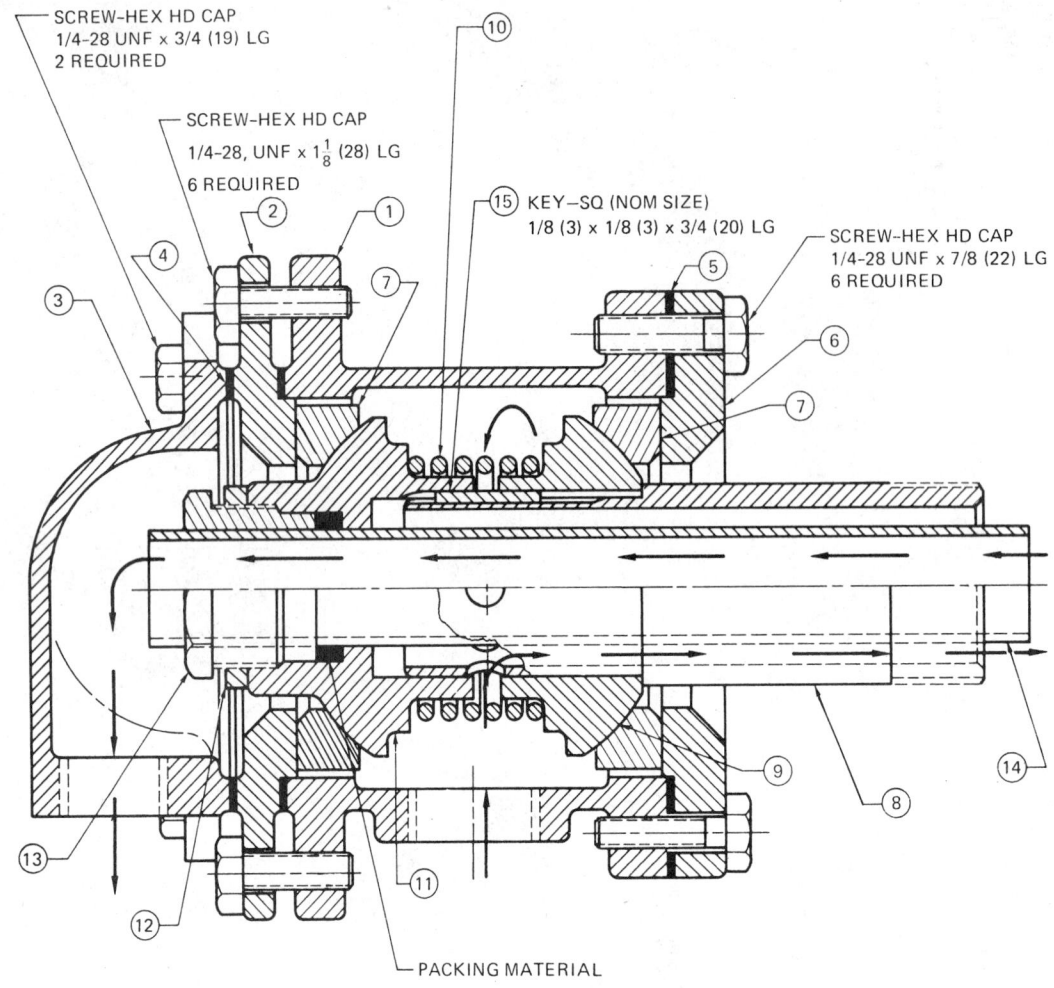

SCREW–HEX HD CAP
1/4–28 UNF x 3/4 (19) LG
2 REQUIRED

SCREW–HEX HD CAP
1/4–28, UNF x $1\frac{1}{8}$ (28) LG
6 REQUIRED

KEY–SQ (NOM SIZE)
1/8 (3) x 1/8 (3) x 3/4 (20) LG

SCREW–HEX HD CAP
1/4–28 UNF x 7/8 (22) LG
6 REQUIRED

PACKING MATERIAL

Figure 5-1   Sectional view

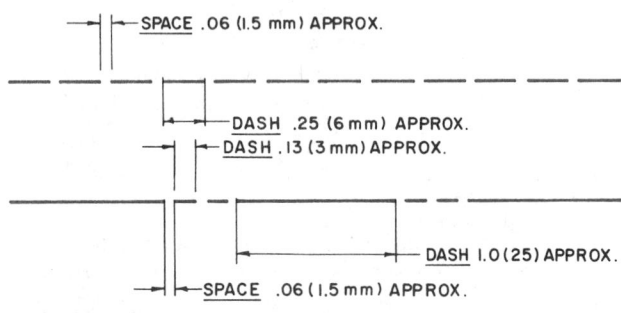

SPACE .06 (1.5 mm) APPROX.

DASH .25 (6 mm) APPROX.
DASH .13 (3 mm) APPROX.

DASH 1.0 (25) APPROX.

SPACE .06 (1.5 mm) APPROX.

Figure 5-2   Cutting-plane line

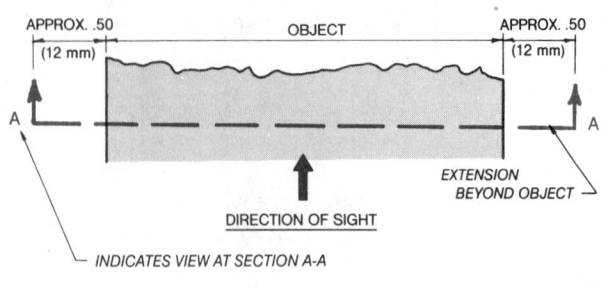

APPROX. .50
(12 mm)

OBJECT

APPROX. .50
(12 mm)

A

A

*EXTENSION
BEYOND OBJECT*

DIRECTION OF SIGHT

INDICATES VIEW AT SECTION A-A

Figure 5-3   The direction of sight

by thin, black lines drawn at 45° to the horizon, unless there is some specific reason for using a different angle. Section lining is spaced by eye from 1/16" (1.5 mm) to 1/4" (6 mm) apart, depending upon the overall size of the object. The average spacing used for most drawings is .13" (3 mm), ***Figure 5-4***. Section lines must be of uniform thickness (thin black) and evenly spaced by eye.

If a cutting plane passes through two parts, each part has section lines using a 45° angle or other principal angle. These section lines should not be aligned in the same direction, ***Figure 5-5***. If the cutting plane passes through more than two parts, the section lining of each individual part must be drawn at different angles. When an angle other than 45° is used, the angle should be 30° or 60°. Section lining should *not* be parallel with the sides of the object to be section lined, ***Figure 5-6***.

In past years, section lining used various symbols to indicate the type of material used to make the object. These symbols used only such general type identifications as cast iron, steel, brass, aluminum, and so forth. Today, because there are so many different kinds of material, section lining symbols have been eliminated, and a single, all-purpose section lining is used (as illustrated in Figure 5-4). Specific information as to the type of material is given in the title block under "material."

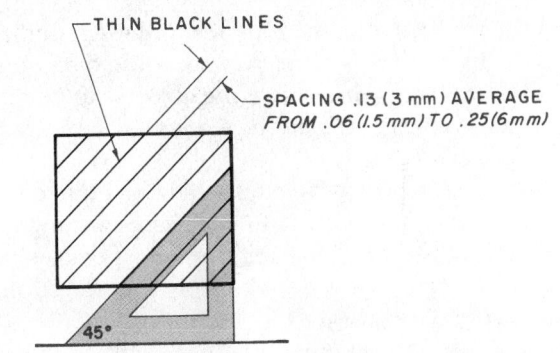

***Figure* 5-4**   Section lining

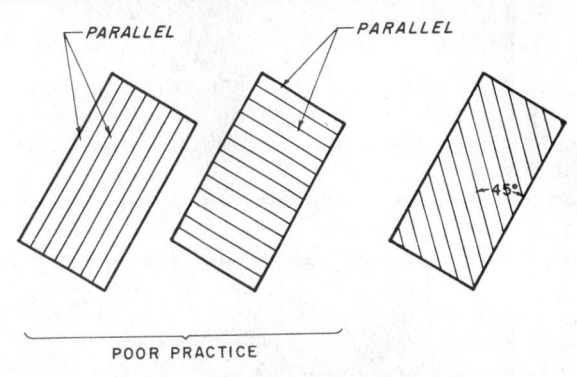

***Figure* 5-6**   Section lining angle

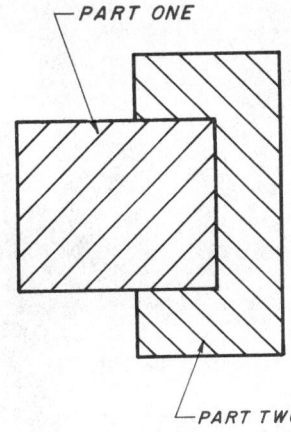

***Figure* 5-5**   Two parts with section lining

Multisection Views

When an object is very complicated and cut in more than one place, each cutting-plane line must be labeled starting with section A-A, followed by B-B, and so forth, ***Figure* 5-7**.

Kinds of Sections

Nine kinds of sections are used today in industry.

- Full section
- Offset section
- Half section
- Broken-out section
- Revolved section
  (rotated section)
- Removed section
- Auxiliary section
- Thinwall section
- Assembly section

Some sections are made up of a combination of the nine kinds of sections. Each is explained in full detail in the following paragraphs.

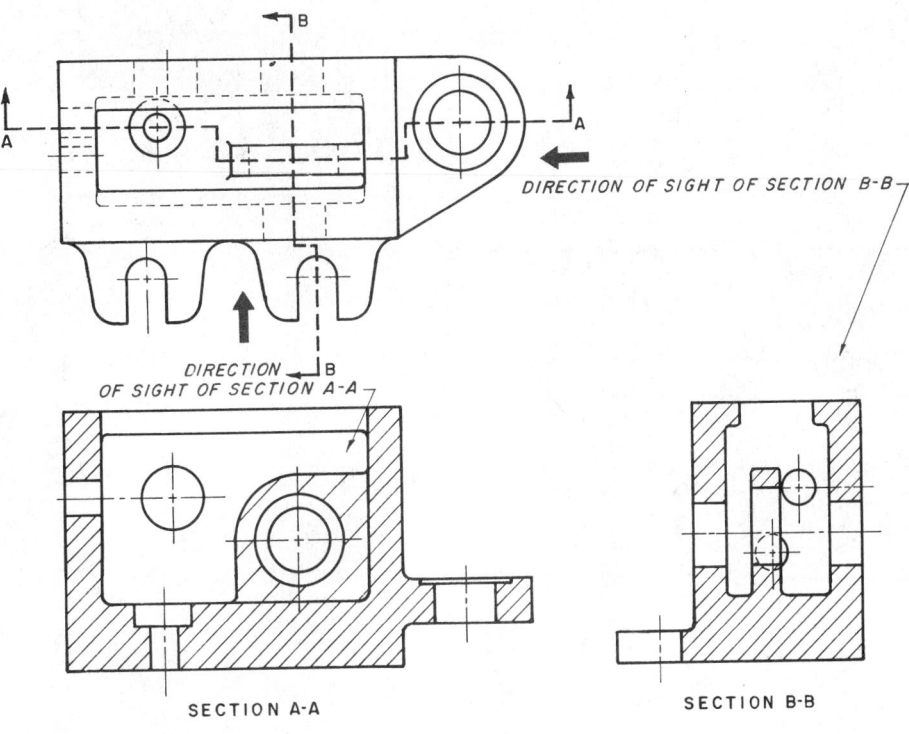

***Figure* 5-7**   Multisection views

Full Section

A *full section* is simply a section of one of the regular multiviews that is sliced or cut completely in two. See the given problem, ***Figure 5-8***, a regular three-view drawing of an object.

Determine which view contains many hard-to-understand hidden lines. In this example, it is the front view. Add a cutting plane to either the top view or right-side view. In this example, the top view is chosen. Indicate how the front view is to be viewed or the direction of sight. After determining where the object is to be sliced or cut and viewed, change the front view into section A-A, ***Figure 5-9***.

Think of the object as a pictorial drawing, ***Figure 5-10***. An imaginary cutting-plane line is passed through the object, ***Figure 5-11***. The front portion is removed and the remaining section is viewed by the direction of sight, ***Figure 5-12***.

Notice that section lining is applied only to the area the imaginary cutting plane passed through. The back side of the hole and the back sides of the notches are *not* section lined.

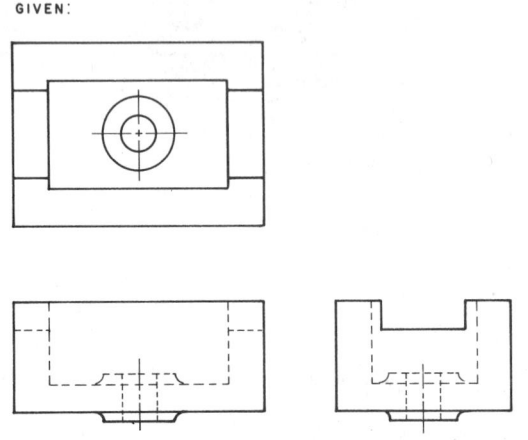

GIVEN:

Figure 5-8   Given: Regular three views of an object

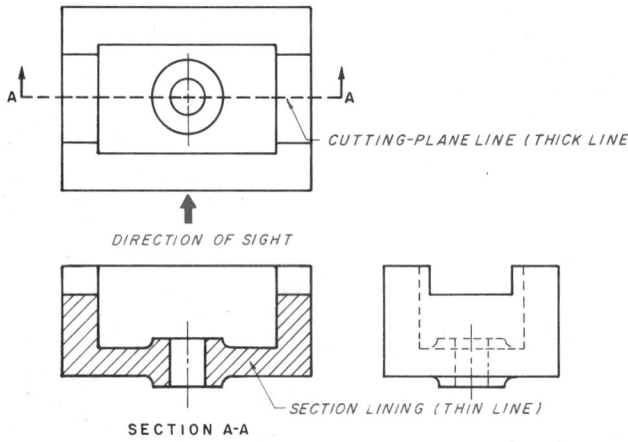

Figure 5-9   Section A-A added

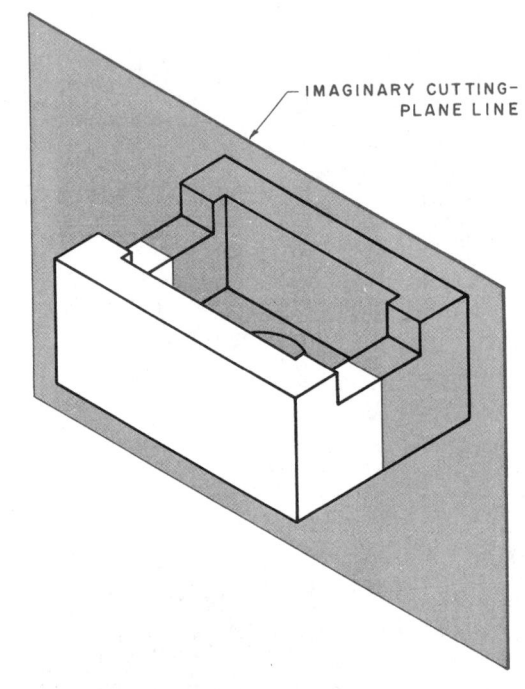

Figure 5-11   Imaginary cutting-plane line added

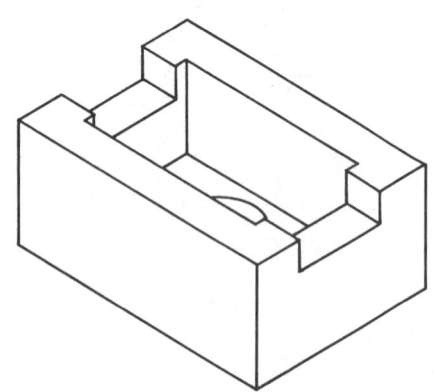

Figure 5-10   Pictorial view of the object

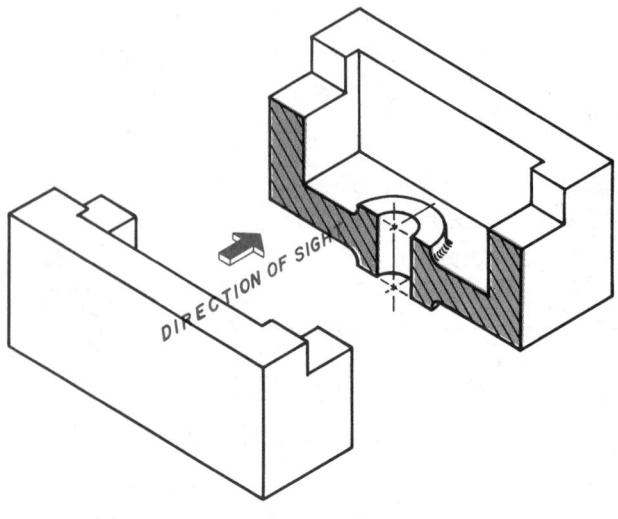

Figure 5-12   Pictorial view of the full section

Offset Section

Many times, important features do not fall in a straight line as they do in a full section. These important features can be illustrated in an *offset section* by bending or offsetting the cutting-plane line. An offset section is very similar to a full section, except that the cutting-plane line is not straight, ***Figure 5-13***.

Note that the features of the countersunk holes A, projection B with its counterbore, and groove C with a shoulder are not aligned with one another. The cutting-plane line is added, and changes of direction (*staggers*) are formed by right angles to pass through these features. An offset cutting-plane line A-A is added to the top view and the material behind the cutting plane is viewed in section A-A, ***Figure 5-14***. The front view is changed into an offset section, similar to a full-sectional view. The actual bends of the cutting-plane lines are omitted in the offset section, ***Figure 5-15***. By using a

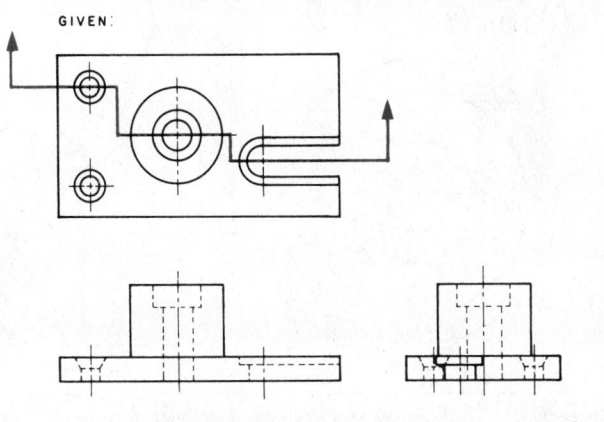

Figure 5-13   Offset section

sectional view, another view often may be omitted. In this example, the right-side view could have been omitted, as it adds nothing to the drawing and takes extra time to draw.

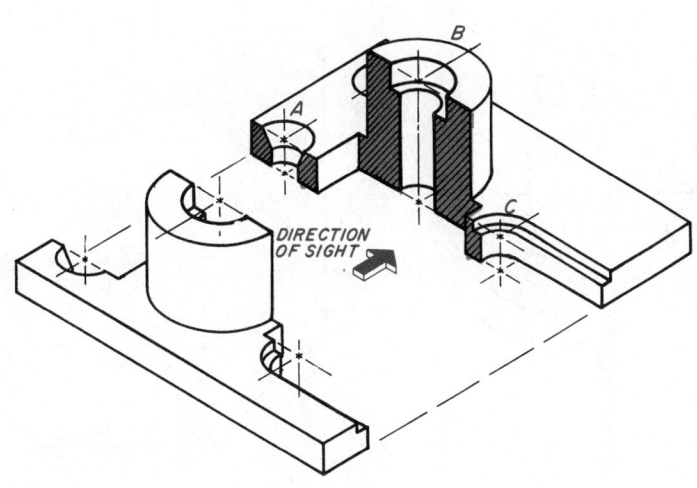

Figure 5-14   Pictorial view of the offset section

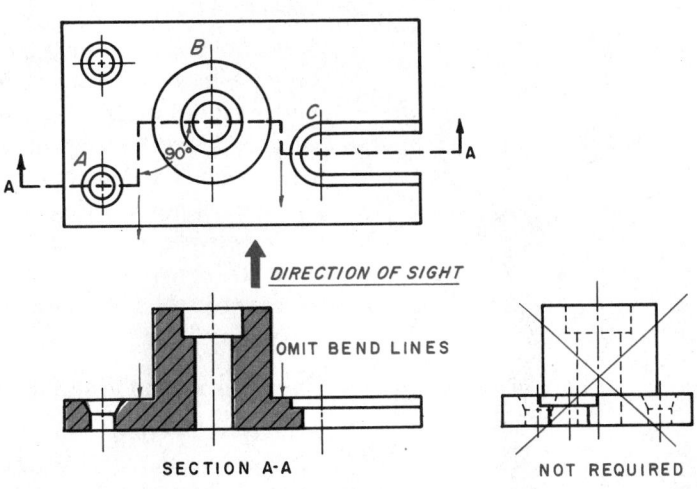

Figure 5-15   Bends omitted from section view

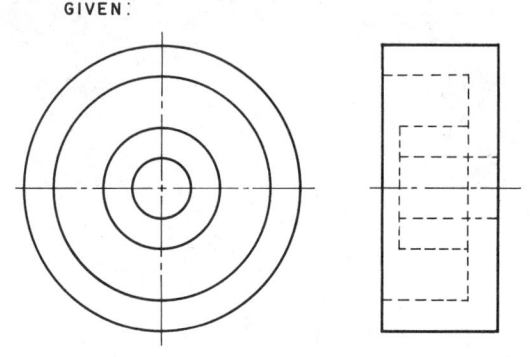

GIVEN:

Figure 5-16   Given: Regular two views of an object

Half Section

In a *half section*, the object is cut only halfway through and a quarter section is removed, *Figure 5-16*. A cutting plane is added to the front view, with only one arrowhead to indicate the viewing direction. Also, a quarter section is removed and, in this example, the right side is sectioned accordingly, *Figure 5-17*. A pictorial view of this half section is illustrated in *Figure 5-18*. The visible half of the object that is not removed shows the *exterior* of the object, and the removed half shows the *interior* of the object. The half of the object not sectioned can be drawn as it would normally be drawn, with the appropriate hidden lines.

Half sections are best used when the object is *symmetrical*, that is, the exact same shape and size on both sides of the cutting-plane line. A half-section view is capable of illustrating both the inside and the outside of an object in the same view. In this example, the top half of the right side illustrates the interior; the bottom half illustrates the exterior. A center line is used to separate the two halves of the half section (refer back to Figure 5-17). A solid line would indicate the presence of a real edge, which would be false information.

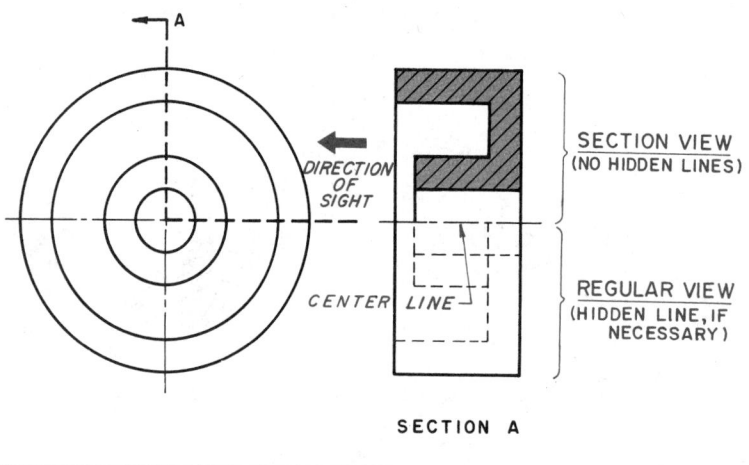

Figure 5-17   Half section

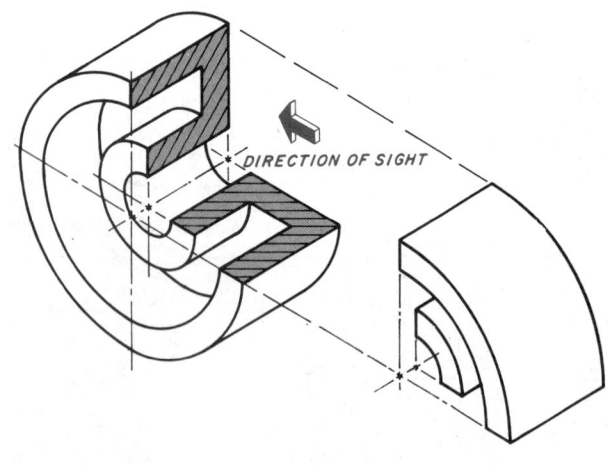

Figure 5-18   Pictorial view of the half section

Broken-out Section

Sometimes, only a small area needs to be sectioned in order to make a particular feature or features easier to understand. In this case, a *broken-out section* is used. Given: *Figure 5-19*. As drawn, the top section is somewhat confusing and could create a question. To clarify this area, a portion is removed, *Figure 5-20*. The finished drawing would be drawn as illustrated in *Figure 5-21*. The broken line is put in freehand, and is drawn as a visible thick line. The actual cutting-plane line is usually omitted.

Revolved Section (Rotated Section)

A *revolved section*, sometimes referred to as a *rotated section*, is used to illustrate the cross section of ribs, webs, bars, arms, spokes or other similar features of an object. *Figure 5-22* is a two-view drawing of an arm.

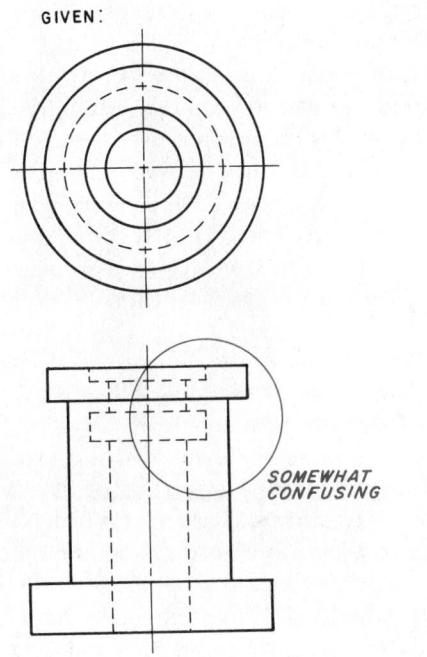

GIVEN:

SOMEWHAT CONFUSING

Figure 5-19   Given: Regular two views of an object

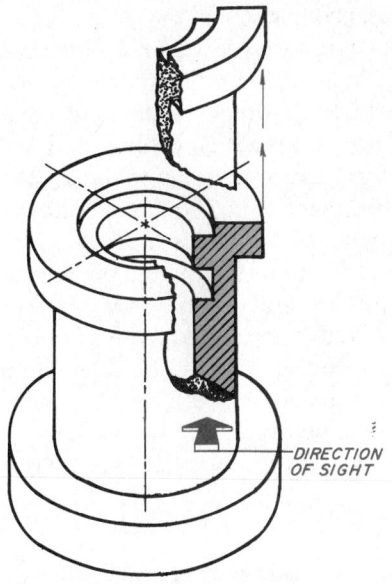

DIRECTION OF SIGHT

Figure 5-20   Pictorial view of the broken-out section

The cross-sectional shape of the center portion of the arm is not defined. In drafting, no feature should remain questionable, and a section through the center portion of the arm would provide the complete information.

A revolved section is made by assuming a cutting plane perpendicular to the axis of the feature of the object to be described, *Figure 5-23*. Note that the rotation point occurs at the cutting-plane location and, theoretically, will be rotated 90°. Rotate the imaginary cutting-plane line about the rotation point of the object, *Figure 5-24*. Notice that dimension X is transferred from the top view to the sectional view of the feature; in this example, the front view. Dimension Y in the top view is also trans-

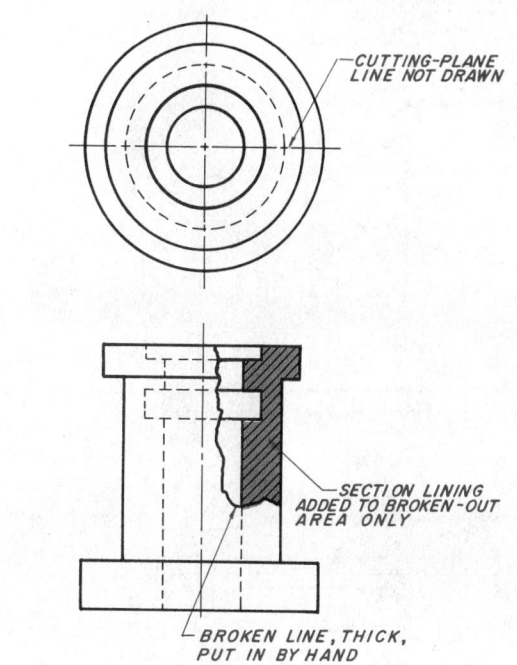

CUTTING-PLANE LINE NOT DRAWN

SECTION LINING ADDED TO BROKEN-OUT AREA ONLY

BROKEN LINE, THICK, PUT IN BY HAND

Figure 5-21   Broken-out section

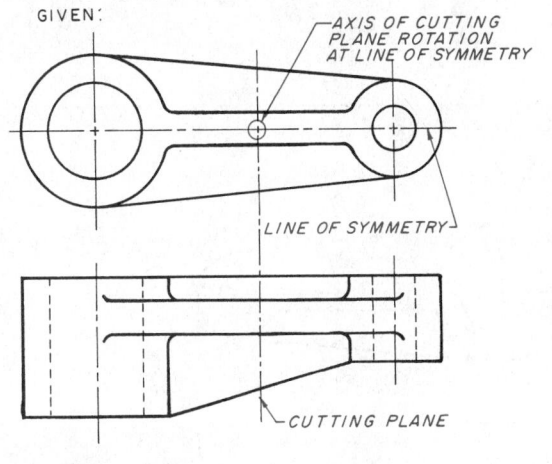

GIVEN:

AXIS OF CUTTING PLANE ROTATION AT LINE OF SYMMETRY

LINE OF SYMMETRY

CUTTING PLANE

Figure 5-22   Given: Regular two views of an object

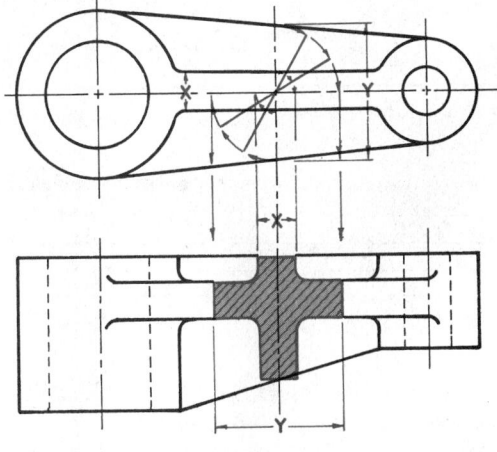

Figure 5-23   Revolved section

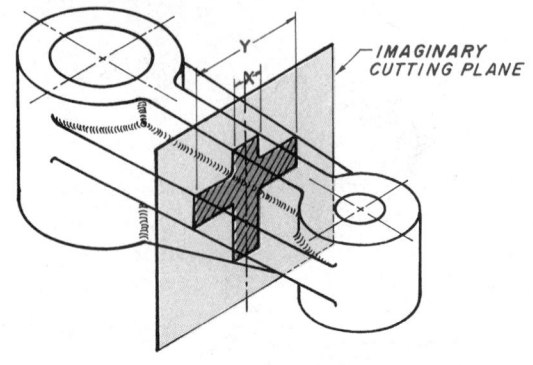

Figure 5-24 Pictorial view of a revolved section

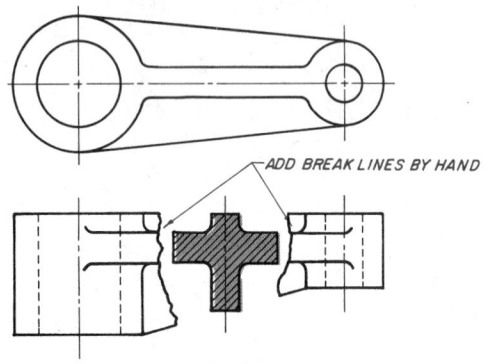

Figure 5-25 Revolved section view

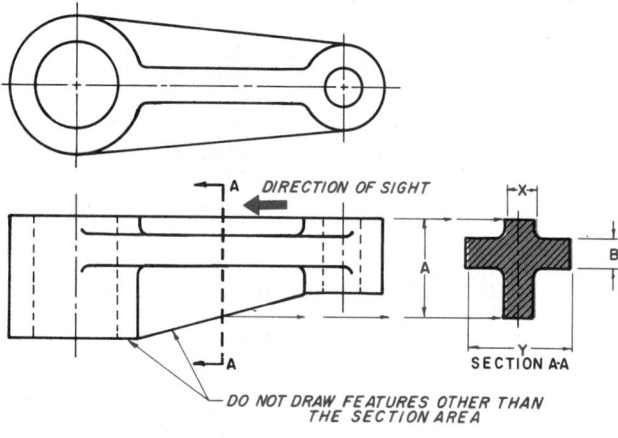

Figure 5-26 Removed section

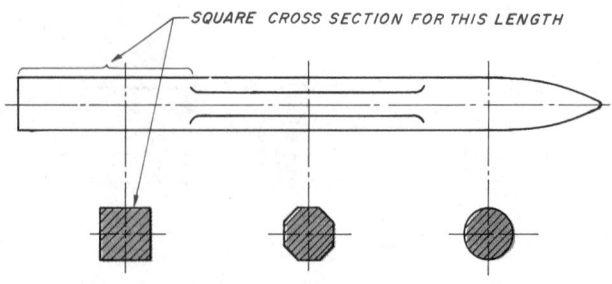

Figure 5-27 Removed section view

ferred to the front view. The section is now drawn in place. The finished drawing is illustrated in Figure 5-25. Note that the break lines in the front view are on each side of the sectional view, and are put in freehand.

The revolved section is not used as much today as it was in past years. Revolved sections tend to be confusing, and often create problems for the people who must interpret the drawings. Today, it is recommended to use a removed section instead of a revolved or rotated section.

Removed Section

A *removed section* is very similar to a rotated section except that, as the name implies, it is drawn removed or away from the regular views, ***Figure 5-26***. The removed section, as with the revolved section, is also used to illustrate the cross section of ribs, webs, bars, arms, spokes or other similar features of an object.

A removed section is made by assuming that a cutting-plane, perpendicular to the axis of the feature of the object, is added through the area that is to be sectioned. (Refer back to Figures 5-23 and 5-24.) Transfer dimensions X and Y to the removed views, exactly as was done in the rotated section. Height features, such as dimensions A and B, are transferred from the front view in this example.

Note that a removed section must identify the cutting-plane line from which it was taken. In the sectional view, do not draw features other than the actual section.

Removed sections are labeled section A-A, section B-B, and so forth, corresponding to the letters at the ends of the cutting-plane line. The sections are usually placed on the drawing in alphabetical order from left to right or from top to bottom, away from the regular views.

Sometimes a removed section is simply drawn on a center line that is extended from the object, ***Figure 5-27***. A removed section can be drawn to an enlarged scale if necessary to illustrate and/or dimension a small feature. The scale of the removed section must be indicated directly below the sectional view, ***Figure 5-28***.

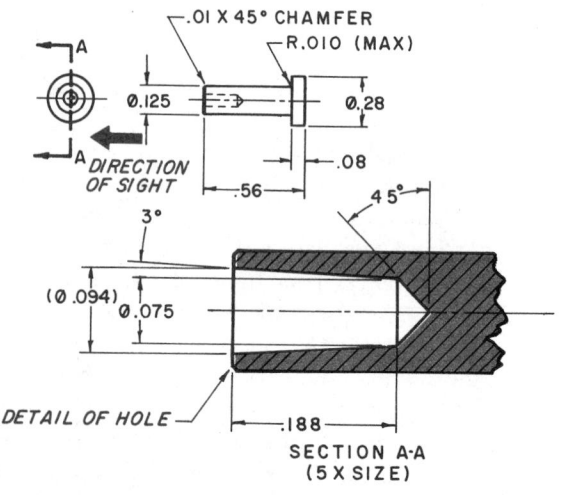

Figure 5-28 Enlarged removed section

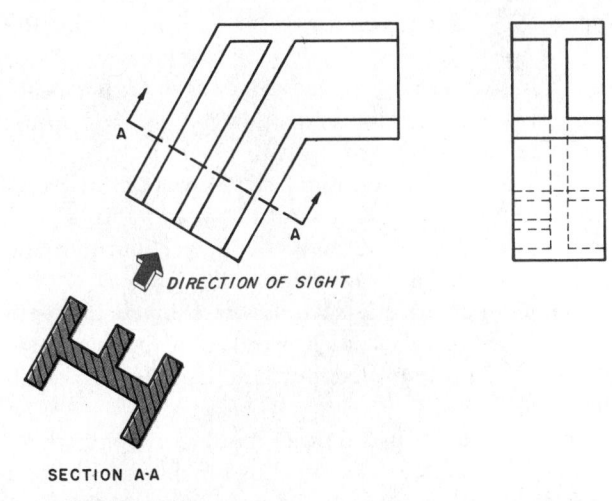

SECTION A-A

Figure 5-29   Auxiliary section

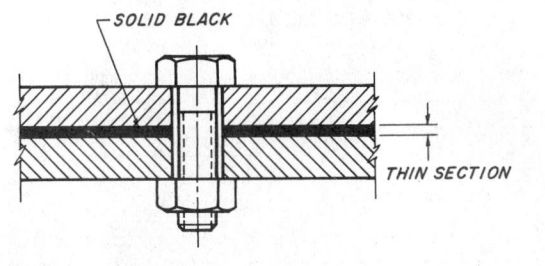

Figure 5-30   Thinwall section

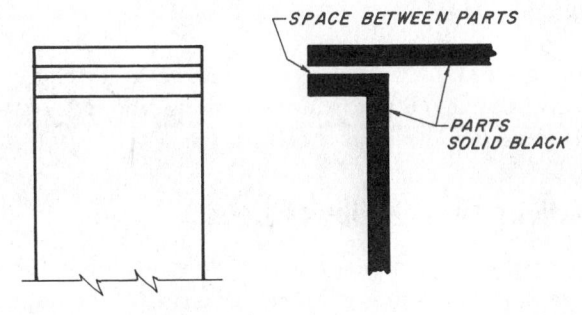

Figure 5-31   Space between thinwall sections

In the field of mechanical drafting, the removed section should be drawn on the same page as the regular views. If there is not room enough on the same page and the removed section is drawn on another page, a page number cross reference must be given as to where the removed section may be found. The page where the removed section is located must refer back to the page from which the section is taken. For example, section A-A on sheet 2 of 4.

Auxiliary Section

If a sectional view of an object is intended to illustrate the true size and shape of an object's boundary, the cutting-plane path must be perpendicular to the axis or surfaces of the object. An *auxiliary section* is projected in the same way as any normal auxiliary view, *Figure 5-29*.

Thinwall Section

Any very thin object that is drawn in section, such as sheet metal, a gasket or a shim, should be filled-in solid black, as it is impossible to show the actual section lining. This is called a *thinwall section*, *Figure 5-30*. If several thin pieces that are filled-in solid black are touching one another, a small white space is left between the solid thinwall section, *Figure 5-31*.

Assembly Section

When a sectional drawing is made up of two or more parts it is called an *assembly section*, *Figure 5-32*. An assembly section can be a full section (as it is in this example), an offset section, a half section or a combination of the various kinds of sectional views. The assembly section shows how the various parts go together.

Each part in the assembly must be labeled with a name, part or plan number, and the quantity required for one complete assembly. If the assembly section does not have many parts, this information is added by a note alongside each part. If the assembly has many parts, and there is not enough room to prevent the drawing from appearing cluttered, each individual part may be identified by a number

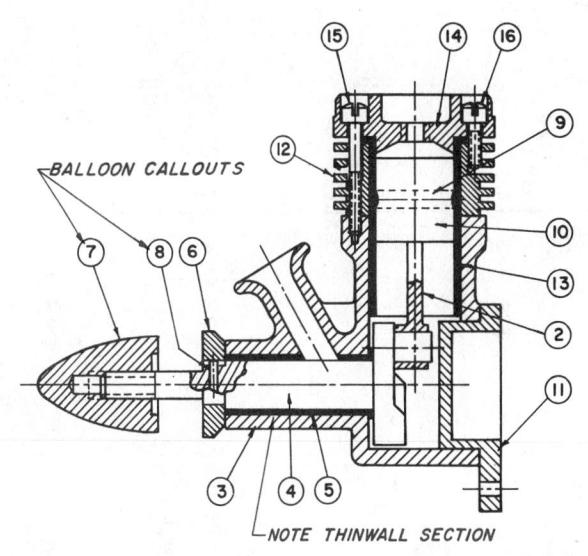

Figure 5-32   Assembly section

within a circle called a *balloon*. The balloon callout system is used in this example. A table must be added to the drawing, listing the name, part or plan number, the quantity required for one complete assembly, and a cross reference to the corresponding balloon number. This is called a *parts list*. The exact form of the list varies from company to company. *Figure 5-33* is an example of a parts list used with the balloon system of callouts. Notice that entries are sometimes listed in reverse (bottom to top) order, as illustrated.

Occasionally drafters deviate from conventional standards for clarity, even though the deviation may violate

┌─── CORRESPONDS TO BALLOON NUMBERS

NO	TITLE	PART NO	NO REQ'D
5	PIVOT PIN	A 9 2 0 0 1	6
4	ARM	B 1 1 6 2 7	1
3	CENTER SHAFT	A 1 1 6 2 6	2
2	BASE	C 1 1 6 2 5	1
1	MAIN FRAME	C 1 1 6 2 4	1

USUAL CALLOUT INFORMATION

Figure 5-33   Example of a parts list

a basic standard. The first part of this chapter covered conventional standards for sectioning; the latter part will focus on deviations for clarity.

Sections through Ribs or Webs

True projection of a sectioned view often produces incorrect impressions of the actual shape of the object.

Figure 5-34 has given a front view and a right-side view. Its pictorial view would look like ***Figure 5-35***. A full section A-A would appear as it does in ***Figure 5-36***. This is a true projection of section A-A, as the cutting-plane line passes through the rib.

However, such a sectional view gives an incorrect impression of the object's actual shape, and is poor drawing practice. It misleads the viewer into thinking the object is actually shaped as it is in ***Figure 5-37***. The conventional practice used to illustrate this section is to draw the section view as illustrated in ***Figure 5-38***, which is not a true projection. Note that the web or rib is not section lined.

Some companies use another method to compensate for this problem. It is somewhat of a middle ground or a combination of true projection and correct representation, ***Figure 5-39***. This is called *alternate section lining*. Section lining, over the rib or web section, is drawn using every *other* section line, and the actual shape is indicated by hidden lines. However, most companies do not use alternate section lining.

Another example of a cutting-plane passing through a

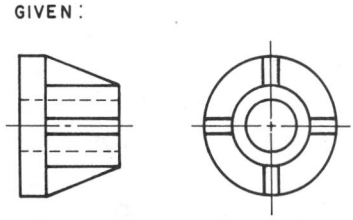

GIVEN:

Figure 5-34   Given: Two-view drawing of an object

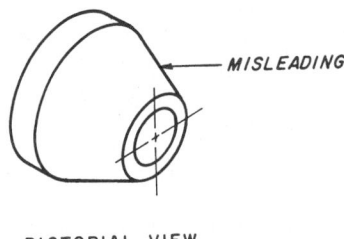

MISLEADING

PICTORIAL VIEW

Figure 5-37   True projection can be misleading

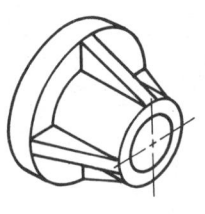

PICTORIAL VIEW

Figure 5-35   Pictorial view of an object

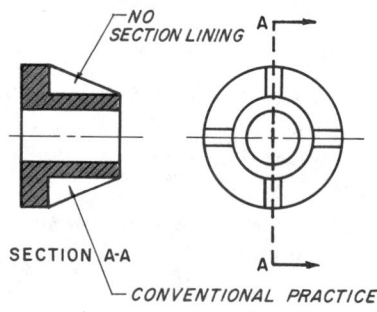

NO SECTION LINING

SECTION A-A

CONVENTIONAL PRACTICE

Figure 5-38   Conventional practice—web or rib *not* sectioned

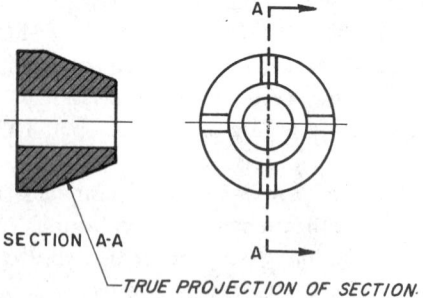

SECTION A-A

TRUE PROJECTION OF SECTION.

Figure 5-36   True projection of an object

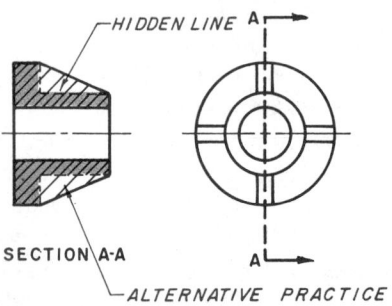

HIDDEN LINE

SECTION A-A

ALTERNATIVE PRACTICE

Figure 5-39   Alternate conventional practice

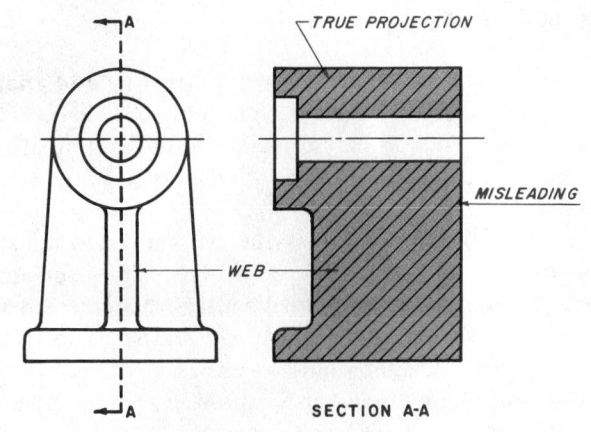

Figure 5-40   Example of true projection

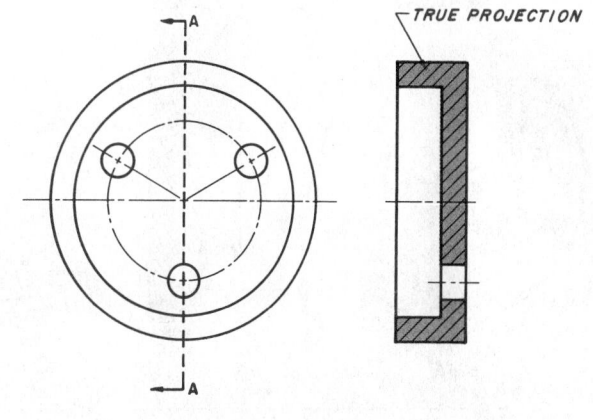

Figure 5-42   Holes using true projection

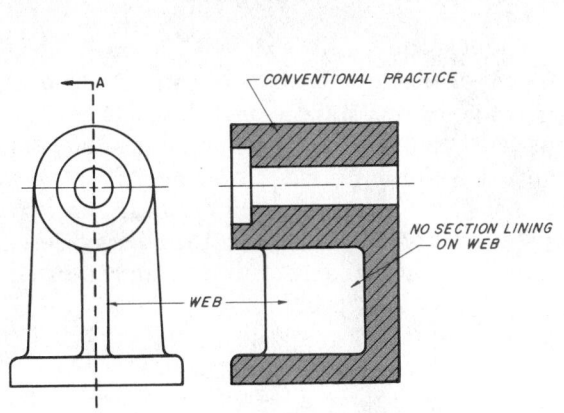

Figure 5-41   Example of conventional practice

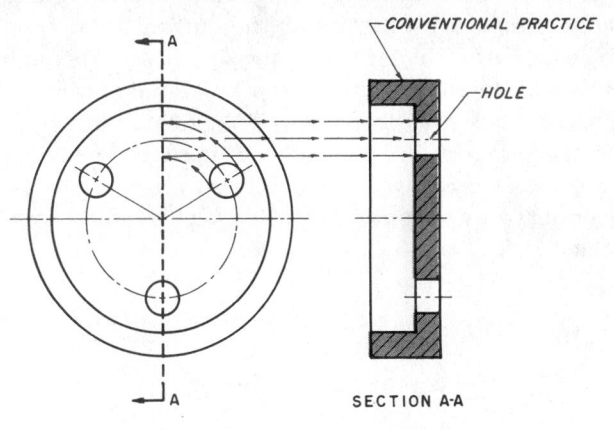

Figure 5-43   Holes using conventional practice (aligning of features)

rib or web is shown in ***Figure 5-40***. This example is a true projection, but it is poor drafting practice, as it gives the impression that the center portion is a thick, solid mass. ***Figure 5-41*** is drawn incorrectly, but does not give the false impression of the object's center portion. This is the *conventional practice* used.

Holes, Ribs and Webs, Spokes and Keyways

Holes located around a bolt circle are sometimes not aligned with the cutting-plane line, ***Figure 5-42***. The cutting plane passes through only one hole. This is a true projection of the object, but poor drafting practice. In actual practice, the top hole is theoretically revolved to the cutting-plane line and projected to the sectional view, ***Figure 5-43***. This practice is called *aligning of features*.

Ribs and Webs

Ribs or webs sometimes do not align with the cutting-plane line, ***Figure 5-44***. The cutting plane passes through only one web and only one hole. This is a true

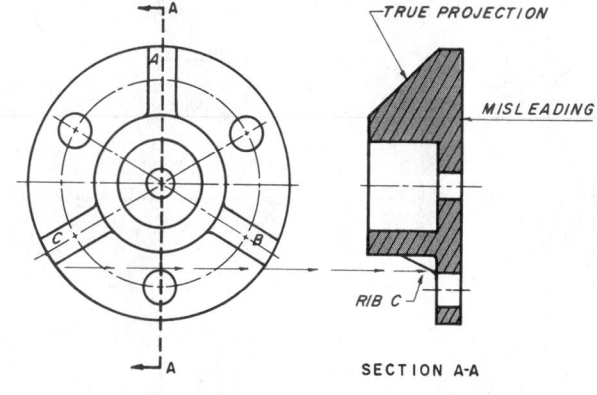

Figure 5-44   Rib or web using true projection

projection of the object, but poor drafting practice. In actual practice, one of the webs is theoretically revolved up to the cutting-plane line and projected to the sectional view, ***Figure 5-45***. Notice that the bottom hole is unaffected and is projected normally. This is another example of aligning of features.

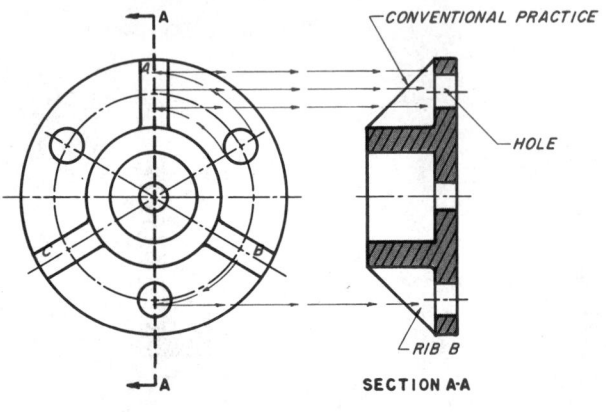

Figure 5-45 Rib or web using conventional practice

Spokes and Keyways

Spokes and keyways and other important features sometimes do not align with the cutting-plane line, ***Figure 5-46***. The cutting-plane line passes through only one spoke and misses the keyway completely. This also is a true projection of the object, but poor drafting practice. In conventional practice, spoke B is revolved to the cutting-plane line and projected to the sectional view, ***Figure 5-47***. The keyway is also projected as illustrated. This is another example of aligning of features.

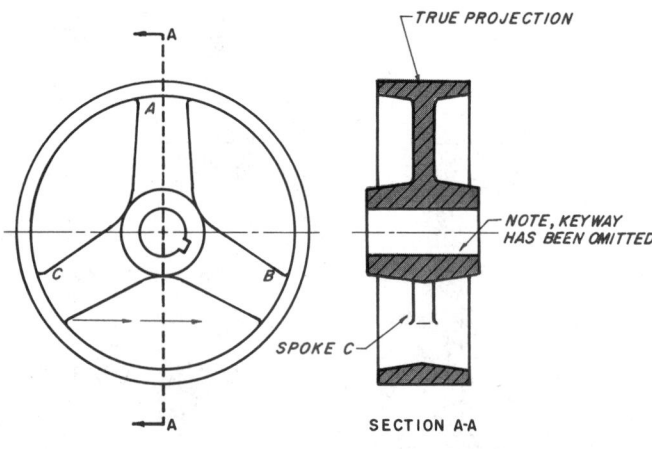

Figure 5-46 Spokes and keyway using true projection

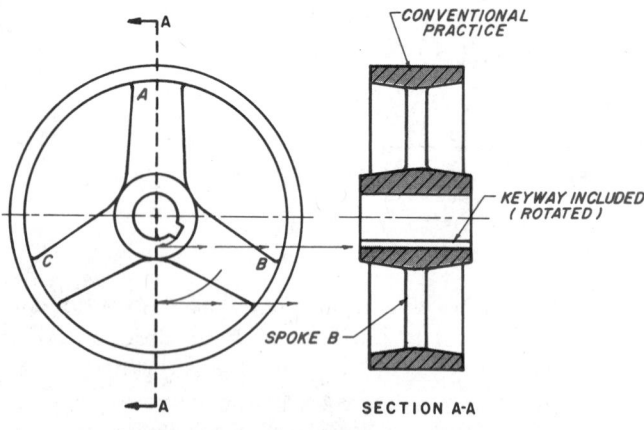

Figure 5-47 Spokes and keyway using conventional practice

Aligned Sections

Arms and other similar features are revolved to alignment in the cutting plane, as were spokes in the preceding section. This procedure is used if the cutting-plane line cannot align completely with the object, as illustrated in ***Figure 5-48***.

The arm or feature is now revolved to the imaginary cutting plane, and projected down to the sectional view, ***Figure 5-49***. The actual cutting-plane line is bent and drawn through the arm or feature and then revolved to a straight, aligned vertical position. Notice that section lining is *not* applied to the arm, and is also omitted from the web area.

Fasteners and Shafts in Section

If a cutting plane passes *lengthwise* through any kind of fastener or shaft, the fastener or shaft is *not* sectioned. Section lining of a fastener or shaft would have no interior detail, thus it would serve no purpose and only add confusion to the drawing, ***Figure 5-50***. The round head machine screw, the hex head cap screw w/nut, and the rivet are not sectioned. The other objects in the figure such as fasteners, ball bearing rollers, and so forth, are also not sectioned.

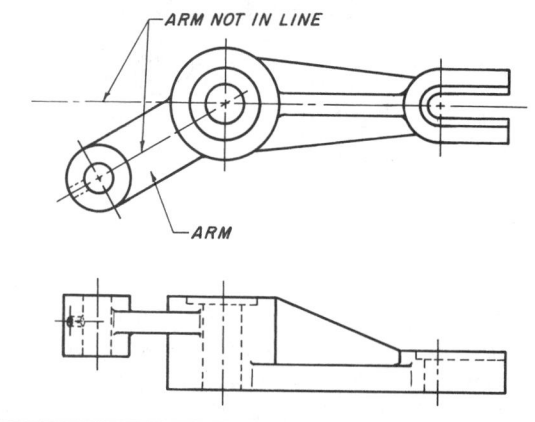

Figure 5-48 Two-view drawing

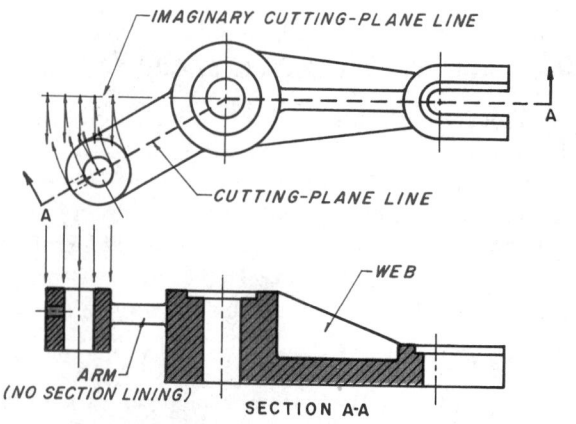

Figure 5-49 Aligned section

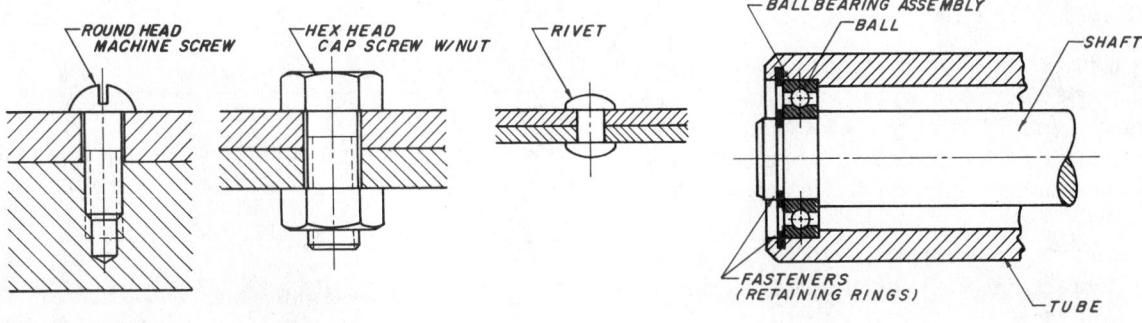

Figure 5-50 Parts *not* sectioned

If a cutting-plane line passes *perpendicularly* through the axis of a fastener or shaft, section lining is added to the fastener or shaft, ***Figure 5-51***. The end view has section lining added as shown.

Intersections in Section

Where an intersection of a small or relatively unimportant feature is cut by a cutting-plane line, it is not drawn as a true projection, ***Figure 5-52***. Since a true projection takes drafting time, it is preferred that it be disregarded, and the feature drawn, using conventional practice, as shown in ***Figure 5-53***. This procedure is much quicker and more easily understood.

Review

1. Explain the difference between true projection and conventional practice. Which is used in a sectional view and why?
2. Explain the difference between a revolved section and a removed section. Which is recommended today?
3. Are hidden lines used in a sectional view? Why?
4. Why is a removed section sometimes drawn at a larger scale?
5. List the nine kinds of sectional views and describe the various features of each.
6. What is alternate section lining? Where is it used?
7. List two major functions of an assembly drawing.
8. Explain the practice used for drawing intersections of small or unimportant features that are cut by a cutting-plane line.
9. What kind of sectional view illustrates both the exterior and interior of the object?
10. What must be done if a removed section is placed on another page other than the page on which the cutting-plane line is placed?
11. Explain the two methods used in regard to a cutting-plane line passing through the long dimension and perpendicularly to the center of a fastener or shaft.
12. What must be included for each part in an assembly section? Explain the two methods used to accomplish this.

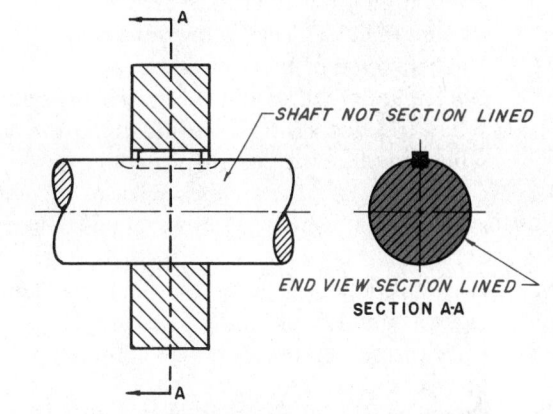

Figure 5-51 Shaft sectioned in end view only

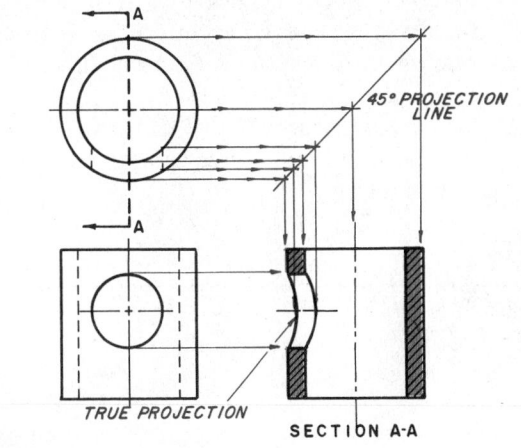

Figure 5-52 Intersection using true projection

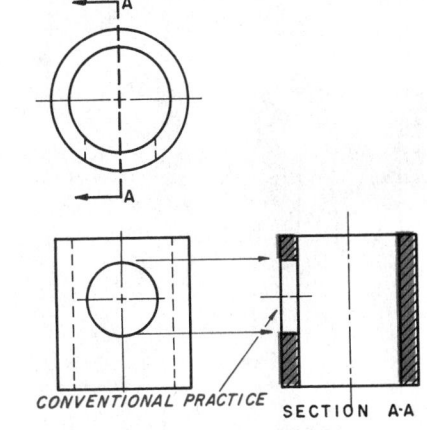

Figure 5-53 Intersection using conventional practice

Chapter Five Problems

The following problems are intended to give the beginning drafter practice in using the various kinds of sectional views used in industry. As these are beginning problems, no dimensions will be used at this time.

The steps to follow in laying out all problems in this chapter are:

Step 1   Study the problem carefully.

Step 2   Choose the view with the most detail as the front view.

Step 3   Position the front view so there will be the least amount of hidden lines in the other views.

Step 4   Make a sketch of all required views.

Step 5   Determine what should be drawn in section, what type of section should be used, and where to place the cutting-plane line.

Step 6   Center the required views within the work area with a 1-inch (25-mm) space between each view.

Step 7   Use light projection lines. Do not erase them.

Step 8   Lightly complete all views.

Step 9   Check to see that all views are centered within the work area.

Step 10   Check to see that there is a 1-inch (25-mm) space between all views.

Step 11   Carefully check all dimensions in all views.

Step 12   Darken in all views using correct line thickness.

Step 13   Add a cutting-plane line and section lining as required.

Step 14   Recheck all work, and, if correct, neatly fill out the title block using light guidelines and neat lettering.

Problem 5-1

Center three views within the work area, and make the front view a full section.

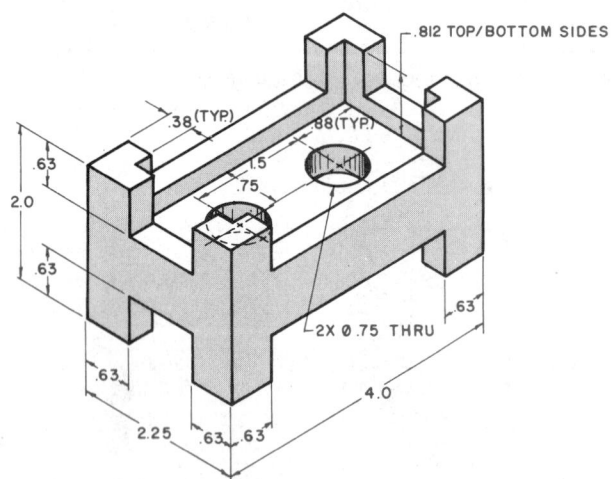

Problem 5-1

Problem 5-2

Center two views within the work area, and make one view a full section. Use correct drafting practices for the ribs.

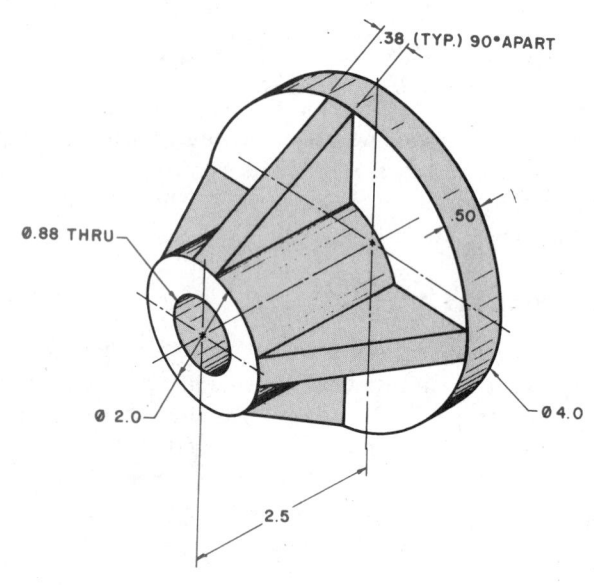

Problem 5-2

Problem 5-3

Center two views within the work area, and make one view a full section. Use correct drafting practices for the holes.

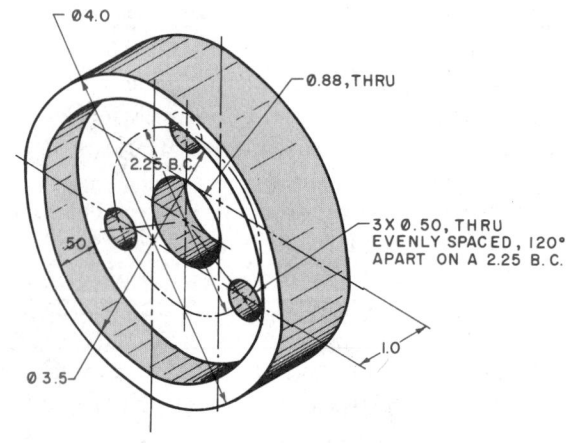

Problem 5-3

Problem 5-4

Center the front view and top view within the work area. Make one view a full section.

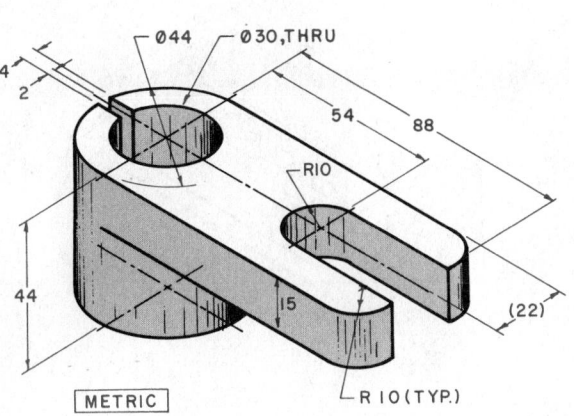

Problem 5-4

Problem 5-5

Center two views within the work area, and make one view a full section. Use correct drafting practices for the arms, horizontal hole, and keyway.

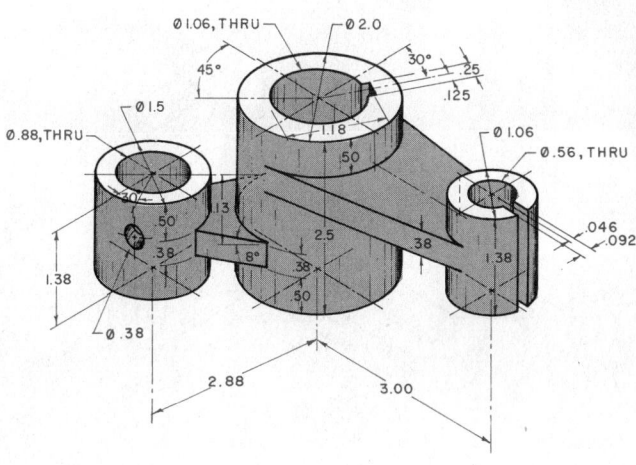

Problem 5-5

Problem 5-6

Center two views within the work area, and make one view a full section. Use correct drafting practices for the keyway, ribs, and holes.

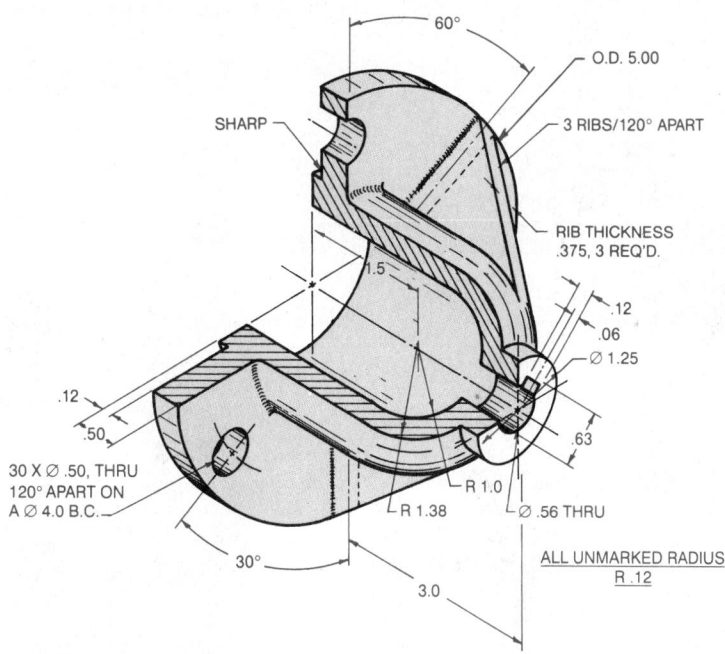

Problem 5-6

Problem 5-7

Center three views within the work area, and make one view an offset section. Be sure to include three major features.

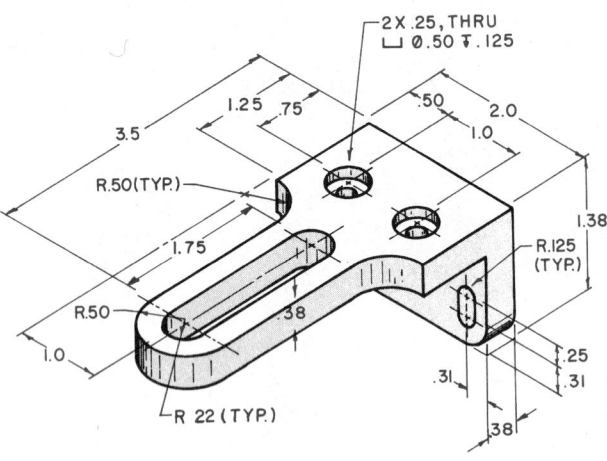

Problem 5-7

Problem 5-9

Center three views within the work area, and make one view an offset section. Be sure to include as many of the important features as possible.

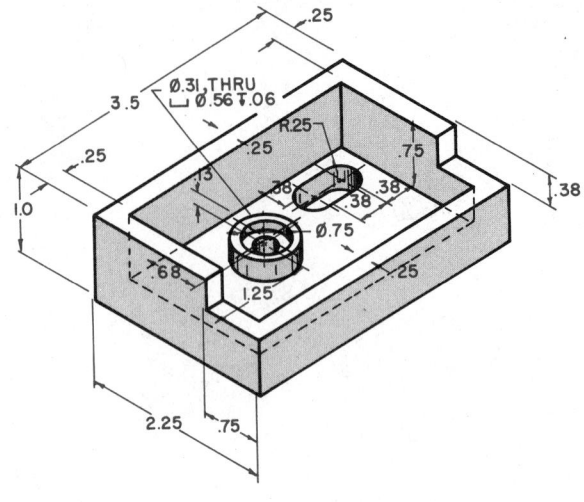

Problem 5-9

Problem 5-8

Center three views within the work area, and make one view an offset section. Be sure to include three major features.

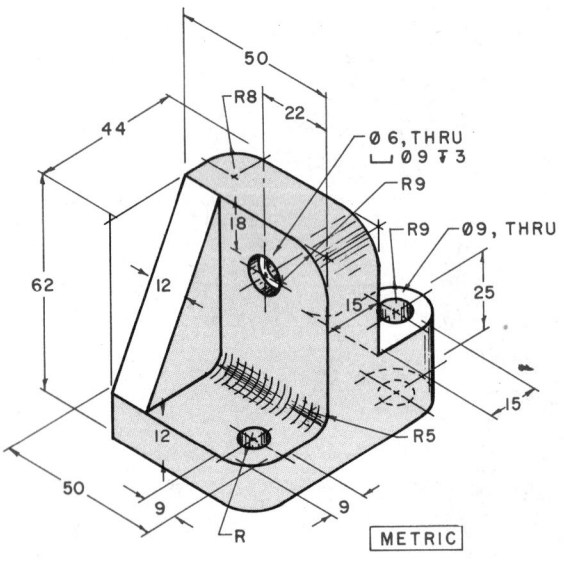

Problem 5-8

Problem 5-10

Center three views within the work area, and make one view an offset section. Be sure to include as many of the important features as possible.

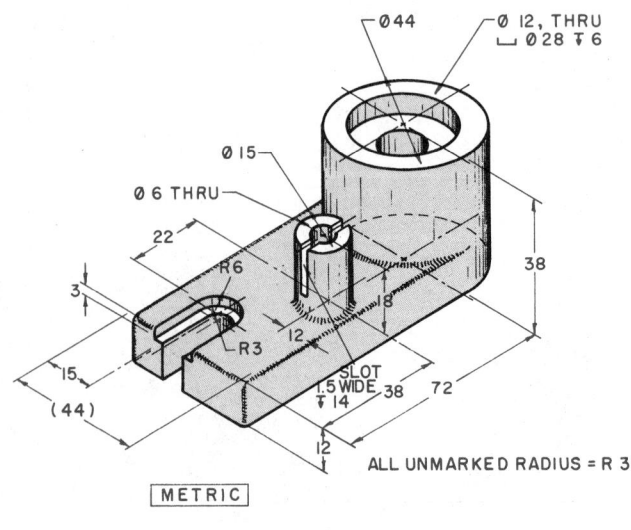

Problem 5-10

Problem 5-11

Center two views within the work area, and make one view an offset section. Be sure to include as many of the important features as possible.

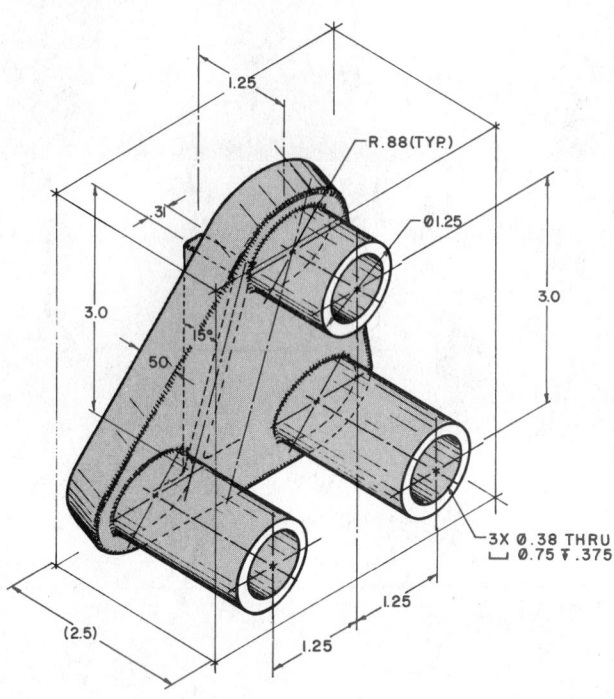

Problem 5-11

Problem 5-12

Center the front view and top view within the work area. Make one view a half section.

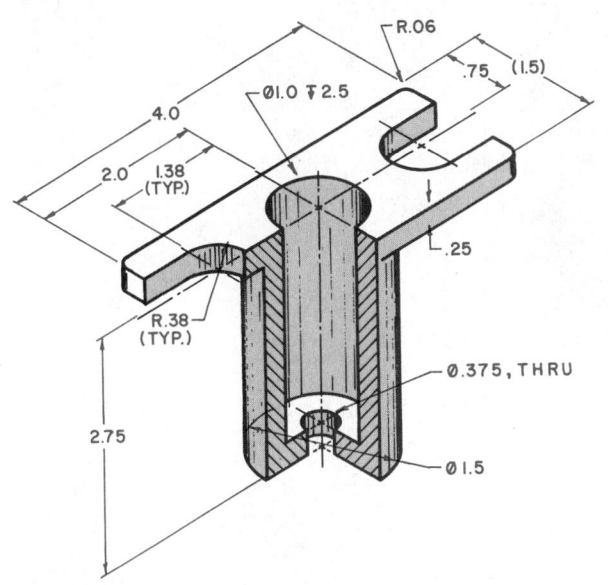

Problem 5-12

Problem 5-13

Center two views within the work area, and make one view a half section.

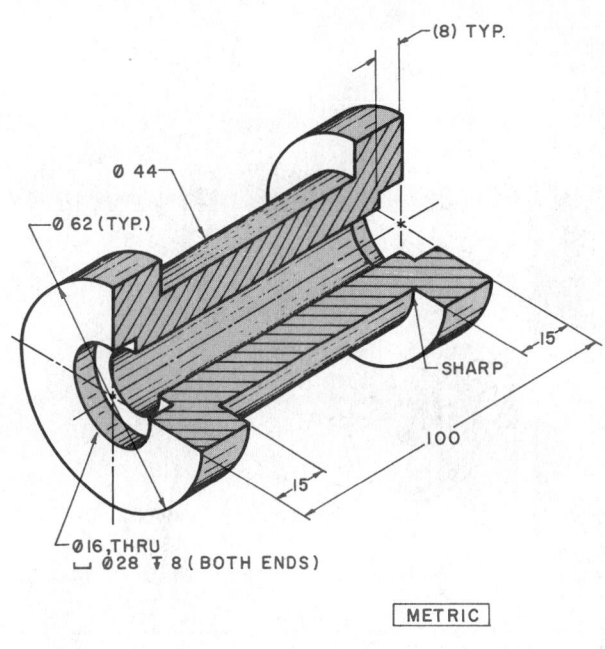

Problem 5-13

Problem 5-14

Center the two views within the work area, and make one view a half section.

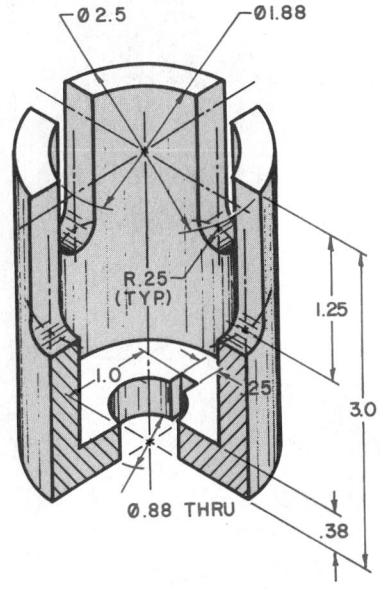

Problem 5-14

Problem 5-15
Center two views within the work area, and make one view a half section.

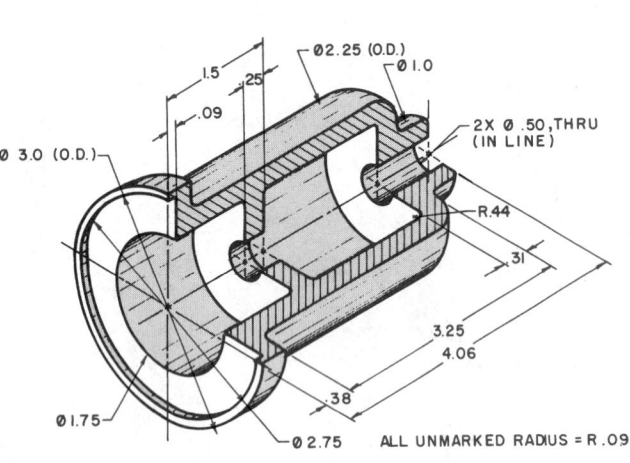

Problem 5-15

Problem 5-16
Center two views within the work area, and make one view a half section.

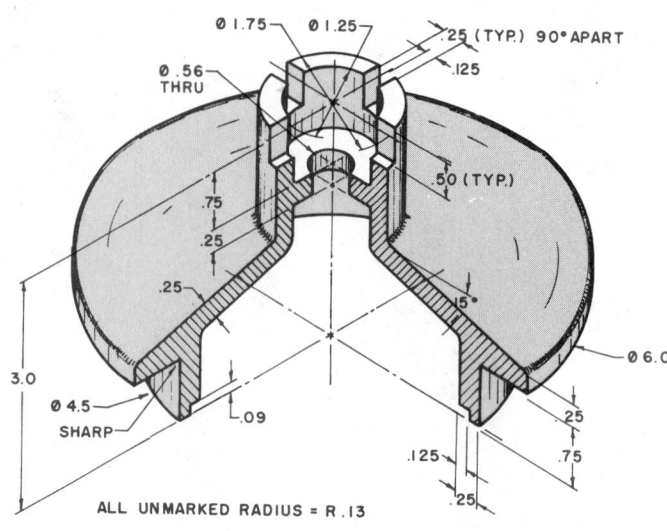

Problem 5-16

Problem 5-17
Center the required views within the work area, and make one view a broken-out section to illustrate the complicated interior area.

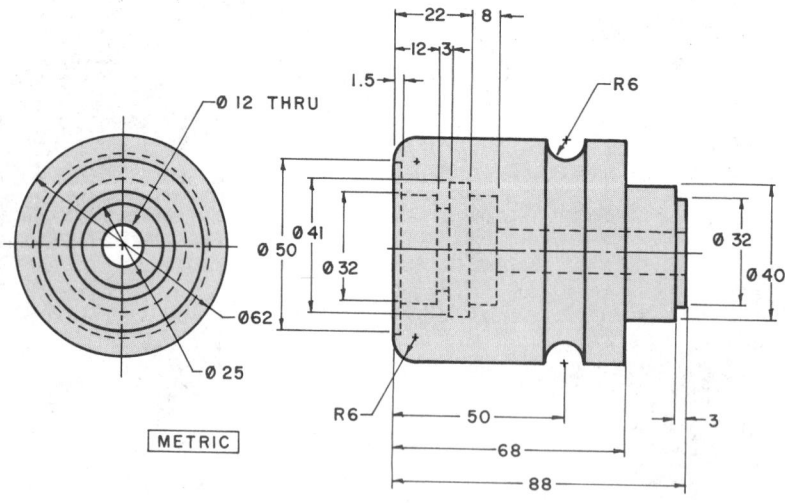

Problem 5-17

Problem 5-18

Center the required views within the work area, and make one view a broken-out section as required.

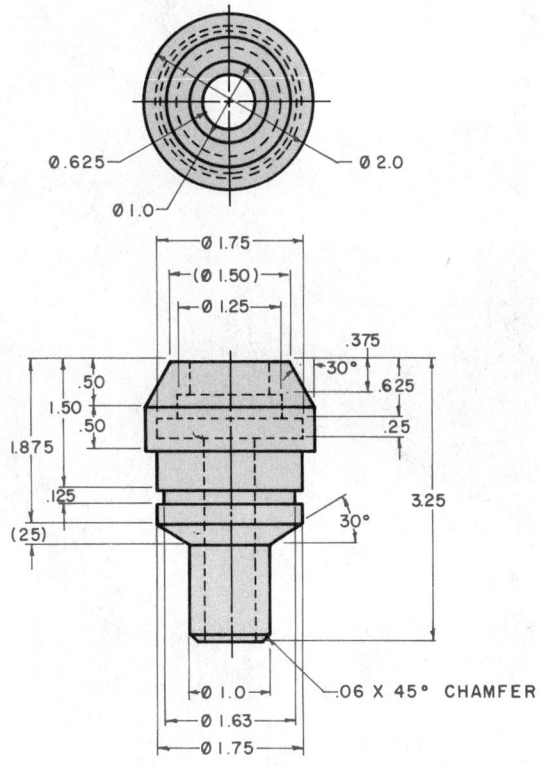

Problem 5-18

Problem 5-19

Center three views within the work area, and add removed sections A-A and B-B.

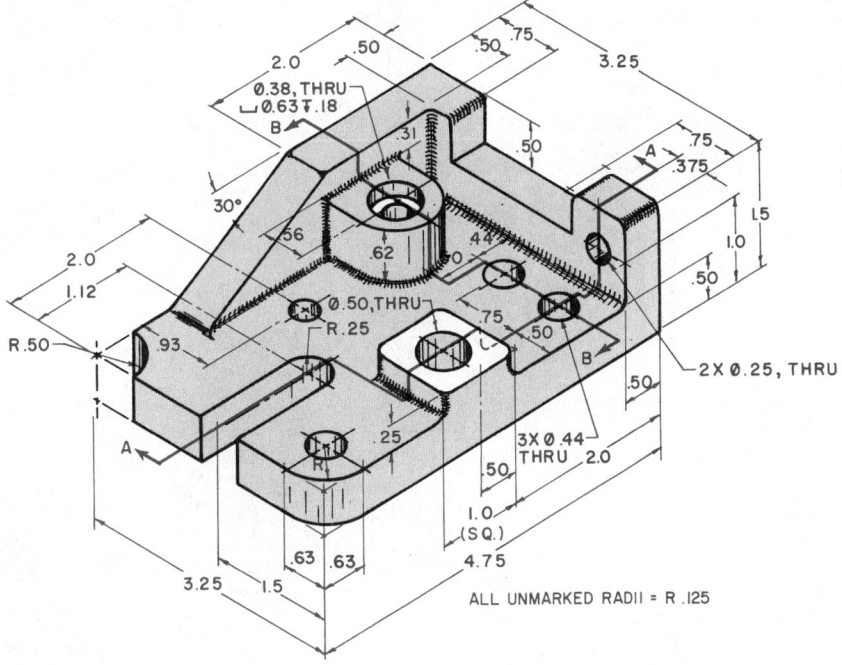

ALL UNMARKED RADII = R .125

Problem 5-19

Problem 5-20

Center two views within the work area, and add removed sections A-A, B-B, and C-C.

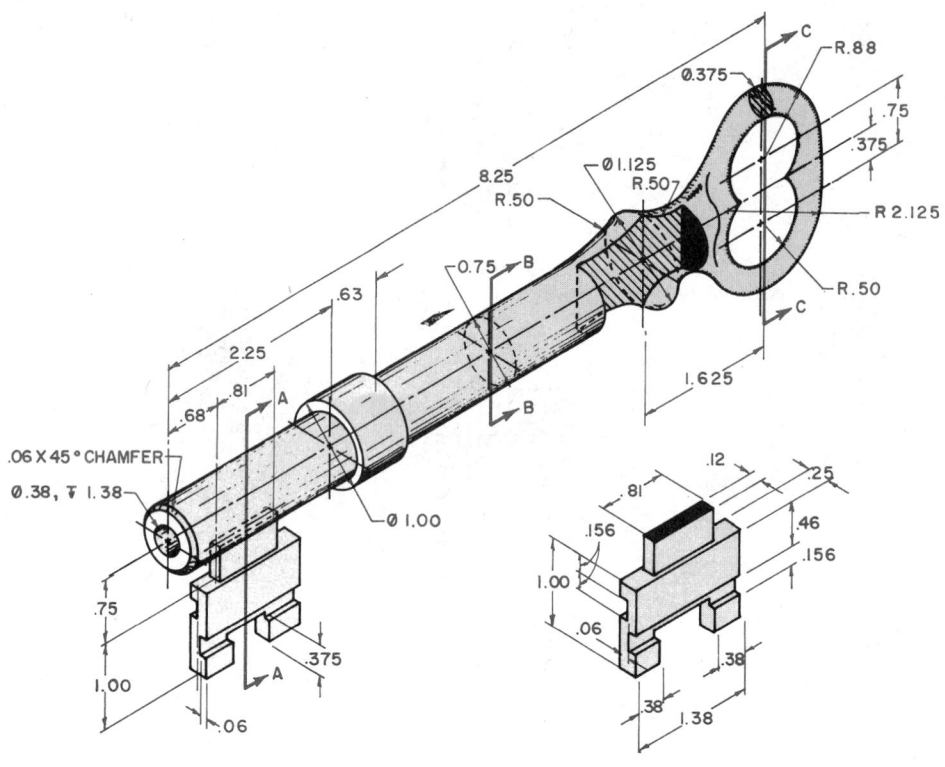

Problem 5-20

Problem 5-21

Center the required views within the work area, and add removed section as required.

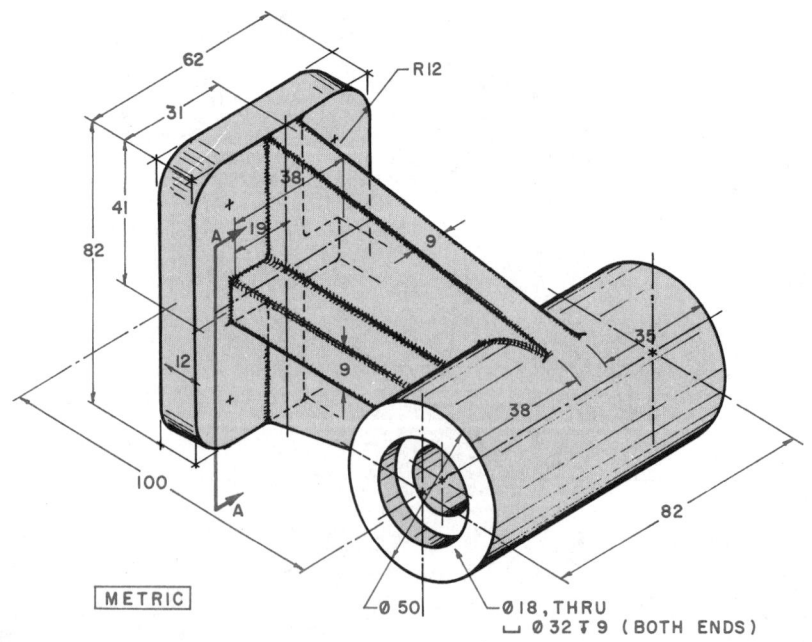

Problem 5-21

Problem 5-22

Center the required views within the work area, and add removed section as required.

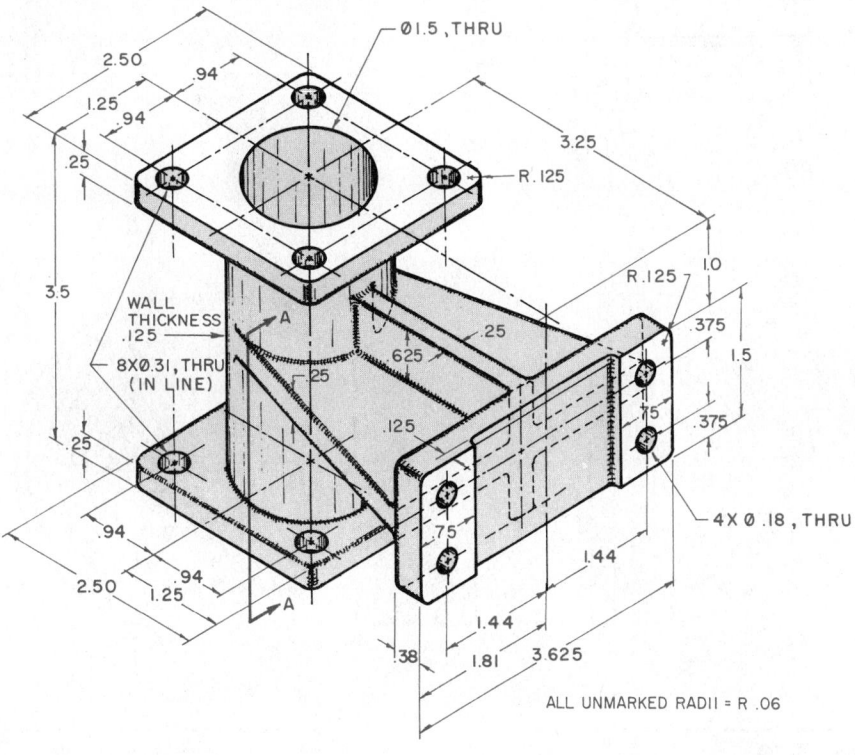

Problem 5-22

Problem 5-23

Center the required views within the work area, and add removed sections A-A and B-B.

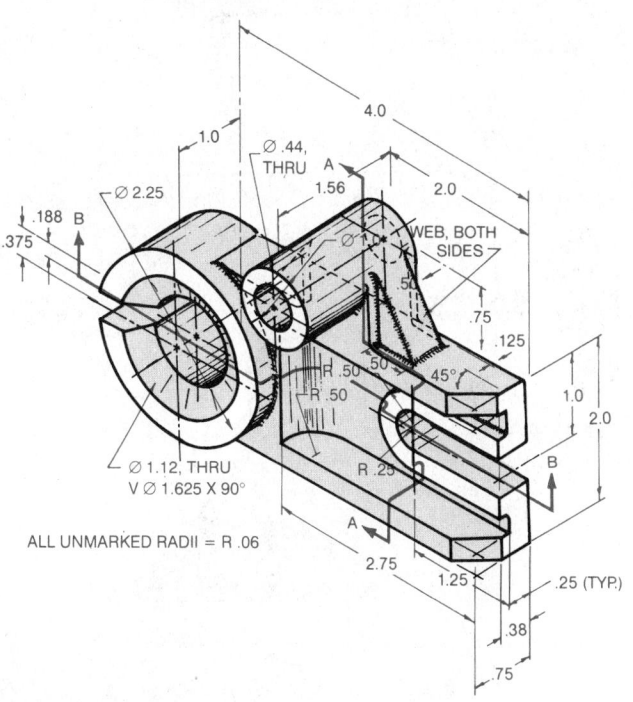

Problem 5-23

Problem 5-24

Center the required views within the work area, and add removed section A-A.

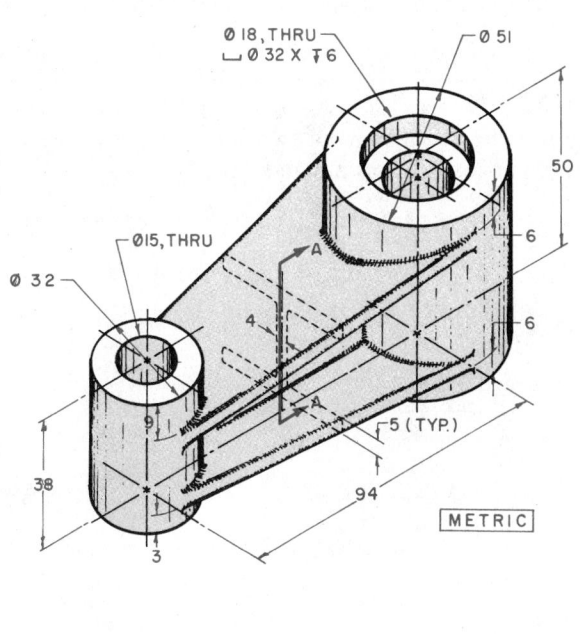

Problem 5-24

Problem 5-25

Center the four views within the work area. Make the top view section A-A and the right-side view section B-B.

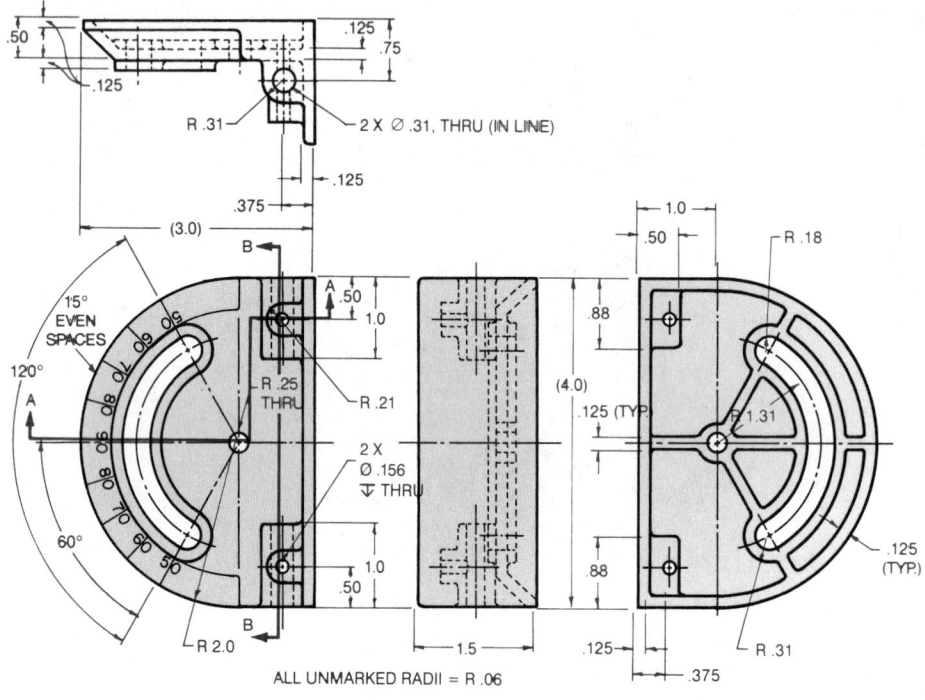

ALL UNMARKED RADII = R .06

Problem 5-25

Problem 5-26

Center the required views within the work area, and add removed sections A-A and B-B.

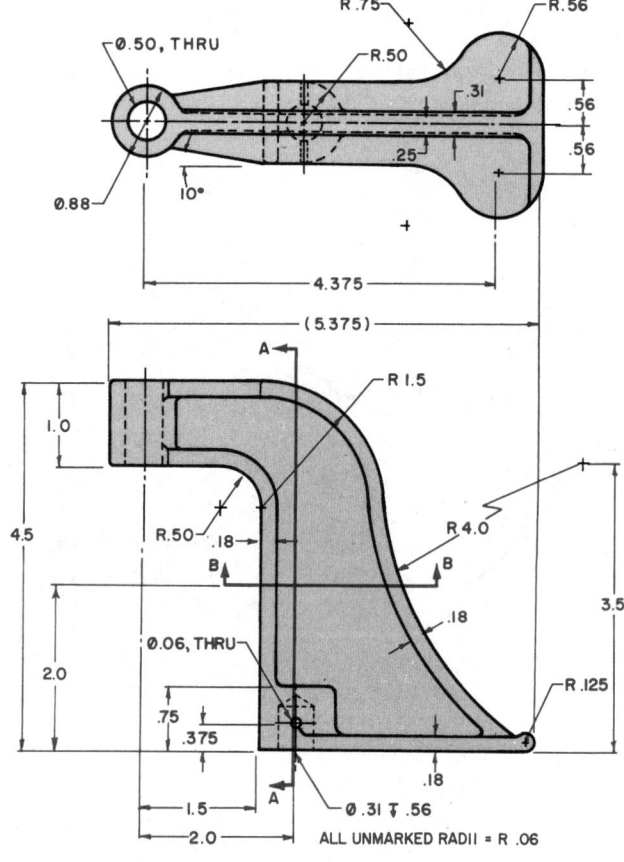

ALL UNMARKED RADII = R .06

Problem 5-26

Problem 5-27

Center the required views within the work area, and add removed sections A-A and B-B.

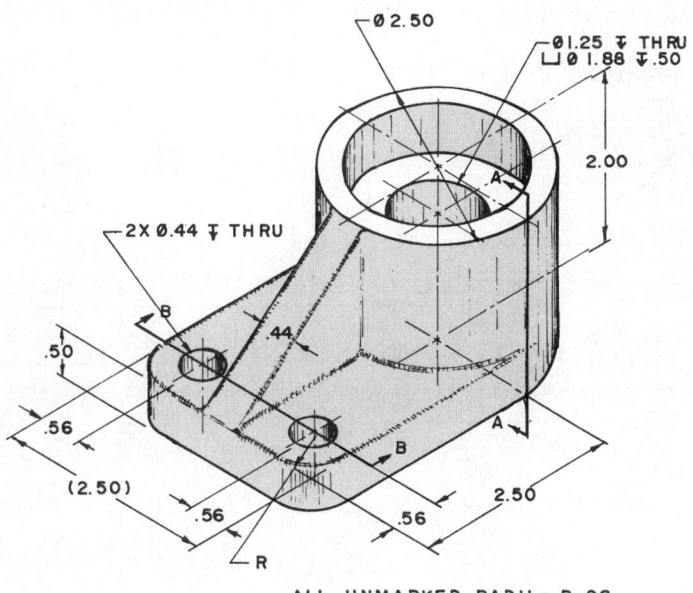

ALL UNMARKED RADII = R.06

Problem 5-27

Problem 5-28

Center the front view, side view and removed sections A-A, B-B, and C-C within the work area.

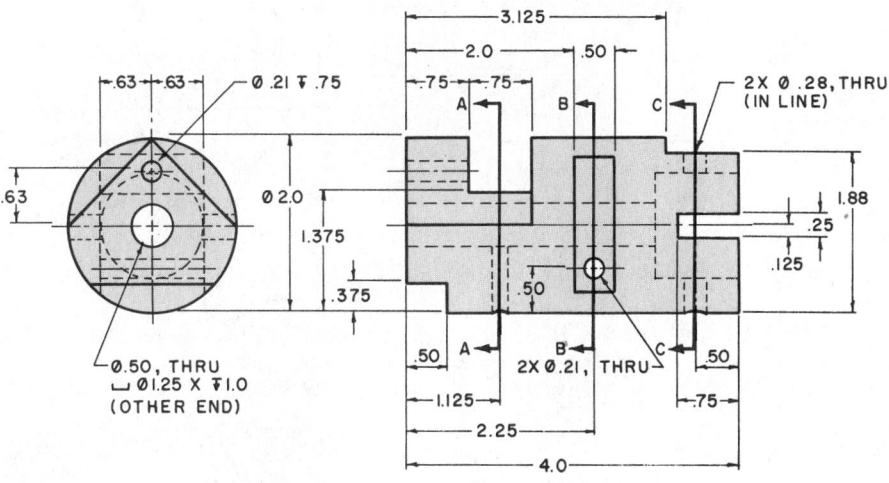

Problem 5-28

Problem 5-29

Make a two-view assembly drawing of parts 1, 2, 3, and 4. Make one view a full-section assembly. Use correct section lining, and all conventional drafting practices.

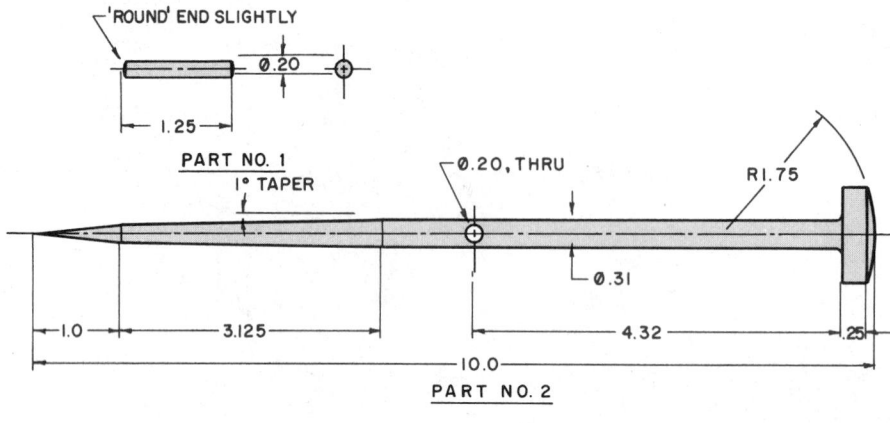

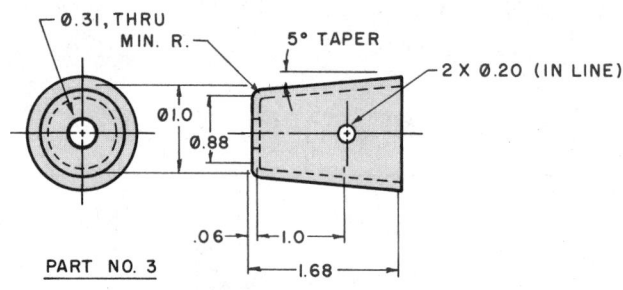

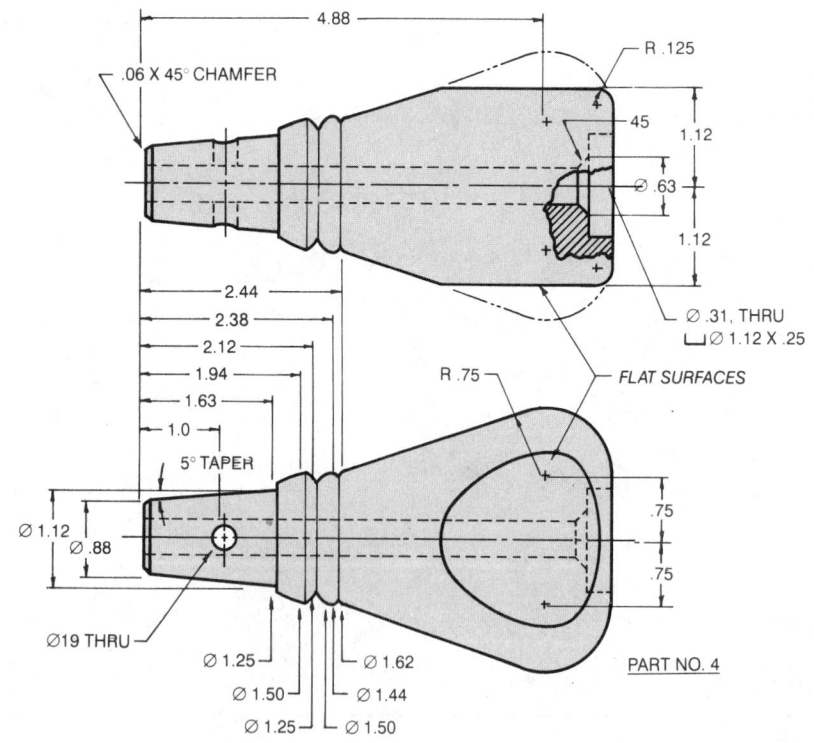

Problem 5-30

Make a two-view assembly drawing of parts 1, 2, and 3. Make one view a full-section assembly. Use correct section lining and all conventional drafting practices.

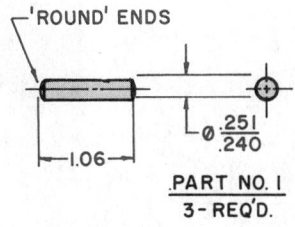

'ROUND' ENDS

$\varnothing \dfrac{.251}{.240}$

1.06

PART NO. 1
3- REQ'D.

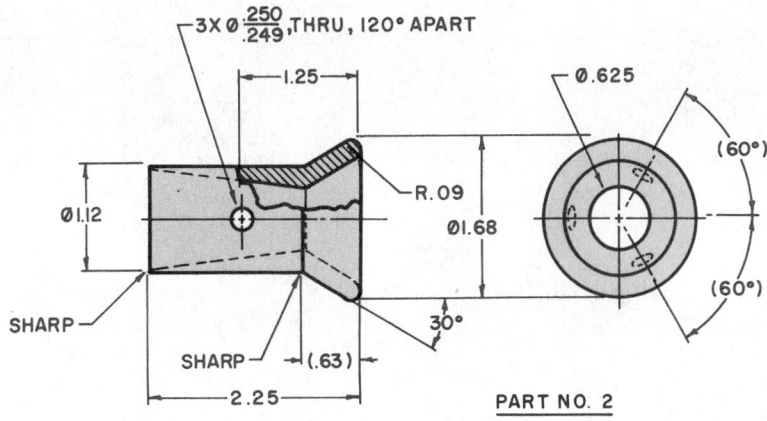

$3 \times \varnothing \dfrac{.250}{.249}$, THRU, 120° APART

1.25

$\varnothing 0.625$

(60°)

R.09

$\varnothing 1.12$

$\varnothing 1.68$

(60°)

30°

SHARP

SHARP (.63)

2.25

PART NO. 2

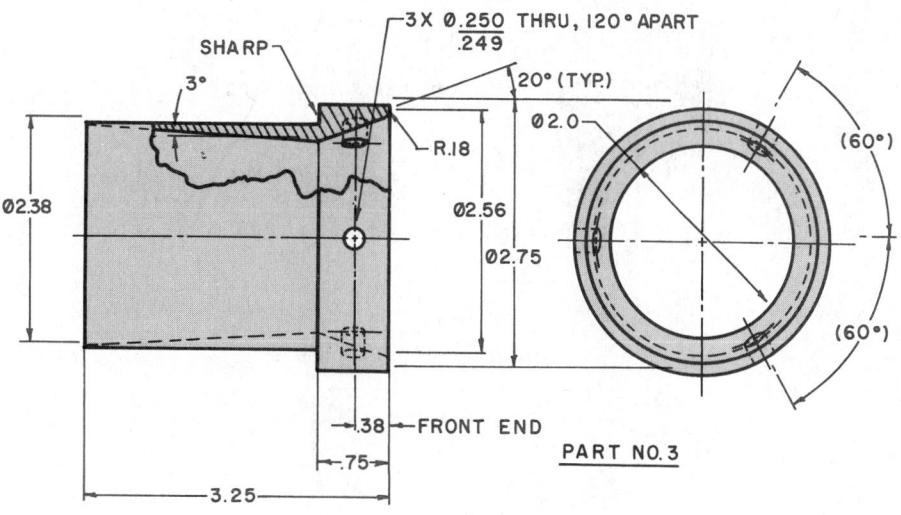

SHARP

$3 \times \varnothing \dfrac{.250}{.249}$ THRU, 120° APART

3°

20° (TYP.)

$\varnothing 2.0$

(60°)

R.18

$\varnothing 2.38$

$\varnothing 2.56$

$\varnothing 2.75$

(60°)

.38 FRONT END

.75

PART NO. 3

3.25

Problem 5-30

Problem 5-31

Make a two-view assembly drawing of parts 1, 2, and 3. Make one view a full-section assembly. Use correct section lining and all conventional drafting practices.

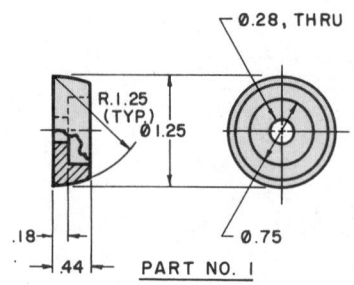

PART NO. 1

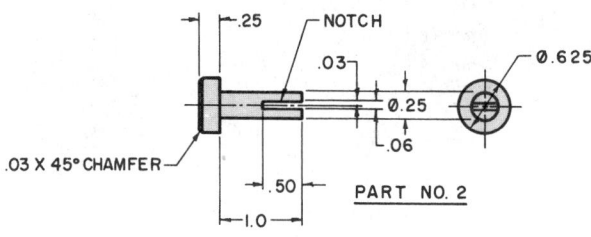

PART NO. 2

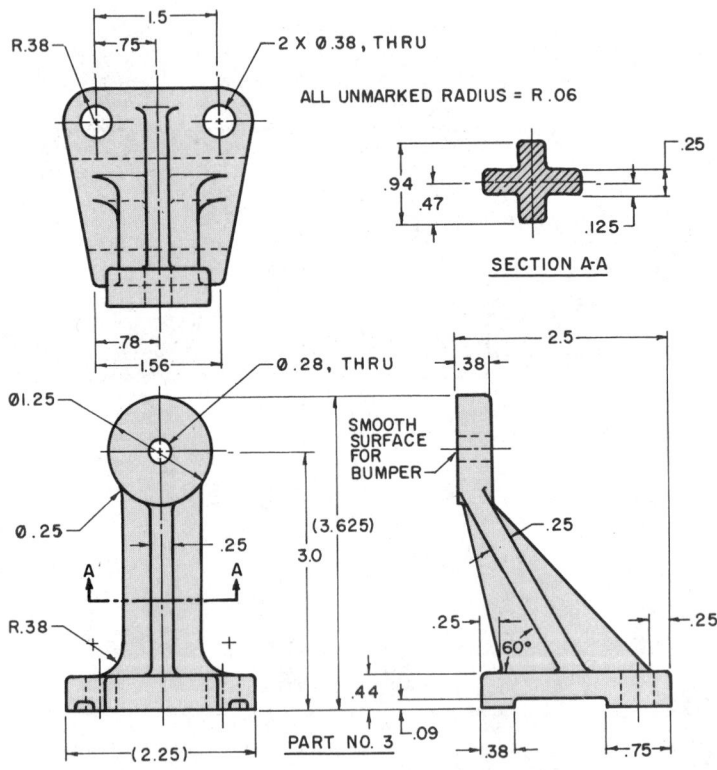

PART NO. 3

Problems 5-32 through 5-37

Center required views within the work area. Leave a 1-inch or 25-mm space between views. Make one view into a section view to fully illustrate the object. Use either a full, half, offset, broken-out, revolved, or removed section. Do not add dimensions.

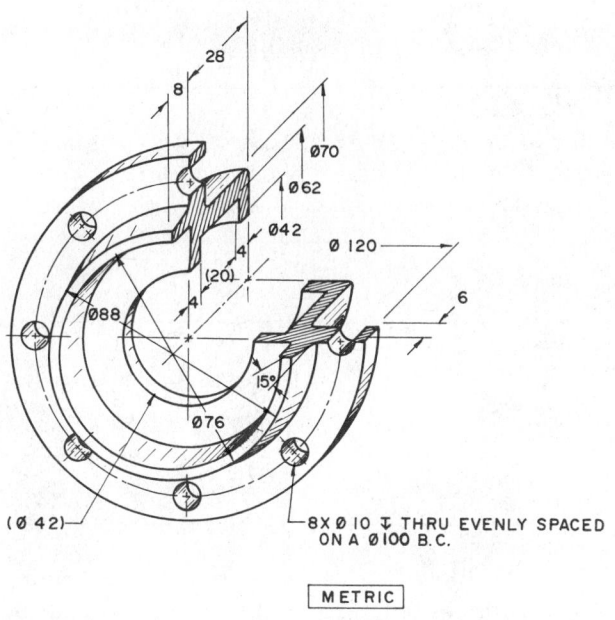

Problem 5-32

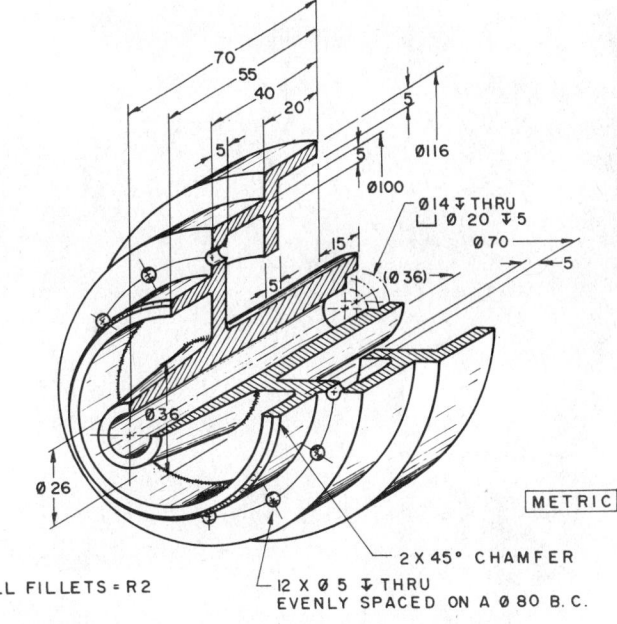

Problem 5-34

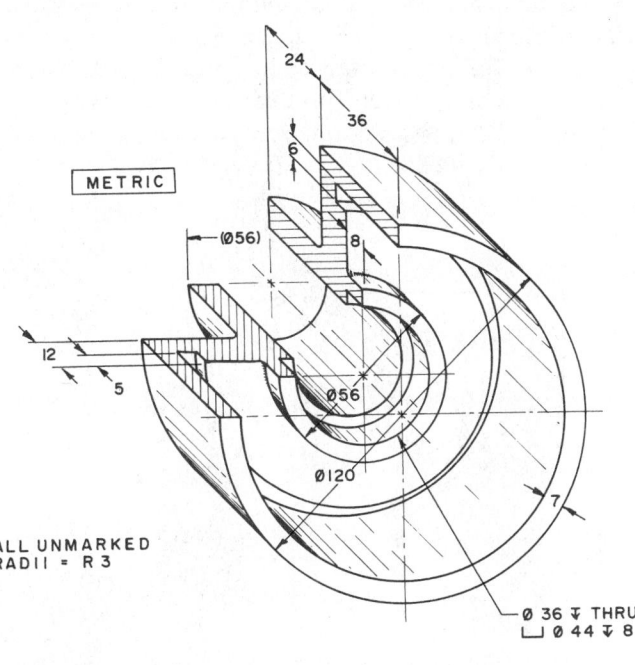

Problem 5-33

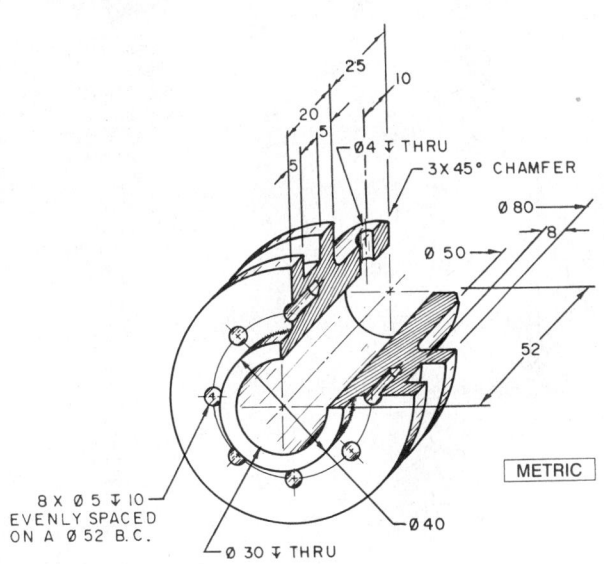

Problem 5-35

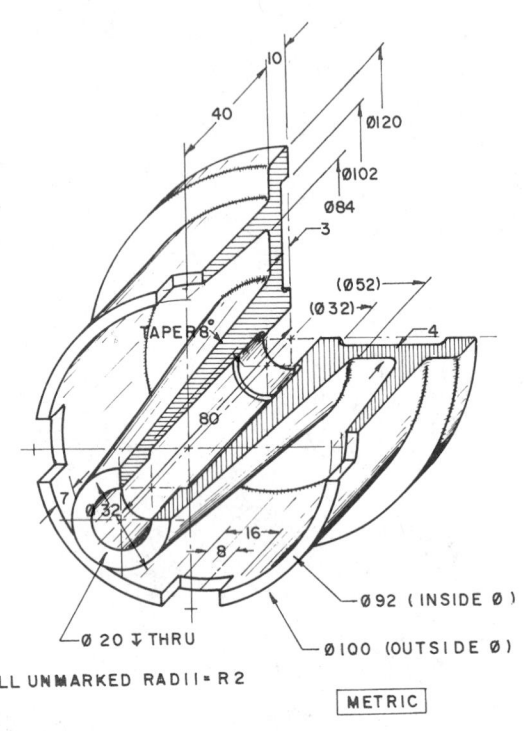

ALL UNMARKED RADII = R2

METRIC

Problem 5-36

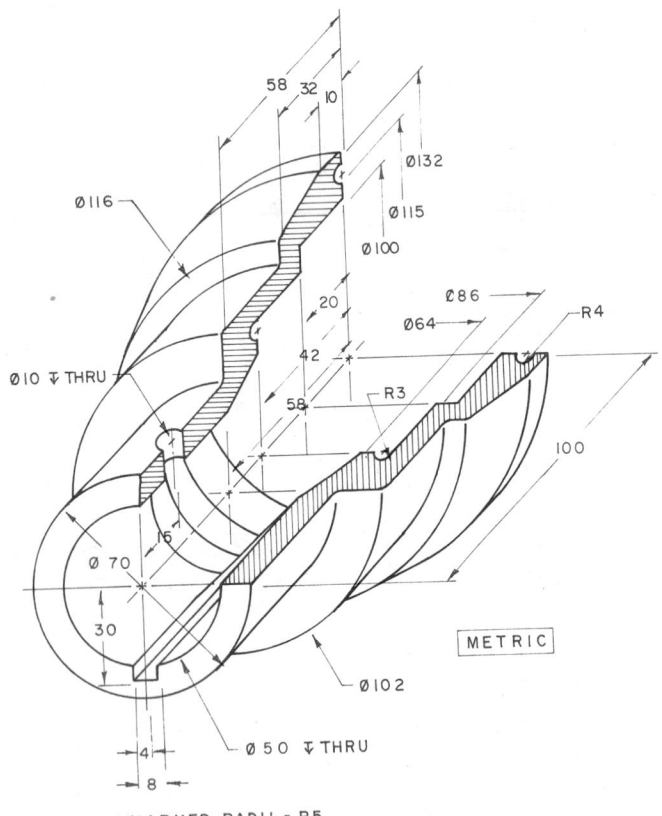

ALL UNMARKED RADII = R5

Problem 5-37

☐ Problem 5-38

Refer to Problem 17-9 and use the full section drawing for information on dimensions to determine the overall distance from the outside of part (1) on the left to the outside of part (1) on the right, along the centerline of the pulley. Prepare a sketch with dimensions for your answer.

☐ Problem 5-39

Refer to Problem 18-52 (assembly) and prepare an assembly section (front-view orientation) of the vise. Indicate approximate dimensions only.

☐ Problem 5-40

Refer to Problem 18-50 and prepare a full-section view of the object.

☐ Problem 5-41

Refer to Problem 9-21 and prepare an offset section view (profile orientation) of the object through the centerline. Include a threaded hole.

☐ Problem 5-42

Select an object having a nonuniform cross section like the scissor half shown and prepare removed section views to illustrate detail of the object. Suggested objects: hammer, pliers, bolt cutters, boat oar, snow ski, water ski, propeller (marine, airplane, contemporary wind machine, etc.), nozzle, handles, spokes, etc.

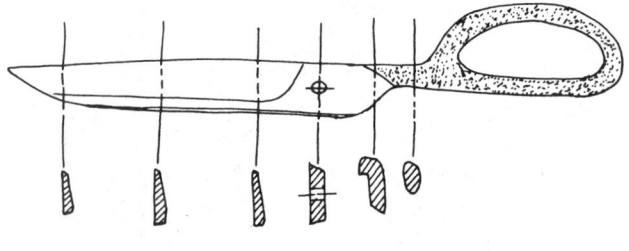

☐ Problem 5-42

☐ Problem 5-43

Refer to Problem 9-12 and prepare an aligned section view of the object. Show the threaded holes and the keyway using conventional practices.

Problem 5-44

Refer to Problem 9-37 and prepare an aligned section view of the object.

Problem 5-45

Refer to Problem 9-43. Prepare a conventional section view, showing ribs of the object.

Problem 5-46

Refer to Problem 17-13 (top view). Prepare a section, drawn to scale, through Part (2) showing Parts (2), (6), (10), (11), and (12) only.

Problem 5-47

Refer to the drawings shown. Prepare an offset section C-C, full size, of the distributor cap. Dimensions are not required.

Problem 5-48

Select one of the following common devices and prepare a section view(s) to best show inside details, and how the device works, for a non-technical person. Simplify the details of the device for clarity. How-things-work books would be useful references. Indicate approximate sizes only.

  a. clock
  b. toaster
  c. internal-combustion-engine cylinder
  d. ski binding
  e. electric drill
  f. car window mechanism (manual or electrical)
  g. electric motor application in an average home including an automobile.

Note: The average home plus an automobile has approximately 35 electric motors.

Problem 5-49

Find material on threaded inserts (usually a threaded collar made of steel, inserted in softer material such as aluminum). Prepare a section drawing showing an application. Indicate approximate dimensions.

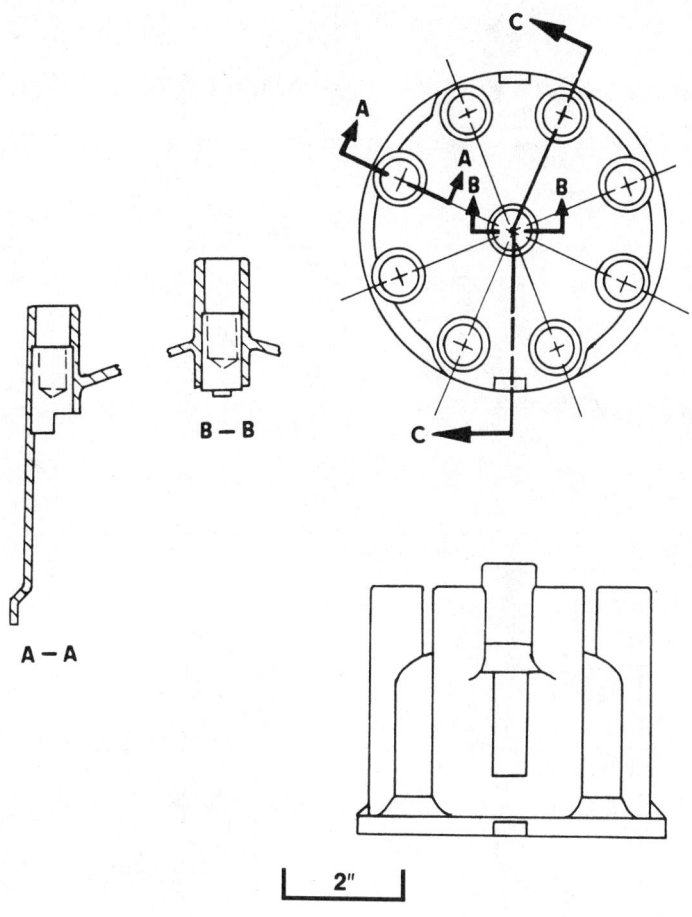

Problem 5-47

❑ Problem 5-50

Select an object from one of the locations or categories listed and prepare a multiview drawing of your redesign ideas for the object. Use section views to replace traditional views where appropriate. (Refer to Figure 5-15.)

- **a.** classroom
- **b.** lab
- **c.** shop
- **d.** garage
- **e.** automobile
- **f.** hobby
- **g.** sports equipment

❑ Problem 5-51

Use the partially completed drawings shown and do the following.

- **a.** Assume that the beam used has a depth of 152 mm and a flange width of 84 mm. Estimate dimensions of the parts which make up dimension X, and show in a freehand sketch dimension X and your assumptions.

b. Make additional estimates: (1) of the thickness of the plates for the trolley wheels, and (2) of the distances from the trolley wheels to the lower bolted connection and to the hole. Prepare an offset section through the trolley wheels, through the bolted connection, and through the hole.

❑ Problem 5-52

Design a lever-type door handle for handicapped individuals who have prosthetics for hands and cannot use knob type door openers. Use removed sections to show the details of the lever arm. Indicate approximate dimensions.

❑ Problem 5-53

Design a desk for a personal computer, keyboard, printer with track-fed paper, a mouse with pad, and a monitor. Prepare a multiview drawing of the desk and show a cross sectional view (profile view orientation) of the desk. A 26" knee standard is a typical standard for computer desks. Indicate approximate dimensions.

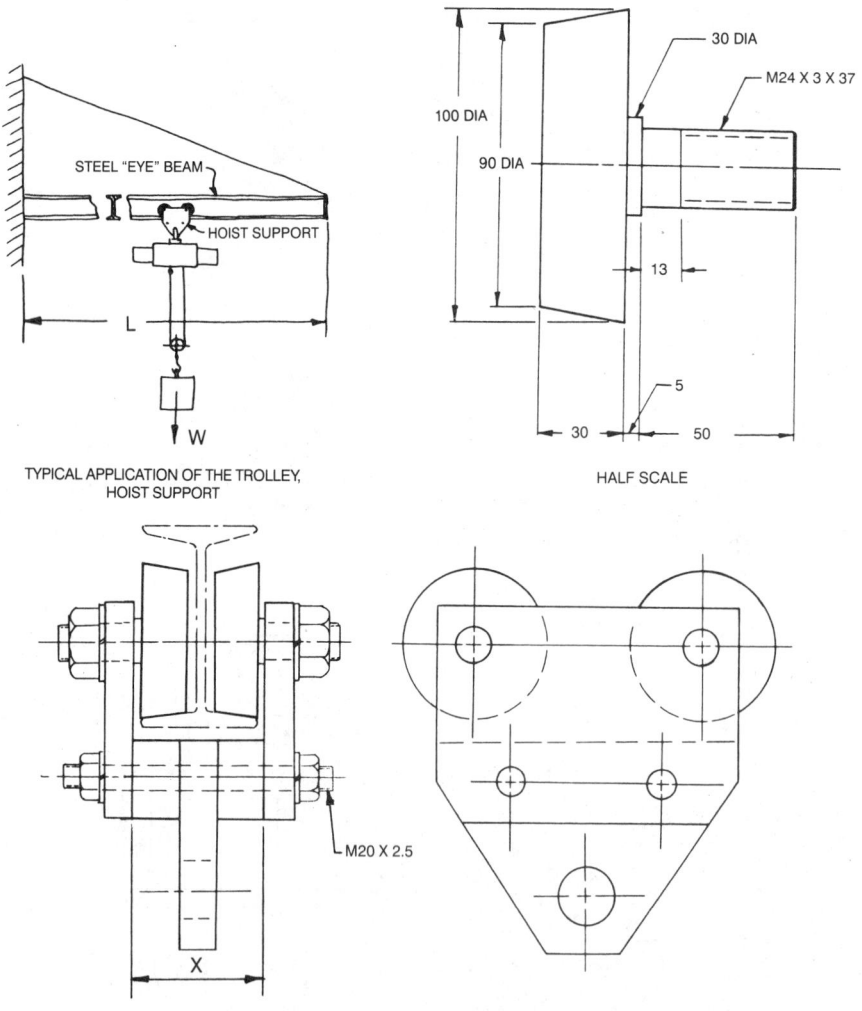

TYPICAL APPLICATION OF THE TROLLEY, HOIST SUPPORT

HALF SCALE

Problem 5-51

Auxiliary Views

CHAPTER 6

Chapter Outline

- Auxiliary Views Defined
- Secondary Auxiliary Views
- Partial Views
- Auxiliary Section
- Half Auxiliary Views
- Review
- Problems

Auxiliary Views Defined

Many objects have inclined surfaces that are not always parallel to the regular planes of projection. For example, in ***Figure 6-1***, the front view is correct as shown, but the top and right-side views do *not* correctly represent the inclined surface. To truly represent the inclined surface and to show its true shape, an auxiliary view must be drawn. An *auxiliary view* has a line of sight that is perpendicular to the inclined surface, as viewed looking directly at the inclined surface. Auxiliary views are always projected 90° from the inclined surface.

An auxiliary view serves three purposes:

- It illustrates the true size of a surface.
- It illustrates the true shape of a surface, including all true angles and/or arcs.
- It is used to project and complete other views.

An auxiliary view can be constructed from any of the regular views. An auxiliary view projected from the front view would appear as it does in Figure 6-1. This is referred to as a *front view auxiliary*. An auxiliary view projected from the top view would appear as it does in

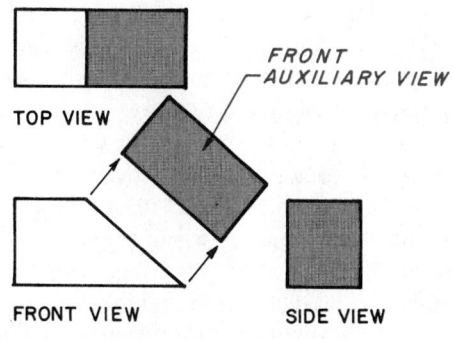

Figure 6-1 Front view auxiliary view

Figure 6-2. This is referred to as a *top view auxiliary*. An auxiliary view projected from the right-side view would appear as it does in ***Figure 6-3***. This is referred to as a *side view auxiliary*. Note in each case that the auxiliary view is projected 90° from the inclined or slanted surface, and is viewed from a line of sight 90° to the inclined or slanted surface, or as viewed looking directly down upon the inclined surface.

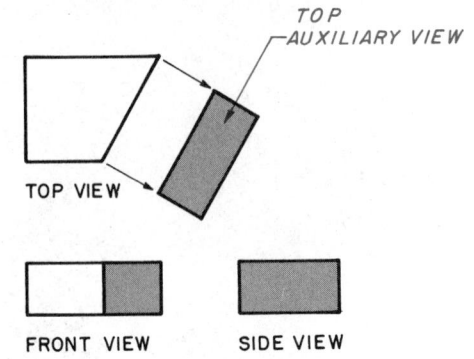

Figure 6-2   Top view auxiliary view

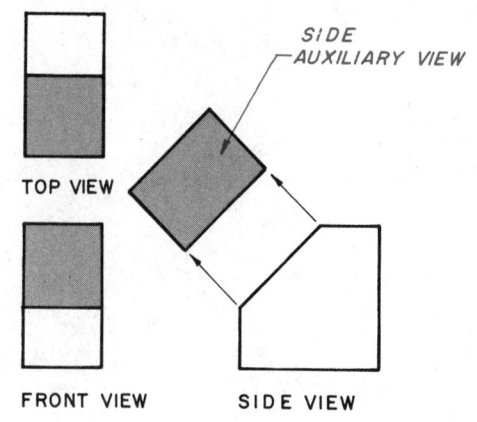

Figure 6-3   Side view auxiliary view

Hidden Lines in an Auxiliary View

Hidden lines should be omitted in an auxiliary view, unless they are needed for clarity. This is the drafter's prerogative or decision.

How To Draw an Auxiliary View

Given: The pictorial view of an object, **Figure 6-4**. Notice the inclined surface. As the inclined surface is on the front view, this will be a front view auxiliary. The usual three views of an object, front view, top view and right-side view, are shown in **Figure 6-5**.

Step 1   Label all important points of the auxiliary view, as illustrated in **Figure 6-6A**.

Step 2   Construct a *reference* line, which is also the edge view of a reference plane, in the right-side view. Always construct a reference line so that it is vertical and passes through as many points as possible. In this example, it passes through points a–d, **Figure 6-6B**.

Step 3   Draw light projection lines 90° from the inclined surface, and construct a reference line *parallel* to the inclined surface at any convenient distance, as shown in **Figure 6-6C**. Label all important points established thus far. Notice points a and d are *on* the reference line.

Step 4   In the right-side view, measure the distance each point is *from* the reference line. Project these distances back to the inclined surface of the front view and up to 90° from the inclined surface to the reference line above. Transfer each distance, and lightly label each point as illustrated in **Figure 6-6D**.

Step 5   Recheck all work. Be sure:
  • The projection is 90° from the inclined surface, in this example the front view.
  • The reference line is *parallel* to the inclined surface of the front view.
  • All distances have been transferred accurately.

If correct, carefully darken in and complete all views. The final finished drawing will appear as it does in **Figure**

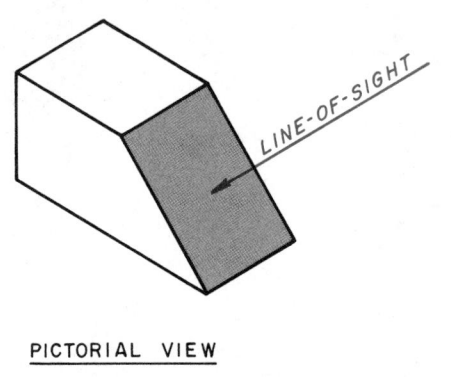

Figure 6-4   Drawing an auxiliary view

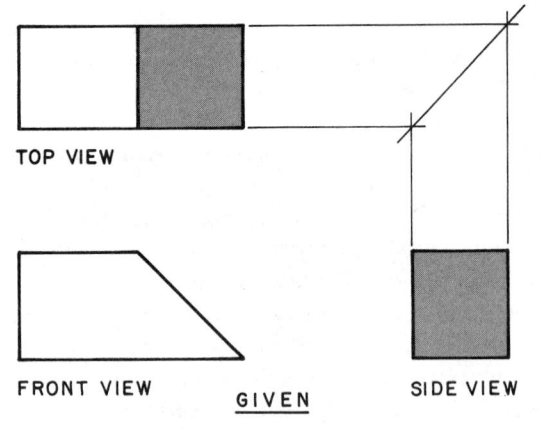

Figure 6-5   Given: Three views of an object

6-6E. Notice that *only* the inclined surface is projected into the auxiliary view. Anything else would be foreshortened and, thus, not of true size or shape, and therefore of no use. Good drafting practice is to project *only* the surface of the inclined line, **Figure 6-7**.

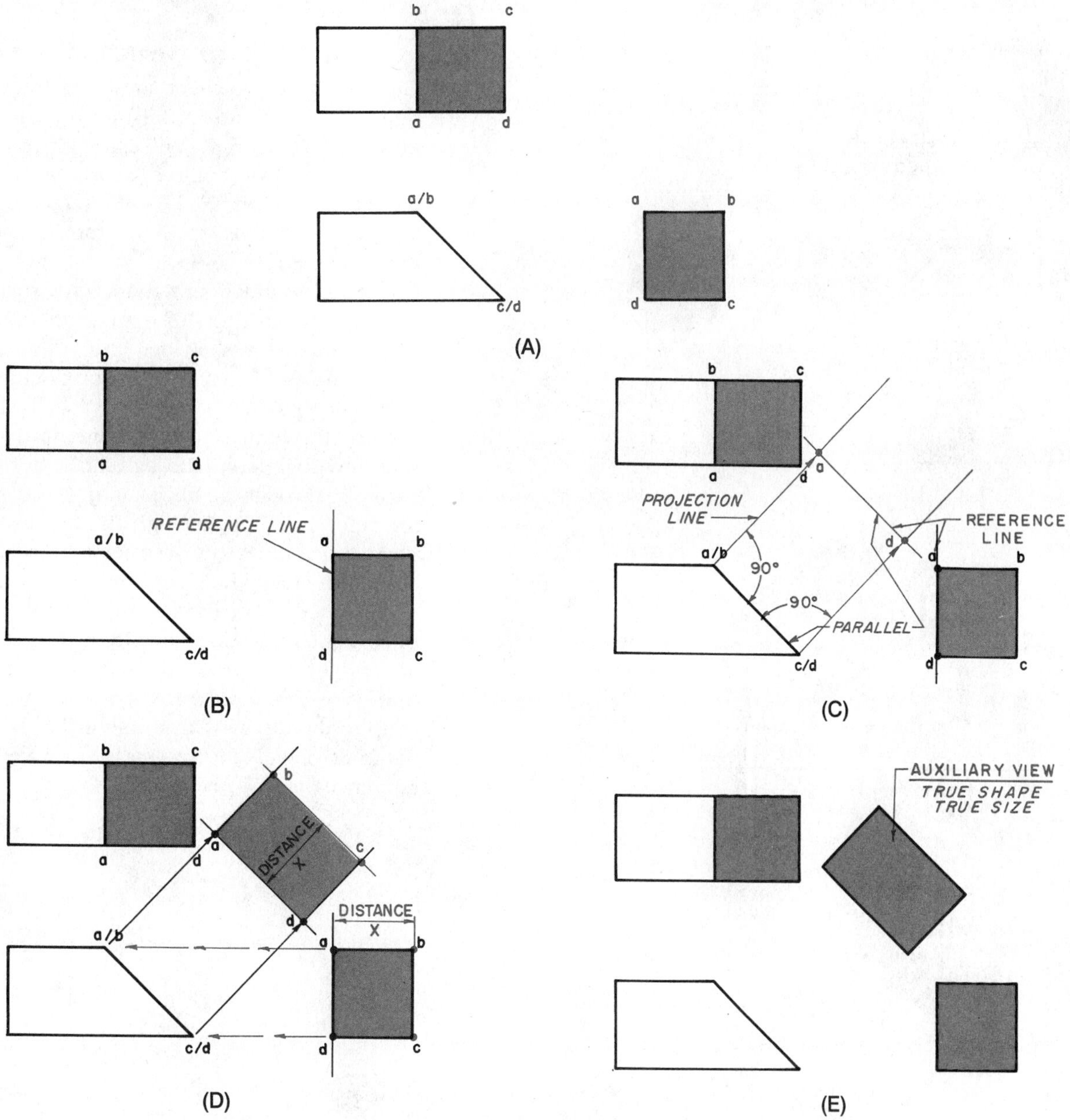

(A)

(B)

REFERENCE LINE

PROJECTION LINE

REFERENCE LINE

90°

90°

PARALLEL

(C)

DISTANCE X

DISTANCE X

(D)

AUXILIARY VIEW
TRUE SHAPE
TRUE SIZE

(E)

Figure 6-6   (A) Step 1;   (B) Step 2;   (C) Step 3;   (D) Step 4;   (E) Step 5 (finished drawing)

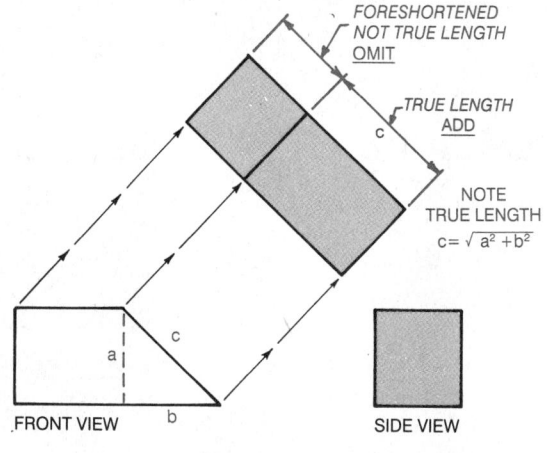

FORESHORTENED
NOT TRUE LENGTH
OMIT

TRUE LENGTH
ADD

NOTE
TRUE LENGTH
$c = \sqrt{a^2 + b^2}$

FRONT VIEW

SIDE VIEW

Figure 6-7   Draw only the inclined surface in the auxiliary view

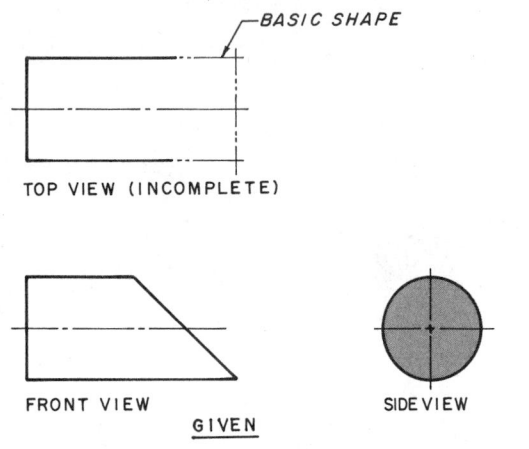

TOP VIEW (INCOMPLETE)

FRONT VIEW   SIDE VIEW

GIVEN

Figure 6-8   Projecting a round surface from an inclined surface

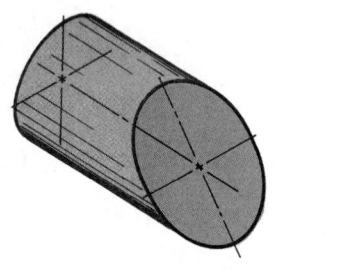

Figure 6-9   Pictorial view of an object

How To Project a Round Surface from an Inclined or Slanted Surface

Given: The usual three views of an object, front view, right-side view and unfinished top view, are shown in ***Figure 6-8***. (The usual 45° projection angle line will be omitted, as a slightly newer projection method will be used to complete the top view.) Refer also to the pictorial drawing of this object, ***Figure 6-9***.

Step 1 Divide the rounded view into equal spaces. In this example, using a 30° triangle, the right-side view is rounded and divided into 12 equal parts, ***Figure 6-10A***. Letter each point clockwise, as shown in the figure.

Step 2 Construct a vertical reference line in the right-side view so it passes through the center; in this example, through points a and g. (Always place the reference line through the center of any symmetrical object.) Construct a reference line in the top view which runs through the center, as illustrated in ***Figure 6-10B***.

Step 3 Draw light projection lines from the 12 points in the right-side view to the inclined edge in the front view. Project these same 12 points directly up to the top view from the inclined edge in the front view. Notice points a and g are *on* the reference line, ***Figure 6-10C***.

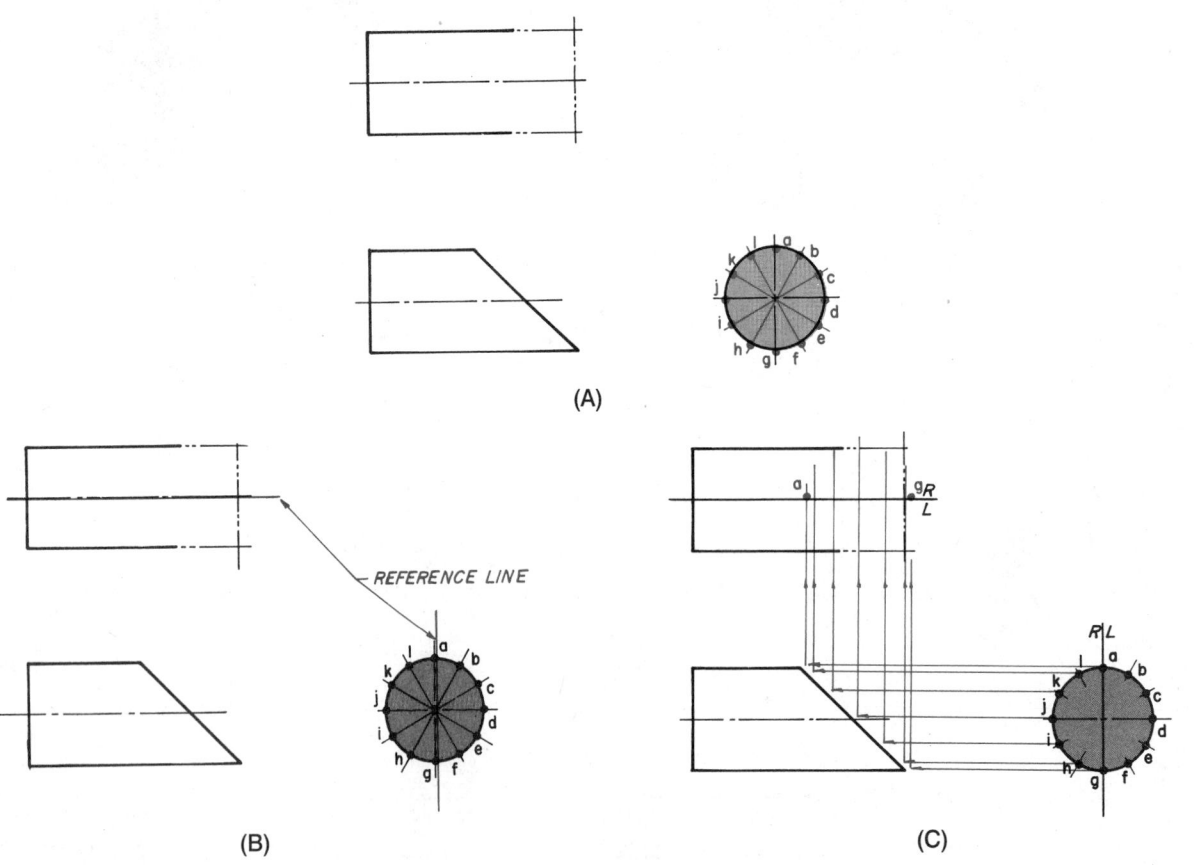

(A)

— REFERENCE LINE

(B)   (C)

Figure 6-10   (A) Step 1;   (B) Step 2;   (C) Step 3

Step 4 In the right-side view, measure the distance each point is *from* the reference line. Transfer each of these distances from the right-side view reference line to the top view reference line. Label each point lightly as each is found, ***Figure 6-10D***.

Step 5 Lightly connect all points and, if correct, darken in all views. This completes the top view, ***Figure 6-10E***. Always darken in the compass and irregular curve layout work first. This completes the top view.

How To Draw an Auxiliary View of a Round Surface

Step 6 Draw light projection lines 90° from the inclined surface of the front view. Construct the reference line *parallel* to the inclined surface at any convenient distance. Label the points that are on the reference line; in this example, a and g, ***Figure 6-10F***.

Step 7 Draw light projection lines from the 12 points in the right-side view to the inclined edge in the front view (Step 3). Project these same 12 points directly up to the auxiliary view from the inclined edge in the front view. Again, notice points a and g are *on* the reference line, ***Figure 6-10G***. In the right-side view, measure the distance each point is from the reference line. Transfer each of these distances from the right-side view reference line to the auxiliary view reference line. Label each point lightly as each is found.

Step 8 Lightly connect all points and, if correct, darken in the auxiliary view. This completes the auxiliary view, ***Figure 6-10H***. Notice that only the inclined surface has been projected into the auxiliary view.

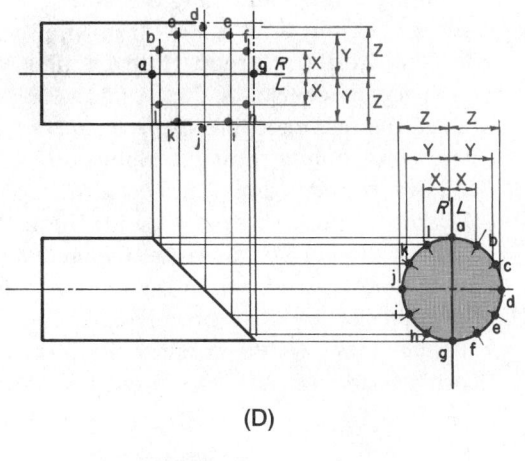

(D)

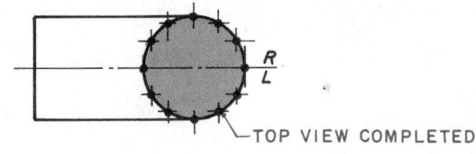

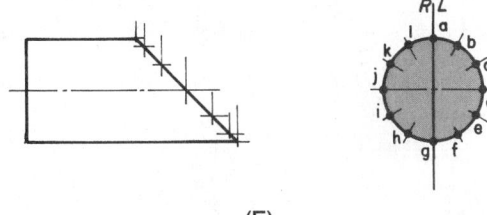

(E)

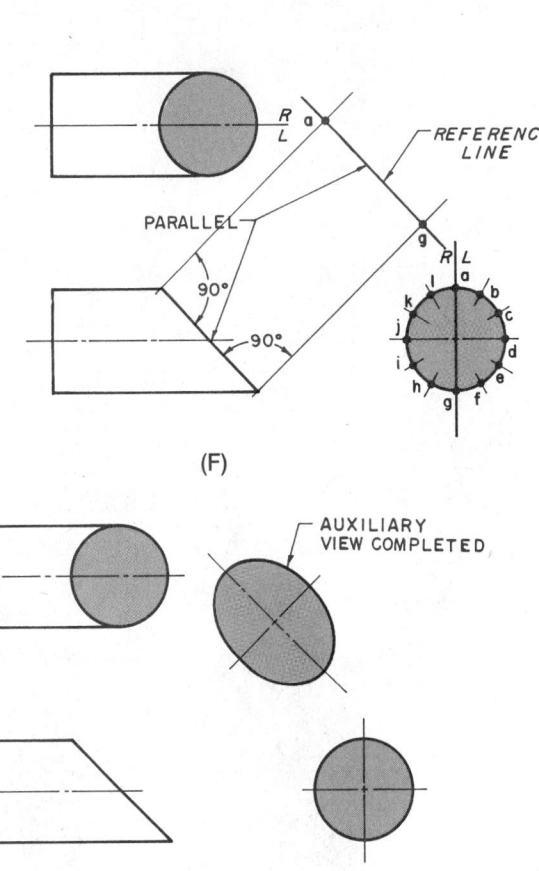

(F)

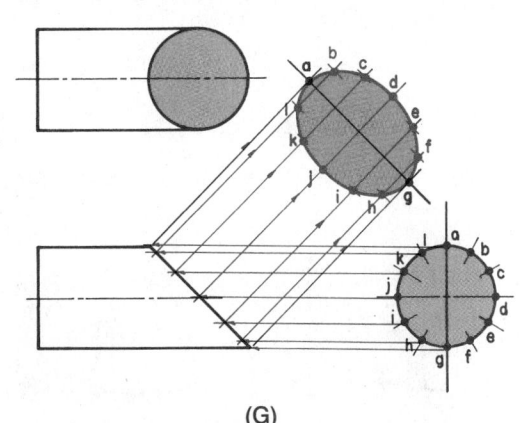

(G)

(H)

Figure 6-10 (Continued) (D) Step 4; (E) Step 5; (F) Step 6; (G) Step 7; (H) Step 8 (finished drawing)

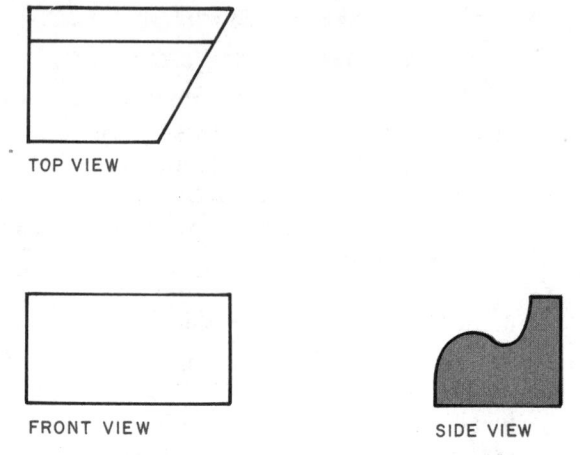

Figure 6-11   Plotting an irregular curved surface

How to Plot an Irregular Curved Surface

Given: The usual three views: front view (incomplete), top view, and the right-side view, *Figure 6-11*. This is an example of a top view auxiliary.

Step 1 Add various points at random along the curved surface. Even spaces are not necessary, but try to choose points that pick up high and low points along the line, *Figure 6-12A*. Always label the points in a clockwise direction.

Step 2 From the right-side view, project these points *up* to the 45° projection line and over to the top view. Where these points intersect with the inclined surface, project directly down into the front view. Project these same points from the right-side view to the front view. Where the points intersect is where the point actually is. Lightly locate and connect all lines and complete the front view, *Figure 6-12B*.

Step 3 Establish a reference line in the right-side view. Construct this reference line so it passes through as many points as possible; in this example, points a and j, *Figure 6-12C*.

Step 4 Add projection lines at 90° from the inclined surface. Add a reference line parallel to the inclined surface. Notice that points a and j again fall *on* the reference line, *Figure 6-12D*.

Step 5 Project all points from the right-side view up and over to the inclined surface of the top view. Where the points intersect with the inclined surface, project 90° from the inclined surface. Transfer all distances from the reference line in the right-side view to the reference line in the auxiliary view. Lightly connect all points and, if correct, darken in all views, *Figure 6-12E*.

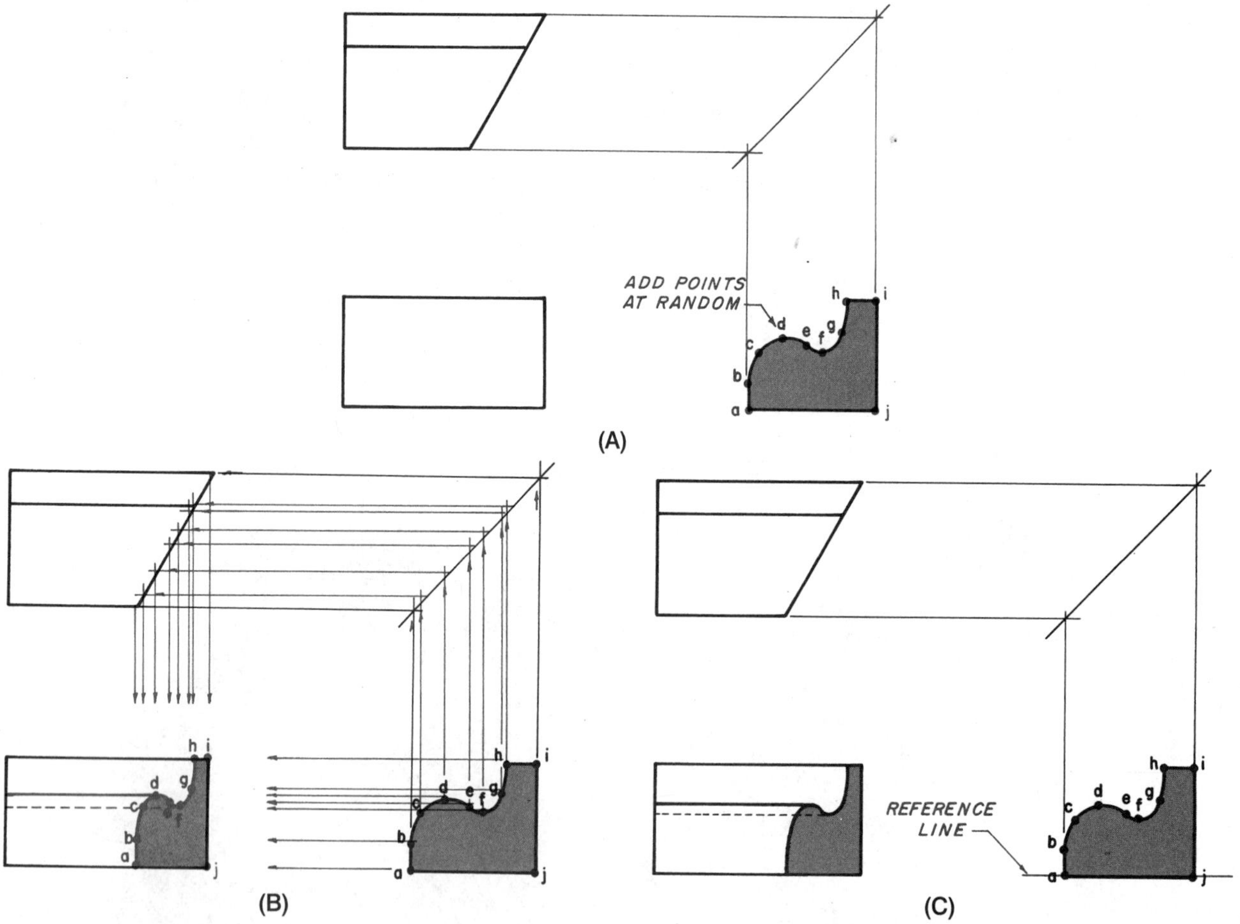

Figure 6-12   (A) Step 1; (B) Step 2; (C) Step 3

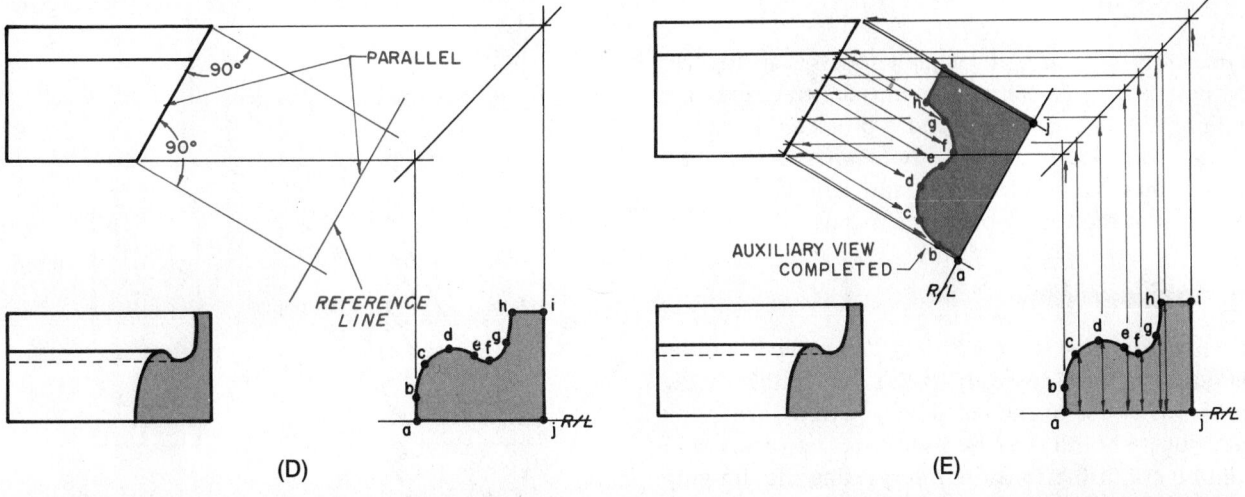

Figure 6-12 (Continued) (D) Step 4; (E) Step 5 (finished drawing)

Secondary Auxiliary Views

Up to this point, primary auxiliary views have been dealt with. A *primary auxiliary view* can be projected from any of the regular views, as has been illustrated thus far.

Sometimes, a primary auxiliary view is not enough to fully illustrate an object; a secondary auxiliary view is needed. A *secondary auxiliary view* is projected directly *from* the auxiliary view. Many times, the auxiliary view and/or some of the other views cannot be fully *completed* until the secondary view has been completed first, ***Figure 6-13***.

Partial Views

The use of a *partial auxiliary view* makes it possible to eliminate one or more of the regular views which, in turn, saves drafting time and cost. ***Figure 6-14*** is an example of a front view, top view, right-side view and auxiliary view. Note, the auxiliary view is always drawn

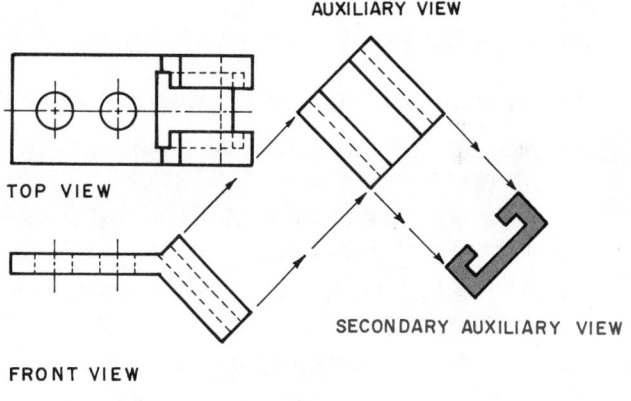

Figure 6-13 Secondary auxiliary view

partial; only the *inclined* surface is drawn on the auxiliary view. By using a partial top and right-side view in conjunction with the particular auxiliary view, the drawing can be simplified without detracting from its clarity; in fact, in most cases, these partial views make the drawing easier to read, ***Figure 6-15***.

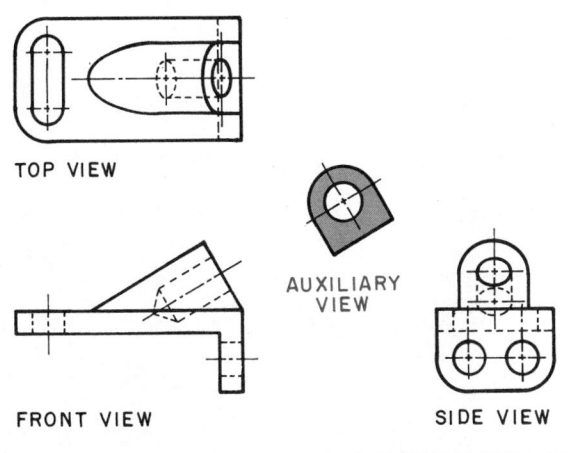

Figure 6-14 Given: Regular three views

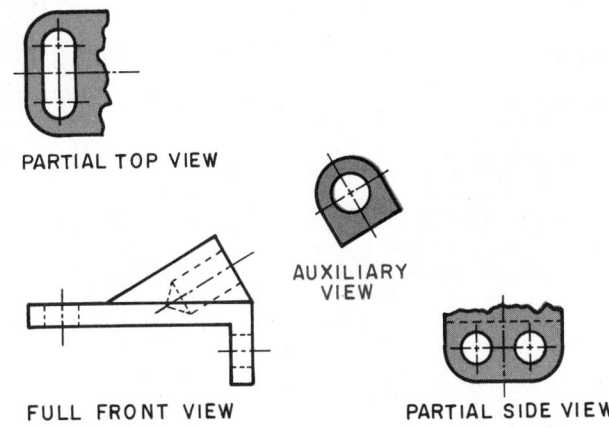

Figure 6-15 Partial auxiliary views

Auxiliary Section

An *auxiliary section*, as its name implies, is an auxiliary view in section. An auxiliary section is drawn exactly as is any removed sectional view, and is projected in exactly the same way as any auxiliary view, *Figure 6-16*. All the usual auxiliary view rules apply, and generally only the surface cut by the cutting-plane line is drawn.

Half Auxiliary Views

If an auxiliary view is symmetrical, and space is limited, it is permitted to draw only half of the auxiliary view, *Figure 6-17*. Use of the half auxiliary view saves some time, but it should only be used as a last resort, as it could be confusing to those interpreting the drawing. Always draw the *nearest* half, as shown in the figure.

Review

1. What three purposes does an auxiliary view serve?

2. Name the three major kinds of auxiliary views.

3. What must be done first if the projected surface is round or has a radius?

4. Explain the use of partial views as used in conjunction with an auxiliary view.

5. What is the practice for the use of hidden lines in an auxiliary view?

6. How should the regular views and the auxiliary view be placed within the work area?

7. Explain the use of a reference line. Where should it be drawn, and at what angle?

8. Projection lines *must* be drawn at what angle from the edge view?

9. When and why is a half auxiliary view used?

10. What is an auxiliary sectional view?

11. Explain the use of a secondary auxiliary view.

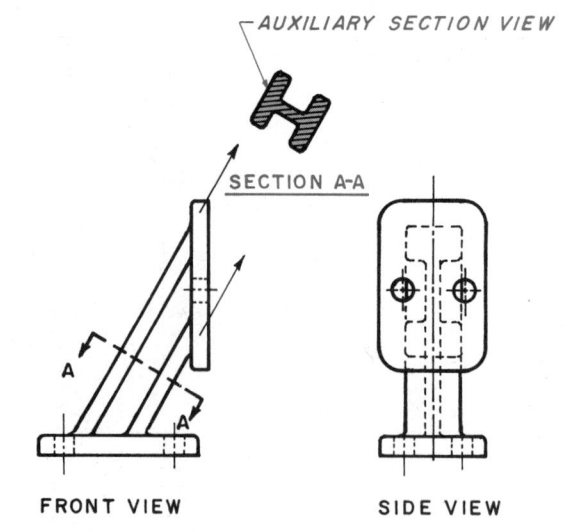

Figure 6-16 Auxiliary section

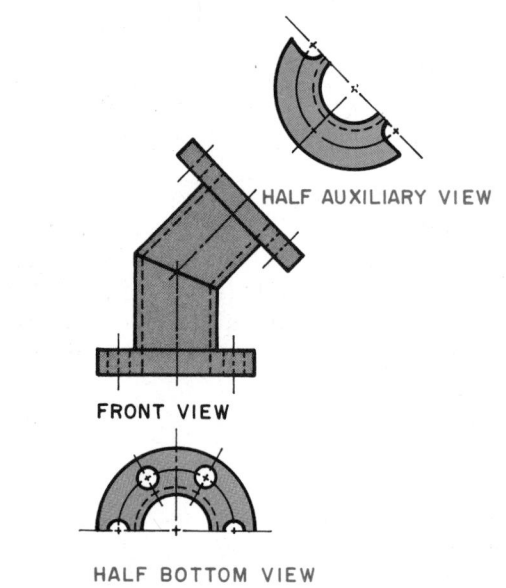

Figure 6-17 Half auxiliary view

Chapter Six Problems

The following problems are intended to give the beginning drafter practice in sketching and laying out multiviews with an auxiliary view.

The steps to follow in laying out any drawing with an auxiliary view are:

Step 1   Study the problem carefully.

Step 2   Choose the view with the most detail as the front view.

Step 3   Position the front view so there will be the least number of hidden lines in the other views.

Step 4   Determine which view from which to project the auxiliary view.

Step 5   Make a sketch of all views, including the auxiliary view.

Step 6   Center the required views within the work area with approximately 1-inch (25-mm) space between the views. Adjust the regular views to accommodate the auxiliary view.

Step 7   Use light projection lines. Do *not* erase them.

Step 8   Lightly complete all views.

Step 9   Check to see that all views are centered within the work area.

Step 10 Carefully check all dimensions in all views.

Step 11 Darken in all views using correct line thickness.

Step 12 Recheck all work, and, if correct, neatly fill out the title block using light guidelines and neat lettering.

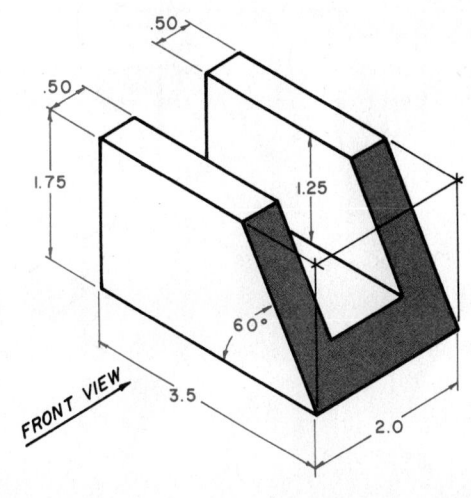

Problem 6-2

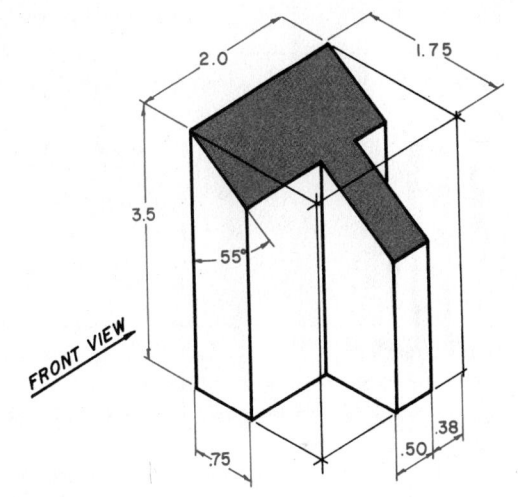

Problem 6-3

Problems 6-1 through 6-4

Draw the front view, top view, right-side view and auxiliary view. Complete all views using the listed steps.

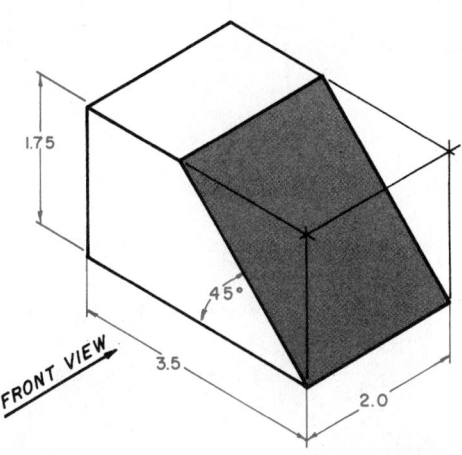

Problem 6-1

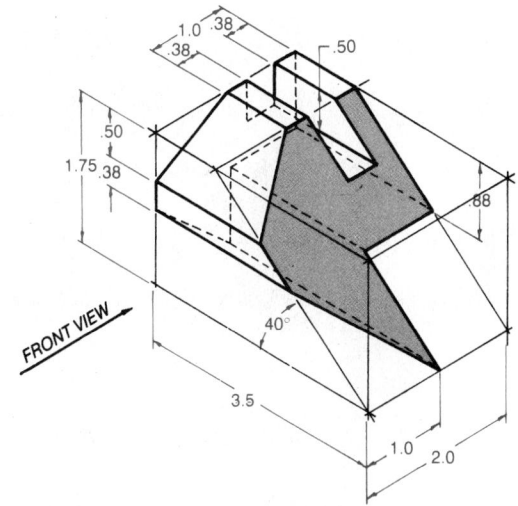

Problem 6-4

Problems 6-5 and 6-6

Draw the front view, top view, right-side view and *two* auxiliary views to illustrate the true size and shape of the slanted surfaces. Complete all views using the listed steps.

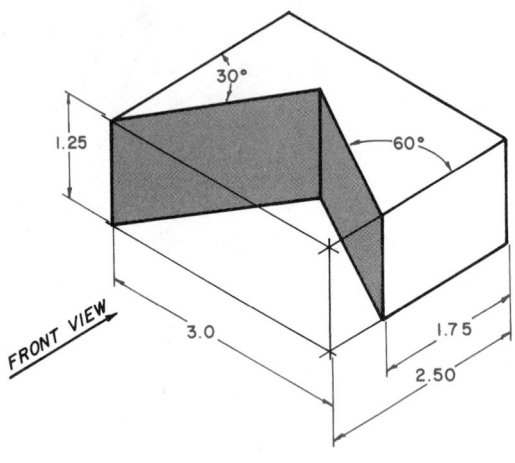

Problem 6-5

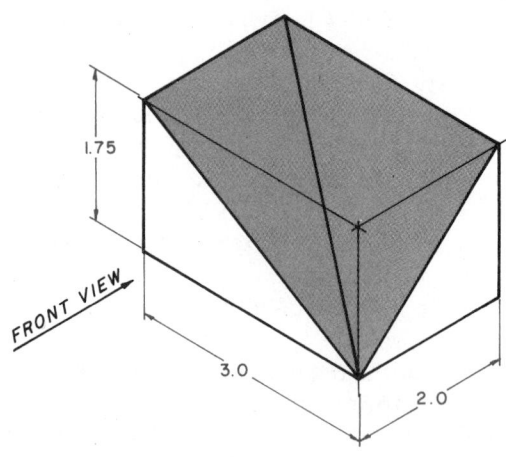

Problem 6-6

Draw the front view, top view, right-side view and auxiliary view. Complete all views using the listed steps.

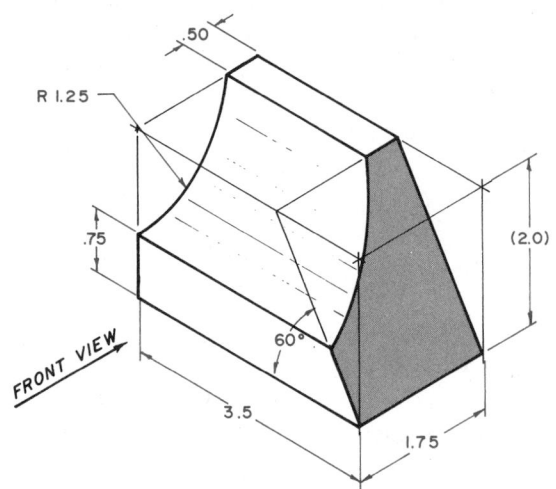

Problem 6-7

Problems 6-7 through 6-13

Draw the front view, top view, right-side view and auxiliary view. Complete all views using the listed steps.

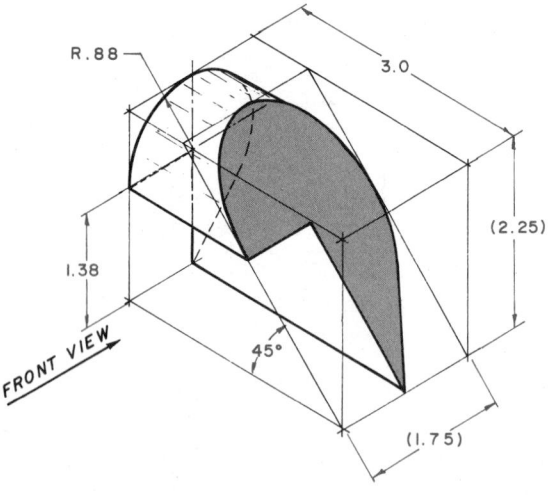

Problem 6-8

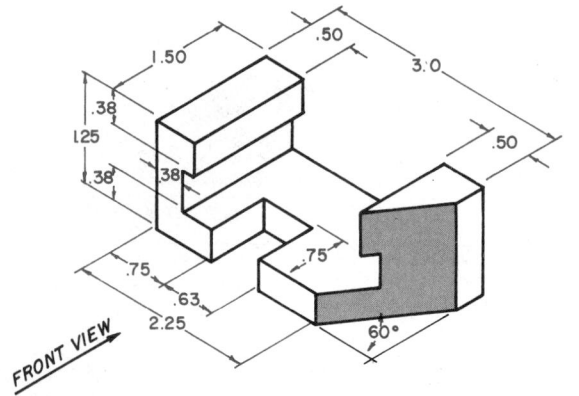

Problem 6-9

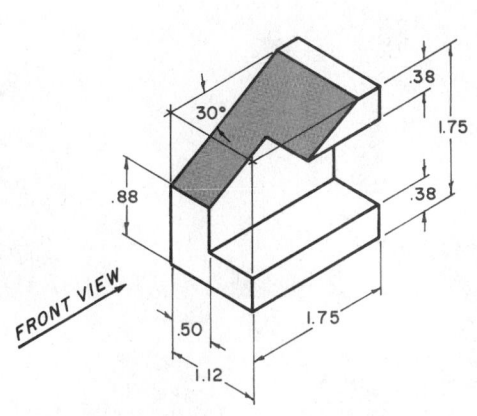

Problem 6-10

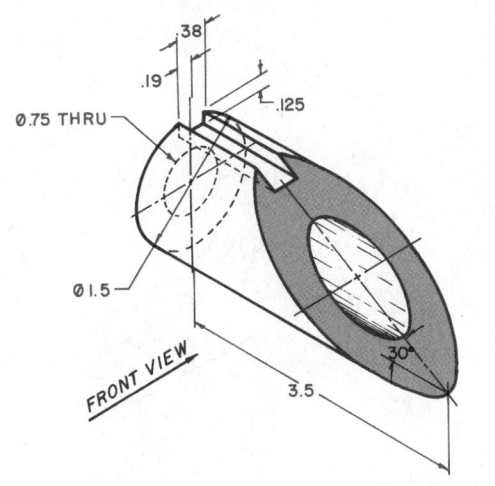

Problem 6-11

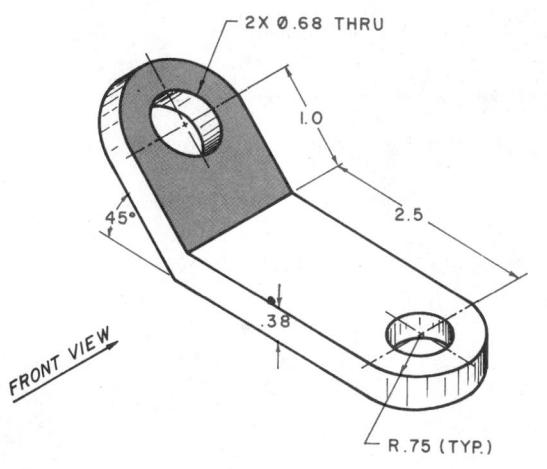

Problem 6-12

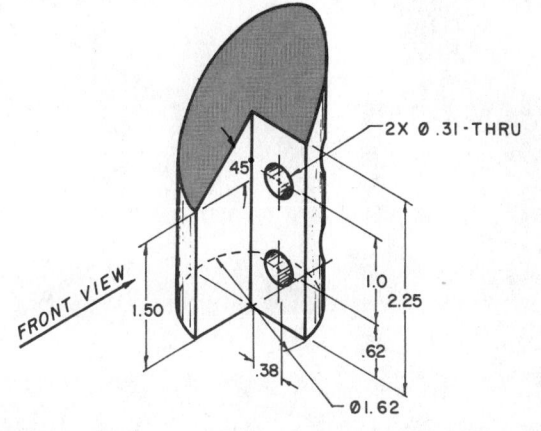

Problem 6-13

Problem 6-14

Draw the front view, top view, right-side view and *two* auxiliary views to illustrate the true size and shape of all surfaces. Complete all views using the listed steps.

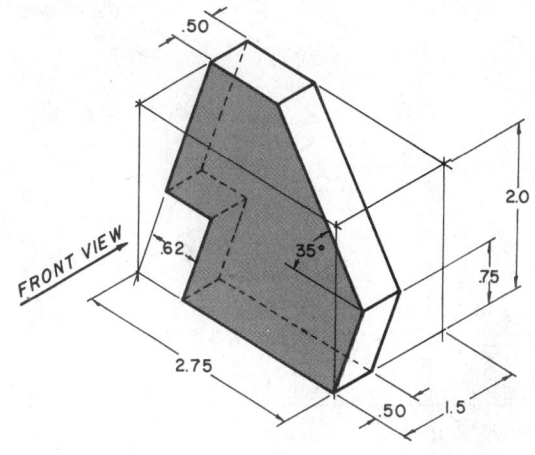

Problem 6-14

Problems 6-15 through 6-22

Draw the required views to fully illustrate each object. Complete all views using the listed steps.

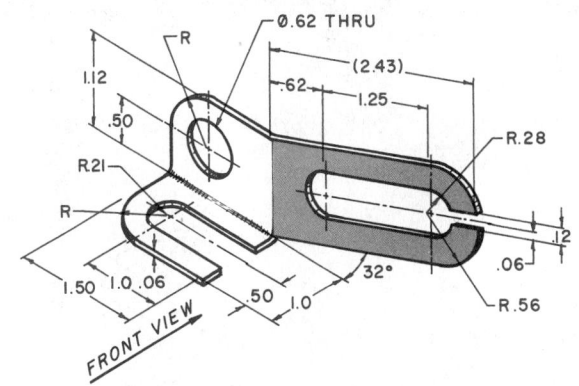

Problem 6-15

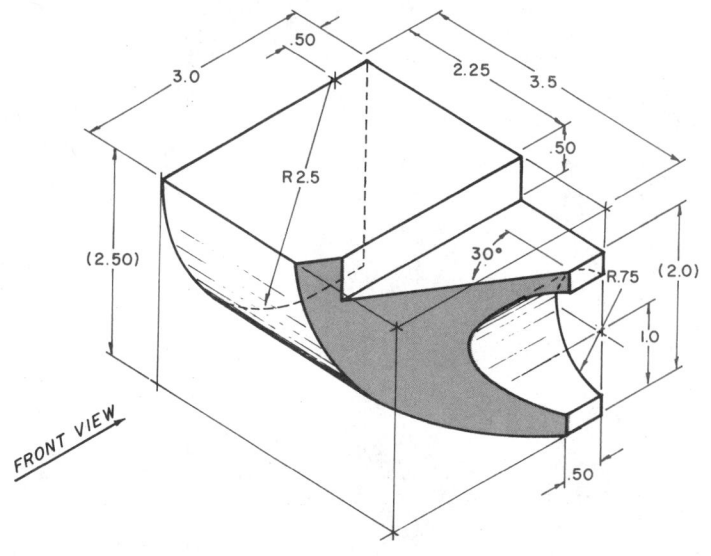

FRONT VIEW

Problem 6-16

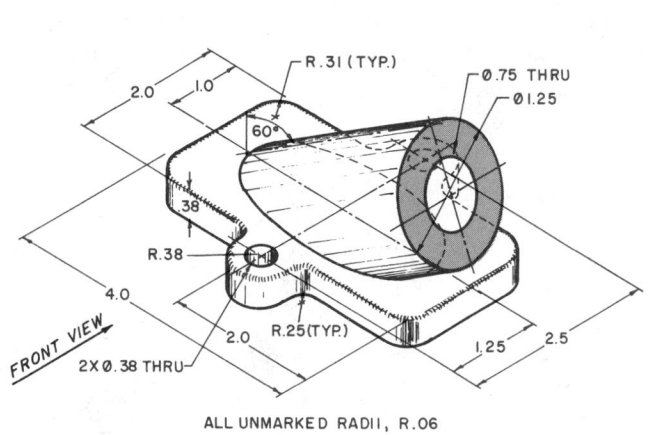

FRONT VIEW

ALL UNMARKED RADII, R.06

Problem 6-17

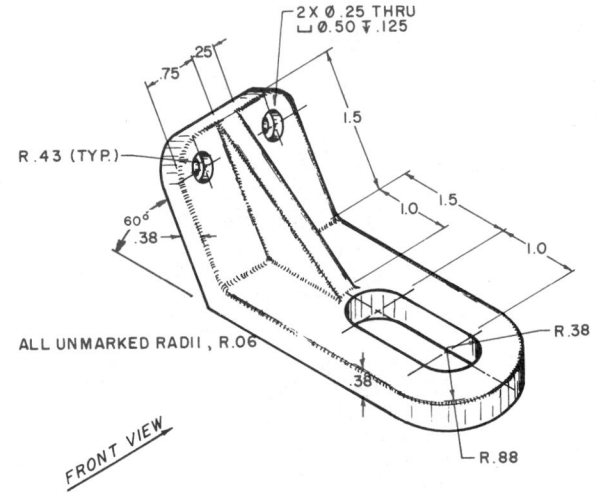

FRONT VIEW

Problem 6-18

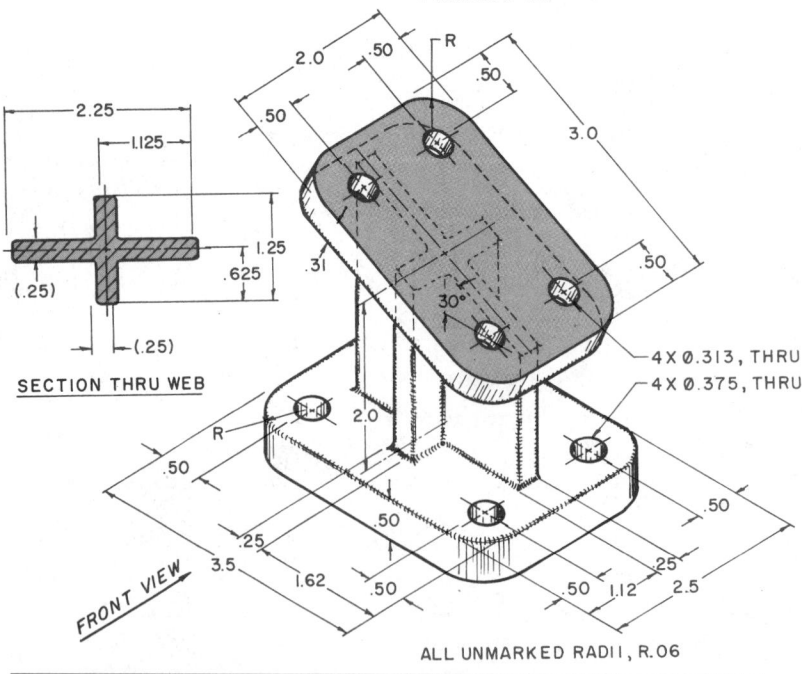

SECTION THRU WEB

FRONT VIEW

ALL UNMARKED RADII, R.06

Problem 6-19

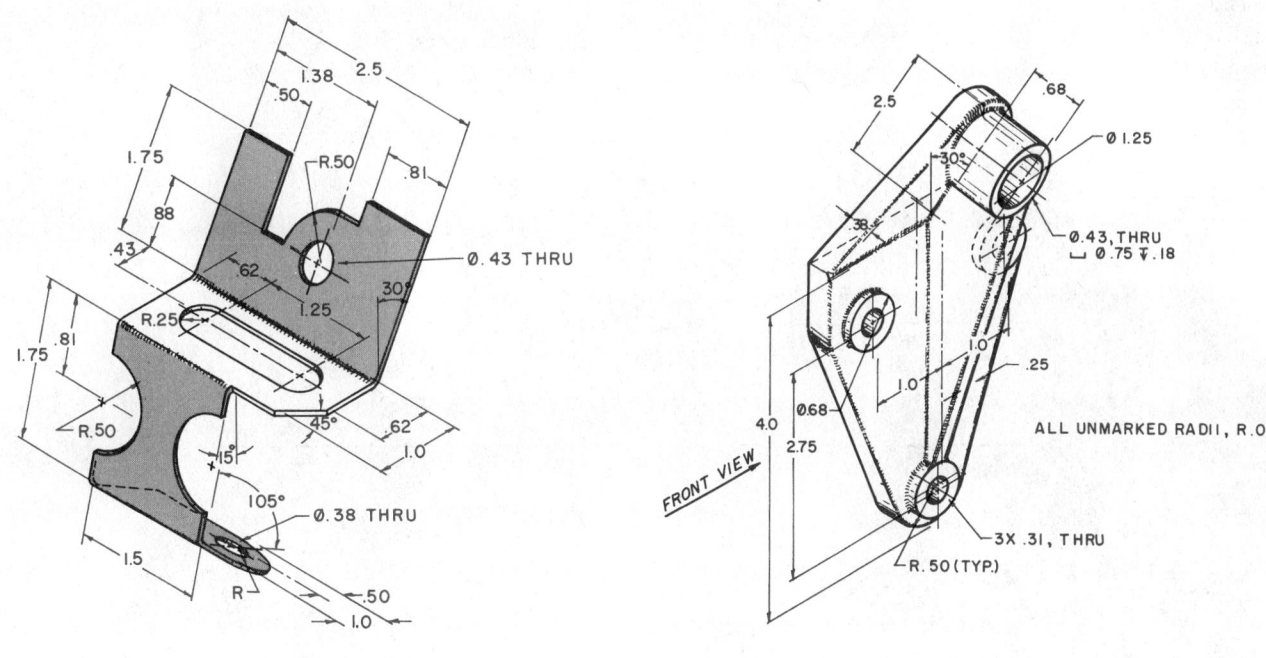

Problem 6-20

Problem 6-21

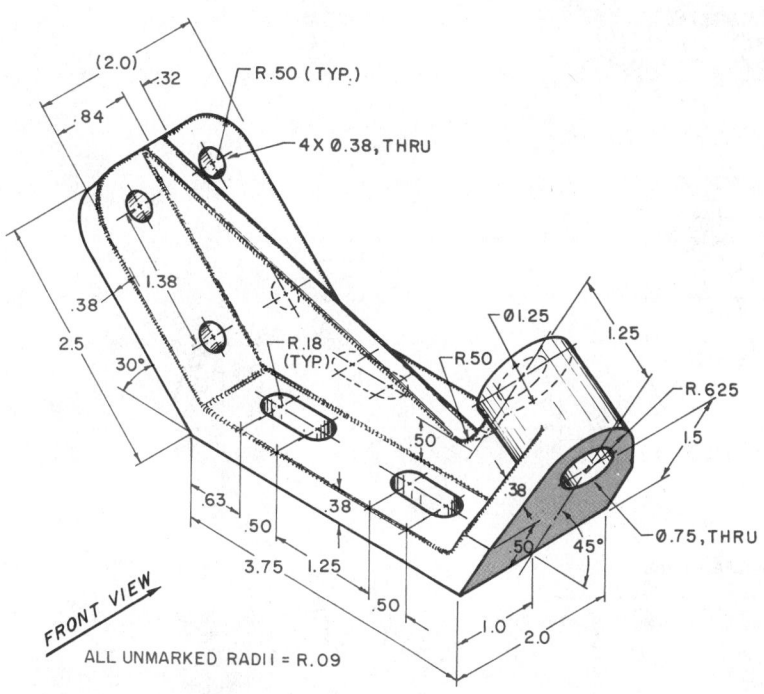

Problem 6-22

Problems 6-23 through 6-29

Make a finished drawing of selected problems as assigned by the instructor. Draw the given front and right-side views, add the top view with hidden lines if required, and add an auxiliary view. If assigned, design your own right-side view consistent with the given front view, and add a complete auxiliary view. Do not add dimensions.

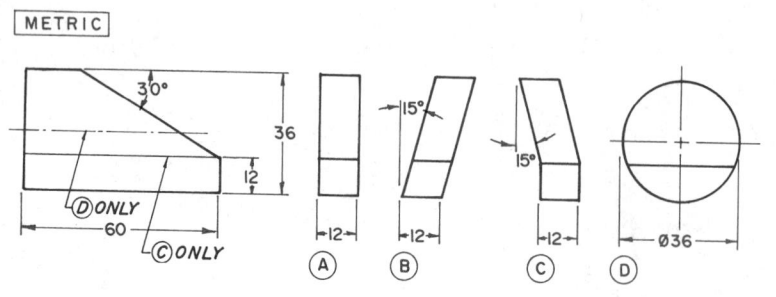

Problem 6-23

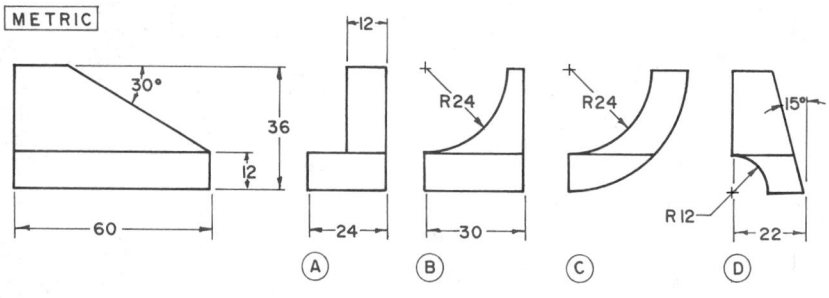

Problem 6-24

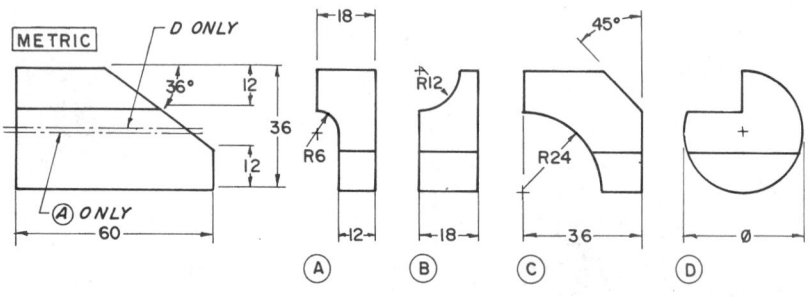

Problem 6-25

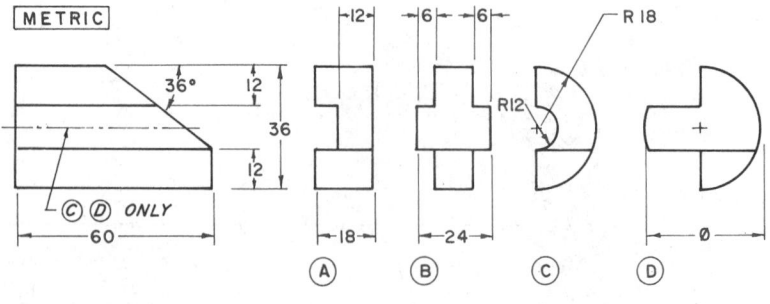

Problem 6-26

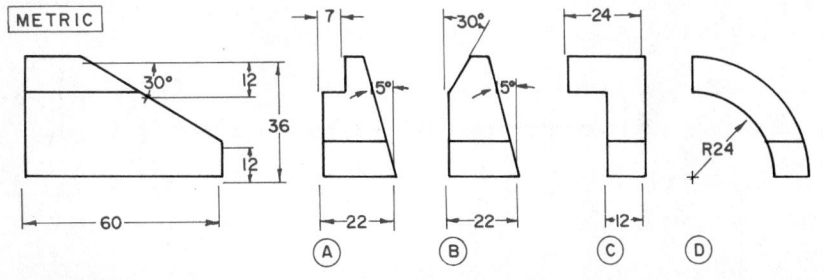

Problem 6-27

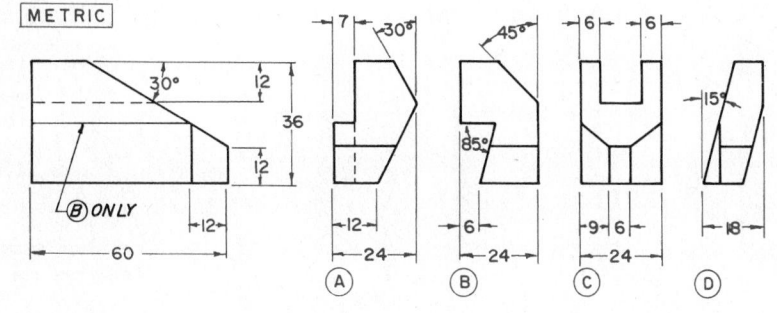

Problem 6-28

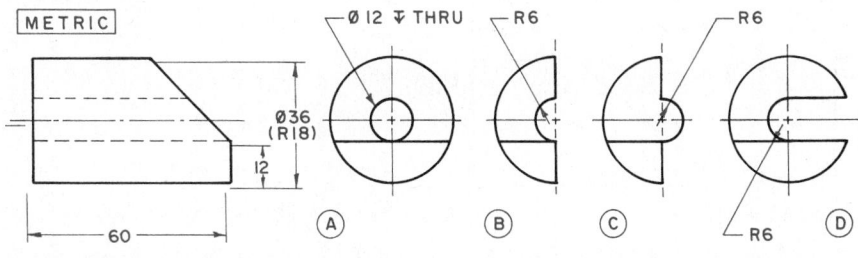

Problem 6-29

Problems 6-30 through 6-37

Draw the given views and add the required auxiliary view. Present the views within the work area so all views have a neat, centered appearance. Do not add dimensions.

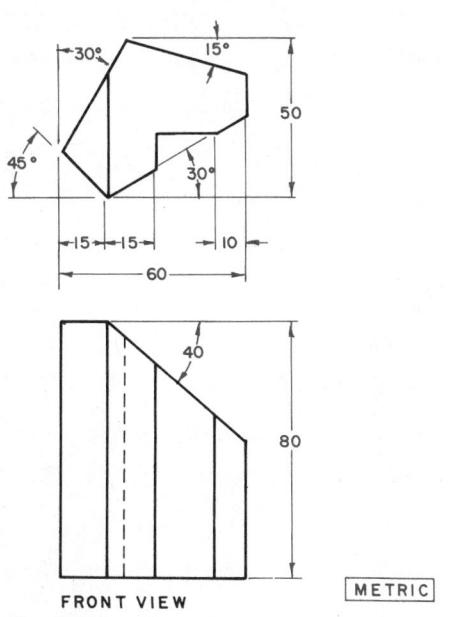

Problem 6-30

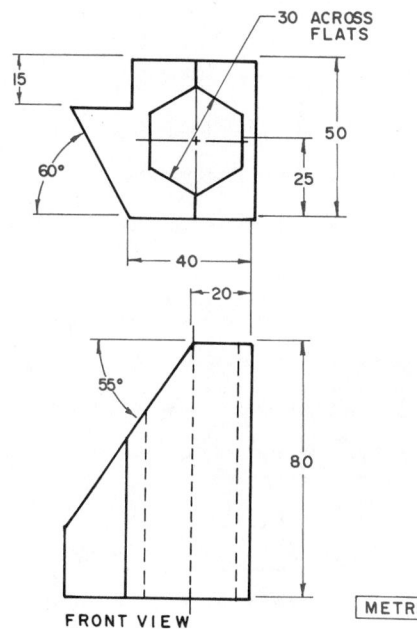

Problem 6-31

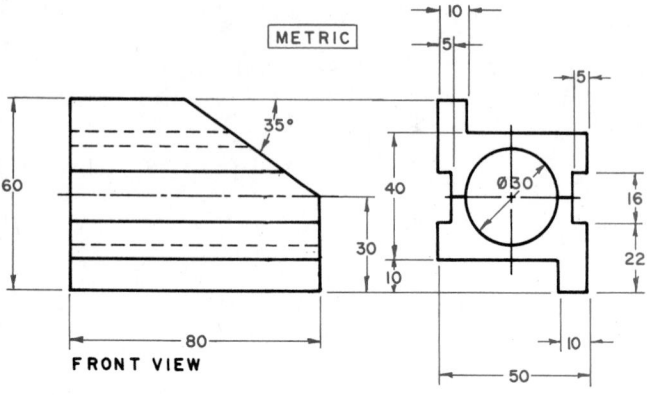

Problem 6-32

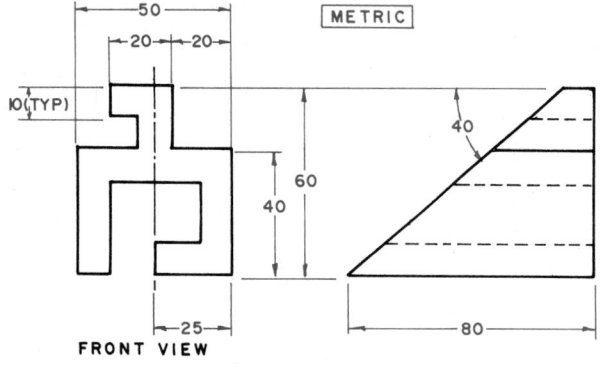

Problem 6-33

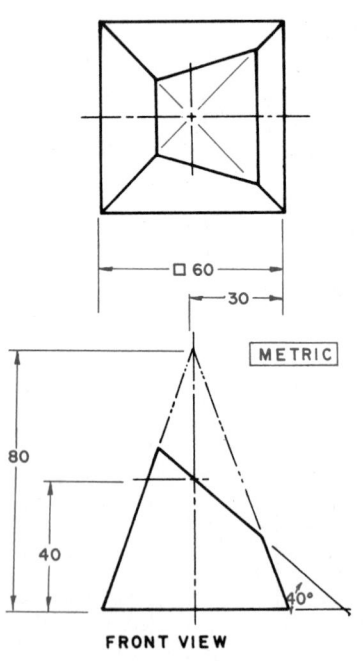

Problem 6-34

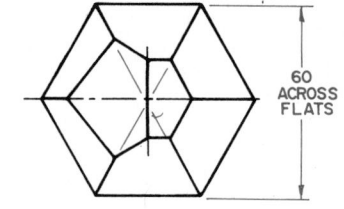

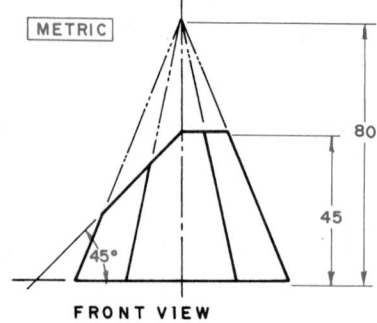

Problem 6-35

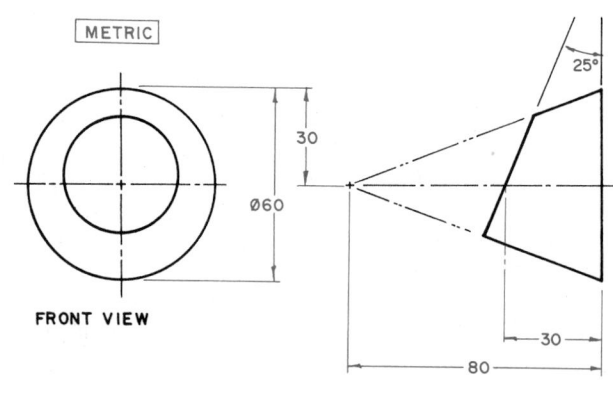

Problem 6-36

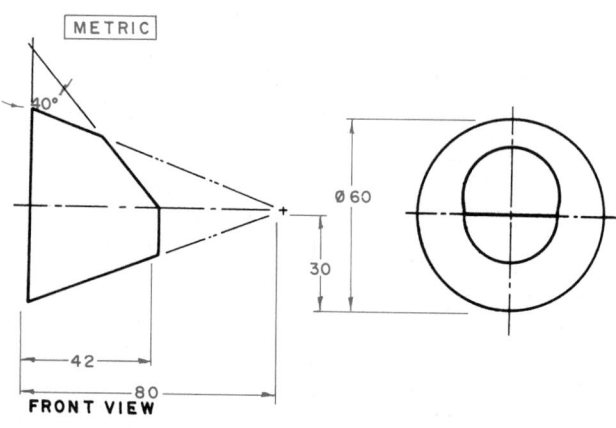

Problem 6-37

❑ Problem 6-38

The partially completed building, in outline form, is to have the name of your city painted on one side of the roof, in letters to be read by individuals in aircraft as high as 5000′ above the roof sign. A rule of thumb for good readability is that the letters on a sign, in a classroom, in an auditorium, or on a roof, should be as high as approximately 1/250 of the distance the viewer is from the sign. Use an auxiliary view of the roof to layout your city's name. Then specify the length of the building to suit your sign. Refer to Figure 2-4 for example fonts.

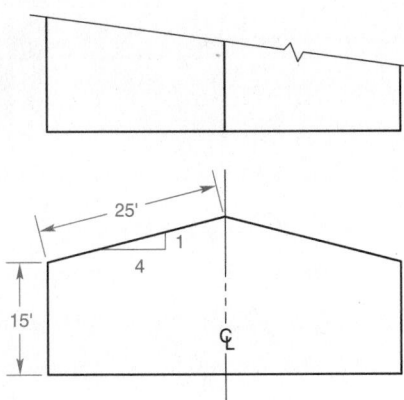

Problem 6-38

❑ Problem 6-40

Panels A, B, and C in the control station shown, satisfy anthropometric requirements. They are 30″ from the operator's head at location O, and the edge of the desk is 10″ from the operator. Panel B needs to be replaced with a new panel that has the 5 new instruments shown, located in a reasonable arrangement in the panel. Prepare an auxiliary view of panel B and locate the 5 instruments, using your own judgment for the arrangement. Note approximate dimensions to their locations. Suggestion: Construct the auxiliary view of B, which will show the panel in true size. Cut out five rectangles of paper to represent the 5 instruments; and make trial arrangements of them on your auxiliary view. Select the best arrangement, in your opinion. Mark their locations and give approximate dimensions to their locations.

❑ Problem 6-39

The 750-mm outside diameter flange for the special 45° elbow shown is to have M24x3.0, in through holes on a 650-mm bolt circle. The recommended distance between fasteners (for strength and to provide space for wrenches) 2.5 to 3.0 times the diameter of the fastener. Decide on the best, even number of bolts, for the 650-mm bolt circle on this flange. Show your answer in an auxiliary view of the flange with the bolt holes.

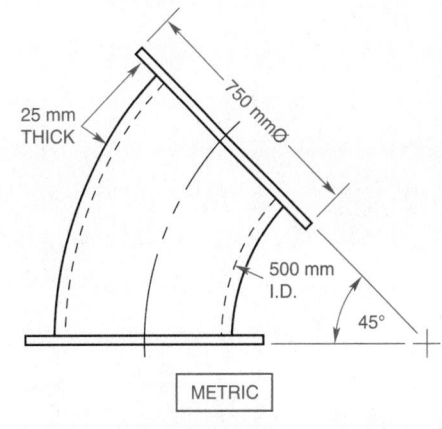

Problem 6-39

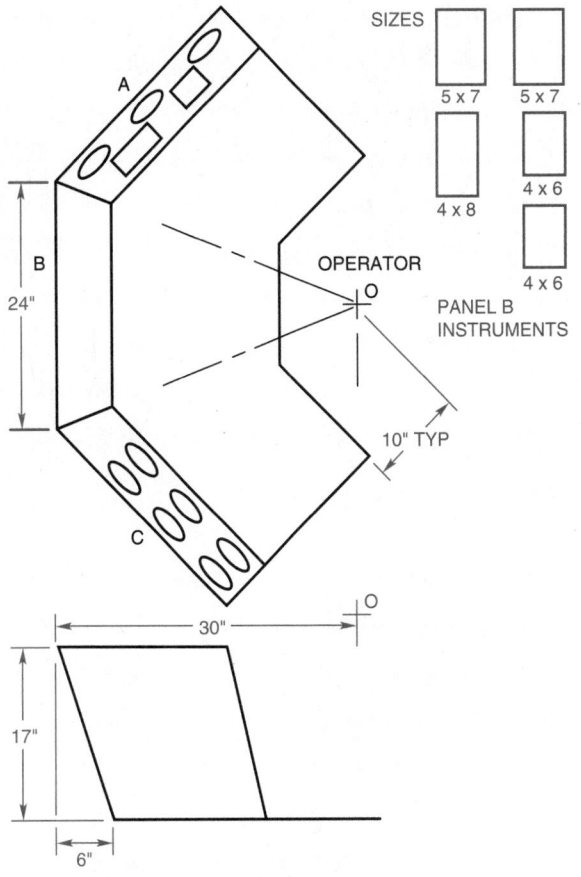

Problem 6-40

❏ Problem 6-41

The hopper shown needs a vibrator attached to the left side to shake down material to the conveyor below. Without the vibrator, some materials tend to "hang up." Prepare an auxiliary view of the left side of the hopper to determine approximate dimensions to locate a mounting plane, that already has threaded holes, shown. (The plate is to be welded to the hopper.) The center of the plate should be approximately 1.9 m above the floor level so maintenance personnel can reach the vibrator. Also, locate the plate to either side of center so the maintenance person won't have to reach over the conveyor.

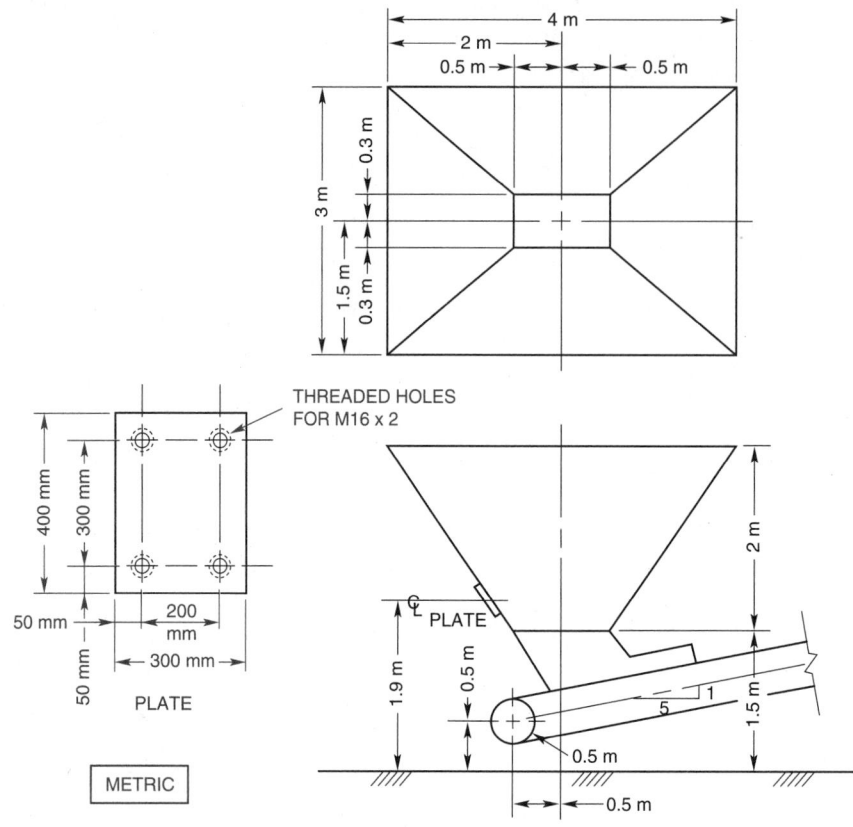

Problem 6-41

Descriptive Geometry

CHAPTER
7

Chapter Outline

Descriptive Geometry Projection

Primary auxiliary views are views that have been projected perpendicular from an inclined surface and viewed looking directly at that surface. As studied in Chapter 6, the auxiliary view shows true size and true shape of the inclined surface and is sometimes used to complete other views. If a surface is oblique, that is, slanting in more than one direction and not 90° from one of the other views, the true shape and true size of the successive surface is based on auxiliary views.

Using the descriptive geometry method of projection not only gives true size and shape, but it also can be used to find intersections, true distances of lines in space, true angles between surfaces, and exact piercing points. Descriptive geometry graphically shows the solution to problems dealing with points, lines and planes, and their relationship in space. In order to be able to apply descriptive geometry to various drafting problems, the drafter must know and understand the various basic steps involved. This chapter explains these basic steps. The basic theories covered in this chapter are:

- Projecting a line into other views
- Projecting points into other views
- Determining the true length of a line
- Determining the point view of a line
- Finding the true distance between a line and a point in space
- Determining the true distance between two lines in space
- Projecting a plane surface in space
- Developing an edge view of a plane surface in space
- Determining the true distance between a plane surface and a point in space
- Determining the true angle between plane surfaces in space

Steps Used

All problems, regardless of their complexity, use the same procedures or steps. These steps ultimately must be done in the same order each time, although some intermediate steps may be selected or are optional. Like climbing a flight of stairs, each step must be executed, and only experience can provide the ability to perform multiple steps simultaneously. An example of the sequential steps needed in descriptive geometry problems is as follows. In order to find an edge view of a plane (flat surface), a true length line must be located in that plane (Step 1); a point view of the true length line must be projected, which yields an edge view of the plane (Step 2); see **Figure 7-1**. By projective means, it would be impossible to find the edge view of an oblique surface from normal views without performing Steps 1 and 2.

Notations

In order to progress through sequential steps it is important to keep track of all views and points in space. (Recall that a straight line is composed of two spatially located end points.) To do this, a system of *notations* or labeling is advisable. Each view and each point should be labeled to provide an accurate identity of each at all times. For purposes of this text, the following notations are used in this chapter.

Each *point* in space is called out in *lowercase letters*.

Example:

a, b, c, d

Each *line* in space is identified by the *two lowercase end points*.

Example:

Line a-b

Each *line* in space is assumed to end at the indicated end points, unless it is specified as an *Extended Line* (continues without end).

Example:

Line p-q and Extended Line j-k

Each *plane* in space, such as a triangle, is called out in lowercase letters.

Example:

Triangle abc

Each *plane* in space is assumed to be limited (stops at the indicated boundaries) unless otherwise defined as an *Unlimited Plane* (limitless extension beyond the indicated boundaries).

Example:

Plane defg and Unlimited Plane mno

Each *viewing plane* is called out in *uppercase letters*.

Example:

F = Frontal viewing plane (a principal plane).
T = Top (horizontal) viewing plane (a principal plane).
R = Right side (profile) viewing plane (principal plane).
L = Left side (profile) viewing plane (principal plane).
A(digit) = an auxiliary viewing plane, with the number of successive projections from a principal viewing plane.
B(digit) = same as A(digit), but using a different series of successive projections.

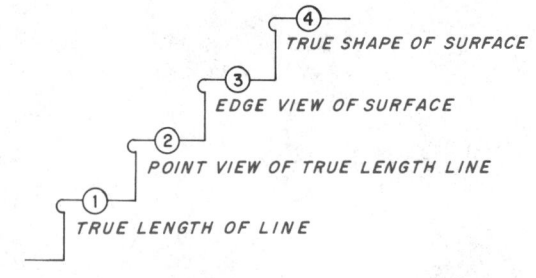

***Figure* 7-1**   Steps used to solve descriptive geometric problems

A combination of lowercase and uppercase letters is used to fully describe each point in space, and the view in which it is located.

Example:

aF would be *a* location as seen in the Frontal viewing plane.

aT-bT would be line *a-b* location as seen in the Top (horizontal) viewing plane.

aR-bR-cR would be plane *a-b-c* as seen in the Right side (profile) viewing plane.

Fold Lines

A fold line is represented by a thin black line, similar to a phantom line. It indicates a 90° intersection of two viewing planes. Each fold line must be labeled using uppercase letters, **Figure** 7-2. In this example, the fold line is placed between the front view and the top view. This placement eliminates the usual 45° projection line, **Figure** 7-3. It is important to add these notations in order to keep track of exactly which viewing plane is being executed.

***Figure* 7-2**   Fold line

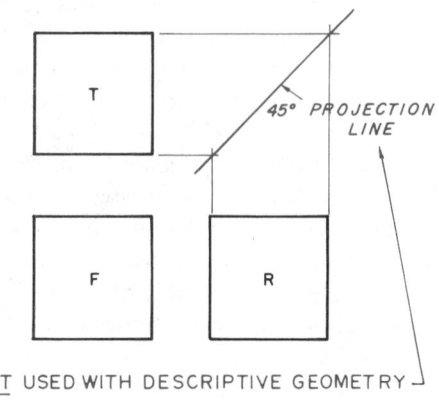

***Figure* 7-3**   Regular multiview drawing using 45° projection line

How To Project a Line into Other Views

Fold lines are placed between successive orthographic views and labeled with uppercase letters, *Figure 7-4*. Any appropriate spacing is permissible, but the distance from the image to the frontal viewing plane fold line *must* be equal at both the top and right-side views. The spatial positioning of a line a-b (surrounded by an imaginary transparent box) is shown in *Figure 7-5*. Point a is closer to the frontal viewing plane than is point b, but is farther below the top (horizontal) viewing plane.

Represented on a normal layout drawing, it would be illustrated and labeled as shown in *Figure 7-6*.

The customary 45° projection line is *not* used in descriptive geometry projection. All points are projected along light projection lines drawn at 90° to the fold lines. All measurements are taken *along* these light projection lines and measured *from* the fold line *to* the point in space, *Figure 7-7*.

An important rule to remember is to *always skip-a-view* between all measurements. In *Figure 7-7*, notice in the right-side view that dimension X (distance from F/R fold line to point aR) is projected into and through the front view and *up* to the top view and transferred into the top view. (Distance from F/T fold line to point aT.) The front view was skipped.

Again, referring to the right-side view of Figure 7-7, dimension Y (distance from F/R fold line to point bR) is

transferred into the top view (distance from F/T fold line to point bT) and the front view was skipped. In each of the instances, the distance is measured as the perpendicular distance from the fold line to the point.

Example:

To project a top view of a line from a given front and right-side view.

Given: Line a-b in the front view and line a-b in the right-side view, *Figure 7-8*. Extend projection lines into the top view from the end points in the front view, aF and bF. Find the distances X and Y from the line end points in the right-side

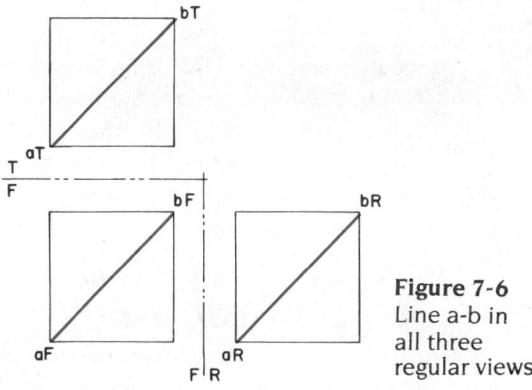

Figure 7-6
Line a-b in all three regular views

Figure 7-6   Line a-b in all three regular views

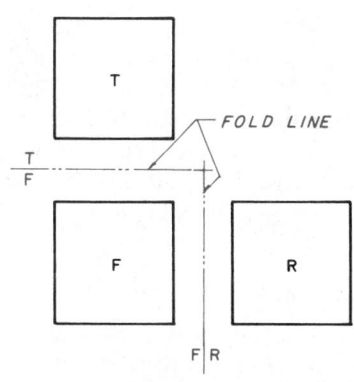

Figure 7-4   Regular multiview drawing using fold lines instead of 45° projection line

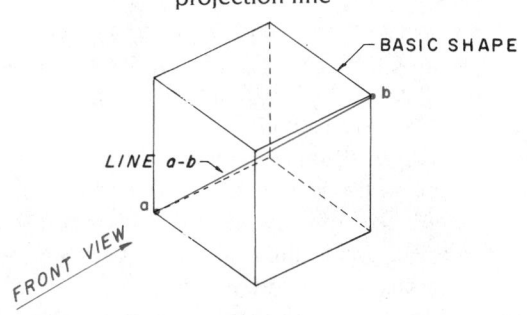

Figure 7-5   Spatial positioning of line a-b

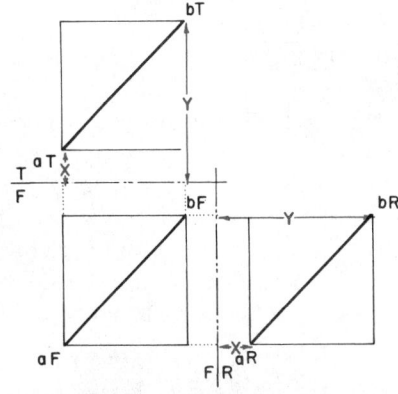

Figure 7-7   Skip-a-view when transferring distances

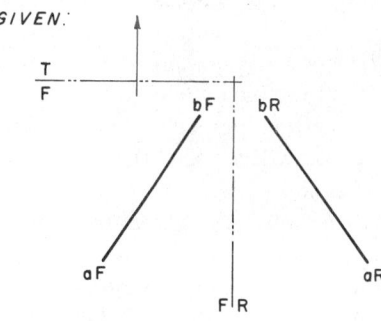

Figure 7-8   Locating line a-b in top view

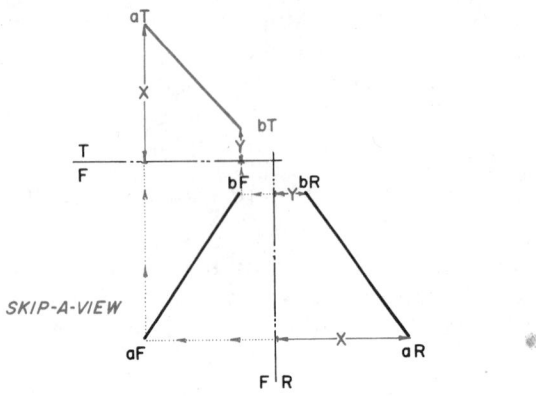

Figure 7-9 Line a-b in the top view using the fold line

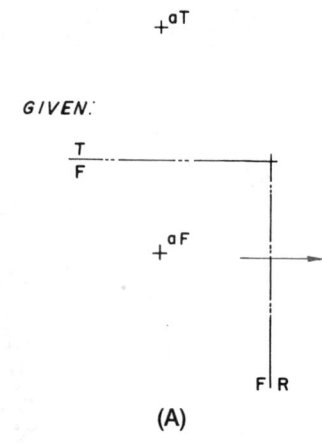

(A)

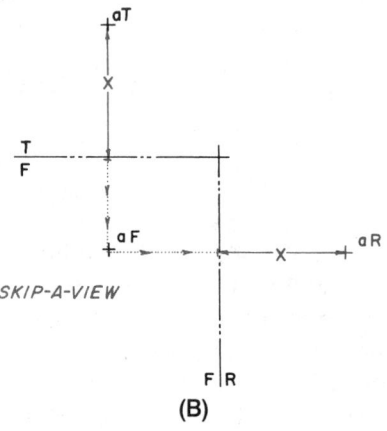

(B)

Figure 7-10 (A) Locating a point in space (right view); (B) Step 1

view aR and bR to the frontal viewing plane at fold line F/R and transfer them into the top view. Label all points and fold lines in all views. Be sure that all projections are made at 90° from the relevant fold lines, *Figure 7-9*.

How To Locate a Point in Space (Right View)

A point in space is projected and measured in exactly the same way as a line in space, except that the point is a line with a single end point, *Figure 7-10A*.

Example:

To project the right-side view of a point from a given top and front view.

Given: Point a in the front view and top view (Figure 7-10A).

Project point aF from the front view into the right-side view. Point aR must lie on this projection line. Find X, which is the distance from fold line F/T to aT in the top view and project it into and through the front view, *Figure 7-10B*. Project it over into the right-side view. Transfer distance X to find point a. Be sure to always label all points and fold lines in all views.

How To Find the True Length of a Line

Any line that is parallel to a fold line will appear in its true length in the next successive view adjoining that fold line.

To find the true length of any line:

Step 1 Draw a fold line parallel to the line of which the true length is required. This can be done at any convenient distance, such as approximately one-half inch.

Step 2 Label the fold line "A" for auxiliary view.

Step 3 Extend projection lines from the end points of the line being projected into the auxiliary view. These must always be at 90° to the fold line.

Step 4 Transfer the end point distances from the fold line in the second preceding view from the one being drawn, to locate the corresponding end points in the view being drawn.

Example:

Refer to *Figure 7-11A*.

Given: Line a-b in the front view, side view, and top view. The problem is to find the true length of line a-b. A true length can be projected from any of the three principal views by placing a fold line parallel to the line in any view. The right-side view is selected in this example.

Step 1 Draw a fold line parallel to line a-b, and label it as shown in *Figure 7-11B*. Extend light projection lines at 90° to the fold line from points a and b into the auxiliary view.

Step 2 Determine distance X and Y from the front view to the near-fold line and transfer them into the auxiliary view, as shown in *Figure 7-11C*. Label all points and fold lines. The result will be the actual true length of line a-b.

Use these steps to find the true length of any line.

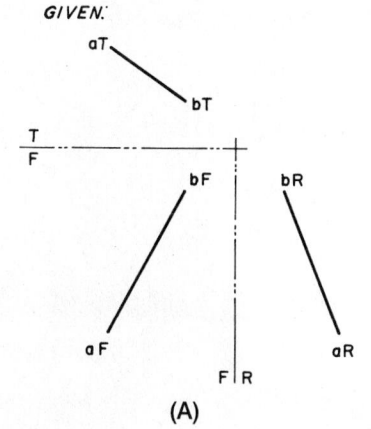

(A)

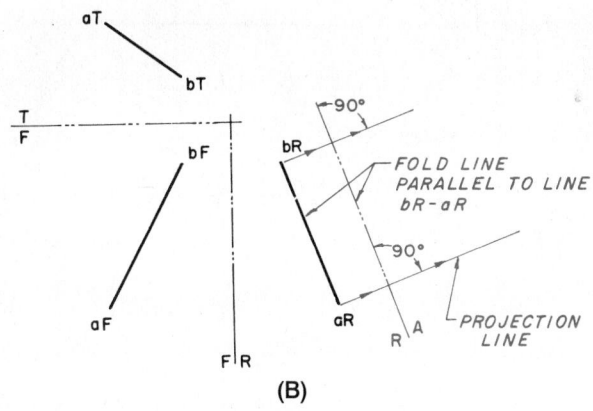

(B)

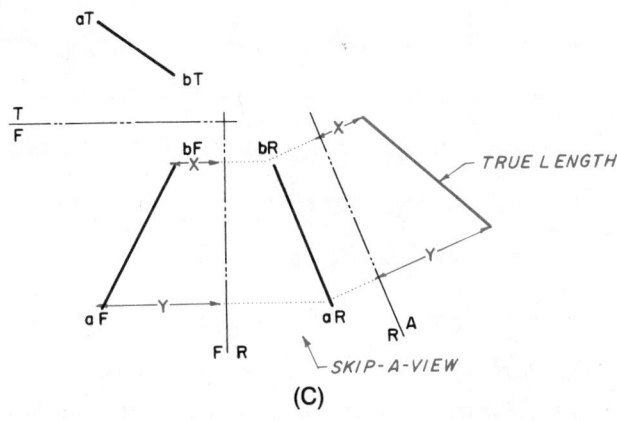

(C)

Figure 7-11   (A) Finding the true length of a line; (B) Step 1; (C) Step 2

How To Construct a Point View of a Line

To construct a point view of a line:

Step 1   Find the true length of the line.

Step 2   Draw a fold line perpendicular to the true length line at any convenient distance from either end of the true length line.

Step 3   Label the fold line A-B (B indicates a secondary auxiliary view).

Step 4   Extend a light projection line from the true length line into the secondary auxiliary view.

Step 5   Transfer the distance of the line end points into the secondary auxiliary view (B) from the corresponding points in the second preceding view.

Example:

Refer to ***Figure 7-12A***.

Given: Line a-b in the front view, side view, and auxiliary view. The true length is located in the auxiliary viewing plane (A), which is projected from the front view. (The true length could have been projected from any of the given views.)

Step 1   At any convenient distance from either end, draw a fold line A/B perpendicular to the true length of line a-b. This will establish a viewing plane that is perpendicular to the direction of the line's path. Extend a projection line aligned with the true length line into this new secondary auxiliary view. The projection lines from both points a and b appear to align in this common projection line. Therefore, both points a and b must be located on this projection line in the secondary auxiliary view. In ***Figure 7-12B***, the fold line A/B was added to the left of the true length line a-b.

Step 2   The front view is the second preceding view to the secondary auxiliary view being constructed.

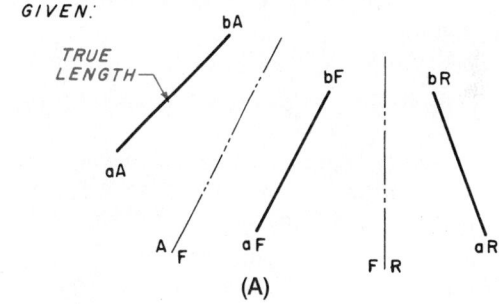

(A)

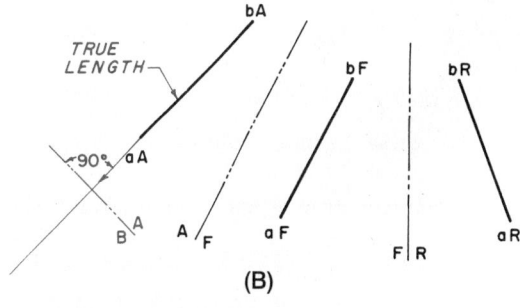

(B)

Figure 7-12   (A) Constructing a point view of a line; (B) Step 1

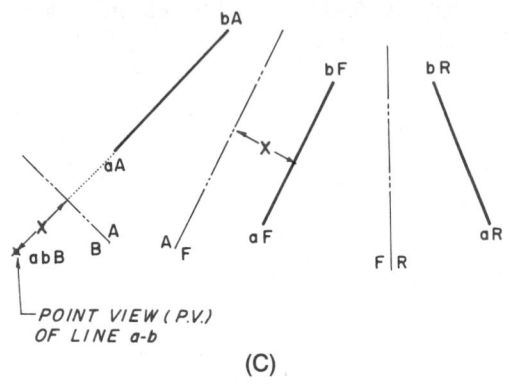

(C)

Figure 7-12 (Continued) (C) Step 2

Therefore, distance X in the front view from the fold line A/F to line a/b is transferred to the secondary auxiliary view from the fold line A/B, **Figure 7-12C**. As points a and b in the front view are both at the same distance from the fold line A/F, and they both lie on the same projection line in the secondary auxiliary view, they will appear to coincide at the same location. This provides evidence that the end view of the line has been achieved.

How To Find the True Distance Between a Line and a Point in Space

The true distance between a line and a point in space will be evident in a view that shows the end point of the line and that point, simultaneously.

Step 1 Project an auxiliary view (viewing plane A) to find the true length of the line, and project the point into the same view.

Step 2 Project a secondary auxiliary view (B) to find the point view of the line, and project the point into the same view.

Step 3 The observable distance between the end view of the line and the point will be the actual distance between them. The location in the line that is nearest to the point is on the path that is perpendicular to the line from the point, but this is not discernible in the secondary auxiliary view.

Example:

Refer to **Figure 7-13A**.

Given: Line a-b and point c in the front view and right-side view.

Step 1 Project the true length of line a-b by placing a fold line parallel to a given view of the line, and project point c into the auxiliary view. Recall that point locations are transferred from the second preceding view. Label all fold lines and all points, **Figure 7-13B**.

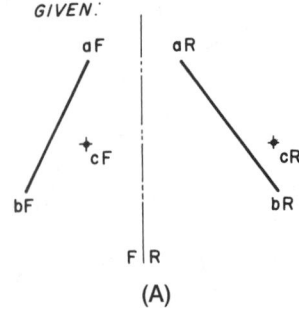

(A)

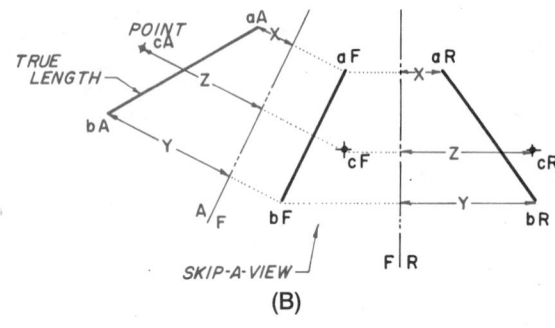

(B)

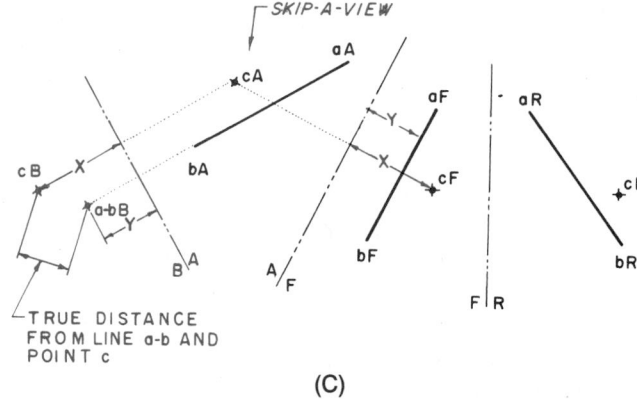

(C)

Figure 7-13 (A) Finding the true distance between a line and a point in space; (B) Step 1; (C) Step 2

Step 2 Draw a fold line perpendicular to the true length of line a-b, and label the fold line. Project line a-b into the secondary auxiliary view (B) to find the point view of line a-b. Project point c along into the second auxiliary view (B). The actual distance between the point view of line a-b/B and point cB is evident, **Figure 7-13C**.

The location in the line a-b that is nearest to point c lies on a path that is perpendicular to line ab and passes through c. Any path that is perpendicular to a line will appear perpendicular in a view where the line is true length. Therefore, the path can be drawn in the preceding view to locate point p on the line. Point p can be projected to its correct location on the original views of line a-b.

How To Find the True Distance Between Two Parallel Lines

If two lines are actually parallel, they will appear parallel in all views. A view is needed to show the end view of both lines simultaneously, where the real distance between them will be apparent.

Step 1   Find the true length of each of the two lines. Projecting a view that will provide the true length of a line will automatically provide the true length of any other parallel line that is projected into the same view.

Step 2   Find the point view of each of the two lines. Projecting a view that will provide the end view of a line will automatically provide the end view of any other parallel line that is projected into the same view.

Step 3   The true distance between the parallel lines is the straight-line path between their end points. There is no single location within the length of the lines where this occurs, as each location in a line has a corresponding closest point location on the other line, each on the path connecting them and perpendicular to their respective lines.

Example:

Refer to *Figure 7-14A*.

Given: The parallel lines a-b and c-d, in a front view, right side view, and a top view.

Step 1   Find the true lengths of line a-b and line c-d in auxiliary view A. The two parallel lines will also be parallel in the auxiliary, if they are actually parallel. See *Figure 7-14B*.

Step 2   Draw a fold line perpendicular to the true length lines, a-b and c-d. Project the lines into the secondary auxiliary view (B) to find the point view of the two lines a-b and c-d. Measure the true distance between the point views of lines a-b/B and c-d/B. See *Figure 7-14C*.

How To Find the True Distance Between Two Nonparallel (or Skewed) Lines

If two lines are not parallel, they will appear nonparallel in at least one view. It is wise to first check two nonparallel (or skewed) lines to determine if they actually intersect, which would make the distance between them to be zero. This can be verified by projecting the apparent point of intersection of a view to the adjoining view. If the apparent points of intersection align, then the lines actually intersect.

To determine the true distance between two nonparallel lines:

Step 1   Project the true length of either one of the two

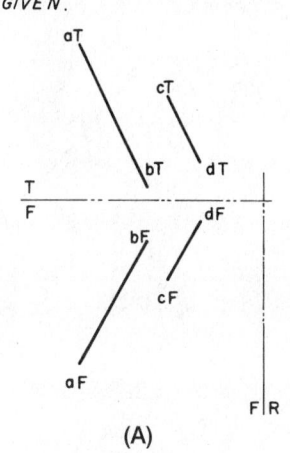

(A)

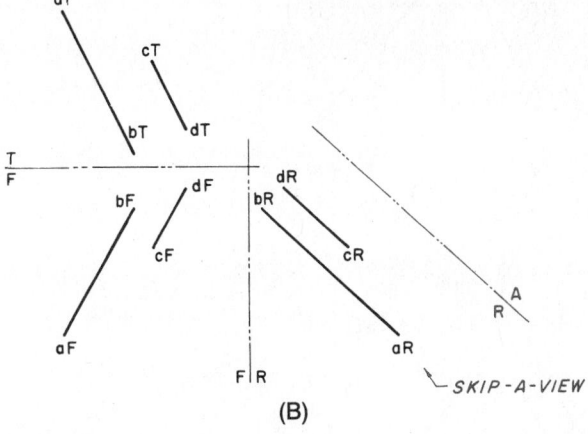

(B)

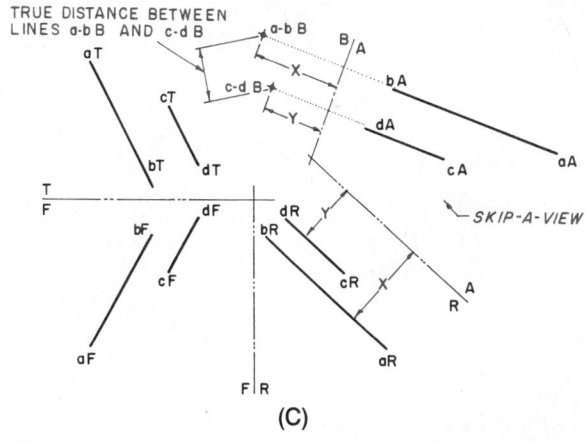

(C)

Figure 7-14   (A) Finding the true distance between two parallel lines; (B) Step 1; (C) Step 2

lines, and project the other line into that auxiliary view (A).

Step 2   Find the point view of the true length line, and project the other line into that secondary auxiliary view (B).

Step 3   Measure the true distance between the point view line at a location that is perpendicular to the other line.

GIVEN:

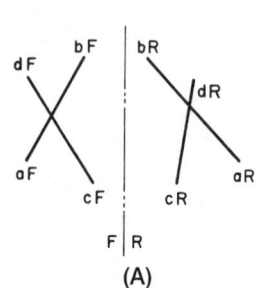

(A)

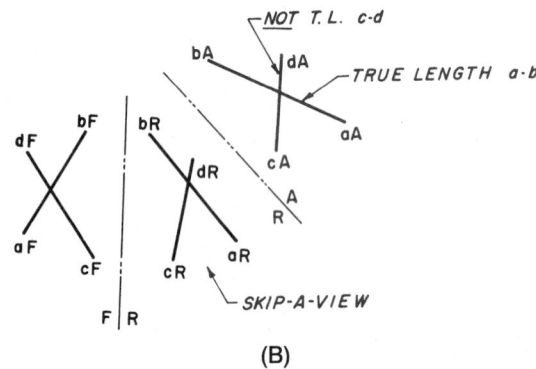

(B)

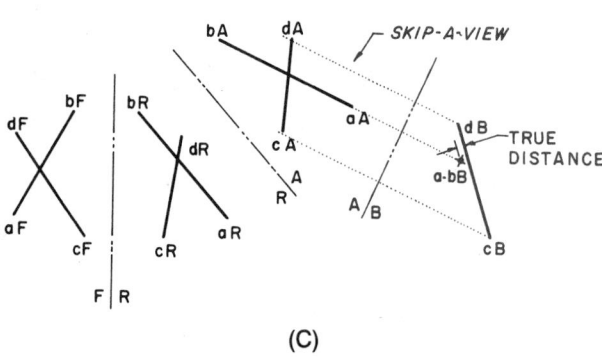

(C)

Figure 7-15   (A) Finding the true distance between two nonparallel (or skewed) lines; (B) Step 1; (C) Step 2

Example:

Refer to **Figure 7-15A**.

Given: Nonparallel lines a-b and c-d in a front view and right-side view.

Step 1   Project the true length of line a-b in an auxiliary view (A), and project line c-d along into that auxiliary view, **Figure 7-15B**. Notice that c-d is not the true length.

Step 2   Project the point view of line a-b into the secondary view (B) and project line c-d into this view. Measure the true distance between the point view of line a-b/B, perpendicular from line c-d/B, **Figure 7-15C**.

How To Project a Plane into Another View

A limited plane is located by points in space joined by straight lines. In order to transfer a plane, find at least three points in the plane and project each point into the next view.

Example:

Refer to **Figure 7-16A**.

Given: Triangular plane abc shown in the top view and front view.

Step 1   In the top view, find the distance from the fold line to points aT, bT, and cT, **Figure 7-16B**.

Step 2   Project these same points from the front view into the right-side view, and transfer the corresponding distances from the top view points to the top view fold line into the right-side view, **Figure 7-16C**. Construct straight lines between points a, b, and c. This transfers the plane surface.

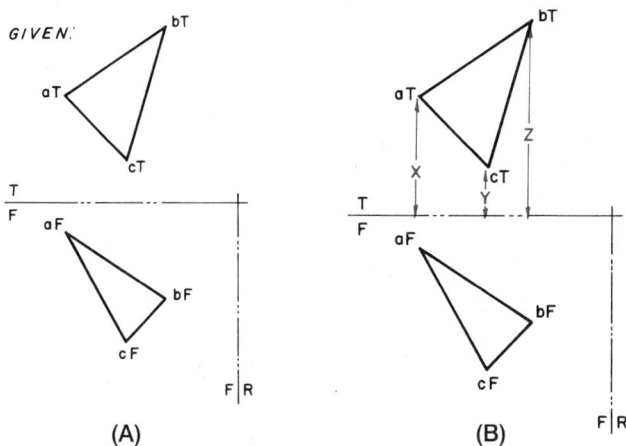

(A)   (B)

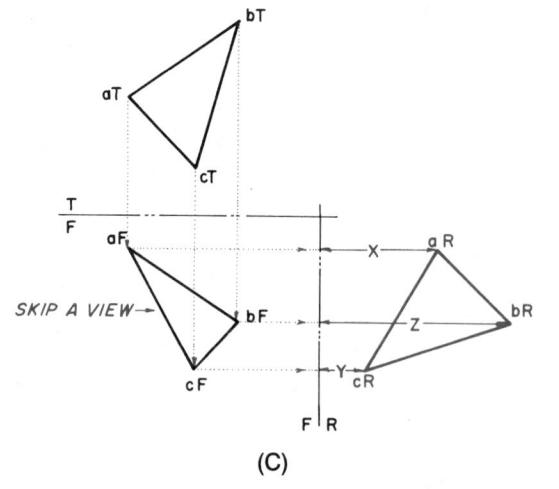

(C)

Figure 7-16   (A) Projecting a plane into another view; (B) Step 1; (C) Step 2

How To Construct an Edge View of a Plane Surface

If the end view of a line that is in a plane is shown, then the edge view of that plane is shown also. Recall that to find the end view of a line, a true length must be found first. When the boundaries of a plane are provided, projecting a view to find the true length of a selected boundary is a simple procedure. It is necessary only to locate a fold line parallel to the boundary to ensure its true length in the next projected view.

A shorter method of securing a true length line in a plane is also used, if a true length line is not present already. A line may be added on the plane in any view, arranged to be parallel to an adjoining fold line, and to have its terminations at the plane boundaries. Projecting the added-line terminations to the corresponding boundaries in the adjoining view relocates the added line in that view. Moreover, it will be a true length line, as it was made to be parallel to the fold line in the preceding view.

Whichever method is used to find a true length line in the plane, all point locations within the plane should be projected to the view where the true length line exists, if not there already. A fold line perpendicular to the true length line will provide a direction for projecting the edge view of the plane. Projecting all the points in the plane will provide evidence of this, as they will align into a single, straight path.

Example:

Refer to *Figure 7-17A*.

Given: Plane surface abc in the top view and front view.

Step 1   Construct a line parallel to fold line F/R and through one or more points if possible. In this example, the line passes through point c. Label this newly constructed line cF/xF, *Figure 7-17B*.

Step 2   Construct the true length of line cF/xF in the next view, *Figure 7-17C*.

Step 3   Construct the point view of the true length line c-R/xR. Project the remaining points of the plane surface into this view. Also, label all points and fold lines. If done correctly, the result should be a straight line which passes through the point view. This straight line is the edge view of the plane, *Figure 7-17D*.

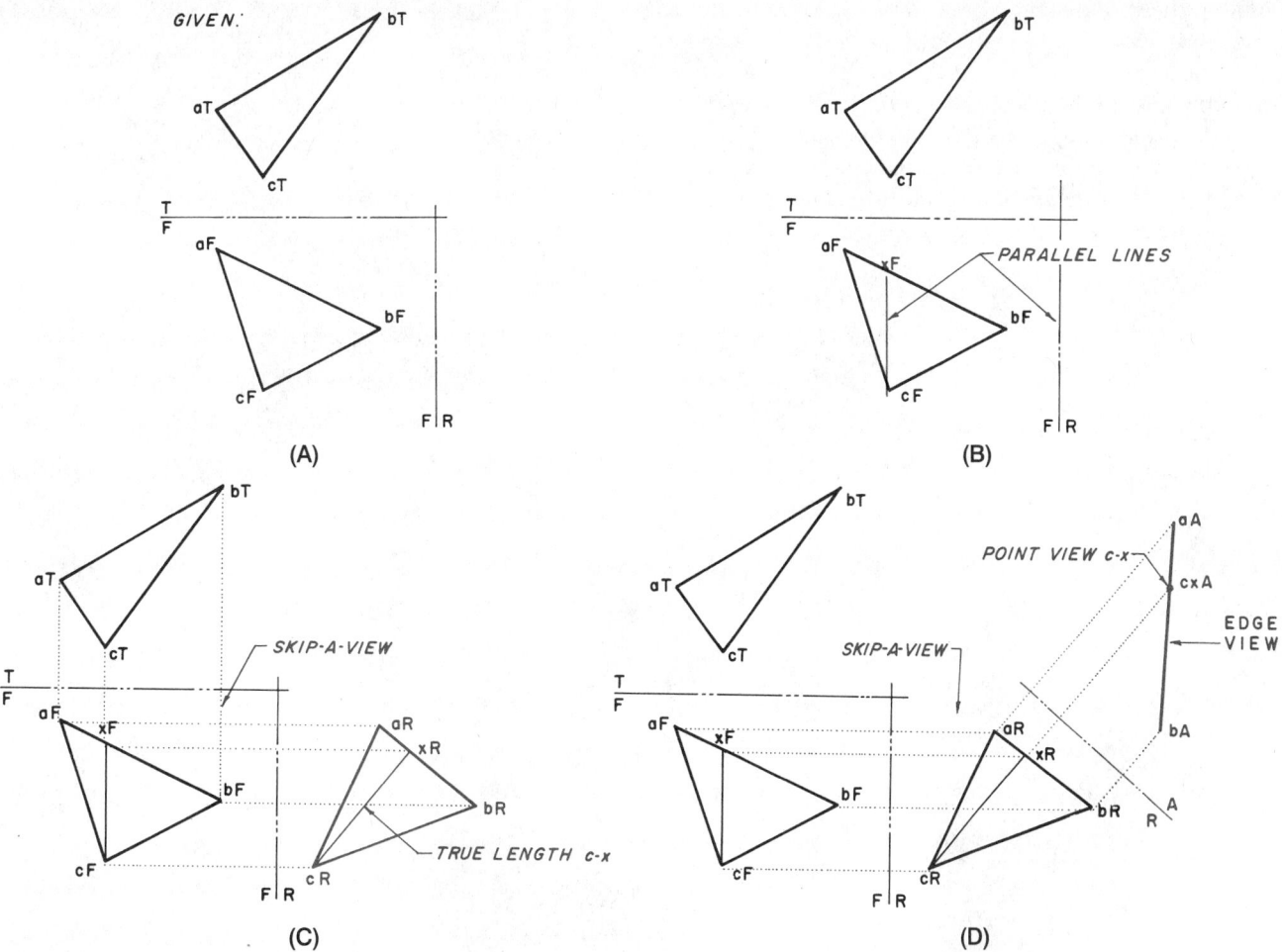

Figure 7-17   (A) Constructing an edge view of a plane surface;   (B) Step 1;   (C) Step 2;   (D) Step 3

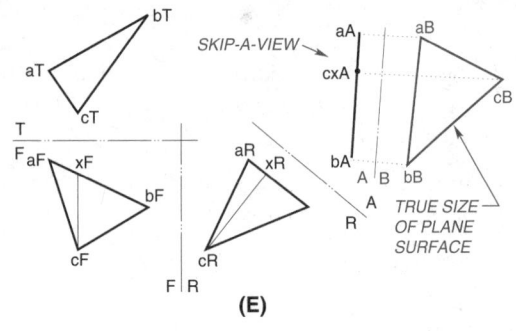

(E)

Figure 7-17 (Continued) (E) Step 4

Step 4 Construct a fold line A/B parallel to plane abc (shown as an edge in view A) and construct the plane surface abc in view B. Plane surface abc is now seen in true size, in view B, *Figure 7-17E*, because plane abc was seen as an edge in the adjacent view A.

How To Find the True Distance Between a Plane Surface and a Point in Space

Finding the true distance between a given point and a plane in space requires a view that shows the edge view of the plane and the point in the same view. The distance from the point to the plane is the path that is perpendicular to the plane, and passing through the point.

Step 1 Add a line on the plane parallel to a fold line, if a true length line is not already available in the plane.

Step 2 Project the added line to an adjoining view, where it will be seen in true length.

Step 3 Project a point view of the added true length line.

Step 4 Project the points of the plane into this view. The result should be a straight line which passes through the added line end view, which is the edge view. Project the point in space into this view. Measure the perpendicular distance from the plane's edge view to the point in space. This is the actual true distance between the plane surface and the point in space.

Example:

Refer to *Figure 7-18A*.

Given: Triangular plane abc and point X in the front view and right-side view.

Step 1 Construct a line parallel to the fold line F/R and through one or more points. Point c is a convenient line termination. Label this newly constructed line cR/zR, *Figure 7-18B*.

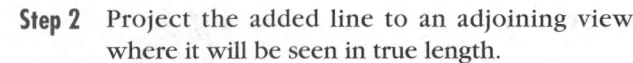

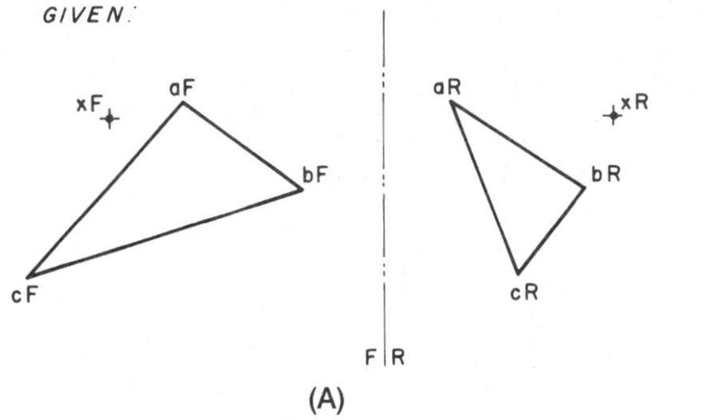

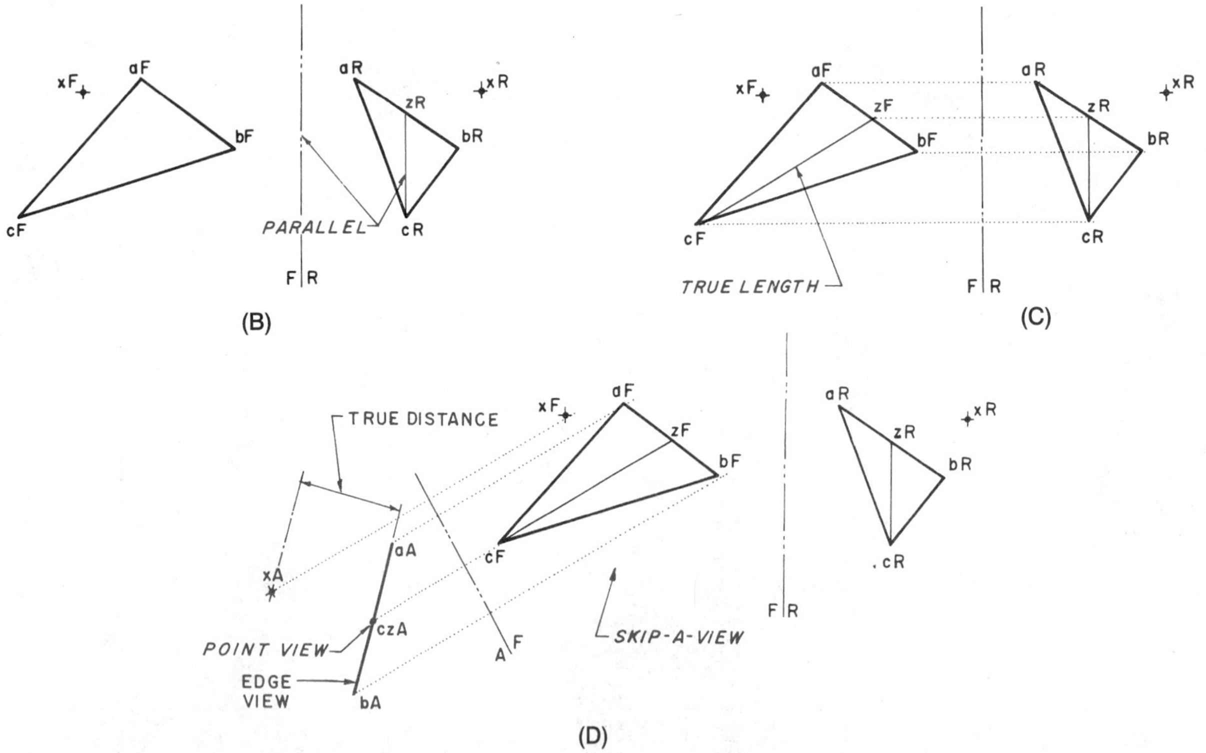

Figure 7-18 (A) Finding the true distance between a plane surface and a point in space; (B) Step 1; (C) Step 2 (D) Step 3

Step 2 Construct the true length of line cR/zR in the front view, and label the line cF/zF, *Figure 7-18C*. Bring point X into this view, also. Label all points and fold lines.

Step 3 Project the end view of the true length line cF/zF in the auxiliary view, *Figure 7-18D*. Project the points of the plane surface into this view, and also point X, projecting 90° from the fold line. Label all points and fold lines. Measure the perpendicular distance from the edge view to the point in space. This is the true distance between the plane surface and the point X in space.

How To Find the True Angle Between Two Planes (Dihedral Angle)

The edge of intersection between two planes is a line that is common to both planes. The end view of that line will provide the edge view of both planes simultaneously, and the angle between them will be evident.

To find the true angle between two surfaces:

Step 1 Construct the true length of the intersection between the two surfaces, and project all other points of both planes into that auxiliary view.

Step 2 Project the end view of the true length edge of intersection line and project the points of both planes into this secondary auxiliary view. Label all points and fold lines. Measure the true angle between the two surfaces.

Example:

Refer to *Figure 7-19A*.

Given: Plane abc and plane abd that intersect one edge, a-b, in the front view, side view, and top view.

Step 1 Project the true length of the edge of intersection a-b, and project all other points into this auxiliary view (A), *Figure 7-19B*.

Step 2 Project the point view of the true length of the edge of intersection a-b and project all other points into this secondary auxiliary view (B), *Figure 7-19C*. Measure the true angle between two surfaces.

How To Determine the Visibility of Lines

To determine which line of an apparent intersection of two lines is closer to the viewer, the following steps are used.

Step 1 From the exact crossover point of the lines, project to an adjoining view.

Step 2 In the adjoining view, determine which of the lines is closest to the fold line between the views, on the projection line from the first view (or the first line that the projection line encounters on its path from the first view).

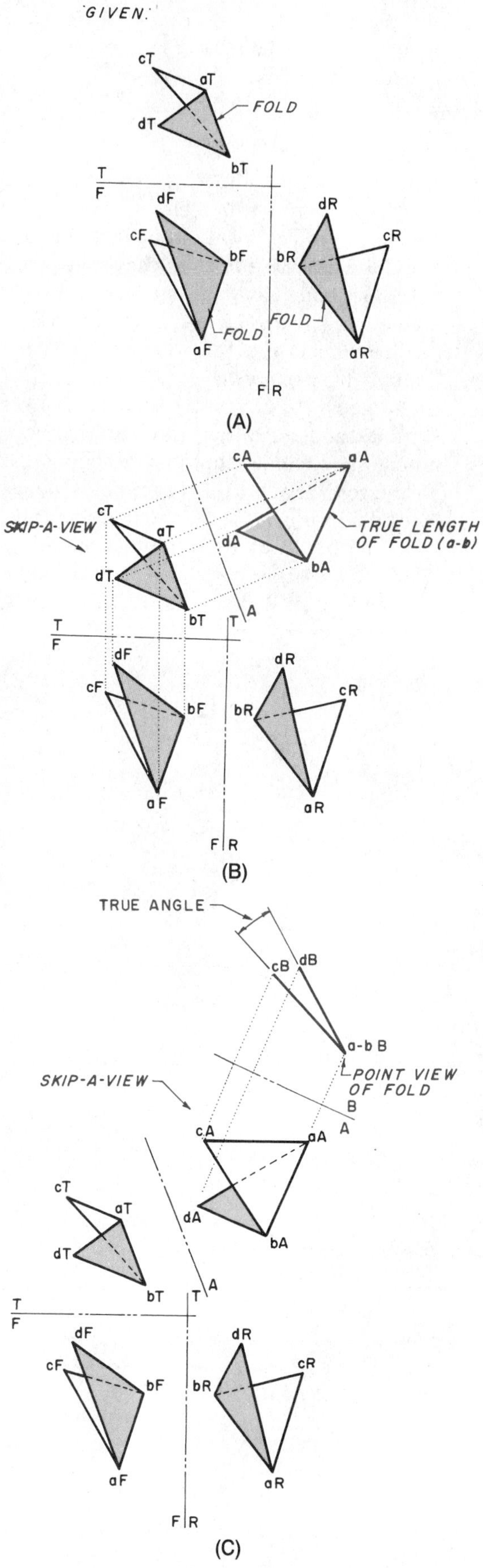

Figure 7-19 (A) Finding the angle between two surfaces; (B) Step 1; (C) Step 2

Step 3 Whichever line is closest to the fold line at that point only is the line that is in front of the other line in the first view.

Example:

Refer to *Figure 7-20A*.

Given: Lines a-b and c-d in the front view and top view.

Step 1 At the exact crossover of lines a-b and c-d in the top view, project down through the fold line to the corresponding lines in the front view. At this exact point along the lines, line c-d is closer to the fold line and line a-b is farther away from the fold line; thus, in the top view, line c-d is in front of line a-b, *Figure 7-20B*.

Step 2 See *Figure 7-20C*. At the exact crossover of lines a-b and c-d in the front view, project up through the fold line to the corresponding lines in the top view. At this exact point along the lines, line a-b is closer to the fold line and line c-d is farther away from the fold line; thus, in the front view, line a-b is in front of line c-d.

Step 3 The end result is drawn (in *Figure 7-20D*) to illustrate this crossover.

How To Determine the Piercing Point by Inspection

The *piercing point* is the exact location of the intersection of a surface and a line. The exact piercing point is determined from the view where the surface appears as an edge view. In *Figure 7-21A*, the top view illustrates the edge view of the surface. The piercing point is established in the top view and projected down into the front view, *Figure 7-21B*. In *Figure 7-22A*, the front view illustrates the edge view of the surface. The piercing point is established in the front view and projected up into the top view, *Figure 7-22B*.

How To Determine the Piercing Point by Construction

To determine the exact piercing point by construction:

Step 1 Construct a true length of a line on the plane surface.

Step 2 Project a point view of the true length line, and an edge view of the plane surface. The edge view of the plane lies in the path of the line in

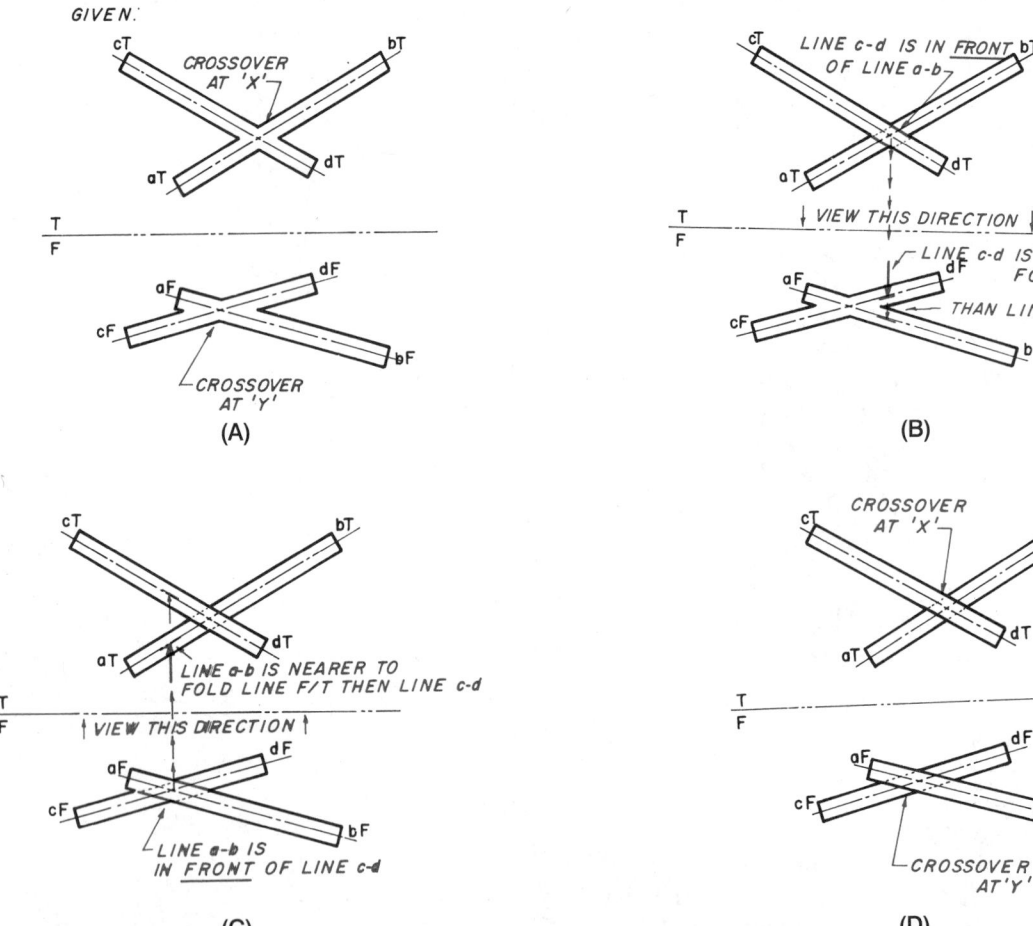

Figure 7-20 (A) Determining the visibility of lines; (B) Step 1; (C) Step 2; (D) Step 3

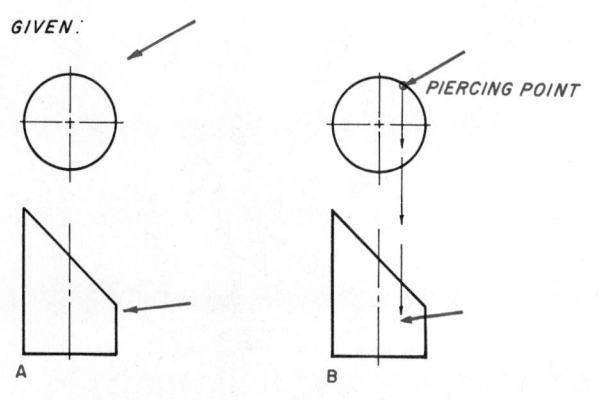

Figure 7-21   Determining the piercing point by inspection

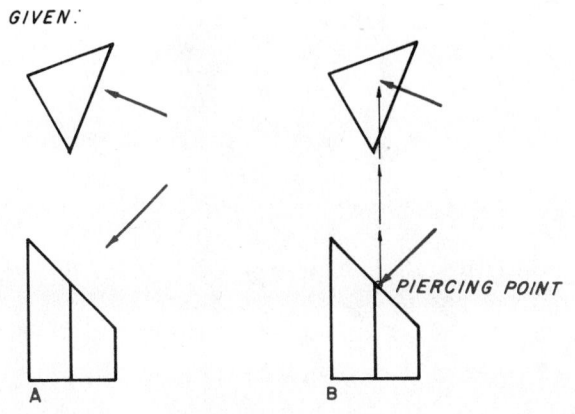

Figure 7-22   Determining the piercing point by inspection

this view, indicating the piercing point location.

Step 3   Transfer the piercing point back into the other views.

Step 4   Determine the visibility of lines to find which part of the line is visible from the viewing direction. Use the fold line to determine visibility, if necessary.

Example:

Refer to *Figure 7-23A*.

Given: Plane surface abc and line x-y, in the top view and front view.

Step 1   In the front view of plane abc, construct a line parallel to fold line (bF-dF) and find its true length in the top view, *Figure 7-23B*.

Step 2   Project an auxiliary view to find the point view of the true length line from the top view, and the edge view of the plane. The line intersection with the edge view of the plane indicates the piercing point location, *Figure 7-23C*.

Step 3   Project the piercing point back into other views, *Figure 7-23D*.

Step 4   In each view that the piercing point is being

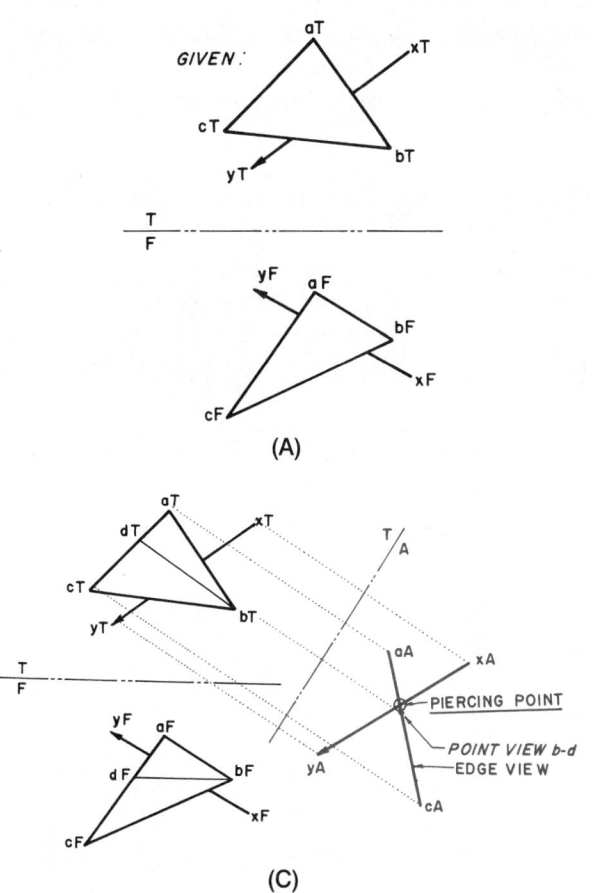

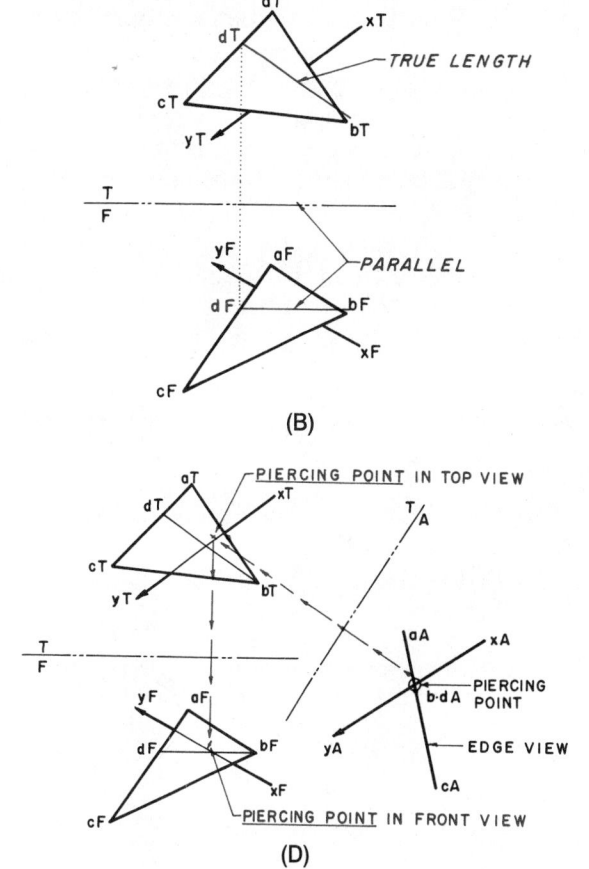

Figure 7-23   (A) Determining the piercing point by construction;   (B) Step 1;   (C) Step 2;   (D) Step 3

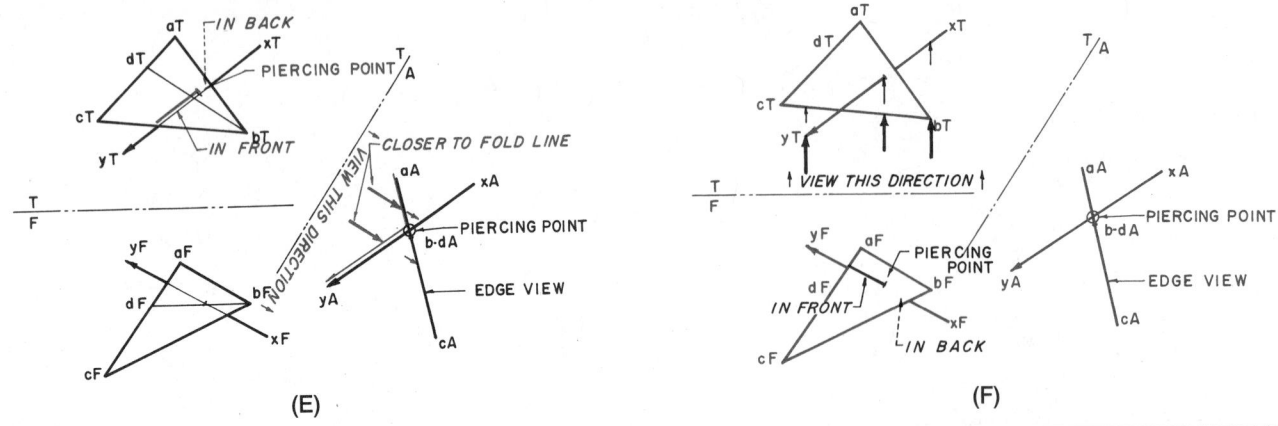

Figure 7-23 (Continued) (E) Step 4; (F) Step 5

projected to, the portion of the line that is visible can be determined by checking its preceding view, *Figure 7-23E*.

Notice in the auxiliary view that line x-y is closer to the fold line from the piercing point to point yA than the edge view of the plane surface from the piercing point to point cA. Therefore, in the top view, line x-y is seen only from the piercing point to yT.

Step 5 By viewing back into the top view, determine the visibility of lines of the plane surface and line x-y in the front view, *Figure 7-23F*. Notice in the top view that line x-y is closer to the fold line from the piercing point to point yT than the edge view of the plane surface cT/bT; therefore, in the front view, line x-y is seen only from the piercing point to yF.

When the edge view does not appear in the regular views, it must be constructed using the preceding steps, *Figure 7-24A*. Once the edge view is constructed, the piercing point can easily be seen. The piercing point is now projected down into the front view and up into the top view, *Figure 7-24B*.

How To Determine the Piercing Point by Line Projection

An alternate method of determining the piercing point simply locates the path of a cutting plane that passes through the line, leaving a "scar" on the surface of the given plane. Since the line lies in the cutting plane, an intersection of the line and the scar on the given plane locates the piercing point. The end points of the scar are found by aligning the cutting plane with the given line in one view, where it crosses the boundary of the given plane in two places. Projecting the cutting plane intersection to the corresponding boundaries of the given plane in the next view, locates the scar end points in that view, and provides a view of the scar that the given line can cross at the piercing point.

Example:

See *Figure 7-25A*.

Given: Skewed surface abc, and a line.

Step 1 In the top view, lightly draw the piercing line and determine points X and Y where they cross

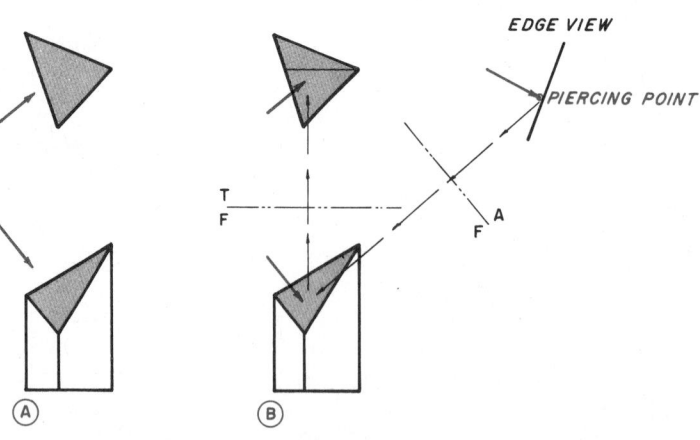

Figure 7-24 Determining the piercing point by construction

GIVEN:

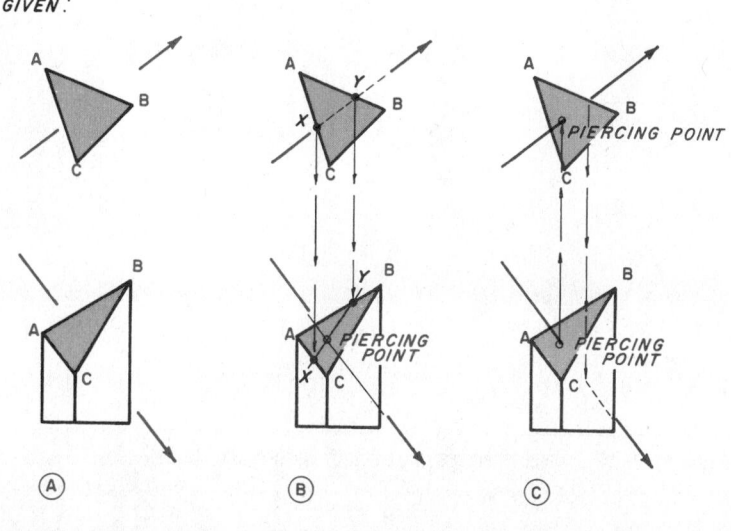

Figure 7-25 Determining piercing points by line projection (skewed surface)

the boundary of plane abc. The piercing point is located somewhere between these points.

Step 2 Project points X and Y down into the front view to the corresponding edges of the plane abc, *Figure 7-25B*. Where line X-Y crosses the given line is the exact piercing point.

Step 3 Project the piercing point to the top view, *Figure 7-25C*. Visibility of the piercing line is determined by inspection.

Figure 7-26A illustrates a cylinder pierced by a line. Lightly draw the piercing line in the top view and find point X. Project point X down into the front view, *Figure 7-26B*, to find piercing point X. Because the piercing line exits the skewed surface, piercing point Y is found on the edge view of the front view, *Figure 7-26C*. Project piercing point Y up into the top view. Visibility of the piercing line is determined by inspection.

If the object is a cone, line segments must be located and drawn from the base of the cone up to the vertex of the cone.

Given: A cone with a piercing line, *Figure 7-27A*.

Lightly draw the piercing line in the top view and find points X and Y, *Figure 7-27B*. Project points X and Y down to the base of the cone. Project light line segments from points X and Y on the base up to the vertex of the cone. Where these lines cross the piercing line is the location of the two piercing points. Visibility of the piercing line is determined by inspection, *Figure 7-27C*.

If the object is spherical, an imaginary flat surface on the sphere must be established along the piercing line.

Given: A sphere with a piercing line, *Figure 7-28A*.

Lightly draw the piercing line in the top view. Where the piercing line intersects with the sphere is the location of the imaginary flat surface, *Figure 7-28B*. This also establishes the diameter of the flat surface. Transfer the imaginary flat surface to the front view. Where the piercing line intersects the imaginary flat surface is the exact piercing point locations. Once the piercing point

GIVEN:

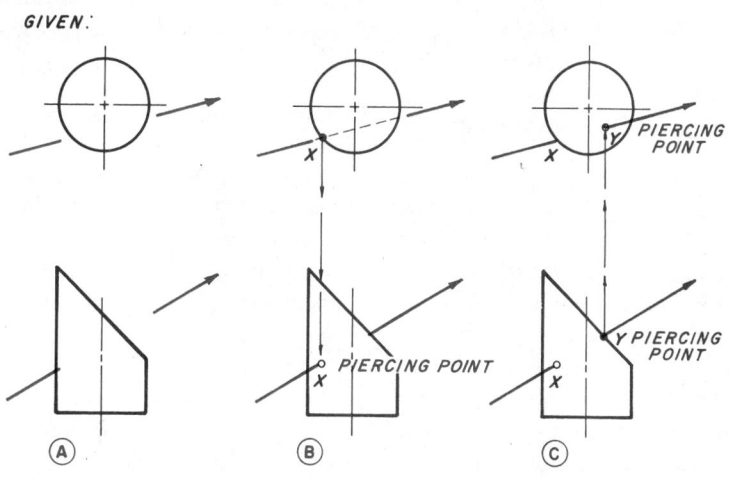

Figure 7-26 Determining piercing points by line projection (cylinder)

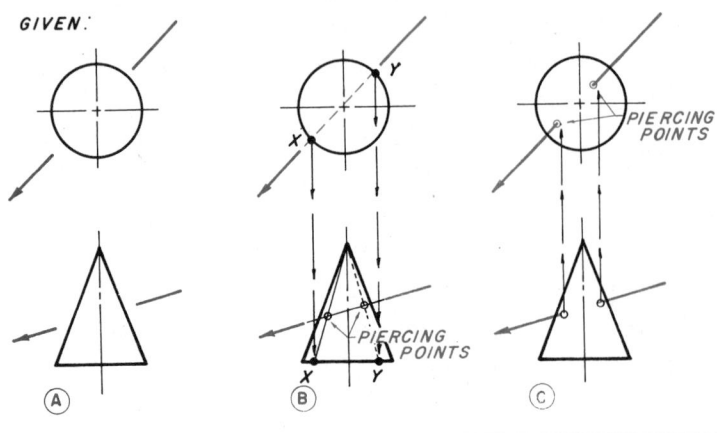

Figure 7-27 Determining piercing points by line projection (cone)

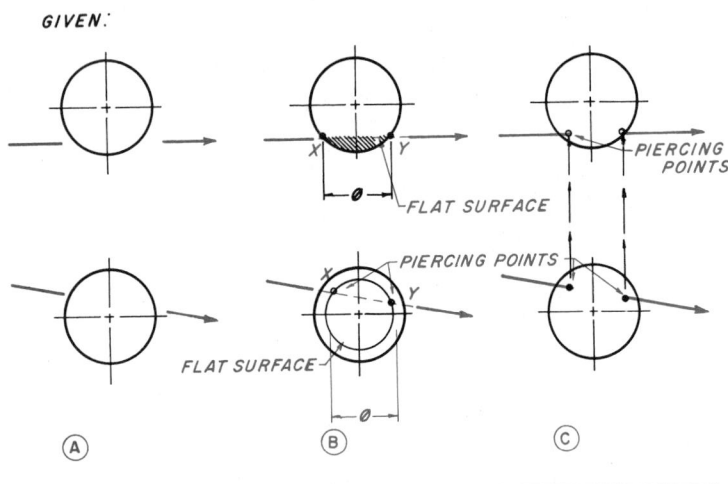

Figure 7-28 Determining piercing points by line projection (sphere)

locations are found they are projected up into the top view, ***Figure 7-28C***. Visibility of the piercing line is determined by inspection.

How To Find the Intersection of Two Planes by Line Projection

The intersection of two planes can be determined by generating an edge view of one of the planes. Another method, somewhat simpler, is the line-projection method.

Given: Two plane surfaces, ***Figure 7-29A***.

In the top view, locate points X and Y on edge ab of one of the planes. Project points X and Y down into the front view as illustrated in ***Figure 7-29B***, and draw a light line from X-Y in the front view. The piercing point is where line X-Y crosses the edge ab. Project points from edge ac of the same plane. Project these points down into the front view. Draw a line connecting these points; where they cross edge ac is the exact piercing point. Complete the views as illustrated in ***Figure 7-29C***.

How To Find the Intersection of a Cylinder and a Plane Surface by Line Projection

The intersection of a cylinder and a plane surface can also be determined by generating an edge view of the plane surface. Another method is the line-projection method.

Given: A cylinder and a plane surface, ***Figure 7-30A***.

Divide the circle into equal parts; in this example, 12 equal parts. Extend these lines out to the edges of the plane surface, ***Figure 7-30B***. The example illustrated uses points 1 o'clock and 7 o'clock to find points X and Y on the edge of the plane surface. Project points X and Y down into the front view. Draw a line from X-Y in the front view. Points 1 o'clock and 7 o'clock are located somewhere on this line. To find their exact locations, project points 1 and 7 from the top view to the line in the front view. Continue around the various points to locate each of the 12 points. Complete the drawing as shown in ***Figure 7-30C***.

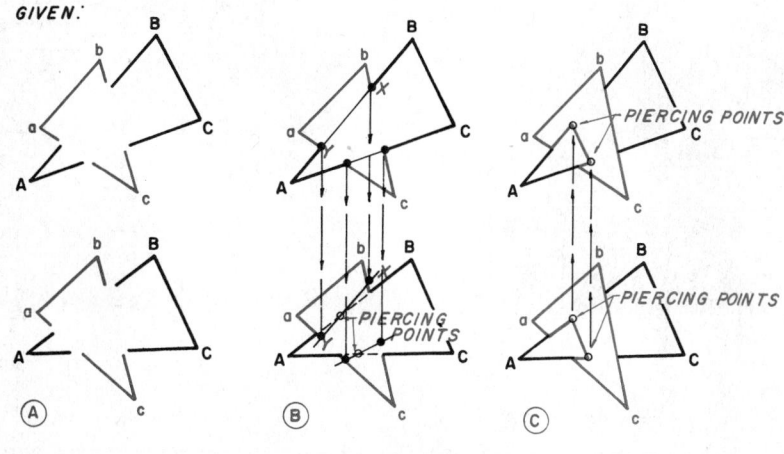

Figure 7-29   Determining the intersection of two planes

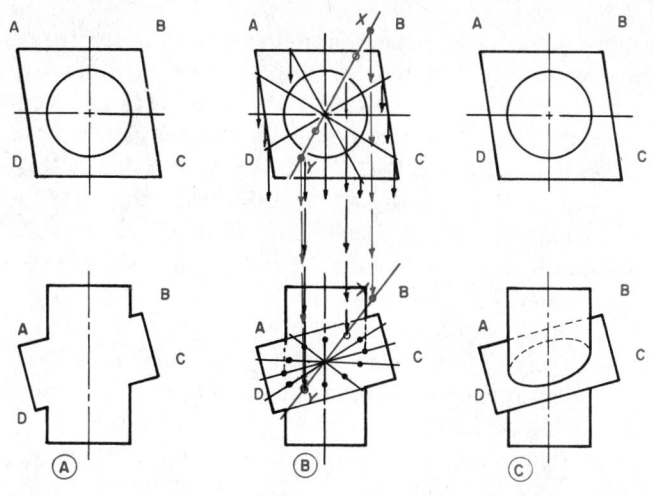

Figure 7-30   Determining the intersection of a cylinder and a plane surface

How To Find the Intersection of a Sphere and a Plane Surface

Given: A sphere and a plane, ***Figure 7-31A***.

Construct an edge view of the plane, ***Figure 7-31B***. Divide the sphere into even spaces as illustrated. Think of each evenly spaced line as an imaginary sliced-off portion of the sphere. Draw each imaginary slice in the top view and where the projection from the corresponding point on the edge view cross is the piercing point. Complete the top view as illustrated. Once these points have been found in the top view, the line-projection method is used to transfer them to the front view, ***Figure 7-31C***.

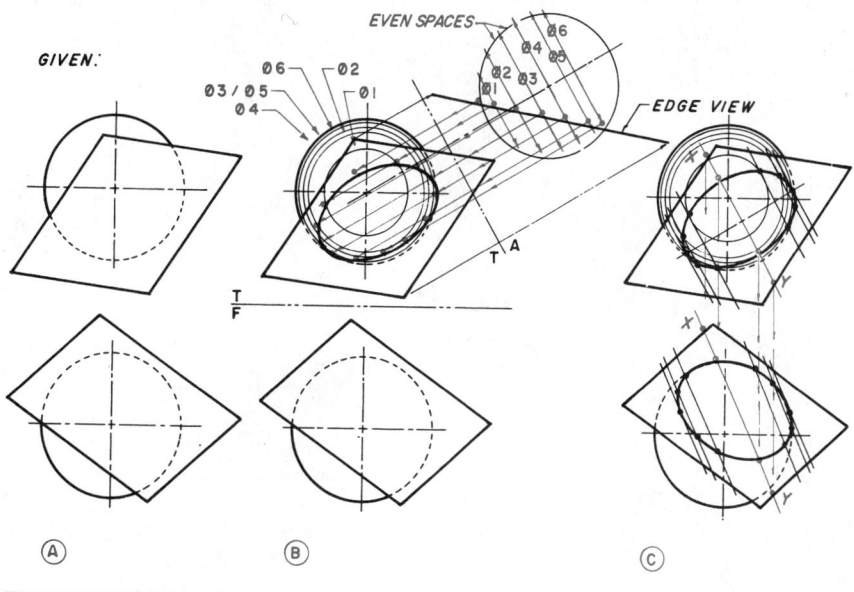

Figure 7-31   Determining the intersection of a sphere and a plane surface

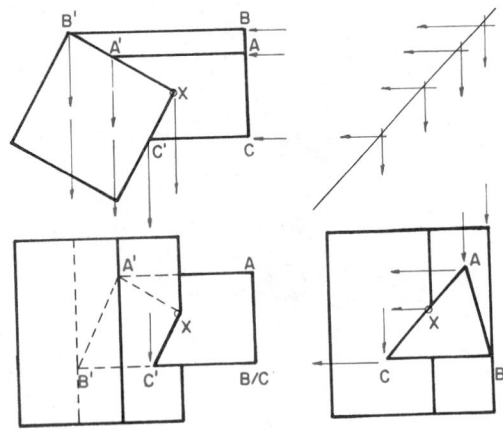

Figure 7-32   Finding the intersection of two prisms (method 1)

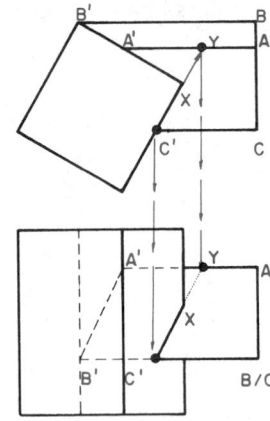

Figure 7-33   Finding the intersection of two prisms (method 2)

How To Find the Intersection of Two Prisms (Method 1)

Given: The top, front, and right-side views of two intersecting prisms, ***Figure 7-32***.

Label the various points as illustrated. Project each point to the 45° projection and into the next view and where they intersect is the location of each point, as illustrated. The only point in question is point X; using the three views, it can be easily found.

How To Find the Intersection of Two Prisms (Method 2)

Using only two views, point X creates a problem to locate point X in the front view. Refer to ***Figure 7-33***. In the top view extend line C'-X to where it crosses line segment A-A'. Project point Y down into the front view to line A-A'. Draw a line from C' to Y to locate point X in the front view.

Bearings, Slope, and Grade

In some fields of drafting the exact position of a line in space is described by its *bearing* and *slope* or by its *bearing* and *grade*.

Bearing of a Line

The *bearing* or *heading* of a line is the direction of that line as it is drawn in the *top view* of a drawing on a map. Using a map of a given area on the earth's surface, a view looking directly down would be an example of how a bearing of a line is used.

There are two methods used to call off a bearing. It is called off by either its *north/south* deviation or by its *azimuth bearing*.

North/South Deviation

A bearing of a line is measured in degrees with respect to the north or the south. It is customary to consider the north as being located from the *top* of the page and

the south as being located from the *bottom* of the page, unless otherwise noted. See ***Figure 7-34***. Bearings are called off in angles *less* than 90° and are always given *from* the north or south, heading *toward* either the east or west. See ***Figure 7-35***. In this example a line, 1-2, is projected 45° *from* the north and *toward* the west. It has a bearing of, and is called off as, N 45° W. A line, 1-2, is projected 30° *from* the south *toward* the east. See ***Figure 7-36***. It has a bearing of, and is called off as, S 30° E. It is important to know *where* the point of origin or the beginning of the line is, and in which direction the line is heading. Refer to ***Figure 7-37***. Line 2-2' could be called off as either N 45° W or as S 45° E, depending on its point of origin. If the line is projected from point 1 to point 2, the bearing would be N 45° W; if from 1' to point 2', the bearing would be S 45° E.

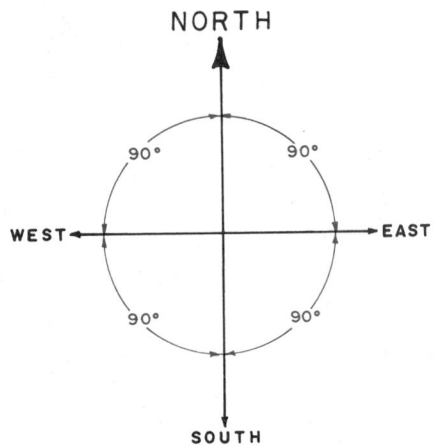

Figure 7-34   North, south, east, west headings; north is usually located at the top of the page

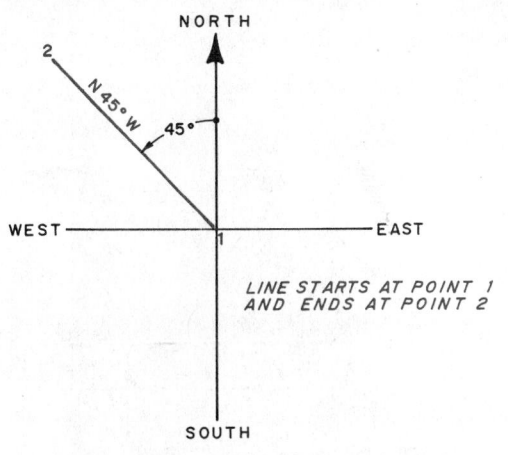

Figure 7-35 A line with a heading of N 45° W; headings are always drawn in the top view

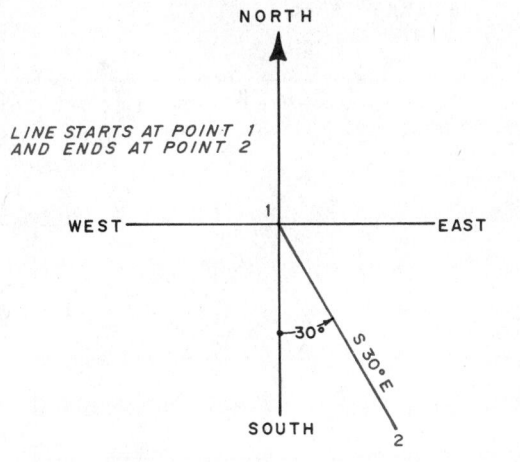

Figure 7-36 A line with a heading of S 30° E

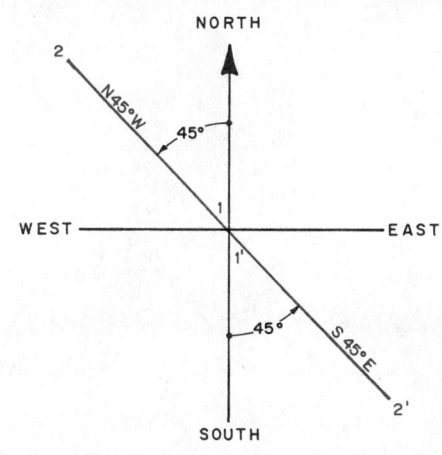

Figure 7-37 This line could be either N 45° W or S 45° W, depending upon the starting point and the direction in which it is projected

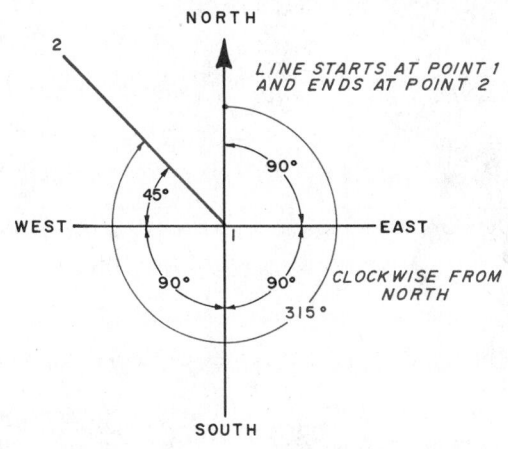

Figure 7-38 Azimuth bearing is indicated from the north clockwise to the line

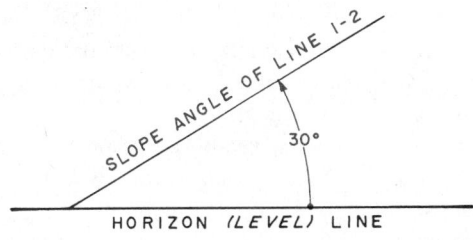

Figure 7-39 Slope angle is the angle in degrees from the horizon

Azimuth Bearing

In using the azimuth bearing method, the *total* angle is used, going *clockwise* from the north. A given line, 1-2, is drawn at 45° from the north and heading toward the west, ***Figure 7-38***. Using the azimuth method and measuring *clockwise* from the north, it has a bearing of 315°, (90° plus 90° plus 90° plus 45°). It is usually understood that the north bearing is the starting point, thus the "N" is usually omitted in verbally describing the bearing; on the drawing, however, it is best to use the "N" or "North" so as to avoid any confusion. Note that *bearing* is sometimes noted as *heading*.

Slope Angle

Using the slope angle is a way to describe the inclination of a line. The slope of a line is the angle, in degrees, that the line makes with a horizontal (level) plane, ***Figure 7-39***. The line is heading *from* point 1 *toward* point 2, and it is rising; therefore, it is a *plus* (+) slope angle.

Note that the *true slope* of a line can only be seen in the view that has a horizon (level) line and in which it is

drawn as a *true length*, ***Figure 7-40***. In this example, the true length can be found from the front view and/or from the top view. However, projecting from the front view will *not* give a horizon (level) line—it is only found by projecting from the top view.

Grade Percent

Another way to describe the inclination of a line in relation to the horizon is by the ratio of the vertical rise (R)

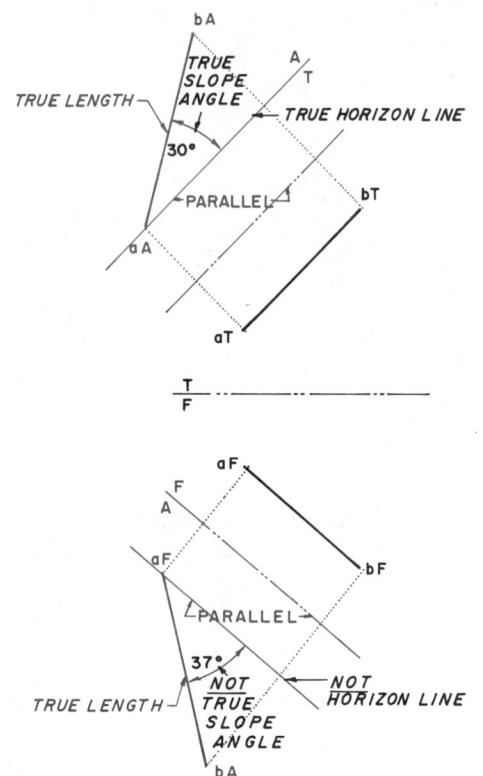

Figure 7-40 True slope angle of a line can only be seen in a view that has a horizon line. It is noted in degrees.

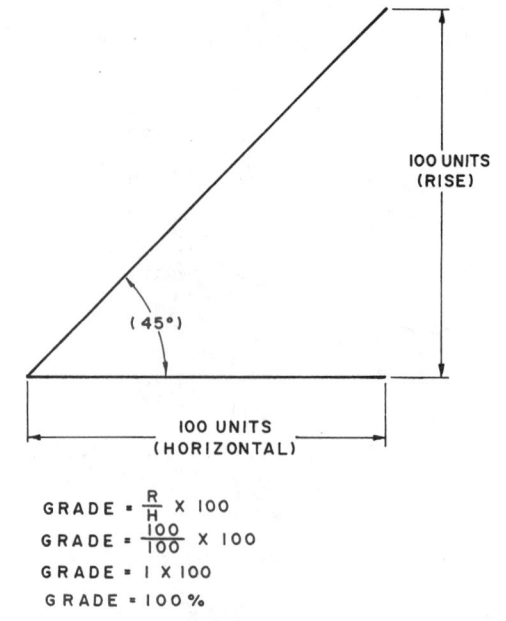

Figure 7-41 Grade percent is the inclination of a line in relation to the horizon, noted in percent

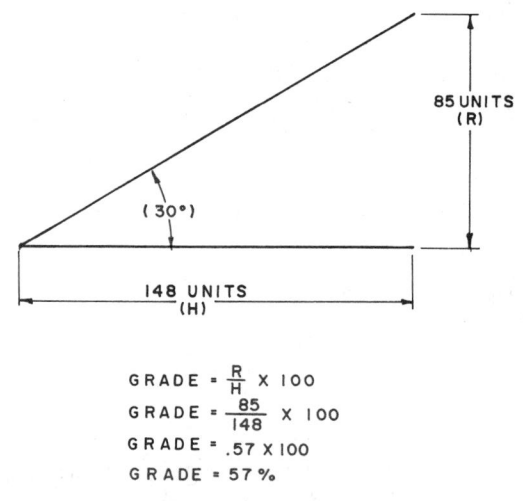

Figure 7-42 Example of a line with a grade of 57%

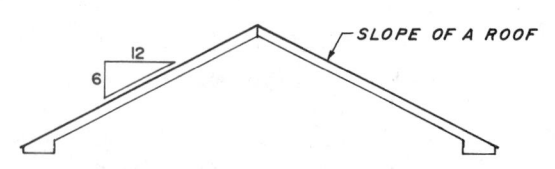

Figure 7-43 Pitch is the slope of a roof and is a ratio of vertical rise for every 12 inches or 12 feet of horizontal span

to the horizontal (H). It is expressed in percent and calculated by the simple formula:

$$\text{PERCENT OF GRADE} = \frac{\text{UNITS OF VERTICAL RISE (R)}}{\text{UNITS OF HORIZON (H)}} \times 100$$

Figure 7-41 illustrates a simple 45° angle line with 100 horizontal units and 100 vertical units. 100 divided by 100 times 100 equals a 100% grade. If the angle were 30°, the grade would be 57%, *Figure 7-42*. The most common method of measuring the horizontal and vertical units is with an engineer's scale using multiples of ten. Percent of grade is another way to express the slope of a line, but, as in laying out grade percents, *the view that has both the true horizon and true length must be used.* Refer back to Figure 7-40.

Other Ways to Call Off the Slope of a Line

Various fields of drafting use slightly different methods to call off the specified inclination of lines. Each method is similar to that of calling off percent of slope and/or percent of grade in that each expresses the relationship of the line to the horizontal (level) plane. Three common methods are *pitch, bevel,* and *batter.*

Pitch. Pitch is usually used in the architectural field of drafting and is used to call off the *slope* of a roof. It is expressed as the ratio of vertical rise to 12 inches or 12 feet of horizontal span, *Figure 7-43*. In each horizontal 12 inches the roof *rises* 6 inches.

Bevel. In structural engineering, where steel members are used in the construction of the building, the slope of a member is called off by the bevel of the beam, *Figure 7-44*. It is very similar to that of the pitch used in architectural drafting.

Batter. In the field of civil engineering the slope of the grade (earth) is used and is referred to as *batter*. It is called off as illustrated by *Figure 7-45*. This means that for each unit of rise (R), there are four horizontal (H) units.

Angles Between Bearings or Headings

If you know the bearings of two lines, the angle between the lines can be determined, *Figure 7-46*

Given are lines 1-2 and 1-3, as illustrated. Both these lines fall within one 90° quadrant; therefore, to find the angle between the lines, subtract one bearing from the other, *Figure 7-47*. Fifteen degrees is subtracted from eighty degrees, and the true angle between the bearings is sixty-five degrees.

If the bearings fall within two or more quadrants, the actual angles in each quadrant must be added or subtracted, as illustrated in *Figures 7-48*, *7-49*, *7-50*, and *7-51*.

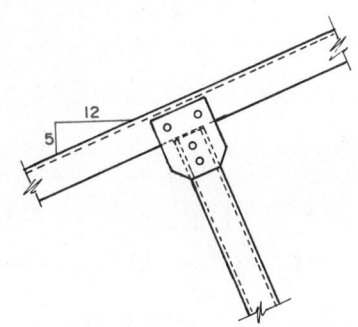

Figure 7-44   Bevel is used in structural engineering to indicate the slope of a structural member

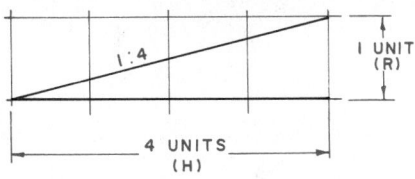

Figure 7-45   Batter is used in the civil engineering field to indicate the slope of a grade

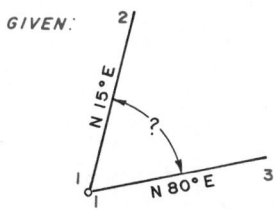

Figure 7-46   Angle between two headings in one quadrant (90°)

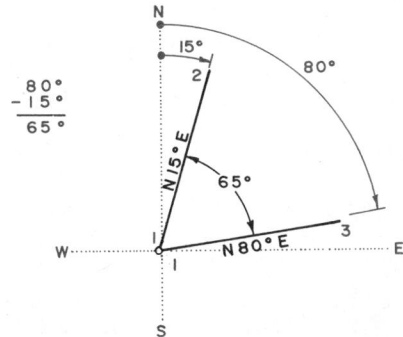

Figure 7-47   How to calculate the angle between headings in one quadrant

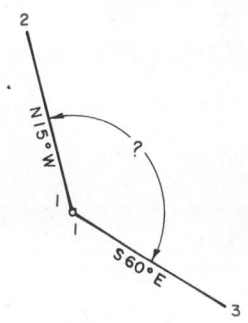

Figure 7-48   Angle between two headings in three quadrants

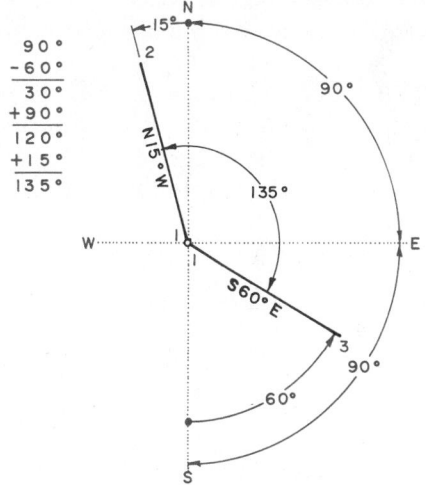

Figure 7-49   How to calculate the angle between headings in three quadrants

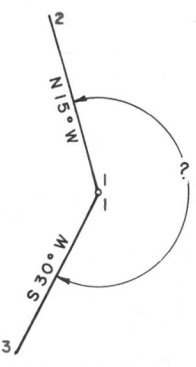

Figure 7-50   Angle between two headings in four quadrants

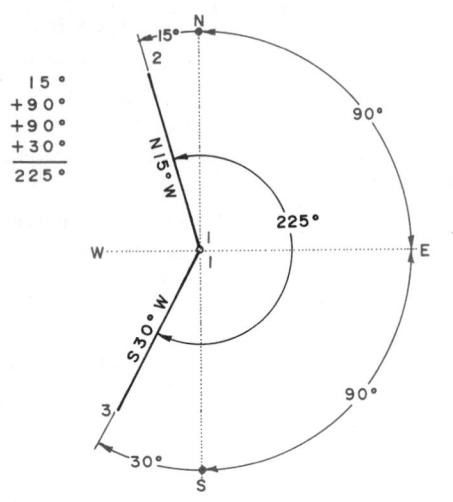

Figure 7-51 How to calculate the angle between headings in four quadrants

How to Construct a Line with a Specified Bearing, Slope Angle, and Length

The bearing is always drawn in the top view, and the true length can be drawn *if* the line is to be *exactly* horizontal. Any line with an incline up (+) or down (–) from the horizon *must* be laid out using the following steps:

Given: The starting point aT and aF of a line with a bearing of N 60° E, with a slope angle of +15°, and 230.0 feet long, *Figure 7-52A*.

Because the line is *not* horizontal and has a slope angle of plus (+) 15°, the true length must

LINE BEARING N 60° E
15° (PLUS) SLOPE
230.0' TRUE LENGTH

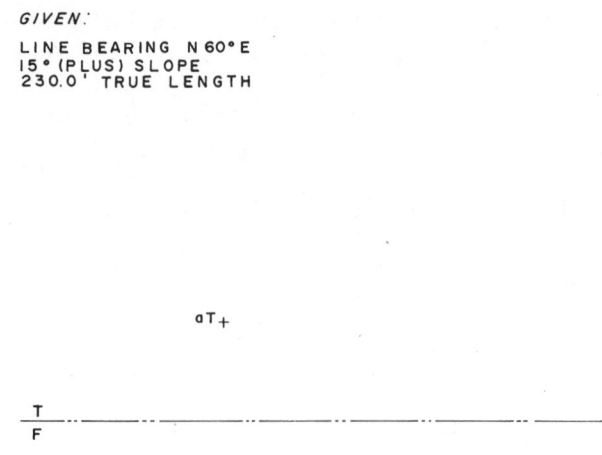

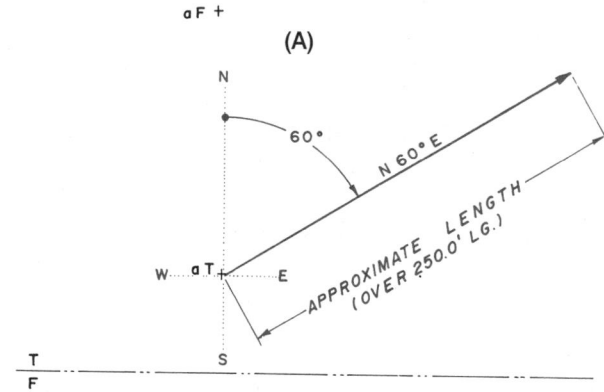

(A)

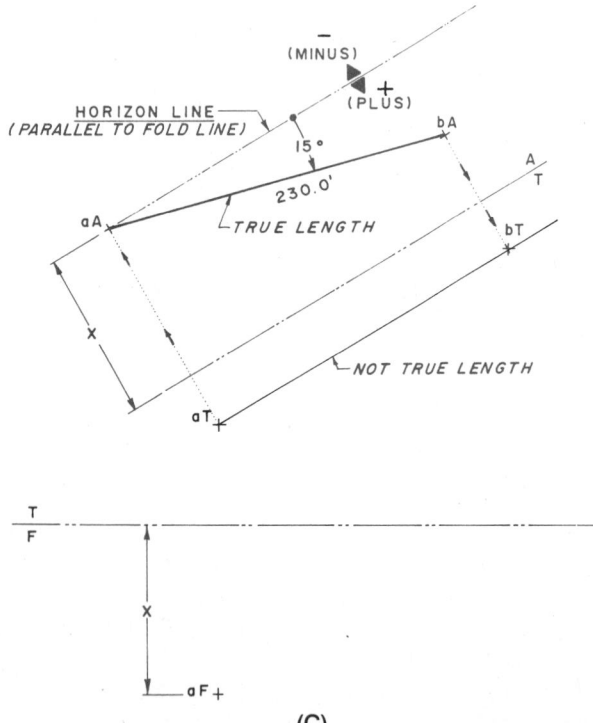

(C)

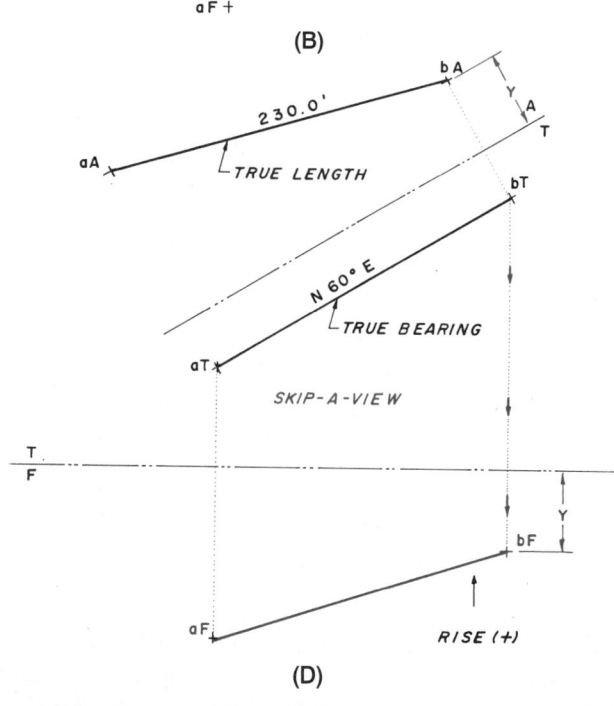

(D)

Figure 7-52 (A) Constructing a line with a specified bearing, slope angle, and length; (B) Step 1; (C) Step 2; (D) Step 3

be constructed *before* the heading length can be drawn in the top view.

Step 1 In the top view, draw a line starting from aT, at 60° from the north and toward the east, ***Figure 7-52B***. Temporarily draw the line actually *longer* than the 230.0 feet required.

Step 2 Find the true length of the line by adding a fold line, T/A, *parallel* to the bearing line aT. Project point a into the auxiliary view to find aA, ***Figure 7-52C***, and refer to dimension X. Don't forget to skip-a-view.

From point aA in the auxiliary view, draw a light horizon line *parallel* to the fold line T/A. From point aA, construct a 15° angle *toward* the fold line as shown.

Note: Draw the line *toward* the fold line for a rise (+) and *away* from the fold line for a drop (–) in elevation. Along *this* line measure the true length, 230.0 feet, to establish point bA. Project bA back into the top view to find point bT. Remember, this line is *not* the true length in the top view.

Step 3 Project point bT down into the front view to find point bF. Use the Y dimension to find point bF, but do not forget to skip-a-view, ***Figure 7-52D***. Construct line aF/bF. Note that the line *rises* from a to b in the front view, which indicates a plus (+) rise.

The true bearing is found in the top view and the true length is found in the auxiliary view. Remember, true length must be found before the other views can be completed.

How to Construct a Line with a Specified Bearing, Percent of Grade, and Length

Approximately the same steps are used in constructing a line with a specified bearing, percent of grade, and length as was used in constructing a line with a specified bearing, slope angle, and length. The bearing is always drawn in the top view, and the true length of the line can be drawn *if* the line is to be *exactly* horizontal. Any line with an incline up (+) or down (–) from the horizon *must* be laid out using the following steps:

Given: The starting point aT and aF of a line with a bearing of N 45° W, with a grade of (–) 40%, and 190.0 feet long, ***Figure 7-53A***. Because the line is *not* horizontal and has a grade of minus (–) 40%, the true length must be constructed *before* the bearing length can be drawn in the top view.

Step 1 In the top view, draw a line starting from aT, at 45° from the north and toward the west, ***Figure 7-53B***. Temporarily draw the line actually *longer* than the 190.0 feet required.

Step 2 Find the true length of the line by adding a fold line, T/A, *parallel* to the bearing line aT. Project

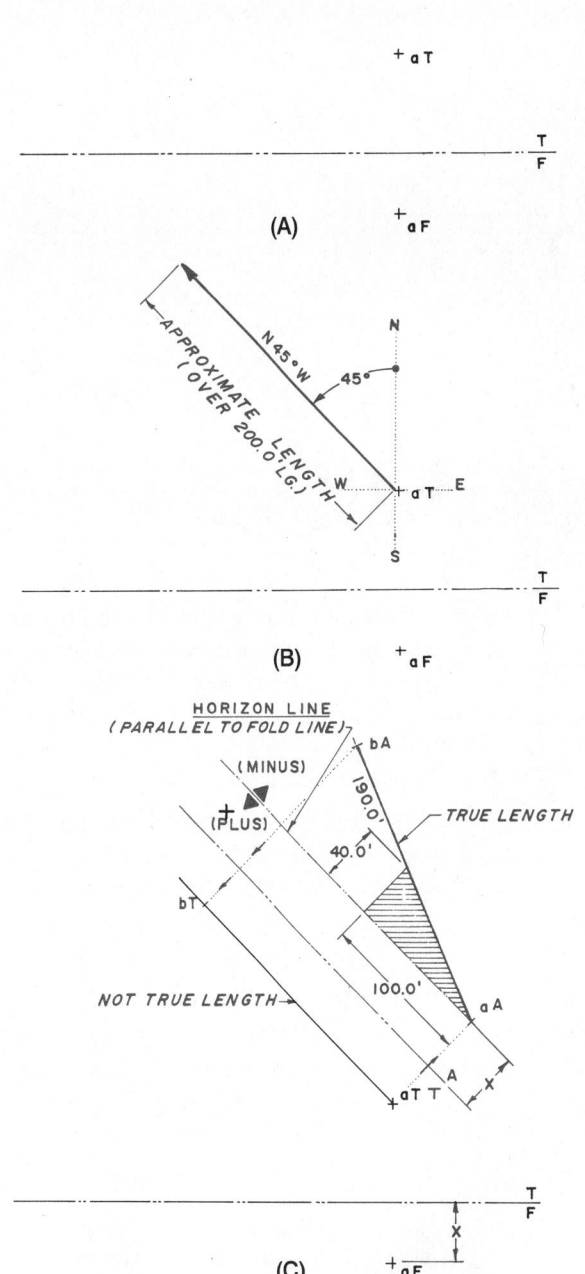

GIVEN:
LINE BEARING N45°W
40% (MINUS) GRADE
190.0' TRUE LENGTH

Figure 7-53 (A) Constructing a line with a specified bearing, percent of grade, and length; (B) Step 1; (C) Step 2

point a into the auxiliary view to find aA, ***Figure 7-53C***, and refer to dimension X. Don't forget to skip-a-view.

From point aA in the auxiliary view, draw a light horizon line *parallel* to the fold line T/A.

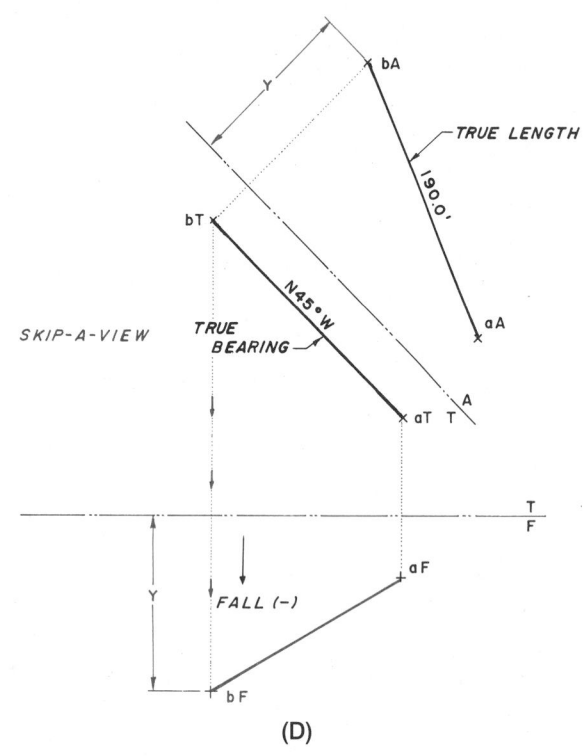

(D)

Figure 7-53 (Continued) (D) Step 3

From point aA, and *along* the horizon line, measure 100 units. From this point, turn 90° *away* from the fold line and measure 40 units, and put a point. Construct a line from aA to this new point and *extend* it out to the 190.0 feet true length to find point bA.

Note: Construct the triangle *toward* the fold line for a rise (+) and *away* from the fold line for a drop (–) in elevation. Project bA back into the top view to find point bT. Remember, this line is *not* the true length in the top view.

Step 3 Project point bT down into the front view to find point bF. Use the Y dimension to find point bF, but do not forget to skip-a-view, *Figure 7-53D*. Construct the line aF/bF. Note that the line *falls* from a to b in the front view, which indicates a minus (–) drop. The true bearing is found in the top view and the true length is found in the auxiliary view. Remember, true length must be found before the other views can be completed.

Applied Descriptive Geometry

The topics in the remainder of this chapter are essentially applications of the descriptive geometry basics already presented. Some of the topics are contours and topographic maps, vectors, forces in trusses, and revolution.

Topographic Topics

Surveys

Compass directions with distances describe real estate boundaries in cities and counties for commercial and domestic property owners. The descriptions usually include references to governmental surveys, such as being located in a government lot, section, township, and range and has specific coordinates so a surveyor can locate the property and stake out the area. Architects and builders have drafters prepare scaled drawings of the areas for planning the locations of buildings and houses.

How To Plot the Boundary of Real Estate

Example:

Given: Compass bearings with distances in feet, outlining a real estate lot, shown in a rough sketch with North in the traditional direction toward the top of the sheet, *Figure 7-54A*.

Step 1 Layout the boundary to scale starting with the S17°W 59′, then the S31°W 47′, and next the S54°30′ 170′ while utilizing the drawing area efficiently, i.e., the direction North may not be toward the top of the sheet, *Figure 7-54B*. Note that the 71°30′ angle shown is an internal angle, and that the sum of all the internal angles equal (number of sides N – 2) x 180°. For the layout shown (N – 2) x 180° = (5 – 2) x 180° = 540°. The 85°30′ angle is called a deflection angle; the smallest angle surveyor must turn the transit to align with the next side. The sum of

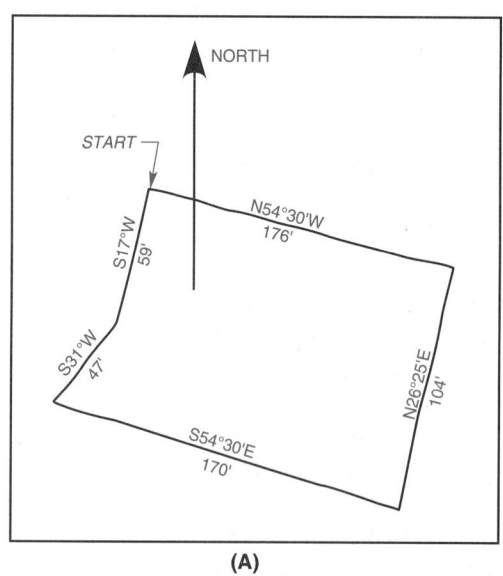

(A)

Figure 7-54 (A) Compass bearings and distances: Real estate lot

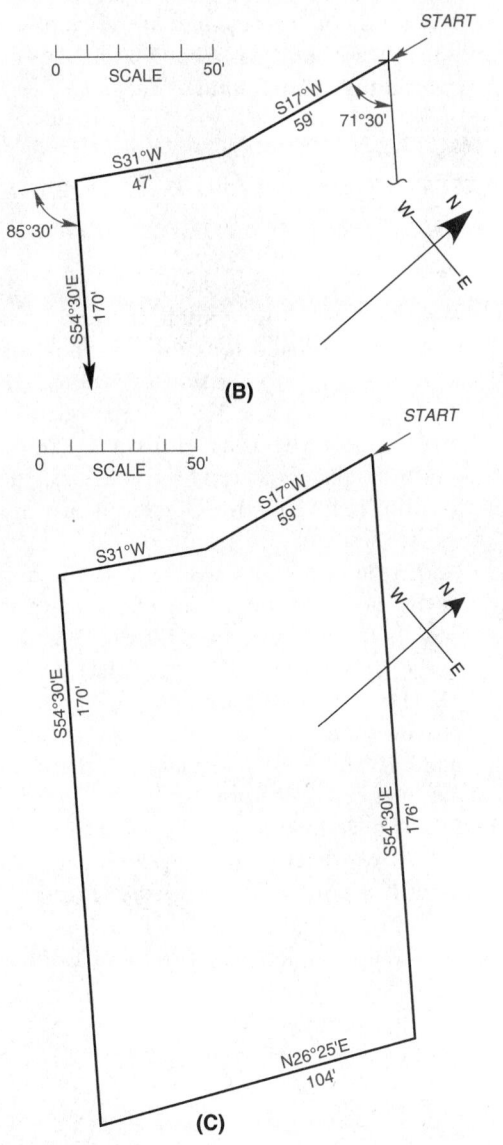

Figure 7-54 (Continued) (B) Step 1; (C) Step 2

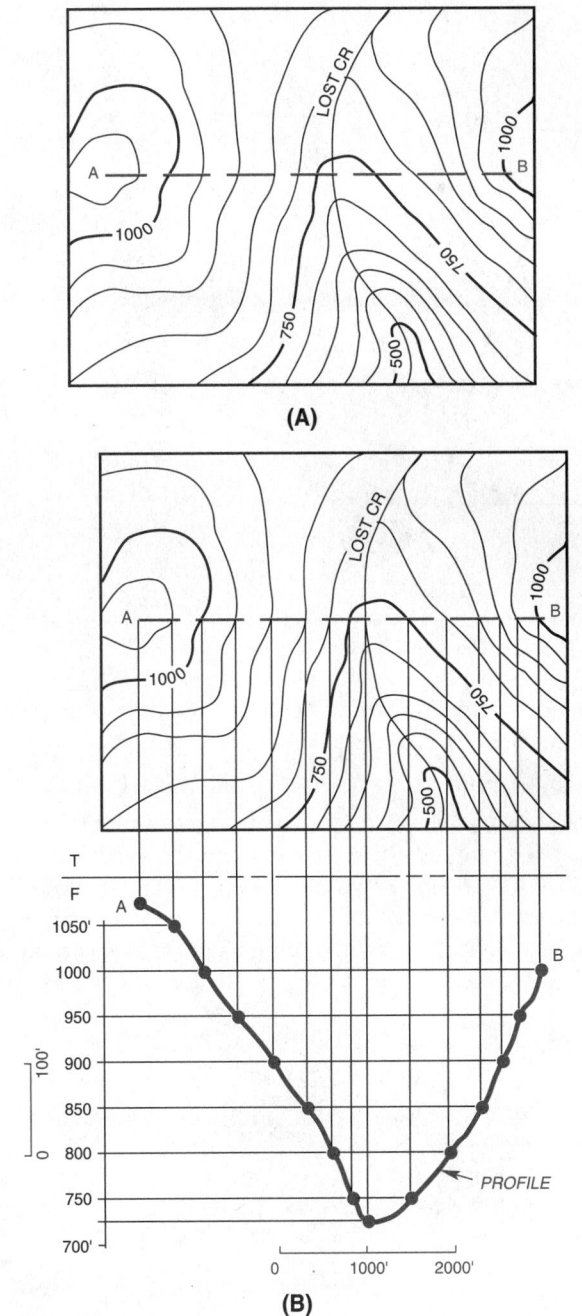

Figure 7-55   (A) Contour map; (B) Profile

all the deflection angles must equal 360° for a complete traverse around the property.

Step 2   Complete the boundary with N26°25' 104' next and N54°30' 176' last. The last segment should close on the starting point, **Figure 7-54C**.

For excavation and landscaping planning, an architect may ask a drafter to draw-in contour lines. This is the topic of the next section.

Contour Lines

A *contour line* on a plot or map is a line at a constant elevation on the surface of the earth, above sea level, **Figure 7-55A**. A *contour interval* is the difference in elevation between neighboring contour lines. It is the same value for any one drawing. For example, 50 feet is the interval in Figure 7-55A due to mountainous terrain. Ten-foot intervals are common; architectural plots may only have one-foot intervals on relatively level terrain for houses or buildings.

Usually, every fifth contour line is heavier and its elevation indicated. Lines close together represent steep terrain, etc.

Profiles

Imagine a vertical plane cutting through a contoured map, and being able to see the cross section at the cut. The cross section would show a profile of the terrain in an elevation view, as illustrated in **Figure 7-55B** for a straight path from point A to B. The fold line is added as a

PROFILES FOR "CUTS AND FILLS"

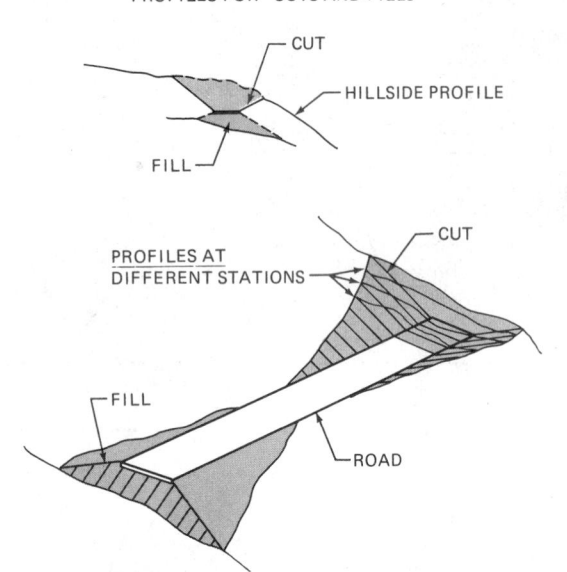

Figure 7-56   Cuts and fills

reminder that all topographic maps are T views and elevations, of course, occur in views projected off a T view.

Cuts and Fills

Highways passing through hilly terrain require that the earth of some hills be excavated and the earth used to fill in valleys—*cuts and fills*, ***Figure 7-56***. The volume of the earth to be cut and filled needs to be determined prior to having an army of construction workers and their equipment arrive at the job site, for efficient use of labor and equipment. The volume of the cut determines how much fill can be completed. Volumes of cuts and fills are determined from boundary maps.

How To Determine the Boundaries of Cuts and Fills from Topographic Maps with Contour Lines

Example:

Given: A portion of a topographic map with contours and the proposed location of a 60-foot wide, level road, at an elevation of 500 feet, ***Figure 7-57A***. The fold line T/F is perpendicular to the direction of the road and the F view shows elevation planes at ten-foot intervals; the topographic map and the elevations are drawn to the same scale.

Step 1   Construct cutting planes at 45° to the horizontal in the elevation view F, to each side of the proposed 60-foot wide road, ***Figure 7-57B***. Where the cutting planes intersect the elevation planes, mark the corresponding intersections on the map. Connect the intersections, freehand, to show the boundaries of the excavation cut, Figure 7-57B.

Step 2   In a similar manner construct cutting planes at 45° to represent the proposed earth fill, and mark the boundaries as before, ***Figure 7-57C***. Note that the angles of cuts and fills vary, e.g., a commonly used fill has a slope of 1.5 horizontal to 1.0 vertical.

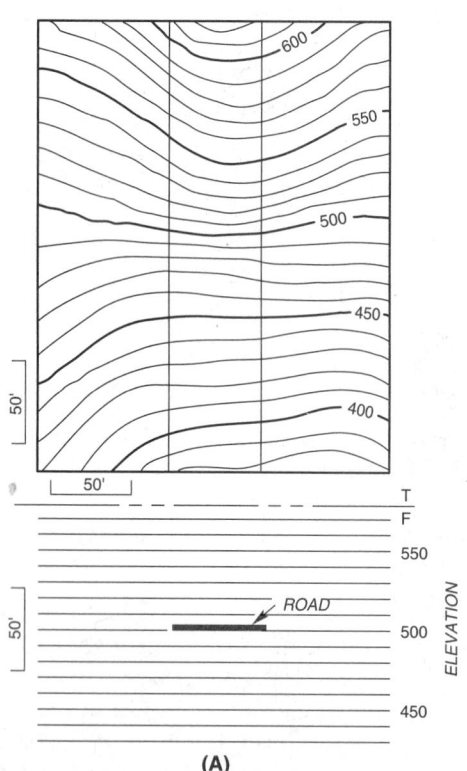

(A)

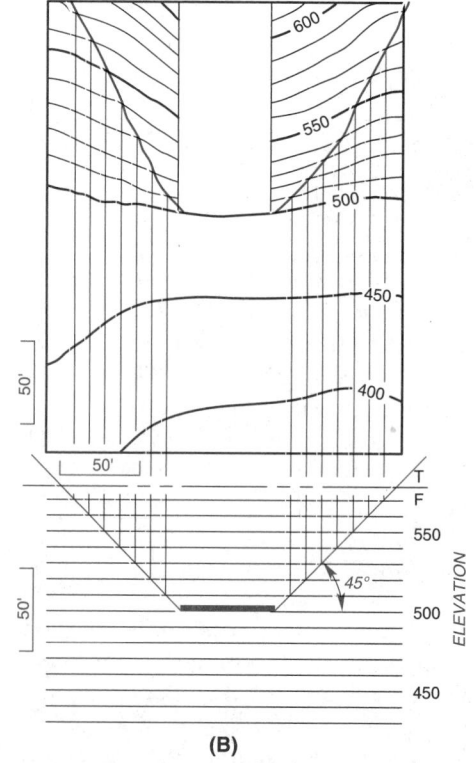

(B)

Figure 7-57   (A) Proposed road; (B) Step 1

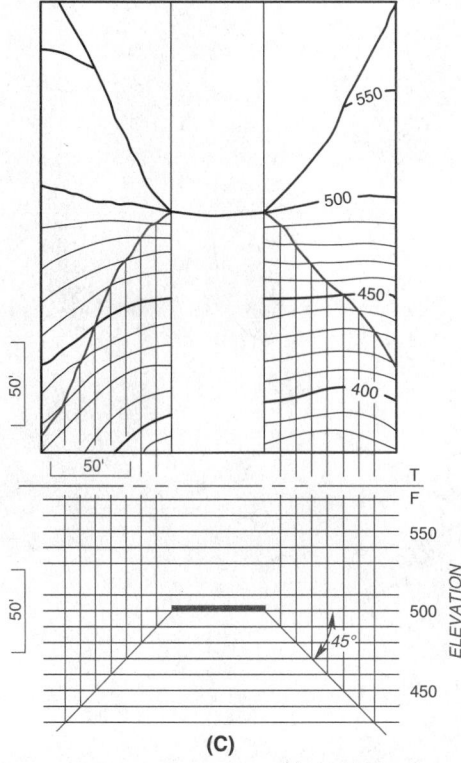

Figure 7-57 (Continued) (C) Step 2

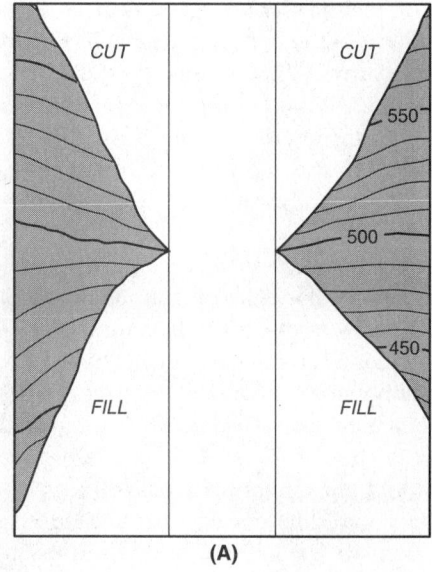

(A)

How To Determine the Volumes of Cuts and Fills (Graphical, Mechanical, and Electronic)

Example:

Given: The cut and fill boundaries from the preceding example shown in a simplified T view, *Figure 7-58A*.

Step 1 Start at the 500-foot elevation. Divide the cut and the fill with vertical planes at 25-foot intervals called stations (STA), *Figure 7-58B*. Estimate the elevations of the points where the vertical planes at each station intersect the boundaries by reading the contour lines.

Step 2 Construct scaled drawings of the profiles at the stations, and determine the area of each profile by using a graphical method for calculating geometric areas, *Figure 7-58C*. Tabulate the results.

Three methods used to calculate geometric areas of cross sections are graphical, mechanical, and electronic. The three are applied to a variety of problems, such as cuts and fills, cross sections of ship and boat hulls, and cross sections of layers of odd shapes for volume calculations.

A *graphical method* for approximating areas was used in Figure 7-58C. The method is to divide the area into convenient shapes, e.g., trapezoids and triangles, scale appropriate lengths, and calculate the areas.

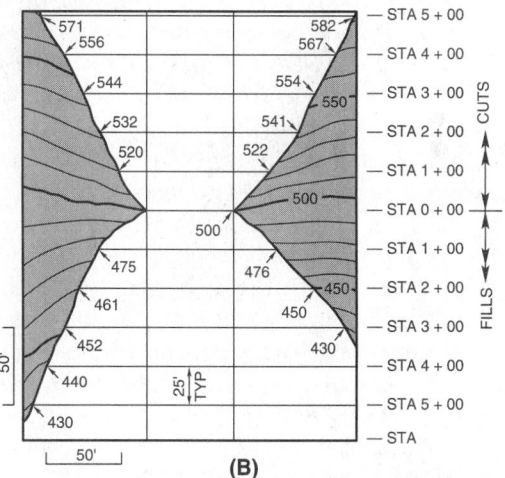

(B)

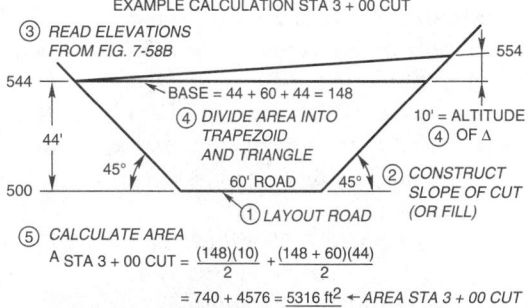

(C)

Figure 7-58 (A) Cut and fill boundaries; (B) Step 1; (C) Step 2 (graphical method)

A *mechanical method* utilizes a planimeter shown in the conceptual sketch, *Figure 7-58D*. Pole arm OA is confined to swivel about point O. Point P is a pointer the operator uses to trace the complete boundary, drawn to scale, causing an indicating wheel to rotate and slide. The wheel's final position gives a reading of the area traversed.

An *electronic method* employs a computer and a digitizer (looks like a computer mouse). A cross-haired window at the bottom of the digitizer is positioned at small increments of distances around the boundary of an area, *Figure 7-58E*. At each increment along the boundary, a key on the digitizer is clicked. This causes x-y coordinates to be stored in a computer. Finally, the computer uses the coordinates to calculate the digitized area.

Step 3 Calculate the amount of earth to excavate (cut) between two stations (STA) by averaging the areas at the stations and multiplying by the 25-foot distance between the stations. For example, the area STA 2+00 of the cut = 3502 square feet; the area at STA 3+00 of the cut = 5316 square feet; the volume between STAs 2 and 3 of the cut is ((3502 + 5316)/2 x (25) = 110,225 cubic feet. The volume in cubic yards (called *yards* in construction) is 110,225/27 = 4,082. Use this calculation procedure to determine the volume of earth between STAs 0+00 and 5+00 of the cut, Figure 7-58B. Get 20,709 yards available for fills. The fill requirements (using the same calculation procedure) for the volume between STAs 0+00 and 3+00 FILLS is only 9,390 yards. The remaining earth from the cut would be used to fill to additional stations.

Strike and Dip

Strike and dip are terms used in mining to describe three attributes of ore veins, *Figure 7-59*. *Strike* is a compass direction of a horizontal line lying in the plane of an ore vein. *Dip* is the downward angle of the vein with respect to a horizontal plane at the site of the ore vein. Dip also includes the general compass direction of the downward slope.

The vein in Figure 7-59 would be described as N64°30′ E, 35°NW, i.e., the compass direction of the strike, and the 35° dip downward toward the NW.

How To Determine the Strike and Dip of a Vein of Ore (or Stratum of Rock)

Example:

Given: The depths of three core-holes, A,B and C, sunk over a discovered ore vein, *Figure 7-60A*.

Step 1 Layout the points A, B and C in a T view to scale, Figure 7-60A. Since the points, which are

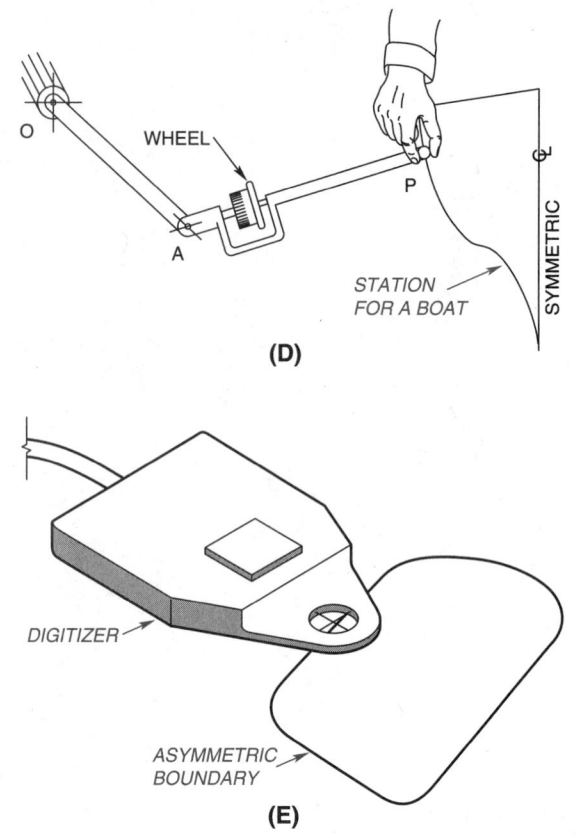

(D)

(E)

Figure 7-58 (Continued) (D) Mechanical method; (E) Electronic method

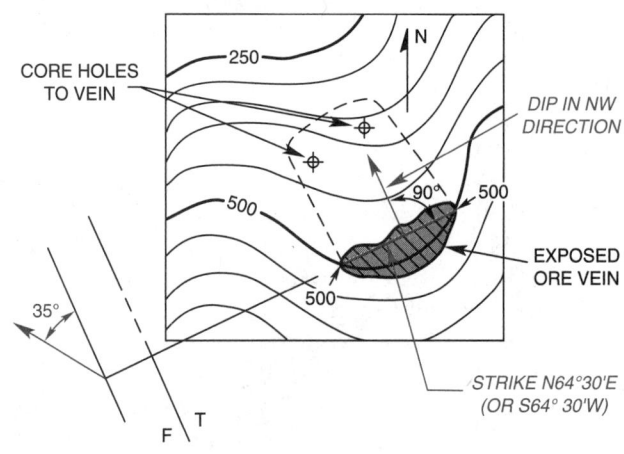

Figure 7-59 Strike and dip attributes

at different depths, do not fall in a straight line, they describe a triangular plane lying in the vein, *Figure 7-60B*.

Step 2 Find plane ABC as an edge in view A by using fold line T/A, *Figure 7-60C* and the technique described in Figures 7-17A through 7-17D in this chapter.

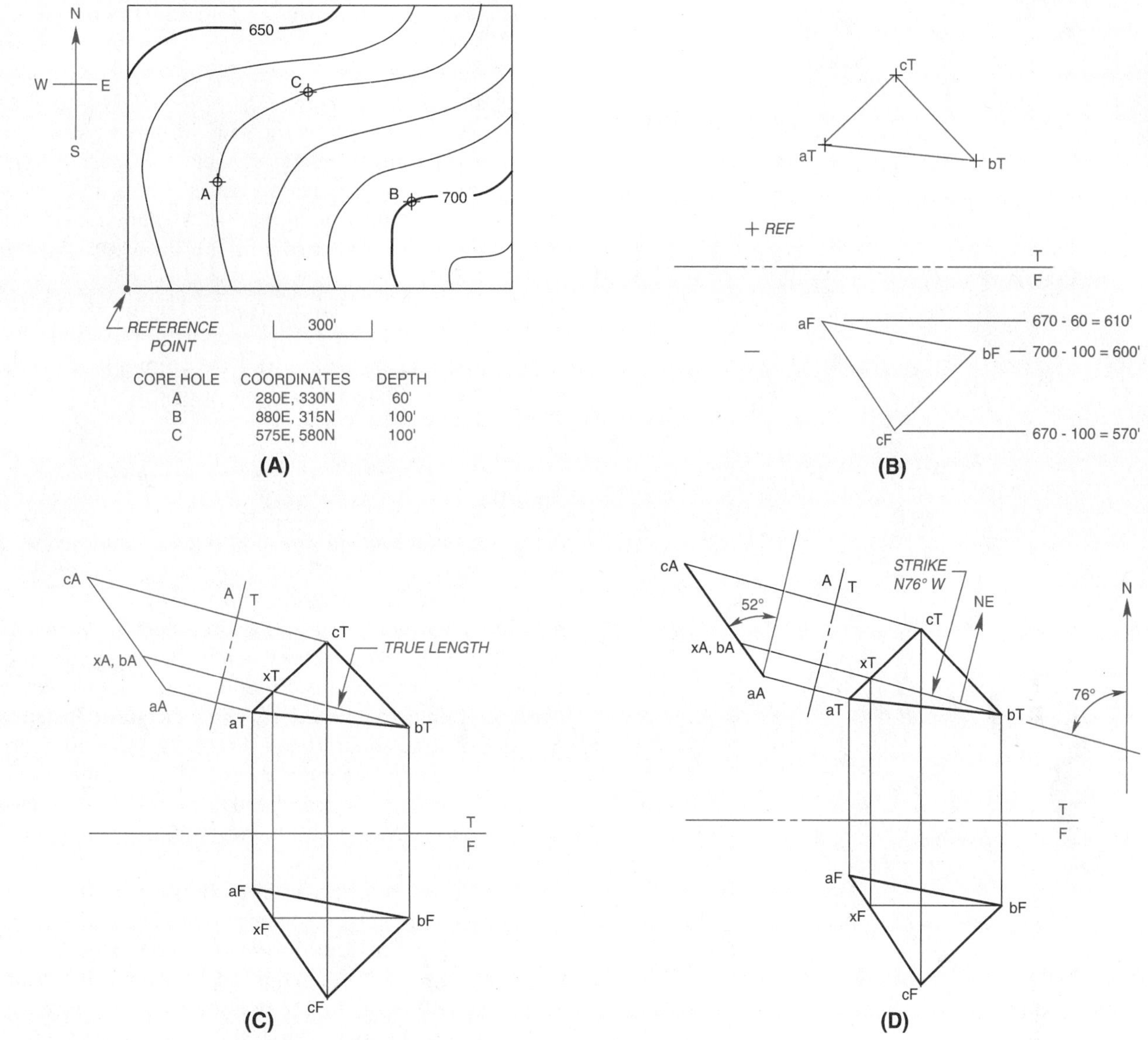

650

C

A

B 700

N
W — E
S

REFERENCE
POINT

| 300' |

CORE HOLE	COORDINATES	DEPTH
A	280E, 330N	60'
B	880E, 315N	100'
C	575E, 580N	100'

(A)

cT

aT   bT

+ REF

_____ T
                                F

aF _____ 670 - 60 = 610'
—   bF — 700 - 100 = 600'

cF _____ 670 - 100 = 570'

(B)

cA

A / T

xA, bA

aA   cT

aT   xT   *TRUE LENGTH*
              bT

_____ T
                           F

aF

xF   bF

cF

(C)

cA

52°   A / T   *STRIKE N76° W*

xA, bA   NE

aA   cT

aT   xT   bT   76°

_____ T
                           F

aF

xF   bF

cF

N

(D)

Figure 7-60   (A) Core holes for vein location;   (B) Step 1;   (C) Step 2;   (D) Step 3

Step 3   In view A measure the dip angle, and in the T view measure the strike bearing, *Figure 7-60D*. Note the general direction of the dip to complete the attributes of the ore vein—N76°W, 52°NE.

How To Determine the Strike and Dip of a Vein given Two Lines in the Plane of the Vein and Intersecting

Example:

Given: Two apparent dip directions (not strikes) and angles of the dips as follows: N61°W, 40° and N22°E, 25° both originating from the same point P, a bore-hole, *Figure 7-61A*.

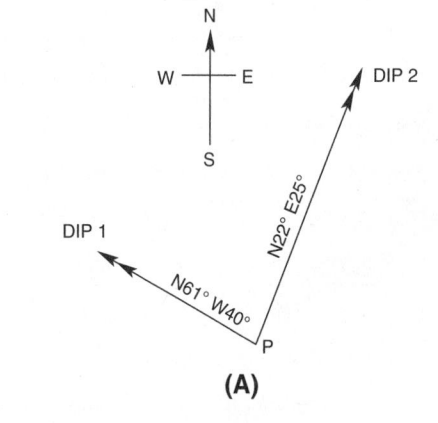

Figure 7-61   (A) Given: Dip directions and angles

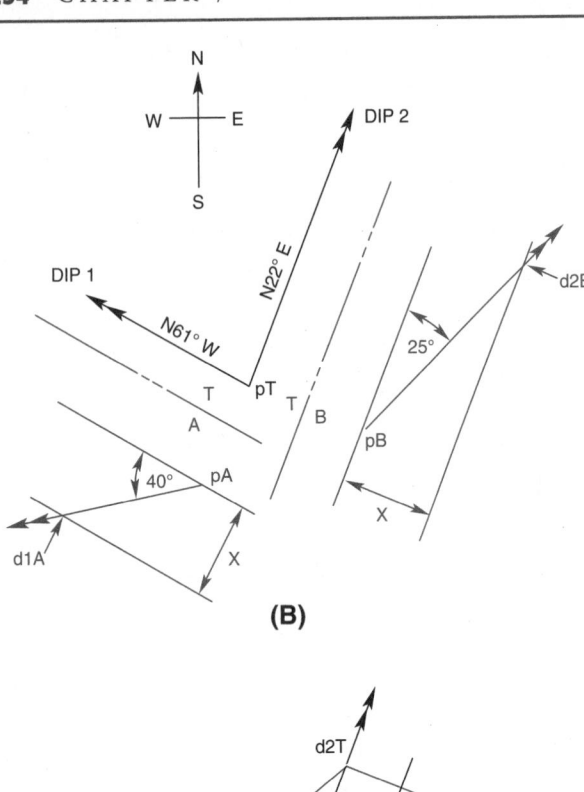

(B)

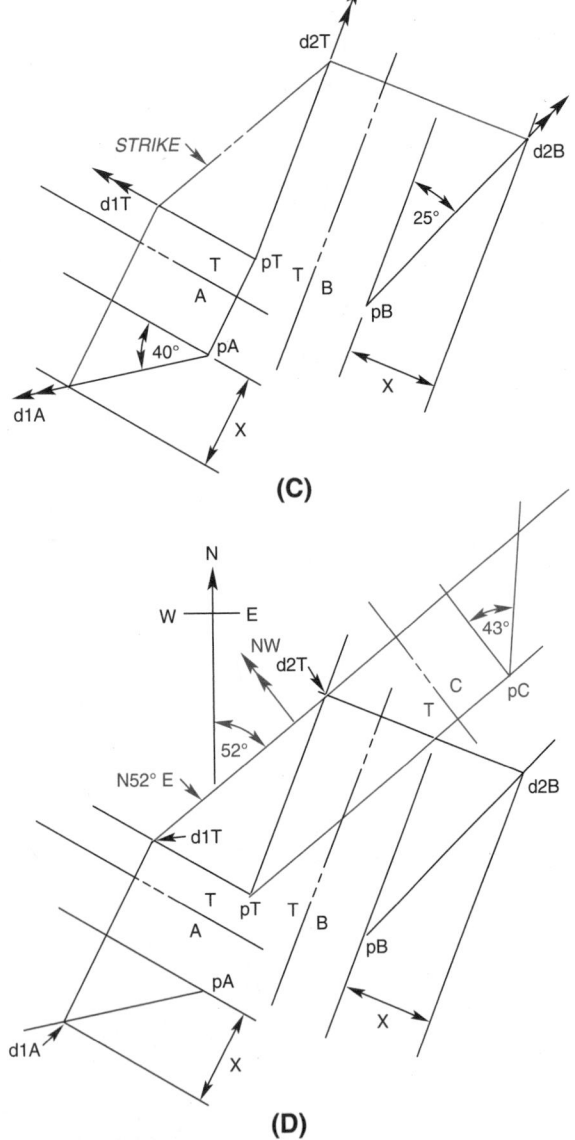

(C)

(D)

Figure 7-61 (Continued) (B) Step 1; (C) Step 2; (D) Step 3

Step 1 Construct view A using fold line T/A parallel to DIP1 to obtain the true view of the dip. Construct view B using T/B parallel to DIP2 to obtain its true view, *Figure 7-61B*. From points pA and pB layout the same distance, x, downward. Label the marks d1A and d2B, respectively.

Step 2 Project the points d1A and d2B into the T view and note that the line d1T-d2T is horizontal and, therefore, the strike of the ore vein, *Figure 7-61C*.

Step 3 Project the triangle (p-d1-d2)T into a view C using fold line T/C perpendicular to the strike line and measure the true dip angle, *Figure 7-61D*. The true strike and dip are N52°E, 43°NW.

Vectors

A vector as defined in engineering, technology, and mathematics must have both magnitude and direction. The vector shown in *Figure 7-62*, drawn to scale and having an arrowhead at one end could represent any one of the following: displacement (length and direction); force (pounds and direction); velocity (distance per unit of time and direction); or acceleration (distance per unit of time squared and direction). As a comparison direction only is a characteristic of the omnipresent straight line with an arrowhead at one end. A straight line with arrowheads at both ends in dimensioning indicates magnitude. These are not vectors but they efficiently convey information in technical material, such as compass bearings, cursor positions, callouts, identifying features, and of course dimensions on orthographic drawings, *Figure 7-63*. Magnitudes without directions are called *scalars*, such as temperature and pressure.

Vector equivalents of physical systems are simplifications used in engineering so that the vectors can be manipulated efficiently in mathematical formulas. For example, bulky objects or portions of a bridge deck, are assumed to be loads concentrated at a point. Internal loads, although acting over an area, are assumed to be point loads acting at the centroid of the area. All parts of an object traveling at the same velocity are represented by one velocity vector acting at the center of mass of the object. (See *Figure 7-64*.) Always keep in mind that these simplifications do not change the physical situation, and that point loadings are only approximate. Thus, graphical solutions employing these simplifica-

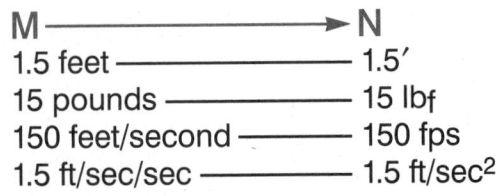

Figure 7-62 Vector

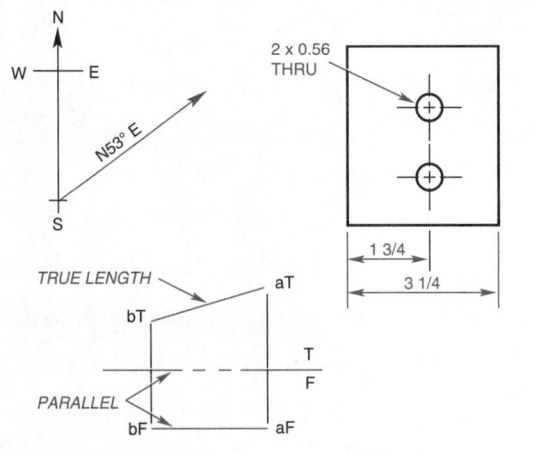

Figure 7-63   Non-vectors

tions are as accurate as mathematical solutions. A number of the vector manipulations that can be done graphically are discussed in this chapter, for two main reasons—to demonstrate the efficacy of graphics and to emphasize that when you can visualize what is going on, you will have a better understanding of the analysis and the answer.

Vectors will be labeled with capital letters, e.g., F(AB) is a force F in the direction of, or in, member AB. An applied load would be W or F. A velocity vector would be V.

Vector Systems

Six possibilities exist for analyses using vectors: coplanar (concurrent, parallel, and nonparallel) and noncoplanar (concurrent, parallel, and nonparallel). Coplanar-concurrent and noncoplanar-concurrent are the categories to be emphasized in this chapter.

Coplanar-Concurrent Force Systems— The Parallelogram Law: Two Vectors

The *parallelogram law* basically states that two vectors, not parallel, drawn to scale, having the same units, and acting at the same point can be added or subtracted as shown in **Figure 7-65A**. Use Vector A plus Vector B to construct a parallelogram. The diagonal of the parallelogram Vector (A+B) is called the resultant R, which is equivalent to the effect of the two vectors, **Figure 7-65B**. Vector A minus Vector B can also be used to construct a parallelogram. The resultant Vector (A-B) is the diagonal, **Figure 7-65C**.

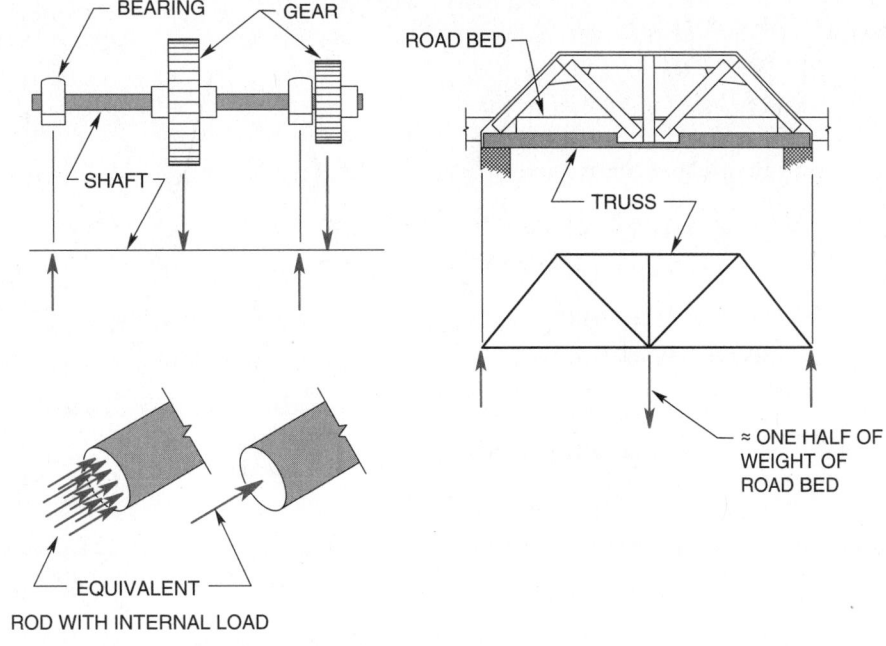

Figure 7-64   Vectors used in engineering simplifications

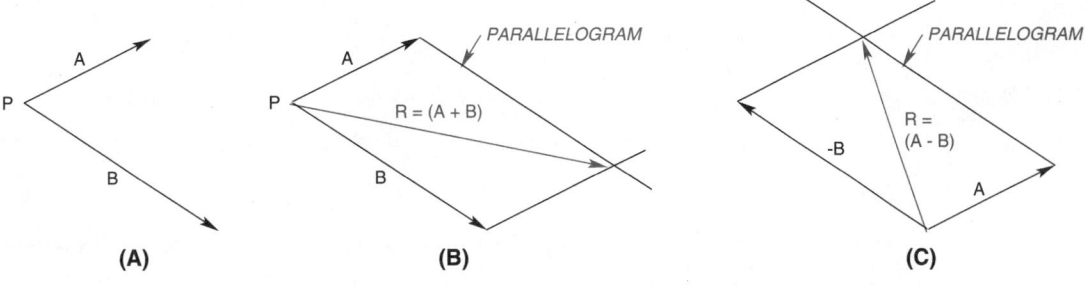

Figure 7-65   (A) Coplanar concurrent vectors;   (B) Adding vectors: Parallelogram law;   (C) Subtracting vectors: Parallelogram law

If the vectors being added or subtracted are force vectors, the equal and opposite reaction to the resultant is defined as the *equilibrant*. The reaction must be present to satisfy a basic law of mechanics, which is the sum of the forces acting on an object that is in equilibrium must add to zero in any one direction, i.e., $\Sigma F = 0$. A force may be defined as the action of one body on another, and it tends to change the state of rest or motion of the body acted on.

A vector force, acting on a body, can be considered to be moved along its line of action without changing its effect on the body.

Example:

Refer to *Figure 7-66A*.

Given: Force vectors P and Q acting on body B. Also, S and T acting on W.

Step 1 Move vectors P and Q along their respective lines of action until they meet and are tail to tail, *Figure 7-66B*. Also, move S and T.

Step 2 The resultant R of vectors P and Q shows the combined effect of P and Q on body B, *Figure 7-66C*. Also, the resultant U of vectors S and T shows the effect on body W.

How To Add More than Two Coplanar-Concurrent Vectors

Two graphical methods can be used to add more than two coplanar-concurrent vectors. One requires adding any two vectors first, and then using the resultant to add to the third, etc. The second, involves adding vectors "tail to head," in any order, to obtain a resultant.

Example: Method 1

Refer to *Figure 7-67A*.

Given: Vectors A, B, and C acting at point P.

Step 1 Add vectors A and B using the parallelogram law to obtain resultant R(1), *Figure 7-67B*.

Step 2 Add vectors R(1) and vector C using the parallelogram law to obtain resultant R(2), *Figure 7-67C*. R(2) is the resultant of the three vectors and would have the same effect as A, B, and C acting together at point P.

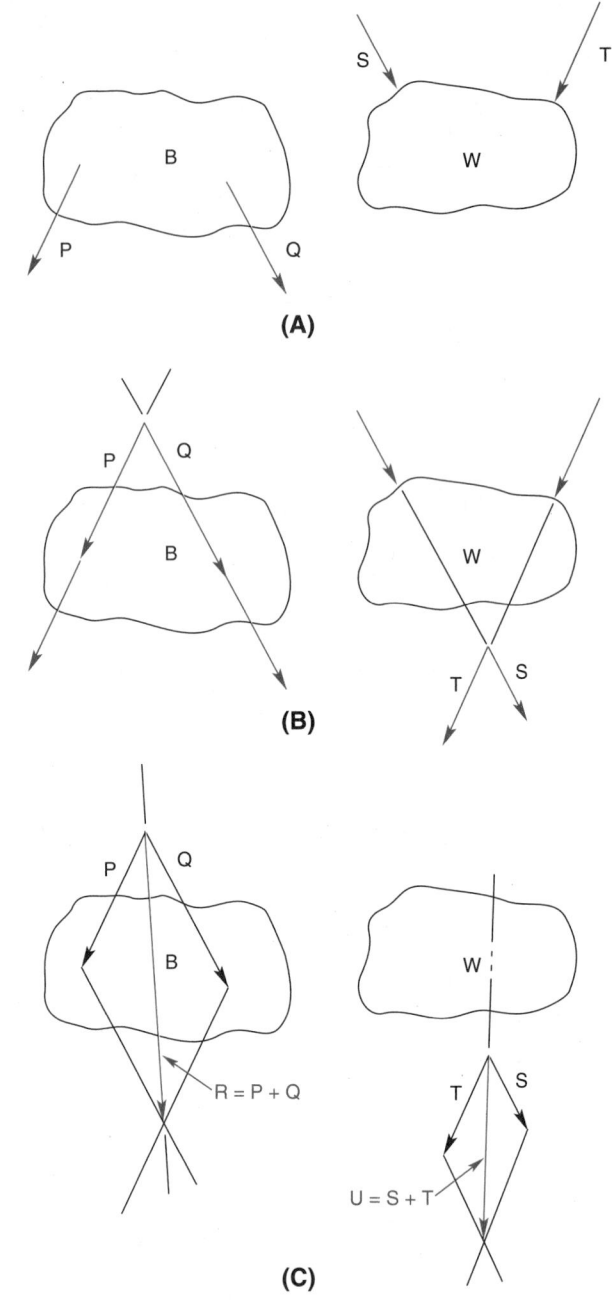

Figure 7-66 (A) Vector line of action; (B) Step 1; (C) Step 2

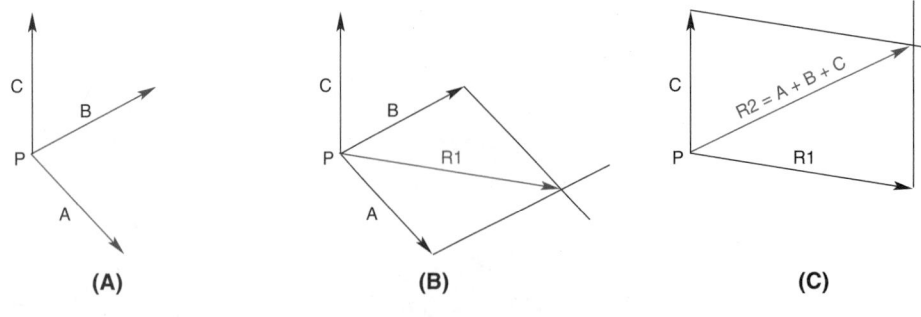

Figure 7-67 (A) Three coplanar concurrent vectors (method 1); (B) Step 1; (C) Step 2

Example: Method 2

Refer to *Figure 7-68A*.

Given: Vectors A, B, and C acting at point P.

Step 1 Move vector B parallel to its given position so that the tail of B is at the head of vector A, *Figure 7-68B*.

Step 2 Locate vector C in a similar manner so that the tail of C is at the head of vector B, *Figure 7-68C*.

Step 3 A vector from point P to the head of vector C is the resultant of the three vectors, i.e., the resultant would have the same effect on point P as vectors A, B, and C, *Figure 7-68D*.

Resolution of Vectors

A given vector can be resolved into two coplanar vector components that act along two axes that intersect at the tail of the given vector.

How To Resolve a Vector onto Perpendicular Axes

Example 1:

Refer to *Figure 7-69A*.

Given: Vector A and axes x and y intersecting at O.

Step 1 From the tip of vector A construct lines parallel to the two axes so that the lines intersect the axes x and y at X(1) and Y(1), *Figure 7-69B*.

Step 2 The two vector components that replace vector A are OX(1) and OY(1) with arrowheads at X(1) and Y(1), respectively, *Figure 7-69C*.

Example 2:

A three-dimensional mechanism problem in dynamics, but simplified to a statics problem at the instant shown. A further simplification presents the components as straight lines. The simplified mechanism and the velocity vector are drawn to scale. Refer to *Figure 7-70A*.

Given: Arm R confined to rotate about point A. Slider S pinned to C of crank BC. Crank BC confined to rotate about point B, 360 degrees. (The apparent interference of arm R with crank BC doesn't occur in the three-dimensional mechanism.) Slider C is free to slide along arm R.

Slider C at the instant shown has a velocity V(C) m/sec = ω(BC) radians/sec x BC m. The direction of the velocity of slider S at the instant shown is perpendicular to the crank BC because slider C is attached to point C.

To Find: The angular velocity ω(ARM)

Step 1 Resolve vector V(C), drawn to scale, into two components, one parallel to ARM(R) and one perpendicular, V(perpendicular), to ARM(R), *Figure 7-70B*.

Step 2 Divide V(perpendicular) m/sec by L m to obtain ω(ARM) radians/sec. (Note that radians/sec = 2 x π x RPS. RPS = revolutions per second).

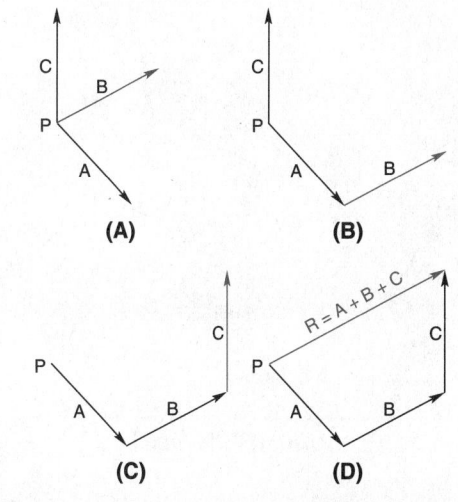

Figure 7-68 Four coplanar concurrent vectors (method 2)

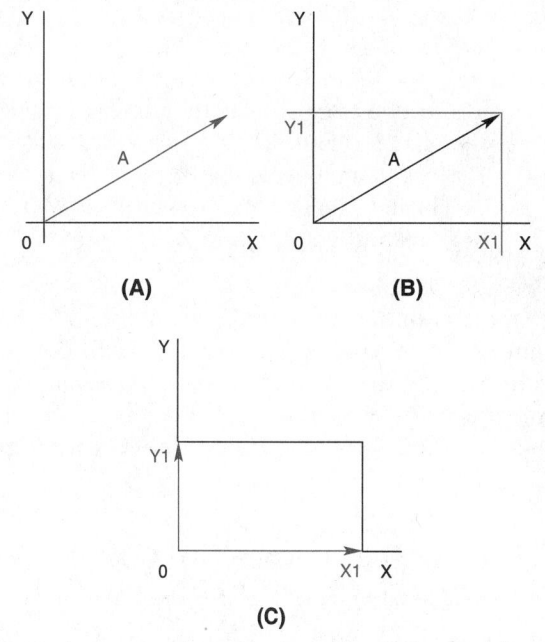

Figure 7-69 (A) Resolution on perpendicular axes; (B) Step 1; (C) Step 2

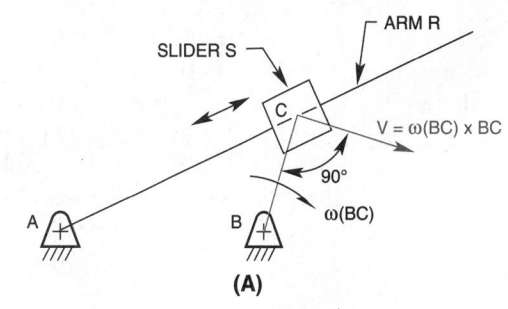

Figure 7-70 (A) Quick-return mechanism

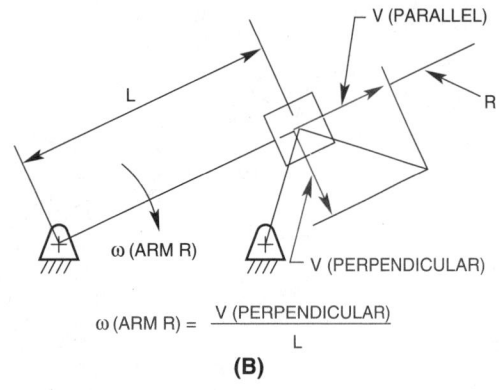

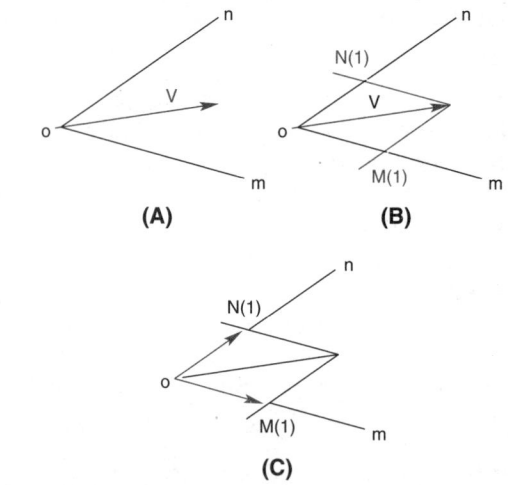

Figure 7-70 (Continued) (B) Steps 1 and 2

Figure 7-71 (A) Resolving a vector onto two axes not perpendicular; (B) Step 1; (C) Step 2

How To Resolve a Vector onto Two Axes not Perpendicular

Example:

Given: Vector V originating at O and axes m and n intersecting at O, *Figure 7-71A*.

Step 1 From the tip of vector V construct lines parallel to the two axes so that they intersect axes m and n at M(1) and N(1), *Figure 7-71B*.

Step 2 The two components that replace vector V are OM(1) and ON(1) with arrowheads at M(1) and N(1), respectively, *Figure 7-71C*.

Two Force Members

Two force members have forces at each end, equal in magnitude, in the opposite direction, and collinear. Therefore, they are in equilibrium. The weight of the member is considered to be negligible due to the usually much greater effect of the forces applied. See *Figure 7-72*.

How To Determine the Forces in the Members of a Pinned Joint

Example:

Given: A space diagram showing weight W supported by a frame in equilibrium, made up of two-force members AB and BC, drawn to scale, and pinned to a wall, *Figure 7-73A*.

Step 1 Isolate pinned joint B by "cutting" the members AB and BC, and redrawing the joint, parallel to the original drawing. An engineering expedient is to represent members of a frame (or truss) as straight lines. The lines of action of the internal forces F(AB) and F(BC) intersect at joint B. Their directions must be tentatively assumed by inspection. The direction of force F(AB) must keep the joint away from the wall. Force F(BC) keeps the joint from dropping. See *Figure 7-73B*.

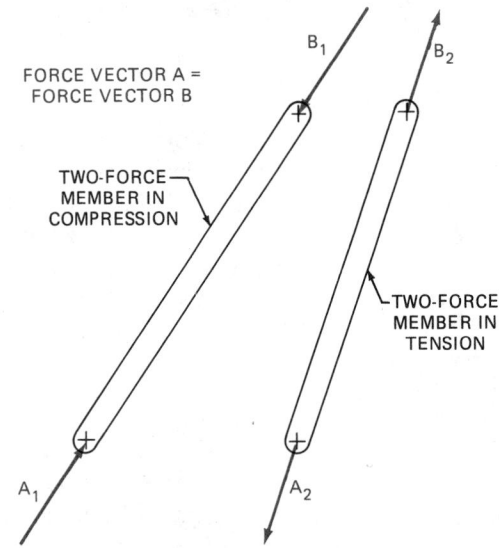

Figure 7-72 Two force members

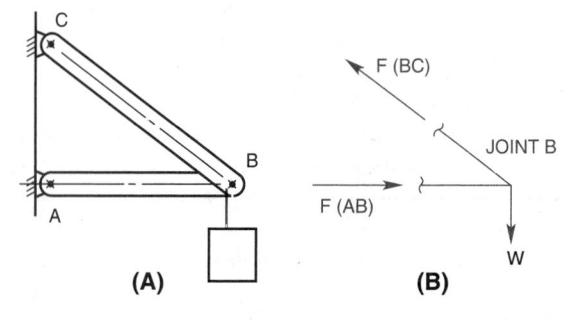

Figure 7-73 (A) Determing forces of a pinned joint; (B) Step 1

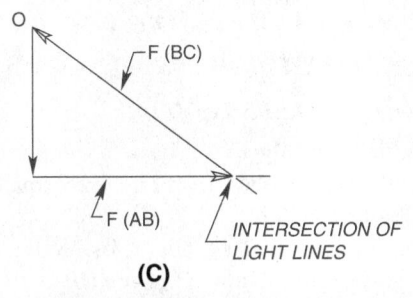

(C)

Figure 7-73 (Continued) (C) Step 2

Step 2 Construct a vector force diagram of the forces at joint B. First, lay out the applied force W, from a point O, vertically downward, to scale. The two remaining forces F(AB) and F(BC) must add—tail to head—to W and end at the starting point O because joint B is in equilibrium. To determine the magnitudes of F(AB) and F(BC) construct a light line through the head of vector W parallel to the direction F(AB) and a light line through the tail of vector W parallel to the direction of F(BC), *Figure 7-73C*. Where the

two lines intersect determines the magnitudes of F(AB) and F(BC). The tail-to-head method of adding vectors determines their directions. The assumed directions are confirmed in the vector force diagram.

Note that only two unknown magnitudes of forces can be determined in these coplanar joint analyses. Their directions are determined in the space diagram.

Trusses

Trusses are frames that are usually two-dimensional and are made of structural shapes, such as angles, tubes, rounds, and channels. They are assembled in triangular patterns for rigidity. The members are considered to be two-force members and the joints are assumed to be pinned even though welded, bolted, or riveted. Observe some features of the truss shown in *Figure 7-74A* and the simplified drawing of the truss in *Figure 7-74B*. Both are drawn to scale.

Trusses seldom stand alone. They are used in pairs or in multiples. For example, some highway bridges and railroad bridges employ pairs. Multiples are used to support roofs, *Figure 7-75A*. In a coplanar analysis, each truss is assigned its share of the roadbed loading or the roof loads at the appropriate joints of the trusses, *Figure 7-75B*.

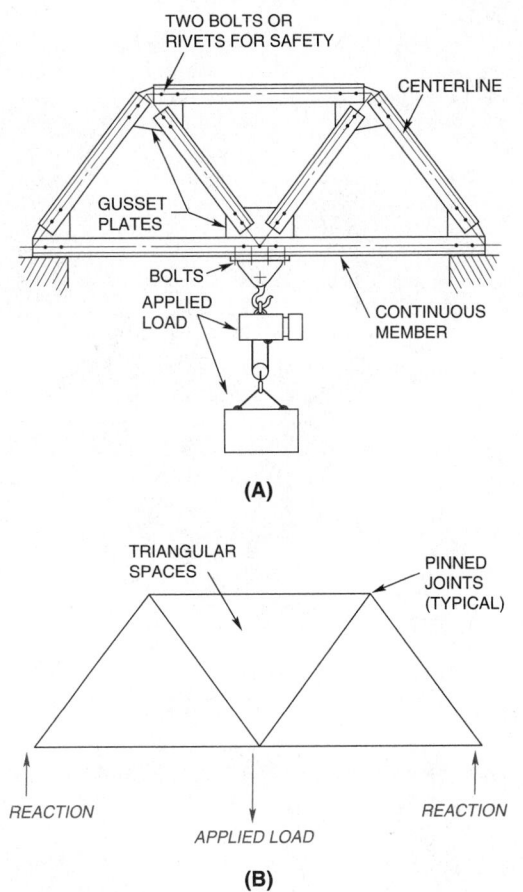

Figure 7-74 (A) Example truss; (B) Simplified drawing of truss

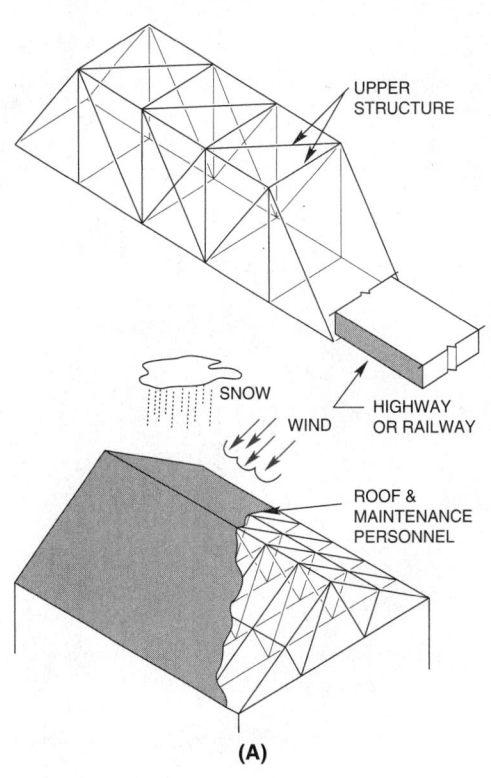

Figure 7-75 (A) Truss applications

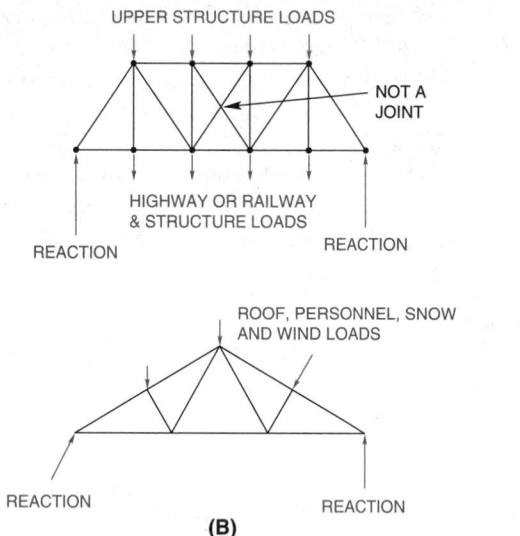

UPPER STRUCTURE LOADS

NOT A JOINT

HIGHWAY OR RAILWAY & STRUCTURE LOADS

REACTION   REACTION

ROOF, PERSONNEL, SNOW AND WIND LOADS

REACTION   REACTION

(B)

Figure 7-75 (Continued) (B) Truss drawings for analyses

How To Determine the Forces in the Members of a Truss

Example 1(a): Method of Joints

Given: Line drawing to scale of a symmetrical truss with one applied load and two equal reactions, *Figure 7-76A*.

Step 1 Select a joint in the truss with only two unknowns. A and C meet this criterion. Use joint A and isolate it in a space diagram, *Figure 7-76B*. Assume the direction of F(AD) by inspection to "go into" the joint, i.e., a *compressive* force. Assume F(AB) "goes away from" the joint, thus, a *tensile* force. Next, layout a force diagram to scale, starting with the known force of 2500N. Thus, follow the procedure described in step 2 of the preceding discussion, Figure 7-73C, where the light lines indicate the lines of action of the two unknown forces. Their inter-

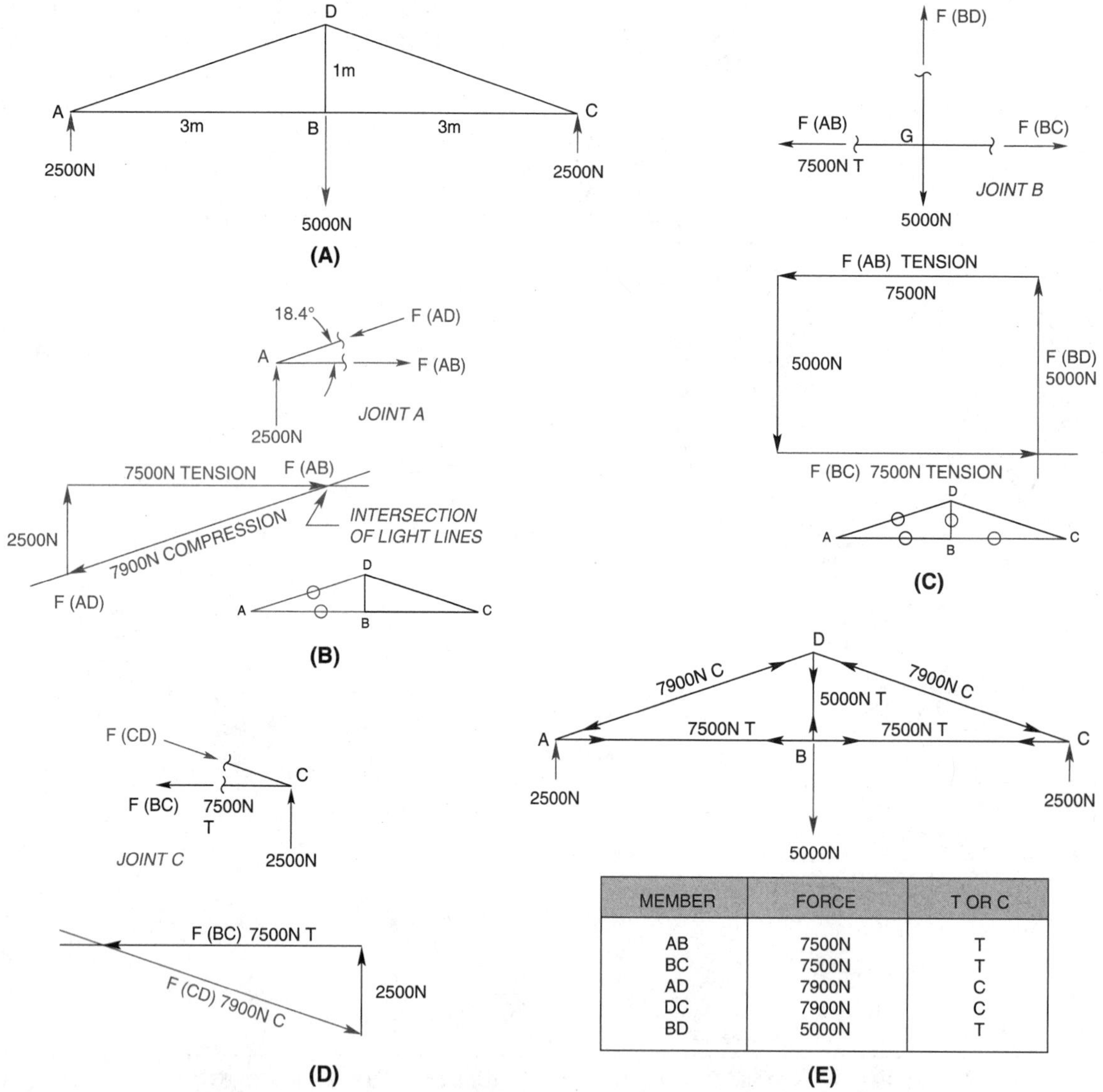

D

1m

A   3m   B   3m   C

2500N   2500N

5000N

(A)

F (BD)

F (AB)   G   F (BC)

7500N T   *JOINT B*

5000N

F (AB) TENSION

7500N

5000N   F (BD) 5000N

F (BC) 7500N TENSION

(C)

18.4°   F (AD)

A   F (AB)

JOINT A

2500N

7500N TENSION   F (AB)

2500N   *INTERSECTION OF LIGHT LINES*

7900N COMPRESSION

F (AD)

(B)

F (CD)

F (BC)   C   7500N T

JOINT C   2500N

F (BC) 7500N T

F (CD) 7900N C   2500N

(D)

D

7900N C   7900N C

5000N T

A   7500N T   7500N T   C

B

2500N   5000N   2500N

MEMBER	FORCE	T OR C
AB	7500N	T
BC	7500N	T
AD	7900N	C
DC	7900N	C
BD	5000N	T

(E)

Figure 7-76 (A) Example 1(a) method of joints; (B) Step 1; (C) Step 2; (D) Step 3; (E) Step 4

section point determines the magnitudes. Scale the magnitudes and put arrowheads on the vectors so that the vectors add head to tail back to the tail of the 2500N force, the starting point. Finally, make a small freehand sketch of the truss and circle the known members, AD and AB.

Step 2 Joints B or D have only two unknowns. Select either one and isolate it in an equilibrium sketch. Use joint B, *Figure 7-76C*. Construct a force diagram, starting with known information, the 5000N applied load, and the 7500N tensile force in AB. Construct the light lines for the forces in members BD and BC. Scale their magnitudes and add arrowheads so the forces add head to tail back to the tail of the 5000N force. On a new freehand sketch, circle the known members, AD, AB, BD and BC.

Step 3 The remaining member DC is known due to the symmetry of the truss, F(CD) = F(AD), and is a compressive force. If the truss or the loading were not symmetrical, joint C would be isolated and a force diagram prepared as shown, *Figure 7-76D*.

Step 4 Prepare a summary of the forces on a sketch of the truss or in a table, *Figure 7-76E*.

Example 1(b): Method: Maxwell Diagram with Bow's Notation

Given: The same truss used for the method of joints, but to be labeled with Bow's notation scheme, *Figure 7-77A*. Label the triangular spaces within the truss with numbers or letters. Create enclosed spaces between the applied forces and reaction forces by drawing a light envelope line

surrounding the truss and passing through each applied force and reaction. Label each "space" with a letter.

Step 1 Construct a force diagram from reading the external spaces in the envelope in a clockwise direction, *Figure 7-77B*. Use a convenient scale to keep the force diagram on the drawing sheet and use the space efficiently. Read spaces a to b for the left reaction, 2500N. Then read spaces b to c for 2500N. Finally, read c to a for the 5000N force, which brings the force diagram back to the starting point. Thus, the truss is in equilibrium.

Step 2 Select any joint with only two unknown members and read spaces clockwise around the joint, *Figure 7-77C*. Construct lines parallel to the two members crossed. For example, start at joint A and go a to b, b to 1, and 1 to a. Point 1 is located where the construction lines cross.

Step 3 Select the next joint with only two unknown members, joint B or D. Around B go a to 1, 1 to 2, 2 to c, and c to a, *Figure 7-77D*. The two light lines for 1 to 2 and 2 to c locate point 2. To finish the force diagram for the truss, go around joint C or joint D. Choose joint C and go b to c, c to 2, and 2 to b.

Step 4 Prepare a summary of the forces on a sketch of the truss, *Figure 7-77E*. Go clockwise around each joint of the truss to determine tensile and compressive forces. For example, reading clockwise around joint A, 1 to a in the force diagram goes in a direction away from A. Therefore, AB is in tension and the magnitude is scaled to be

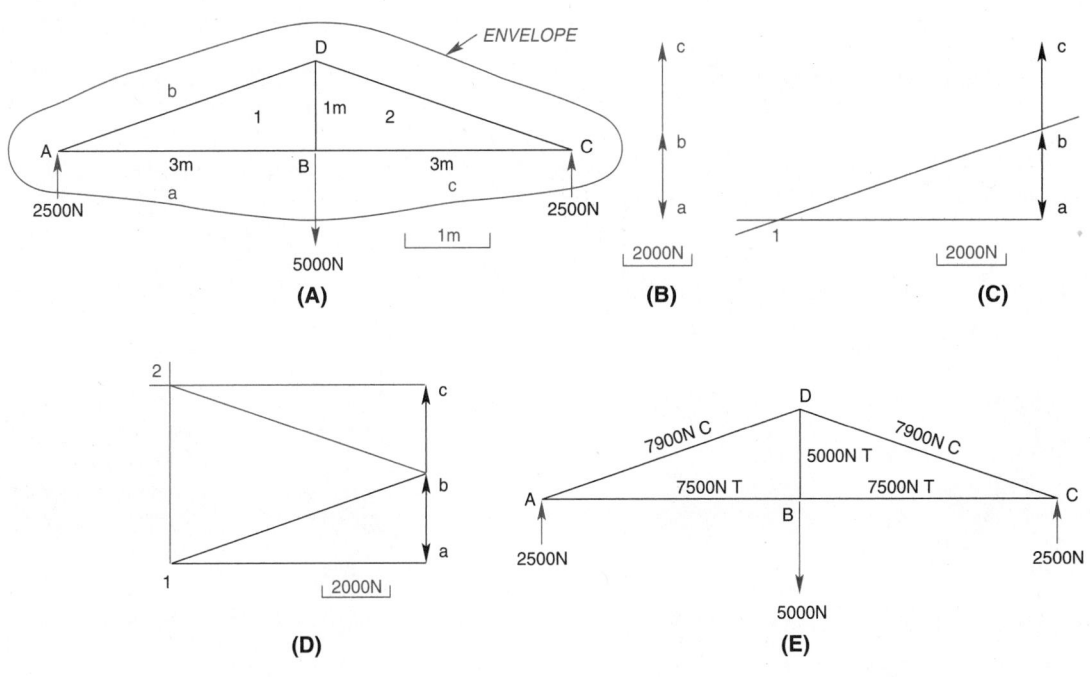

Figure 7-77 (A) Example 1(b) Maxwell diagram with Bow's notation; (B) Step 1; (C) Step 2; (D) Step 3; (E) Step 4

7500N. Going b to 1 goes into joint A; thus a compressive force of 7900N. In a similar clockwise pattern, determine the type and magnitude of the force for each member.

The next example focuses on a more complicated truss with asymmetrical loading.

Example 2(a): Method of Joints

Given: A freehand sketch and a scaled line drawing of a symmetrical roof truss, but with asymmetrical loading due to anticipated wind and snow conditions, *Figure 7-78A*. The reactions at A and D were determined analytically using the fundamental equations of statics, $\Sigma F = 0$ and $\Sigma M = 0$. At joint A, the small freehand sketch shows a

typical symbol for a pinned support, i.e., it provides support vertically and horizontally. The symbol at the right end indicates support in one direction only, normal to the surface of support.

Step 1 Start at a joint with only two unknowns, A or D. Isolate joint A, *Figure 7-78B*, and construct a force diagram to find F(AB) and F(AG). (Use the technique discussed with respect to Figure 7-73C.) Then, to keep a handy record, circle the known members AB and AG in a miniature sketch of the truss. Circle additional known members as they are found.

Step 2 Isolate joint G, the next joint with only two unknowns, *Figure 7-78C*. Construct a force diagram starting with known forces F(AG) and

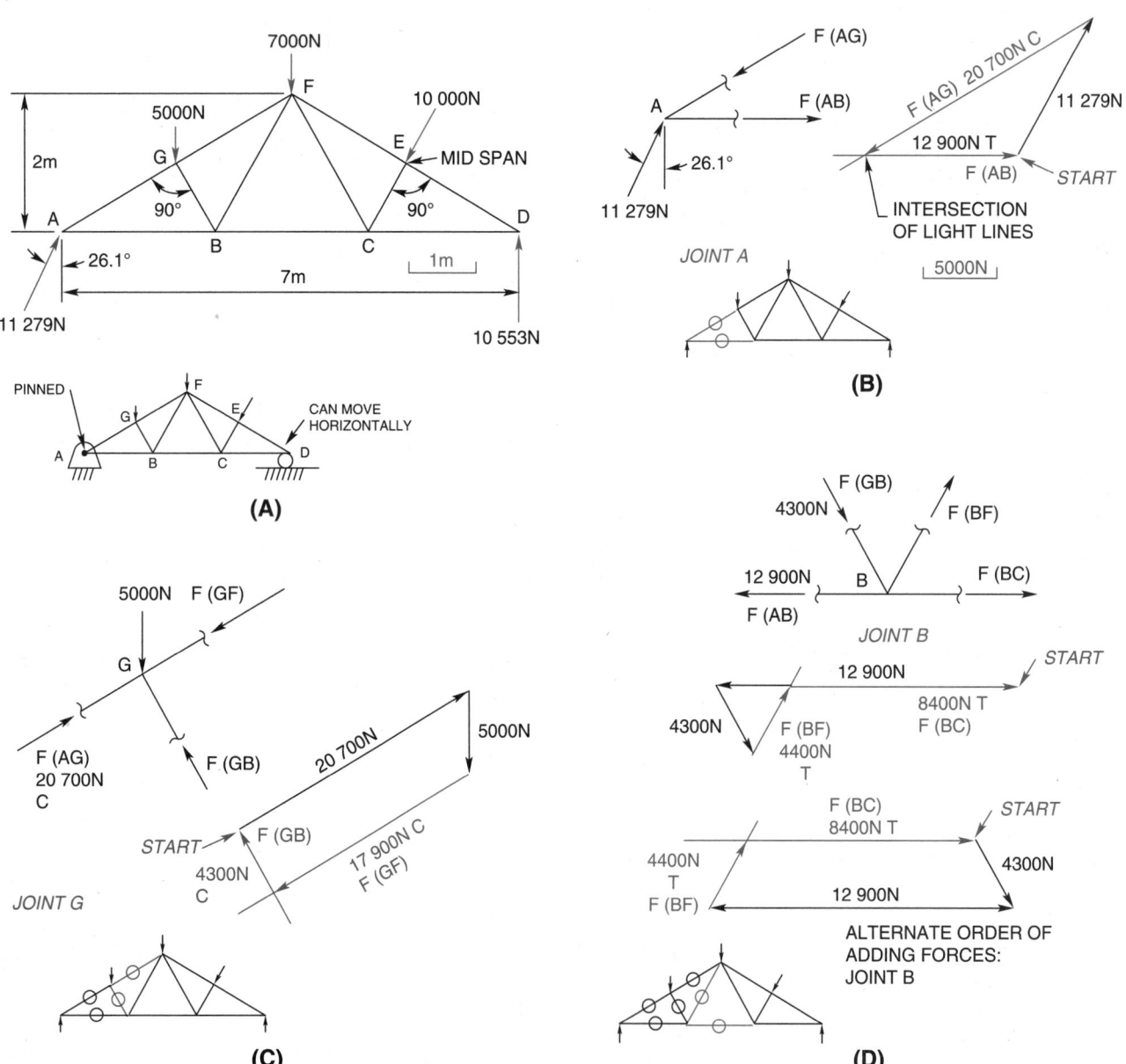

Figure 7-78 (A) Example 2(a) method of joints; (B) Step 1; (C) Step 2; (D) Step 3

the 5000N applied force. Find F(GF) and F(GB) and circle those members in a miniature sketch.

Step 3 Isolate joint B next, *Figure 7-78D*, and construct a force diagram starting with F(AB), then F(GB) to find F(BF) and F(BC). Note that the light line for F(BC) coincides with F(AB) because of the arbitrary clockwise selection of forces around joint B. However, any order of taking the forces around a joint can be used for the method of joints; the final magnitudes and directions would be identical. Circle the known members, in a miniature sketch.

Step 4 Isolate joint F, *Figure 7-78E*. Construct a force diagram by using any order of forces. A clockwise order was used for the figure, starting with known F(FB), then F(FG) followed by the 7000N applied force to determine F(FE) and F(FC). Add these to a miniature sketch.

    Determine F(CE) by inspection to be 10000N. A force acting normal to a joint that has members at 90 degrees to each other can only transmit force to the truss member along the same line as the force. Circle this member in the sketch.

Step 5 Isolate the last joint D, *Figure 7-78F*, to determine F(DC) and F(DE). Observe that force F(DE) should equal F(EF) due to the 90-degree angles between member DE and CE, and between DE and the 10000N force. Circle the last two members to complete the analysis.

Step 6 Prepare a table of the forces in the truss members, *Figure 7-78G*.

Example 2(b): Method: Maxwell Diagram with Bow's Notation

Given: The same symmetrical truss, drawn to scale, with asymmetrical loads, used in the preceding method of joints example, *Figure 7-79A*. Label the triangular spaces within the truss with numbers. Draw a light line envelope surrounding the truss and passing through each applied force and reaction. Label these spaces with letters.

Step 1 Start a force diagram (to a scale to keep the diagram on the drawing sheet) by reading the external spaces in a clockwise direction, a to b, b to c, c to d, d to e, and e to a. Plot each applied force and reaction, *Figure 7-79B*.

    Either clockwise or counterclockwise will work, but once a pattern is started it must be used throughout the complete problem for both the Maxwell diagram and determining whether a member is in tension or compression.

Step 2 Start at a joint with only two unknowns, A or D. At joint D, go clockwise a to 1, 1 to e, and e to a to locate point 1, *Figure 7-79C*.

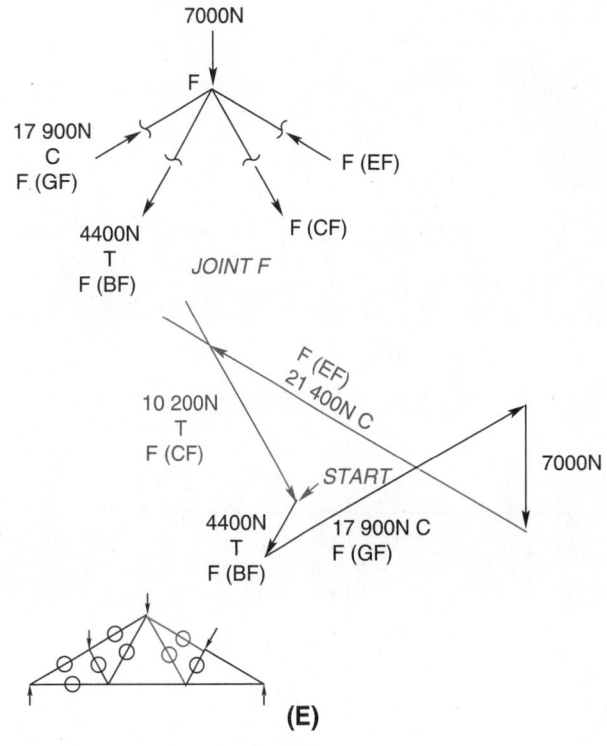

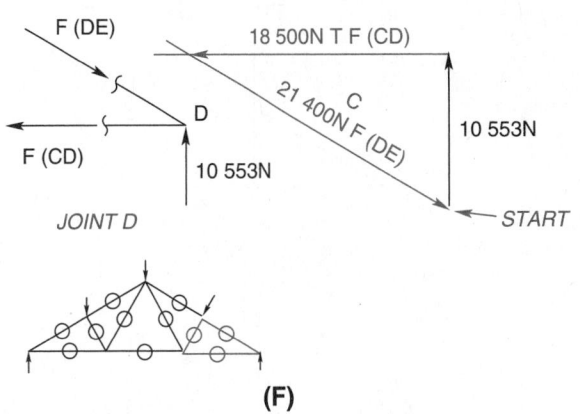

MEMBER	FORCE N	T OR C
AB	12900	T
BC	8400	T
CD	18500	T
DE	21400	C
EF	21400	C
FG	17900	C
GA	20700	C
BG	4300	C
BF	4400	T
CF	10200	T
CE	10000	C

(G)

Figure 7-78 (Continued) (E) Step 4; (F) Step 5; (G) Step 6

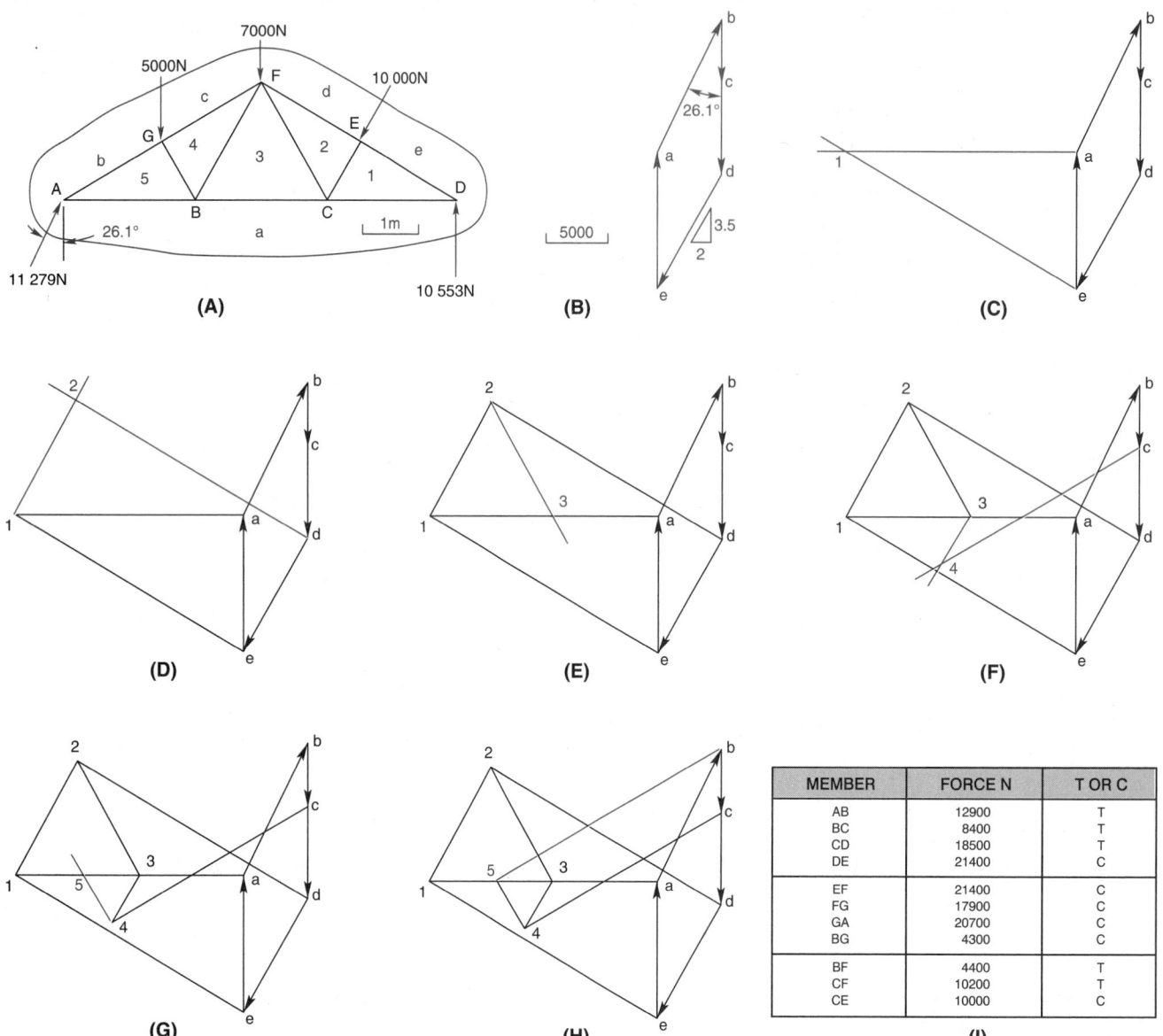

Figure 7-79 (A) Example 2(b) Maxwell diagram with Bow's notation; (B) Step 1; (C) Step 2; (D) Step 3; (E) Step 4; (F) Step 5; (G) Step 6; (H) Step 7; (I) Step 8

MEMBER	FORCE N	T OR C
AB	12900	T
BC	8400	T
CD	18500	T
DE	21400	C
EF	21400	C
FG	17900	C
GA	20700	C
BG	4300	C
BF	4400	T
CF	10200	T
CE	10000	C

Step 3 Now joint E has only two unknowns so continue d to e, e to 1, 1 to 2, and 2 to d to locate point 2, *Figure 7-79D*.

Step 4 For joint C go a–3–2–1–a to locate point 3, *Figure 7-79E*.

Step 5 Joint F has only two unknowns so go d–2–3–4–c–d to locate point 4, *Figure 7-79F*.

Step 6 Around point B go a–5–4–3–a to locate point 5, *Figure 7-79G*.

Step 7 For the last joint, A, go a–b–5–a, *Figure 7-79H*. The closing line in direction 5–b should inter-

sect point b on the Maxwell diagram for a correct solution.

Step 8 Prepare a table of the forces in each member; scale the diagram for magnitudes and use the clockwise pattern to determine whether the force goes into the joint (compression) or goes away (tension), *Figure 7-79I*. For example, around joint C for member CF, 3–2 goes away from the joint; thus, CF is in tension. Scale the magnitude to be 10000N. Around joint F, 2–3 goes away from the joint F as expected because CF is a tension member.

Noncoplanar Concurrent Force Systems

How To Determine the Forces in the Members of a Noncoplanar-Concurrent Force System

A noncoplanar-concurrent force system could be described as a "tripod" with a force or forces applied at the apex of the tripod, i.e., three two-force members, usually rigid, meeting at a point and supporting a load, *Figure 7-80*. Only three unknown forces may be present in order to solve for their magnitudes using basic statics; again the governing equation is $\Sigma F = 0$. Furthermore, the vectors, representing the forces in the three members, plus the applied load(s) must add "tail to head" starting at the apex and ending at the apex because the system is in equilibrium, Figure 7-80.

Graphical solutions of three-member, noncoplanar-concurrent force systems incorporate two fundamental descriptive geometry techniques; one, finding the piercing point of a line in a plane and two, finding the true length of a line. In a space diagram, the direction line of a force vector of one member, pierces the plane made by the other two vectors. In a force diagram the true lengths of the three vectors representing the forces in the members are found. The space diagram and the force diagram are drawn to scale and are usually superimposed for convenience.

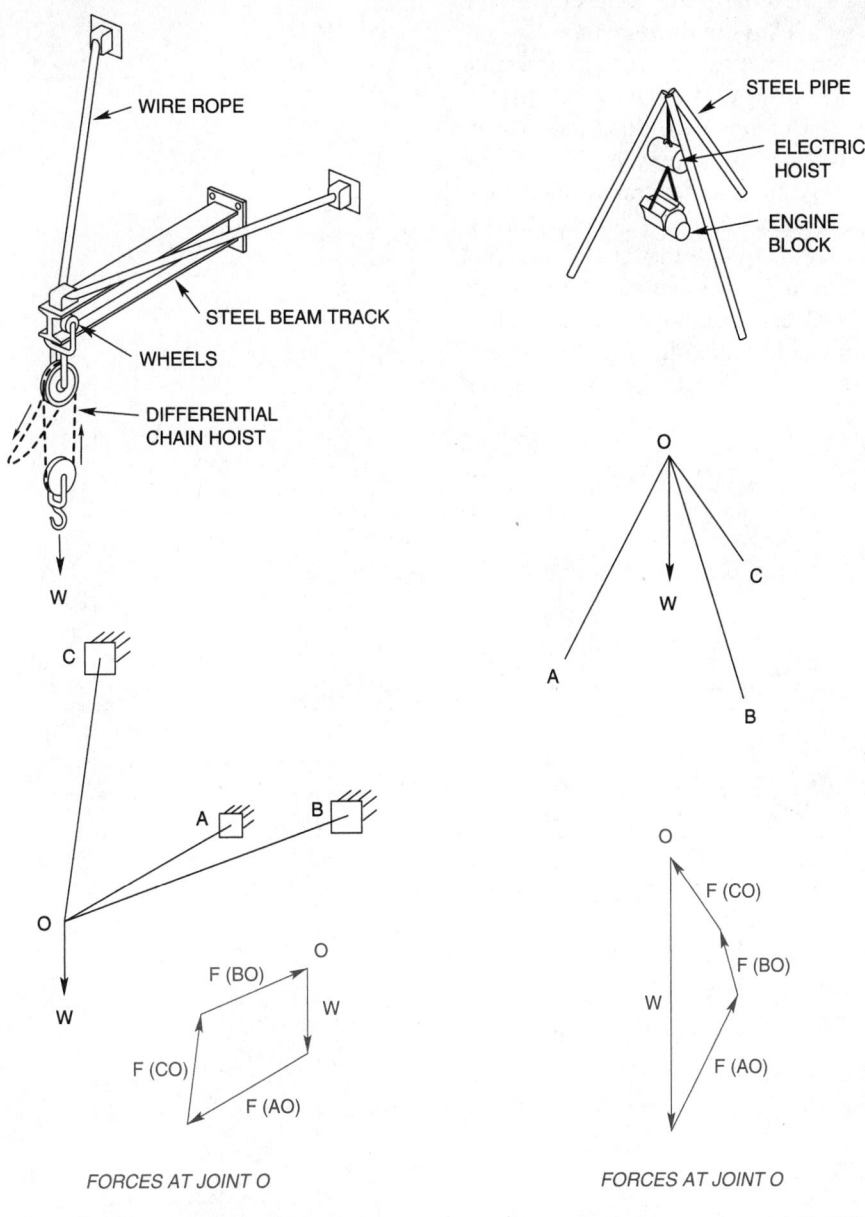

Figure 7-80 Noncoplanar concurrent force systems

Example 1:

Given: A track beam for a hoist, attached to a wall, held up by two wire ropes, and supporting a load at the end as illustrated in the simplified line drawing, *Figure 7-81A*. The line drawing of the beam and wire ropes is drawn to scale. The applied load of 40000N is drawn to scale.

Step 1 Draw a direction line, originating at the tip of the applied load vector, and parallel to any one of the three members, *Figure 7-81B*. Find the piercing point made by this line, in the plane made by the other two members.

Select this first direction line for convenience in arriving at a solution. For example, the plane made by (b-o)T and (c-o)T in view T shows as an edge in view F. Having the edge view expedites finding a piercing point. Therefore, construct a line parallel to the third member (a-o)F, starting the line at the tip of vector F and note where the line pierces the (b-o-c)F plane, which actually extends beyond the apparent boundaries of (b-o-c)F. (Keep in mind that a willingness to "go beyond arbitrary boundaries" is a step toward being a creative individual.)

Step 2 Find F(BO) and F(CO) now that F(AO) has been established. The force vectors parallel to members OB and OC, and lying in the plane of BOC extended, must add tail to head to F(AO) and end at the origin of the force vector F, *Figure 7-81C*.

Step 3 Determine the magnitudes of the three unknown forces and prepare a summary sketch, *Figure 7-81D*. Use the principle of revolution (to be discussed in the next section of this chapter) to determine the magnitudes. The sketch of isolated joint O with directions of the three forces found in the force diagram, indicate whether each member is in tension or compression. Of course, cables supporting a load must be in tension.

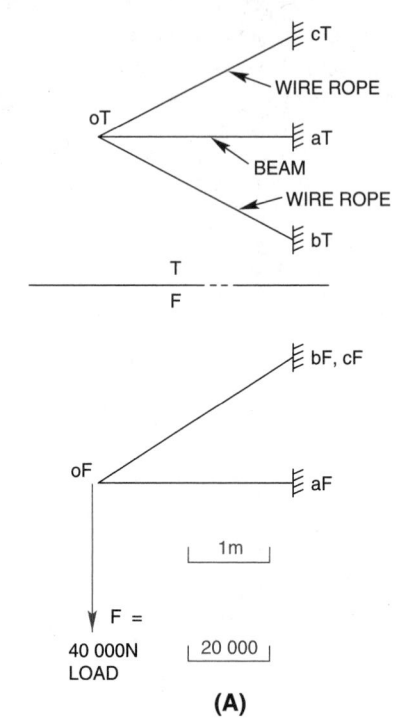

(A)

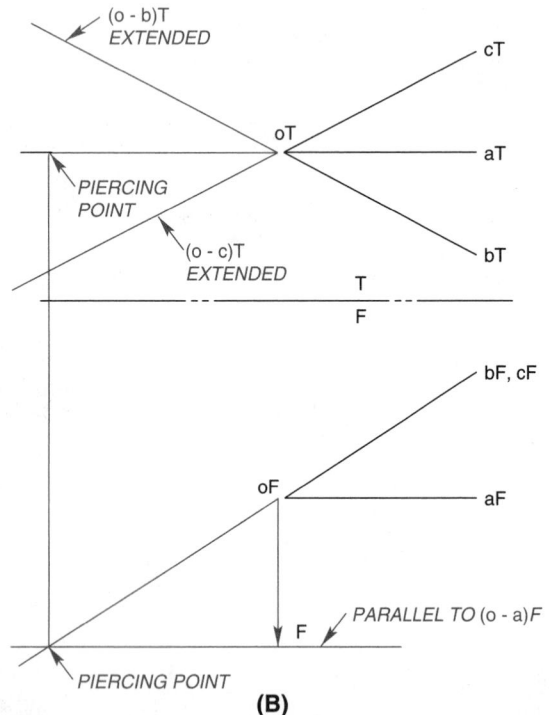

(B)

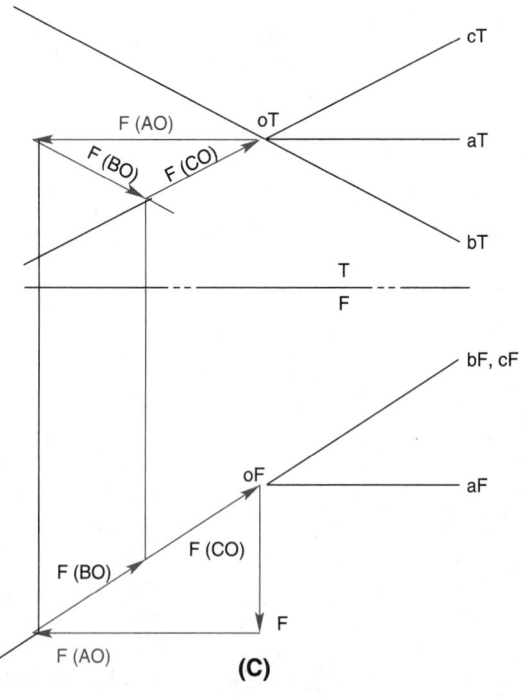

(C)

Figure 7-81 (A) Determining the forces of a noncoplanar concurrent force system—example 1; (B) Step 1; (C) Step 2

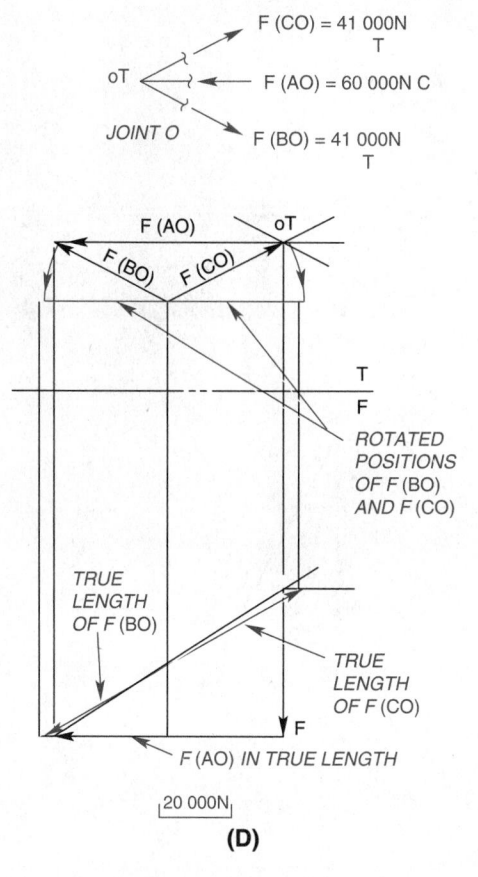

F (CO) = 41 000N
T

oT

F (AO) = 60 000N C

JOINT O

F (BO) = 41 000N
T

F (AO)   oT

F (BO)   F (CO)

T
F

ROTATED
POSITIONS
OF F (BO)
AND F (CO)

TRUE
LENGTH
OF F (BO)

TRUE
LENGTH
OF F (CO)

F

F (AO) IN TRUE LENGTH

20 000N

(D)

Figure 7-81 (Continued) (D) Step 3

Example 2:

Given: A tripod on a sloping terrain supporting a 12000N load, *Figure 7-82A*. The tripod and the applied load are drawn to scale.

Step 1 Select a plane made by two of the members, to expedite the solution, and construct a view showing this plane as an edge, *Figure 7-82B*. Use members (a–o)F and (c–o)F to establish the plane using auxiliary line (a–x)F, and project the plane (a–o–c)T into view A. Use the technique shown in Figure 7-17, discussed earlier in this chapter, for finding an edge view of a plane surface.

Step 2 Find the piercing point in plane (a–o–c)A of the vector representing the force in member (o–b)A by constructing a direction line in view A, originating at the tip of the 12000N force and parallel to member (o–b)A. Label the piercing point pA and show F(OB), *Figure 7-82C*. Project a line from point pA into view T and locate pT by noting where extended line (o–b)T crosses the projected line from pA in view A.

Step 3 In the T view construct a line from the piercing point pT, parallel to (o–a)T, until it intersects (o–c)T. This establishes F(OA) and F(OC), *Figure 7-82D*. Project the point of intersection of F(OA) with (o–c)T into view A and indicate both vectors F(OA) and F(OC) in view A.

Step 4 Determine the magnitudes of the three unknowns—F(OB), F(OA) and F(OC), and list

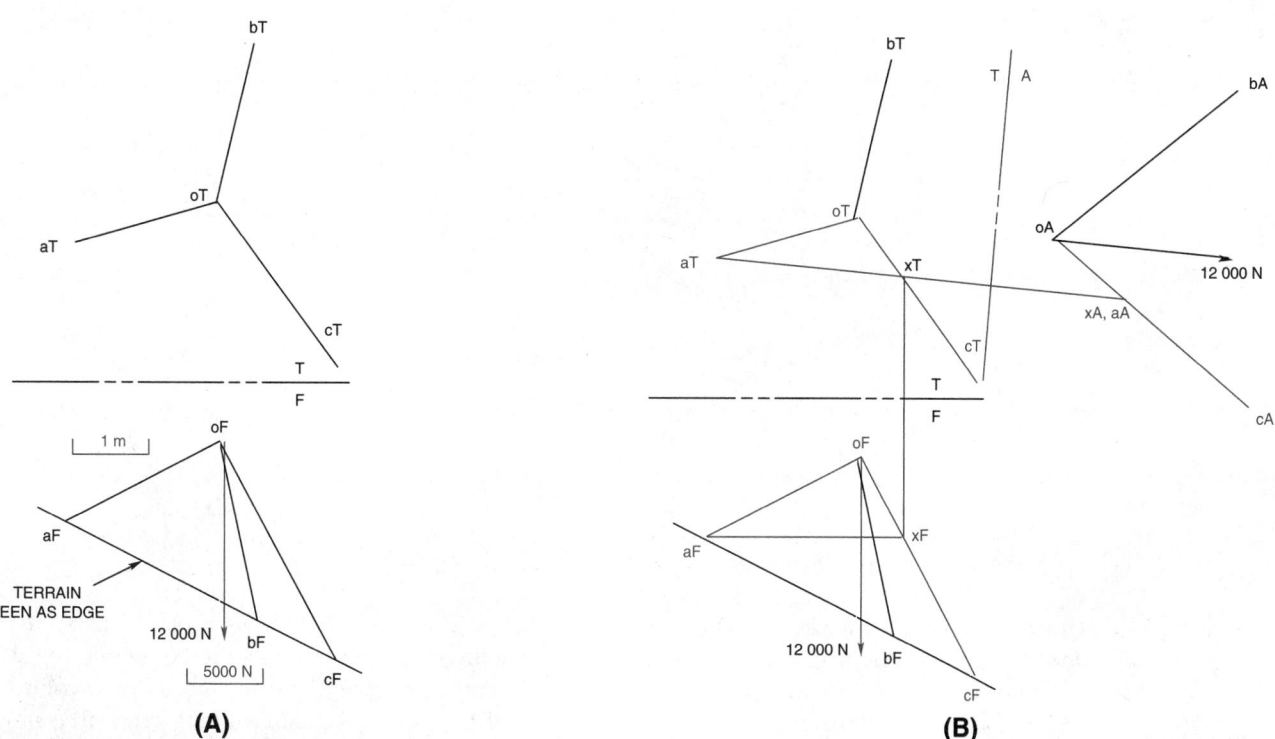

(A)

(B)

Figure 7-82   (A) Determining the forces of a noncoplanar concurrent force system—example 2;   (B) Step 1

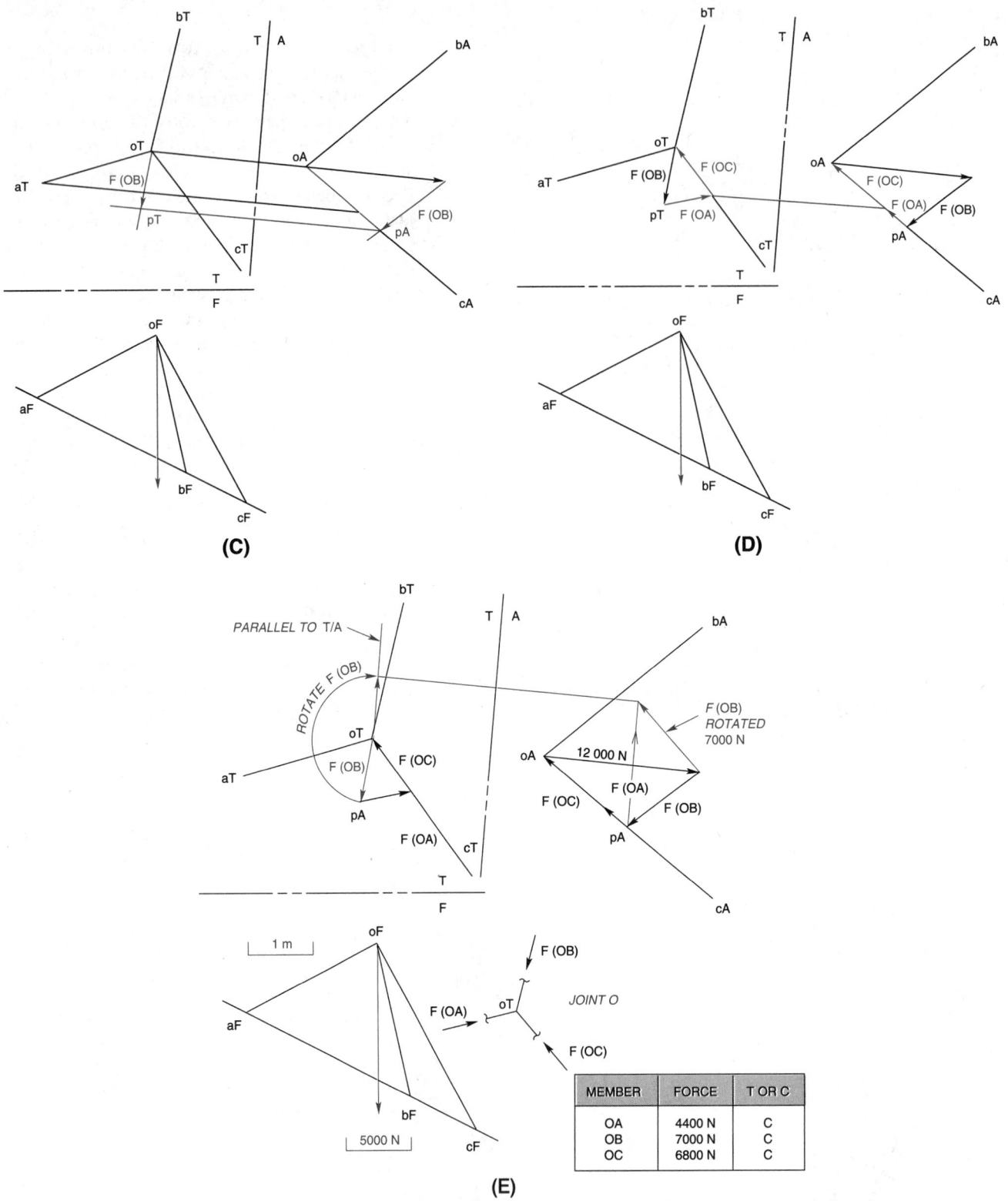

Figure 7-82 (Continued) (C) Step 2; (D) Step 3; (E) Step 4

them in a table, *Figure 7-82E*. All three unknown forces could be determined by the true-length-of-a-line method. For example, F(OA) and F(OC) could be determined in an auxiliary view B, by making a fold line A/B parallel to the edge view of plane (a–o–c)A. Then, F(OA) and F(OC) would be in true size in view B.

A different method, however, will be used for the three unknowns—the technique of revolution. For clarity, the technique of revolution will be shown for determining one force only, force F(OB). Revolution will be discussed in the next section.

Comment: An alternative method of finding the piercing point described in step 2 would be to employ the line projection method, Figure 7-25, discussed earlier in this chapter. The method is shown in views F and T, ***Figure 7-83***.

Revolution

The revolution technique in graphics saves time and space. The technique can be used to solve many true length and true angle problems, while requiring less views.

Consider the basic attributes of the technique by observing the straight member, a-b, rigidly attached to an axis m-n, (which can rotate 360 degrees), shown in the T and F views, ***Figure 7-84A***. The true length of member AB is seen twice in the F view as the axis makes one revolution, ***Figure 7-84B***. The end bT of member a-b traces a circular path in the T view and bF traces a straight line in the F view. The circle made by bT is in true size in the T view and bF's path shows as an edge parallel to the fold line T/F in the F view.

The revolution technique works only if the axis, whether real or imaginary, shows as point in one view and in true length in any adjacent view.

How To Determine the True Length of a Line

Example:

Given: The T and F views from Figure 7-11A are redrawn here, ***Figure 7-85A***. In the T view, assume that an axis (m-n)T goes through bT; but it only shows as a point, mT,nT. Then, in the F view axis mF-nF shows in true length.

Step 1 In the T view, rotate member (a-b)T about axis mT, nT until (a-b)T is parallel to the fold line

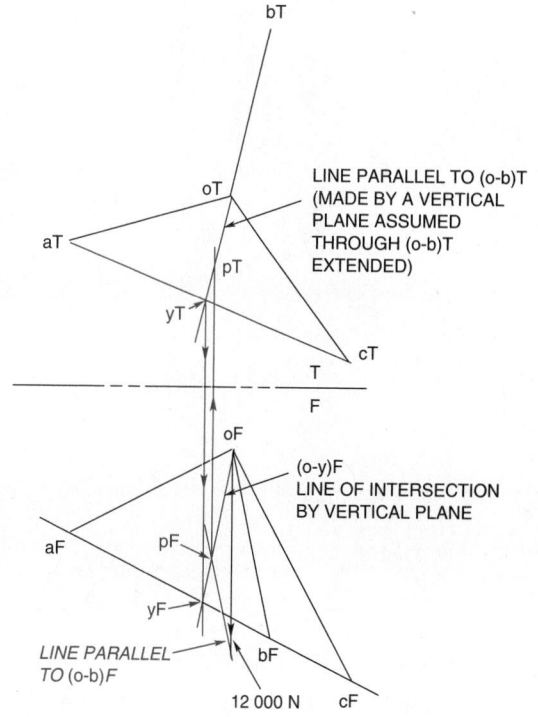

Figure 7-83   Piercing point by line projection method

T/F, ***Figure 7-85B***. Next, in the F view, move point aF parallel to the fold line T/F until it lines up with the projection line from aT's new position in the T view. Redraw (a-b)F in the revolved position where it is now in true length.

Comment: Revolution techniques also save time in laying out developments, Chapter 8.

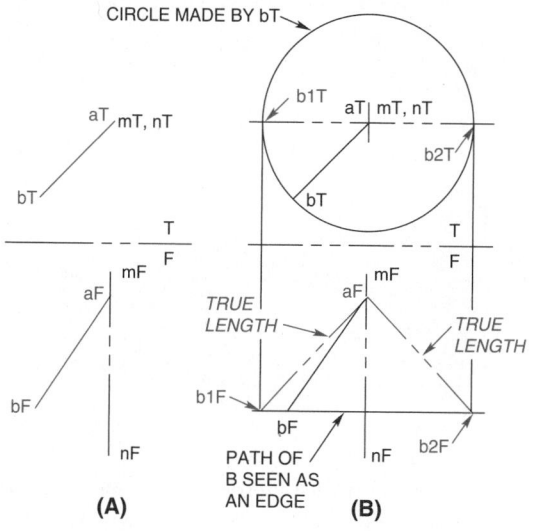

(A)   **(B)**

Figure 7-84   (A) Revolution attributes; (B) True length by revolution

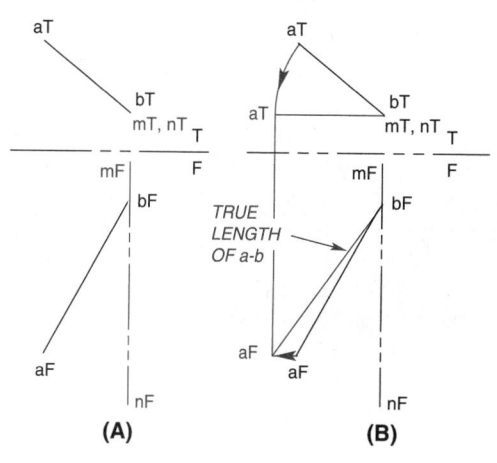

(A)   **(B)**

Figure 7-85   (A) Determining the true length of a line—example 1;
(B) Step 1

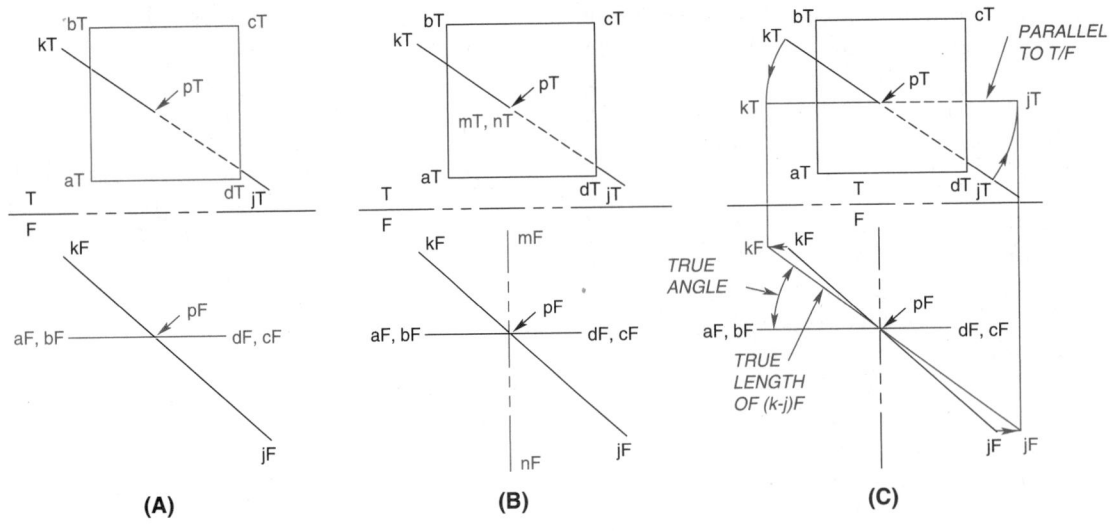

Figure 7-86 (A) Determining the true angle; (B) Step 1; (C) Step 2

How To Determine the True Angle a Line makes with a Plane

Example 1:

Given: Plane (a-b-c-d)T in true size and line (k-j)T piercing the plane at point pT, in the T view, *Figure 7-86A*. In the F view, the line (k-j)F pierces plane (a-b-c-d)F, which shows as an edge.

Step 1 Construct an auxiliary axis m-n which shows as point mT,nT in the T view, and goes through piercing point pT, *Figure 7-86B*. Construct the axis (m-n)F as a true length line in the F view, perpendicular to the T/F fold line.

Step 2 Rotate line (k-j)T about the axis, in the direction of the smallest angle, until line (k-j)T is parallel to the fold line T/F, *Figure 7-86C*. Construct projection lines from the new positions of kT and jT into the F view. Then locate the new positions of kF and jF by moving them parallel to the fold line T/F. The point kF moves to the left to the projection line from kT, and point jF

moves to the right to the projection from jT. Draw (k-j)F in its revolved position and observe that it is now in true length. Measure the true angle between the line (k-j)F and the plane (a-b-c-d)F.

In summary, the true angle a line makes with a plane must be seen in a view where the plane shows as an edge and the line is in true length.

Example 2:

Given: Views F,T and A from Figure 7-23F and redrawn here, *Figure 7-87A*.

This example shows that with only one more view, view B, the true angle between the piercing line (x-y)A, and the plane (a-b-c)A can be found easily using revolution.

Construct a fold line A/B parallel to the edge view of plane (a-b-c)A shown in the A view, *Figure 7-87B*. Project the plane and the piercing line into the B view. The plane (a-b-c)B will be in true size, but the line (x-y)B is not.

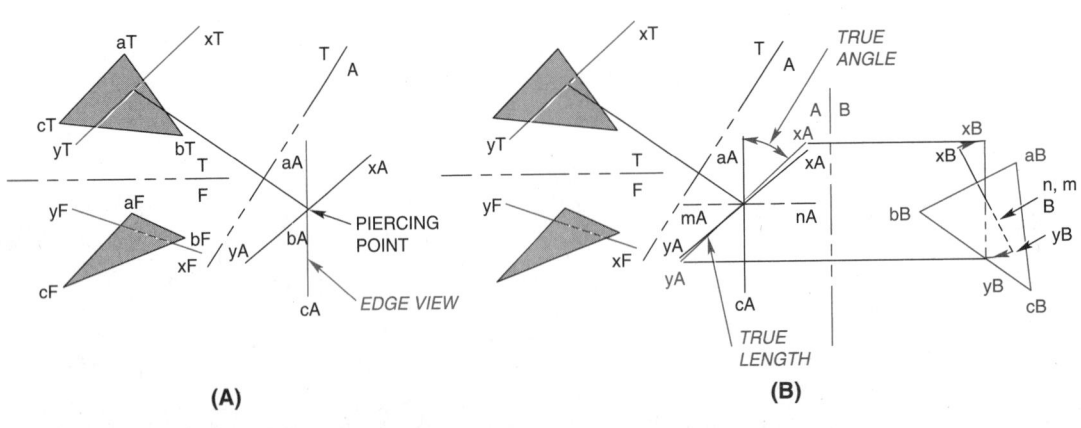

Figure 7-87 (A) Determining the true angle—example 2 (Refer to Figure 7-23F); (B) True angle

Therefore, in view A, construct a true length axis (m-n)A going through the piercing point and perpendicular to plane (a-b-c)A. The axis will show as a point (m-n)B in view B, at the piercing point. Rotate the line (x-y)B about the axis in view B until the line is parallel to the fold line A/B. Then in view A, move point xA parallel to fold line A/B to coincide with the new position of xB in view B. In view A draw the line (x-y)A in its rotated position, where it now shows in true length. In view A measure the true angle between the line (x-y)A and the edge view of plane (a-b-c)A.

How To Determine Clearances Using Revolution

Example:

Given: The top and front views of a proposed installation of a vertical wind turbine and turbine blade details, *Figure 7-88A*. Requirement: Determine the vertical distance to raise anchor B so cable (o-b)F will clear any portion of the turbine blades by at least .3 meter. Cables (o-a)F and (o-c)F are assumed to meet this requirement, but should be checked.

Step 1 Rotate (o-b)T in the T view until it is parallel to the fold line T/F, *Figure 7-88B*. Note that the vertical axis of the turbine qualifies as the axis of revolution because it is in true length in the F view and shows as a point in the adjacent view, view T. Move point bF to the right in the F view as shown. Then, lay out the true size of the turbine blade in the F view.

Construct a straight line from oF, representing (o-b)F, which clears the turbine blade by at least .3 meter. Find the intersection point of this straight line (o-b)F with a vertical line directly above anchor B.

The vertical distance down from the intersection point to the original elevation of anchor B is the required distance to raise anchor B.

Review

1. Why is descriptive geometry used by drafters, designers, and engineers?
2. Explain the basic steps involved in laying out the true shape of a plane.
3. What are notations and why are they so important?
4. What is a point view of a line?
5. Explain the basic steps involved in laying out the true distance between two parallel lines in space.
6. Explain the basic steps involved in laying out the true distance between two nonparallel lines.
7. What is an edge view?

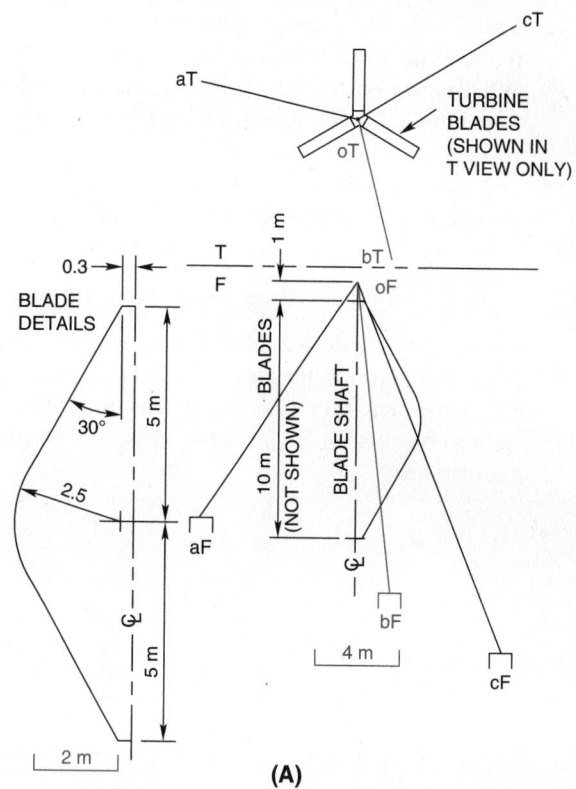

(A)

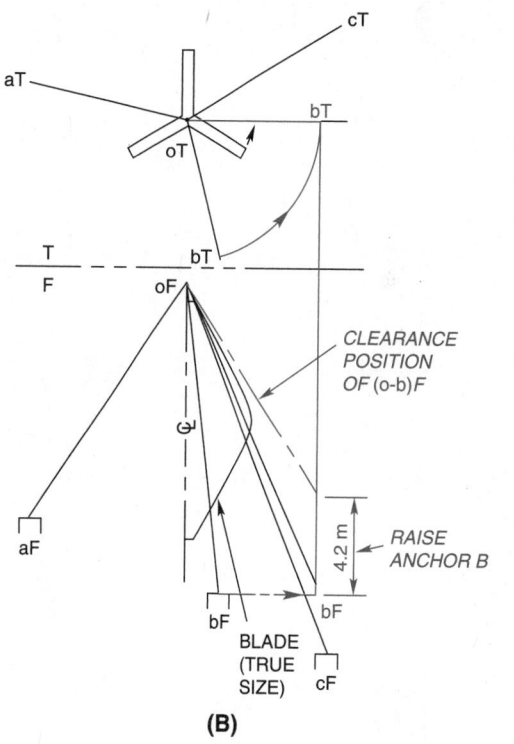

(B)

Figure 7-88 (A) Vertical wind turbine; (B) Step 1

8. How is a point called-out in a given view?
9. What is an important rule to remember in projecting from one view to another?
10. Explain the basic steps involved to find the true length of any straight line.
11. What is a fold line and what does it represent?

12. Explain the basic steps to find the point view of a line as projected from its true length.

13. What is the basic concept of visibility regarding two lines in space drawn in a front view and a top view?

14. Explain how to find a piercing point by line projection in cones, spheres, and cylinders.

15. What is meant by the bearing and grade, or bearing and slope of a line?

16. Given the bearing, slope and length of a line, how is the line constructed in a drawing?

17. What is a contour line and what is the significance of contour lines being close together on a map?

18. What are cuts and fills? How are their volumes determined?

19. Compare "strike and dip" to "bearing and grade."

20. Define a vector. What are some uses of vectors in engineering and technology?

21. What is the parallelogram law? Compare the law to resolution of vectors.

22. What is a two-force member?

23. What is a truss?

24. How are space diagrams and force diagrams used in truss analyses?

25. What is the method of joints?

26. Explain the use of Maxwell diagram and Bow's notation in determining the forces in the members of a truss.

27. Describe how to find the true length of a line using revolution.

Chapter Seven Problems

The following problems are intended to give the beginning drafter practice in using the various principles of descriptive geometry. Problems 1 through 13 deal with each of the various principles used to solve actual design problems. Problems 14 through 32 apply these principles to develop required views of various objects.

The steps to follow in laying out problems 1 through 13 and 23 through 32 are:

1. On an 8½ x 11 sheet of paper with a sharp 4-H lead, locate all lines, points, and fold lines per the given dimensions.
2. Complete the problem per the given instructions. Project in the direction of the large arrow.

The steps to follow in laying out problems 14 through 27 are:

Step 1 Study the problem carefully.

Step 2 Using the given front view, make a sketch of all required views.

Step 3 Center the required views within the work area with a 1-inch (25-mm) space.

Step 4 Use light projection lines. Do *not* erase them.

Step 5 Lightly complete all views.

Step 6 Check to see that all views are centered within the work area.

Step 7 Check that there is a 1-inch (25-mm) space between all views.

Step 8 Carefully check all dimensions in all views.

Step 9 Darken in all views using correct line thickness.

Step 10 Recheck all work, and, if correct, neatly fill out the title block using light guidelines and neat lettering.

Problem 7-1

Locate line a-b and fold lines per given dimensions. Locate line a-b in the right-side view. Label all points in all views.

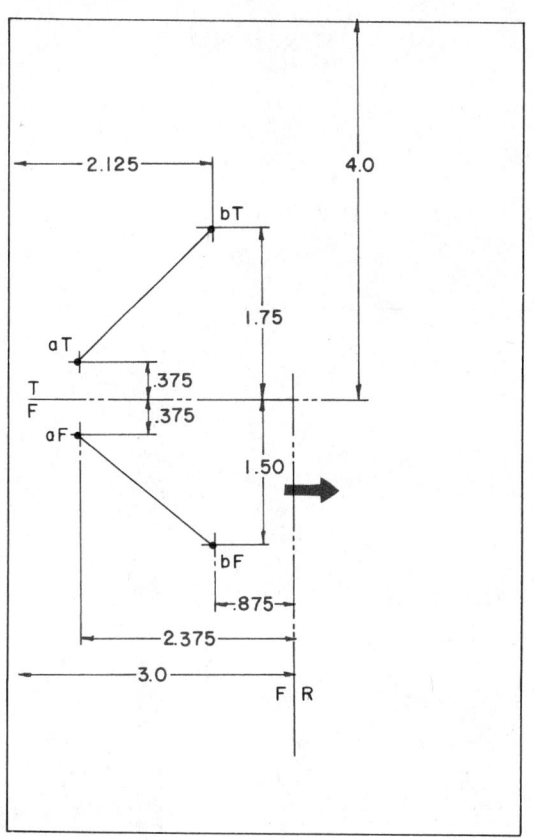

Problem 7-1

Problem 7-2

Locate line a-b and fold lines per given dimensions. Locate line a-b in the top view. Locate point c on line a-b in all views. Label all points in all views.

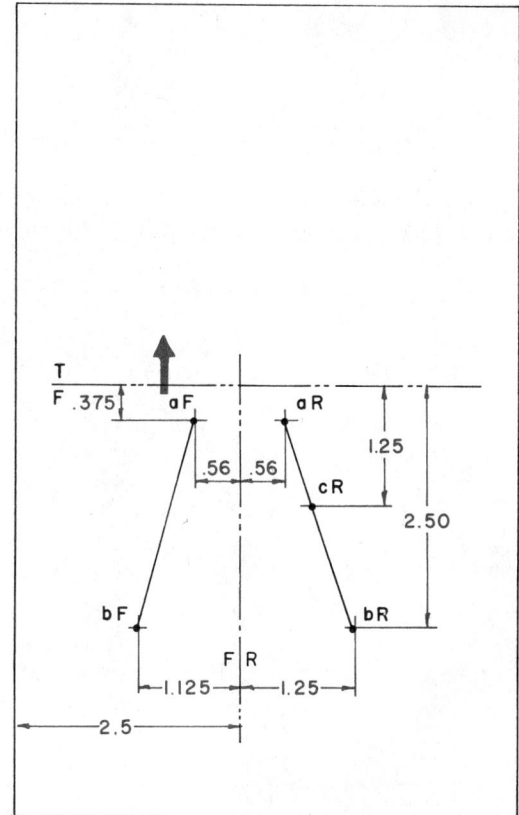

Problem 7-2

Problem 7-3

Locate line a-b and fold lines per given dimensions. Locate line a-b in the right view. Find the true length of line a-b, projecting from the right side. Label all points in all views.

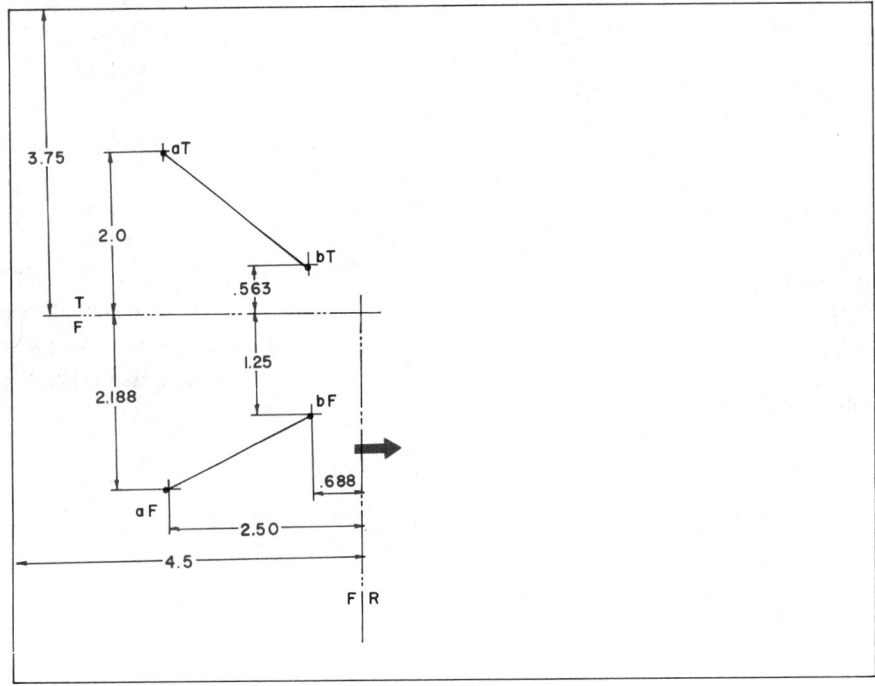

Problem 7-3

Problem 7-4

Locate the line a-b and fold lines per given dimensions. Locate line a-b in the top view. Find the point view of line a-b, projecting from the top view. Label all points in all views.

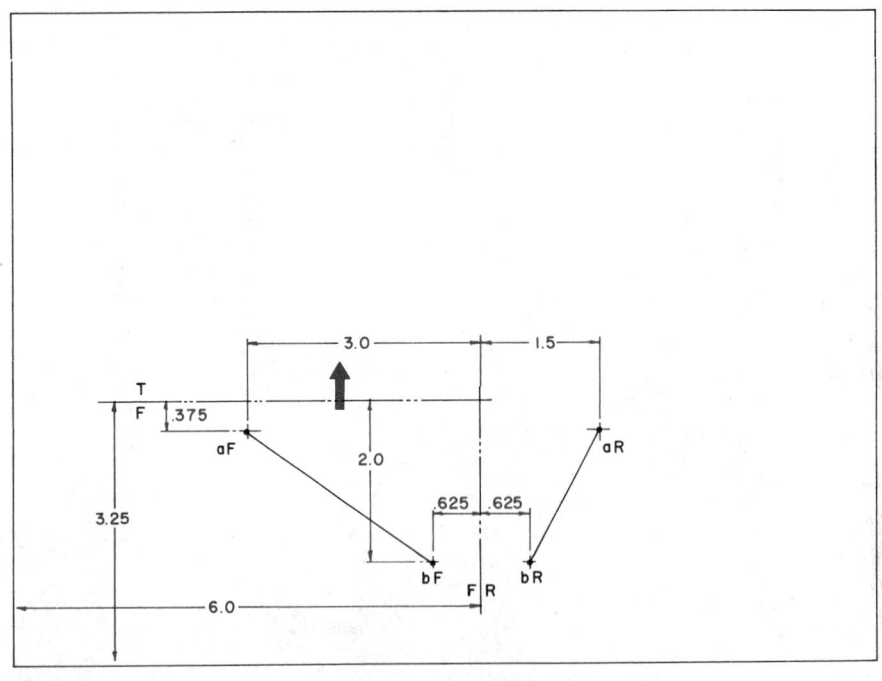

Problem 7-4

Problem 7-5

Locate line a-b, point c, and fold lines per given dimensions. Find the true distance from line a-b to point c.

Project from the right view. Label all points in all views.

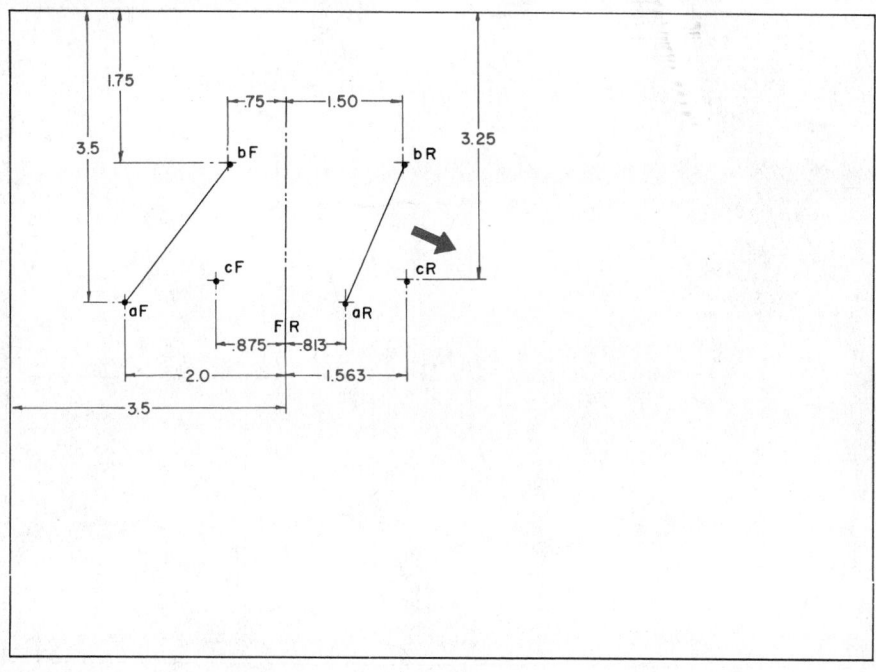

Problem 7-5

Problem 7-6

Locate parallel lines a-b and c-d, and fold lines per given dimensions. Find the true distance between lines a-b and c-d. Project from the front view. Label all points in all views.

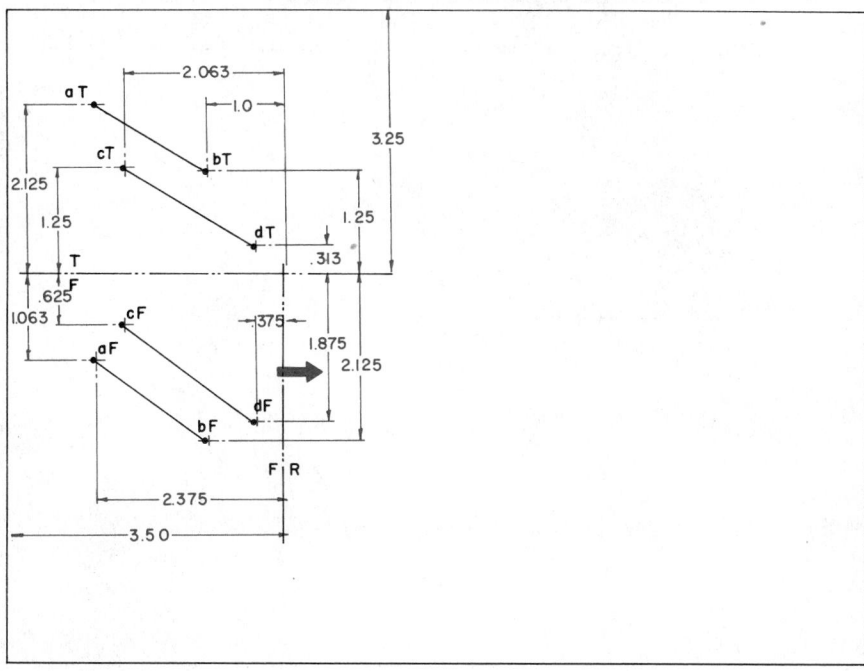

Problem 7-6

Problem 7-7

Locate lines a-b and c-d, and fold lines per given dimensions. Find the true distance between lines a-b and c-d.

Project from the top view. Label all points in all views.

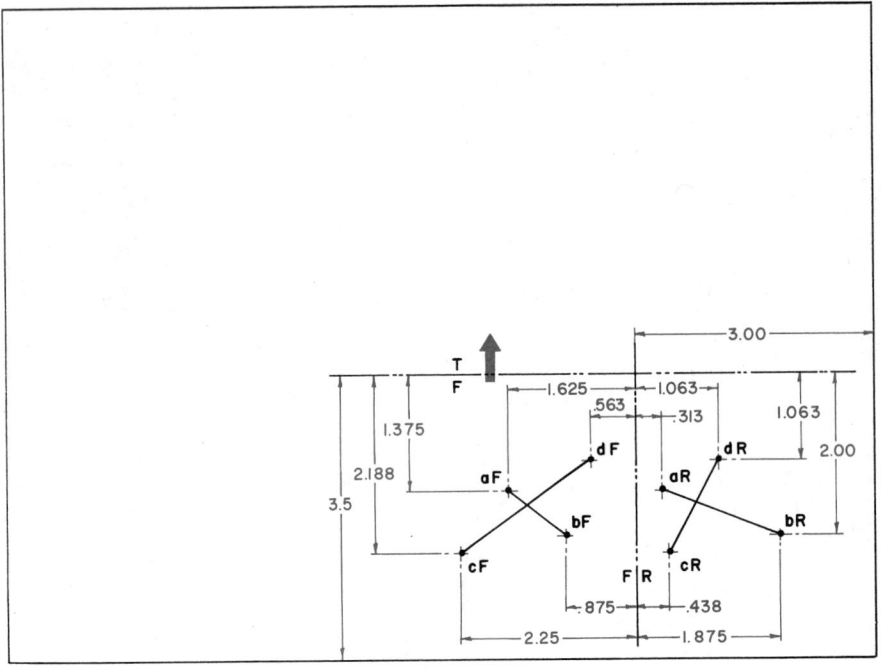

Problem 7-7

Problem 7-8

Locate points a, b, and c, and fold lines per given dimensions. Connect points a, b, and c to form a plane.

Project plane abc into the top view. Label all points in all views.

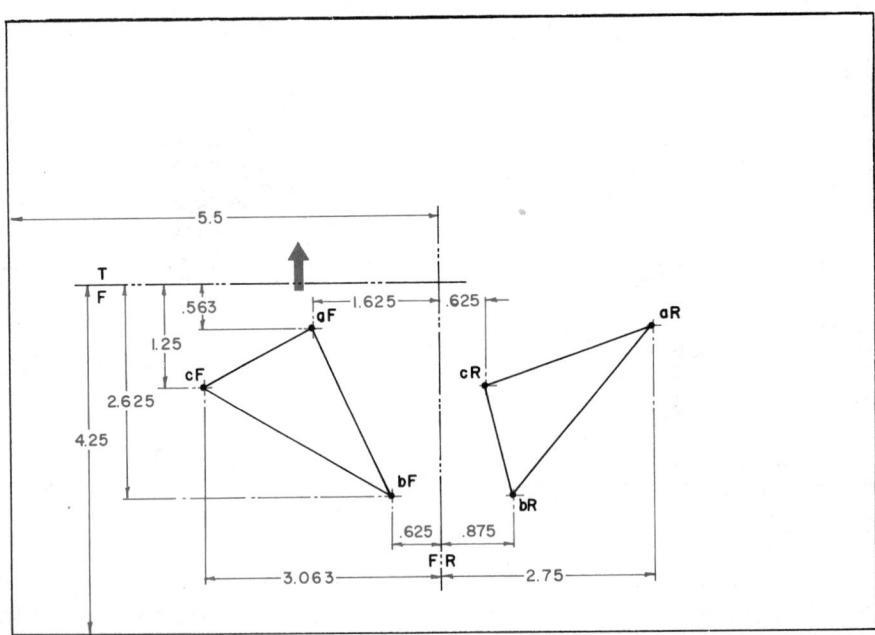

Problem 7-8

Problem 7-9

Locate points a, b, c, and d, point x, and fold lines per given dimensions. Connect points a, b, c, and d to form a plane. Project plane abcd into the right view; locate point x in all views. Label all points in all views.

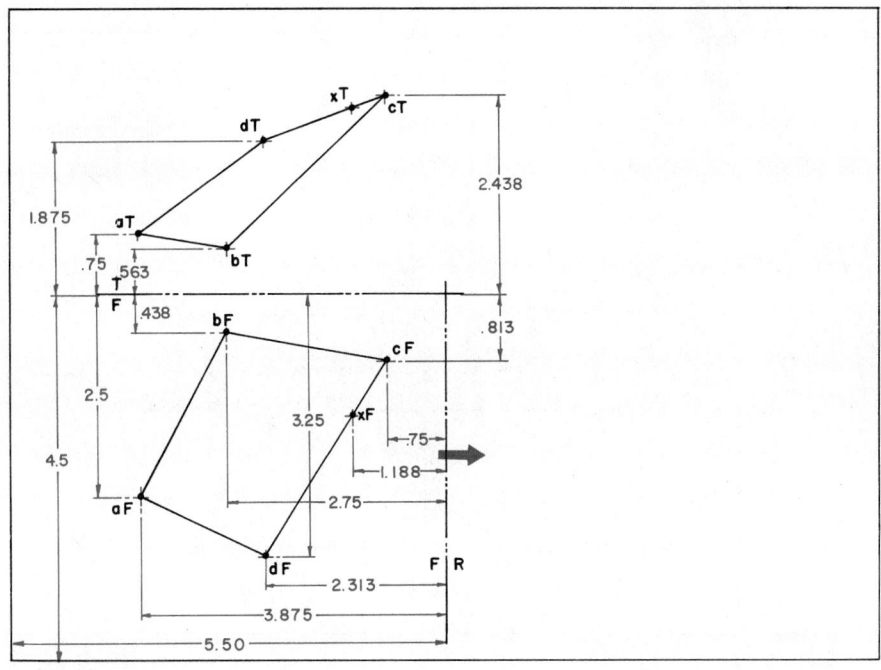

Problem 7-9

Problem 7-10

Locate points a, b, c, and fold lines per given dimensions. Connect points a, b, and c to form a plane. Draw a line from point cF, parallel to fold line F-R; label this line cF-xF. Find the edge view of plane surface abc. Project from the right view. Label all points in all views.

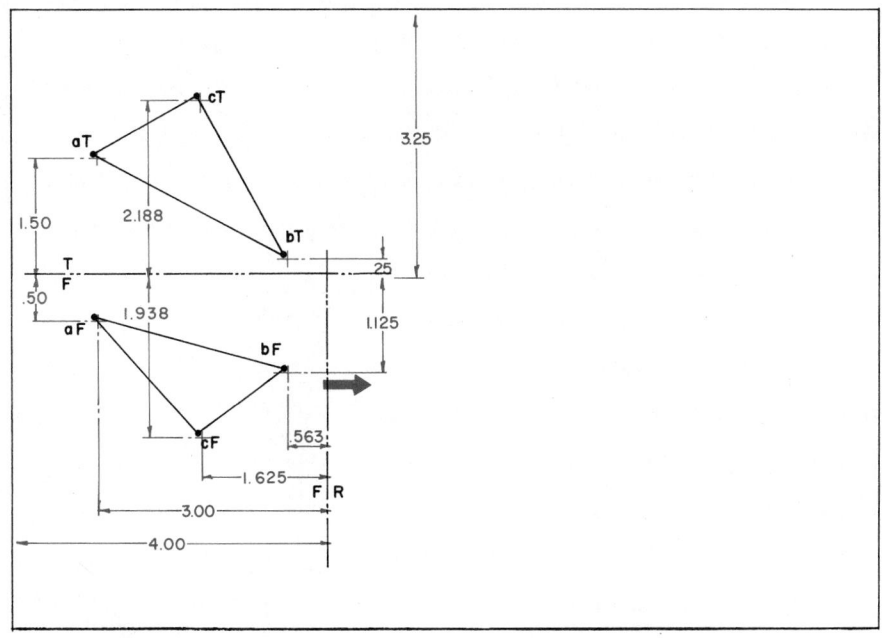

Problem 7-10

Problem 7-11

Locate points a, b, and c, and fold lines per given dimensions. Connect points a, b, and c to form a plane. Draw a line from point bF, parallel to fold line F-T; label this line bF-xF. Find the edge view of the plane abc and project from the front view. label all points in all views.

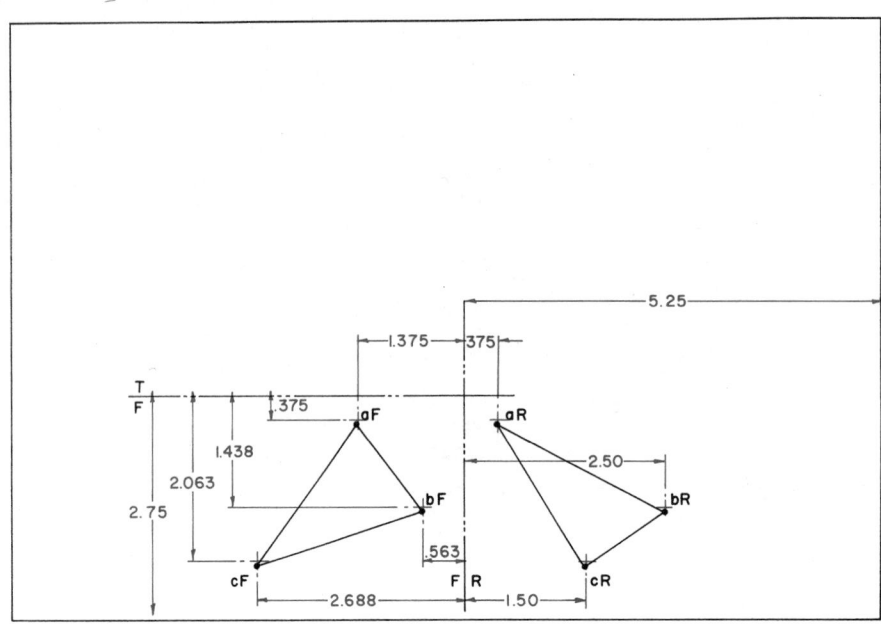

Problem 7-11

Problem 7-12

Locate points a, b, and c, point x, and fold lines per given dimensions. Connect points a, b, and c to form a plane. Find the true distance between plane surface abc, and point x. Project from the front view. Label all points in all views.

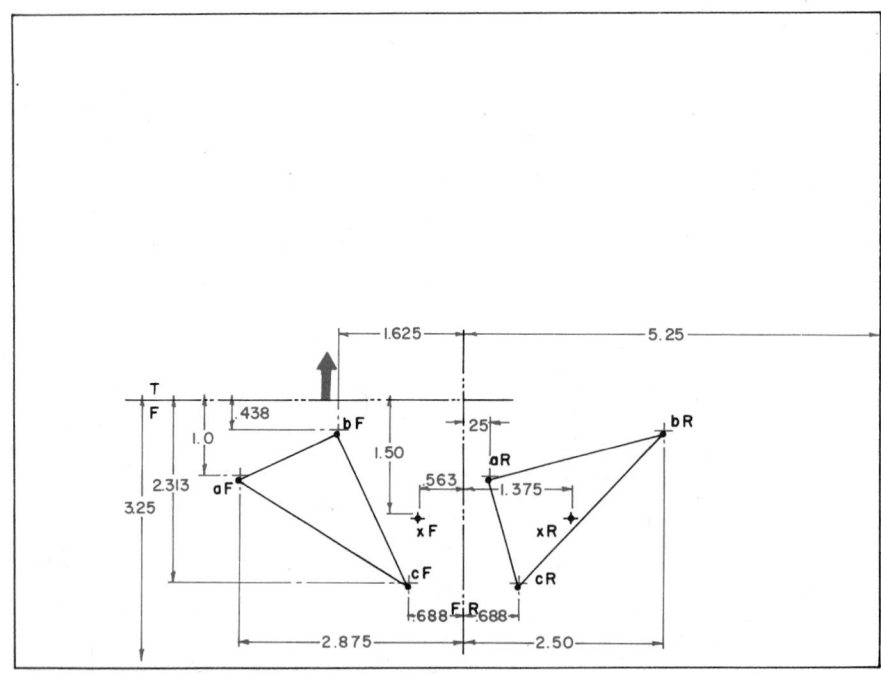

Problem 7-12

Problem 7-13

Locate points a, b, c, and d, and fold lines per given dimensions. Connect points a, b, c, and d to form a folded object. Find the true angle between the plane surfaces as viewed directly along the fold, a-b. Project from the right view. Label all points in all views.

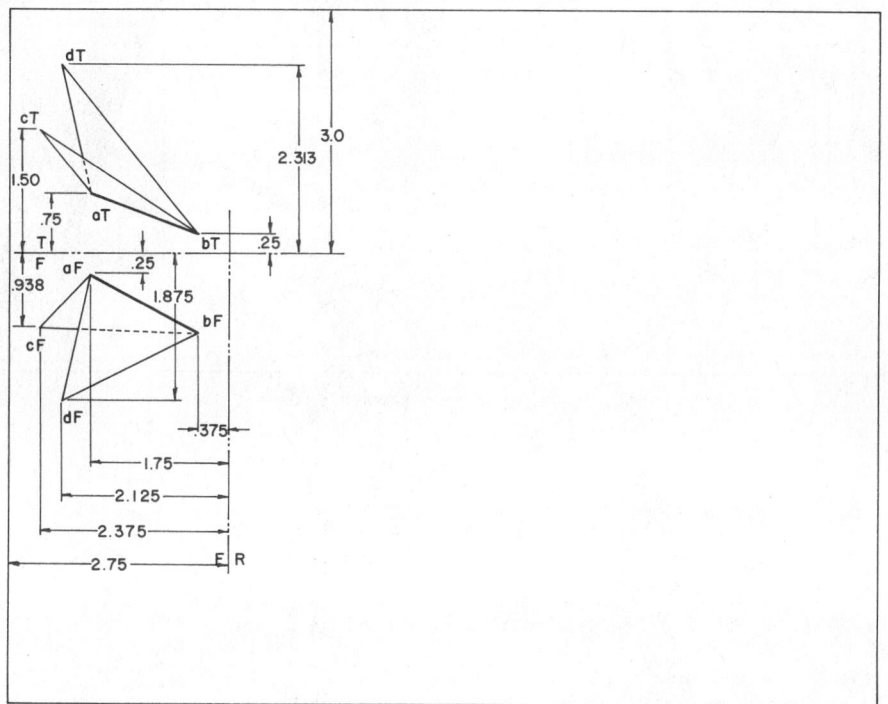

Problem 7-13

Problems 7-14 through 7-22

Draw the front view, top view, right view, and auxiliary view per the listed steps.

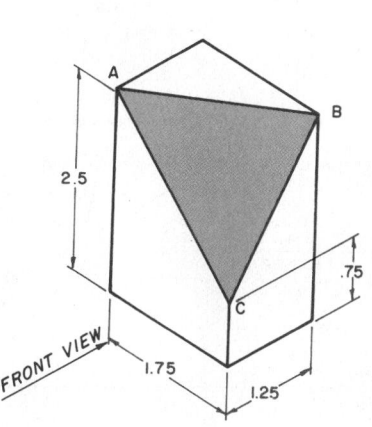

Problem 7-14

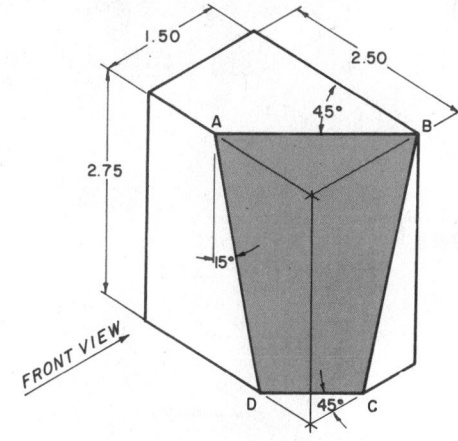

Problem 7-15

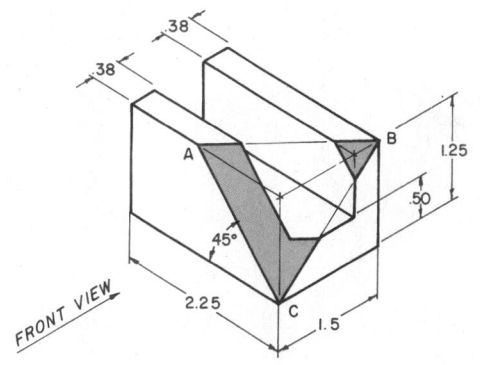

Problem 7-16

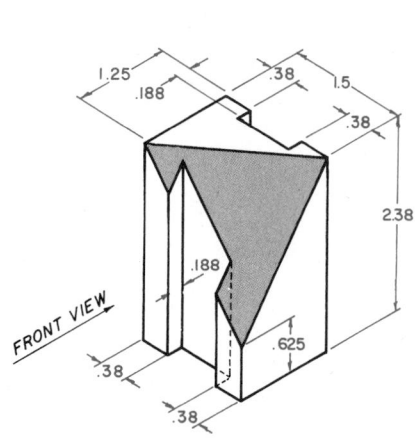

Problem 7-17

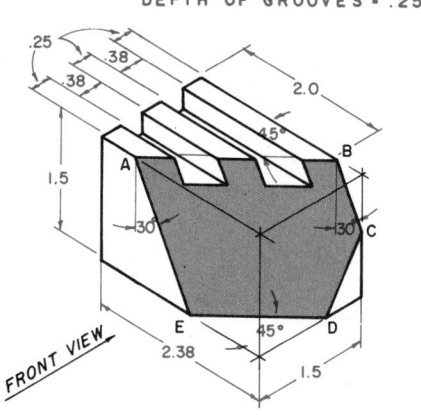

Problem 7-18

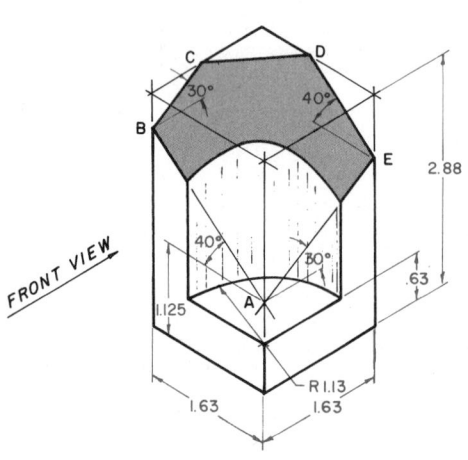

Problem 7-19

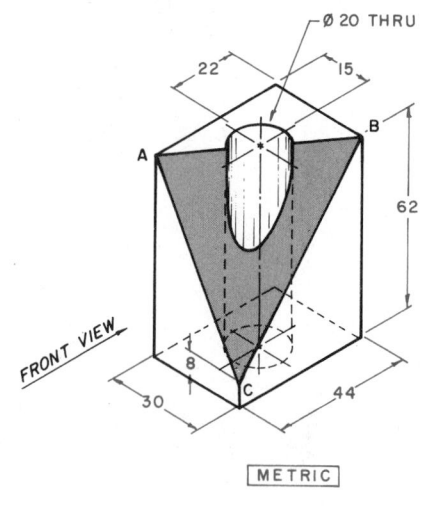

Problem 7-20

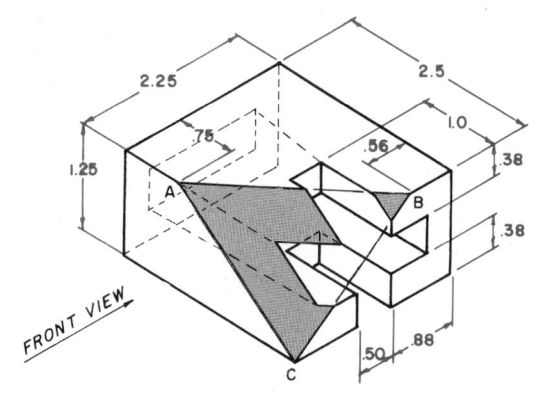

Problem 7-21

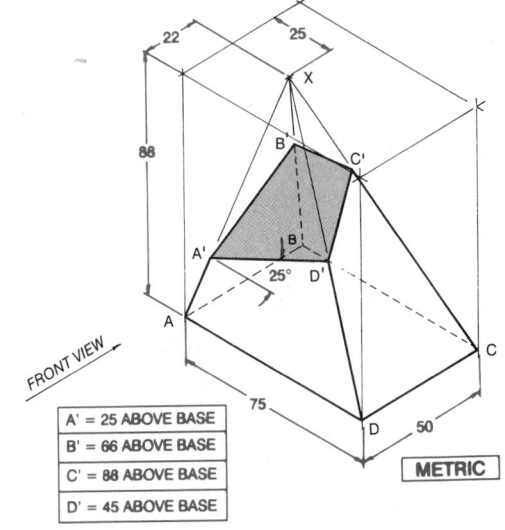

Problem 7-22

Problem 7-23

Draw the front view, top view, right view, and required auxiliary views per the listed steps. Find the true angle at abc, as viewed along the fold.

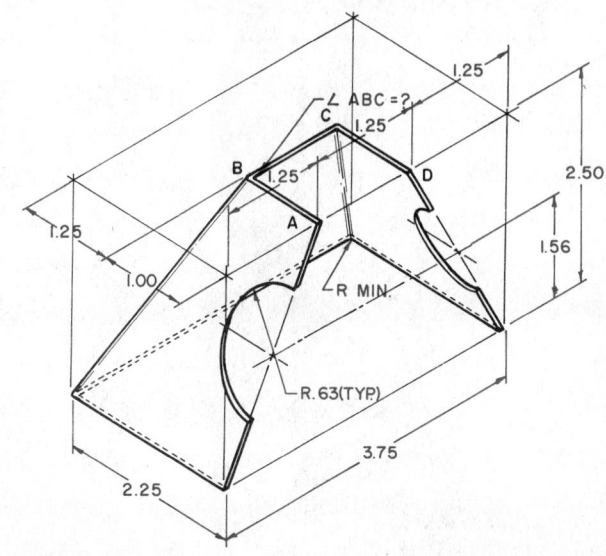

Problem 7-23

Problems 7-24 and 7-25

Draw the front view, top view, right view, and auxiliary view per the listed steps.

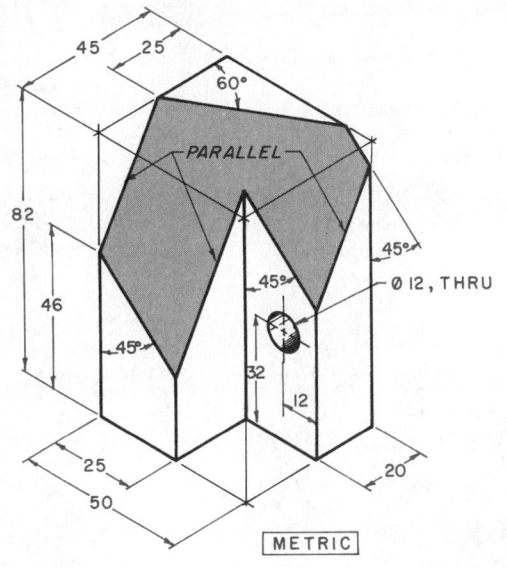

Problem 7-24

Problems 7-26 and 7-27

Draw the front view, top view, right view, and required auxiliary views per the listed steps. Find the true angle between the various surfaces as viewed along the fold.

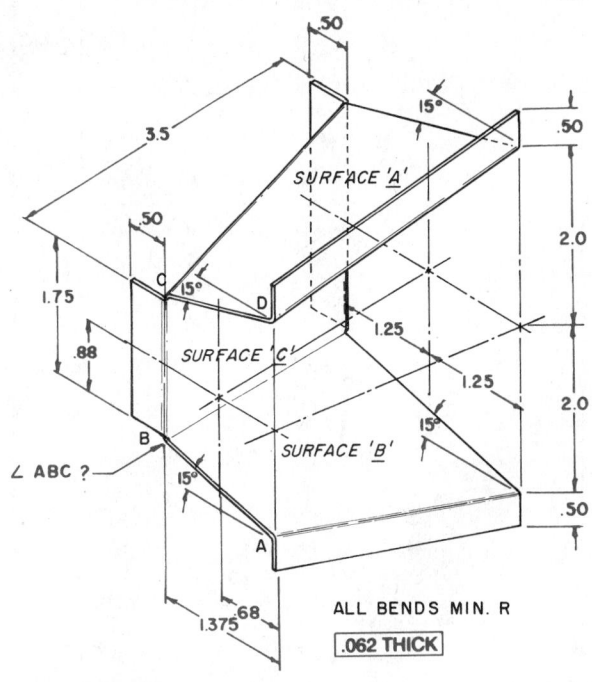

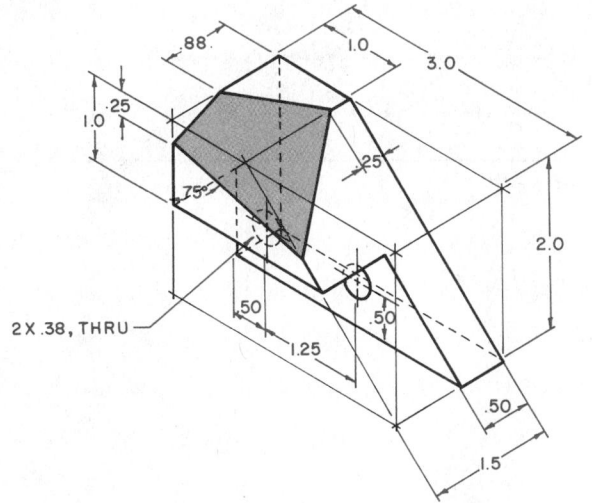

Problem 7-25

Problem 7-26

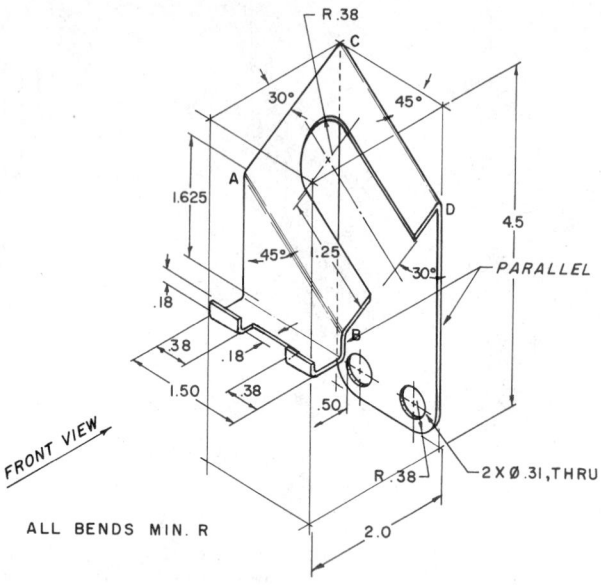

R .38
C
30° 45°
A
1.625
45° .25
30°
4.5
PARALLEL
.18
B
.38
.18
1.50 .38 .50
R .38 2 X Ø .31,THRU

FRONT VIEW

ALL BENDS MIN. R
2.0

Problem 7-27

Problem 7-28

Complete the front view using the line-projection method. Draw the correct visibility of all lines in the front view.

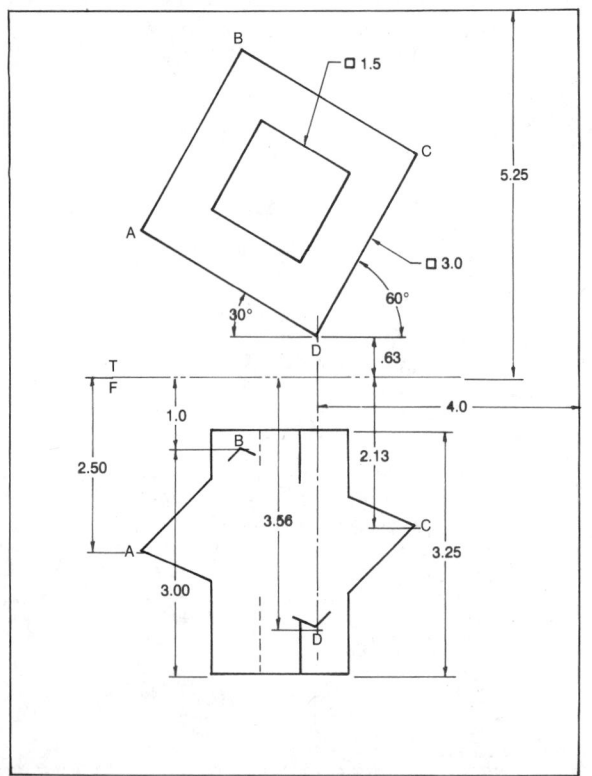

Problem 7-28

Problem 7-29

Complete the top and front views using the line-projection method. Draw the correct visibility of all lines by inspection.

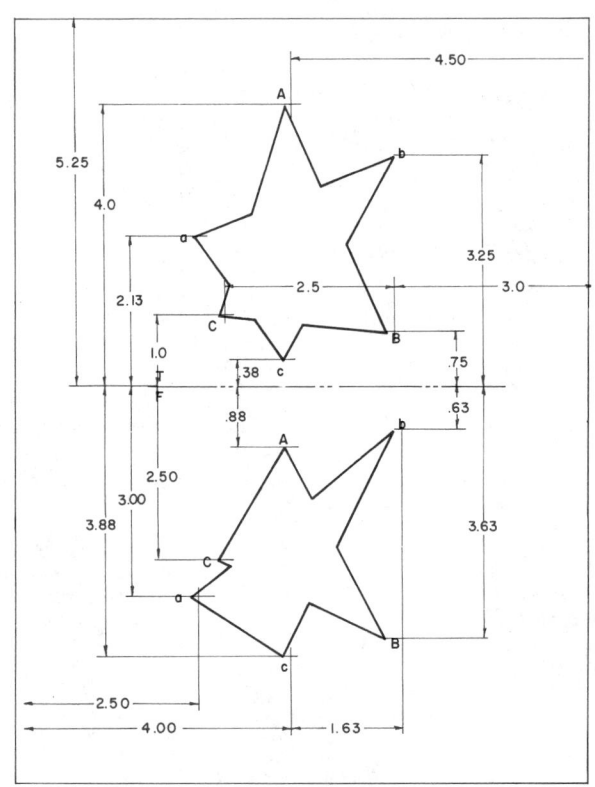

Problem 7-29

Problem 7-30

Complete the front view by inspection and projection. Draw the correct visibility of all lines.

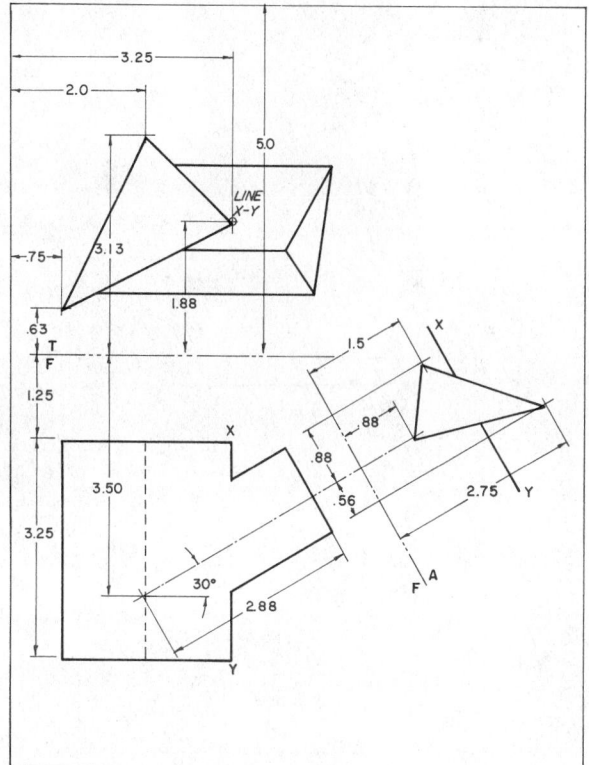

Problem 7-30

Problem 7-31

Complete the front view using the line-projection method. Draw the correct visibility of all lines.

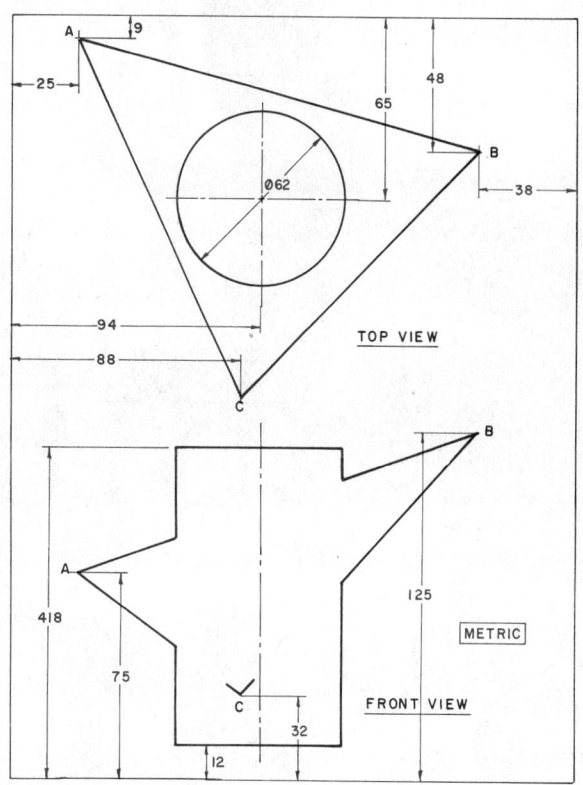

Problem 7-31

Problem 7-32

Complete the top and front views using the line-projection method. Do not use the right-side view for solving the problem. Draw the correct visibility of all lines.

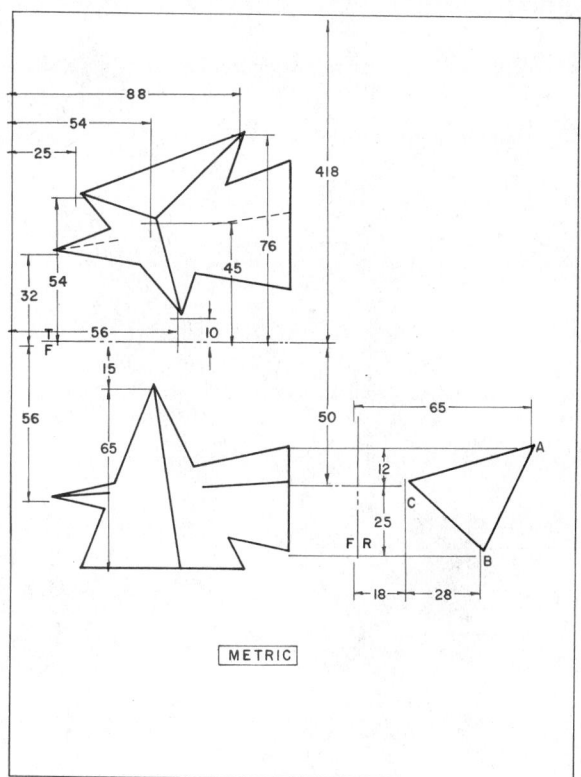

Problem 7-32

Problems 7-33 through 7-42

Lightly lay out and draw the two views and the fold line on an A size sheet of vellum using the given dimen-sions. Complete all views using correct line thickness—add all hidden lines.

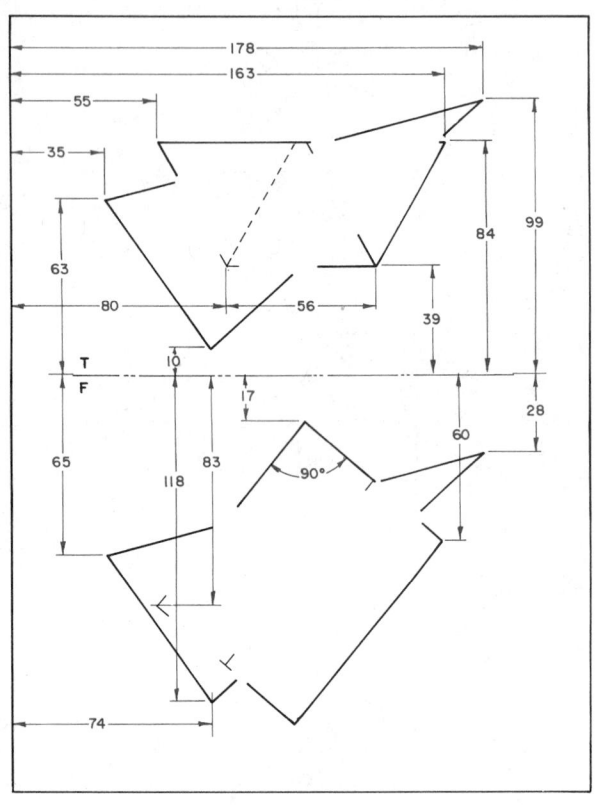

Problem 7-33

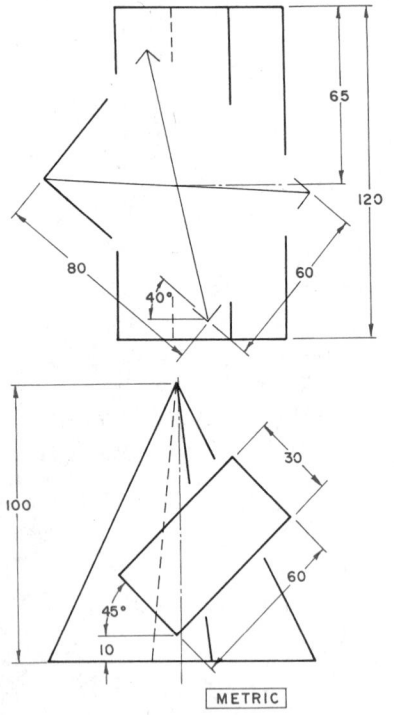

Problem 7-34

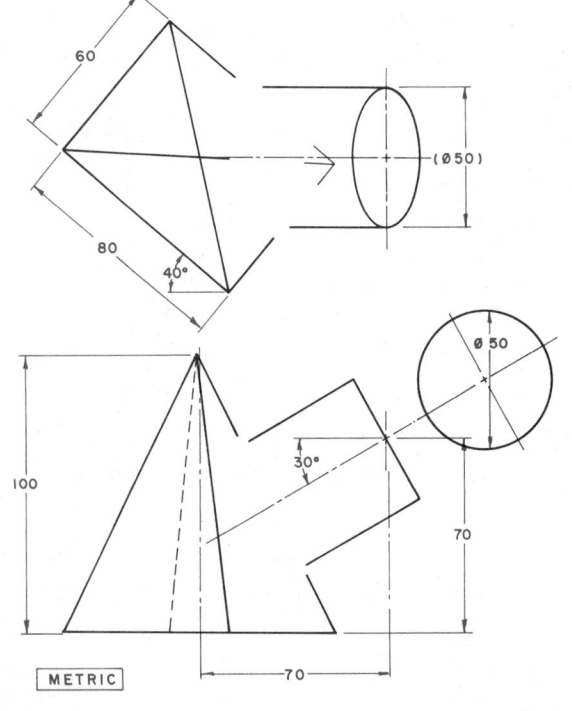

Problem 7-35

Problem 7-36

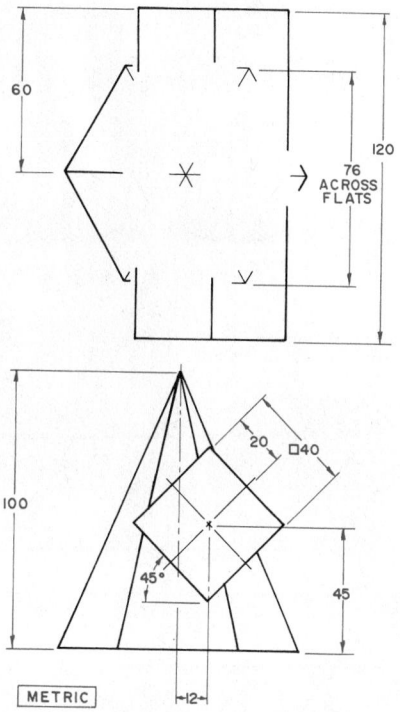

Problem 7-37

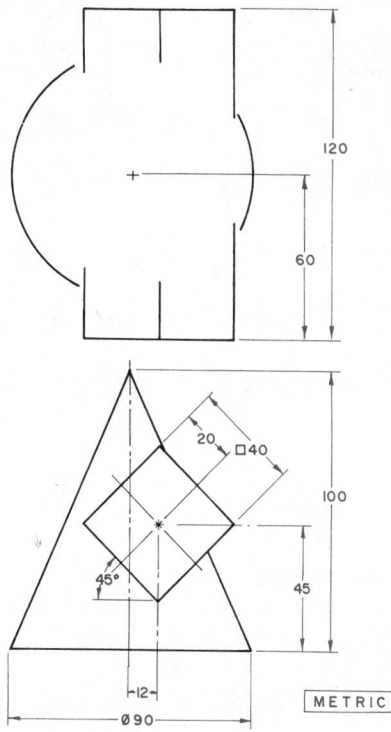

Problem 7-38

Problem 7-39

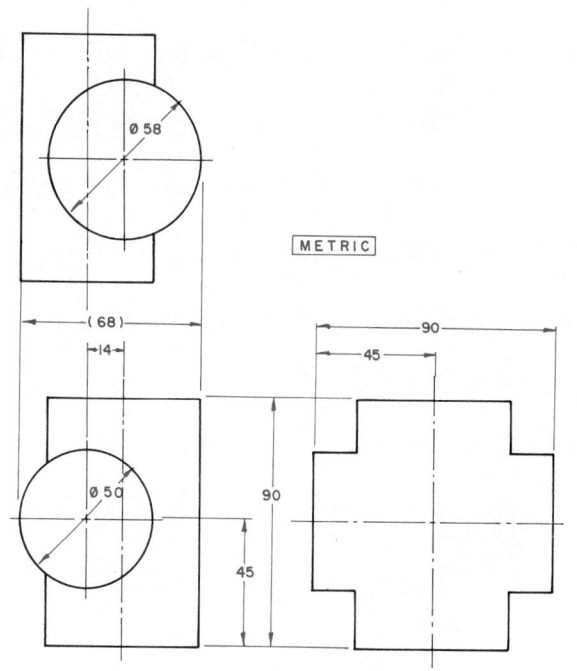

Problem 7-40

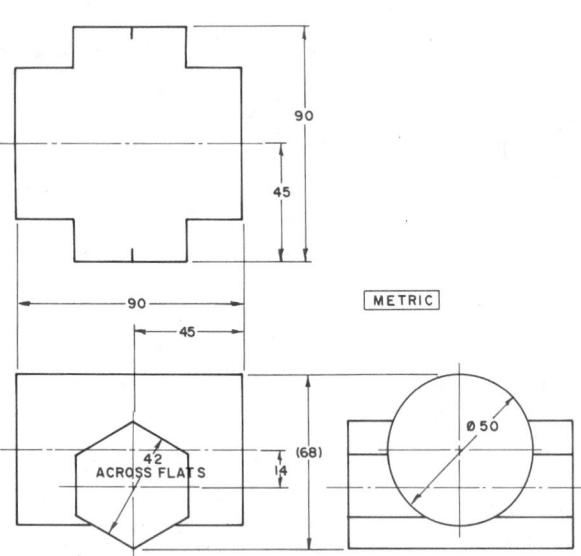

Problem 7-41

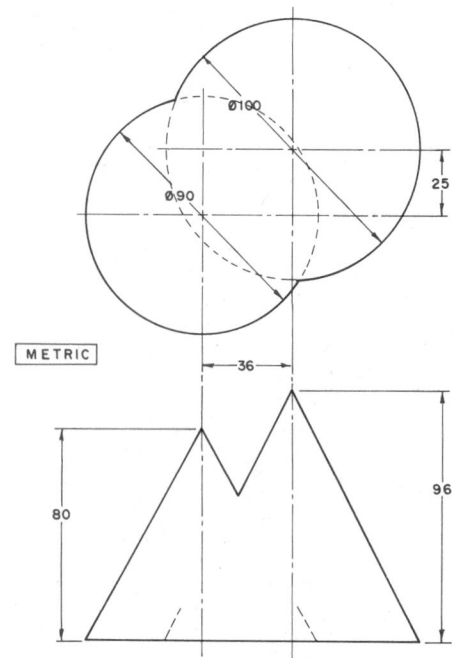

Problem 7-42

Problem 7-44

On a B size paper, draw the *top view* and *front view* of a straight roadway (a-b) 280.0 feet long, with an azimuth bearing of N 288°, and a grade of minus (–) 20%. Use a 1 inch = 50.0 feet scale. Show all light construction work.

Given: Location of fold line F/T and location of point aT and aF.

Problem 7-43

Calculate the angles using the given bearings for A through F. Show all math work.

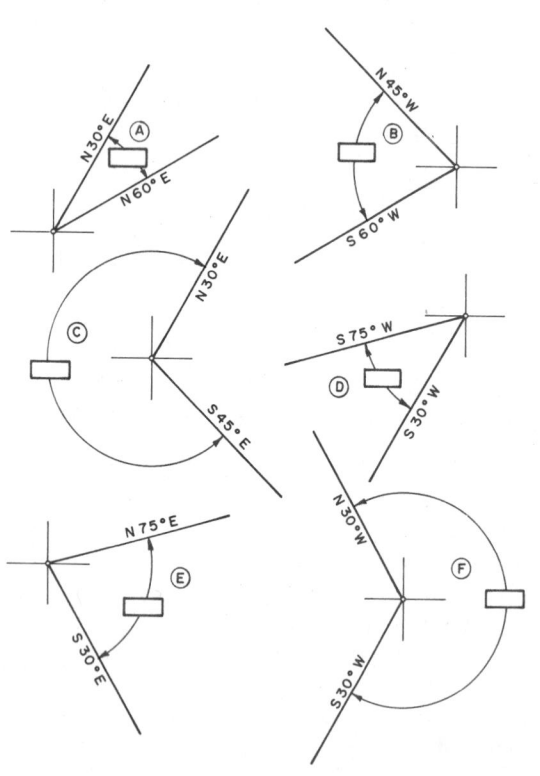

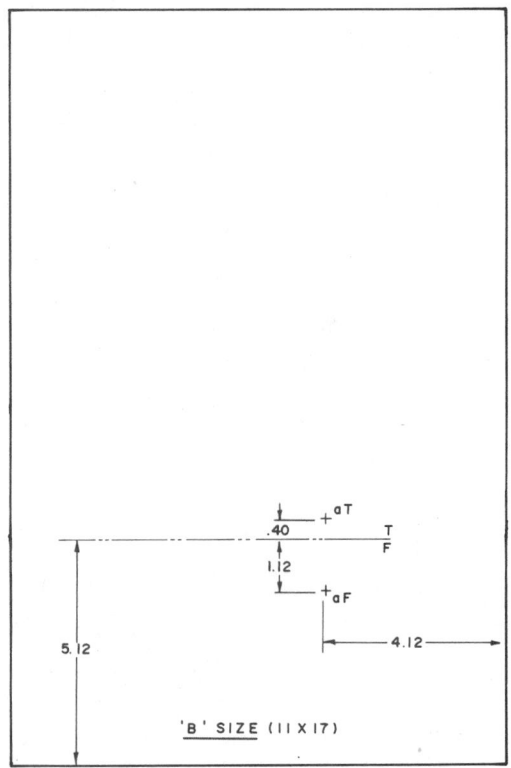

Problem 7-43

Problem 7-44

Problem 7-45

On a B size paper, lay out the *top view* and *front view* of a radio tower. Use a 1 inch = 50.0 feet scale. Show all light construction work. Answer the following questions:

1. What is the grade percent for the guide wire a/b?
2. What is the slope angle of the guide wire?
3. What is the true length of the wire?

Given: Location of the radio tower, guide wires, and fold line F/T.

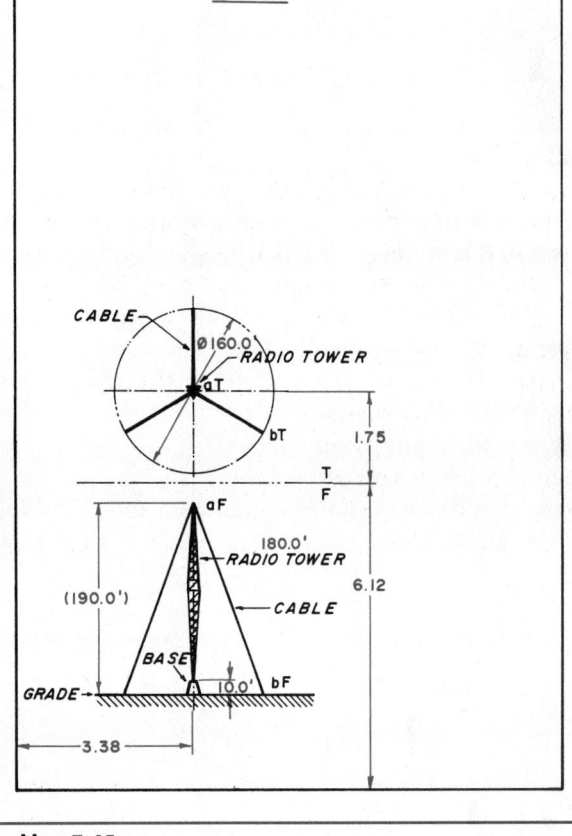

Problem 7-45

BELT	BEARING	TRUE LG.	GRADE %
a – b	N 30° W	180.0'	40%
b – c	N 80° W	140.0'	30%
c – d	S 45° W	145.0'	15%
ANSWER			
a – d			

Problem 7-46

On a B size paper, draw an existing conveyor belt system with the given specifications (see specifications in box). The conveyor belt system is to be redesigned to go from point a directly to point d. Use a 1 inch = 50.0 feet scale. Show all light construction work. Answer the following questions:

1. What is the new redesigned bearing from points a to d?
2. What is the new redesigned true length?
3. What is the new grade percent?
4. How much shorter is the new design than the old design?

Given: Location of grade, starting point a, and fold line F/T.

Problem 7-46

Problem 7-47

On a B size paper, construct two tunnels, a and b. Tunnel a starts at point a and has a bearing of N 70° W. It is 240.0 feet long and has a plus (+) slope angle of 34°. Tunnel b starts at point b and has a bearing of N 60° E. It is 200.0 feet long and has a plus (+) slope angle of 19°. Use a 1 inch = 50 feet scale. Show all light construction work. Project from the front view. Answer the following questions:

1. Do the tunnels ever intersect?
2. If *not*, how far apart are they?

Given: Locations of tunnel entrances a and b and fold line F/T.

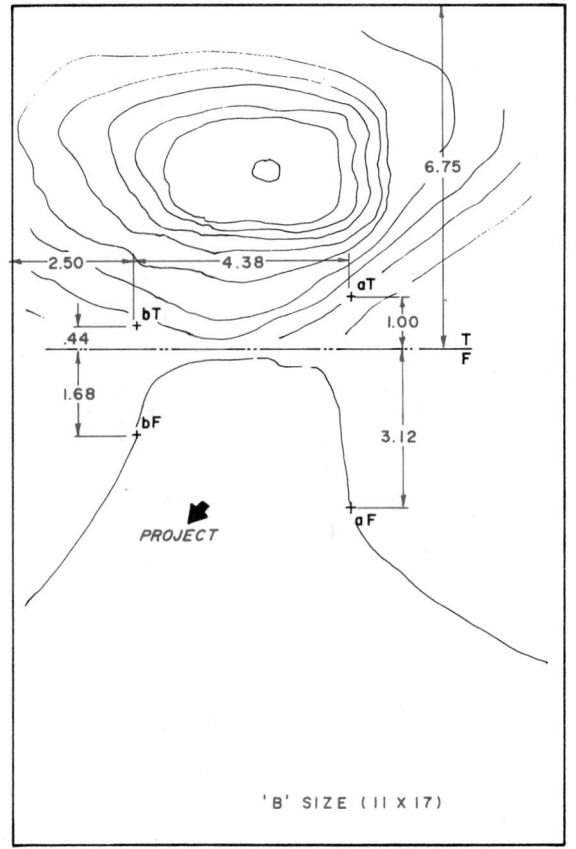

'B' SIZE (11 X 17)

Problem 7-47

Problem 7-48

On a B size paper, construct two mining shafts into a mountain from two different directions and elevations. Both are cut straight into the mountain at different percents of grades. One shaft starts at point a and the other starts at point b. Shaft a has a bearing of S 60° W, with a grade of plus (+) 35%, and is 165.0 feet long. Shaft b has a bearing of S 45° E, with a grade of plus (+) 40%, and is 176.0 feet long. Use a 1 inch = 50.0 feet scale. Show all light construction work. Answer the following questions:

1. Do the shafts intersect?
2. What is the true distance from the entrance of shaft a to the entrance of shaft b?
3. What is the bearing and percent of grade from the two entrances?

Given: Locations of the shaft entrances a and b, fold line F/T, and mountain.

'B' SIZE (11 X 17)

Problem 7-48

❑ Problem 7-49

A 40′ wide, level road, at elevation 150′, is to be run through a hilly terrain as shown on the contour map. Trace the map and use the accompanying scale to determine cut and fill boundaries for a slope of 1:1 for both the cuts and the fills.

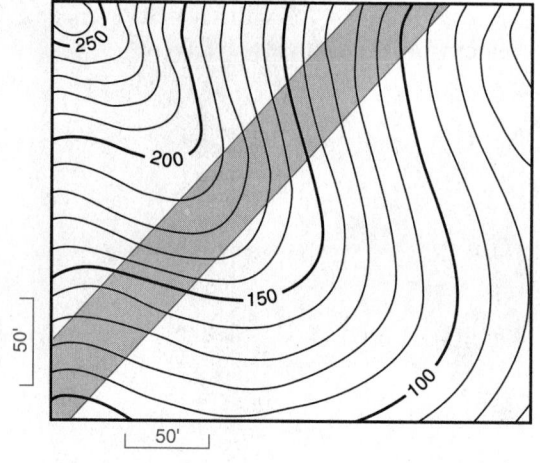

Problem 7-49

❑ Problem 7-50

Vehicle A is traveling due North at 150 kph (90 mph); vehicle B, due West at 110 kph (66 mph). Vehicle B is 40 kilometers North and 40 kilometers East of vehicle A, at the instant shown. What is the shortest distance between the vehicles A and B when they pass each other? An example velocity vector diagram illustrates the consequences of A and B going at the same speed. What should vehicle B's speed be so that the shortest distance between A and B is 15 kilometers when they pass each other?

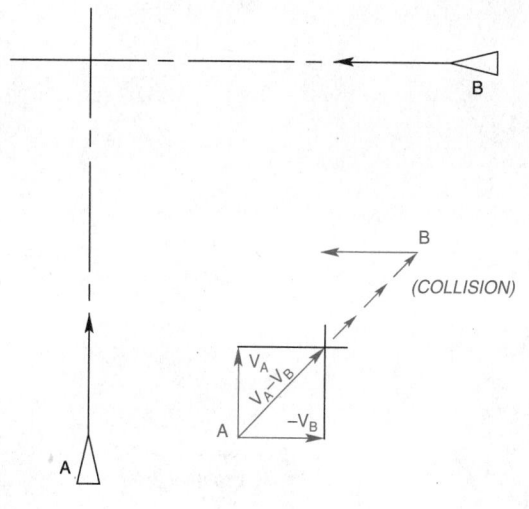

Problem 7-50

❑ Problem 7-51

An ore outcropping is shown on the topographic contour map. Points A and B are on the upper surface (called the hanging wall) of the vein. A vertical bore hole at point C indicates that the hanging wall is at 877′ elevation. The core hole extended at C indicates that the bottom surface of the ore vein (foot wall) is at 841′ elevation. For the information given for the locations of points A, B, and C and the vein surfaces, determine the following:

1. The strike and dip of the ore vein.
2. The true thickness of the ore vein.

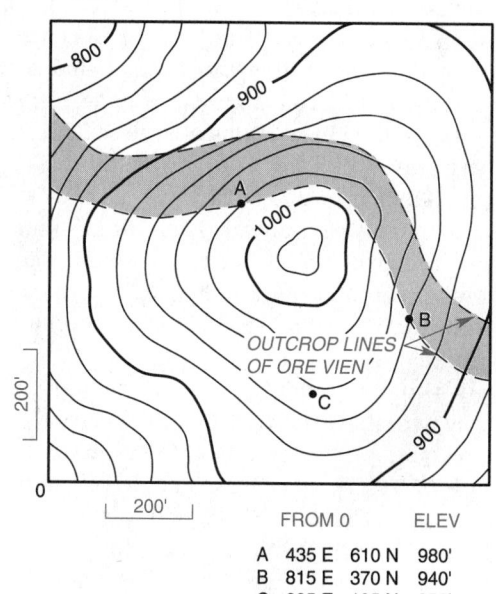

	FROM 0		ELEV
A	435 E	610 N	980′
B	815 E	370 N	940′
C	605 E	195 N	955′

Problem 7-51

❑ Problem 7-52

The simplified drawing of the 4-bar linkage shown (3 links and the rigid support) has link A-B rotate 360° around pinned connection A. Link A-B's motion causes link C-D to oscillate about pinned connection D. At the instant shown, a velocity analysis has link C-D rotating CCW at 18.3 radians per second. Note that the velocity of point B of link A-B is V(B) = ω(AB) x (0.03m) = 1.8 m per second. The component of V(B), which is coincident with member B-C, causes link B-C to move to the left with a velocity of V(BC) = 1.2 m per second as found graphically. Next, a component of V(BC) perpendicular to C-D, causes C-D to rotate CCW about point D with an angular velocity of ω(CD) = (1.19)/0.065 = 18.3 radians per second. Determine the following:

1. The instantaneous angular velocities of ω(CD) when A-B is at 90, 180, and 270 degrees.
2. At what values of θ will C-D have an angular velocity of zero? Why?

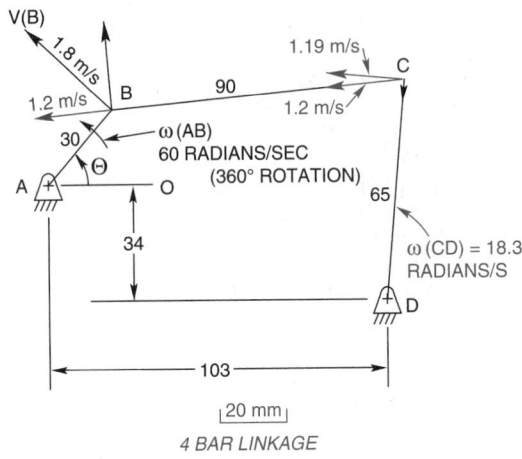

4 BAR LINKAGE

Problem 7-52

❑ Problem 7-53

The cam and cam follower shown represent a camshaft cam and valve follower in an internal combustion engine. The cam rotates at a constant rate of 2,000 rpm, which is 209 radians per second (2000/60 = 33.33 RPS; and 33.33 x 2 x π = 209 radians per second). The cam has rotated 225° at the instant shown and is transferring a velocity of 120 inches per second to the follower as determined by a graphical analysis. The velocity of the point in contact T at this instant is V(T) = ϖ(cam) x 0.98 inch = (209 radians per second) x (0.98 inch) = 205 inches per second. Then the resolution of the 205 in./sec gives 120 in./sec to the follower and the other component indicates a sliding velocity parallel to the bottom surface of the valve. Hence, good lubrication is required.

Lay out the cam from the dimensions given and graphically determine the velocity of the valve throughout one revolution, starting at θ = 0°. Record the results in a graph of V(follower) versus angle of rotation θ. Suggestions: (1) From θ = 180° to 360° determine V(follower) at 15° intervals. (2) Lay out the cam on one sheet of drawing paper and use an overlay sheet of drawing paper for the intervals.

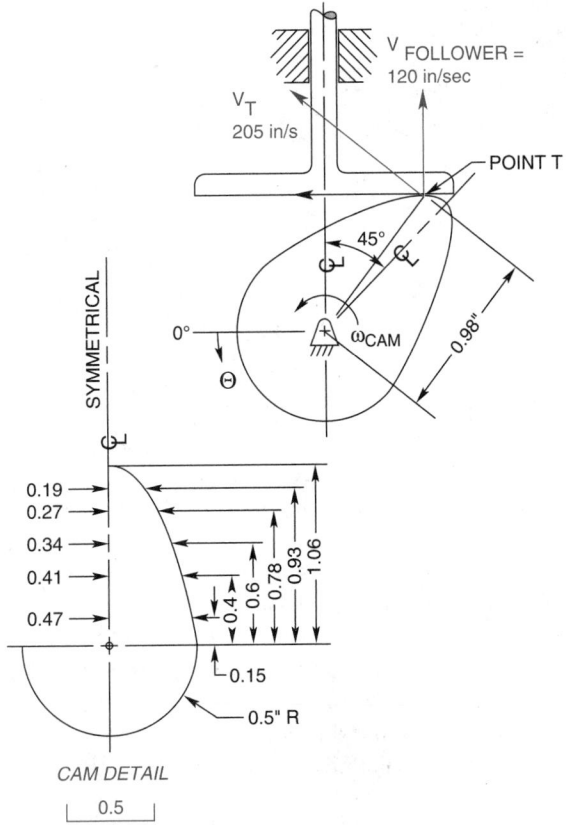

CAM DETAIL

Problem 7-53

❏ Problem 7-54

The laser beam a-b shines on the metal door shown. As the door swings open the laser beam traces a path on the door's surface. Determine the path for the door opening 150° and show the path in view F.

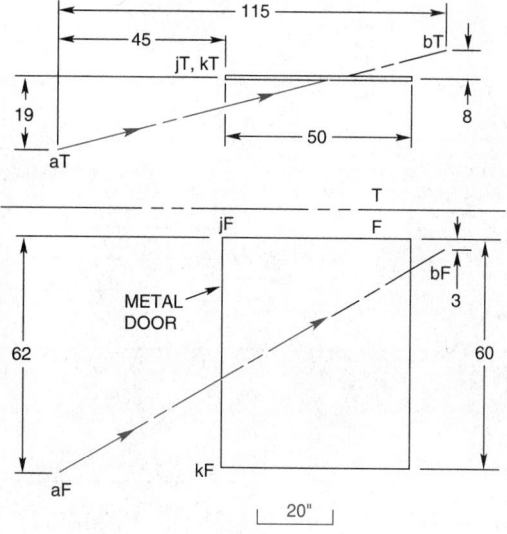

Problem 7-54

❏ Problem 7-55

A basic toggle mechanism is used to generate a large force from a smaller force, for punching holes in metal. For the position shown in the simplified mechanism, what force is generated at punch A, vertically downward? When the 40000N force is at B2, what force at A is generated, vertically downward?

❏ Problem 7-56

For the coplanar-concurrent system shown, determine the forces in members AB and BC using the method of joints.

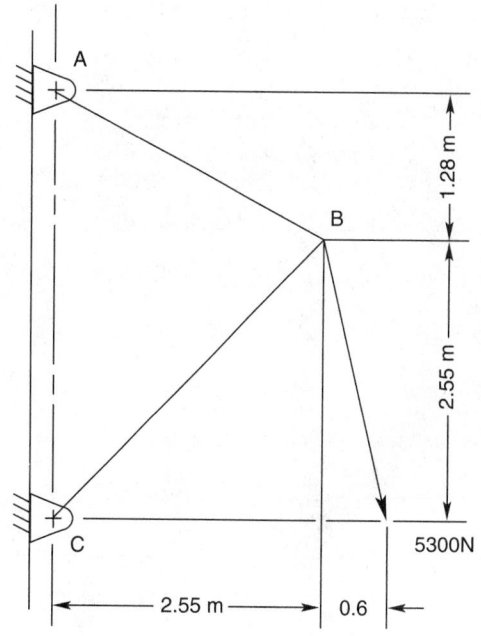

Problem 7-55

Problem 7-56

❑ Problem 7-57

Use the method of Maxwell diagram with Bow's notation to determine the forces in each member of the truss shown. Tabulate the magnitudes and whether the members are in tension or compression.

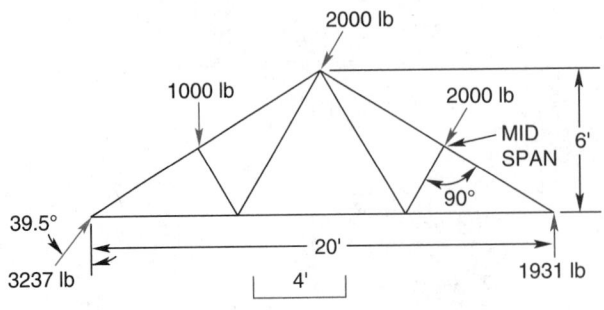

Problem 7-57

❑ Problem 7-58

Redraw the tripod and the 1500 lb applied force starting at the lower left-hand area of a drawing sheet. Determine the force in each member by the method shown in this chapter, and use revolution to finally determine the magnitude of each force. Suggestion: Use members o-b and o-c to form a plane.

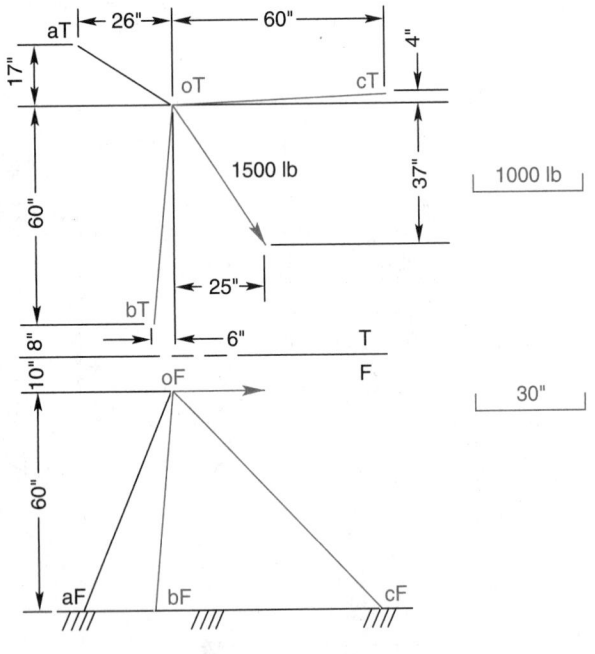

Problem 7-58

❑ Problem 7-59

From the given information, lay out the four noncoplanar-concurrent force vectors and add them "tail to head" in any order (the order can differ in each view also) to determine the equivalent resultant of the four forces. Then, find the magnitude of the resultant by revolution. Also, determine the bearing and percent slope of the resultant.

The coordinates of the four forces originating from point o are:

a (10,5,5)   **c** (–8,–8,8)
b (4,–13,–10)   **d** (–15,5,–10)

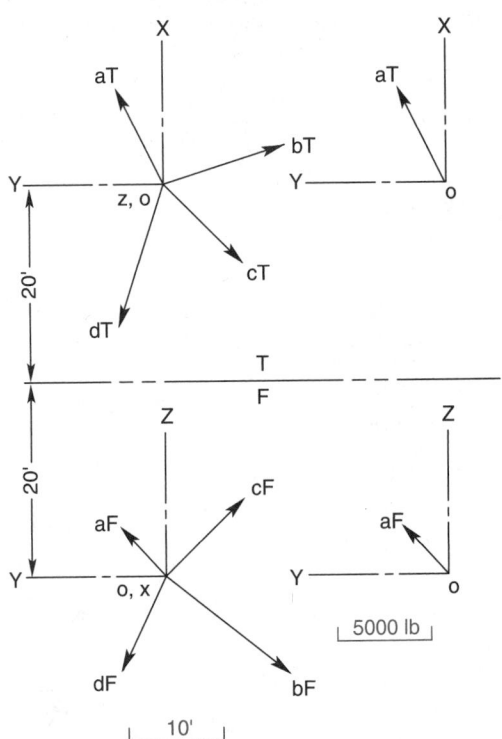

Problem 7-59

❑ Problem 7-60

A ham radio operator needs to know the lengths of three guy wires for his antenna installation, and where to locate guy wire o-c. At the lower left corner of a sheet of drawing paper, construct a layout of the given roof information and solve the ham's guy wire problem using rotation. Guy wire o-c is to clear corner f of the chimney by 1 foot. Show all work and tabulate the answers.

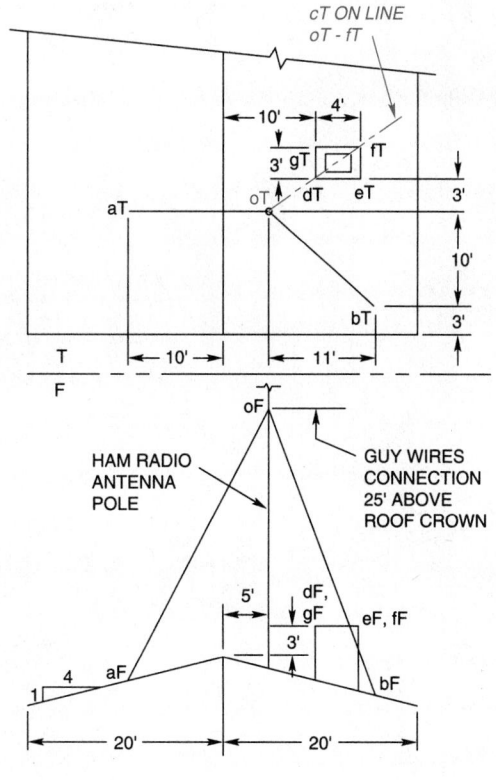

Problem 7-60

❑ Problem 7-61

A tension spring a-b constrains a 30″ wide door. The spring shown in simplified schematic form has a spring constant of 5 lb/in., i.e., 5 lb will stretch the spring 1 inch. The door shown in the closed position pivots about a hinge located at point c. At the position shown the spring has an initial tension of 6 lb. What will be the total tension in the spring when the door is opened 90°? Solve the problem using revolution.

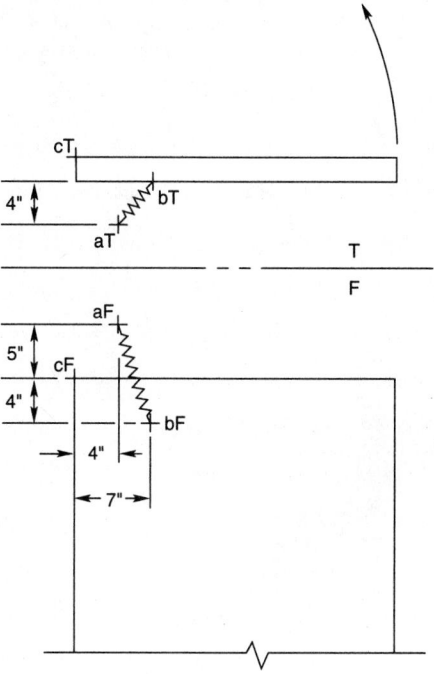

Problem 7-61

❏ Problem 7-62

The simplified drawing shows a personnel lift, powered by an hydraulic cylinder, DC. At different angles θ the force F(DC) required by the cylinder is calculated by dividing the moment due to the 1000-lb load (1000-lb load x the shortest distance to the line of action of the 1000 lb load from point A), by the shortest distance to the line of action of the cylinder's force F(DC). The shortest distances are called moment arms. Redraw the positions of the lift as specified in the table, locate the cylinder, measure moment arms, and do the calculations to finish the table.

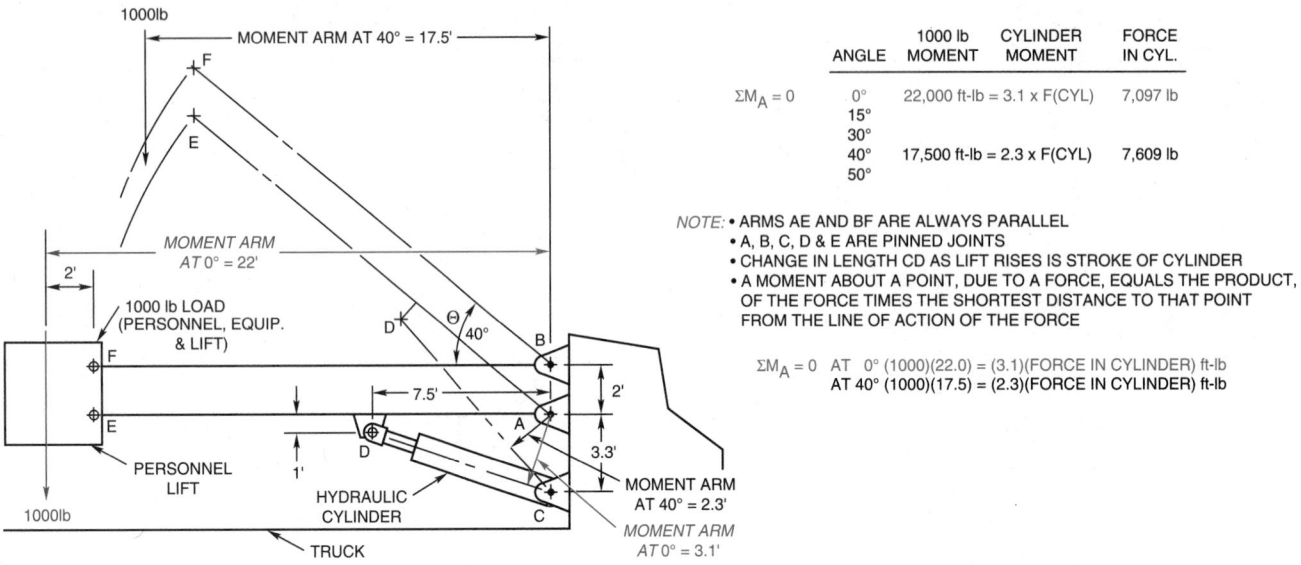

	ANGLE	1000 lb MOMENT	CYLINDER MOMENT	FORCE IN CYL.
$\Sigma M_A = 0$	0°	22,000 ft-lb =	3.1 x F(CYL)	7,097 lb
	15°			
	30°			
	40°	17,500 ft-lb =	2.3 x F(CYL)	7,609 lb
	50°			

NOTE: • ARMS AE AND BF ARE ALWAYS PARALLEL
• A, B, C, D & E ARE PINNED JOINTS
• CHANGE IN LENGTH CD AS LIFT RISES IS STROKE OF CYLINDER
• A MOMENT ABOUT A POINT, DUE TO A FORCE, EQUALS THE PRODUCT, OF THE FORCE TIMES THE SHORTEST DISTANCE TO THAT POINT FROM THE LINE OF ACTION OF THE FORCE

$\Sigma M_A = 0$ AT 0° (1000)(22.0) = (3.1)(FORCE IN CYLINDER) ft-lb
AT 40° (1000)(17.5) = (2.3)(FORCE IN CYLINDER) ft-lb

Problem 7-62

❏ Problem 7-63

A floating golf-course green, supported by a concrete float, is moved by changing the lengths of four wire ropes powered by winches inside the float. From the position shown, the green is moved 130 yd West and 20 yd North. Use revolution to determine the change in length of each cable, A, B, C, and D. Tabulate the changes. (Ask your golfing friends if they could hit the green at the new position.)

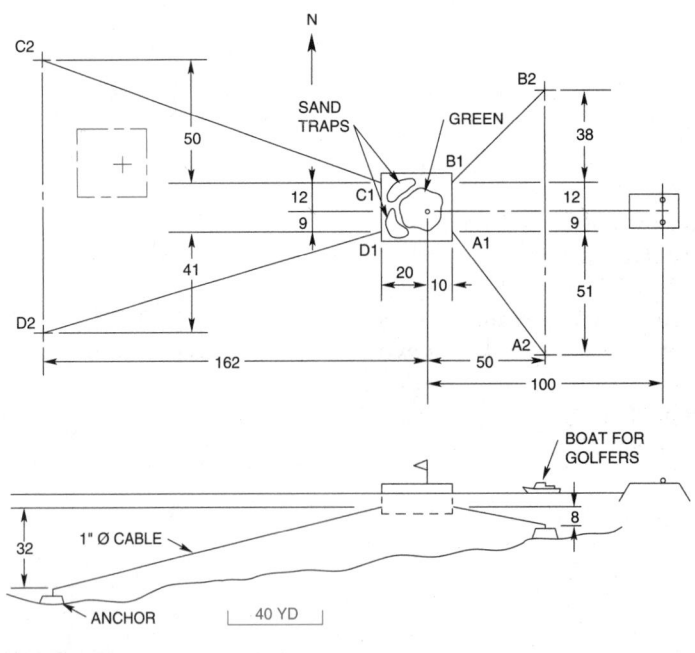

Problem 7-63

Patterns and Developments

Developments

A *development* is the pattern or template of a shape that is laid out in a single flat plane in preparation for the bending or folding of a material to a required shape. Surface developments are used in many different industries. Some examples of objects requiring developments are cereal boxes, tool boxes, funnels, air-conditioning ducts, and simple mail boxes.

Three major kinds of surface developments are parallel line developments, radial line developments, and triangulation developments. *Parallel line developments* are used for objects having parallel fold lines. *Radial line developments* are those whose fold lines radiate from one point. See *Figure 8-1*. *Triangulation developments* are the development process of breaking up an object into a series of triangular plane surfaces, *Figure 8-2*. Each kind of development is explained here in full.

Surfaces

A *development surface* is the exterior and/or interior of the sheet material used to form an object. The various

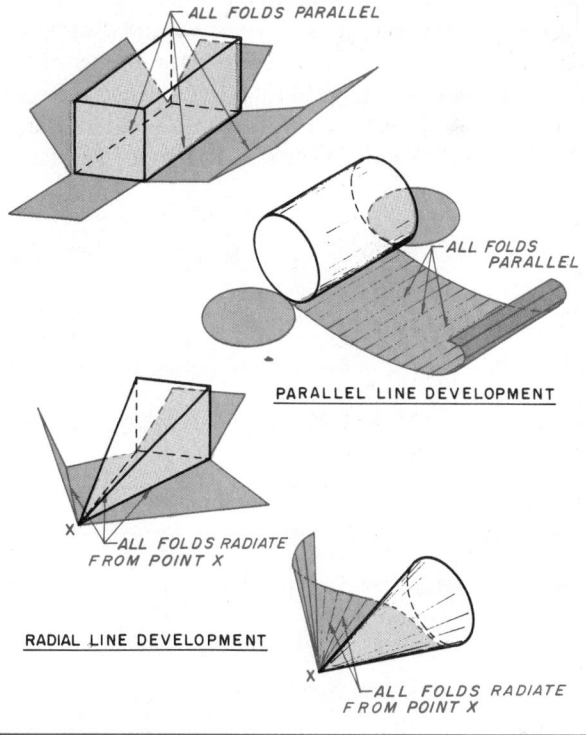

Figure 8-1   Parallel line development and radial line development

295

FOLD LINES DEVELOP
A SERIES OF TRIANGULAR
PLANE SURFACES

RADIAL LINE DEVELOPMENT

Figure 8-2   Triangular development

kinds of surfaces include plane surface, single-curved surface, and double-curved surface.

If any two points anywhere on a surface are connected to form a straight line, and that line rests upon the surface, it is a *plane surface*. If all points on a surface can be interconnected to form straight lines without exception, it is a *flat-plane surface*. The top of a drawing board is an example of a flat-plane surface. A flat-plane surface can have three or more straight edges. Such objects as a cube or a pyramid are bounded by plane surfaces. If a surface can be unrolled to form a plane, it is a *single-curved surface*. A cylinder or a cone is an example of a *single-curved surface*.

A surface that cannot be developed because it is neither a plane surface nor a curved surface is a *warped surface*. An automobile fender is an example of a warped surface. This kind of surface is usually stamped or pressed into shape. An object fully formed by curved lines with no straight lines is a *double-curved surface*. A sphere is an example of a double-curved surface. This type of surface

cannot be developed exactly by using flat patterns; only an approximate development can be made.

Laps and Seams

Extra material must be provided for laps and seams. Many kinds of seams are available, ***Figure 8-3***. When making a choice, the drafter must take into account the thickness of the metal so that crowding of the metal at the joints is not a problem. Allowance must be made for the gluing, soldering, riveting or welding processes for joints. The method of fastening joints together varies with the material and, accordingly, the elimination of rough or sharp edges. Tabs for the drawing problems at the end of this chapter should be designed to support the joining surface, ***Figure 8-4***.

Design Practices

The drafter should lay out developments according to the dimensions of stock materials for economy and the best use of materials and labor. The stock material area required to cut a pattern should be kept to the smallest convenient size. It is good practice to put the seam at the shortest joint, and to attach tops and bottoms along the longest possible seam or bend to reduce the length requiring soldering, riveting or welding. It is assumed that the inside surface of the final object is the side that the pattern defines. (The important dimension sizes are usually the inside surfaces.) Fold or bend locations in the material are shown on the pattern with thin solid lines, and are locations that are also assumed to occur on the inside surface of the final object.

Thickness of Material

The actual thickness of sheet metal is specified by gage numbers. Each gage number indicates a particular thickness of material. ***Figure 8-5*** lists the gage number and gage sizes for sheet and plate steel. These are nominal

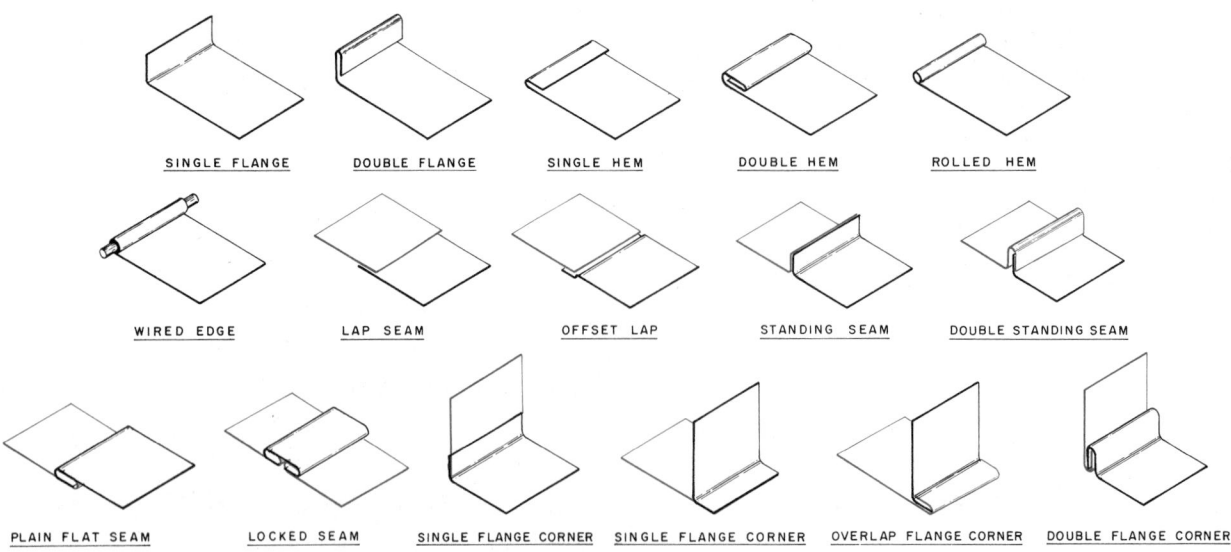

SINGLE FLANGE   DOUBLE FLANGE   SINGLE HEM   DOUBLE HEM   ROLLED HEM

WIRED EDGE   LAP SEAM   OFFSET LAP   STANDING SEAM   DOUBLE STANDING SEAM

PLAIN FLAT SEAM   LOCKED SEAM   SINGLE FLANGE CORNER   SINGLE FLANGE CORNER   OVERLAP FLANGE CORNER   DOUBLE FLANGE CORNER

Figure 8-3   **Many types of seams are available**

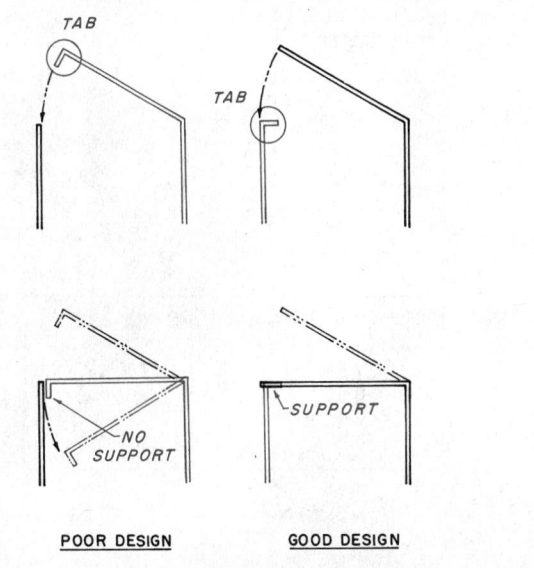

Figure 8-4 Tabs should support the joining surfaces

POOR DESIGN   GOOD DESIGN

thicknesses, subject to permissible tolerances. Today, there is considerable confusion when using the gage numbering system.

In 1893, Congress established the United States Standard Gage, and it was primarily a *weight* gage rather than a *thickness* gage. It was derived from the weight of wrought iron. At that time, the weight of wrought iron was calculated at 480 pounds per cubic foot; thus, a plate 12 inches square and 1 inch thick weighs 40 pounds. A No. 3 U.S. gage represents a wrought iron plate weighing 10 pounds per square foot. Therefore, if a weight per square foot 1 inch thick is 40 pounds, the plate thickness for a No. 3 gage equals $10 \div 40 = 0.25$ inch, which is the original thickness equivalent for a No. 3 U.S. gage. Since this and all other gage numbers were based on the weight of wrought iron, they are not correct for steel. To add to the confusion, there is considerable variation in the gage thickness for different kinds of material. For example, a gage

SHEET METAL

GAGE	ALUMINUM		BRASS		STEEL	
	THICKNESS	WT./SQ. FT.	THICKNESS	WT./SQ. FT.	THICKNESS	WT./SQ. FT.
8	.1285	1.812	.1285	5.662	.1644	6.875
9	.1144	1.613	.1144	5.041	.1494	6.250
10	.1019	1.440	.1019	4.490	.1345	5.625
11	.0907	1.300	.0907	3.997	.1196	5.000
12	.0808	1.160	.0808	3.560	.1046	4.375
13	.0720	1.020	.0720	3.173	.0897	3.750
14	.0641	.907	.0641	2.825	.0747	3.125
15	.0571	.805	.0571	2.516	.0673	2.812
16	.0508	.720	.0508	2.238	.0598	2.500
17	.0453	.639	.0453	1.996	.0538	2.250
18	.0403	.580	.0403	1.776	.0478	2.000
19	.0359	.506	.0359	1.582	.0418	1.750
20	.0320	.461	.0320	1.410	.0359	1.500
21	.0285	.402	.0285	1.256	.0329	1.375
22	.0253	.364	.0253	1.119	.0299	1.250
23	.0226	.318	.0226	.996	.0269	1.125
24	.0201	.289	.0201	.886	.0239	1.000
25	.0179	.252	.0179	.789	.0209	.875
26	.0159	.224	.0159	.700	.0179	.750
27	.0142	.200	.0142	.626	.0164	.688
28	.0126	.178	.0126	.555	.0149	.625
29	.0113	.159	.0113	.498	.0135	.562
30	.0100	.141	.0100	.441	.0120	.500
31	.0089	.126	.0089	.392	.0105	.438
32	.0080	.112	.0080	.353	.0097	.406
33	.0071	.100	.0071	.313	.0090	.375
34	.0063	.089	.0063	.278	.0082	.344
35	.0056	.079	.0056	.247	.0075	.312
36	.0050	.071	.0050	.220	.0067	.281
37	.0045	.063	.0045	.198	.0064	.266
38	.0040	.056	.0040	.176	.0060	.250
39	.0035	.050	.0035	.154	-	-
40	.0031	.044	.0031	.137	-	-

Figure 8-5 Thickness of material

used for such nonferrous materials as brass and copper is sometimes also used to specify a thickness for steel or vice versa.

Today, to help eliminate the problems, the decimal or metric system of indicating gage size is now replacing the older gage size numbering system.

Parallel Line Development

In parallel line developments all fold lines are parallel. (Refer back to Figure 8-1.)

A three-view drawing of a simple 2-inch cube is shown in *Figure 8-6*. A cube is a specific kind of prism. Notice that all the fold lines that form the lateral surfaces of a prism are parallel, *Figure 8-7A*. If the cube were to be unfolded, it would appear as it does in *Figures 8-7B*, *8-7C*, and *8-7D*. The finished development is shown in *Figure 8-7E*.

Notice that the cube was developed or laid out from a baseline, sometimes referred to as the *stretchout line*. Fold lines are always oriented 90° from the baseline, as illustrated. If the pattern were to be cut from a flat sheet of material, tabs would be needed to fasten the cube together. Tabs must be positioned so as to make

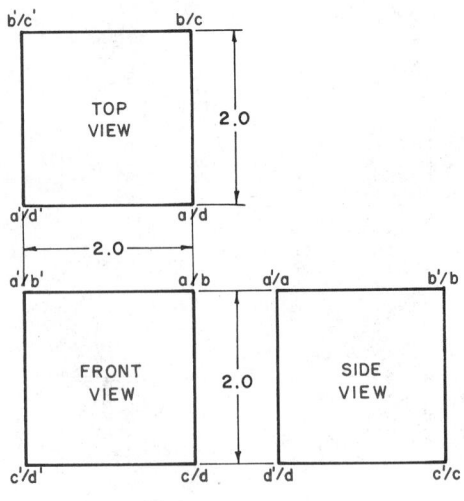

Figure 8-6   Simple multiview drawing of a cube

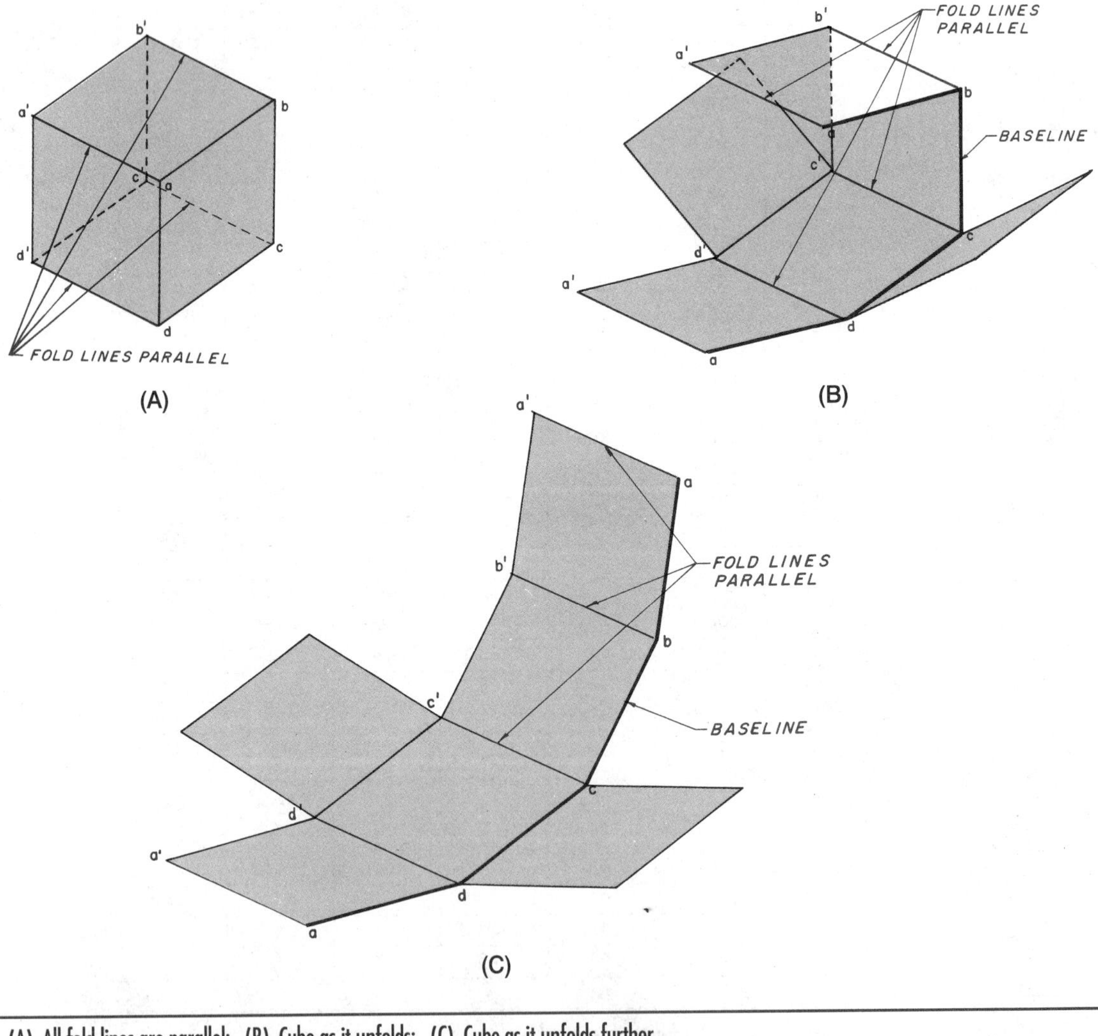

each meet with an untabbed edge, *Figure 8-7F*. Notice that there are seven tabs and seven blank edges. Each tab folds to meet a particular blank edge.

(A)

(B)

(C)

Figure 8-7   (A) All fold lines are parallel;   (B) Cube as it unfolds;   (C) Cube as it unfolds further

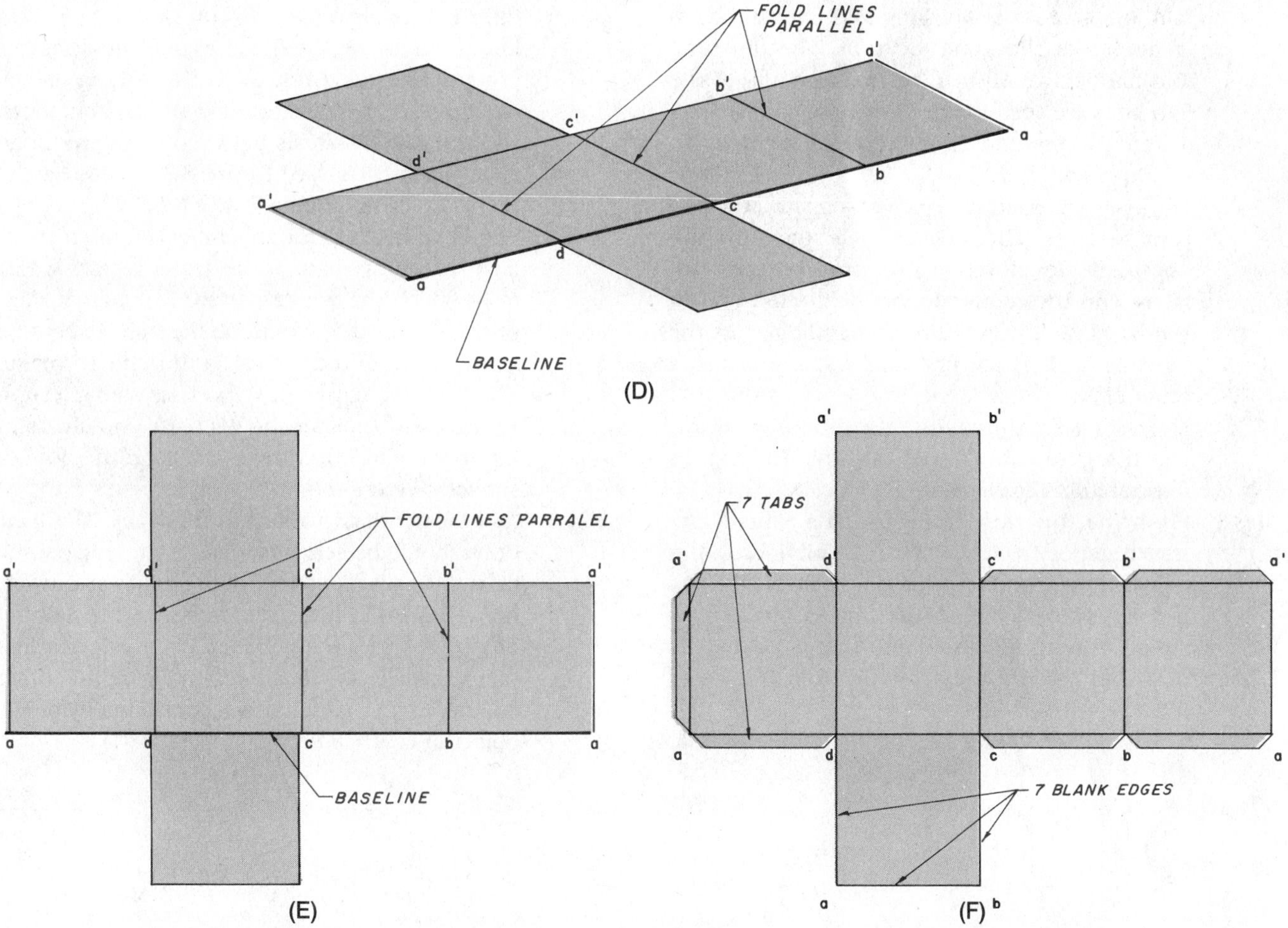

FOLD LINES
PARALLEL

BASELINE

(D)

FOLD LINES PARRALEL

BASELINE

(E)

7 TABS

7 BLANK EDGES

(F)

Figure 8-7 (Continued) (D) Cube flattened out; (E) Cube in the finished development; (F) Cube flattened out with tabs added

How To Develop a Truncated Prism Using Parallel Line Development

Given: A prism with a front view, top view, and auxiliary view, ***Figure 8-8A***.

Step 1 Locate the baseline, which must always be 90° from the parallel fold lines. Label each fold line

clockwise in alphabetical or numerical order. Always start with the shortest fold line, ***Figure 8-8B***. Either fold line a-a' or b-b' would be eligible, and a-a' is selected. Notice that the starting

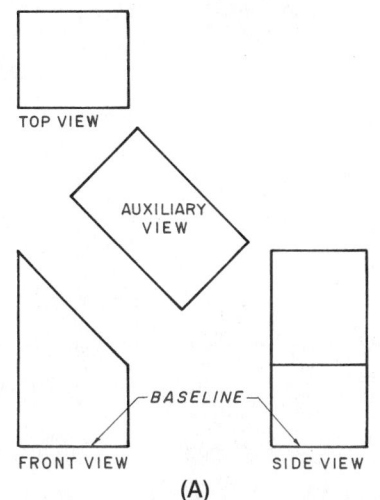

TOP VIEW

AUXILIARY VIEW

BASELINE

FRONT VIEW   SIDE VIEW

(A)

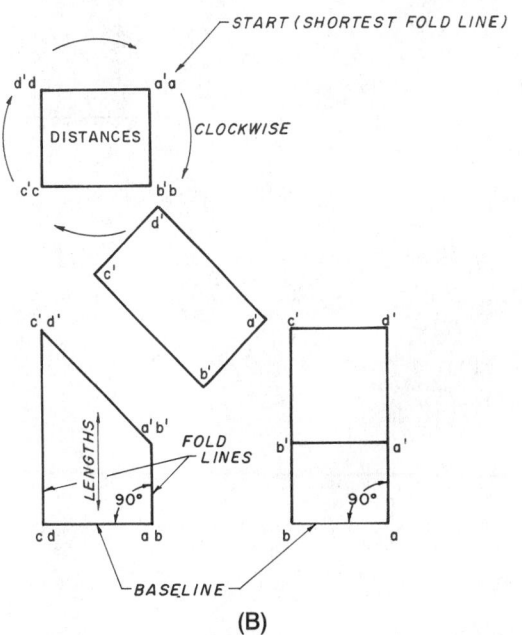

START (SHORTEST FOLD LINE)

DISTANCES

CLOCKWISE

LENGTHS

FOLD LINES

90°   90°

BASELINE

(B)

Figure 8-8 (A) Development of a truncated prism using parallel line development; (B) Step 1

fold line is also the finishing fold line, and each is labeled at the same location. The finishing fold line is a-a'. All real (or true) lengths of the fold lines are seen in the front view and the real distances between their locations are seen in the top view.

Step 2 On a separate sheet of paper, construct a pattern baseline, allowing enough space for the complete development, top and bottom surfaces, and the required tabs. At the left end of the baseline, draw a line perpendicular to the baseline. This is the first fold line and is also a seam edge. Transfer the true length from fold line a-a' (as measured from the three-view drawing in Figure 8-8A). Label this line a-a' also, as illustrated in *Figure 8-8C*.

Step 3 From the top view of Figure 8-8A, obtain the true distance from the first fold line location at a-a' to the second fold line location at b-b'. On the pattern baseline, transfer this distance, measuring from the first fold line a-a', and draw the next fold line parallel to the first fold line. Label it b-b', *Figure 8-8D*.

Step 4 Repeat the first three steps, alternating true fold

line lengths and true distances between fold lines until all the true lengths and true distances are transferred. Each of the true lengths a-a' through e-e' are transferred from the front view, and the true distances between them are found from the top view of Figure 8-8A. Connect the points as illustrated in *Figure 8-8E*.

Step 5 The edge selected for attachment of the top surface should be made to keep the rectangle size from which the whole pattern is cut as small as possible. Line a'b' would be the best, except an acute angle cutout c'b'a' is difficult to make. The fold line a'd' is next best, as shown. From Figure 8-8A, transfer the true size and shape of the top surface (auxiliary view) and the bottom surface, *Figure 8-8F*.

Step 6 Add tabs enough to fasten the edges. The locations should be selected so as not to enlarge the material area needed for the pattern cutout, *Figure 8-8G*. Check to verify that the number of tabs equals the number of blank edges in the same manner as is illustrated in Figure 8-7F. This completes the parallel line development of the object in Figure 8-8A.

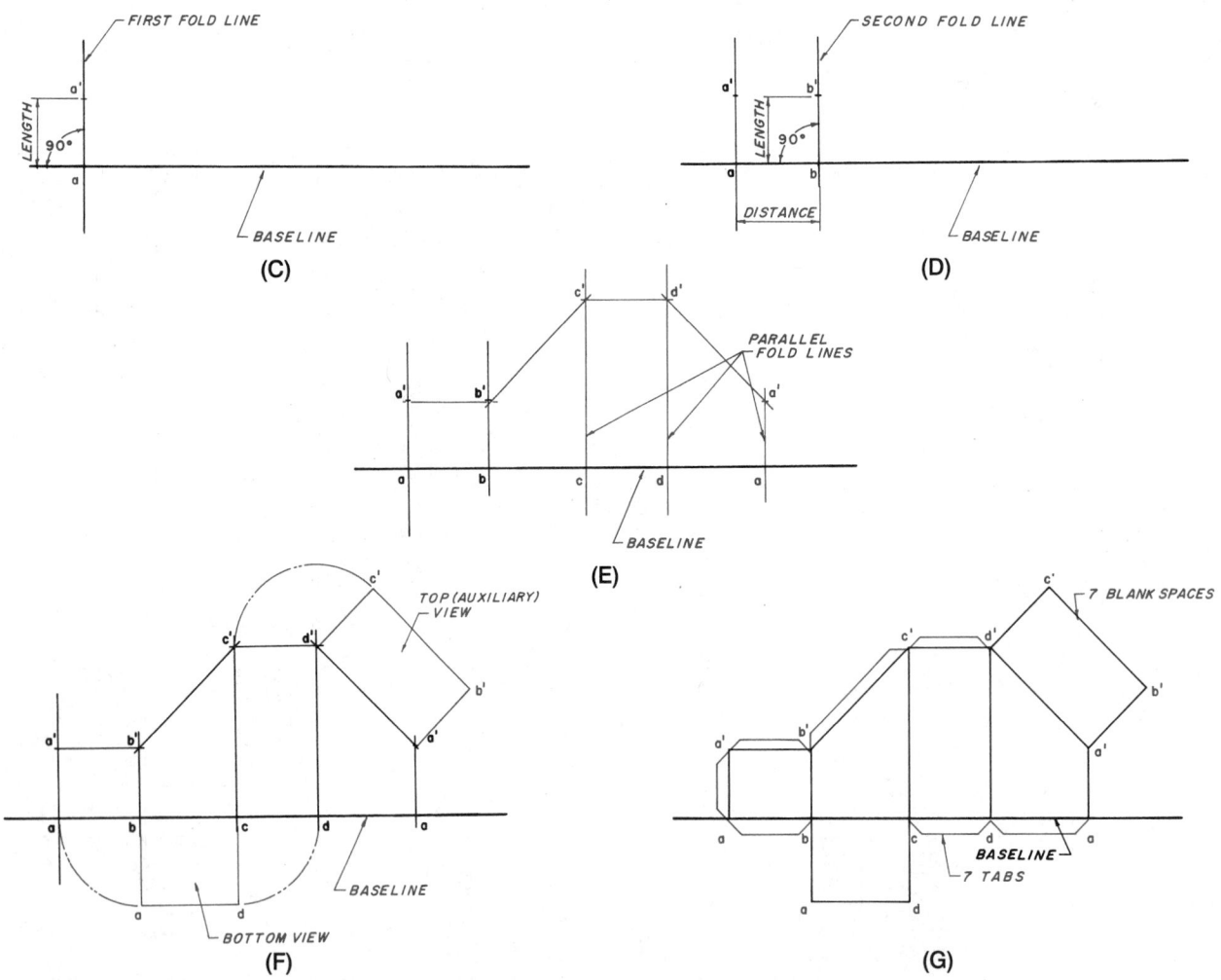

Figure 8-8 (Continued) (C) Step 2; (D) Step 3; (E) Step 4; (F) Step 5; (G) Step 6 (completed development of the prism)

How To Develop a Truncated Cylinder Using Parallel Line Development

Given: A cylinder with a front view, top view, and auxiliary view, *Figure 8-9A*. In developing a cylinder, the exact procedure is used as when developing a prism. Because the cylinder has a rounded surface, line segments called *station lines* must be assumed or chosen on the lateral surface. This is explained in Step 1.

Step 1 Locate a baseline which must always be 90° from the fold lines, sometimes referred to as *parallel station lines*. The bottom corner of the lateral surface is selected as a convenient location because it already exists. In the view where the cylinder appears as a diameter (the top view in this example), divide the diameter into equal divisions, *Figure 8-9B*. This example uses 12 equal divisions, but any number of appropriate equal divisions would do. These divisions locate fold line positions, which are used only for construction and layout purposes. Starting with the shortest fold line, consecutively label each divi-

sion; in this example, a through 1. Note that a-a' marks the beginning and ending of the same line which meets from opposite directions. The true "distances" between station lines can be approximated in this top view. Project each point into the next (front) view in order to find the true "lengths" of the fold lines. It should be reemphasized that the fold lines must be 90° from the baseline.

Step 2 On a separate sheet of paper, construct a pattern baseline allowing enough space for the complete development, both ends, and the required tabs. At the left end of the pattern baseline draw a line perpendicular to the baseline. This is the first fold line location a-a'. Transfer the true length a-a' from the front view of Figure 8-9B. Label this line as illustrated in *Figure 8-9C*.

Step 3 Two methods are used for finding the distances between station lines. Method A is an approximate method that assumes the cylinder to be a prism, whereas Method B is mathematically correct.

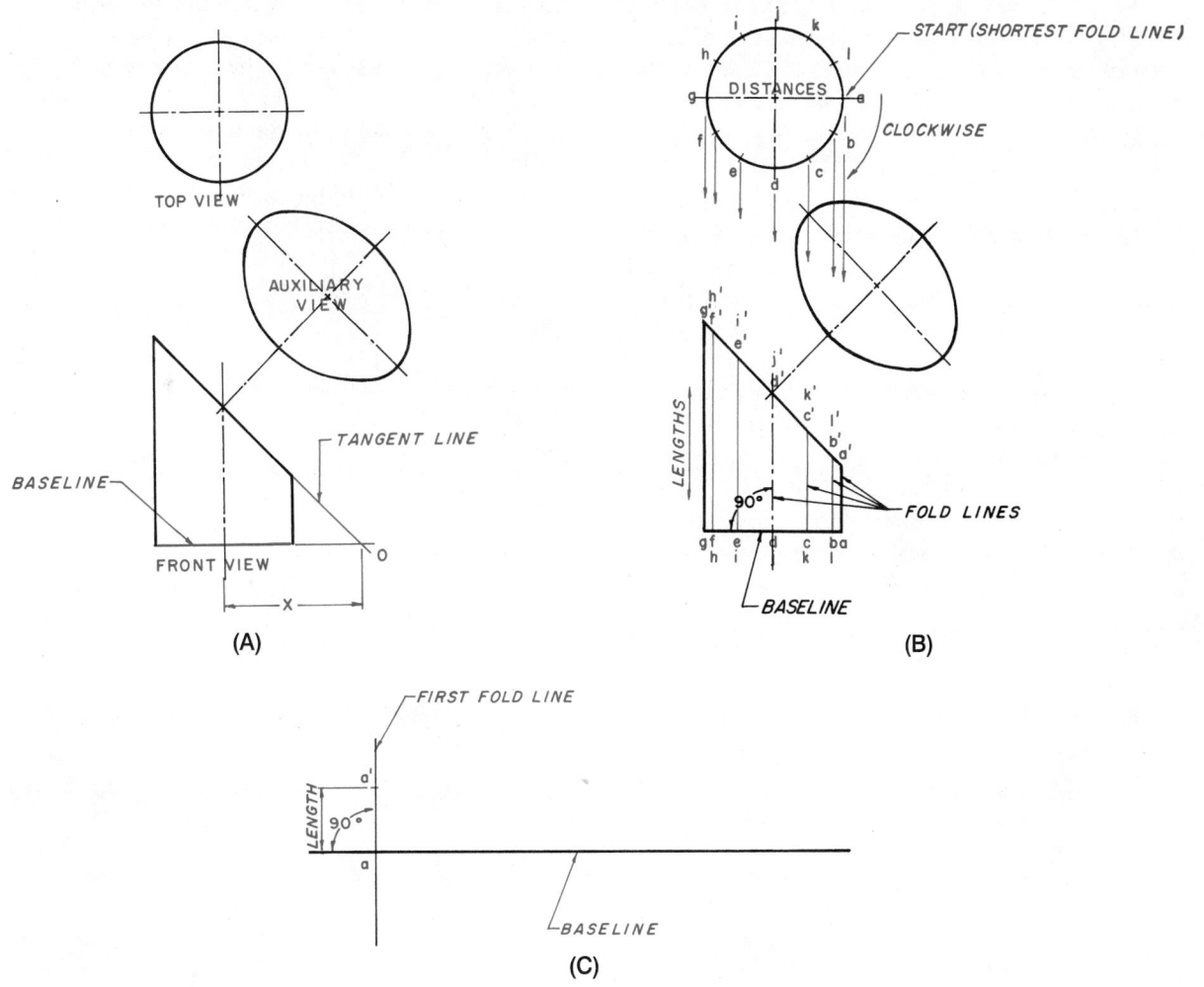

Figure 8-9 (A) Development of a truncated cylinder using parallel line development; (B) Step 1; (C) Step 2

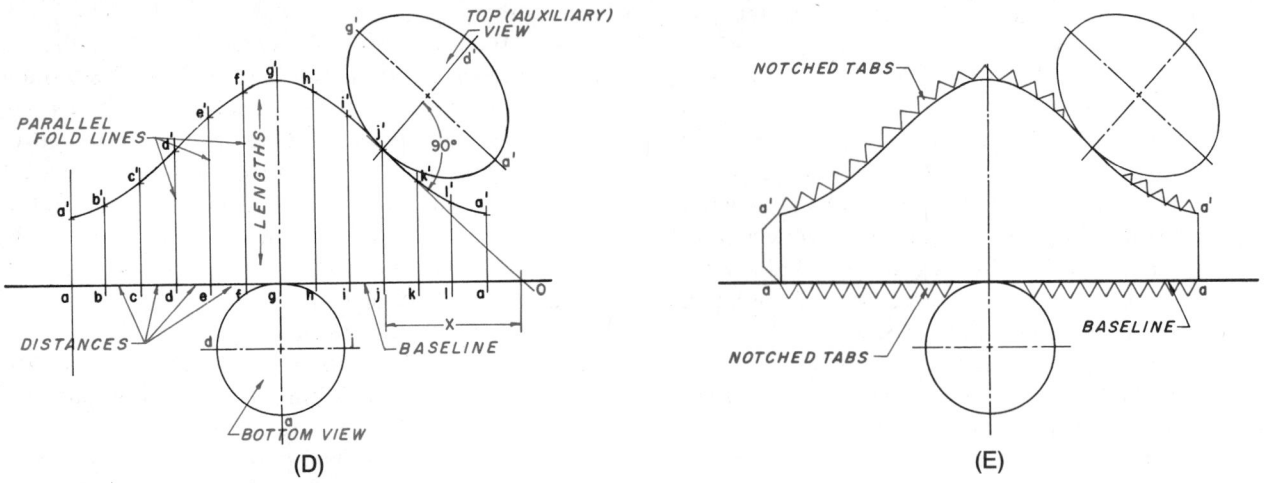

Figure 8-9 (Continued) (D) Steps 3 and 4; (E) Step 5 (completed development of the cylinder)

Method A: From the top view of Figure 8-9B, obtain the direct chordal distance from the first station line a-a' to the second fold line b-b'. On the pattern baseline, draw the second fold line parallel to the first station line a-a' at the chordal distance found from the top view, and label it b-b', *Figure 8-9D*. Repeat these steps, alternating chordal lengths and true distances until all station lines have been located and drawn. Note that the last fold line length a-a' is a duplicate of the first.

Method B: The required length to form the cylinder's circumference can be calculated using the following formula: circumference = cylinder diameter x pi. (Pi = 3.1416, approximately.) The baseline is extended to this length, and divided into the same number of divisions as was done at Step 1 in the top view of the cylinder, Figure 8-9B. The division can best be performed as a construction exercise, rather than mathematically. Refer back to Chapter 3, "Geometric Construction." Each of the 13 division locations (from 12 spaces) is the correct location of the required station lines, and can be successively labeled.

Step 4 Add the true size and shape of the bottom surface (bottom view), and the top surface (auxiliary view). The attachment locations of these surfaces are particularly difficult, as the mathematical point of tangency is not easily isolated on the pattern. The top surface location is chosen to allow the widest access of cutting to this point. The bottom surface can be added to any of the fold lines. In order to locate the top surface, the distance X from center to point 0 is transferred from Figure 8-9B to the development, and projected 90° as illustrated in Figure 8-9D.

Step 5 Add the tabs necessary for edge attachments. In cylinders, a notched tab must be used in order

to accommodate crowding by surface curvature. A tab must also be added at fold line a-a' in order to attach edge a-a' to the other edge a-a'; the top (auxiliary) and bottom fold to meet the notched tabs, *Figure 8-9E*.

Regardless of the shape, whether it be a cylinder or a prism, parallel line developments are used with essentially the same procedure. *Figure 8-10* shows a prism truncated at both ends, which necessitates creating a new baseline other than a part edge to ensure that all fold lines (instead of station lines) are at 90° to the baseline. *Figure 8-11* is a cylindrical prism, truncated at each end, which requires the selection of station lines instead of using existing fold lines. *Figure 8-12* shows a series of truncated cylinders, similar in construction to the preceding views, whose patterns are used to form an elbow.

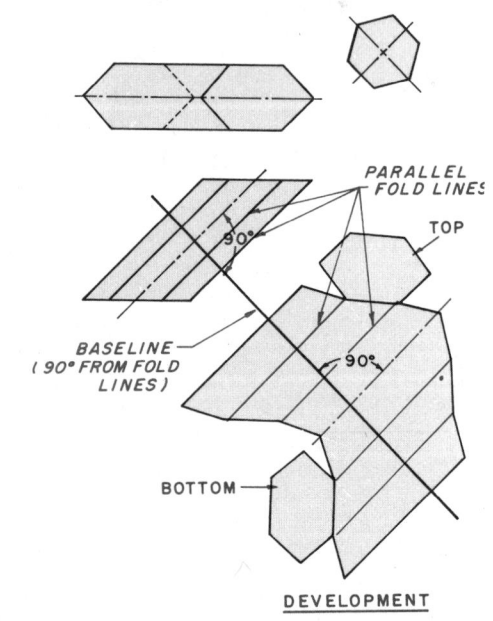

Figure 8-10 Creating a new baseline

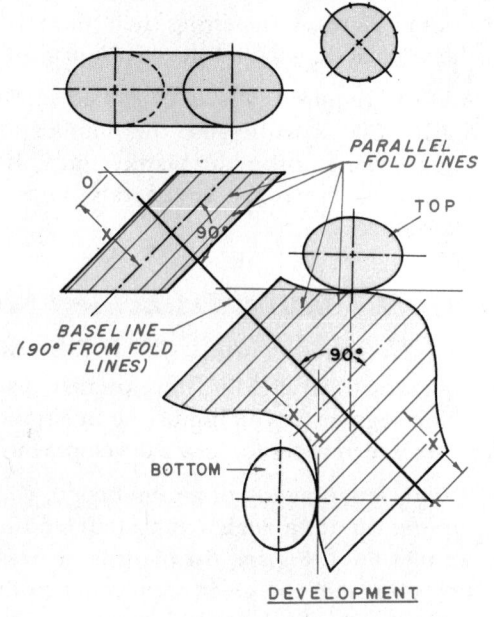

Figure 8-11 Using station lines instead of fold lines

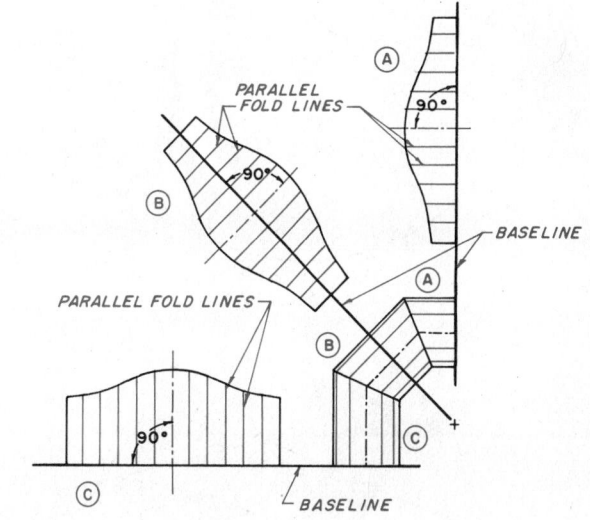

Figure 8-12 Series of patterns used to form an elbow

Radial Line Development

Radial line development is different from parallel line development in that all fold lines or line segments radiate from one point. (Review Figure 8-1.) As in all patterns, true lengths and true distances must be used in laying out the developments.

How To Develop a Truncated Pyramid Using Radial Line Development

Given: A two-view drawing, and an auxiliary view of a pyramid with the top cut off at an angle (truncated), *Figure 8-13A*. Label points clockwise

starting with the shortest fold line. In this example, either a-1 or b-2 could be used. Line a-1 is selected, but the fold line location a-1 is also the meeting corner of edges a-1 and a-1. True distances between the end points of the fold lines are evident in the top view. For example, a-b, b-c, c-d, and d-a are true length.

Step 1 To find the true length of the fold lines, they must be rotated to a position that is parallel to the frontal viewing plane, *Figure 8-13B*. In the top view, point a is rotated as shown, and projected into the front view. This revolved line from a' to 0 is its true length. Similarly, point c in the top view is rotated and projected into the front view, giving the true length of c'-0. The

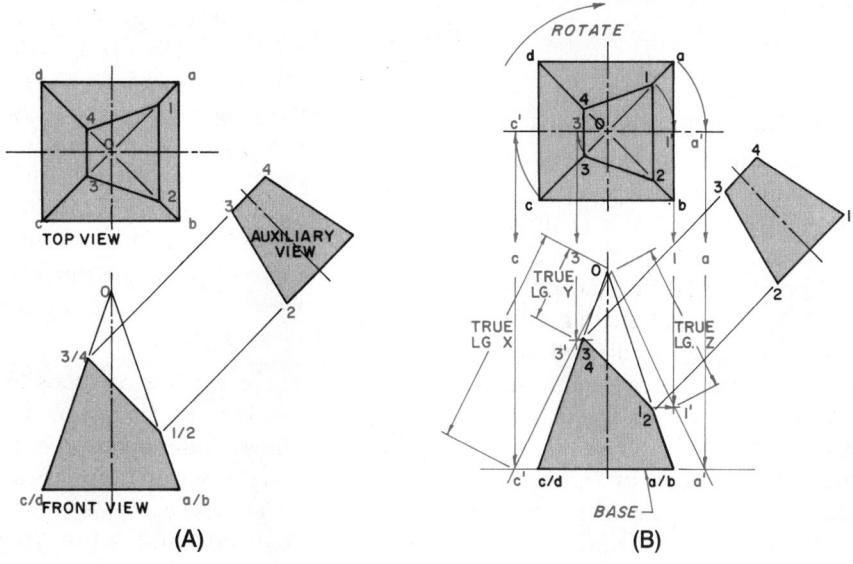

(A)   (B)

Figure 8-13 (A) Development of a pyramid using radial line development; (B) Step 1

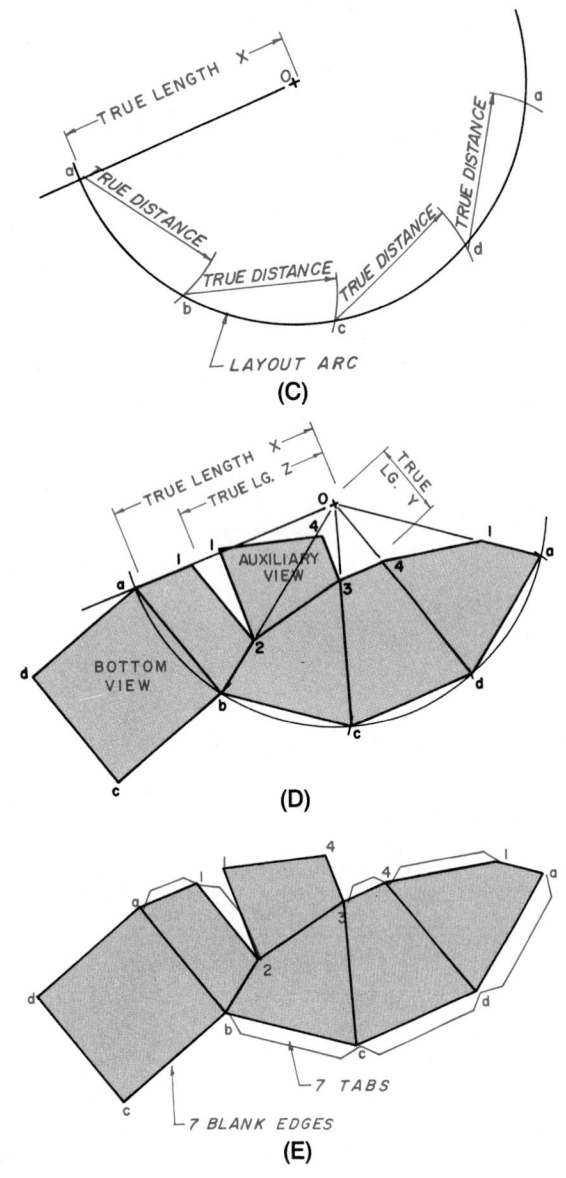

Figure 8-13 (Continued) (C) Step 2; (D) Step 3;
(E) Step 4 (completed development of a pyramid)

bottom views, transferring them directly from the original drawing 8-13A. See *Figure 8-13D*.

Step 4 Add the required tabs as illustrated in *Figure 8-13E*. Check to be sure the number of tabs equals the number of blank edges. In this example, there are seven tabs and seven blank spaces.

How To Develop a Truncated Cone Using Radial Line Development

As with cylindrical parallel line developments, station lines are line segments which must be positioned on the lateral surface in order to draw a development.

Given: A two-view drawing of a round cone, with the top cut off at an angle (truncated), and having no fold lines, *Figure 8-14A*. It is necessary to first complete the given views, and to draw a true view of the inclined flat surface. To project the flat surface to the top view, the front view is sliced at points A through G. These seven slices are projected into the top view and appear as circles, correspondingly labeled A through G. The points of intersection of the edge view with each slice is projected into the top view. To construct the auxiliary view, project from the seven point intersections perpendicular from the edge view of the flat surface to any convenient distance for the auxiliary view. Draw and use the center line as a reference line, and transfer all point-to-center line distances from the top view to the auxiliary view; refer to reference distances X.

Step 1 To construct a development of this object, station lines need to be established, and their true lengths and true distances must be determined. In the top view of *Figure 8-14B*, divide the base circle into 12 equal parts, and label each point. In this example, numbers 1 through 12 are arbitrarily used, positioned as illustrated. The station line 0-3 must have two identities as two edges will meet at this point. The true distances between station lines are located between their end points on the cone base, seen in the top view. Project the 12 points down into the base of the front view. These are the end points of the station lines in the front view. Draw a line from each of these end points up to point 0.

Step 2 There are two methods of drawing the pattern outline. Each begins with swinging an arc whose radius is the true length of the lateral distance from the apex to the base, or length 0-3 or 0-9. The compass radius can be set to this distance on Figure 8-14B. This distance can also be derived mathematically if the diameter of the base of the cone is known, and the altitude

same procedure is used to find the true lengths 3'-0 and 1'-0. To construct a development of this object, continue with the following steps.

Step 2 Using the true length from 0 to a (a'-0), swing the arc as shown in *Figure 8-13C*. On this arc, mark off as chord lengths the true distances between the fold line end points a to b, b to c, c to d, and d to e, as transferred from the top view of Figure 8-13B. Label all points as shown. Connect the fold lines a to 0, b to 0, c to 0, d to 0, and a to 0.

Step 3 Using the true lengths found in Figure 8-13B, locate points 1, 2, 3, 4, and 1 on the respective fold lines 0-a, 0-b, 0-c, and 0-a. Note that 0-1 is the same length as both 0-2 and 0-1, and 0-3 is the same length as 0-4. Add the auxiliary and

(perpendicular distance from base to apex) is known. The formula is:

RADIUS = THE SQUARE ROOT OF [(½ CONE BASE DIAMETER)^2 + (ALTITUDE)^2]

After swinging this arc, choose Method A, which approximates the cone as a pyramid, or Method B, which is mathematically correct.

Method A: In the top view of Figure 8-14B, the direct chordal distance between any two station line end points, say 5 and 6 (they are all equal), is an approximation of the arc length between them. On the 0-3 radius just drawn, strike off this chordal distance between station line end points repeatedly to equal the same number of spaces on the arc as the top view of the cone was divided into in Step 1, and ending at point 3, *Figure 8-14C*. Connect the station line end points from point 0 to each of these arc intersections, and proceed to Step 3.

Method B: The central angle of the pattern is determined by a pattern arc length needed to equal the circumference of the base of the cone. The ratio of the central angle (A°) to a full-circle 360° is the same as the circumference of the base of the cone is to the full circumference generated by the pattern's radius.

$$\frac{A°}{360°} = \frac{\text{CIRCUMFERENCE OF BASE OF CONE}}{\text{CIRCUMFERENCE OF PATTERN}}$$

$$\frac{A°}{360°} = \frac{\text{PI} \times \text{CONE DIAMETER AT BASE}}{\text{PI} \times \text{PATTERN DIAMETER}}$$

$$\frac{A°}{360°} = \frac{\text{CONE BASE DIAMETER}}{2 \times \text{PATTERN RADIUS}}$$

$$A° = \frac{180 \times \text{CONE BASE DIAMETER}}{\text{PATTERN RADIUS (A'} - 0)}$$

After the central pattern radius A° is found, the angle between successive station lines (a°) is found by dividing the same number of conic divisions done in Step 1 into A°. The chordal lengths (D) of these divisions is found by using the formula:

D = 2 x PATTERN RADIUS x [SIN (½ A°)]

This chordal distance can then be struck off on the pattern radius to form the same number of spaces as the conic divisions in Step 1, and labeled accordingly. See Figure 8-14C. Connect these station line end points from point 0 to each of the intersections, and proceed to Step 3.

Step 3 To draw the edge of intersection that the cone's lateral surfaces makes with the upper flat surface, true lengths from the apex to this edge at each station line must be determined. Only 3'-0 and 9'-0 are visible as true lengths in the front view of Figure 8-14B. To find the other true lengths, each segment must be rotated to a posi-

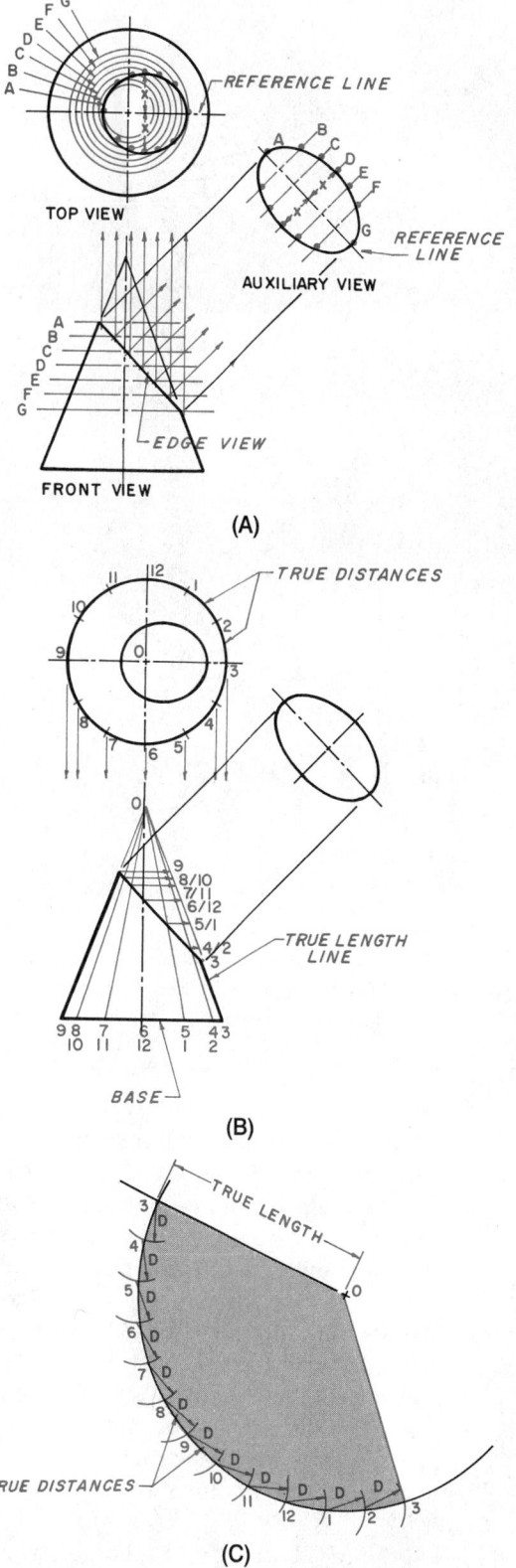

Figure 8-14 (A) Development of a truncated cone using radial line development; (B) Step 1; (C) Step 2

tion parallel to the frontal viewing plane; in this example, on line 3-0. Project the intersection of each station line at the truncated surface edge to line 3-0 on the cone profile in the front view (on line 3-0).

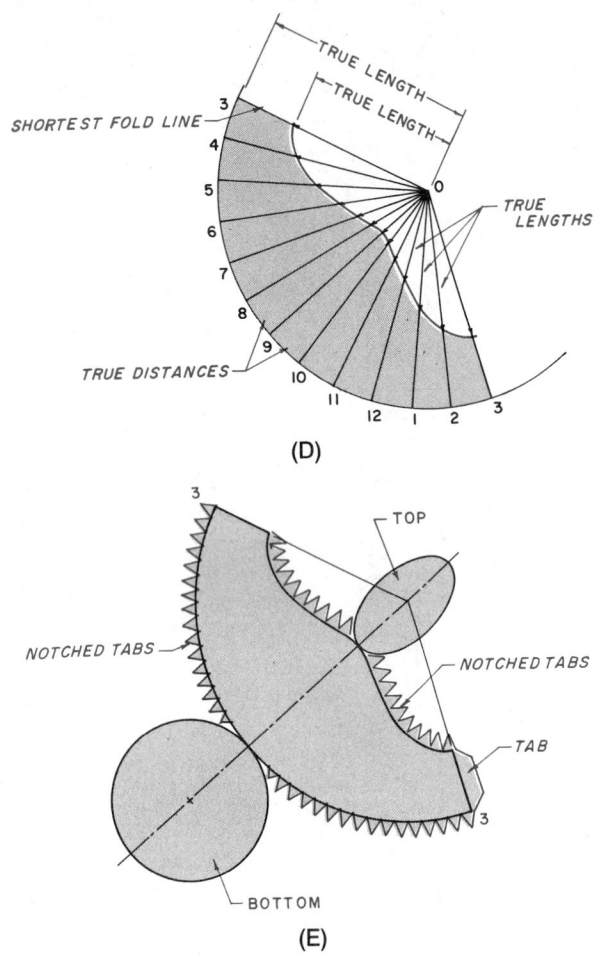

(D)

(E)

Figure 8-14 *(Continued)* (D) Step 3; (E) Step 4 (completed development of a cone)

Starting with the shortest line segments of the cone, swing true lengths arc 0-3' on the corresponding pattern station lines, as illustrated in ***Figure 8-14D***. Swing the apex-to-edge true length found for each station point, at the corresponding station line on the pattern. Repeat for all 13 stations, 3 through 12 and back to 3. Label each point as illustrated.

Draw light line segments between each of the station line intersection positions found on the corresponding pattern station lines. This completes the true contour of the top edge, Figure 8-14D.

Step 4 Add the top, bottom, and split tabs, as illustrated in ***Figure 8-14E***.

Triangulation Development

Triangulation is the third major method used to lay out a surface development. *Triangulation* is a method of

dividing a surface into a number of triangles and then transferring each triangle's true size and shape to the development. (Review Figure 8-2.) As with parallel line and radial line developments, *true lengths* and *true distances* must be used exclusively in pattern constructions.

True-Length Diagram

Before any layouts can be started, true lengths and true distances of the object's boundary edges must be determined. A true-length diagram is usually used to develop true lengths and true distances of these edges. A true-length diagram is often a more rapid method to obtain the needed projections than by descriptive geometry methods.

How To Develop a Transitional Piece Using Triangulation Development

Given: ***Figure 8-15A*** describes a transitional piece using a top view, front view, and isometric view. Each of the object's corners are labeled as illustrated. Note that the true distances A to B, B to C, C to D, D to A, 1 to 2, 2 to 3, 3 to 4, and 4 to 1 can be measured directly from the top view. As drawn, the lengths A-1, B-1, B-2, C-2, C-3, D-3, D-4, and A-4 are not shown in their true lengths. A true-length diagram must be used to find these.

Step 1 A true-length diagram is a combination of two views. In this example, it is a combination of the top view and the front view, ***Figure 8-15B***. The height between points A and 1 is shown projected from the front view to the true-length diagram. The distance X between corresponding line end points A and 1 is found in the top view, and transferred to the true-length diagram. The illustrated diagonal line, drawn in the true-length diagram, is the true length of A-1. As all the heights and top-view distances between end points are the same for each of the lateral edges, then A-1 is also the same length as B-1, B-2, C-2, C-3, D-3, D-4, and A-4. These true lengths are needed to lay out the development.

Step 2 Starting from line A-1 and using true lengths and true distances, reconstruct each triangle representing a surface of the object, as illustrated in ***Figure 8-15C***. (Review Chapter 2 for aid in transferring a triangle.) Notice in this example that all true distances are transferred from the top view, and all true distances are transferred from the true-length diagram. A tab is added to join A-1 to A'-1'. This completes the transitional piece development as drawn in Figure 8-15A.

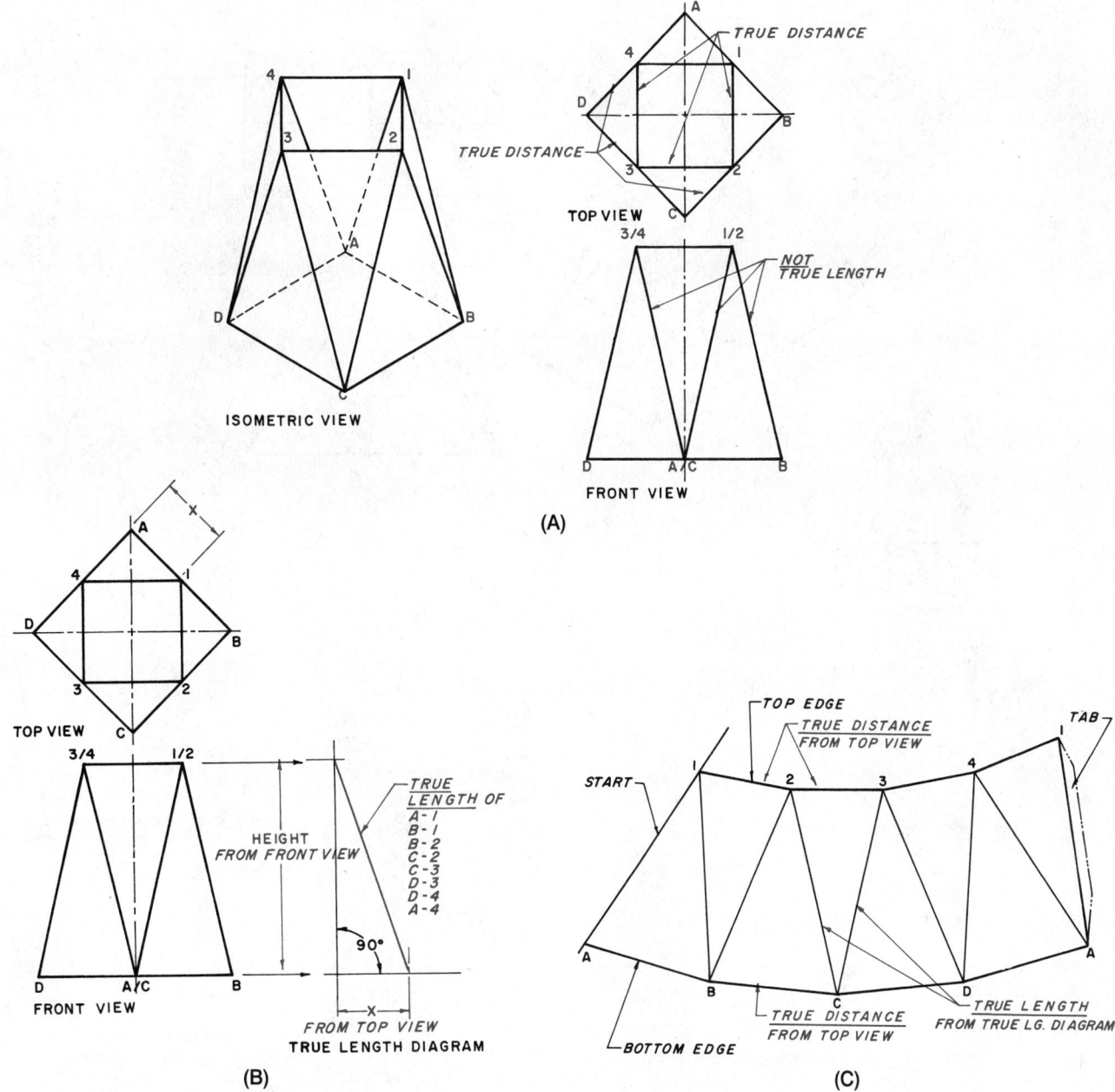

Figure 8-15 (A) Development of a transitional piece using triangulation development; (B) Step 1 (true-length diagram); (C) Step 2

How To Develop a Transitional Piece with a Round End Using Triangular Development

A transitional piece that is square or rectangular at one end and round at the other is laid out using a procedure very similar to the one outlined previously. The only difference is that station lines must be positioned on the lateral surface of the round-to-corner transitions. *Figure 8-16A* illustrates a two-view transitional piece with a round top end and a rectangular bottom end. In this example, the top end is divided into 12 equal spaces. Each point is numbered clockwise as illustrated.

The bottom four corners making up the bottom rectangle are labeled A through D, *Figure 8-16B*. The true distances from A-B, B-C, C-D, and D-A' are found in the top view. Station lines W, X, Y, and Z are not true lengths, but the true lengths must be determined with a true-length diagram. Note that in the true-length diagram, the true length of line W, from A' to 0, is also the same lengths for B-6, C-6, and D-12. Lines X, Y, and Z are also used four times at corresponding locations. Segment the object into triangles and develop the pattern using true lengths and true distances to reconstruct each adjoining segment, *Figure 8-16C*.

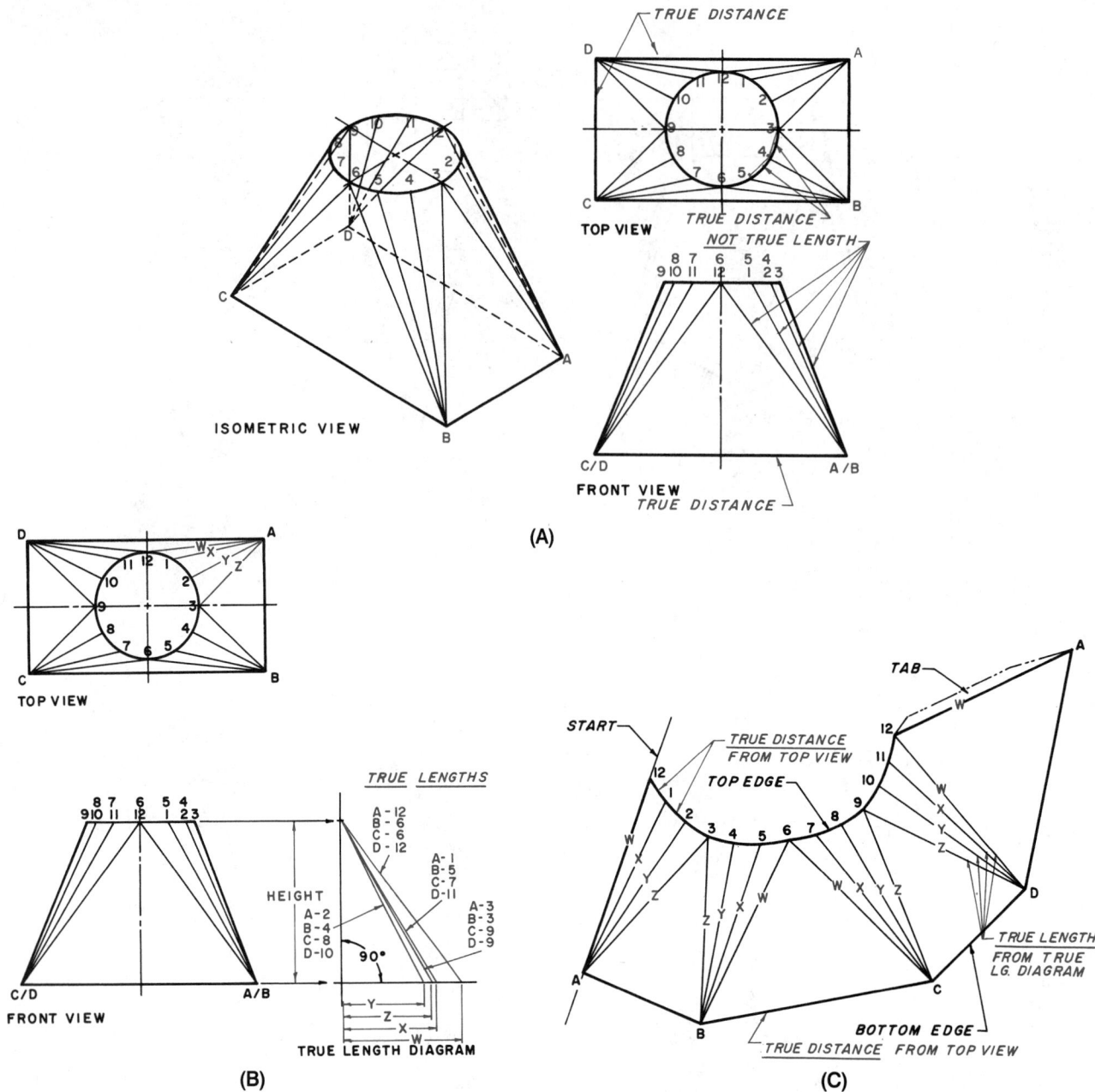

Figure 8-16 (A) Development of a transitional piece with a round end using triangulation; (B) True-length diagram; (C) Completed development of a transitional piece with a round end

How To Develop a Transitional Piece with No Fold Lines Using Triangular Development

Figure 8-17A shows a transitional piece with no evident fold lines or line segments. It is developed in the same manner as any other triangulation development.

Step 1   Divide the top and bottom surfaces into equal spaces as illustrated in **Figure 8-17B**. In this example, the top edge is divided into 12 equal spaces, 0 through 12. The bottom edge is divided into the exact same number of equal spaces, and lettered A through L to A.

Step 2   Connect points 12 to A, 1 to B, 2 to C, and con-

secutively to points L to 11 with solid lines, as illustrated in **Figure 8-17C**. Dash lines are added to segment the object into various triangles.

Step 3   True distances from 1 to 2, 2 to 3, 3 to 4, and so on through 11 to 12, and the true distances from A to B, B to C through to L to A' are found in the top view, Figure 8-17C. The other fold lines, connecting the top and bottom are not true lengths and, therefore, true-length diagrams must be made, one for the solid lines and the other for the dash lines, **Figure 8-17D**. The development is laid out by constructing connecting triangles using the lengths of each leg of each triangle, **Figure 8-17E**.

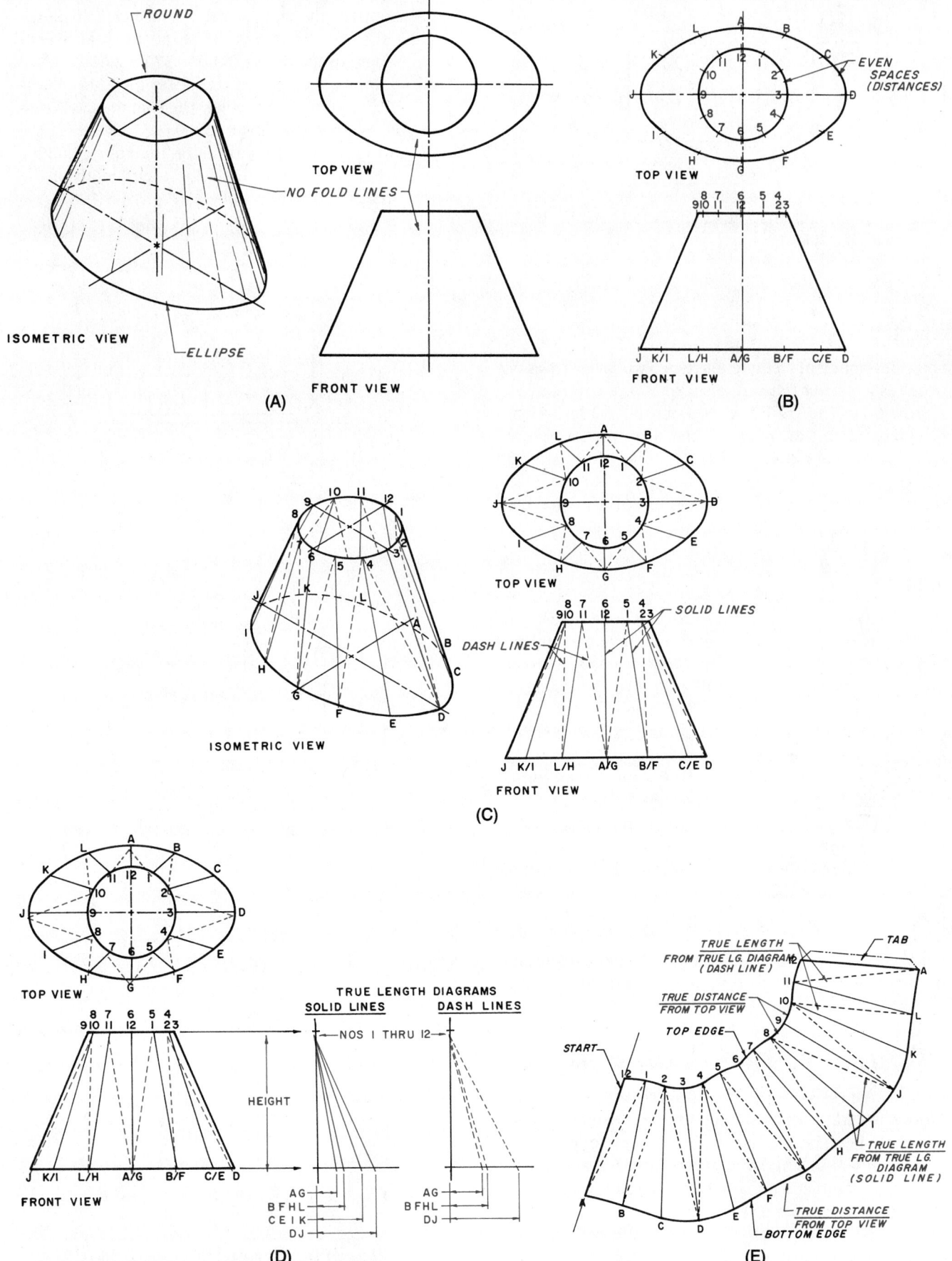

Figure 8-17 (A) Development of a transitional piece with no fold lines; (B) Step 1; (C) Step 2; (D) Step 3 (true-length diagram); (E) Completed development of a transitional piece

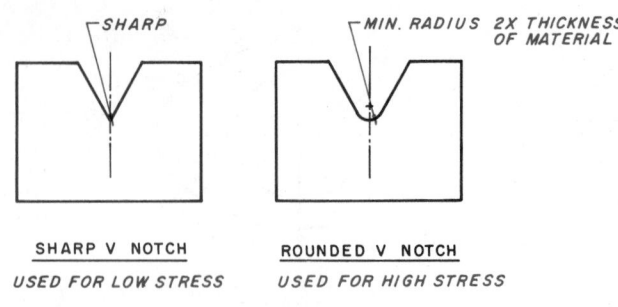

Figure 8-18   Notches

Notches

Some developments require notches. Two major types of notches are usually used: a sharp V or a rounded V, **Figure 8-18**. The sharp V is used only if a minimal force is tending to part the material. The sharp point of the V will tear or crack under stress. The rounded V is used where parts would be under greater stress. The radius of the vertex of the rounded V should be at least twice the thickness of the metal used, and larger if possible.

Bends

When bending sheet metal to form a corner, rib, or design, the outer surface stretches and the inner surface compresses. The length of material needed by each bend must be calculated and added to the straight unbent portions of the pattern. The material needed by each bend is the length of the neutral axis within the material, where neither compression nor stretching occurs. This is calculated either by formulas or by using various charts available for this purpose.

Bend allowance charts are included in the Appendix of this text, and are much faster to use than a formula. Two basic kinds of charts are used to calculate bend allowance, in both the English and metric systems. One type is used for 90° bends; the other for bends from 1° through 180°. The total length of a pattern is called the *developed length*. To determine the developed length, the stretched-out flat pattern must include all straight sides, plus the calculated bend allowances.

How To Find All Straight Sides of a 90° Bend

Figure 8-19A shows a simple 90° bend with an inside radius of .25, a sheet metal thickness of .125, and legs of 2.0 and 3.0 (the English system inch is used in this example). Bend radii are always measured from the surface closest to the bend radius center.

Step 1   Locate the tangent points at the ends of the straight sides.

Step 2   Add the thickness of the sheet metal to the bend radius .25 (.125 + .25 = .375).

Step 3   Subtract the sum of the sheet metal thickness .125 and the radius .25 from the 3.00 overall length of the object (3.00 – .375 = 2.625).

Step 4   Subtract the sum of the sheet metal thickness .125 and the bend radius .25 from the 2.00 overall height of the object (2.00 – .375 = 1.625).

Step 5   Add the two straight sides together (2.625 + 1.625 = 4.250). This is the total length of the straight sides of this object, **Figure 8-19B**.

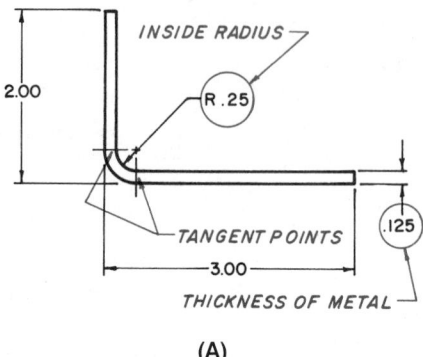

(A)

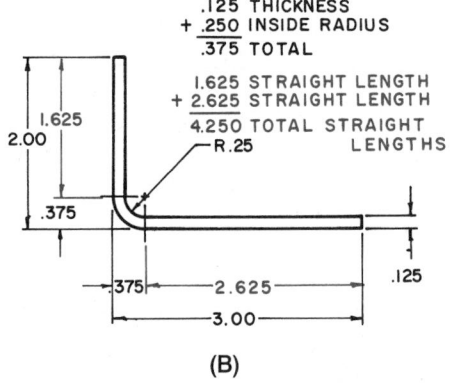

(B)

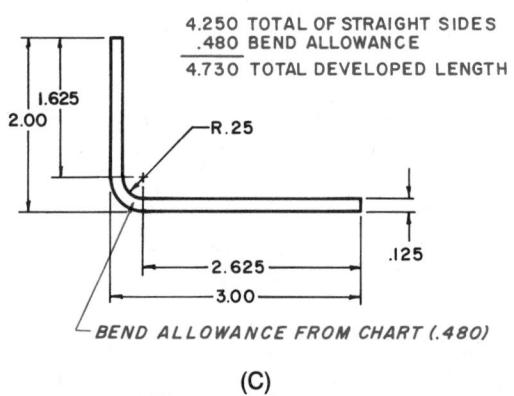

(C)

Figure 8-19   (A) Bend allowance of 90° bend;   (B) Total of straight lengths;   (C) Total length including straight lengths and bend allowance

How To Find the Bend Allowance of a 90° Bend

Review Figure 8-19A.

Step 1 Note the metal thickness, in this example .125, and the inside radius, in this example .25.

Step 2 Refer to the bend allowance chart in inches for 90° bends in the Appendix at the end of this text.

Step 3 The left-hand column gives various metal thicknesses. Go down the left-hand column until the required size or the closest size is found. In this example .125.

Step 4 Along the top of the chart is listed various inside radii; go across the top of the chart to the required size or the closest radius, in this example .25.

Step 5 From the .125 number in the left-hand column project across to the right; from the .25 number along the top of the chart, project down to where the two columns intersect. Given is the bend allowance for material .125 thick with a .25 radius. In this example the bend allowance is .480, **Figure 8-19C**.

Step 6 The stretched-out dimension is found by adding the straight sides to the bend allowance; in this example, 4.250 + .480 = 4.730 (see Figure 8-19C).

How To Find the Total Straight-Side Length Adjoining a Bend Other Than 90°

Figure 8-20A shows a simple 30° bend with an inside radius of 6.35 and a sheet metal thickness of 3.175 (the metric system is used in this example).

Step 1 Locate the tangent points at the ends of the straight sides.

Step 2 Determine the length of the straight sides and add them together (64.0 + 50.0 = 114.0). This is the total length of the straight sides of this object.

How To Find the Bend Allowance of Other Than a 90° Bend

Review Figure 8-20A.

Step 1 Note the metal thickness, in this example 3.175, and the inside radius, in this case R6.35.

Step 2 Refer to the bend allowance chart in millimetres for 1° bends in the Appendix at the end of this text.

Step 3 The left-hand column gives various metal thicknesses. Go down the left-hand column until the required size or the closest size is found. In this example 3.175.

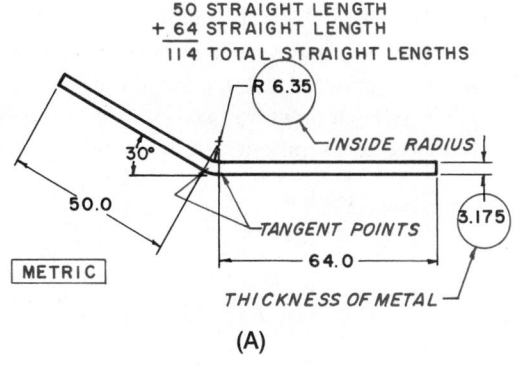

(A)

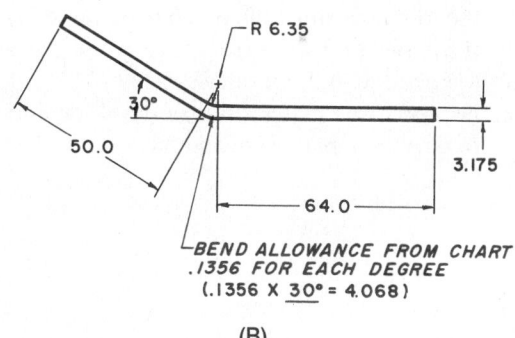

(B)

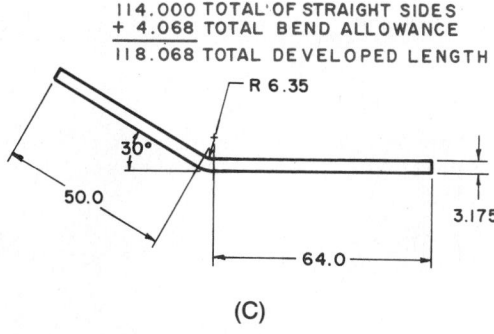

(C)

Figure 8-20 (A) Bend allowance of a bend other than 90°—total of straight lengths; (B) Total length of bend allowance; (C) Total length including straight lengths and bend allowance

Step 4 Along the top of the chart is listed various inside radii. Go across the top of the chart to the required size or closest radius, in this example 6.35.

Step 5 From the 3.175 number of the left-hand column, project across to the right; from the 6.35 number along the top of the chart, project down to where the two columns intersect. Given is the factor used to calculate the bend allowance. In this example .1356.

Step 6 Multiply this factor times the actual degrees in the bend, from a straight 180° line, *Figure 8-20B*. This is the bend allowance (.1356 x 30° = 4.068).

Step 7 Add the straight-side lengths to the bend allowance to get the total developed length: 114.0 + 4.068 = 118.068, *Figure 8-20C*.

Note: A 30° bend dimension as illustrated in *Figure 8-21A* is actually 150° (180° – 30° = 150°). See *Figure 8-21B*. The total bend must be calculated from a straight piece that is actually 180° before bending.

Review

1. What is a transitional piece and which kind of development would be used to develop a pattern?
2. What are the two major kinds of notches used in a development?
3. What does a gage number represent?
4. In parallel line development, at which angle to the fold lines must the baseline be projected?
5. Why is it important to develop the pattern or template with the inside surface up?
6. List the three major kinds of developments used to develop a pattern of an object.

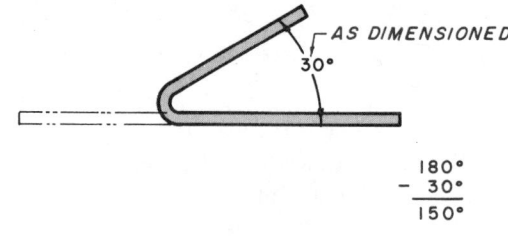

(A)

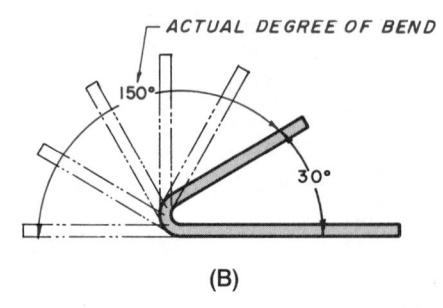

(B)

Figure 8-21   (A) Total bend must be calculated from straight line dimension—given is 30°; (B) As dimensioned 30° is actually a 150° bend from a straight line

7. Why are tabs used?
8. What is bend allowance, and why is it used?
9. Explain what must be done if a transitional piece does not have fold lines.
10. What is a true-length diagram and why is it used?
11. Why is the gage system of calling out the thickness of a material being phased out?
12. What two elements must be located or laid out before a development can be actually started?

Chapter Eight Problems

The following problems are intended to give the beginning drafter practice in sketching, laying out auxiliary views if required, and drawing the stretched-out development of various objects. These problems will use one or more of the three standard methods of development: *parallel line developments*, *radial line developments* and *triangulation developments*. The student will also practice calculating developed lengths using various charts to determine full length before bending.

The steps to follow in laying out any object that is to be developed are:

Step 1   Study the problem carefully.

Step 2   Determine which method or methods of development will be used (parallel line/radial line/triangulation).

Step 3   On a scrap sheet of paper, draw to scale enough views as required to lay out *all* true lengths and true distances. Add the end view and auxiliary views, in scale, if necessary. Draw with a *sharp* 4-H lead, and work as accurately as possible.

Step 4   Label each point, if necessary, in order to keep track of progress.

Step 5   Check all lengths, and auxiliary views if included.

Step 6   Starting from a baseline, lightly lay out the development. Use *true lengths* and *true distances* for all measurements. (Be sure to *start* the seam at the *shortest* fold line possible.)

Step 7   Label each point, if necessary, in order to keep track of progress.

Step 8   Use *phantom lines* to represent all *fold lines*.

Step 9   Add tabs as required, and check to see that there are enough but without duplications.

Step 10   Recheck all true lengths and true distances.

Step 11   Darken in all lines.

Problems 8-1 through 8-18

Using the parallel line development method, develop each object starting from the given *seam*. Label each point clockwise; add ends and/or auxiliary views as necessary to develop a complete object. Add tabs .125 (3) x 45° to suit. Use phantom lines for all fold lines.

Extra assignment(s): Cut out the development(s) and glue or tape it together to prove its accuracy.

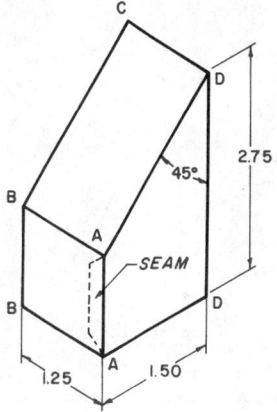

Problem 8-2

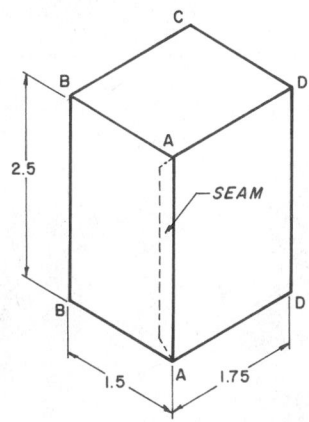

Problem 8-1

Problem 8-3

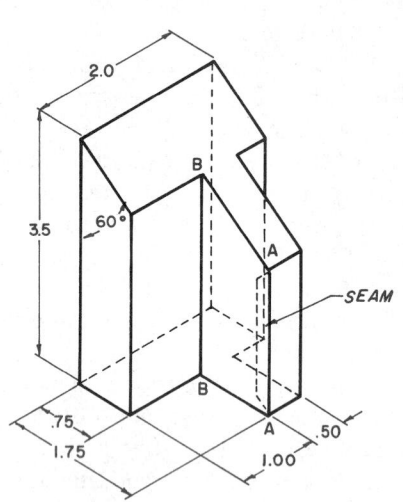

Problem 8-4

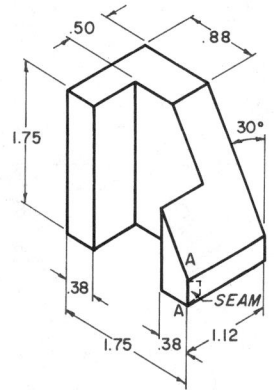

Problem 8-5

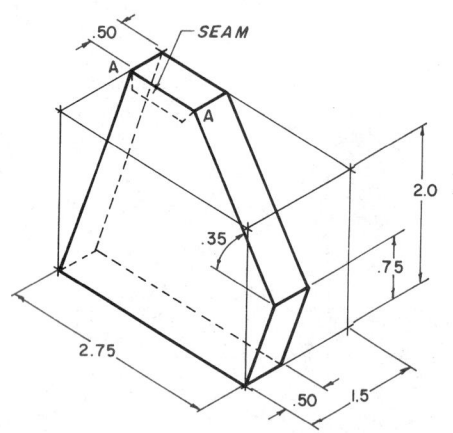

Problem 8-6

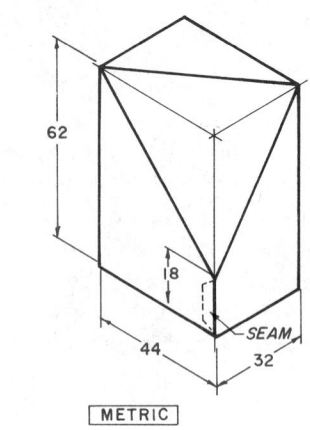

Problem 8-7

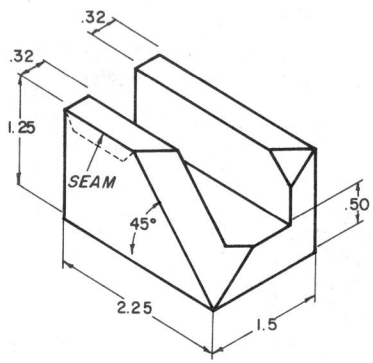

Problem 8-8

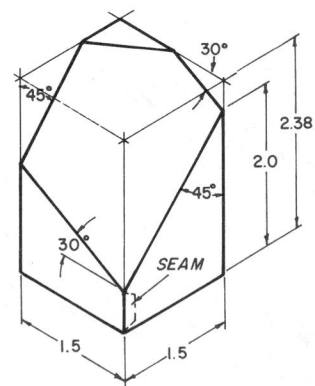

Problem 8-9

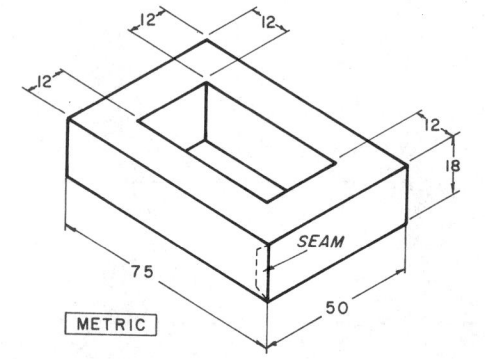

Problem 8-10

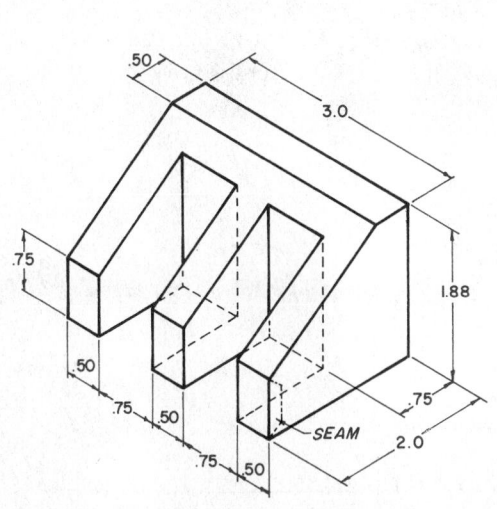

Problem 8-11

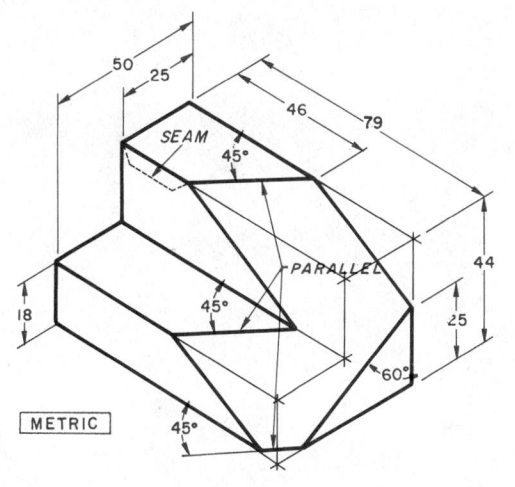

METRIC

Problem 8-12

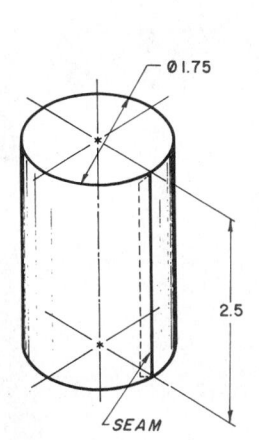

Problem 8-13

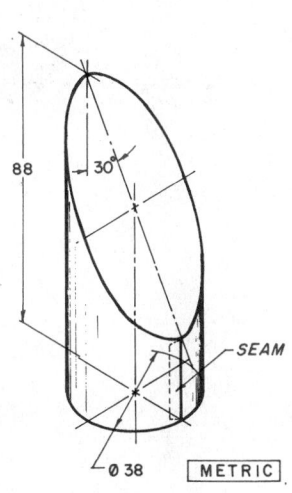

METRIC

Problem 8-14

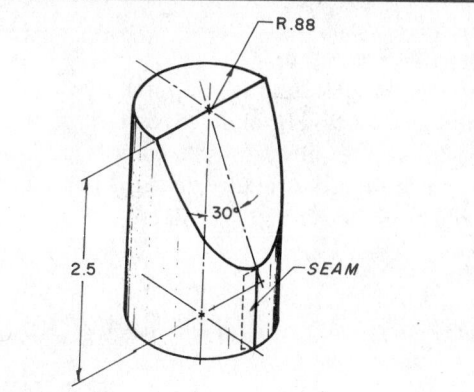

Problem 8-15

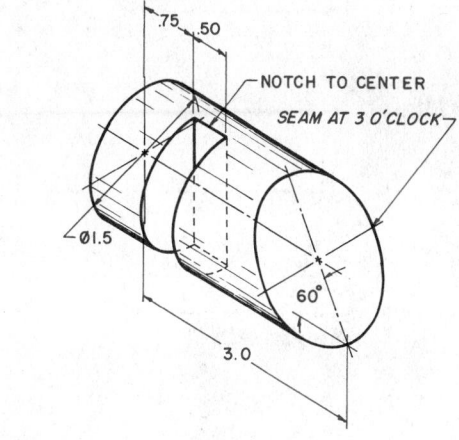

Problem 8-16

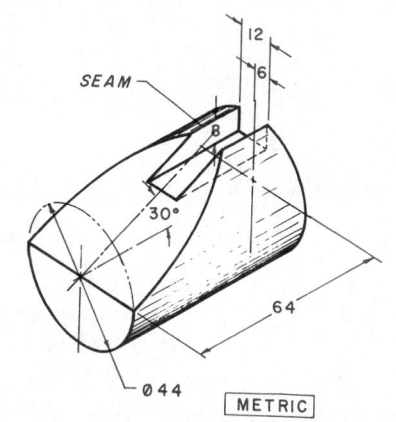

METRIC

Problem 8-17

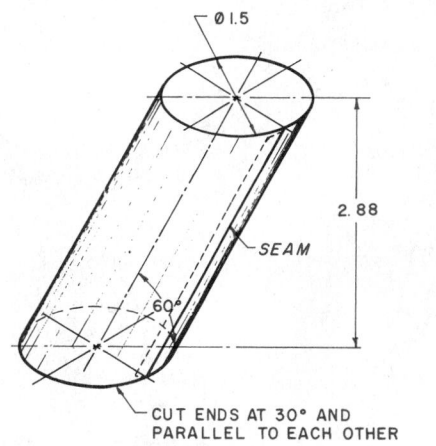

Problem 8-18

Problems 8-19 through 8-23

Using the parallel line development method, develop each object starting from the given *seams*. Add .125 (3) x 45° tabs as required to hold parts together. Design parts so there are as many identical parts as possible. Use phantom lines for all fold lines.

Extra assignment(s): Cut out the development(s) and glue or tape it together to prove its accuracy.

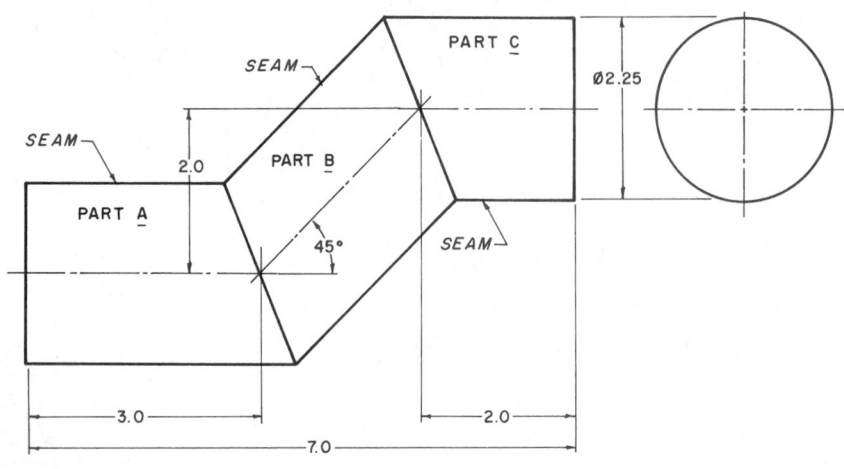

Problem 8-19

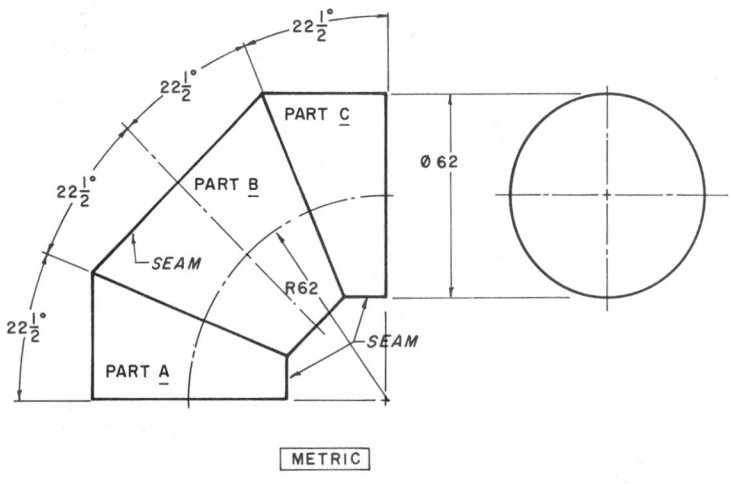

METRIC

Problem 8-20

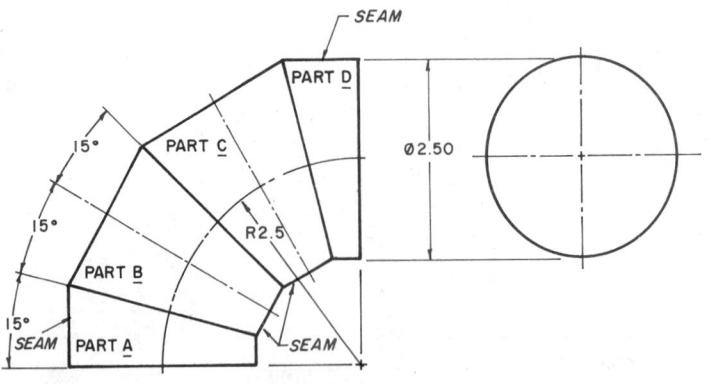

Problem 8-21

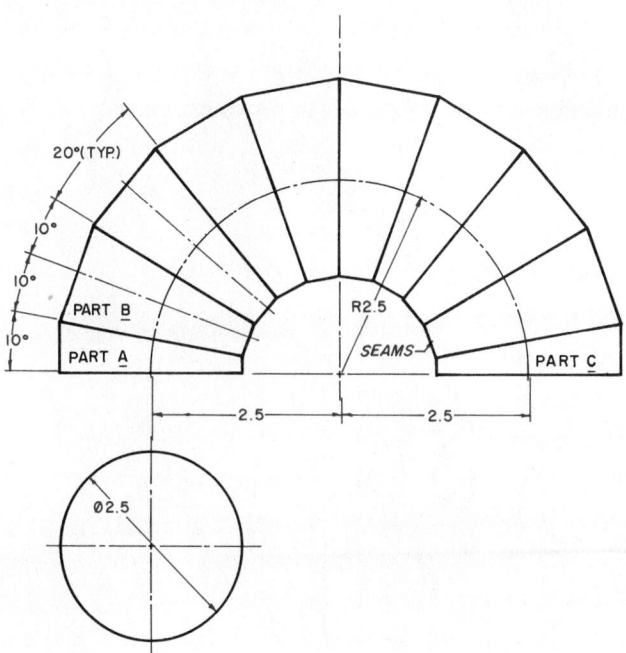

Problem 8-22

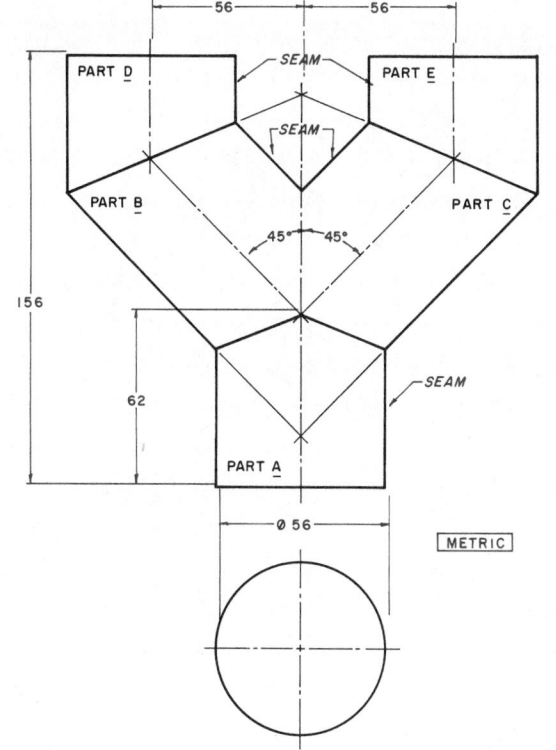

Problem 8-23

Problems 8-24 through 8-26

Develop the objects, using given dimensions. Using *bend allowance charts*, design layouts to include material for bends.

Extra assignment(s): Cut out the development(s) and glue or tape it together to prove its accuracy.

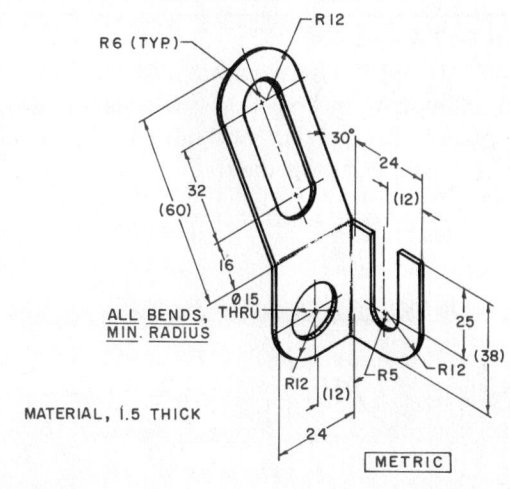

Problem 8-24

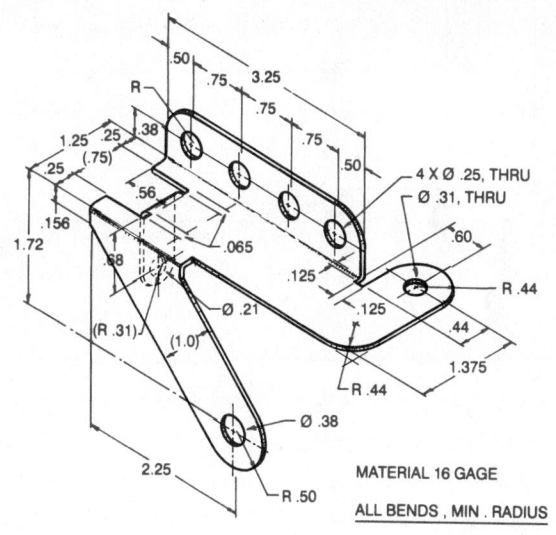

Problem 8-25

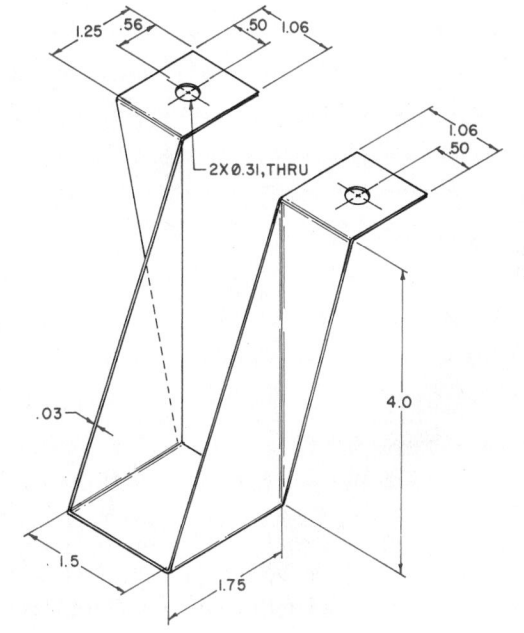

Problem 8-26

Problems 8-27 through 8-30

Carefully lay out a full-size multiview drawing of the object, include all intersection lines. Make a full-size development of the object. Add tabs .125 (3) x 45° as required to hold parts together.

Extra assignment(s): Cut out the development(s) and glue or tape it together to prove its accuracy.

ALL DIMENSIONS ARE TO OUTSIDE
LIMITS OF PARTS

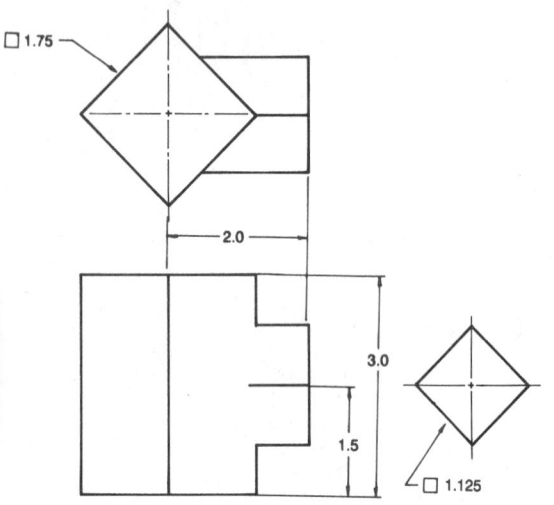

Problem 8-27

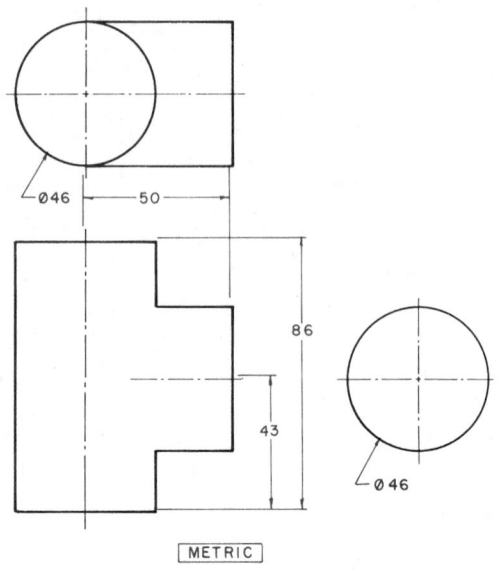

METRIC

Problem 8-29

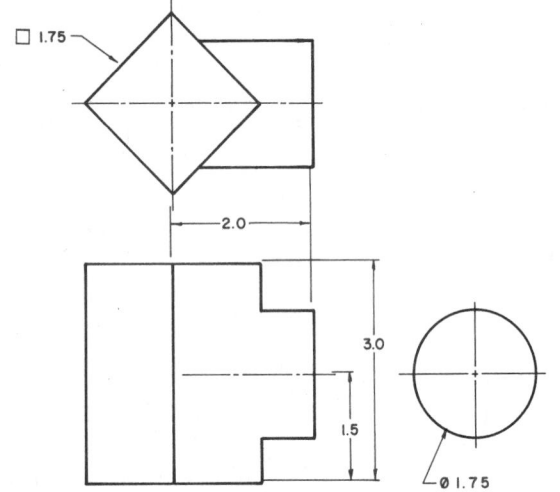

Problem 8-28

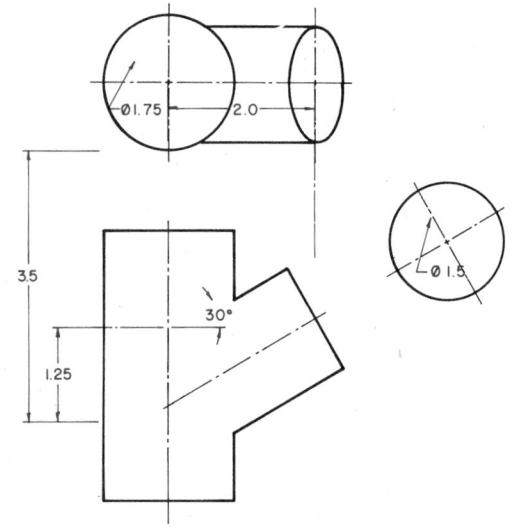

Problem 8-30

Problems 8-31 through 8-42

Using the radial line development method, develop each object starting from the given *seam*. Label each point clockwise; add ends and/or auxiliary views as necessary to develop a complete object. Add tabs .125 (3) x 45° to suit. Use phantom lines for all fold lines.

Extra assignment(s): Cut out the development(s) and glue or tape it together to prove its accuracy.

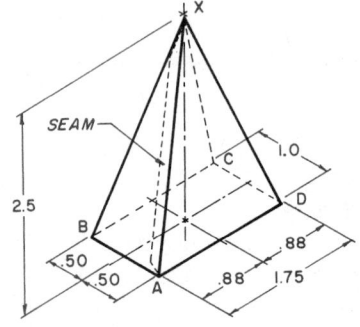

Problem 8-31

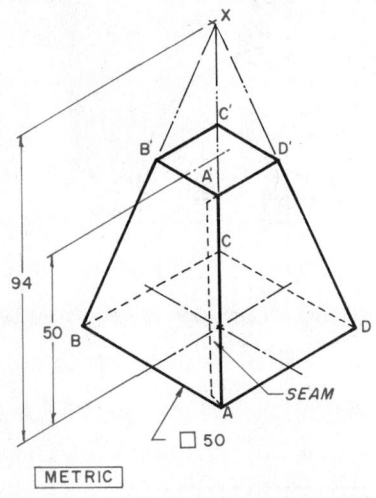

Problem 8-32

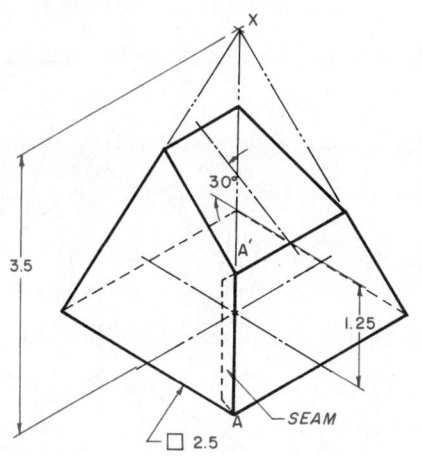

Problem 8-33

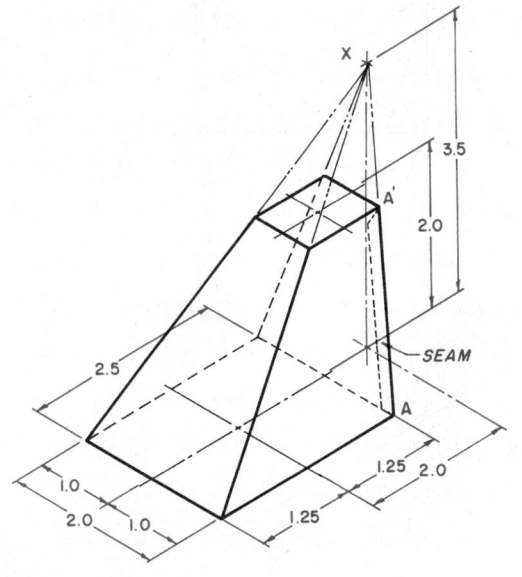

Problem 8-34

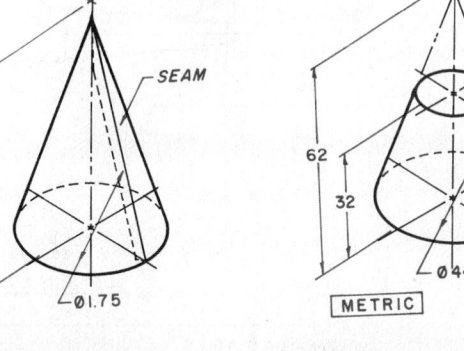

Problem 8-35　　　**Problem 8-36**

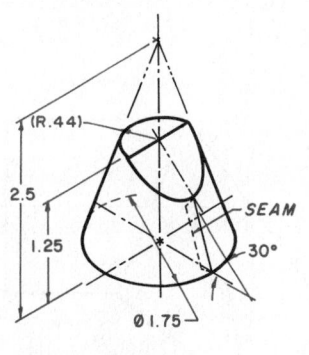

Problem 8-37

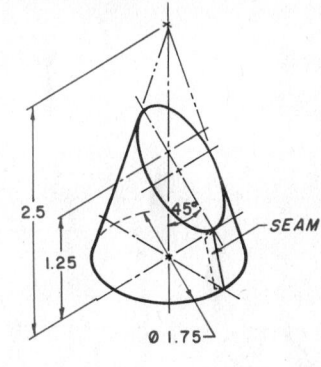

Problem 8-38

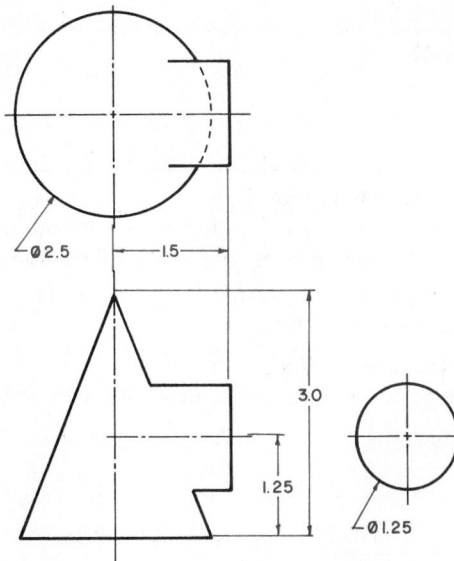

Problem 8-39

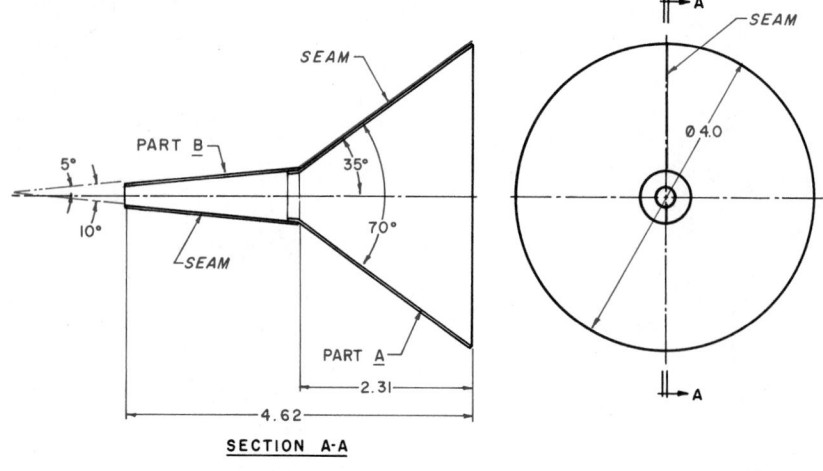

SECTION A-A

Problem 8-40

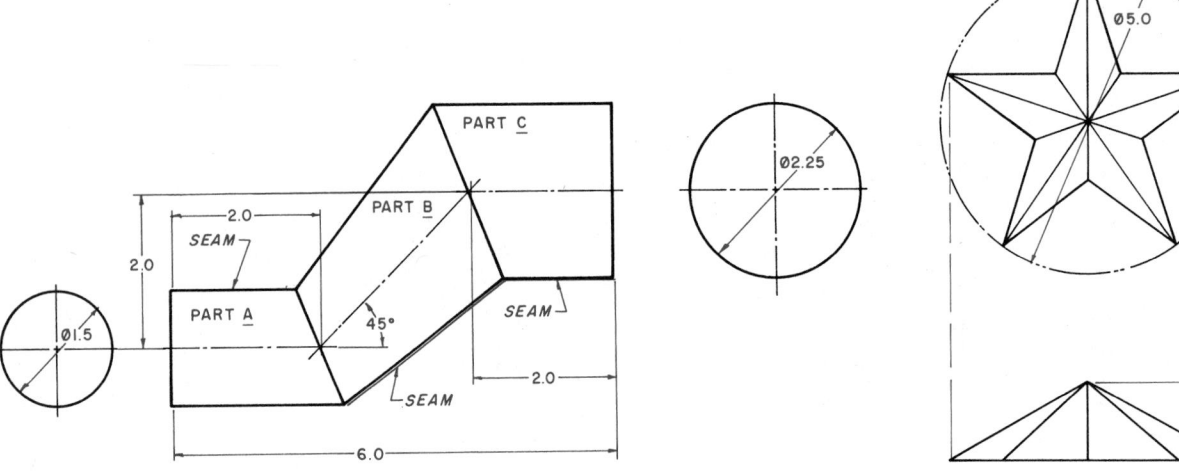

Problem 8-41

Problem 8-42

Problems 8-43 through 8-46

Using the triangulation development method, develop each object starting from the given *seam*. Label each point clockwise: add tabs .125 (3) x 45° to suit. Use phantom lines for all fold lines.

Extra assignment(s): Cut out the development(s) and glue or tape it together to prove its accuracy.

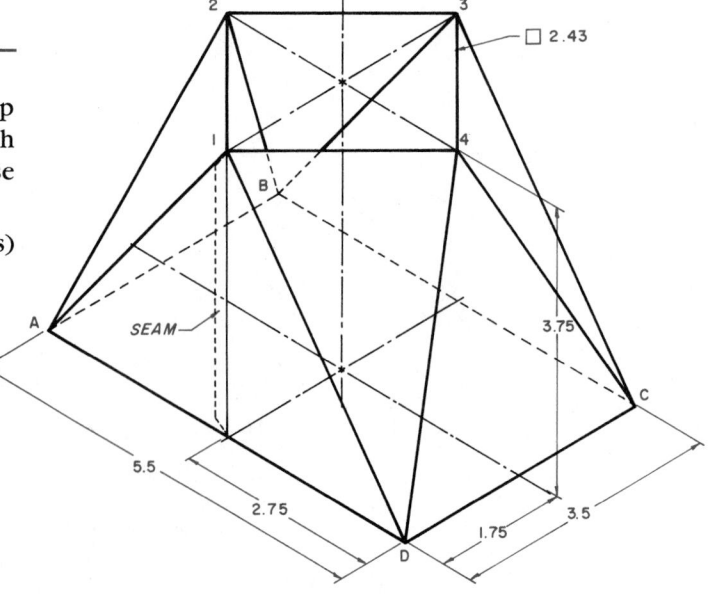

Problem 8-43

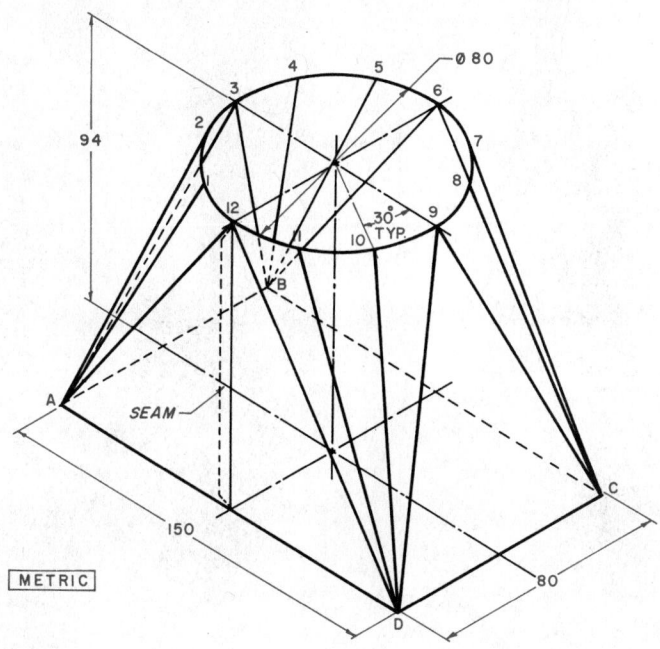

Problem 8-44

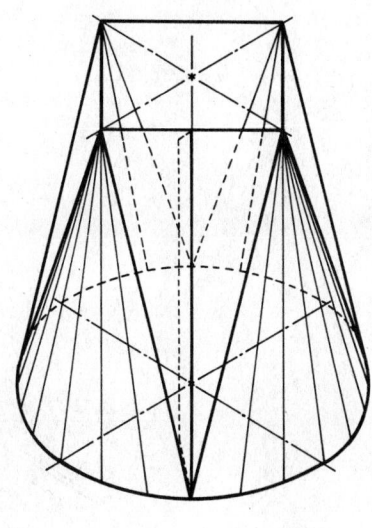

Problem 8-45

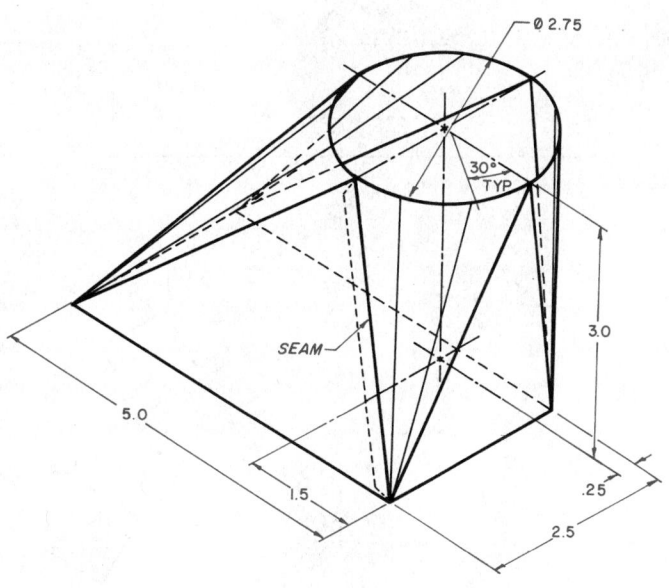

Problem 8-46

❏ Problems 8-47 through 8-49

Using development charts, calculate the true developed length of each object. Round the answer to the nearest three places. Recheck all calculations.

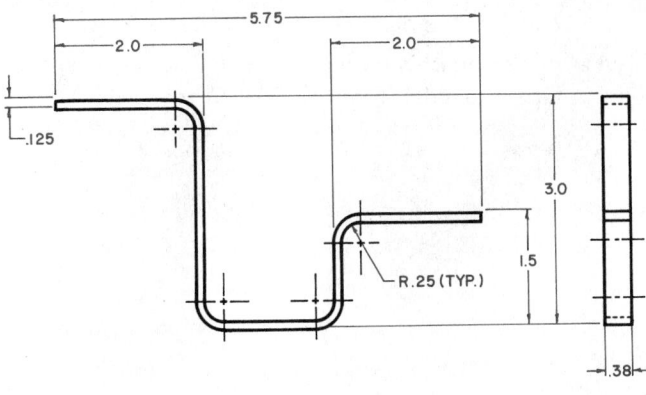

Problem 8-47

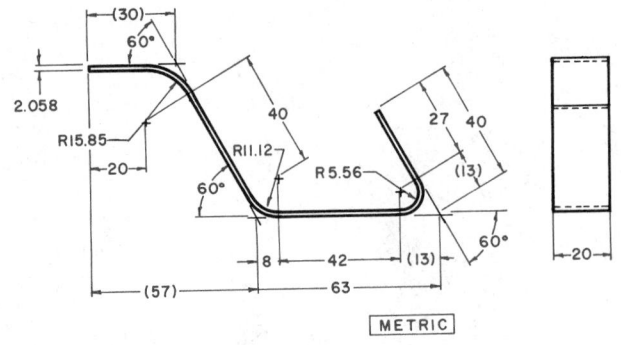

Problem 8-48

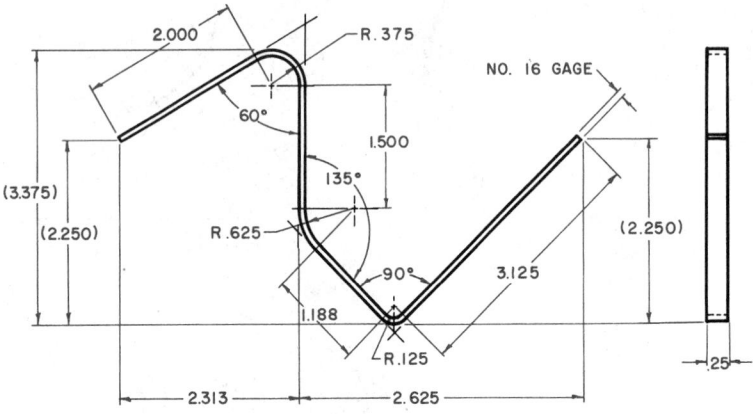

Problem 8-49

❏ Problems 8-50 through 8-54

Using either the parallel line, the radial line, or the triangulation development method, develop each object. Be sure to leave a seam and to add all required tabs to suit as required. Use phantom lines for all fold lines.

Extra assignment(s): cut out the development and glue or tape it together to prove its accuracy.

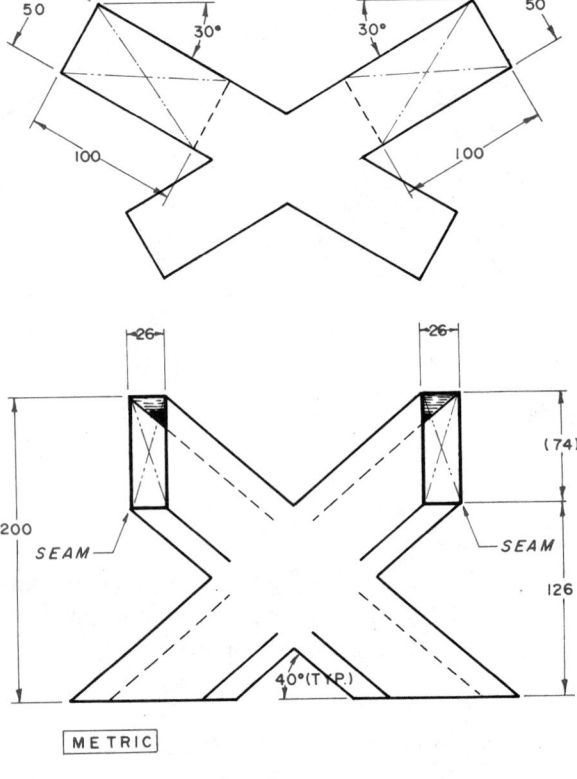

Problem 8-50

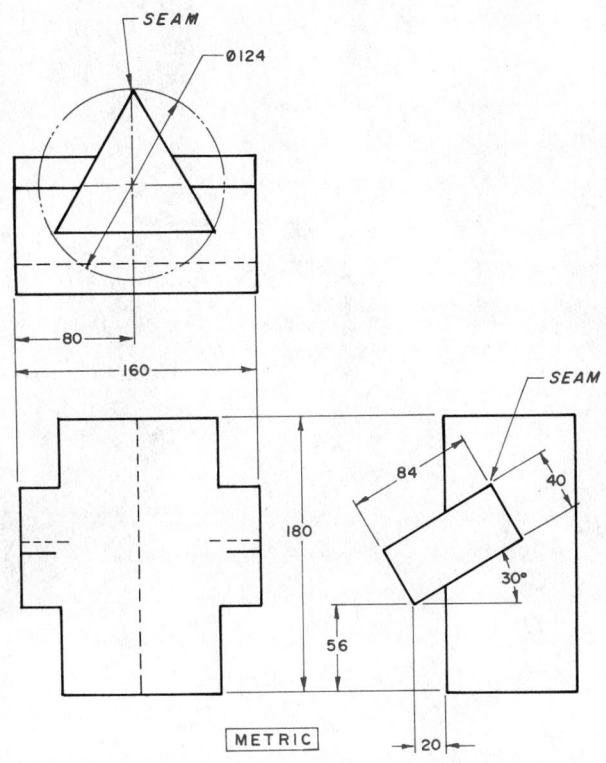

Problem 8-51

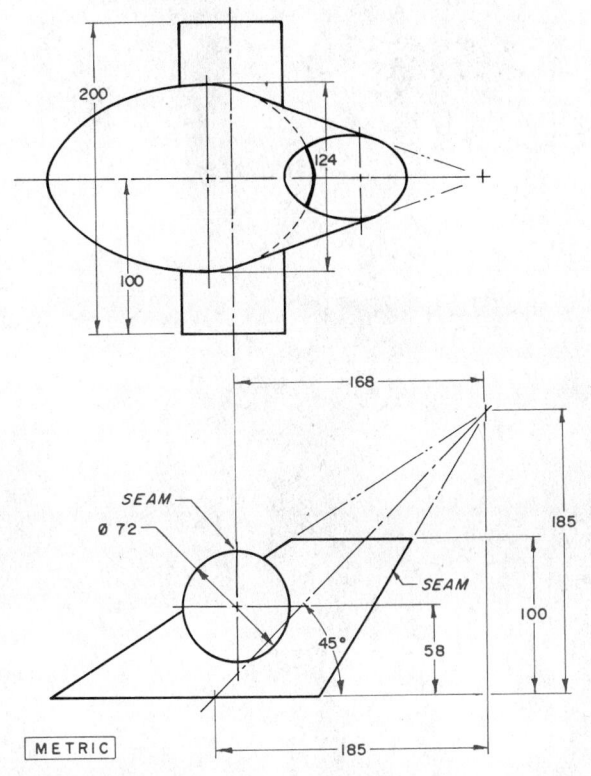

Problem 8-53

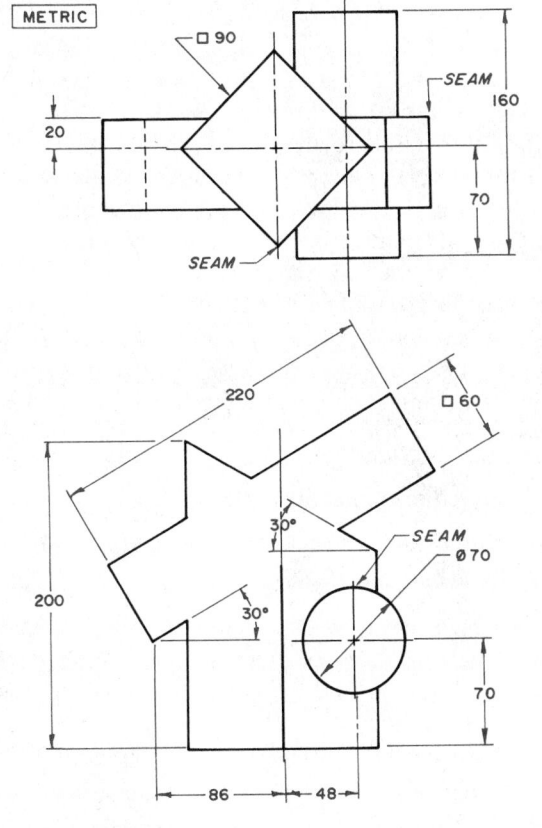

Problem 8-52

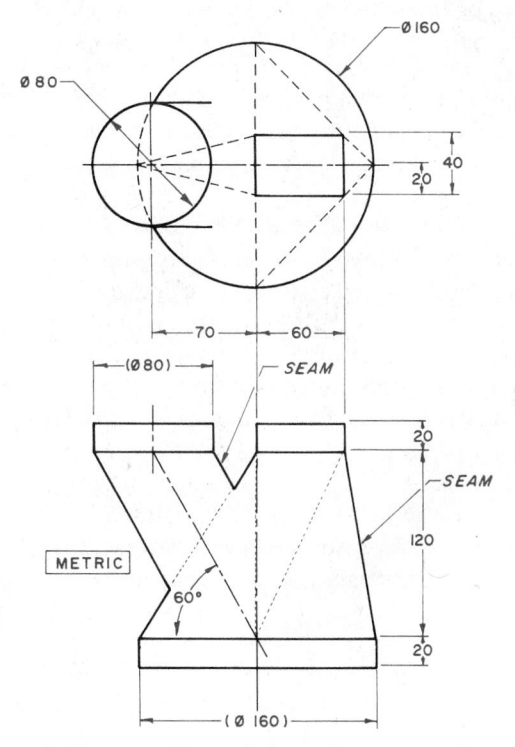

Problem 8-54

Dimensioning and Notation

Chapter Outline

One of the most fundamental drafting tasks is to meet the requirements of the engineering definition of the part while providing for the most economical production process and interchangeability considerations. All of this is accomplished by the use of proper dimensions and notation on drawings. *Dimensioning* is the process whereby size and location data for the subject of a technical drawing are provided. *Notation* is the process whereby needed information not covered by dimensions is placed on a technical drawing.

It is critical that drafters, designers, and engineers be proficient in standard dimensioning practices. The most widely accepted dimensioning standard is American National Standards Institute document Y14.5M-1988R (ANSI Y14.5M-1988R). Similar standards are produced by the International Standards Organization (ISO). However, unless otherwise specified, ANSI Y14.5M-1988R is the standard used for guiding dimensioning practices.

Modern dimensioning practices described in ANSI Y14.5M-1988R apply in most instances in which interchangeability of parts is a major consideration. The concept dictates that parts produced from a drawing at one manufacturing site should be interchangeable with those produced at another manufacturing site. Automotive parts are an excellent example of production for interchangeability. Some parts are manufactured in America, some in Europe, and some in Japan, but they must all fit together in one car during assembly. Although interchangeability is not a factor with all parts that are produced, the drafter should still use the basic dimensioning principles of ANSI Y14.5M-1988R. This is particularly important when the parts will be produced using such ever-increasing automated, semiautomated, or integrated processes as numerical control, computer-aided manufacturing (CAM), or computer-integrated manufacturing (CIM).

Dimensioning Systems

Three dimensioning systems are used on technical drawings in the United States. Metric dimensioning, decimal-inch dimensioning, and fractional dimensioning. Certain rules of practice with which drafters should be familiar pertain to each of these dimensioning systems.

UNIT	MULTIPLE OF A METRE
METRE	1.0
DECIMETRE	0.1
CENTIMETRE	0.01
MILLIMETRE	0.001

Figure 9-1   Metric linear measurements

Metric Dimensioning

The standard metric unit of measurement for use on technical drawings is the millimeter (.001 meter) or .039 inch. **Figure 9-1** is a chart of various metric units of measurement of less than a meter that shows where the millimeter fits in.

When using metric dimensioning, several general rules should be observed. When a dimension is less than 1 millimeter, a zero must be placed to the left of the decimal point, **Figure 9-2A**. When a metric dimension is a whole number, neither the zero nor the decimal is required, **Figure 9-2B**. When a metric dimension consists of a whole number and a decimal portion of another millimeter, it is written as follows: whole number first, decimal point second, and finally, the decimal part of the number. The decimal part of the number is not followed by a zero in metric dimensioning, **Figure 9-2C**. Individual digits in metric dimensions are not separated by commas or spaces, **Figure 9-3**. Drawings prepared with metric dimensions are identified with the word "metric" contained in a small rectangle below the part.

Decimal-Inch Dimensioning

Decimal-inch dimensioning is frequently used in the dimensioning of technical drawings. It is a much less

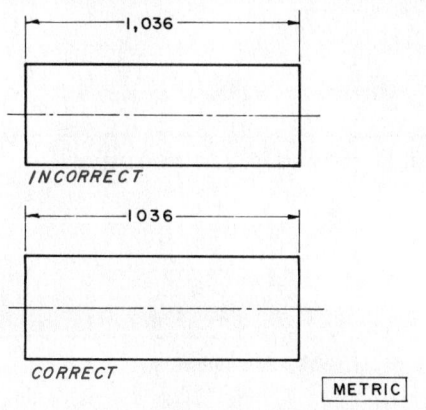

Figure 9-3   No commas in metric dimensions

cumbersome system for mechanical drawings than is the fractional system, and it is still used more than the metric system. When using the decimal-inch dimensioning system, several rules should be observed. If a dimension is less than one inch, only a decimal point and the numbers of the decimal fraction are required. A zero is not required to precede the decimal point, **Figure 9-4A**. The number of places beyond the decimal point that a decimal-inch dimension is carried is determined by the specified tolerance for the part in question, **Figure 9-4B**. In this figure, a tolerance of .002 (±.001), three places to the right of the decimal, is specified. Consequently, the dimension of 1.637 is carried out three places to the right of the decimal.

There are no specified sizes for decimal points, but they should be made dark enough and large enough to be seen and to reproduce through any normal reproduction process (diazo, photocopy, or microfilm).

Fractional Dimensioning

Fractional dimensioning is not frequently used on mechanical technical drawings. Its primary use is on architectural and structural engineering drawings. However, since it is occasionally still used on mechanical drawings, drafters should be familiar with this sys-

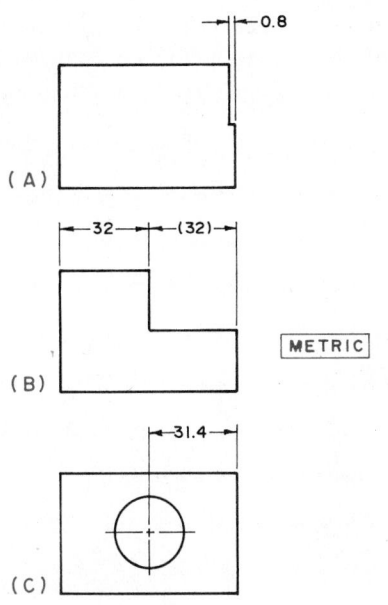

Figure 9-2   Metric dimensioning

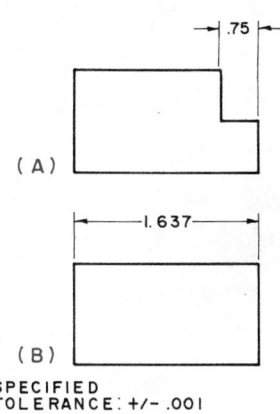

Figure 9-4   Decimal-inch dimensioning

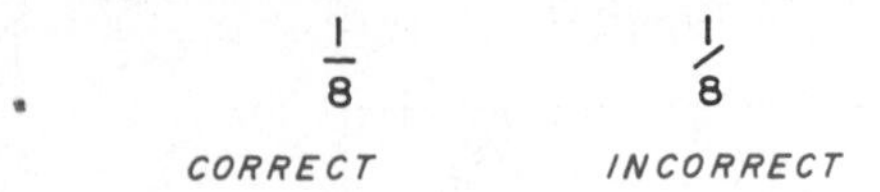

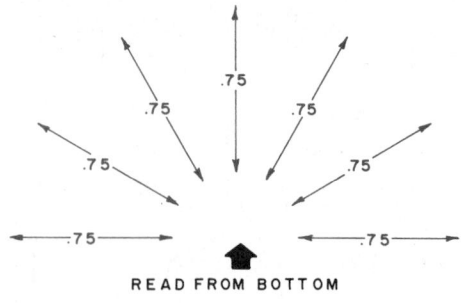

Figure 9-5   Horizontal line is the correct method

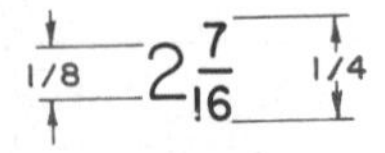

Figure 9-6   Proportions for fractions

Figure 9-8   Unidirectional dimensioning system

tem. When using fractional dimensions on mechanical drawings, several rules should be observed. The line separating the numerator and denominator of a fraction should be a horizontal line, not an inclined line, ***Figure 9-5***. Full-inch dimensions should be a minimum of one-eighth inch in height. The combined height of the numerator, denominator, and horizontal line of a fraction should be one-quarter inch, ***Figure 9-6***, for A, B, and C size drawings, and a minimum of five-sixteenths inch for D, E, and F sizes. Also, be sure to leave a visible gap between the numerator, denominator, and the horizontal line of the fraction.

Orienting Dimensions

Regardless of whether you are dimensioning in the metric, decimal-inch, or fractional dimensioning system, there are two subsystems for orienting dimensions. These orientation subsystems are known as the aligned system and the unidirectional system. The aligned dimensioning system is illustrated in ***Figure 9-7***. In this system the dimensions are aligned with the dimension line so that they can be read either from the bottom of the drawing sheet or from the right-hand side. Notice in Figure 9-7 the area which has been marked off to be avoided if possible. Dimensions written in this area are difficult to read from either the bottom or the right-hand side of the drawing sheet and, therefore, should be avoided. The unidirectional dimensioning system is illustrated in ***Figure 9-8***. In this system all dimensions are written so that they can be read from the bottom of the drawing sheet. This means that the guidelines used

in entering the dimensions should always be parallel with the bottom of the page.

Dimension Components

Several components are common to all dimensioning systems. These include extension lines, dimension lines, arrowheads, leader lines, and the actual numbers or dimensions. Drafters and engineers should be knowledgeable in the proper use of these components.

Extension Lines

An *extension line* is a thin, solid line that extends from the object in question or from some feature of the object. Several rules should be observed when placing extension lines on technical drawings. There should be a small but visible gap between the object or object feature and the beginning of an extension line, ***Figure 9-9A***. Extension lines should extend uniformly beyond dimension lines for a distance of approximately one-eighth inch, ***Figure 9-9B***. Extension lines that originate on the object, such as center lines, may cross visible lines with no gap required, ***Figure 9-10***. ***Figure 9-11*** is a fully dimensioned mechanical drawing which summarizes the basic rules drafters and engineers should remember about spacing of extension lines, dimension lines, and dimensions. Notice in this figure that when

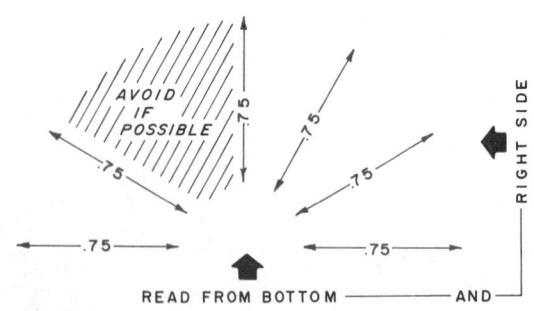

Figure 9-7   Aligned dimensioning system

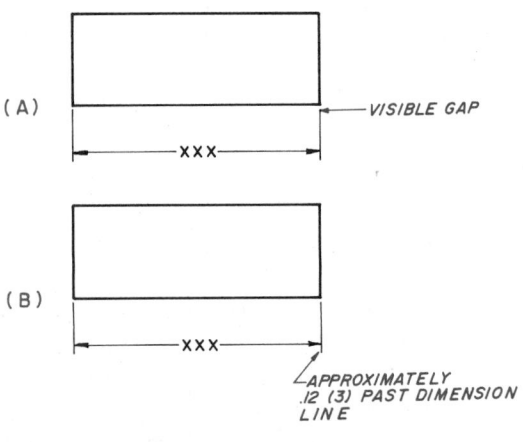

Figure 9-9   Drawing extension lines

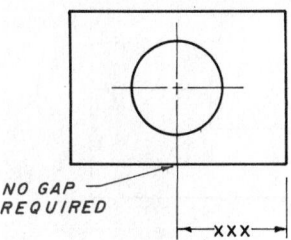

Figure 9-10   Center lines as extension lines

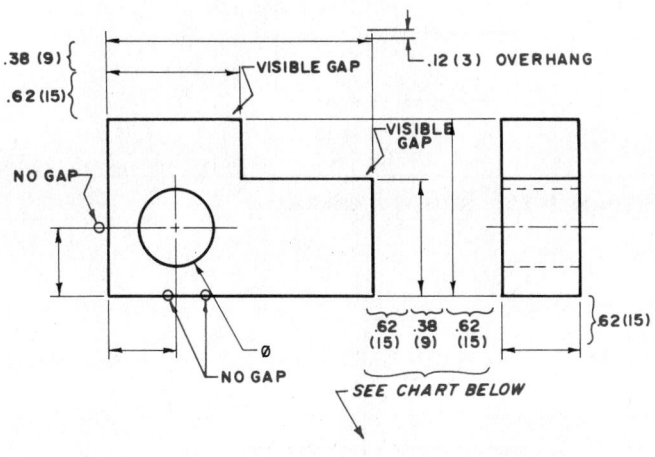

Figure 9-11   Spacing for extension lines, dimension lines, and dimensions

REQUIRED DIMENSIONS BETWEEN VIEWS	SUGGESTED DISTANCE BETWEEN VIEWS	
1	1.25	30
2	1.62	39
3	2.00	48
4	2.38	57
5	2.75	66
6 MAX.	3.12	75

center lines become extension lines there is no gap required as they cross object lines. Notice also that leader lines are not broken when they cross object lines. Note the visible gap between the object itself and the beginning of extension lines. Note the .12 or approximately one-eighth inch overhang of the extension line beyond dimension lines. Suggested spacing of dimension lines and distances between views are summarized in the chart that accompanies the drawing in Figure 9-11.

Dimension Lines

A *dimension line* is a thin, solid line used to indicate graphically the linear distance being dimensioned. Dimension lines are normally broken for placement of the dimension, ***Figure 9-12A***. If a horizontal dimension line is not broken, the dimension is placed above the dimension line with guidelines parallel to it, ***Figure 9-12B***. ***Figure 9-13*** illustrates the proper methods to be used for locating dimensions when space is limited on a drawing. *Notice from this illustration that one*

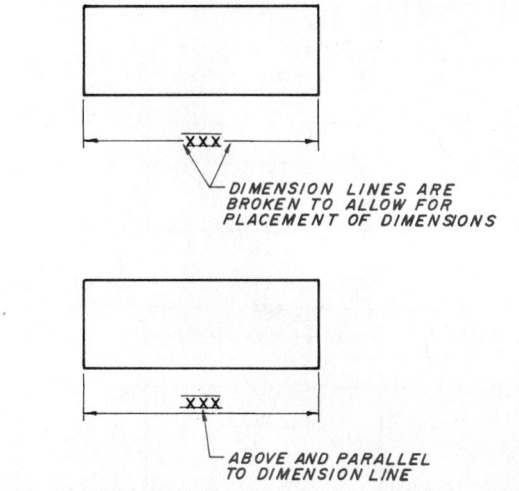

Figure 9-12   Placement of dimensions

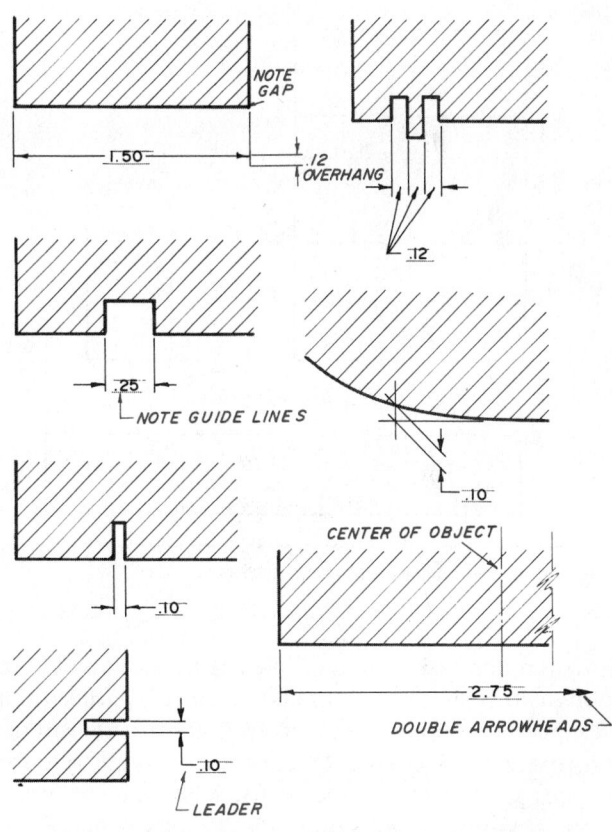

Figure 9-13   Placement of dimensions when space is limited

dimension can be written and then leaders can be used to point to where the dimension applies. A dimension may be written within the extension lines with dimension lines extending inward rather than outward. Dimension lines may be bent at a 90° angle; and extension lines, when necessary, can be drawn at an oblique angle. Double arrowheads can be used on a dimension line when a long or short break line has been used.

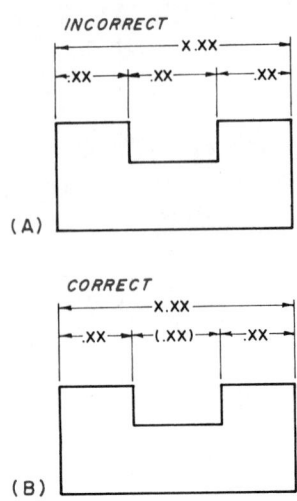

Figure 9-14 Proper placement of dimensions when dimensioning multiple features

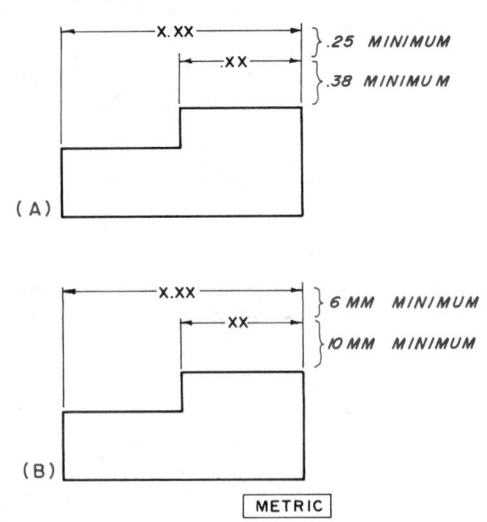

Figure 9-16 Successive dimension lines

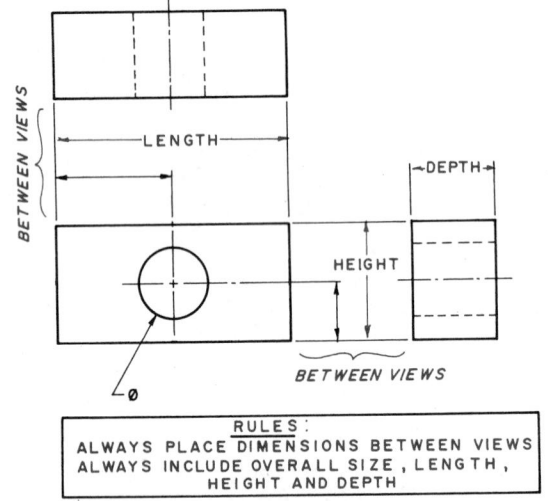

RULES:
ALWAYS PLACE DIMENSIONS BETWEEN VIEWS
ALWAYS INCLUDE OVERALL SIZE, LENGTH, HEIGHT AND DEPTH

Figure 9-15 Dimensioning between views

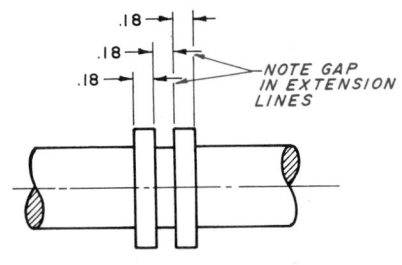

Figure 9-17 Successive dimension lines in close areas

When dimensioning multiple features of an object, dimensions should be aligned uniformly rather than staggered or randomly scattered about the object, *Figure 9-14*. *Figure 9-15* illustrates another rule for proper placement of dimensions. When dimensioning several views of the same object, always place the dimensions between the views, and always include the overall size, length, height, and depth of the object.

Dimension lines are drawn parallel to the direction of measurement. Sufficient distance between the object and the dimension lines, and between successive dimension lines, is important so that cramped and crowded dimensions do not result. First dimension lines should be at least .62 inch from the object, *Figure 9-16A*. Successive dimension lines should be at least .25 inch apart. If using metric dimensions, the first dimension lines should be at least ten millimeters away from the object. Successive lines should have at least six mil-

limeters between them, *Figure 9-16B*. *Figure 9-17* illustrates the methods to be used when successive dimensions are called for in close areas. Notice how the dimension lines and corresponding dimensions are offset, and notice also the gap left in the extension lines to accommodate crossover of dimension lines.

When the shape of an object requires a series of parallel lines, the breaks and dimensions should be slightly staggered to make it easier to read the dimensions, *Figure 9-18*. *Figure 9-19* illustrates the proper way of stacking a successive group of dimension lines. Drafters and engineers should always place the shortest dimensions closest to the objects, and always keep dimensions outside and away from the object, as illustrated in Figure 9-19.

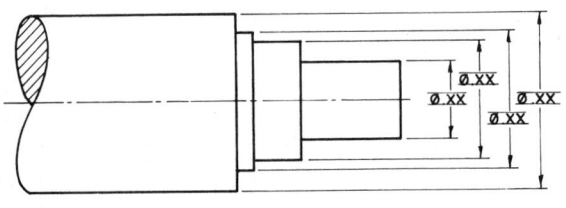

Figure 9-18 Staggering dimensions

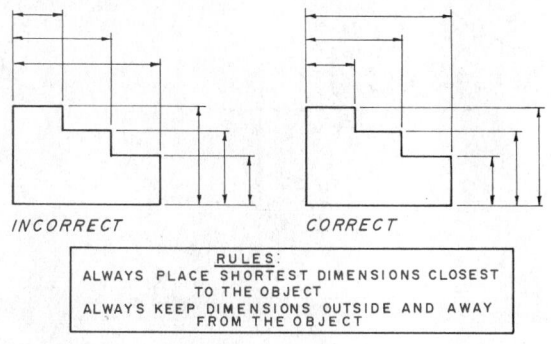

Figure 9-19   Stacking dimension lines

Leader Lines

A *leader line* is a thin line that begins horizontally, breaks at an angle, and terminates in an arrowhead, or, on occasion, a dot, ***Figure 9-20***. Leaders which terminate at an edge or at some other specific point on a drawing should have an arrowhead. Leaders that terminate inside an object on a flat surface should end in a

dot. The preferred angle of a leader line is from a minimum of 30° to a maximum of 60°, ***Figure 9-21***. The horizontal portion of a leader should be from .125 inch to .25 inch in length, Figure 9-21. To avoid confusion, leader lines should not be drawn parallel to extension lines or dimension lines. When placing a dimension and/or note at the end of a leader line, it is easier to letter from the leader than into the leader, ***Figure 9-22***.

When using a leader line to direct a dimension to its appropriate feature on a drawing, one dimension for each leader is the preferred method, ***Figure 9-23***. More than one leader line extending from the same dimension can create a confusing situation that is difficult to interpret. Leader lines pointing to the center of a circle should be directed toward but not extended into the circle center, ***Figure 9-24***. ***Figure 9-25*** illustrates the proper use of leader lines in a variety of dimensioning situations.

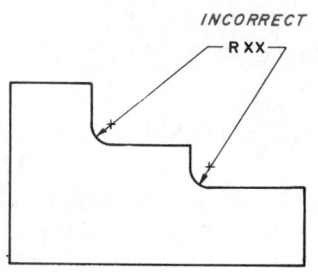

Figure 9-20   Leader lines

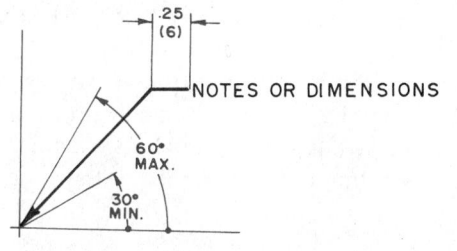

Figure 9-21   Angle of a leader line

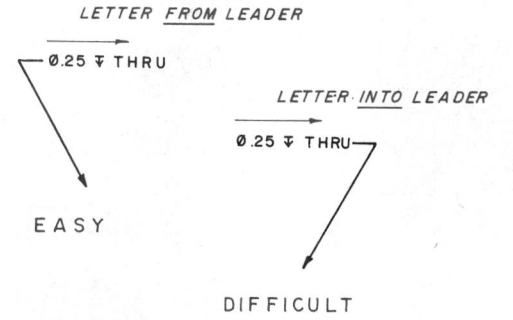

Figure 9-22   Lettering and leaders

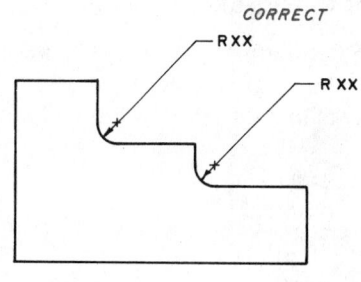

Figure 9-23   One dimension per leader line is preferred

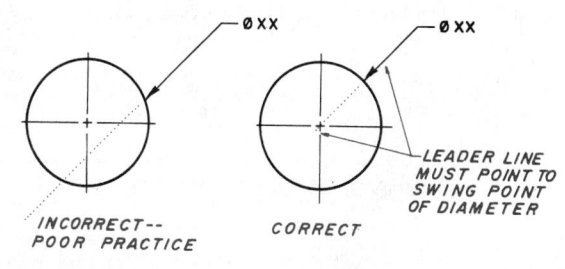

Figure 9-24   Leader Lines point at the center of a circle

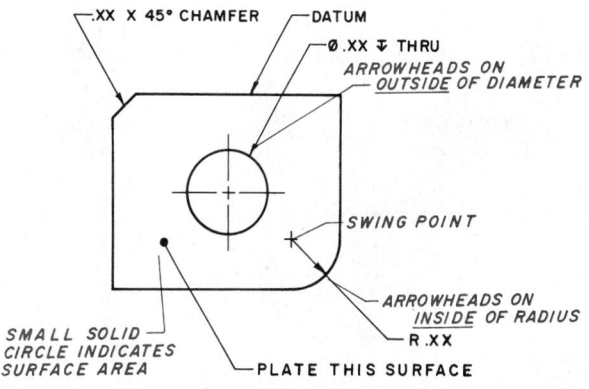

Figure 9-25   Proper use of leader lines

Arrowheads

An *arrowhead* is the most commonly used termination symbol for dimension and leader lines. Arrowheads should be approximately three times as long as they are wide. They should be large enough to be seen but small enough that they do not detract from the appearance of the drawing. A commonly accepted length for arrowheads is .12 inch. Arrowheads may be slightly larger or smaller than this, but, regardless of their size, they should be uniform throughout a drawing, *Figure 9-26*. Although the same standard applies to drawings prepared on a CAD system, arrowheads do not have to be filled in on drawings prepared using automated processes. Some CAD systems fill in arrowheads and some do not. Both open and filled-in arrowheads have been considered acceptable since the advent of CAD.

Figure 9-27 illustrates the three steps used in drawing arrowheads.

Laying Out Dimensions

Figure 9-28 illustrates the four steps in laying out dimensions for an object. First, place light layout lines on the drawing sheet. Second, draw in light guidelines for the

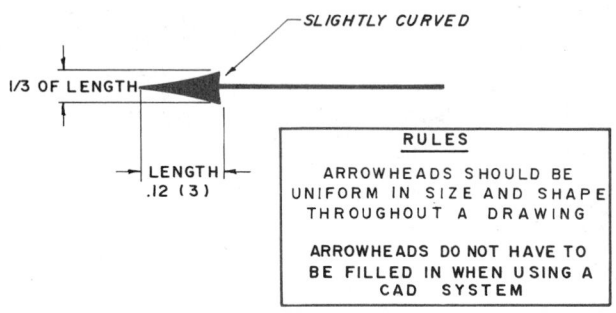

Figure 9-26   Proper size and shape of arrowheads

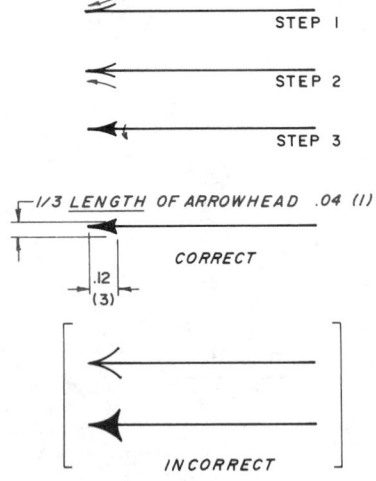

Figure 9-27   Drawing arrowheads

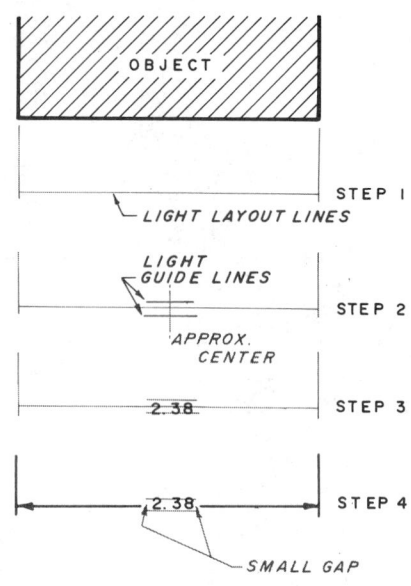

Figure 9-28   Laying out dimensions

dimensions. Third, darken your extension lines, dimension lines, and dimensions. Fourth, add arrowheads.

Steps in Dimensioning

There are two basic steps in dimensioning objects, regardless of the type of object.

Step 1   Apply the size dimensions. These are dimensions which indicate the overall size of the object and the various features which make up the object.

Step 2   Apply the locational dimensions. Locational dimensions are dimensions which locate various features of an object from some specified datum or surface. *Figure 9-29* gives examples of size and location dimensions.

In order to properly dimension an object, drafters and engineers must be able to mentally break the object down into component parts and subelements. *Figure 9-30* illustrates graphically how this is done. On the left

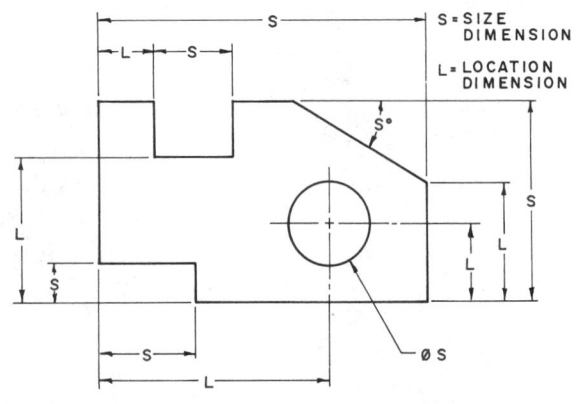

Figure 9-29   Size and location dimensions

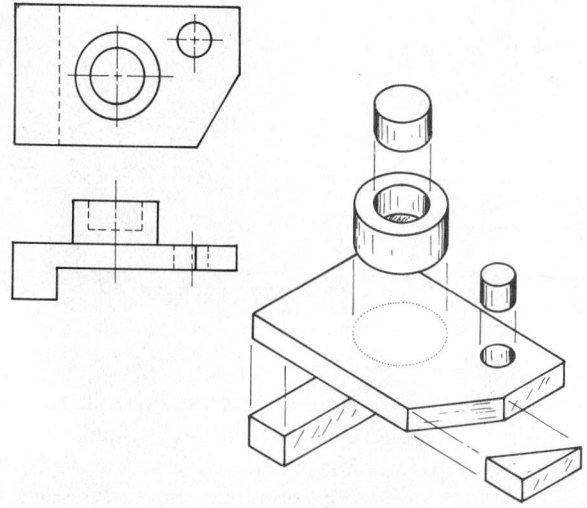

Figure 9-30   Geometric breakdown of an object

R	RADIUS
Ø	DIAMETER
2 X	NUMBER OF TIMES - PLACES
▽	DEEP OR DEPTH
∨	COUNTERSINK
⊔	COUNTERBORE OR SPOTFACE
SF	SPOTFACE
□	SQUARE
( )	REFERENCE DIMENSION
.25	NOT TO SCALE DIMENSION
◺	SLOPE
▷	CONICAL TAPER
⌒	ARC LENGTH
SR	SPHERICAL RADIUS
SØ	SPHERICAL DIAMETER
-x-	BASIC DIMENSION

Figure 9-31   Dimensioning symbols

side of the figure is a two-dimensional representation of the object containing a top and front view. On the right side of the figure is a three-dimensional geometric breakdown of the object. This three-dimensional breakdown shows the drafter and engineer what geometric shapes must be sized using dimensions and which must be located using dimensions.

Specific Dimensioning Techniques

All of the information on dimensioning so far has been of a general nature. The following sections deal with techniques used for applying dimensions to specific situations that are recurrent in drafting. With a thorough knowledge of the general information presented earlier, and the specific information presented in these sections, drafters and engineers will be able to dimension any situation confronted on technical drawings.

Dimensioning Symbols

There are a number of symbols used in different dimensioning situations with which drafters and engineers should be familiar. These symbols are summarized in *Figure 9-31*. Review these symbols carefully and commit them and their uses to memory before proceeding.

Dimensioning Chords, Arcs, and Angles

Chords, arcs, and angles are dimensioned in a similar manner. When dimensioning a chord, the dimension line should be perpendicular and the extension lines parallel to the chord. When dimensioning an arc, the dimension line runs concurrent with the arc curve, but the extension lines are either vertical or horizontal. An arc symbol is placed above the dimension. When dimensioning an angle, the extension lines extend from the sides forming the angle, and the dimension line forms an arc. These methods are illustrated in *Figure 9-32*. *Figures 9-33* and *9-34* contain additional information relative to dimensioning angles.

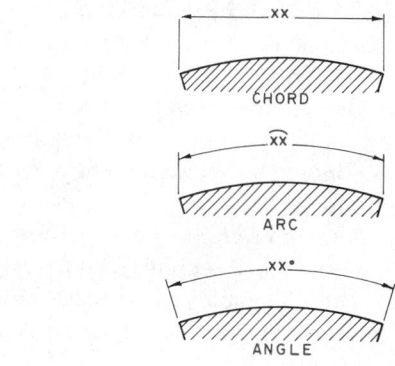

Figure 9-32   Dimensioning chords, arcs, and angles

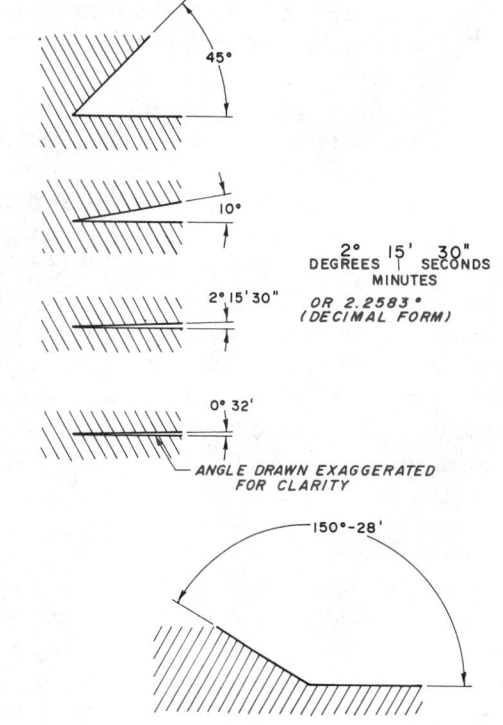

Figure 9-33   Dimensioning angles

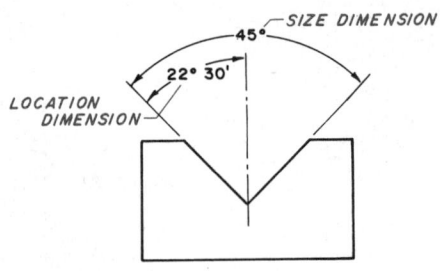

Figure 9-34 Dimensioning angles

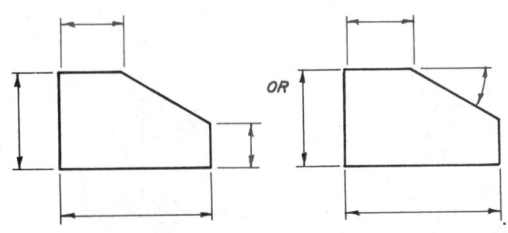

Figure 9-35 Using normal linear dimensions for effecting the proper angle

Notice in Figure 9-33 that angles are normally written in degrees, minutes, and seconds. The symbols used to depict degrees, minutes, and seconds are also shown in this figure. Angular measurements may also be stated in decimal form. This is particularly advantageous when they must be entered into an electronic digital calculator. The key to converting angular measurements to decimal form is in knowing that each degree contains 60 minutes, and each minute contains 60 seconds. Therefore to convert a measurement stated in degrees, minutes, and seconds into decimal form is a two-step process. Consider the example of the angular measurement 2 degrees, 15 minutes, and 30 seconds. This measurement would be converted to decimal form as follows:

Step 1 The seconds are converted to decimal form by dividing by 60. Thirty seconds divided by 60 equals .50. The 15 minutes are added to this so that the minutes are expressed in decimal form or 15.50.

Step 2 The minutes stated in decimal form are converted to decimal degrees by dividing by 60. The number 15.50 divided by 60 equals .25833. The 2 degrees are added to this number to have the measurement stated in decimal form or 2.2583 degrees.

Figure 9-34 illustrates how angles can be dimensioned for size and location. The size dimension gives the overall size of the angular cut. However, it does not locate the angular cut in the object. This is done by using a locational dimension off of the center line of the object. ***Figure 9-35*** shows how normal linear dimensions can be used for effecting the proper angle. When this method is used in the example on the left, all surfaces are dimensioned except the angular surface. When this method is used the angular surface is defined by default. In the method on the right, three linear dimensions and one angular dimension define the angular surface.

Dimensioning a Radius

Dimension lines used to specify a radius have one arrowhead, normally at the arc end. An arrowhead should not be used at the center of the arc. Where space permits, the arrowhead should be placed between the arc and the arc center with the arrowhead touching the arc. If space permits, the dimension is placed between the arc center and the arrowhead. If

space is not available, the dimension may be placed outside the arc by extending the dimension line into a leader. The arc center of a radius is denoted with a small cross. ***Figure 9-36*** illustrates the preferred method for dimensioning a radius.

Dimensioning a Foreshortening Radius

On occasion the center of an arc radius will fall outside the drawing or will be so far removed from the drawing as to interfere with other views. When this is the case, the radius dimension line should be foreshortened and the radius center located using coordinate dimensions. This is done by relocating the arc center and placing a zig-zag in the radius dimension line, as shown in ***Figure 9-37***. When this method is used it is important that the arc center actually lie upon the real center line of the arc.

Dimensioning Curved Surfaces

Curved surfaces containing two or more arcs are dimensioned by showing the radii of all arcs and locating the

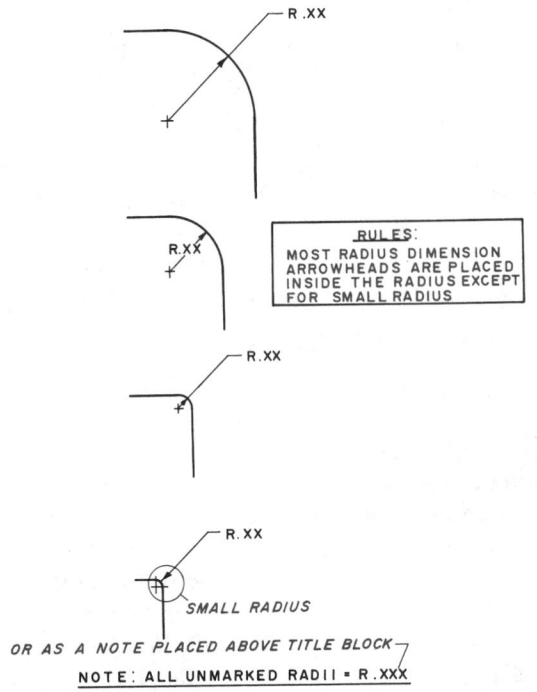

Figure 9-36 Dimensioning a radius

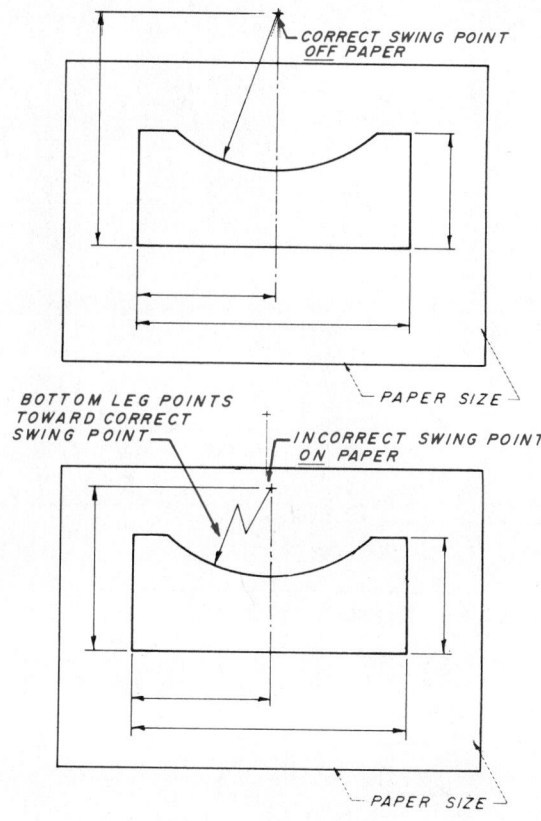

Figure 9-37 Dimensioning a foreshortening radius

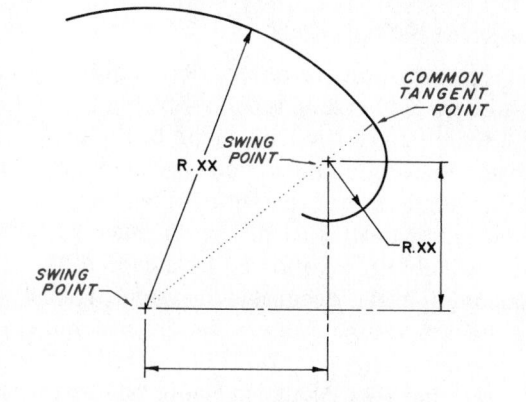

Figure 9-38 Dimensioning curved surfaces

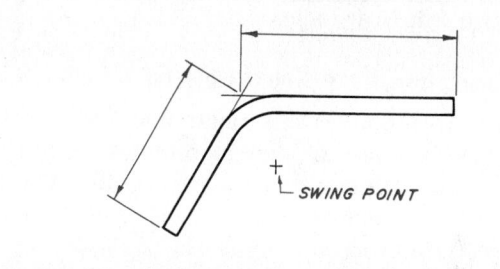

Figure 9-39 Dimensioning offsets

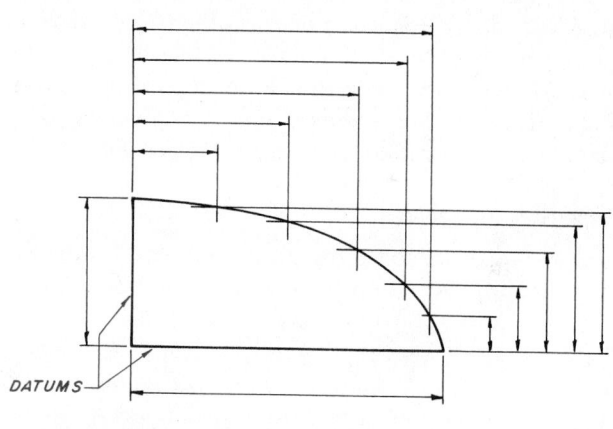

Figure 9-40 Dimensioning irregular curves

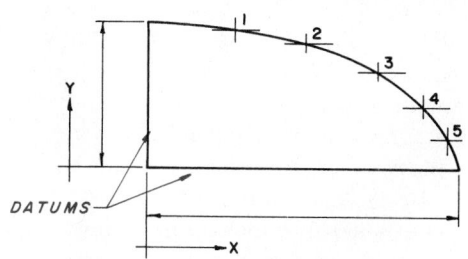

STATION	1	2	3	4	5
X	1.12	2.12	3.00	3.62	4.00
Y	1.75	1.50	1.12	.75	.31

Figure 9-41 Dimensioning contours not defined as arcs

arc centers using coordinate dimensions. Other radii may be located using their points of tangency. This method is illustrated in *Figure 9-38*.

Dimensioning Offsets

Offsets are dimensioned from the points of intersection of the tangents along one side of the object, *Figure 9-39*. The distance from one end of the offset to the intersection is specified with a coordinate dimension. The distance from the other end to the offset is also specified with a coordinate dimension, as shown in Figure 9-39.

Dimensioning Irregular Curves

You have already seen in Figure 9-38 how to dimension curved surfaces using arc centers, radius dimensions, and tangent points. Irregular curves may also be dimensioned using coordinate dimensions from a specified datum. When this is the case the coordinate dimensions extend from a common datum to specified points along the curve, *Figure 9-40*.

Dimensioning Contours Not Defined as Arcs

Contours not defined as arcs can be dimensioned by indicating X, Y coordinates at points along the surface of the contour. Each of these points, sometimes referred to as stations, is numbered. The X and Y coordinates for each station are tabulated and placed in table form under the drawing, *Figure 9-41*.

Dimensioning Multiple Radii

When dimensioning an object that requires several radii, arcs should be dimensioned showing the radius in a view which gives the true shape of the curve. The dimension lines for a radius should be drawn as a radial line at an angle, rather than horizontally or vertically. Only one arrowhead is used. The dimensional value of the radius should be followed by a capital "R" when dimensioning in the decimal-inch system. The "R" precedes the dimensional value when dimensioning in the metric system. This method is illustrated in *Figures 9-42*, *9-43*, and *9-44*. Notice in Figure 9-44 that where a radius is dimensioned in a view that does not show the true radius, a note should be used to indicate that the true radius is not shown, and a separate note used to indicate what the true radius is.

Dimensioning by Offset (Round Objects)

Another way to dimension a round object is the offset method. In this method, dimension lines are used as extension lines. Dimension lines are distributed across the object perpendicular to the center line of the object and spaced using coordinate dimensions, *Figure 9-45*.

Dimensioning by Offset (Flat Objects)

Flat objects may also be dimensioned using the offset method. Again, dimension lines on the object become extension lines. Dimension lines are spaced across the object perpendicular to the center line of the datum and extended to become extension lines for the coordinate dimensions which space them, *Figure 9-46*.

Dimensioning Spheres

Figure 9-47 illustrates the proper method for dimensioning spheres. When the diameter method is used, a leader points to the center of the sphere, and the diameter note is preceded with a capital "S" to indicate that it is a spherical diameter. When the radius method is used, a dimension extends from the arc center to the arc and is extended on with a leader, and the note is preceded by a capital "SR" to indicate a spherical radius.

Dimensioning Round Holes

Round holes are dimensioned in the view in which they appear as circles, *Figure 9-48*. Holes may be dimensioned using a leader which points toward the center of the hole in which the note gives the diameter, or extension lines may be drawn from the circle with a dimension that also indicates the diameter. Larger circles are dimensioned with a dimension line drawn across the circle through its center at an angle with the diameter dimension shown. Except for very large holes, the arrowhead and the dimensional value are placed outside the hole. It is important when dimensioning holes to call off the diameter, not the radius.

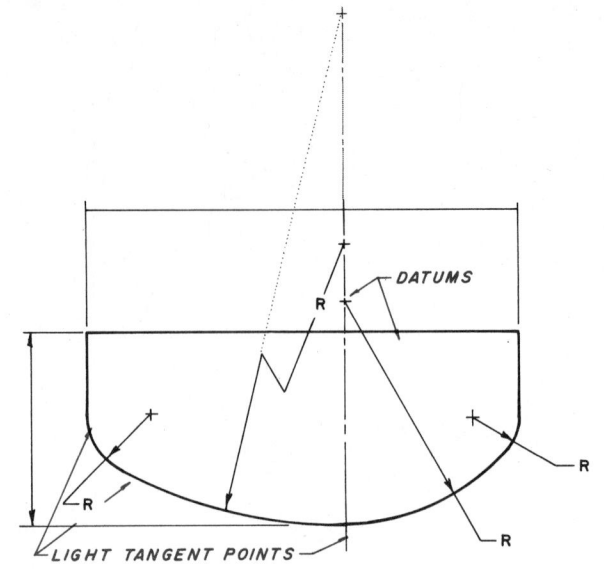

Figure 9-42   Dimensioning multiple radii

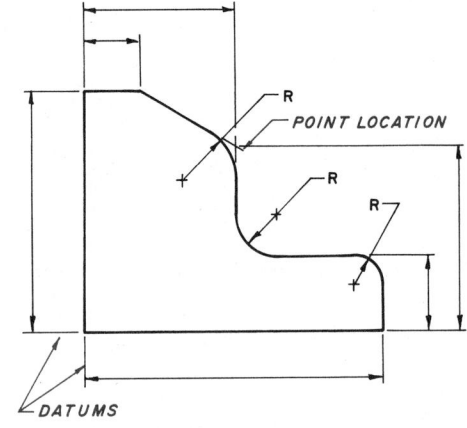

Figure 9-43   Dimensioning multiple radii

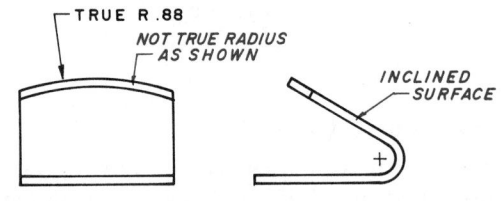

Figure 9-44   Dimensioning multiple radii

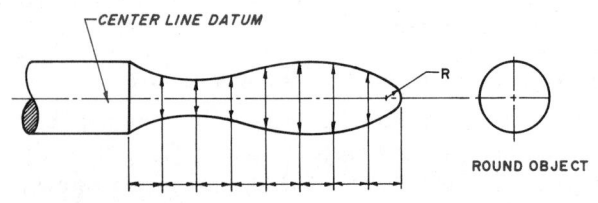

Figure 9-45   Dimensioning by offset (round object)

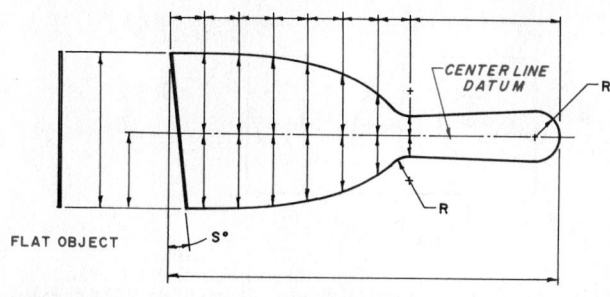

Figure 9-46 Dimensioning by offset (flat object)

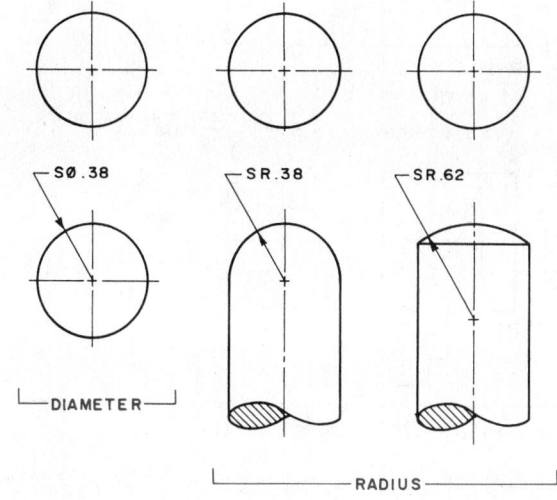

Figure 9-47 Dimensioning spheres

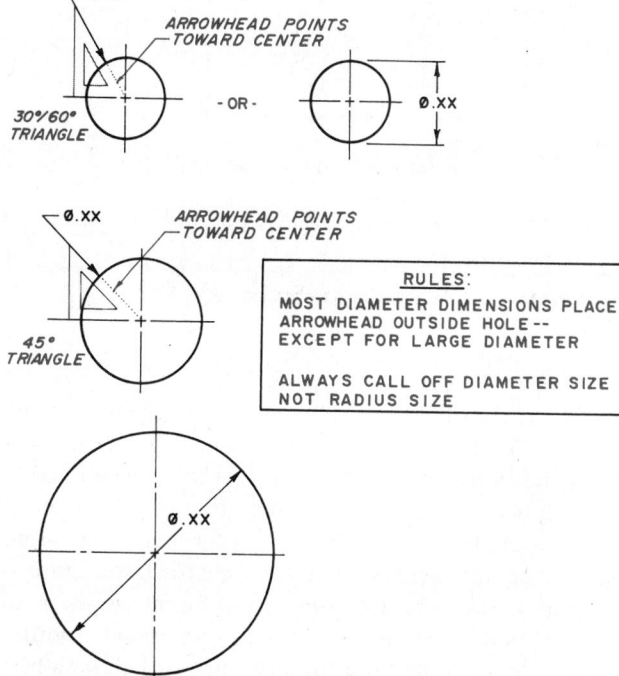

Figure 9-48 Dimensioning round holes

Simple Hole Callouts

Drafters and engineers need to know how to apply simple callouts to both through-holes and blind-holes. A through-hole is one which passes all the way through the object. A blind-hole is one which cuts into but does not pass through the object. Both types of holes, and the callout used for each, are illustrated in *Figure 9-49*. Machine holes are generally drilled to make the rough hole and then reamed to refine the hole. Figure 9-49 shows the difference between a drill and a ream. No hole is only reamed. A hole must be drilled before it can be reamed.

A through-hole callout has a leader line extending toward the center of the hole in the view in which the hole appears as a circle. The note attached to the leader gives the diameter of the hole, the depth symbol, and the word "thru" to indicate that the hole passes through an object. Blind-hole callouts are similar to through-hole callouts, except that the depth symbol is followed by the actual depth of the hole.

Drill Size Tolerancing

Holes are not drilled to the exact size specified on a drawing. This is because there are several factors which mitigate against a perfectly sized hole. The accuracy of the actual drill, the tolerance level of the machine, and the qualifications of the machinist all have an impact on the actual size of a hole once it's drilled. It is accepted in manufacturing that no hole, even with the added accuracy provided by computer-aided manufacturing, is going to be drilled exactly the size specified. Therefore, drafters and engineers need to know how much variation to expect in a hole so that they can decide what limits to give the hole and whether the actual drilled hole will give the fit required in a given situation.

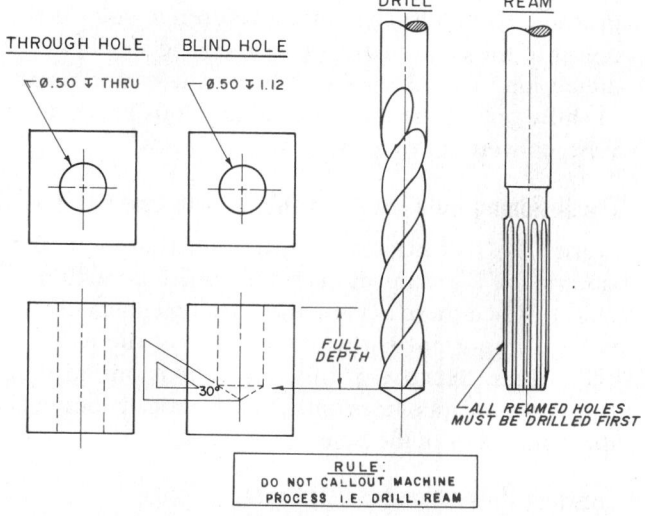

Figure 9-49 Simple hole callouts

Drill size tolerance charts have been developed to assist drafters and engineers in determining the expected upper and lower limits of a drilled hole. Such charts are contained in Appendix B as Table 16. Turn to these tables and you will notice that the standard drill size is given a number/letter, a fractional designation, a decimal designation, and a metric designation. To the right-hand side of the table the tolerance for each drill size is given in decimals. The left-hand column is the plus tolerance, and the right-hand column is the minus. To use these tables, apply the following steps:

1. Find the letter or number of the drill in question on the left-hand side of the table.
2. Find the corresponding size for the drill in decimal form.
3. Find the plus tolerance for that size drill and add it to the decimal drill size. This will give you the upper limit for the hole size.
4. Find the minus tolerance for that drill size and subtract it from the decimal drill size. This will give you the lower limit for the hole.
5. Write the upper and lower limits with the smaller number on top separated from the larger number by a horizontal line. For example, an H-size drill has a decimal diameter of 0.2660. To find the upper limits for a hole using this drill size, add the plus tolerance of 0.0064 to this decimal drill size to get 0.2724 as the upper limit for the hole. To find the lower limit, subtract the minus tolerance of .002 from the decimal drill size to get a lower limit of 0.2640.

Dimensioning Hole Locations

In addition to dimensioning the size of a hole, drafters and engineers must also dimension the locations of holes. *Figure 9-50* illustrates the proper methods for dimensioning the location of a hole. The preferred method is to dimension the location of the hole in the view where the hole appears as a circle. It is common practice to dimension from a reference side of the object to the center line of the hole. One should never dimension to a hidden line, however. If it is necessary to show one of the locational dimensions in a hidden view, convert the view into a sectional view.

Dimensioning Hole Locations (More Than One Hole)

Figure 9-51 illustrates the proper method of dimensioning the locations of more than one hole within an object. When there is more than one hole in an object, what is important is the relationship of the holes to each other. Because of this, the dimensions applied show the distance horizontally and vertically between the center lines of the holes.

Locating Holes About a Bolt Center

A common dimensioning situation is to have holes spaced about a bolt center. When this is the case the holes may be dimensioned using coordinate dimensions

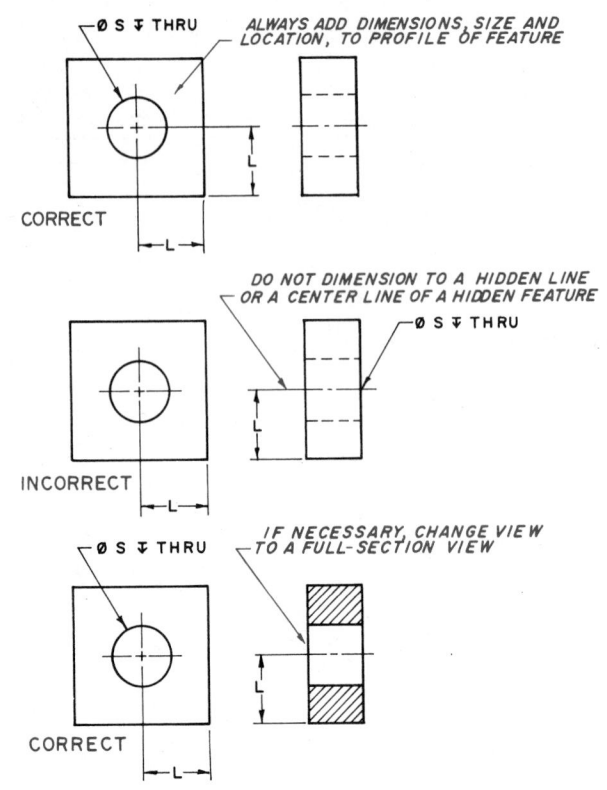

Figure 9-50   Dimensioning hole locations

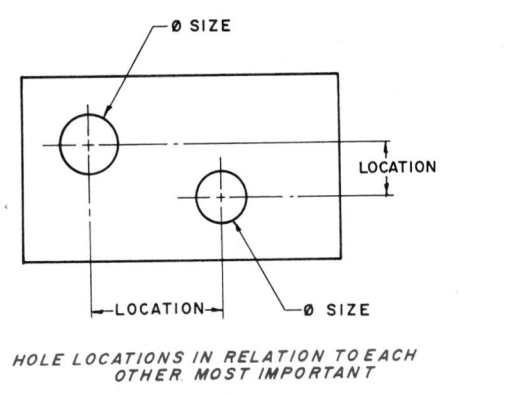

Figure 9-51   Dimensioning hole locations (more than one hole)

or angular dimensions, *Figure 9-52*. For the greatest accuracy, coordinate dimensions can be given from the center line of the bolt's center as well as center-to-center dimensions of the holes. This method is illustrated in the top half of Figure 9-52. Notice that the diameter of the bolt center is specified as a reference dimension. Reference dimensions are given only for informational purposes. They are not intended to be measured or to be used in dictating manufacturing operations. Another method that can be used for locating holes about a bolt center is illustrated in the bottom half of Figure 9-52. With this method the diameter of the bolt circle is shown and a note specifying the number of holes, their diameters, and the indication "equally spaced" is given.

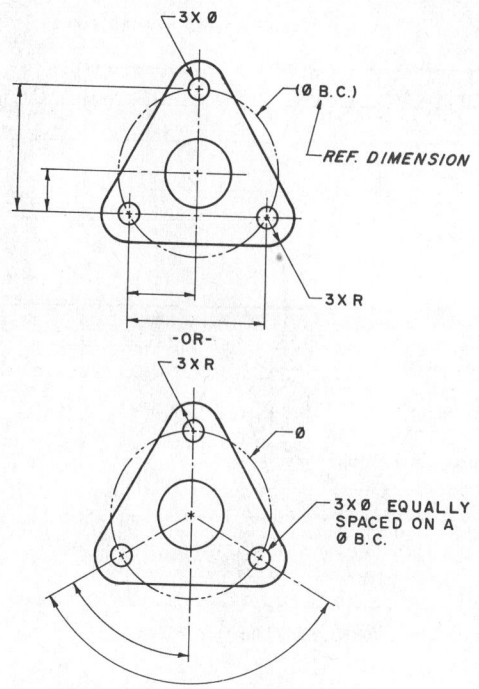

Figure 9-52 Locating holes about a bolt center

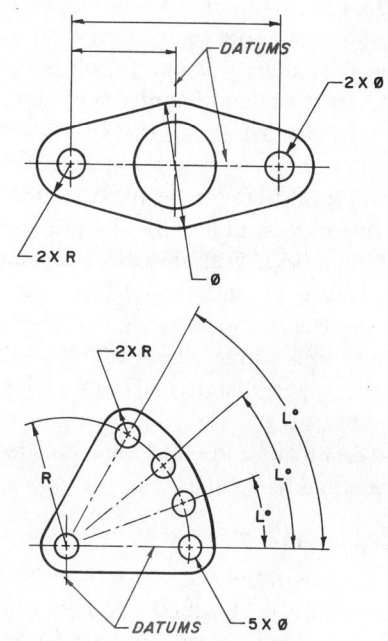

Figure 9-53 Locating holes on center lines and concentric arcs

If the holes are unequally spaced, the same method can be used except that the note "equally spaced" is dropped and an angular measurement with reference to one of the center lines is indicated.

Locating Holes on Center Lines and Concentric Arcs

When locating holes on common center lines, the center lines become datums from which coordinate measurements can be made, as illustrated in the top half of *Figure 9-53*. All three holes in the object in this figure share the same horizontal center line. Consequently, the only dimensions required are dimensions from center to center of the circles. This is accomplished by providing a coordinate dimension from the center of a large circle to the center of one of the smaller circles, and the overall coordinate dimension between the centers of the two small circles, as illustrated. When centers are located on concentric arcs, the polar system of dimensioning can be used, as illustrated in the bottom half of Figure 9-53. When this is the case the radial distances from the point of concurrency and the angular measurements between the holes are used to locate the centers of the holes. The distance from the point of concurrency to the center of each hole is indicated by the radius, as shown.

Dimensioning Multiple Holes Along the Same Center Line

Figure 9-54 shows the proper method used for dimensioning multiple holes along the same center line. The method shown at the top of Figure 9-54 is known as "chain dimensioning." The center line of the first hole is dimensioned from a reference datum which is normally one side of the object. Then each successive center line

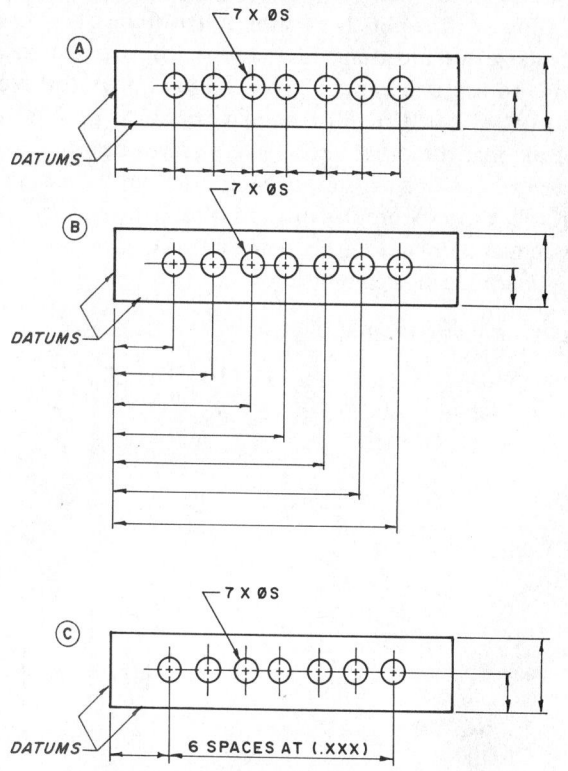

Figure 9-54 Dimensioning multiple holes

is dimensioned from the immediate preceding center line. A problem with this type of dimensioning is that if one of the features in the chain is improperly located, all successive features will be improperly located. As a result, another method of dimensioning multiple holes or features is often used in place of chain dimensioning.

This dimensioning concept is known as "datum," or "baseline" dimensioning. As is shown in the middle part of Figure 9-54, each successive center line is dimensioned from a baseline or datum. In this way any one of the multiple features can be improperly located by the machinist without affecting the others. When this happens, the error can sometimes be corrected.

Another method that can be used for dimensioning multiple features within an object is illustrated at the bottom of Figure 9-54. With this method the center line of the first feature is dimensioned from a given datum, usually the edge of the object. Remaining features are dimensioned using a note which indicates the number of spaces at a given distance. This method does not have the disadvantage of relying on an accurate placement of the first feature in order for each successive feature to be accurately located.

Dimensioning Repetitive Features

Figure 9-55 illustrates the proper methods used for dimensioning an object which contains repetitive features. Repetitive features may be located on a drawing using a note which specifies the number of times a given dimension is repeated, and giving one particular dimension and the total length as a reference. When dealing with repetitive features it is necessary to use both size and locational dimensions. The size dimension for one feature can be given followed by the word "TYPICAL" or the shortened version "TYP," which means that the other successive features have the same size dimensions as the one single feature that is completely dimensioned. Note the dimensions that have been given in Figure 9-55. Some are size dimensions and some are location dimensions.

Callouts for Tapered Holes

Occasionally the design of a manufactured part will require a tapered hole. *Figure 9-56* illustrates the proper

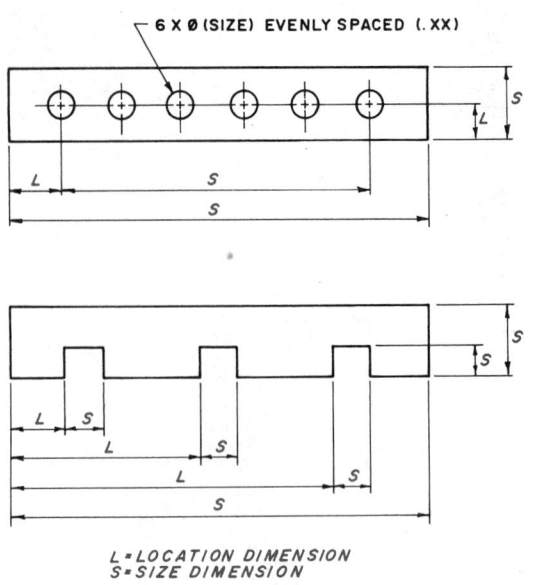

Figure 9-55 Dimensioning repetitive features

method for annotating a tapered hole and shows an example of a 4° tapered bit. Callouts for tapered holes are similar to those for regular holes except that the note must also contain the taper symbol and the degree and the length of taper. The callout for the tapered hole in Figure 9-56 means that the subject hole has a .50 diameter at the top. It is drilled through and tapered at 4° for the full length of the hole.

Callouts for Countersunk Holes

Countersinking is a machining process to form a conical head so that a fastener can fit flush with the top of a part. The callout for a countersunk hole should contain the diameter of the countersink hole, which is the maximum diameter on the surface, and the angle of the countersink. The proper method for calling out a

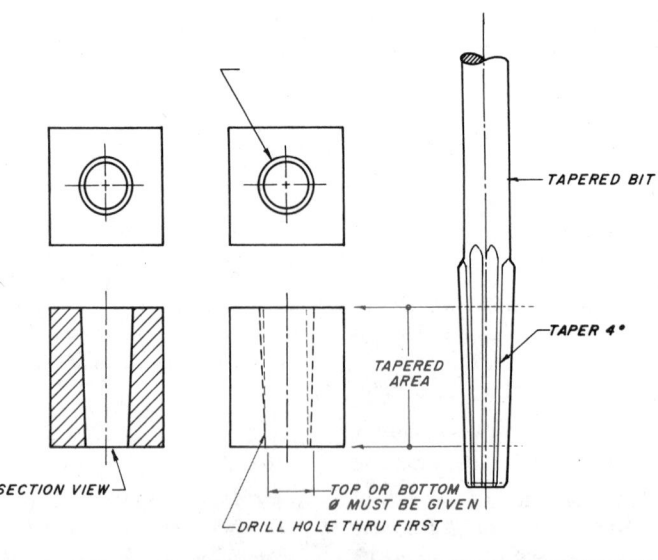

Figure 9-56 Callouts for tapered holes

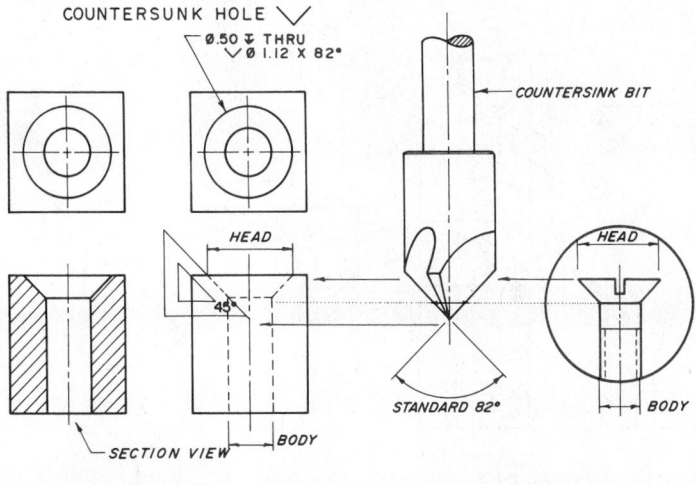

Figure 9-57 Callouts for countersunk holes

countersunk hole is illustrated in ***Figure 9-57***. First a .50 diameter hole is drilled through the part. Then, using a countersink bit, the top of the hole is countersunk at 82° until the head of the countersink has a diameter of 1.12.

Callouts for Counterbored Holes

On occasion a hole in a part is counterbored to allow the head of a bolt or another type fastener to be recessed. The bottom of a counterbore is flat rather than angled as in a countersink. The callout for a counterbored hole shows the diameter of the drilled hole, the counterbore symbol, the diameter of the counterbore itself, and the depth of the counterbore. When a hole is to be counterbored, the hole itself is drilled first and then the counterbore is applied using a special counterbore bit, as shown in ***Figure 9-58***.

Callouts for Spotfaces

A spotface is a machining process used to finish the surface around the top of a hole so that a bolt or other type of threaded fastener can feed properly. A spotface resembles a very shallow counterbore. The callout for a spotface contains the diameter of the actual hole, the capital letters "SF," and the diameter of the spotface. The depth of a spotface is not normally shown. When a hole is to be spotfaced, the hole itself is drilled first and then the spotface is applied with a special spotface bit, as illustrated in ***Figure 9-59***.

Dimensioning a Cylinder

Figure 9-60 shows the proper way for dimensioning a cylinder. Notice from this figure that there are two methods which can be used. In the first method the longitudinal view and the end view of the cylinder are given. When this method is used, the length of the cylinder is shown and the diameter dimension is applied between the longitudinal view and the end view. In the second method, the end view is left off. Consequently, the diameter dimension is applied to the longitudinal view. When dimensioning a cylinder, always specify the diameter rather than the radius.

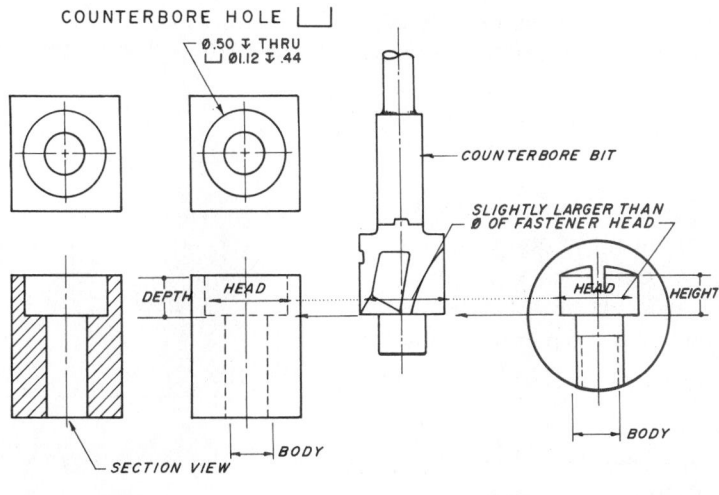

Figure 9-58 Callouts for counterbored holes

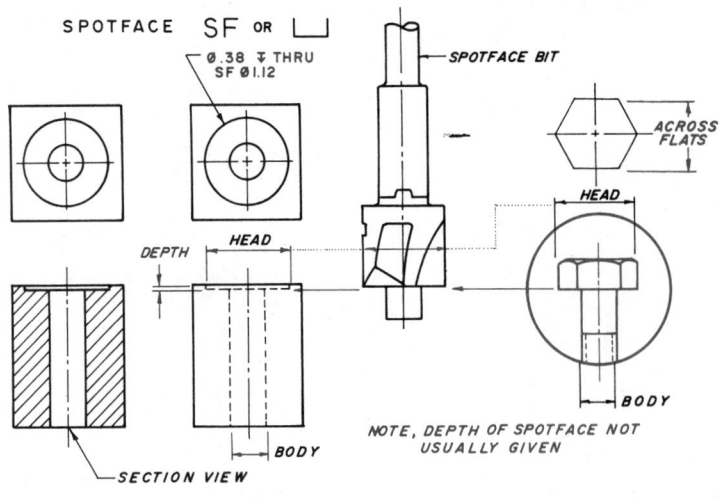

Figure 9-59 Callouts for spotfaces

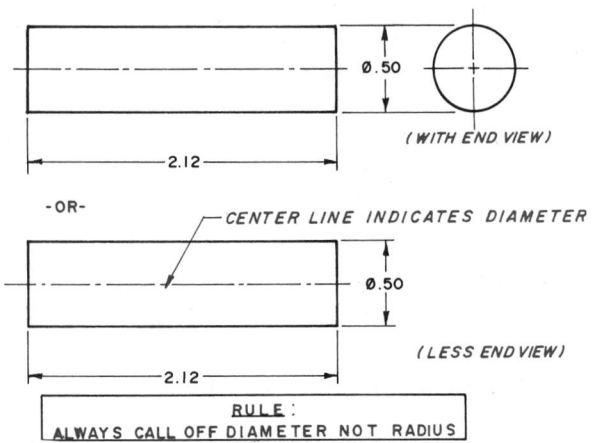

Figure 9-60 Dimensioning a cylinder

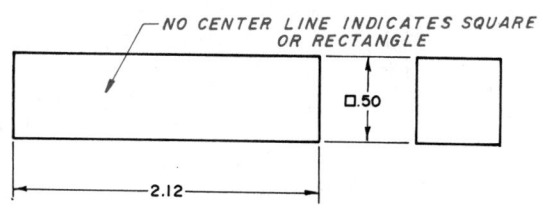

Figure 9-61 Dimensioning a square

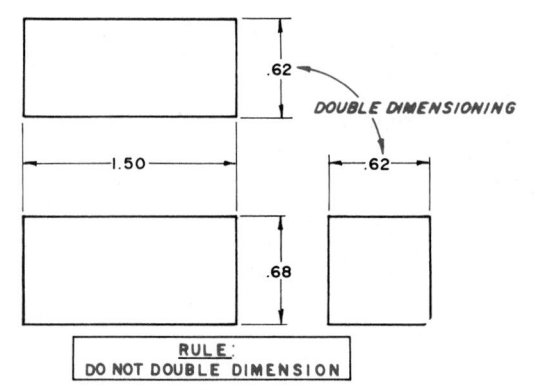

Figure 9-62 Double dimensioning

Dimensioning a Square

Figure 9-61 illustrates the proper method for dimensioning an object that is square in cross section. No center line on the object indicates that the object is square rather than round in cross section. This type of object requires a length dimension on the longitudinal view and a height dimension placed between the longitudinal view and the end view, and preceded by the square symbol.

Double Dimensioning

Figure 9-62 illustrates the concept of double dimensioning. This is sometimes referred to as "superfluous" dimensioning. Drafters and engineers should avoid double dimensioning. It is redundant and can lead to confusion.

Reference Dimensioning

Figure 9-63 illustrates the concept of reference dimensioning. In the object shown, the .62 dimension in parentheses is a reference dimension. Reference

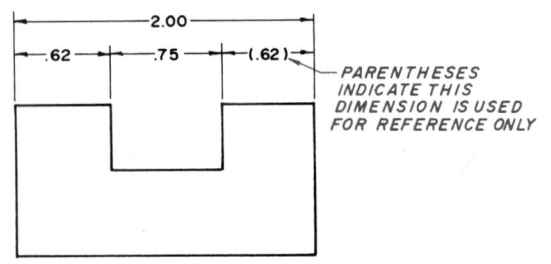

Figure 9-63 Reference dimensioning

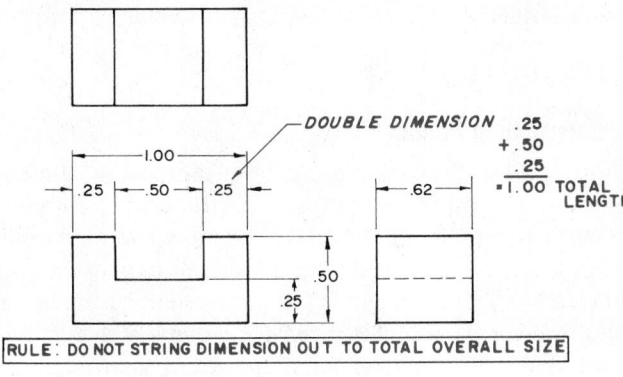

Figure 9-64   Reference dimensions

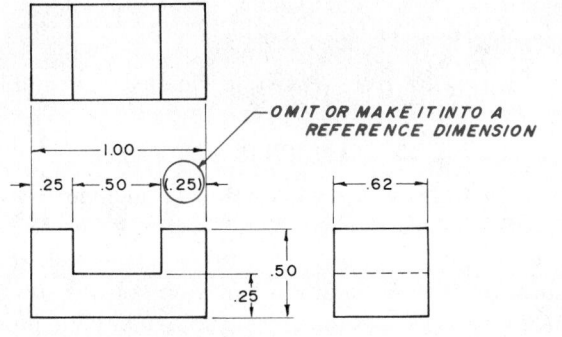

Figure 9-65   Reference dimensions

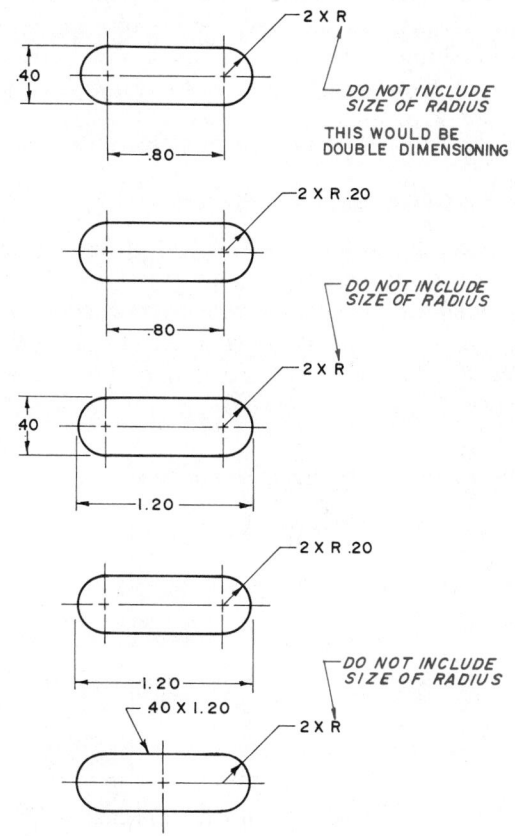

Figure 9-66   Dimensioning internal slots

dimensions are not required. They are occasionally given for information purposes only. The old method used the letters "REF" following the dimension to indicate a reference dimension. The newer and accepted method is to place reference dimensions in parentheses. *Figures 9-64* and *9-65* further illustrate the concept of reference dimensions. In Figure 9-64, the second .25 dimension in the front view of the object is a double dimension; it is superfluous. In Figure 9-65, the problem has been solved by placing the second dimension in parentheses, thereby making it a reference dimension. The drafter or engineer had two options in this case for solving the problem: omit the dimension altogether or make it a reference dimension.

Dimensioning Internal Slots

Internal slots represent a common manufacturing situation. Drafters and engineers approach internal slots as two partial holes separated by a space. There are a variety of methods for dimensioning internal slots. The preferred methods are illustrated in *Figure 9-66*. Notice from these examples that when the width of the slot is dimensioned, there is no need to indicate the size of the radius. To do so would be double dimensioning. All of the methods shown in Figure 9-66 are acceptable methods and can be substituted for each other based on the needs of the individual drawing and local preferences.

Figures 9-67 and *9-68* are examples of how to dimension internal slots for two other manufacturing

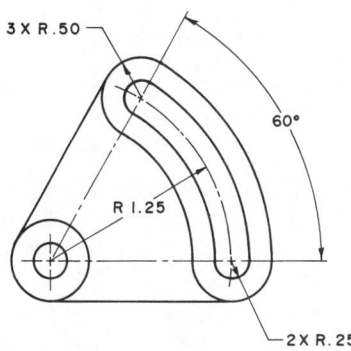

Figure 9-67   Dimensioning internal slots

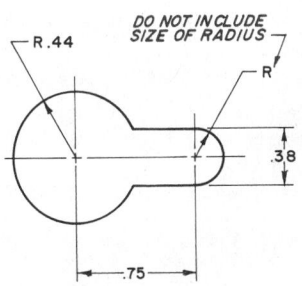

Figure 9-68   Dimensioning internal slots

situations. Note in Figure 9-67 that the distance from the concurrent center point to the center line of the slot is indicated with a radius dimension. In addition, the angle formed by the extension of the 2-hole center lines to their intersection at the converging point must also be dimensioned as an angle. Figure 9-68 illustrates the proper method for dimensioning an internal slot when one of the holes is larger than the other. If a width dimension is provided, a radius value is not provided. If a width dimension is not provided, the radius dimension must be.

Dimensioning Rounded Ends

Figure 9-69 illustrates the proper method for dimensioning the three most common situations involving external rounded ends. Parts with fully rounded ends should be dimensioned with an overall dimension and the radius should be indicated but not dimensioned, as indicated at the top of Figure 9-69. Figures with partially rounded ends, such as the exmaple in the middle of Figure 9-69, require a radius specification. For parts with rounded ends and holes, overall dimensions can be given as reference dimensions, as shown in the bottom example of Figure 9-69.

Dimensioning and Hidden Lines

There is only one rule relative to dimensioning and hidden lines:

DO NOT DIMENSION TO HIDDEN LINES

Figure 9-70 illustrates what drafters and engineers can do when faced with a situation where it is necessary to dimension hidden lines. When this is the case, and it cannot be avoided in any other way, convert the hidden-line view to a sectional view so that you are dimensioning to a solid line.

Not to Scale (NTS) Dimensions

On occasion it is necessary to enter a dimension not to scale. When this is the case, the not-to-scale dimension should be underlined with a solid straight line, as illustrated in *Figure 9-71*. The old method used a wavy line under the dimension. This is no longer acceptable practice.

Nominal Dimensions

A nominal dimension is one which represents a general classification size given to a part or product. It may not, and usually does not, express the actual size of the object dimensioned. There is usually a small amount of difference between the actual size of a part and the nominal size, as illustrated in *Figure 9-72*.

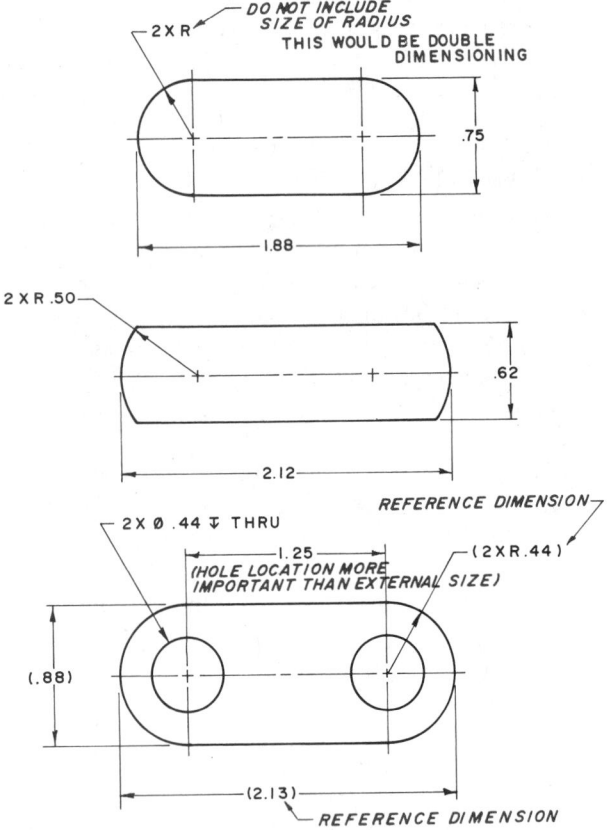

Figure 9-69   Dimensioning external rounded ends

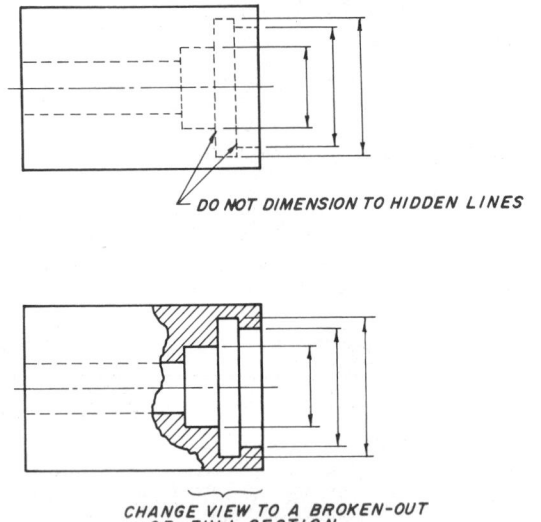

Figure 9-70   Do not dimension to hidden lines

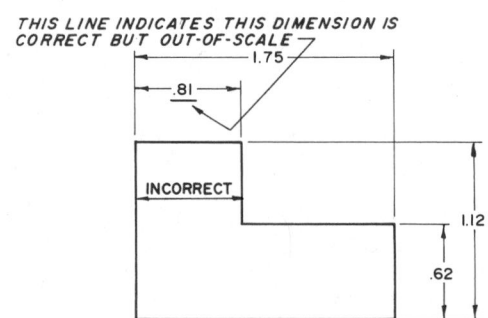

Figure 9-71   Not to scale (NTS) dimensions

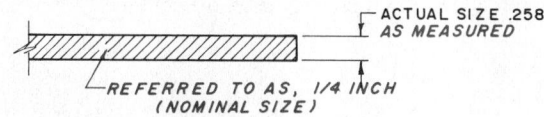

Figure 9-72   Nominal dimensions

Dimensioning External Chamfers

A chamfer is a beveled edge normally applied to cylindrical parts and a variety of different types of fasteners. Chamfers are used to eliminate rough edges that might break off or impede the assembly of parts. ***Figure 9-73*** illustrates three accepted procedures for dimensioning external chamfers. The illustration at the top shows how to dimension a chamfer using a leader and note. This is frequently done when the chamfer is a 45° chamfer. The middle illustration shows how to dimension larger chamfers at angles other than 45°. Notice that the length of the chamfer is dimensioned, as is the angle of the chamfer. The bottom illustration shows how to dimension a 45° chamfer that is very small. The dimension is called out and followed by the term "MAX." This tells the machinist that a 45° chamfer is to be applied but its length is to be no more than .010.

Dimensioning Internal Chamfers

Figure 9-74 illustrates two accepted methods for dimensioning internal chamfers. The example on the left illustrates how to call out a 45° chamfer. The example on the right illustrates how to call out a chamfer other than 45°. Notice that when the chamfer is other than 45° the dimension of the top of the hole after the chamfer has been applied must be given, as must the angle of the chamfer.

Staggered Dimensions

Figure 9-75 illustrates the proper method for applying staggered dimensions. A succession of parallel dimensions can become confusing and difficult to read. As a result, the accepted practice is to stagger the series of parallel dimensions as shown in the bottom half of Figure 9-75. Staggered dimensions are usually easier to read than aligned parallel dimensions.

Dimensioning Necks and Undercuts

Necks and undercuts are recessed cuts into cylindrical parts. They are generally used where cylinders or other geometric shapes come together. ***Figure 9-76*** illustrates the proper method for dimensioning necks and undercuts. Notice from the three examples on the left side of Figure 9-76 that necks and undercuts may be dimensioned using a leader and accompanying note. The three examples on the right side of Figure 9-76 show that necks and undercuts may also be dimensioned using a combination of notes, angular dimensions, and size dimensions.

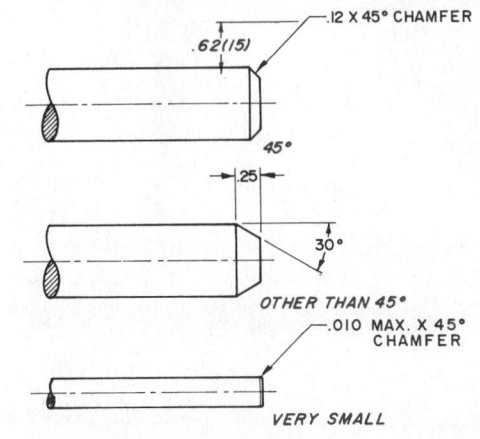

Figure 9-73   Dimensioning external chamfers

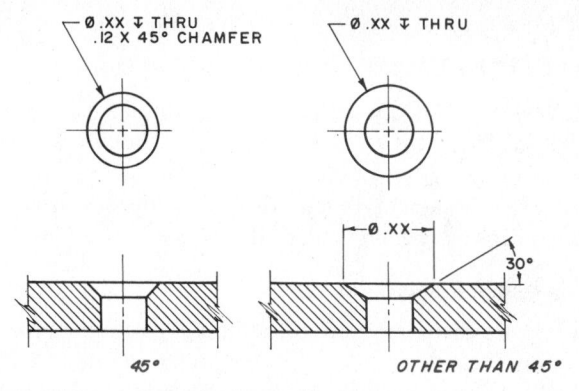

Figure 9-74   Dimensioning internal chamfers

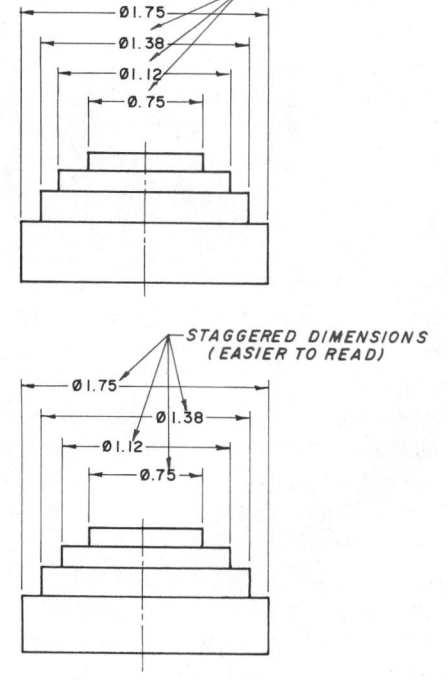

Figure 9-75   Staggered dimensions

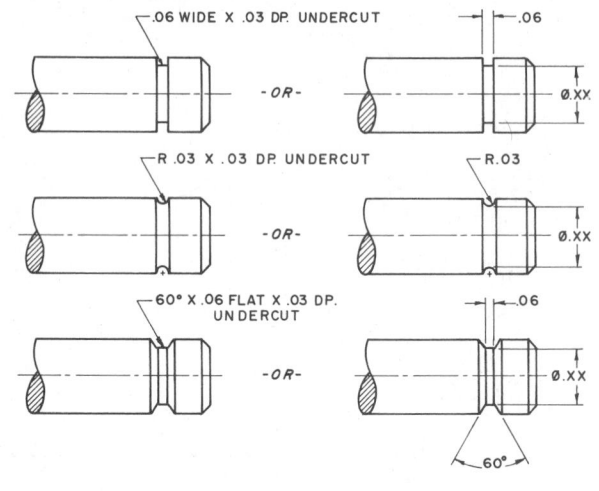

Figure *9-76*   Dimensioning necks and undercuts

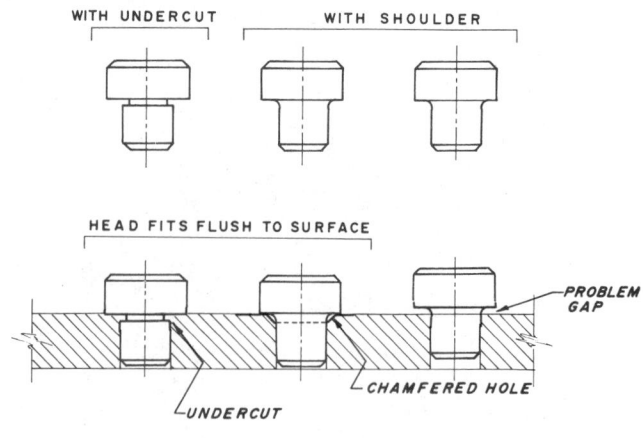

Figure *9-77*   Headfits for undercuts and chamfers

Head Fits for Undercuts and Chamfers

Figure 9-77 illustrates the proper method for showing head fits for parts with undercuts and shoulders. With an undercut, the head fits flush with the surface. With a shoulder, the head will only fit flush with the surface if the surface has been chamfered.

Dimensioning Knurls

Knurls are diamond-shaped or parallel patterns cut into cylindrical surfaces to improve gripping, to improve the bonding between parts for permanent press fits, and, sometimes, for decoration. The accepted method for dimensioning both parallel and diamond-shaped knurls are illustrated in *Figure 9-78*. Notice in this figure that when knurls are fully dimensioned on a drawing, there is no need to actually show them. The callouts for a knurl should include the type, pitch, and diameter. The initials "DP" in the knurl callouts stand for diametrical pitch. The most commonly used diametrical pitches for knurling are 64 DP, 96 DP, 128 DP, and 160 DP. In addition to the knurling callout, the length of the knurl and the diameter of the cylinder that will receive the knurls must be shown.

Dimensioning Keyways

Figure 9-79 illustrates the preferred methods for dimensioning both exterior and interior keyways. Notice that both size and location dimensions are required to completely dimension a keyway. The size dimension gives the overall width of the keyway, and the location dimension locates the keyway on the object in question from a specified datum such as the center line of the object. With an exterior keyway, the depth is shown using a dimension which runs from the bottom of the keyway to the bottom of the part in question and parallel to the center line or axis of the shaft. With an interior keyway, the depth of the keyway is specified using a dimension which runs from the top of the keyway to the bottom of the hole and parallel to the center line of the hole.

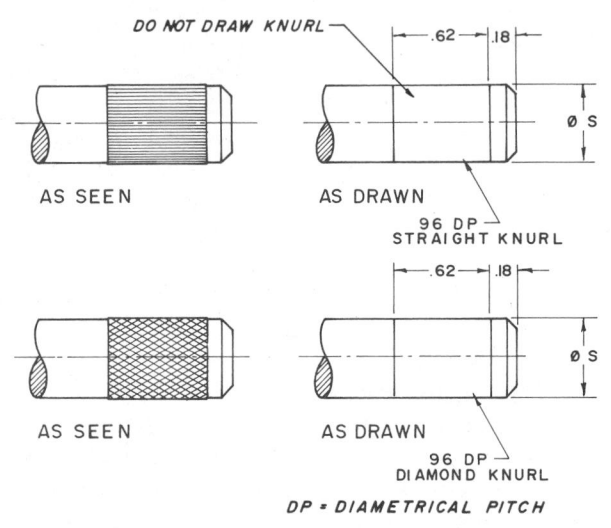

Figure *9-78*   Dimensioning knurls

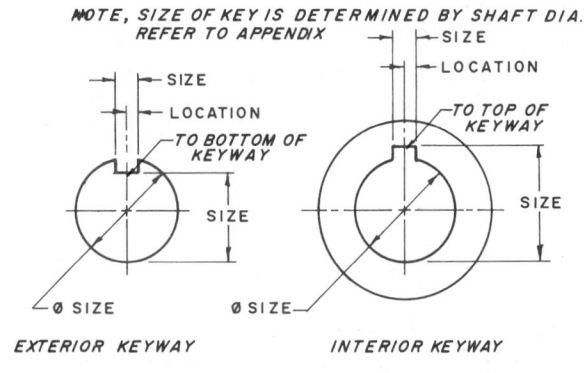

Figure *9-79*   Dimensioning keyways

Figure 9-80 illustrates the preferred method for dimensioning a woodruff keyway. A woodruff keyway requires both size and location dimensions. The size dimensions specify the width and depth of the keyway. The location dimension from a specified datum such as a center line locates the keyway in the part as shown in Figure 9-80.

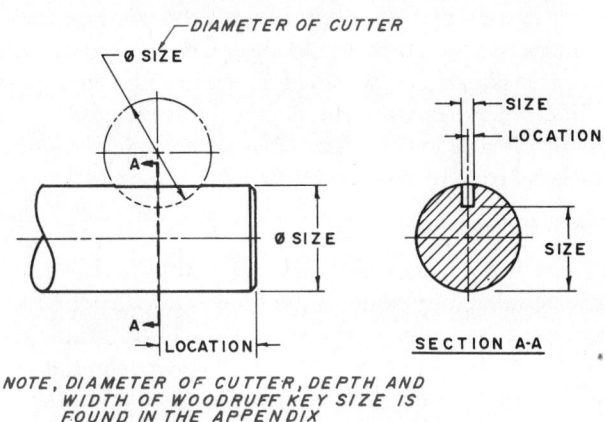

NOTE, DIAMETER OF CUTTER, DEPTH AND
WIDTH OF WOODRUFF KEY SIZE IS
FOUND IN THE APPENDIX

Figure 9-80   Dimensioning a woodruff keyway

In addition, the location of the center of the keyway from one end of the shaft must also be shown. The diameter of the shaft and the woodruff cutter must also be dimensioned, as shown in Figure 9-80.

Dimensioning Flat Tapers

Figure 9-81 illustrates the preferred method for dimensioning flat tapers. The method on the left uses 2-size dimensions and a leader with a taper symbol attached. In addition, the taper ratio must be shown. The 5:1 taper ratio called out for the part on the left means that the part tapers one unit in the vertical direction for every five units traveled in the horizontal. The example on the right illustrates another acceptable method for dimensioning flat tapers. In this method, 3-size dimensions provide all of the information needed to manufacture the part with the proper taper.

Dimensioning Round Tapers

Figure 9-82 illustrates two acceptable methods for dimensioning round tapers. Round tapers can be dimensioned by providing the following dimensional data: (a) the diameter or width at both ends of the taper; (b) the length of the tapered feature; and (c) the rate of taper. The example on the right shows the diameter at both ends of the taper. The example on the left shows the length of the tapered feature and is in the taper ratio of 10:1, meaning that the part is tapered one unit for every ten units parallel to the horizontal center line of the shaft.

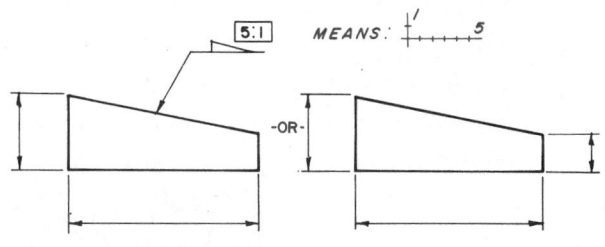

Figure 9-81   Dimensioning flat tapers

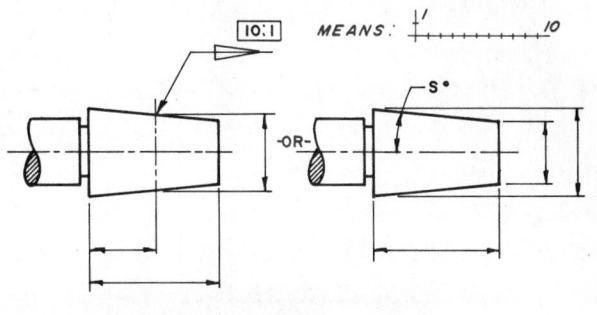

Figure 9-82   Dimensioning round tapers

Dimensioning Threads

Figure 9-83 shows the preferred methods for dimensioning threaded fasteners. Notice that it is not necessary to attempt to draw an actual representation of the thread. Rather, the schematic approach shown in the middle and bottom examples is preferred. The threads themselves are called out in a thread note attached to a leader. In addition, the nominal diameter of the shaft being threaded and the longitudinal distance of the threading must also be dimensioned, as shown in Figure 9-83. A chamfer call out should be shown with the leader line to indicate how the end of the threaded fastener is to be chamfered.

Dimensioning a Pad and a Boss

Figure 9-84 illustrates the preferred method for dimensioning a pad and a boss. A boss is a round projection from a part. A pad is a projection of any shape other than round. Notice from Figure 9-84 that the boss requires a diameter dimension, a dimension showing how far the boss projects out from the part, and a radius indicator. The pad requires 2-size dimensions: one for length and one for width. It also requires a projection dimension and a radius indicator.

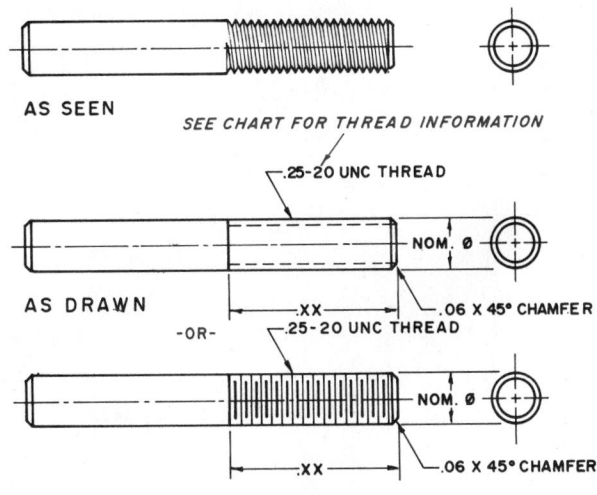

Figure 9-83   Dimensioning threads

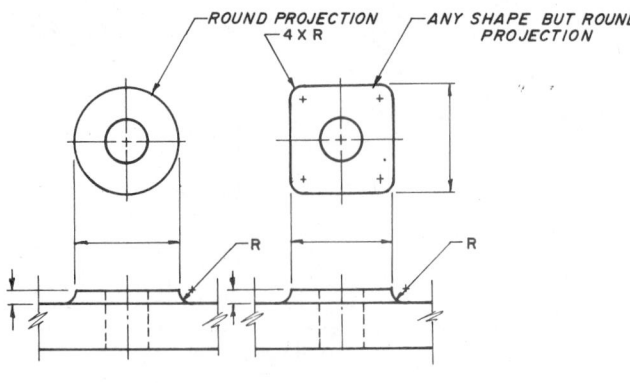

Figure 9-84 Dimensioning a pad and a boss

Dimensioning Sheet Metal Bends

Figure 9-85 illustrates the different methods for dimensioning sheet metal bends. When dimensioning a sheet metal object, allowances must be made for the bends. The intersection of the surfaces adjacent to a bend is called the "mold line." This is the line which is used for dimensioning purposes. Notice in Figure 9-85 that each bend has an inside mold line and an outside mold line. Dimensions begin at the intersection of the outside mold line, as shown.

Dimensioning Sectional Views

Figure 9-86 illustrates the rule for dimensioning sectional views. The dimension should, whenever possible, be placed outside the object. Placing dimensions inside an object causes them to conflict with the section lines. When there is no other way to apply a dimension than to place it inside of a sectioned area, remove enough of the section lining to permit the dimension to be easily read and the dimension lines to be easily seen, as shown in the right-hand example in Figure 9-86.

Dimensioning Pyramids

Figure 9-87 illustrates the acceptable method for dimensioning a pyramid. A pyramid is dimensioned by showing the height in the front view and the dimensions of the base and the center of the vertex in the top view. If the pyramid base is square, it is only necessary to show the base dimension on one side of the base in the top view and to indicate that the base is square by using the appropriate symbol.

Dimensioning a Cone

Figure 9-88 illustrates the preferred methods for dimensioning a cone. A cone is dimensioned by showing its altitude in the view where the cone appears as a triangle and showing the diameter of the base between the top view and front view. The frustum of a cone is dimensioned by giving the diameter between the top and front view, and both heights in the front view where the cone would appear as a triangle. The example on the right-hand side of Figure 9-88 illustrates how to dimension a cone that is cut off at an angle. The diameter of the cone's base is shown between the top and front views. The height of the cone before it is cut and the height of the cone at the center line of the cone after the cut are shown in

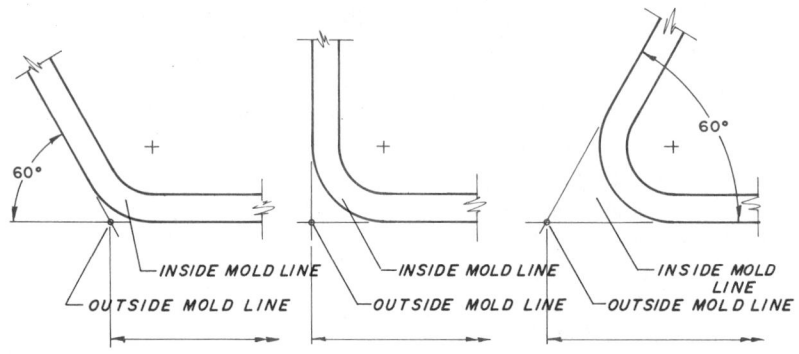

Figure 9-85 Dimensioning sheet metal bends

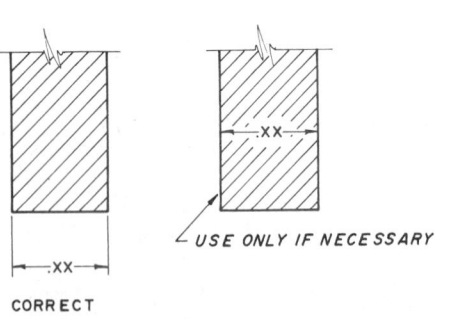

Figure 9-86 Dimensioning sectional views

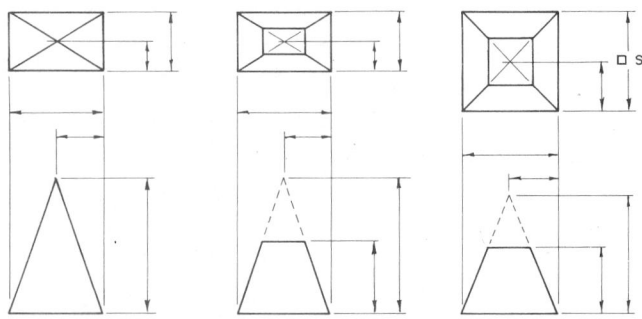

Figure 9-87 Dimensioning a pyramid

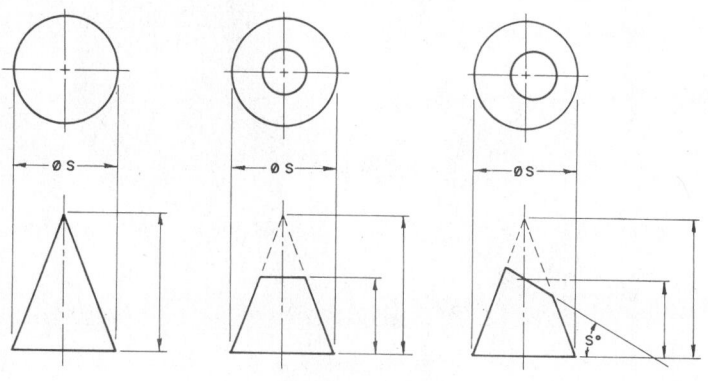

Figure 9-88 Dimensioning a cone

the front view where the cone would appear as a triangle. The angle of the cut also is dimensioned in this view.

Dimensioning Concentric and Nonconcentric Shafts

Figure 9-89 illustrates the various methods that can be used for dimensioning concentric and nonconcentric shafts and cylinders. The example at the top uses the chain dimensioning concept. The example in the center uses the datum or baseline dimensioning concept. The example at the bottom uses the datum dimensioning concept combined with a reference dimension.

Rectangular Coordinate Dimensioning

Figure 9-90 illustrates the proper methods for applying rectangular coordinate dimensioning. Rectangular coordinate dimensioning is a system which locates features of the drawing relative to each other, individually or as a group, from a datum or series of datums. Round holes and other similar features are located by dimensioning distances to their center points. Linear dimensions in the rectangular coordinate dimension system specify distances from two or three mutually perpendicular planes which are used as datums. Coordinate dimensions should clearly indicate what features of the part being dimensioned establish these datums.

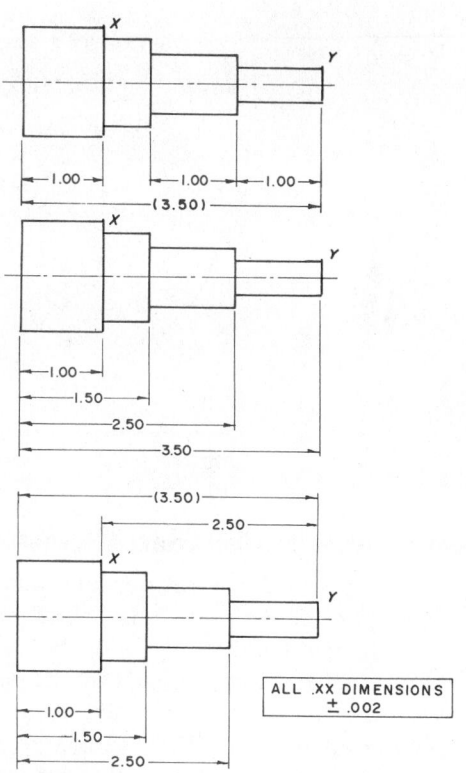

Figure 9-89 Dimensioning concentric and nonconcentric shafts

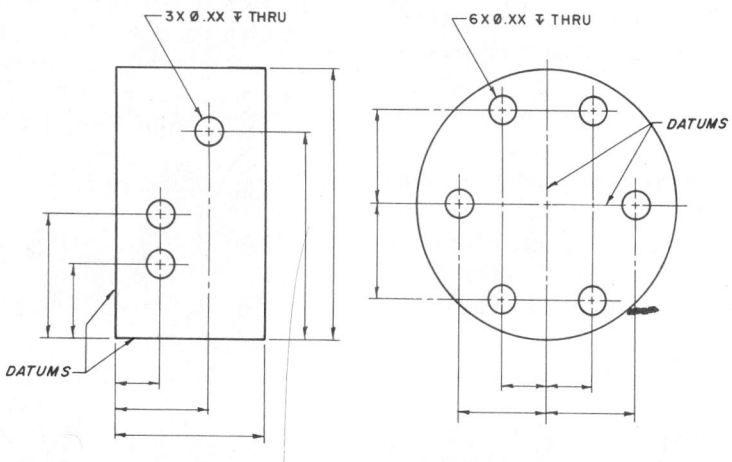

Figure 9-90 Rectangular coordinate dimensioning

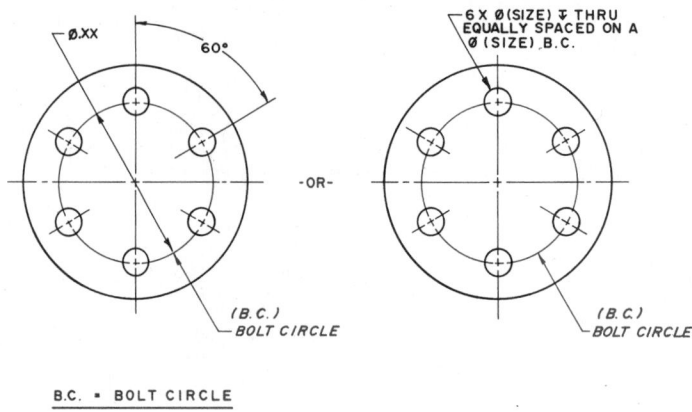

Figure 9-91   Dimensioning holes on a bolt center diameter

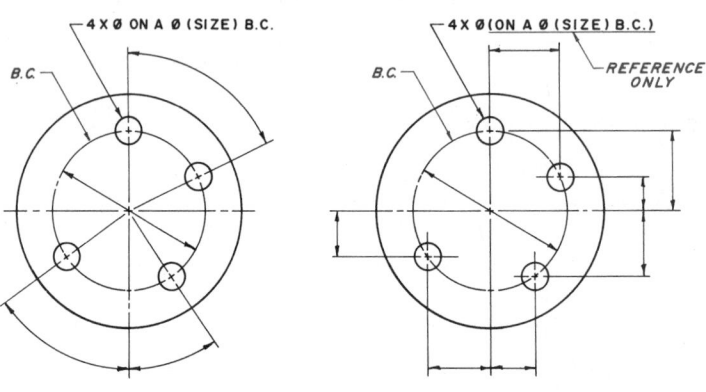

Figure 9-92   Dimensioning holes around a center

Dimensioning Holes on a Bolt Center Diameter

Figure 9-91 illustrates the proper methods used in dimensioning holes on a bolt center diameter. In the example on the left, the diameter of the bolt's circle is given and one angle is shown to indicate the number of degrees between the center lines of the six circles. This is sufficient information for the machinist to properly drill the holes. In the example on the right, a note is used to accomplish the same thing. Notice that the note specifies the number and size of the holes. It explains that the holes are drilled through the object and equally spaced on a bolt center. Either of these methods may be used for dimensioning holes around a center. *Figure 9-92* shows two other methods which are used for dimensioning holes around a center. These methods are used when the holes around the bolt center are not equally spaced. The example on the left gives a dimension for the bolt center, a note containing the number of holes and their size, and angles turned off of a reference center line for locating the center lines of the other three holes. The example on the right substitutes linear dimensions for angular dimensions.

Finish Marks

Figure 9-93 illustrates three different systems for applying finish marks to drawings. The traditional system and the general system have been replaced by the new system. However, engineers should be familiar with all three systems so that they are prepared to interpret finish marks applied to drawings produced in years past. Note the sizes specified when applying finish marks. A finish mark in the new system, which is the type of finish mark that should be used on all drawings now, resembles a mechanically produced check mark that forms a 60° "v." Finish marks are used for specifying the type of surface finish to be applied to any machined surface. Finish marks should always appear where the surface in question appears as an edge view. Finish marks are not used when dimensioning notes are given for such machining processes as drilling and boring, as they would be redundant. Finish marks are also omitted when the part in question is made of machined stock. When all surfaces on a part are finished, a note stating "Finish all over" or the letters "FAO" may be substituted.

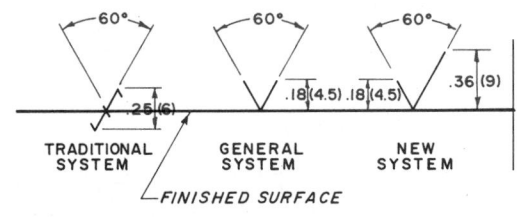

Figure 9-93   Finish marks

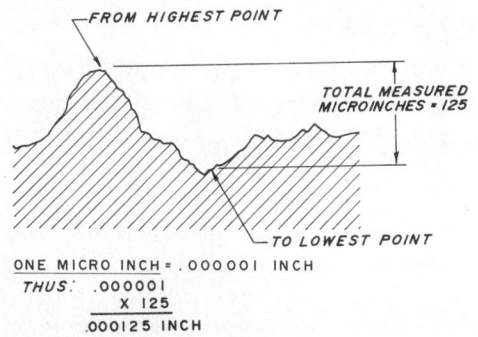

Figure 9-94   Microinches

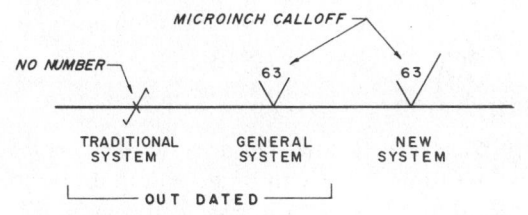

Figure 9-95   Microinch callouts

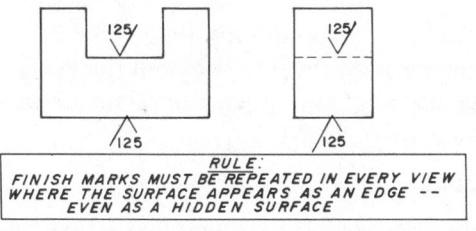

RULE:
FINISH MARKS MUST BE REPEATED IN EVERY VIEW
WHERE THE SURFACE APPEARS AS AN EDGE --
EVEN AS A HIDDEN SURFACE

Figure 9-96   Applying finish marks to drawings

MICROMETRES AA RATING	MICROINCHES AA RATING	APPLICATION
25.2	1000	Rough, low grade surface resulting from sand casting, torch or saw cutting, chipping, or rough forging. Machine operations are not required because appearance is not objectionable. This surface, rarely specified, is suitable for unmachined clearance areas on rough construction items.
12.5	500	Rough, low grade surface resulting from heavy cuts and coarse feeds in milling turning, shaping, boring, and rough filing, disc grinding and snagging. It is suitable for clearance areas on machinery, jigs, and fixtures. Sand casting or rough forging produces this surface.
6.3	250	Coarse production surfaces, for unimportant clearance and cleanup operations, resulting from coarse surface grind, rough file, disc grind, rapid feeds in turning, milling, shaping, drilling, boring, grinding, etc., where tool marks are not objectionable. The natural surfaces of forgings, permanent mold castings, extrusions, and rolled surfaces also produce this roughness. It can be produced economically and is used on parts where stress requirements, appearance, and conditions of operations and design permit.
3.2	125	The roughest surface recommended for parts subject to loads, vibration, and high stress. It is also permitted for bearing surfaces when motion is slow and loads light or infrequent. It is a medium commercial machine finish produced by relatively high speeds and fine feeds taking light cuts with sharp tools. It may be economically produced on lathes, milling machines, shapers, grinders, etc., or on permanent mold castings, die castings, extrusion, and rolled surfaces.
1.6	63	A good machine finish produced under controlled conditions using relatively high speeds and fine feeds to take light cuts with sharp cutters. It may be specified for close fits and used for all stressed parts, except fast rotating shafts, axles, and parts subject to severe vibration or extreme tension. It is satisfactory for bearing surfaces when motion is slow and loads light or infrequent. It may also be obtained on extrusions, rolled surfaces, die castings and permanent mold castings when rigidly controlled.
0.8	32	A high-grade machine finish requiring close control when produced by lathes, shapers, milling machines, etc., but relatively easy to produce by centerless, cylindrical, or surface grinders. Also, extruding, rolling or die casting may produce a comparable surface when rigidly controlled. This surface may be specified in parts where stress concentration is present. It is used for bearings when motion is not continuous and loads are light. When finer finishes are specified, production costs rise rapidly; therefore, such finishes must be analyzed carefully.
0.4	16	A high quality surface produced by fine cylindrical grinding, emery buffing, coarse honing, or lapping, it is specified where smoothness is of primary importance, such as rapidly rotating shaft bearings, heavily loaded bearing and extreme tension members.
0.2	8	A fine surface produced by honing, lapping, or buffing. It is specified where packings and rings must slide across the direction of the surface grain, maintaining or withstanding pressures, or for interior honed surfaces of hydraulic cylinders. It may also be required in precision gauges and instrument work, or sensitive value surfaces, or on rapidly rotating shafts and on bearings where lubrication is not dependable.
0.1	4	A costly refined surface produced by honing, lapping, and buffing. It is specified only when the design requirements make it mandatory. It is required in instrument work, gauge work, and where packing and rings must slide across the direction of surface grain such as on chrome plated piston rods, etc. where lubrication is not dependable.
0.05	2	Costly refined surfaces produced only by the finest of modern honing, lapping, buffing, and superfinishing equipment. These surfaces may have a satin or highly polished appearance depending on the finishing operation and material. These surfaces are specified only when design requirements make it mandatory. They are specified on fine or sensitive instrument parts or other laboratory items, and certain gauge surfaces, such as precision gauge blocks.
0.025	1	

Figure 9-97   Typical surface roughness height applications

familiarize themselves with these ratings and their corresponding applications.

X-Y-Z Coordinates

Figure 9-98 illustrates the concept of X-Y-Z coordinates. In order to be able to fully understand the various dimensioning systems, one must understand the concept of X-Y-Z coordinates. This is particularly true in the case of tabular dimensioning. Note from this illustration that X coordinates move from an origin in a horizontal direction, Y coordinates move from an origin in a vertical direction, and Z coordinates move from an origin in a direction indicating depth.

Surface imperfections of a part are usually so small that they may be barely visible to the human eye. Consequently, they are measured in microinches. *Figure 9-94* illustrates the concept of the microinch.

Figure 9-95 illustrates the proper method for applying microinch callouts to finish marks. Notice that microinch callouts are not used with the traditional finish mark system.

Figure 9-96 illustrates the rule for applying finish marks to drawings. Note that finish marks are to be repeated in every view where the surface in question appears as an edge. Table 46 in the Appendix shows the surface roughness textures obtainable by common production methods. Engineers should be familiar with the information in this table. The various manufacturing processes are listed along the left side of the table. The typical applications are listed along the bottom of the table, and the corresponding roughness height rating for each process/application is listed horizontally in a bar-graph format. *Figure 9-97* is a table of typical surface roughness height applications. Microinch ratings are listed in the left-most column of the table. The corresponding micrometer ratings are listed in the next column. Applications are then explained horizontally beside each rating. Engineering students should

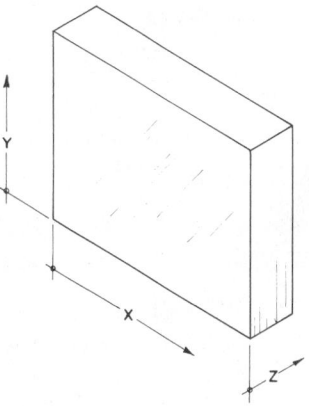

Figure 9-98   X-Y-Z coordinates

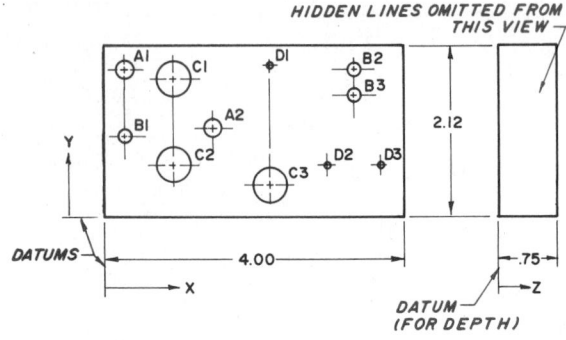

HIDDEN LINES OMITTED FROM THIS VIEW

DATUMS

DATUM (FOR DEPTH)

HOLE	Ø	X	Y	Z	NOTE
A1	.25	.30	1.75	THRU	–
A2		1.38	1.06	.38	–
B1	.18	.30	.94	THRU	–
B2		3.25	1.75	THRU	–
B3		3.25	1.50	THRU	–
C1	.38	.88	1.68	.38	–
C2		.88	.62	.25	–
C3		1.12	.38	.38	–
D1	.08	2.12	1.81	THRU	–
D2		3.00	.62	.38	–
D3		3.62	.62	.38	–

Figure 9-99   Tabular dimensioning

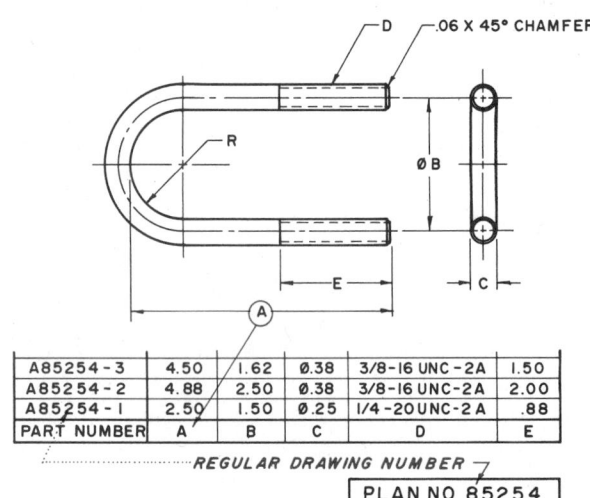

.06 X 45° CHAMFER

A85254-3	4.50	1.62	Ø.38	3/8-16 UNC-2A	1.50
A85254-2	4.88	2.50	Ø.38	3/8-16 UNC-2A	2.00
A85254-1	2.50	1.50	Ø.25	1/4-20 UNC-2A	.88
PART NUMBER	A	B	C	D	E

REGULAR DRAWING NUMBER

PLAN NO 85254

Figure 9-100   Tabular drawing

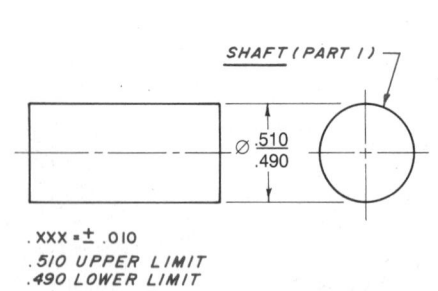

SHAFT (PART 1)

$\varnothing \begin{smallmatrix} .510 \\ .490 \end{smallmatrix}$

.XXX = ± .010
.510 UPPER LIMIT
.490 LOWER LIMIT

Figure 9-101   Shaft limits

Tabular Dimensioning

Figure 9-99 illustrates the concept of tabular dimensioning. Tabular dimensioning is a system in which size and location dimensions are given in a table rather than on the drawing itself. Notice that the only dimensions applied to the drawing itself are the overall length, width, and depth dimensions. Each hole is given an alphanumeric designation. Holes with the same letter designation have the same diameter. Holes with the same letter designation but a different numerical suffix have the same diameter but a different depth. The holes are located in the part using X, Y, and Z coordinates from the specified datums. In certain cases such as the one illustrated in Figure 9-99, tabular dimensioning is a more convenient, less confusing approach.

Tabular Drawing

When parts have similar features but different dimensions, the dimensions can be given in tabular form to decrease the amount of drawing that must be done. With a tabular drawing, one drawing of the part is made with all appropriate dimension lines applied. However, instead of dimension notations, numerical notations are substituted. Then a table of dimensions is prepared showing the part numbers for all of the parts that the drawing applies to and the dimensions for each letter notation, *Figure 9-100*.

Tolerancing

Even the best machinist operating the most accurate machine tool is not able to produce a part as precisely and accurately as specified on a drawing. For this reason it is necessary to apply tolerances when dimensioning parts. A tolerance is the total amount of variation permitted from the design size of a part. A basic rule of thumb when applying tolerances is to make them as large as possible and still be able to produce a usable part. Tolerances may be expressed as limits, as shown in *Figures 9-101* and *9-102* or as the design size followed by the tolerance (12.382 ± 0.003). The design size of a part is also known as the nominal size. This is the dimension from which the limits are calculated. To calculate the upper limit of a toleranced dimension,

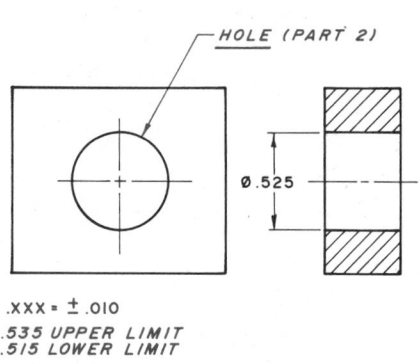

HOLE (PART 2)

Ø .525

.XXX = ± .010
.535 UPPER LIMIT
.515 LOWER LIMIT

Figure 9-102   Hole limits

add the allowable tolerance to the nominal size. To calculate the lower limits for such dimensions, subtract the allowable tolerance from the nominal size, Figure 9-101 and 9-102. There are several definitions with which drafters and engineers should be familiar before applying tolerances on drawings. The most important of these are:

Actual Size. Also called the produced size, this is the actual size of the part after it is produced.

Allowance. The amount of room between two main parts, usually considered the tightest possible fit between the two parts.

Bilateral Tolerance. A bilateral tolerance is one in which variation from the nominal dimension is allowed in both the plus and the minus direction (.500 ± .002).

Datum. A theoretically perfect surface, plane, axis, center plane, or point from which dimensions for related features are established.

Least Material Condition (LMC). The condition which exists when the least material is available for the features being dimensioned. LMC is the lower limit for an external feature and the upper limit for an internal feature.

Limits. The largest and smallest possible size in which a part may be produced and still be usable. The limits are calculated by applying the specified tolerance to the nominal dimension.

Maximum Material Condition (MMC). The condition of a part in which the most material is present. MMC is the upper limit for an external feature and the lower limit for an internal feature.

Nominal Dimension. This is the dimension from which the limits are calculated. By applying the specified tolerance to the nominal dimension one can calculate the upper and lower limits.

Tolerance. The total permissible variation in size for a part. The tolerance is the difference between the upper and lower limits for a part.

Unilateral Tolerance. A tolerance that may vary in only one direction; either in the plus or minus direction but not both:

$$.500 \begin{array}{c} +.000 \\ -.002 \end{array} \qquad .250 \begin{array}{c} +.002 \\ -.000 \end{array}$$

In addition to these tolerance-related definitions, drafters and engineers should also be familiar with the types of fits that are used with mating parts. The following definitions relate specifically to fits between mating parts.

Fit. Fit refers to the amount of tightness or looseness between mating parts. There are three types of fits: clearance, interference, and transition.

Clearance Fit. A clearance fit is one in which there is space remaining after mating parts have been assembled.

Interference Fit. The type of fit which occurs when there is a negative allowance, or in other words, when the shaft is larger than the collar with which it must mate.

Transition Fit. The type of fit in which there may be either a clearance or an interference fit after mating parts have been assembled. This occurs when the smallest shaft size within the allowable tolerance will fit within the largest hole size, resulting in a clearance fit, and when the largest shaft size within the allowable tolerance will interfere with the smallest hole within an allowable tolerance, creating an interference fit.

Shaft Limits

Figure 9-101 illustrates the concept of shaft limits. In this figure the shaft has a nominal dimension of .500. The tolerance specified is ±.010. To determine the upper limit for the shaft, the tolerance of .010 is added to the nominal dimension of .500, resulting in an upper limit of .510. To determine the lower limit of the shaft, the tolerance of .010 is subtracted from the nominal dimension of .500, resulting in a lower limit of .490.

Hole Limits

Figure 9-102 illustrates the concept of hole limits. The hole through the part shown has a nominal diameter of .525. The specified tolerance is ±.010. To determine the upper limit for the hole, the tolerance of .010 is added to the nominal dimension of .525, resulting in an upper limit of .535. The lower limit is determined by subtracting the tolerance of .010 from the nominal dimension of .525, resulting in a lower limit of .515.

Allowance

Figure 9-103 illustrates the concept of allowance or the tightest fit between mating parts. Part 1 is a shaft that, when assembled, will mate with Part 2. To determine the allowance between Part 1 and Part 2, the maximum material condition of the shaft (the largest allowable size) is subtracted from the maximum material condition of the hole (the smallest size hole). The maximum material condition of the shaft is .510. The maximum material condition of the hole is .515. Therefore, the allowance between Part 1 and Part 2 is .005.

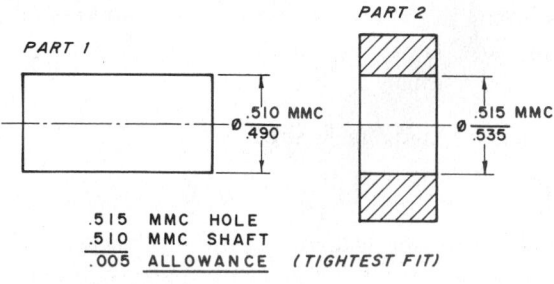

Figure 9-103 Allowance

Clearance

Figure 9-104 illustrates the concept of clearance. The shaft, Part 1, is to mate with the collar, Part 2, during assembly. To determine the clearance between Part 1 and Part 2, the least material condition of the shaft is subtracted from the least material condition of the hole. The least material condition of the shaft is .490. The least material condition of the hole is .535. Therefore, the clearance between Part 1 and Part 2 is .045.

Interference Fit

Figure 9-105 illustrates the concept of interference fit. The smallest shaft in this example is larger than the largest corresponding hole. To determine the amount of interference, you can calculate the allowance and the clearance. The allowance, in the case of an interference fit, represents the maximum amount of interference. The clearance, in an interference fit, represents the smallest amount of interference. You can see from the calculations in Figure 9-105 that the maximum amount of interference during assembly will be .007, or the allowance. The minimum amount of interference will be .004, or the clearance.

Transition Fit

Figure 9-106 illustrates the concept of transition fit. In this example there can be both a clearance fit and an interference fit. The largest shaft in this example would interfere with the smallest hole. However, the smallest shaft will fit through the largest hole. This condition is illustrated in the calculations.

Size Limits

Figure 9-107 illustrates the concept of size limits. The cylindrical part shown in this figure will be usable if produced at any size between the upper and lower length limits, and at any diameter between the upper and lower diameter limits, inclusive. Applying size limits such as those illustrated in Figure 9-107 cuts down on waste, requires less time to produce the part, and, therefore, cuts the cost of production and the corresponding costs of the part.

Design Size

Figure 9-108 illustrates the concept of design size. Design size is another way of stating nominal size. The design size represents the ideal the drafter or engineer is seeking. Ideally, the produced size will match the design size. However, since this is not practical, tolerances are normally applied.

Maximum and Minimum Sizes

Figure 9-109 illustrates the concept of maximum and minimum sizes. These are the sizes between which the actual length and diameter of the produced part must fall. After all machining processes have been completed, the inspection department will check the part. If the actual produced size falls within the established maxi-

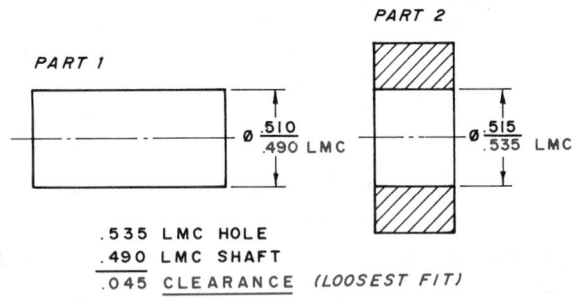

Figure 9-104 Clearance

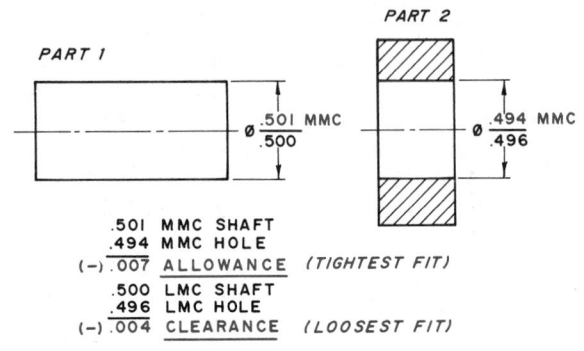

Figure 9-105 Interference fit

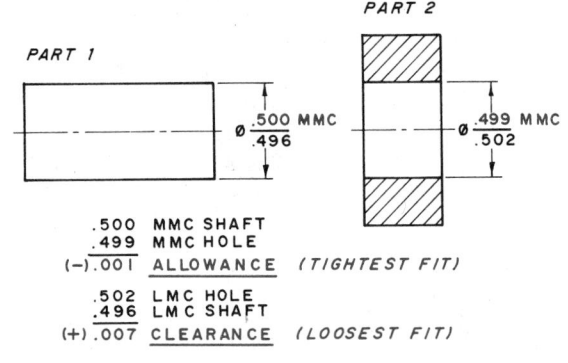

Figure 9-106 Transition fit

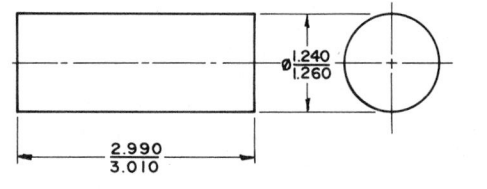

Figure 9-107 Size limits

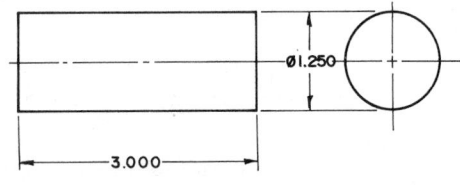

Figure 9-108 Design Size

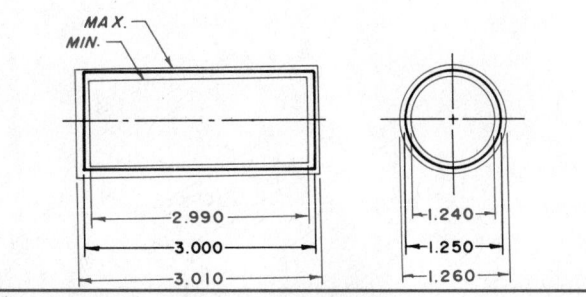

Figure 9-109 Maximum and minimum sizes

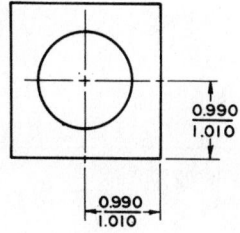

Figure 9-110 Location limits

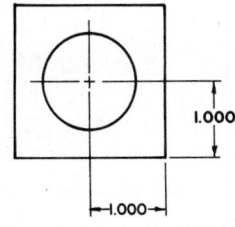

Figure 9-111 Design location

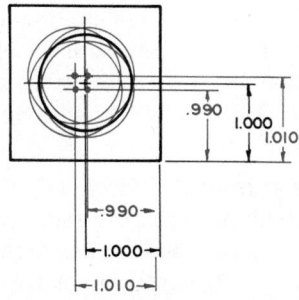

Figure 9-112 Maximum and minimum locations

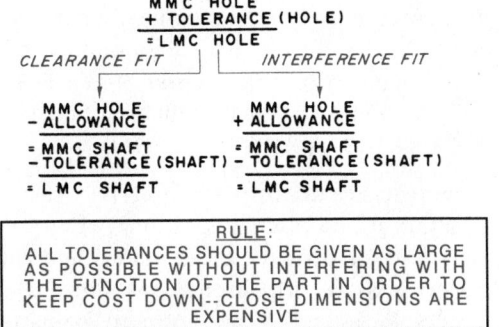

Figure 9-113 Calculating fits

mum and minimum sizes (inclusive), the part will be passed as usable. If it is smaller in any dimension than the minimum size, the part will be scrapped as waste. If it is larger in any dimension than the maximum size, it might be salvaged through additional machining.

Location Limits

Just as limits are applied to size dimensions, they are also applied to location dimensions, and for the same reasons. Machinists are not able to locate features on a part precisely to the design dimension. Consequently, limits must be applied to location dimensions.

Figure 9-110 illustrates how limits are applied to location dimensions. The part shown in Figure 9-110 will be accepted if the central line of the hole falls within the limits shown.

Design Location

Figure 9-111 illustrates the concept of design location. The design location is the location the drafter or engineer actually wants. It is the theoretically ideal location of the feature. The dimensions shown on this figure are the design dimensions, sometimes referred to as nominal dimensions. The specified tolerance is applied to these nominal dimensions to determine the upper and lower limits for locating the feature.

Maximum and Minimum Locations

Figure 9-112 illustrates the concept of maximum and minimum locations. By applying the specified tolerance to the nominal dimensions, the maximum and minimum locations of the center line of the feature in this figure can be determined. The tolerances applied to the nominal dimension give the machinist latitude in both the X and Y directions. This establishes a square tolerance zone within which the actual center point must fall. So long as the center of the feature falls anywhere within the prescribed tolerance zone, the parts will be accepted as usable.

Calculating Fits

Figure 9-113 illustrates the mathematics involved in calculating fits. Take special note of the rule that states "All dimensions should be given as large as possible without interfering with the function of the part in order to keep costs down—close dimensions are expensive." The tighter the fit, the more difficult the part is to manufacture. The more difficult the part is to manufacture, the more it costs to manufacture. Therefore, an engineer and a drafter should never assign a tighter fit than is absolutely necessary for the part to properly function.

Matching Parts

Figure 9-114 illustrates the concept of matching parts. During assembly, Part A must fit over Part B. You can see from this drawing that the hole in Part A must be large enough to accommodate the size and location tolerances of the pins in Part B. The parts must be able to fit if the holes in Part A are bored and located on the low side of the tolerance, and the pins in Part B are made and located on the high side of the tolerance. Obviously, determining the dimensions to be used for matching parts takes a good deal of careful planning. It is critical that features on mating parts be located from the same datum as shown in this figure. Some additional factors to consider when planning the dimensions for matching parts are:

1. The length of time the parts are to be engaged
2. The speed at which mating parts will move after assembly
3. The load of the parts after assembly
4. The type of lubrication used if lubrication is necessary
5. The average temperature at which parts will operate
6. The level of humidity during operation
7. The materials of which parts are made
8. The required functional lifespan of the part
9. The capabilities of machine tools and the people that operate them in the manufacturing plant
10. Overall costs of the part

Standard Fits

Figure 9-115 shows the different types of standard fits. The American National Standards Institute document governing fits is ANSI B4.1-1967, R1987. This standard covers the five types of fits listed in Figure 9-115. Each of these fits is explained below.

Running or Sliding Clearance Fits (RC). Running or sliding fits are used to provide a similar running or sliding performance, with an allowance for suitable lubrication, for all sizes within the specified range. *Figure 9-116* is a table of American National Standard Running and Sliding Fits (ANSI B4.1-1967, R1987). From this table you can see that there are nine classes of running and sliding fits: Classes RC1, RC2, RC3, RC4, RC5, RC6, RC7, RC8, and RC9.

Class RC1 is used with close sliding fits for accurately locating parts that must fit together without any play. Class RC2 is used for sliding fits where accurate location is important, but where greater maximum clearance than is provided by class RC1 fits is necessary. Parts manufactured to this fit do not run freely, but they do move and turn easily. Class RC3 is used for precision running fits. This class ensures the closest possible fit that can be expected to run freely. This fit is used for precision work that will operate at slow speeds under light pressures, but is not suitable where significant temperature differences are likely to occur.

Class RC4 is used for close running fits on accurate machinery with medium surface speeds and pressure when accurate location and very little play is desired. Classes RC5 and RC6 are used for medium running fits on accurate machinery with higher running speeds

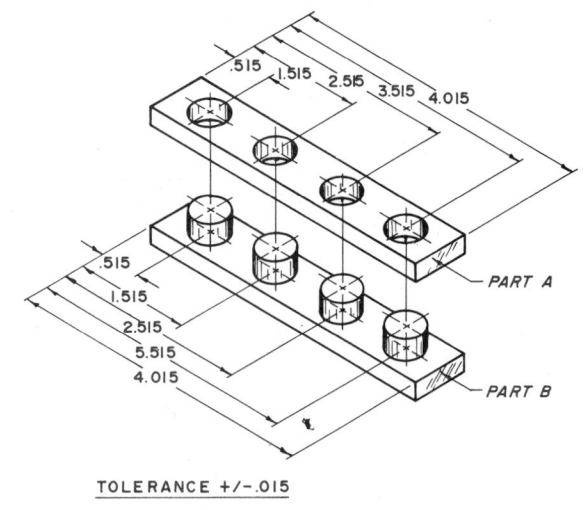

TOLERANCE +/-.015

Figure 9-114 Matching parts

STANDARD FITS	
RC	RUNNING AND SLIDING FITS
LC	CLEARANCE LOCATIONAL FITS
LT	TRANSITION LOCATIONAL FITS
LN	INTERFERENCE LOCATIONAL FITS
FN	FORCE AND SHRINK FITS

Figure 9-115 Standard fits

and/or heavier pressures. Class RC7 is used for free running fits where accuracy is less important and/or where significant temperature variations are likely. Classes RC8 and RC9 are used for loose running fits when large tolerances are necessary, together with an allowance on the external parts.

Clearance Locational Fits (LC). Clearance locational fits are used with parts that are stationary but must be able to be freely assembled or disassembled. *Figure 9-117* is a table of American National Standard Clearance Locational Fits from ANSI 4.1-1967, R1987.

Transition Locational Fits (LT). Transition locational fits fall on a continuum between clearance and interference fits, and are used in applications where locational accuracy is important but a small amount of clearance or interference is allowable. *Figure 9-118* is a table of American National Standard Transition Locational Fits from ANSI B4.1-1978, R1987.

Interference Locational Fits (LN). Interference locational fits are used when locational accuracy is very important and for parts requiring rigidity and alignment but no special requirements for bore pressure. Interference locational fits are not used for parts which

American National Standard Running and Sliding Fits (ANSI B4.1-1967, R1987)

Tolerance limits given in body of table are added or subtracted to basic size (as indicated by + or – sign) to obtain maximum and minimum sizes of mating parts.

Nominal Size Range, Inches (Over / To)	Class RC 1 Clearance*	Class RC 1 Hole H5	Class RC 1 Shaft g4	Class RC 2 Clearance*	Class RC 2 Hole H6	Class RC 2 Shaft g5	Class RC 3 Clearance*	Class RC 3 Hole H7	Class RC 3 Shaft f6	Class RC 4 Clearance*	Class RC 4 Hole H8	Class RC 4 Shaft f7
			Values shown below are in thousandths of an inch									
0– 0.12	0.1 / 0.45	+0.2 / 0	−0.1 / −0.25	0.1 / 0.55	+0.25 / 0	−0.1 / −0.3	0.3 / 0.95	+0.4 / 0	−0.3 / −0.55	0.3 / 1.3	+0.6 / 0	−0.3 / −0.7
0.12– 0.24	0.15 / 0.5	+0.2 / 0	−0.15 / −0.3	0.15 / 0.65	+0.3 / 0	−0.15 / −0.35	0.4 / 1.12	+0.5 / 0	−0.4 / −0.7	0.4 / 1.6	+0.7 / 0	−0.4 / −0.9
0.24– 0.40	0.2 / 0.6	+0.25 / 0	−0.2 / −0.35	0.2 / 0.85	+0.4 / 0	−0.2 / −0.45	0.5 / 1.5	+0.6 / 0	−0.5 / −0.9	0.5 / 2.0	+0.9 / 0	−0.5 / −1.1
0.40– 0.71	0.25 / 0.75	+0.3 / 0	−0.25 / −0.45	0.25 / 0.95	+0.4 / 0	−0.25 / −0.55	0.6 / 1.7	+0.7 / 0	−0.6 / −1.0	0.6 / 2.3	+1.0 / 0	−0.6 / −1.3
0.71– 1.19	0.3 / 0.95	+0.4 / 0	−0.3 / −0.55	0.3 / 1.2	+0.5 / 0	−0.3 / −0.7	0.8 / 2.1	+0.8 / 0	−0.8 / −1.3	0.8 / 2.8	+1.2 / 0	−0.8 / −1.6
1.19– 1.97	0.4 / 1.1	+0.4 / 0	−0.4 / −0.7	0.4 / 1.4	+0.6 / 0	−0.4 / −0.8	1.0 / 2.6	+1.0 / 0	−1.0 / −1.6	1.0 / 3.6	+1.6 / 0	−1.0 / −2.0
1.97– 3.15	0.4 / 1.2	+0.5 / 0	−0.4 / −0.7	0.4 / 1.6	+0.7 / 0	−0.4 / −0.9	1.2 / 3.1	+1.2 / 0	−1.2 / −1.9	1.2 / 4.2	+1.8 / 0	−1.2 / −2.4
3.15– 4.73	0.5 / 1.5	+0.6 / 0	−0.5 / −0.9	0.5 / 2.0	+0.9 / 0	−0.5 / −1.1	1.4 / 3.7	+1.4 / 0	−1.4 / −2.3	1.4 / 5.0	+2.2 / 0	−1.4 / −2.8
4.73– 7.09	0.6 / 1.8	+0.7 / 0	−0.6 / −1.1	0.6 / 2.3	+1.0 / 0	−0.6 / −1.3	1.6 / 4.2	+1.6 / 0	−1.6 / −2.6	1.6 / 5.7	+2.5 / 0	−1.6 / −3.2
7.09– 9.85	0.6 / 2.0	+0.8 / 0	−0.6 / −1.2	0.6 / 2.6	+1.2 / 0	−0.6 / −1.4	2.0 / 5.0	+1.8 / 0	−2.0 / −3.2	2.0 / 6.6	+2.8 / 0	−2.0 / −3.8
9.85–12.41	0.8 / 2.3	+0.9 / 0	−0.8 / −1.4	0.8 / 2.9	+1.2 / 0	−0.8 / −1.7	2.5 / 5.7	+2.0 / 0	−2.5 / −3.7	2.5 / 7.5	+3.0 / 0	−2.5 / −4.5
12.41–15.75	1.0 / 2.7	+1.0 / 0	−1.0 / −1.7	1.0 / 3.4	+1.4 / 0	−1.0 / −2.0	3.0 / 6.6	+2.2 / 0	−3.0 / −4.4	3.0 / 8.7	+3.5 / 0	−3.0 / −5.2
15.75–19.69	1.2 / 3.0	+1.0 / 0	−1.2 / −2.0	1.2 / 3.8	+1.6 / 0	−1.2 / −2.2	4.0 / 8.1	+2.5 / 0	−4.0 / −5.6	4.0 / 10.5	+4.0 / 0	−4.0 / −6.5

See footnotes at end of table.

Nominal Size Range, Inches (Over / To)	Class RC 5 Clearance*	Class RC 5 Hole H8	Class RC 5 Shaft e7	Class RC 6 Clearance*	Class RC 6 Hole H9	Class RC 6 Shaft e8	Class RC 7 Clearance*	Class RC 7 Hole H9	Class RC 7 Shaft d8	Class RC 8 Clearance*	Class RC 8 Hole H10	Class RC 8 Shaft c9	Class RC 9 Clearance*	Class RC 9 Hole H11	Class RC 9 Shaft
				Values shown below are in thousandths of an inch											
0– 0.12	0.6 / 1.6	+0.6 / 0	−0.6 / −1.0	0.6 / 2.2	+1.0 / 0	−0.6 / −1.2	1.0 / 2.6	+1.0 / 0	−1.0 / −1.6	2.5 / 5.1	+1.6 / 0	−2.5 / −3.5	4.0 / 8.1	+2.5 / 0	−4.0 / −5.6
0.12– 0.24	0.8 / 2.0	+0.7 / 0	−0.8 / −1.3	0.8 / 2.7	+1.2 / 0	−0.8 / −1.5	1.2 / 3.1	+1.2 / 0	−1.2 / −1.9	2.8 / 5.8	+1.8 / 0	−2.8 / −4.0	4.5 / 9.0	+3.0 / 0	−4.5 / −6.0
0.24– 0.40	1.0 / 2.5	+0.9 / 0	−1.0 / −1.6	1.0 / 3.3	+1.4 / 0	−1.0 / −1.9	1.6 / 3.9	+1.4 / 0	−1.6 / −2.5	3.0 / 6.6	+2.2 / 0	−3.0 / −4.4	5.0 / 10.7	+3.5 / 0	−5.0 / −7.2
0.40– 0.71	1.2 / 2.9	+1.0 / 0	−1.2 / −1.9	1.2 / 3.8	+1.6 / 0	−1.2 / −2.2	2.0 / 4.6	+1.6 / 0	−2.0 / −3.0	3.5 / 7.9	+2.8 / 0	−3.5 / −5.1	6.0 / 12.8	+4.0 / 0	−6.0 / −8.8
0.71– 1.19	1.6 / 3.6	+1.2 / 0	−1.6 / −2.4	1.6 / 4.8	+2.0 / 0	−1.6 / −2.8	2.5 / 5.7	+2.0 / 0	−2.5 / −3.7	4.5 / 10.0	+3.5 / 0	−4.5 / −6.5	7.0 / 15.5	+5.0 / 0	−7.0 / −10.5
1.19– 1.97	2.0 / 4.6	+1.6 / 0	−2.0 / −3.0	2.0 / 6.1	+2.5 / 0	−2.0 / −3.6	3.0 / 7.1	+2.5 / 0	−3.0 / −4.6	5.0 / 11.5	+4.0 / 0	−5.0 / −7.5	8.0 / 18.0	+6.0 / 0	−8.0 / −12.0
1.97– 3.15	2.5 / 5.5	+1.8 / 0	−2.5 / −3.7	2.5 / 7.3	+3.0 / 0	−2.5 / −4.3	4.0 / 8.8	+3.0 / 0	−4.0 / −5.8	6.0 / 13.5	+4.5 / 0	−6.0 / −9.0	9.0 / 20.5	+7.0 / 0	−9.0 / −13.5
3.15– 4.73	3.0 / 6.6	+2.2 / 0	−3.0 / −4.4	3.0 / 8.7	+3.5 / 0	−3.0 / −5.2	5.0 / 10.7	+3.5 / 0	−5.0 / −7.2	7.0 / 15.5	+5.0 / 0	−7.0 / −10.5	10.0 / 24.0	+9.0 / 0	−10.0 / −15.0
4.73– 7.09	3.5 / 7.6	+2.5 / 0	−3.5 / −5.1	3.5 / 10.0	+4.0 / 0	−3.5 / −6.0	6.0 / 12.5	+4.0 / 0	−6.0 / −8.5	8.0 / 18.0	+6.0 / 0	−8.0 / −12.0	12.0 / 28.0	+10.0 / 0	−12.0 / −18.0
7.09– 9.85	4.0 / 8.6	+2.8 / 0	−4.0 / −5.8	4.0 / 11.3	+4.5 / 0	−4.0 / −6.8	7.0 / 14.3	+4.5 / 0	−7.0 / −9.8	10.0 / 21.5	+7.0 / 0	−10.0 / −14.5	15.0 / 34.0	+12.0 / 0	−15.0 / −22.0
9.85–12.41	5.0 / 10.0	+3.0 / 0	−5.0 / −7.0	5.0 / 13.0	+5.0 / 0	−5.0 / −8.0	8.0 / 16.0	+5.0 / 0	−8.0 / −11.0	12.0 / 25.0	+8.0 / 0	−12.0 / −17.0	18.0 / 38.0	+12.0 / 0	−18.0 / −26.0
12.41–15.75	6.0 / 11.7	+3.5 / 0	−6.0 / −8.2	6.0 / 15.5	+6.0 / 0	−6.0 / −9.5	10.0 / 19.5	+6.0 / 0	−10.0 / −13.5	14.0 / 29.0	+9.0 / 0	−14.0 / −20.0	22.0 / 45.0	+14.0 / 0	−22.0 / −31.0
15.75–19.69	8.0 / 14.5	+4.0 / 0	−8.0 / −10.5	8.0 / 18.0	+6.0 / 0	−8.0 / −12.0	12.0 / 22.0	+6.0 / 0	−12.0 / −16.0	16.0 / 32.0	+10.0 / 0	−16.0 / −22.0	25.0 / 51.0	+16.0 / 0	−25.0 / −35.0

All data above heavy lines are in accord with ABC agreements. Symbols H5, g4, etc. are hole and shaft designations in ABC system. Limits for sizes above 19.69 inches are also given in the ANSI Standard.

* Pairs of values shown represent minimum and maximum amounts of clearance resulting from application of standard tolerance limits.

Figure 9-116   ANSI Standard Running and Sliding Fits *(Courtesy ASME)*

American National Standard Clearance Locational Fits (ANSI B4.1-1967, R1987)

Tolerance limits given in body of table are added or subtracted to basic size (as indicated by + or – sign) to obtain maximum and minimum sizes of mating parts.

Nominal Size Range, Inches Over To	Class LC 1 Clear-ance*	Standard Tolerance Limits Hole H6	Shaft h5	Class LC 2 Clear-ance*	Standard Tolerance Limits Hole H7	Shaft h6	Class LC 3 Clear-ance*	Standard Tolerance Limits Hole H8	Shaft h7	Class LC 4 Clear-ance*	Standard Tolerance Limits Hole H10	Shaft h9	Class LC 5 Clear-ance*	Standard Tolerance Limits Hole H7	Shaft g6
							Values shown below are in thousandths of an inch								
0– 0.12	0 0.45	+ 0.25 0	0 – 0.2	0 0.65	+ 0.4 0	0 – 0.25	0 1	+ 0.6 0	0 – 0.4	0 2.6	+ 1.6 0	0 – 1.0	0.1 0.75	+ 0.4 0	– 0.1 – 0.35
0.12– 0.24	0 0.5	+ 0.3 0	0 – 0.2	0 0.8	+ 0.5 0	0 – 0.3	0 1.2	+ 0.7 0	0 – 0.5	0 3.0	+ 1.8 0	0 – 1.2	0.15 0.95	+ 0.5 0	– 0.15 – 0.45
0.24– 0.40	0 0.65	+ 0.4 0	0 – 0.25	0 1.0	+ 0.6 0	0 – 0.4	0 1.5	+ 0.9 0	0 – 0.6	0 3.6	+ 2.2 0	0 – 1.4	0.2 1.2	+ 0.6 0	– 0.2 – 0.6
0.40– 0.71	0 0.7	+ 0.4 0	0 – 0.3	0 1.1	+ 0.7 0	0 – 0.4	0 1.7	+ 1.0 0	0 – 0.7	0 4.4	+ 2.8 0	0 – 1.6	0.25 1.35	+ 0.7 0	– 0.25 – 0.65
0.71– 1.19	0 0.9	+ 0.5 0	0 – 0.4	0 1.3	+ 0.8 0	0 – 0.5	0 2	+ 1.2 0	0 – 0.8	0 5.5	+ 3.5 0	0 – 2.0	0.3 1.6	+ 0.8 0	– 0.3 – 0.8
1.19– 1.97	0 1.0	+ 0.6 0	0 – 0.4	0 1.6	+ 1.0 0	0 – 0.6	0 2.6	+ 1.6 0	0 – 1	0 6.5	+ 4.0 0	0 – 2.5	0.4 2.0	+ 1.0 0	– 0.4 – 1.0
1.97– 3.15	0 1.2	+ 0.7 0	0 – 0.5	0 1.9	+ 1.2 0	0 – 0.7	0 3	+ 1.8 0	0 – 1.2	0 7.5	+ 4.5 0	0 – 3	0.4 2.3	+ 1.2 0	– 0.4 – 1.1
3.15– 4.73	0 1.5	+ 0.9 0	0 – 0.6	0 2.3	+ 1.4 0	0 – 0.9	0 3.6	+ 2.2 0	0 – 1.4	0 8.5	+ 5.0 0	0 – 3.5	0.5 2.8	+ 1.4 0	– 0.5 – 1.4
4.73– 7.09	0 1.7	+ 1.0 0	0 – 0.7	0 2.6	+ 1.6 0	0 – 1.0	0 4.1	+ 2.5 0	0 – 1.6	0 10.0	+ 6.0 0	0 – 4	0.6 3.2	+ 1.6 0	– 0.6 – 1.6
7.09– 9.85	0 2.0	+ 1.2 0	0 – 0.8	0 3.0	+ 1.8 0	0 – 1.2	0 4.6	+ 2.8 0	0 – 1.8	0 11.5	+ 7.0 0	0 – 4.5	0.6 3.6	+ 1.8 0	– 0.6 – 1.8
9.85–12.41	0 2.1	+ 1.2 0	0 – 0.9	0 3.2	+ 2.0 0	0 – 1.2	0 5	+ 3.0 0	0 – 2.0	0 13.0	+ 8.0 0	0 – 5	0.7 3.9	+ 2.0 0	– 0.7 – 1.9
12.41–15.75	0 2.4	+ 1.4 0	0 – 1.0	0 3.6	+ 2.2 0	0 – 1.4	0 5.7	+ 3.5 0	0 – 2.2	0 15.0	+ 9.0 0	0 – 6	0.7 4.3	+ 2.2 0	– 0.7 – 2.1
15.75–19.69	0 2.6	+ 1.6 0	0 – 1.0	0 4.1	+ 2.5 0	0 – 1.6	0 6.5	+ 4 0	0 – 2.5	0 16.0	+ 10.0 0	0 – 6	0.8 4.9	+ 2.5 0	– 0.8 – 2.4

See footnotes at end of table.

Nominal Size Range, Inches Over To	Class LC 6 Clear-ance*	Std. Tolerance Limits Hole H9	Shaft f8	Class LC 7 Clear-ance*	Std. Tolerance Limits Hole H10	Shaft e9	Class LC 8 Clear-ance*	Std. Tolerance Limits Hole H10	Shaft d9	Class LC 9 Clear-ance*	Std. Tolerance Limits Hole H11	Shaft c10	Class LC 10 Clear-ance*	Std. Tolerance Limits Hole H12	Shaft	Class LC 11 Clear-ance*	Std. Tolerance Limits Hole H13	Shaft
							Values shown below are in thousandths of an inch											
0– 0.12	0.3 1.9	+ 1.0 0	– 0.3 – 0.9	0.6 3.2	+ 1.6 0	– 0.6 – 1.6	1.0 2.0	+ 1.6 0	– 1.0 – 2.0	2.5 6.6	+ 2.5 0	– 2.5 – 4.1	4 12	+ 4 0	– 4 – 8	5 17	+ 6 0	– 5 – 11
0.12– 0.24	0.4 2.3	+ 1.2 0	– 0.4 – 1.1	0.8 3.8	+ 1.8 0	– 0.8 – 2.0	1.2 4.2	+ 1.8 0	– 1.2 – 2.4	2.8 7.6	+ 3.0 0	– 2.8 – 4.6	4.5 14.5	+ 5 0	– 4.5 – 9.5	6 20	+ 7 0	– 6 – 13
0.24– 0.40	0.5 2.8	+ 1.4 0	– 0.5 – 1.4	1.0 4.6	+ 2.2 0	– 1.0 – 2.4	1.6 5.2	+ 2.2 0	– 1.6 – 3.0	3.0 8.7	+ 3.5 0	– 3.0 – 5.2	5 17	+ 6 0	– 5 – 11	7 25	+ 9 0	– 7 – 16
0.40– 0.71	0.6 3.2	+ 1.6 0	– 0.6 – 1.6	1.2 5.6	+ 2.8 0	– 1.2 – 2.8	2.0 6.4	+ 2.8 0	– 2.0 – 3.6	3.5 10.3	+ 4.0 0	– 3.5 – 6.3	6 20	+ 7 0	– 6 – 13	8 28	+ 10 0	– 8 – 18
0.71– 1.19	0.8 4.0	+ 2.0 0	– 0.8 – 2.0	1.6 7.1	+ 3.5 0	– 1.6 – 3.6	2.5 8.0	+ 3.5 0	– 2.5 – 4.5	4.5 13.0	+ 5.0 0	– 4.5 – 8.0	7 23	+ 8 0	– 7 – 15	10 34	+ 12 0	– 10 – 22
1.19– 1.97	1.0 5.1	+ 2.5 0	– 1.0 – 2.6	2.0 8.5	+ 4.0 0	– 2.0 – 4.5	3.6 9.5	+ 4.0 0	– 3.0 – 5.5	5.0 15.0	+ 6 0	– 5.0 – 9.0	8 28	+ 10 0	– 8 – 18	12 44	+ 16 0	– 12 – 28
1.97– 3.15	1.2 6.0	+ 3.0 0	– 1.0 – 3.0	2.5 10.0	+ 4.5 0	– 2.5 – 5.5	4.0 11.5	+ 4.5 0	– 4.0 – 7.0	6.0 17.5	+ 7 0	– 6.0 – 10.5	10 34	+ 12 0	– 10 – 22	14 50	+ 18 0	– 14 – 32
3.15– 4.73	1.4 7.1	+ 3.5 0	– 1.4 – 3.6	3.0 11.5	+ 5.0 0	– 3.0 – 6.5	5.0 13.5	+ 5.0 0	– 5.0 – 8.5	7 21	+ 9 0	– 7 – 12	11 39	+ 14 0	– 11 – 25	16 60	+ 22 0	– 16 – 38
4.73– 7.09	1.6 8.1	+ 4.0 0	– 1.6 – 4.1	3.5 13.5	+ 6.0 0	– 3.5 – 7.5	6 16	+ 6 0	– 6 – 10	8 24	+ 10 0	– 8 – 14	12 44	+ 16 0	– 12 – 28	18 68	+ 25 0	– 18 – 43
7.09– 9.85	2.0 9.3	+ 4.5 0	– 2.0 – 4.8	4.0 15.5	+ 7.0 0	– 4.0 – 8.5	7 18.5	+ 7 0	– 7 – 11.5	10 29	+ 12 0	– 10 – 17	16 52	+ 18 0	– 16 – 34	22 78	+ 28 0	– 22 – 50
9.85–12.41	2.2 10.2	+ 5.0 0	– 2.2 – 5.2	4.5 17.5	+ 8.0 0	– 4.5 – 9.5	7 20	+ 8 0	– 7 – 12	12 32	+ 12 0	– 12 – 20	20 60	+ 20 0	– 20 – 40	28 88	+ 30 0	– 28 – 58
12.41–15.75	2.5 12.0	+ 6.0 0	– 2.5 – 6.0	5.0 20.0	+ 9.0 0	– 5 – 11	8 23	+ 9 0	– 8 – 14	14 37	+ 14 0	– 14 – 23	22 66	+ 22 0	– 22 – 44	30 100	+ 35 0	– 30 – 65
15.75–19.69	2.8 12.8	+ 6.0 0	– 2.8 – 6.8	5.0 21.0	+ 10.0 0	– 5 – 11	9 25	+ 10 0	– 9 – 15	16 42	+ 16 0	– 16 – 26	25 75	+ 25 0	– 25 – 50	35 115	+ 40 0	– 35 – 75

All data above heavy lines are in accordance with American-British-Canadian (ABC) agreements. Symbols H6, H7, s6, etc. are hole and shaft designations in ABC system. Limits for sizes above 19.69 inches are not covered by ABC agreements but are given in the ANSI Standard.
* Pairs of values shown represent minimum and maximum amounts of interference resulting from application of standard tolerance limits.

Figure 9-117   ANSI Standard Clearance Locational Fits *(Courtesy ASME)*

ANSI Standard Transition Locational Fits (ANSI B4.1-1967, R1987)

Nominal Size Range, Inches	Class LT 1			Class LT 2			Class LT 3			Class LT 4			Class LT 5			Class LT 6		
		Std. Tolerance Limits			Std. Tolerance Limits			Std. Tolerance Limits			Std. Tolerance Limits			Std. Tolerance Limits			Std. Tolerance Limits	
Over To	Fit*	Hole H7	Shaft js6	Fit*	Hole H8	Shaft js7	Fit*	Hole H7	Shaft k6	Fit*	Hole H8	Shaft k7	Fit*	Hole H7	Shaft n6	Fit*	Hole H7	Shaft n7
	Values shown below are in thousandths of an inch																	
0– 0.12	−0.12 +0.52	+0.4 0	+0.12 −0.12	−0.2 +0.8	+0.6 0	+0.2 −0.2							−0.5 +0.15	+0.4 0	+0.5 +0.25	−0.65 +0.15	+0.4 0	+0.65 +0.25
0.12– 0.24	−0.15 +0.65	+0.5 0	+0.15 −0.15	−0.25 +0.95	+0.7 0	+0.25 −0.25							−0.6 +0.2	+0.5 0	+0.6 +0.3	−0.8 +0.2	+0.5 0	+0.8 +0.3
0.24– 0.40	−0.2 +0.8	+0.6 0	+0.2 −0.2	−0.3 +1.2	+0.9 0	+0.3 −0.3	−0.5 +0.5	+0.6 0	+0.5 +0.1	−0.7 +0.8	+0.9 0	+0.7 +0.1	−0.8 +0.2	+0.6 0	+0.8 +0.4	−1.0 +0.2	+0.6 0	+1.0 +0.4
0.40– 0.71	−0.2 +0.9	+0.7 0	+0.2 −0.2	−0.35 +1.35	+1.0 0	+0.35 −0.35	−0.5 +0.6	+0.7 0	+0.5 +0.1	−0.8 +0.9	+1.0 0	+0.8 +0.1	−0.9 +0.2	+0.7 0	+0.9 +0.5	−1.2 +0.2	+0.7 0	+1.2 +0.5
0.71– 1.19	−0.25 +1.05	+0.8 0	+0.25 −0.25	−0.4 +1.6	+1.2 0	+0.4 −0.4	−0.6 +0.7	+0.8 0	+0.6 +0.1	−0.9 +1.1	+1.2 0	+0.9 +0.1	−1.1 +0.2	+0.8 0	+1.1 +0.6	−1.4 +0.2	+0.8 0	+1.4 +0.6
1.19– 1.97	−0.3 +1.3	+1.0 0	+0.3 −0.3	−0.5 +2.1	+1.6 0	+0.5 −0.5	−0.7 +0.9	+1.0 0	+0.7 +0.1	−1.1 +1.5	+1.6 0	+1.1 +0.1	−1.3 +0.3	+1.0 0	+1.3 +0.7	−1.7 +0.3	+1.0 0	+1.7 +0.7
1.97– 3.15	−0.3 +1.5	+1.2 0	+0.3 −0.3	−0.6 +2.4	+1.8 0	+0.6 −0.6	−0.8 +1.1	+1.2 0	+0.8 +0.1	−1.3 +1.7	+1.8 0	+1.3 +0.1	−1.5 +0.4	+1.2 0	+1.5 +0.8	−2.0 +0.4	+1.2 0	+2.0 +0.8
3.15– 4.73	−0.4 +1.8	+1.4 0	+0.4 −0.4	−0.7 +2.9	+2.2 0	+0.7 −0.7	−1.0 +1.3	+1.4 0	+1.0 +0.1	−1.5 +2.1	+2.2 0	+1.5 +0.1	−1.9 +0.4	+1.4 0	+1.9 +1.0	−2.4 +0.4	+1.4 0	+2.4 +1.0
4.73– 7.09	−0.5 +2.1	+1.6 0	+0.5 −0.5	−0.8 +3.3	+2.5 0	+0.8 −0.8	−1.1 +1.5	+1.6 0	+1.1 +0.1	−1.7 +2.4	+2.5 0	+1.7 +0.1	−2.2 +0.4	+1.6 0	+2.2 +1.2	−2.8 +0.4	+1.6 0	+2.8 +1.2
7.09– 9.85	−0.6 +2.4	+1.8 0	+0.6 −0.6	−0.9 +3.7	+2.8 0	+0.9 −0.9	−1.4 +1.6	+1.8 0	+1.4 +0.2	−2.0 +2.6	+2.8 0	+2.0 +0.2	−2.6 +0.4	+1.8 0	+2.6 +1.4	−3.2 +0.4	+1.8 0	+3.2 +1.4
9.85–12.41	−0.6 +2.6	+2.0 0	+0.6 −0.6	−1.0 +4.0	+3.0 0	+1.0 −1.0	−1.4 +1.8	+2.0 0	+1.4 +0.2	−2.2 +2.8	+3.0 0	+2.2 +0.2	−2.6 +0.6	+2.0 0	+2.6 +1.4	−3.4 +0.6	+2.0 0	+3.4 +1.4
12.41–15.75	−0.7 +2.9	+2.2 0	+0.7 −0.7	−1.0 +4.5	+3.5 0	+1.0 −1.0	−1.6 +2.0	+2.2 0	+1.6 +0.2	−2.4 +3.3	+3.5 0	+2.4 +0.2	−3.0 +0.6	+2.2 0	+3.0 +1.6	−3.8 +0.6	+2.2 0	+3.8 +1.6
15.75–19.69	−0.8 +3.3	+2.5 0	+0.8 −0.8	−1.2 +5.2	+4.0 0	+1.2 −1.2	−1.8 +2.3	+2.5 0	+1.8 +0.2	−2.7 +3.8	+4.0 0	+2.7 +0.2	−3.4 +0.7	+2.5 0	+3.4 +1.8	−4.3 +0.7	+2.5 0	+4.3 +1.8

All data above heavy lines are in accord with ABC agreements. Symbols H7, js6, etc. are hole and shaft designations in ABC system.

* Pairs of values shown represent maximum amount of interference (−) and maximum amount of clearance (+) resulting from application of standard tolerance limits.

Figure 9-118 ANSI Standard Transition Locational Fits *(Courtesy ASME)*

are designed to transmit frictional loads to other parts through the tightness of the fits. These types of applications are covered by force fits. ***Figure 9-119*** is a table of American National Standard Interference Locational Fits from ANSI B4.1-1967, R1987.

Force and Shrink Fits (FN). Force or shrink fits are a special type of interference fit used when maintenance of constant bore pressure for all sizes within the specified range is important. The amount of interference varies with the diameter, and the difference between minimum and maximum values is small to ensure maintenance of the resulting pressures within reasonable limits. ***Figure 9-120*** is a table of American National Standard Force and Shrink Fits from ANSI B4.1-1967, R1987.

Using Fit Tables

Figures 9-116 through 9-120 are fit tables taken from ANSI B4.1-1967, R1987. It is important for engineers and drafters to know how to use tables such as these. Look at the table in Figure 9-116. The nine classes explained earlier are arranged across the top of the table. Under each class column there are three other columns: one giving the clearance, one giving the standard tolerance limits for the hole, and one giving the standard tolerance limits for the shaft. On the extreme left-hand side of the chart the nominal size range in inches is given. To use this table, as well as the others contained in this chapter, select the appropriate class across the top of the table and locate the nominal size at the extreme left-hand side of the table. Moving across the row that contains the appropriate nominal size range to the class column you have selected, you will

ANSI Standard Interference Locational Fits (ANSI B4.1-1967, R1987)

Tolerance limits given in body of table are added or subtracted to basic size (as indicated by + or − sign) to obtain maximum and minimum sizes of mating parts.

Nominal Size Range, Inches	Class LN 1			Class LN 2			Class LN 3		
	Limits of Interference	Standard Limits		Limits of Interference	Standard Limits		Limits of Interference	Standard Limits	
Over To		Hole H6	Shaft n5		Hole H7	Shaft p6		Hole H7	Shaft r6
	Values shown below are given in thousandths of an inch								
0– 0.12	0 0.45	+0.25 0	+0.45 +0.25	0 0.65	+0.4 0	+0.65 +0.4	0.1 0.75	+0.4 0	+0.75 +0.5
0.12– 0.24	0 0.5	+0.3 0	+0.5 +0.3	0 0.8	+0.5 0	+0.8 +0.5	0.1 0.9	+0.5 0	+0.9 +0.6
0.24– 0.40	0 0.65	+0.4 0	+0.65 +0.4	0 1.0	+0.6 0	+1.0 +0.6	0.2 1.2	+0.6 0	+1.2 +0.8
0.40– 0.71	0 0.8	+0.4 0	+0.8 +0.4	0 1.1	+0.7 0	+1.1 +0.7	0.3 1.4	+0.7 0	+1.4 +1.0
0.71– 1.19	0 1.0	+0.5 0	+1.0 +0.5	0 1.3	+0.8 0	+1.3 +0.8	0.4 1.7	+0.8 0	+1.7 +1.2
1.19– 1.97	0 1.1	+0.6 0	+1.1 +0.6	0 1.6	+1.0 0	+1.6 +1.0	0.4 2.0	+1.0 0	+2.0 +1.4
1.97– 3.15	0.1 1.3	+0.7 0	+1.3 +0.8	0.2 2.1	+1.2 0	+2.1 +1.4	0.4 2.3	+1.2 0	+2.3 +1.6
3.15– 4.73	0.1 1.6	+0.9 0	+1.6 +1.0	0.2 2.5	+1.4 0	+2.5 +1.6	0.6 2.9	+1.4 0	+2.9 +2.0
4.73– 7.09	0.2 1.9	+1.0 0	+1.9 +1.2	0.2 2.8	+1.6 0	+2.8 +1.8	0.9 3.5	+1.6 0	+3.5 +2.5
7.09– 9.85	0.2 2.2	+1.2 0	+2.2 +1.4	0.2 3.2	+1.8 0	+3.2 +2.0	1.2 4.2	+1.8 0	+4.2 +3.0
9.85–12.41	0.2 2.3	+1.2 0	+2.3 +1.4	0.2 3.4	+2.0 0	+3.4 +2.2	1.5 4.7	+2.0 0	+4.7 +3.5
12.41–15.75	0.2 2.6	+1.4 0	+2.6 +1.6	0.3 3.9	+2.2 0	+3.9 +2.5	2.3 5.9	+2.2 0	+5.9 +4.5
15.75–19.69	0.2 2.8	+1.6 0	+2.8 +1.8	0.3 4.4	+2.5 0	+4.4 +2.8	2.5 6.6	+2.5 0	+6.6 +5.0

All data in this table are in accordance with American-British-Canadian (ABC) agreements.

Limits for sizes above 19.69 inches are not covered by ABC agreements but are given in the ANSI Standard.

Symbols H7, p6, etc. are hole and shaft designations in ABC system.

* Pairs of values shown represent minimum and maximum amounts of interference resulting from application of standard tolerance limits.

Figure 9-119 ANSI Standard Interference Locational Fits *(Courtesy ASME)*

ANSI Standard Force and Shrink Fits (ANSI B4.1-1967, R1987)

Nominal Size Range, Inches Over To	Class FN 1 Inter-ference*	Standard Tolerance Limits Hole H6	Shaft	Class FN 2 Inter-ference*	Standard Tolerance Limits Hole H7	Shaft s6	Class FN 3 Inter-ference*	Standard Tolerance Limits Hole H7	Shaft t6	Class FN 4 Inter-ference*	Standard Tolerance Limits Hole H7	Shaft u6	Class FN 5 Inter-ference*	Standard Tolerance Limits Hole H8	Shaft x7
					Values shown below are in thousandths of an inch										
0–0.12	0.05 0.5	+0.25 0	+0.5 +0.3	0.2 0.85	+0.4 0	+0.85 +0.6				0.3 0.95	+0.4 0	+0.95 +0.7	0.3 1.3	+0.6 0	+1.3 +0.9
0.12–0.24	0.1 0.6	+0.3 0	+0.6 +0.4	0.2 1.0	+0.5 0	+1.0 +0.7				0.4 1.2	+0.5 0	+1.2 +0.9	0.5 1.7	+0.7 0	+1.7 +1.2
0.24–0.40	0.1 0.75	+0.4 0	+0.75 +0.5	0.4 1.4	+0.6 0	+1.4 +1.0				0.6 1.6	+0.6 0	+1.6 +1.2	0.5 2.0	+0.9 0	+2.0 +1.4
0.40–0.56	0.1 0.8	+0.4 0	+0.8 +0.5	0.5 1.6	+0.7 0	+1.6 +1.2				0.7 1.8	+0.7 0	+1.8 +1.4	0.6 2.3	+1.0 0	+2.3 +1.6
0.56–0.71	0.2 0.9	+0.4 0	+0.9 +0.6	0.5 1.6	+0.7 0	+1.6 +1.2				0.7 1.8	+0.7 0	+1.8 +1.4	0.8 2.5	+1.0 0	+2.5 +1.8
0.71–0.95	0.2 1.1	+0.5 0	+1.1 +0.7	0.6 1.9	+0.8 0	+1.9 +1.4				0.8 2.1	+0.8 0	+2.1 +1.6	1.0 3.0	+1.2 0	+3.0 +2.2
0.95–1.19	0.3 1.2	+0.5 0	+1.2 +0.8	0.6 1.9	+0.8 0	+1.9 +1.4	0.8 2.1	+0.8 0	+2.1 +1.6	1.0 2.3	+0.8 0	+2.3 +1.8	1.3 3.3	+1.2 0	+3.3 +2.5
1.19–1.58	0.3 1.3	+0.6 0	+1.3 +1.0	0.8 2.4	+1.0 0	+2.4 +1.8	1.0 2.6	+1.0 0	+2.6 +2.0	1.5 3.1	+1.0 0	+3.1 +2.5	1.4 4.0	+1.6 0	+4.0 +3.0
1.58–1.97	0.4 1.4	+0.6 0	+1.4 +1.0	0.8 2.4	+1.0 0	+2.4 +1.8	1.2 2.8	+1.0 0	+2.8 +2.2	1.8 3.4	+1.0 0	+3.4 +2.8	2.4 5.0	+1.6 0	+5.0 +4.0
1.97–2.56	0.6 1.8	+0.7 0	+1.8 +1.3	0.8 2.7	+1.2 0	+2.7 +2.0	1.3 3.2	+1.2 0	+3.2 +2.5	2.3 4.2	+1.2 0	+4.2 +3.5	3.2 6.2	+1.8 0	+6.2 +5.0
2.56–3.15	0.7 1.9	+0.7 0	+1.9 +1.4	1.0 2.9	+1.2 0	+2.9 +2.2	1.8 3.7	+1.2 0	+3.7 +3.0	2.8 4.7	+1.2 0	+4.7 +4.0	4.2 7.2	+1.8 0	+7.2 +6.0
3.15–3.94	0.9 2.4	+0.9 0	+2.4 +1.8	1.4 3.7	+1.4 0	+3.7 +2.8	2.1 4.4	+1.4 0	+4.4 +3.5	3.6 5.9	+1.4 0	+5.9 +5.0	4.8 8.4	+2.2 0	+8.4 +7.0
3.94–4.73	1.1 2.6	+0.9 0	+2.6 +2.0	1.6 3.9	+1.4 0	+3.9 +3.0	2.6 4.9	+1.4 0	+4.9 +4.0	4.6 6.9	+1.4 0	+6.9 +6.0	5.8 9.4	+2.2 0	+9.4 +8.0

See footnotes at end of table.

Nominal Size Range, Inches Over To	Class FN 1 Inter-ference*	Standard Tolerance Limits Hole H6	Shaft	Class FN 2 Inter-ference*	Standard Tolerance Limits Hole H7	Shaft s6	Class FN 3 Inter-ference*	Standard Tolerance Limits Hole H7	Shaft t6	Class FN 4 Inter-ference*	Standard Tolerance Limits Hole H7	Shaft u6	Class FN 5 Inter-ference*	Standard Tolerance Limits Hole H8	Shaft x7
					Values shown below are in thousandths of an inch										
4.73–5.52	1.2 2.9	+1.0 0	+2.9 +2.2	1.9 4.5	+1.6 0	+4.5 +3.5	3.4 6.0	+1.6 0	+6.0 +5.0	5.4 8.0	+1.6 0	+8.0 +7.0	7.5 11.6	+2.5 0	+11.6 +10.0
5.52–6.30	1.5 3.2	+1.0 0	+3.2 +2.5	2.4 5.0	+1.6 0	+5.0 +4.0	3.4 6.0	+1.6 0	+6.0 +5.0	5.4 8.0	+1.6 0	+8.0 +7.0	9.5 13.6	+2.5 0	+13.6 +12.0
6.30–7.09	1.8 3.5	+1.0 0	+3.5 +2.8	2.9 5.5	+1.6 0	+5.5 +4.5	4.4 7.0	+1.6 0	+7.0 +6.0	6.4 9.0	+1.6 0	+9.0 +8.0	9.5 13.6	+2.5 0	+13.6 +12.0
7.09–7.88	1.8 3.8	+1.2 0	+3.8 +3.0	3.2 6.2	+1.8 0	+6.2 +5.0	5.2 8.2	+1.8 0	+8.2 +7.0	7.2 10.2	+1.8 0	+10.2 +9.0	11.2 15.8	+2.8 0	+15.8 +14.0
7.88–8.86	2.3 4.3	+1.2 0	+4.3 +3.5	3.2 6.2	+1.8 0	+6.2 +5.0	5.2 8.2	+1.8 0	+8.2 +7.0	8.2 11.2	+1.8 0	+11.2 +10.0	13.2 17.8	+2.8 0	+17.8 +16.0
8.86–9.85	2.3 4.3	+1.2 0	+4.3 +3.5	4.2 7.2	+1.8 0	+7.2 +6.0	6.2 9.2	+1.8 0	+9.2 +8.0	10.2 13.2	+1.8 0	+13.2 +12.0	13.2 17.8	+2.8 0	+17.8 +16.0
9.85–11.03	2.8 4.9	+1.2 0	+4.9 +4.0	4.0 7.2	+2.0 0	+7.2 +6.0	7.0 10.2	+2.0 0	+10.2 +9.0	10.0 13.2	+2.0 0	+13.2 +12.0	15.0 20.0	+3.0 0	+20.0 +18.0
11.03–12.41	2.8 4.9	+1.2 0	+4.9 +4.0	5.0 8.2	+2.0 0	+8.2 +7.0	7.0 10.2	+2.0 0	+10.2 +9.0	12.0 15.2	+2.0 0	+15.2 +14.0	17.0 22.0	+3.0 0	+22.0 +20.0
12.41–13.98	3.1 5.5	+1.4 0	+5.5 +4.5	5.8 9.4	+2.2 0	+9.4 +8.0	7.8 11.4	+2.2 0	+11.4 +10.0	13.8 17.4	+2.2 0	+17.4 +16.0	18.5 24.2	+3.5 0	+24.2 +22.0
13.98–15.75	3.6 6.1	+1.4 0	+6.1 +5.0	5.8 9.4	+2.2 0	+9.4 +8.0	9.8 13.4	+2.2 0	+13.4 +12.0	15.8 19.4	+2.2 0	+19.4 +18.0	21.5 27.2	+3.5 0	+27.2 +25.0
15.75–17.72	4.4 7.0	+1.6 0	+7.0 +6.0	6.5 10.6	+2.5 0	+10.6 +9.0	9.5 13.6	+2.5 0	+13.6 +12.0	17.5 21.6	+2.5 0	+21.6 +20.0	24.0 30.5	+4.0 0	+30.5 +28.0
17.72–19.69	4.4 7.0	+1.6 0	+7.0 +6.0	7.5 11.6	+2.5 0	+11.6 +10.0	11.5 15.6	+2.5 0	+15.6 +14.0	19.5 23.6	+2.5 0	+23.6 +22.0	26.0 32.5	+4.0 0	+32.5 +30.0

All data above heavy lines are in accordance with American-British-Canadian (ABC) agreements. Symbols H6, H7, s6, etc. are hole and shaft designations in ABC system. Limits for sizes above 19.69 inches are not covered by ABC agreements but are given in the ANSI standard.
* Pairs of values shown represent minimum and maximum amounts of interference resulting from application of standard tolerance limits.

Figure 9-120   ANSI Standard Force and Shrink Fits *(Courtesy ASME)*

find the tolerance limits you need. These tolerance limits are added to or subtracted from the basic size to obtain maximum and minimum sizes of mating parts. Each tolerance is preceded by either a plus or a minus to indicate whether it is added to or subtracted from the basic size. Values must be changed to inches before adding and subtracting.

Example:

Using fit tables:

Values are limits in thousandths of an inch.

Basic size: .0156

Nominal size range: .0–0.12

Limits of clearance: $0.1 = 000.1 = .0001$
$0.45 = 000.45 = .00045$

Standard limits: (Move decimal 3 places to the left to obtain values in inches.)

Hole: $+0.2 = 000.2 = +.0002$
$-0 = -0$

Shaft: $-0.1 = -000.1 = -.0001$
$-0.25 = -000.25 = -.00035$

Calculating for tolerance dimensions:

Hole:	.0156	basic hole size
	+.0002	standard limits in thousands
	.0158	upper limit for hole

	.0156	basic hole size
	-0	standard limits in thousands
	.0156	lower limit for hole

Shaft:	.0156	basic hole size
	-.0001	standard limits for shaft
	.0155	upper limit for shaft

	.0156	basic hole size
	-.00025	standard limit for shaft
	.01535	lower limit for shaft

Tolerance dimensions for basic size of .0156:

| Hole: | .01580 | Shaft: | .01550 |
| | .01560 | | .01535 |

To check calculations:

Subtract:	.01580	upper limit of hole
	-.01535	lower limit of shaft
	.00045	limit of clearance

	.01560	lower limit of hole
	-.01550	upper limit of shaft
	.0001	limit of clearance

Limit of Clearance:

$.0001 = .1$ from chart
$.00045 = .45$ from chart

Manufacturing Precision

Figure 9-121 is a table showing the levels of tolerances engineers and drafters can expect from various common manufacturing processes. It is important for engineers and drafters to know what to expect from the manufacturing processes that will be used in converting their design into an actual working part. It will serve no purpose to specify a tolerance greater than that which can be expected of a given manufacturing process.

Summary of Dimensioning Rules

This section summarizes most of the dimensioning rules with which engineers and drafters should be familiar. This section can be used as a checklist for ensuring that drawings you do throughout this book are properly dimensioned. Use this summary as a checklist until you are so familiar with these rules that the checklist is no longer necessary.

1. Give all dimensions necessary to accurately communicate the design to manufacturing personnel, but only those dimensions necessary.

2. Apply dimensions in such a way that they cannot be misinterpreted.

3. Show dimensions between surfaces or points that have a functional relationship.

4. Do not dimension to or from rough surfaces. Always attempt to dimension from a finished surface.

5. Give complete dimensions, so that manufacturing personnel do not have to interrupt their work to calculate or otherwise determine a dimension they need.

6. Apply dimensions in the views where the shape of the part is shown most accurately.

7. Apply dimensions in those views where the part being dimensioned appears true size and true shape.

8. Do not dimension to hidden lines. If it becomes necessary to dimension to hidden lines, change the view to a sectional view.

9. Attempt to place dimensions which apply to two adjacent views between the views.

10. Avoid crossing extension lines with dimension lines.

11. Clearly dimension and annotate all features. Never expect manufacturing personnel to assume anything.

12. Do not apply a complete chain of detail dimensions. Omit the last one or indicate that the last dimension in the chain is a reference dimension.

13. Avoid placing dimensions on the part itself.

14. Always place dimension lines uniformly throughout the drawing.

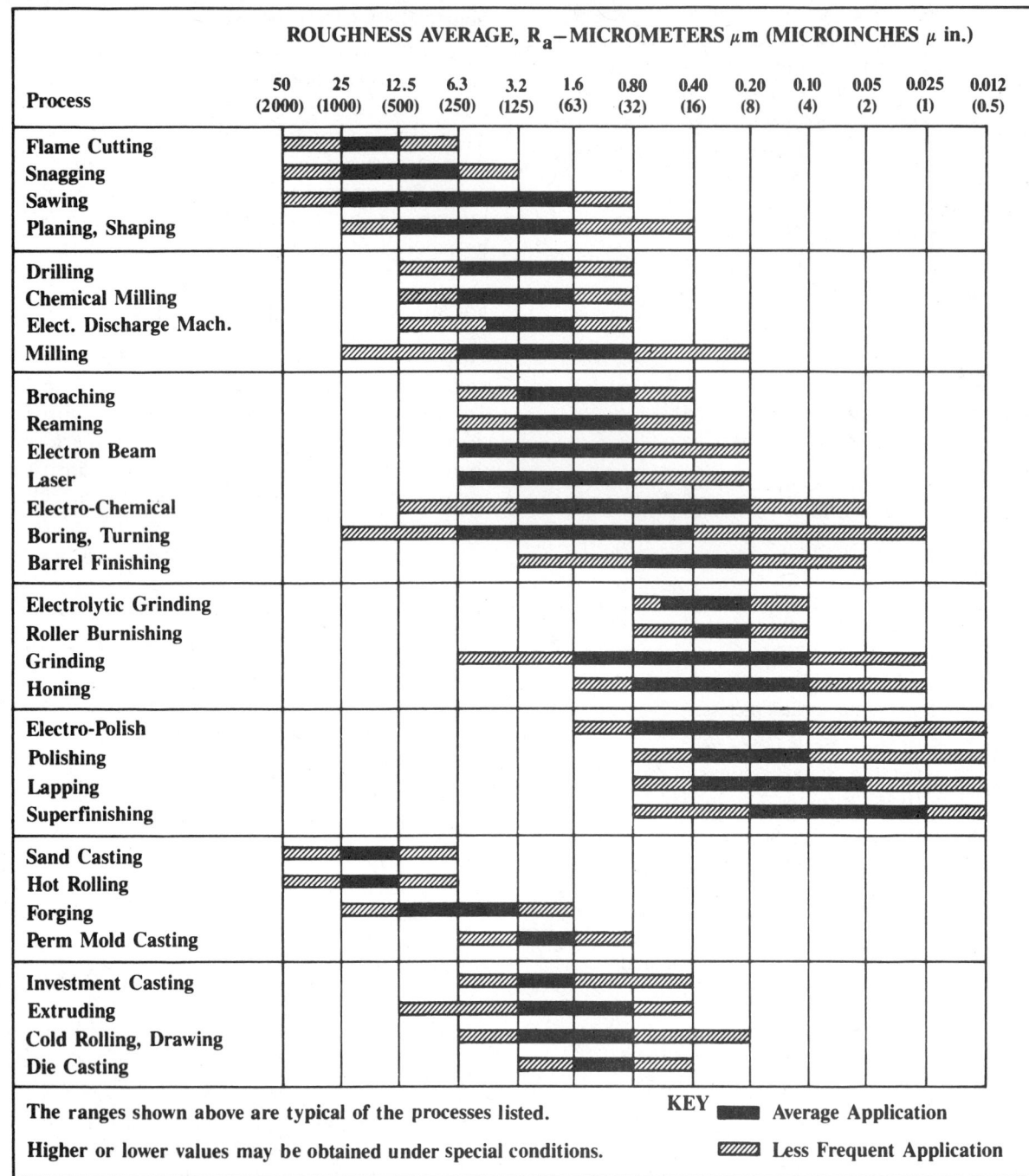

Process	ROUGHNESS AVERAGE, R_a – MICROMETERS µm (MICROINCHES µ in.)												
	50 (2000)	25 (1000)	12.5 (500)	6.3 (250)	3.2 (125)	1.6 (63)	0.80 (32)	0.40 (16)	0.20 (8)	0.10 (4)	0.05 (2)	0.025 (1)	0.012 (0.5)

Figure 9-121   Manufacturing precision *(From Machinery's Handbook, 23rd Edition, © 1988 by Industrial Press Inc. Used with permission.)*

15. Never use a line that constitutes a portion of the drawing of a part as a dimension line.
16. Avoid crossing dimension lines wherever possible.
17. Extension lines that cross other extension lines or object lines should not be broken.
18. A center line may be used as an extension line.
19. Center lines should not extend from view to view.
20. Leader lines should point to the center of a hole or circular feature.
21. Leader lines should be straight and broken at precise angles. They should never curve.

22. Leader lines should begin approximately at the center of a note height and extend either from the beginning or end of the note.
23. If it becomes necessary to letter within a section-lined area, erase enough of the section lines to leave a clear space so that the dimension can be easily read.
24. Fraction bars should be horizontal rather than angled.
25. The numerator and denominator of a fraction should not touch the horizontal fraction bar.

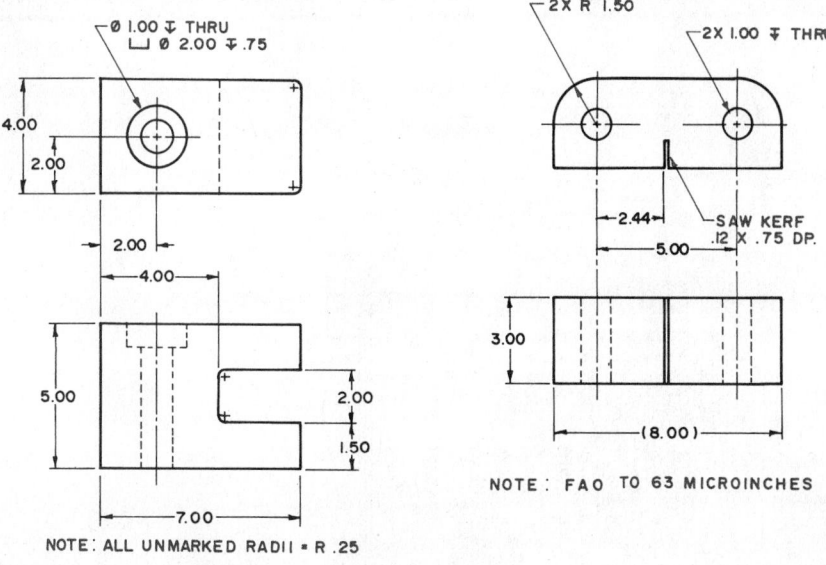

Figure 9-122   Notes on drawings improve communication

26. Finish marks should be applied on views in which the surface appears as an edge.
27. Finish marks are not necessary on parts made from rolled stock.
28. If a part is finished all over, finish marks are not used. Substitute the letters FAO or apply a note that says "Finish All Over."
29. Dimension a cylinder by showing the diameter and length on the rectangular view or by showing the diameter as a diagonal dimension in the circular view.
30. Always specify drill sizes in decimals.
31. A metric diameter dimension should be preceded by the diameter symbol.
32. A metric radius dimension should be preceded by a capital "R."
33. Dimensions that are not to scale should be underlined with a straight line.
34. Decimal dimensioning is preferred for mechanical and manufacturing related drawings.

Notation

A good drawing may be defined as one that contains all of the information required by the various design and manufacturing people who will use it in producing the subject part. Most of this information can be conveyed graphically, using standard dimensioning practices. However, it is not uncommon to encounter a situation in which all of the needed information cannot be communicated graphically. In these cases, notes are used to communicate or clarify the designer's intent.

Notes are brief, carefully worded statements placed on drawings to convey information not covered or not adequately explained using graphics, *Figure 9-122*. Notes should be clearly worded so as to allow only one correct interpretation.

There are no ANSI standards specifically governing the use of notes on technical drawings. However, several rules of a general nature should be observed. These rules apply to both general notes and the more specific detail notes.

Rules for Applying Notes on Drawings

Notes may be lettered freehand, entered using a keyboard in a computer-aided drafting (CAD) system, or through any one of several mechanical lettering processes. Sample notes are included on the drawings in *Figures 9-123* and *9-124*. In any case, regardless of how they are put on the drawing, notes should be oriented horizontally on the drafting sheet, *Figure 9-125*. General notes should be located directly above the title block, *Figure 9-126*. When using manual processes to apply general notes such as those in Figure 9-126, the first note is placed directly above the title block, the second is placed on top of it and so on up the line. This allows notes to be added as needed without renumbering. However, when using a CAD system, this is not necessary since one of the advantages of CAD

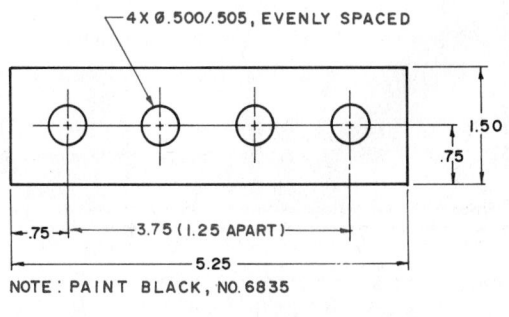

Figure 9-123   Sample note lettered mechanically

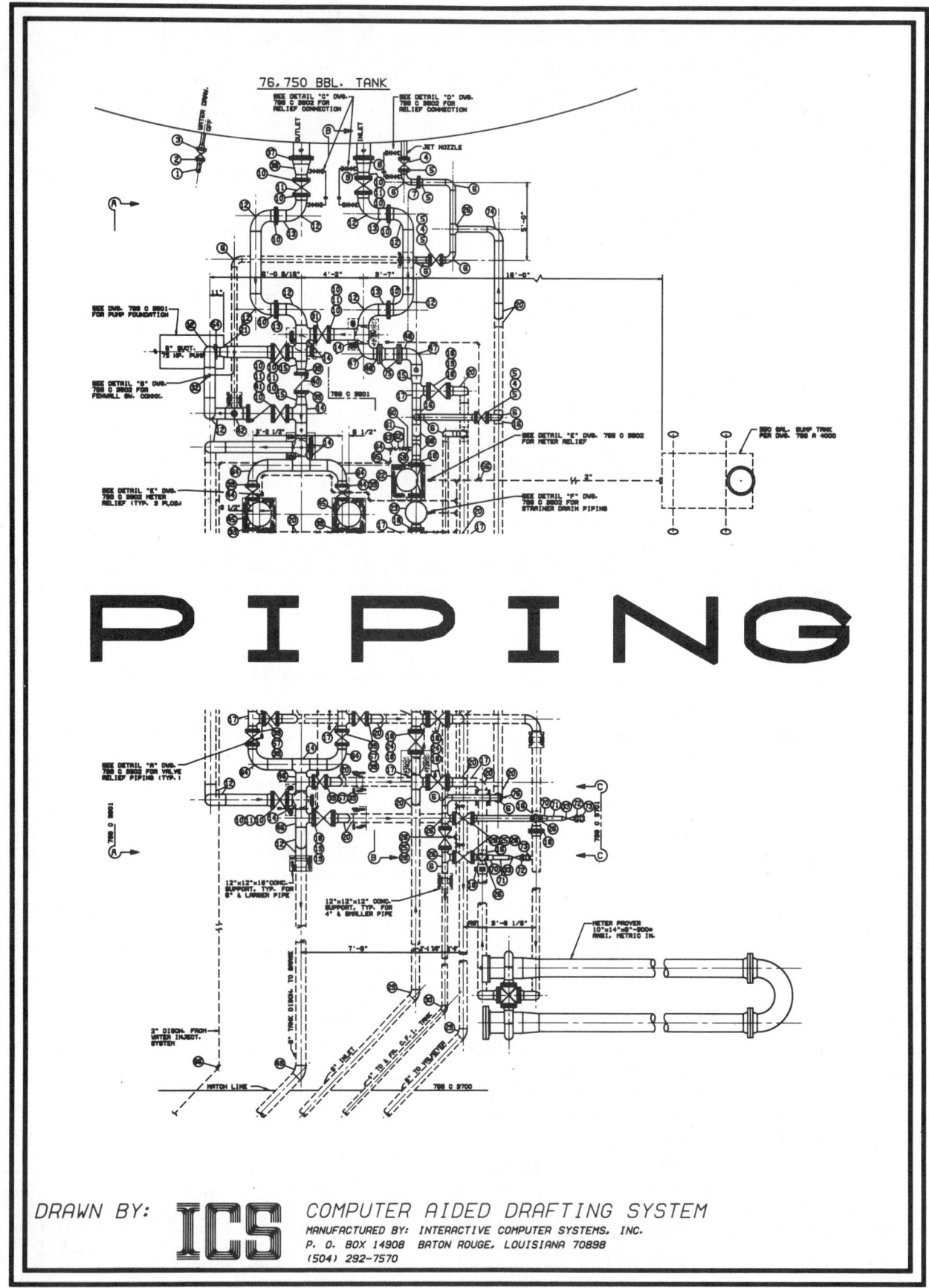

Figure 9-124 Notes lettered on a CAD system *(Courtesy Interactive Computer Systems Inc.)*

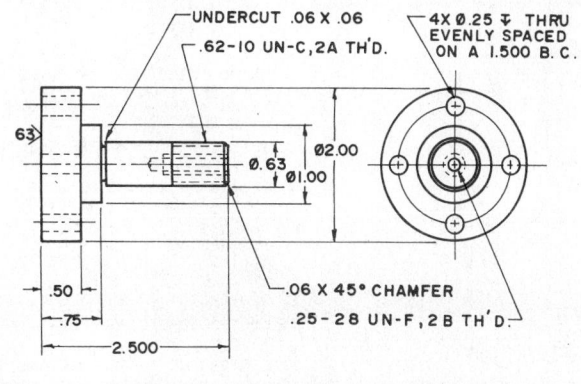

Figure 9-125   Notes are placed horizontally

is that notes can be renumbered and rearranged automatically. Detail notes should be located as close as possible to the detail they are describing, and connected to it by a leader line, *Figure 9-127*.

Notes should be applied to drawings after all graphics have been completed. This prevents notes and graphics from overlapping, and minimizes other technique problems such as smudging the worksheet and frequent erasing of notes as changes are required on a drawing.

Two basic types of notes are used on technical drawings: general notes and detail notes. They serve different purposes and, therefore, must be examined separately.

General Notes

General notes are broad items of information which have a job- or project-wide application rather than relating to just one single element of a project or a part. They are usually placed immediately above the title block on the drawing sheet and numbered sequentially.

Information placed in general notes includes such characteristics of a product as finish specifications; standard sizes of fillets, rounds, and radii; heat treatment specifications; cleaning instructions; general tolerancing data; hardness testing instructions; and stamping specifications. *Figures 9-128* and *9-129* are examples of drawings containing general notes.

Detail Notes

Detail notes are specific notes that pertain to one particular element or characteristic of a part. They are placed as near as possible to the characteristic to which they apply, and are connected using a leader line, *Figure 9-130*.

Detail notes should not be placed on views, *Figure 9-131*. The only exception to this rule is in cases where a great deal of open space exists on a view, but very little around it, *Figure 9-132*. Notes should never be superimposed over other data such as dimensions, lines or symbols, *Figure 9-133*.

Common sense is the best rule to follow in applying detail notes. Since detail notes are used to more completely communicate or to clarify intent, they should be placed as close as possible to the element to which they pertain and in such a way as to be easily read.

Writing Notes

The written word lends itself to interpretations that may vary. Thus, notes used on drawings must convey the exact intent of the designer. Consequently, drafters must be especially careful in the preparation of notes.

The first step is to place in a legend all abbreviations used in the notes, *Figure 9-134*. This ensures that all readers of the notes interpret the abbreviations consistently and correctly.

Another technique which will limit misinterpretation of notes is punctuation or indentation. Notes containing more than one sentence should be properly punctuated so that readers know where one sentence leaves off and the next one begins, *Figure 9-135*. Indentation is used when the length of a note requires more than one line. The second line is indented so that readers know it is part of the first line, *Figure 9-136*.

Another technique is particularly useful for beginning drafters. It is called *verification*. Notes to be placed on drawings are first jotted down, legibly, on a print of the drawing. Another drafter, preferably one with experience, is then asked to read the notes to ensure that they are not open to multiple interpretations.

Note Specifications

A number of specifications should be observed when writing notes. The most widely accepted lettering style for notes is uppercase (all capital) block Gothic letters in a vertical format. Some companies accept uppercase block Gothic lettering in an inclined or slanted format. The proper lettering height is one-quarter inch for titles such as GENERAL NOTES, and one-eighth to three-sixteenths inch for actual notes, *Figure 9-137*.

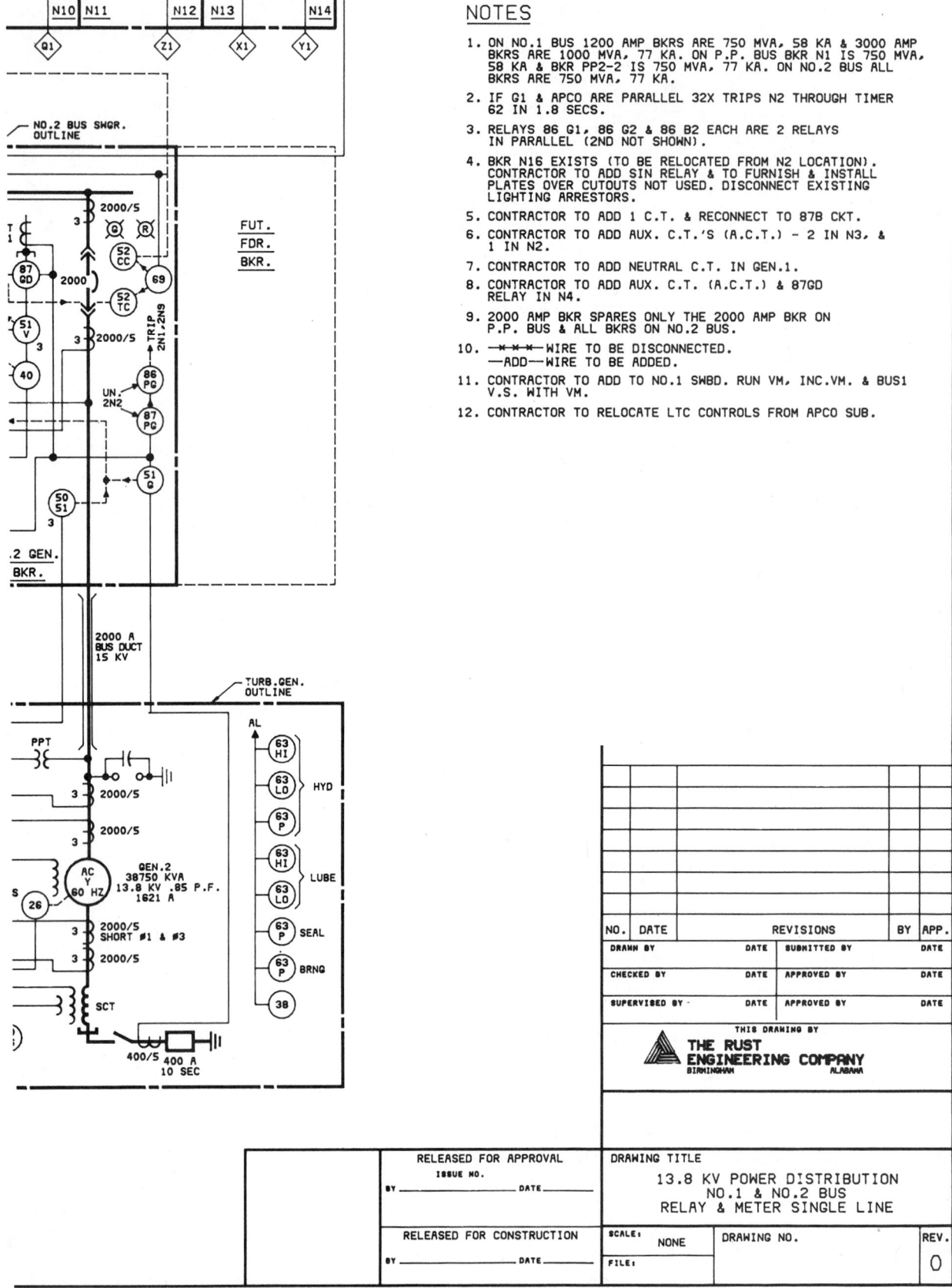

NOTES

1. ON NO.1 BUS 1200 AMP BKRS ARE 750 MVA, 58 KA & 3000 AMP BKRS ARE 1000 MVA, 77 KA. ON P.P. BUS BKR N1 IS 750 MVA, 58 KA & BKR PP2-2 IS 750 MVA, 77 KA. ON NO.2 BUS ALL BKRS ARE 750 MVA, 77 KA.

2. IF G1 & APCO ARE PARALLEL 32X TRIPS N2 THROUGH TIMER 62 IN 1.8 SECS.

3. RELAYS 86 G1, 86 G2 & 86 B2 EACH ARE 2 RELAYS IN PARALLEL (2ND NOT SHOWN).

4. BKR N16 EXISTS (TO BE RELOCATED FROM N2 LOCATION). CONTRACTOR TO ADD SIN RELAY & TO FURNISH & INSTALL PLATES OVER CUTOUTS NOT USED. DISCONNECT EXISTING LIGHTING ARRESTORS.

5. CONTRACTOR TO ADD 1 C.T. & RECONNECT TO 87B CKT.

6. CONTRACTOR TO ADD AUX. C.T.'S (A.C.T.) - 2 IN N3, & 1 IN N2.

7. CONTRACTOR TO ADD NEUTRAL C.T. IN GEN.1.

8. CONTRACTOR TO ADD AUX. C.T. (A.C.T.) & 87GD RELAY IN N4.

9. 2000 AMP BKR SPARES ONLY THE 2000 AMP BKR ON P.P. BUS & ALL BKRS ON NO.2 BUS.

10. ─×─×─×─ WIRE TO BE DISCONNECTED.
    ─ADD─ WIRE TO BE ADDED.

11. CONTRACTOR TO ADD TO NO.1 SWBD. RUN VM, INC.VM. & BUS1 V.S. WITH VM.

12. CONTRACTOR TO RELOCATE LTC CONTROLS FROM APCO SUB.

NO.	DATE	REVISIONS	BY	APP.

DRAWN BY	DATE	SUBMITTED BY	DATE
CHECKED BY	DATE	APPROVED BY	DATE
SUPERVISED BY	DATE	APPROVED BY	DATE

THIS DRAWING BY

THE RUST ENGINEERING COMPANY
BIRMINGHAM   ALABAMA

RELEASED FOR APPROVAL	DRAWING TITLE
ISSUE NO.	13.8 KV POWER DISTRIBUTION NO.1 & NO.2 BUS RELAY & METER SINGLE LINE
BY _____ DATE _____	
RELEASED FOR CONSTRUCTION	SCALE: NONE	DRAWING NO.	REV. 0
BY _____ DATE _____	FILE:		

CDD868

Figure 9-126 General notes are located over the titleblock *(Courtesy The Rust Engineering Co.)*

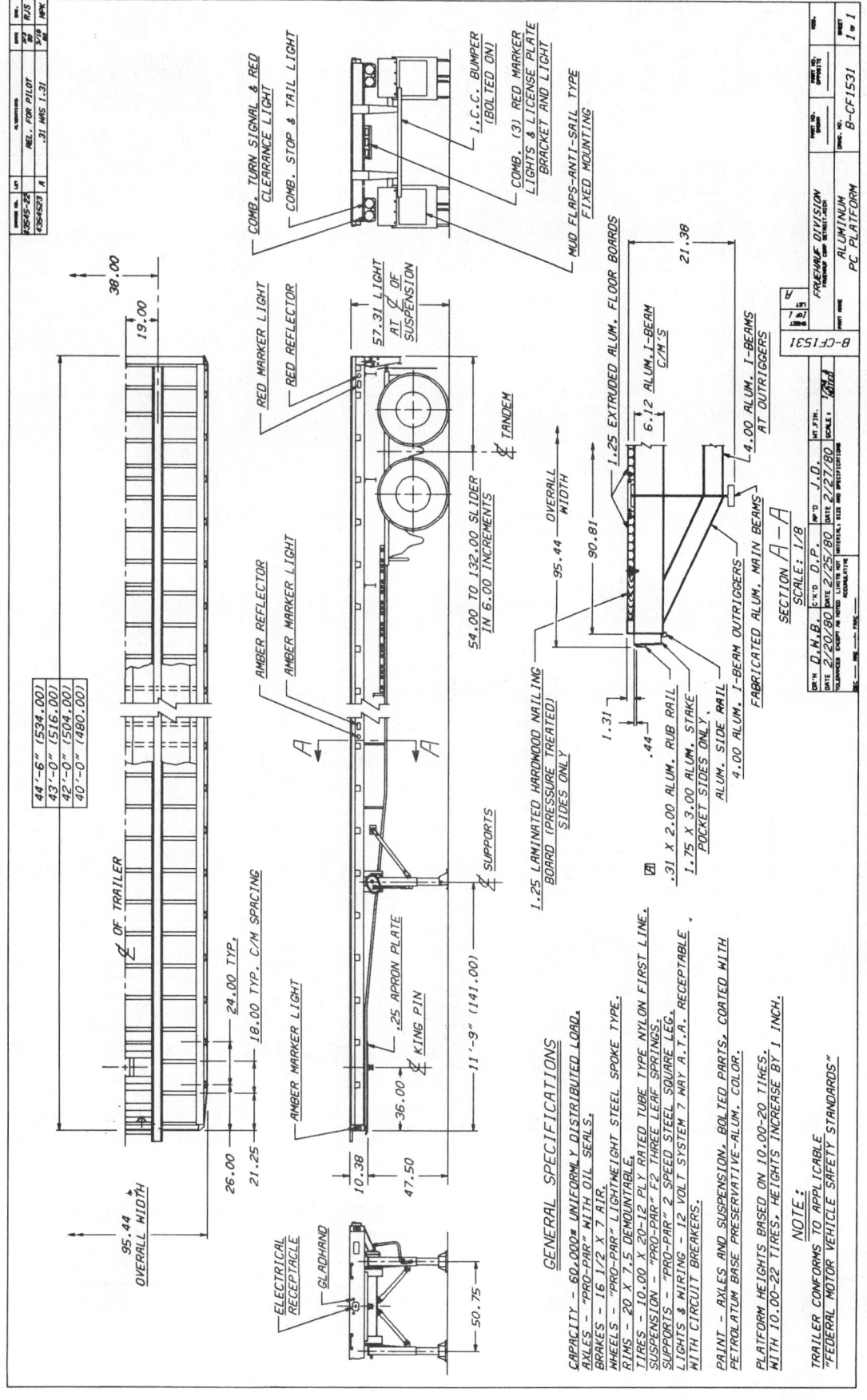

Figure 9-127 Detail notes using leader lines *(Courtesy Fruehauf Division, Fruehauf Corp.)*

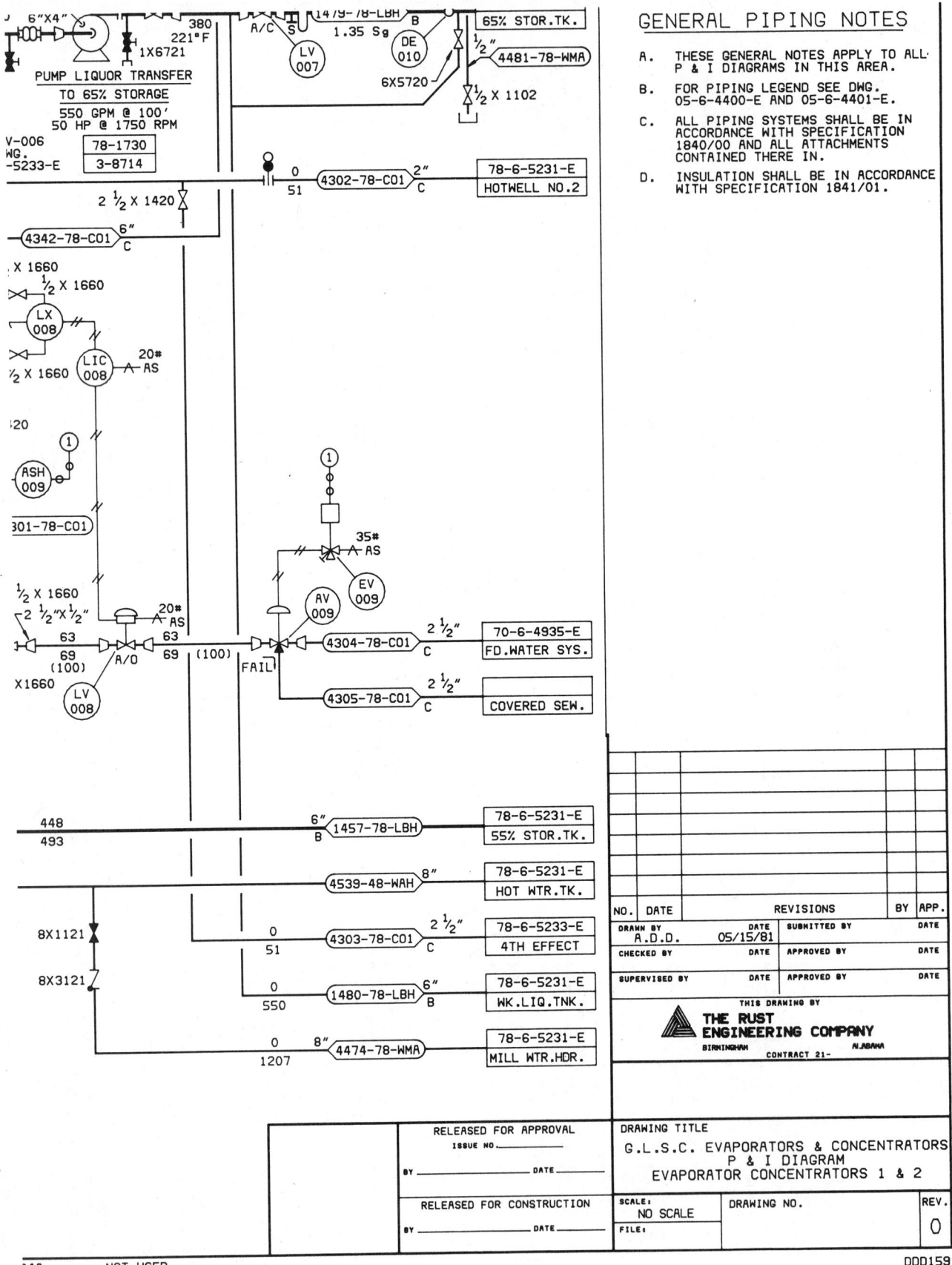

Figure 9-128 General notes on drawings *(Courtesy The Rust Engineering Co.)*

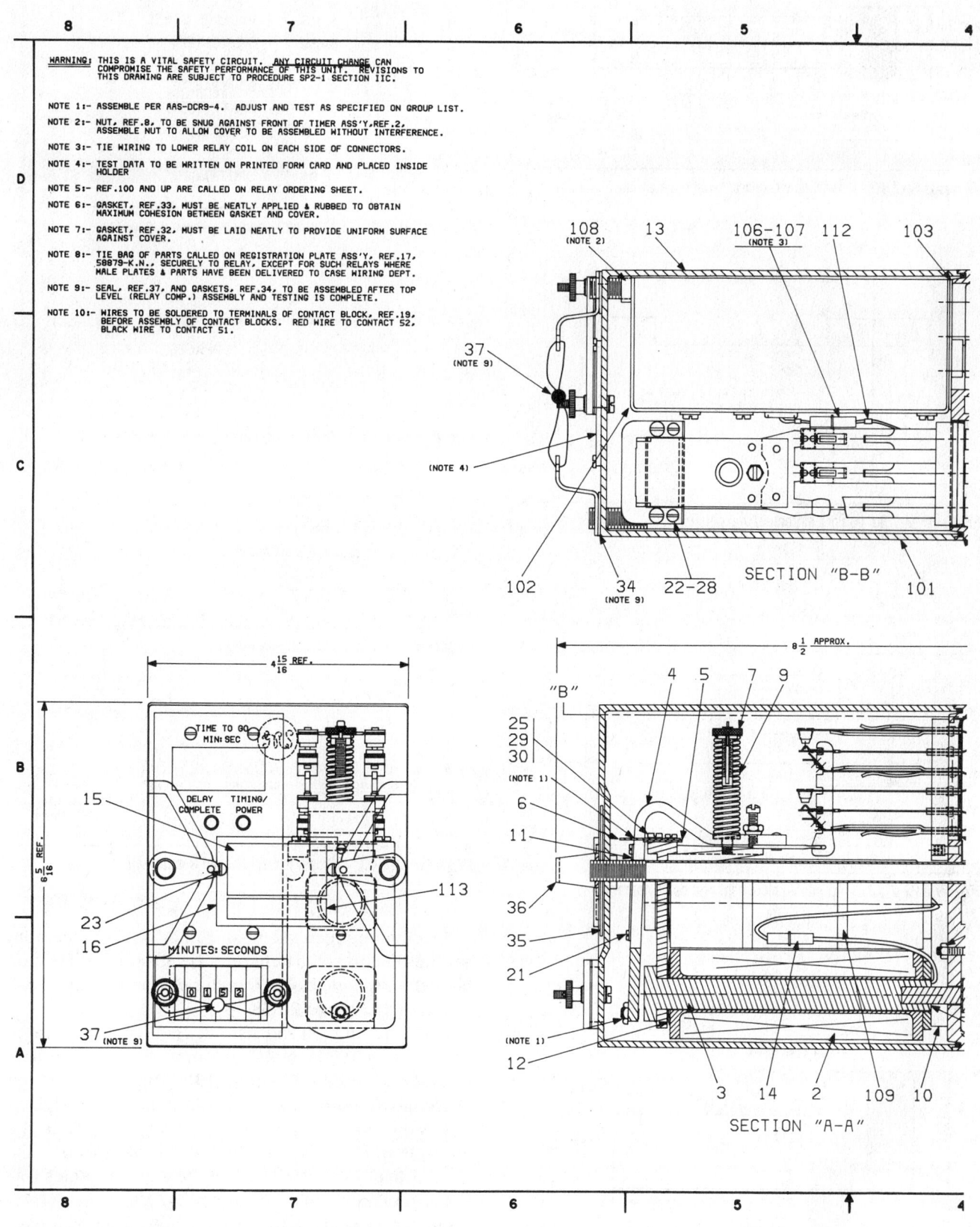

WARNING: THIS IS A VITAL SAFETY CIRCUIT. ANY CIRCUIT CHANGE CAN COMPROMISE THE SAFETY PERFORMANCE OF THIS UNIT. REVISIONS TO THIS DRAWING ARE SUBJECT TO PROCEDURE SP2-1 SECTION IIC.

NOTE 1:- ASSEMBLE PER AAS-DCR9-4. ADJUST AND TEST AS SPECIFIED ON GROUP LIST.

NOTE 2:- NUT, REF.8, TO BE SNUG AGAINST FRONT OF TIMER ASS'Y, REF.2, ASSEMBLE NUT TO ALLOW COVER TO BE ASSEMBLED WITHOUT INTERFERENCE.

NOTE 3:- TIE WIRING TO LOWER RELAY COIL ON EACH SIDE OF CONNECTORS.

NOTE 4:- TEST DATA TO BE WRITTEN ON PRINTED FORM CARD AND PLACED INSIDE HOLDER

NOTE 5:- REF.100 AND UP ARE CALLED ON RELAY ORDERING SHEET.

NOTE 6:- GASKET, REF.33, MUST BE NEATLY APPLIED & RUBBED TO OBTAIN MAXIMUM COHESION BETWEEN GASKET AND COVER.

NOTE 7:- GASKET, REF.32, MUST BE LAID NEATLY TO PROVIDE UNIFORM SURFACE AGAINST COVER.

NOTE 8:- TIE BAG OF PARTS CALLED ON REGISTRATION PLATE ASS'Y, REF.17, 58879-K.N., SECURELY TO RELAY, EXCEPT FOR SUCH RELAYS WHERE MALE PLATES & PARTS HAVE BEEN DELIVERED TO CASE WIRING DEPT.

NOTE 9:- SEAL, REF.37, AND GASKETS, REF.34, TO BE ASSEMBLED AFTER TOP LEVEL (RELAY COMP.) ASSEMBLY AND TESTING IS COMPLETE.

NOTE 10:- WIRES TO BE SOLDERED TO TERMINALS OF CONTACT BLOCK, REF.19, BEFORE ASSEMBLY OF CONTACT BLOCKS. RED WIRE TO CONTACT 52, BLACK WIRE TO CONTACT 51.

SECTION "B-B"

SECTION "A-A"

Figure 9-129 General notes on drawings *(Courtesy General Railway Signal Corp.)*

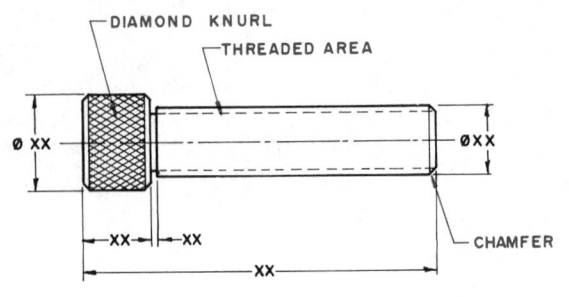

Figure 9-130 Detail note connected with a leader line

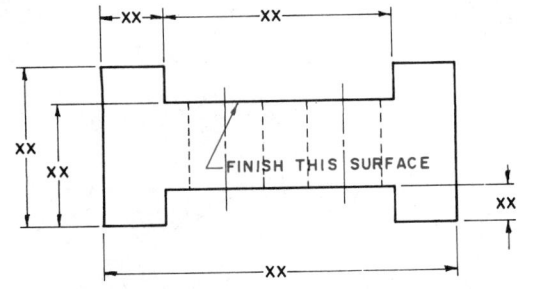

Figure 9-131 Poor placement of a detail note

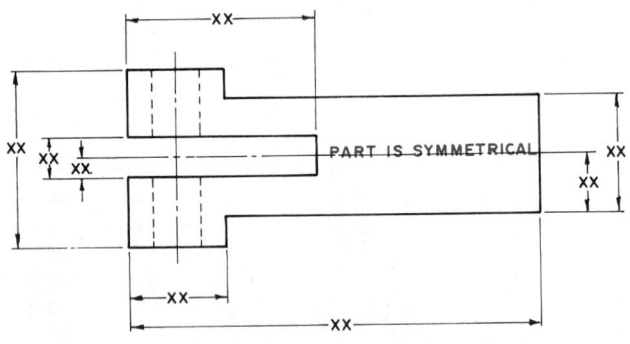

Figure 9-132 Acceptable placement of a detail note when there is space

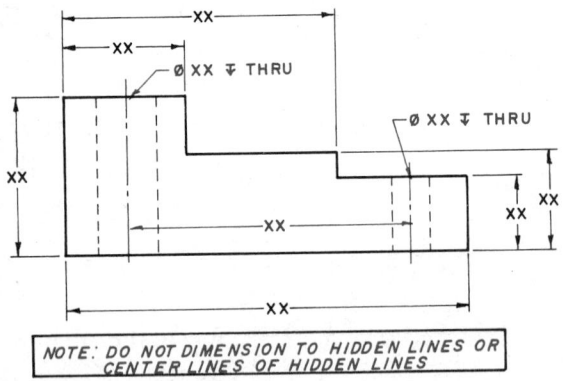

Figure 9-133 Superimposing a note is poor practice

LEGEND
1. ASSY = ASSEMBLY
2. FAO = FINISH ALL OVER
3. MS = MACHINE STEEL
4. FAB = FABRICATE
5. CARB = CARBURIZE
6. DVTL = DOVETAIL
7. HT TR = HEAT TREAT
8. KST = KEYSEAT
9. TPR = TAPER
10. SF = SPOTFACED

Figure 9-134 Abbreviations are explained in a legend

NOTE

POWER BRUSH ALL SURFACES. APPLY
THREE COATS PAINT. STAMP PART
WITH NUMBER 01929612.

Figure 9-135 Proper punctuation of notes is important

NOTE

ALL DIMENSIONS AND CONDITIONS
SHALL BE FIELD CHECKED AND
VERIFIED BY PROPER TRADES

Figure 9-136 Indenting for clarity

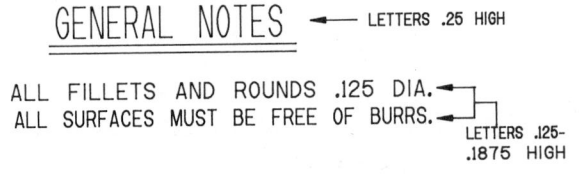

Figure 9-137 Proper lettering height is important

Spacing between successive lines of the same note should be approximately one-sixteenth inch. Spacing between separate notes should be approximately one-eighth inch, ***Figure 9-138***.

There are no hard and fast rules governing the length of a line of notation, but from four to six words per line is widely accepted, ***Figure 9-139***. When a note string will contain more than six words, it should be divided into more than one line. The number of words in the multiple line note should be divided so as to balance the finished note. One long line followed by a drastically shorter line is bad form, ***Figure 9-140***.

On occasion, the last word in a note string will have to be hyphenated. There are rules governing the dividing of words, according to syllables. If the proper point of division is not obvious from the makeup of the word, consult a dictionary. Do not hyphenate abbreviated words.

NOTES

1. SPACING BETWEEN LINES OF THE SAME NOTE SHOULD BE .0625.

2. SPACING BETWEEN SEPARATE NOTES SHOULD BE .1250.

Figure 9-138   Spacing of notes

NOTES

NOTES SHOULD BE LIMITED TO 4 TO 6 WORDS PER LINE

FOUR TO SIX WORDS

Figure 9-139   Number of words per line limitations are important

NOTES

AVOID LONG LINE IN A NOTE FOLLOWED BY ONE SHORT LINE

Figure 9-140   Lines in notes should be approximately equal

Sample Notes

The following is a list of sample notes of the types frequently used on technical drawings:

- All fillets and rounds R1/8
- 45 degree chamfer all edges
- Surfaces A & B parallel
- Stock .125 thick
- Heat treat
- All internal radii R.0625
- FAO
- Ream for 1/4 dowels
- All bends R5/32
- CBORE from bottom
- Identical bolts—both sides
- 3/32 x 1/16 oil groove
- 87 CSK 3/4 dia—4 holes
- Rounds .125 R unless otherwise specified
- 4 holes equally spaced
- 1/32 x 45 chamfer
- #7 Drill .75 deep
- Machine steel—4 reqd.
- Power brush all ext surfaces
- Sandblast before painting

Notice that these sample notes are very brief, concise, and to the point. Words such as "a," "an," "the," and "are" are used only sparingly. The notes are not always complete sentences in terms of proper grammar, but they are complete thoughts in terms of communication. Notice also the use of abbreviations such as FAO

(finish all over), CBORE (counterbore), and CSK (countersink). Abbreviations can be used frequently to cut down on the length of notes and the time required to letter them. However, when used, abbreviations should always be placed in a legend to ensure consistency of interpretation.

Group Technology

Group technology is a manufacturing concept in which similar parts are grouped together in parts groups or families. Parts may be alike in two ways:

1. In their design characteristics
2. In the manufacturing processes required to produce them

By grouping similar parts into families, manufacturing personnel can improve efficiently. Such improvements are the result of advantages gained in such areas as set-up time, standardization of processes and scheduling.

Group technology can also improve the productivity of design personnel by decreasing the amount of work and time involved in designing a new part. Chances are a new part will be similar to an existing part in a given family. When this is the case, the new part can be developed by simply modifying the design of the already existing part. Design modifications tend to require less time and work than new design. This is especially true in the age of CAD. The advantages of group technology are dealt with at greater length later in this chapter.

Historical Development

Ever since the industrial revolution, manufacturing and engineering personnel have been searching for ways to optimize manufacturing processes. There have been numerous developments over the years since the industrial revolution mechanized production. Mass production and interchangeability of parts in the 1800s were major steps forward in optimizing manufacturing.

However, even with mass production and assembly lines, most manufacturing is done in small batches ranging from one workpiece to two or three thousand. In fact, even today over 70 percent of manufacturing involves batches of less than three thousand workpieces. Historically, less has been done to optimize small batch production than has been done for assembly line work.

There have been attempts to standardize a variety of design tasks and some working queuing and sequencing in manufacturing, but until recent years design and manufacturing in small batch settings has been somewhat random.

The underlying problem that has historically prevented significant improvements to small batch manufacturing is that any solution must apply broadly to general production processes and principles rather than to a specific product. This is a difficult problem because the various workpieces in a small batch can be so random and different.

When manufacturing entered the age of automation and computerization, such developments as scheduling software, sequencing software, and materials requirements planning (MRP) systems became available to improve production of both small and large batches. Even these developments have not optimized the production of small batch manufacturing lots. The problem has become even more critical because, since the end of World War II, the trend has been toward more small batch and less production.

In recent years the problems of small batch production have finally begun to receive the attention necessary to bring about improvements. A major step is the ongoing development of group technology.

Part Families

It has already been stated that parts may be similar in design (size and shape) and/or in the manufacturing processes used to produce them. A group of such parts is called a *part family*. It is possible for parts in the same family to be very similar in design yet radically different in the area of production requirements. The opposite may also be true.

Figure 9-141 contains examples of two parts from the same family. These parts were placed in the same family based on design characteristics. They have exactly the same shape and size. However, you will notice that they differ in the area of production processes.

Part 1, after it is drilled, will go to a painting station for two coats of primer. Its dimensions must be held to a tolerance of plus or minus .125 inch. Part 2, after it is drilled, will go to a finishing station for sanding and buffing. Its dimensions require more restrictive tolerances of plus or minus .003 inch. The parts differ in material. The material for Part 1 is cold-rolled steel; Part 2 is aluminum.

Figure 9-141 contains examples of two parts from the same family. Although the design characteristics of these two parts are drastically different (i.e., different sizes and shapes), a close examination will reveal that they are similar in the area of production processes. Part 1 is made of stainless steel. Its dimensions must be held to a tolerance of plus or minus .002 inch, and three holes must be drilled through it. Part 2, in spite of the differences in its size and shape, has exactly the same manufacturing characteristics.

The parts shown in Figure 9-141, because of their similar design characteristics, were grouped in the same family. Such a family is referred to as a design part family. Those in *Figure 9-142* were grouped together because of similar manufacturing characteristics. Such a family is referred to as a *manufacturing part family*. The characteristics used in classifying parts are referred to as *attributes*.

By grouping parts into families, manufacturing personnel can cut down significantly on the amount of materials handled and movement wasted in producing

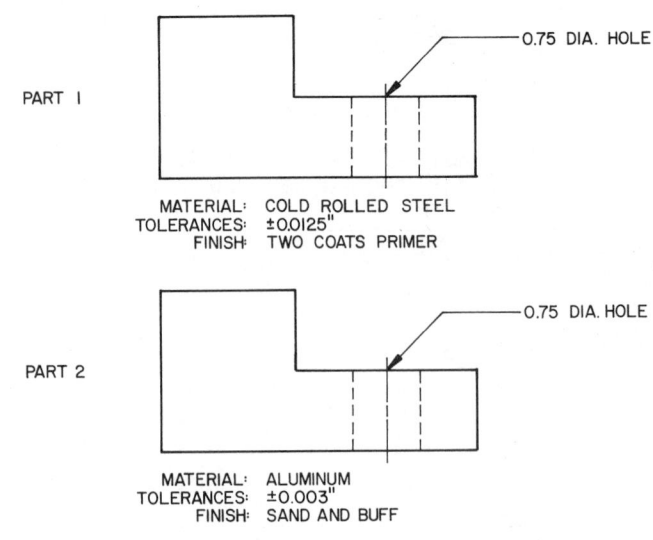

MATERIAL: COLD ROLLED STEEL
TOLERANCES: ±0.0125"
FINISH: TWO COATS PRIMER

MATERIAL: ALUMINUM
TOLERANCES: ±0.003"
FINISH: SAND AND BUFF

Figure 9-141 Two parts from the same family

them. This is because manufacturing machines can be correspondingly grouped into specialized work cells instead of the traditional arrangement of machines according to function (i.e., mills together, lathes together, drills together, etc.).

Each work can be specially configured to produce a given family of parts. When this is done, the number of setups, the amount of materials handling, the length of lead time, and the amount of in-process inventory are all reduced.

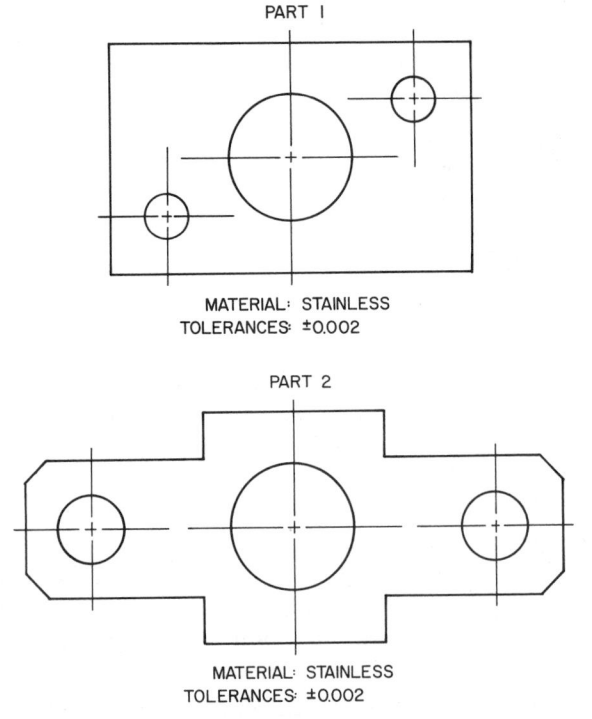

MATERIAL: STAINLESS
TOLERANCES: ±0.002

MATERIAL: STAINLESS
TOLERANCES: ±0.002

Figure 9-142 Two parts from the same family

Grouping Parts into Families

Part grouping is not as simple a process as one might think. You already know the criteria used: design similarities and manufacturing similarities. But how does the actual grouping take place?

There are three methods that can be used for grouping parts into families (*Figure 9-143*).

1. Sight inspection
2. Route sheet inspection
3. Parts classification and coding

All three methods require the expertise of experienced manufacturing personnel.

Sight inspection is the simplest, least sophisticated method. It involves looking at parts, photos of parts, or drawings of parts. Through such an examination, experienced personnel are able to identify similar characteristics and group the parts accordingly. This is the easiest approach, especially for grouping parts by design attributes, but it is also the least accurate of the three methods.

The second method involves inspecting the routing sheets used to route the parts through the various operations to be performed. This can be an effective way to group parts into manufacturing part families, provided the routing sheets are correct. If they are, this method is more accurate than the sign inspection approach. This method is sometimes referred to as the PFA or production flow analysis method.

The most widely used method for grouping parts is the third method: parts classification and coding. This is also the most sophisticated, most difficult, and most time-consuming method. Parts classification and coding is complex enough to require a more in-depth treatment than the other two methods.

Parts Classification and Coding

Parts classification and coding is a method in which the various design and/or manufacturing characteristics of a part are identified, listed, and assigned a code number. Recall that these characteristics are referred to as *attributes*. This is a general approach used in classifying and coding parts. There are many different systems that have been developed for actually carrying out the process, none of which has emerged as the standard.

The many different classification and coding systems that have been developed all fall into one of three groups (*Figure 9-144*).

1. Design attribute group
2. Manufacturing attribute group
3. Combined attribute group

Students of CAD/CAM should be familiar with how systems are grouped.

Design Attribute Group. Classification and coding systems that use design attributes as the qualifying criteria fall into this group. Commonly used design attributes include: dimensions, tolerances, shape, finish, and material.

Manufacturing Attribute Group. Classification and coding systems that use manufacturing attributes as the qualifying criteria fall into this group. Commonly used manufacturing attributes include: production processes, operational sequence, production time, tools required, fixtures required, and batch size.

Combined Attribute Group. There are advantages in using design attributes and advantages in using manufacturing attributes. Systems that fall into the design attribute group are particularly advantageous if the goal is

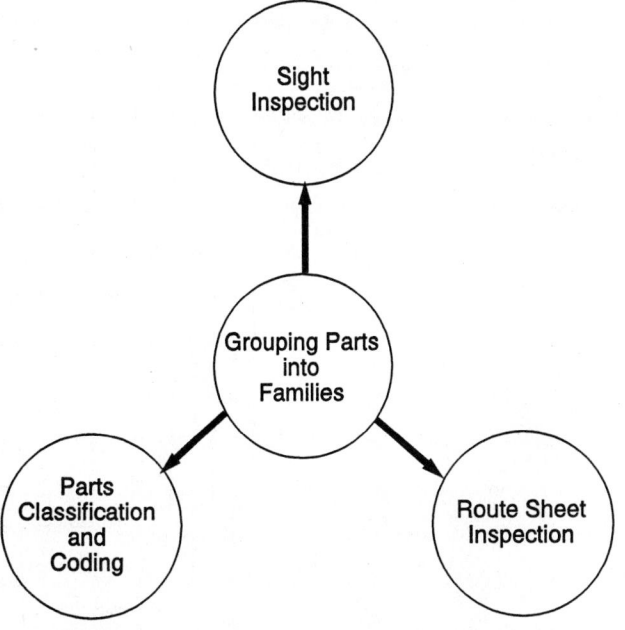

Figure 9-143 Methods of grouping parts into families

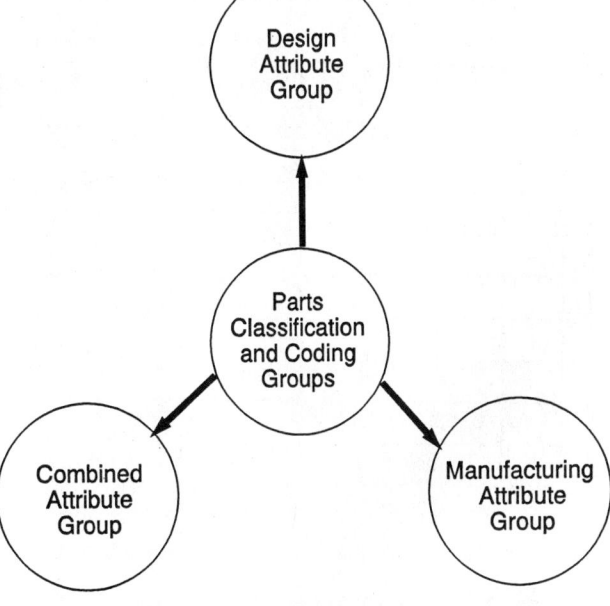

Figure 9-144 Parts classification and coding groups

design retrieval. Those in the manufacturing group are better if the goal is any of a number of production-related functions. However, there is a need for systems that combine the best characteristics of both. Such systems are both design and manufacturing attributes.

Sample Parts Classification and Coding System. Some companies develop their own parts classification and coding system. But this can be an expensive and time-consuming approach. The more widely used approach is to purchase a commercially prepared system. There are several such systems available. However, the most widely used of these is the *Opitz system*. It is a good example of a parts classification and coding system and how one works.

Opitz System

Classification and coding systems use alphanumeric symbols to represent the various attributes of a part. One of the many classification and coding systems is the Opitz system. This system uses characters in 13 places to code the attributes of parts, and hence, to classify them. These digit places are represented as follows:

12345 6789 ABCD

The first five digits (12345) code the major design attributes of a part. The next four digits (6789) are for coding manufacturing-related attributes and are called the supplementary code. The letters (ABCD) code the production operation and sequence.

The alphanumeric characters shown represent places. For example, the actual numeral used in each place can be 0–9. The numeral used in the "1" place indicates the length-to-diameter ratio of the part. The numeral used in the "2" place indicates the external shape of the part. The numeral used in the "3" place indicates the internal shape of the part. The numeral used in the "4" place indicates the type of surface machining. The numeral used in the "5" place indicates gear teeth and auxiliary holes. With such a system, a part might be coded as follows:

20801

The "2" means that the part has a certain length-to-diameter ratio. The first "0" means the part has no external shape elements. The "8" means the part has an internal thread. The second "0" means no surface machining is required. The "1" means the part is axial, not on pitch.

Figure 9-145 is a chart that shows the basic overall structure of the Opitz system for parts classification and coding. *Figure 9-146* is a chart used for assigning an Opitz code to rotational parts. *Figure 9-147* is a drawing of an actual part that has been assigned an Opitz code of 15400. This code was arrived at as follows:

Step 1   The total length of the part is divided by the overall diameter:

$$\frac{1.75(L)}{1.25(D)} = 1.40$$

Since 1.40 is greater than .5 but less than 3, the first digit in the code is "1."

Step 2   An overall description of the external shape of the part would read "…a rotational part that is stepped on both ends with one stepped end threaded." Consequently, the most appropriate second digit in the code is "5."

Step 3   A description of the internal shape of the part would read "…a through hole." Consequently, the third digit in the code is "4."

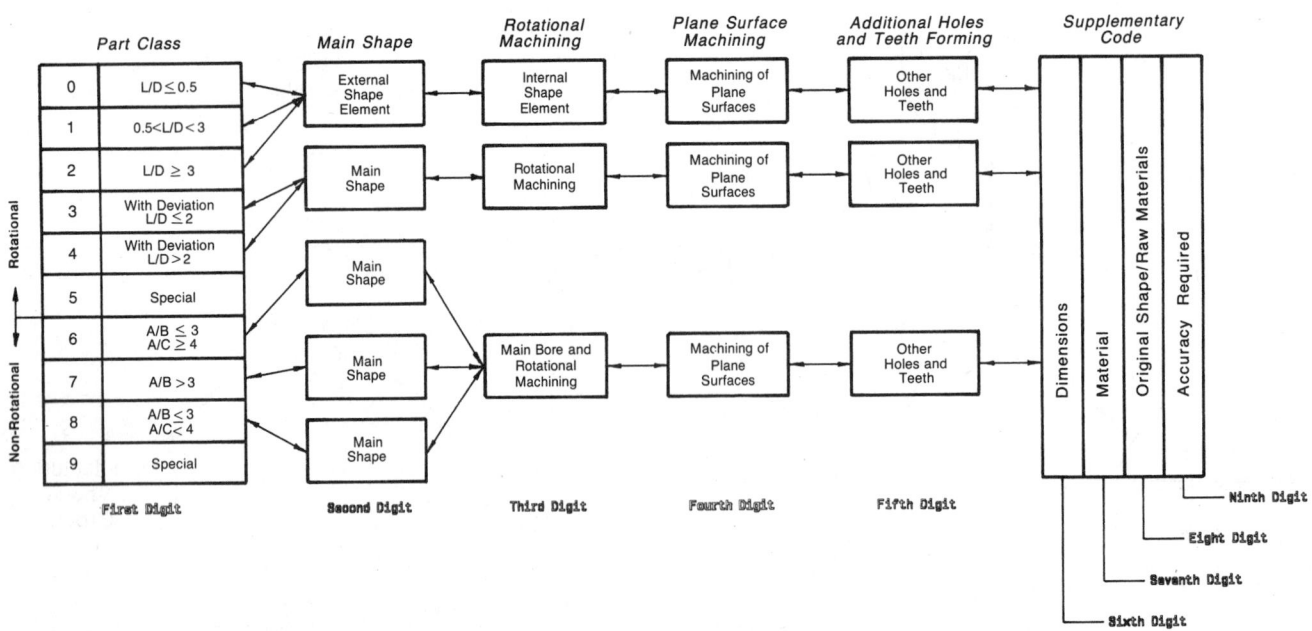

Figure 9-145   Basic structure of the Opitz system

	PART CLASS		MAIN SHAPE EXTERNAL	ROTATIONAL MACHINING INTERNAL	PLANE SURFACE MACHINING	ADDITIONAL HOLES AND TEETH
Rotational		$L/D \le 0.5$	Smooth, no shape elements	No hole, no breakthrough	No surface machining	No auxiliary hole
Rotational	1	$0.5 < L/D < 3$	Stepped to one end or smooth: No shape elements	Smooth or stepped to one end: No shape elements	Surface plane and/or curved in one direction, external	No gear teeth: Axial, not on pitch circle diameter
Rotational	2	$L/D \ge 3$	Stepped to one end or smooth: Thread	Smooth or stepped to one end: Thread	External plane surface related by graduation around a circle	No gear teeth: Axial on pitch circle diameter
Rotational	3	—	Stepped to one end or smooth: Functional groove	Smooth or stepped to one end: Functional groove	External groove and/or slot	No gear teeth: Radial, not on pitch circle diameter
Rotational	4	—	Stepped to both ends: No shape elements	Stepped to both ends: No shape elements	External spline (polygon)	No gear teeth: Axial and/or radial and/or other direction
Rotational	5	—	Stepped to both ends: Thread	Stepped to both ends: Thread	External plane surface and/or slot, external spline	No gear teeth: Axial and/or radial on PCD and/or other directions
Non-Rotational	6	—	Stepped to both ends: Functional groove	Stepped to both ends: Functional groove	Internal plane surface and/or slot	With gear teeth: Spur gear teeth
Non-Rotational	7	—	Functional cone	Functional cone	Internal spline (ploygon)	With gear teeth: Bevel gear teeth
Non-Rotational	8	—	Operating thread	Operating thread	Internal and external polygon, groove and/or slot	With gear teeth: Other gear teeth
Non-Rotational	9	—	All others	All others	All others	All others
	First Digit		Second Digit	Third Digit	Fourth Digit	Fifth Digit

Figure 9-146 Assigning a code using the Opitz system

Step 4 By examining the part it can be seen that no surface machining is required. Therefore, the fourth digit in the code is "0."

Step 5 By examining the part it can be seen that no auxiliary holes or gear teeth are required. Therefore, the fifth digit in the code is also "9."

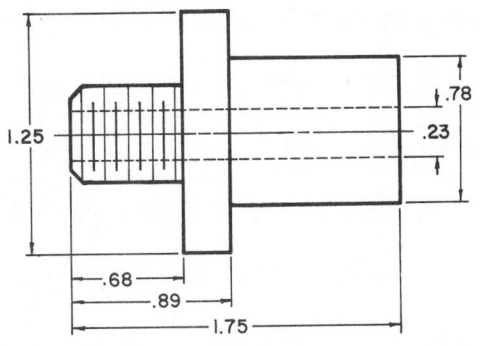

Figure 9-147 Sample part that can be coded using the Opitz system

Advantages and Disadvantages of Group Technology

There are several advantages that can be realized through the application of group technology. These advantages include:

1. improved design
2. enhanced standardization
3. reduced materials handling
4. simplified production scheduling
5. improved quality control

Improved Design

Group technology allows designers to use their time more efficiently and productively by decreasing the amount of new design work required each time a part is to be designed. When a new part is needed, its various attributes can be listed. Then, an existing part with as many of these attributes as possible can be identified and retrieved. The only new design required is that which relates to attriburtes of the new part not contained in the existing part. Because this characteristic of group technology tends to promote design standardization, additional design benefits accrue.

Enhanced Standardization

Parts are classified into groups according to their similarities. The more similarities, the better. Consequently, enhanced standardization is promoted by group technology. As you have already seen, design factors account for part of the enhanced standardization. Setups and tooling also become more standardized since similar parts require similar setups and similar tooling.

Reduced Materials Handling

One of the major elements of group technology is the arrangement of machines into specialized work cells, each cell producing a given family of parts. Traditionally, machines are arranged by function (i.e., lathes together in one group, mills together in another, etc.). A functional arrangement causes excessive movement of parts from machines to machine and back. Specialized work cells reduce such movement to a minimum and, in turn, reduce materials handling.

Simplified Production Scheduling

With machines grouped into specialized work cells and arrangement of parts into families, there is less work center scheduling requiring less scheduling overall. Group technology involves less sophisticated difficult scheduling.

Improved Quality Control

Quality control is improved through group technology because each work cell is responsible for a specific family of parts. This does two things, both of which tend to improve the quality of parts: 1) it promotes pride in the work among employees assigned to each work cell, and 2) it makes production errors and problems easier to trace to a specific source. In traditional manufacturing shops, each work center performs certain tasks, but no one center is responsible for an entire part. Such an approach does not promote quality control.

In spite of these advantages, there are several disadvantages associated with group technology. Transitioning from machines in a traditional shop into specialized work cells can cause a major disruption. Such disruptions translate quickly into monetary losses. As a consequence, some managers are reluctant to pursue group technology.

Another problem with group technology is that parts classification coding is a time-consuming, difficult, and expensive process, especially in the initial stages. This causes some managers to balk at the expense.

Review

1. What is the most frequently used dimensioning standard?
2. How does a drafter indicate on a drawing that a dimension is NTS?
3. Of the three dimensioning systems used, which is most frequently used on mechanical technical drawings?
4. Illustrate how to label a metric dimension that is less than one millimeter.
5. How should the line separating the numerator and denominator of a fraction be drawn?
6. List the five components of all dimensioning systems.
7. How far should an extension line extend beyond the dimension line?
8. What is the recommended length/width ratio of arrowheads?
9. What is the keyword to remember in making arrowheads, regardless of the size?
10. What is unidirectional dimensioning?
11. Define the term *notes*.
12. Describe the proper location of general notes.
13. How should notes be oriented on the drafting sheet?
14. Define the term *general notes*.
15. What is the proper height for lettering note titles?
16. Define the term *verification* as it relates to notes.
17. Explain the significance of balance in terms of notation.
18. What is group technology?
19. How are parts grouped into families?
20. Explain the three methods used for grouping parts into families.
21. What is an attribute?
22. List five commonly applied design attributes.
23. List five commonly applied manufacturing attributes.
24. List and explain five advantages of group technology.
25. What are the disadvantages of group technology?

Chapter Nine Problems

The following problems are intended to give beginning drafters and engineers practice in applying the various dimensioning techniques required on technical drawings in industry. Students should follow the general instructions which apply to all problems, and any additional instructions provided for special problems. All problems should be dimensioned in strict accordance with the rules set forth in this chapter.

General Instructions

1. Study each problem carefully.
2. Make a complete sketch of the problem, applying all dimensions on the sketch. Note on the sketch the proper spacing between views, between the object and first dimension lines, and between successive dimension lines.
3. Compare your fully dimensioned sketch against all applicable sections of this chapter and "The Summary of Dimensioning Rules."

Note: These instructions do not apply to Problems 9-47 through 9-50.

Problems 9-1 through 9-24

1. Use the grid to determine all dimensions.
2. Redraw the views shown, and fully dimension your drawing.

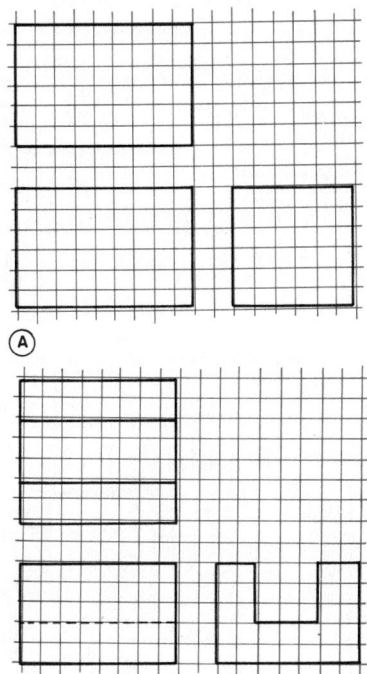

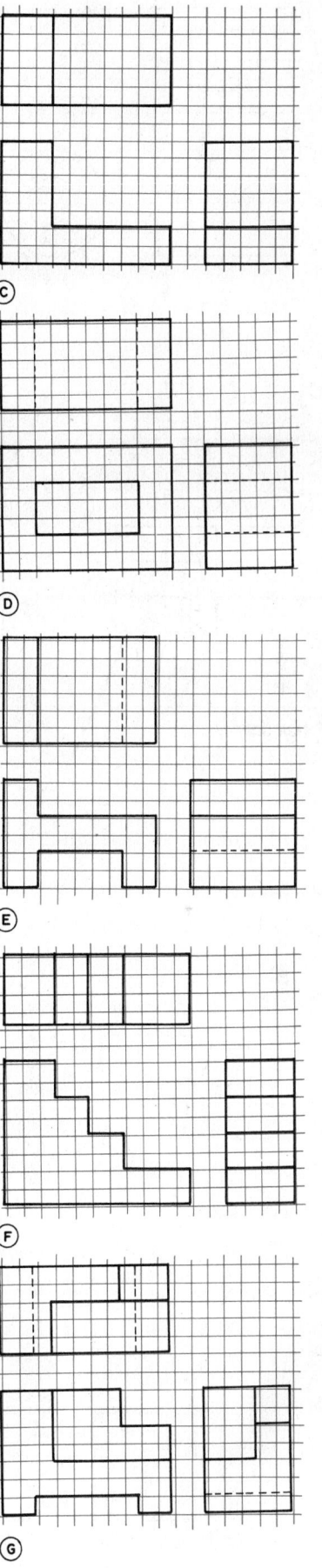

Problem 9-1

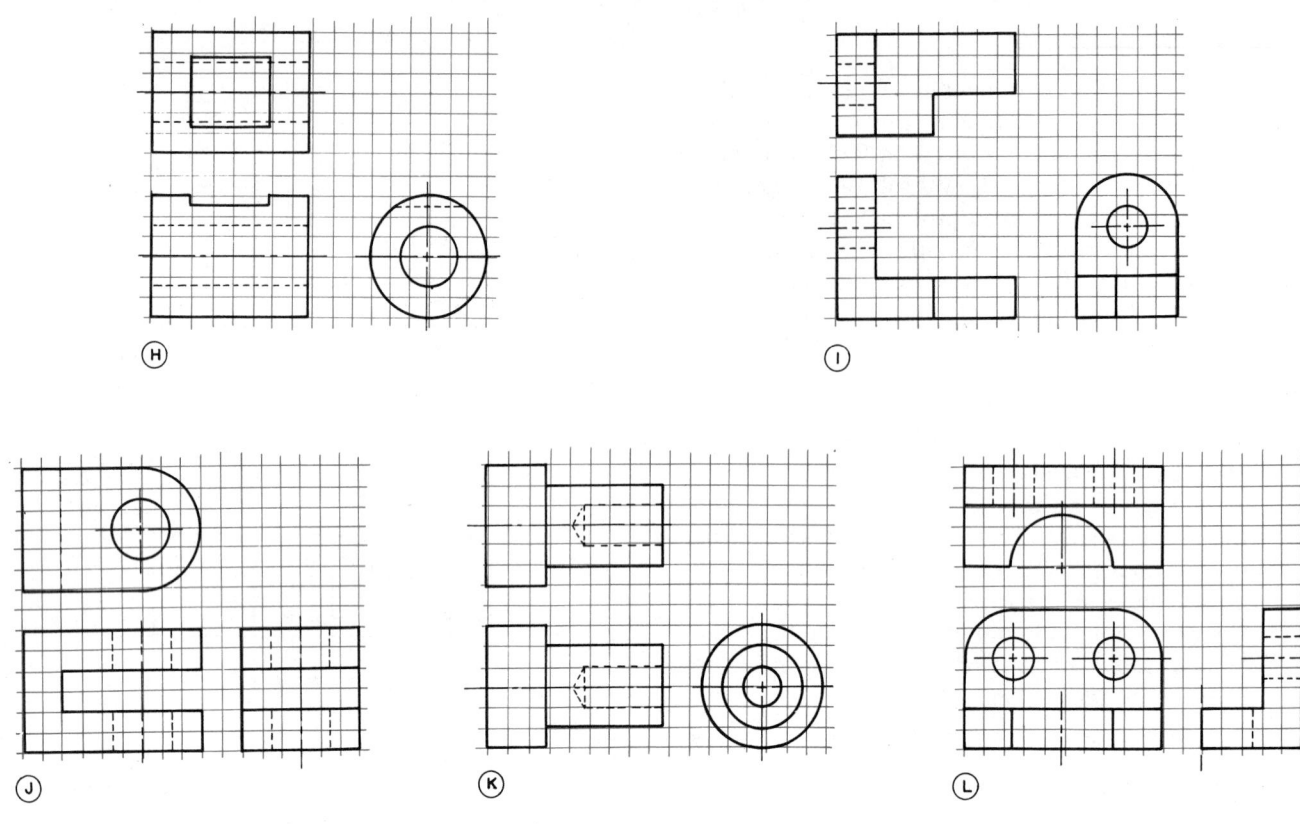

Problem 9-1 (Continued)

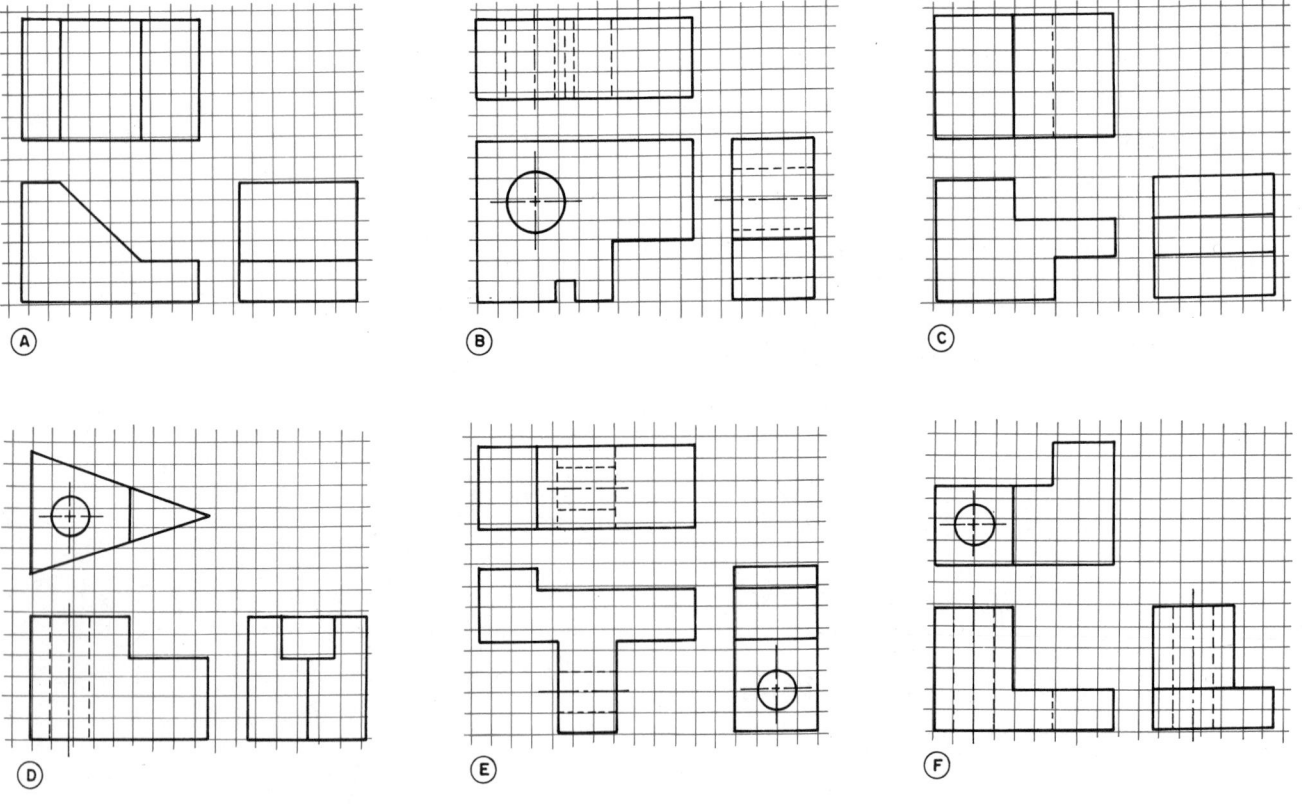

Problem 9-2

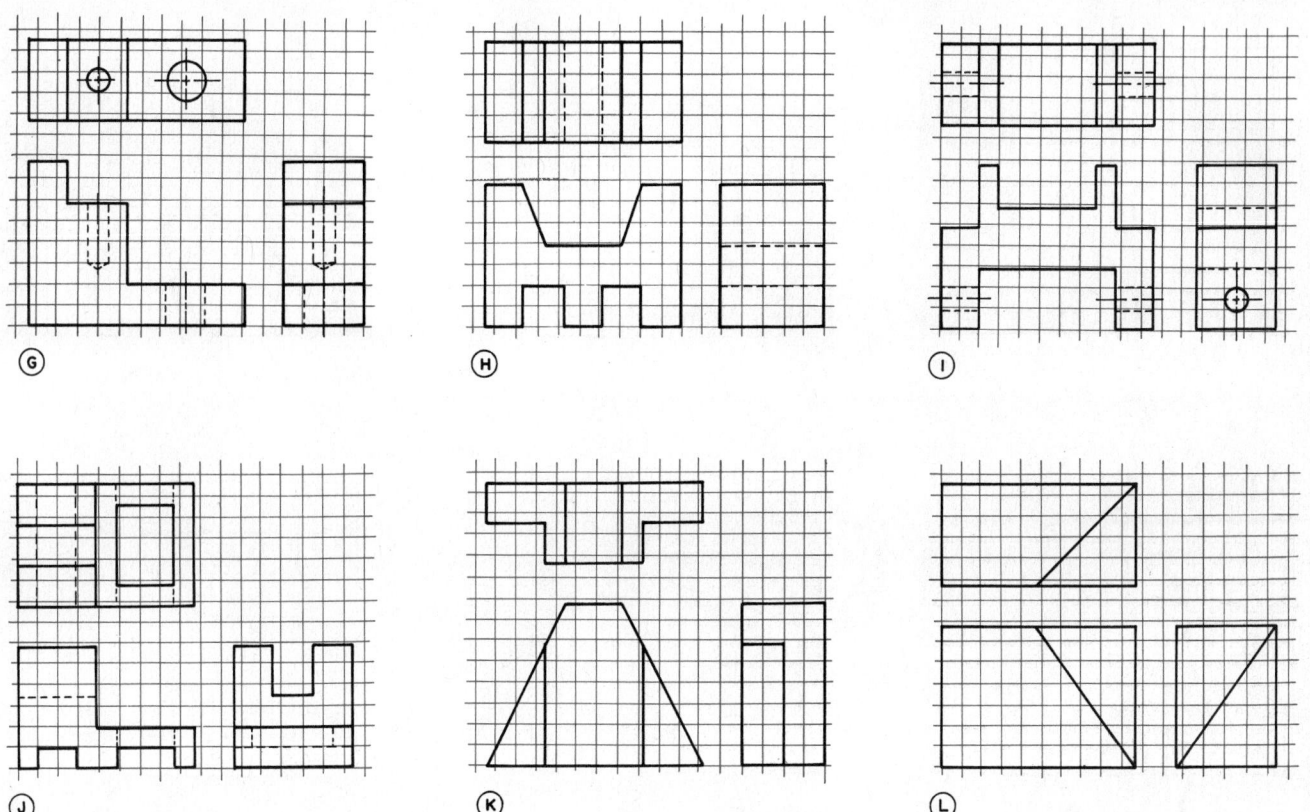

Problem 9-2 *(Continued)*

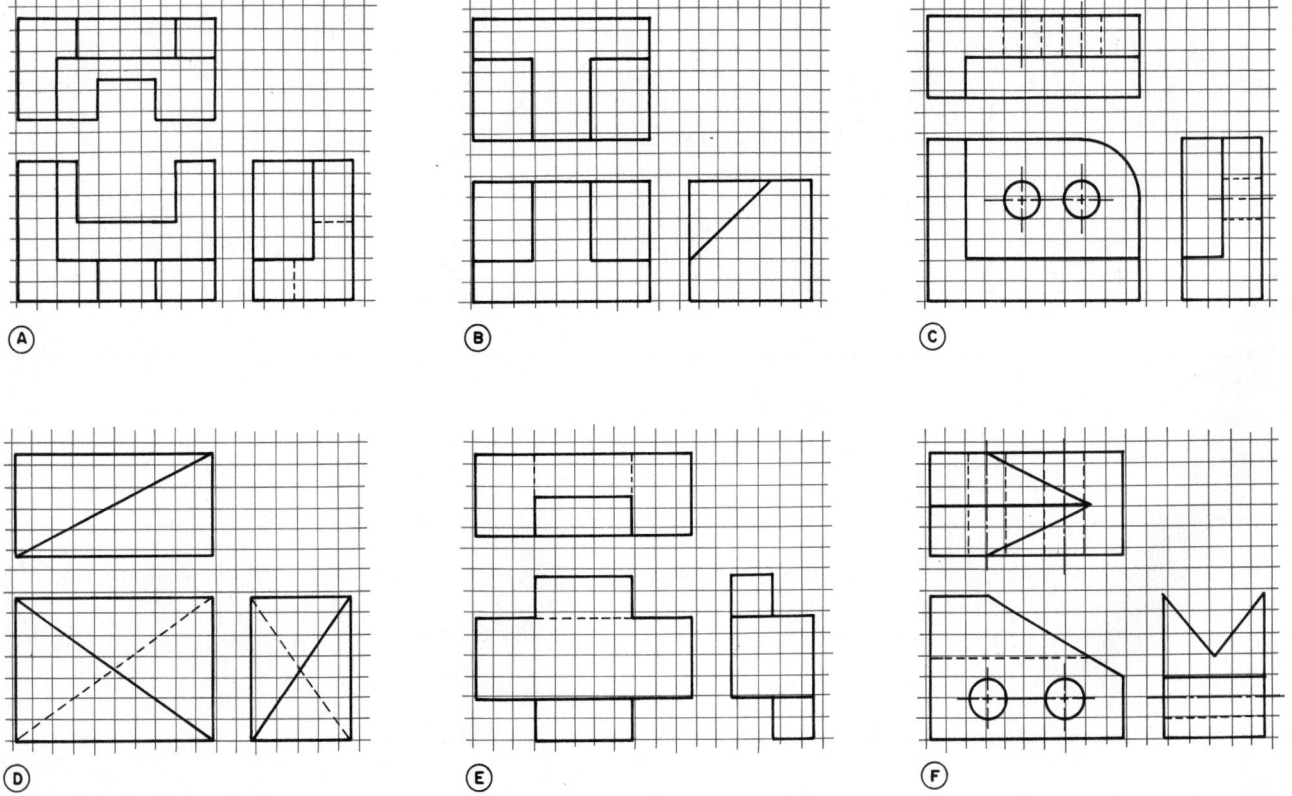

Problem 9-3

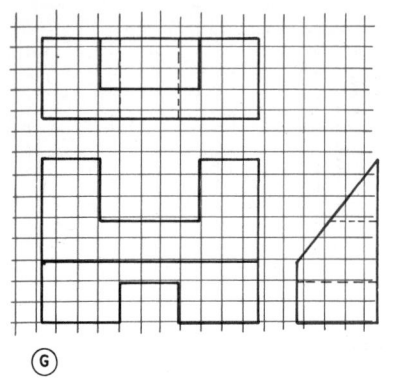

(G)

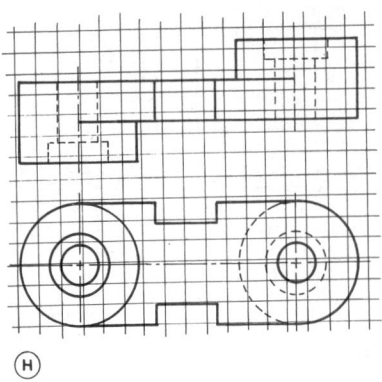

(H)

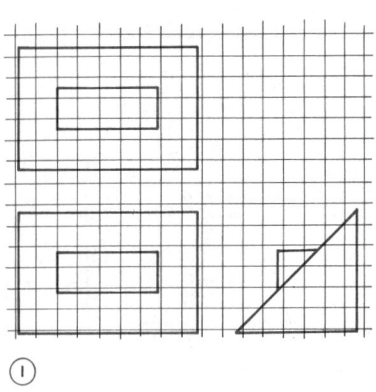

(I)

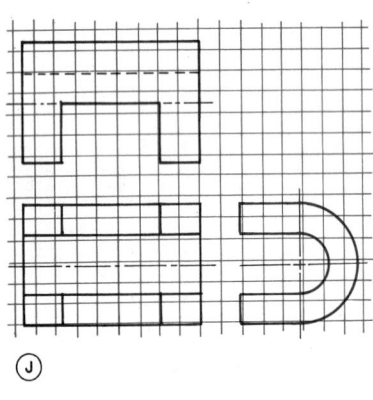

(J)

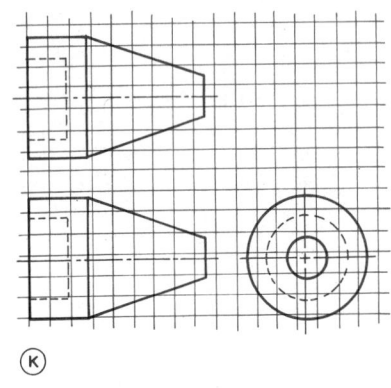

(K)

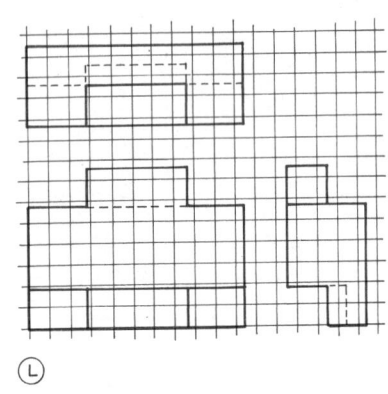

(L)

Problem 9-3 (Continued)

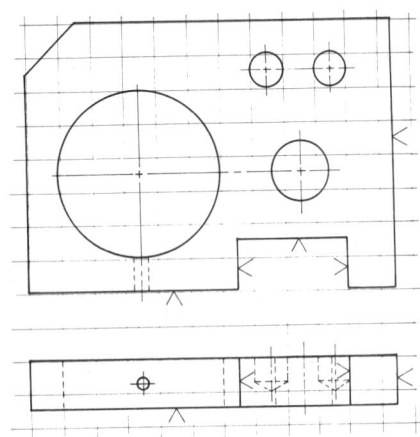

Problem 9-4

Problem 9-6

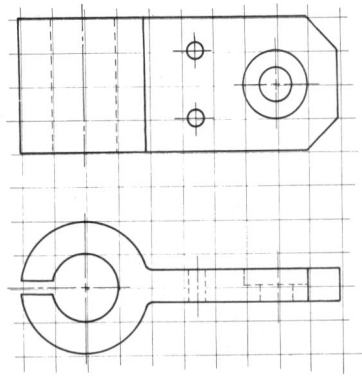

Problem 9-5

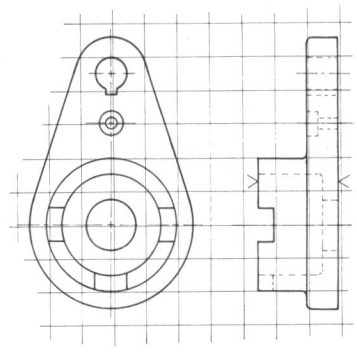

Problem 9-7

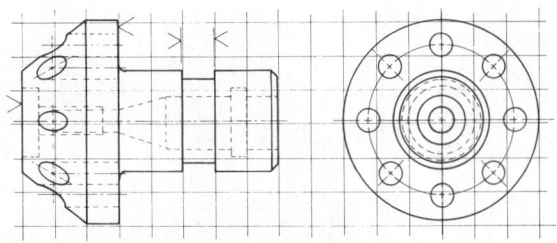

Problem 9-8

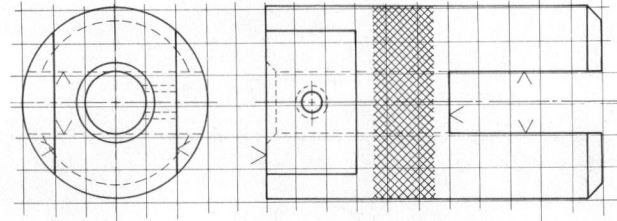

Problem 9-9

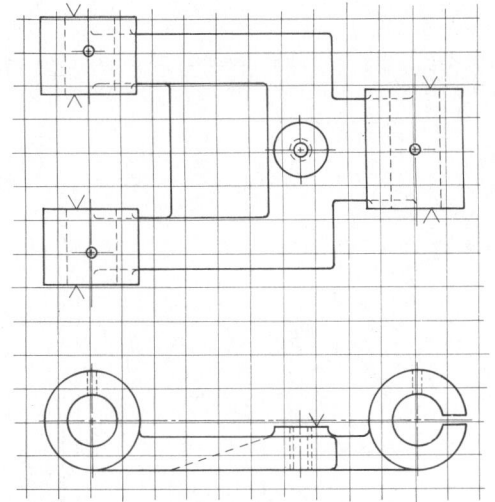

Problem 9-10

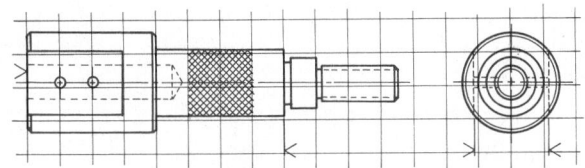

Problem 9-11

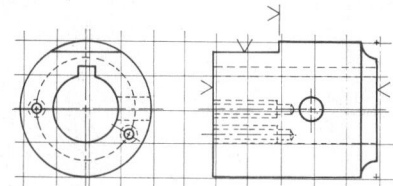

Problem 9-12

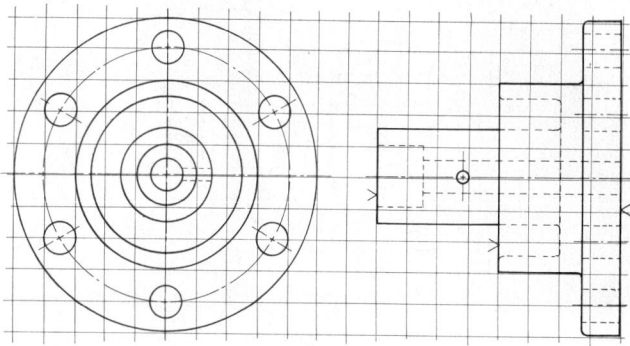

Problem 9-13

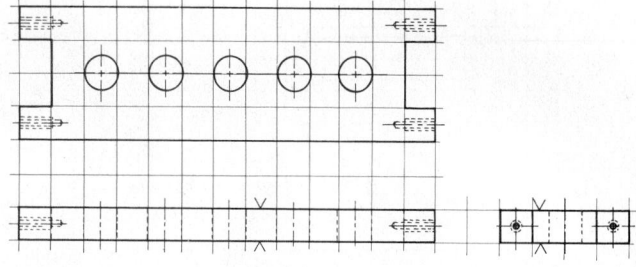

Problem 9-14

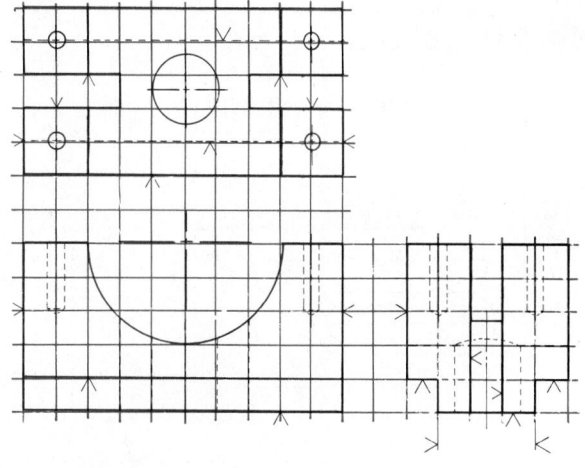

Problem 9-15

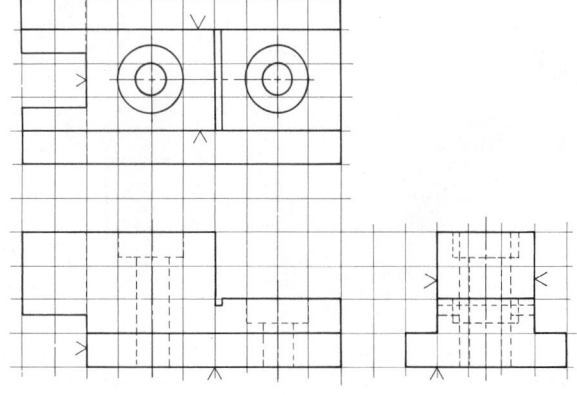

Problem 9-16

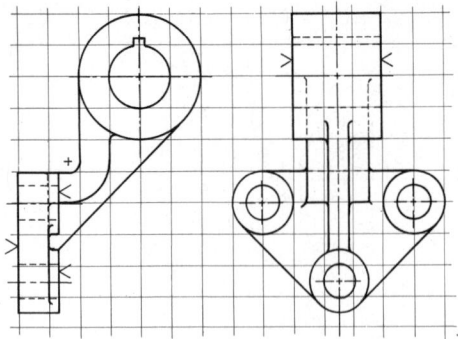

Problem 9-17

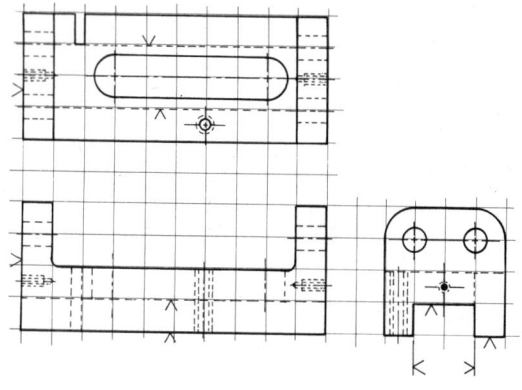

Problem 9-18

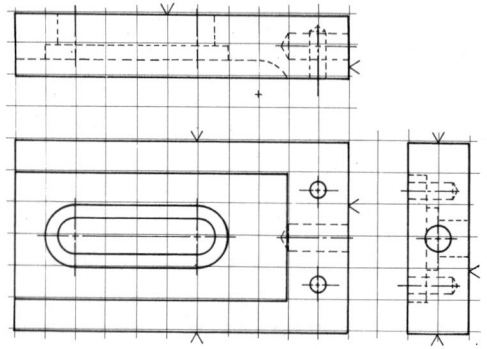

Problem 9-19

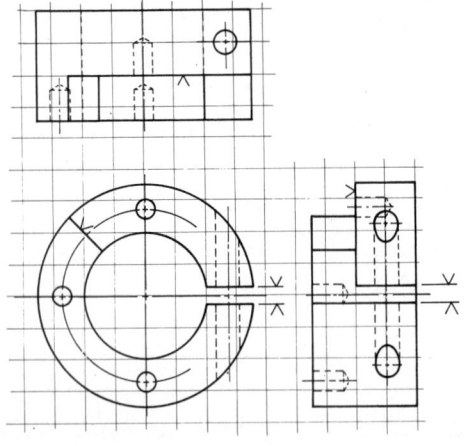

Problem 9-20

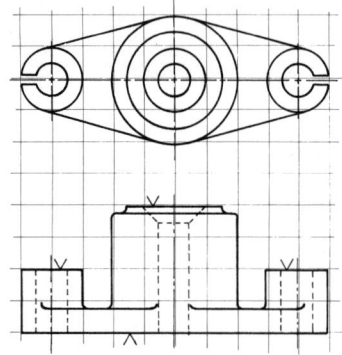

Problem 9-21

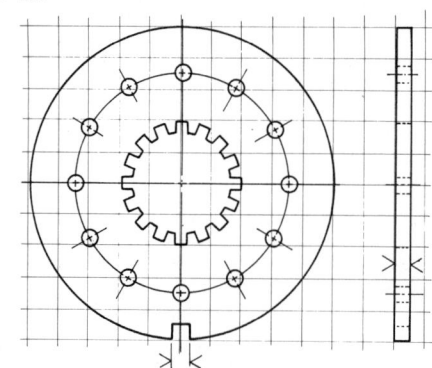

Problem 9-22

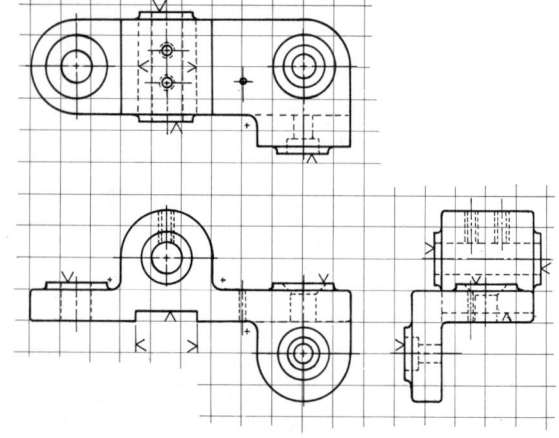

Problem 9-23

Problem 9-24

Problems 9-25 through 9-46

1. Convert the isometric drawings provided into orthographic drawings, showing as many views and/or sections as necessary to communicate the design.
2. Fully dimension your drawing.

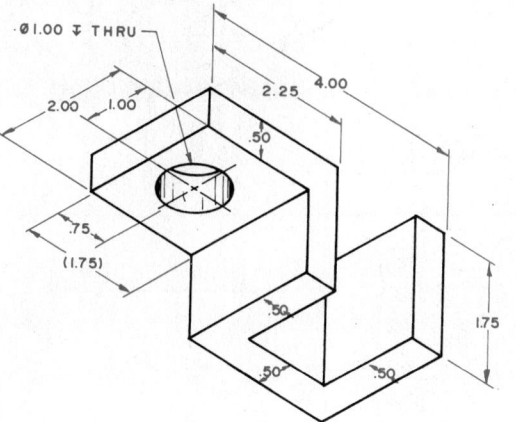

Problem 9-25

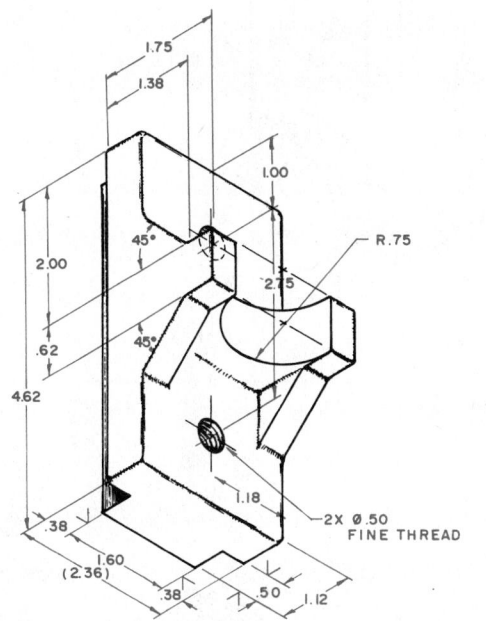

Problem 9-26

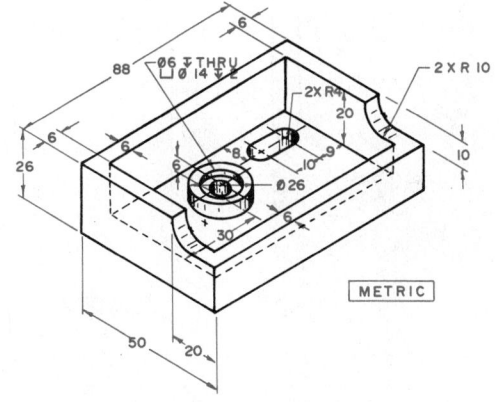

Problem 9-27

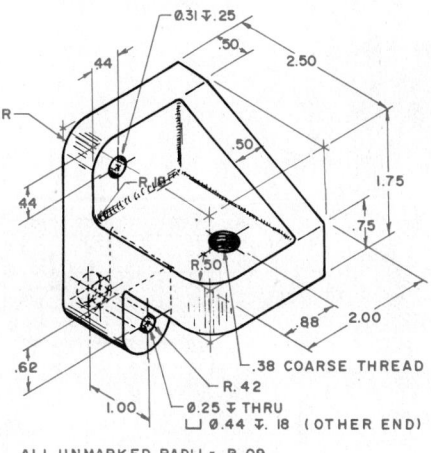

Problem 9-28

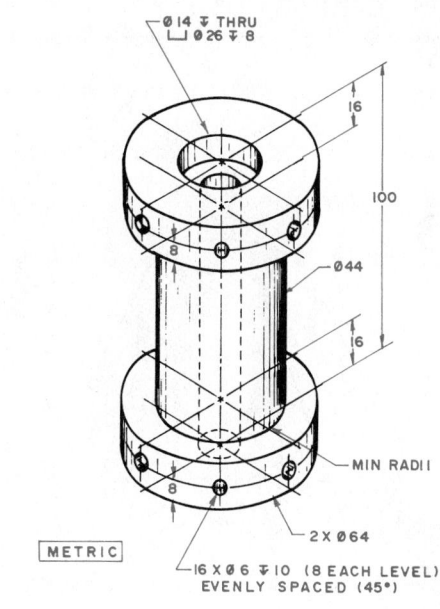

Problem 9-29

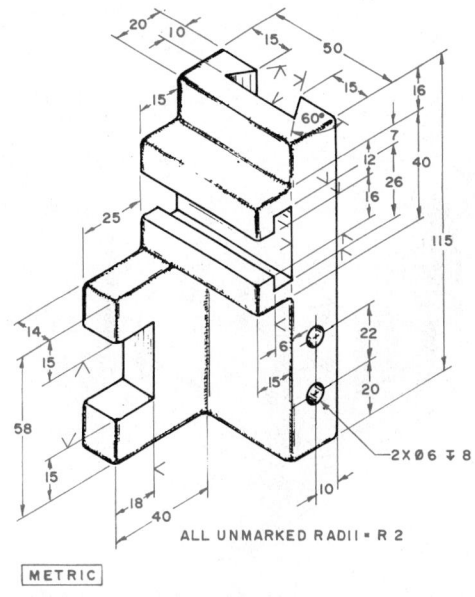

Problem 9-30

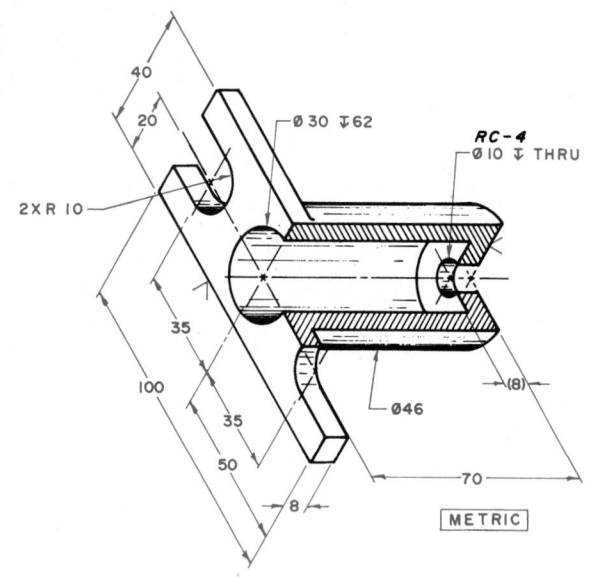

Problem 9-31

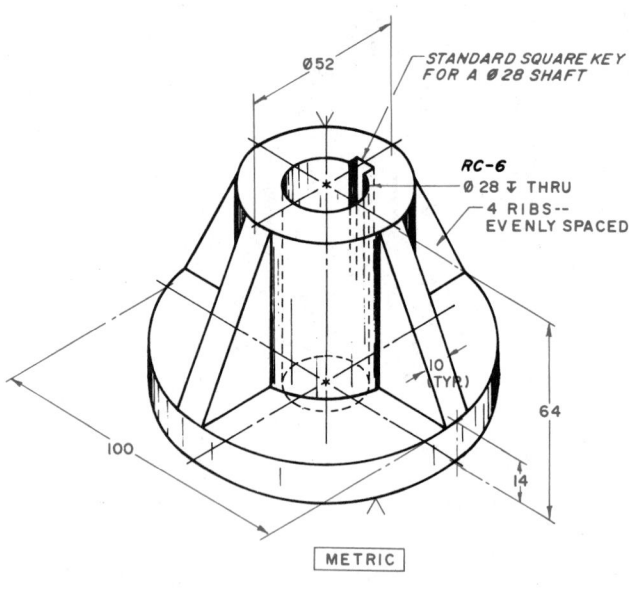

Problem 9-32

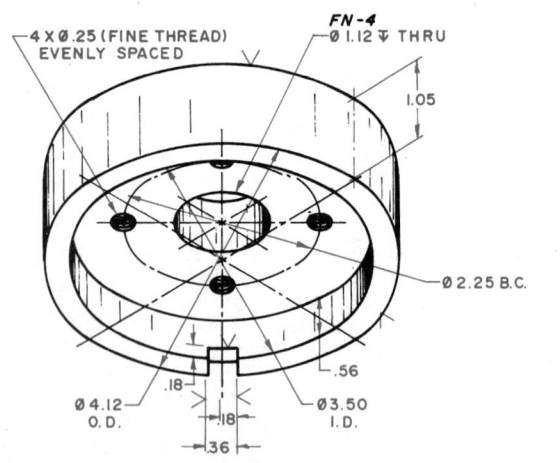

Problem 9-33

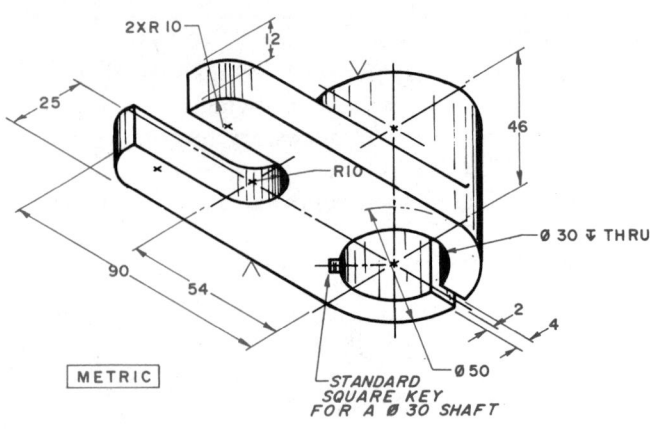

Problem 9-34

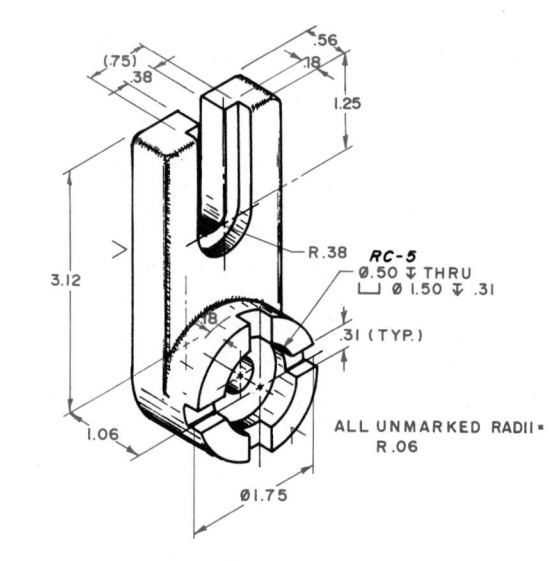

Problem 9-35

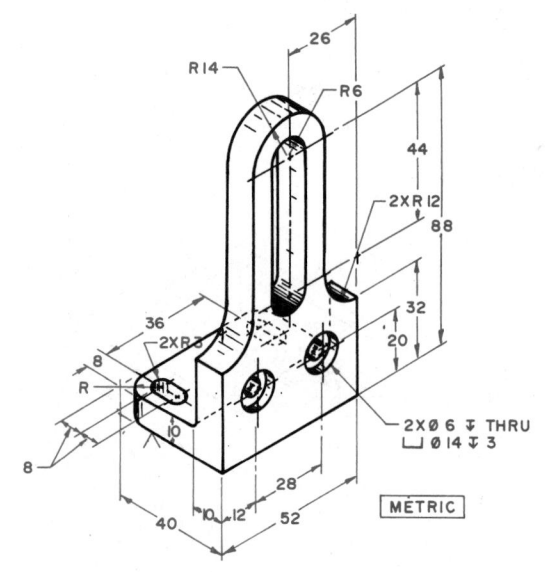

Problem 9-36

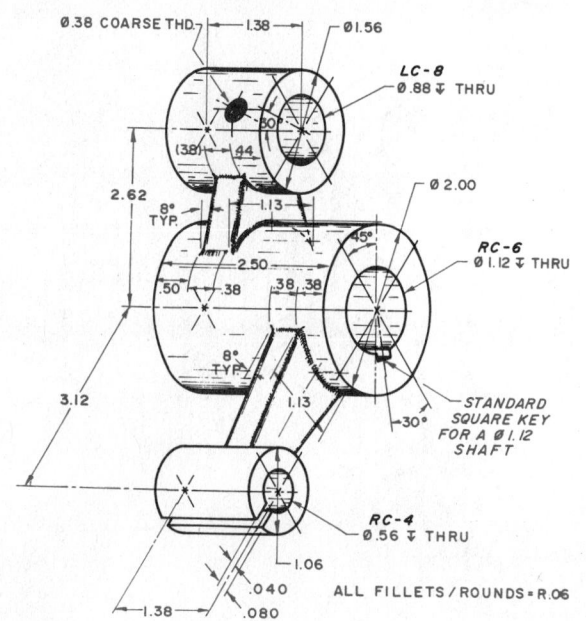

Ø.38 COARSE THD.
1.38
Ø1.56
LC-8
Ø.88 ⟱ THRU
(38) 44
2.62
8° TYP
1.13
Ø2.00
2.50
45°
RC-6
Ø1.12 ⟱ THRU
50 .38 .38 .38
8° TYP
3.12
1.13
30°
STANDARD SQUARE KEY FOR A Ø1.12 SHAFT
RC-4
Ø.56 ⟱ THRU
1.06
.040
1.38
.080
ALL FILLETS/ROUNDS = R.06

Problem 9-37

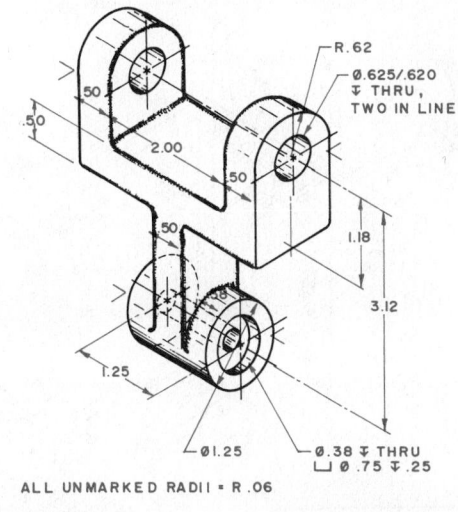

R.62
Ø.625/.620
⟱ THRU, TWO IN LINE
50
.50
2.00
50
50
1.18
3.12
1.25
Ø1.25
Ø.38 ⟱ THRU
□ Ø.75 ⟱ .25
ALL UNMARKED RADII = R.06

Problem 9-40

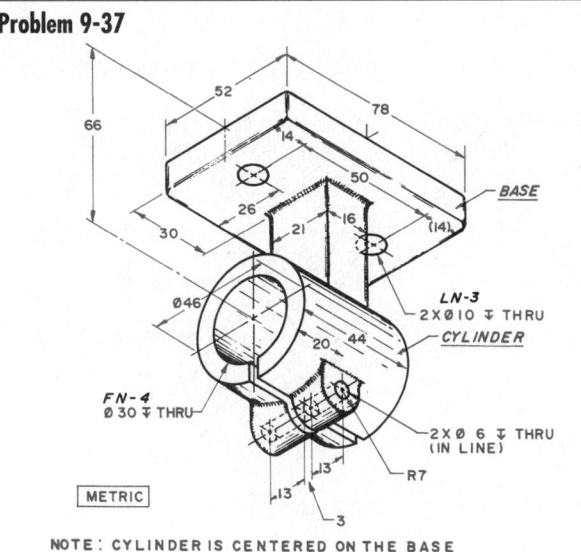

52
66
78
14
50
BASE
26
16
30
21
(14)
LN-3
2 X Ø10 ⟱ THRU
Ø46
CYLINDER
20
44
FN-4
Ø30 ⟱ THRU
2 X Ø6 ⟱ THRU (IN LINE)
13
R7
13
3
METRIC

NOTE: CYLINDER IS CENTERED ON THE BASE

Problem 9-38

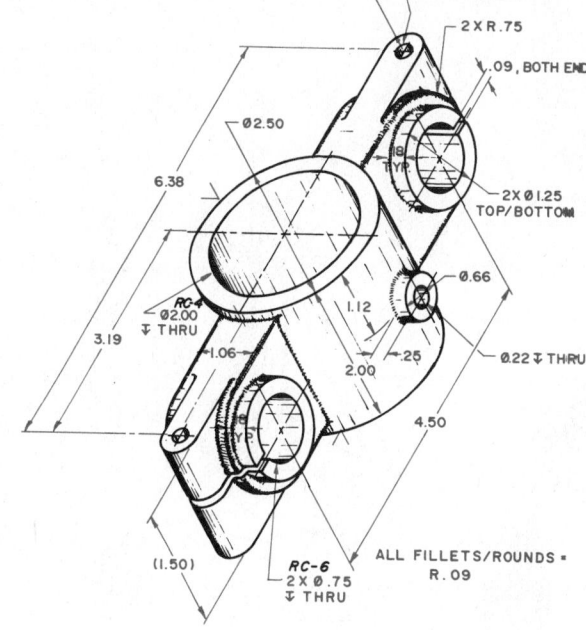

2 X R.52
2 X Ø.75 ⟱ THRU (BOTH ENDS)
2 X R.75
.09, BOTH ENDS
Ø2.50
6.38
TYP
2 X Ø1.25 TOP/BOTTOM
RC-4
Ø2.00 ⟱ THRU
1.12
Ø.66
3.19
1.06
.25
Ø.22 ⟱ THRU
2.00
4.50
(1.50)
RC-6
2 X Ø.75 ⟱ THRU
ALL FILLETS/ROUNDS = R.09

Problem 9-41

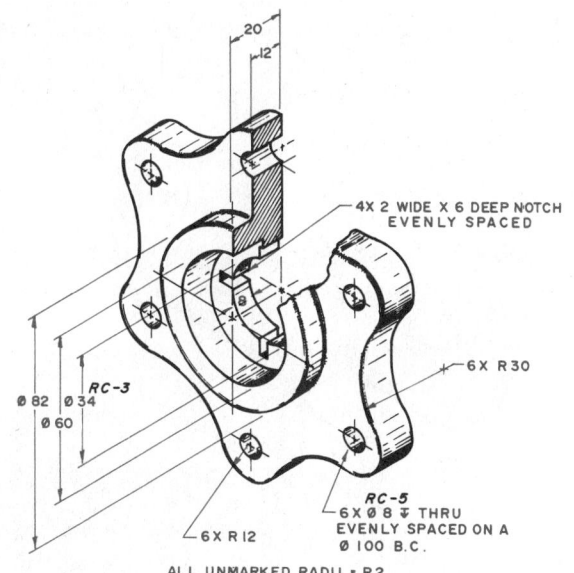

20
12
4X 2 WIDE X 6 DEEP NOTCH EVENLY SPACED
6X R30
Ø82 Ø34
RC-3
Ø60
RC-5
6 X Ø8 ⟱ THRU EVENLY SPACED ON A Ø100 B.C.
6X R12
ALL UNMARKED RADII = R2
METRIC

Problem 9-39

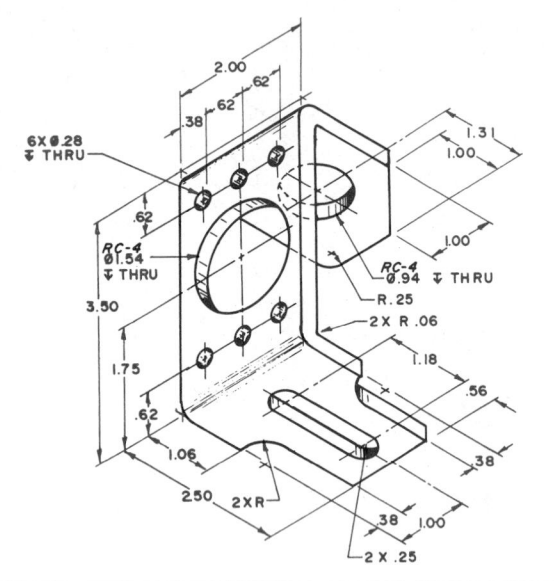

Problem 9-42

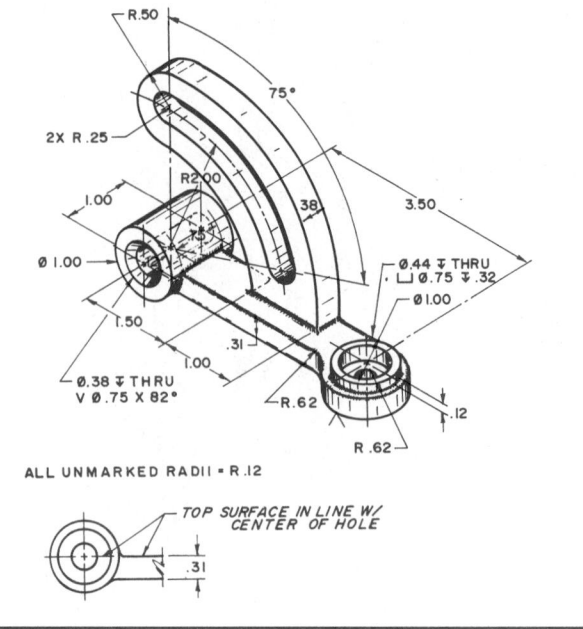

ALL UNMARKED RADII = R.12

TOP SURFACE IN LINE W/
CENTER OF HOLE

Problem 9-44

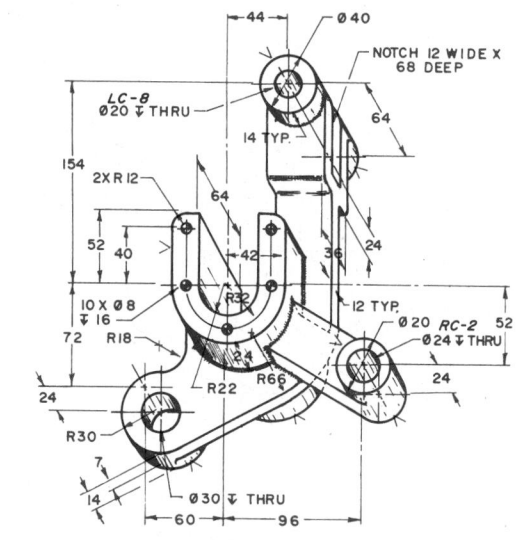

Problem 9-45

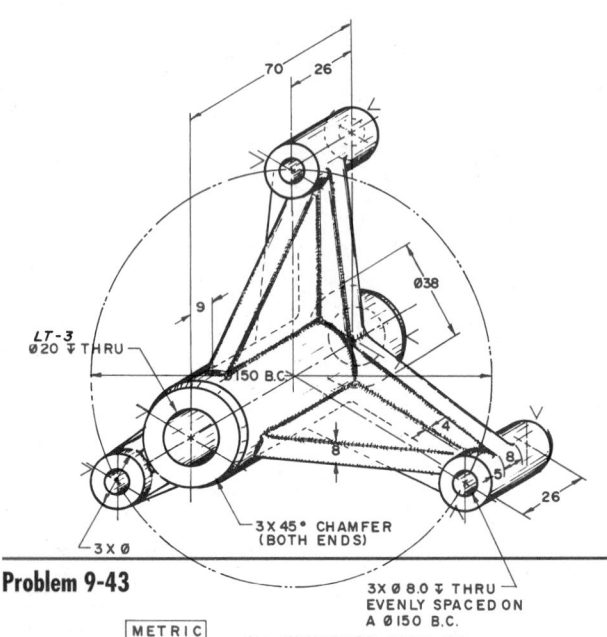

Problem 9-43

METRIC

3X Ø 8.0 ⌄ THRU
EVENLY SPACED ON
A Ø150 B.C.
ALL UNMARKED RADII = R2

Problem 9-46

❏ Problem 9-47

Convert Problem 9-47 into a tabular drawing.

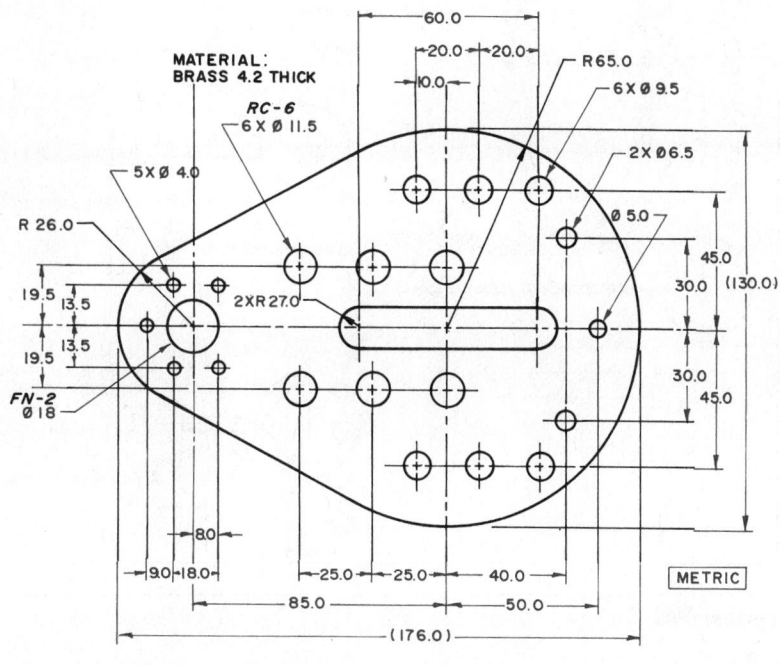

Problem 9-47

❏ Problem 9-48

Convert Problem 9-48 into a standard coordinate dimensioning system drawing.

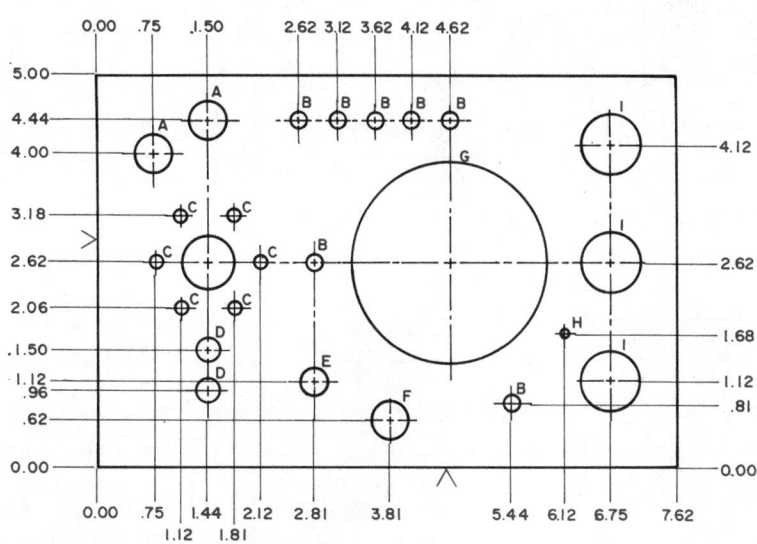

Problem 9-48

❏ Problem 9-49

Convert Problem 9-49 into a tabular drawing.

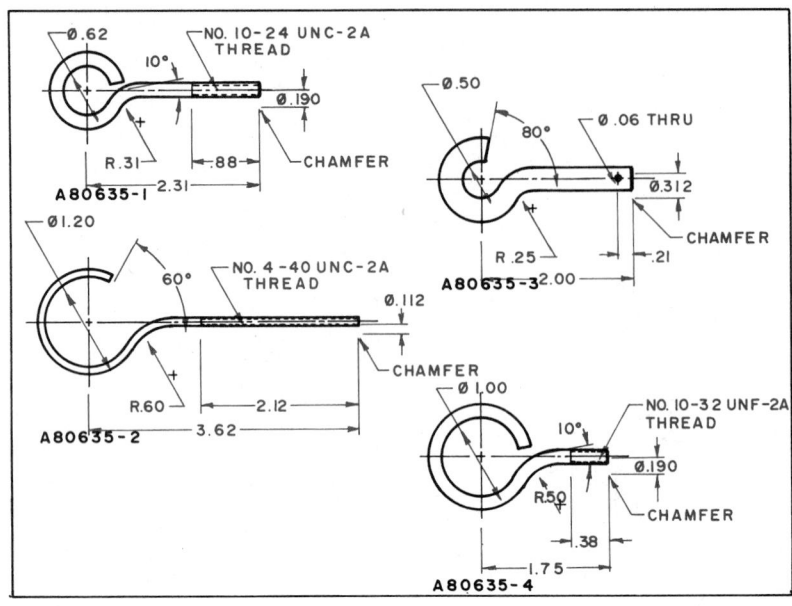

Problem 9-49

❏ Problem 9-50

Draw parts AO608-5 and AO608-8 full size.

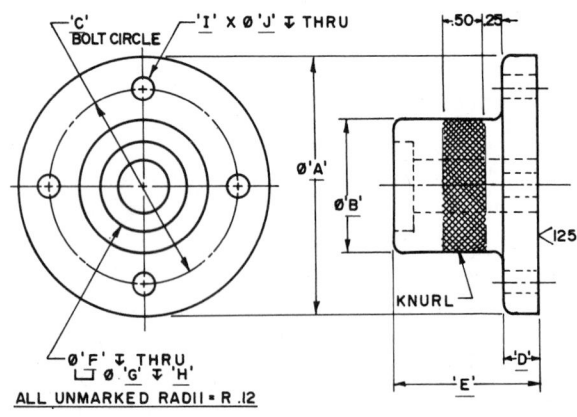

ALL UNMARKED RADII = R .12

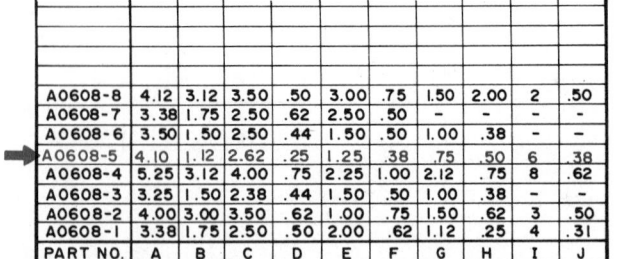

PART NO.	A	B	C	D	E	F	G	H	I	J
A0608-8	4.12	3.12	3.50	.50	3.00	.75	1.50	2.00	2	.50
A0608-7	3.38	1.75	2.50	.62	2.50	.50	–	–	–	–
A0608-6	3.50	1.50	2.50	.44	1.50	.50	1.00	.38	–	–
A0608-5	4.10	1.12	2.62	.25	1.25	.38	.75	.50	6	.38
A0608-4	5.25	3.12	4.00	.75	2.25	1.00	2.12	.75	8	.62
A0608-3	3.25	1.50	2.38	.44	1.50	.50	1.00	.38	–	–
A0608-2	4.00	3.00	3.50	.62	1.00	.75	1.50	.62	3	.50
A0608-1	3.38	1.75	2.50	.50	2.00	.62	1.12	.25	4	.31

Problem 9-50

❏ Problems 9-51 and 9-52

Carefully fill in all blank lines.

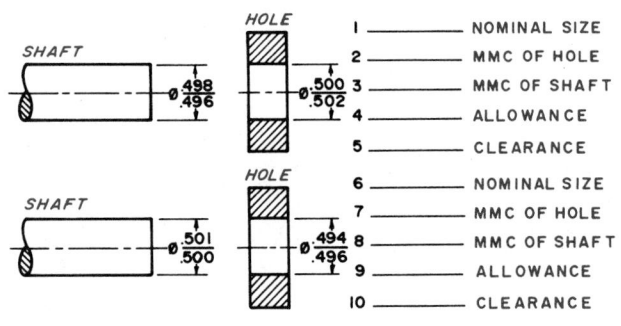

1 _____ NOMINAL SIZE
2 _____ MMC OF HOLE
3 _____ MMC OF SHAFT
4 _____ ALLOWANCE
5 _____ CLEARANCE

6 _____ NOMINAL SIZE
7 _____ MMC OF HOLE
8 _____ MMC OF SHAFT
9 _____ ALLOWANCE
10 _____ CLEARANCE

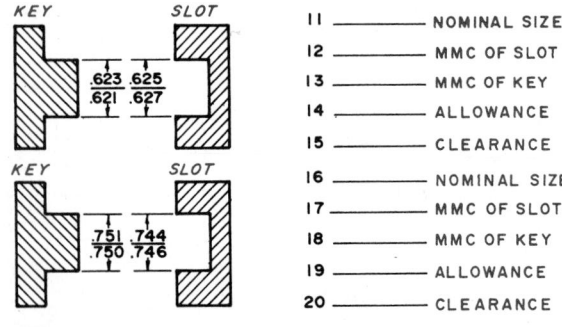

11 _____ NOMINAL SIZE
12 _____ MMC OF SLOT
13 _____ MMC OF KEY
14 _____ ALLOWANCE
15 _____ CLEARANCE

16 _____ NOMINAL SIZE
17 _____ MMC OF SLOT
18 _____ MMC OF KEY
19 _____ ALLOWANCE
20 _____ CLEARANCE

Problem 9-51

PROB	NOM SIZE	BASIC SIZE	DIMENSION OF HOLE	DIMENSION OF SHAFT	HOLE TOLERANCE	SHAFT TOLERANCE	ALLOWANCE	CLEARANCE
1			.625 / .630	—		.005	+.002	
2	1.25		—	—	.001	.002	+.001	
3			.312 / .314	.316 / .314				
4	.75		—	—	.003	.002	+.001	
5			.812 / .813	.815 / .814				
6			22.2 / 22.3	22.5 / 22.4				
7	20		—	—	0.4	0.3	+0.2	
8			9.6 / 9.4	9.8 / 9.4				
9	30		—	—	0.2	0.4	+0.2	
10			15.6 / 15.4	—		0.6	+0.4	

(Rows 1–6 labeled INCH; rows 7–10 labeled METRIC)

Problem 9-52

❑ Problem 9-53

Refer to the mounting guide drawing, No. 392450, and answer the following questions:

1. What is dimension A?
2. What tolerance is expected for two-place dimensions?
3. What are the dimension limits (maximum and minimum) of dimension B?
4. What size tap hole is specified for the 5/8 threaded hole?
5. What thread series is indicated for the 5/8 threaded hole?
6. When was the drawing last revised?
7. What are the dimension limits (maximum and minimum) of dimension C?
8. What is the maximum dimension of F?
9. For one assembly of the mounting guide two stud bolts that are supposed to go through the two holes D are found to be spaced too far apart. The two stud bolts have a center-to-center distance of 1.625″ and are .373″ in diameter. The .422″ D holes are 1.563″ apart. One remedy for this type of problem is to enlarge the mating holes. Therefore, what is the minimum diameter for the two holes D that would permit the assembly?

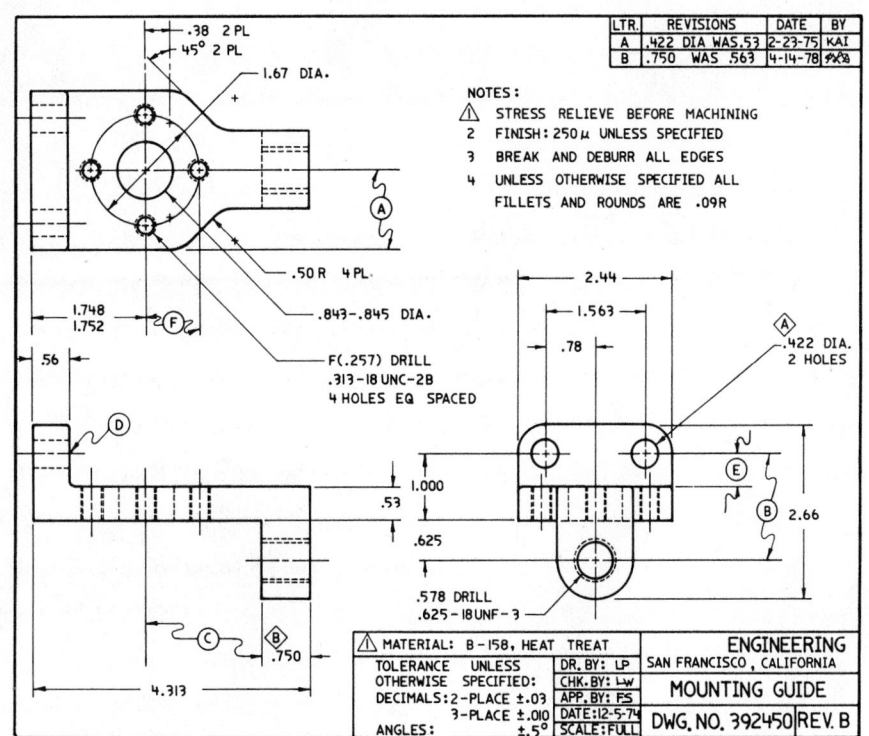

Problem 9-53

Computer-Aided Drafting

Career Profile: Raymond Rickman

Upon graduation from high school, Raymond Rickman entered the Air Force where he intended to get into the electronics field and become a computer technician. He chose this field because he had worked for IBM as a trainee during his senior year in high school and had enjoyed it very much. To his surprise, he found himself unable to pass the color vision test. As such, he spent his time in the Air Force as an administrative clerk.

Near the end of his enlistment, Raymond decided to enroll in Okaloosa-Walton Community College (OWCC). He completed his associate of science degree in drafting. Here, Raymond was taught a very valuable life-long skill: "learn how to learn." This skill became valuable when he gained employment as a mechanical drafter at Orlando Technology Inc. (OTI) prior to his graduation from OWCC. Although Raymond's drafting education was quite broad, it did not cover every type of drafting. At OTI, Raymond's responsibilities included many types of drawing—mechanical, electrical, civil, architectural, structural, and piping. He had to apply his ability to learn on his own in order to complete numerous assigned tasks.

Raymond accepted a teaching position at OWCC. Going to OWCC with no prior teaching experience, he found it very demanding at first to prepare for classes. He had ten days to prepare to teach seven different courses, three of which were taught at two different time frames. This was difficult in itself; however, the courses were also multi-leveled. That is, as many as five courses were simultaneously taught in the same time frame. Courses included basic mechanical drawing, engineering graphics, descriptive geometry, technical illustration, and geometric tolerancing.

During Raymond's first semester as an instructor, he had advanced students that had taken the majority of their specialty courses with the former instructor. He had to adjust his grading policy and teaching strategies to meet their needs.

Raymond soon learned that the life of an instructor is a never-ending learning experience. Keeping his technical skills current is a constant concern. Staying abreast of changes in technology and modifying lesson plans, course outlines, and syllabi to keep instructional material current is very time consuming. Additionally, the constant barrage of meetings and workshops that generally must be attended often interfere with instructional concerns. These are the age-old problems that every instructor encounters.

The average semester consists of 21 credit hours of instruction in as many as 8 different courses. Raymond's classes usually begin at 7:30 a.m. and his day generally ends around 6 p.m. Normally, between teaching and office hours he puts in about 50 hours per week. This does not include the time devoted to grading papers and drawing assignments, which takes another 10 to 15 hours per week.

Teaching at OWCC is more than just a job to Raymond. He sees it as a rewarding challenge. He enjoys being challenged to perform under abnormal conditions that have become the norm. He takes pride in overcoming obstacles that are constantly placed in his path and that every instructor encounters. But most of all, he enjoys imparting the knowledge needed to acquire employment in the drafting field. The life-long learning skills that he learned as a student are passed on to his students. He has a sincere desire to help the students achieve their goals. It is very rewarding for Raymond to see students that he has taught become successful.

CAD/CAE: Technology and Systems

Chapter Outline

This chapter provides an in-depth treatment of modern computer-aided drafting (CAD) and computer-aided engineering (CAE). These topics are covered from a conceptual perspective as are the technology and systems by which they are implemented.

Overview of CAD

The dawning of the age of computers brought significant changes to the fields of design, drafting, engineering, and the various other fields related to these areas. You learned in the Introduction that the fundamental nucleus of these fields—the design process—was altered by computers in two ways: 1) the fourth step in the traditional design process, the making and testing of models or prototypes, was replaced by the developing and testing of three-dimensional computer models; and 2) the tools and techniques used in accomplishing each step of the process were significantly changed.

During the 1960s, shortly after the development of the integrated circuit (IC), the computer began to be used to save time, work, and money in hundreds of different fields. One of the more innovative uses of the computer was in the area of design and drafting, *Figure 10-1*.

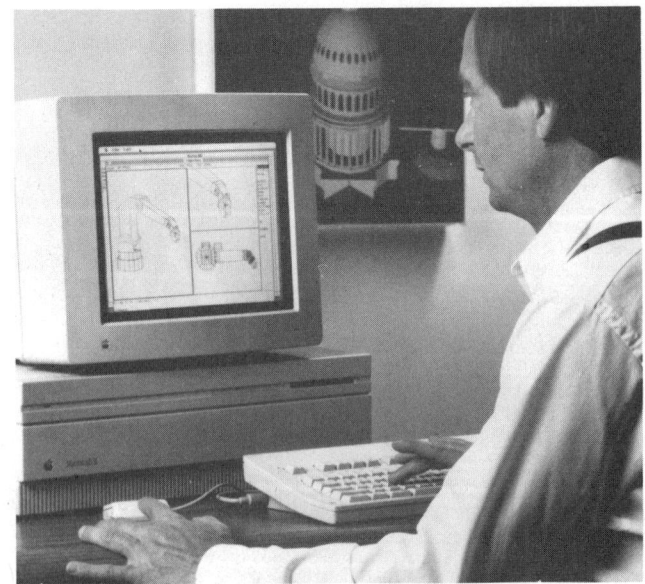

Figure 10-1   CAD system; screen shows AutoCAD Release 10 *(Courtesy Autodesk Inc.)*

Defining Terms

The rapid growth of the use of computers in the world of design and drafting brought forth a deluge of new, and often confusing, terms. Such terms as CAD, CADD, CG, CAE, AD and hundreds of others spawned by the computer revolution began to be heard in drafting and design departments. As a result, drafting and design practitioners found themselves confused by it all.

The reader should be conversant in the language of computers as it applies to design and drafting. The first step is to develop an understanding of the term "computer-aided drafting" or CAD.

Before attempting to form a definition for CAD, the definition of "drafting" should be thoroughly understood. Most people think of drafting as simply the drawing of plans. True, drafters do draw plans. However, this is a very limited definition because drafters do much more than draw plans.

To understand drafting in its broadest sense, one must begin with the design process. The *design process* is an organized, systematic procedure used to accomplish a design that is needed to solve a problem or meet a need, *Figure 10-2*. Each step in the design process requires various types of documentation, such as sketches, preliminary drawings, working drawings, calculations, bills of material, parts lists, and schedules.

Producing or "drafting" this documentation is the job of the drafter. Therefore, *drafting* means producing the documentation required in support of the design process. With this understanding of drafting as a concept, one may now begin to develop an understanding of CAD. *Computer-aided drafting* or CAD means using the computer and peripheral devices in producing the documentation needed in support of the design process. Put another way, CAD means using the computer and peripheral devices to do drafting.

A well-informed drafter should be able to distinguish CAD from among the various other related terms frequently heard in modern drafting settings, such as AD, CADD, CAE, CG, and CAM.

CAD is usually taken to mean computer-aided drafting, but it can also be used to mean *computer-aided design*. It is up to the person using the term to make the distinction. AD means *automated drafting*, and it is synonymous with CAD or computer-aided drafting. Originally, AD was used when referring to the use of the computer to do drafting, and CAD was used when referring to the use of the computer as a design tool. However, AD did not catch on because it is a less distinctive term than CAD. Consequently, CAD began to be used to mean both computer-aided drafting and computer-aided design.

CADD is the acronym for *computer-aided design and drafting*. This is a broad concept in which a computer system is used for both design and drafting.

CAE is the acronym for *computer-aided engineering*. This is a newer term that is replacing the term CAD when it is used to mean computer-aided design. The

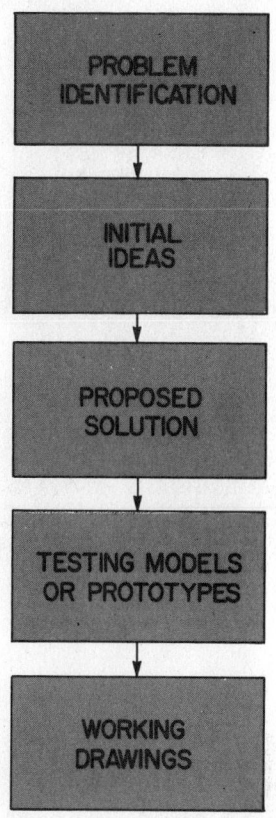

Figure 10-2 The modern design process

term CAE serves to eliminate the computer-aided design/computer-aided drafting confusion when using the term CAD. CAE is covered later in this Chapter.

CG is short for *computer graphics*. This is a term that is used and misused a great deal. It is frequently used in place of CAD (drafting). However, in reality, it is a much broader term. CG means using the computer and peripheral devices for producing any type of graphic image, technical or artistic.

None of these terms has been formally standardized. However, there is a definite trend toward greater use of CAD when speaking of drafting, and CAE when speaking of design and engineering.

The manufacturing equivalent of CAD is *computer-aided manufacturing* or CAM. CAM refers to computer-controlled automation of the various manufacturing processes used in modern industry. CAD/CAM links together the processes of design, drafting, and manufacturing. The CAD/CAM process involves designing a product on the computer, using the computer to produce necessary documentation, and using the data base stored in the computer from the design and drafting phases to issue the manufacturing instructions to automated machines and industrial robots.

Differences between Manual Drafting and CAD

The techniques used by manual drafters in documenting a design include freehand lettering, manual scaling, and mechanical line work. CAD technicians accomplish

Figure 10-3 Tools of the CAD technician *(Courtesy VersaCAD Corporation)*

Figure 10-4 Storage on magnetic tape *(Courtesy Clinton Corn Processing Co.)*

Figure 10-5 Magnetic disk storage *(Courtesy Autodesk Inc.)*

these same tasks by interacting with a keyboard, terminal display, digitizer, and function menu, *Figure 10-3*.

Another important difference between manual drafting and CAD is in the methods used for storing design documentation. Manual drafters store drawings in large flat files. Calculations, parts lists, bills of material, and schedules are normally stored in folders in filing cabinets. CAD technicians store all design documentation on magnetic tapes or disks, *Figures 10-4* and *10-5*.

CAD Systems

CAD system is one of the most frequently heard terms in modern drafting rooms. People who use this term are usually referring to a configuration of computer hardware or the tools of the CAD technician. However, this is a misuse of the term. CAD hardware is only one of three components which must be present in order to have a CAD system.

A CAD system actually consists of hardware, software, and users, *Figure 10-6*. A collection of CAD hardware is properly referred to as a *hardware configuration*. In order to turn a hardware configuration into a CAD system, software and well-trained users must be added.

CAD Hardware

Hardware is the term given to the computer's physical equipment or devices. Many different companies manufacture CAD hardware. Consequently, numerous configurations are on the market. However, a typical hardware configuration consists of a terminal display, a keyboard, a digitizer equipped with either a light pen or a puck, function menus, a plotter, and a processor, *Figure 10-7*.

The Terminal Display

The *terminal display* is an output defice resembling a television that displays the data on which the CAD technician is working. As a drawing or any other type of documentation is created, it is displayed on the screen of the terminal, *Figure 10-8*.

There are three common types of terminal displays: 1) refresh displays, 2) raster displays, and 3) storage tube displays. In a *refresh display*, the image is traced on the back of the screen by an electron beam. This image must be constantly retraced or "refreshed." *Raster displays* create images by illuminating *pixels* (picture elements) on the screen. The more pixels available on the screen for illumination, the higher the quality of the image. *High resolution* produces a high-quality

CAD SYSTEM=

HARDWARE + SOFTWARE + USERS

Figure 10-6 Components of a CAD system

Figure 10-7   Typical hardware configuration (Courtesy Micro Control Systems Inc.)

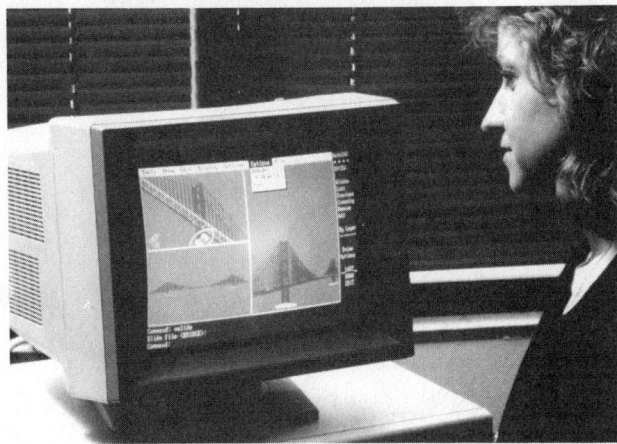

Figure 10-8   Viewing the display on a terminal (Courtesy Autodesk Inc.)

image. *Storage tube displays* create images by tracing them on the back of the display screen with an electron beam like refresh displays. However, unlike refresh displays, storage tube images are stored and do not have to be refreshed. Graphics displays may be color or monochrome (amber or green) devices.

The Keyboard

The keyboard is one of the most frequently used input devices in a CAD system. *Keyboards* are used for entering text, system commands, X-Y or X-Y-Z coordinates, and any other type of alphanumeric input.

Most CAD systems have keyboards which contain both the standard alphanumeric keys and special auxiliary keys, *Figure 10-9*. The special auxiliary keys may be part of a separate numeric pad or a group of display control keys.

Keying is the process used for entering text on drawings, entering certain system commands, entering

dimensions not entered automatically, and for logging-on to the system.

The Text Display

The *text display* is an output device with a screen for displaying user prompts, instructions, and other alphanumeric data. Most text displays are noncolor cathode-ray tube (CRT) devices.

The Digitizer

The *digitizer* is a special electromechanical input device that resembles an electronic tablet. In fact, the digitizer is frequently referred to as the *tablet*. Some CAD systems have digitizers that are as small as tablets, and some have digitizers that are as large as conventional drafting tables, *Figures 10-10* and *10-11*.

Digitizing is a process through which graphic data, such as a sketch or a drawing, is converted to digital data as it is input. A digitizer can be used for a number of different functions. It can be used in conjunction with a light pen or a puck to control the screen cursor, Figure 10-10. A *cursor* is the small crosshairs symbol on the graphics display used in creating, locating, and manipulating graphic and alphanumeric data.

A digitizer can be used as a place to mount menus for giving system commands or calling up stored symbols from memory. It can also be used for electronically tracing (digitizing) graphic data so that they can be entered into a CAD system.

Function Menus

Menus are lists of system commands and stored data, or overlays that serve the same purpose, which correspond with storage locations in a CAD system's memory. They are used as a means of speeding and simplifying human interaction with the CAD system. One of the fastest, easiest ways to enter a system command is to have the command on a menu. Just as you need a menu when you sit down in a restaurant so that you know what

Figure 10-9   Typical keyboard (Courtesy Micro Control Systems Inc.)

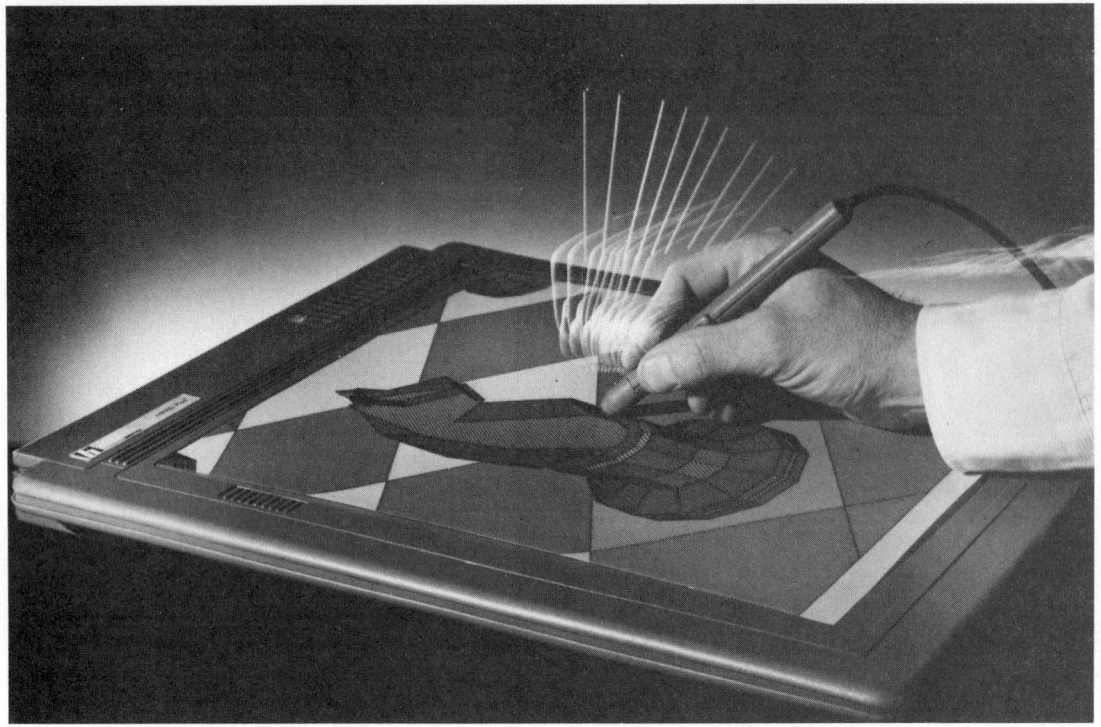

Figure 10-10   An example of using a digitizer as a drawing tablet *(Courtesy Houston Instrument Division, AMETEK Inc.)*

Figure 10-11   Complete hardware configuration with large digitizing table *(Courtesy Auto-Trol Technology Corp.)*

options you have for ordering, you need a menu when you sit down at a CAD system so that you know what options you have for commanding the system.

The most frequently used types of menus are *screen-displayed menus* and *tablet-mounted menus*. A command on a screen-displayed menu may be activated by touching it with a light pen, moving the screen cursor to its location on the screen, or pressing a key that corresponds with the command. For example, the menu in ***Figure 10-12*** lists seventeen options. To activate the ARC option, the technician would key the letter "A" on

```
ARC
CIRCLE
INSERT:
LINE:
PLINE:
POINT:
SHAPE:
SKETCH:
SOLID:
TEXT:
TRACE:

DISPLAY
EDIT
LAYERS
MODES

LASTMENU
ROOTMENU
```

Figure 10-12   Sample menu

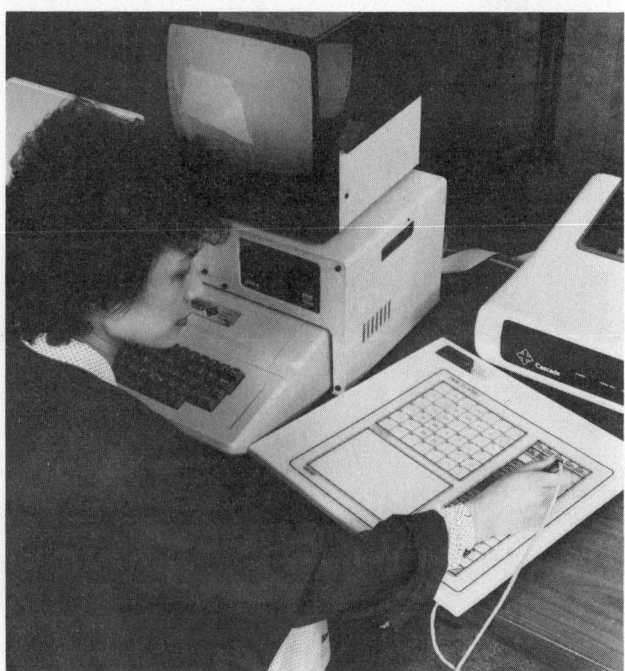

Figure 10-13 Activating menu commands *(Courtesy Cascade Development Corp.)*

the keyboard and, then, press either the RETURN or ENTER key. Pressing the RETURN or ENTER key is a necessary activation step in most cases when entering commands from a keyboard. Tablet-mounted menu commands are activated by touching the appropriate position on the menu with a light pen or the sight of the puck, ***Figure 10-13***. To activate a tablet-mounted menu command, the technician would align the sight of the puck with the menu position occupied by the option in question, and then press the puck button.

In addition to system commands, menus may contain frequently used symbols or items of graphic data. If a certain symbol is used frequently, it may be created once, saved, and assigned to a specific position on a menu. Then, when needed, it can be called up from storage in a matter of milliseconds rather than having to be completely redrawn each time it is needed on a drawing.

The Plotter

The *plotter* is the output device which actually draws (plots) drawings and other types of documentation. Three types of plotters are used in CAD systems: pen plotters, electrostatic plotters, and photoplotters.

Pen plotters may be bed plotters or drum plotters. They create drawings by converting into lines X-Y coordinates received from the processor in the form of electrical impulses. Pen plotters are rated according to their accuracy, repeatability, resolution, and plotting speed. Most pen plotters can plot drawings accurately to plus or minus .001″ or better.

Repeatability in plotting is the ability to redraw lines without creating a double image. This is important because on many plotters line thickness or density is increased by retracing lines until the desired density is achieved.

Resolution in plotting is judged by the shortest line segment the plotter is able to plot. This becomes a critical factor when plotting circles, arcs, and curves, all of which are formed by a series of short, straight-line segments. The shorter the line segments, the better the resolution. Resolution of .001″ is common for pen plotters. A *plotting speed* of 40 inches per second is a common, and frequently exceeded, mean plotting speed with modern plotters.

Electrostatic plotters are much faster than pen plotters, but they are not as accurate and they require special electrostatic paper. This type of plotting paper is not dimensionally stable; a factor which limits the applications of electrostatic plotters.

Photoplotters are used in those situations requiring extremely close tolerances and dimensional stability, such as in the production of printed circuit board (PCB) artwork. Of the three types of plotters, the pen plotter is the one most frequently used.

The Processor

The *processor* is the computer in a computer-aided drafting system. As in any computer, processors have a memory component and a logic component. In addition, many CAD system processors have an added feature: a graphics processor, ***Figure 10-14***.

A processor's memory is rated according to the amount of data it can hold. Individual memory location sizes are rated in *bits* or binary digits. This is why some processors are referred to as 8-, 16-, or 32-bit processors. An 8-bit processor is slower than a 16-bit or 32-bit processor. The 32-bit processor is the fastest.

Overall memory size is rated in *bytes* (the commonly accepted ratio is 1 byte = 8 bits). Because a byte represents a relatively small amount of data, memory sizes are usually given in kilobytes. A 256-kilobyte memory would not be uncommon in a CAD system's processor.

The term *kilo* normally means 1000. However, when used in conjunction with computers, it means 1024. To determine the actual size of a 256-kilobyte processor, multiply 1024 x 256 = 262,144. To convert this to bits, multiply 262.144 x 8 = 2,097,152.

The processor console also houses a disk or tape drive unit for secondary storage devices such as disks, diskettes or magnetic tapes.

CAD Software

Software is the name given to the various programs that actually make the CAD system work, and any printed materials such as instruction sheets or operator's manuals, for instance. There are three basic types of software for CAD systems: 1) operational software; 2) application software; and 3) user-defined software.

Operational software consists of programs which

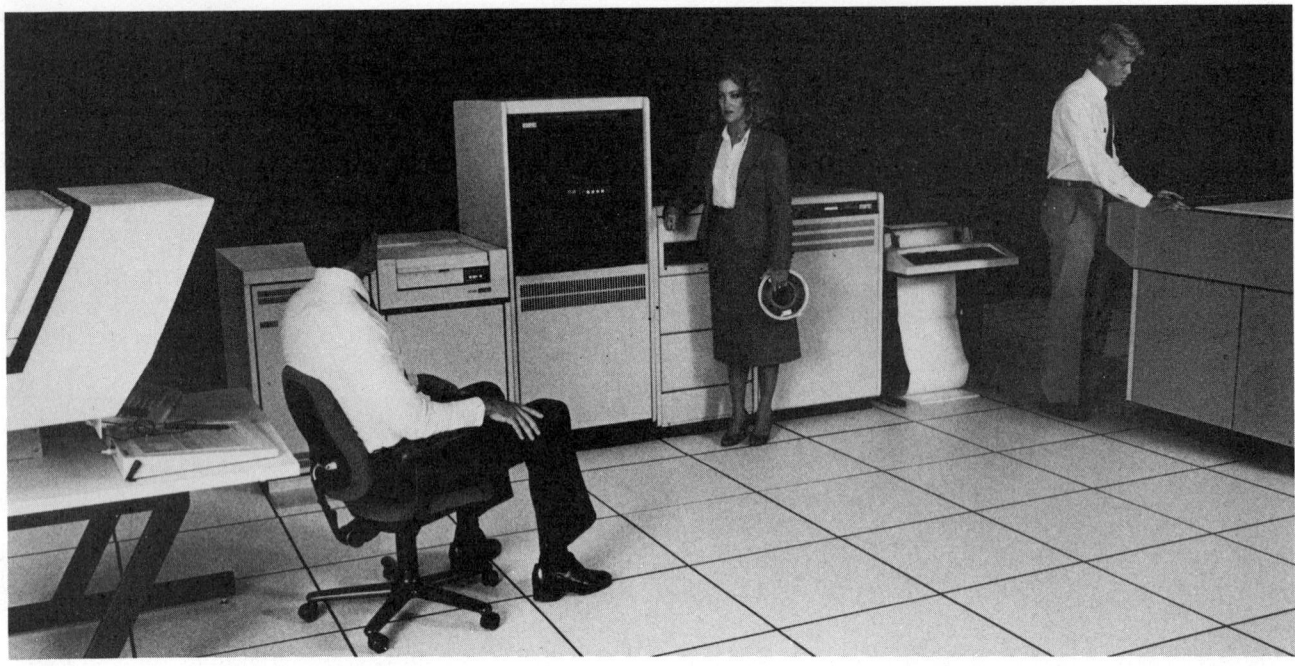

Figure 10-14   One type of graphics processor *(Courtesy Calma Co.)*

make the system perform general operational tasks such as accepting commands, storing data, retrieving data, and so forth.

Application software consists of programs that command the CAD system to perform drafting functions in such specific application areas as architectural, mechanical, structural, electronic, and civil drafting.

User-defined software usually is restricted to such various user-specific needs as symbols that are added to a symbols library and a menu. Users do not normally write the code for CAD software.

CAD Users

Users of CAD systems can be placed in two categories: 1) CAD system operators, and 2) CAD technicians. CAD *operators* are people who are able to operate a CAD system in terms of inputting, manipulating, and outputting data. However, they have little or no background in design, drafting or engineering.

CAD *technicians*, on the other hand, have both system operational skills and strong backgrounds in design, drafting or engineering. For CAD technicians, a CAD system is simply a tool to help them do their jobs faster and better.

Modern CAD System Configurations

There are more than 100 turnkey CAD systems on the market now, and the list will probably continue to increase. Some are general-purpose systems which can be used in a number of different drafting fields. Others are dedicated systems designed for one specific application.

Modern CAD systems range from systems based on

large mainframe computers to small microcomputer-based configurations. *Figure 10-15* shows an example of a modern CAD system.

Advantages of CAD

The applications of computer-aided drafting grew rapidly because of its many advantages over manual drafting. CAD is being used in a variety of settings in both large and small companies. When compared with manual drafting, CAD is more accurate, faster, neater, more consistent, and better in terms of storing, correcting, and revising drawings, *Figure 10-16*.

Figure 10-15   Modern CAD system *(Courtesy Autodesk Inc.)*

MORE ACCURATE

FASTER

NEATER

MORE CONSISTENT

BETTER STORAGE

EASIER CORRECTIONS

EASIER REVISIONS

Figure 10-16 Advantages of CAD

CAD hardware is consistently accurate to better than .001 inch. Manual scaling and drawing cannot match this figure consistently, even under the most positive conditions. In fact, inaccuracy is a built-in human characteristic. Such typical drafting tasks as lettering and line work are slow and tedious for manual drafters; but, for CAD technicians, they are fast and easy because they are accomplished by simply pressing buttons and executing commands.

The computer usually does the work for the person and it does it very fast. Using the computer, a drafting technician can produce lettering as much as 25 times faster than by doing it manually. A drafting technician with typing skills can improve on this figure considerably.

Whereas a manual drafter might draw at an average rate of four to eight inches per minute, it is not uncommon for a CAD system's plotter to create drawings at a rate of 40 inches per *second*.

Neatness and consistency in line work and lettering are the major weaknesses of manual drafting and the major strengths of CAD. Since documentation in CAD is produced electromechanically, it is not subject to the negative impact of such factors as fatigue, boredom or a lack of skills. *Figures 10-17* and *10-18* illustrate the neatness and consistency that is characteristic of documentation prepared on a CAD system.

Some of the most important advantages of CAD can be found in the area of corrections and revisions. On the job, a great deal of time is devoted to correcting and revising drawings and other types of documentation. Time spent correcting and revising documentation is nonproductive time. Because of the drawing manipulation capabilities of CAD systems (for instance, such command functions as MOVE, COPY, DELETE, ROTATE, and ZOOM), correcting and revising time can be cut by as much as 75 percent in many cases. The time and work-saving potential of CAD in terms of corrections and revisions is even more important than that associated with the original creation of design documentation.

An additional advantage of CAD over manual drafting involves storage of documentation. In CAD, one 5.25-inch floppy disk might hold as much documentation as a large file drawer. With CAD, years of drafting work can be stored in a container the size of a shoe box. Backup disks or copies of file disks can be made quickly, easily, and inexpensively, and stored in protective containers. Thus, the deterioration problems normally associated with traditional drafting media are eliminated.

MicroCAD

In the early 1980s a development known as "microCAD" began to revolutionize the fields of design, engineering, and drafting. MicroCAD sometimes referred to as PC-CAD means design and drafting accomplished using microcomputers. Both terms, microCAD and PC-CAD, are acceptable. However, caution is in order when using these terms. The problem stems from the fact that although PCs (personal computers) are all microcomputers, not all microcomputers are PCs. Further, not all PCs are capable of design and drafting applications.

Most microcomputers used in design and drafting applications have more memory and processing power. They also have high-resolution monitors and dual-disk-drive configurations. Many are configured with a hard disk and one floppy drive.

MicroCAD in the Beginning

The rapid growth of microCAD is relatively new, but microCAD itself is not. There have been microCAD systems in use since the late 1970s. However, these earlier systems were seriously lacking in real design and drafting capabilities. Early microcomputers had neither the processing power nor the graphic capabilities needed in CAD. Consequently, microCAD got off to a bad start.

Early microCAD systems were slow, limited to only the most basic two-dimensional drafting tasks, and able to produce documentation of only marginal quality. Such systems were thought of as being "toys," and, for the most part, were rejected by the private sector.

For the most part, the problems associated with early microCAD systems grew out of problems with early microcomputers. Before microCAD could become a real alternative for companies that wished to convert to CAD but could not afford traditional systems, two things had to happen.

1. Development of the microcomputer had to reach the point where more memory and processing power were available.
2. Development of the microcomputer had to reach the point where better graphic capabilities were available.

By the early 1980s both of the developments had taken place. More powerful 16-bit microcomputers were available. Microcomputers with 512K bytes of RAM became commonplace. Dual-disk-drive configurations or a combination hard disk with one floppy disk drive also became the norm. High-resolution color monitors became readily available at affordable prices. These developments paved the way for the acceptance of microCAD.

Modern microCAD computers may have over 32 megabytes of RAM and the trend toward more memory

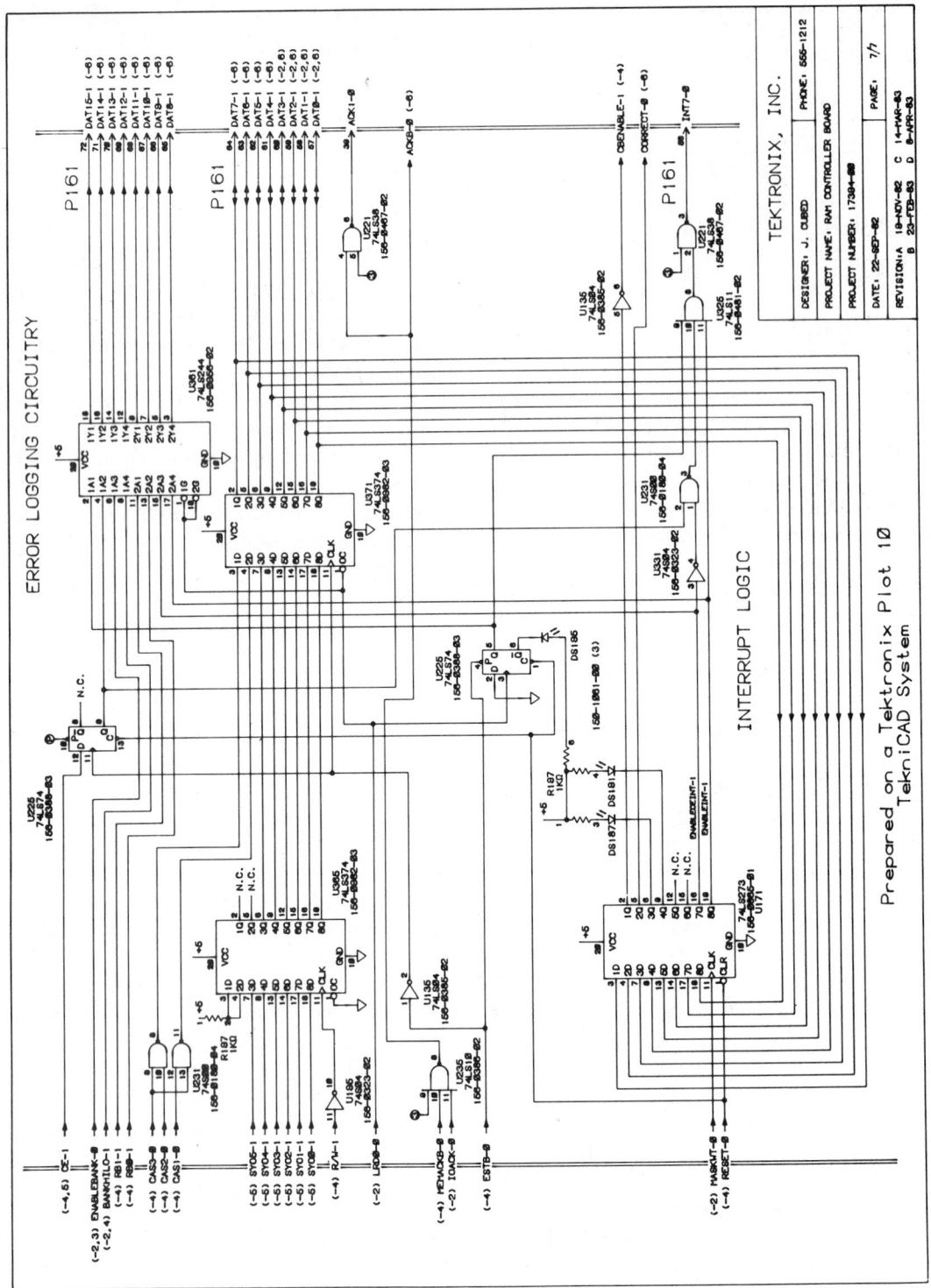

Figure 10-17   Drawing produced on a CAD system *(Courtesy Tektronix Inc.)*

and more processing power continues. Math co-processors are now added to microcomputers, particularly in CAD settings, to speed the time required to process complex drawings and three-dimensional solid models.

Advantages of MicroCAD

MicroCAD can be compared with both manual design and drafting and traditional CAD. When making such comparisons, one will find that microCAD offers users five major advantages.

1. Increased productivity

2. Low cost

3. Easy learning and use

4. Convenience

5. Versatility

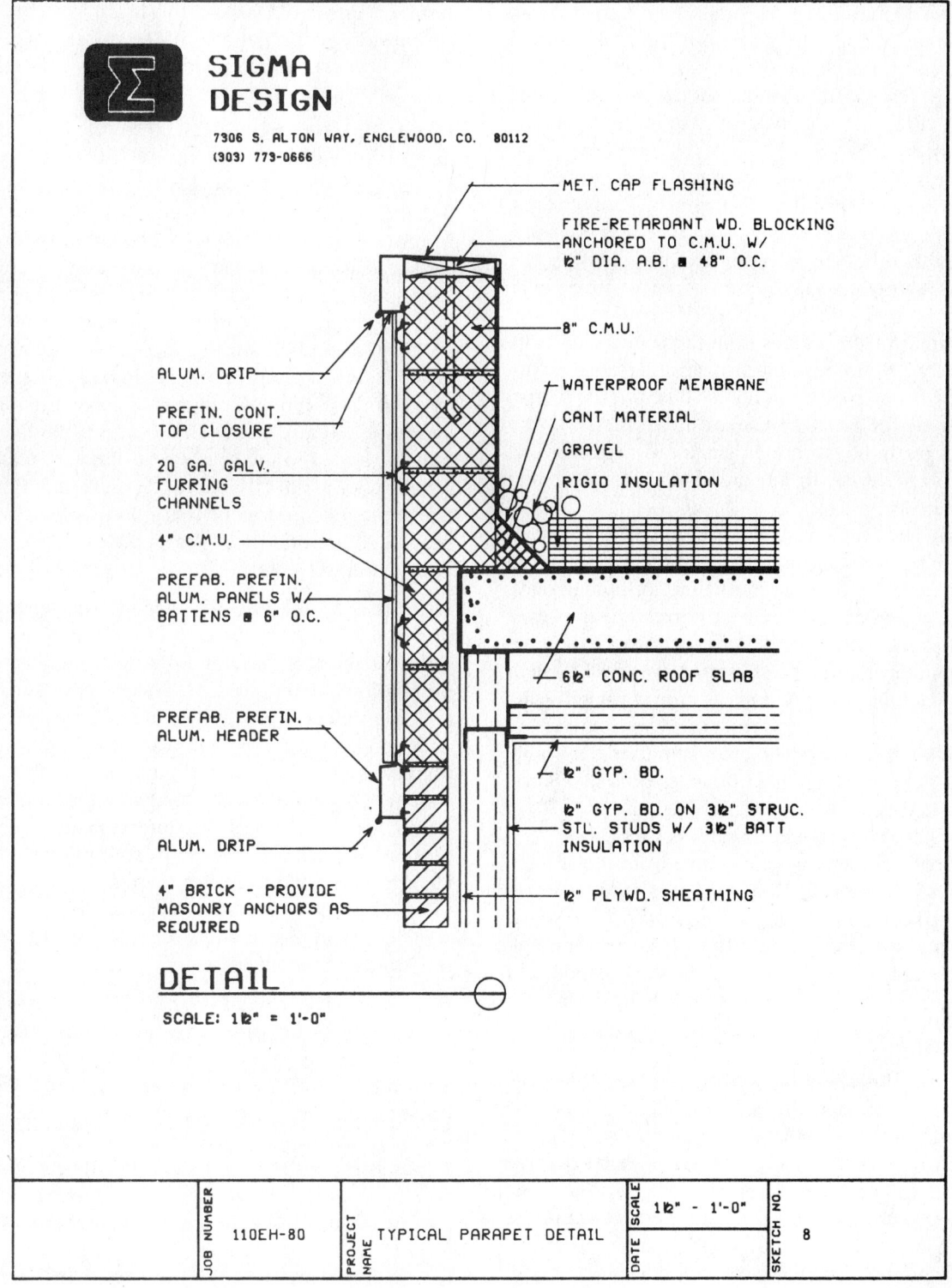

SIGMA DESIGN

7306 S. ALTON WAY, ENGLEWOOD, CO. 80112
(303) 773-0666

MET. CAP FLASHING

FIRE-RETARDANT WD. BLOCKING ANCHORED TO C.M.U. W/ ½" DIA. A.B. @ 48" O.C.

8" C.M.U.

ALUM. DRIP

PREFIN. CONT. TOP CLOSURE

20 GA. GALV. FURRING CHANNELS

4" C.M.U.

PREFAB. PREFIN. ALUM. PANELS W/ BATTENS @ 6" O.C.

WATERPROOF MEMBRANE

CANT MATERIAL

GRAVEL

RIGID INSULATION

6½" CONC. ROOF SLAB

PREFAB. PREFIN. ALUM. HEADER

½" GYP. BD.

½" GYP. BD. ON 3½" STRUC. STL. STUDS W/ 3½" BATT INSULATION

ALUM. DRIP

4" BRICK - PROVIDE MASONRY ANCHORS AS REQUIRED

½" PLYWD. SHEATHING

DETAIL

SCALE: 1½" = 1'-0"

JOB NUMBER		PROJECT NAME		DATE	SCALE		SKETCH NO.
	110EH-80		TYPICAL PARAPET DETAIL		1½" - 1'-0"		8

Figure 10-18 Drawing produced on a CAD system *(Courtesy Sigma Design)*

Increased Productivity

If one compares microCAD with manual design and drafting techniques, increased productivity is probably its principal and most important advantage. The degree of improvement in productivity of microCAD over manual design and drafting depends on a number of factors.

Such factors as the application area and the skills and creativity of the user must be considered, of course. However, speaking in general terms, one can expect productivity ratios (of microCAD to manual drafting) from 3:1 to 5:1 in such applications as mechanical, manufacturing, construction, or architecture. If the application area is electronic, electrical, survey, civil engineering, or any other single-line symbol or schematic-oriented type of design and drafting, productivity ratios of 5:1 or even higher can be expected.

There are a number of reasons why microCAD is a more productive approach to design and drafting than manual techniques. Some of the more important reasons are: (1) faster data creation; (2) faster data manipulation; (3) faster, more convenient data storage; and (4) faster data output.

Some of the reasons why documentation can be created faster on a CAD system than manually are obvious. For example, even a slow typist can type text considerably faster than he or she could produce neat, legible freehand lettering. The same is true of linework. However, the greatest advantage with regard to creating graphic data probably comes from the concept of "nonrepeatability." Nonrepeatability means that once a symbol or any other type of geometric character is produced, it never has to be produced again. Once produced, it can be stored in primary storage or a secondary storage device, or in a menu, then called up and used repeatedly.

The principal time-saving factor in microCAD as compared with manual drafting comes in the manipulation of graphic data. Two of the most time-consuming tasks in manual design and drafting are correcting and revising the documentation for designs. Because of the editing capabilities of a microCAD system—as well as such standard manipulation functions as mirror, rotate, scale, move, and copy—corrections and revisions can be made quickly and easily. The manual drafter is not able to call up standard symbols or other graphic configurations, rotate them, mirror them, move them, scale them up or down, or copy them a number of times.

Additional productivity gains come from the storage advantages of a microCAD system. It is much simpler and faster to call up a file containing a drawing or drawings from computer storage than it is to rummage through file drawers, locate the drawings in question, and pull them out of the drawer without damaging them.

Lower Costs

When low cost is used as an advantage of microCAD, the comparison is generally between microCAD and the larger traditional CAD systems. MicroCAD systems are less expensive than the traditional large systems that are based on mini- and mainframe computers. The most obvious reason for the difference in prices, of course, is the lower initial cost of a microCAD system compared with a traditional system. However, as indicated below, lower initial costs represent only one of the reasons.

1. Lower initial cost
2. Lower maintenance costs
3. Shorter payback period
4. Fewer facility adaptations
5. Less user training

Lower Initial Cost.   The price of a traditional CAD system might range from as low as $25,000 to as high as $3 million, or even higher. From $50,000 to $150,000 would be the medium range for the typical traditional CAD system. The medium range for a typical microCAD

system is only $10,000. It is this initial low cost that represents the principal advantage of microCAD in the minds of most people who must justify an investment in CAD.

Lower Maintenance Costs.   Like the initial costs, the maintenance costs for microCAD are lower than those associated with traditional CAD systems. Users typically purchase maintenance agreements on CAD systems from the vendor. In this way the vendor is responsible for keeping the system in good operating condition and for repairing it quickly when breakdowns occur.

Maintenance agreements vary significantly among vendors, depending on the application and a variety of other factors. However, a traditional CAD system that costs in the $250,000 range may have a maintenance agreement that costs as much as $50,000. This cost must be added to the yearly cost of the system. On the other hand, a $10,000 microCAD system might have a maintenance agreement that costs as little as $500 a year. A figure of 15% of the purchase price is a good rule of thumb.

Shorter Payback Period.   One of the principal costs of doing business is the cost of money, or the interest rate that must be paid on borrowed money. Whether purchasing a traditional CAD system or a microCAD system, a business is probably going to borrow the money and, consequently, pay interest on a loan. In this case the advantage of microCAD is that the payback period is shorter. Consequently, the total amount of interest paid on the loan will be less. The interest rate itself may also be less as a result of the shorter payback period. Managers who must justify a CAD conversion are particularly appreciative of this advantage because it represents considerably less risk on their part than would a traditional CAD conversion.

Fewer Facility Adaptations.   Most microCAD systems will fit on top of the average desk and occupy very little space. This is not true of traditional CAD systems. Typically, traditional CAD systems will require a number of facility adaptations in order to be properly accommodated. These facility adaptations include such things as the installation of additional electrical outlets, built-up flooring, climate control, and dust control adaptations. MicroCAD systems, on the other hand, rarely require facility adaptations. Although there are vendors that manufacture special furniture for microcomputers and microCAD systems, special furniture is not a necessity when converting to microCAD.

Less User Training.   It can take from six months to a year to become proficient in the use of a traditional CAD system. The training time required to learn to operate a traditional system will vary because the systems vary in size and degree of complexity. However, traditional CAD systems are more sophisticated, more complex, and hence more difficult to learn to operate than microCAD systems. This means (1) that users spend more time in a learning mode than in a productive mode during the initial phase of a CAD conversion, and (2) that the initial productivity curve will probably go down to a level below that of the manual format.

MicroCAD systems, on the other hand, tend to be simple and easy to learn and operate. An experienced engineer, architect, designer, or drafter can expect to be reasonably skilled in operating a microCAD system within a matter of a few days or a week. This shortens the amount of time users spend in a learning mode and, in turn, the amount of time that the productivity curve will remain below the level prior to the conversion.

Convenience

MicroCAD systems are more convenient than traditional CAD systems in several ways. The first and possibly most important way is that they are more convenient to purchase. Whereas most traditional CAD systems must be purchased directly from the specific vendors that manufacture and market them, microCAD systems may be purchased directly from vendors, indirectly through computer stores, or even by mail order. Frequently, a microCAD software package may be run on a microcomputer that already exists in a firm. Whereas a traditional CAD system requires a separate workstation for the operator, a microCAD system can fit easily and conveniently on most office-size desks. MicroCAD systems are light and portable. Consequently, they are easily moved and accommodated.

Versatility

One of the most important advantages of a microCAD system is its versatility. A microCAD system may be used for standard design and drafting tasks, but it is not limited to these tasks. MicroCAD systems can be used for all of the many other purposes to which microcomputers are often put. Of course, the most important and most frequent use of microCAD systems is in performing such traditional design and drafting functions as design analysis, solid modeling, finite-element analysis, kinematics, and two- and three-dimensional drafting.

In addition to the design and drafting uses of microCAD systems, there are other uses, such as project management, office automation, electronic spreadsheets, decision support, and data base management. All of these capabilities are within the realm of what a microCAD system can do for users.

Computer-Aided Engineering (CAE)

Computer-aided engineering (CAE) is any use of a computer to assist in engineering tasks, analysis testing, calculating, design review, and experimentation. While there is some overlap between computer-aided design (CAD) and CAE, there are also distinctions. While CAD focuses on those tasks necessary to design a product (see Chapter 24), CAE focuses more on analyzing and reviewing the design.

CAE workstations look just like CAD workstations,

Figure 10-19 CAE workstation *(Courtesy Auto-Trol Technology Corp.)*

Figure 10-19. The noticeable differences are in software rather than hardware.

The driving force behind the evolution of CAE has been the need to improve products while simultaneously reducing the time—and subsequently the cost—associated with analysis, engineering calculations, testing, and experimentation. In other words, the driving force has been and is competitiveness.

The most widely practiced uses of CAE are engineering analysis and design review. These two uses are explained in the following sections.

Engineering Analysis

When designing a manufactured product, there are various types of analyses that must be done. Stress-strain analysis and heat-transfer analysis are two of the most common types. CAE can increase the accuracy of these and other types while simultaneously reducing the amount of time and effort required to conduct an analysis.

Software is readily available that allows engineers to analyze the mass properties of a part and to analyze finite or individual sections of the part. The former process is known as mass-properties analysis. The latter is known as finite-element analysis. The two CAE-related processes are explained in the following paragraphs:

Mass Properties Analysis

This process involves performing analysis computations concerning such mass properties as weight, volume, surface area, moment of inertia, and center of gravity. The computer can perform such calculations quickly and accurately. The results of such calculations

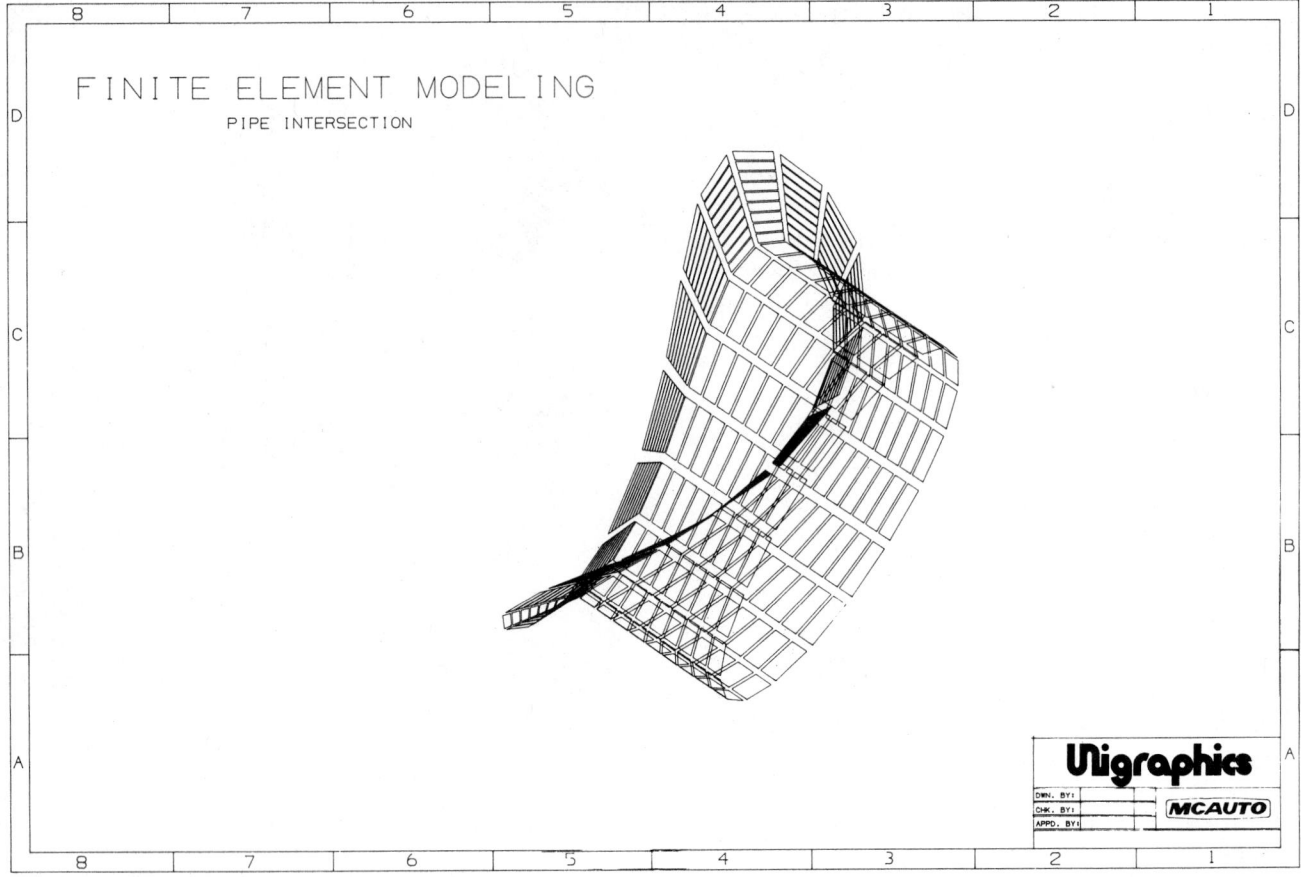

Figure 10-20   Finite element analysis *(Courtesy McAuto)*

are then used in conducting additional calculations, tests, and experiments to determine if the proposed design represents the optimum design.

Finite Element Analysis

This process involves dividing the geometric computer model of a part into many small inter-connected elements, *Figure 10-20*. Each element is typically rectangular or triangular in shape. The computer is then able to analyze each individual element, and, thereby, the entire part. A particularly helpful aspect of finite-element analysis is the ability of a computer to display the entire object as shown in Figure 10-20 and to highlight problem elements that might fail or fail to perform in the intended manner.

Review

1. Explain how CAD changed the design process.
2. Define the term *drafting*.
3. Define the term *computer-aided drafting*.
4. What do each of the following acronyms stand for? CAD, AD, CADD, CAE, CG, CAM.
5. What are the most obvious differences between CAD and manual drafting?
6. List the typical tools of the CAD technician.
7. How do CAD technicians store design documentation?
8. What three components must be present in order to have a CAD system?
9. What are the components in a typical CAD hardware configuration?
10. What are the three types of graphics displays?
11. Name two categories of keys found on CAD system keyboards.
12. What does CRT stand for?
13. Explain three uses of a digitizer.
14. What is a function menu?
15. Name two different types of function menus.
16. Name three different types of plotters used in CAD systems.
17. What is the processor? Name its two most important components.
18. Name three types of CAD software.
19. List five advantages of CAD over manual drafting.
20. How will CAD affect your future as a drafting technician?
21. Define the term *computer-aided engineering (CAE)*.
22. What is mass properties analysis?
23. Describe the process of finite element analysis.

Engineering Visualization: Computational Design and Analysis

Chapter Outline

V*isualization* is the pictorial representation of data for the purpose of more clearly and concisely communicating information. (See *Figure 11-1*.) The computer has made it possible to create pictorial representations of data that are virtually impossible with previous graphics techniques. The purpose of this chapter is to define and describe existing and emerging techniques for creating and analyzing data by means of com-

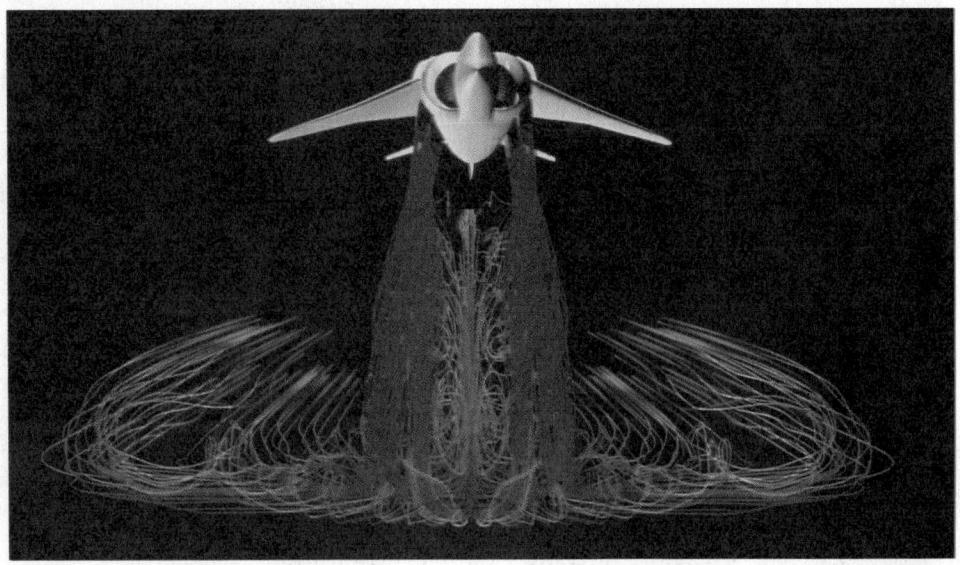

Figure 11-1 Harrier jet *(Courtesy NASA)*

puter-generated visualization. The old saying that a picture is worth a thousand words has never been more true. Today, however, the saying is more appropriately stated as: "A picture is worth 786,432 (1,024 x 768) pixels."

Visualization makes it possible to compress an engineering problem consisting of multiple variables, any one of which has massive amounts of related information, into a variety of types of images that more clearly and concisely convey critical concepts and findings.

The technology and techniques for solving engineering problems have changed dramatically in a very short period of time. The pervasive use of computers has brought with it the ability to resolve problems with far greater accuracy. Computers have also made it possible to solve problems that were previously too complex to resolve. One very important result of computer technology is the massive amount of information generated in solving any given problem. A particular problem may generate stacks of paper printouts with so much data that interpreting the results becomes impossible.

One example of this is the tremendous amount of information obtained by the Voyager 1 and Voyager 2 satellites during their flights past Jupiter. *Figure 11-2* illustrates a composite of true color and filtered images of Jupiter that have been extracted from an animation of the planet rotating on its axis. These images represent digital data transmitted by the Voyager satellites that were used to render a three-dimensional model that was then animated to rotate on the planet's axis. The ability to view side-by-side animations of the real and filtered images greatly improved research on cloud composition compared with the previous method of viewing individual static images of small portions of the planet.

During March and July of 1979 there were approximately 13,600 images captured by the Voyager 1 and Voyager 2 space satellites. Each image contains the equivalent of 640,000 bytes of information. The total amount of data equates to more than 8,700,000,000 bytes of information. It would require more than 6,000 high density 3.5" disks (1.44 megabytes) to store the data acquired by the Jupiter flybys.

Figure 11-2   Jupiter: True color and spectral classification image (*Courtesy Planetary Arts and Sciences*)

To date, researchers have quantitatively looked at approximately sixty images. These sixty images represent less than one-half of one percent (.44 percent) of the total information available. Who knows what enlightening and beneficial information is sitting in a computer bank just waiting to be discovered?

Increasingly complex engineering problems generate corresponding increases in the amount of information produced. The improvements in products made possible by more accurate data justifies the large amounts of data to be utilized. In many cases the amount of data to be analyzed is so great that it is impossible to comprehend in numerical form. Extremely large amounts of data are one of the major reasons for utilizing visualization techniques. Stacks of printouts consisting of numerical data are frequently impossible to comprehend without the benefit of visualization.

Advantages and Disadvantages of Using Visualization

Several factors are leading to increased use of visualization techniques and output.

- More complex engineering problems are generating increasingly large amounts of information.
- Increased computational speed makes it possible to perform visualization tasks on workstations and personal computers instead of being limited to supercomputers.
- Increasing performance/price ratio makes visualization hardware and software affordable for more people and companies who can benefit from visualization.
- Improved user interfaces make the operation of systems and production of visualization output easier to perform. Rather than being so complex and unique that only a specialist can use a system to create visualization output, it has become increasingly easier for greater numbers of people to productively utilize visualization technology and techniques.

Even with tremendous improvements in performance and ease of use, major barriers to the adoption of visualization techniques still exist. Among the more critical factors are the following:

- The complexity of visualization techniques and technology continue to make it difficult to learn and use, particularly for more complex data. Both hardware and software have varying degrees of complexity depending on the type and level of visualization being produced.
- Visualization generally involves three-dimensional data. This represents a major increase in complexity from two-dimensional data and requires a solid understanding of three-dimensional information.
- An unlimited number of ways to represent any given data set presents complex choices. The decision of how to represent the data is subjective. Frequently, no standard rules exist to govern selection.

3-Dimensional Imaging Techniques

Data can be represented by a variety of techniques. As a general rule, the more detailed or complex the representation, the more time-consuming and costly the production is. If the information being conveyed can be clearly represented by a two-dimensional, monochrome image, it is not necessary, timely, or cost effective to produce a 3-D animation of the data. It is the task of the person creating the visualization image to determine how to best represent the information to be communicated. Considering the number of variables and the variety of means available, this at times can be a very complex decision-making process. It is a task that requires personal expertise, training, and comprehension.

The numerous kinds of information to be communicated necessitate a variety of ways in which the information may be represented. In fact, any one problem can be represented in multiple ways. The more variables in any given problem, the greater the number of possibilities for the ways to represent the data. Each of these kinds of information requires different considerations regarding the type of image that most clearly represents the data while at the same time is the most cost efficient.

The following types of images provide a complete range of techniques for representing data. They are arranged from least complex to most complex. Complexity typically has a direct correlation with cost of production.

There are three basic methods available for representing engineering visualization models. The methods are wireframe, surface, and solid modeling. Each of these methods has its own advantages and disadvantages. The selection of which method of representation to use is based on the particular design project and requirements. (See *Figures 11-3* through *11-7*).

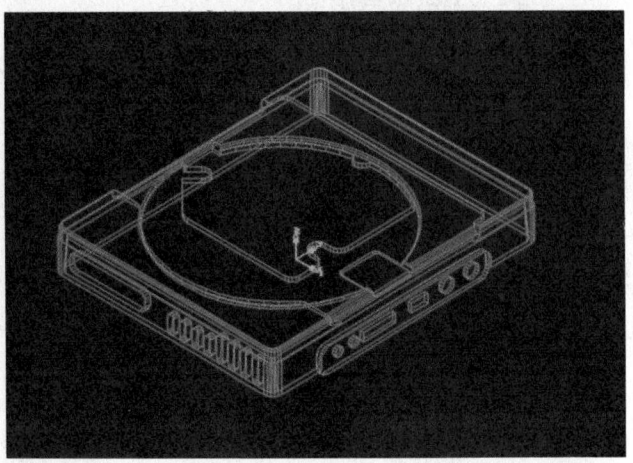

Figure 11-3 Wireframe *(Courtesy Matra Datavision Inc.)*

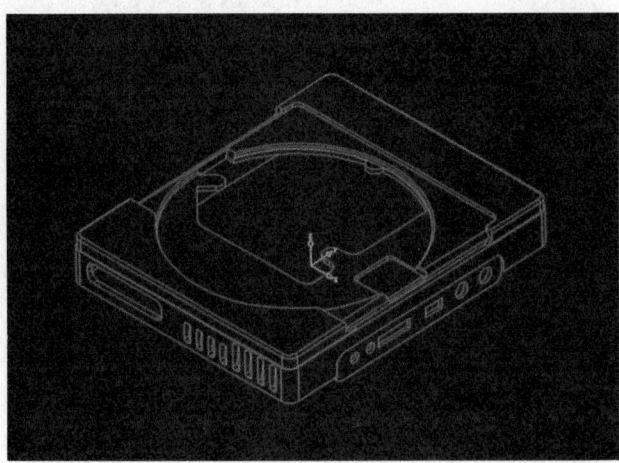

Figure 11-4 Wireframe with hidden lines removed *(Courtesy Matra Datavision Inc.)*

• Least time-consuming computationally

• Least intrinsic information

•Least descriptive image: frequently confusing, especially without hidden lines removed

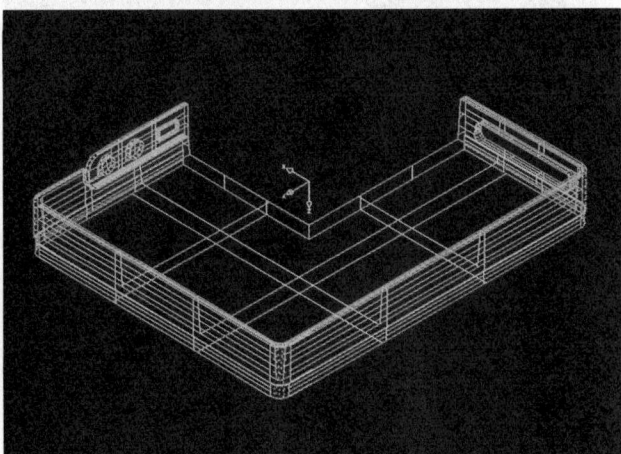

Figure 11-5 Surface model made by EUCLID CAD/CAM *(Courtesy Matra Datavision Inc.)*

• Moderate to highly time-consuming computationally

• Descriptive surfaces: the interior of the object is hollow

• Works well with certain automated machining processes

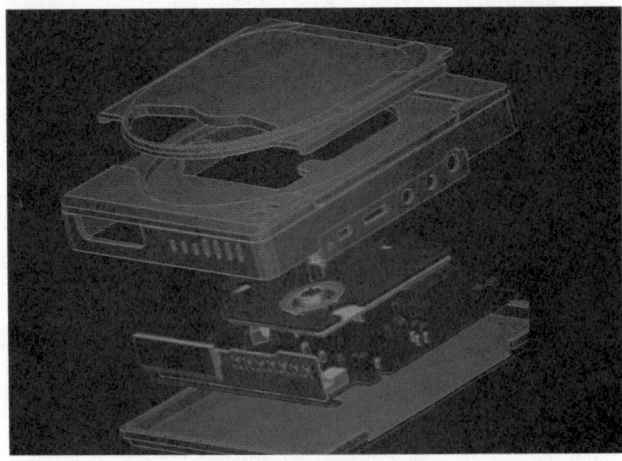

Figure 11-6   Solid *(Courtesy Matra Datavision Inc.)*

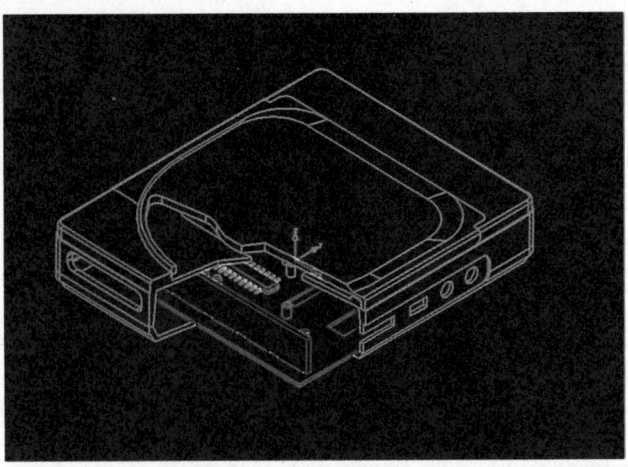

Figure 11-7   Sectioned solid *(Courtesy Matra Datavision Inc.)*

- Highly time consuming computationally
- Provides realistic views of both internal and external detail
- Contains object information such as properties and characteristics

As computers increase in speed and capability, design methodology will change. The use of 3-D technology will become more prevalent than 2-D technology. Systems are emerging that make it feasible and productive to design in 3-D using solid models. When solid models are as easy to create and manipulate as is now the case with 2-D methods, they will provide increased clarity and intrinsic information that permit additional analysis and evaluation.

Types of Visualization Images

Still Images

Single-frame, still images remain one of the most prevalent types of images used for visualization. The explod-

ed view of a fully shaded lifting block assembly in *Figure 11-8* presents a clear means of communicating product details. The ability to separate product components demonstrates an additional benefit of 3-D modeling. This same 3-D database may be used to generate multiple views of the product by simply designating an alternate viewing location.

Transparency

The ability to view the internal detail of a product is frequently necessary. Current technology makes it possible to readily assign varying degrees of transparency to multiple objects within an assembled product. The lifting block in *Figure 11-9* illustrates how transparency can be used to clearly show the internal detail of assembled products.

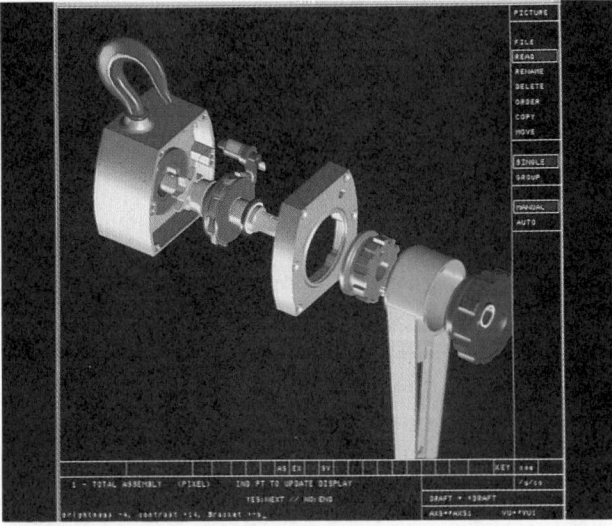

Figure 11-8   Lifting block: Exploded assembly *(Courtesy IBM© 1992 by IBM Corporation. All rights reserved.)*

Figure 11-9   Lifting block: Transparent *(Courtesy IBM© 1992 by IBM Corporation. All rights reserved.)*

Figure 11-10   Truck with digital image background *(Courtesy Paul Maxwell, the Goodyear Tire and Rubber Company with Silicon Graphics Inc. computer system)*

Photorealism

Systems now exist that also make it practical to render realistic images. Rendering an object adds lighting sources, shades and shadows, and surface colors and textures. Rendered images may also incorporate digital images of real objects with computer-generated models. The computer-generated model of the truck in *Figure 11-10* has been superimposed on a digital image of real-world surroundings.

Animation

Animation provides a means of representing time-dependant variables and active visualization of objects or events. Products consisting of moving parts and active functions often cannot be clearly represented by still images. Animation is becoming increasingly useful for a variety of innovative applications. In addition to playing a programmed animation it is now possible to move through 3-D models interactively. Advancements in visualization technology will continue to increase the reality with which 3-D models can be viewed and toured electronically.

The NASA Ames Research Center utilizes animation for a variety of design and analysis purposes. One project uti-lizing animation involves the flowfield about a delta wing equipped with thrust-reverser jets. Powered-lift flight in close vicinity to the ground can have serious implications for aircraft functionality. The three images in *Figure 11-11* have been extracted from an animated sequence to show the progression of the jet and the formation of a ground vortex. Particle traces are colored by time at release. Although the three images give a general understanding of the flowfield, they do not communicate the continuous and real time progression of a live video.

Stereo Images

Stereo images are a special type of 3-D model imaging. Stereo images consist of pairs of images that replicate views according to what the left and right eyes see. Since there is a slightly different angle of view between left and right eyes, it is necessary to create two images with slightly different views of an object. These matched sets of images must be viewed in a manner that limits the left eye to the left view image and the right eye to the right view image. The benefit of stereo images is their ability to recreate a view conveying depth similar to what would be seen in our true 3-D world.

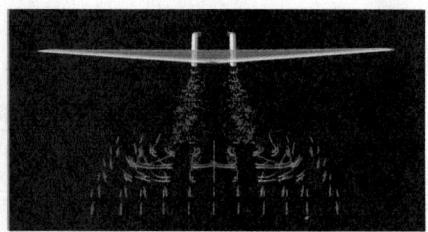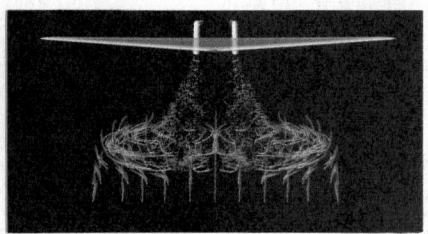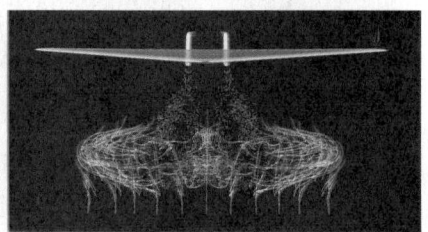

Figure 11-11   Delta wing animation series (three views) *(Courtesy NASA)*

Figure 11-12 showing two images side by side of an F18 jet aircraft.

Figure 11-12   Stereo pair: F18 jet aircraft (left and right) *(Courtesy NASA)*

The images in *Figure 11-12* represent a stereo pair of an F-18 jet aircraft. They may be viewed with a stereo viewer to create a stereo image. In place of a stereo viewer, the images may be viewed by placing a business-size envelope between the images with the long dimension projecting out from the images. Place the end of the envelope against your nose and relax your eyes until the images merge. There are numerous methods of creating stereo images. A very basic way of viewing 3-D stereo images is with the use of red and blue lens glasses similar to those used with 3-D movies. More sophisticated methods include computer and glasses combinations that can also create 3-D stereo animations.

Holography

Holography produces images that simulate the effect of seeing the three-dimensionality of an object from multiple viewpoints. Holographic images have become fairly common for a variety of applications. Perhaps the most prevalent application has been the use of holographic images on credit cards. In the past, holographic images have been limited to physical objects that could be photographed in a manner that permits compositing multiple images into a holographic image.

It is now possible to produce holographic images from computer-generated 3-D models. In their most basic form, these images are static images on sheet material. One advancement over a static image is the combination of computer and holographic technology to create holographic images in 3-D space. *Figure 11-13* illustrates a system that integrates a computer-generated 3-D model of an automobile with holography equipment to produce an interactive hologram. The holographic image may be seen from different viewpoints by rotating the 3-D model within the computer system. *Figure 11-14* represents a series of wireframe views of the automobile ending on the right with a surfaced model.

The complexity and computational requirements involved in generating a holographic image in this manner are extremely demanding. For this reason, the model is initially viewed as a wireframe and then rendered as a surfaced model when the desired viewpoint

Figure 11-13   Holography: System *(Courtesy MIT: Media Lab)* ©*Hiroshi Nishikawa*

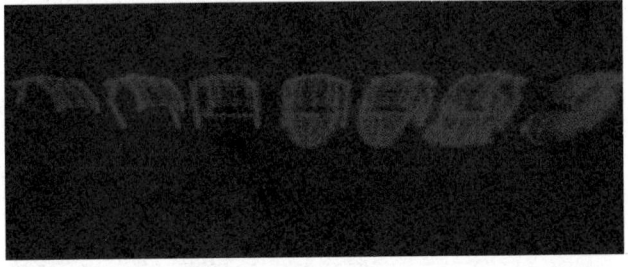

Figure 11-14   Holography: Multiple views of vehicle *(Courtesy MIT: Media Lab)* ©*Quesada/Burke Studio, NY, Courtesy Scientific American*

is established. As computational speeds increase, it will be possible to generate increasingly realistic images. Increased size and the addition of color now exist for computer-generated holograms. Faster computational speeds will also improve the capability to interact with the model in real time.

Virtual Reality

Virtual reality is a technology designed to simulate either a physical or hypothetical environment. While

Figure 11-15   Virtual reality *(Courtesy NASA)*

the emphasis in this chapter is on visualization, it is important to note that virtual reality is intended to involve all the natural senses. Virtual reality systems can incorporate stimulus for all the five basic senses: sight, sound, touch, smell, and taste.

Virtual reality has a wealth of applications. It can create an environment that permits a sense of how a particular engineering design may feel or function. By being able to experience such an environment, the researcher can test and analyze alternative designs before investing in physical prototypes. Where multiple senses are important to a design project, virtual reality provides a more complete and realistic simulation.

Virtual reality also has great potential for hypothetical and hazardous environments. These environments either don't exist, are too dangerous for humans to enter physically, or may be beyond the reach of humans. The virtual reality system illustrated in *Figure 11-15* is designed to provide the user with a simulated experience of walking on the planet Mars. The view the operator sees is replicated on the computer screen in order that observers may see what the operator is seeing. The operator is equipped with sensors that signal locational and relationship data to the simulation. All of these components work together to make it possible for the operator to interact with the computational world in a manner that simulates being in the real environment.

Methods of Representing Variables

Engineering projects involve a large variety of variables. Visualization may be called upon to represent either single or multiple variables in a given visualization image. Numerous means of representing product characteristics and data are available. The methods of representing variables may be used either individually or in combination depending on the project requirements.

Color is one means of representing variables. It has been used extensively for a long time. Generally understood associations with colors have evolved over time and now serve as de facto standards. The color blue is generally associated with cold temperatures, while red is associated with hot temperatures. Other means are

achieving similar standardization as visualization evolves. There will always be new or alternative meanings associated with standard representations depending on specific projects and contemporary practices. The following means of representation comprise some of the more common types in use.

3-Dimensional Images

Three-dimensional modeling is rapidly becoming the norm for product design. The technology and techniques available to produce 3-D models now justify a transition from 2-D drawings for many companies. In addition to providing improved clarity for visualizing an object, a 3-D model possesses object characteristics and information essential for analysis purposes. *Figure 11-16* is a 3-D model of a Buell motorcycle frame. The figure illustrates the ability to use a 3-D model to generate 2-D orthographic drawings as well as pictorial views. Future development of computer systems to decrease the learning curve and increase interactivity will further promote the transition from 2-D drawings to 3-D modeling.

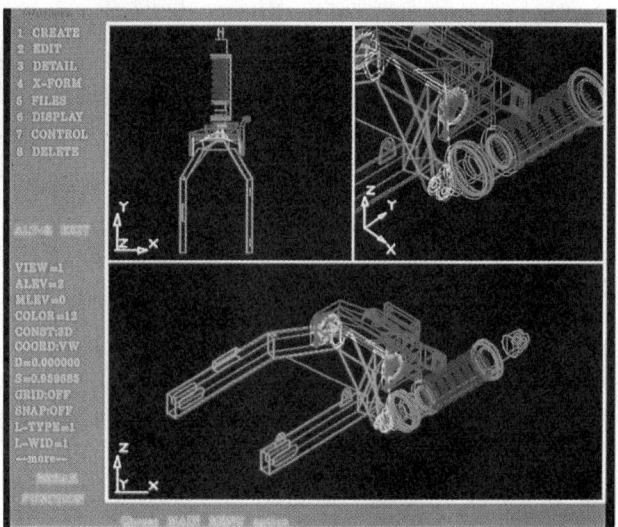

Figure 11-16   3-D Model of frame of Buell Motorcycle, Model RS 1200, designed with Cadkey *(Courtesy Cadkey Inc.)*

Color

Color is one of the most common and useful means of representing data with visualization. It may be assigned to numerous types of variables including temperature, function, and pressure. *Figure 11-17* illustrates a hypersonic vehicle and related flight pressures. Pressures are shown for both the vehicle and a cross section of the shock layer. The colors range from magenta to blue and represent highest to lowest pressures respectively.

It should be noted that some applications are better visualized by using multiple degrees of gray scales. In certain instances it is easier to determine minor variations in physical characteristics by viewing gradual changes from black to white.

Shape

Shape may be used to indicate both visible and invisible variables. This can be anything from pressure levels to molecular bonding forces. An object may in reality be physically deformed or may be represented as deformed to emphasize force. Images may also represent forces that are not visible to the human eye.

The perspective view in *Figure 11-18* represents the convective thermal field computed for a cross section (two-dimensional) of the thermal printhead geometry in an ink jet printing system. The model incorporates composite material structures, convective flow of the ink, and selective heating levels from the thermal device. The problem cross section is outlined in the lower area. The temperature field is mapped into the third (vertical) dimension. This image also uses color and is mapped so that blue-yellow-red corresponds to a range of low- to high-temperature values. In this case the surface plot clearly emphasizes the jump discontinuities in temperature gradient where crossing material interfaces with differing conductivities.

Symbols

Numerous types of symbols have been developed to represent variable characteristics. Some symbols have become standards and have very logical applications. Arrows are one example of a symbol with standard applications. They may be used to clearly indicate vectors and direction.

The types of symbols available for visualization and their possible combinations are infinite. Common types of variable symbols include points, lines, (+) and (−) symbols, as well as 2-D and 3-D polygons and shapes. The challenge of visualization is to develop the most clear and intuitive representations possible to convey the information being described. A variety of symbols are illustrated in examples throughout this chapter.

Relationships

There are a variety of relationships between variable symbols that may be visually represented to assist with engineering visualization. Similar to the infinite types of

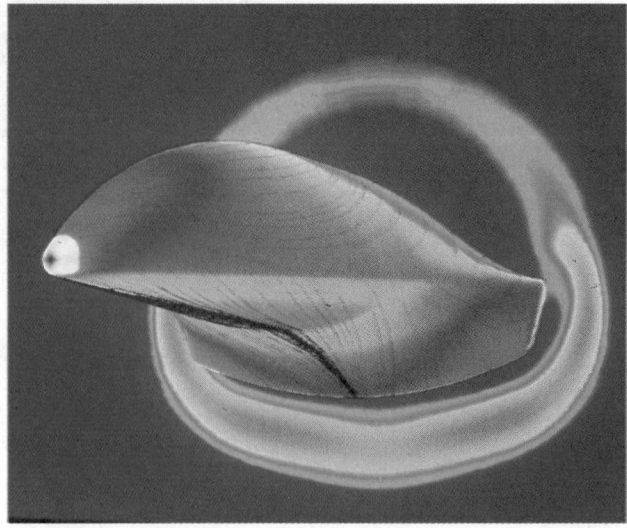

Figure 11-17   Hypersonic vehicle *(Courtesy NASA)*

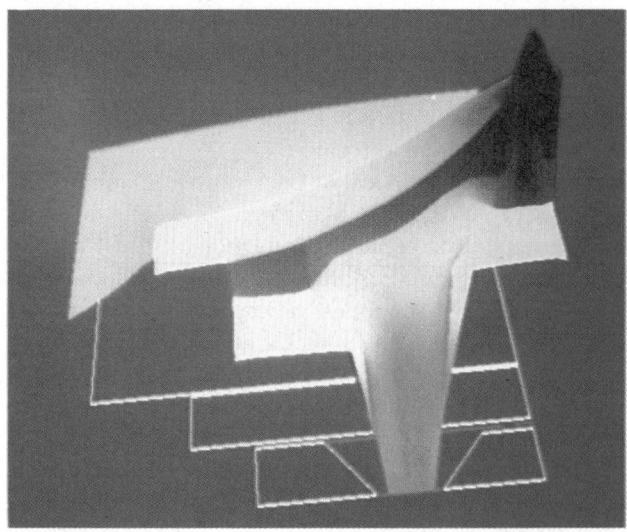

Figure 11-18   Convective thermal field *(Courtesy Xerox Corporation)*

variable symbols, there are numerous ways of representing relationships. Three common types of relationships are size, direction, and density.

Direction

The visualization of the aircraft hull illustrated in *Figure 11-19* utilizes arrows to indicate load distributions. The arrows indicate directional load forces on the hull.

Size

The same arrows used to indicate directional load forces in Figure 11-19 also provide an indication of the amount of load forces on the hull at specific locations. This is achieved by varying the length of the arrows so that longer arrows indicate greater amounts of load force.

Density

Another common method of representing a variable is by the density of symbols. In this case the arrows could

Pressure Loading on the Fuselage of Finite Element Model

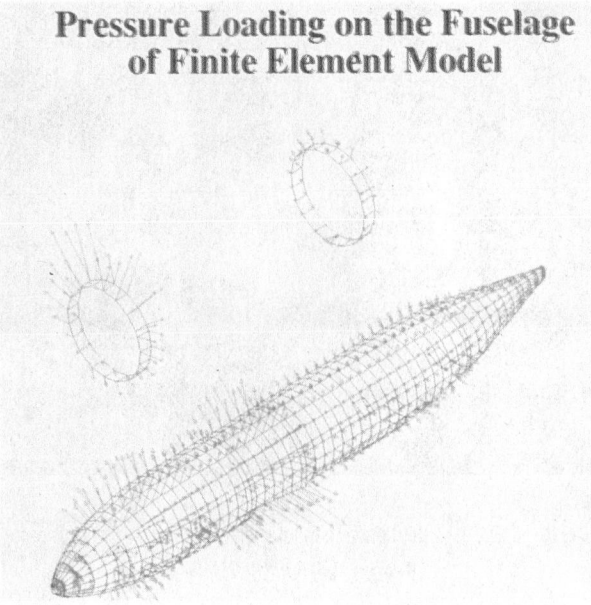

Figure 11-19   Boeing aircraft vectors: Direction and size (Courtesy Boeing)

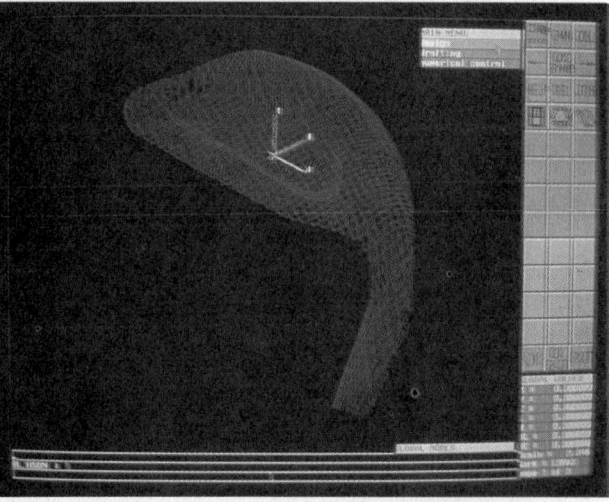

Figure 11-20   Finite element analysis (Courtesy Karsten Manufacturing Corporation/Ping Golf Clubs)

all be the same length but more closely packed where greater amounts of load forces exist.

The decision of how to best represent variables is very subjective. The person responsible for visually representing the data being viewed must select the method of representation based on experience. The more knowledge a designer has about the alternative visualization techniques available, the better that person will be able to communicate necessary information.

Motion

Many engineering applications involve movement. Movement may be either moving parts or various types of flowing matter. The use of animation to represent moving entities has become increasingly practical in terms of both cost and production. Motion is vitally important to a vast number of engineering visualization applications. Aerodynamics is a prime example of an application that must include an understanding of moving matter, specifically air. The ability to portray airflow has dramatically improved the means of evaluating and optimizing the aerodynamics of aircraft. As an example, the image of the YAV-8B Harrier jet in Figure 11-1 may be viewed as an animation with the airflow progressing through simulated paths.

Engineering Visualization Categories

Engineering visualization involves numerous types of design and research information. Each of these types of information requires unique and specific means of representation in order to most clearly and concisely convey selected information. The following examples represent major types of engineering visualization.

Finite Element Analysis

Finite element analysis (FEA) is a vital means of designing, modifying, and analyzing objects and products. The process of FEA begins with the creation of a *finite element model (FEM)*. This 3-D model is created with points throughout the object called *nodes*. The golf club in *Figure 11-20* has been created as a finite element model. Each node contains specific information about the golf club characteristics at that particular point. All of the nodes of a model are connected to create a FEA mesh. In this example the FEA mesh is being analyzed to determine the structural and impact characteristics of the golf club.

Thermal Analysis

Thermal analysis is a means of evaluating the thermal characteristics of a part or product. Thermal analysis is vital to numerous types of design applications. A clear understanding of a product's thermal characteristics can improve both efficiency and longevity. It may, in fact, prevent product failure.

Thermal analysis may be used for applications where thermal characteristics are the only design variable. It may also be used for design projects where heat or cold is just one of many variables, and where each of the multiple variables has a direct impact on related variables. The heat may be flowing through a conduit or around the exterior of an object. Other products may be solid objects involving critical thermal transfer through a specific material or materials.

One very common application for thermal analysis is the design of heat ducts. These same principles apply to a variety of similar types of heating and cooling systems. The cross sectional shape and internal volume of a duct significantly impact thermal transfer characteristics.

Thermal analysis systems employ visualization techniques which produce color-coded images that clearly

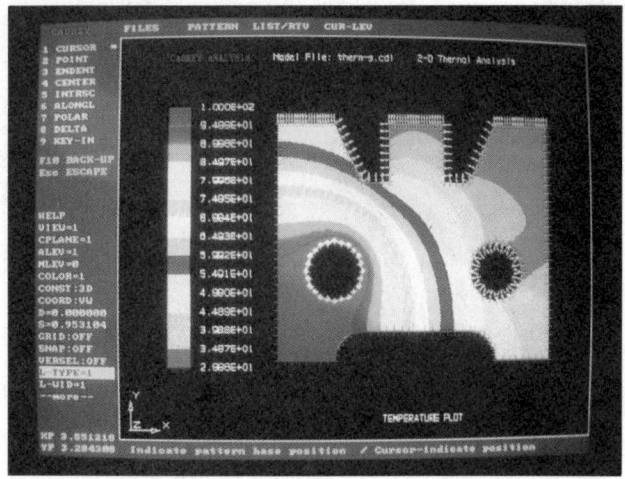

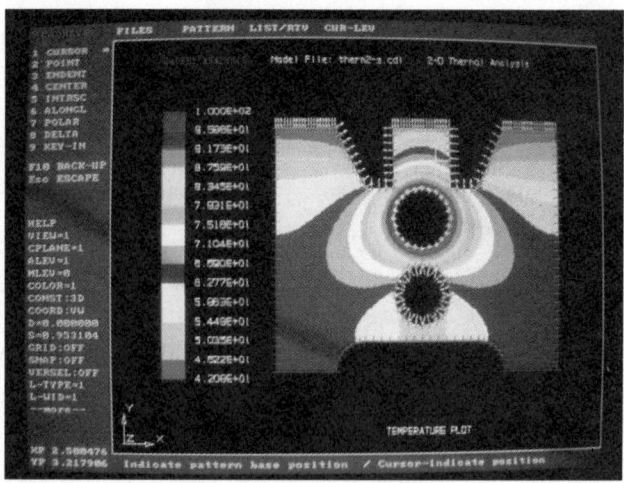

Figure 11-21   Heat duct: Thermal analysis *(Courtesy Colorado State University, Department of Industrial Sciences)*

Figure 11-22   Heat duct: Thermal analysis *(Courtesy Colorado State University, Department of Industrial Sciences)*

illustrate temperature distributions. These simulations permit quick and easy evaluation of alternative conduit designs.

The two heat duct designs illustrated in **Figures 11-21** and **11-22** depict two design configurations. The thermal color coding shows distinctly different thermal characteristics. Depending on the desired thermal characteristics, one design may provide significantly better results.

Stress Analysis

Stress is a critical factor in engineering designs of all types. Designs where stress is a factor should be analyzed for functionality in order to prevent failure. Stress occurs where force is exerted on an object. It may be concentrated in a small area or distributed over a large span. The ability to create visualizations that graphically represent stress has become commonplace and promotes optimum design solutions.

Computer systems are ideal for computational testing of product stresses. Stress analysis techniques may vary dramatically depending on the type of object being tested. This requires designers to be familiar with and select appropriate stress analysis systems. The object illustrated in **Figure 11-23** uses color to represent stress distribution in a beam. The beam in this case demonstrates even stress distribution. In the event that stress analysis should indicate a problem like localized stress at a corner, the design may be reworked to resolve the problem, perhaps by incorporating a fillet or round.

Computational Fluid Dynamics

Computational fluid dynamics (*CFD*) is a means of using computers to mathematically simulate physical events. This involves the study of fluids and their characteristics under a variety of conditions. Typical types of fluids consist of air, gases, and liquids. Computer programs and algorithms have evolved to a level that make them highly accurate for predicting fluid flow

characteristics for mechanical and electrical products, architectural structures, medicine, and the automotive and aerospace industries.

Aerodynamics

Aerodynamic visualization deals with the flow of air around and through objects. The ability to interact with alternative designs and to observe their resulting characteristics and impact permits much more timely and optimized products.

CFD plays a vital role in the aerospace industry. The design of airplanes depends upon CFD to produce efficient planes that also perform optimally. **Figure 11-24** illustrates the use of CFD to determine airflow over an F-16A aircraft. The particle traces illustrate the simulated flowfield about the F-16A. The colors are used to indicate height in relation to the wing. The red color is closest to the horizontal center plane of the wing and

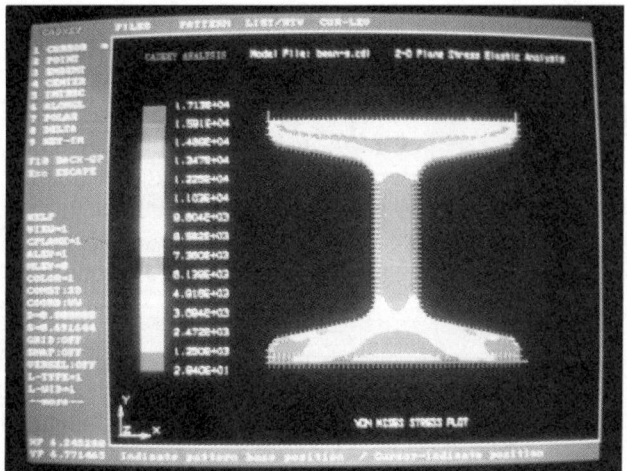

Figure 11-23   Stress analysis *(Courtesy Colorado State University, Department of Industrial Sciences)*

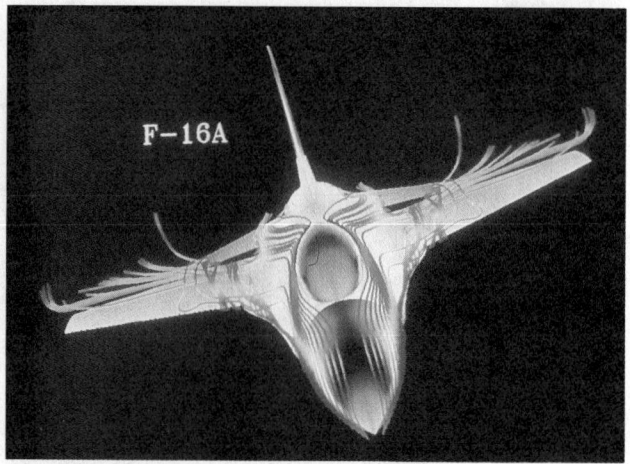

Figure 11-24   F-16 aircraft *(Courtesy NASA)*

red is highest above the plane. While it may not eliminate the need for testing a physical prototype in a wind tunnel, it significantly reduces the number of physical prototypes that must be built and tested.

In addition to the aerospace industry there are growing numbers of other industries discovering innovative applications for aerodynamic visualization. Applications in which aerodynamic visualization has strong potential include:

1. Sports equipment and clothing
2. Architecture: airflow between high-rise buildings in large cities with concentrated buildings
3. Automobiles: friction reduction for improved mileage and reduced interior noise levels
4. Mechanical and electrical products: proper cooling

Liquids

Medical science is developing numerous innovative applications for visualization. One such application is the use of CFD to design artificial hearts. This requires an understanding of liquid fluid flows and involves several variables. The heart illustrated in *Figure 11-25* is being studied to determine vorticity. Excessive vorticity may result in high shear regions within the heart. This is critical since high shear may damage blood cells as they travel through the heart.

In this example, vorticity magnitude is represented by colors with blue being low magnitude and white being high magnitude. The green cubes represent blood cell locations that have reached a specified threshold of vorticity.

Structural Analysis

Virtually every product involves structural factors that are critical to their success. Structural factors may include pressure loads and interference/clearance checking. Boeing® utilizes CFD to determine the surface pressure characteristics of aircraft configurations. *Figure 11-26* shows how color is used to represent varying surface pressures on aircraft. Knowledge of the pressures and their locations is then used to assure structural integrity throughout the aircraft.

Commercial aircraft incorporate many diverse systems that must work in unison. Physically fitting these systems and their components into an aircraft is a monumental task. Not only must the systems fit into limited space, they are also constrained by aerodynamic considerations for the shape of the aircraft. Visualization techniques make it possible to locate potential system interferences during the design stage and eliminate costly revisions.

Space allocation in the area beneath the flight deck is a complex task involving several groups within Boeing. Computer technology and visualization provide a common database that can be used interactively by multiple groups to resolve problems in the early stages of design. This concurrent engineering procedure saves a tremendous amount of time and resources. Equally as important, it results in optimum solutions to design criteria.

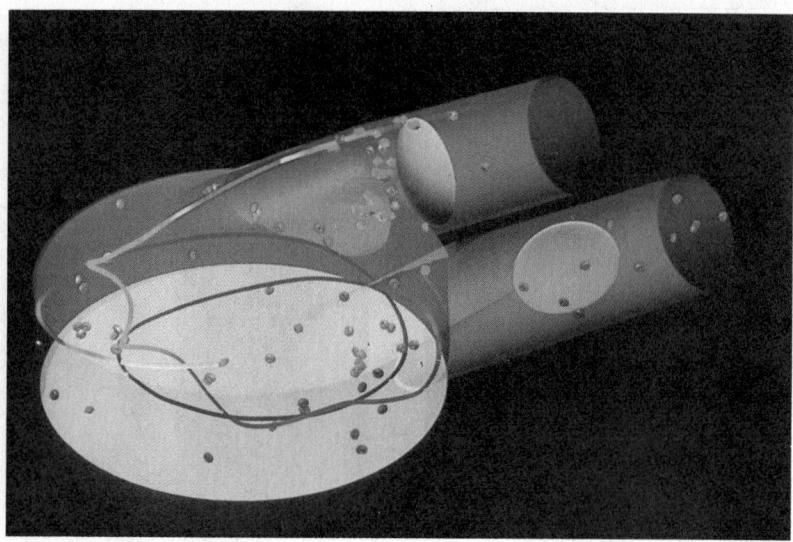

Figure 11-25   Artificial heart *(Courtesy NASA)*

Figure 11-26   Boeing aircraft: Pressures *(Courtesy Boeing)*

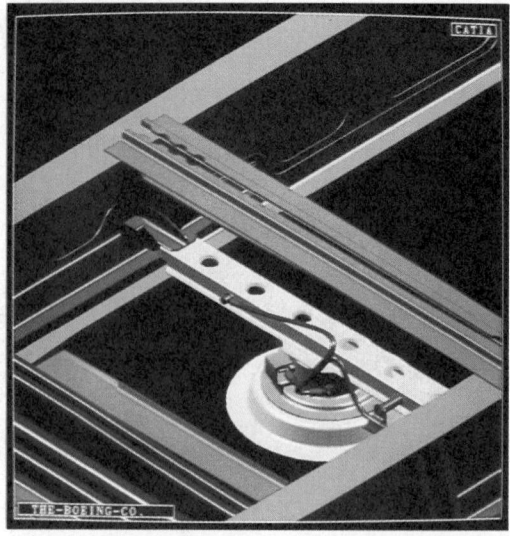

Figure 11-27   Boeing structural: Interference *(Courtesy Boeing)*

Figure 11-27 illustrates an interference in the cargo compartment located beneath the passenger compartment. The interference occurs where the left end of the cream-colored channel meets the green mounting bracket. The channel serves as a mounting device for a light fixture used to illuminate the cargo compartment.

Configuration management software developed by Boeing is used to manage and "assemble" individual 3-D solid modeled parts. The computational assembly of 3-D components provides a computational prototype equivalent of the actual physical airplane.

The interference was corrected by replacing the original channel with a shorter channel. The correctly fitted channel is illustrated in *Figure 11-28*. By detecting the interference early in the design process, it was possible to eliminate the interference before getting into physical prototyping and additional expenses.

Simulation

Simulation is a means of conducting product design alternatives and evaluation without costly physical prototypes. It has become common practice to use simulation technology in computational building products of all kinds. These computer models contain data that may be used to predict the physical characteristics of a product to be designed. The accuracy of simulations has become sufficient to justify computational simulations as a means of reducing the need for physical prototypes. Animation is another advantage available in order to further evaluate design criteria and solutions.

The image of the space shuttle in *Figure 11-29* illustrates how computational and physical prototypes may be compared. The left half of the shuttle image is the result of a computational simulation representing surface pressures with color. The right half of the image illustrates the corresponding surface pressures obtained with a physical model in a wind tunnel. The simulation approximates the physical results with sufficient accuracy to justify the use of simulation for design purposes.

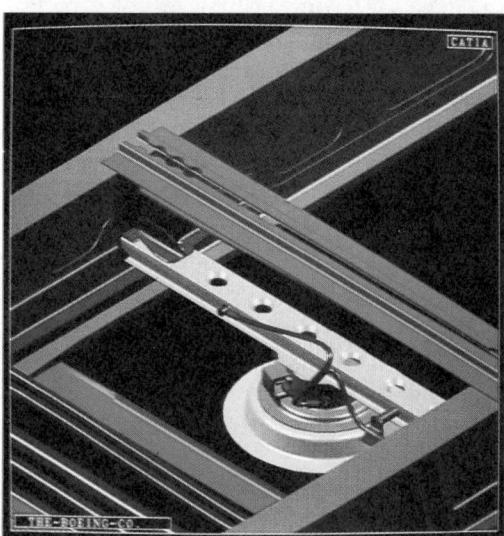

Figure 11-28   Boeing structural: Clearance *(Courtesy Boeing)*

Figure 11-29   Shuttle: Simulated/physical *(Courtesy NASA)*

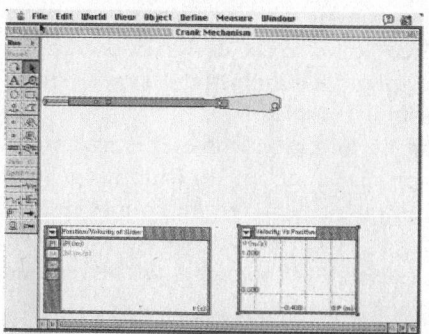 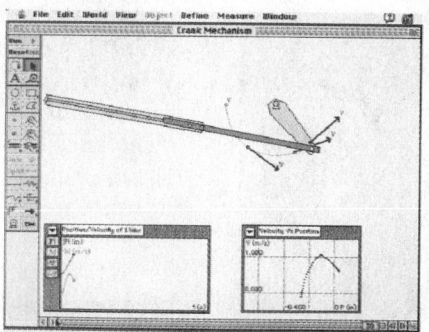 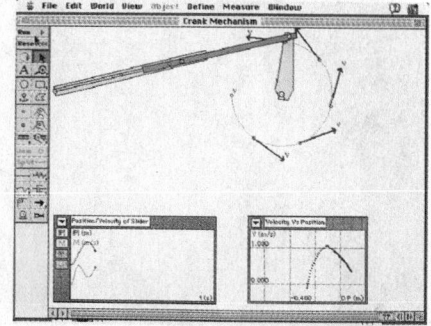

Figure 11-30   Kinematic sequence *(Courtesy Knowledge Revolution)*

Kinematic Simulations

Kinematics is an excellent application for visualization and simulation. Linkages and mechanisms are present in numerous products that we use daily. Computers are ideal for solving the mathematical calculations necessary. Once the calculations have been performed, images may be generated that clearly illustrate resulting designs. Technology exists that makes it easy to computationally build kinematic systems. Once a system has been built, it may easily be animated and tested for performance. ***Figure 11-30*** illustrates three stages of an animated sequence representing a linkage system and relevant system characteristics.

Automotive Crash Simulations

The automotive industry is using visualization in computational simulations to better understand structural damage resulting from crashes. This crash information is vital to improving the safety of vehicles. In the past, such information was only available from physically crashing real vehicles. ***Figure 11-31*** illustrates computer-simulated crash results. These three images have been extracted from an animation sequence that permits the study of progressive stages of crash impact. The ability to interactively study computational crashes permits a better understanding of the crash results. This information may then be used to optimize designs first with simulations. This reduces the need for crash testing real vehicles.

Case Studies

Engineering visualization is one aspect of product design. In reality there are multiple phases that occur between product conception and the final product. The following two examples illustrate how visualization is utilized to improve products, their production, and their cost effectiveness. These two examples also consist of several types of visualization images and considerations that need to be addressed.

Biomechanical Visualization

Biomechanical technology is incorporating ever increasing use of engineering and visualization techniques to improve the quality and capabilities of work in this field. The following example describes work being done with hip and knee implant replacement techniques. In addition to demonstrating benefits of modern technology, it exemplifies very real benefits on a personal level. Prior to the use of this technology, a typical hip replacement had a life expectancy of 4–5 years. With the increased information and accuracy of visualization and related technologies, the life expectancy of the replacement has increased to 10 years and continues to improve. In simple terms, it reduces return surgery to one half the previous frequency.

There are several steps involved in producing implant prosthetics. The first step in the process involves obtain-

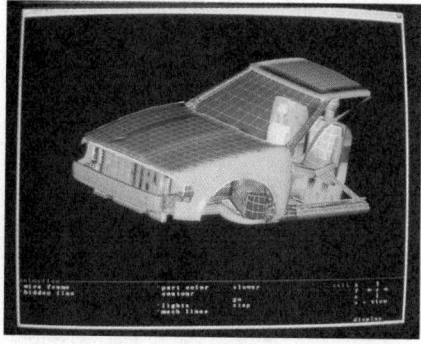 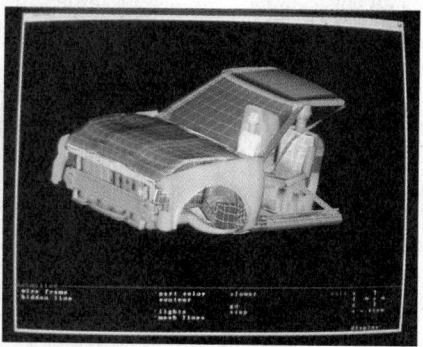 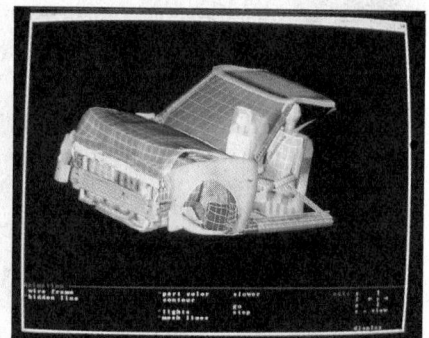

Figure 11-31   Vehicle crash testing: Real and simulated *(Courtesy Altair Engineering with CRAY XMS Supercomputer)*

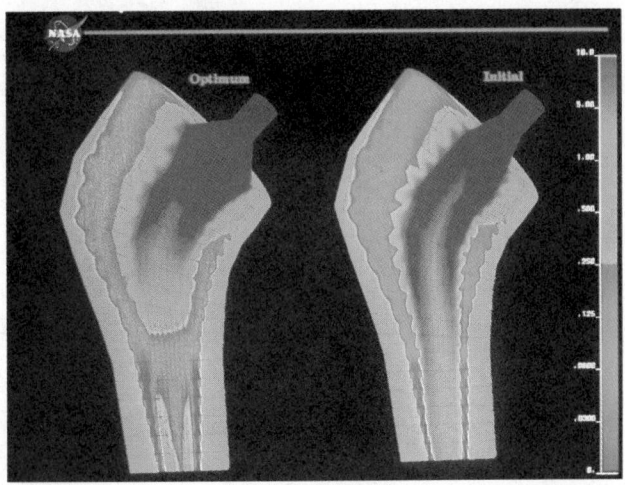

Figure 11-32 Hip implant: Optimal/initial *(Courtesy NASA and Case Western Reserve University)*

ing three-dimensional data, which can be used to create a 3-D FEM of the problem area. This information is typically obtained by techniques such as magnetic resonance imaging (MRI). The 3-D FEM may be used for a variety of design and analysis purposes.

The hip implant illustrated in ***Figure 11-32*** is being optimized relative to stresses in the region of the bone near the implant interface. The FEM on the right is the initial prosthetic design and the FEM on the left is the optimized prosthetic design. The material selected for this particular implant was titanium. Implant material selection is critical to hip replacement functionality. This type of surgery involves several critical factors including bone material stress and growth characteristics. Understanding stress and growth characteristics has proven to be one of the major advancements related to increased prosthetic implant life expectancy.

Orthopedic knee implants involve techniques similar to hip implants. Since a knee is subjected to both symmetrical and asymmetrical loading, it is necessary to evaluate the effects of both types of loading. ***Figure***

11-33 illustrates equivalent stress distributions in the cortical bone at three points in the design evolution for symmetric loads. ***Figure 11-34*** illustrates similar stress distributions for asymmetrical loads.

The knee implant in this case involves multiple materials in order to optimize functionality and longevity. ***Figure 11-35*** illustrates idealized morphology and material regions for a typical adult male tibia with an implant. The materials and the numbers of the regions they correspond to are as follows:

1 Polythene
2 Titanium
3 Cancellous HD
4 Cancellous LD
5 Cortical Shell

Once the 3-D FEM has been optimized for fit and functionality, it must next be produced. It is here that the relationship between design and production becomes evident. Typical knee implant components are illustrated in ***Figure 11-36*** and consist of (from left to right): femoral component, patellar component, and tibial component. The production of implant components is now possible with automated machining equipment. The same 3-D model that was used for design purposes may be used to generate code to control machining operations. The automated machining equipment may then produce the prosthetic implant.

Product Design

Product design techniques are changing dramatically. Competition is demanding faster design cycles and improved design solutions. The following example describes how Xerox® utilizes computer technology for product design. In addition to visualization techniques, methods of conducting real time interactive design with variable controls on the computer screen are discussed. The goal is improved product quality and design time.

The use of physical prototypes is both time consuming and expensive. One means of significantly improving design solutions and time requirements is to perform initial design using computational prototypes. Computational prototypes are computer-generated models that

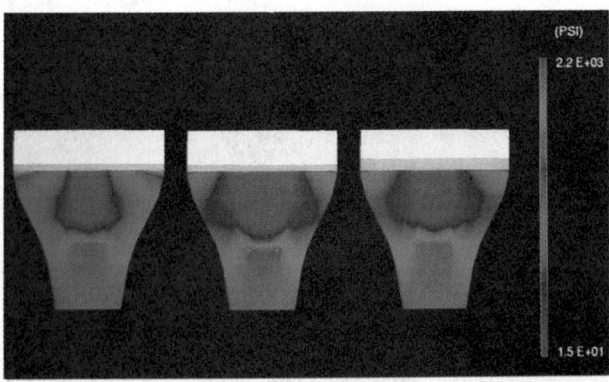

Figure 11-33 Knee implant: Symmetrical loads *(Courtesy NASA and Case Western Reserve University)*

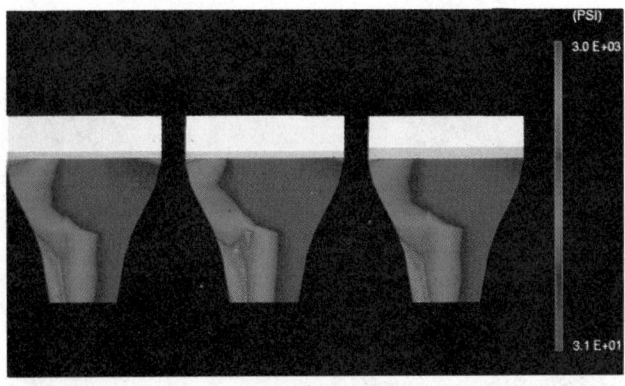

Figure 11-34 Knee implant: Unsymmetrical loads *(Courtesy NASA and Case Western Reserve University)*

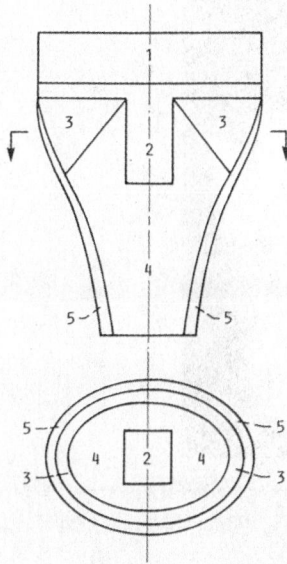

Figure 11-35 Knee implant: Composite materials *(Courtesy NASA and Case Western Reserve University)*

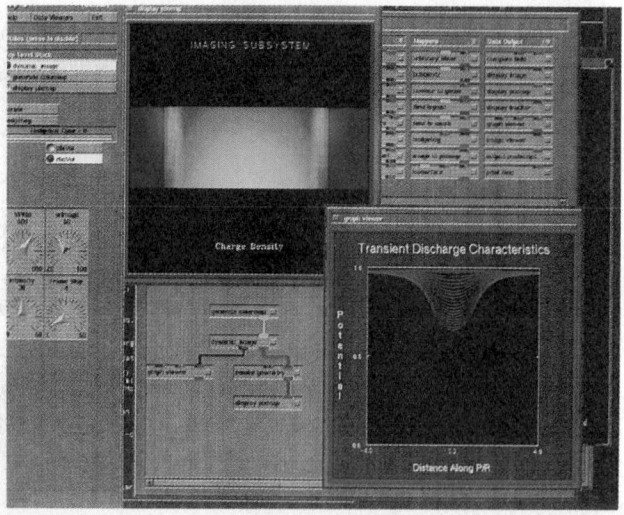

Figure 11-37 Xerox interactive display *(Courtesy Xerox Corporation)*

incorporate the physical characteristics of a physical model. A computational prototype may be evaluated and modified more quickly than a physical model. This permits ease of testing design alternatives and promotes investigating more designs than traditional physical prototyping methods. Testing more designs provides increased information from which to select the optimum design solution.

Xerographic printing and copying is pervasive throughout the world. The process of xerography is a very complex process demanding exacting standards. The design of xerography systems is accordingly complicated and presents an important case for the need to develop computational prototyping techniques. It is equally important that researchers and designers are provided fully integrated computer systems with highly functional interactive user interfaces.

One critical aspect of the xerography process is understanding the effects of laser exposure on the photoreceptor. *Figure 11-37* is a screen-captured image from the designer's workstation. It represents an interactive session for analyzing the xerography imaging process. The upper window shows a cross section of the photoreceptor with space charge densities color mapped as blue-yellow-red to indicate zero, moderate, and high levels respectively. The lower right window shows the evolution of the surface voltage on the photoreceptor. As the image discharges subsequent to the laser illumination, the photo-generated charge from the bottom moves upward to lower the surface voltage. Development of the image is voltage sensitive.

The designer's screen illustrates built-in capabilities permitting interactive changing of variables. On the left are four dials that can be used to change photoreceptor velocity, dielectric constants, backplate voltage, and wire voltage. These dials let the designer computational test varying levels and combinations of critical variables.

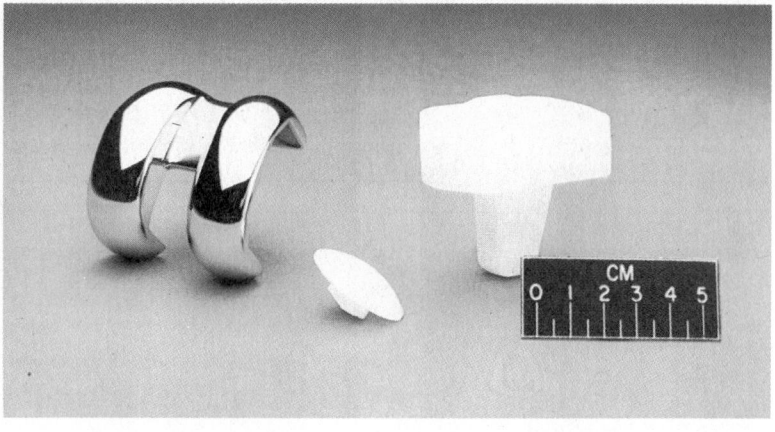

Figure 11-36 Knee implant: Components *(Courtesy NASA and Case Western Reserve University)*

Figure 11-38   Xerox laser exposure of photoreceptor
(*Courtesy Xerox Corporation*)

Further clarification of the effects of the laser exposure on the photoreceptor are illustrated in *Figure 11-38*. This model graphically shows the scattering and attenuation of plane waves representing the laser illumination as it moves through the photoreceptor. In this instance, intensity is color mapped so that blue-yellow-red indicates zero, moderate, and high levels of intensity respectively. The resulting model indicates that light is a wave-like phenomenon. The plane waves are incident on the left and become compressed and wrapped around the toner particles on the surface of the photoreceptor. It may be noted that the plane waves are similar to water waves as they flow around poles placed in water.

Review

1. Write a description of visualization and state why it is beneficial.
2. How have computer technology and visualization changed the way products are designed?
3. List the three basic methods of representing engineering visualization models. Describe one advantage and one disadvantage for each of these methods.
4. List six types of visualization images. Describe an application for three of these image types and state why it is appropriate for the image type selected.
5. Name six methods of representing visualization variables.
6. Describe a visualization application that uses two different types of variable symbols. State why the symbols are appropriate.
7. List the five engineering visualization categories described in this chapter. Describe an application for each of three selected categories. The three applications should be different than the examples in this chapter.
8. What expertise must a person have in order to create effective visualization images?
9. List and describe three advantages of using visualization.
10. List and describe three disadvantages of using visualization.

Design Drafting
Applications

Career Profile: Arlene Frank

As a recent graduate from the University of Utah with a degree in civil engineering, Arlene Frank accepted a position as a civil engineer with the division of structures construction in the California Department of Transportation (CAL-TRANS). Arlene enjoys the challenge and the responsibility of overseeing employees' and contractors' progress and procedures in the field, documenting daily field work, changing orders, making structural calculations, paying estimates, and making decisions on project changes with the appropriate approval.

Each week begins with a safety meeting because Arlene's responsibilities and the decisions she makes are not only for the moment, but for the life of the structure and the safety of other employees and the public.

When Arlene is not in the field, she prepares for the different phases of her job by reviewing the daily plans and reading the standards specifications that will apply to the next day's work. She also enjoys opportunities to consult with her coworkers and supervisor who are always willing to help.

Arlene's other responsibilities include preparing pay estimates. This is a payment made to the contractor for the work performed. It requires the reference of a daily diary. From the diary, Arlene determines the payment for material and labor.

The experience Arlene is gaining is part of the company's three-year engineering rotation program. Once Arlene completes her nine-month assignment in structures construction, she will rotate to the structural design section, where she will gain experience designing transportation-related structures such as pumping plant and rest area facilities, maintenance stations, toll facilities, parking structures, transit stations and platforms, and transit-related pedestrian bridges.

After Arlene completes her rotation assignment in the structural design section, she will choose an elective assignment in structural and seismic analysis, hydrology and hydraulics, externally financed projects, local assistance and programming, or one of several other areas.

As a new civil engineer employee, Arlene is gaining valuable experience that will help her obtain her registration as a professional engineer. Once she has her professional engineer's license, she will be able to increase her responsibilities and rise to new challenges; in addition she will have the opportunity to compete in her company's promotional examination for associate transportation engineer.

As a woman engineer, Arlene hopes to be a model for other minorities pursuing engineering degrees. Arlene feels it is important to keep young students informed of their potential futures in transportation and engineering by encouraging them to stay in school, study math and science, and to discover the possibilities.

Arlene is excited about the work she is doing, the experience she is gaining, the challenge of the job, and unlimited future potential.

Geometric Dimensioning and Tolerancing

Summary of Geometric Dimensioning and Positional Tolerance Terms

Actual Size.   Actual measured size of a feature.

Angularity.   Tolerancing of a feature at a specified angle from another feature.

Basic Dimension.   A theoretically "perfect" dimension similar to a reference or nominal dimension. It is used to identify the exact location, size, or shape of a feature.

Bilateral Tolerances.   Tolerances that are applied to a nominal dimension in the positive and negative directions.

Circular Runout.   A tolerance that identifies an infinite number of elemental tolerance zones on a feature when the feature is rotated 360° for each element.

Circularity.   A tolerance that controls elemental tolerance zones around a circular feature that is independent of other features.

Clearance Fit.   A condition between mating parts in which there is always a clearance assembly.

Concentricity.   A tolerance in which the axis of a feature must be coaxial to a specified datum regardless of the datum's and the feature's size.

Cylindricity.   A tolerance that simultaneously controls the surface of revolution for straightness, parallelism, and circularity of a feature, and is independent of any other features on a part.

Datums.   Reference points, lines, planes, cylinders, and axes which are assumed to be exact and perfect.

Datum Feature.   A feature which is used to establish a datum.

Datum Identification Symbol.   A special rectangular box which contains the datum reference letter.

Feature.   A component of a part such as a hole, slot, surface, or boss.

Fit.   A term used to describe the range of assembly that results from tolerances on two mating parts.

Flatness.   A tolerance that controls the amount of variation from the perfect plane on a feature independent of any other features on the part.

Form Tolerance. A tolerance that specifies the allowable variation of a feature from perfect form.

Least Material Condition (LMC). A condition of a feature in which it contains the least amount of material.

Limit Dimensions. A tolerancing method showing only the maximum and minimum dimensions which establish the limits.

Limits. The maximum and minimum allowable sizes of a feature.

Location Tolerance. A tolerance which specifies the allowable variation from the perfect location.

Maximum Material Condition (MMC). A condition in which the feature contains the maximum amount of material.

Modifier. The application of MMC or RFS to alter the normally implied interpretation of a tolerance specification.

Parallelism. A tolerance that controls interdependent surfaces and axes which must be equal distance from a datum plane or axis.

Perpendicularity. A tolerance that controls surfaces and axes which must be at right angles with a datum axis.

Position Tolerance. A tolerance that controls the position of a feature as related to a datum or datums.

Profile of a Line. A tolerance that controls the allowable variation along an elemental tolerance zone with regard to a basic profile.

Profile of a Surface. A tolerance that controls the allowable variation of a surface form a basic profile or configuration.

Projected Tolerance Zone. A tolerance zone that applies to the location of an axis above the surface of the feature being controlled.

Reference Dimension. A non-toleranced dimension used for information purposes only which does not govern production or inspection operations.

Regardless of Feature Size (RFS). A condition of a tolerance in which the tolerance must be met regardless of the size of the feature.

Runout. The composite variation from the desired form of a part surface of revolution during full rotation of the part on a datum axis.

Size Tolerance. A tolerance that specifies how far individual features may vary from the desired size.

Straightness. A tolerance that controls the allowable variation of a surface or an axis from a theoretically perfect plane or line.

Tolerance. The variation or allowance of a dimension.

Total Runout. A tolerance that simultaneously controls a surface related to a datum in all directions.

Transition Fit. A condition in which two parts are toleranced so that either a clearance or an interference can result when the parts are assembled.

True Position. The theoretically exact location of a feature.

Unilateral Tolerance. A tolerance which allows variations in only one direction.

Virtual Condition. A condition of a feature in which all tolerance of location, size, and form are considered.

General Tolerancing

The industrial revolution created a need for mass production; assembling interchangeable parts on an assembly line to turn out great quantities of a given finished product. Interchangeability of parts was the key. If a particular product was composed of 100 parts, each individual part could be produced in quantity, checked for accuracy, stored, and used as necessary.

Since it was humanly and technologically impossible to have every individual part produced exactly alike (it still is), the concept of geometric and positional tolerancing was introduced. *Tolerancing* means setting acceptable limits of deviation. For example, if a mass produced part is to be 4″ in length under ideal conditions, but is acceptable as long as it is not less than 3.99″ and not longer than 4.01″, there is a tolerance of plus or minus .01″, *Figure 12-1*. This type of tolerance is called a *size tolerance*.

There are three different types of size tolerances: unilateral and bilateral, shown in *Figure 12-2*, and limit dimensioning. When a *unilateral tolerance* is applied

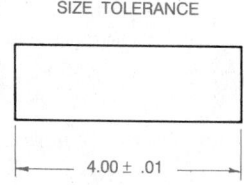

SIZE TOLERANCE

4.00 ± .01

Figure 12-1 Size tolerance

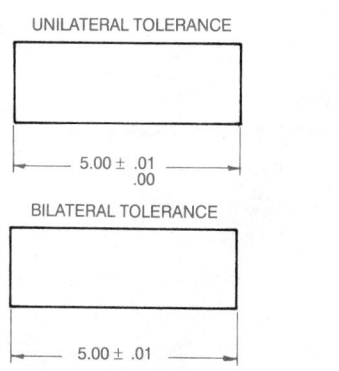

UNILATERAL TOLERANCE

5.00 ± .01
.00

BILATERAL TOLERANCE

5.00 ± .01

Figure 12-2 Two types of tolerances

to a dimension, the tolerance applies in one direction only (for example, the object may be larger but not smaller, or it may be smaller but not larger). When a *bilateral tolerance* is applied to a dimension, the tolerance applies in both directions, but not necessarily evenly distributed. In *limit dimensioning*, the high limit is placed above the low value. When placed in a single line, the low limit precedes the high limit and the two are separated by a dash.

Tolerancing size dimensions offers a number of advantages. It allows for acceptable error without compromises in design, cuts down on unacceptable parts, decreases manufacturing time, and makes the product less expensive to produce. However, it soon became apparent that in spite of advantages gained from size tolerances, tolerancing only the size of an object was not enough. Other characteristics of objects also needed to be toleranced, such as location of features, orientation, form, runout, and profile.

In order for parts to be acceptable, depending on their use, they need to be straight, round, cylindrical, flat, angular, and so forth. This concept is illustrated in *Figure 12-3*. The object depicted is a shaft that is to be manufactured to within plus or minus .01 of 1.00 inch in diameter. The finished product meets the size specifications but, since it is not straight, the part might be rejected.

The need to tolerance more than just the size of objects led to the development of a more precise system of tolerancing called geometric dimensioning and positional tolerancing. This new practice improved on conventional tolerancing significantly by allowing designers to tolerance size, form, orientation, profile, location, and runout, *Figure 12-4*. In turn, these are the characteristics that make it possible to achieve a high degree of interchangeability.

Geometric Dimensioning and Positional Tolerancing Defined

Geometric dimensioning and positional tolerancing is a dimensioning practice which allows designers to set

FOR INDIVIDUAL FEATURES	FORM
FOR INDIVIDUAL RELATED FEATURES	PROFILE
FOR RELATED FEATURES	ORIENTATION LOCATION RUNOUT

Figure 12-4   Types of tolerances

tolerance limits not just for the size of an object, but for all of the various critical characteristics of a part. In applying geometric dimensioning and tolerancing to a part, the designer must examine it in terms of its function and its relationship to mating parts.

Figure 12-5 is an example of a drawing of an object that has been geometrically dimensioned and toleranced. It is taken from the dimensioning standards manual produced by the American National Standards Institute or ANSI Y14.5M-R1988. This manual is a necessary reference for drafters and designers involved in geometric dimensioning and positional tolerancing.

The key to learning geometric dimensioning and positional tolerancing is to learn the various building blocks which make up the system, as well as how to properly apply them. *Figure 12-6* contains a chart of the building blocks of the geometric dimensioning and tolerancing system. In addition to the standard building blocks shown in the figure, several modifying symbols are used when applying geometric tolerancing, as discussed in detail in upcoming paragraphs.

Another concept that must be understood in order to effectively apply geometric tolerancing is the concept of datums. For skilled, experienced designers, the geometric building blocks, modifiers, and datums blend together as a single concept. However, for the purpose of learning, they are dealt with separately, and undertaken step-by-step as individual concepts. They are presented now in the following order: modifiers, datums, and geometric building blocks.

ANSI's dimensioning standards manual (Y14.5 series) changes from time to time as standards are updated. For example, the Y14.5 manual became Y14.5M in 1982 to accommodate metric dimensioning. Revised again in 1988, it became Y14.5M-R1988. The process is on-going. This chapter helps students learn the basics of geometric dimensioning and positional tolerancing so they will be able to apply latest standards set forth by ANSI at any point in time and in accordance with any edition of the manual that is specified. Students should not use this chapter as a reference in place of the ASNI standard. Always refer to the latest edition of the standard for specifics that go beyond the basics covered herein.

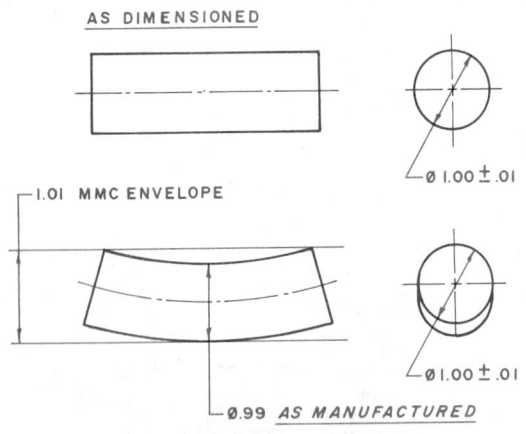

AS DIMENSIONED

1.01 MMC ENVELOPE

Ø 1.00 ± .01

Ø 1.00 ± .01

0.99 *AS MANUFACTURED*

Figure 12-3   Tolerance of form

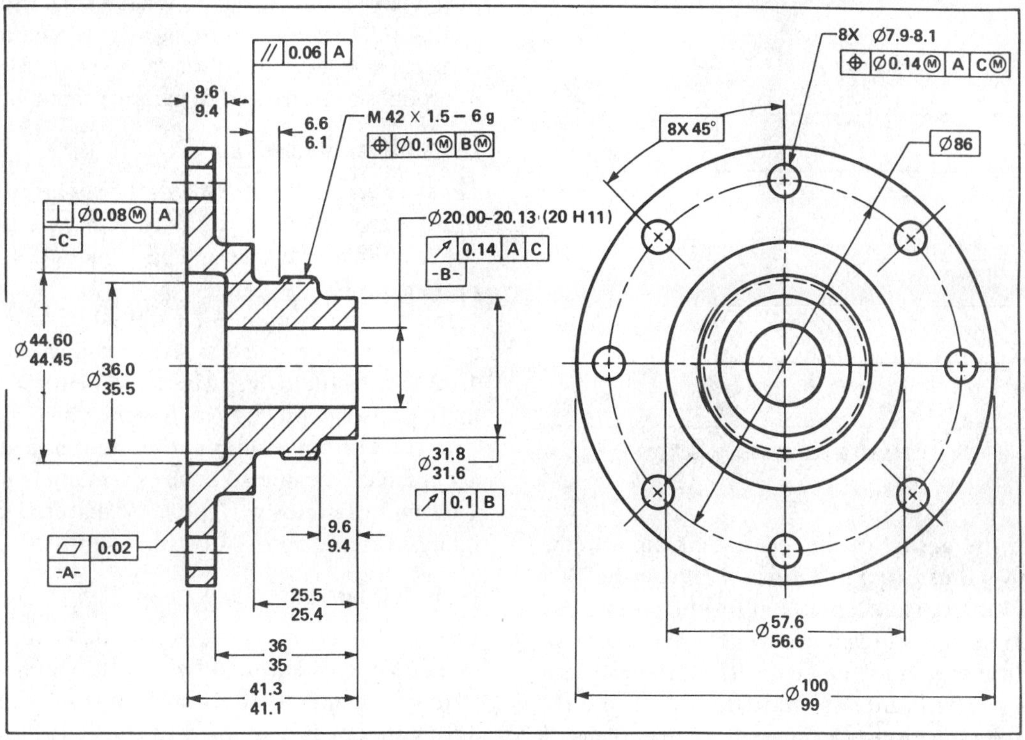

Figure 12-5 Geometrically dimensioned and toleranced drawing *(From ANSI Y14.5M-R 1988)*

SYMBOL	CHARACTERISTIC	GEOMETRIC TOLERANCE
—	STRAIGHTNESS	FORM
⏥	FLATNESS	
○	CIRCULARITY	
⌭	CYLINDRICITY	
⌒	PROFILE OF A LINE	PROFILE
⌓	PROFILE OF A SURFACE	
∠	ANGULARITY	ORIENTATION
⊥	PERPENDICULARITY	
//	PARALLELISM	
⌖	TRUE POSITION	LOCATION
◎	CONCENTRICITY	
⌯	SYMMETRY	
↗	CIRCULAR RUNOUT	RUNOUT
⌰	TOTAL RUNOUT	

THE SYMBOL FOR SYMMETRY IS SOMETIMES ⌖

Figure 12-6 Building blocks

Ⓜ	MAXIMUM MATERIAL CONDITION
Ⓛ	LEAST MATERIAL CONDITION
Ⓢ	REGARDLESS OF FEATURE SIZE
Ⓟ	PROJECTED TOLERANCE ZONE

Figure 12-7 Modifiers used when applying geometric tolerancing

Modifiers

Modifiers are symbols that can be attached to the standard geometric building blocks to alter their application or interpretation. The proper use of modifiers is fundamental to effective geometric tolerancing. Four modifiers are frequently used: maximum material condition, regardless of feature size, least material condition, and projected tolerance zone, ***Figure 12-7***.

Maximum Material Condition

Maximum material condition, abbreviated MMC, is the condition of a characteristic when the most material exists. For example, MMC of the external feature in ***Figure 12-8*** is .77 inch. This is the MMC because it represents the condition where the most material exists on the part being manufactured. The MMC of the internal feature in the figure is .73 inch. This is the MMC because the most material exists when the hole is produced at the smallest allowable size.

In using this concept, the designer must remember that the MMC of an internal feature is the smallest allowable size. The MMC of an external feature is the largest allowable size within specified tolerance limits inclusive. A rule of thumb to remember is that MMC means *most material*.

Regardless of Feature Size

Regardless of feature size, abbreviated RFS, is a modifier which tells machinists that a tolerance of form or position or any characteristic must be maintained,

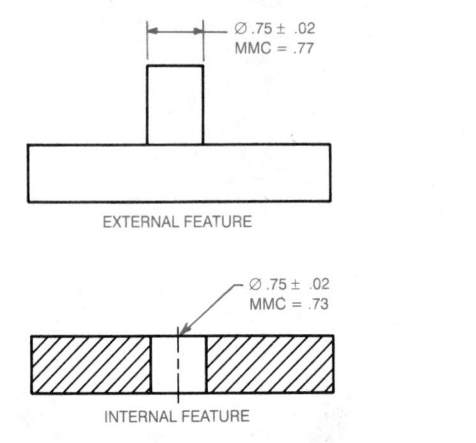

Figure 12-8 MMC of an external and an internal feature

regardless of the actual produced size of the object. This concept is illustrated in ***Figure 12-9***. In the RFS example, the object is acceptable if produced in sizes from 1.01 inches to .99 inch inclusive. The form control is axis straightness to a tolerance of .02 inch regardless of feature size. This means that the .02 inch axis straightness tolerance must be adhered to, regardless of the produced size of the part.

Contrast this with the MMC example. In this case, the produced sizes are still 1.01 inches, 1.00 inch, and .99 inch. However, because of the MMC modifier, the .02 inch axis straightness tolerance applies only at MMC or 1.01 inches.

If the produced size is smaller, the straightness tolerance can be increased proportionally. Of course, this makes the MMC modifier more popular with machinists for several reasons: 1) it allows them greater room for error without actually increasing the tolerance, 2) it decreases the number of parts rejected, 3) it cuts down on unacceptable parts, 4) it decreases the number of

inspections required, and 5) it allows the use of functional gaging. All of these advantages translate into substantial financial savings while, at the same time, making it possible to produce interchangeable parts at minimum expense.

Least Material Condition

Least material condition, abbreviated LMC, is the opposite of MMC. It refers to the condition in which the least material exists. This concept is illustrated in ***Figure 12-10***.

In the top example, the external feature of the part is acceptable if produced in sizes ranging from .98 inch to 1.02 inches inclusive. The least material exists at .98 inch. Consequently, .98 inch is the LMC.

In the bottom example, the internal feature (hole) is acceptable if produced in sizes ranging from .98 inch to 1.02 inches inclusive. The least material exists at 1.02 inches. Consequently, 1.02 represents the LMC.

Projected Tolerance Zone

Projected tolerance zone is a modifier that allows a tolerance zone established by a locational tolerance to be extended a specified distance beyond a given surface. This concept is discussed further later in this chapter under the heading "True Position."

Datums

Datums are points, lines, axes, surfaces or planes used for referencing features of an object. Datums are identified on drawings by datum "flags" or symbols, ***Figure 12-11***. This figure shows several ways in which datum feature symbols are placed on drawings.

In Part A of the figure, the datum symbol is attached to an extension line. In this example, DATUM A would be the plane upon which the part is resting. In Parts B, C, and D, the datum symbol is associated with a dimension by a dimension line, a leader line or by appearing under a dimension. In these cases, the symbol applies to the entire feature rather than just one surface as in object A.

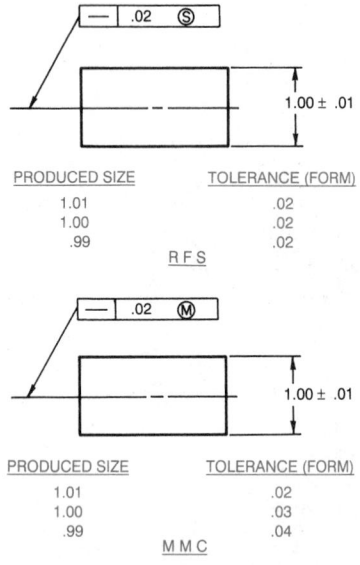

Figure 12-9 Regardless of feature size (RFS)

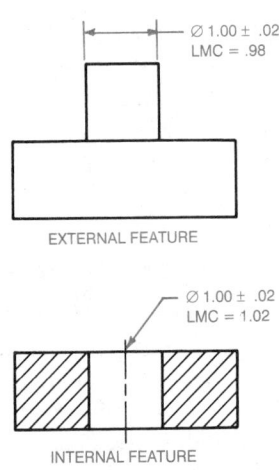

Figure 12-10 Least material condition (LMC)

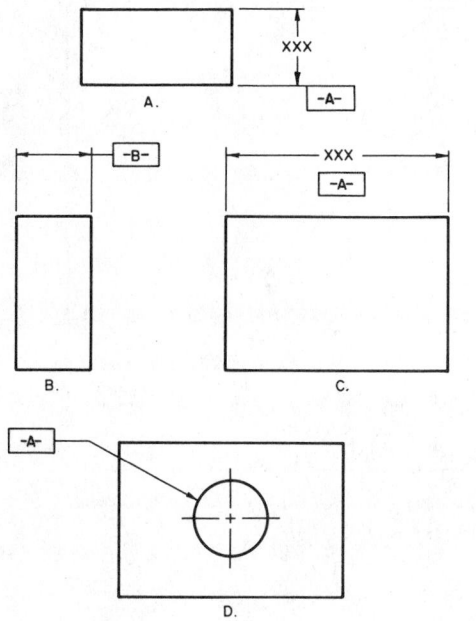

Figure 12-11   Datum flags

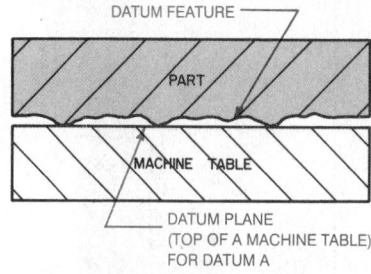

Figure 12-14   Establishing datums

Figure 12-12 shows the top, front, and right-side views of an object on which the front surface has been designated Datum A. Any depth dimension for this object can be referenced to Datum A. **Figure 12-13** shows the proper method for calling out the diameter of a cylindrical object as a datum.

Establishing Datums

In establishing datums, designers must consider the function of the part, the manufacturing processes that will be used in producing the part, how the part will be inspected, and the part's relationship to other parts after assembly. Designers and drafters must also under-

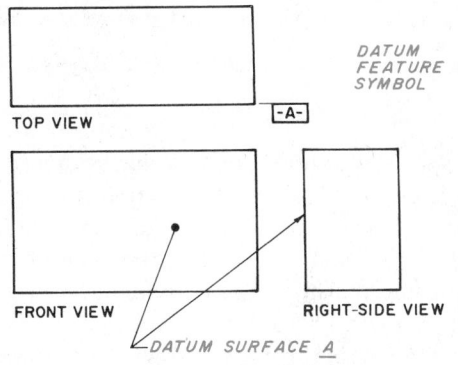

Figure 12-12   Datum callout on a flat object

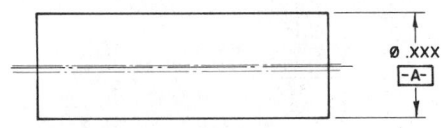

Figure 12-13   Datum callout on a diameter

stand the difference between a datum feature and a datum surface.

A *datum plane* is a theoretically perfect plane from which measurements are made. A *datum surface* is the inexact corresponding surface of the object. This concept is illustrated in **Figure 12-14**.

Notice the irregularities on the datum surface. The high points on the datum surface actually establish the datum plane, which, in this case, is the top of a machine table. All measurements referenced to DATUM A are measured from the theoretically perfect datum plane. High point contact is used for establishing datums when the entire surface in question will be a machined surface. On rougher more irregular surfaces, such as those associated with castings, specified point contacts are used for establishing datums. In these cases, only specified points on surfaces are employed to establish datum features.

When specified point contact is used for establishing datums, a minimum of three points is required for the primary datum, a minimum of two for the secondary, and a minimum of one for the tertiary, **Figure 12-15**. These points are normally located with basic dimensions and identified with datum target symbols, **Figure 12-16**. A *basic dimension* is a theoretically exact dimension enclosed in a rectangular box.

In Figure 12-16, DATUM A is the top of the object and it is established by points A1, A2, and A3. The characters in each datum target symbol identify the datum by letter and target numbers. DATUM B is the front of the object and DATUM C is the right side.

Notice that a datum is called out on a drawing in the view where the surface in question appears as an edge. Notice also that the secondary datum must be perpen-

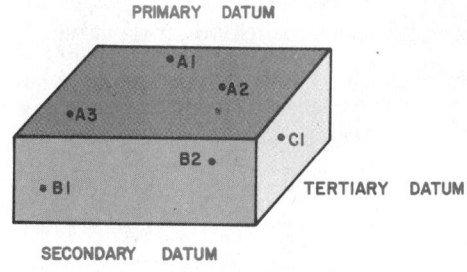

Figure 12-15   Specified point contact for establishing datums

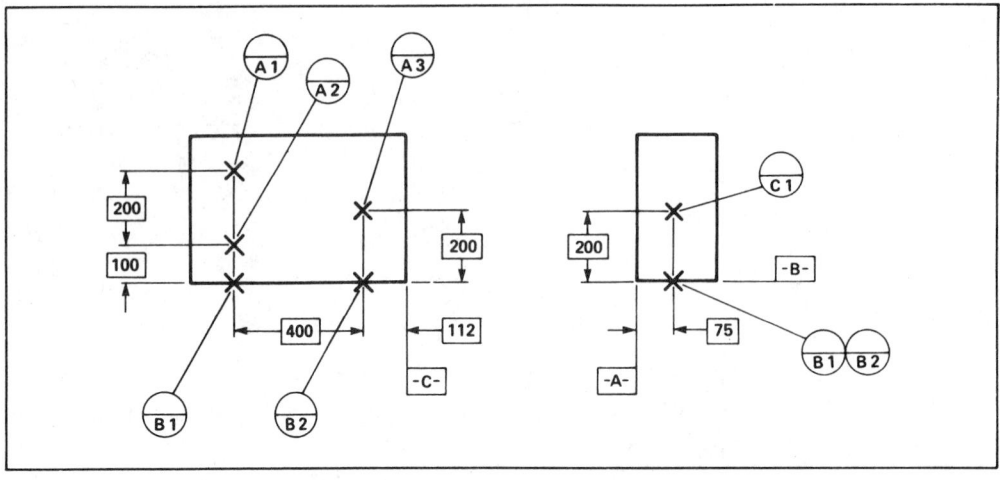

Figure 12-16   Datum target symbols with basic dimensions *(From ANSI Y14.5M-R 1988)*

dicular to the first, and the tertiary datum must be perpendicular to both the primary and secondary datums. These three datums establish what is called the datum frame. The *datum frame* is a hypothetical three-dimensional box into which the object being produced fits and from which measurements can be made.

Figure 12-17 is an example of a "basic dimension." A basic dimension is a theoretically perfect dimension, much like a nominal or design dimension, that is used to locate or specify the size of a feature. Basic dimensions are enclosed in rectangular boxes, as shown in Figure 12-17.

Feature Control Symbol

The *feature control symbol* is a rectangular box in which all data referring to the subject feature control are placed, including: the symbol, datum references, the feature control tolerance, and modifiers. These various feature control elements are separated by vertical lines. (Figure 12-5 contains a drawing showing how feature control symbols are actually composed.)

The order of the data contained in a feature control frame is important. The first element is the feature control symbol. Next is the zone descriptor, such as a diameter symbol where applicable. Then, there is the feature control tolerance, modifiers when used, and datum references listed in order from left to right, *Figure 12-18*.

Figures 12-19 through 12-23 illustrate how feature control symbols are developed for a variety of design

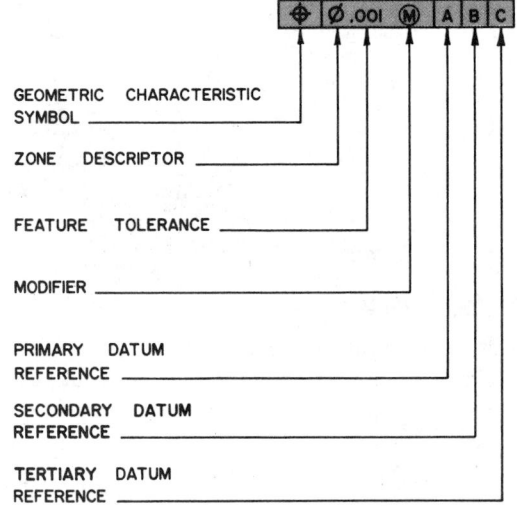

Figure 12-18   Order of elements in a feature control symbol

situations. Figure 12-19 is a feature control symbol which specifies a .005 tolerance for symmetry and no datum reference. Figure 12-20 specifies a tolerance of .005 for the true position of a feature relative to Datum A. Figures 12-21 and 12-22 show the proper methods for constructing feature control symbols with two and three datum references, respectively. Figure 12-23 illustrates a feature control symbol with a modifier and a controlled datum added.

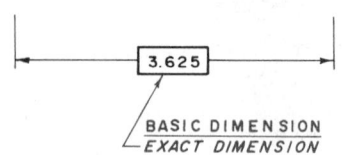

Figure 12-17   Basic dimensions

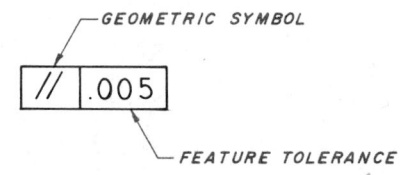

Figure 12-19   Feature control symbol with no datum reference

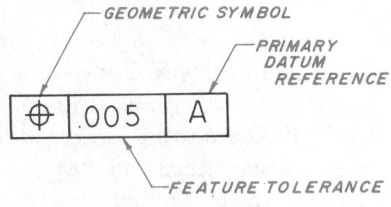

Figure 12-20   Feature control symbol with one datum reference

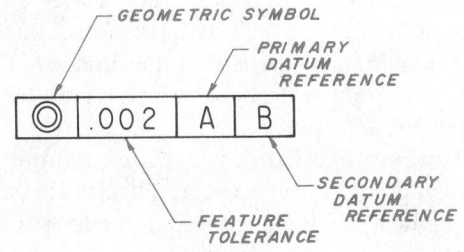

Figure 12-21   Feature control symbol with two datum references

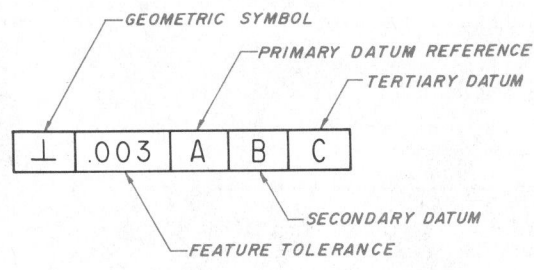

Figure 12-22   Feature control symbol with three datum references

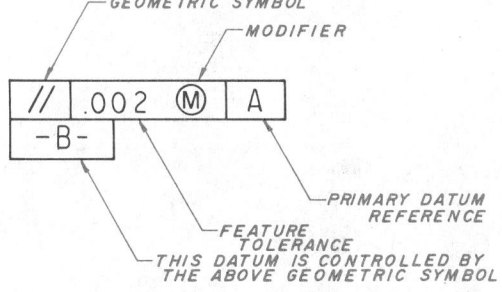

Figure 12-23   Feature control symbol with a modifier

True Position

True position is the theoretically exact location of the center line of a product feature such as a hole. The tolerance zone created by a position tolerance is an imaginary cylinder, the diameter of which is equal to the stated position tolerance. The dimensions used to locate a feature, that is to have a position tolerance, must be basic dimensions.

Figure 12-24 contains an example of a part with two holes drilled through it. The holes have a position toler-

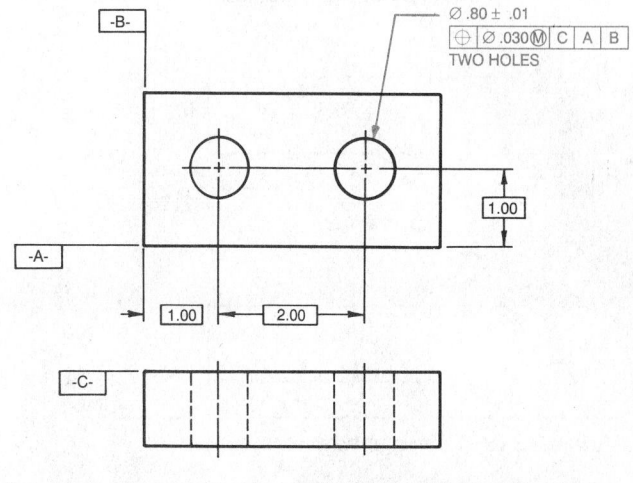

Figure 12-24   True position

ance relative to three datums: A, B, and C. The holes are located by basic dimensions. The feature control frame states that the positions of the center lines of the holes must fall within cylindrical tolerance zones having diameters of .030 inch at MMC relative to DATUMS A, B, and C. The modifier indicates that the .030 inch tolerance applies only at MMC. As the holes are produced larger than MMC, the diameter of the tolerance zones can be increased correspondingly.

Figure 12-25 illustrates the concept of the cylindrical tolerance zone from Figure 12-24. The feature control frame is repeated showing a .030 inch diameter tolerance zone. The broken-out section of the object from Figure 12-24 provides the interpretation. The cylindrical tolerance zone is shown in phantom lines. The center line of the hole is acceptable as long as it falls anywhere within the hypothetical cylinder.

Using the Projected Tolerance Zone Modifier

As stated previously, a *projected tolerance zone* is a tolerance zone that extends beyond the surface of a part. When a projected tolerance zone modifier is used, the surface in question must be identified as a datum and the length of the projected tolerance zone specified.

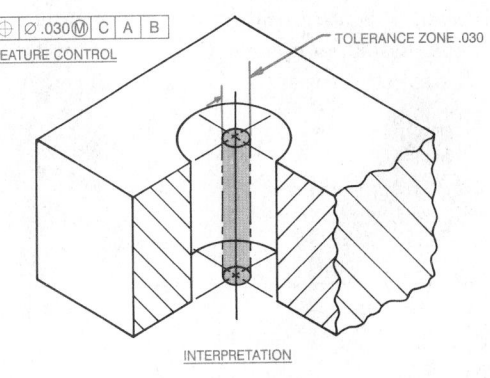

Figure 12-25   Cylindrical tolerance zone

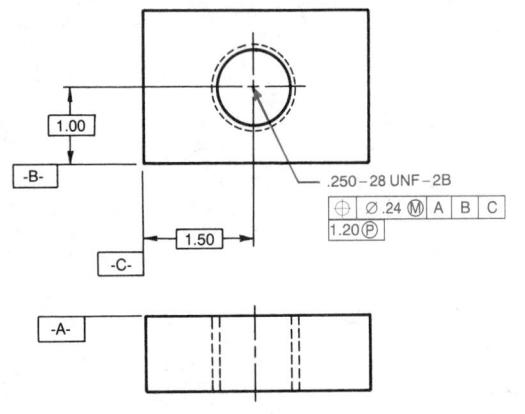

Figure 12-26 Projected tolerance zone

The tolerance zone illustrated in ***Figure 12-26*** has a depth equal to that of the part through which it extends. On occasion, there is a need to extend a tolerance zone beyond the surface of a part. ANSI recommends the use of the projected tolerance zone concept when the variation in perpendiculars of threaded or press-fit holes could cause fasteners, such as screws, studs, or pins, to interfere with mating parts. In these cases, the projected tolerance zone concept is used.

Flatness

Flatness is a feature control of a surface which requires all elements of the surface to lie within two hypothetical parallel planes. When flatness is the feature control, a datum reference is neither required nor proper.

Figure 12-27 shows how flatness is called out in a drawing and the effect such a callout has on the produced part. The surface indicated must be flat within a tolerance zone of .002, as shown in Figure 12-27 (right).

Flatness is specified when size tolerances alone are not sufficient to control the form and quality of the surface and when a surface must be flat enough to provide a stable base or a smooth interface with a mating part.

Flatness is inspected for a full indicator movement (FIM) using a dial indicator. FIM is the newer term which has replaced the older "total indicator movement" or TIR. FIM means one full swing of the indicator needle. The dial indicator is set to run parallel to a surface table which is a theoretically perfect surface. The dial indicator is mounted on a stand or height gauge. The machined surface is run under it, allowing the dial indicator to detect irregularities that fall outside of the tolerance zone.

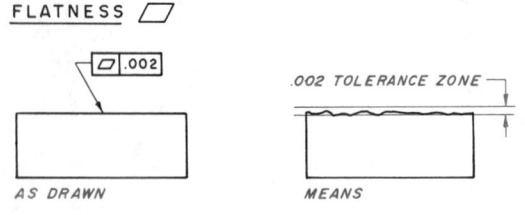

Figure 12-27 Flatness

Straightness

Straightness is a feature control of one single element of a surface that must be a straight line to within the stated tolerance. It differs from flatness in that flatness covers an entire surface rather than just one element on a surface. A straightness tolerance yields a tolerance zone of specified width between which all points on the line in question must lie. Straightness is generally applied to longitudinal elements.

Figure 12-28 shows how a straightness tolerance is applied on a drawing. The feature control frame states that any longitudinal element of the surface in the direction indicated must lie between two parallel straight lines that are .002 inch apart.

Straightness, like flatness, does not require a datum reference. Straightness is not additive to the size tolerance and must be contained within the limits of the size tolerance.

Figure 12-29 illustrates the relationship between a straightness tolerance and a size tolerance of a part.

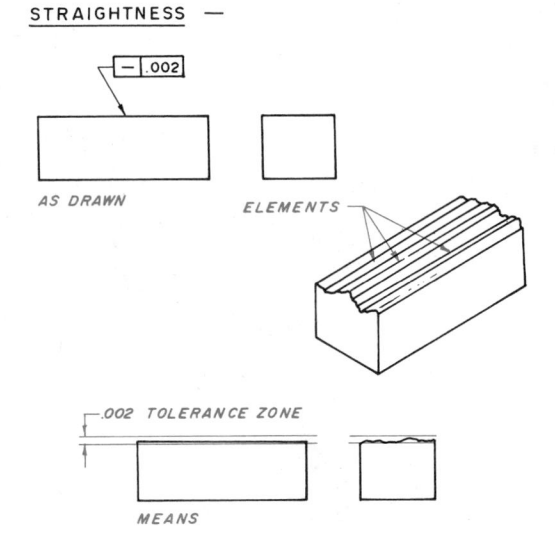

Figure 12-28 Straightness

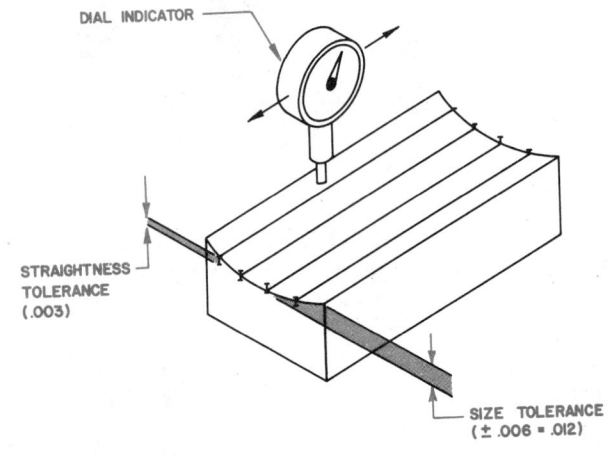

Figure 12-29 Straightness interpreted

Each element surface must stay within the specified straightness tolerance zone and within the size tolerance envelope. Straightness is effected by running the single-line elements of a surface under a dial indicator for a full indicator movement (FIM).

Figures 12-30 through 12-33 further illustrate the concept of straightness. Figure 12-30 shows a part with a size tolerance, but no feature control tolerance. For each acceptable size in which the part can be produced,

it must be straight. Figure 12-31 is the same part with a straightness tolerance of .002 at maximum material condition applied. The drawing at the top of the figure illustrates how the part would be drawn. The five illustrations below the part as drawn illustrate the actual

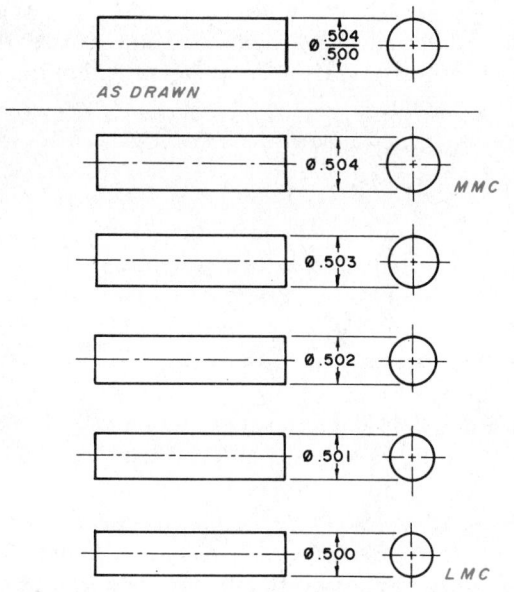

Figure 12-30 Object with no feature control symbol

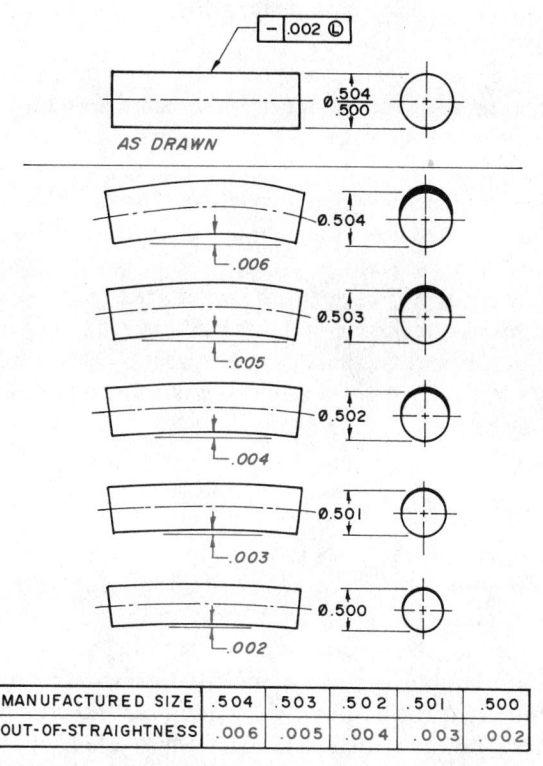

MANUFACTURED SIZE	.504	.503	.502	.501	.500
OUT-OF-STRAIGHTNESS	.006	.005	.004	.003	.002

Figure 12-32 Straightness at LMC

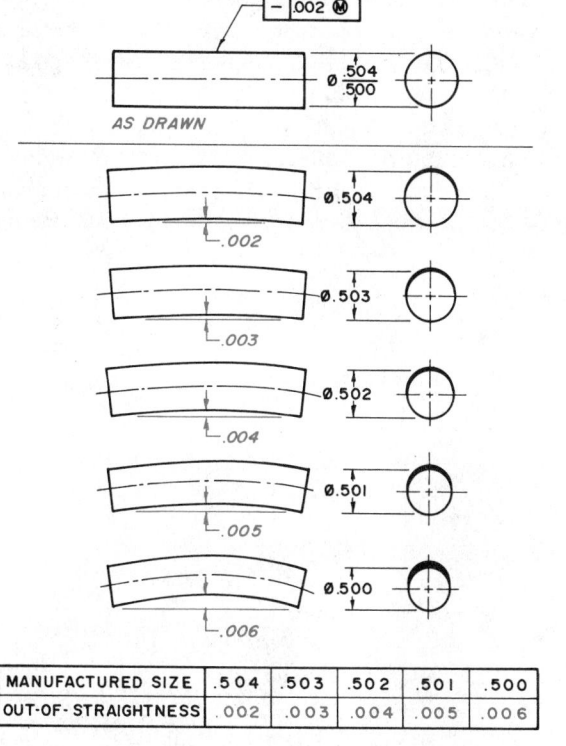

MANUFACTURED SIZE	.504	.503	.502	.501	.500
OUT-OF-STRAIGHTNESS	.002	.003	.004	.005	.006

Figure 12-31 Straightness at MMC

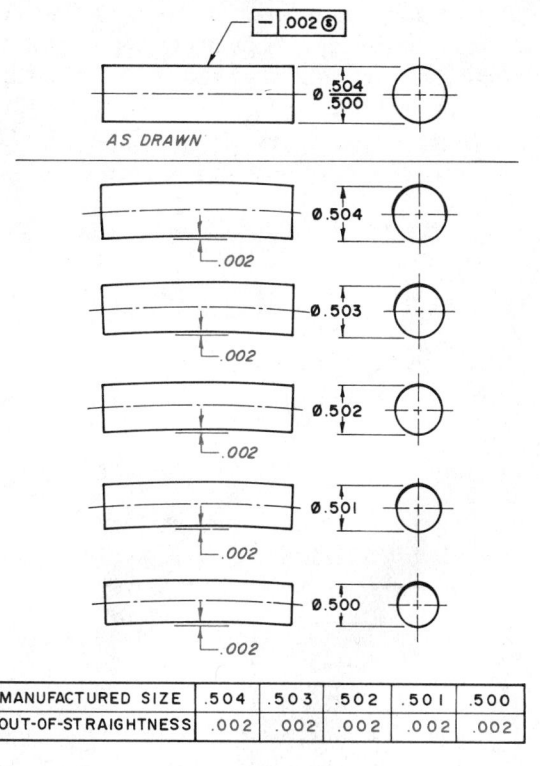

MANUFACTURED SIZE	.504	.503	.502	.501	.500
OUT-OF-STRAIGHTNESS	.002	.002	.002	.002	.002

Figure 12-33 Straightness at RFS

shape of the object with each corresponding produced size. Since the .002 straightness tolerance applies at maximum material condition, the amount that the part can be out of straightness increases correspondingly as the produced size decreases. The table at the bottom Figure 12-31 summarizes the manufactured sizes and the corresponding amounts that the part can be out of straightness for each side.

Figure 12-32 is an example of the same part with a .002 straightness tolerance at least material condition. This results in the opposite effect of what occurred in Figure 12-31. Notice that the .002 straightness tolerance applies at the least material condition. As the actual produced size increases, the amount of out of straightness allowed increases correspondingly.

Figure 12-33 illustrates the same part from a .002 straightness tolerance and a regardless of feature size modifier. Notice in this example that the .002 straightness tolerance applies regardless of the actual produced size of the part.

Circularity (Roundness)

Circularity, sometimes referred to as *roundness*, is a feature control for a surface of revolution (cylinder, sphere, cone, and so forth). It specifies that all points of a surface must be equidistant from the center line or axis of the object in question. The tolerance zone for circularity is formed by two concentric and coplanar circles between which all points on the surface of revolution must lie.

Figure 12-34 illustrates how circularity is called-out on a drawing and provides an interpretation of what the circularity tolerance actually means. At any selected cross section of the part, all points on the surface must fall within the zone created by the two concentric circles. At any point where circularity is measured, it must fall within the size tolerance. Notice that a circularity tolerance cannot specify a datum reference.

Circularity establishes elemental single-line tolerance zones that may be located anywhere along a surface. The tolerance zones are taken at any cross section of the feature. Therefore, the object may be spherical, cylindrical, tapered, or even hourglass shaped so long as the cross-section for inspection is taken at 90° to the nominal axis of the object. A circularity tolerance is inspected using a dial indicator and making readings relative to the axis of the feature. In measuring a circularity tolerance, the full indicator movement (FIM) of the dial indicator should not be any larger than the size tolerance, and there should be several measurements made at different points along the surface of the diameter. All measurements taken must fall within the circularity tolerance.

Cylindricity

Cylindricity is a feature control in which all elements of a surface of revolution form a cylinder. It gives the effect of circularity extended the entire length of the object, rather than just a specified cross section. The tolerance zone is formed by two hypothetical concentric cylinders.

Figure 12-35 illustrates how cylindricity is called-out on a drawing. Notice that a cylindricity tolerance does not require a datum reference.

Figure 12-35 also provides an illustration of what the cylindricity tolerance actually means. Two hypothetical concentric cylinders form the tolerance zone. The outside cylinder is established by the outer limits of the object at its produced size within specified size limits. The inner cylinder is smaller (on radius) by a distance equal to the cylindricity tolerance.

Cylindricity requires that all elements on the surface fall within the size tolerance and the tolerance established by the feature control.

A cylindricity tolerance must be less than the size tolerance and is not additive to the maximum material condition of the feature. Cylindricity is inspected by passing the tolerance object through a gauge. The

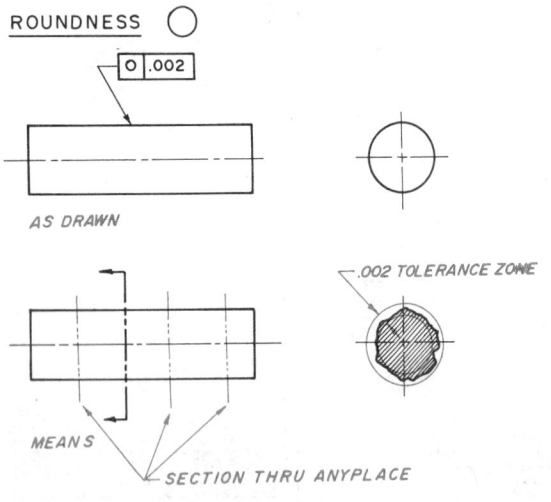

Figure 12-34 Circularity (roundness)

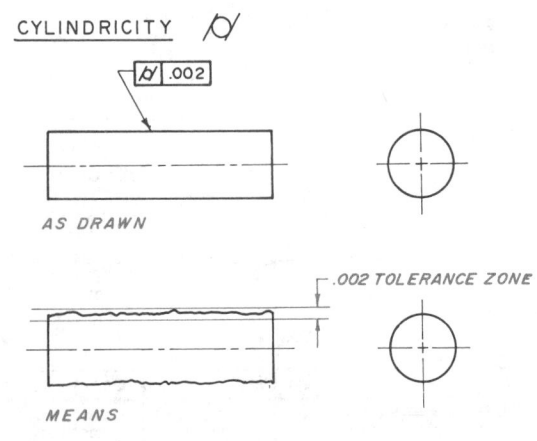

Figure 12-35 Cylindricity

object should pass through a gauge that is equal to or greater than the diameter of the external envelope establishing the cylindrical tolerance zone. It should not pass through a gauge that is slightly smaller than the internal envelope, which establishes the cylindrical tolerance zone.

Angularity

Angularity is a feature control in which a given surface, axis, or center plane must form a specified angle with a datum other than 90°. Consequently, an angularity tolerance requires a datum reference. The tolerance zone formed by an angularity callout consists of two hypothetical parallel planes which form the specified angle with the datum. All points on the angular surface or along the angular axis must lie between these parallel planes.

Figure 12-36 illustrates how an angularity tolerance is called out on a drawing. Notice that the specified angle is enclosed in a BASIC box. This is required when applying an angularity tolerance. Figure 12-36 also provides an interpretation of what the angularity tolerance actually means. Notice that the outside plane of the tolerance zone is established by the outermost extremities of the angular surface. The inner plane is measured in a distance equal to the angularity tolerance. All elements of the toleranced surface must lie within the size tolerance limits.

Parallelism

Parallelism is a feature control that specifies that all points on a surface, center plane, axis or line must be equidistant from a datum. Consequently, a parallelism tolerance must have a datum reference. A parallelism tolerance zone is formed by two hypothetical planes that are parallel to a specified datum. They are spaced apart at a distance equal to the parallelism tolerance.

Figure 12-37 illustrates how a parallelism is called-out on a drawing and provides an interpretation of what the parallelism tolerance actually means. Notice that all elements of the toleranced surface must fall within the size limits.

Notice in Figure 12-37 that the .002 parallelism tolerance is called out relative to Datum A. You must specify a datum when calling out a parallelism tolerance.

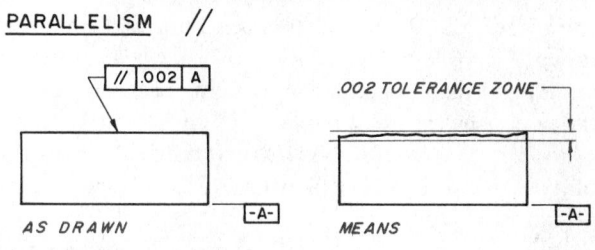

Figure 12-37 Parallelism

Parallelism should be specified when features such as surfaces, axes, and planes are required to lie in a common orientation. Parallelism is inspected by placing the part on an inspection table and running a dial indicator a full indicator movement across the surface of the part.

Perpendicularity

Perpendicularity is a feature control which specifies that all elements of a surface, axis, center plane, or line form a 90° angle with a datum. Consequently, a perpendicularity tolerance requires a datum reference. A perpendicularity tolerance is formed by two hypothetical parallel planes.

Figure 12-38 illustrates how a perpendicularity tolerance is called-out on a drawing and provides an interpretation of what the perpendicularity tolerance actually means. The elements of the toleranced surface must fall within the size limits and between two hypothetical parallel planes that are a distance apart equal to the perpendicularity tolerance.

The perpendicularity of a part such as the one shown in Figure 12-38 could be inspected by clamping the part to an inspection angle. The datum surface should rest against the inspection angle. Then a dial indicator should be passed over the entire surface for a full indicator movement to determine if the perpendicularity tolerance has been complied with.

Profile

Profile is a feature control that specifies the amount of allowance variance of a surface or line elements on a surface. There are three different variations of the pro-

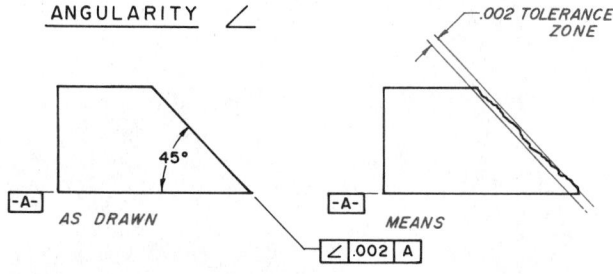

Figure 12-36 Angularity

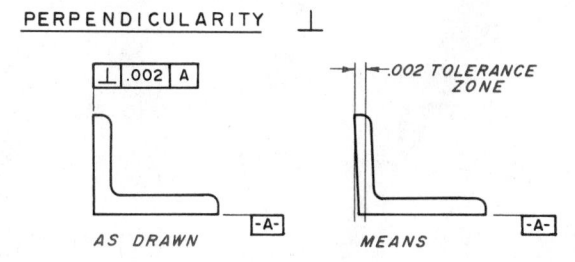

Figure 12-38 Perpendicularity

file tolerance: bilateral, unilateral up, and unilateral down, *Figures 12-39*, *12-40* and *12-41*. A profile tolerance is normally used for controlling arcs, curves, and other unusual profiles not covered by the other feature controls. It is a valuable feature control for use on objects that are so irregular that other feature controls do not easily apply.

When applying a profile tolerance, the symbol used indicates whether the designer intends profile of a line or profile of a surface, *Figures 12-42* and *12-43*. *Profile of a line* establishes a tolerance for a given single element of a surface. *Profile of a surface* applies to the entire surface. The difference between profile of a line and profile of a surface is similar to the difference between circularity and cylindricity.

When using a profile tolerance, drafters and designers should remember to use phantom lines to indicate whether the tolerance is applied unilaterally up or unilaterally down. A bilateral profile tolerance requires no phantom lines. An ALL AROUND symbol should also be placed on the leader line of the feature control frame to specify whether the tolerance applies ALL AROUND or between specific points on the object, *Figure 12-44*.

Figure 12-45 provides an interpretation of what the BETWEEN A & B profile tolerance in Figure 12-44 actually means. The rounded top surface, and only the top surface, of the object must fall within the specified tolerance zone. *Figure 12-46* provides an interpretation

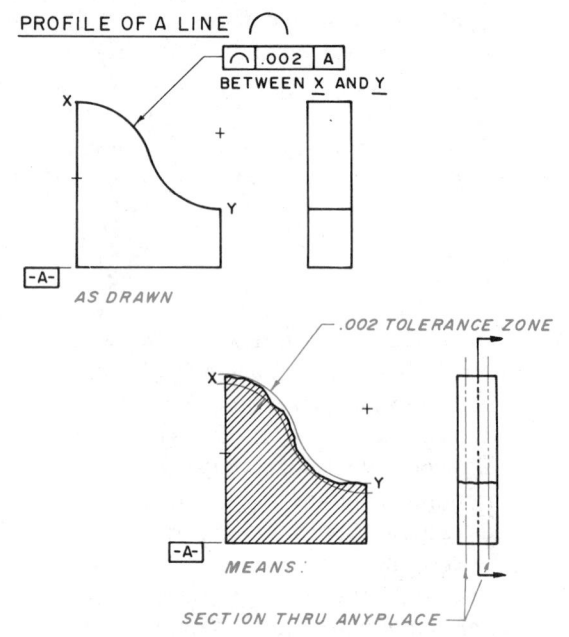

Figure 12-42   Profile of a line

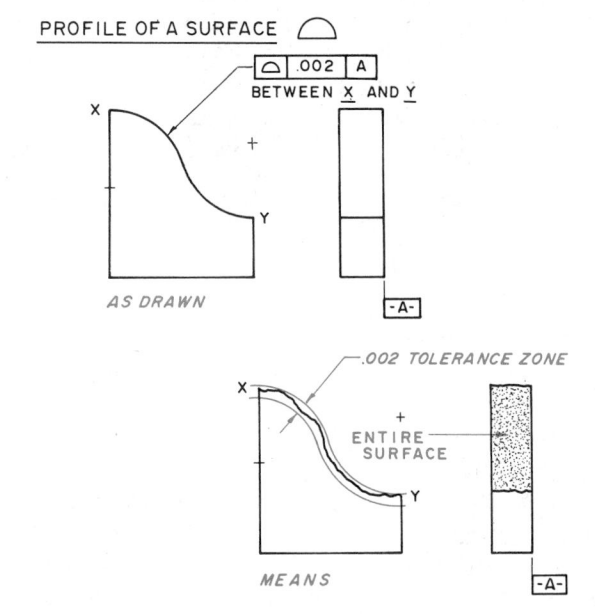

Figure 12-43   Profile of a surface

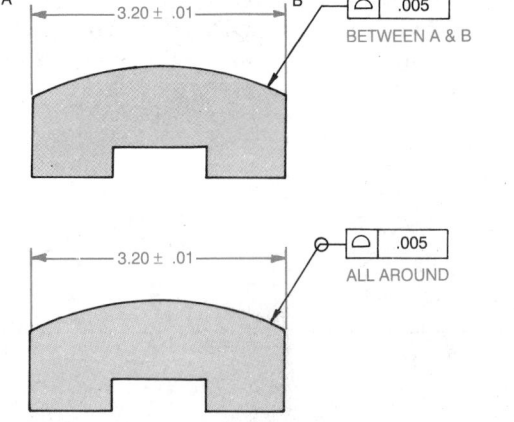

Figure 12-44   Profile "ALL AROUND"

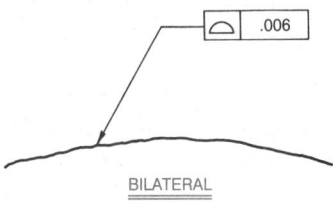

Figure 12-39   Bilateral profile tolerance

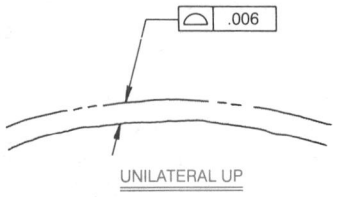

Figure 12-40   Unilateral up profile

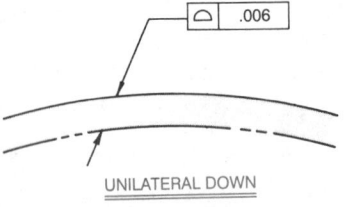

Figure 12-41   Unilateral down profile

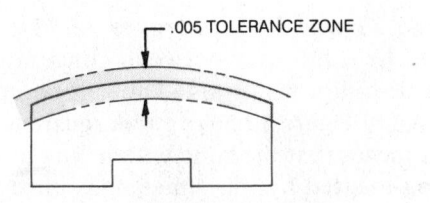

Figure 12-45 Interpretation of "BETWEEN A & B"

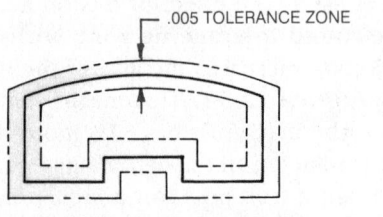

Figure 12-46 Interpretation of "ALL AROUND"

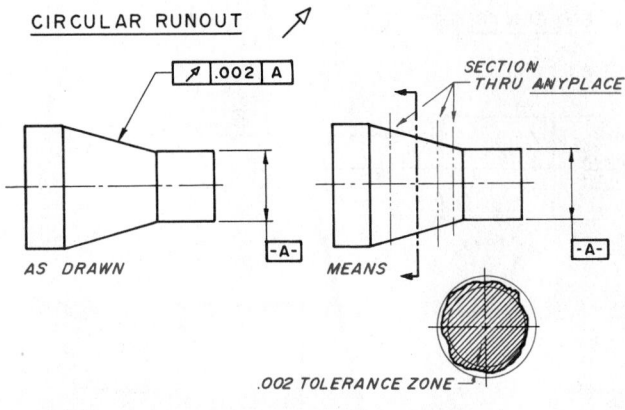

Figure 12-47 Circular runout

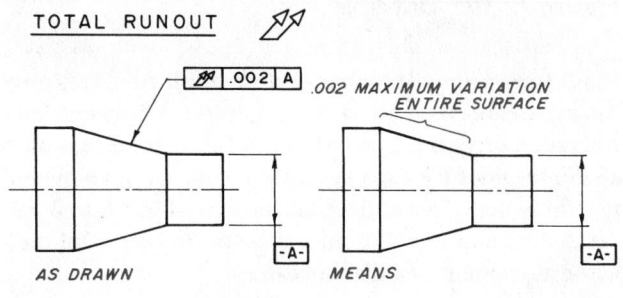

Figure 12-48 Total runout

of what the ALL AROUND profile tolerance in Figure 12-44 actually means. The entire surface of the object, all around the object, must fall within the specified tolerance zone.

Profile tolerances may be inspected using a dial indicator. However, because the tolerance zone must be measured at right angles to the basic true profile and perpendicular to the datum, the dial indicator must be set up to move and read in both directions. Other methods of inspecting profile tolerances are becoming more popular, however. Optical comparators are becoming widely used for inspecting profile tolerances. An optical comparator magnifies the silhouette of the part and projects it onto a screen where it is compared to a calibrated grid or template so that the profile and size tolerances may be inspected visually.

Runout

Runout is a feature control that limits the amount of deviation from perfect form allowed on surfaces or rotation through one full rotation of the object about its axis. Revolution of the object is around a datum axis. Consequently, a runout tolerance does require a datum reference.

Runout is most frequently used on objects consisting of a series of concentric cylinders and other shapes of revolution that have circular cross sections; usually, the types of objects manufactured on lathes, *Figures 12-47* and *12-48*.

Notice in Figures 12-47 and 12-48 that there are two types of runout: circular runout and total runout. The circular runout tolerance applies at any single-line element through which a section passes. The total runout tolerance applies along an entire surface, as illustrated in Figure 12-48. Runout is most frequently used when the actual produced size of the feature is not as important as the form, and the quality of the feature must be related to some other feature. Circular runout is inspected using a dial indicator along a single fixed position so that errors

are read only along a single line. Total runout requires that the dial indicator move in both directions along the entire surface being toleranced.

Concentricity

It is not uncommon in manufacturing to have a part made up of several subparts all sharing the same center line or axis. Such a part is illustrated in *Figure 12-49*. In such a part it is critical that the center line for each subsequent subpart be concentric with the center lines of the other subparts. When this is the case, a concentricity tolerance is applied. A concentricity tolerance locates the axis of a feature relative to the axis of a datum. A concentricity tolerance deals only with the center-line relationship. It does not affect the size, form, or surface quality of the part. Concentricity deals only with axial relationships. Regardless of how large or

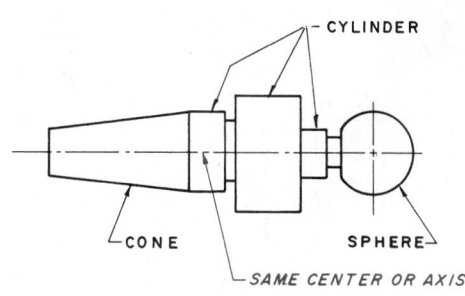

Figure 12-49 Part with concentric subparts

CONCENTRICITY ◎

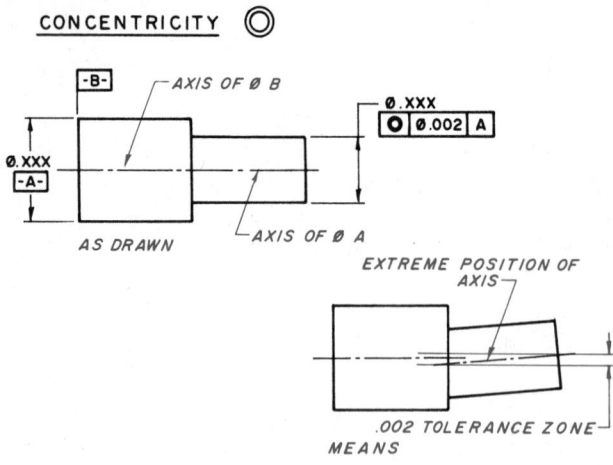

Figure 12-50   Concentricity

small the various subparts of an overall part are, only their axes are required to be concentric. A concentricity tolerance creates a cylindrical tolerance zone in which all center lines for each successive subpart of an overall part must fall. This concept is illustrated in **Figure 12-50**. A concentricity tolerance is inspected by a full indicator movement of a dial indicator.

Symmetry

Parts that are symmetrically disposed about the center plane of a datum feature are common in manufacturing settings. If it is necessary that a feature be located symmetrically with regard to the center plane of a datum feature, a symmetry tolerance may be applied, **Figure 12-51**. The part in Figure 12-51 is symmetrical about a center plane that is perpendicular to Datum A. To ensure that the part is located symmetrically with respect to the center plane, a .002 symmetry tolerance is applied. This creates a .002 tolerance zone within which the center plane in question must fall, as illustrated on the right-hand side of Figure 12-51.

True Positioning

True position tolerancing is used to locate features of parts that are to be assembled and mated. True position is symbolized by a circle overlaid by a large plus sign or cross. This symbol is followed by the tolerance, a modifier when appropriate, and a reference

SYMMETRY ═

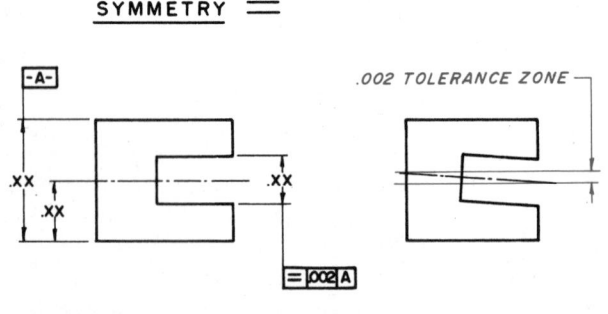

Figure 12-51   Symmetry

datum, **Figure 12-52**. **Figures 12-53** and **12-54** illustrate the difference between conventional and true position dimensioning. The tolerance dimensions shown in Figure 12-53 create a square tolerance zone. This means that the zone within which the center line being located by the dimensions must fall takes the shape of a square. As you can see in Figure 12-54, the tolerancing zone is round when true position dimensioning is used. The effect of this on manufacturing is that the round tolerancing zone with true position dimensioning increases the size of the tolerance zone by 57%, **Figure 12-55**. This means that for the same tolerance the machinist has 57% more room for error without producing an out-of-tolerance part.

When using true position dimensioning, the tolerance is assumed to apply regardless of the feature size unless modified otherwise. **Figure 12-56** illustrates the effect of modifying a true position tolerance with a maximum material condition modifier. In this exam-

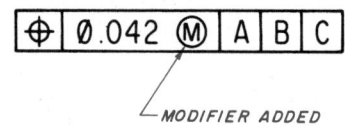

Figure 12-52   True position symbology

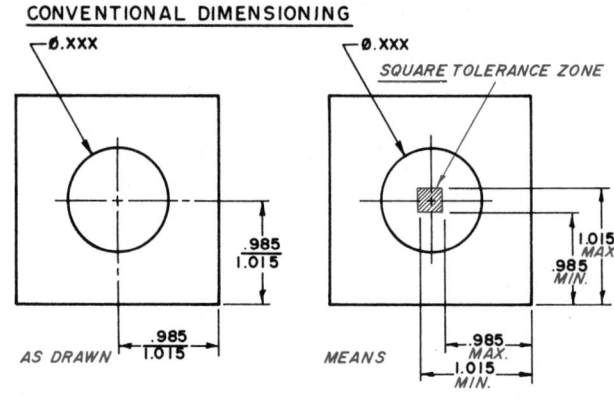

Figure 12-53   Conventional dimensioning

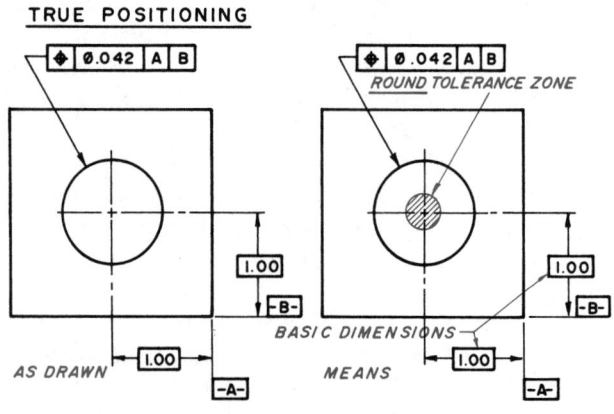

Figure 12-54   True position dimensioning

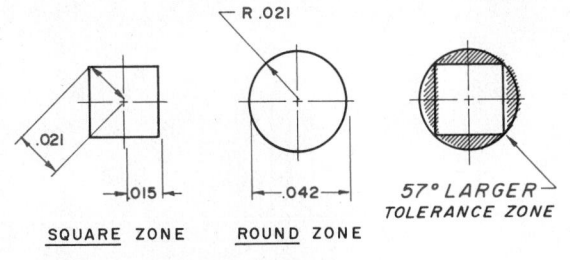

Figure 12-55 Comparison of tolerance zones

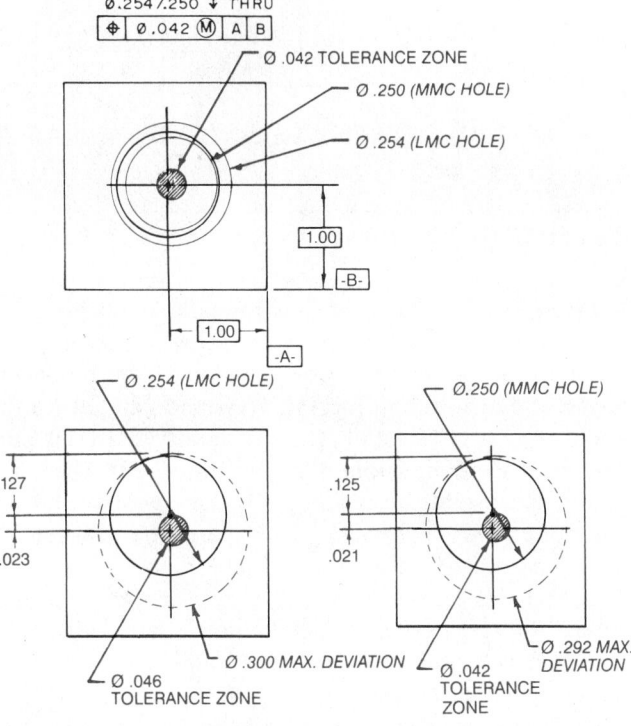

Figure 12-56 True positioning at MMC

ple, a hole is to be drilled through a plate. The maximum diameter is .254 and the minimum diameter is .250. Therefore, the maximum material condition of the part occurs when the hole is drilled to a diameter of .250. Notice from this example that as the hole increases, the positional tolerance increases. At maximum material condition (.250 diameter), the tolerance zone has a diameter of .042. At least material condition (.254 diameter), the tolerance zone increases to .046 diameter. The tolerance zone diameter increases correspondingly as the hole size increases.

Projected Tolerance Zone

Occasionally when working with mating parts it becomes necessary to control the perpendicularity of a surface of a part to ensure ease of assembly. When this is the case, a designer can specify a projected tolerance zone. This means that the tolerance zone is projected above the surface for a specified distance.

PROJECTED TOLERANCE ZONE

Figure 12-57 Projected tolerance zone symbol

Figure 12-57 illustrates the symbol used for specifying a projected tolerance zone.

Figure 12-58 illustrates how a projected tolerance zone symbol is attached to a normal feature control box and what doing so means for manufacturing personnel. In the example in Figure 12-58, a .25 diameter threaded hole is to be placed in a part. The hole is located using true position dimensioning with a positional tolerance of .042 at maximum material condition. The designer has specified that the tolerance zone is to project above the surface of the part for a distance of .50 of an inch. This is illustrated in Figure A. Figures B and C illustrate what such a callout actually means.

Review of Datums

Fundamental to an understanding of geometric dimensioning and tolerancing is an understanding of datums. Since many engineering and drafting students find the concept of datums difficult to understand, this section will review the concept in depth. It is important to understand datums because they represent the starting point for referencing dimensions to various features on parts and for making calculations relative to those dimensions. Datums are usually physical components. However, they can also be invisible lines, planes, axes, or points that are located by calculations or as they relate to other features. Features such as diameters, widths, holes and slots, are frequently specified as datum features.

Datums are classified as being a primary, secondary, or tertiary datum, **Figure 12-59**. Three points are required to establish a primary datum. Two points are required to establish a secondary datum. One point is required to establish a tertiary datum, **Figure 12-60**. Each point used to establish a datum is called off by a

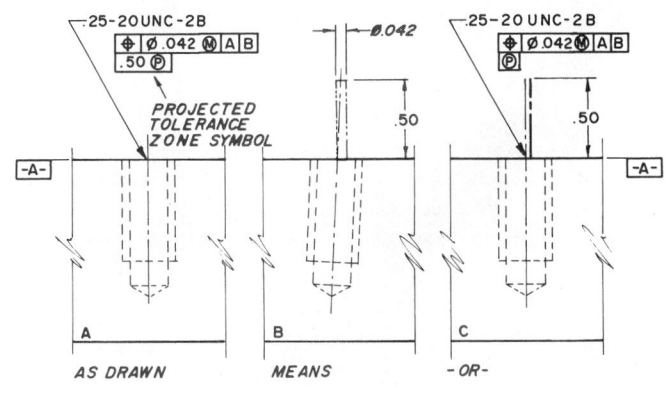

Figure 12-58 Projected tolerance zone

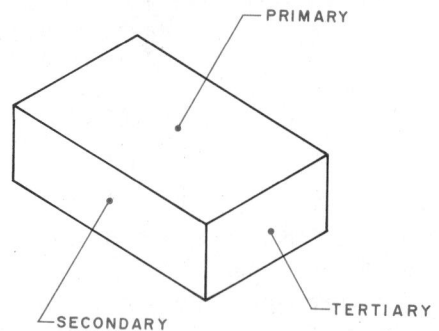

Figure 12-59 Datums

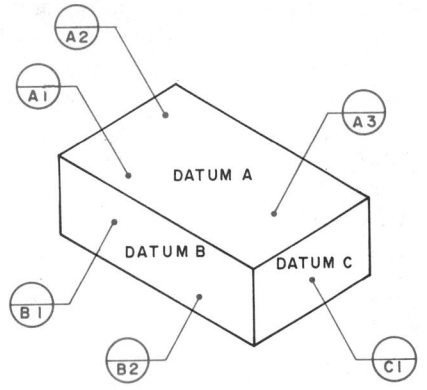

Figure 12-60 Establishing datums

datum target symbol, ***Figure 12-61***. The letter designation in the datum target symbol is the datum identifier. For example, the letter A in ***Figure 12-62*** is the datum designator for Datum A. The number 2 in ***Figure 12-63*** is the point designator for Point 2. Therefore, the complete designation of "A2" means Datum A-Point 2.

Figure 12-61 Datum target symbol

Figure 12-62 Datum designation

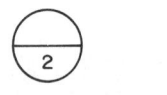

Figure 12-63 Point designator

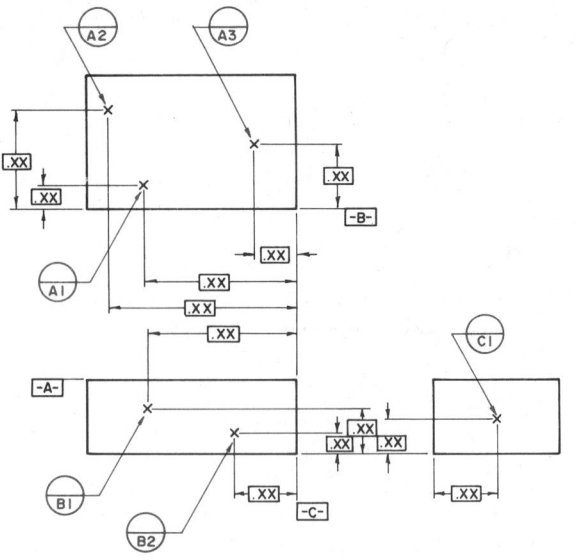

Figure 12-64 Dimensioning datum points

Figure 12-64 illustrates how the points which establish datums should be dimensioned on a drawing. In this illustration, the three points which establish Datum A are dimensioned in the top view and labeled using the datum target symbol. The two points that establish Datum B are dimensioned in the front view. The one point that establishes Datum C is dimensioned in the right-side view. ***Figure 12-65*** illustrates the concept of datum plane and datum surface. The theoretically perfect plane is represented by the top of the machine table. The less perfect actual datum surface is the bottom surface of the part. ***Figure 12-66*** shows how the differences between the perfect datum plane and the actual datum surface are reconciled. The three points protruding from the machine table correspond with the three points which establish Datum A. Once this difference has been reconciled, inspections of the part can be carried out.

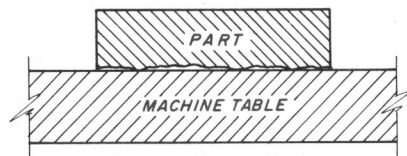

Figure 12-65 Datum plane versus datum surface

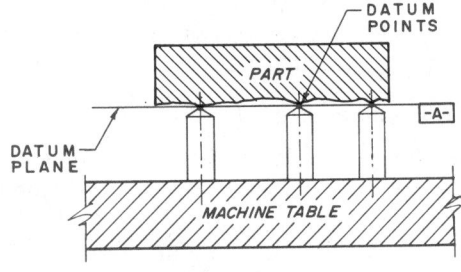

Figure 12-66 Reconciling the datum surface to the datum plane

Review

1. Define the term *tolerancing*.
2. What are the two types of size tolerances?
3. What led to the development of geometric dimensioning and tolerancing?
4. Define the term *geometric dimensioning and tolerancing*.
5. Identify the ANSI standard that pertains to geometric dimensioning and tolerancing.
6. Sketch the symbols for the following:

   a. Flatness   **e.** Perpendicularity
   b. Circularity   **f.** Parallelism
   c. Straightness   **g.** Angularity
   d. True position

7. Explain the term *maximum material condition*.
8. Explain the term *regardless of feature size*.
9. Explain the term *least material condition*.
10. What is a datum?
11. How is a datum established on a machined surface?
12. How is a datum established on a cast surface?
13. Sketch a sample feature control symbol that illustrates the proper order of elements.
14. Which feature controls *do not* require a datum reference?
15. Which feature controls *must* have a datum reference?

Chapter Twelve Problems

The following problems are intended to give beginning drafters practice in applying the principles of geometric dimensioning and tolerancing.

The steps to follow in completing the problems are:

Step 1 Study the problem carefully.

Step 2 Make a checklist of tasks you will need to complete.

Step 3 Center the required view or views in the work area.

Step 4 Include all dimensions according to ANSI Y14.5M-R1988.

Step 5 Re-check all work. If it's correct, neatly fill out the title block using light guidelines and freehand lettering.

Note: These problems do not follow current drafting standards. You are to use the information shown here to develop properly drawn, dimensioned, and toleranced drawings.

Problem 12-1

Apply tolerances so that this part is straight to within .004 at MMZ.

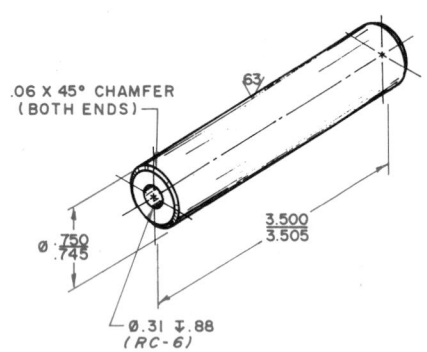

Problem 12-1

Problem 12-4

Apply tolerances to locate the holes using true position and basic dimensions relative to datums A-B-C.

Problem 12-2

Apply tolerances so that the top surface of this part is flat to within .001 and the two sides of the slot are parallel to each other within .002 RFS.

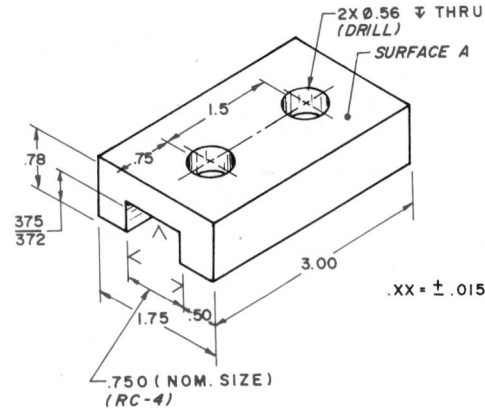

Problem 12-2

Problem 12-3

Apply tolerances so that the smaller diameter has a cylindricity tolerance of .005 and the smaller diameter is concentric to the larger diameter to within .002. The shoulder must be perpendicular to the axis of the part to within .002.

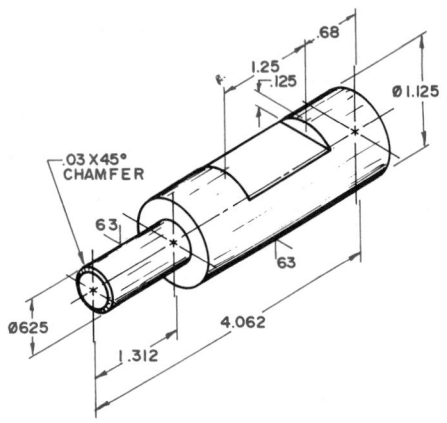

Problem 12-3

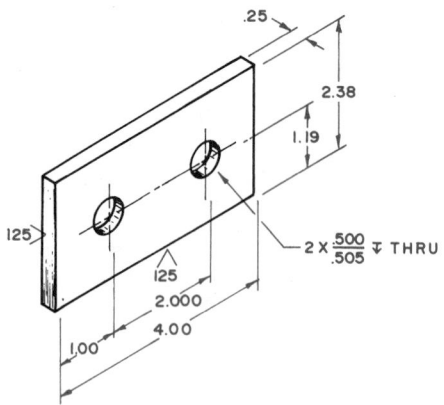

Problem 12-4

Problem 12-5

Apply angularity, true position, and parallelism tolerances of .001 to this part. Select the appropriate datums. The parallelism tolerances should be applied to the sides of the slot.

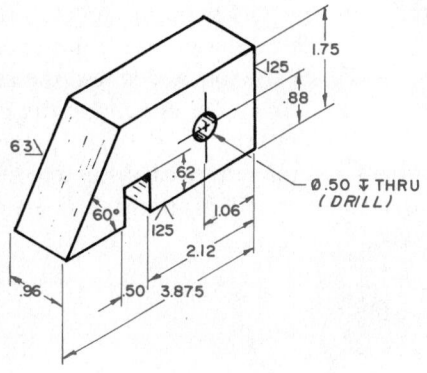

Problem 12-5

Problem 12-6

Apply tolerances so that the outside diameter of the part is round to within .004 and the ends are parallel to within .001 at maximum material condition.

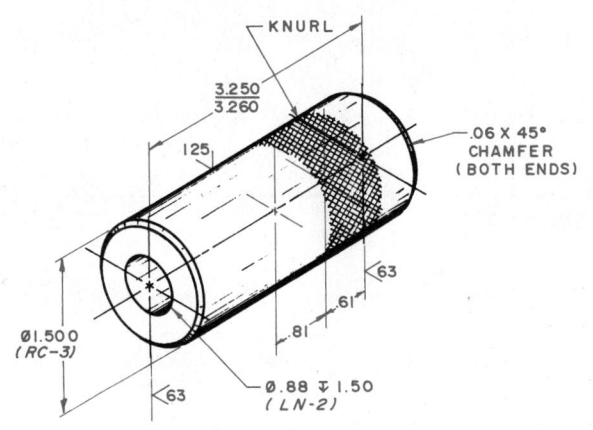

Problem 12-6

Problem 12-9

Select datums and apply tolerances in such a way as to ensure that the slot is symmetrical to within .002 with the .50 diameter hole, and the bottom surface is parallel to the top surface to within .004.

Problem 12-7

Apply a line profile tolerance to the top of the part between points X and Y of .004. Apply true position tolerances to the holes of .021, and parallelism tolerances of .001 to the two finished sides.

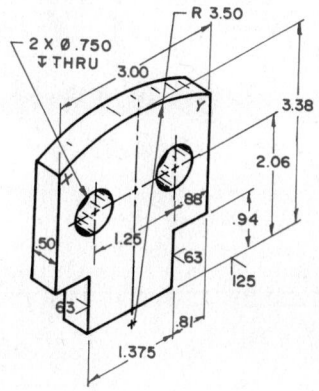

Problem 12-7

Problem 12-8

Use the bottom of the part as datum A and the right side of the part as datum B. Apply surface profile tolerances of .001 to the top of the part between points X and Y.

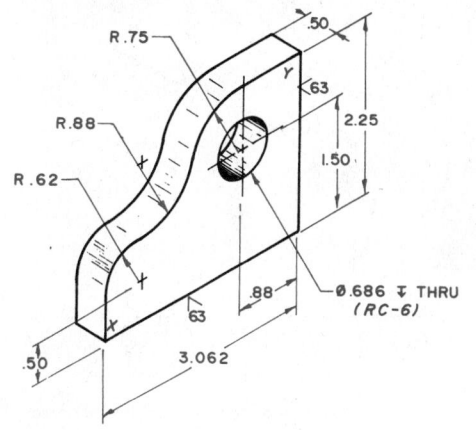

Problem 12-8

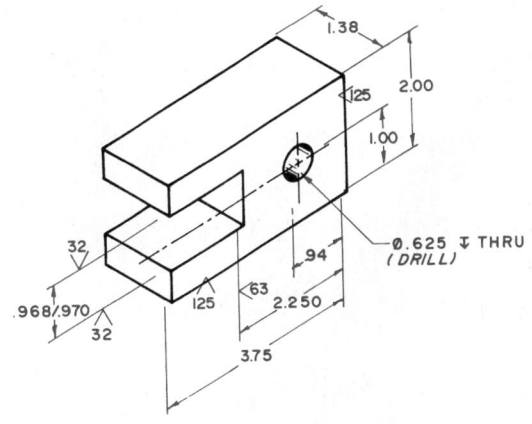

Problem 12-9

Problem 12-10

Apply tolerances to this part so that the tapered end has a total runout of .002.

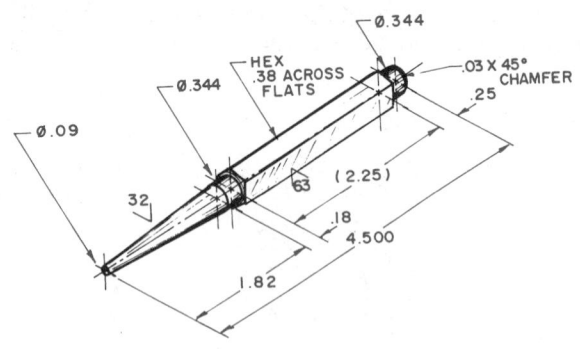

Problem 12-10

Problem 12-11

Apply tolerances to this part so that diameters X and Z have a total runout of .02 relative to datum A (the large diameter of the part) and line runout of .004 to the two tapered surfaces.

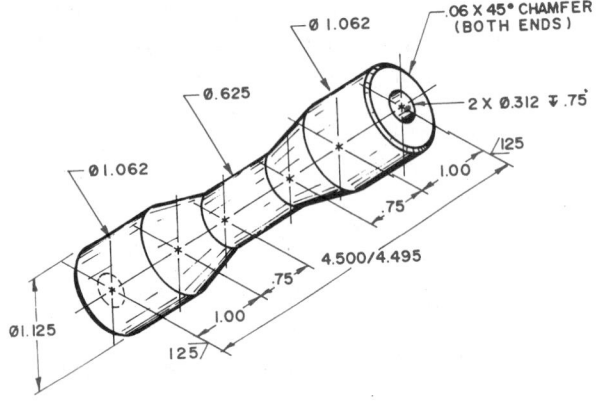

Problem 12-11

Problem 12-12

Select datums and apply a true position tolerance of .001 at MMC to the holes, and a perpendicularity tolerance of .003 to the vertical leg of the angle.

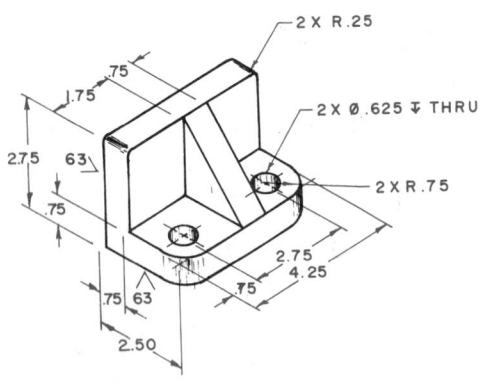

Problem 12-12

Problems 12-13 through 12-30

For each of the remaining geometric dimensioning and tolerancing problems, examine the problem closely with an eye to the purpose that will be served by the part. Then select datums, tolerances, and feature controls as appropriate, and apply them properly to the parts. In this way you will begin to develop the skills required of a mechanical designer. Do not overdesign. Remember, the closer the tolerances, and the more feature control applied, the more expensive the part. Try to use the rule of thumb that says: "Apply only as many feature controls and tolerances as absolutely necessary to ensure that the part will properly serve its purpose after assembly."

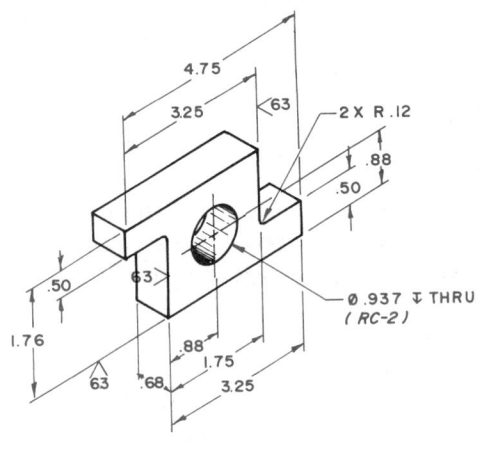

Problem 12-13

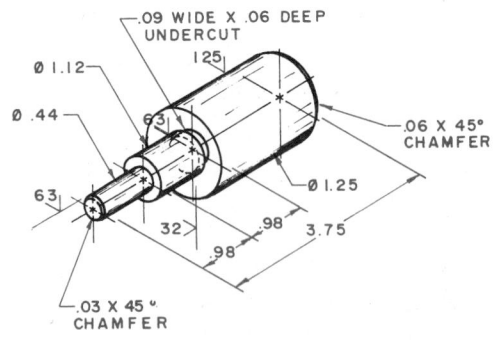

Problem 12-14

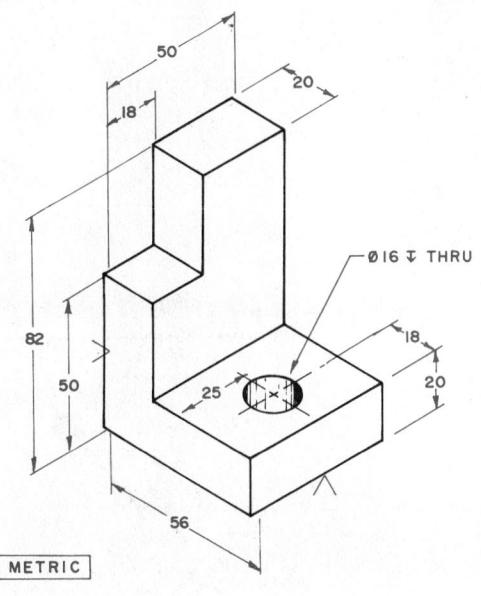

50
20
18
82
50
25
Ø16 ⵌ THRU
18
20
56

METRIC

Problem 12-15

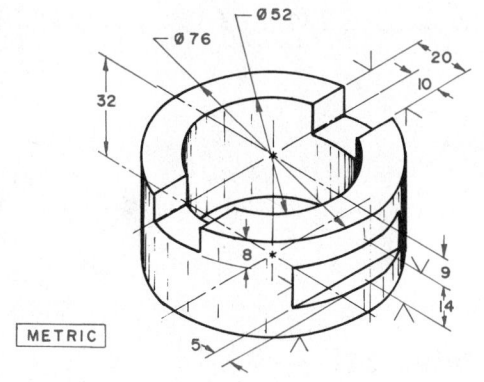

Ø52
Ø76
20
10
32
8
9
4
5

METRIC

Problem 12-16

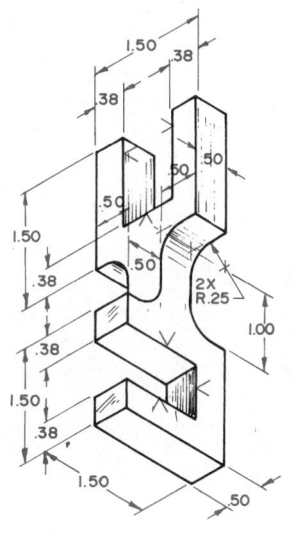

1.50
.38
38
.50
.50
.50
1.50
.50
.38
2X R.25
1.00
.38
.38
1.50
.38
1.50
.50

Problem 12-17

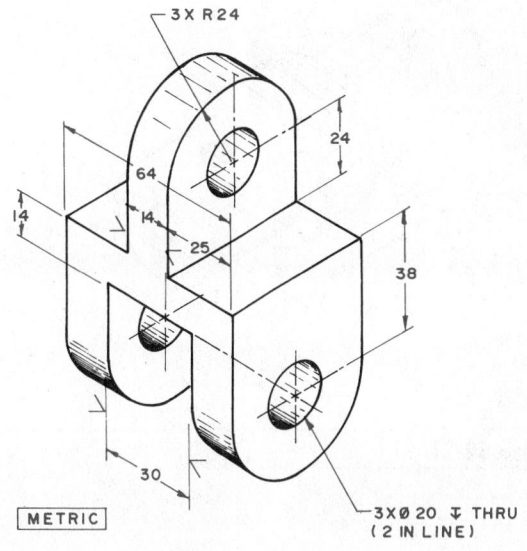

3X R24
24
64
14
14
25
38
30
3XØ20 ⵌ THRU
(2 IN LINE)

METRIC

Problem 12-18

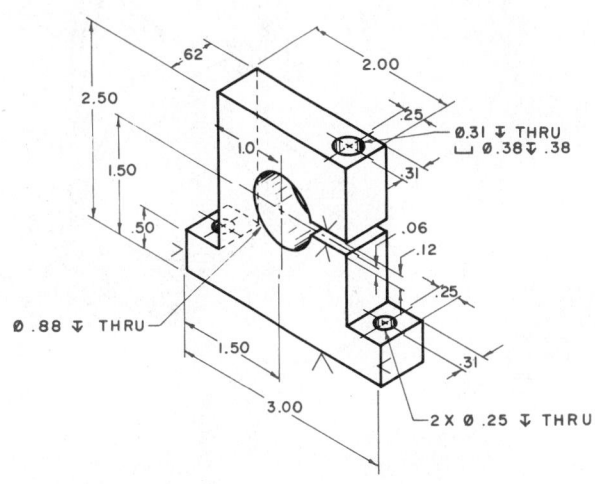

62
2.00
2.50
.25
Ø.31 ⵌ THRU
⌴ Ø.38ⵌ.38
1.50
1.0
.31
.50
.06
.12
Ø.88 ⵌ THRU
.25
1.50
.31
3.00
2X Ø.25 ⵌ THRU

Problem 12-19

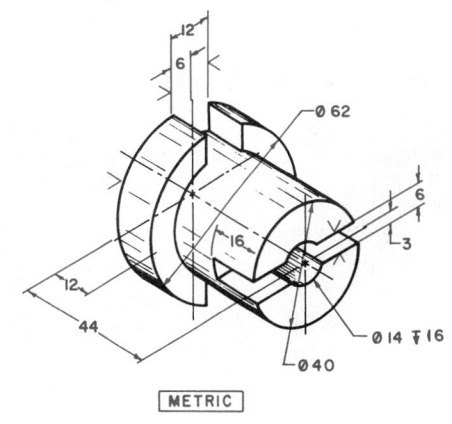

12
6
Ø62
6
16
3
12
44
Ø14 ⵌ16
Ø40

METRIC

Problem 12-20

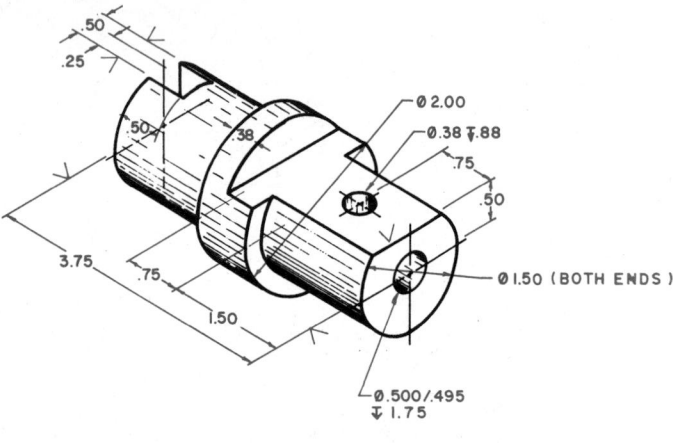

.50
.25
Ø 2.00
0.38 ↧.88
.75
.50
.50
.38
3.75
.75
Ø 1.50 (BOTH ENDS)
1.50
Ø.500/.495
↧ 1.75

Problem 12-21

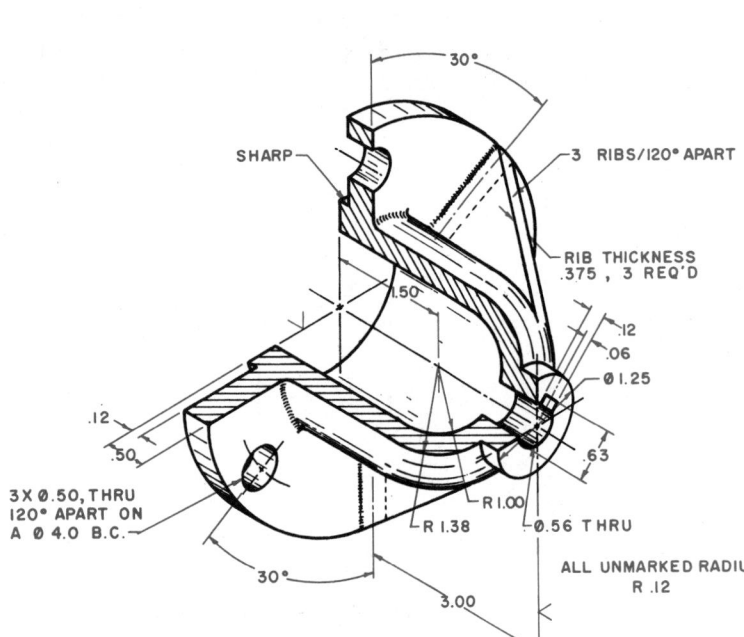

30°
SHARP
3 RIBS/120° APART
RIB THICKNESS
.375 , 3 REQ'D
1.50
.12
.06
Ø 1.25
.12
.50
.63
3X Ø.50, THRU
120° APART ON
A Ø 4.0 B.C.
R 1.00
R 1.38
Ø.56 THRU
30°
3.00
ALL UNMARKED RADIUS
R .12

Problem 12-22

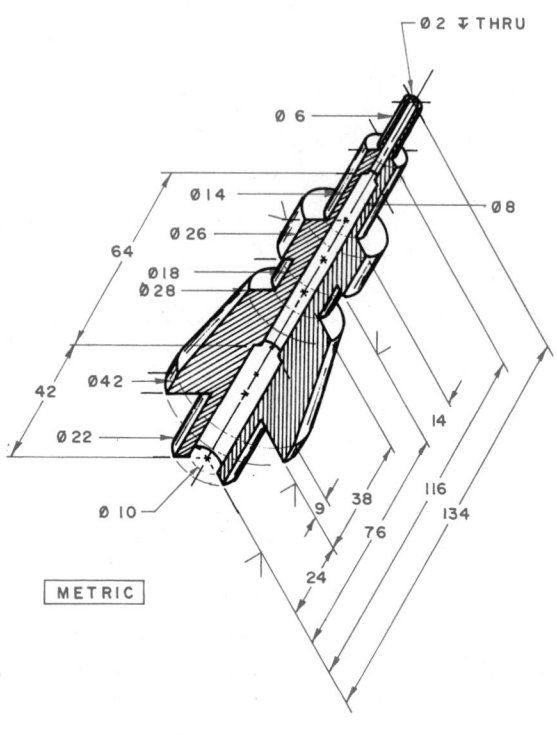

Ø 2 ↧ THRU
Ø 6
Ø 14
Ø 8
Ø 26
64
Ø 18
Ø 28
42
Ø 42
14
Ø 22
116
Ø 10
9 38 134
24 76

METRIC

Problem 12-23

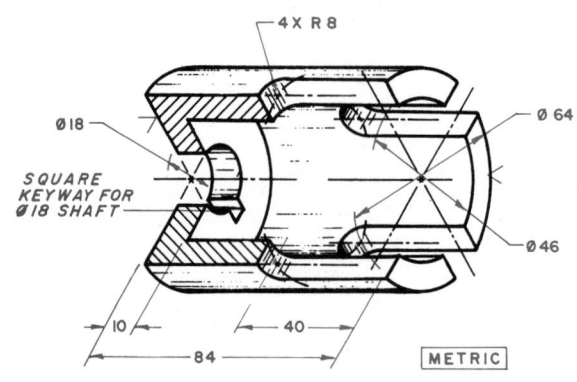

4X R 8
Ø 64
Ø 18
SQUARE
KEYWAY FOR
Ø 18 SHAFT
Ø 46
10
40
84

METRIC

Problem 12-24

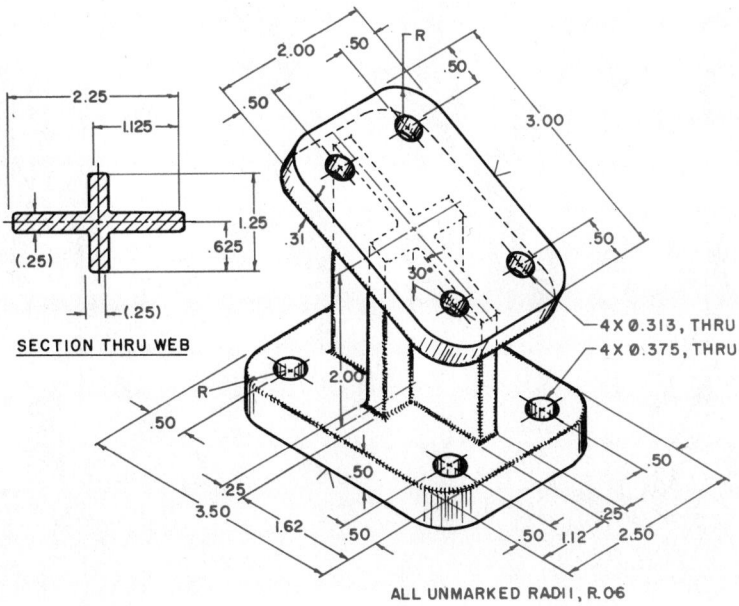

2.25
1.125

1.25
.625
(.25)
(.25)

SECTION THRU WEB

2.00 .50 R
.50 .50
.50
3.00
.31
30°
.50

4X Ø.313, THRU
4X Ø.375, THRU
2.00
R
.50
.25
3.50
1.62 .50
.50
.50 1.12 2.50
.25
.50

ALL UNMARKED RADII, R.06

Problem 12-25

(2.00)
.32
.84
R.50 (TYP.)
4X Ø.38, THRU
1.38
.38
R.18
(TYP.)
Ø1.25
1.25
R.50
2.50
30°
.50
R.625
.63
.38
1.5
.50
.38
3.75 1.25
.50
.50
45°
Ø.75, THRU
1.00 2.00

ALL UNMARKED RADII = R.09

❏ **Problem 12-26**

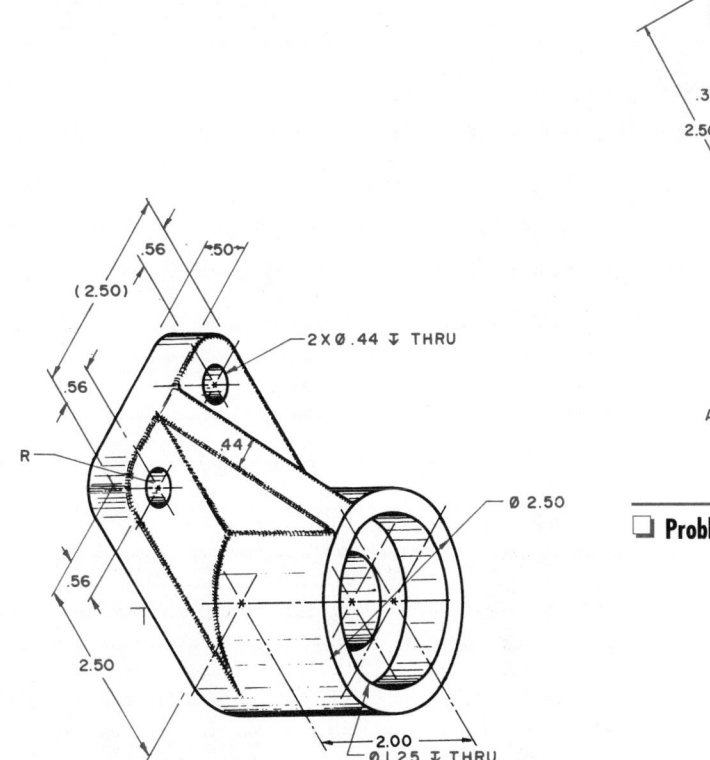

.56 .50
(2.50)
.56
2X Ø.44 ⤓ THRU
R
.44
.56
Ø 2.50
2.50
2.00
Ø1.25 ⤓ THRU
⌴ Ø 1.88 ⤓.50
ALL UNMARKED RADII = R.06

❏ **Problem 12-27**

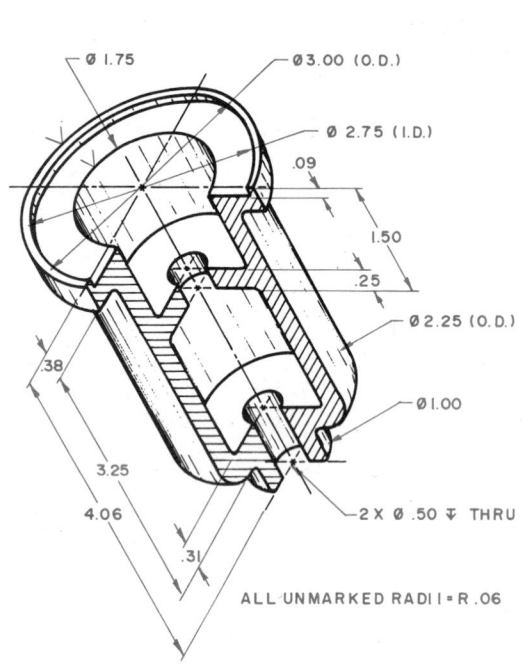

❏ Problem 12-28

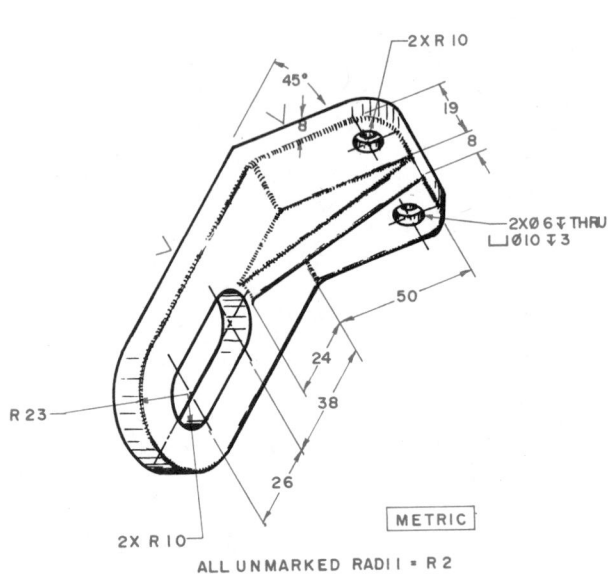

❏ Problem 12-29

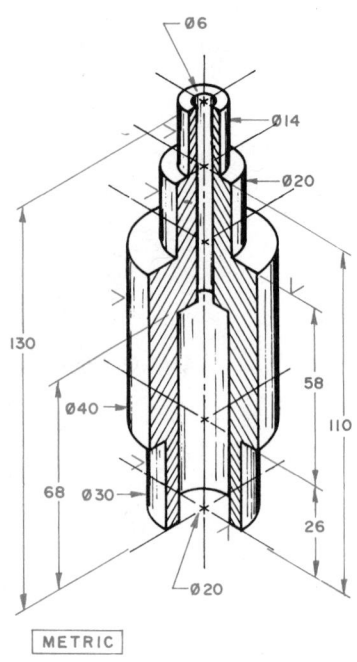

METRIC

❏ Problem 12-30

❏ Problem 12-31

This problem deals with feature control symbols. In items 1–19 explain what each symbol means. In items 20–30, draw the required symbols.

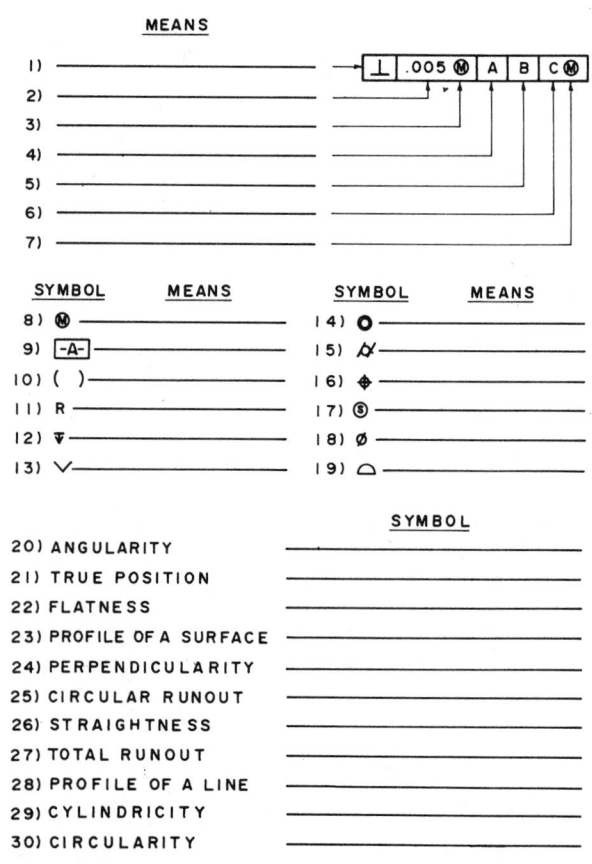

20) ANGULARITY _____
21) TRUE POSITION _____
22) FLATNESS _____
23) PROFILE OF A SURFACE _____
24) PERPENDICULARITY _____
25) CIRCULAR RUNOUT _____
26) STRAIGHTNESS _____
27) TOTAL RUNOUT _____
28) PROFILE OF A LINE _____
29) CYLINDRICITY _____
30) CIRCULARITY _____

Problem 12-31

Fasteners

Classifications of Fasteners

As a new product is developed, determining how to fasten it together is a major consideration. The product must be assembled quickly, using standard, easily available, low-cost fasteners. Some products are designed to be taken apart easily—others are designed to be permanently assembled. Many considerations are required as to what kind, type, and material of fastener to be used. Sometimes the stress load upon a joint must be considered. There are two major classifications of fasteners: permanent and temporary. *Permanent* fasteners are used when parts will not be disassembled. *Temporary* fasteners are used when the parts will be disassembled at some future time.

Permanent fastening methods include welding, brazing, stapling, nailing, gluing, and riveting. Temporary fasteners include screws, bolts, keys, and pins.

Many temporary fasteners include threads in their design. In early days, there was no such thing as standardization. Nuts and bolts from one company would not fit nuts and bolts from another company. In 1841, Sir Joseph Whitworth worked toward some kind of standardization through England. His efforts were finally

accepted, and England came up with a standard thread form called Whitworth Threads.

In 1864, the United States tried to develop a standardization of its own but, because it would not interchange with the English Whitworth Threads, it was not adopted at that time. It was not until 1935 that the United States adopted the American Standard Thread. It was actually the same 60° V-thread form proposed back in 1864. Still, there was no standardization between countries. This created many problems, but nothing was done until World War II, which changeability of parts that, in 1948, the United States, Canada, and the United Kingdom developed the Unified Screw Thread. It was a compromise between the newer American Standard Thread and the old Whitworth Threads.

Today, with the changeover to the metric system, new standards are being developed. The International Organization for Standardization (ISO) was formed to develop a single international system using metric screw threads. This new ISO standard will be united with the American National Standards Institute (ANSI) standards. At the present time, we are in a transitional period and a combination of both systems is still being used.

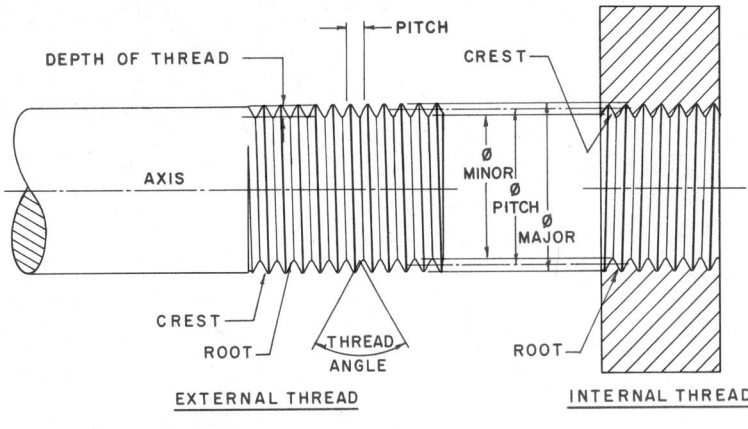

Figure 13-1   Thread terms

Threads

Threads are used for four basic applications:

1. to fasten parts together, such as a nut and a bolt.
2. for fine adjustment between parts in relation to each other, such as the fine adjusting screw on a surveyor's transit.
3. for fine measurement, such as a micrometer.
4. to transmit motion or power, such as an automatic screw threading attachment on a lathe or a house jack.

There are many types and sizes of fasteners, each designed for a particular function. Permanently fastening parts together by welding or brazing is discussed in Chapter 19. Although screw threads have other important uses, such as adjusting parts and measuring and transmitting power, only their use as a fastener and only the most used kinds of fasteners are discussed in this chapter.

Thread Terms

Refer to *Figure 13-1* for the following terms.

- *External thread*—Threads located on the outside of a part, such as those on a bolt.
- *Internal thread*—Threads located on the inside of a part, such as those on a nut.
- *Axis*—A longitudinal center line of the thread.
- *Major diameter*—The largest diameter of a screw thread, both external and internal.
- *Minor diameter*—The smallest diameter of a screw thread, both external and internal.
- *Pitch diameter*—The diameter of an imaginary diameter centrally located between the major diameter and the minor diameter.
- *Pitch*—The distance from a point on a screw thread to a corresponding point on the next thread, as measured parallel to the axis.
- *Root*—The bottom point joining the sides of a thread.
- *Crest*—The top point joining the sides of a thread.
- *Depth of thread*—The distance between the crest and the root of the thread, as measured at a right angle to the axis.

- *Angle of thread*—The included angle between the sides of the thread.
- *Series of thread*—A standard number of threads per inch (TPI) for each standard diameter.

Screw Thread Forms

The *form* of a screw thread is actually its profile shape. There are many kinds of screw thread forms. Seven major kinds are discussed next.

Unified National Thread Form

The Unified National thread form has been the standard thread used in the United States, Canada, and the United Kingdom since 1948, *Figure 13-2A*. This thread form is used mostly for fasteners and adjustments.

ISO Metric Thread Form

The ISO metric thread form is the new standard to be used throughout the world. Its form or profile is very similar to that of the Unified National thread, except that the thread depth is slightly less, *Figure 13-2B*. This thread form is used mostly for fasteners and adjustments.

Square Thread Form

The square thread's profile is exactly as its name implies; that is, square. The faces of the teeth are at right angles to the axis and, theoretically, this is the best

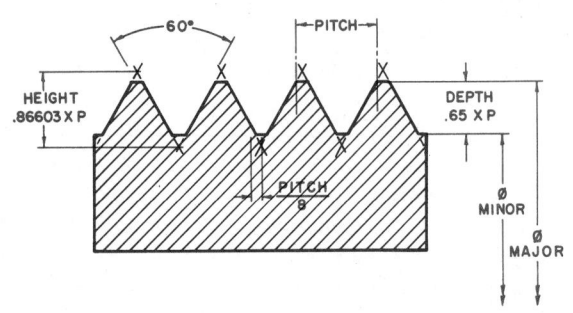

Figure 13-2A   Unified national thread form (UN)

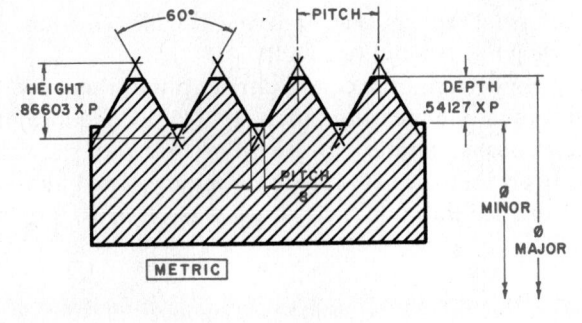

Figure 13-2B   ISO metric thread form

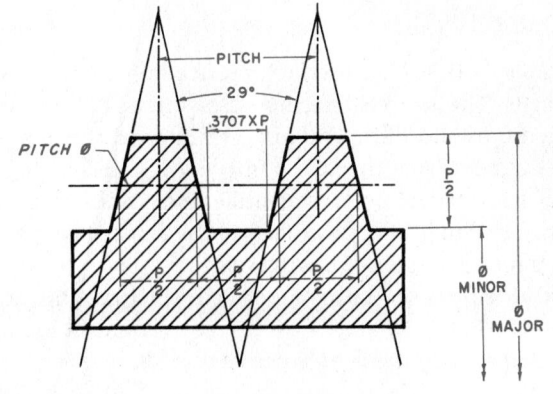

Figure 13-2D   Acme thread form

thread to transmit power, **Figure 13-2C**. Because this thread is difficult to manufacture, it is being replaced by the Acme thread.

Acme Thread Form

The Acme thread is a slight modification of the square thread. It is easier to manufacture and is actually stronger than the square thread, **Figure 13-2D**. It, too, is used to transmit power.

Worm Thread Form

The worm thread is similar to the Acme thread, and is used primarily to transmit power, **Figure 13-2E**.

Knuckle Thread Form

The knuckle thread is usually rolled from sheet metal and is used, slightly modified, in electric light bulbs, electric light sockets, and sometimes for bottle tops. The knuckle thread is sometimes cast, **Figure 13-2F**.

Buttress Thread Form

The buttress thread has certain advantages in applications involving exceptionally high stress along its axis in *one* direction only. Examples of applications are the breech assemblies of large guns, airplane propeller hubs, and columns for hydraulic presses, **Figure 13-2G**.

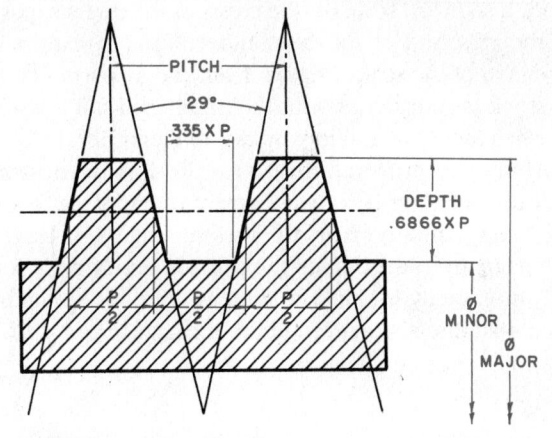

Figure 13-2E   Worm thread form

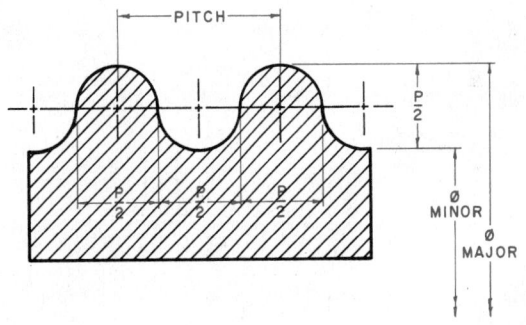

Figure 13-2F   Knuckle thread form

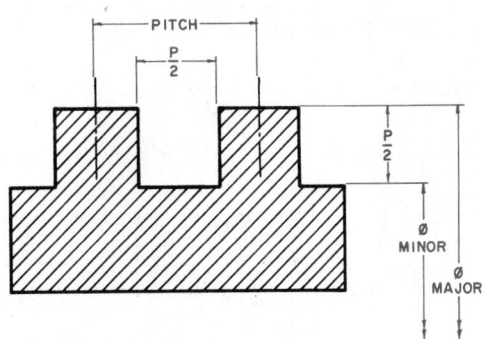

Figure 13-2C   Square thread form

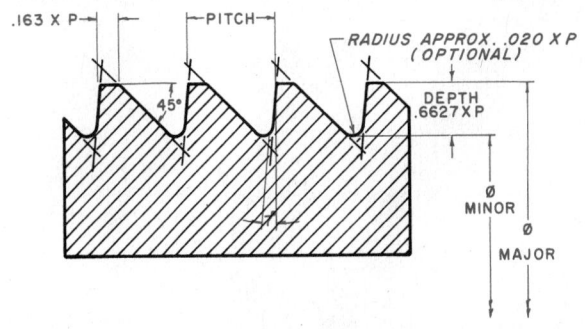

Figure 13-2G   Buttress thread form

Tap and Die

Various methods are used to produce inside and outside threads. The simplest method uses thread-cutting tools called taps and dies. The *tap* cuts internal threads; the *die* cuts external threads, **Figure 13-3**. In making an internal threaded hole, a tap-drilled hole must be drilled first. This hole is approximately the same diameter as the minor diameter of the threads.

Notice how the tap is tapered at the end; this taper allows the tap to start into the tap-drilled hole. This tapered area contains only partial threads.

Threads per Inch (TPI)

One method of measuring threads per inch (TPI) is to place a standard scale on the crests of the threads, parallel to the axis, and count the number of full threads within one inch of the scale, **Figure 13-4**. If only part of an inch of stock is threaded, count the number of full threads in one half inch and multiply by two to determine TPI.

A simple, more accurate method of determining threads per inch is to use a screw thread gage, **Figure 13-5**. By trial and error, the various fingers or leaves of the gage are placed over the threads until one is found that fits exactly into all the threads. Threads per inch are then read directly on each leaf of the gage, **Figure 13-6**.

Pitch

The *pitch* of any thread, regardless of its thread form or profile, is the distance from one point on a thread to the corresponding point on the adjacent thread as mea-

sured parallel to its axis, **Figure 13-7**. Pitch is found by dividing the TPI into one inch.

In this example, a coarse thread pitch, there are 10 threads in one measured inch; 10 TPI divided into one inch equals a pitch of 10. In a fine thread of the same diameter there are 20 threads in one measured inch; 20 TPI divided into one inch equals a pitch of 20, **Figure 13-8**.

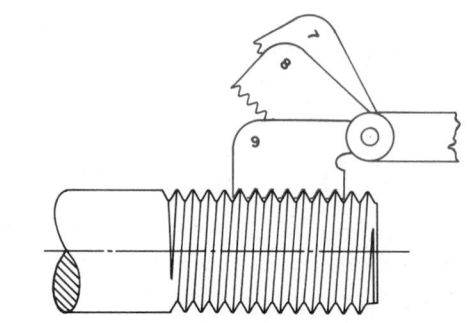

Figure 13-5   Use of a screw pitch gage

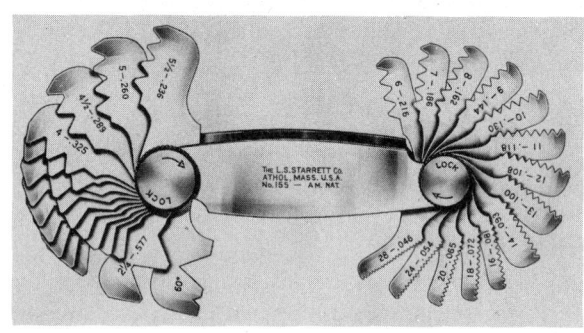

Figure 13-6   Reading screw pitch gage *(Courtesy The L. S. Starrett Co.)*

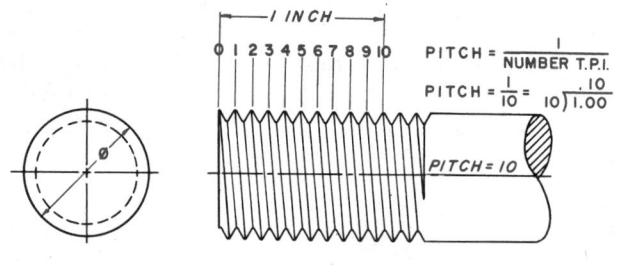

Figure 13-7   Coarse thread pitch

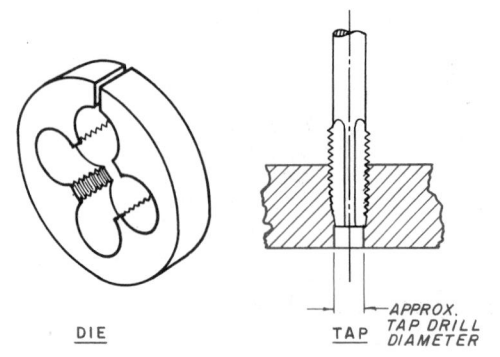

Figure 13-3   Tap and die

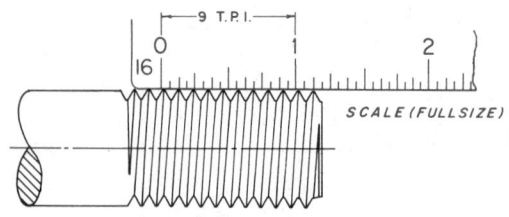

Figure 13-4   Use of a scale to calculate threads per inch (TPI)

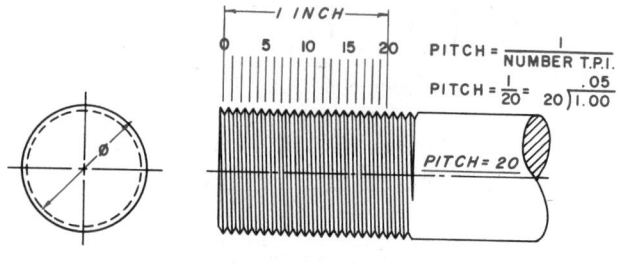

Figure 13-8   Fine thread pitch

For metric threads, the pitch is specified in millimeters. Pitch for a metric thread is included in its call-off designation. For example: M10 x 1.5. The 1.5 indicates the pitch; therefore, it does not as a rule have to be calculated.

Single and Multiple Threads

A *single thread* is composed of one continuous ridge. The lead of a single thread is equal to the pitch. *Lead* is the distance a screw thread advances axially in one full turn. Most threads are single threads.

Multiple threads are made up of two or more continuous ridges following side-by-side. The lead of a double thread is equal to twice the pitch. The lead of a triple thread is equal to three times the pitch, *Figure 13-9*.

Multiple threads are used when speed or travel distance is an important design factor. A good example of a double or triple thread is found in an inexpensive ballpoint pen. Take a ball-point pen apart and study the end of the external threads. There will probably be two or three ridges starting at the end of the threads. Notice how fast the parts screw together. The speed, not power, is the characteristic of multiple threads.

Right-Hand and Left-Hand Threads

Threads can be either right-handed or left-handed. To distinguish between a right-hand and a left-hand thread, use this simple trick. A right-hand thread winding tends

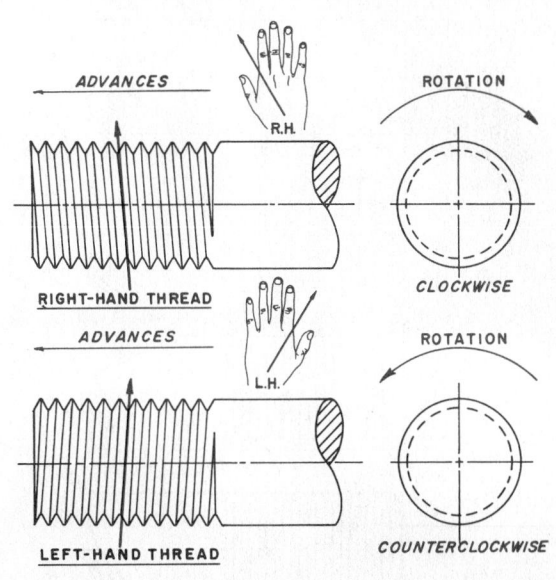

Figure 13-10   Right-hand and left-hand threads

to lean toward the left. If the thread leans toward the left, the right-hand thumb points in the same direction. If the thread leans to the right, *Figure 13-10*, the left-hand thumb leans in that direction indicating that it is a left-hand thread.

Thread Representation

The top illustration of *Figure 13-11* shows a normal view of an external thread. To draw a thread exactly as it will actually look takes too much drafting time. To help speed up the drawing of threads, one of two basic systems is used and each is described and illustrated. The schematic system of representing threads was developed approximately in 1940, and is still used somewhat today. The simplified system of representing threads was developed 15 years later, and is actually quicker and in greater use today.

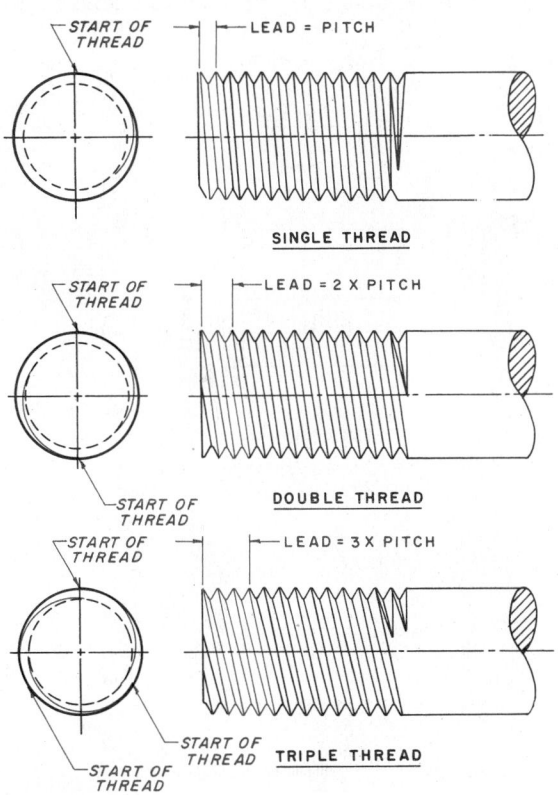

Figure 13-9   Single and multiple threads

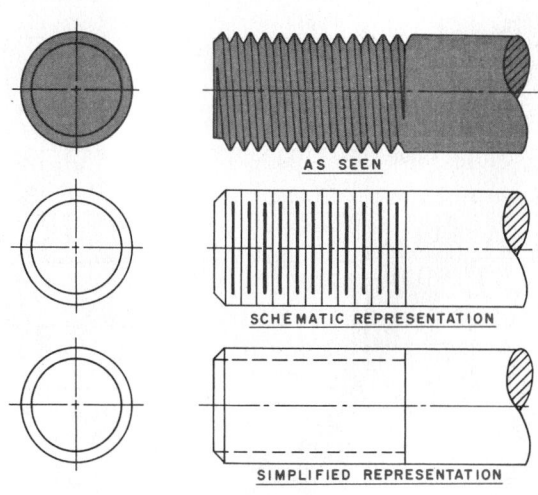

Figure 13-11   Thread representation

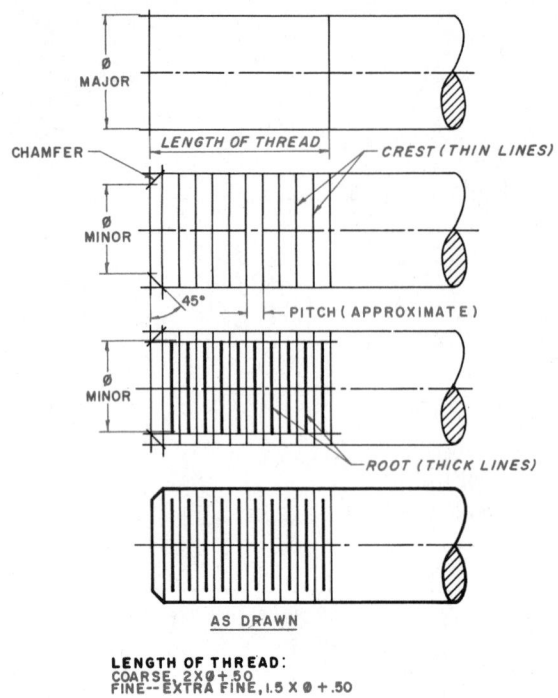

Figure 13-12   How to draw threads using the schematic system

How To Draw Threads Using the Schematic System

Step 1   Refer to *Figure 13-12*. Lightly draw the major diameter, and locate the approximate length of full threads.

Step 2   Lightly locate the minor diameter and draw the 45° chamfered ends as illustrated. Draw lines to represent the crest of the threads spaced approximately equal to the pitch.

Step 3   Draw slightly thicker lines centered between the crest lines to the minor diameter. These lines represent the root of the threads.

Step 4   Check all work and darken in. Notice the crest lines are *thin* black lines and the root lines are *thick* black lines.

How To Draw Threads Using the Simplified System

Step 1   Refer to *Figure 13-13*. Lightly draw the major diameter and locate the approximate length of full threads.

Step 2   Lightly locate the minor diameter and draw the 45° chamfered ends as illustrated. Draw dash lines along the minor diameter. This represents the root of the threads.

Step 3   Check all work and darken in. The dash lines are *thin* black lines.

Standard External Thread Representation

The most recent standard to illustrate external threads using either the schematic or simplified system is illustrated in *Figure 13-14*. Note how section views are illustrated using schematic and simplified systems.

Standard Internal Thread Representation

There are two major kinds of interior holes: through holes and blind holes. A *through hole*, as its name implies, goes completely through an object. A *blind hole* is a hole that does *not* go completely through an object. In the manufacture of a blind hole, a tap drill must be drilled into the

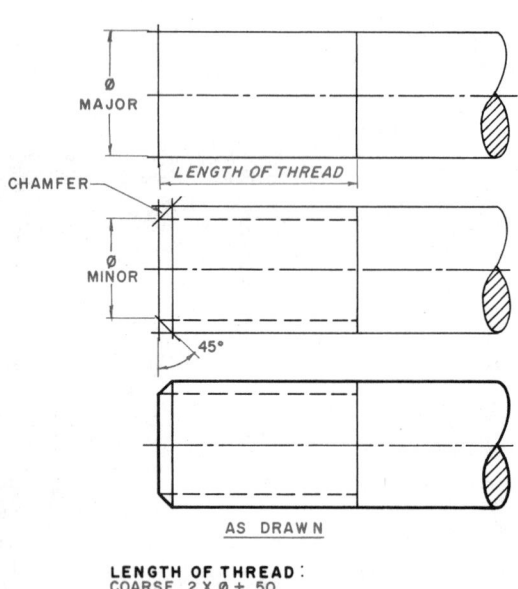

Figure 13-13   How to draw threads using the simplified system

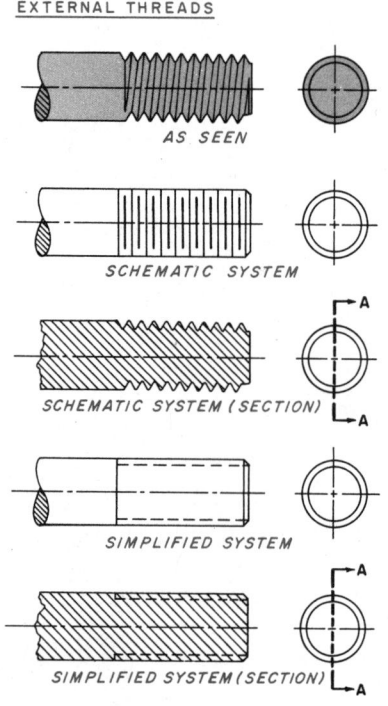

Figure 13-14   Standard external thread representation

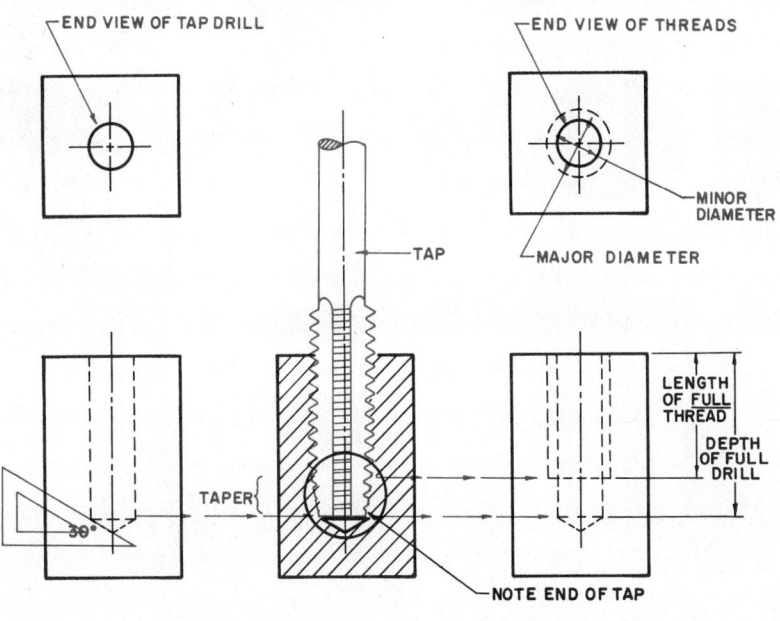

Figure 13-15   Standard internal thread representation

part first, **Figure 13-15**. To illustrate a tap drill, use the 30°–60° triangle. This is *not* the actual angle of a drill point but is close enough for illustration. The tap is now turned into the tap drill hole. Because of the taper on the tap, *full* threads do not extend to the bottom of the hole (refer back to Figure 13-3). The drafter illustrates the tap drill and the full threaded section as shown to the right in Figure 13-15.

Using the schematic system to represent a through hole is illustrated in **Figure 13-16**. A blind hole and a section view are drawn as illustrated in **Figure 13-17**.

Using the simplified system to represent a through hole is illustrated in **Figure 13-18**. A blind hole and a section view are drawn as illustrated in **Figure 13-19**.

Thread Relief (Undercut)

On exterior threads, it is impossible to make perfectly uniform threads up to a shoulder; thus, the threads tend to run out, as illustrated in **Figure 13-20**. Where mat-

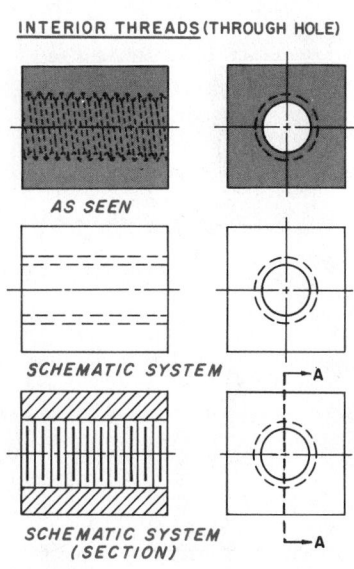

Figure 13-16   Standard internal thread representation for a through hole (schematic system)

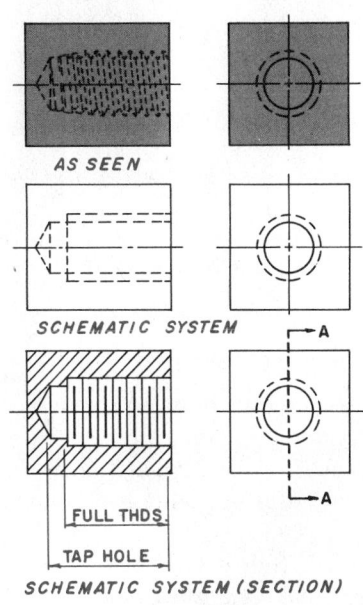

Figure 13-17   Standard internal thread representation for a blind hold (schematic system)

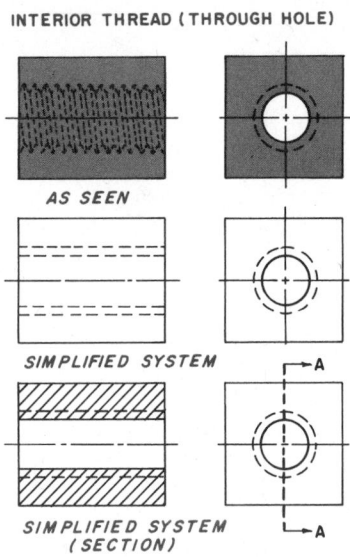

Figure 13-18 Standard internal thread representation for a through hole (simplified system)

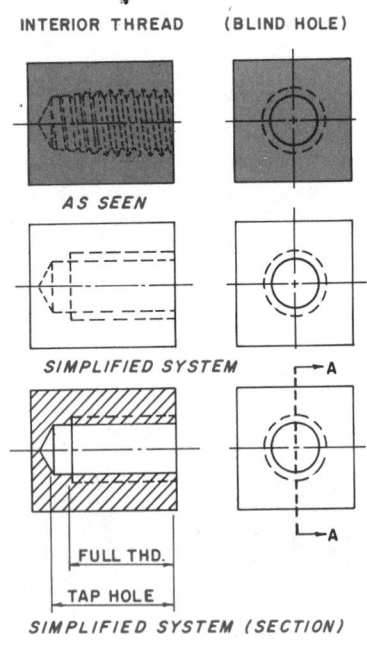

Figure 13-19 Standard internal thread representation for a blind hold (simplified system)

ing parts must be held tightly against the shoulder, the last one or two threads must be removed or *relieved*. This is usually done no farther than to the depth of the threads so as not to weaken the fastener. The simplified system of thread representation is illustrated at the bottom of Figure 13-20.

Full interior threads cannot be manufactured to the end of a blind hole. One way to eliminate this problem is to call-off a thread relief or undercut, as illustrated in *Figure 13-21*. The bottom illustration is as it would be drawn by the drafter.

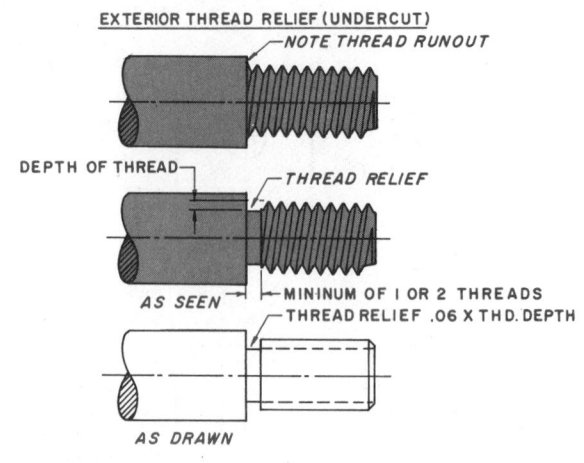

Figure 13-20 External thread relief (undercut)

Screw, Bolt, and Stud

Figure 13-22 illustrates and describes a screw, a bolt, and a stud. A *screw* is a fastener that does not use a nut and is screwed directly into a part.

A *bolt* is a fastener that passes directly through parts to hold them together, and uses a nut to tighten or hold the parts together.

A *stud* is a fastener that is a steel rod with threads at both ends. It is screwed into a blind hole and holds other parts together by a nut on its free end. In general practice, a stud has either fine threads at one end and coarse threads at the other, or Class 3-fit threads at one end and Class 2-fit threads at the other end. Class of fit is fully explained later in this chapter under "Classes of Fit."

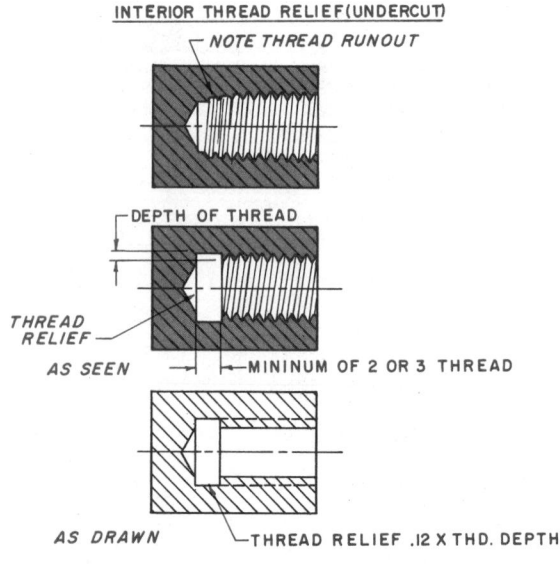

Figure 13-21 Internal thread relief (undercut)

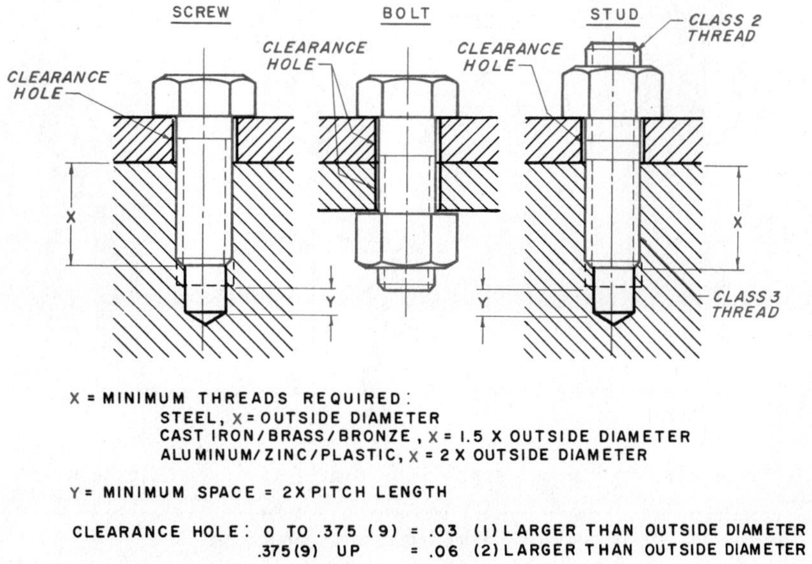

SCREW   BOLT   STUD — *CLASS 2 THREAD*

CLEARANCE HOLE —
CLEARANCE HOLE —
CLEARANCE HOLE —

X   X   X

Y   Y

— *CLASS 3 THREAD*

X = MINIMUM THREADS REQUIRED:
STEEL, X = OUTSIDE DIAMETER
CAST IRON/BRASS/BRONZE, X = 1.5 X OUTSIDE DIAMETER
ALUMINUM/ZINC/PLASTIC, X = 2 X OUTSIDE DIAMETER

Y = MINIMUM SPACE = 2 X PITCH LENGTH

CLEARANCE HOLE: 0 TO .375 (9) = .03 (1) LARGER THAN OUTSIDE DIAMETER
.375 (9) UP   = .06 (2) LARGER THAN OUTSIDE DIAMETER

Figure 13-22   Screw, bolt, and stud

The minimum full thread length for a screw or a stud is:

In steel: equal to the diameter.

In cast iron, brass, bronze: equal to 1.5 times the diameter.

In aluminum, zinc, plastic: equal to 2 times the diameter.

The clearance hole for holes up to .375 (9) diameter is approximately .03 oversize; for larger holes, .06 oversize.

Machine Screws

Machine screw sizes run from .021 (.3) to .750 (20) in diameter. There are eight standard head forms. Four major kinds are illustrated in *Figure 13-23*. *Machine screws* are used for screwing into thin materials. Most machine screws are threaded within a thread or two to the head. Although these are screws, machine screws sometimes incorporate a hex-head nut to fasten parts together.

The length of a machine screw is measured from the bottom of the head to the end of the screw (refer again to Figure 13-23).

Cap Screws

Cap screw sizes run from .250 (6) and up. There are five standard head forms, *Figure 13-24*. A *cap screw* is usually used as a true screw, and it passes through a clearance hole in one part and screws into another part.

How To Draw a Machine Screw or Cap Screw

The exact dimensions of machine screws and cap screws are given in the Appendix of the text but, in actual practice, they are seldom used for drawing purposes. See *Figures 13-25* and *13-26*, which show the various sizes as they are proportioned in regard to the diameter of the fastener. Various fastener templates are now available to further speed up drafting time.

Set Screws

A *set screw* is used to prevent motion between mating parts, such as the hub of a pulley on a shaft. The set screw is screwed into and through one part so that it applies pressure against another part, thus preventing motion. Set screws are usually manufactured of steel, and are hardened to make them stronger than the average fastener.

Set screws have various kinds of heads and many kinds of points. *Figure 13-27* illustrates a few of the

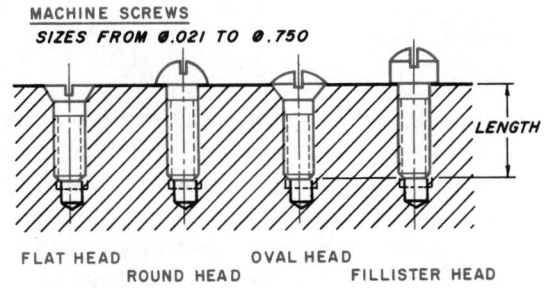

MACHINE SCREWS
SIZES FROM 0.021 TO 0.750

LENGTH

FLAT HEAD   OVAL HEAD
ROUND HEAD   FILLISTER HEAD

Figure 13-23   Machine screws

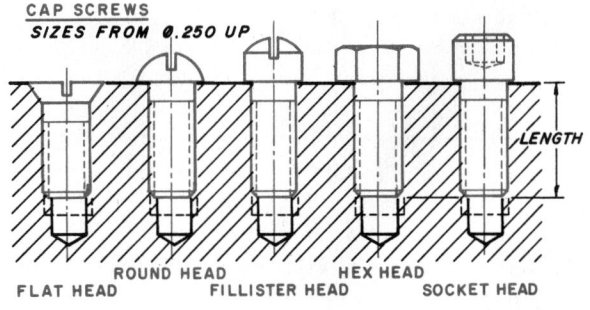

CAP SCREWS
SIZES FROM 0.250 UP

LENGTH

ROUND HEAD   HEX HEAD
FLAT HEAD   FILLISTER HEAD   SOCKET HEAD

Figure 13-24   Cap screws

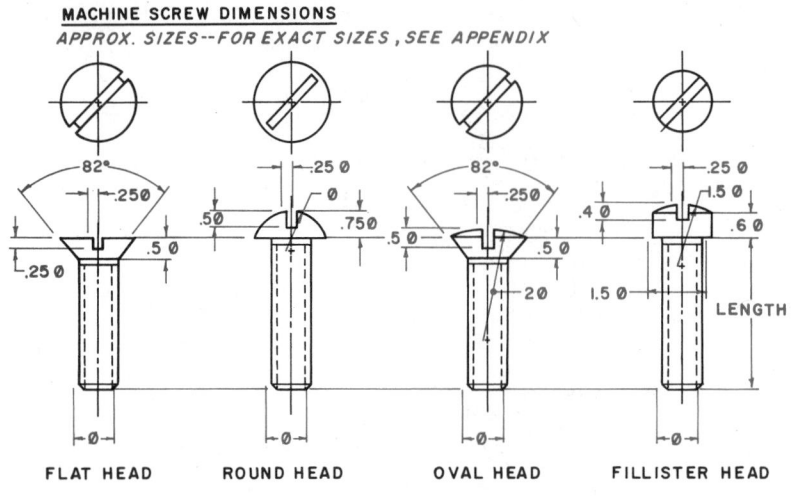

Figure 13-25 Approximate sizes for machine screws or cap screws

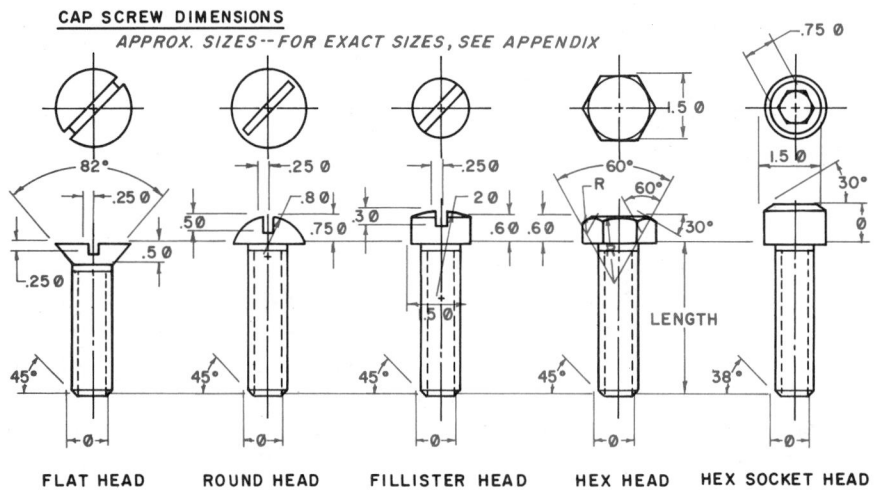

Figure 13-26 Approximate sizes for machine screws or cap screws

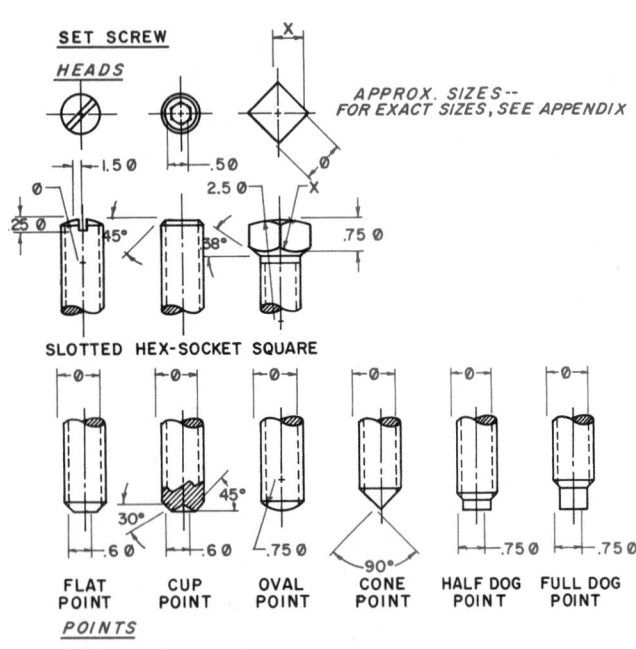

Figure 13-27 Approximate sizes for set screws and set screw points

more common kinds of set screws. Set screws are manufactured in many standard lengths of very small increments, so almost any required length is probably "standard." Exact sizes and lengths can be found in the Appendix. As with machine screws and cap screws, in actual practice, the actual drawing of set screws is done using their proportions in relationship to their diameters.

How To Draw Square- and Hex-Head Bolts

Exact dimensions for square- and hex-head bolts are given in the Appendix of the text, but, in actual practice, they are drawn using the proportions as given in **Figures 13-28** and **13-29**. Notice that the heads are shown in the profile so three surfaces are seen in the front view. In the event a square- or hex-head bolt must be illustrated 90°, the proportions as illustrated in **Figure 13-30** are used.

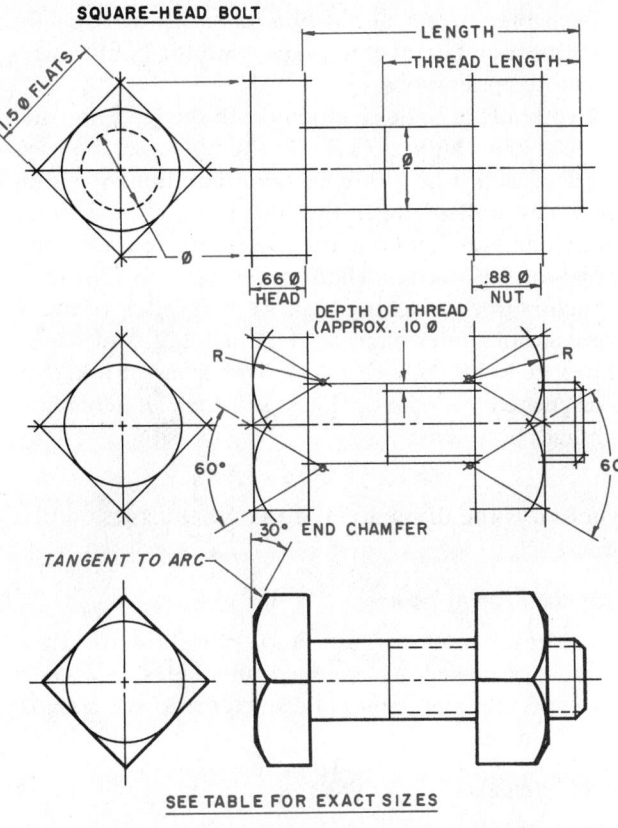

Figure 13-28 Approximate sizes for a square-head bolt

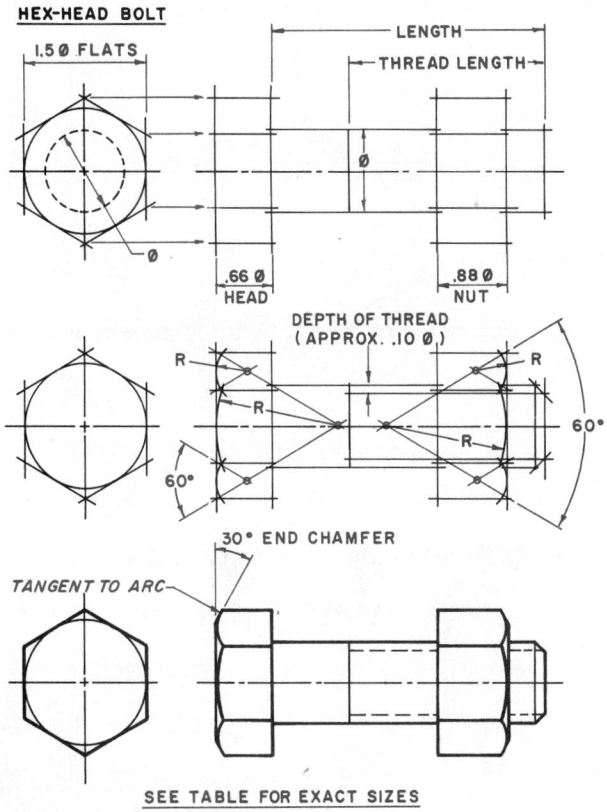

Figure 13-29 Approximate size for a hex-head bolt

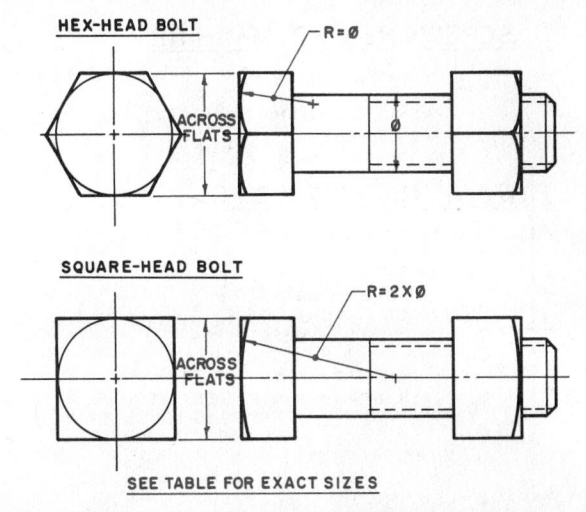

Figure 13-30 Side view of hex- and square-head bolts

Nuts, Bolts, and Other Fasteners in Section

If the cutting plane passes through the axis of any fastener, the fastener is *not* sectioned. It is treated exactly as a shaft and drawn exactly as it is viewed. Refer to *Figure 13-31*. The illustration at the left is drawn correctly. The figure at the right is drawn incorrectly (notice how difficult it is to understand, especially the nut).

Thread Call-offs

Although not all companies use the exact same call-offs for various fasteners, it is important that all drafters within one company use the same method. One method used to call-off fasteners is illustrated in *Figure 13-32*. Regardless of which system is used, the first line contains the fastener's general identification, type of head, and classification. The second line contains all the exact detailed information. All threads are assumed to be right hand (R.H.), unless otherwise noted. If a thread is to be left hand (L.H.), it is noted at the end of the second line.

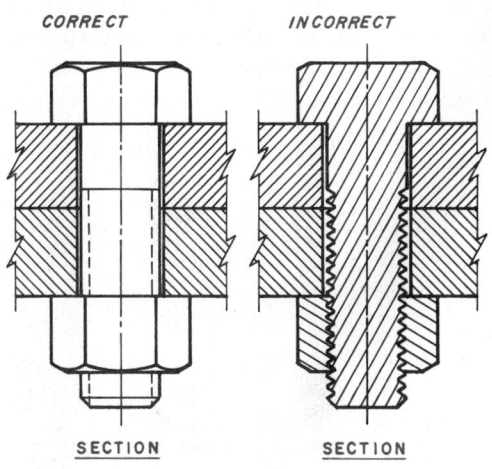

Figure 13-31 Fasteners in section

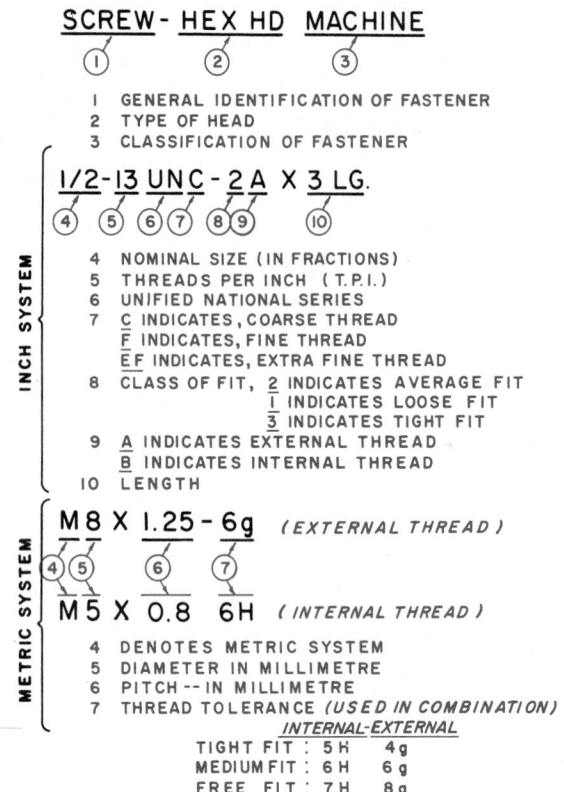

Figure 13-32 Thread call-off

Various Kinds of Heads

Many different kinds of screw heads are used today. *Figure 13-33* illustrates a few of the standard heads.

Rivets

Rivets are permanent fasteners, usually used to hold sheet metal together. Most rivets are made of wrought iron or soft steel and, for aircraft and space missiles, copper, aluminum, alloy or other exotic metals.

Riveted joints are classified by applications, such as pressure vessels, structural and machine members. For data concerning joints for pressure vessels refer to such sources as ASME boiler codes. For data concerning larger field structural rivets, such as bridges, buildings and ships, see ANSI standards or a *Machinery's Handbook*.

This chapter covers information for small-size rivets for machine-member riveted joints used for lighter mass produced applications.

Two kinds of basic rivet joints are the lap joint, and the butt joint, *Figure 13-34*. In the *lap joint*, the parts overlap each other, and are held together by one or more rows of rivets. In the butt joint, the parts are butted, and are held together by a cover plate or butt strap which is riveted to both parts.

Factors to consider are: type of joint, pitch of rivets, type and diameter of rivet, rivet material, and size of clearance holes, *Figure 13-35*. The diameter of a rivet is calculated from the thickness of metal, and commonly ranges between

$$d = 1.2 \sqrt{t} \text{ AND } d = 1.4 \sqrt{t}$$

where d is the diameter, and t is the thickness of the plate.

Size and Type of Hole

Rivet holes must be punched, punched and then reamed, or drilled. As a general rule, holes are usually made .06 (1.5 mm) larger in diameter than the nominal rivet diameter.

Rivet Symbols

Rivets applied in mass produced applications are represented in *Figure 13-36A*, illustrating the kind of rivet, to which side it is applied, and if it is to be countersunk, and so forth.

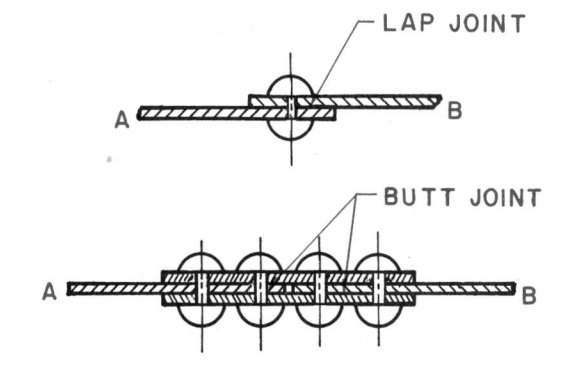

Figure 13-34 Two basic rivet joints (lap joint and butt joint)

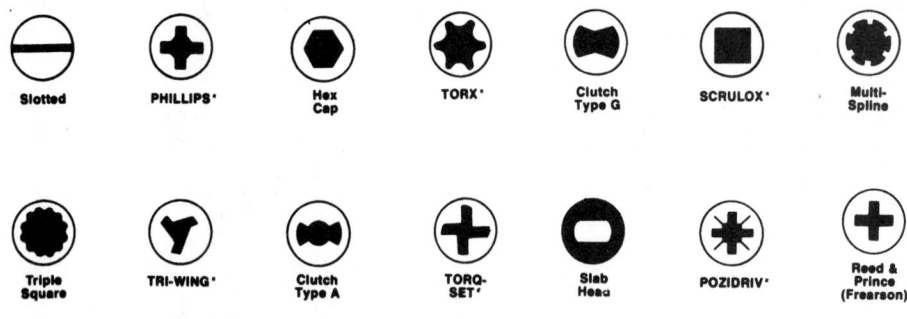

Figure 13-33 Kinds of screw heads

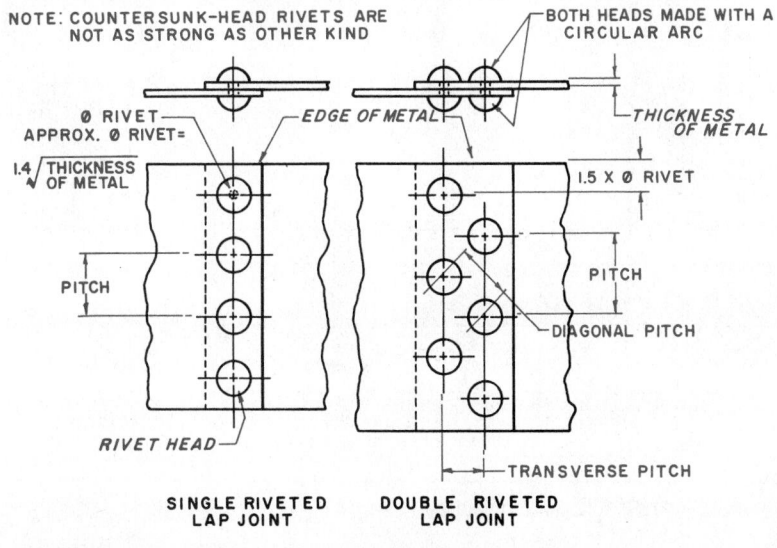

Figure 13-35 Factors to consider in rivet joints

Kinds of Rivets

There are many different kinds of small-size rivets. The five major kinds are truss head, button head, pan head, countersunk head, and flat head. The countersunk-head rivet is not as strong as the other kinds of rivets; therefore, more rivets must be used to gain strength equal to the other types.

Drawing of Rivets

American standard small solid rivets are shown in their approximate standard proportions in Figure 13-36A. These sizes are close enough to be used for drawing the rivet if necessary. For exact sizes, data must be obtained from other sources.

End Points

The end of a standard rivet is usually cut off straight. If a point is required, a standard size point is illustrated at the lower end of a countersunk head (refer back to Figure 13-36A).

The drafter must indicate what kind of rivet is to be used, and on which side the head is to be positioned. To illustrate this, the drafter uses a code such as that illustrated in ***Figure 13-36B***.

Keys and Keyseats

Key

A *key* is a demountable part that provides a positive means of transferring torque between a shaft and a hub.

Keys are used to prevent slippage and to transmit torque between a shaft and a hub. There are many kinds of keys. The five major kinds used in industry today are illustrated in ***Figure 13-37***: square key, flat key, gib head key, Pratt & Whitney key and Woodruff key. Where a lot of torque is present, a double key and keyseat is often used. In extreme conditions, a spline is machined into the shaft and into the hub. For spline information, refer to a *Machinery's Handbook*.

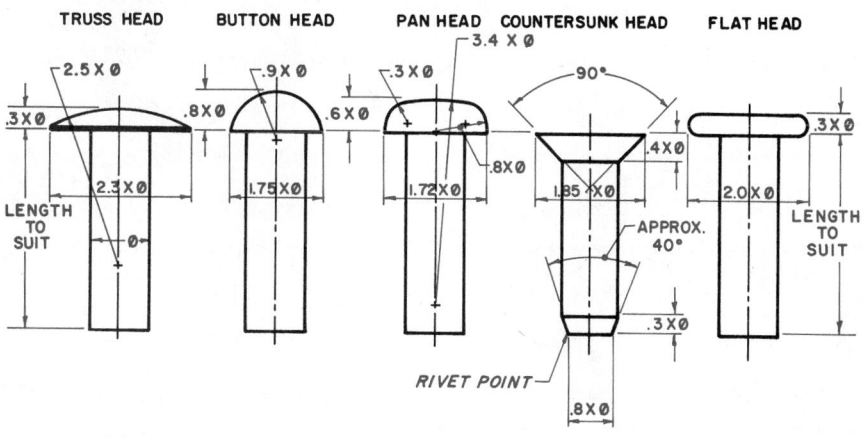

Figure 13-36A Approximate sizes for standard small rivets

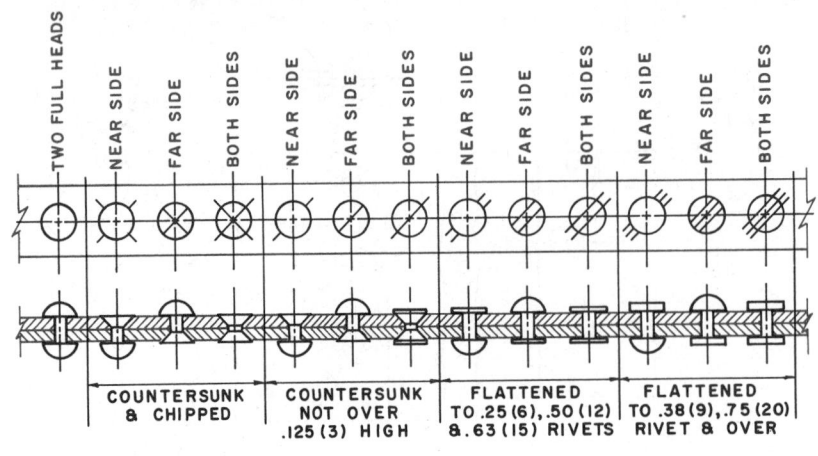

Figure 13-36B Illustrated rivet code

Keyseat

A *keyseat* is an auxiliary-located rectangular groove machined into the shaft and/or hub to receive the key.

Classes of Fit

There are three classifications of fit:

Class 1 A side surface clearance fit obtained by using bar stock key and keyseat tolerances. This is a relatively free fit.

Class 2 A possible side surface interference or side surface clearance fit obtained by using bar stock key and keyseat tolerances. This is a relatively tight fit.

Class 3 A side surface interference fit obtained by inter-ference fit tolerances. This is a very tight fit and has not been generally standardized.

Key Sizes

For a general rule, the key width is about one-fourth the nominal diameter of the shaft. For exact recommended key sizes, refer to the Appendix in the text or a *Machinery's Handbook*.

Dimensioning Keyseats

Methods of dimensioning a stock key are shown in ***Figure 13-38***.

For dimensioning a Woodruff keyseat, the key number must be included, ***Figure 13-39***. Refer to the key size in the Appendix to obtain the exact sizes. To dimension a Pratt & Whitney keyseat, see ***Figure 13-40***.

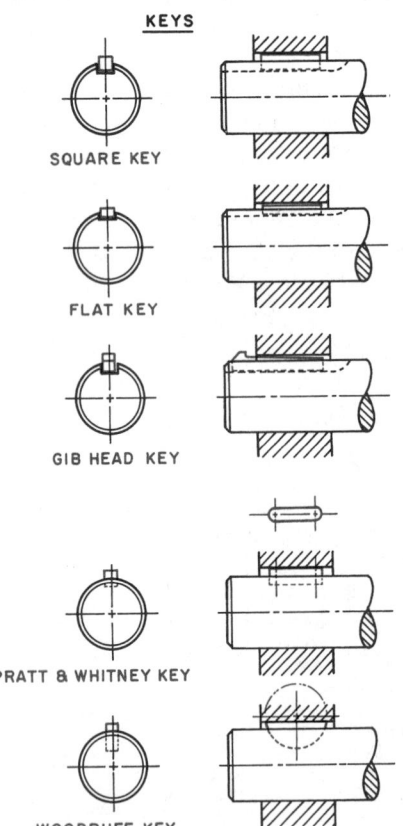

KEYS

SQUARE KEY

FLAT KEY

GIB HEAD KEY

PRATT & WHITNEY KEY

WOODRUFF KEY

Figure 13-37 Keys and keyseats

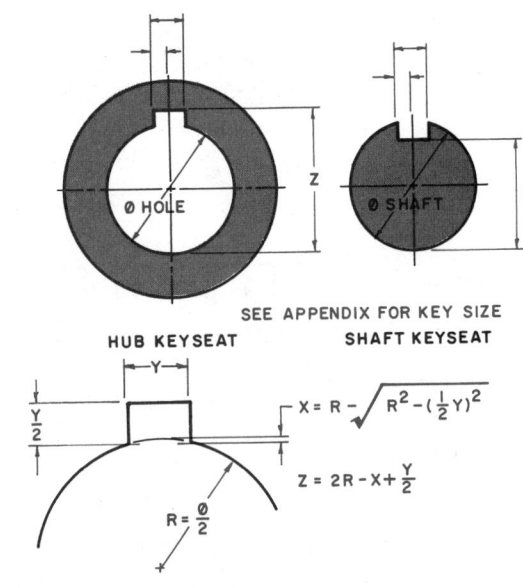

Ø HOLE

Ø SHAFT

SEE APPENDIX FOR KEY SIZE

HUB KEYSEAT   **SHAFT KEYSEAT**

$$X = R - \sqrt{R^2 - (\tfrac{1}{2}Y)^2}$$

$$Z = 2R - X + \frac{Y}{2}$$

$$R = \frac{\emptyset}{2}$$

Figure 13-38 Dimensioning keyseats

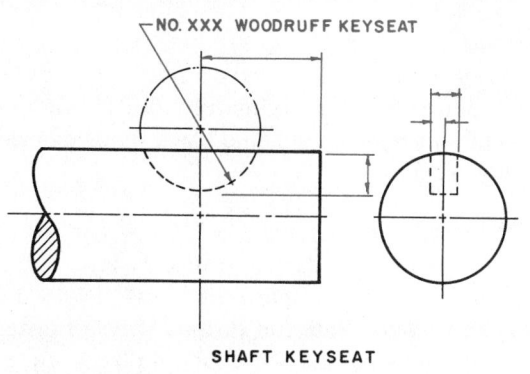

SHAFT KEYSEAT

Figure 13-39 Woodruff keyseat

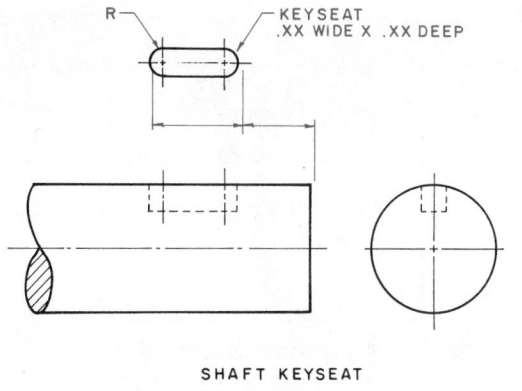

SHAFT KEYSEAT

Figure 13-40 Pratt & Whitney keyseat

Grooved Fasteners

Many types of fasteners are used in industry, each with its own application. Threaded fasteners, such as nuts and bolts, are used to hold parts in tension. *Grooved*

fasteners are used to solve metal-to-metal pinning needs with shear application, *Figure 13-41*.

Grooved fasteners have great holding power and are resistant to shock, vibration, and fatigue. They are available in a wide range of types, sizes, and materials. A grooved fastener often has a better appearance than most other methods of fastening. This can be important to the overall design, if the fastener is visible.

Grooved fasteners have three parallel grooves, equally spaced, impressed longitudinally on their exterior surface. To make these grooves, a grooving tool is pressed *below* the surface to displace a carefully determined amount of material. *Nothing* is removed. The metal is displaced to each side, forming a raised portion or flute extending along each side of the groove, *Figure 13-42*. The crest of the flute constitutes the *expanded diameter* (Dx). The expanded diameter (Dx) is a few thousandths larger than the nominal diameter (D) of the stock.

Grooved fasteners can be custom manufactured in order to meet most any application. For example, a custom grooved pin can be made with a cross-drilled hole, a groove for a snap ring, or a threaded hole in one end. The groove length can also be varied and placed anywhere along the length of the pin.

Installation

The grooved fastener is forced into a drilled hole slightly larger than the nominal or specified diameter of the pin, *Figure 13-43*. The crest or flutes are forced back into the grooves when the fastener is driven into the hole. The resiliency of the metal forced back into the grooves creates powerful radial forces against the hole wall.

In many cases, grooved fasteners are lower in cost than knurled pins, taper pins, pins with cotter pins, rivets, set screws, keys or other methods used to fasten metal-to-metal parts together. The installation cost of

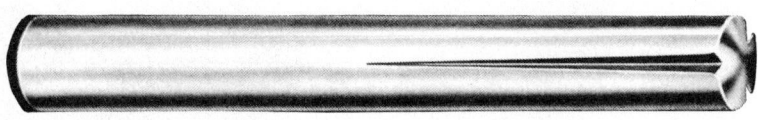

Figure 13-41 Grooved fastener

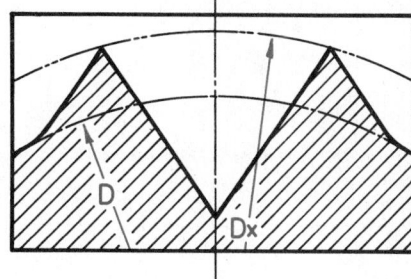

Figure 13-42 Enlarged sectional view of one of the grooves before inserting fastener

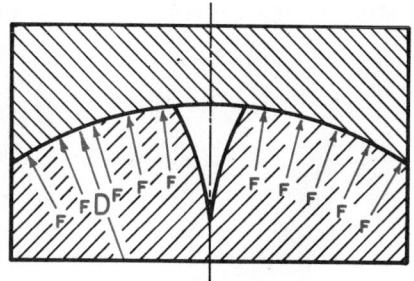

Figure 13-43 The same view after insertion. A powerful radial thrust is obtained.

grooved pins is invariably lower because of the required hole tolerances, and no special guides are required at assembly.

Material

Standard grooved fasteners are made of cold-drawn, low-carbon steel. The physical properties of this material are more than enough for ordinary applications. Alloy steel, hard brass, silicon bronze, stainless steel, and other exotic metals may also be specially ordered. These special materials are usually heat treated for optimum physical qualities.

Finish

Standard grooved fasteners have a finish of zinc electroplated, deposited approximately .00015 inch deep. Chromate, brass, cadmium, and black oxide can also be specially ordered.

Standard Types

Study the various types of grooved fasteners and the related technical data in *Figures 13-44*, *13-45*, *13-46*, *13-47*, and *13-48*. Note the various types of fasteners, how each functions, and what each replaces. For example, Type A is used in place of taper pins, rivets, set screws, and keys.

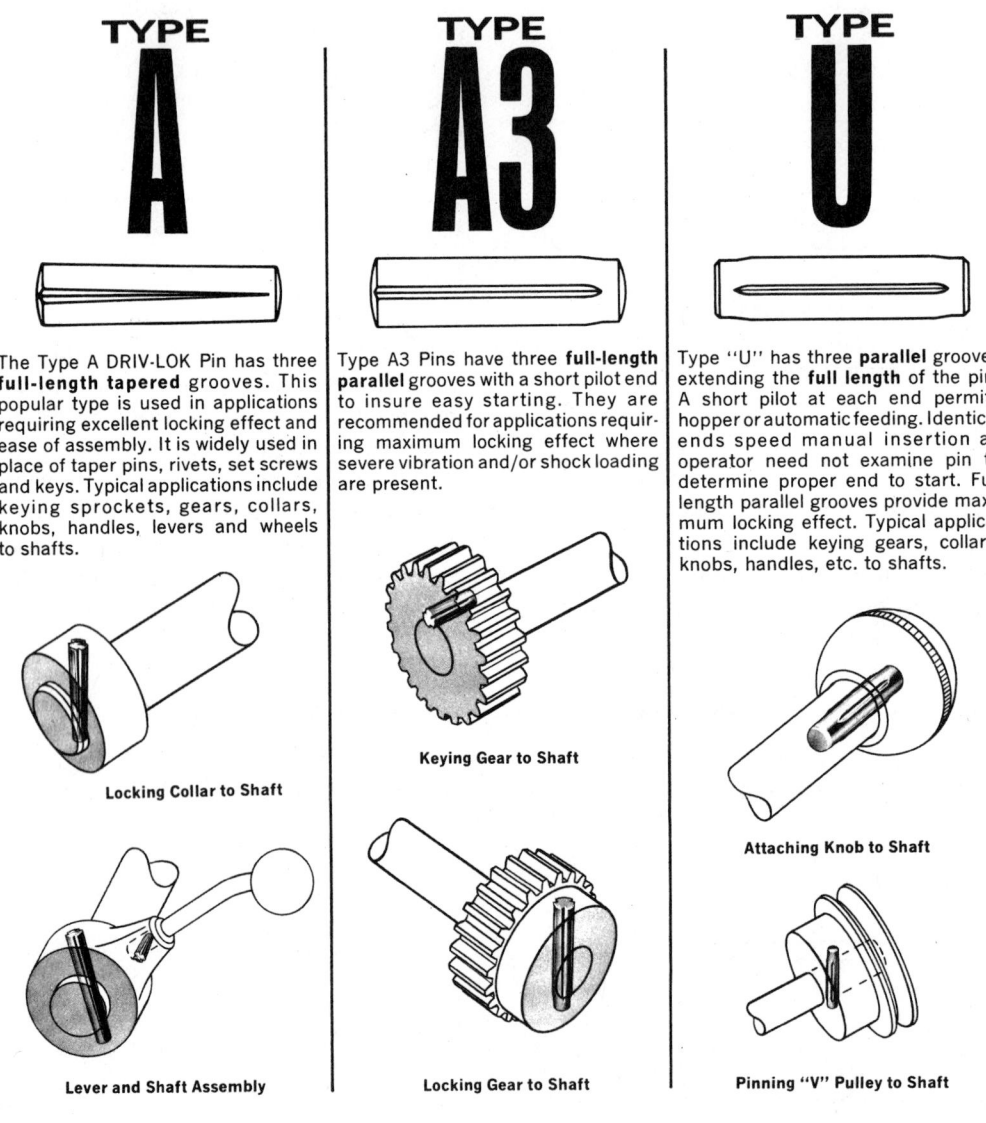

TYPE A

The Type A DRIV-LOK Pin has three **full-length tapered** grooves. This popular type is used in applications requiring excellent locking effect and ease of assembly. It is widely used in place of taper pins, rivets, set screws and keys. Typical applications include keying sprockets, gears, collars, knobs, handles, levers and wheels to shafts.

Locking Collar to Shaft

Lever and Shaft Assembly

TYPE A3

Type A3 Pins have three **full-length parallel** grooves with a short pilot end to insure easy starting. They are recommended for applications requiring maximum locking effect where severe vibration and/or shock loading are present.

Keying Gear to Shaft

Locking Gear to Shaft

TYPE U

Type "U" has three **parallel** grooves extending the **full length** of the pin. A short pilot at each end permits hopper or automatic feeding. Identical ends speed manual insertion as operator need not examine pin to determine proper end to start. Full length parallel grooves provide maximum locking effect. Typical applications include keying gears, collars, knobs, handles, etc. to shafts.

Attaching Knob to Shaft

Pinning "V" Pulley to Shaft

Figure 13-44　Typical applications for grooved pin types A, A3, and U

TYPE B

Type B Pins have three **tapered** grooves extending **one-half** the length of the Pin. This type is widely used as a hinge or pivot Pin. Driven or pressed into a straight drilled hole, the grooved portion locks in one part, while the ungrooved portion will remain free. Also excellent for dowel and locating applications.

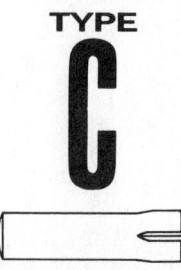

TYPE C

The Type C Pin has three **parallel** grooves extending **one-quarter** its over-all length. It is ideally suited for linkage or pivot applications, especially where a relatively short locking section and longer free length are required. Widely used in certain types of hinge applications. The long lead permits easy insertion.

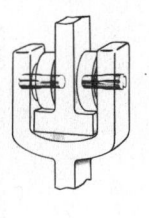

Roller Pins

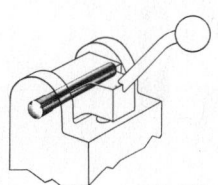

Control Valve Hinge Assembly

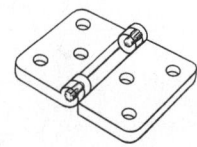

Hinge Pins

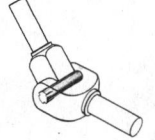

Linkage or Hinge Pin

Figure 13-45 Typical applications for grooved pin types B and C

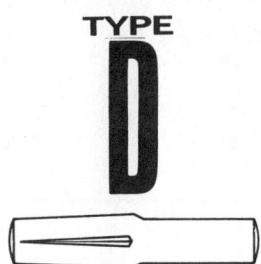

TYPE D

Type D has three **reverse** taper grooves extending **one-half** the length of the pin. Recommended for use in blind holes as a stop pin, roller pivot, dowel, or for certain hinge or linkage applications. Reverse taper grooves permit easy insertion in blind holes.

TYPE E

Type E has three **parallel half-length** grooves located equidistant from each end. Widely used as T handle on valves and tools. Also used as a cross pin, cotter pin, pivot pin, etc. where center locking is required.

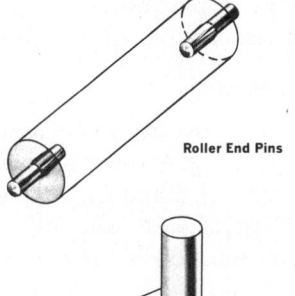

Roller End Pins

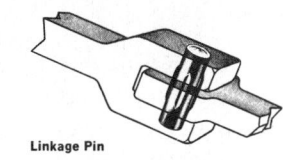

Linkage Pin

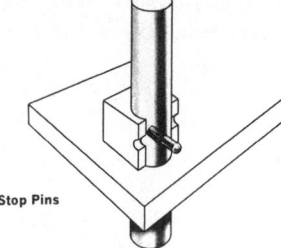

Stop Pins

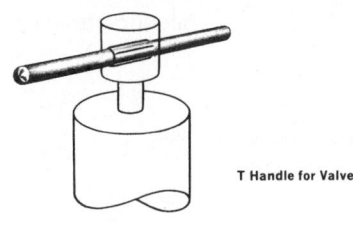

T Handle for Valve

Figure 13-46 Typical applications for grooved pin types D and E

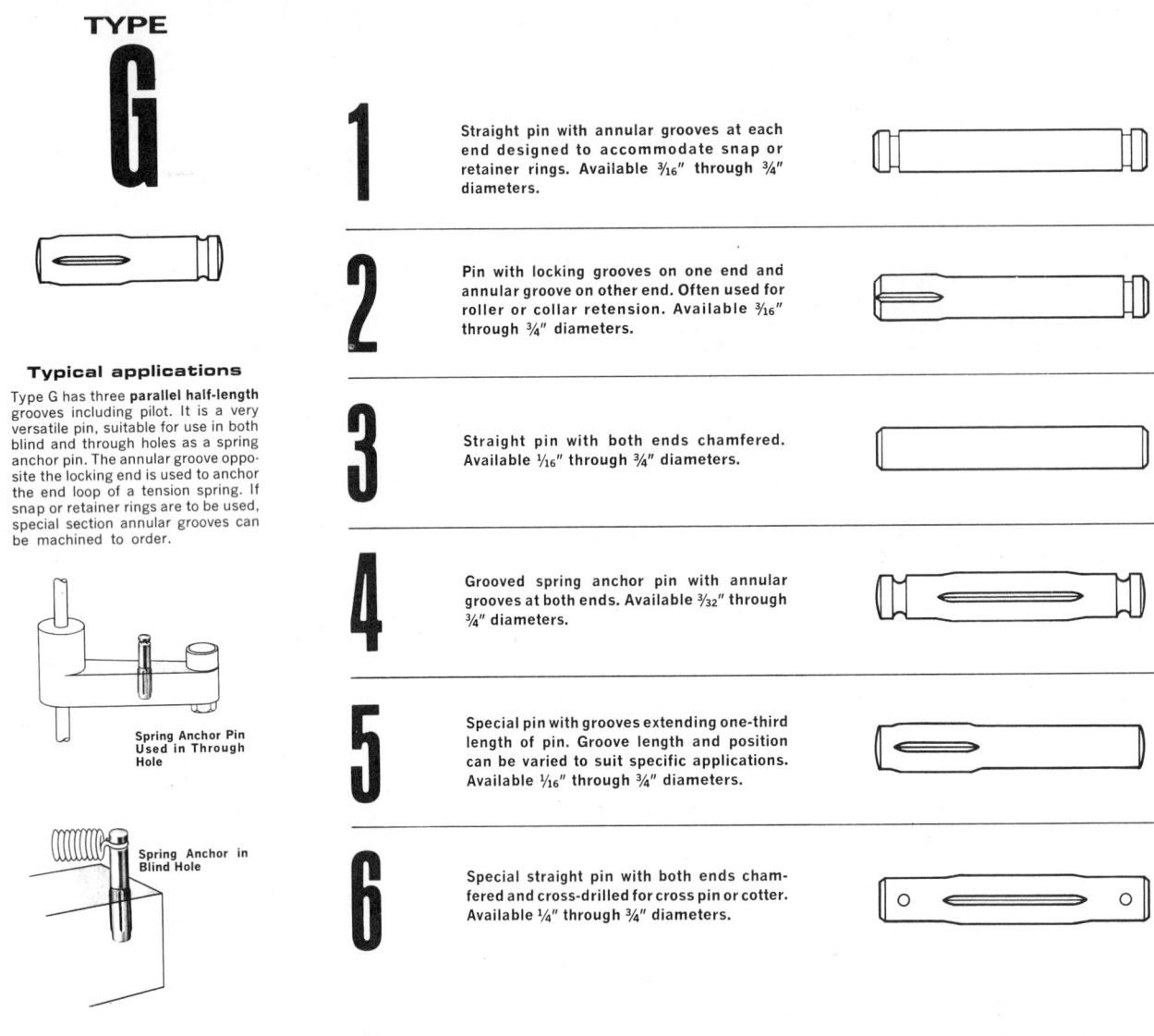

TYPE G

Typical applications

Type G has three **parallel half-length** grooves including pilot. It is a very versatile pin, suitable for use in both blind and through holes as a spring anchor pin. The annular groove opposite the locking end is used to anchor the end loop of a tension spring. If snap or retainer rings are to be used, special section annular grooves can be machined to order.

Spring Anchor Pin Used in Through Hole

Spring Anchor in Blind Hole

1 Straight pin with annular grooves at each end designed to accommodate snap or retainer rings. Available ³⁄₁₆″ through ¾″ diameters.

2 Pin with locking grooves on one end and annular groove on other end. Often used for roller or collar retension. Available ³⁄₁₆″ through ¾″ diameters.

3 Straight pin with both ends chamfered. Available ¹⁄₁₆″ through ¾″ diameters.

4 Grooved spring anchor pin with annular grooves at both ends. Available ³⁄₃₂″ through ¾″ diameters.

5 Special pin with grooves extending one-third length of pin. Groove length and position can be varied to suit specific applications. Available ¹⁄₁₆″ through ¾″ diameters.

6 Special straight pin with both ends chamfered and cross-drilled for cross pin or cotter. Available ¼″ through ¾″ diameters.

Figure 13-47 Typical application for grooved pin type G

Figure 13-48 Special pins

Standard Sizes

Refer to the standard size charts in **Figures 13-49**, **13-50**, **13-51**, and **13-52**. Across the top of each chart are the nominal sizes from 1/16-inch diameter to 1/2-inch diameter. At the left side of each chart are listed the standard lengths from 1/4-inch long to 4 1/2-inches long. Various other technical information can be derived from the charts. For drilling procedure, hole tolerances, application data, and high-alloy pin applications, see **Figures 13-53**, **13-54**, **13-55**.

Grooved Studs

Grooved studs are widely used for fastening light metal or plastic parts to heavier members or assemblies, **Figures 13-56** and **13-57**. They replace screws, rivets, peened pins, and many other types of fasteners. The grooves function as any grooved fastener.

TYPE A   # TYPE A3   # TYPE U

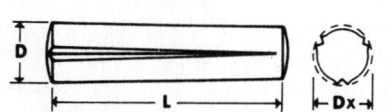
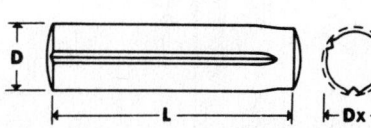

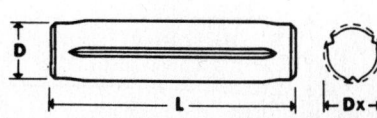

STANDARD SIZES

Nominal diameter and recommended drill sizes	1/16	5/64	3/32	7/64	1/8	5/32	3/16	7/32	1/4	5/16	3/8	7/16	1/2
Dec. Equivalents	.0625	.0781	.0938	.1094	.1250	.1563	.1875	.2188	.2500	.3125	.3750	.4375	.5000
Crown Height, In.	.0065	.0087	.0091	.0110	.0130	.0170	.0180	.0220	.0260	.0340	.0390	.0470	.0520
Radius, In. ±.010	5/64	3/32	1/8	9/64	5/32	3/16	1/4	9/32	5/16	3/8	15/32	17/32	5/8
Pilot Length, In. (Ref.)	1/32	1/32	1/32	1/32	1/32	1/16	1/16	1/16	1/16	3/32	3/32	3/32	3/32
Chamfer Length, In. (Type U Only)	1/64	1/64	1/64	1/64	1/64	1/32	1/32	1/32	1/32	3/64	3/64	3/64	3/64

"Dx" EXPANDED DIAMETER—CAN BE DETERMINED ACCURATELY ONLY WITH RING GAGES

LENGTH OF PIN IN INCHES

Length of pin (in.)	1/16	5/64	3/32	7/64	1/8	5/32	3/16	7/32	1/4	5/16	3/8	7/16	1/2
¼ (.250)	.068	.084	.101	.117	.134								
⅜ (.375)	.068	.084	.101	.117	.134	.166	.198						
½ (.500)	.068	.084	.101	.117	.134	.166	.198	.230	.263				
⅝ (.625)	.068	.084	.101	.117	.134	.166	.198	.230	.263	.329			
¾ (.750)	.068	.084	.101	.116	.134	.166	.198	.230	.263	.329	.394		
⅞ (.875)	.068	.084	.101	.116	.133	.165	.198	.230	.263	.329	.394	.459	
1 (1.000)	.068	.084	.101	.115	.133	.165	.198	.230	.263	.329	.394	.459	.525
1¼ (1.250)			.101	.115	.132	.164	.197	.230	.263	.329	.394	.459	.525
1½ (1.500)					.132	.164	.197	.230	.263	.329	.394	.459	.525
1¾ (1.750)						.163	.197	.229	.262	.328	.393	.459	.525
2 (2.000)						.163	.196	.229	.262	.328	.393	.458	.525
2¼ (2.250)							.196	.229	.262	.328	.393	.458	.524
2½ (2.500)								.228	.261	.327	.393	.458	.524
2¾ (2.750)								.228	.261	.327	.393	.458	.524
3 (3.000)								.227	.260	.327	.392	.457	.523
3¼ (3.250)									.260	.326	.392	.457	.523
3½ (3.500)										.326	.391	.456	.522
3¾ (3.750)											.391	.456	.522
4 (4.000)											.390	.455	.521
4¼ (4.250)											.390	.455	.521
4½ (4.500)												.454	.520

Nom. Diam. (D)	Exp. Diam. (Dx) reduced by
1/16	.002
5/64	.002
3/32	.002
7/64	.002
1/8	.002
5/32	.002
3/16	.003
7/32	.003
1/4	.003
5/16	.004
3/8	.005
7/16	.006
1/2	.006

TOLERANCES:

On Nominal Diameter "D"
+.000—.001 up to 7/64" diameter
+.000—.002 7/64" and above

On Expanded Diameter "Dx"
±.001 up to 7/64" diameter
±.002 7/64" and above

On over-all Length "L"
±.010 for all diameters

◄ For stainless steels and other special materials, the expanded diameters shown in above table are reduced by amounts shown at left.

Note: Intermediate pin lengths, pin diameters up to ¾", groove lengths, and groove positions to order as specials. When ordering, specify type, diameter, length, and special finishes.

Figure 13-49   Standard size chart of grooved pin types A, A3, and U

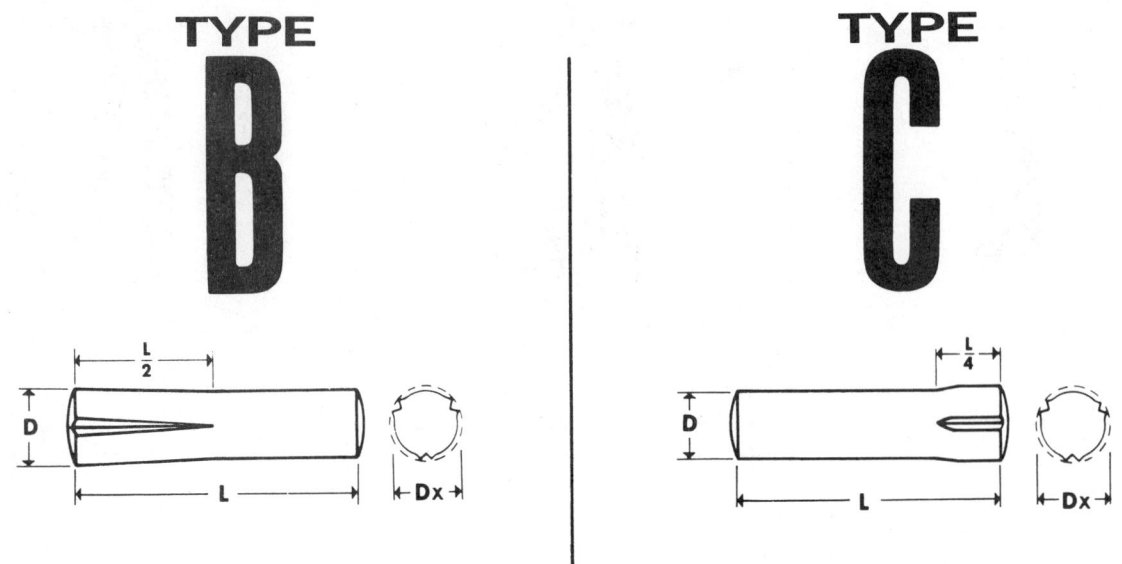

STANDARD SIZES

Nominal diameter and recommended drill sizes	1/16	5/64	3/32	7/64	1/8	5/32	3/16	7/32	1/4	5/16	3/8	7/16	1/2
Dec. Equivalents	.0625	.0781	.0938	.1094	.1250	.1563	.1875	.2188	.2500	.3125	.3750	.4375	.5000
Crown Height, In.	.0065	.0087	.0091	.0110	.0130	.0170	.0180	.0220	.0260	.0340	.0390	.0470	.0520
Radius, In. ±.010	5/64	3/32	1/8	9/64	5/32	3/16	1/4	9/32	5/16	3/8	15/32	17/32	5/8

"Dx" EXPANDED DIAMETER—CAN BE DETERMINED ACCURATELY ONLY WITH RING GAGES

LENGTH OF PIN IN INCHES

Length	(dec)	1/16	5/64	3/32	7/64	1/8	5/32	3/16	7/32	1/4	5/16	3/8	7/16	1/2
1/4	(.250)	.068	.084	.101	.117	.134								
3/8	(.375)	.068	.084	.101	.117	.134	.166	.198						
1/2	(.500)	.068	.084	.101	.117	.134	.166	.198	.230	.263				
5/8	(.625)	.068	.084	.101	.117	.134	.166	.198	.230	.263	.329			
3/4	(.750)	.068	.084	.101	.117	.134	.166	.198	.230	.263	.329	.394		
7/8	(.875)	.068	.084	.101	.117	.134	.166	.198	.230	.263	.329	.394	.459	
1	(1.000)	.068	.084	.101	.117	.134	.166	.198	.230	.263	.329	.394	.459	.525
1¼	(1.250)			.101	.117	.134	.166	.198	.230	.263	.329	.394	.459	.525
1½	(1.500)					.134	.166	.198	.230	.263	.329	.394	.459	.525
1¾	(1.750)						.165	.198	.230	.263	.329	.394	.459	.525
2	(2.000)						.165	.198	.230	.263	.329	.394	.459	.525
2¼	(2.250)							.197	.230	.263	.329	.394	.459	.525
2½	(2.500)								.230	.263	.329	.394	.459	.525
2¾	(2.750)								.229	.262	.329	.394	.459	.525
3	(3.000)								.229	.262	.329	.394	.459	.525
3¼	(3.250)									.262	.328	.393	.459	.525
3½	(3.500)										.328	.393	.459	.525
3¾	(3.750)											.393	.458	.525
4	(4.000)											.393	.458	.525
4¼	(4.250)											.393	.458	.524
4½	(4.500)												.458	.524

Nom. Diam. (D)	Exp. Diam. (Dx) reduced by
1/16	.002
5/64	.002
3/32	.002
7/64	.002
1/8	.002
5/32	.002
3/16	.003
7/32	.003
1/4	.003
5/16	.004
3/8	.005
7/16	.006
1/2	.006

TOLERANCES:

On Nominal Diameter "D"
+.000—.001 up to 7/64" diameter
+.000—.002 7/64" and above

On Expanded Diameter "Dx"
±.001 up to 7/64" diameter
±.002 7/64" and above

On over-all Length "L"
±.010 for all diameters

◄ For stainless steels and other special materials, the expanded diameters shown in above table are reduced by amounts shown at left.

Note: Intermediate pin lengths, pin diameters up to ¾", groove lengths, and groove positions to order as specials. When ordering, specify type, diameter, length, and special finishes.

Figure 13-50   Standard size chart of grooved pin types B and C

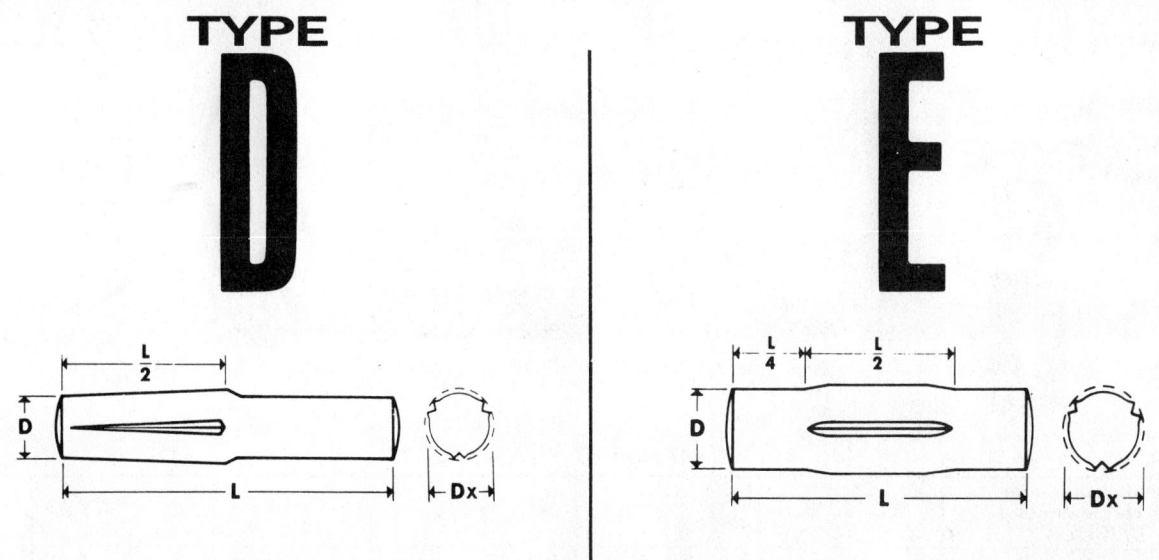

STANDARD SIZES

Nominal diameter and recommended drill sizes	¹⁄₁₆	⁵⁄₆₄	³⁄₃₂	⁷⁄₆₄	¹⁄₈	⁵⁄₃₂	³⁄₁₆	⁷⁄₃₂	¹⁄₄	⁵⁄₁₆	³⁄₈	⁷⁄₁₆	¹⁄₂
Dec. Equivalents	.0625	.0781	.0938	.1094	.1250	.1563	.1875	.2188	.2500	.3125	.3750	.4375	.5000
Crown Height, In.	.0065	.0087	.0091	.0110	.0130	.0170	.0180	.0220	.0260	.0340	.0390	.0470	.0520
Radius, In. ± .010	⁵⁄₆₄	³⁄₃₂	¹⁄₈	⁹⁄₆₄	⁵⁄₃₂	³⁄₁₆	¹⁄₄	⁹⁄₃₂	⁵⁄₁₆	³⁄₈	¹⁵⁄₃₂	¹⁷⁄₃₂	⁵⁄₈

"Dx" EXPANDED DIAMETER—CAN BE DETERMINED ACCURATELY ONLY WITH RING GAGES

LENGTH OF PIN IN INCHES

¼	(.250)	.068	.084	.101	.117	.134								
⅜	(.375)	.068	.084	.101	.117	.134	.166	.198						
½	(.500)	.068	.084	.101	.117	.134	.166	.198	.230	.263				
⅝	(.625)	.068	.084	.101	.117	.134	.166	.198	.230	.263	.329			
¾	(.750)	.068	.084	.101	.117	.134	.166	.198	.230	.263	.329	.394		
⅞	(.875)	.068	.084	.101	.117	.134	.166	.198	.230	.263	.329	.394	.459	
1	(1.000)	.068	.084	.101	.117	.134	.166	.198	.230	.263	.329	.394	.459	.525
1¼	(1.250)		.101	.117	.134	.166	.198	.230	.263	.329	.394	.459	.525	
1½	(1.500)			.134	.166	.198	.230	.263	.329	.394	.459	.525		
1¾	(1.750)				.165	.198	.230	.263	.329	.394	.459	.525		
2	(2.000)					.165	.198	.230	.263	.329	.394	.459	.525	
2¼	(2.250)						.197	.230	.263	.329	.394	.459	.525	
2½	(2.500)							.230	.263	.329	.394	.459	.525	
2¾	(2.750)							.229	.262	.329	.394	.459	.525	
3	(3.000)								.229	.262	.329	.394	.459	.525
3¼	(3.250)									.262	.328	.393	.459	.525
3½	(3.500)										.328	.393	.459	.525
3¾	(3.750)											.393	.458	.525
4	(4.000)											.393	.458	.525
4¼	(4.250)											.393	.458	.524
4½	(4.500)												.458	.524

Nom. Diam. (D)	Exp. Diam. (Dx) reduced by
¹⁄₁₆	.002
⁵⁄₆₄	.002
³⁄₃₂	.002
⁷⁄₆₄	.002
¹⁄₈	.002
⁵⁄₃₂	.002
³⁄₁₆	.003
⁷⁄₃₂	.003
¹⁄₄	.003
⁵⁄₁₆	.004
⅜	.005
⁷⁄₁₆	.006
½	.006

TOLERANCES:

On Nominal Diameter "D"
+.000—.001 up to ⁷⁄₆₄" diameter
+.000—.002 ⁷⁄₆₄" and above

On Expanded Diameter "Dx"
± .001 up to ⁷⁄₆₄" diameter
± .002 ⁷⁄₆₄" and above
On over-all Length "L"
± .010 for all diameters

◄ For stainless steels and other special materials, the expanded diameters shown in above table are reduced by amounts shown at left.

Note: Intermediate pin lengths, pin diameters up to ¾", groove lengths, and groove positions to order as specials. When ordering, specify type, diameter, length, and special finishes.

Figure 13-51 Standard size chart of grooved pin types D and E

TYPE G

(combined former F and G)

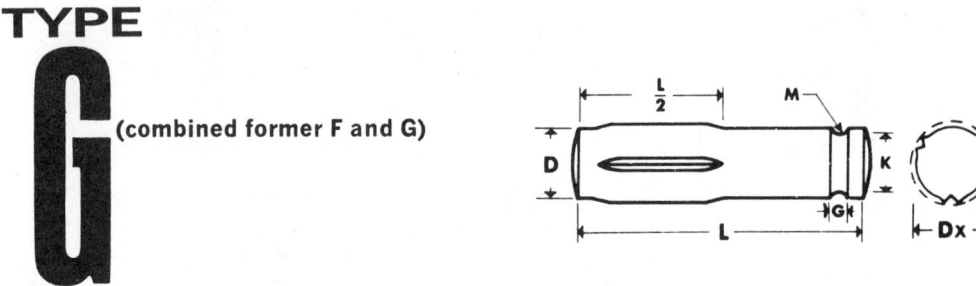

STANDARD SIZES

Nominal diameter and recommended drill sizes	³⁄₃₂	⁷⁄₆₄	⅛	⁵⁄₃₂	³⁄₁₆	⁷⁄₃₂	¼	⁵⁄₁₆	⅜	⁷⁄₁₆	½
Dec. Equivalents	.0938	.1094	.1250	.1563	.1875	.2188	.2500	.3125	.3750	.4375	.5000
Crown Height, In.	.0091	.0110	.0130	.0170	.0180	.0220	.0260	.0340	.0390	.0470	.0520
Radius, In., ±0.010	⅛	⁹⁄₆₄	⁵⁄₃₂	³⁄₁₆	¼	⁹⁄₃₂	⁵⁄₁₆	⅜	¹⁵⁄₃₂	¹⁷⁄₃₂	⅝
Pilot Length, In. (Ref.)	¹⁄₃₂	¹⁄₃₂	¹⁄₃₂	¹⁄₁₆	¹⁄₁₆	¹⁄₁₆	¹⁄₁₆	³⁄₃₂	³⁄₃₂	³⁄₃₂	³⁄₃₂
M Neck Radius (Ref.)	¹⁄₆₄	¹⁄₆₄	¹⁄₃₂	¹⁄₃₂	¹⁄₃₂	³⁄₆₄	³⁄₆₄	¹⁄₁₆	¹⁄₁₆	³⁄₃₂	³⁄₃₂
G Neck Width (Ref.)	¹⁄₃₂	¹⁄₃₂	¹⁄₁₆	¹⁄₁₆	¹⁄₁₆	³⁄₃₂	³⁄₃₂	⅛	⅛	³⁄₁₆	³⁄₁₆
Shoulder Width +.010—.000	¹⁄₃₂	¹⁄₃₂	¹⁄₃₂	³⁄₆₄	³⁄₆₄	¹⁄₁₆	¹⁄₁₆	³⁄₃₂	⅛	⅛	⅛
K Neck Diameter ±.005	.062	.078	.083	.104	.125	.146	.167	.209	.250	.293	.312

"Dx" EXPANDED DIAMETER—CAN BE DETERMINED ACCURATELY ONLY WITH RING GAGES

LENGTH OF PIN IN INCHES

		³⁄₃₂	⁷⁄₆₄	⅛	⁵⁄₃₂	³⁄₁₆	⁷⁄₃₂	¼	⁵⁄₁₆	⅜	⁷⁄₁₆	½
⅜	(.375)	.101	.117	.134	.166	.198						
½	(.500)	.101	.117	.134	.166	.198	.230	.263				
⅝	(.625)	.101	.117	.134	.166	.198	.230	.263	.329			
¾	(.750)	.101	.117	.134	.166	.198	.230	.263	.329	.394		
⅞	(.875)	.101	.117	.134	.166	.198	.230	.263	.329	.394	.459	
1	(1.000)	.101	.117	.134	.166	.198	.230	.263	.329	.394	.459	.525
1¼	(1.250)	.101	.117	.134	.166	.198	.230	.263	.329	.394	.459	.525
1½	(1.500)			.134	.166	.198	.230	.263	.329	.394	.459	.525
1¾	(1.750)				.165	.198	.230	.263	.329	.394	.459	.525
2	(2.000)				.165	.198	.230	.263	.329	.394	.459	.525
2¼	(2.250)					.197	.230	.263	.329	.394	.459	.525
2½	(2.500)						.230	.263	.329	.394	.459	.525
2¾	(2.750)							.229	.262	.329	.394	.459	.525
3	(3.000)							.229	.262	.329	.394	.459	.525
3¼	(3.250)								.262	.329	.393	.459	.525
3½	(3.500)									.328	.393	.459	.525
3¾	(3.750)									.328	.393	.458	.525
4	(4.000)										.393	.458	.525
4¼	(4.250)										.393	.458	.524
4½	(4.500)											.458	.524

Nom. Diam. (D)	Exp. Diam. (Dx) reduced by
³⁄₃₂	.002
⁷⁄₆₄	.002
⅛	.002
⁵⁄₃₂	.002
³⁄₁₆	.003
⁷⁄₃₂	.003
¼	.003
⁵⁄₁₆	.004
⅜	.005
⁷⁄₁₆	.006
½	.006

TOLERANCES:

On Nominal Diameter "D"
+.000—.001 up to ⁷⁄₆₄″ diameter
+.000—.002 ⁷⁄₆₄″ and above

On Expanded Diameter "Dx"
±.001 up to ⁷⁄₆₄″ diameter
±.002 ⁷⁄₆₄″ and above

On over-all Length "L"
±.010 for all diameters

◄ For stainless steels and other special materials, the expanded diameters shown in above table are reduced by amounts shown at left.

Note: Intermediate pin lengths, pin diameters up to ¾″, groove lengths, and groove positions to order as specials. When ordering, specify type, diameter, length, and special finishes.

Figure 13-52   Standard size chart of grooved pin type G

Pin Diameter	Decimal Equivalent	Recommended Drill Size	Hole Tolerances ADD To Nominal Diameter
1/16"	.0625	1/16"	.002"
5/64	.0781	5/64	.002"
3/32	.0938	3/32	.003"
7/64	.1094	7/64	.003"
1/8	.1250	1/8	.003"
5/32	.1563	5/32	.003"
3/16	.1875	3/16	.004"
7/32	.2188	7/32	.004"
1/4	.2500	1/4	.004"
5/16	.3125	5/16	.005"
3/8	.3750	3/8	.005"
7/16	.4375	7/16	.006"
1/2	.5000	1/2	.006"

Tolerances for drilled holes shown in table are based on depth to diameter ratio of approximately 5 to 1. Higher ratios may cause these figures to be exceeded, but in no case should they be exceeded by more than 10%. Specifications for holes having a depth to diameter ratio of 1 to 1 or less should be held extremely close; 60% of figures shown in table is recommended.

Undersized drills should never be used to produce holes for Driv-Lok Pins. This malpractice results from the false assumption that the pins will "hold better." Instead, they bend, damage the hole wall, crack castings, and peel their expanded flutes, thus reducing their retaining characteristics and preventing their reuse.

Holes made in hardened steel or cast iron are recommended to have a slight chamfer at the entrance. This eliminates shearing of the flutes as the pins are forced in.

Care should be exercised in all drilling for Driv-Lok Pins. Drills used should be new or properly ground with the aid of an approved grinding fixture. The drilling machine spindle must be in good condition and operated at correct speeds and feeds for the metal being drilled, and suitable coolant is always recommended. Drill jigs with accurate bushings always facilitate good drilling practice.

Figure 13-53   Drilling procedure and hole tolerance

RECOMMENDED PIN DIAMETER FOR VARIOUS SHAFT SIZES AND TORQUE TRANSMITTED BY PIN IN DOUBLE SHEAR

Shaft Size	Pin Diameter	Torque Inch Lbs.	H.P. at 100 R.P.M.	Shaft Size	Pin Diameter	Torque Inch Lbs.	H.P. at 100 R.P.M.
3/16	1/16	4.6	.007	7/8	1/4	347	.555
7/32	5/64	8.4	.013	15/16	5/16	580	.927
1/4	3/32	13.7	.022	1	5/16	618	.990
5/16	7/64	23.6	.038	1 1/16	5/16	657	1.05
3/8	1/8	37.2	.060	1 1/8	3/8	1010	1.61
7/16	5/32	67.6	.108	1 3/16	3/8	1065	1.70
1/2	5/32	77.2	.124	1 1/4	3/8	1120	1.79
9/16	3/16	125.0	.200	1 5/16	7/16	1590	2.55
5/8	3/16	139.0	.222	1 3/8	7/16	1670	2.67
11/16	7/32	207.0	.332	1 7/16	7/16	1740	2.79
3/4	1/4	297.0	.476	1 1/2	1/2	2380	3.81
13/16	1/4	322	.516				

This table is a guide in selecting the proper size Driv-Lok Grooved Pin to use in keying machine members to shafts of given sizes and for specific load requirements. Torque and horsepower ratings are based on pins made of cold finished, low carbon steel and a safety factor of 8 is assumed.

MINIMUM SINGLE SHEAR VALUES (LBS.) OF DRIV-LOK PINS OF VARIOUS MATERIALS

DRIV-LOK PIN DIAM.	Cold Finished 1213 Steel	Shear-Proof® ALLOY STEEL R.C. 40-48	Brass	Silicon Bronze	Heat Treated Stainless Steel
1/16	200	363	124	186	308
5/64	312	562	192	288	478
3/32	442	798	272	408	680
7/64	605	1091	372	558	933
1/8	800	1443	492	738	1230
5/32	1240	2236	764	1145	1910
3/16	1790	3220	1100	1650	2750
7/32	2430	4386	1495	2240	3740
1/4	3190	5753	1960	2940	4910
5/16	4970	8974	3060	4580	7650
3/8	5810	12960	4420	6630	11050
7/16	7910	17580	6010	9010	15000
1/2	10300	23020	7850	11800	19640

Figure 13-54   Grooved pin application/engineering data

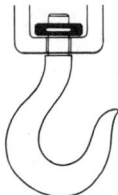

MATERIAL HANDLING EQUIPMENT—The Type E pin provides positive locking with a half-length groove in the center of the Pin. Extreme shear is exerted in this application, yet the Shear-Proof Pin is used with complete safety for both men and materials. Type E is a special Pin.

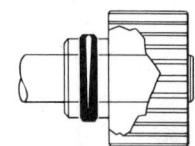

HEAVY-DUTY GEAR AND SHAFT ASSEMBLY—Type A Shear-Proof Pin as specified for this application to give maximum locking power over the entire pin and gear hub area. The Type A Pin, with grooves the full length of the Pin, is the standard stock Pin which meets most applications.

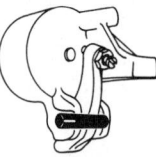

AUTOMATIC TRANSMISSION IN AUTOMOBILES—Special Type C was selected as a shaft in this transmission servo to replace a cross drilled shaft with a cross pin for holding shaft in position. This eliminated a costly drilling operation and the cross pin.

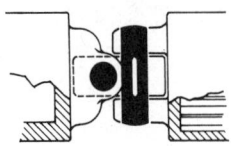

UNIVERSAL JOINTS IN HAND TOOLS—Special Type E Pin with center groove eliminates costly staking and grinding operations and improves product appearance. This Pin is easily installed, fits flush and permits plating before assembly.

EYE BOLT HINGE PIN — Type C Pin, with quarter-length grooves, provides maximum ease of assembly. There is no interference until three-fourths of the pin is in position. The high safety factor inherent in Shear-Proof Pins makes them practical and efficient for such constant shear applications. Type C is a special Pin.

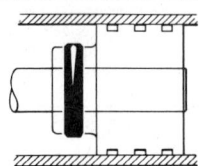

HIGH-PRESSURE PISTON AND ROD ASSEMBLY—Type B Pin was used here because the half-length grooves simplified the job of starting the pin into the hole. Ease of assembly was matched with sufficient locking power even when subjected to continuous, strong reciprocating forces. Type B is a special Pin.

Figure 13-55 High alloy shear-proof pins

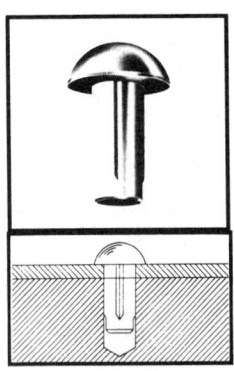

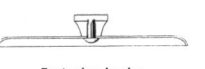

Fastening knobs
handles, etc.

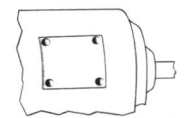

Attaching nameplates,
instruction panels

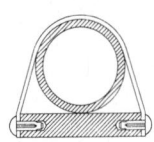

Widely used for
fastening brackets

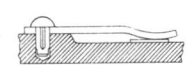

Fastening spring assemblies
or control arms

STANDARD SIZES and SPECIFICATIONS

Stud Number	Nominal Shank Diameter	Recom-mended Drill Size	Head Dia. Max.	Head Dia. Min.	Head Height Max.	Head Height Min.
0	.067	51	.130	.120	.050	.040
2	.086	44	.162	.146	.070	.059
4	.104	37	.211	.193	.086	.075
6	.120	31	.260	.240	.103	.091
7	.136	29	.309	.287	.119	.107
8	.144	27	.309	.287	.119	.107
10	.161	20	.359	.334	.136	.124
12	.196	9	.408	.382	.152	.140
14	.221	2	.457	.429	.169	.156
16	.250	¼"	.472	.443	.174	.161

Maximum Expansion—Standard Lengths

Stud No.	⅛"	3/16"	¼"	5/16"	⅜"	½"
0	.074	.074	.074			
2	.096	.096	.095			
4		.115	.113	.113		
6			.132	.130	.130	
7				.147	.147	.144
8					.155	.153
10					.173	.171
12						.206
14						.234
16						.263

TOLERANCES:
On length - .010
On Exp. Diameter - .002
On Nominal Diameter +.000
-.002

Note: The expanded diameter can be determined accurately only with ring gages.

Figure 13-56 Standard studs

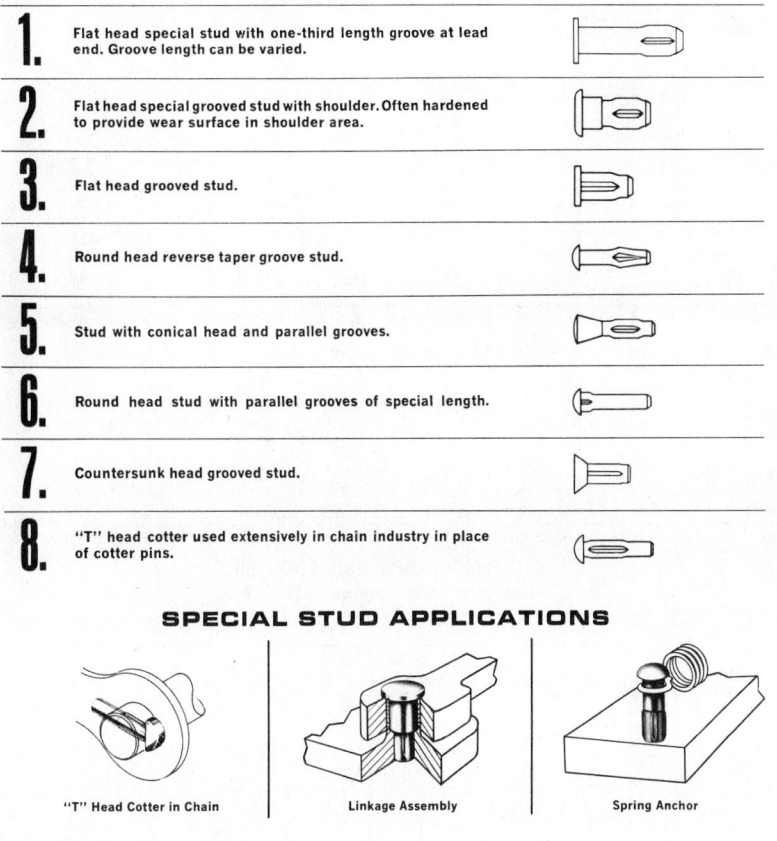

1. Flat head special stud with one-third length groove at lead end. Groove length can be varied.

2. Flat head special grooved stud with shoulder. Often hardened to provide wear surface in shoulder area.

3. Flat head grooved stud.

4. Round head reverse taper groove stud.

5. Stud with conical head and parallel grooves.

6. Round head stud with parallel grooves of special length.

7. Countersunk head grooved stud.

8. "T" head cotter used extensively in chain industry in place of cotter pins.

SPECIAL STUD APPLICATIONS

"T" Head Cotter in Chain

Linkage Assembly

Spring Anchor

Figure 13-57 Special studs

Spring Pins

Another type of fastener is the *spring pin*, **Figure 13-58**. Spring pins are manufactured by cold forming strip metal in a progressive roll forming operation. After forming, the pins are broken off and deburred to eliminate any sharp edges. They are then heat treated to a R/C 46/53. This develops spring qualities or resiliency in the metal.

In the free state, the pins are larger in diameter than the hole into which they are to inserted. The pins compress themselves as they are driven into the hole, thus exerting radial forces around the entire circumference of the hole.

Spring pins are made from a high carbon steel—1074, stainless steel, either 420 heat treated or 300 series cold worked. Some spring pins are also made from brass or beryllium copper. Spring pins meet the demand of many industrial applications. **Figures 13-59** and **13-60** list some suggested applications.

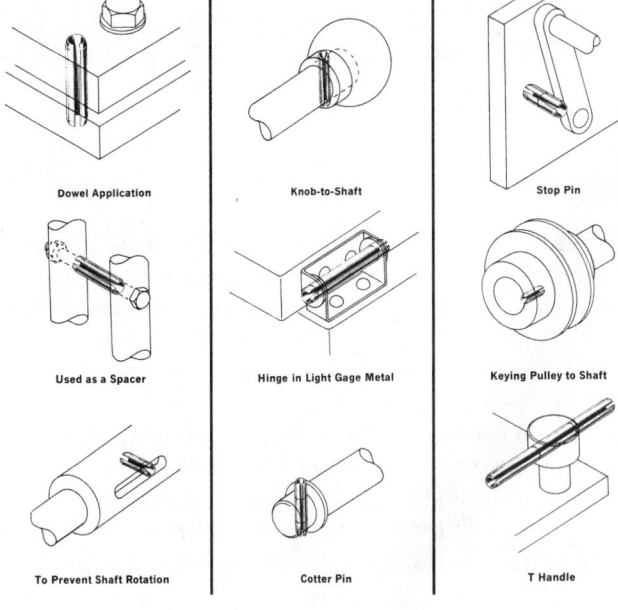

Dowel Application

Knob-to-Shaft

Stop Pin

Used as a Spacer

Hinge in Light Gage Metal

Keying Pulley to Shaft

To Prevent Shaft Rotation

Cotter Pin

T Handle

Figure 13-59 Spring pin application

Figure 13-58 Spring pin

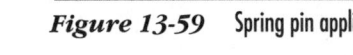

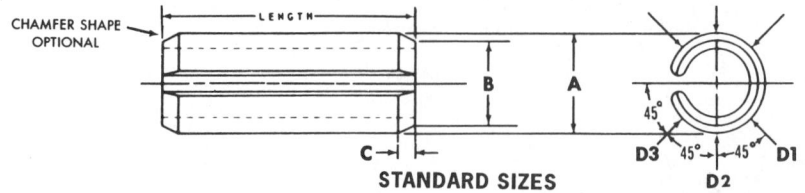

STANDARD SIZES

A			B Max.	C		WALL THICK- NESS	RECOMMENDED HOLE SIZE		MINIMUM DOUBLE SHEAR STRENGTH POUNDS Carbon Steel and Stainless Steel
Nominal	Minimum ⅓(D₁+D₂+D₃)	Maximum (Go Ring Gage)		Min.	Max.		Min.	Max.	
.062	.066	.069	.059	.007	.028	.012	.062	.065	425
.078	.083	.086	.075	.008	.032	.018	.078	.081	650
.094	.099	.103	.091	.008	.038	.022	.094	.097	1,000
.125	.131	.135	.122	.008	.044	.028	.125	.129	2,100
.156	.162	.167	.151	.010	.048	.032	.156	.160	3,000
.187	.194	.199	.182	.011	.055	.040	.187	.192	4,400
.219	.226	.232	.214	.011	.065	.048	.219	.224	5,700
.250	.258	.264	.245	.012	.065	.048	.250	.256	7,700
.312	.321	.328	.306	.014	.080	.062	.312	.318	11,500
.375	.385	.392	.368	.016	.095	.077	.375	.382	17,600
.437	.448	.456	.430	.017	.095	.077	.437	.445	20,000
.500	.513	.521	.485	.025	.110	.094	.500	.510	25,800

All dimensions listed on this page are in accordance with National Standards. Wall thicknesses within the Spring Pin industry are standard.

SPRING PIN WEIGHT PER 1000 PIECES
Material—Steel · Nominal Diameter

LENGTH	.062	.078	.094	.125	.156	.187
0.187	.10	0.18				
0.250	.15	0.22	0.33			
0.312	.19	0.28	0.41			
0.375	.23	0.34	0.50	0.89		
0.437	.27	0.40	0.58	1.00	1.50	
0.500	.30	0.46	0.66	1.20	1.70	2.50
0.562	.34	0.51	0.75	1.30	1.90	2.90
0.625	.38	0.57	0.83	1.50	2.10	3.20
0.687	.42	0.63	0.92	1.60	2.40	3.50
0.750	.46	0.69	1.00	1.80	2.60	3.80
0.812	.49	0.74	1.10	1.90	2.80	4.10
0.875	.53	0.80	1.20	2.10	3.00	4.50
0.937	.57	0.86	1.30	2.20	3.20	4.80
1.000	.61	0.92	1.40	2.40	3.40	5.10
1.125		1.00	1.50	2.70	3.80	5.70
1.250		1.20	1.70	3.00	4.30	6.40
1.375		1.30	1.80	3.30	4.70	7.00
1.500		1.40	2.00	3.60	5.10	7.60
1.625				3.90	5.50	8.30
1.750				4.20	6.00	8.90
1.875				4.50	6.40	9.60
2.000				4.70	6.80	10.0
2.250					7.80	12.0
2.500					8.60	13.0

LENGTH	.219	.250	.312	.375	.437	.500
0.562	4.00					
0.625	4.50	5.30				
0.687	4.90	5.90				
0.750	5.30	6.30	9.90	15.0		
0.812	5.80	6.90	11.0			
0.875	6.20	7.40	12.0			
0.937	6.70	8.00	12.0			
1.000	7.00	8.50	13.0	20.0	24.0	
1.125	7.90	9.50	14.0			
1.250	8.80	11.0	16.0	24.0	30.0	41.0
1.375	9.70	12.0	18.0			
1.500	11.0	13.0	19.0	29.0	36.0	49.0
1.625	12.0	14.0	21.0			
1.750	12.0	15.0	22.0	34.0	42.0	57.0
1.875	13.0	16.0	24.0			
2.000	14.0	17.0	25.0	38.0	48.0	65.0
2.250	16.0	19.0	29.0	43.0	54.0	73.0
2.500	18.0	21.0	32.0	48.0	60.0	81.0
2.750	19.0	23.0	35.0	53.0	66.0	89.0
3.000	21.0	25.0	38.0	58.0	72.0	97.0
3.250		28.0	41.0	62.0	77.0	105.0
3.500		31.0	44.0	67.0	83.0	114.0
3.750			48.0	72.0	89.0	122.0
4.000			51.0	77.0	95.0	130.0

Figure 13-60   Spring pin data

Fastening Systems

Fastening systems play a critical role in most product design. They often do more than position and secure components. In many cases, fastening systems have a direct effect upon the product's durability, reliability, size, and weight. They affect the speed with which the product may be assembled and disassembled, both during manufacturing and later in field service. Fastening systems also affect cost; not only for the fasteners, but for the machining and assembly operations they require.

Unless a drafter has had a great deal of experience in using different fastening devices and techniques, it is often difficult to choose a fastener that combines optimum function with maximum economy. A fastening system that is best for one product may not be desirable for another.

Retaining Rings

Retaining rings are precision-engineered fasteners. They provide removable shoulders for positioning or limiting the movement of parts in an assembly, *Figure 13-61*. Applications range from miniaturized electronic systems to massive earth-moving equipment. Retaining rings are used in automobiles, business machines, and complex components for guided missiles. They are found in such commonplace items as doorknobs to sophisticated underwater seismic cable connectors.

Typical ring applications are shown in *Figures 13-62 through 13-66*. From these figures, the drafter may examine the design of various types of rings, determine their purpose and function, and decide how they may be used to the best advantage.

Most retaining rings are made of materials that have

Figure 13-61   Retaining rings *(Courtesy Koh-I-Noor Rapidograph)*

Figure 13-62   Ring application in a precision differential *(Courtesy Koh-I-Noor Rapidograph)*

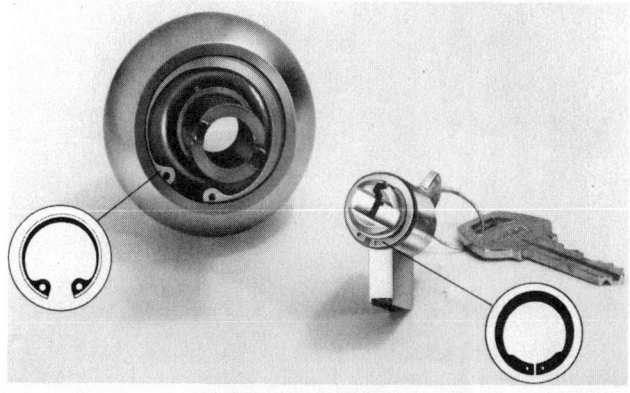

Figure 13-63   Ring application in a cylindrical lockset *(Courtesy Koh-I-Noor Rapidograph)*

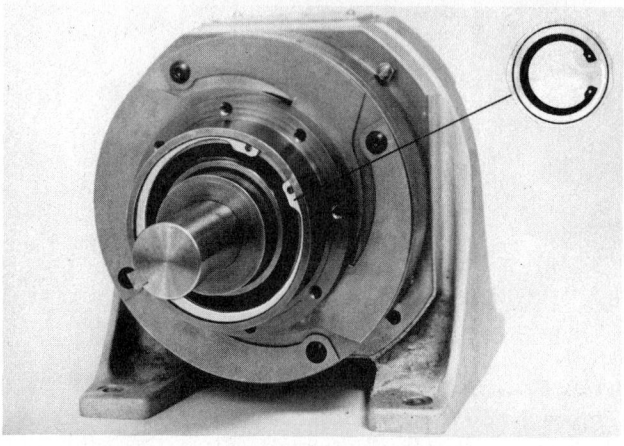

Figure 13-64   Ring application in an electromagnetic clutch brake *(Courtesy Koh-I-Noor Rapidograph)*

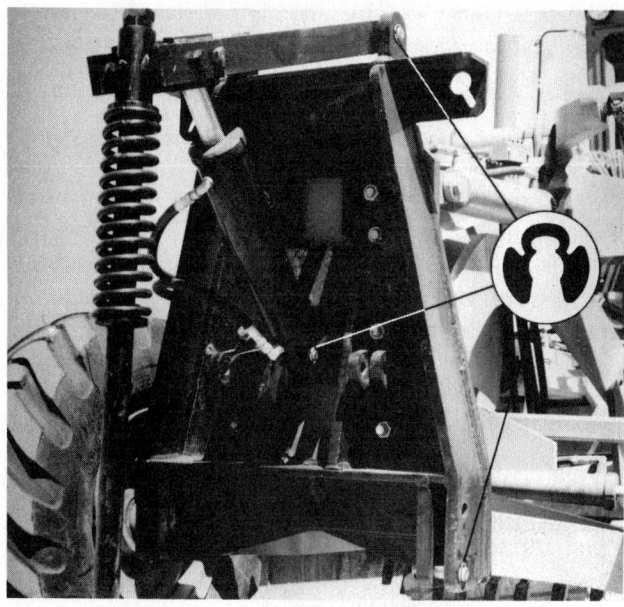

Figure 13-65   Ring application in a road grader *(Courtesy Koh-I-Noor Rapidograph)*

good spring properties. This permits the rings to be deformed elastically to a substantial degree, yet still spring back to their original shape during assembly and disassembly. This allows most rings to function in one of two ways: 1) they may be sprung into a groove or other recess in a part, or 2) they may be seated on a part in a deformed condition so that they grip the part by frictional means. In either case, the rings form a fixed shoulder against which other components may be abutted and prevented from moving.

Unlike wire-formed rings which have a uniform section height, stamped rings have a tapered radial width. The width decreases symmetrically from the center section to the free ends. The tapered section permits the rings to remain circular after they have been compressed for insertion into a bore or expanded for assembly over a shaft. Most rings are designed to be seated in grooves. This constant circularity assures maximum

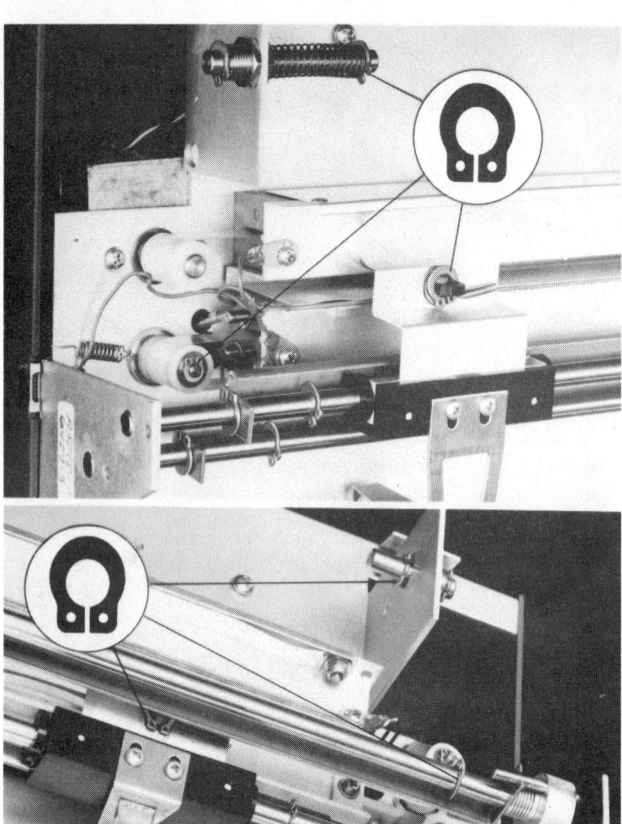

Figure 13-66 Ring application in strip chart recorder (Courtesy Koh-I-Noor Rapidograph)

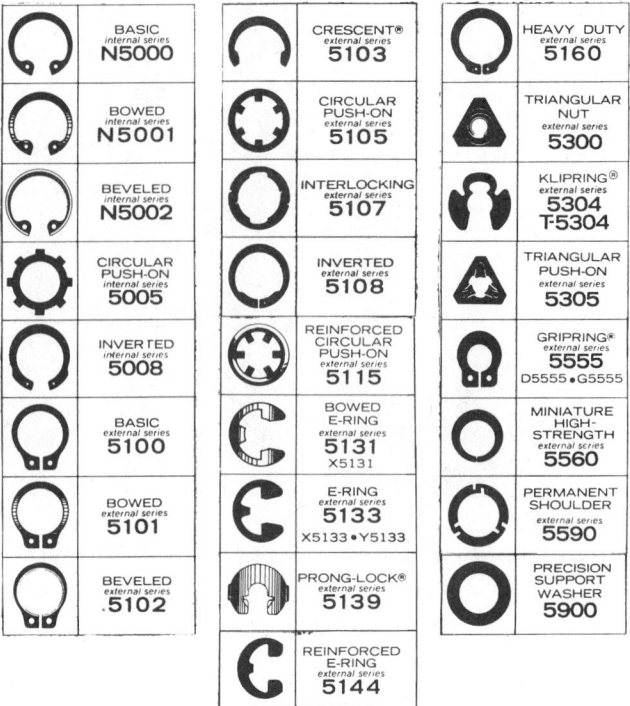

Figure 13-67 Standard retaining rings series (Courtesy Koh-I-Noor Rapidograph)

Figure 13-68 Internal ring pliers (Courtesy Koh-I-Noor Rapidograph)

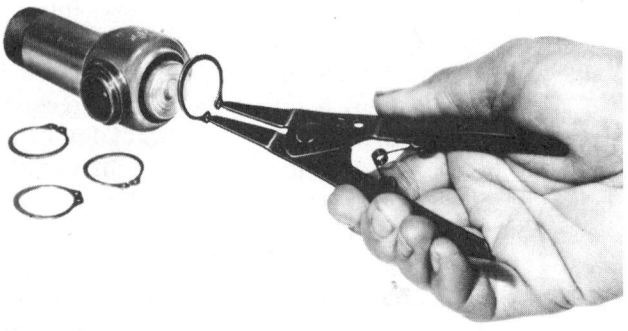

Figure 13-69 External ring pliers (Courtesy Koh-I-Noor Rapidograph)

contact surface with the bottom of the groove. It is also an important factor in achieving high static and dynamic thrust load capacities.

Types of Retaining Rings

A great number of fastening requirements are involved in product design. This factor has led to the development of many different types of retaining rings. Standard rings are shown in *Figure 13-67*.

Limited space prevents describing in detail all the different ring types available. In general, however, retaining rings can be grouped into two major categories.

Internal rings for axial assembly are compressed for insertion into a bore or housing. They generally have a large gap and holes in the lugs, located at the free ends, for pliers which are used to grasp the rings securely during installation and removal, *Figure 13-68*.

External rings for axial assembly are expanded with pliers, *Figure 13-69*, so they may be slipped over the end of a shaft, stud, or similar part. They have a small gap and the lug position is reversed from that of the internal ring. Radially assembled external rings do not have holes for pliers. Instead, the rings have a large gap and are pushed into the shaft directly in the plane of the groove with a special application tool, *Figure 13-70*.

Figure 13-70 Applicator and dispenser for retaining rings *(Courtesy Koh-I-Noor Rapidograph)*

In addition to the tools shown here, retaining rings can be installed with automatic equipment. This equipment can be designed for specific, high-speed, automatic assembly lines. Retaining ring grooves serve two purposes: 1) they assure precise seating of the ring in the assembly, and 2) they permit the ring to withstand heavy thrust loads. The grooves must be located accurately and precut in the housing or shaft before the rings are assembled. Shaft grooves often can be made at no additional cost during the cut-off and chamfering operations.

Self-locking rings do not require any grooves because they exert a frictional hold against axial displacement. They are used mainly as positioning and locking devices where the ring will be subjected to only moderate or light loading.

Bowed rings differ from conventional types in that they are bowed around an axis perpendicular to the diameter bisecting the gap. The bowed construction permits the rings to function as springs as well as fasteners. This provides resilient end play take-up in the assembly.

Beveled rings have a 15-degree bevel on the groove-engaging edge. They are installed in grooves having a comparable bevel on the load-bearing wall. When the ring is seated in the groove, it acts as a wedge against the retained part. Sometimes play develops between the ring and retained part because of accumulated toler-

ances or wear in the assembly. If play develops, the spring action of the ring causes the fastener to seat more deeply in its groove and move in an axial direction, automatically taking up the end play. (Because self-locking rings can be seated at any point on a shaft or in a bore, they too can be used to compensate for tolerances and eliminate end play.)

Materials and Finishes

As indicated previously, retaining rings are made of materials having good spring properties. Some also have high tensile and yield strengths. They must also have adequate ratio or ultimate tensile strength to elasticity. This permits the required deformation without too much permanent set. A ratio of 1:100 is satisfactory for most rings having the tapered-section design.

Standard material for most rings is carbon spring steel (SAE 1060-1090). For special applications, rings are also available in stainless steel (PH 15-7 Mo), beryllium copper (Alloy #25), and aluminum (Alclad 7075-T6). Rings are normally phosphate coated. Cadmium, zinc, and other platings and finishes are used for assemblies where extra corrosion resistance is needed or if the rings must withstand other unusual environmental conditions. Selection of the ring material and finish for a specific product design should be based upon the operating conditions under which the ring must function. These may include temperature, the presence of corrosive elements, thrust loads, and other factors.

Selection Considerations

Selecting the best ring for a product load capacity is a critical factor in some product designs. There are other factors the drafter should consider, however, before selecting specific ring types for a given product.
- Will there be adequate clearance to assemble the ring with pliers or other tools?
- Must the ring take up accumulated tolerances, either resiliently or rigidly?
- Is it possible to machine a ring groove on the shaft or inside a bore?
- Should the ring be adjustable to several positions on a shaft?

Making the Choice

After considering all these conditions, the drafter may find that more than one ring type is suitable. How, then, can the drafter make the best choice?

The most important design criterion is the ability of the ring to do the fastening job required. Before a final selection of ring type is made, the drafter should consider savings which may be possible in various parts of the assembly. These include:
- The cost of installing the ring.
- Whether or not a groove is required.
- If the ring can be installed *permanently* or if it may have to be removed for field service.

A self-locking ring, for example, eliminates the need for the ring grooves. If the ring will be subjected to only moderate loading, this may be ideal for the assembly. But most self-locking rings must be destroyed for removal. If field service is anticipated, another style of ring should be adopted.

The ideal ring is the one which will function adequately and provide the most economical means of fastening. Retaining rings are designed primarily as shoulders for positioning and retaining machine components on shafts and in housings and bores. Different rings have been developed and manufactured to meet specific fastening needs and problems.

To ensure correct selection of the proper type for any individual application, rings have been grouped according to their basic function. The selector guides, *Figures 13-71* and *13-72*, provide a visual index to all standard types.

DESIGN FEATURES

RING TYPES FOR AXIAL ASSEMBLY

Series N5000, 5100: Tapered section assures constant circularity and groove pressure. Secure against heavy thrust loads and high rotational speeds.

Series 5008, 5108: Lugs inverted to abut groove bottom. Rings form high circular shoulder, concentric with bore or shaft. Good for parts having large corner radii or chamfers.

Series 5160: Heavy-duty ring resists high thrust, impact loads. Eliminates spacer washers in bearing assemblies.

Series 5560: New miniature, high-strength ring. Forms tamper-proof shoulder on small diameter shafts subject to heavy thrust loads.

Series 5590: Permanent-shoulder ring for small diameter shafts. When compressed into groove, notches deform to close gaps, reducing both I.D. and O.D.

RING TYPES FOR RADIAL ASSEMBLY

Series 5103: Forms narrow, uniformly concentric shoulder. Excellent for assemblies where clearance is limited.

Series 5133: Provides large shoulder on small diameter shafts. Installed in deep groove for added thrust capacity.

Series 5144: Reinforced to provide five times greater gripping strength, 50% higher rpm limits than conventional E-rings. Secure against rotation.

Series 5107: Two-part ring balanced to withstand high rpm's, heavy thrust loads, relative rotation between parts.

Series 5304: New high-strength ring for large bearing surface. Can be installed quickly with pliers or mallet, removed with ordinary screw driver.

Series T5304: Thinner model of 5304. Can be seated in same width grooves as E-rings, has more gripping power. Good for cast or molded grooves.

RING TYPES FOR TAKING UP END-PLAY

Series N5001, 5101: Bowed cylindrically to accommodate large tolerances, provide resilient end-play take-up.

Series N5002, 5102: Rings beveled 15° on groove-engaging edge for use in groove with similar bevel. Wedge action provides rigid end-play take-up.

Series 5131: Provides large shoulder on small diameter shafts. Bowed for resilient end-play take-up.

Series 5139: Bowed ring designed for use as shoulder against rotating parts. Prongs lock against shaft, prevent ring from being forced from groove.

SELF-LOCKING TYPE RINGS (No groove required)

Series 5115: Push-on type fastener for ungrooved shafts and studs. Has arched rim for extra strength, long prongs for wide shaft tolerances.

Series 5105, 5005: Flat rim, shorter prongs, smaller O.D. than 5115. For flat contact surface, better clearance.

Series 5555: Secure against axial displacement from either direction. No groove needed. Adjustable, reusable.

Series 5305: Dished body, three heavy prongs lock on shaft under spring tension. Withstands heavy thrust loads.

Series 5300: Free-spinning nut. Dished body flattens under torque, eliminating need for separate lock washers.

INTERNAL	BASIC **N5000** For housings and bores	Size Range .250—10.0 in. 6.4—254.0 mm.	EXTERNAL	BOWED **5101** For shafts and pins	Size Range .188—1.750 in. 4.8—44.4 mm.	EXTERNAL	REINFORCED **5115** For shafts and pins	Size Range .094—1.0 in. ●	EXTERNAL	TRIANGULAR NUT **5300** For threaded parts	Size Range 6-32 and 8-32 10-24 and 10-32 1/4-20 and 1/4-28
INTERNAL	BOWED **N5001** For housings and bores	Size Range .250—1.750 in. 6.4—44.4 mm.	EXTERNAL	BEVELED **5102** For shafts and pins	Size Range 1.0—10.0 in. 25.4—254.0 mm.	EXTERNAL	BOWED E-RING **5131** For shafts and pins	Size Range .110—1.375 in. 2.8—34.9 mm.	EXTERNAL	KLIPRING **5304** T-5304 For shafts and pins	Size Range .156—1.000 in. 4.0—25.4 mm.
INTERNAL	BEVELED **N5002** For housings and bores	Size Range 1.0—10.0 in. 25.4—254.0 mm.	EXTERNAL	CRESCENT® **5103** For shafts and pins	Size Range .125—2.0 in. 3.2—50.8 mm.	EXTERNAL	E-RING **5133** For shafts and pins	Size Range .040—1.375 in. 1.0—34.9 mm.	EXTERNAL	TRIANGULAR **5305** For shafts and pins	Size Range .062—.438 in. ●
INTERNAL	CIRCULAR **5005** For housings and bores	Size Range .312—2.0 in. ●	EXTERNAL	CIRCULAR **5105** For shafts and pins	Size Range .094—1.0 in. ●	EXTERNAL	PRONG-LOCK® **5139** For shafts and pins	Size Range .092—.438 in. ●	EXTERNAL	GRIPRING® **5555** For shafts and pins	Size Range .079—.750 in. 2.0—19.0 mm.
INTERNAL	INVERTED **5008** For housings and bores	Size Range .750—4.0 in. 19.0—101.6 mm.	EXTERNAL	INTERLOCKING **5107** For shafts and pins	Size Range .469—3.375 in. 11.9—85.7 mm.	EXTERNAL	REINFORCED E-RING **5144** For shafts and pins	Size Range .094—.562 in. 2.4—14.3 mm.	EXTERNAL	HIGH-STRENGTH **5560** For shafts and pins	Size Range .101—.328 in.
EXTERNAL	BASIC **5100** For shafts and pins	Size Range .125—10.0 in. 3.2—254.0 mm.	EXTERNAL	INVERTED **5108** For shafts and pins	Size Range .500—4.0 in. 12.7—101.6 mm.	EXTERNAL	HEAVY-DUTY **5160** For shafts and pins	Size Range .394—2.0 in. 10.0—50.8 mm.	EXTERNAL	PERMANENT SHOULDER **5590** For shafts and pins	Size Range .250—.750 6.4—19.0 mm.

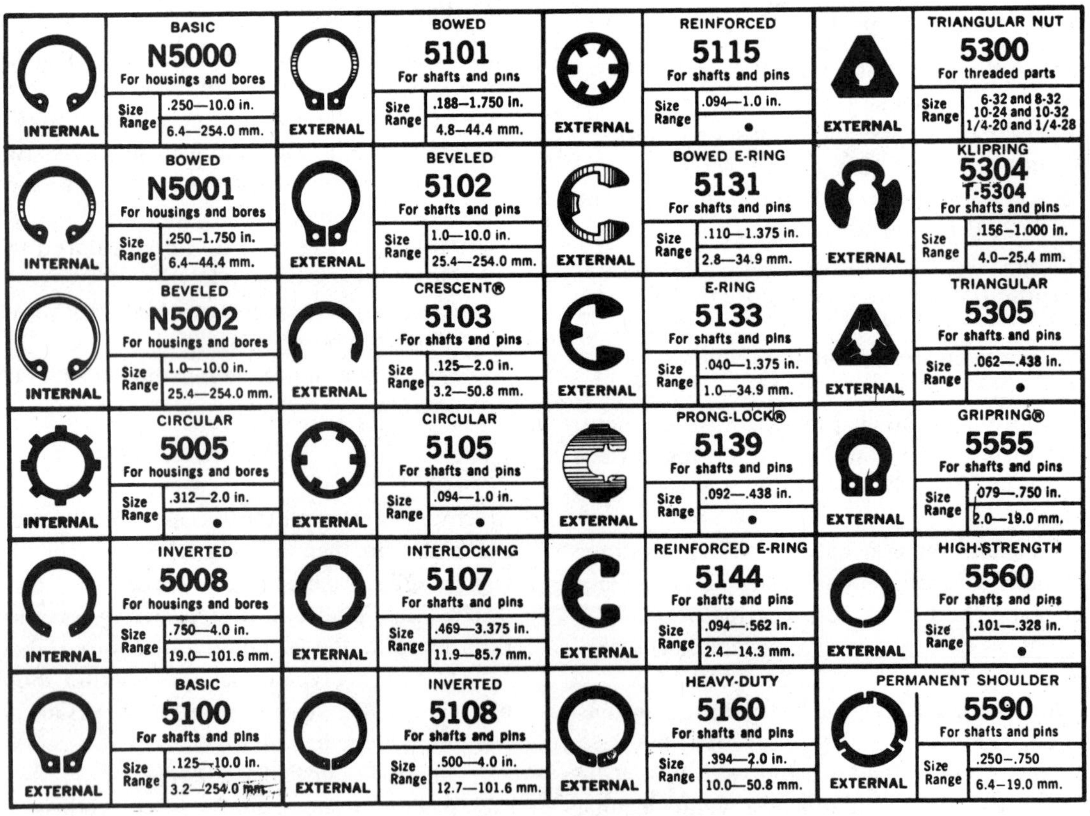

Figure 13-71 Selector guide: Standard ring series *(Courtesy Koh-I-Noor Rapidograph)*

The symbols listed below are used in the data charts for various ring types. Ring, groove and retained part dimensions are in inches; allowable thrust loads are in pounds.

SYMBOL		DEFINITION	RING SERIES WHERE APPLICABLE	SYMBOL		DEFINITION	RING SERIES WHERE APPLICABLE
A	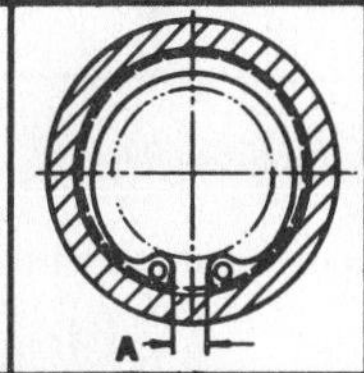	Minimum gap width: internal ring installed in groove	N5000, N5001, N5002, 5008	Ch max.		Maximum allowable chamfer height of retained part	N5000, N5001, N5002, 5008, 5100, 5101, 5102, 5103, 5107, 5108, 5131, 5133, 5144, 5160, 5555, 5560
B	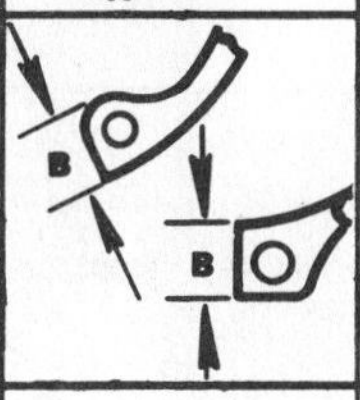	Lug height	N5000, N5001, N5002, 5100, 5101, 5102, 5160, 5555	D		Free diameter	N5000, N5001, N5002, 5008, 5100, 5101, 5102, 5103, 5107, 5108, 5131, 5133, 5144, 5160, 5304, 5555, 5560, 5590
b	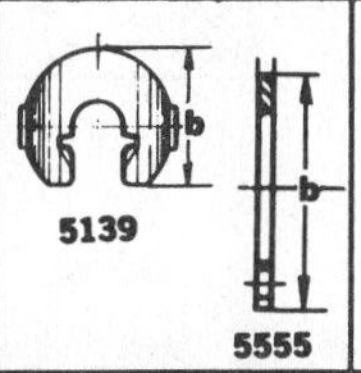	Ring height	5139, 5555, 5305, 5300, 5304	d		Nominal groove depth	**ALL RINGS USED IN GROOVES**
C	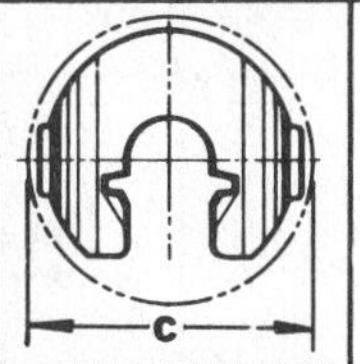	Clearance diameter	5139, 5590	E		Large section height	N5000, N5001, N5002, 5008, 5100, 5101, 5102, 5103, 5107, 5108, 5160, 5560
C_1	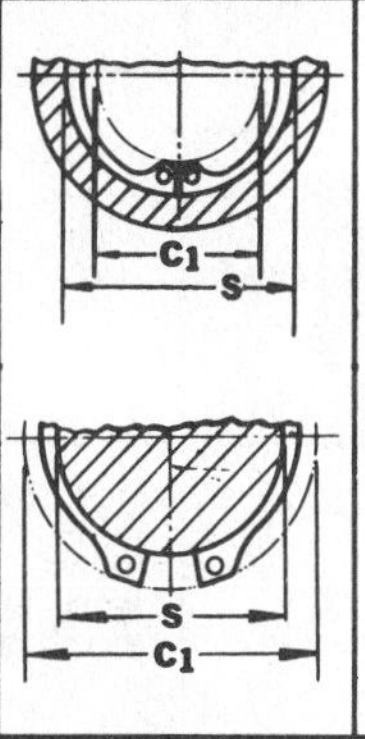	Clearance diameter: ring sprung into housing or over shaft, prior to installation in groove	N5000, N5001, N5002, 5008, 5100, 5101, 5102, 5108, 5160, 5555, 5560, 5590	e REF.		Distance from center of ring to outer edge (Reference)	5139
				G		Groove diameter	**ALL RINGS USED IN GROOVES**
C_2	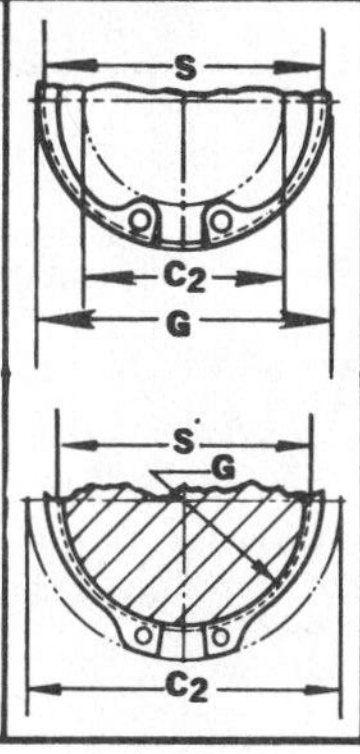	Clearance diameter: ring installed in groove	N5000, N5001, N5002, 5008, 5100, 5101, 5102, 5103, 5107, 5108, 5131, 5133, 5144, 5160, 5560	H	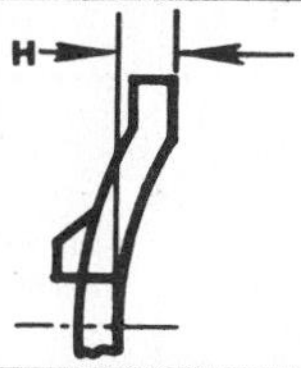	Bow height	5139
				h		Ring height	5115

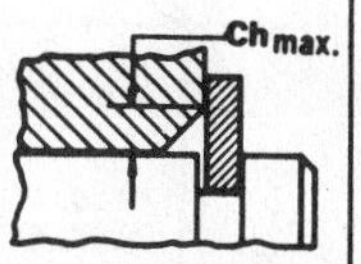

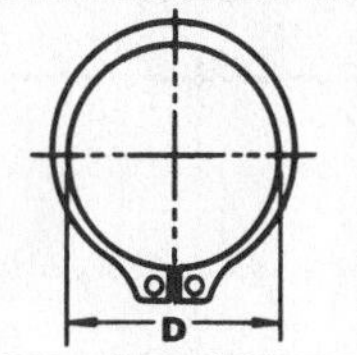

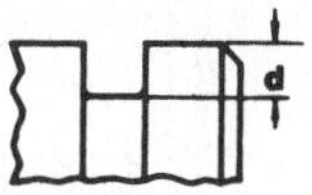

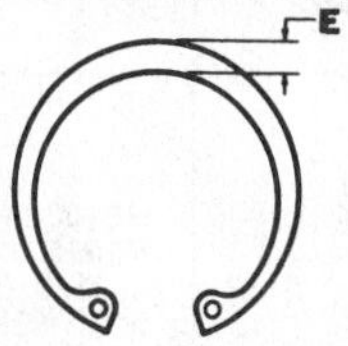

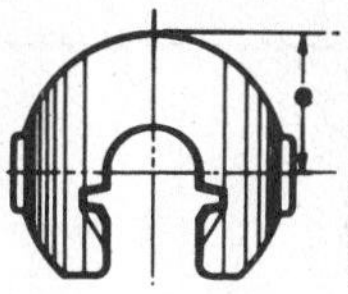

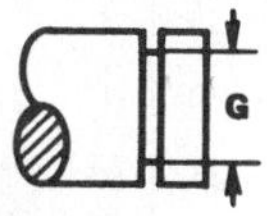

Figure 13-72   Definition symbols *(Courtesy Koh-I-Noor Rapidograph)*

SYMBOL		DEFINITION	RING SERIES WHERE APPLICABLE	SYMBOL		DEFINITION	RING SERIES WHERE APPLICABLE
J	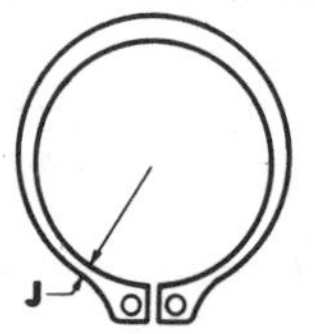	Small section height	N5000, N5001, N5002, 5008, 5100, 5101, 5102, 5108, 5160, 5560	S	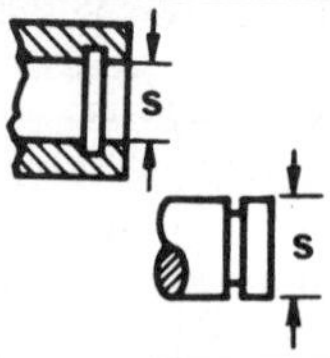	Shaft or housing diameter	ALL RINGS
K	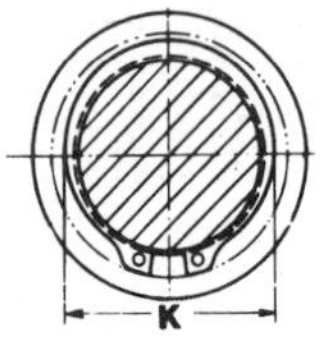	Maximum gaging diameter: ring installed in groove	5100, 5101, 5102, 5108, 5160, 5560	t		Ring thickness	ALL RINGS
k	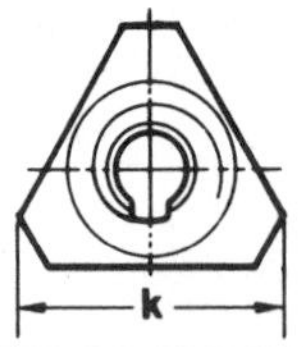	Overall ring width	5305, 5300	U	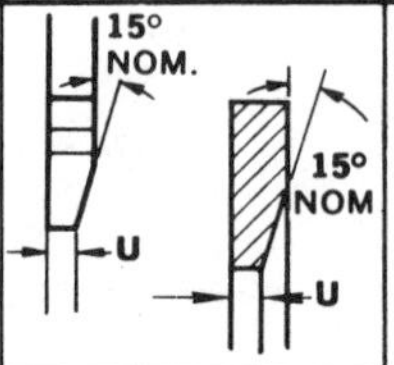	Ring thickness at beveled edge	N5002, 5102
L M	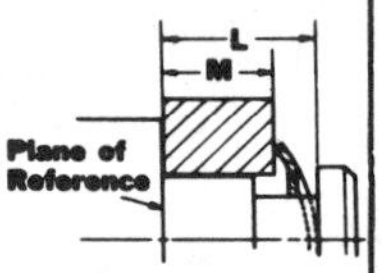	L: Location of outer groove wall from plane of reference M: Width of retained part	N5001, N5002, 5101, 5102, 5131, 5139	V	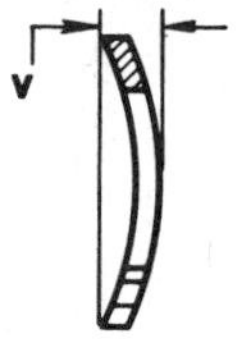	Overall bow height	N5001, 5101, 5131
P	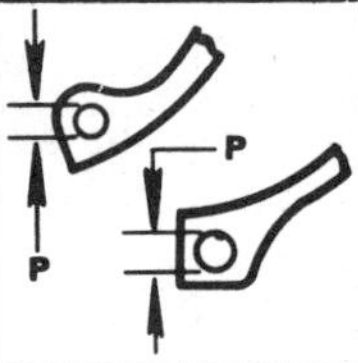	Pliers hole diameter	N5000, N5001, N5002, 5008, 5100, 5101, 5102, 5108, 5160, 5555	W	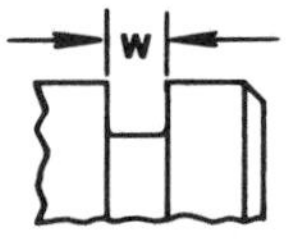	Groove width	ALL RINGS USED IN GROOVES
p	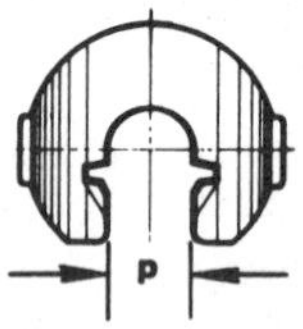	Gap width	5139	X	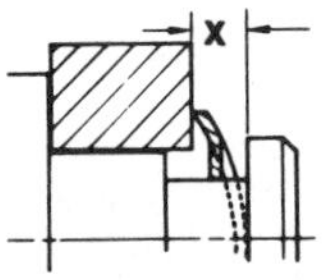	Distance from outer groove wall to face of retained part	N5001, 5101, 5131, 5139
P_r P'_r P_g	Allowable thrust load for ring (lbs.) Allowable assembly load with maximum corner radius or chamfer Allowable thrust load for groove (lbs.)		ALL RINGS USED IN GROOVES	Y	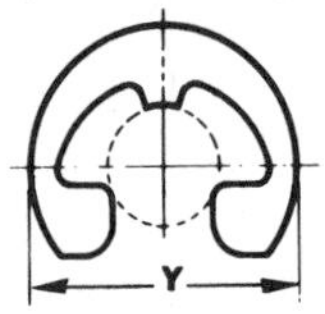	Free outside diameter	5103, 5133, 5144, 5131, 5304
R	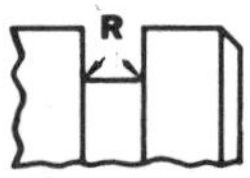	Radius of groove bottom	ALL RINGS USED IN GROOVES	Z	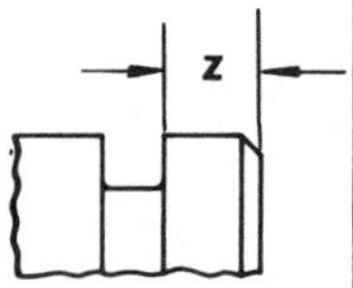	Edge margin	ALL RINGS USED IN GROOVES
$R_{max.}$	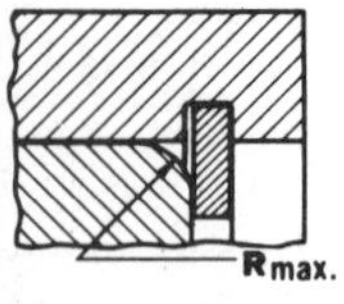	Maximum allowable corner radius of retained part	N5000, N5001, N5002, 5008, 5100, 5101, 5102, 5103, 5107, 5108, 5131, 5133, 5144, 5160, 5555, 5560	Z_1	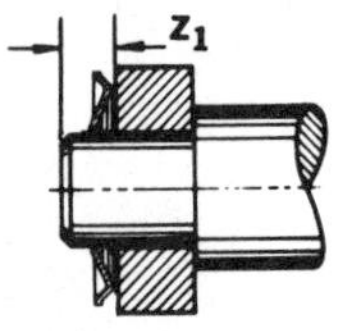	Minimum distance from face of retained part to end of shaft or housing	5115, 5005, 5105, 5305

Figure 13-72 (Continued)

Figures 13-73 and ***13-74*** are from the latest *Truarc Technical Manual.* Figure 13-73 is a sample of an internal series (N5000); Figure 13-74 is from an external series (5100).

Example (Refer to Figure 13-73):

A shaft of 1.000 (24.4) diameter would use a No. N5000-100.

- Free diameter = 1.111 $\pm\,^{.015}_{.010}$
- Thickness = .042 ± .002
- Lug size = .155 ± .005
- Plier hole diameter = .062

The required groove must be:

- Diameter = 1.066 ± .003 (.004 TIR)
- Width = .046 $\pm\,^{.003}_{.000}$

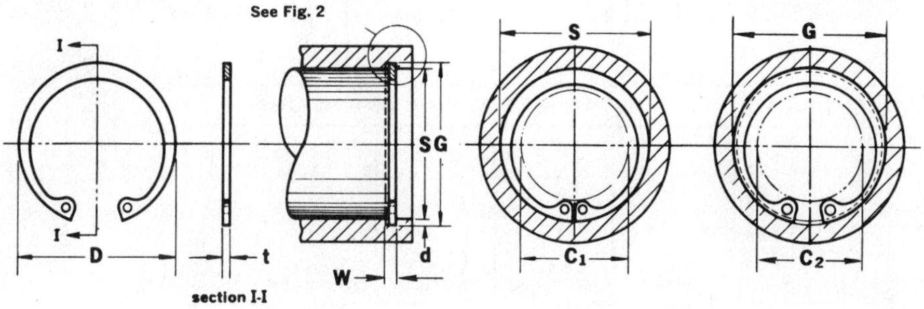

HOUSING DIA.			MIL-R-21248 MS 16625 INTERNAL SERIES **N5000**	TRUARC RING DIMENSIONS: Thickness t applies only to unplated rings. For plated and stainless steel (Type H) rings, add .002" to the listed maximum thickness. Maximum ring thickness will be at least .0002" less than the listed minimum groove width (W).				Approx. weight per 1000 pieces	GROOVE DIMENSIONS: T.I.R. (total indicator reading) is the maximum allowable deviation of concentricity between groove and housing.				Nominal groove depth	APPLICATION DATA: CLEARANCE DIAMETER		ALLOW. THRUST LOAD (lbs.) Sharp corner abutment	
Dec. equiv. inch	Approx. fract. equiv. inch	Approx. mm.		FREE DIA.		THICKNESS			DIAMETER		WIDTH			When sprung into housing S	When sprung into groove G	RINGS (Standard material) Safety factor = 4	GROOVES (Cold rolled steel bores and housings) Safety factor = 2
S	S	S	size—no.	D	tol.	t	tol.	lbs.	G	tol.	W	tol.	d	C_1	C_2	P_r	P_g
.250	1/4	6.4	N5000-25	.280		.015		.08	.268	±.001	.018	+.002	.009	.115	.133	420	190
.312	5/16	7.9	N5000-31	.346		.015		.11	.330	.0015 T.I.R.	.018	−.000	.009	.173	.191	530	240
.375	3/8	9.5	N5000-37	.415		.025		.25	.397	± .002	.029		.011	.204	.226	1050	350
.438	7/16	11.1	N5000-43	.482		.025		.37	.461	.002	.029		.012	.23	.254	1220	440
.453	29/64	11.5	N5000-45	.498		.025		.43	.477	T.I.R.	.029		.012	.25	.274	1280	460
.500	1/2	12.7	N5000-50	.548	+.010	.035		.70	.530		.039		.015	.26	.29	1980	510
.512	—	13.0	N5000-51	.560	−.005	.035		.77	.542		.039		.015	.27	.30	2030	520
.562	9/16	14.3	N5000-56	.620		.035		.86	.596	± .002	.039		.017	.275	.305	2220	710
.625	5/8	15.9	N5000-62	.694		.035		1.0	.665	.004	.039		.020	.34	.38	2470	1050
.688	11/16	17.5	N5000-68	.763		.035		1.2	.732	T.I.R.	.039	+.003	.022	.40	.44	2700	1280
.750	3/4	19.0	N5000-75	.831		.035		1.3	.796		.039	−.000	.023	.45	.49	3000	1460
.777	—	19.7	N5000-77	.859		.042		1.7	.825		.046		.024	.475	.52	4550	1580
.812	13/16	20.6	N5000-81	.901		.042		1.9	.862		.046		.025	.49	.54	4800	1710
.866	—	22.0	N5000-86	.961		.042		2.0	.920		.046		.027	.54	.59	5100	1980
.875	7/8	22.2	N5000-87	.971	+.015	.042		2.1	.931	± .003	.046		.028	.545	.60	5150	2080
.901	—	22.9	N5000-90	1.000	−.010	.042	±.002	2.2	.959	.004 T.I.R.	.046		.029	.565	.62	5350	2200
.938	15/16	23.8	N5000-93	1.041		.042		2.4	1.000		.046		.031	.61	.67	5600	2450
1.000	1	25.4	N5000-100	1.111		.042		2.7	1.066		.046		.033	.665	.73	5950	2800
1.023	—	26.0	N5000-102	1.136		.042		2.8	1.091		.046		.034	.69	.755	6050	3000
1.062	1 1/16	27.0	N5000-106	1.180		.050		3.7	1.130		.056		.034	.685	.75	7450	3050
1.125	1 1/8	28.6	N5000-112	1.249		.050		4.0	1.197		.056		.036	.745	.815	7900	3400
1.181	—	30.0	N5000-118	1.319		.050		4.3	1.255		.056		.037	.79	.86	8400	3700
1.188	1 3/16	30.2	N5000-118	1.319		.050		4.3	1.262		.056		.037	.80	.87	8400	3700
1.250	1 1/4	31.7	N5000-125	1.388		.050		4.8	1.330		.056		.040	.875	.955	8800	4250
1.259	—	32.0	N5000-125	1.388	+.025	.050		4.8	1.339		.056		.040	.885	.965	8800	4250
1.312	1 5/16	33.3	N5000-131	1.456	−.020	.050		5.0	1.396	±.004	.056		.042	.93	1.01	9300	4700
1.375	1 3/8	34.9	N5000-137	1.526		.050		5.1	1.461	.005	.056	+.004	.043	.99	1.07	9700	5050
1.378	—	35.0	N5000-137	1.526		.050		5.1	1.464	T.I.R.	.056	−.000	.043	.99	1.07	9700	5050
1.438	1 7/16	36.5	N5000-143	1.596		.050		5.8	1.528		.056		.045	1.06	1.15	10200	5500
1.456	—	37.0	N5000-145	1.616		.050		6.4	1.548		.056		.046	1.08	1.17	10300	5700
1.500	1 1/2	38.1	N5000-150	1.660		.050		6.5	1.594		.056		.047	1.12	1.21	10550	6000
1.562	1 9/16	39.7	N5000-156	1.734		.062		8.9	1.658		.068		.048	1.14	1.23	13700	6350
1.575	—	40.0	N5000-156	1.734	+.035	.062		8.9	1.671	± .005	.068		.048	1.15	1.24	13700	6350
1.625	1 5/8	41.3	N5000-162	1.804	−.025	.062	±.003	10.0	1.725	.005	.068		.050	1.15	1.25	14200	6900
1.653	—	42.0	N5000-165	1.835		.062		10.4	1.755	T.I.R.	.068		.051	1.17	1.27	14500	7200
1.688	1 11/16	42.9	N5000-168	1.874		.062		10.8	1.792		.068		.052	1.21	1.31	14800	7450

Figure 13-73 Internal series N5000 *(Courtesy Koh-I-Noor Rapidograph)*

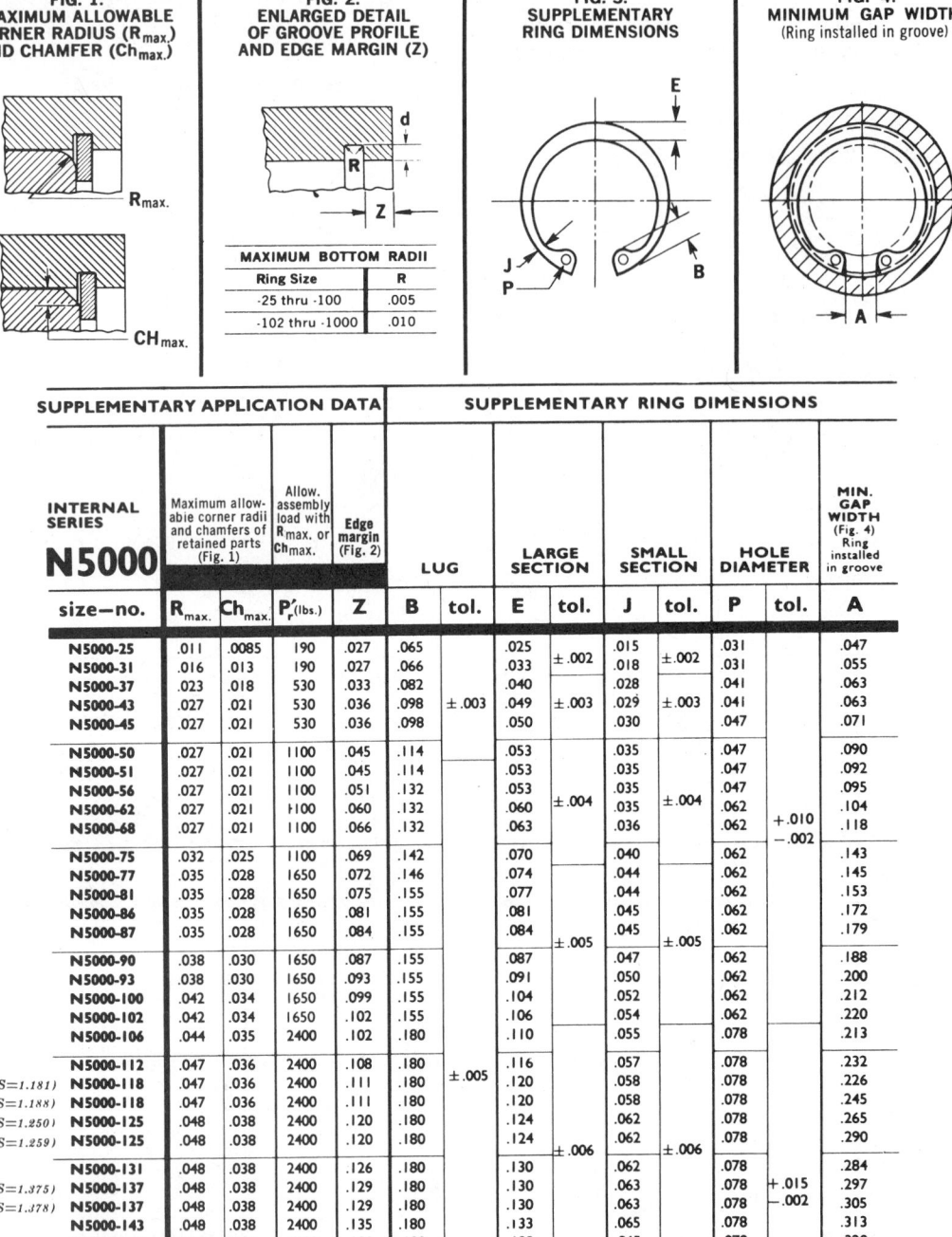

FIG. 1:
MAXIMUM ALLOWABLE
CORNER RADIUS (R$_{max.}$)
AND CHAMFER (Ch$_{max.}$)

FIG. 2:
ENLARGED DETAIL
OF GROOVE PROFILE
AND EDGE MARGIN (Z)

FIG. 3:
SUPPLEMENTARY
RING DIMENSIONS

FIG. 4:
MINIMUM GAP WIDTH
(Ring installed in groove)

MAXIMUM BOTTOM RADII

Ring Size	R
-25 thru -100	.005
-102 thru -1000	.010

	SUPPLEMENTARY APPLICATION DATA				SUPPLEMENTARY RING DIMENSIONS								
INTERNAL SERIES **N5000**	Maximum allowable corner radii and chamfers of retained parts (Fig. 1)		Allow. assembly load with R$_{max.}$ or Ch$_{max.}$ (Fig. 2)	Edge margin (Fig. 2)	LUG		LARGE SECTION		SMALL SECTION		HOLE DIAMETER		MIN. GAP WIDTH (Fig. 4) Ring installed in groove
size—no.	R$_{max.}$	Ch$_{max.}$	P′$_r$(lbs.)	Z	B	tol.	E	tol.	J	tol.	P	tol.	A
N5000-25	.011	.0085	190	.027	.065		.025	±.002	.015	±.002	.031		.047
N5000-31	.016	.013	190	.027	.066		.033		.018		.031		.055
N5000-37	.023	.018	530	.033	.082		.040		.028		.041		.063
N5000-43	.027	.021	530	.036	.098	±.003	.049	±.003	.029	±.003	.041		.063
N5000-45	.027	.021	530	.036	.098		.050		.030		.047		.071
N5000-50	.027	.021	1100	.045	.114		.053		.035		.047		.090
N5000-51	.027	.021	1100	.045	.114		.053		.035		.047		.092
N5000-56	.027	.021	1100	.051	.132		.053		.035		.047		.095
N5000-62	.027	.021	⊦100	.060	.132		.060	±.004	.035	±.004	.062	+.010 −.002	.104
N5000-68	.027	.021	⊦100	.066	.132		.063		.036		.062		.118
N5000-75	.032	.025	1100	.069	.142		.070		.040		.062		.143
N5000-77	.035	.028	1650	.072	.146		.074		.044		.062		.145
N5000-81	.035	.028	1650	.075	.155		.077		.044		.062		.153
N5000-86	.035	.028	1650	.081	.155		.081		.045		.062		.172
N5000-87	.035	.028	1650	.084	.155		.084	±.005	.045	±.005	.062		.179
N5000-90	.038	.030	1650	.087	.155		.087		.047		.062		.188
N5000-93	.038	.030	1650	.093	.155		.091		.050		.062		.200
N5000-100	.042	.034	1650	.099	.155		.104		.052		.062		.212
N5000-102	.042	.034	1650	.102	.155		.106		.054		.062		.220
N5000-106	.044	.035	2400	.102	.180		.110		.055		.078		.213
N5000-112	.047	.036	2400	.108	.180	±.005	.116		.057		.078		.232
(S=1.181) N5000-118	.047	.036	2400	.111	.180		.120		.058		.078		.226
(S=1.188) N5000-118	.047	.036	2400	.111	.180		.120		.058		.078		.245
(S=1.250) N5000-125	.048	.038	2400	.120	.180		.124		.062		.078		.265
(S=1.259) N5000-125	.048	.038	2400	.120	.180		.124	±.006	.062	±.006	.078		.290
N5000-131	.048	.038	2400	.126	.180		.130		.062		.078		.284
(S=1.375) N5000-137	.048	.038	2400	.129	.180		.130		.063		.078	+.015 −.002	.297
(S=1.378) N5000-137	.048	.038	2400	.129	.180		.130		.063		.078		.305
N5000-143	.048	.038	2400	.135	.180		.133		.065		.078		.313
N5000-145	.048	.038	2400	.138	.180		.133		.065		.078		.320
N5000-150	.048	.038	2400	.141	.180		.133		.066		.078		.340
(S=1.562) N5000-156	.064	.050	3900	.144	.202		.157		.078		.078		.338
(S=1.575) N5000-156	.064	.050	3900	.144	.202		.157		.078		.078		.374
N5000-162	.064	.050	3900	.150	.227		.164	±.007	.082	±.007	.078		.339
N5000-165	.064	.050	3900	.153	.227		.167		.083		.078		.348
N5000-168	.064	.050	3900	.156	.227		.170		.085		.078		.357

Figure 13-73 (Continued)

Example (Refer to Figure 13-74):

A shaft of 4.000 (101.6) diameter would use a No. 5100-400.

- Free diameter = 3.700 $\pm \begin{smallmatrix} .020 \\ .030 \end{smallmatrix}$

- Thickness = .109 ± .003

- Lug size = .352 ± .005
- Plier hole size = .125 $\pm \begin{smallmatrix} .015 \\ .002 \end{smallmatrix}$

The required groove must be:

- Diameter = 3.792 ± .006 (.006 TIR)
- Width = .120 $\pm \begin{smallmatrix} .005 \\ .000 \end{smallmatrix}$

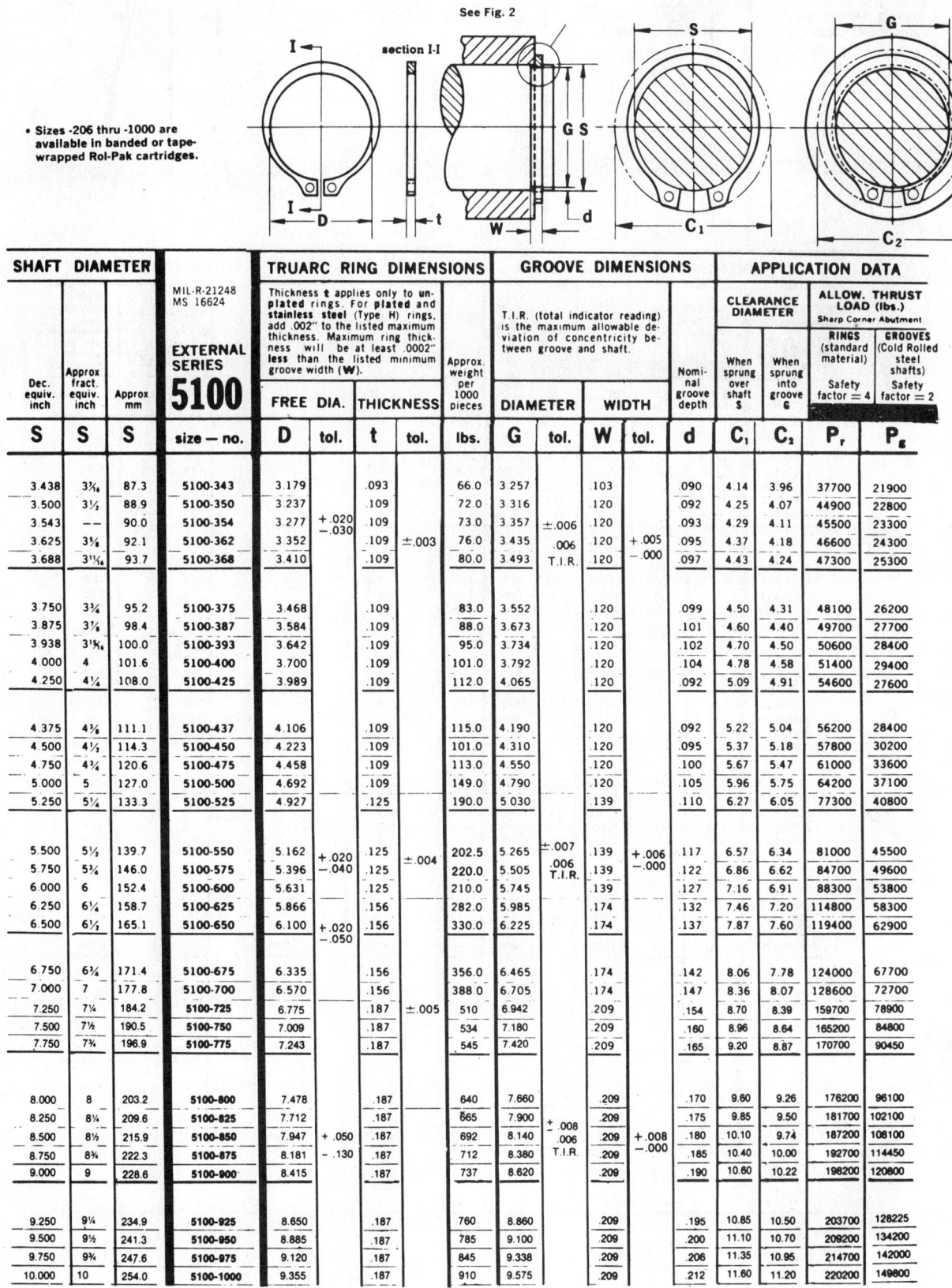

- Sizes -206 thru -1000 are available in banded or tape-wrapped Rol-Pak cartridges.

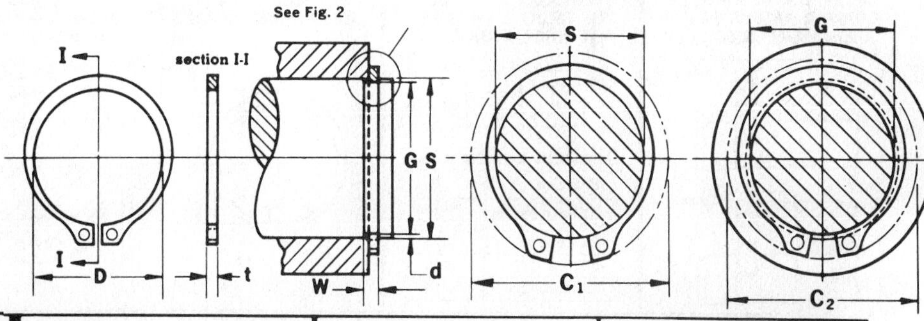
See Fig. 2
section I-I
I
I
D
t
W
G S
d
S
C1
G
C2

SHAFT DIAMETER: Dec. equiv. inch	SHAFT DIAMETER: Approx fract. equiv. inch	SHAFT DIAMETER: Approx mm	MIL-R-21248 MS 16624 EXTERNAL SERIES 5100	TRUARC RING DIMENSIONS (Thickness t applies only to unplated rings. For plated and stainless steel (Type H) rings, add .002" to the listed maximum thickness. Maximum ring thickness will be at least .0002" less than the listed minimum groove width (W).): FREE DIA.	FREE DIA.	THICKNESS	THICKNESS	Approx. weight per 1000 pieces	GROOVE DIMENSIONS (T.I.R. (total indicator reading) is the maximum allowable deviation of concentricity between groove and shaft.): DIAMETER	DIAMETER	WIDTH	WIDTH	Nominal groove depth	APPLICATION DATA: CLEARANCE DIAMETER: When sprung over shaft S	CLEARANCE DIAMETER: When sprung into groove G	ALLOW. THRUST LOAD (lbs.) Sharp Corner Abutment: RINGS (standard material) Safety factor = 4	ALLOW. THRUST LOAD (lbs.): GROOVES (Cold Rolled steel shafts) Safety factor = 2
S	S	S	size — no.	D	tol.	t	tol.	lbs.	G	tol.	W	tol.	d	C_1	C_2	P_r	P_g
3.438	3 7/16	87.3	5100-343	3.179		.093		66.0	3.257		.103		.090	4.14	3.96	37700	21900
3.500	3 1/2	88.9	5100-350	3.237		.109		72.0	3.316		.120		.092	4.25	4.07	44900	22800
3.543	--	90.0	5100-354	3.277	+.020 −.030	.109		73.0	3.357	±.006	.120		.093	4.29	4.11	45500	23300
3.625	3 5/8	92.1	5100-362	3.352		.109	±.003	76.0	3.435	.006	.120	+.005 −.000	.095	4.37	4.18	46600	24300
3.688	3 11/16	93.7	5100-368	3.410		.109		80.0	3.493	T.I.R.	.120		.097	4.43	4.24	47300	25300
3.750	3 3/4	95.2	5100-375	3.468		.109		83.0	3.552		.120		.099	4.50	4.31	48100	26200
3.875	3 7/8	98.4	5100-387	3.584		.109		88.0	3.673		.120		.101	4.60	4.40	49700	27700
3.938	3 15/16	100.0	5100-393	3.642		.109		95.0	3.734		.120		.102	4.70	4.50	50600	28400
4.000	4	101.6	5100-400	3.700		.109		101.0	3.792		.120		.104	4.78	4.58	51400	29400
4.250	4 1/4	108.0	5100-425	3.989		.109		112.0	4.065		.120		.092	5.09	4.91	54600	27600
4.375	4 3/8	111.1	5100-437	4.106		.109		115.0	4.190		.120		.092	5.22	5.04	56200	28400
4.500	4 1/2	114.3	5100-450	4.223		.109		101.0	4.310		.120		.095	5.37	5.18	57800	30200
4.750	4 3/4	120.6	5100-475	4.458		.109		113.0	4.550		.120		.100	5.67	5.47	61000	33600
5.000	5	127.0	5100-500	4.692		.109		149.0	4.790		.120		.105	5.96	5.75	64200	37100
5.250	5 1/4	133.3	5100-525	4.927		.125		190.0	5.030		.139		.110	6.27	6.05	77300	40800
5.500	5 1/2	139.7	5100-550	5.162	+.020 −.040	.125	±.004	202.5	5.265	±.007 .006 T.I.R.	.139	+.006 −.000	.117	6.57	6.34	81000	45500
5.750	5 3/4	146.0	5100-575	5.396		.125		220.0	5.505		.139		.122	6.86	6.62	84700	49600
6.000	6	152.4	5100-600	5.631		.125		210.0	5.745		.139		.127	7.16	6.91	88300	53800
6.250	6 1/4	158.7	5100-625	5.866		.156		282.0	5.985		.174		.132	7.46	7.20	114800	58300
6.500	6 1/2	165.1	5100-650	6.100	+.020 −.050	.156		330.0	6.225		.174		.137	7.87	7.60	119400	62900
6.750	6 3/4	171.4	5100-675	6.335		.156		356.0	6.465		.174		.142	8.06	7.78	124000	67700
7.000	7	177.8	5100-700	6.570		.156		388.0	6.705		.174		.147	8.36	8.07	128600	72700
7.250	7 1/4	184.2	5100-725	6.775		.187	±.005	510	6.942		.209		.154	8.70	8.39	159700	78900
7.500	7 1/2	190.5	5100-750	7.009		.187		534	7.180		.209		.160	8.96	8.64	165200	84800
7.750	7 3/4	196.9	5100-775	7.243		.187		545	7.420		.209		.165	9.20	8.87	170700	90450
8.000	8	203.2	5100-800	7.478		.187		640	7.660		.209		.170	9.60	9.26	176200	96100
8.250	8 1/4	209.6	5100-825	7.712		.187		665	7.900		.209		.175	9.85	9.50	181700	102100
8.500	8 1/2	215.9	5100-850	7.947	+.050 −.130	.187		692	8.140	±.008 .006 T.I.R.	.209	+.008 −.000	.180	10.10	9.74	187200	108100
8.750	8 3/4	222.3	5100-875	8.181		.187		712	8.380		.209		.185	10.40	10.00	192700	114450
9.000	9	228.6	5100-900	8.415		.187		737	8.620		.209		.190	10.60	10.22	198200	120800
9.250	9 1/4	234.9	5100-925	8.650		.187		760	8.860		.209		.195	10.85	10.50	203700	128225
9.500	9 1/2	241.3	5100-950	8.885		.187		785	9.100		.209		.200	11.10	10.70	209200	134200
9.750	9 3/4	247.6	5100-975	9.120		.187		845	9.338		.209		.206	11.35	10.95	214700	142000
10.000	10	254.0	5100-1000	9.355		.187		910	9.575		.209		.212	11.60	11.20	220200	149800

Figure 13-74 External series 5100 *(Courtesy Koh-I-Noor Rapidograph)*

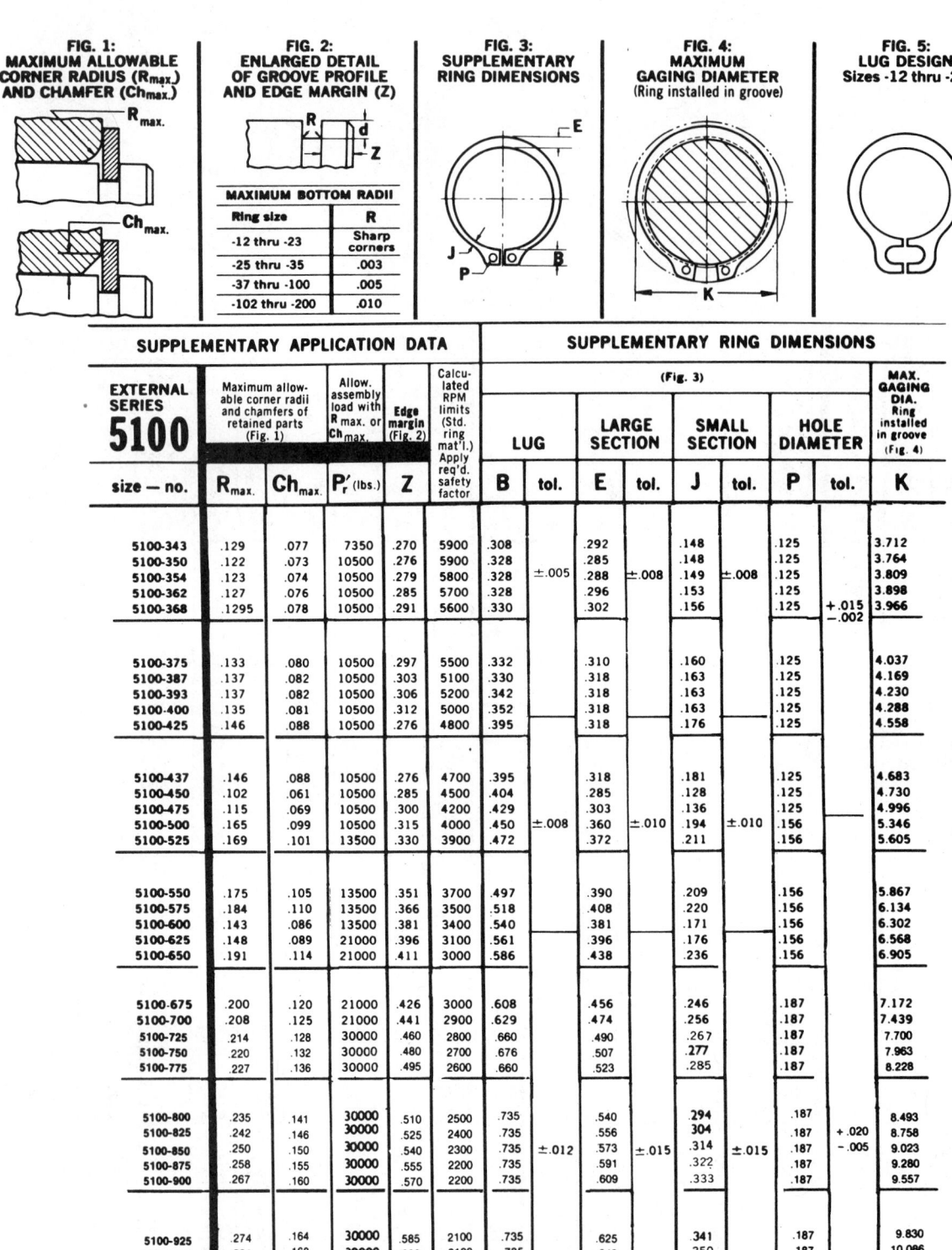

FIG. 1: MAXIMUM ALLOWABLE CORNER RADIUS (R max.) AND CHAMFER (Ch max.)

FIG. 2: ENLARGED DETAIL OF GROOVE PROFILE AND EDGE MARGIN (Z)

MAXIMUM BOTTOM RADII

Ring size	R
-12 thru -23	Sharp corners
-25 thru -35	.003
-37 thru -100	.005
-102 thru -200	.010

FIG. 3: SUPPLEMENTARY RING DIMENSIONS

FIG. 4: MAXIMUM GAGING DIAMETER (Ring installed in groove)

FIG. 5: LUG DESIGN Sizes -12 thru -23

EXTERNAL SERIES 5100	SUPPLEMENTARY APPLICATION DATA					SUPPLEMENTARY RING DIMENSIONS								MAX. GAGING DIA.
	Maximum allowable corner radii and chamfers of retained parts (Fig. 1)		Allow. assembly load with R max. or Ch max.	Edge margin (Fig. 2)	Calculated RPM limits (Std. ring mat'l.) Apply req'd. safety factor	LUG		LARGE SECTION		SMALL SECTION		HOLE DIAMETER		Ring installed in groove (Fig. 4)
size — no.	R max.	Ch max.	P'r (lbs.)	Z		B	tol.	E	tol.	J	tol.	P	tol.	K
5100-343	.129	.077	7350	.270	5900	.308		.292		.148		.125		3.712
5100-350	.122	.073	10500	.276	5900	.328	±.005	.285	±.008	.148	±.008	.125		3.764
5100-354	.123	.074	10500	.279	5800	.328		.288		.149		.125		3.809
5100-362	.127	.076	10500	.285	5700	.328		.296		.153		.125		3.898
5100-368	.1295	.078	10500	.291	5600	.330		.302		.156		.125	+.015 −.002	3.966
5100-375	.133	.080	10500	.297	5500	.332		.310		.160		.125		4.037
5100-387	.137	.082	10500	.303	5100	.330		.318		.163		.125		4.169
5100-393	.137	.082	10500	.306	5200	.342		.318		.163		.125		4.230
5100-400	.135	.081	10500	.312	5000	.352		.318		.163		.125		4.288
5100-425	.146	.088	10500	.276	4800	.395		.318		.176		.125		4.558
5100-437	.146	.088	10500	.276	4700	.395		.318		.181		.125		4.683
5100-450	.102	.061	10500	.285	4500	.404		.285		.128		.125		4.730
5100-475	.115	.069	10500	.300	4200	.429		.303		.136		.125		4.996
5100-500	.165	.099	10500	.315	4000	.450	±.008	.360	±.010	.194	±.010	.156		5.346
5100-525	.169	.101	13500	.330	3900	.472		.372		.211		.156		5.605
5100-550	.175	.105	13500	.351	3700	.497		.390		.209		.156		5.867
5100-575	.184	.110	13500	.366	3500	.518		.408		.220		.156		6.134
5100-600	.143	.086	13500	.381	3400	.540		.381		.171		.156		6.302
5100-625	.148	.089	21000	.396	3100	.561		.396		.176		.156		6.568
5100-650	.191	.114	21000	.411	3000	.586		.438		.236		.156		6.905
5100-675	.200	.120	21000	.426	3000	.608		.456		.246		.187		7.172
5100-700	.208	.125	21000	.441	2900	.629		.474		.256		.187		7.439
5100-725	.214	.128	30000	.460	2800	.660		.490		.267		.187		7.700
5100-750	.220	.132	30000	.480	2700	.676		.507		.277		.187		7.963
5100-775	.227	.136	30000	.495	2600	.660		.523		.285		.187		8.228
5100-800	.235	.141	30000	.510	2500	.735		.540		.294		.187		8.493
5100-825	.242	.146	30000	.525	2400	.735		.556		.304		.187	+.020 −.005	8.758
5100-850	.250	.150	30000	.540	2300	.735	±.012	.573	±.015	.314	±.015	.187		9.023
5100-875	.258	.155	30000	.555	2200	.735		.591		.322		.187		9.280
5100-900	.267	.160	30000	.570	2200	.735		.609		.333		.187		9.557
5100-925	.274	.164	30000	.585	2100	.735		.625		.341		.187		9.830
5100-950	.281	.168	30000	.600	2100	.735		.642		.350		.187		10.086
5100-975	.287	.172	30000	.618	2000	.735		.658		.358		.187		10.340
5100-1000	.294	.176	30000	.636	2000	.735		.675		.367		.187		10.610

Figure 13-74 (Continued)

Review

1. In addition to fastening parts together, what are two other uses for threads?

2. What number 5100 series retaining ring would be recommended for a shaft diameter of 203.2 mm?

3. List four factors to consider before selecting any type of retaining ring, other than load capacity.

4. Explain in full what ⅜ – 11 UNC – 2A X 2½ LG means.

5. What is the recommended hole size for a .312-inch (8) diameter spring pin?

6. If you have a ¼ – 20 UNC threaded screw and rotate it 10 full turns, how far will the end travel?

7. What series retaining ring would be recommended to provide rigid end play take-up?

8. How is a screw called-out that is 3 inches long, ⅜-inch in diameter, coarse, right-hand threads, average fit, and round-head style?

9. What is the free diameter of an external 5100-775 retaining ring? What is the tolerance of the dimension?

10. A 2-inch (50.8) diameter shaft uses what size square key?

11. What is the pliers hole diameter and tolerance for an internal N5000-131 retaining ring?

12. How is the depth of thread figured?

Chapter Thirteen Problems

The following problems are intended to give the beginning drafter practice using various size charts of fastener and, by using these charts, practice in laying out and drawing the many fasteners used.

The steps to follow in laying out fasteners are:

Step 1 Study all required specifications.

Step 2 Using the appropriate size chart, for size dimension, *lightly* lay out fastener in place.

Step 3 Check each dimension.

Step 4 Darken in the fastener, starting with diameters and arcs first to correct line thickness. Use simplified thread representation.

Step 5 Neatly add all required call-off specifications, using the latest drafting standard.

Problem 13-1

On an A-size, 8½ x 11 sheet of paper, lay out the center lines per given dimensions.

Calculate standard thread lengths.

Draw the following fasteners on the center lines as illustrated.

A. Flat-head cap screw — ⅜ - 16 UNC - 2A x 2.0 lg.

B. Round-head cap screw — ½ - 13 UNC - 2A x 2.25 lg.

C. Fillister-head cap screw — ⅝ - 11 UNC - 2A x 2.5 lg.

D. Oval-head cap screw — ⅜ - 24 UNF - 2A x 1.75 lg.

E. Hex-head cap screw with hex nut — 1 - 8 UNC - 2A x 2.5 lg.

F. Socket-head cap screw — ½ - 20 UNF - 2A x 2.0 lg.

G. Cotter pin — ⅜ dia. x 3.5 lg.

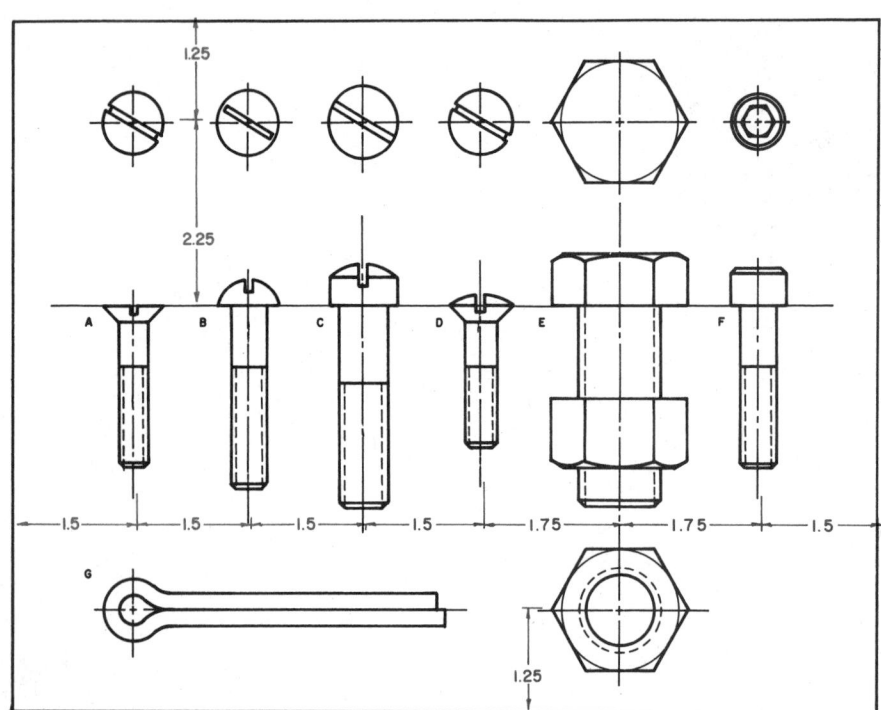

Problem 13-1

Problem 13-2

On an A-size, 8½ x 11 sheet of paper, lay out the center lines per given dimensions.

Calculate standard thread lengths.

Draw the following fasteners on the center lines as illustrated.

A. Hex-head bolt — 1⅜ - UNC x 3.0 lg.

B. Lock washer (For a ⅞ dia. cap screw).

C. Square nut — 1 - 12 UNF 2B.

D. Square-head cap screw — ¾ - 10 UNC x 3.0 lg.

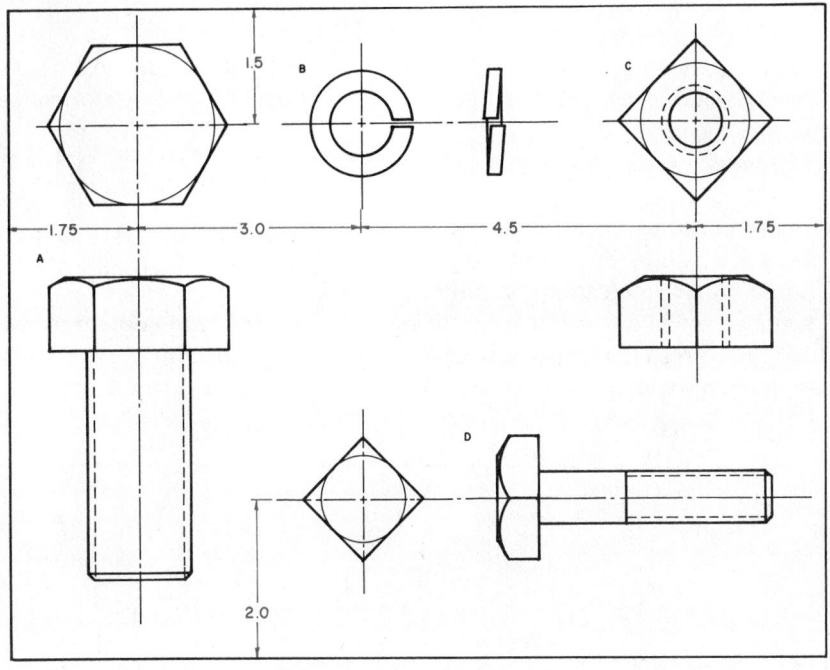

Problem 13-2

Problem 13-3

On an A-size, 8½ x 11 sheet of paper, lay out the center lines per given dimensions.

A. Calculate the required size of square key needed for a 2.25 ID collar. Draw collar and keyway, and dimension per drafting standards.

B. Calculate the required size of square key needed for the matching shaft of the collar. Draw the end view of the shaft and dimension per drafting standards.

C. Calculate the *minimum* length hex-head cap screw, ¾ – 10 UNC – 2A required to safely secure part X to part Y (material: aluminum). Illustrate and call-off the required pilot drill size and recommended depth into part Y. Illustrate and call-off the required size and depth of *full* threads for the hex-head cap screw.

Illustrate and call-off the diameter for clearance hole in part X.

D. Draw a square-head set screw with a full dog point, ⅜ – 18 UNF x 3.0 lg.

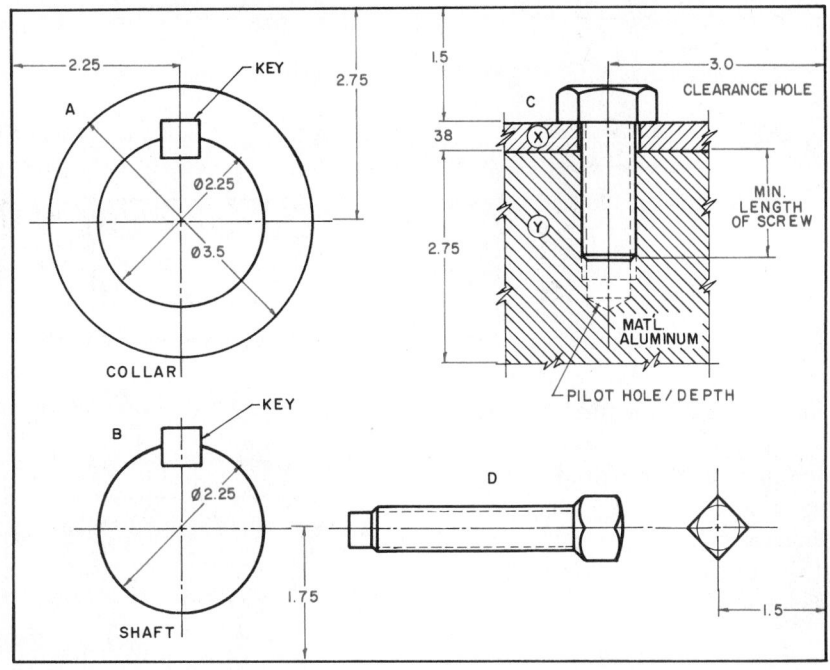

Problem 13-3

Problem 13-4

On an A-size, 8½ x 11 sheet of paper, lay out the center lines, part X, and part Y per given dimensions.

Calculate the standard thread lengths.

Draw the following fasteners on the center lines as illustrated.

Give *all* specific hole call-off information for each fastener in the space above each as illustrated (include clearance hole sizes, counterbore specifications, countersink specifications, spotface specifications, required tap drill size and depth for 75% thread; and tap size and required depth per latest drafting standards.

A. Socket-head cap screw — ⅞ - 9 UNC x 1.75 lg (calculate c'bore information).

B. Oval-head cap screw — ¾ - 10 UNC x 2.5 lg (calculate c'sink information).

C. Hex-head cap screw — 1 - 12 UNF x 2.5 lg (calculate S.F. information).

D. Fillister-head cap screw with plain washer and hex-head nut — ¾ - 16 UNF x length to suit. (Calculate clearance hole required.)

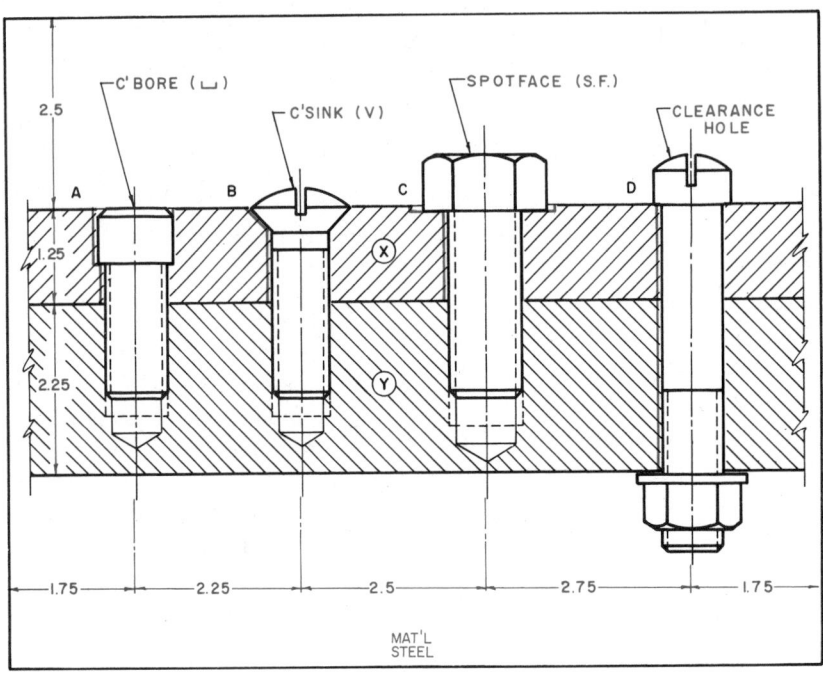

Problem 13-4

❑ Problem 13-5

Given, an illustration of an old design (Figure 13-5A) that uses a washer and cotter pin to hold the .3543 dia. shaft in place. Using this design, it was necessary to locate and drill a hole for the cotter pin. The new design incorporated the use of one retaining ring to achieve the exact same function at a much lower cost. Choose the correct type and size retaining ring and fill in the required dimensions at A, B and C (see Figure 13-6A). Neatly letter all retaining ring data below the dimensioned shaft as illustrated.

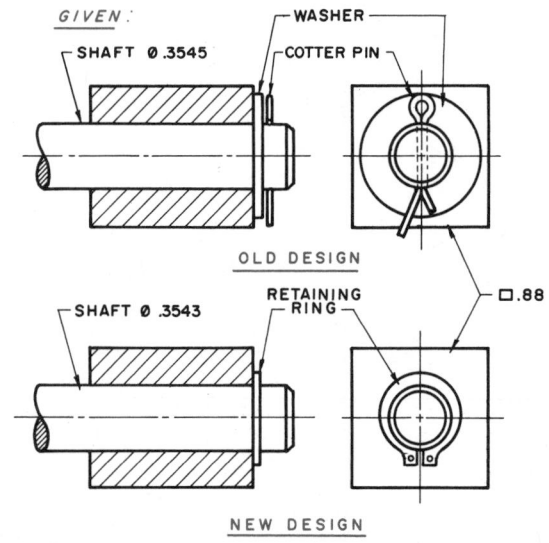

Problem 13-5A

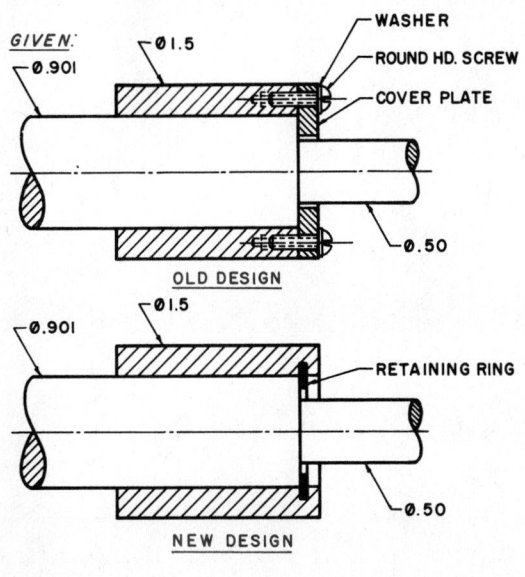

Problem 13-5B

❏ Problem 13-6

Given, an illustration of an old design (Figure 13-5B) that uses a cover plate with 4 holes for 4 washers and 4 round-head machine screws to hold the shaft in place. Using this design required drilling 4 blind tap-drill holes which had to be tapped for the 4 round-head screws.

The new design incorporated the use of one retaining ring to achieve the exact same function at a much lower cost. Choose the correct type and size retaining ring and fill in the required dimensions at A, B and C (see Figure 13-6B). Neatly print all retaining ring data below the dimensioned collar as illustrated.

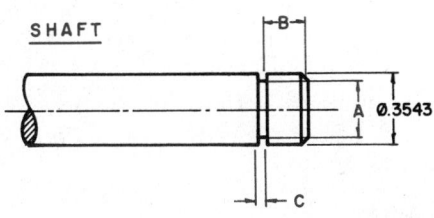

DATA:
RETAINING RING NUMBER _____
FREE DIAMETER _____
THICKNESS _____
LUG SIZE _____
LARGE SECTION _____
SMALL SECTION _____
PLIER HOLE SIZE _____
MAX. GAGING DIA. WHEN INSTALLED
_____

Problem 13-6A

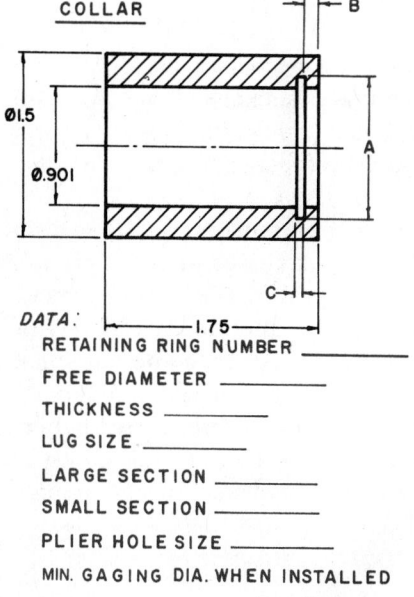

DATA:
RETAINING RING NUMBER _____
FREE DIAMETER _____
THICKNESS _____
LUG SIZE _____
LARGE SECTION _____
SMALL SECTION _____
PLIER HOLE SIZE _____
MIN. GAGING DIA. WHEN INSTALLED
_____

Problem 13-6B

Springs

Chapter Outline

A *spring* is a mechanical device that is used to store and apply mechanical energy. A spring can be designed and manufactured to apply a pushing action, a pulling action, a torque or twisting action, or a simple power action.

Product manufacturers usually use standard-size springs that are purchased from companies that specialize in making springs. These springs are usually mass produced and are, therefore, relatively inexpensive. Occasionally, however, a special spring must be designed to perform a special function. To be able to design special-function springs, the drafter or designer must know the many terms associated with springs, and how to design and construct special springs.

Spring Classification

A spring is generally classified as either a helical spring or a flat spring. *Helical springs* are usually cylindrical or conical; the *flat springs*, as their name implies, are usually flat.

Helical Springs

Three types of helical springs are compression springs, extension springs, and torsion springs, *Figures 14-1A*, *14-1B*, and *14-1C*.

Compression Spring

A *compression spring* offers resistance to a compressive force or applies a pushing action. In its free state, the coils of the compression spring do not touch. The types of ends of a compression spring are illustrated and described as follows.

Plain open end—*Figure 14-2A*. The plain open end spring is very unstable, has only point contact, and the ends are *not* perpendicular to the axis of the spring.

Plain closed end—*Figure 14-2B*. The plain closed end spring is a little more stable, and provides a round parallel surface contact perpendicular to the axis of the spring.

Ground open end—*Figure 14-2C*. The ground open end spring is much more stable, and provides a flat surface contact perpendicular to the axis of the springs.

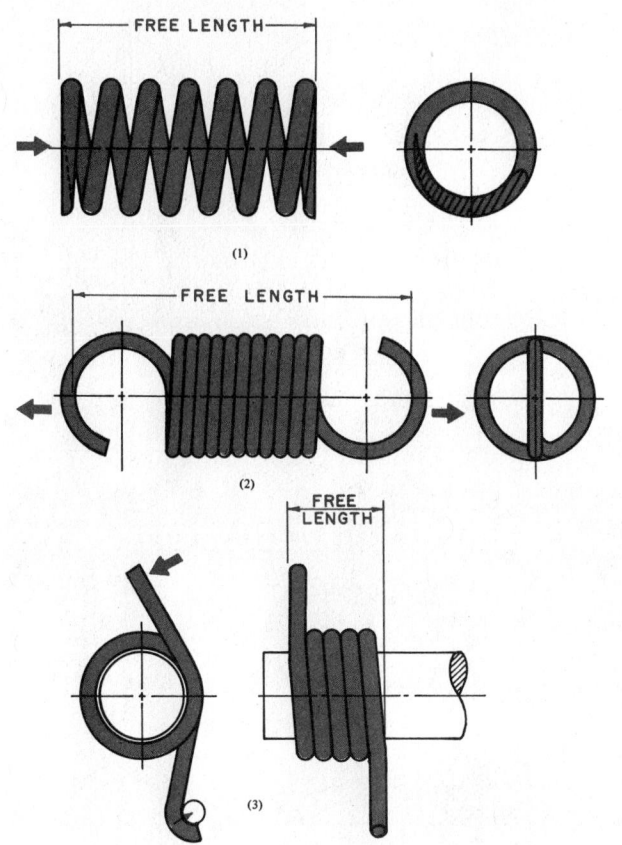

Figure 14-1A   Types of helical springs (1) compression spring, (2) extension spring, and (3) torsion spring

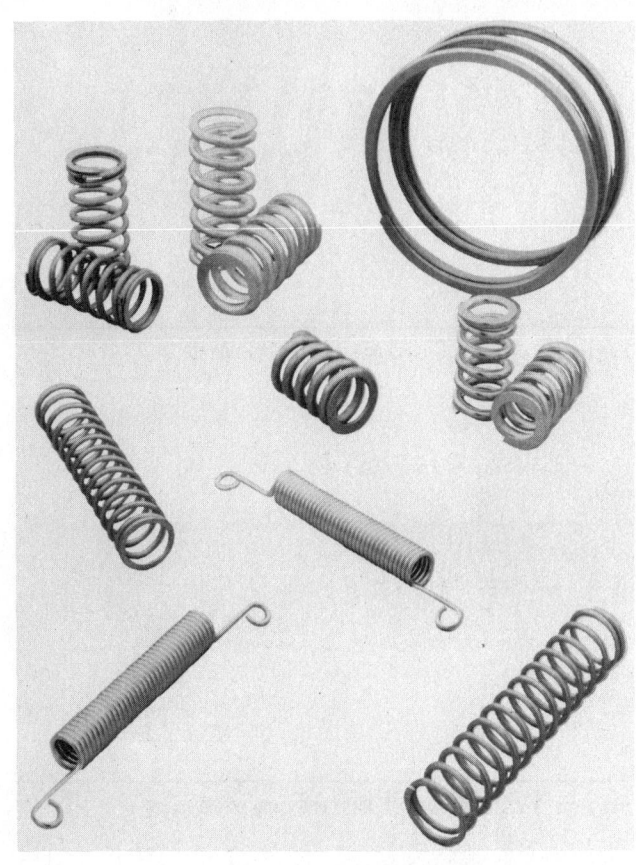

Figure 14-1B   Varieties of compression and extension springs (Courtesy AMETEK Inc., Hunter Spring Division)

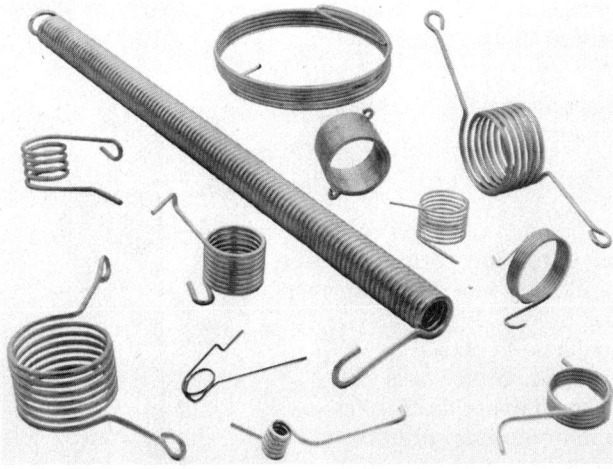

Figure 14-1C   Varieties of torsion springs (Courtesy AMETEK Inc., Hunter Spring Division)

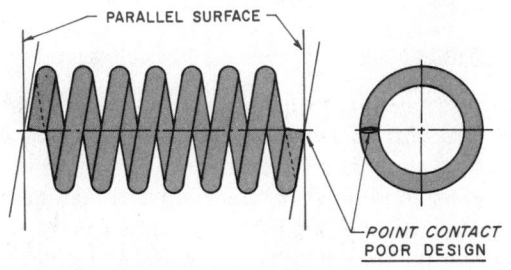

Figure 14-2A   Plain open end compression spring

Figure 14-2B   Plain closed end compression spring

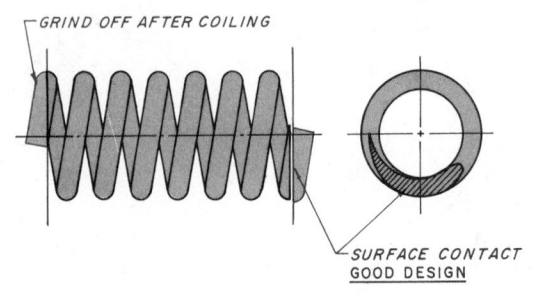

Figure 14-2C Ground open end compression spring

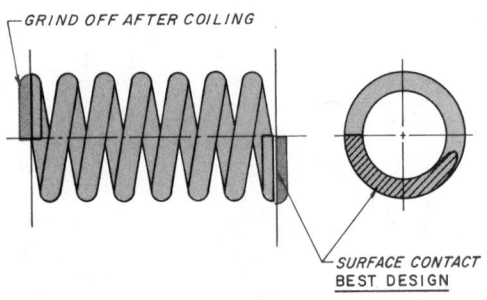

Figure 14-2D Ground closed end compression spring

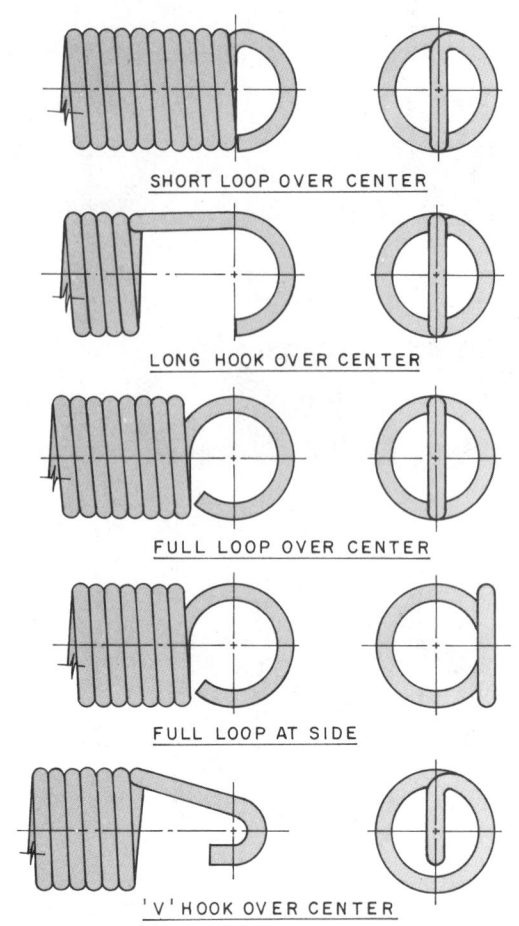

Figure 14-3 Ends of extension spring

Ground closed end—*Figure 14-2D*. The ground closed end spring is the most stable, and the best design. This design provides a large flat surface contact perpendicular to the axis of the spring.

A good example of a compression spring is the kind that is used in the front end of an automobile.

Extension Spring

An *extension spring* offers resistance to a pulling force. In its free state, the coils of the extension spring are usually either touching or very close. The ends of an extension spring are usually a hook or loop, *Figure 14-3*. Illustrated are a few of the many kinds of ends that are used. The loop or hook can be over the center or at the side. Each end is designed for a specific application or assembly. Regardless of the shape of the loop or hook, the overall length of an extension spring is measured from the *inside* of the loops or hook. A good example of an extension spring is the kind that is used to counterbalance a garage door.

Torsion Spring

A *torsion spring* offers resistance to a torque or twisting action. In its free state, the coils of a torsion spring are usually either touching or very close. The ends of a torsion spring are usually specially designed to fit a particular mechanical device, *Figure 14-4*. A good example of a torsion spring is the kind that is used to return a doorknob back to its original position.

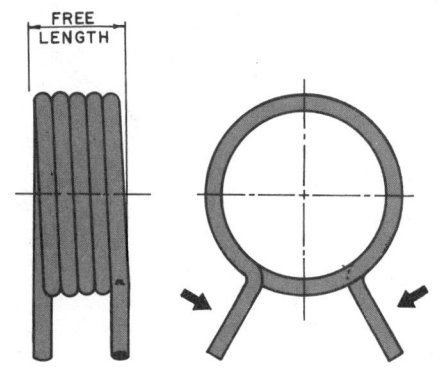

Figure 14-4 Torsion spring

Flat Springs

The other classification or type of spring is the flat spring. The *flat spring* is made of spring steel and is designed to perform a special function. Flat springs are not standard, and must be designed and manufactured to fit each particular need. The flat spring is considered a power spring, and it resists a pressure. A good example of a flat spring is the kind that is used for a door latch on a cabinet, *Figure 14-5*, or a leaf spring as used in a truck, *Figure 14-6*. A flat spiral spring used to power a windup clock or toy is another example, *Figure 14-7*.

Terminology of Springs

It is important to know and fully understand the various terms associated with springs and their design, *Figure 14-8*.

Free Length

Free length is the overall length of the spring when it is in its free state of unloaded condition. Free length of a compression spring is measured from the extreme ends of the spring. In an extension spring, the free length is measured inside the loops or hooks.

Solid Length

Solid length of a compression spring is the overall length of the spring when *all* coils are compressed together so they touch. This can be mathematically calculated. If the wire diameter is .500, the overall solid length is 3.25.

Outside Diameter

Outside diameter (O.D.) is the outside diameter of the spring (2 x wire size *plus* inside diameter).

Wire Size

Wire size is the diameter of the wire used to make up the spring.

Inside Diameter

Inside diameter (I.D.) is the inside diameter of the spring (2 x wire size *minus* outside diameter).

Active Coils

Active coils in a compression spring are usually the total coils minus the two end coils. In a compression spring, the coil at each end is considered nonfunctional. In an extension spring, active coils include *all* coils.

Loaded Length

Loaded length is the overall length of the spring with a special given or designed load applied to it.

Mean Diameter

Mean diameter is the theoretical diameter of the spring measured to the center of the wire diameter. This theoretical diameter is used by the drafter to lay out and draw a spring. To mathematically calculate the mean diameter, subtract the wire diameter from the outside diameter.

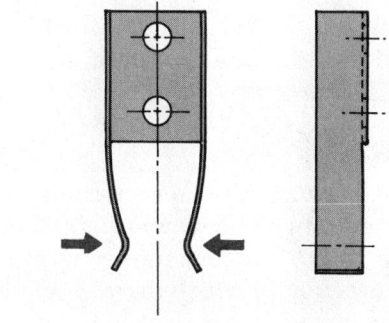

Figure 14-5 Flat spring

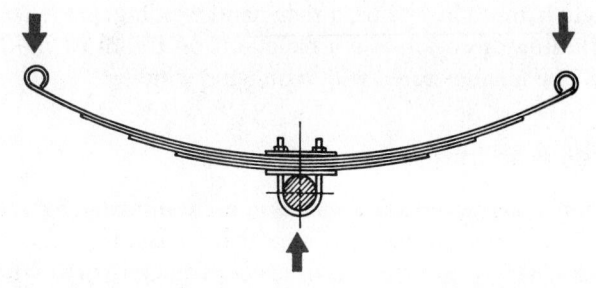

Figure 14-6 Leaf spring

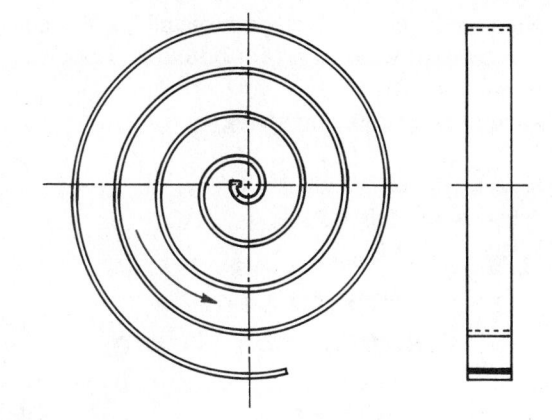

Figure 14-7 Special spring

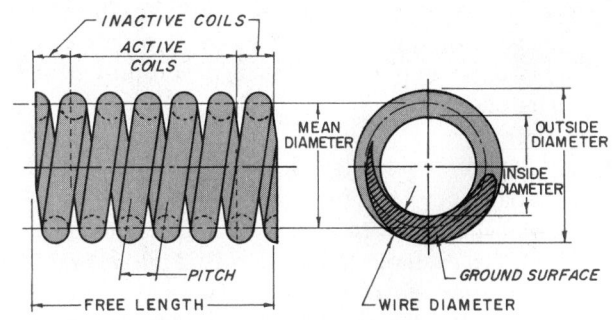

Figure 14-8 Spring terminology

Coil

Coil or *turn* is one full turn or 360° of the wire about the center axis, *Figure 14-9*.

Direction of Winding

Direction of winding is the direction in which the spring is wound. It can be wound right hand or left hand, *Figure 14-10*. The thumbs of your hands point toward the direction in which the coil windings are leaning. If the coil windings slant to the right, as the example to the left illustrates, it agrees with the direction of your left-hand thumb; thus, a left-hand winding. If the coil windings slant to the left as the example to the right illustrates, it agrees with the direction of your right-hand thumb; thus, a right-hand winding. If the coil winding direction is not called-off on the drawing, it will be manufactured with right-hand winding.

Required Spring Data

Each drawing must include complete dimensions and specific data. Besides the regular dimensions, i.e., outside diameter (O.D.) and/or inside diameter (I.D.) wire size; free length; and solid length (compression spring), the following data should be called-off at the lower right side of the work area:

- Material
- Number of coils (including active and inactive coils)
- Direction of winding (if not noted, it is assumed to be right hand)
- Torque data (if torsion spring)
- Finish
- Heat treatment specification
- Any other required data

How To Draw a Compression Spring

Given: Plain open ends, 1.50 outside diameter, .25 diameter wire size, 4.00 free length, oil tempered spring steel wire, 8 total coils (6 active coils), left-hand winding, heat treatment: heat to relieve coiling stresses, finish: black paint. Make a rough sketch of the spring using the given specifications, *Figure 14-11A*. A rough sketch is extremely important in developing a new spring design.

Step 1 Measure and construct two vertical lines the required overall length (front view). Divide the space between the two vertical lines into 17 equal spaces (front view) 2 x total coils plus 1; in this example, 2 x 8 + 1 = 17, *Figure 14-11B*. Measure and construct the mean diameter. To calculate the mean diameter, subtract the wire diameter from the outside diameter (end view).

Step 2 Determine the direction of windings; refer to sketch and lightly construct the wire diameter accordingly, (front view). The wire diameters

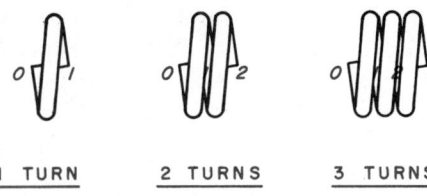

Figure 14-9 Coils or turns

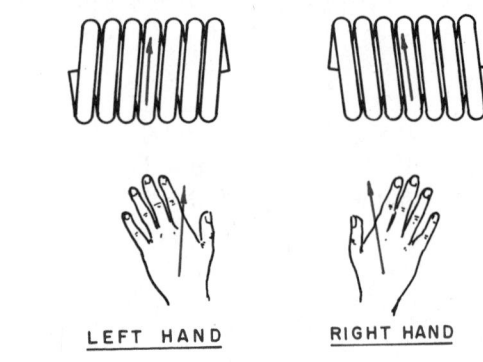

Figure 14-10 Direction of winding

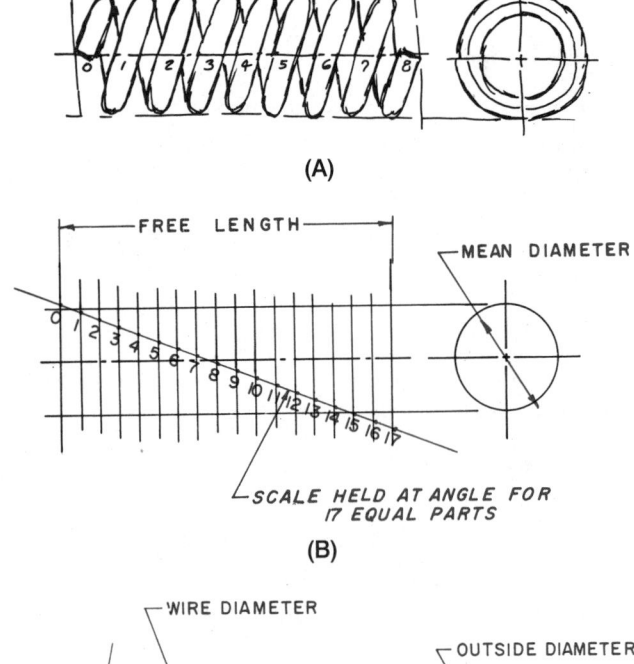

(A)

(B)

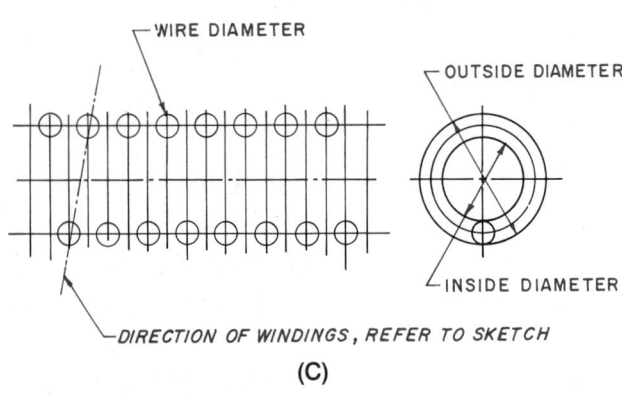

(C)

Figure 14-11 (A) How to draw a compression spring with plain open ends; (B) Step 1; (C) Step 2

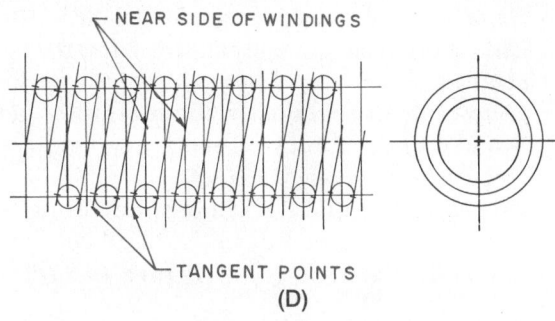

NEAR SIDE OF WINDINGS

TANGENT POINTS

(D)

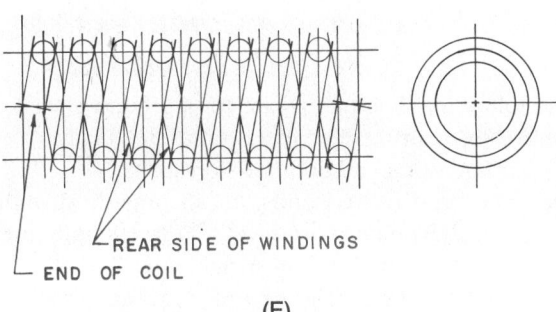

REAR SIDE OF WINDINGS
END OF COIL

(E)

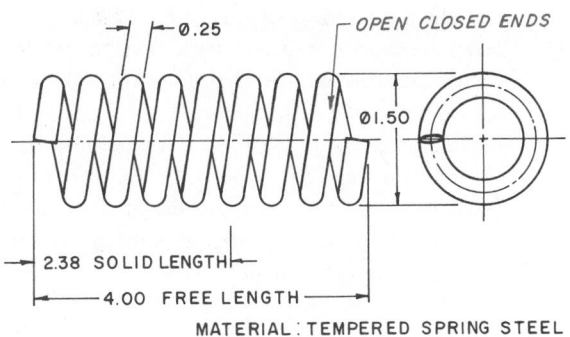

Ø.25

OPEN CLOSED ENDS

Ø1.50

2.38 SOLID LENGTH

4.00 FREE LENGTH

MATERIAL: TEMPERED SPRING STEEL

8 TOTAL COILS (6 ACTIVE)

LEFT-HAND COILS

HEAT TO RELIEVE COILING STRESSES

FINISH: BLACK

(F)

Figure 14-11 (Continued) (D) Step 3; (E) Step 4; (F) Step 5

are constructed on the mean diameter. Lightly draw the outside diameter and inside diameter (end view), *Figure 14-11C*.

Step 3 Lightly construct the near side of the spring. Add short, light tangent points, *Figure 14-11D*.

Step 4 Lightly construct the rear side of the spring. Add ends of spring; in this example, plain open ends. Check all work against given requirements, *Figure 14-11E*.

Recheck all work. Darken in spring using the latest drafting standards. Add end of spring, right end, dimensions and add all required spring data, *Figure 14-11F*.

Notice: outside diameter, wire size, free length, and solid length are given as regular dimensions.

The material, number of coils, direction of winding, heat treatment specifications and finish requirements are listed below and to the right as a note. In this example, the outside diameter is the controlling dimension; that is, the outside dimension is more important than the inside dimension. Both inside and outside dimensions should not be added in the same drawing.

How To Draw a Compression Spring (Plain Closed Ends)

Given: Plain closed ends, 20 mm inside diameter, 5 mm diameter wire size, 100 mm free length, material: hard drawn steel spring wire, 5 active coils (total 7 coils), left-hand winding, heat treatment: heat to relieve coiling stresses, finish: zinc plate. Make a rough sketch of the spring using the given specifications, *Figure 14-12A*.

Step 1 Measure and construct two vertical lines the required overall length (front view). Draw a line *inside* the two vertical lines equal to half the wire diameter from each end. Divide the space between the two vertical lines into 13 equal spaces (front view) 2 x total coils minus 1; in this example, 2 x 7 – 1 = 13, *Figure 14-12B*. Measure and construct the mean diameter. To calculate the mean diameter, add the wire diameter to the inside diameter (end view).

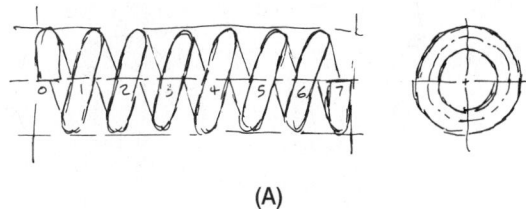

(A)

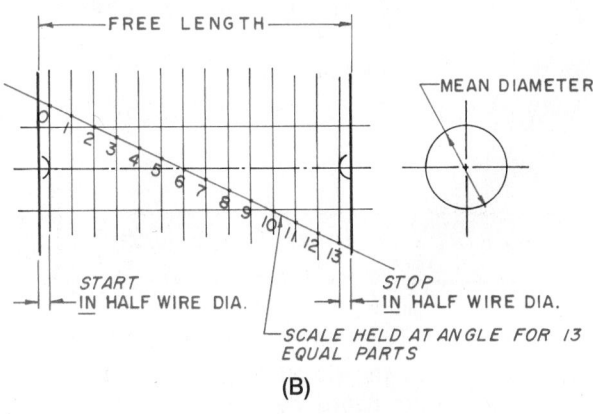

FREE LENGTH

MEAN DIAMETER

START
IN HALF WIRE DIA.

STOP
IN HALF WIRE DIA.

SCALE HELD AT ANGLE FOR 13 EQUAL PARTS

(B)

Figure 14-12 (A) How to draw a compression spring with plain closed ends; (B) Step 1

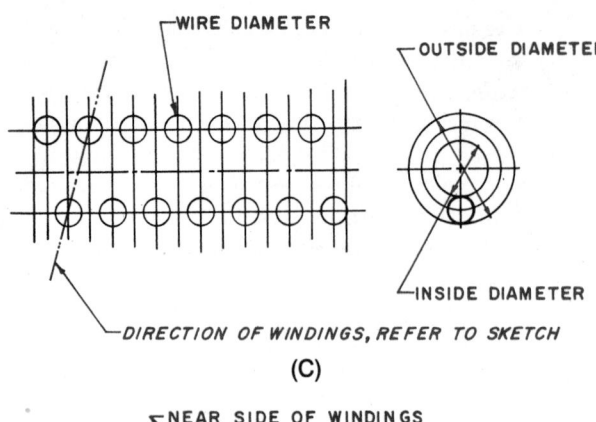

(C)

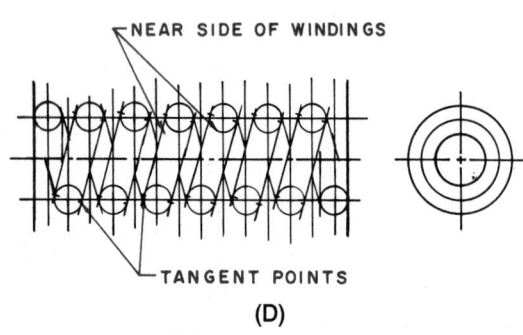

(D)

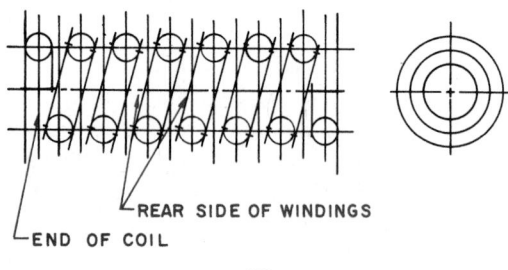

(E)

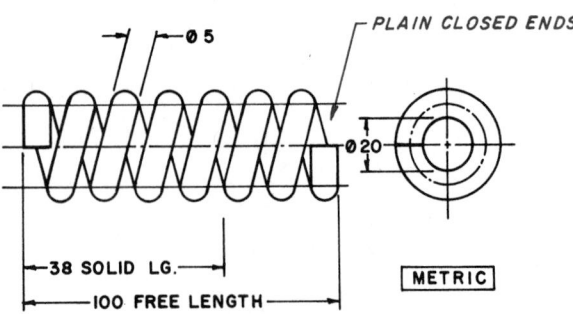

(F)

Figure 14-12 (Continued) (C) Step 2; (D) Step 3; (E) Step 4; (F) Step 5

Step 2 Determine the direction of winding; refer to the sketch and lightly construct the wire diameter accordingly (front view). Lightly draw the outside diameter and inside diameter (end view), ***Figure 14-12C***.

Step 3 Lightly construct the near side of the spring. Add short tangent points, ***Figure 14-12D***.

Step 4 Lightly construct the rear side of the spring. Add ends of springs. In this example, plan closed ends. Check all work against given requirements, ***Figure 14-12E***.

Step 5 Recheck all work. Darken in spring using the latest drafting standards. Add all required dimensions and spring data, ***Figure 14-12F***.

How To Draw a Compression Spring (Ground Closed Ends)

(The following illustrations are for a ground closed end compression spring. The exact same steps can be used for a ground open end compression spring.)

Given: Ground closed ends, 1.125 outside diameter, .188 diameter wire size, 3.00 free length, material: hard drawn steel spring wire, 7 total coils (5 active coils), right-hand winding, heat treatment: heat to relieve coiling stresses, finish: zinc plate. Make a rough sketch of the spring using the given specifications, ***Figure 14-13A***.

Step 1 Measure and construct two vertical lines the required free length (front view). Divide the space between the two vertical lines into 13 equal spaces (front view), 2 x total coils minus 1; in this example, 2 x 7 – 1 = 13, ***Figure 14-13B***. Measure and construct the mean diameter. To calculate the mean diameter, subtract the wire diameter from the outside diameter (end view).

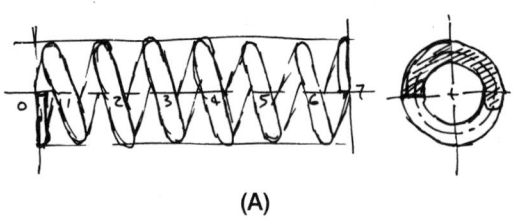

(A)

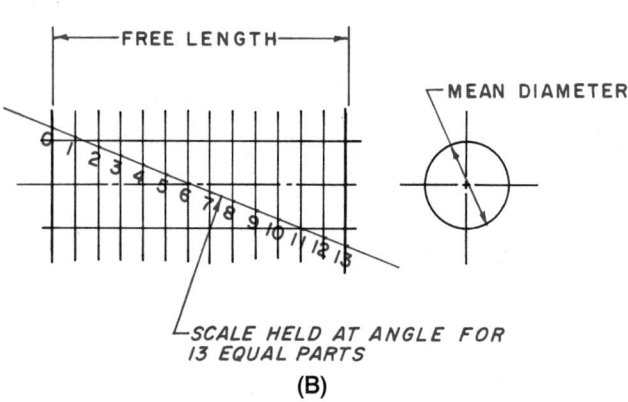

(B)

Figure 14-13 (A) How to draw a compression spring with ground closed ends; (B) Step 1

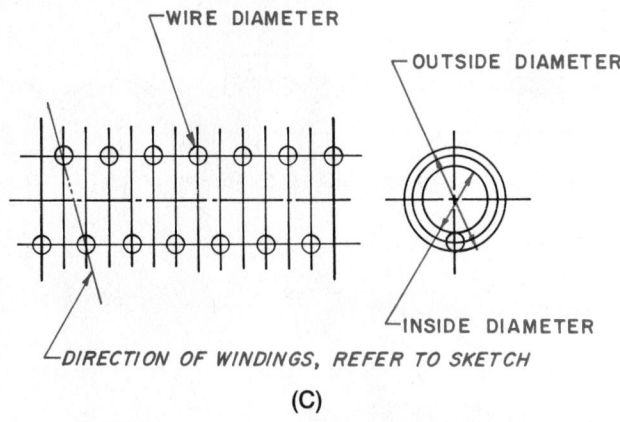

WIRE DIAMETER
OUTSIDE DIAMETER
INSIDE DIAMETER
DIRECTION OF WINDINGS, REFER TO SKETCH

(C)

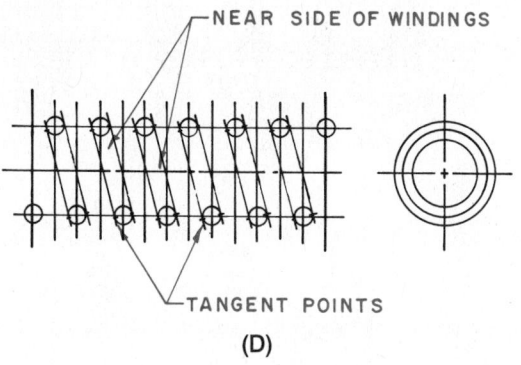

NEAR SIDE OF WINDINGS
TANGENT POINTS

(D)

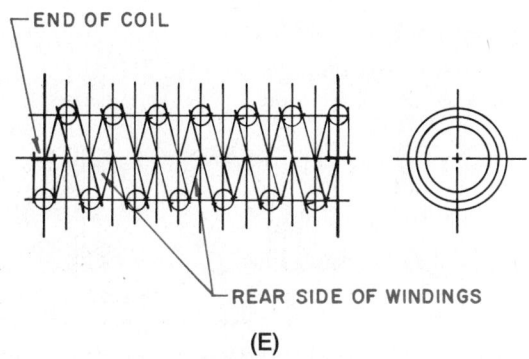

END OF COIL
REAR SIDE OF WINDINGS

(E)

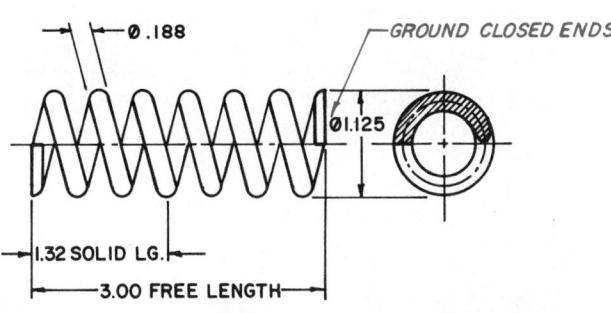

Ø.188
GROUND CLOSED ENDS
Ø1.125
1.32 SOLID LG.
3.00 FREE LENGTH

MATERIAL: HARD DRAWN SPRING STEEL WIRE
7 TOTAL COILS (5 ACTIVE COILS)
HEAT TO RELIEVE COILING STRESSES
FINISH: ZINC PLATE

(F)

Figure 14-13 (Continued) (C) Step 2; (D) Step 3; (E) Step 4; (F) Step 5

Step 2 Determine the direction of winding; refer to the sketch and lightly construct the wire diameter accordingly (front view). Lightly draw the outside diameter and inside diameter (end view), ***Figure 14-13C***.

Step 3 Lightly construct the rear side of the spring. Add short tangent points, ***Figure 14-13D***.

Step 4 Lightly construct the rear side of the spring. Add ends of spring; in this example, ground closed end. Check all work against given requirements, ***Figure 14-13E***.

Step 5 Recheck all work. Darken in spring using the latest drafting standards. Add all required dimensions and spring data, ***Figure 14-13F***.

How To Draw an Extension Spring

Given: Full loop, over center each end 1.625 outside diameter, .188 diameter wire size, 4.750 approximate free length, material: hard drawn spring steel wire, heat treatment: heat to relieve coiling stresses, finish: black paint, windings as tight as possible. Make a rough sketch of the spring using the given specifications, ***Figure 14-14A***.

Step 1 Draw the outside diameter, inside diameter and mean diameter (end view). Lightly draw the required free length. In the design of an extension spring, it is almost impossible to arrive at the *exact* free length. In designing an extension spring, it is best to design it slightly *shorter* than actually required. Construct the end loops with the *inside* diameter within the specified free length, ***Figure 14-14B***. Note in this example the loops are the exact diameter as the spring itself.

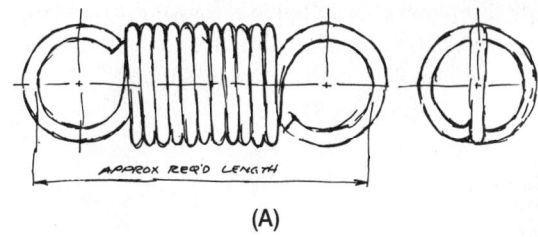

APPROX REQ'D LENGTH

(A)

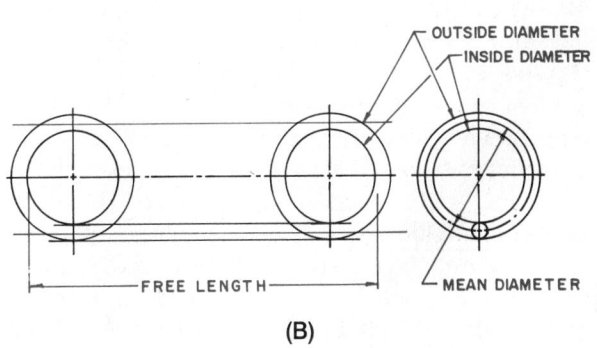

OUTSIDE DIAMETER
INSIDE DIAMETER
FREE LENGTH
MEAN DIAMETER

(B)

Figure 14-14 (A) How to draw an extension spring with full loop ends; (B) Step 1

Step 2 Starting at the intersection of the mean diameter of the right end loop, and the mean diameter of the spring, lower side, locate and draw the first lower wire diameter, *Figure 14-14C*. Project directly up to the mean diameter of the spring, upper side, and draw the first upper wire diameter *over* to the left, half a wire diameter as illustrated. Lay out, side-by-side, wire diameters touching, as illustrated; in this example, 10½ coils. Notice the coils must end on the mean diameter, left end. This may mean adjusting slightly the location of the left end loop. It is best to bring it *in* slightly, if necessary.

Step 3 Locate points X and Y, as illustrated in *Figure 14-14D*, and adjust the drafting machine at this angle and lock on this angle. Draw the wire loops.

Step 4 Lightly draw the loop in the right-side view. To calculate just what the loop will actually look like, number various points along the loop in the front view; in this example, point 1, starting point; point 2, up; point 3, over; and point 4, around. Project these points to the end view, as illustrated in *Figure 14-14E*.

Step 5 Recheck all work. Darken in spring using the latest drafting standards. Add all required dimensions and spring data, *Figure 14-14F*.

Other Spring Design Layout

When drawing any specially designed spring for a particular function, it is recommended to make a rough sketch of it, including all required specifications. A torsion spring is developed in a manner very similar to an extension spring with its tight coils. Usually, the required torque pressure is included in the given spring data. The actual designed deflection of the torsion spring is illustrated by phantom lines, and the angle is noted.

Standard Drafting Practices

Some company standards specify that the drafter not draw the complete spring because of the time and cost involved. A shortcut method to draw a spring is shown in *Figure 14-15*. At top is the conventional method of representing a compression spring; below it is the same spring drawn by the schematic drawing system.

Another method of drawing a spring, especially a long spring, is illustrated in *Figure 14-16*. The incompleted coils are illustrated by phantom lines. If the company does not have a standard method of drawing springs it is the drafter's decision as to which method of representation is used. In most cases, it is best to draw the object, in this case the spring, in such a way that there is no question whatsoever as to exactly what is required. Of course, regardless of which system is used, full dimensions and specification data must be included.

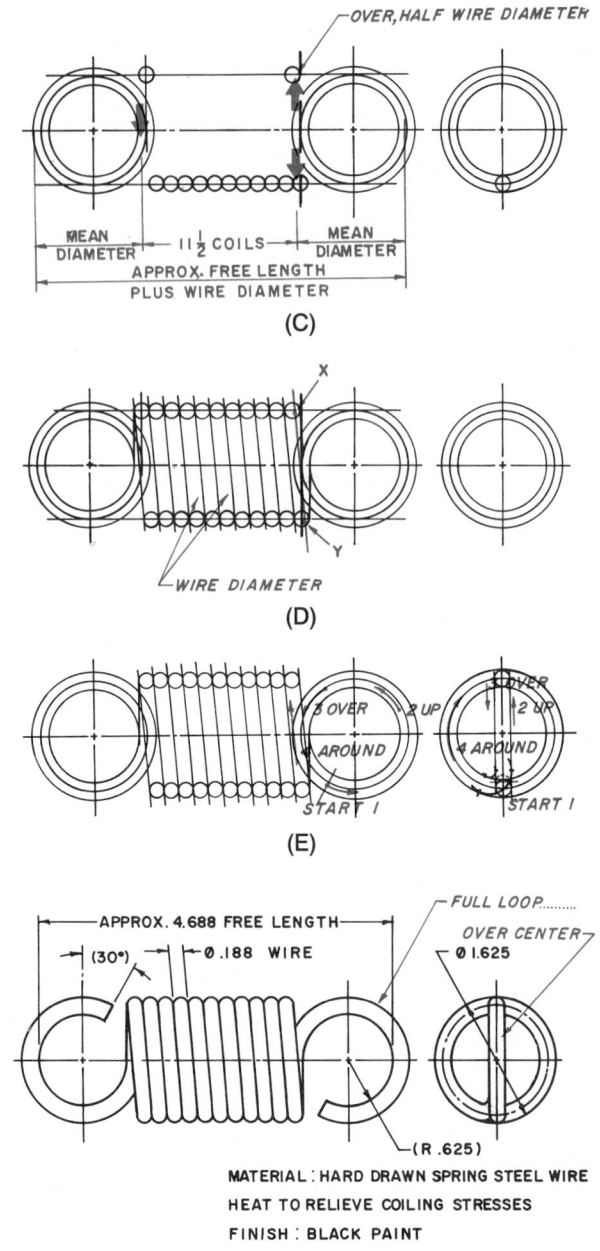

Figure 14-14 (Continued) (C) Step 2; (D) Step 3; (E) Step 4; (F) Step 5

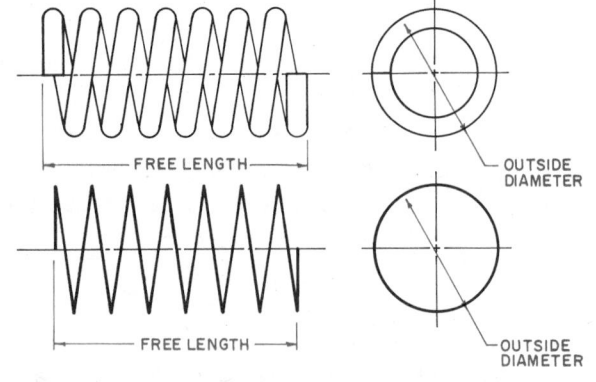

Figure 14-15   Schematic system of representing a compression spring

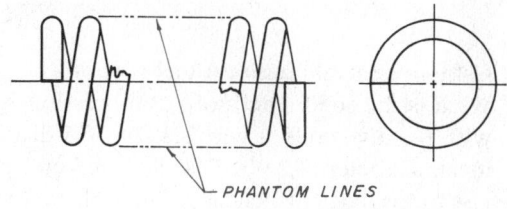

Figure 14-16 Incompleted coils with phantom lines

Section View of a Spring

If a cutting plane line passes through the axis of a spring, the spring is drawn in one of two ways. A small spring is drawn with the back coils showing, as illustrated in *Figure 14-17A*, with the section lining of the coils filled in solid. A larger spring is drawn the same way, but uses standard section lining, *Figure 14-17B*. Notice that both illustrations are right-hand springs, but, because the front half has been removed, only the rear half is seen which appears as left hand.

Isometric Views

A template, such as the one shown in *Figure 14-18*, makes it quick and simple to draw isometric views

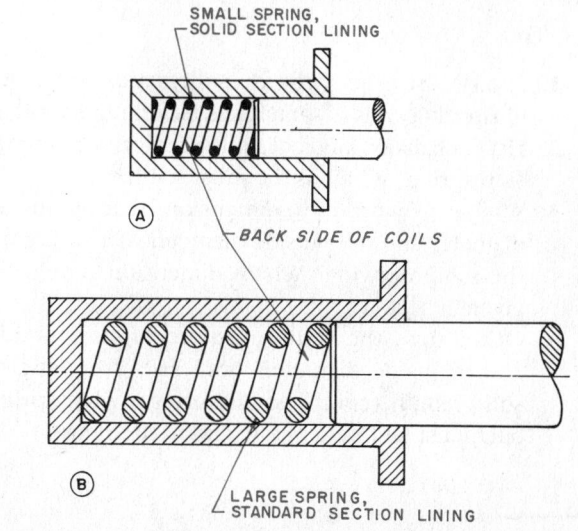

Figure 14-17 Section view of a spring

of light, medium, or heavyweight open springs. Lightweight springs are slightly smaller than the dimension printed on the template because they are drawn with the two smallest cutouts in any set. The thick lower edge and pinpoint centers ensure precise positioning to repeat each loop for any length spring.

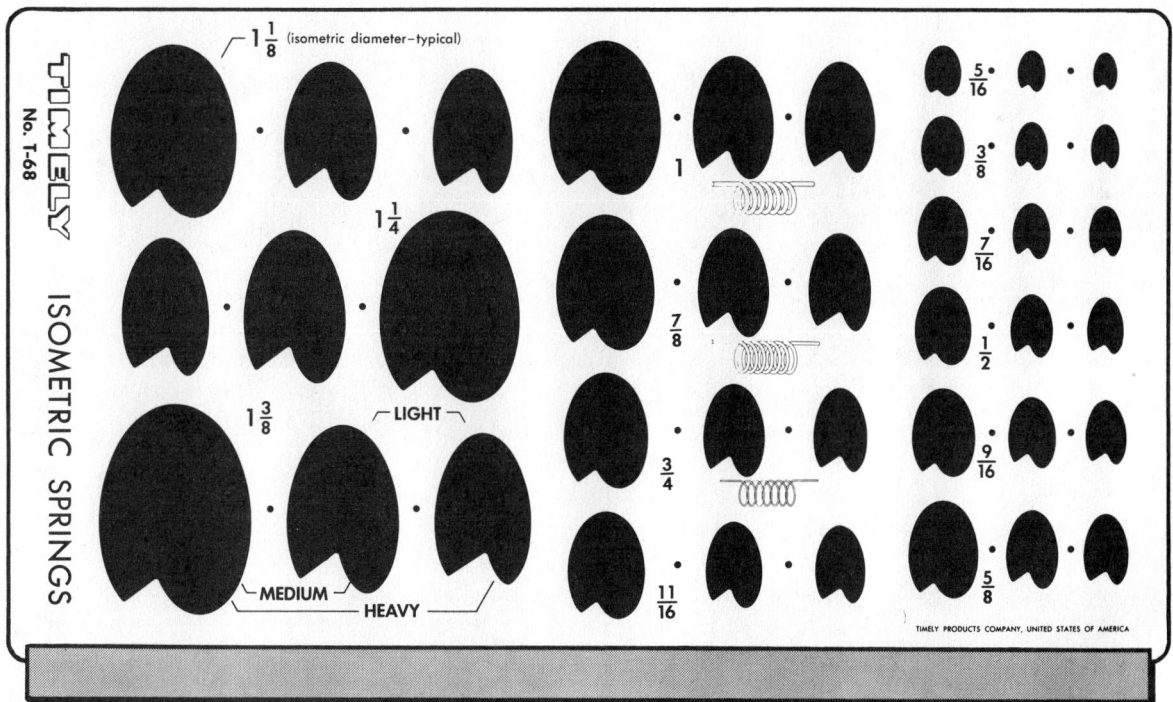

This template makes it quick and simple to draw isometric views of light, medium, or heavy weight open springs. Lightweight springs are slightly smaller than dimension printed on template because they are drawn with the two smallest cutouts in any set. The thick lower edge and pin point centers insure precise positioning to repeat each loop — for any length spring.

Figure 14-18 Isometric spring template *(Courtesy Timely Products Co.)*

Review

1. List two reasons why the schematic method of illustrating coils is sometimes used in industry.
2. List four major kinds of ends used on a compression spring. Which is the most stable?
3. Why is it incorrect to dimension both the inside diameter and the outside diameter of a spring on the same drawing? Which dimension should be given?
4. Other than the required dimensions, outside and/or inside diameter, wire size, free length, solid length (compression spring), what spring data must be noted below and to the right?
5. List two general classifications of springs.
6. What is the solid length of a compression spring with 8 active coils, a wire size of .375 diameter, mean diameter of 2.00, closed ground ends?
7. List three types of helical springs. Make a rough sketch of each illustrating how pressure is applied to each.
8. How is the section lining illustrated in a small spring where the cutting plane line passes through the center axis of the spring?
9. What is meant by the free length of a spring?
10. List three common uses for a torsion spring.

Chapter Fourteen Problems

The following problems are intended to give the beginning drafter practice in drawing various kinds of springs with special required specifications.

The steps to follow in laying out springs are:

Step 1 Study all required specifications.

Step 2 Make a rough sketch of the spring, incorporating *all* required specifications and required dimensions.

Step 3 Calculate the size of the basic shape of the front and end views.

Step 4 Center the two views within the work area with correct spacing for dimensions between views.

Step 5 Starting with the end view, or the view that contains the diameter, lightly draw the spring adhering to *all* specific required specifications. Take care to show correct direction of winding, correct type of ends, and total count of windings.

Step 6 Check to see that both views are centered within the work area.

Step 7 Check all dimensions in both views.

Step 8 Darken in both views, starting first with all diameters and arcs.

Step 9 Neatly add all dimensioning as required to fully describe the object using the latest drafting standards.

Step 10 Neatly add all required specifications under the two views. The drawing must include the following:
>   Free length
>   Wire size
>   Total number of coils
>   Type of ends
>   Solid length
>   Outside diameter
>   Inside diameter
>   Direction of windings
>   Material (usually, hard drawn steel, spring wire)
>   Required heat treating process (usually, heat to relieve coiling stresses)
>   Any other special requirements

Step 11 For torsion-type springs, illustrate *working location* and travel with phantom lines and dimension.

Problems 14-1 through 14-7

Construct a 2-view drawing of each spring using the listed steps. Center the views on an A-size, 8½ x 11 sheet of paper with a 1″ (25 mm) space between views. Use correct line thicknesses and all drafting standards. Add all required dimensions and specifications per the latest drafting standards.

Compression-type spring
Free length 7.00
Wire size .25
12 active coils
(14 *total* coils)
Plain open ends
O.D. 2.00/I.D. 1.50
R.H. winding
Calculate solid length

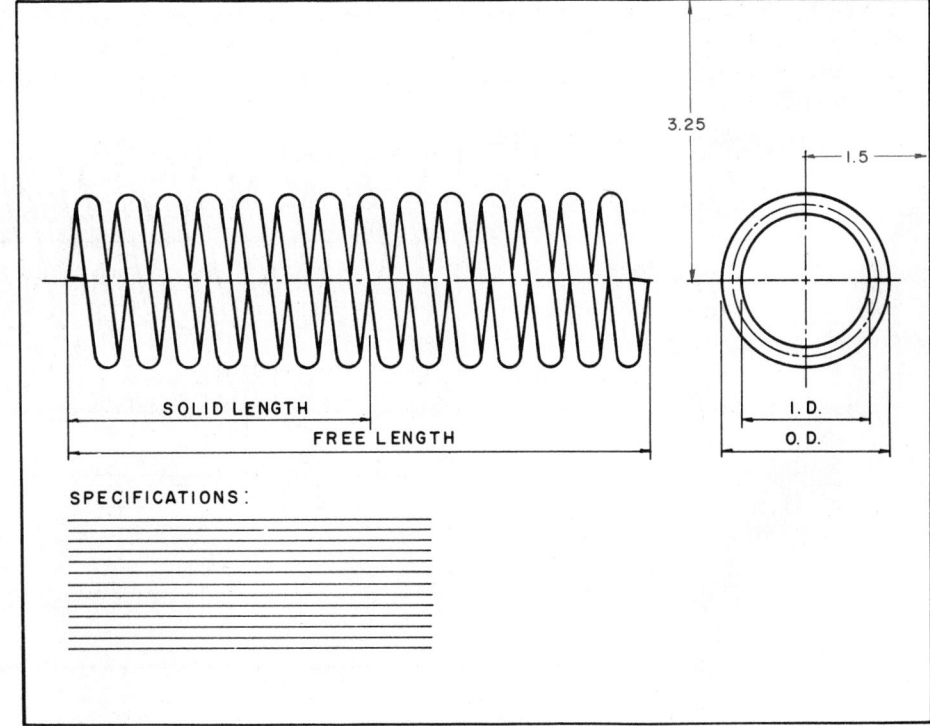

Problem 14-1

Compression-type spring
Free length 5.25
Wire size .375
4 active coils
(6 *total* coils)
Ground open ends
O.D. 2.75/I.D. 2.00
L.H. winding
Calculate solid length

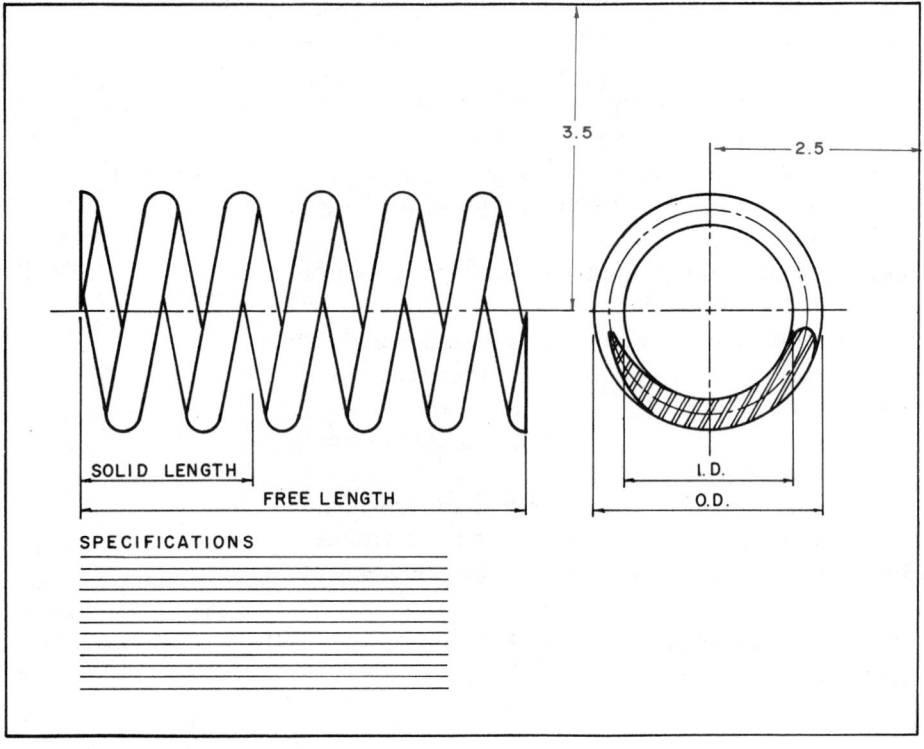

Problem 14-2

Compression-type spring
Free length 6.25
Wire size .375
9 total coils
(7 *active* coils)
Plain closed ends
O.D. 2.00/I.D. 1.25
L.H. winding
Calculate solid length

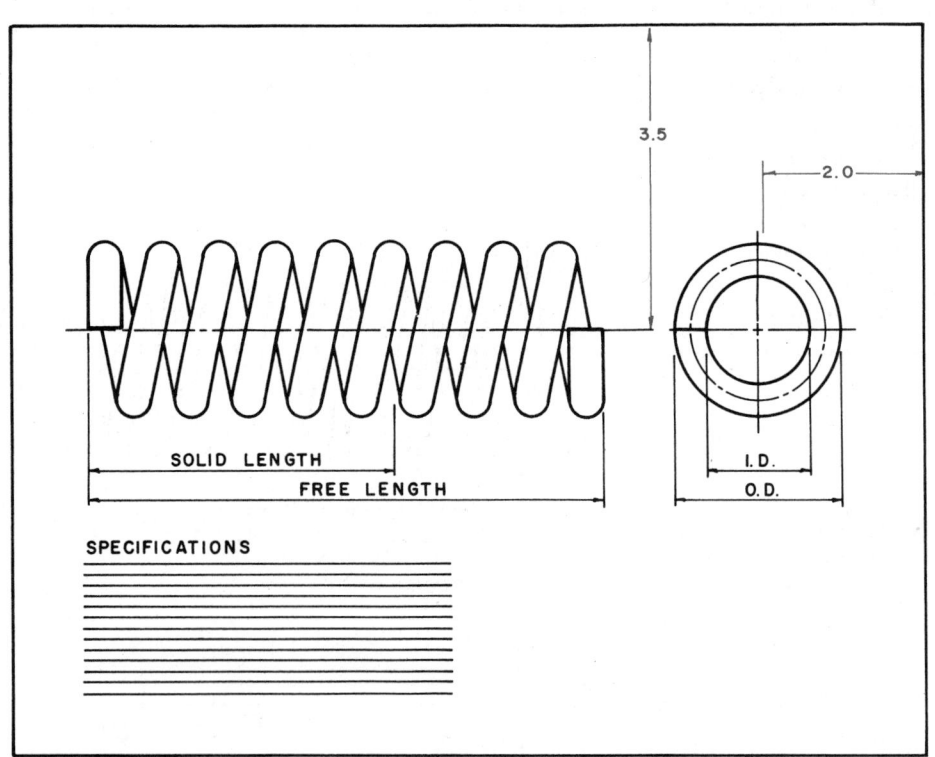

Problem 14-3

(Metric)
Compression-type spring
Free length 90
Wire size 12
4 total coils
   (2 active coils)
Ground closed ends
O.D. 96/I.D. 72
R.H. winding
Calculate solid length

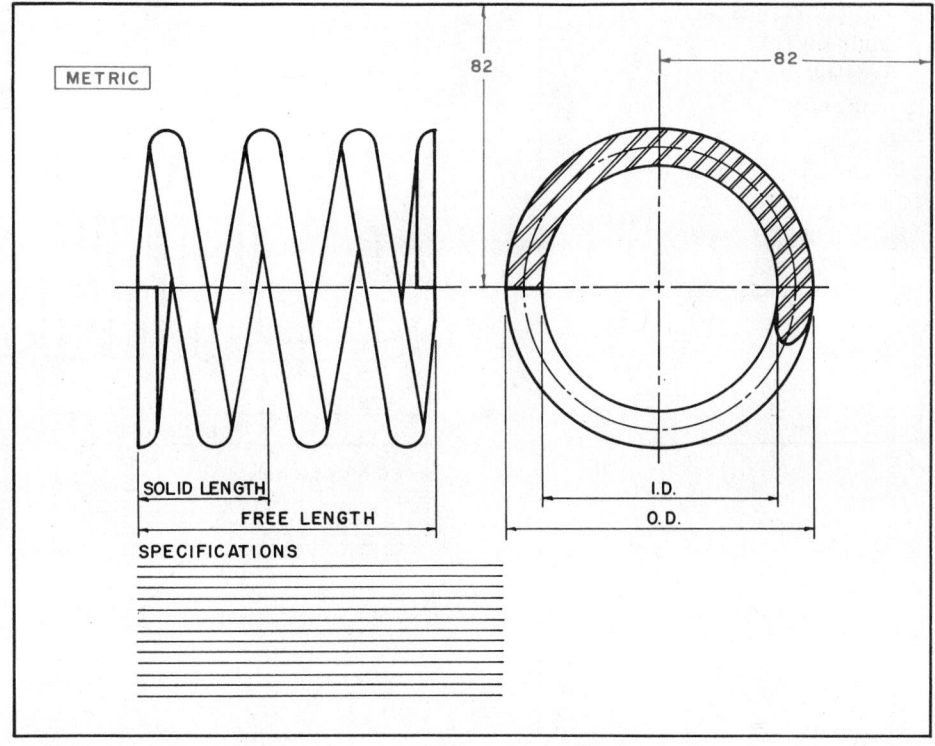

Problem 14-4

Extension-type spring
Full loop over center,
   both ends
Approx. free length 7.0
Wire size .25
O.D. 2.0/I.D. 1.50
R.H. winding (This is *standard*
   for all extension springs)

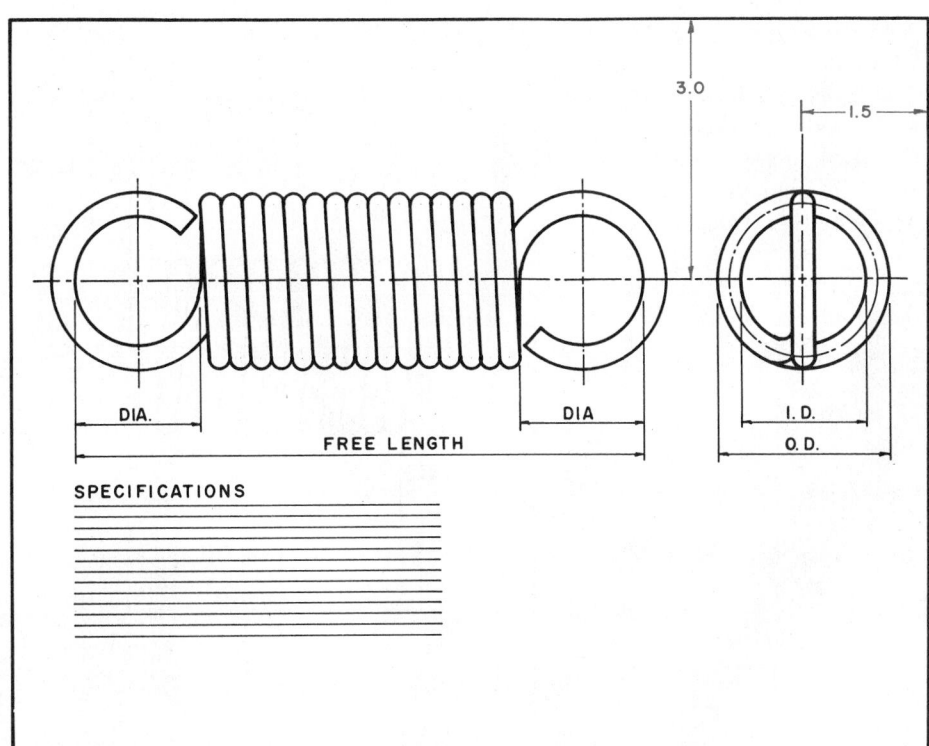

Problem 14-5

(Metric)
Extension-type spring
Full loop over center
   (right end)
Long hook over center
   (left end)
Free length 150
Wire size 5
12 total coils
65 coil length
O.D. 42/I.D. 32
Standard R.H. winding

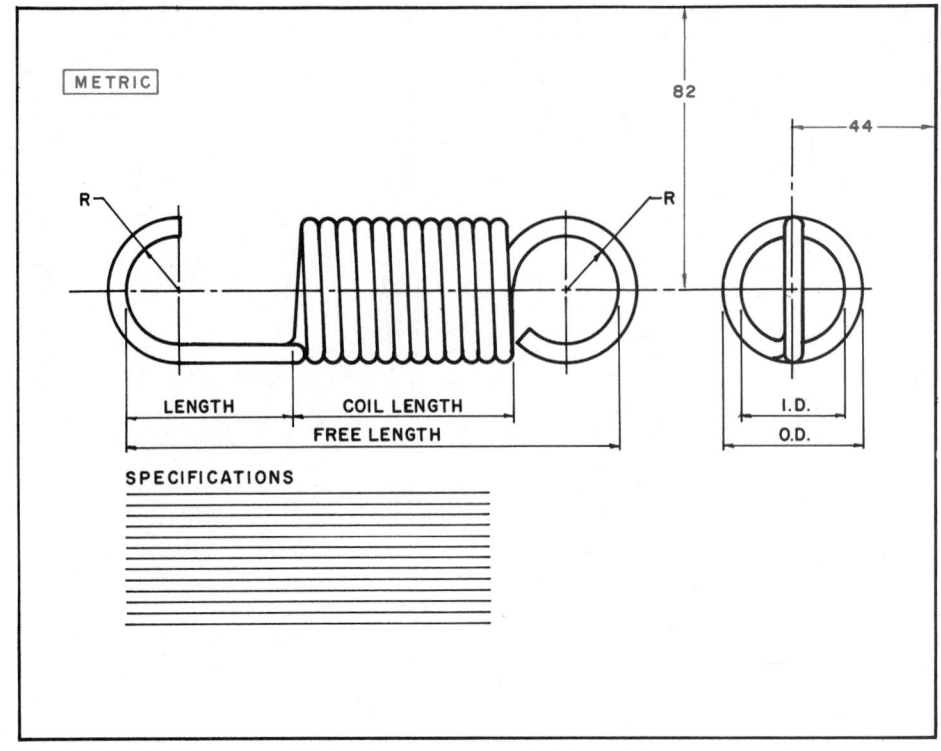

Problem 14-6

Torsion-type spring
Free length 3.625
Wire size .25
13 total coils
O.D. 2.0/I.D. 1.50
R.H. standard winding
90° maximum working flex
   (counterclockwise)
Arm lengths 2.0 (as shown)

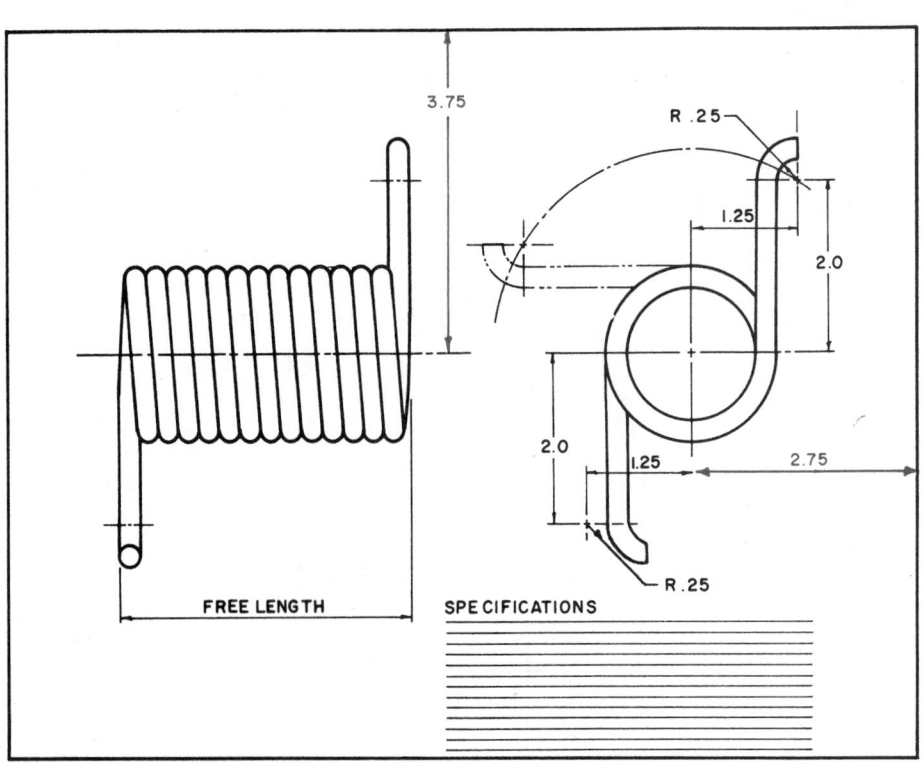

❏ **Problem 14-7**

Cams

CHAPTER 15

Chapter Outline

Cam Principle

A *cam* produces a simple means to obtain irregular or specified predictable designed motion. These motions would be very difficult to obtain in any other way.

Figure 15-1 illustrates the basic principle and terms of a cam. In this example, a rotating shaft has an irregularly shaped disc attached to it. This disc is the cam. The follower, with a small roller attached to it, pushes against the cam. As the shaft is rotated the roller follows the irregular surface of the cam, rising or falling according to the profile of the cam. The roller is held tightly against the cam either by gravity or a spring.

Figure 15-2 illustrates a simple cam in action. Notice how the rotation of the shaft converts into an up and down motion of the follower. A cam using a flat-faced follower is shown in ***Figure 15-3***. This type of cam, for example, is used to raise and lower the valve in an automobile engine. A modified cam follower is used to

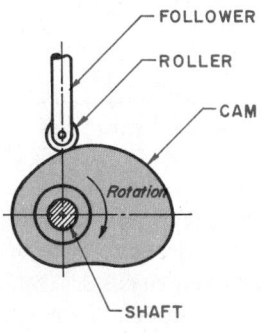

Figure 15-1 Basic cam principle and terms

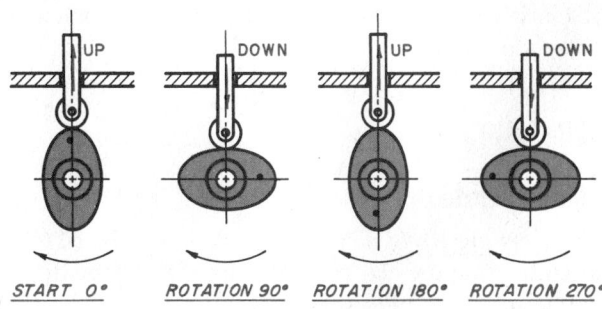

Figure 15-2 Simple cam in action

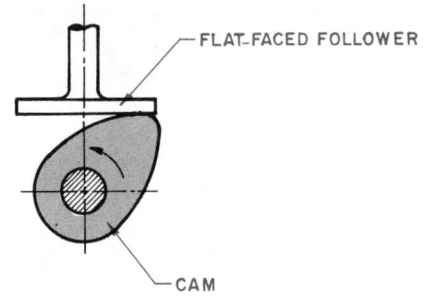

Figure 15-3 Flat-faced follower

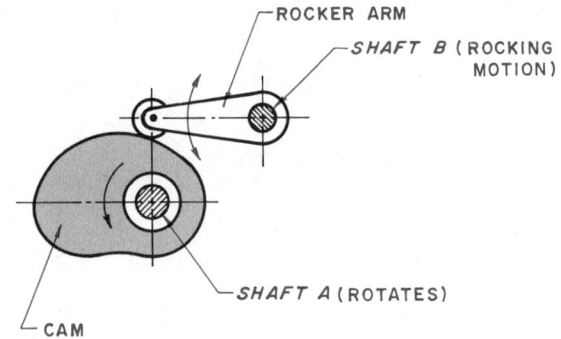

Figure 15-4 Rocker arm follower with wheel

change the rotary motion of shaft A into an up and down of the rocker arm, and then into a rocking motion of shaft B, ***Figure 15-4***.

Basic Types of Followers

Speed of rotation and the actual load applied upon the lifter determines the type of follower to be used. There are various basic types of followers: roller, pointed, flat faced and spherical faced, ***Figure 15-5***.

Cam Mechanism

Two major kinds of cams are used in industry: radial arm design and cylindrical design. The *radial arm* design changes a rotary motion into either an up and down motion or a rocking action as discussed previously. In the *cylindrical* design, a shaft rotates exactly as it does in the radial design, but the action or direction of the follower differs greatly. In the radial arm design, the follower operates *perpendicular* to the cam shaft. In the cylindrical design, the follower operates *parallel* to the cam shaft. See ***Figure 15-6*** for a simplified example of a cylindrical design cam. Shaft A has an irregular groove cut into it. As the cam rotates, the follower traces along the groove. The follower is directly attached to a shaft that moves to the left and right. Notice, this shaft does *not* rotate; the motion is *parallel* to the axis of the rotating cam shaft.

Cams produce one of three kinds of motion: uniform velocity, harmonic, or uniform acceleration. Each is discussed in full later.

Cam Terms

Working Circle

The *working circle* is considered a distance equal to the distance from the center of the cam shaft to the highest point on the cam, ***Figure 15-7***.

Displacement Diagram

The *displacement diagram* is a designed layout of the required motion of the cam. It is laid out on a grid, and its length represents one complete revolution of the cam.

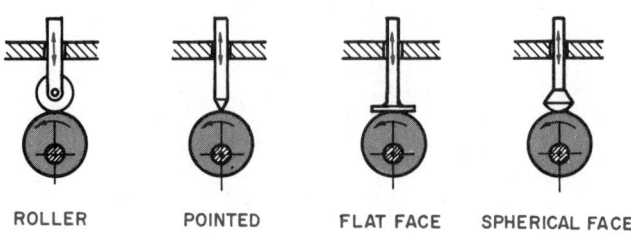

Figure 15-5 Basic types of followers

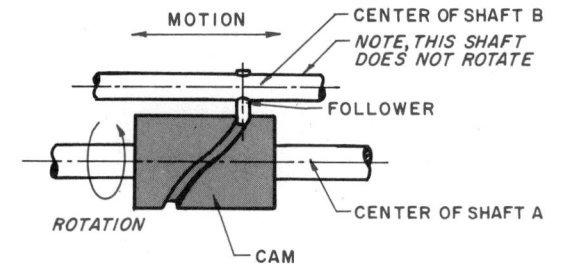

Figure 15-6 Cylindrical cam mechanism

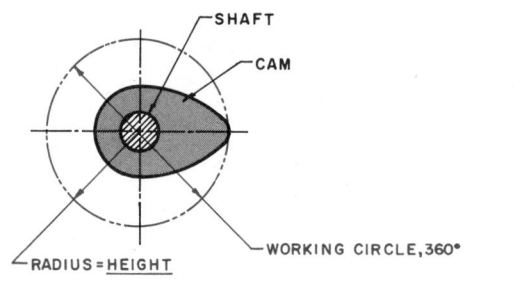

Figure 15-7 Cam terms and working circle

The length of the displacement diagram is usually drawn equal in length to the circumference of the working circle. This is not absolutely necessary, but it will give an in-scale idea of the cam's profile. The height of the displacement diagram is equal to the radius of a working circle, ***Figure 15-8***. The bottom line of the displacement diagram is the *baseline*. All dimensions should be measured upward from the baseline. On the cam itself, think of the center of the cam as the baseline.

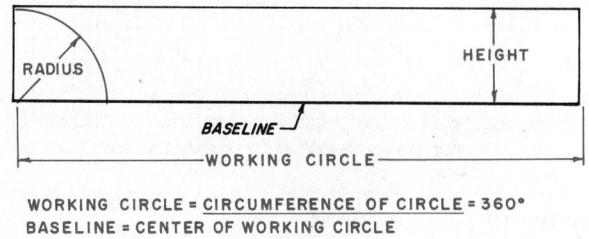

WORKING CIRCLE = CIRCUMFERENCE OF CIRCLE = 360°
BASELINE = CENTER OF WORKING CIRCLE

Figure 15-8 Displacement diagram (basic layout)

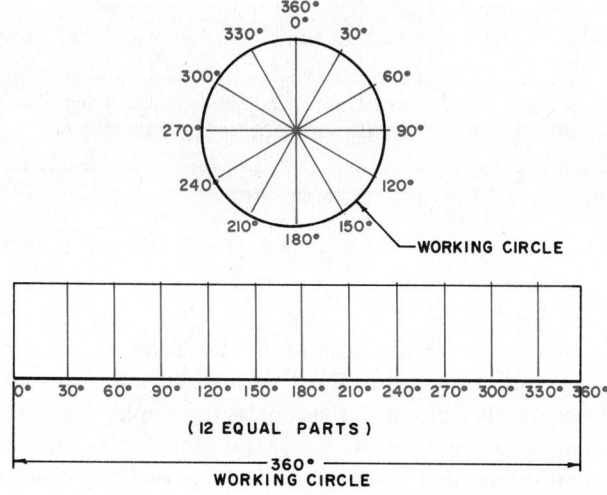

Figure 15-9 Displacement diagram with 30° increments

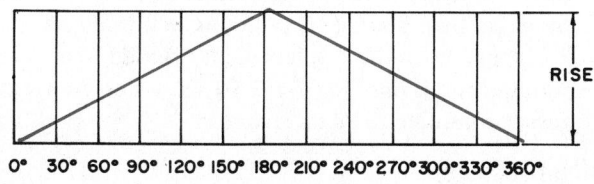

Figure 15-10 Uniform velocity cam

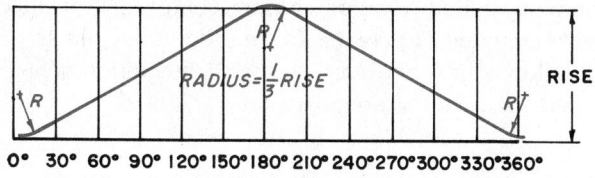

Figure 15-11 Modified uniform velocity cam

The length of the displacement diagram is divided into equal lines or grid, each of which represents degrees around the cam. These divisions can be 30°, 15° or even 10°. The finer the divisions, the more accurate is the final cam profile, *Figure 15-9*.

Dwell

Dwell is the period of time during which the follower does not move. This is shown on the displacement dia-

gram by a straight horizontal line throughout the dwell angle. On the cam, the dwell is drawn by a radius.

Time Interval

The *time interval* is the time it takes the cam to move the follower to the designed height.

Rotation

Rotation of the cam is either clockwise or counter-clockwise. The actual cam profile is laid out *opposite* of the rotation.

Base Circle

The *base circle* is used to lay out an offset follower which is illustrated in detail later. The base circle is a circle with a radius equal to the distance from the center of the shaft to the center of the follower wheel *at its lowest position*. On an offset follower, the base circle replaces the working circle.

Cam Motion

Four major types of curves are usually employed. Special irregular curves other than the four major types can be designed to meet a specific movement, if necessary.

The four types are: uniform velocity, modified uniform velocity, harmonic motion, and uniform acceleration. For comparison, the following four examples use the same working circle, same shaft, and same rise.

Uniform Velocity

In this type of motion, the cam follower moves with a *uniform velocity*; that is, it rises and falls at a constant speed, but the start and stop is very abrupt and rough, *Figure 15-10*.

Modified Uniform Velocity

Because of the abrupt and rough start and stop of the uniform velocity, it is modified slightly. *Modified uniform velocity* smoothes out the roughness slightly by adding a radius at the ends of the high and low points. This radius is equal to 1/3 the rise or fall, *Figure 15-11*. This radius smoothes out the start and stop somewhat, and is good for slow speed.

Harmonic Motion

Harmonic motion is very smooth, but the speed is not uniform. Harmonic motion has a smooth start and stop, and is good for fast speed.

To lay out a harmonic motion on the displacement diagram, draw a semicircle whose diameter is equal to the designed rise or fall. Divide the semicircle into equal divisions; in this example, 30°. Divide the overall horizontal distance of the rise or fall equally into increments of the same number as the semicircle; in this example, 6 equal divisions. Projection points on the semicircle are projected horizontally to the corresponding vertical lines. For example, the first point on the semicircle is projected to the first vertical line, the second point on the semicircle is projected to the second vertical line, and so forth. Connect the curve with an irregular curve. This completes the harmonic curve, *Figure 15-12*.

Uniform Acceleration

Uniform acceleration is the smoothest motion of all cams, and its speed is constant throughout the cam travel. The uniform acceleration curve is actually a parabolic curve; the first half of the curve is exactly the reverse of the second half. This form is best for high speed.

To lay out a uniform acceleration motion on the displacement diagram, divide the designed rise or fall into 18 equal parts. To do this, place the edge of a scale with its "0" on the starting level of the rise or fall, and equally space 18 units of measure on the high or low elevation of the rise or fall, *Figure 15-13*. Mark off points 1, 4, 9, 4 (14), and 1 (17), and draw horizontal lines. Note that 14 is actually 4, and 17 is actually 1. Divide the overall horizontal distance of the rise or fall into 12 equal increments. From points 1, 4, 9, 4 (14), and 1 (17), project points to the corresponding vertical lines. For example, the first point is from point 1 on the division line to the first vertical line, and then from point 4 on the division line to the second vertical, and so forth. Connect the curve with an irregular curve. This completes the uniform acceleration curve.

Combination of Motions

The illustrations thus far have covered the four major types of cams: uniform velocity, modified uniform velocity, harmonic motion, and uniform acceleration. Any combinations of these can be designed for a particular function or requirement. *Figure 15-14* illustrates a displacement diagram with a 90° fall using a harmonic motion, followed by a 30° dwell. The cam then falls again 90° using a modified uniform velocity followed by a 30° dwell. The cam then rises using uniform acceleration motion followed by another 30° dwell.

Laying Out the Cam from the Displacement Diagram

Regardless of which type of motion is used, the layout process is exactly the same. The working circle is

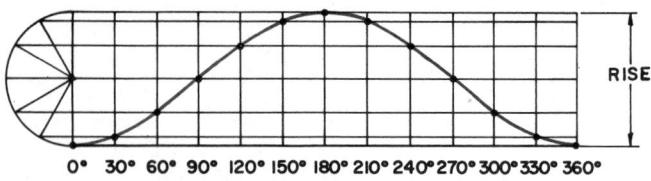

Figure 15-12   Harmonic motion cam

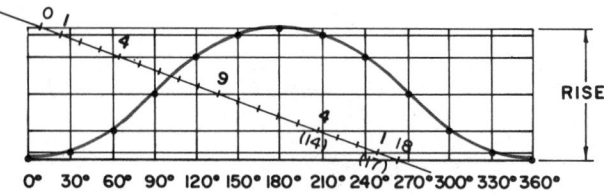

Figure 15-13   Uniform acceleration cam

drawn equal to the height of the displacement diagram. The circle of the cam must be divided into the exact same number of equal divisions as the displacement diagram, *Figure 15-15A*. Label the increments on both the displacement diagram and the working circle of the cam; in this example, harmonic motion is illustrated.

Notice that the even increment on the cam layout is labeled *opposite* of the rotation of the cam. In this example, rotation is clockwise; therefore, the labeling of the radial lines should be counterclockwise. Transfer each distance from the displacement diagram to the corresponding radial line, *Figure 15-15B*. Using an irregular curve, complete the cam layout.

Offset Follower

The cam follower is usually in line with the same center line as the cam (see Figure 15-1). Occasionally, however, when position or space is a problem, the follower can be designed to be on another center line, or offset as illustrated in *Figure 15-16*. In order to lay out an offset follower, it is necessary to use slightly different steps than those used for a regular cam.

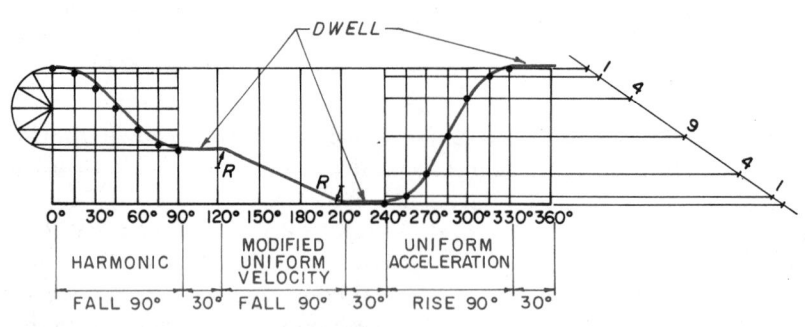

Figure 15-14   Combination of motions

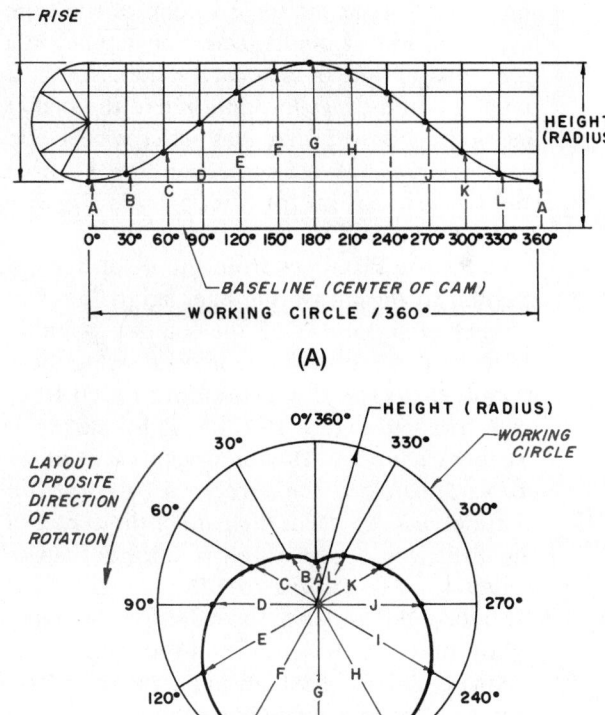

(A)

(B)

Figure 15-15 (A) Displacement diagram example; (B) Heights transferred from displacement diagram

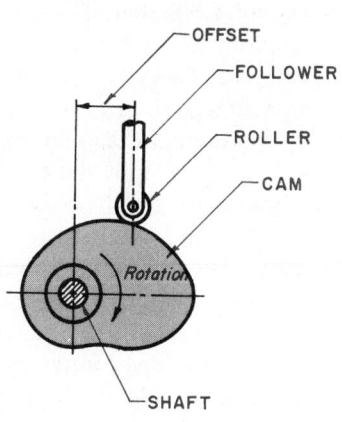

Figure 15-16 Associated cam terms

How To Draw a Cam with an Offset Follower

Given: Cam data: harmonic motion cam, follower offset .625, base circle .88 radius, rise of 1.50, shaft diameter .375, direction of rotation: counter-clockwise.

Follower data: Follower width .250, roller .375 diameter.

Step 1 Design and lay out a displacement diagram exactly as with any cam, *Figure 15-17A*.

Step 2 From the center of the shaft of the cam, construct the offset circle. The offset circle has a radius equal to the required follower offset. From the same swing point, lay out the base circle. The radius of the base circle is equal to the distance from the center of the shaft to the center of the follower *wheel* at its *lowest* position, *Figure 15-17B*.

Step 3 The point where it crosses a constructed vertical line from the center line of the follower is point A on the displacement diagram, *Figure 15-17C*. This represents the *lowest* position of the follower wheel. This vertical line is the measuring line and will be used to lay out the other points around the cam.

Step 4 Divide the offset circle into the same equal divisions as used on the displacement diagram; in this example, 12 equal parts. Lightly, consecutively, letter or number the divisions on the off-

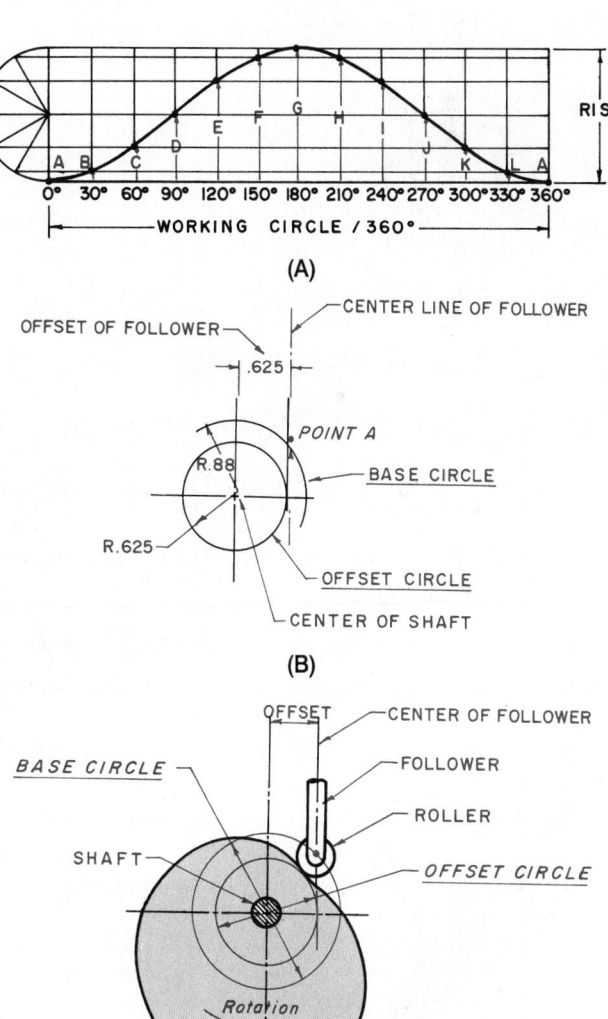

(A)

(B)

(C)

Figure 15-17 (A) Step 1 (drawing a cam with an offset follower design layout); (B) Step 2; (C) Step 3

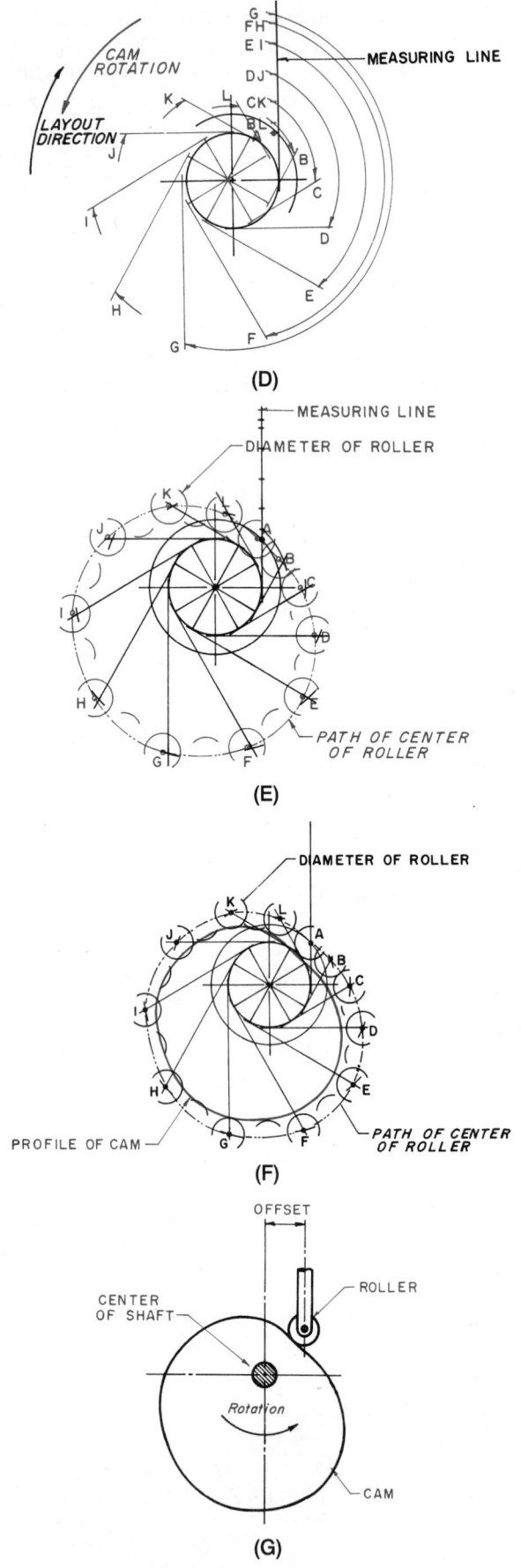

(D)

(E)

(F)

(G)

Figure 15-17 (Continued) (D) Step 4; (E) Step 5; (F) Step 6; (G) Step 7

set circle. Start at the lowest point of the cam; in this example, point A. Letter or number in a direction *opposite* of the cam rotation. Note that point A is the exact location where the center line of the follower is tangent with the offset circle. Construct a layout projection line 90° from the tangent point of the circle, **Figure 15-17D**.

Step 5 Transfer the distances from the displacement diagram to the measuring line, lightly letter or number each point. With the compass point on the center of the shaft of the cam, swing these distances to the corresponding radial lines (refer back to Figure 15-17D). Lightly letter or number each point. Do not forget to *lay out the cam opposite to the direction of rotation*. These points represent the path of the center of the follower wheel. Lightly connect these points together, **Figure 15-17E**.

Step 6 With the compass set at the radius for the roller, lightly draw the roller around the path of the follower. The inside, high points of these arcs represent the actual profile of the cam, **Figure 15-17F**.

Step 7 Check all construction work. If correct, darken and complete the cam, **Figure 15-17G**.

How To Draw a Cam with a Flat-faced Follower

Given: Cam data: harmonic motion cam, base circle .75 radius, rise of 1.50, shaft diameter .375, direction of rotation: clockwise.

Follower data: flat-faced follower, 1.25 diameter distance across the flat surface. Design and lay out a displacement diagram exactly as with any cam. In this example, the displacement diagram in Figure 15-17A will be used.

Step 1 Construct the base circle. The base circle is drawn with a radius equal to the distance from the center of the shaft to the face of the follower in its *lowest* position. Divide the base circle into the same equal divisions as used on the displacement diagram. Lightly consecutively letter or number the divisions on the base circle, **Figure 15-18A**.

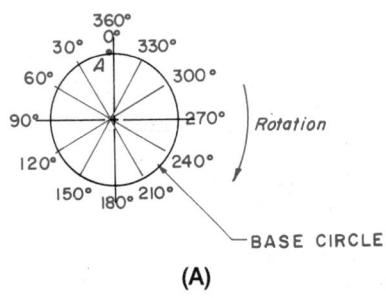

(A)

Figure 15-18 (A) Step 1 (drawing a cam with a flat-faced follower)

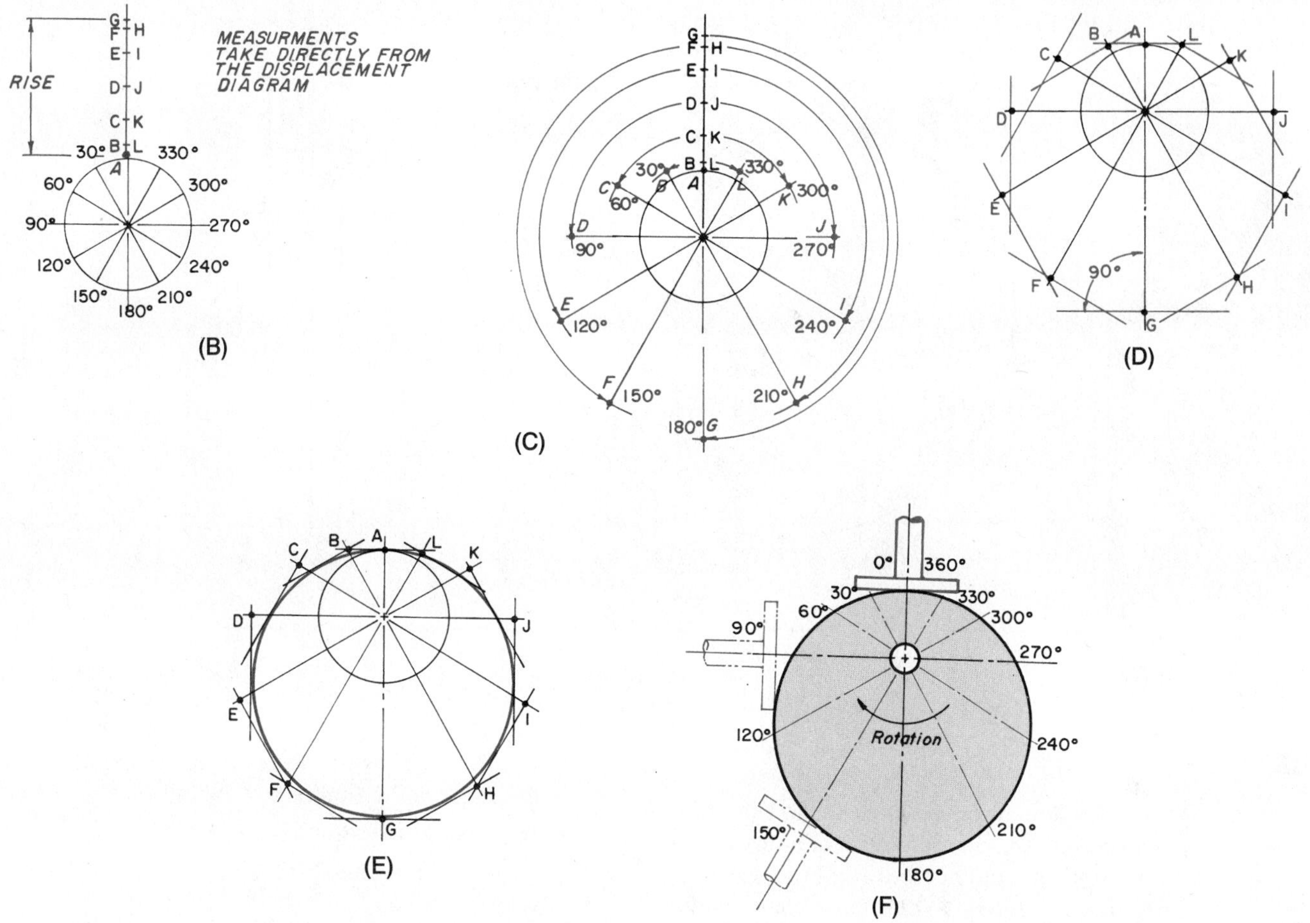

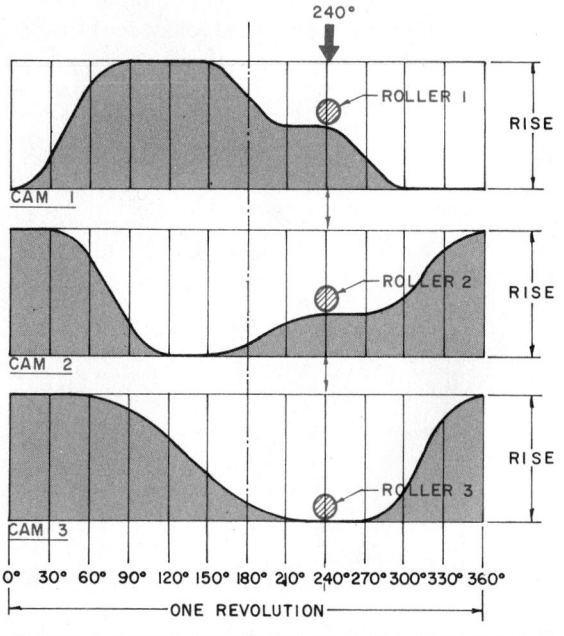

Figure 15-18 (Continued) (B) Step 2; (C) Step 3; (D) Step 4; (E) Step 5; (F) Step 6

Step 2 Locate point A which represents the location of the flat-faced follower at its lowest point. Construct a measuring line starting from the base circle and transfer the distances from the displacement diagram to the measuring line. Letter or number each point, *Figure 15-18B*. Do not forget to lay out the *cam opposite to the direction of rotation*.

Step 3 Swing these distances to the corresponding radial lines, *Figure 15-18C*.

Step 4 From these distances, construct lines perpendicular from these radial lines at the point of intersection, *Figure 15-18D*.

Step 5 Lightly construct the cam within these perpendicular lines, *Figure 15-18E*.

Step 6 Check all construction work. If correct, darken and complete the cam, *Figure 15-18F*. Notice the position of the flat-faced follower at the 0°, 90°, and 150° positions.

Timing Diagram

Many times, more than one cam is attached to the same shaft. In this case, each cam works independently but must function in relation to the other cams. In order to be able to visualize graphically and study the interaction of the cams on the same shaft, place the various displacement diagrams in a column with the starting points and ending points in line, as illustrated in *Figure 15-19*. This represents one full revolution and is referred to as a *timing diagram*. For example, at 240° it is easy to visualize the exact position of each cam roller.

Figure 15-19 Timing diagram

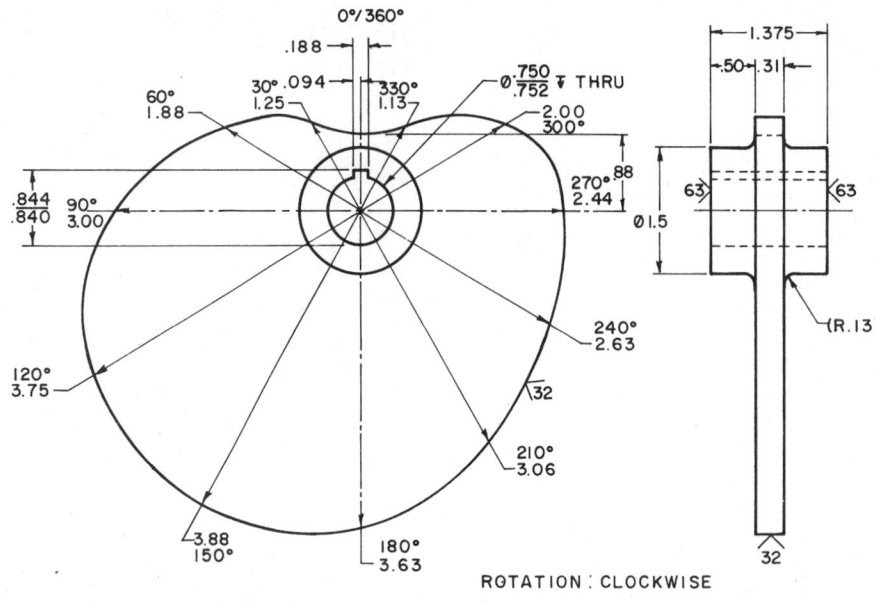

Figure 15-20   Dimensioning a cam

Dimensioning a Cam

A cam must be fully dimensioned. The cam size dimensions are constructed radiating from the center of the shaft by both radii length and degrees from the starting point, as illustrated in *Figure 15-20*. Other important dimensions pertain to the shaft hole size and keyway size, and are dimensioned according to most recent drafting standards.

Try not to place dimensions on the cam itself; place them around the perimeter, if possible.

Review

1. List three examples of uses of a simple cam.
2. What determines the type of follower to be used?
3. List the two major kinds of cams used in industry. Explain the function of each.
4. What is the working circle?
5. Explain what the displacement diagram is and why it is used.
6. What do the length and height of the displacement diagram represent on the actual cam?
7. List the four major types of curves usually used in the design and layout of the cam displacement diagram. Which one gives the smoothest motion?
8. What is meant by an offset follower?
9. Explain why a timing diagram is used.
10. Why is the cam laid out opposite of the direction of rotation?

Chapter Fifteen Problems

The following problems are intended to give the beginning drafter practice in designing and layout out cam displacement diagrams according to the required specifications to do a particular function. The cam displacement diagram will then be transferred to an actual cam profile layout. The drafter will practice laying out uniform velocity, harmonic motion, and uniform acceleration motions.

The steps to follow in laying out cams are:

Step 1   Study all required specifications.

Step 2   Make a rough sketch of the cam displacement diagram incorporating *all* required specifications and required dimensions.

Step 3   Lightly lay out the displacement diagram. Space out horizontally as far as possible.

Step 4   Break down the area into equal spaces that correspond to degrees around the cam. *Label* each line indicating the degrees.

Step 5   Following the required specifications, lightly develop the cam's travel on the displacement diagram. (Show all light layout work—do not erase.)

Step 6   Check all work.

Step 7   Darken in displacement diagram.

Step 8   To lay out the cam profile, locate the center of the cam.

Step 9   Draw the shaft, hub, working circle, and key, if required, in place.

Step 10   Divide the working circle into equal degrees as required.

Step 11   Note *direction of travel*. Lightly letter the degrees around the working circle *opposite* of direction of travel.

Step 12   Locate the starting elevation and draw the cam follower *from* the starting point.

Step 13   Lightly draw in the rest of the cam, following details as required.

Step 14   Transfer all distances *from* the cam displacement diagram to the corresponding degree line *around* the cam layout.

Step 15   Lightly connect all points using an irregular curve.

Step 16   Check all work.

Step 17   Darken in all work.

Step 18   Neatly add all required specifications and dimensions to the drawing.

The drawing must include the following:

Shaft diameter
Hub size
Key or set screw size
Direction of rotation
Working circle size
All dimensions according to the most recent drafting standards. (For these problems, dimension only the front view.)

Problems 15-1A and 15-1B

Construct a cam displacement diagram on a A-size, 8½ x 11 sheet of paper per dimensions given.

Lay out a cam with the following specifications:
1. Working circle 5.0 dia.   5. Rotation: counterclockwise
2. Shaft dia. .75
3. Hub dia. 1.25   6. Follower dia. — 1.0
4. Square key .18   7. Start 1.0 from center

Required motion:
1. Rise 90°, modified uniform velocity 1.5
2. Dwell 60°
3. Fall 60°, modified uniform velocity .75
4. Dwell 30°
5. Fall 60°, modified uniform velocity .75
6. Dwell 60°

Construct the cam and follower on an A-size, 8½ x 11 sheet of paper per dimensions given, and transfer all dimensions from the displacement diagram to the cam layout.

Check all work and add all required dimensions and specifications according to the most recent drafting standards.

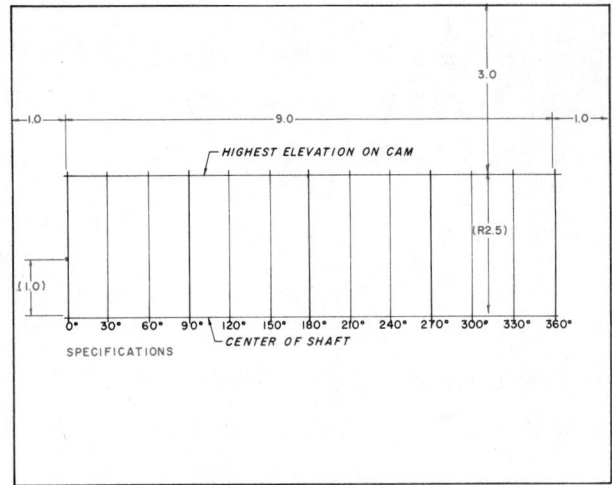

Problem 15-1A

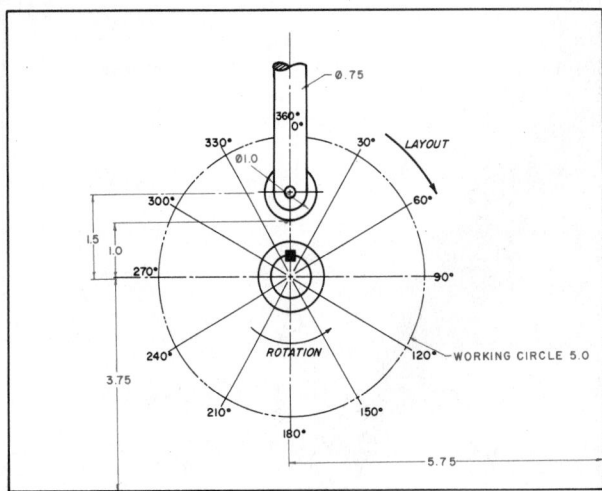

Problem 15-1B

Problems 15-2A and 15-2B

Construct a cam displacement diagram on an A-size, 8½ x 11 sheet of paper per dimensions given.

Lay out a cam with the following specifications:

1. Working circle 5.0 dia.   5. Rotation: clockwise
2. Shaft dia. .75   6. Follower dia. 1.0
3. Hub dia. 1.25   7. Start 1.0 from center
4. Square key .18

Required motion:

1. Rise 90°, harmonic motion 1.00
2. Dwell 30°
3. Rise 90°, harmonic motion .50
4. Dwell 30°
5. Fall 90°, harmonic motion 1.50
6. Dwell 30°

Construct the cam and follower on an A-size, 8½ x 11 sheet of paper per dimensions given, and transfer all dimensions from the displacement diagram to the cam layout.

Check all work and add all required dimensions and specifications according to the most recent drafting standards.

Problems 15-3A and 15-3B

Construct a cam displacement diagram on a A-size, 8½ x 11 sheet of paper per dimensions given.

Lay out a cam with the following specifications:

1. Working circle 5.0   5. Rotation: clockwise
2. Shaft dia. .63   6. Follower dia. .75
3. Hub dia. 1.0   7. Start 2.5 from center
4. Square key to suit

Required motion:

1. Fall 180°, uniform acceleration 1.50
2. Dwell 45°
3. Rise 90°, uniform acceleration 1.50
4. Dwell 45°

Construct the cam and follower on an A-size, 8½ x 11 sheet of paper per dimensions given, and transfer all dimensions from the displacement diagram to the cam layout.

Check all work and add all required dimensions and specifications according to the most recent drafting standards.

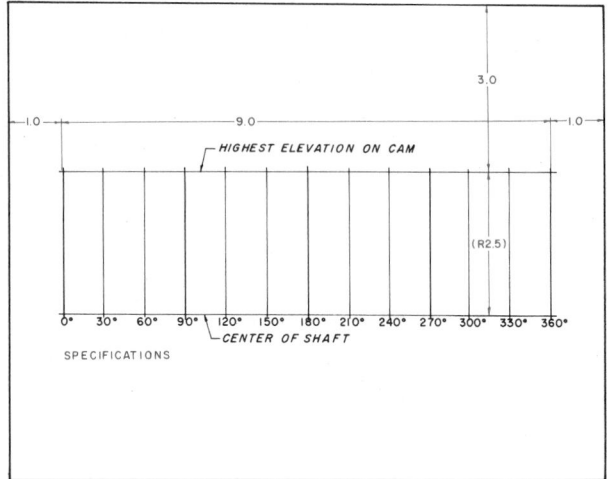

Problem 15-2A

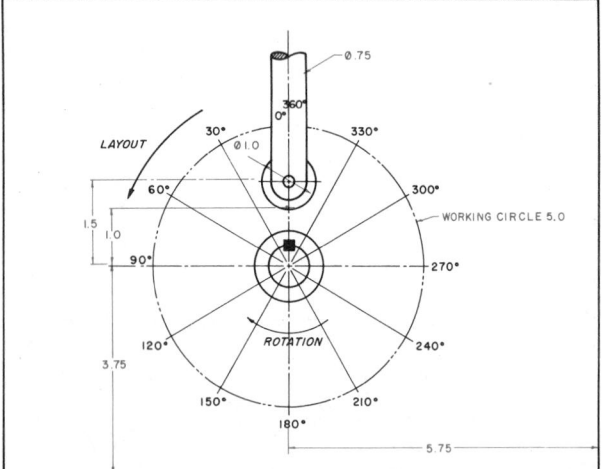

Problem 15-2B

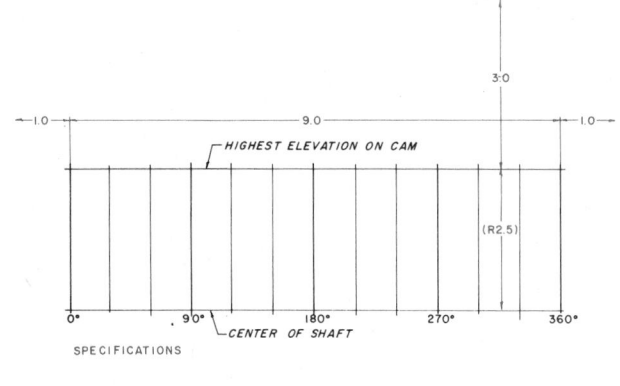

Problem 15-3A

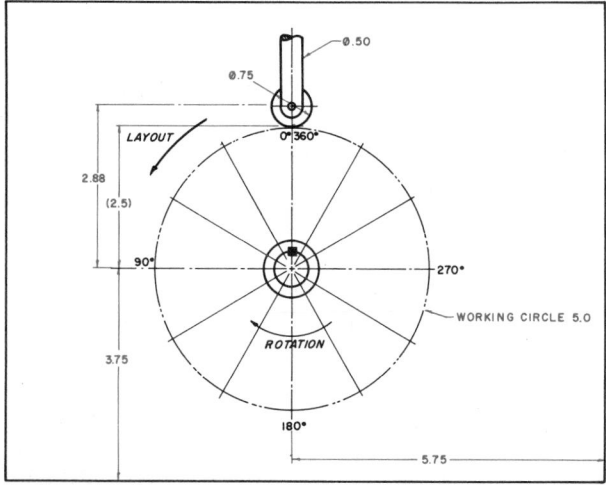

Problem 15-3B

Problems 15-4A and 15-4B

Construct a cam displacement diagram on a A-size, 8½ x 11 sheet of paper per dimensions given.

Lay out a cam with the following specifications (metric):

1. Working Circle 245
2. Shaft dia. 28
3. Hub dia. 44
4. Square key to suit
5. Rotation: clockwise
6. Follower dia. 22
7. Start 32 from center

Required motion:

1. Rise 60°, harmonic motion 25
2. Dwell 15°
3. Rise 90°, harmonic motion 38
4. Dwell 15°
5. Fall 45°, harmonic motion 44
6. Dwell 15°
7. Rise 30°, harmonic motion 18
8. Dwell 15°
9. Fall to starting level 75°, harmonic motion 38

Construct the cam and follower on an A-size, 8½ x 11 sheet of paper per dimensions given, and transfer all dimensions from the displacement diagram to the cam layout.

Check all work and add all required dimensions and specifications according to the most recent drafting standards.

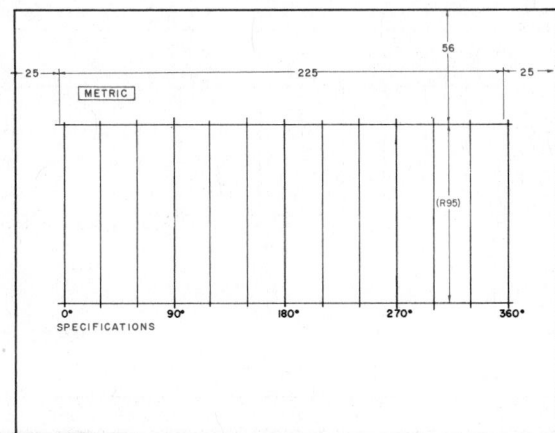

Problem 15-4A

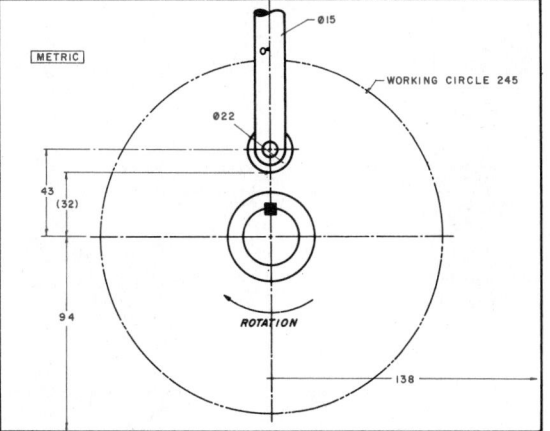

Problem 15-4B

Problems 15-5A and 15-5B

Construct a cam displacement diagram on a A-size, 8½ x 11 sheet of paper per dimensions given.

Lay out a cam with the following specifications:

1. Working circle 7.0
2. Shaft dia. .625
3. Hub dia. 1.0
4. Square key to suit
5. Rotation: counterclockwise
6. Follower dia. 1.0
7. Start 1.0 from center

Required motion:

1. Rise 120°, modified uniform velocity 2.5
2. Dwell 60°
3. Fall 30°, modified uniform velocity .75
4. Dwell 30°
5. Fall to starting level 90°, modified uniform velocity 1.75
6. Dwell 30°

Construct the cam and follower on an A-size, 8½ x 11 sheet of paper per dimensions given, and transfer all dimensions from the displacement diagram to the cam layout.

Check all work and add all required dimensions and specifications according to the most recent drafting standards.

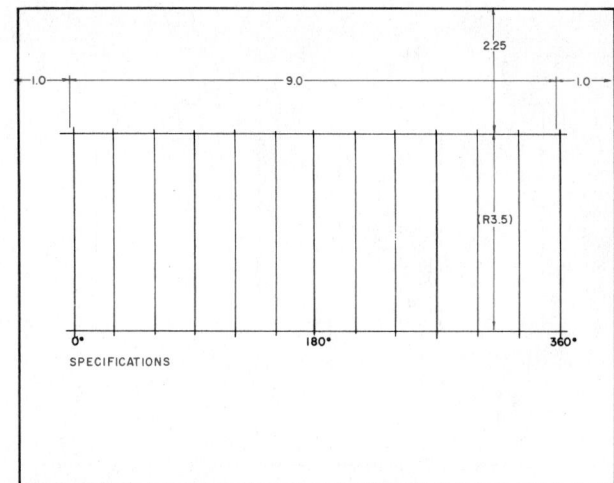

Problem 15-5A

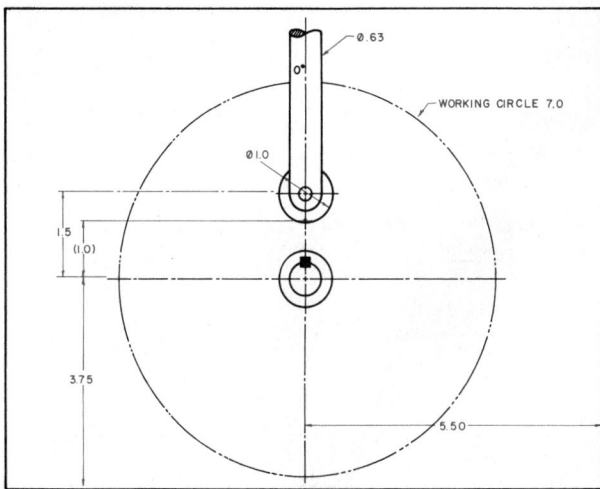

Problem 15-5B

Problems 15-6A and 15-6B

Construct a cam displacement diagram on a A-size, 8½ x 11 sheet of paper per dimensions given.

Lay out a cam with the following specifications:

1. Working circle 7.75
2. Shaft dia. .56
3. Hub dia. .88
4. Square key to suit
5. Rotation: clockwise
6. Follower dia. .375
7. Start 1.625 from center

Required motion:

1. Fall 90°, uniform acceleration 3.0
2. Dwell 15°
3. Rise 60°, uniform acceleration 2.0
4. Dwell 105°
5. Rise 90°, uniform acceleration to starting level (1.0)

Construct the cam and follower on an A-size, 8½ x 11 sheet of paper per dimensions given, and transfer all dimensions from the displacement diagram to the cam layout.

Check all work and add all required dimensions and specifications according to the most recent drafting standards.

❑ Problems 15-7A and 15-7B

Construct a metric cam displacement diagram per dimensions given.

Lay out a cam with the following specifications:

1. Rise/fall 56
2. Shaft dia. 16
3. Hub dia. 32
4. Square key to suit
5. Rotation: clockwise
6. Follower: *flat-faced (34 dia.)
7. Base circle 26 radius
8. Harmonic motion/one full turn

Construct the cam and flat-faced follower per dimensions given.

Check all work and add required dimensions and specifications according to the most recent drafting standards and using the metric system.

*Design the flat-faced follower to suit.

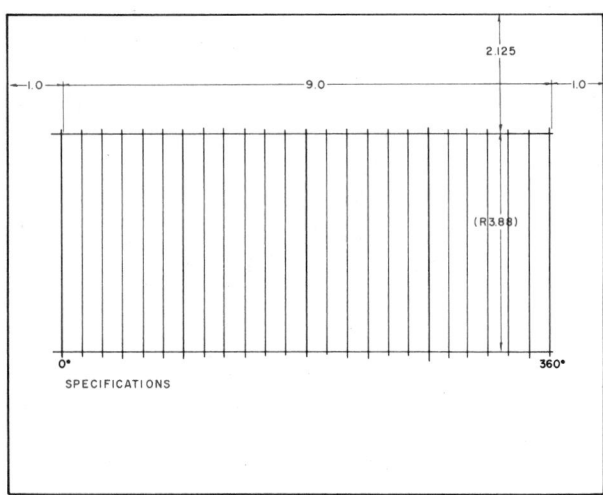

Problem 15-6A

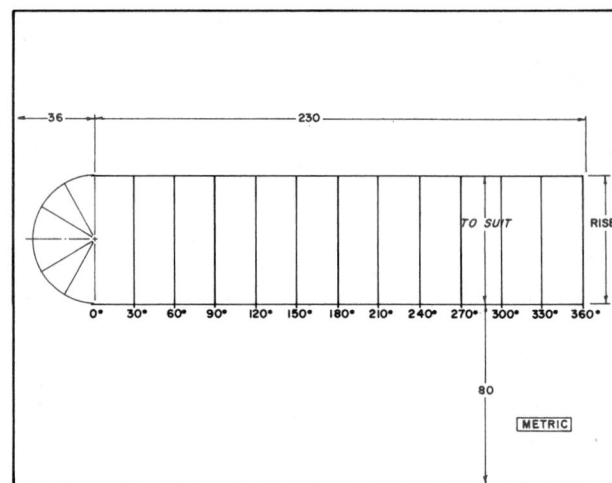

Problem 15-7A

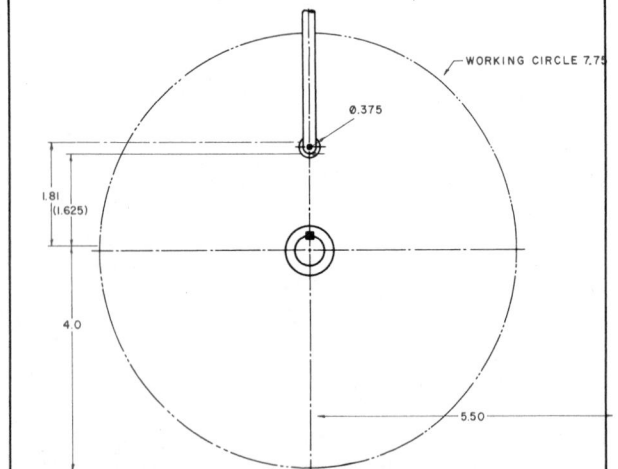

Problem 15-6B

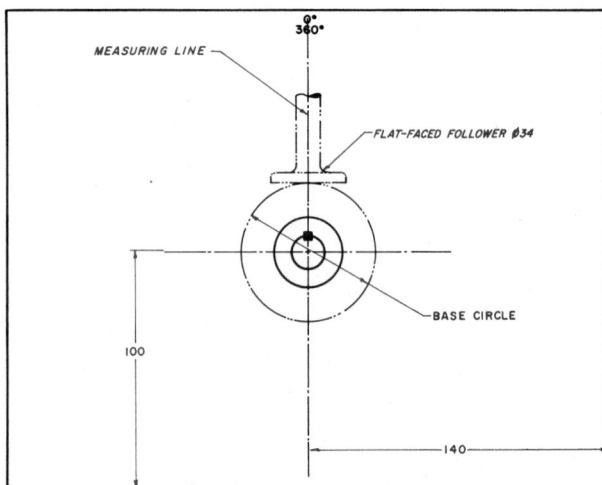

Problem 15-7B

❏ Problems 15-8A and 15-8B

Construct a cam displacement diagram per dimensions given.

Lay out an offset cam with the following specifications:

1. Offset circle dia. 2.00
2. Shaft dia. .75
3. Hub dia. 1.25
4. Square key to suit
5. Rotation: clockwise
6. Follower dia. .75
7. Follower offset 1.00
8. Base circle 1.375 radius

Required motion:

1. Fall 75°, uniform acceleration 1.75
2. Dwell 10°
3. Rise 35°, harmonic motion .75
4. Dwell 5°
5. Fall 70°, modified uniform velocity 1.25
6. Dwell 15°
7. Rise 45°, harmonic motion .75
8. Dwell 15°
9. Rise 60°, uniform acceleration 1.50
10. Dwell 30°

Construct the cam and offset follower per dimensions given, and transfer all dimensions from the displacement diagram to the cam layout.

Check all work and add all required dimensions and specifications according to the most recent drafting standards.

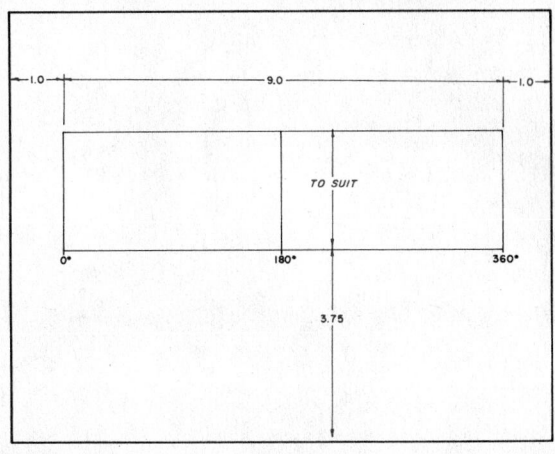

Problem 15-8A

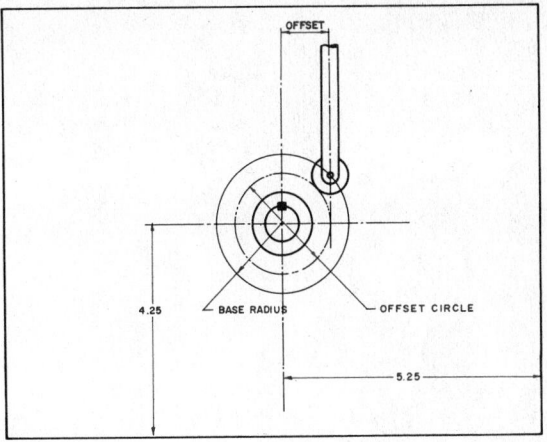

Problem 15-8B

Gears

Gears transmit or transfer rotary motion from one shaft to another shaft. Gears can change the direction of rotation, speed up or slow down rotation, transmit power, and change rotary motion into reciprocating motion. There are various kinds of gears, each with their own function, *Figure 16-1*.

The drafter must be able to identify each kind of gear, know the various functions of each and be able to design and draw the various gears using correct terminology associated with gears.

Kinds of Gears

Gears are usually classified by the position or location of the shafts they connect. A spur gear, pinion gear or helical gear is usually used to connect shafts that are parallel to each other. Intersecting shafts at 90° are usually connected by a beveled gear or angle gear. Shafts that are not parallel to each other and that do not intersect use a worm and worm gear. In order to connect rotary motion into a reciprocating or back and forth motion, a rack gear and pinion would be used.

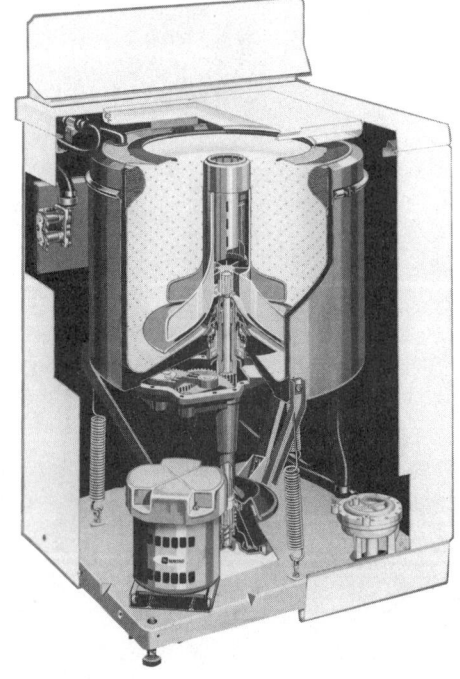

Figure 16-1   Example of gears in use *(Courtesy The Maytag Co.)*

Spur Gear

The *spur gear* is the most commonly used gear, *Figure 16-2*. It is cylindrical in form, with teeth that are cut straight across the face of the gear. All teeth are parallel to the axis of the shaft. The spur gear is usually considered the *driven gear*.

Pinion Gear

The *pinion gear* is exactly like a spur gear but it is usually smaller, and has fewer teeth, Figure 16-2. The pinion gear is normally considered the *drive gear*.

Rack Gear

The *rack gear* is a type of spur gear, but its teeth are in a straight line or flat instead of in a cylindrical form, *Figure 16-3*. The rack gear is used to transfer circular motion into straight-line motion.

Ring Gear

The *ring gear* is similar to the spur, pinion, and rack gears, except that the teeth are internal, *Figure 16-4*.

Bevel Gear

A bevel gear is another gear commonly used, *Figure 16-5*. A *bevel gear* is cone shaped in form with straight teeth that are on an angle to the axis of the shaft. Bevel gears are used to transmit power and motion between intersecting shafts that are at 90° to each other.

Angle Gear

The *angle gear* is similar to a bevel gear, except that the angles are at other than 90° to each other.

Miter Gear

The *miter gear* is exactly the same as a bevel gear, except that both mating gears have the same number of teeth. The shafts are at 90° to each other, *Figure 16-6*.

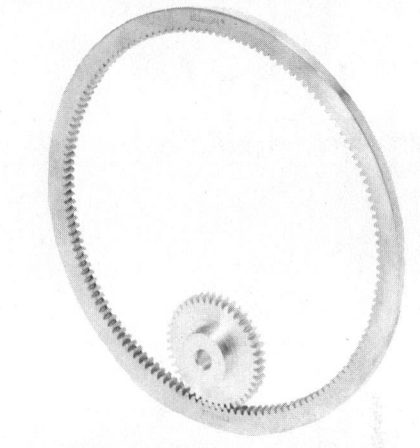

Figure 16-4 Ring gear *(Courtesy Boston Gear)*

Figure 16-2 Spur gear/pinion gear *(Courtesy Boston Gear)*

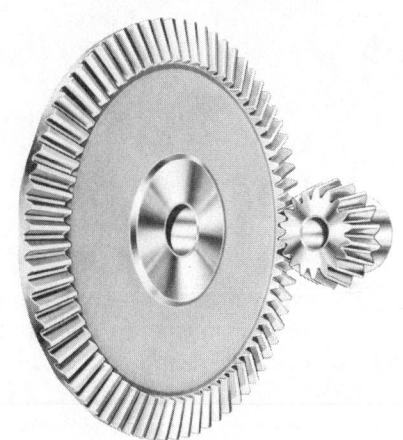

Figure 16-5 Bevel gear *(Courtesy Boston Gear)*

Figure 16-3 Rack gear *(Courtesy Boston Gear)*

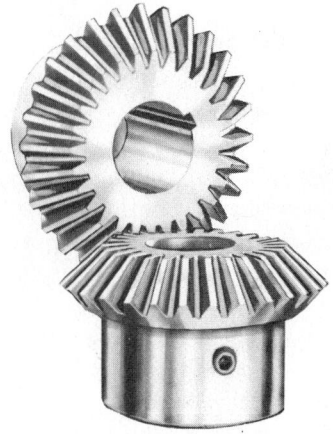

Figure 16-6 Miter gear *(Courtesy Boston Gear)*

Figure 16-7   Spiral bevel gear *(Courtesy Boston Gear)*

Figure 16-8   Worm gear *(Courtesy Boston Gear)*

Spiral Bevel or Miter Gears

Any bevel gears with curved teeth are called *spiral bevel gears*, *Figure 16-7*.

Worm Gear

Worm gears are used to transmit power and motion at a 90° angle between nonintersecting shafts, *Figure 16-8*. They are normally used as a speed reducer. The worm gear is round like a wheel. The worm is shaped like a screw, with threads (or teeth) wound around it. Because one full turn of the worm is required to advance the worm gear one tooth, a high-ratio speed reduction is achieved. The worm drives the larger worm gear.

Figure 16-9   Chain and sprockets *(Courtesy Boston Gear)*

Chain and Sprockets

A *chain and sprockets* are used to transmit motion and power to shafts that are parallel to each other, *Figure 16-9*. Sprockets are similar to spur and pinion gears, as the teeth are cut straight across the face of the sprocket and are parallel to the shaft. There are many types of chains and sprockets but all use the same terms, formulas, ratios, and so forth.

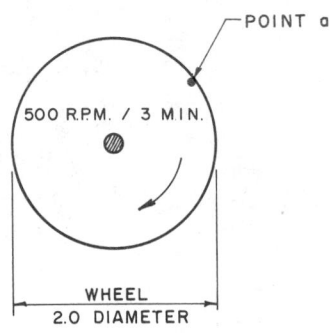

Figure 16-10   Relationship of diameter of a wheel to given revolutions per minute and time

Velocity of Feet Per Minute (F.P.M.)

A gear is very similar to a simple wheel. In the field of horology (the science of measuring time), a gear is referred to as a wheel. In making calculations for a gear, it is sometimes easier to think of it as a wheel. *Figure 16-10* illustrates the relationship of the diameter of a wheel to given revolutions per minute (R.P.M.) and time. Velocity is calculated by recording the distance that a given point on a wheel or gear travels during a certain period of time. To calculate velocity of a gear, the point is usually assumed to be located on the pitch circle. The pitch circle is discussed in detail later.

The formula is:

$$\pi \text{DIA. x R.P.M. x TIME = DISTANCE}$$

In this example, point a on the wheel will travel 785 feet.

$$\pi 2.0 \text{ Dia. x 500 R.P.M. x 3 Min. = Distance}$$
$$6.28 \text{ x } 500 \text{ x } 3 = 9420 \text{ inches}$$
$$9420 \text{ inches} = 785 \text{ feet}$$

If two wheels of the exact same diameter are placed together, as illustrated in *Figure 16-11*, and assuming there is no slippage, point a on wheel A will travel the exact same distance as point b on wheel B. Notice, wheel a is rotating clockwise, and wheel B is rotating counterclockwise. If wheel A is rotated at 500 R.P.M. for 3 minutes point a will travel 785 feet.

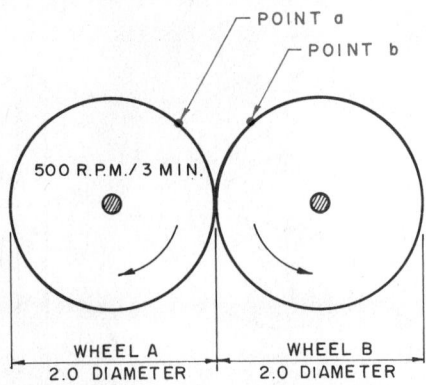

Figure 16-11 Relationship of two wheels of the exact same diameters

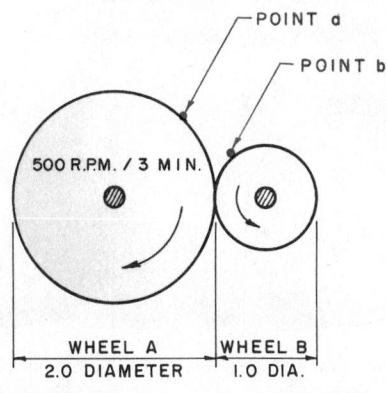

Figure 16-12 Relationship of two wheels of different diameters

Gear Ratio

If two friction wheels of different diameters are placed together, **Figure 16-12**, and assuming there is no slippage, point a on wheel A will travel the same distance as point b on wheel B. In this example, wheel A is the driver.

$$\pi\text{Diameter x revolutions per minute}$$
$$\text{x time = distance}$$

WHEEL A	**WHEEL B**
πDia. x R.P.M. x time =	πDia. x R.P.M. x time
π2.0 Dia. x 500 R.P.M.	π1.0 Dia. x ? R.P.M.
x 3 min. =	x 3 min.
6.28 x 500 x 3 = 9420 =	3.14 x 3 = 9.42

$$9420 \div 9.42 = 1000 \text{ R.P.M. of Wheel B}$$

Wheel A has a 2.0 diameter and turns at 500 R.P.M. Wheel B has a diameter of half of wheel A, but turns twice as fast, thus point a and point b travel the exact same distance. Note that both wheel A and wheel B travel at the same velocity.

The *ratio* between gears is:

$$\frac{\text{Diameter, wheel A}}{\text{Diameter, wheel B}} = \frac{2}{1} = 2{:}1$$

The ratio of 2:1 means that wheel B rotates two times each time wheel A rotates once.

The ratio of one gear to another can be calculated by using any of three different methods: the number of teeth of corresponding gears, pitch diameters of corresponding gears, or the R.P.M. between the gears.

Method 1

To find the ratio, divide the number of teeth on the spur gear by the number of teeth on the pinion gear:

$$\frac{\text{Number of teeth (spur gear)}}{\text{Number of teeth (pinion gear)}} = \text{Ratio}$$

Example:

If a spur gear has 60 teeth and a pinion gear has 30 teeth, the ratio would be:

$$\frac{\text{Number of teeth (spur gear)}}{\text{Number of teeth (pinion gear)}} = \frac{60}{30} = 2{:}1$$

The pinion gear will rotate two times for each full revolution of the spur gear.

Method 2

To find the ratio, divide the pitch diameter (D) of the spur gear by the pitch diameter (D) of the pinion gear:

$$\frac{\text{Pitch Diameter (D) (spur gear)}}{\text{Pitch Diameter (D) (pinion gear)}} = \text{Ratio}$$

Example:

If a spur gear has a pitch diameter (P.D.) of 4.000 and a pinion gear has a pitch diameter (P.D.) of 1.000, the ratio would be:

$$\frac{\text{P.D. (spur gear)}}{\text{P.D. (pinion gear)}} = \frac{4.000}{1.000} = 4{:}1$$

The pinion gear will rotate four times for each full revolution of the spur gear.

Method 3

To find the ratio, divide the revolutions per minute (R.P.M.) of the pinion gear by the R.P.M. of the spur gear:

$$\frac{\text{R.P.M. (pinion gear)}}{\text{R.P.M. (spur gear)}} = \text{Ratio}$$

Example:

If a pinion gear rotates at 1000 R.P.M. and a spur gear rotates at 200 R.P.M., the ratio would be:

$$\frac{\text{R.P.M. (pinion gear)}}{\text{R.P.M. (spur gear)}} = \frac{1000}{200} = 5{:}1$$

The pinion gear will rotate five times for each full revolution of the spur gear.

Pitch Diameter

Think of the pitch diameter of a gear as the outside diameter of the friction wheel, **Figure 16-13**. All ratios as discussed so far with the friction wheel apply also to gears. To apply formulas, substitute the outside diameter of the friction wheel for the pitch diameter of the gear.

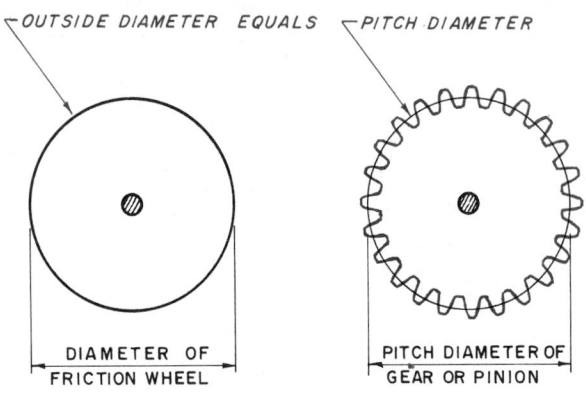

Figure 16-13 Pitch diameter of a gear

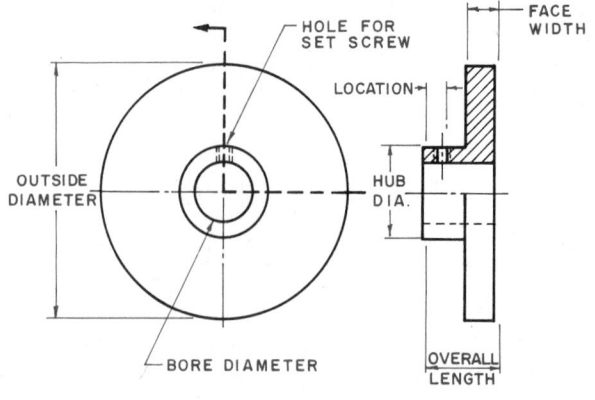

Figure 16-14 Gear blank

Gear Blank

Gears are usually cut from a gear blank. The gear blank must allow sufficient space for the gears and a method to attach the gear to the shaft. Illustrated in ***Figure 16-14*** is a common gear blank with an outside diameter and face width for the gears to be cut into, a hub, and a hole for the set screw to secure the gear blank to the shaft. A half-section is used to illustrate the gear blank in this example. Two complete sets of dimensions must be applied to the gear drawing; those relating to the gear blank, as illustrated, and those relating to the teeth. Gear dimensions do not hold tight tolerancing, but all gear tooth dimensions must hold very tight tolerancing.

Backlash

Backlash is the amount by which the gear tooth is less than the tooth space. It is measured by locking one gear and rocking the other back and forth, and measuring that rock at a known radius, usually the pitch diameter. The amount of backlash is the shortest distance between mating teeth as measured between the nondriving surfaces of adjacent teeth, ***Figure 16-15***.

Basic Terminology

The drafter must know and understand all major terms associated with various kinds of gears used today in industry. Illustrated in ***Figure 16-16*** are the basic terms associated with most gears, regardless of type.

Pitch Diameter (D)

The *pitch diameter* is the theoretical circle on which the teeth of mating gear mesh. Think of the pitch diameter on the gear as being the same as the outer diameter on the friction wheel. Note: The pitch diameter is considered the *nominal* size of the gear.

Root Diameter (DR)

The *root diameter* is the measurement over the extreme inner edges of teeth. It is equal to the pitch diameter minus 2 times the dedendum (b).

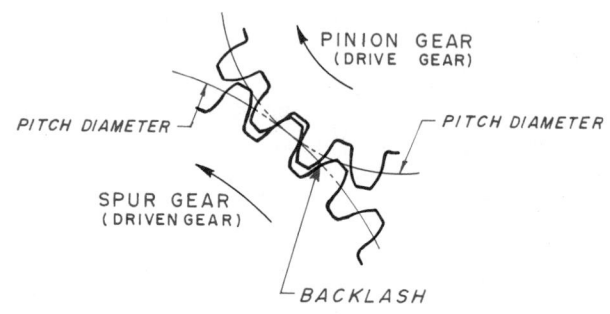

Figure 16-15 Backlash

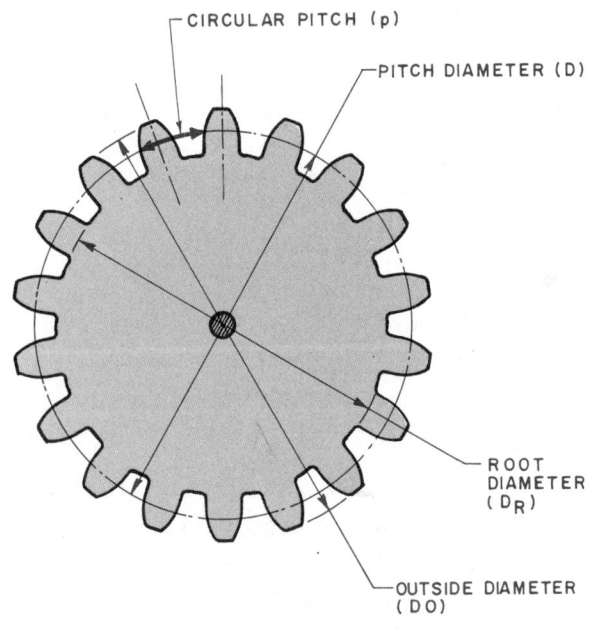

Figure 16-16 Basic gear terminology

Outside Diameter (OD)

The *outside diameter* is the measurement over the extreme outer edge of teeth. It is equal to the pitch diameter, plus 2 times the addendum (a).

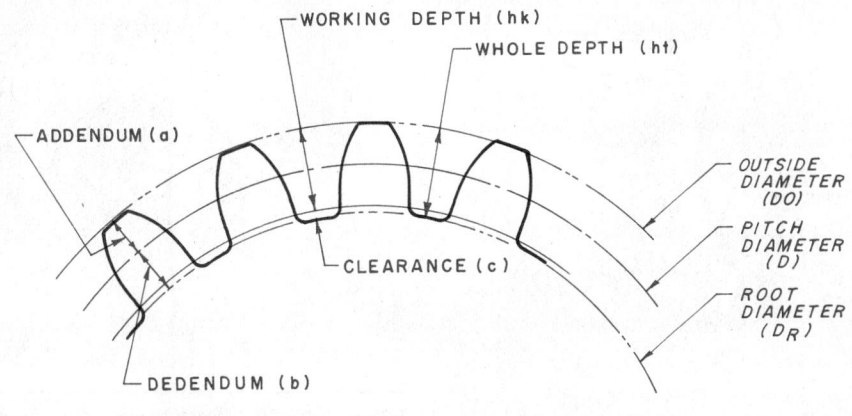

Figure 16-17   Major gear terms

Circular Pitch (p)

Circular pitch is the distance between corresponding points of adjacent teeth measured on the circumference of the pitch diameter.

Figure 16-17 illustrates the major terms associated with individual gears. There are many more technical terms associated with the actual form and cutting of gear teeth, but the following terms are the ones the drafter should be familiar with.

Working Depth (hk)

The *working depth* is the distance that a tooth projects into the mating space. It is equal to gear addendum (a) plus pinion addendum (a).

Addendum (a)

The *addendum* is the radial distance from the pitch diameter to the top of the tooth. It is equal to $\frac{1}{P}$.

Dedendum (b)

The *dedendum* is the radial distance from the pitch diameter to the bottom of the tooth. It is equal to $\frac{1.157}{P}$.

Clearance (c)

Clearance is the space between the working depth and the whole depth. It is equal to dedendum (b) minus addendum (a).

Whole Depth (ht)

The *whole depth* is the total depth of a tooth space. It is equal to the addendum (a), plus the dedendum (b), plus the clearance (c).

Note: All notations in parentheses ( ) correspond to those used in the most recent edition of *Machinery's Handbook* so that they agree if a more detailed formula is required by the drafter.

Diametral Pitch (P)

Diametral pitch (P) is a ratio equal to the number of teeth on the gear per inch of pitch diameter (D).

$$\text{Diametral pitch (P)} = \frac{\text{Number of teeth (N)}}{\text{Pitch diameter (D)}}$$

Example:

A gear with 48 teeth on a 3.0 pitch diameter (D) would have a diametral pitch (P) of 16.

$$P = \frac{N}{D} = \frac{48}{3.0} = 16$$

Gear Template

A quick and efficient method of illustrating teeth is by using a gear template, *Figure 16-18*. When using a gear template, the drafter must first calculate the diametral pitch (P). The numbers next to each tooth indicate the diametral pitch. The *lower* part of the template opening is used to draw spur gear *teeth*. The *top* portion of the opening is used to draw teeth of a *rack* or *worm*.

Pressure Angle

The pressure angle determines the tooth form, *Figure 16-19*. The *pressure angle* is the angle at which pressure from the driver gear tooth passes to the tooth of the driven gear. The standard pressure angle is either 14½° or 20°. The 14½° pressure angle is still popular,

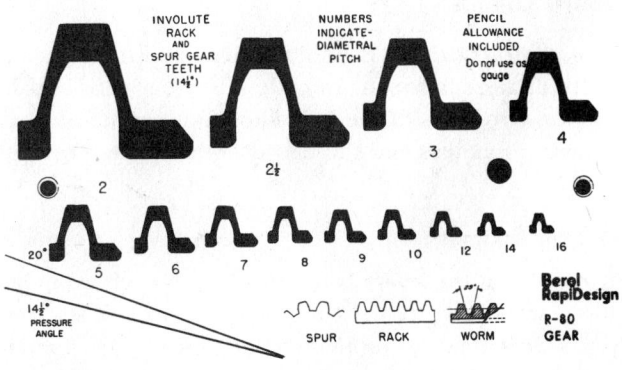

Figure 16-18   Gear template—spur/rack/worm *(Courtesy Modern School Supply)*

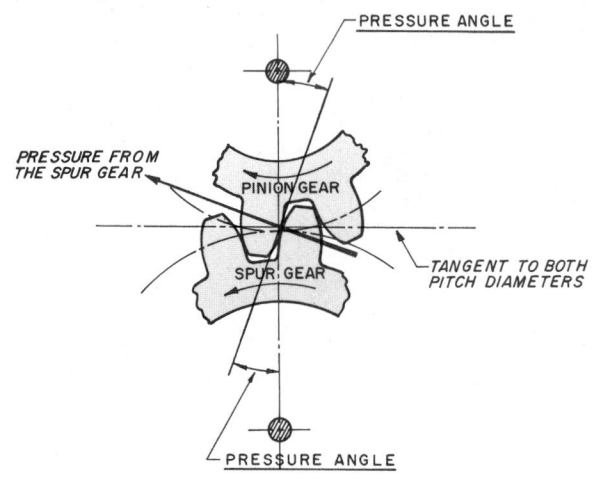

Figure 16-19   Pressure angle

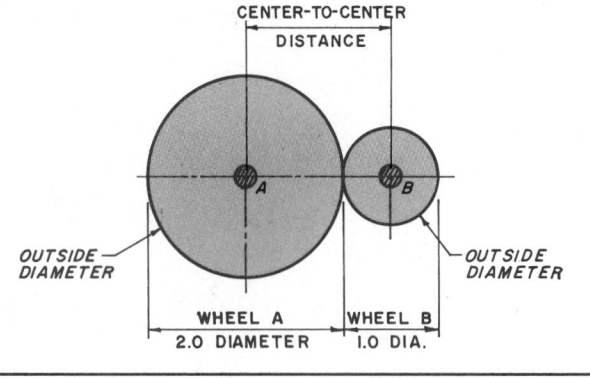

Figure 16-20   Center-to-center distances of two wheels

especially when replacing old machinery, and because it is quiet. The 20° pressure angle is stronger, due to the heavier tooth section at the base of the tooth form.

Center-to-Center Distances

Calculating the center-to-center distance between two friction wheels is a simple math problem, *Figure 16-20*.

$$\text{Center-to-center distance} = \frac{\text{Outside diameter of wheel A} + \text{Outside diameter of wheel B}}{2}$$

In this example, $\dfrac{2.0 + 1.0}{2} = 1.5$ center-to-center distance

The exact same formula applies to a gear and pinion, except that the pitch diameter (D) is used in place of the outside diameter (OD). Refer to *Figure 16-21*.

$$\text{Center-to-center distance} = \frac{\text{Pitch diameter of gear} + \text{Pitch diameter of pinion}}{2}$$

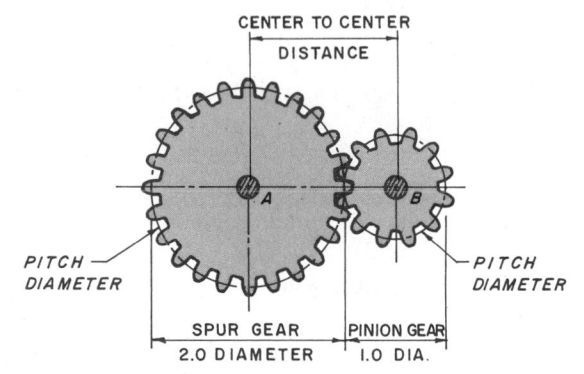

Figure 16-21   Center-to-center distances of two gears

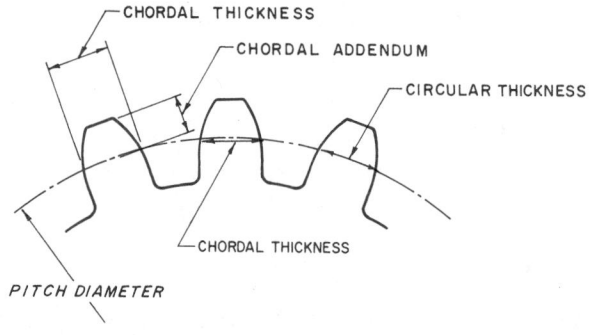

Figure 16-22   Measurements required to use a gear tooth caliper

Measurements Required to Use a Gear Tooth Caliper

A *gear tooth caliper* is used to check and measure an individual gear tooth. In order to use a gear tooth caliper, two parts of the individual tooth are used: the chordal thickness and the chordal addendum, *Figure 16-22*.

Chordal Thickness

The *chordal thickness* is the length of the chord measured straight across at the pitch diameter. It is measured straight across tooth, not along the pitch diameter as is the circular thickness.

$$\text{Chordal thickness} = \text{Sin}\left(\frac{90°}{N}\right) \times \text{Pitch dia.}$$

Chordal Addendum

The *chordal addendum* is the distance from the top of the tooth to the point at which the chordal thickness is measured.

$$\text{Chordal addendum} = \left[1 - \text{Cos}\left(\frac{90°}{N}\right)\right] \times \frac{\text{Pitch dia.}}{2} + \text{addendum}$$

Figures 16-23 and *16-24* show a gear tooth vernier caliper and how it is used.

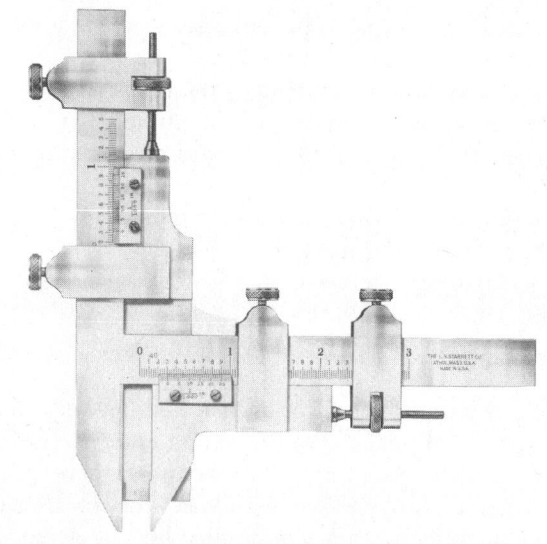

Figure 16-23 Gear tooth caliper *(Courtesy L. S. Starrett Co.)*

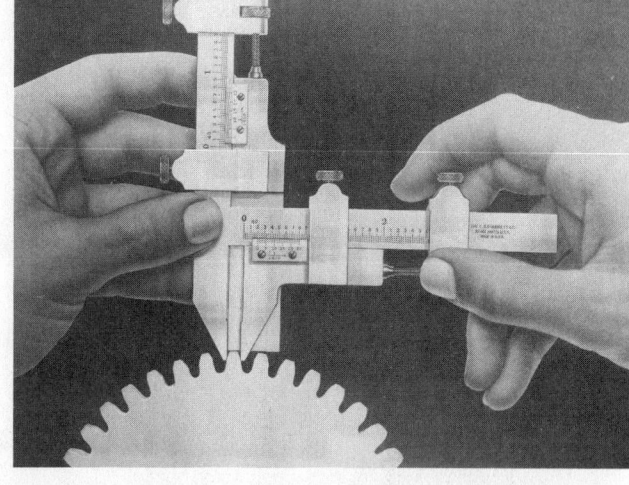

Figure 16-24 Using a gear tooth caliper *(Courtesy L. S. Starrett Co.)*

Required Tooth-Cutting Data

Overall gear blank dimensions are shown on the detail drawing (as illustrated in Figure 16-15) and can hold loose tolerancing. The actual gear dimensions must hold tight tolerancing.

Twelve essential items of information must be calculated and listed on the gear detail drawing. This list is usually located at the lower right-hand side of the drawing, and generally consists of the data in ***Figure 16-25***, but these items may vary slightly from company to company. The required material, heat-treating process, and other important information must also be included, usually in the title block.

In actual practice, the gear teeth are *not* drawn as it takes too much drawing time. A simplified method is used to illustrate the actual gear tooth, similar to that used to illustrate threads of a fastener, ***Figure 16-26***. The outside diameter and the root diameter are illustrated by a thin phantom line. The pitch diameter is illustrated by a thin center line. Actual teeth are not drawn, except to illustrate some special feature in relationship to a tooth, a spline, a keyway or a locating point. In this example, all dimensions in color are related to the gear blank only. All gear dimensions are called-off in the cutting data box underneath.

Rack

A *rack* is simply a gear with teeth formed on a flat surface, ***Figure 16-27***. A rack changes rotary motion into reciprocating motion. All terms and formulas associated with spur and pinion gears apply to the rack. The sides of the teeth are *straight*, not involute as on a spur gear. The teeth are inclined at the same angle as the pressure angle of the mating pinion gear. Note that the circular pitch of the pinion gear is the same as the linear pitch of the rack. All dimensions for heights of depth, pitch,

ITEM	TO FIND:	HAVING	FORMULA
1	Number of teeth (N)	D & P	$D \times P$
		DO & P	$(DO \times P) - 2$
2	Diametral pitch (P)	p	$\dfrac{3.1416}{p}$
		D & N	$\dfrac{N}{D}$
		DO & N	$\dfrac{N+2}{DO}$
3	Pressure angle (ø)	–	20° STANDARD 14°-30' OLD STANDARD
4	Pitch diameter (D)	N & P	$\dfrac{N}{P}$
		N & DO	$\dfrac{N \times DO}{N+2}$
		DO & P	$DO - \dfrac{2}{P}$
5	Whole depth (ht)	a & b	$a + b$
		P	$\dfrac{2.157}{P}$
6	Outside diameter (DO)	N & P	$\dfrac{N+2}{P}$
		D & P	$D + \dfrac{2}{P}$
		D & N	$\dfrac{(N+2) \times D}{N}$
7	Addendum (a)	P	$\dfrac{1}{P}$
8	Working depth (hk)	a	$2 \times a$
		P	$\dfrac{2}{P}$
9	Circular thickness (t)	P	$\dfrac{3.1416}{2 \times P}$
10	Chordal thickness	N, D & a	$\sin\left(\dfrac{90°}{N}\right) \times D$
11	Chordal addendum	N, D & a	$\left[1-\cos\left(\dfrac{90°}{N}\right)\right] \times \dfrac{D}{2} + a$
12	Dedendum (b)	P	$\dfrac{1.157}{P}$

REQUIRED CUTTING DATA

Figure 16-25 Required tooth-cutting data for spur and pinion gears

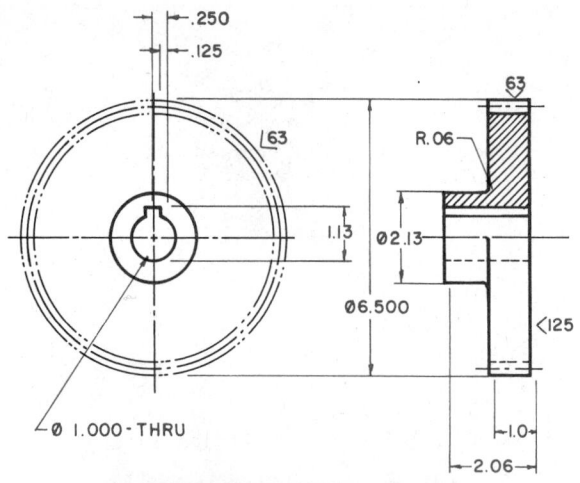

CUTTING DATA	
NUMBER OF TEETH	24
DIAMETRAL PITCH	4
PRESSURE ANGLE	20°
PITCH DIAMETER	6.000
WHOLE DEPTH	0.5393
OUTSIDE DIAMETER	6.500
ADDENDUM	.250
WORKING DEPTH	.500
CIRCULAR THICKNESS	.3925
CHORDAL THICKNESS	.2566
CHORDAL ADDENDUM	.3924
DEDENDUM	.289

Figure 16-26   Example of a spur gear detail drawing

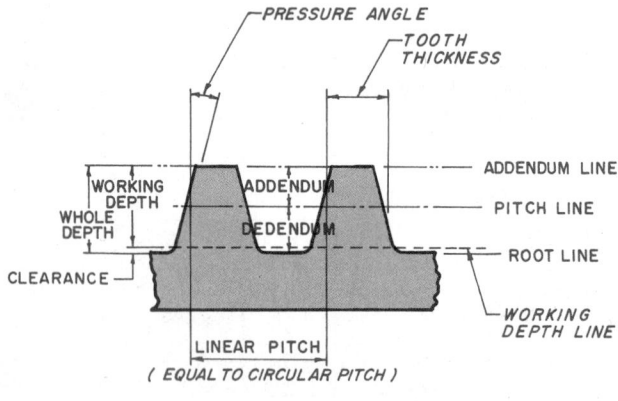

Figure 16-27   Gear terminology for a rack

and addendum lines are calculated from a datum or reference line. The rack is usually manufactured out of rectangular stock but, occasionally, is manufactured out of round stock to meet a specific design.

Two methods can be used to obtain full meshing of the pinion and the rack:

1. The rack is cut down so that it has less backlash.
2. The outside diameter of the pinion is increased slightly so that the pinion is *not* a standard size. If a spur gear is meshed with the oversized pinion gear, its outside diameter is manufactured slightly undersize.

Both methods maintain the standard shaft center-to-center distances between the gear and the pinion. Standard

tabulated charts are available to calculate these amounts of increased or decreased outside diameters.

The required tooth-cutting data for a rack are:
- Pressure angle (same as mating pinion gear)
- Tooth thickness at pitch line (same as mating pinion gear)
- Whole depth (same as mating pinion gear)
- Maximum allowance pitch variation
- Accumulated pitch error max. (over total length)

Bevel Gear

Bevel gears transmit power and motion between intersecting shafts at right angles to each other. *Figure 16-28* illustrates the various terms associated with bevel gears. The required tooth-cutting data for bevel gears are similar to spur/pinion gears. These must be calculated and listed on the detail drawing. *Figure 16-29* shows the order in which the data and formulas should be listed on the drawing. Bevel gears must be designed and drawn in pairs to ensure a perfect fit.

Worm and Worm Gear

The *worm* and *worm gear* is used to transmit power between nonintersecting shafts at right angles to each other. When using the worm and worm gear, a large speed ratio is possible as one revolution of a single-thread worm turns the worm gear only one tooth and one space.

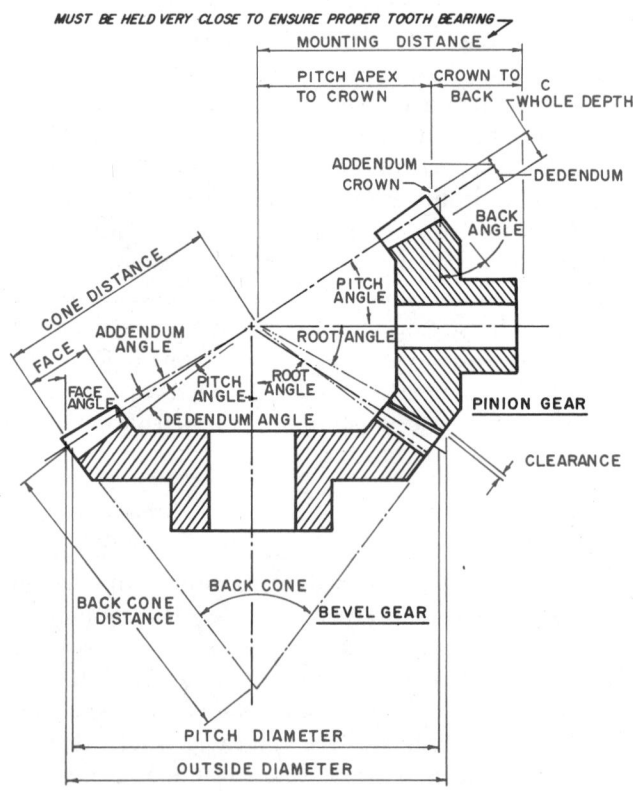

Figure 16-28   Gear terminology for bevel gears

The worm's thread is similar in form to a rack tooth. The worm gear is similar in form to a spur gear, except that the teeth are twisted slightly and curved to fit the curvature of the worm.

When drawing the worm and worm gear, an approxi-mate representation is used, *Figure 16-30*. Cutting data for the worm and worm gear must be listed on the drawing in the lower right side in the order illustrated in *Figure 16-31* for the worm gear, and in *Figure 16-32* for the worm.

ITEM	TO FIND:	HAVING	FORMULA SPUR	FORMULA PINION
	REQUIRED CUTTING DATA			
1	Number of teeth (N)	–	AS REQ'D.	
2	Diametral pitch (P)	p	$\dfrac{3.1416}{p}$	
3	Pressure angle (ø)	–	20° STANDARD 14°-30′ OLD STANDARD	
4	Cone distance (A)	D & d	$\sin d \sqrt{\dfrac{D}{2}}$	
5	Pitch distance (D)	p	$\dfrac{N}{p}$	
6	Circular thickness (t)	p	$\dfrac{1.5708}{p}$	
7	Pitch angle (d)	N & d (of pinion)	90°–d(pinion)	$\tan d \dfrac{N \text{ pinion}}{N \text{ gear}}$
8	Root angle (γ R)	d & δ	$d - \delta$	
9	Addendum (a)	p	$\dfrac{1}{p}$	
10	Whole depth (ht)	p	$\dfrac{2.188}{p} + .002$	
11	Chordal thickness (C)	D & d	$\dfrac{1}{2}\left(\dfrac{D}{\cos d}\right)$	$1 - \cos\left(\dfrac{90°}{N}\right) + a$

ITEM	TO FIND:	HAVING	FORMULA SPUR	FORMULA PINION
	REQUIRED CUTTING DATA			
12	Chordal addendum (aC)	d	$\sin\left(\dfrac{\frac{90°}{N}}{\cos d}\right)$	
13	Dedendum (bC)	P	$\dfrac{2.188}{P} - a\text{(pinion)}$	$\dfrac{2.188}{P} - a\text{(gear)}$
14	Outside diameter (DO)	D, a & d	$D + (2 \times a) \times \cos. d$	
15	Face	A	1/3 A (max.)	
16	Circular pitch (p)	p & N	$\dfrac{3.1416 \times p}{N}$	
17	Ratio	N gear & N pinion	$\dfrac{N \text{ gear}}{N \text{ pinion}}$	
18	Back angle (γ O)	–	SAME AS PITCH ANGLE	
19	Angle of shafts	–	90°	
20	Part number of mating gear	–	AS REQ'D.	
21	Dedendum angle δ	A & b	$\dfrac{b}{A} = \tan \delta$	

Figure 16-29   Required tooth-cutting data for bevel gears

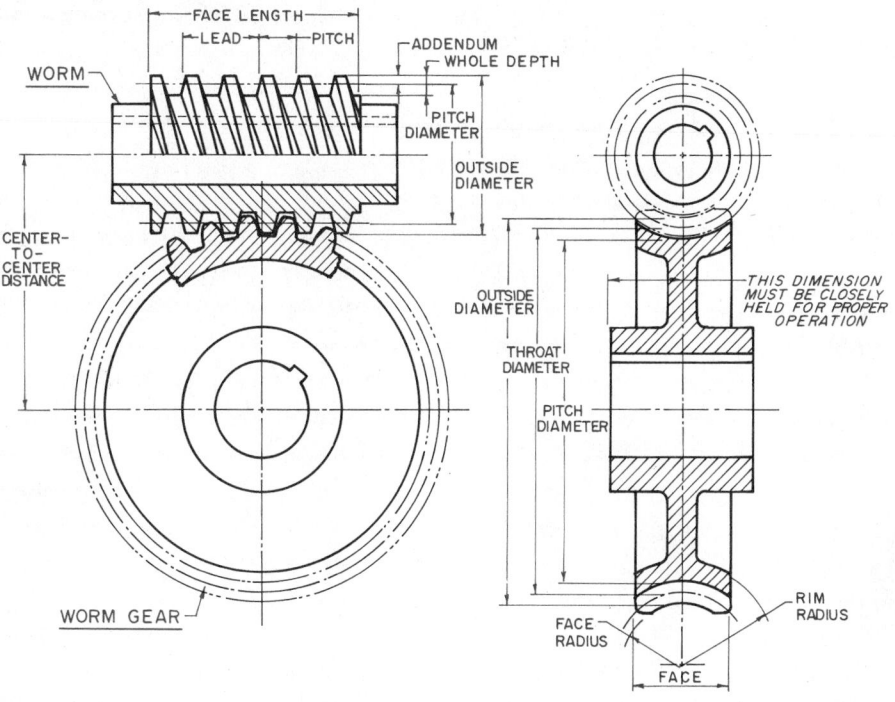

Figure 16-30   Example of worm and worm gear data

REQUIRED CUTTING DATA			
ITEM	TO FIND:	HAVING	FORMULA
1	Number of teeth (N)	–	AS REQ'D.
2	Pitch diameter (D)	N & p	$\dfrac{N \times p}{3.1416}$
3	Addendum (a)	p	p x .3181
		P	$\dfrac{1}{P}$
4	Whole depth (ht)	p	p x .6866
		P	$\dfrac{2.157}{P}$
5	Lead (L) Right-Left	p & N	p x N
6	Worm part no.	–	AS REQ'D.
7	Pressure angle ø	–	20° STANDARD 14°-30' OLD STANDARD
8	Outside diameter (DO)	Dt & Pa	Dt + .4775 x Pa
9	*Circular pitch (p)	P	$\dfrac{3.1416}{P}$
		L & N	$\dfrac{L}{N}$
10	Diametral pitch (P)	p	$\dfrac{3.1416}{p}$
11	Throat diameter (Dt)	D & Pa	D + .636 x Pa
12	Ratio of worm/ worm gear	N worm & N worm gear	$\dfrac{N \text{ worm gear}}{N \text{ gear}}$
13	Center to center distance between worm & worm gear	D worm & D worm gear	$\dfrac{D \text{ worm} + D \text{ worm gear}}{2}$

*Circular pitch (p) must be same as worm Axial pitch (Pa)

Figure 16-31   Required tooth-cutting data for a worm gear

REQUIRED CUTTING DATA			
ITEM	TO FIND:	HAVING	FORMULA
1	Number of teeth (N)	P	$\dfrac{3.1416}{P}$
2	Pitch diameter (D)	Pa	(2.4 x Pa) + 1.1
		DO & a	DO – (2 x a)
3	*Axial pitch (Pa)	–	Distance from a point on one tooth to same point on next tooth
4	Lead (L) Right or Left	p & N	p x N
5	Lead angle (La)	L & D	$\dfrac{L}{3.1416 \times D} = \tan La$
6	Pressure angle (ø)	–	20° STANDARD 14°-30' OLD STANDARD
7	Addendum (a)	p	p x .3183
		P	$\dfrac{1}{P}$
8	Whole depth (ht)	Pa	.686 x Pa
9	Chordal thickness	N, D & a	$\left[1\text{-}\cos\left(\dfrac{90°}{N}\right)\right] \times \dfrac{D}{2} + a$
10	Chordal addendum	N, D & a	$\sin\left(\dfrac{90°}{N}\right) \times D$
11	Outside diameter (DO)	D & a	D + (2 x a)
12	Worm gear part no.	–	AS REQ'D.

*Axial pitch (Pa) must be same as worm gear circular pitch (p)

Figure 16-32   Required tooth-cutting data for a worm

Note: It is important that the mounting of a worm and gear set ensures that the central plane of the gear passes essentially through the axis of the worm. This may be accomplished by adjusting the gear axially at assembly by means of shims. When properly mounted and lubricated, worm gear sets will become more efficient after the initial breaking-in period.

Other information that must be included on a worm/ worm gear detail drawing includes:
- Gear blank information
- Tooth-cutting data
- Reference to mating part

Gear Train

A *gear train* is two or more gears used to achieve a designed R.P.M. The ratio of the R.P.M. of the first gear to the R.P.M. of the final gear is called the *value* of the gear train.

Example:

$$\frac{\text{R.P.M. of last shaft}}{\text{R.P.M. of first shaft}} = \frac{\text{Value of}}{\text{gear train}}$$

Study the example on gear trains in Figure 16-33:
- Shafts 1 and 2 are connected by gear A and pinion B
- Shafts 2 and 3 are connected by gear C and pinion D
- Shafts 3 and 4 are connected by gear E and pinion F
- Gear A has 60 teeth
- Pinion B has 15 teeth
- Gear C has 40 teeth
- Pinion D has 20 teeth
- Gear E has 45 teeth
- Pinion F has 9 teeth

Problem:

The R.P.M. of shaft 1 is 200 R.P.M. What is the R.P.M. of shaft 4? What is the gear train value?

Each shaft taken individually:

$$\text{Ratio} = \frac{\text{N spur gear (driver gear)}}{\text{N pinion gear (drive gear)}}$$

R.P.M. of driven pinion = Ratio x R.P.M. driver

Step 1   a. Shaft 1 rotates at 200 R.P.M.
   b. Gear A has 60 teeth; pinion B has 15 teeth.
   c. This ratio is:

$$\frac{\text{N spur gear}}{\text{N pinion gear}} = \frac{60}{15} = \frac{4}{1} \text{ or 4:1 ratio}$$

d. R.P.M. of driven gear (shaft 2) = Ratio x
R.P.M. driver = 4 x 200 = 800 R.P.M.

Step 2   **a.** Shaft 2 rotates at 800 R.P.M.
b. Gear C has 40 teeth; pinion D has 20 teeth.
c. This ratio is:

$$\frac{\text{N spur gear}}{\text{N pinion gear}} = \frac{40}{20} = \frac{2}{1} \text{ or } 2:1 \text{ ratio}$$

d. R.P.M. of driven gear (shaft 3) = Ratio x
R.P.M. driver = 2 x 800 = 1600 R.P.M.

Step 3   **a.** Shaft 3 rotates at 1600 R.P.M.
b. Gear E has 45 teeth; pinion F has 9 teeth.
c. This ratio is:

$$\frac{\text{N spur gear}}{\text{N pinion gear}} = \frac{45}{9} = \frac{5}{1} \text{ or } 5:1 \text{ ratio}$$

d. R.P.M. of driven gear (shaft 4) = Ratio x
R.P.M. driver = 5 x 1600 = 8000 R.P.M.

Step 4

$$\frac{\text{Value of}}{\text{gear train}} = \frac{\text{R.P.M. last shaft}}{\text{R.P.M. first shaft}} = \frac{8000}{200} = 40 \text{ value}$$

To calculate the value of a *complete* gear train use the following formula:

The product of the number of teeth on all the drivers (spur gears) divided by the product of the number of teeth on all followers (pinion gears) equals the value of a complete gear train.

Example:

$$\frac{\text{N gear A x N gear C x N gear E}}{\text{N pinion B x N pinion D x N pinion F}}$$

$$\frac{60 \times 40 \times 45}{15 \times 20 \times 9} = \frac{108,000}{2700} = 40$$

In *Figure 16-33*, the gear train has been sketched with the gears in a neat row. This is done for the purpose of showing the sketch or basic design. In the final assem-bly, gears are actually clustered closer together to save space, and they are almost never in a row as shown.

Materials

Gears are made of many materials, such as brass, cast iron, steel, and plastic to mention but a few. Many bevel gears are forged, some are stamped from thin material, some are cast as a blank and machined, and others are die-cast to the exact size and shape.

Each application must be carefully analyzed. Metal gears have been used for years, but, if the load is not excessive, plastic gears have many advantages. Plastic runs quieter, has a self-lubricating effect, weighs much less, and costs much less. In addition, complicated multiple gears can be molded into a single piece which further reduces costs.

Design and Layout of Gears

The initial design of gears starts with the nominal size of the pitch diameters. The required speed ratio, loading, space limitations, and center-to-center distances are also important factors to consider. Mating gears and pinions must have equal diametral pitch in order to correctly mesh; therefore, the diametral pitch should be one of the first considerations. A complete analysis and design of a gear or a complete gear chain are very complex, and far beyond the scope of this text. Most designers try to use standard gears from a company that specializes in gears, gear chains, and gear design. Most gear manufacturing companies can be of assistance in designing gears for special applications. These companies can usually manufacture gears at a lower price than if they were made in-house.

Further analysis, in-depth study, and design data can be found in the most recent edition of *Machinery's Handbook*.

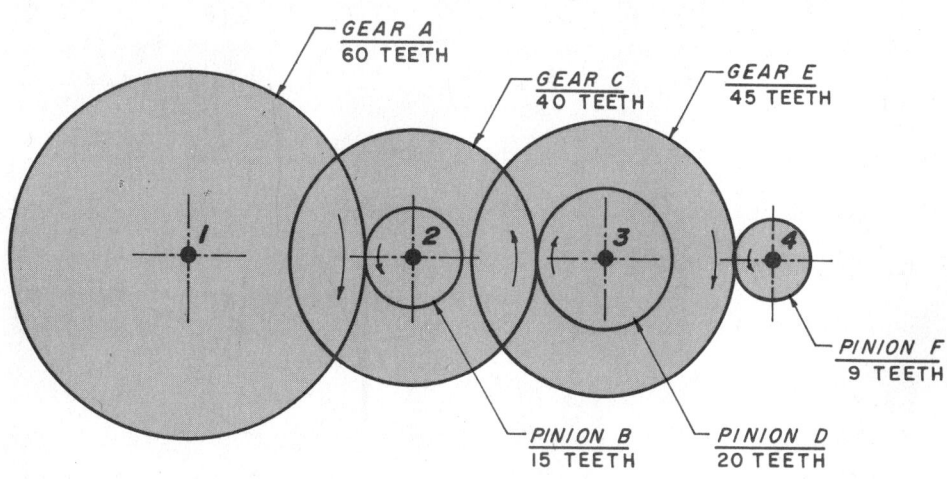

Figure 16-33   Gear train

Review

Calculate the following math problems; include all math work to illustrate how the answers were derived.

1. What is the outside diameter of a spur gear having a pitch diameter of 1.500 and 48 teeth?

2. What is the addendum and dedendum of a spur gear having a pitch diameter of 3.000 and 48 teeth?

3. With the addendum, dedendum, and pitch diameter of Problem 2, what is the root diameter and outside diameter?

4. How many teeth are on a spur gear having a pitch diameter of 1.750, diametral pitch of 20, and an outside diameter of 1,850?

5. What is the pitch diameter of a spur gear having a diametral pitch of 24, 78 teeth, and an outside diameter of 3.333?

6. What is the diametral pitch of a spur gear having 80 teeth and a 2.500 pitch diameter?

7. What is the root diameter of a spur gear having a pitch diameter of 2.000 and 32 teeth?

8. What is the pitch diameter of a spur gear having 40 teeth and a diametral pitch of 20?

9. What is the diametral pitch of a pinion gear having 75 teeth and 3.208 outside diameter? Use the formula $P = \dfrac{N}{D}$

10. How many teeth are required for a pinion gear with a mating spur gear having a P of 48, a 2.000 pitch diameter, and a 4:1 ratio?

11. What is the whole depth of a spur gear with the following specifications: 70 teeth, .7291 pitch diameter, .750 outside diameter, and 96 diametral pitch?

12. What is the circular pitch of a bevel pinion having a 16 diametral pitch, and 20 teeth?

13. A spur gear has a pitch diameter of 3.000 and a pinion gear has a pitch diameter of 1.500. If an R.P.M. 250 is needed at the shaft of the pinion gear, how fast should the spur gear be rotated?

14. What is the ratio between a spur gear with an R.P.M. of 175 and a pinion gear with an R.P.M. of 1050?

15. The ratio between a pair of bevel gears at a 90° angle and 48 P is:
    Gear: D = .9375, 45 teeth
    Pinion: D = .6250, 30 teeth

16. What is the outside diameter of a bevel gear with the following specifications: a 2.500 pitch diameter, a 48 diametral pitch, a 4:1 ratio, and 120 teeth?

17. What is the shaft center distance between a worm and a worm gear with the following specifications:
    Worm: Pitch diameter .500
            Diametral pitch 24
            Single thread
    Worm Gear: Pitch diameter 4.000
            Diametral pitch 24
            (N) Teeth 100

18. What is the gear ratio of a spur gear and pinion gear with the following specifications? (Calculate the answer using three different methods.)

Spur gear:	Pinion gear:
75 teeth	15 teeth
3.208 DO	.6250 D
3.125 D	.708 DO
725 R.P.M.	24 P
24 P	3/16 face
3/16 face	3625 R.P.M.

Chapter Sixteen Problems

The following problems are intended to give the beginning drafter practice in using the many formulas associated with gears, and practice in designing and laying out finished professional detail drawings of many major kinds of gears.

The steps to follow in laying out gears are:

Step 1 Study the problem carefully.

Step 2 Make a sketch if necessary

Step 3 Do all math required for each problem. Keep all math work for rechecking.

Step 4 Center the required views within the work area.

Step 5 Include all dimensioning according to the most recent drafting standards.

Step 6 Add all required gear cutting specifications in the lower, right side of the paper.

Step 7 Recheck all work, and, if correct, neatly fill out the title block using light guidelines and neat lettering.

Problem 16-1

Use the following specifications to lay out a single-view drawing of a 2:1 ratio spur gear and pinion gear *in mesh*.

Spur gear:
88 teeth
pressure angle 20°
pitch diameter 5.500
hub diameter 2.625
bore of 2.000 diameter

Pinion gear:
hub diameter 1.625
bore of 1.000 diameter
pressure angle 20°

Use all standard drafting methods to illustrate the outside diameter, pitch diameter, and root diameter. Calculate and add the center-to-center distance between the shafts.

Problem 16-2

Draw a spur gear having the following specifications: 56 teeth, pressure angle 20°, pitch diameter 3.500, hub diameter 1.00 gear blank with overall size (width) 1.00, face size .50, bore .5625; use a ⅛″ set screw.

Complete two views using the half-section method of representing gears. Do not show the gear teeth; use the conventional method of illustrating teeth. Dimension per all standard practices. Calculate and add all standard required cutting data to the lower right side of the work area. Material S.A.E. #1320 cast steel. Heat treatment: carburize .015–.020 deep.

Problem 16-3

Complete a two-view detail drawing of a pinion gear having the following specifications: calculate and add all standard cutting data: 30 teeth, pressure angle 20°, pitch diameter 6.000, hub diameter 2.000, heat treatment: carburize .050 deep, gear blank with overall size (width) 3.500, face size 1.250, bore 1.000, material S.A.E. #4620 steel.

Problem 16-4

Make a detail drawing, completely dimensioned, of a spur gear with the following specifications: diametral pitch 16, pressure angle 20°, pitch diameter 5.500, whole depth .1348, outside diameter 5.625, addendum .0625, working depth .125, circular thickness .098, chordal thickness .0633, chordal addendum .0985, whole depth .1348, and dedendum .0723, material cast brass.

Design a simple gear blank with a set screw to fasten gear to a .75 dia. shaft.

Problem 16-5

Design and draw a detailed drawing, completely dimensioned, of a pair of bevel gears having teeth of a 4.0 diametral pitch; the gear with 25 teeth, the pinion gear with 13 teeth, face width of 1.00, gear shaft 1.25 diameter, and pinion shaft .875 diameter. (Calculate for FN-2 fits with the shaft.)

Design the hub diameter to be approximately twice the diameter of the shaft diameter. Backing for the gear 1.375 and for the pinion .75. Design and dimension the remaining portion to suit, using standard drafting practices. Add all cutting data to the lower right side of the work area. *Material*: S.A.E. #3120 cast steel, heat treatment: carburize .015/.020 deep.

Problem 16-6

Make a design layout of a worm and worm gear with the following specifications: shaft diameter 1.25, single-thread worm, lead .75, worm gear with 28 teeth. Add all dimensions and cutting data as required.

Problem 16-7

Sketch a clock gear chain with the following specifications:

First gear (main wheel): 84 teeth
Second gear: pinion gear 8 teeth
    spur gear 60 teeth
Third gear (center shaft): pinion gear 12 teeth
    spur gear 68 teeth
Fourth gear: pinion gear 7 teeth
    spur gear 66 teeth
Fifth gear (escapement gear): pinion gear 7 teeth
    spur gear 33 teeth

Note: In actual practice, a clock's first gear (main wheel) is usually driven by a spring. The minute hand is connected to the shaft of the third gear (center shaft) and the fifth gear is the escapement gear.

Problem 16-8

Using the sketch of a clock gear chain in Problem 16-7, answer the following questions:

1. How many times will the second gear rotate for one full revolution of the first gear?
2. How many turns will the third gear rotate for one full revolution of the second gear?
3. How many teeth of the main wheel are required to turn the third gear shaft one complete revolution? (On a clock, one revolution of the third gear equals one hour. This wheel turns 24 times for one day.)
4. What is the required revolution of the first gear (main wheel) for one day's running time (24 turns of the center wheel)?
5. What is the value of the complete gear train?
6. As this is a clock gear chain in revolutions, how many will the fifth gear (escapement) make in twelve hours? (The third gear will turn 12 times.)

Problem 16-9

Design and develop a design layout two-view drawing of a gear train with the following specifications. Design a compact clustered arrangement of the gears within an 8.0 x 10.0 supporting plates area.

First shaft: 42 teeth/4.0 outside diameter-gear
Second shaft: 6 teeth/1.0 outside diameter-pinion
Third shaft: 6 teeth/.75 outside diameter-pinion
        30 teeth/2.5 outside diameter-gear
Fourth shaft: 6 teeth/.75 outside diameter-pinion
        30 teeth/2.0 outside diameter-gear
Shaft sizes = .375 dia/supporting plates 1.5 apart

Problem 16-10

Locate a windup clock or some other such small gear chain device and make a design layout similar to that in Problem 16-9.

Assembly and Detail Drawings for Design

Chapter Outline

ACRONYMS FOR DESIGN, ASSEMBLY, AND DETAIL DRAWINGS	
DFM	Design for Manufacturing
DFA	Design for Assembly
REP	Request for Proposal
ECO	Engineering Change Order
ECR	Engineering Change Request
CFR	Code of Federal Regulations

Technical Drawings and the Engineering Department

A representative engineering department and a chain of commands for the department provide a framework for discussing technical drawings—assembly, detail, and layout—and related engineering practices. Design procedures, master parts lists, engineering revisions, invention agreements, personal technical files, designing for manufacturing and assembly, and quotation proposals make up the list of engineering practices discussed in this chapter. In problems at the end of the chapter, the student drafter-designer is given the opportunity to do basic designing using the procedures, drawings, and practices discussed, and is encouraged to incorporate topics from preceding chapters, such as dimensioning standards, fasteners, tolerances to ensure the function of mating parts, and machine elements.

Engineering departments vary from company to company, but most of them have functions in common. An example of an engineering functional chart is illustrated in *Figure 17-1*.

An engineering department consists of a variety of sections, usually directed by a chief engineer. The various sections associated with a medium-size engineering department would include design, application, and quotation. The design section is responsible for new-design products and design improvement of existing products. The application engineering section handles special order modifications of existing products. For example, a scale manufacturing company would customize a scale platform to meet a customer's special need. The quotation department estimates costs and prepares quotations for new or modified products and systems. In

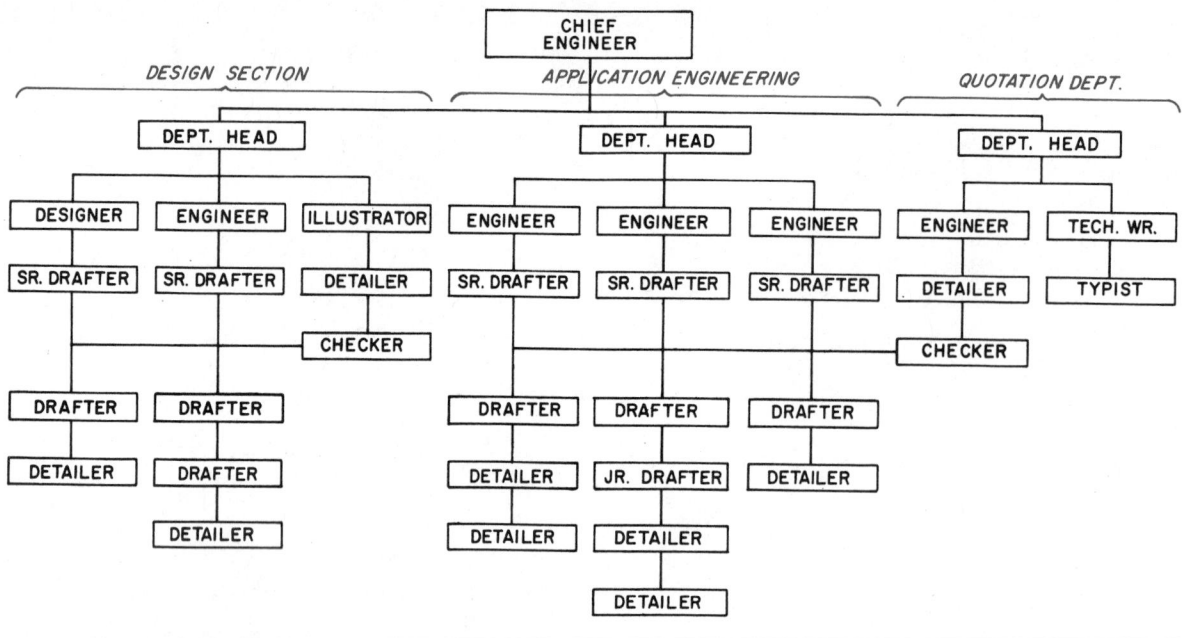

Figure 17-1   Chain of command in an engineering department

larger engineering departments, these various sections may be divided into fields of engineering such as mechanical, civil, electrical, electronic, structural, materials, agricultural, petroleum, and so forth.

Directly under the chief engineer are the section heads, usually engineers or designers with extensive design-engineering experience plus an ability to lead teams of skilled personnel. The teams could consist of engineers, designers, drafters, detailers, checkers, technical writers, illustrators, and typists. Although most engineering departments have definite departmental structures, many shift drafters, engineers, and designers move back and forth among the various sections as needed. So, new personnel should fully understand the departmental structure and operation of their organization.

One means to gain an understanding of an engineering organization is to follow paper documents along the chains of command as indicated in the flow chart, Figure 17-1. For illustrative purposes in this chapter, two hypothetical engineering organizations are discussed. The first company produces "one-of-a-kind" systems to meet specific needs of customers who generate electrical power. The second is a company that mass produces a device for positioning a small cross-hair cursor or a pointer on a computer screen, the ubiquitous mouse.

Systems Company

Assume that a systems company has received a request for proposal (RFP) from a solid-waste handling company that needs to convert burnable waste into electrical power. The proposal is to be submitted to the solid-waste handling company in two months. The president of the systems company sends a memo to his chief engineer with directions to process the RFP. Consequently,

the chief calls a meeting of the design, application, and quotation department heads. They decide that the design section will need to design a new furnace and one new conveyor for handling solid waste coming from a shredder. The application engineering group will need to modify several designs of existing conveyors. A steam turbine and an electrical generator will only require small changes. The quotation department has the task of preparing a proposal to be submitted to the waste-handling company before the two-month deadline. The proposal will include the selling price of the complete energy-generating system, the delivery dates, and descriptions of the system components. Then, if the waste-handling company accepts the proposal and signs it, the proposal becomes a binding contract on both companies.

Design Section

The design section will be discussed first to illustrate the uses of the drawings and the related engineering practices listed earlier, which include, assembly, detail and layout drawings, design procedures, master parts lists, engineering revisions, invention agreements, personal technical files, and quotations. Application engineering will be discussed later.

Design Procedures

A design process or design procedure used by the engineering department acts as a general guideline for a team of designers, engineers and drafters formed into a team, ***Figure 17-2***. Their preliminary design tasks are to design a new conveyor plus create a new furnace, and to prepare sufficient information on these components so that the quotation department can prepare estimates of the conveyor and furnace costs.

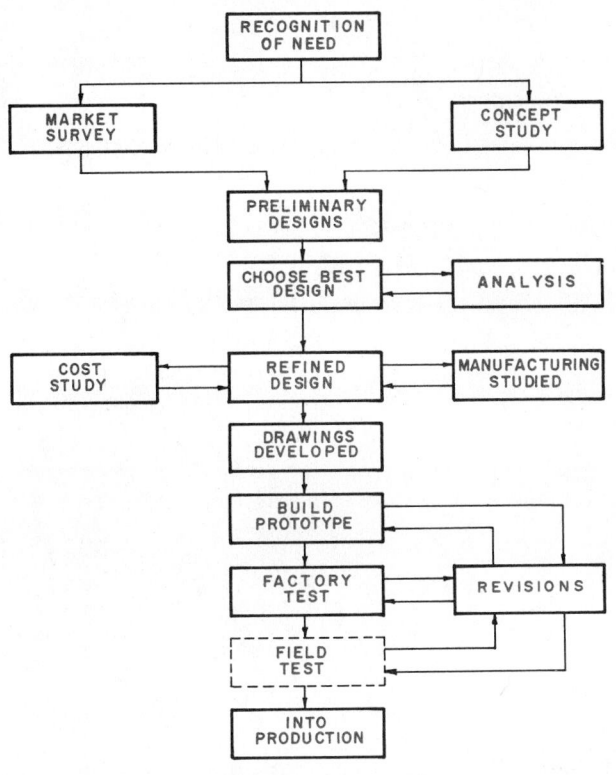

Figure 17-2 Design procedure

No one design process serves an individual designer, design team, or organization. Each individual, team, and organization develops a design process to suit their respective needs. (Similar-but-different design processes also are discussed in chapters 1, 10, and 24.)

The actual process of developing a new product from its established need (the RFP in this example) to its final production, consumes many hours and involves a variety of highly trained, skilled personnel. The flow chart in Figure 17-2 shows the overall process to production; however, in this preliminary RFP phase, the design team only goes through to the cost study stage to prepare information for quoting purposes. Later, if the proposal is accepted by the customer, the design section team would develop the conveyor and furnace through to the production stage.

Nevertheless, to continue describing the design procedure followed by the design section team for the new conveyor and new furnace, concept-study sketches would be done first. (Note that a market survey would not be required for this project because the RFP stated the need.) Sketches would be made of conveyor ideas for moving the waste from a shredder to the furnace, *Figure 17-3*. Next, where and how to burn the waste in the furnace and where to utilize the hot gases to generate steam would be considered. During these studies, many ideas are presented, usually in sketches. Designers are encouraged to generate and sketch ideas without restraint; that is, without being influenced by existing designs, in order to arrive at a fresh and perhaps unique solution. The more ideas that are proposed in the time

available, the greater the chances of developing the best solution to the problem. These brainstorming sessions are conducted in an atmosphere of "no criticism" so members will not feel inhibited. Analysis of the ideas are scheduled after the free-wheeling sessions.

The design section team employs two effective design approaches. In *conceptual design*, much use is made of known technical information from such sources as reference books, technical journals, handbooks, and manufacturers' catalogs. (Refer to chapters 13, 21 and 24 for examples.) All members of the team should be aware of the latest technical information available. In *scientific design*, use is made of concepts from physics, mechanics, materials science, chemistry, and mathematics.

During the analysis phase the best ideas are identified and developed further in sketches. Suitable engineering materials are investigated. Loads on the structures are analyzed and energy balances (output versus input) are estimated.

Layout Drawings

The design section department head follows the team's progress and soon directs the team to prepare *layout drawings* for a preliminary design review. The sketches of the best ideas are redrawn to scale in layout drawings. *Figure 17-4* shows the team's sketch of a preferred drag type conveyor for handling burnable shredded-waste to the furnace. *Figure 17-5* shows portions of the preferred conveyor in a design development layout drawing. Layout drawings in which the views are drawn to scale are informal and contain all the pertinent information. Standard dimensioning practices are encouraged, but not required, and clear handwritten notes are acceptable. The drawings are usually done by a designer or an engineer who also specifies standard materials and standard manufacturing processes where possible. A design layout may not be completely dimensioned, but it has important overall dimensions and has important center lines located. Tolerances are noted, especially for mating parts. Nominal sizes are indicated and special requirements are listed. A title block is optional. (Refer to chapter 24 for further discussions of design development layout drawings. Kinematic layout drawings are discussed, also.)

During the design review, the layout drawings help the team to decide on the final designs to consider. The quotation department assists the team in preliminary costs analyses. Finally, the conveyor design and a new furnace concept that best meet the requirements, are selected. Manufacturing personnel who also have participated in these discussions recommend appropriate manufacturing processes for the conveyor and furnace.

Assume now that the quotation department has received all the technical data and specifications they need from the design section and application engineering departments to prepare a cost estimate to satisfy the RFP. A proposal is written and presented to the waste han-

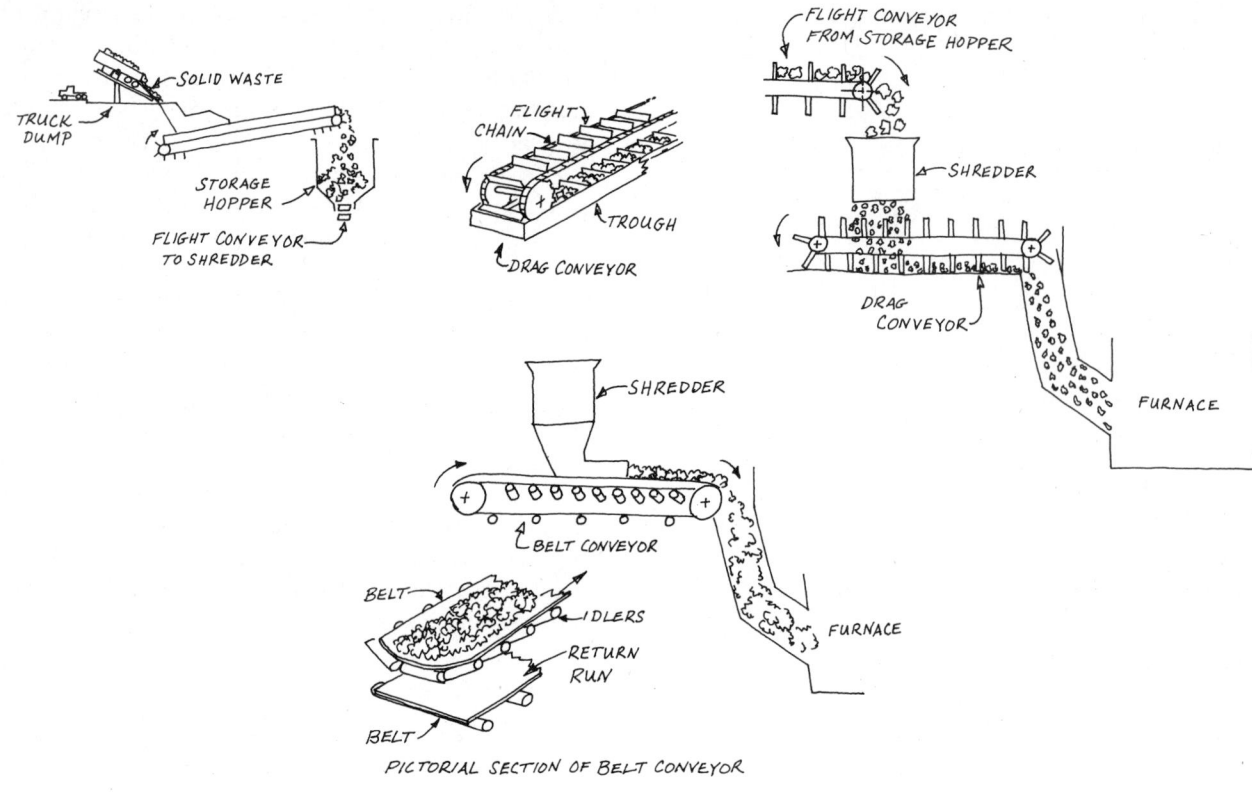

Figure 17-3   Sketches of conveyor ideas for conveying waste to the furnace

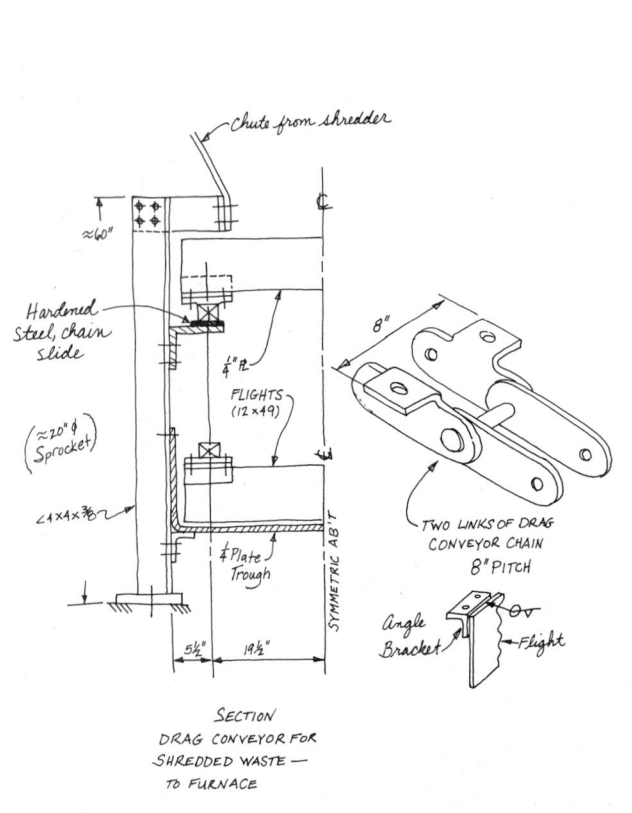

Figure 17-4   Sketch of preferred conveyor for conveying waste to the furnace

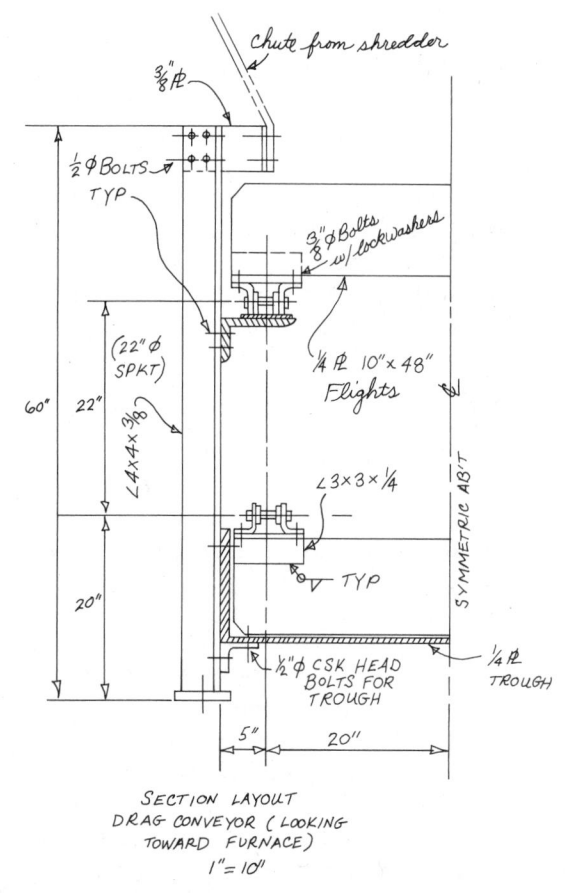

Figure 17-5   Layout drawing of preferred conveyor

dling company just before the two-month deadline. In addition to the written proposal and several systems site-drawings, two types of illustrative material may be presented for better visualization of the system. An illustrator may prepare a watercolor perspective drawing of the whole system. A second type that might be presented is an architectural type scale model of the whole system showing dumping sites, chutes, conveyors, the furnace and boiler, and the electricity-generating equipment.

The waste-handling company representatives may ask for further information and explanations, such as brief descriptions of system components and when specific components will be delivered, installed, and operating. Finally, after further comments and questions, the proposal is accepted and signed. It is now a binding, legal contract on both companies. ***Figure 17-6*** shows an example signature page from a proposal-contract form.

Working Drawings: Detail and Assembly

The systems company president notifies the chief engineer that the proposal is now a contract. The chief directs the designers and engineers to forward the design development layout drawings of the new conveyor and the furnace to drafters and detailers who use these drawings to create working drawings; i.e., detail and assembly drawings for manufacturing. There are a number of different kinds of working drawings associated with the development and production of a product, so the drafter must be familiar with each kind, know the company standards, and be able to draw each kind. The major kinds of working drawings are:

- Assembly drawings
- Sub-assembly drawings
- Detail drawings
- Modified purchased drawings

Assembly Drawing

Any product that has more than one part must have an assembly drawing. The *assembly drawing* illustrates how a product is assembled when completed. The assembly drawing can have one, two, three, or more views as needed. Often, a full section view is used to show how the parts are assembled.

Assembly drawings usually are not dimensioned, except for general overall dimensions or to indicate the capabilities of the assembly. For example, a clamp assembly drawing might have a dimension to illustrate how wide it opens. (Refer to Problem 17-17.) Assembly drawings do not contain hidden lines unless they are absolutely necessary to illustrate some important feature that the reader might otherwise miss.

An assembly drawing must call-off each part that is used to make up the assembly. Each part must be called-off by its part number and title, and the total number required to make up one complete assembly. Companies most often use one of two methods to call-

COMPANY

20—**APPROVAL:** There are no agreements or oral understandings outside of this proposal, which shall become a contract only when accepted by you, ______________________

______________________, and approved in writing by a plant manager or an executive officer of Company.

Prepared and submitted to you for Company

by ______________________.

Accepted this ________ day of ____________

19______ for

____

by ______________________

Title ______________________

Approved for Company at

____

this ________ day of ____________ 19______

by ______________________.

Title ______________________

Figure 17-6 Proposal-contract signature page

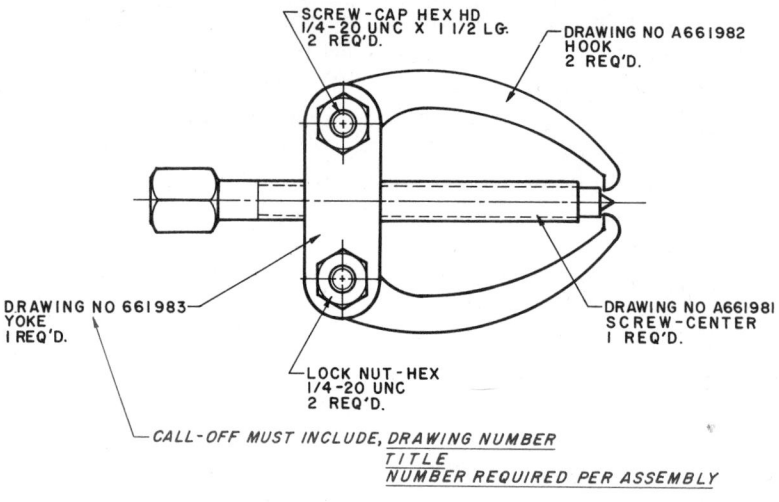

SCREW-CAP HEX HD
1/4-20 UNC X I 1/2 LG.
2 REQ'D.

DRAWING NO A661982
HOOK
2 REQ'D.

DRAWING NO 661983
YOKE
I REQ'D.

DRAWING NO A661981
SCREW-CENTER
I REQ'D.

LOCK NUT-HEX
1/4-20 UNC
2 REQ'D.

CALL-OFF MUST INCLUDE, *DRAWING NUMBER*
TITLE
NUMBER REQUIRED PER ASSEMBLY

Figure 17-7   Assembly call-off (method 1)

off the parts. In method one, each part has a leader line with the part number, title and number required, *Figure 17-7*. In method two, each part has a leader line with a balloon call-off number inside, *Figure 17-8*. This method uses a tabulation chart to list the part number, title, and number required.

Subassembly Drawing
A subassembly is composed of two or more parts permanently fastened together. A subassembly drawing is very similar to an assembly drawing. All standard practices associated with assembly drawings apply to subassembly drawings. That is, full section views, few overall dimensions, hidden lines only if necessary, and call-offs using either method one or two, are employed.

Subassemblies sometimes require machining operations after assembly, *Figure 17-9*. For instance, a hole is to be drilled through the parts after assembly. Drilling each individual part and then aligning the holes at

assembly would be much more difficult and costly. Thus, the required dimensions to drill the hole are added to the subassembly drawing.

There can be any number of subassemblies in the final assembly. For example, in an automobile, the engine is a subassembly, the alternator is a subassembly, the transmission is a subassembly, and so forth. These subassemblies, together with various single parts, make up the final assembly of the automobile. Note that subassemblies are often purchased, stocked, and stored for later final assemblies.

Detail Drawings
Every manufactured part must have its own fully dimensioned detail drawing, with its own drawing number and title block, *Figure 17-10*. All information needed to manufacture the part is contained in the detail draw-

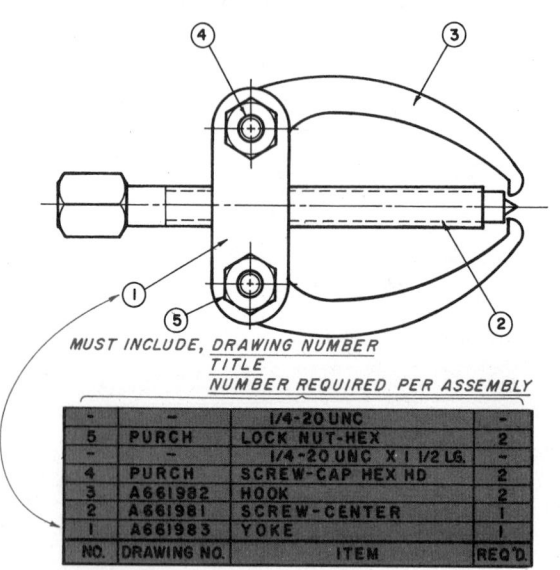

MUST INCLUDE, *DRAWING NUMBER*
TITLE
NUMBER REQUIRED PER ASSEMBLY

NO.	DRAWING NO.	ITEM	REQ'D.
-	-	1/4-20 UNC	-
5	PURCH	LOCK NUT-HEX	2
-	-	1/4-20 UNC X I 1/2 LG.	-
4	PURCH	SCREW-CAP HEX HD	2
3	A661982	HOOK	2
2	A661981	SCREW-CENTER	I
1	A661983	YOKE	I

Figure 17-8   Assembly call-off (method 2)

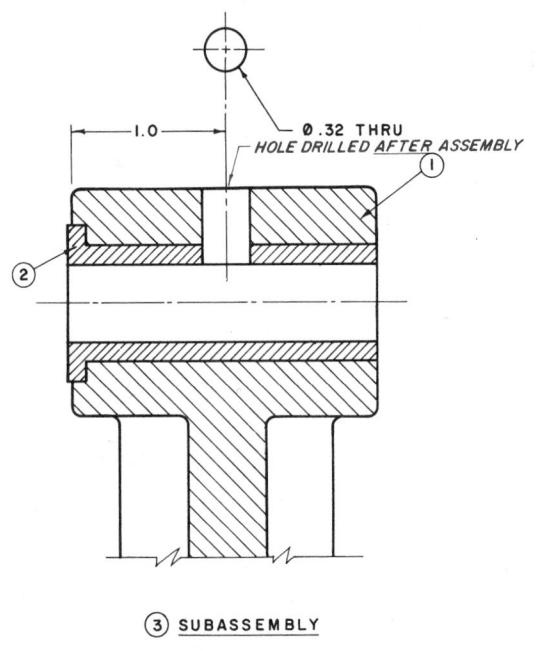

Ø.32 THRU
HOLE DRILLED *AFTER ASSEMBLY*

1.0

③ SUBASSEMBLY

Figure 17-9   Subassembly drawing

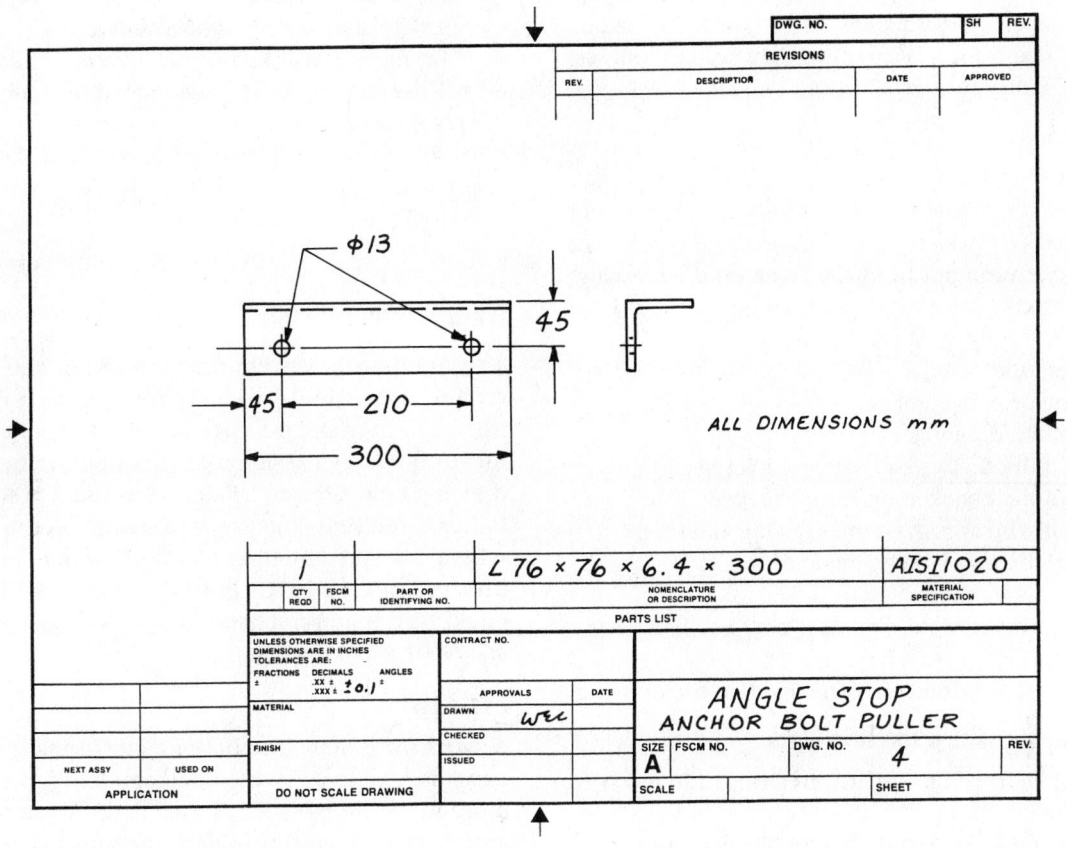

Figure 17-10   Detail drawing

ing, including as many views as needed to fully illustrate and dimension the parts.

The general practice in technical drawing is to draw only one part to one sheet of paper, regardless of how small the part is, because individual parts are often made on different machines or by different processes, and the bookkeeping plus costing is easier for one part per drawing.

Most companies use detail drawings as just described. However, some companies separate the detail drawings into various manufacturing processes, such as:

- Pattern detail drawings for castings
- Casting detail drawings
- Machining detail drawings
- Forging detail drawings
- Welding detail drawings
- Stamping detail drawings

Modified Purchased Parts

A purchased part that needs to be modified, however slightly, must have the modification fully described and dimensioned as with any detailed part, except that the purchased part size and description is listed in the title block under "material." *Figure 17-11* shows a standard fastener, M20 x 2.5-6g x 70 long hex head machine screw, which is to have a tip specially turned and a 4-mm diameter hole drilled through the tip. The standard fastener is drawn with the modifications added and dimensioned. Then it would be called-out as a stan-

dard metric fastener in the title block under material. This directs the machinist to make the part from a standard fastener.

Purchased Parts

Most manufacturing companies cannot make such purchased items as electric motors, bearings, shafting, bolts, nuts, screws, and washers as cheaply as companies that specialize in making these items. Therefore, designers should specify these standard items whenever possible. Purchased parts are drawn in assembly drawings, usually in outline form, to show their function and relationship to the product. All specific information for a purchased part should be listed on the drawing or

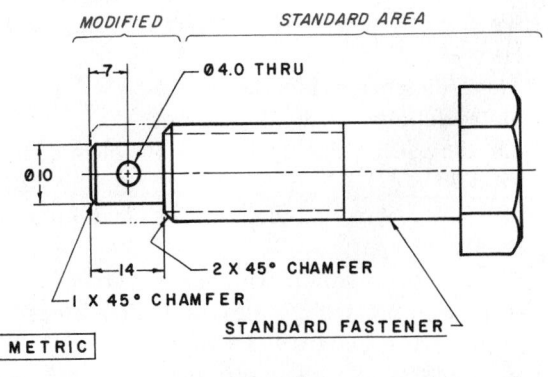

Figure 17-11   Drawing of modified purchased part

parts list, including size, material, special features, vendor part number, performance, and so forth. For example, specifications for a DC motor might read as follows: "12 volt DC motor, GE no. 776113 or equal."

Title Block

Title blocks vary from company to company, although most companies follow the ANSI standard illustrated in Figure 17-10. The primary function of a title block is to provide information not normally given on the drawing, such as:

- Name and address of the company
- Drawing title
- Drawing or part number
- Scale of the drawing
- Name of the drafter and date completed
- Name of the checker and date checked
- Name of who approved the drawing and date
- Material the part is to be made of
- Tolerances/limits of dimensions
- Heat treatment requirements if not specified on the drawing
- Finish requirements if not specified on the drawing

Size of Lettering within the Title Block

- General information, .125 (3) freehand/.120 mechanical
- Drawing title, .250 (6) freehand/.240 mechanical
- Drawing number, .312 (8) freehand/.350 mechanical

Checking Drawings

The importance of absolute accuracy in engineering drawings cannot be stressed enough. The slightest error could cause large and unnecessary expenses. In some fields of engineering, such as in aircraft and space vehicles, an error could cause tremendous consequences—lost lives! Therefore, every precaution should be taken to avoid errors. In most engineering departments, checkers are employed to verify all dimensions and to see that standard materials and features (holes, threads, key ways, etc.) are used wherever possible. The checker's responsibility is to check the entire drawing and sign it before the drawing is finally approved (usually by a chief engineer) and released.

Conscientious drafters check their drawings before submitting them to the checkers to ensure accuracy and to save the checker's time. A partial checklist for drafters is as follows:

- Are all dimensions included?
- Are there unnecessary dimensions?
- Can the part be manufactured in the most economical way?
- Are dimensions and instructions clear and understandable?
- Will the part assemble with mating parts?
- Have all limits, tolerances, and allowances been properly analyzed for all moving parts?
- Have undesirable accumulations of tolerances been analyzed?

- Are all notes added?
- Are finish-texture symbols added?
- Are the material and treatment for each part specified?
- What is the drawing's general appearance? (legibility, neatness, arrangement of views)
- Does it follow all drawing and company standards?
- Is the title block complete? (Title, part number, drafter's name, etc.)

Numbering System

The numbering system for identifying and recording technical drawings varies from company to company. Although no standard system is used by all companies, most companies assign a sequential number for each drawing as it is finished. Some also add a prefix letter to indicate the drawing size. A drawing on a C-size sheet with an assigned number of 114937 would be indicated on the drawing by C-114937. The drawing number could be done in freehand, 0.312 (8) high, or mechanically lettered, 350 (9) high.

Parts List

A *parts list* is actually a bill of material that itemizes the parts needed to assemble one complete full assembly of a product. The parts list is usually on A-size pages separate from the assembly drawing and usually has the same number as the corresponding assembly drawing. For example, an assembly drawing with a drawing number D-114937 (D-size sheet) would have a corresponding parts list number A-114937 (A-size sheet).

The parts list contains the assembly drawing number, all subassembly drawing numbers, all detail drawing numbers, all purchased parts, material of each part, and the quantity of individual parts needed for one complete assembly.

A drafter usually prepares the parts list while following company guidelines. Some companies list the parts in order of size, some in the order of importance in the manufacturing process, and others list the parts in order of assembly. The assembly method is preferred due to the time savings later in the assembly area. An assembly method parts list is shown in *Figure 17-12*, for a machine vise. The parts listed in the order used for assembling the machine vise.

Note that the purchased parts do not have a drawing number; therefore, they are listed under the column titled "drawing number" as PURCH., indicating that these are standard parts to be purchased, not made, by the company. Additional typical purchases for most companies are motors, bearings, switches, gears, chains, belts, and controls. Complete specifications must be listed for purchased parts so that there is no question as to what must be purchased. The purchasing agent for the company may not have a technical background, and therefore, will rely on the parts list specifications for purchasing "buy-outs." In this example, items numbered 7, 8, 17, 23, and 24 are purchased parts.

MASTER PARTS LIST				
NO	DRAWING NO.	DESCRIPTION	MATERIAL	QUAN
1	D77942	VISE ASSEMBLY - MACHINE	AS NOTED	1
2				
3	C77947	BASE-VISE	AS NOTED	1
4	B77952	BASE-LOWER	C.I.	1
5	A77951	BASE-UPPER	C.I.	1
6	A77946	SPACER-BASE	STEEL	1
7	PURCH.	BOLT 1/2-13 UNC -2.0 LG.	STEEL	1
8	PURCH.	NUT 1/2-13 UNC	STEEL	1
9				
10	C77955	JAW-SLIDING	STEEL	1
11				
12	A77954	SCREW-VISE	STEEL	1
13	A77953	ROD-HANDLE	STEEL	1
14	A77956	BALL-HANDLE	STEEL	2
15				
16	A77961	PLATE-JAW	STEEL	2
17	PURCH.	SCREW 1/4-20 UNC-1.0LG.	STEEL	4
18				
19	A77962	COLLAR	STEEL	1
20				
21	A77841	KEY-SPECIAL VISE	STEEL	2
22				
23	PURCH.	BOLT 1/2-13 UNC-4.0LG.	STEEL	4
24	PURCH.	NUT 1/2-13 UNC	STEEL	4

INDENT 1 = ASSEMBLY DRAWING
INDENT 2 = SUB-ASSEMBLY DRAWINGS
INDENT 3 = DETAIL DRAWINGS
INDENT 4 = PURCHASED PARTS

JAN ENGINEERING PETERBOROUGH, N.H. — Model No. 160 — Parts Lister NELSON — Date 6 AUG 86
Title VISE ASSEMBLY-MACHINE — PAGE 1 OF 1 PAGES — Drawing No. A1198891

Figure 17-12   Parts list

Some companies use a system of indents to list the parts, Figure 17-12. There are four indents, indicating the various kinds of drawings. The first indent indicates the main assembly drawing. The second indent indicates all subassembly drawings used to make up the assembly. The third indent indicates all detail drawings and modified purchase parts with drawing numbers to make up the assembly. The fourth indent indicates all purchased parts to make up the assembly.

Not all companies use this system, but using the indent system gives a quick indication of which parts are used to make up the various subassemblies. Note on the example list that item number 3 is indented one place, indicating a subassembly. Item numbers 4, 5, and 6 are detail drawings used to make up the subassembly item number 3. Item numbers 7 and 8 are purchased parts used to complete the subassembly item number 3.

Drawing Revisions: ECRs and ECOs

After any technical drawing has been "signed-off" (drawn, checked, and approved) and incorporated into the company's system, a change in the drawing is usually accomplished through ECRs and ECOs. Anyone having a suggestion or need to improve a part or assembly within the product, or an error to correct on the drawing, must bring it up in an engineering change request (ECR) procedure meeting, *Figure 17-13*, and request an engineering change order (ECO). No change can be

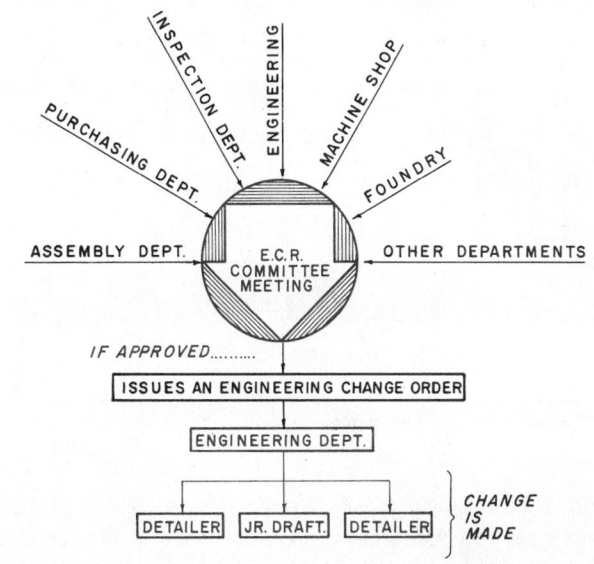

Figure 17-13   Engineering change request order procedure

made in a drawing after it has been released (signed-off) to manufacturing, even by the drafter or designer who drew it. All departments concerned must agree to the change, Figure 17-13. The ECR and ECO explain what change is to be made, what it was before, why the change is to be made, whether it affects other parts, and who suggested the change. When the engineering change request has been approved, the committee issues the engineering change order. The ECR and ECO list the departments in the company that will be directly affected by the change.

Beginning drafters, detailers, designers, or newly-hired engineering graduates may "get their feet wet" in the company by being assigned the task of incorporating ECOs in the company's technical drawings. These assignments are made primarily to introduce the employees, new to the company, to technical documents that flow through the company and how the company functions. Introducing new employees to the flow of technical documents illustrates an admonition of a commencement speaker who said to the graduates, "Get dirt under your fingernails so you'll know what's going on!"

Minor changes on original drawings may be just erasures or additions made directly on the original, whether on paper or in a computer. If a dimension is not noticeably affected by the change, the dimension is simply corrected and underlined with a heavy black line. This indicates that the dimension is correct, but slightly out of scale, *Figure 17-14*.

If the change is extensive, the drawing must be redrawn. The original drawing is not destroyed: it is stamped OBSOLETE and includes the note "SUPERSEDED BY (the number of the corrected drawing)." The new drawing must have a note referring back to the superseded drawing. The change on the drawing must be clearly marked where the change was made, and clearly describe what

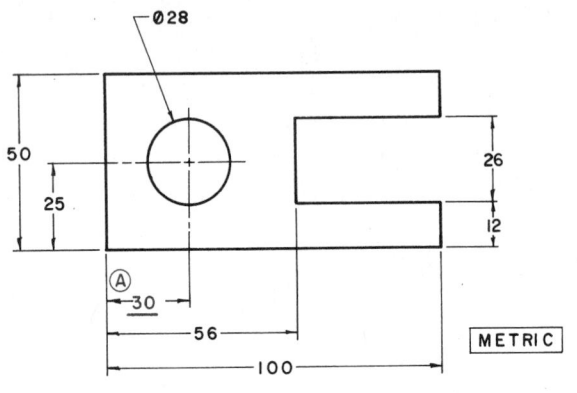

Figure 17-14   Balloon change call-off

was changed. Each company has its own standards to indicate this information. The most common method is to place a letter or number in a small balloon near a changed dimension or dimensions, or near the change made to the drawing. The letter or number must be listed in the revision block, *Figure 17-15*. The revision block, located either to the left of the title block or on the top right corner of the drawing sheet, shows the letter or number of the change, what was changed, the date of the change, and who made it. If the change is extensive, only the ECO number is listed. Anyone needing to know why the change was made can refer back to the ECO or the ECR.

Some companies add the last change letter to the drawing number, which, in effect, actually changes the part number. For example, part number C–65498 would become C–65498–A for the first change. Then, if changed again at a later date, it would become C–65498–B, and so on.

In certain cases, appropriate administrative changes may be made without an engineering change order. Such simple changes as misspelled words or incorrect references are non-engineering changes and probably would not affect other departments.

The next section in the system company's structure, application engineering, primarily processes existing designs.

Application Engineering

Preliminary activity in the application engineering section proceeds to the cost studies step in the design procedure, Figure 17-2, in a manner similar to the design

section. The procedure acts as a guide to an application engineering team of engineers, designers, and drafters who consider designs of conveyors, steam turbines, and electrical generators the company has produced in the past. Also, the team considers concepts that already exist in industry. The team's primary task is to modify and incorporate these existing concepts to meet the design requirements for the waste-burning system. For example, a scale is modified to monitor the amount of waste being delivered. A platform over the scale must be modified to accommodate large trucks. After being weighed, the trucks would dump the waste into a chute going to a flight conveyor, a standard product of the company, which would carry the waste to a hopper where the waste would be stored until conveyed to a shredder. From the shredder the waste would be conveyed to the furnace on the new drag conveyor created in the design section, Figures 17-4 and 17-5. Drawings of a steam turbine would be reviewed in an ECR meeting where ECOs would be written to modify the existing specifications and drawings of the turbine. Drawings of an electrical generator system also would be modified from ECOs. Drafters and detailers would process the ECOs.

Several new designs are developed in application engineering, buy they are minor and are based on the designer's previous experience. For example, sketches and a layout drawing to determine overall dimensions of the chute previously mentioned are shown in *Figure 17-16*.

Finally, specifications and appropriate layout drawings are forwarded to the quotation department for their use in the cost study.

Quotation Department

During the design and development of the waste-burning furnace and during the processing of the conveyors, steam turbine, and electrical generator, the quotation department closely followed the design section's progress and the application engineering's activities. Thus an estimate of the selling-price for the entire system was nearly completed for the RFP by the time design and application finished their respective tasks. The selling price is the major dollar figure submitted to the customer, but subtotals for component systems, such as the furnace, the steam turbine and the generator may in included. The selling price is prepared by the quotation department in steps illustrated in *Figure 17-17*. Basic costs include labor and engineering materials. Basic labor (sometimes labeled direct labor) includes estimates of hours for engineering, drafting, manufacturing, assembling, and installing. Engineering materials include steel plates, structural steel shapes, fasteners, welding rods, switches, motors, controls, and so forth. Manufacturing expenses, such as depreciation of machines and maintenance, add to the basic costs to make up manufacturing costs. Sales salaries and expenses plus general overhead costs (electricity, water, office supplies, etc.) are added to manufacturing costs to arrive at the total cost to pro-

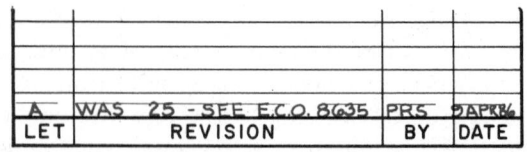

Figure 17-15   Revision block *(Courtesy Bishop Graphics Accupress)*

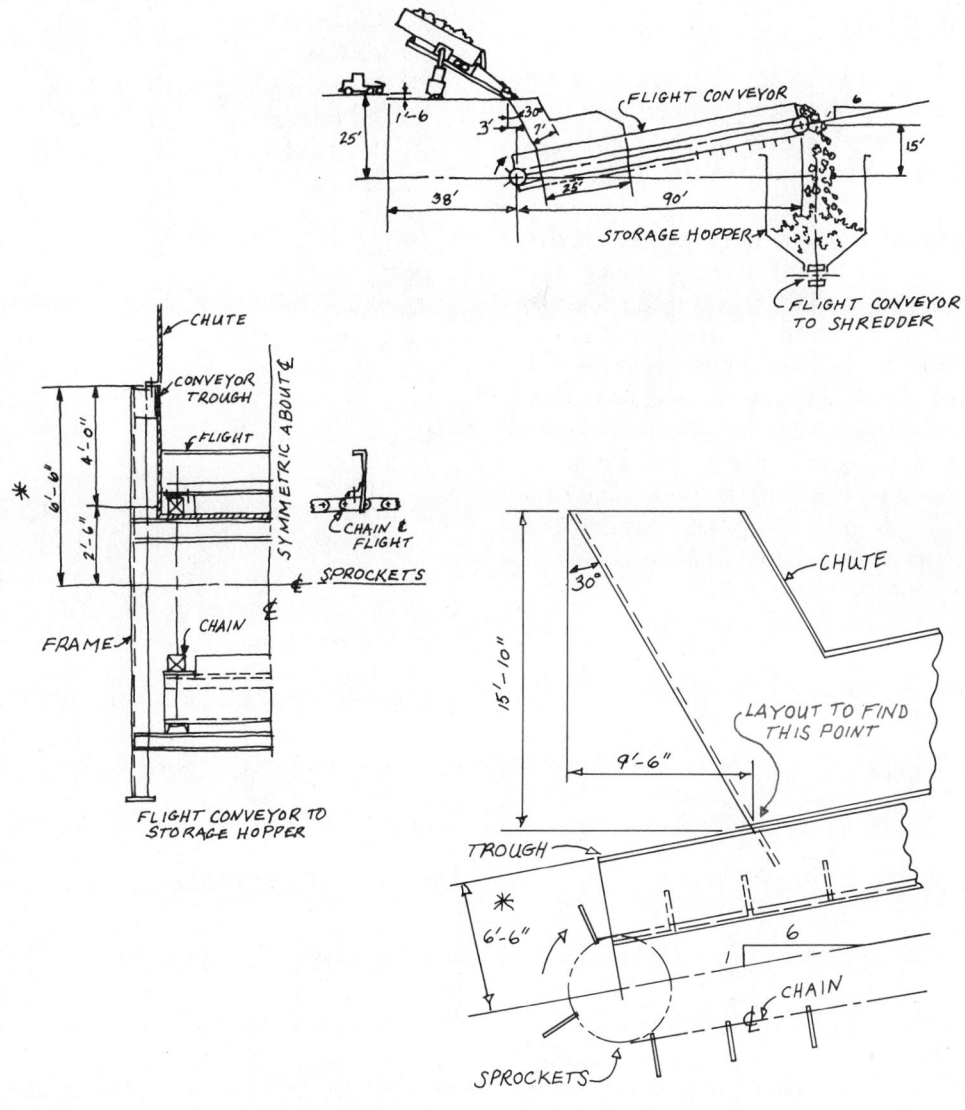

Figure 17-16   Sketches and layout drawing of chute

duce the energy system. Profit added to total cost establishes the selling price.

The selling price, specifications, delivery schedules, and the "boiler plate" make up the proposal. A typical boiler plate would be stored in a computer ready for use in any new proposal. It includes introductory comments and sections intended to "sell" the company to the prospective customer, such as descriptions of past systems designed and installed and summaries of the expertise of the personnel who would handle the job if the proposal were accepted. Another section of the boiler plate protects the company with legal statements referring to late delivery due to strikes, acts of God (earthquakes, tornadoes, etc.), unavailability of materials, and so forth.

A cover letter with a bound copy of the proposal would be mailed to the prospective customer before the due date, or delivered in person if an oral presentation was arranged. Oral presentations give the recipient an opportunity to ask questions, suggest changes, and correct any misunderstandings. When the customer accepts the proposal and signs it, it becomes a binding legal contract on both parties. It also becomes a document that immediately generates continuing work for the design and application sections of the systems company.

This continuing work for the design section was discussed earlier. Working drawings were prepared, checked, signed, and sent to manufacturing. In a similar manner application engineering would produce working drawings, have them checked, signed, and sent on to manufacturing.

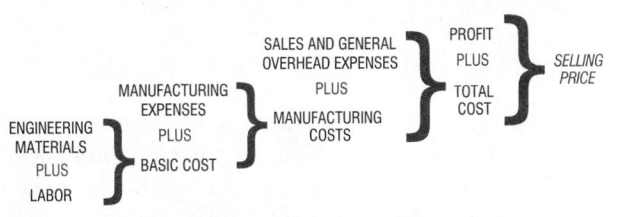

Figure 17-17   Selling price components

Personal Technical File

During the various design and development activities each engineer, designer, detailer, checker, technical writer, and illustrator should develop and maintain a technical file to enhance their involvement in the company. The objective is to minimize the time spent locating technical information pertinent to their respective jobs. Usually a company manufactures products and performs engineering in specific areas of technology, so files can be concentrated in these areas for efficient retrieval of information. An efficient technical filing system uses an alphabetical-numerical scheme. For example, the first part of filing scheme for a drafter-designer in a company specializing in conveyors would be an alphabetical listing of technical topics with numerical references. This file should be kept in a three-ringed binder or a computer so that additional information can be added easily. Example entries would be:

Section A			
Aluminum Products	112	Channel Sections	134
ANSI Standards	410	Conveyors	500
Section B		Codes and Standards	410
Bar Stock	132	Controls	210
Beams	133	*Section D*	
Belt Conveyors	510	Drives	600
Belt Materials	122	*Section E*	
Bucket Elevators	530	Electrical Components	200
Section C		Controls	210
Castings	140	Switches	230
Chain Conveyors	520	And so on…	

The numerical part of the file, kept in a file cabinet, would contain tab-numbered dividers, numbered folders, and numbered materials in the folders. Example tabs, folders and materials would be numbered as follows. (The letters in parentheses are for this discussion purposes only: T = Tab number for a divider, F = Folder number, and M = Material labeled with a number.)

100	Engineering Materials	(T)
110	Metals	(F)
111	Ferrous Based	(M)
112	Aluminum	(M)
120	Non-metals	(F)
121	Ceramics	(M)
122	Rubber (belts)	(M)
130	Structural Shapes	(F)
131	Angles	(M)
132	Bar Stock	(M)
133	Beams	(M)
134	Channels	(M)
200	Electrical Components	(T)
210	Controls	(F)
220	Motors	(F)
221	AC Motors	(M)
222	DC Motors	(M)
230	Switches	(F)

300	Materials Handling	(T)
310	Solid Waste	(F)
400	Codes and Standards	(T)
410	ANCI Standards	(F)
411	ANSI Y14.5M–1982	(M)
500	Conveyors	(T)
510	Belt	(F)
520	Chain	(F)
530	Bucket	(F)
600	Drives	(T)
610	Gear Box	(F)
620	Roller Chain and Sprockets	(F)
630	V-Belt and Pulleys	(F)
And so on…		

Start a technical file now, as a student, and expand it later, to suit your technical jobs in industry.

Occasionally, a drafter, designer, or engineer may be required by an employer to keep a bound-page notebook for patent development. The pages are bound together, numbered, and each one signed by the individual inventor. Also, each page needs to be signed by at least two witnesses who understand the area of technology of the potential patent. Furthermore, the individual inventor may be required to sign an invention agreement.

Invention Agreement

Anyone who is creative and working for a company, such as a drafter, designer, or engineer, usually must sign an *invention agreement* form giving the company the right to any new invention designed while working for the company. The form specifies that the employee will not reveal any of the company's discoveries or projects. The invention agreement is binding on the employee from six months to two years after the employee leaves the company, so that the employee will not invent something, quit, and patent the invention themselves.

A company is not obligated to, so it usually does not give extra pay for an invention developed by an employee. This is what the employee is paid to do. However, companies recognize talent and will reward that talent with a promotion, stocks, bonds, or possibly a raise. Signing an employee/employer invention agreement is a normal request in any company. Of course, prospective employees should read the agreement carefully and understand it before signing. An applicant who refuses to sign the agreement may be asked to look elsewhere for employment.

Patent Drawings

All patent applications for a new idea or an invention must include a *patent drawing* to illustrate and explain its function. Patent drawings must be mechanically correct, and must adhere to strict patent drawing regulations, *Figure 17-18*. For example, patent drawings are

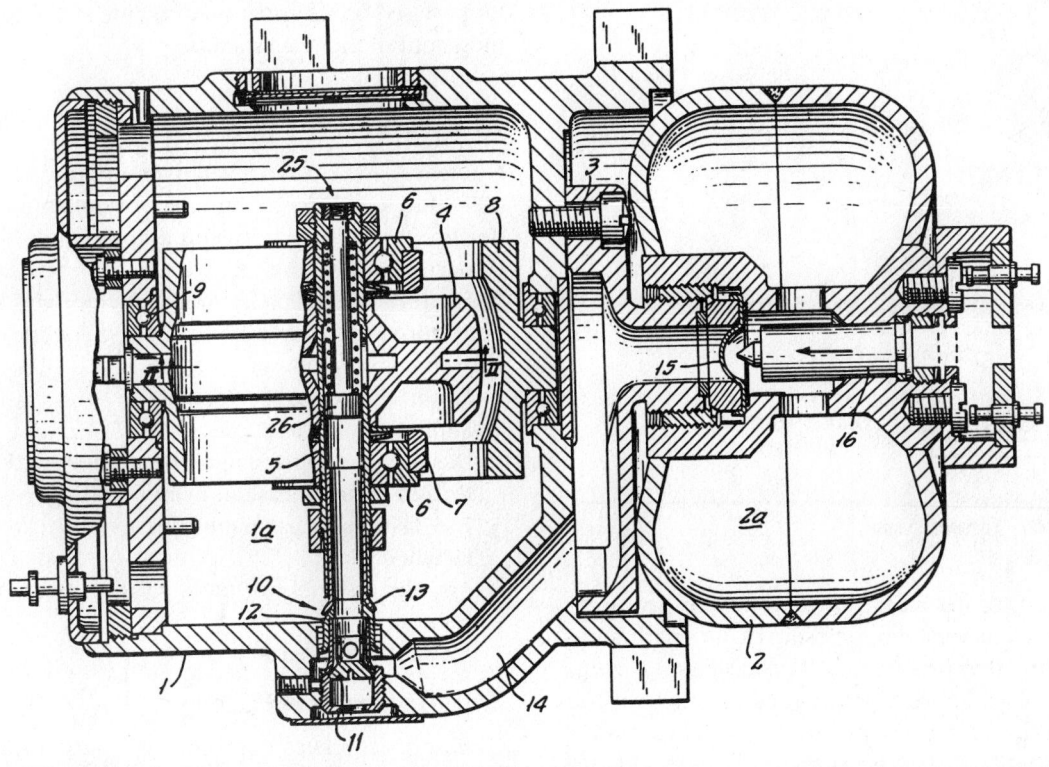

Figure 17-18 Patent drawing *(Courtesy Timex Corp.)*

more pictorial and explanatory than regular detail or assembly drawings. Center lines, dimensions, and notes are omitted. All features and parts are identified by numbers that refer to written explanations and descriptions. Line shading is used to improve readability.

All patent drawings must be made so that they reproduce clearly, which essentially means that the drawings be done in ink. Current requirements specify that India ink or its equivalent be used, and that the drawings be done on strong, white paper, 8.5 x 13 (21.6 x 33.1 cm) or 8.5 x 14 (21.6 x 35.6 cm). A precise "sight" area (20.3 x 29.8 cm) must be used for the drawing itself, on the strong white paper. (For further requirements, refer to *37 CFR* Ch 1 7-1-91) edition.) The publication, *A Guide for Patent Drawings*, may be obtained from the Superintendent of Documents, U. S. Government Printing Office, Washington, DC 20402.

These comments on patent drawings end the discussion of the systems company; the mass production company is next.

Mass Production Company

A *mass production company* that produces large quantities of a product would have an engineering departmental structure similar to the systems company discussed earlier, Figure 17-1; however, the mass production company would have a somewhat different approach to producing a product, ***Figure 17-19***. The company's design, application-engineering, quotation, and manufacturing departments would follow this

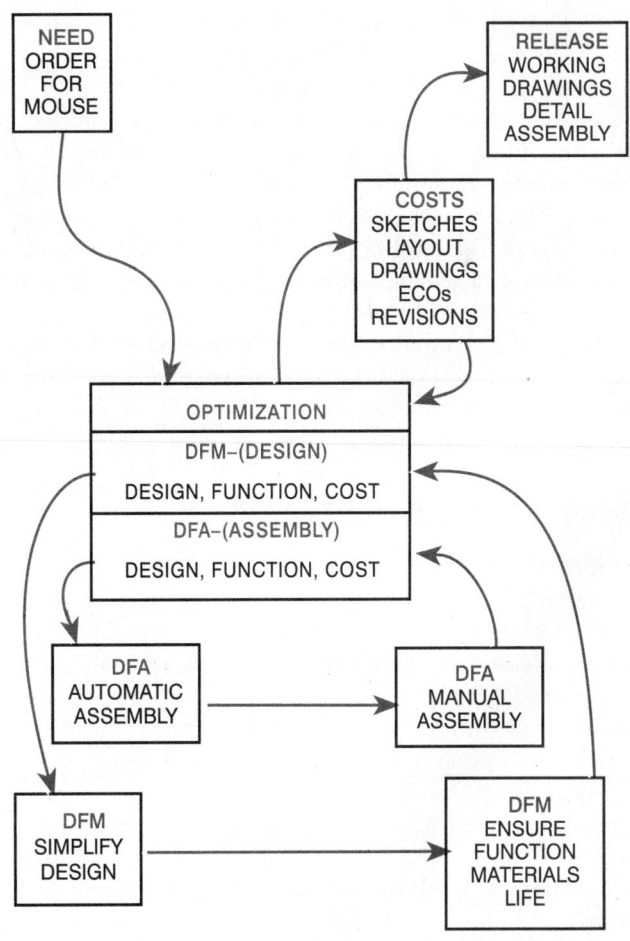

Figure 17-19 DFM-DFA procedures

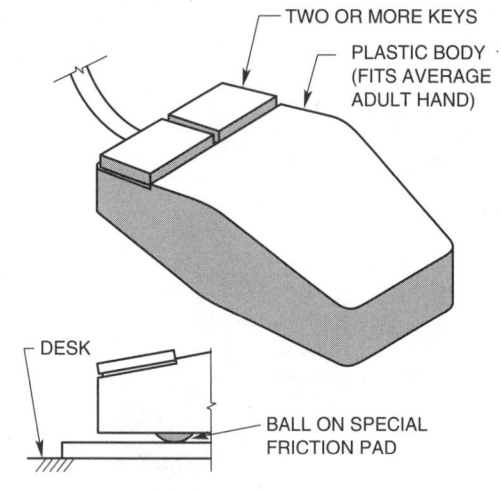

Figure 17-20   Computer mouse

design procedure for mass producing "one product" while working together more closely than departments in the company that produced a "one-of-a-kind" system. To illustrate these different procedures, assume that the mass production company primarily produces mouse units, *Figure 17-20*, for use with a computer. Essentially, a mouse provides a quick way for computer operators to position a small cross-hair cursor or a small pointer on a video screen by moving the mouse across a special desk pad. Then once the cross-hair or pointer has been positioned, one key atop the mouse is clicked to initiate a basic computer operation.

Assume that the mass production company has received an order for two million mouse units and that each unit requires more functions than their standard mouse. The major challenge to the company is to design and mass produce these units as efficiently as possible and still have a reliable product. The company's design efforts are focused on two important areas of mass production. The first, DFM (design for manufacturability) emphasizes design of the parts that make up a product. The second, DFA (design for assembly) concentrates on assembly of the parts that make up a product.

Design for Manufacturability (DFM)

The *DFM* procedure is an iterative loop and acts as a guide for a design team, Figure 17-19. The primary goals of the DFM team are to develop the new mouse design, for more efficient manufacturing, at the least cost, within a reasonable time, and ensure its function for a reasonable life. The team made up of engineers, designers, drafters, and manufacturing personnel pursue these goals as illustrated by the iterative path shown. Start with optimizing (design, function, and cost), then go on to simplifying (DFM Guidelines), ensuring (function, life, materials), and back to optimizing. One factor not shown in Figure 17-19 is the time constraint for making decisions. This can affect the success of the DFM process. Decisions must be made at each step in the

loop, based on information available at the time and the judgment of the team members.

DFM Guidelines

Guidelines for DFM evolved over a period of time essentially due to an increasing need for collaboration on the design of products between a company's engineering department and manufacturing. To be competitive, companies needed to be more efficient and to produce better products. A number of DFM guidelines from DFM literature are as follows:

Minimize the number of parts.

Minimize part variations.

Design parts to have more than one function or use.

Design parts for ease of manufacturing.

Use fasteners that handle easily and can be readily installed.

Arrange assembly of parts so they assemble in layer fashion.

Avoid flexible parts.

Minimize Number of Parts

One approach is to combine two parts into one, thus eliminating fasteners. An obvious example is shown in *Figure 17-21*.

Minimize Part Variations

If fasteners are required, use all the same size to save storage space and sorting-and-assembly costs. Use standard "off-the-shelf" parts where possible.

Design Parts to be Multi-functional

For example, a leaf spring designed to exert a small force could also conduct an electrical current. A structural part could have ridges for guiding a second part into position. A spacer post could be used for a fasten-

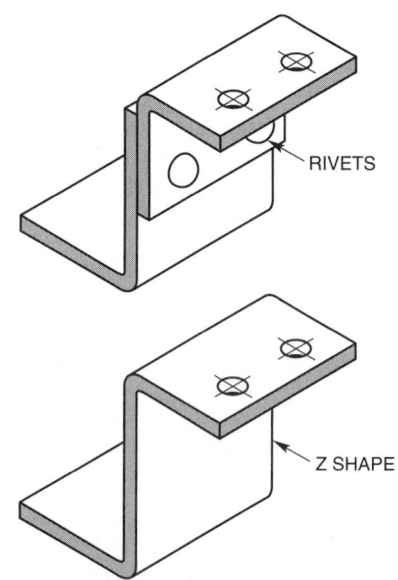

Figure 17-21   Combining two parts into one

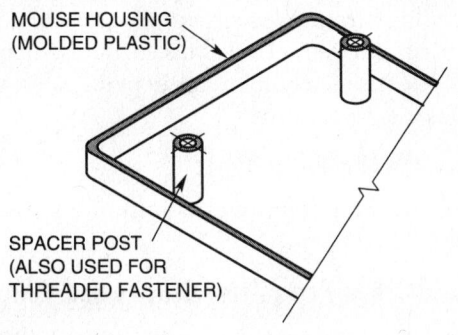

Figure 17-22   Post used for spacer and fastener

er, *Figure 17-22*. One shaft of a specific length and diameter could be used in a variety of products. One-size parts such as bearings, springs, or gears could be used in a variety of products in a company.

Design Parts for Ease of Manufacturing

Precision cast gears or stamped gears need little or no machining or surface finishing. Basic materials with relatively expensive finishes, such as anodized aluminum, may ultimately save painting, polishing, and machining later in the manufacturing processes, thus keeping labor costs down.

Use Fasteners Easily Handled and Installed

Snap-in fasteners save assembly and screw-driver time, *Figure 17-23*.

Arrange Parts Assembly for Layering

Assembly of the mouse shown in *Figure 17-24* illustrates layering; that is, essentially, parts are stacked from the bottom, up.

Avoid Flexible Parts

Use circuit boards instead of flexible, difficult-to-assemble wires for electrical circuits where possible. Of course, the mouse, Figure 17-20, needs one flexible connector for easy movement on a desk pad.

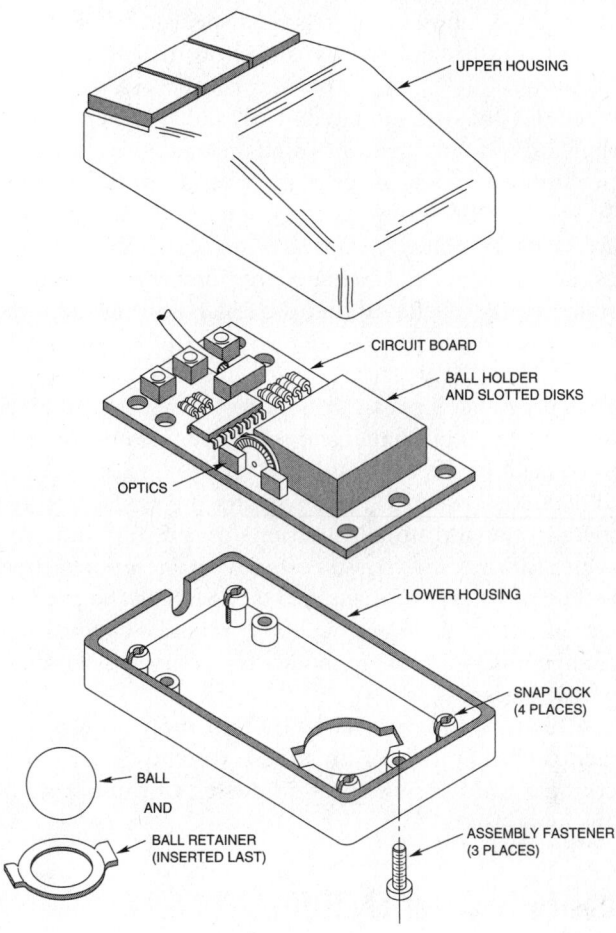

Figure 17-24   Layering for mouse assembly

The next step in the loop, Figure 17-19, is to ensure function and life, and select the best materials. Most mouse units utilize a printed circuit on a board with electronic components, to integrate the digital signals generated by light passing through rotating slotted disks inside the mouse, *Figure 17-25*. The shafts of the two slotted disks are driven by a small ball, mostly inside the

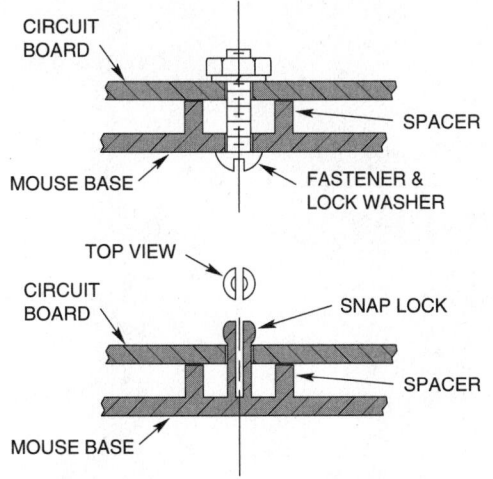

Figure 17-23   Snap-in fastener for circuit board

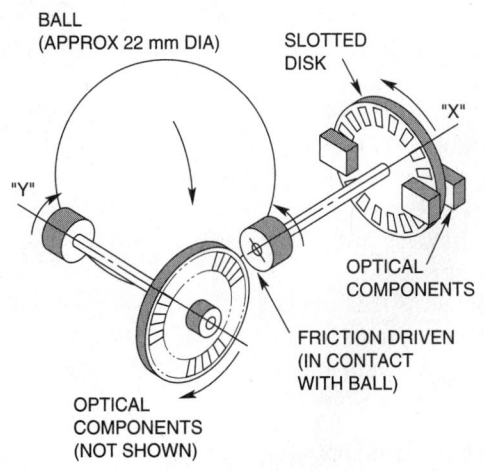

Figure 17-25   Slotted disks inside a mouse

body of the mouse, as it rolls on a special desk pad in a direction determined by the computer user, who observes a cross-hair or pointer on the video screen. Since the disks are mounted in "x" and "y" orientations, the sum of their rotations (digital signals), are used to position the cross-hair or pointer on the video screen. The body of the mouse and a number of the inside parts are made of molded plastic. Resistors, capacitors, optical components, and switches are purchased "buyouts." Stainless steel shafts support the disks; bearing supports for the shafts are part of the plastic body, thus no lubrication is required. They keys for initiating basic operations are molded plastic with narrow portions acting as hinges. The only maintenance required for the mouse is to occasionally remove the ball (approximately 22 cm in diameter), wash it in soapy water, rinse, dry, and re-install. The additional functions ordered by the customer for the new mouse design were incorporated by adding one key, one switch, and expanding the printed circuit. These additions, of course, required engineering change orders (ECOs) to modify the company's existing mouse design.

After the team was satisfied that their designs for manufacturing parts of the mouse were optimum, they concentrated their efforts on design of the parts for assembly (DFA).

Design for Assembly (DFA)

The DFA procedure is an iterative loop that occurs simultaneously with the DFM loop, Figure 17-19, but the emphasis is on assembly of parts vis a vis the design and manufacturing of parts. The team has similar goals in the DFA loop—develop the best design of parts for assembly of the mouse, at the least cost, within a reasonable time, and ensure its function. They follow the DFA loop shown: first, consider which parts should be assembled automatically, and second, consider which parts need to be assembled manually.

DFA Guidelines

A partial list of DFA guidelines from DFA literature are as follows:

Use "pick-and-place" robots for simple, routine placement of parts.
Keep the number of parts to a minimum (DFM Guideline, also).
Use gravity and vibration for moving parts into assembly areas.
Avoid parts that "tangle" (hang up on each other).
Use tapers and chamfers to guide parts into position.
Avoid time-consuming manual operations such as screwdriving, soldering, and riveting.

Use Robots for Simple Operations

Robots in chapter 20, Figures 20-66, 20-67, and 20-70 can be programmed to pick electronic components off

a conveyor or from a feeder and place them accurately into a circuit board for the mouse. A mouse may require 25 to 50 electronic components depending on the basic functions it is to perform.

Minimize the Number of Parts

Combine several components into one molded part, *Figure 17-26*.

Move and Orient Parts Using Gravity and Vibration

The sketch in *Figure 17-27* shows one method of orientating spacers, and steps to avoid tangling of lock washers, and tangling of rivets.

Use Tapers and Chamfers for Positioning

An example of tapered fasteners, and a receptacle with a tapered hole, both for ease of assembly are shown in *Figure 17-28*.

Avoid Time-Consuming Operations

The snap-locks in the body shown in Figure 17-23 allow for quick assembly and accurate positioning of the circuit board.

Finally, after the DFA team members agree on the best designs for manufacturability and the best designs for assembly for the new mouse, and the DFM team concludes their optimization efforts, costs for producing the new mouse are checked. If the cost analysis results in further modifications, sketches and layout drawings may be prepared and ECOs may be issued. To continue the design procedure, the new mouse design is formally released. Designers and drafters prepare working drawings. Eventually, all the drawings are released to manufacturing supervisors who have anticipated the drawings because they had worked closely with the design teams during the development of the new mouse design.

As the new mouse goes into production, the personnel of the mass production company realize that even a small savings in cost resulting from following DFM and DFA Guidelines is worthwhile when millions of units

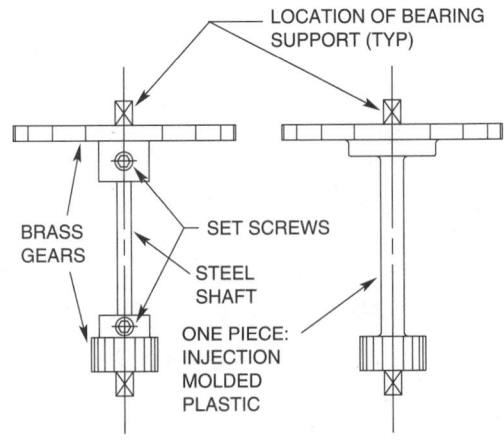

Figure 17-26   Combine several parts into one

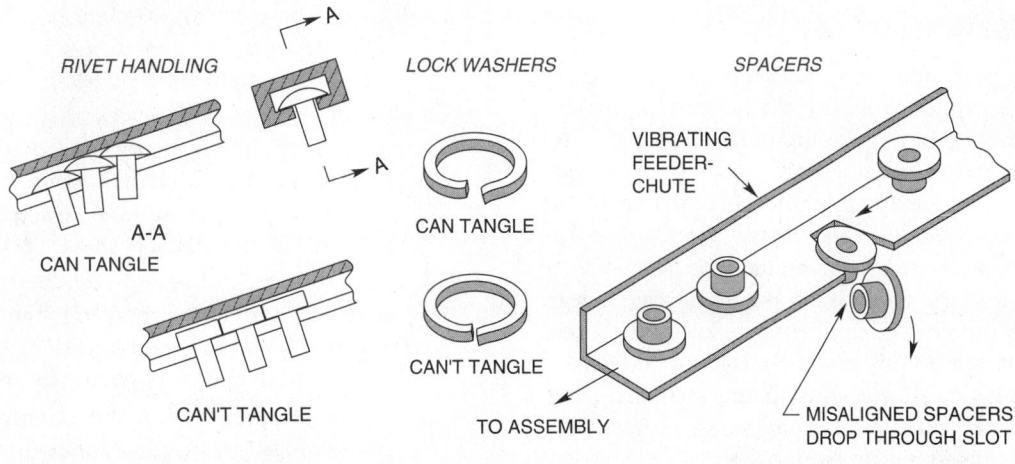

Figure 17-27 Spacers handling and anti-tangling techniques

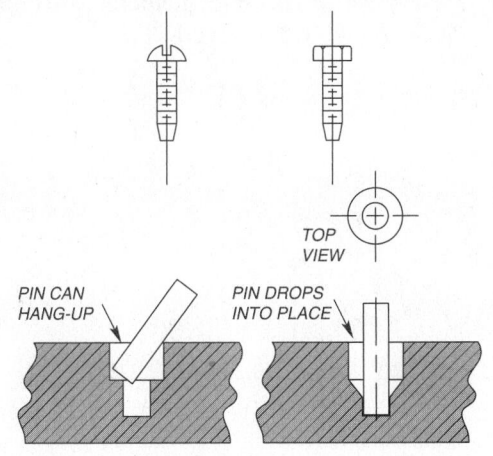

Figure 17-28 Tapered fasteners and a tapered hole for ease of assembly

are involved. Moreover, the company finds that following the guidelines, not only results in cost savings, but the design and function of products are invariably improved.

Review

1. Describe the functional differences between the design section and the application engineering section in a typical engineering department.
2. List the possible members of a design team. Write one sentence describing each member's role.
3. Give several examples of "one-of-a-kind" systems. Give several examples of items that might be produced by a mass production company.
4. What is an RFP?
5. How is a freehand sketch related to a layout drawing? Who usually draws layout drawings?
6. What is a design procedure and how is it used by a design team?
7. Describe the differences between conceptual design and scientific design.
8. What are two types of illustrations used by a systems company to help a customer visualize the overall system?
9. Describe the differences between an assembly drawing and a detail drawing. What is a subassembly drawing?
10. Why does a company use purchased parts, i.e., parts made by another company?
11. List the information usually included in a title block.
12. List the important parts of a drawing that should be checked by a drafter before submitting the drawings to a checker.
13. What information would you expect to find in a parts list?
14. Explain the differences between an ECR and ECO.
15. How is the phrase, "get dirt under your fingernails," used in this chapter?
16. What are the steps in preparing a selling price?
17. What is the "boiler plate" in a quotation?
18. Describe the alphabetical and numerical parts of the filing system discussed in this chapter.
19. What is an invention agreement?
20. Define the acronyms DFM and DFA and describe each one briefly.
21. Compare the design procedure and the goals for the systems company to the design procedures for the mass production company. What are the differences, similarities?

Chapter Seventeen Problems

The following problems are intended to give the beginning drafter knowledge of the many drafting procedures, and practice in developing detail drawings, subassembly drawings, and assembly drawings. Students should practice calculating tolerances for various fits of mating parts, and filling out parts lists.

The steps to follow in developing these problems are:

Step 1 Using a sharp 4-H lead, lay out a full scale design layout drawing of the problem, using the basic overall dimensions provided. Lay out and design any missing dimensions, using standard sizes and materials. (Use as many views as necessary to fully illustrate the assembly.)

Step 2 Calculate all required fits and tolerances as necessary on a separate sheet of paper.

Step 3 Center a fully dimensioned detail drawing of each part using the latest drafting standards. Apply tolerances and limits to mating parts.

Assign the drawing number per the numbering provided on the design layout drawing, and include the material to be used.

Step 4 Develop *subassembly drawings* wherever two or more parts are assembled, *before* the final assembly, as needed. List each part and part number, using standard drafting practice. (Material is to be listed "AS NOTED.")

Step 5 Develop a final assembly drawing and list each part and part number, using standard drafting practice. (Material is to be listed "AS NOTED.") Use standard fasteners wherever possible.

Step 6 Develop a parts list of the assembly. List parts per standard drafting practices, in the recommended order of assembly.

Step 7 Recheck all drawings and related calculations for accuracy.

Step 8 Make a list of recommendations as to how the design could be improved.

Problems 17-1 through 17-8

Follow the eight listed steps to develop Problems 17-1 through 17-8.

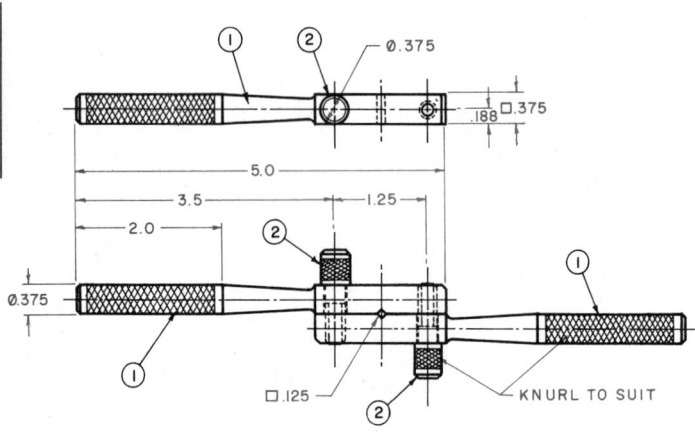

Problem 17-1

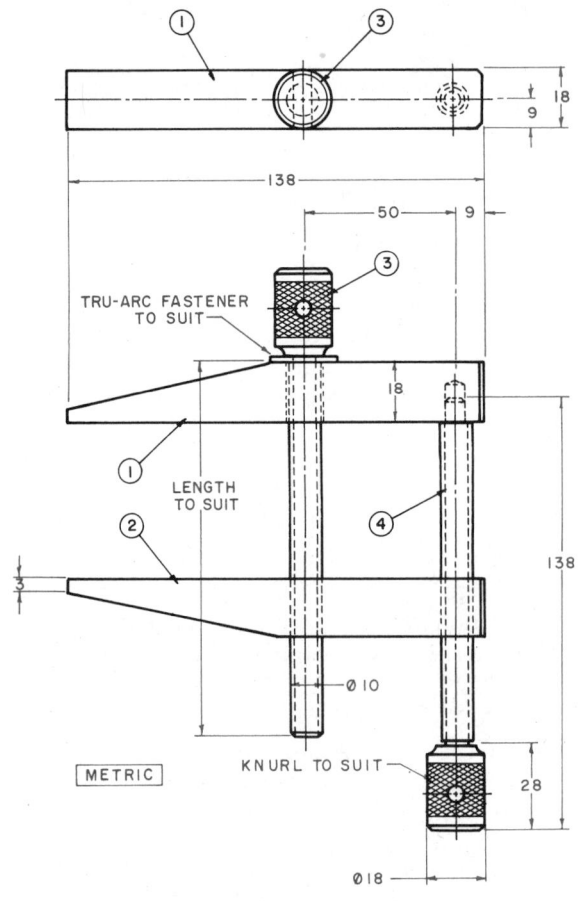

Problem 17-2

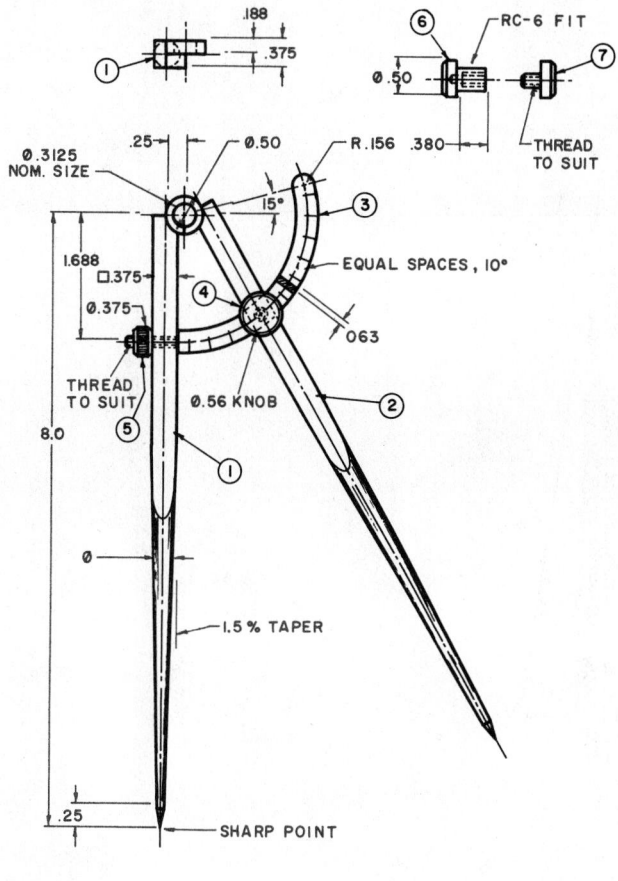

Problem 17-3

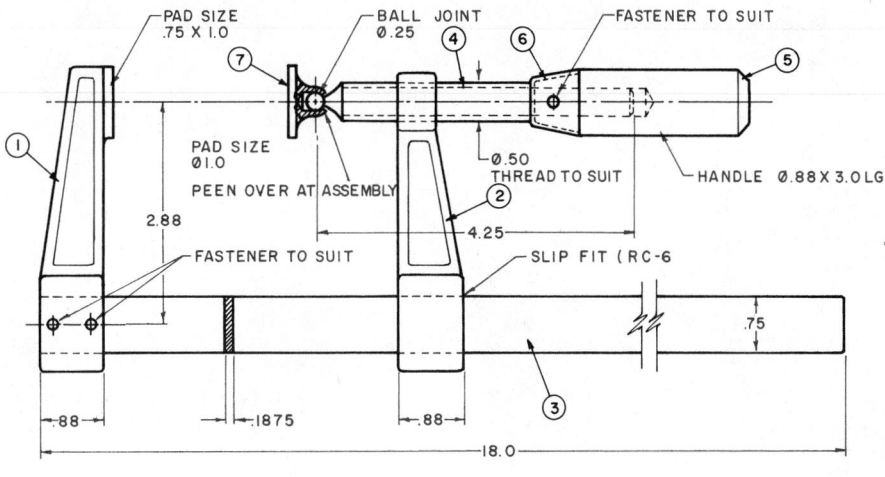

Problem 17-4

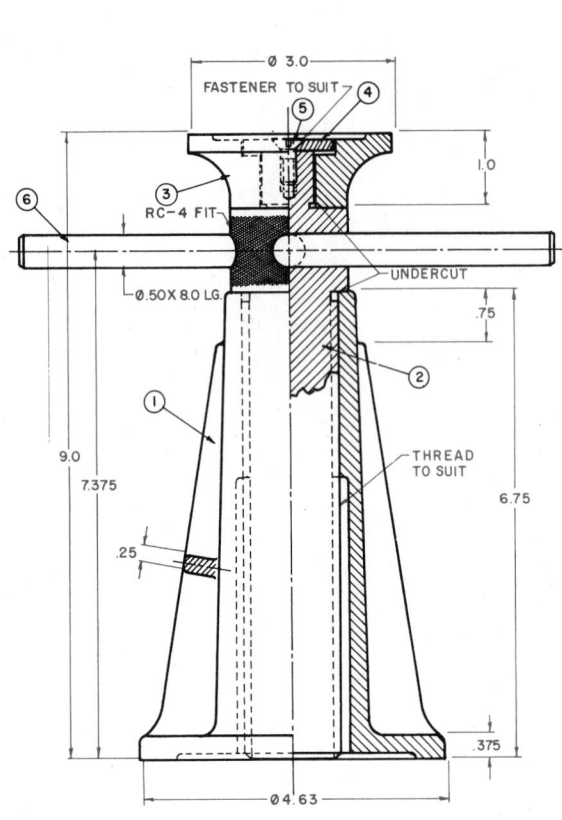

Problem 17-5

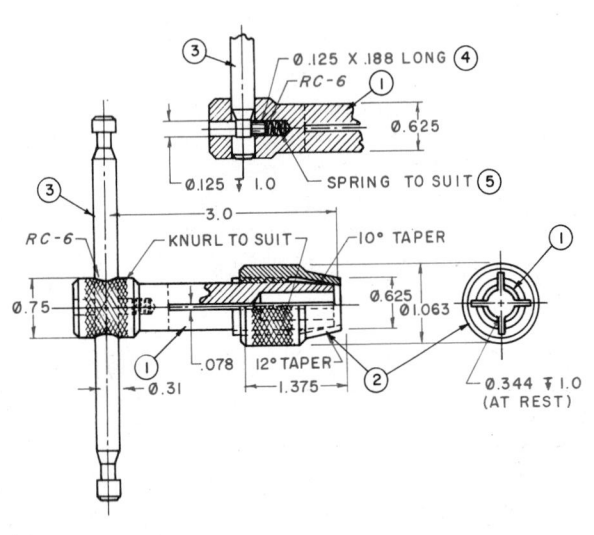

Problem 17-6

Problem 17-7

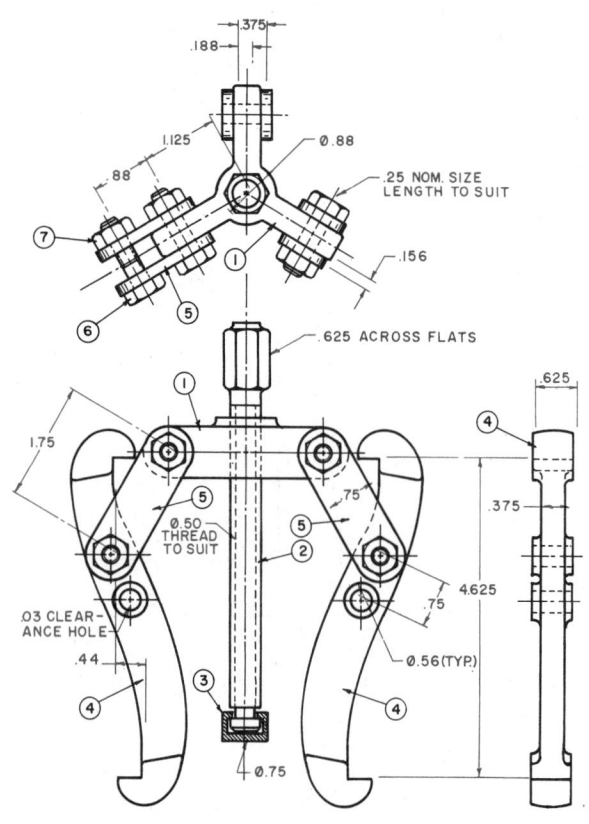

Problem 17-8

Problem 17-9

Follow the eight listed steps to develop Problem 17-9. Redesign the layout to include standard ball bearings in place of the bushings, part No. 4.

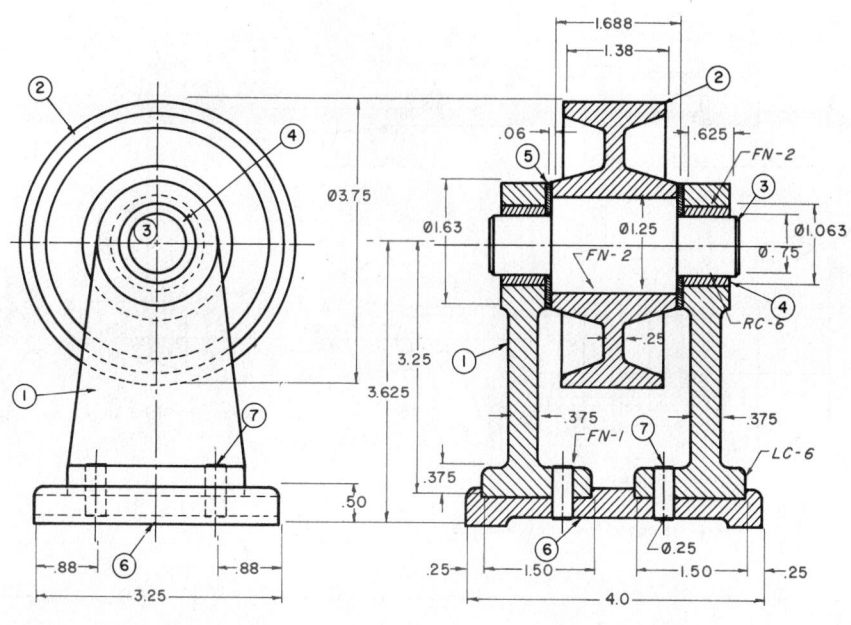

Problem 17-9

Problems 17-10 through 17-12

Follow the eight listed steps to develop Problems 17-10 through 17-12.

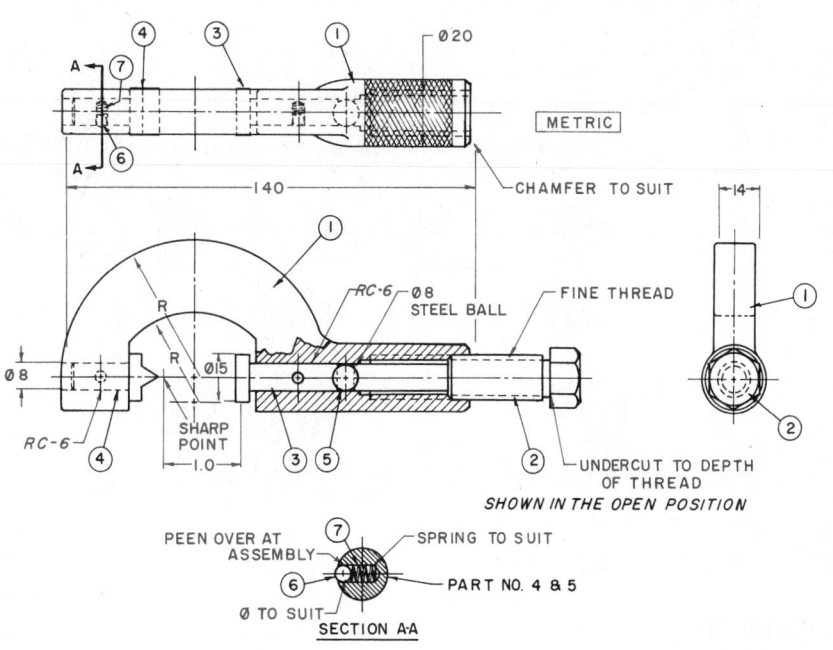

Problem 17-10

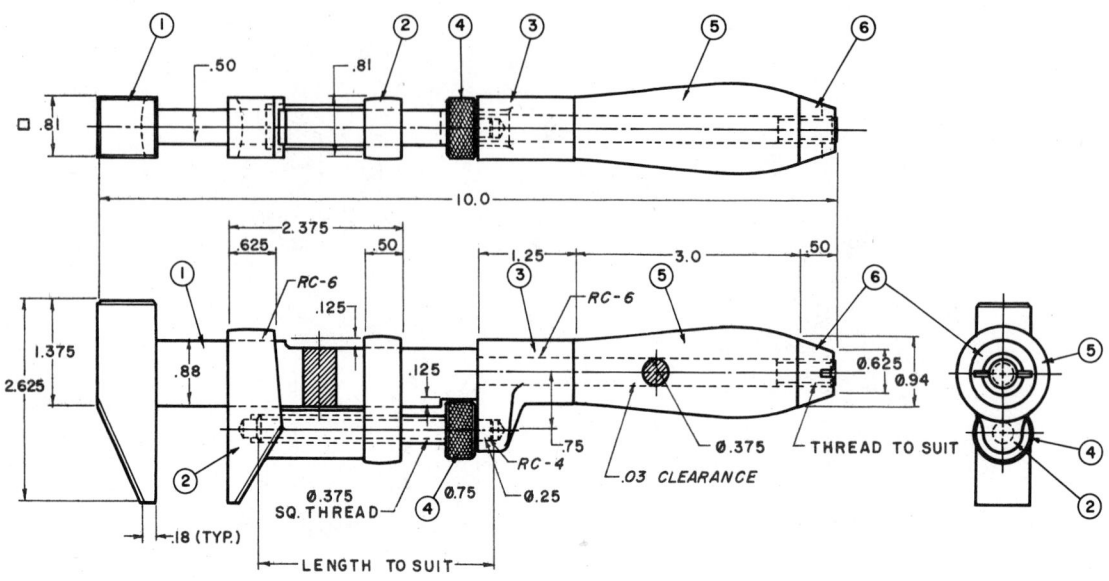

Problem 17-11

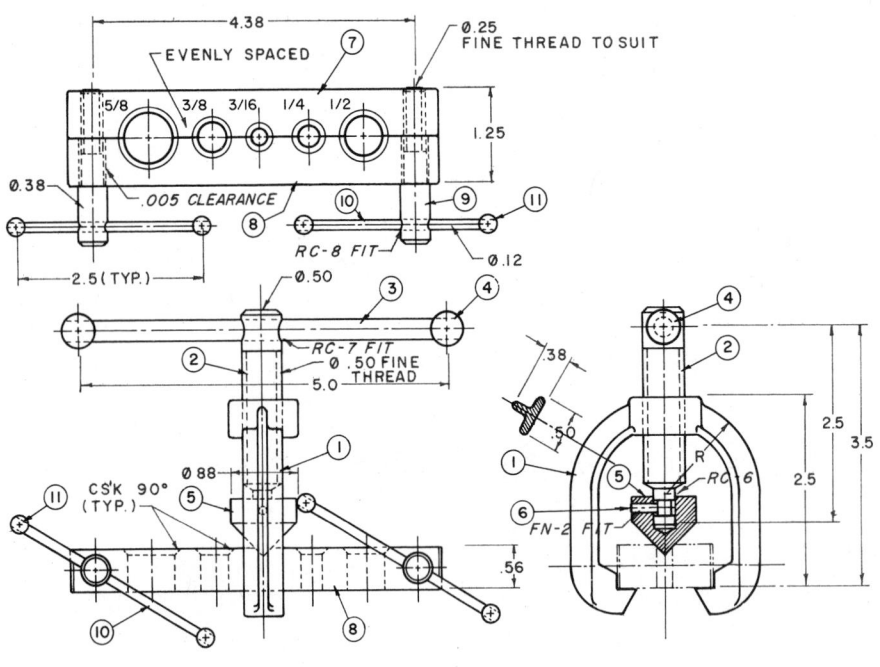

Problem 17-12

Problem 17-13

Follow the eight listed steps to develop Problem 17-13.
Redesign the layout to secure the round head locking
screw, part No. 10, from turning.

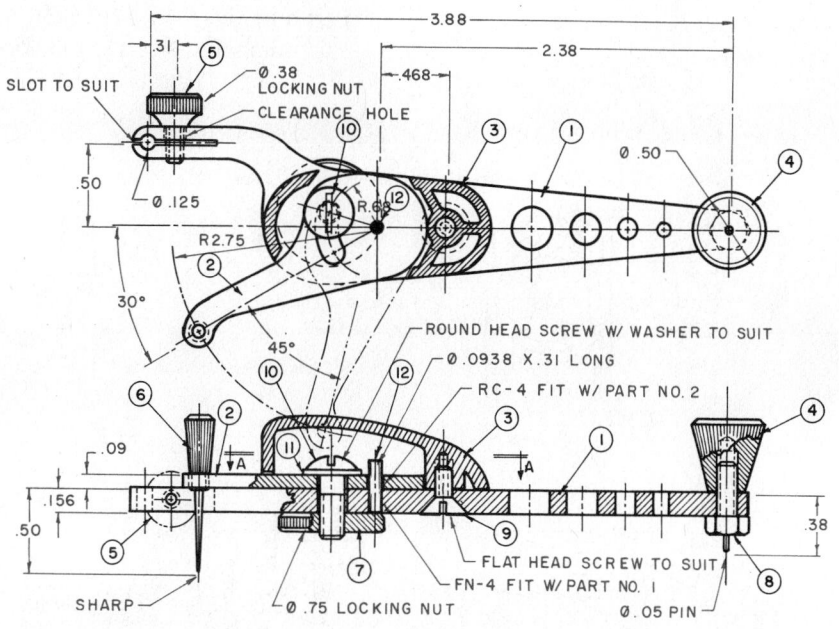

Problem 17-13

Problem 17-14

Follow the eight listed steps to develop Problem 17-14.

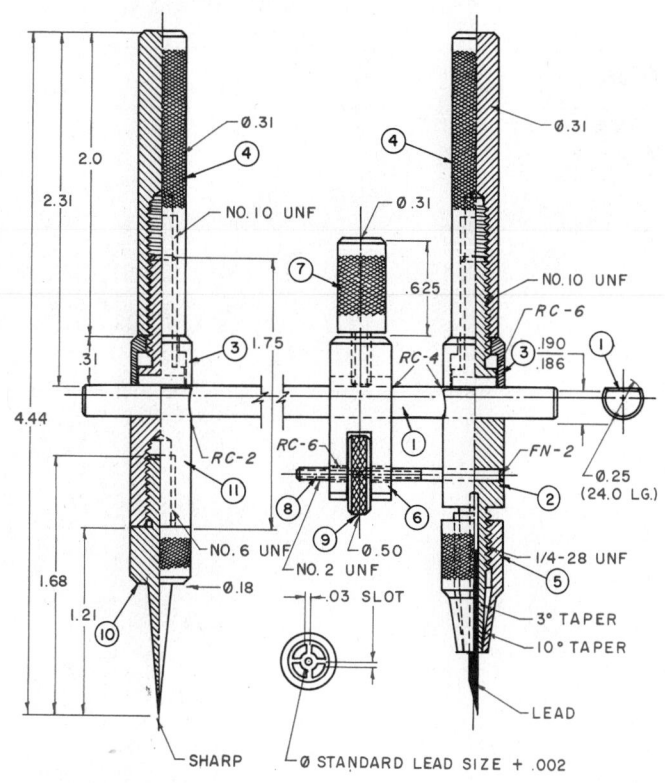

Problem 17-14

Problem 17-15

Using detail drawings 17-15A through 17-15O, lay out a three-view assembly drawing. Call off parts per standard drafting practices. Calculate the following fits using each of the 15 detail drawings. Show all calculations.

Part numbers 1-12, FN-1/1.062
Part numbers 1-2, FN-2/.688

Part numbers 1-11, LC-4/1.50
Part numbers 3-4, RC-5/.562
Part numbers 4-5, RC-5/.375
Part numbers 4-8, LC-2/.125
Part numbers 9-14, FN-2/.250
Part numbers 13-14-15, RC-6/.938

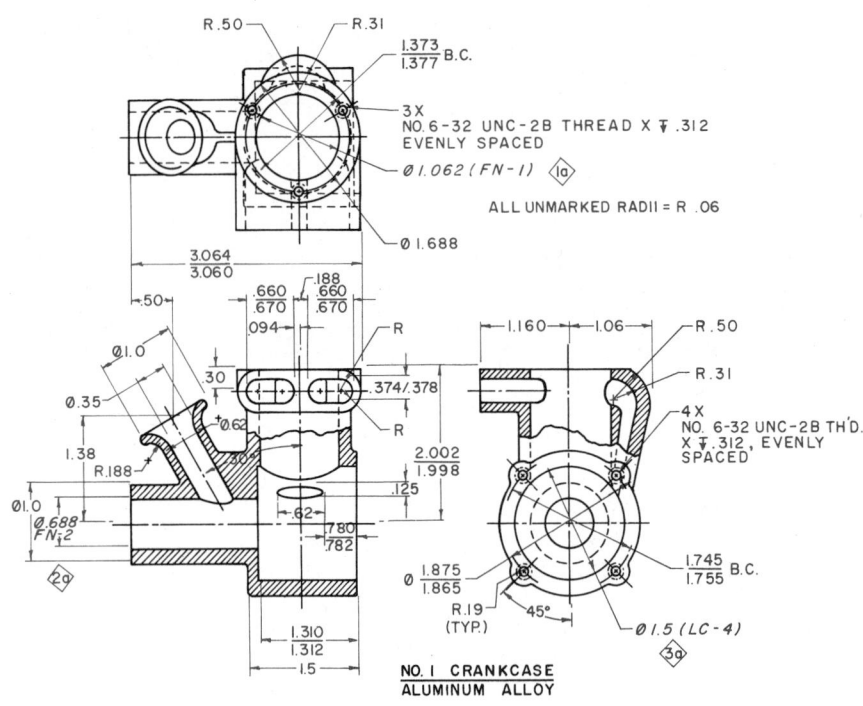

NO. I CRANKCASE
ALUMINUM ALLOY

Problem 17-15A

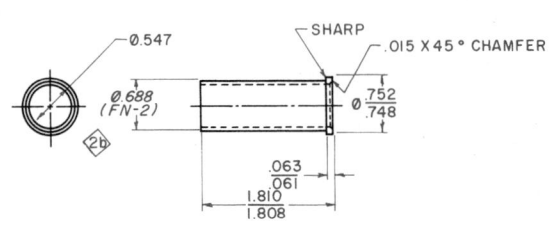

NO.2 BUSHING CRANKCASE
LEADED BRONZE

Problem 17-15B

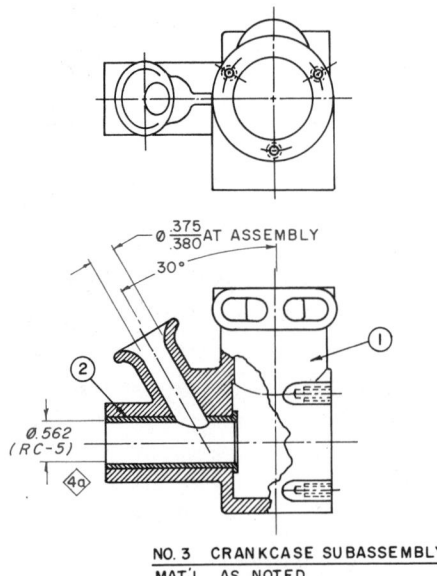

NO. 3 CRANKCASE SUBASSEMBLY
MAT'L. AS NOTED

Problem 17-15C

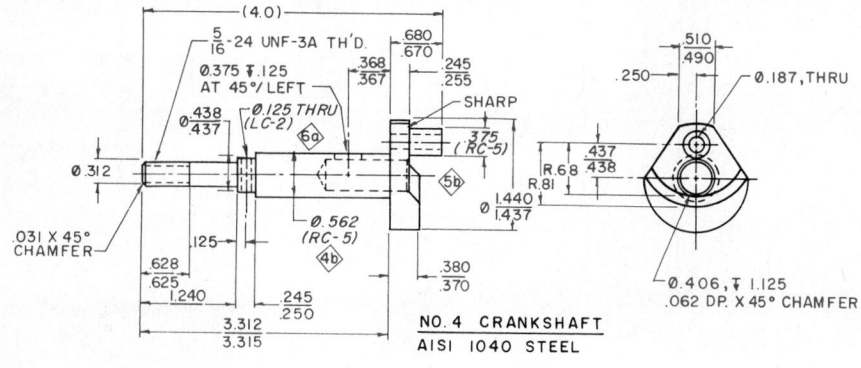

NO. 4 CRANKSHAFT
AISI 1040 STEEL

Problem 17-15D

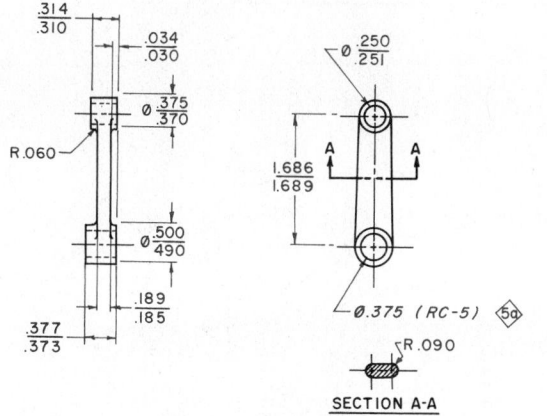

SECTION A-A

NO. 5 ROD-CONNECTING
ALUMINUM ALLOY 7075-T6

Problem 17-15E

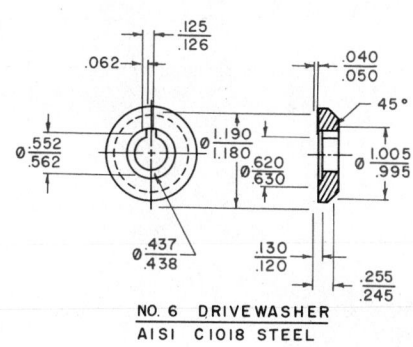

NO. 6 DRIVEWASHER
AISI C1018 STEEL

Problem 17-15F

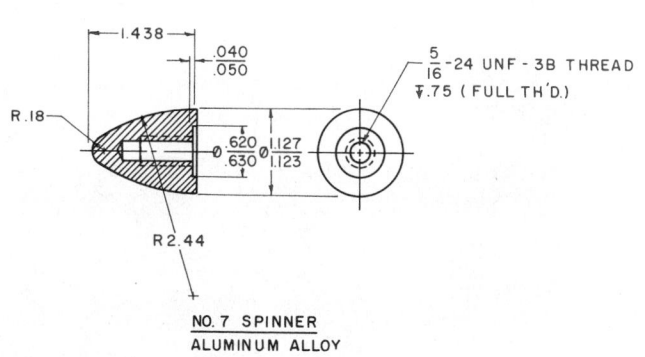

NO. 7 SPINNER
ALUMINUM ALLOY

Problem 17-15G

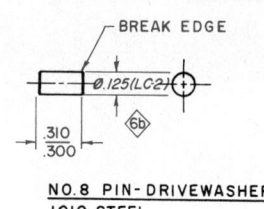

NO. 8 PIN-DRIVEWASHER
1010 STEEL

Problem 17-15H

Ø.250
(FN-2)
.877
.873

NO. 9 PIN PISTON
1010 STEEL

Problem 17-15I

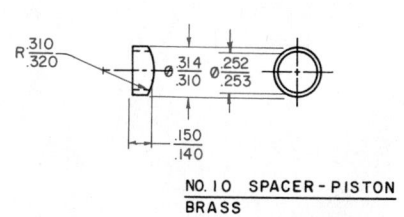

NO. 10 SPACER-PISTON
BRASS

Problem 17-15J

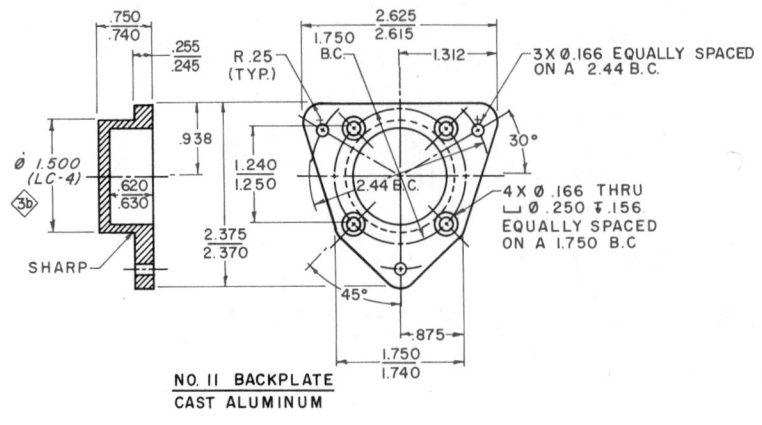

NO. 11 BACKPLATE
CAST ALUMINUM

Problem 17-15K

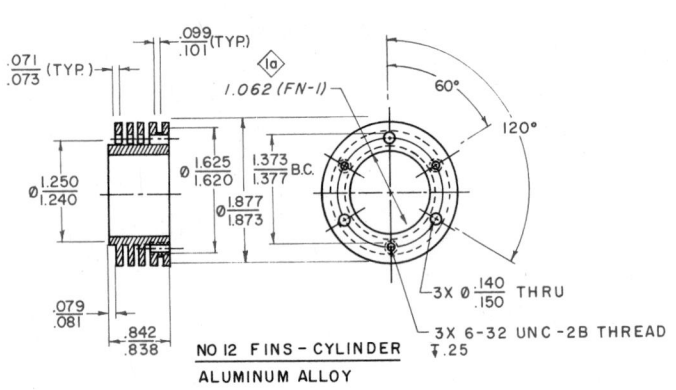

NO 12 FINS – CYLINDER
ALUMINUM ALLOY

Problem 17-15L

NO. 14 PISTON
DUCTILE IRON

Problem 17-15N

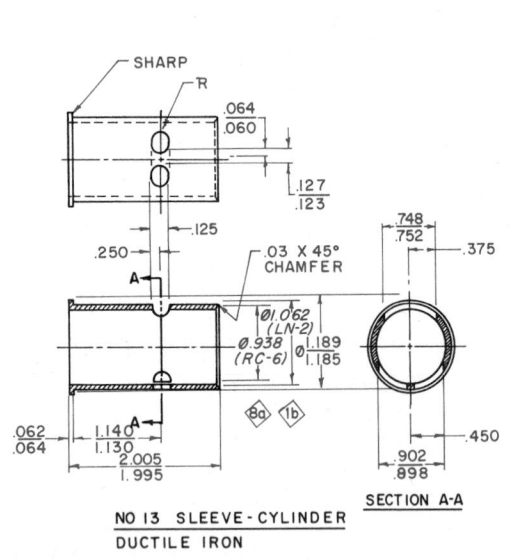

NO 13 SLEEVE – CYLINDER
DUCTILE IRON

SECTION A-A

Problem 17-15M

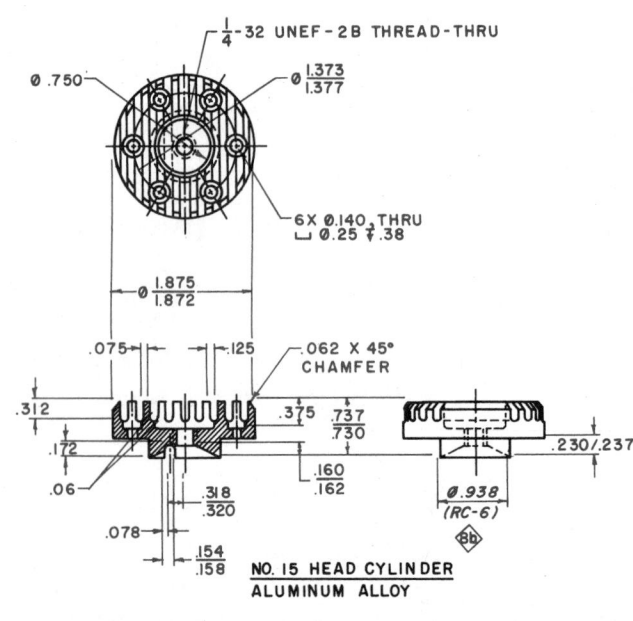

NO. 15 HEAD CYLINDER
ALUMINUM ALLOY

Problem 17-15O

Problem 17-16

Using detail drawings 17-16A through 17-16G, lay out a one-view assembly drawing. Call off parts per standard drafting practices. Design special extra-fine threads where required.

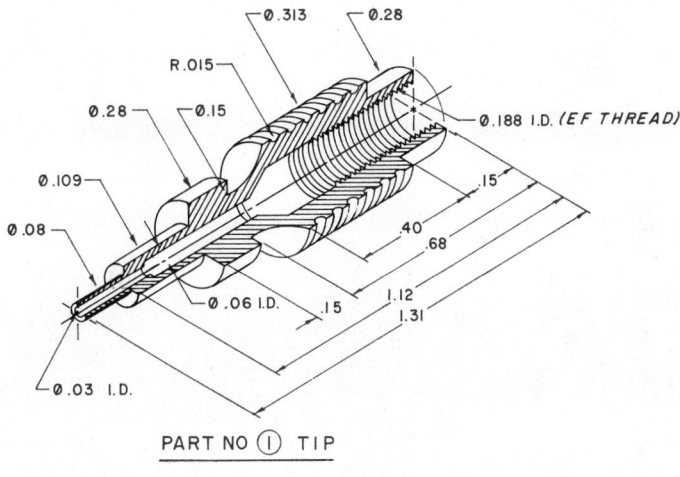

Problem 17-16A

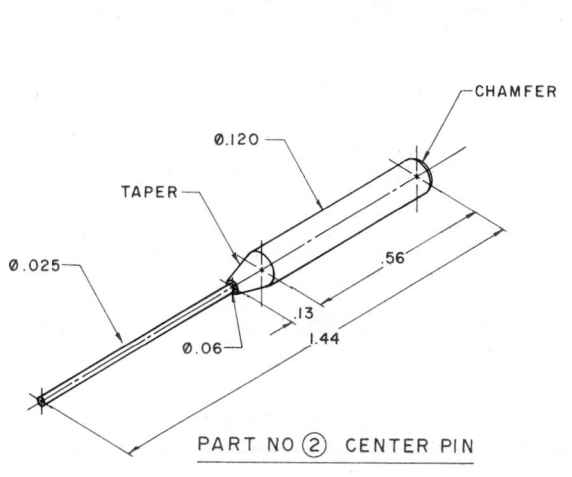

Problem 17-16B

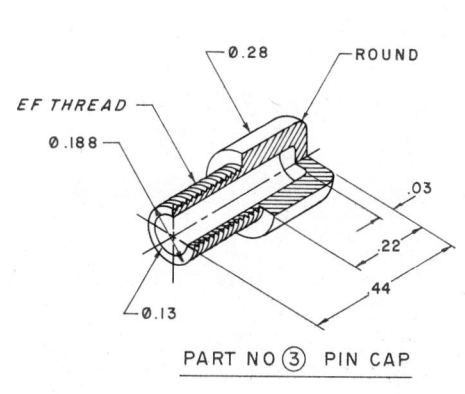

Problem 17-16C

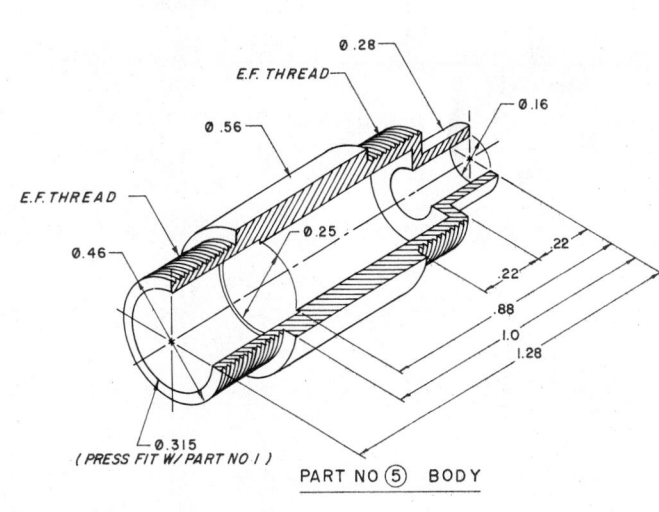

Problem 17-16D

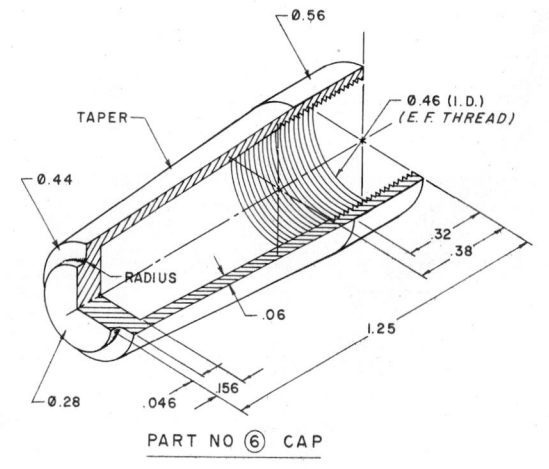

Problem 17-16E

Problem 17-16F

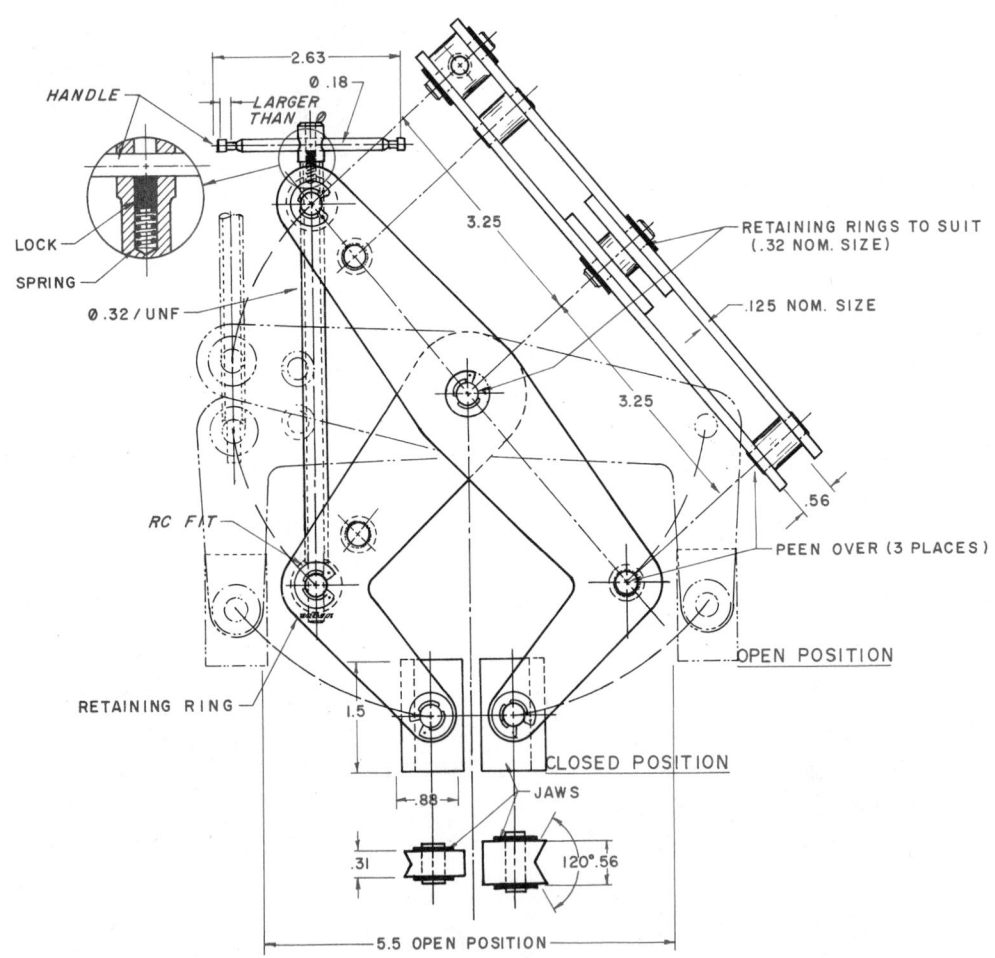

Ø.44 ── Ø.28 ── ROUND

Ø.06

TAPER

.25

.06 (TYP)

Ø.56

3.5

.06

PART NO⑦ END

.63

Ø.46 (I.D.)
E.F. THREAD

Problem 17-16G

❏ Problem 17-17

Follow the eight listed steps to develop Problem 17-17.

HANDLE

2.63

Ø .18

LARGER
THAN Ø

LOCK

SPRING

Ø.32 / UNF

RETAINING RINGS TO SUIT
(.32 NOM. SIZE)

3.25

.125 NOM. SIZE

3.25

.56

RC FIT

PEEN OVER (3 PLACES)

OPEN POSITION

RETAINING RING

1.5

CLOSED POSITION

.88

JAWS

.31

120° .56

5.5 OPEN POSITION

Problem 17-17

❏ **Problem 17-18**

Follow the eight listed steps to develop Problem 17-18.

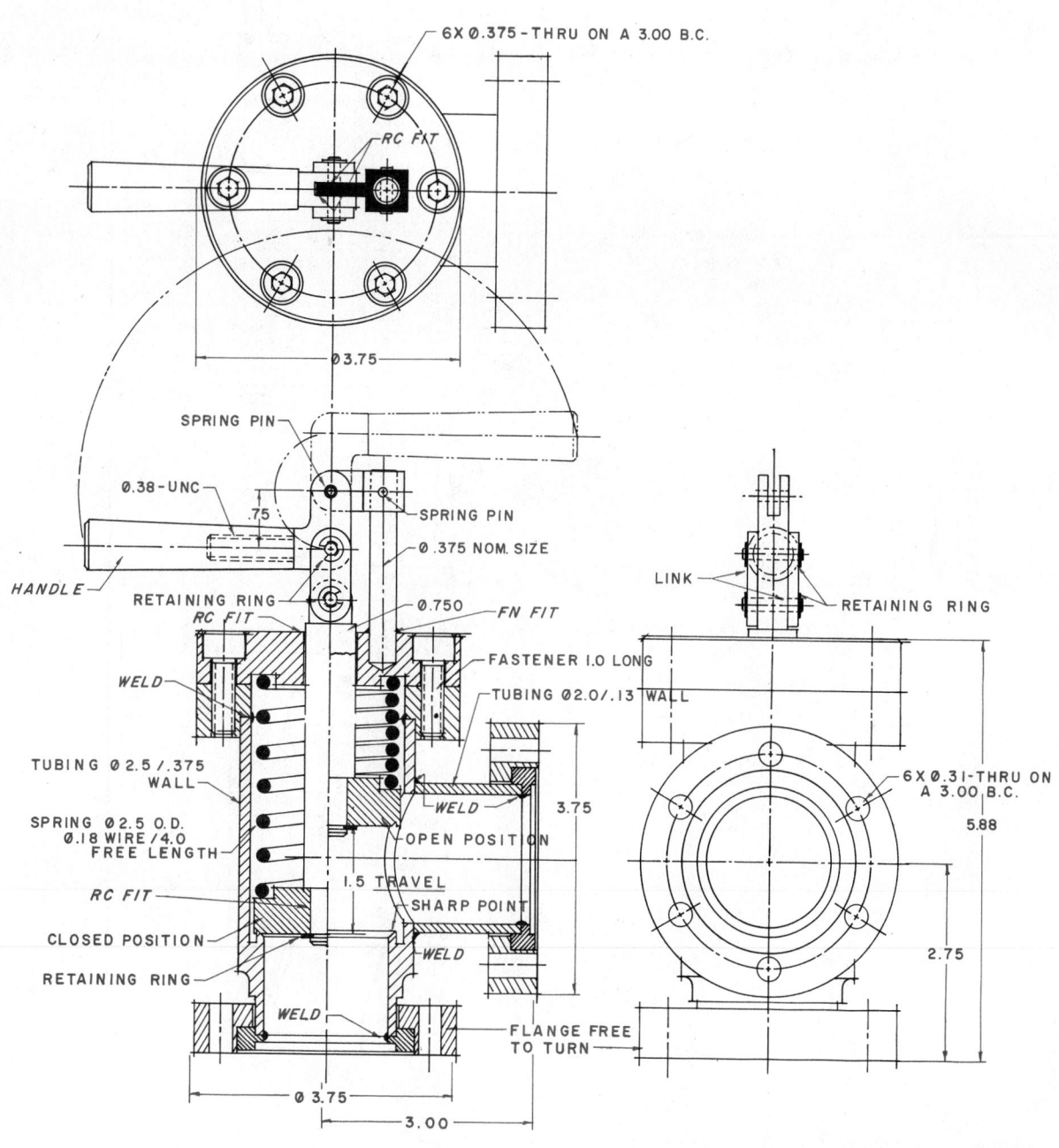

Problem 17-18

❏ Problem 17-19

Follow the eight listed steps to develop Problem 17-19. Draw the clamping device in the position shown, and draw arm A at 30° and 60° to the left as shown. Design a complete pneumatic cylinder with a length to suit based on .75 extra interior space at each end of the cycle as shown. Improve the basic designs where necessary.

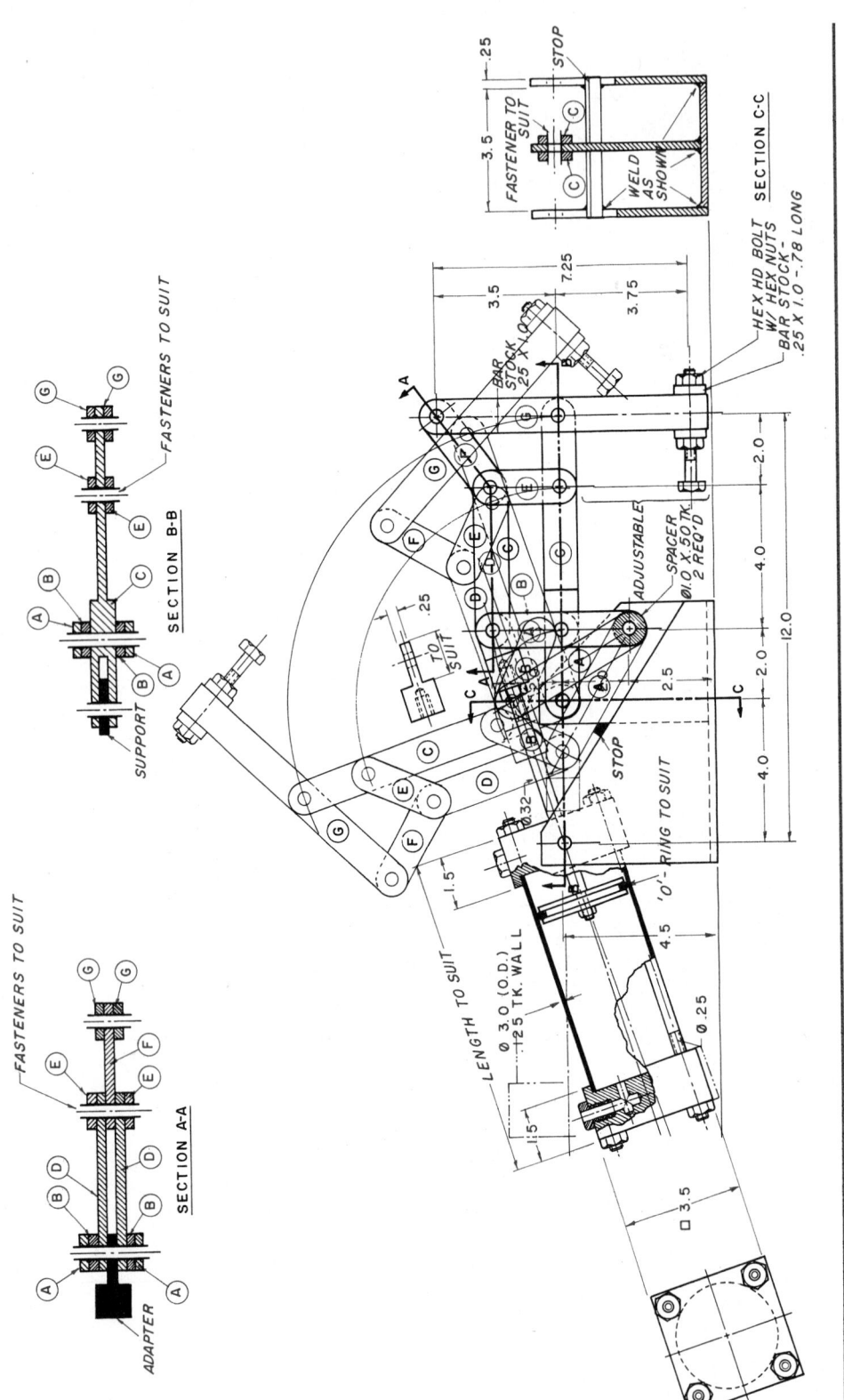

Problem 17-19

❏ Problem 17-20

Follow the eight steps listed to develop Problem 17-20. For this problem, use the information from the figures shown in Figures 17-7 and 17-8 in the text to prepare freehand sketches of each part. Then, follow the eight steps. Assume an overall length of approximately 9 inches for the assembly of the yoke, screw center and hooks as shown. Make each part proportional to that overall length. Make the yoke and hooks of medium carbon steel, AISI 1045. (See Table 47 in the Appendix.) Make the hooks at least 1/4″ thick, and use UNF–20, 1/2″ diameter threads or larger for the screw center.

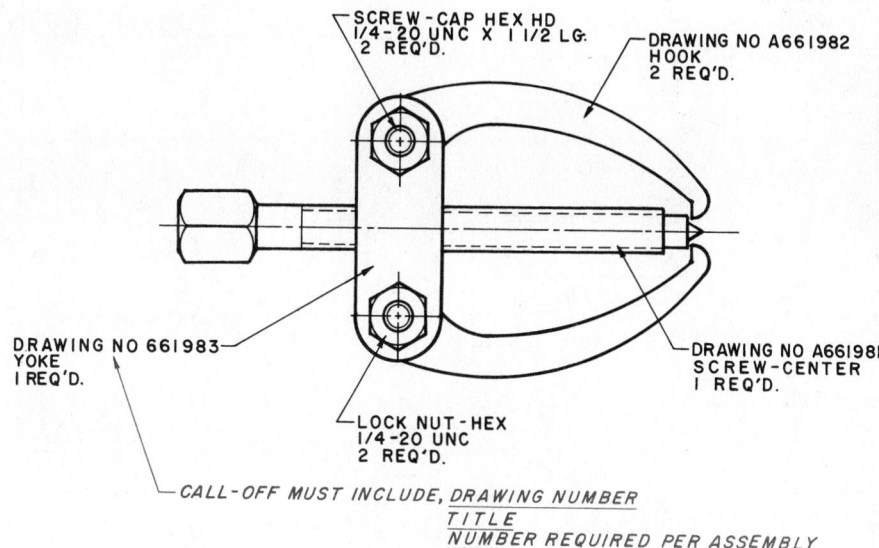

SCREW-CAP HEX HD
1/4-20 UNC X 1 1/2 LG.
2 REQ'D.

DRAWING NO A661982
HOOK
2 REQ'D.

DRAWING NO 661983
YOKE
1 REQ'D.

LOCK NUT-HEX
1/4-20 UNC
2 REQ'D.

DRAWING NO A661981
SCREW-CENTER
1 REQ'D.

CALL-OFF MUST INCLUDE, DRAWING NUMBER
TITLE
NUMBER REQUIRED PER ASSEMBLY

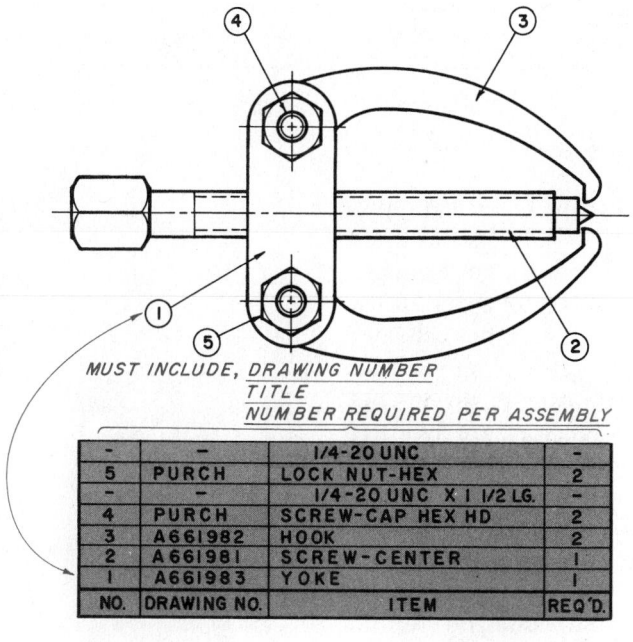

MUST INCLUDE, DRAWING NUMBER
TITLE
NUMBER REQUIRED PER ASSEMBLY

NO.	DRAWING NO.	ITEM	REQ'D.
-	-	1/4-20 UNC	-
5	PURCH	LOCK NUT-HEX	2
-	-	1/4-20 UNC X 1 1/2 LG.	-
4	PURCH	SCREW-CAP HEX HD	2
3	A661982	HOOK	2
2	A661981	SCREW-CENTER	1
1	A661983	YOKE	1

Problem 17-20

❏ Problem 17-21

The angle stop shown in the drawing is to be changed according to ECO Number C159, originated by GNC on Sept. 11, and justified because of a new model of anchor-bolt puller to be manufactured. The description of change on a typical ECO form indicates that the 210-mm dimension changes to 250 and the holes change from a diameter of 13 to 19. An ECO committee determined that the change was necessary and that the change would not affect performance, reliability, or safety. Each member of the committee signed-off the ECO (see Figure 17-13). Make a machine copy of the drawing as it is shown. Make the changes. (Refer to Figure 17-14.) Record the changes in the revisions block (upper right-hand corner).

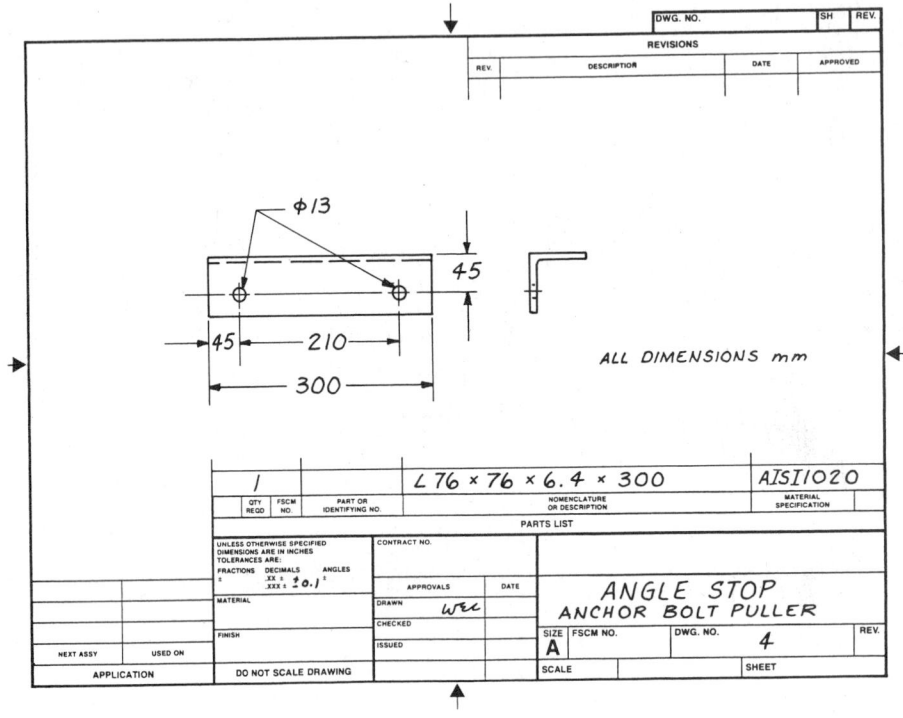

Problem 17-21

❏ Problem 17-22

Limit switches, which are basically electrical on-off switches, limit travel of machine parts, act as interlocks, and indicate positions of devices. For example, the machines in the photographs in chapter 20 use limit switches to limit travel. Limit switches activated by open doors prevent an elevator from moving. Positions of remote devices in hostile environments, such as control-rods in nuclear reactors, are indicated safely by limit switches. Rotating antennas can be stopped at pre-set positions by limit switches to keep wires from tangling.

Use information in the figure to do a layout drawing to determine the dimensions of an actuating shoe that would cause the limit switch to operate and stop the rotation of the antenna. Also, do a layout drawing of the support-plate for the actuating shoe. Include a hub and set screw. Hub diameters may average approximately 2 times shaft diameters, and are about 2 times shaft diameters long. The hub is to be welded to the support plate. Use low-carbon steel. (Refer to Table 47 in the Appendix.)

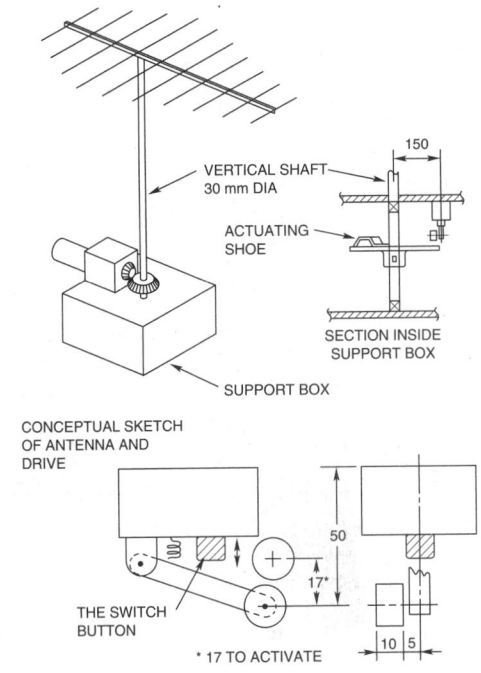

Problem 17-22

❏ Problem 17-23

The pressure transducer shown indicates pressure by the distortion of the piezoelectric crystal that generates a voltage proportional to the distortion. A digital voltmeter is used to read the voltage. Assembly of the transducer could be simplified if the tubular spacer were part of the hex-shaped fitting with the 1/2″–UNF threads. Prepare a freehand sketch of your idea to simplify the assembly. Then, follow the eight listed steps to develop the transducer. The piezoelectric crystal is a purchased item. Use a rust-resistant steel. (Refer to Table 47 in the Appendix.)

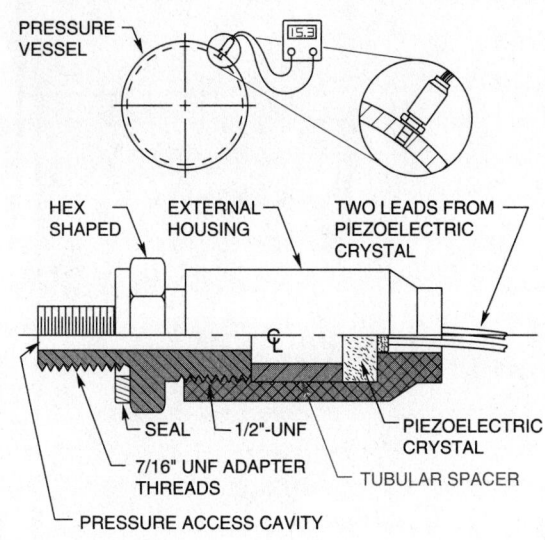

Problem 17-23

❏ Problem 17-24

Manufacturing and assembly of the adjustable roller support shown could be improved. Prepare freehand sketches of your ideas for improvement and then follow the eight listed steps to develop the adjustable roller support. Note that slotted holes are often used for adjustments.

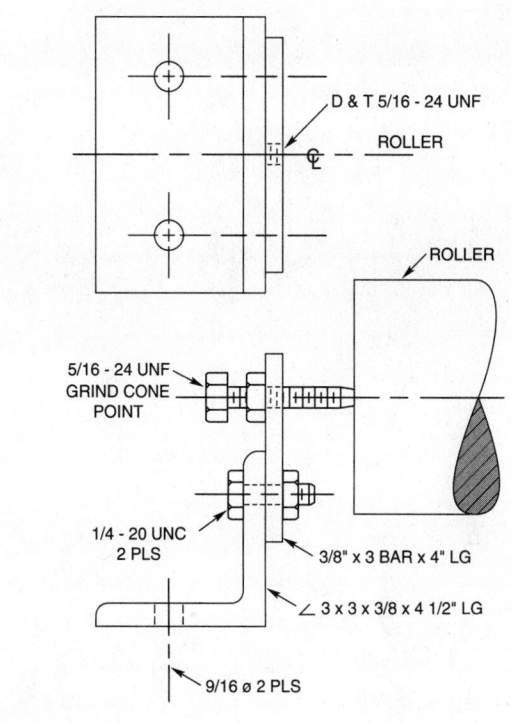

Problem 17-24

❏ Problem 17-25

Choose an existing mechanical or electrical-mechanical device that you believe could be improved in one or more of the following ways:

    a. Better use of standard materials

    b. Function

    c. Manufacturability

    d. Assembly

Prepare freehand sketches of your ideas and then follow the eight listed steps to develop them.

Pictorial Drawings

Chapter Outline

Pictorial Drawings

Oblique Drawings

Isometric Drawings

Perspective Drawings

Axonometric: Drawing and Projection

Other Oblique Drawings

Pictorials by Computers

Review

Problems

Pictorial Drawings

Pictorial drawings facilitate communication of technical information during all phases of the design process—from early sketching of product ideas for future uses, to talking through concepts with colleagues, and finally, to the more formal pictorial drawings (for selling products, maintenance manuals, catalogs, assembly instructions, etc). This chapter begins with the three most common types of pictorial drawings used by drafters, designers, and engineers—oblique, isometric, and perspective. Sketching these three types was introduced in chapter 2.

After discussing oblique, isometric and perspective drawings, more complete discussions will follow on additional types of pictorial drawings and projections.

Oblique Drawings

Oblique drawing is the easiest of the three to draw. In oblique drawing, one surface of the object, usually the front view, is drawn exactly as it would appear in a multiview drawing. Circles are true circles and rectangles parallel to the frontal plane are in true size, *Figure 18-1*. Circles are drawn with a compass or circle template, rectangles have square corners.

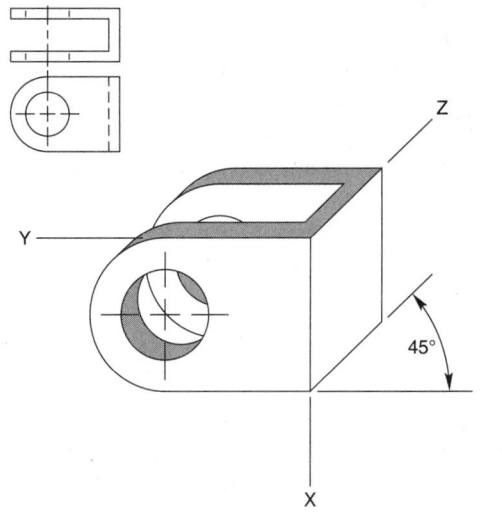

Figure 18-1   Axes for oblique drawings

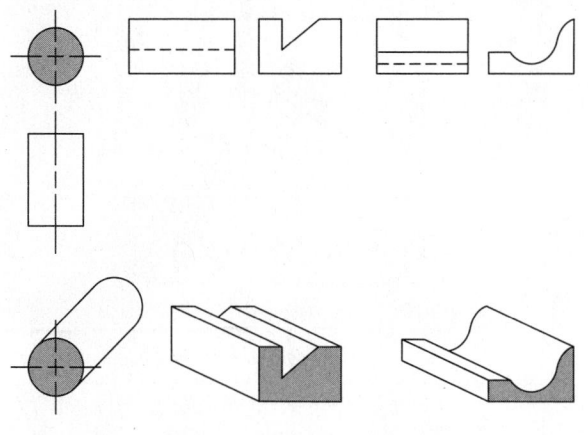

Figure 18-2 Advantages of oblique

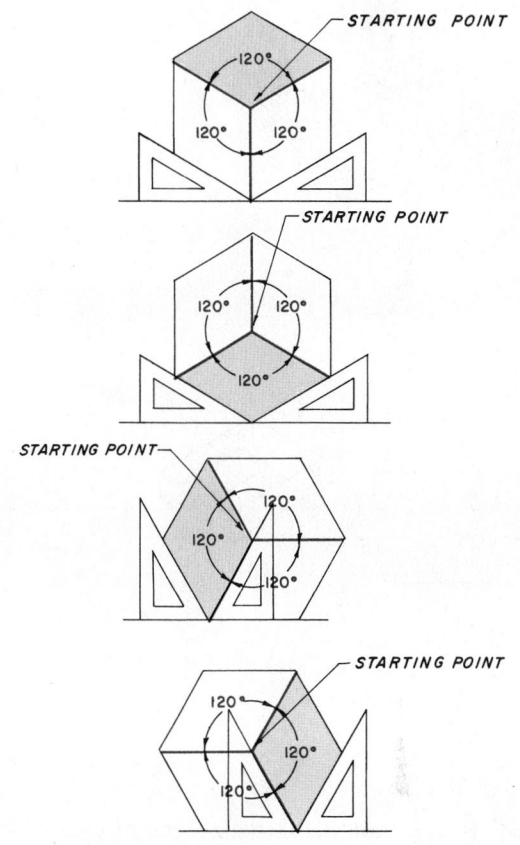

Figure 18-3 Isometric principles

Constructing Oblique Drawings

Oblique drawings use three axes; two at right angles (for example, X and Y in Figure 18-1) and one, the Z axis, usually at an angle of the convenient 30° or 45° standard triangle. Measurements are taken from the multiview drawing; X and Y dimensions match those in front views, and Z dimensions, depth, are laid off parallel to the third axis.

Cylinders, Angles, and Curves in Oblique Drawings

Figure 18-2 has further illustrations of the primary advantages of oblique drawings, especially starting with the front view. Cylinders are round, angles are in true size, and irregular curves are not distorted.

Isometric Drawings

Isometric drawings are the most used of the three common pictorials. They are easier to draw than perspective drawings and they illustrate objects almost as well as perspective drawings.

Isometric drawings use three axes, sometimes called principal edges, at 120° to each other, *Figure 18-3*.

How to Make an Isometric Drawing

Given: A multiview drawing of an object, *Figure 18-4A*.

Step 1 Locate the starting point and lightly draw the three principal edges at 102° using a 30°–60° triangle, *Figure 18-4B*.

Step 2 Transfer the depth, height, and width directly from the multiview drawing, Figure 18-4C, to the three principal edges. Measure full size directly along each edge. Lightly construct an isometric, basic outline, or "box," of the object, *Figure 18-4C*.

GIVEN: MULTIVIEW PROJECTION

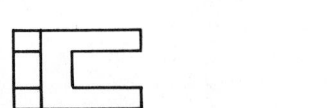

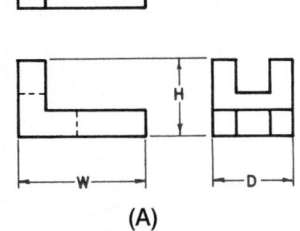

(A)

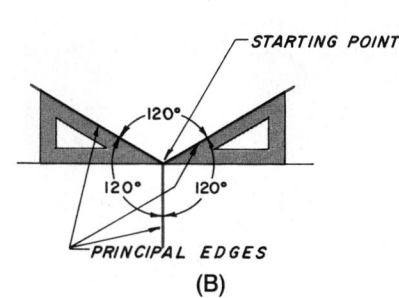

(B)

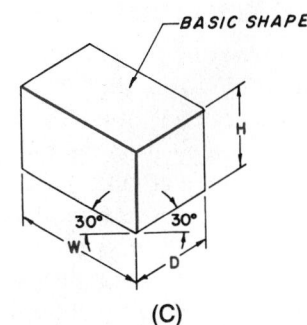

(C)

Figure 18-4 (A) How to draw an isometric drawing; (B) Step 1; (C) Step 2

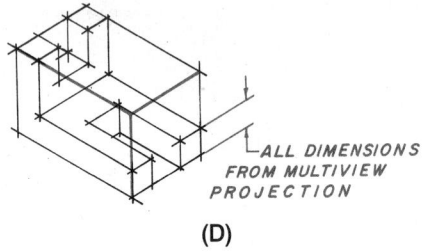

(D)

ISOMETRIC PROJECTION

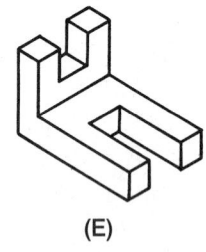

(E)

Figure 18-4 (Continued) (D) Step 3; (E) Completed isometric view

GIVEN: MULTIVIEW DRAWING

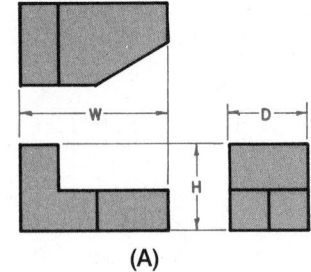

(A)

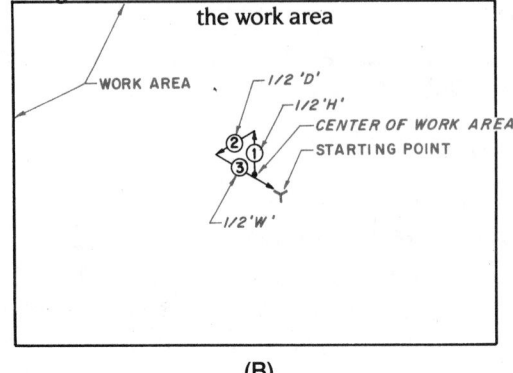

(B)

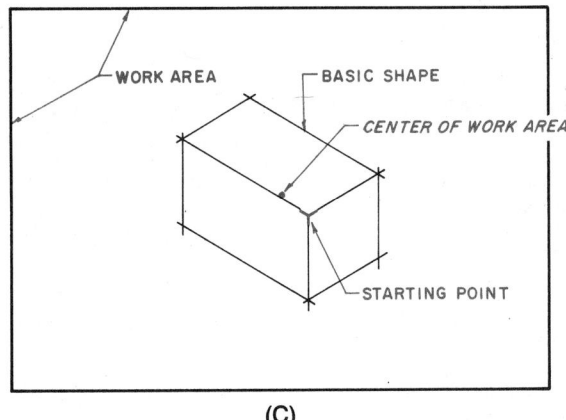

(C)

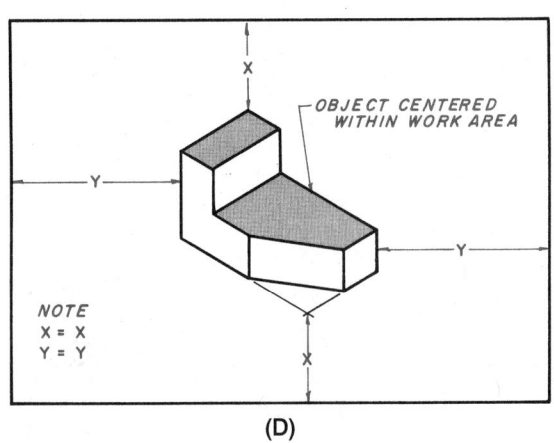

(D)

Figure 18-5 (A) Centering an isometric view within the work area; (B) Step 1; (C) Step 2; (D) Completed isometric view

Step 3 Transfer true size lengths, lengths parallel to the principal axes, from the various features of the object, *Figure 18-4D*.

Step 4 Check all lines, correct if necessary, and darken in the object using correct line thickness, *Figure 18-4E*.

How to Center an Isometric Object

Save time in locating isometric drawings in or near the center of a drawing space by following these steps.

Given: A multiview of an object, *Figure 18-5A*.

Step 1 Locate the center point of the work area and do the following: go North 1/2 H, go S60°W a distance of 1/2 D, and go S60°E, 1/2 W to establish a starting point, *Figure 18-5B*.

Step 2 From the starting point, draw the basic shape of the object, *Figure 18-5C*.

Step 3 Construct the object within the basic shape, *Figure 18-5D*.

Offset Measurements

Offset measurements are used to locate one feature in relationship to another. Measure offsets parallel or perpendicular to one of the three standard planes—frontal, top, or profile—from the multiview of the object being drawn, *Figure 18-6A*.

Develop the isometric drawing as before; first, the basic shape or "box" with dimensions W, D, and H, and then, the offset measurements A, B, and C parallel or perpendicular to the frontal, top, or profile planes, *Figure 18-6B*. Note lines that are truly parallel in the multiview drawing must be parallel in the isometric drawing.

GIVEN: MULTIVIEW PROJECTION

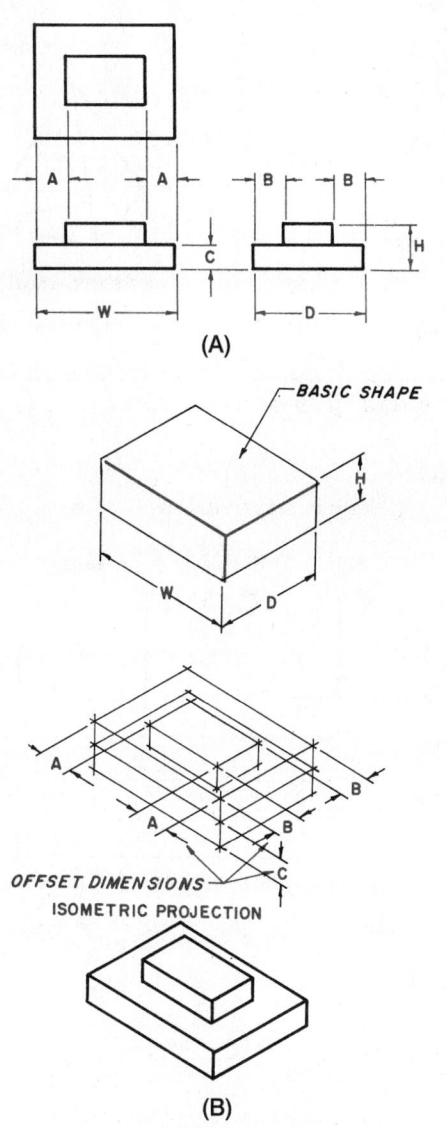

Figure 18-6 (A) Multiview projection; (B) Offset measurements

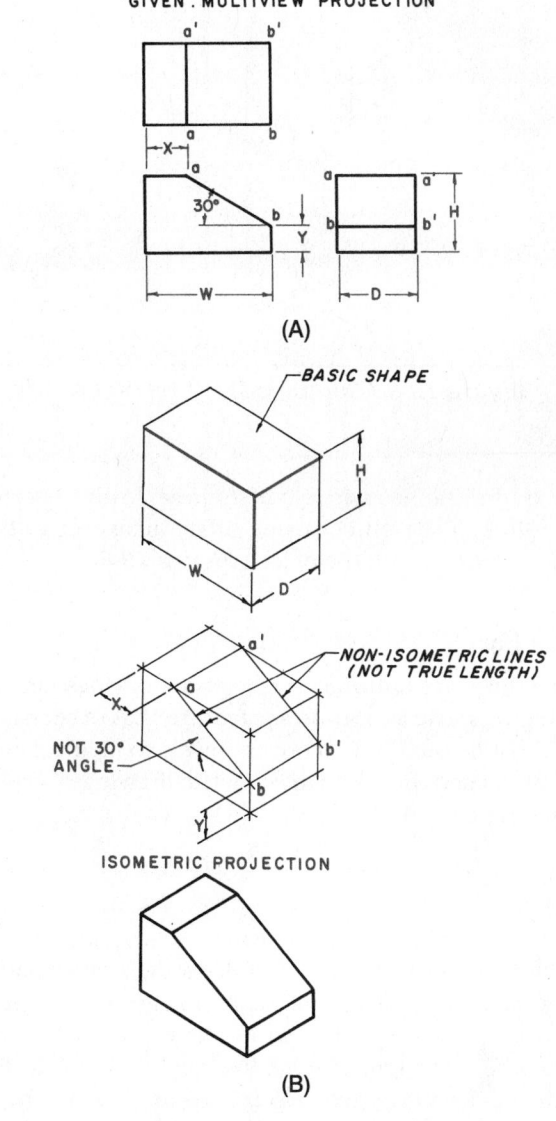

Figure 18-7 (A) Nonisometric lines; (B) Locating points

Nonisometric Lines

If a line is not parallel or perpendicular to any of the three principal edges in an isometric drawing, it is a *nonisometric line*. *Figure 18-7A* shows a multiview drawing of an object with a 30° inclined surface, labeled as shown with a, b, a' and b'.

Construct the isometric drawing as before; "box" first, then the feature using the X and Y dimensions, and finally, heavy in the lines of the object, *Figure 18-7B*. Line a-b is not in the true length and the 30° angle is distorted.

Box Construction for Irregular Objects

Some objects, *Figure 18-8A*, have lines that do not line up with the isometric axes, so use a box or basic shape to establish offset dimensions for drawing the object, *Figure 18-8B*.

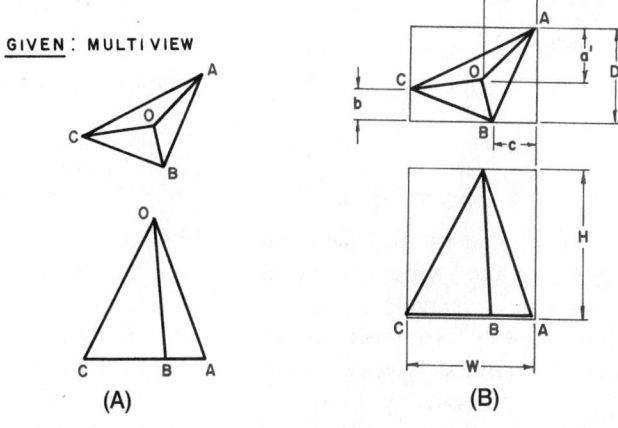

Figure 18-8 (A) Box construction; (B) Basic shape

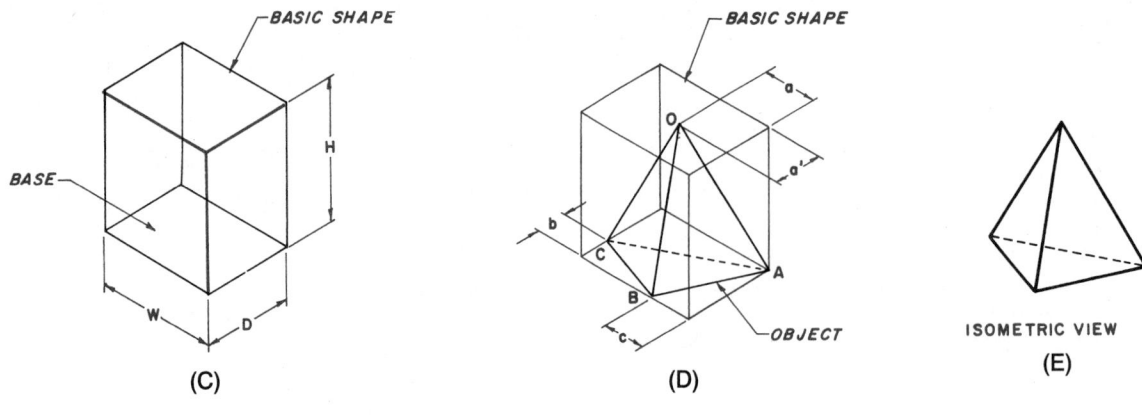

Figure 18-8 (Continued) (C) Isometric basic shape; (D) Locating points of object; (E) Isometric view completed

Next, draw an isometric box, *Figure 18-8C*. Locate key points of the object using offset measurements, *Figure 18-8D*. Finish the object, *Figure 18-8E*.

Hidden Lines

Hidden lines are not used in isometric drawings unless needed to show an important feature that otherwise would not be seen. Try to orient objects so hidden lines are not needed, not only in isometric drawings, but in all pictorials.

Isometric Curves

Construct isometric curves in isometric drawings using offsets or a grid. For a series of offset distances, use a number of evenly spaced lines, all parallel to one of the principal edges, *Figure 18-9A*. Transfer these lines to the isometric layout and transfer each distance, respectively, to locate points on the curve, *Figure 18-9B*. Give depth to the drawing by projecting D distances at 30°, which is parallel to an isometric axis, from each point on the curve. Use a French curve to connect the points.

For transferring a curve using a grid, choose a grid spacing that will give a sufficient number of points along the curve to give as much detail of the curve as needed, *Figure 18-10A*.

Construct the basic shape of the object using the W, H, and D dimensions. Transfer the grid, *Figure 18-10B*. Finally, transfer the irregular curve, square by square. Use a French curve to connect the points on the curve.

Irregularly Shaped Objects

Transfer and draw irregularly shaped objects by dividing the object into a series of convenient sections. Draw each section as an isometric, *Figure 18-11A*. The two-view drawing of the boat has been simplified for this example. Divide the object into a series of sections, A through H. Use a logical base line (in this case a center line). Locate and draw each section, *Figure 18-11B*. Connect all the sections and complete the boat, *Figure 18-11C*.

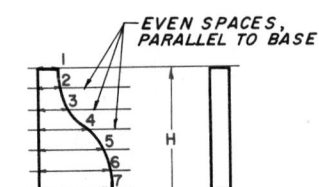

Figure 18-9A Drawing an isometric curve (offset method)

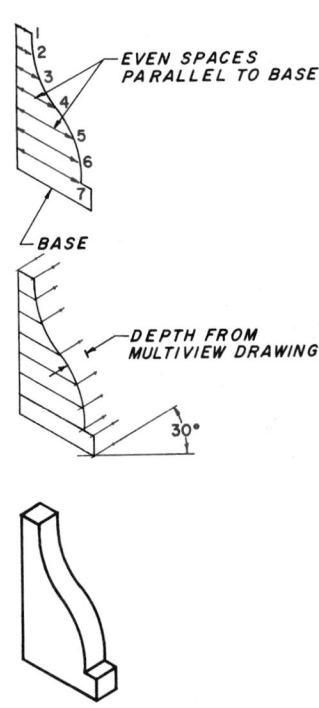

Figure 18-9B Drawing an isometric curve (offset distances and depth)

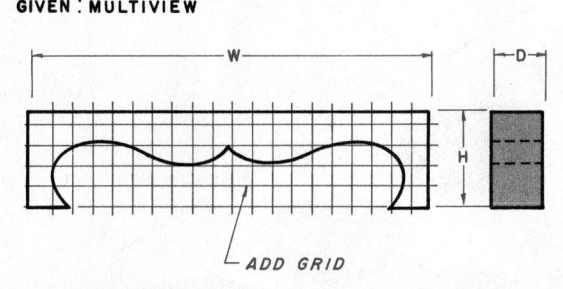

Figure 18-10A Drawing an isometric curve (grid method)

Isometric Circles and Arcs

Drawing isometric circles and arcs, without using templates, takes too much time for isometric drawing. Use an isometric ellipse template, **Figure 18-12**. The template has eight hash marks printed around the perimeter of each ellipse. Four indicate the isometric center lines of the circle or arc. For example, an ellipse for a one-inch circle would be labeled number –1– and it would measure 1″ across the ellipse along the isometric center line. These same hash marks also indicate the four tangent points of the isometric circle with the isometric basic shape of the circle. (See **Figures 18-13A** and **18-13B**.) The other four hash marks line up with the axes of the circle; two are coincident with the axis of the circle and are lined up across the narrow part of the ellipse, and

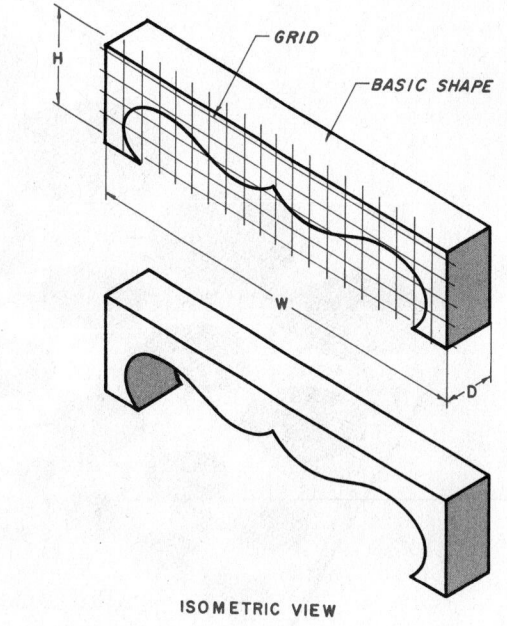

Figure 18-10B Isometric grid layout

two are perpendicular to the axis and are lined up across the longest part of the ellipse, **Figure 18-13C**. The longest distance, across the one-inch ellipse, actually measures more than one inch due to the way the basic shape is constructed for isometric drawing.

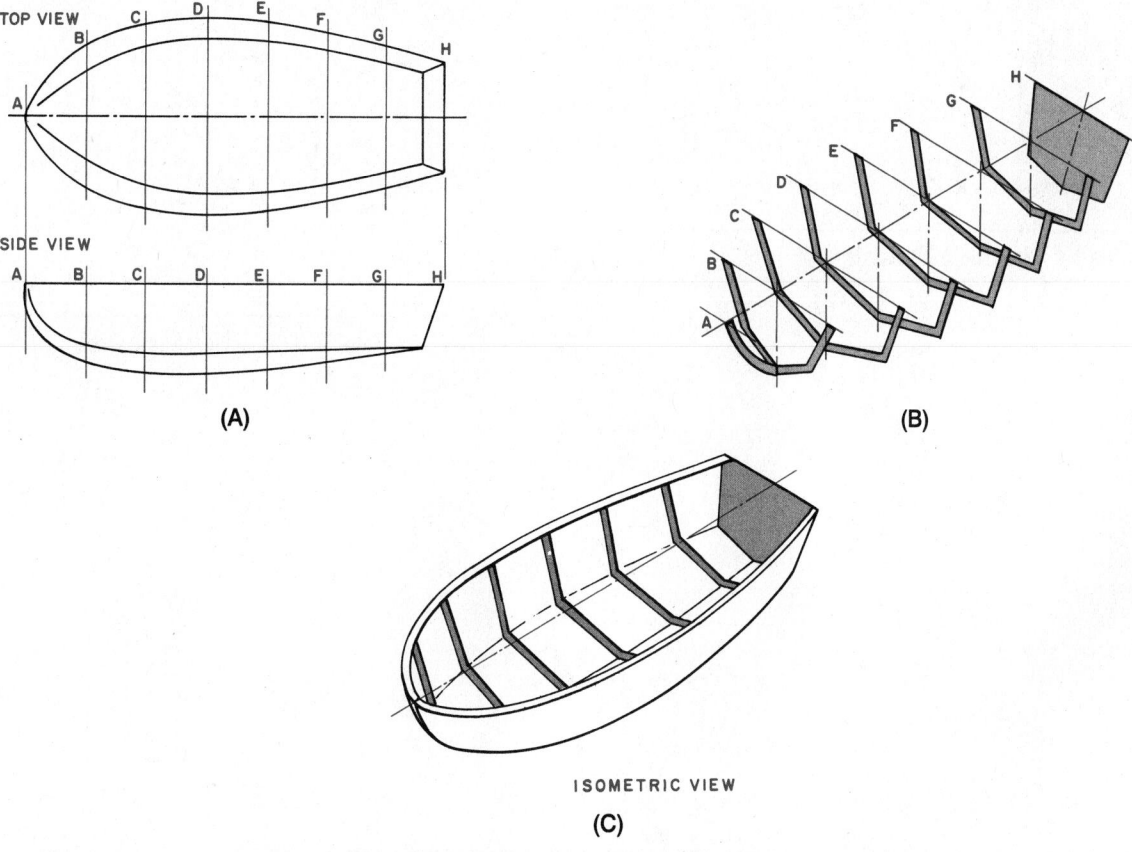

Figure 18-11 (A) Irregularly shaped object; (B) Drawing each section; (C) Completed object

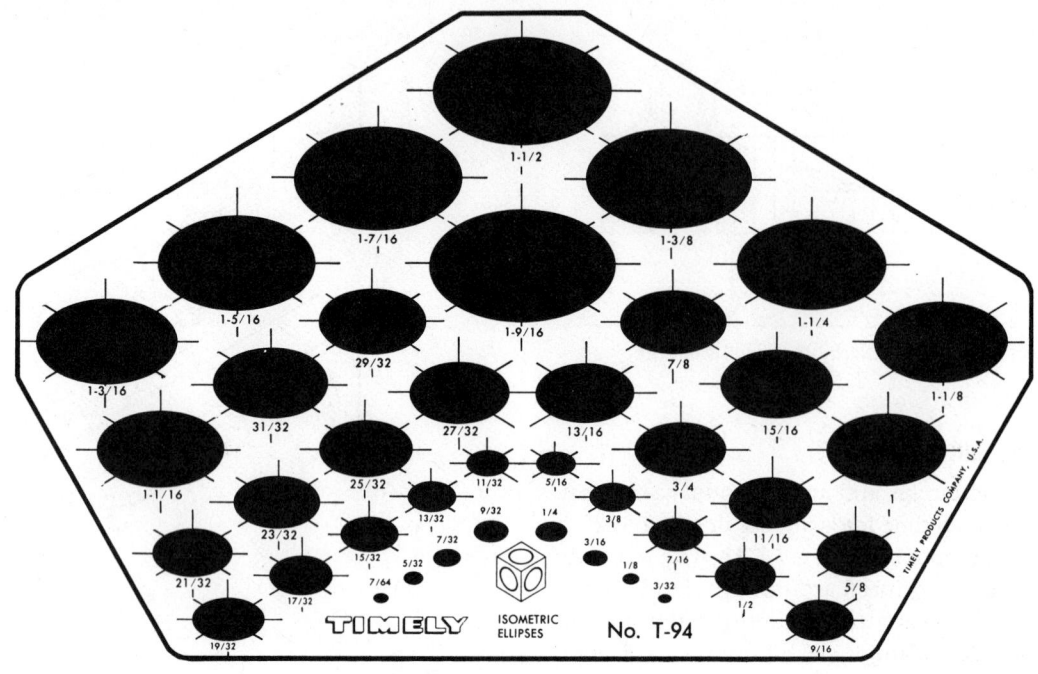

Figure 18-12   Isometric ellipse template *(Courtesy Timely Products Co.)*

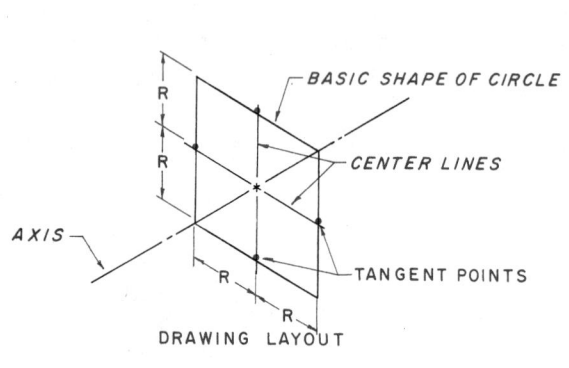

Figure 18-13A   Basic shape of the circle

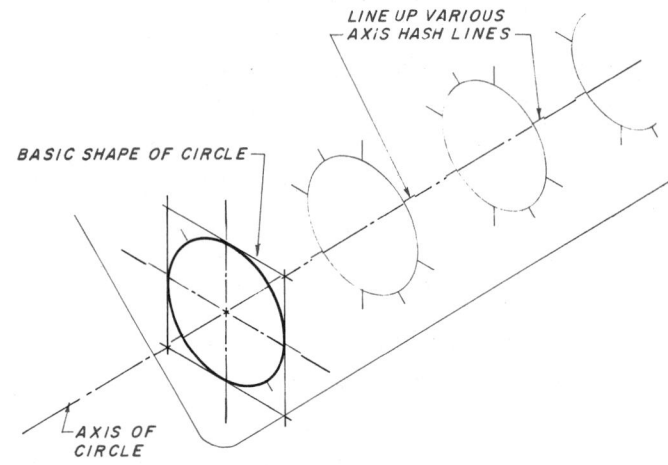

Figure 18-13B   Draw the ellipse within the basic shape

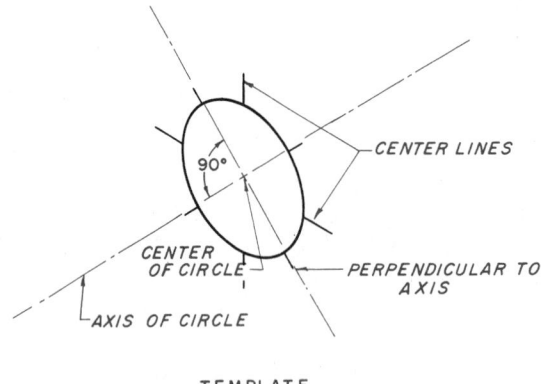

Figure 18-13C   Isometric ellipse template aligned on axis of circle

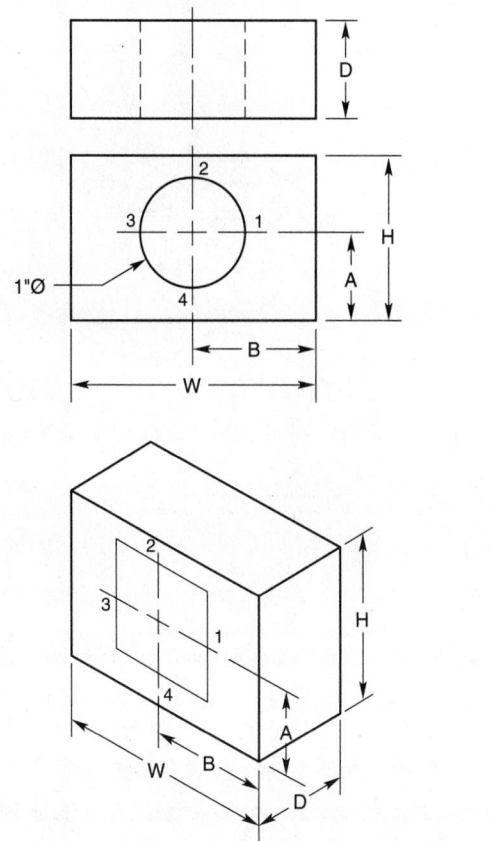

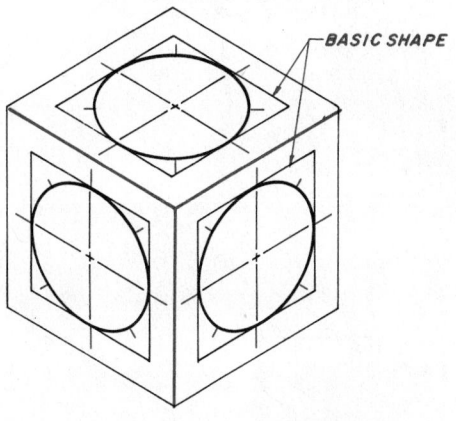

Figure 18-15 Isometric circles in various planes

Figure 18-14A How to use an isometric ellipse template

How to Use Isometric Ellipse Templates for Isometric Drawings

Step 1 For the object given, *Figure 18-14A*, locate the center of the circle or arc to be drawn using offset measurements, and construct the basic shape for the circle so that the sides of the basic shape are parallel to the two appropriate principal edges.

Step 2 Line up the 1"-ellipse template so the the isometric center line hash marks touch the tangent points 1, 2, 3, and 4 of the basic shape, *Figure 18-14B*. Trace the ellipse. If some doubt exists as to which hash marks to line up, refer to *Figure 18-15* for guidance.

How to Draw True Isometric Circles

Assume that a relatively accurate isometric drawing of a circle is required, but a template is not available for the size needed.

Given: A regular circle in its basic shape, and the same circle as it would be drawn in an isometric drawing, in its basic shape, *Figure 18-16A*. In both drawings the tangent points are given "clock" numbers.

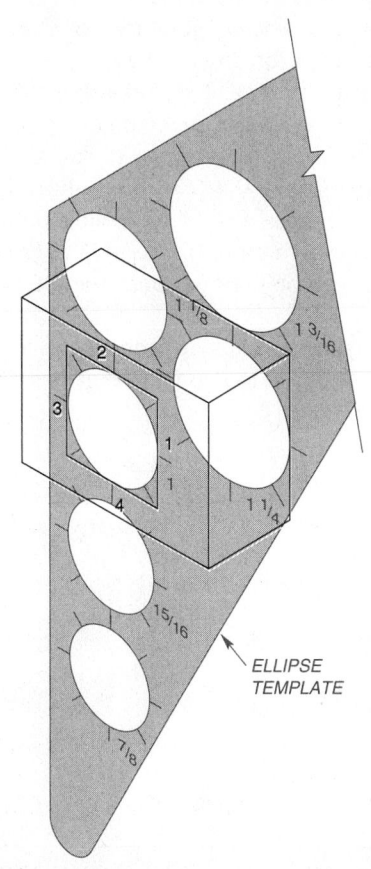

Figure 18-14B Line up hash marks on tangent points

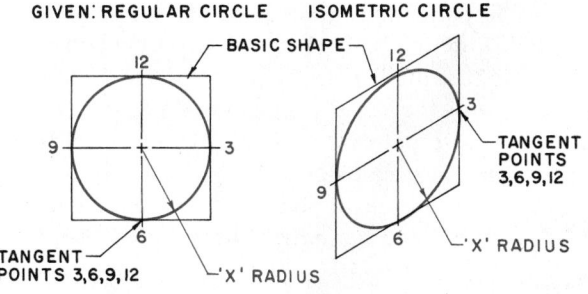

Figure 18-16 (A) Regular circle and isometric circle

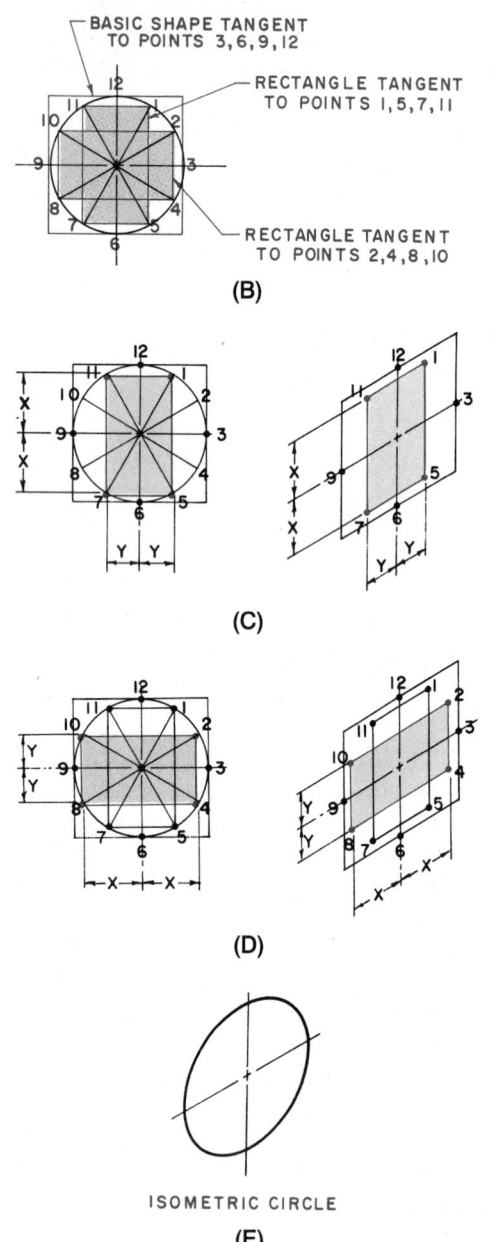

(B)

(C)

(D)

ISOMETRIC CIRCLE

(E)

Figure 18-16 (Continued) (B) Step 1; (C) Step 2;
(D) Step 3; (E) Completed isometric circle

Step 1 Use the circle in the basic shape, a square, with the tangent points 3, 6, 9, and 12. Use the remaining clock numbers and divide the circle into the rectangles shown, *Figure 18-16B*.

Step 2 Draw the basic shape of the circle and use the offset method to locate points 1, 5, 7, and 11 using the X and Y dimensions, *Figure 18-16C*.

Step 3 Locate points 2, 4, 8, and 10 in a similar manner, *Figure 18-16D*.

Step 4 Connect the 12 points with a French curve and finish the isometric drawing of the circle, *Figure 18-16E*.

To draw an arc of a circle, use these steps and use the portion of the circle needed, *Figure 18-17*.

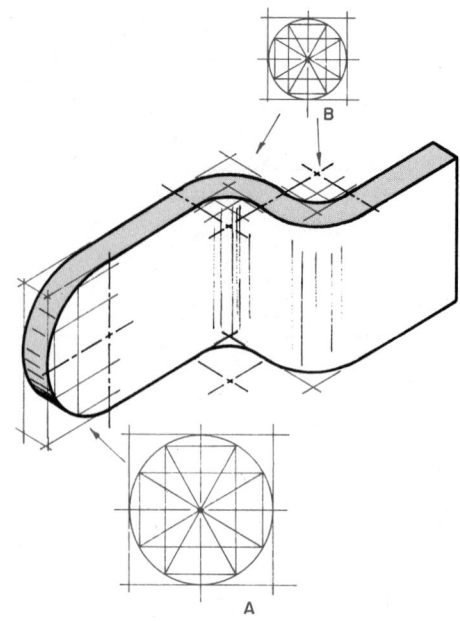

Figure 18-17 How to draw an isometric arc (offset method)

How to Draw Approximate Isometric Circles

Assume that an approximate isometric circle is needed in an isometric drawing, but a template is not available with the size required. The four-center method described in the following steps can be used.

Given: A regular circle with a radius X, Figure 18-16A.

Step 1 Draw the basic shape of the circle in the appropriate location in the drawing. Label the four tangent points, 1, 2, 3, and 4, *Figure 18-18A*.

Step 2 From each of the four tangent points, draw a line at 90° to the opposite sides of the basic shape, *Figure 18-18B*. These lines intersect at four locations called the swing points or centers. Label them A, B, C, and D, *Figure 18-18C*. A line can be used to check the location centers A and B.

Step 3 Construct a radius at each of the four centers, A, B, C, and D as shown in *Figures 18-18D* and *18-18E*.

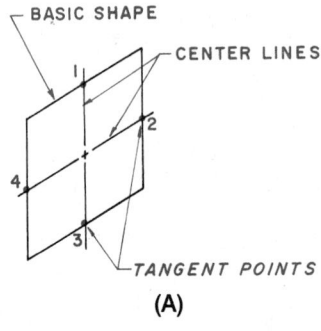

(A)

Figure 18-18 (A) Step 1 (how to draw an approximate isometric circle)

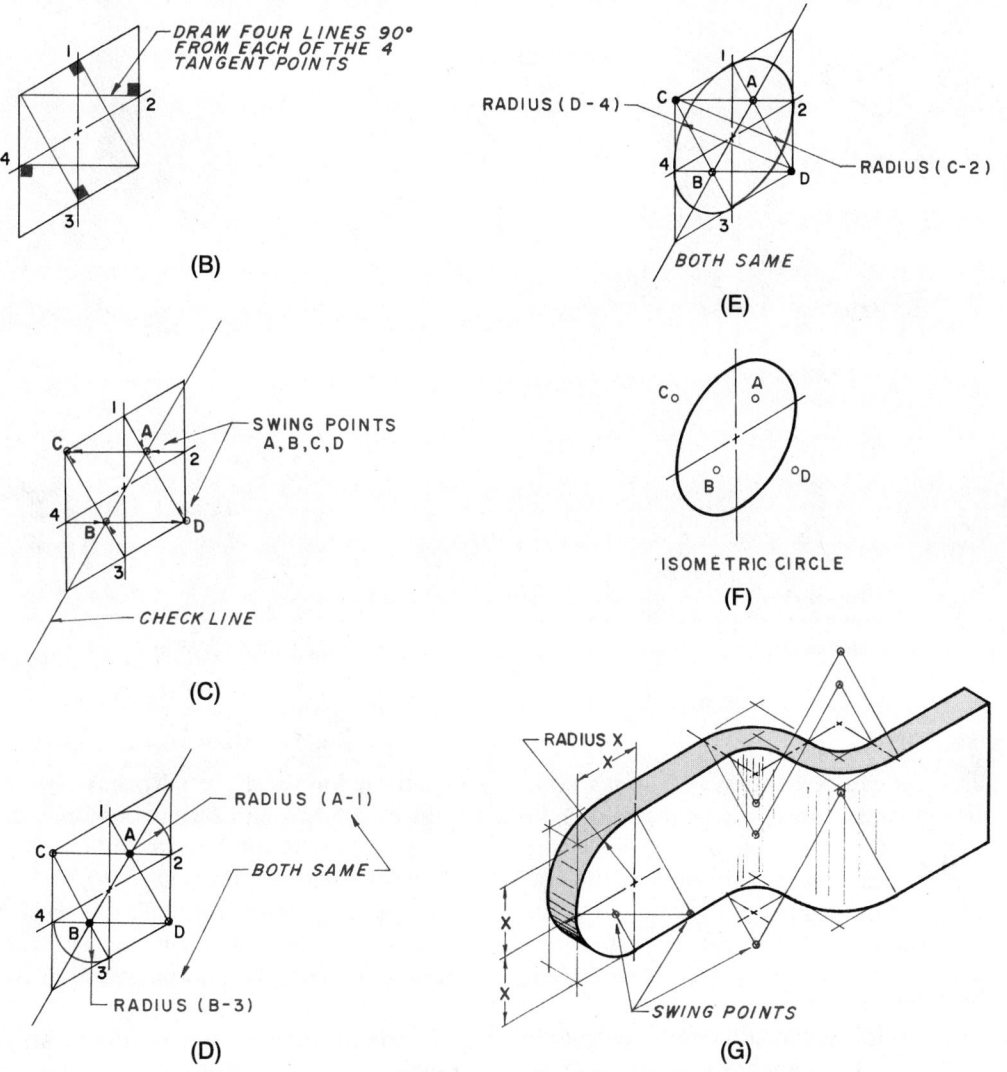

Figure 18-18 (Continued) (B) and (C) Step 2; (D) and (E) Step 3; (F) Completed isometric circle; (G) Drawing isometric arcs (four-center method)

Step 4 Darken in the completed isometric circle and center lines, ***Figure 18-18F***.

Construct arcs using the four-center method and the radius X, discussed in these four steps, ***Figure 18-18G***.

How to Draw an Ellipse on an Inclined Plane

Given: A two-view drawing of a cylinder cut at an angle, ***Figure 18-19A***.

Step 1 Divide the view of the cylinder, showing as a circle, in 12 equally spaced arcs (1 through 12). Project the points into the view showing the base of the cylinder as an edge. Locate the true length line segments from each point, Figure 18-19A.

Step 2 Construct an isometric circle of the base and locate each of the 12 points using the offset technique, ***Figure 18-19B***.

Step 3 Project each true length line segment up from the base and connect the end points to form the boundary of the elliptically shaped surface. Use a French curve.

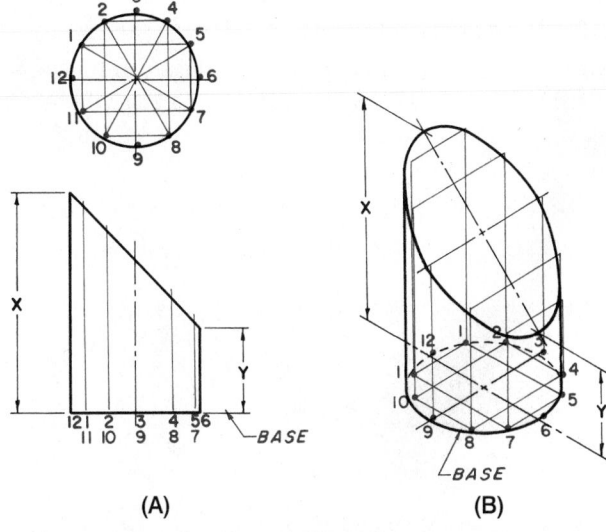

Figure 18-19 (A) How to draw an ellipse on an inclined plane—given: two views; (B) An isometric ellipse on an inclined plane

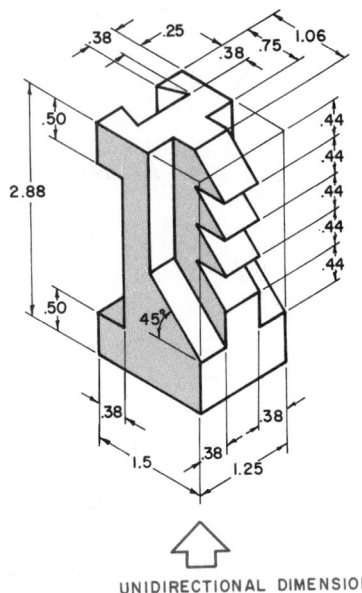

UNIDIRECTIONAL DIMENSIONS

Figure 18-20   Isometric dimensioning

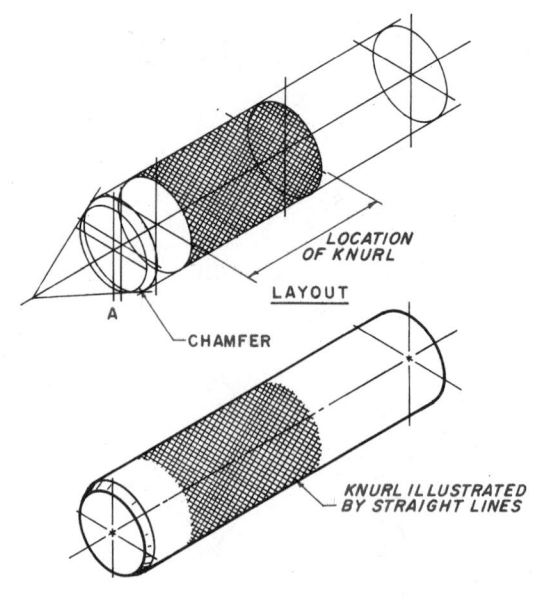

Figure 18-22   Isometric knurl

Isometric Intersections

An isometric drawing can be constructed from a completed multiview drawing, showing an intersection(s), by following the steps outlined in this chapter. First, construct a basic shape, box, and then transfer a sufficient number of points from the multiview drawing, using the offset method, to complete the isometric drawing.

Isometric Dimensioning

Use the same dimensioning standards, rules and methods discussed in chapter 9 to dimension isometric drawings. Furthermore, use unidirectional dimensioning for clarity and ease of placing the dimensions, *Figure 18-20*. Note that extension lines and dimension lines are parallel to one of the three axes.

Isometric Rounds and Fillets

Rounds and fillets in isometric drawings can be shown three ways, *Figure 18-21 A*, *B*, and *C*. Use whichever way, or combinations of ways, that best shows the fillets or rounds, and where they are located.

Isometric Knurls, Screw Threads, and Spheres

Isometric knurls are usually drawn with straight lines, *Figure 18-22A* and *B*. Part A shows the location of a proposed knurl and how to indicate a chamfer. Part B shows the finished drawing.

Isometric threads are readily represented by a series of ellipses drawn along the center line of the fastener, parallel to each other, and approximating the thread spacing, *Figure 18-23*.

Isometric spheres require three ellipses to be drawn, *Figure 18-24*—one along each axis, and one circumscribing the first two. If only a portion of a sphere is wanted, then the first two ellipses can be used to outline a half sphere, *Figure 18-25*; a quarter sphere, *Figure 18-26*; or a three-quarter sphere, *Figure 18-27*.

Based on the two-view drawing in *Figure 18-28* draw a flat surface in a sphere by following the implied steps in *Figure 18-29*.

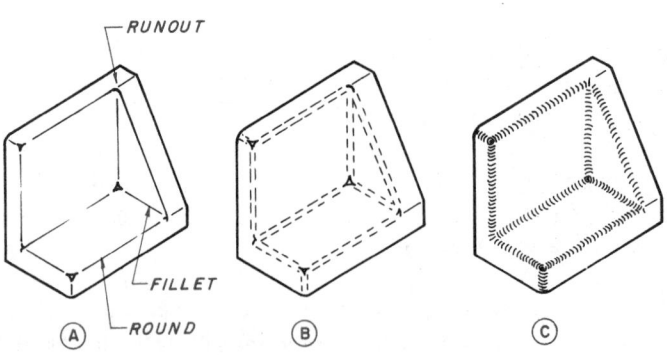

Figure 18-21   Isometric rounds and fillets

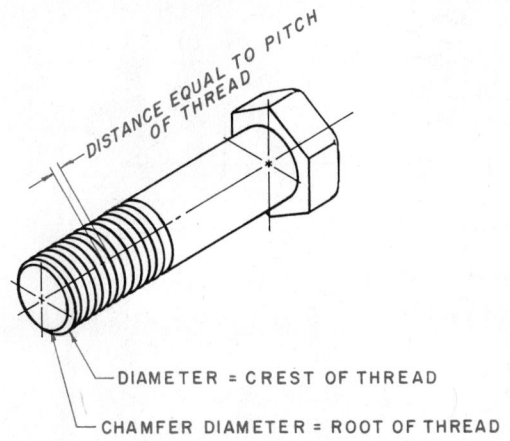

Figure 18-23   Isometric threads

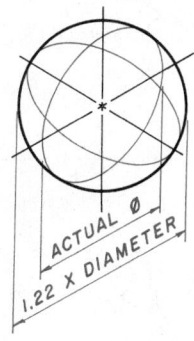

Figure 18-24   Isometric sphere

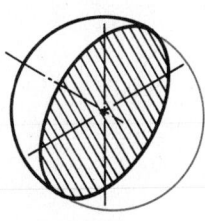

Figure 18-25   Isometric half sphere

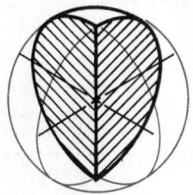

Figure 18-26   Isometric quarter sphere

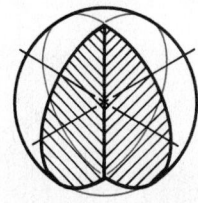

Figure 18-27   Isometric three-quarter sphere

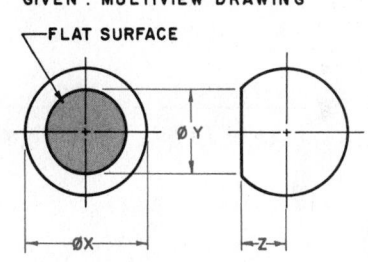

Figure 18-28   Two-view drawing of sphere

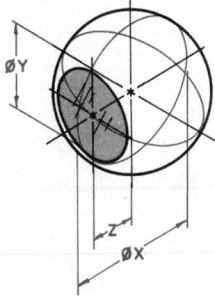

Figure 18-29   Flat surface on a sphere

Perspective Drawings

A *perspective drawing* is a three-dimensional drawing of an object or scene that shows visible lines almost as a person's eyes would perceive them from one particular viewing location. A perspective drawing is somewhat like a photograph, *Figure 18-30*. The upper drawing shows the lines of a building from the outside; the lower one shows an inside view of a room. Both views are drawn in two-point perspective like several of the sketches in chapter 2.

Perspective Drawing Terms

Learn the terms used in perspective drawing while studying the figures and procedures that follow.

Station Point (SP).   The exact location from which the observer views the object, *Figure 18-31*.

Ground Line (GL).   The edge view of the surface upon which the object to be viewed rests, Figure 18-31.

Horizon (H).   A line at the level of the viewer's eyes, Figure 18-31, and the line where the vanishing points are located.

Outside view in two-point perspective

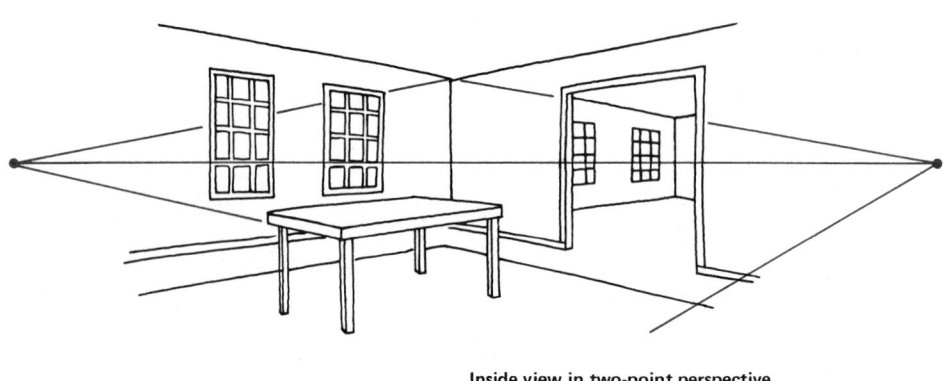

Inside view in two-point perspective

Figure 18-30 Outside and inside perspective views

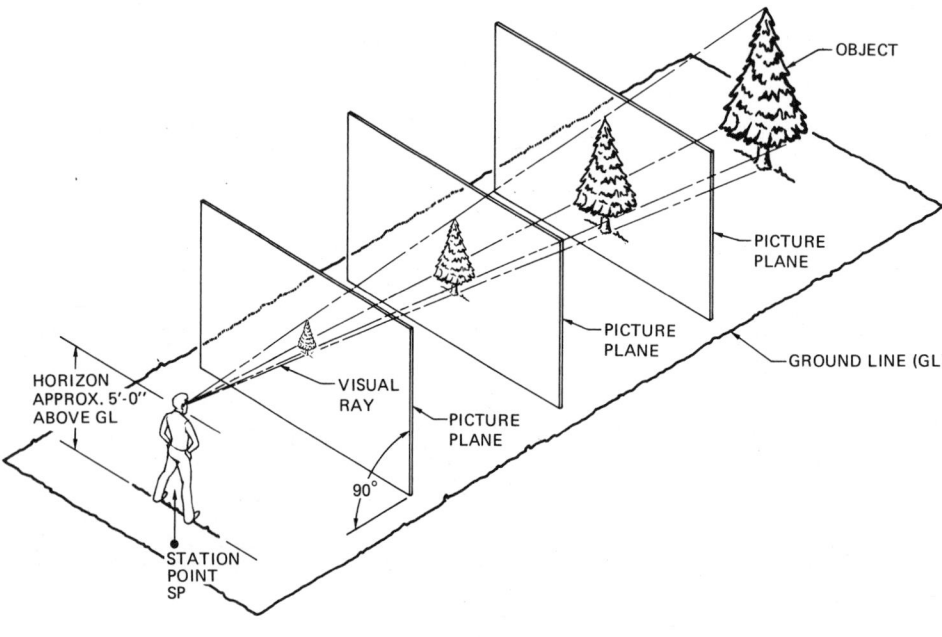

Figure 18-31 Perspective terms

Vanishing Point (VP). A point on the horizon line to which all other lines are projected, ***Figure 18-32***. Six cubes, A through F, are used to show which sides of an object would be seen for various left and right positions of the cubes in perspective drawings. Note that if the dimension X shown is too great, a two-point perspective drawing would be needed.

Picture Plane (PP) (like the projection plane in multiview drawings). An imaginary plane perpendicular to the ground line, Figure 18-31.

Visible Rays (VR). Lines from the eyes of the viewer to each and every point on the object to be viewed. Where the visible rays intersect the picture plane creates the perspective drawing, Figure 18-31.

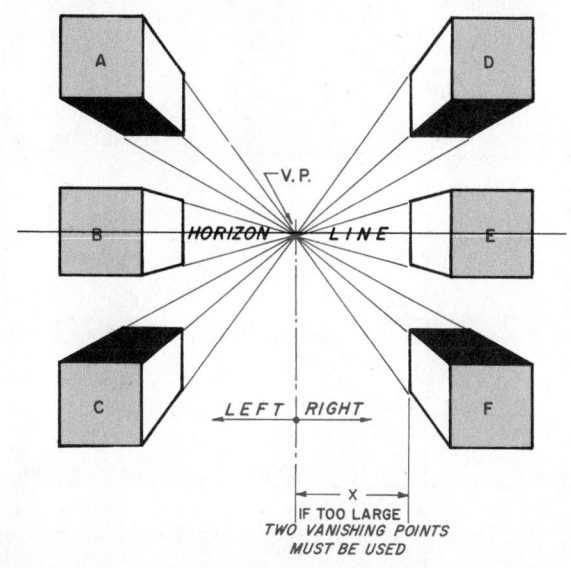

Figure 18-32   Vanishing point and horizon line

Measuring Line (M).   Actually used as a vertical line, but is appears horizontal because of the way it is established. The line originates at the projection plane and is 90° to it, *Figure 18-33D*. *Figures 18-33G* and *H* show the measuring line as a vertical line used to measure true heights.

Types of Perspective Views

Three types of perspective views are shown in *Figure 18-34*; one-, two- and three-point, respectively. Only two-point perspective is discussed in this chapter because it is the type most often used in industry.

Pre-sketching for Perspective Drawings

Make simple, quick sketches of an object to be drawn in two-point perspective to determine the best arrangement of the horizon, station point, ground line, and vanishing points, *Figures 18-33A* and *B*. Two of the multiviews are required for the two-point perspective (usually a top view and a front or side view).

How to Draw a Two-Point Perspective

Given: Multiview drawing *Figure 18-33A* and the sketches of the object in *Figure 18-33B*. Sketch C is selected as the best position to show the features of the object.

Step 1   On the upper portion of a drawing sheet, draw the top view at any orientation, but attempt to match the best position from the sketches, *Figure 18-33C*. Locate the front edge of the object tangent to the picture plane.

Step 2   From the front edge of the object, where it is tangent to the picture plane line, project the measuring line at 90° to the picture plane, *Figure 18-33D*. Locate the station point on the

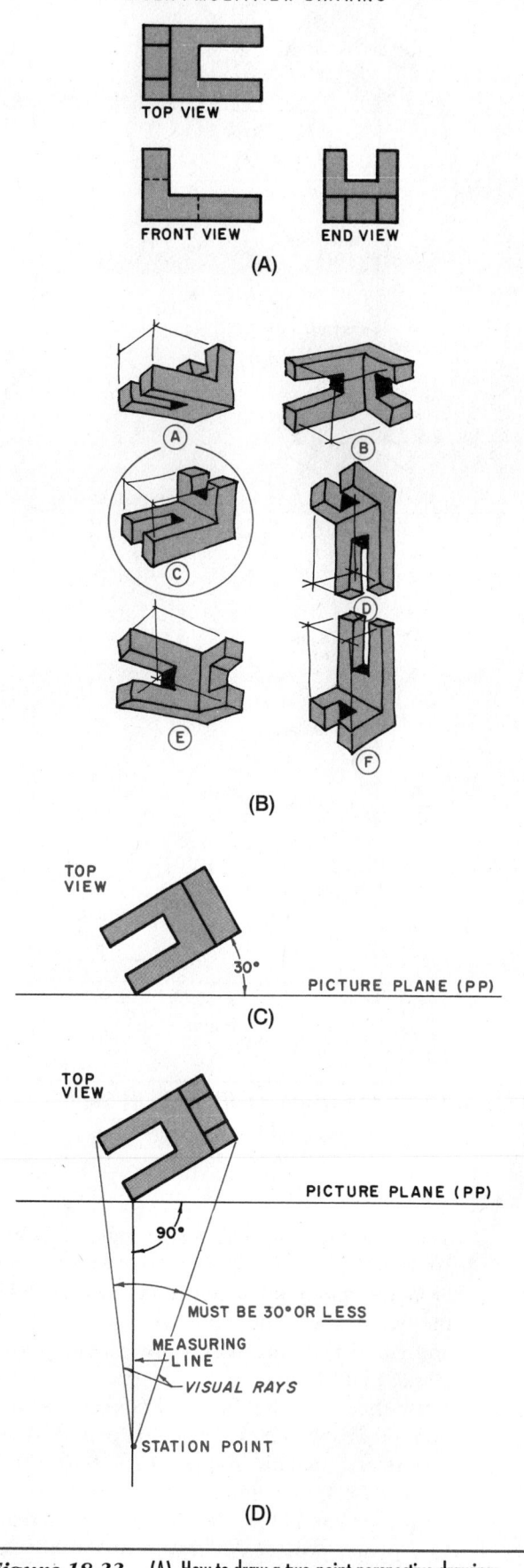

Figure 18-33   (A) How to draw a two-point perspective drawing; (B) Sketch of object in various positions; (C) Step 1; (D) Step 2

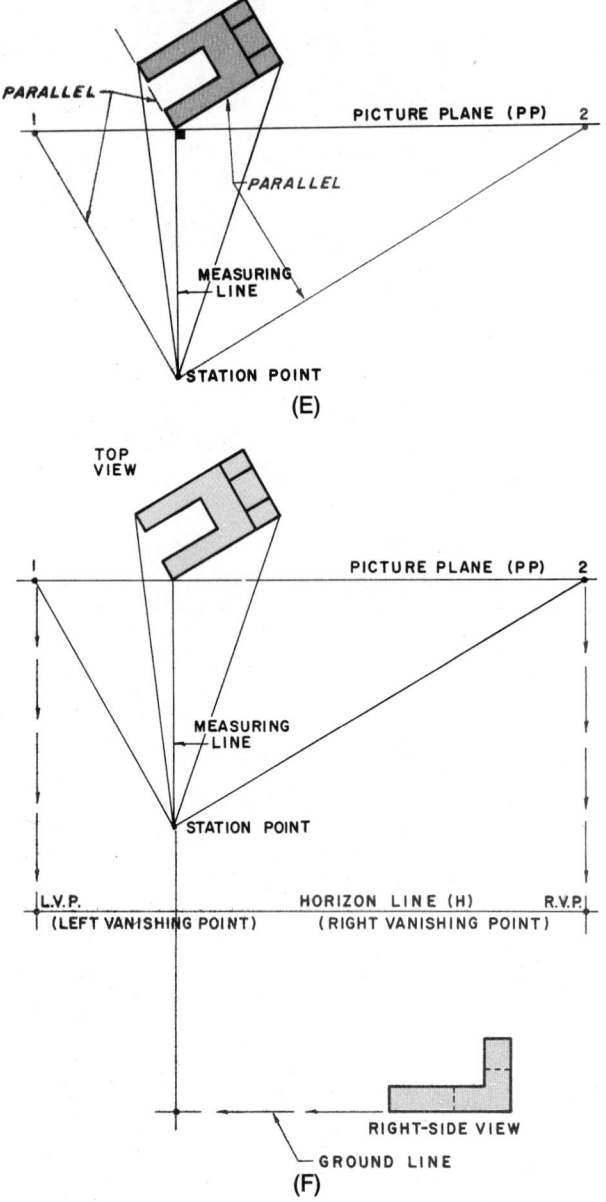

Figure 18-33 (Continued) (E) Step 3; (F) Step 4; (G) Step 5; (H) Step 6; (I) Step 7

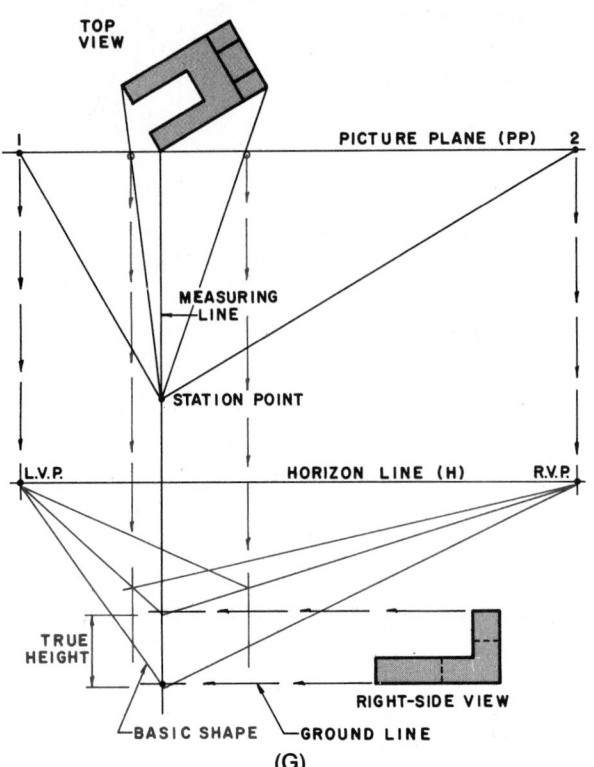

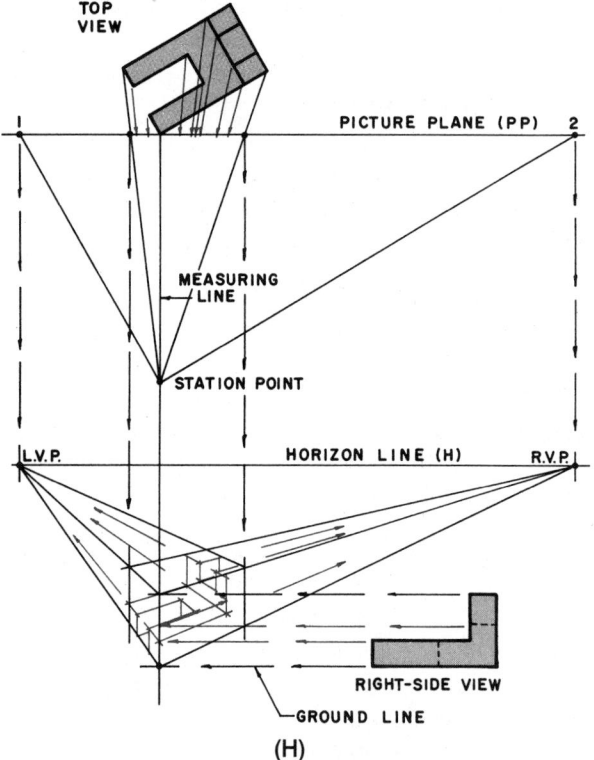

PERSPECTIVE VIEW

(I)

measuring line so the inclusive angle of view is 30° or less.

Step 3 From the station point construct lines parallel to the two edges of the top view, *Figure 18-33E*, and label their intersections with the picture plane, 1 and 2.

Step 4 Draw the horizontal line at a convenient location, preferably below the station point to avoid a cluttered drawing, *Figure 18-33F*. (Refer to *Figure 18-35* for an alternative position of the horizontal line.) From points 1 and 2 on the picture plane, project downward to the horizon line to locate the left and right vanishing points. Locate a ground line and draw a side or front view on the line as shown.

Step 5 From the view selected for the ground line, in this example the right-side view, project the true height of the object to the measuring line, Figure 18-33G. From the true height, project lines to the left and right vanishing points, and then, construct a perspective basic shape of the object. Note how the vertical boundaries of the basic shape are established by the top view.

Step 6 Project the true heights of each feature from the right-side view to the measuring line and back toward the appropriate vanishing point, to a point where the same feature intersects the picture plane line, Figure 18-33H. Develop the various features of the object in light lines until satisfied that the drawing is correct.

Step 7 Finally, darken the lines of the object, *Figure 18-33I*. Hidden lines are not usually drawn except to show an important feature that otherwise would not be seen.

Perspective Circles and Arcs

Divide the circle (or arc) into equal segments, *Figure 18-36*. Follow the steps from the previous example to construct the perspective drawing.

Perspective Irregular Curves

Use a multiview drawing and a grid as was done for the isometric drawing. Then follow the steps in the previous example, to construct a perspective drawing, Figure 18-36.

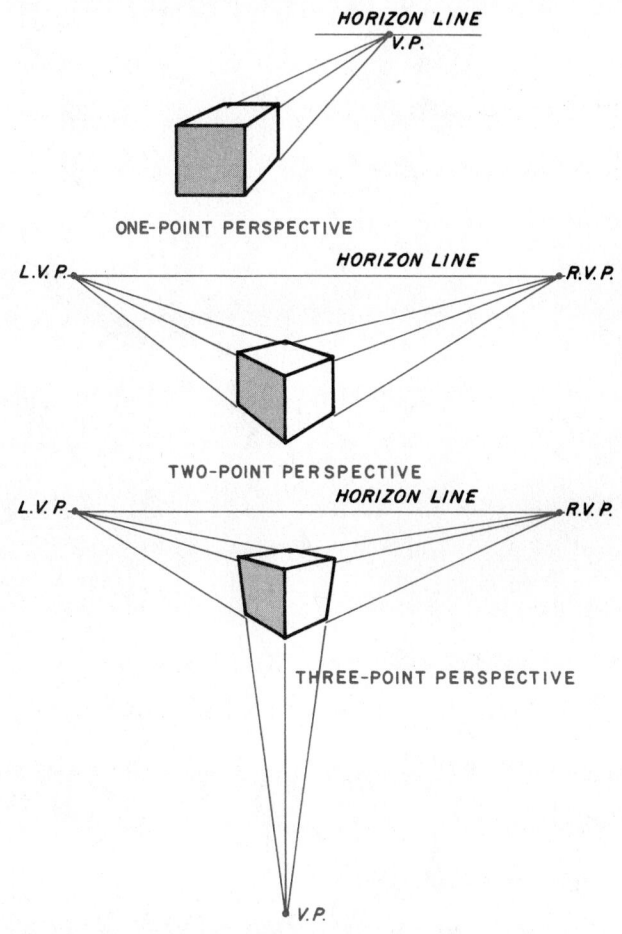

Figure 18-34 Kinds of perspective drawings

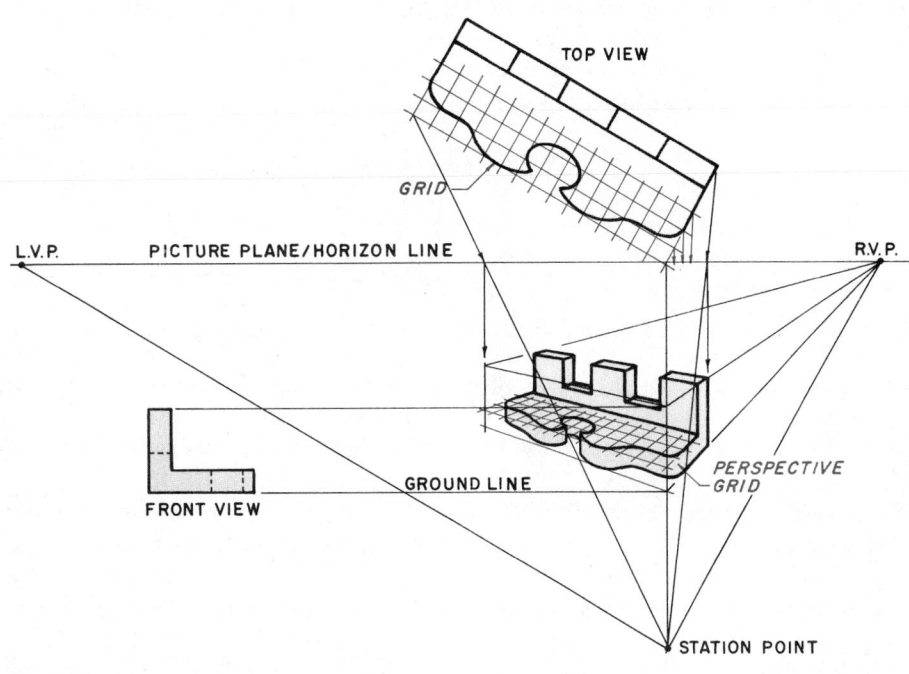

Figure 18-35 Perspective irregular curves

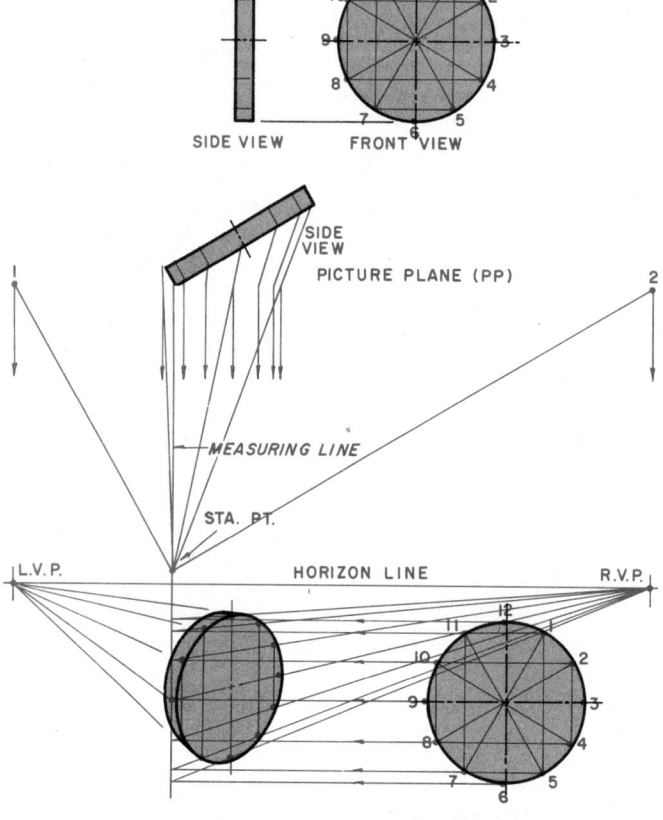

Figure 18-36 Perspective circles or arcs

Axonometric: Drawing and Projection

Drafters, designers, and engineers use the pictorial drawings—oblique, isometric and perspective—discussed in the first part of this chapter because approximations of these drawings are easy to sketch, and relatively accurate drawings of these types are easy to construct. Additional pictorial drawings and several projections will be considered next.

Isometric Projection

Isometric projection causes dimensions along the three principal edges to be foreshortened to approximately 82 percent of their true length, *Figure 18-37*. Note that the diagonal of the 1″ cube is seen as a point in the A view, the three edges 3-2, 3-4, and 3-7 equal 0.816″, and the angles between edges 3-2, 3-4, and 3-7 are all equal to 120°.

Isometric Assemblies

Isometric assembly drawings appear in two popular formats, exploded and assembled.

Exploded isometric drawings show all the components of an assembly as individual isometric drawings, including fasteners; however, the components are aligned in space the identical way they are to be assembled. See problem 18-52 for an exploded-assembly

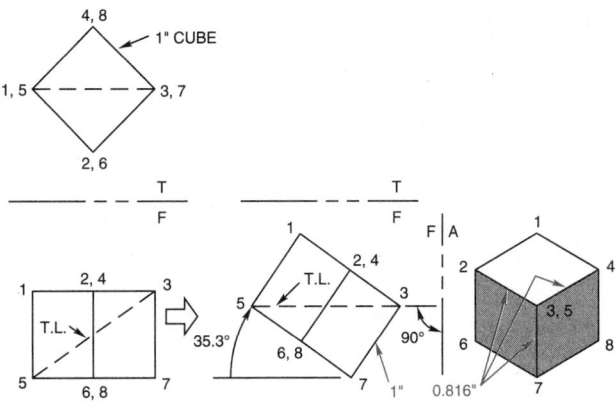

Figure 18-37 Isometric projection

sketch. Exploded-assembly instructional drawings usually accompany products that are to be assembled after the buyer purchases the product. Exploded-assembly drawings are used extensively in automotive maintenance manuals.

Assembled isometric assembly drawings show how a single product looks when assembled. For example, refer to the isometric drawing of a robot in chapter 20. Other isometric assembly drawings may show a large number of individual components, for example, a CIM system (computer-integrated manufacturing) in chapter 20, and a piping-system drawing in chapter 21.

Dimetric Drawing

Dimetric drawing differs from isometric drawing in that only two angles are equal. *Figure 18-38* shows a number of orientations for drawing dimetric views; lines of an object that are parallel to the first two axes are drawn full size, lines parallel to the third are drawn to a different scale. Because two different scales are used, less distortion is apparent and an object looks more realistic. One disadvantage, however, is that circles are difficult to draw in dimetric views.

Trimetric Drawing

Trimetric drawing uses axes that are rotated so that each axis is drawn at a different angle, and each axis uses a different scale, *Figure 18-39*. Trimetric drawings can appear more realistic than any of the other types, but they are the most time consuming to draw, so aren't used by drafters.

Other Oblique Drawings

Cavalier Drawing

Cavalier drawings have a receding axis drawn between 30° and 60° to the horizon, and the receding distances are drawn full size, *Figures 18-40* and *18-41*.

Cabinet Drawing

Cabinet drawings have the same range of receding axes as cavalier drawings, but the receding distances are drawn to half size, *Figure 18-42*.

General Oblique Drawing

General oblique drawing is similar to cabinet drawing except that the lines drawn along the direction of the receding may be at any scale the drafter chooses, *Figure 18-43*.

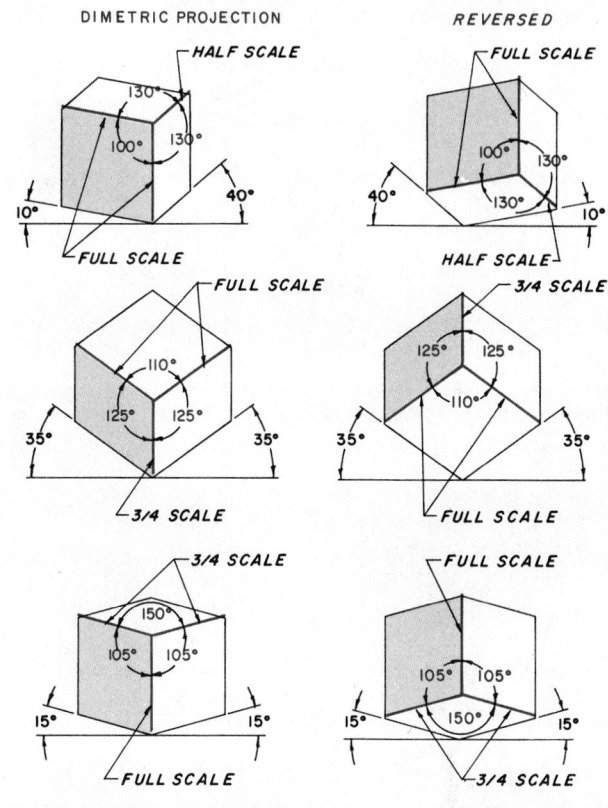

Figure 18-38   Dimetric drawing

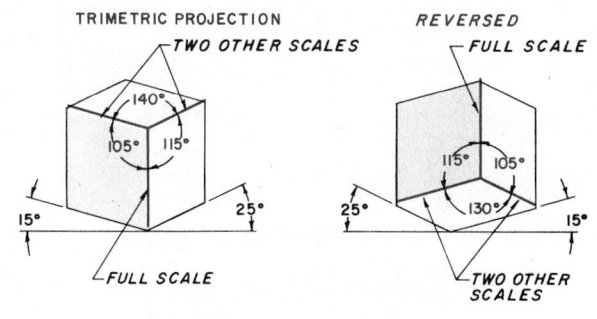

Figure 18-39   Trimetric drawing

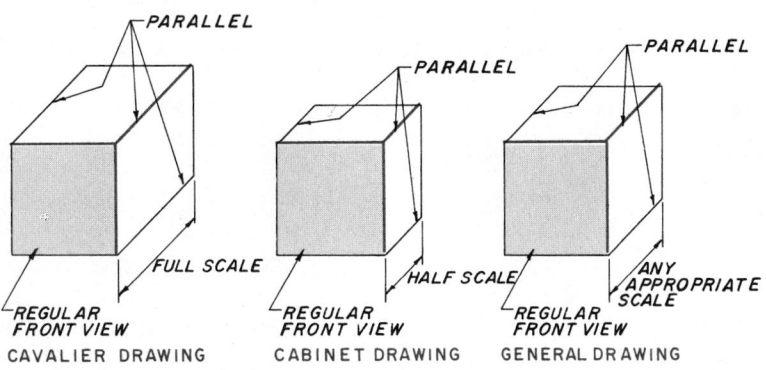

Figure 18-40   Oblique drawing

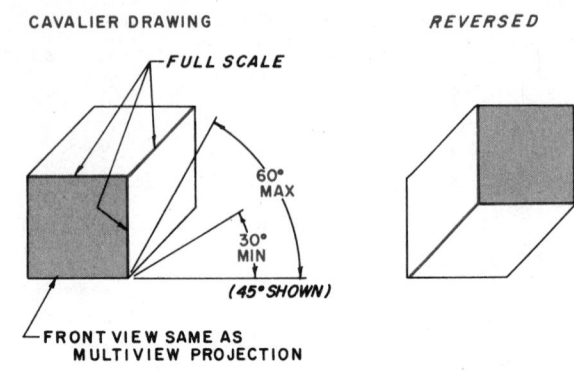

Figure 18-41   Cavalier drawing

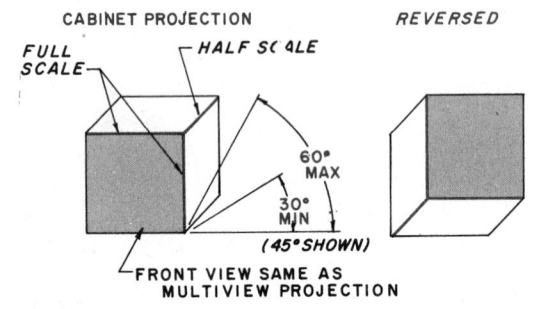

Figure 18-42   Cabinet drawing

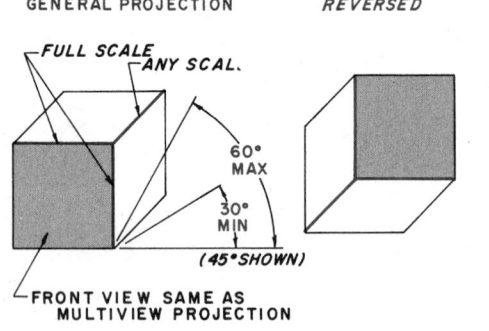

Figure 18-43   General oblique

Pictorials by Computers

Most of the pictorial drawings discussed in this chapter are done on computers—obliques, isometrics, and perspectives. Some computer programs allow the operator to "walk" through a perspective of a ship, a building, or an airport. Other computer programs show solid models of objects. Several of these techniques are discussed elsewhere in the text.

Review

1. Explain when the true isometric circle method is preferred over the four-center isometric circle method, even though the true method is much slower.
2. Explain how an isometric drawing is dimensioned.
3. Explain the used of box construction and why it is used.
4. Under what classification do isometric, dimetric, and trimetric drawings fall?
5. List the three types of pictorial drawings most used in industry.
6. What are the principal edges of a pictorial drawing?
7. Which kind of pictorial drawing best illustrates the object as it would be viewed by the eye?
8. Describe two ways to draw an irregular curve in an isometric drawing.
9. What is meant by a nonisometric line?
10. How are pictorial drawings used today?
11. List three kinds of oblique drawings. Explain their differences.
12. When are hidden lines used in pictorial drawings?
13. What is the difference between an isometric drawing and an isometric projection?
14. Describe how to construct a two-point perspective drawing. Use an example object.

Chapter Eighteen Problems

The following problems are intended to give the beginning drafter practice in laying out and developing isometric and perspective renderings of various objects.

The steps to follow in laying out the drawings in this chapter are:

Step 1 Study the problem carefully.
Step 2 Position the object so it *best* illustrates the most features.
Step 3 Make a pictorial sketch of the object.
Step 4 Calculate the basic shape size.
Step 5 Find the center of the work area.
Step 6 Lightly center and lay out the basic shape within the work area.
Step 7 Lightly develop the object *within* the basic shape.
Step 8 Check to see that the completed object is centered within the work area.
Step 9 Check all size dimensions, and check for accuracy of object.
Step 10 Darken in the object.
Step 11 Recheck all work, and, if correct, neatly fill out the title block using light guidelines and neat lettering.

Problems 18-1 through 18-27

Construct an isometric drawing of the object in Problems 18-1 through 18-27. (These problems have only *straight* lines.)

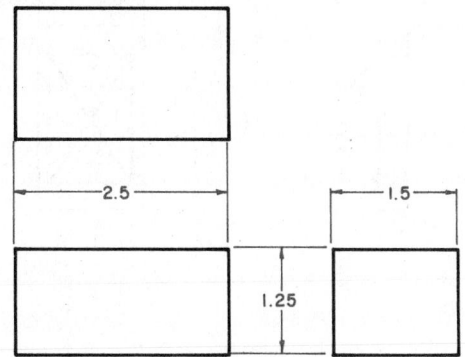

Problem 18-1

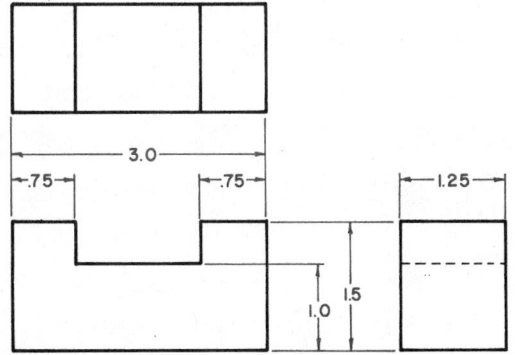

Problem 18-2

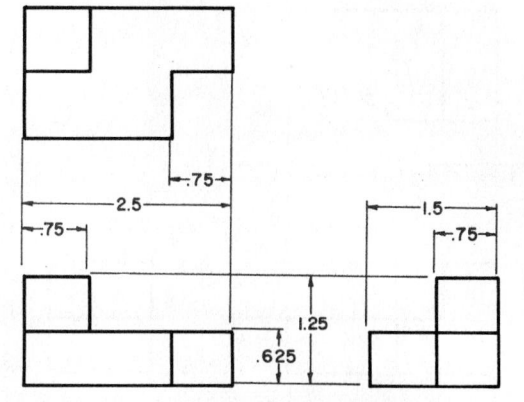

Problem 18-3

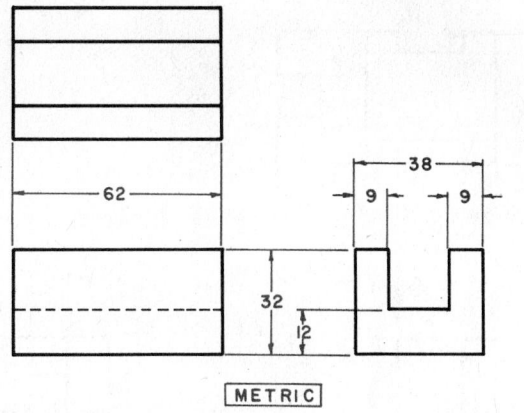

Problem 18-4

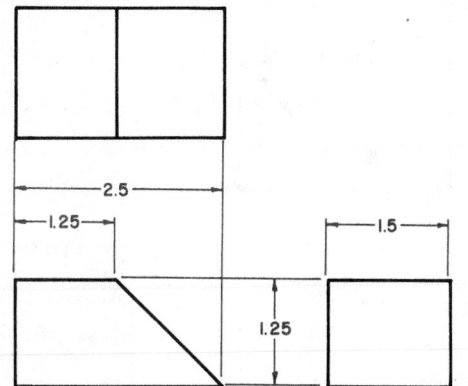

Problem 18-5

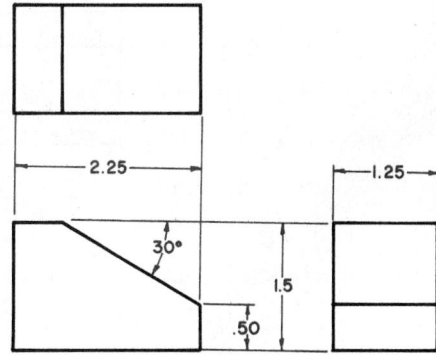

Problem 18-6

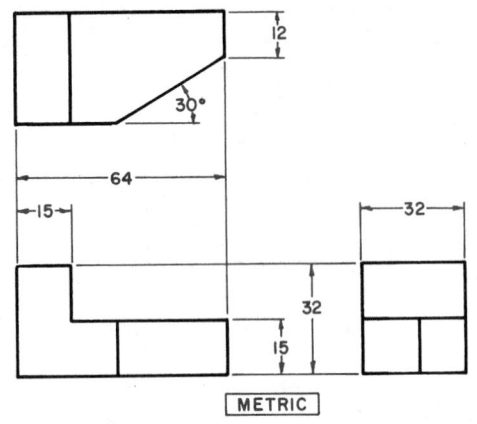

METRIC

Problem 18-7

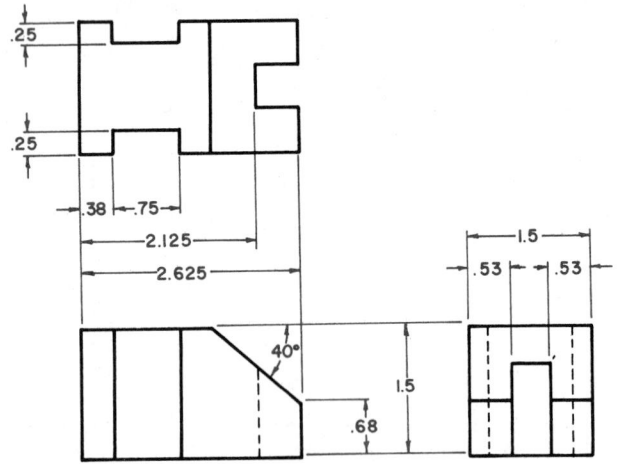

Problem 18-8

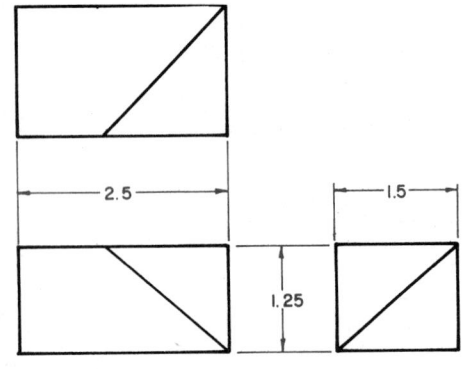

Problem 18-9

Problem 18-11

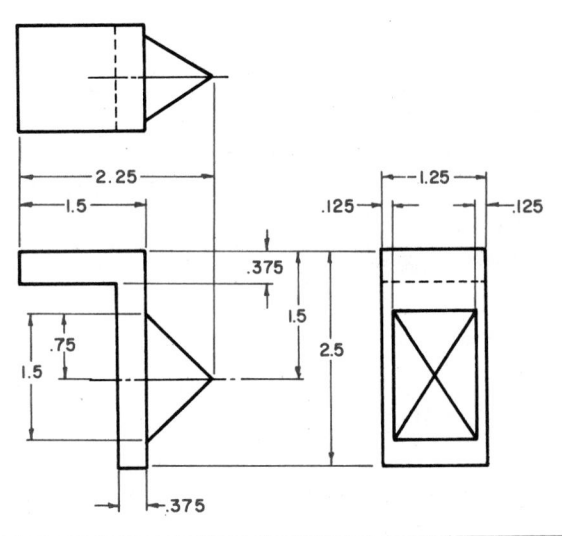

Problem 18-12

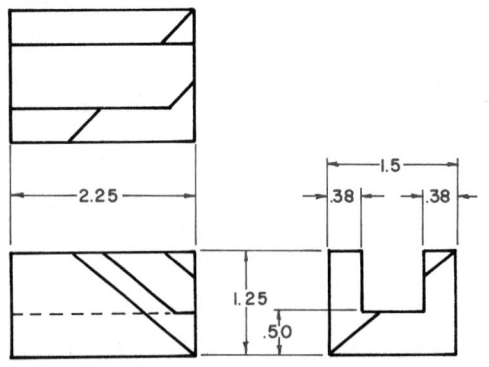

Problem 18-10

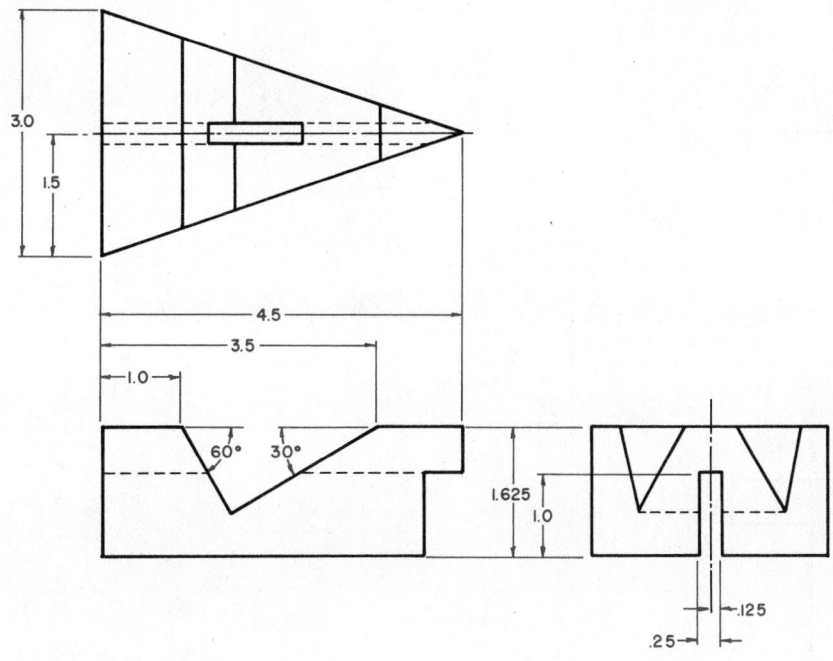

Problem 18-13

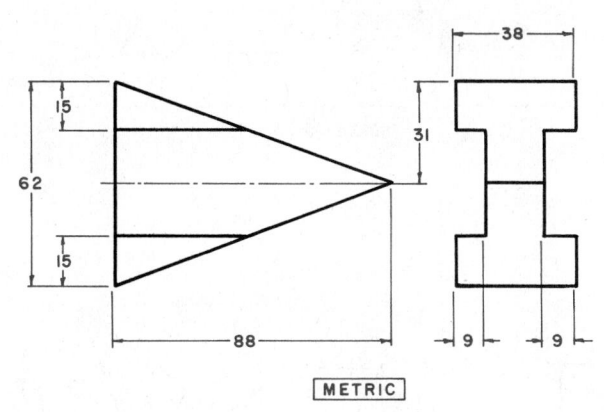

METRIC

Problem 18-14

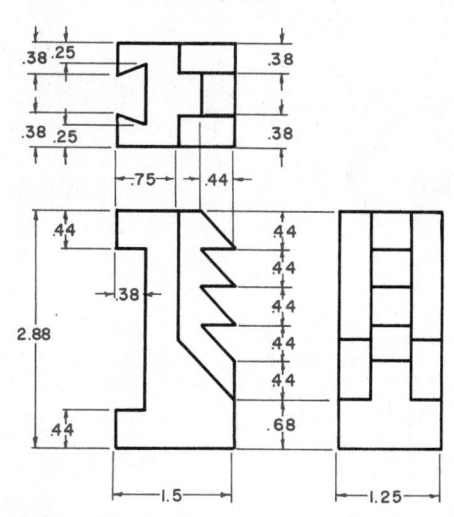

Problem 18-15

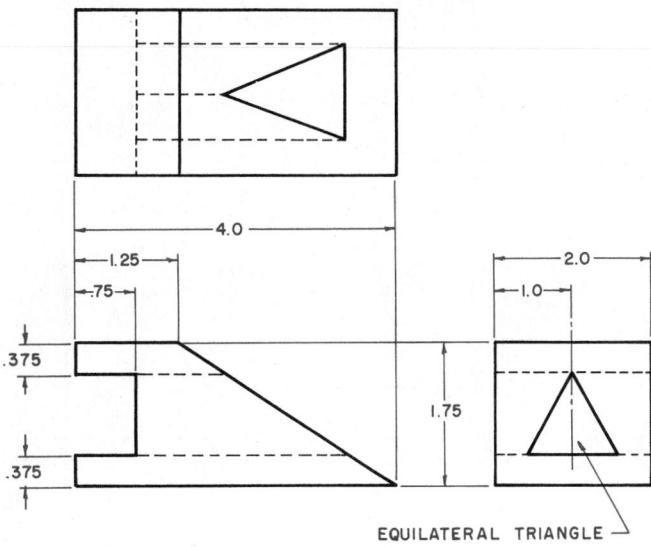

EQUILATERAL TRIANGLE

Problem 18-16

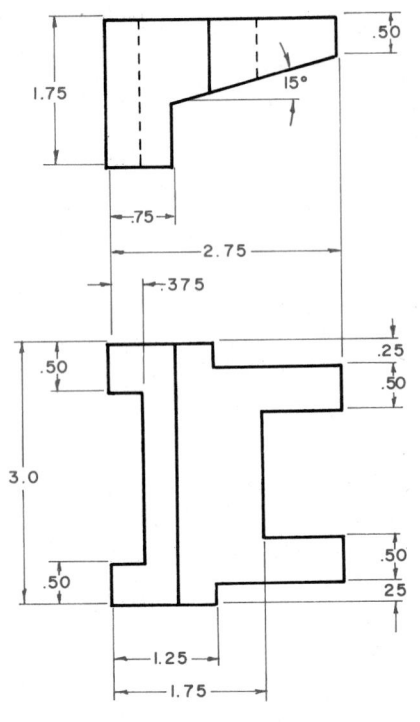

Problem 18-17

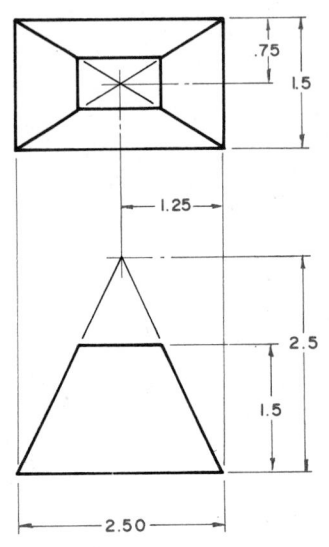

Problem 18-19

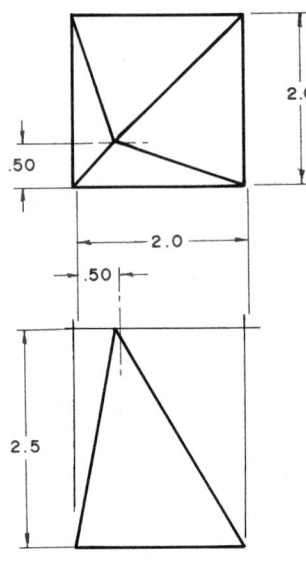

Problem 18-21

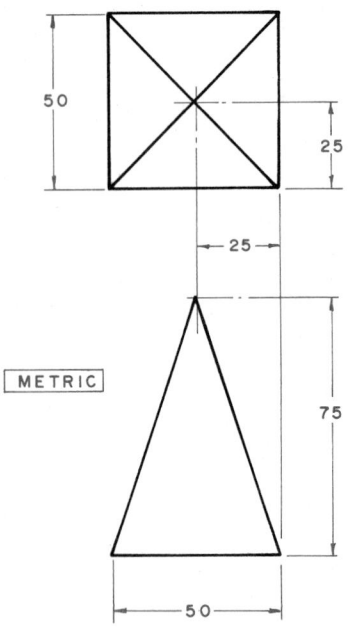

Problem 18-18

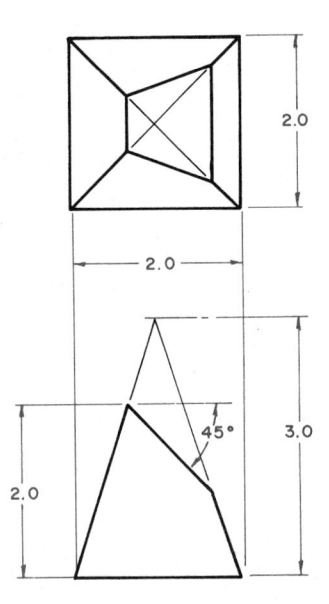

Problem 18-20

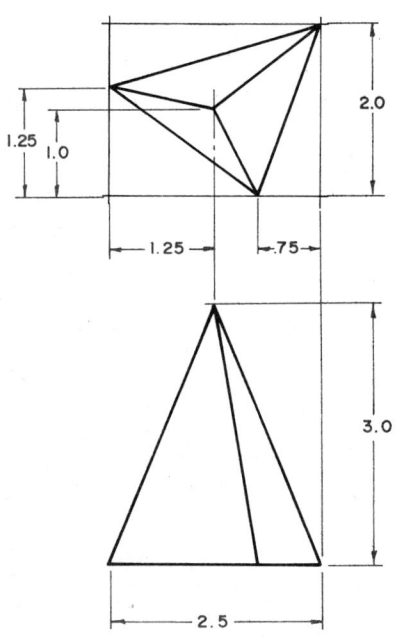

Problem 18-22

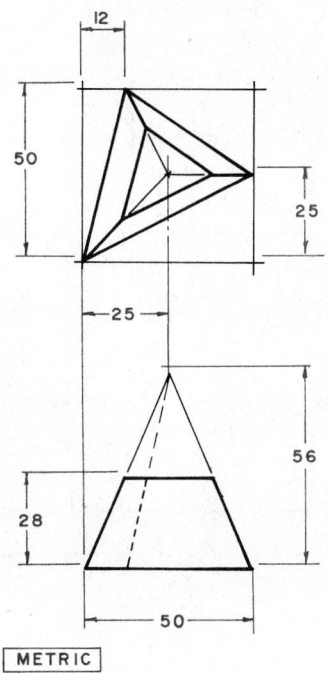

METRIC

Problem 18-23

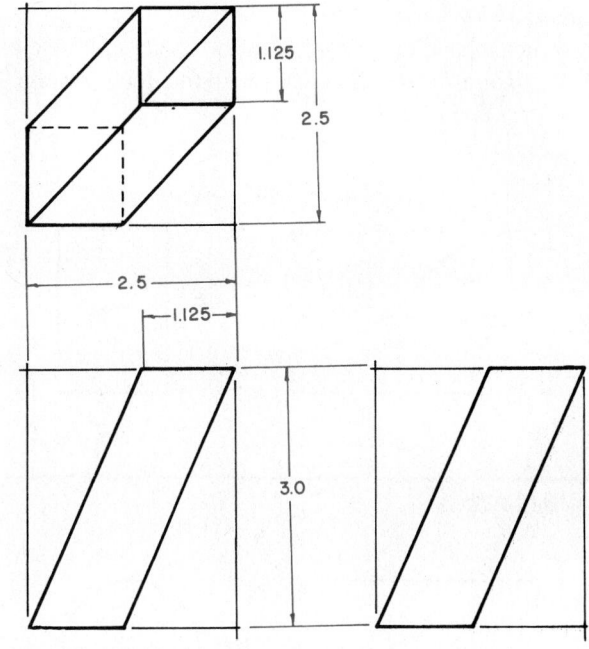

Problem 18-25

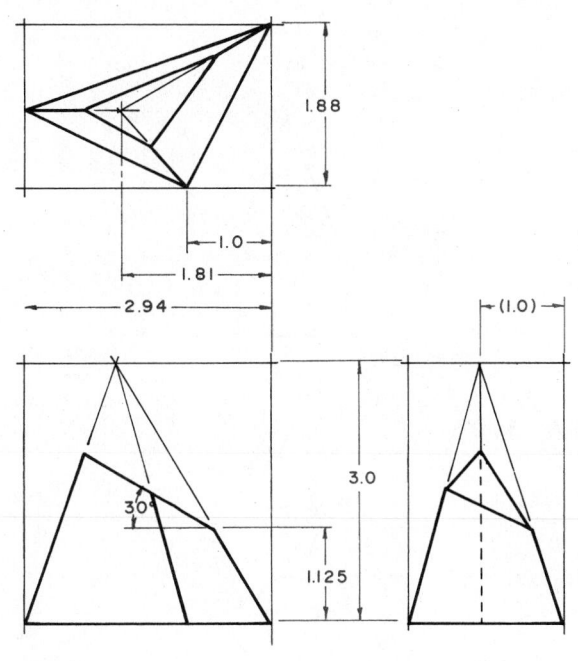

Problem 18-24

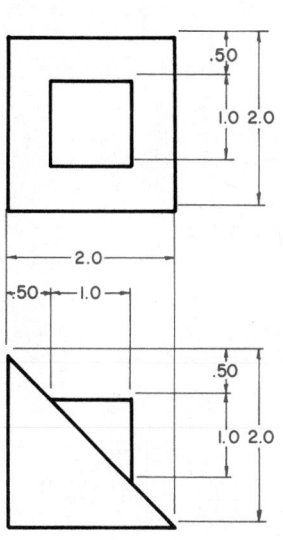

Problem 18-26

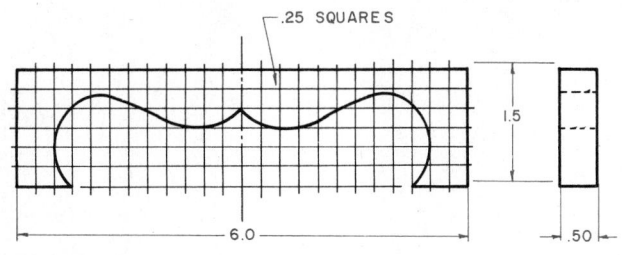

Problem 18-27

Problems 18-28 through 18-51

Construct an isometric drawing of the object in Problems 18-28 through 18-51. (These problems have *straight lines*, *arcs* and *circles*.)

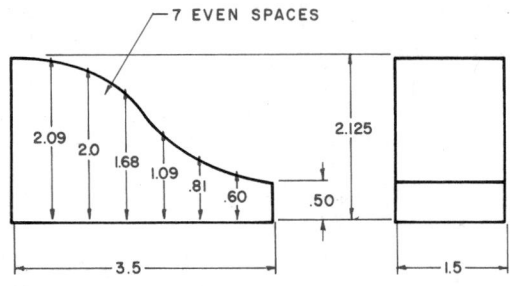

Problem 18-28

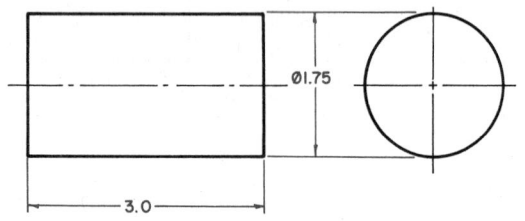

Problem 18-29

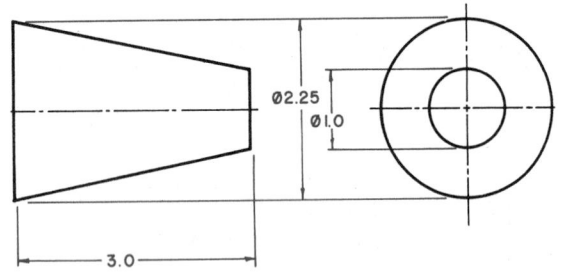

Problem 18-30

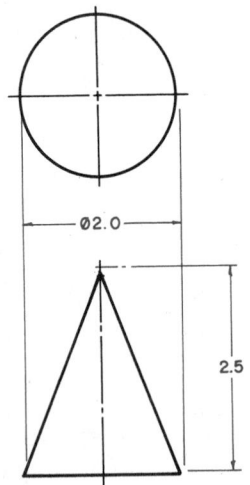

Problem 18-31

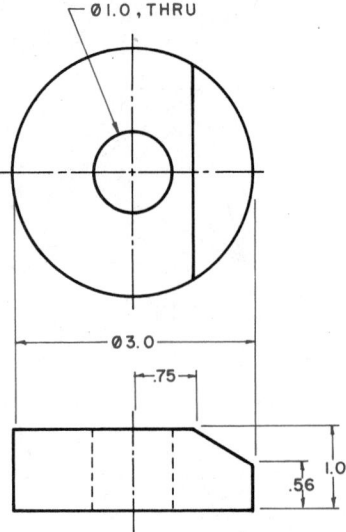

Problem 18-32

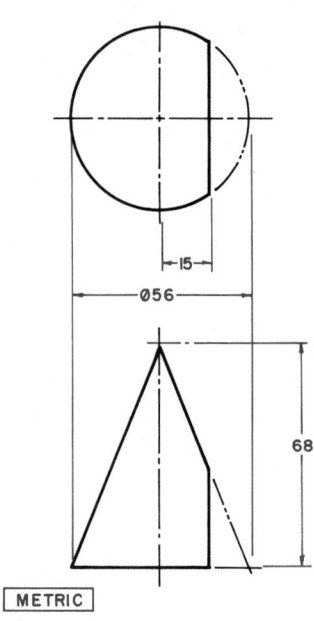

Problem 18-33

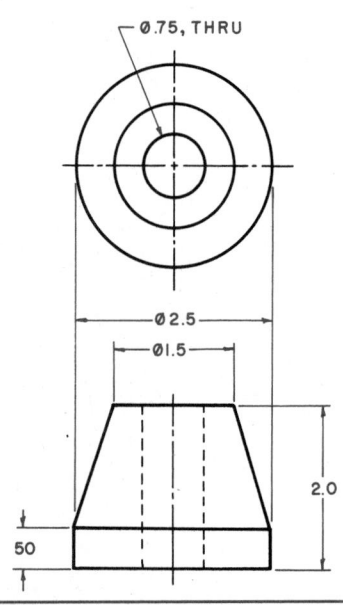

Problem 18-34

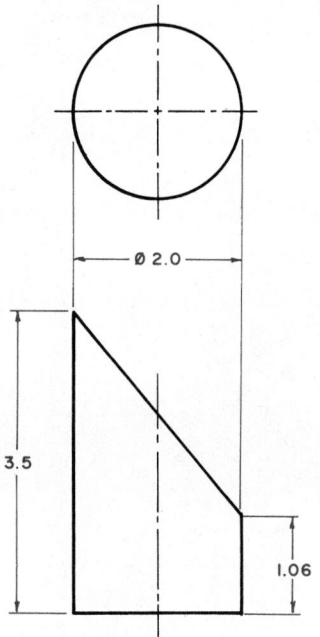

Problem 18-35

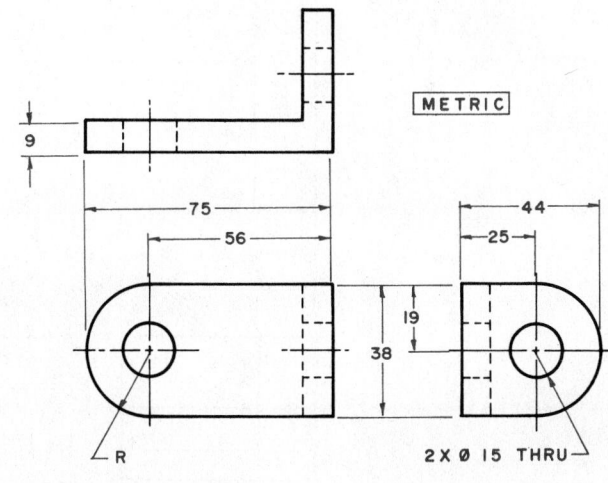

Problem 18-38

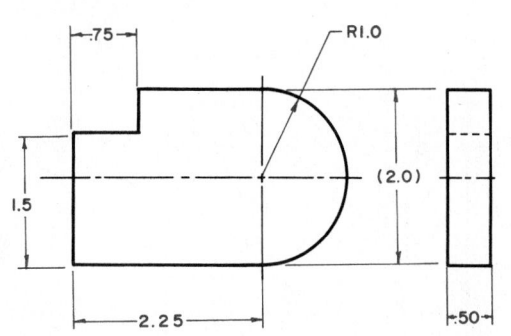

Problem 18-36

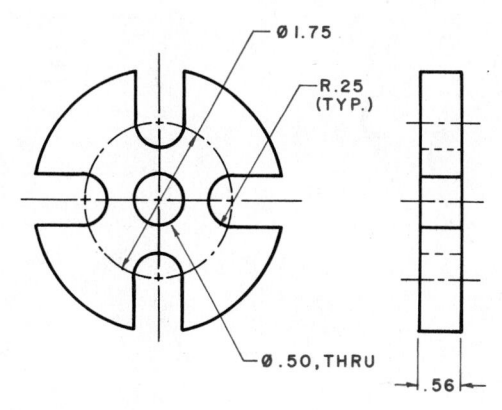

Problem 18-39

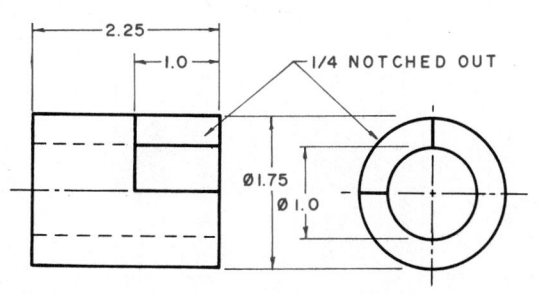

Problem 18-37

Problem 18-40

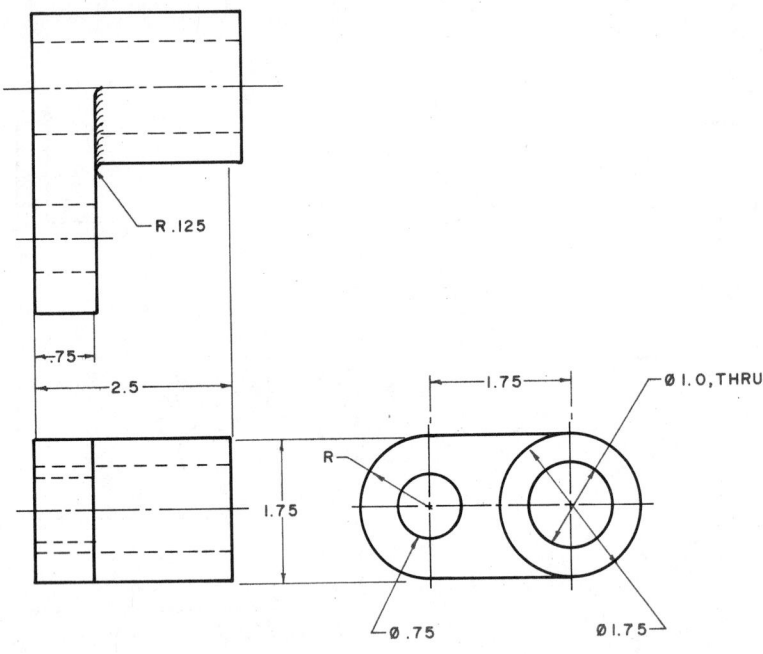

Problem 18-41

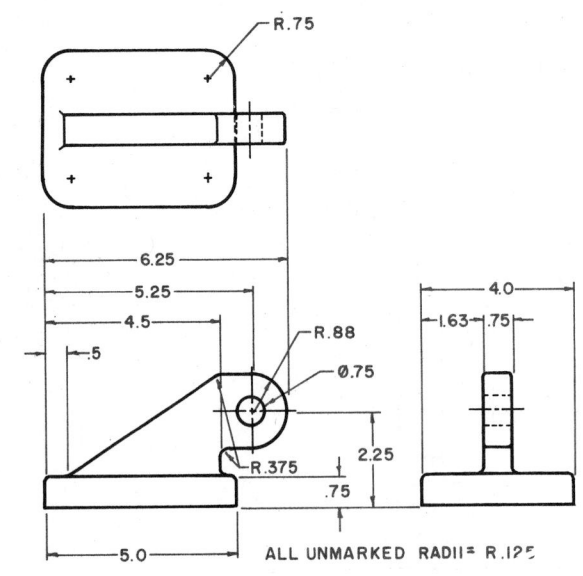

Problem 18-42

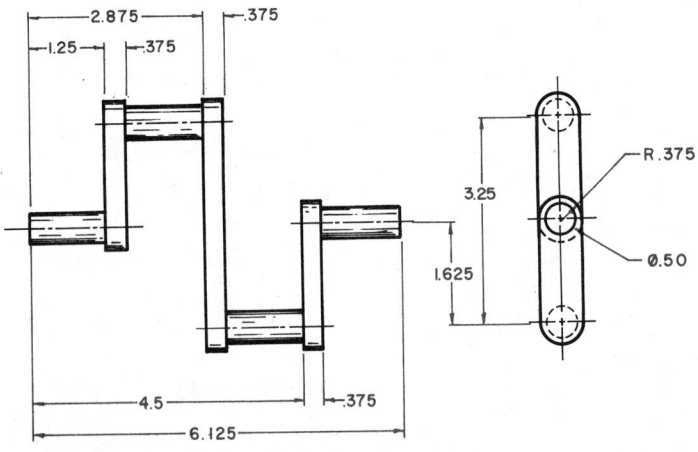

Problem 18-43

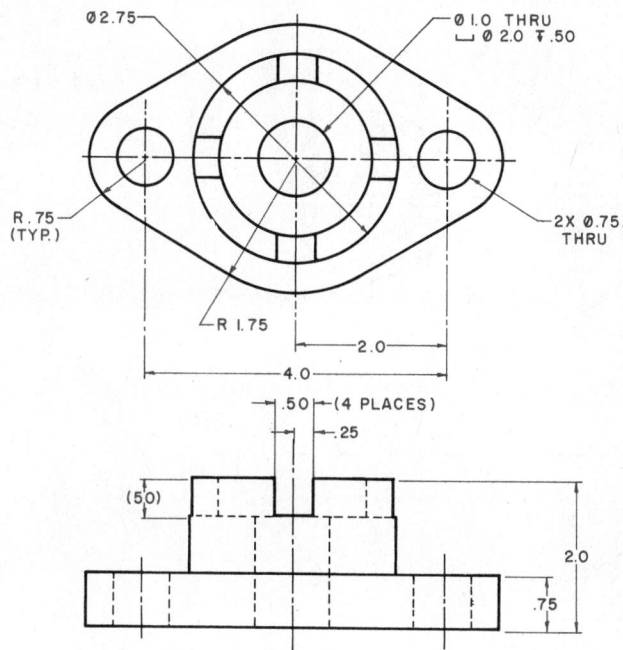

Ø2.75

Ø1.0 THRU
⌴ Ø2.0 ⊤.50

R.75
(TYP.)

2X Ø.75,
THRU

R 1.75

2.0

4.0

.50 (4 PLACES)
.25

(.50)

2.0

.75

Problem 18-44

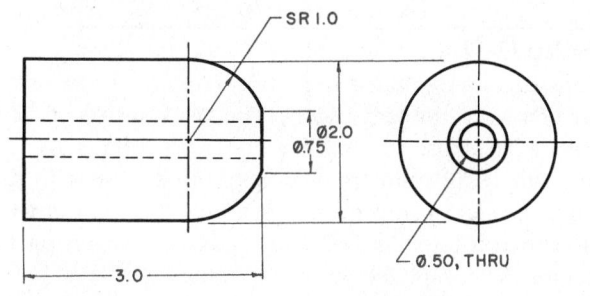

SR 1.0

Ø2.0
Ø.75

Ø.50, THRU

3.0

Problem 18-45

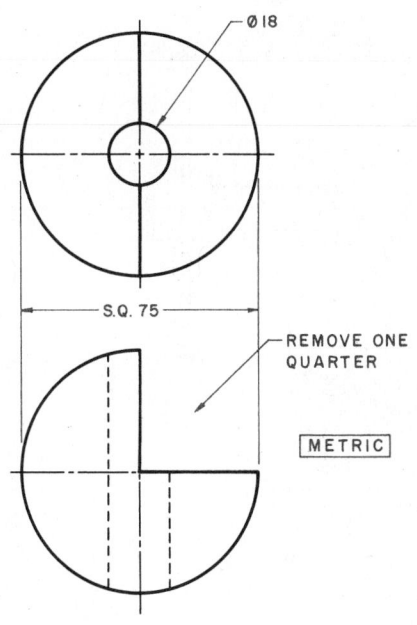

Ø18

S.Q. 75

REMOVE ONE
QUARTER

METRIC

Problem 18-46

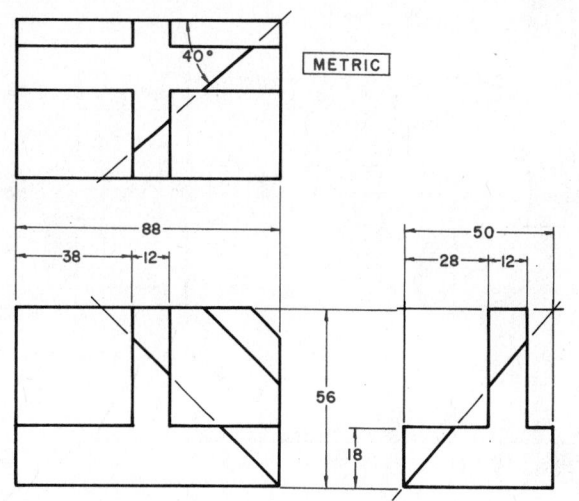

2.0

1.0

2.0

1.0

1.0

2.0

3X 1.5, THRU

Problem 18-47

40°

METRIC

88
38
12

56

18

50
28
12

Problem 18-48

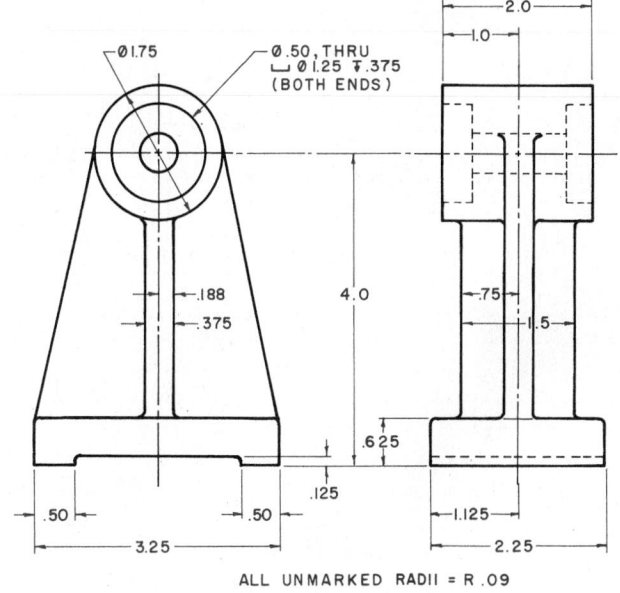

Ø1.75

Ø.50, THRU
⌴ Ø1.25 ⊤.375
(BOTH ENDS)

2.0

1.0

4.0

.188
.375

.75

1.5

.625

.125

.50

.50

1.125

3.25

2.25

ALL UNMARKED RADII = R .09

Problem 18-49

Problem 18-50

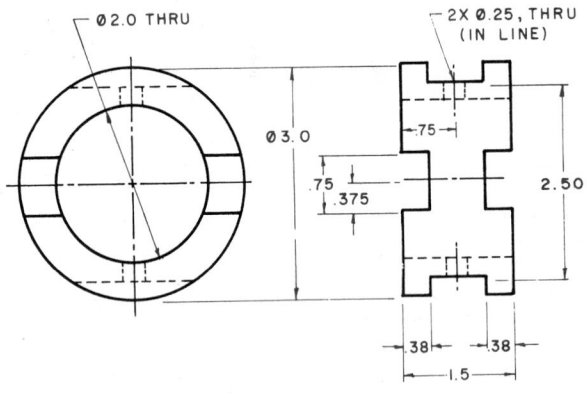

Problem 18-51

Problem 18-52

Construct an exploded isometric drawing of the assembly of a vise, Problem 18-52A. The vise consists of 7 different parts and various fasteners, Problems 18-52B through 18-52I. Illustrate the exploded view with the parts in relationship to its assembly. Space parts with an approximate 1″ (25 mm) space between parts. Center the completed view within the work area.

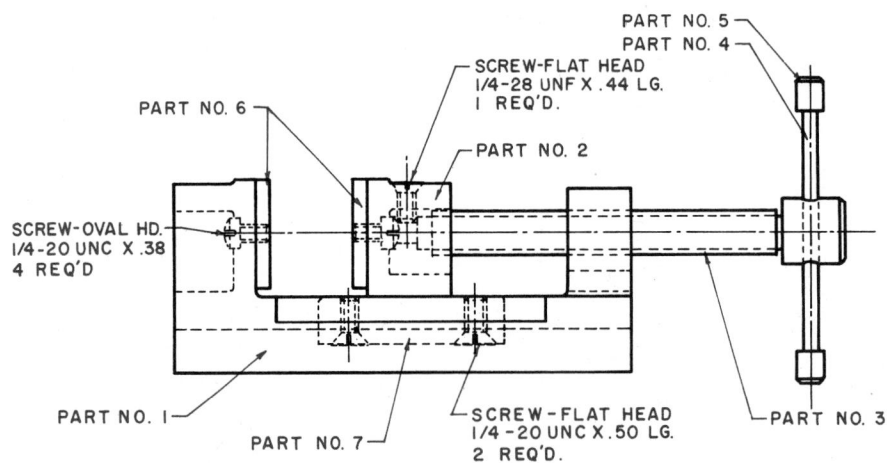

Problem 18-52A

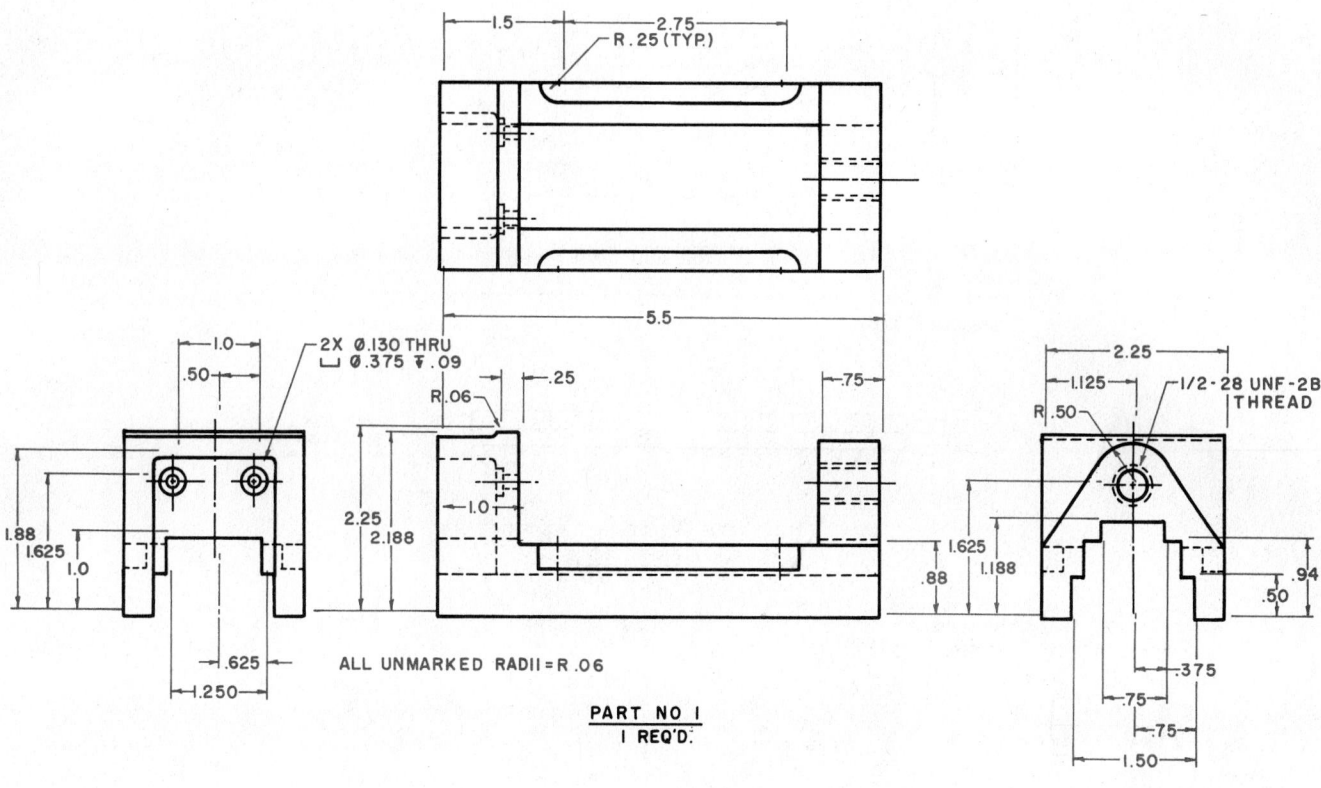

ALL UNMARKED RADII = R .06

PART NO I
I REQ'D.

Problem 18-52B

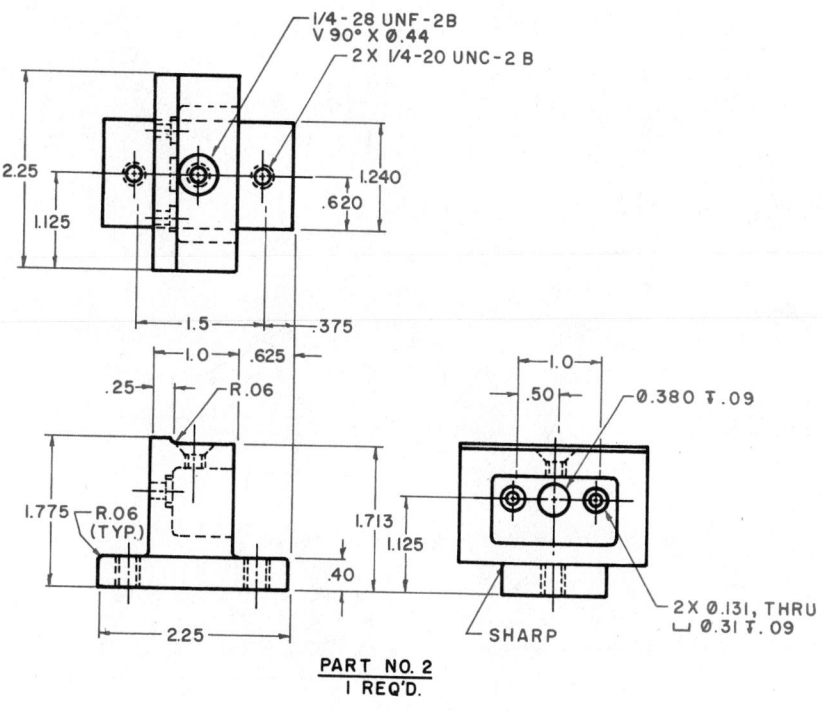

PART NO. 2
I REQ'D.

Problem 18-52C

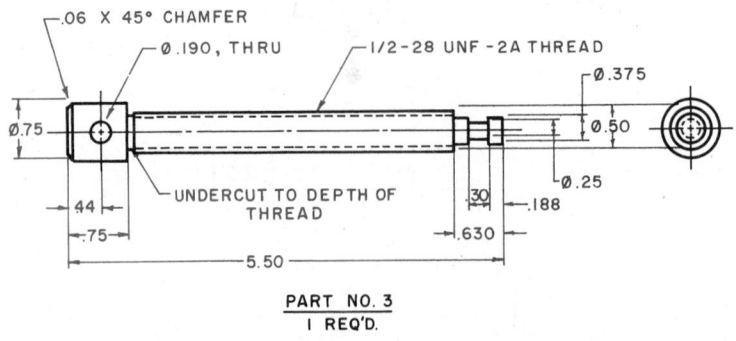

.06 X 45° CHAMFER
Ø.190, THRU
1/2-28 UNF-2A THREAD
Ø.375
Ø.50
Ø.75
Ø.25
UNDERCUT TO DEPTH OF THREAD
44
30
.188
.75
.630
5.50

PART NO. 3
1 REQ'D.

Problem 18-52D

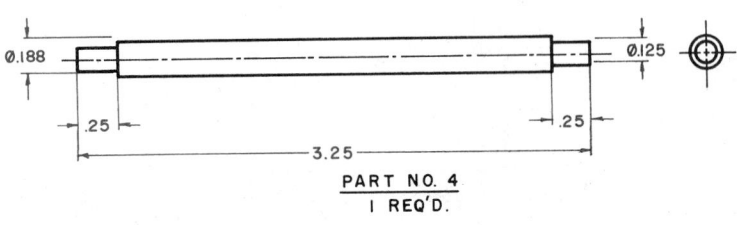

Ø.188
Ø.125
.25
.25
3.25

PART NO. 4
1 REQ'D.

Problem 18-52E

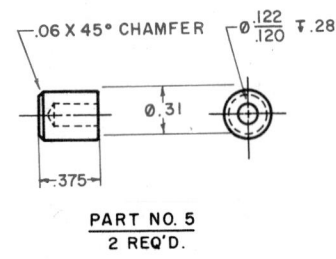

.06 X 45° CHAMFER
Ø.122/.120 ⊤.28
Ø.31
.375

PART NO. 5
2 REQ'D.

Problem 18-52F

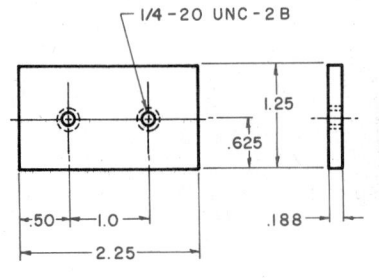

1/4-20 UNC-2B
1.25
.625
.50
1.0
.188
2.25

Problem 18-52G

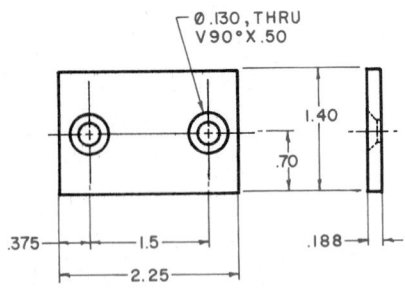

Ø.130, THRU
V90°X.50
1.40
.70
.375
1.5
.188
2.25

Problem 18-52H

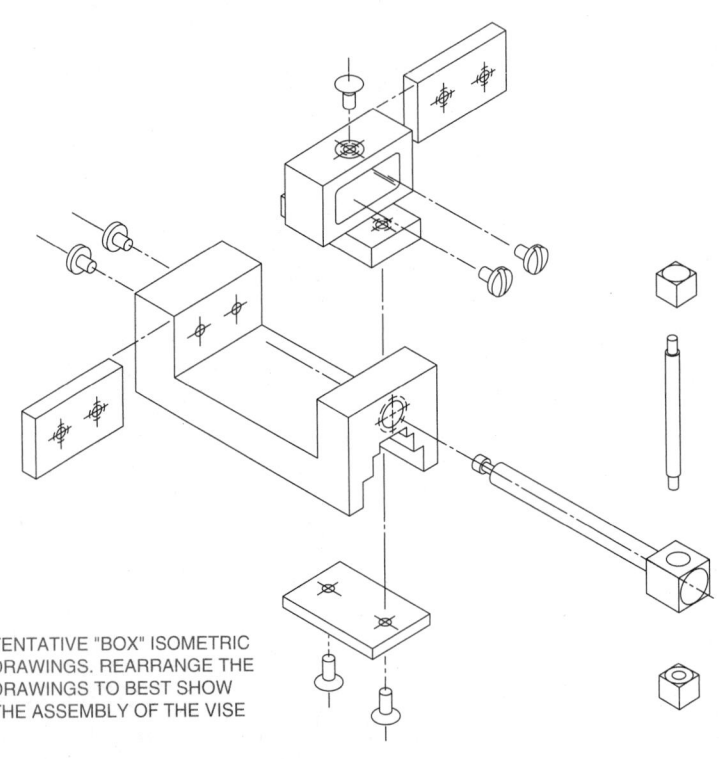

TENTATIVE "BOX" ISOMETRIC DRAWINGS. REARRANGE THE DRAWINGS TO BEST SHOW THE ASSEMBLY OF THE VISE

Problem 18-52I

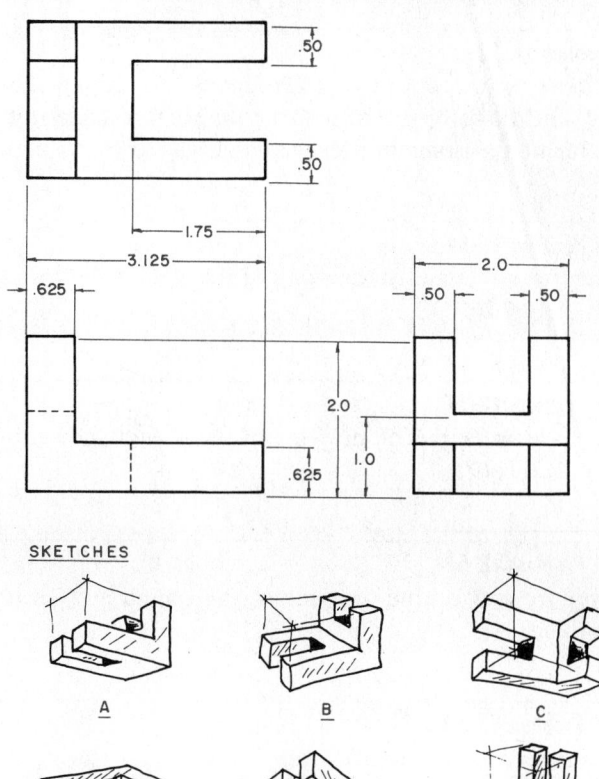

Problem 18-53

Using the given multiview drawing in Problem 18-53, develop a perspective drawing in the positions as illustrated by sketches A, B, C, D, E and F.

SKETCHES

Problem 18-53

Problem 18-54

Using the given multiview drawing in Problem 18-54, develop a perspective drawing in the positions as illustrated by sketches A and B.

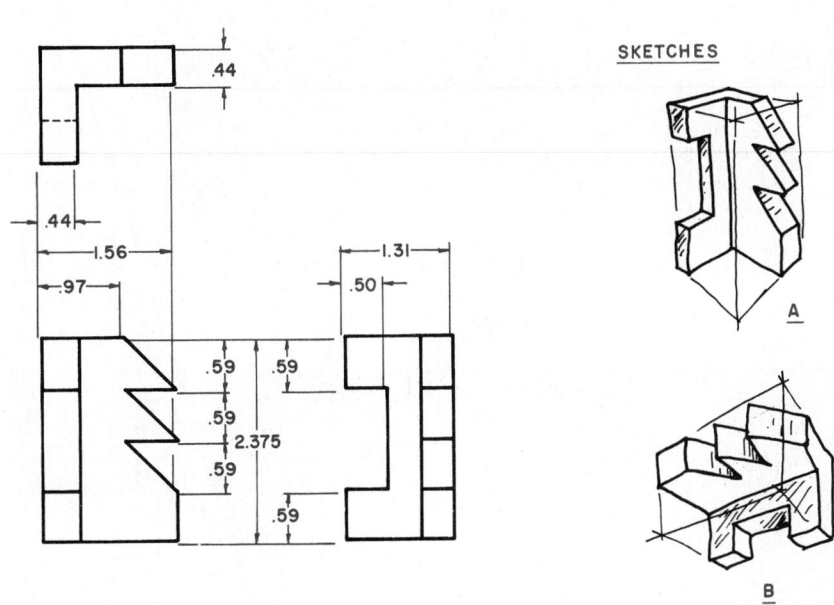

SKETCHES

Problem 18-54

Problem 18-55

Choose various objects from Problems 18-1 through 18-51, and develop them into a perspective drawing. Compare the isometric view with the isometric drawing.

❏ **Problem 18-56**

Prepare an isometric drawing of the object shown in Problem 5-26.

❏ **Problem 18-57**

Prepare an isometric drawing of the object shown in Problem 8-49.

❏ **Problem 18-58**

Prepare an oblique drawing of the object shown in Problem 9-13.

❏ **Problem 18-59**

Prepare an oblique drawing of the object shown in Problem 9-40.

❏ **Problem 18-60**

Redesign parts 2 and 3 of the vise in problem 18-52. Propose a more positive means for connecting part 2 to part 3. Show the redesign in any combination of the following formats: multiview drawings, section drawings, isometric drawings, oblique drawings, perspective drawings, etc.

❏ **Problem 18-61**

Select an assembled product (one that can be disassembled easily, or if necessary destroyed during disassembly) and draw an exploded, approximate isometric drawing of the components. The objective is to show how the product works. Suggestions: household items, automotive components, office devices (pencil sharpener, file drawer mechanism…), sports equipment, weight-training apparatus, laboratory equipment, classroom demonstrations, etc.

❏ **Problem 18-62**

Design a means, made of simple parts, to raise water, one meter, from one irrigation canal to another, for a third-world country. Employ animal power or a windmill-type apparatus. Prepare an exploded isometric of the components that make up the design.

❏ **Problem 18-63**

Design a wooden toy (but metal for axles, fasteners, etc.) to be made out of simple shapes; for example, a train engine made of blocks, cylinders, circles, cones, etc. Prepare an exploded isometric showing how the pieces are to go together and how they are to be joined.

Related Technology

Career Profile: Kapriel Karagozyan

Kapriel Karagozyan is a senior applications engineer for DSP Products at Analog Devices. DSP applications engineers provide technical support on DSP processor and related development tools such as assemblers, compilers, simulators, development boards, and emulators.

The role of an applications engineer is a combination of many types of responsibilities. There is really no "typical" day for Kapriel but rather a number of different types of days. Kapriel works with design engineers, product engineers, marketing engineers, sales people, and customers. His work day is usually dictated by what type of project he is involved in at the moment or for what type of technical support he is responsible. He provides technical support in both a pre- and post-sales environment as well as in a product development environment.

Pre-sales support consists of accompanying sales people into a prospective customer site to make technical presentations about the company's products—features and capabilities. Often competitors are also under consideration by the potential customer and Kapriel is expected to compare and contrast technical approaches based on competing opponents as well. Usually, Kapriel spends some time preparing for the visit by gathering background information about the other company, strategizing with our marketing people, and preparing some related applications information.

In meetings with the prospective client company, Kapriel finds out what their engineers are trying to accomplish, what the end product will be, and what is the schedule for production. Once he has that information, he works with the engineers to sketch and propose several technical alternatives for their design. Post-sales support consists of keeping the customers happy once they have decided to use DSP's processors.

Applications engineers take turns signing up for telephone duty on the applications assistance line, which customers turn to for help with design prob-lems. Sometimes, the customer's situation needs to be recreated in the lab to analyze the problem and test out various potential solutions.

On days when Kapriel is in the office, most of his time is spent in the lab, working either on a circuit board or in front of a pc writing or debugging a DSP code.

Application notes serve as a "how to" guide and show how a DSP processor can be used in a particular application such as speech, telecommunications, audio, or video. Kapriel is responsible for both hardware and software aspects of the application note, which can include program listings for DSP applications codes as well as schematics of a relevant circuit. Writing an application note involves researching the application, implementing the application in the lab, documenting what is done, and finally publishing the application note.

Kapriel participates in several technical project and product development teams that include various hardware/software engineers and members of the marketing staff. They meet to decide how to set specifications for a new product, how to better promote existing products, and to discuss issues related to the release of a new product. Support needs of a particular product or tactics for a key customer may also be discussed.

What interests Kapriel most about his job is that it allows him exposure to a diverse range of activities and technologies.

Welding

Welding Processes

Pieces of metal can be fastened together with mechanical fasteners as was discussed in Chapter 13, "Fasteners." They can also be held together by soldering or brazing, and some are fastened together by an adhesive. A permanent way to join pieces of metal together is by *welding*.

One method used to weld parts together is to heat the edges of the pieces to be joined until they melt and join or fuse together. When the pieces cool, they become one homogeneous mass, permanently joined together. A filler rod is sometimes used to mix with the molten metal to make the joint stronger. When a lot of heat is used to melt and fuse the joint, the weld is called a *fusion weld*. Heat for this process can come from a torch, burning gasses or with a high electric current. Because of the extremely high temperatures involved in welding, the part or parts may be distorted; thus, all machining is usually done after welding.

Another kind of weld used in years past was the pressure weld. In this process, the parts to be welded together are heated to a plastic state, and forced together by

pressure or by hammering. This was done by the local blacksmith. This old method was called *forge welding*. Today, faster and better methods are used.

Welding assemblies are usually built-up from stock forms, such as plate steel, square bars, tubing, angle iron and the like. These parts are cut to shape and welded together. The various welding processes are illustrated in *Figure 19-1*. Welded assemblies are much less expensive and more satisfactory that the casting process, especially in a case where only one or only a few identical parts are required.

Welding is used on large structures that would be difficult or impossible to build in a manufacturing plant. Large structures such as building frames, bridges, and ships are welded into one large assembly.

Major welding methods classified by the American Welding Society include: brazing, gas welding, arc welding, resistance welding, and fusion welding. *Brazing* is the process of joining metals together with a nonferrous filler rod. A temperature above 1000°F, but just below the melting point, is used to join the parts. The most commonly used *gas* welding is oxyacetylene welding. Resistance welding and fusion welding, the two major

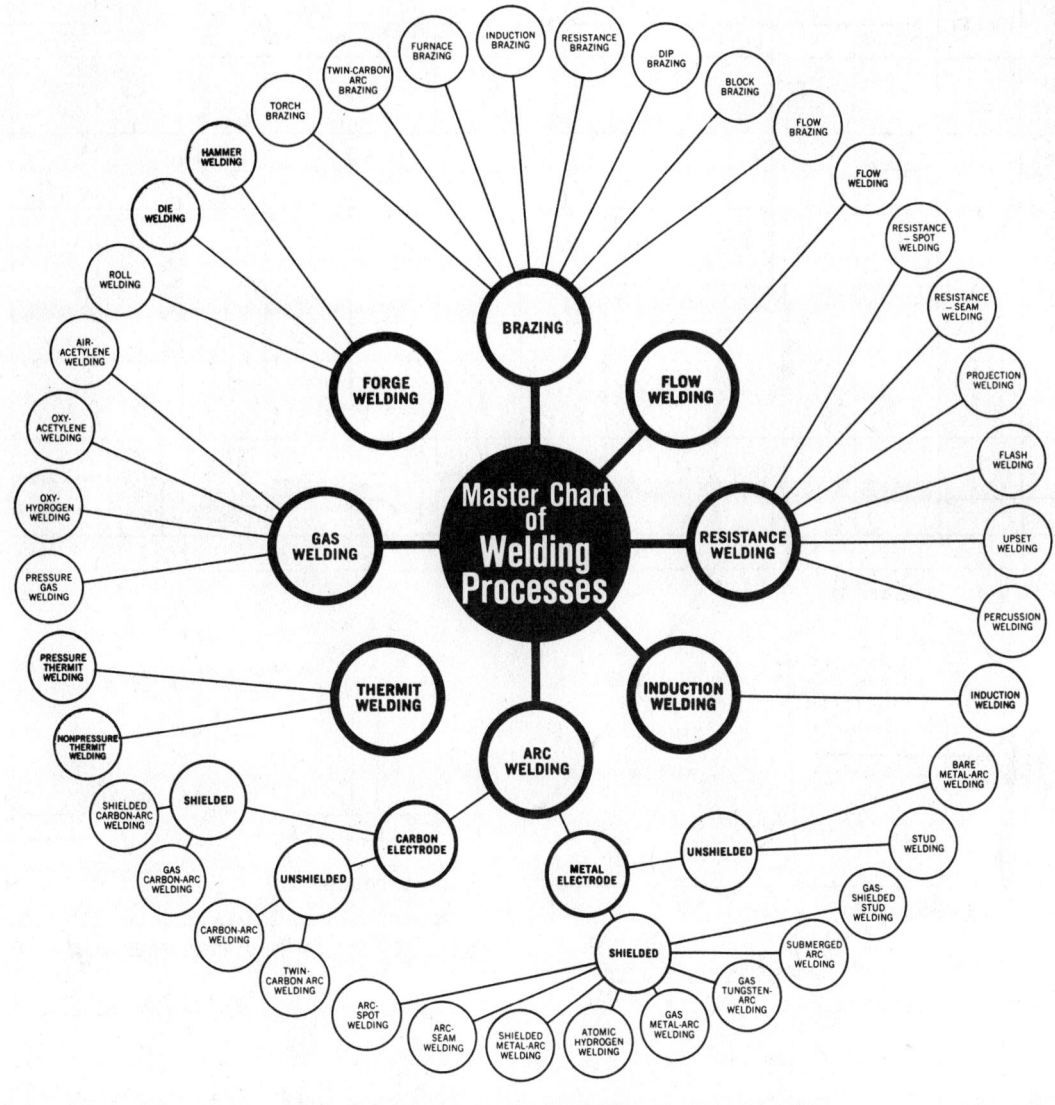

Figure 19-1 Master chart of welding processes *(Courtesy American Welding Society)*

processes, are explained in this text. Detailed information regarding the other types of welding processes can be obtained by writing to the American Welding Society, P.O. Box 351040, Miami, Florida 33125.

Resistance welding is the process of passing an electric current through the exact location where the parts are to be joined. This is usually done under pressure. The combination of the pressure and the heat generated by the electric current welds the parts together. This process is usually done on thin sheet metal parts.

Fusion welding is usually used for large parts. The fusion welding process uses many standard types of welds, *Figure 19-2*. Each is explained in full.

The next few figures use the fillet welding symbol as an example, but the same method applies, regardless of which welding symbol is used.

Basic Welding Symbol

The *basic welding symbol* consists of a reference line, a leader line and arrow, and, if needed, a tail, *Figure 19-3*.

The tail is added for specific information or notes in regard to welding specifications, processes or reference information. The reference line of the basic welding symbol is usually drawn horizontally. Any welding symbol placed on the upper side of the reference line indicates WELD OPPOSITE SIDE, *Figure 19-4*. Any welding symbol placed on the lower side of the reference line indicates WELD ARROW SIDE, Figure 19-4. The direction of the leader line and arrow had no significance whatsoever to the reference line.

A fillet weld symbol (see Figure 19-2) is added above or below the reference line with the left leg always drawn vertical, *Figure 19-5*. A welding symbol placed *above* the reference line means to weld the OPPOSITE side, as illustrated in *Figure 19-6*. A weld symbol placed *below* the reference line means to weld ARROW side, as illustrated in Figure 19-6. The positions as drawn and as welded are shown in *Figures 19-7* and *19-8*. A weld symbol placed above and below the reference line means to weld both the OPPOSITE side and the ARROW side, as illustrated in *Figure 19-9*.

TYPE OF WELD	FILLET	GROOVE							BACK OR BACKING	PLUG OR SLOT	SURFAC-ING	FLANGE WELD	
		SQUARE	V	BEVEL	U	J	FLARE V	FLARE BEVEL				EDGE	CORNER
SYMBOL													
WELD ARROW SIDE													
WELD OPPOSITE SIDE									V GROOVE WELD / U GROOVE WELD				
WELD BOTH SIDES													
NO ARROW/ OPPOSITE SIDE SIGNIFICANCE													
EXAMPLE													

Figure 19-2 Fusion welding

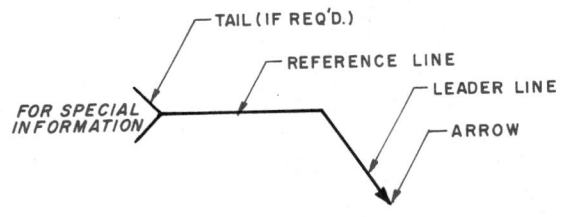

Figure 19-3 Basic welding symbol

(TAIL (IF REQ'D.), FOR SPECIAL INFORMATION, REFERENCE LINE, LEADER LINE, ARROW)

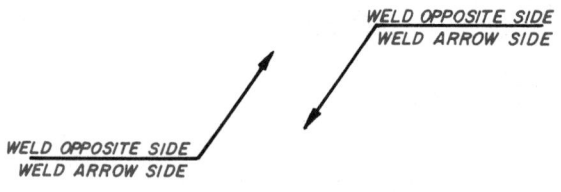

WELD OPPOSITE SIDE / WELD ARROW SIDE

WELD OPPOSITE SIDE / WELD ARROW SIDE

Figure 19-4 Any welding symbol placed on the upper side of the reference line indicates weld OPPOSITE side. Any welding symbol placed on the lower side of the reference line indicates weld ARROW side.

LEFT LEG ALWAYS DRAWN VERTICAL

Figure 19-5 Left leg of welding symbol is always drawn vertical

INDICATES, WELD OPPOSITE SIDE

INDICATES, WELD ARROW SIDE

Figure 19-6 Welding symbol placed above the reference line means to weld OPPOSITE side. Welding symbol placed below the reference line means to weld ARROW side.

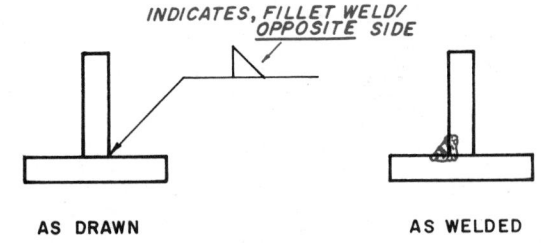

INDICATES, FILLET WELD/ OPPOSITE SIDE

AS DRAWN   AS WELDED

Figure 19-7 Position as drawn and as welded

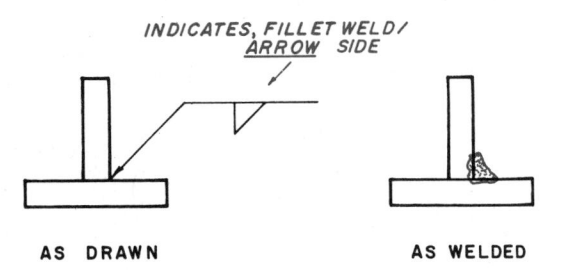

INDICATES, FILLET WELD/ ARROW SIDE

AS DRAWN   AS WELDED

Figure 19-8 Position as drawn and as welded

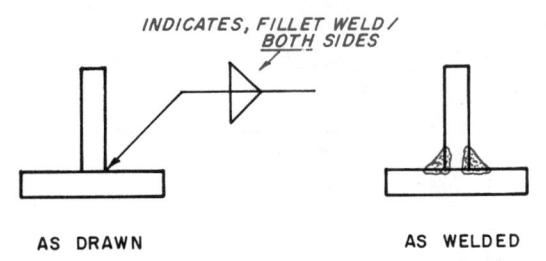

INDICATES, FILLET WELD/ BOTH SIDES

AS DRAWN   AS WELDED

Figure 19-9 Welding symbol placed above and below the reference line means to weld both sides

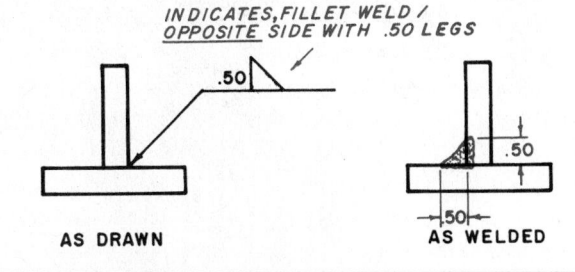

Figure 19-10   Dimension to the left of welding symbol indicates length of leg or side of weld

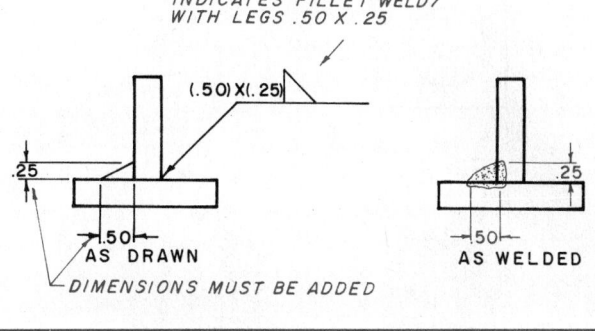

Figure 19-12   Welds with legs or side of different sizes

Size of Weld

The weld must be fully dimensioned so that there is no question whatsoever as to its intended and designed size. The size of a weld refers to the length of the leg or side of the weld. The size is placed directly to the left of the welding symbol. The two legs are assumed to be equal in size unless otherwise dimensioned, **Figure 19-10**. If a double weld is needed, the size of each weld must be included as illustrated in **Figure 19-11**.

If the legs or sides of the weld are to be different, the dimensions of the sides must be indicated to the left of the welding symbol in parentheses, as reference only dimensions, **Figure 19-12**. Because the symbol does not indicate which is the .50 leg and which is the .25 leg, the detail drawing must include the dimensions.

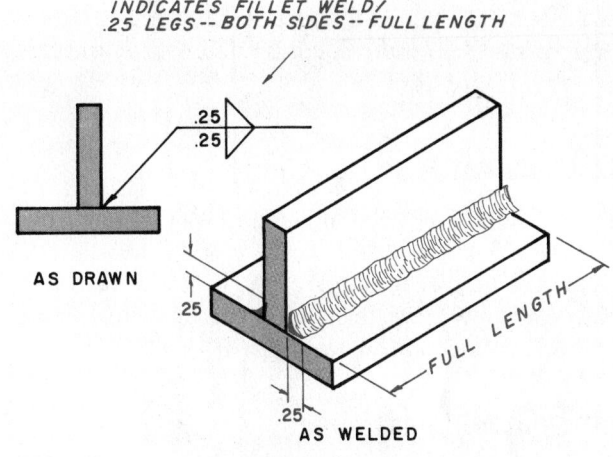

Figure 19-13   If weld length is not specified, it is assumed to be continuous full length

Length of Weld

When a weld length is not specified it is assumed to be continuous the full length, **Figure 19-13**. If a weld must be made to a special length, it must be indicated. This is done by a dimension directly to the right of the weld symbol, **Figure 19-14**.

When a weld is to be made continuously all around an object, it is indicated on the welding symbol by adding a small circle between the reference line and the leader line, **Figure 19-15**.

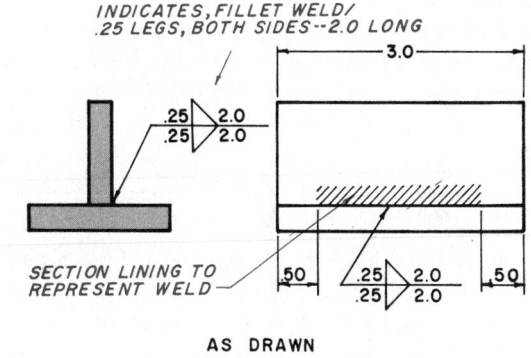

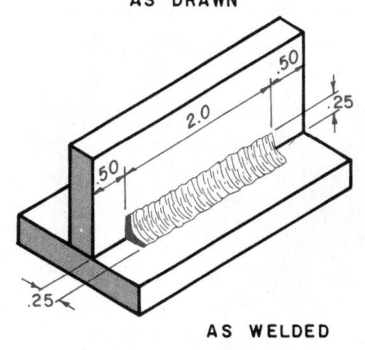

Figure 19-14   Length of weld must be noted to the right of the welding symbol

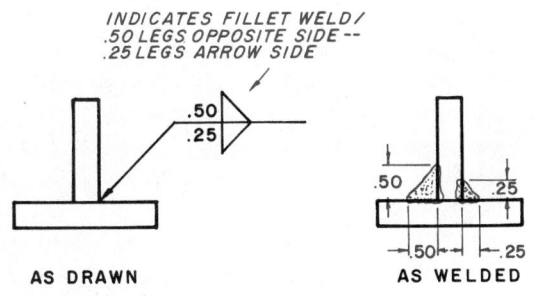

Figure 19-11   Size of each weld is required

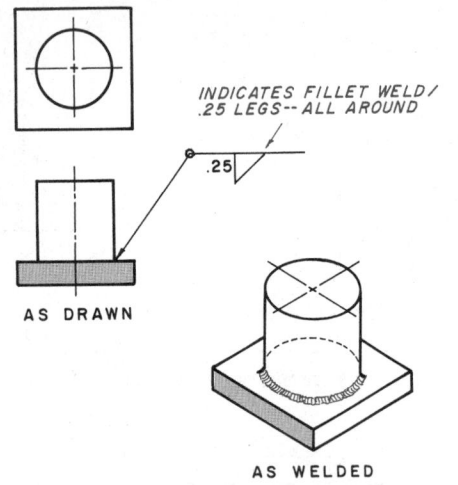

Figure 19-15 A small circle indicates weld continuously around an object

Placement of Weld

When a weld is not continuous and is needed in only a few areas, *section lining* is used to indicate where the weld is to be placed, **Figure 19-16**. If a weld is to be placed in a few areas, this is indicated by the use of multiple leader lines and arrows, **Figure 19-17**.

Intermittent Welds

Intermittent welds are a series of short welds. When this type of weld is needed, an extra dimension is added

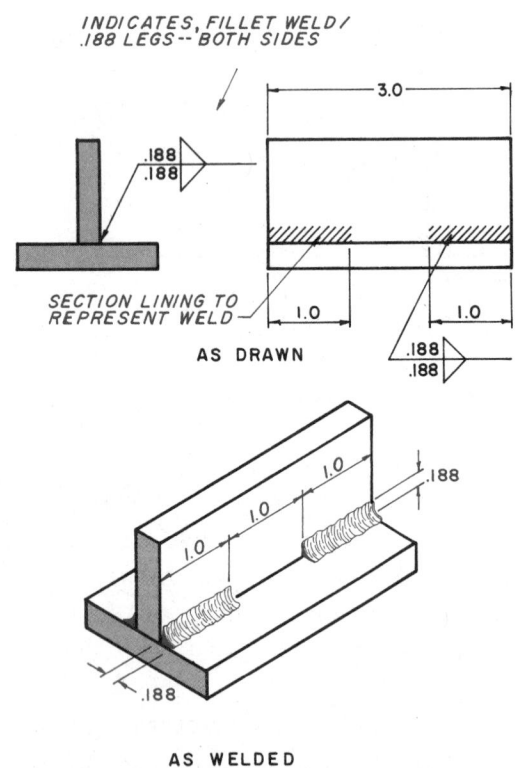

Figure 19-16 Section lining indicates location of weld

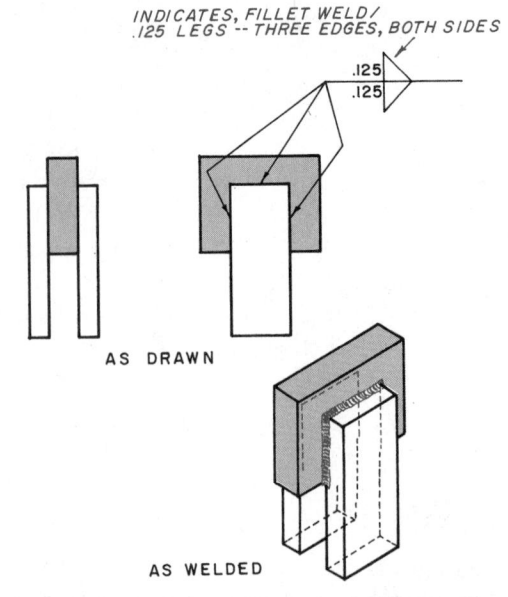

Figure 19-17 Leader line and arrows indicate location of weld

to the welding symbol, **Figure 19-18**. The usual size of leg is noted to the left of the symbol, and the length of the weld is indicated to the right of the welding symbol followed by a dash line and the pitch of the intermittent welds. See **Figure 19-19**. An *increment* is the length of the weld; the *pitch* is the center-to-center distance between increments.

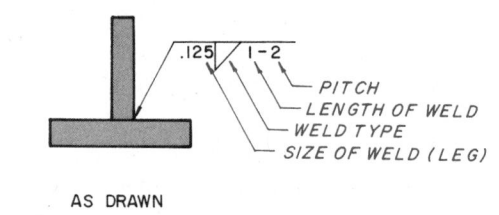

Figure 19-18 Intermittent welds are noted by dimensions to the right of the welding symbol

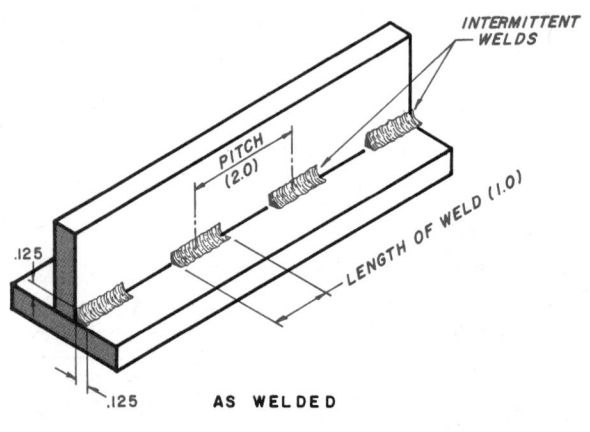

Figure 19-19 Pitch is the center-to-center distance between welds

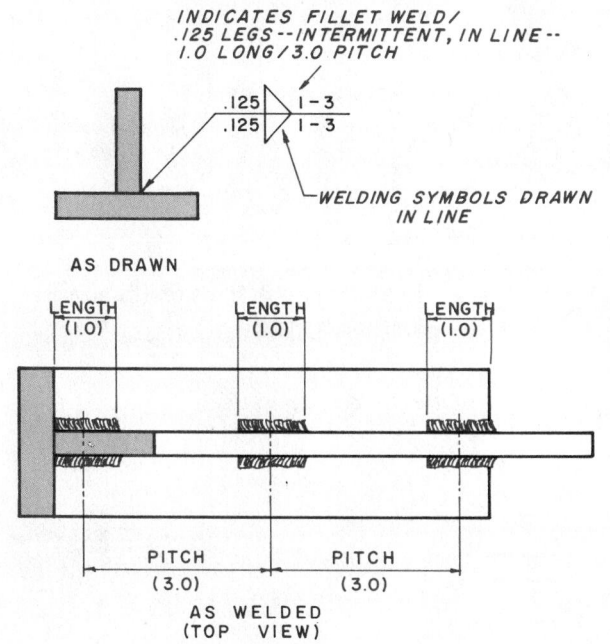

Figure 19-20   Chain intermittent welds are applied directly opposite to each other

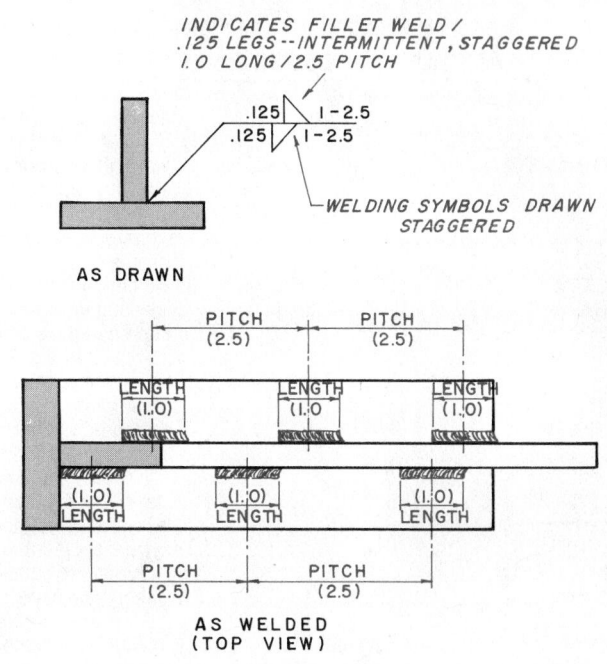

Figure 19-21   Staggered intermittent welds are placed staggered opposite each other

Chain Intermittent Weld

A *chain intermittent weld* is a weld where each weld is applied directly opposite to each other on opposite sides of the joint, *Figure 19-20*. Note the welding symbol is applied above and below the reference line in line with each other.

Staggered Intermittent Weld

A *staggered intermittent weld* is similar to the chain intermittent weld except the welds are staggered directly opposite to each other on opposite sides, *Figure 19-21*. Note the welding symbol is applied above and below the reference line staggered to each other.

Process Reference

There are many welding processes developed by the American Welding Society. Each process has a standard abbreviation designation, *Figure 19-22*. Letter designations are used to specify a method used to obtain a particular contour of a weld, *Figure 19-23*. These abbreviations are taken from the latest American Welding Society's standard #AWS-2.4-79,71.

The standard process abbreviations or letter designations are usually added to the tail of the basic welding symbol as required, *Figure 19-24*. Sometimes, they are called-off in a note somewhere on the drawing, *Figure 19-25*.

Contour Symbol

If a special finish must be made to a weld, a contour symbol must be added to the welding symbol. There are

DESIGNATION OF WELDING AND ALLIED PROCESSES BY LETTERS

AAC	air carbon arc cutting
AAW	air acetylene welding
ABD	adhesive bonding
AB	arc brazing
AC	arc cutting
AHW	atomic hydrogen welding
AOC	oxygen arc cutting
AW	arc welding
B	brazing
BB	block brazing
BMAW	bare metal arc welding
CAC	carbon arc cutting
CAW	carbon arc welding
CAW-G	gas carbon arc welding
CAW-S	shielded carbon arc welding
CAW-T	twin carbon arc welding
CW	cold welding
DB	dip brazing
DFB	diffusion brazing
DFW	diffusion welding
DS	dip soldering
EASP	electric arc spraying
EBC	electron beam cutting
EBW	electron beam welding
ESW	electroslag welding
EXW	explosion welding
FB	furnace brazing
FCAW	flux cored arc welding
FCAW-EG	flux cored arc welding–electrogas
FLB	flow brazing
FLOW	flow welding
FLSP	flame spraying
FOC	chemical flux cutting
FOW	forge welding
FRW	friction welding
FS	furnace soldering
FW	flash welding

Figure 19-22   Designation of welding and allied processes by letters (abbreviations)

DESIGNATION OF WELDING AND ALLIED PROCESSES BY LETTERS

GMAC ....gas metal arc cutting
GMAW ....gas metal arc welding
GMAW-EG..gas metal arc welding–electrogas
GMAW-P ....gas metal arc welding–pulsed arc
GMAW-S ....gas metal arc welding–short circuiting arc
GTAC...gas tungsten arc cutting
GTAW ....gas tungsten arc welding
GTAW-P ....gas tungsten arc welding–pulsed arc
HFRW...............................high frequency resistance welding
HPW ....hot pressure welding
IB ....induction brazing
INS ....iron soldering
IRB ....infrared brazing
IRS ....infrared soldering
IS ....induction soldering
IW ....induction welding
LBC ....laser beam cutting
LBW ....laser beam welding
LOC ....oxygen lance cutting
MAC ....metal arc cutting
OAW ....oxyacetylene welding
OC ....oxygen cutting
OFC...oxyfuel gas cutting
OFC-A ....oxyacetylene cutting
OFC-H ....oxyhydrogen cutting
OFC-N ....oxynatural gas cutting
OFC-P ....oxypropane cutting
OFW ....oxyfuel gas welding
OHW ....oxyhydrogen welding
PAC ....plasma arc cutting
PAW ....plasma arc welding
PEW ....percussion welding
PGW ....pressure gas welding
POC ....metal powder cutting
PSP ....plasma spraying
RB ....resistance brazing
RPW ....projection welding
RS ....resistance soldering
RSEW ....resistance seam welding
RSW ....resistance spot welding
ROW ....roll welding
RW ....resistance welding
S ....soldering
SAW ....submerged arc welding
SAW-S ....series submerged arc welding
SMAC ....shielded metal arc cutting
SMAW ....shielded metal arc welding
SSW...solid state welding
SW ....stud arc welding
TB ....torch brazing
TCAB ....twin carbon arc brazing
TW ....thermit welding
USW ....ultrasonic welding
UW ....upset welding

Figure 19-22 (Continued) Designation of welding and allied processes by letters (abbreviations)

LETTER	METHOD
C	CHIPPING
G	GRINDING
H	HAMMERING
R	ROLLING
M	MACHINING

Figure 19-23 Abbreviation of standard method used to obtain a particular contour

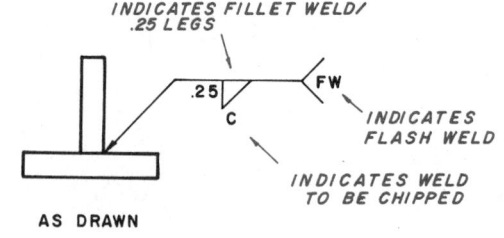

Figure 19-24 Process abbreviations are placed in the tail of the basic welding symbol

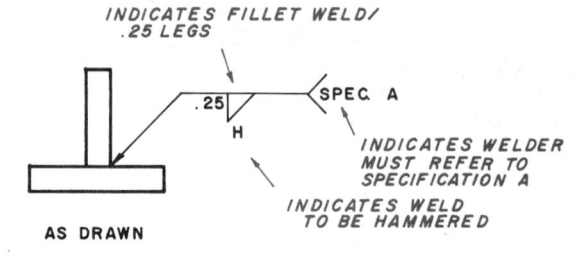

Figure 19-25 Sometimes a note is indicated somewhere on the drawing

three kinds of contour symbols used: flush, convex, or concave, ***Figure 19-26***. The contour symbol is added directly above or below the welding symbol. If the welding symbol is *above* the reference line, the contour symbol is placed *above* the welding symbol. If the welding symbol is placed *below* the reference line, the contour symbol is placed *below* the welding symbol, ***Figure 19-27***. The degree of finish is not usually included, but, if a specific finish is desired, a finish number designation must be added, ***Figure 19-28***.

Figure 19-26 Contour symbol

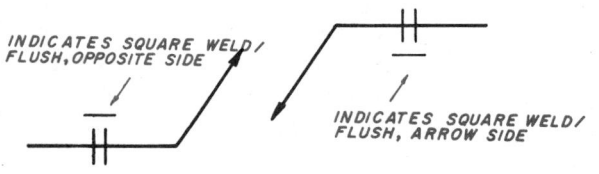

Figure 19-27 Contour symbol is added to the welding symbol

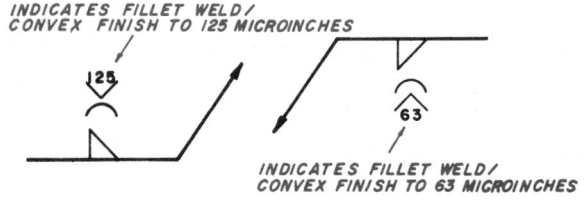

Figure 19-28 Finish number specifies finish required

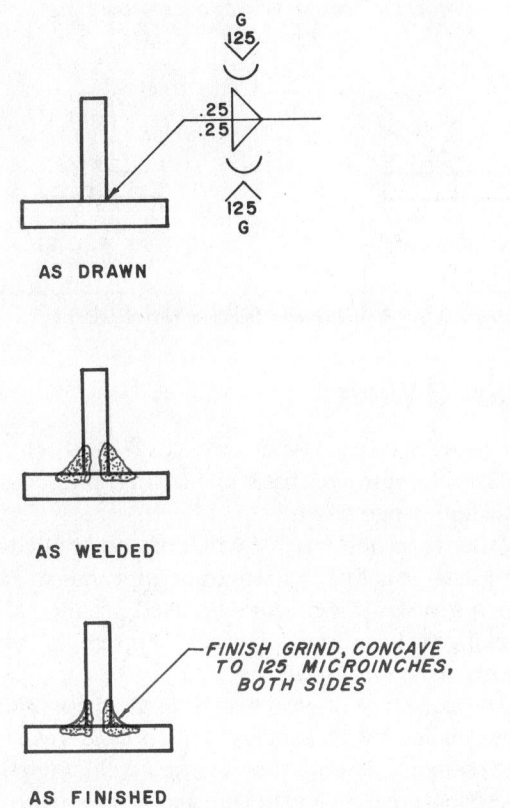

AS DRAWN

AS WELDED

AS FINISHED

FINISH GRIND, CONCAVE
TO 125 MICROINCHES,
BOTH SIDES

Figure 19-29 Finish weld is to be concave and ground to a
125-microinch surface finish

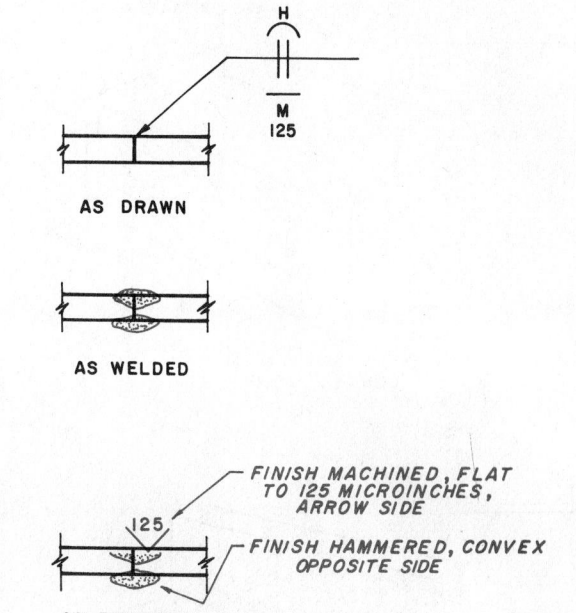

AS DRAWN

AS WELDED

FINISH MACHINED, FLAT
TO 125 MICROINCHES,
ARROW SIDE

FINISH HAMMERED, CONVEX
OPPOSITE SIDE

AS FINISHED

Figure 19-30 Finish weld is to be machined flat on top surface to
125-microinch and convex on the bottom surface with
a hammered finish

Figure 19-29 represents that the finish weld must be concave and ground to a 125-microinch surface finish. ***Figure 19-30*** represents that the finish weld must be machined flat on the top surface to a surface finish of 125 microinches. The weld must be convex on the bottom surface with a hammered finish.

Field Welds

Any weld not made in the factory, that is, it is to be made at a later date, perhaps at final assembly on site, is called a *field weld*. Its symbol is added to the basic weld symbol by a filled-in flag, located between the reference line and the leader line and drawn using a 30°–60° triangle, ***Figure 19-31***.

Welding Joints

There are five basic kinds of welding joints and they are classified according to the position of the parts that are being joined, ***Figures 19-32A*** through ***19-32E***. There are many types of welds used to weld these joints together. Considerations as to which type of weld to use are based on the particular application, thickness of material, required strength of the joint, and available welding equipment, among others.

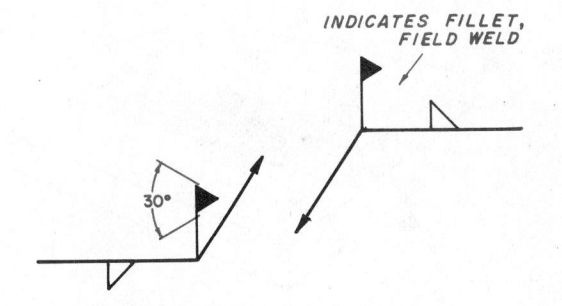

INDICATES FILLET,
FIELD WELD

Figure 19-31 Field weld

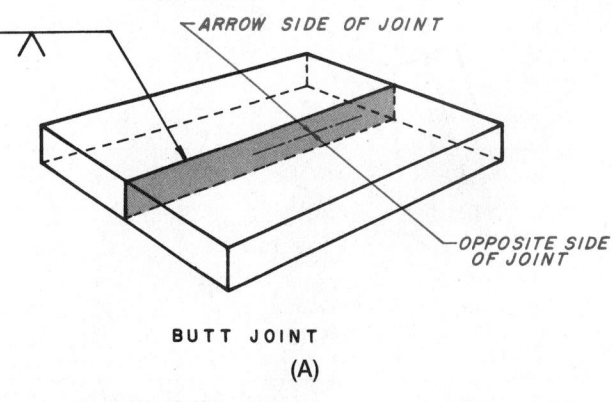

ARROW SIDE OF JOINT

OPPOSITE SIDE
OF JOINT

BUTT JOINT
(A)

Figure 19-32 (A) Butt joint

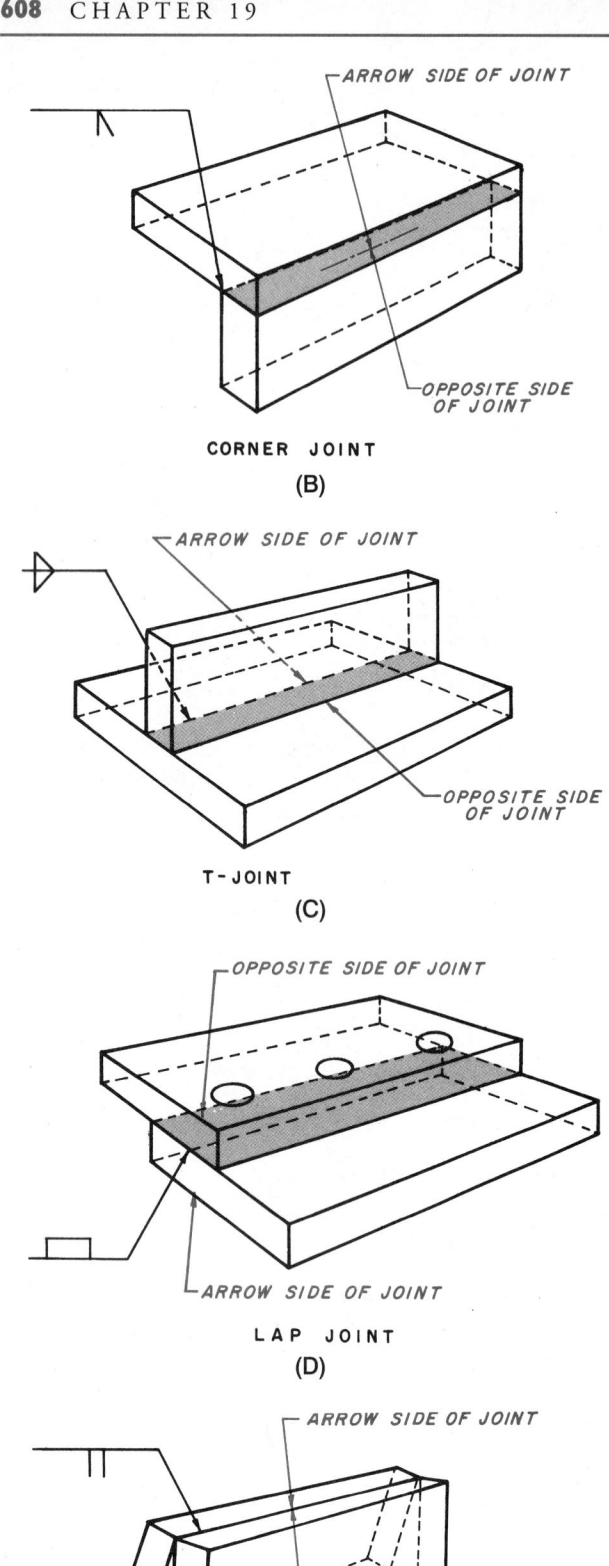

Figure 19-32 (Continued) (B) Corner joint; (C) T-joint; (D) Lap joint; (E) Edge joint

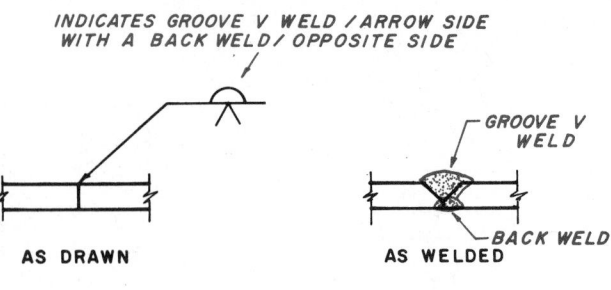

Figure 19-33 Multiwelds (different type of welds)

Types of Welds

There are six major types of welds (refer back to Figure 19-2): fillet, groove, back or backing, plug or slot, surface, and flange welds.

More than one type of weld may be applied to a single joint—usually for strength or appearance. For example, a groove V weld may be used on one side, and a backing weld on the other side, *Figure 19-33*. This is known as a *multiweld*.

The same type of weld may be used on opposite sides of a single joint. Such a joint is called a *double weld*; for example, double V, double U or double J weld, *Figure 19-34*. These too, are used for strength and/or appearance.

Fillet Welds

Up to this point, the fillet weld and its symbol have been used as an example. The other five major types of welding symbols used are very similar.

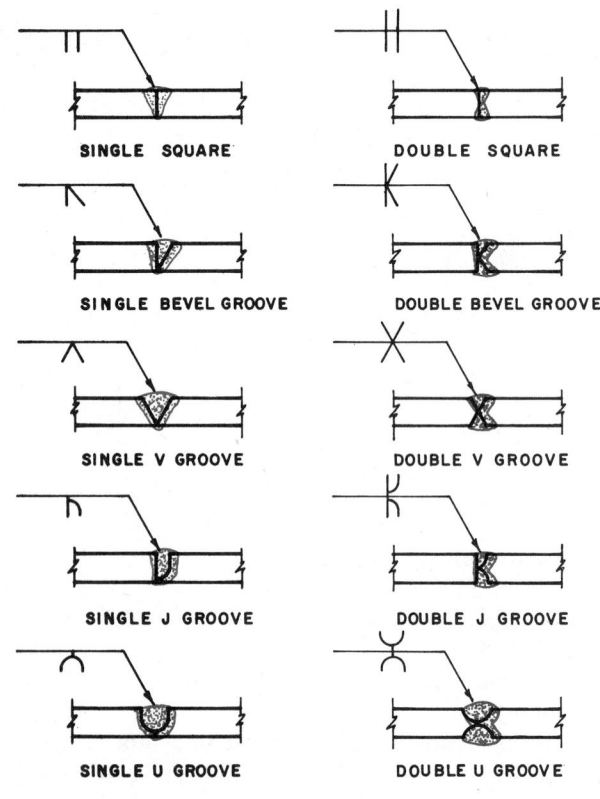

Figure 19-34 Double welds (same type of welds)

GROOVE WELDS		
TYPE	SYMBOL	AS SEEN
SQUARE	‖	
V	V	
BEVEL GROOVE	V	
U	∩	
J	P	
FLARE V	⟩⟨	
FLARE BEVEL	I⟨	

Figure 19-35 Groove welds

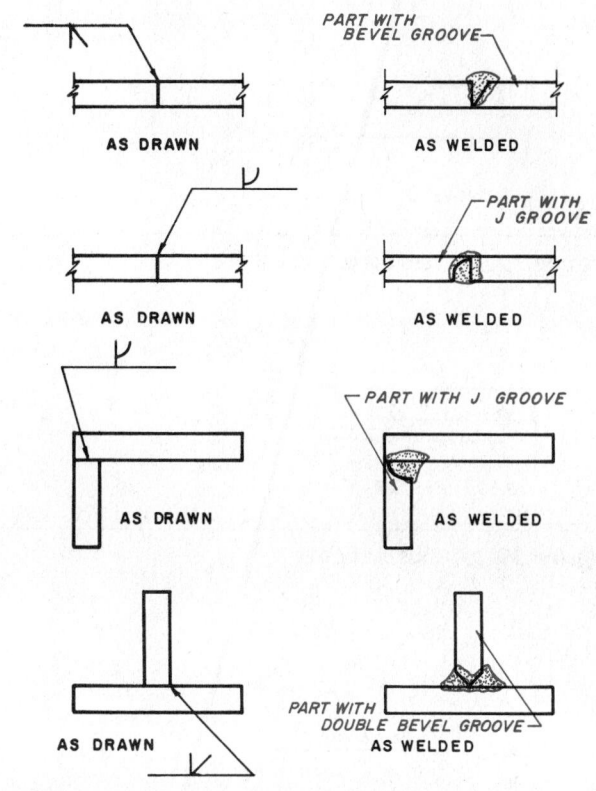

Figure 19-36 Groove weld illustrations

Groove Weld

Seven types of welds are considered to be *groove welds*: square, V, bevel groove, J, U, flare V, and flare bevel. Each type has its own welding symbol, *Figure 19-35*. In the bevel groove and J welds, only one part is actually chamfered or grooved. To do this, the leader line and arrow point toward the part that has the bevel groove or J groove, *Figure 19-36*. Note that the welding symbol is placed above or below the reference line, as usual, to indicate which side the chamfer or groove is located.

Size of Groove Weld.   In the groove weld, the dimension refers to the actual depth of the chamfer or groove *not* the size of the weld. With the fillet weld, the size of the chamfer or groove is located directly to the left of the welding symbol and on the same side of the reference line, *Figure 19-37*. If a V groove weld symbol is shown without a size dimension, the size or depth is assumed to be equal to the thickness of the pieces.

Depth or Actual Size of Weld.   The depth or actual size of the weld includes the penetration of the complete weld. This is illustrated in *Figure 19-38*. Note in this example that the depth or actual size of the weld is deeper than the depth of the chamfer. Occasionally, the depth or actual size of the weld is less than the size of the chamfer, *Figure 19-39*.

In order to call-off the depth or actual size of the weld *and* the size of the chamfer or groove, a dual dimension is used and both are added to the *left* of the welding symbol. In order to differentiate between the two, the size of the chamfer or groove is added to the extreme left of the welding symbol. The size of the

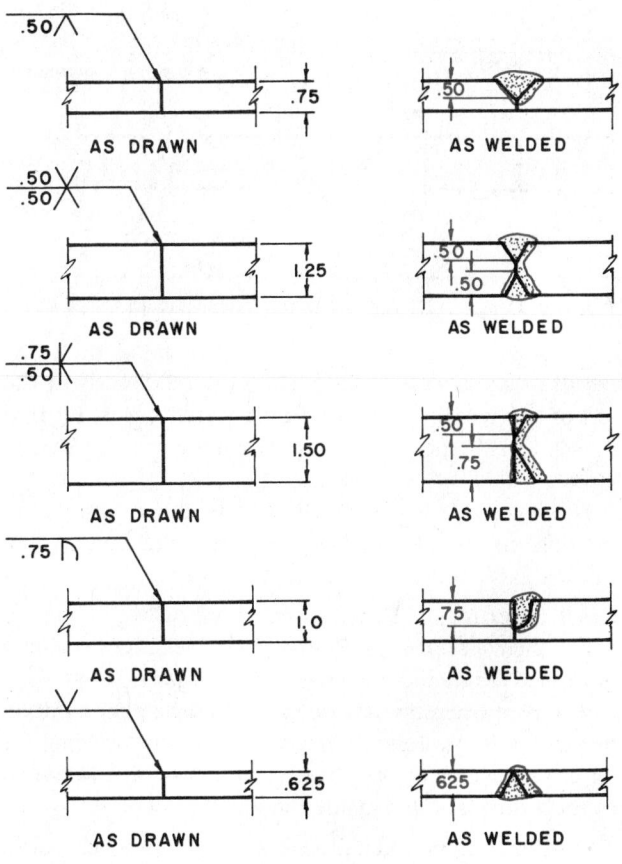

Figure 19-37 Groove welds with dimensions

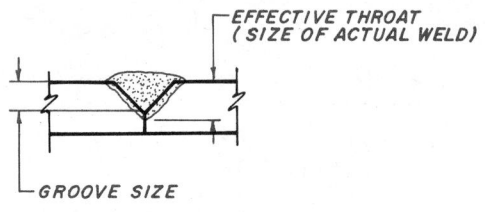

Figure 19-38   Depth or actual size of weld

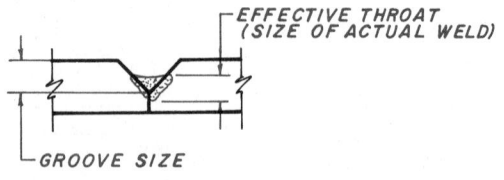

Figure 19-39   Size of weld only

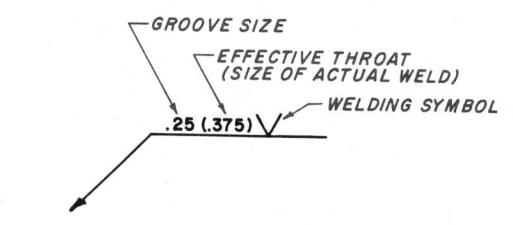

Figure 19-40   Size of weld is added to the left of the welding symbol in parentheses

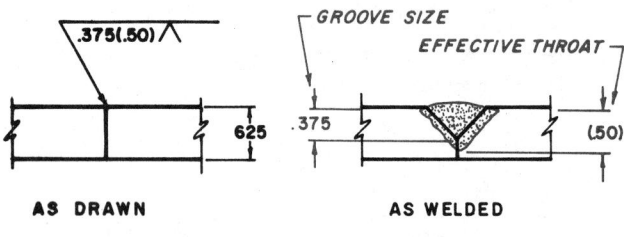

Figure 19-41   Example of effective throat-weld larger than groove

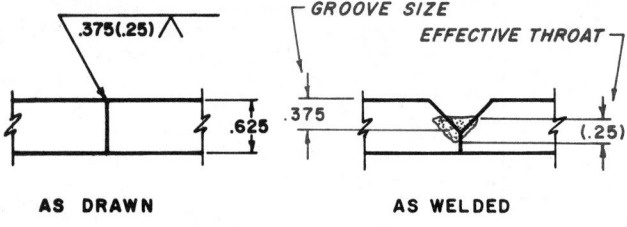

Figure 19-42   Example of effective throat-weld smaller than groove

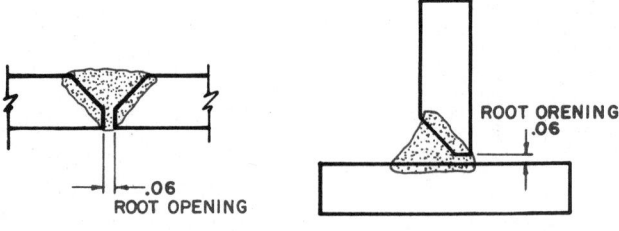

Figure 19-43   Root opening

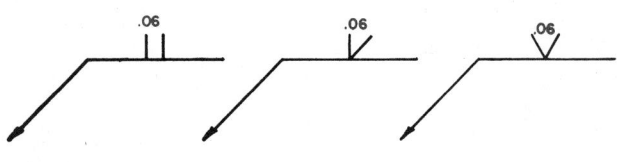

Figure 19-44   Size of root opening

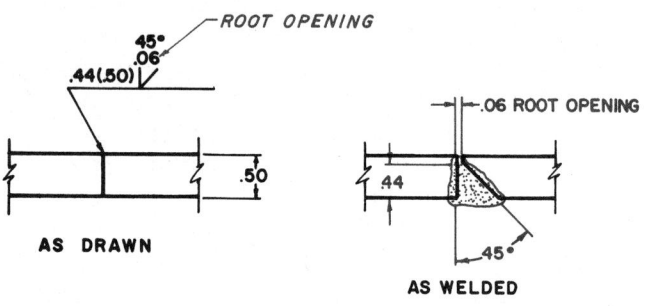

Figure 19-45   Chamfered angle

depth or actual size of the weld is added directly to the left of the welding symbol, but in parentheses, ***Figure 19-40***. Usually the weld is larger than the groove, ***Figure 19-41***.

If the weld is less than the size of the chamfer, the dimensions are added to the left of the welding symbol in the exact same manner, ***Figure 19-42***.

Root Opening.   The allowed space between parts is called the *root opening*, ***Figure 19-43***. The root opening is applied *inside* the welding symbol, ***Figure 19-44***.

The root opening may be called-off as a general note. For example, "unless otherwise noted, root opening for all groove welds is .06." If there is *no* opening between parts, a zero is added inside the welding symbol.

Chamfer Angle.   In a groove weld, a V or bevel groove must be held to tight tolerances. The angle must also be included with the size dimension in order to obtain the

exact required geometric size and shape of all manufactured parts. The angle is also added inside the welding symbol, ***Figure 19-45***. Note that the leader line and arrow point toward the part with the chamfer. Because the welding symbol is placed *above* the reference line, the chamfer and weld are on the side opposite from the arrow.

Groove Radius.   If a J or U groove weld must be held to tight tolerances, the angle and radius must be included with the size dimensions in order to obtain the exact required geometric size and shape of all manufactured parts. The angle is also added inside the welding symbol and the required radius is called-off as a note, ***Figure 19-46***. Note that the leader line and arrow point toward the part with the groove. Because the welding symbol is placed *below* the reference line, the groove and weld is on the side of the arrow.

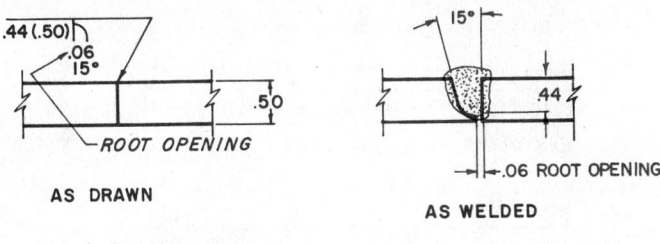

Figure 19-46   Required radius

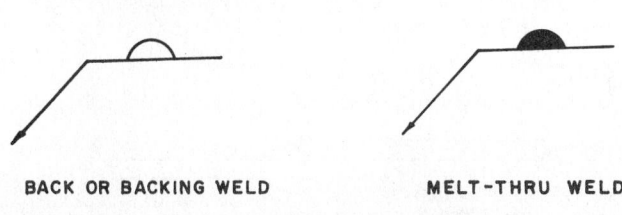

Figure 19-47   Back weld symbol/melt-thru weld symbol

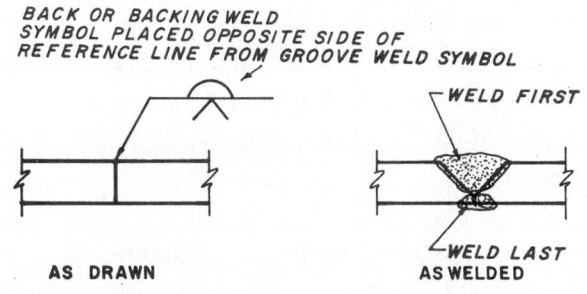

Figure 19-48   Illustration of back weld symbol

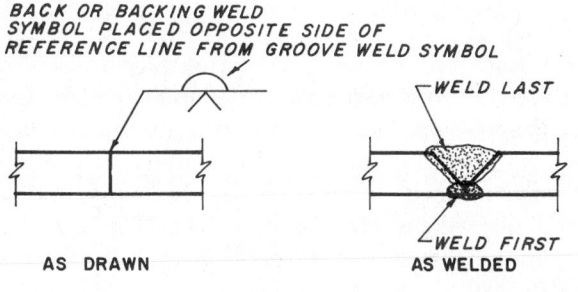

Figure 19-49   Illustration of back weld symbol

Back Weld — Backing Weld

A *back weld* is a weld applied to the opposite side of the joint *after* the major weld has been applied. A backing weld is a weld applied *first*, followed by the major weld. Both use the same welding symbol and both are used to strengthen a weld (refer back to Figure 19-2). The back weld is also used for appearance.

If a welding melt-through is required, the same symbol is used, except it is filled in solid, *Figure 19-47*. When the back weld or backing weld symbol is used, it is placed on the opposite side of the reference line from the groove weld symbol, *Figures 19-48* and *19-49*.

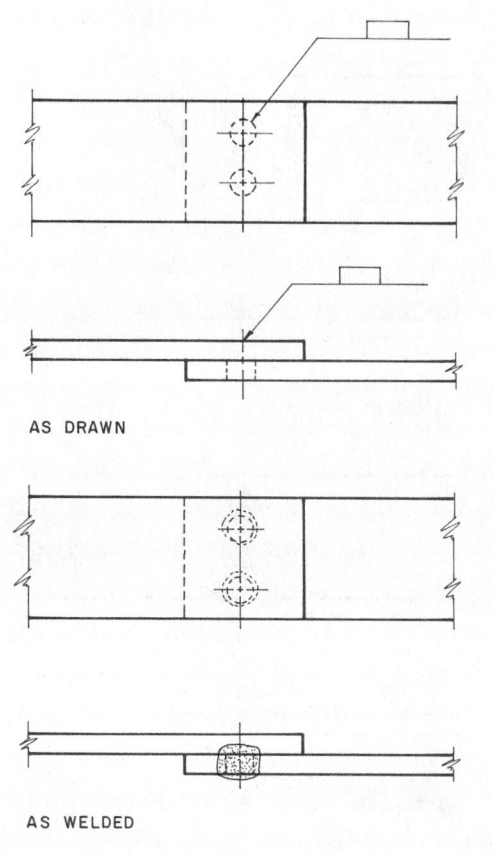

Figure 19-50   Plug and slot welds

Plug and Slot Welds

The *plug* and *slot welds* both use the same welding symbol. The only difference between the two welds is the shape of the hole through which the weld is applied. A *plug weld* is made through a round hole. A *slot weld* is made through an elongated hole (refer back to Figure 19-2).

The plug or slot welding symbol is applied to the basic welding symbol and interpreted in exactly the same way. If the welding symbol is applied *above* the reference line, the hole or slot is on the *opposite* side of the arrow. If *below* the reference line, the hole or slot is on the *same side* as the arrow, *Figure 19-50*.

Size of Plug or Slot Weld.   The given dimension for a plug or slot weld refers to the diameter of the plug or slot at the *base* of the weld. The required dimension is placed directly to the left of the plug or slot weld symbol, *Figure 19-51*. The size of the slot weld is drawn and dimensioned directly on the detail drawing.

Depth of Weld.   Unless noted, the plug or slot weld hole is completely filled by the weld. If filling the hole is not necessary, the full depth of weld must be called-off. The required depth of weld is added *inside* the welding symbol, *Figure 19-52*.

Tapered Plug or Slot.   If the plug or slot hole is tapered, the taper inclusive angle is added directly above or below the welding symbol, *Figure 19-53*.

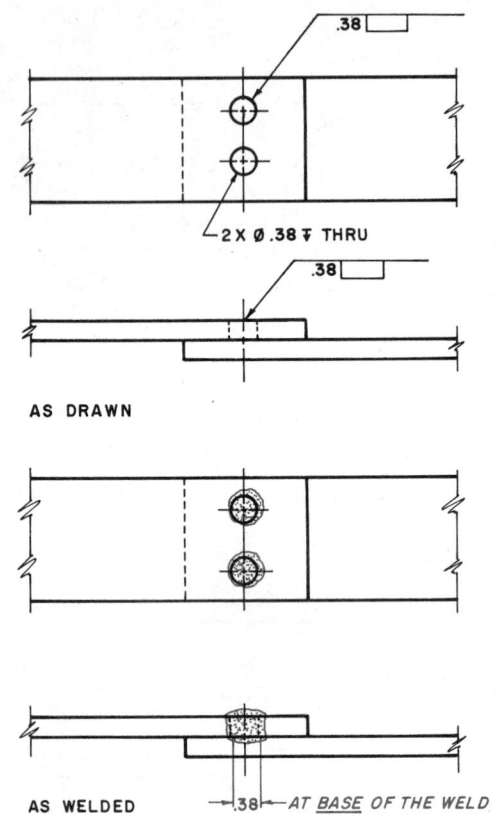

Figure 19-51 Size of plug or slot weld

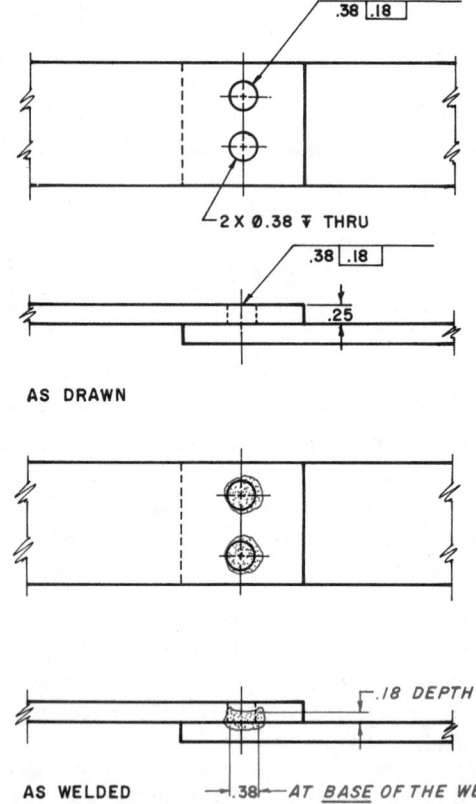

Figure 19-52 Depth of weld

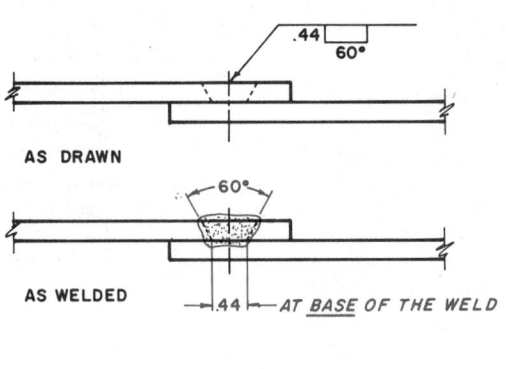

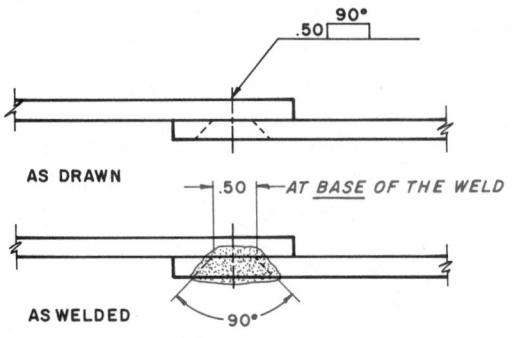

Figure 19-53 Tapered plug or slot

Surface Weld

If a surface must have material added to it or built-up, a *surface weld* is added (refer back to Figure 19-2). This welding symbol is *not* used to indicate the joining of parts together; therefore, the welding symbol is simply added below the reference line, **Figure 19-54**.

Size of Surface Weld. If the surface is to be built-up and a special height is *not* required, the welding symbol is drawn as illustrated in Figure 19-54. If the surface is to be built-up and a *specific* height is required, the required height is added directly to the left of the welding symbol, **Figure 19-55**. If only a *portion* of the surface is to be built-up, that portion must be illustrated and dimensioned on the detail drawing, **Figure 19-56**. The actual weld is represented by section lining.

Flange Weld

A *flange weld* is used to join thin metal parts together. Instead of welding to the surfaces of the parts to be

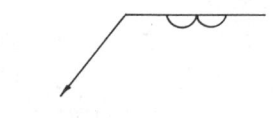

Figure 19-54 Surface weld

Figure 19-55 Size of surface weld

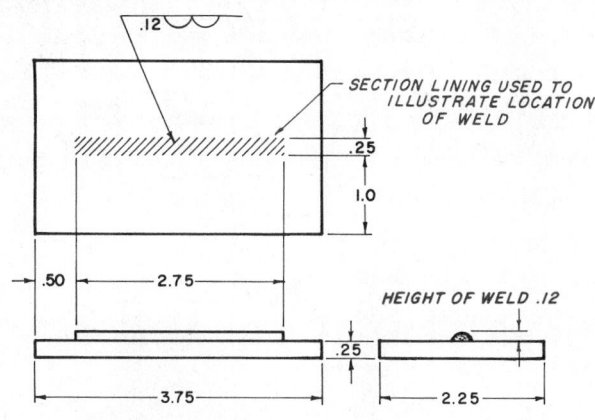

Figure 19-56 Illustration of a surface weld

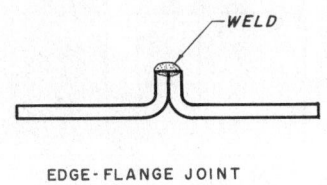

Figure 19-57 Edge-flange weld

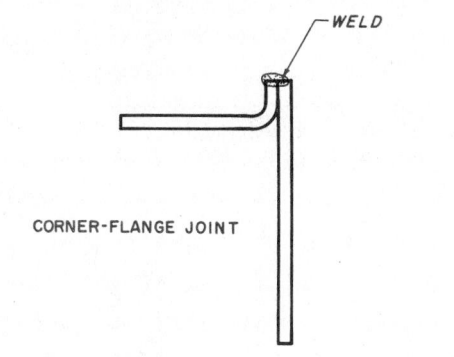

Figure 19-58 Corner-flange weld

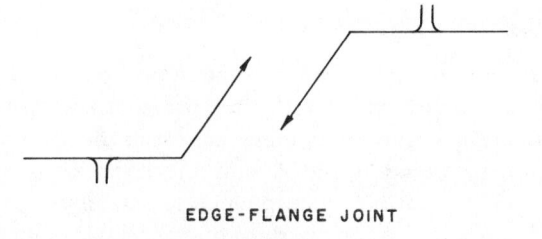

Figure 19-59 Edge-flange weld symbol

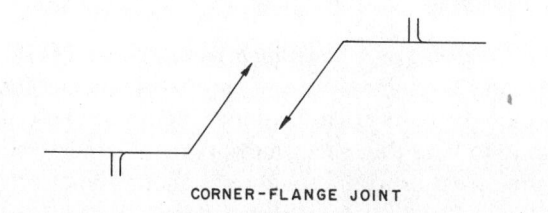

Figure 19-60 Corner-flange weld

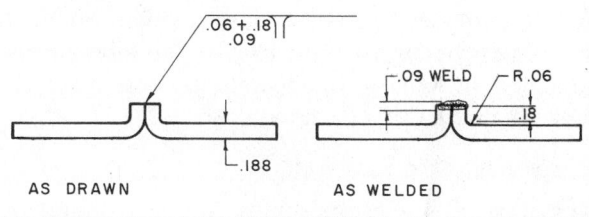

Figure 19-61 Dimensioning a flange weld

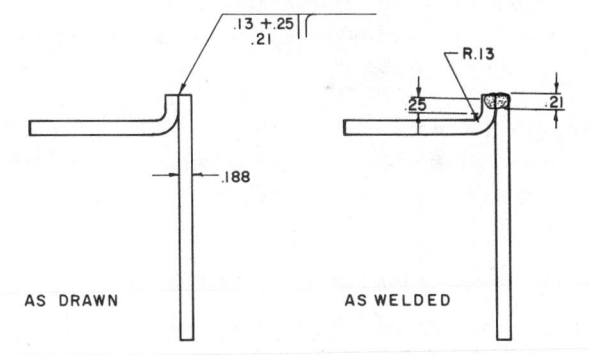

Figure 19-62 Dimensioning a corner-flange weld

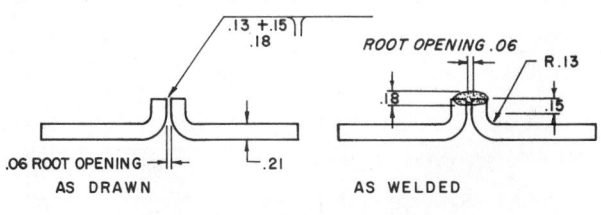

Figure 19-63 Root opening

joined, the weld is applied to the edges of thin material. A weld applied to thin metal could burn right through.

There are two distinct flange weld symbols (refer back to Figure 19-2): the first is called an edge-flange weld, *Figure 19-57*; the other is called a corner-flange weld, *Figure 19-58*. The straight line of this welding symbol is always drawn to the left of the partially curved line, regardless of the actual joint being illustrated. The edge-flange or corner-flange weld symbol is placed above or below the reference line to indicate OPPOSITE side or ARROW side, but never on both sides, *Figure 19-59* and *19-60*.

Dimensioning a Flange Weld. It requires three dimensions to fully dimension a flange weld: the radius of the flange, the height above the point of tangency between parts, and the size of the weld itself, *Figures 19-61* and *19-62*. The three dimensions are added to the welding symbol as indicated previously. The welding symbol should appear at the peak of the joint as illustrated. If there is to be a root opening, that is, a space between parts, the required space must be indicated on the drawing, *not* on the welding symbol, *Figure 19-63*.

Multiple Reference Line

If there is more than one operation on a particular joint, these operations are indicated on *multiple reference lines*. The first reference line closest to the arrow is for the first operation, the second reference line is for the second operation or supplemental data, and the third reference line is for the third operation or test information, *Figure 19-64*.

Spot Weld

Spot welding is a resistance welding process done by passing an electric current through the exact location where the parts are to be joined. Spot welding is usually done to hold thin sheet metal parts together. The resistance weld process uses three standard types of welds, *Figure 19-65*. Each is explained in full.

The symbol for a spot weld, as illustrated in *Figure 19-66*, is located tangent to or touching the reference line, and uses the same added supplemental data symbols and dimensions as are illustrated with the fusion welding process, *Figure 19-67*. The tail is *always* included, in order to indicate the process required to make the weld (refer back to Figure 19-22).

Dimensioning the Spot Weld

The diameter size of the spot weld is added directly to the left of the weld symbol, *Figure 19-68*. Sometimes the diameter size is omitted, and a required shear strength is inserted in its place. In this condition, an explanation of the value is added to the drawing by a note, *Figure 19-69*.

The spacing of the spot weld is called-off by a number directly to the right of the weld symbol. This number is the pitch, and refers to the center-to-center dis-

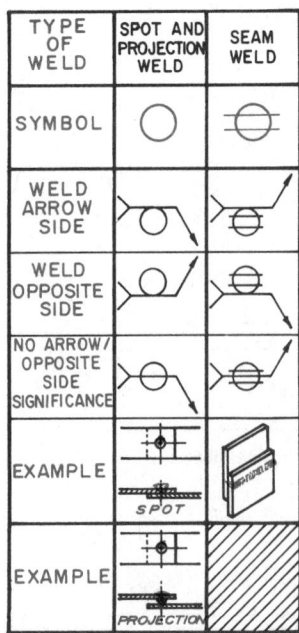

TYPE OF WELD	SPOT AND PROJECTION WELD	SEAM WELD
SYMBOL		
WELD ARROW SIDE		
WELD OPPOSITE SIDE		
NO ARROW/OPPOSITE SIDE SIGNIFICANCE		
EXAMPLE	*SPOT*	
EXAMPLE	*PROJECTION*	

Figure 19-65   Spot welds

SPOT WELD SYMBOL

Figure 19-66   Spot weld symbol

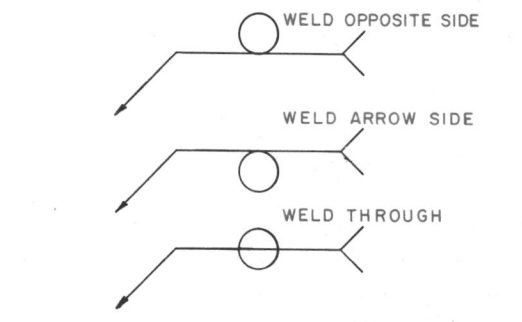

WELD OPPOSITE SIDE

WELD ARROW SIDE

WELD THROUGH

Figure 19-67   Spot weld symbol used with reference line

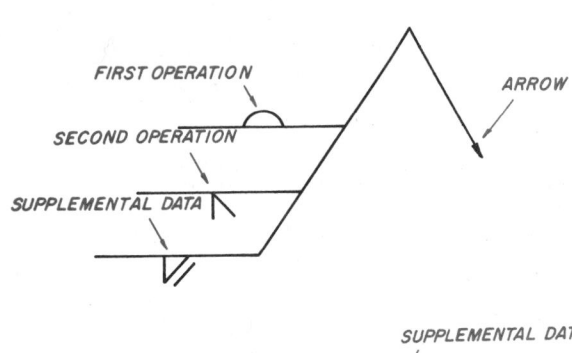

FIRST OPERATION

SECOND OPERATION

SUPPLEMENTAL DATA

ARROW

Figure 19-64   Multiple reference line

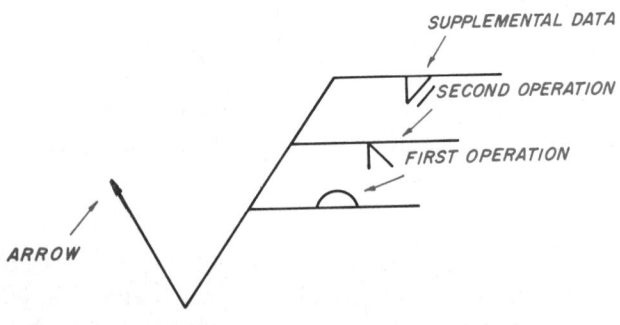

SUPPLEMENTAL DATA

SECOND OPERATION

FIRST OPERATION

ARROW

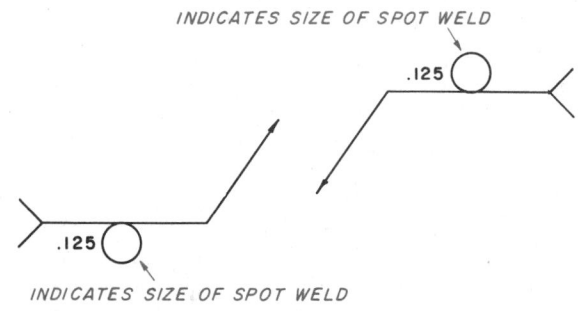

INDICATES SIZE OF SPOT WELD

.125

.125

INDICATES SIZE OF SPOT WELD

Figure 19-68   Diameter of spot weld

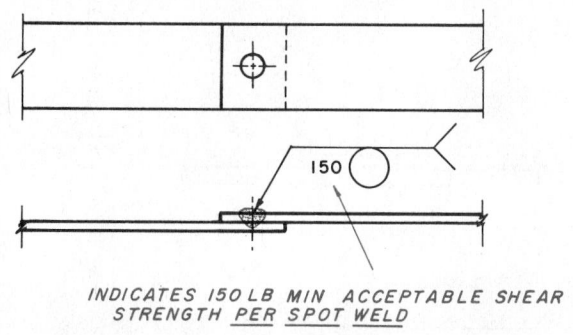

Figure 19-69 Shear strength of spot weld

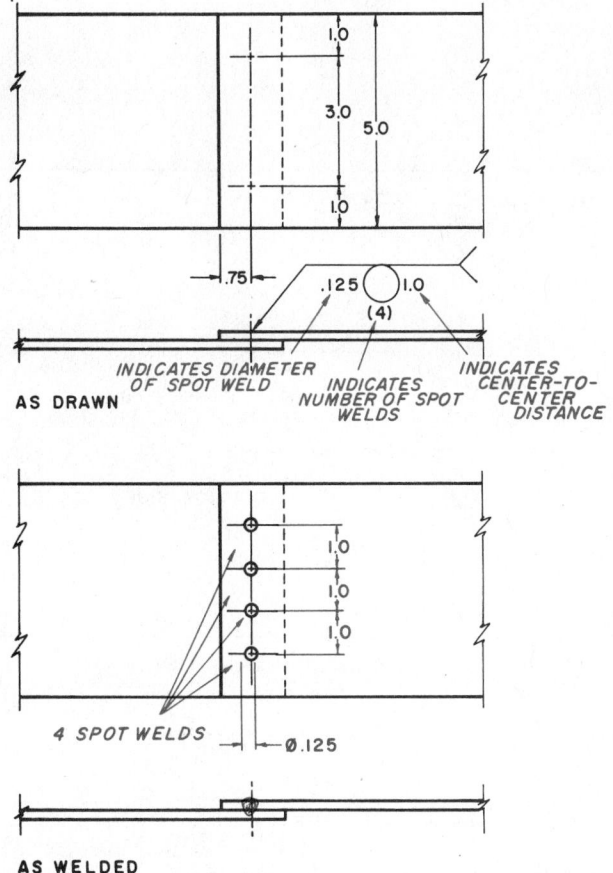

Figure 19-70 Pitch of spot welds

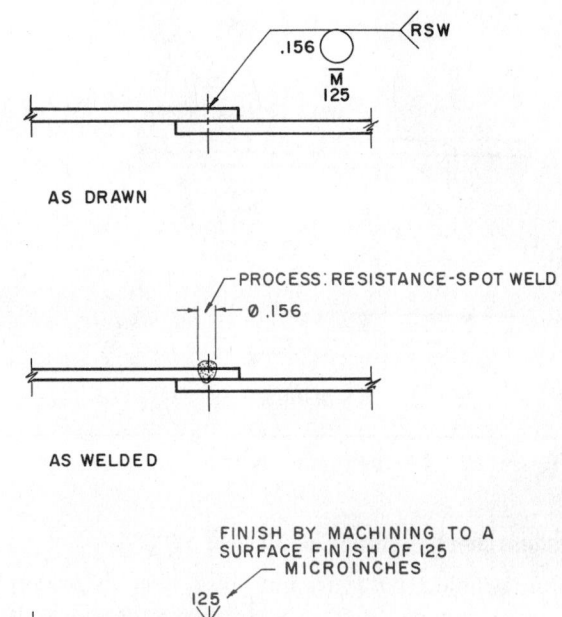

Figure 19-71 Contour and finish symbols

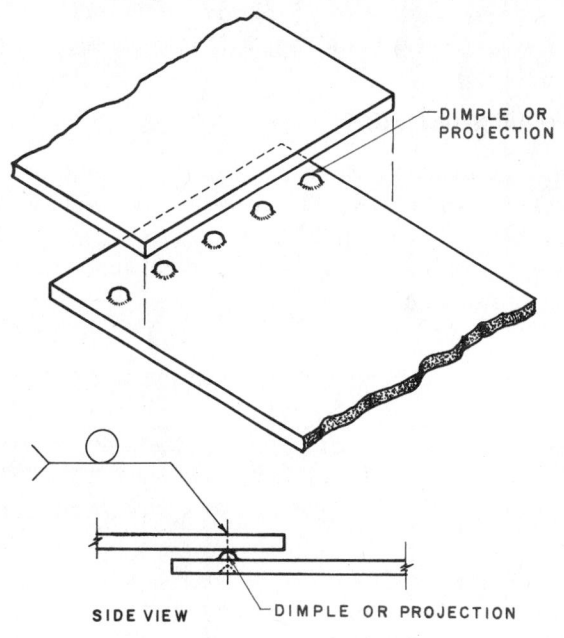

Figure 19-72 Projection weld

tance between spots, *Figure 19-70*. The number of welds for a particular joint is added above or below the welding symbol in parentheses.

Contour and Finish Symbols

Contour and finish symbols are added to the welding symbol exactly as is done in the fusion welding symbol, *Figure 19-71*.

Projection Weld

A *projection weld* is similar to a spot weld and uses the exact same welding symbol. One of the two parts has a

series of evenly spaced dimples stamped into it. Each stamped dimple is the location of an individual spot weld, *Figure 19-72*. A projection weld allows more penetration and, as a result, a much better weld. Note, the dimples are *not* illustrated on the detail drawing.

The reference line (OPPOSITE side-ARROW side significance) is changed slightly as used for a projection weld. The OPPOSITE side-ARROW side indicates which of the two parts to be joined has the dimples on it, *Figures 19-73* and *19-74*.

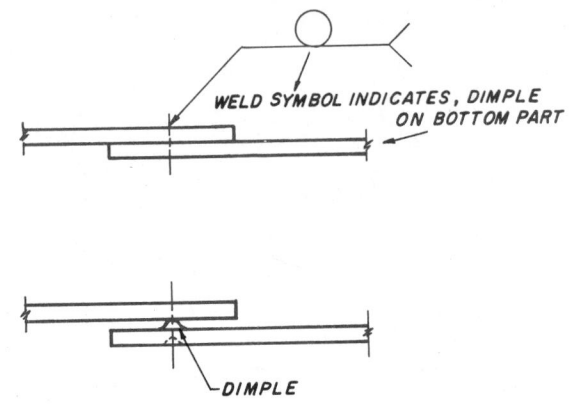

Figure 19-73   Projection weld symbol

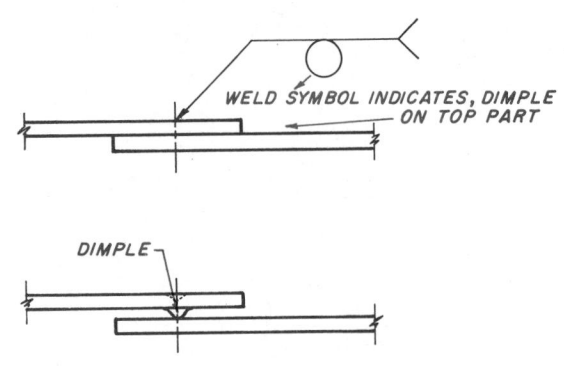

Figure 19-74   Projection weld symbol

Dimensioning, Contour, and Finish Symbols

Dimensioning, contour, and finish symbols are applied to projection welds exactly as they are for spot welds.

Figure 19-75   Seam weld symbol

Seam Weld

A *seam weld* is like a spot weld, except that the weld is actually continuous from start to finish.

The welding symbol is modified slightly to indicate a seam weld, *Figure 19-75*. *Figure 19-76* indicates how the seam weld symbol is applied to the reference line.

Welding Template

In order to simplify and speed up the drawing time for welding symbols, welding templates can be used. The welding template standardizes the size of each welding symbol and provides a quick, ready reference for welding symbols, *Figure 19-77*.

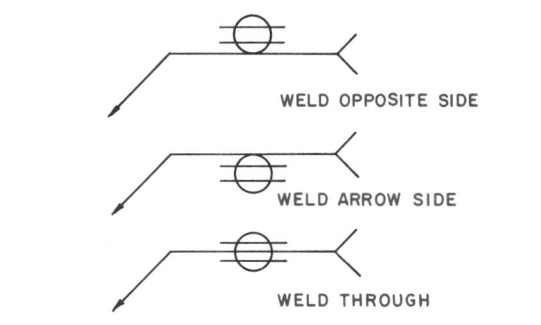

Figure 19-76   Seam weld symbol used with reference line

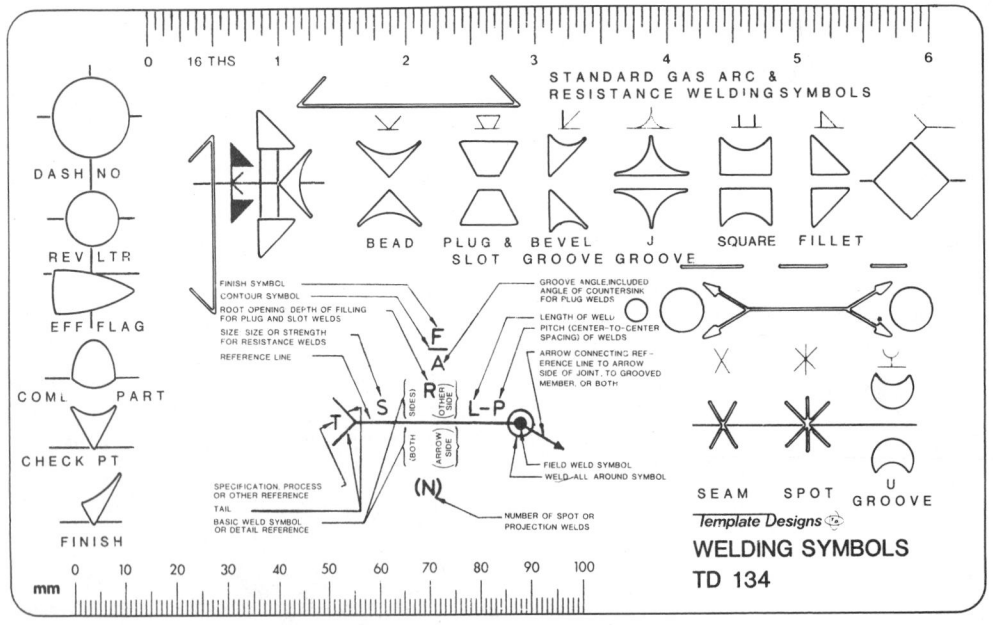

Figure 19-77   Welding template

Design of Weldments

Welded items are held together by fixtures during the actual welding process. This maintains the proper relationships between the parts. However, once the parts are released from the fixture, the potential for warping and other problems is introduced. This problem can be solved for the most part through proper design practices. Designers should keep the following objectives in mind when designing weldments.

1. design for the convenience of the welder or welding machine/robot
2. design for proper heat control in the area of the weld
3. design in appropriate clamping mechanisms to prevent warping
4. design in outlets for the welding atmosphere
5. design in the appropriate clearances to accommodate for filler material
6. consider ease of operation after the weld is completed

Review

1. What is the difference between a plug weld and a slot weld?
2. Make a sketch of a welding symbol for a fillet weld with legs of .250 and a weld 4.5 long and weld on OPPOSITE side of arrowhead.
3. List six major types of welds.
4. Explain what the *root opening* is.
5. Describe an intermittent weld.
6. What are the two major welding processes and generally, where is each used?
7. List the five basic kinds of welding joints. How are they classified?
8. Describe the basic welding symbol and explain the significance of welding symbols being placed above and below the reference line.
9. Where and why is a flange weld used?
10. What is meant by a process reference and where is it located on the welding symbol?
11. What is the advantage of using a projection weld over a spot weld?
12. Explain in full the difference between a back weld and a backing weld. How is the melt-through designated to these symbols?
13. List the seven types of groove welds.
14. How is a field weld indicated on the welding symbol?
15. How is a weld that is to be continuously around an object indicated on the welding symbol?

Chapter Nineteen Problems

The following problems are intended to give the beginning drafter practice in studying the many factors involved in dimensioning and calling-off welding processes. The student will sharpen skills in centering the views within the work area, line weight, dimensioning, and speed.

The steps to follow in laying out the following problems are:

1. Study the problem carefully.
2. Choose the view with the most detail as the front view.
3. Make a sketch of the required views.
4. Add required dimensions to the above sketch per latest drafting standards.
5. Add welding symbols as required to the sketch per the latest welding standards.
6. Lightly lay out all required views and check that they are centered within the work area.
7. Check that all dimensions are added using the latest drafting standards.
8. Darken in all views using correct line thickness and add all dimensioning using light guidelines.
9. Recheck all work, and, if correct, neatly fill out the title block using light guidelines and neat lettering.
10. Recheck that all required welding symbols are added correctly.

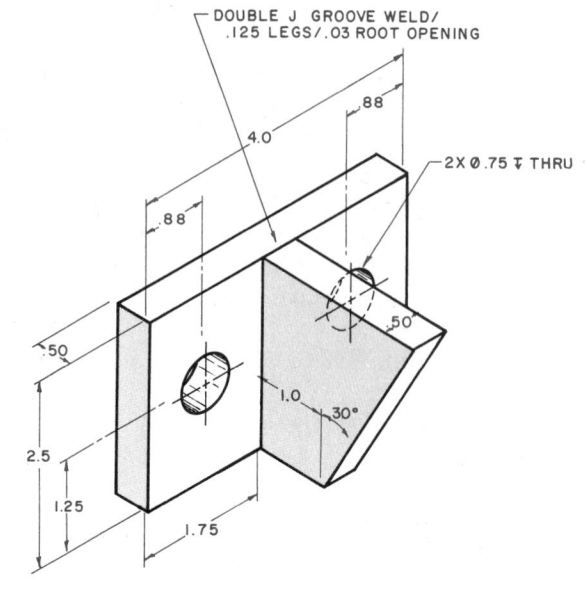

Problem 19-2

Problems 19-1 through 19-11

Follow the listed steps, using the latest drafting standards.

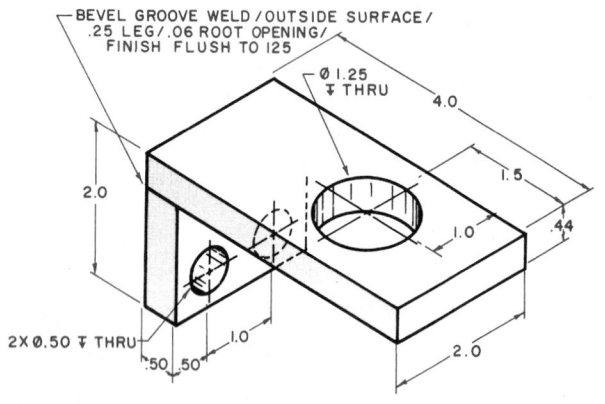

Problem 19-1

Problem 19-3

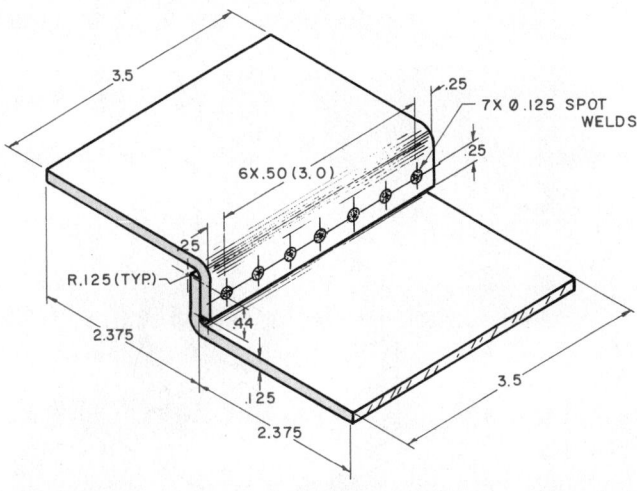

Problem 19-4

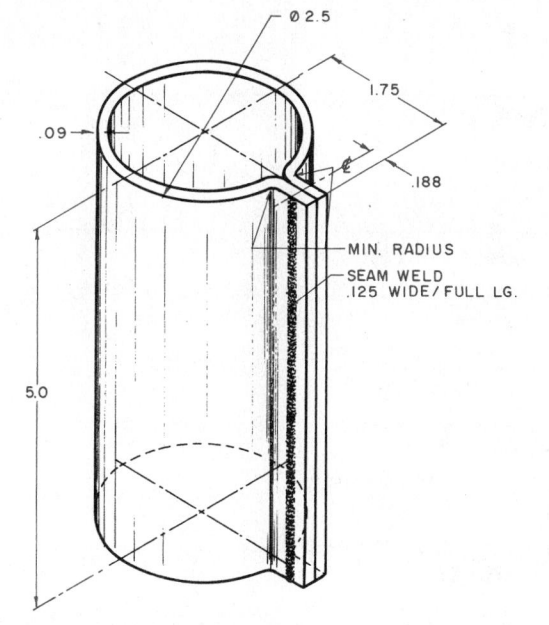

Problem 19-5

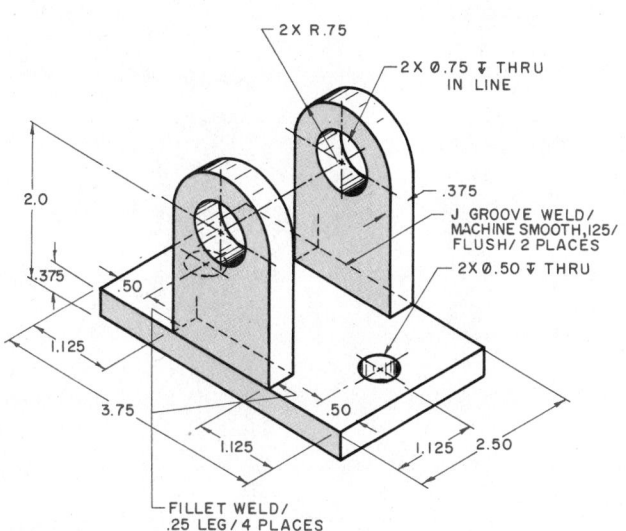

Problem 19-6

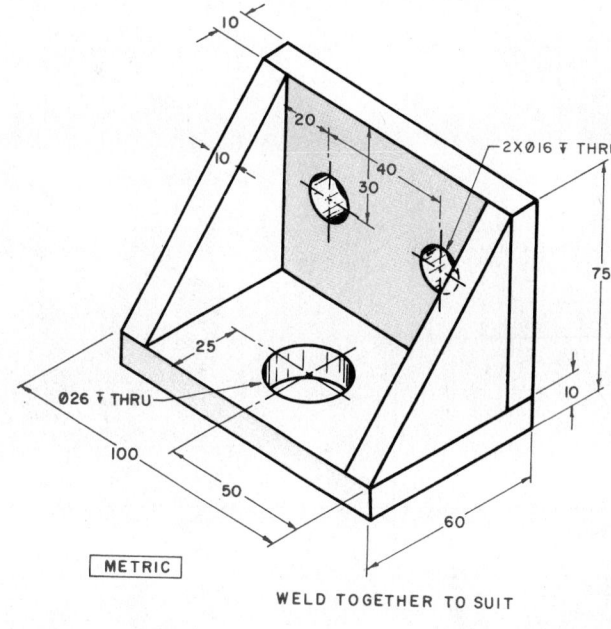

METRIC

WELD TOGETHER TO SUIT

Problem 19-7

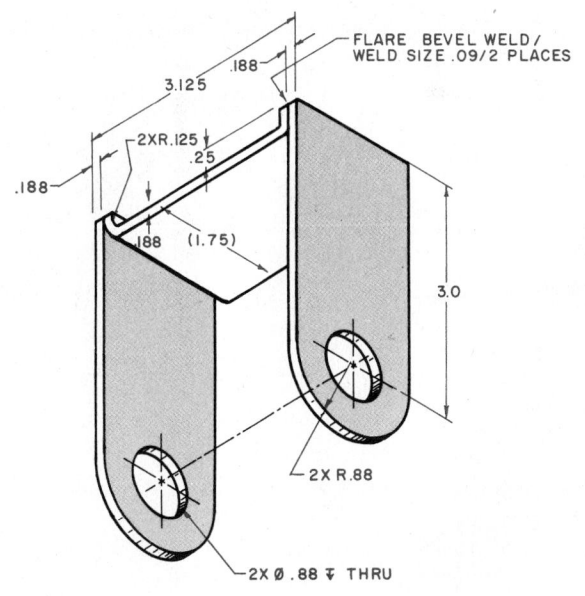

Problem 19-8

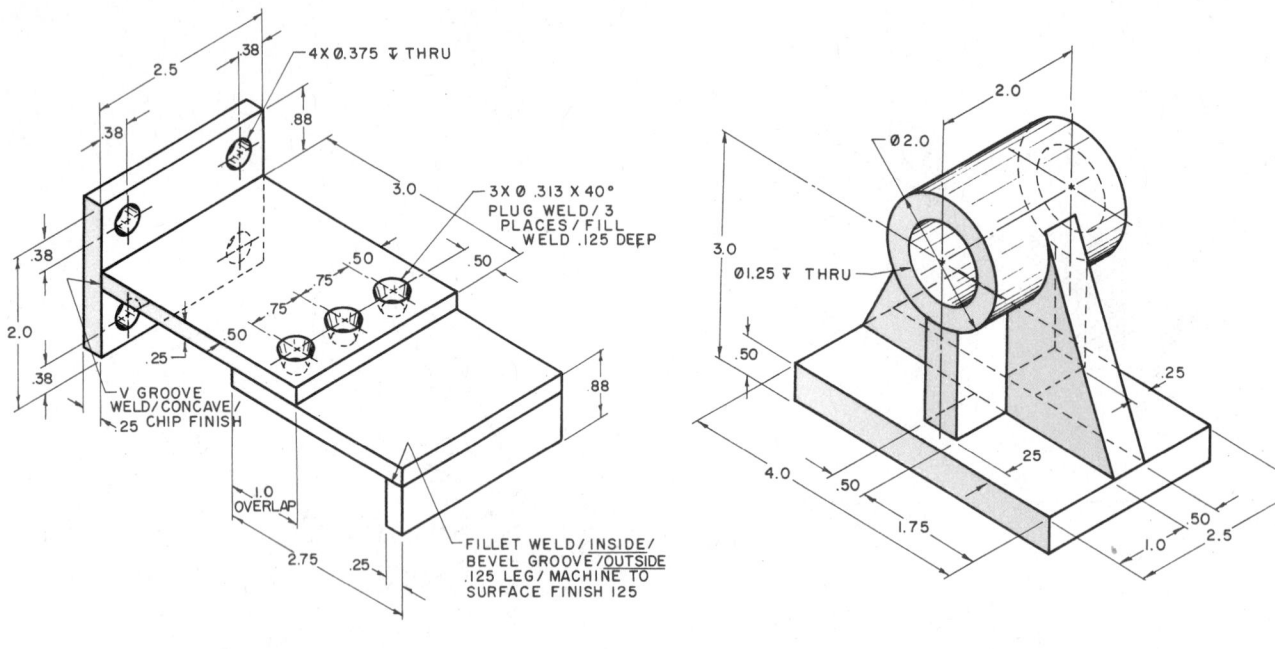

Problem 19-9

❏ **Problem 19-11**

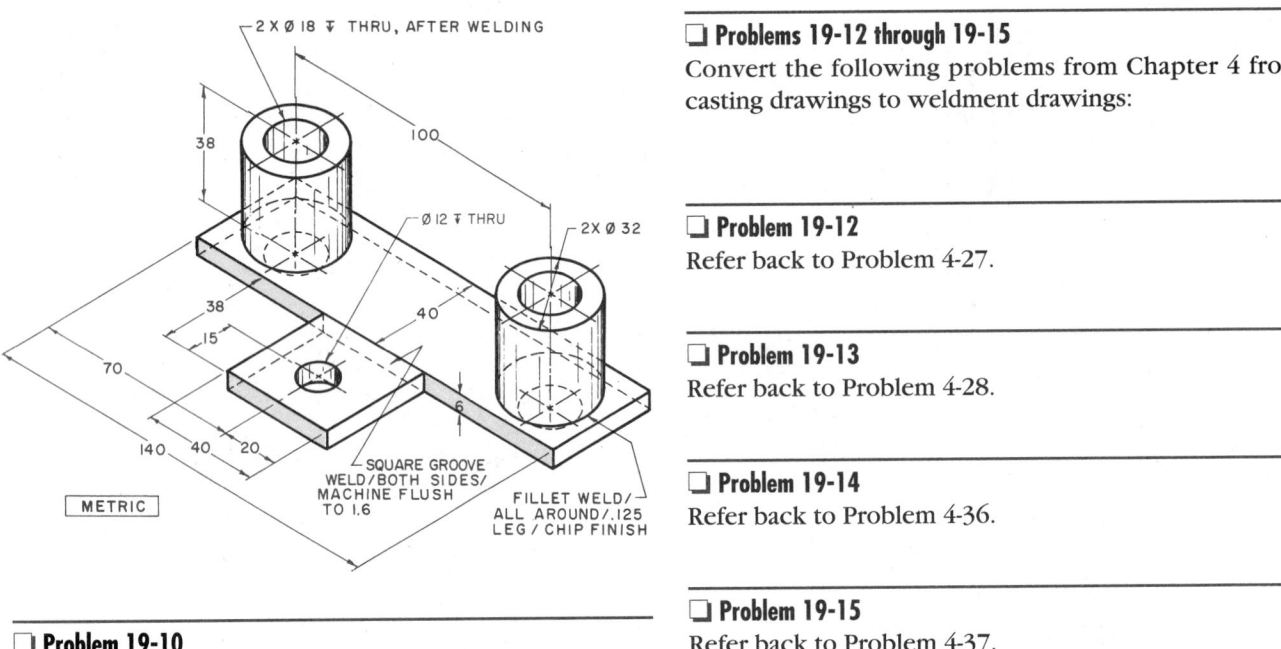

❏ **Problems 19-12 through 19-15**

Convert the following problems from Chapter 4 from casting drawings to weldment drawings:

❏ **Problem 19-12**

Refer back to Problem 4-27.

❏ **Problem 19-13**

Refer back to Problem 4-28.

❏ **Problem 19-14**

Refer back to Problem 4-36.

❏ **Problem 19-15**

Refer back to Problem 4-37.

❏ **Problem 19-10**

Modern Manufacturing: Materials, Processes, and Automation

Chapter Outline

Engineering Materials

Manufacturing Processes

Casting

Forging

Extruding

Stamping

Machining

Special Workholding Devices

Heat Treatment of Steels

Nontraditional Machining Processes

Automation and Integration (CAM and CIM/FMS)

Computer-Aided Manufacturing (CAM)

Industrial Robots

CIM/FMS

Just-in-Time Manufacturing (JIT)

Manufacturing Resource Planning (MRPII)

Statistical Process Control (SPC)

Review

This chapter describes the processes used in a modern manufacturing setting to produce metal and non-metal products as well as the engineering materials from which these products can be made. It also describes how these processes have been automated and integrated in an effort to improve quality and increase productivity.

The ultimate purpose of any engineering drawing is to provide the information necessary to make or fabricate an object. To achieve this purpose an engineering drawing must completely describe and detail the desired object. The drawing must show the specific size and geometric shape of the object as well as furnishing the related information concerning the material specifications, finish requirements, and any special treatments required.

To properly construct an engineering drawing, the designer must be thoroughly familiar with the materials and methods used to transform the drawn image into the actual object specified in the drawing. By knowing the materials and basic processes used to manufacture an object, the designer can design the object with the manufacturing processes in mind. This will not only make manufacture easier but will many times reduce the cost of a manufactured product. Therefore, the purpose of this chapter is to acquaint the new designer with the basic processes used to fabricate objects, as well as the capabilities and limitations of the machine tools used to machine these objects to their final form.

Engineering Materials

Manufacturing is the enterprise through which raw materials are converted into finished products. This conversion adds value to the raw materials. In a modern engineering and manufacturing setting these materials are referred to as *engineering materials*. Engineering materials can be divided into two broad categories: metals and non-metals. Metals include steels, alloy steels, and a variety of high-performance alloys. Non-metals include plastics, composites, elastomers, and ceramics.

Metals

Metals are widely used in modern manufacturing. The most frequently used metals are carbon steels; alloy steels; high-strength, low-alloy steels; stainless steels; maraging steels; cast steels; cast irons; high-performance alloys; tungsten, molybdenum, and titanium; aluminum; copper; magnesium; lead; tin; zinc; and powdered metals. These various metals are described in the following paragraphs.

Carbon Steels

The main elements in steels are iron and carbon. Carbon steels are divided into three groups based on their carbon content. Low-carbon steels contain from .001 to .30 percent carbon. Medium-carbon steels contain from .70 to 1.30 percent carbon. High-carbon steels contain from .70 to 1.30 percent carbon. Certain grades of carbon steels also contain boron to improve hardenability and aluminum to control grain size and promote deoxidation. The following types of products are made of carbon steels: bar, sheet, strip, plate, wire, tubing, and structural shapes.

Alloy Steels

An *alloy steel* is any steel that contains one or more alloying elements added to product characteristics not available in unalloyed carbon steels. Widely used alloying elements include manganese, silicon, copper, aluminum, chromium, cobalt, columbium, molybdenum, nickel, titanium, tungsten, vanadium, and zirconium. Such alloys are added to improve the mechanical and physical properties of steels.

High-Strength, Low-Alloy Steels

These are steels that are significantly stronger than carbon steels as a result of the addition of small amounts of alloying elements and special processing. These alloying elements include manganese, silicon, phosphorus, copper, aluminum, chromium, niobium, vanadium, titanium, molybdenum, nickel, zirconium, nitrogen, and calcium. These steels are categorized according to composition. The different categories vary as to strength, toughness, formability, weldability, and corrosion resistance.

Stainless Steels

Stainless steels are steels that contain 10.5 percent or more chromium. The special appearance of stainless steels is the result of a chromium-based oxide film that coats the metal's surface. Stainless steels also contain a number of other elements that are added to improve their machinability and corrosion resistance. These elements include nickel, molybdenum, copper, titanium, silicon, manganese, columbium, aluminum, nitrogen, and sulfur. There are almost sixty standard grades of stainless steel, but there are five main types. These types are austenitic, ferritic, martensitic, prescription-hardening, and duplex.

Maraging Steels

Maraging steels are a special class of high-strength steels in which nickel is the primary alloy. The name *maraging* is derived from the term *martensite age hardening*, which is a process of age hardening of a matrix comprised of low-carbon, iron-nickel martensite. The process used to produce maraging steels is a double-vacuum melting process. The first process involves vacuum-induction melting. The second process involves vacuum arc remelting. This double-vacuum process maintains a high level of purity and reduces residual elements. Maraging steels are produced in bar, plate, and sheet form.

Cast Steels

Cast steels are steels that are produced in ingot form by melting all component elements together and pouring the molten metal into a mold. This is an intermediate step. Cast steel ingots are subsequently hot or cold worked into another shape. Alloying elements include nickel, chromium, vanadium, and copper. The main types of cast steels are carbon, low-alloy, corrosion-resistant, and heat-resistant.

Cast Irons

Cast irons are metals in a family of cast ferrous metals. Cast irons are distinguished from steels by the relative carbon content (i.e., less than 2 percent in steels and over 2 percent in cast irons). In addition, cast irons also contain from 1 percent to 3 percent silicon. These differences are the basis of the corresponding differences in how steels and cast irons can be used. The metallurgical properties of cast irons make them easy to melt and cast. These are the following basic types of cast irons: white, malleable, gray, ductile, and compacted-graphite iron.

High-Performance Alloys

High-performance alloys are a group of metals that have high strength and/or high corrosion resistance over a broad range of temperatures. High-performance alloys are primarily iron-nickel, nickel, cobalt, or iron-based. They typically have a complex composition with at least three and sometimes over ten component elements. Also known as the super-alloys, their most common elements include chromium, molybdenum, tungsten, columbium, aluminum, cobalt, titanium, carbon, and nitrogen. The most common applications of high-performance alloys are in aircraft and aerospace settings.

Tungsten, Molybdenum, and Titanium

Tungsten and molybdenum are known as *refractory* metals because of their exceptional ability to withstand high temperatures. They also have the following additional characteristics: low coefficient of expansion, high modulus of elasticity, high resistance to thermal shock, high conductivity, and excellent hardness properties at high temperatures. Unlike tungsten and molybdenum, titanium is not a refractory metal, but its properties are similar to such metals. The processes used to produce these three metals are similar. They are extracted from ore, processed into chemicals, and processed further into powders or sponges. Typical applications of tungsten and molybdenum are the automotive, aerospace, mining, chemical, processing, electrical, nuclear, petroleum processes, ordnance, and metal-processing industries. Titanium is used primarily in the aerospace, chemical, and medical industries.

Aluminum

Aluminum is a bright silvery metal that is very light. A given amount of ferrous metal will weigh three times as much as the same amount of aluminum. Aluminum is an abundant metal accounting for approximately 8 percent of the earth's outer surface. Pure aluminum is ductile, soft, and weak compared with other metals. However, with the addition of alloying elements such as copper, magnesium, manganese, and zinc it can be made much harder and stronger. Aluminum is light, an excellent conductor, and able to resist corrosion. Consequently, it is widely used in both manufacturing and construction.

Copper

Copper is a heavy metal that is salmon pink in its pure form and abundant. It is non-magnetic and has excellent conductivity. Only silver is a better conductor. There are over 300 different types of copper and copper alloys. They are produced in a variety of forms including strip, plate, sheet, pipe, tube, rod, forgings, wire, foil, extrusions, and castings. Copper is widely used in both manufacturing and construction.

Magnesium

Magnesium is a soft silvery white metal that lacks the strength needed for structural uses. However, it is extremely light and machinable. To increase its strength, magnesium is alloyed with aluminum, manganese, lithium, thorium, zinc, and zirconium. The result is a metal that can be machined at high speeds, has good conductivity, a high level of fatigue, and is dent resistant.

Lead

Lead is a dark grey metal that has the following characteristics: high density, low melting point, corrosion resistance, chemical stability, malleability, lubricity, and electrical properties. Lead can be easily produced in a variety of forms. Most lead is used for paint pigments, gasoline additives, and batteries.

Tin

Tin is a silvery white metal that is soft and ductile. Its high level of malleability allows tin to be drawn into wire or hammered into extremely thin sheets. Tin is too weak to be used in its pure form, but alloyed with antimony, copper, lead, zinc, or silver, tin becomes useful. Its most common use is in coating steel containers used to preserve food. Tin is also used in solder alloys, bronzes, fusible alloys, type metals, pewter, and dental amalgams. Tin chemical compounds are used in the electroplating, ceramics, plastics, pesticide, and antifungal industries.

Zinc

Zinc is a bluish-white metal that is widely used in manufacturing. Only iron, aluminum, and copper are more widely used than zinc. Zinc is produced in castings and it is widely used for coating other materials. Zinc alloys offer several advantages over other die-cast metals including ease of casting, strength, ductility, and the ability to hold close tolerances and to be cast in thin sections.

Powdered Metals

Powdered metals are exactly what the name says. They are produced by atomization, electrolysis, chemical reduction, or oxide reduction. Various powdered metals are mixed and sintered to produce alloys with an almost endless number of compositions. Sintering produces bonds among the various powdered particles. The properties of products made from powdered metals depend on which metals are mixed, in what amounts, and by what processes. These characteristics can be controlled by varying particle size, particle shape, size distribution, sieve analysis, bulk density of the powder, rate of flow into the die, and the compressibility of the powder in the die.

Non-Metals

Non-metals are widely used in modern manufacturing. The most widely used non-metals are plastics, composites, ceramics, and elastomers. These various non-metals are described in the following paragraphs.

Plastics

Plastics are non-metal materials that are made from either natural or synthetic resins. Most plastics used in modern manufacturing are made from synthetic resins. There are many different kinds of plastics including polystyrene, high-impact polystyrene, acrylics, polycarbonate, ABS, acetal, nylon, polypropylene, polyethylene, epoxy, and phenolic. Most plastics used in an engineering or manufacturing setting are made from thermoplastic resins. Such plastics have the following characteristics: a) thermal, mechanical, chemical, corrosion resistance, and fabricability; b) ability to sustain high mechanical loads; and c) predictable performance.

Composites

Composites are materials created by combining two or more materials. Perhaps the most readily recognizable composite is plywood. Fiberglass is another example of a composite. Composite technology is divided into four broad areas: organ matrix composites (resins), metal matrix composites, carbon-carbon composites, and ceramic matrix composites. Composites can be divided into two categories in terms of how they are produced; laminates and sandwiches. Laminates consist of two or more layers bonded together. Sandwiches consist of thick, low-density layers (i.e., honeycomb or foam materials) pressed between thin layers of higher strength, higher density material. Applications of composites are extensive and varied. They include aircraft and aerospace applications, athletic equipment, automobiles, prosthetic devices, printed circuit boards, boats, and many others.

Ceramics

Ceramics are materials made by combining metallic and non-metallic elements such as oxides, carbides, and nitrites. Glass, bricks, tile, and porcelain are all common ceramics. Refractory ceramics are important in modern manufacturing because of their usefulness in electronics applications. They are used in thermistors, rectifiers, capacitors, transducers, and for high-voltage insulation. Characteristics of ceramics include the fact that they are hard, brittle, have a high melting point, have low conductivity, are chemically and thermally stable, have high compressive strength, and have high creep resistance.

Elastomers

The most recognizable form of elastomer is natural rubber. *Elastomers* are linear polymers that are able to stretch and return to their original shape and size (i.e., a rubber band). Widely used elastomers include polyacrylate, ethylene propylene, neoprene, polysulfide, silicone, and urethane. Common applications of elastomers include oil hoses, o-rings, electric insulation, belts, gaskets, seals, value disks, and foam padding.

Manufacturing Processes

The term manufacturing processes refers to the basic methods used by the shop to make the object described in the engineering drawing. The specific processes used to make the object depend on the object iself. In some cases the part may be cast, while other parts may be forged, extruded, or stamped. In many instances the part, once fabricated into its basic form, must also be machined to maintain a specific degree of accuracy or to produce a feature not possible with other manufacturing processes.

When the manufacturing shop receives the engineering drawing, in the form of a part print, it is first reviewed to ensure all pertinent data and information necessary to make the object are contained in the drawing. During this review, the shop personnel will consider several factors necessary to determine how the part must be made. These factors include: the type and condition of the material used to make the object, the overall size and shape of the part, the types of operations required, the required accuracy of the part, and number of parts to be made.

The type and condition of the material used to make the part are important considerations. Some parts may be made from solid bar stock while other parts must be extruded, cast, forged, or stamped. In most cases, parts made from bar stock require the least lead time. The *lead time* is the interval from the time the shop receives the drawing until production begins. Most shops maintain a sufficient supply of bar stock to begin production as soon as the drawing is received. However, when a part must be extruded, cast, forged, or stamped a longer lead time is required to make the necessary molds or dies to fabricate the object.

The size and shape of the object must be considered to determine if the object is within the capabilities of the shop to make. In addition, the size and shape may also determine the size of the machine tools required as well as the datum, or reference, surfaces used to locate the part during manufacture.

The types of operations required determine the types of equipment and machine tools needed to make the part. If, for example, the part requires holes, a drill press, vertical milling machine, or other machine tool may be used. The next consideration, the required accuracy of the part, also determines how the operations are performed. A hole with a required accuracy of .002″ (0.05mm), for example, would require reaming, while a hole with a required accuracy of .020″ (0.5mm) could be drilled.

The number of parts to be made will frequently determine if the part should be cast, forged, extruded, or machined from solid stock. Likewise, the number of parts will also determine if any special workholders are to be made. Larger production runs will normally justify more sophisticated tools and processes since the cost can be spread over a larger number of parts. Smaller production runs normally demand the parts be made at the lowest possible cost with little or no investment in special molds, dies, or workholders.

While each of these factors has been separated for the purpose of this discussion, in practice each is considered as part of the other. These factors are so closely related in production that they frequently overlap and one cannot be considered without the others.

The primary processes used to fabricate manufactured products are casting, forging, extruding, stamping, and machining. To use these processes to their best advantage, the designer must be familiar with the strengths and weaknesses of each process as well as the fundamental aspects of each process.

Casting

Casting is basically a process of pouring molten metal into a mold that contains the desired shape in the form of a cavity. The principal types of casting used in manufacturing today are sand casting, investment casting, centrifugal casting, and die casting.

Sand Casting

Sand casting is the most common type of casting method. The major components of the molds used to make sand castings are shown in *Figure 20-1* and include the *flask*, *pattern*, and *green sand*. The flask is a two-part box, or frame, used to contain the sand. The top half of the flask is called the *cope* and the bottom half is called the *drag*. Occasionally, a third section may be installed between the cope and drag. This section is called a *cheek*, and is used where a deep or complex shape must be cast. In sand casting the mold is prepared by ramming the green sand around a model of a part, called the *pattern*. The model is then removed to form the cavity for the molten metal. The plane of division between the cope and drag is called the *parting line*. In many castings, the parting line occurs at the approximate

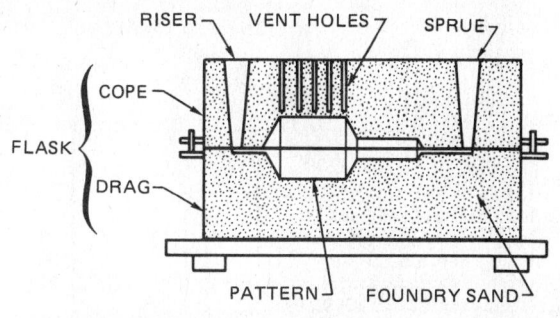

Figure 20-1   Components of a sand casting mold *(From Materials Processing, B. R. Thode, Delmar Publishers Inc.)*

middle of the part. The parting line can be seen on most castings by a ragged line that is usually ground off.

The molten metal enters the mold through the *sprue hole* and is directed to the cavity by one or more *gates*. The sprue hole is formed by installing a sprue peg in the cope. This peg is then removed after the final ramming of the cope. The *riser* is used to vent the mold and to allow gases to escape. The riser also acts as a small reservoir to keep the cavity full as the metal begins to shrink during the cooling process.

Once the cope and drag are rammed and the pattern is removed, a solid part could be poured in the mold. However, in some cases a hollow part, or one which has large holes, must be poured. To reduce the amount of material needed to fill the cavity and to reduce the time necessary to machine the part a *sand core* may be installed in the cavity. When sand cores are intended to be installed in a cavity, *core prints* must also be provided to locate and anchor the sand cores during the casting process. Core prints are normally a simple extension of the pattern.

Making Patterns.   When patterns are made for casting two important factors must always be considered. The first is the draft. *Draft* is the slope or taper of the sides of a pattern which permits it to be removed from the cope and drag without disturbing the cavity. The draft also permits the cast part to be removed easily from some molds. The amount of draft necessary will normally depend on the part being cast, but will normally be about one degree.

The second consideration is shrinkage. When metals are cast a certain amount of *shrinkage* occurs as the metal cools. The specific amount of shrinkage depends on the metal being cast. Steel, for example, shrinks at a rate of approximately three-sixteenths of an inch per foot, while cast iron shrinks about one-eighth of an inch per foot. To allow for this shrinkage and to make sure the final part is the correct size the pattern must be made slightly larger than the final size of the desired part.

When making a pattern, the pattern maker will normally use a shrink rule to compensate for the shrinkage. A *shrink rule*, *Figure 20-2*, is a standard steel rule which has the graduations marked for a specific amount of shrinkage. A shrink rule used for cast iron has an extra one-eighth inch added to each foot, while a shrink rule for steel has an additional three-sixteenths added.

In most instances, when a pattern must be made, the pattern maker will receive the engineering drawing which contains all the final sizes of the part. The pattern maker will then make all the necessary calculations needed to make the oversized pattern. But, occasionally, the pattern maker will be given a drawing with all the calculations already made to produce the pattern. In either case, the patterns must be made to suit the material being cast.

Investment Casting

Investment casting produces parts with great detail and accuracy, while at the same time allowing very thin cross sections to be cast. In this process, the pattern is made by casting wax in the desired form. The pattern is then placed in a sand mold and the mold is fired to melt out the wax pattern. This process is also referred to as *lost wax casting*. The molten metal is then fed into the mold to produce the cast part.

Centrifugal Casting

Centrifugal casting, *Figure 20-3*, is a process of pouring a measured amount of molten metal into a rotating mold. This process can be used for a single mold or multiple

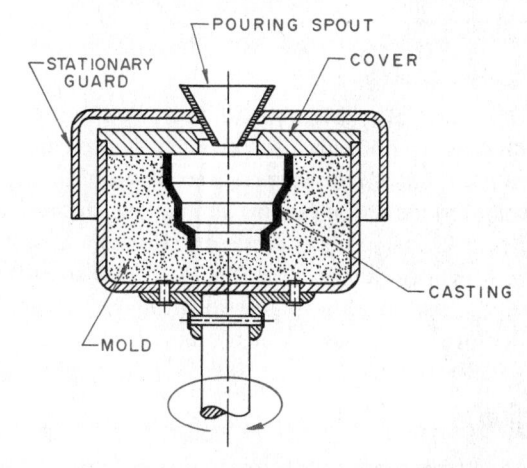

Figure 20-3   Centrifugal casting

Figure 20-2   Shrink rule *(Courtesy L. S. Starrett Co.)*

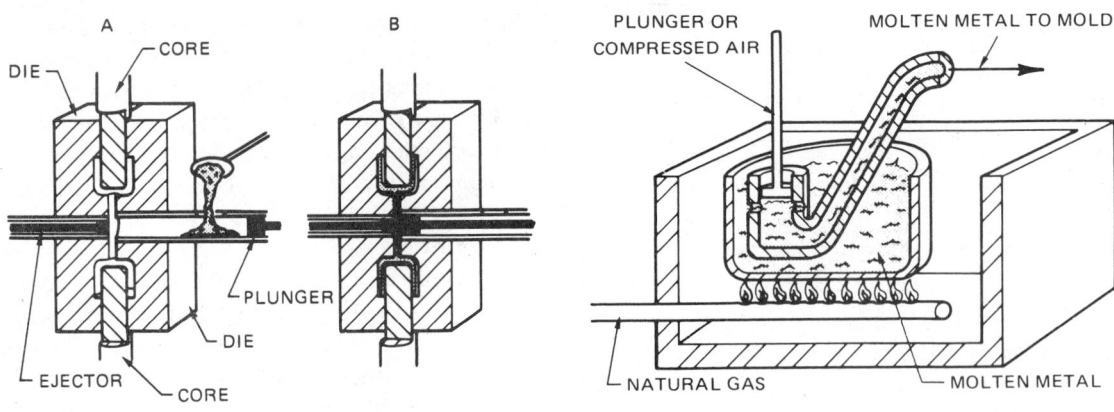

Figure 20-4   Die casting *(From Materials Processing, B. R. Thode, Delmar Publishers Inc.)*

molds. The centrifugal force created by the rotation forces the molten metal to fill the cavity. This same centrifugal force, along with the measured amount of material, also controls the wall thickness of the cast part and results in a less porous cast surface than is possible with sand casting. The molds used for this purpose are usually permanent molds made from metal rather than sand. These molds are used repeatedly and produce highly accurate parts requiring very little machining. However, due to their cost, permanent molds are normally used only for high-volume production.

Die Casting

Die casting, *Figure 20-4*, is a process where molten metal is forced into metal dies under pressure. This process is very well suited for such materials as zinc alloys, aluminum alloys, copper alloys, and magnesium alloys. Parts produced by die casting are superior in appearance and accuracy and require little or no machining to final size.

Powder Metallurgy

Powder metallurgy, *Figure 20-5*, while not an actual casting process, does have some similarities to casting. In the *powder metallurgy* process, metal particles, or powder, are blended and mixed to achieve the desired composition. The powder is then forced into a die of the desired form under pressure from 15,000 to 100,000 pounds per square inch. The resulting heat fuses the powder into a solid piece which can be machined.

Forging

Forging is the process of forming metals under pressure using a variety of different processes. The most common forms of forging are drop forging and press forging. Other variations of forging include rolling and upsetting.

Drop Forging

Drop forging is the process of forming a heated metal bar, or billet, in dies. In practice, the heated metal is placed on a lower portion of a forging die and struck repeatedly with the upper die portion. This forces the metal into the

shape of the cavity of the dies. The pressure required to form the metal is produced by a drop hammer.

Most drop forge dies contain at least four different stations, or dies, to complete the part, *Figure 20-6*. The first station is called the *fuller*. This station is used to rough-form the part to fit the other cavities. The second station, when used, is called the *breakdown*, or *bender*. This station forms the part into any special contours required by the next station, the roughing impression. The roughing impression rough-forms the part to the desired form. The final station is called the *finishing impression* and is used to finish the part to the desired shape and size. Additional stations, such as a trimmer to remove the fins and a cutoff to sever the forged part from the bar, may also be included on the die if they are necessary.

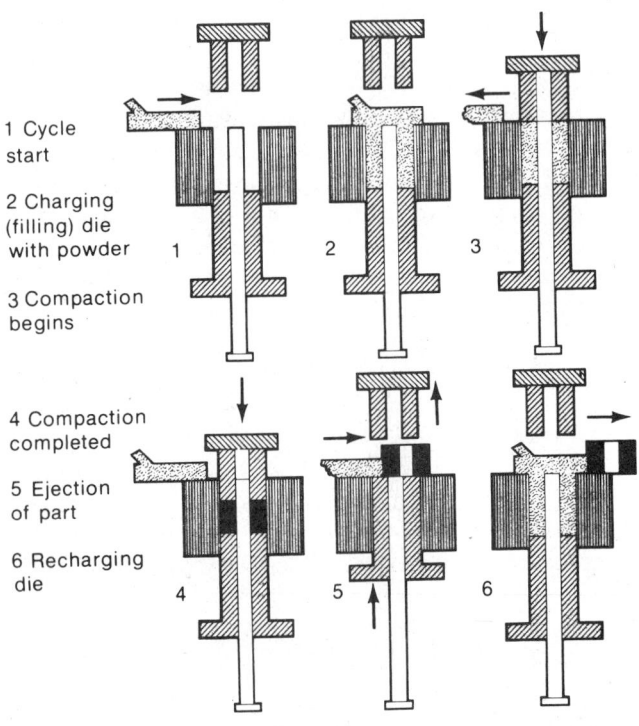

1 Cycle start

2 Charging (filling) die with powder

3 Compaction begins

4 Compaction completed

5 Ejection of part

6 Recharging die

Figure 20-5   Producing parts with powder metallurgy *(Courtesy Metal Powder Industry Foundation)*

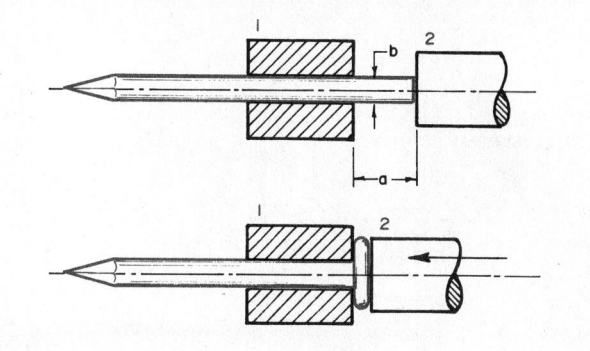

Figure 20-7 Upsetting

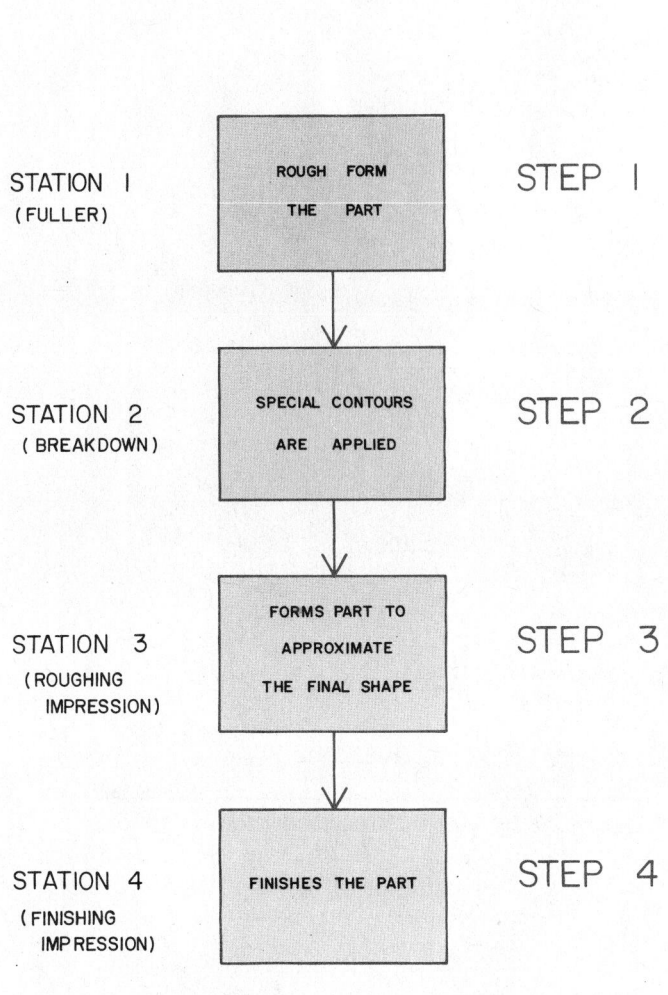

STATION 1
(FULLER)

ROUGH   FORM
THE   PART

STEP 1

STATION 2
(BREAKDOWN)

SPECIAL   CONTOURS
ARE   APPLIED

STEP 2

STATION 3
(ROUGHING
IMPRESSION)

FORMS PART TO
APPROXIMATE
THE FINAL SHAPE

STEP 3

STATION 4
(FINISHING
IMPRESSION)

FINISHES   THE   PART

STEP 4

Figure 20-6 Drop forging process

Press Forging

Press forging is a single-step process in which the heated metal bar or billet is forced into a single die and is completed in a single press stroke. The press normally used for press forging is hydraulic.

Rolling

Rolling is a process where the heated metal bar is formed by passing through rollers having the desired form impressed on their surfaces. Rolling is frequently used to flatten and thin out thick sections.

Upsetting

Upsetting, **Figure 20-7**, is a process used to enlarge selected sections of a metal part. Bolt heads, for example, are normally produced by an upsetting process.

Extruding

Extruding is a process of forcing metal through a die of a desired form and cross section. The bars produced are then cut to the required lengths. Typical examples of extruded parts include parts for aluminum windows and doors. Extrusions provide a near final shape, as shown in **Figure 20-8** and, in many cases, only need to be cut off and machined slightly to complete a finished part.

Stamping

Stamping is a process of using dies to cut or form metal sheets or strips into a desired form. The main tool used for metal stamping is a die. The term *die* has a double

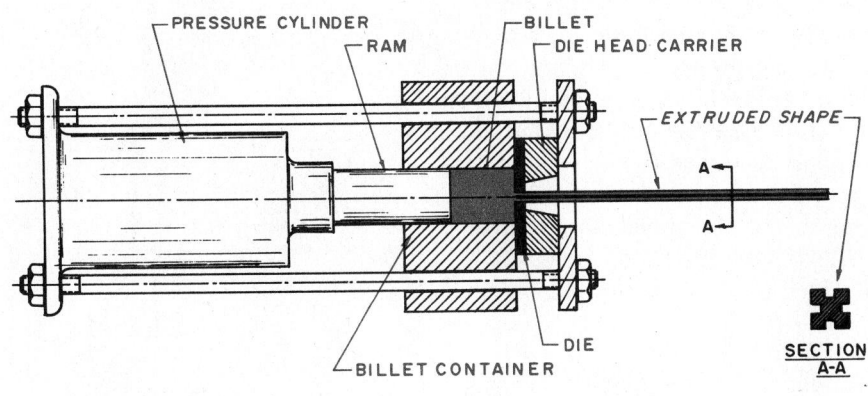

Figure 20-8 Extruding

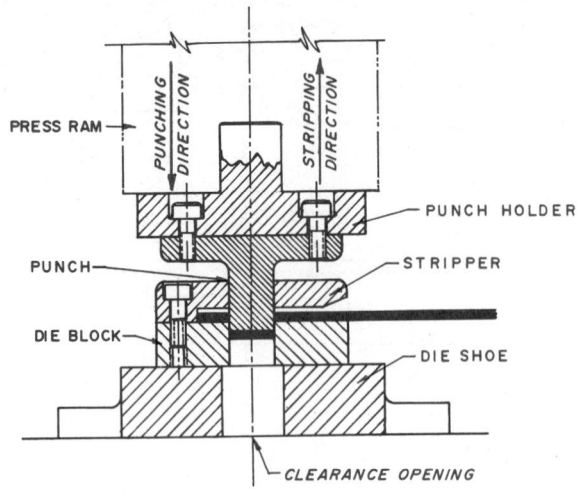

Figure 20-9 Metal stamping die

meaning. It can be used to describe the entire assembled tool or the lower cutting part of the tool. The exact meaning of the term can usually be determined from the context of its use. The principal parts of a die are shown in ***Figure 20-9***. The upper die shoe is used to mount the punch. The die is mounted on the lower die shoe. The guide pins and guide pin bushings are used to maintain the alignment between the punch and die. The stripper is designed to strip the stock material from the punch after the cutting stroke of the die.

When more than one operation takes place on a part during a single stroke, blanking and punching for example, the operation is said to be *compound*. However, when several operations take place sequentially in a die the die is called a *progressive die*.

The principal stamping operations normally performed are shearing, cutting off, parting, blanking, punching, piercing, perforating, trimming, slitting, shaving, bending, forming, drawing, coining, and embossing.

Operations that Produce Blanks

The operations that produce blanks are shearing, cutting off, parting, and blanking, ***Figure 20-10***. *Shearing* is a cutting action performed along a straight line. *Cutting off* is a cutting operation which is performed on a part to produce an edge other than a straight edge. In many cases, cutting off is used to finish the edges of a part. Like shearing, cutting off does not produce any scrap. *Parting*, on the other hand, is also an operation used to finish the edges of a part, but, unlike the other operations, parting does produce scrap.

Blanking is an operation that produces a part cut completely by the punch and die. In blanking operations, the piece that falls through the die is the desired part. The scrap is the skeleton left on the stock strip.

Operations that Produce Holes

The operations that produce holes are punching, piercing, and perforating. *Punching* is an operation that cuts a hole in a metal sheet or strip, ***Figure 20-11A***. In

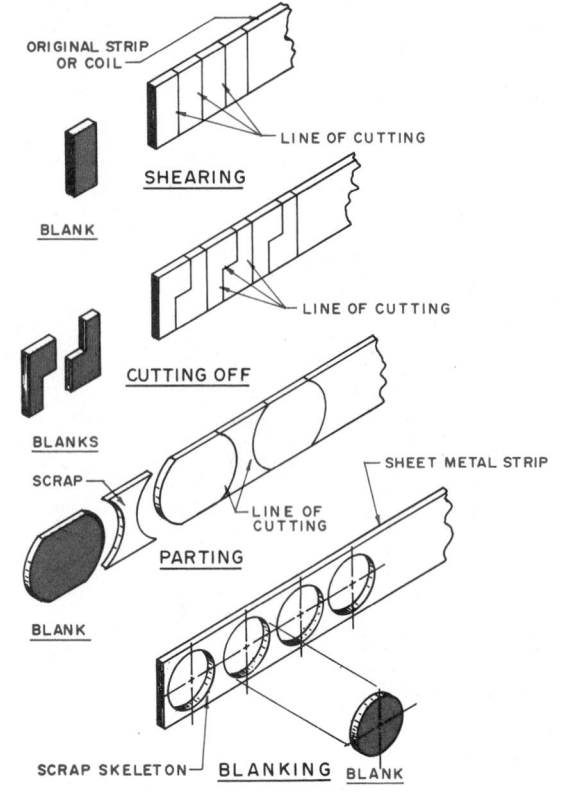

Figure 20-10 Operations that produce blanks

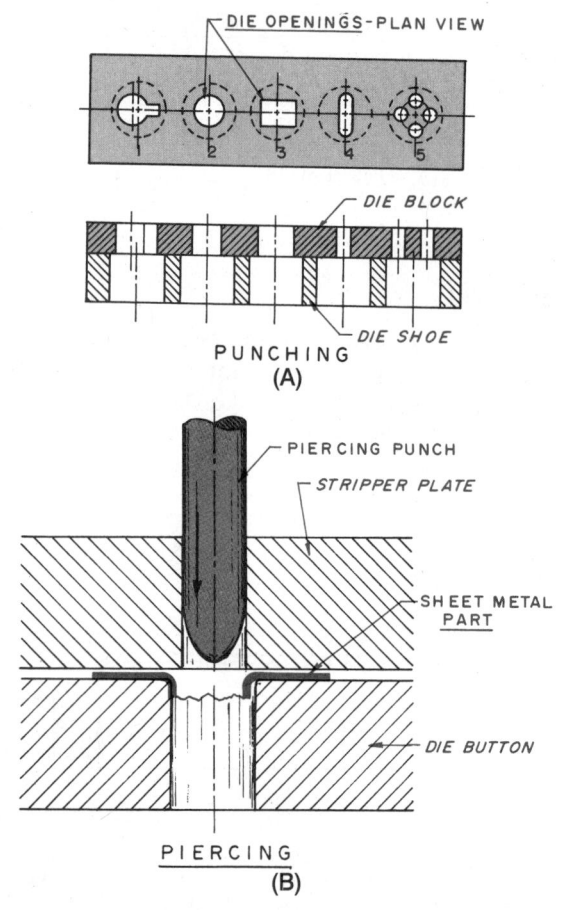

Figure 20-11 (A) Punching operation to produce holes;
(B) Piercing operation to produce holes

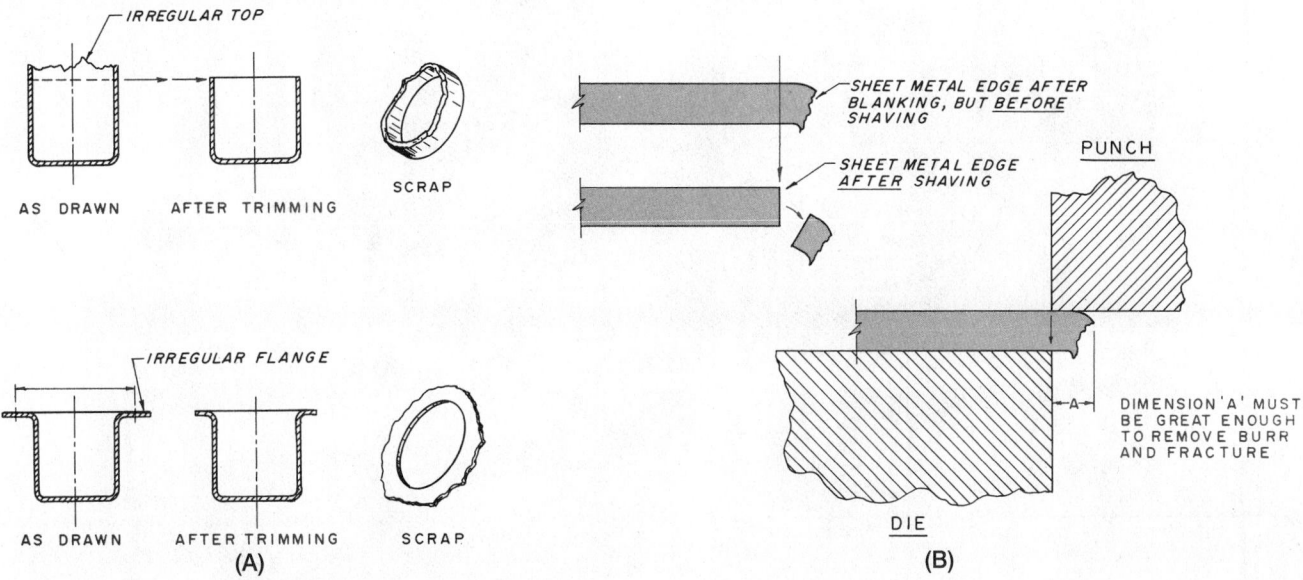

Figure 20-12   (A) Operation that controls size–trimming; (B) Operation than controls size–shaving

punching, the piece that falls through the die is scrap, and the area on the strip is the desired part. Punching is normally performed on parts that are to be blanked to provide holes for bolts, screws, or similar parts.

Piercing is an operation similar to punching, except that in piercing no scrap is produced, ***Figure 20-11B***. A pierced hole is sometimes used to increase the thickness of metal around a hole for tapping threads in a stamped part. *Perforating* is simply a punching operation performed on sheets to produce either a uniform hole pattern or a decorative form in the sheet.

Operations that Control Size

The operations used to control size are trimming, slitting, and shaving, ***Figures 20-12A*** and ***20-12B***. *Trimming* is an operation performed on formed or drawn parts to remove the ragged edge of the blank. *Slitting* is performed on large sheets to produce thin stock strips. Slitting may be performed by rollers or by straight blades. *Shaving* is an operation performed on a blanked part to produce an exact dimensional size or to square a

blanked edge. Shaving produces a part with a close dimensional size, and a cut edge which is almost perpendicular to the top and bottom surfaces of the blanked part.

Operations that Bend or Form Parts

The operations used to bend or form parts are bending, forming, and drawing, ***Figure 20-13***. *Bending* is an operation in which a metal part is simply bent to a desired angle. *Forming*, on the other hand, is an operation in which a part is bent or formed into a complex shape. Forming is also a catchall word frequently used to describe any bending operation that does not fall into one of the other categories. *Drawing* is a process of stretching and forming a metal sheet into a shape similar to a cup or top hat.

Machining

Machining is the process of removing metal with machine tools and cutters to achieve a desired form or

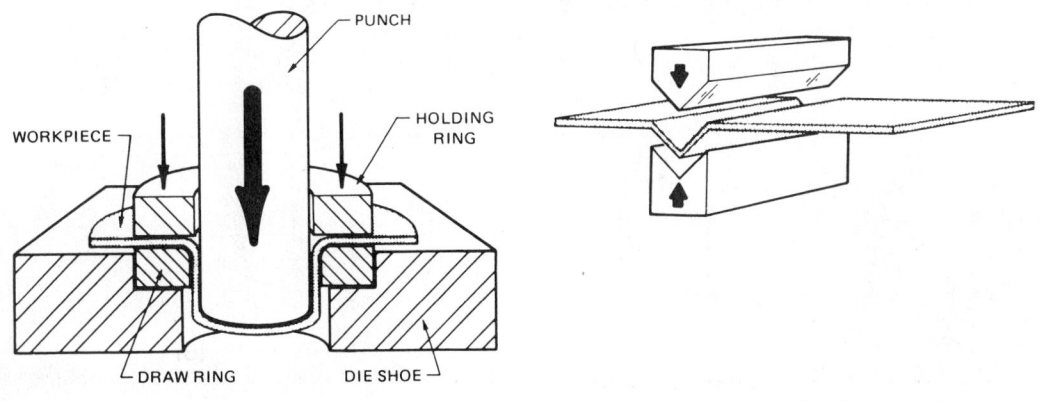

Figure 20-13   Operations that bend or form parts *(From Materials Processing, B. R. Thode, Delmar Publishers Inc.)*

Figure 20-14 Lathe *(Courtesy Lodge & Shipley Co.)*

feature. The principal types of machine tools used to perform these operations are lathes, milling machines, drill presses, saws, and grinders. Other machines that are variations of the basic machines used in the machine shop include shapers and planners, boring mills, broaching machines, and numerically controlled machines. The specific type of machine used to perform a particular machining operation is determined by the type of operation and the degree of accuracy required.

Lathes

A lathe, ***Figure 20-14***, is one of the most commonly used and versatile machine tools in the shop. Typically, a *lathe* performs such operations as straight and taper turning, facing, straight and taper boring, drilling, reaming, tapping, threading, and knurling, ***Figure 20-15***.

In operation, the workpiece is held in the headstock and supported by the tailstock. The cutting tool is mounted in the tool post and is traversed past the rotating workpiece to machine the desired form, ***Figure***

20-16. The most common method used to hold parts in a lathe is with a chuck. A *chuck* is a device much like a vise, having movable jaws that grip and hold the workpiece. The most popular type of chuck is the three-jaw chuck, but there are also two-jaw, four-jaw, and six-jaw chucks as well. Other types of devices used to hold and drive a workpiece in a lathe include collets, face plates, drive plates, and lathe dogs.

Milling Machines

Milling machines are used to machine workpieces by feeding the part into a rotating cutter. The two basic variations of the milling machine are the vertical milling machine and the horizontal milling machine, ***Figure 20-17***.

In operation, the milling cutter is mounted in either the spindle or arbor, and the workpiece is held on the machine table. The most commonly used device to mount the workpiece for milling is the milling machine vise, but the workpiece may also be held directly on the machine table or in other workholding devices. The operations most frequently

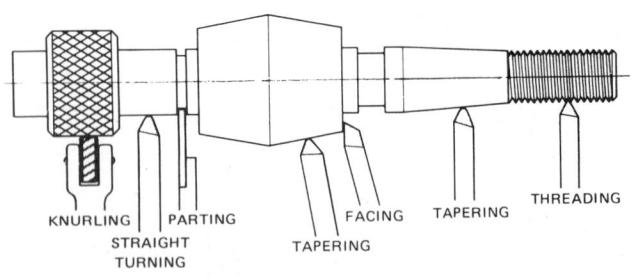

Figure 20-15 Basic lathe operations *(From Materials Processing, B. R. Thode, Delmar Publishers Inc.)*

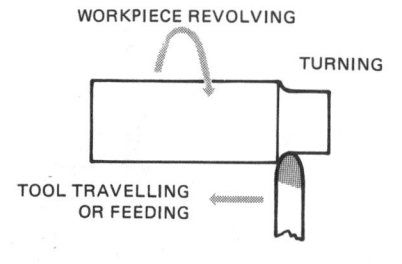

Figure 20-16 Lathe setup for straight turning *(From Turning Technology, S. F. Krar and J. W. Oswald, Delmar Publishers Inc.)*

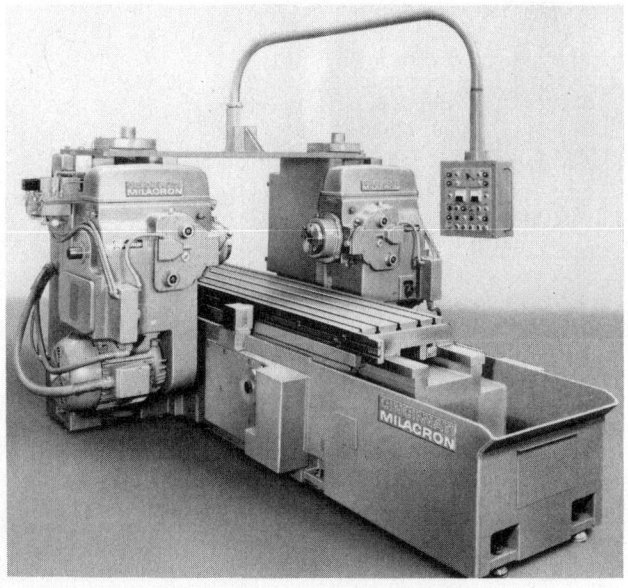

Figure 20-17   Milling machines *(Courtesy Cincinnati Milacron)*

CONVENTIONAL MILLING   CLIMB MILLING

Figure 20-18   Basic milling machine operations *(From Materials Processing, B. R. Thode, Delmar Publishers Inc.)*

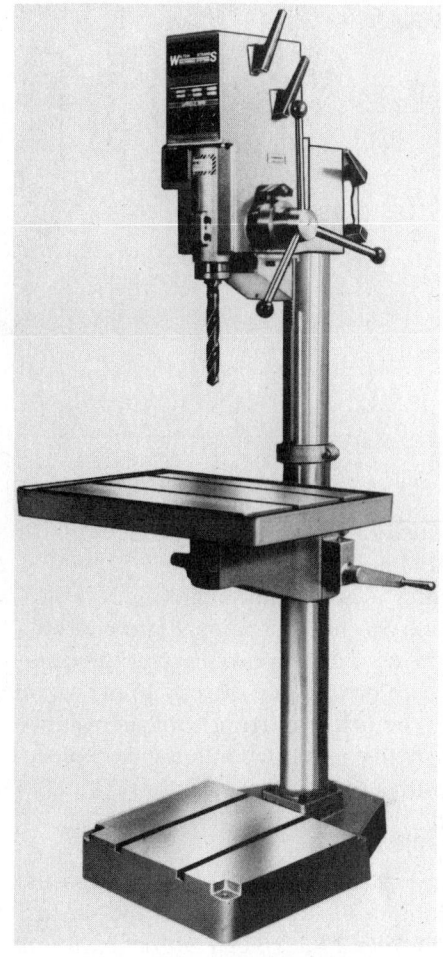

Figure 20-20   Drill press *(Courtesy Wilton Corp.)*

performed on a milling machine include plain milling, face milling, end milling, straddle milling, gang milling, form milling, keyseat milling, and gear cutting, *Figure 20-18*. *Figures 20-19A* and *20-19B* show two types of cutters commonly used for milling operations.

Drill Presses

A *drill press*, *Figure 20-20*, is a machine that is used mainly for producing holes. The principal types of drill presses used in the shop are the radial drill press and the sensitive drill press.

(A)

(B)

Figure 20-19   (A) and (B) Standard milling cutters *(Courtesy Cincinnati Milacron)*

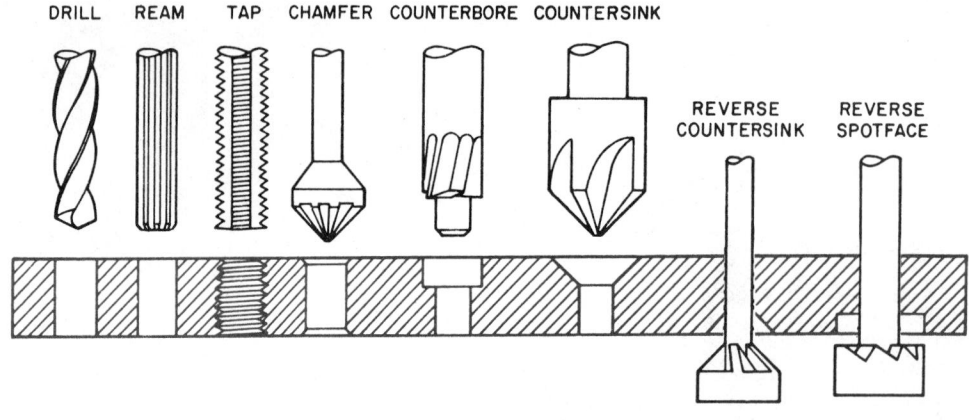

DRILL   REAM   TAP   CHAMFER   COUNTERBORE   COUNTERSINK

REVERSE COUNTERSINK   REVERSE SPOTFACE

Figure 20-21   Basic drill press operations

The operations normally performed on drill presses include drilling, reaming, tapping, chamfering, spotfacing, counterboring, countersinking, reverse countersinking, and reverse spotfacing, ***Figure 20-21***.

When using a drill press, the workpiece is normally held in a vise or clamped directly to the machine table. The drill, or other cutting tool, is mounted in the machine spindle and is fed into the workpiece either by hand or with a mechanical feed unit.

Power Saws

The *power saws* used most often in the machine shop are the contour band machine or band saw and the cutoff saw, ***Figure 20-22***. The *contour band machine* or *band saw* is used primarily for sawing intricate or detailed shapes; the *cutoff saw* is used to cut rough bar stock to lengths suitable for machining in other machine tools.

Precision Grinders

Precision grinders are available in several styles and types to suit their many and varied applications. The principal types of precision grinder used in the machine shop are the surface grinder and the cylindrical grinder, among others.

The *surface grinder*, ***Figure 20-23***, is mainly used to produce flat, angular or special contours on flat workpieces. The most popular variation of this machine consists of a grinding wheel mounted on a horizontal spindle, and a reciprocating table that traverses back and forth under the grinding wheel. One other variation of the surface grinder frequently found in the machine shop uses a vertical spindle and a round, rotating table, ***Figure 20-24***. The workpiece is generally mounted and held on a magnetic chuck during the grinding operation on both styles of surface grinders.

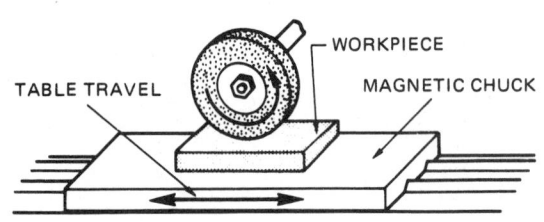

TABLE TRAVEL   WORKPIECE   MAGNETIC CHUCK

Figure 20-23   Surface grinder *(From Materials Processing, B. R. Thode, Delmar Publishers Inc.)*

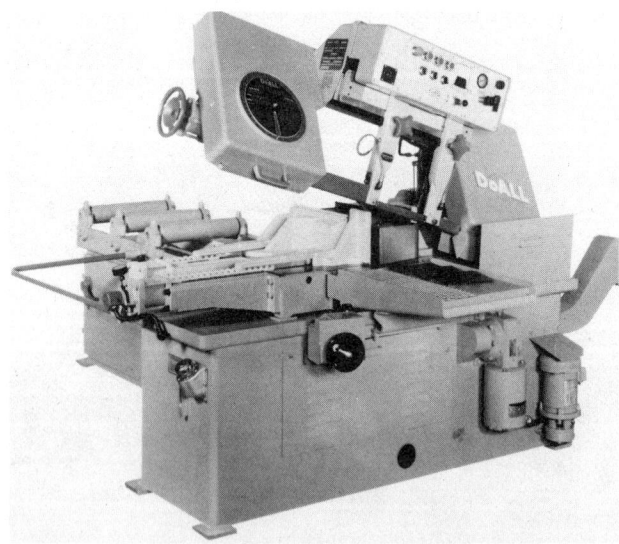

Figure 20-22   Power saw *(Courtesy DoALL Co.)*

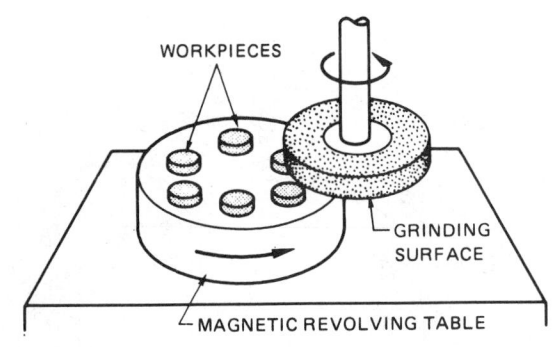

WORKPIECES

GRINDING SURFACE

MAGNETIC REVOLVING TABLE

Figure 20-24   Vertical spindle—rotating table surface grinder *(From Materials Processing, B. R. Thode, Delmar Publishers Inc.)*

Cylindrical grinders are used to precisely grind cylindrical or conical workpieces. When grinding, the workpiece is mounted either between centers or in a precision chuck, much like a lathe. The grinding wheel is mounted behind the workpiece on a horizontal spindle. The table of the grinder traverses the workpiece past the grinding wheel in a reciprocating motion, while the workpiece rotates at a preset speed.

Shapers and Planers

Shapers and *planers* are machine tools that use single-point cutting tools to perform their cutting operations. The shaper uses a reciprocating ram to drive the cutter. The cutting tool on this machine tool is mounted on the end of the ram in a unit called the *clapper box*. The clapper box can be positioned vertically or at an angle, depending on the work to be performed. The depth of cut is adjusted by lowering the vertical slide of the clapper box or by raising the position of the table. The workpiece is mounted on the table or vise and is fed past the reciprocating cutter by the table feed. The length and position of the stroke are regulated by the position of the ram and are easily adjusted to suit the size of the workpiece.

Planers operate in a manner similar to the shaper. The major differences between these two machines are their size and method of cutting. The planer is normally much larger than the shaper and the work, rather than the cutter, moves on the planer. In operation, the workpiece is mounted on the table of the planer and the table reciprocates under the stationary tool. As the table reciprocates, the tool is moved across the workpiece by the feed unit. The depth of cut is determined by the height of the tool from the table and is adjusted by lowering or raising the cross beam. The length and position of the cut are determined by the position of the table.

These processes are seldom used today.

Boring Mills

A *boring mill*, **Figure 20-25**, is a machine tool normally used to machine large workpieces. The two common

Figure 20-25 Boring mill *(Courtesy Cincinnati Milacron)*

variations of these machines are horizontal and vertical. The distinction between the two is determined by the position of the spindle. *Horizontal boring mills* perform a wide variety of different machining tasks normally associated with a milling machine but on a much larger scale. *Vertical boring mills* are commonly used to turn, bore, and face large parts in much the same way as a lathe. Another variation of the horizontal boring mill sometimes found in the machine shop is the *vertical turret lathe*. This machine serves the same basic function as the vertical boring mill but, rather than using a single tool, a vertical turret lathe uses a turret arrangement to mount and position the cutting tools.

Broaching Machines

A *broaching machine*, **Figure 20-26**, is used to modify the shape of a workpiece by pulling tools called *broaches* across or through the part. Both internal and external forms can be broached. *Internal broaching* can produce

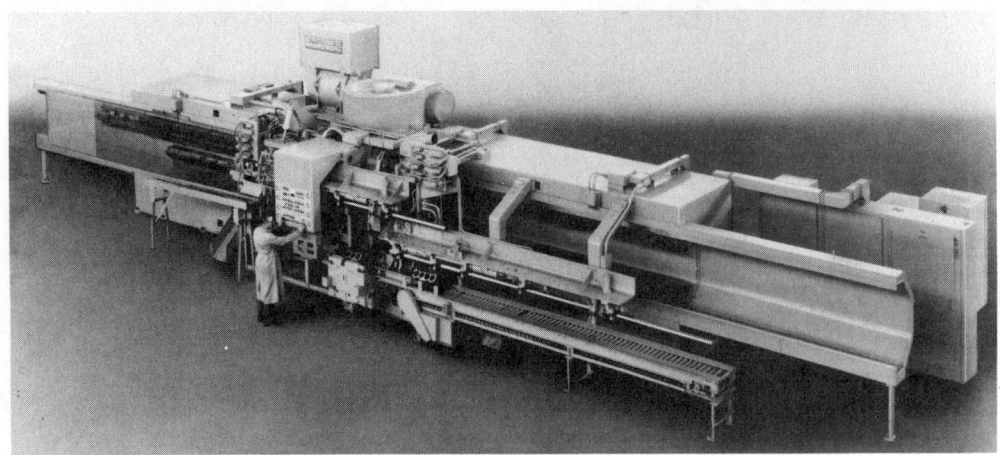

Figure 20-26 Broaching machine *(Courtesy Cincinnati Milacron)*

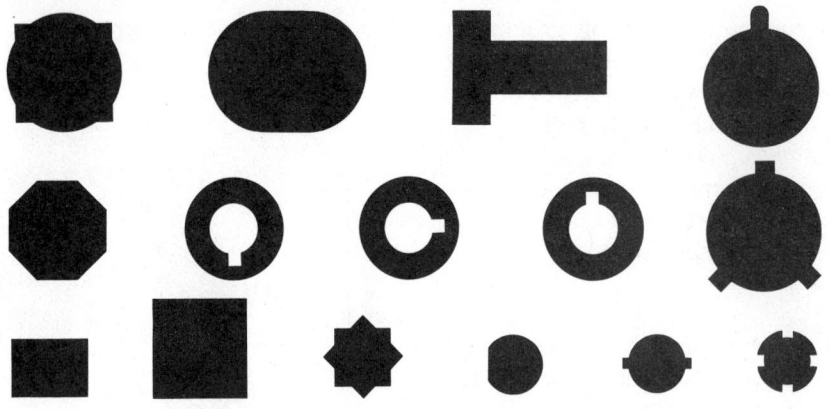

Figure 20-27 Typical internal broached shapes *(Courtesy The DuMont Corp.)*

holes with a wide variety of different forms, *Figure 20-27*. *External broaching* is typically used to produce some gear teeth, plier jaws, and other similar details.

Another type of internal broaching operation is broached in an *arbor press* using a push-type, rather than a pull-type broach. This type of broaching is often used in the top to produce keyways or other simple shapes.

Numerically Controlled Machines

Numerically controlled machine tools represent one of the newer innovations in machine tool design in wide use today. These machine tools are operated by either punched tapes or by computers, and virtually eliminate the errors caused by human operators.

The two basic variations of these machines in use today are the point-to-point and the continuous path machines. The principal difference between the two is in the movements of the tool with reference to the workpiece. *Point-to-point* machines operate on a series of pre-

programmed coordinates to locate the position of the tool. When the tool finishes at one point, it automatically goes to the next point by the shortest route. This type of control is very useful for drilling machines, but machines such as milling machines require greater control of the tool movement. So, *continuous path* controls are used to control the movement of the tool throughout the cutting cycle. For example, if a circle or radius were to be milled, the operator or programmer would need only to program a few points along the arc. The computer would compute the remaining points and guide the cutting tool throughout the complete cycle.

These controls are frequently used for a wide range of different machine tools, from drill presses and milling machines to lathes and grinders. Due to the advent of these controls, a whole new form of machine tool has evolved: the machining center, *Figure 20-28*. These machines can take a rough, unmachined part and completely machine the whole part without removing it from the machine or changing a setup.

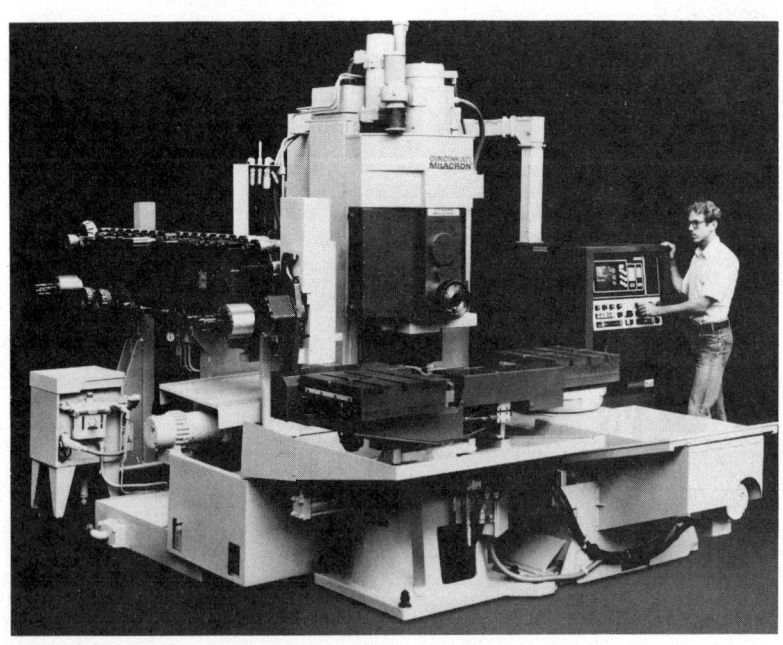

Figure 20-28 Machining center *(Courtesy Cincinnati Milacron)*

Special Workholding Devices

Special workholding devices are frequently used to produce parts in high-volume production runs. The major categories of special workholding devices commonly used in the shop are jigs and fixtures. The primary function of *jigs* and *fixtures* is to transfer the required accuracy and precision from the operator to the tool. This permits duplicate parts to be produced within the specified limits of size without error. Both jigs and fixtures hold, support and locate the workpiece; the principal difference between these tools is the method used to control the relationship of the tool to the workpiece. *Jigs* guide the cutting tool through a hardened drill bushing during the cutting cycle. *Fixtures*, on the other hand, reference the cutting tool by means of a set block, *Figure 20-29*.

Classification of Jigs

Jigs are normally classified by the type of operation they perform and their basic construction. Typically, jigs are used to drill, ream, tap, countersink, chamfer, counterbore, and spotface. Jigs are also divided into two general construction categories: open and closed. *Open jigs* are jigs that cover only one side of the part, and are used for relatively simple operations. *Closed jigs* are jigs that enclose the part on more than one side, and are intended to machine the part on several sides without removing the part from the jig.

The most common types of jigs include plate jigs, angle plate jigs, box jigs, and indexing jigs. While there are several other distinct styles of jigs, these represent the most common forms.

Plate jigs, *Figure 20-30*, are the most common form of jig. These jigs consist of a simple plate which contains the required drill bushings, locators, and clamping elements. Typical variations of the basic plate jig include the *table jig*, *Figure 20-31A*, and the *sandwich jig*, *Figure 20-31B*. Another and even simpler version of the plate jig is the *template jig*, *Figure 20-32*. These

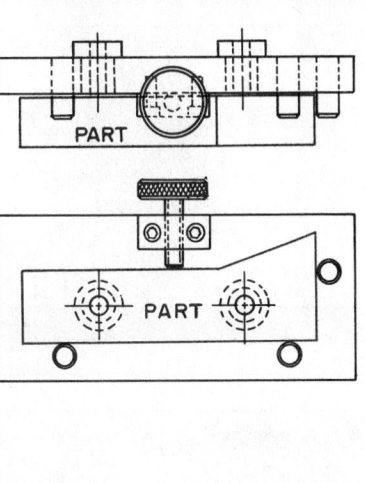

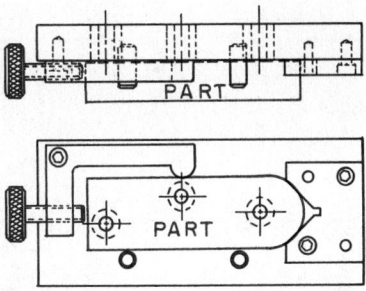

Figure 20-30   Plate jig

jigs are used where accuracy and not speed is the prime consideration. Template jigs may or may not have drill bushings, and do not normally have a clamping device. This form of jig is frequently used for light machining or for layout work.

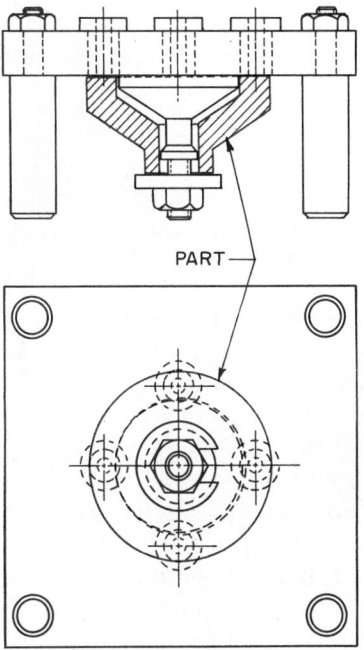

Figure 20-31A   Table jig

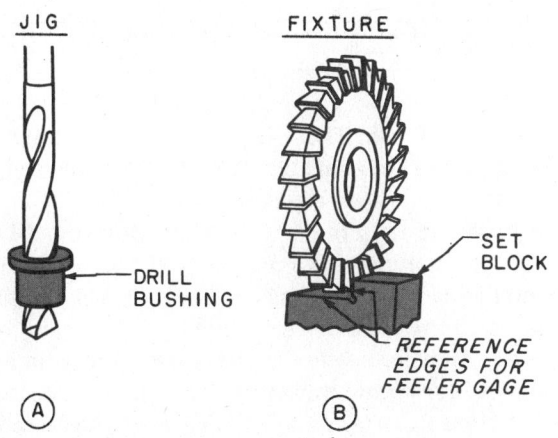

Figure 20-29   Referencing the tool to the workpiece

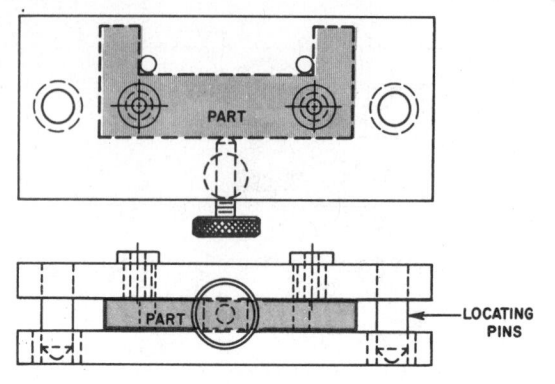

Figure 20-31B Sandwich jig

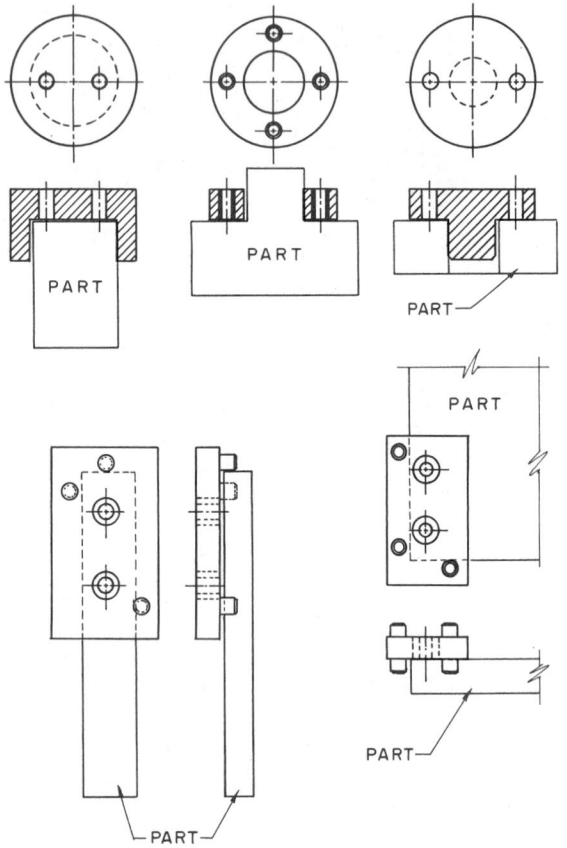

Figure 20-32 Template jigs

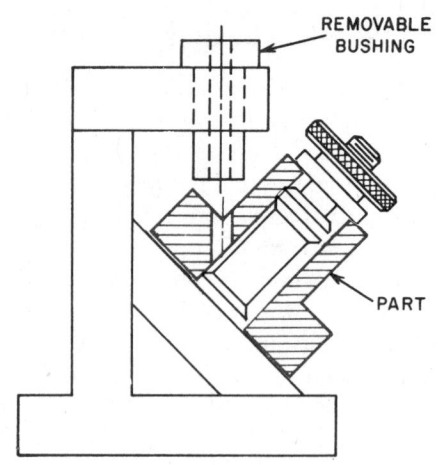

Figure 20-33 Angle plate jig

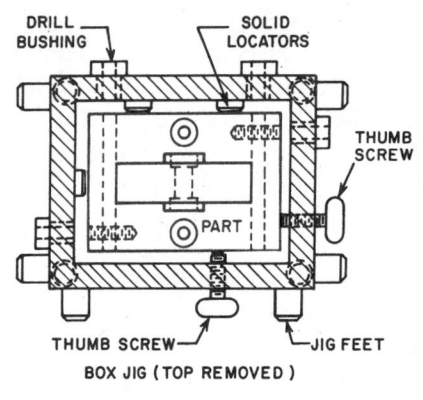

BOX JIG (TOP REMOVED)

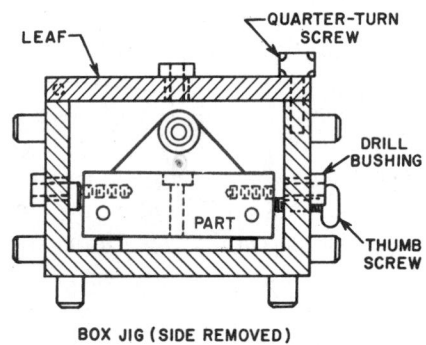

BOX JIG (SIDE REMOVED)

Figure 20-34 Box jig

An *angle plate jig*, *Figure 20-33*, is a modified form of a plate jig in which the surface to be machined is perpendicular to the locating surface. This type of jig is often used to machine pulleys, gears, hubs or similar parts. Another variation of this type of jig is the *modified angle plate jig*. These jigs are used to machine parts at angles other than 90°.

A *box jig*, *Figure 20-34*, is designed to be used for parts that require machining on several sides. With these jigs, the part is mounted in the jig and clamped with a leaf or door. Other variations of the basic box jig include the *channel jig*, *Figure 20-35A*, and the *leaf*

jig, *Figure 20-35B*. These jigs are similar in design to the box jig but they machine the part on only two or three sides, rather than all six sides.

An *indexing jig*, *Figure 20-36*, is used primarily to machine parts which have machined details at intervals around the part. Drilling four holes 90° apart is a typical example of the type of work this jig is best suited to perform. Another type of jig which uses an indexing arrangement to locate the jig, rather than the part, is the *multistation jig*, *Figure 20-37*. These jigs are used to machine several parts at one time.

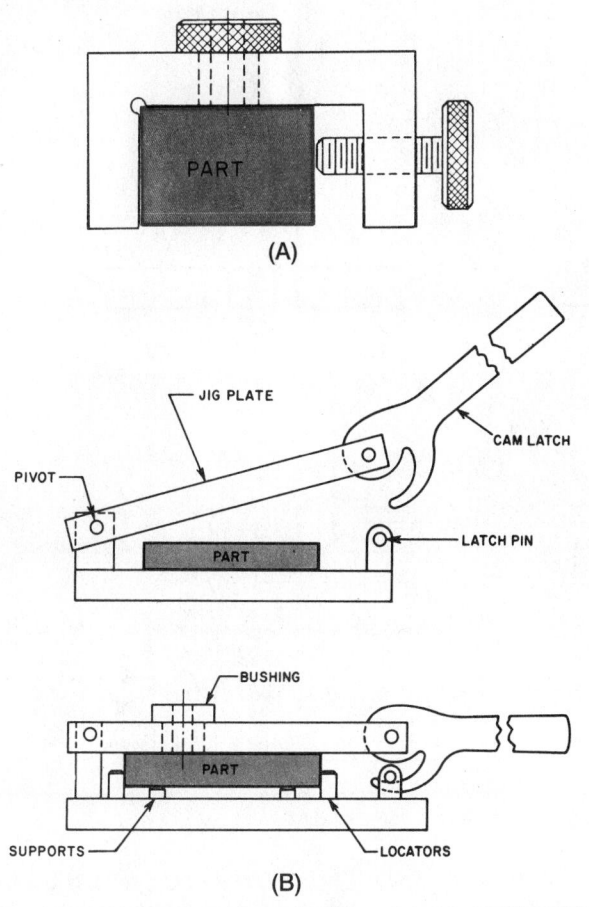

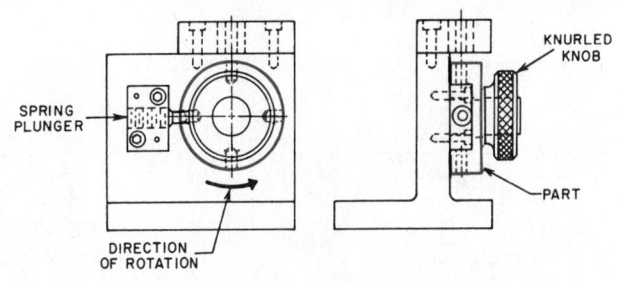

Figure 20-36 Indexing jig

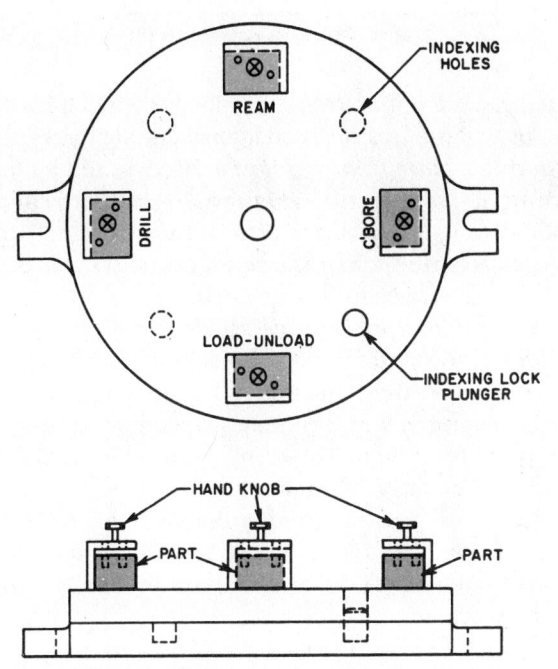

Figure 20-35 (A) Channel jig; (B) Leaf jig

Classification of Fixtures

Fixtures are classified by the type of machine they are used on, the type of operation performed, and by their basic construction features. The principal types of fixtures normally used in the shop include plate fixtures, angle plate fixtures, vise jaw fixtures, and indexing fixtures. Typically, fixtures are used for milling, turning, sawing, grinding, inspecting, and several other varied operations.

Figure 20-37 Multistation jig

A *plate fixture*, **Figure 20-38**, is the most common type of fixture. Like plate jigs, plate fixtures are simply a plate containing the locators, set blocks, and clamping devices necessary to locate and hold the workpiece and to reference the cutter. While similar in design to a

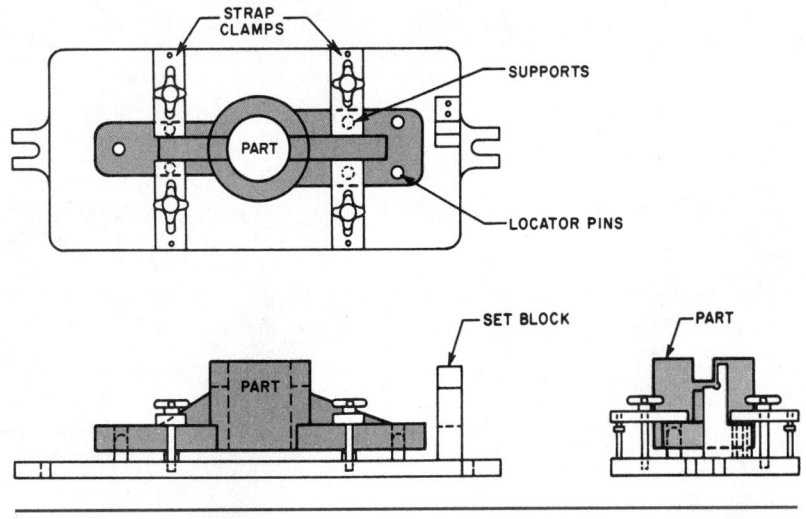

Figure 20-38 Plate fixture

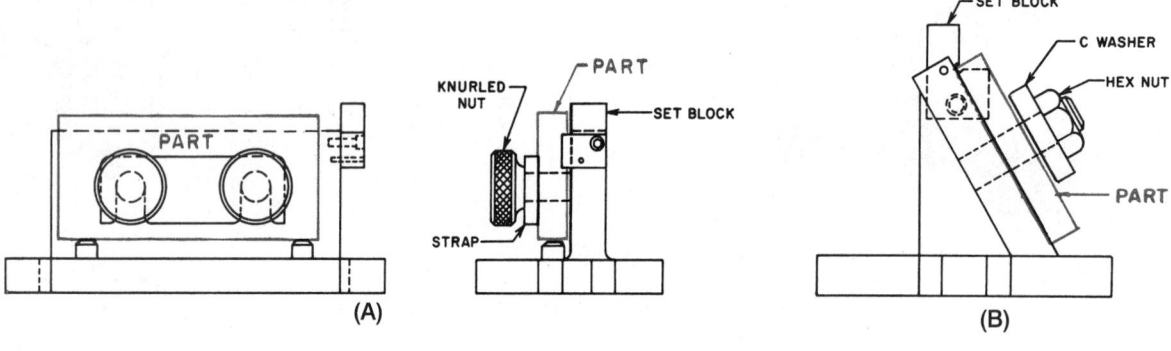

Figure 20-39 (A) Angle plate fixture; (B) Modified angle plate fixture

plate jig, plate fixtures are normally made much heavier than plate jigs to resist the additional cutting forces.

An *angle plate fixture*, **Figure 20-39A**, and a *modified angle plate fixture*, **Figure 20-39B**, are simple modifications of the basic plate fixture design. These fixtures are used when the reference surface is at an angle to the surface to be machined.

A *vise jaw fixture*, **Figure 20-40**, is a useful modification to the standard milling machine vise. With this type of fixture, the standard jaws of a milling machine vise are replaced with specially shaped jaws to suit the part to be machined. The result is an accurate fixture which can be made at a minimal cost. Since the clamping device is contained within the vise, this fixture is very cost effective. Likewise, one vise can be used for a countless number of different fixtures by simply changing the jaws.

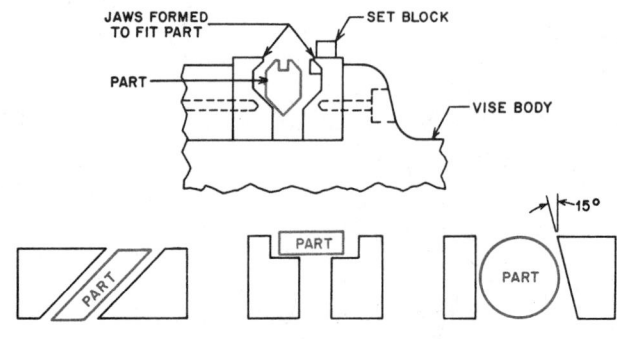

Figure 20-40 Vise jaw fixture

Indexing fixtures, **Figure 20-41**, are mainly used to machine parts which have a repeating part feature. Typical examples of the types of parts which are machined in an indexing fixture are shown in **Figure 20-42**. Here

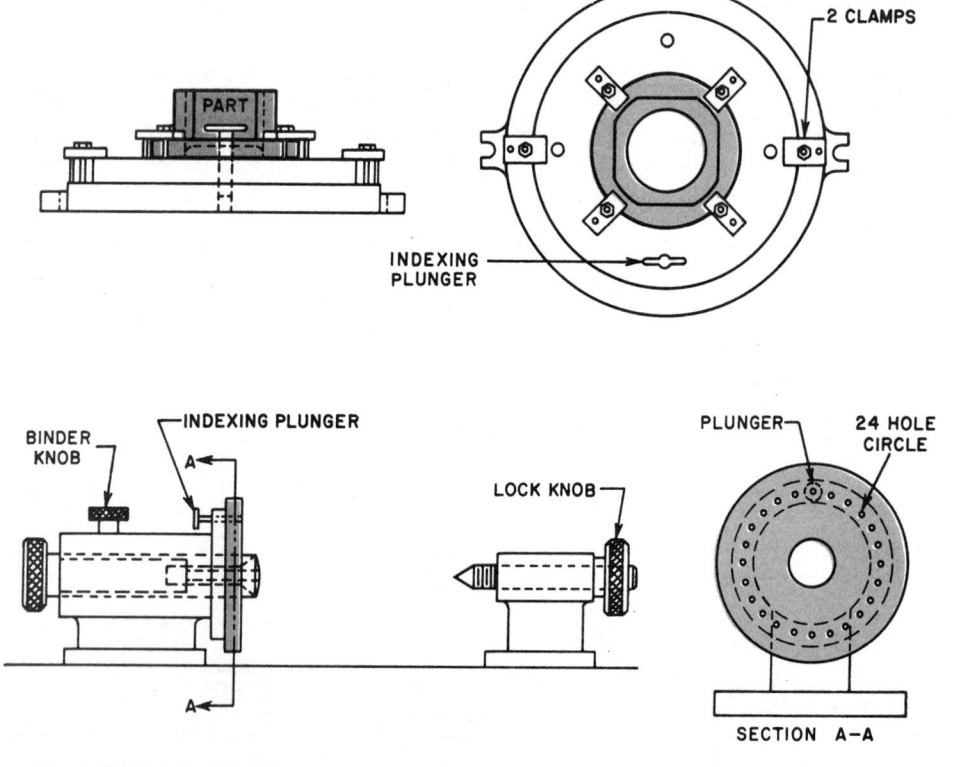

Figure 20-41 Indexing fixture

Figure 20-42   Typical parts machined in an indexing fixture

again, the indexing feature may also be used to position the tool as well as the part. The *duplex fixture* shown in *Figure 20-43* shows a method of indexing the fixture to machine two parts. In use, the first part is machined while the second is loaded. The fixture is then rotated and the second is unloaded and a fresh part loaded. This process is continued throughout the production run.

Locating Principles

To properly machine any part in a jig or fixture, the part must first be located correctly. The first rule of locating is repeatability. *Repeatability* is the feature of a jig or fixture which permits parts to be loaded in the tool in the same position part after part. When selecting locators for a part, the prime considerations must be position, foolproofing, tolerance, and elimination of duplicate or redundant locators.

Locators must be *positioned* as far apart as practical, and should contact the workpiece on a reliable surface to ensure repeatability. The locators must also be designed to minimize the effect of chips or dirt. *Figure 20-44* shows a few methods generally used to relieve locators to prevent interference from chips.

Foolproofing is simply a method used to prevent the part from being loaded incorrectly. A simple pin, *Figure 20-45*, is normally enough to ensure proper loading of

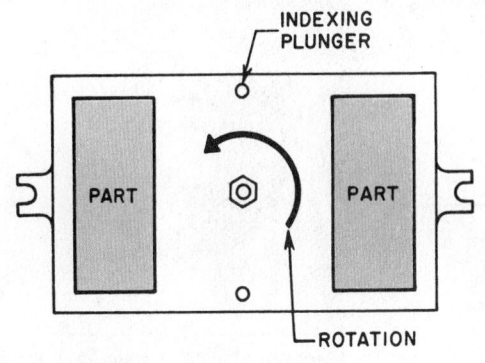

Figure 20-43   Duplex fixture

every part. The *tolerance* of a locator is determined by the specific size of the part to be machined. As a general rule, the tolerance of jigs and fixtures should be approximately 30 percent to 50 percent of the part tolerance. For example, if a hole were to be located within .010″ (.25 mm) from an edge, the locators in the jig or fixture should be positioned within .003″ to .005″ (.08 mm to .13 mm) to make sure the part is properly positioned. An overly tight tolerance only adds cost, not quality, to a tool.

Finally, locators should never duplicate any location. As shown in *Figure 20-46*, a part should be located on

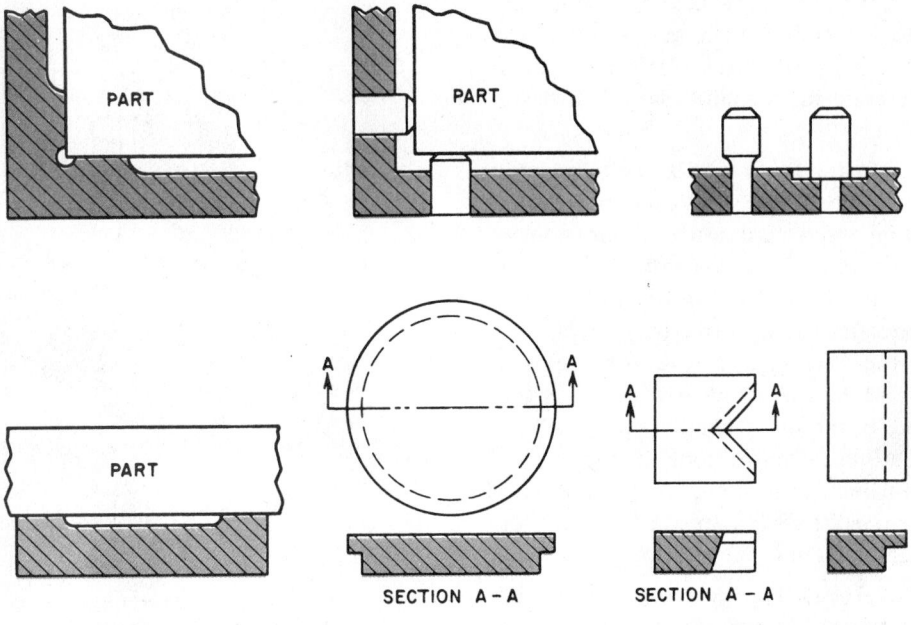

Figure 20-44   Relieving locators

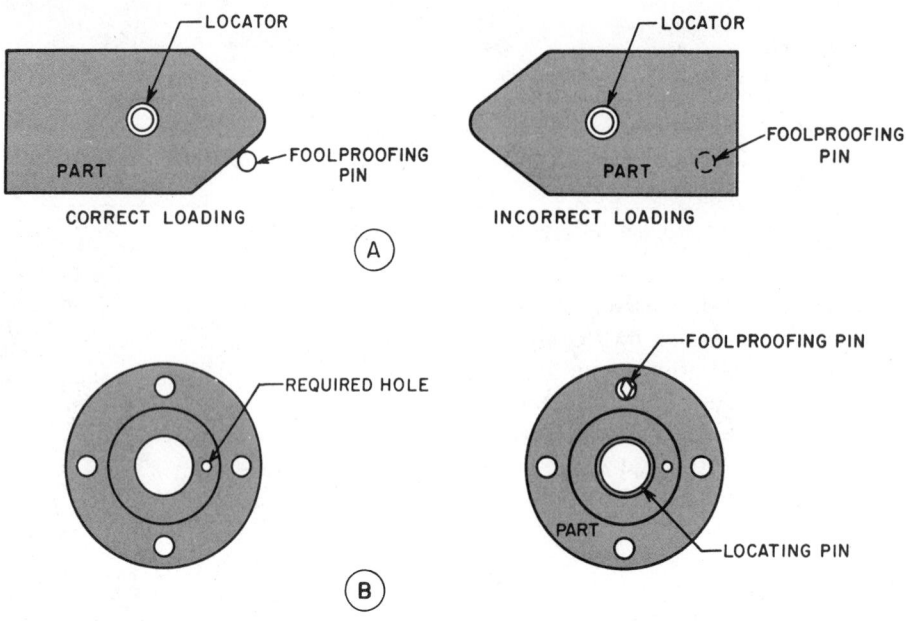

Figure 20-45 Foolproofing a workholder

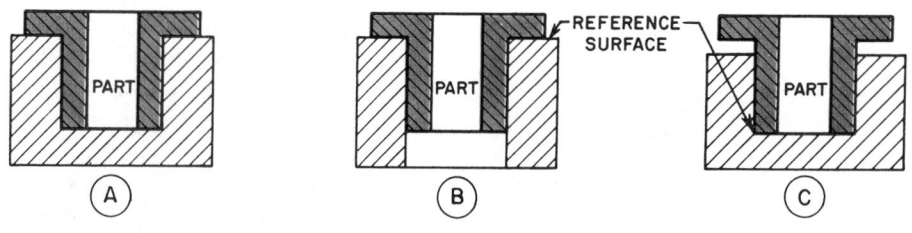

Figure 20-46 Duplicate locating

only one surface. Locating a part on two parallel surfaces only serves to reduce the effectiveness of the location, and could improperly locate the part. First determine the reference surface, and use only that surface for locating.

Restricting Movement. Every part is free to move in a limitless number of directions if left unrestricted. But, for the purpose of jig and fixture design, the number of directions in which a part can move has been limited to twelve: six axial and six radial, **Figure 20-47**. The methods used to restrict these twelve movements normally depend on the part itself, but the examples in **Figures 20-48A** and **20-48B** show two methods that are frequently used. In the first example, three-pin base restricts five directions of movement. In the second example, a five-pin base restricts eight directions of movement. In **Figure 20-48C**, a six-pin base restricts nine directions of movement.

Types of Locators. Locators are commercially available in many styles and types. **Figures 20-49A** through **20-49F** show several of the most common types of locators.

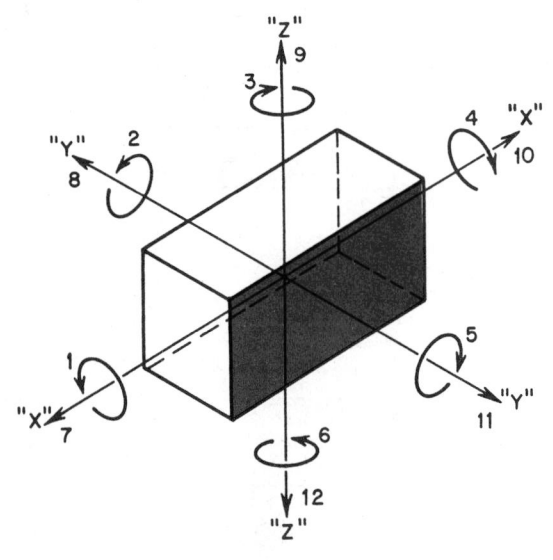

Figure 20-47 Planes of movement

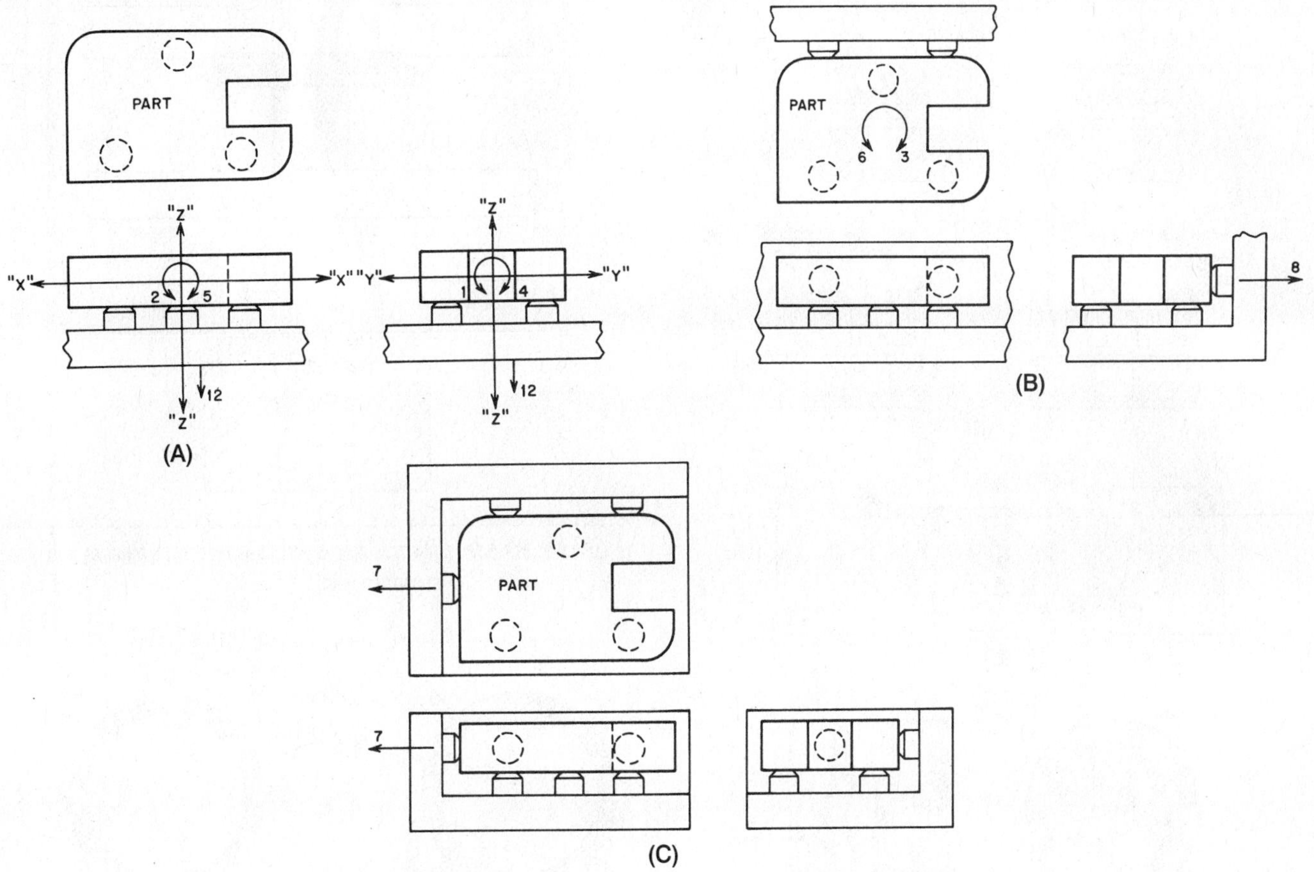

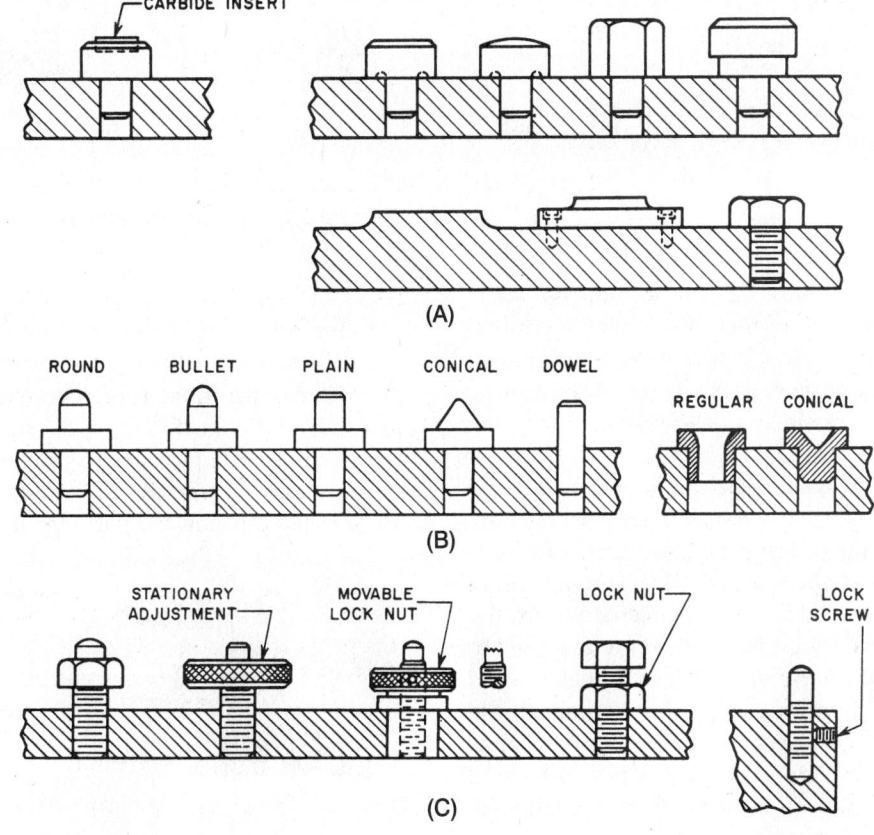

Figure 20-48   (A), (B), and (C) Restricting part movement

Figure 20-49   (A–C) Types of locators

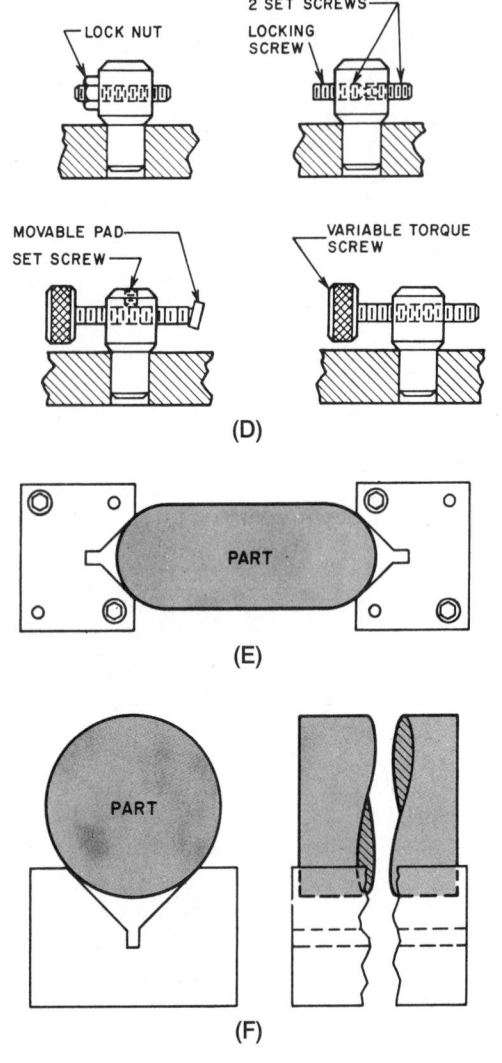

(D)

(E)

(F)

Figure 20-49 (Continued) (D–F) Types of locators

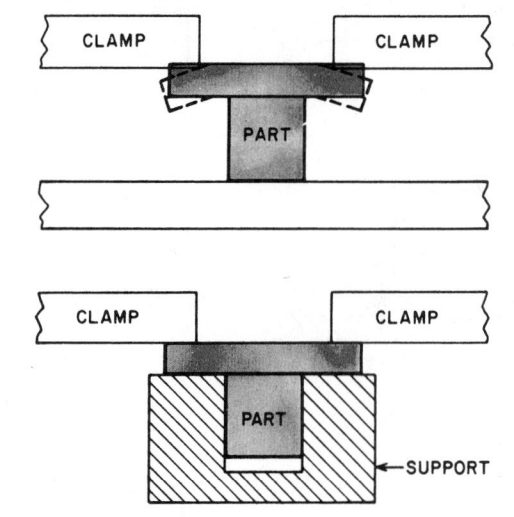

Figure 20-50 Always clamp a part at its most rigid point or a supported point

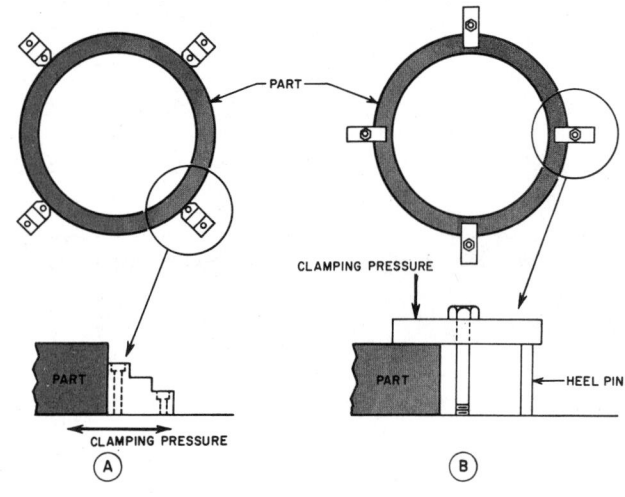

Figure 20-51 Clamping forces

Clamping Principles

In addition to locators, most jigs and fixtures use some type of clamping device to restrict the directions of movement not contained by the locators. When designing a clamping arrangement, several factors should be considered. These include position of the clamps, clamping forces, and tool forces.

Clamps should always be positioned to contact the part at either its most rigid point or a supported point. Clamping a part at any other point could bend or distort the part, as shown in ***Figure 20-50***. The clamping forces used to hold a part should always be directed toward the most solid part of the tool. Clamping a part as shown in ***Figure 20-51A***, will normally result in an egg-shaped part because the clamping forces are directed only toward each other. However, by clamping the part as shown in ***Figure 20-51B***, the part is not only held securely, but the chance of distortion is greatly reduced. As a rule, use only enough clamping force to hold the part against the locators. The locators should always

resist the bulk of the tool thrust, not the clamps.

When designing a jig or fixture, remember to direct the tool forces toward the locators, not the clamps. *Tool forces* are those forces generated by the cutting tool during the machining cycle. In most cases, the tool forces can be used to advantage when holding any part. As shown in ***Figure 20-52***, the downward tool forces are actually pushing the part into the tool. The rotation-

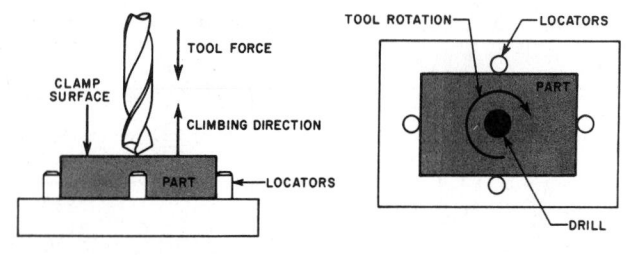

Figure 20-52 Tool forces

al tool forces are contained by the locators. The only force the clamps need to hold are the forces generated by the drill as the point breaks out the opposite side of the part, and these forces are only a fraction of the cutting forces. So, when designing a jig or fixture, always direct the tool forces so that they act to hold the part in the tool. Use the clamps only to hold the part against the locators.

Types of Clamps. Clamps, like locators, are commercially available in many styles and types. *Figures 20-53A* through *20-53D* show some of the more common types of clamps used for jigs and fixtures. The specific type of clamp you should select for any workholding application will normally be determined by the part, and the type of holding force desired.

Basic Construction Principles

When designing any jig or fixture, the first consideration is normally the tool body. Tool bodies are generally made in any of three ways: cast, welded or built-up, *Figure 20-54*. *Cast tool* bodies are generally the most expensive, and require the longest lead time to make. They do, however, offer such advantages as good material distribution, good stability, and good vibration-damping qualities. *Welded tool* bodies offer a faster lead time, but must be machined frequently to remove any distortion caused by the heat of welding. The most popular and common type of tool body in use today is the built-up. *Built-up* tool bodies are made from pieces of preformed stock, such as precision-ground flat stock, ground rod or plain cold-rolled sections which are

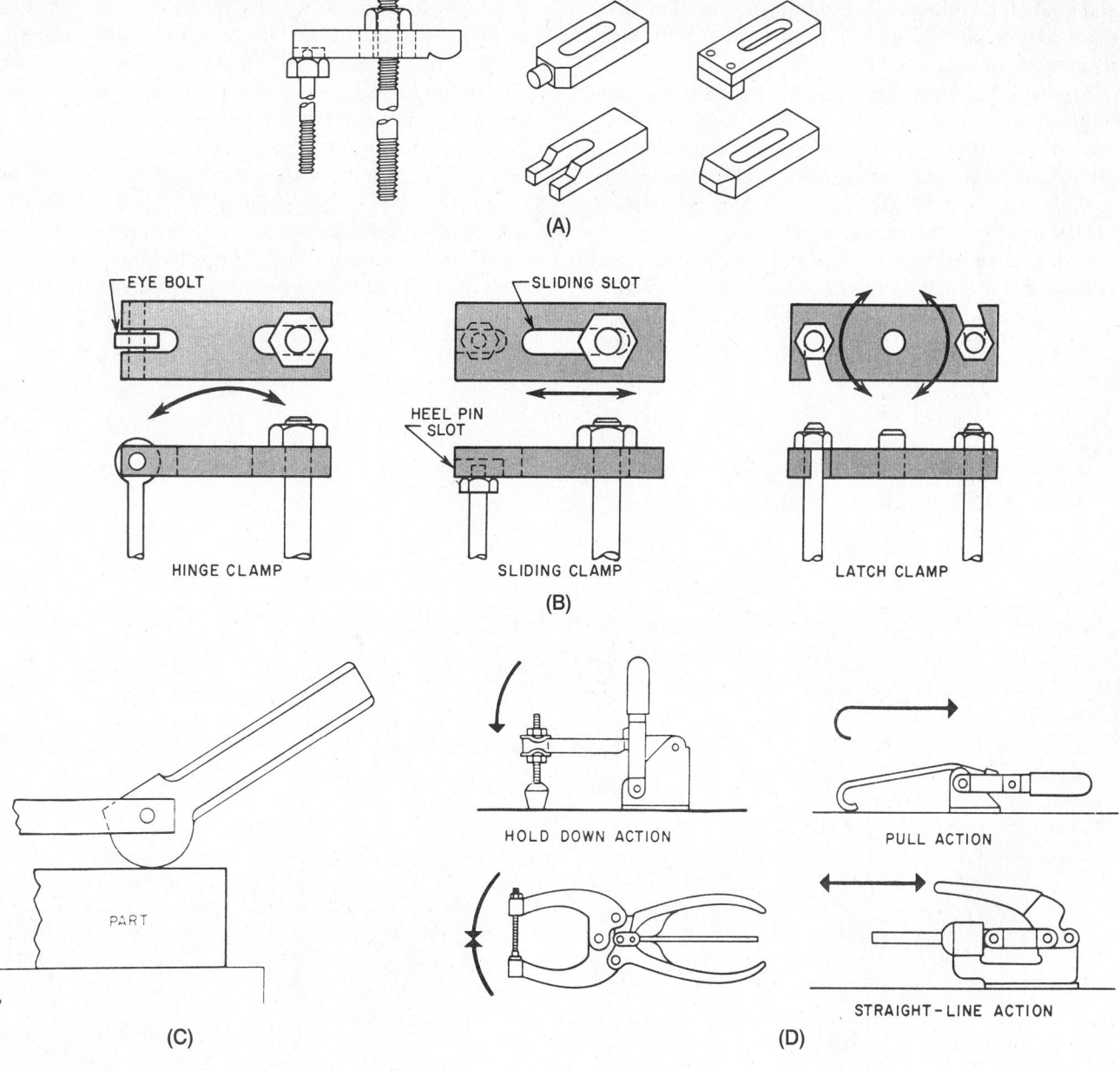

Figure 20-53 (A–D) Types of clamps

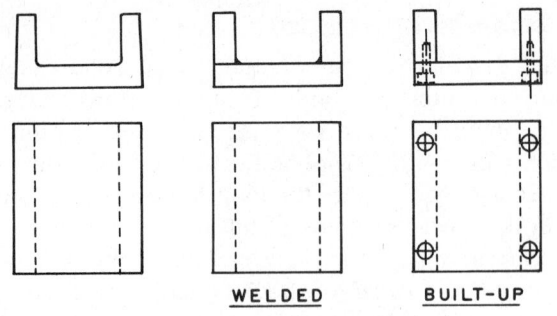

Figure 20-54 Tool bodies

Figure 20-55 Press-fit drill bushings

pinned and bolted together. The principle advantages of using this type of construction are fast lead time, minimal machining, and easy modification.

Drill Bushings. *Drill bushings* are used to position and guide the cutting tools used in jigs. The three basic types of drill bushing used for jigs are press fit bushings, renewable bushings, and liner bushings.

Press-fit bushings, *Figure 20-55*, as their name implies, are pressed directly into the jig plate. These bushings are useful for short-run jigs or in applications where the bushings are not likely to be replaced frequently. The two principle types of press fit bushings are the head type and headless type.

Renewable bushings are used in high-volume applications or where the bushings must be changed to suit different cutting tools used in the same hole. The two types

of renewable bushings that are commercially available are the slip-renewable type, *Figure 20-56A*, and the fixed-renewable type, *Figure 20-56B*.

Slip-renewable bushings, *Figure 20-56C*, are used where the bushing must be changed quickly. Typical applications include holes which must be drilled, countersunk, and tapped. The bushings are held in place with a lock screw and need only to be turned counterclockwise and lifted out of the hole. The next bushing is inserted in the hole and turned clockwise to lock it in place.

Fixed-renewable bushings are used for applications where the bushings do not need to be changed often, but when they are changed more often than press fit bushings. High-volume production is a typical example where fixed renewable bushings are often used.

Liner bushings, *Figure 20-56D*, are used in conjunction with renewable bushings and provide a hardened, wear-resistant mount for renewable bushings. Liner bushings are actually press-fit bushings with a large-diameter hole. Like press-fit bushings, liner bushings are

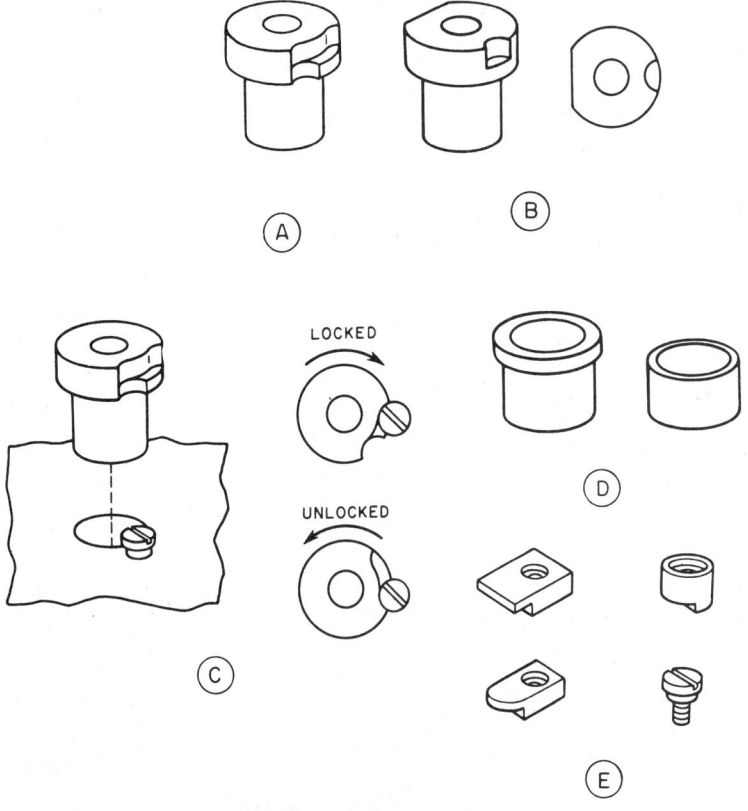

Figure 20-56 (A–E) Renewable drill bushings

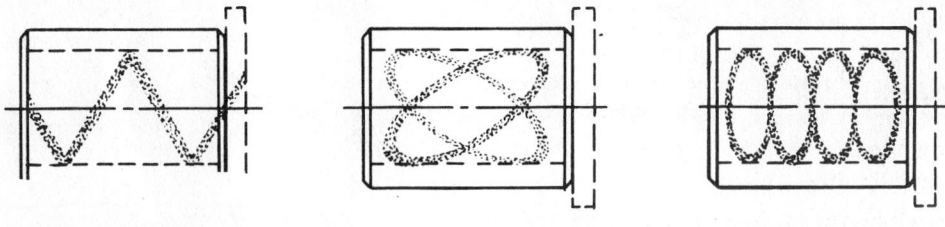

Figure 20-57   Oil-groove bushings

available in both a head type and a headless style. The clamps normally used to hold the fixed-renewable bushing in the jig plate are shown in *Figure 20-56E*. The screw is used to mount slip renewable bushings, whereas the other three styles are used to secure the fixed-renewable bushings.

Several other variations of these basic bushing styles include special purpose bushings, serrated and knurled bushings, and oil-groove bushings. Oil-groove bushings are shown in *Figure 20-57*.

Mounting Drill Bushings.   When installing a drill bushing in the jig plate, the specific size of the jig plate and spacing from the part are important factors the designer must consider. As shown in *Figure 20-58A*, the jig plate should be between one to two times the tool diameter. If a thinner jig plate must be used, a

head-type bushing can be used to achieve the added thickness needed to support the cutting tool. The space between the jig plate and the workpiece should be one to one and one half the tool diameter for drilling, *Figure 20-58B*, and one quarter to one half the tool diameter for reaming, *Figure 20-58C*.

In those cases where both drilling and reaming operations are to be performed in the same hole, an arrangement similar to the one shown in *Figure 20-58D* may be used. Here the jig plate is properly positioned for drilling, and the bushing used for reaming is made longer to achieve the desired distance for reaming. When special shapes or contours must be drilled, the ends of the bushings can be modified to suit the contour and maintain the proper support for the cutting tool, *Figures 20-58E* and *20-58F*.

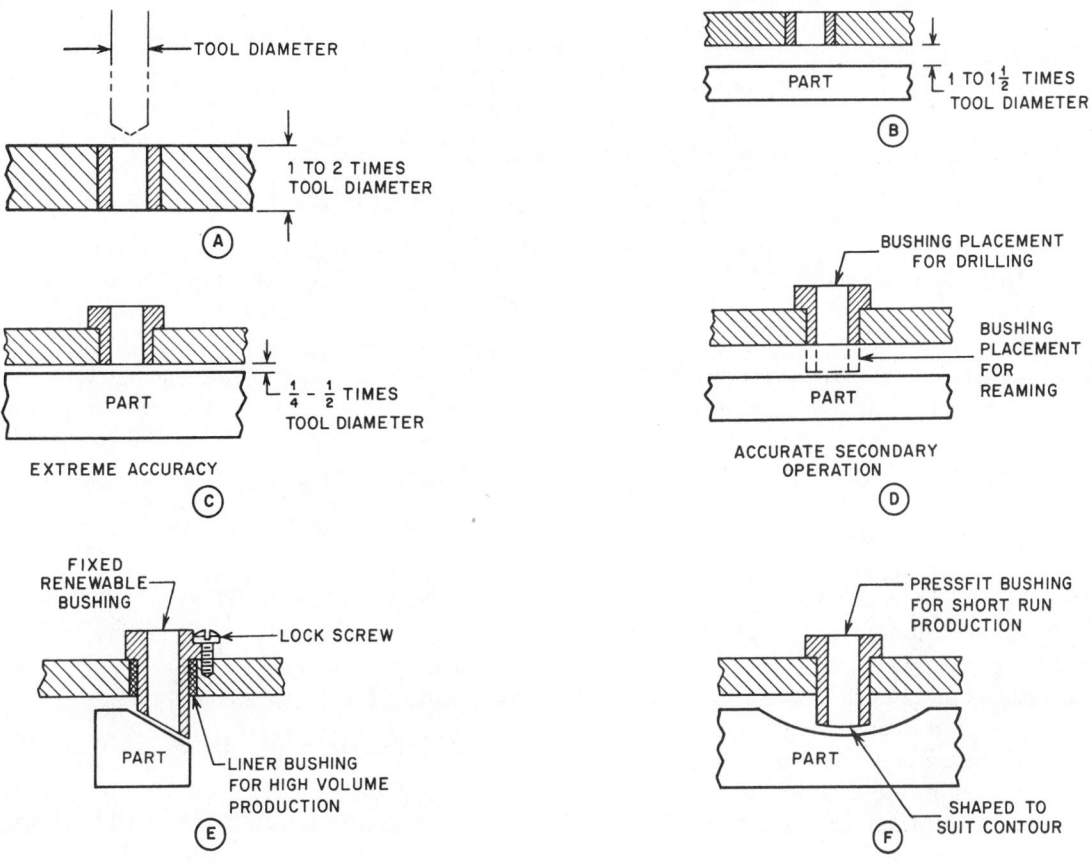

Figure 20-58   (A–F) Mounting drill bushings

Set Blocks. Set blocks are used along with thickness gages to properly position the cutting tool with a fixture. The specific design of the set block is determined by the part and shape to be machined. The basic set block designs shown in *Figure 20-59* are typical of the styles found on many fixtures.

Fastening Devices. The most common fastening devices used with jigs and fixtures are dowel pins and sockethead cap screws. Other fasteners sometimes used with these tools include the nuts and washers shown in *Figure 20-60*.

Heat Treatment of Steels

Heat treatment is a series of processes used to alter or modify the existing properties in a metal to obtain a specific condition required for a workpiece. The specific properties normally changed by heat treatment include hardness, toughness, brittleness, malleability, ductility, wear resistance, tensile strength, and yield strength. The five standard heat treating operations normally performed on steel parts are hardening, tempering, annealing, normalizing, and case hardening.

Hardening is the process of heating a metal to a predetermined temperature, allowing the part to soak until thoroughly heated, and cooling rapidly in a cooling material called a *quench*. The most common quench media are air, oil, water, and a water and salt mixture called brine. Hardening is mainly used to increase the hardness, wear resistance, toughness, tensile strength, and yield strength of a workpiece.

Tempering is the process of heating a hardened workpiece to a temperature below the hardening temperature, allowing the part to soak for a specific time period, and cooling. Tempering is mainly used to reduce the hardness so the toughness is increased and brittleness is decreased. Almost every hardened part is tempered to control the amount of hardness in the finished part.

Annealing is the process of heating a part, soaking it until it is thoroughly heated, and slowly cooling it by turning off the furnace. Annealing is mainly used to completely remove all hardness in a metal. Frequently, hardened parts are annealed to be remachined or modified and then rehardened.

Normalizing is the process used to remove the effects of machining or cold-working metals. In this process, the metal is heated, allowed to soak, and cooled in still air. This process produces a uniform grain size, and eliminates almost all stresses in the metal.

Case hardening is the process of hardening the outside surface of a part to a reselected depth. In most case-hardening operations, carbon is added to the surface of the part by packing the part in a carbonous material, and then heating it so that the carbon is transferred into the surface of the part. Once the carbon is added to the surface of the part, the part is then hardened to produce a hardened shell around a normalized core.

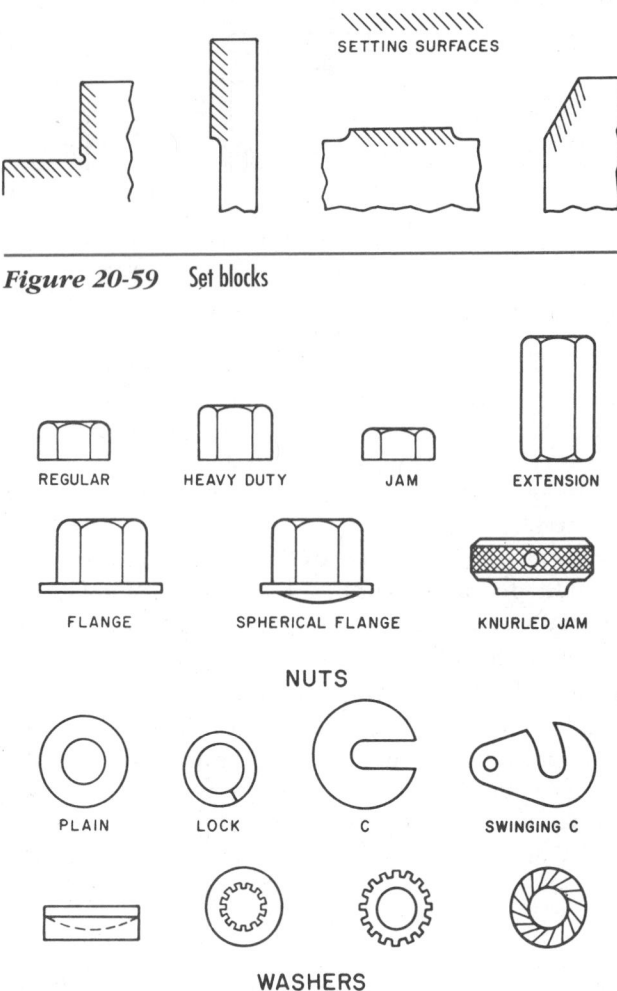

Figure 20-59 Set blocks

Figure 20-60 Nuts and washers frequently used with jigs and fixtures

Nontraditional Machining Processes

The term *nontraditional machining* encompasses a number of advance machining processes that have been developed in response to the demands of modern manufacturing for more effective, more efficient ways to machine the many new materials that have been developed and are still developing. The most widely used nontraditional machining processes are explained in the following paragraphs.

Hydrodynamic Machining (HDM)

Hydrodynamic machining (HDM) removes material using a high-velocity, high-pressure stream of liquid solution. It is used primarily for slitting and cutting nonmetallic materials such as wood, paper, plastics, gypsum, leather, felt, rubber, and fiberglass.

Ultrasonic Machining (USM)

Ultrasonic machining (USM) removes material using an abrasive slurry driven by a special tool vibrating at a high-frequency along a longitudinal axis. It is used primarily for producing blind holes, through holes, slots, and irregular shapes.

Rotary Ultrasonic Machining (RUM)

Rotary ultrasonic machining (RUM) is similar to USM except that: 1) the grinding motion is rotary instead of longitudinal; 2) RUM uses diamond tools instead of steel or Monel; and 3) RUM uses the diamond of the tool to remove material while USM uses an abrasive slurry. RUM is used primarily in the development of prototypes. It is also used for machining ceramics, quartz, glass, ferrite computer parts, and composites.

Ultrasonically Assisted Machining (UAM)

Ultrasonically assisted machining (UAM) involves assisting such traditional machining processes as drilling and turning by coupling the tool with an external source of vibrating energy. It is used in most drilling and turning applications and is effective in reducing the amount of subsurface tearing and plastic flow. It also evens out tool wear.

Electromechanical Machining (EMM)

Electromechanical machining (EMM) is a process that improves the performance of traditional machining processes. Metal removal is accomplished in the traditional manner (i.e., drilling, milling, turning, etc.) with the exception that the workpiece is electrochemically polarized by applying a controlled voltage across the connection point between the workpiece and an electrolyte. This changes the surface of the workpiece making it easier to machine.

Electrochemical Discharge Grinding (ECDG)

Electrochemical discharge grinding (ECDG) is a combination of two processes: electrochemical grinding (ECG) and electrical discharge grinding (EDG). In ECDG alternating current or pulsating direct current moves from a bonded graphite wheel (conductor) to a positively charged workpiece. No mechanical contact is made between the workpiece and the wheel. Instead, an electrolyte is pumped into the gap and it serves as the connecting agent. Material removal occurs as follows: 1) the ECG process converts the surface of the metal into an oxide film; and 2) the EDG process removes the oxide film.

Electrochemical Grinding (ECG)

Electrochemical grinding (ECG) combines an electrochemical reaction that occurs on the surface of the metal with abrasion. Subsequently, it is applied to the metal's surface to remove material. The process is typically used on tough, hard metals. The abrasion is applied by a rotary grinding wheel that grinds off oxidized surface metal.

Electrochemical Honing (ECH)

Electrochemical honing (ECH) is a process similar to ECG. It combines electrochemical oxidation of surface metal and abrasion. The primary difference is that honing is a process of removing material inside of a drilled hole. Consequently, the abrasion is applied using non-conductive honing stones attached to a tool that rotates and reciprocates.

Electrochemical Machining (ECM)

Electrochemical machining (ECM) is a broad concept that applies to a number of process variations that all use an electrolytic action in removing metal. With ECM electricity is combined with a chemical to form a reaction that is the opposite of metal plating. ECM can be used with any conductive workpiece but is used most often with hard metals. Its best application is external shaping of materials.

Electrochemical Turning (ECT)

Electrochemical turning (ECT) is a variation of ECM. The same basic principles apply except that the workpiece rotates. The primary application of ECT is the machining of large disc-shaped forgings.

Electron Beam Machining (EBM)

Electron beam machining (EBM) is a thermal energy process. This means it uses thermal energy to remove metal. The thermal energy comes in the form of a pulsating stream of high-speed electrons that are focused on a very small area of the workpiece by electrostatic and electro-magnetic fields. Material removal occurs as the result of melting or evaporation. EBMT is used for drilling small wire-drawing dies, metering holes, and holes for spinnerets used in textile manufacturing.

Electrical Discharge Machining (EDM)

Electrical discharge machining (EDM) is a process that removes metal using a series of rapidly recurring electrical discharges between an electrically charged cutting tool and the workpiece submerged in a dielectric fluid. EDM is widely used in the aerospace, electronics, and tool and die industries.

Electrical Discharge Wire Cutting (EDWC)

Electrical discharge wire cutting (EDWC) is a process that is similar to the bandsawing process. Material removal occurs as the result of an electrical discharge as a wire electrode is drawn through the workpiece. EDWC can be used only with conductive workpieces. Primary applications of EDWC include the production of stamping dies, prototypes, molds, lathe tools, special form inserts, pot and wobble broaches, extrusion dies, and templates.

Electrical Discharge Grinding (EDG)

Electrical discharge grinding (EDG) is similar to EDM in that it employs rapidly recurring electrical discharges between an electrode and the workpiece in a dielectric fluid. The difference is that with EDG the electrode is a rotating wheel made of graphite or brass. Primary applications include the grinding of steel and carbide at the same time, thin sections, and brittle and fragile parts.

Laser Beam Machining (LBM)

Laser beam machining (LBM) is a process that uses lasers for performing operations that are traditionally performed with cutting tools. These operations include cutting, drilling, slotting, and scribing. The basic principle of LBM is the conversion of electrical energy into coherent light. This coherent light or laser beam is focused on the workpiece where it is converted into thermal energy. This thermal energy melts or vaporizes the material. LBM has broad applications, but it is particularly effective at cutting and drilling.

Plasma Arc Machining (PAM)

Plasma arc machining (PAM) is a process that makes use of a plasma arc to remove material. *Plasma* is a gas that has been heated to the point that it ionizes and becomes electrically conductive. With PAM, a stream of ionized particles directed through a nozzle against the workpiece removes material. The primary applications of PAM are hole piercing, stack cutting, gouging, grooving, and bevel cutting.

Chemical Milling

Chemical milling is a process etches or shapes workpieces to close tolerances by chemically induced material removal. Primary applications include metal removal from irregularly shaped workpieces, reduction of web thicknesses, removal of the decarburized layer from low-alloy steel forgings, improved surface finish, and removal of defects.

Photochemical Machining

Photochemical machining, also known as chemical blanking, involves producing parts by chemical action. It is accomplished by placing an exact image of the part to be produced on a sheet of metal (the prototype must be chemical-resistant) and immersing them both in a chemical. The chemical action dissolves all of the metal except the desired part. Most photochemically machined parts are thin and flat.

Water Jet Machining (WJM)

Unlike the nontraditional processes presented so far, *water jet machining (WJM)* is used with non-metal workpieces. It involves cutting plastics, non-metal composites, ceramics, glass, and wood by directing a thin, pressurized jet of water against the workpiece.

Abrasive Water Jet Machining (AWJM)

Abrasive water jet machining (AWJM) is similar to WJM except that an abrasive is added to the water jet after it exits the nozzle. This allows AWJM to be used with brittle and/or heat-sensitive materials such as glass, composites, and plastics.

Automation and Integration (CAM and CIM/FMS)

There have been four phases in the development of shop and manufacturing processes:

1. Manual phase
2. Mechanization phase
3. Automation phase (CAM)
4. Integration phase (CIM/FMS)

Of course, there has been and continues to be a good deal of overlap among these phases. Manual manufacturing did not suddenly stop when the age of mechanization began, nor did mechanized manufacturing stop when the automation phase began. Correspondingly, the beginning of the integration phase did not bring automated manufacturing to a sudden stop.

It takes many years for a given stage in the development of shop and manufacturing processes to phase out completely. For example, by the year 1900, mechanized manufacturing had replaced manual manufacturing as the dominant approach. However, manual manufacturing was still widely practiced at that time. By the year 2000, automated manufacturing will be the dominant approach; nevertheless, mechanized manufacturing will still be widely practiced, and even manual manufacturing will still be practiced in very isolated cases.

At present, we are seeing the introduction of the integration phase, which will eventually dominate but not completely replace—at least not for a long time—the other three approaches.

The manual phase preceded the Industrial Revolution. During these early times, manufactured products were produced by hand, using a variety of manual tools. The Industrial Revolution ushered in the mechanization phase. The fundamental component of this phase was the machine. Early machines were steam powered, while later machines, of course, were dependent on electricity. In this phase, machines such as mills, drills, saws, and lathes did the work, but people provided the control.

The automation phase has changed the way machines are controlled. It has also lessened the amount of human involvement in manufacturing. With the automation approach, manufacturing machines are controlled by computer. This has come to be known as CAM or computer-aided manufacturing. CAM machines are operated automatically according to computer programs. People still write the programs, set up the machines, monitor operations, and unload machines.

With CAM, individual operations such as milling, drilling, turning, and cutting are automated, but they still operate independently of each other. The most advanced manufacturing approach to date—computer-integrated manufacturing or CIM—solves this problem by linking automated manufacturing processes with other processes such as materials handling, warehousing, contracts and bids, scheduling, and design. CIM represents a major step toward the wholly automated factory. These two concepts—CAM and its eventual successor CIM—deserve special attention.

Computer-Aided Manufacturing (CAM)

Through CAM, such manufacturing processes as machining and materials handling can be automated. CAM encompasses any manufacturing process that is automated through the use of the computer, but the best-known examples of CAM are computer numerical control (CNC) of machine tools and robotics.

Computer Numerical Control (CNC)

Numerical control is a method of controlling machine tools by using coded programs. These programs consist of numbers, letters, and special characters that define the path of the machine tool in accomplishing specific tasks. When the job changes, the program must be rewritten to accommodate the change. The programmability feature is the key to the growth of NC. Machine tools that are programmable are more flexible than their nonprogrammable counterparts.

NC is a broad term that encompasses the traditional approach to NC as well as the more modern outgrowths of computer numerical control (CNC) and direct numerical control (DNC).

The first NC machines used punched cards as the programming medium. Later, punched tape was substituted, which was an improvement over manual control but did have some disadvantages. Paper tape was easy to damage and difficult to correct or edit.

Mylar tape, as a less fragile medium, began to replace paper tape. This solved the problem of frequent damage but did not make punched tape any easier to edit. Such problems, coupled with advances in microelectronic technology, led to the use of computers as the means for programming machine tools. Computers solved the damage and editing problems, and this gave birth to both CNC and DNC. CNC involves using a computer in writing, storing, and editing NC programs. DNC involves using a computer as the controller for one or more NC machines, **Figure 20-61**. A more advanced form of DNC is called "distributive numerical control," **Figure 20-62**.

In this concept there is a host coupler and several intermediate computers which are tied to NC machines, robots, and other NC manufacturing devices. The main host computer is the central repository for all programs for all jobs. Specific programs for specific jobs are "dumped" from the host computer to the intermediate computers. This cuts down on the amount of time required to get the programmed instructions from the computer to the machine tool. CNC machines are available for most machining processes, including cleaning and finishing, inspection and quality control, pressing and forming, and material removal.

Figure 20-63 is a Bridgeport Series II CNC Interact 520 bed-type vertical machining center.

Figure 20-64 is a Bridgeport Interact I CNC milling machine.

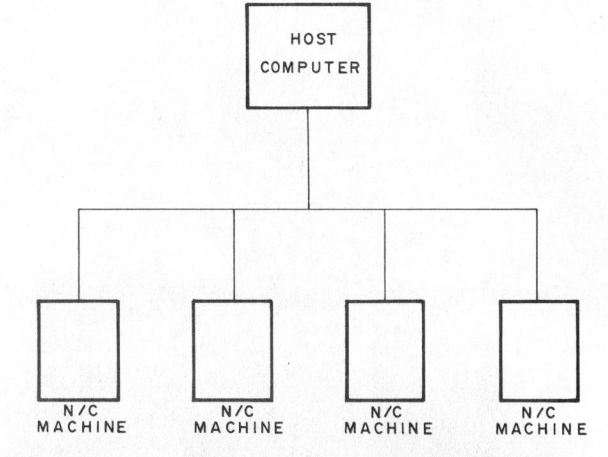

Figure 20-61 Direct numerical control (*Adapted from* Computer Numerical Control, *by Warren S. Seames, Copyright 1986 by Delmar Publishers Inc.*)

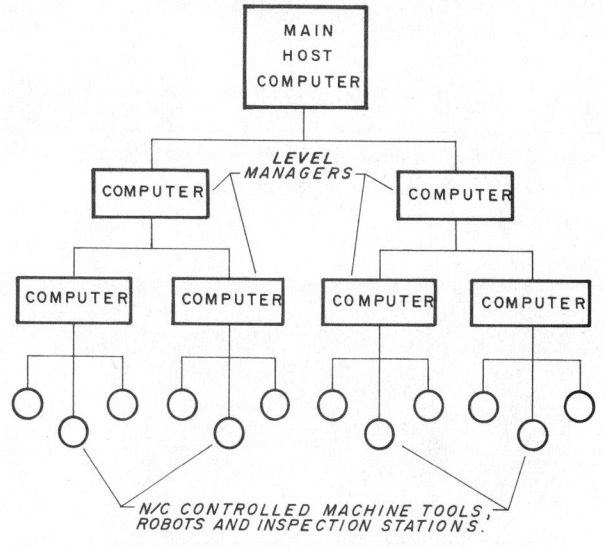

Figure 20-62 Distributive numerical control (*Adapted from* Computer Numerical Control, *by Warren S. Seames, Copyright 1986 by Delmar Publishers Inc.*)

Figure 20-65 is a Cincinnati Milacron 2EF Cinternal multi-surface grinding machine.

All of the CNC machines shown in Figures 20-63 through 20-65 have controllers. There are as many different controllers as there are CNC machines. Regardless of the type of controller, there must be some means of storing and inputting programs which instruct the machines. These means are called input media.

The most frequently used types of input media are punched tape and magnetic tape. Punched tape may be either the paper or mylar (plastic) variety. Mylar is more widely used because it is less prone to damage.

With earlier CNC machines a tape punch was used to put holes in a tape that was fed through it. The holes

Figure 20-63 Series II CNC INTERACT *(Courtesy Bridgeport Machines, Division of Textron Inc.)*

Figure 20-64 INTERACT I CNC milling machine *(Courtesy Bridgeport Machines, Division of Textron Inc.)*

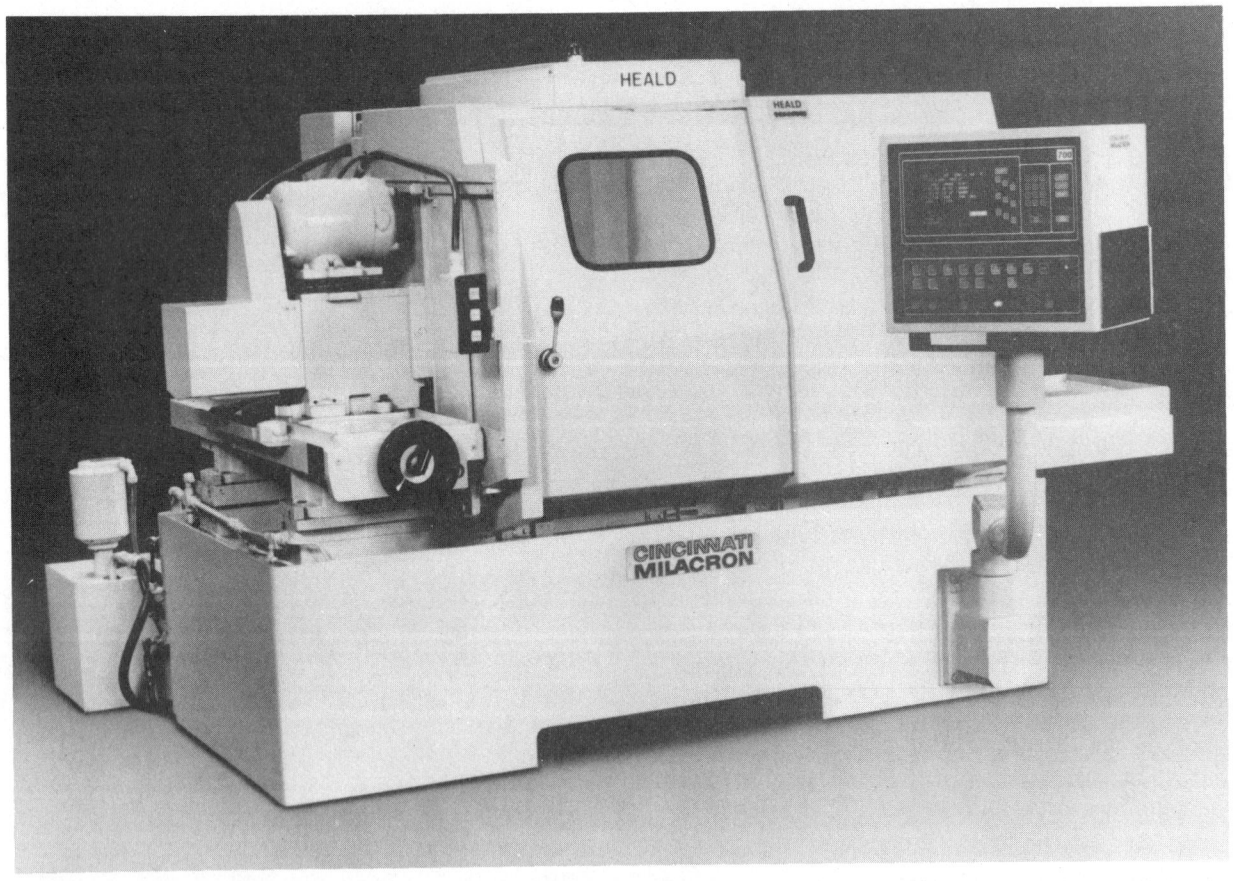

Figure 20-65 CINTERNAL grinding machine *(Courtesy Cincinnati Milacron Inc.)*

represented the program code. The punched tape was run through a tape reader that sensed the holes and fed the code into the controller. A problem with this type of input media is that it is easy to make an error when punching the tape and difficult to correct it.

An improvement to this early, cumbersome method of input is to interface a tape punch directly with a microcomputer. The code is typed via the microcomputer's keyboard. As the code is entered, it is stored so that it can be checked for accuracy and edited. Once correct, it is fed to the tape punch, and the tape is prepared. This solves the error correction problem.

Magnetic tape is now more popular than punched tape. With this type of medium the program code is affixed to the tape as magnetic spots rather than as holes punched through it. Because this type of medium is becoming so popular, standards for format and coding have been developed by the Electronics Industries Association.

CNC is the most modern and most technologically advanced method of controlling manufacturing machines. However, it is not the best control method in every case. Like any technological development, CNC has its advantages and disadvantages. There are times when the traditional manual approach is better. CNC is indicated when one or both of the following factors is important:

- Increased productivity
- Decreased labor and production costs

When CNC is indicated, there are advantages and disadvantages with which students of CAD/CAM should be familiar. The advantages include:

1. Better production and quality control
2. Increased productivity, flexibility, accuracy, and uniformity
3. Reduced labor, production, tool, and fixture costs
4. Less parts handling and tool storage

There are other advantages of CNC; however, these are the most important. There are also disadvantages associated with CNC. Those most frequently stated are:

1. High "up-front" costs
2. Higher operating costs
3. Retraining costs
4. Potential personnel problems

The advantages of CNC outweigh the disadvantages. But the list of disadvantages does make the point that it is not always the appropriate choice.

CNC Applications

CNC applications can be viewed at two levels. First, there is the industry level. At this level, all of the various manufacturing industries use CNC machines. These include:

- Aerospace manufacturers
- Electronics manufacturers
- Automobile, truck, and bus manufacturers
- Appliance manufacturers
- Tooling manufacturers
- Locomotive/train manufacturers

The second application level to consider is the machine level. At this level, CNC applications include:

- Cleaning and finishing applications
- Material removal applications
- Presswork and forming applications
- Inspection/quality control applications

Cleaning and finishing machines perform such tasks as washing, degreasing, blasting, lapping, grinding, and deburring. There are currently several CNC cleaning and finishing machines on the market.

Material removal machines include mills, lathes, grinders, drills, and saws. All such machines may be CNC controlled. In fact, this is the machine level application area most strongly associated with CNC.

Presswork and forming machines perform such tasks as pressing, stamping, blanking, punching, bending, and swagging. There are currently several CNC presswork and forming machines on the market.

Inspection and quality control machines perform such tasks as measuring, gaging, testing, and weighing. There are several CNC inspection and quality control machines on the market.

Industrial Robots

As is sometimes the case with new and emerging technological developments, there are a variety of definitions used for the term "robot." Depending on the definition used, the number of robot installations in this country and others will vary widely. There are a variety of single-purpose machines used in manufacturing plants which, to the lay person, would appear to be robots. These machines are hardwired to perform one function. They cannot be reprogrammed to perform a different function. These single-purpose machines do not fit the definition of industrial robots that is coming to be widely accepted. This definition is the one developed by the Robot Institute of America:

> A robot is a reprogrammable multifunctional manipulator designed to move material, parts, tools, or specialized devices through variable programmed motions for the performance of a variety of tasks.

Notice that the RIA's definition contains the words "reprogrammable" and "multifunctional." It is these two characteristics that separate the true industrial robot from the various single-purpose machines used in modern manufacturing firms. The term "reprogrammable" implies, first, that the robot operates according to a written program, and, second, that this program can be rewritten to accommodate a variety of manufacturing tasks. The term "multifunctional" means that the robot is able, through reprogramming and the use of a variety of end effectors, to perform a number of different manufacturing tasks. Definitions written around these two critical characteristics are becoming the accepted definitions among manufacturing professionals. *Figures*

20-66 and *20-67* are examples of modern industrial robots that fit this definition.

The microprocessor was the enabling device with regard to the wide-scale development and use of industrial robots. But major technological developments do not take place simply because of a new capability. Something must provide the impetus for taking advantage of the new capability. In the case of industrial robots, the impetus was economics.

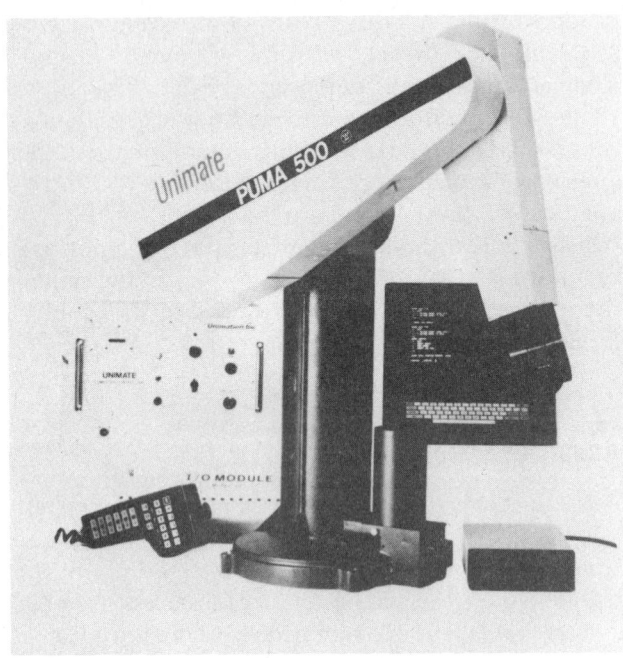

Figure 20-66 Industrial robot *(Courtesy UNIMATION Inc., a Westinghouse Co.)*

Figure 20-67 Industrial robot *(Courtesy UNIMATION Inc., a Westinghouse Co.)*

In the 1970's it became imperative for the U.S. to produce better products at lower costs in order to be competitive with foreign manufacturers. Other factors, such as the need to find better ways of performing dangerous manufacturing tasks, contributed to the development of industrial robots. However, the principle rationale has always been, and is now, improved productivity.

Industrial robots offer a number of benefits that account for the rapid current and projected growth of industrial robot installations.

1. Increased productivity
2. Improved product quality
3. More consistent product quality
4. Reduced scrap and waste
5. Reduced reworking costs
6. Reduced raw goods inventory
7. Direct labor cost savings
8. Savings in related costs, such as lighting, heating, and cooling
9. Savings in safety-related costs
10. Savings from correctly forecasting production schedules

The Robot System

Work in a manufacturing setting is not accomplished by a robot. Rather, it is accomplished by a robot system. A robot system has four major components: the controller, the robot arm or manipulator, the end-of-arm tools, and the power sources, *Figure 20-68*. These components, coupled with the various other pieces of equipment and tools needed to perform the job for which a robot is programmed, are called the robot's work cell.

Figure 20-69 contains a schematic drawing of the work cell for the Cincinnati Milicron Corporation's T3-726 robot, which is used for TIG welding. Notice that the work cell contains not only the robot system but an index table, an operator's safety shield, and special welding equipment.

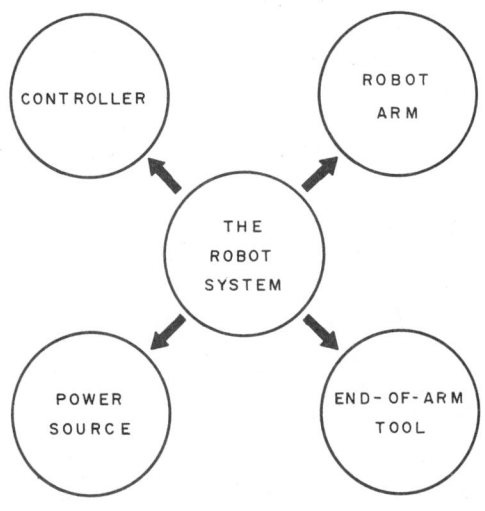

Figure 20-68 The robot system

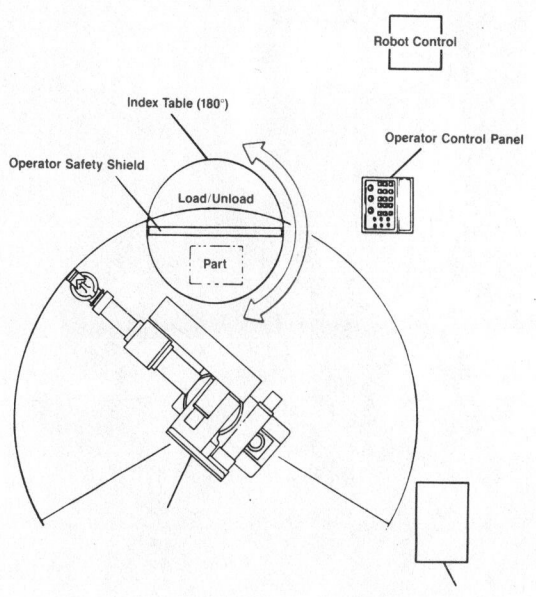

Figure 20-69   Robot work cell *(Courtesy Cincinnati Milacron)*

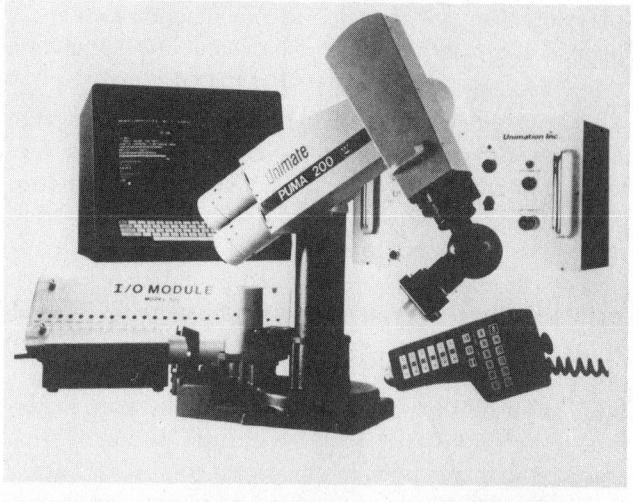

Figure 20-70   Puma 200 industrial robot *(Courtesy UNIMATION Inc., a Westinghouse Co.)*

The contents of a robot's work cell will vary according to the application of the robot. However, the one constant in a robot's work cell is the robot system.

The Controller.   A robot is a special-purpose device similar to a computer. As such, it has all the normal components of a computer, including the central processing unit—made up of a control section and an arithmetic/logic section—and a variety of input and output devices.

The controller for a robot system does not look like the microcomputer one is used to seeing on a desk. It must be packaged differently so as to be able to withstand the rigors of a manufacturing environment. Typical input/output devices used in conjunction with a robot controller include teach stations, teach pendants, a display terminal, a controller front panel, and a permanent storage device.

Teach terminals, teach pendants, or front panels are used for interacting with the robot system. These devices allow humans to turn the robot on, write programs, and key in commands to the robot system. Display terminals give operators a soft copy output source. The permanent storage device is a special device on which reusable programs can be stored. *Figure 20-70* is a photograph of a robot system. In the background you will notice a display terminal with a special keyboard, an input/output module, and a teach pendant.

Mechanical Arm.   The mechanical arm, or manipulator, is the part of a robot system with which most people are familiar. It is the part that actually performs the principle movements in doing manufacturing-oriented work. Mechanical arms are classified according to the types of motions of which they are capable. The basic categories of motion for mechanical arms are rectangular, cylindrical, and spherical. *Figure 20-71* is a drawing of the mechanical arm for the Unimate Puma Series 700 robot.

End-of-Arm Tooling.   The human arm by itself can perform no work. It must have a hand, which in turn, grips a tool. The mechanical arm of a robot, like the human arm, can perform no work. It, too, must have a hand, a tool, or a device which combines both functions. On industrial robots, the function of the wrist is performed by the tool plate. The tool plate is a special device to which end-of-arm tools are attached. The tools themselves vary according to the type of tasks the robot will

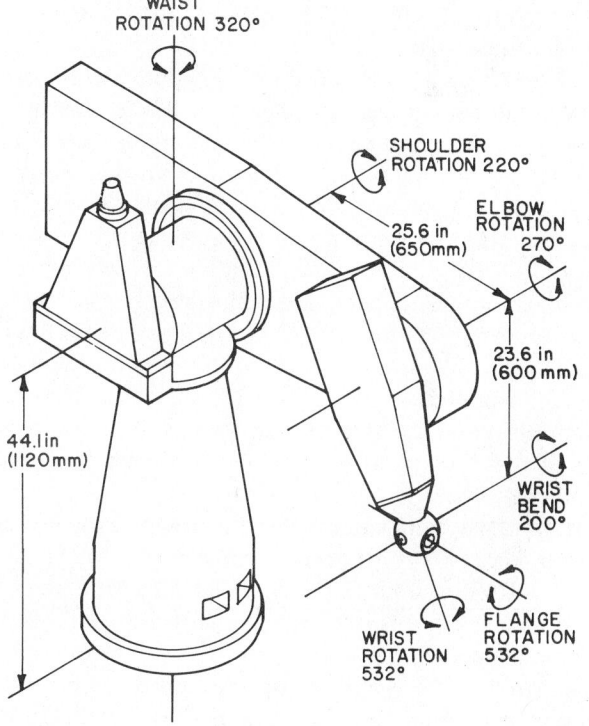

Figure 20-71   Robot mechanical arm *(Courtesy UNIMATION Inc., a Westinghouse Co.)*

perform. Any special device or tool attached to the tool plate to allow the robot to perform some specialized task is classified as an end-of-arm tool or an end effector.

The Power Source.   A robot system can be powered by three different types of power. The power sources used in a robot system are electrical, pneumatic, or hydraulic. The controller, of course, is powered by electricity, as is any computer. The mechanical arm and end-of-arm tools may be powered by either pneumatic or hydraulic power. Some robot systems will use all three types of power. For example, a given robot might use electricity to power the controller, hydraulic power to manipulate the arm, and pneumatic power to manipulate the end-of-arm tool. Hydraulic power is fluid based. Pneumatic power comes from compressed gas.

Figure 20-72 shows a widely used industrial robot with a gripper device as an end-of-arm tool. The tool plate is the rectangular plate immediately behind the gripper.

CIM/FMS

Computer-integrated manufacturing is the most modern, most automated form of production. It involves tying different phases of production together into one wholly integrated system. The term "flexible manufacturing system" is sometimes used synonymously with computer-integrated manufacturing system. Actually, however, a flexible manufacturing system is one type of CIM system designed for medium-range production volumes and moderate flexibility. Other types of CIM systems are special systems and manufacturing cells. There are other terms that have been used to describe the same concept, but CIM and FMS are the two that are most widely used today.

There are a number of different types of computer-integrated manufacturing systems. This is because each system is designed to meet the specific needs of the individual manufacturing setting where it will be used. However, in general, a CIM system is any computerized manufacturing system in which numerically controlled machines are joined together and connected by some form of automated material handling system.

Figure 20-73 shows CNC machining centers linked together in an automated cell. In this system, two CNC machining centers are linked together by a cart-type material handling system. The cart system is further linked to four part load/unload stations to permit untended operation.

In a CIM system such as this, the computer is used in several ways: 1) it is used to control CNC machines; 2) it is used to control the materials handling system; and 3) it is used for monitoring all production tasks accomplished by the system.

Figure 20-74 is a photograph of a flexible manufacturing machine/robot cell. In this CIM system, rough castings are machined and inspected. The system contains two CNC machines, gaging devices, a conveyor, and two robots. The rough castings are loaded onto the

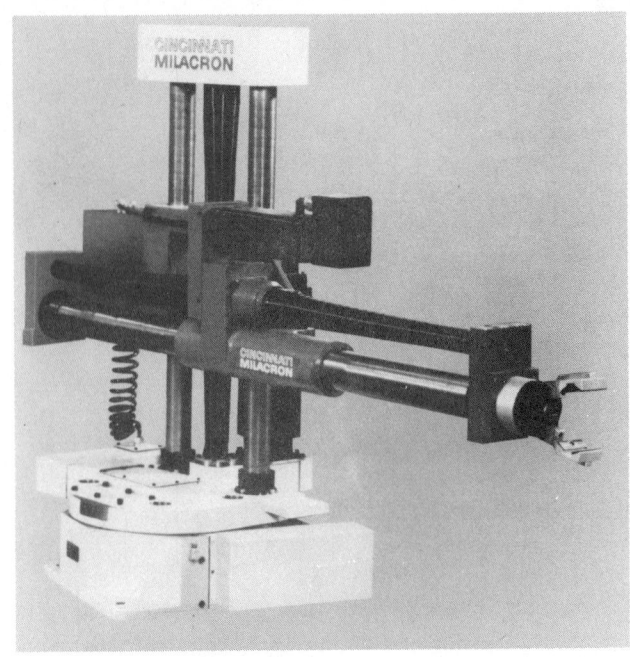

Figure 20-72   End-of-arm gripper *(Courtesy Cincinnati Milacron)* (Note: Safety equipment may have been removed or opened to clearly illustrate products and must be in place prior to operation.)

conveyor by a robot, and finished castings are unloaded by a robot.

Figure 20-75 is a photograph of a composite tape laying (CTL) system. The CTL system automatically dispenses tape at speeds of 100 feet per minute. Tape can be automatically laid in 3-inch, 6-inch, or 12-inch widths, debulked, cut, and overlaid ply-on-ply. Such systems are used in the manufacture of aircraft and space products.

In spite of advances in manufacturing automation, there is human involvement with a CIM system. Typically, this involvement falls into six broad categories: 1) loading raw stock and materials onto the system for processing; 2) unloading processed work pieces from the system; 3) changing tools on machines within the system; 4) setting tools on machines within the system; 5) continuous maintenance of the system; and 6) occasional repair of the system when there is a breakdown or malfunction.

Stand-alone CNC machines are used in low-volume manufacturing applications that require a high degree of flexibility in order to produce a wide variety of parts. They represent one extreme on the manufacturing spectrum. At the other extreme are transfer lines. Transfer lines are used in high-volume manufacturing applications where all parts produced are identical. Transfer lines are not flexible; they cannot accommodate variety.

Transfer lines are a major component of what is sometimes referred to as "Detroit Automation." A transfer line consists of several workstations linked together

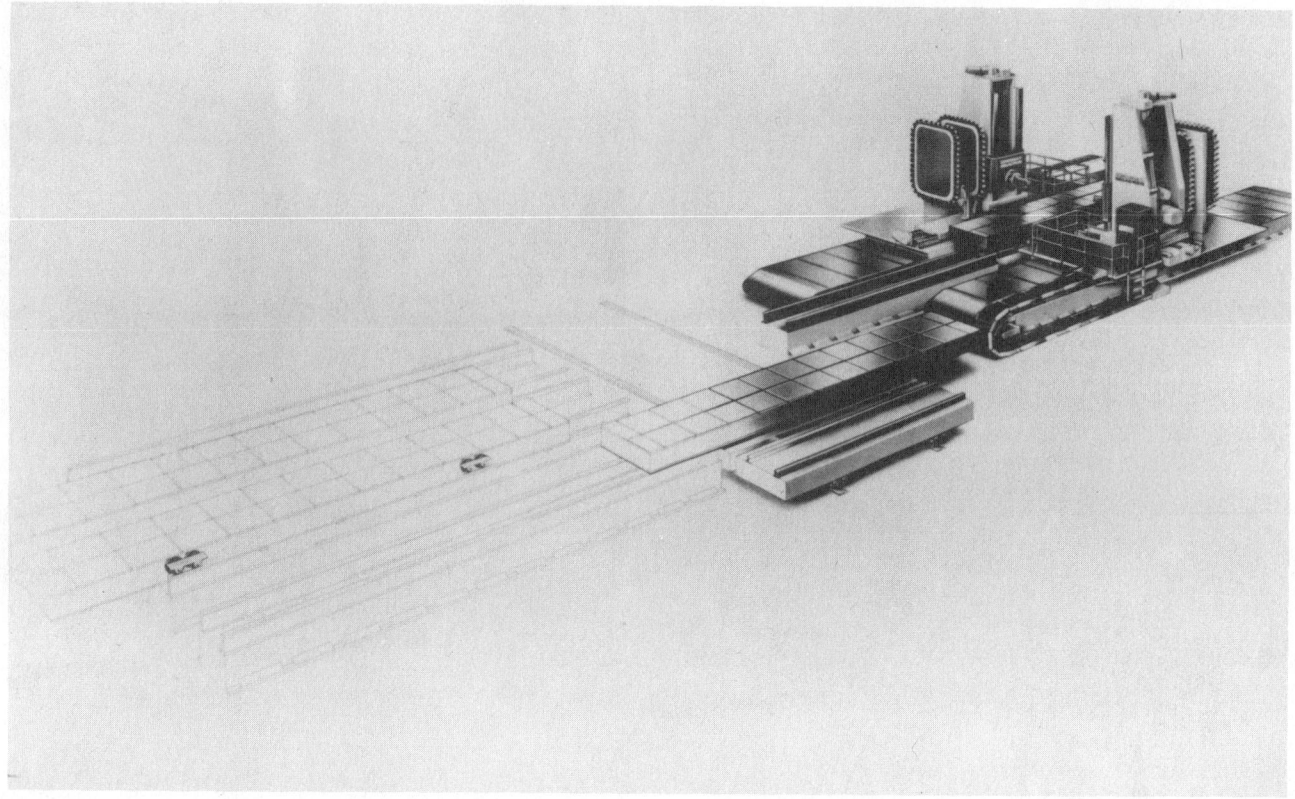

Figure 20-73   Example of a CIM system *(Courtesy Cincinnati Milacron Inc.)*

by materials handling devices that transfer workpieces from station to station.

The first station holds the raw material. The last station is a bin for collecting the finished parts. Each station in-between performs some type of operation on the parts as

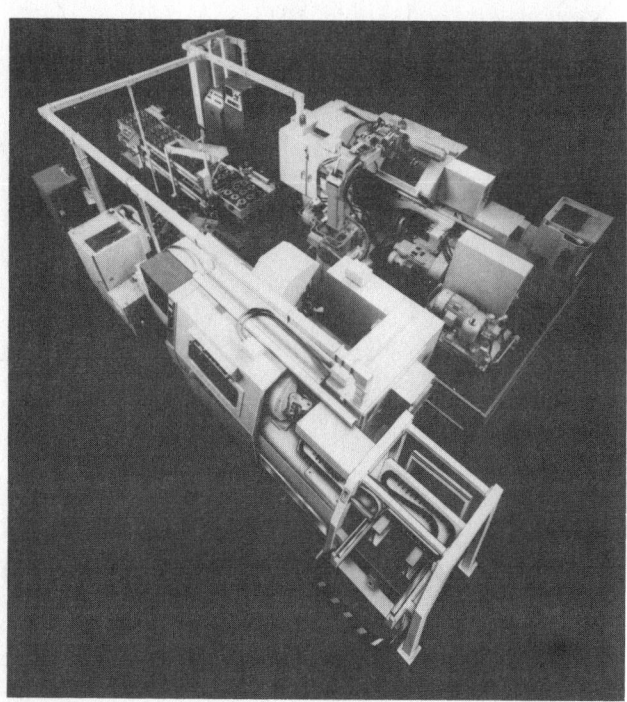

Figure 20-74   FMS *(Courtesy Cincinnati Milacron Inc.)*

they pass through. The transfer of parts from station to station, and the work performed on them, are automatic. Transfer lines are appropriate in situations that involve high-volume production of identical parts.

There has always been a gap between these two extremes, and there has always been a need to fill this gap. The medium-volume, moderate flexibility manufacturing situation has long been a problem in need of a solution. CIM systems are such a solution. CIM fills the void between stand-alone CNC machines and transfer lines, *Figure 20-76*.

CIM offers a number of advantages that, when taken together, form the rationale for this modern approach to medium-volume, flexible production. The most important of these are:

1. Produces families of parts
2. Accommodates the random introduction of parts
3. Requires less lead-time
4. Allows a closer relationship between parts to be produced and workpieces loaded onto the system
5. Allows better machine utilization
6. Requires less labor

CIM systems reduce the amount of direct and indirect labor costs associated with finished workpieces. This is because CIM systems require less human involvement in producing them. Ten to 12 traditional CNC machines require 10 to 12 operators. A CIM system with 10 to 12 CNC machines might require as few as 4 people. This means less direct labor. Indirect labor costs resulting from such tasks as materials handling are also reduced

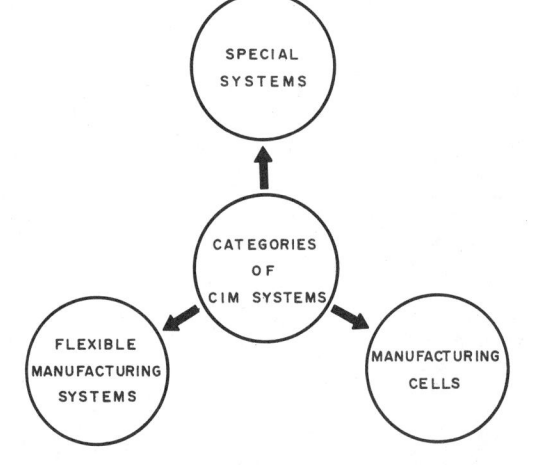

Figure 20-75 CTL system *(Courtesy Cincinnati Milacron Inc.)*

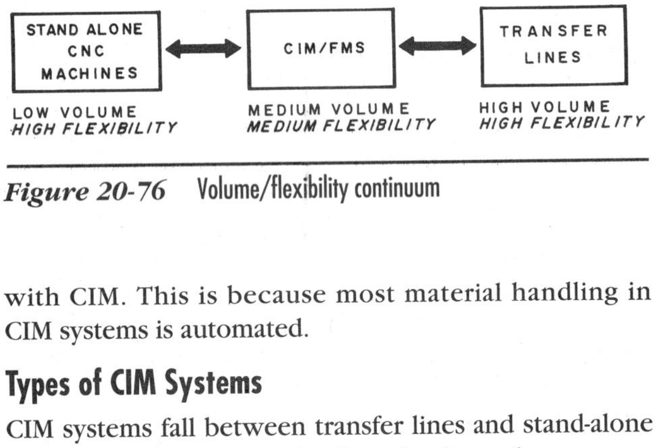

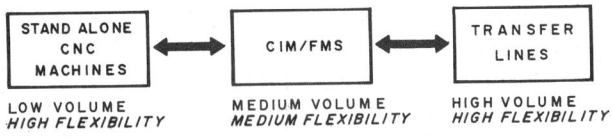

Figure 20-76 Volume/flexibility continuum

with CIM. This is because most material handling in CIM systems is automated.

Types of CIM Systems

CIM systems fall between transfer lines and stand-alone CNC machines on a graph of production volume versus part variety flexibility. By applying these same criteria—production volume and part variety—CIM systems can be divided into three categories, *Figure 20-77*.

1. Special systems
2. Flexible manufacturing systems
3. Manufacturing cells

The systems above are listed in order from the least flexible (special systems) to the most flexible (manufacturing cells). Of all CIM systems, those classified as special systems are capable of the most volume production and the least flexibility. Special systems might produce as many as 15,000 parts per year, but they are limited to less than 10 different part types.

Figure 20-77 Categories of CIM systems

Flexible manufacturing systems represent the middle group in CIM systems. An FMS might have a production volume as high as 2,000 parts per year and a capability of handling as many as 200 different parts. *Figure 20-78* is a diagram of a FMS produced by Cincinnati Milacron.

The most flexible of CIM systems are manufacturing cells, which, in turn, are capable of the lowest produc-

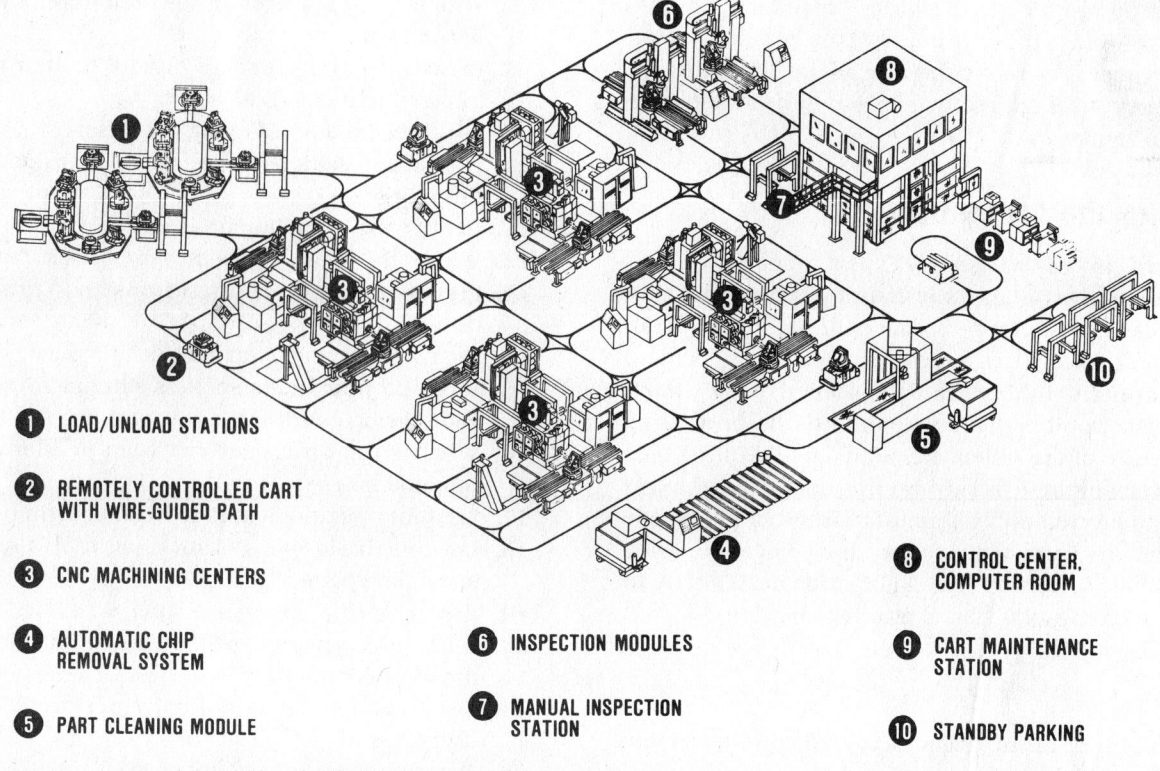

1 **LOAD/UNLOAD STATIONS**

2 **REMOTELY CONTROLLED CART WITH WIRE-GUIDED PATH**

3 **CNC MACHINING CENTERS**

4 **AUTOMATIC CHIP REMOVAL SYSTEM**

5 **PART CLEANING MODULE**

6 **INSPECTION MODULES**

7 **MANUAL INSPECTION STATION**

8 **CONTROL CENTER, COMPUTER ROOM**

9 **CART MAINTENANCE STATION**

10 **STANDBY PARKING**

Figure 20-78 CIM system *(Courtesy Cincinnati Milacron Inc.)*

tion volumes. A manufacturing cell might have a production volume as low as 500 parts per year but a capability of handling as many as 500 different part types.

Just-in-Time Manufacturing (JIT)

Just-in-time manufacturing (JIT) is an approach to manufacturing that seeks to completely eliminate waste. The term *waste* is used here in the broadest sense and encompasses rejected parts, wasted personnel time, wasted machine time, excessive inventory, and any other situation that wastes times, material, or any other manufacturing resource. JIT has the following basic characteristics.

Process Synchronization

This characteristic is what gives JIT its name. *Process synchronization* means that all processes involved in producing a product are in balance and synchronized so that production flows smoothly and continuously. Traditionally, production processes have been operated on a *hurry-up-and-wait* basis. While one process is hurrying to complete its task, the next one in succession is idle. Process synchronization smoothes this out so that each successive process receives its material just-in-time to be processed.

Simplicity

This characteristic is fundamental to JIT. The view being that the simpler things are, the easier JIT will be to accomplish. With JIT a continuous effort is made to

simplify processes so that they can be performed with a continually decreasing amount of resources.

Wholistic View

This characteristic means that JIT cannot be accomplished by any individual unit operating in isolation. The entire company—all units—must be included in the effort. All processes, from purchasing to accounting and production, must be synchronized in order for JIT to work properly.

Manufacturing Resource Planning (MRPII)

Manufacturing resource planning is often referred to as MRPII to distinguish it from its predecessor, which was material requirements planning (MRP). MRP was a way to reduce inventory and waste through systematic planning for material requirements. MRPII is a broader concept that allows companies to systematically plan for all resources needed in manufacturing (i.e., material, capacity of production systems, personnel, time, etc.).

An MRPII system is computer-based and typically consists of at least the following four components: 1) control database; 2) planning database; 3) functional modules; and 4) integrated modules. The control database contains inventory data and a master file on workcenters. The planning database contains a bill of materials file and a process routing file. Functional modules include production scheduling, capacity planning, material planning, and job cost reporting. Integrated modules or modules that might be integrated include sales, accounting, customer order processing, payroll, etc. Using MRPII properly simplifies the implementation of JIT because it helps synchronize all processes that are either directly or indirectly involved in producing the products in question.

Statistical Process Control (SPC)

Statistical process control (SPC) is a process that uses statistical control charts in monitoring production processes in an attempt to reduce the amount of scrap and rework. Tolerances are transformed into upper and lower control limits that are placed on a chart. Periodically, points that correspond to the processing performance of the system are plotted on this chart. The plots are monitored visually or electronically. As long as all points plotted fall within established limits on the chart, the process is under control. If a point falls outside of the limits, corrections should be made immediately to prevent further production of unacceptable parts.

Review

1. List four of the major factors to consider before any part is made.
2. What is meant by the term *lead time*?
3. List four primary methods used to fabricate manufactured products.
4. What terms are used to describe the top and bottom sections of the flask used for casting?
5. What two important factors must be considered when making a pattern for casting?
6. What is the approximate rate of shrinkage for steel? For cast iron?
7. Which casting process is also referred to as *lost wax casting*?
8. In which casting process is the molten metal forced into metal dies?
9. List three common forging processes.
10. How many stations are normally contained in a drop forge die?
11. Explain the extruding process.
12. What is the principal tool used for metal stamping?
13. List two stamping operations that produce blanks.
14. List two stamping operations that produce holes.
15. What do operations such as slitting, trimming, and shaving control?
16. What is the difference between notching and seminotching?
17. List four operations used to bend or form parts.
18. List four basic types of machine tools used for machining parts.
19. Explain the operation of a lathe.
20. What two variations of the milling machine are the most common?
21. List five operations normally performed on a drill press.
22. Describe the process known as JIT.
23. What is MRPII?
24. Explain the concept known as SPC.
25. Define the following nontraditional machining processes:
    - HDM
    - USM
    - ECM
    - EDM
    - PAM

Drafting Applications: Pipe, Structural, and Architectural

This is an application chapter that allows students to use the design and drafting skills they have learned in developing real-world projects in the areas of pipe, structural, and architectural drafting. Projects developed in this chapter relate directly to specific design and drafting career fields.

Types of Pipe

The wide variety of fluids used by modern society requires a number of different types of pipe. The most widely used of these are:

- Steel pipe
- Cast iron pipe
- Brass and copper pipe
- Copper tubing
- Plastic pipe

Steel Pipe

Steel pipe is well suited for high-pressure and high-temperature applications. It is frequently used in piping systems that transport water, oil, petroleum, and steam. Steel pipe comes in several cross-sectional configurations.

The most widely used of these are the standard, extra strong, and double extra strong configurations, *Figure 21-1*. There are actually ten different cross-sectional configurations for steel pipe.

Steel pipe is specified by a nominal diameter callout. The actual diameter will vary slightly from the nominal. The American National Standard Institute (ANSI) specifies pipe in ten different schedules. Each schedule corresponds to a wall thickness. For example, standard and extra strong pipe are Schedules 40 and 80, respectively. The nominal diameter for pipe up to 12 inches refers to the inside diameter. Callouts are given in inches or millimeters. The nominal diameter for pipe over 12 inches refers to the outside diameter.

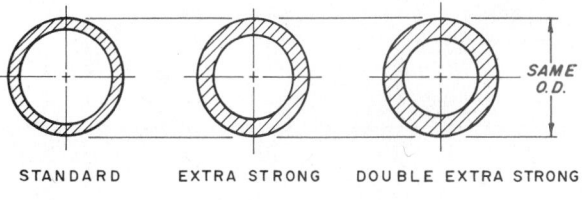

STANDARD   EXTRA STRONG   DOUBLE EXTRA STRONG

Figure 21-1   Schedule of pipe

Cast Iron Pipe

Cast iron pipe is used most frequently for underground applications—to transport water, gas, and sewage. It is well suited for low-pressure steam connections also.

Brass and Copper Pipe

Brass and copper pipe are used in applications where corrosion will be a problem. Because brass and copper are able to withstand corrosion, the expected lifetime of a piping system in a high-corrosive setting will be longer if brass or copper pipe are used.

Copper Tubing

Copper tubing is used extensively in applications where vibration and misalignment are important factors. It is used in hydraulic and pneumatic applications, such as industrial settings involving robots, and in automotive settings.

Plastic Pipe

Plastic pipe is used extensively in modern piping settings. It is highly resistant to corrosion and chemical degradation. Plastic pipe is flexible and can be easily installed. However, it is not used where heat or pressure are factors. Plastic pipe does not have the chemical makeup to withstand heat or the wall strength to withstand high pressures.

Types of Joints and Fittings

Each kind of pipe explained in the previous section comes in straight length. However, piping systems require turns and branches and changes of size. In every instance where a pipe must change directions or size there is a joint. Joints are accomplished using fittings. There are three broad classifications of fittings.

1. Screwed
2. Welded
3. Flanged

Figure 21-2 illustrates these three types of fittings.

Engineers and drafters need to be able to specify pipe fittings. A given pipe fitting is specified by stating the nominal pipe size, the name of the fitting, and the material out of which the fitting is made. Any fitting that is used to connect different sizes of pipes is referred to as a "reduc-

ing fitting." When specifying reducing fittings, you must state the nominal pipe sizes for both the large and small end of the fitting. The large size is stated first. There are enough differences among screwed, welded, and flanged fittings that each must be studied separately.

Screwed Fittings

Screwed fittings are generally used in applications requiring small diameter pipe of 2.5 inches or less. The threaded end of the pipe and the internal threads on the fitting are usually coated with a special lubricant to seal the joint and to ease the connection process. *Figures 21-3*, *21-4*, *21-5*, and *21-6* contain information on the most commonly used threaded type fittings. Figure 21-3 contains information on the 90°elbow, 90°street elbow, 45° elbow, 45° street elbow, and 90° reducing elbow. The table accompanying each illustration gives the type sizes with which that particular fitting can be used, the various dimensions in which it is available, and the weight of the fitting itself. Figure 21-4 contains information for blind flange, screwed flange, and flat band cap fittings, as well as bushings. Figure 21-5 contains information on union, coupling, cross, Tee, and reducing Tee fittings. Figure 21-6 contains information on plug, return bend, and lock nut fittings, as well as reducers and nipples.

There are two types of American Standard pipe threads: tapered and straight. Tapered threads are more common. Straight threads are usually used only for special applications. The ANSI Standard for Pipe Threads is ANSI/ASME B1.20.1-1983. *Figure 21-7* illustrates the conventions used for drawing pipe threads. *Figure 21-8* illustrates the American National Standard taper pipe thread notation methods. Both types of threads have the same number of threads per inch.

Welded Fittings and Joints

Some piping applications, such as high-pressure and high-temperature systems, require permanent fittings and joints. When this is the case, welded fittings are used. To accommodate the welding process, the connection ends of welded fittings as well as the ends of adjoining pipe are usually beveled. Welded fittings usually weigh less than flanged or screwed fittings, and they are easier to insulate.

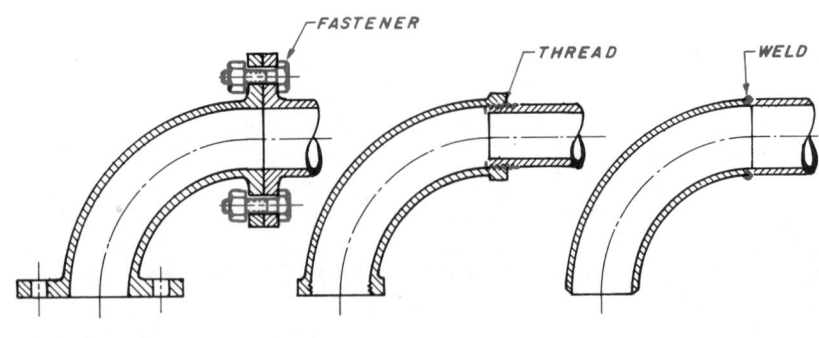

Figure 21-2   Pipe connections

AVAILABLE STYLES AND SIZES

90° ELBOW

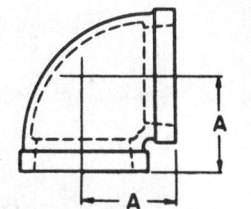

REFERENCE	PIPE SIZE, INCHES														
	⅛	¼	⅜	½	¾	1	1¼	1½	2	2½	3	3½	4	5	6
A	¹¹⁄₁₆	¹³⁄₁₆	¹⁵⁄₁₆	1⅛	1⁵⁄₁₆	1⁷⁄₁₆	1¾	1¹⁵⁄₁₆	2¼	2¹¹⁄₁₆	3⅛	3⁷⁄₁₆	3¾	4½	5⅛
Weight	.055	.060	.095	.145	.210	.355	.705	.790	1.180	1.670	2.590	3.250	4.065	6.900	9.800

90° STREET ELBOW

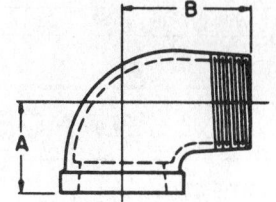

REFERENCE	PIPE SIZE, INCHES										
	⅛	¼	⅜	½	¾	1	1¼	1½	2	2½	3
A	¹¹⁄₁₆	¹³⁄₁₆	¹⁵⁄₁₆	1⅛	1⁵⁄₁₆	1⁷⁄₁₆	1¾	1¹⁵⁄₁₆	2¼	2¹¹⁄₁₆	3⅛
B	1⅛	1⁵⁄₁₆	1⁷⁄₁₆	1⅝	1⅞	2⅛	2½	2¹¹⁄₁₆	3³⁄₁₆	3¹³⁄₁₆	4½
Weight	.025	.045	.085	.140	.180	.205	.495	.750	1.250	1.850	2.900

45° ELBOW

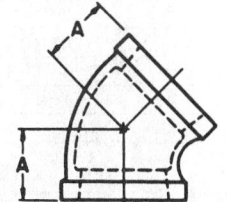

REFERENCE	PIPE SIZE, INCHES														
	⅛	¼	⅜	½	¾	1	1¼	1½	2	2½	3	3½	4	5	6
A	¹¹⁄₁₆	¾	¹³⁄₁₆	⅞	1	1⅛	1⁵⁄₁₆	1⁷⁄₁₆	1¹¹⁄₁₆	1¹⁵⁄₁₆	2³⁄₁₆	2⅜	2⅝	3¹⁄₁₆	3¹⁵⁄₃₂
Weight	.040	.040	.075	.115	.200	.260	.455	.605	.970	1.420	1.925	2.530	3.335	5.650	8.700

45° STREET ELBOW

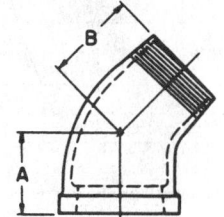

REFERENCE	PIPE SIZE, INCHES									
	⅛	¼	⅜	½	¾	1	1¼	1½	2	2½
A	—	—	—	1¹⁄₁₆	1³⁄₁₆	1⅜	1¹⁷⁄₃₂	1⁵⁄₁₆	1⅝	2¹⁄₁₆
B	—	—	—	1³⁄₁₆	1⅜	1¹⁷⁄₃₂	1²³⁄₃₂	2¼	2⅜	2½
Weight	—	—	—	.140	.180	.205	.495	.700	1.000	1.625

90° REDUCING ELBOW

Size	Weight	Size	Weight	Size	Weight
½" x ¼"	.115	1½" x 1"	.560	4" x 3"	3.065
½" x ⅜"	.120	1½" x 1¼"	.720		
¾" x ½"	.190	2" x 1¼"	1.000		
1" x ½"	.260	2" x 1½"	1.030		
1" x ¾"	.300	2½" x 2"	1.745		
1¼" x ¾"	.405	3" x 2"	2.205		
1¼" x 1"	.470	3" x 2½"	2.315		

LATERALS — .1. 45° Y

    a. Size Range - ½'' through 3''

    b. Dimensional Standard - F-52618-C (Revision)

NOTE: Any reducing fittings not specifically listed can be produced on a Special Order basis and will be subject to a Special Order charge of 25% of the listed retail price.

(For additional fittings see Page 8)

Figure 21-3 Pipe fittings (Courtesy Latrobe Foundry Machine and Supply Co.)

AVAILABLE STYLES AND SIZES

FLANGE - Blind

REFERENCE	PIPE SIZE, INCHES										
	½	¾	1	1¼	1½	2	2½	3	3½	4	6
A	3½	3⅞	4¼	4⅝	5	6	7	7½	8½	9	11
B	7/16	7/16	7/16	½	9/16	⅝	11/16	¾	13/16	15/16	1
C	2⅜	2¾	3⅛	3½	3⅞	4¾	5½	6	7	7½	9½
D	⅝	⅝	⅝	⅝	⅝	¾	¾	¾	¾	¾	⅞
E	4	4	4	4	4	4	4	4	8	8	8
Weight	.275	.385	.560	.765	1.015	1.640	2.440	3.075	4.730	5.285	9.250

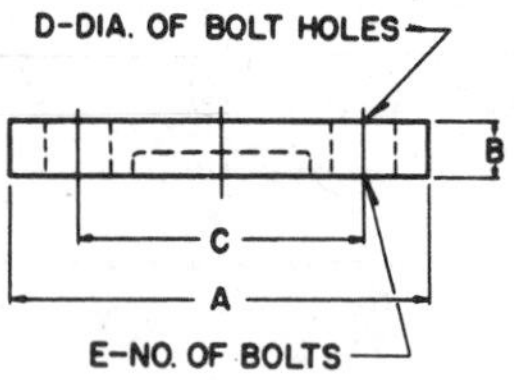
D-DIA. OF BOLT HOLES
B
C
A
E-NO. OF BOLTS

FLANGE - Screwed

FOR REDUCING, SLIP-ON, FLOOR AND WELDING NECK FLANGES SEE PAGE 8.

REFERENCE	PIPE SIZE, INCHES											
	½	¾	1	1¼	1½	2	2½	3	3½	4	5	6
A	3½	3⅞	4¼	4⅝	5	6	7	7½	8½	9	10	11
B	⅝	⅝	11/16	13/16	⅞	1	1 3/16	1¼	1¼	1 5/16	1 7/16	1 9/16
C	2⅜	2¾	3⅛	3½	3⅞	4¾	5½	6	7	7½	8½	9½
D	⅝	⅝	⅝	⅝	⅝	¾	¾	¾	¾	¾	⅞	⅞
E	4	4	4	4	4	4	4	4	8	8	8	8
Weight	.255	.375	.550	.740	.970	1.530	2.385	2.835	4.250	4.565	5.650	6.850

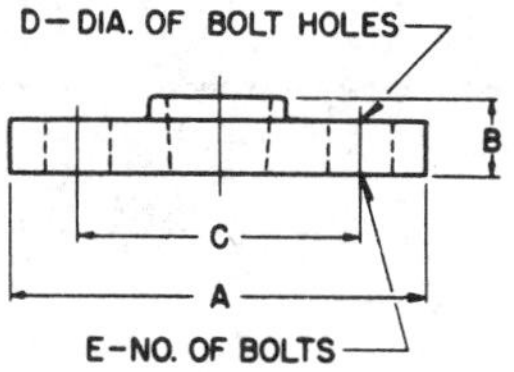
D-DIA. OF BOLT HOLES
B
C
A
E-NO. OF BOLTS

CAP - Flat Band

REFERENCE	PIPE SIZE, INCHES														
	⅛	¼	⅜	½	¾	1	1¼	1½	2	2½	3	3½	4	5	6
A	19/32	25/32	27/32	1 1/16	1 5/32	1¼	1 5/16	1 15/32	1 9/16	2 1/32	2 1/16	2 3/32	2 3/16	2⅜	2⅝
Weight	.015	.025	.050	.075	.100	.150	.290	.350	.450	.760	1.435	1.825	3.080	4.950	6.500

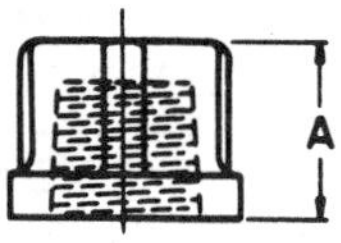
A

BUSHINGS

Size	Weight	Size	Weight	Size	Weight	Size	Weight
¼" x ⅛"	.010	1" x ¾"	.070	2" x 1¼"	.325	3½" x 3"	.750
⅜" x ⅛"	.020	1¼" x ⅜"	.170	2" x 1½"	.285	4" x 1½"	1.860
⅜" x ¼"	.015	1¼" x ½"	.165	2½" x 1"	.785	4" x 2"	1.720
½" x ⅛"	.045	1¼" x ¾"	.145	2½" x 1¼"	.755	4" x 2½"	1.535
½" x ¼"	.030	1¼" x 1"	.130	2½" x 1½"	.715	4" x 3"	1.255
½" x ⅜"	.030	1½" x ½"	.215	2½" x 2"	.605	5" x 2"	2.600
¾" x ¼"	.055	1½" x ¾"	.210	3" x 1¼"	1.095	5" x 3"	2.350
¾" x ⅜"	.050	1½" x 1"	.205	3" x 1½"	1.015	5" x 4"	2.000
¾" x ½"	.045	1½" x 1¼"	.125	3" x 2"	.895	6" x 3"	3.950
1" x ¼"	.110	2" x ½"	.370	3" x 2½"	.705	6" x 4"	3.600
1" x ⅜"	.105	2" x ¾"	.370	3½" x 2"	1.120	6" x 5"	3.250
1" x ½"	.090	2" x 1"	.365	3½" x 2½"	.920		

Figure 21-4 Pipe fittings *(Courtesy Latrobe Foundry Machine and Supply Co.)*

AVAILABLE STYLES AND SIZES

UNION

REFERENCE	PIPE SIZE, INCHES												
	1/8	1/4	3/8	1/2	3/4	1	1 1/4	1 1/2	2	2 1/2	3	3 1/2	4
A	1 9/16	1 13/16	1 15/16	2	2 1/8	2 3/4	3	3	3 1/4	3 5/8	4 1/8	4 7/8	5
Weight	.145	.130	.170	.255	.330	.505	.790	.815	1.250	1.745	2.865	5.000	5.800

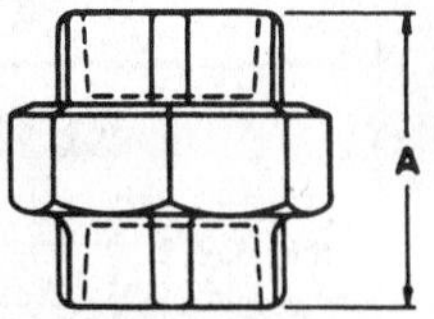

COUPLING   FOR HALF COUPLINGS AND HEAVY WALL COUPLINGS SEE PAGE 8

REFERENCE	PIPE SIZE, INCHES														
	1/8	1/4	3/8	1/2	3/4	1	1 1/4	1 1/2	2	2 1/2	3	3 1/2	4	5	6
A	19/32	3/4	29/32	1 1/16	1 11/32	1 5/8	1 31/32	2 15/64	2 23/32	3 5/16	3 15/16	4 7/16	4 15/16	6 1/16	7 3/16
B	15/16	1 1/32	15/32	15/16	1 9/16	1 13/16	2 1/16	2 5/16	2 9/16	2 7/8	3 1/16	3 7/16	3 7/16	4 1/8	4 1/8
Weight	.020	.025	.035	.060	.100	.135	.230	.310	.500	.730	1.015	1.740	2.040	3.300	4.500

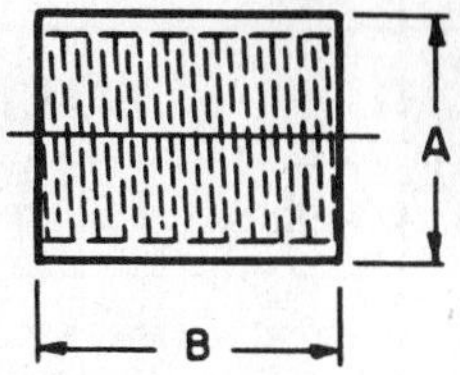

CROSS

REFERENCE	PIPE SIZE, INCHES												
	1/8	1/4	3/8	1/2	3/4	1	1 1/4	1 1/2	2	2 1/2	3	3 1/2	4
A	11/16	13/16	15/16	1 1/8	1 5/16	1 7/16	1 3/4	1 15/16	2 1/4	2 11/16	3 1/8	3 7/16	3 3/4
Weight	.050	.085	.155	.215	.340	.565	.825	1.115	2.115	2.730	4.000	5.200	6.210

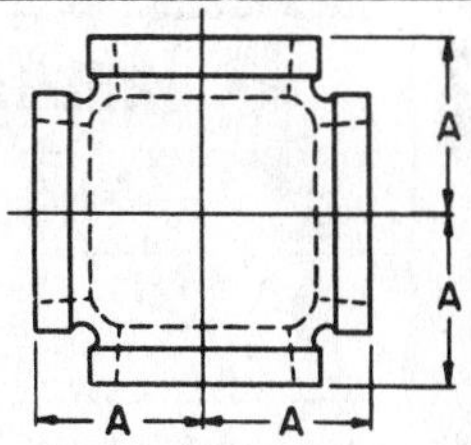

TEE   FOR LATERALS SEE PAGE 2

REFERENCE	PIPE SIZE, INCHES													
	1/8	1/4	3/8	1/2	3/4	1	1 1/4	1 1/2	2	2 1/2	3	3 1/2	4	5	6
A	11/16	13/16	15/16	1 1/8	1 5/16	1 7/16	1 3/4	1 15/16	2 1/4	2 11/16	3 1/8	3 7/16	3 3/4	4 1/2	5 1/8
Weight	.060	.070	.130	.180	.285	.435	.745	1.060	1.760	2.250	3.035	4.135	5.580	9.510	13.000

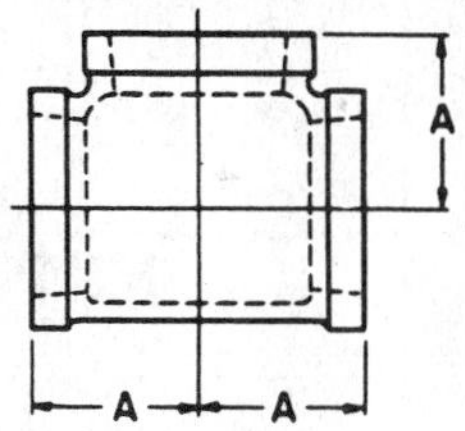

REDUCING TEE   NOTE: In the listing of reducing tees, the last dimension shown is the size of the branch.

Size	Wt.	Size	Wt.	Size	Wt.	Size	Wt.
1/2" x 1/2" x 1/4"	.195	1" x 1" x 1/2"	.485	1 1/4" x 1" x 1 1/2"	1.180	2" x 2" x 1 1/4"	1.325
1/2" x 1/2" x 3/8"	.190	1" x 1" x 3/4"	.465	1 1/2" x 1 1/4" x 1"	1.230	2" x 2" x 1 1/2"	1.470
1/2" x 1/2" x 3/4"	.320	1" x 1" x 1 1/4"	.850	1 1/2" x 1 1/4" x 1 1/4"	1.170	2" x 2" x 2 1/2"	1.910
3/4" x 1/2" x 1/2"	.320	1" x 1" x 1 1/2"	1.300	1 1/2" x 1 1/2" x 1/2"	1.245	2 1/2" x 2" x 2"	1.925
3/4" x 1/2" x 3/4"	.305	1 1/4" x 3/4" x 1 1/4"	.830	1 1/2" x 1 1/2" x 3/4"	1.220	2 1/2" x 2 1/2" x 2"	2.095
3/4" x 3/4" x 3/8"	.315	1 1/4" x 1" x 3/4"	.885	1 1/2" x 1 1/2" x 1"	1.180	3" x 3" x 2"	2.690
3/4" x 3/4" x 1/2"	.305	1 1/4" x 1" x 1"	.855	1 1/2" x 1 1/2" x 1 1/4"	1.115	4" x 4" x 2"	4.015
3/4" x 3/4" x 1"	.465	1 1/4" x 1 1/4" x 1/2"	.850	1 1/2" x 1 1/2" x 2"	1.280	4" x 4" x 3"	5.095
1" x 1/2" x 1/2"	.190	1 1/4" x 1 1/4" x 3/4"	.850	2" x 1 1/2" x 1 1/2"	1.315	6" x 6" x 4"	10.850
1" x 3/4" x 1/2"	.515	1 1/4" x 1 1/4" x 1"	.800	2" x 1 1/2" x 2"	1.755		
1" x 3/4" x 3/4"	.495	1 1/4" x 1 1/4" x 1 1/2"	1.170	2" x 2" x 1/2"	0.995		
1" x 3/4" x 1"	.465	1 1/2" x 3/4" x 1 1/2"	1.220	2" x 2" x 3/4"	1.290		
1" x 1" x 3/8"	.495	1 1/2" x 1" x 1"	1.300	2" x 2" x 1"	1.300		

Figure 21-5   Pipe fittings (Courtesy Latrobe Foundry Machine and Supply Co.)

AVAILABLE STYLES AND SIZES

PLUG

COUNTERSUNK PLUGS

REFERENCE	PIPE SIZE, INCHES														
	⅛	¼	⅜	½	¾	1	1¼	1½	2	2½	3	3½	4	5	6
A	⅝	¹¹⁄₁₆	¹³⁄₁₆	⅞	1	1³⁄₁₆	1⁷⁄₁₆	1⅜	1⁹⁄₁₆	1¹¹⁄₁₆	1¹³⁄₁₆	2	2³⁄₁₆	2⅜	2⁹⁄₁₆
Weight	.010	.015	.020	.030	.055	.090	.150	.215	.315	.560	.755	.910	1.500	2.150	3.500

COUNTERSUNK PLUGS - Square Head   Size Range 3/8'' through 4''
Dimension Standard - ANSI   B16.14-1949

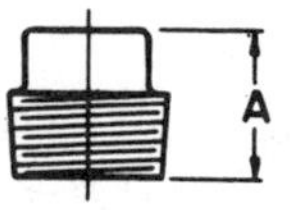

RETURN BEND - Close

REFERENCE	PIPE SIZE, INCHES								
	⅛	¼	⅜	½	¾	1	1¼	1½	2
A	–	–	–	1¼	1½	1¾	2¼	2½	3¼
B	–	–	–	2⅝	3⁵⁄₃₂	3²⁹⁄₃₂	4²¹⁄₃₂	5⁷⁄₃₂	6¹⁹⁄₃₂
Weight	–	–	–	.205	.275	.440	.780	1.000	1.840

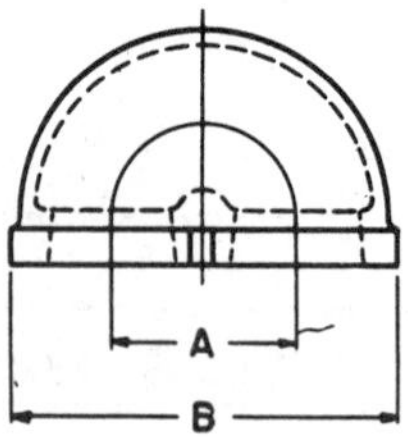

LOCK NUT

REFERENCE	PIPE SIZE, INCHES										
	⅛	¼	⅜	½	¾	1	1¼	1½	2	2½	3
A	–	⁵⁄₃₂	³⁄₁₆	¼	⁵⁄₁₆	⅜	⁷⁄₁₆	½	⅝	¾	¹⁵⁄₁₆
Weight	–	.015	.020	.025	.045	.080	.105	.175	.345	.425	.740

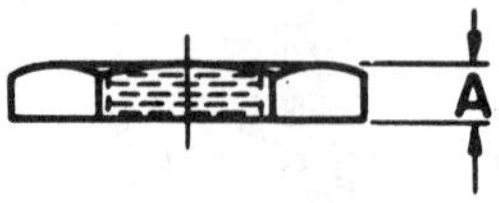

REDUCERS

Size	Weight		Size	Weight		Size	Weight
¼" x ⅛"	.040		1" x ¾"	.240		2" x 1½"	.825
⅜" x ⅛"	.060		1¼" x ¾"	.330		2½" x 2"	1.180
⅜" x ¼"	.060		1¼" x 1"	.340		2½" x 1½"	1.200
½" x ¼"	.100		1½" x ¾"	.480		3" x 2"	1.790
½" x ⅜"	.095		1½" x 1"	.490		3" x 2½"	1.750
¾" x ¼"	.150		1½" x 1¼"	.560		4" x 3"	3.375
¾" x ⅜"	.130		2" x 1"	.710		5" x 4"	4.75
¾" x ½"	.160		2" x 1¼"	.745		6" x 4"	5.00
1" x ½"	.215					6" x 5"	5.00

NIPPLES - Close   ADDITIONAL NIPPLES *

Pipe Size	Length	Weight	Pipe Size	Length	Weight	Pipe Size	Length	Weight	Pipe Size	Length	Weight	Pipe Size	Length	Weight
⅛"	¾"	.005	¾"	1⅜"	.045	1½"	1¾"	.135	3"	2⅝"	.570
¼"	⅞"	.010	1"	1½"	.075	2"	2"	.210	3½"	2¾"	.720
⅜"	1"	.015	1¼"	1⅝"	.105	2½"	2½"	.415	4"	2⅞"	.895
½"	1⅛"	.030									

* NIPPLES 1. Long Nipples a. Size Range — ⅛" through 4"
b. Length — through 12"

Figure 21-6   Pipe fittings (Courtesy Latrobe Foundry Machine and Supply Co.)

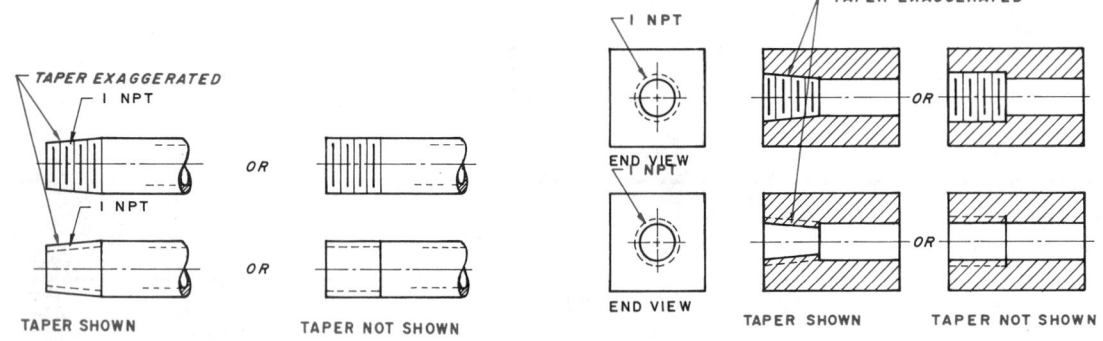

Figure 21-7   Pipe thread conventions

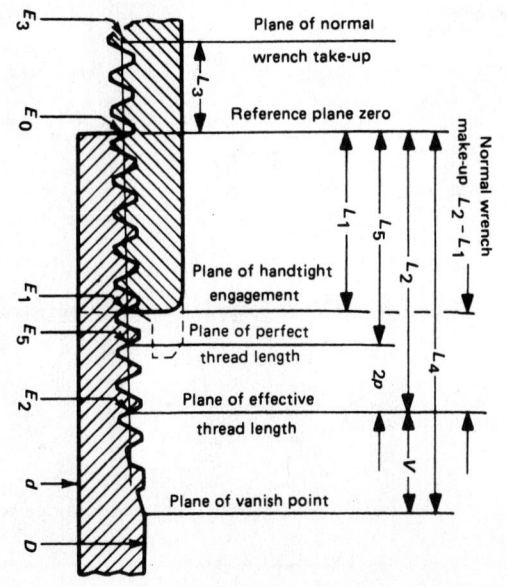

Figure 21-8   Pipe thread notation *(Reprinted from ANSI/ASME Standard B1.20.1—1983. Courtesy ANSI/ASME)*

Flanged Fittings and Joints

Some piping applications require that the piping systems occasionally be disassembled. When this is the case, flanged fittings are appropriate. Flanged fittings and adjoining pipes are bolted together. On occasion, flanged fittings and pipe may be welded together. However, they are normally bolted or glued together. *Figure 21-9* contains size, weight, and dimensional data for the most common type of flanged fittings: 90° elbows, 45° elbows, tees, and reducers.

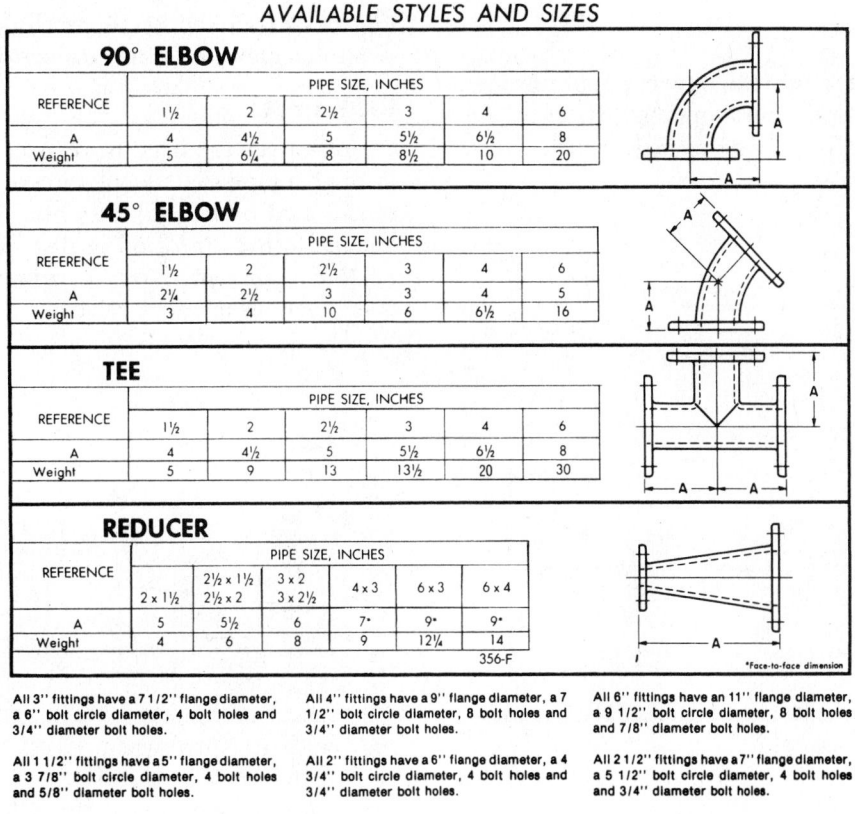

Figure 21-9   Pipe fittings *(Courtesy Latrobe Foundry Machine and Supply Co.)*

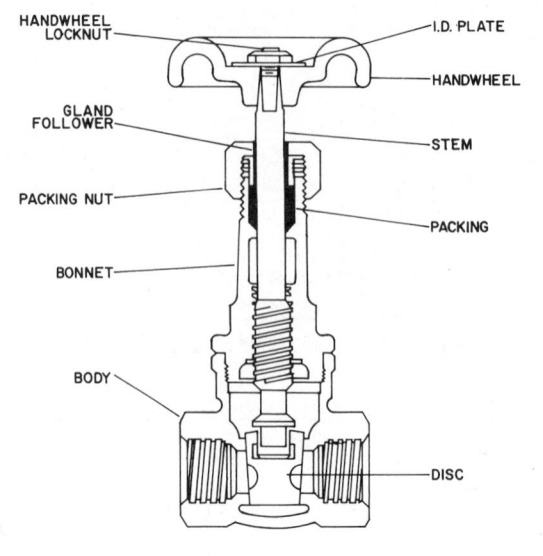

Figure 21-10   Gate valve

Types of Valves

Fluids and gases do not just flow freely through piping systems. They must be regulated, and at certain points stopped. There are a number of different types of valves, including gate, globe, jet, swiveled joint, ball, check, and butterfly. The most frequently used of these are gate, globe, and check valves. These commonly used types of valves are illustrated in *Figures 21-10*, *21-11*, and *21-12*.

Gate Valves

Gate valves are used to turn the flow of liquids through a pipe on or off without restricting the flow of the liquids through the valve or with as little restriction as possible. Gate valves are not meant to be used to regulate the degree of flow. *Figures 21-13* and *21-14* contain dimensional data for two commonly used gate valve configurations.

Globe Valves

Globe valves are used to turn the flow of liquids through a pipe on and off, and they are used also to regulate the flow of fluids through the valve to the desired level. *Figures 21-15* and *21-16* contain dimensional data for two commonly used configurations of globe valves.

Check Valves

Check valves are used to restrict the flow of liquids through a pipe in only one direction. A backward flow is checked by the valve, which is activated by any change in the direction of flow. *Figures 21-17* and *21-18* contain dimensional data for two commonly used configurations of check valves.

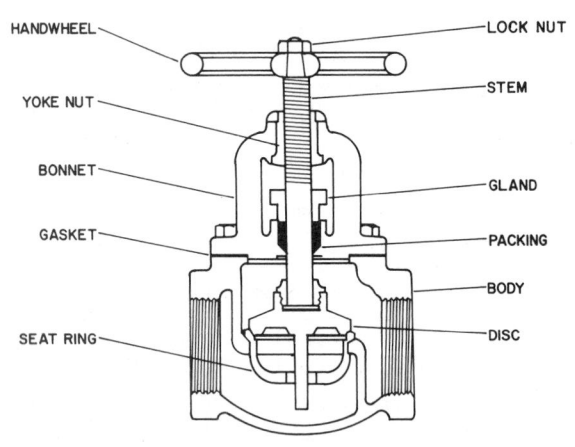

Figure 21-11   Globe valve

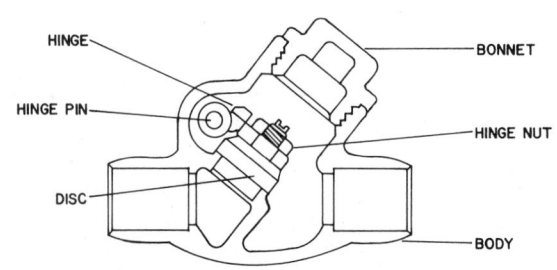

Figure 21-12   Check valve

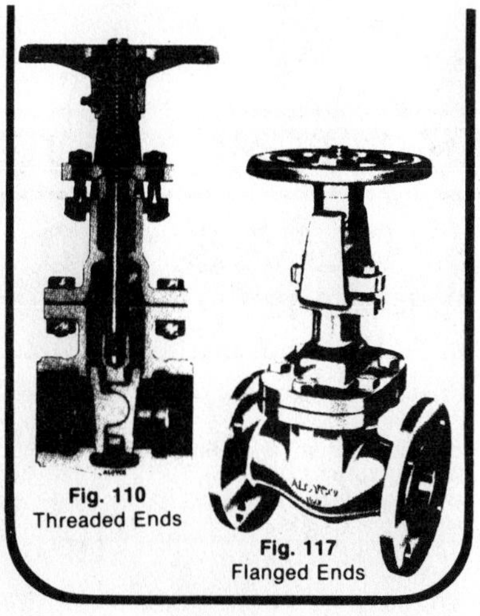
Fig. 110
Threaded Ends
Fig. 117
Flanged Ends

STAINLESS STEEL GATE

CLASS 150

FIGURES: 110 (½″ to 2″), 114 (½″ to 2″), 116 (1½″ to 12″), 117 (½″ to 12″)

DESIGN DESCRIPTION:

Outside Screw and Yoke	Socket Weld Ends (Fig. 114)
Bolted Bonnet	Buttwelding Ends (Fig. 116)
Rising Stem	Flanged Ends (Fig. 117)
Non-Rising Handwheel	Male-Female Bonnet Joint Sizes ½″ to 1″ Incl.
Integral Seats	Double Disc Ball-and-Socket Type Wedge
Threaded Ends (Fig 110)	Flat Faced Flanged Valves, Specify Fig. 117-FF

PARTS AND MATERIAL LIST:

DESCRIPTION	ALOYCO 18-8 SMO	DESCRIPTION	ALOYCO 18-8 SMO
1 body*	A351 GR CF8M	15 yoke bushing†	wrought B16 or cast B584-C83600
2 bonnet	A351 GR CF8M		
3 male disc	{wrought-type 316 or	16 handwheel	malleable iron
4 female disc	{cast A351 GR CF8M	17 yoke nut set screw	type 303
5 disc arm	A351 GR CF8M	18 disc arm pin	type 316
6 gasket†	comp. asbestos or teflon	19 handwheel key	steel
7 stem	type 316	20 yoke bushing nut†	wrought B16 or cast B584-C83600
8 bonnet bolt	A193 GR B8		
9 bonnet bolt nut	A194 GR 8F	21 grease fitting	steel
10 gland plate	A351 GR CF8M	22 yoke	CF8
11 gland follower	wrought-type 316 or cast A351 GR CF8M	23 yoke bolt	A193 GR B8
		24 yoke bolt nut	A194 GR 8F
12 packing†	braided asbestos or teflon	25 gland stud	A193 GR B8
13 gland bolt	A193 GR B8	26 gland stud nut	A194 GR 8F
14 gland bolt nut	A194 GR 8F		

Remarks: *Body material on threaded and welding end valves is ELC grade. Other Alloys available. (Refer to Introduction)

†Unless otherwise specified, these valves may be supplied at the manufacturer's option with: PTFE Gasket and Packing; Type 303 yoke bushing and nut. Such valves are limited to a temperature of 550°F and are so tagged.

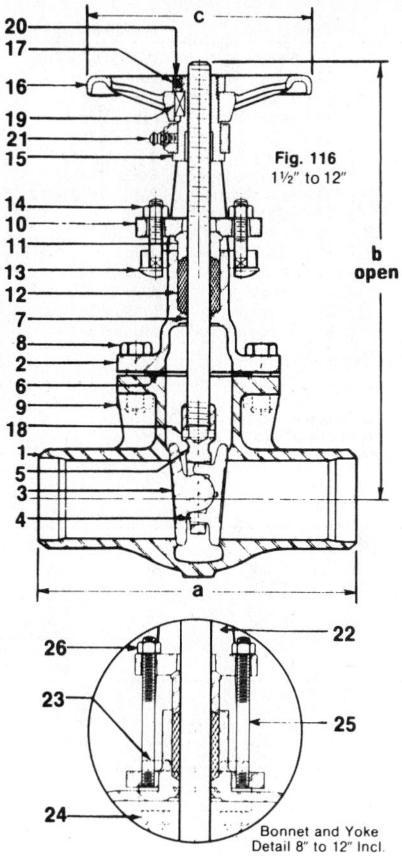
c
20
17
16
19
21
15
Fig. 116
1½″ to 12″
14
10
11
13
12
7
8
2
6
9
18
1
5
3
4
b
open
a
26
22
23
25
24
Bonnet and Yoke
Detail 8″ to 12″ Incl.

DIMENSIONS IN MM AND INCHES:

SIZE	MM	15	20	25	32	40	50	65	80	100	150	200	250	300
	INCH	½	¾	1	1¼	1½	2	2½	3	4	6	8	10	12
a (110, 114)	mm	70	80	89	108	114	121	—	—	—	—	—	—	—
	inch	2¾	3	3½	4¼	4½	4¾							
a (116)	mm	—	—	—	—	165	216	241	283	305	403	419	457	502
	inch					6½	8½	9½	11⅞	12	15⅞	16½	18	19¾
a (117)	mm	108	118	127	—	165	178	191	203	229	267	292	330	356
	inch	4¼	4⅝	5		6½	7	7½	8	9	10½	11½	13	14
b (110, 114)	mm	210	210	238	277	286	337	—	—	—	—	—	—	—
	inch	8¼	8¼	9⅜	10⅞	11¼	13¼							
b (116, 117)	mm	210	210	235	—	291	337	394	432	527	781	1016	1219	1524
	inch	8¼	8¼	9¼	—	11 7/16	13¼	15½	17	20¾	30¾	40	48	60
c	mm	89	100	100	127	127	150	178	191	250	305	356	406	457
	inch	3½	4	4	5	5	6	7	7½	10	12	14	16	18
d (114)	mm	13	13	13	13	13	16	—	—	—	—	—	—	—
	inch	½	½	½	½	½	⅝							
d (117)	mm	11	11	11	—	14	16	18	19	24	25	29	30	32
	inch	7/16	7/16	7/16	—	9/16	⅝	11/16	¾	15/16	1	1⅛	1 3/16	1¼

WEIGHTS IN KG AND POUNDS:

110, 114	kg	2.8	2.9	4.0	5.9	7.0	9.0	—	—	—	—	—	—	—
	lb	6.3	6.5	9	13	15.5	20							
116	kg	—	—	—	—	7.0	10	14	16	26	56	91	151	223
	lb					15.5	21	30	35	63	124	201	332	492
117	kg	3.6	4.3	5.4	—	9.9	14	19	24	40	72	124	206	305
	lb	8	9.5	12		22	30	42	51	88	158	275	455	675

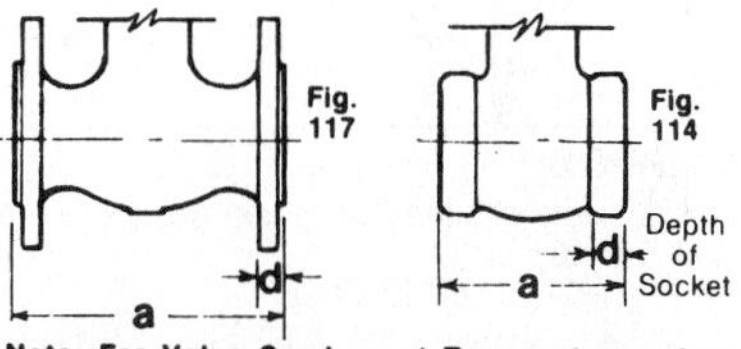
Fig. 117
Fig. 114
Depth of Socket
a
d
a
d

APPLICABLE CLASS 150 STANDARDS:

End Flanges, ANSI B16.5	SW Ends (Bore and Depth), ANSI B16.11
Pipe Threads, ANSI B2.1	BW Ends (Schedule 40), ANSI B16.25 and B16.10
Design, API 603	Face-to-Face, ANSI B16.10
Wall Section, ANSI B16.34	Pressure-Temperature Ratings, ANSI B16.34-1977

Note: For Valve Services at Temperatures above 700°F, the Aloyco Engineering Department should be consulted for materials and features.

For Engineering Specifications and Data, see Engineering Section of Catalog.

Figure 21-13 Gate valve data (*Courtesy Crane-Aloyco Inc.*)

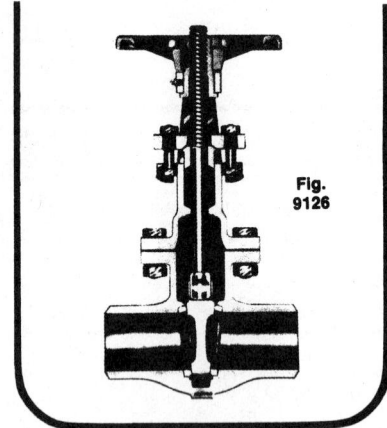

Fig. 9126

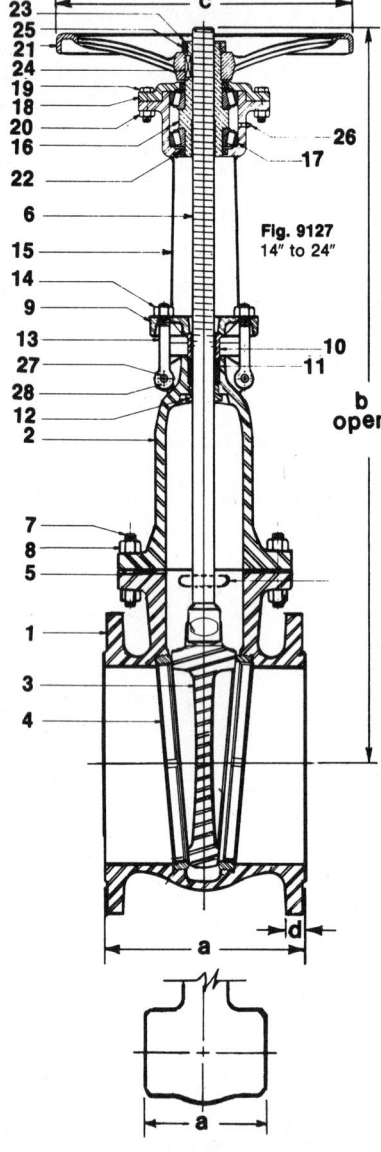

Fig. 9126

STAINLESS STEEL GATE
CLASS 150
FIGURES: 9126, 9127 (14″ to 24″)

DESIGN DESCRIPTION:

Outside Screw and Yoke
Bolted Bonnet
Rising Stem
Non-rising Handwheel
Solid Wedge

For Flat Faced Flanged Valves, Specify Fig. 9127-FF
Roller Bearing Yoke
Renewable Seat Rings
Buttwelding Ends (Fig. 9126)
Flanged Ends (Fig. 9127)

PARTS AND MATERIAL LIST:

DESCRIPTION	ALOYCO 18-8 SMO	DESCRIPTION	ALOYCO 18-8 SMO
1 body*	A351 GR CF8M	16 yoke bushing	B584 alloy C86200
2 bonnet	A351 GR CF8M	17 roller bearing	steel
3 wedge	A351 GR CF8M	18 yoke cap	CF8
4 seat ring	A351 GR CF8M	19 yoke cap bolt	A193 GR B8
5 gasket	comp. asbestos	20 yoke cap bolt nut	A194 GR 8F
6 stem	type 316	21 handwheel	malleable iron
7 bonnet studbolt	A193 GR B7	22 oil seal	flax
8 bonnet studbolt nut	A194 2H	23 yoke bushing nut	B584 alloy C86200
9 gland plate	A351 GR CF8M	24 handwheel key	steel
10 gland follower	A351 GR CF8M	25 yoke nut set screw	type 303
11 packing	braided asbestos	26 grease fitting	steel
12 stem hole bushing	A351 GR CF8M	27 hinge bolt	A193 GR B8
13 gland eyebolt	A193 GR B8	28 hinge bolt nut	A294 GR 8F
14 gland eyebolt nut	A194 GR 8F	29 yoke studbolt**	A193 GR B8
15 yoke	CF8	30 yoke studbolt nut**	A194 GR 8F

Remarks: *Body material on welding end valves to be ELC grade.
**Not shown.
Other Alloys Available. Refer to Introduction.

**Note: For Valve Services at Temperatures above
700°F, the Aloyco Engineering Department should be
consulted for materials and features.**

DIMENSIONS IN MM AND INCHES:

SIZE	MM	350	400	450	500	600
	INCH	14	16	18	20	24
a(9126)	mm	571	610	660	711	813
	inch	22½	24	26	28	32
a(9127)	mm	381	406	432	457	508
	inch	15	16	17	18	20
b	mm	1765	1870	2140	2369	2753
	inch	69½	73⅜	84¼	93¼	108⅜
c	mm	508	559	610	660	762
	inch	20	22	24	26	30
d(9127) thickness of flange	mm	35	37	40	43	48
	inch	1⅜	1⁷/₁₆	1⁹/₁₆	1¹¹/₁₆	1⅞

WEIGHTS IN KG AND POUNDS:

9126	kg	342	474	535	708	1043
	lb	754	1046	1180	1560	2300
9127	kg	419	559	644	848	1234
	lb	924	1232	1420	1870	2720

APPLICABLE CLASS 150 STANDARDS

End Flanges, ANSI B16.5
Design, ANSI B16.34
Wall Section, API 600

Buttwelding Ends (Schedule 40), ANSI B16.25 and B16.10
Face-to-Face, ANSI B16.10
Pressure-Temperature Ratings, ANSI B16.34-1977

Figure 21-14   Gate valve data *(Courtesy Crane-Aloyco Inc.)*

STAINLESS STEEL GLOBE
CLASS 150
FIGURES: 422 (½" to 2"), 427 (½" to 6")

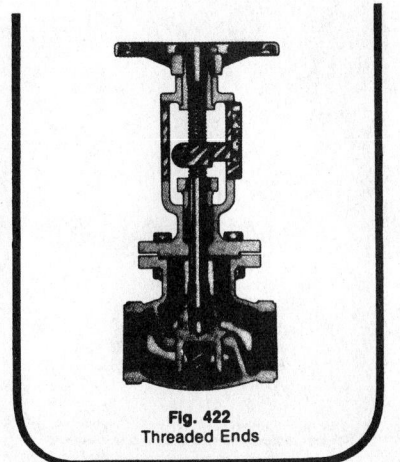

Fig. 422
Threaded Ends

DESIGN DESCRIPTION:

"V" Port Throttling
Bolted Bonnet
Outside Screw and Yoke
Position Indicator
Rising Stem

Non-rising Handwheel
Threaded Ends (Fig. 422)
Flanged Ends (Fig. 427)
When Flat Face End Flanges are
  Required Order as 427-FF

PARTS AND MATERIAL LIST:

DESCRIPTION	ALOYCO 18-8 SMO
1 body	A351 GR CF8M
2 bonnet	A351 GR CF8M
3 disc	A351 GR CF8M
4 stem	type 316
5 gasket	PTFE
6 bonnet bolts	A193 GR B8
7 bonnet bolt nuts	A194 GR 8F
8 gland plate	CF8M
9 gland follower	type 316
10 packing	PTFE
11 gland bolts	A193 GR B8
12 gland bolt nuts	A194 GR 8F

DESCRIPTION	ALOYCO 18-8 SMO
13 yoke bushing	cast B584 alloy 836 or wrought B16 half hard
14 handwheel	malleable iron
15 indicator	CF-16F
16 indicator bolt	A193 GR B8
17 escutcheon pins	type 304
18 indicating plate	type 304
19 yoke bushing nut	cast B584 alloy 836 or wrought B16 half hard
20 handwheel key	steel
21 disc pin	type 316
22 yoke nut set screw	type 303

Other Alloys available. (Refer to Introduction)

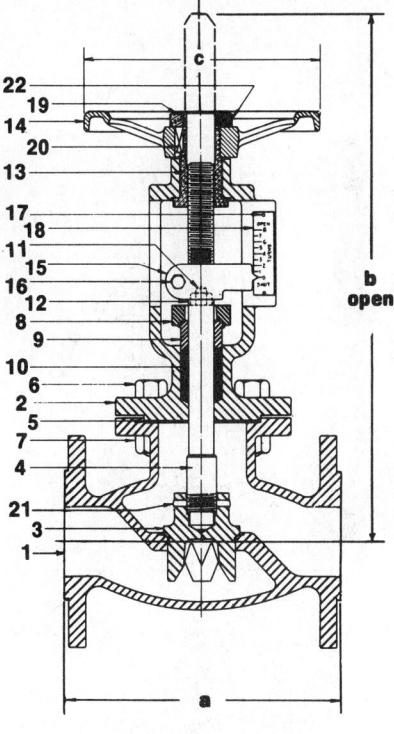

Fig. 427
Sizes ½" to 6"

DIMENSIONS IN MM AND INCHES:

SIZE		15	20	25	40	50	65	80	100	150
	MM	15	20	25	40	50	65	80	100	150
	INCH	½	¾	1	1½	2	2½	3	4	6
a	mm	86	95	108	140	165	—	—	—	—
	inch	3⅜	3¾	4¼	5½	6½				
a (F)	mm	108	117	127	165	203	216	241	292	406
	inch	4¼	4⅝	5	6½	8	8½	9½	11½	16
b	mm	180	235	259	283	308	330	368	419	584
	inch	7¹¹/₁₆	9¼	10³/₁₆	11⅛	12⅛	13	14½	16½	23
c	mm	102	127	127	152	178	191	254	305	356
	inch	4	5	5	6	7	7½	10	12	14

WEIGHTS IN KG AND POUNDS:

422	kg	1.8	2.7	4.5	7.2	9.5	—	—	—	—
	lb	4	6	10	16	21				
427	kg	2.7	4	5.9	9.5	13	20	26	43	78
	lb	6	9	13	21	29	43	58	95	173

APPLICABLE CLASS 150 STANDARDS:

End Flanges, ANSI B16.5
Wall Section, ANSI B16.34
Face-to-Face, ANSI B16.10
Pressure-Temperature Ratings, ANSI B16.34-1977
Pipe Threads, ANSI B2.1

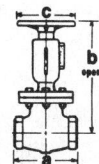

Fig. 422 (½" to 2")

Figure 21-15   Globe valve data *(Courtesy Crane-Aloyco Inc.)*

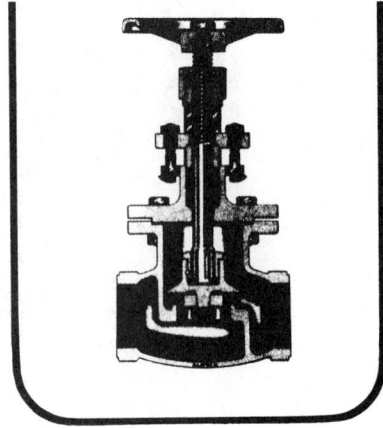

STAINLESS STEEL GLOBE
CLASS 150
FIGURES: 502 (½″ to 2″), 504 (½″ to 2″), 507 (½″ to 10″)

DESIGN DESCRIPTION:

Outside Screw and Yoke
Bolted Bonnet
Renewable PTFE Disc
Retained Gasket
Rising Stem and Handwheel
For Flat Faced Flanged Valve
 Specify Fig. 507 FF

Integral Seat
Threaded Ends (Fig. 502)
Socket Weld Ends (Fig. 504)
Flanged Ends (Fig. 507)
Valves in Sizes 8″ and 10″ Have
 Disc Guide Below Seat

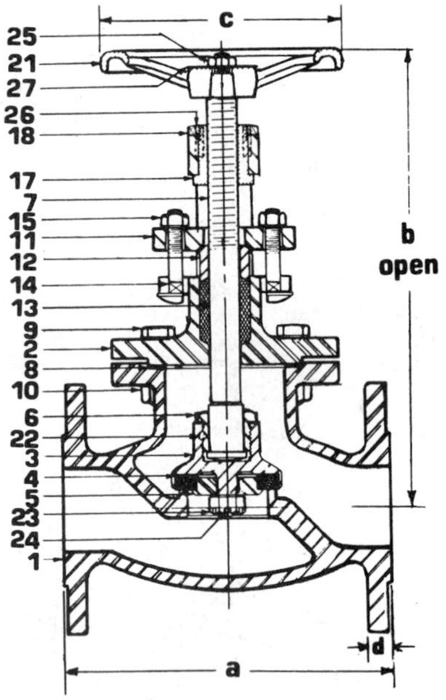

PARTS AND MATERIAL LIST:

DESCRIPTION	ALOYCO 18-8 SMO	DESCRIPTION	ALOYCO 18-8 SMO
1 body*	A351 GR CF8M	14 gland bolt	A193 GR B8
2 bonnet	A351 GR CF8M	15 gland bolt nut	A194 GR 8F
3 disc holder	cast-CF8M, wrought-type 316	16 yoke ***	CF8
4 disc	PTFE	17 yoke bushing	B16 half hard
5 disc holder plate	cast-CF8M, wrought-type 316	18 yoke bushing nut	B16 half hard
6 swivel nut		19 yoke bolt ***	A193 GR B8
7 stem	type 316	20 yoke bolt nut ***	A194 GR 8F
8 gasket	PTFE	21 handwheel	malleable
9 bonnet bolt	A193 GR B8	22 swivel nut pin	type 316
10 bonnet bolt nut	A194 GR 8F	23 disc holder plate nut**	type 316
11 gland plate	CF8M	24 disc holder nut pin	type 316
12 gland follower	cast-CF8M, wrought-type 316	25 handwheel nut	type 303
13 packing	PTFE	26 yoke nut pin	type 303
		27 I.D. plate	type 304

Remarks: *Body material on threaded and welding end valves is ELC grade.
 **Sizes ½″ to 2″ incl.
 ***Sizes 8″ and 10″ only.
 Other Alloys Available. Refer to Introduction.
 Note: For Valve Services at Temperatures above 700°F, the Aloyco Engineering Department should be consulted for materials and features.

DIMENSIONS IN MM AND INCHES:

SIZE	MM INCH	15 ½	20 ¾	25 1	40 1½	50 2	65 2½	80 3	100 4	150 6	200 8	250 10
a(502,504)	mm	86	95	108	140	165	—	—	—	—	—	—
	inch	3⅜	3¾	4¼	5½	6½						
a(507)	mm	108	117	127	165	203	216	241	292	406	495	622
	inch	4¼	4⅝	5	6½	8	8½	9½	11½	16	19½	24½
b(502,504)	mm	165	210	248	283	305	—	—	—	—	—	—
	inch	6½	8¼	9¾	11⅛	12						
b(507)	mm	165	210	251	283	305	327	359	425	533	622	822
	inch	6½	8¼	9⅞	11⅛	12	12⅞	14⅛	16¾	21	24½	32⅜
c	mm	67	76	102	127	152	178	203	254	305	406	457
	inch	2⅝	3	4	5	6	7	8	10	12	16	18
d(504) Depth of Socket	mm inch	10 ⅜	13 ½	13 ½	13 ½	16 ⅝	—	—	—	—	—	—
d(507) Flange Thickness	mm inch	11 ⁷/₁₆	11 ⁷/₁₆	11 ⁷/₁₆	14 ⁹/₁₆	16 ⅝	17 ¹¹/₁₆	19 ¾	24 ¹⁵/₁₆	25 1	29 1⅛	30 1³/₁₆

WEIGHTS IN KG AND POUNDS:

502,504	kg	2.7	5	5	7.3	11	—	—	—	—	—	—
	lb	6	10	10	16	24						
507	kg	5	6.4	6.4	10	15	20.8	26	43	78	128.8	188.7
	lb	10	14	14	22	34	46	58	95	173	284	416

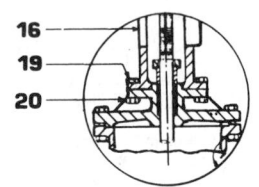
Separable Yoke Detail
Sizes 8″ and 10″ only

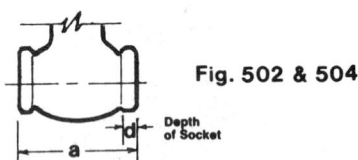
Fig. 502 & 504
Depth of Socket

APPLICABLE CLASS 150 STANDARDS:

End Flanges, ANSI B16.5
Wall Section, ANSI B16.34
Face-to-Face, ANSI B16.10
Pipe Threads, ANSI B2.1

Socket Weld Ends (Bore and Depth), ANSI B16.11
Pressure-Temp. Ratings, ANSI B16.34-1977
 Maximum Service Temperature, 450°F

Figure 21-16 Globe valve data *(Courtesy Crane-Aloyco Inc.)*

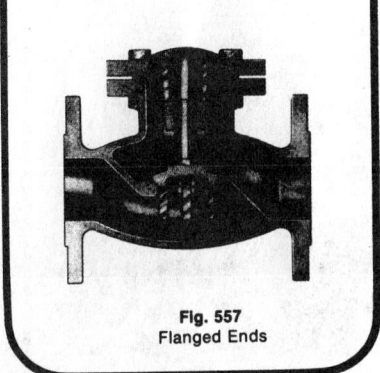

Fig. 557
Flanged Ends

STAINLESS STEEL
LIFT CHECK VALVES
CLASS 150

**FIGURES: 550 (¼″ to 2″), 554 (¼″ to 2″),
556 (½″ to 6″), 557 (½″ to 6″)***

DESIGN DESCRIPTION:

Horizontal Type
Bolted Cover
Retained Gasket
Regrinding Disc
Integral Seat

Threaded Ends (Fig. 550)
Socket Weld Ends (Fig. 554)
Buttwelding Ends (Fig. 556)
Flanged Ends (Fig. 557)

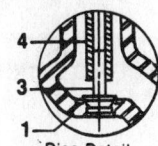

Disc Detail
Sizes ¼″ to ¾″ incl.

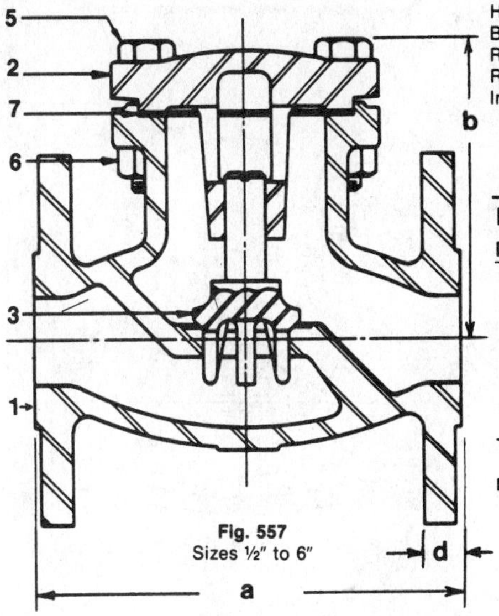

Fig. 557
Sizes ½″ to 6″

PARTS AND MATERIAL LIST:

DESCRIPTION	ALOYCO 18-8 SMO
1 body*	A351 GR CF8M
2 cover	A351 GR CF8M
3 disc	cast CF8M or wrought type 316
4 disc guide**	type 316
5 cover bolt	A193 GR B8
6 cover bolt nut	A194 GR 8F
7 gasket	comp. asbestos

Remarks: *Body material on threaded and welding end valves is ELC grade.
Other Alloys available. (Refer to Introduction)
**Sizes ¼″ to ¾″ incl.
†Unless otherwise specified, these valves may be supplied at the manufacturer's option with: PTFE Gasket.

Note: For Valve Services at Temperatures above 700°F, the Aloyco Engineering Department should be consulted for materials and features.

DIMENSIONS IN MM AND INCHES:

SIZE	MM	6	10	15	20	25	40	50	65	80	100	150
	INCH	¼	⅜	½	¾	1	1½	2	2½	3	4	6
a (550,554)	mm	79	79	86	95	108	140	165	—	—	—	—
	inch	3⅛	3⅛	3⅜	3¾	4¼	5½	6½				
a (556,557)	mm	—	—	108	117	127	165	203	216	241	292	406
	inch			4¼	4⅝	5	6½	8	8½	9½	11½	16
b	mm	67	67	67	76	86	102	116	133	146	168	200
	inch	2⅝	2⅝	2⅝	3	3⅜	4	4⁹/₁₆	5¼	5¾	6⅝	7⅞
d (554)	mm	10	10	10	13	13	13	16	—	—	—	—
	inch	⅜	⅜	⅜	½	½	½	⅝				
d (557)	mm	—	—	11	11	11	14	16	17	19	24	25
	inch			⁷/₁₆	⁷/₁₆	⁷/₁₆	⁹/₁₆	⅝	¹¹/₁₆	¾	¹⁵/₁₆	1

WEIGHTS IN KG AND POUNDS:

		6	10	15	20	25	40	50	65	80	100	150
550,554	kg	1.4	1.4	1.8	2.3	2.9	5.9	7.3	—	—	—	—
	lb	3	3	4	5	6.3	13	16				
556	kg	—	—	2	2.7	3.4	6.8	10	12	18.6	28	51
	lb			4.5	6	7.5	15	22	26	41	61	112
557	kg	—	—	2.3	3	4.1	8	12.7	14	22	34	68
	lb			5	7	9	17.5	28	31	48	75	150

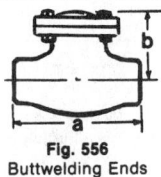

Fig. 556
Buttwelding Ends
½″ to 6″

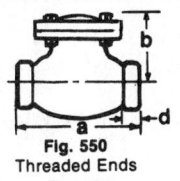

Fig. 550
Threaded Ends
Fig. 554
Socket Weld Ends

APPLICABLE CLASS 150 STANDARDS:

End Flanges, ANSI B16.5
Wall Section, ANSI B16.34
Face-to-Face, ANSI B16.10
Pipe Threads, ANSI B2.1

SW Ends (Bore and Depth), ANSI B16.11
BW Ends (Sch. 40), ANSI B16.25 and B16.10
Pressure-Temp. Ratings, ANSI B16.34-1977

Figure 21-17   Check valve data *(Courtesy Crane-Aloyco Inc.)*

STAINLESS STEEL
LIFT CHECK VALVES
CLASS 600

FIGURES: 4550-A (½″ to 2″)
        4554-A (½″ to 2″)
        4557-A (½″ to 2″)

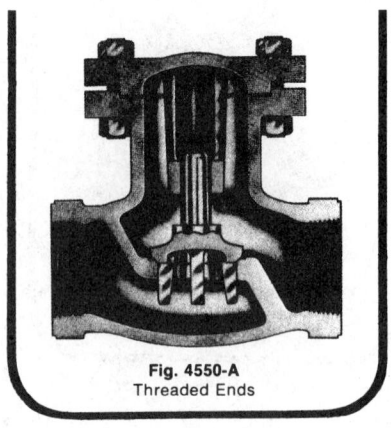

Fig. 4550-A
Threaded Ends

DESIGN DESCRIPTION:

Bolted Cover
Integral Seat
Horizontal Type
Threaded Ends (Fig. 4550-A)
Socket Weld Ends (Fig. 4554-A)
Buttwelding Ends (Fig. 4557-A)

PARTS AND MATERIAL LIST:

DESCRIPTION	ALOYCO 18-8 SMO
1 body	A351 GR CF8M
2 cover	A351 GR CF8M
3 disc	type 316
4 disc guide (½″ and ¾″)	type 316
5 cover studbolt	A193 GR B8
6 cover studbolt nut	A194 GR 8F
7 gasket	comp. asbestos

Other Alloys Available. Refer to Introduction.

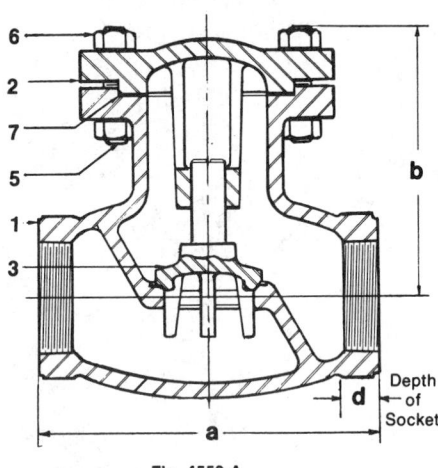

Fig. 4550-A
Sizes ½″ to 2″

DIMENSIONS IN MM AND INCHES:

SIZE	MM	15	20	25	40	50
	INCH	½	¾	1	1½	2
a (4550-A, 4554-A)	mm	95	108	127	165	191
	inch	3¾	4¼	5	6½	7½
a (4557-A)	mm	165	191	216	241	387
	inch	6½	7½	8½	9½	15¼
b	mm	94	94	97	148	165
	inch	3¹¹/₁₆	3¹¹/₁₆	3¹³/₁₆	5¹³/₁₆	6½
d (4554-A)	mm	10	13	13	13	16
	inch	⅜	½	½	½	⅝
d (4557-A)	mm	14	16	17	22	25
	inch	⁹/₁₆	⅝	¹¹/₁₆	⅞	1

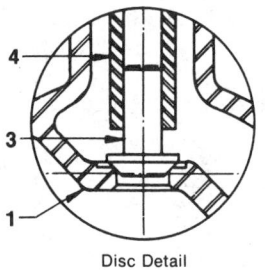

Disc Detail
Sizes ½″ and ¾″

WEIGHTS IN KG AND POUNDS:

4550-A, 4554-A	kg	2.3	3	5	10	14
	lb	5	7	11	22	30
4557-A	kg	5	6.3	10	15	22
	lb	11	14	22	32	48

APPLICABLE CLASS 600 STANDARDS:

End Flanges, ANSI B16.5
Wall Section, ANSI B16.34
Face-to-Face, ANSI B16.10
Buttwelding Ends (Schedule 40),
    ANSI B16.25 and B16.10

Pipe Threads, ANSI B2.1
Pressure-Temperature Rating, ANSI B16.34-1977
Socket Weld Ends (Bore and Depth), ANSI B16.11

**Note: For Valve Services at Temperatures above
700°F, the Aloyco Engineering Department should be
consulted for materials and features.**

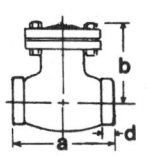   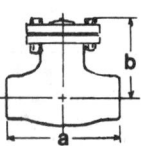

Fig. 4554-A
Socket Weld Ends
½″ to 2″

Fig. 4557-A
Buttwelding Ends
½″ to 2″

Figure 21-18   Check valve data *(Courtesy Crane-Aloyco Inc.)*

Pipe Drawings

Pipe drawings are used like any other design drawing, to convey the intentions of the designer and engineer to the trades people who will actually construct the piping system. There are two types of pipe drawings: single-line and double-line. *Figure 21-19* is an example of a double-line drawing. *Figure 21-20* is an example of the single-line version of the same drawing. Single-line drawings, since they are schematic, can be drawn much faster. However, drafters and engineers should be famil-

iar both with double-line and single-line versions of pipe drawings, since both have applications on the job.

Single-Line Drawings

Single-line pipe drawings consist of schematic, symbolic representations of valve fittings connected by a single line that represents the center line of the pipe. *Figure 21-21* contains all of the various schematic symbols used for representing pipe fittings and valves on single-line pipe drawings. *Figure 21-22* contains the graphic symbols used for representing piping on single-line pipe drawings.

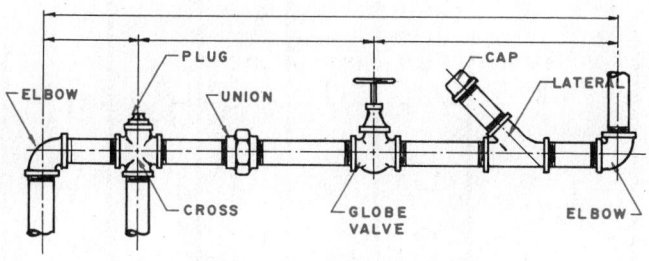

Figure 21-19   Double-line pipe drawing

Figure 21-20   Single-line pipe drawing

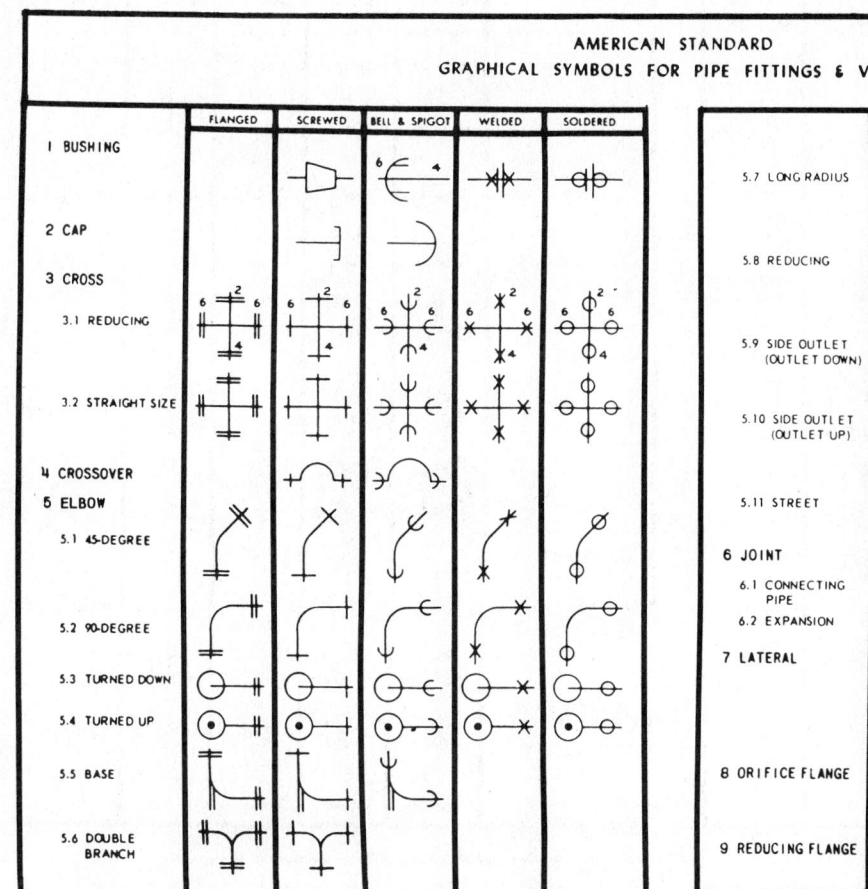
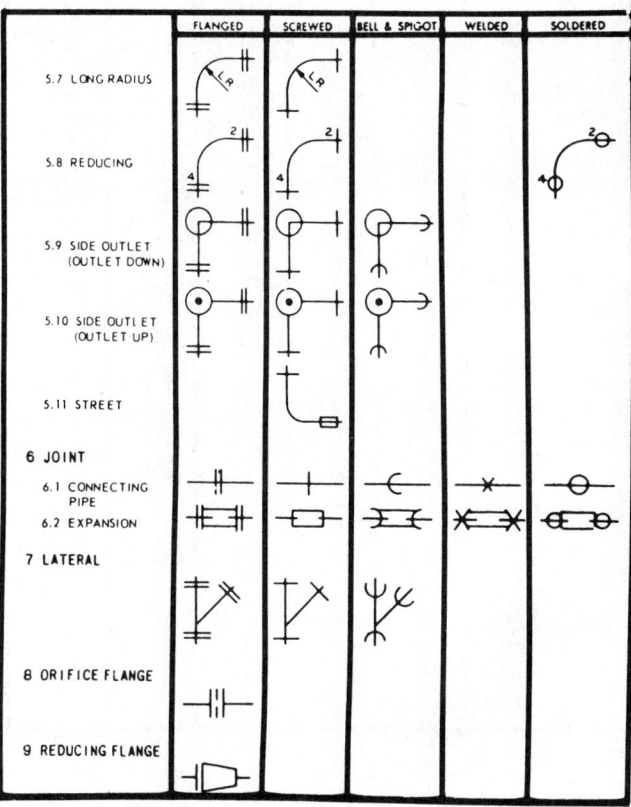

Figure 21-21   Symbols for pipe fittings and valves *(Reprinted from ASA Z32.2.3—1949. Courtesy ASME)*

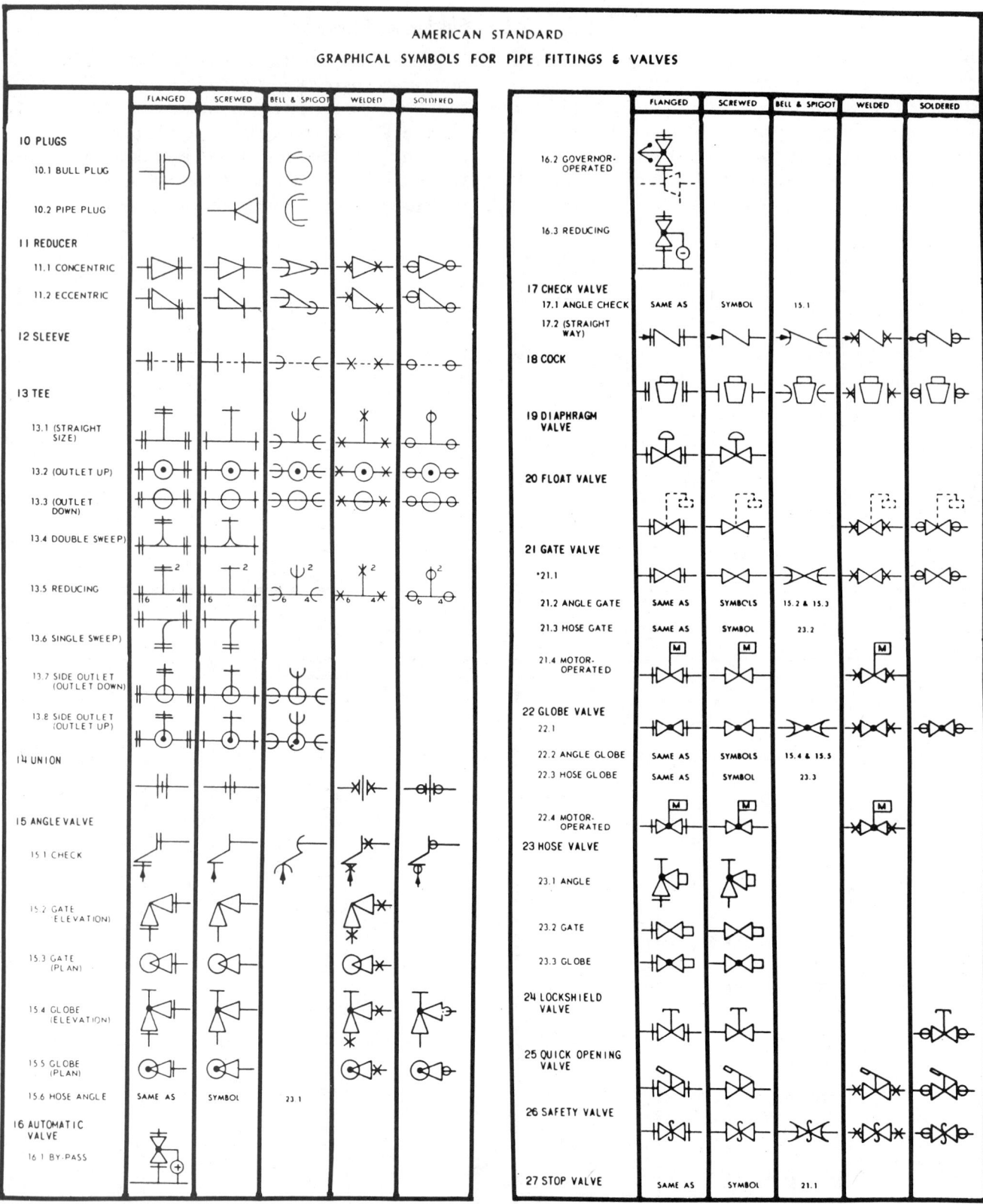

Figure 21-21 (Continued) Symbols for pipe fittings and valves (Reprinted from ASA Z32.2.3—1949. Courtesy ASME)

AMERICAN STANDARD
GRAPHICAL SYMBOLS FOR PIPING

AIR CONDITIONING
28 BRINE RETURN
29 BRINE SUPPLY
30 CIRCULATING CHILLED OR HOT-WATER FLOW
31 CIRCULATING CHILLED OR HOT-WATER RETURN
32 CONDENSER WATER FLOW
33 CONDENSER WATER RETURN
34 DRAIN
35 HUMIDIFICATION LINE
36 MAKE-UP WATER
37 REFRIGERANT DISCHARGE
38 REFRIGERANT LIQUID
39 REFRIGERANT SUCTION
HEATING
40 AIR-RELIEF LINE
41 BOILER BLOW OFF
42 COMPRESSED AIR
43 CONDENSATE OR VACUUM PUMP DISCHARGE
44 FEEDWATER PUMP DISCHARGE
45 FUEL-OIL FLOW
46 FUEL-OIL RETURN
47 FUEL-OIL TANK VENT
48 HIGH-PRESSURE RETURN
49 HIGH-PRESSURE STEAM
50 HOT-WATER HEATING RETURN
51 HOT-WATER HEATING SUPPLY

52 LOW-PRESSURE RETURN
53 LOW-PRESSURE STEAM
54 MAKE-UP WATER
55 MEDIUM PRESSURE RETURN
56 MEDIUM PRESSURE STEAM
PLUMBING
57 ACID WASTE
58 COLD WATER
59 COMPRESSED AIR
60 DRINKING-WATER FLOW
61 DRINKING-WATER RETURN
62 FIRE LINE
63 GAS
64 HOT WATER
65 HOT-WATER RETURN
66 SOIL, WASTE OR LEADER (ABOVE GRADE)
67 SOIL, WASTE OR LEADER (BELOW GRADE)
68 VACUUM CLEANING
69 VENT
PNEUMATIC TUBES
70 TUBE RUNS
SPRINKLERS
71 BRANCH AND HEAD
72 DRAIN
73 MAIN SUPPLIES

Figure 21-22   Symbols for piping *(Reprinted from ASA Z32.2.3—1949. Courtesy ASME)*

Like mechanical drawings, pipe drawings may be drawn using orthographic or isometric projection. Single-line orthographic projection is the recommended form of representation in most cases. ***Figure 21-23*** is an example of single-line orthographic projection of a pipe drawing. Single-line isometric projection, as illustrated in ***Figure 21-24***, is often used for assembly and layout work because the drawing is easier for trades people to understand. Regardless of whether you are using orthographic or isometric projection, there are certain drawing conventions with which you should be familiar. The most important of these are crossings, connections, fittings, and machine/devices.

Crossings.   When lines representing pipe on a drawing cross but do not intersect or make connections, they are usually drawn without breaks. However, on occasion it will be necessary to show breaks so that trades people using the plans will be absolutely sure as

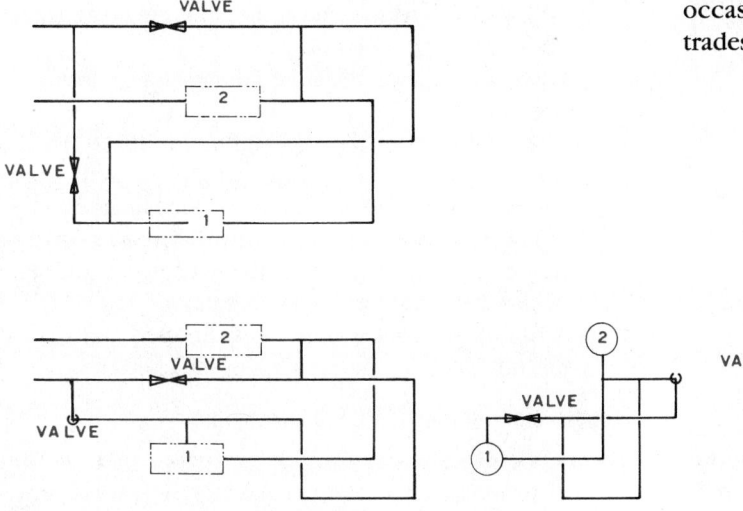

Figure 21-23   Orthographic single-line pipe drawing

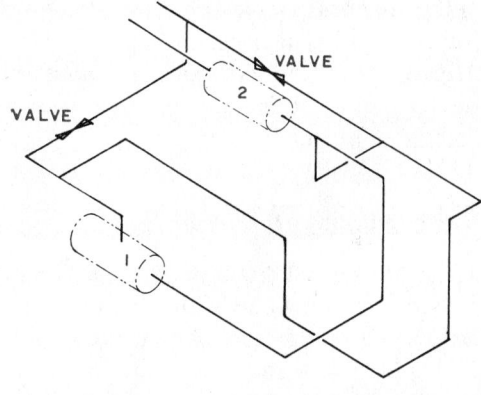

Figure 21-24   Isometric single-line pipe drawing

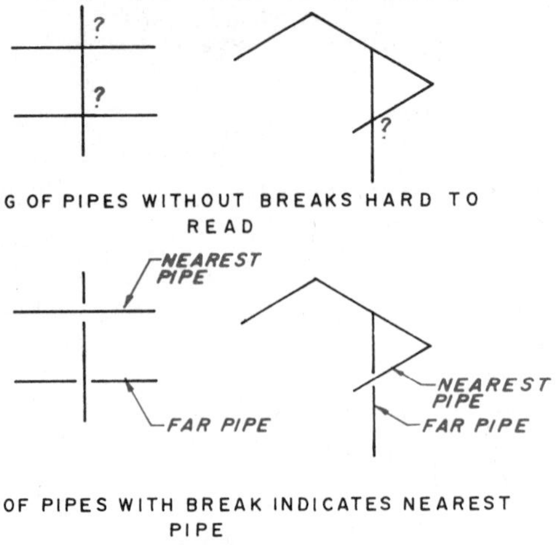

CROSSING OF PIPES WITHOUT BREAKS HARD TO READ

CROSSING OF PIPES WITH BREAK INDICATES NEAREST PIPE

Figure 21-25 Drawing crossing pipe

to which pipe is nearest to them and which is farthest away. When such a need arises, breaks can be used as illustrated in **Figure 21-25**. When using a break, it should be wide enough to be easily seen but not so wide as to create an inordinate gap. A widely used rule of thumb is to make the break from five to ten times the thickness of the lines representing the pipe.

Connections. All of the various means of making connections at joints and fittings fall into two categories: permanent and detachable. On single-line pipe drawings, permanent connections are indicated by using a heavy dot. Detachable connections are indicated by a single thick line. Both of these methods are illustrated in **Figure 21-26**.

Fittings. When drawing fittings on single-line drawings it is sometimes necessary to be able to indicate whether a pipe is coming toward the viewer or going away from the viewer. It is also necessary at times to indicate whether the viewer is looking at a front view of the fitting or a rear view. All of the various different types of single-line drawing situations that you may confront when drawing fittings are covered in Figure 21-21. Refer to Figure 21-21 when drawing any type of single-line fitting symbol.

Machines and Devices. Most piping systems will tie into machines, devices, and other types of apparatus, **Figure 21-27**. When this is the case, the machines, devices, and other types of apparatus can be drawn using thin phantom lines.

Dimensioning Pipe Drawings

There are several dimensioning rules that apply specifically to pipe drawings with which you should be familiar. The most important of these are summarized below.

- Dimensions for pipes and pipe fittings should be shown from center to center of pipe and to the outside face of the pipe end or flange.

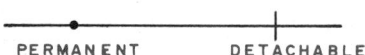

Figure 21-26 Drawing pipe connections

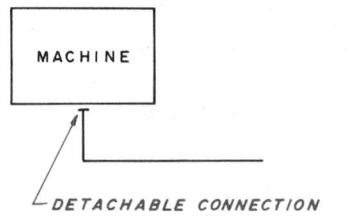

Figure 21-27 Drawing adjoining machinery

- The length of pipe is not shown on the drawing except in rare cases.
- Pipe sizes are shown on the drawing using leader lines or as a general note.
- Fitting sizes are shown on the drawing using leader lines or as a general note.
- Pipes with bends should be dimensioned from vertex to vertex.
- The radii of bent pipe should be indicated using a leader line. Supplementary angles of bent pipes should be dimensioned in the normal manner.
- Outside diameters and wall thicknesses of pipe may be indicated on the drawing using a leader line or may be made part of a general note.
- A bill of material should be developed and placed directly on the pipe drawing or attached to it.

Single-Line Isometric Piping Drawings

Isometric piping drawings are prepared in a manner similar to mechanical piping drawings. The lines representing pipes are drawn parallel to either the X,Y, or Z axis, **Figure 21-28**. Flanges are represented using short

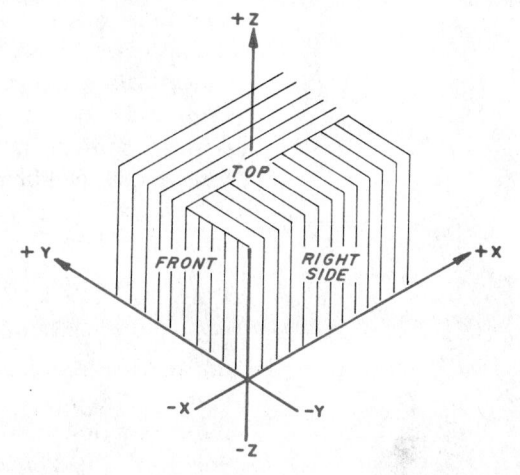

Figure 21-28 Isometric axes

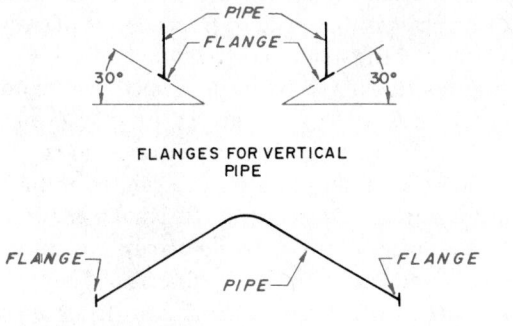

Figure 21-29 Drawing flanges

strokes of equal thickness, **Figure 21-29**. Notice that flanges for horizontal pipe are drawn vertically and flanges for vertical pipe are drawn at the appropriate 30° angle to the horizontal.

Figure 21-30 illustrates the various methods used for drawing valves. The unidirectional dimensioning illustrated in **Figure 21-31** should be used in isometric piping drawings.

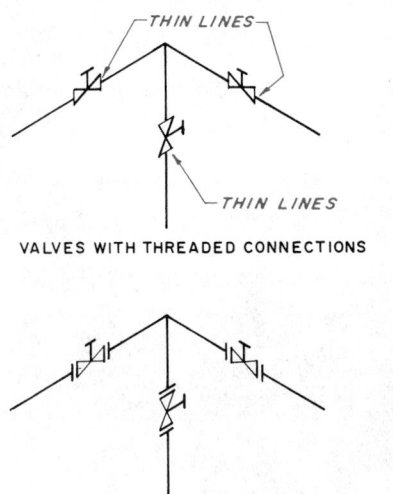

Figure 21-30 Drawing valves

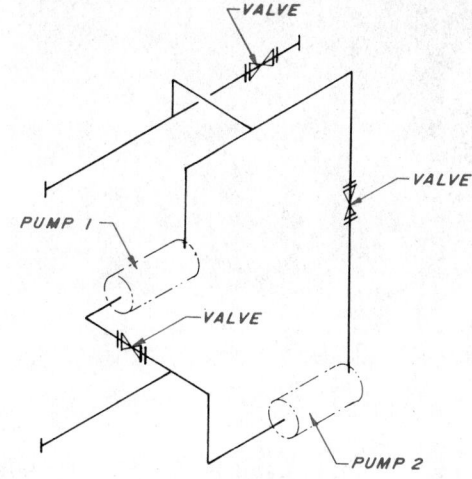

Figure 21-31 Isometric pipe drawing

Structural Drafting

A *structure* is anything constructed of parts. In heavy construction this encompasses commercial buildings, industrial buildings, bridges, towers, parking decks, stadiums, and numerous other types of structures. **Figures 21-32** and **21-33** are examples of modern structures. Structures such as those shown in these figures are constructed of many parts. The overall structure must be designed and the design must be documented by engineering drawings. In addition, each individual part must be designed and the design must be documented by shop drawings. Structural drafting consists of preparing the documentation for the design of a structure so that the individual parts can be manufactured and the overall structure erected.

Documentation prepared by structural drafters consists primarily of engineering drawings and shop drawings. Engineering drawings convey structural design information (i.e., what types of products are to be used and in what sizes, how the various parts fit together, how the individual parts are connected, etc.). They con-

Figure 21-32 Cleveland state office building *(Courtesy Bethlehem Steel Corp.)*

sometimes referred to as detail drawings. They consist of the necessary views of individual members with sectional views added when necessary and a bill of material. *Figure 21-35* is an example of a structural steel shop drawing showing all the information necessary to fabricate several steel beams. It also contains a bill of material showing all of the various materials needed to fabricate the beams shown on this shop drawing.

Drawing, Checking, Correcting, and Revising

Structural drafters prepare original drawings—engineering and/or shop drawing—using either CAD or manual drafting techniques. The drawings are then checked by a more experienced drafter, known as a checker, and an engineer. Checkers and engineers make sure the drawings conform to the specifications of the original material used to prepare them and that the drafter has correctly interpreted engineering calculations/sketches/instructions. Corrections that need to be made are marked on a checkprint and the drafter corrects the original.

Drawings that have been produced, checked, and corrected must also be changed on occasion. Such changes are known as revisions. A revision is typically required as the result of an engineering or architectural change whereas a correction is usually made as the result of a drafting error. Revisions are typically specified on a document called an Engineering Change Order (ECO). *Figure 21-36* is an example of an ECO. Normally the cost of making revisions is charged against the company or department that ordered them.

Structural Products

The products that become the individual components of a structure are standardized for the most part. They are steel and concrete and, to a lesser extent, wood products. *Figure 21-37* illustrates the most widely used structural steel shapes. They are W shapes, S shapes, M shapes, HP shapes, C shapes, MC shapes, equal-legged angles, unequal-legged angles, square tubing, rectangular tubing, circular tubing, pipe, and structural tees.

The various steel shapes shown in Figure 21-37 may be used to form the vertical and horizontal members in the structural skeleton of a building, stadium, tower, or

sist of various types of framing plans (i.e., anchor bolt, column, beam, floor, and roof framing plans); sections; and connection details. *Figure 21-34* is an example of an engineering drawing. It is a framing plan showing the placement of steel beams and joists for a commercial building. The original material used for preparing structural engineering drawings is typically a set of architectural plans and/or a set of engineering sketches and calculations.

Shop drawings convey all of the information needed to manufacture the individual parts of a structure and are

Figure 21-33 Rock Island municipal parking structure *(Courtesy Prestressed Concrete Institute)*

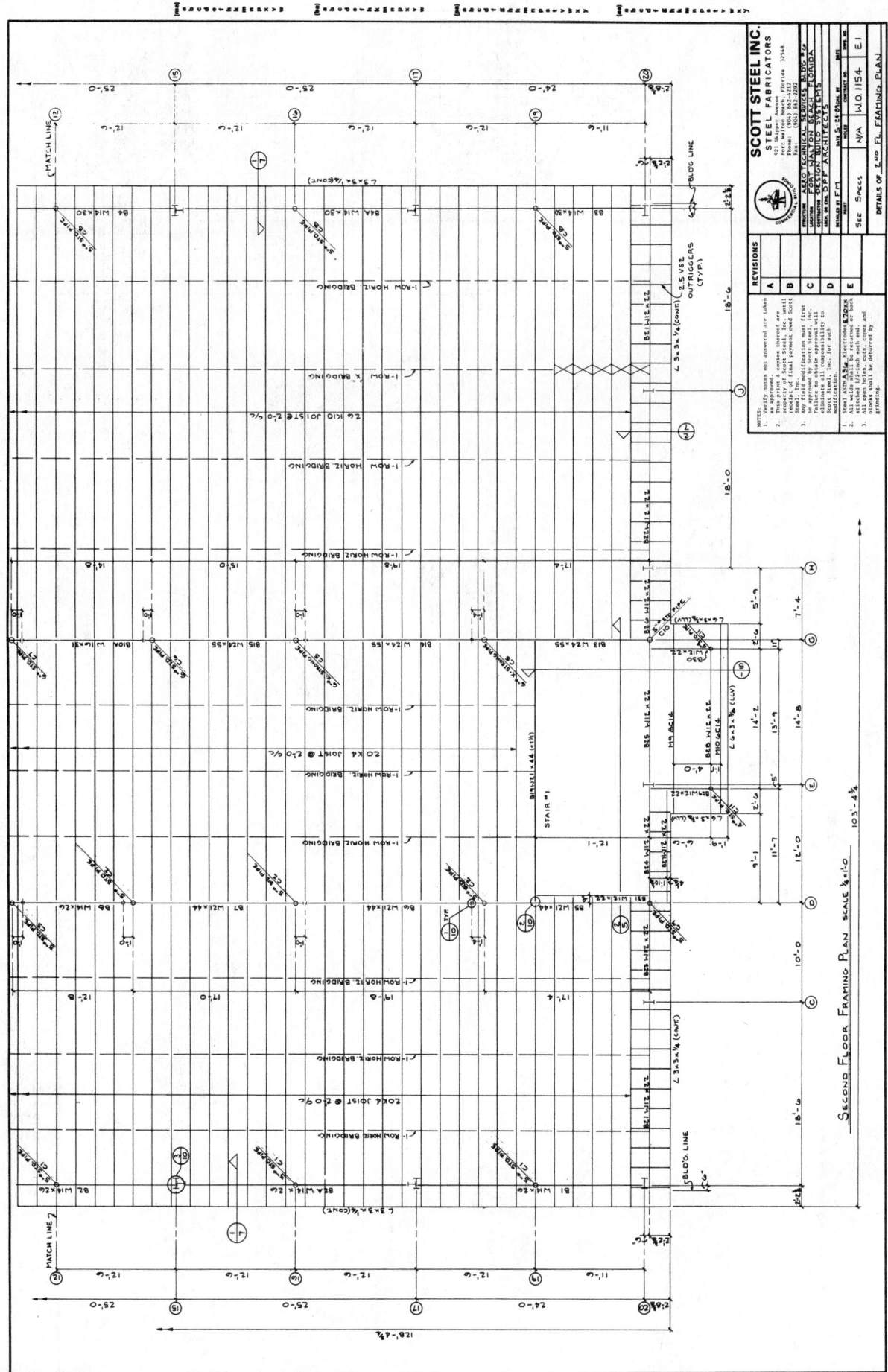

Figure 21-34 Engineering drawing *(Courtesy Scott Steel Inc.)*

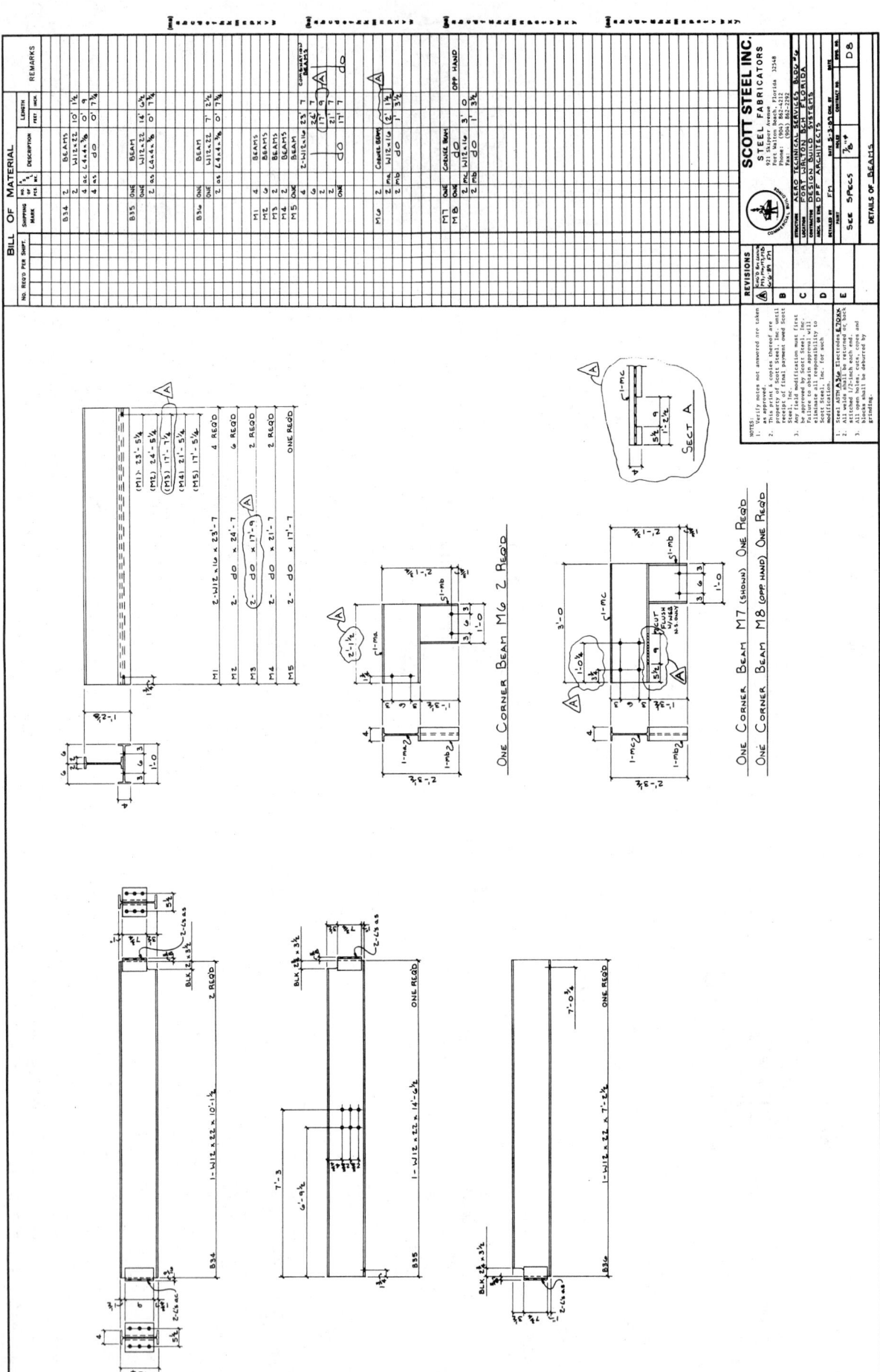

Figure 21-35 Structural steel shop drawing *(Courtesy Scott Steel Inc.)*

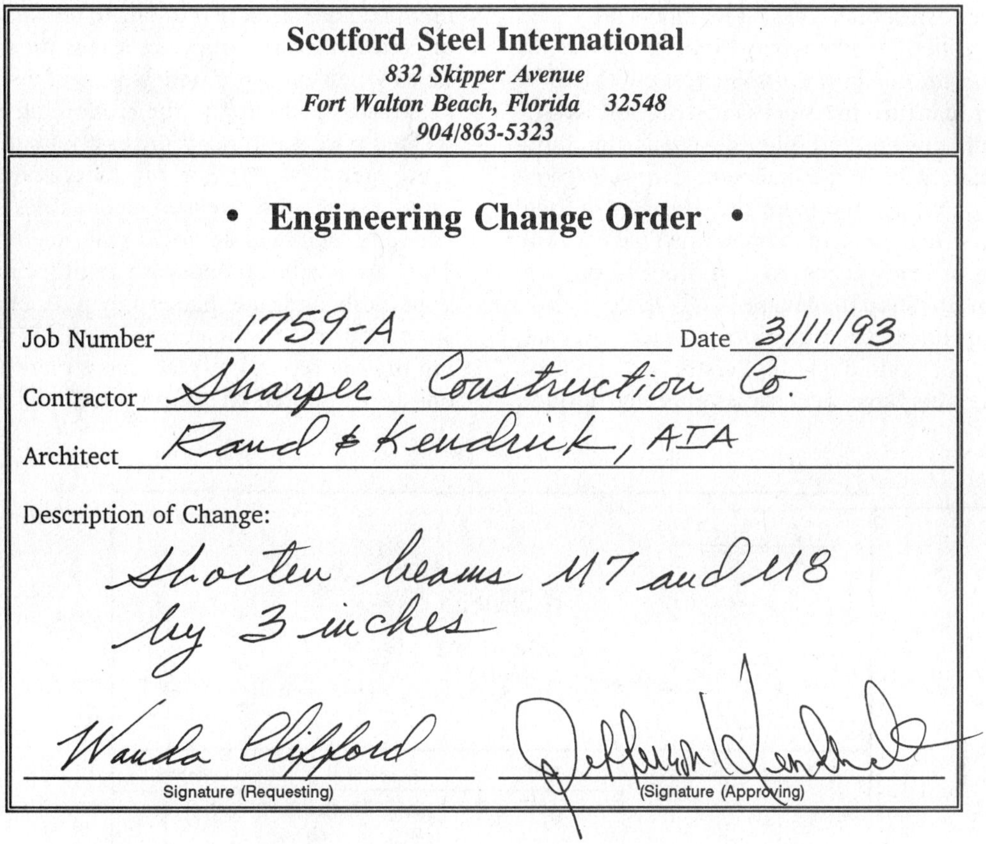

Scotford Steel International
832 Skipper Avenue
Fort Walton Beach, Florida   32548
904/863-5323

• Engineering Change Order •

Job Number __*1759-A*__   Date __*3/11/93*__

Contractor __*Sharper Construction Co.*__

Architect __*Rand & Kendrick, AIA*__

Description of Change:

*Shorten beams M7 and M8
by 3 inches*

__*Wanda Clifford*__   __*Jefferson Kendrick*__
Signature (Requesting)   (Signature (Approving)

Figure 21-36   Engineering drawing *(Courtesy Scott Steel Inc.)*

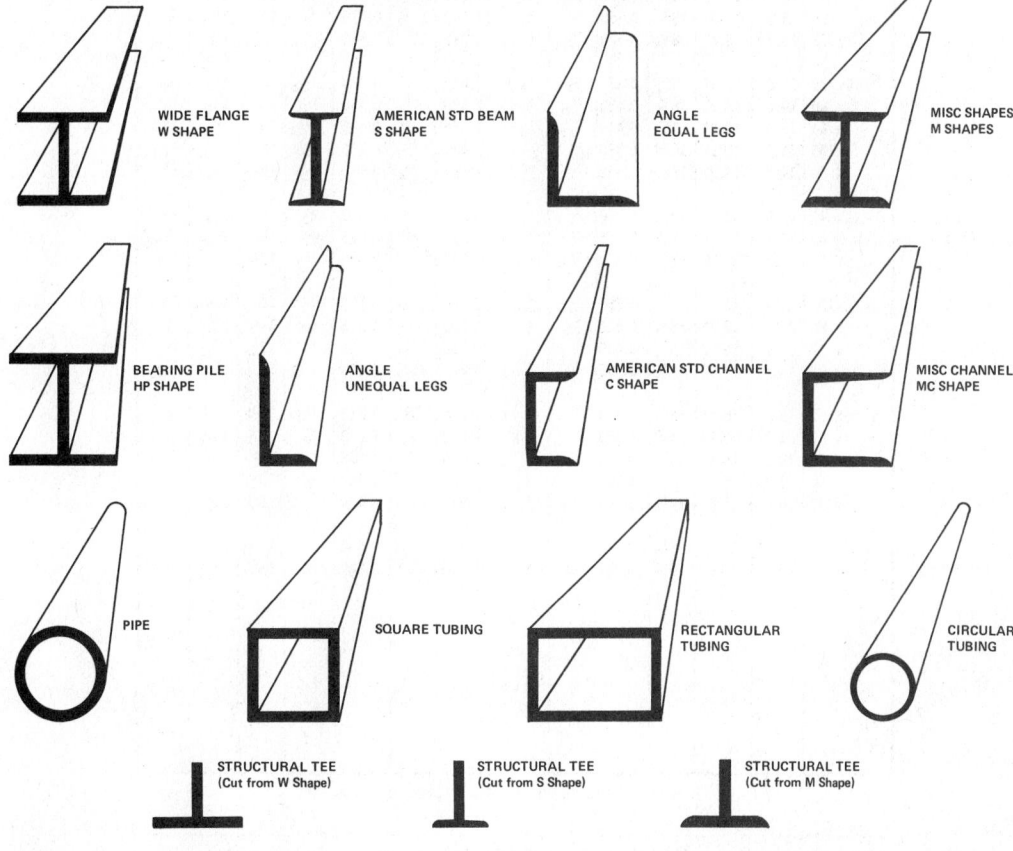

Figure 21-37   Structural steel shapes *(From Structural Drafting by Goetsch, Delmar Publishers Inc.)*

any other type of structure. Vertical members are called columns. Horizontal members may be beams, joists, or girders. The *Manual of Steel Construction* published by the American Institute for Steel Construction (AISC) contains the dimensions and other design and drafting-related information for structural steel shapes. *Figure 21-38* is a page excerpted from the *Manual of Steel Construction.* Structural steel drafters must have a copy of this manual or ready access to it in order to prepare engineering and/or shop drawings.

Concrete products are either prestressed, precast, post-tensioned, or poured-on-site. Prestressed and post-tensioned products have steel cables running through them that are stretched to create tension that is equal to or greater than the opposite forces they will be subjected to when incorporated as part of a structure. These cables are in addition to the reinforcing bars, plates, and welded wire mesh normally cast into all structural concrete members. Precast products contain all of these items except the stressed steel cables. Poured-on-site concrete members do not contain the tensioned cables. They are strengthened with reinforcing bars, welded wire mesh, and steel plates only. Precast and poured-on-site members are the same except for where the pouring of concrete takes place; in a plant for precast and on-site for poured-on-site products.

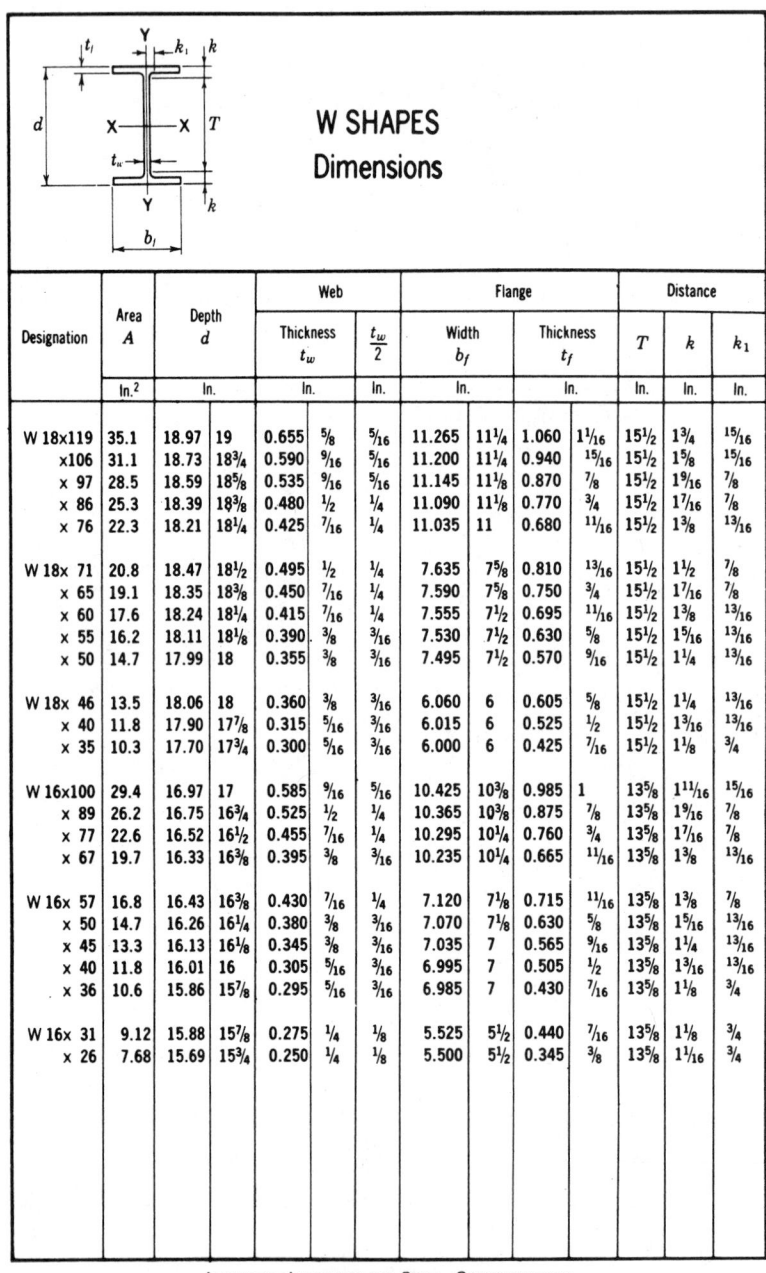

Designation	Area A	Depth d		Web			Flange				Distance		
				Thickness t_w	$\frac{t_w}{2}$		Width b_f		Thickness t_f		T	k	k_1
	In.²	In.		In.	In.		In.		In.		In.	In.	In.
W 18x119	35.1	18.97	19	0.655	⅝	⁵⁄₁₆	11.265	11¼	1.060	1¹⁄₁₆	15½	1¾	¹⁵⁄₁₆
x106	31.1	18.73	18¾	0.590	⁹⁄₁₆	⁵⁄₁₆	11.200	11¼	0.940	¹⁵⁄₁₆	15½	1⅝	¹⁵⁄₁₆
x 97	28.5	18.59	18⅝	0.535	⁹⁄₁₆	⁵⁄₁₆	11.145	11⅛	0.870	⅞	15½	1⁹⁄₁₆	⅞
x 86	25.3	18.39	18⅜	0.480	½	¼	11.090	11⅛	0.770	¾	15½	1⁷⁄₁₆	⅞
x 76	22.3	18.21	18¼	0.425	⁷⁄₁₆	¼	11.035	11	0.680	¹¹⁄₁₆	15½	1⅜	¹³⁄₁₆
W 18x 71	20.8	18.47	18½	0.495	½	¼	7.635	7⅝	0.810	¹³⁄₁₆	15½	1½	⅞
x 65	19.1	18.35	18⅜	0.450	⁷⁄₁₆	¼	7.590	7⅝	0.750	¾	15½	1⁷⁄₁₆	⅞
x 60	17.6	18.24	18¼	0.415	⁷⁄₁₆	¼	7.555	7½	0.695	¹¹⁄₁₆	15½	1⅜	¹³⁄₁₆
x 55	16.2	18.11	18⅛	0.390	⅜	³⁄₁₆	7.530	7½	0.630	⅝	15½	1⁵⁄₁₆	¹³⁄₁₆
x 50	14.7	17.99	18	0.355	⅜	³⁄₁₆	7.495	7½	0.570	⁹⁄₁₆	15½	1¼	¹³⁄₁₆
W 18x 46	13.5	18.06	18	0.360	⅜	³⁄₁₆	6.060	6	0.605	⅝	15½	1¼	¹³⁄₁₆
x 40	11.8	17.90	17⅞	0.315	⁵⁄₁₆	³⁄₁₆	6.015	6	0.525	½	15½	1³⁄₁₆	¹³⁄₁₆
x 35	10.3	17.70	17¾	0.300	⁵⁄₁₆	³⁄₁₆	6.000	6	0.425	⁷⁄₁₆	15½	1⅛	¾
W 16x100	29.4	16.97	17	0.585	⁹⁄₁₆	⁵⁄₁₆	10.425	10⅜	0.985	1	13⅝	1¹¹⁄₁₆	¹⁵⁄₁₆
x 89	26.2	16.75	16¾	0.525	½	¼	10.365	10⅜	0.875	⅞	13⅝	1⁹⁄₁₆	⅞
x 77	22.6	16.52	16½	0.455	⁷⁄₁₆	¼	10.295	10¼	0.760	¾	13⅝	1⁷⁄₁₆	⅞
x 67	19.7	16.33	16⅜	0.395	⅜	³⁄₁₆	10.235	10¼	0.665	¹¹⁄₁₆	13⅝	1⅜	¹³⁄₁₆
W 16x 57	16.8	16.43	16⅜	0.430	⁷⁄₁₆	¼	7.120	7⅛	0.715	¹¹⁄₁₆	13⅝	1⅜	⅞
x 50	14.7	16.26	16¼	0.380	⅜	³⁄₁₆	7.070	7⅛	0.630	⅝	13⅝	1⁵⁄₁₆	¹³⁄₁₆
x 45	13.3	16.13	16⅛	0.345	⅜	³⁄₁₆	7.035	7	0.565	⁹⁄₁₆	13⅝	1¼	¹³⁄₁₆
x 40	11.8	16.01	16	0.305	⁵⁄₁₆	³⁄₁₆	6.995	7	0.505	½	13⅝	1³⁄₁₆	¹³⁄₁₆
x 36	10.6	15.86	15⅞	0.295	⁵⁄₁₆	³⁄₁₆	6.985	7	0.430	⁷⁄₁₆	13⅝	1⅛	¾
W 16x 31	9.12	15.88	15⅞	0.275	¼	⅛	5.525	5½	0.440	⁷⁄₁₆	13⅝	1⅛	¾
x 26	7.68	15.69	15¾	0.250	¼	⅛	5.500	5½	0.345	⅜	13⅝	1¹⁄₁₆	¾

AMERICAN INSTITUTE OF STEEL CONSTRUCTION

Figure 21-38   Page from the Manual of Steel Construction *(Courtesy American Institute of Steel Construction)*

With prestressed concrete products, the steel cables are tensioned before the concrete is poured into the form or before it is extruded in the case of cored flat slabs. With post-tensioned products the cables are placed in the form encased in a special conduit and tensioned after the concrete has cured.

Figures 21-39 and *21-40* illustrate some of the more widely used prestressed and post-tensioned concrete products. They include double-tee members, single-tee members, solid flat slabs, cored flat slabs, L-shaped beams, inverted T-shaped beams, rectangular beams, tee joists, and keystone joists.

Connecting Structural Members

The two most widely used methods for connecting individual structural members are bolting and welding. Structural steel construction uses both widely. Concrete construction relies more on welding, but does use some bolted connections. The information contained in Chapter 13 applies to the bolts that are used to fasten structural members. *Figure 21-41* shows a structural steel pipe column that will be bolted at the bottom (base plate) and at the top (top plate).

The information structural drafters need to know about welding and the proper use of welding symbols is covered in Chapter 19. *Figure 21-42* shows both steel and prestressed concrete members connected by welding. *Figure 21-43* shows the welding types and their corresponding symbols as approved by the American Institute of Steel Construction (AISC).

Structural Steel Drawings

The point was made earlier that structural steel drawings consist of both engineering drawings and shop or detail drawings. Engineering drawings are used to communicate what products are to be used; in what sizes and shapes; and how they are to be spaced, connected, and otherwise put together to form a structure. Engi-

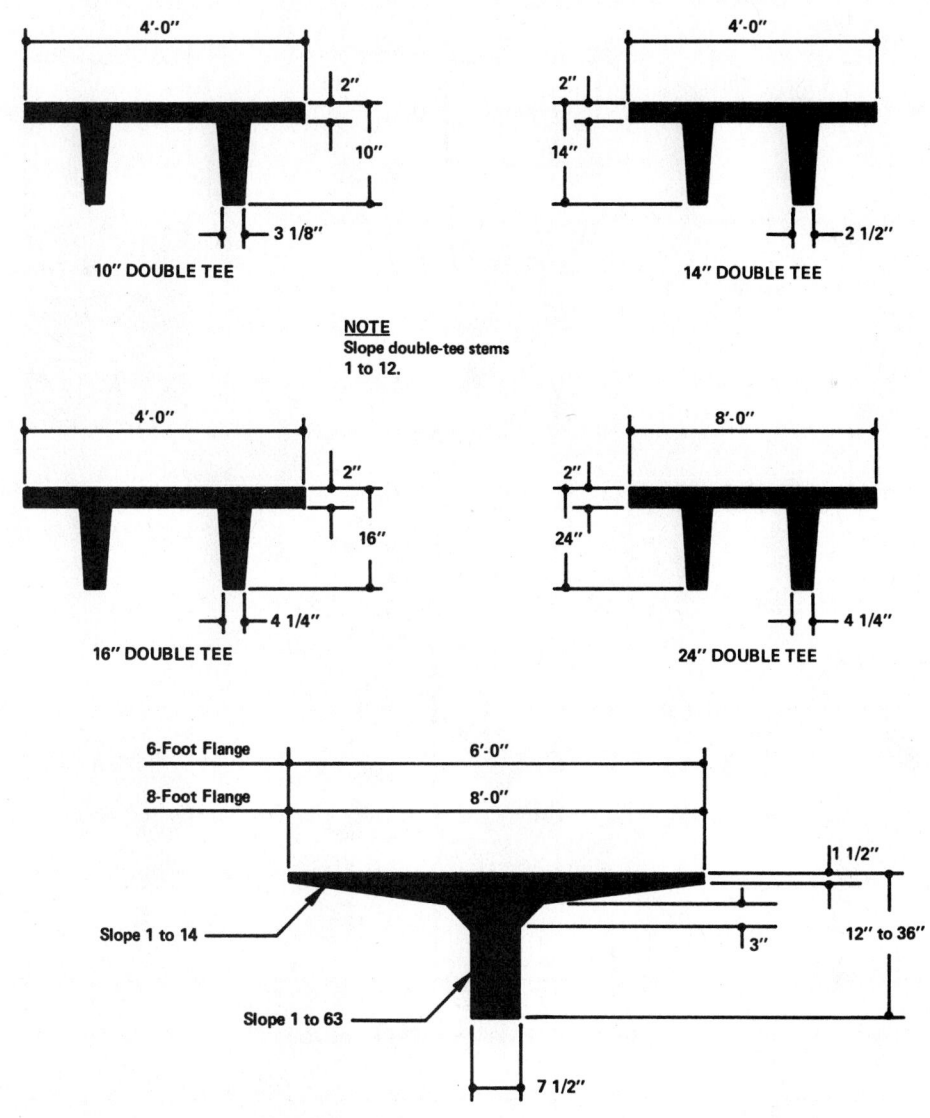

Figure 21-39   Standard precast concrete products *(From Structural Drafting by Goetsch, Delmar Publishers Inc.)*

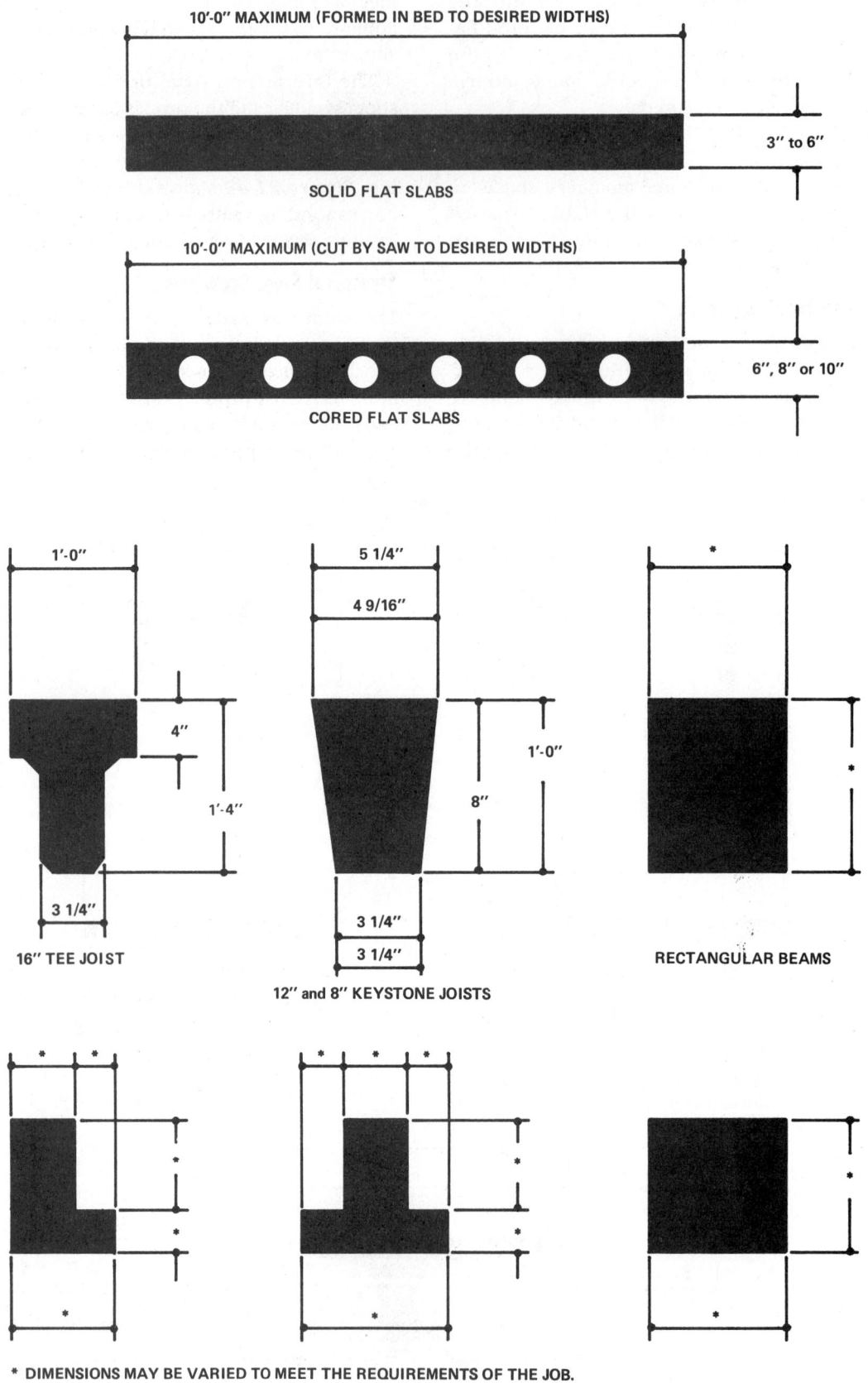

10'-0" MAXIMUM (FORMED IN BED TO DESIRED WIDTHS)

3" to 6"

SOLID FLAT SLABS

10'-0" MAXIMUM (CUT BY SAW TO DESIRED WIDTHS)

6", 8" or 10"

CORED FLAT SLABS

1'-0"

4"

1'-4"

3 1/4"

16" TEE JOIST

5 1/4"

4 9/16"

1'-0"

8"

3 1/4"

3 1/4"

12" and 8" KEYSTONE JOISTS

*

*

*

RECTANGULAR BEAMS

*

*

*

*

*

*

*

*

*

*

*

*

*

*

*

* DIMENSIONS MAY BE VARIED TO MEET THE REQUIREMENTS OF THE JOB.

Figure 21-40   Standard precast concrete products *(From* Structural Drafting *by Goetsch, Delmar Publishers Inc.)*

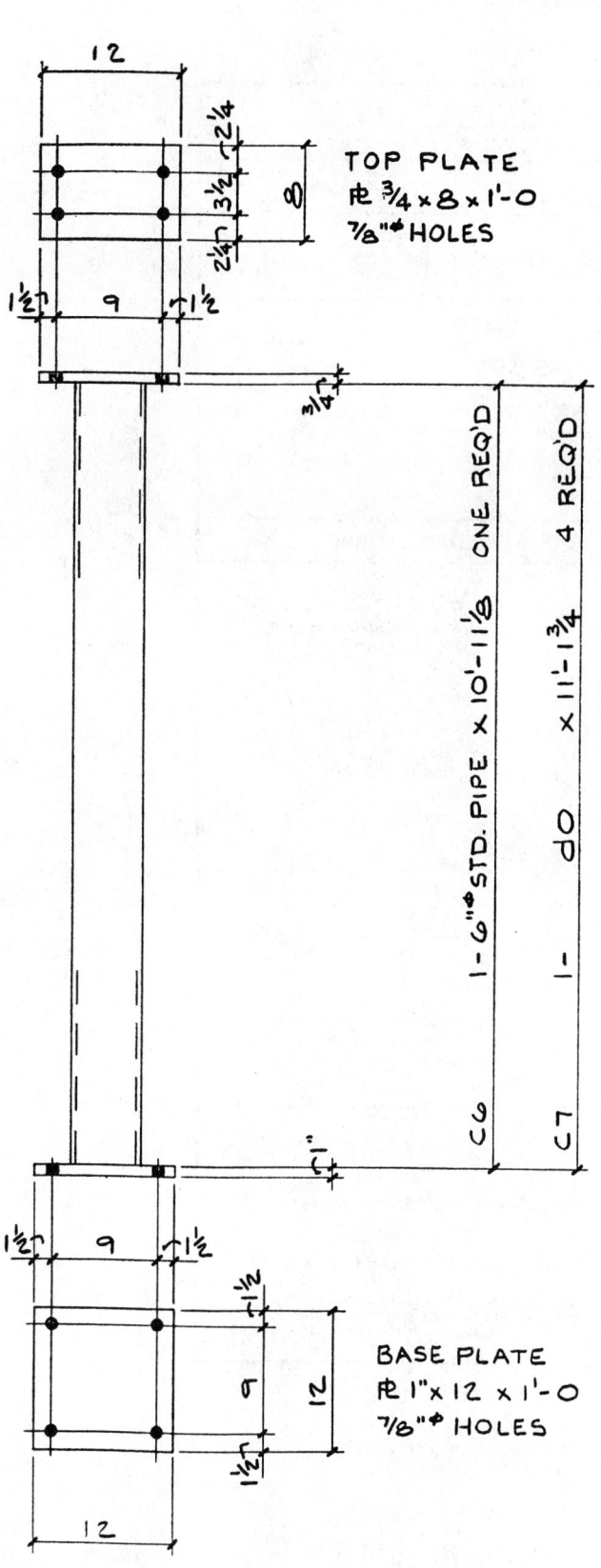

Figure 21-41 Structural steel pipe fabricated as a column *(Courtesy Scott Steel Inc.)*

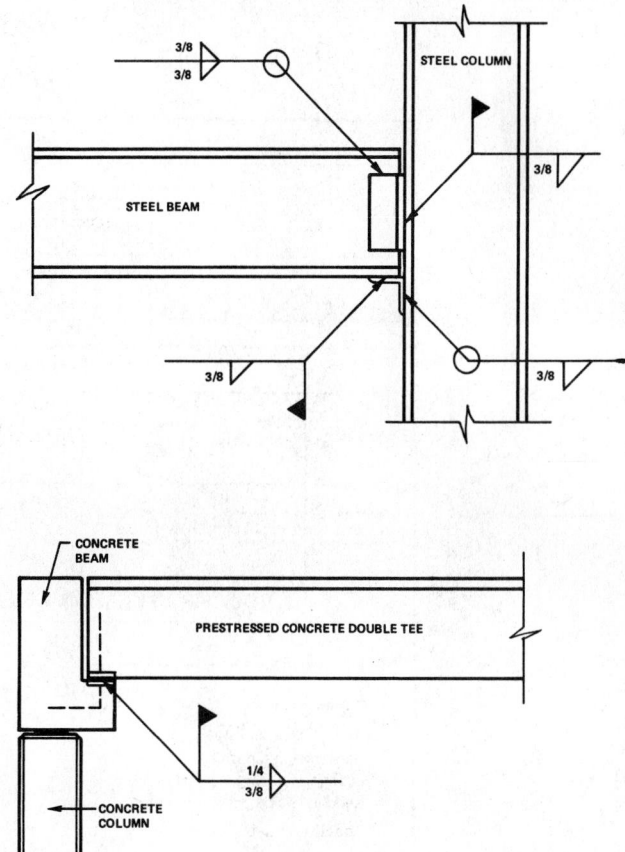

Figure 21-42 Welded connections *(From Structural Drafting by Goetsch, Delmar Publishers Inc.)*

neering drawings for structural steel construction consist of framing plans, sections (as needed), and connection details. ***Figure 21-44*** is an example of a second floor framing plan for a commercial building. ***Figure 21-45*** is an example of connection details for such a building.

Shop or detail drawings are used to fabricate the individual structural members that are laid out on the engineering drawings. They are very detailed and typically are accompanied by a bill of material. ***Figure 21-46*** is a shop drawing showing the fabrication details for several structural steel beams. The shop drawing contains a bill of material listing all of the material needed to fabricate each beam.

Structural Concrete Drawings

Structural concrete products may be poured on-site, precast, prestressed, or post-tensioned. Regardless of the approach, engineering and shop drawings are needed. In the case of concrete members that are poured on site, shop drawings are known as *placing drawings*. Placing drawings show the placement of reinforcing bars in the on-site form. Shop drawings and placing drawings serve the same purpose. The difference is in where the work occurs, in a shop or on-site.

WELDED JOINTS
Standard symbols

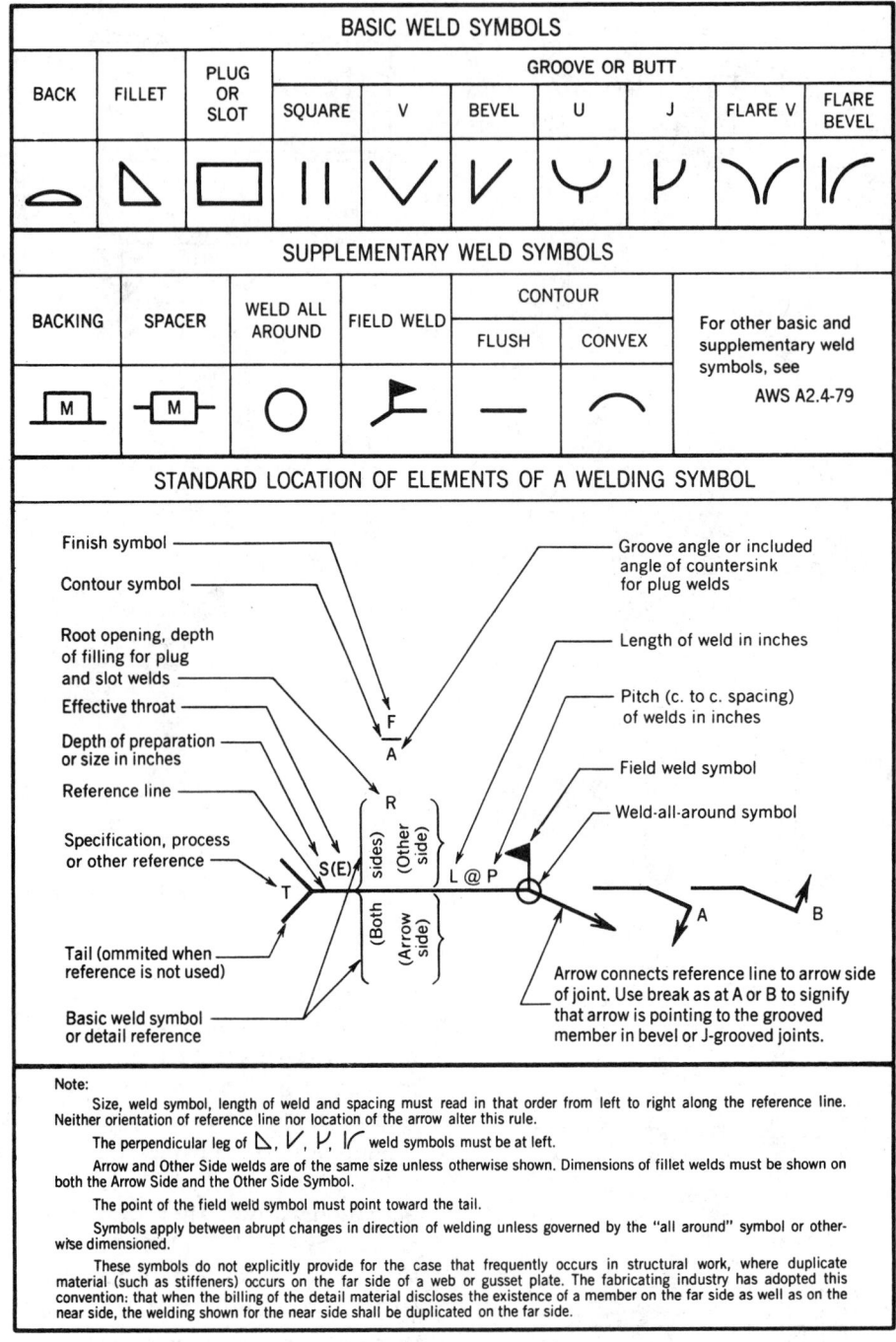

Figure 21-43   Welding types of symbols *(Courtesy American Institute of Steel Construction)*

Engineering drawings are used to communicate what products are to be used, in what sizes and shapes, and how they are to be spaced, connected, and otherwise put together to form a structure. Engineering drawings for structural concrete consist of framing plans, sections, (as needed), and connection details. *Figure 21-47* is an example of an engineering drawing for a small industrial building. The drawing contains a roof framing plan and sections. The sections also double as connection details. The roof is to be constructed of 4'0" wide cored flat slabs.

Shop and/or placement drawings are used to fabricate the individual structural members that are laid out on the engineering drawings. They are very detailed and

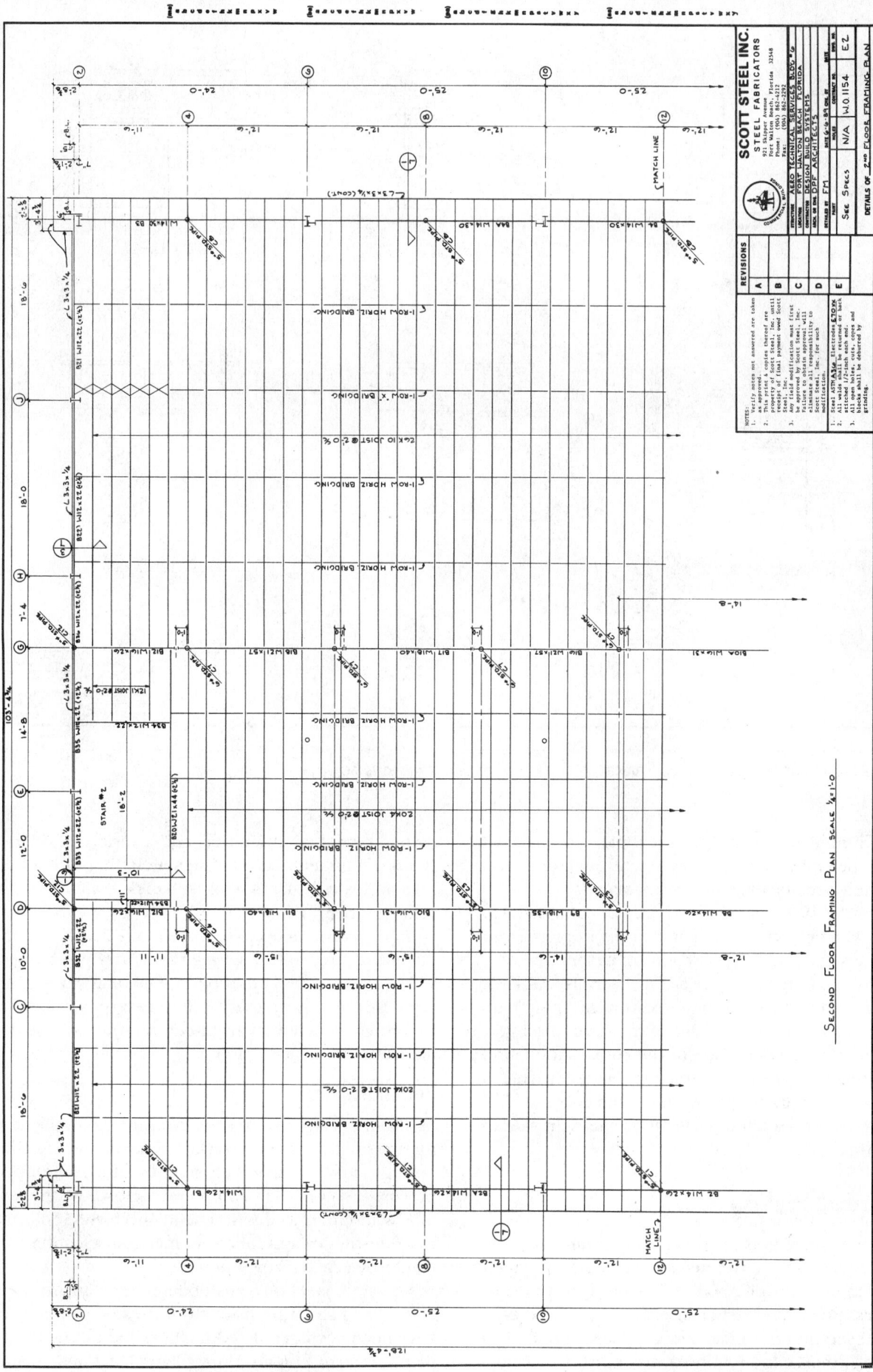

Figure 21-44   Second floor framing plan *(Courtesy Scott Steel Inc.)*

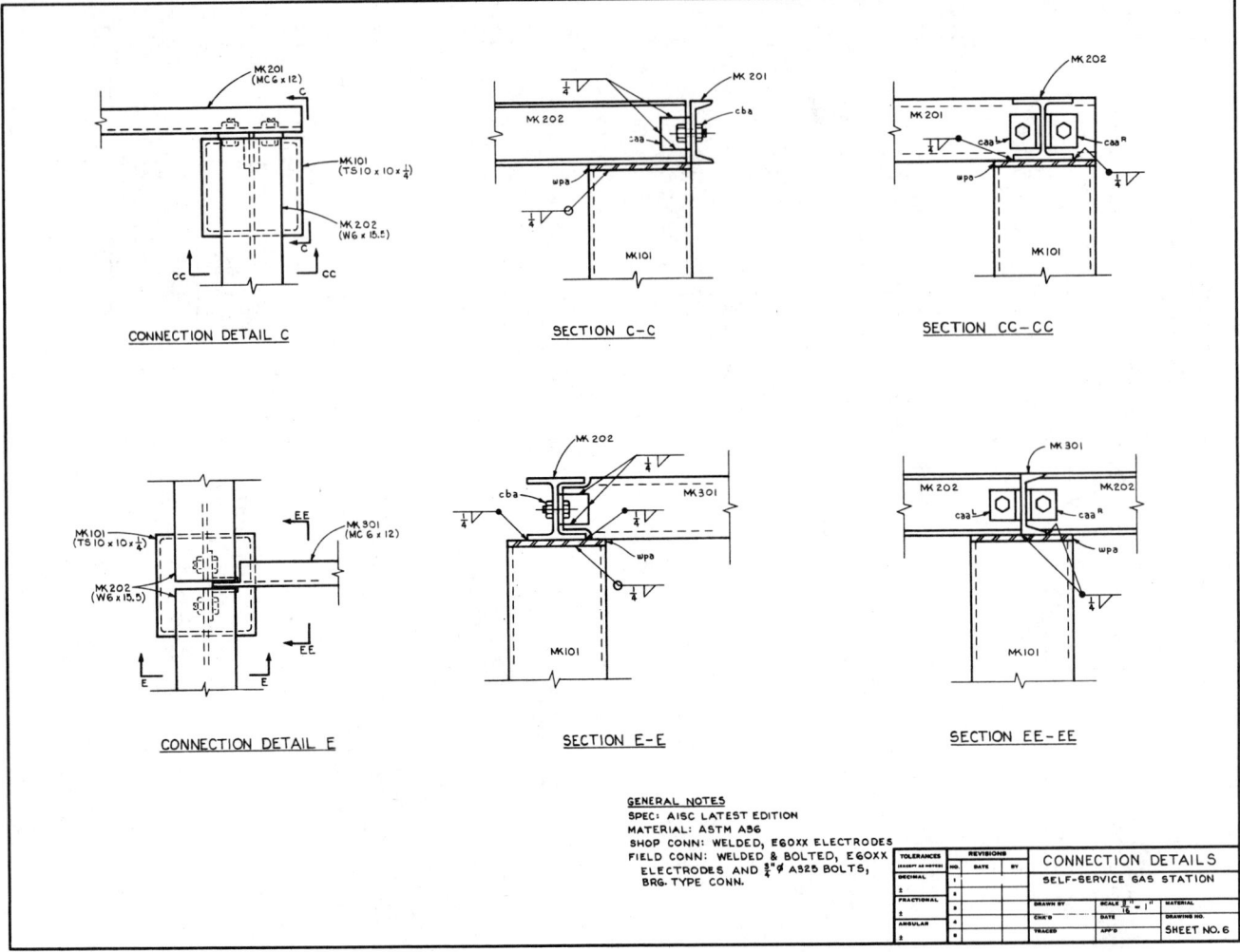

Figure 21-45   Connection details *(From Structural Drafting by Goetsch, Delmar Publishers Inc.)*

must communicate every individual piece of material that is to be embedded in the concrete. Such drawings are typically accompanied by a bill of material.

Figure 21-48 is a shop drawing for a solid flat slab member with one side beveled. It contains 17 prestressed steel strands that run lengthwise and are wrapped by reinforcing bars spaced lengthwise at 1'0" intervals (MK443). There are lifting hooks made of bent reinforcing bars (lha) for lifting the member out of the form, placing it on a truck for transportation to the construction site, and erecting it as part of the structure. Plates for welded connections (swpl) are placed along the edges at 7'0" intervals. There is a bill of material on the right side of the drawing.

Architectural Drafting

Architectural drafting is the process of documenting the designs developed by architects. Architects are called on to plan and design a variety of different types of structures. However, most of an architect's work is typically focused on residential and commercial buildings including single-family houses, town houses, apart-ments, model communities, and a variety of different types of commercial buildings including office buildings, industrial plants, stores, banks, malls, hospitals, schools, libraries, and so on.

Architectural drafters prepare plans that convey and document an architect's design for any of these types of structures as well as other types. The best way to understand this field is to divide it into two broad categories as follows: residential architectural drafting and commercial architectural drafting.

Residential Plans

Residential design involves the development of the design for a residential structure (i.e., single-family dwelling, cottage, town house, apartments, etc.). The design process involves such considerations as how the structure will fit into its environment, space and room planning, aesthetics and function, material selection, code compliance, and many others.

A complete set of plans documenting an architect's design and showing how all of these considerations have been accommodated includes the following: plot plan; foundation plan; floor plan; elevations; heating,

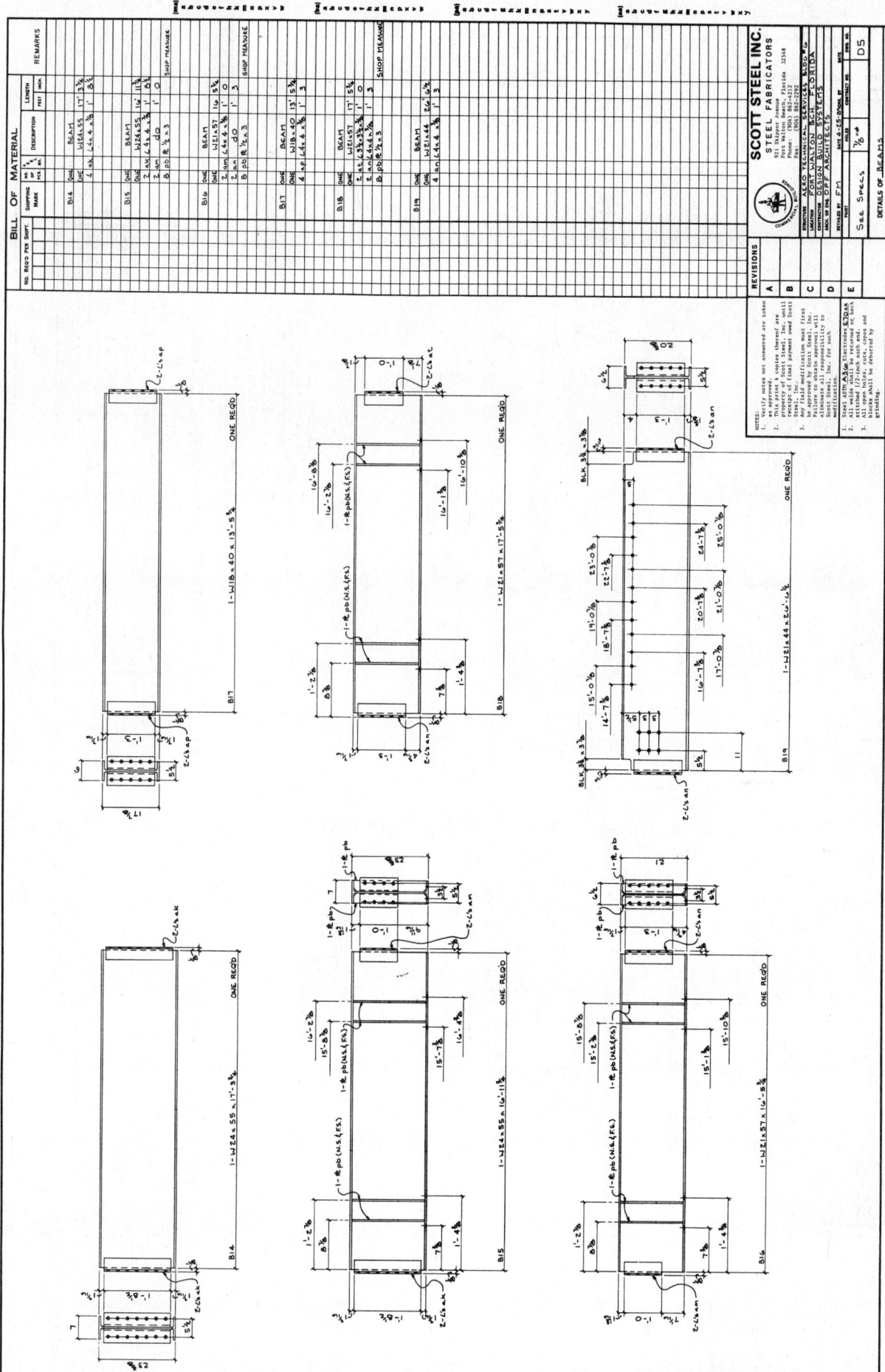

Figure 21-46 Structural steel shop drawing *(Courtesy Scott Steel Inc.)*

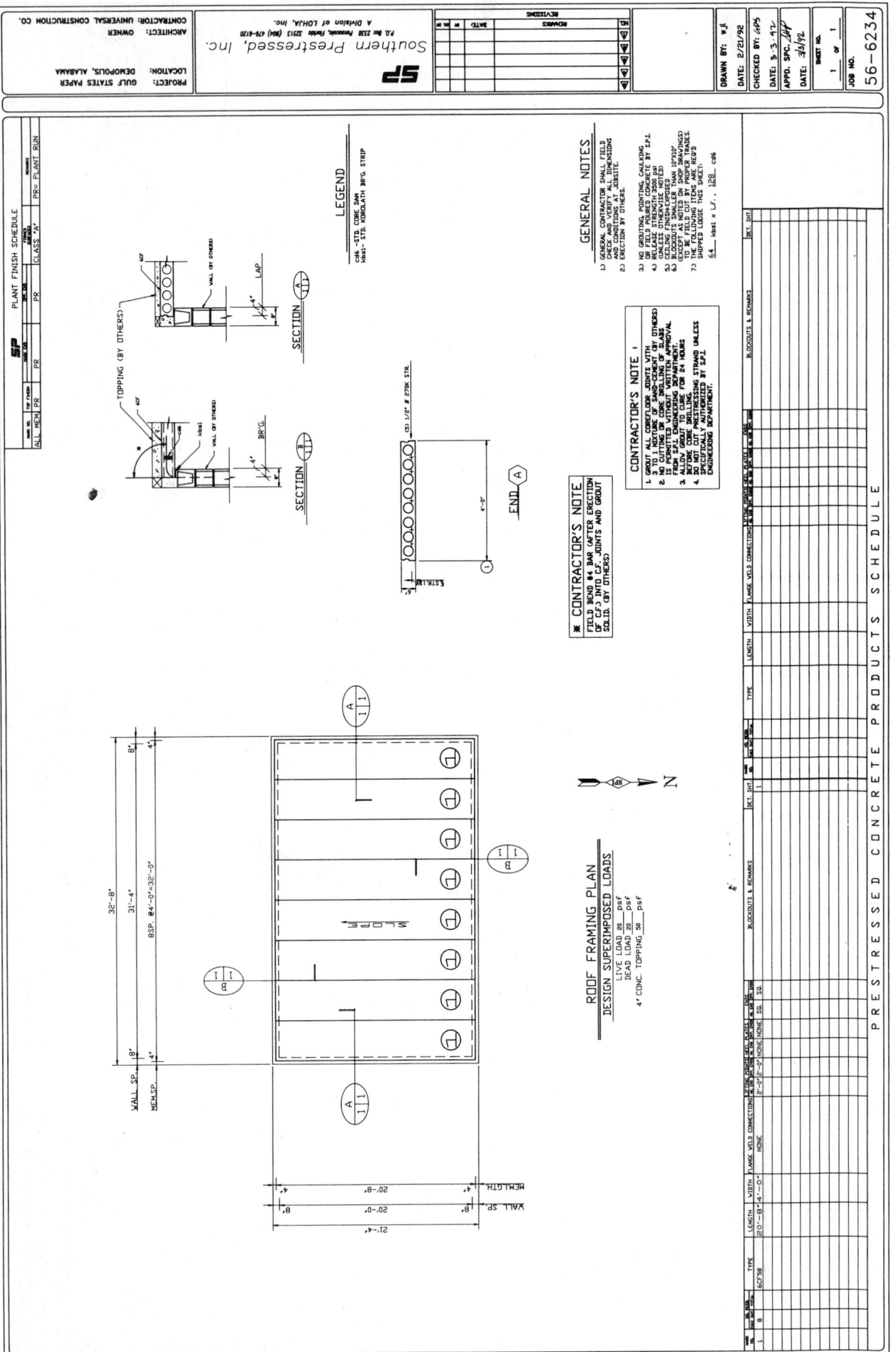

Figure 21-47 Engineering drawing (*Courtesy Southern Prestressed Inc.*)

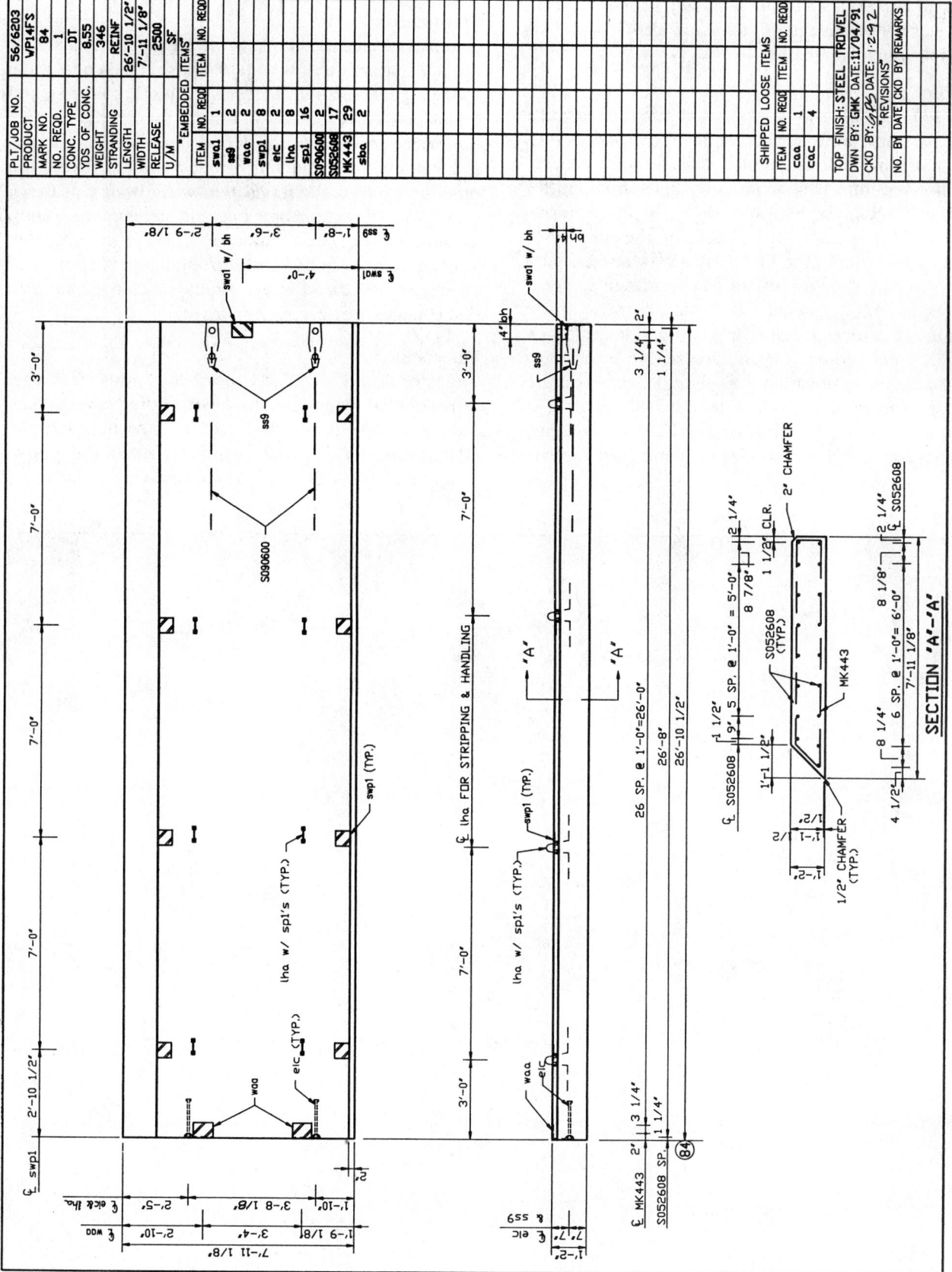

Figure 21-48 Shop drawing *(Courtesy Southern Prestressed Inc.)*

ventilation, and air conditioning (HVAC) plan; electrical plan; and framing plans (for roofs, floors, or both). The floor plan also doubles as the plumbing plan in modern construction. The contents of each component is described in the following paragraphs.

Plot Plan

A *plot plan* is a plan view of the building site showing where the structure is located on the lot. Surveyors provide the original material for plot plans. It should contain the following information: property lines showing the compass bearing and length of each; shape, location, and size of the building or buildings on the site; elevation of each corner of the site; north arrow; streets, sidewalks, driveways, patios; utilities and easements; wells, septic tanks, and drain fields (as applicable); scale of the drawing and a property description (i.e., lot and block number, street address, or legal description from a survey). Additional information that might be shown on the drawing but is not required includes the following: fences and retaining walls; trees, bushes, shrubs, rivers/streams, and other natural components; and contour lines. *Figure*

21-49 is an example of a plot plan that contains the necessary information.

Foundation Plan

A *foundation plan* is a plan view in section that shows the understructure of the building. It should contain the following information: footings for piers, columns, and foundation walls; foundation walls; columns; piers; retention walls; partition walls, doors, and plumbing fixtures in houses with a basement; windows, vents, doors, and other openings in the foundation walls; beams; pilasters; floor joists (direction, spacing, and size); drains, sump, air-conditioning line; dimension; scale of the drawing; applicable notes; and footing/foundation sections and details. *Figure 21-50* is an example of a foundation plan that contains the necessary information.

Floor Plan

The floor plan is the most important component in a set of residential plans. It is the heart of the plans. A *floor plan* is a plan view of the structure in section with the ceiling and roof removed. It should contain the follow-

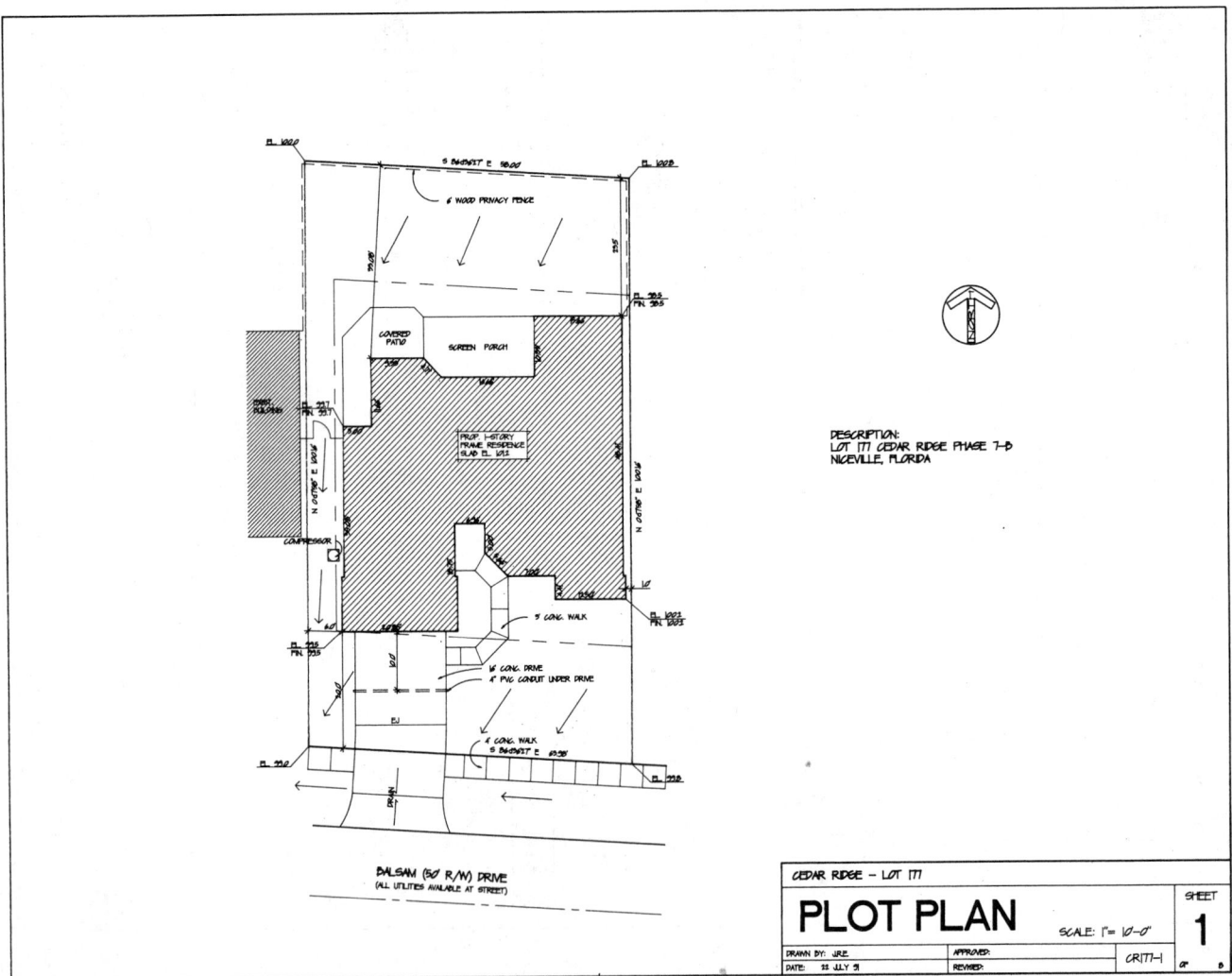

Figure 21-49 Plot plan *(Courtesy Barton Homes Inc.)*

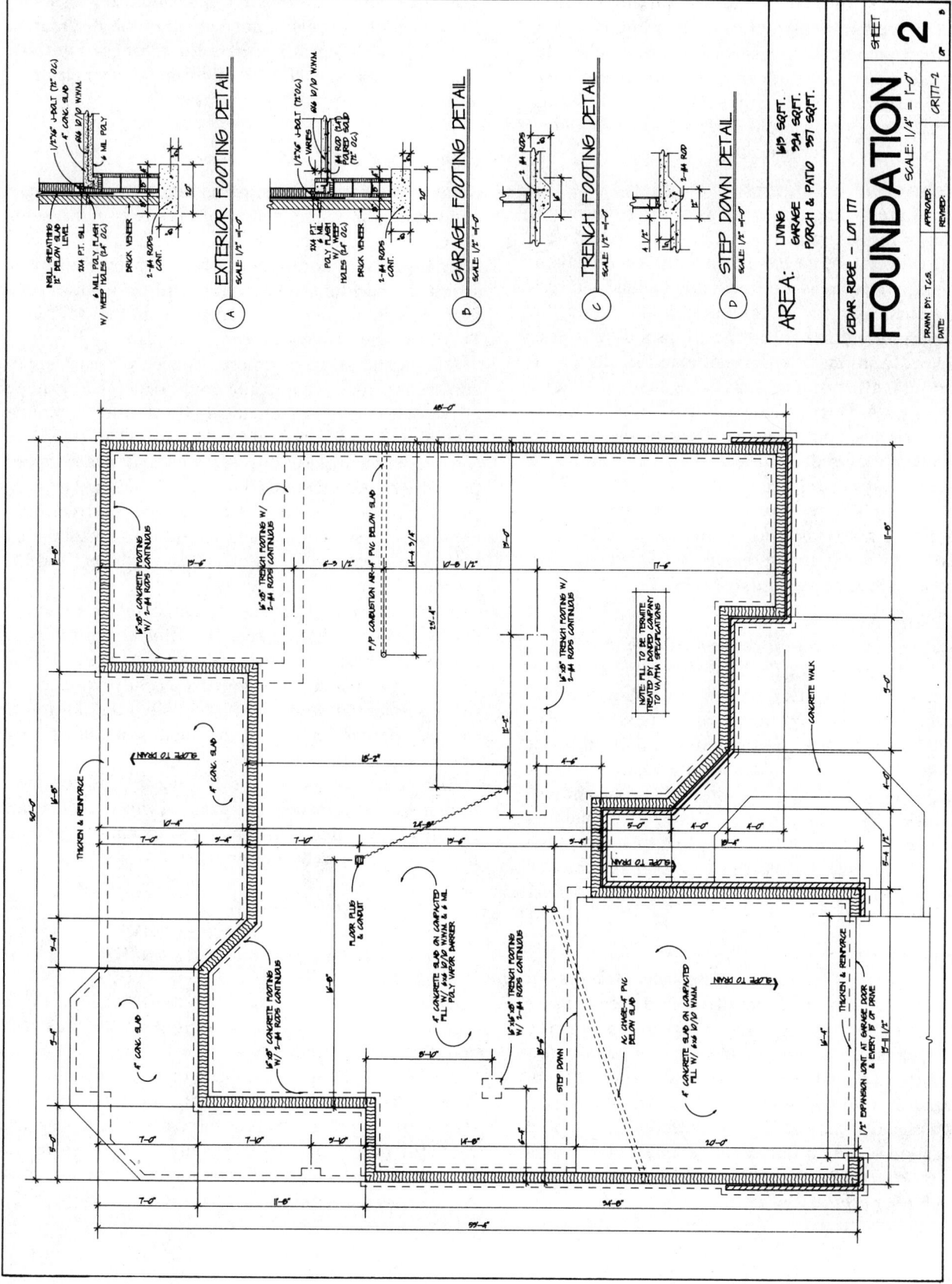

Figure 21-50   Foundation plan *(Courtesy Barton Homes Inc.)*

ing information: all necessary dimensions; exterior and interior walls; doors, windows, and other openings; built-in cabinets; appliances; plumbing fixtures; stairs; fireplaces; freestanding structures (i.e., garage); decks, patios, and porches; room labels; notes; door schedule; window schedule; area information (i.e., living area, porch area, garage, etc.); and scale of the drawing. *Figure 21-51* is an example of a floor plan that contains the necessary information.

Elevations

Elevations are orthographic views of the front, back, and sides of the structure. Most elevations are exterior views. However, interior elevations of the kitchen, bathrooms, and utility areas are frequently included in residential plans. Exterior elevations should contain the following information: grade lines, finished floor line, ceiling line, locations of corners of exterior walls, windows, doors, all roof features (i.e., gables, dormers, chimneys, etc.), vertical dimensions, porches, decks, patios, sun-rooms, applicable material symbols, and special details as required. *Figure 21-52* is an example of elevations that contain the necessary information. Notice that the kitchen, bath, and utility area interior elevations are not contained on this sheet. As is often the case, they were drawn on another sheet that had more room (see Sheet 5, *Figure 21-53*).

HVAC Plan

The *heating-ventilation-air-conditioning plan (HVAC)* shows the duct work for the house's heating and cooling system superimposed on the floor plan without dimensions. The location of interior and exterior heating/cooling units, size of duct work, direction of air flow, notes, details as needed, hot water heater, thermostat, registers, inlets, and vents should also be shown. Figure 21-53 is an example of an HVAC plan containing the necessary information. Notice that this sheet also contains interior elevations. This is a common practice to cut down on the number of sheets in the overall set of plans.

Electrical Plans

The *electrical plan* shows the house's electrical system superimposed on a floor plan without dimensions. It should contain the following information: service entrance capacity, meter and distribution panel, location and types of switches, location and types of lighting fixtures, number and types of electrical circuits, details as needed, notes, electrical symbol legend, and an indication of which fixtures are connected to which switches. *Figure 21-54* is an example of an electrical plan with the necessary information.

Roof Plan

The *roof plan* is a plan view showing the location and spacing of rafters and/or trusses. It should also contain a truss schedule, details as needed, and truss diagrams as needed. It is superimposed on the floor plan without dimensions. *Figure 21-55* is an example of an electrical plan with the necessary information.

Commercial Plans

Commercial architectural plans differ somewhat from residential plans. The major difference is in who produces the various components of the plans. With residential plans all components can be produced by an architect, residential designer, and/or architectural drafter including the plot, electrical, and HVAC plans. This is not the case with commercial plans.

A complete set of commercial plans will usually contain architectural, civil engineering, structural (as needed), electrical, and mechanical (HVAC and plumbing) drawings. The architectural component of the plans (i.e., floor plans, elevations, sections, and details) is prepared by an architect or architectural drafter. The component relating to the site is prepared by a civil engineer, the structural component by a structural engineer, the electrical by an electrical engineer, and the mechanical by a mechanical engineer.

Architectural drafters and designers in a commercial setting prepare floor plans, elevations, sections, and details. The basic contents of these plans are similar to those in residential plans. *Figure 21-56* is an example of a commercial floor plan. Notice that in addition to the plan view in sections of the building it contains several sectional details.

Each detail has a designator that, when translated, gives its numerical designation, the sheet on which it is cut, and the sheet on which it is drawn. For example, sectional detail "1-A2-A2" can be interpreted as follows:

1 = Detail number (1)
A2 = Sheet on which the detail is cut (A2)
A2 = Sheet on which the detail is drawn (A2)

This floor plan also contains a symbols legend, material legend, equipment legend, keynotes, a scale indicator, and titleblock information.

Figure 21-57 is an example of commercial elevations. In addition to the elevations themselves, it contains a roof plan, two full sections cut through the building, and a roof plan. *Figure 21-58* contains numerous other miscellaneous sections and details commonly found in a set of commercial plans plus their corresponding notes.

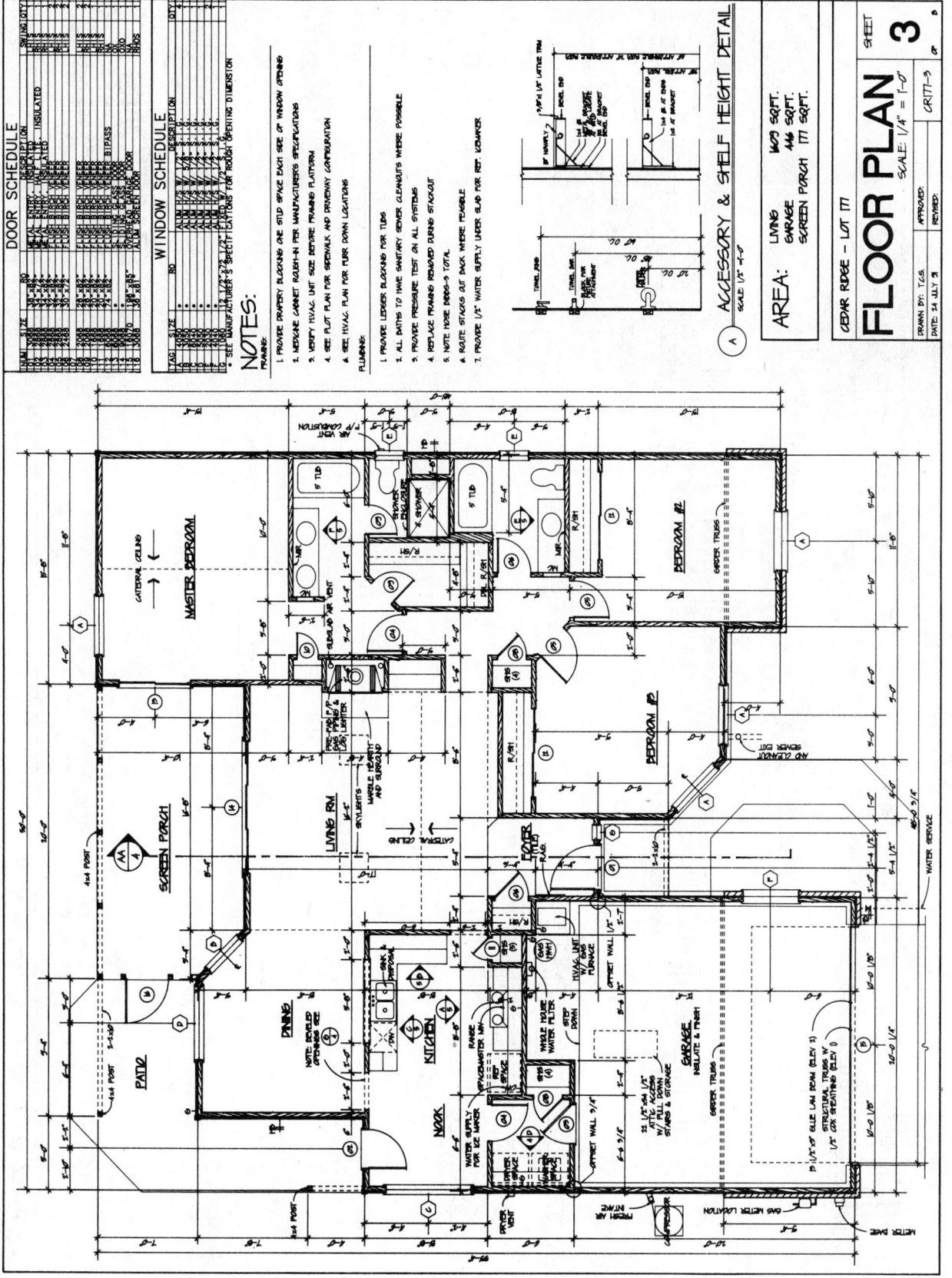

Figure 21-51   Floor plan *(Courtesy Barton Homes Inc.)*

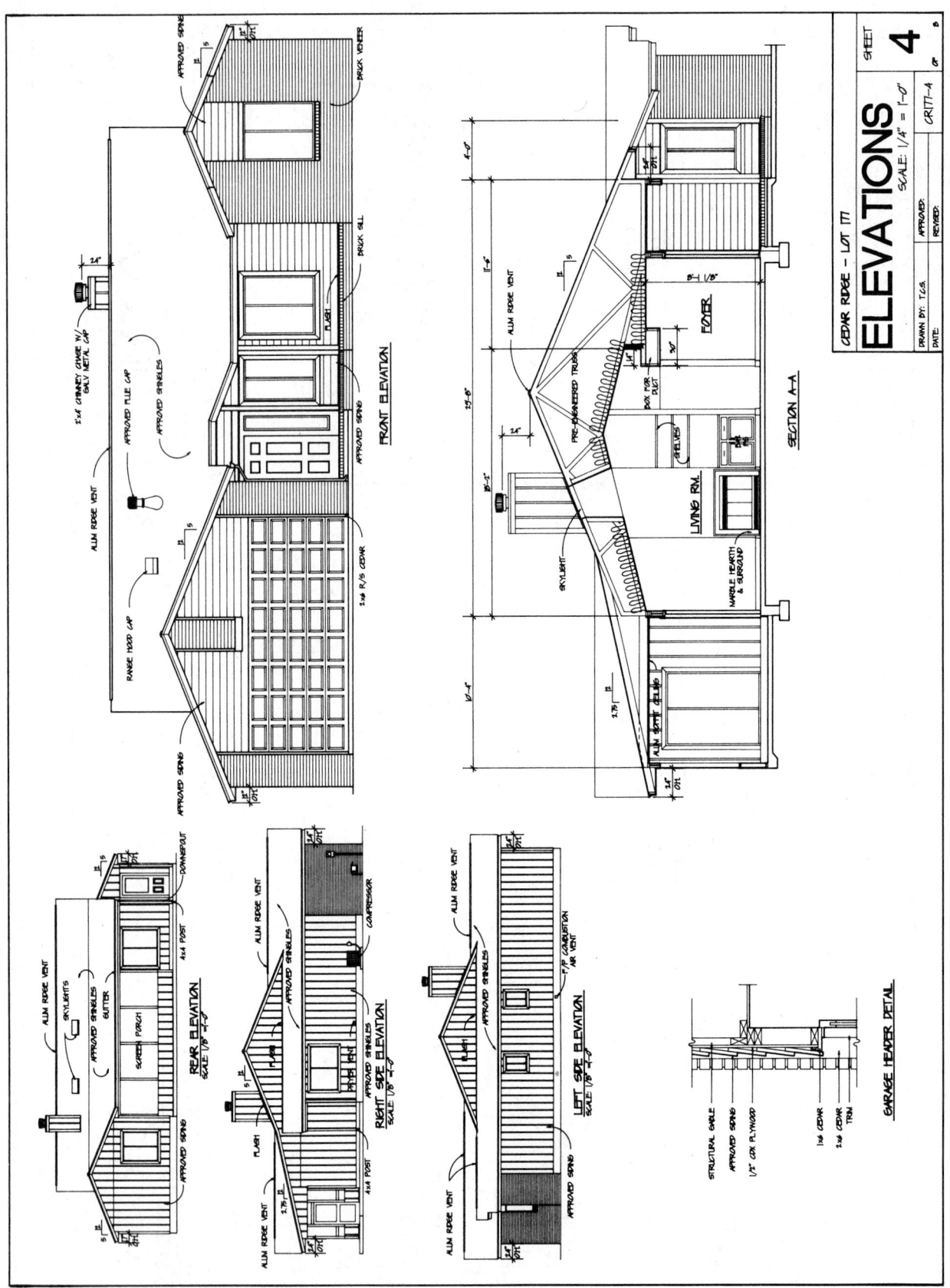

Figure 21-52 Elevations *(Courtesy Barton Homes Inc.)*

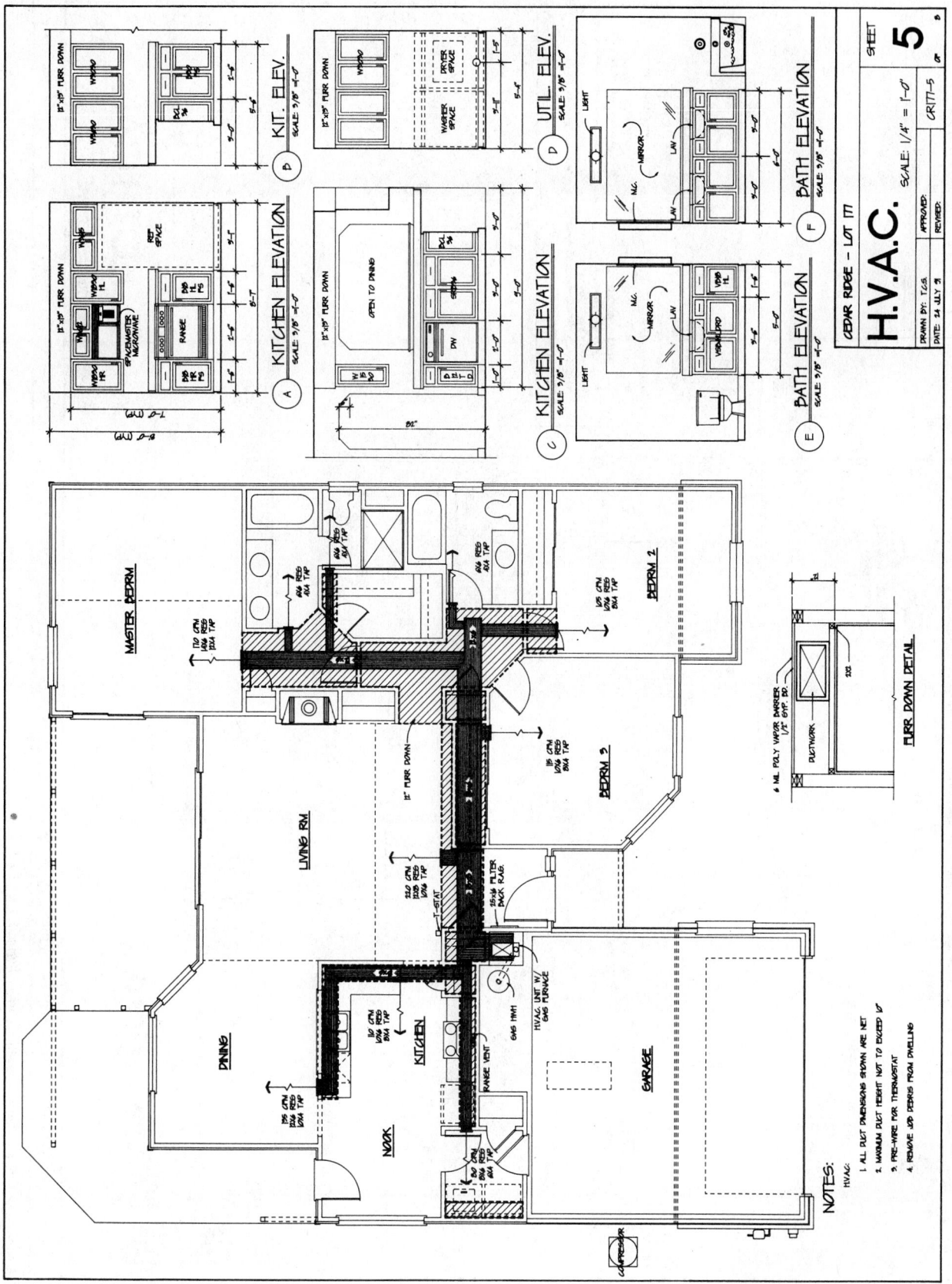

Figure 21-53 HVAC plan (Courtesy Barton Homes Inc.)

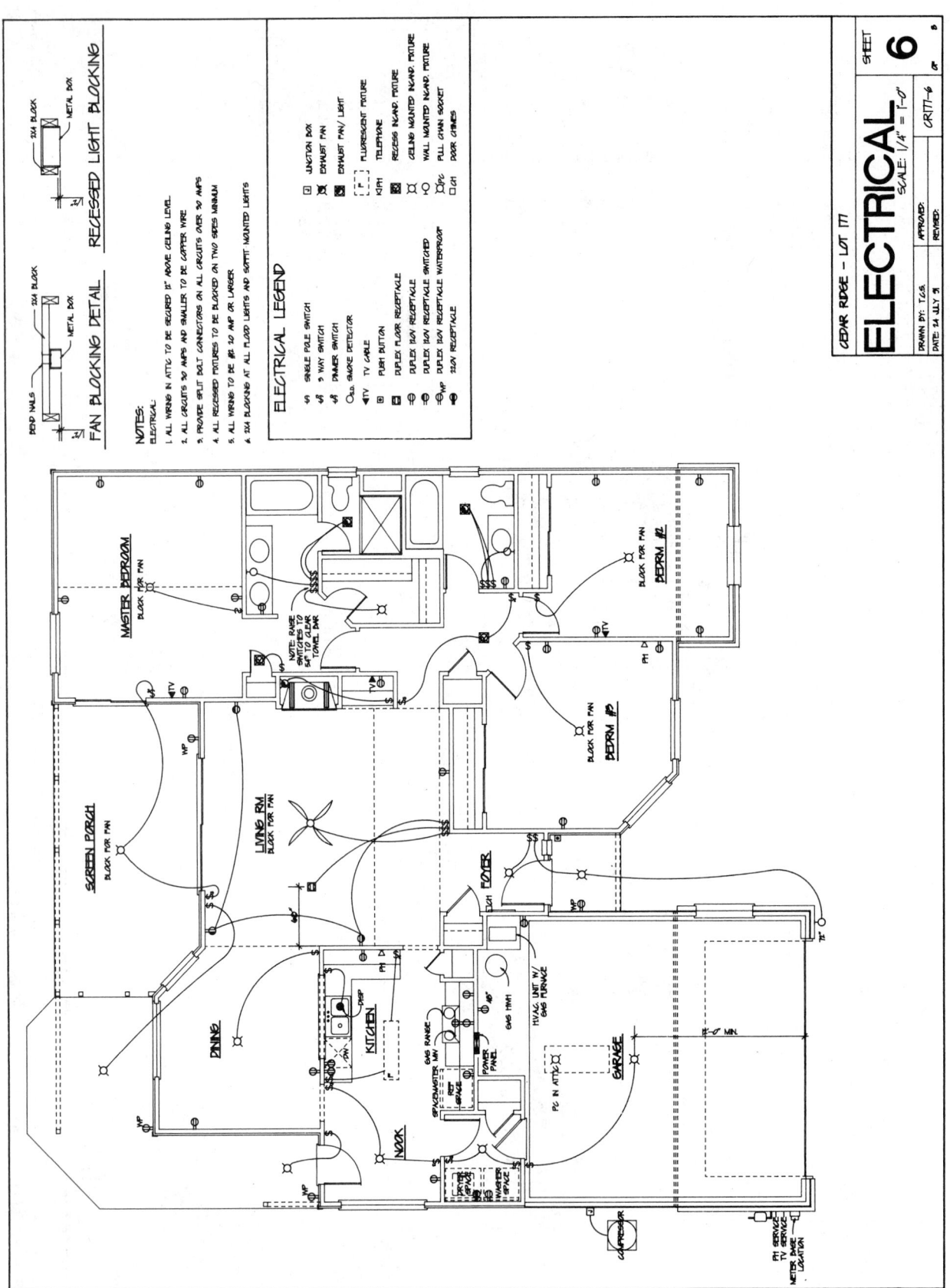

Figure 21-54 Electrical plan (Courtesy Barton Homes Inc.)

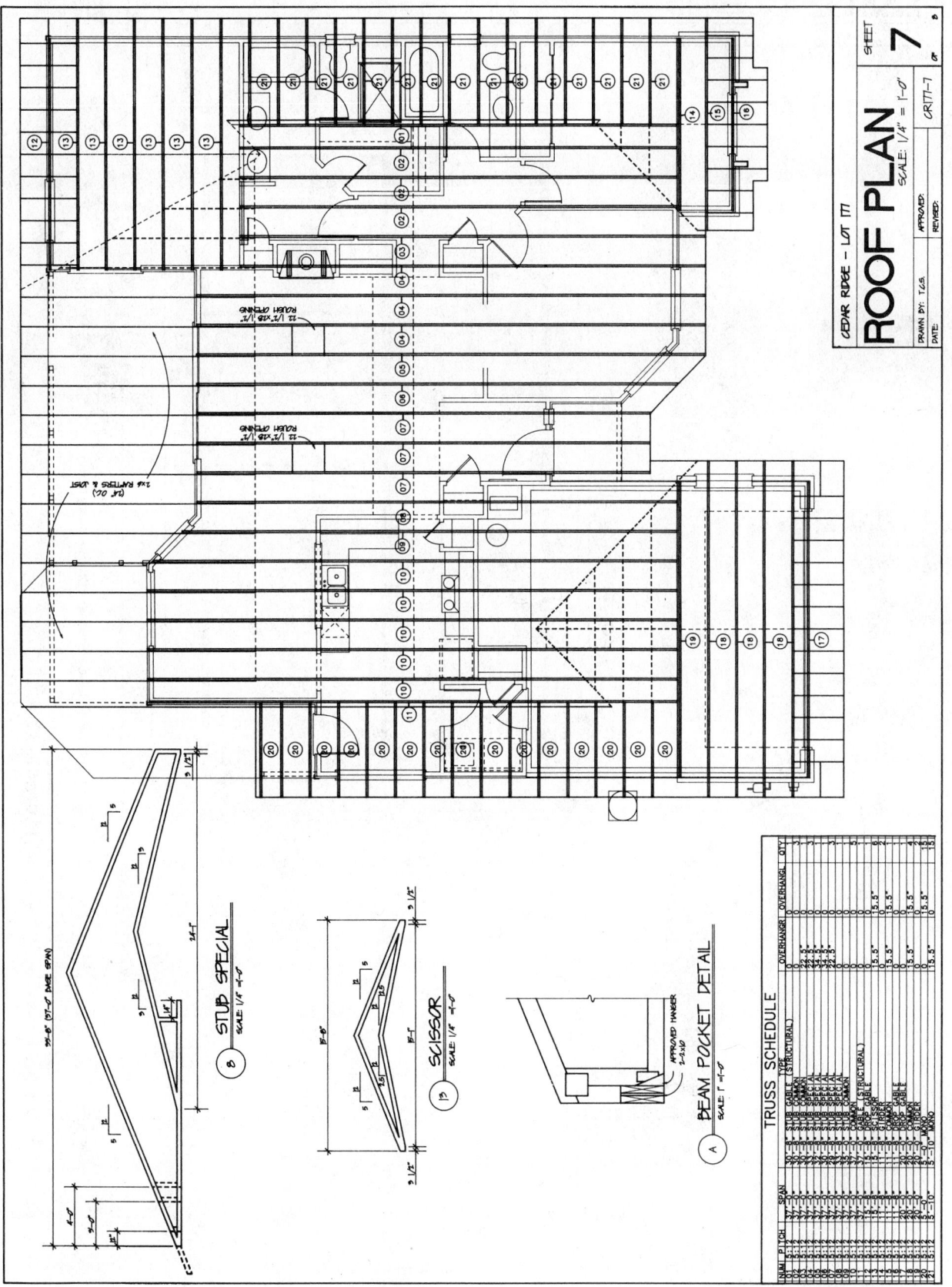

Figure 21-55 Roof plan *(Courtesy Barton Homes Inc.)*

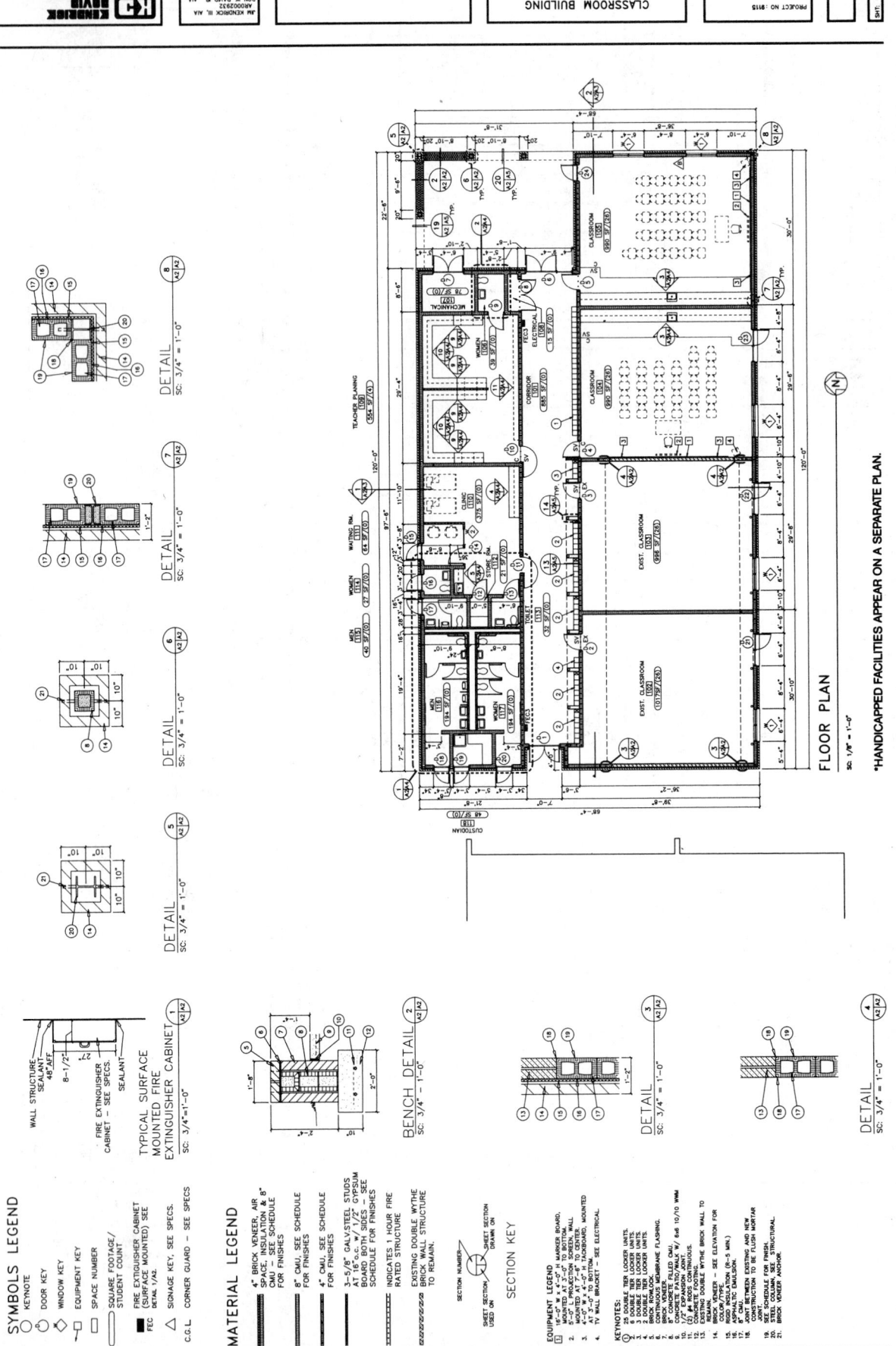

Figure 21-56 Commercial floor plan *(Courtesy Kendrick, David, and Dowling Architects Inc.)*

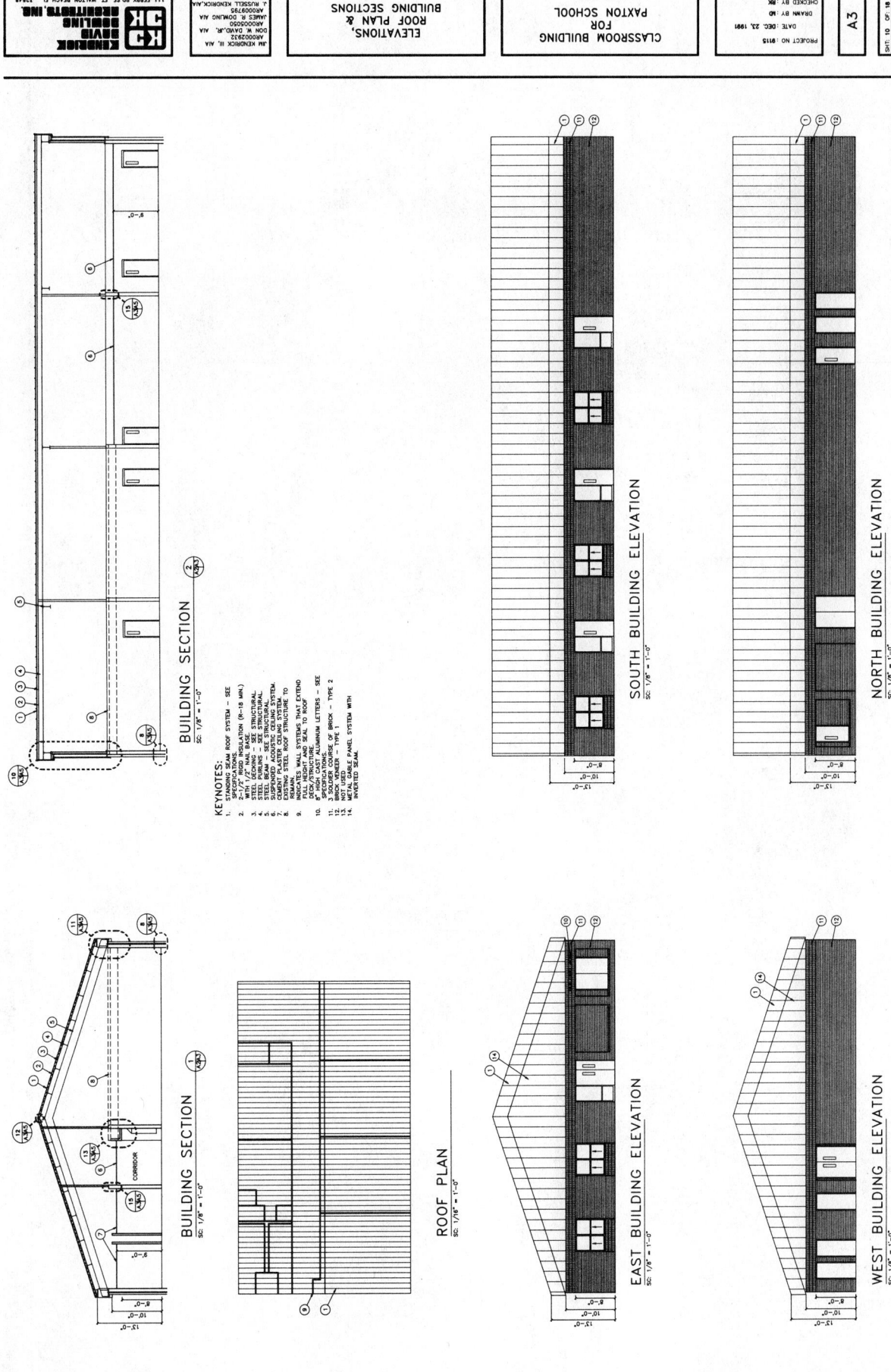

Figure 21-57 *Commercial elevations (Courtesy Kendrick, David, and Dowling Architects Inc.)*

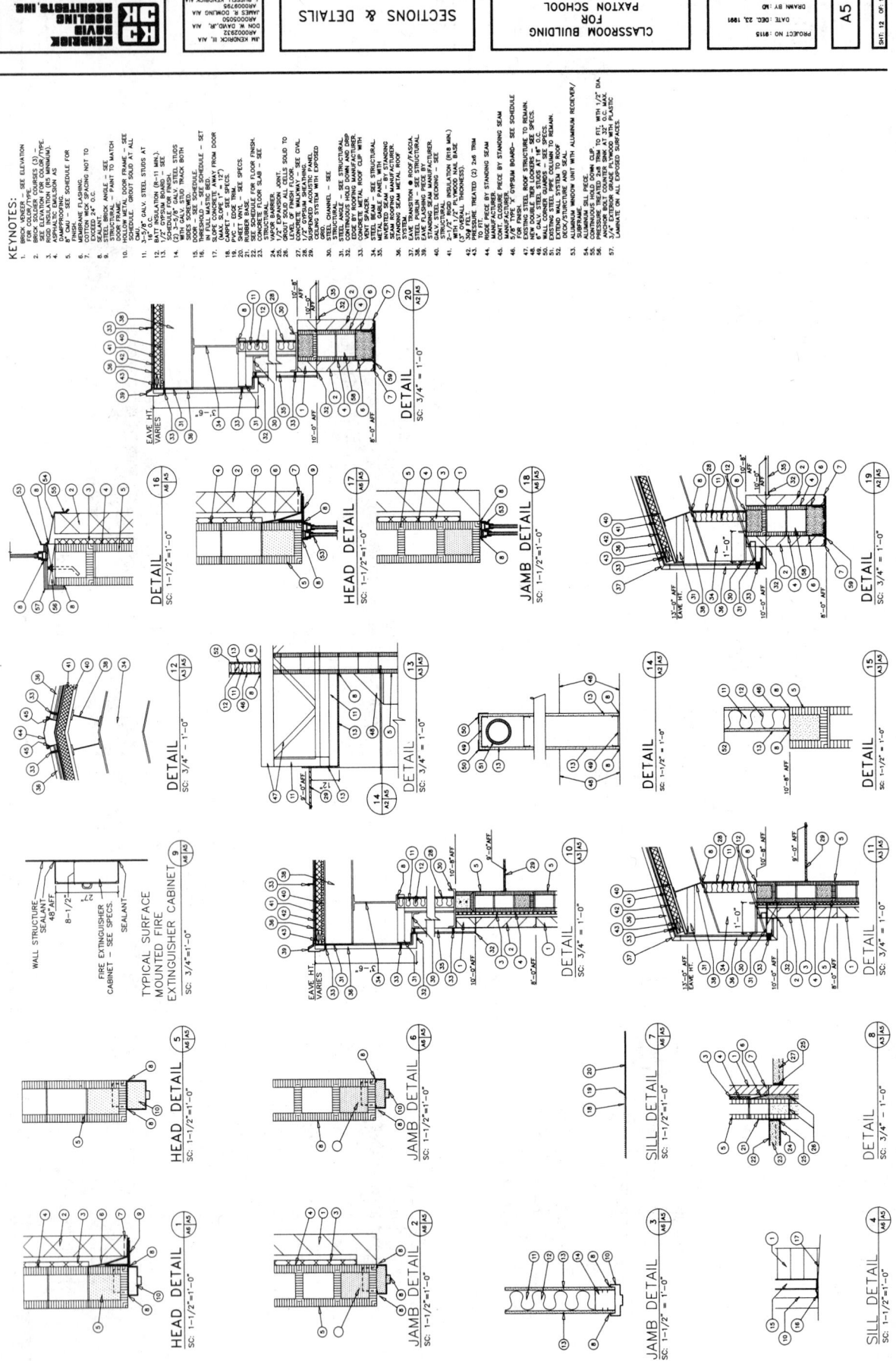

Figure 21-58 Commercial details and sections *(Courtesy Kendrick, David, and Dowling Architects Inc.)*

Review

1. Explain why engineering and drafting students should be proficient in the preparation of pipe drawings.
2. What are the five most widely used types of pipe?
3. What are the three broad classifications of pipe fittings?
4. What are the three most widely used categories of valves?
5. Sketch the symbols for the following:
   Flanged lateral
   Screwed sleeve
   Welded union
   Flanged gate valve
   Screwed glove valve
   Fuel-oil flow
   Hot-water heating return
   Low-pressure steam
   Cold-water line
   Gas line
6. What is the difference between a correction and a revision?
7. What are the two most widely used methods for connecting structural members?
8. Define and distinguish between engineering drawings and shop drawings.
9. List the typical components in a set of residential architectural plans.
10. Explain how residential and commercial architectural plans differ.

Chapter Twenty-One Problems

Problem 21-1

Problem 21-1 is a single-line isometric pipe drawing. Convert this drawing into a single-line orthographic drawing showing the following views of the system: top, front, right side, and left side.

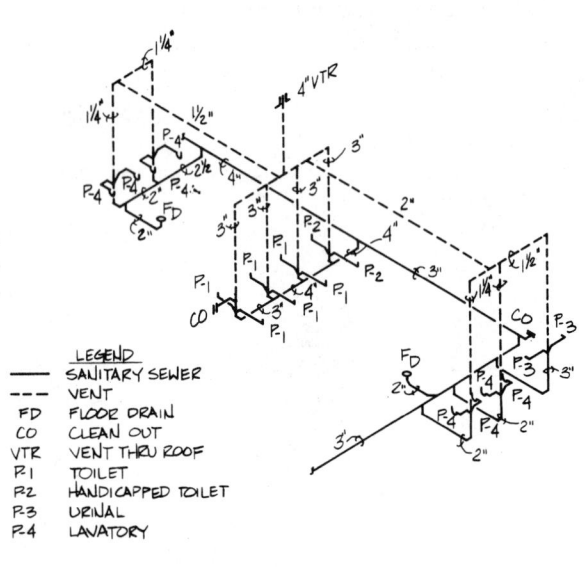

SANITARY SEWER RISER DIAGRAM
NTS

Problem 21-1

Problem 21-2

Problem 21-2 is a double-line orthographic pipe drawing. Convert it into a single-line orthographic drawing.

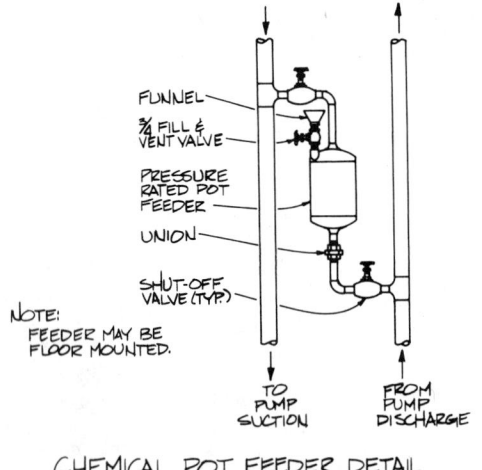

CHEMICAL POT FEEDER DETAIL
N.T.S.

Problem 21-2

Problem 21-3

Problem 21-3 is a double-line orthographic pipe drawing. Convert it into a single-line orthographic drawing.

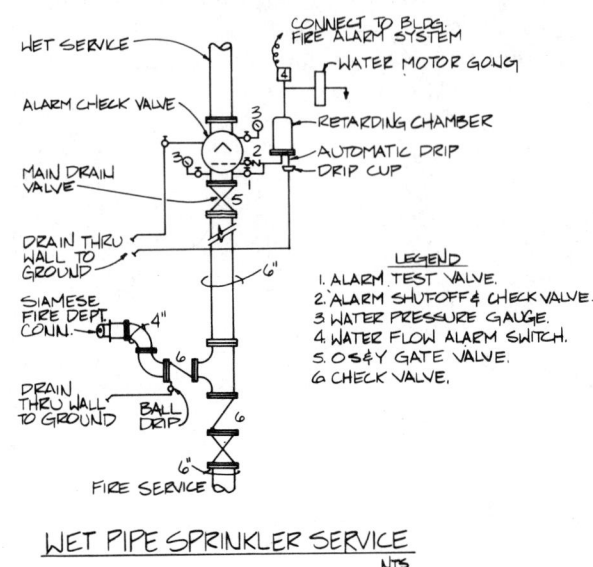

WET PIPE SPRINKLER SERVICE
NTS

Problem 21-3

Problem 21-4

Problem 21-4 is a double-line orthographic pipe drawing. Convert it into a single-line orthographic drawing.

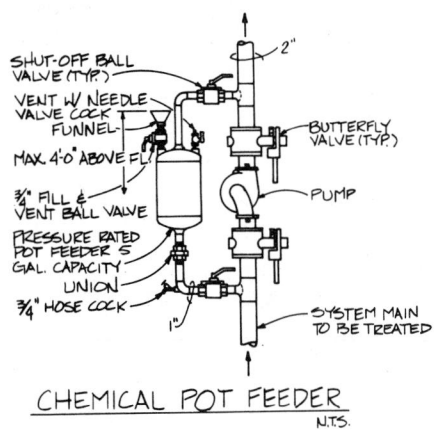

CHEMICAL POT FEEDER
N.T.S.

Problem 21-4

Problem 21-5

Problem 21-5 is a double-line representation of a piping system. Reconstruct the drawing as shown in single-line representation.

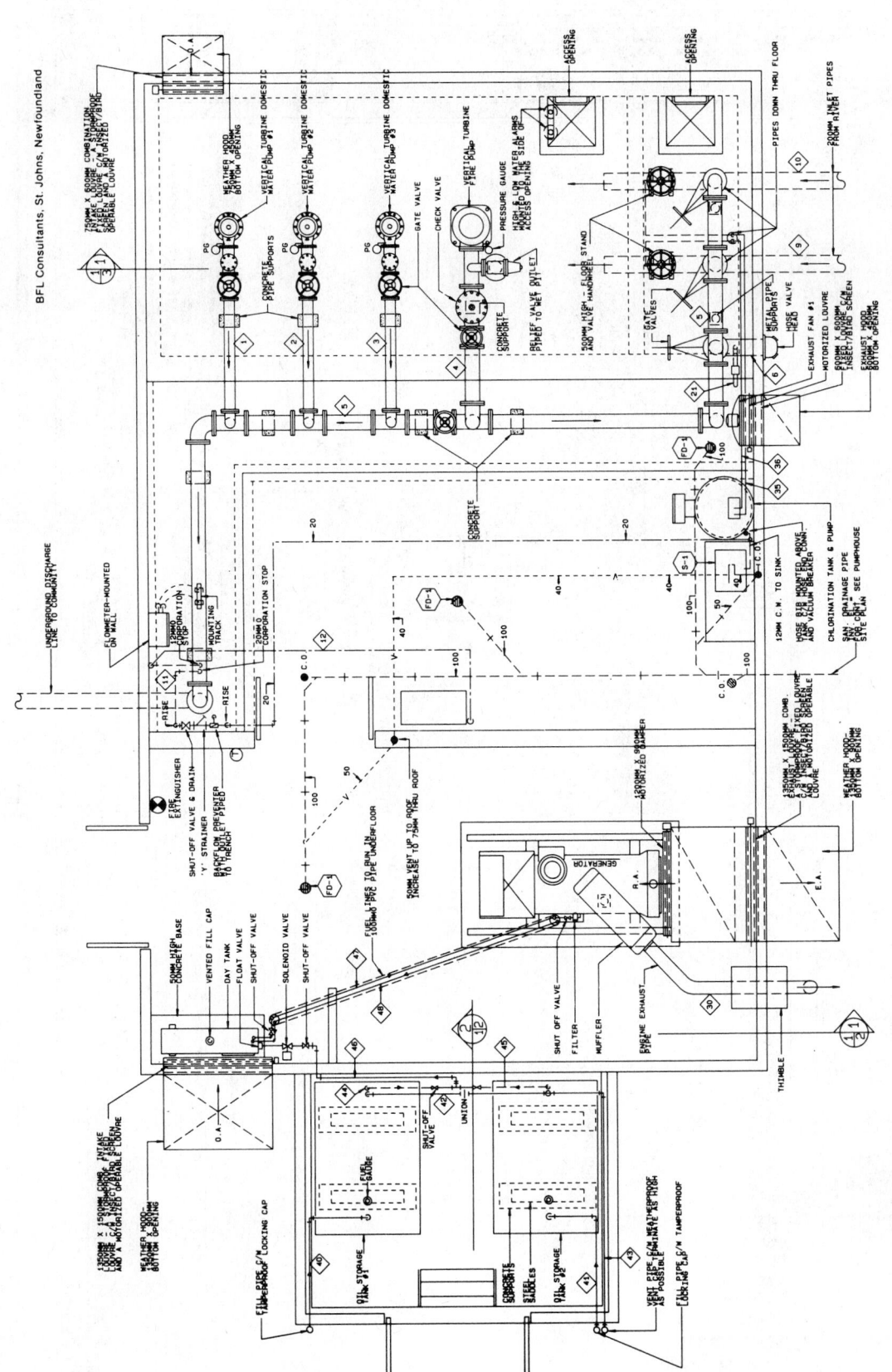

Problem 21-5   *(Courtesy VersaCAD Inc.)*

❏ Problem 21-6

Problem 21-6 is an engineering drawing for a set of bleachers. Reconstruct the entire drawing making the following changes: a) add 21′6″ to the *Out-to-Out*

dimension and adjust all others accordingly; and b) add one more row of bleachers at the bottom.

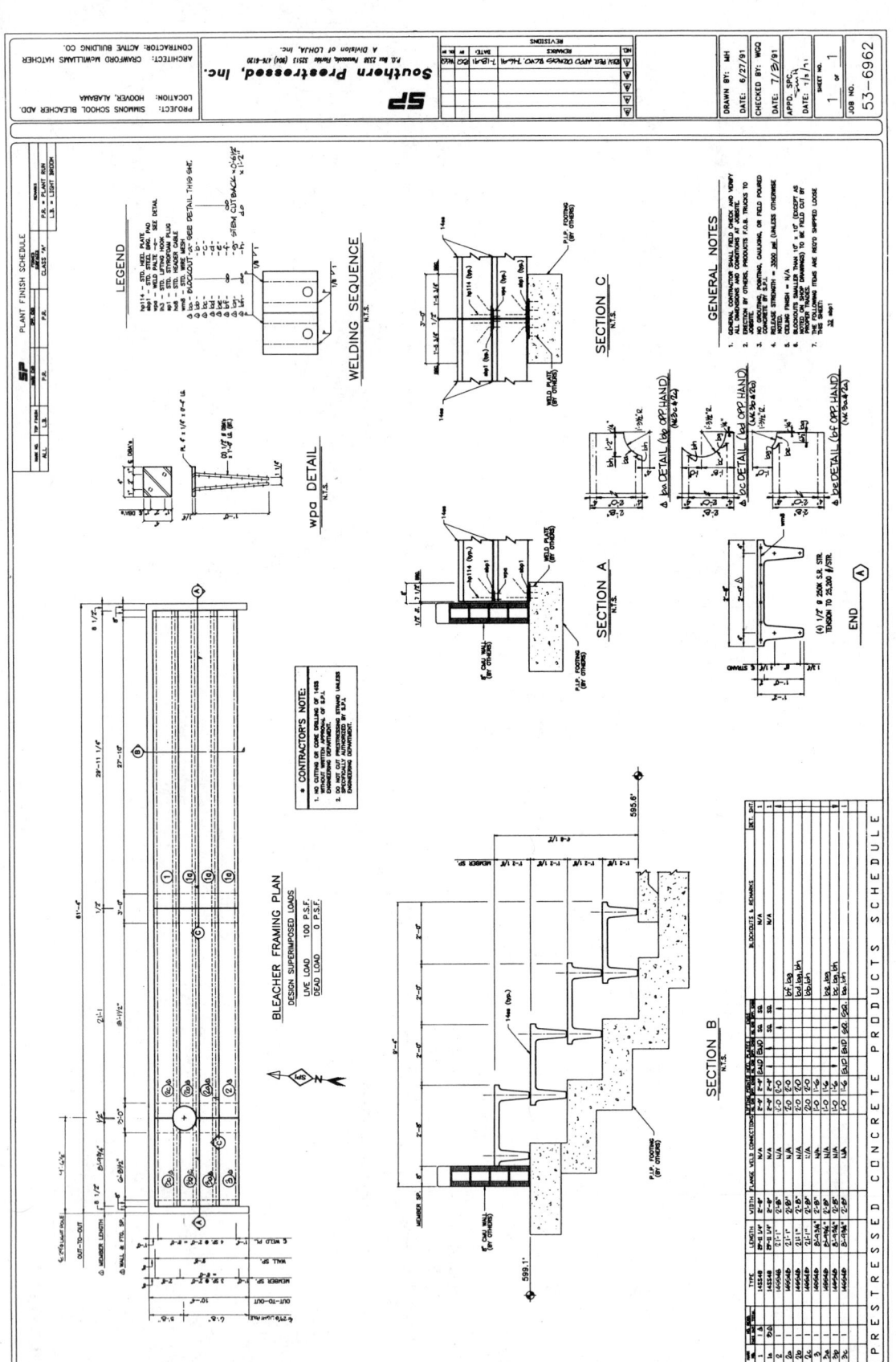

Problem 21-6

☐ Problem 21-7

Problem 21-7 is an engineering drawing for a pedestrian bridge. Reconstruct the entire drawing making the following change: add one more 33′2″ bay to the bridge and adjust all dimensions accordingly.

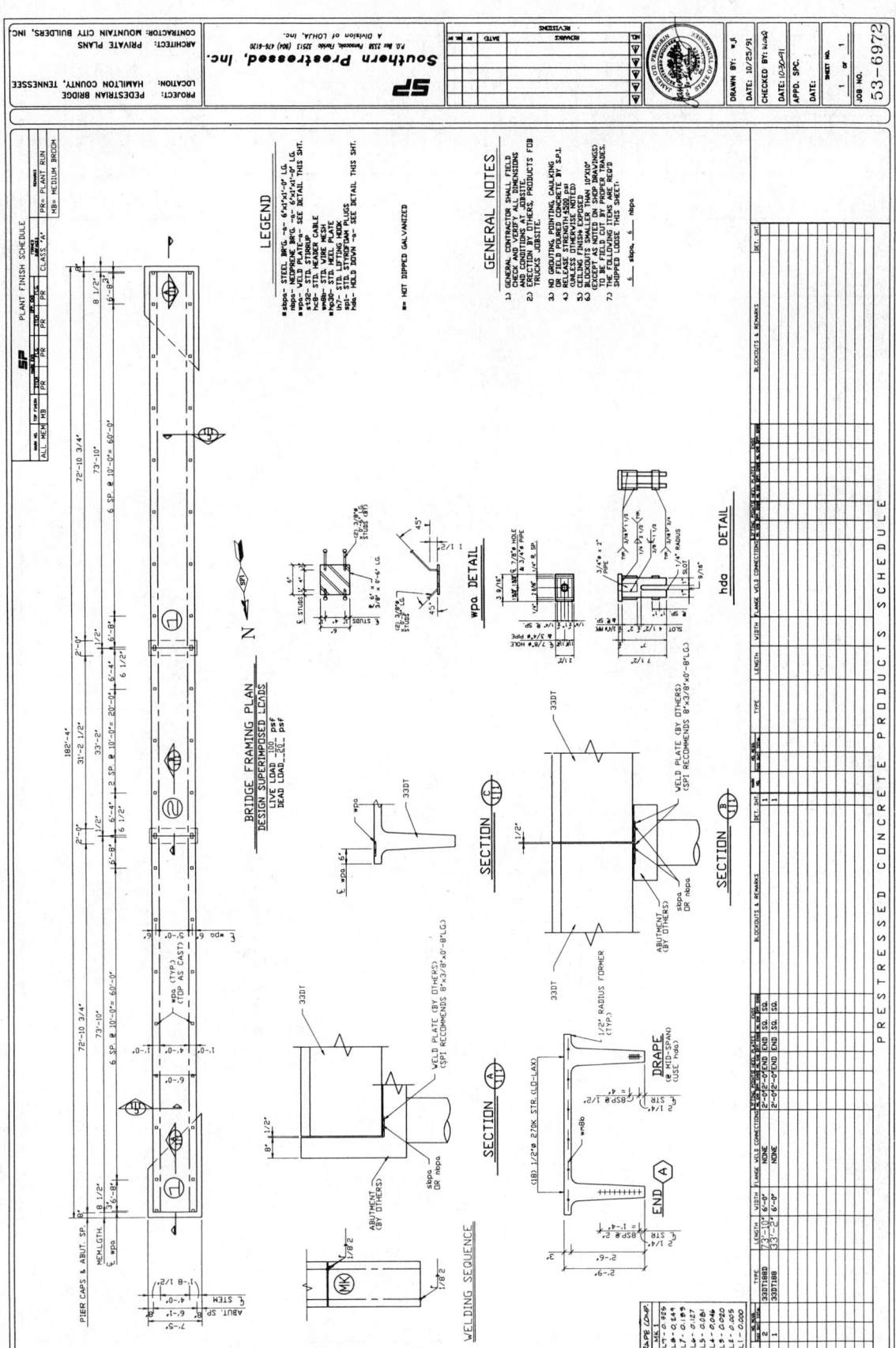

Problem 21-7

❏ Problem 21-8

Problem 21-8 is a shop drawing for a prestressed concrete beam. Reconstruct the entire drawing making the following change: change the depth of the beam to 1′6″ and adjust accordingly.

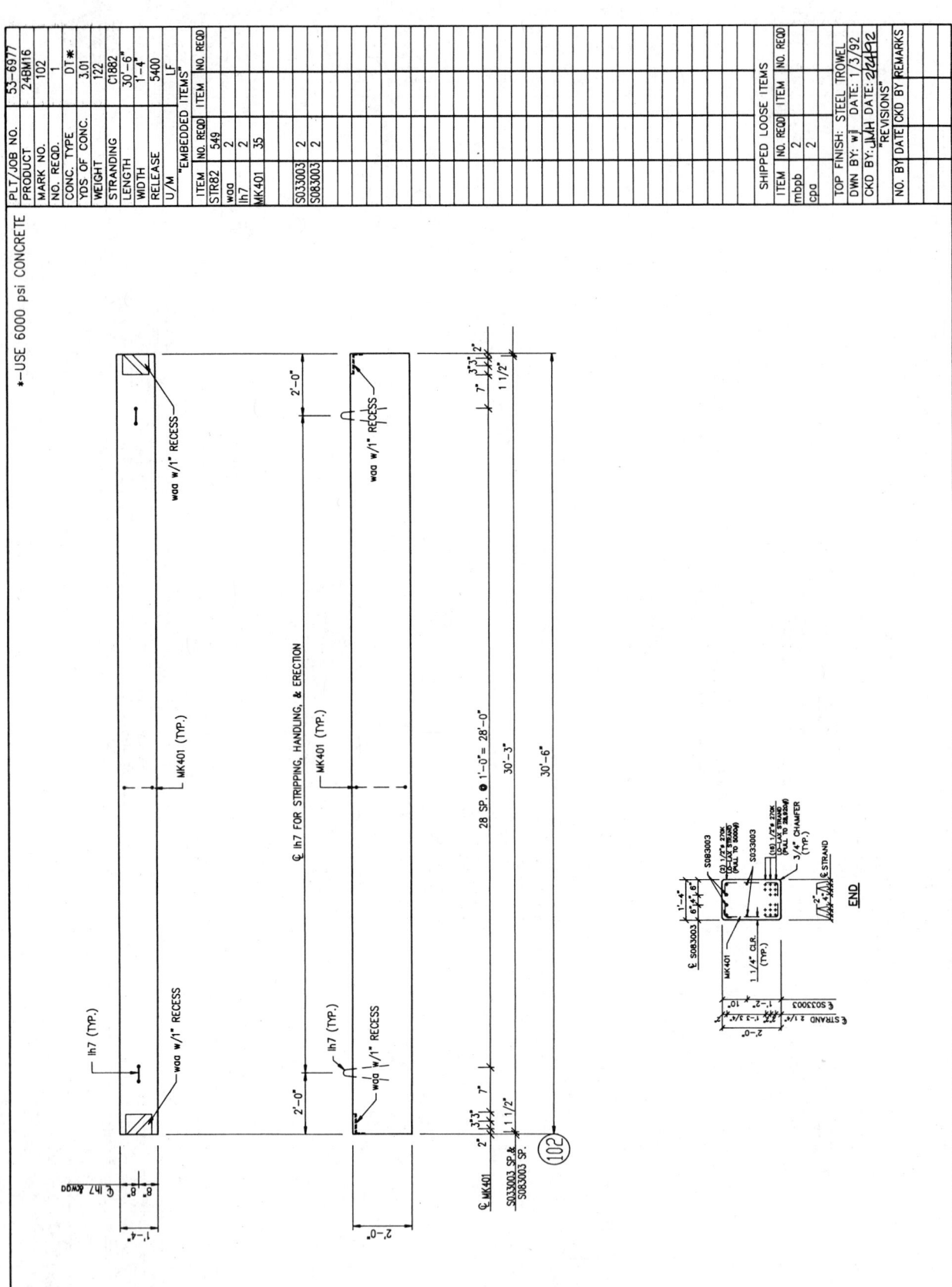

❏ Problem 21-9

Problem 21-9 is a shop drawing for a precast concrete flat slab. Reconstruct the entire drawing making the following change: decrease the width by 1′1″ and the length by 9″ and adjust accordingly.

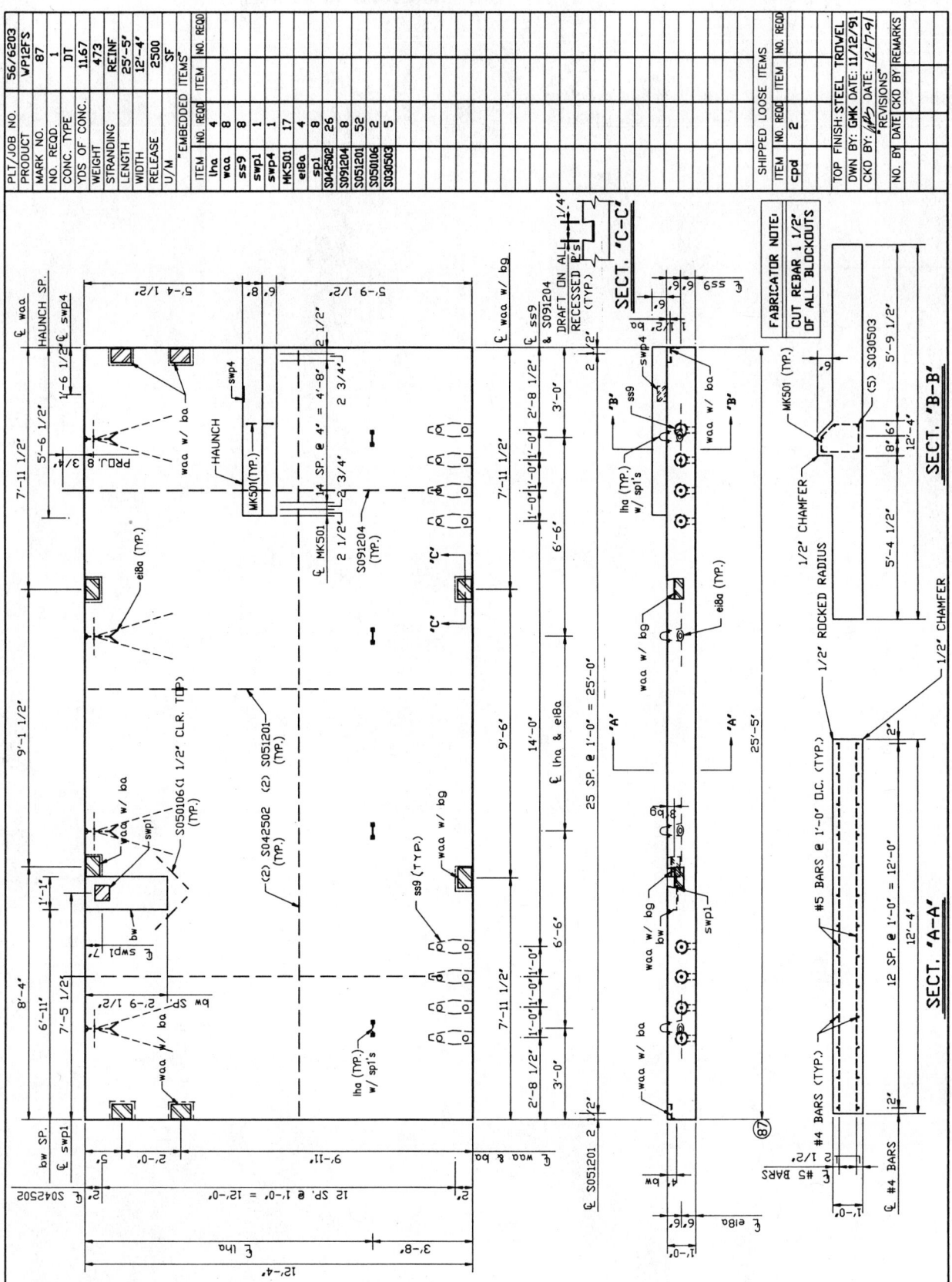

Problem 21-9

❑ **Problem 21-10**

Figure 21-51 in this chapter is a floor plan for a residential dwelling. Subtract 11″ from the overall length and width of this figure. Now make all necessary adjustments to the floor plan.

❑ **Problem 21-11**

Using the floor plan developed in the previous problem as the nucleus, develop a complete set of residential plans.

❑ **Problem 21-12**

Figures 21-56, 21-57, and 21-58 in this chapter contain the basic architectural plans for a classroom building. Add one additional classroom (106) and revise the entire set of plans accordingly.

Drafting Applications: Electronics and Printed Circuit Boards

Electronics Symbols

Electronic products are made of numerous different types of electronic components. Each of these components can be represented symbolically and pictorially. The most frequently used components are antennas, batteries, capacitors, diodes, grounds, inductors, meters, resistors, switches, transformers, and transistors. Other less frequently used components include amplifiers, bells, buzzers, circuit breakers, connectors, crystals, delays, envelopes, fuses, headsets, lamps, motors, magnets, phase shifters, pickup heads, rectifiers, relays, relay coils, repeaters, safety interlocks, speakers, terminal boards, thermal elements, and thermocouples. *Figure 22-1* illustrates the symbols for these electronic components.

Schematic Diagrams

Schematic diagrams are used to show what components are contained on a circuit board and which components connect with which, *Figure 22-2*. It should be noted that the actual location of a component on a circuit board cannot be determined by looking at a schematic

diagram. Schematics show components and connections, but they do not show their actual locations.

Rules for Drawing Schematics

Drafters prepare finished schematic diagrams from sketches provided by engineers. *Figure 22-3* is an example of an engineer's sketch and the corresponding finished schematic diagram. The general rules which should be applied in drawing schematic diagrams are:

1. Inputs go left.
2. Outputs go right.
3. Break connectors and jumble pins.
4. Grounds point downward.
5. Curved side of capacitors points to ground.
6. Arrange relay symbols.
7. Remove doglegs.
8. Avoid crossovers.
9. Don't cramp and crowd components.
10. Group leads.
11. Remove four-way ties.
12. Number symbols from left to right and top to bottom.

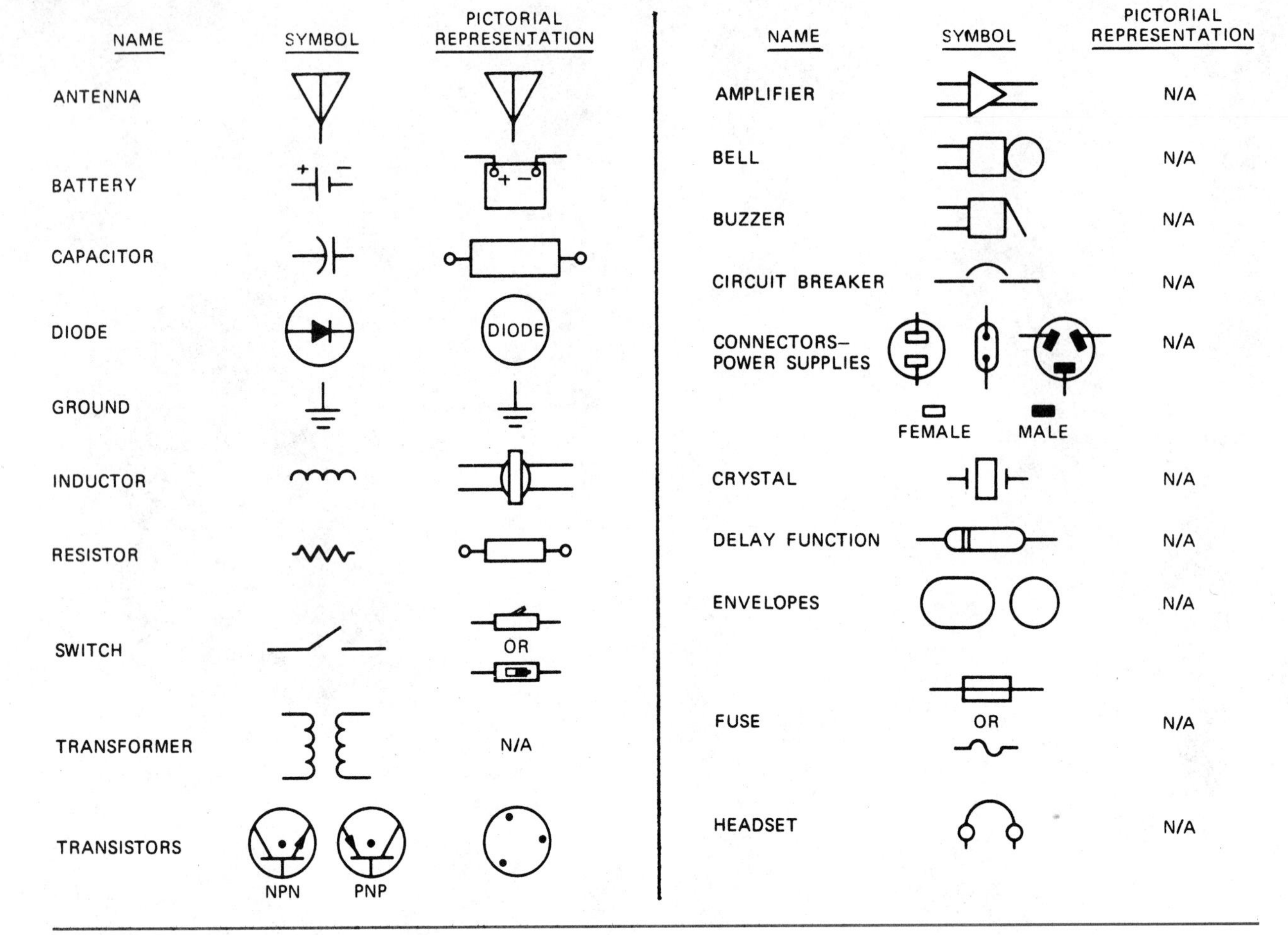

NAME
SYMBOL
PICTORIAL REPRESENTATION
ANTENNA
BATTERY
CAPACITOR
DIODE
DIODE
GROUND
INDUCTOR
RESISTOR
SWITCH
OR
TRANSFORMER
N/A
TRANSISTORS
NPN
PNP
NAME
SYMBOL
PICTORIAL REPRESENTATION
AMPLIFIER
N/A
BELL
N/A
BUZZER
N/A
CIRCUIT BREAKER
N/A
CONNECTORS—POWER SUPPLIES
N/A
FEMALE
MALE
CRYSTAL
N/A
DELAY FUNCTION
N/A
ENVELOPES
N/A
FUSE
OR
N/A
HEADSET
N/A

Figure 22-1 Electronics drafting symbols

R7 1K
R6 1K
U32
LS04
Y1
12/16MHZ
C13 .01UF
C14 300PF
C15 150 PF
U32
LS04
+5V
VCC
U42
LS92/93
W8
6/18MHZ
12 MHZ XTAL
2 MHZ (SH1,2)
C12 33PF
R4 220
R8 22
R5 1K
Q1 2N3906
+5V
U32
LS04
U32
Ø (SH2)
* NOTE:
USE 74LS92 WITH 12 MHZ XTAL (FOR 6MHZ OPERATION)
AND 74LS93 WITH 16 MHZ XTAL (FOR 4 & 8 MHZ OPERATION)

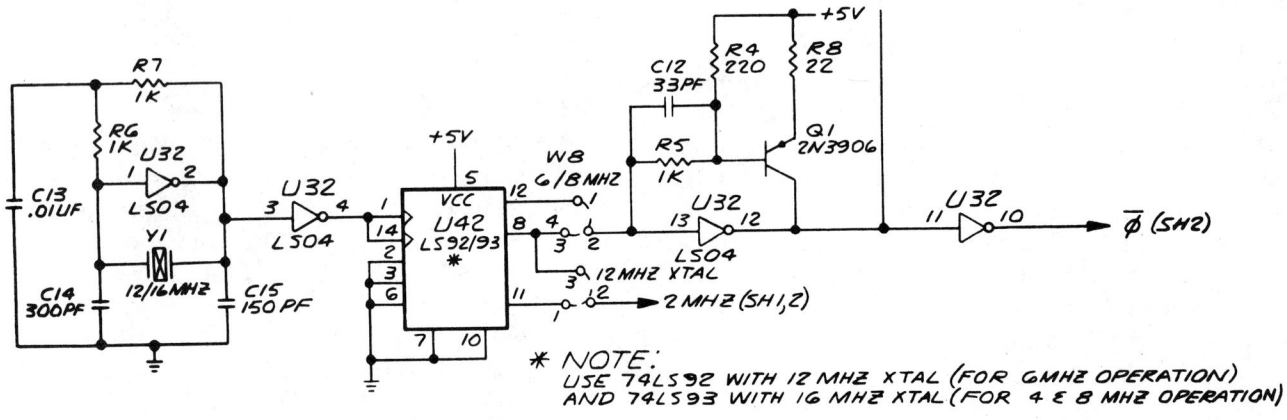

Figure 22-2 Schematic diagram (From Electronic Drafting and Printed Circuit Board Design, 2nd Edition, by James M. Kirkpatrick, ©1989 by Delmar Publishers Inc.)

Figure 22-4 is an example of a schematic diagram drawn according to the rules stated previously.

Connection Diagrams

Connection diagrams are used to show how the components in an electronic system are connected. They are used as a guide in assembling electronic systems and maintaining electronic equipment. You might have used a connection diagram in connecting a new stereo, VCR, or compact disk system. There are four types of connection diagrams:

1. Point-to-point diagrams
2. Baseline diagrams
3. Highway diagrams
4. Lineless diagrams

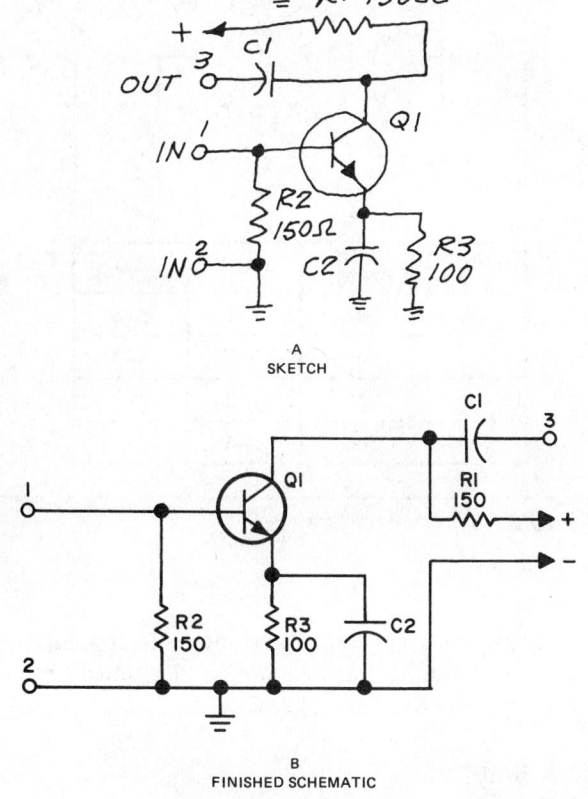

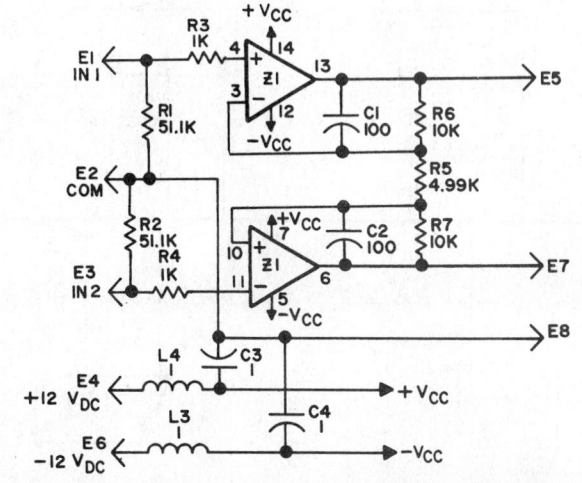

Figure 22-4 Schematic diagram *(From Electronic Drafting and Printed Circuit Board Design, 2nd Edition, by James M. Kirkpatrick, ©1989 by Delmar Publishers Inc.)*

Figure 22-3 Sketch with finished schematic *(From Electronic Drafting and Printed Circuit Board Design, 2nd Edition, by James M. Kirkpatrick, ©1989 by Delmar Publishers Inc.)*

Point-to-Point Diagrams

This type of connection diagram shows the various terminal connection locations and the routing paths of every wire in an electronic system. Point-to-point diagrams are used primarily with simple circuits. They can be difficult to read when circuits are complex and contain many wires. *Figure 22-5* is an example of a point-to-point diagram. The steps in drawing such a diagram are:

1. Draw the various electronic components in the actual locations they hold in the circuit.
2. Draw the wire paths.
3. Label the wire colors.

Baseline Diagrams

This type of connection diagram shows all wire paths running into a single baseline. Baseline diagrams are easier to read and follow than point-to-point diagrams, but they can be misleading because they do not show the electronic components in their proper positions. *Figure 22-6* is an example of a baseline diagram. The following steps are used in drawing such a diagram:

1. Draw a thick single line that will serve as the baseline.
2. Draw the electronic components placing half above the baseline and half below it.
3. Draw lines from each connection point to the baseline. The lines must be perpendicular to the baseline.
4. Label each wire path with a destination code.

Highway Diagrams

This type of connection diagram collects wires that run along similar paths and combines them into groups of wires called "highways." As with point-to-point dia-

Figure 22-5 Point-to-point diagram *(From Electronic Drafting and Printed Circuit Board Design, 2nd Edition, by James M. Kirkpatrick, ©1989 by Delmar Publishers Inc.)*

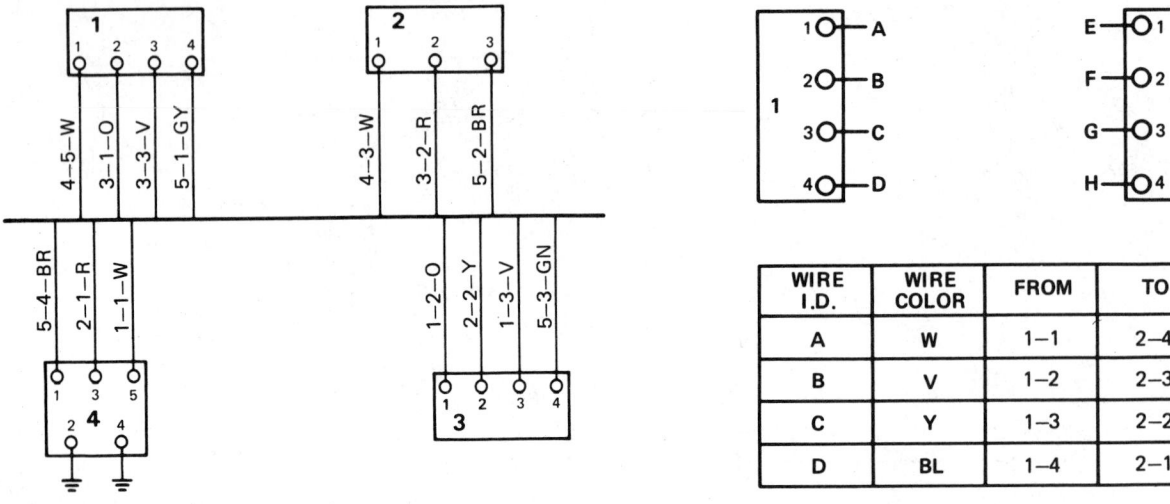

Figure 22-6   Baseline diagram

Figure 22-8   Lineless diagram

WIRE I.D.	WIRE COLOR	FROM	TO
A	W	1—1	2—4
B	V	1—2	2—3
C	Y	1—3	2—2
D	BL	1—4	2—1

grams, the components are located in the positions they will hold in the actual circuit. ***Figure 22-7*** is an example of a highway diagram. The following steps are used in drawing such a diagram:

1. Draw the electronic components in their proper positions.
2. Lightly sketch in the wire paths to determine where potential highways exist.
3. Darken the highways, the lines from the connection points to the highways, and the components.
4. Label the wire paths with a destination code, component number, and wire color.

Lineless Diagrams

This type of connection diagram shows the electronic components with the connection points labeled, but it does not show wires. Instead of wires, the components are accompanied by a table which contains a designation, color, and destination code for each wire. ***Figure 22-8*** is an example of a lineless diagram. The following steps are used in drawing such a diagram:

1. Draw the electronic components in the approximate locations they will hold in the actual circuit.
2. Construct the wiring table.

Block Diagrams

Block diagrams are the most widely used types of diagrams in electronics engineering settings. They are also used a great deal in other applications. There are three types of block diagrams.

1. Organizational charts
2. Process flow charts
3. Functional block diagrams

Figures 22-9, ***22-10***, and ***22-11*** illustrate the three types of block diagrams. Block diagrams are drawn according to the following steps:

1. Lightly lay out the boxes at an approximate size.
2. Lightly lay out connecting lines.
3. Darken all horizontal lines.
4. Darken all vertical lines.
5. Draw circles (when required).

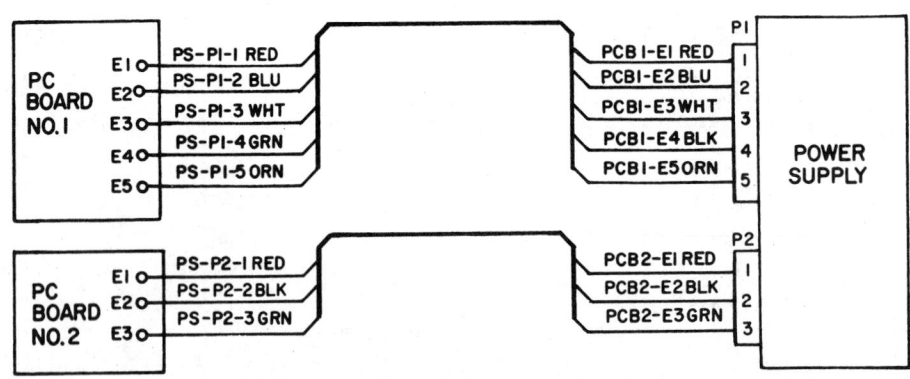

Figure 22-7   Highway diagram *(From Electronic Drafting and Printed Circuit Board Design, 2nd Edition, by James M. Kirkpatrick, ©1989 by Delmar Publishers Inc.)*

ORGANIZATIONAL CHART
QUALITY ASSURANCE AND HUMAN RESOURCES

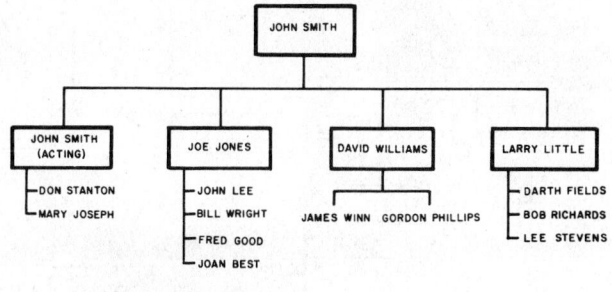

Figure 22-9   Organizational chart

6. Draw arrowheads.
7. Letter text moving from left to right.

Rules for Drawing Block Diagrams

There are several rules of thumb which should be observed where drawing block diagrams. These rules lend a degree of standardization to block diagrams and make them easier to read.

1. Whenever possible, draw all boxes the same size.
2. Whenever possible, show the flow from left to right (**Figure 22-12**).
3. When a left-to-right flow is not possible, top-to-bottom flow is acceptable.
4. Crossovers and doglegs should be avoided (**Figure 22-13**).

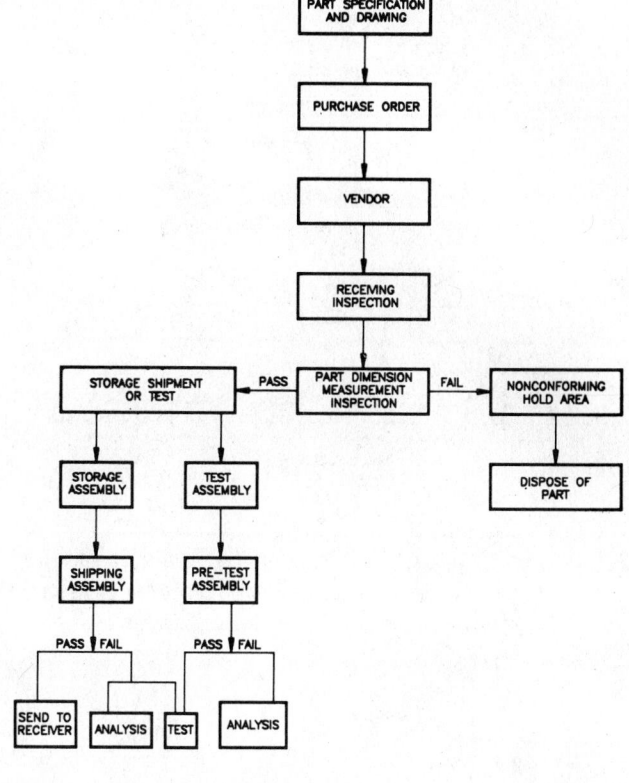

Figure 22-10   Process flow chart

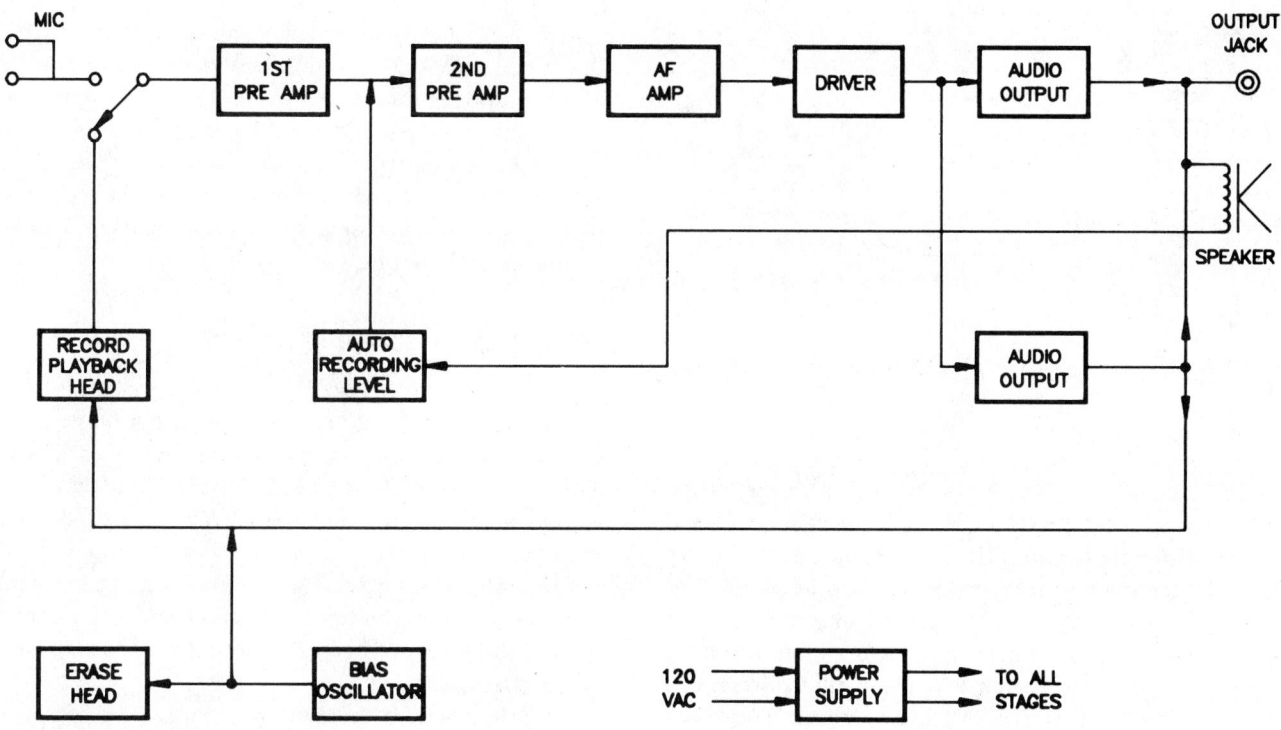

Figure 22-11   Functional block diagram

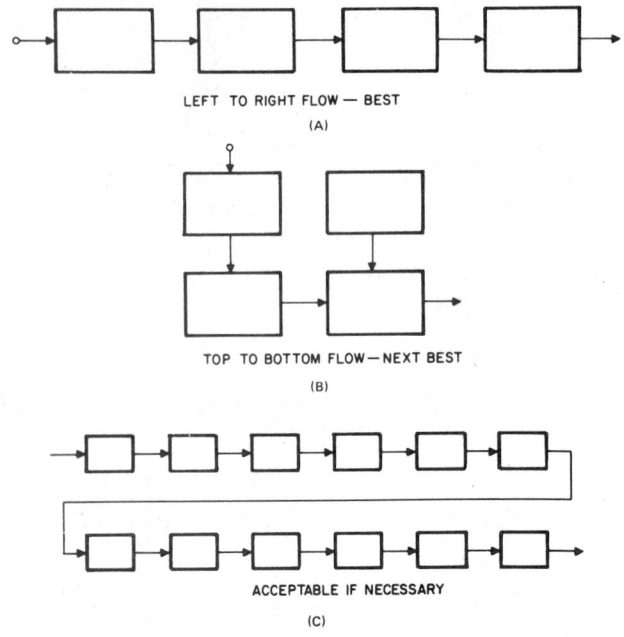

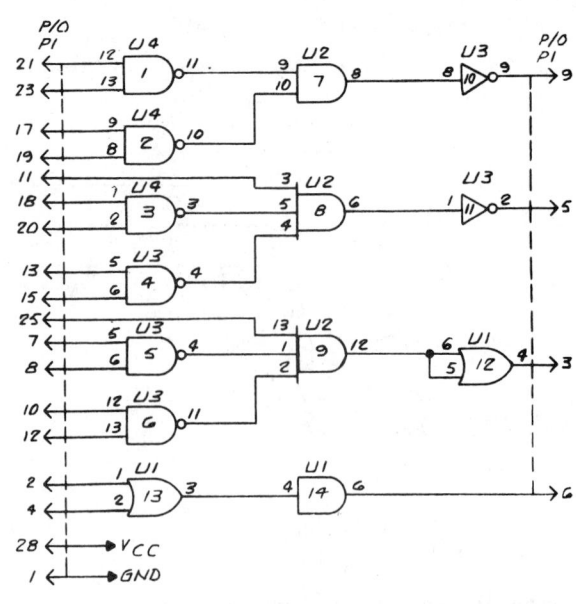

Figure 22-12 Acceptable flow patterns *(From Electronic Drafting and Printed Circuit Board Design, 2nd Edition, by James M. Kirkpatrick, ©1989 by Delmar Publishers Inc.)*

Figure 22-14 Logic diagram *(From Electronic Drafting and Printed Circuit Board Design, 2nd Edition, by James M. Kirkpatrick, ©1989 by Delmar Publishers Inc.)*

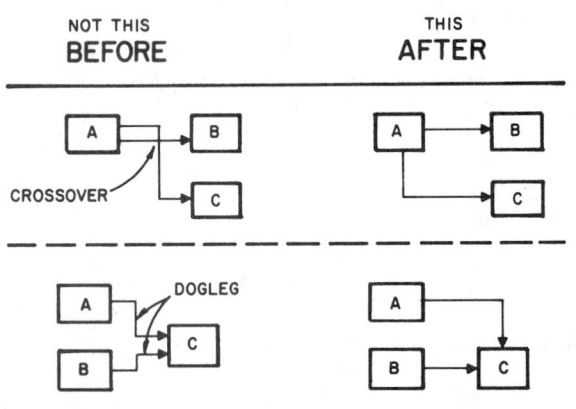

Figure 22-13 Avoiding crossovers and doglegs *(From Electronic Drafting and Printed Circuit Board Design, 2nd Edition, by James M. Kirkpatrick, ©1989 by Delmar Publishers Inc.)*

1. AND gate
2. OR gate
3. NAND gate (negative AND gate)
4. NOR gate (negative OR gate)

Other symbols that are frequently used with these gates are those for AMPLIFIER, FLIP FLOP, DELAY, SINGLE SHOT, GENERAL and SCHMITT TRIGGER. These symbols are also illustrated in Figure 22-15.

The number and letter tables which accompany each gate symbol in Figure 22-15 are called "truth tables." Truth tables are used for analyzing the various outputs that a particular gate will produce as a result of different combinations of inputs. For example, an AND gate such as the one shown in Figure 22-15 can be analyzed using its accompanying truth table. An input of "O" at "A" and "I" at "B" will produce an output of "O" at "F." Logic diagrams are drawn using special templates or symbols menus in the case of CADD.

Logic Diagrams

Logic diagrams are used in conjunction with the design of integrated circuits (ICs). They are most widely used with electronic circuits that are based on the binary numbering concept, such as those found in computers. *Figure 22-14* is an example of a logic diagram.

The principle components in logic diagrams are called "gates." A gate is a miniature circuit that performs a specific function within a larger circuit. The most widely used gates are illustrated in *Figure 22-15*. The gates are:

Printed Circuit Board Drawings

A printed circuit board is a special laminated board upon which electronic components are mounted. The components are connected by metal "traces" which run along the surface of the board. *Figure 22-16* is an illustration of a printed circuit board. This is a single-sided printed circuit board. In the top view, the electronics components can be seen as mounted on the board. In the bottom view, the traces which connect the components can be seen. There are three main types of printed circuit boards:

1. Single-sided boards
2. Double-sided boards
3. Multilayered boards

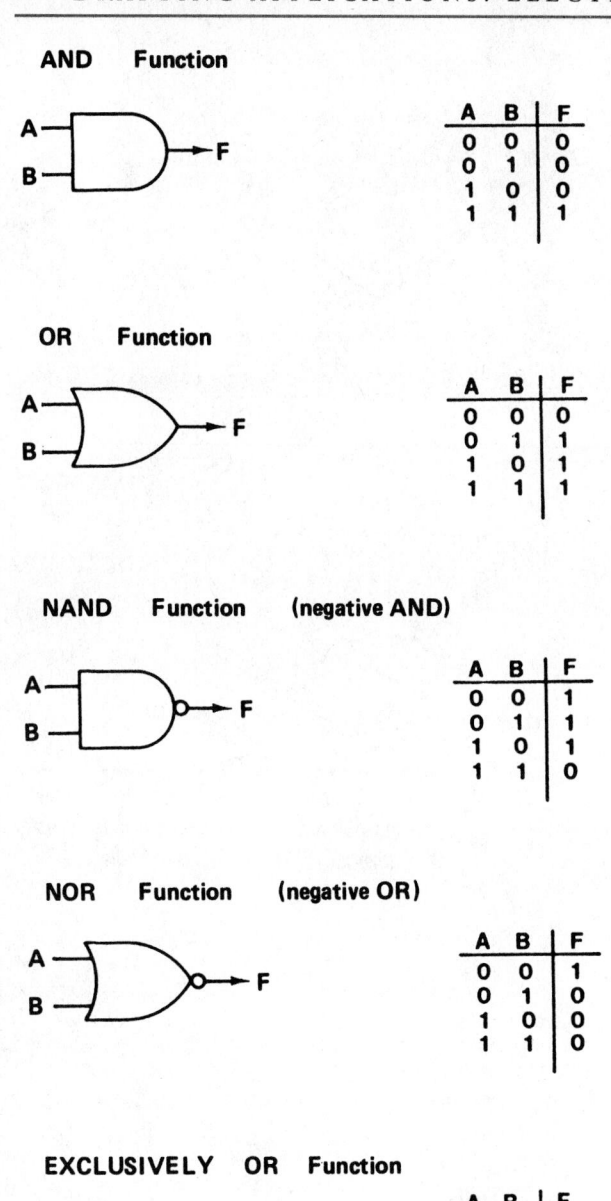

AND Function

A	B	F
0	0	0
0	1	0
1	0	0
1	1	1

OR Function

A	B	F
0	0	0
0	1	1
1	0	1
1	1	1

NAND Function (negative AND)

A	B	F
0	0	1
0	1	1
1	0	1
1	1	0

NOR Function (negative OR)

A	B	F
0	0	1
0	1	0
1	0	0
1	1	0

EXCLUSIVELY OR Function

A	B	F
0	0	0
0	1	1
1	0	1
1	1	0

OTHER SYMBOLS

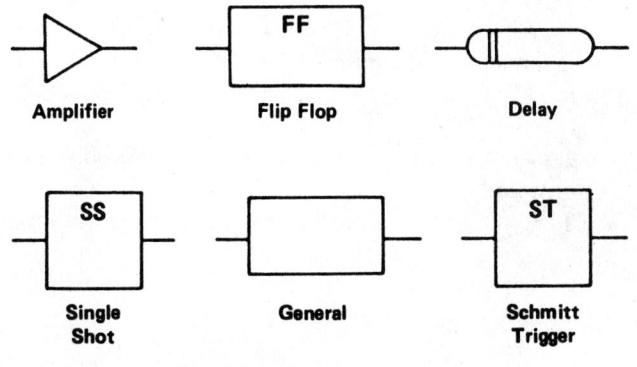

Amplifier   Flip Flop   Delay

Single Shot   General   Schmitt Trigger

Figure 22-15   Logic symbols

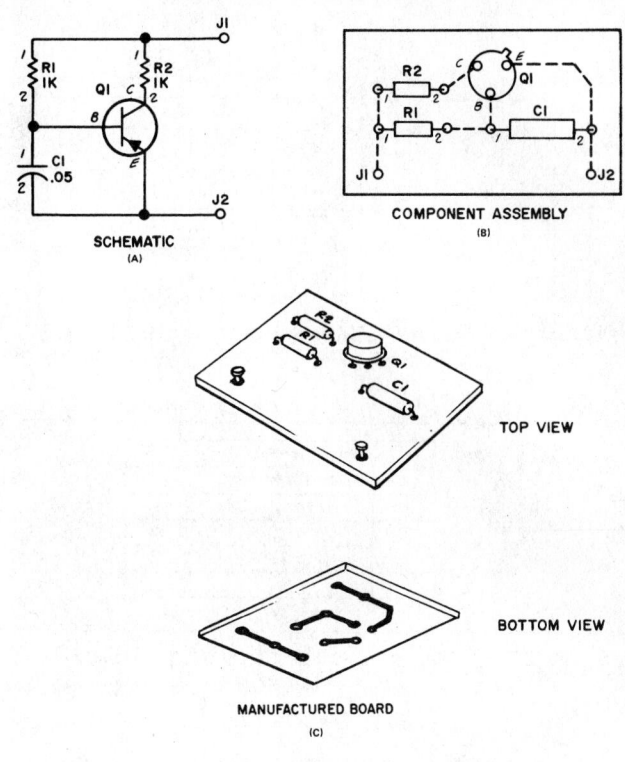

Figure 22-16   Printed circuit boards *(From Electronic Drafting and Printed Circuit Board Design, 2nd Edition, by James M. Kirkpatrick, ©1989 by Delmar Publishers Inc.)*

Single-sided boards have the electronic components mounted on one side and the connecting circuit etched on the other side. Double-sided boards have components mounted on one side also, but the circuits are etched on both sides. Multilayered boards are made up of two or more single- and/or double-sided boards.

The most frequently made types of printed circuit board drawings are the drill plan and artwork. The drill plan is a typical mechanical drawing that shows where holes are to be drilled through the board. The drill plan includes a hole schedule that contains at least three components.

1. Hole designation (usually a letter)
2. Hole size or description
3. Number of holes of each designation

Figure 22-17 is an example of a drill plan for a printed circuit board.

Artwork for a printed circuit board may be produced on a photoplotter, CAD system, or with tape. The artwork shows the circuit that will be etched on the printed circuit board, ***Figure 22-18***. Artwork is usually layed out at least twice the size of the actual board, and reduced photographically. Figure 22-18 also shows the component side of the same board. ***Figure 22-19*** shows how components are actually mounted on a printed circuit board.

The process of preparing printed circuit board drawings involves four steps.

1. Study the schematic and make a list of all electronic components it contains.

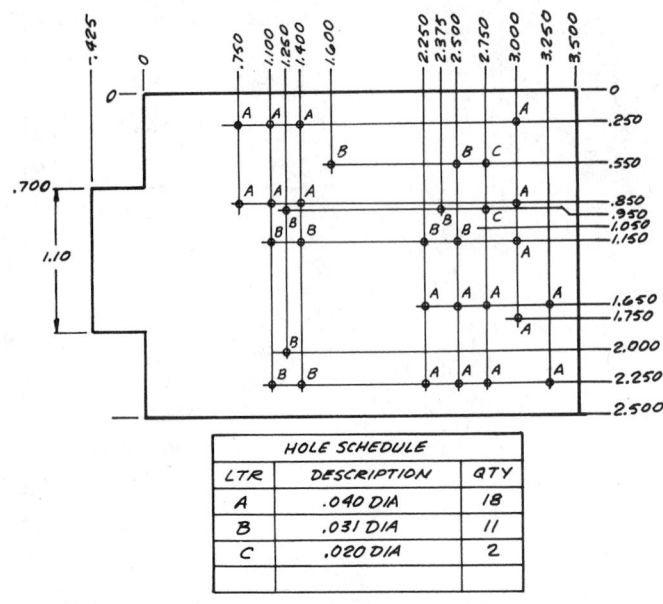

Figure 22-17 Drill plan *(From Electronic Drafting and Printed Circuit Board Design, 2nd Edition, by James M. Kirkpatrick, ©1989 by Delmar Publishers Inc.)*

LTR	DESCRIPTION	QTY
A	.040 DIA	18
B	.031 DIA	11
C	.020 DIA	2

HOLE SCHEDULE

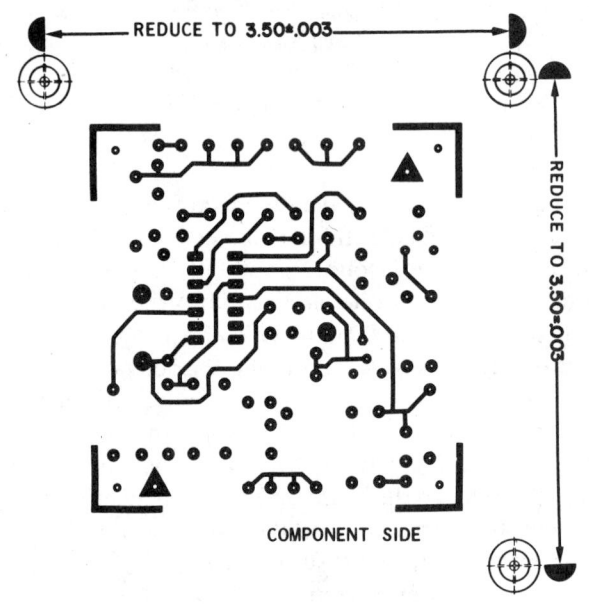

Figure 22-18 Artwork

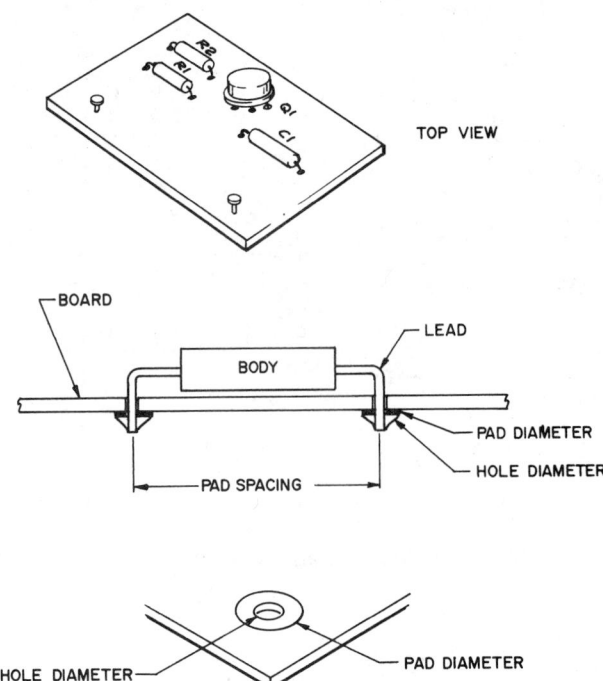

Figure 22-19 Mounting components *(From Electronic Drafting and Printed Circuit Board Design, 2nd Edition, by James M. Kirkpatrick, ©1989 by Delmar Publishers Inc.)*

Review

1. Draw the symbols for the following:
   - Capacitor
   - Diode
   - Resistor
   - Transistor
   - Delay
   - Rectifier
2. What are schematic diagrams used for?
3. On a schematic diagram, what are the rules pertaining to inputs, outputs, and grounds?
4. What are the four types of connection diagrams?
5. What are the three types of block diagrams?
6. What is the preferred direction of flow on a block diagram?
7. What is the second preferred direction of flow on a block diagram?
8. What are the four most widely used gates in logic diagrams?
9. What are the three main types of printed circuit boards?
10. What are the components of a drill plan for a printed circuit board?

2. Redraw the schematic, substituting pictorial representations of each symbol. You will have to consult manufacturers catalogs for the dimensions of components.
3. Rearrange the components in such a way as to produce the best conductor paths for the circuit. Avoid arrangements that will cause sharp turns, acute angles, and doglegs in conductor paths. Keep the paths as short as possible
4. Prepare the printed circuit board drawings at a scale of 2, 3, 4, or 5 to 1.

Chapter Twenty-Two Problems

Problems 22-1 and 22-2

Below are engineer's sketches containing several logic gates and other electronic components. Convert the sketches into finished schematics.

a. Redraw the schematics using pictorial representations.

b. Use the pictorial schematics as guides in developing drill plans and printed circuit board artwork.

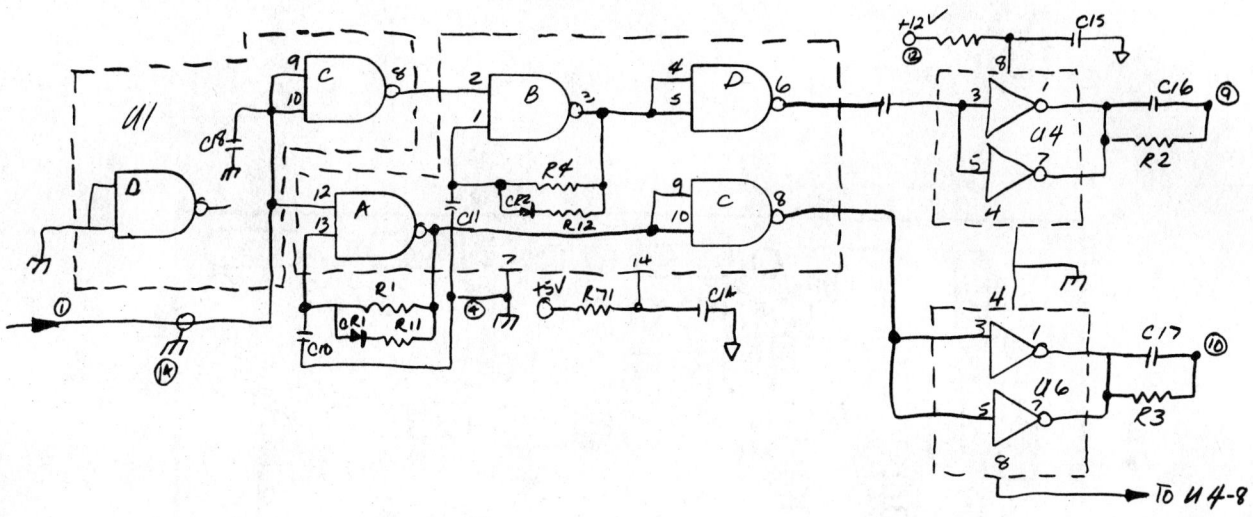

Problem 22-1

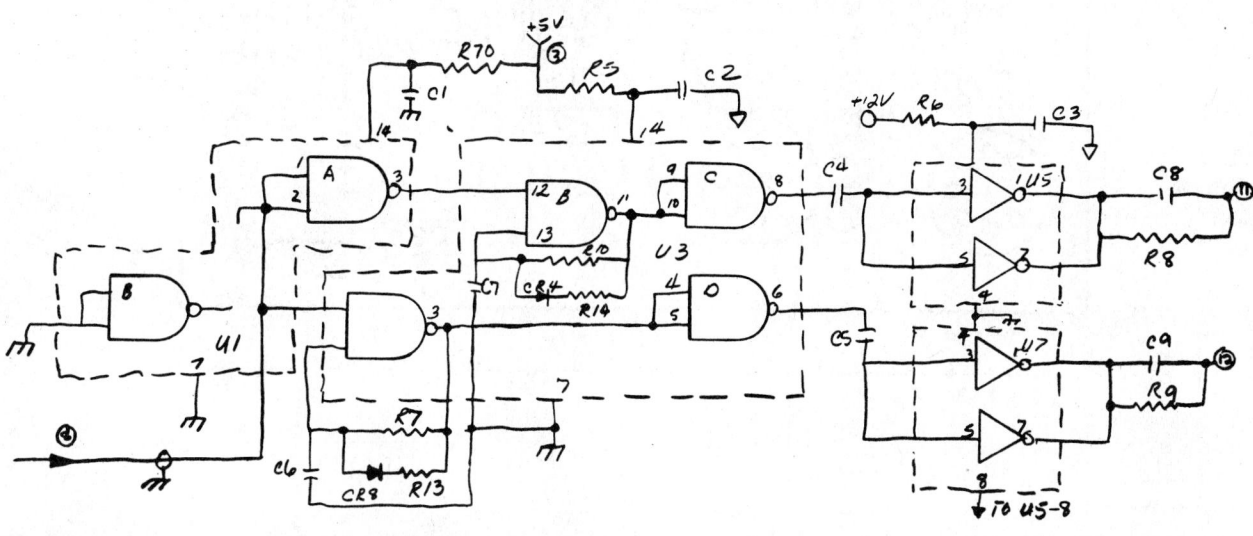

Problem 22-2

Problem 22-3

Problem 22-3 is a sketch of a pictorial schematic. Using it as a guide, complete the following tasks:

    a. Develop the symbolic schematic plan from which such a pictorial schematic might have been drawn.

    b. Make a drill plan and printed circuit board artwork using the pictorial schematic as a guide.

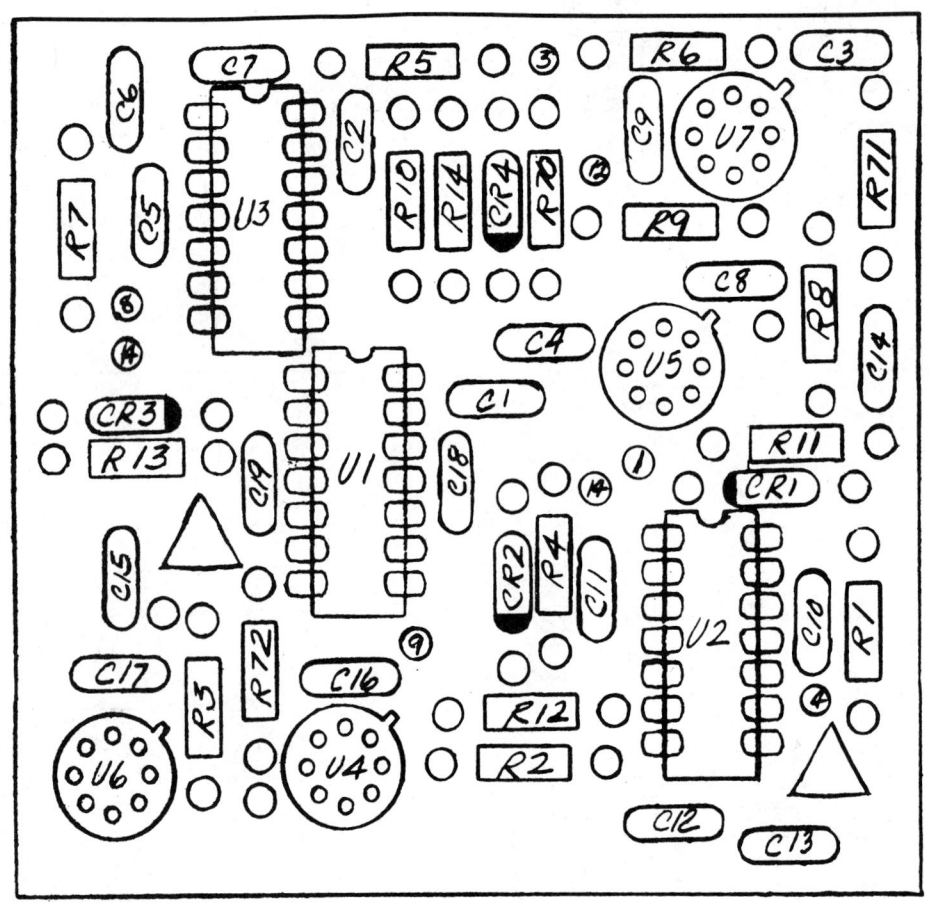

Problem 22-3

Problem 22-4

Following the rules and steps set forth in this chapter, lay out and draw finished diagrams from the sketches provided.

a. Block diagram

b. Schematic diagram

c. Schematic diagram

d. Logic diagram

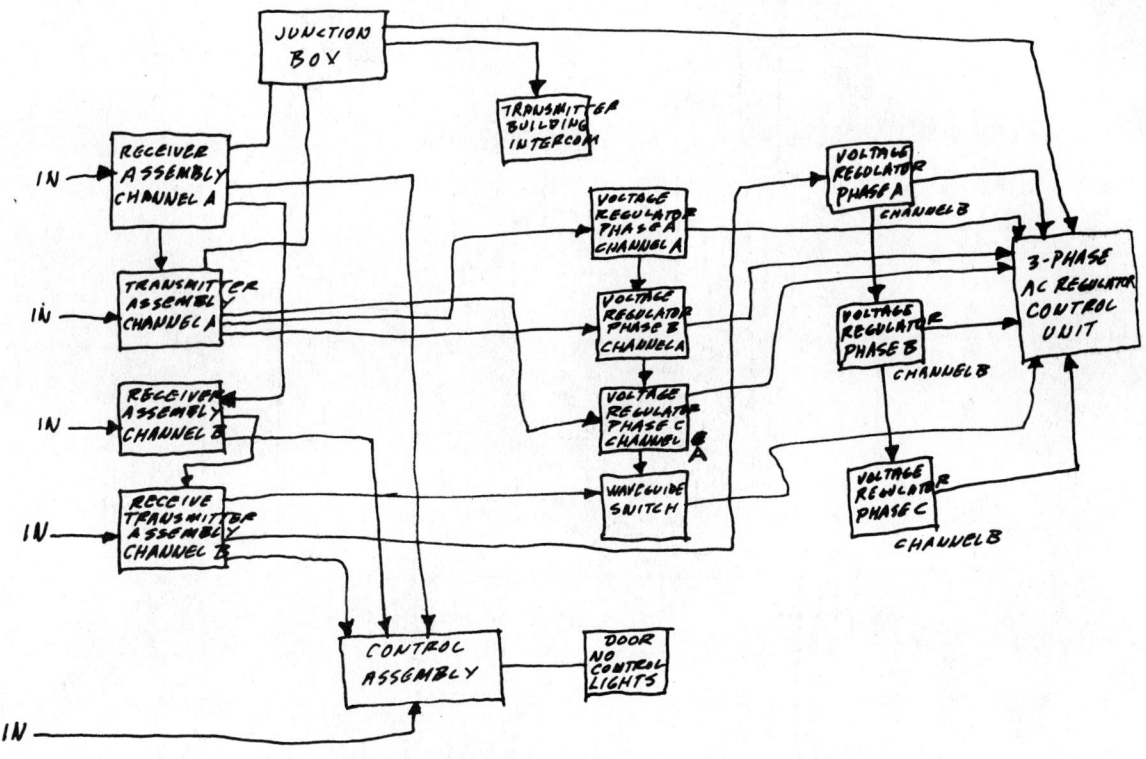

Problem 22-4A *(From Electronic Drafting and Printed Circuit Board Design, 2nd Edition, by James M. Kirkpatrick, ©1989 by Delmar Publishers Inc.)*

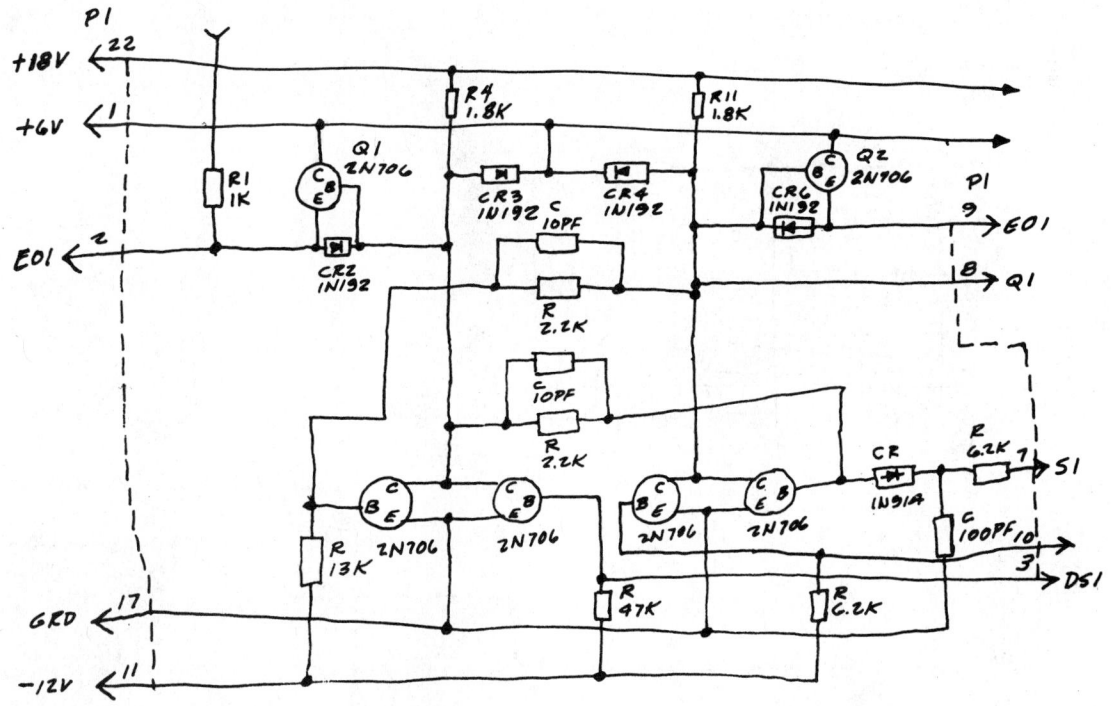

Problem 22-4B *(From Electronic Drafting and Printed Circuit Board Design, 2nd Edition, by James M. Kirkpatrick, ©1989 by Delmar Publishers Inc.)*

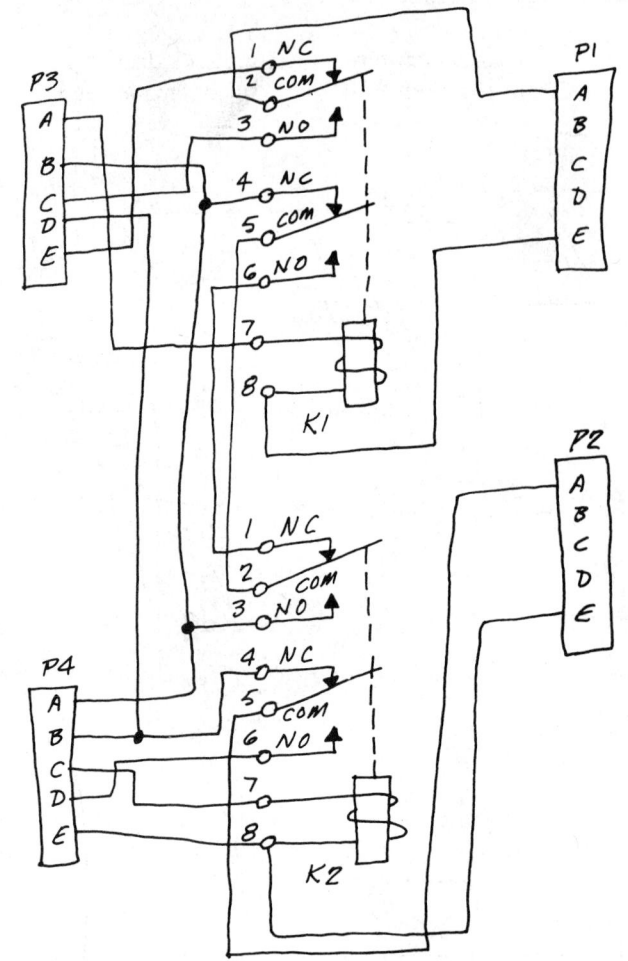

❑ **Problem 22-4C** *(From Electronic Drafting and Printed Circuit Board Design, 2nd Edition, by James M. Kirkpatrick, ©1989 by Delmar Publishers*

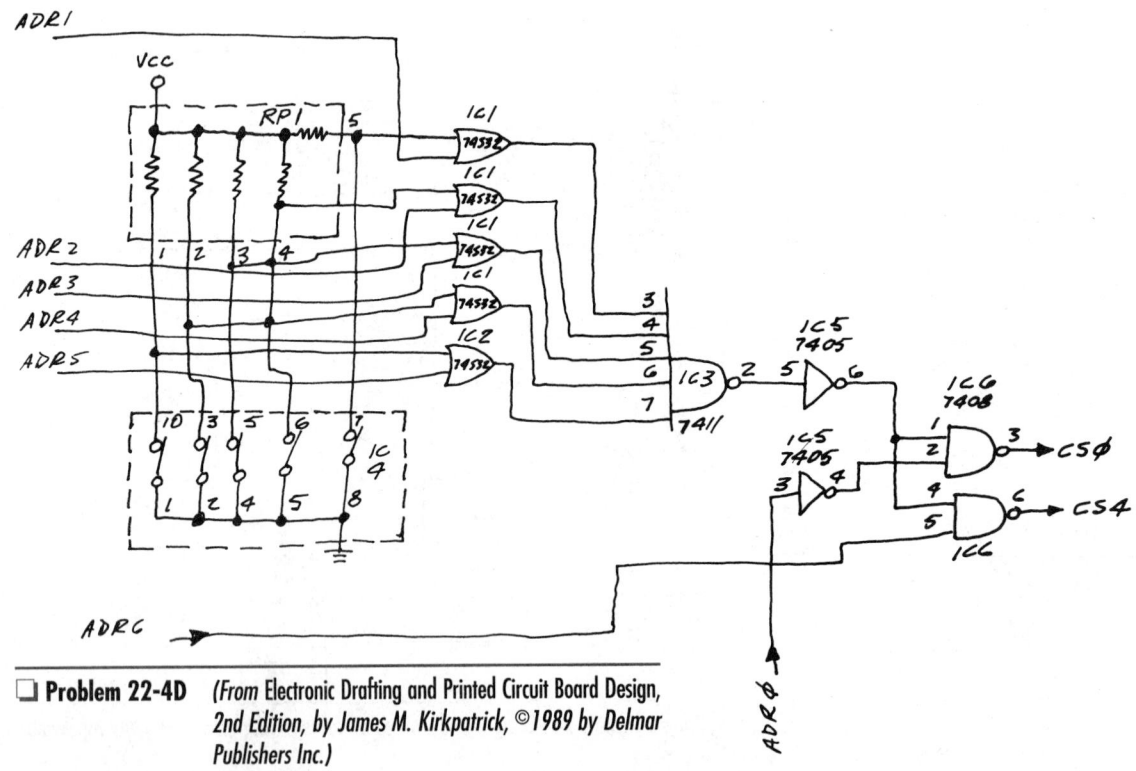

❑ **Problem 22-4D** *(From Electronic Drafting and Printed Circuit Board Design, 2nd Edition, by James M. Kirkpatrick, ©1989 by Delmar Publishers Inc.)*

Charts and Graphs

Functional Classes: An Overview

The two major classes of charts and graphs to be discussed are those used to communicate information and those used to generate information.

Charts and Graphs Used to Communicate Information

Pie Charts. Pie charts look like circular pies and usually represent 100% of a budget, *Figure 23-1*. Hence, the cliche "That's how they slice the (money) pie." School newspapers show these annually. A pie chart appears later in this chapter in conjunction with spreadsheets.

Bar Charts. Bar charts, either horizontal or vertical, most often show relative amounts, *Figure 23-2A*; however, they are frequently used to show the relative timing of activities in the form of planning chart, *Figure 23-2B*.

Pictorial Charts. Pictorial charts employing varying sizes of figures seem to be favored by newspapers as a means of illustrating relative amounts to the general public, *Figures 23-3A* and *23-3B*.

Profile Charts. Material science texts use profile charts to illustrate the roughness of the surface of a metal by slicing the metal at a shallow angle and show-

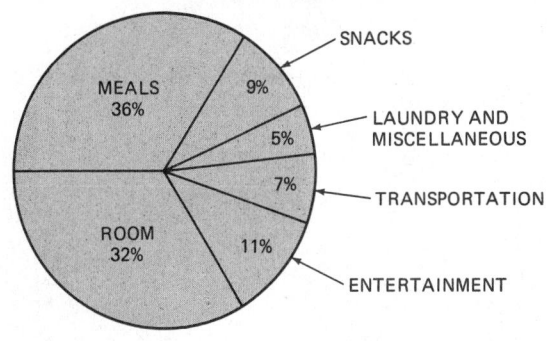

MONTHLY BUDGET:
STUDENT LIVING AWAY FROM HOME

Figure 23-1   Pie chart

ing the cut surface in a magnified plot, *Figure 23-4A*. A cross section of a portion of a continent, with condensed scales, is used to show terrain characteristics in a profile chart. Civil engineers prepare profile charts to determine how much earth to remove or replace ("cuts and fills") when constructing a highway through a hilly terrain, *Figure 23-4B*.

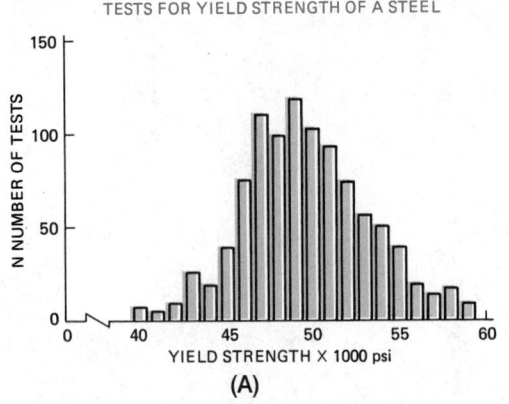

TYPICAL DISTRIBUTION OF TENSILE
TESTS FOR YIELD STRENGTH OF A STEEL

(A)

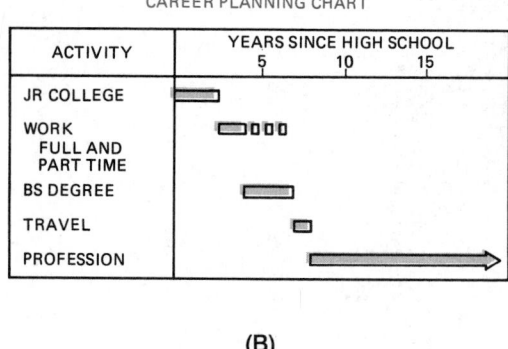

CAREER PLANNING CHART

(B)

Figure 23-2   (A) Bar chart;   (B) Planning chart

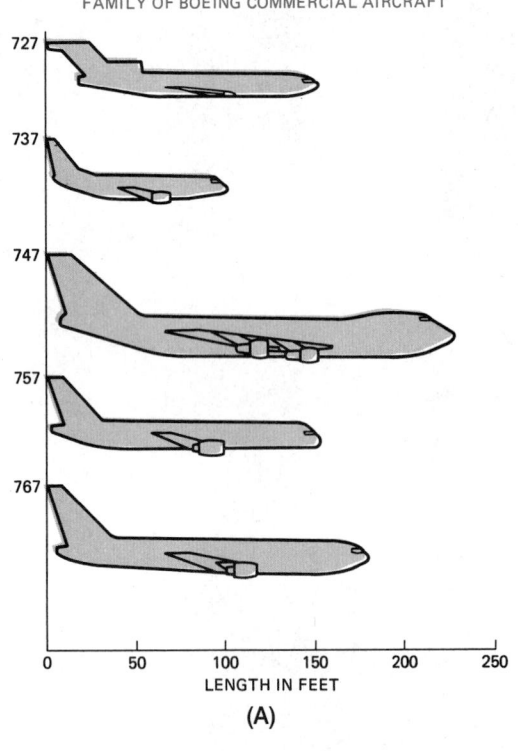

FAMILY OF BOEING COMMERCIAL AIRCRAFT

(A)

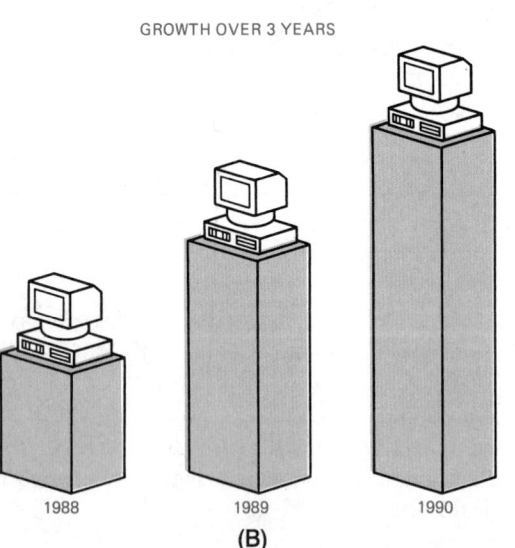

GROWTH OVER 3 YEARS

(B)

Figure 23-3   (A) and (B) Pictorial bar charts

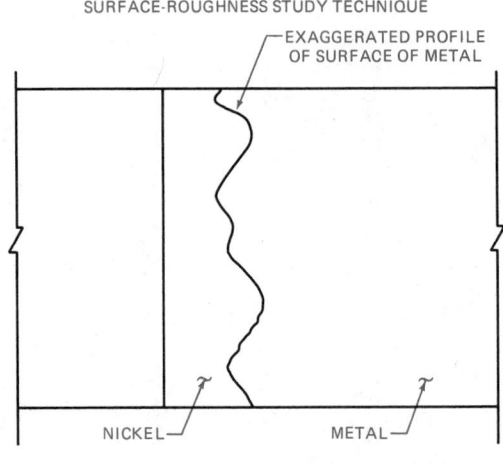

SURFACE-ROUGHNESS STUDY TECHNIQUE

(A)

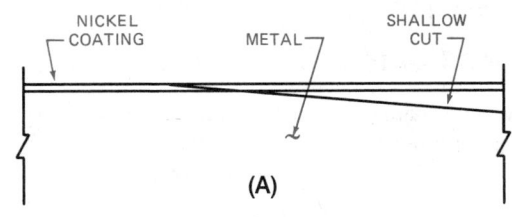

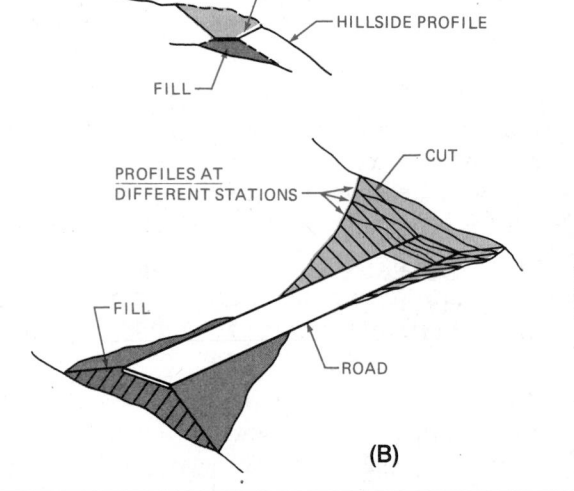

PROFILES FOR "CUTS AND FILLS"

(B)

Figure 23-4   (A) Profile chart: Material science engineering;   (B)
Cuts and fills profiles: Civil engineering

Schematic Charts. Electrical and electronic schematics of circuits are the most common, ***Figure 23-5A***; however, there are others, such as piping, hydraulic power (***Figure 23-5B***), logic, fault tree, and failure tree schematics.

Flow Charts. Students encounter flow charts in computer programming courses, ***Figure 23-6A***. Other uses include flow of information in an organization, flow of materials in a manufacturing plant, and flow of energy uses in an agricultural facility, ***Figure 23-6B***.

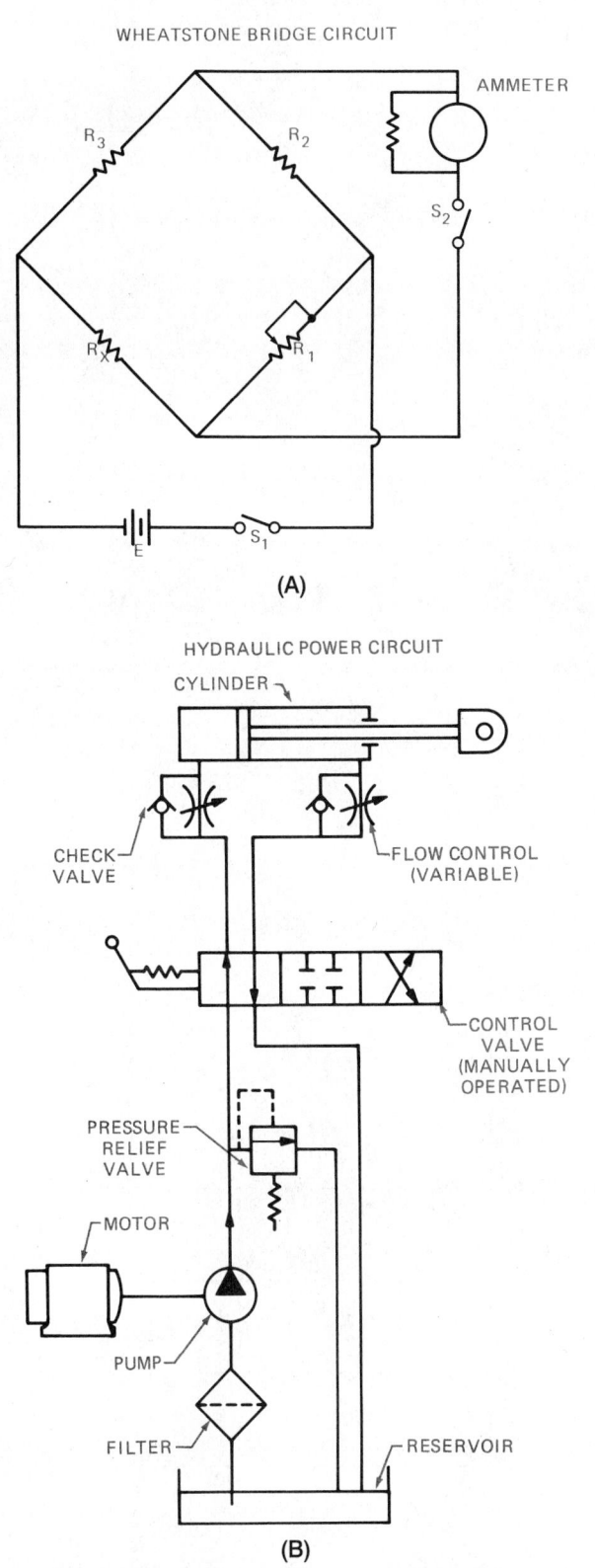

Figure 23-5 (A) Electrical schematic; (B) Hydraulic power schematic

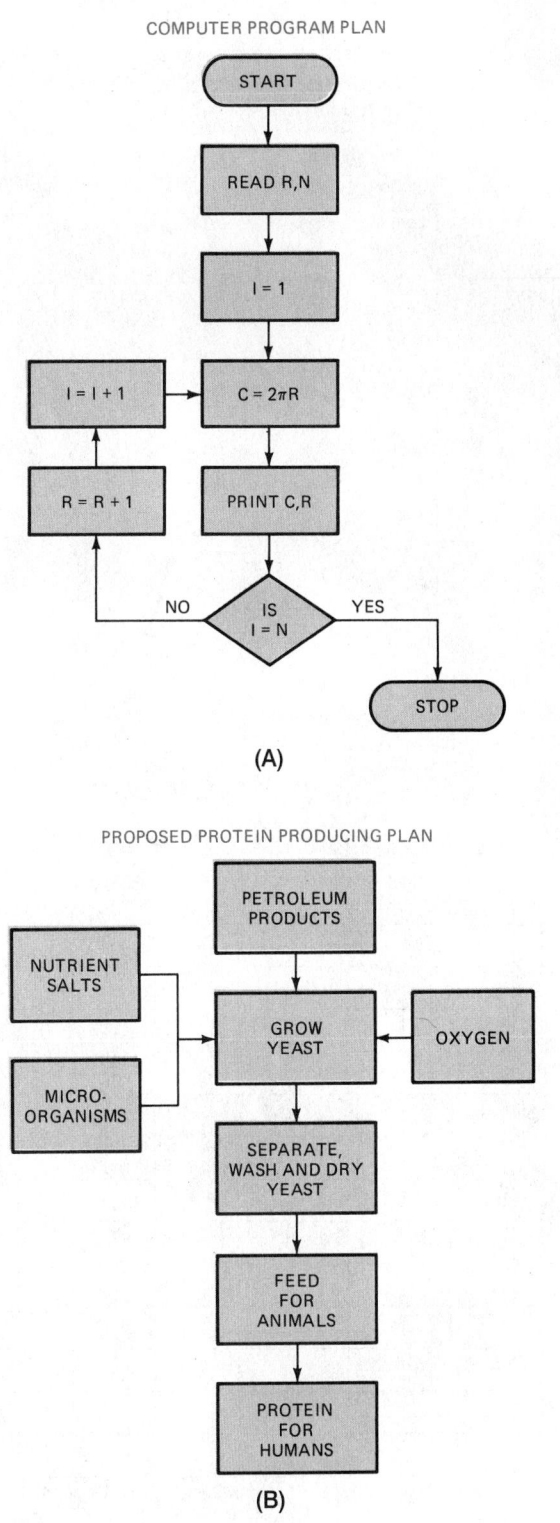

Figure 23-6 (A) Flow chart: Computer program; (B) Flow chart: Agricultural process

Linear Coordinate Graphs. Basic science (mathematics, physics, biology, and chemistry) and engineering science (statics, dynamics, kinematics, strength of materials, materials, and thermodynamics) textbooks save thousands of words by using linear coordinate graphs, in both two-dimensional and three-dimensional formats, *Figures 23-7A* and *23-7B*. Newspapers use them to save space and to communicate effectively. One important use in technology is to show "ranges of values" or "error bars" at data points, *Figures 23-7C* and *23-7D*. The range of the Dow Jones Average during any one day is an excellent illustration. Error bars on data points representing an experiment show the ranges of accuracy that can be expected for certain conditions.

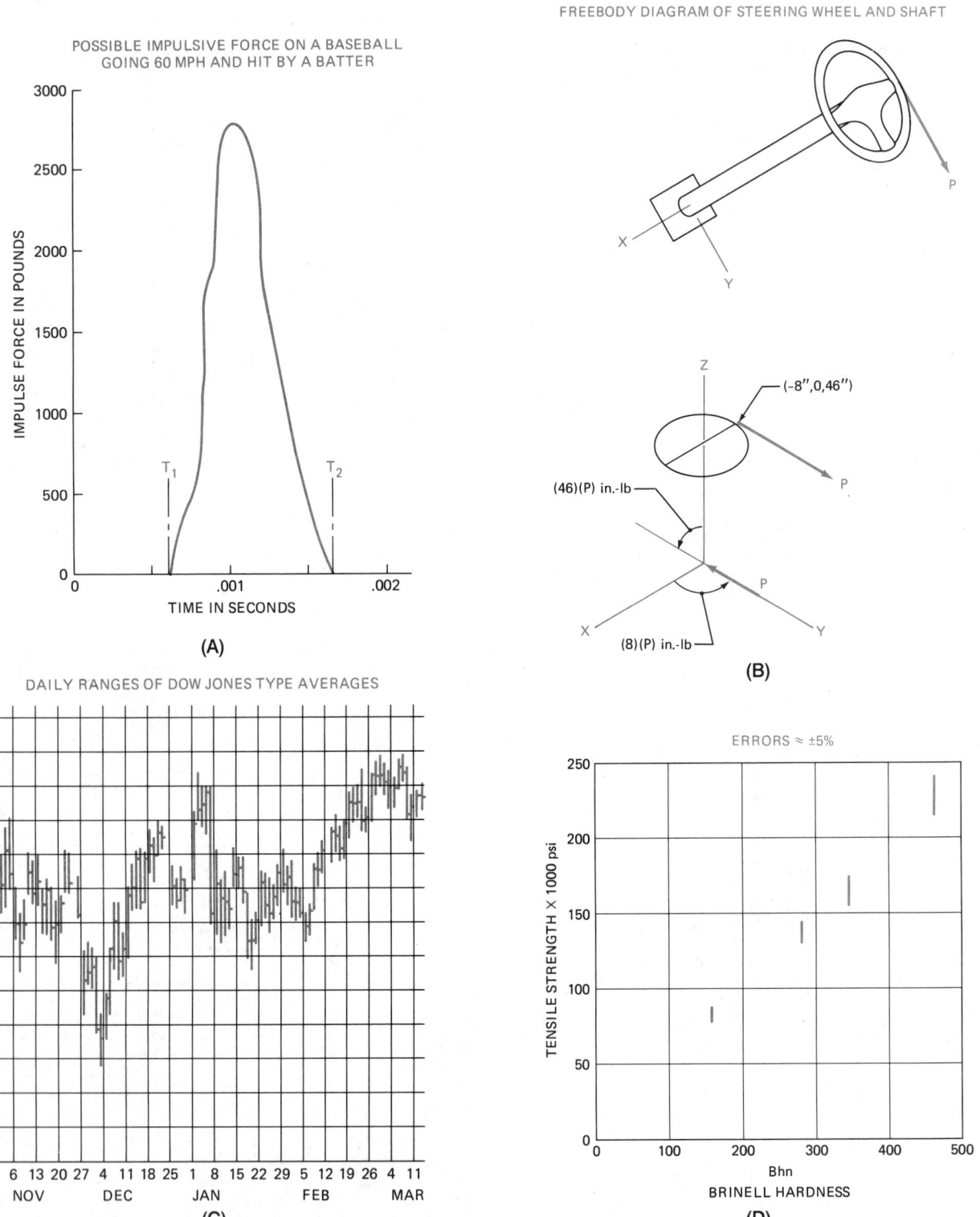

Figure 23-7 Linear coordinates: (A) Impulsive force vs. time; (B) Three-dimensional; (C) Ranges of values; (D) Error bars

Polar Graphs.   Polar graphs are 360° maps usually used to show the effect of a point source of energy (heat, light, electromagnetic sound) at varying distances from the source. Radio and TV stations use these to determine their local coverage. The use of polar graph paper saves hours in preparing these plots, *Figure 23-8*.

Charts and Graphs Used to Generate Information

Trilinear Graphs.   Metallurgical and chemical personnel use these equilaterally shaped graphs to illustrate how differing percentages of three elements of a mixture behave or should behave. For example, the sum of the three perpendiculars, one to each side, from a point equal the altitude of the triangle which would be 100% in all cases. Commercial grids are available, *Figure 23-9*.

Linear Graphs.   Grids for linear graphs are ruled uniformly along the horizontal and the vertical axes. This type is used here to introduce the general equation of a straight line, y = mx + b, where m is the slope, x the independent variable, y the dependent variable, and b the intercept along the vertical axis. Furthermore, y = mx + b appears here because it is used in conjunction with graphs to generate information, *Figures 23-10A*, *23-10B*, and *23-10C*. A more complete discussion appears later in this chapter in a section on empirical equations. Log-linear and log-log applications are also included.

Log-Linear Graphs.   Log-linear or semilogarithmic grids are linear, or uniformly ruled, along the horizontal axis and logarithmic, or ruled proportionately to the logarithms of numbers, along the vertical axis, *Figure 23-11*. Curves with equations y = b(e**mx) plot as straight lines on log-linear graphs. Therefore, if data from an experiment is plotted on a log-linear grid, and the data forms a straight line, then an equation for the data may be generated in the form y = b(e**mx). The variables are defined in the linear graph discussion. The number e = 2.718282.

Log-Log Graphs.   Log-log or logarithmic grids are ruled proportionately to the logarithms of numbers along both the horizontal and vertical axes. Curves with

LUMENS/FT² FOR A LIGHT FIXTURE @ 3 METERS

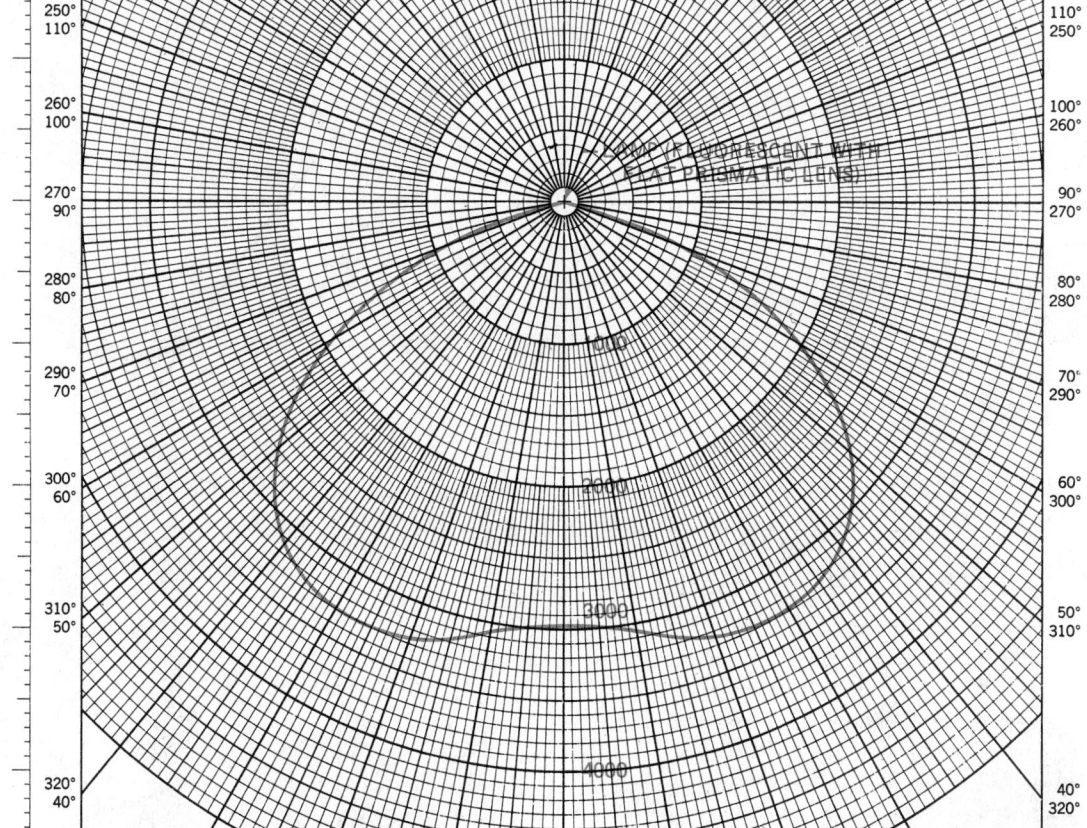

Figure 23-8   Polar coordinates graph

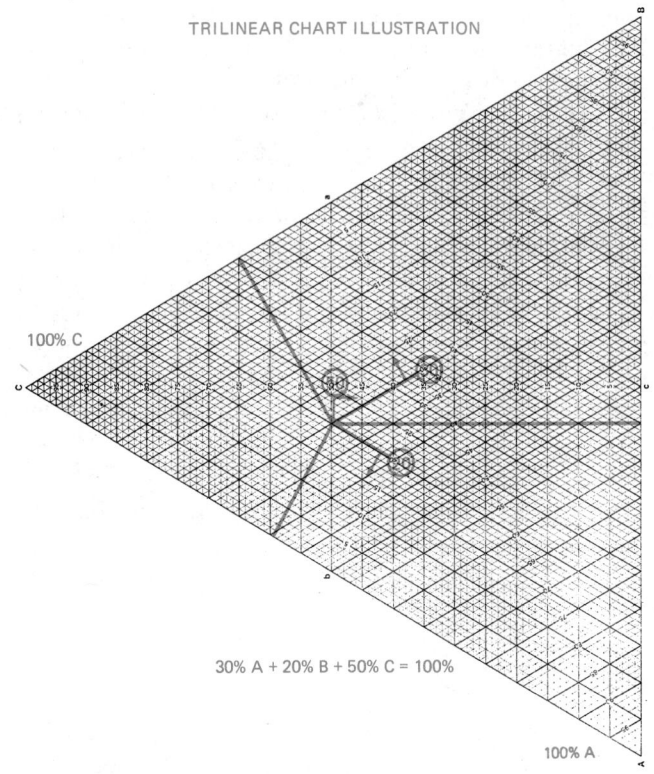

TRILINEAR CHART ILLUSTRATION

100% C

30% A + 20% B + 50% C = 100%

100% A

Figure 23-9   Trilinear graph

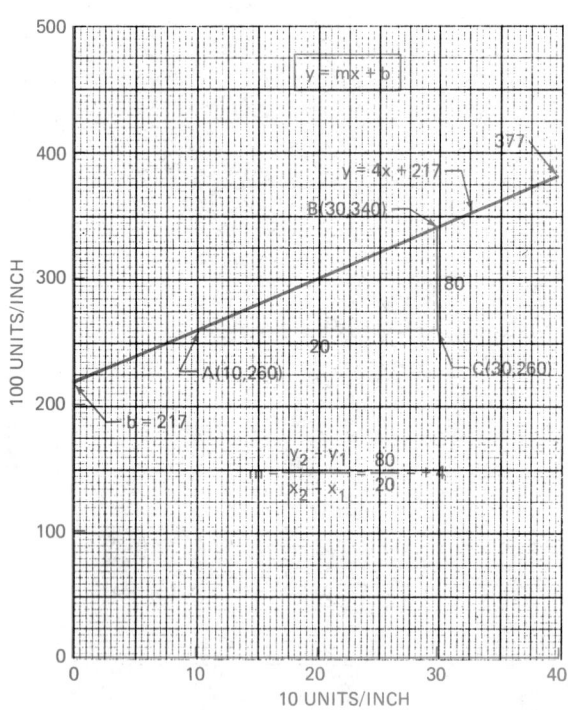

Figure 23-10A   Straight line equation: y = mx + b

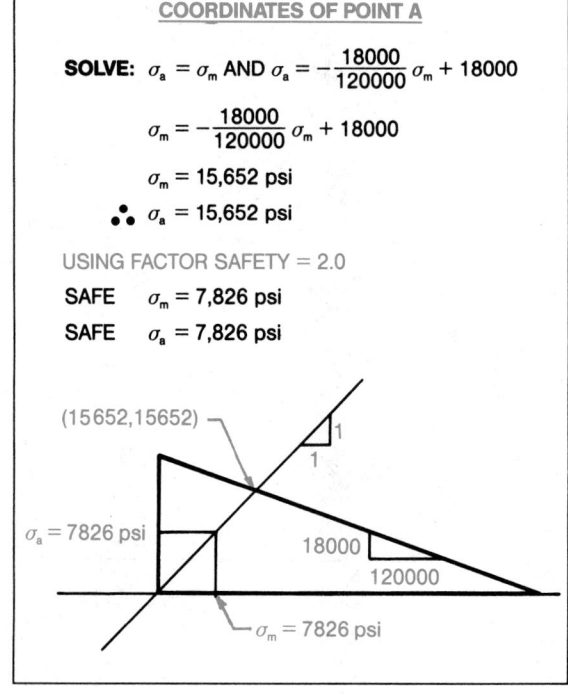

Figure 23-10B   Straight line equations: Fatigue analysis

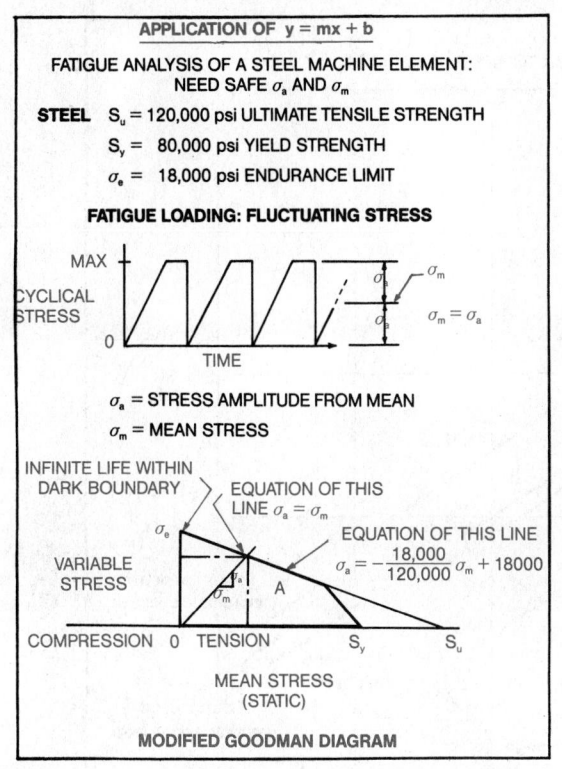

Figure 23-10C Straight line equations: Fatigue analysis

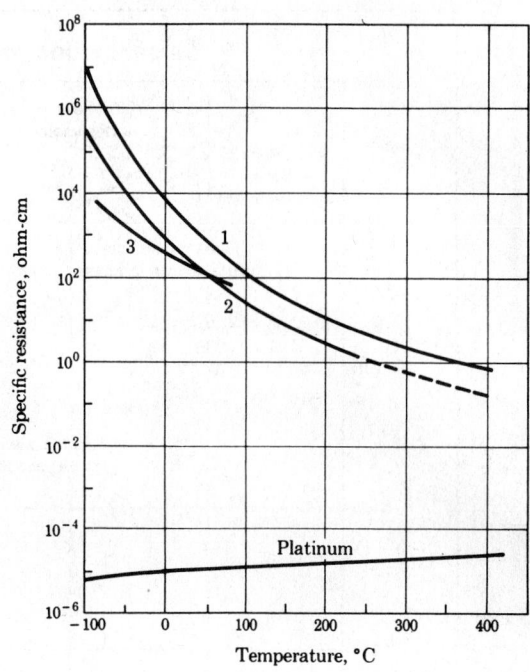

Figure 23-11 Log-linear graph (semilogarithmic) (*From Holman, J.P., Experimental Methods for Engineers, ©1966, McGraw-Hill Book Co. Used with permission.*)

equations y = b(x**m), plot as straight lines on log-log grid sheets, **Figures 23-12A** and **23-12B**. Therefore, if data from an experiment plots as a straight line on log-

log paper, an equation for the data may be generated in the form y = b(x**m). The variables are defined in the linear graph discussion.

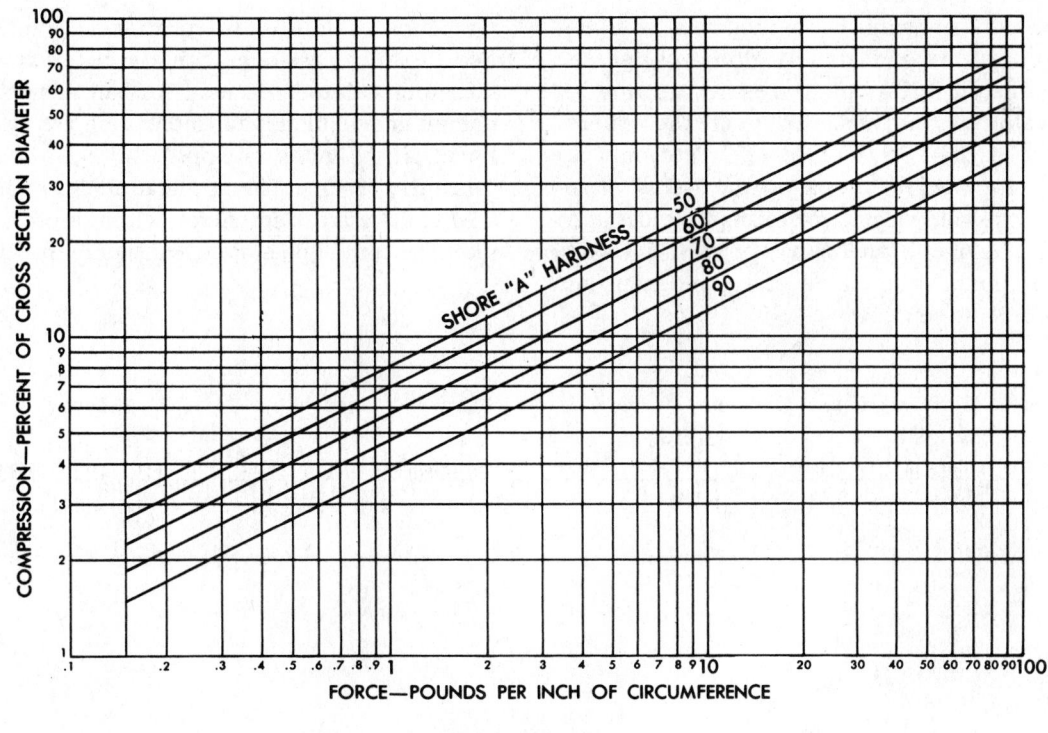

.139 Cross Section

Figure 23-12A Logarithmic graph: O-ring compression (*Courtesy Parker Seal*)

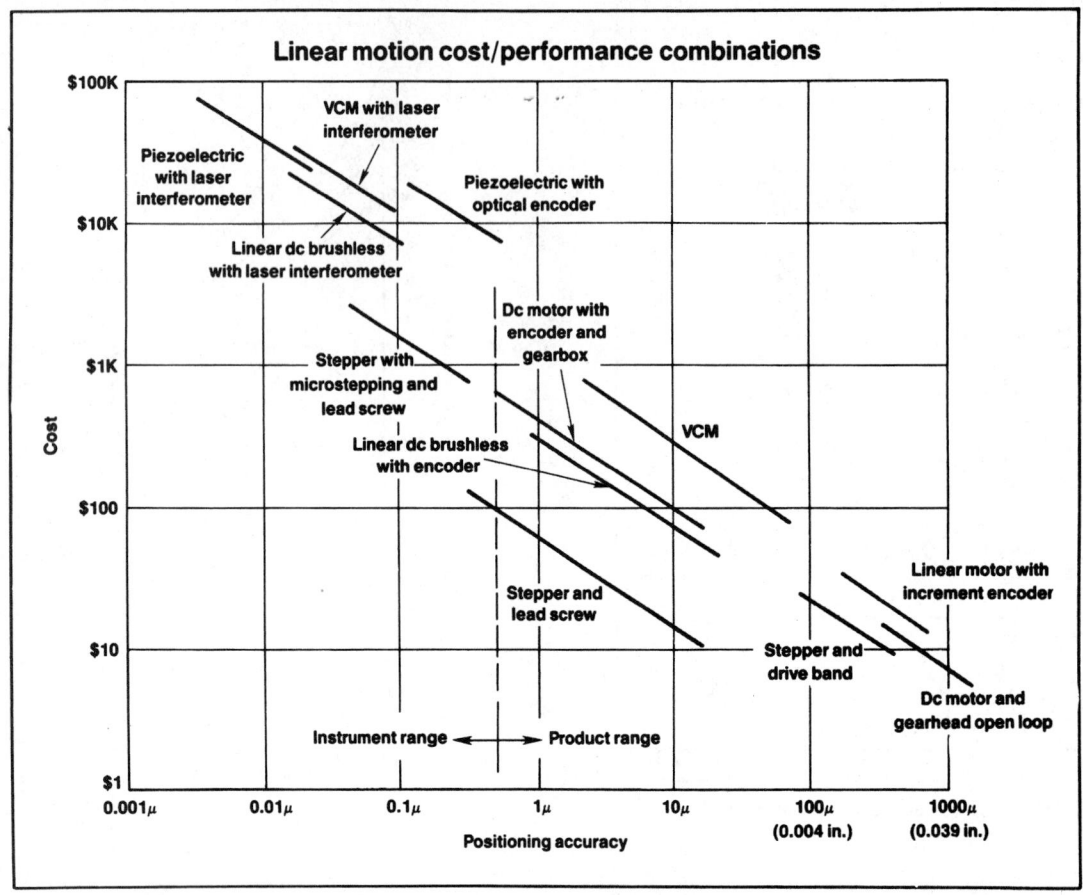

Figure 23-12B   Logarithmic graph: Cost vs. accuracy *(Reprinted from Machine Design, Oct. 27, 1987, ©1987, Penton Publishing Inc.)*

Nomographs.   Nomographs or nomograms are graphical representations of formulas along straight or curved, graduated lines which are used to find an unknown value of one variable, given the others, *Figure 23-13A*, *23-13B*, and *23-13C*. Not only do nomographs save time, but they allow the user to visualize ranges of possible values. For example, a designer can generate a number of alternatives quickly, and work toward an optimum solution. Several nomograms are discussed later in this chapter.

Graphical Calculus: Integration.   Graphical integration provides designers with a visual, reasonably accurate, approach to integration. Analytical, mechanical, and computer techniques for integrating are also available; however, the graphical approach helps the designer to "see" the results. The application, shown in *Figure 23-14*, generates the energy within a pressure-volume plot. The technique is discussed later in this chapter.

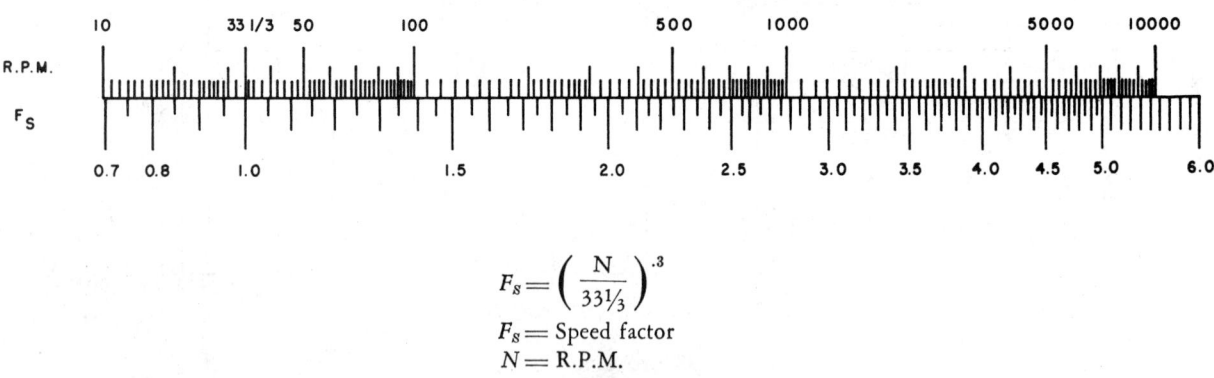

$$F_s = \left(\frac{N}{33\frac{1}{3}} \right)^{.3}$$

F_s = Speed factor
N = R.P.M.

Figure 23-13A   Adjacent alignment chart *(Courtesy McGill Manufacturing Co. Inc.)*

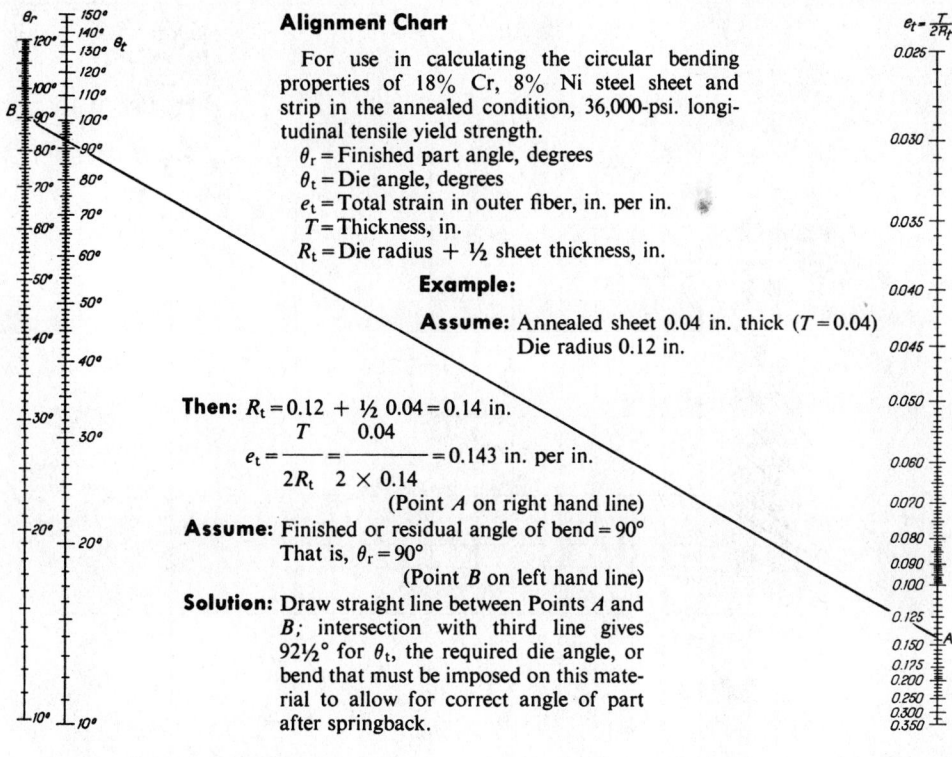

Alignment Chart

For use in calculating the circular bending properties of 18% Cr, 8% Ni steel sheet and strip in the annealed condition, 36,000-psi. longitudinal tensile yield strength.

θ_r = Finished part angle, degrees
θ_t = Die angle, degrees
e_t = Total strain in outer fiber, in. per in.
T = Thickness, in.
R_t = Die radius + ½ sheet thickness, in.

Example:

Assume: Annealed sheet 0.04 in. thick ($T = 0.04$)
Die radius 0.12 in.

Then: $R_t = 0.12 + ½\ 0.04 = 0.14$ in.

$$e_t = \frac{T}{2R_t} = \frac{0.04}{2 \times 0.14} = 0.143 \text{ in. per in.}$$

(Point A on right hand line)

Assume: Finished or residual angle of bend = 90°
That is, $\theta_r = 90°$

(Point B on left hand line)

Solution: Draw straight line between Points A and B; intersection with third line gives 92½° for θ_t, the required die angle, or bend that must be imposed on this material to allow for correct angle of part after springback.

Circular bends on annealed 18-8.

Figure 23-13B   Parallel alignment chart: Three variables *(Courtesy Allegheny Ludlum Corp.)*

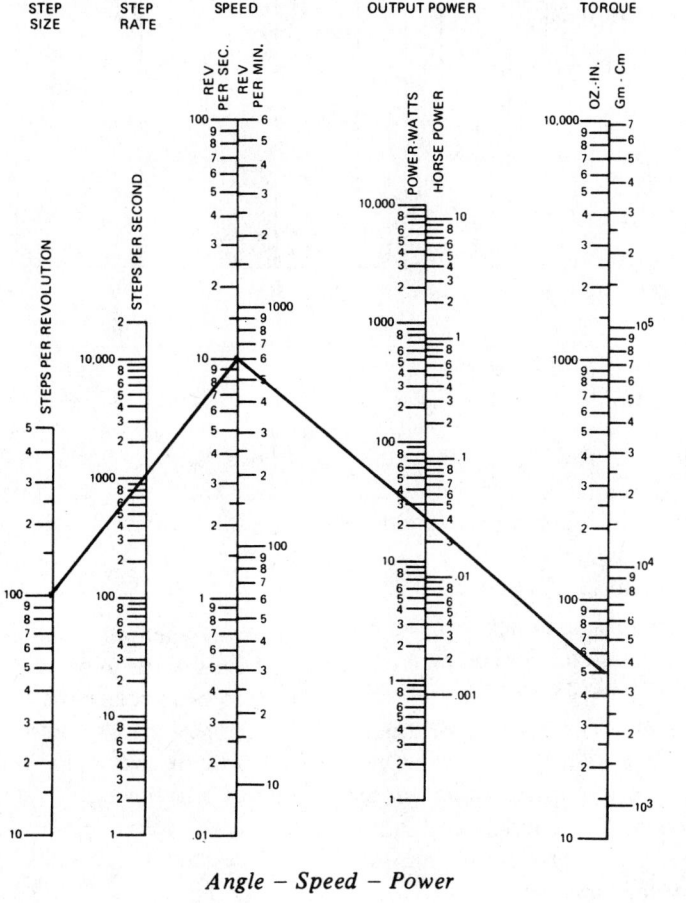

Angle – Speed – Power

Figure 23-13C   Parallel alignment chart: Four variables *(Courtesy Pacific Scientific)*

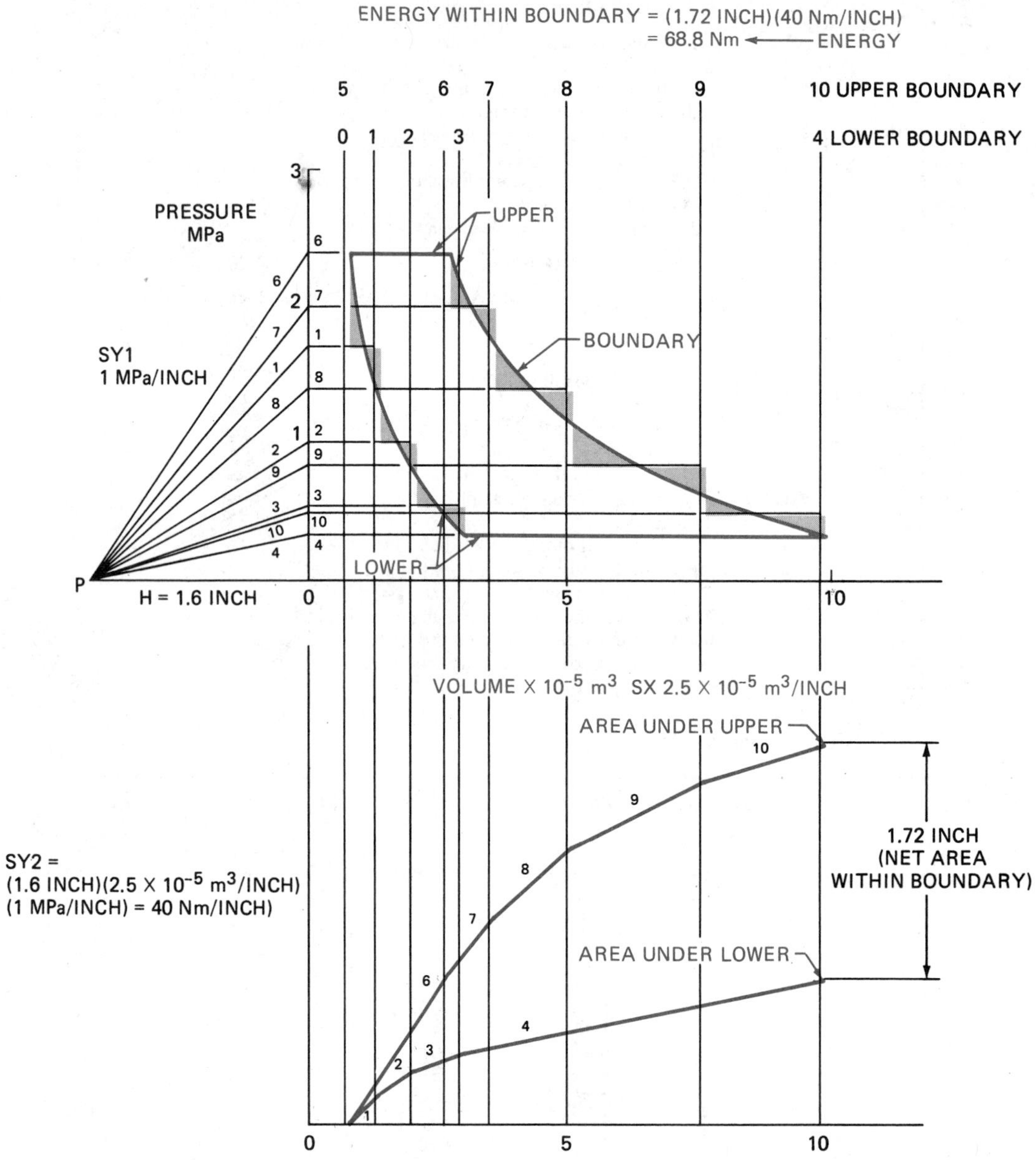

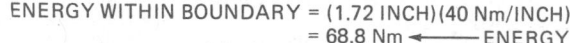

ENERGY WITHIN BOUNDARY = (1.72 INCH)(40 Nm/INCH)
= 68.8 Nm ← ENERGY
10 UPPER BOUNDARY
4 LOWER BOUNDARY
PRESSURE
MPa
UPPER
BOUNDARY
SY1
1 MPa/INCH
LOWER
P
H = 1.6 INCH
VOLUME X 10-5 m3 SX 2.5 X 10-5 m3/INCH
AREA UNDER UPPER
AREA UNDER LOWER
1.72 INCH
(NET AREA
WITHIN BOUNDARY)
SY2 =
(1.6 INCH)(2.5 X 10-5 m3/INCH)
(1 MPa/INCH) = 40 Nm/INCH)

Figure 23-14 Graphical calculus: Integration

Graphical Calculus: Differentiation. The technique for doing graphical differentiation, and the advantages to a designer are similar to graphical integration; however, obtaining reasonably accurate results requires more skill than integrating graphically, ***Figure 23-15***.

Timing Graphs. Timing graphs aid designers in coordinating electrical and electronics components in a system driven by clock pulses, ***Figure 23-16***.

Topographic Maps (Charts). These speak for themselves, ***Figure 23-17***. The pilots of space, air, land, sea, and undersea vehicles use maps and sophisticated devices to locate their positions. Civil, agricultural, mining, and petroleum engineers, generate information for their use and explorations from maps.

Force Maps. Force polygons drawn on force maps (same scale in all directions) are included here because of their utility in science and technology courses, ***Figure 23-18***. Force polygons are discussed later in reference to an elementary structures design.

Beam Diagrams. Beams of varying shapes, materials, and sizes occur in so many designs that a brief introduction to beam diagrams in this chapter seems appropriate, ***Figure 23-19***. These diagrams are discussed later.

Knowing the major functions of these graphs and charts, and having the skill and knowledge to prepare them, will enhance your ability to read and use them in courses and in industry.

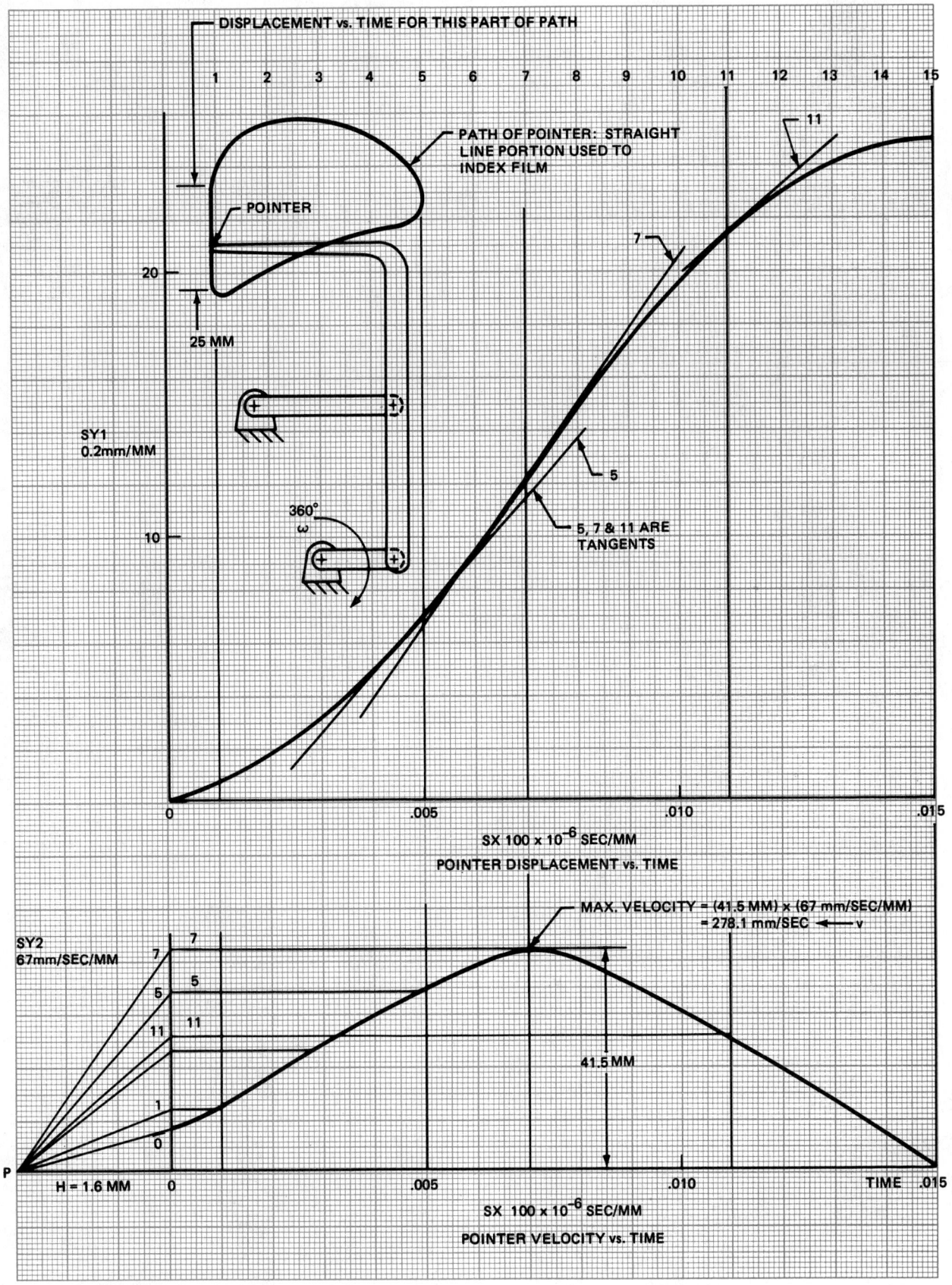

Figure 23-15   Graphical calculus: Differentiation

INPUT TO LOGIC CIRCUIT FOR DRIVING A STEPPING MOTOR

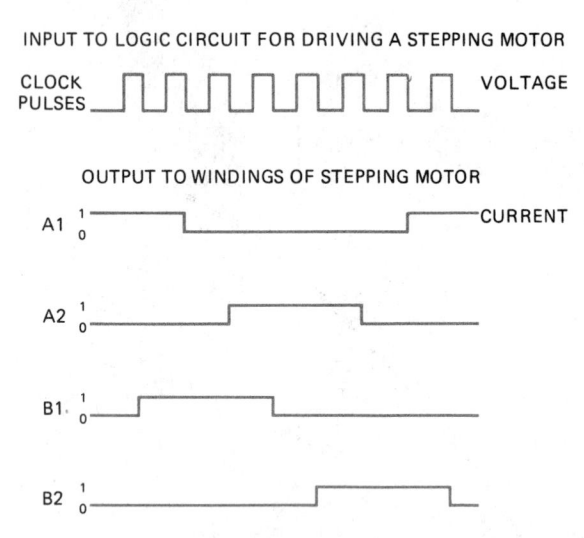

Figure 23-16 Timing graphs

ABBREVIATED TOPOGRAPHIC MAP

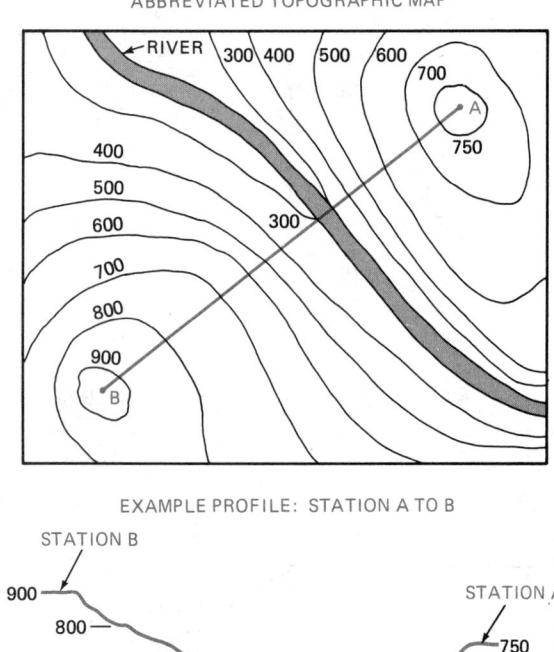

EXAMPLE PROFILE: STATION A TO B

VERTICAL SCALE 1″ = 250′
HORIZONTAL SCALE 1″ = 1000′

Figure 23-17 Topographic map

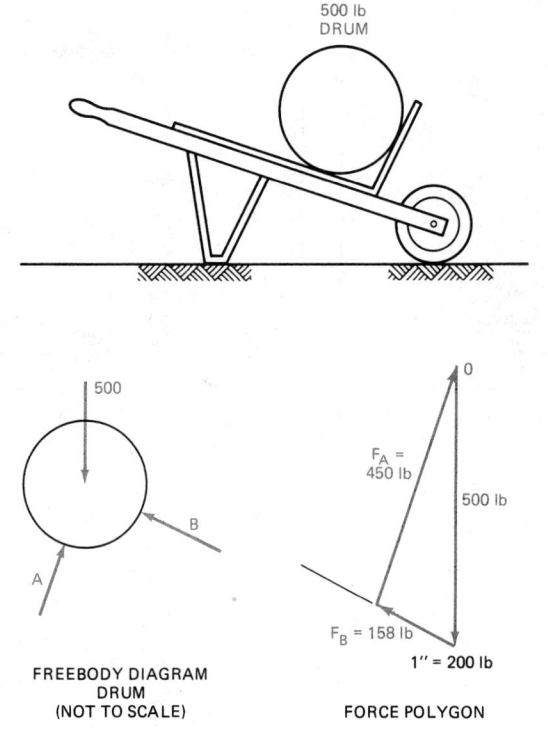

Figure 23-18 Force maps

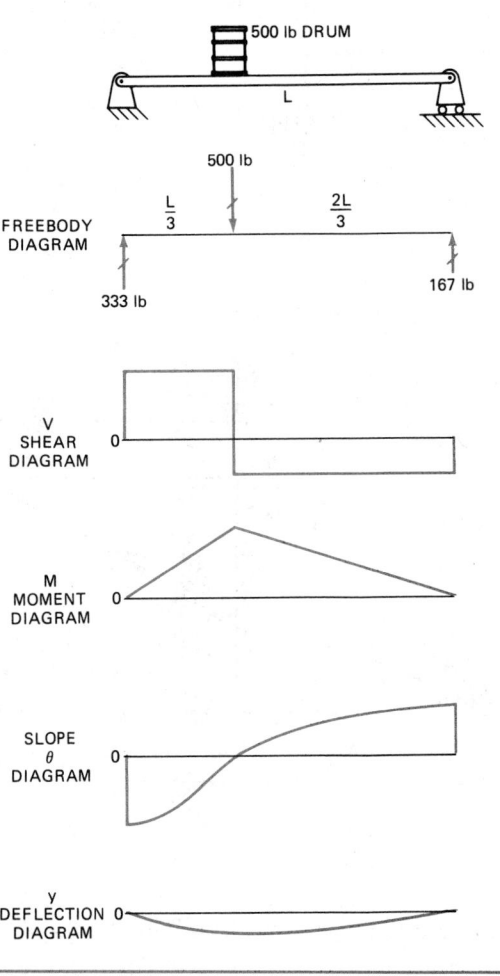

Figure 23-19 Beam diagrams

Five Basic Components

First, consider the five basic components of almost all charts and graphs.

1. The grid
2. The scales
3. The scales' labels
4. The plot
5. The title

These are identified in *Figure 23-20*. Diligently including all five components will make your instructor extremely happy! Selecting and developing all five requires a number of steps that act as a checklist to ensure professional charts and graphs.

The Steps

Assuming that the data to be plotted has been collected, and that the function of the chart or graph has been decided, do the following:

1. Examine the data to determine the ranges to be plotted.
2. Use scales to present the data for optimum communication or generation of information, *Figure 23-21*. To help users interpolate between scale numbers, use scales that are multiples of 1, 2, 5, or 10 (unless the intervals are topical, such as days, weeks, months, years, or geographical locations), *Figure 23-22*. Use horizontal scales for independent variables, and vertical scales for dependent variables.
3. Select commercial grid sheets to save time. Grids familiar to readers save them time. Moreover, commercial grids were developed for specific, frequently needed, functions. Your experiences and observations will help you to select the best formats for presenting data.
4. Plots usually start at the lower left unless the points are negative or the function of the chart or graph requires a different format. Figure 23-20 shows a typical plot. *Figure 23-23* shows a different format.

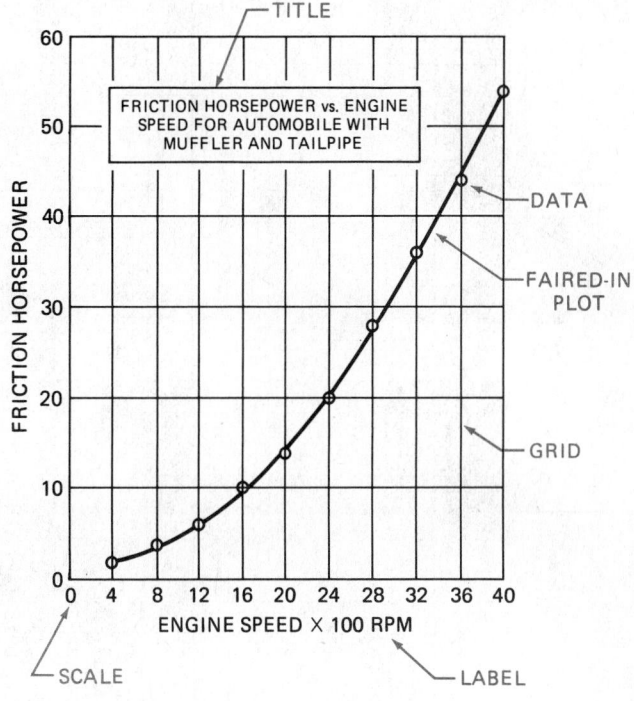

Figure 23-20   Basic components

5. Use symbols for specific points when plotting data from laboratory experiments. Circles, squares, and triangles are common. Curves do not go through the symbols. See Figure 23-20.
6. Draw all curves lightly before finally darkening them in. Use smooth curves, without symbols, for equations. "Best" fit curves through points (symbols) for experimental data may miss some of the points. See Figure 23-20. Employ line segments for connecting discontinuous data, *Figure 23-24*. *Note:* curves representing changing physical phenomena (mechanical, electrical, and chemical) usually appear as smooth plots.
7. Clearly identify multiple plots on the same grid by labels and by a variety of different lines, *Figure 23-25*.

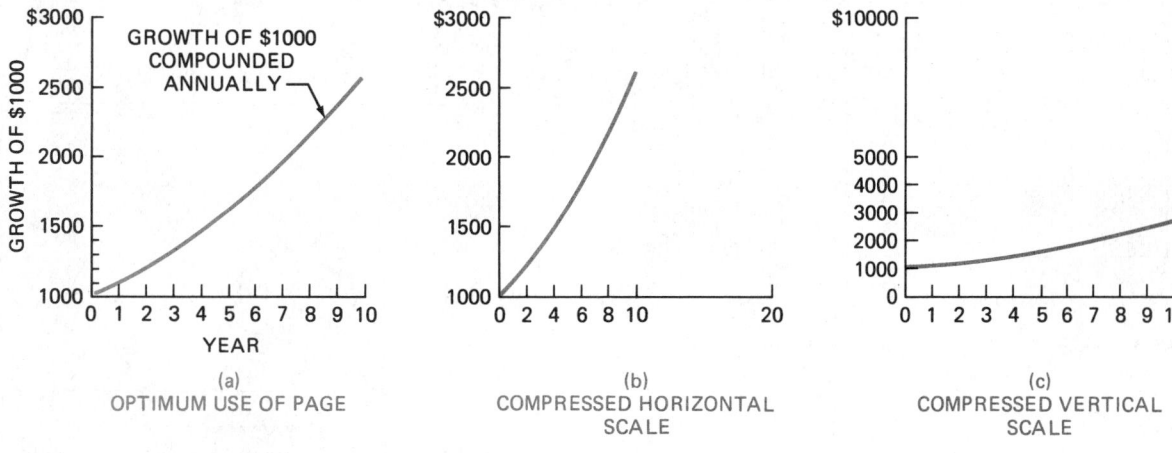

Figure 23-21   Optimum use of page area

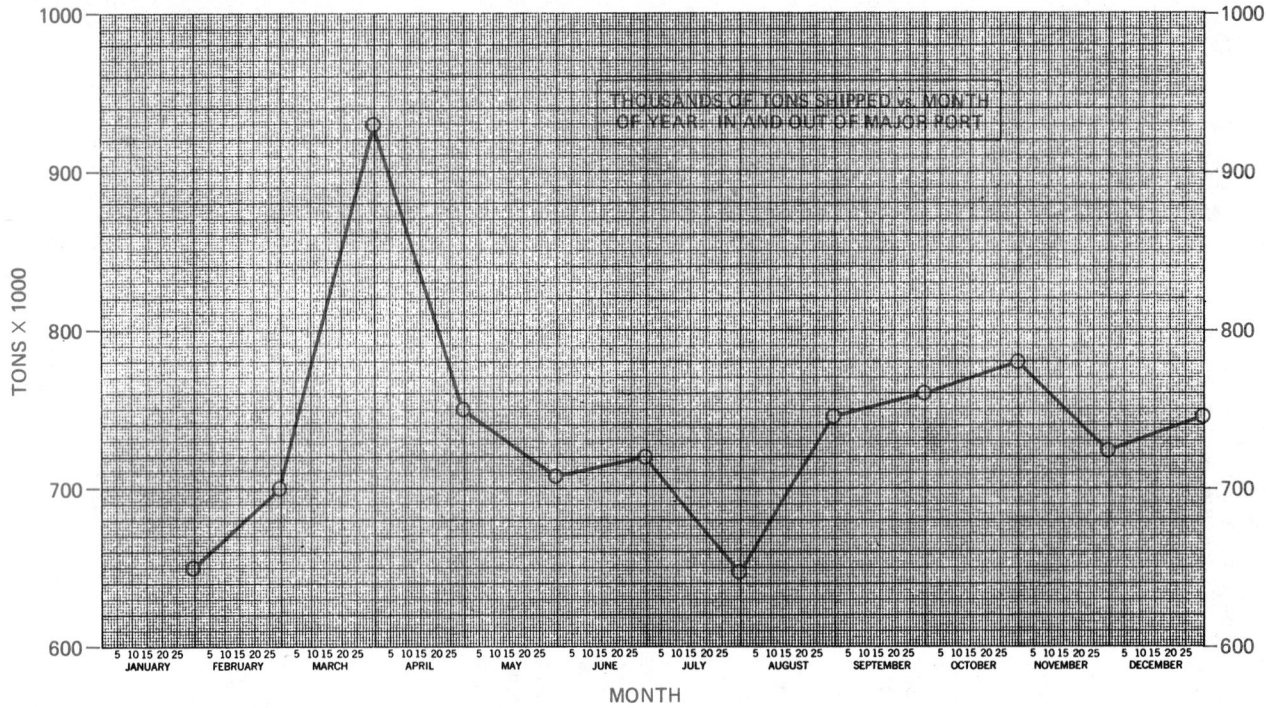

Figure 23-22 Topical scales

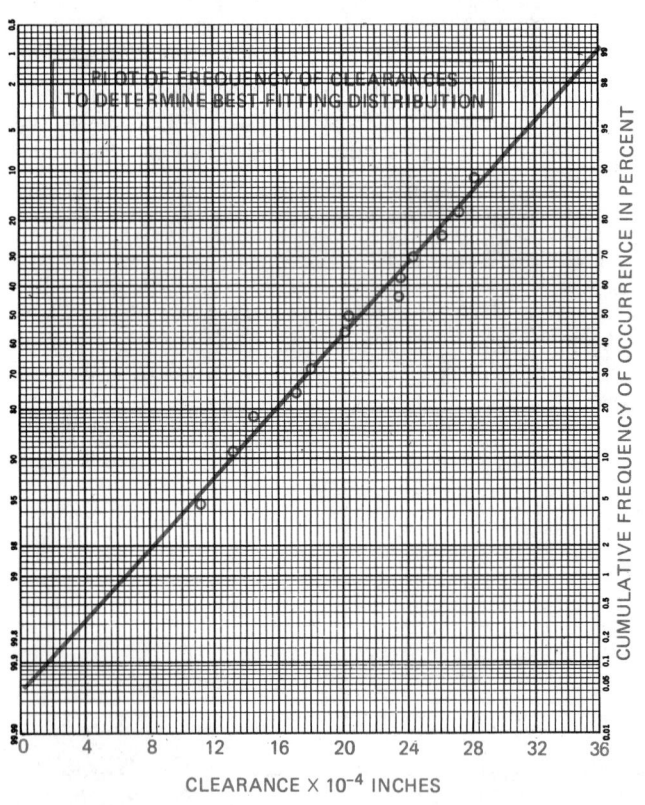

Figure 23-23 Different format

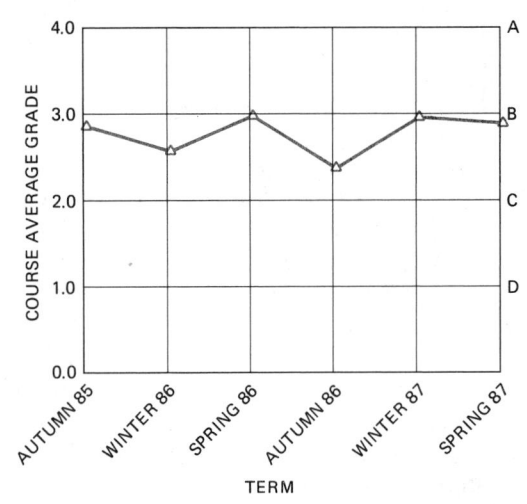

Figure 23-24 Discontinuous data

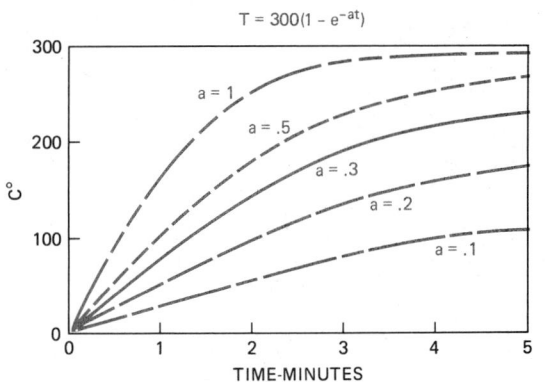

$$T = 300(1 - e^{-at})$$

Figure 23-25 Variety of lines

8. Label each axis by describing the variable and the unit of measurement. See Figure 23-20.

9. As a minimum, include in the title the dependent variable as a function of the independent variable. Also include the source of the data, the date, and your name. Arrange all labels, titles, and notes to be read either parallel to the bottom of the page or from the right side. Use a cleared space for the title if it appears in the grid area. Titles may be placed outside the grid area.

10. Step back and ask if the final form of the chart or graph communicates effectively or assists in generating the intended information. Successful charts and graphs are three orders of magnitude ($10^{**}3 = 1000$) better at conveying information than words!

The preceding discussion on preparing graphs and charts is summarized in the flow chart in *Figure 23-26*.

Further examples of uses and instructions for preparing specific charts and graphs constitute the remainder of the chapter. The examples illustrate student activities, classroom assignments in technical courses, and industrial applications.

Specific Charts and Graphs

In the preceding overview of the functions and formats of charts and graphs, six types were designated for further discussions.

- Pie Charts and Bar Charts: generated from a spreadsheet in a personal computer.
- Force Polygons: used in an optimization investigation.
- Nomograms: several alignment charts and concurrency charts to graphically represent equations.
- Empirical Equations: $y = mx + b$, $y = b(e^{**}mx)$, and $y = b(x^{**}m)$ for matching equations to data.
- Graphical Integration and Differentiation: Pole and Ray method.
- Beam Diagrams: free-body, shear, moment, slope, and deflection diagrams.

Pie Charts and Bar Charts from a Spreadsheet

Assume that a student in an orientation-to-higher-education class has prepared a weekly activity schedule for the first term as shown in *Figure 23-27*. The student's anticipated schedule of eating, sleeping, playing, attending class, studying, and everything else has been typed into a spreadsheet format in a personal computer. The format resembles "pigeon holes" for stuffing information into, arranged in columns and rows. One column lists each of the twenty-four hours; seven columns list the activities for each day of the week, coded in categories as follows: (ST) study time, (CL) class time, (SL) sleep time, (PL) playtime, (EA) eating time, and (MI) miscellaneous—transportation, laundry, club meetings, and church.

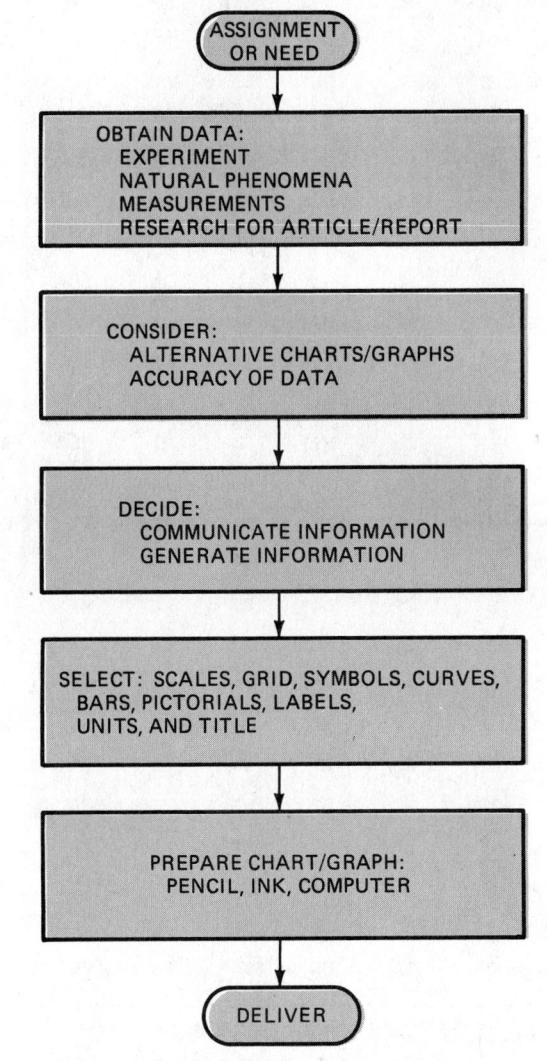

Figure 23-26 Design procedure: Charts and graphs

A program generated by the student, or one already contained in software, sorts through the seven days and totals the number of hours for each category. Then, instructions are typed or selected from a menu to generate and plot the pie chart and bar chart shown in *Figure 23-28*. Later, or for the next term, the student can, with a push of a button, easily adjust the schedule and have new plots made.

Force Polygons in an Optimization Investigation

Force polygons, vector diagrams, and vector polygons all refer to the same graphical technique. *Force polygons* are used in this discussion. However, before proceeding to the optimization investigation, several assumptions and definitions need to be stated. This discussion assumes that the reader has had a brief introduction to forces and vectors but will appreciate a condensed review of some basic concepts. Only solid bodies at rest with coplanar, concurrent (all in one plane going through one point) vectors are considered. Furthermore, the weights of the bodies, or members,

HOUR	SUN	MON	TUE	WED	THU	FRI	SAT
0000	SL	SL	SL	SL	SL	SL	SL
0100	SL	SL	SL	SL	SL	SL	SL
0200	SL	SL	SL	SL	SL	SL	SL
0300	SL	SL	SL	SL	SL	SL	SL
0400	SL	SL	SL	SL	SL	SL	SL
0500	SL	SL	SL	SL	SL	SL	SL
0600	SL	SL	SL	SL	SL	SL	SL
0700	SL	MI	MI	MI	MI	MI	SL
0800	SL	EA	EA	EA	EA	EA	SL
0900	MI	CL	CL	CL	CL	CL	MI
1000	EA	CL	MI	CL	MI	CL	EA
1100	MI	CL	CL	CL	CL	CL	ST
1200	MI	EA	EA	EA	EA	EA	ST
1300	EA	ST	CL	ST	CL	MI	EA
1400	ST	ST	CL	ST	CL	MI	PL
1500	ST	ST	CL	ST	CL	PL	PL
1600	ST	ST	MI	ST	MI	PL	PL
1700	ST	MI	MI	MI	MI	PL	PL
1800	EA	EA	EA	EA	EA	EA	PL
1900	EA	MI	ST	MI	ST	PL	EA
2000	PL	ST	ST	ST	ST	PL	PL
2100	ST	ST	ST	ST	ST	PL	PL
2200	ST	ST	ST	ST	ST	PL	PL
2300	ST	SL	SL	SL	SL	PL	PL

Figure 23-27   Week activities: spreadsheet

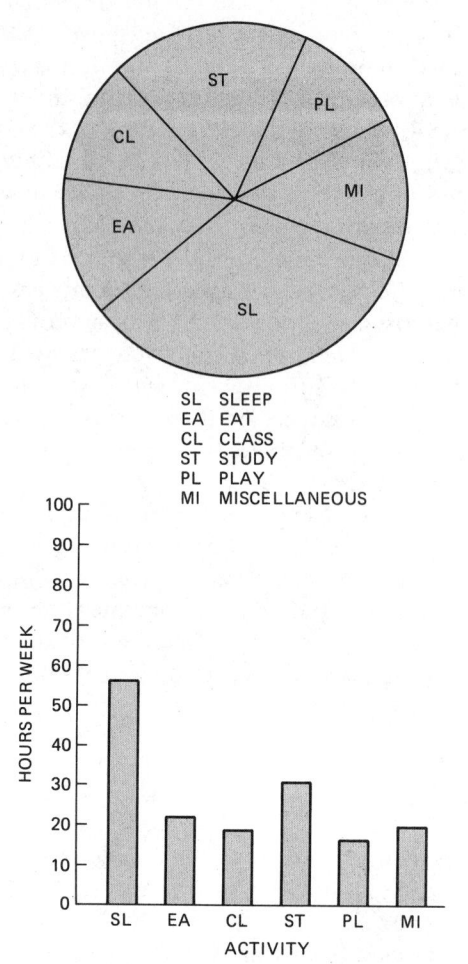

Figure 23-28   Pie and bar charts from spreadsheet

are considered to be negligible (zero) compared to the forces applied.

1. **Vector:**   A vector is a graphical symbol that represents direction by means of an arrowhead at one end of a straight line whose scaled length represents a *magnitude*.

2. **Physical Simplifications:**   A physical simplification describes the essentials of a real phenomenon. For example, a vector is a graphical model which can represent a weight, a force, a velocity, or an acceleration of a body.

3. **Force:**   A force is a directed action of one body on a second one that tends to change the state of motion, or rest, of the second body; therefore, a force is a vector quantity.

4. **Equilibrium:**   The sum of forces acting on a solid body, in any direction, add to zero on a body in equilibrium that is either at rest or moving at a constant velocity. The following equation of mechanics states this principle succinctly: $\Sigma F = 0$.

5. **Two-Force Members:**   The majority of two-force members are found in structures, frames, trusses, and flexible members in tension. The two forces are equal and opposite in direction, colin-

ear, and applied at the ends of the member. See ***Figure 23-29*** for two examples, one in tension and one in compression.

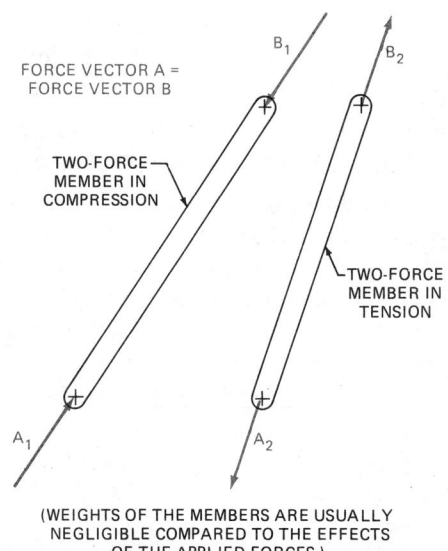

Figure 23-29   Two-force members

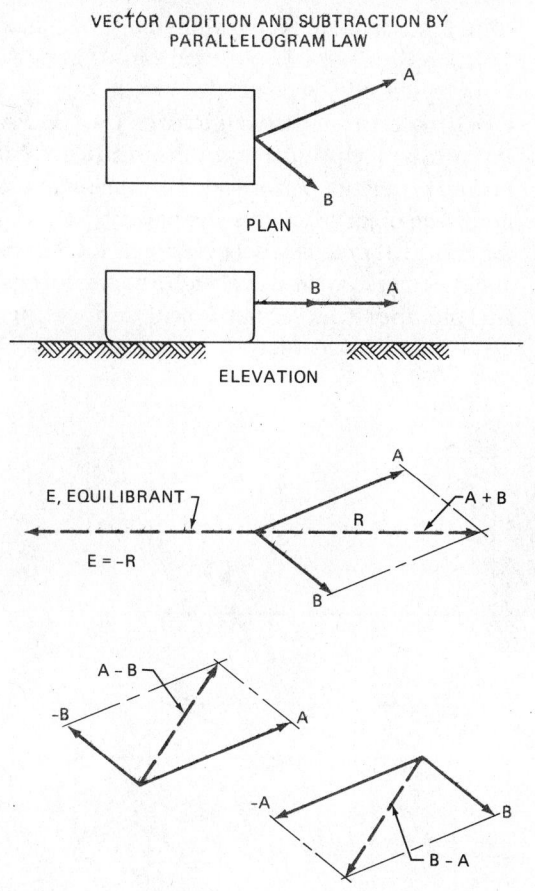

Figure 23-30 Parallelogram law: Vector addition and subtraction

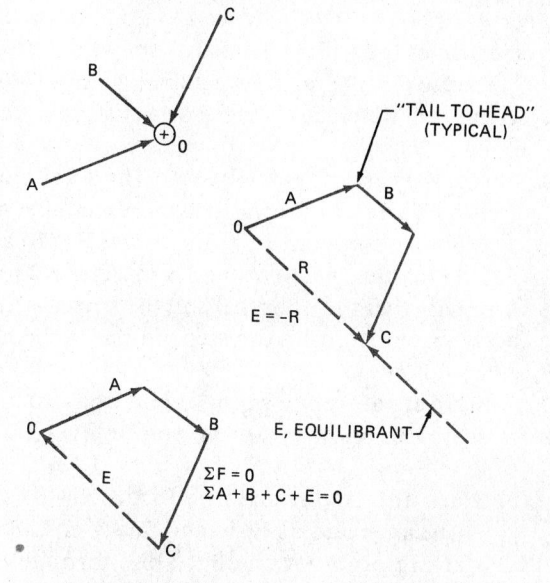

Figure 23-31 Addition: Three or more vectors

6. Vector Addition and Subtraction: Two vectors that are not parallel and not colinear but are acting on the same body and going through the same point may be added by the parallelogram-law technique. Refer to *Figure 23-30* for an example showing the addition of two vectors A and B to obtain a single vector R, which is equivalent to the effect of the two vectors and is called a *resultant*. Figure 23-30 also shows the subtraction of vector B from vector A. A vector labeled *equilibrant* is the reaction to the resultant vector R and is equal and opposite to R, and its line of action goes through the same point. Of course, the sum of the resultant R and the equilibrant equals zero. (Note: the term "equilibrant" is used mostly in definitions, but not much in actual problems in mechanics.)

More than two vectors—such as A, B, and C—acting on a body and going through a common point add, "tail to head," in any order to create a resultant and an equilibrant, *Figure 23-31*. Figure 23-31 also illustrates that the sum of vectors A, B, and C plus the equilibrant equals zero. Moreover, when all the force vectors (three or more) acting through a common point on a body that is in equilibrium are added "tail to head," the sum must equal zero.

7. Free-Body Diagrams (FBD): A body in equilib-

rium, free from its supports, or a portion of the body shown separately, is a free-body diagram when all the forces acting on the body are shown. *Figure 23-32* shows a complete body in equilibrium, and it is drawn as an FBD. The isolated pinned joint at B is also drawn as an FBD. Isolating joints saves space and time and allows for both external forces and internal forces to be shown.

The forces on the isolated joint B are assumed to go through the "pin" at the joint. Joints in trusses and frames are usually considered to be pinned even though the joint may be welded, riv-

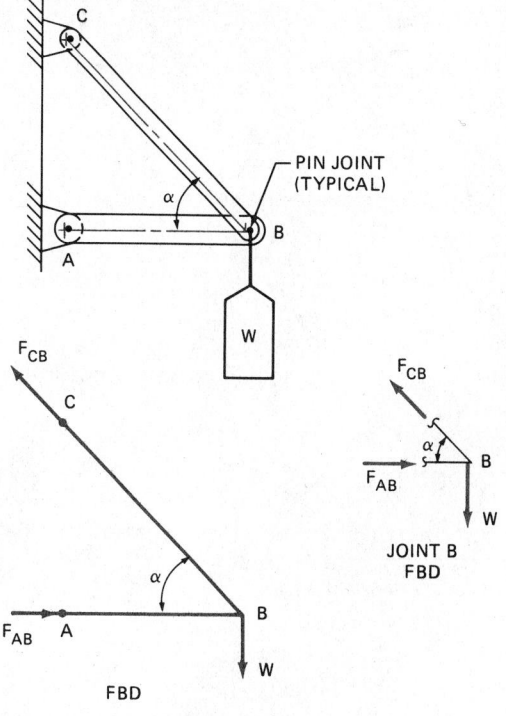

Figure 23-32 FBD (freebody diagrams)

eted, or bolted. This practice is a simplification in mechanics to facilitate joint analyses with FBD's.

Having the skill to draw accurate Free-body diagrams will significantly increase a student's grade point average!

8. **Force Polygon for Analysis:** The force polygon in *Figure 23-33* was used to determine the direction and magnitude of the forces F (CB) and F (AB) for the isolated joint B in Figure 23-32. Assuming that the boom AB and the strut CB have been drawn to scale, the steps in the procedure are as follows:

a. Start at reference point O and construct to scale the force vector W pointing vertically downward.

b. Since the two remaining forces F(CB) and F(AB) must add on to W and finally end at the starting point O, one light line through the head of W is drawn parallel to the strut CB. A second line is drawn through point O parallel to boom AB. The intersection point of the two lines determines the magnitudes of F(CB) and F(AB); also, the "tail to head" pattern determines their directions. (Only two unknown forces may be determined using this technique in coplanar problems).

c. The direction of F(CB) is "away" from the joint B; therefore, F(CB) is a tensile force. The direction of F(AB) "into" the joint indicates a compressive force. W is an externally applied force, so it is neither tensile nor compressive. F(AB) could have been considered first in step b and the results would be identical. Now that definitions and assumptions have been stated, the optimization investigation utilizing force polygons comes next.

9. **Optimization Investigation Using Force Polygons:** Optimizing a design usually entails finding maximum performance or minimum cost or both. The objective in this elementary example is to determine the angle theta between the boom AB and the strut CB that will give the minimum cost for the member CB, with CB still supporting the 10,000-pound load safely. Refer to *Figure 23-34A*.

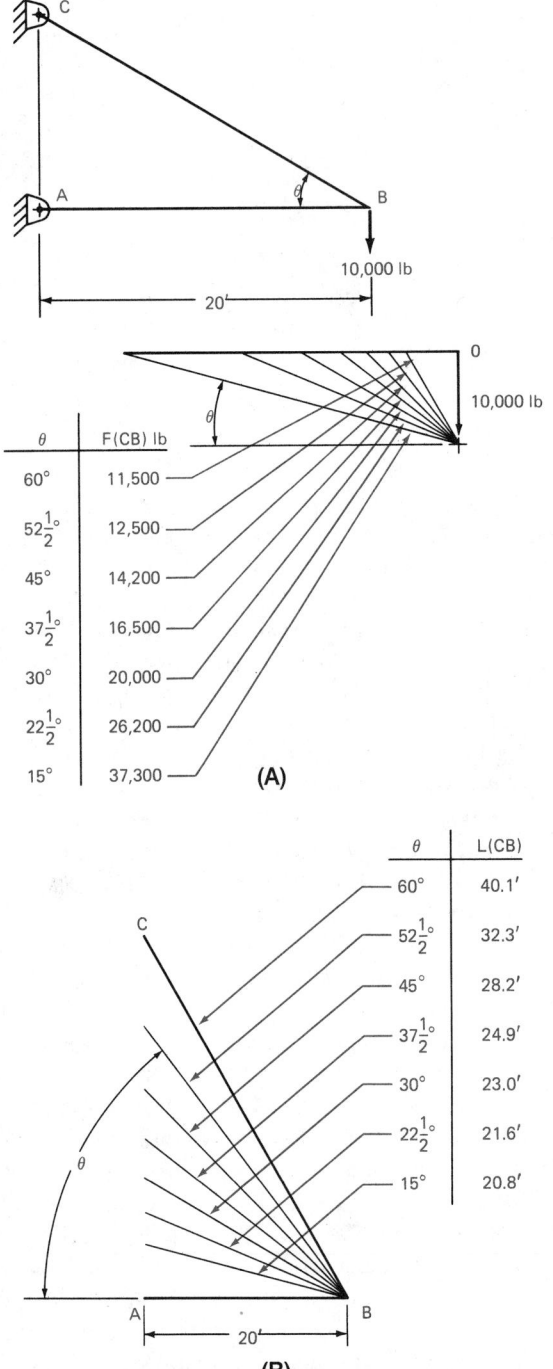

θ	F(CB) lb
60°	11,500
$52\frac{1}{2}°$	12,500
45°	14,200
$37\frac{1}{2}°$	16,500
30°	20,000
$22\frac{1}{2}°$	26,200
15°	37,300

(A)

θ	L(CB)
60°	40.1'
$52\frac{1}{2}°$	32.3'
45°	28.2'
$37\frac{1}{2}°$	24.9'
30°	23.0'
$22\frac{1}{2}°$	21.6'
15°	20.8'

(B)

Figure 23-34 (A) Force polygons for F(BC); (B) Space diagrams for L(BC)

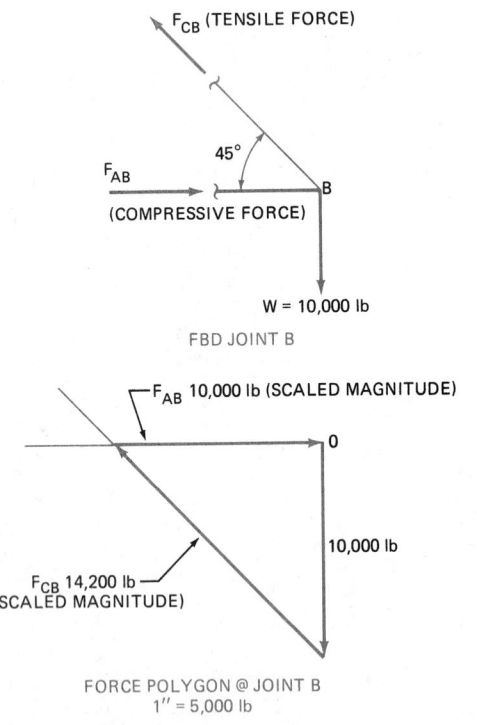

F_{CB} (TENSILE FORCE)

F_{AB}
(COMPRESSIVE FORCE)

45°

B

W = 10,000 lb

FBD JOINT B

F_{AB} 10,000 lb (SCALED MAGNITUDE)

O

10,000 lb

F_{CB} 14,200 lb
(SCALED MAGNITUDE)

FORCE POLYGON @ JOINT B
1" = 5,000 lb

Figure 23-33 Force polygon application

a. Member CB is to be an equal leg, structural angle made of mild steel. Boom AB is 20 feet long and is assumed to be satisfactory for any angle theta between 15° and 60°. Weights of the members are negligible.

b. Force polygons shown in Figure 23-34A determine the tensile forces F(CB) in member CB for different angle thetas. Space diagrams (same linear scale in all directions) in *Figure 23-34B* give the required lengths L(CB) for member CB for the same angles.

c. Next, the forces are entered in a stress equation from the theory of mechanics of materials to select the equal leg structural angles. The equation S = (F)/(A) states that:

normal stress in pounds per square inch
(psi) =

$$\frac{\text{pounds Force}}{\text{cross-sectional area in square inches}}$$

The *force* is assumed to be normal to the cross-sectional area, and each fiber in the entire member is stressed the same. One type of steel used for structural members has a *yield strength* of 36,000 psi. This means that the member could withstand tensile forces, in a linear manner (see *Figure 23-35*), without excessive yielding (stretching), up to 36,000 psi. Therefore, a designer would divide the yield stress by a *factor* of *safety* to obtain an *allowable* or *working stress* and specify that the stress in the member must not exceed it. A typical factor of safety for this application would be 2.0, so the allowable stress would be 18,000 psi; that is, halfway down the linear plot in Figure 23-35. This allowable stress is used in the stress equation.

The stress equation may now be rearranged to solve for area, the third unknown.

Cross-sectional Area =
pounds force/18,000 psi

d. *Figure 23-36* contains an excerpt from an AISC Handbook which lists equal leg structural angles, cross-sectional areas in square inches, and weight per foot of each angle. The following example-calculation for an angle theta of 30° summarizes several of the steps already described and adds several more steps to arrive at cost.

   (i)   L(CB) is found in a space diagram to be 23 feet long for theta = 30°.

   (ii) F(CB) is found in a force polygon to be 20,000 pounds for theta = 30°.

   (iii) A(CB), the required cross-sectional area to limit the stress to 18,000 psi, is found from the equation:

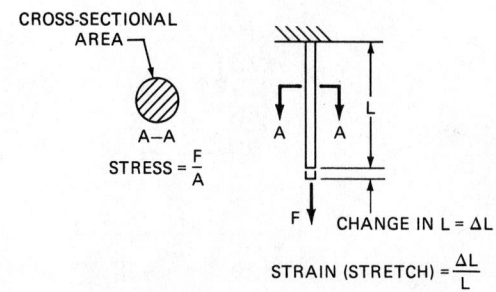

STRESS-STRAIN CURVE FOR MILD STEEL

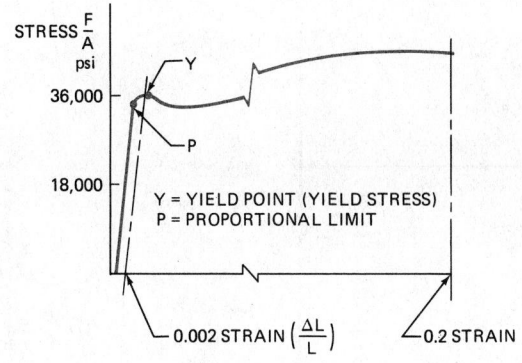

Figure 23-35   Stress vs. strain

$$A(CB) = \frac{F(CB)}{18,000} = \frac{20,000}{18,000}$$

=1.111 square inches.

   (iv) From the AISC Handbook data, the nearest angle with a cross-sectional area of 1.111 or more is Angle 2-½ x 2-½ x ¼ with an area of 1.19 and a weight of 4.1 pounds/foot. Select the lightest member of a group in the handbook, because it is usually the one readily available in warehouses.

   (v)   Finally, assuming that the cost is $2.00 per pound, this angle would cost: (23 feet) (4.1 pounds/foot) (2.00/pound) = to the nearest dollar, $189.00.

   (vi) The rest of the structural angles were selected in a similar manner and tabulated in *Figure 23-37A*.

   (vii) *Figure 23-37B* shows three curves, each a function of angle theta. The optimizing curve, $(CB) vs. theta, shows the lowest (optimum) cost at approximately 48°; therefore, the best angle to specify is angle 2 x 2 x 3/16 x 30 feet long.

Nomograms: Alignment Charts and Concurrency Charts

Nomograms represent equations graphically as shown in Figure 23-13. *Alignment Charts*, one class of nomograms, appear in periodicals, vendor catalogs, and textbooks. These charts are designed to help the reader

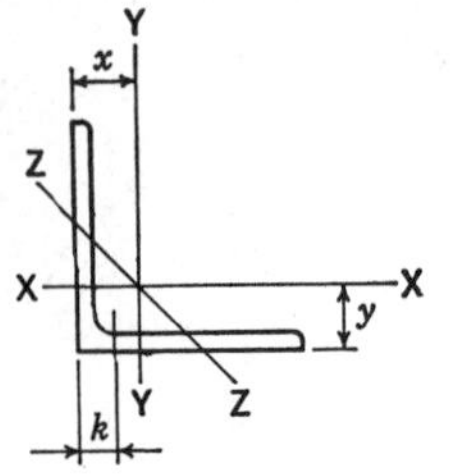

ANGLES
Equal legs

Properties for designing

Size and Thickness	k	Weight per Foot	Area	AXIS X-X AND AXIS Y-Y				AXIS Z-Z
				I	S	r	x or y	r
In.	In.	Lb.	In.²	In.⁴	In.³	In.	In.	In.
L 3½ × 3½ × ½	⅞	11.1	3.25	3.64	1.49	1.06	1.06	.683
⁷⁄₁₆	¹³⁄₁₆	9.8	2.87	3.26	1.32	1.07	1.04	.684
⅜	¾	8.5	2.48	2.87	1.15	1.07	1.01	.687
⁵⁄₁₆	¹¹⁄₁₆	7.2	2.09	2.45	.976	1.08	.990	.690
¼	⅝	5.8	1.69	2.01	.794	1.09	.968	.694
L 3 × 3 × ½	¹³⁄₁₆	9.4	2.75	2.22	1.07	.898	.932	.584
⁷⁄₁₆	¾	8.3	2.43	1.99	.954	.905	.910	.585
⅜	¹¹⁄₁₆	7.2	2.11	1.76	.833	.913	.888	.587
⁵⁄₁₆	⅝	6.1	1.78	1.51	.707	.922	.865	.589
¼	⁹⁄₁₆	4.9	1.44	1.24	.577	.930	.842	.592
³⁄₁₆	½	3.71	°1.09	.962	.441	.939	.820	.596
L 2½ × 2½ × ½	¹³⁄₁₆	7.7	2.25	1.23	.724	.739	.806	.487
⅜	¹¹⁄₁₆	5.9	1.73	.984	.566	.753	.762	.487
⁵⁄₁₆	⅝	5.0	1.46	.849	.482	.761	.740	.489
¼	⁹⁄₁₆	4.1	1.19	.703	.394	.769	.717	.491
³⁄₁₆	½	3.07	.902	.547	.303	.778	.694	.495
L 2 × 2 × ⅜	¹¹⁄₁₆	4.7	1.36	.479	.351	.594	.636	.389
⁵⁄₁₆	⅝	3.92	1.15	.416	.300	.601	.614	.390
¼	⁹⁄₁₆	3.19	.938	.348	.247	.609	.592	.391
³⁄₁₆	½	2.44	.715	.272	.190	.617	.569	.394
⅛	⁷⁄₁₆	1.65	.484	.190	.131	.626	.546	.398
L 1¾ × 1¾ × ¼	½	2.77	.813	.227	.186	.529	.529	.341
³⁄₁₆	⁷⁄₁₆	2.12	.621	.179	.144	.537	.506	.343
⅛	⅜	1.44	.422	.126	.099	.546	.484	.347
L 1½ × 1½ × ¼	⁷⁄₁₆	2.34	.688	.139	.134	.449	.466	.292
³⁄₁₆	⅜	1.80	.527	.110	.104	.457	.444	.293
⁵⁄₃₂	⅜	1.52	.444	.094	.088	.461	.433	.295
⅛	⁵⁄₁₆	1.23	.359	.078	.072	.465	.421	.296
L 1¼ × 1¼ × ¼	⁷⁄₁₆	1.92	.563	.077	.091	.369	.403	.243
³⁄₁₆	⅜	1.48	.434	.061	.071	.377	.381	.244
⅛	⁵⁄₁₆	1.01	.297	.044	.049	.385	.359	.246
L 1 × 1 × ¼	⅜	1.49	.438	.037	.056	.290	.339	.196
³⁄₁₆	⁵⁄₁₆	1.16	.340	.030	.044	.297	.318	.195
⅛	¼	.80	.234	.022	.031	.304	.296	.196

AMERICAN INSTITUTE OF STEEL CONSTRUCTION

Figure 23-36 Equal leg structural angles *(Courtesy American Institute of Steel Construction)*

θ	L(CB) ft	F(CB) lb	A(CB) Calculated	Angle ∠	A(CB) Handbook	lb/ft	Wt(CB) lb	$(CB)
60°	40.1	11,500	0.639	2 x 2 x ³⁄₁₆	0.715	2.44	97.8	196
52½°	32.3	12,500	0.694	2 x 2 x ³⁄₁₆	0.715	2.44	78.8	158
45°	28.2	14,200	0.789	2½ x 2½ x ³⁄₁₆	0.902	3.07	86.6	173
37½°	24.9	16,500	0.917	3 x 3 x ³⁄₁₆	1.09	3.71	92.4	184
30°	23.0	20,000	1.111	2½ x 2½ x ¼	1.19	4.1	94.3	189
22½°	21.6	26,200	1.456	2½ x 2½ x ⁵⁄₁₆	1.46	5.0	108.0	216
15°	20.8	37,200	2.067	3½ x 3½ x ⁵⁄₁₆	2.09	7.2	149.8	300

Figure 23-37A Table: Optimization investigation

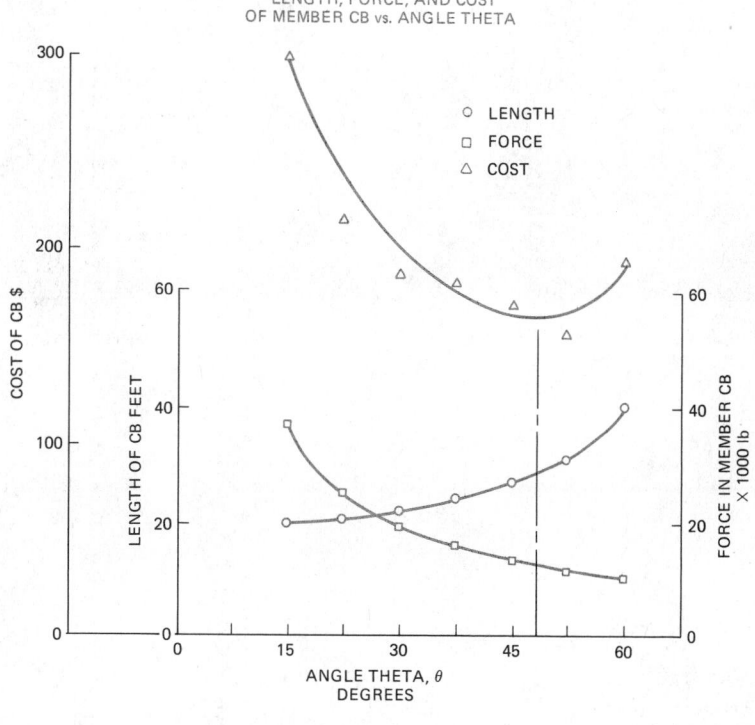

Figure 23-37B   Graphs: Optimization investigation

visualize the information to be generated, and to obtain the information quickly and accurate enough for most design calculations. As a further aid for the reader, the charts contain an example or two showing how to use them. Thus, designers can use the charts easily without having to know how the chart was constructed.

However, if designers know the theory and how to construct alignment charts, they will use them more effectively and confidently. Moreover, these designers will be able to decide whether the effort and time to prepare alignment charts will ultimately save time for themselves or their colleagues. Some may need to make alignment charts for reports and manuals.

First, consider a few reminders regarding mathematics, followed by a few definitions; then, a number of charts.

1. **Common Logarithm.**   A common logarithm is an exponent of the base number 10, so that when 10 is raised to that exponent a specific number is generated. For example, the common logarithm of the number 25, written log 25, is equal to 1.397940; that is, $10^{**}1.397940 = 25$. The table in ***Figure 23-38*** shows an interesting comparison of "the old way" of obtaining logs and the current source, hand calculators.

2. **Function of a Variable: y = f(x).**   In the equation $y = x^{**}2 + 3x + 10$, y is a function of the variable x, written in short form as $y = f(x)$. Occasionally, in alignment chart derivations a single variable, such as u, that may have a range of numerical values is written as $f(u) = u$.

3. **Straight-Line Equation y = mx + b.**   The straight-line equation, $y = mx + b$, specifies that a

N Number	Log N Characteristic from Five-Place Table and Mantissa by observation		Log N Complete Log from Hand Calculator	
	mantissa+characteristic			
1	0.	.00000	0.000000	00
7560	3.	.87852	3.878522	00
756	2.	.87852	2.878522	00
75.6	1.	.87852	1.878522	00
7.56	0.	.87852	8.875218	–01
.756	7.56/10 0.87852 – 1.00000 = –.12148		–1.214782	–01
.0756	7.56/100 0.87852 – 2.00000 = –1.12148		–1.121478	00

Note:
(a) logarithms obey exponent laws:

$$\frac{10^m}{10^n} = 10^{m-n} \text{ AND } 10^m \cdot 10^n = 10^{m+n}$$

(b) mantissa: one less than number of digits before decimal

Figure 23-38   Logarithms

dependent variable y is a function of an independent variable x, m is the slope of the straight line, and b is the intercept along the vertical axis. Assume that the coordinate axes follow a tradi-

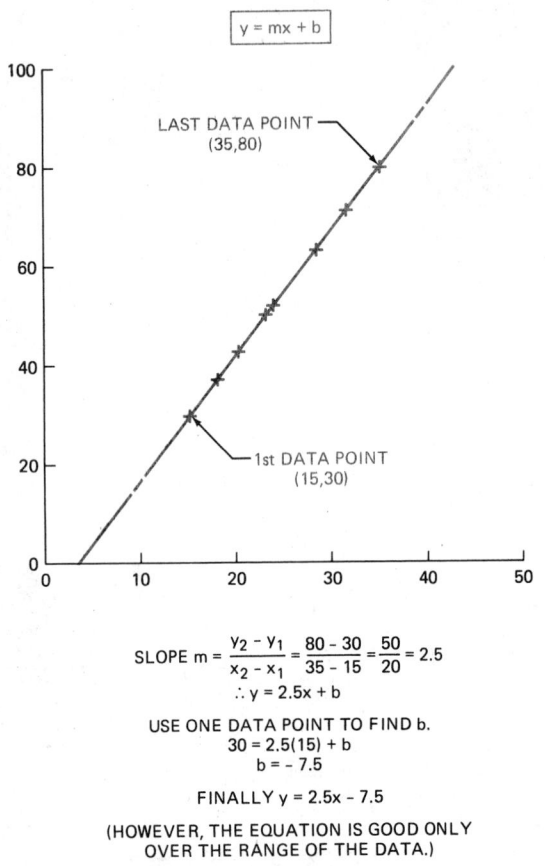

Figure 23-39   y = mx + b: Straight-line equation

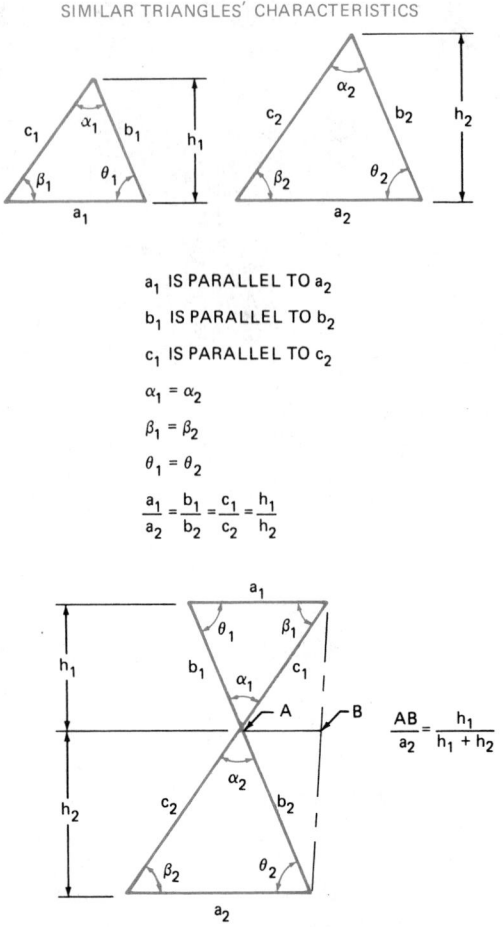

Figure 23-40   Similar triangles

tional pattern: y is plotted in the vertical direction (ordinate axis) and x is plotted in the horizontal direction (abscissa), as shown in *Figure 23-39*.

4. **Similar Triangles.**   Similar triangles form the basis of almost all alignment chart designs. Some properties of the triangles are summarized in *Figure 23-40*.

5. **Alignment Chart.**   An *alignment chart* is a graphical representation of the relationship of two or more variables, such as y = log X, u + v = w, uv = w, and u = v/w.

6. **Scale Equation L = mf(x).**   The *scale equation* fits functions to the page area. In the equation: L = m[f(X_n) - f(X_0)], L = the distance between the final value X_n and the initial value X_0. Designers select distance L, arbitrarily: m = scale modulus which makes the scale fit the page. The quantity f(X_n) = the value of the function of X (when X_n is the largest number of the given range of numbers) and f(X_0) = the lowest value of the function of X.

7. **Functional Scale for y = f(X).**   A *functional scale* is a line graduated and labeled according to a functional relationship. Distances to the labels are calculated using the scale equation, but the distances are not shown on the scale.

The functional scale in *Figure 23-41* is labeled for the function y = log X. The distances to the labels were calculated using the following steps:

a. Since X ranges from X = 1 to X = 100, and 5.5 inches was arbitrarily selected to fit the scale on the page, the scale modulus was calculated.

$$L = m[f(X_n) - f(X_0)]$$
$$5.5 = m[\log 100 - \log 1]$$
$$m = \frac{5.5}{(2-0)}$$
= 2.75 inches per difference in f(X).

b. Next, distances to each label were calculated; for example, when X = 30:

$$L = (2.75)(\log 30 - \log 1) = 4.062''$$

This distance and several others are shown in Figure 23-41.

8. **Adjacent Alignment Chart for f(u) = f(v)**

a. Two related variables plotted along the same line, but on opposite sides, make up an adjacent alignment chart. A temperature conversion chart for relating degrees Celsius to degrees Fahrenheit is an excellent example. Assume that 8 inches is arbitrarily designated as the scale length and do

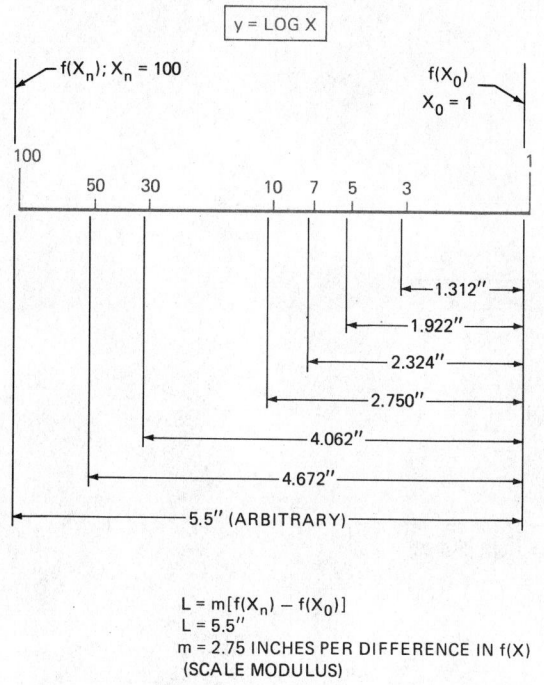

Figure 23-41   Functional scales: y = f(x)

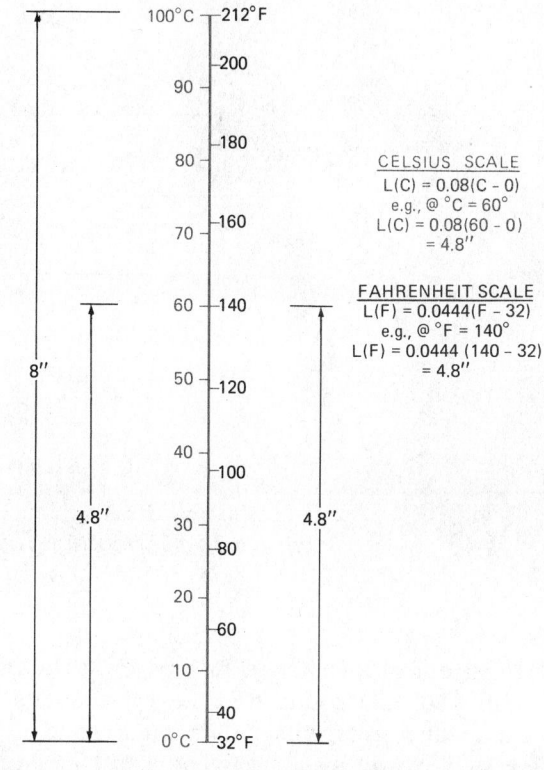

Figure 23-42   Adjacent alignment chart: f(u) = f(v)

the following to create the conversion scale shown in *Figure 23-42*:

(i)   L(Celsius) = L(C) = m(C) (100 – 0)

8 = m(c) (100)

m(C) = 0.08 inches per difference in degrees Celsius

thus, L(C) = 0.08C is the scale equation.

(ii)   L(Fahrenheit) = L(F) = m(F) $\frac{5}{9}$ (F – 32)

8 = m(F) $\frac{5}{9}$ (F – 32)

when F = 212 degrees

m(F) = $\dfrac{8}{\frac{5}{9}(212 - 32)}$ = 0.08

and L(F) = (0.08) $\frac{5}{9}$ (F – 32),

the scale equation.

b.   Another adjacent alignment chart is the speedometer in most automobiles which have MPH and kmPH graduated along the same curved line.

c.   A third example solves a designer's needs. The stress equation used in the *Force Polygons* discussed can be plotted along one line, as shown in *Figure 23-43*. The steps to produce the adjacent plots for the equation "stress = Force/Area" are as follows:

(i)   Assume an arbitrary scale length of 6 inches and the equation restated to read

Area = Force/18,000

(ii)   L(area) = m(A) [(F/18,000) – (0)]

Assume that the largest value of F will be 45,000 pounds force.

Then m(A) = 6/[(45,000/18,000) – (0)]

m(A) = 2.4

L(A) = 2.4[(F/18,000) – (0)]

L(A) = (F/7500 – 0)

(iii)   Since F was found from the force polygons and used in the equation, rewrite the stress equation for F to be F = 18,000A

Then L(F) = m(F) (18,000A)

The maximum value of A is

A = 45,000/18,000 = 2.5

Thus, m(F) = $\dfrac{6}{(18,000)\,(2.5)}$ = $\dfrac{1}{7500}$

L(F) = $\dfrac{1}{7500}$ (18,000A)

Finally, L(F) = 2.4A

(iv)   Calculate and plot distances L(A) for a number of values of force between 0 and 45,000 pounds.

(v)   Calculate and plot distances L(F) for a number of values of area between 0 and 2.5. Compare several values of area opposite forces, from the chart with the tabulated values in Figure 23-37A.

9.   Non-Adjacent Alignment Chart for f(u) = f(v). Refer to the adjacent alignment chart discussion and visualize that the two scales are separated by

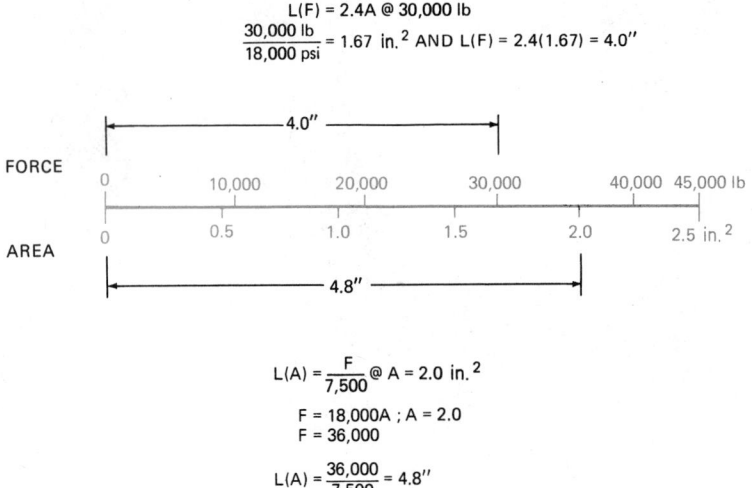

L(F) = 2.4A @ 30,000 lb

$$\frac{30,000\ lb}{18,000\ psi} = 1.67\ in.^2\ \text{AND}\ L(F) = 2.4(1.67) = 4.0''$$

$$L(A) = \frac{F}{7,500} @ A = 2.0\ in.^2$$

F = 18,000A ; A = 2.0
F = 36,000

$$L(A) = \frac{36,000}{7,500} = 4.8''$$

Figure 23-43   Adjacent alignment chart: Stress = (force)/(area)

a convenient distance to fit a page. Rotate one scale 180° and connect the lower value end of each with a straight line. All that remains to do to create a non-adjacent alignment chart is to locate a pivot point K to "hinge" straight lines called "isopleths." "Isopleths" are the lines used to locate values in a nomogram.

a. The steps for deriving the technique for preparing nonadjacent alignment charts are as follows:

(i) Locate two parallel lines at an arbitrary distance apart to fit the page. Select L(u) and L(v) as convenient lengths also to fit the page. Construct the similar triangles as shown in ***Figure 23-44*** and label intersections of the lines. Assume that the initial values of u and v are equal to zero—to facilitate the derivation.

(ii) From similar triangles
$$\frac{L(u)}{L(v)} = \frac{BK}{KC} = \frac{a}{b}$$

From the scale equation
$$\frac{L(u)}{L(v)} = \frac{m(u)\ f(u)}{m(v)\ f(v)} = \frac{a}{b}$$

And since f(u) = f(v)
$$\frac{m(u)}{m(v)} = \frac{a}{b}$$

(iii) Determine m(u) and m(v) so the ratio a/b can be found, which is used as a guide to divide the arbitrary distance between the scales in proportion to a and b. Point K locates the division and also acts as the pivot point for isopleths.

b. Use these steps to prepare a non-adjacent alignment chart for the equation C = 2πR where circumference is a function of radius R, or f(C) = f(R). R varies from 0 to 20 inches,

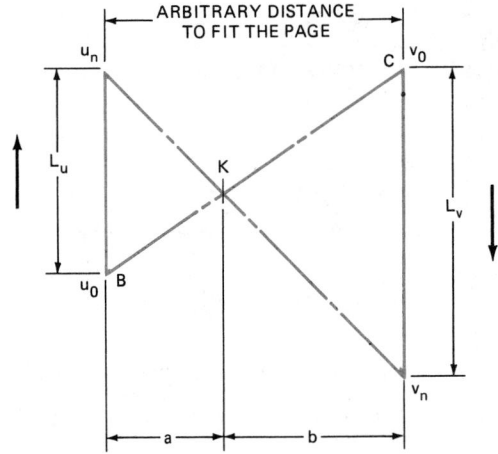

NOTE: L_u IN THIS FIGURE IS THE L(u) IN THE TEXT, ETC.

Figure 23-44   Non-adjacent scales

and L(R) is arbitrarily set at 4 inches. See ***Figure 23-45***.

(i) Determine m(R) from
$$L(R) = m(R)\ [f(20) - f(O)]$$

$$m(R)\ \frac{4}{20} = 0.2$$

Thus, L(R) = 0.2R

(ii) Assume that L(C) is arbitrarily 6 inches long and that L(C) and L(R) are separated by 4 inches, (L(R) on the left side), to fit the page. Now determine:

m(C) using R = C/2π because the values of R are specified.

$$L(C) = m(C)\ [f(C(20) - f(C(O))]$$

$$= m(C)\ \frac{2\pi 20}{2\pi} - \frac{2\pi}{\pi}$$

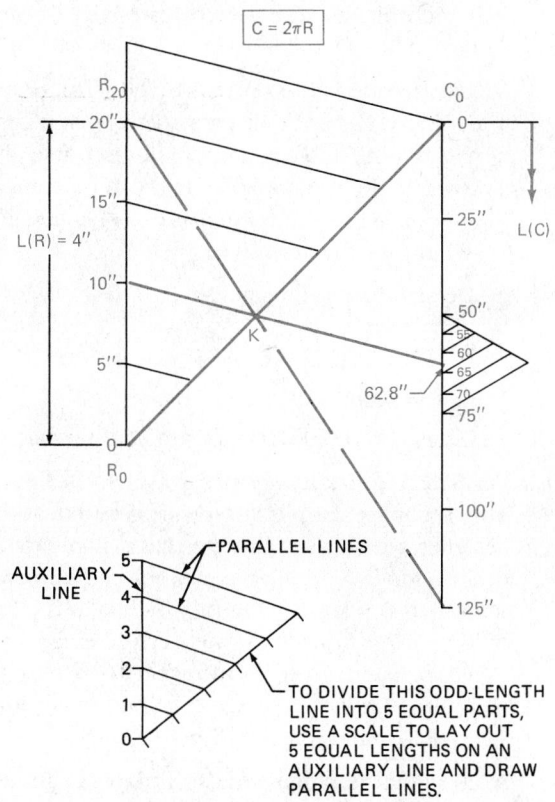

Figure 23-45 $C = 2(3.14)R$

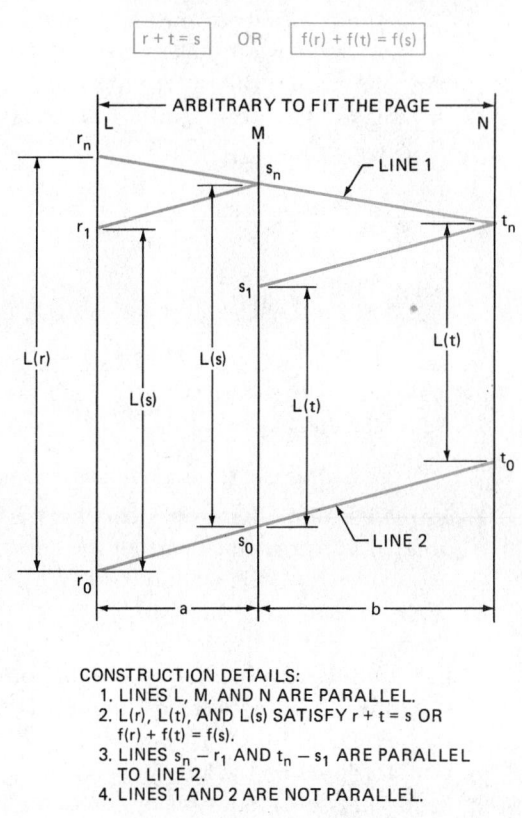

CONSTRUCTION DETAILS:
1. LINES L, M, AND N ARE PARALLEL.
2. L(r), L(t), AND L(s) SATISFY r + t = s OR
   f(r) + f(t) = f(s).
3. LINES $s_n - r_1$ AND $t_n - s_1$ ARE PARALLEL
   TO LINE 2.
4. LINES 1 AND 2 ARE NOT PARALLEL.

Figure 23-46 Parallel alignment chart: f(r) + f(t) = f(s)

$$L(C) = m(C)(20)$$

$$m(C) = \frac{6}{20} = 0.3$$

Thus, $L(C) = 0.3\dfrac{C}{2\pi}$

or $L(C) = 0.0478C$

(iii) Locate point K from

$$\frac{m(R)}{m(C)} = \frac{0.2}{0.3} = \frac{2}{3}$$

Divide line R(O) – C(O) into five parts and locate K a distance of 2 parts from R(O).

(iv) Graduate the two scales, L(R) and L(C), using scale equations. A helpful graphical technique for dividing lines is shown in the Figure 23-45.

(v) Use several isopleths to verify the chart, e.g., when R = 10, C should equal 62.8 inches.

10. Parallel Alignment Chart for f(r) + f(t) = f(s)

a. The procedure for preparing a set of three parallel scales for a three-variable equation, r + t = s, or more generally f(r) + f(t) = f(s), is derived using similar triangles and scale equations. The derivation aims at obtaining relationships from two of the scales so that the third can be located and graduated.

b. The following steps in the derivation are illustrated in *Figure 23-46*.

(i) Preliminary construction requirements and assumptions are summarized in the Figure 23-46.

(ii) To facilitate the derivation assume that:

—Initial values of the variables are equal to zero: r(O) = t(O) = s(O) = 0
—The scales L(r) and L(t) have been graduated and that m(r) and m(t) are known.

$$L(r) = m(r) f(r)$$
$$L(t) = m(t) f(t)$$

To be determined:

$$L(s) = m(s) f(s)$$

(iii) From similar triangles:

$$\frac{L(r) - L(s)}{a} = \frac{L(s) - L(t)}{b}$$

Substitute scale equations:

$$\frac{m(r) f(r) - m(s) f(s)}{a} = \frac{m(s) f(s) - m(t) f(t)}{b}$$

Cross multiply, collect terms and get
m(r)f(r) + a/b m(t)f(t) = (a + b)/b m(s)f(s). Then since f(r) + f(t) = f(s),

the coefficients must all be equal; that is, $m(r) = a/b\ m(t) = (a + b)/b\ m(s)$. From step (iii): $m(r)/m(t) = a/b$ which is used to locate the third scale, $L(s)$.

Also from step (iii):

$$m(r) = \frac{a + b}{b}\ m(s) = (1 + \frac{a}{b})\ m(s)$$

Since $\dfrac{a}{b} = \dfrac{m(r)}{m(t)}$

$$m(r) = (1 + \frac{m(r)}{m(t)})\ m(s)$$

Finally, $m(s) = \dfrac{m(r)\ m(t)}{m(r) + m(t)}$

which is the scale modulus for $L(s)$.

11. **Example Parallel Alignment Chart for r + t = s.** *Figure 23-47* shows the chart for the equation $r + t = s$ where $r = 0$ to 20 and $t = 0$ to 50. Do the following to construct the chart:
   a. Draw the parallel scales for r and t to fit the page and leave room between them for scale s. Arbitrarily select $L(r) = 4$ inches, $L(t) = 5$ inches. Locate their zero values at the bottom. $L(r)$ goes on the left.
   b. Calculate scale moduli for $m(r)$ and $m(t)$.

   $L(r) = m(r)\ (20 - 0)$

   $m(r) = \dfrac{4}{20} = 0.2$

   $L(t) = m(t)\ (50 - 0)$

   $m(t) = \dfrac{5}{50} = 0.1$

   c. Locate a vertical line for the center scale, $L(s)$, from

   $\dfrac{a}{b} = \dfrac{m(r)}{m(t)} = \dfrac{0.2}{0.1} = \dfrac{2}{1}$

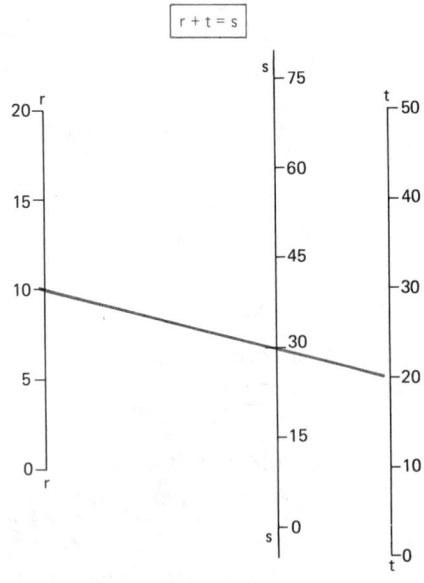

$$\boxed{r + t = s}$$

Figure 23-47   Parallel alignment chart: r + t = s

which places $L(s)$ two-thirds of the distance between $L(r)$ and $L(t)$, from $L(r)$

   d. Locate one value on $L(s)$ by constructing an isopleth between any two values of r and t, e.g., $r = 10$ and $t = 20$, which of course gives $s = 30$. The remainder of $L(s)$ can be completed using the scale equation $L(s) = m(s)f(s)$, where s varies from 0 to 70.
   e. Determine $m(s)$ from

   $$m(s) = \frac{m(r)\ m(t)}{m(r) + m(t)} = \frac{(.2)\ (.1)}{(.2 + .1)}$$

   $m(s) = 0.067$

   Thus, $L(s) = 0.067f(s)$ for graduating scale s.

12. **Parallel Alignment Chart for f(r) • f(t) = f(s).** This second example is typical of an equation a designer would use. Assume that a nomogram is desired for $E = 1R$. The equations can be rewritten as $\log E = \log 1 + \log R$, which is in the form of $f(r) + f(t) = f(s)$, which was derived earlier.

   To construct the nomogram for this basic equation, (volts) = (amps) (ohms), follow the steps below and refer to *Figure 23-48*.

   a. Assume the ranges for the variables I and R; 1 = 0.1 to 10 amps and R = 1 to 1000 ohms. Select $L(I) = 5$ inches and $L(R) = 5$ inches: place them 4 inches apart, vertical, parallel, and with the smaller values at the bottom. Locate $L(I)$ on the left side. Graduate them using the scale equations:

   $L(I) = m(I) • f(I)$ and $L(R) = m(R) • f(R)$.

   The $y = \log X$ scale in Figure 23-41 is a useful graphical aid in graduating logarithmic scales as illustrated in Figure 23-48.

   b. Calculate scale moduli $m(I)$ and $m(R)$ to use in determining a/b and $m(E)$.

   $L(I) = m(I)\ (\log 10 - \log 0.1)$

   $$m(I) = \frac{5}{(1.0 - (-1.0))} = 2.5$$

   Thus, $L(I) = 2.5\ f(I)$

   $L(R) = m(R)\ (\log 900 - \log 1)$

   $$m(R) = \frac{5}{2.954243 - 0} = 1.6925$$

   Thus, $L(R) = 1.6925\ f(R)$

   c. $\dfrac{a}{b} = \dfrac{m(I)}{m(R)} = \dfrac{2.5}{1.6925} = 1.477$

   Thus, $a = 1.477\ b$ and $a + b = 4.00$

   and $1.477\ b + b = 4.00$

   so, $b = 1.615$ inches from $L(R)$; $a = 2.385$ inches from $L(I)$.

   d. $m(E) = \dfrac{m(I)\ m(R)}{m(I) + m(R)} = \dfrac{(2.5)\ (1.6925)}{(2.5 + 1.6925)}$

   $m(E) = 1.009$

   Thus, $L(E) = 1.009\ f(E)$

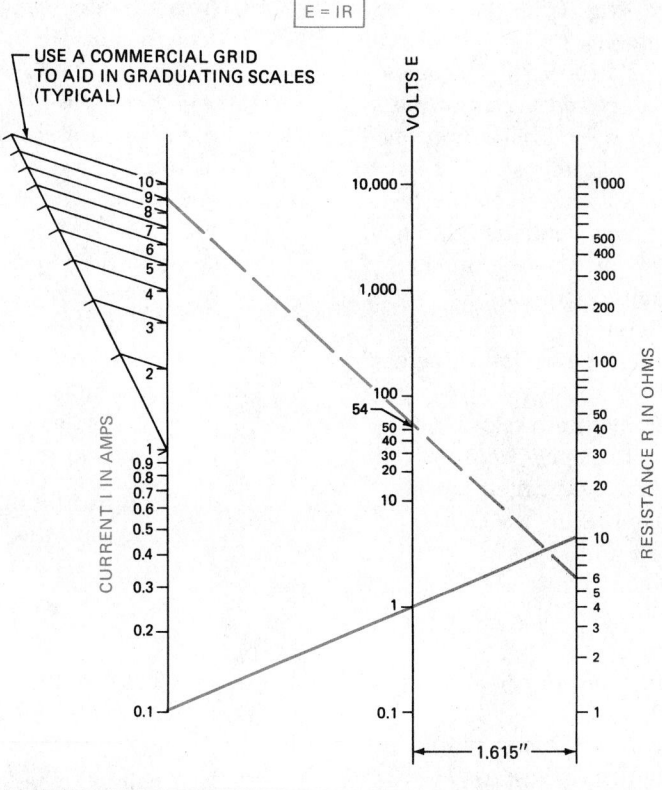

Figure 23-48 $E = IR$

so, L(E) = 1.009 (log 9000 – log 0.1)

= 1.009 times (3.954243 – (–1))

L(E) = 4.9988 inches

e. Construct an isopleth for any value of E from IR, e.g., (0.1 amps) x (10 ohms) = 1 volt, which locates the scale L(E).

f. Graduate L(E) using the scale equation and the graphical aid, y = log x.

13. Short Method for a Parallel Alignment Chart for E = IR. After constructing a number of alignment charts and learning the theories plus graphical techniques, users will appreciate more practical, faster methods. One method is shown in **Figure 23-49** for the steps below.

a. On a sheet of commercial 3-cycle, log-linear (semi-logarithmic) grid paper:

Locate 1 = 0.1, 1.0, and 10 amps on the left side of the sheet. Locate R = 1, 10, 100, and 1,000 ohms in a similar manner at a convenient distance to the right; say 5 or 6 inches.

b. Do two different calculations for E, but make $E_1 = E_2$, e.g.,

$E_1 = (10)\ (10) = 100$ volts
$E_2 = (1)\ (100) = 100$ volts

c. Connect the I and R values, respectively, with isopleths to locate the E scale where the two isopleths intersect. Label the intersection, 100.

d. Finish the E scale as before.

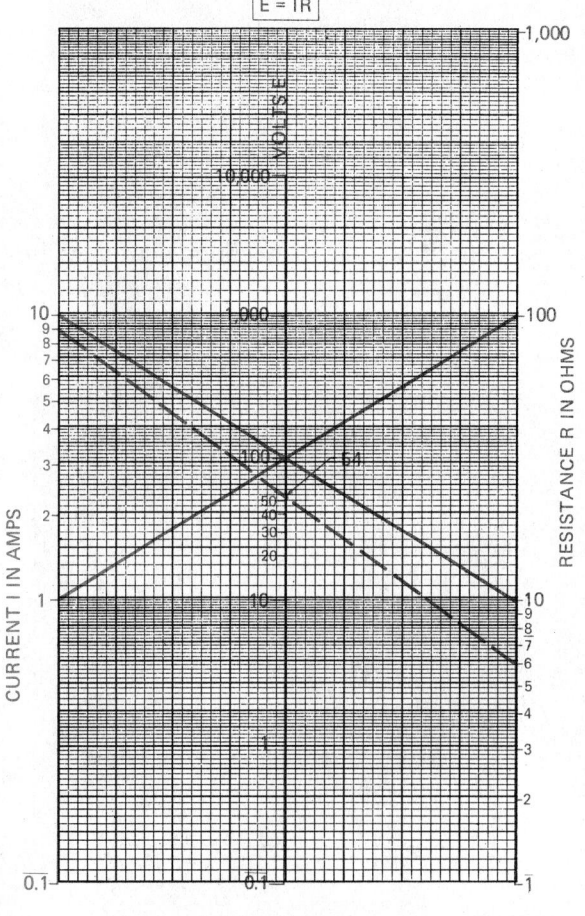

Figure 23-49   Shortcut: Parallel alignment chart

14. Concurrency Charts for f(r) + f(t) = f(s).

Concurrency charts provide designers with an additional graphical format to visualize equations with three or more variables, and to generate information. The basic format is the familiar horizontal and vertical cartesian coordinate system. This first concurrency chart example has two objectives: one is to illustrate the format, and the other is to point out that designers may have a number of choices of formats for any one equation.

Note that the equation form r + t = s, or f(r) + f(t) = f(s), has been discussed earlier for parallel alignment charts.

Assume that the variables have values: r = 0 to 12, and t = 0 to 10. *Figure 23-50* shows the concurrency chart for finding the third variable, s. Follow the steps used to prepare the chart:

a. Observe that the equation s = r + t is a straight-line equation, where s is y, r is x, and t is b in the general straight-line equation y = mx + b. The slope m = 1.

b. Divide the horizontal axis in equal segments for the variable r, 0 to 12.

c. Graduate the vertical axis to account for the minimum and maximum values of s, 0 to 22.

d. Write equations to account for the intercept t in s = (1)r + t. For example,

When t = 0, s = r
  t = 1, s = r + 1
  t = 2, s = r + 2
    ○
    ○
    ○
  t = 10, s = r + 10

e. Plot the equations which all have the same slope of 1, and label them with their appropriate t value.

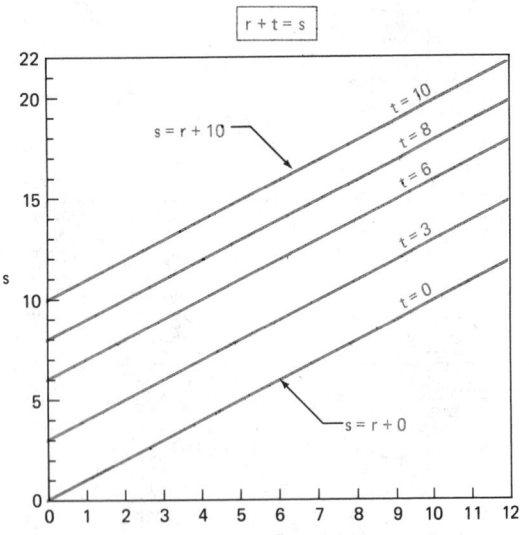

15. Concurrency Charts for f(r) • f(t) = f(s).

A second example of a concurrency chart, shown in *Figure 23-51*, is for the equation E = IR, which was used earlier. The equation can be rewritten in the form log E = log I + log R, as before. This equation, also, is patterned after the familiar straight-line equation, because log E is y, log I is x, and log R is b. The slope m = 1. The steps to prepare the chart are similar to those for the preceding chart, except that the horizontal and vertical scales are graduated in the logarithmic format. The variables I and R have the following ranges for this example:

I = 0 to 50 amps and R = 1 to 10 ohms.
E, then will be 0 to 500 volts.

a. Straight-line equations can be written to account for the intercept log R, as follows:

When R = 1
log E = log I + log 1
and E = I

When R = 10
log E = log I + log 10
and E = 10 I

b. Finally, the curves are drawn with a slope of 1, as shown in Figure 23-51.

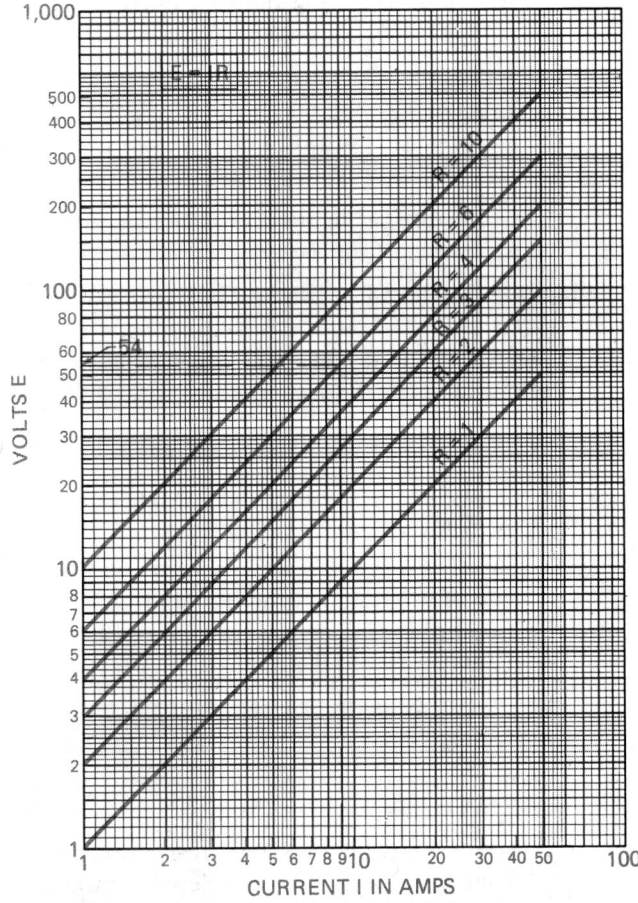

Figure 23-50 Concurrency chart: f(r) + f(t) = f(s)

Figure 23-51 Concurrency chart: f(r) • f(t) = f(s)

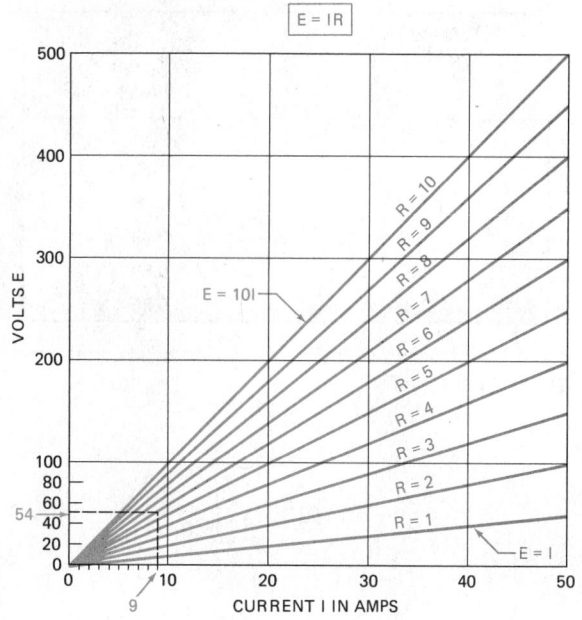

Figure 23-52   Concurrency chart: f(r) • f(t) = f(s)

16. **Concurrency Charts for f(r) • f(t) = f(s).**
This chart is an alternative format to the preceding discussion. The same equation, E = IR, is used but the steps are different and the chart is different, as seen in *Figure 23-52*. Do the following for this alternative:

a. Observe that E = IR can be compared to the straight-line equations in a slightly different way. In the equations y = mx + b and E = RI:

E is y, I is x, and R is the slope of the line.

b. Let the two variables have the same ranges as before, I = 0 to 50 amps and R = 1 to 10 ohms. Next, the following equations can be written:

When R = 1
    E = 1

When R = 2
    E = 21

    °

    °

    °

When R = 10
    E = 101

c. Graduate the horizontal and vertical scales in a linear manner for I and E, respectively.

The foregoing definitions and discussions of alignment charts and concurrency charts, plus the steps to prepare them, provide sufficient theory and construction guidelines to help designers use these charts effectively to obtain information or to prepare their own. However, further discussions and instructions are available for those who need them. An excellent reference was written by A. S. Levens, NOMOGRAPHY, 1959, John Wiley and Sons, Inc.

Empirical Equations: y = mx + b, y = b(e**mx) and y = b(x**m): and Other Curve Matching Techniques

Consider two problems involving the need to analytically describe curves, either plotted from data in an experiment or developed during the design of a surface.

1. **Data Problem.** Data involving two changing variables, one independent and the other dependent, has been recorded for an experiment in a lab course. An equation is required to describe the relationship of the two variables so additional data points can be calculated within the range of the data already recorded. What steps should be taken next? How can an equation be generated?

2. **Curve Problem.** This problem differs slightly from the preceding one because the data points are to be part of a design development to generate curved lines for an automobile body surface. Equations of the lines involve the use of a personal computer; however, the problem is still graphical. What steps should be taken? How are the curves generated? Equations?

The Data Problem will be discussed first, primarily from a graphics approach. Then the *curve problem* will be considered from a more analytical basis and involving personal computers. Both problems use physical and graphical illustrations and explanations.

Data Problem. Empirical Equations for Experimental Data. The flow chart in *Figure 23-53* depicts the overall approach for deriving and verifying empirical equations. The steps contained in the chart are followed while considering three sets of data.

First Set of Data

1. Assume that the data tabulated in *Figure 23-54* has been collected during a lab assignment involving surface factors for various steels, and that the next task is to plot the data on linear coordinate graph paper.

2. Before plotting the data, consider the three types of curves that are generally produced from experiments or from observing natural phenomena: (a) Straight-Line, (b) Power, and (c) Exponential.

   a. Data plotting in a *straight line* on linear coordinate paper would obviously indicate a linear relationship and an equation would be easily derived; however, be certain to indicate that the line applies only within the range of the data taken.

   b. Plots of data tending toward a *power curve* y = B(x**m) would show patterns similar to those in *Figure 23-55*. When m = 1, a unique straight line results which goes through zero and has a slope equal to B. See curve 1 in Figure 23-55. When m equals a positive number greater than one, curve 2 would be typical. Positive m's less than one look like curve 3,

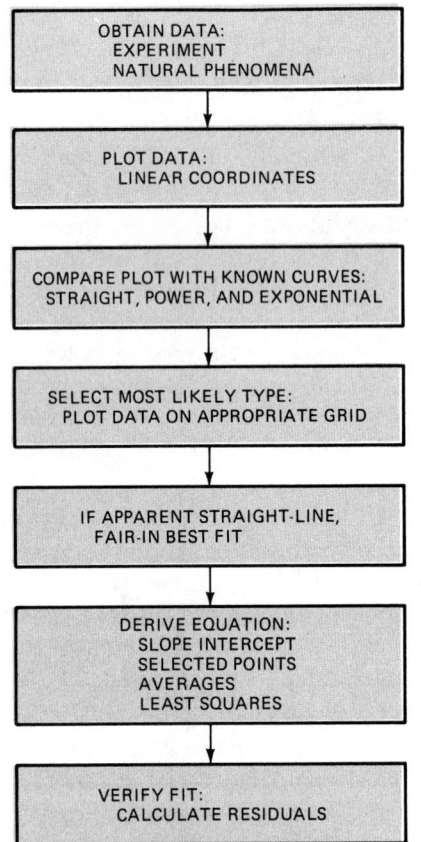

Figure 23-53 Empirical equations: Procedure

ULTIMATE TENSILE STRENGTH S x 1000 psi	SURFACE FACTOR k(a)
100	0.53
120	0.50
140	0.46
160	0.43
180	0.38
200	0.36

Figure 23-54 Data

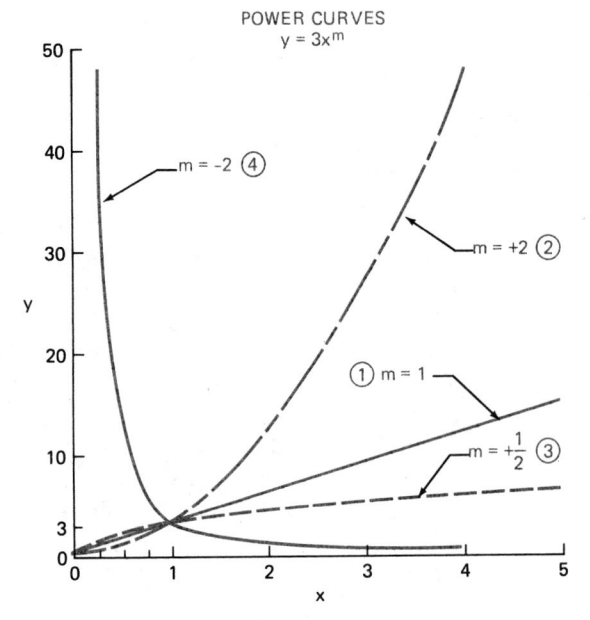

Figure 23-55 Power curves on linear coordinates

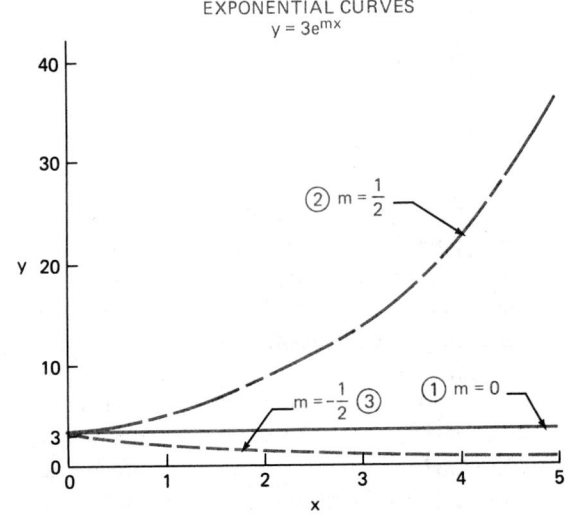

Figure 23-56 Exponential curves on linear coordinates

which continues in an upward direction. Negative values of m give the hyperbolic shape of curve 4. Curves 1, 2, and 3 go through zero when x = 0.

c. *Exponential curves*, in general, look like those in *Figure 23-56*. When m = 0 in the equation y = B(e**mx), the plot is the horizontal line 1 parallel to the x axis. Positive values of m give curve 2, which intercepts the y axis at the value of B and continues upward. Curve 3 starts downward from B on the y axis and asymptotically approaches the x-axis. Curve 3 is repeated in *Figure 23-57*, because 3 and 4 are the most common types. These two represent phenomena where the rate of change in a variable is proportional to the variable itself. For example, curve 3 illustrates the decay of a radioactive substance N. The equation N(at any time t) = N(originally) (e** – at) can be derived from data. The constant, a, represents 0.693/half life. Curve 4 illustrates the change in temperature of an object being heated. The temperature approaches a maximum value asymptotically as described by T(actual at time t) = T(max) (1 – e** – mt).

3. This first set of data plots almost as a straight line of the form y = mx + b on linear coordinate axes,

TYPICAL EXPONENTIAL PHENOMENA

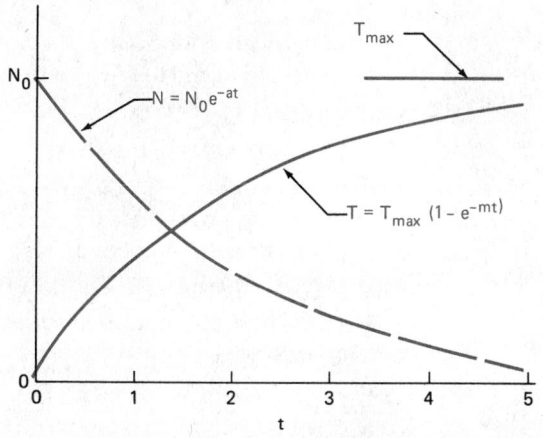

Figure 23-57   Exponential curves on linear coordinates

SURFACE FACTOR k_a
vs.
ULTIMATE STRENGTH OF HOT ROLLED STEEL

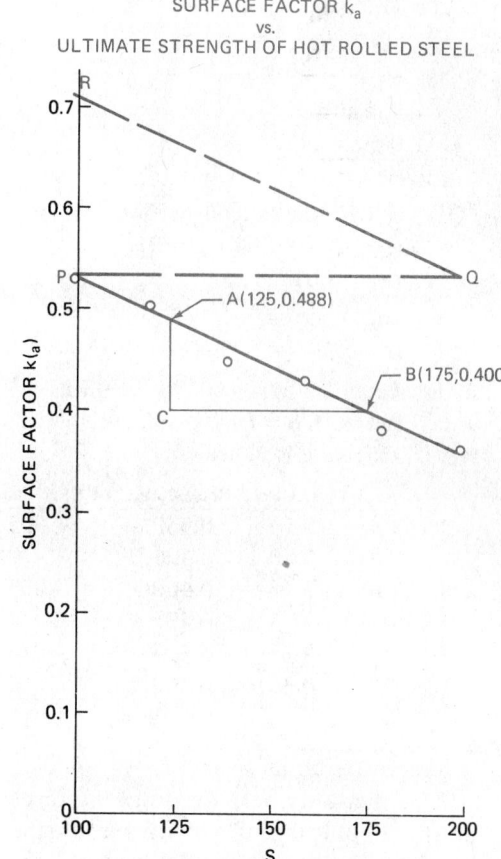

Figure 23-58   Data on linear coordinates

as shown in **Figure 23-58**. A "best" fit straight line is drawn "by eye" through most of the points; that is, a straight line has been "faired-in."

4. Next, derive an equation of the form y = mx + b. Of the many approaches available, the four listed here are the ones most often found in graphics texts. In increasing order of difficulty and time involvement, they are:

   a. Method of Slope-Intercept
   b. Method of Selected Points
   c. Method of Averages
   d. Method of Least Squares

   All four methods are used for this first set of data, and the curve from each method is verified (checked) by the method of residuals.

5. The methods.

   a. Slope-Intercept Method

   (i)   Select two points, e.g., A and B in Figure 23-58.

   (ii) Graphically, determine m, negative because of the downward slope, using the distance AC divided by CB. The slope must be compatible with the equation; therefore AC = 0.088 surface factor units, BC = 50 kpsi, and AC/BC = m = −.088/50 = −.00176 surface factor units/kpsi.

   (iii) Thus, the equation is k(a) = −.00176 S + b. Next, graphically determine the intercept b, but note that the S values only go down to 100 kpsi. The intercept must be found when S = O. The graphical method for finding b when the independent variable range does not include zero, is to simply project backward to zero, as shown with line PQ and QR in Figure 23-58. The intercept is found to be 0.71.

   (iv) Finally the equation is k(a) = −.00176 S + 0.71. The 0.71 value is the intercept at S = O; however, remember to state that the equation is good only between S = 100 kpsi and S = 200 kpsi.

   (v) Calculate and sum the residuals, which are equal to observed values of k(a) − calculated values of k(a), using the derived equation, at the six values of S.

S	k(a) obs.	k(a) calc.	Residuals R
100	0.53	0.528	+.002
120	0.50	0.493	+.007
140	0.46	0.458	+.002
160	0.43	0.422	+.008
180	0.38	0.387	−.007
200	0.36	0.352	+.008
		SUM R =	+.020

   A positive sum indicates that the curve could be moved upward slightly for a better fit.

   b. Selected Points Method

   (i)   Select two pairs of coordinate points from the "best" fit line and substitute them into the basic form of the straight-line equation. Then solve simultaneously for m and b.

The pair of points are:
k(a) = 0.5, S = 120
k(a) = 0.36, S = 200

Substitute:
0.50 = m(120) + b (q)
0.36 = m(200) + b (p)

(ii) Solve simultaneously for m and b.

(p) − (q) = −.14 + m(80)

m = −.00175

and 0.36 = −.00175(200) + b

b = 0.71

(iii) Finally, k(a) = −.00175 S + 0.71
between S = 100 kpsi and 200 kpsi

(iv) Calculate Residuals R

S	k(a) obs.	k(a) calc.	Residuals R
100	0.53	0.530	+.005
120	0.50	0.496	0
140	0.46	0.462	−.005
160	0.43	0.428	0
180	0.38	0.394	−.015
200	0.36	0.360	0
		SUM R =	−.015

c. Averages Method

This method doesn't require the direct use of the graphical plot; however, the data must be plotted to decide which type of curve the data points more closely match.

The expression SUM(y − mx − b) = 0 says that for the best fit line, the differences between the observed values of y and the calculated values of mx + b add to zero.

(i) Divide the data into two equal groups, substitute the data into equations y − mx − b = 0, and add them.

.53 − 100m − b
.50 − 120m − b
.46 − 140m − b
1.49 − 360m − 3b = 0   (r)

.43 − 160m − b
.38 − 180m − b
.36 − 200m − b
1.17 − 540m − 3b = 0   (s)

Solve (r) and (s) simultaneously and find
m = −.00178 and b = 0.71

(ii) The equation of the line becomes k(a) = −.00178 S + 0.71 for S from 100 kpsi to 200 kpsi.

(iii) Calculate the Residuals in the same way as before and find that SUM R = +.003

d. Least Squares Method

This method consumes the most time, but it gives the best accuracy of the four methods. The expression SUM(*y − mx − b*)² = *a minimum* says that for the best fit, the differences between the observed data for y and the cal-

culated data mx + b, squared, should be a minimum.

(i) Set the derivatives of SUM(y − mx − b)² with respect to m and to b, equal to zero:

With respect to m:
SUM [2(y − mx − b) (−x)] = 0

With respect to b:
SUM [2(y − mx − b) (−1)] = 0

(ii) Multiply and rewrite to obtain:

SUM (x) (y) = b SUM(x) + m SUM(x**2)
SUM (y) = b SUM(number of observations) + m SUM(x)

(x) (y)	(b) (x)	M(x**2)
100 (.53)	=b100	+m(100)**2
120 (.50)	=b120	+m(120)**2
140 (.46)	=b140	+m(140)**2
160 (.43)	=b160	+m(160)**2
180 (.38)	=b180	+m(180)**2
200 (.36)	=b200	+m(200)**2
386.6	=b900	+m142,000 (t)

Y	b	mx
.53	=b	+m100
.50	=b	+m120
.46	=b	+m140
.43	=b	+m160
.38	=b	+m180
.36	=b	+m200
2.66	=6b	+m900 (w)

(iv) Solve (t) and (w) simultaneously and find

m = −.00177
b = 0.709

so, k(a) = −.00177S + 0.709

(v) Calculate Residuals R and get SUM R = −.001, which is the best fit of the four methods summarized here.

a. Slope-Intercept	SUM = +0.014
b. Selected Points	SUM = −0.015
c. Averages	SUM = +.003
d. Least Squares	SUM = −.001

Second Set of Data. To calibrate a Weir, the following data was taken:

H(feet)	2.	3.	4.	5.	6.
Q (cu. ft/min)	14.2	41.5	80.0	135.2	225.3

H(feet)	7.	8.	9.	10.
Q (cu. ft/min)	330.5	445.5	620.5	790.

A Weir is a rigid plate used to measure the flow of water in open channels (e.g., irrigation channels). The rigid plate goes crosswise in the channel, as shown in **Figure 23-59**, so the water must spill over it. Height of the water, H, above the bottom of the notch indicates the rate of flow, Q.

1. Follow the general empirical-equations approach depicted in Figure 23-53 and plot the Weir data on linear coordinate paper to ascertain the type of

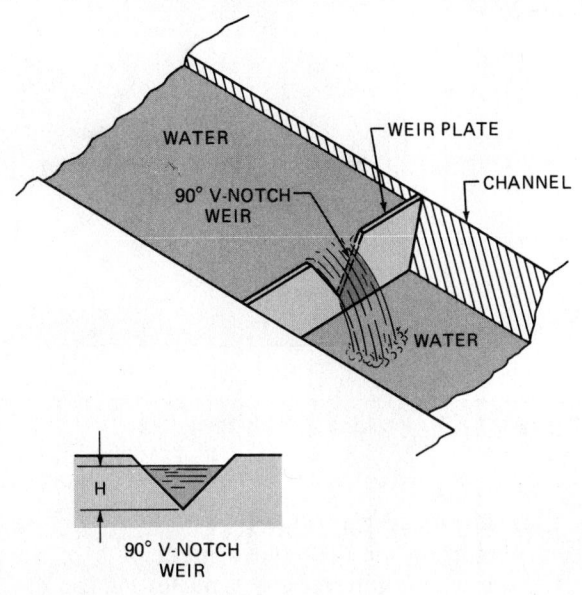

Figure 23-59 Weir

a. Slope-Intercept Method

   (i) Select two points on the best fit line, A and D in Figure 23-61.

   (ii) Determine slope m from

$$m = (\log A - \log C)/(\log C - \log D);$$

$$(\log 450 - \log 80) = 0.7501$$

and $(\log 8 - \log 4) = .301$

so $m = +(.7501)/(.301)$
= 2.492 cubic feet per second per foot.

  (iii) The logarithmic equation changes to log Q = log B + 2.492 log H. Next, graphically determine the slope-intercept. Use an offset technique to find an intercept for H = 0, as shown in Figure 23-61. Upward from P to R and then parallel to the original line from R to B, in effect adds an additional cycle so the intercept can be found: it is B = 2.5. Finally, the

curve — straight, power, or exponential. The plot of the Weir data in *Figure 23-60* appears similar to the one in Figure 23-55, which seems to be increasing upward as H gets larger. Also, the plot of Q would go through zero for H = O. Thus, a power type curve is indicated.

2. Try a logarithmic 2 x 2 cycle commercial grid for obtaining a straight line. *Figure 23-61* indeed shows that the data forms a linear relationship and will fit the logarithmic form of the general power equation, log y = log B + m log x.

3. Fair-in a best fit straight line and prepare to derive a power equation from log Q = log B + m log H.

4. The Slope-Intercept and Selected Points methods are used for this set of data.

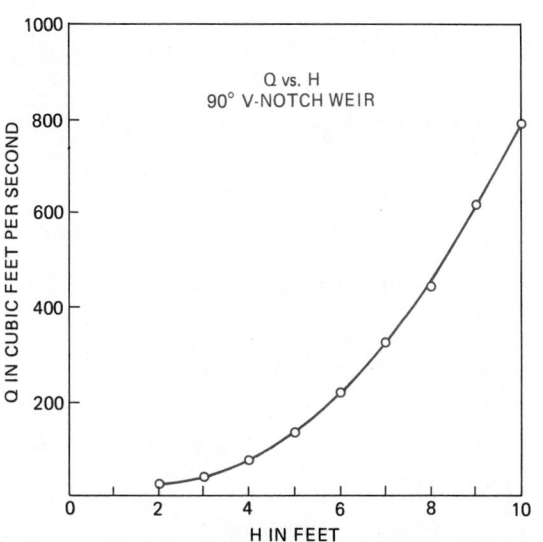

Figure 23-60 Weir data on linear coordinates

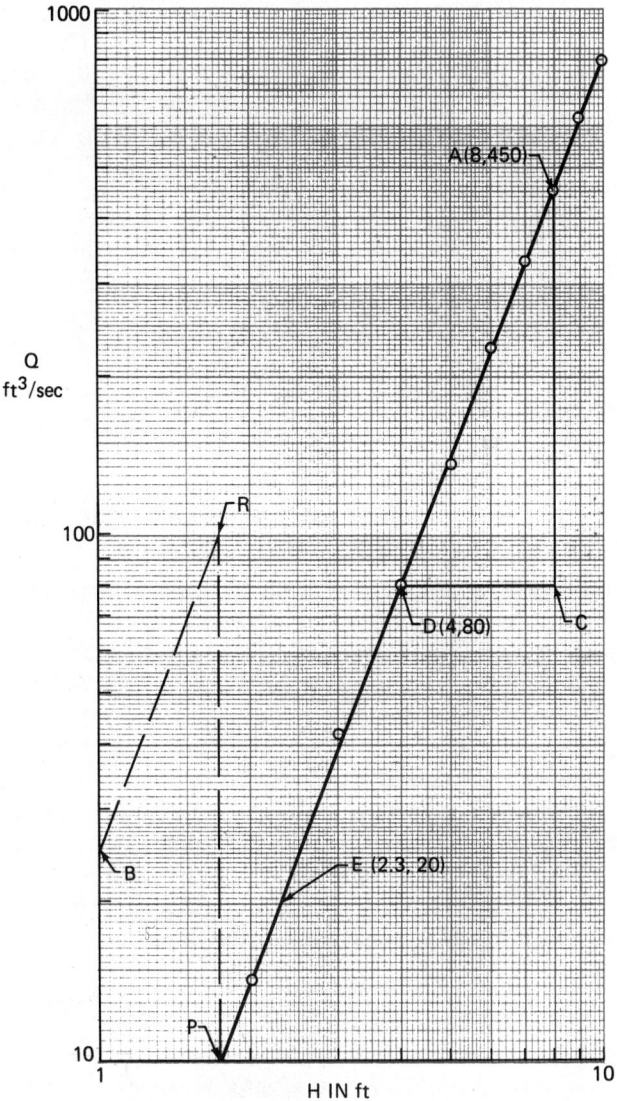

Figure 23-61 Weir data on logarithmic grid

equation can be written as Q = 2.5(H**2.492) for values of H between 2 feet and 10 feet only. Further measurements should be taken below 2 feet.

(iv) Calculate and SUM the residuals.

H	Q (Observed)	Q (Calculated)	Residual R
2	14.2	14.06	+0.14
3	41.5	38.63	+2.87
4	80.0	79.12	+0.88
5	135.2	137.97	-2.77
6	225.3	217.32	+7.98
7	330.5	319.10	+11.40
8	445.5	445.08	+0.42
9	620.5	596.92	+23.50
10	790.6	776.14	+14.46
		SUM R =	+58.88

The positive sum indicates that the faired-in straight line should be moved slightly upward for a better fit.

b. Selected Points Method

(i) Select two pairs of coordinates from the best fit line and substitute them into the equation log Q = log B + m log H. From the line, obtain A(450, 8) and E(20, 2.3) to write log 450 = log B + m log 8 and log 20 = log B + m log 2.3

(ii) Solve simultaneously:

log 450 – log 20 = m(log 8 – log 2, 3)
m = 2.498

and then, log 450 = log B + 2.498 log 8, so B = 2.496.

(iii) The equation finally is:

log Q = log 2.496 + 2.498 log H

and Q = 2.496 (H**2.498)

(iv) Calculate residuals to verify the equation

Third Set of Data. Picture a researcher operating a remote device used to place small canisters containing material into a nuclear reactor used for research. The material is lowered into a flux of neutrons to be irradiated. The material absorbs the neutron energy, but later gives up the energy by emitting gamma rays hundreds of times per second. A graph showing the material's pattern of energy change, E vs. time t, is in **Figure 23-62**. The detention period starting at time T1 helps the researcher obtain gamma ray data to be used to determine the half-life of the irradiated material. The data recorded from a gamma ray detector was:

N (counts per second)	2000	210	40	7
t (time in seconds)	0	900	1640	2450

For this set of data, the researcher knew that the decay rate (gamma rays emitted) would be directly proportional to the energy level of the irradiated material. Physical phenomena of this type behave exponentially as y = B(e**mx),

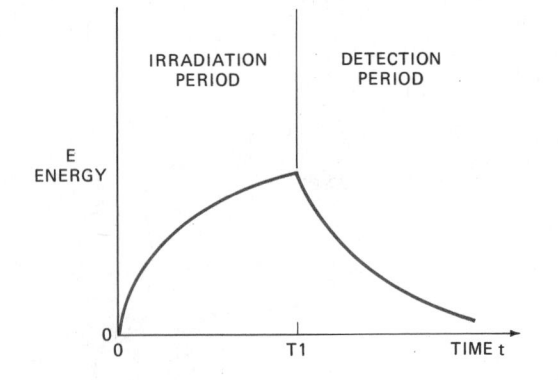

Figure 23-62 Irradiation of sample: Energy level

so a log-linear grid was selected without bothering to check the data on linear coordinate paper.

1. Follow the general approach in Figure 23-53, starting at the straight-line plotting on a log-linear grid. Use a commercial grid, semi-logarithmic, 4 cycles. *Figure 23-63* shows the plot plus a faired-in best fit straight line. The form of the equation of this line is log y = log B + m log e x, where the coefficient of x is m log e.

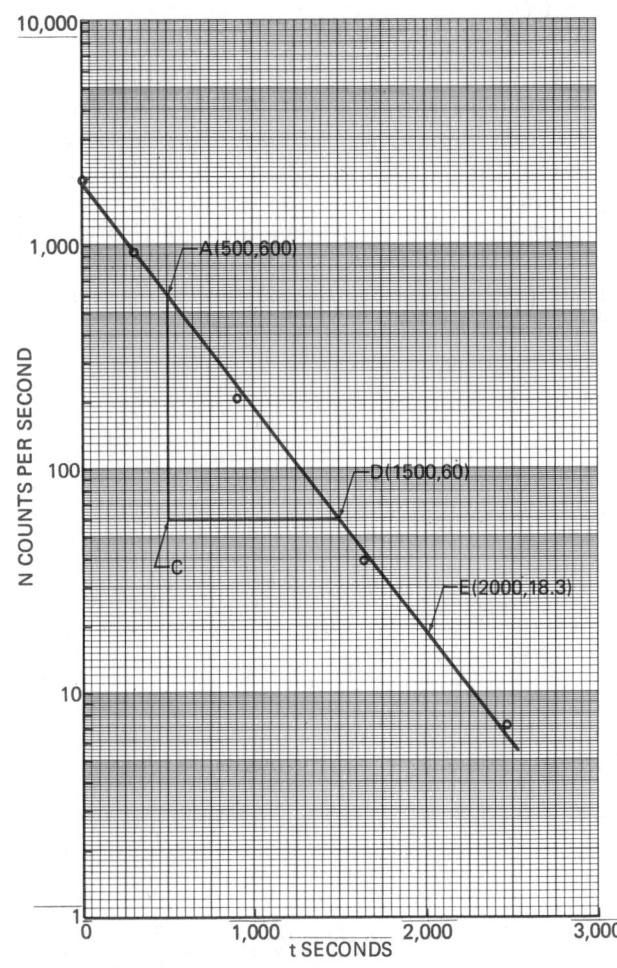

Figure 23-63 Irradiation data on semilogarithmic grid

2. Use the Slope-Intercept and Selected Points method for matching an equation to the plot of the data. The straight-line equation for the variables N and t is log N = log B + m log e t.

   a. Slope-Intercept Method

   (i) Select points A(500, 600) and D(1500, 60) on the best fit line.

   (ii) Determine the slope m from

   m log e =
   (log 600 – log 60)/(500 – 1500)

   m(.4343) = (2.77815 – 1.77815)/(–1000)

   m = –(1)/((1000) (.4343))
   = –.002303 counts per second.

   (iii) Then the equation is log N = log B – .002303 (log e)t. The B intercept from the graph is 1900, so the final equation is N = 1900(e** – .002303 t).

   (iv) The time elapsed to arrive at the half-life t(HL) value of the counts can now be calculated by noting that one-half of the initial count equals one-half of 1900; therefore,

   1900/(2) = 1900 e** – .002303t(HL)

   and ½ = e** – .002303 t(HL).

   Thus, log 1 – log 2 =
   –.002303(log e)t (HL)

   and t(HL) = 300.9 seconds.

   The straight-line plot confirms this value of 950 counts per second at approximately 300 seconds.

   (v) Calculate and sum the residuals.

t seconds	N observed	N calculated	Residuals R
0	2000	1900	+100
900	210	239.1	–29.1
1640	40	53.5	–3.7
2450	7	6.7	+.3
		SUM R =	67.5

   b. Selected Points Method

   (i) Two pairs of coordinate points from the best fit line are A(500, 600) and E(2000, 18.3).

   (ii) Substitute them in log N = 1900 – .002303(.4343)t and solve simultaneously:

   log 600 = log B + m (.4343) (500)

   log 18.3 = log B + m (.4343) (2000)

   so m = –.00233 counts per second.

   Then, to solve for B:

   log 600 = log B – .00233 (log e) 500,
   B = 1923.6 counts per second.

   (iii) Finally

   log N = log 1923.6 – .00233 (log e) t

   so N = 1923.6(e** – .00233t)

   (iv) Calculate Residuals next to verify the derived equation.

Note how the straight-line equation formed the basis for handling all three sets of data. Additional plotting techniques and analytical approaches are discussed next.

Curve Problem. The goal in the Data Problem was to find an equation to fit all of the data of an experiment, over the range of the data. The equations that were derived showed the relationships of dependent to independent variables such as k(a) = –.00176S + .71, Q = 2.5(H**2.492), and N = 1900(e** – .002303t) – linear, power, and exponential equations, respectively. Finding the proper coordinate grids so that the data plotted in approximately a straight line was the key to deriving the equations.

A number of additional approaches may be used to match equations to curves. One is similar to the Selected Points technique already discussed, but the basic equation could be third order, such as y = ax**3 + bx**2 + cx + d, which requires four sets of "selected points." Sometimes the entire curve may be described by a number of equations, piecewise, in series. Still another method describes a curve that may touch the first and last points, but only goes near the others. These types of curves are introduced in this section for three main reasons.

1. They require graphical-physical explanations to visualize the concepts.
2. Personal computers make it possible for students and designers to do this type of curve matching and curve generating in a relatively short time.
3. The trend in one type of manufacturing uses these approaches to generate curved surfaces and the related equations, which are then used to program numerically controlled milling machines to cut the surfaces.

Two physical models help to visualize the following curve types:

1. Splines

   a. *Figure 23-64* shows the technique traditionally used by ship builders (before computers) to draw smooth contour lines through a series of points. The "ducks" shown in the figure weigh about 5 pounds, are cast iron with felt glued to their bottoms, and are used to apply pressure at specified points along a thin metal strip called a *spline*. These pressure points are placed over the points that define a contour line of a ship. Because of the stiffness of the spline, the curve going into a point is exactly in-line with the curve leaving the point. Distances between points are called spans.

   b. *Cubic splines* can be generated which in effect take the place of the metal spline. These equa-

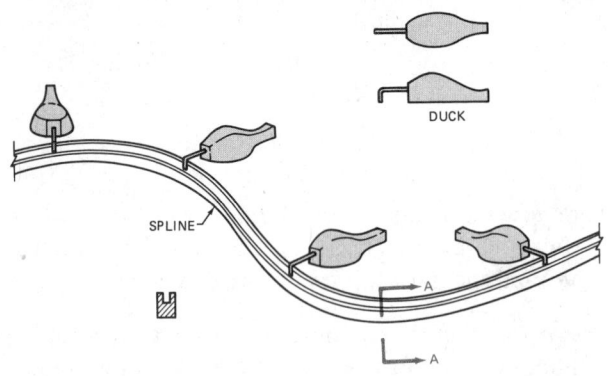

Figure 23-64 Spline

tions, one for each span, not only describe the curve over the span but ensure that the span segments are "in line." Software is available to generate cubic splines on personal computers.

2. Bezier (pronounced bay-zee-ay) and B-Splines
   a. Bezier Curves: Assume that the thin metal strip shown in *Figure 23-65* is magnetic and without the "ducks." The strip is placed on a flat surface between a number of "point" magnets that pull on the strip but never touch it. The strip is anchored to the end points of a series of points: all points except the ends are the magnets. At the ends, the strip is tangent to a line from the end point to the nearest point. Also, if one magnet along the strip moves, the whole strip may move slightly.

   Designers use BEZIER curves in conjunction with a computer screen to generate pleasing curves for the design of automobile bodies. Once the curve satisfies the designer, an equation is automatically generated to match the curve. Ultimately surfaces are generated using curves in horizontal and vertical planes.

   One disadvantage of the BEZIER curve generating procedure is that when one point moves, the entire curve can be affected. Usually all but a portion of a curve is OK, and just changing a single portion is desirable.

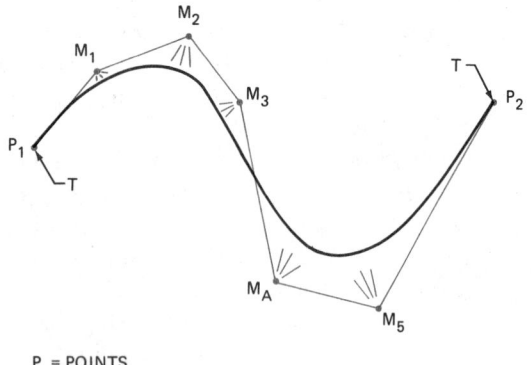

P = POINTS
M = MAGNETS
T = TANGENT POINTS

Figure 23-65 "B-spline" type curve

b. B-Splines: The B-Spline curve-making technique allows a single portion of a Bezier-type curve to be modified without affecting the entire curve. Therefore, automobile body designers find that the B-Spline technique saves considerable time in designing contour lines and surfaces. Software is also available for generating B-Splines on personal computers.

One application of these curve-generating techniques illustrates a trend in some machining operations. Shoe "lasts" (full scale models of the left and right feet for a given shoe size) are completely described by spline-type equations. The equations are automatically coded into language that instructs numerically controlled machines to cut the lasts. Thus, no drawings are needed to communicate the information from the designer to the machine.

Conclusion. This curve problem discussion was included in empirical equations to emphasize that the computer is not only affecting the process of making drawings, in CADD systems, but is also being used to assist designers to generate graphical curves, surfaces, and models.

The next section contains discussions on graphical calculus, integrating, and differentiating that is also useful to designers.

Graphical Integration and Differentiation: Pole and Ray Method

The advantages of knowing how to do graphical calculus, integrating, and differentiating are twofold: (a) they enhance insight and understanding of analytical integrating and differentiating, and (b) they provide a means of handling data and some equations, which would be too difficult and expensive to do analytically. Furthermore, graphical calculus, as do other graphical techniques, provides designers with relatively rapid visual techniques for designing, analyzing, and checking. Therefore, this section focuses on the graphical techniques for integrating and differentiating using the *Pole* and *Ray* method.

1. First, consider two fundamental ideas, one for integrating, the other for differentiating.
   a. *Figure 23-66* illustrates a graphical explanation of the process of integrating and the resulting integral curve.
      (i) The area between the curve and the horizontal axis in the Y1 vs. X graph is divided into segments by vertical division lines 1 and 2.
      (ii) The cross-hatched area between the division lines 1 and 2 represents the change of the integral curve in the vertical direction between the same two division lines. The area shown is above the x axis, so it is positive; therefore, the

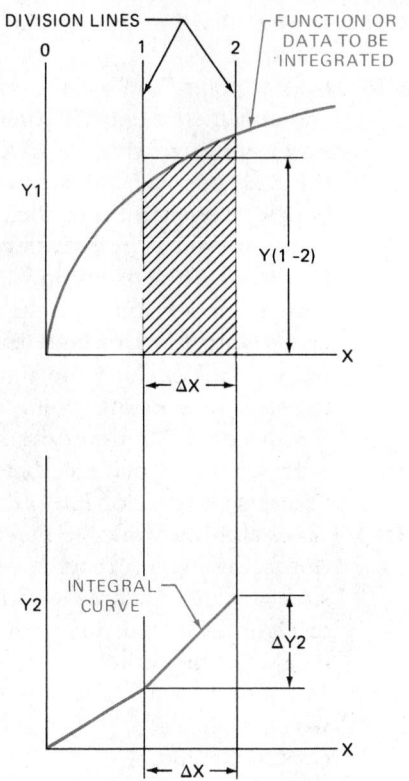

Figure 23-66   Graphical integration: Fundamentals

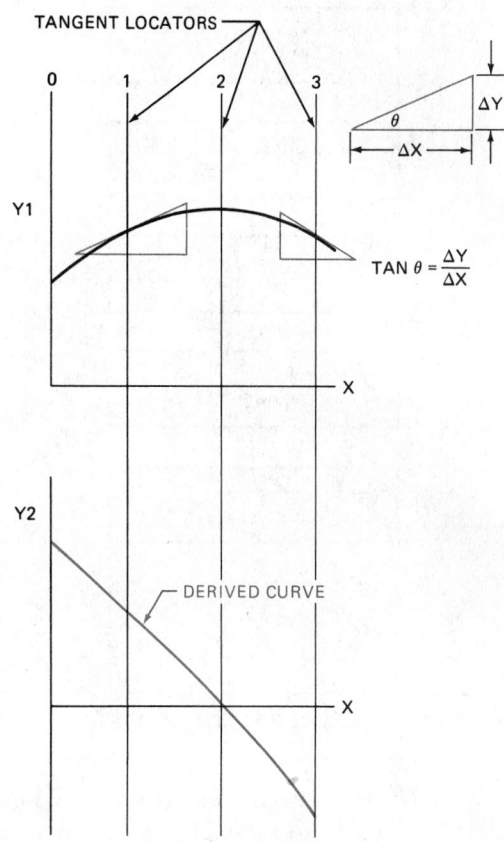

Figure 23-67   Graphical differentiation: Fundamentals

change in the integral curve is upward between the same two division lines.

    (iii) Items (1) and (2) above restated in the symbols shown would be: the average height of the area, Y(1 – 2), times the distance between division lines 1 and 2, Δx, equals ΔY2 of the integral curve between the same two division lines in the Y2 and X graph.

    (iv) Vertical scales Y1 and Y2 are different, but in-line: both horizontal scales are identical.

b. ***Figure 23-67*** shows a graphical explanation of differentiation and the resulting derived curve.

    (i) Tangents to the curve in the Y1 vs. X graph are constructed "by eye" at the three tangent locator points, and the slope of the tangents are determined by a ΔY/ΔX calculation.

    (ii) The positive, zero, and negative slopes from points 1, 2, and 3, respectively, are plotted in the Y2 vs. X graph.

    (iii) Items (1) and (2) restated in the symbols shown would be: the slopes ΔY/ΔX at specific points in the Y1 curve are represented by proportionate vertical distances in the Y2 vs. X graph.

    (iv) Vertical scales Y1 and Y2 are different but in line: the horizontal scales are identical.

2. Graphical Integration. The overall procedure to do graphical integration follows the flow chart in

Figure 23-68. Most of the procedure entails graphical construction steps: the last step, deriving a scale, is both analytical and graphical.

a. Refer to ***Figure 23-69*** and follow the steps to do the graphical construction. (*Note:* (i) is the circled number 1 in the figure, etc.)

    (i) Plan space for two sets of axes, the upper space for SY1 vs. SX and the lower for SY2 vs. SX. For illustration purposes, data from an aircraft launch is used. The following data for acceleration in g's vs. time t, during the launching cycle for an airplane from an aircraft carrier was recorded in a control center in the carrier.

g's Meters, per sec./per sec.	t seconds
0.00	0.0
0.50	0.2
1.00	0.3
2.55	0.5
4.60	0.6
5.20	0.7
5.40	0.8
5.26	1.0
4.95	1.3
4.60	1.6
4.00	2.0

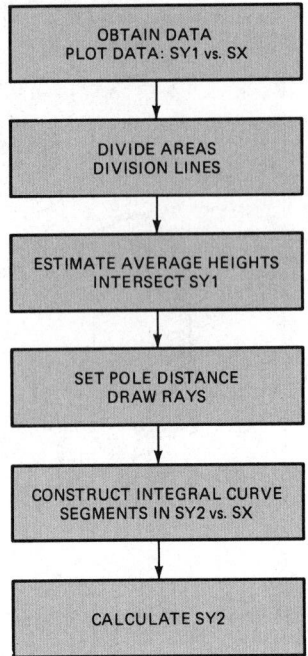

Figure 23-68   Graphical integration: Procedures

(ii) Plot the data on a commercial grid 10 x 10 to the inch, and fair-in a best fit curve. Use vertical scale SY1 equal to 2 g's/inch, and the horizontal scale SX equal to 0.5 seconds/INCH. (The capital letters, INCH, represent the basic unit of distance on the grid.)

(iii) Use division lines to divide the area between the curve and the horizontal axis. Use more lines during curving portions of the curve for better accuracy. Label the lines 1, 2, 3, 4, 5, and 6, for this plot.

(iv) Estimate "by eye" the average height of each division of area. Sketch A in Figure 23-69 shows that a line for average height is approximated when the negative area is about equal to the positive area. Extend a line to the SY1 axis for each average height, and label each extension.

(v) Set the *Pole* distance H equal to 1.5 INCHES as an extension of the horizontal axis SX, to the left, to point P. The distance H affects the slope of the derived curve: short *Pole* distances cause the curve to be steep; long distances, flat.

(vi) Construct *Rays* from the extension-lines intersections on axis SY1 to the *Pole* point P.

(vii) Construct and label the second set of axes, SY2 vs. SX, so SY2 is aligned with SY1. The calculation for SY2 is dis-

cussed after the steps for the graphical construction.

(viii) Start at point T on the SY2 vs. SX axes and construct a line TK parallel to the *Ray* associated with the area between the SY1 axis and division line 1. The height of point K above the horizontal axis represents the area between the SY1 axis and division line 1. From K draw a line KJ parallel to the *Ray* associated with the area between division lines 1 and 2. Continue similarly for the remaining *Rays* to point W. Finally, the distance WZ times the scale SY2 represents the total area between the SY1 axis and division line 6.

(ix) To establish a workable *Pole* distance, the following approach works. Construct a trial H distance, say to P1, shown in Figure 23-69. Estimate the average height of the entire area between the SY1 axis and division line 6, which would be line RQ. Draw *Ray* R-P1 and construct a line TU parallel to R-P1 and construct a line TU parallel to R-P1 in the SY2 vs. SX graph. Point U indicates the approximate location of where the integral curve will probably terminate; that is, where point W will be.

(x) Calculate the scale SY2 for the integral curve graph using the equation SY2 = (H) (SX) (SY1), which is derived as follows:

- *Figure 23-70* shows SY1 vs. SX with SY2 vs. SX directly below. The first curve is plotted and the area is divided by division lines 1 and 2. FD represents the average height of the area between the SY1 axis and division line 1. FP is the *Ray* associated with the area. H is the *Pole* distance.

- Line OK in the SY2 vs. SX graph is constructed parallel to ray FP, and KJ = ΔY.

- The desired result of this graphical integration construction is that (ΔY) (SY2) = (FD) (SY1) (ΔX) (SX), which says generally that the area between any two division lines represents the change in height of the integral curve between the same two division lines. Note how the scales must be used in the equation to obtain the correct values.

- Geometrically from similar triangles:

$$\frac{FD}{PD} = \frac{KJ}{OJ} = \frac{\Delta Y}{\Delta X} \text{ and } PD = H$$

so, $\frac{FD}{H} = \frac{\Delta Y}{\Delta X}$; then $\Delta Y = \frac{(\Delta X)(FD)}{H}$

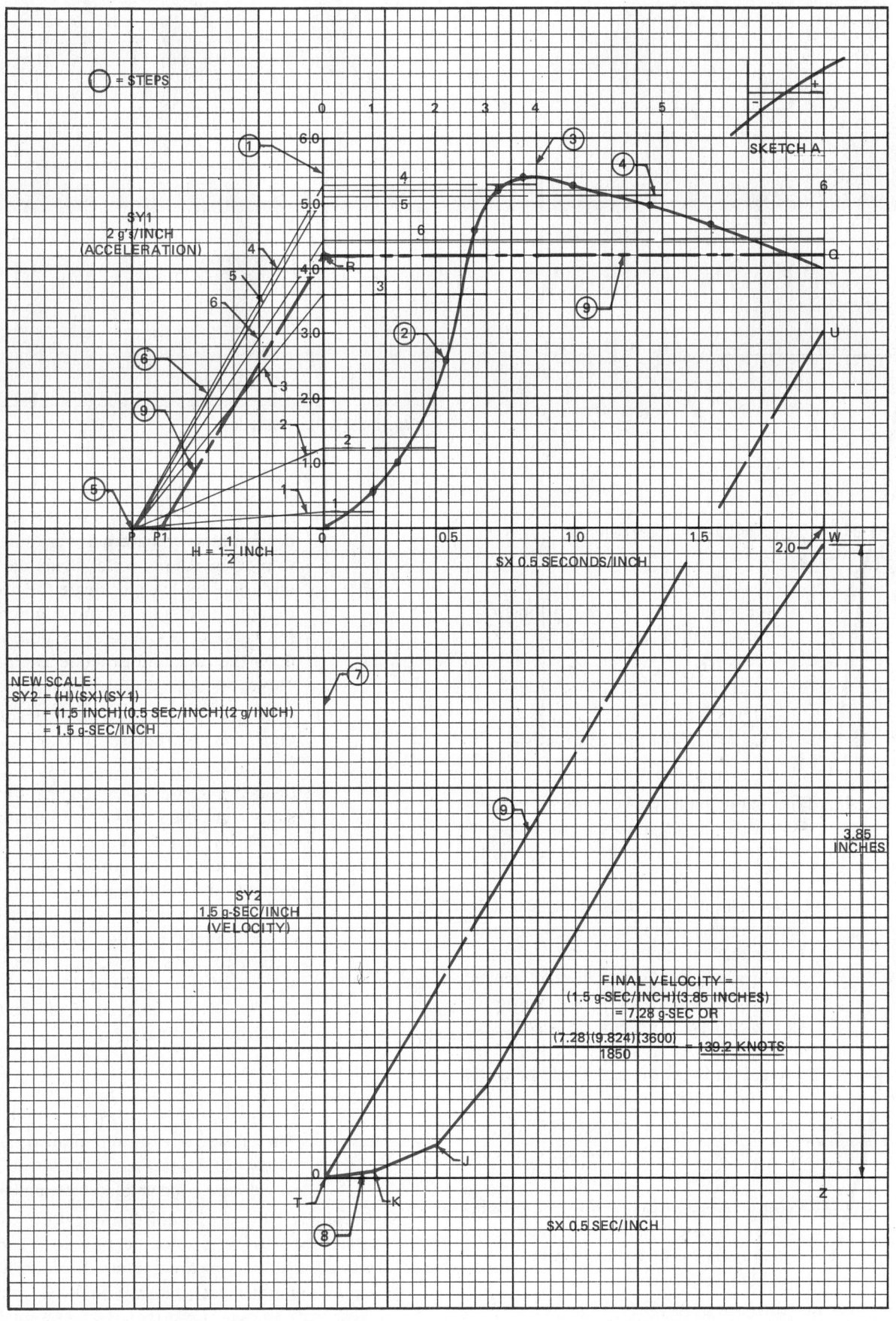

Figure 23-69 Graphical integration: Steps

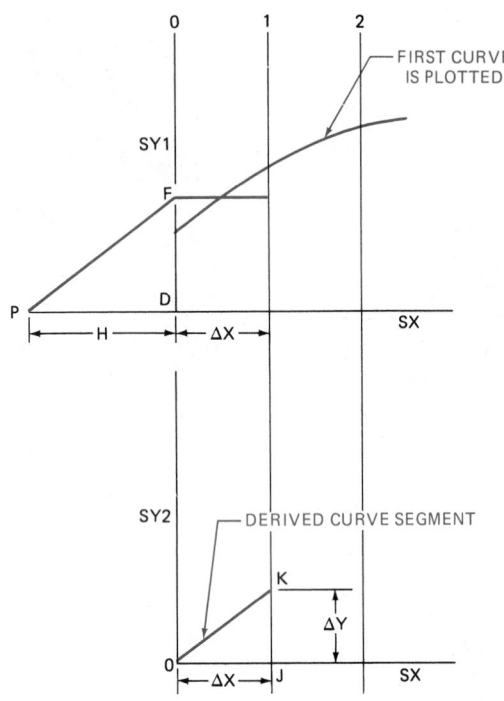

Figure 23-70 Graphical integration: Scale derivation

Substitute ΔY in the equation from the preceding step to get

$$\frac{(\Delta X)\,(FD)}{H} \cdot (SY2)=$$

(FD) (SY1) (ΔX) (SX)

Cancel like quantities and obtain

SY2 = (H) (SX) (SY1)

b. An additional capability of graphical integration is to divide odd-shaped areas and volumes into equal portions. The following example demonstrates this capability.

   (i) A right-circular-cone shaped water tank is used for this example to convince the reader that the method works, since the graphical results can be checked analytically. After following this example it should be easy to imagine an odd-shaped fuel tank nestled in a wing section of an airplane and how the tank could be graduated into equal volumes.

   (ii) *Figure 23-71* shows the water tank that needs to be graduated and marked at ¾, ½, and ¼ FULL.

  (iii) Cross-sectional areas of the 15-foot tall, 10-foot diameter tank were calculated at 3-foot intervals and recorded here:

Station S in Feet	0.	3.	6.	9.	12.	15.
Area A (square feet)	0.	3.14	12.6	28.3	50.2	78.5

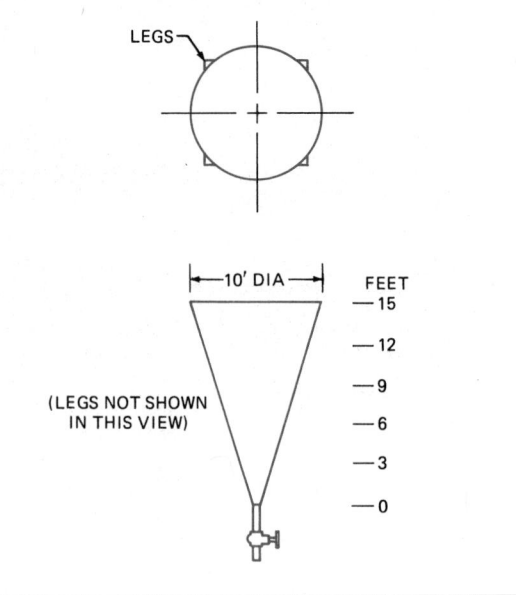

Figure 23-71 Water tank

   (iv) *Figure 23-72* shows how the steps to do graphical integration were followed to plot the data to scale, divide the areas with division lines, estimate the average heights, and transfer those heights to the vertical axis, locate the pole point, draw the rays, and generate the integral curve.

   (v) Distance WZ times SY2 scale of 125 cubic feet/INCH equals 390 cubic feet.

  (vi) To obtain the graduations, WZ was divided into four equal lengths: the divisions were then transferred horizontally to the integral curve and, finally, these intersections were projected vertically down to the horizontal axis.

     The intercepts on the horizontal axis represent the location along the tank height to mark the ¼, ½, and ¾ FULL, 9.5′, 12.2′, and 13.5′, respectively. Now the following comparison can be made with the analytical solution.

	¼	½	¾	FULL
GRAPHICAL	9.5′	12.2′	13.5′	15′
ANALYTICAL	9.45′	11.91′	13.63′	15′

The close correlation demonstrates the efficacy of this technique to graduate areas or volumes using graphical integration.

3. Graphical Differentiation. Graphical differentiation uses slopes at selected points in a curve instead of adding the areas under the curve as was done in graphical integration. Use the steps in the flow chart shown in *Figure 23-73* as a guide for following the graphical differentiation example discussed next.

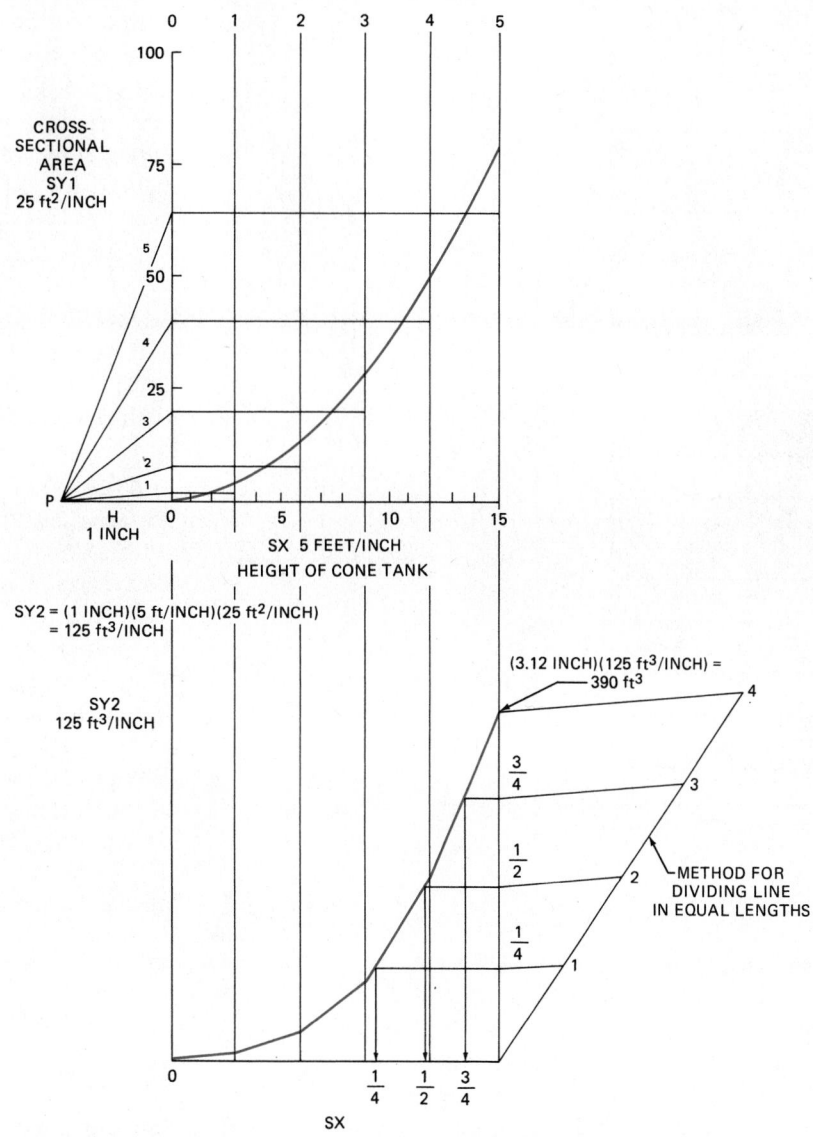

Figure 23-72 Graduating volumes

a. The cam shown in *Figure 23-74* provides an excellent example of the application of graphical differentiation to visualize the operation of the cam and follower, and to check the cam's operation. Cams convert rotary motion of a cam shaft to reciprocating motion for the follower. If the cam profile changes too rapidly, the follower may separate from the cam surface because of excessive acceleration — an undesirable condition. Therefore, designers need to check their cam profile designs for accelerations problems. Refer to Figure 23-74 for the steps to do a graphical differentiation of the proposed displacement of the follower vs. angular positions of the cam. The first differentiation will generate a curve showing the velocity of the follower at any time during the first half of one rotation: the second half is to be a mirror image of the first half. Differentiation of the velocity curve produces an acceleration curve for the follower.

(i) Prepare space on a commercial metric grid for three sets of axes, the first set in the upper space, for follower displacement vs. time. Follower displacement was calculated at every 15° rotation of the cam, which rotated at a constant speed of one revolution every 4 seconds. The time elapsed at each 15° was also calculated. The results were as follows:

Vertical Displacement

S in mm	0	.35	1.39	3.13	5.56	8.68
Time in fractions of a second	0	1/6	1/3	1/2	2/3	5/6

12.5	16.32	19.44	21.87	23.60	24.65	25
1	1-1/6	1-1/3	1-1/2	1-2/3	1-5/6	2

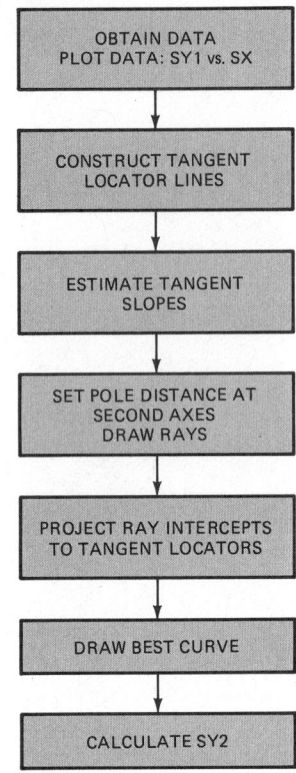

Figure 23-73 Graphical differentiation: Procedure

(ii) Plot the results and fair-in a best fit curve. Use a vertical scale SY1 equal to 0.5 mm/MM, and a horizontal scale SX equal to 1/60 second/MM. (The capital letters MM represent the basic unit of distance on the metric grid.)

(iii) Use tangent locators, vertical lines, to intersect the plotted curve at selected points; place more locators where the curve changes rapidly, for better accuracy. Label them 1, 2, 3, 4, 5, and 6, for this example.

(iv) Estimate "by eye" the slope of the curve, a tangent, at each locator intersection. Label the tangents, also.

(v) Set a *Pole* distance H = 30 MM as an extension of the horizontal axis SX of the second set of axes. The distance H affects the height of the derived curve: longer *Pole* distances cause higher curves; shorter distances, lower curves.

(vi) Construct *Rays* parallel to each tangent, from point P to the vertical axis SY2. Label the *Rays*.

(vii) Extend the tangent locators into the area of SY2 vs. SX. Then, extend horizontal lines from the *Ray* intercepts on the SY2 axis, to intersect the related tangent locator.

(viii) Fair-in a best fit curve, or line, through the intersections on the tangent locators.

(ix) To help establish a workable *Pole* distance, try the following. Determine the steepest slope of the curve being differentiated, and use the *Ray* parallel to it to set a *Pole* distance. Tangent 3 is the steepest in this example. A trial *Pole* location of H = 60 MM causes the *Ray* PR to go too high. A second trial H = 30 MM was workable.

(x) Calculate the SY2 scale using the equation which is derived as follows:

$$SY2 = \frac{SY1}{(H)(SX)}$$

- *Figure 23-75* shows SY1 vs. SX, with SY2 vs. SX directly below. The curve is plotted on axes SY1 vs. SX, and then vertical, tangent locators are drawn.
- The slope, tangent to the curve, at location 1 is calculated by $\Delta Y / \Delta X$ from the triangle shown.
- *Ray* PF is drawn parallel to the slope at locator 1, and a horizontal line is drawn from F to locator 1, point G.
- The desired result of this graphical-differentiation construction is that

$$(GK)(SY2) = \frac{(\Delta K)(SY1)}{(\Delta X)(SX)}$$

From similar triangles:

$$\frac{\Delta Y}{\Delta X} = \frac{FD}{H} = \frac{GK}{H}$$

Substitute and factor.

$$(GK)(SY2) = \frac{(GK)}{(H)} \cdot \frac{(SY1)}{(SX)}$$

$$SY2 = \frac{SY1}{(H)(SX)}$$

which is the derivation of the new scale.
SY2 for this example is

$$SY2 = \frac{(.5mm/MM)}{(30\ MM)(1/60\ sec/MM)}$$

SY2 = 1 mm/second/MM
which is a velocity scale as expected.

(xi) To finish the example, i.e., derive the acceleration curve, repeat the procedures: set axes SY3 vs. SX, set a *Pole* distance, determine the slopes, draw *Rays* parallel to the slopes, fair-in the curve, and calculate SY3.

$$SY3 = \frac{1mm/sec/MM}{(40\ MM)(1/60\ sec/MM)}$$

SY3 = 1.5mm/sec²/MM

an acceleration as expected.

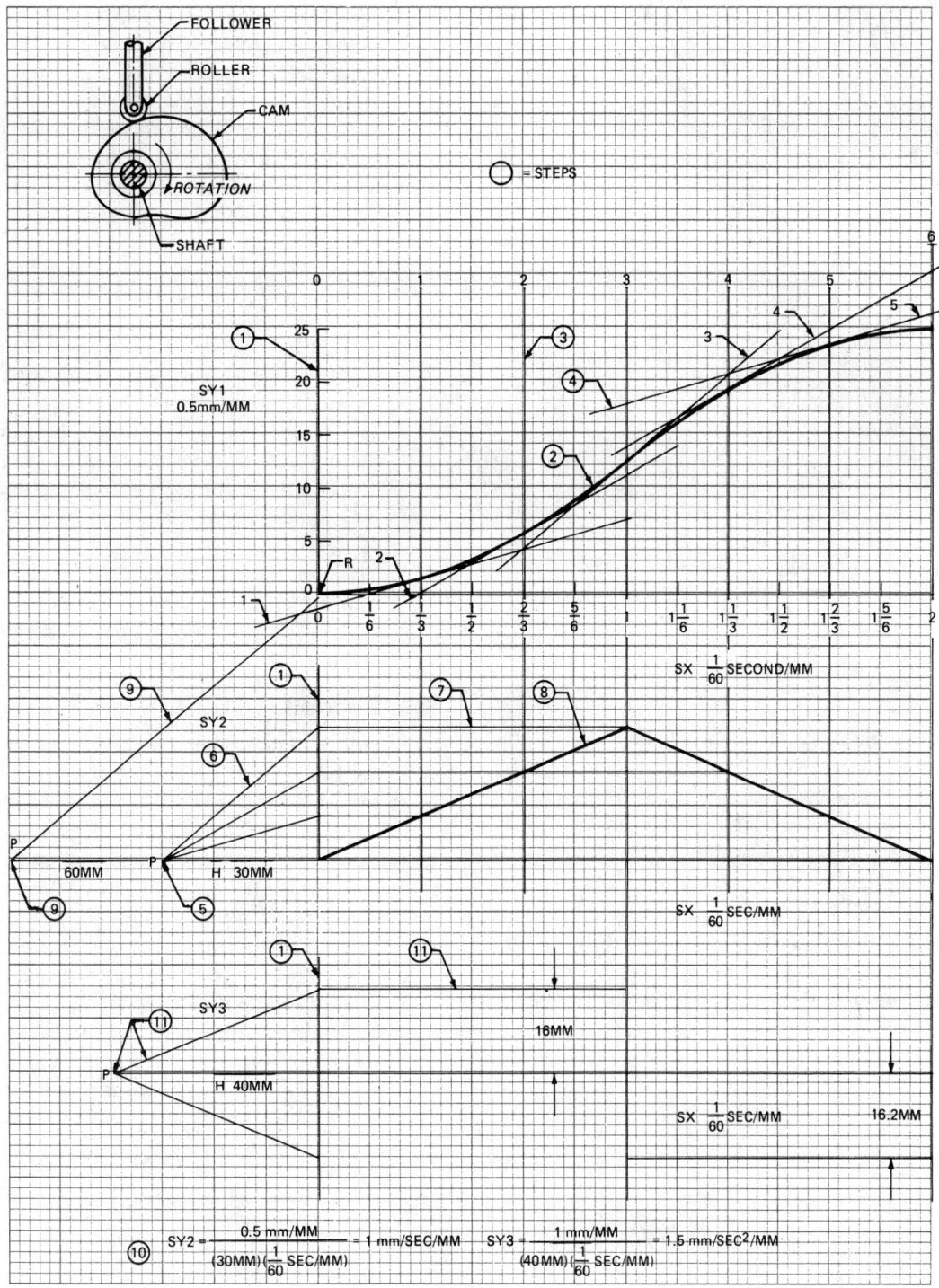

Figure 23-74   Graphical differentiation: Steps

(xii) The cam profile does cause constant acceleration and deceleration, as shown in the Figure 23-75. Their magnitudes are calculated from measurements on the graph to be

acceleration =

$(16\text{MM}) (1.5\text{mm/sec}^2/\text{MM}) =$ 24mm/sec^2

deceleration =

$(16.2 \text{ MM}) (1.5\text{mm/sec}^2/\text{MM}) =$ 24.3mm/sec^2

b. Graphical differentiation is more difficult to do accurately than graphical integration because of the possibility of errors in determining the slopes of curves "by eye." However, doing and understanding graphical differentiation helps

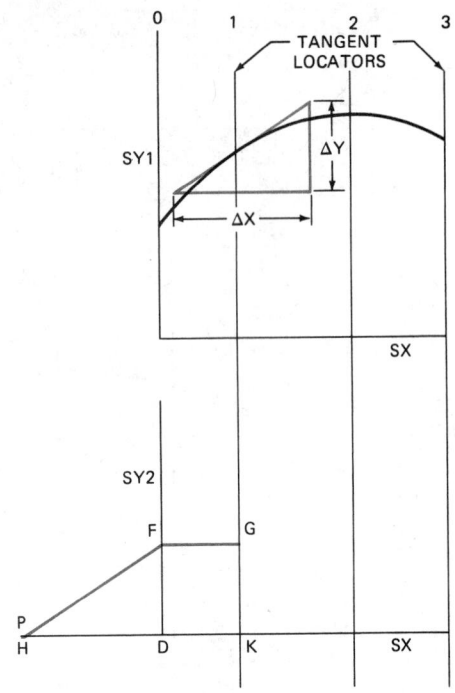

Figure 23-75 Graphical differentiation: Scale derivation

to visualize that the slope of the higher-order curve at a specific locator is represented by the height from the horizontal axis to the derived lower-order curve at the same locator.

c. The conceptual relationships between graphical differentiation and integration are summarized in **Figure 23-76**. Curve #1 in the upper left was differentiated, and the derived curve #2 was placed just below at the lower left. Then the derived curve #2 was reproduced at the lower

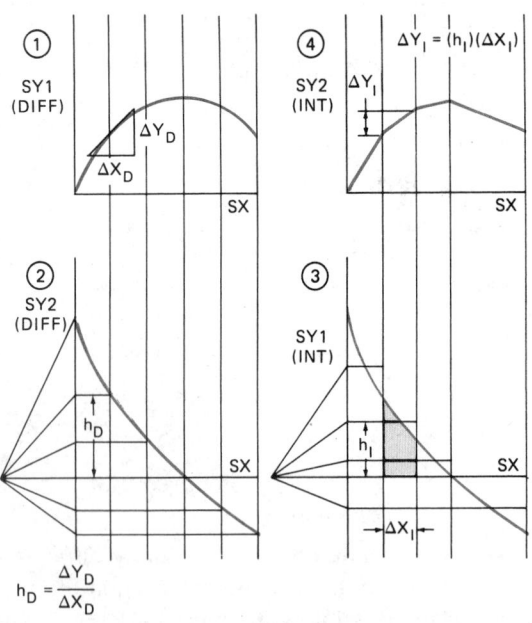

Figure 23-76 Comparison: Graphical integration vs. differentiation

right and labeled #3. Curve #3 was integrated to obtain the curve in the upper right, which should be identical to curve #1. In any case, vertical distances in #2 represent the slopes in #1: sections of area in #3 represent changes in the height of #4 between the same division lines.

4. The following are concluding comments on the usefulness of graphical integration.

   a. Some functions such as

   $$\int_0^\pi \sqrt{1 + \cos^2 X}\, dx$$

   that have no analytical integral can be integrated using graphical integration. Numerical methods can also be used.

   b. Profiles of cuts through hilly terrains may be impractical to express analytically, so volumes of earth can be calculated. Graphical integration is used. Planimeters, mechanical integrators, are also used.

5. The concepts of graphical differentiation and integration provide a unique insight into analyzing beam diagrams, which are considered in the next section.

Beam Diagrams

Most students using *Technical Drawing and Design* will have had only a brief introduction to the mechanics of materials, so they may not have been introduced to beam diagrams. However, a brief introduction in this section to beam diagrams related to graphical differentiation and integration should help them understand beam theories in future courses. Review the graphical calculus concepts in the preceding section, particularly Figure 23-76, before perusing this section.

1. Several preliminary concepts:

   a. The centrally loaded beam in **Figure 23-77A** has supports at ends A and E. At A, a symbol shows that the beam is pinned to a support

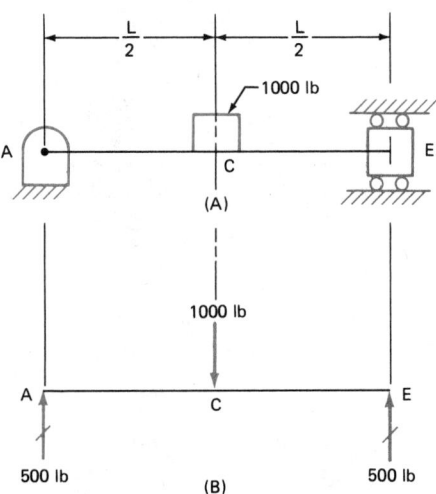

Figure 23-77 (A-B) Centrally loaded beam

that is fixed to a rigid frame or structure. "Whiskers" represent the rigid frame or structure.

b. The FBD (free-body diagram) in *Figure 23-77B* is drawn in the traditional simplified-model form used in beam analyses. In this example, the weight of the beam is considered negligible compared to the effect of the applied load, 1000 pounds. Vectors representing the reactions at A and E, and the applied load at C, are all parallel: Their algebraic sum must equal zero for the beam·to be in equilibrium.

2. The 1000-pound load causes the beam to deflect: the maximum deflection occurring at C is shown in *Figure 23-78A*. This deflection curve is usually the last beam diagram to be derived in beam analyses. However, the deflection diagram is considered first here to illustrate how graphical differentiation principles apply in deriving beam diagrams.

a. Differentiate the deflection curve to obtain the slope curve shown in *Figure 23-78B*. From left to right, the slope curve indicates a large negative slope at A, less negative at B, zero at C, positive at D, and larger positive at E. The deflection anywhere along the beam is represented by the variable y; slopes are represented by dy/dx.

b. The next derived curve is called a moment diagram, *Figure 23-78C*. The moment diagram shows that the slope curve has changing positive slopes throughout, and a maximum slope

at C. The maximum moment in this beam also occurs at C. A moment about a point is defined as *the force times the shortest distance from the point to the line of action of the force*. Therefore, the varying moment, zero at A and E, and increasing to the maximum at C, varies as a function of the distance from the reactions at A and E. Moment diagrams are second-order derivatives.

c. Finally, in *Figure 23-78D* the shear diagram appears as the derivative of the moment diagram. Since the moment diagram has two linear portions, one positive sloped and one negative, the shear diagram appears as two horizontal lines, one above the axis and one below. Shear diagrams are third-order derivatives.

3. Refer to Figure 23-76 again to review the comparisons of graphical differentiation and integration, because the beam diagrams are integrated next.

4. Follow the progression of diagrams in *Figure 23-79* for the same beam; however, this time the shear diagram is considered first, which is the usual progression in beam analyses.

a. Before constructing the shear diagram, look at the FBD. Upward forces are positive; downward, negative. Start at A and consider the beam from left to right. B shows the shearing tendency caused by the reaction force at A. This tendency, for this beam, remains constant along the beam until a change in loading occurs at C. From C onward, the effect of the reaction at A plus the larger downward load at C causes the tendency for shear to be downward, and constant through to E.

b. The moment diagram in C, going from left to right, shows the moment increasing as X increases. The conceptual sketch at right isolates a portion of the beam and shows it in equilibrium, because the 500X is balanced by the reaction force of 500 pounds times the moment arm X.

c. Consider the order of the graphical curves compared to the analytical derivatives.

   (i) The moment diagram is the integral of the shear diagram. Since the shear diagram has a zero-order slope, the moment diagram is a first-order curve.

   (ii) Analytically, the shear diagram, always a third derivative, integrates to a second derivative.

d. D shows the slope diagram, integral of the moment diagram. The slope diagram starts with an initial value of slope at A. This value must be found either analytically or graphically. The graphical method is shown in step 6, below. Slope diagrams are first-order derivatives.

e. Finally, E shows the deflection curve, the inte-

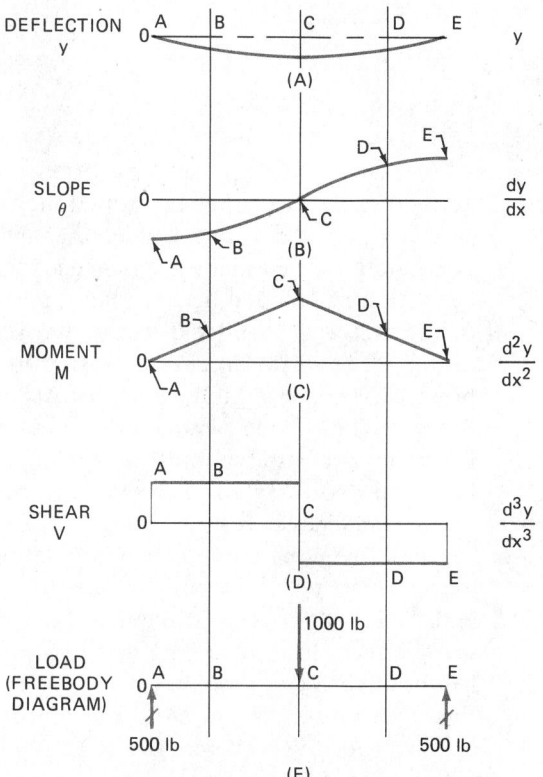

Figure 23-78   (A-E) Centrally loaded beam: Beam diagrams

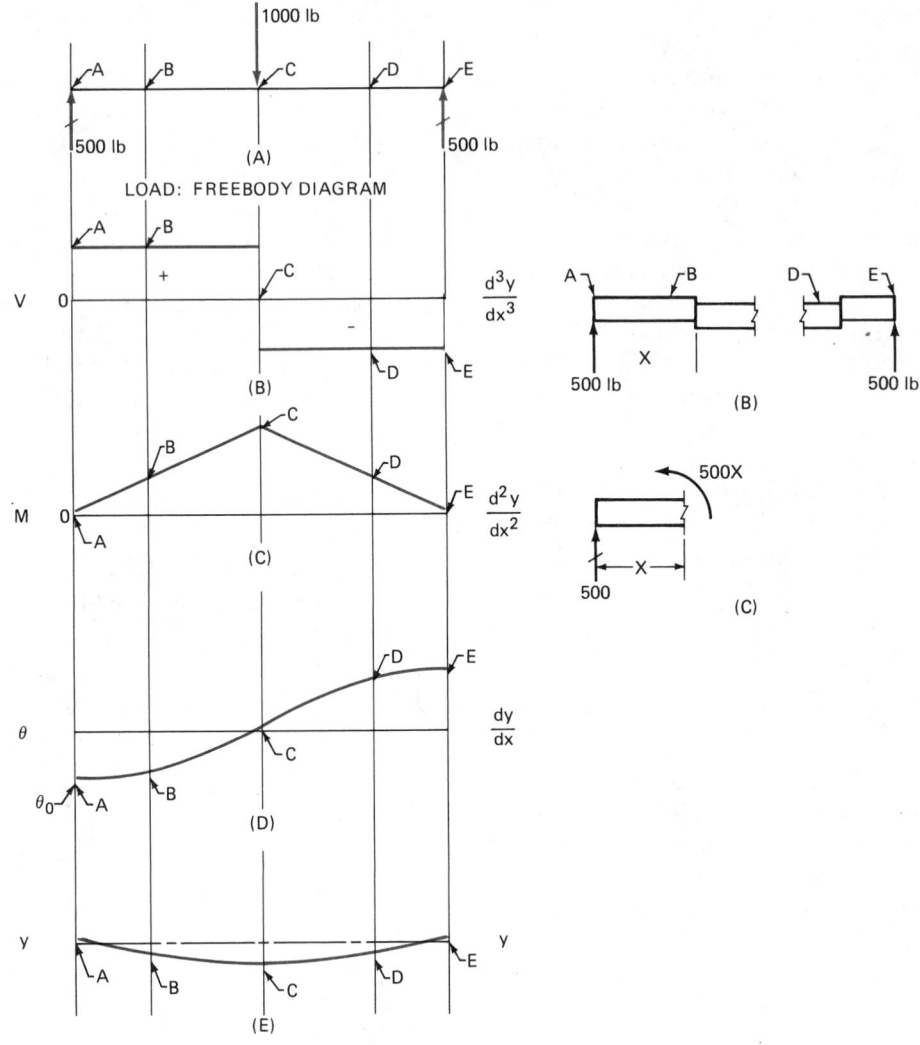

Figure 23-79 (A-E) Centrally loaded beam: Beam diagrams

gral of the slope curve. (Moment curves and deflection curves are the ones designers use most: the slope curve is just the intermediate step.) The deflection curve looks as the real beam would with a concentrated load at the center.

5. Compare the beam diagrams shown in *Figures 23-80* and *23-81* with the preceding discussion. Shear and moment diagrams are shown, but deflections are given as equations only. These are typical of the diagrams that appear in handbooks and texts.

6. *Figure 23-82* is the simply loaded beam again, but with the actual application of graphical integration.
   a. The steps summarized in Figure 23-68 to do graphical integration were followed to prepare the diagrams. Scale calculations are shown at right. SYI was arbitrarily selected; the other values were derived using the equation SY2 = (H) (SX) (SYI). The deflection at C equals the

derived scale SY4 times the actual measurement at C, divided by the product of E times I.
   b. The modulus of elasticity E was needed in the deflection calculation because E is a measure of the material's ability to resist deforming under loads. E for steel equals 30,000,000 pounds per square inch. (SI Units: 210 Giga Pascals) The second parameter I, moment of inertia in inches to the fourth power, is a geometrical property and is a measure of the beam's tendency to resist bending.
   c. Refer to the slope curve for an illustration of the graphical technique used to determine the initial value of the slope. Note that the derived slope curve (integral of the moment curve) begins at zero and then, because of the positive area under the moment curve, moves upward (positively) across the beam. Next, a new horizontal axis is established "by eye" so that the area under the new horizontal axis approximately equals the area above. If these

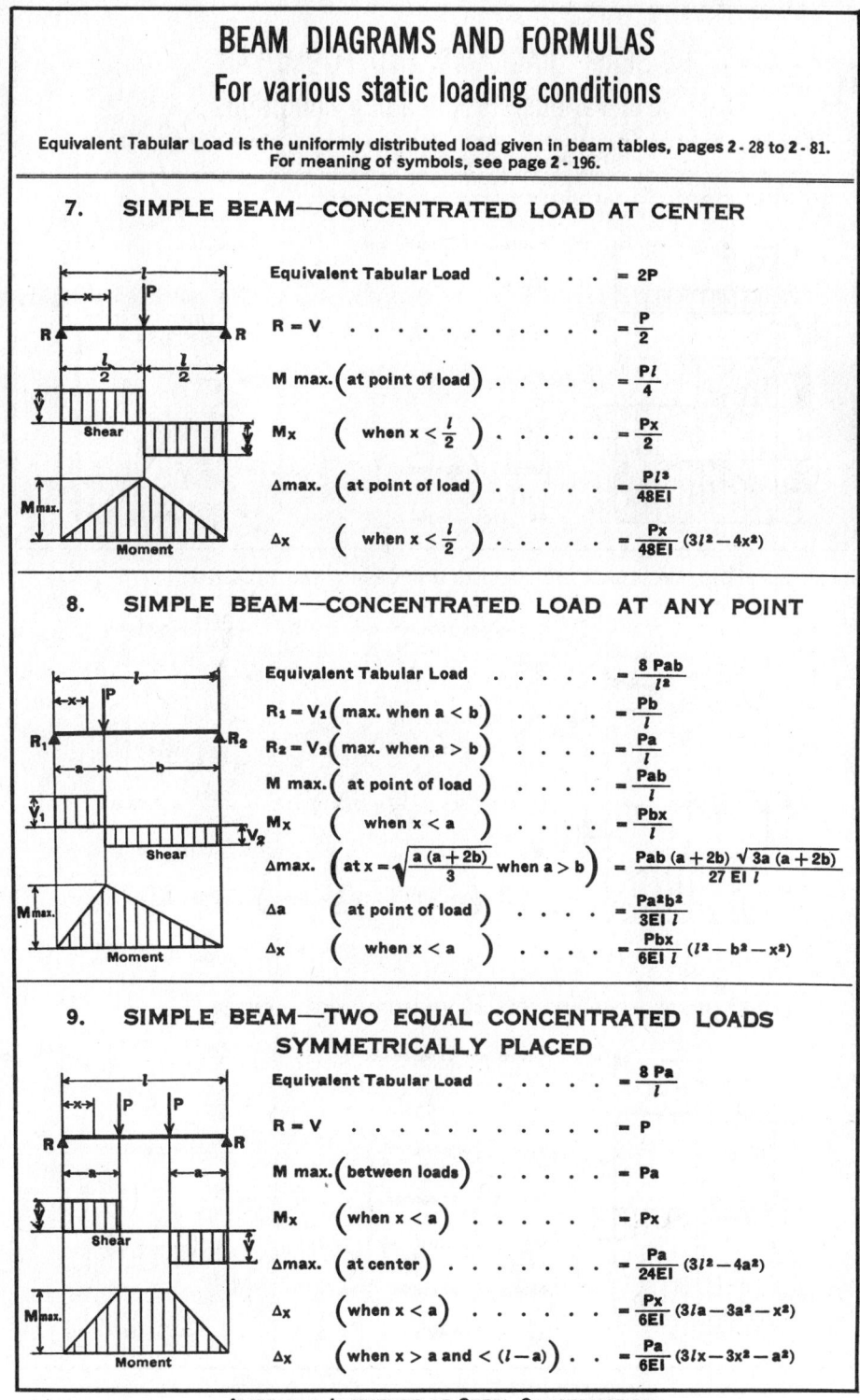

BEAM DIAGRAMS AND FORMULAS
For various static loading conditions

Equivalent Tabular Load is the uniformly distributed load given in beam tables, pages 2 - 28 to 2 - 81.
For meaning of symbols, see page 2 - 196.

7. SIMPLE BEAM—CONCENTRATED LOAD AT CENTER

Equivalent Tabular Load . . . . . . $= 2P$

$R = V$ . . . . $= \dfrac{P}{2}$

M max. $\left(\text{at point of load}\right)$ . . . . $= \dfrac{Pl}{4}$

M_x $\left(\text{when } x < \dfrac{l}{2}\right)$ . . . . $= \dfrac{Px}{2}$

Δmax. $\left(\text{at point of load}\right)$ . . . . . $= \dfrac{Pl^3}{48EI}$

Δ_x $\left(\text{when } x < \dfrac{l}{2}\right)$ . . . . . $= \dfrac{Px}{48EI}(3l^2 - 4x^2)$

8. SIMPLE BEAM—CONCENTRATED LOAD AT ANY POINT

Equivalent Tabular Load . . . . . $= \dfrac{8\,Pab}{l^2}$

$R_1 = V_1 \left(\text{max. when } a < b\right)$ . . . . $= \dfrac{Pb}{l}$

$R_2 = V_2 \left(\text{max. when } a > b\right)$ . . . . $= \dfrac{Pa}{l}$

M max. $\left(\text{at point of load}\right)$ . . . . $= \dfrac{Pab}{l}$

M_x $\left(\text{when } x < a\right)$ . . . . $= \dfrac{Pbx}{l}$

Δmax. $\left(\text{at } x = \sqrt{\dfrac{a(a+2b)}{3}} \text{ when } a > b\right) = \dfrac{Pab\,(a+2b)\,\sqrt{3a\,(a+2b)}}{27\,EI\,l}$

Δa $\left(\text{at point of load}\right)$ . . . . $= \dfrac{Pa^2 b^2}{3EI\,l}$

Δ_x $\left(\text{when } x < a\right)$ . . . . $= \dfrac{Pbx}{6EI\,l}(l^2 - b^2 - x^2)$

9. SIMPLE BEAM—TWO EQUAL CONCENTRATED LOADS SYMMETRICALLY PLACED

Equivalent Tabular Load . . . . . $= \dfrac{8\,Pa}{l}$

$R = V$ . . . . $= P$

M max. $\left(\text{between loads}\right)$ . . . . . $= Pa$

M_x $\left(\text{when } x < a\right)$ . . . . . $= Px$

Δmax. $\left(\text{at center}\right)$ . . . . $= \dfrac{Pa}{24EI}(3l^2 - 4a^2)$

Δ_x $\left(\text{when } x < a\right)$ . . . . . $= \dfrac{Px}{6EI}(3la - 3a^2 - x^2)$

Δ_x $\left(\text{when } x > a \text{ and } < (l-a)\right)$ . . $= \dfrac{Pa}{6EI}(3lx - 3x^2 - a^2)$

AMERICAN INSTITUTE OF STEEL CONSTRUCTION

Figure 23-80 Beam diagrams: Handbook *(Courtesy American Institute of Steel Construction)*

negative and positive areas are indeed equal, then the deflection curve will start and end at the level of the horizontal reference line for the derived deflection curve. Returning to the same level confirms the new position of the horizontal axis. Furthermore, the initial value of the slope curve is also confirmed.

d. Suppose that the new horizontal axis was located too low. What would happen? ***Figure 23-83*** shows the slope curve redrawn with the new horizontal axis located too low; that is, the positive area exceeds the negative, so the deflection curve ends too high. However, the magnitude of the deflection can be easily

BEAM DIAGRAMS AND FORMULAS
For various static loading conditions

Equivalent Tabular Load is the unifomly distributed load given in beam tables, pages 2 - 28 to 2 - 81.
For meaning of symbols, see page 2 - 196.

1. SIMPLE BEAM—UNIFORMLY DISTRIBUTED LOAD

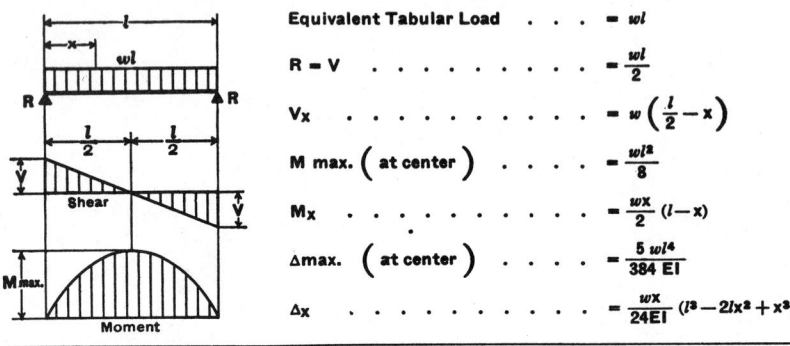

Equivalent Tabular Load . . . . $= wl$

$R = V$ . . . . $= \dfrac{wl}{2}$

V_x . . . . $= w\left(\dfrac{l}{2} - x\right)$

M max. (at center) . . . . $= \dfrac{wl^2}{8}$

M_x . . . . $= \dfrac{wx}{2}(l - x)$

Δmax. (at center) . . . . $= \dfrac{5\,wl^4}{384\,EI}$

Δ_x . . . . $= \dfrac{wx}{24EI}(l^3 - 2lx^2 + x^3)$

2. SIMPLE BEAM—LOAD INCREASING UNIFORMLY TO ONE END

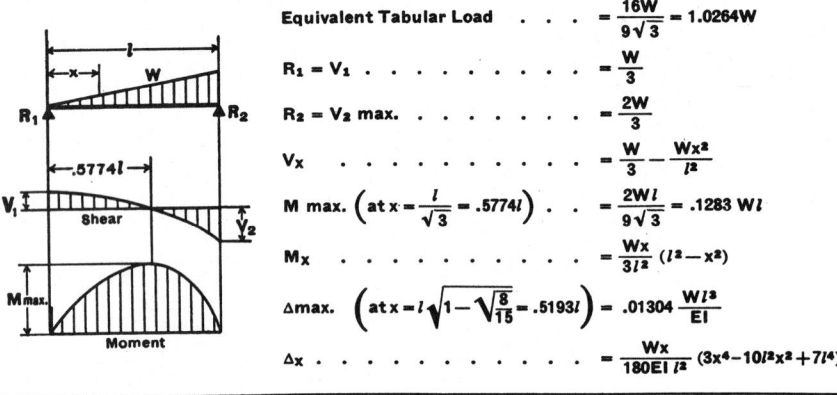

Equivalent Tabular Load . . . $= \dfrac{16W}{9\sqrt{3}} = 1.0264W$

$R_1 = V_1$ . . . . $= \dfrac{W}{3}$

$R_2 = V_2$ max. . . . . $= \dfrac{2W}{3}$

V_x . . . . $= \dfrac{W}{3} - \dfrac{Wx^2}{l^2}$

M max. $\left(\text{at } x = \dfrac{l}{\sqrt{3}} = .5774l\right)$ . . $= \dfrac{2Wl}{9\sqrt{3}} = .1283\,Wl$

M_x . . . . $= \dfrac{Wx}{3l^2}(l^2 - x^2)$

Δmax. $\left(\text{at } x = l\sqrt{1 - \sqrt{\dfrac{8}{15}}} = .5193l\right) = .01304\,\dfrac{Wl^3}{EI}$

Δ_x . . . . $= \dfrac{Wx}{180EI\,l^2}(3x^4 - 10l^2x^2 + 7l^4)$

3. SIMPLE BEAM—LOAD INCREASING UNIFORMLY TO CENTER

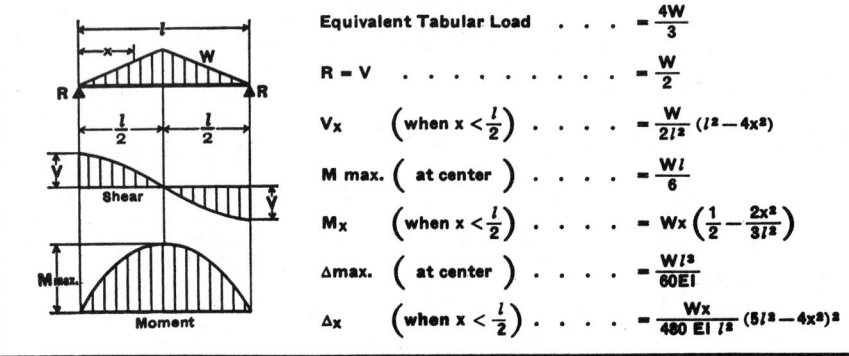

Equivalent Tabular Load . . . $= \dfrac{4W}{3}$

$R = V$ . . . . $= \dfrac{W}{2}$

V_x $\left(\text{when } x < \dfrac{l}{2}\right)$ . . . . $= \dfrac{W}{2l^2}(l^2 - 4x^2)$

M max. (at center) . . . . $= \dfrac{Wl}{6}$

M_x $\left(\text{when } x < \dfrac{l}{2}\right)$ . . . . $= Wx\left(\dfrac{1}{2} - \dfrac{2x^2}{3l^2}\right)$

Δmax. (at center) . . . . $= \dfrac{Wl^3}{60EI}$

Δ_x $\left(\text{when } x < \dfrac{l}{2}\right)$ . . . . $= \dfrac{Wx}{480\,EI\,l^2}(5l^2 - 4x^2)^2$

AMERICAN INSTITUTE OF STEEL CONSTRUCTION

Figure 23-81 Beam diagrams: Handbook *(Courtesy American Institute of Steel Construction)*

determined by a simple graphical correction. Line AE is drawn first. Then FG is constructed tangent to the deflection curve and parallel to AE. The deflection is measured from the point of tangency in the *vertical* direction to the line AE. Finally, the deflection is calculated by multiplying the measured distance 0.56 *inches* times SY4 and dividing by EI to obtain 0.63″, as before.

7. Graphical integration can be used for most beam analyses; however, the best use is for beams which have a number of different cross sections. These beams are difficult to describe analytically but can readily be integrated using graphical inte-

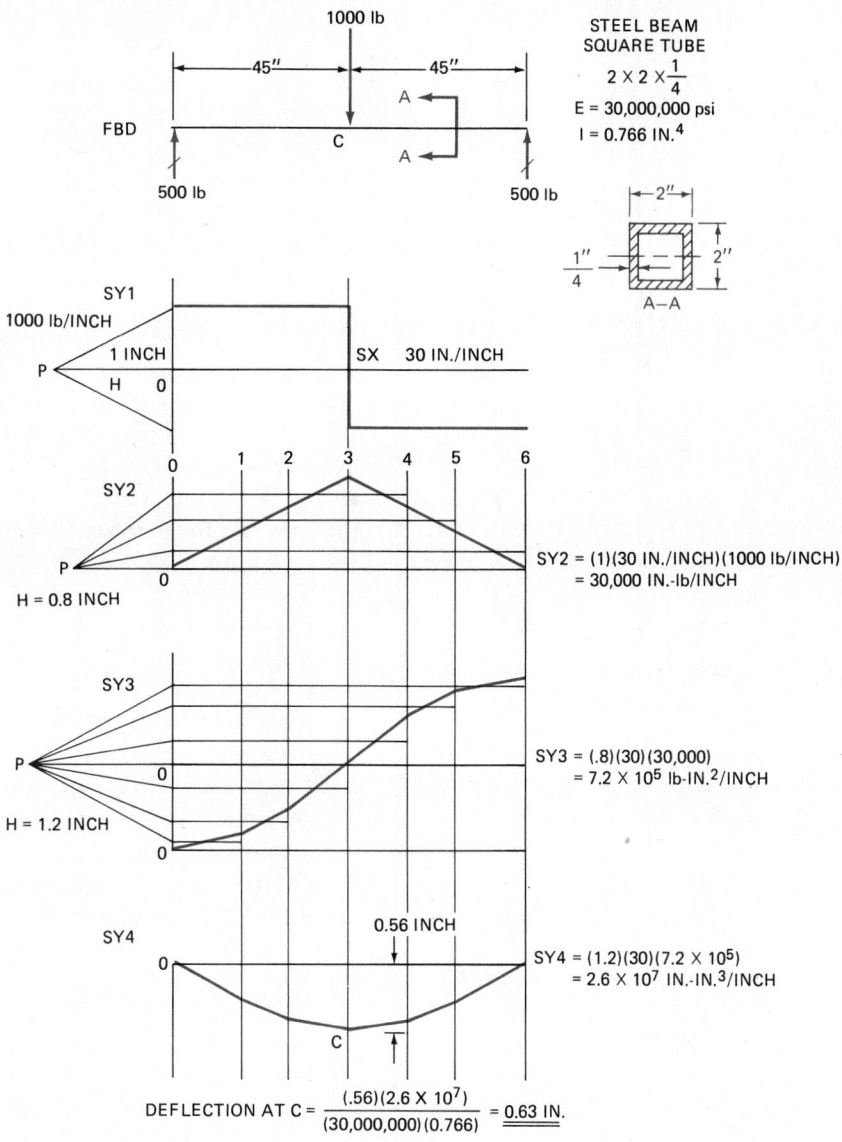

Figure 23-82   Centrally loaded beam: Graphical integration

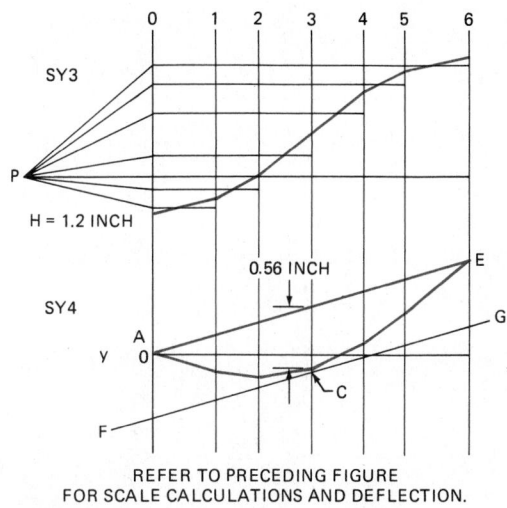

REFER TO PRECEDING FIGURE
FOR SCALE CALCULATIONS AND DEFLECTION.

Figure 23-83   Centrally loaded beam: Initial slope

gration. **Figure 23-84** shows a complete set of diagrams, without calculations, to demonstrate how graphical integration is used effectively when the cross-section changes. The significant difference in these diagrams comes after the moment diagram. Selected values of moments are divided by appropriate values of EI to obtain a new curve, M/EI, plotted in a new set of axes. The M/EI diagram comes after the moment diagram to incorporate the properties, E and I, to resist deformation and bending, respectively. At this point graphical integration procedures start anew.

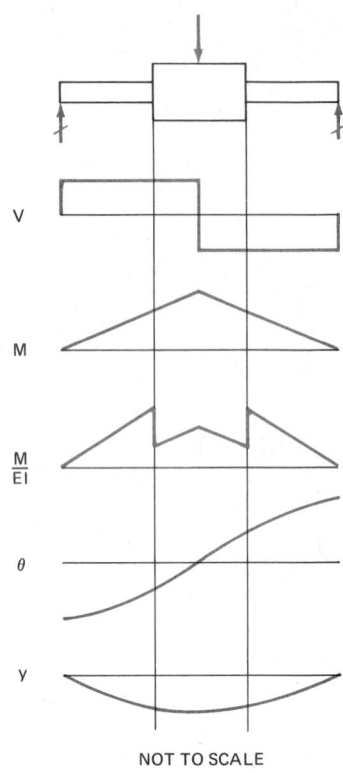

NOT TO SCALE

Figure 23-84   Stepped beam: Graphical integration

Conclusion

After studying this chapter most readers should agree that communicating information is the popular use of *charts* and *graphs* in a variety of media, including textbooks, periodicals, newspapers, and television. However, generating information is the most useful function of *charts* and *graphs* for designers.

Review

1. With respect to *charts* and *graphs*, what was meant by communicating information as compared to generating information?

2. List the five basic components of *charts* and *graphs*.
3. Why were commercial grids, such as logarithmic grids, developed?
4. What is a spreadsheet?
5. Define and describe a vector. How are they added and subtracted?
6. Define equilibrium as used in the study of mechanics.
7. What is a 2-force member?
8. Why are free-body diagrams (FBD) important in the study of mechanics?
9. What is meant by optimization?
10. What does yield strength mean?
11. Define scale modulus.
12. What is a logarithm?
13. Describe the variables in the following equations:
    a. $y = mx + b$
    b. $y = B(x^{**}m)$
    c. $y = B(e^{**}mx)$
14. What determines the length of an alignment chart?
15. Why is the straight-line equation so useful in matching equations to data?
16. How were similar triangles used in this chapter?
17. Describe the short method for designing a parallel alignment chart.
18. List and describe the four graphical techniques used in this chapter to derive equations once the data is plotted in an approximate straight line.
19. What is a spline?
20. Describe the steps to do:
    a. Graphical integration.
    b. Graphical differentiation.
21. Explain the derivations using graphical sketches of:
    a. $SY2 = (H)(SX)(SYI)$.
    b. $SY2 = SY1/((H)(SX))$.
22. What affect does the distance H have on the next derived curve in:
    a. Graphical integration?
    b. Graphical differentiation?
23. What is an isopleth?

Chapter Twenty-Three Problems

Problems for this chapter are divided into two main categories. To Communicate Information and To Generate Information. A third group contains a miscellaneous collection of related problems. Refer to appropriate sections of the chapter for guidelines and suggested steps to prepare the charts and graphs.

To Communicate Information

Basic Graphs (Refer to the Five Basic Components in the chapter)

Problem 23-1

Prepare plots of BHP (brake horsepower) and T (torque), both as a function of RPM (revolutions per minute), for the following data:

RPM	BHP	T
500	15	190
1000	40	210
1500	65	230
2000	90	232
2200		240
2500	115	234
3000	130	230
3500	145	220
4000	150	200
4500	145	170

Problem 23-2

Electronic parts fail over a period of time in a frequency pattern shaped like a "bath tub." Plot the given failure-rate data with failure-rate ratios as a function of time.

Period	Failure-Rate Ratio	Time (hours)
Break-in Period		
Initial Failure Rate	2.00	0
	1.56	100
Constant Failure Rate	1.40	200
	1.28	300
	1.19	400
	1.12	500
	1.05	700
	1.00	1000
Typical Operating Period		
Constant Failure Rate	1.00	1000
		to 5000
Wear Out Period		
Wear Out Failure Rate	1.00	50,000
	1.05	51,000
Constant Failure Rate	1.18	52,000
	1.40	53,000
	1.72	54,000
	2.50	55,000

Problem 23-3

Plot Average Cost in $ vs. Weight of Steel Forgings in pounds for the following:

Cost $	4.20	7.50	15.00	30.00	54.00	115.00
Weight lb	2.0	4.0	10.0	20.0	40.0	100.0

Problem 23-4

Prepare a graph of % Power as a function of Air-Fuel-Ratio for the data representing a typical Internal Combustion Engine.

% Power	90	100	95	87	75	60	43
Air-Fuel-Ratio	10:1	12:1	14:1	16:1	18:1	20:1	22:1

Problem 23-5

During a tension test of a steel bar, the following data was calculated and recorded; prepare a graph of Stress in psi vs. Strain (stretching) in inches per inch.

Stress lbs/square inch (psi)	Strain inches/inch
00000	0.00000
3000	0.00010
5000	0.00020
10,000	0.00030
15,000	0.00050
20,000	0.00070
25,000	0.00080
30,000	0.00100
36,000	0.00120

Problem 23-6

A convenient expression in strength of materials calculations is the relationship of G (shear modulus) as a function of E (modulus of elasticity). The equation is:

$$G = \frac{E}{2(1 + 0.3)} = \frac{E}{2.6}$$

where the 0.3 is Poisson's Ratio.

Prepare a plot of the following data for a variety of engineering materials:

Material	E psi	G psi
Plastic	1,000,000	400,000
Magnesium	2,300,000	6,100,000
Aluminum	3,800,000	10,000,000
Zinc	5,000,000	12,000,000
Copper	6,000,000	16,000,000
Palladium	6,300,000	17,000,000
Platinum	9,300,000	24,000,000
Steel	11,600,000	30,000,000

Pie Charts

Problem 23-7

Prepare a weekly schedule of your activities for 24 hours per day. Show the different activities in a pie chart.

Problem 23-8

Use a pie chart to show the distribution of percentages of the sources of electrical power generation as follows: Coal, 56; Nuclear, 17; Hydro-electric, 12; Gas, 10; and Oil, 5.

Problem 23-9

Show in a pie chart the relative importance of the qualifications of a successful, professional, technical person: Character, 40%; Judgement, 18%; Efficiency, 15%; Understanding People, 14%; and Technical Knowledge, 13%.

Problem 23-10

Select a pie or bar chart from your local newspaper or national business publication and revise it based on the ideas presented in a book by E.R. Tufte titled *The Visual Display of Quantitative Information*. Mr. Tufte contends that charts can, and should, be made much simpler and still convey the required information. Some of his suggestions and observations are:

1. Since desktop computers now have the ability to fill in bound spaces with a variety of patterns, pie charts and bar charts are filled with these patterns—because they are there! Mr. Tufte believes that the patterns only clutter up the charts.

2. Scales should only go as far as the data, as shown in Figure Problem 23-10. The location of zero values are implied.

3. Pleasing "to the eye" dimensions for the overall chart space are approximately equal to the proportions of the "golden rectangle": the height to width ratio of 1:1.618. (Most ancient Greek and Roman structures seemed to fit these proportions.)

4. Try to be creative in presenting the data; for example, Mr. Tufte displayed a photograph of people lined up in columns of various lengths to represent a statistical histogram to approximate a normal distribution.

Bar Charts

Problem 23-11

A local newspaper published data relating to three alternate choices for handling city sewage. The data appeared in tabular form as shown below. Prepare a bar chart to better represent the data.

Monthly Household Sewer Rates (Dollars)

	1995 Utility Est.	1995 City Est.	2005 Utility Est.	2005 City Est.	2030 Utility Est.	2030 City Est.	45 Yr. Avg. Utility Est.	45 Yr. Avg. City Est.
Alternative 1	14.95	14.75	9.75	11.45	4.60	6.00	8.85	9.45
Alternative 2	19.70	18.30	11.10	11.45	4.35	6.10	10.45	10.35
Alternative 3	17.65	17.00	10.40	11.20	4.55	6.30	9.75	10.00

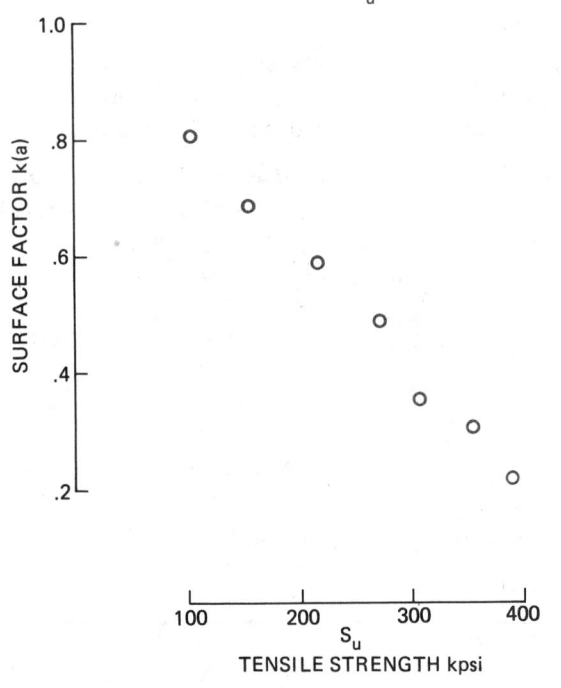

SURFACE FACTOR k(a) vs. S_u TENSILE STRENGTH

Problem 23-10

Problem 23-12

Prepare a bar chart to represent the population of the United States in Millions vs. Year as listed:

Population in Millions	Year
5.3	1800
7.2	1810
9.6	1820
12.9	1830
17.1	1840
23.2	1850
31.4	1860
38.6	1870
56.2	1880
63.0	1890
76.0	1900
97.0	1910
105.7	1920
122.8	1930
131.7	1940
151.1	1950
179.3	1960
211.4	1970
255.0	1980
300.0 (est)	1990
360.0 (est)	2000

Problem 23-13

A recent newspaper article compared a number of baseball pitchers' fastball speeds to speeds of a sneeze, horses, automobiles (freeway speed), jaguars, and horses in MPH. Prepare a creative bar chart which shows the following:

Pitcher/Other	Speed in MPH
A	100
B	98
C	98
D	97
E	96
F	96
G	95
H	95
I	94
J	94
K	92
L	88
M	88
N	87
O	86
Sneeze	103
Automobile	65
Jaguar	63
Horse	43

Planning Charts

Problem 23-14

Construct a planning chart to show the activities required to do an experiment in an eight-hour day. A report also is required for this experiment. The activities necessary to do the experiment and the estimated time for each activity are listed below. Some activities will overlap.

Hours

1. Study experiment.................................1 hour
2. Set up Apparatus 1-½
3. Take data 3
4. Plot data 4-½
5. Write first draft.................................1
6. Edit and rewrite on word processor 1-½
7. Collect parts for report and assemble 1
8. Put the report in a folder and submit it 1-½

Problem 23-15

Prepare a planning chart for the activities required to make a float for a parade in your school or community.

Problem 23-16

Prepare a planning chart for any one of the following:

1. A major repair to an automobile or other major appliance.

2. A field trip for 50 students to a local industry.

3. Manufacturing a product which has several components.

Flow Charts

Problem 23-17

Redraw the rough sketch in Figure Problem 23-17, which shows an example Fault Tree Analysis flow chart. Flow is upward on the page. Lines are either horizontal or vertical, are drawn with a straight edge, and change direction with 90° corners.

Problem 23-18

Construct a flow chart of a computer program used in your classes.

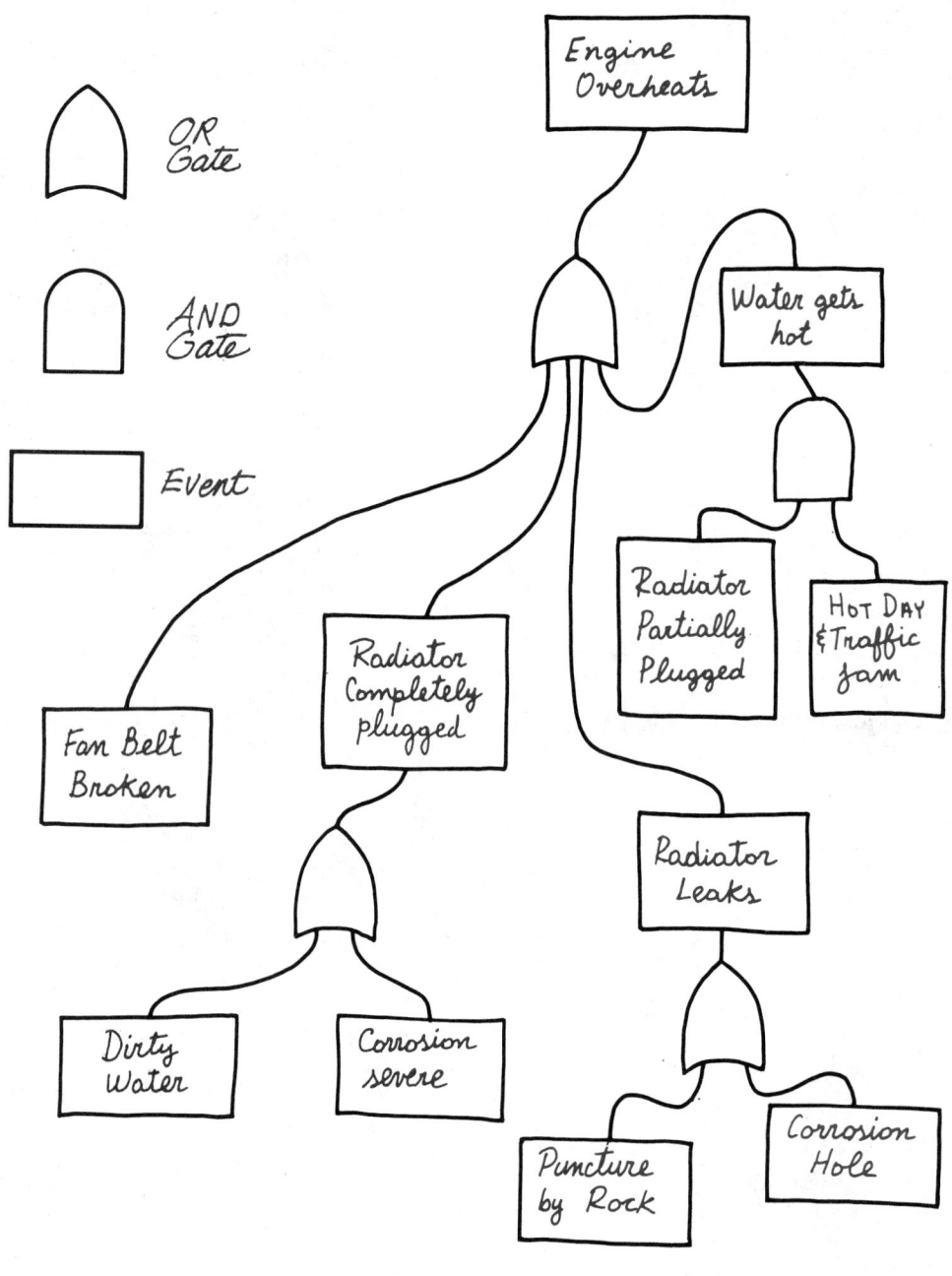

Schematics

Problem 23-19

Using a straight edge and a circle template, redraw the rough hydraulic power schematic shown in Figure Problem 23-19 and label each component. Show the directional valve in the position for pushing the piston to the right. Lines in schematics are either horizontal or vertical, and change direction in 90° corners.

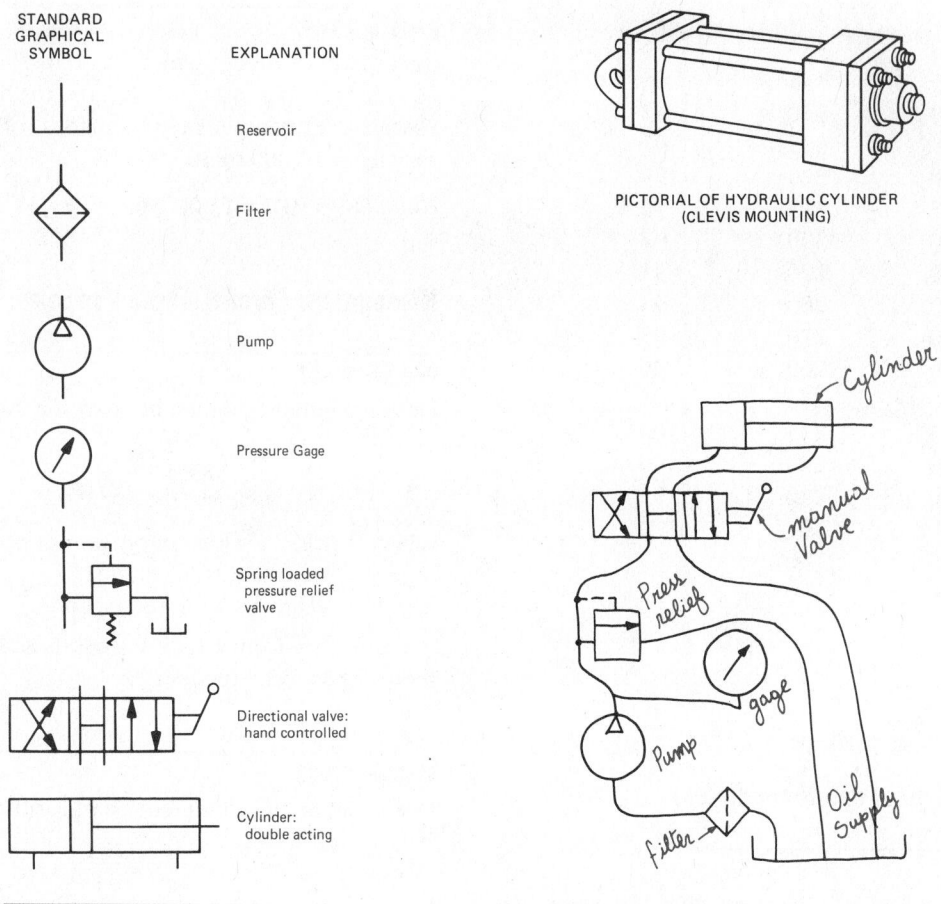

Problem 23-19

Problem 23-20

Using a straight edge, redraw the rough schematic of an overtone oscillator shown in Figure Problem 23-20. Generally, signals (voltages, current, and information) travel from left to right; positive voltages are in the upper areas and negative voltages are in the lower.

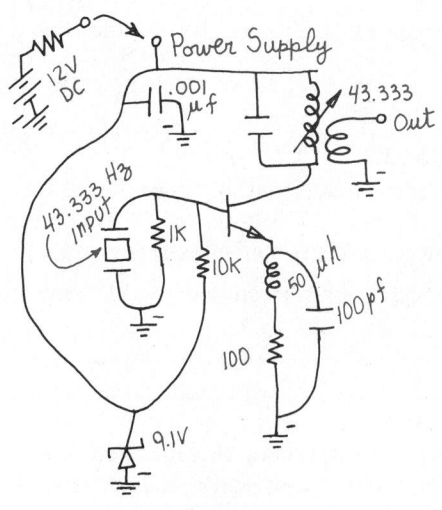

Problem 23-20

Polar Charts

Problem 23-21

Use a commercial polar grid to plot noise-level values surrounding a jet engine: decibels vs. angle.

Angle (Zero Degrees is North)		Decibels
0	360	20
20	340	76
25	335	85
30	330	93
35	325	97
40	320	99
45	315	96
60	300	92
70	290	91
80	280	94
90	270	95
100	260	96
110	250	95
125	235	100
140	220	90
150	210	95
160	200	90
170	190	82
180	180	65

To Generate Information

Nomograms: Adjacent and Non-adjacent

Problem 23-22

1. Design an adjacent alignment chart for the equation E = 33,000 HP, where E is in foot-pounds per minute and HP is horsepower than varies from 2 to 20.
2. Design a non-adjacent chart for the same equation.

Problem 23-23

1. Design an adjacent alignment chart for the equation kW = .746 HP, where HP is horsepower and kW is kilowatts. Let HP vary from 1 to 100.
2. Design a non-adjacent chart for the same equation.

Problem 23-24

1. Design an adjacent alignment chart for the equation 57.3 R = N, where R is in radians and N is number of degrees. Let R vary from 0 to 6.28.
2. Design a non-adjacent chart for the same equation.

Problem 23-25

Design adjacent alignment charts for any or all of the following useful conversions:

1. kilograms to pounds
2. pounds to Newtons
3. Pascals to psi
4. Newton-meters to foot-pounds or pound-inches

Problem 23-26

Design an adjacent alignment chart for the conversion of *stress* in pounds per square inch to MPa, mega-Pascals (Newton per meter squared = Pascal). Let stress vary from 10,000 to 300,000 psi.

Note: 6895 MPa = 1,000 psi.

Nomograms: Parallel, Three Variables

Problem 23-27

Design a parallel alignment chart for the equation

$$KE = \frac{(W)(V^{**}2)}{(2)(g)}$$

where
KE = kinetic energy in foot-pounds
W = weight in pounds
V = velocity in feet/second

Let W vary from 1 to 500 pounds and V from 1 to 100 feet/sec. g = 32.2 feet/sec**2

Problem 23-28

Design a parallel alignment chart for the equation

$$S = \frac{PD}{4T}$$

where
S = Circumferential Stress in a spherical tank
P = internal gas pressure, psi
D = diameter in inches = 20″
T = thickness of shell

Let P vary from 0 to 150 psi and T from .1″ to .5″

Problem 23-29

Design a parallel alignment chart for the following helical spring equation:

$$y = \frac{8F(D^{**}3)N}{(d^{**}4)G}$$

where
y = deflection in inches
F = force in lb applied
D = mean diameter of coil in inches = .6″
d = wire diameter in inches = .1″
G = shear modulus 11,500,000 psi
N = number of coils

Let F vary from 0 to 50 pounds and N from 3 to 15.

Problem 23-30

Design a parallel alignment chart for the following shaft design equation:

$$S = \frac{16T}{(3.14)\, d^{**}3}$$

where   S = shear stress, psi
T = torque, inch-pounds (2,000 to 10,000 in-lb)
d = diameter, inches (1 inch to 5 inches)

Note: For a factor of safety of 2, S should not exceed ½ the yield strength of the metal being considered.

Problem 23-31

Design a parallel alignment chart for the following moment of inertia equation for a rectangular cross-sectional member:

$$I = \frac{b\, h^{**}3}{12}$$

where   I = moment of inertia
(or actually second moment of area)-----in**4
b = width, inches (0.5 to 3 inches)
h = height, inches (1.5 to 9 inches)

(I is a measure of a members ability to resist bending about an axis parallel to b and perpendicular to h.)

Nomograms: Concurrency Charts

Problem 23-32

The "learning curve" premise says that as workers gain experience on producing a new product they can produce more products in a given length of time. An equation that expresses this improvement in production as learning increases is the following:

$$T = E\, C^{**}m.$$

where   T = hours of production time to produce the cumulated total of batches of products (10 to 1000)
C = cumulated total of batches of products (1 to 100,000)
E = initial effort in hours to produce the first batch of products (1000)
m = an exponent which depends on the improvement in learning

For example, assume that the cumulative total of batches of products has doubled, and that the time used for the second half of the double is only 75% of the time used for the first half. Then the exponent m is calculated as follows:

$$\frac{.75T}{T} = \frac{E(2C)^{**}m}{E(C)^{**}m} \text{ so } .75 = (2)^{**}m$$

and from log .75 = m log 2, m = −.415. Then the equation for the learning curve becomes

$$T = E\, C^{**}(-.415) \text{ and } \log T = \log E - .415\log C$$

which indicates that a straight-line plot could be done on a log-log (logarithmic) grid with a slope of −.415.

Prepare a concurrency chart of percentage learning curves on a 3 x 5 cycles logarithmic grid for 65%, 70%, 75%, 80%, and 85%. Check one of the equations by plotting the curve on linear coordinates until the cumulative total doubles several times.

Problem 23-33

Prepare a concurrency chart of the equation for determining the deflection at the center of a beam that has a concentrated load at the center and is simply supported at the ends. The equation is:

$$y = \frac{W\, L^{**}3}{48\, EI}$$

where   y = deflection in inches
W = the applied concentrated load (1000 to 5000 pounds)
L = length between supports (30″ to 100″)
EI = (30,000,000psi for steel) x (.766 in**4 for a 2 x 2 x ¼ inch tube) = (2.3) (10**7) lb-in**2

Problem 23-34

Prepare a parallel alignment chart of the equation in Problem 23-33.

Problem 23-35

Prepare a concurrency chart for an equation from one of your courses.

Empirical Equations: Semilogarithmic

Problem 23-36

Derive an equation for the following data taken in a grassy field downwind from a source of radioactive Iodine, over a period of one month.

Activity (counts/second)	Day
102	1
48	5
29	10
13	15
7	20
3	25
2	31

Problem 23-37

Derive an equation for the following data for the change in electrical resistance in a ceramic because of a change in temperature.

Resistance R (ohms)	Temperature (degrees Fahrenheit)
240.0	750
120.0	840
54.0	930
32.0	1020
15.0	1100
7.0	1200
3.7	1300
1.9	1370
1.0	1450

Problem 23-38

Derive an equation that describes the voltage across a resistor in a circuit where a capacitor is being charged, using the following data:

Voltage	Time (tenths of a second)
10.0	0
5.4	2
3.0	4
1.5	6
1.0	8
.6	10

Problem 23-39

Derive an equation that describes the relative response of an electric recording device as a function of frequency.

Relative Response (decibels)	Frequency (Hertz)
–38	10
–20	100
–12	200
–7	300
–5	400
–3	500
–2	600
–1	700

Empirical Equations: Logarithmic

Problem 23-40

Verify the *Weir equation* that was derived in the discussion of Empirical Equations in the chapter. The data is repeated here for convenience.

Flow Q (cubic feet per minute)	Head H (feet)
14.2	2
41.5	3
80.0	4
135.2	5
225.3	6
330.5	7
445.5	8
620.5	9
790.6	10

❏ Problem 23-41

Derive an equation that describes data taken from film showing height vs. time of an object dropped by students from an airplane flying at 10,000 feet elevation.

Airplane Height Minus Object Height (feet)	Time (seconds)
4	.5
17	1.0
120	3.0
400	5.0
1500	10.0
3700	15.0
6500	20.0

❏ Problem 23-42

During a laboratory exercise students measured the deflection of a cantilevered beam due to a concentrated load applied at the end vs. the length of the cantilevered beam. Derive an equation from their data to show them how deflection varies as the length is increased for the same cross section of beam.

Deflection (inches)	Beam Length (inches)
.01	10
.09	20
.25	30
.60	40
1.50	50
3.40	70
6.00	80
10.00	100

Miscellaneous

The data for the next seven problems should be plotted first on linear-coordinate paper to determine the type of equation—linear, power, or exponential—the data matches. Then, appropriate grids can be used to derive an equation for the data.

❏ Problem 23-43

In the Saturated Steam: Pressure Table, Specific Volume V(g) is tabulated as a function of Absolute Pressure P in psi. Derive an equation with Specific Volume vs. Absolute Pressure for the following steam table data:

P	V
1.000	333.60
5.000	73.53
10.000	38.42
14.696	26.80
15.000	26.29
20.000	20.09
30.000	13.74
40.000	10.50

❏ Problem 23-44

Iron pipe without insulation loses heat to the atmosphere at various rates, depending on the diameter of the pipe. Determine an equation for the following data relating standard pipe sizes to measured heat loss:

Pipe Size P (see std tables for dimensions)	Heat Loss (Btu/foot/hour)
1	79
1-½	114
2	143
2-½	173
3	211
3-½	240
4	271
4-½	300
5	334
6	398
8	518
10	647

❏ Problem 23-45

A typical centrifugal fan creates a pressure differential across the fan which causes air to flow. Derive an equation to relate *air flow* to *revolutions per minute* of this squirrel-cage type fan, from the following data:

Air Flow (cubic feet per min.)	500	800	1050	1400	1900	2100	2550	3000
RPM R	110	290	510	700	1010	1190	1520	1800

❏ Problem 23-46

A sample exposed to neutron radiation became radioactive. The decay in energy of the sample was recorded hourly. Derive an equation for counts per minute as a function of time. (*Note:* If this data is exponential, a tangent to the beginning of the curve, assuming that the curve is plotted on linear-coordinate paper, will intersect the horizontal axis at a time when the counts equal approximately 36% of the initial value of counts per minute. Try this check when it is believed that the data is exponential.)

Counts Per Minute CPM	Time (minutes)
450	60
215	125
138	175
110	250
95	300
90	355
80	550
75	650

❏ Problem 23-47

The following measurements of diameter vs. station (distance in the horizontal direction) were found in a design notebook. Plot the measurements and derive an equation relating diameter to station distance.

Diameter D (mm)	Distance to Station (mm)
3.93	10
3.18	20
2.42	30
1.78	40
1.12	50
0.43	60

❏ Problem 23-48

During a recent consumer product research project an observer noted that the level of popcorn in a box descended rapidly at first, and then the rate of descent decreased slowly. The level of the popcorn was noted by the observer at one-minute intervals. Use the data to derive an equation for popcorn level vs. time.

Depth of Popcorn H (mm)	Time (minutes)
250	0
130	1
100	2
70	3
40	4
26	5
17	6
11	7
8	8
5	9
3	10

Graphical Differentiation

❏ Problem 23-49

During a broadcast of a space launch an armchair observer made notes on the speed of the launch vs. time. Plot the observer's data, fair-in a best fit curve, and differentiate it to determine a curve of acceleration vs. time.

Time (minutes)	Speed (kilometers/hour)
0	0
1	1700
2	5800
3	11,500
4	15,000

❏ Problem 23-50

Plot the following data related to a field test of an automobile and fair-in a best fit curve through the data. Since the data is velocity vs. time, integrating the curve will give distance vs. time, and differentiating the curve will give acceleration vs. time. Do both for the data:

Velocity (mph)	Time (sec)
0	0
20	3
29	4
34	5
48	10
58	15
63	20
66	25
67	30
68	35

❏ Problem 23-51

Occasionally, known quantities, known equations, and known results are used to build confidence in techniques new to an individual, such as graphical differentiation and integration. The data listed below represent the x and y values for the equation $y = (2)(x^{**}2) - (3)(x) + 10$. Plot the points, draw a curve through the points, and then differentiate the curve. Check your results analytically. Where does the curve have zero slope?

x	y
0	10
1	9
2	12
3	19
4	30

One technique for assisting "by-eye" drawing of tangents to curves while doing graphical differentiation is to use a small mirror so that the plane of the mirror is perpendicular to the curve at the point of tangency. When the reflected curve appears "smooth" and "in-line," use the mirror to construct the perpendicular to the curve. Then draw the tangent at 90°.

Graphical Integration

❏ Problem 23-52

Some equations cannot be integrated analytically, so other methods are used, such as numerical integration and graphical integration. Use graphical integration to integrate the expression $y = (5\sin X)/(X^{**}2)$.

❏ Problem 23-53

The objective in this problem is to use graphical integration to graduate the volume in a water tank. The tank looks like a light bulb upside down. Use the diameters at the given elevations to calculate cross-sectional areas, then plot area vs. elevations. Integrate and divide the tank into 1/4, 1/3, 1/2, 2/3, and 3/4 full.

Diameter (feet)	Elevation (feet)
10.0	0
10.0	10
10.0	20
10.2	30
11.8	32
13.0	34
15.0	36
18.4	38
26.4	40
32.8	42
37.0	44
39.0	46
40.2	48
40.0	50
39.0	52
36.6	54
31.0	56
26.0	58
0.0	60

❏ Problem 23-54

Do the Aircraft Launch example discussed in the chapter and add the speed of the Aircraft Carrier (24 knots) to the speed of the aircraft attained by graphical integration. Will the aircraft stay airborne?

❏ Problem 23-55

Use the same beam shown in Figure 23-82, but assume that the 1000-pound load is applied 30″ from the left end instead of 45″. The reaction at the left end will be 667 pounds, and at the right, 333 pounds. Use graphical integration to do the diagrams, and determine the deflection of the beam at the point where the 1000-pound force is applied. Check your answer using the beam deflection calculation equation in Figure 23-80:

$$\text{delta (a)} = y = \frac{P\,(a^{**}2)\,(b^{**}2)}{3\,E\,I\,L}$$

where P = 1000 lbs
  a = 30″
  b = 60″
  E = 30,000,000 psi
  I = .766 in**4
  L = 90″

❏ Problem 23-56

Use Figure 23-82 as a guide for graphically integrating to obtain the deflection at the center of a beam that is 90″ long, that is loaded at the center with a 2500-pound concentrated force, and that is an S beam ("eye" beam) S 3 x 5.7 with a value of I = 2.52 in**4. Use the equation in Figure 23-80 to check your results:

$$\text{delta (max)} = y = \frac{P\,(L^{**}3)}{48\,E\,I}$$

where: P = 2500 lbs
  L = 90″
  E = 30,000,000 psi
  I = 2.52 in**4

Force Polygons

❏ Problem 23-57

Refer to Figure Problem 23-57 and do the following:

1. Draw the figure to scale using only single lines for members AB and BC.

2. Show a free-body diagram of joint B.

3. Use a force polygon, drawn to scale, to graphically determine the internal forces F (AB) and F (BC).

4. Label the values of F(AB) and F(BC) on the force polygon and indicate also whether these internal forces are going into joint B (compression) or away, (tension).

❏ Problem 23-58

Figure Problem 23-58 shows a Bell Crank with forces F(A) and F(B) equal to 60 lbs and 100 lbs, respectively. The two applied forces and the reaction at C must go through a common point P for equilibrium conditions. Do the following:

1. Draw the Bell Crank, to scale, using single lines for AC and BC.

2. Extend the lines of action of the forces until they intersect; label the intersection point P.

3. From P, construct a line through the pivot point C.

4. Construct a force polygon to scale to determine the magnitude of the reaction at C.

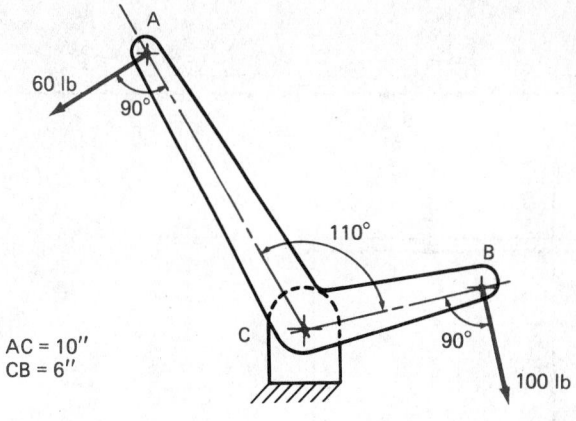

Problem 23-58

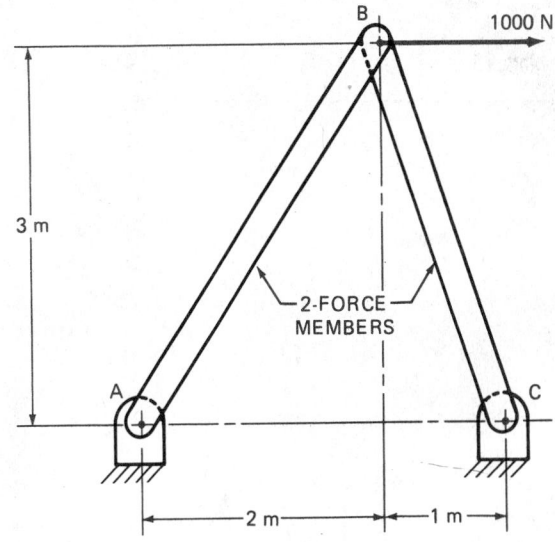

Problem 23-57

❏ **Problem 23-59**

Figure Problem 23-59 shows an excited rider in a carnival-ride chair. The rider plus the chair weigh 750N, and the angle beta of the cable supporting the chair is 35°. Use a force polygon drawn to scale to determine the tensile force in the cable and the centrifugal force.

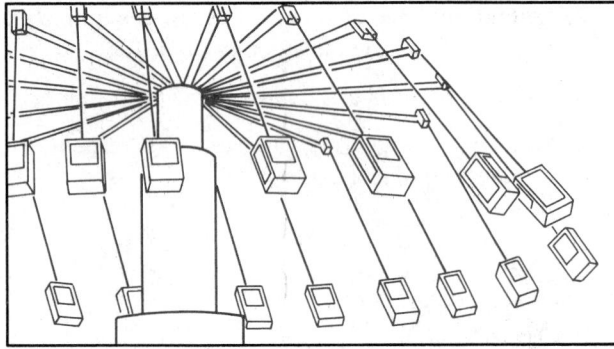

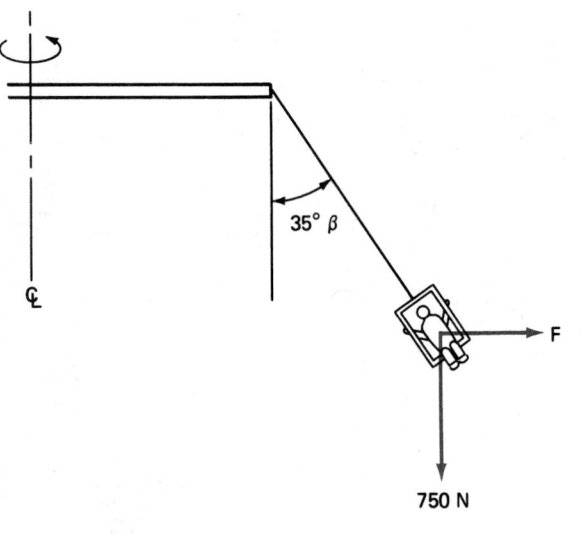

Problem 23-59

Miscellaneous

❏ **Problem 23-60**

Optimization problem. The two equations below give costs for insulating steam pipes over a period of a year.

$ (Fixed Cost) = 50t + 50

$ (Heat Loss) = 150/t, where t = insulation thickness

Plot both equations plus their sum as a function of insulation thickness t, and determine the optimum thickness to give the least cost.

❏ **Problem 23-61**

Curve matching technique. There are a number of curve matching techniques other than graphical ones; however, to visualize whether the results of a particular technique has indeed matched the curve, a plot of the results is usually done. One technique for matching a cyclic curve—that is, one which repeats periodically—is called a Fourier Series. The following series was derived to match a cyclic, square wave, function that varied from 1 to 2 every 3.14 radians.

$$y = \frac{3}{2} - \frac{2}{3.14}\left(\sin X + \frac{\sin 3X}{3} + \frac{\sin 5X}{5} + \ldots\right)$$

Calculate and plot values of y for X varying from zero to 10 radians. One radian = 57.6 degrees.

❏ **Problem 23-62**

Curve matching technique. A quadratic equation is used sometimes to match an equation to a curve. Try 3 sets of data from the Weir problem discussed in the chapter (H, Q: 4, 80; 6, 225.3; 8, 445.5) in the following equation: $Q = A(H^{**}2) + B(H) + C$. Substitute the 3 sets of data to obtain 3 equations and solve them simultaneously for A, B, and C. Then check the derived equation with additional data, e.g.,

H = 5 and Q = 135.2.

❏ **Problem 23-63**

Investigate ANSI (American National Standards Institute) Standard ANSI Y15.1M-1979 entitled "Illustrations for Publication and Preparation" and prepare a report for classmates, telling why the standard is useful to designers and showing several illustrations.

The Design Process

CHAPTER 24

Chapter Outline

The *design process* is a plan of action for reaching a goal. The plan, sometimes labeled problem-solving strategy, is used by engineers, designers, drafters, scientists, technologists, and a multitude of professionals. This strategy, the *design process*, occurs in five *phases* during a product design, and each *phase* has a number of *steps*. The *phases* listed below occur sequentially with some overlapping, as shown in ***Figure 24-1***.

Phase I	Recognize and Define the Problem
Phase II	Generate Alternative Solutions: SYNTHESIS
Phase III	Evaluate Solution Ideas: ANALYSIS
Phase IV	Document, Specify, and Communicate the Solution
Phase V	Develop, Manufacture, and Deliver

The *steps*, not shown in Figure 24-1, may occur sequentially; however, each *phase* has a different number of steps depending on the complexity of the problem to be solved and the time available.

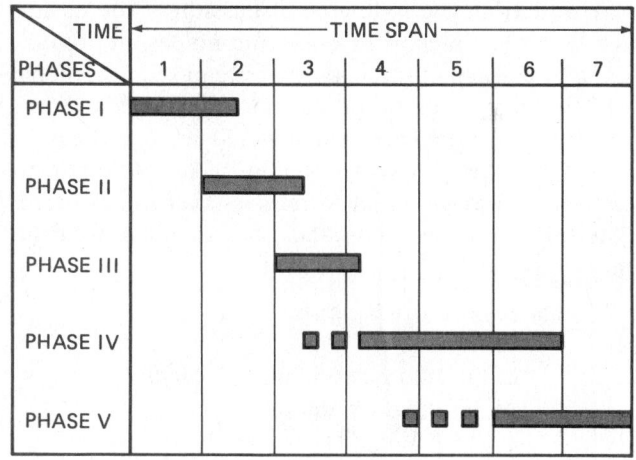

Figure 24-1 The phases in the design process

Time

The available *time*, whether hours, days, months, or years, controls the starting and stopping of the phases during the design process. Therefore, successive problem solving requires careful planning. And since the phases must occur within the time span, they must be scheduled as needed to reach the required goal.

Learning The Design Process

Learn the design process by doing it! That is, follow the phases and steps, as a guide, to do at least two design projects. First, individually, do a routine design project. Next, in a group, do a non-routine design project. Use these design experiences to augment and expand your design skills, and to acquire an understanding that the design process can be modified to suit the level and complexity of the problem to be solved. Start with this chapter, which provides the following.

1. An outline and discussion of the phases and steps in the design process (see *Figure 24-2*)

2. Examples from a non-routine design project (done by a student group)

3. A discussion of routine design projects (the type usually assigned to a neophyte drafter, designer, or engineer as their first task in a company)

4. A discussion of non-routine design projects (the type assigned to more experienced drafters, designers, and engineers as they assume more responsibility in their respective jobs)

5. A collection of design hints to enhance your problem-solving skills and design solutions

6. Review questions

7. Exercises to complement the design process

8. A list of design project ideas—routine and non-routine

Throughout the following discussion of the design process, examples from a non-routine design project, done by a group of students, illustrate most of the steps. The problem, given to the students from a medical doctor, was to design a device that would exercise the muscles of a patient's jaw, continuously day and night, to accelerate healing of the jaw muscles after an operation. The following figures contain the examples from the student's project:

Figure 24–3   Planning chart
      24–4   Log
      24–5   Information from the client
      24–6   Progress report
     24–18   Letter of transmittal
     24–19   Title page and abstract
     24–20   Table of contents
     24–21   Introduction
     24–22   Power pack assembly sketch
     24–23   Alternatives discarded
     24–24   Final recommendations
     24–25   Appendix: materials, electrical schematic, and costs

Follow the phases and steps in Figure 24-2, generally in the sequence shown. Use the steps as a checklist.

DESIGN PROCESS: PHASES AND STEPS

PHASE I
RECOGNIZE AND DEFINE THE PROBLEM

STEP 1   Establish Plans for the Phases, Steps, and Records

STEP 2   Gather Information
     a. Primary Sources
     b. Secondary Sources

STEP 3   Combine Information Into a Preliminary Definition
     a. Constraints
     b. Specifications
     c. Preliminary Solution Ideas
     d. Iterate

STEP 4   Prepare A Progress Report
     a. Internal Client
     b. External Client

PHASE II
GENERATE ALTERNATIVE SOLUTIONS: SYNTHESIS

STEP 5   Generate Solution Ideas
     a. Use Systematic Techniques
     b. Use Creative Techniques and Attitudes

PHASE III
EVALUATE ALTERNATIVE SOLUTION IDEAS: ANALYSIS

STEP 6   Sift, Combine, and Prepare Solution Ideas for Analysis
     a. Preliminary Preparation
     b. Specific Preparation

STEP 7   Compare Solution Alternatives to Constraints and Specifications
     a. Iterate
     b. Persevere
     c. Qualitative Evaluation
     d. Quantitative Evaluation

STEP 8   Conduct a Design Review
     a. In-House Design Reviews
     b. Outside Design Reviews

STEP 9   Refine the Final Solution

PHASE IV
DOCUMENT, SPECIFY, AND COMMUNICATE THE SOLUTION

STEP 10   Prepare Drawings and a Written Report

STEP 11   Give an Oral Presentation

PHASE V
DEVELOP, MANUFACTURE, AND DELIVER

STEP 12   Prepare Technical Detail Drawings of Each Component To Be Manufactured

STEP 13   Schedule Assembly, Final Testing, Packaging, and Delivery

Figure 24-2   Phases and steps in the design process

The Design Process: Phases and Steps

Phase I Recognize and Define the Problem

Step 1   Establish Plans for the Phases, Steps, and Records.   Make a *planning chart*, which is an excellent visual display of when phases and steps are scheduled to begin and end, *Figure 24-3*. Note that over each bar is the name of the individual responsible for that activity, whether the individual does all of the work or not.

Start and keep a *log* of activities and the time spent. Each individual in the group should keep a *log*. Newspaper-style entries, who—where—what—when—why, are useful because they tell specifically what was done. Supervisors use the information for checking progress and for accounting purposes. But more importantly, the phone numbers, names, dates, and facts may be useful on future projects. Instructors use the log information for checking progress and for grading purposes. See *Figure 24-4* for a sample log from a student's notebook.

Some companies may require selected individuals to keep a more detailed log-design notebook for legal purposes or for patent development. For patent development, each page is bound, numbered, and then arranged to be signed by a knowledgeable witness.

Start a *design file* with every page dated to agree with the log, and file the pages chronologically for easy reference. Each individual keeps a *design file* which includes freehand sketches, machine copies of technical information, copies of related drawings, summaries of telephone conversations and interviews, and vendor brochures. Samples from a student's *design file* are in *Figures 24-5A*, *24-5B*, and *24-5C*. A group log and file may be required by a supervisor in industry.

Step 2   Gather Information.

a. *Primary Sources.* Contact the source of the problem, who will provide information to establish most of the project's constraints and specifications, verbally or in writing. Later, in STEP 4, confirm any verbal decisions made by the source, in a progress report.

Next, review similar products and experiences for information because most designs are actually new combinations of existing ideas.

b. *Secondary Sources.* Investigate secondary sources, which include a variety of people and references. The following checklist can be expanded to fit the needs of specific problems.

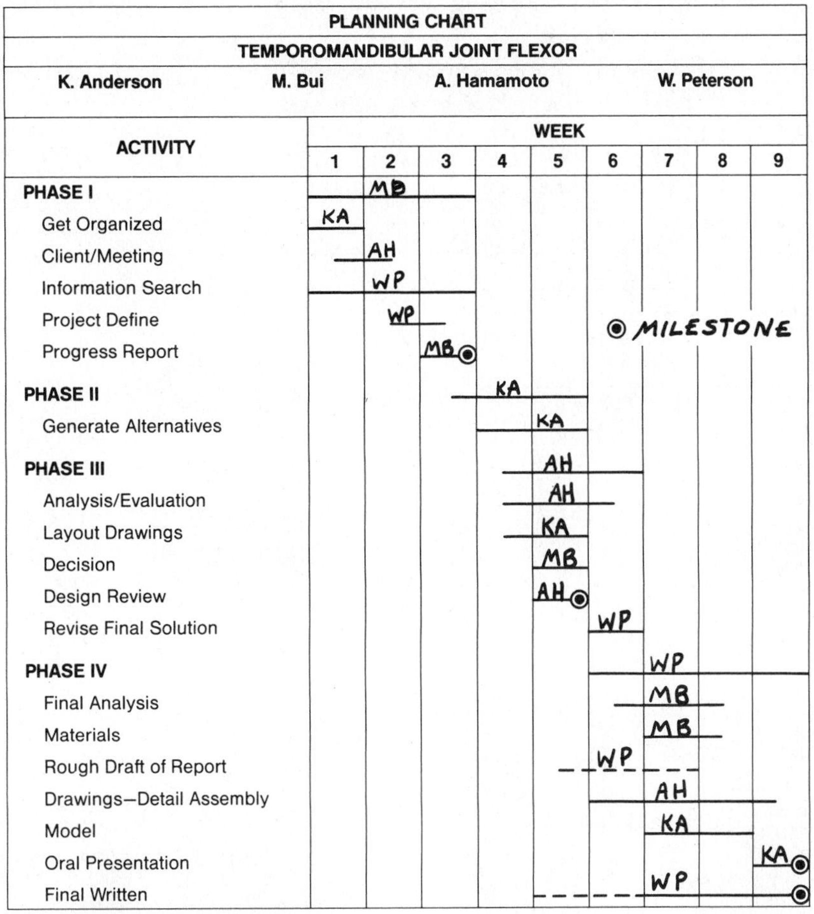

Figure 24-3   Planning chart for student group

TIME LOG	W. PETERSON		
	TEMPOROMANDIBULAR JOINT FLEXOR		
DATE	ACTIVITY	TIME HR	CUM. TIME HR
10-2	Met with Anderson, Bui, Hamamoto- -Exchanged phone numbers -Copied Dr. P's proposal	1/2	1/2
10-8	MET - KA MB & AH - discussed proposal - made up questions for Dr. P Discussed planning chart	2 3/4	3 1/4
10-10	Questions for Dr. P MET at Dr. P's OFFICE IN MED SCHOOL	2 1/2	5 3/4
10-15	Discussed electrical circuits w/ MB	1/4	6
10-17	MET w/ group, checked drawings/ Went to hobby shop for solenoids Worked on progress report w/MB Typed progress report	2 1/4	8 1/4
10-20	Talked to Mike Wieggand re: valves Group meeting	2	10 1/4
10-30	Contacted vendors for bellows information Prepared transparencies for DESIGN REVIEW - worked with Hamamoto	3	13 1/4
11-1	Trip to surplus yard- found plastics	2 1/2	15 3/4

Figure 24-4   Log

PRELIMINARY DESIGN CONSIDERATIONS

1. SIZE   Date 10-10

2. POWER
   –BELT POWER PACK
   –CHARGER

3. CONTROLS
   –FREQUENCY
   –OPENING
   –PATIENT OR DR.?

4. EXPECTED PRODUCTION
   –REUSABLE?

5. COST

6. MATERIALS
   –STAINLESS
   –PLASTIC
   –FDA RESTRICTIONS

7. NOISE LEVEL

8. FORCES REQUIRED
   –OPENING
   –CLOSING

9. SAFETY OVERDRIVE

10. JAW DIMENSIONS

11. ACCURACY

Figure 24-5A   File: Preliminary design considerations

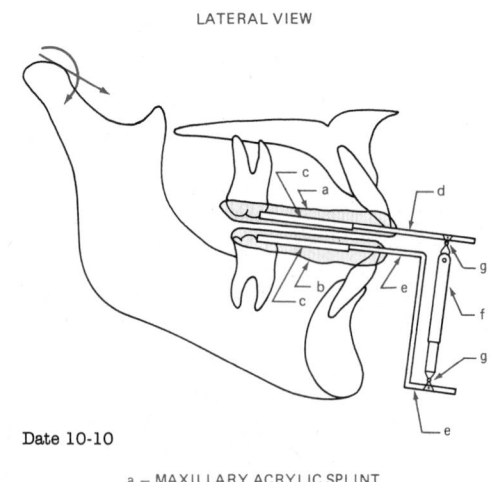

LATERAL VIEW

Date 10-10

a — MAXILLARY ACRYLIC SPLINT
b — MANDIBULAR ACRYLIC SPLINT
c — BUCCAL TUBES
d — UPPER MEMBER
e — LOWER MEMBER
f — PNEUMATIC PISTON
g — UNIVERSAL JOINTS

Figure 24-5B   File: Information from clients

FRONTAL VIEW

Date 10-10

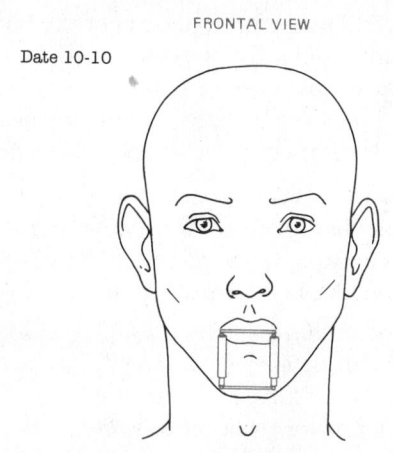

Figure 24-5C File: Information from clients

People: shop personnel (machine, structural, electronic, electrical, packaging, shipping, welding...)

Practicing professionals: (engineers, lawyers, doctors, technologists, scientists...)

City services: (fire, police, waste, utility, library.)

Most helpful for students: (Instructors, Librarians, and Shop Personnel.)

Patents: Note that only 20% of the information in Patents appears elsewhere; that is, 80% is only available in Patents!

References:

  Books (textbooks on specific topics):

    Best listing of references found to date: Kolb, John and Ross, Steven, PRODUCT SAFETY AND RELIABILITY, 1980, McGraw-Hill, Inc. (It also contains more than 45 pages of checklists for Designers and Installers.)

  Handbooks: Each engineering discipline has a handbook, e.g., aeronautical, chemical, civil, electrical, mechanical, materials...

Standards and codes (a small sampling):

  ANSI American National Standards Institute

  ASTM American Society for Testing Metals

  Mil Military Standard

  UL Underwriters Laboratory

Catalogs:

  Vendor (products, hardware, electrical, electronic, structural shapes...)

  Thomas Register (listing of companies, what they make, their address, sample catalogs...)

  McMaster-Carr Supply Co. (Most complete hardware catalog found to date, P.O. Box 54960, Los Angeles, CA 90054, 213/945-2811.)

Step 3   Combine Information into a Preliminary Definition.   Prepare a preliminary definition of the problem so that all parties concerned can see the direction and course of the project. The definition mainly consists of two types of statements, *constraints* and *specifications*. These are written as specifically as possible even at this early stage in the project. For example, a vague specification might say that the product is to be safe to use. A more specific specification would say that the product will be safe to use by a 12-year-old child who can read and understand simple directions, and that the directions will be affixed to the product.

a. *Constraints.* List the more restrictive *constraints*, such as cost, delivery date, availability of materials, codes, and standards. Then, list the less restrictive ones, such as shop capability, available personnel, product life, and aesthetics.

b. *Specifications*. Divide specifications into two main categories, *performance* and *design. Performance* specifications describe what the design is to do, what it is to accomplish, what it is to produce, who is to use it, what environment it will operate in, how it will be maintained, how safe it is, and how easy it is to use. *Design* specifications indicate what materials are to be used and how they are to be fabricated and finished. Purchased products also are included.

Well-written specifications are stated so that they can be tested or measured. For example, the chemistry and strength of a material can be measured and tested, respectively. Resistance, frequency, volume, temperature, and weight can all be measured.

c. *Preliminary Solution Ideas.* Collect *preliminary solution ideas* which occur while the definition, constraints, and specifications are being developed and written. File these for use in Phase II.

d. *Iterate.* Return to STEP 2 to gather further information and to revise the definition. Returning and revising is a *iterative* process that occurs continually throughout the design process.

Step 4   Prepare a Progress Report.   Prepare a *progress report* to the *client* who is usually the primary contact during the project. By now, approximately one-fourth of the project time will have been used, so a report to the client should be made preferably in writing. Include the current definition — constraints and specifications — so that if there are any misunderstandings, they can be resolved early in the project. Decisions made verbally should be summarized. Indicate immediate plans and ask for confirmation as soon as possible. The *client*, a person or an organization, wants to know how the project is progressing and if any problems have arisen.

For the purposes of this discussion, clients are divided into two groups.

a. *Internal Clients* are instructors in classrooms and supervisory persons in industry. And even with this close proximity to the client, important communications should be in writing.

b. *External Clients* comprise a large variety of people and organizations, both in classrooms and in industry.

In classroom situations, an *external client* could be another instructor, a student, a department, a design contest, a handicapped person, a rehabilitation department, a local government, or a national organization. In industry, the *external client* could be another company; an individual; a department in a city, state, or national government; the military; or even another nation.

Figure 24-6 shows a sample *progress report* from a student project. A copy was sent to their instructor.

Phase II Generate Alternative Solutions: Synthesis

Step 5   Generate Solution Ideas.   Review the solution ideas already collected and realize that additional effort and discipline need to be exerted to generate more. Frequently, an idea for a solution occurs early in the process, and there is a tendency—especially among students, because of their time constraints—to select the first idea as the final solution. Successful problem solvers know that additional ideas can be generated beyond the initial obvious ones by continuing to focus on the design problem and employing two types of idea-generating techniques: systematic techniques and creative techniques. Having a creative attitude also helps.

a. *Use Systematic Techniques.* *Verb lists* provide a systematic checklist for generating new ideas, primarily building on existing ones. Ask the following questions regarding and idea being considered:

- Can it be *modified*?
- Can it be made *larger*? *Smaller*?
- Can it be used for *something else*?
- Can other materials be *substituted*?
- Can components be *rearranged*?
- Can it be *combined* with another idea?

Divide an idea into *smaller parts* so that each part can be systematically examined and developed.

Date 10-10

                    Department of Mechanical Engineering
                    (School)
                    (City, State, Zip Code)

Dr.
University Hospital
(City, State, Zip Code)

Dear Dr.,

    This letter is to inform you of our design team's progress and
decisions regarding your CPM (continuous passive motion) device. We
have discussed the design criteria set forth in your proposal and also
examined the additional criteria we discussed with you at our last
meeting. Presently, we are considering the feasibility of a pneumatic
device with an external power pack. We are currently evaluating this
concept and its ability to meet the following criteria:

    1) Weight: less than 4 oz
    2) Accurate and adjustable movement
    3) Frequency in the range of 2-3 cycles per minute with a maximum
       range of 1-4 cycles per minute
    4) Device must be safe for use by a 10-year-old and older for 24
       hours at a time
    5) Operation must be simple (few moving parts and adjustable by
       anyone who can read and understand simple directions)
    6) Low cost (less than $500, including labor and materials)
    7) Parts must be readily available (most from local sources)

Based on our pending evaluation of the above, we will pick what we
feel is the best design, and prepare a report outlining our choice.
Included in the report will be a cost analysis, detail and assembly
drawings, and an overall evaluation. This report will be presented to
you at a date and time decided by our instructor.

    If you have any questions, please contact us through our
instructor.

                    Sincerely,

                    W. Peterson

                    K. Anderson

                    M. Bui

                    A. Hamamoto

Figure 24-6   Progress report

Try a *180-degree approach*. For instance, instead of breaking a walnut shell with a hammer, poke a hollow needle inside and explode the shell with air pressure.

List the *obvious attributes* of an object to uncover ideas.

- Describe the size using dimensioned sketches.
- List the intended functions of the object.
- Outline how it was probably manufactured.
- List the materials used to make it.
- Describe any other features, such as aesthetics, weight, and color.

Use attribute listing, not only to generate ideas, but also as a checklist for the design of parts on technical drawings. That is, check for complete dimensions to describe the geometry of each part and dimensions to locate each feature, such as bolt holes, slots, special surfaces, datum, etc. Review appropriate dimensions to ensure the function of the part, for example, clearances, fits, etc. Check the material specified for the part, and whether or not the manufacturability of the part was considered during the design phase. Finally, check all remaining attributes, such as how easily the part can be assembled, surface finishes, heat treatments, aesthetics, etc. Also list the *not-so-obvious attributes*, such as other uses which were not intended functions. Developing the habit of proposing these "not-so-obvious" uses for objects will help to generate new ideas for future designs.

Figure 24-7 shows the obvious attributes of a paper clip, and the not-so-obvious ones are listed in *Figure 24-8*.

Browse through *technical texts, periodicals,* and *handbooks* that contain numerous ideas for components, mechanisms, and applications of existing products. The partial list that follows includes the three types.

Texts:

Herkimer, Herbert, THE ENGINEERS ILLUSTRATED THESAURUS, 1952, Chemical Publishing Co.

Schwartz, Otto B./Grafstein, Paul, PICTORIAL HANDBOOK OF TECHNICAL DEVICES, 1971, Chemical Publishing Co.

McCormick, Ernest J., HUMAN FACTORS ENGINEERING, 1970 McGraw-Hill, Inc.

Chapanis, A., MAN-MACHINE-ENGINEERING, 1965, Wadsworth Publishing Co.

Periodicals (found in most libraries):

Machine Design
Product Engineering
Test and Measurement World
Control Engineering
Medical Product Manufacturing News
Hydraulics and Pneumatics
Research and Development
Engineer's Digest

Handbooks:

Shigley, Joseph E., STANDARD HANDBOOK OF MACHINE DESIGN, 1986, McGraw-Hill, Inc.

Laughner, V.H./Hargan, A.D., HANDBOOK OF FASTENING AND JOINING OF METAL PARTS, 1956, McGraw-Hill, Inc.

Horger, A.J., ASME HANDBOOK METALS ENGINEERING DESIGN, 1965, ASME Publisher

b. *Use Creative Techniques and Attitudes.* A clear understanding of the *creative process* will help to generate ideas. The process—as described by technical people, composers, writers, builders, doctors, hobbiests, and many others—proceeds as follows:

1. *Accumulate* all the information about the problem being attacked; that is, thoroughly and conscientiously get involved in the problem, even emotionally if necessary. This involvement, according to creative individuals, causes the subconscious portion of the brain to be alerted that a problem needs solving. Next, relax and let the information collected...

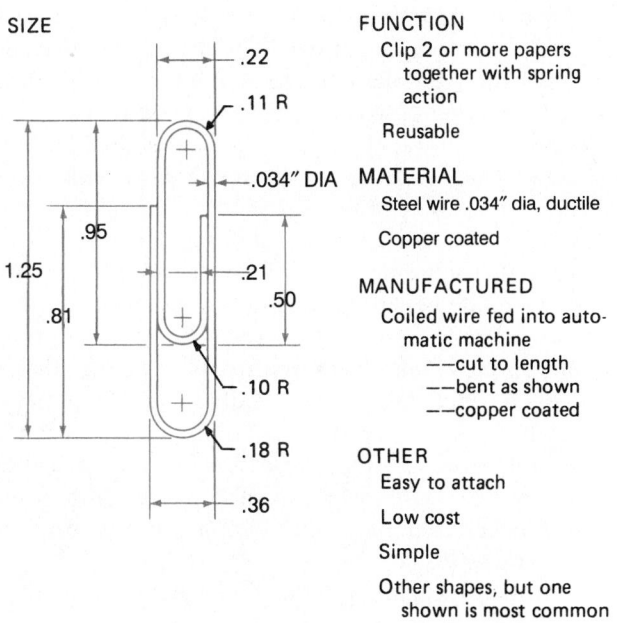

EXAMPLE OF OBVIOUS ATTRIBUTES

SIZE

.22
.11 R
.034" DIA
.95
1.25
.81
.21
.50
.10 R
.18 R
.36

FUNCTION
Clip 2 or more papers together with spring action
Reusable

MATERIAL
Steel wire .034" dia, ductile
Copper coated

MANUFACTURED
Coiled wire fed into automatic machine
——cut to length
——bent as shown
——copper coated

OTHER
Easy to attach
Low cost
Simple
Other shapes, but one shown is most common

Figure 24-7 Attributes of a paper clip

EXAMPLES OF NOT-SO-OBVIOUS ATTRIBUTES	
AS IS	**CHANGE**
–dig out ear wax	–fish hook
–make a chain	–cake tester
–balance for paper airplane	–electrical conductor
–clip other items	–chess pieces
–sinker for fishing	–lock pick
–checkers	–artistic shape
–poker chips	–therapy–bend for relaxing
–hanger for small picture	–melt many to make casting
–standard of weight	–fatigue failure demo

Figure 24-8 Not-so-obvious attributes

2. *Incubate* with other ideas that the subconscious brain retrieves from memory locations. Take a walk, jog. Plop both feet upon a desk and lean back while contemplating the ceiling. (Supervisors may need to be informed that the "process" is working.) Soon, one hopes an idea will surface in the conscious portion of the brain; that is, the idea...

3. *Illuminates* its presence and causes exclamations like, "Eureka, I've got it!" Finally, the idea needs to be...

4. *Verified*. Does it actually solve the problem? Meet the constraints? Satisfy the specifications?

Develop a *creative attitude* and *behavior* by adopting characteristics of creative people. Several are listed here.

1. *Keep an open mind*. Be willing to *accept change* and *new* ideas.

2. *Look for similarities and unusual relationships*.

3. *Think positive. Avoid* or ignore *negative comments* such as: "It's been tried before." "It's not logical." "What a dumb idea." "It'll never work."

4. *Be persistent*.

5. ALWAYS CARRY A NOTEBOOK!!!!!!! Ideas come when least expected and may be lost if not recorded.

Participate in exercises designed to revive creative talents, buried during years of being told what to do and to conform. For example, the late Professor John Arnold, M.I.T. and Stanford University, proposed a fictitious planet, "Arcturus IV", populated with creatures who were unusually tall and slow moving, had reaction times one-tenth as fast as humans, had three fingers on each hand, had three eyes, one of which had x-ray vision, and breathed methane. Professor Arnold asked his students to design an automobile-type vehicle for these inhabitants. Tackle this problem before reading on, and compare your design ideas with other members of your class. For example, what speed limit should be required due to the slow reaction time of the inhabitants?

Build a *simple model* related to the design problem being developed. Building a model, or a prototype, almost always produces new ideas because of the simultaneous involvement of the hands, eyes, and brain, plus the three-dimensional insight.

Do *exercises* to develop imagination, for the same reason that muscles are exercised: for growth and to avoid atrophying. One convenient source of creative exercises is the daily newspaper, which contains word puzzles, crossword puzzles, and picture puzzles.

Try the following exercises to help develop a creative attitude.

1. Use analogies. For example, pretend to be one of the components in a machine and imagine how each part would function if designed a certain way. What would happen if the design were changed?

2. Systematically propose that a product be changed in any one of the following ways: made bigger, smaller, longer, shorter, colorful, drab, nontradi-tional, expensive, cheap, etc.

Read books devoted to developing creativity and try some of their recommendations. Several are:

1. Osborne, A.F., APPLIED IMAGINATION, 1953, Charles Scribners Publisher.

2. Harrisberger, Lee. ENGINEERSMANSHIP, 1966, Brooks/Cole Publishing Co.

3. Adams, James. CONCEPTUAL BLOCKBUSTING, 1974. W.H. Freeman and Co.

4. Gordon, J.J., SYNECTICS, 1961 Harper and Row.

5. Raudsepp, Eugene. CREATIVE GROWTH GAMES, 1977, Jove Publications, Inc.

A person's creative potential can be enhanced by taking courses in areas outside of engineering and technology. Take courses to obtain different points of view in other disciplines, such as biology, geology, agriculture, environment, political science, art, philosophy, oceanography, etc.

Brainstorm according to the ideas of Alex Osborne in his APPLIED IMAGINATION. The procedure is simple to follow. Meet with a group who are willing to abide by the following guidelines:

1. Have one person *record ideas* as they are generated, on a surface visible to all participants. New ideas often stem from the visible ones.

2. Agree that *no one is to criticize* any ideas while the brainstorming is in progress, no matter how ridiculous an idea may seem. Frequently, viable ideas stem from less obvious ones.

3. Meet in as *pleasant surroundings* as practical, and include refreshments. Several one-hour sessions produce better results than one long session.

4. Finally, meet to *critique* the ideas for possible solutions.

Brainstorming can also be done individually.

Develop *observation skills* through freehand sketching by looking at a component while imagining how a sketch of the component would be made. Of course, do the sketch if possible! Looking at an object while imagining how a sketch would be made causes the viewer to see more about the object than he or she would if only a casual glance were given. Try sketching assemblies of components, also.

Assume now that a number of design ideas have been generated. The group usually wants to continue in this creative mode because creating is a satisfying activity. However, since *time* controls the schedule, the group needs to move on. Students involved in the design process for the first time reluctantly leave this synthesis phase believing that if they had more *time* they could do a better job. Creative professionals know that they must do the best they can in each phase, in the time available, and then move on to complete their projects within the schedule.

So now is the time to focus on evaluating instead of generating.

Phase III Evaluate Alternative Solution Ideas: Analysis

Step 6   Sift, Combine, and Prepare Solution Ideas for Analysis.   *Sift* through all the collected ideas and keep those which seem to have a potential as a part solution or as a complete solution. File the remaining ideas for possible use in the future.

a. *Preliminary Preparation.* Expand each potential idea in more detail. Clarify, if possible, vague statements and generalities. Ask each member of the group to develop *preliminary statements* and *sketches* of those of his or her ideas that survived the initial sifting and culling.

Do further research, if needed, to expand the ideas.

b. *Specific Preparation.* Prepare *layout development drawings* of the preliminary sketches. Layout drawings are essentially a designer's freehand sketch redrawn to scale with instruments. Designers use two types, *design development* and *kinematic development* (the latter is used primarily for mechanical products). Schematic layouts, using lines and symbols, but not to scale, are used for electronic circuits, electrical circuits, piping diagrams, routing paths, processing steps, chemical plants, and logic circuits.

A *design development layout drawing* is made from a freehand sketch of a design idea for an anchor bolt puller shown in *Figure 24-9A*. The design idea is drawn to scale in the *design development layout drawing, Figure 24-9B*. Notes and comments may be in writing or in casual lettering. Compare this informal approach to the formal technical detail and assembly drawings for shop use, which have standard lettering, lines, and symbols made on standard size drawing sheets, complete with title blocks.

Use the design development layouts to establish the locations and sizes of parts to be fabricated so as to ensure the functions of the final product. Also use this layout to verify that purchased products will fit where required.

In industry, designers forward their *design development layout drawings* to drafters who prepare the formal technical detail and assembly drawings. Students in classrooms do both.

A *kinematic development layout drawing* is shown in *Figure 24-10*. This type of layout shows the relative motions of assembled components, verifies the function of the assembly, checks for interferences, and does velocity plus acceleration analyses. The kinematic analyses are usually done in a two-dimensional format even though the assemblies are, of course, three dimensional.

A useful complement to the *kinematic development*

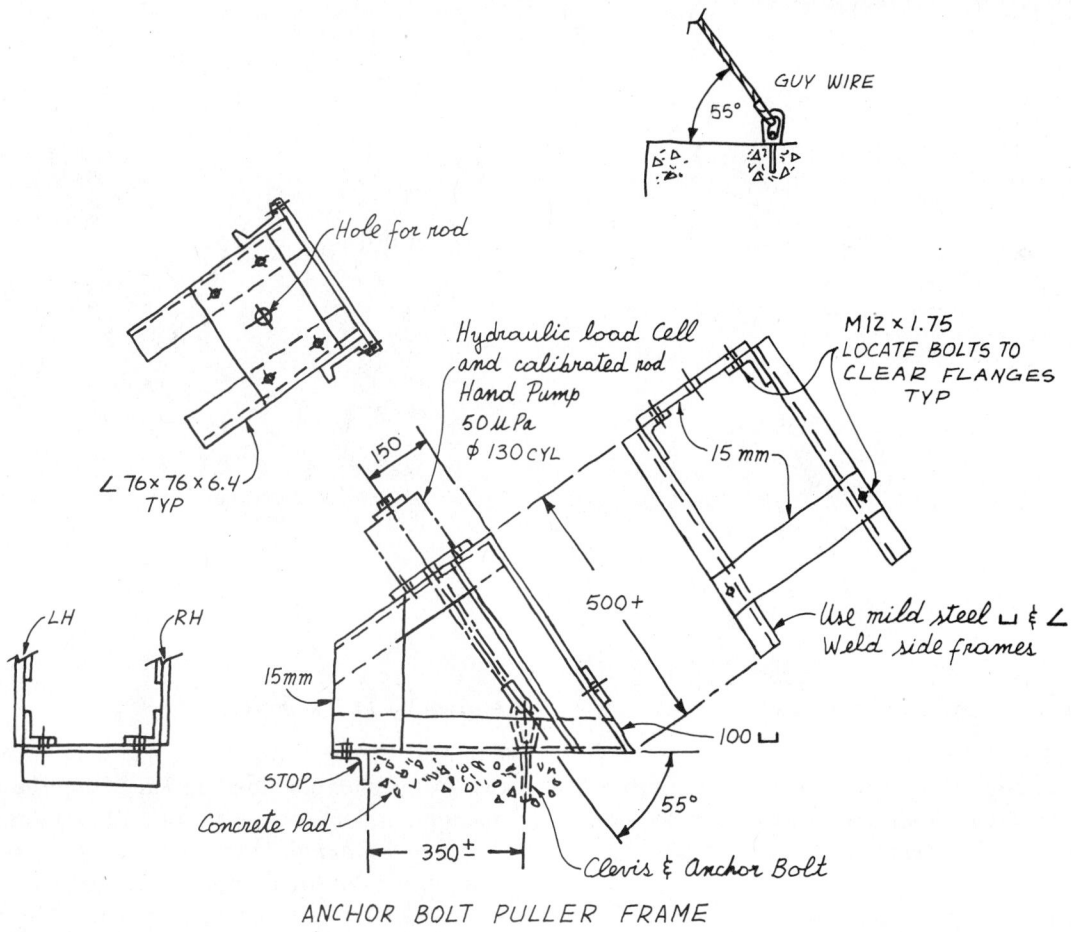

Figure 24-9A   Freehand sketch of a design idea

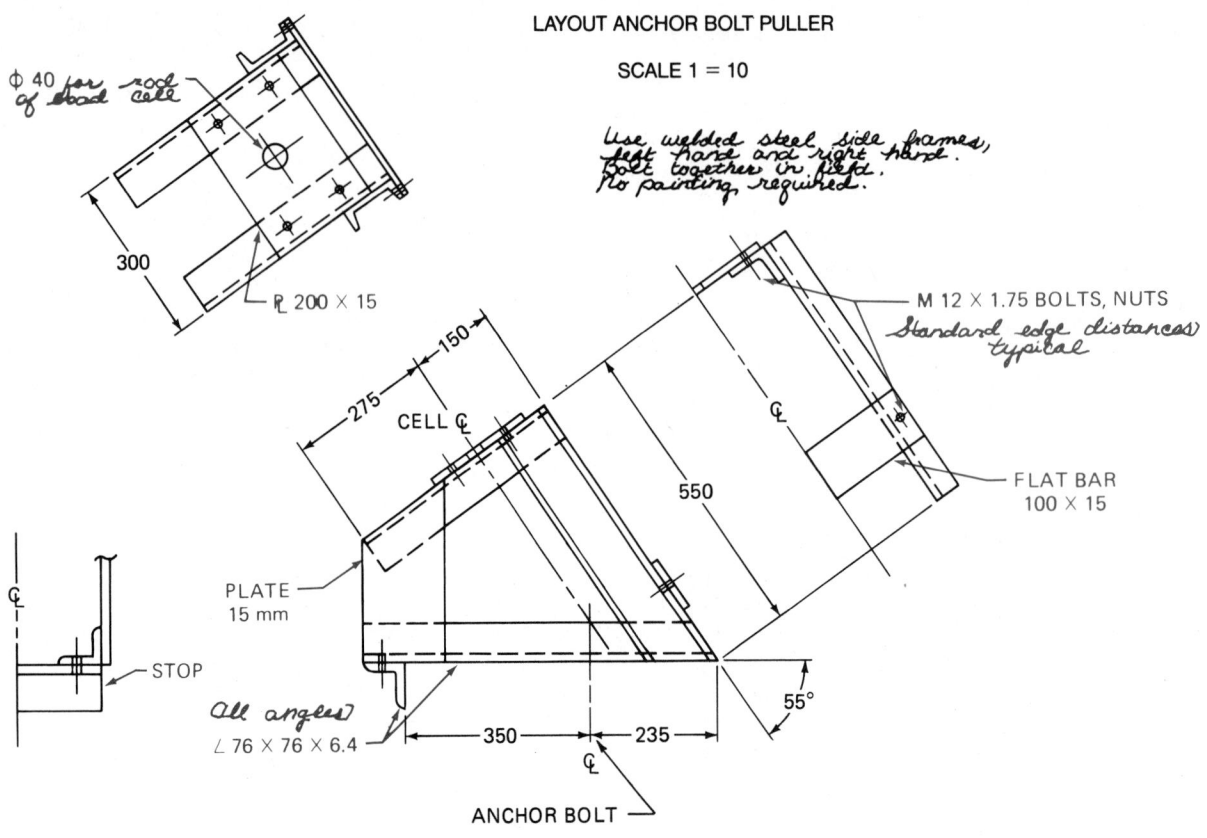

LAYOUT ANCHOR BOLT PULLER

SCALE 1 = 10

Use welded steel side frames,
left hand and right hand.
Bolt together in field.
No painting required.

φ 40 for rod
of load cell

300

℄ 200 × 15

M 12 × 1.75 BOLTS, NUTS

Standard edge distances
typical

FLAT BAR
100 × 15

150

275

CELL ℄

550

PLATE
15 mm

STOP

All angles
∠ 76 × 76 × 6.4

350   235

55°

ANCHOR BOLT

Figure 24-9B   Design development layout

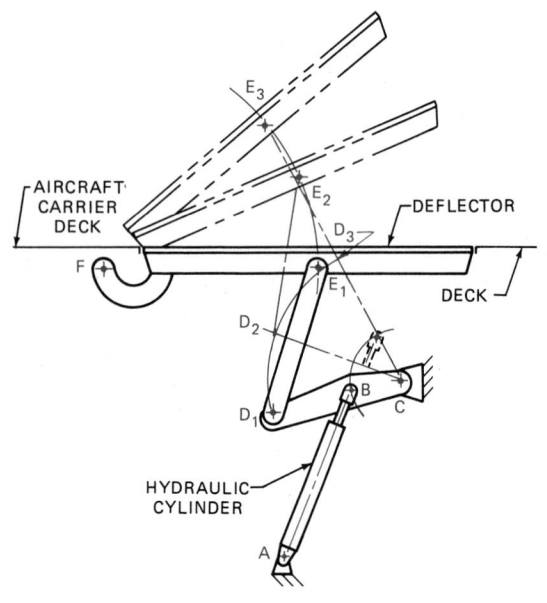

AIRCRAFT
CARRIER
DECK

E₃

E₂

D₃

DEFLECTOR

DECK

F

E₁

D₂

D₁   B   C

HYDRAULIC
CYLINDER

A

Figure 24-10   Kinematic development layout

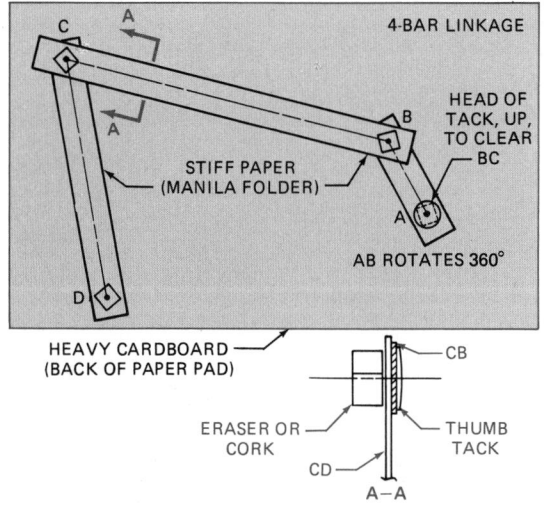

4-BAR LINKAGE

C

A

A

B

HEAD OF
TACK, UP,
TO CLEAR
BC

STIFF PAPER
(MANILA FOLDER)

A

D

AB ROTATES 360°

HEAVY CARDBOARD
(BACK OF PAPER PAD)

ERASER OR
CORK

CB

CD

THUMB
TACK

A–A

Figure 24-11   Kinematic model

layout drawing is a simple cardboard-thumbtack model, as illustrated in **Figure 24-11**. CADD systems with programs for kinematic analyses are also useful and available.

Use *"cardboard and glue" models* (noted earlier) to further complement the layout drawings. Students, designers, drafters and practicing engineers attest that

two major phenomenon happen during the process of making a model of a portion or all of a design from a *design development layout*. One, they discover that the components in the three-dimensional model may not work as planned, so adjustments and refinements are required. And two, new solution ideas surface, ideas no one had considered. They believe that the combination of using their hands and eyes while shaping and assembling materials triggers an untapped area in their subconscious mind. Furthermore, since ideas evolve as pictures in the mind, sketching, drawing, and model making seem to enhance the forming of these pictures.

Two additional advantages of cardboard models are the relative low cast (compared to prototypes) and the fact that the model may be used later in a design review.

The *solution ideas* which have been expanded and prepared for further analyses can now be considered as *solution alternatives*.

Step 7 Compare Solution Alternatives to Constraints and Specifications.

a. *Iterate.* Review Phases I and II for possible additions or deletions to the constraints and specifications. New information may have been discovered. The client may want changes made.

b. *Persevere.* Be cognizant of a *morale* problem that may occur at about this time in the design process. Imagine that a group, or an individual, has started a project with high hopes and confidence that solution can be found. However, even though the first five or six STEPS have been conscientiously done, no real viable solution ideas have been generated. The individual or all the participants find themselves at a low point in the *morale* curve shown in *Figure 24-12*. They discover that there is more to the problem than they thought, and useful information is difficult to find. They become discouraged. However, experienced problem solvers know that perseverance will get them through the slump, and that satisfactory results eventually will follow. They know that toward the end of Phase IV their morale and self-confidence for attacking problems will be higher than when they started.

c. *Qualitative Evaluation.* Compare solution alternatives with what-has-been-done-before. If a solution alternative is similar to existing designs, either within the company or in industry, the feasibility of it is relatively easy to determine.

Use a *range* technique for evaluating preliminary alternatives. Intuition, experience, and judgment help to determine if the size or thickness of a member is within a reasonable range. For example, most students could evaluate design alternatives for home-use ladders, based on their own experiences. Similar *range* approaches have names like *back-of-the-envelope calculations* and *order-of-magnitude analyses.* Interestingly, all of these approaches are useful in designing as well as in evaluating.

d. *Quantitative Evaluations.* Build a *prototype* of part or all of a solution alternative. For example, a student group had the problem of compressing a 2' x 2' x 4' bale of hay into a 2' x 2' x 2' cube so that their farmer client could ship more hay per shipload from Seattle, Washington, to a port in northern Alaska. The hay compressor was to be part of a complete hay-baler machine. One key factor that they needed for their design was the force required to compress a bale to half size. So they made a crude, but effective, set of rectangular welded-steel boxes, one to telescope within the other, to hold a bale in a testing machine that could generate up to 60,000 pounds of compressive force. See *Figure 24-13* for a conceptual sketch showing the two containers and the bale of hay. The students learned that to compress the bale to half size required 14,000 pounds of force. A hydraulic cylinder was selected to provide the force.

Another type of prototype building is a process called *stereolithography*. Actual solid models are created layer by layer by an electronic-laser process in a special liquid solution. A focused beam causes the liquid to solidify over exact areas at accurate locations. It adheres to the previous layer, until a complete component is built up.

Two alternatives to actually building a prototype of a proposed design can be done on a computer: wire frame drawing and solid modeling.

Perspective, wire frame drawings (line drawings in perspective of objects, no hidden lines) on expensive computer systems that can rotate views and zoom into or out of buildings, or marine vessels, are used by architects and marine engineers to visualize what the internal structure would look like.

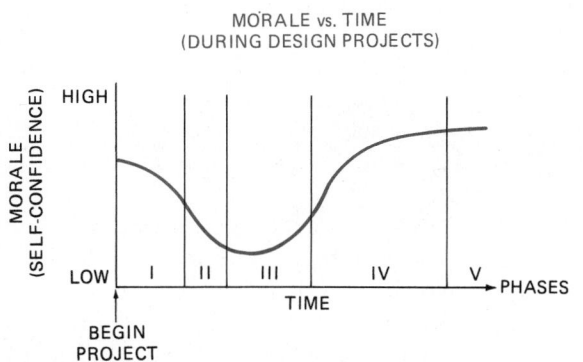

Figure 24-12 Morale curve

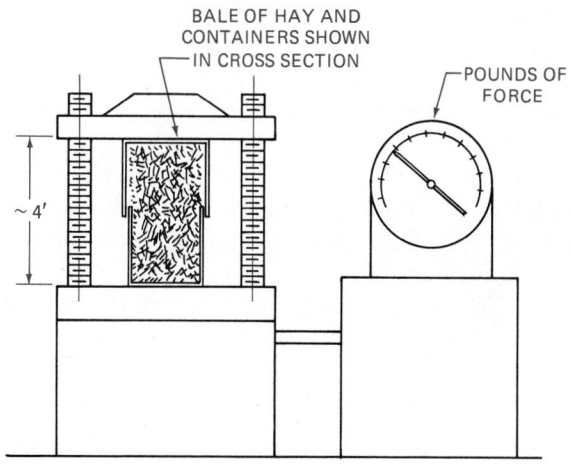

Figure 24-13 Prototype test

Solid modeling on a computer screen can be helpful in evaluating a design for shape, structure, and interferences. Also, the solid model can be used as the basis for a finite element analysis for checking the strength of an object.

Consider the *feasibility* of each alternative by asking questions such as: How much will it cost? Can the shop make it? Are engineering materials available? Are finished components available? Are sufficient labor hours available? Can the delivery date be met?

Prepare a *decision table* to systematically compare each solution alternative with every other alternative as to how well each one satisfies the constraints and specifications from the definition of the problem in STEP 3.

An example *decision table* is shown in *Figure 24-14*. Although the example is simple, and was made up for illustration purposes, the procedure is the same for more complex analyses. The example is based on the question to a class, "Which alternative for transportation to and from school for a distance of five miles (10 miles round trip) is the best one?" The class was asked to create a *decision table* for a quantitative evaluation of four transportation alternatives, walk, bike, bus, and car. First, a list of criteria was generated and tabulated in the left-hand column. Next, each criterion was compared to every other criterion and was given a *weighting factor* for its relative importance according to the students.

The four alternatives then were systematically compared against each other for each criterion. First, the expense criterion was considered. Walking was the least expensive so it rated a score of 10 as compared to a car, which was given a 4. Biking and bussing scored 7 and 6, respectively.

After the four alternatives were compared in relation to each criterion, products of the weighting factors and the scores were totaled for each alternative. Walking scored the highest; therefore, walking would be the alternative to investigate and develop further. The bus and car are close contenders and should also be looked over once more.

To evaluate solution alternatives in a design project, list the *constraints* and *specifications* in the left-hand column and give them weighting factors. Then list the alternatives across the top. Compare and rate every alternative with the others for each constraint and specification. Finally, compute the products of the weighting factors and the alterative scores, total the products for each alternative, and select the alternatives with the higher totals for further study.

Finally, select the best one, realizing that it is the best for the time available, the information collected, and the judgment of the group. Even though it is difficult to do, a *decision* must be made so that the project stays on schedule.

Schedule a *design review* next.

Step 8   Conduct a Design Review.

a. *In-House Design Reviews.* Prepare drawings, transparencies, slides, models and appropriate documents to

DECISION TABLE					
Criteria	Weighting Factor	Walk	Bike	Bus	Car
$	10	10 100	7 70	6 60	4 40
Protection from weather	7	3 21	2 14	7 49	8 56
Time spent	6	2 12	5 30	4 24	6 36
Safety	3	7 21	6 18	9 27	8 24
Reliability	8	9 72	8 64	6 48	7 56
		226	196	208	212

Figure 24-14   Decision table

effectively and efficiently communicate the design solution to the attendees.

In a *classroom*, present sufficient information so that the students in other groups will ask thoughtful questions and make helpful comments. The goal of the review is to benefit from the collective expertise of classmates, and to be sure that all facets of the design have been considered.

The instructor may be the only non-student present, as clients usually do not attend. Also, the instructor may schedule more than one design review throughout the project or may have conferences with each group.

During the *design review* have someone record the questions and comments for modifying or reassessing the solution soon afterwards.

An in-house *design review* in *industry* is more comprehensive because, in addition to technical personnel, representatives from accounting, finance, sales, management, manufacturing, shipping, and installation may be present. Furthermore, the design solution discussed may involve a variety of products or an entire complex system.

b. *Outside Design Reviews.* An *outside design review* provides information to the client at a location selected by the client. Often, the main objective of an *outside design review* is to "sell" ideas to a client, whether the client is a private citizen, a private company, or a governmental organization. Companies and governmental organizations publish requests for bids or proposals on a variety of projects, such as aircraft, spacecraft, dams, highways, buildings, chemical processes, electronic devices, electrical system, power plants, vehicles, military devices, ships, submarines, and numerous products. The bidding individual or proposing organization prepares an *outside design review*, and if the bid or proposal is accepted and a contract is signed, the bidder starts a new *design process* in more detail than the one leading to the *outside review*.

Students may give *outside design reviews* to a client as progress reports; however, most classroom projects require only *in-house design reviews* that lead to the final solution.

Step 9   Refine the Final Solution.   Review the questions, comments, and suggestions recorded during the in-house review and decide which ideas to incorporate in design changes or additions. Revise previous steps if necessary; however, time constraints dictate that the next phase should be started soon.

Phase IV Document, Specify, and Communicate the Solution

In classroom projects, Phase IV requires about one-half of the time available. Moreover, most classroom projects end with Phase IV because of limited time, limited shop facilities, and limited funds.

In industry, Phase IV and Phase V overlap, and together they consume approximately one-half of the time available.

Step 10   Prepare Drawings and a Written Report. Use the informal development layout drawings for preparing the formal *technical detail* and *assembly* drawings in standard views, lines, dimensions, and symbols. Only those parts which require shop processes are detailed in *technical drawings*. Parts—such as fasteners, keys, and pins—that do not require shop processes only appear in the parts list of the assembly drawings. Purchased components such as motors, switches, and bearings appear in assembly drawings in the parts list and are usually drawn in outline form.

Four examples of *technical detail* drawings of components to be made in a shop are shown in *Figure 24-15* and *Figures 24-16 A*, *B*, and *C*. Figure 24-15 is a subassembly of structural steel parts, welded together for use in the anchor bolt puller. Figures 24-16 A, B, and C show several individual components. *Figure 24-17A* is an *assembly* drawing with a parts list (*Figure 24-17B*) for the complete anchor bolt puller. Note, that the subassembly from Figure 24-15 is given only one part number in the assembly drawing parts list.

Technical detail and *assembly* drawings in industry are similar to the anchor bolt puller drawings, but with several differences. Complex drawing-numbering systems tell the user which ones are details, where individual components are used, which ones are assemblies, and which job or contract used the drawings. These systems are as varied as the number and size of industries and organizations. (Refer to Chapter 17) Usually the drawings are on larger sheets. Also, each individual part needing to be processed in a shop is detailed separately to accommodate accounting procedures and to merge into the flow of information within the company.

Prepare a *written report* for the client. Although each client's needs in a report differ, the parts of reports are similar and all the parts should be included. The order is arranged to suit the information to be communicated.

Use the following discussion as a guide for preparing professional, typed project-reports to clients.

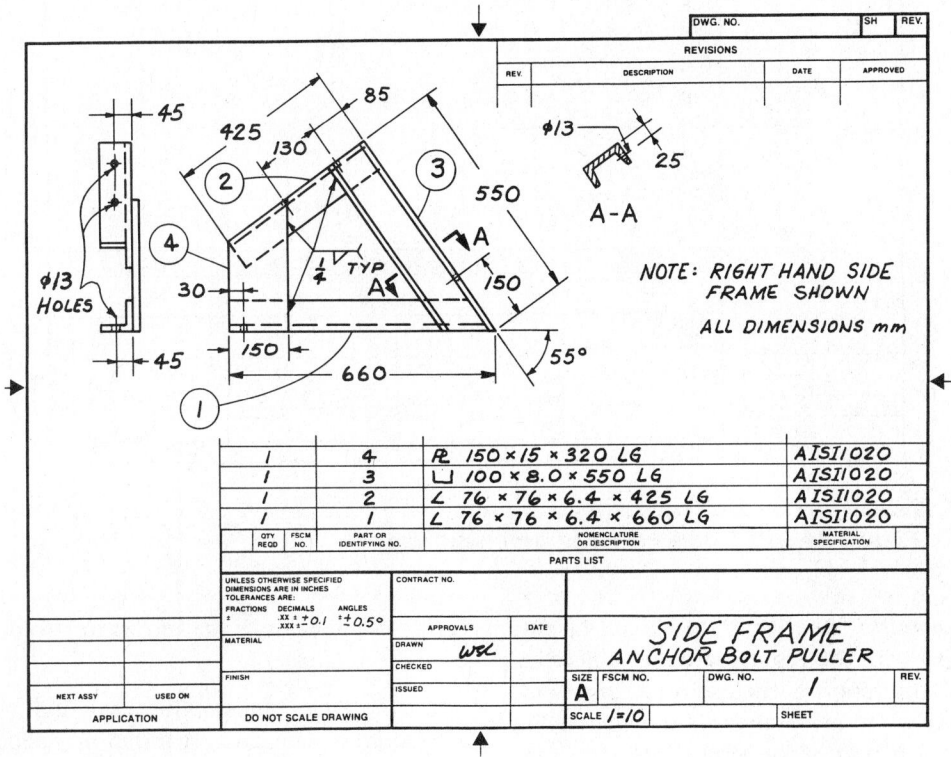

Figure 24-15   Technical detail drawing of sub-assembly

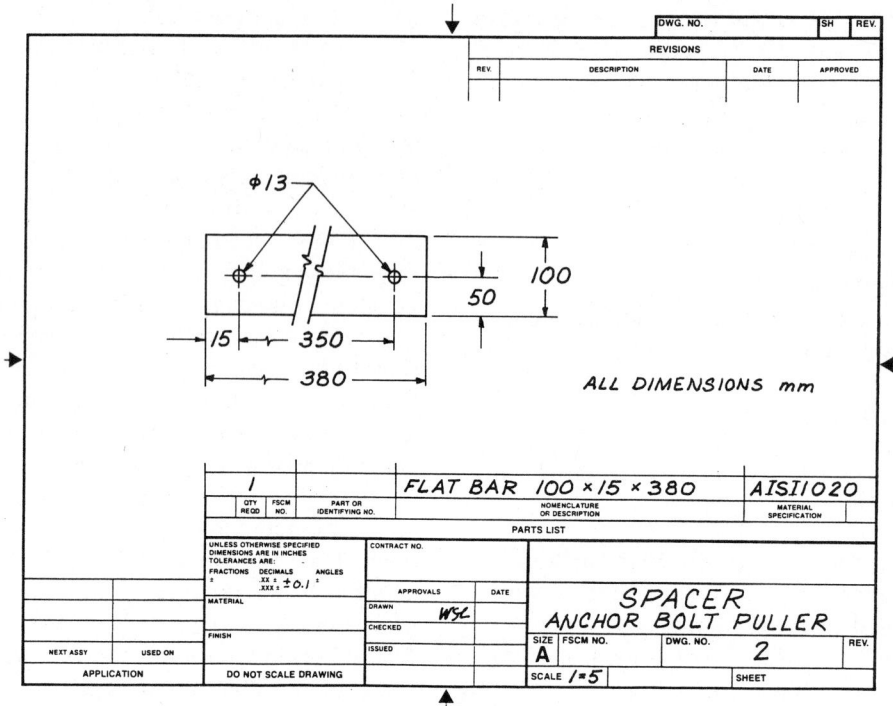

Figure 24-16A Technical detail drawing of a spacer

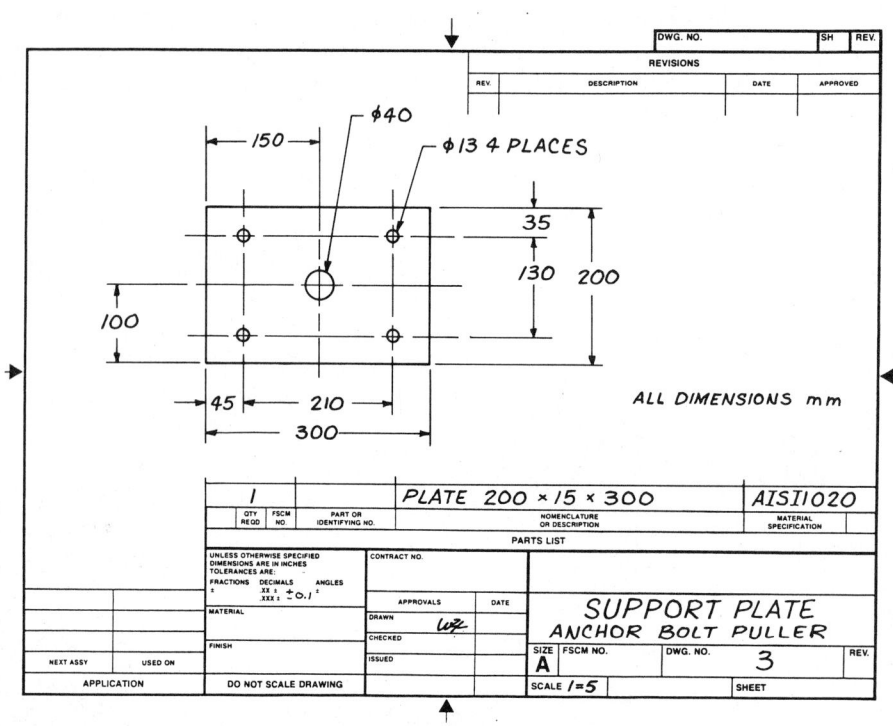

Figure 24-16B Technical detail drawing of a support plate

a. *Letter of Transmittal.* A letter to the client, separate from but enclosed with the report, says the project has been completed, the report is enclosed, and that working with the client on future projects is anticipated. Fee statements may be a part of the letter. Include phone number(s) where the client can call for questions, and an address for mailing fees.

b. *Cover.* Enclose the report in a folder, unless otherwise directed.

c. *Title Page.* Include, title, by whom written, to whom submitted, date, and organization (college or company).

d. *Abstract.* Include one or two sentences about the content of the report to capture a potential reader's

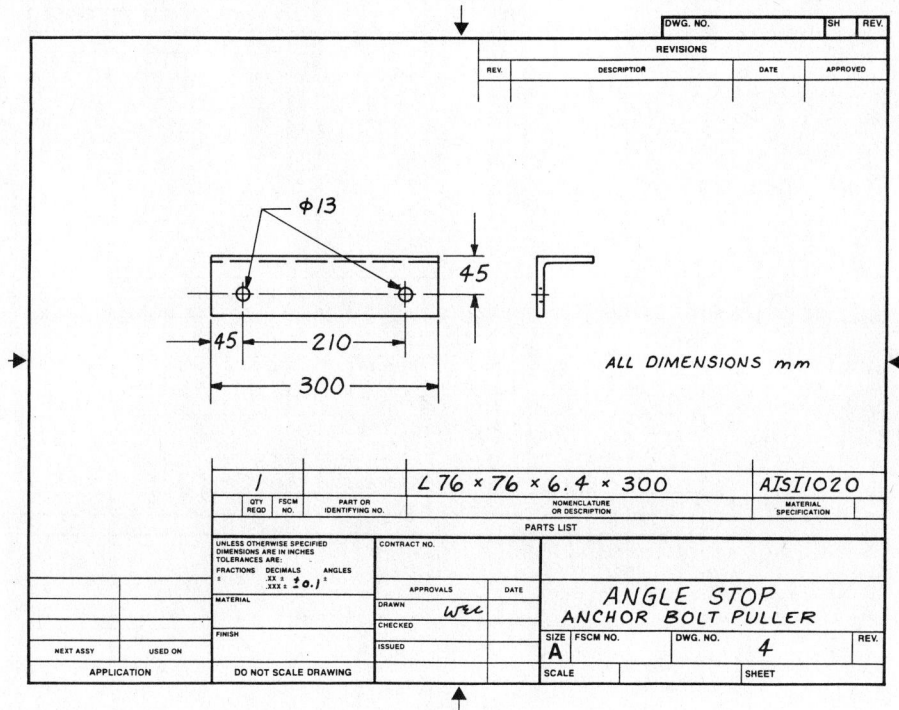

Figure 24-16C   Technical detail drawing of an angle stop

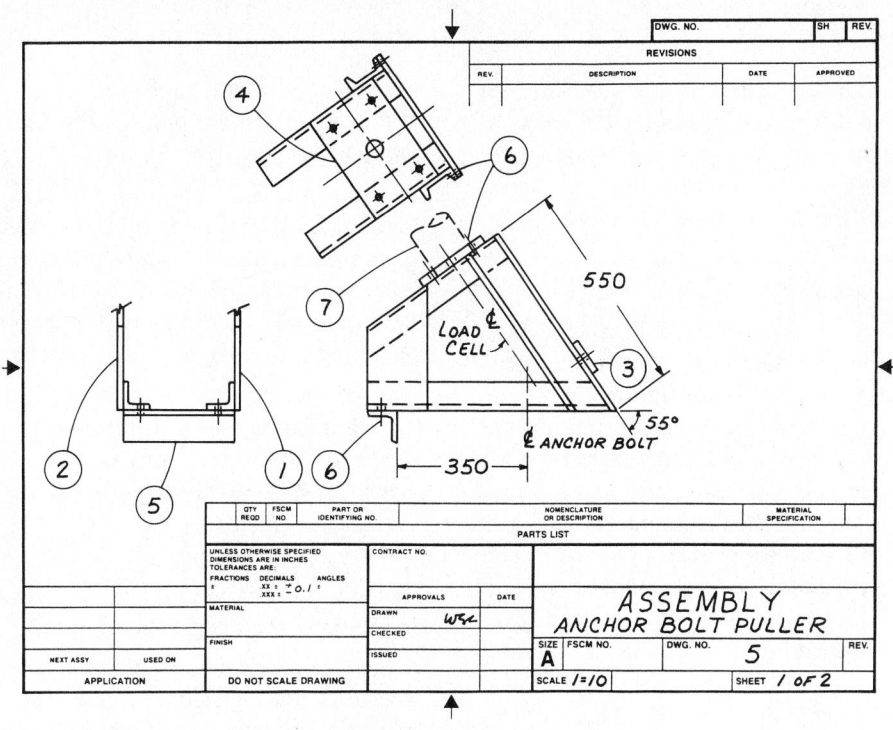

Figure 24-17A   Assembly drawing of anchor bolt puller frame

interest. For product designs, describe the product, specify its performance, and give the cost or price.

e. *Table of Contents.* List the main headings and the subheadings, with appropriate page numbers. List the illustrations, drawings, and tables, plus an appendix if used.

f. *Introduction.* Describe briefly—what, who, when, where, and why—the origin of the project/problem, because even though the client knows this information, other members of the client's organization may read the report.

g. *Solution Description.* A *written* portion should use a liberal number of headings and subheadings, and should

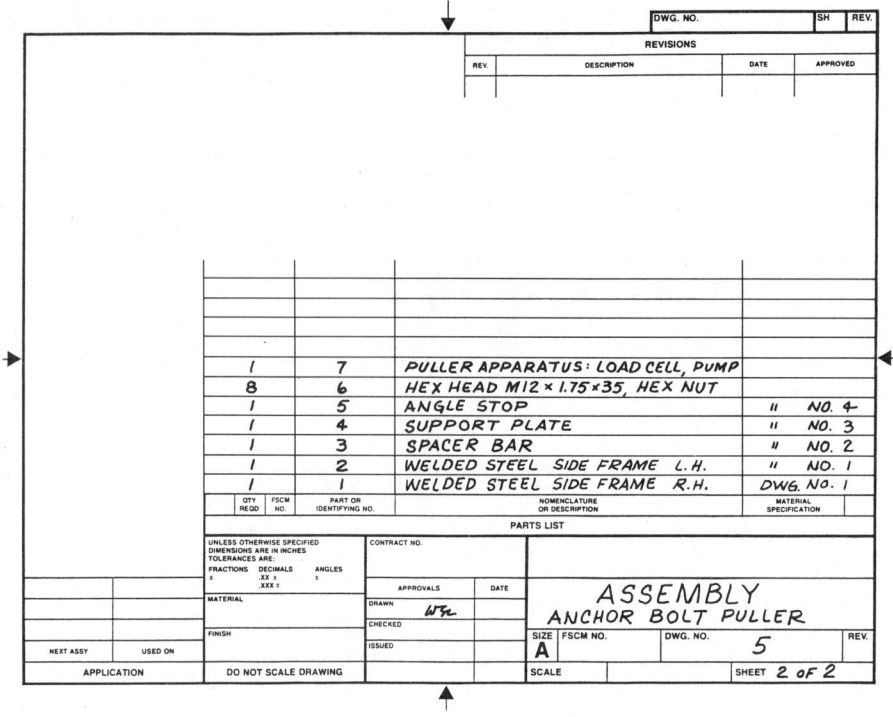

The parts list table reads:

QTY REQD	FSCM NO.	PART OR IDENTIFYING NO.	NOMENCLATURE OR DESCRIPTION	MATERIAL SPECIFICATION
1		7	PULLER APPARATUS: LOAD CELL, PUMP	
8		6	HEX HEAD M12 × 1.75 × 35, HEX NUT	
1		5	ANGLE STOP	" NO. 4
1		4	SUPPORT PLATE	" NO. 3
1		3	SPACER BAR	" NO. 2
1		2	WELDED STEEL SIDE FRAME L.H.	" NO. 1
1		1	WELDED STEEL SIDE FRAME R.H.	DWG. No. 1

PARTS LIST

ASSEMBLY
ANCHOR BOLT PULLER

SIZE **A** FSCM NO. DWG. NO. **5** REV.
SCALE SHEET 2 OF 2

Figure 24-17B   Parts list for anchor bolt puller

---

describe the final solution, including the performance specifications and the design specifications. List the constraints, costs, materials, and special features. Also include the delivery date. Refer to illustrations, figures, charts, graphs, and drawings for efficient communication.

h. *Analyses.* Summarize the major assumptions and calculations. Place detailed analyses in an appendix.

i. *Costs.* For a classroom project, which usually ends with Phase IV, list the basic direct costs that the client would have to outlay to have the design made. These costs could include, materials, shop labor, and a number of purchased components—such as fasteners, pins, shafts, switches, controls, electronic circuits, motors, and similar hardware. Even if the classroom project continues through Phase V and a prototype is built, the client will most likely be required to reimburse the students only for their actual expenses incurred. The excellent learning experience is payment for the students' labor!

In industry, only the selling price would be given in the report, because the basis for arriving at the selling price (labor rates, material costs, overhead, sales, and profit margin) would be proprietary information.

j. *Drawings.* Include the technical assembly drawings in the main report, and the technical detail drawings in an appendix. *Pictorial* drawings (isometric, oblique, or perspective) may also be appropriate, depending upon the client's technical experience and background.

k. *Illustrations, Figures, Charts,* and *Graphs.* Check the illustrations, figures, charts, and graphs for com-

plete information. Each one should have a legend. In addition, charts and graphs should have clearly marked and labeled axes.

l. *Alternatives.* Convince the client that a thorough investigation was done by the design group by including a brief summary of the alternatives that were considered but discarded. Use partially dimensioned sketches, drawings, and pictorials with appropriate comments as to why the alternatives were not selected.

m. *Recommendations.* Discuss briefly any recommended further developments or investigations that relate to the project. For example, one of the discarded alternatives may have some potential and should be investigated further.

n. *References.* List references and the specific pages that were used. Number the references, and use these numbers in the text of the report.

o. *Appendices.* Place information in the appendices that does not logically fit in the main report. Such information includes extensive calculations, computer printouts (separated, of course), vendor brochures, catalogs, machine copies of referenced material, and technical detail drawings.

In industry, when a company or organization wants to "sell" its expertise to clients it includes a roster of individuals, indicating each individual's technical skills, education, and experience. A summary of projects completed by the company may be included.

Selected parts of a final report prepared by students are shown in *Figures 24-18 through 24-25*.

Date 12-6

```
                    Department of Mechanical Engineering
                    (School)
                    (City, State, Zip Code)


Dr.
University Hospital
(City, State, Zip Code)


Dear Dr.,
    This letter is to inform you that we have completed our design for
a CPM device to be used in your research project. The enclosed report
details the design, giving complete drawings, a cost analysis, and
future design considerations. The design is now ready for the
prototype stage.

    We are pleased to have had the opportunity to work on this project
for you, and we hope we will be able to work with you more in the
future.


                              Sincerely,


                              W. Peterson

                              K. Anderson

                              M. Bui

                              A. Hamamoto
```

Figure 24-18   Letter of transmittal

```
DESIGN PROJECT: TEMPOROMANDIBULAR JOINT FLEXOR


SUBMITTED TO: DOCTOR

DEPARTMENT OF ORTHODONTICS
UNIVERSITY MEDICAL SCHOOL
(CITY, STATE, ZIP CODE)
Date 12-6



BY: W. PETERSON
    K. ANDERSON
    M. BUI
    A. HAMAMOTO
```

```
    ABSTRACT

    This report contains a description of a continuous passive
motion (CPM) device to exercise the temporomandibular joint (TMJ).
Briefly, the design incorporates the use of two compressible bellows
that are connected by an air hose. While one of the bellows is
compressed at a controlled rate, the other bellows expands at a
controlled rate. The system is based on user feedback and provides
for excellent emergency override, which is essential for safe use.
The device costs less than $200.
```

Figure 24-19   Title page and abstract

Written reports in industry include parts similar to the students' report; however, most organizations have an established format for their reports. Furthermore, a number of companies and organizations use computers to store "boiler plate" information that appears in all of their reports. Standard paragraphs and phrases appear at the beginning and at the end of their reports–hence, the "wrap-around" image of a boiler. New material fills the blanks.

Rewrite the report text at least two (2) times, whether a classroom report is being prepared or one in industry!! Check for completeness. Check grammar, punctuation, and style. Avoid *clichés*. Eliminate unnecessary words.

Figure 24-20   Table of contents

INTRODUCTION

This project stems from a graduate research project in the department of orthodontics. Studies have shown that continuous passive motion (CPM) on joints such as knees, elbows, and fingers have proven to be quite successful in alleviating post-operative pain, swelling, and adhesions. It is the Doctor's hopes as well as ours that a similar application of this research on the temporo-mandibular joint will prove to be equally successful. Our task was to develop a device that would work the jaw up and down to a controlled distance in a slow cyclic manner and at the same time provide a maximum in safety. It was also necessary to use feedback switches to control the inflating and deflating of a bellows that opened and closed the jaw of the patient so that any irreg-ularities or overwhelming resistance to CPM by the individual (sneezing, coughing, yawning, etc.) will immediately be sensed by the system and the necessary override would be provided automatically.

Figure 24-21   Introduction

Step 11   Give an Oral Presentation.   Prepare an *oral presentation* of the design solution to the client. The following guidelines for planning a presentation apply to student presentations as well as those done in industry.

a. For a simple, but effective presentation, first tell the audience what is going to be presented. Then, present the material. Concluding, tell the audience what was presented.

b. Use transparencies on an overhead projector in relatively small classrooms (30 persons) and boardrooms. Cardboard borders for mounting transparencies are useful for notes, key words, and phrases.

c. Use slides in larger rooms, lecture halls, and auditoriums.

d. Make projected words and numbers large enough so everyone can read them. A conservative guide is to make the letters on the screen as high as 1/250 of the distance to the person the farthest away from the screen. Thus a 25-meter distance requires letters on the screen to be 0.1 meter

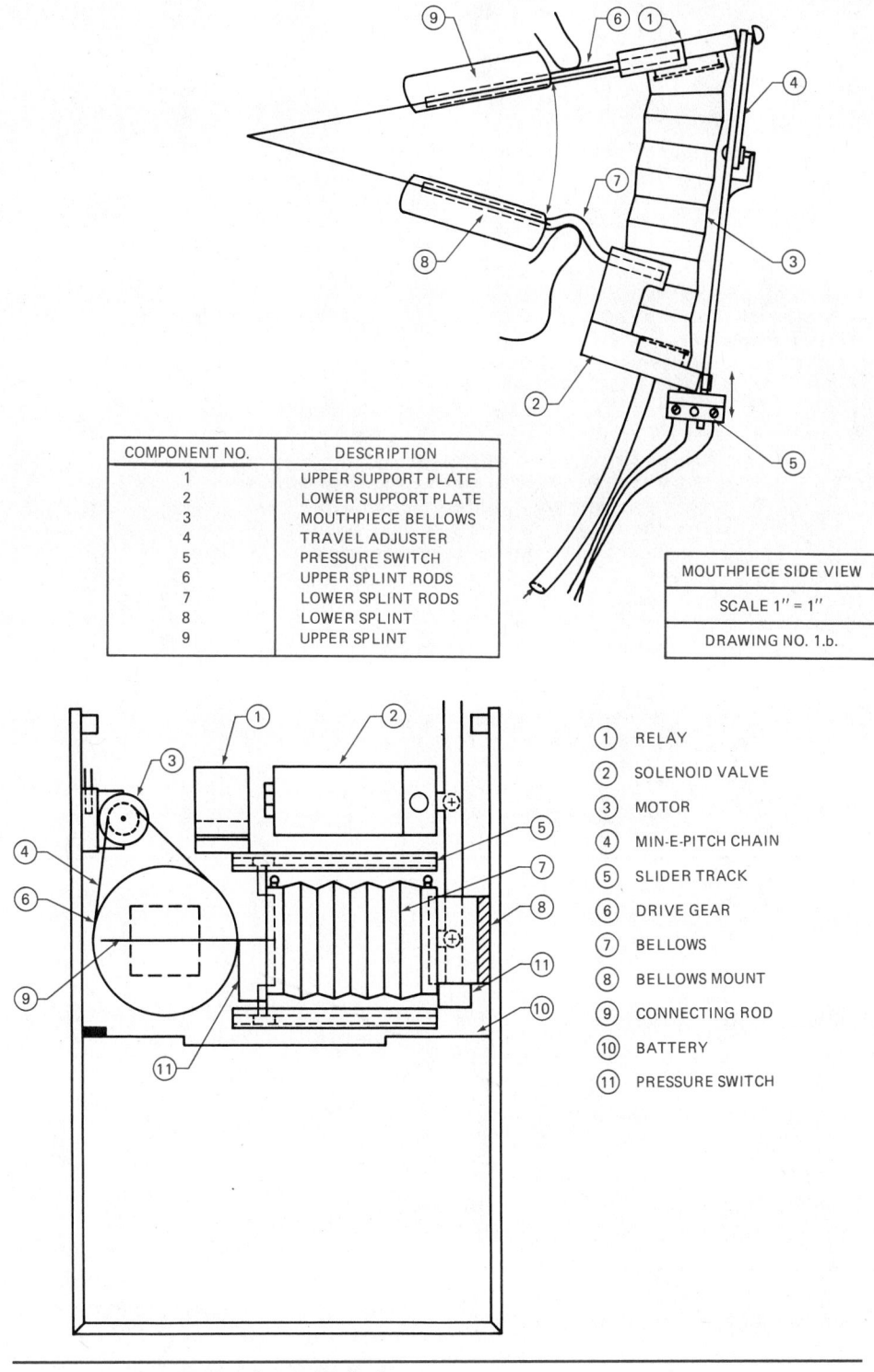

COMPONENT NO.	DESCRIPTION
1	UPPER SUPPORT PLATE
2	LOWER SUPPORT PLATE
3	MOUTHPIECE BELLOWS
4	TRAVEL ADJUSTER
5	PRESSURE SWITCH
6	UPPER SPLINT RODS
7	LOWER SPLINT RODS
8	LOWER SPLINT
9	UPPER SPLINT

MOUTHPIECE SIDE VIEW

SCALE 1″ = 1″

DRAWING NO. 1.b.

① RELAY
② SOLENOID VALVE
③ MOTOR
④ MIN-E-PITCH CHAIN
⑤ SLIDER TRACK
⑥ DRIVE GEAR
⑦ BELLOWS
⑧ BELLOWS MOUNT
⑨ CONNECTING ROD
⑩ BATTERY
⑪ PRESSURE SWITCH

Figure 24-22 Power pack assembly sketch

high. Typed material reproduced on transparencies is usually unsatisfactory.

e. Restrict the number of different ideas on a visual to five or less. If more are on the transparency, distribute a copy to the audience beforehand.

f. Use a pointer.

g. Use models large enough to be seen easily.

h. PRACTICE at least once. One or more persons can do the presentation.

i. Allow time for questions.

In industry, a final *oral presentation* made to a client at the end of Phase IV would be done by a consulting-type firm whose final "product" consists of engineering drawings and a written report.

```
ALTERNATIVES CONSIDERED AND DISCARDED

1. Pneumatic Cylinders as Actuators
   Advantages:
                ..small in size
   Disadvantages:
                ..limited control of expansion rate
                ..higher working pressures, noise
                ..difficult to find stroke and size in standard
                  products
                ..lubrication may be a problem
                ..allows minimal user resistance

2. Source of Compressed Air: Bottle and Intricate Valving
   Advantages:
                ..power pack simple and light-weight
                ..no batteries
                ..volume of air easily varied
   Disadvantages:
                ..complicated valving
                ..supplying air bottles cumbersome
                ..compressor needed to charge bottles
                ..noise
                ..difficult to get emergency override

3. Motor Driven Cam and Linkage
   Advantages:
                ..simple to construct
   Disadvantages:
                ..difficult to interrupt
                ..excessive weight
                ..bulky, physically restrictive
                ..little adjustment
                ..little freedom of jaw movement (patients need rota-
                  tion, translation, and lateral movements)
```

Figure 24-23   Alternatives discarded

Phase V Develop, Manufacture, and Deliver

Arrange for the necessary paperwork, drawings, materials manufacturing processes, personnel, and funds to develop the device or system to meet the client's requirements.

These requirements may range from modifying an existing device to producing a "state-of-the-art" system. Thousands of different combinations of requirements occur in industry, so describing them all here is impractical. Consequently, a more appropriate tack for this chapter is to present the steps that are common to most industries.

Step 12   Prepare Technical Detail Drawings of Each Component to Be Manufactured.   These drawings "get-the-wheels-turning" within a company. A typical drawing enclosed in a clear-plastic folder may follow the component through a manufacturing area from casting or forging, to machining, to heat treating, to finishing, to assembly and delivery. Additional areas could include checking, quality control, subassembly, and temporary storage.

Construct and test a prototype of the final solution if a prototype was scheduled. Then, revise any part of the design as needed.

Step 13   Schedule Assembly, Final Testing, Packaging, and Delivery.   Plan the final processes prior to sending the product to the client. Compare appropriate subassemblies and complete assemblies with the specifications, particularly performance specifications. For example, an electronic device may be tested in a variety of environments before being released for shipment.

Design packaging and shipping containers so that the product will survive shipping without damage. Some devices and assemblies must endure more severe loading enroute than after being installed—for example, large frames, large cylindrical-shaped housings, and pressure vessels.

Write maintenance and service manuals to help prolong the life of the product. Provide instructions for proper and safe use of the product.

Step 14   Keep Records.   Document and file all aspects of the product's development and use. Record the sources and quality of materials used. File the detail and assembly drawings. Summarize the manufacturing processes, and prototype testing results. Add feedback comments from the client. Request cost, accounting, purchasing and related records from appropriate personnel.

FINAL RECOMMENDATIONS

ASSUMPTIONS

It is important to note that any state-of-the-art design is subject to change since so many variables, having no cut and dry answers, must be dealt with, and values must sometimes be assumed using good engineering judgment. Some of the variables in this design project include average mouth dimensions, forces to open and close the jaw, allowable weights for the mouthpiece and powerpack, influence of an oversize solenoid valve on battery life, and so on.

PROTOTYPE

The presence of these types of variables strongly implies a need for prototype testing to see if the actual design concept is valid. A second type of testing might be called human factor testing, which involves sizing and comfort factors. Are the dimensions on the mouthpiece satisfactory to accommodate an average-size person? Is the size of the powerpack cumbersome? Are weights unbearable? Are jaw movements restricted?

SOLENOID

It was previously mentioned that the specific solenoid valve was oversized. At this point in time the valve is the smallest one made. Because of its power consumption, which is excessive (7 watts max), the battery must be extraordinarily large. It is hoped that eventually a smaller solenoid valve could be purchased or even designed to be much more economical. This could drastically increase battery life and decrease battery size, ultimately decreasing the size of the powerpack.

CNC LATHE

From a manufacturing viewpoint it is recommended that the prototype have all its plastic parts machined on the CNC lathe, since molds are expensive as used in injection molding. However, if the device is ever mass produced, it is recommended that injection molding be the way to go, since production would be much faster, decreasing labor costs per unit.

Figure 24-24   Final recommendations

APPENDIX NO. 1

DELRIN

The material chosen to use for the mouthpiece is Delrin. Delrin is a trade name for an acetal homopolymer. It was chosen because it has many desirable characteristics. Some of the advantages of Delrin are:

1) Good chemical resistance, especially to organic compounds

2) Desired regulatory status:
   a) FDA approved for intermittent contact with aqueous or alcohol-based foods
   b) USDA approved for meat and blood

3) Can be machined like soft brass

4) Can be welded or adhesively bonded

5) Forms good mechanical joints

6) Has good structural integrity

7) Tensile strength (at break) 9700 psi

8) Yield strength 9500-1200 psi

9) High impact resistance

10) Rockwell hardness 92-94

Figure 24-25A   Appendix: Material

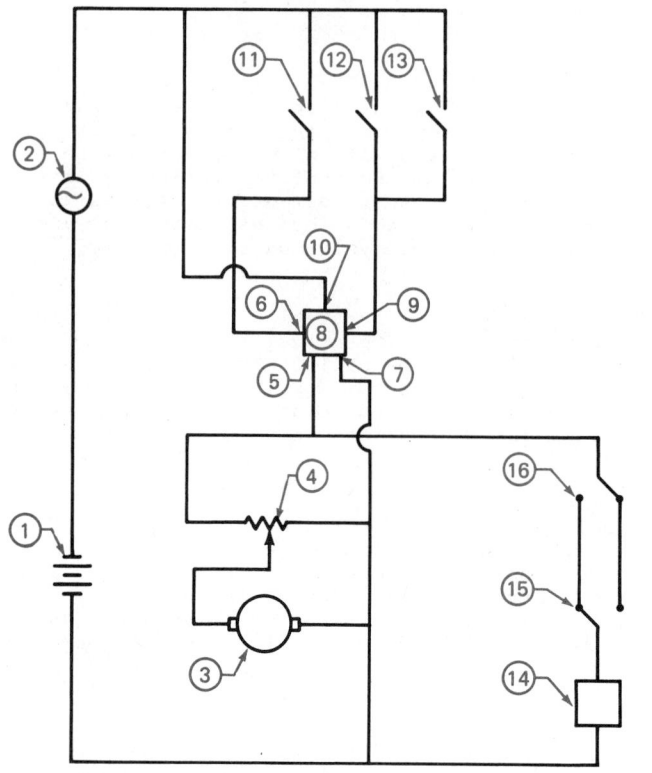

Figure 24-25B   Appendix: Electrical schematic

QUANTITY	PART NO.	DESCRIPTION	COST
1	NP8-6	Yuasa Battery	$29.55
1	BC-3B	Yuasa Charger	$5.39
1	B11DK 040	Skinner Solenoid Valve	$33.00
1	1615 900-1	Mini Motor (9rpm) 6v DC	$50.00
1	Rel CY6-5DC-SCO	Relay (Lafayette)	$4.88
2	80511-E	SPST Momentary switch (Lafayette)	$2.94
1	ZC-10B	Fuse Holder (Lafayette)	$1.20
1	3MP10A-10	10-Teeth pinion	$6.34
1	3MP10A-40	40-Teeth gear	$8.59
1	3CCF-50-E	Min-E-Pitch Chain 55 link	$3.24
1	E-62-10K	Unimax Switch	$3.08
1	4PTN-50K	Potentiometer 50 Kohm	$1.98
1	CH3-R10	Charging Jack	$2.99
		Miscellaneous Parts, Tubing, Fittings, Screws	$30.00
		TOTAL	$183.18

Figure 24-25C   Appendix: Costs

These records are useful for future use if the client wants replacement parts or another complete product. Also, other customers may purchase the product.

But more importantly, keep good records and documentation in case the product becomes involved in a product liability suit.

The *design process* discussed in the preceding pages may be used for a variety of projects, from relatively short ones to large, complex ones. Adapt the *process* to fit the project.

Two types of projects for students who are learning the *design process* are discussed next.

Design Projects: Routine and Non-Routine

Routine Design Problems

Routine design projects and problems are used to introduce new members of an organization to the various people in the company and to the routes design information needs to flow through. Beginning engineers, designers, technologists, and drafters need to know how jobs are initiated, documented, processed and completed. An effective way to start these new employees is to assign *routine* projects and problems to them.

These first assignments usually are assigned to an individual and are characterized as follows:

  a. The problem may be a modification to an existing product line, a change in the company's plant facilities, or just a drawing change.
  b. The problem difficulty matches the individual's abilities and skills.
  c. Information is communicated in a memorandum in a typical industrial format, such as an ECO (engineering change order) or an EDO (engineering design order).
  d. Requirements are clearly stated and the information is nearly complete; however, some new data may be needed.
  e. An analysis may be required.
  f. For guidance, there are similar designs, knowledgeable colleagues, the company library, and vendor catalogs.
  g. Technical drawings are required; they become part of the documentation of the modification or change.
  h. Labor and materials costs are always present.
  i. A condensed *design process* is followed; however, all the essential steps are followed, including *planning*, keeping a *log*, keeping a *file*, understanding the problem, *synthesis*, *analysis*, *comparing* alternatives, *reviewing* designs with supervisors, *documenting* the project in *drawings* and *text*, *manufacturing*, *assembly* and *installation*.

An example first assignment that a newly graduated engineering student received in an EDO was to design a support for a tank containing a liquid for a paper-making process. But before resolving the technical parts of the support, the new engineer, escorted by the chief engineer, was introduced to personnel in drafting, purchasing, the shop, and the paper-making area where the tank was to be installed. Further introductions included individuals in general supplies, copy service, and even those who made coffee.

The technical challenges involved load calculation, sizing beams and columns, and finally, making a *design development layout* drawing. The chief engineer was consulted before the layout was handed to a drafter. The drafter questioned a few items and then produced *technical detail drawings* plus an *assembly* drawing.

The new engineer checked, approved, and signed the drawings. Then he followed them through the shop and ultimately used the drawings at the installation site.

The last last step was to document the project. The drawings and vendor information were filed. A short written report summarized the need for the tank support, and a copy of the EDO was included. Cost information from the accounting department completed the file.

Non-Routine Design Problems

Non-routine design problems are assigned to experienced engineers, designers, technologists, and drafters—their complexity and level of difficulty in direct proportion to the experience, skills, and ability of the recipients. Usually a team of technical personnel attack *non-routine design* problems.

Non-routine design problems are characterized as follows:

  a. The problem originates outside the company, and it stems from the needs of a client who may be an individual, another company, or a governmental organization.
  b. A team of individuals with diverse backgrounds or a group with similar backgrounds is formed, depending on the experience, skills, and abilities required.
  c. The problem, not clearly defined, must be researched thoroughly so that constraints and specifications can be written.
  d. A complete *design process*, phases and steps, is required, from the initial planning to the final delivery and installation. More alternatives are generated than for routine problems, and, therefore, more decisions are made.
  e. The projects require a longer time span than routine projects.
  f. Detailed planning is required for personnel assignments and hours, materials procuring, budget allocations, production processes, and delivery/installation.
  g. More engineering *trade-offs* occur. A *trade-off* is a compromise, usually between costs and materials. For example, an expensive material (titanium) may cost more but weigh less than a less costly material (aluminum) that weighs more.

Since *non-routine design* problems vary as widely as the number and size of the sources (individuals to organizations), there are no typical non-routine problems. However, the manner in which they are attacked—*the design process*—is typical. Therefore, follow the phases and steps presented in this chapter to obtain an insight into the efficacy of this process. Then design you own *design process* as needed.

Design Hints

Use *design hints* during all phases of the *design process* as extensions to the design hints already included. These *design hints* help in initial designing and become checklists as the project progresses. Keep a file during design problem experiences to expand these lists, and remember: CARRY A NOTEBOOK AT ALL TIMES!!!

Attribute Listing for Checking Drawings. Use the technique of *attribute listing* as a systematic guide for checking the completeness of technical detail drawings. For example, ask the following questions:

a. What is the *function*? The function dictates the dimensioning plan.

b. What *size* is it? Are all geometrical shapes dimensioned? Are all the features located?

c. What materials are used?

d. Can it be made? How will it be manufactured? What processes will be used?

e. What other facets need to be checked? Surface finish? Weight? Vibration? Safety? Aesthetics?

"Rules of Thumb". Use *"rules of thumb,"* practical guides based on experience, to establish or check "ballpark" (within + or –10%) quantities. Use them as qualitative checks.

a. *Tolerance.* As a beginning problem solver in technical fields, keep in mind that the tighter the tolerance on the dimensions of an object, the more the object costs. Tolerances should only be as tight (small) as needed to ensure the function of the part. Use as a conceptual reference the thickness of the paper this sentence is written on: It is .003″ to .004″ or 0.08 mm to 0.10 mm thick. Most machinists and machines can easily work to .001″, but .0001″ takes more time, and costs more.

b. *Rigidity.* "When in doubt make it stout." This statement is just a reminder that not all designs need to be based on an extensive stress analysis and strength of a material. Rigidity and accurate locations may be more important than strength. For example, a surveyor's transit needs to be rigid to keep the lenses aligned. Instrument panels need to be rigid to support the dials and meters, and to prevent vibration. Another practical application of the statement refers to sizing members when budgets are low. That is, a brief stress analysis producing a conservative size (stout) is less expensive than an extensive analysis and a more exact size.

c. *Murphy's Law.* "If an event can happen, it will." Use this guide to evaluate how a designed part or assembly might behave. A more important application is to ask, "How could someone misuse the part or assembly?"

d. *Weld Size.* Fillet and butt welds constitute approximately 85% of all welds used in industry. And if the thinnest cross-section dimension of the weld is at least as thick as the thinnest member joined, the weld will be as strong as the members, assuming that the weld is done by a skilled welder.

e. *Threads.* The depth of threads for steel threaded fasteners should be equal to the diameter of the fastener. Use 1.5 to 2.0 times the diameter for the depth of threads for non-ferous threaded fasteners, or steel fasteners in non-ferous metals.

f. *Intuitive Ranges.* Early in a design, establish sizes by asking what would be a reasonable range for this size? Establish performance, weights, and function in a similar manner. During an analysis, ask questions such as: Is this a reasonable size? Weight? Function? Performance?

Human Factors. There are published tabulations of body dimensions of average males, females and children. This statistical data may be found in textbooks and handbooks; however, for classroom designs use a classmate or yourself for measurements, since most of the adult population is close to the average size. Use children if needed.

For a design file, observe or measure objects designed for average humans. Typical designs include chairs, tables, desks, computer stations, work counters, door knobs, ladder rungs, switches, handles, and numerous objects encountered daily.

Also, consider and observe the various techniques that are used to "interface" machines with human operators. A variety of sounds such as voices speaking messages, a number of visual devices using a multitude of colors, numerous tactile devices such as clicks during dialing, and different shapes of knobs are all used.

Manufacturing Process Costs. The following sample list is headed by the least costly and ends with the more costly manufacturing process. Costs are also related to the roughness of the surfaces of the materials. The roughness is a result of the manufacturing or production processes.

Sawing	Rough Surfaces
Cutting (torch)	(lower cost)
Drilling	
Turning	
Punching	
Milling	Average Surface Roughness
Reaming	(Average Cost)
Broaching	
Grinding	
Polishing	
Honing	
Lapping	Smooth Surface
	(Higher Cost)

Relative Costs of Materials. Use relative cost numbers as an aid to evaluating the feasibility of a design alternative, and for establishing approximate costs. Mild steel is given a factor of 1.00; all other engineering materials are given relative numbers. Find the current cost of mild steel, and use the relative numbers to establish costs of the other materials. Revise the numbers as needed.

Material	Relative Cost (By Weight)
Mild steel	1.00
Spring steel	2.00
Tool steel	11.00
Cast iron	3.00
Aluminum	6.00
Stainless steel	7.00
Brass	7.00
Copper	7.50
Titanium	130.00
Teflon	67.00
Nylon	19.00
Rubber	4.00
Wood	1.00

Filing System. Create a design file that uses an alphabetical-numerical cross referencing system for efficient retrieval of design information.

a. The alphabetical part includes the subject and their numerical references. For example:

A	Aluminum	111.2
	Angles	112.4
B	Bar Stock	121.1
C	Casting	230
	Ceramics	112.1

b. The numerical file contains the numbered folders in a file cabinet. For example:

100	Engineering Materials
110	Categories
111	Metals
111.1	Ferrous Based
111.2	Aluminum
112	Non-Metals
112.1	Ceramics
112.2	Plastics
120	Shapes
121	Rounds
122	Flats
123	Plate
124	Angles
200	Manufacturing
300	Mechanical Designs
400	Power Transmission
500	Electrical
600	Electronic
700	Hydraulic Power
800	Pneumatic Power
900	Codes and Standards

etc.

Create new categories and subcategories as needed.

Vendor Catalogs. Vendors provide customers with convenient guides to select products and materials from their catalogs.

a. *Machine Elements* such as V-Belt drives, ball and roller bearings, roller chain drives, clutches, brakes, couplings, gearmotors (motor and gear reduction box as one unit), shafting, motors (AC and DC), and controls, all have similar guidelines for selecting them. These guidelines include formulas, tables, and charts to convert most customer applications and environmental conditions to numbers or letters in their catalogs.

For example, assume that the belt conveyor shown in *Figure 24-26* needs a roller chain drive to transmit power from the gearmotor's 200 RPM output shaft to the headshaft of the conveyor. Calculations indicate 14 horsepower to operate the conveyor at 500 FPM (feet per minute).

The roller chain drive, sprockets and chain, can be selected from a vendor's catalog using the tables in *Figures 24-27 A-E*. Figure 24-27A contains general guidelines for selecting and applying roller chain drives. To select a roller chain drive, follow the five steps in Figure 24-27B. The first step determines the service factor to be 1.25, as indicated in Figure 24-27C, assuming heavy duty, not uniformly fed and operating less than 10 hours per day. Step two says that the design horsepower is the calculated horsepower times the service factor, (14) (1.25) = 17.5. The third step is to use the design horsepower in the selection table in Figure 24-27B to determine chain size. Chain size No. 120 with a 17-tooth smaller sprocket will work. The next step (using Figure 24-27D) indicates that an 11-tooth smaller sprocket will theoretically work; however, good practice dictates the use of a 17-tooth sprocket according to comments in Figure 24-27A. Finally, for the fifth step use Figure 24-27E to find the larger sprocket to give a drive

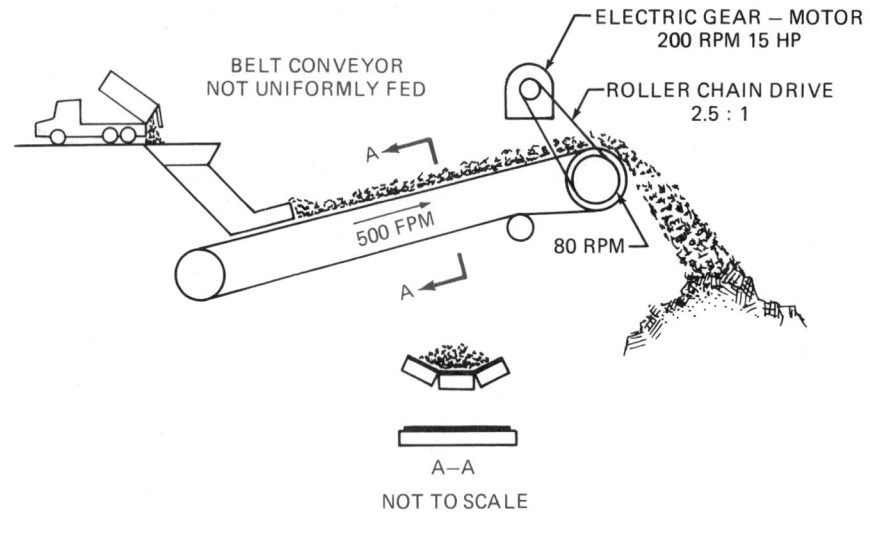

BELT CONVEYOR
NOT UNIFORMLY FED

ELECTRIC GEAR — MOTOR
200 RPM 15 HP

ROLLER CHAIN DRIVE
2.5 : 1

500 FPM

80 RPM

A—A

NOT TO SCALE

Figure 24-26   Belt conveyor system schematic

ROLLER CHAIN DRIVE SELECTION

The following considerations are very important in the selection and application of roller chain drives:

HORSEPOWER RATINGS—This catalog lists Horsepower Ratings for ANSI Series, single pitch, single strand chains No. 25 through No. 160 (and lightweight machinery series No. 41 and No. 43). Ratings are listed for various numbers of teeth and speeds of the smaller sprocket. Ratings for intermediate numbers of teeth or RPM may be determined by interpolation. The ratings reflect a service factor of 1, a chain length of approximately 100 pitches, the use of recommended lubrication methods and a drive arrangement where two aligned sprockets are mounted on parallel horizontal shafts. For maximum service life, sprockets with small numbers of teeth, operating at moderate to high speeds or near the rated horsepower should have hardened teeth. Approximately 15,000 hours of service life at full load operation may be expected under these conditions.

NO. OF TEETH—It is good practice to select a pinion sprocket with no less than 17 Teeth, to assure 120° of chain wrap and minimize overhung load. However, certain conditions, i.e., space limitations, light loads, intermittent duty, etc. will permit the use of smaller pinions.

RATIO—Sprocket ratios should not exceed about 6 to 1 for normal chain life.

HARDENED TEETH—Boston Gear steel sprockets can be hardened. Consult the factory for recommended procedure.

CENTER DISTANCE—The correct center distance is very important. In designing chain drives, it is important that the Center Distance should be long enough to provide at least 120° of chain wrap on the smaller sprocket.

RELATIVE SHAFT LOCATIONS—It is desirable that the line between the two shaft centers be as nearly horizontal as possible. If this line is more than 60° from the horizontal, special precautions should be taken.

Figure 24-27A   Roller chain drive selection considerations *(Courtesy Boston Gear Incom International Inc.)*

ratio of 2.50. A 42-tooth sprocket combined with a 17-tooth gives a ratio of 2.47, which is close enough. The center distance between the two sprockets using a 64-inch long chain will be 16.777 inches.

b. *Engineering materials and shapes* can be selected from vendor catalogs.

   (i) The catalogs published by vendors of materials list materials and their typical uses to aid in their selection. To illustrate, one catalog describes 6061 aluminum as being heat treatable, having good formability (will not crack during bending), having good weldability, and being corrosion resistant. Also,

6061 aluminum is used for structural application, boats, and transportation equipment.

   (ii) Some materials are selected primarily on the basis of their shape; for example, steel and aluminum are rolled into plates, bars, channels, beams, and angles, and are extruded into pipe, square tubes, and rectangular tubes. Square and rectangular shapes are easier to join in fabricating than pipe sections. Pipe sections are equally strong in all directions.

   (iii) Materials may be selected on the basis of their weight, rigidity, and coefficient of expansion due to temperature changes. INVAR is used in instrumentation because of its low

ROLLER CHAIN DRIVE SELECTION (Continued)

A roller chain consists essentially of numerous small bearings operating under high pressures and requires adequate lubrication. There are four basic types of lubrication suggested for chain drives, depending upon the chain speed and the power transmitted. The Horsepower Rating Tables indicate the type of lubrication recommended.

TYPE I — MANUAL LUBRICATION

Manual lubrication is accomplished by applying oil with a brush or spout can to the inside of the chain at the edges of the side plates. Volume and frequency should be determined by periodic inspection.

TYPE II — DRIP LUBRICATION

Oil is directed between link plate edges from a drip lubricator. Only enough oil to keep the chain moist is necessary and a light metal splash guard will keep the floor and surroundings clean.

TYPE III — BATH OR DISC LUBRICATION

With bath lubrication, the lower strand of the chain runs through a sump of oil. The oil level should reach the pitch line of the chain at its lowest point while operating. With disc lubrication, the chain operates above the oil level. The disc picks up oil from the sump and deposits it on the chain, usually by means of a trough. The disc diameter should be such as to produce rim speeds from 600 minimum to 8000 maximum FPM. This type of lubrication requires that the drive be enclosed in an oil tight chain case.

TYPE IV — OIL STREAM LUBRICATION

The lubricant is uaually supplied by a circulating pump capable of supplying the chain drive with a continuous stream of oil. The oil should be applied inside the chain loop evenly across the chain width, and directed at the lower strand. This type of lubrication requires that the drive be enclosed in an oil tight chain case.

Recommended lubricant viscosities for various ambient temperatures are listed in the following table:

Temp. Degrees F.	Lubricant	Temp. Degrees F.	Lubricant
20-40	SAE20	100-120	SAE40
40-100	SAE30	120-140	SAE50

SURROUNDING CONDITIONS — Abrasive, corrosive, or high temperature conditions can shorten chain life. If adverse conditions exist, special precautions should be taken. It may be advisable to use a drive with higher capacity than normal, stainless steel chain, etc.

Roller chain drives may be selected with the following procedure:

a.   From Table #1 of the Application Classification Chart on Page A94. determine the Service Factor.

b.   Multiply the Application HP by the Service Factor to obtain a Design HP. *

c.   The Selection Table below may be used to select an appropriate chain size using a sprocket of 17 teeth or larger.

d.   From the appropriate horsepower rating table (pages B6-B8) determine the minimum size sprocket needed to provide, at the required speed, a rating equal to (or greater than) the Design horsepower.

e.   The tables on pages B9-B11 may then be used to select number of sprocket teeth, shaft center distance and chain length of a drive suitable for the application.

*For Stainless Steel Chains, operating under wet or dry conditions, the Design Horsepower must be multiplied by a Factor (see Table below) for selection purposes.

NOTE: Standard Steel Chains are not recommended for wet or dry applications.

Application Conditions	Factor
Wet (Moisture)	2.0
Dry (Unlubricated)	5.0

Horsepower ratings of Multiple Strand chain may be obtained by multiplying the Single Strand rating by the proper Factor from the following table:

MULTIPLE STRAND RATING FACTORS

Number of Strands	Double	Triple	Quadruple
Rating Factor	1.7	2.5	3.3

*These Horsepower Ratings are based on certain operating conditions, see Page B4.

SELECTION TABLE

RPM of Smaller Sprocket	DESIGN HORSEPOWER												
	1/2	1	1-1/2	2	3	4	5	7-1/2	10	15	20	25	30
	CHAIN NUMBER												
1800	25	25	35	35	35	40	40	40	50	80	60−2	80−2	—
1500	25	25	35	35	35	40	40	40	60	60	80	60−2	80−2
1200	25	35	35	35	40	40	40	50	60	60	60	80	100
1000	25	35	35	35	40	40	40	50	60	60	80	80	80
800	25	35	35	40	40	40	50	50	60	60	80	80	80
700	25	35	35	40	40	50	50	50	60	80	80	80	80
600	35	35	35	40	40	50	50	60	60	80	80	80	100
500	35	35	40	40	50	50	50	60	80	80	80	100	100
400	35	35	40	40	50	50	60	60	80	80	100	100	100
350	35	40	40	40	50	50	60	80	80	80	100	100	100
300	35	40	40	50	50	60	60	80	80	100	100	100	120
250	35	40	40	50	50	60	60	80	80	100	100	120	120
200	35	40	50	50	60	60	80	80	80	100	120	120	120
175	40	40	50	50	60	80	80	80	100	100	120	120	140
150	40	50	50	60	60	80	80	80	100	120	120	120	140
125	40	50	50	60	80	80	80	100	100	120	120	140	140
100	40	50	60	60	80	80	80	100	100	120	140	140	160
80	40	50	60	80	80	80	100	100	120	140	140	160	160
70	50	60	60	80	80	80	100	120	120	140	160	160	
60	50	60	80	80	80	100	100	120	120	140	160		
50	50	60	80	80	80	100	100	120	140	160	160		
40	50	60	80	80	100	100	120	120	140	160			
30	60	80	80	100	100	120	120	140	160				
25	60	80	80	100	120	120	140	140	160				
20	60	80	100	100	120	120	140	160					
15	80	100	100	120	120	140	160						
10	80	100	120	120	140	140							

*Based on 17 Tooth Sprocket. For other numbers of teeth refer to the applicable Horsepower Table, pages B6-B8.

Figure 24-27B   Roller chain drive selection procedure *(Courtesy Boston Gear Incom International Inc.)*

APPLICATION CLASSIFICATION FOR VARIOUS LOADS

Type of Machine To Be Driven	Chart I For all Drives		
	Service Factor		
	Loading		
	Not More Than 15 Mins. In 2 Hrs.	Not More Than 10 Hrs. per Day	More Than 10 Hrs. Per Day
AGITATORS			
Pure Liquid	0.80	1.00	1.25
Semi-Liquids, Variable Density	1.00	1.25	1.50
BLOWERS			
Centrifugal and Vane	0.80	1.00	1.25
Lobe	1.00	1.25	1.50
BREWING AND DISTILLING			
Bottling Machinery	0.80	1.00	1.25
Brew Kettles—Continuous Duty	—	—	1.25
Cookers—Continuous Duty	—	—	1.25
Mash Tubs—Continuous Duty	—	—	1.25
Scale Hopper—Frequent Starts	—	1.25	1.50
CAN FILLING MACHINES	—	1.00	—
CANE KNIVES	—	1.50	—
CAR DUMPERS	—	1.75	—
CAR PULLERS	—	1.25	—
CLARIFIERS	—	1.00	1.25
CLASSIFIERS	—	1.25	1.50
CLAY WORKING MACHINERY			
Brick Press & Briquette Machine	—	1.75	2.00
Extruders and Mixers	1.00	1.25	1.50
COMPRESSORS			
Centrifugal	—	1.00	1.25
Lobe—Reciprocating, Multi-Cycle	—	1.25	1.50
Reciprocating—Single Cycle	—	1.75	2.00
CONVEYORS— UNIFORMLY LOADED & FED			
Apron	—	1.00	1.25
Assembly-Belt—Bucket or Pan	—	1.00	1.25
Chain—Flight	—	1.00	1.25
Oven—Live Roll—Screw	—	1.00	1.25
CONVEYORS—HEAVY DUTY NOT UNIFORMLY FED			
Apron		1.25	1.50
Assembly-Belt—Bucket or Pan	—	1.25	1.50
Chain—Flight	—	1.25	1.50
Live Roll	—	—	—
Oven—Screw	—	1.25	1.50
Reciprocating—Shaker	—	1.75	2.00
CRANES AND HOISTS			
Main Hoists			
Bridge and Trolley Drive	*	1.00	1.25
CRUSHER			
Ore, Stone	—	1.75	2.00
Sugar	—	1.50	1.50
ELEVATORS			
Bucket—Uniform Load	—	1.00	1.25
Bucket—Heavy Load	—	1.25	1.50
Centrifugal Discharge	—	1.25	1.50
Freight	—	1.25	1.50
Gravity Discharge	—	1.00	1.25
ELEVATORS			
Bucket—Uniform Load	—	1.00	1.25
Bucket—Heavy Load	—	1.25	1.50
Centrifugal Discharge	—	1.25	1.50
Freight	—	1.25	1.50
Gravity Discharge	—	1.00	1.25
FANS			
Centrifugal—Light (Small Diam.)	—	1.00	1.25
Large Industrial	—	1.25	1.50

Type of Machine To Be Driven	Chart I For All Drives		
	Service Factor		
	Loading		
	Not More Than 15 Mins. In 2 Hours	Not More Than 10 Hours per Day	More Than 10 Hours Per Day
FEEDERS			
Apron—Belt—Screw	—	1.25	1.50
Disc	—	1.00	1.25
Reciprocating	—	1.75	2.00
FOOD INDUSTRY			
Beet Slicer	—	1.25	1.50
Cereal Cooker	—	1.00	1.25
Dough Mixer—Meat Grinder	—	1.25	1.50
GENERATORS (NOT WELDING)	—	1.00	1.25
HAMMER MILLS	—	1.75	2.00
HOISTS			
Heavy Duty	—	1.75	2.00
Medium Duty and Skip Type	—	1.25	1.50
LAUNDRY TUMBLERS	—	1.25	1.50
LINE SHAFTS			
Uniform Load	—	1.00	1.25
Heavy Load	—	1.25	1.50
MACHINE TOOLS			
Auxiliary Drive	—	1.00	1.25
Main Drive—Uniform Load	—	1.25	1.50
Main Drive—Heavy Duty	—	1.75	2.00
METAL MILLS			
Draw Bench Carriers & Main Drive	—	1.25	1.50
SLITTERS	—	1.25	1.50
TABLE CONVEYORS— NON REVERSING			
Group Drives	—	1.25	1.50
Individual Drives	—	1.75	2.00
Wire Drawing, Flattening or Winding	—	1.25	1.50
MILLS ROTARY TYPE BALL & ROD			
Spur Ring Gear and Direct Connected	—	—	2.00
Cement Kilns, Pebble	—	—	1.50
Dryers and Coolers	—	—	1.50
Plain and Wedge Bar	—	—	1.50
Tumbling Barrels	—	—	2.00
MIXERS			
Concrete—Continuous	—	1.25	1.50
Concrete—Intermittent	—	1.25	1.50
Constant Density	—	1.00	1.25
Semi-Liquid	—	1.25	1.50
OIL INDUSTRY			
Oil Well Pumping	—	—	*
Chillers, Paraffin Filter Press	—	1.25	1.50
Rotary Kilns	—	1.25	1.50
PAPER MILLS			
Agitator (Mixer)	—	1.25	1.50
Agitator—Pure Liquids	—	1.00	1.25
Barking Drums—Mechanical Barkers	—	1.75	2.00
Bleacher	—	1.00	1.25
Beater	—	1.25	1.50
Calender Heavy Duty	—	—	2.00
Calender Anti-Friction Brgs.	—	1.00	1.25
Cylinders	—	1.25	1.50
Chipper	—	—	2.00
Chip Feeder	—	1.25	1.50
Coating Rolls—Couch Rolls	—	1.00	1.25
Conveyors—Chips—Bark— Chemical	—	1.00	1.25
Conveyors—Log and Slab	—	—	2.00
Cutter	—	—	2.00
Cylinder Molds, Dryers (Anti-Friction Brg.)	—	—	1.25
Felt Stretcher	—	1.25	1.50
Screens—Chip and Rotary	—	1.25	1.50
Thickener (AC)	—	1.25	1.50
Washer (AC)	—	1.25	1.50
Winder—Surface Type	—	—	1.25

Figure 24-27C Service factors for various applications (Courtesy Boston Gear Incom International Inc.)

Small Sprocket		HP RATINGS—STANDARD SINGLE * STRAND ROLLER CHAIN—NO. 120—1-1/2″ PITCH																		
RPM→		10	20	30	50	75	100	125	150	200	250	300	400	500	600	700	800	900	1000	1200
Teeth	P.D.																			
11	5.324″	1.37	2.56	3.68	5.82	8.40	10.9	13.3	15.6	20.3	24.8	29.2	37.8	46.3	54.5	46.3	37.9	31.8	27.1	20.6
13	6.268	1.64	3.06	4.41	6.97	10.1	13.0	15.9	18.7	24.3	29.7	35.0	45.3	55.4	65.3	59.5	48.7	40.8	34.9	26.5
15	7.215	1.91	3.57	5.14	8.13	11.7	15.2	18.6	21.9	28.3	34.7	40.8	52.9	64.6	76.1	73.8	60.4	50.6	43.2	32.9
17	8.164	2.19	4.09	5.88	9.31	13.4	17.4	21.3	25.0	32.4	39.7	46.7	60.5	74.0	87.2	89.0	72.8	61.0	52.1	39.6
19	9.114	2.47	4.61	6.64	10.5	15.2	19.6	24.0	28.2	36.5	44.8	52.7	68.2	83.4	98.3	105	86.1	72.1	61.6	46.8
21	10.064	2.75	5.13	7.39	11.7	16.9	21.8	26.7	31.4	40.7	49.8	58.7	76.0	93.0	110	122	100	83.8	71.6	54.4
Lubrication #		Type I		Type II					Type III				Type IV							

Figure 24-27D Horsepower ratings for No. 120 roller chain (Courtesy Boston Gear Incom International Inc.)

SPEED RATIOS – CENTER DISTANCES – CHAIN LENGTHS

RATIO
CENTER DISTANCE – IN PITCHES
CHAIN LENGTH – IN PITCHES
To obtain corresponding values in INCHES, multiply by the appropriate Chain Pitch.

Teeth on DriveN Sprocket	Teeth on DriveR Sprocket														
	11	12	13	14	15	16	17	18	19	20	21	22	23	24	25
15	1.36 6.469 26	1.25 7.235 28	1.15 6.993 28	1.07 7.748 20	1.00 7.500 30										
16	1.45 7.207 28	1.33 6.971 28	1.23 7.736 30	1.14 7.494 30	1.07 8.249 32	1.00 8.000 32									
17	1.55 7.943 30	1.42 7.710 30	1.31 7.473 30	1.21 8.237 32	1.13 7.994 32	1.06 8.749 34	1.00 8.500 34								
18	1.64 7.669 30	1.50 8.446 32	1.38 8.212 32	1.29 7.975 32	1.20 8.737 34	1.13 8.495 34	1.06 9.249 36	1.00 9.000 36							
19	1.73 8.404 32	1.58 8.174 32	1.46 8.949 34	1.36 8.714 34	1.27 8.477 34	1.19 9.238 36	1.12 8.995 36	1.06 9.749 38	1.00 9.500 38						
20	1.82 8.124 32	1.67 8.909 34	1.54 8.679 34	1.43 9.452 36	1.33 9.216 36	1.25 8.978 36	1.18 9.739 38	1.11 9.495 38	1.05 10.249 40	1.00 10.000 40					
21	1.91 8.857 34	1.75 8.632 34	1.61 9.414 36	1.50 9.183 36	1.40 9.955 38	1.31 9.718 38	1.24 9.479 38	1.17 10.239 40	1.11 9.995 40	1.05 10.749 42	1.00 10.500 42				
22	2.00 9.590 36	1.83 9.365 36	1.69 9.139 36	1.57 9.918 38	1.47 9.686 38	1.37 10.457 40	1.29 10.220 40	1.22 9.980 40	1.16 10.740 42	1.10 10.496 42	1.05 11.249 44	1.00 11.000 44			
23	2.09 9.304 36	1.92 10.098 38	1.77 9.872 38	1.64 9.645 38	1.53 10.422 40	1.44 10.189 40	1.35 10.959 42	1.28 10.721 42	1.21 10.481 42	1.15 11.240 44	1.10 10.996 44	1.05 11.749 46	1.00 11.500 46		
24	2.18 10.037 38	2.00 9.815 38	1.85 10.605 40	1.72 10.378 40	1.69 10.150 40	1.50 10.926 42	1.41 10.692 42	1.33 11.461 44	1.26 11.222 44	1.20 10.982 44	1.14 11.741 46	1.09 11.496 46	1.04 12.249 48	1.00 12.000 48	
25	2.27 9.744 38	2.08 10.547 40	1.92 10.324 40	1.79 11.112 42	1.67 10.884 42	1.56 10.654 42	1.47 11.429 44	1.39 11.195 44	1.31 11.963 46	1.25 11.723 46	1.19 11.483 46	1.14 12.241 48	1.09 11.996 48	1.04 12.750 50	1.00 12.500 50
30	2.72 11.345 44	2.50 12.161 46	2.31 11.943 46	2.14 12.746 48	2.00 12.522 48	1.88 12.299 48	1.76 13.087 50	1.67 12.858 50	1.58 13.638 52	1.50 13.406 52	1.43 13.172 52	1.36 13.942 54	1.30 13.705 54	1.25 14.469 56	1.20 14.228 56
32	2.91 12.812 48	2.66 12.597 48	2.46 12.379 48	2.28 13.188 50	2.14 12.967 50	2.00 13.765 52	1.88 13.539 52	1.78 13.314 52	1.68 14.099 54	1.60 13.869 54	1.52 14.646 56	1.45 14.413 56	1.39 14.178 56	1.33 14.946 58	1.28 14.708 58
35	3.18 13.976 52	2.92 13.761 52	2.69 13.546 52	2.50 14.361 54	2.33 14.141 54	2.19 13.921 54	2.06 14.721 56	1.94 14.497 56	1.84 15.288 58	1.75 15.061 58	1.67 14.833 58	1.59 15.613 60	1.52 15.382 60	1.46 16.155 62	1.40 15.921 62
36	3.27 13.668 52	3.00 14.495 54	2.77 14.279 54	2.57 14.063 54	2.40 14.874 56	2.25 14.653 56	2.12 14.433 56	2.00 15.230 58	1.89 15.006 58	1.80 15.795 60	1.71 15.567 60	1.64 15.338 60	1.56 16.117 62	1.50 15.886 62	1.44 16.658 64
40	3.64 15.561 58	3.34 15.349 58	3.08 15.136 58	2.86 15.961 60	2.67 15.746 60	2.50 15.528 60	2.35 16.339 62	2.22 16.119 62	2.10 16.920 64	2.00 16.697 64	1.90 16.473 64	1.82 17.262 66	1.74 17.035 66	1.67 17.818 68	1.60 17.588 68
42	3.82 15.983 60	3.50 15.773 60	3.23 16.605 62	3.00 16.391 62	2.80 16.177 62	2.62 16.994 64	2.47 16.777 64	2.34 16.557 64	2.21 17.364 66	2.10 17.142 66	2.00 17.939 68	1.91 17.714 68	1.83 17.489 68	1.75 18.275 70	1.68 18.047 70
45	4.09 17.139 64	3.75 16.930 64	3.46 16.719 64	3.22 17.553 66	3.00 17.340 66	2.81 18.161 68	2.65 17.945 68	2.50 17.728 68	2.37 18.536 70	2.25 18.317 70	2.14 18.096 70	2.04 18.895 72	1.96 18.671 72	1.88 19.463 74	1.80 19.237 74

Figure 24-27E   Speed ratios–center distances–chain lengths

coefficient of expansion of 1.6 x 10^{-6} inches per inch per degree Fahrenheit.

(iv) Additional bases for selecting materials are corrosion resistance, machinability, strength, and electrical properties.

(v) Combinations of all of the above, plus aesthetics and prior experiences, may be used for selecting materials.

Call or visit a local vendor of materials or machine elements, and request one of their catalogs. Usually, they are eager to have their catalogs in the hands of future buyers, so they are happy to donate one or more.

Furthermore, since most vendors handle a variety of products, a call or a visit may result in a small library!

Logical Dimensioning. Design components so they can be measured readily and logically—for two main reasons. One, shop personnel need to locate points, planes, and features such as holes, from reference planes or surfaces. Two, the function of the component needs to be ensured.

Shop personnel use a large number of measuring devices, some very sophisticated, some relatively simple, but all measure from a reference plane or surface that is usually machined. Some of the more common measuring devices and their measuring capabilities are:

a. Flat, stainless steel scale: to 0.010″
b. Micrometer: inside and outside: to 0.0002″
c. Vernier Calipers: inside and outside: to 0.001″
d. Lathes, milling machines, shapers, boring machines, and gear cutters with calibrated dials to indicate the movement of the cutting tool or the part being machined: to 0.0005″
e. Some newer milling machines and numerically controlled machines with electronic readouts: to 0.0001″

Engineers, designers, and drafters can facilitate these shop processes by dimensioning components from obvious reference planes or surfaces. The plane or surface may need to be created in the manufacturing process.

Ensure the function of a component or an assembly of components by dimensioning to common reference planes or surfaces. An example to illustrate reference planes is shown in **Figure 24-28**. The two identical L-shaped brackets are cast iron. Surface No. 1 is milled to establish a plane of reference to locate vertical surface No. 2 to be perpendicular to surface No. 1. Then, the hole at No. 3 is machined perpendicular to surface No. 2 (and hence parallel to surface No. 1) so the shaft for the roller will be horizontal. The two holes in the bracket would probably be located with respect to vertical surface No. 2 and centered about the centerline of the machined hole, surface No. 3.

A more exact dimensioning technique called positional tolerancing and form tolerancing (or geometrical tolerancing) establishes datum surfaces for referencing

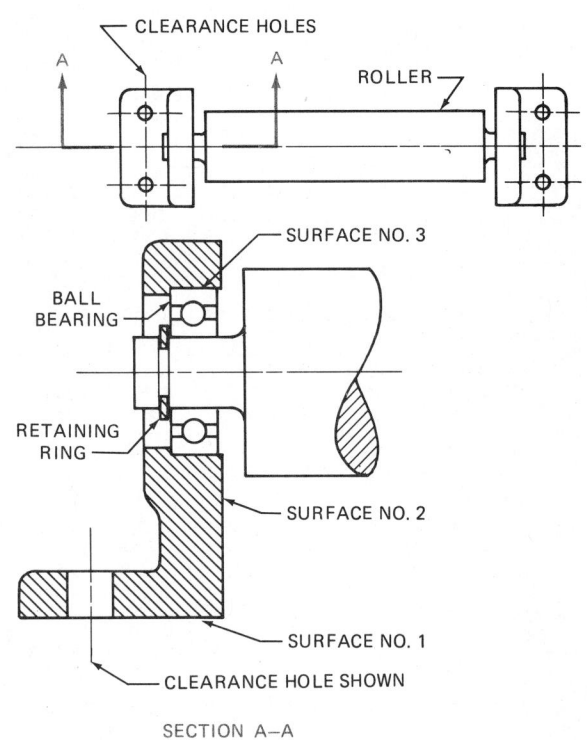

Figure 24-28 Reference plane for dimensioning

dimensions and features. The technique requires more skill and time to apply; however, time and money are saved because less parts are scrapped. They fit better and their function is assured.

This tolerancing technique has grown as machines and systems have become more complex and expensive. Furthermore, when one company manufactures components that will be assembled together with components from another company, the dimensioning and features need to be accurately located so the components will go together.

Oral Presentations. According to William L. Trippe in GRADUATING ENGINEER, January 1988, 90% of respondents in a survey indicated that they regularly gave oral presentations several times per week. More than 90% considered public speaking skills to be an important part of the job.

During the design process discussed in this chapter, oral presentations are scheduled at least two times, the design review and the final presentation. As a guide for giving these two presentations, and others, use the following three practical steps:

1. Tell them what you're going to tell them.
2. Tell them.
3. Tell them what you've told them.

Further guidelines are in the exercises at the end of the chapter.

Safe Designs. Use the following checklist to provide safe designs, safe installations, and protection against negligence suits.

Document the design to show evidence of the following:

a. Accurate calculations
b. Proper use of codes and standards
c. Use of state-of-the-art technology
d. Proper use of materials and hardware
e. Having run tests, if appropriate
f. Use of proper manufacturing processes, proper fasteners, and appropriate heat treatments
g. A quality control program
h. Preparation of instructions for safe use
i. Addition of warning signs to the product
j. Having told future owners of the proper safety devices and practices needed, such as guards around any exposed moving parts of machines, and painted stripes to show danger areas
k. Thorough exploration of all possible uses of the product other than those for which it was designed

Access. Locate fasteners and components for installation and servicing so personnel can have access to them with their hands and tools. Also, provide access to weld areas for welders.

Vibration. Use locking devices such as lock washers and lock nuts on threaded fasteners subjected to vibration. Some high-speed engines have wires passing through holes in the nuts to prevent the nuts from "backing off" because of vibrating conditions.

Fatigue. Avoid sharp changes of cross-section and small fillet radii on parts subjected to varying loads. See *Figure 24-29* for examples of good and poor practices. The sharp corners and radii (and scored surfaces) cause stress concentrations and therefore, possible fatigue failure.

Cleaning/Corrosion. Provide sloping surfaces wherever standing fluids could cause corrosion. For example, tank bottoms should slope toward a drain hole so that chemicals and cleaning fluids can be completely drained. Upper surfaces should slope to prevent standing rainwater or melting snow.

Temperature. Avoid stringent tolerances on dimensions of parts subjected to extreme temperature changes. Furthermore, check the strength and ductility of materials at extreme temperatures. Also, the temperature of a part may need to be specified for verifying some critical dimensions at that temperature if tolerance on the dimension is less than ±0.0002″.

Design for Manufacturability (DFM) and Design for Assembly (DFA).
Refer to Chapter 17 for a detailed discussion of these two topics. Design hints for DFM include goals, such as the following:

• Minimize the number of parts
• Minimize part variations
• Design parts for more than one function or use
• Design parts for ease of manufacturing

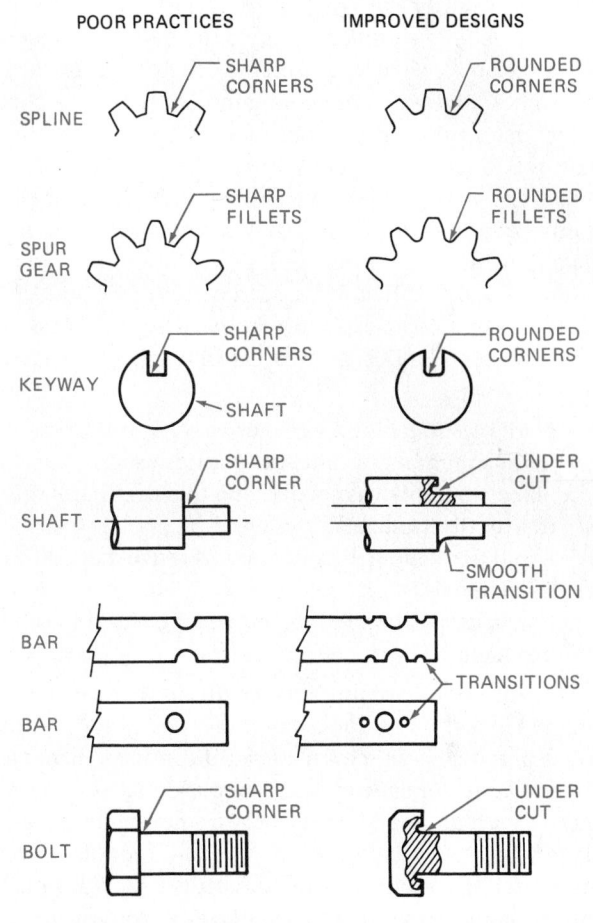

Figure 24-29 Fatigue considerations

• Use fasteners that handle easily and can be readily installed
• Arrange assembly of parts so they assemble in layer fashion
• Avoid flexible parts

These goals are discussed early in the design process in meetings where not only designers, drafters, and engineers attend, but manufacturing personnel participate also. Design hints for DFA include goals such as the following:

• Use "pick-and-place" robots for simple, routine placement of parts
• Keep the number of parts to a minimum (DFM guideline, also)
• Use gravity and vibration for moving parts into assembly areas
• Avoid parts that may "tangle" (hang up on each other)
• Use tapers and chamfers to guide parts into position
• Avoid time-consuming manual operations such as screwdriving, soldering, and riveting

These goals are also discussed early in the design process in the same engineering and manufacturing meetings previously noted.

Designing for Recyclability (DFR) and Designing for Disassembly (DFD).

Recently, a large manufacturing company that produces an automobile, several companies that produce large appliances, and one large company that produces computers, have initiated programs for DFR and DFD. Their goals are to recover as much material as feasible, mainly metals and plastic, to use again. Not only are their programs economically sound, they acknowledge the fact that landfills are becoming less available, and some basic materials are becoming scarce. Both DFR and DFD programs are considered early in the design process, as are DFM and DFA. Design hints for DFR and DFD include the following:

- Mark plastic molded parts for easy identification
- Reduce number of different plastics used
- Make assembly easy with snap-together parts (also easy to disassemble)
- Avoid permanent attachment (adhesives) of dissimilar materials
- Use materials that can be recycled or used in other products

The automobile manufacturer found that recycling, engine blocks, alternators, starter motors, water pumps, and differentials cut expenses significantly. A manufacturer of large appliances marks all plastic parts (over 50 grams) with an SAE (Society of Automotive Engineers) identification, which gives the type and recyclability of the parts. The computer manufacturer recycles glass and phosphorous from CRTs, plastics from housings (ground up and made into shingles), and metals from parts (steel, aluminum, and copper).

Review

1. What are the five phases of the design process? Since the design process is a problem-solving strategy, describe how it might apply to a non-technical problem, such as:
   a. borrowing the car from parents
   b. generating money for a student organization
   c. writing a paper for a non-technical class

2. Why is a planning chart used?

3. Why keep a log and a design file?

4. What sources of technical information are available in your school, and where can they be found?

5. What are the two major categories of specifications?

6. Explain the term iteration and how it applies to the design process.

7. What is the main function of a progress report to the client?

8. How can attribute listing help a designer to generate ideas?

9. What are the four steps in the creative process?

10. Why should designers carry a notebook at all times?

11. What are some characteristics of a creative person?

12. What is a design layout drawing? How does it differ from a technical detail drawing?

13. What are several advantages of making "cardboard-glue" models?

14. Describe a decision table.

15. Why do a design review?

16. Why are only the first four phases usually done for classroom design projects?

17. Why is the ability to communicate usually rated higher than technical skills by companies looking for new technical personnel?

18. Who are some of the people (occupations) you would most likely contact while doing the design process in school? In industry?

19. How do records kept for patent development differ from a design file?

20. What are some differences between routine and non-routine design projects? Similarities?

21. How does attribute listing help as a checklist for technical detail drawing?

22. Explain briefly what the acronyms DFM, DFA, DFR, and DFD mean.

Chapter Twenty-Four Problems

The following problems complement the design process and provided practice to understand and learn the steps before attacking routine and non-routine design projects.

Problem 24-1

Prepare a 5-minute oral presentation on one of the sources of information listed below. Use transparencies for an overhead projector. Suggestions for transparencies (**Tx**):

Tx 1. Name of source, its location and name of speaker.

Tx 2. Outline of the talk:
  a. description of the source and its specific location
  b. why source is useful to designers
  c. example of use of source
  d. conclusion

Tx 3. Description of source and its location.

Tx 4. Why source is useful to designers.

Tx 5. Example of use of source.

Tx 6. Conclusion.

Suggested Sources:
1. ASTI (Applied Science and Technology Index)
2. ASTM (American Society for Testing Materials)
3. ANSI (American National Standards Institute)
4. Patents
5. UL (Underwriter's Laboratory)
6. Machinery handbook
7. Handbook (instructor's or your choice)
8. Vendor's catalog

Problem 24-2

Prepare performance and design specifications for one of the items listed below. Well-written specifications can be verified by testing. For example, instead of specifying that a bumper jack will be easy to use, lightweight, and made of metal, well-written "specs" would be the following: easy to use by a licensed driver who is not handicapped, a maximum of 20 pounds of force to operate, weighs less than 10 pounds, and is made of SAE 1035 steel. Include a sketch with notes and dimensions.

1. Screwdriver for use in average household
2. Light switch in average household
3. Can opener for kitchen use
4. Knife for kitchen use
5. Pliers for electrician, or average household
6. Hammer for carpenter, or average household
7. Wood saw
8. Metal saw (hack saw)

Problem 24-3

Prepare a decision table using criteria and weighting factors for the situations listed. Generate at least three alternatives for each situation.

1. Vehicle for town driving
2. Vehicle for country driving
3. Vehicle for backwoods driving
4. Vacation
5. College to attend
6. Job choice in one city
7. Job choice for different locations (West Coast, East Coast, South, North, etc.)
8. Your choice

Problem 24-4

Prepare a freehand sketch of a relatively uncomplicated design and then prepare a design development layout drawing of the sketched design. Draw the object to scale, but add notes and dimensions in an informal manner. No title block is needed. Complete information is required. Design suggestions:

1. Wall bracket to support object of your choice. Instructor's choice. Use welded assemblies.
2. Minor modification of a component from:
   a. This text (instructor's choice)
   b. Your choice
   c. Instructor's choice

Problem 24-5

Prepare a kinematic development layout drawing of one of the assemblies drawn below or of an assembly assigned by your instructor.

Suggestions for Kinematic Analyses:

1. Redraw the figures as single straight lines to represent the linkages.
2. Construct radii about fixed points of rotation.
3. Trace, on a separate sheet of paper, the lines representing the linkages whose motions are to be analyzed.
4. Use the lines as underlays to locate the new positions.

Problem 24-5A

The figure below shows the linkages and hydraulic cylinder to operate a jet-blast deflector EF. The hydraulic cylinder AB causes linkage CD to rotate CCW (counterclockwise) about point C, which causes linkage DE to move the deflector EF in a CW (clockwise) direction. When EF is in the maximum CW position it acts as a shield for personnel standing behind jet aircraft when the jets are warming up for launching from an aircraft carrier. Determine the maximum CW angle that EF rotates and the change in length (stroke) of cylinder AB to obtain the maximum angle. Draw lines EF, DE, CD, and AB when they are in the position for the maximum CW rotation of EF.

Problem 24-5B

One application of a 4-Bar Linkage is to control the movement of the hood of an automobile: Actually a pair of 4-Bar Linkages support the hood. Linkage GH is attached to the hood as shown in the figure. When the hood closes, GH rotates and translates to a horizontal position. Determine the total CCW (counterclockwise) rotation of the hood from the open to the closed position. Also determine the change in length of the spring SP from the open to closed positions. Draw the lines representing the linkages and the spring in the closed position.

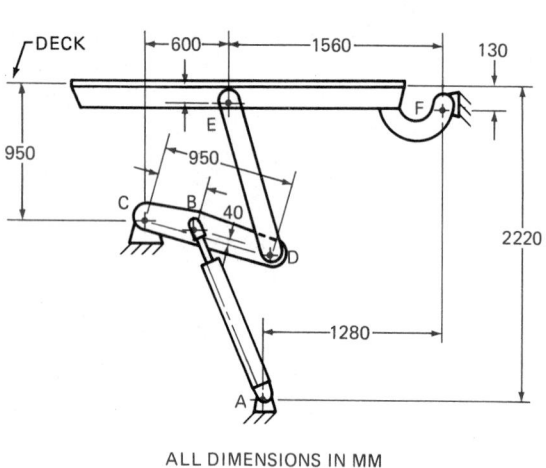

ALL DIMENSIONS IN MM

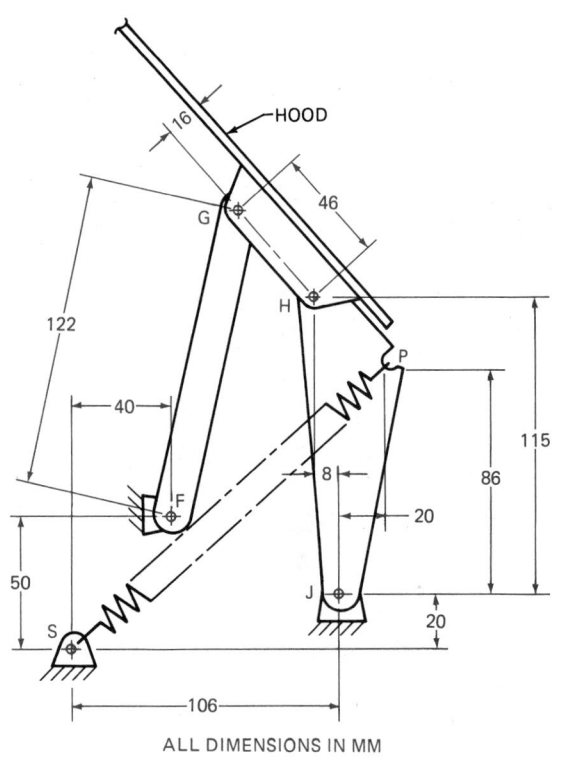

ALL DIMENSIONS IN MM

Problem 24-5A

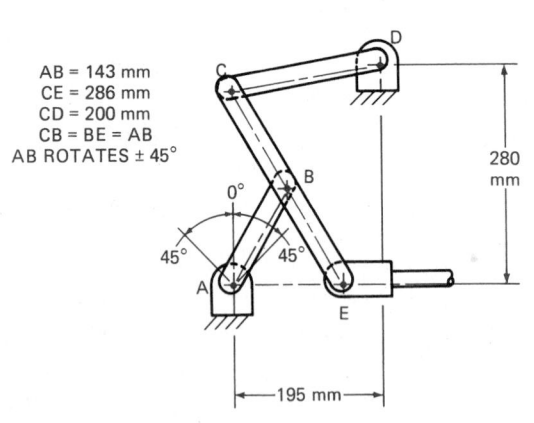

AB = 143 mm
CE = 286 mm
CD = 200 mm
CB = BE = AB
AB ROTATES ± 45°

Problem 24-5C

Problem 24-5B

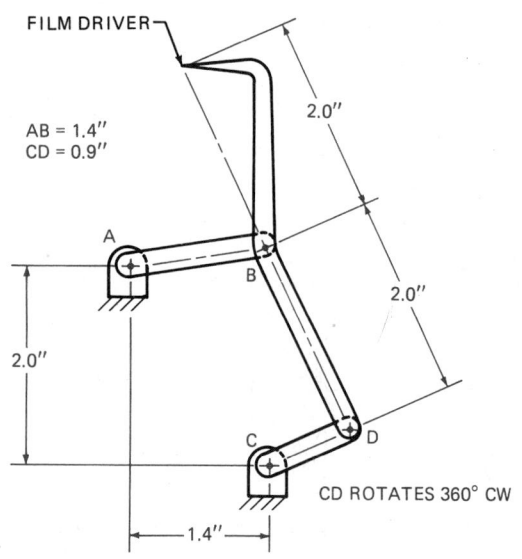

FILM DRIVER

AB = 1.4″
CD = 0.9″

CD ROTATES 360° CW

Problem 24-5D

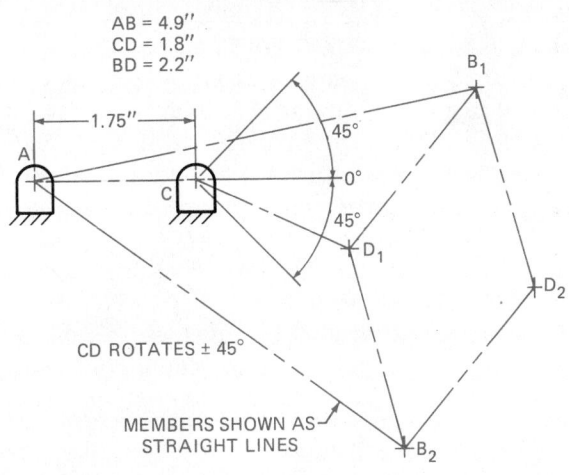

AB = 4.9″
CD = 1.8″
BD = 2.2″

1.75″

A

C

45°

0°

45°

B_1

D_1

D_2

CD ROTATES ± 45°

MEMBERS SHOWN AS
STRAIGHT LINES

B_2

Problem 24-5E

Problem 24-6

The 19-step Kraft Mill Engineering Roadmap listed below summarizes a design process from a company for improving or changing the company's plant facilities. A recent graduate from an engineering program sent the summary to one of her professors to show what she was doing during the first few months on the job.

Kraft Mill Engineering Roadmap

1. PROJECT SCOPE: Define the problem to be resolved, the changes that are required, and the objective to be achieved.

2. SCOPE REVIEW MEETING: Verify that both Maintenance and Operations approve of the project scope.

3. PRELIMINARY DESIGN: Select several options for consideration. Prepare rough cost estimates, sketch general layouts/schematics, and estimate R01/payback for each option.

4. PRELIMINARY DESIGN REVIEW MEETING: Select one or two options to pursue. Obtain Management's approval to proceed. Determine if project is to be capitalized or expensed.

5. SUBMIT AR OR WO: Write a cost estimate. Use class ±10%, ±20%, or ±40%, as required. Write and submit an AR (Authorization Request) or a WO (work order).

6. RECEIVE APPROVED AR OR WO: No money may be spent until approval is received.

7. DETAILED DESIGN: Make detailed layout drawings and flow schematics. Do energy balance. Develop schedule; make time-bar diagram for larger projects. Determine equipment requirements.

8. PROJECT DESIGN REVIEW MEETING: Obtain "buy off" from Operations and Maintenance of the design and layout. Verify schedule, shutdown periods, etc. Decide whether contractors or the mill will provide labor.

9. BIDS: Prepare bid package for material and for labor. Draw up a bidders list. Send to Purchasing Dept. and go out for bids.

10. REVIEW BIDS: Select the best one or two vendors for each portion of the project. Run background check on contractors. Check references of equipment manufacturers.

11. VENDOR APPROVAL MEETING: Obtain approval to purchase equipment and/or hire contractors.

12. PURCHASE ORDERS: Award bids. Write Purchase Orders (P.O.'s).

13. EQUIPMENT TRACKING: Track equipment as it comes into mill. Store in safe place until installation.

14. CONSTRUCTION: Install equipment.

15. PRE-STARTUP: Obtain and file vendor drawings, guarantees, and instruction files. Set up spare parts in the storeroom. Write an operating manual. Develop a start-up plan.

16. PRE-STARTUP: Review startup plan. Set a startup date.

17. STARTUP: Put system into operation.

18. PROJECT REVIEW: Either hold meeting or circulate memo. Restate original project scope. Review the project. Determine that the desired objectives were (or were not) met. Discuss any problems. State final project cost.

19. PROJECT CLOSEOUT: Do necessary paperwork to close out project. Sign the WO as complete, or fill out an AR Capital Job Closure Form.

Note: Some of these events overlap. Depending on the size and type of job, some events may be deleted or some more may need to be added.

Study the 19 steps and do all or any one of the following as assigned by your instructor:

1. Prepare a planning chart assuming that a 6-month period is required for a particular plant modification, and that you are doing the job.

2. Relate the 19 steps to the design process discussion in this chapter.

3. List and comment on the different people most likely contacted during the 19 steps.

4. Comment on the apparent meaning of some of the terms and acronyms: for example, scope, R01, capitalized, expensed, AR, WO, ±10% class, ±20% class, ±40% class, time-bar diagram, flow schematic, "buy-off", bid package, background check, PO, tracking, and AR Capital Job Closure Form.

Problem 24-7

1. Contact local vendors of material to obtain current prices of common ones, such as steel shapes (angles, plates, bars, channels, and beams). Note that prices are usually given in dollars per 100 pounds and that price reductions are given for purchasing large quantities (e.g., a full railroad car).

2. Contact vendors of finished components–such as motors, switches, controls, bearings, chain, and similar products–to obtain costs to several categories of buyers.

   a. Private individuals usually pay the full retail price.

   b. Companies that include the finished components in a kit may get a 15% to 30% discount from the vendors because the companies sell the kits to customers.

   c. Companies that include the finished components as an integral part of an assembled machine are called original equipment manufacturers (OEM). They may receive even larger discounts.

Problem 24-8

Frequently, professionals are asked to make quick estimates during conversations over lunch or while phoning a client. This type of quick thinking is also an excellent approach to use for checking final results, whether it is homework in a engineering science course or a system in industry. Ask question such as, does this look right, does this look reasonable, or does the output seem reasonable for the input?

For practicing "back-of-the-envelope" thinking and calculations, do the following in a time period assigned by your instructor and record the assumptions made:

1. Estimate the quantities of food and the cost of lunch for all of the people in your classroom.

2. Estimate the number of gallons of water used by your school in one day. (Perhaps maintenance personnel can verify the accuracy of the estimate.)

3. Estimate the cost of transporting a college football team 1000 miles, and expenses for a 3-day road trip.

4. Estimate the selling price of a new automobile, truck, van, sports car, or motorcycle based on the price per pound of existing vehicles.

5. Estimate selling prices of other products using the approach in 4.

Problem 24-9

List the manufacturing processes required to make the parts shown in the following figures.

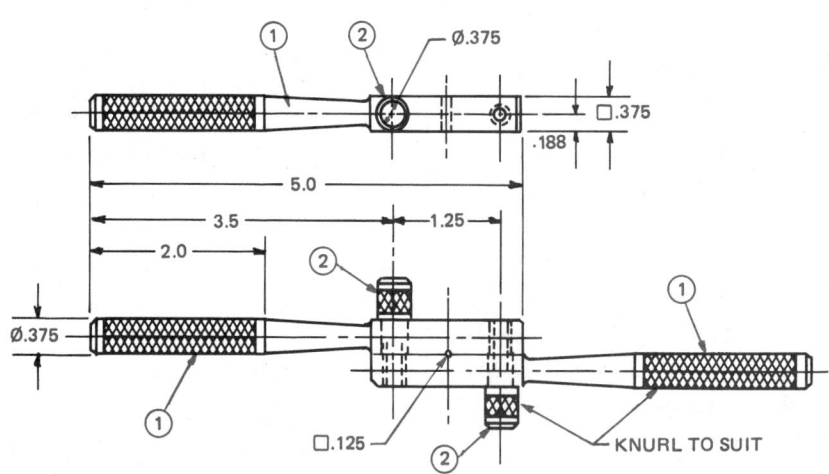

Problem 24-9A

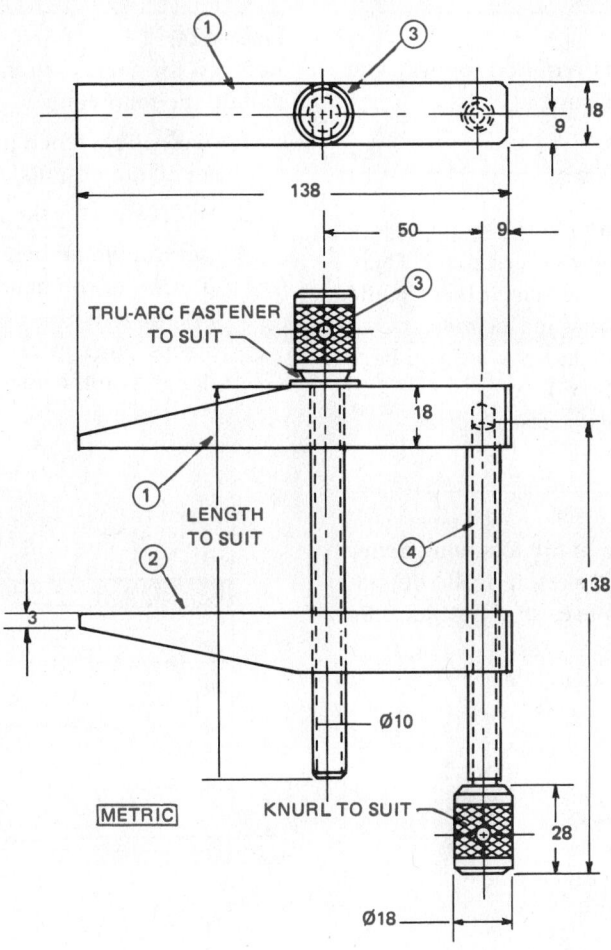

Problem 24-9B

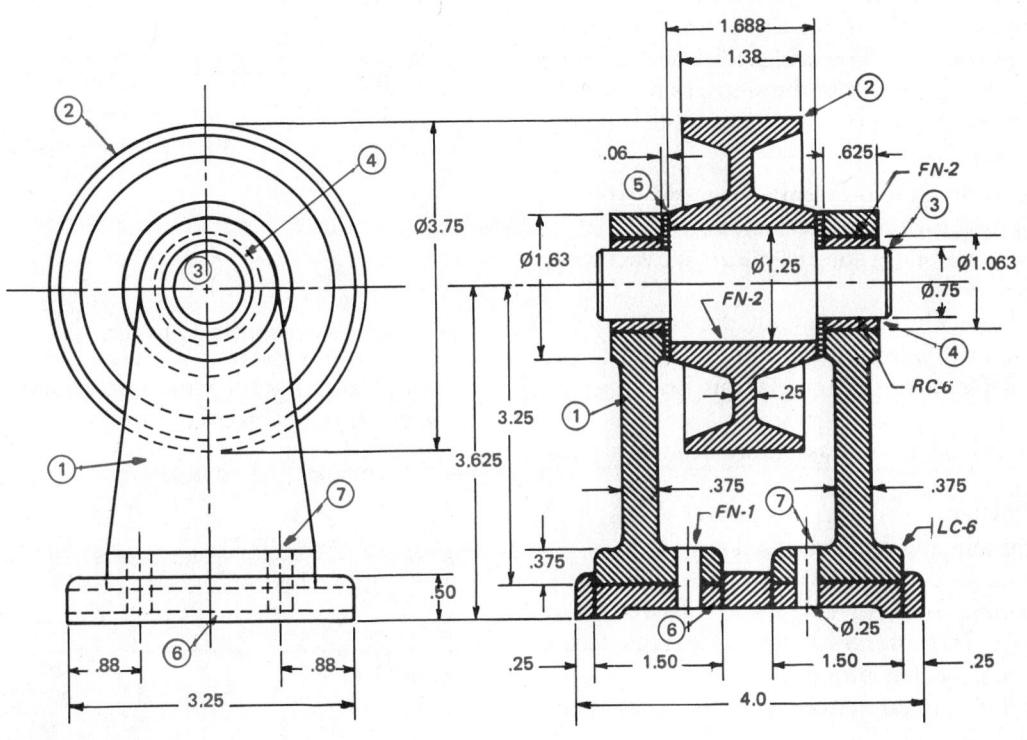

Problem 24-9C

Problem 24-9D

List the manufacturing processes required to make the part or parts assigned by your instructor.

Problem 24-10

From memory, make a freehand sketch of an object shown to you for a limited time by your instructor. Then, do a second freehand sketch of the same object while you have a much longer time to view and examine it.

This practice of viewing an object knowing you have to sketch it from memory will train you to see and remember more than you would at a casual glance.

Problem 24-11

List the obvious attributes of one of the following items. Include a freehand dimensioned sketch of the object. Then list all the not-so-obvious uses of the object. Be creative.

1. Kitchen table
2. Clothespin
3. Bookshelves
   a. Wood
   b. Metal
4. Cardboard milk carton
5. Plastic container of 35 mm film
6. Screwdriver
7. Nail

Routine Design Problems

The routine design problems listed here are intended to give students experience in the design process while doing problems similar to those assigned to new employees in technical industries. Use the steps, as needed, in the design process to produce the following short report (printed or typed):

1. A statement of the problem in your own words
2. A list of information sources used
3. Freehand sketch(es) or preliminary ideas of what needs to be done
4. Design development layout drawings of the ideas in the freehand sketches
5. Kinematic development layout drawings, if required
6. Complete detail and assembly drawings with parts lists
7. Cost analysis
8. Conclusions and recommendations
9. Appendices
   a. Machine copies of 3, 4, 5, and 6 above (These copies will show how well drawings reproduce.)
   b. Vendor catalog information
   c. Any additional materials: work order, photographs, etc.
10. A title page and a folder for the report

Problem 24-12

Refer to the torque wrench assembly in Problem 24-12 and do the following:

1. Make the wrench just two times as large as shown for all dimensions.
2. Select steel for the handles and the fasteners.
3. Determine the bending stress in the necked-down portion of the handle. Specify as large a radius as practical at the necked-down location to avoid stress concentration. The bending stress will depend on the assumed force F and the length L.

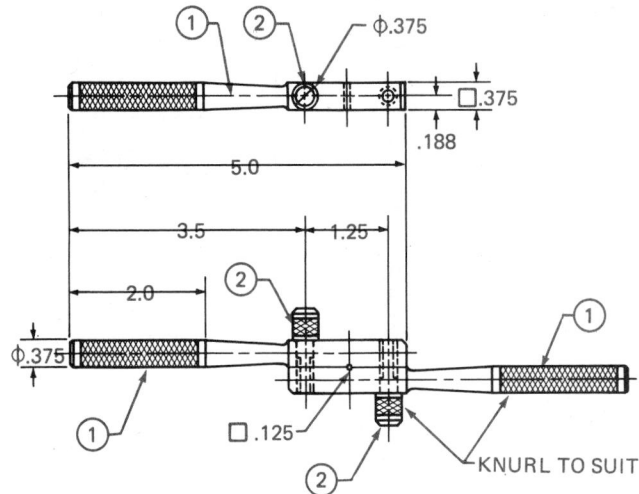

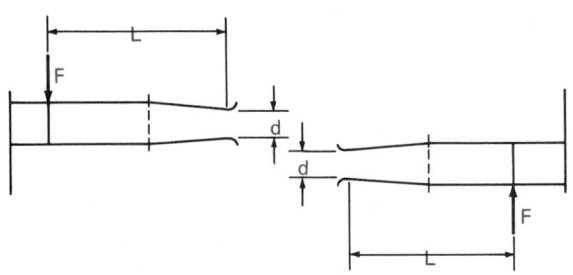

F = FORCE BY OPERATOR IN POUNDS
(NEED TO ASSUME)

L = ASSUME DISTANCE FROM F TO d IN INCHES

d = DIAMETER OF NECKED DOWN PORTION IN INCHES

STRESS DUE TO MOMENT F X L AT d IS

$$\text{STRESS} = \frac{(F \times L)\,(d/2)}{(\pi d^4)/64} \quad \text{IN POUNDS PER SQUARE INCH}$$

Problem 24-12

Problem 24-13

Refer to the clamp shown in Problem 24-13 and assume that parts No. 1 and No. 2 are to be aluminum, and parts No. 3 and No. 4 are steel. When threads in aluminum may be too weak because of the softer metal, a threaded-steel insert is pressed into the aluminum. Use an insert with appropriate threads inside and a nominal outside diameter of 15 mm, but dimensioned for an FN-2 medium drive fit. Increase the 18mm dimension for parts No. 1 and No. 2 to 25 mm.

Problem 24-14

The clamp shown in Problem 24-14 needs to be modified to clamp rectangular assemblies 50 x 300 mm between the two threaded members instead of between the tapered extensions. The thread size should be increased 100% in diameter. Make any other changes that seem appropriate. Select materials. (*Note:* part No. 4 must function like part No. 3, and the tapered portions of parts No. 1 and No. 2 are no longer needed.)

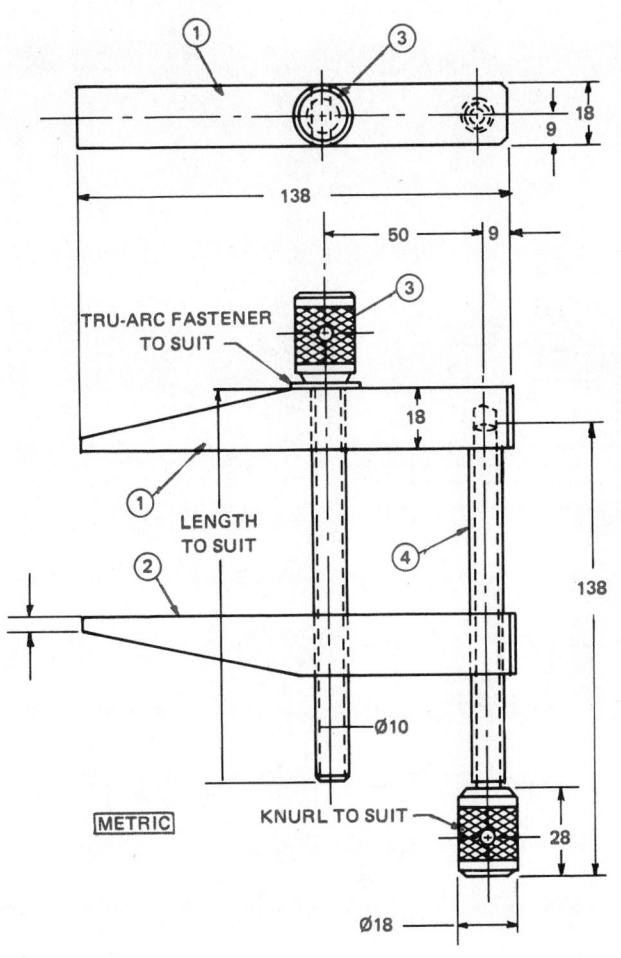

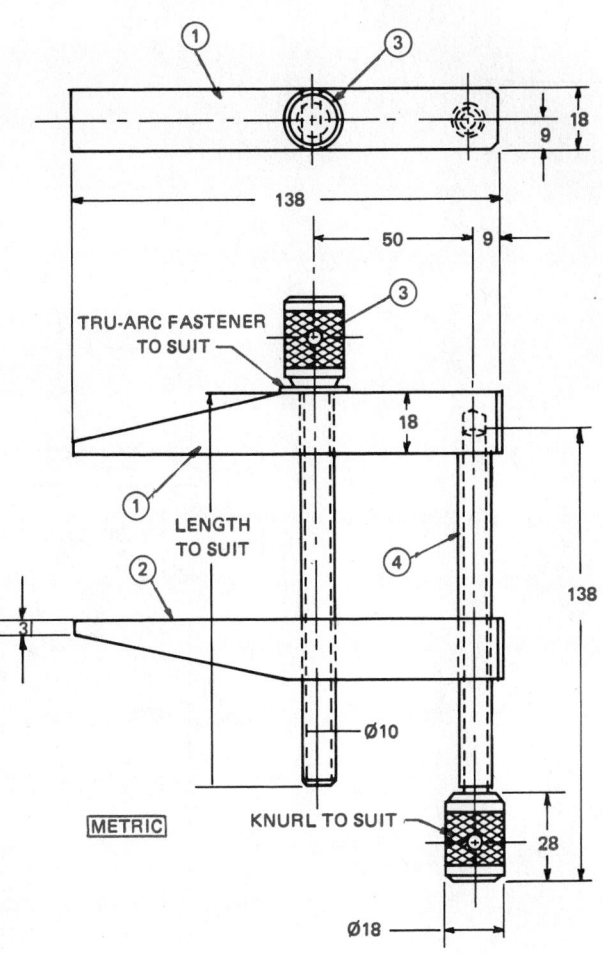

Problem 24-13   **Problem 24-14**

Problem 24-15

Modify the screw jack shown in Problem 24-15 to have an 8-inch vertical movement instead of the apparent 4-inch vertical one.

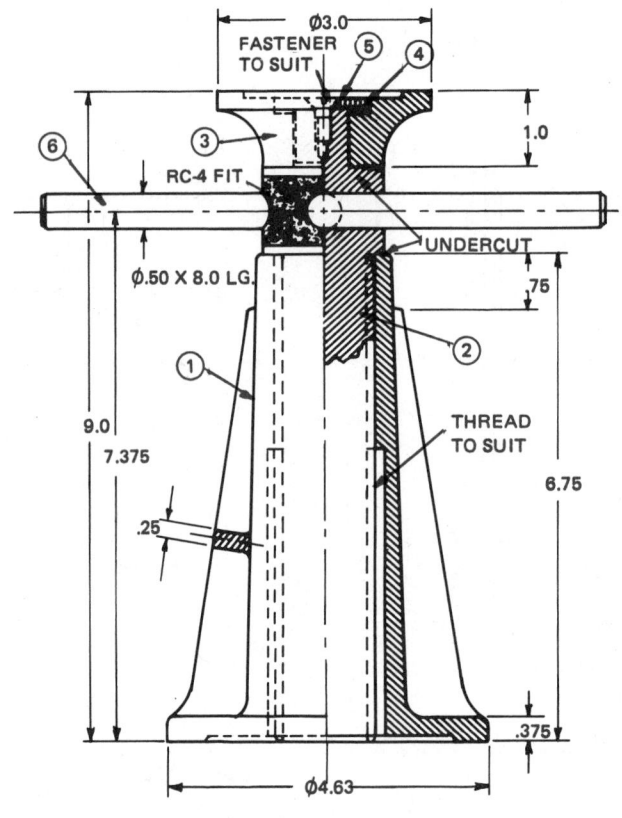

Problem 24-15

Problem 24-16

Modify the roller assembly shown in Problem 24-16 as follows:

1. Use standard ball bearings in place of the bushings in part No. 1. Use an FN-2 medium drive fit for the bearings pressed into part No. 1.

2. Change parts No. 1 and No. 6 as indicated in the sketch. Omit part No. 7.

3. Make part No. 2 cast aluminum.

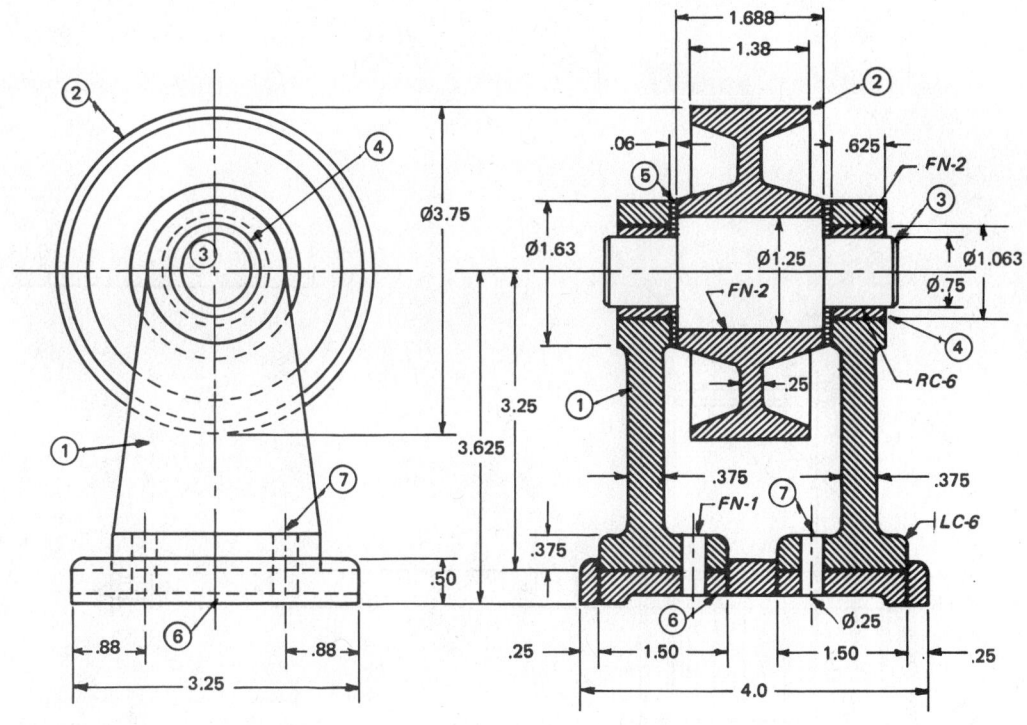

Problem 24-16

Problem 24-17

Change the vise shown in Problem 24-17 to be 50% longer and 75% wider. In a kinematic layout drawing, show the extreme open and closed positions.

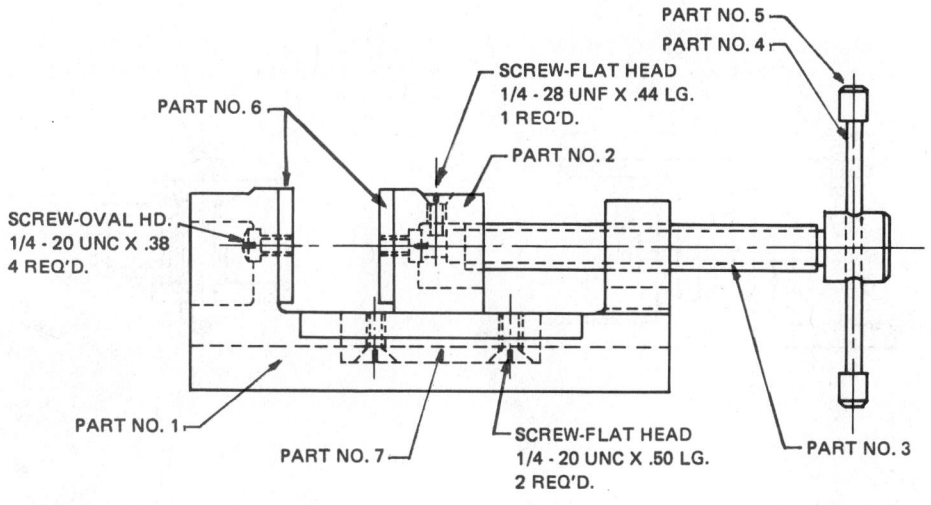

Problem 24-17

Problem 24-18

Develop the concept of the beverage can crusher shown in Problem 24-18. Use standard materials and parts as indicated. Include a kinematic layout drawing showing extreme positions of the crusher head.

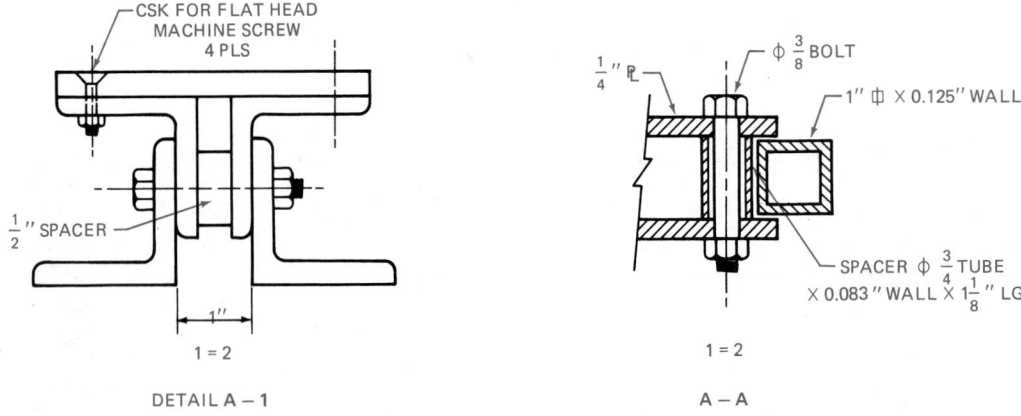

BEVERAGE-CAN CRUSHER
ALL ALUMINUM EXCEPT FOR FASTENERS

Problem 24-19

Use press-fit drill-bushings with heads to make a leaf type jig, for the hole pattern shown in the part to be drilled.

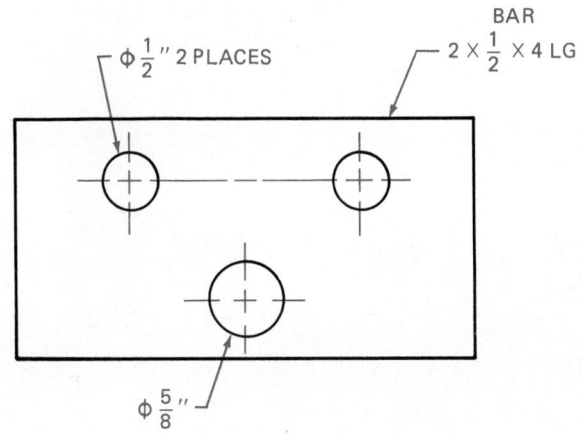

Problem 24-19

❏ Non-Routine Design Problems

The non-routine design problems listed or described here are intended to give students experience in the design process at least in Phases I-IV, which include Steps 1–11. Where possible, students are encouraged to build prototypes and discuss the other activities in Phase V, which includes Steps 12–14.

The following list includes the general sources for finding real problems.

1. School
   Classroom demonstrators of basic principles of chemistry, physics, statics, dynamics, mechanics of materials, engineering sciences, and manufacturing processes.

2. Handicapped
   a. Designs to help individuals confined to a wheelchair to:
      (i) Reach objects and prepare meals in a kitchen
      (ii) Obtain exercise
      (iii) Obtain access to heights higher than curbs
      (iv) Have access to classroom work surfaces
      (v) Have work surfaces on the wheelchair
   b. Designs to continually flex injured elbows and knee joints to accelerate healing.
   c. Designs to aid communications for individuals who have speech difficulties.

3. Hospitals
   a. Games or devices for children who are confined to a bed and have limited use of their arms.

   b. Devices to help injured individuals to perform simple functions such as dress, read, play games, and eat.

4. Civic Organizations
   a. Designs for YWCA or YMCA
   b. Designs to assist scouting programs such as neighborhood improvement projects

5. Local Industry
   a. Projects for the industry
   b. Projects for the industry's client

6. Private individuals (designs for)
   a. Home improvements
   b. Hobbies
   c. Inventions
   d. Invalid care

7. City (designs for)
   a. Better parking meters
   b. Playgrounds
   c. Park facilities
   d. Signs
   e. Junior Achievement

8. State (designs for)
   a. Signs
   b. National Parks
   c. Recreational facilities

9. National (designs for)
   a. Design contests (Alcoa, Lincoln Arc Welding, Engineering Societies…)
   b. Environment

10. International (designs for)
    a. Third World countries
    b. To help Peace Corp volunteers

Specific suggestions for non-routine design problems are as follows:

1. Design an inexpensive fruit dryer.

2. Design a device to assist an individual who has no legs, but full use of two arms, to get into a wheelchair from the floor, and back to the floor.

3. Design a device to lift a 150-pound handicapped person out of a swimming pool.

4. Design a means to get rid of moisture in beverage cans that come in truckloads to a recycling center. The driver usually needs to be paid within one-half hour, a realistic price based on the actual metal turned in.

5. Design a motorized vehicle to move one or more employees throughout a large industrial manufacturing plant or warehouse. The vehicle should be cost effective; that is, it should save enough time (money) to justify its use.

Mechanical Drafting Mathematics

Chapter Outline

Mathematics for Drafters

The drafter is often required to solve problems by using formulas obtained from technical handbooks. Geometry is fundamental to mechanical technology: An engineering drawing is an example of applied geometry. Geometric principles are the bases of many dimension calculations and problem solutions. Trigonometry is used to compute unknown angles and sides of triangles. The drafter and designer often work with triangular configurations. Working dimensions must be determined in which parts of complex shapes are broken down into a series of triangles.

Problems encountered in fully defining details of parts or assemblies to be manufactured are often solved by using a combination of algebra, geometry, and trigonometry.

It is essential that the drafter develop the ability to analyze a problem and relate the mathematical principles that are involved in its solution. Then the problem must be solved in clear, orderly steps based on mathematical fact.

Rounding Decimal Fractions

When working with decimals, the computations and answers may contain more decimal places than are required. The number of decimal places needed depends on the degree of precision required. The degree of precision depends on how the decimal value is going to be used. *Rounding a decimal* means expressing the decimal with fewer decimal places.

Procedure: To round a decimal fraction

- Determine the number of decimal places required in an answer.

- If the digit directly following the last decimal place required is less than 5, drop all digits that follow the required number of decimal places.

- If the digit directly following the last decimal place required is 5 or larger, add one to the last required digit and drop all digits that follow the required number of decimal places.

Example 1:   Round 0.68247 in to three decimal places.

The digit following the third decimal place is 4.
0.684④7 in.
Because 4 is less than 5, drop all digits after the third decimal place.
0.682 in Ans

Example 2:   Round 12.3876 mm to two decimal places.

The digit following the second decimal place is 7.
12.38⑦6 mm
Because 7 is greater than 5, add 1 to the 8.
12.39 mm Ans

Example 3:   Round 3.918256 in to four decimal places.

The digit following the fourth decimal place is 5.
3.9182⑤6 in
Therefore, add 1 to the 2.
3.9183 in Ans

Expressing Common Fractions as Decimal Fractions

Given or computed common = fraction values are often converted to decimal = fraction dimensions.

A common fraction is an indicated division. For example, 7/16 is the same as 7 ÷ 16. Because the numerator and the denominator of a common fraction are both whole numbers, expressing a common fraction as a decimal fraction requires division with whole numbers.

Procedure: To express a common fraction as a decimal fraction divide the numerator by the denominator.

Example: Express 7/8 in as a decimal.

Divide the numerator 7 by the denominator 8.

$$\frac{0.875}{8\overline{)7.000}}$$

Decimal Equivalent Tables

Generally, fractional engineering drawing dimensions are given in multiples of 64ths of an inch. Decimal equivalent tables are widely used in the manufacturing industry. Many of the equivalents are memorized after using the tables for a period of time. Some of the decimals listed in the table are given to six places. A decimal is rounded to the degree of precision required for a particular application. The amount of computation and the chances of error can be reduced by using the decimal equivalent table shown in *Figure A-1*.

DECIMAL EQUIVALENT TABLE

Fraction	Decimal	Fraction	Decimal	Fraction	Decimal	Fraction	Decimal
1/64	.015625	17/64	.265625	33/64	.515625	49/64	.765625
1/32	.03125	9/32	.28125	17/32	.53125	25/32	.78125
3/64	.046875	19/64	.296875	35/64	.546875	51/64	.796875
1/16	.0625	5/16	.3125	9/16	.5625	13/16	.8125
5/64	.078125	21/64	.328125	37/64	.578125	53/64	.828125
3/32	.09375	11/32	.34375	19/32	.59375	27/32	.84375
7/64	.109375	23/64	.359375	39/64	.609375	55/64	.859375
1/8	.125	3/8	.375	5/8	.625	7/8	.875
9/64	.140625	25/64	.390625	41/64	.640625	57/64	.890625
5/32	.15625	13/32	.40625	21/32	.65625	29/32	.90625
11/64	.171875	27/64	.421875	43/64	.671875	59/64	.921875
3/16	.1875	7/16	.4375	11/16	.6875	15/16	.9375
13/64	.203125	29/64	.453125	45/64	.703125	61/64	.953125
7/32	.21875	15/32	.46875	23/32	.71875	31/32	.96875
15/64	.234375	31/64	.48438	47/64	.734375	63/64	.984375
1/4	.250	1/2	.500	3/4	.750	1	1.000

Figure A-1   Decimal equivalent table

The following examples illustrate the use of the decimal equivalent table.

Example 1:   Find the decimal equivalent of 7/32 in. The decimal equivalent is shown directly to the right of the common fraction.

7/32 in = 0.21875 in Ans

Example 2:   Find the fractional equivalent of 0.8125 in. The fractional equivalent is shown directly to the left of the decimal fraction.

0.8125 in = 13/16 in Ans

Example 3:   Find the nearer fractional equivalent of 0.757 in. The decimal 0.757 lies between 0.750 and 0.765625. The difference between 0.757 and 0.750 is 0.007. The difference between 0.757 and 0.765625 is 0.008625. Since 0.007 is less than 0.008625, the 0.750 value is closer to 0.757. The nearer fractional equivalent of 0.757 in is therefore 3/4 in.

Millimetre-Inch Equivalents (Conversion Factors)

Since the English and the metric systems are both used in this country, it is sometimes necessary to express equivalents between systems. Metric-English equivalents other than millimetre-inch are seldom used in mechanical technology.

The relationship between English decimal inch units and metric millimetre units is shown by comparing the scales in *Figure A-2*.

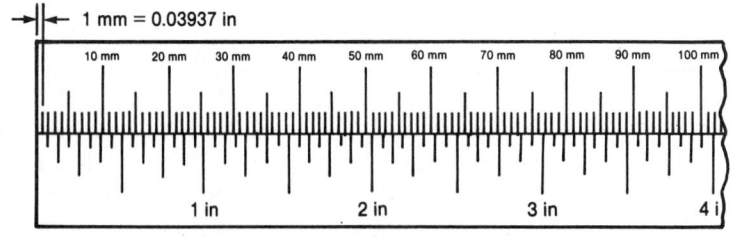

Figure A-2   Decimal inch-millimetre scales

1 inch (in) = 25.4 millimetres (mm)
1 millimetre (mm) = 0.03937 inch (in)

Procedure: To express a dimension given in inches as an equivalent dimension in millimetres, multiply the inch value by 25.4 mm.

Example: Express 1.278 in millimetres. Round the answer to two decimal places.

Multiply 1.278 by 25.4 mm.
1.278 x 25.4 mm = 32.46 mm Ans

Procedure: To express a dimension given in millimetres as an equivalent dimension in inches, multiply the millimetre value by 0.03937 inch.

Example: Express 68.74 mm in inches. Round the answer to three decimal places.

Multiply 68.74 by 0.03937 in.
68.74 x 0.03937 in = 2.706 in Ans

Example: The template shown in **Figure A-3** is dimensioned in millimetres. Determine, in inches, the total length of the template. Round the answer to three decimal places. Do *not* express each of the dimensions as inches. Add the dimensions as they are given and express the sum in inches.

87.63 mm + 102.34 mm + 98.75 mm = 288.72 mm
Multiply 288.72 by 0.03937 in:
288.72 x 0.03937 in = 11.367 in Ans

Evaluating Formulas

By letters and other symbols, generalizations called *formulas* can be stated mathematically. A formula uses symbols to show the relationship between quantities. Problems are often solved by using formulas found in technical manuals and other reference materials.

Procedure: To solve (evaluate) a formula substitute the numerical values for the letter values and solve by following the order of operations of arithmetic. The order of operations follows:

• Do all operations within the grouping symbol first. Parentheses, the fraction bar, and the radical symbol are used to group numbers. If an expression con-

tains parentheses within parentheses or brackets, do the work within the innermost parentheses first.

• Do powers and roots next. The operations are performed in the order in which they occur. If a root consists of two or more operations within the radical symbol, perform all the operations within the radical symbol, then extract the root.

• Do multiplication and division next in the order in which they occur.

• Do addition and subtraction last in the order in which they occur.

Example 1: Determine the outside diameter of a gear with 16 teeth and a circular pitch of 0.7854 in. Use the formula

$$D_o = \frac{P_c(N + 2)}{3.1416}$$

where D_o = outside diameter
P_c = circular pitch
N = number of teeth

Substitute the given numerical values for the letter values.

$$D_o = \frac{0.7854\,(16 + 2)}{3.1416}$$

Do the work in parentheses.

$$D_o = \frac{0.7854\,(18)}{3.1416}$$

Multiply.

$$D_o = \frac{14.1372}{3.1416}$$

Divide.

$$D_o = 4.5000 \text{ in Ans}$$

Example 2: The dimensions in **Figure A-4** are given in inches. Refer to the figure and determine slant height(s) by using the formula

$$S = \sqrt{(R - r)^2 + h^2}.$$

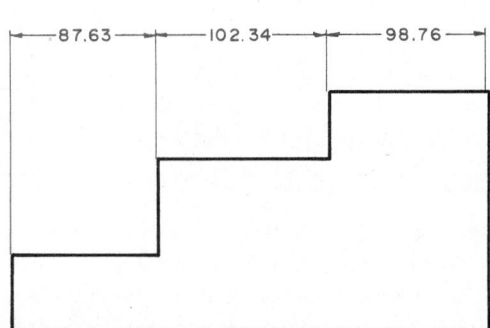

Figure A-3 Template

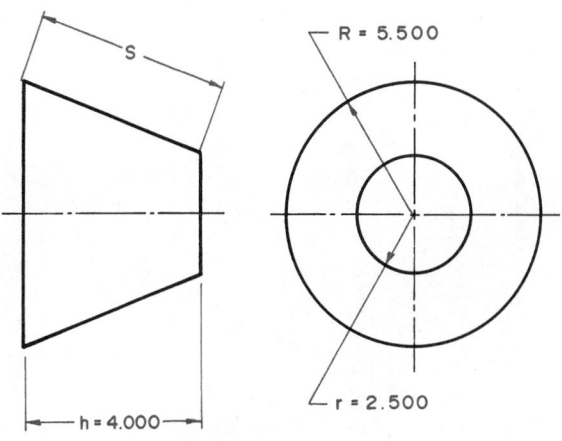

Figure A-4

Substitute numerical values for letter values.

$S = \sqrt{(5,500 - 2.500)^2 + 4.000^2}$

Do the work within the grouping (radical) symbol.
Do the work in parentheses.

$S = \sqrt{3.000^2 + 4.000^2}$

Compute powers within the radical symbol.

$S = \sqrt{9.000 + 16.000}$

Add the values within the radical symbol.

$S = \sqrt{25.000}$

Extract the root.

$S = 5.000$ in Ans

Ratio and Proportion

Ratios

Ratio and proportion are widely used in manufacturing applications, such as computing gear sizes and speeds.

Ratio is the comparison of two *like* quantities. The terms of a ratio are the two numbers that are compared. Both terms of a ratio must be expressed in the same units of measure. Ratios are generally expressed with a colon between the two terms, such as 3:8 (which is read 3 to 8) or as a fraction: ⅜.

The terms of a ratio must be compared in the order in which they are given. The first term is the numerator of a fraction, and the second is the denominator.

Examples:

1. The ratio 2 to 3 = 2 ÷ 3 = ⅔.
2. The ratio 3 to 2 = 3 ÷ 2 = 3⁄2.

   Generally, a ratio should be expressed in lowest fractional terms.

Examples: A pulley is shown in *Figure A-5.*

1. The ratio of the 2-in diameter pulley to the 6-in diameter pulley is 2 to 6 = 2⁄6 = ⅓.
2. The ratio of the 6-in pulley to the 3-in pulley is 6 to 3 = 6⁄3 = 2⁄1.

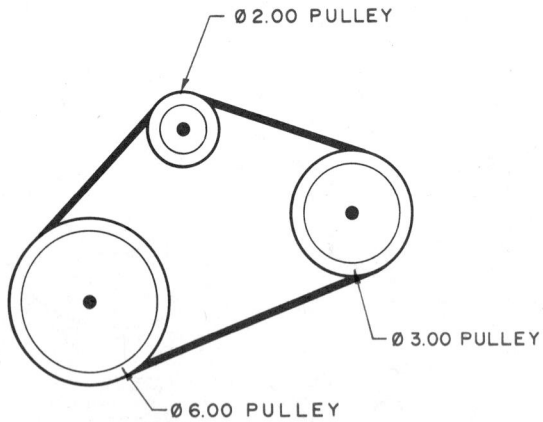

Figure A-5   Pulley system

Proportions

A *proportion* is an expression that states the equality of two ratios. Proportions are expressed in the following two ways:

1. 3:4 :: 6:8, which is read: "3 is to 4 as 6 is to 8."
2. ¾ = 6⁄8. This form is the usual way to write a proportion.

A proportion consists of four terms. The first and the fourth terms are called *extremes*, and the second and third terms are called the *means*.

Example: 7⁄16 = 14⁄32. 7 and 32 are the extremes; 16 and 14 are the means.

In a proportion the product of the means equals the product of the extremes. If the terms are cross multiplied, their products are equal.

Example: ¾ = 6⁄8

Cross Multiply. ¾ x 6⁄8

$3 \times 8 = 4 \times 6$

$24 = 24$

The method of cross multiplying is used in solving proportions that have an unknown term. After the terms have been cross multiplied, the division principle of equality is applied.

Example 1: $\dfrac{X}{7.5} = \dfrac{23.4}{20}$. Solve for the value of X.

Cross multiply.

$20X = 7.5 (23.4)$

$20X = 175.5$

Apply the division principle of equality. Divide both sides of the equation by 20.

$$\frac{20X}{20} = \frac{175.5}{20}$$

$X = 8.775$ Ans

Example 2: $\dfrac{80 \text{ mm}}{150 \text{ mm}} = \dfrac{70 \text{ mm}}{X}$. Solve for the value of X.

Cross multiply.

$80X = 150(70 \text{ mm})$

$80X = 10,500 \text{ mm}$

Apply the division principle of equality. Divide both sides of the equation by 80.

$$\frac{80X}{80} = \frac{10,500 \text{ mm}}{80}$$

$X = 131.25 \text{ mm}$ Ans

Arithmetic Operations on Angles Expressed in Degrees, Minutes, and Seconds

The *degree* is the basic unit of angular measure. The symbol ° means degree. A radius rotated one revolution makes a complete circle or 360°. In the English system, computations and measurements are in degrees and minutes. Degrees, minutes, and seconds are used for applications requiring precise angular measurement.

DECIMAL INCH SCALE

1 circle	= 360 degrees	1 degree	= $\frac{1}{360}$ circle
1 degree	= 60 minutes	1 minute	= $\frac{1}{60}$ degree
1 minute	= 60 seconds	1 second	= $\frac{1}{60}$ minute

Figure A-6 Relationship of degrees, minutes, and seconds

A degree is divided into 60 equal parts called *minutes*. The symbol for minute is ′. A minute is divided into 60 equal parts called *seconds*. The symbol for seconds is ″. The relationship of degrees, minutes, and seconds is shown in ***Figure A-6***.

Adding Angles Expressed in Degrees, Minutes, and Seconds

Example 1: Determine ∠1 in ***Figure A-7***.

$$∠1 = 43°50' + 81°37'$$

$$\begin{array}{r} 43°50' \\ + \quad 81°37' \\ \hline 124°87' = 125°27' \text{ Ans} \end{array}$$

Note: 87′ = 60′ + 27′ = 1°27′; therefore, 124°87′ = 125°27′.

Example 2: Determine ∠2 in ***Figure A-8***.

$$∠2 = 78°43'27'' + 29°38'52''$$

$$\begin{array}{r} 78°43'27'' \\ + \quad 29°38'52'' \\ \hline 107°81'79'' = 107°82'19'' = 108°22'19'' \text{ Ans} \end{array}$$

Note: 79″ = 60″ + 19″ = 1′19″; therefore, 107°81′79″ = 107°82′19″.
82′ = 60′ + 22′ = 1°22′; therefore, 107°82′19″ = 108°22′19″.

Subtracting Angles Expressed in Degrees, Minutes, and Seconds

Example 1: Determine ∠1 in ***Figure A-9***.

$$∠1 = 98°26' - 44°37'$$

$$\begin{array}{r} 98°26' = 97°86' \\ - \quad 44°37' = 44°37' \\ \hline 53°49' \text{ Ans} \end{array}$$

Note: Since 37′ cannot be subtracted from 26′, express 1° as 60′.
98°26′ = 97° + 1° + 26′ = 97° + 60′ + 26′ = 97°86′.

Example 2: Determine ∠2 in ***Figure A-10***.

$$∠2 = 57°13'28'' - 44°19'88''$$

$$\begin{array}{r} 57°13'28'' = 56°73'28'' = 56°72'88'' \\ - \quad 44°19'42'' = 44°19'42'' = 44°19'42'' \\ \hline 12°53'46'' \text{ Ans} \end{array}$$

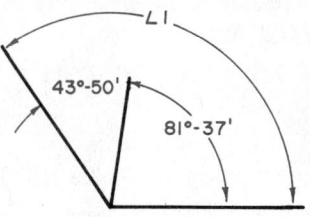

Figure A-7

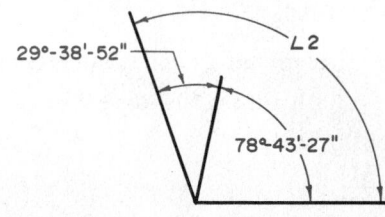

Figure A-8

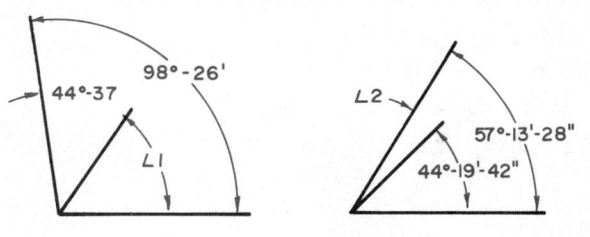

Figure A-9   ***Figure A-10***

Note: Since 19′ cannot be subtracted from 13′, and 42″ cannot be subtracted from 28″, express 1° as 60′ and 1′ as 60″.
57°13′28″ = 56° + 1° + 13′ + 28″ = 56°73′28″ = 56°72′ + 1′ + 28″ = 56°72′88″.

Multiplying Angles Expressed in Degrees, Minutes, and Seconds

Example 1: Determine ∠1 in ***Figure A-11***.

$$∠1 = 2(62°47')$$

$$\begin{array}{r} 62°47' \\ \times \quad 2 \\ \hline 124°94' = 125°34' \text{ Ans} \end{array}$$

Note: 94′ = 1°34′.

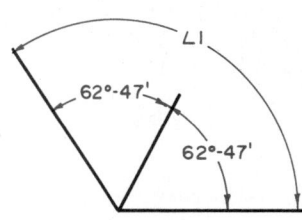

Figure A-11

Example 2: Refer to *Figure A-12*. Determine $\angle 2$ when $x = 41°27'42''$.

$$\angle 2 = 5x = 5(41°27'42'')$$

$$
\begin{array}{r}
41°27'42'' \\
\times \quad 5 \\
\hline
205°135'210''
\end{array}
= 205°138'30'' = 207°18'30'' \text{ Ans}
$$

Note: $210'' = 3'30''$ and $138' = 2°18'$.

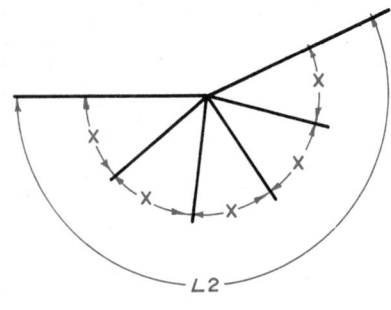

Figure A-12

Dividing Angles Expressed in Degrees, Minutes, and Seconds

Example 1: Determine $\angle 1$ and $\angle 2$ in *Figure A-13*.

$$\angle 1 = \angle 2 = 103°26' \div 2$$

Divide $103°$ by 2.

$$
\begin{array}{r}
51° \\
2\overline{)103°} \\
102° \\
\hline
1°
\end{array}
$$

$103 \div 2 = 51°$ plus a remainder of $1°$

Add the $1°$ ($60'$) to the $26'$.

$60' + 26' = 86'$.

Divide $86'$ by 2.

$$
\begin{array}{r}
43' \\
2\overline{)86'}
\end{array}
$$

$86' \div 2 = 43'$

Combine degrees and minutes.

$51° + 43' = 51°43'$ Ans

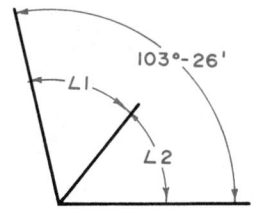

Figure A-13

Example 2: Determine $\angle 1$, $\angle 2$, and $\angle 3$ in *Figure A-14*.

$$\angle 1 = \angle 2 - \angle 3 = 128°37'21'' \div 3$$

Divide $128'$ by 3.

$$
\begin{array}{r}
42° \\
3\overline{)128°} \\
126° \\
\hline
2°
\end{array}
$$

$128 \div 3 = 42°$ plus a remainder of $2°$.

Add the $2°$ ($120'$) to the $37'$.

$120' + 37' = 157'$

Divide $157'$ by 3.

$$
\begin{array}{r}
52' \\
3\overline{)157'} \\
156' \\
\hline
1'
\end{array}
$$

$157' \div 3 = 52'$ plus a remainder of $1'$.

Add the $1'$ ($60''$) to the $21''$.

$60'' + 21'' = 81''$.

Divide $81''$ by 3.

$$
\begin{array}{r}
27'' \\
3\overline{)81''}
\end{array}
$$

$81'' \div 3 = 27''$

Combine.

$42° + 52' + 27'' = 42°52'27''$ Ans

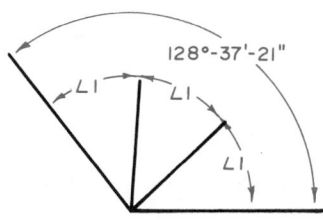

Figure A-14

Degrees, Minutes, Seconds— Decimal Degree Conversion

In the metric system the preferred method of angular measure is decimal degrees. However, degrees, minutes, and seconds are also used. It is important to be able to compute and express angular measurements in both the English and the metric systems and to express angular measure between systems.

Expressing Decimal Degrees As Degrees, Minutes, and Seconds

The measure of an angle given in the form of decimal degrees, such as $41.1938°$, must often be expressed as degrees, minutes, and seconds.

Procedure: To express decimal degrees as degrees, minutes, and seconds

- Multiply the decimal part of the degrees by $60'$ in order to obtain minutes.
- If the number of minutes obtained is not a whole number, multiply the decimal part of the minutes

by 60″ in order to obtain seconds. Round to the nearer whole second if necessary.
- Combine degrees, minutes, and seconds.

Example: Express 47.1938° as degrees, minutes, and seconds.

Multiply 0.1938 by 60′ to obtain minutes.
  60′(0.1938) = 11.6280′
Multiply 0.6280 by 60″ to obtain seconds.
  60″(0.6280) = 38″
Round to the nearer whole second.
Combine degrees, minutes, and seconds.
  47° + 11′ + 38″ = 47°11′38″ Ans

Expressing Degrees, Minutes, and Seconds As Decimal Degrees

Often an angle given in degrees and minutes is to be expressed as decimal degrees.

Procedure: To express degrees and minutes as decimal degrees

- Divide the minutes by 60.
- Combine whole degrees and decimal degrees. Round the answer to two places.

Example: Express 76°29′ as decimal degrees.

Divide 29 by 60 to obtain decimal degrees.
  29 ÷ 60 = 0.48°
Combine whole degrees with decimal degrees.
  76° + 0.48 = 76.48° Ans

When working with English and metric units of measure, it may be necessary to express angles given in degrees, minutes, and seconds as angles in decimal degrees.

Procedure: To express degrees, minutes, and seconds as decimal degrees

- Divide the seconds by 60 in order to obtain decimal minutes.
- Combine whole minutes and decimal minutes.
- Divide the total minutes by 60 in order to obtain decimal degrees.
- Combine whole degrees and decimal degrees. Round the answer to four decimal places.

Example: Express 23°18′44″ as decimal degrees.

Divide 44 by 60 to obtain decimal minutes.
  44 ÷ 60 = 0.7333′
Combine whole minutes and decimal minutes.
  18′ + 0.7333′ = 18.7333′
Divide 18.7333 by 60 to obtain decimal degrees.
  18.7333 ÷ 60 = 0.3122°
Combine whole degrees with decimal degrees.
  23° + 0.3122° = 23.3122° Ans

Types of Angles

An *acute angle* is an angle that is less than 90°. Angle 1 in *Figure A-15* is acute.

A *right angle* is an angle of 90°. Angle A in *Figure A-16* is a right angle.

An *obtuse angle* is an angle greater than 90° but less than 180°. Angle ABC in *Figure A-17* is an obtuse angle.

A *straight angle* is an angle of 180°. A straight line is a straight angle. Line EFG in *Figure A-18* is a straight angle.

A *reflex angle* is an angle greater than 180° and less than 360°. Angle 3 in *Figure A-19* is a reflex angle.

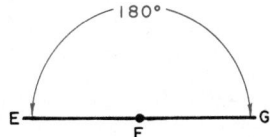

Figure A-18   Straight angle

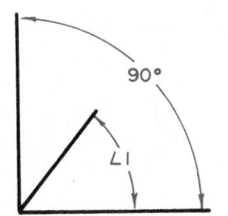

Figure A-15   Acute angle

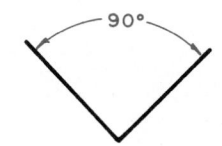

Figure A-16   Right angle

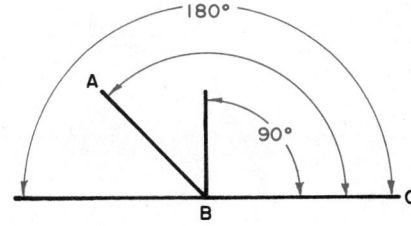

Figure A-17   Obtuse angle

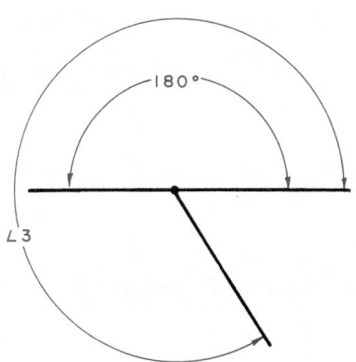

Figure A-19   Reflex angle

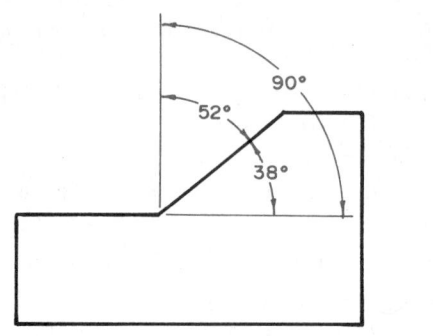

Figure A-20   Complementary angles

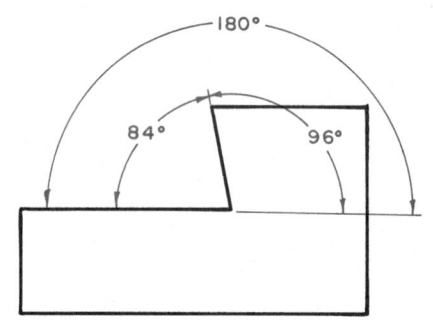

Figure A-21   Supplementary angles

Angles Formed by a Transversal

A *transversal* is a line that intersects (cuts) two or more lines. Refer to ***Figure A-24***. Line EF is a transversal since it cuts lines AB and CD.

Alternate interior angles are pairs of interior angles on opposite sides of the transversal. The angles have different vertices. For example, angles 3 and 5 and angles 4 and 6 are alternate interior angles.

Corresponding angles are pairs of angles, one interior and one exterior, on the same side of the transversal. The angles have different vertices. For example, angles 1 and 5, 2 and 6, 3 and 7, and 4 and 8 are corresponding angles.

If two parallel lines are intersected by a transversal, then the alternate interior angles are equal.

Refer to ***Figure A-25***. Given AB ∥ CD. (The symbol ∥ means parallel.) Conclusion: $\angle 3 = \angle 5$; therefore $\angle 5 = 62°$. $\angle 4 = \angle 6$; therefore $\angle 6 = 118°$.

If two lines are intersected by a transversal and a pair of alternate interior angles are equal, then the lines are parallel. Refer to ***Figure A-26***. Given: $\angle 1 = 70°$, $\angle 2 = 70°$; $\angle 1 = \angle 2$. Conclusion: AB ∥ CD.

If two parallel lines are intersected by a transversal, then the corresponding angles are equal. Refer to ***Figure A-27***. Given: AB ∥ CD. Conclusion: $\angle 1 = \angle 5 = 57°$, $\angle 2 = \angle 6 = 123°$, $\angle 3 = \angle 7 = 57°$, $\angle 4 = \angle 8 = 123°$.

Two angles are *complementary* when their sum is 90°. For example, $38° + 52° = 90°$. Therefore, 38° is the complement of 52°, and 52° is the complement of 38°. Refer to ***Figure A-20***.

Two angles are *supplementary* when their sum is 180°. For example, $84° + 96° = 180°$. Therefore, 84° is the supplement of 96°, and 96° is the supplement of 84°. Refer to ***Figure A-21***.

Two angles are *adjacent* if they have a common side and a common vertex. A *vertex* is the point where the two lines forming the angle meet. Angles 1 and 2 in ***Figure A-22*** are adjacent since they both contain the common side BC and the common vertex B. Angles 3 and 4 in ***Figure A-23*** are *not* adjacent because they do not have a common vertex.

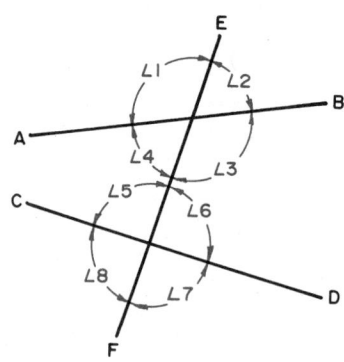

Figure A-24

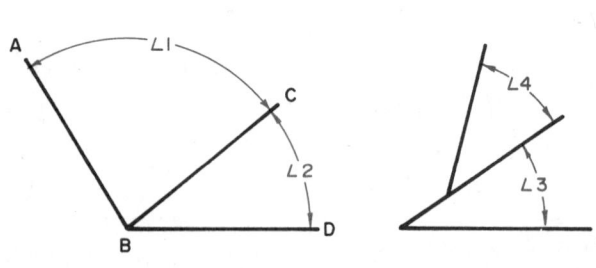

Figure A-22   ***Figure A-23***

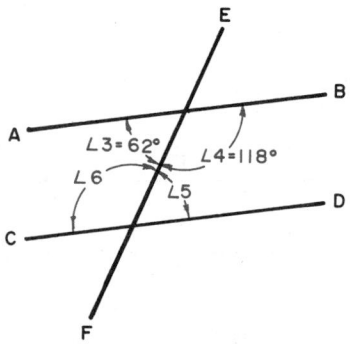

Figure A-25

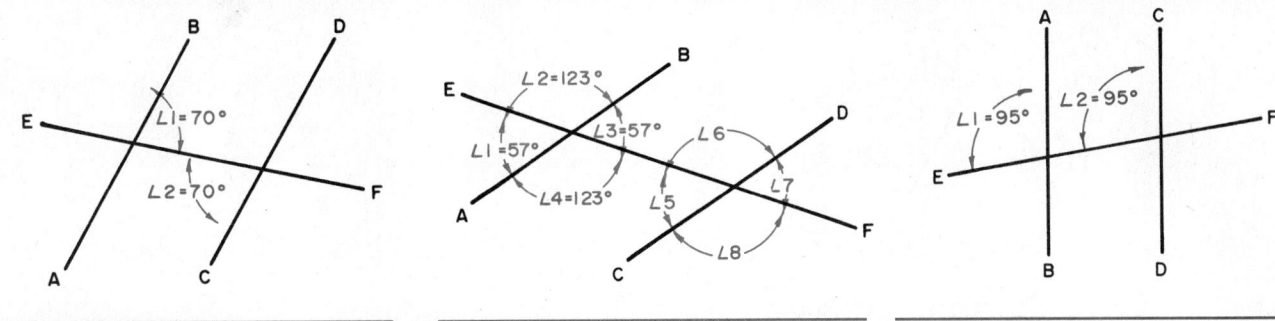

Figure A-26

Figure A-27

Figure A-28

If two lines are intersected by a transversal and a pair of corresponding angles are equal, then the lines are parallel. Refer to **Figure A-28**. Given: ∠1 = 95°, ∠2 = 95°;∠1 = ∠2. Conclusion: AB ∥ CD.

Types of Triangles

A *polygon* is a closed plane figure formed by three or more straight-line segments.

A *triangle* is a three-sided polygon; it is the simplest kind of polygon.

The sum of the three angles of any triangle equals 180°.

Scalene Triangle

A *scalene triangle* has three unequal sides. It also has three unequal angles. Triangle ABC shown in **Figure A-29** is scalene. Sides AB, AC, and BC are unequal, and angles A, B, and C are unequal.

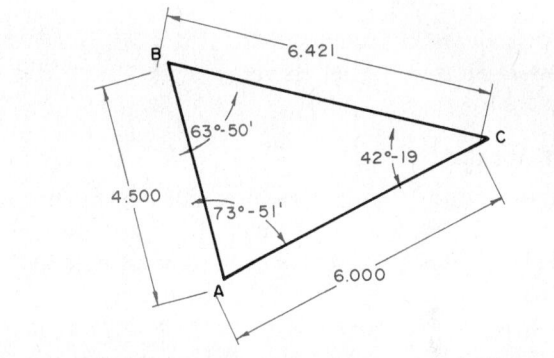

Figure A-29   Scalene triangle

Isosceles Triangle

An *isosceles triangle* has two equal sides, called *legs*, and two equal *base angles* (the angles opposite the legs). In isosceles triangle EFG in **Figure A-30**, leg EF = leg FG = 27.60 mm and base ∠E base ∠G = 40°.

In an isosceles triangle an altitude to the base bisects the base and the vertex angle.

An *altitude* is a line drawn from a vertex perpendicular to the opposite side.

To *bisect* means to divide into two equal parts.

Isosceles triangle ABC in **Figure A-31** is formed by the intersection of center lines between holes A, B, and C. Altitude CD bisects base AB; AD = DB = 6.840 ÷ 2 = 3.420 in. Altitude CD bisects vertex angle ACB; ∠1 = ∠2 = 71°38′ ÷ 2 = 35°49′.

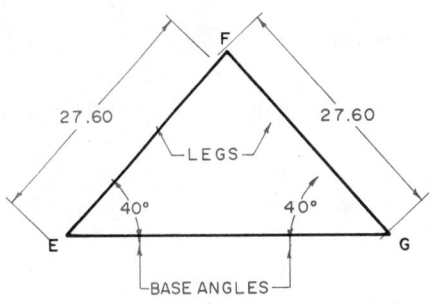

Figure A-30   Isosceles triangle

Equilateral Triangle

An *equilateral triangle* has three equal sides and three equal angles, each equal to 60°. In equilateral triangle DEF in **Figure A-32**, sides DE = DF = EF = 37.86 mm and ∠D = ∠E = ∠F = 60°.

In an equilateral triangle an altitude to any side bisects the side and the vertex angle.

Equilateral triangle HJK in **Figure A-33** is formed by the intersection of center lines between holes H, J, and

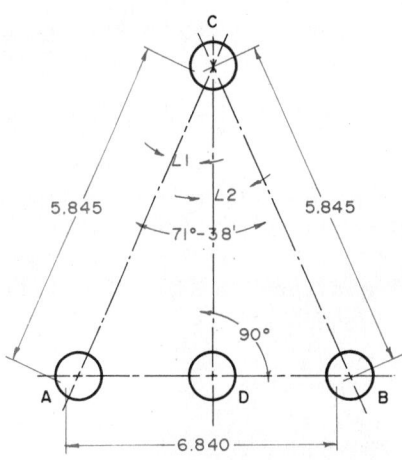

Figure A-31

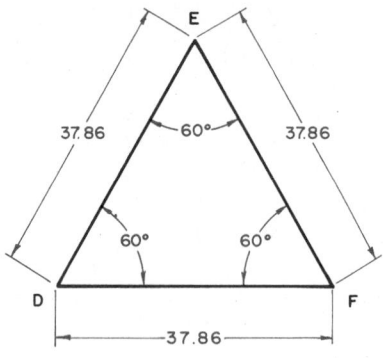

Figure A-32   Equilateral triangle

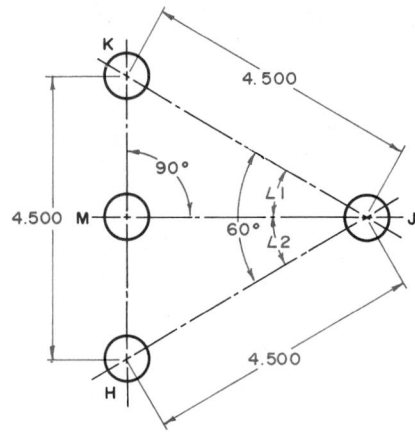

Figure A-33

K. Altitude MJ bisects side HK; HM = MK = 4.500 ÷ 2.250 in. Altitude MJ bisects vertex angle HJK; ∠1 = ∠2 = 60 ÷ 2 = 30°.

Right Triangle

A *right triangle* has one right or 90° angle. The side opposite the right angle is called the *hypotenuse*. The other two sides are called *legs*. In right triangle EFG in *Figure A-34*, ∠E = 90°, and FG is the hypotenuse.

In a right triangle the square of the hypotenuse is equal to the sum of the squares of the other two sides. This principle, called the *Pythagorean Theorem*, is often used for solving drafting problems. As applied to the right triangle with sides a, b, and c (hypotenuse) shown in *Figure A-35*, the Pythagorean Theorem formula is expressed as $c^2 = a^2 + b^2$.

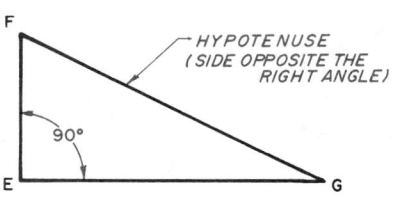

Figure A-34   Right triangle

Example 1:   Solve for side c when side a = 60.00 mm and side b = 80.00 mm, as shown in Figure A-35.

$c^2 = a^2 + b^2$

Substitute the given values and solve for c.

$c^2 = (60.00 \text{ mm})^2 + (80.00 \text{ mm})^2$

$c^2 = 3600 \text{ mm}^2 + 6400 \text{ mm}^2$

$c^2 = 10,000 \text{ mm}^2$

Add terms. Apply the root principle of equality. Extract the square root of both sides of the equation.

$\sqrt{c^2} = \sqrt{10,000 \text{ mm}^2}$

$c^2 = 100.00 \text{ mm Ans}$

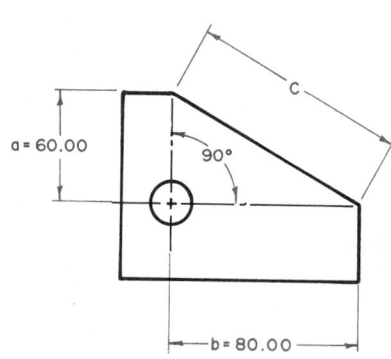

Figure A-35

Example 2:   In right triangle EFG in *Figure A-36*, side g = 5.800 in and hypotenuse e = 7.200 in.

Determine side f.

$e^2 = f^2 + g^2$

Substitute the given values and solve for f.

$(7.200 \text{ in})^2 = f^2 + (5.800 \text{ in})^2$

$51.840 \text{ sq in} = f^2 + 33.640 \text{ sq in}$

Apply the subtraction principle of equality. Subtract 33.640 sq in from both sides of the equation.

$51.840 \text{ sq in} = f^2 + 33.640 \text{ sq in}$

$\underline{-33.640 \text{ sq in} \qquad -33.640 \text{ sq in}}$

$18.200 \text{ sq in} = f^2$

Extract the square root.

$\sqrt{18.200 \text{ sq in}} = \sqrt{f^2}$

$4.266 \text{ in} = f^2$

$f^2 = 4.266 \text{ in Ans}$

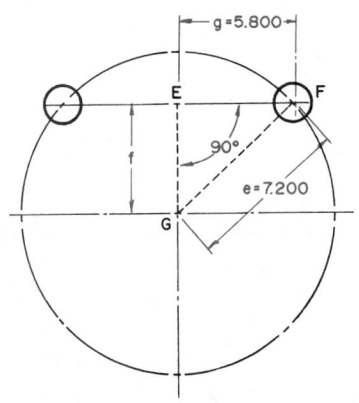

Figure A-36

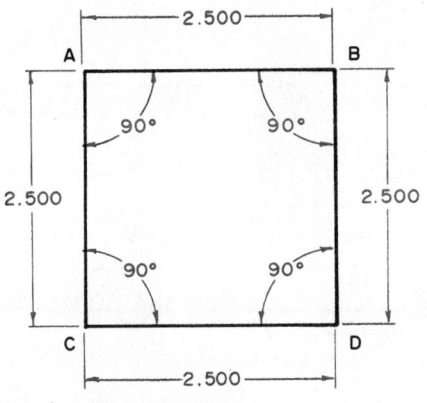

Figure A-37   Square

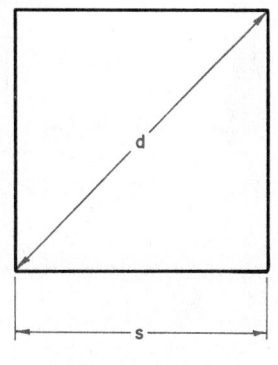

Figure A-38

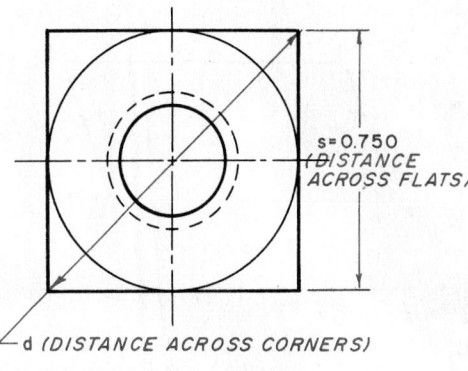

Figure A-39

Common Polygons

The types of polygons most widely used in drafting in addition to triangles are squares, rectangles, parallelograms, and regular hexagons. A *regular polygon* is one with equal sides and equal angles.

In addition to the description and properties of commonly used polygons, formulas are given that simplify computations when working with polygons.

Squares

A *square* is a regular four-sided polygon. Each angle equals 90°. In square ABCD in **Figure A-37**, AB = BC = CD = AD = 2.500 in, and ∠A = ∠B = ∠C = ∠D = 90°.

The following formulas are used with squares. Refer to **Figure A-38**.

$a = 1.414s$ where d = diagonal
$s = 0.7071\,d$ s = side
$A = s^2$ or $0.5d^2$ A = area

Example: A square nut is shown in **Figure A-39**. Given S (distance across flats) = 0.750 in, determine d (distance across corners). Substitute 0.750 for S in the formula $d = 1.414$S and solve.

$d = 1.414$S
$d = 1.414(0.750$ in)
$d = 1.061$ in Ans

Rectangles

A rectangle is a four-sided polygon with opposite sides parallel and equal. Each angle is 90°. In rectangle EFGH in **Figure A-40**, EF ∥ GH, EH ∥ FG; EF = GH = 40.85 mm, EH = FG = 75.20 mm; ∠E = ∠F = ∠G = ∠H = 90°.

The following formulas are used with rectangles. Refer to **Figure A-41**.

$d = \sqrt{l^2 + w^2}$ where d = diagonal
$l = \sqrt{d^2 - w^2}$ l = length
$w = \sqrt{d^2 - l^2}$ w = width
$A = lw$ A = area

Example: A rectangular sheet of aluminum is 30 in wide and 42 in long.

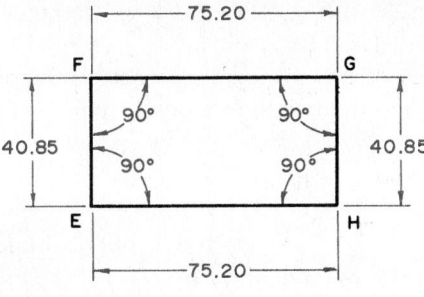

Figure A-40   Rectangle

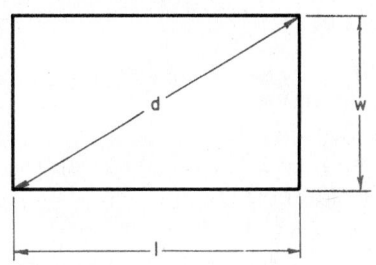

Figure A-41

Determine:
   a. the distance across opposite corners, diagonal d.
   b. the area or the sheet in square feet.
   a. Substitute the given values in the formula
      $d = \sqrt{l^2 + w^2}$ and solve.
      $d = \sqrt{l^2 + w^2}$
      $d = \sqrt{(42\text{ in})^2 + (30\text{ in})^2}$
      $d = \sqrt{1764\text{ sq in} + 900\text{ sq in}}$
      $d = \sqrt{2664\text{ sq in}}$
      $d = 51.614$ in Ans
   b. Substitute the values in the formula A = lw
      and solve. Since 1 sq ft = 144 sq in, 1260 sq in is divided by 144 to obtain square feet.
      $A = lw$
      $A = 42$ in (30 in)
      $A = 1260$ sq in
      1260 sq in ÷ 144 = 8.75 sq ft Ans

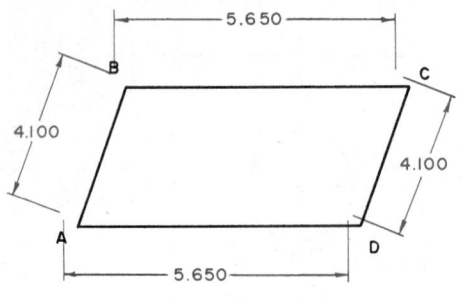

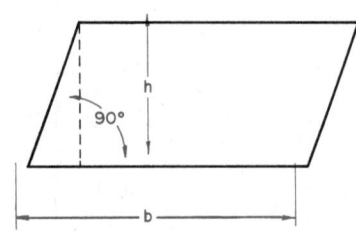

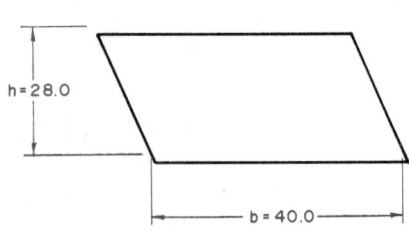

Figure A-42 Parallelogram

Figure A-43

Figure A-44

Parallelograms

A *parallelogram* is a four-sided polygon with opposite sides parallel and equal and opposite angles equal. In parallelogram ABCD in *Figure A-42*, AB ∥ CD, AD ∥ BC; AB = CD = 4.100 in, AD = BC = 5.650 in; ∠A = ∠C = 70°, ∠B = ∠D = 110°.

The area of a parallelogram is equal to the product of the base and the height or altitude. Refer to *Figure A-43*.

$$A = bh \qquad \text{where} \qquad A = \text{area}$$
$$b = \text{base}$$
$$h = \text{height or altitude.}$$

Note that h is perpendicular to b.

Example: A sheet-metal detail in the shape of a parallelogram is shown in *Figure A-44*. Determine the area in square feet. Substitute values in the formula $A = bh$ and solve.

$A = bh$
$A = 28$ in (40 in)
$A = 1120$ sq in

Since 1 sq ft. = 144 sq in, 1120 sq in is divided by 144 to obtain square feet.

1120 sq in ÷ 144 = 7.78 sq ft Ans

Regular Hexagons

A *regular hexagon* is a six-sided figure with all sides equal and all angles equal. Each angle equals 120°. In the regular hexagon ABCDEF shown in *Figure A-45*, AB = BC = CD = DE = EF = AF = 50.25 mm, and ∠A = ∠B = ∠C = ∠D = ∠E = ∠F = 120°.

The following formulas are used with regular hexagons. Refer to *Figure A-46*.

$$c = 2s \qquad \text{where} \quad s = \text{side}$$
$$s = 0.5c \qquad\qquad c = \text{distance across corners}$$
$$f = 1.732s \qquad\qquad f = \text{distance across flats}$$
$$s = 0.577f \qquad\qquad A = \text{area}$$
$$c = 1.55f$$
$$f = 0.866c$$
$$A = 2.598s^2 = 0.650c^2 = 0.866f^2$$

Example: A hexagonal nut is shown in *Figure A-47*. Given f (distance across flats) determine:
a. s (length of side)
b. c (distance across corners)
　a. Substitute 0.875 for f in the formula $s = 0.577f$ and solve.
　$s = 0.577f$
　$s = 0.577(0.875$ in)
　$s = 0.505$ in Ans
　b. Substitute 0.875 for f in the formula $c = 1.555f$ and solve.
　$c = 1.155f$
　$c = 1.155(0.875$ in)
　$c = 1.011$ in Ans
　$c = 1.155f$

The following principle is widely applied when determining unknown angles in any type of polygon.

The sum of the interior angles of a polygon of N sides is equal to (N − 2) times 180°.

Example: Determine ∠1 in the polygon in *Figure A-48*. Count the number of sides. The polygon has six

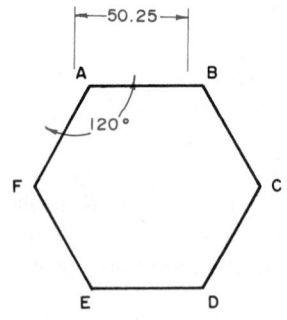

Figure A-45 Regular hexagon

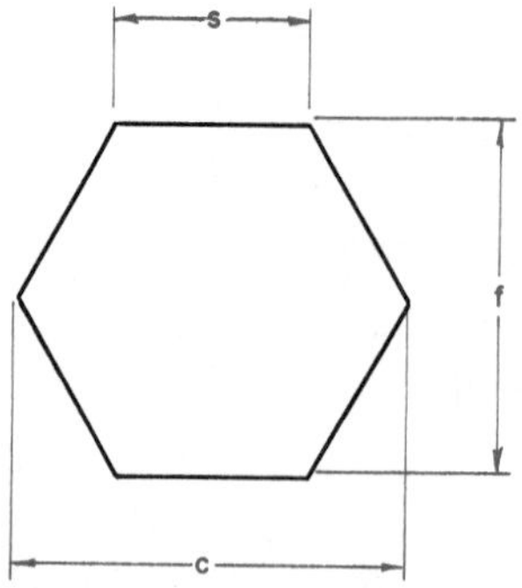

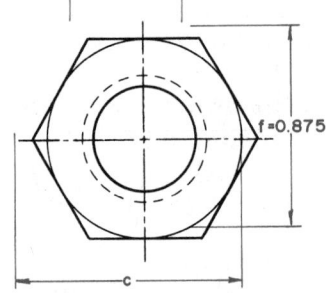

Figure A-46

Figure A-47

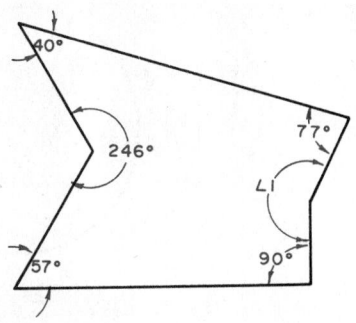

Figure A-48

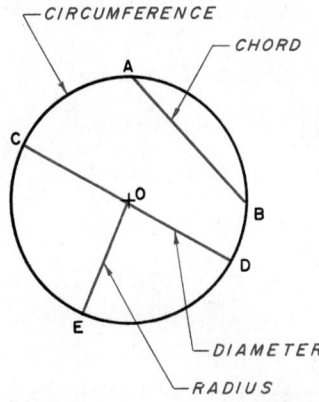

sides: N = 6. The sum of the six angles is (N – 2) 180°
= (6 – 2) 180° = 4 (180°) = 720°.

Add the five known interior angles and subtract from
720° to find ∠1.
∠1 = 720° – (57° + 246° + 40° + 77° + 90°)
∠1 = 720° – 510°
∠1 = 210° Ans

Definitions of Properties of Circles

The following terms are commonly used to describe the
properties of circles. These properties are often applied
in mechanical drafting and design situations.

A *circle* is a closed curve of which every point on the
curve is equally distant from a fixed point called the center.

Refer to **Figure A-49** for the following definitions.

A *circumference* is the length of the curved line
which forms the circle.

A *chord* is a straight-line segment that joins two points
on the circle. AB is a chord.

A *diameter* is a chord that passes through the center
of a circle. CD is a diameter.

A *radius* (plural radii) is a straight-line segment that
connects the center of the circle with a point on the circle. The radius is one-half the diameter. OE is a radius.

Refer to **Figure A-50** for the following definitions.

An *arc* is that part of a circle between any two points
on the circle. The symbol ⌒ written above the letters
means arc. A͡B is an arc.

A *tangent* to a circle is a straight line that touches the
circle at one point only. The point on the circle
touched by the tangent is called the *point of tangency*
or *tangent point*. CD is a tangent, P is a tangent point.

A *secant* is a straight line that passes through a circle
and intersects the circle at two points. EF is a secant.

Refer to **Figure A-51** for the following definitions.

A *segment* is that part of a circle bounded by a chord
and its arc. The shaded figure ABC is a segment.

A *sector* is that part of a circle bounded by two radii
and the arc intercepted by the radii. The shaded figure
EOF is a sector.

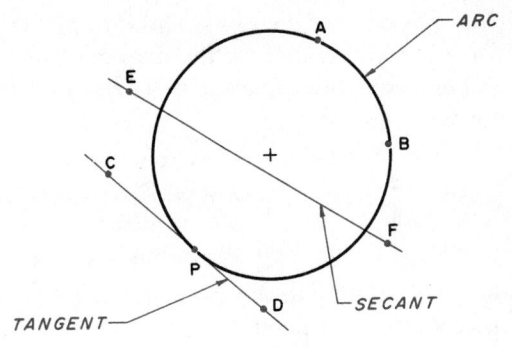

Figure A-50

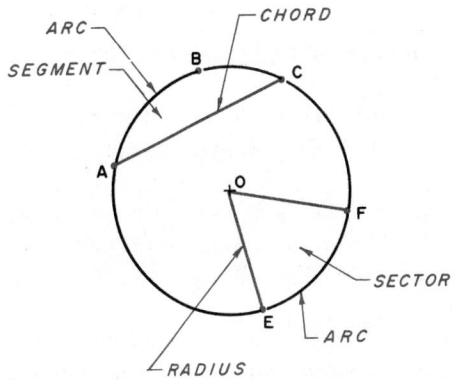

Figure A-51

Refer to **Figure A-52** for the following definitions.

A *central angle* is an angle whose vertex is at the
center of a circle and whose sides are radii. Angle MON
is a central angle.

An *inscribed angle* is an angle whose vertex is on the
circle and whose sides are chords. Angle SRT is an
inscribed angle.

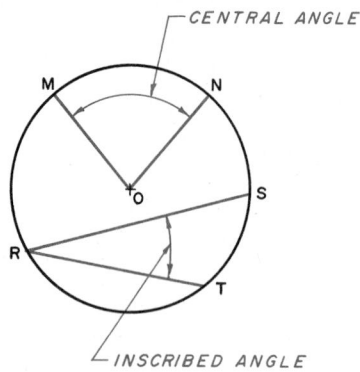

Figure A-52

Geometric Principles of Circle Circumference, Central Angles, Arcs, and Tangents

The circumference of a circle is equal to pi (π) times the diameter. Generally, for the degree of precision required in machining applications, a value of 3.1416 is used for π.

$$C = \pi d \quad \text{where} \quad C = \text{circumference}$$
$$\text{or} \qquad\qquad\qquad \pi = \text{pi}$$
$$C = 2\pi r \qquad\qquad d = \text{diameter}$$
$$r = \text{radius}$$

Example 1: Compute the circumference of a circle with a 50.70-mm diameter.

$$C = \pi d = 3.1416(50.70\text{mm}) = 159.28 \text{ mm Ans}$$

Example 2: Determine the radius of a circle with a circumference of 14.860 in.

$$C = 2\pi r$$
$$14.860 \text{ in} = 2(3.1416)r$$
$$r = 2.365 \text{ in Ans}$$

In the same circle or in equal circles, equal chords cut off equal arcs.

Refer to *Figure A-53*. Given: Circle A = Circle B and chords CD = EF = GH = MS. Conclusion: CD = EF = GH = MS.

In the same circle or in equal circles, equal central angles cut off equal arcs. Refer to *Figure A-54*. Given: Circle D and ∠1 = ∠2 = ∠3 = ∠4. Conclusion: AB = FG = HK = MP.

In the same circle or in equal circles, two central angles have the same ratio as the arcs that are cut off by the angles.

Example: Refer to *Figure A-55*. Circle E = Circle F. If ∠COD = 90°, ∠EOF = 50°, CD = 1.400″, and GH = 2.100″, determine (a) the length of EF and (b) the size of ∠GOH. All dimensions are in inches.

a. Set up a proportion between CD and EF with their respective central angles. Solve for EF.

$$\frac{\angle COD}{\angle EOF} = \frac{CD}{EF}$$

$$\frac{90°}{50°} = \frac{1.400″}{EF}$$

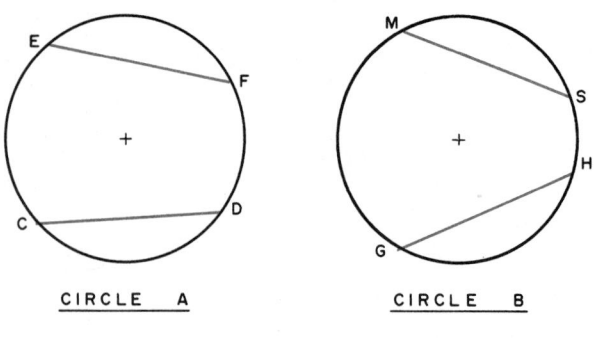

Figure A-53

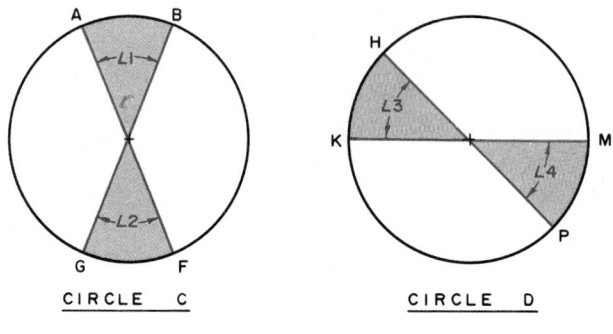

Figure A-54

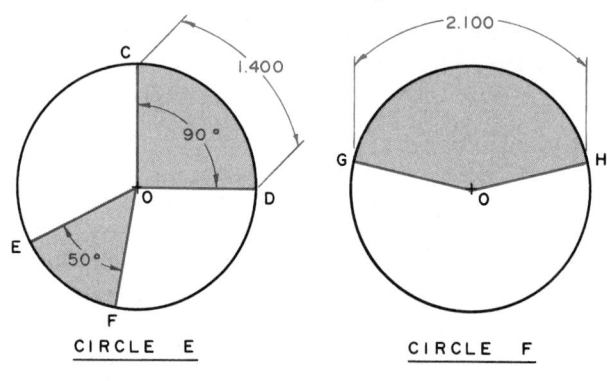

Figure A-55

$$9EF = 5(1.400″)$$
$$EF = \frac{5(1.400″)}{9}$$

$$EF = 0.778″ \text{ Ans}$$

b. Set up a proportion between CD and GH with their central angles. Solve for ∠GOH.

$$\frac{\angle COD}{\angle GOH} = \frac{CD}{GH}$$

$$\frac{90°}{\angle GOH} = \frac{1.400″}{2.100″}$$

$$1.400(\angle GOH) = 90°(2.100)$$
$$\angle GOH = \frac{90°(2.100)}{1.400}$$

$$\angle GOH = 135° \text{ Ans}$$

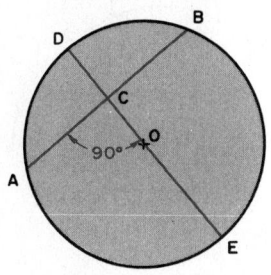

Figure A-56

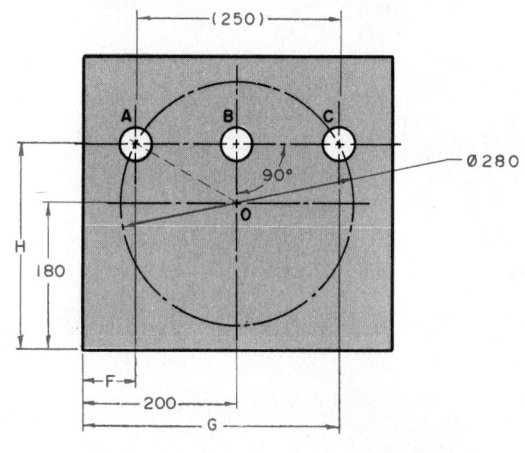

Figure A-57

A line drawn from the center of a circle perpendicular to a chord bisects the chord and the arc cut off by the chord. The perpendicular bisector of a chord passes through the center of a circle.

Refer to **Figure A-56**. Given: Diameter DE is perpendicular to chord AB. Conclusion: AC = BC and AD = BD and AE = BE.

The use of this principle with the Pythagorean Theorem has wide practical application in mechanical technology.

Example: Holes A, B, C are to be drilled in the plate shown in **Figure A-57**. The centers of holes A and C lie on a 280-mm diameter circle. The center of hole B lies on the intersection of chord AC and segment OB, which is perpendicular to AC. Compute working dimensions F, G, and H. All dimensions are in millimetres.

Compute dimension F: Applying the principle, AC is bisected by OB.

$$AB = BC = 250 \text{ mm} \div 2 = 125 \text{ mm}$$
$$F = 200 \text{ mm} - 125 \text{ mm} = 75 \text{ mm Ans}$$

Compute dimension G.

$$G = 200 \text{ mm} + 125 \text{ mm} = 325 \text{ mm Ans}$$

Compute dimension H. In right triangle ABO, AB = 125 mm, AO = 280 mm ÷ 2 = 140 mm. Compute OB by applying the Pythagorean Theorem.

$$AO^2 = OB^2 + AB^2$$
$$(140 \text{ mm})^2 = OB^2 + (125 \text{ mm})^2$$
$$OB = 63.05 \text{ mm}$$

Hence, H = 180 mm + 63.05 mm = 243.05 mm Ans

A line perpendicular to a radius at its extremity is tangent to the circle. A tangent is perpendicular to a radius at its tangent point.

Refer to **Figure A-58**.

Example 1:

Given: Line AB is perpendicular to a radius CO at point C. Conclusion: Line AB is a tangent.

Example 2:

Given: Tangent DE passes through point F of radius FO. Conclusion: Tangent DE is perpendicular to radius FO.

Two tangents drawn to a circle from a point outside the circle are equal. The angle at the outside

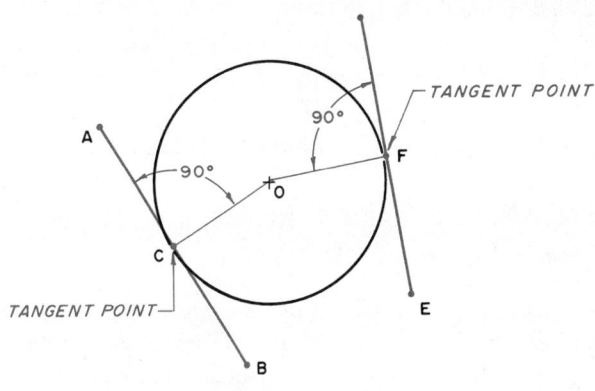

Figure A-58

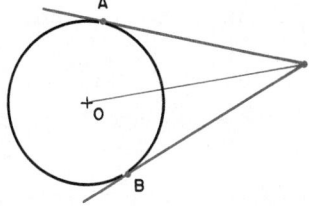

Figure A-59

point is bisected by a line drawn from the point to the center of the circle.

Refer to **Figure A-59**.

Example 1:

Given: Tangents AP and BP are drawn to the circle from point P. Conclusion: AP = BP

Example 2:

Given: Line OP extending from outside point P to center O. Conclusion: ∠APO = ∠BPO

If two chords intersect inside a circle, the product of the two segments of one chord is equal to the product of the two segments of the other chord.

Refer to **Figure A-60**.

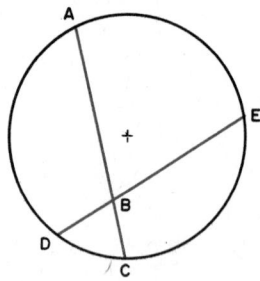

Figure A-60

Example 1:

Given: Chords AC and DE intersect at point B.
Conclusion: AB(BC) = BD(BE).

Example 2: If AB = 7.5 in, BC = 2.8 in, and BD = 2.1 in, determine the length of BE.

$$AB(BC) = BD(BE)$$
$$7.5(2.8) = 2.1(BE)$$
$$21.0 = 2.1BE$$
$$BE = 10.0 \text{ in Ans}$$

Geometric Principles of Angles Formed Inside a Circle

A central angle is equal to its intercepted arc. (An intercepted arc is an arc cut off by a central angle.) Refer to *Figure A-61*. Given AB = 78°. Conclusion: ∠AOB = 78°.

An angle formed by two chords intersecting inside a circle is equal to one-half the sum of its two intercepted arcs.

Refer to *Figure A-62*.

Example 1: Given: Chords CD and EF intersect at point P.
Conclusion: ∠EPD = ½(CF + DE).

Example 2: If CF = 106° and ED = 42°, determine ∠EPD. ∠EPD = ½(106°) + 42° = 74° Ans

An inscribed angle is equal to one-half of its intercepted arc. Refer to *Figure A-63*. Given: AC = 105°. Conclusion: ∠ABC = ½AC = ½(105°) = 52°30'.

Arc Length Formula

Consider a complete circle as an arc of 360°. The ratio of the number of degrees of an arc to 360° is the frac-

tional part of the circumference that is used to find the length of an arc. *The length of an arc equals the ratio of the number of degrees of the arc to 360° times the circumference.*

$$\text{Arc Length} = \frac{\text{Arc Degrees}}{360°}(2\pi r)$$

or

$$\text{Arc Length} = \frac{\text{Central Angle}}{360°}(2\pi r)$$

Example: Refer to *Figure A-64*. ABC = 130° and the radius is 120 mm. Determine the arc length ABC.

$$\text{Arc Length} = \frac{\text{Arc Degrees}}{360°}(2\pi r)$$
$$= \frac{130°}{360°}[2(3.1416)(120 \text{ mm})]$$
$$= 272.271 \text{ mm Ans}$$

Geometric Principles of Angles Formed on a Circle and Angles Formed Outside a Circle

An angle formed by a tangent and a chord at the tangent point is equal to one-half of its intercepted arc.

Example 1: Refer to *Figure A-65*. Tangent CD meets chord AB at tangent point A and AEB = 110°. Determine ∠CAB.

$$\angle CAB = ½AEB = ½(110°) = 55° \text{ Ans}$$

Example 2: Refer to *Figure A-66*. The centers of three holes lie on line ABC. Line ABC is tangent to

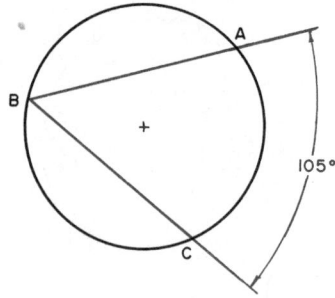

Figure A-63

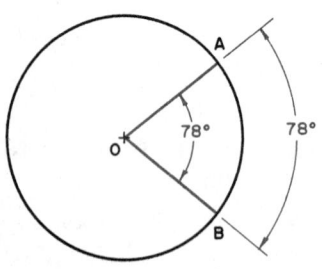

Figure A-61   *Figure A-62*

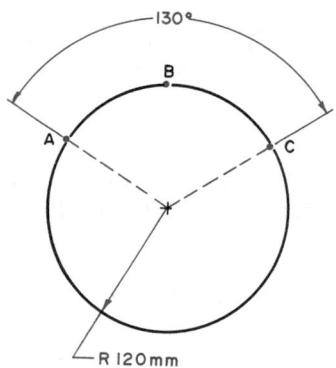

Figure A-64

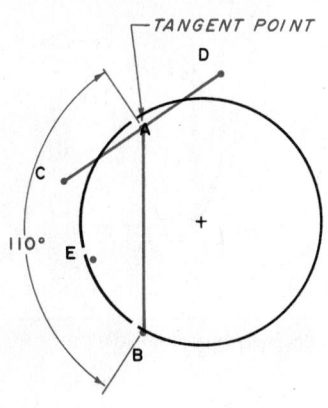

Figure A-65

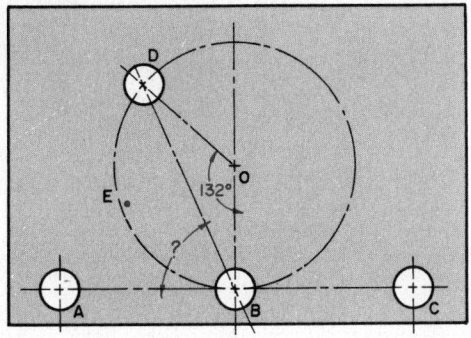

Figure A-66

circle O at hole-center B. The hole-center D, of a fourth hole, lies on the circle. Determine ∠ABD. A central angle is equal to its intercepted arc.

DEB = ∠DOB = 132°

An angle formed by a tangent and a chord is equal to one-half of its intercepted arc.

∠ABD = ½DEB = ½(132°) = 66° Ans

An angle formed at a point outside a circle by two secants, two tangents, or a secant and a tangent is equal to one-half the difference of the intercepted arcs.

Two Secants

Refer to **Figure A-67**.

Example 1:

Given: Secants AP and DP meet at point P and intercept BC and AD. Conclusion: ∠P = ½(AD – BC).

Example 2: If AD = 85°40′ and BC = 39°17′, find ∠P.

∠P = ½(AD – BC) = ½(85°40′ – 39°17′) = ½(46°23′) = 23°11′30″ Ans

Two Tangents

Refer to **Figure A-68**.

Example 1:

Given: Tangents DP and EP meet at point P and intercept DE and DCE. Conclusion: ∠P = ½(DCE – DE).

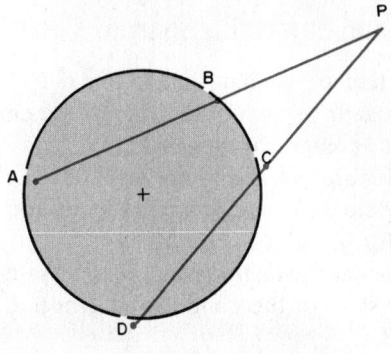

Figure A-67

Example 2: If DCE = 253°37′ and DE = 106°23′, determine ∠P.

∠P = ½(DCE – DE) = ½(253°37′ – 106°23′) = ½(147°14′) = 73°37′ Ans

A Tangent and a Secant

Example 1: Refer to **Figure A-69**. Given: Tangent AP and secant CP meet at point P and intercept AC and AB. Conclusion: ∠P = ½(AC – AB).

Example 2: If AC – 126°38′ and AB = 68°58′, determine ∠P.

∠P = ½(AC – AB) = ½(126°38′ – 68°58′) = ½(57°40′) = 28°50′ Ans

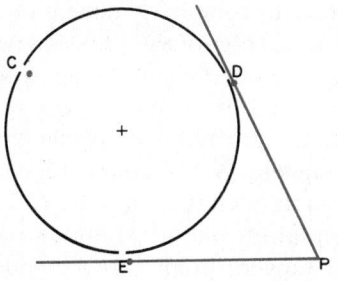

Figure A-68

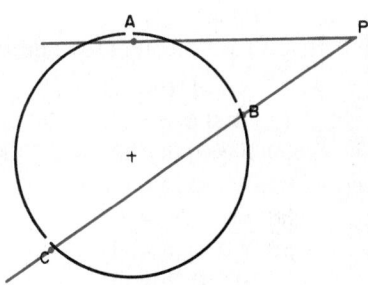

Figure A-69

Internally and Externally Tangent Circles

Two circles that are tangent to the same line at the same point are tangent to each other. Circles are either internally tangent or externally tangent.

Two circles are *internally tangent* if both circles are on the same side of the common tangent line. Refer to *Figure A-70*.

Two circles are *externally tangent* if the circles are on opposite sides of the common tangent line. Refer to *Figure A-71*.

If two circles are either internally tangent or externally tangent, a line connecting the centers of the circles passes through the point of tangency and is perpendicular to the tangent line.

Internally Tangent Circles

Example 1: Refer to *Figure A-72*. Given: Circle D and Circle E are internally tangent at point C. D is the center of Circle D, and E is the center of Circle E. Line AB is tangent to both circles at point C. Conclusion: An extension of line DE passes through tangent point C and line CDE is perpendicular to tangent line AB.

This principle is often used as the basis for computing dimensions of parts on which two or more radii blend to give a smooth curved surface. This type of application is illustrated by the following example.

Example 2: A part is to be drawn as shown in *Figure A-73*. The proper locations of the two radii will result in a smooth curve from point A to point B. Note that the curve from A to B is not an arc of one circle, but is made up of two circles of different sizes. In order to completely dimension the part the location to the center of the 12.000-in radius (dimension x) must be determined. Compute x. All dimensions are in inches. Refer to *Figure A-74*. The 12.000-in radius arc and the 25.000-in radius arc are internally tangent. A line connecting arc centers F and H passes through tangent point C. Tangent point C is the endpoint of the 25.000-in radius: (FH = 25.000-in). Tangent point C is the endpoint of the 12.000-in radius: CF = 12.000-in.

$$FH = 25.000\text{-in} - 12.000\text{-in} = 13.000\text{-in}.$$

Since BFE is vertical and AEH is horizontal, ∠FEH = 90°. In right triangle FEH, FH = 13.000-in, FE = 21.000-in − BF = 9.000-in.

Apply the Pythagorean Theorem to compute EH.

$$FH^2 = EH^2 + FE^2$$
$$(13.000 \text{ in})^2 = EH^2 + (9.000 \text{ in})^2$$
$$169.000 \text{ sq in} = EH^2 + 81.000 \text{ sq in}$$
$$169.000 \text{ sq in} - 81.000 \text{ sq in} = EH^2$$
$$88.000 \text{ sq in} = EH^2$$
$$\sqrt{88.000} \text{ sq in} = EH$$
$$9.381 \text{ in} = EH$$
$$x = EH,$$
$$x = 9.381 \text{ in Ans}$$

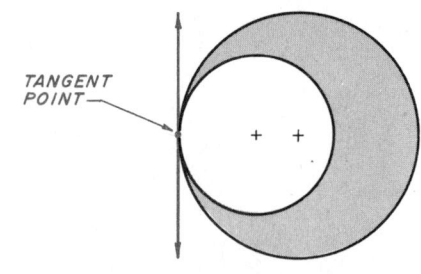

Figure A-70   Internally tangent circles

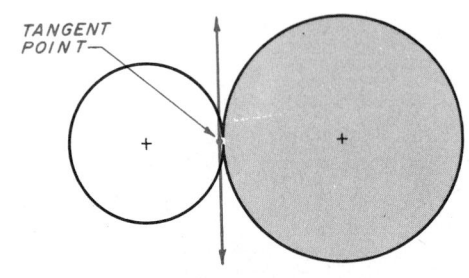

Figure A-71   Externally tangent circles

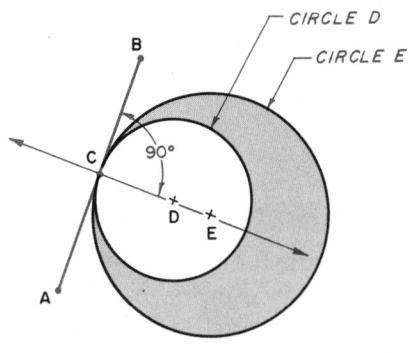

Figure A-72

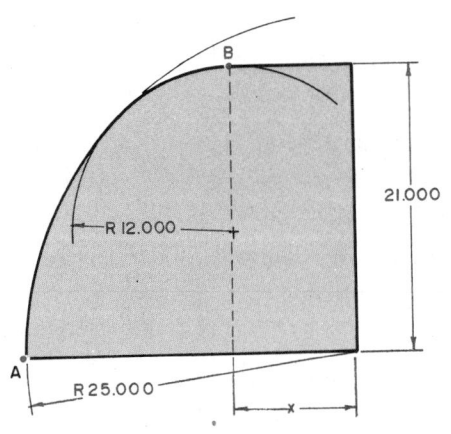

Figure A-73

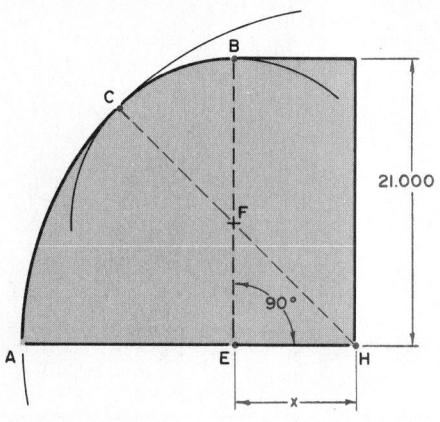

Figure A-74

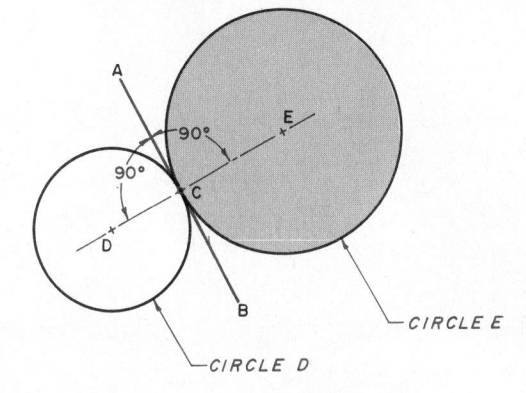

Figure A-75

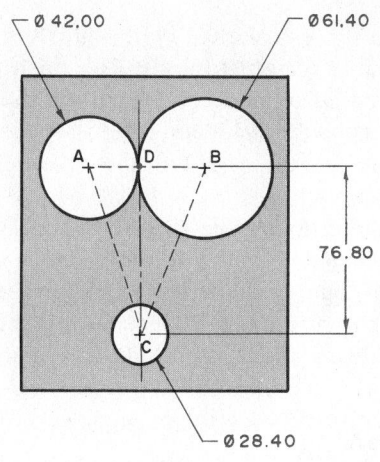

Figure A-76

Externally Tangent Circles

Example 1: Refer to *Figure A-75*. Given: Circle D and Circle E are externally tangent at point C. D is the center of Circle D, and E is the center of Circle E. Line AB is tangent to both circles at point C. Conclusion: Line DE passes through tangent point C, and line DE is perpendicular to tangent line AB at point C.

Example 2: Three holes are to be bored in the steel plate of *Figure A-76*. The 42.00-mm and 61.40-mm diameter holes are tangent at point C. CD is the common tangent line. Determine the distances between hole centers (AB, AC, and BC). All dimensions are in millimetres. Compute AB. AB connects the centers of two tangent circles; AB passes through tangent point D.

$$AB = AD + DB = 21.00 \text{ mm} + 30.70 \text{ mm} =$$
$$51.70 \text{ mm Ans}$$

Compute AC and BC. Since AB connects the centers of two tangent circles, AB is perpendicular to tangent line DC. Triangle ADC and triangle BDC are right triangles. Apply the Pythagorean Theorem.

In right triangle ADC, AD = 21.00 mm and DC = 76.80 mm.

$$AC^2 = AD^2 + DC^2$$
$$AC^2 = (21.00 \text{ mm})^2 + (76.80 \text{ mm})^2$$
$$AC^2 = 441.00 \text{ mm}^2 + 5898.24 \text{ mm}^2$$
$$AC^2 = 6339.24 \text{ mm}^2$$
$$AC = 79.62 \text{ mm Ans}$$

In right triangle BDC, DB = 30.70 mm and DC = 76.80 mm.

$$BC^2 = DB^2 + DC^2$$
$$BC^2 = (30.70 \text{ mm})^2 + (76.80 \text{ mm})^2$$
$$BC^2 = 942.49 \text{ mm}^2 + 5898.24 \text{ mm}^2$$
$$BC^2 = 6840.73 \text{ mm}^2$$
$$BC = \sqrt{6840.73} \text{ mm}^2$$
$$BC = 82.71 \text{ mm Ans}$$

Trigonometry: Trigonometric Functions

Trigonometry is the branch of mathematics used to compute unknown angles and sides of triangles. Many drafting problems that cannot be solved by the use of geometry alone are easily solved by trigonometry.

Ratio of Right Triangle Sides

In a right triangle the ratio of two sides of the triangle determines the sizes of the angles, and the angles determine the ratio of two sides. Refer to the triangles in *Figures A-77 through A-79*. The size of angle A is determined by the ratio of side *a* to side *b*. When side *a* = 1 in and side *b* = 2 in (Figure A-77), the ratio of *a* to *b* is 1:2 or 1/2. If side *a* is increased to 2 in and side *b* remains 2 in (Figure A-78), the ratio of *a* to *b* is 1:1 or 1/1. Observe the increase in angle A (Figure A-79) as the ratio changed from 1/2 to 1/1.

Note: The symbol for a right angle is ∟, which is shown at the vertex of the angle.

Identifying Right Triangles by Name

The sides of a right triangle are named opposite side, adjacent side, and hypotenuse. The *hypotenuse* (hyp) is

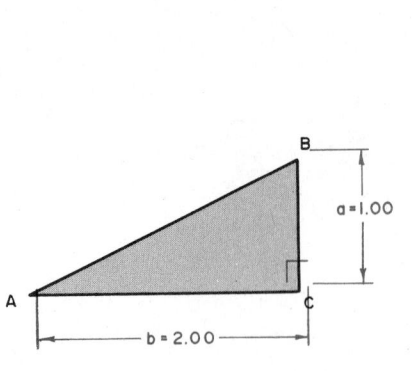

Figure A-77

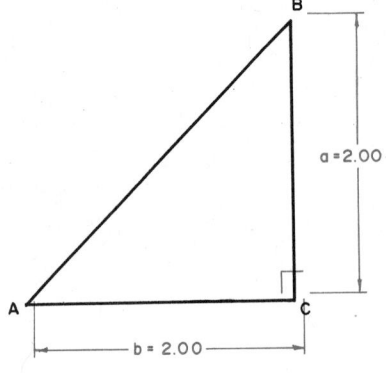

Figure A-78

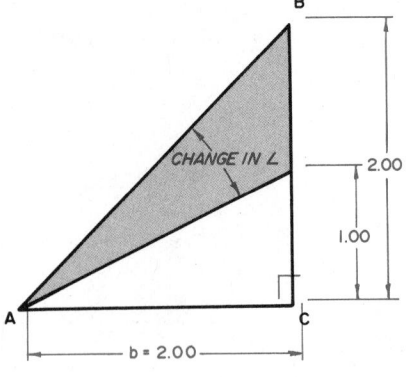

Figure A-79

always the side opposite the right angle. It is always the longest side of a right triangle. The positions of the opposite and adjacent sides depend on the reference angle. The *opposite side* (opp side) is opposite the reference angle, and the *adjacent side* (adj side) is next to the reference angle.

In the triangle in *Figure A-80* showing ∠A as the reference angle, side *b* is the adjacent side, and side *a* is the opposite side. In the triangle in *Figure A-81* showing ∠B as the reference angle, side *b* is the opposite side, and side *a* is the adjacent side. It is important to be able to identify the opposite and adjacent sides of right triangles with reference to any angle regardless of the positions of the triangles.

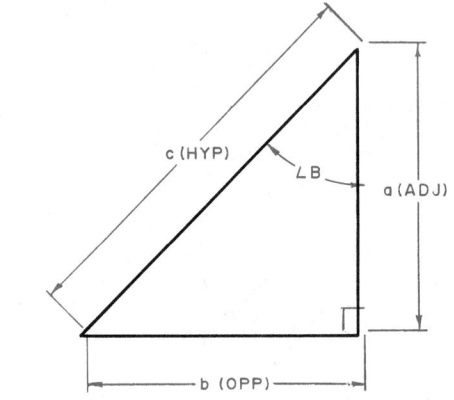

Figure A-81

Trigonometric Functions: Ratio Method

There are two methods of defining trigonometric functions: the unity or unit circle method, and the ratio method. Only the ratio method is presented here.

Since a triangle has three sides and a ratio is the comparison of any two sides, there are six different ratios. The names of the ratios are *sine, cosine, tangent, cotangent, secant,* and *cosecant.*

The six trigonometric functions are defined in the table in *Figure A-82* in relation to the triangle shown in

FUNCTION	SYMBOL	DEFINITION OF FUNCTION
sine of angle A	sine A	$\sin A = \dfrac{\text{opp side}}{\text{hyp}} = \dfrac{a}{c}$
cosine of angle A	cos A	$\cos A = \dfrac{\text{adj side}}{\text{hyp}} = \dfrac{b}{c}$
tangent of angle A	tan A	$\tan A = \dfrac{\text{opp side}}{\text{adj side}} = \dfrac{a}{b}$
cotangent of angle A	cot A	$\cot A = \dfrac{\text{adj side}}{\text{opp side}} = \dfrac{b}{a}$
secant of angle A	sec A	$\sec A = \dfrac{\text{hyp}}{\text{adj side}} = \dfrac{c}{b}$
cosecant of angle A	csc A	$\csc A = \dfrac{\text{hyp}}{\text{opp side}} = \dfrac{c}{a}$

Figure A-82

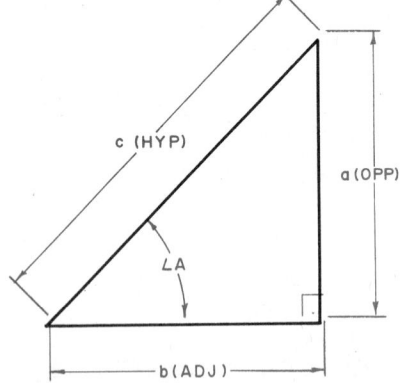

Figure A-80

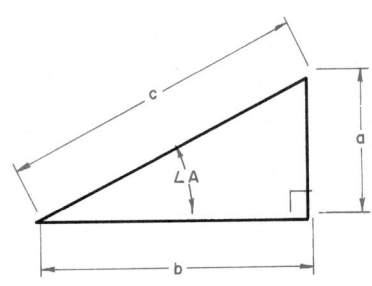

Figure A-83

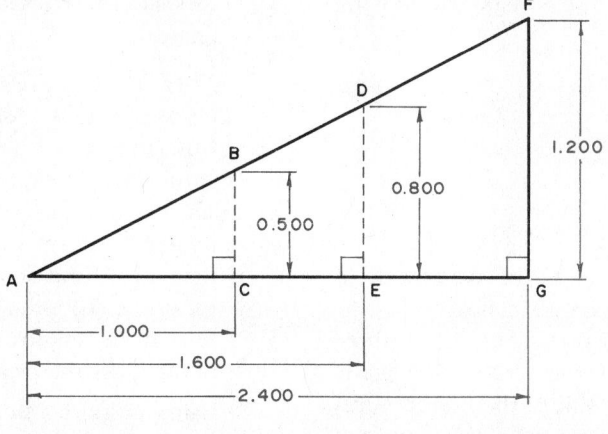

Figure A-84

Figure A-83. The reference angle is A, and adjacent side is *b*, the opposite side is *a*, and the hypotenuse is *c*.

To properly use trigonometric functions it is essential to know that the function of an angle depends upon the ratio of the sides and not upon the size of the triangle. Similar triangles are alike in shape but different in size. The functions of similar triangles are the same regardless of the sizes of the triangles, since the sides of similar triangles are proportional. For example, in the similar triangles shown in *Figure A-84*, the functions of angle A are the same for the three triangles. The equality of the tangent function is shown. Each of the other five functions has equal values for the three similar triangles.

Note: The symbol △ means triangle.

$$\text{In} \triangle ABC, \tan\angle A = \frac{0.500}{1.000} = 0.500$$

$$\text{In} \triangle ADE, \tan\angle A = \frac{0.800}{1.600} = 0.500$$

$$\text{In} \triangle AFG, \tan\angle A = \frac{1.200}{2.400} = 0.500$$

In a right triangle the reference angle is greater than 0° and less than 90°. Values for the six trigonometric functions from 0° to 90° are available in degree-minute and decimal-degree tables. Scientific electronic calculators provide a quick and convenient determination of trigonometric functions. It is assumed that users of this text will use calculators; therefore, trigonometric function tables are not provided. However, if a calculator is not used, supplementary trigonometric function tables are generally readily available. Conversion between decimal-minutes and decimal-degree angular measure may be required. If necessary, refer to the section Degrees, Minutes, Seconds—Decimal Degree Conversion.

Trigonometry: Basic Calculations of Angles

In order to solve for an unknown angle of a right triangle when neither acute angle is shown, at least two sides must be known. An understanding of the procedures required for solving for unknown angles is essential to the drafter.

Procedure: *To determine an unknown angle when two sides of a right triangle are known*

- Identify two given sides as adjacent, opposite, or hypotenuse in relation to the desired angle.
- Determine the functions that are ratios of the sides identified in relation to the desired angle.

Note: Two of the six trigonometric functions are ratios of the two known sides. Either of the two functions can be used. Both produce the same value for the unknown, except for cotangents, secants, and cosecants of angles less than 15° and tangents, secants, and cosecants of angles greater than 75°.

- Choose one of the two functions, substitute the given sides in the ratio, and divide.
- Using either a calculator or a trigonometric function table, determine the angle that corresponds to the quotient obtained.

Example 1: Determine ∠B of the triangle in *Figure A-85*. All dimensions are in millimetres.

In relation to ∠B, the 290.000-mm side is the adjacent side and the 575.00-mm side is the hypotenuse. Determine the two functions whose ratios consist of the adjacent side and the hypotenuse:

$$\text{secant } \angle B = \frac{\text{hypotenuse}}{\text{adjacent side}},$$
and
$$\text{cosine } \angle B = \frac{\text{adjacent side}}{\text{hypotenuse}}.$$

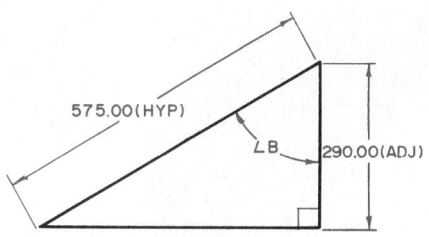

Figure A-85

Either function can be used. Choosing the cosine function:

$$\cos \angle B = \frac{\text{adj side}}{\text{hyp}}$$

$$\cos \angle B = \frac{290.00}{575.00}$$

$$\cos \angle B = 0.50435$$

$$\angle B = 59.72° \text{ or } 59°43' \text{ Ans}$$

Example 2: Determine ∠1 and ∠2 of the triangle in *Figure A-86*. All dimensions are in inches.

Calculate either ∠1 or ∠2. Choose any two of the three given sides. In relation to ∠1, the 3.420-in side is the opposite side. The 5.845-in side is the hypotenuse. Determine the two functions whose ratios consist of the opposite side and the hypotenuse:

$$\angle 1 = \frac{\text{opposite side}}{\text{hypotenuse}},$$

and

$$\text{cosecant } \angle 1 = \frac{\text{hypotenuse}}{\text{opposite side}}.$$

Either function can be used. Choosing the sine function:

$$\sin \angle 1 = \frac{\text{opp side}}{\text{hyp}}$$

$$\sin \angle 1 = \frac{3.420}{5.845}$$

$$\sin \angle 1 = 0.58512$$

$$\angle 1 = 35°49' \text{ Ans}$$

$$\angle 2 = 90° - 35°49'$$

$$\angle 2 = 54°11' \text{ Ans}$$

Trigonometry: Basic Calculations of Sides

Procedure: *To determine an unknown side when an acute angle and one side of a right triangle are known*

- Identify the given side and the unknown side as adjacent, opposite, or hypotenuse in relation to the given angle.
- Determine the trigonometric functions that are ratios of the sides identified in relation to the given angle.

Note: Two of the six functions will be found as ratios of the two identified sides. Either of the two functions can be used. Both produce the same value for the unknown, except for cotangents, secants, and cosecants of angles less than 15° and tangents, secants, and cosecants of angles greater than 75°. If the unknown side is made the numerator of the ratio, the problem is solved by multiplication. If the unknown side is made the denominator of the ratio, the problem is solved by division.

- Using either a calculator or a trigonometric function table, find the function of the given angle and substitute this value.
- Solve as a proportion for the unknown side.

Example: Determine the side of the triangle in *Figure A-87*. All dimensions are in inches.

In relation to the 63°20' angle, the 8.100-in side is the adjacent side and side *x* is the opposite side. Determine the two functions whose ratios consist of the adjacent and opposite sides:

$$\text{tangent } 63°20' = \frac{\text{opposite side}}{\text{adjacent side}},$$

and

$$\text{cotangent } 63°20' = \frac{\text{adjacent side}}{\text{opposite side}}.$$

Either function can be used. Choosing the tangent function:

$$\tan 63°20' = \frac{\text{opp side}}{\text{adj side}}$$

$$\tan 63°20' = \frac{x}{8.100}$$

$$1.9912 = \frac{x}{8.100}$$

Solve as a proportion.

$$\frac{1.9912}{1} = \frac{x}{8.100}$$

$$x = 1.9912 \,(8.100)$$

$$x = 16.129 \text{ in Ans}$$

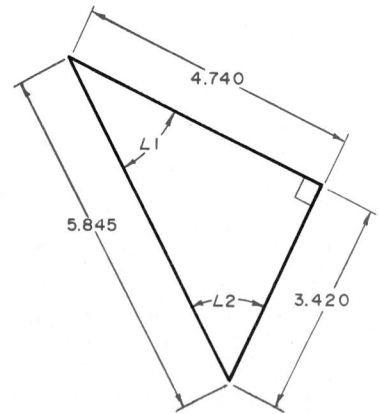

Figure A-86

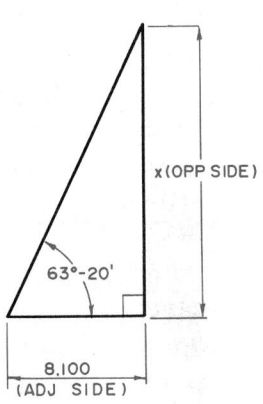

Figure A-87

Trigonometry: Common Drafting Applications

Method of Solution

The following examples are simple practical applications of right-angle trigonometry, although they may not be given directly in the form of right triangles. To solve the examples it is necessary to project auxiliary lines to produce a right triangle. The unknown, or a dimension required to compute the unknown, is part of the triangle. The auxiliary lines may be projected between given points or from given points. The lines may be projected parallel or perpendicular to center lines, tangents, or other reference lines.

It is important to study carefully the procedures and the use of auxiliary lines as they are applied to the following examples. The same basic method is used in solving many similar drafting problems. A knowledge of both geometric principles and trigonometric functions and the ability to relate and apply them to specific situations are required in solving many applied problems.

Tapers and Bevels

Example 1: Determine the included taper angle of the shaft in *Figure A-88*. All dimensions are in inches.

The problem must be solved by using a figure in the form of a right triangle. Therefore, project line AB from point A parallel to the center line. Right $\triangle$ABC is formed in which $\angle$BAC is one-half the included taper angle.

$$\text{Side AB} = 10.500''$$
$$\text{Side BC} = \frac{1.800'' - 0.700''}{2} = 0.550''$$

Using sides AB and BC, solve for $\angle$BAC.

$$\tan \angle\text{BAC} = \frac{\text{BC}}{\text{AB}} = \frac{0.500}{10.500} = 0.05238$$
$$\angle\text{BAC} = 3°0'$$

The included taper angle = $2(3°0') = 6°0'$ Ans

Example 2: Determine diameter x of the part shown in *Figure A-89*. All dimensions are in millimetres. Project line DE from point D parallel to the center-line in order to form right $\triangle$DEF.

Side DE = 21.80 mm – 7.50 mm = 14.30 mm.
$\angle$EDF = 32.5°

Using side DE and $\angle$EDF, solve for side EF.

$$\tan \angle\text{EDF} = \frac{\text{EF}}{\text{DE}}$$
$$\tan 32.5° = \frac{\text{EF}}{14.30}$$
$$\text{EF} = 0.63707(14.30)$$
$$\text{EF} = 9.11 \text{ mm}$$

Dia x = 26.25 mm – 2(9.11 mm) = 8.03 mm Ans

Isosceles Triangle Applications: Distance Between Holes and V-Slots

The solutions to many practical trigonometry problems are based on recognizing figures as isosceles triangles. In an isosceles triangle an altitude to the base bisects the base and the vertex angle.

Example 1: In *Figure A-90* five holes are equally spaced on a 5.200-in diameter circle. Determine the straight-line distance between centers of two consecutive holes.

Project radii from center O to hole centers A and B.

Project a line from A to B. $\angle$AOB = $\dfrac{360°}{5}$ = 72°.

Since OA = OB, $\triangle$AOB is isosceles. Project line OC perpendicular to AB from center O. Line OC bisects

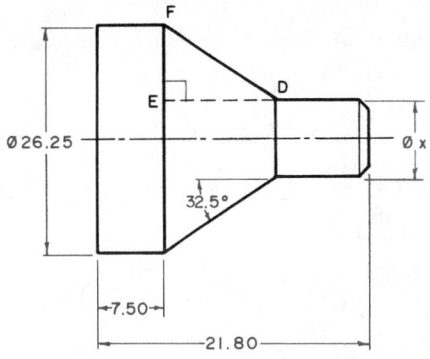

Figure A-89

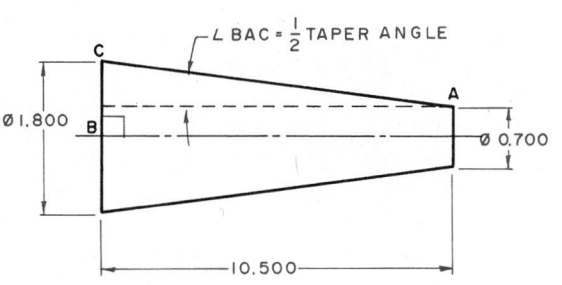

Figure A-88

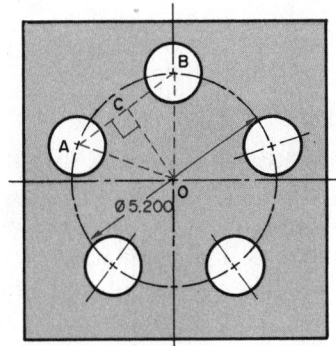

Figure A-90

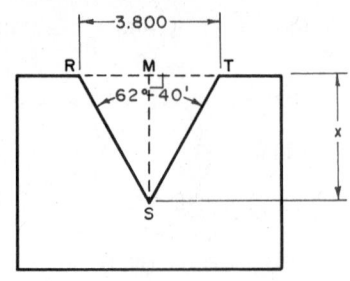

Figure A-91

∠AOB and side AB. In right △ AOC, ∠AOC = $\frac{72°}{2}$ = 36°. AO = $\frac{5.200 \text{ in}}{2}$ = 2.600 in.

Solve for side AC.

$$\sin \angle AOC = \frac{AC}{AO}$$

$$\sin 36° = \frac{AC}{2.600}$$

$$0.58779 = \frac{AC}{2.600}$$

$$AC = 0.58779(2.600)$$

$$AC = 1.528 \text{ in}$$

$$AB = 2(1.528 \text{ in}) = 3.056 \text{ in Ans}$$

Example 2: Determine dimension x of the V-slot in ***Figure A-91***. All dimensions are in inches.

Connect a line between points R and T.

Sides RS = TS; therefore, △ RST is isosceles.

Project line SM from point S perpendicular to RT. Side RT and ∠RST are bisected. In right △ RSM,

$$\angle RSM = \frac{62°40'}{2} = 31°20'.$$

$$RM = \frac{3.800 \text{ in}}{2} = 1.900 \text{ in}.$$

Solve for distance MS.

$$\cot \angle RSM = \frac{MS}{RM}$$

$$\cot \angle 31°20' = \frac{MS}{1.900}$$

$$1.6426 = \frac{MS}{1.900}$$

$$MS = 1.6426(1.900)$$

$$MS = 3.121 \text{ in}$$

$$x = MS = 3.121 \text{ in Ans}$$

Tangents to Circles Applications: Angle Cuts and Dovetails

Example: An internal dovetail is shown in ***Figure A-92***. Two pins or balls are used to check the dovetail for both location and angular accuracy. Calculate check dimension x. All dimensions are in inches.

Project line HO from point H to the pin center O; HO bisects the 72°20′ angle. Project a radius from point O to the point of tangency K; ∠HKO is a right

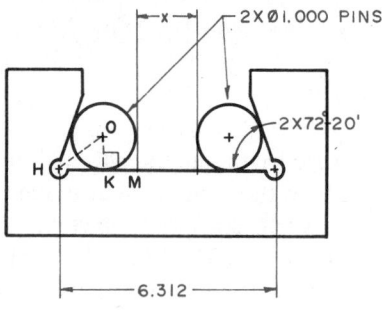

Figure A-92

angle, since a radius is perpendicular to a tangent at the point of tangency. In right △ HOK, ∠KHO = $\frac{72°20'}{2}$ = 36°10′, KO = $\frac{1.000 \text{ in}}{2}$ = 0.500 in

Solve for side HK.

$$\cot \angle KHO = \frac{HK}{KO}$$

$$\cot 36°10' = \frac{HK}{0.500}$$

$$1.3680 = \frac{HK}{0.500}$$

$$HK = 0.500(1.3680)$$

$$HK = 0.684 \text{ in}$$

KM = pin radius = 0.500 in

HM = HK + KM = 0.684 in + 0.500 in = 1.184. in x = 6.312 in − 2(HM) = 6.312 in − 2(1.184 in) = 3.944 in Ans

Trigonometry: Oblique Triangles— Law of Sines and Law of Cosines

An *oblique triangle* is one that does not contain a right angle. Drafting and design problems often involve oblique triangles. These problems can be reduced to a series of right triangles, but the process can be cumbersome and time consuming. Two formulas, the *law of sines* and the *law of cosines*, can be used to simplify such computations. In order to use either formula, three parts of an oblique triangle must be known; at least one part must be a side.

An oblique triangle can contain an angle greater than 90°, such as oblique triangle ABC in ***Figure A-93***. Therefore, sine and cosine functions of angles between 90° and 180° must be determined.

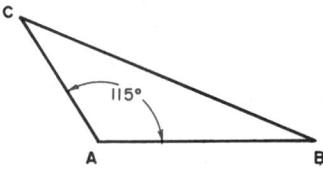

Figure A-93

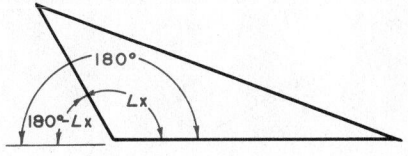

Figure A-94

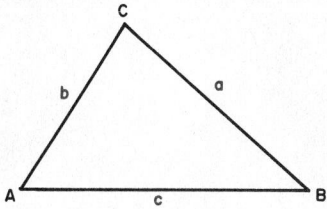

Figure A-95

Sine Functions of Angles Between 90° and 180°

The sine of an angle greater than 90° and less than 180° equals the sine of the supplement of the angle. If the value of ∠x is between 90° and 180°, as shown in *Figure A-94*, sin ∠x = sin(180° – ∠x).

Examples:

1. sin 115° = sin (180° – 115°) = sin 65° = 0.90631 Ans
2. sin 90°42′ = sin (180° – 90°42′) = sin 89°18′ = 0.99992 Ans
3. sin 176.25° = sin(180° – 176.25°) = sin 3.75° = 0.06540 Ans

Law of Sines

The law of sines states that, in any triangle, the sides are proportional to the sines of the opposite angles. That is (see *Figure A-95*)

$$\frac{a}{\sin A} = \frac{b}{\sin B} = \frac{c}{\sin C}$$

The law of sines is used to solve the following kinds of problems:

- Problems where any two angles and any one side of an oblique triangle are known.
- Problems where any two sides and an angle opposite one of the given sides of an oblique triangle are known.

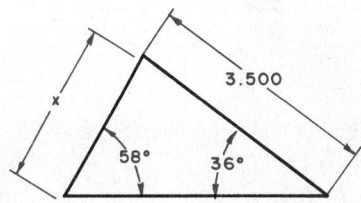

Figure A-96

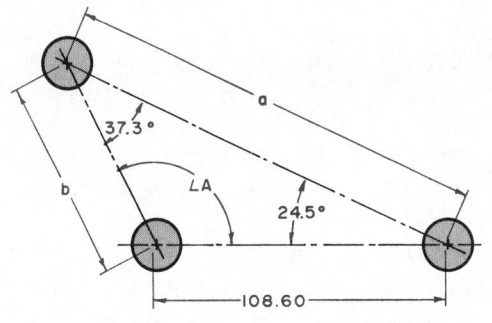

Figure A-97

Solving Oblique Triangle Problems Given Two Angles and a Side

Example 1: Given two angles and a side, determine side *x* of the oblique triangle in *Figure A-96*. (All dimensions are in inches.)

Set up a proportion and solve for *x*.

$$\frac{x}{\sin 36°} = \frac{3.500}{\sin 58°}$$

$$\frac{x}{0.58779} = \frac{3.500}{0.84805}$$

$$0.84805x = 3.500(0.58779)$$

$$x = \frac{3.500(0.58779)}{0.84805}$$

$$x = 2.426 \text{ in Ans}$$

Example 2: Given two angles and a side of the oblique triangle formed by the intersection of hole centerlines shown in *Figure A-97*. All dimensions are in millimetres.

a. Determine ∠A.
b. Determine side *a*.

a. Determine ∠A.

∠A = 180° – (37.3° + 24.5°) = 118.2° Ans

b. Determine side *a*. Set up a proportion and solve for side *a*.

$$\frac{a}{\sin 118.2°} = \frac{108.60}{\sin 37.3°}$$

$$\sin 118.2° = \sin(180°-118.2°) = \sin 61.8° = 0.88130$$

$$\sin 37.3° = 0.60599$$

$$\frac{a}{0.88130} = \frac{108.60}{0.60599}$$

$$a = 157.94 \text{ mm Ans}$$

Solving Oblique Triangle Problems Given Two Sides and an Angle Opposite One of the Given Sides

Example 1: Given two sides and an opposite angle of the oblique triangle formed by the intersection of

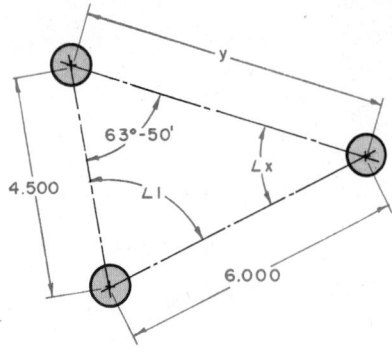

Figure A-98

hole center lines shown in **Figure A-98**. (All dimensions are in inches.)

    a. Determine $\angle x$.

    b. Determine side y.

a. Determine $\angle x$.

$$\frac{4.500}{\sin \angle x} = \frac{6.000}{\sin 63°50'}$$

$$\frac{4.500}{\sin \angle x} = \frac{6.000}{0.89752}$$

$$6.000(\sin \angle x) = 4.500(0.89752)$$

$$\sin \angle x = \frac{4.500(0.89752)}{6.000}$$

$$\sin \angle x = 0.67314$$

$$\angle x = 42°18' \text{ Ans}$$

b. Determine side y.

$$\angle 1 = 180° - (63°50' + \angle x) =$$
$$180° - (63°50' + 42°18') = 180° - 106°8' = 73°52'$$

$$\frac{6.000}{\sin 63°50'} = \frac{y}{\sin 73°52'}$$

$$\frac{6.000}{0.89752} = \frac{y}{0.96062}$$

$$0.89752y = 6.000(0.96054)$$

$$y = \frac{6.000(0.96054)}{0.89752}$$

$$y = 6.422 \text{ in Ans}$$

Example 2: Given two sides and an opposite angle, determine $\angle x$ of the oblique triangle shown in **Figure A-99**. (All dimensions are in millimetres.)

$$\frac{140.00}{\sin 28.17°} = \frac{275.00}{\sin \angle x}$$

$$\frac{140.00}{0.47209} = \frac{275.00}{\sin \angle x}$$

$$\sin \angle x = 0.92731$$

The angle that corresponds to the sine function 0.92731 is 68°. Because $\angle x$ is greater than 90°. $\angle x =$ the supplement of 68°. $\angle x = 180° - 68° = 112°$. Ans

Cosine Functions of Angles Between 90° and 180°

The cosine of an angle greater than 90° and less than 180° equals the negative cosine of the supplement of

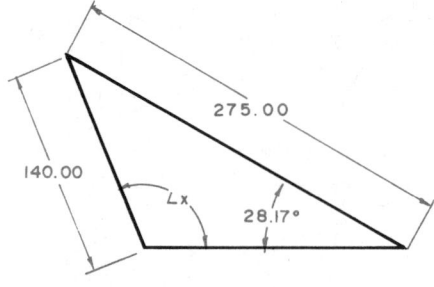

Figure A-99

the angle. If the value of $\angle x$ is between 90° and 180°, $\cos \angle x = -\cos(180° - \angle x)$. A negative function of an angle does *not* mean that the angle is negative; it is a negative function of a positive angle. For example, $-\cos$ 65° does *not* mean $\cos(-65°)$.

Examples:

    1. $\cos 115° = -\cos(180° - 115°) = -\cos 65° = -0.42262$.

Note: Determine the cosine of 65° and prefix a negative sign:

    $\cos 65° = 0.42262$; $-\cos 65° = 0.42262$.

    2. $\cos 90°42' = -\cos(180° - 90°42') = -\cos 89°18' = -0.01222$.

    3. $\cos 176.25° = -\cos(180° - 176.25°) = -\cos 3.75° = -0.99786$.

Law of Cosines

The law of cosines states that, in any triangle, the square of any side is equal to the sum of the squares of the other two sides minus twice the product of these two sides multiplied by the cosine of their included angle. That is (see **Figure A-100**)

$$a^2 = b^2 + c^2 - 2bc(\cos A)$$
$$b^2 = a^2 + c^2 - 2ac(\cos B)$$
$$c^2 = a^2 + b^2 - 2ab(\cos C)$$

The equations can be rearranged in this form:

$$\cos A = \frac{b^2 + c^2 - a^2}{2bc}$$

$$\cos B = \frac{a^2 + c^2 - b^2}{2ac}$$

$$\cos C = \frac{a^2 + b^2 - c^2}{2ab}$$

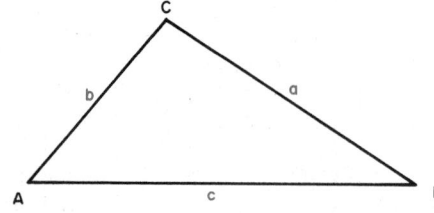

Figure A-100

The law of cosines is used to solve the following kinds of problems:

- Problems where two sides and the included angle of an oblique triangle are known.
- Problems where three sides of an oblique triangle are known.

Solving Oblique Triangle Problems Given Two Sides and the Included Angle

Example 1: Internally and externally tangent circles are shown in *Figure A-101*. Determine dimension y, the distance between hole centers A and C.

All dimensions are in inches.

In oblique $\triangle$ABC:

$$AB = \frac{3.000}{2} - \frac{0.930}{2} = 1.035$$

$$BC = \frac{3.000}{2} + \frac{1.500}{2} = 2.250$$

The 82° angle is the included angle between AB and BC. Substitute values in their appropriate places in the formula and solve for dimension y.

$y^2 = 1.035^2 + 2.250^2 - 2(1.035)(2.250)(\cos 82°)$
$y^2 = 1.0712 + 5.0625 - 2(1.035)(2.250)(0.13917)$
$y^2 = 6.1337 - 0.6482$
$y^2 = 5.4855$
$y = 2.342$ in Ans

Example 2: Given two sides and the included angle, determine side x of the oblique triangle in *Figure A-102*. All dimensions are in inches.

Substitute values to solve for x.

$x^2 = 3.800^2 + 4.100^2 - 2(3.800)(4.100)(\cos 123°)$

Since the given angle is greater than 90°, the cosine of the angle is equal to the negative cosine of its supplement. Therefore,

$\cos 123° = -\cos(180° - 123°) = -\cos 57° = -0.54464$

This negative value must be used in computing side x.

$x^2 = 3.800^2 + 4.100^2 - 2(3.800)(4.100)(-0.54464)$
$x^2 = 31.250 - (-16.971)$

Recall that subtracting a negative value is the same as adding a positive value.

$$x^2 = 31.250 + 16.971$$
$$x^2 = 48.221$$
$$x = 6.944 \text{ in Ans}$$

Solving Oblique Triangle Problems Given Three Sides

Example 1: Four pins are located on a fixture at A, B, C, and D as shown in *Figure A-103*. Compute $\angle 3$.

All dimensions are in millimetres.

$\angle 3 = \angle 1 + \angle 2$. Compute $\angle 1$ in the right $\triangle$ADC. Compute $\angle 2$ in oblique $\triangle$ABC. In right $\triangle$ADC, AC = 75.00, DC = 73.00.

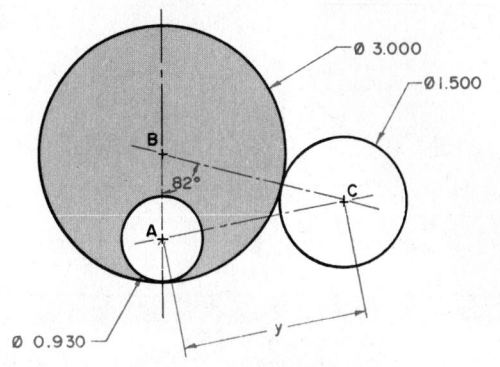

Figure A-101

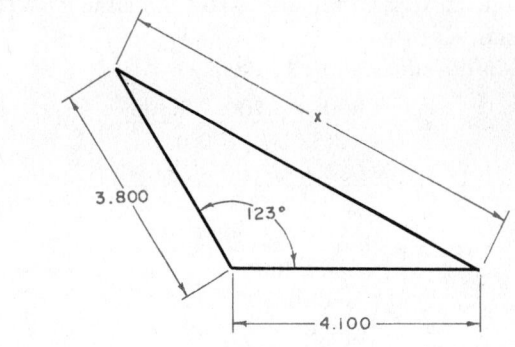

Figure A-102

$$\sin \angle 1 = \frac{DC}{AC}$$

$$\sin \angle 1 = \frac{73.00}{75.00} = 0.97333$$

$$\angle 1 = 76.73°$$

In oblique $\triangle$ABC, AC = 75.00, AB = 80.00, BC = 48.00. Substitute values in their appropriate places in the formula.

$$\cos \angle 2 = \frac{75.00^2 + 80.00^2 - 48.00^2}{2(75.00)(80.00)}$$

$$\cos \angle 2 = \frac{5625 + 6400 - 2304}{12,000} = 0.81008$$

$$\angle 2 = 35.90°$$

$$\angle 3 = \angle 1 + \angle 2 = 76.73° + 35.90° = 112.63° \text{ Ans}$$

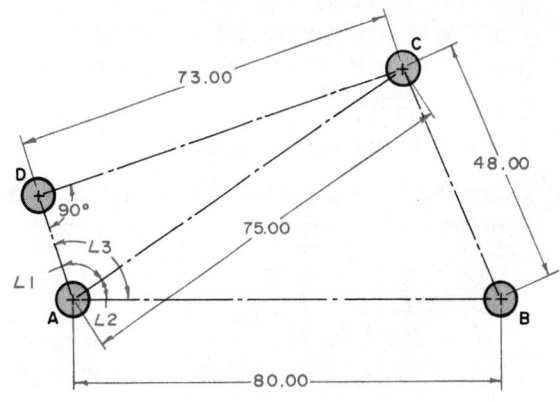

Figure A-103

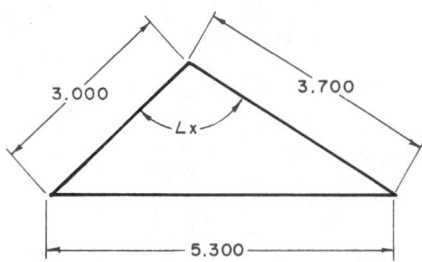

Figure A-104

Example 2: Given three sides, determine $\angle x$ of the oblique triangle in **Figure A-104**. All dimensions are in inches.

Substitute values in the formula.

$$\cos \angle x = \frac{3.000^2 + 3.700^2 - 5.300^2}{2(3.000)(3.700)}$$

$$\cos \angle x = \frac{9.000 + 13.690 - 28.090}{22.200}$$

$$\cos \angle x = \frac{22.690 - 28.090}{22.200}$$

$$\cos \angle x = \frac{-5.400}{22.200}$$

$$\cos \angle x = -0.24324$$

Since $\cos \angle x$ is a negative value, $\angle x$ is equal to the supplement of the angle found. The angle corresponding to the cosine function 0.24324 is 75°55'. Therefore, the angle whose cosine function is −0.24324 is equal to 180° − 75°55' = 104°05'. Therefore, $\angle x = 104°05'$ Ans

Review

Solutions to the following problems require the application of mathematical operations, procedures, and principles presented in this appendix.

1. Round 4.02463 inches to three decimal places.

2. Round 59.6841 mm to two decimal places.

3. Using the decimal equivalent table,
   a. find the decimal equivalent of 57/64 in;
   b. find the fractional equivalent of 0.4375 in;
   c. find the nearer fractional equivalent of 0.706 in.

4. Express 0.983 in in millimetres. Round the answer to two decimal places.

5. Express 70.15 mm in inches. Round the answer to three decimal places.

   Problems 6–11 involve working with formulas that have been obtained from technical handbooks. For each problem substitute the given numerical values for letters in the formula and solve for the unknown. Where necessary, round answers to two decimal places.

6. $A = \pi(ab - cd)$. Solve for A when $\pi = 3.1416$, $a = 5.000$ in, $b = 3.000$ in, $c = 4.000$ in, and $d = 2.000$ in.

7. $A = \pi \dfrac{(R^2 - r^2)}{2}$. Solve for A when $\pi = 3.1416$, R = 7.000 in, and $r = 3.000$ in.

8. $V = \dfrac{(2a + c)bh}{6}$. Solve for V when $a = 9.000$ in, $b = 7.000$ in, $c = 11.000$ in, and $h = 8.000$ in.

9. All dimensions in millimetres.

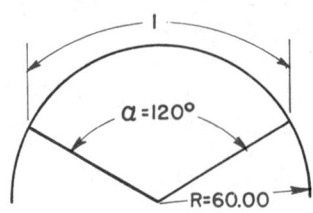

   a. Find the length of this arc (l).
   $$l = \frac{3.1416Ra}{180°}$$
   b. Find the area of this sector (A).
   $$A = \frac{1}{2}Rl$$

10. All dimensions are in millimetres.

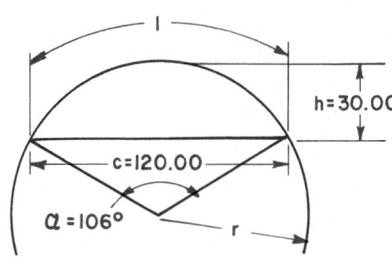

   a. Find the radius of this circle.
   $$r = \frac{c^2 + 4h^2}{8h}$$
   b. Find the length of the arc (l).
   $$l = 0.175ra$$

11. All dimensions are in inches. Find the belt length on the pulleys.
   $$\text{Length of belt} = 2C + \frac{11D + 11d}{7} + \frac{(D - d)^2}{4C}$$

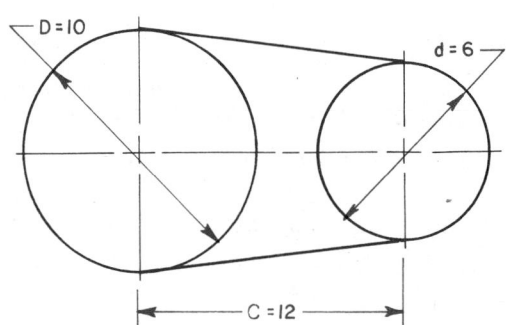

12. Of two gears that mesh, the one which has the greatest number of teeth is called the gear, and the one which has the fewer teeth is called the pinion. The proportion that expresses the relationship of gear teeth and speeds is

$$\frac{\text{Number of teeth on gear}}{\text{Number of teeth on pinion}} = \frac{\text{Speed of pinion}}{\text{Speed of gear}}$$

For each problem, a–e, determine the unknown value x. Where necessary, round answer to one decimal place.

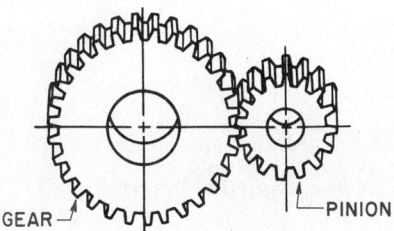

GEAR ⌐   ⌐PINION

	Number of Teeth on Gear	Number of Teeth on Pinion	Speed of Gear (rpm)	Speed of Pinion (rpm)
a	48	20	120	x
b	32	24	x	210
c	35	x	160	200
d	x	15	150	250
e	54	28	80	x

13. Add the angles in each of these problems.
  a. $23°43' + 71°12' + 19°27'$
  b. $9°33'46'' + 37°12'19''$

14. Subtract the angles in each of these problems.
  a. $63°23' - 32°58'$
  b. $54°32'13'' - 19°13'42''$

15. Multiply the angles in each of these problems.
  a. $2(43°43')$
  b. $5(22°10'13'')$

16. Divide the angles in each of these problems.
  a. $105°20' \div 4$
  b. $110°51'5'' \div 5$

17. In the figure,
  $\angle 1 = \angle 2 = \angle 3 = \angle 4 = \angle 5 = 53°41'$
  Determine $\angle 6$.

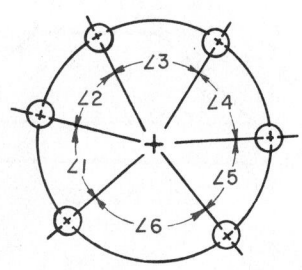

18. In the figure,
  $\angle 1 = \angle 2 = \angle 3 = \angle 4$
  Determine $\angle 1$.

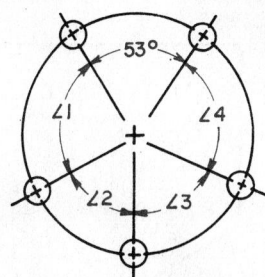

19. Express each of the following decimal-degrees as degrees and minutes.
  a. $76.95°$
  b. $117.70°$

20. Express the following decimal-degrees as degrees, minutes, and seconds.
  a. $7.9250°$
  b. $37.9365°$

21. Express each of the following degrees and minutes as decimal-degrees. Round the answer to two decimal places.
  a. $93°18'$
  b. $79°59'$

22. Express each of the following degrees, minutes, and seconds as decimal-degrees. Round the answer to four decimal places.
  a. $53°10'45''$
  b. $176°27'18''$

23. Centerlines AB ∥ CD and EF ∥ GH. Determine the values of $\angle 1 - \angle 2$.

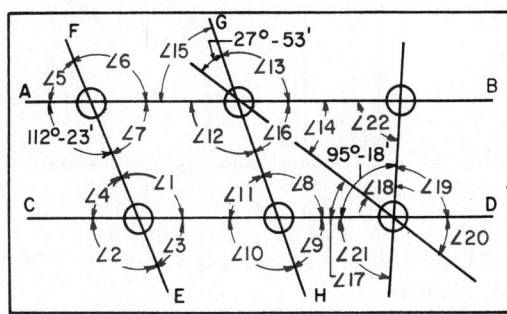

24. Determine dimension y. All dimensions are in millimetres.

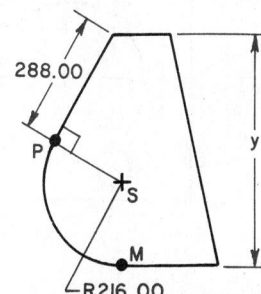

25. Determine dimension *x*. All dimensions are in inches.

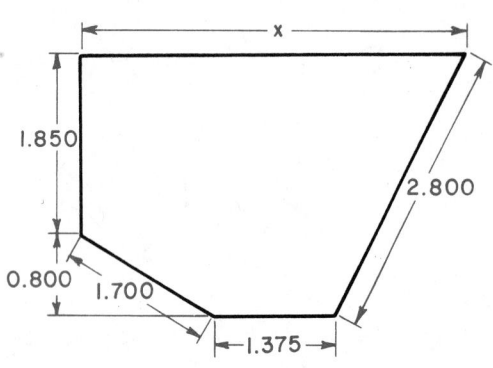

26. Determine ∠1.

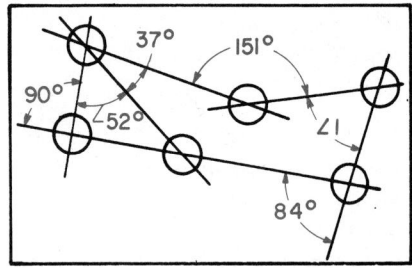

27. Point A is a tangent point. All dimensions are in inches. Determine dimension *x*.

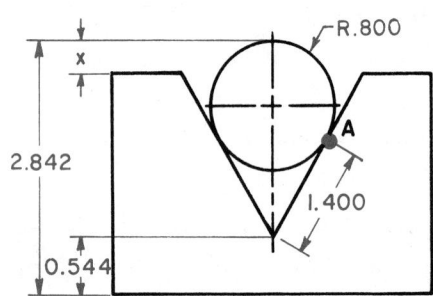

28. Determine dimension *y*. All dimensions are in millimetres.

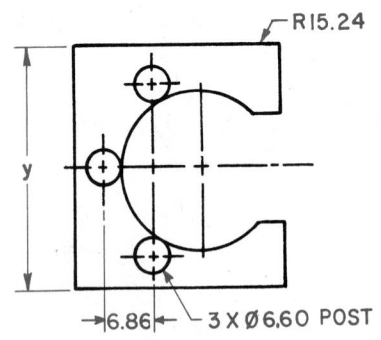

29. Determine dimension *x*. All dimensions are in inches.

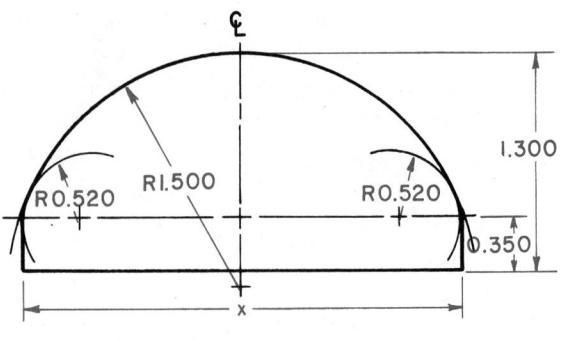

30. AC is a diameter. Determine ∠1.

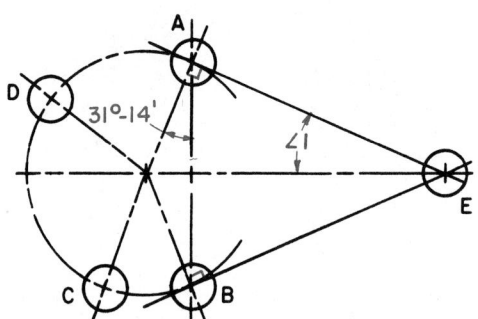

31. Determine the included taper angle *x*. All dimensions are in inches.

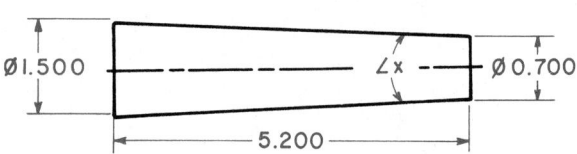

32. Determine diameter *x*. All dimensions are in inches.

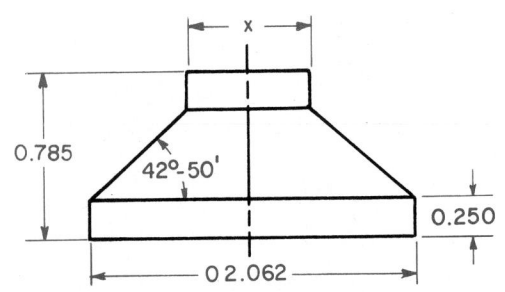

33. Determine center distance *y*. All dimensions are in millimetres.

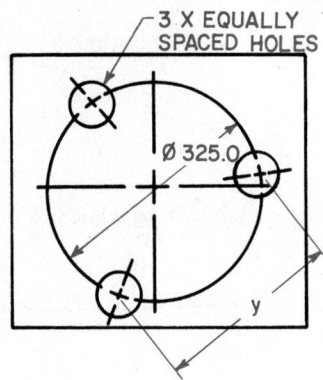

34. Determine depth of cut *x*. All dimensions are in inches.

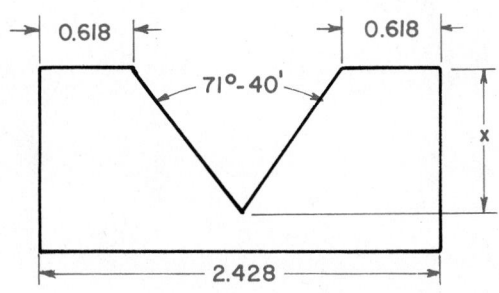

35. Determine ∠*x*. All dimensions are in inches.

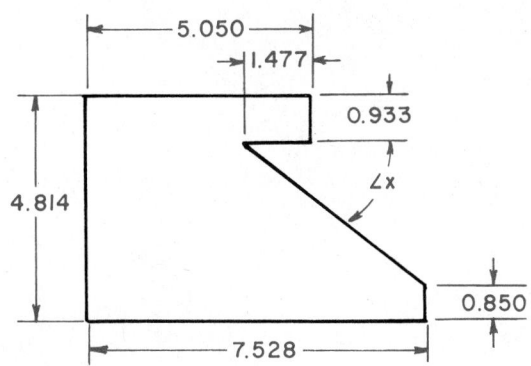

36. Determine gage dimension *x*. All dimensions are in inches.

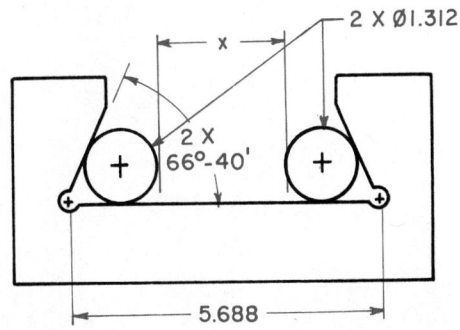

37. Determine dimension *y*. All dimensions are in millimetres.

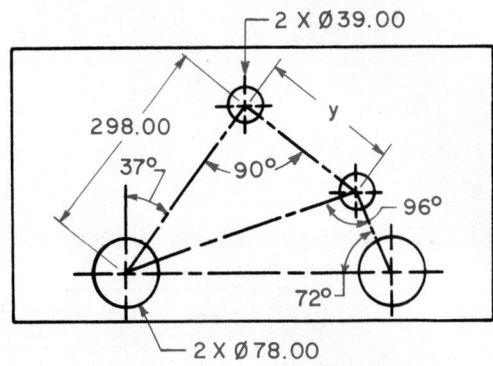

38. Determine ∠*x*. All dimensions are in millimetres.

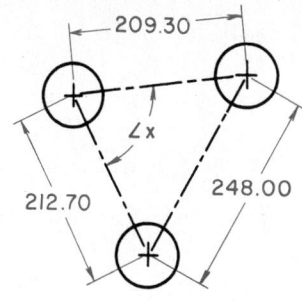

39. Determine dimension *y*. All dimensions are in inches.

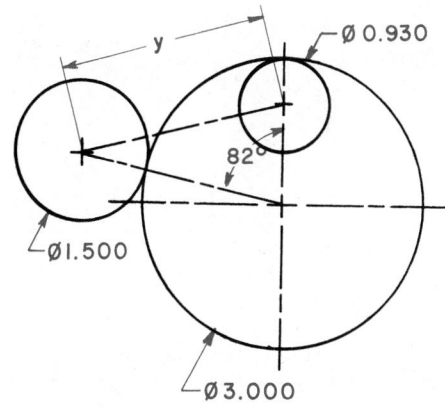

40. Determine dimension *x*. All dimensions are in inches.

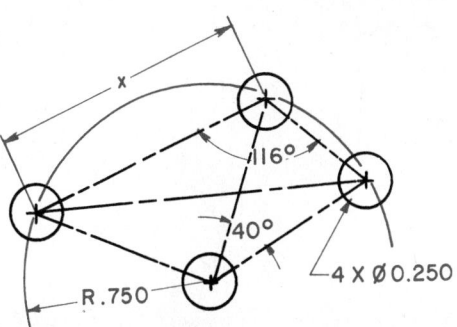

41. Determine ∠x. All dimensions are in inches.

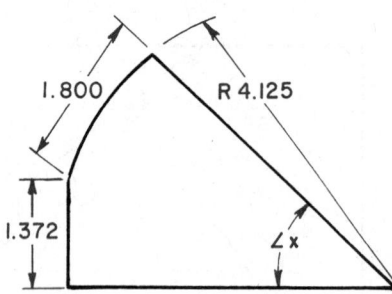

42. Determine dimension y. All dimensions are in millimetres.

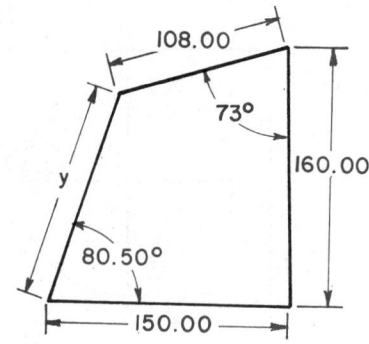

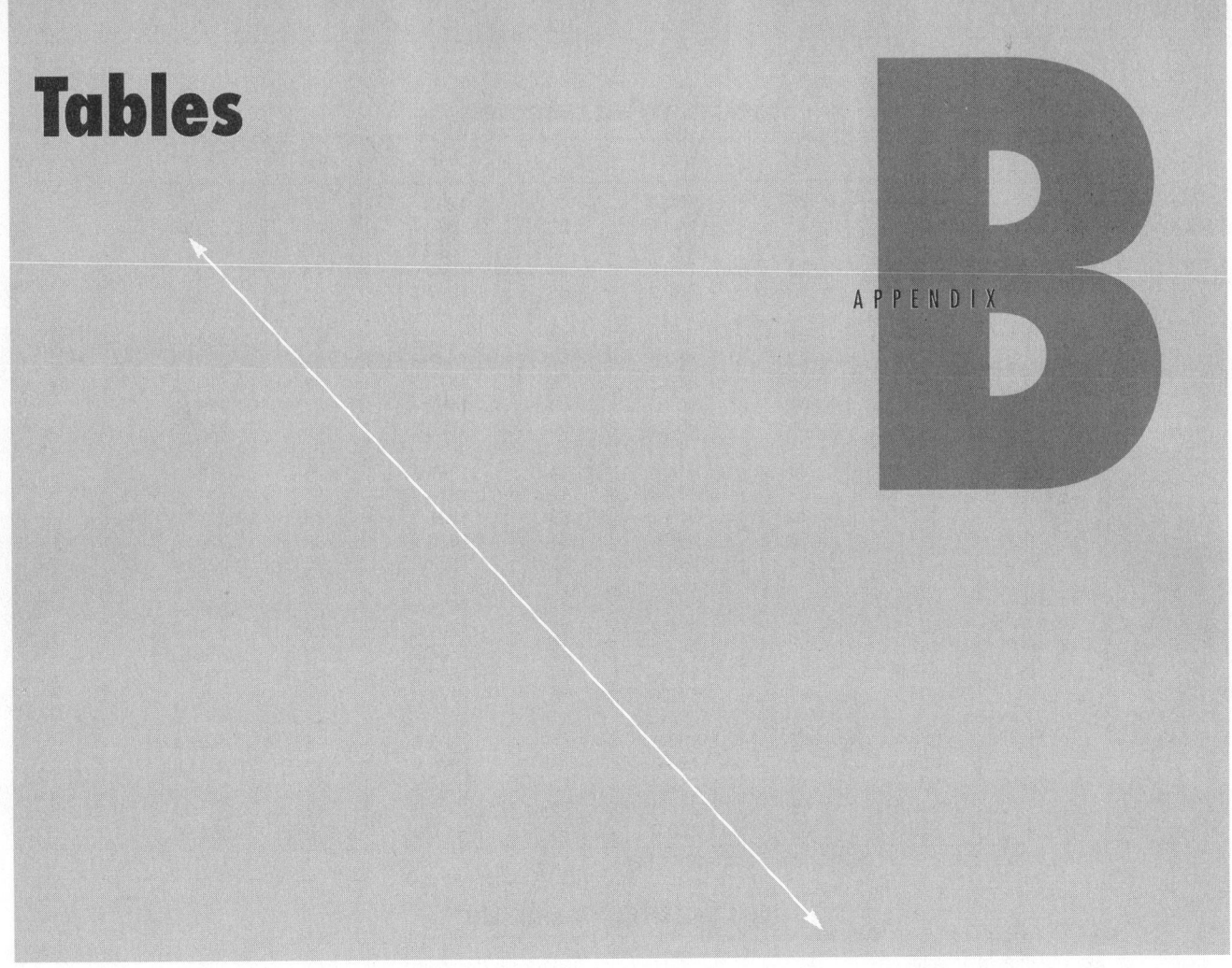

Tables

APPENDIX

B

Contents

INCHES TO MILLIMETRES

in.	mm	in.	mm	in.	mm	in.	mm
1	25.4	26	660.4	51	1295.4	76	1930.4
2	50.8	27	685.8	52	1320.8	77	1955.8
3	76.2	28	711.2	53	1346.2	78	1981.2
4	101.6	29	736.6	54	1371.6	79	2006.6
5	127.0	30	762.0	55	1397.0	80	2032.0
6	152.4	31	787.4	56	1422.4	81	2057.4
7	177.8	32	812.8	57	1447.8	82	2082.8
8	203.2	33	838.2	58	1473.2	83	2108.2
9	228.6	34	863.6	59	1498.6	84	2133.6
10	254.0	35	889.0	60	1524.0	85	2159.0
11	279.4	36	914.4	61	1549.4	86	2184.4
12	304.8	37	939.8	62	1574.8	87	2209.8
13	330.2	38	965.2	63	1600.2	88	2235.2
14	355.6	39	990.6	64	1625.6	89	2260.6
15	381.0	40	1016.0	65	1651.0	90	2286.0
16	406.4	41	1041.4	66	1676.4	91	2311.4
17	431.8	42	1066.8	67	1701.8	92	2336.8
18	457.2	43	1092.2	68	1727.2	93	2362.2
19	482.6	44	1117.6	69	1752.6	94	2387.6
20	508.0	45	1143.0	70	1778.0	95	2413.0
21	533.4	46	1168.4	71	1803.4	96	2438.4
22	558.8	47	1193.8	72	1828.8	97	2463.8
23	584.2	48	1219.2	73	1854.2	98	2489.2
24	609.6	49	1244.6	74	1879.6	99	2514.6
25	635.0	50	1270.0	75	1905.0	100	2540.0

The above table is exact on the basis: 1 in. = 25.4 mm

MILLIMETRES TO INCHES

mm	in.	mm	in.	mm	in.	mm	in.
1	0.039370	26	1.023622	51	2.007874	76	2.992126
2	0.078740	27	1.062992	52	2.047244	77	3.031496
3	0.118110	28	1.102362	53	2.086614	78	3.070866
4	0.157480	29	1.141732	54	2.125984	79	3.110236
5	0.196850	30	1.181102	55	2.165354	80	3.149606
6	0.236220	31	1.220472	56	2.204724	81	3.188976
7	0.275591	32	1.259843	57	2.244094	82	3.228346
8	0.314961	33	1.299213	58	2.283465	83	3.267717
9	0.354331	34	1.338583	59	2.322835	84	3.307087
10	0.393701	35	1.377953	60	2.362205	85	3.346457
11	0.433071	36	1.417323	61	2.401575	86	3.385827
12	0.472441	37	1.456693	62	2.440945	87	3.425197
13	0.511811	38	1.496063	63	2.480315	88	3.464567
14	0.551181	39	1.535433	64	2.519685	89	3.503937
15	0.590551	40	1.574803	65	2.559055	90	3.543307
16	0.629921	41	1.614173	66	2.598425	91	3.582677
17	0.669291	42	1.653543	67	2.637795	92	3.622047
18	0.708661	43	1.692913	68	2.677165	93	3.661417
19	0.748031	44	1.732283	69	2.716535	94	3.700787
20	0.787402	45	1.771654	70	2.755906	95	3.740157
21	0.826772	46	1.811024	71	2.795276	96	3.779528
22	0.866142	47	1.850394	72	2.834646	97	3.818898
23	0.905512	48	1.889764	73	2.874016	98	3.858268
24	0.944882	49	1.929134	74	2.913386	99	3.897638
25	0.984252	50	1.968504	75	2.952756	100	3.937008

The above table is approximate on the basis: 1 in. = 25.4 mm, 1/25.4 = 0.039370078740+

Table 1

| | | INCH/METRIC — EQUIVALENTS | | | | | |

Fraction	Decimal Equivalent		Fraction	Decimal Equivalent	
	Customary (in.)	Metric (mm)		Customary (in.)	Metric (mm)
1/64 —— .015625		0.3969	33/64 —— .515625		13.0969
1/32 ———— .03125		0.7938	17/32 ———— .53125		13.4938
3/64 —— .046875		1.1906	35/64 —— .546875		13.8906
1/16 ———————— .0625		1.5875	9/16 ———————— .5625		14.2875
5/64 —— .078125		1.9844	37/64 —— .578125		14.6844
3/32 ———— .09375		2.3813	19/32 ———— .59375		15.0813
7/64 —— .109375		2.7781	39/64 —— .609375		15.4781
1/8 ———————————— .1250		3.1750	5/8 ———————————— .6250		15.8750
9/64 —— .140625		3.5719	41/64 —— .640625		16.2719
5/32 ———— .15625		3.9688	21/32 ———— .65625		16.6688
11/64 —— .171875		4.3656	43/64 —— .671875		17.0656
3/16 ———————— .1875		4.7625	11/16 ———————— .6875		17.4625
13/64 —— .203125		5.1594	45/64 —— .703125		17.8594
7/32 ———— .21875		5.5563	23/32 ———— .71875		18.2563
15/64 —— .234375		5.9531	47/64 —— .734375		18.6531
1/4 ———————————— .250		6.3500	3/4 ———————————— .750		19.0500
17/64 —— .265625		6.7469	49/64 —— .765625		19.4469
9/32 ———— .28125		7.1438	25/32 ———— .78125		19.8438
19/64 —— .296875		7.5406	51/64 —— .796875		20.2406
5/16 ———————— .3125		7.9375	13/16 ———————— .8125		20.6375
21/64 —— .328125		8.3384	53/64 —— .828125		21.0344
11/32 ———— .34375		8.7313	27/32 ———— .84375		21.4313
23/64 —— .359375		9.1281	55/64 —— .859375		21.8281
3/8 ———————————— .3750		9.5250	7/8 ———————————— .8750		22.2250
25/64 —— .390625		9.9219	57/64 —— .890625		22.6219
13/32 ———— .40625		10.3188	29/32 ———— .90625		23.0188
27/64 —— .421875		10.7156	59/64 —— .921875		23.4156
7/16 ———————— .4375		11.1125	15/16 ———————— .9375		23.8125
29/64 —— .453125		11.5094	61/64 —— .953125		24.2094
15/32 ———— .46875		11.9063	31/32 ———— .96875		24.6063
31/64 —— .484375		12.3031	63/64 —— .984375		25.0031
1/2 ———————————— .500		12.7000	1 ———————————— 1.000		25.4000

From Drafting for Trades and Industry—Basic Skills. Nelson. Delmar Publishers Inc.

Table 2

METRIC EQUIVALENTS

LENGTH

U.S. to Metric

1 inch = 2.540 centimetres
1 foot = .305 metre
1 yard = .914 metre
1 mile = 1.609 kilometres

Metric to U.S.

1 millimetre = .039 inch
1 centimetre = .394 inch
1 metre = 3.281 feet or 1.094 yards
1 kilometre = .621 mile

AREA

$1 \text{ inch}^2 = 6.451 \text{ centimetre}^2$
$1 \text{ foot}^2 = .093 \text{ metre}^2$
$1 \text{ yard}^2 = .836 \text{ metre}^2$
$1 \text{ acre}^2 = 4,046.873 \text{ metre}^2$

$1 \text{ millimetre}^2 = .00155 \text{ inch}^2$
$1 \text{ centimetre}^2 = .155 \text{ inch}^2$
$1 \text{ metre}^2 = 10.764 \text{ foot}^2 \text{ or } 1.196 \text{ yard}^2$
$1 \text{ kilometre}^2 = .386 \text{ mile}^2 \text{ or } 247.04 \text{ acre}^2$

VOLUME

$1 \text{ inch}^3 = 16.387 \text{ centimetre}^3$
$1 \text{ foot}^3 = .028 \text{ metre}^3$
$1 \text{ yard}^3 = .764 \text{ metre}^3$
1 quart = .946 litre
$1 \text{ gallon} = .003785 \text{ metre}^3$

$1 \text{ centimetre}^3 = 0.61 \text{ inch}^3$
$1 \text{ metre}^3 = 35.314 \text{ foot}^3 \text{ or } 1.308 \text{ yard}^3$
1 litre = .2642 gallons
1 litre = 1.057 quarts
$1 \text{ metre}^3 = 264.02 \text{ gallons}$

WEIGHT

1 ounce = 28.349 grams
1 pound = .454 kilogram
1 ton = .907 metric ton

1 gram = .035 ounce
1 kilogram = 2.205 pounds
1 metric ton = 1.102 tons

VELOCITY

1 foot/second = .305 metre/second
1 mile/hour = .447 metre/second

1 metre/second = 3.281 feet/second
1 kilometre/hour = .621 mile/second

ACCELERATION

$1 \text{ inch/second}^2 = .0254 \text{ metre/second}^2$
$1 \text{ foot/second}^2 = .305 \text{ metre/second}^2$

$1 \text{ metre/second}^2 = 3.278 \text{ feet/second}^2$

FORCE

N (newton) = basic unit of force, $kg\text{-}m/s^2$. A mass of one kilogram (1 kg) exerts a gravitational force of 9.8 N (theoretically 9.80665 N) at mean sea level.

Table 3

MULTIPLIERS FOR DRAFTERS

Multiply	By	To Obtain	Multiply	By	To Obtain
Acres	43,560	Square feet	Degrees/sec.	0.002778	Revolutions/sec.
Acres	4047	Square metres	Fathoms	6	Feet
Acres	1.562×10^{-3}	Square miles	Feet	30.48	Centimetres
Acres	4840	Square yards	Feet	12	Inches
Acre–feet	43,560	Cubic feet	Feet	0.3048	Metres
Atmospheres	76.0	Cms. of mercury	Foot–pounds	1.286×10^{-3}	British Thermal Units
Atmospheres	29.92	Inches of mercury	Foot–pounds	5.050×10^{-7}	Horsepower–hrs.
Atmospheres	33.90	Feet of water	Foot–pounds	3.241×10^{-4}	Kilogram–calories
Atmospheres	10,333	Kgs./sq. metre	Foot–pounds	0.1383	Kilogram–metres
Atmospheres	14.70	Lbs./sq. inch	Foot–pounds	3.766×10^{-7}	Kilowatt–hrs.
Atmospheres	1.058	Tons/sq. ft.	Foot–pounds/min.	1.286×10^{-3}	B.T.U./min.
Board feet	144 sq. in. × 1 in.	Cubic inches	Foot–pounds/min.	0.01667	Foot–pounds/sec.
British Thermal Units	0.2520	Kilogram–calories	Foot–pounds/min.	3.030×10^{-5}	Horsepower
British Thermal Units	777.5	Foot–lbs.	Foot–pounds/min.	3.241×10^{-4}	Kg.–calories/min.
British Thermal Units	3.927×10^{-4}	Horsepower–hrs.	Foot–pounds/min.	2.260×10^{-5}	Kilowatts
British Thermal Units	107.5	Kilogram–metres	Foot–pounds/sec.	7.717×10^{-2}	B.T.U./min.
British Thermal Units	2.928×10^{-4}	Kilowatt–hrs.	Foot–pounds/sec.	1.818×10^{-3}	Horsepower
B.T.U./min.	12.96	Foot–lbs./sec.	Foot–pounds/sec.	1.945×10^{-2}	Kg.–calories/min.
B.T.U./min.	0.02356	Horsepower	Foot–pounds/sec.	1.356×10^{-3}	Kilowatts
B.T.U./min.	0.01757	Kilowatts	Gallons	3785	Cubic centimetres
B.T.U./min.	17.57	Watts	Gallons	0.1337	Cubic feet
Cubic centimetres	3.531×10^{-5}	Cubic feet	Gallons	231	Cubic inches
Cubic centimetres	6.102×10^{-2}	Cubic inches	Gallons	3.785×10^{-3}	Cubic metres
Cubic centimetres	10^{-6}	Cubic metres	Gallons	4.951×10^{-3}	Cubic yards
Cubic centimetres	1.308×10^{-6}	Cubic yards	Gallons	3.785	Litres
Cubic centimetres	2.642×10^{-4}	Gallons	Gallons	8	Pints (liq.)
Cubic centimetres	10^{-3}	Litres	Gallons	4	Quarts (liq.)
Cubic centimetres	2.113×10^{-3}	Pints (liq.)	Gallons–Imperial	1.20095	U.S. gallons
Cubic centimetres	1.057×10^{-3}	Quarts (liq.)	Gallons–U.S.	0.83267	Imperial gallons
Cubic feet	2.832×10^{4}	Cubic cms.	Gallons water	8.3453	Pounds of water
Cubic feet	1728	Cubic inches	Horsepower	42.44	B.T.U./min.
Cubic feet	0.02832	Cubic metres	Horsepower	33,000	Foot–lbs./min.
Cubic feet	0.03704	Cubic yards	Horsepower	550	Foot–lbs./sec.
Cubic feet	7.48052	Gallons	Horsepower	1.014	Horsepower (metric)
Cubic feet	28.32	Litres	Horsepower	10.70	Kg.–calories/min.
Cubic feet	59.84	Pints (liq.)	Horsepower	0.7457	Kilowatts
Cubic feet	29.92	Quarts (liq.)	Horsepower	745.7	Watts
Cubic feet/min.	472.0	Cubic cms./sec.	Horsepower–hours	2547	B.T.U.
Cubic feet/min.	0.1247	Gallons/sec.	Horsepower–hours	1.98×10^{6}	Foot–lbs.
Cubic feet/min.	0.4720	Litres/sec.	Horsepower–hours	641.7	Kilogram–calories
Cubic feet/min.	62.43	Pounds of water/min.	Horsepower–hours	2.737×10^{5}	Kilogram–metres
Cubic feet/sec.	0.646317	Millions gals./day	Horsepower–hours	0.7457	Kilowatt–hours
Cubic feet/sec.	448.831	Gallons/min.	Kilometres	10^{5}	Centimetres
Cubic inches	16.39	Cubic centimetres	Kilometres	3281	Feet
Cubic inches	5.787×10^{-4}	Cubic feet	Kilometres	10^{3}	Metres
Cubic inches	1.639×10^{-5}	Cubic metres	Kilometres	0.6214	Miles
Cubic inches	2.143×10^{-5}	Cubic yards	Kilometres	1094	Yards
Cubic inches	4.329×10^{-3}	Gallons	Kilowatts	56.92	B.T.U./min.
Cubic inches	1.639×10^{-2}	Litres	Kilowatts	4.425×10^{4}	Foot–lbs./min.
Cubic inches	0.03463	Pints (liq.)	Kilowatts	737.6	Foot–lbs./sec.
Cubic inches	0.01732	Quarts (liq.)	Kilowatts	1.341	Horsepower
Cubic metres	10^{6}	Cubic centimetres	Kilowatts	14.34	Kg.–calories/min.
Cubic metres	35.31	Cubic feet	Kilowatts	10^{3}	Watts
Cubic metres	61.023	Cubic inches	Kilowatt–hours	3415	B.T.U.
Cubic metres	1.308	Cubic yards	Kilowatt–hours	2.655×10^{6}	Foot–lbs.
Cubic metres	264.2	Gallons	Kilowatt–hours	1.341	Horsepower–hrs.
Cubic metres	10^{3}	Litres	Kilowatt–hours	860.5	Kilogram–calories
Cubic metres	2113	Pints (liq.)	Kilowatt–hours	3.671×10^{5}	Kilogram–metres
Cubic metres	1057	Quarts (liq.)	Lumber Width (in.) ×		
Degrees (angle)	60	Minutes	$\dfrac{\text{Thickness (in.)}}{12}$	Length (ft.)	Board feet
Degrees (angle)	0.01745	Radians			
Degrees (angle)	3600	Seconds	Metres	100	Centimetres
Degrees/sec.	0.01745	Radians/sec.	Metres	3.281	Feet
Degrees/sec.	0.1667	Revolutions/min.	Metres	39.37	Inches

Table 4

MULTIPLIERS FOR DRAFTERS (cont'd)

Multiply	By	To Obtain	Multiply	By	To Obtain
Metres	10^{-3}	Kilometres	Pounds (troy)	373.24177	Grams
Metres	10^3	Millimetres	Pounds (troy)	0.822857	Pounds (avoir.)
Metres	1.094	Yards	Pounds (troy)	13.1657	Ounces (avoir.)
Metres/min.	1.667	Centimetres/sec.	Pounds (troy)	3.6735×10^{-4}	Tons (long)
Metres/min.	3.281	Feet/min.	Pounds (troy)	4.1143×10^{-4}	Tons (short)
Metres/min.	0.05468	Feet/sec.	Pounds (troy)	3.7324×10^{-4}	Tons (metric)
Metres/min.	0.06	Kilometres/hr.	Quadrants (angle)	90	Degrees
Metres/min.	0.03728	Miles/hr.	Quadrants (angle)	5400	Minutes
Metres/sec.	196.8	Feet/min.	Quadrants (angle)	1.571	Radians
Metres/sec.	3.281	Feet/sec.	Radians	57.30	Degrees
Metres/sec.	3.6	Kilometres/hr.	Radians	3438	Minutes
Metres/sec.	0.06	Kilometres/min.	Radians	0.637	Quadrants
Metres/sec.	2.237	Miles/hr.	Radians/sec.	57.30	Degrees/sec.
Metres/sec.	0.03728	Miles/min.	Radians/sec.	0.1592	Revolutions/sec.
Microns	10^{-6}	Metres	Radians/sec.	9.549	Revolutions/min.
Miles	5280	Feet	Radians/sec./sec.	573.0	Revs./min./min.
Miles	1.609	Kilometres	Radians/sec./sec.	0.1592	Revs./sec./sec.
Miles	1760	Yards	Reams	500	Sheets
Miles/hr.	1.609	Kilometres/hr.	Revolutions	360	Degrees
Miles/hr.	0.8684	Knots	Revolutions	4	Quadrants
Minutes (angle)	2.909×10^{-4}	Radians	Revolutions	6.283	Radians
Ounces	16	Drams	Revolutions/min.	6	Degrees/sec.
Ounces	437.5	Grains	Square yards	2.066×10^{-4}	Acres
Ounces	0.0625	Pounds	Square yards	9	Square feet
Ounces	28.349527	Grams	Square yards	0.8361	Square metres
Ounces	0.9115	Ounces (troy)	Square yards	3.228×10^{-7}	Square miles
Ounces	2.790×10^{-5}	Tons (long)	Temp. (°C.) + 273	1	Abs. temp. (°C.)
Ounces	2.835×10^{-5}	Tons (metric)	Temp. (°C.) + 17.78	1.8	Temp. (°F.)
Ounces (troy)	480	Grains	Temp. (°F.) + 460	1	Abs. temp. (°F.)
Ounces (troy)	20	Pennyweights (troy)	Temp. (°F.) − 32	5/9	Temp. (°C.)
Ounces (troy)	0.08333	Pounds (troy)	Watts	0.05692	B.T.U./min.
Ounces (troy)	31.103481	Grams	Watts	44.26	Foot—pounds/min.
Ounces (troy)	1.09714	Ounces (avoir.)	Watts	0.7376	Foot—pounds/sec.
Ounces (fluid)	1.805	Cubic inches	Watts	1.341×10^{-3}	Horsepower
Ounces (fluid)	0.02957	Litres	Watts	0.01434	Kg.—calories/min.
Ounces/sq. inch	0.0625	Lbs./sq. inch	Watts	10^{-3}	Kilowatts
Pounds	16	Ounces	Watt—hours	3.415	B.T.U.
Pounds	256	Drams	Watt—hours	2655	Foot—pounds
Pounds	7000	Grains	Watt—hours	1.341×10^{-3}	Horsepower—hrs.
Pounds	0.0005	Tons (short)	Watt—hours	0.8605	Kilogram—calories
Pounds	453.5924	Grams	Watt—hours	367.1	Kilogram—metres
Pounds	1.21528	Pounds (troy)	Watt—hours	10^{-3}	Kilowatt—hours
Pounds	14.5833	Ounces (troy)	Yards	91.44	Centimetres
Pounds (troy)	5760	Grains	Yards	3	Feet
Pounds (troy)	240	Pennyweights (troy)	Yards	36	Inches
Pounds (troy)	12	Ounces (troy)	Yards	0.9144	Metres

Table 4 (Continued)

CIRCUMFERENCES AND AREAS OF CIRCLES
From 1/64 to 50, Diameter

Dia.	Circum.	Area	Dia.	Circum.	Area	Dia.	Circum.	Area	Dia.	Circum.	Area
1/64	.04909	.00019	8	25.1327	50.2655	17	53.4071	226.980	26	81.6814	530.929
1/32	.09818	.00077	8 1/8	25.5254	51.8485	17 1/8	53.7998	230.330	26 1/8	82.0741	536.047
1/16	.19635	.00307	8 1/4	25.9181	53.4562	17 1/4	54.1925	233.705	26 1/4	82.4668	541.188
1/8	.39270	.01227	8 3/8	26.3108	55.0883	17 3/8	54.5852	237.104	26 3/8	82.8595	546.355
3/16	.58905	.02761	8 1/2	26.7035	56.7450	17 1/2	54.9779	240.528	26 1/2	83.2522	551.546
1/4	.78540	.04909	8 5/8	27.0962	58.4262	17 5/8	55.3706	243.977	26 5/8	83.6449	556.761
5/16	.98175	.07670	8 3/4	27.4889	60.1321	17 3/4	55.7633	247.450	26 3/4	84.0376	562.002
3/8	1.1781	.11045	8 7/8	27.8816	61.8624	17 7/8	56.1560	250.947	26 7/8	84.4303	567.266
7/16	1.3744	.15033	9	28.2743	63.6173	18	56.5487	254.469	27	84.8230	572.555
1/2	1.5708	.19635	9 1/8	28.6670	65.3967	18 1/8	56.9414	258.016	27 1/8	85.2157	577.869
9/16	1.7671	.24850	9 1/4	29.0597	67.2007	18 1/4	57.3341	261.587	27 1/4	85.6084	583.207
5/8	1.9635	.30680	9 3/8	29.4524	69.0292	18 3/8	57.7268	265.182	27 3/8	86.0011	588.570
11/16	2.1598	.37122	9 1/2	29.8451	70.8822	18 1/2	58.1195	268.803	27 1/2	86.3938	593.957
3/4	2.3562	.44179	9 5/8	30.2378	72.7597	18 5/8	58.5122	272.447	27 5/8	86.7865	599.369
13/16	2.5525	.51849	9 3/4	30.6305	74.6619	18 3/4	58.9049	276.117	27 3/4	87.1792	604.806
7/8	2.7489	.60132	9 7/8	31.0232	76.5886	18 7/8	59.2976	279.810	27 7/8	87.5719	610.267
15/16	2.9452	.69029				19	59.6903	283.529	28	87.9646	615.752
1	3.1416	.78540	10	31.4159	78.5398	19 1/8	60.0830	287.272	28 1/8	88.3573	621.262
1 1/8	3.5343	.99402	10 1/8	31.8086	80.5156	19 1/4	60.4757	291.039	28 1/4	88.7500	626.797
1 1/4	3.9270	1.2272	10 1/4	32.2013	82.5159	19 3/8	60.8684	294.831	28 3/8	89.1427	632.356
1 3/8	4.3197	1.4849	10 3/8	32.5940	84.5408	19 1/2	61.2611	298.648	28 1/2	89.5354	637.940
1 1/2	4.7124	1.7671	10 1/2	32.9867	86.5902	19 5/8	61.6538	302.489	28 5/8	89.9281	643.548
1 5/8	5.1051	2.0739	10 5/8	33.3794	88.6641	19 3/4	62.0465	306.354	28 3/4	90.3208	649.181
1 3/4	5.4978	2.4053	10 3/4	33.7721	90.7626	19 7/8	62.4392	310.245	28 7/8	90.7135	654.838
1 7/8	5.8905	2.7612	10 7/8	34.1648	92.8856	20	62.8319	314.159	29	91.1062	660.520
2	6.2832	3.1416	11	34.5575	95.0332	20 1/8	63.2246	318.099	29 1/8	91.4989	666.226
2 1/8	6.6759	3.5466	11 1/8	34.9502	97.2053	20 1/4	63.6173	322.062	29 1/4	91.8916	671.957
2 1/4	7.0686	3.9761	11 1/4	35.3429	99.4020	20 3/8	64.0100	326.051	29 3/8	92.2843	677.713
2 3/8	7.4613	4.4301	11 3/8	35.7356	101.623	20 1/2	64.4027	330.064	29 1/2	92.6770	683.493
2 1/2	7.8540	4.9087	11 1/2	36.1283	103.869	20 5/8	64.7954	334.101	29 5/8	93.0697	689.297
2 5/8	8.2467	5.4119	11 5/8	36.5210	106.139	20 3/4	65.1881	338.163	29 3/4	93.4624	695.127
2 3/4	8.6394	5.9396	11 3/4	36.9137	108.434	20 7/8	65.5808	342.250	29 7/8	93.8551	700.980
2 7/8	9.0321	6.4918	11 7/8	37.3064	110.753	21	65.9735	346.361	30	94.2478	706.858
3	9.4248	7.0686	12	37.6991	113.097	21 1/8	66.3662	350.496	30 1/8	94.6405	712.761
3 1/8	9.8175	7.6699	12 1/8	38.0918	115.466	21 1/4	66.7589	354.656	30 1/4	95.0332	718.689
3 1/4	10.2102	8.2958	12 1/4	38.4845	117.859	21 3/8	67.1516	358.841	30 3/8	95.4259	724.640
3 3/8	10.6029	8.9462	12 3/8	38.8772	120.276	21 1/2	67.5442	363.050	30 1/2	95.8186	730.617
3 1/2	10.9956	9.6211	12 1/2	39.2699	122.718	21 5/8	67.9369	367.284	30 5/8	96.2113	736.618
3 5/8	11.3883	10.3206	12 5/8	39.6626	125.185	21 3/4	68.3296	371.542	30 3/4	96.6040	742.643
3 3/4	11.7810	11.0447	12 3/4	40.0553	127.676	21 7/8	68.7223	375.825	30 7/8	96.9967	748.693
3 7/8	12.1737	11.7932	12 7/8	40.4480	130.191	22	69.1150	380.133	31	97.3894	754.768
4	12.5664	12.5664	13	40.8407	132.732	22 1/8	69.5077	384.465	31 1/8	97.7821	760.867
4 1/8	12.9591	13.3640	13 1/8	41.2334	135.297	22 1/4	69.9004	388.821	31 1/4	98.1748	766.990
4 1/4	13.3518	14.1863	13 1/4	41.6261	137.886	22 3/8	70.2931	393.203	31 3/8	98.5675	773.139
4 3/8	13.7445	15.0330	13 3/8	42.0188	140.500	22 1/2	70.6858	397.608	31 1/2	98.9602	779.311
4 1/2	14.1372	15.9043	13 1/2	42.4115	143.139	22 5/8	71.0785	402.038	31 5/8	99.3529	785.509
4 5/8	14.5299	16.8002	13 5/8	42.8042	145.802	22 3/4	71.4712	406.493	31 3/4	99.7456	791.731
4 3/4	14.9226	17.7206	13 3/4	43.1969	148.489	22 7/8	71.8639	410.972	31 7/8	100.1383	797.977
4 7/8	15.3153	18.6655	13 7/8	43.5896	151.201	23	72.2566	415.476	32	100.5310	804.248
5	15.7080	19.6350	14	43.9823	153.938	23 1/8	72.6493	420.004	32 1/8	100.9237	810.543
5 1/8	16.1007	20.6290	14 1/8	44.3750	156.699	23 1/4	73.0420	424.557	32 1/4	101.3164	816.863
5 1/4	16.4934	21.6476	14 1/4	44.7677	159.485	23 3/8	73.4347	429.134	32 3/8	101.7091	823.208
5 3/8	16.8861	22.6906	14 3/8	45.1604	162.295	23 1/2	73.8274	433.736	32 1/2	102.1018	829.577
5 1/2	17.2788	23.7583	14 1/2	45.5531	165.130	23 5/8	74.2201	438.363	32 5/8	102.4945	835.971
5 5/8	17.6715	24.8505	14 5/8	45.9458	167.989	23 3/4	74.6128	443.014	32 3/4	102.8872	842.389
5 3/4	18.0642	25.9672	14 3/4	46.3385	170.873	23 7/8	75.0055	447.689	32 7/8	103.2799	848.831
5 7/8	18.4569	27.1085	14 7/8	46.7312	173.782	24	75.3982	452.389	33	103.6726	855.299
6	18.8496	28.2743	15	47.1239	176.715	24 1/8	75.7909	457.114	33 1/8	104.0653	861.791
6 1/8	19.2423	29.4647	15 1/8	47.5166	179.672	24 1/4	76.1836	461.863	33 1/4	104.4580	868.307
6 1/4	19.6350	30.6796	15 1/4	47.9094	182.654	24 3/8	76.5763	466.637	33 3/8	104.8507	874.848
6 3/8	20.0277	31.9191	15 3/8	48.3020	185.661	24 1/2	76.9690	471.435	33 1/2	105.2434	881.413
6 1/2	20.4204	33.1831	15 1/2	48.6947	188.692	24 5/8	77.3617	476.258	33 5/8	105.6361	888.003
6 5/8	20.8131	34.4716	15 5/8	49.0874	191.748	24 3/4	77.7544	481.106	33 3/4	106.0288	894.618
6 3/4	21.2058	35.7847	15 3/4	49.4801	194.828	24 7/8	78.1471	485.977	33 7/8	106.4215	901.257
6 7/8	21.5985	37.1223	15 7/8	49.8728	197.933	25	78.5398	490.874	34	106.8142	907.920
7	21.9912	38.4845	16	50.2655	201.062	25 1/8	78.9325	495.795	34 1/8	107.2069	914.609
7 1/8	22.3839	39.8712	16 1/8	50.6582	204.216	25 1/4	79.3252	500.740	34 1/4	107.5996	921.321
7 1/4	22.7765	41.2825	16 1/4	51.0509	207.394	25 3/8	79.7179	505.711	34 3/8	107.9923	928.058
7 3/8	23.1692	42.7183	16 3/8	51.4436	210.597	25 1/2	80.1106	510.705	34 1/2	108.3850	934.820
7 1/2	23.5619	44.1787	16 1/2	51.8363	213.825	25 5/8	80.5033	515.724	34 5/8	108.7777	941.607
7 5/8	23.9546	45.6636	16 5/8	52.2290	217.077	25 3/4	80.8960	520.768	34 3/4	109.1704	948.417
7 3/4	24.3473	47.1730	16 3/4	52.6217	220.353	25 7/8	81.2887	525.836	34 7/8	109.5631	955.253
7 7/8	24.7400	48.7069	16 7/8	53.0144	223.654						

Table 5

TRIGONOMETRIC FORMULAS

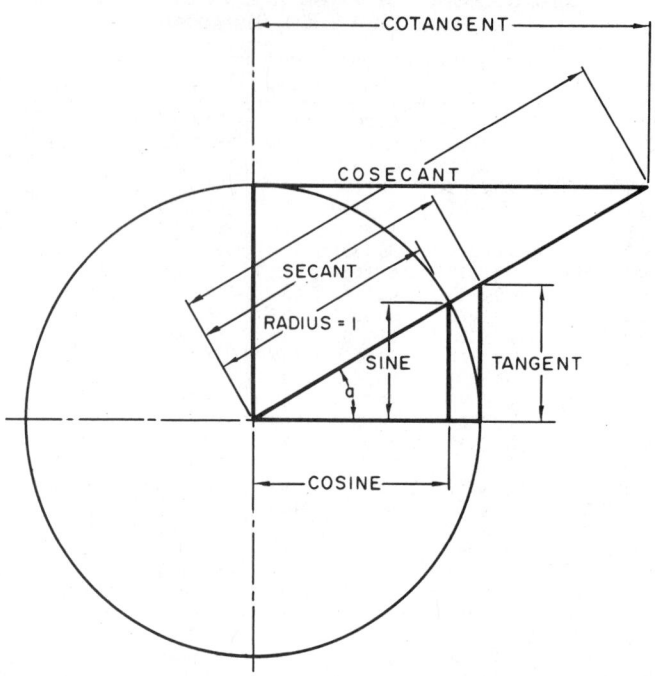

FORMULAS FOR FINDING FUNCTIONS OF ANGLES	
$\dfrac{\text{Side opposite}}{\text{Hypotenuse}}$ = SINE	
$\dfrac{\text{Side adjacent}}{\text{Hypotenuse}}$ = COSINE	
$\dfrac{\text{Side opposite}}{\text{Side adjacent}}$ = TANGENT	
$\dfrac{\text{Side adjacent}}{\text{Side opposite}}$ = COTANGENT	
$\dfrac{\text{Hypotenuse}}{\text{Side adjacent}}$ = SECANT	
$\dfrac{\text{Hypotenuse}}{\text{Side opposite}}$ = COSECANT	

FORMULAS FOR FINDING THE LENGTH OF SIDES FOR RIGHT-ANGLE TRIANGLES WHEN AN ANGLE AND SIDE ARE KNOWN	
Length of side opposite	Hypotenuse × Sine Hypotenuse ÷ Cosecant Side adjacent × Tangent Side adjacent ÷ Cotangent
Length of side adjacent	Hypotenuse × Cosine Hypotenuse ÷ Secant Side opposite × Cotangent Side opposite ÷ Tangent
Length of Hypotenuse	Side opposite × Cosecant Side opposite ÷ Sine Side adjacent × Secant Side adjacent ÷ Cosine

Table 6

RIGHT-TRIANGLE FORMULAS

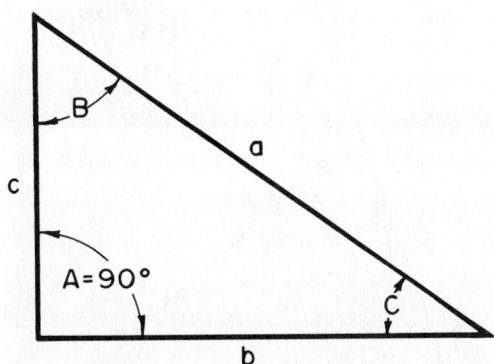

TO FIND ANGLES	FORMULAS	
C	$\dfrac{c}{a}$ = Sine C	90° − B
C	$\dfrac{b}{a}$ = Cosine C	90° − B
C	$\dfrac{c}{b}$ = Tan. C	90° − B
C	$\dfrac{b}{c}$ = Cotan. C	90° − B
C	$\dfrac{a}{b}$ = Secant C	90° − B
C	$\dfrac{a}{c}$ = Cosec. C	90° − B
B	$\dfrac{b}{a}$ = Sine B	90° − C
B	$\dfrac{c}{a}$ = Cosine B	90° − C
B	$\dfrac{b}{c}$ = Tan. B	90° − C
B	$\dfrac{c}{b}$ = Cotan. B	90° − C
B	$\dfrac{a}{c}$ = Secant B	90° − C
B	$\dfrac{a}{b}$ = Cosec. B	90° − C

TO FIND SIDES	FORMULAS	
a	$\sqrt{b^2 + c^2}$	
a	c × Cosec. C	$\dfrac{c}{\text{sine C}}$
a	c × Secant B	$\dfrac{c}{\text{Cosine B}}$
a	b × Cosec. B	$\dfrac{b}{\text{Sine B}}$
a	b × Secant C	$\dfrac{b}{\text{Cosine C}}$
b	$\sqrt{a^2 - c^2}$	
b	a × Sine B	$\dfrac{a}{\text{Cosecant B}}$
b	a × Cos. C	$\dfrac{a}{\text{Secant C}}$
b	c × Tan. B	$\dfrac{c}{\text{Cotangent B}}$
b	c × Cot. C	$\dfrac{c}{\text{Tangent C}}$
c	$\sqrt{a^2 - b^2}$	
c	a × Cos. B	$\dfrac{a}{\text{Secant B}}$
c	a × Sine C	$\dfrac{a}{\text{Cosecant C}}$
c	b × Cot. B	$\dfrac{b}{\text{Tangent B}}$
c	b × Tan C	$\dfrac{b}{\text{Cotangent C}}$

Table 7

OBLIQUE-ANGLED TRIANGLE FORMULAS

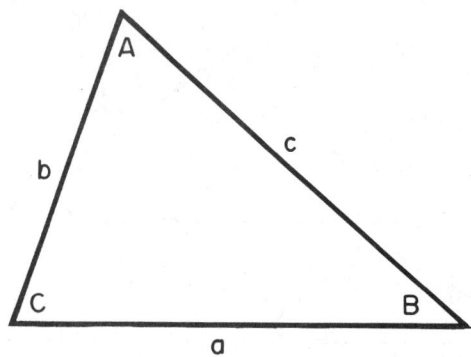

TO FIND	KNOWN	SOLUTION
C	A-B	$180° - (A + B)$
b	a-B-A	$\dfrac{a \times \text{Sin. B}}{\text{Sin. A}}$
c	a-A-C	$\dfrac{a \times \text{Sin. C}}{\text{Sin. A}}$
Tan. A	a-C-b	$\dfrac{a \times \text{Sin. C}}{b - (a \times \text{Cos. C})}$
B	A-C	$180° - (A + C)$
Sin. B	b-A-a	$\dfrac{b \times \text{Sin. A}}{a}$
A	B-C	$180° - (B + C)$
Cos. A	a-b-c	$\dfrac{b^2 + C^2 - a^2}{2bc}$
Sin. C	c-A-a	$\dfrac{c \times \text{Sin. A}}{a}$
Cot. B	a-C-b	$\dfrac{a \times \text{cscC}}{b} - \text{Cot. C}$
c	b-C-B	$b \times \text{Sin. C} \times \text{cscB}$

Table 8

RUNNING AND SLIDING FITS

VALUES IN THOUSANDTHS OF AN INCH

Nominal Size Range Inches		Class RC1 Precision Sliding			Class RC2 Sliding Fit			Class RC3 Precision Running			Class RC4 Close Running			Class RC5 Medium Running		
		Hole Tol. GR5	Minimum Clearance	Shaft Tol. GR4	Hole Tol. GR6	Minimum Clearance	Shaft Tol. GR5	Hole Tol. GR7	Minimum Clearance	Shaft Tol. GR6	Hole Tol. GR8	Minimum Clearance	Shaft Tol. GR7	Hole Tol. GR8	Minimum Clearance	Shaft Tol. GR7
Over	To	-0		+0	-0		+0	-0		+0	-0		+0	-0		+0
0	.12	+0.15	0.10	-0.12	+0.25	0.10	-0.15	+0.40	0.30	-0.25	+0.60	0.30	-0.40	+0.60	0.60	-0.40
.12	.24	+0.20	0.15	-0.15	+0.30	0.15	-0.20	+0.50	0.40	-0.30	+0.70	0.40	-0.50	+0.70	0.80	-0.50
.24	.40	+0.25	0.20	-0.15	+0.40	0.20	-0.25	+0.60	0.50	-0.40	+0.90	0.50	-0.60	+0.90	1.00	-0.60
.40	.71	+0.30	0.25	-0.20	+0.40	0.25	-0.30	+0.70	0.60	-0.40	+1.00	0.60	-0.70	+1.00	1.20	-0.70
.71	1.19	+0.40	0.30	-0.25	+0.50	0.30	-0.40	+0.80	0.80	-0.50	+1.20	0.80	-0.80	+1.20	1.60	-0.50
1.19	1.97	+0.40	0.40	-0.30	+0.60	0.40	-0.40	+1.00	1.00	-0.60	+1.60	1.00	-1.00	+1.60	2.00	-1.00
1.97	3.15	+0.50	0.40	-0.30	+0.70	0.40	-0.50	+1.20	1.20	-0.70	+1.80	1.20	-1.20	+1.80	2.50	-1.20
3.15	4.73	+0.60	0.50	-0.40	+0.90	0.50	-0.60	+1.40	1.40	-0.90	+2.20	1.40	-1.40	+2.20	3.00	-1.40
4.73	7.09	+0.70	0.60	-0.50	+1.00	0.60	-0.70	+1.60	1.60	-1.00	+2.50	1.60	-1.60	+2.50	3.50	-1.60
7.09	9.85	+0.80	0.60	-0.60	+1.20	0.60	-0.80	+1.80	2.00	-1.20	+2.80	2.00	-1.80	+2.80	4.50	-1.80
9.85	12.41	+0.90	0.80	-0.60	+1.20	0.80	-0.90	+2.00	2.50	-1.20	+3.00	2.50	-2.00	+3.00	5.00	-2.00
12.41	15.75	+1.00	1.00	-0.70	+1.40	1.00	-1.00	+2.20	3.00	-1.40	+3.50	3.00	-2.20	+3.50	6.00	-2.20

Nominal Size Range Inches		Class RC6 Medium Running			Class RC7 Free Running			Class RC8 Loose Running			Class RC9 Loose Running		
		Hole Tol. GR9	Minimum Clearance	Shaft Tol. GR8	Hole Tol. GR9	Minimum Clearance	Shaft Tol. GR8	Hole Tol. GR10	Minimum Clearance	Shaft Tol. GR9	Hole Tol. GR11	Minimum Clearance	Shaft Tol. GR10
Over	To	-0		+0	-0		+0	-0		+0	-0		+0
0	.12	+1.00	0.60	-0.60	+1.00	1.00	-0.60	+1.60	2.50	-1.00	+2.50	4.00	-1.60
.12	.24	+1.20	0.80	-0.70	+1.20	1.20	-0.70	+1.80	2.80	-1.20	+3.00	4.50	-1.80
.24	.40	+1.40	1.00	-0.90	+1.40	1.60	-0.90	+2.20	3.00	-1.40	+3.50	6.00	-2.20
.40	.71	+1.60	1.20	-1.00	+1.60	2.00	-1.00	+2.80	3.50	-1.60	+4.00	6.00	-2.80
.71	1.19	+2.00	1.60	-1.20	+2.00	2.50	-1.20	+3.50	4.50	-2.00	+5.00	7.00	-3.50
1.19	1.97	+2.50	2.00	-1.60	+2.50	3.00	-1.60	+4.00	5.00	-2.50	+6.00	8.00	-4.00
1.97	3.15	+3.00	2.50	-1.80	+3.00	4.00	-1.80	+4.50	6.00	-3.00	+7.00	9.00	-4.50
3.15	4.73	+3.50	3.00	-2.20	+3.50	5.00	-2.20	+5.00	7.00	-3.50	+9.00	10.00	-5.00
4.73	7.09	+4.00	3.50	-2.50	+4.00	6.00	-2.50	+6.00	8.00	-4.00	+10.00	12.00	-6.00
7.09	9.85	+4.50	4.00	-2.80	+4.50	7.00	-2.80	+7.00	10.00	-4.50	+12.00	15.00	-7.00
9.85	12.41	+5.00	5.00	-3.00	+5.00	8.00	-3.00	+8.00	12.00	-5.00	+12.00	18.00	-8.00
12.41	15.75	+6.00	6.00	-3.50	+6.00	10.00	-3.50	+9.00	14.00	-6.00	+14.00	22.00	-9.00

VALUES IN MILLIMETRES

Nominal Size Range Millimetres		Class RC1 Precision Sliding			Class RC2 Sliding Fit			Class RC3 Precision Running			Class RC4 Close Running			Class RC5 Medium Running		
		Hole Tol. H5	Minimum Clearance	Shaft Tol. g4	Hole Tol. H6	Minimum Clearance	Shaft Tol. g5	Hole Tol. H7	Minimum Clearance	Shaft Tol. f6	Hole Tol. H8	Minimum Clearance	Shaft Tol. f7	Hole Tol. H8	Minimum Clearance	Shaft Tol. e7
Over	To	-0		+0	-0		+0	-0		+0	-0		+0	-0		+0
0	3	+0.004	0.003	-0.003	+0.006	0.003	-0.004	+0.010	0.008	-0.006	+0.015	0.008	-0.010	+0.015	0.015	-0.010
3	6	+0.005	0.004	-0.004	+0.008	0.004	-0.005	+0.013	0.010	-0.008	+0.018	0.010	-0.013	+0.018	0.020	-0.013
6	10	+0.006	0.005	-0.004	+0.010	0.005	-0.006	+0.015	0.013	-0.010	+0.023	0.013	-0.015	+0.023	0.025	-0.015
10	18	+0.008	0.006	-0.005	+0.010	0.006	-0.008	+0.018	0.015	-0.010	+0.025	0.015	-0.018	+0.025	0.030	-0.018
18	30	+0.010	0.008	-0.006	+0.013	0.008	-0.010	+0.020	0.020	-0.013	+0.030	0.020	-0.020	+0.030	0.040	-0.020
30	50	+0.010	0.010	-0.008	+0.015	0.010	-0.010	+0.030	0.030	-0.015	+0.040	0.030	-0.030	+0.040	0.050	-0.030
50	80	+0.013	0.010	-0.008	+0.018	0.010	-0.013	+0.030	0.030	-0.020	+0.050	0.030	-0.030	+0.050	0.060	-0.030
80	120	+0.015	0.013	-0.010	+0.023	0.013	-0.015	+0.040	0.040	-0.020	+0.060	0.040	-0.040	+0.060	0.080	-0.040
120	180	+0.018	0.015	-0.013	+0.025	0.015	-0.018	+0.040	0.040	-0.030	+0.060	0.040	-0.040	+0.060	0.090	-0.040
180	250	+0.020	0.015	-0.015	+0.030	0.015	-0.020	+0.050	0.050	-0.030	+0.070	0.050	-0.050	+0.070	0.110	-0.050
250	315	+0.023	0.020	-0.015	+0.030	0.020	-0.023	+0.050	0.060	-0.030	+0.080	0.060	-0.050	+0.080	0.130	-0.050
315	400	+0.025	0.025	-0.018	+0.036	0.025	-0.025	+0.060	0.080	-0.040	+0.090	0.080	-0.060	+0.090	0.150	-0.060

Nominal Size Range Millimetres		Class RC6 Medium Running			Class RC7 Free Running			Class RC8 Loose Running			Class RC9 Loose Running		
		Hole Tol. H9	Minimum Clearance	Shaft Tol. e8	Hole Tol. H9	Minimum Clearance	Shaft Tol. d8	Hole Tol. H10	Minimum Clearance	Shaft Tol. e9	Hole Tol. GR11	Minimum Clearance	Shaft Tol. gr10
Over	To	-0		+0	-0		+0	-0		+0	-0		+0
0	3	+0.025	0.015	-0.015	+0.025	0.025	-0.015	+0.041	0.064	-0.025	+0.060	0.100	-0.040
3	6	+0.030	0.015	-0.018	+0.030	0.030	-0.018	+0.046	0.071	-0.030	+0.080	0.110	-0.050
6	10	+0.036	0.025	-0.023	+0.036	0.040	-0.023	+0.056	0.076	-0.036	+0.070	0.130	-0.060
10	18	+0.040	0.030	-0.025	+0.040	0.050	-0.025	+0.070	0.090	-0.040	+0.100	0.150	-0.070
18	30	+0.050	0.040	-0.030	+0.050	0.060	-0.030	+0.090	0.110	-0.050	+0.130	0.180	-0.090
30	50	+0.060	0.050	-0.040	+0.060	0.080	-0.040	+0.100	0.130	-0.060	+0.150	0.200	-0.100
50	80	+0.080	0.060	-0.050	+0.080	0.100	-0.050	+0.110	0.150	-0.080	+0.180	0.230	-0.120
80	120	+0.090	0.080	-0.060	+0.090	0.130	-0.060	+0.130	0.180	-0.090	+0.230	0.250	-0.130
120	180	+0.100	0.090	-0.060	+0.100	0.150	-0.060	+0.150	0.200	-0.100	+0.250	0.300	-0.150
180	250	+0.110	0.100	-0.070	+0.110	0.180	-0.070	+0.180	0.250	-0.110	+0.300	0.380	-0.180
250	315	+0.130	0.130	-0.080	+0.130	0.180	-0.080	+0.200	0.300	-0.130	+0.300	0.460	-0.200
315	400	+0.150	0.150	-0.090	+0.150	0.250	-0.090	+0.230	0.360	-0.150	+0.360	0.560	-0.230

From Drafting for Trades and Industry—Mechanical and Electronic. Nelson. Delmar Publishers Inc.

Table 9

LOCATIONAL CLEARANCE FITS

VALUES IN THOUSANDTHS OF AN INCH

Nominal Size Range Inches		Class LC1			Class LC2			Class LC3			Class LC4			Class LC5			Class LC6		
Over	To	Hole Tol. GR6 (-0)	Minimum Clearance	Shaft Tol. GR5 (+0)	Hole Tol. GR7 (-0)	Minimum Clearance	Shaft Tol. GR6 (+0)	Hole Tol. GR8 (-0)	Minimum Clearance	Shaft Tol. GR7 (+0)	Hole Tol. GR10 (-0)	Minimum Clearance	Shaft Tol. GR9 (+0)	Hole Tol. GR7 (-0)	Minimum Clearance	Shaft Tol. GR6 (+0)	Hole Tol. GR9 (-0)	Minimum Clearance	Shaft Tol. GR8 (+0)
0	.12	+0.25	0	-0.15	+0.4	0	-0.25	+0.6	0	-0.4	+1.6	0	-1.0	+0.4	0.10	-0.25	+1.0	0.3	-0.6
.12	.24	+0.30	0	-0.20	+0.5	0	-0.30	+0.7	0	-0.5	+1.8	0	-1.2	+0.5	0.15	-0.30	+1.2	0.4	-0.7
.24	.40	+0.40	0	-0.25	+0.6	0	-0.40	+0.9	0	-0.6	+2.2	0	-1.4	+0.6	0.20	-0.40	+1.4	0.5	-0.9
.40	.71	+0.40	0	-0.30	+0.7	0	-0.40	+1.0	0	-0.7	+2.8	0	-1.6	+0.7	0.25	-0.40	+1.6	0.6	-1.0
.71	1.19	+0.50	0	-0.40	+0.8	0	-0.50	+1.2	0	-0.8	+3.5	0	-2.0	+0.8	0.30	-0.50	+2.0	0.8	-1.2
1.19	1.97	+0.60	0	-0.40	+1.0	0	-0.60	+1.6	0	-1.0	+4.0	0	-2.5	+1.0	0.40	-0.60	+2.5	1.0	-1.6
1.97	3.15	+0.70	0	-0.50	+1.2	0	-0.70	+1.8	0	-1.2	+4.5	0	-3.0	+1.2	0.40	-0.70	+3.0	1.2	-1.8
3.15	4.73	+0.90	0	-0.60	+1.4	0	-0.90	+2.7	0	-1.4	+5.0	0	-3.5	+1.4	0.50	-0.90	+3.5	1.4	-2.2
4.73	7.09	+1.00	0	-0.70	+1.6	0	-1.00	+2.5	0	-1.6	+6.0	0	-4.0	+1.6	0.60	-1.00	+4.0	1.6	-2.5
7.09	9.85	+1.20	0	-0.80	+1.8	0	-1.20	+2.8	0	-1.8	+7.0	0	-4.5	+1.8	0.60	-1.20	+4.5	2.0	-2.8
9.85	12.41	+1.20	0	-0.90	+2.0	0	-1.20	+3.0	0	-2.0	+8.0	0	-5.0	+2.0	0.70	-1.20	+5.0	2.2	-3.0
12.41	15.75	+1.40	0	-1.00	+2.2	0	-1.40	+3.5	0	-2.2	+9.0	0	-6.0	+2.2	0.70	-1.40	+6.0	2.5	-3.5

Nominal Size Range Inches		Class LC7			Class LC8			Class LC9			Class LC10			Class LC11		
Over	To	Hole Tol. GR10 (-0)	Minimum Clearance	Shaft Tol. GR9 (+0)	Hole Tol. GR10 (-0)	Minimum Clearance	Shaft Tol. GR9 (+0)	Hole Tol. GR11 (-0)	Minimum Clearance	Shaft Tol. GR10 (+0)	Hole Tol. GR12 (-0)	Minimum Clearance	Shaft Tol. GR11 (+0)	Hole Tol. GR13 (-0)	Minimum Clearance	Shaft Tol. GR12 (+0)
0	.12	+1.6	0.6	-1.0	+1.6	1.0	-1.0	+2.5	2.5	-1.6	+4.0	4.0	-2.5	+6.0	5.0	-4.0
.12	.24	+1.8	0.8	-1.2	+1.8	1.2	-1.2	+3.0	2.8	-1.8	+5.0	4.5	-3.0	+7.0	6.0	-5.0
.24	.40	+2.2	1.0	-1.4	+2.2	1.6	-1.4	+3.5	3.0	-2.2	+6.0	5.0	-3.5	+9.0	7.0	-6.0
.40	.71	+2.8	1.2	-1.6	+2.8	2.0	-1.6	+4.0	3.5	-2.8	+7.0	6.0	-4.0	+10.0	8.0	-7.0
.71	1.19	+3.5	1.6	-2.0	+3.5	2.5	-2.0	+5.0	4.5	-3.5	+8.0	7.0	-5.0	+12.0	10.0	-8.0
1.19	1.97	+4.0	2.0	-2.5	+4.0	3.6	-2.5	+6.0	5.0	-4.0	+10.0	8.0	-6.0	+16.0	12.0	-10.0
1.97	3.15	+4.5	2.5	-3.0	+4.5	4.0	-3.0	+7.0	6.0	-4.5	+12.0	10.0	-7.0	+18.0	14.0	-12.0
3.15	4.73	+5.0	3.0	-3.5	+5.0	5.0	-3.5	+9.0	7.0	-5.0	+14.0	11.0	-9.0	+22.0	16.0	-14.0
4.73	7.09	+6.0	3.5	-4.0	+6.0	6.0	-4.0	+10.0	8.0	-6.0	+16.0	12.0	-10.0	+25.0	18.0	-16.0
7.09	9.85	+7.0	4.0	-4.5	+7.0	7.0	-4.5	+12.0	10.0	-7.0	+18.0	16.0	-12.0	+28.0	22.0	-18.0
9.85	12.41	+8.0	4.5	-5.0	+8.0	7.0	-5.0	+12.0	12.0	-8.0	+20.0	20.0	-12.0	+30.0	28.0	-20.0
12.41	15.75	+9.0	5.0	-6.0	+9.0	8.0	-6.0	+14.0	14.0	-9.0	+22.0	22.0	-14.0	+35.0	30.0	-22.0

VALUES IN MILLIMETRES

Nominal Size Range Millimetres		Class LC1			Class LC2			Class LC3			Class LC4			Class LC5			Class LC6		
Over	To	Hole Tol. H6 (-0)	Minimum Clearance	Shaft Tol. h5 (+0)	Hole Tol. H7 (-0)	Minimum Clearance	Shaft Tol. h6 (+0)	Hole Tol. H8 (-0)	Minimum Clearance	Shaft Tol. h7 (+0)	Hole Tol. H10 (-0)	Minimum Clearance	Shaft Tol. h9 (+0)	Hole Tol. H7 (-0)	Minimum Clearance	Shaft Tol. g6 (+0)	Hole Tol. H9 (-0)	Minimum Clearance	Shaft Tol. f8 (+0)
0	3	+0.006	0	-0.004	+0.010	0	-0.006	+0.015	0	-0.010	+0.041	0	-0.025	+0.010	0.002	-0.006	+0.025	0.008	-0.015
3	6	+0.008	0	-0.005	+0.013	0	-0.008	+0.018	0	-0.013	+0.046	0	-0.030	+0.013	0.004	-0.008	+0.030	0.010	-0.018
6	10	+0.010	0	-0.006	+0.015	0	-0.010	+0.023	0	-0.015	+0.056	0	-0.036	+0.015	0.005	-0.010	+0.036	0.013	-0.023
10	18	+0.010	0	-0.008	+0.018	0	-0.010	+0.025	0	-0.018	+0.070	0	-0.040	+0.018	0.006	-0.010	+0.041	0.015	-0.025
18	30	+0.013	0	-0.010	+0.020	0	-0.013	+0.030	0	-0.020	+0.090	0	-0.050	+0.020	0.008	-0.013	+0.050	0.020	-0.030
30	50	+0.015	0	-0.010	+0.025	0	-0.015	+0.041	0	-0.025	+0.100	0	-0.060	+0.025	0.010	-0.015	+0.060	0.030	-0.040
50	80	+0.018	0	-0.013	+0.030	0	-0.018	+0.046	0	-0.030	+0.110	0	-0.080	+0.030	0.010	-0.018	+0.080	0.030	-0.050
80	120	+0.023	0	-0.015	+0.036	0	-0.023	+0.056	0	-0.036	+0.130	0	-0.080	+0.036	0.013	-0.023	+0.090	0.040	-0.060
120	180	+0.025	0	-0.018	+0.041	0	-0.025	+0.064	0	-0.041	+0.150	0	-0.100	+0.041	0.015	-0.025	+0.100	0.040	-0.060
180	250	+0.030	0	-0.020	+0.046	0	-0.030	+0.071	0	-0.046	+0.180	0	-0.110	+0.046	0.015	-0.030	+0.110	0.050	-0.070
250	315	+0.020	0	-0.023	+0.051	0	-0.030	+0.076	0	-0.051	+0.200	0	-0.130	+0.051	0.018	-0.030	+0.130	0.060	-0.080
315	400	+0.036	0	-0.025	+0.056	0	-0.036	+0.089	0	-0.056	+0.230	0	-0.150	+0.056	0.018	-0.036	+0.150	0.060	-0.090

Nominal Size Range Millimetres		Class LC7			Class LC8			Class LC9			Class LC10			Class LC11		
Over	To	Hole Tol. H10 (-0)	Minimum Clearance	Shaft Tol. e9 (+0)	Hole Tol. H10 (-0)	Minimum Clearance	Shaft Tol. d9 (+0)	Hole Tol. H11 (-0)	Minimum Clearance	Shaft Tol. c10 (+0)	Hole Tol. GR12 (-0)	Minimum Clearance	Shaft Tol. gr11 (+0)	Hole Tol. GR13 (-0)	Minimum Clearance	Shaft Tol. gr12 (+0)
0	3	+0.041	0.015	-0.025	+0.041	0.025	-0.025	+0.064	0.06	-0.041	+0.10	0.10	-0.06	+0.15	0.13	-0.10
3	6	+0.046	0.020	-0.030	+0.046	0.030	-0.030	+0.076	0.07	-0.046	+0.13	0.11	-0.08	+0.18	0.15	-0.13
6	10	+0.056	0.025	-0.036	+0.056	0.041	-0.036	+0.089	0.08	-0.056	+0.15	0.13	-0.09	+0.23	0.18	-0.15
10	18	+0.070	0.030	-0.040	+0.070	0.050	-0.040	+0.100	0.09	-0.070	+0.18	0.15	-0.10	+0.25	0.20	-0.18
18	30	+0.090	0.040	-0.050	+0.090	0.060	-0.050	+0.130	0.11	-0.090	+0.20	0.18	-0.13	+0.31	0.25	-0.20
30	50	+0.100	0.050	-0.060	+0.100	0.090	-0.060	+0.150	0.13	-0.100	+0.25	0.20	-0.15	+0.41	0.31	-0.25
50	80	+0.110	0.060	-0.080	+0.110	0.100	-0.080	+0.180	0.15	-0.110	+0.31	0.25	-0.18	+0.46	0.36	-0.31
80	120	+0.130	0.080	-0.090	+0.130	0.130	-0.090	+0.230	0.18	-0.130	+0.36	0.28	-0.23	+0.56	0.41	-0.36
120	180	+0.150	0.090	-0.100	+0.150	0.150	-0.100	+0.250	0.20	-0.150	+0.41	0.31	-0.25	+0.64	0.46	-0.41
180	250	+0.180	0.100	-0.110	+0.180	0.180	-0.110	+0.310	0.25	-0.180	+0.46	0.41	-0.31	+0.71	0.56	-0.46
250	315	+0.200	0.110	-0.130	+0.200	0.180	-0.130	+0.310	0.31	-0.200	+0.51	0.51	-0.31	+0.76	0.71	-0.51
315	400	+0.230	0.130	-0.150	+0.230	0.200	-0.150	+0.360	0.36	-0.230	+0.56	0.56	-0.36	+0.89	0.76	-0.56

Table 10

LOCATIONAL TRANSITION FITS

VALUES IN THOUSANDTHS OF AN INCH

Nominal Size Range Inches		Class LT1			Class LT2			Class LT3			Class LT4			Class LT5			Class LT6		
		Hole Tol. GR7	Maximum Interference	Shaft Tol. GR6	Hole Tol. GR8	Maximum Interference	Shaft Tol. GR7	Hole Tol. GR7	Maximum Interference	Shaft Tol. GR6	Hole Tol. GR8	Maximum Interference	Shaft Tol. GR7	Hole Tol. GR7	Maximum Interference	Shaft Tol. GR6	Hole Tol. GR8	Maximum Interference	Shaft Tol. GR7
Over	To	-0		+0	-0		+0	-0		+0	-0		+0	-0		+0	-0		+0
0	.12	+0.4	0.10	-0.25	+0.6	0.20	-0.4	+0.4	0.25	-0.25	+0.6	0.4	-0.4	+0.4	0.5	-0.25	+0.6	0.65	-0.4
.12	.24	+0.5	0.15	-0.30	+0.7	0.25	-0.5	+0.5	0.40	-0.30	+0.7	0.6	-0.5	+0.5	0.6	-0.30	+0.7	0.80	-0.5
.24	.40	+0.6	0.20	-0.40	+0.9	0.30	-0.6	+0.6	0.50	-0.40	+0.9	0.7	-0.6	+0.6	0.8	-0.40	+0.9	1.00	-0.6
.40	.71	+0.7	0.20	-0.40	+1.0	0.30	-0.7	+0.7	0.50	-0.40	+1.0	0.8	-0.7	+0.7	0.9	-0.40	+1.0	1.20	-0.7
.71	1.19	+0.8	0.25	-0.50	+1.2	0.40	-0.8	+0.8	0.60	-0.50	+1.2	0.9	-0.8	+0.8	1.1	-0.50	+1.2	1.40	-0.8
1.19	1.97	+1.0	0.30	-0.60	+1.6	0.50	-1.0	+1.0	0.70	-0.60	+1.6	1.1	-1.0	+1.0	1.3	-0.60	+1.6	1.70	-1.0
1.97	3.15	+1.2	0.30	-0.70	+1.8	0.60	-1.2	+1.2	0.80	-0.70	+1.8	1.3	-1.2	+1.2	1.5	-0.70	+1.8	2.00	-1.2
3.15	4.73	+1.4	0.40	-0.90	+2.2	0.70	-1.4	+1.4	1.00	-0.90	+2.2	1.5	-1.4	+1.4	1.9	-0.90	+2.2	2.40	-1.4
4.73	7.09	+1.6	0.50	-1.00	+2.5	0.80	-1.6	+1.6	1.10	-1.00	+2.5	1.7	-1.6	+1.6	2.2	-1.00	+2.5	2.80	-1.6
7.09	9.85	+1.8	0.60	-1.20	+2.8	0.90	-1.8	+1.8	1.40	-1.20	+2.8	2.0	-1.8	+1.8	2.6	-1.20	+2.8	3.20	-1.8
9.85	12.41	+2.0	0.60	-1.20	+3.0	1.00	-2.0	+2.0	1.40	-1.20	+3.0	2.2	-2.0	+2.0	2.6	-1.20	+3.0	3.40	-2.0
12.41	15.75	+2.2	0.70	-1.40	+3.5	1.00	-2.2	+2.2	1.60	-1.40	+3.5	2.4	-2.2	+2.2	3.0	-1.40	+3.5	3.80	-2.2

VALUES IN MILLIMETRES

Nominal Size Range Millimetres		Class LT1			Class LT2			Class LT3			Class LT4			Class LT5			Class LT6		
		Hole Tol. H7	Maximum Interference	Shaft Tol. js6	Hole Tol. H8	Maximum Interference	Shaft Tol. js7	Hole Tol. H7	Maximum Interference	Shaft Tol. k6	Hole Tol. H8	Maximum Interference	Shaft Tol. k7	Hole Tol. H7	Maximum Interference	Shaft Tol. n6	Hole Tol. H8	Maximum Interference	Shaft Tol. n7
Over	To	-0		+0	-0		+0	-0		+0	-0		+0	-0		+0	-0		+0
0	3	+0.010	0.002	-0.006	+0.015	0.005	-0.010	+0.010	0.006	-0.006	+0.015	0.010	-0.010	+0.010	0.013	-0.006	+0.015	0.016	-0.010
3	6	+0.013	0.004	-0.008	+0.018	0.006	-0.013	+0.013	0.010	-0.008	+0.018	0.015	-0.013	+0.013	0.015	-0.008	+0.018	0.020	-0.013
6	10	+0.015	0.005	-0.010	+0.023	0.008	-0.015	+0.015	0.013	-0.010	+0.023	0.018	-0.015	+0.015	0.020	-0.010	+0.023	0.025	-0.015
10	18	+0.018	0.005	-0.010	+0.025	0.008	-0.018	+0.018	0.013	-0.010	+0.025	0.020	-0.018	+0.018	0.023	-0.010	+0.025	0.030	-0.018
18	30	+0.020	0.006	-0.013	+0.030	0.010	-0.020	+0.020	0.015	-0.013	+0.030	0.023	-0.020	+0.020	0.028	-0.013	+0.030	0.036	-0.020
30	50	+0.025	0.008	-0.015	+0.041	0.013	-0.025	+0.025	0.018	-0.015	+0.041	0.028	-0.025	+0.025	0.033	-0.015	+0.041	0.044	-0.025
50	80	+0.030	0.008	-0.018	+0.046	0.015	-0.030	+0.030	0.020	-0.018	+0.046	0.033	-0.030	+0.030	0.038	-0.018	+0.046	0.051	-0.030
80	120	+0.036	0.010	-0.023	+0.056	0.018	-0.036	+0.036	0.025	-0.023	+0.056	0.038	-0.036	+0.036	0.048	-0.023	+0.056	0.062	-0.036
120	180	+0.041	0.013	-0.025	+0.064	0.020	-0.041	+0.041	0.028	-0.025	+0.064	0.044	-0.041	+0.041	0.056	-0.025	+0.064	0.071	-0.041
180	250	+0.046	0.015	-0.030	+0.071	0.023	-0.046	+0.046	0.036	-0.030	+0.071	0.051	-0.046	+0.046	0.066	-0.030	+0.071	0.081	-0.046
250	315	+0.051	0.015	-0.030	+0.076	0.025	-0.051	+0.051	0.036	-0.030	+0.076	0.056	-0.051	+0.051	0.066	-0.030	+0.076	0.086	-0.051
315	400	+0.056	0.018	-0.036	+0.089	0.025	-0.056	+0.056	0.041	-0.036	+0.089	0.062	-0.056	+0.056	0.076	-0.036	+0.089	0.096	-0.056

From Drafting for Trades and Industry—Mechanical and Electronic. Nelson. Delmar Publishers Inc.

Table 11

LOCATIONAL INTERFERENCE FITS

VALUES IN THOUSANDTHS OF AN INCH

Nominal Size Range Inches		Class LN1 Light Press Fit			Class LN2 Medium Press Fit			Class LN3 Heavy Press Fit			Class LN4			Class LN5			Class LN6		
		Hole Tol. GR6	Maximum Interference	Shaft Tol. GR5	Hole Tol. GR7	Maximum Interference	Shaft Tol. GR6	Hole Tol. GR7	Maximum Interference	Shaft Tol. GR6	Hole Tol. GR8	Maximum Interference	Shaft Tol. GR7	Hole Tol. GR9	Maximum Interference	Shaft Tol. GR8	Hole Tol. GR10	Maximum Interference	Shaft Tol. GR9
Over	To	-0		+0	-0		+0	-0		+0	-0		+0	-0		+0	-0		+0
0	.12	+0.25	0.40	-0.15	+0.4	0.65	-0.25	+0.4	0.75	-0.25	+0.6	1.2	-0.4	+1.0	1.8	-0.6	+1.6	3.0	-1.0
.12	.24	+0.30	0.50	-0.20	+0.5	0.80	-0.30	+0.5	0.90	-0.30	+0.7	1.5	-0.5	+1.2	2.3	-0.7	+1.8	3.6	-1.2
.24	.40	+0.40	0.65	-0.25	+0.6	1.00	-0.40	+0.6	1.20	-0.40	+0.9	1.8	-0.6	+1.4	2.8	-0.9	+2.2	4.4	-1.4
.40	.71	+0.40	0.70	-0.30	+0.7	1.10	-0.40	+0.7	1.40	-0.40	+1.0	2.2	-0.7	+1.6	3.4	-1.0	+2.8	5.6	-1.6
.71	1.19	+0.50	0.90	-0.40	+0.8	1.30	-0.50	+0.8	1.70	-0.50	+1.2	2.6	-0.8	+2.0	4.2	-1.2	+3.5	7.0	-2.0
1.19	1.97	+0.60	1.00	-0.40	+1.0	1.60	-0.60	+1.0	2.00	-0.60	+1.6	3.4	-1.0	+2.5	5.3	-1.6	+4.0	8.5	-2.5
1.97	3.15	+0.70	1.30	-0.50	+1.2	2.10	-0.70	+1.2	2.30	-0.70	+1.8	4.0	-1.2	+3.0	6.3	-1.8	+4.5	10.0	-3.0
3.15	4.73	+0.90	1.60	-0.60	+1.4	2.50	-0.90	+1.4	2.90	-0.90	+2.2	4.8	-1.4	+4.0	7.7	-2.2	+5.0	11.5	-3.5
4.73	7.09	+1.00	1.90	-0.70	+1.6	2.80	-1.00	+1.6	3.50	-1.00	+2.5	5.6	-1.6	+4.5	8.7	-2.5	+6.0	13.5	-4.0
7.09	9.85	+1.20	2.20	-0.80	+1.8	3.20	-1.20	+1.8	4.20	-1.20	+2.8	6.6	-1.8	+5.0	10.3	-2.8	+7.0	16.5	-4.5
9.85	12.41	+1.20	2.30	-0.90	+2.0	3.40	-1.20	+2.0	4.70	-1.20	+3.0	7.5	-2.0	+6.0	12.0	-3.0	+8.0	19.0	-5.0
12.41	15.75	+1.40	2.60	-1.00	+2.2	3.90	-1.40	+2.2	5.90	-1.40	+3.5	8.7	-2.2	+6.0	14.5	-3.5	+9.0	23.0	-6.0

VALUES IN MILLIMETRES

Nominal Size Range Millimetres		Class LN1 Light Press Fit			Class LN2 Medium Press Fit			Class LN3 Heavy Press Fit			Class LN4			Class LN5			Class LN6		
		Hole Tol. GR6	Maximum Interference	Shaft Tol. gr5	Hole Tol. H7	Maximum Interference	Shaft Tol. p6	Hole Tol. H7	Maximum Interference	Shaft Tol. t6	Hole Tol. GR8	Maximum Interference	Shaft Tol. gr7	Hole Tol. GR9	Maximum Interference	Shaft Tol. gr8	Hole Tol. GR10	Maximum Interference	Shaft Tol. gr9
Over	To	-0		+0	-0		+0	-0		+0	-0		+0	-0		+0	-0		+0
0	3	+0.006		-0.004	+0.010	0.016	-0.006	+0.010	0.019	-0.006	+0.015	0.030	-0.010	+0.025	0.046	-0.015	+0.041	0.076	-0.025
3	6	+0.008		-0.005	+0.013	0.020	-0.008	+0.013	0.023	-0.008	+0.018	0.038	-0.013	+0.030	0.059	-0.018	+0.046	0.091	-0.030
6	10	+0.010		-0.006	+0.015	0.025	-0.010	+0.015	0.030	-0.010	+0.023	0.046	-0.015	+0.036	0.071	-0.023	+0.056	0.112	-0.036
10	18	+0.010		-0.008	+0.018	0.028	-0.010	+0.018	0.036	-0.010	+0.025	0.056	-0.018	+0.041	0.086	-0.025	+0.071	0.142	-0.041
18	30	+0.013		-0.010	+0.020	0.033	-0.013	+0.020	0.044	-0.013	+0.030	0.066	-0.020	+0.051	0.107	-0.030	+0.089	0.178	-0.051
30	50	+0.015		-0.010	+0.025	0.041	-0.015	+0.025	0.051	-0.015	+0.041	0.086	-0.025	+0.064	0.135	-0.041	+0.102	0.216	-0.064
50	80	+0.018		-0.013	+0.030	0.054	-0.018	+0.030	0.059	-0.018	+0.046	0.102	-0.030	+0.076	0.160	-0.046	+0.114	0.254	-0.076
80	120	+0.023		-0.015	+0.036	0.064	-0.023	+0.036	0.074	-0.023	+0.056	0.122	-0.036	+0.102	0.196	-0.056	+0.127	0.292	-0.102
120	180	+0.025		-0.018	+0.041	0.071	-0.025	+0.041	0.089	-0.025	+0.064	0.142	-0.041	+0.114	0.221	-0.064	+0.152	0.343	-0.114
180	250	+0.030		-0.020	+0.046	0.081	-0.030	+0.046	0.107	-0.030	+0.071	0.168	-0.046	+0.127	0.262	-0.071	+0.178	0.419	-0.127
250	315	+0.030		-0.023	+0.051	0.086	-0.030	+0.051	0.119	-0.030	+0.076	0.191	-0.051	+0.152	0.305	-0.076	+0.203	0.483	-0.152
315	400	+0.036		-0.025	+0.056	0.099	-0.036	+0.056	0.150	-0.036	+0.089	0.221	-0.056	+0.152	0.368	-0.089	+0.229	0.584	-0.152

From Drafting for Trades and Industry—Mechanical and Electronic. Nelson. Delmar Publishers Inc.

Table 12

FORCE AND SHRINK FITS

VALUES IN THOUSANDTHS OF AN INCH

Nominal Size Range Inches		Class FN1 Light Drive Fit			Class FN2 Medium Drive Fit			Class FN3 Heavy Drive Fit			Class FN4 Shrink Fit			Class FN5 Heavy Shrink Fit		
		Hole Tol. GR6	Maximum Interference	Shaft Tol. GR5	Hole Tol. GR7	Maximum Interference	Shaft Tol. GR6	Hole Tol. GR7	Maximum Interference	Shaft Tol. GR6	Hole Tol. GR7	Maximum Interference	Shaft Tol. GR6	Hole Tol. GR8	Maximum Interference	Shaft Tol. GR7
Over	To	-0		+0	-0		+0	-0		+0	-0		+0	-0		+0
0	.12	+0.25	0.50	-0.15	+0.40	0.85	-0.25				+0.40	0.95	-0.25	+0.60	1.30	-0.40
.12	.24	+0.30	0.60	-0.20	+0.50	1.00	-0.30				+0.50	1.20	-0.30	+0.70	1.70	-0.50
.24	.40	+0.40	0.75	-0.25	+0.60	1.40	-0.40				+0.60	1.60	-0.40	+0.90	2.00	-0.60
.40	.56	+0.40	0.80	-0.30	+0.70	1.60	-0.40				+0.70	1.80	-0.40	+1.00	2.30	-0.70
.56	.71	+0.40	0.90	-0.30	+0.70	1.60	-0.40				+0.70	1.80	-0.40	+1.00	2.50	-0.70
.71	.95	+0.50	1.10	-0.40	+0.80	1.90	-0.50				+0.80	2.10	-0.50	+1.20	3.00	-0.80
.95	1.19	+0.50	1.20	-0.40	+0.80	1.90	-0.50	+0.80	2.10	-0.50	+0.80	2.30	-0.50	+1.20	3.30	-0.80
1.19	1.58	+0.60	1.30	-0.40	+1.00	2.40	-0.60	+1.00	2.60	-0.60	+1.00	3.10	-0.60	+1.60	4.00	-1.00
1.58	1.97	+0.60	1.40	-0.40	+1.00	2.40	-0.60	+1.00	2.80	-0.60	+1.00	3.40	-0.60	+1.60	5.00	-1.00
1.97	2.56	+0.70	1.80	-0.50	+1.20	2.70	-0.70	+1.20	3.20	-0.70	+1.20	4.20	-0.70	+1.80	6.20	-1.20
2.56	3.15	+0.70	1.90	-0.50	+1.20	2.90	-0.70	+1.20	3.70	-0.70	+1.20	4.70	-0.70	+1.80	7.20	-1.20
3.15	3.94	+0.90	2.40	-0.60	+1.40	3.70	-0.90	+1.40	4.40	-0.70	+1.40	5.90	-0.90	+2.20	8.40	-1.40

VALUES IN MILLIMETRES

Nominal Size Range Millimetres		Class FN1 Light Drive Fit			Class FN2 Medium Drive Fit			Class FN3 Heavy Drive Fit			Class FN4 Shrink Fit			Class FN5 Heavy Shrink Fit		
		Hole Tol. GR6	Maximum Interference	Shaft Tol. gr5	Hole Tol. H7	Maximum Interference	Shaft Tol. s6	Hole Tol. H7	Maximum Interference	Shaft Tol. t6	Hole Tol. GR8	Maximum Interference	Shaft Tol. gr7	Hole Tol. H8	Maximum Interference	Shaft Tol. t7
Over	To	-0		+0	-0		+0	-0		+0	-0		+0	-0		+0
0	3	+0.006	0.013	-0.004	+0.010	0.216	-0.006				+0.010	0.024	-0.006	+0.015	0.033	-0.010
3	6	+0.007	0.015	-0.005	+0.013	0.025	-0.007				+0.013	0.030	-0.007	+0.018	0.043	-0.013
6	10	+0.010	0.019	-0.006	+0.015	0.036	-0.010				+0.015	0.041	-0.010	+0.023	0.051	-0.015
10	14	+0.010	0.020	-0.008	+0.018	0.041	-0.010				+0.018	0.046	-0.010	+0.025	0.058	-0.018
14	18	+0.010	0.023	-0.008	+0.018	0.041	-0.010				+0.018	0.046	-0.010	+0.025	0.064	-0.018
18	24	+0.013	0.028	-0.010	+0.020	0.048	-0.013				+0.020	0.053	-0.013	+0.030	0.076	-0.020
24	30	+0.013	0.030	-0.010	+0.020	0.048	-0.013	+0.020	0.053	-0.013	+0.020	0.058	-0.013	+0.030	0.084	-0.020
30	40	+0.015	0.033	-0.010	+0.025	0.061	-0.015	+0.025	0.066	-0.015	+0.025	0.079	-0.015	+0.041	0.102	-0.025
40	50	+0.015	0.036	-0.010	+0.025	0.061	-0.015	+0.025	0.071	-0.015	+0.025	0.086	-0.015	+0.041	0.127	-0.025
50	65	+0.018	0.046	-0.013	+0.030	0.069	-0.018	+0.030	0.082	-0.018	+0.030	0.107	-0.018	+0.046	0.157	-0.030
65	80	+0.018	0.048	-0.013	+0.030	0.074	-0.018	+0.030	0.094	-0.018	+0.030	0.119	-0.018	+0.046	0.183	-0.030
80	100	+0.023	0.061	-0.015	+0.035	0.094	-0.023	+0.035	0.112	-0.023	+0.036	0.150	-0.023	+0.056	0.213	-0.036

From Drafting for Trades and Industry—Mechanical and Electronic. Nelson. Delmar Publishers Inc.

Table 13

UNIFIED STANDARD SCREW THREAD SERIES

Sizes		Basic Major Diameter	Series with graded pitches			Series with constant pitches								Sizes
Primary	Secondary		Coarse UNC	Fine UNF	Extra fine UNEF	4UN	6UN	8UN	12UN	16UN	20UN	28UN	32UN	
0		0.0600	–	80	–	–	–	–	–	–	–	–	–	0
	1	0.0730	64	72	–	–	–	–	–	–	–	–	–	1
2		0.0860	56	64	–	–	–	–	–	–	–	–	–	2
	3	0.0990	48	56	–	–	–	–	–	–	–	–	–	3
4		0.1120	40	48	–	–	–	–	–	–	–	–	–	4
5		0.1250	40	44	–	–	–	–	–	–	–	–	–	5
6		0.1380	32	40	–	–	–	–	–	–	–	–	UNC	6
8		0.1640	32	36	–	–	–	–	–	–	–	–	UNC	8
10		0.1900	24	32	–	–	–	–	–	–	–	–	UNF	10
	12	0.2160	24	28	32	–	–	–	–	–	–	UNF	UNEF	12
¼		0.2500	20	28	32	–	–	–	–	–	UNC	UNF	UNEF	¼
5/16		0.3125	18	24	32	–	–	–	–	–	20	28	UNEF	5/16
⅜		0.3750	16	24	32	–	–	–	–	UNC	20	28	UNEF	⅜
7/16		0.4375	14	20	28	–	–	–	–	16	UNF	UNEF	32	7/16
½		0.5000	13	20	28	–	–	–	–	16	UNF	UNEF	32	½
9/16		0.5625	12	18	24	–	–	–	UNC	16	20	28	32	9/16
⅝		0.6250	11	18	24	–	–	–	12	16	20	28	32	⅝
	11/16	0.6875	–	–	24	–	–	–	12	16	20	28	32	11/16
¾		0.7500	10	16	20	–	–	–	12	UNF	UNEF	28	32	¾
	13/16	0.8125	–	–	20	–	–	–	12	16	UNEF	28	32	13/16
⅞		0.8750	9	14	20	–	–	–	12	16	UNEF	28	32	⅞
	15/16	0.9375	–	–	20	–	–	–	12	16	UNEF	28	32	15/16
1		1.0000	8	12	20	–	–	UNC	UNF	16	UNEF	28	32	1
	1 1/16	1.0625	–	–	18	–	–	8	12	16	20	28	–	1 1/16
1⅛		1.1250	7	12	18	–	–	8	UNF	16	20	28	–	1⅛
	1 3/16	1.1875	–	–	18	–	–	8	12	16	20	28	–	1 3/16
1¼		1.2500	7	12	18	–	–	8	UNF	16	20	28	–	1¼
	1 5/16	1.3125	–	–	18	–	–	8	12	16	20	28	–	1 5/16
1⅜		1.3750	6	12	18	–	UNC	8	UNF	16	20	28	–	1⅜
	1 7/16	1.4375	–	–	18	–	6	8	12	16	20	28	–	1 7/16
1½		1.5000	6	12	18	–	UNC	8	UNF	16	20	28	–	1½
	1 9/16	1.5625	–	–	18	–	6	8	12	16	20	–	–	1 9/16
1⅝		1.6250	–	–	18	–	6	8	12	16	20	–	–	1⅝
	1 11/16	1.6875	–	–	18	–	6	8	12	16	20	–	–	1 11/16
1¾		1.7500	5	–	–	–	6	8	12	16	20	–	–	1¾
	1 13/16	1.8125	–	–	–	–	6	8	12	16	20	–	–	1 13/16
1⅞		1.8750	–	–	–	–	6	8	12	16	20	–	–	1⅞
	1 15/16	1.9375	–	–	–	–	6	8	12	16	20	–	–	1 15/16
2		2.0000	4½	–	–	–	6	8	12	16	20	–	–	2
	2⅛	2.1250	–	–	–	–	6	8	12	16	20	–	–	2⅛
2¼		2.2500	4½	–	–	–	6	8	12	16	20	–	–	2¼
	2⅜	2.3750	–	–	–	–	6	8	12	16	20	–	–	2⅜
2½		2.5000	4	–	–	UNC	6	8	12	16	20	–	–	2½
	2⅝	2.6250	–	–	–	4	6	8	12	16	20	–	–	2⅝
2¾		2.7500	4	–	–	UNC	6	8	12	16	20	–	–	2¾
	2⅞	2.8750	–	–	–	4	6	8	12	16	20	–	–	2⅞
3		3.0000	4	–	–	UNC	6	8	12	16	20	–	–	3
	3⅛	3.1250	–	–	–	4	6	8	12	16	–	–	–	3⅛
3¼		3.2500	4	–	–	UNC	6	8	12	16	–	–	–	3¼
	3⅜	3.3750	–	–	–	4	6	8	12	16	–	–	–	3⅜
3½		3.5000	4	–	–	UNC	6	8	12	16	–	–	–	3½
	3⅝	3.6250	–	–	–	4	6	8	12	16	–	–	–	3⅝
3¾		3.7500	4	–	–	UNC	6	8	12	16	–	–	–	3¾
	3⅞	3.8750	–	–	–	4	6	8	12	16	–	–	–	3⅞
4		4.0000	4	–	–	UNC	6	8	12	16	–	–	–	4
	4⅛	4.1250	–	–	–	4	6	8	12	16	–	–	–	4⅛
4¼		4.2500	–	–	–	4	6	8	12	16	–	–	–	4¼
	4⅜	4.3750	–	–	–	4	6	8	12	16	–	–	–	4⅜
4½		4.5000	–	–	–	4	6	8	12	16	–	–	–	4½
	4⅝	4.6250	–	–	–	4	6	8	12	16	–	–	–	4⅝
4¾		4.7500	–	–	–	4	6	8	12	16	–	–	–	4¾
	4⅞	4.8750	–	–	–	4	6	8	12	16	–	–	–	4⅞
5		5.0000	–	–	–	4	6	8	12	16	–	–	–	5
	5⅛	5.1250	–	–	–	4	6	8	12	16	–	–	–	5⅛
5¼		5.2500	–	–	–	4	6	8	12	16	–	–	–	5¼
	5⅜	5.3750	–	–	–	4	6	8	12	16	–	–	–	5⅜
5½		5.5000	–	–	–	4	6	8	12	16	–	–	–	5½
	5⅝	5.6250	–	–	–	4	6	8	12	16	–	–	–	5⅝
5¾		5.7500	–	–	–	4	6	8	12	16	–	–	–	5¾
	5⅞	5.8750	–	–	–	4	6	8	12	16	–	–	–	5⅞
6		6.0000	–	–	–	4	6	8	12	16	–	–	–	6

Table 14

DRILL AND TAP SIZES

DECIMAL EQUIVALENTS AND TAP DRILL SIZES
OF WIRE GAGE LETTER AND FRACTIONAL SIZE DRILLS
(TAP DRILL SIZES BASED ON 75% MAXIMUM THREAD)

FRACTIONAL SIZE DRILLS INCHES	WIRE GAGE DRILLS	DECIMAL EQUIVALENT INCHES	TAP SIZES SIZE OF THREAD	TAP SIZES THREADS PER INCH
	80	.0135		
	79	.0145		
1/64	.....	.0156		
	78	.0160		
	77	.0180		
	76	.0200		
	75	.0210		
	74	.0225		
	73	.0240		
	72	.0250		
	71	.0260		
	70	.0280		
	69	.0292		
	68	.0310		
1/32	.....	.0312		
	67	.0320		
	66	.0330		
	65	.0350		
	64	.0360		
	63	.0370		
	62	.0380		
	61	.0390		
	60	.0400		
	59	.0410		
	58	.0420		
	57	.0430		
	56	.0465		
3/64	.....	.0469	0	80
	55	.0520		
	54	.0550		
	53	.0595	1	64
1/16	.....	.0625		72
	52	.0635		
	51	.0670		
	50	.0700	2	56
	49	.0730		64
	48	.0760		
5/64	.....	.0781		
	47	.0785	3	48
	46	.0810		
	45	.0820	3	56
	44	.0860		
	43	.0890	4	40
	42	.0935	4	48
3/32	.....	.0937		
	41	.0960		
	40	.0980		
	39	.0995		
	38	.1015	5	40
	37	.1040	5	44
	36	.1065	6	32
7/64	.....	.1094		
	35	.1100		
	34	.1110		
	33	.1130	6	40
	32	.1160		
	31	.1200		
1/8	.....	.1250		
	30	.1285		
	29	.1360	8	32
	28	.1405		36

FRACTIONAL SIZE DRILLS INCHES	WIRE GAGE DRILLS	DECIMAL EQUIVALENT INCHES	TAP SIZES SIZE OF THREAD	TAP SIZES THREADS PER INCH
9/64	.....	.1406		
	27	.1440		
	26	.1470		
	25	.1495	10	24
	24	.1520		
	23	.1540		
5/32	.....	.1562		
	22	.1570		
	21	.1590	10	32
	20	.1610		
	19	.1660		
	18	.1695		
11/64	.....	.1719		
	17	.1730		
	16	.1770	12	24
	15	.1800		
	14	.1820	12	28
	13	.1850		
3/16	.....	.1875		
	12	.1890		
	11	.1910		
	10	.1935		
	9	.1960		
	8	.1990		
	7	.2010	1/4	20
13/64	.....	.2031		
	6	.2040		
	5	.2055		
	4	.2090		
	3	.2130	1/4	28
7/32	.....	.2187		
	2	.2210		
	1	.2280		
	A	.2340		
15/64	.....	.2344		
	B	.2380		
	C	.2420		
	D	.2460		
1/4	E	.2500		
	F	.2570	5/16	18
	G	.2610		
17/64	.....	.2656		
	H	.2660		
	I	.2720	5/16	24
	J	.2770		
	K	.2810		
9/32	.....	.2812		
	L	.2900		
	M	.2950		
19/64	.....	.2969		
	N	.3020		
5/16	.....	.3125	3/8	16
	O	.3160		
	P	.3230		
21/64	.....	.3281		
	Q	.3320	3/8	24
	R	.3390		
11/32	.....	.3437		
	S	.3480		
	T	.3580		

FRACTIONAL SIZE DRILLS INCHES	WIRE GAGE DRILLS	DECIMAL EQUIVALENT INCHES	TAP SIZES SIZE OF THREAD	TAP SIZES THREADS PER INCH
23/64	.....	.3594		
	U	.3680	7/16	14
3/8	.....	.3750		
	V	.3770		
	W	.3860		
25/64	.....	.3906	7/16	20
	X	.3970		
	Y	.4040		
13/32	.....	.4062		
	Z	.4130		
27/64	.....	.4219	1/2	13
7/16	.....	.4375		
29/64	.....	.4531	1/2	20
15/32	.....	.4687		
31/64	.....	.4844	9/16	12
1/2	.....	.5000		
33/64	.....	.5156	9/16	18
17/32	.....	.5312	5/8	11
35/64	.....	.5469		
9/16	.....	.5625		
37/64	.....	.5781	5/8	18
19/32	.....	.5937		
39/64	.....	.6094		
5/8	.....	.6250		
41/64	.....	.6406		
21/32	.....	.6562	3/4	10
43/64	.....	.6719		
11/16	.....	.6875	3/4	16
45/64	.....	.7031		
23/32	.....	.7187		
47/64	.....	.7344		
3/4	.....	.7500		
49/64	.....	.7656	7/8	9
25/32	.....	.7812		
51/64	.....	.7969		
13/16	.....	.8125	7/8	14
53/64	.....	.8281		
27/32	.....	.8437		
55/64	.....	.8594		
7/8	.....	.8750	1	8
57/64	.....	.8906		
29/32	.....	.9062		
59/64	.....	.9219		
15/16	.....	.9375	1	12
61/64	.....	.9531		
31/32	.....	.9687		
63/64	.....	.9844	1 1/8	7
1	.....	1.0000		

Table 15

DRILLED HOLE TOLERANCE (UNDER NORMAL SHOP CONDITIONS)

STANDARD DRILL SIZE				TOLERANCE IN DECIMALS	
DRILL SIZE				PLUS	MINUS
Number	Fraction	Decimal	Metric (MM)		
80		0.0135	0.3412	0.0023	
79		0.0145	0.3788	0.0024	
—	1/64	0.0156	0.3969	0.0025	
78		0.0160	0.4064	0.0025	
77		0.0180	0.4572	0.0026	
76		0.0200	0.5080	0.0027	
75		0.0210	0.5334	0.0027	
74		0.0225	0.5631	0.0028	
73		0.0240	0.6096	0.0028	
72		0.0250	0.6350	0.0029	
71		0.0260	0.6604	0.0029	
70		0.0280	0.7112	0.0030	.0005
69		0.0292	0.7483	0.0030	
68		0.0310	0.7874	0.0031	
—	1/32	0.0312	0.7937	0.0031	
67		0.0320	0.8128	0.0031	
66		0.0330	0.8382	0.0032	
65		0.0350	0.8890	0.0032	
64		0.0360	0.9144	0.0033	
63		0.0370	0.9398	0.0033	
62		0.0380	0.9652	0.0033	
61		0.0390	0.9906	0.0033	
60		0.0400	1.0160	0.0034	
59		0.0410	1.0414	0.0034	
58		0.0420	1.0668	0.0034	
57		0.0430	1.0922	0.0035	
56		0.0465	1.1684	0.0035	
—	3/64	0.0469	1.1906	0.0036	
55		0.0520	1.3208	0.0037	
54		0.0550	1.3970	0.0038	
53		0.0595	1.5122	0.0039	
—	1/16	0.0625	1.5875	0.0039	
52		0.0635	1.6002	0.0039	
51		0.0670	1.7018	0.0040	
50		0.0700	1.7780	0.0041	
49		0.0730	1.8542	0.0041	.001
48		0.0760	1.9304	0.0042	
—	5/64	0.0781	1.9844	0.0042	
47		0.0785	2.0001	0.0042	
46		0.0810	2.0574	0.0043	
45		0.0820	2.0828	0.0043	
44		0.0860	2.1844	0.0044	
43		0.0890	2.2606	0.0044	
42		0.0935	2.3622	0.0045	
—	3/32	0.0937	2.3812	0.0045	
41		0.0960	2.4384	0.0045	
40		0.0980	2.4892	0.0046	
39		0.0995	2.5377	0.0046	
38		0.1015	2.5908	0.0046	
37		0.1040	2.6416	0.0047	
36		0.1065	2.6924	0.0047	
—	7/64	0.1094	2.7781	0.0047	

Table 16

DRILLED HOLE TOLERANCE (UNDER NORMAL SHOP CONDITIONS)

STANDARD DRILL SIZE				TOLERANCE IN DECIMALS	
DRILL SIZE				PLUS	MINUS
No./Letter	Fraction	Decimal	Metric (MM)		
35		0.1100	2.7490	0.0047	
34		0.1110	2.8194	0.0048	
33		0.1130	2.8702	0.0048	
32		0.1160	2.9464	0.0048	
31		0.1200	3.0480	0.0049	
–	1/8	0.1250	3.1750	0.0050	
30		0.1285	3.2766	0.0050	
29		0.1360	3.4544	0.0051	
28		0.1405	3.5560	0.0052	
–	9/64	0.1406	3.5719	0.0052	
27		0.1440	3.6576	0.0052	
26		0.1470	3.7338	0.0052	
25		0.1495	3.7886	0.0053	
24		0.1520	3.8608	0.0053	
23		0.1540	3.9116	0.0053	
–	5/32	0.1562	3.9687	0.0053	
22		0.1570	3.9878	0.0053	
21		0.1590	4.0386	0.0054	
20		0.1610	4.0894	0.0054	
19		0.1660	4.2164	0.0055	
18		0.1695	4.3180	0.0055	
–	11/64	0.1719	4.3656	0.0055	
17		0.1730	4.3942	0.0055	
16		0.1770	4.4958	0.0056	.001
15		0.1800	4.5720	0.0056	
14		0.1820	4.6228	0.0057	
13		0.1850	4.6990	0.0057	
–	3/16	0.1875	4.7625	0.0057	
12		0.1890	4.8006	0.0057	
11		0.1910	4.8514	0.0057	
10		0.1935	4.9276	0.0058	
9		0.1960	4.9784	0.0058	
8		0.1990	5.0800	0.0058	
7		0.2010	5.1054	0.0058	
–	13/64	0.2031	5.1594	0.0058	
6		0.2040	5.1816	0.0058	
5		0.2055	5.2070	0.0059	
4		0.2090	5.3086	0.0059	
3		0.2130	5.4102	0.0059	
–	7/32	0.2187	5.5562	0.0060	
2		0.2210	5.6134	0.0060	
1		0.2280	5.7912	0.0061	
A		0.2340	5.9436	0.0061	
–	15/64	0.2344	5.9531	0.0061	
B		0.2380	6.0452	0.0061	
C		0.2420	6.1468	0.0062	
D		0.2460	6.2484	0.0062	
E	1/4	0.2500	6.3500	0.0063	
F		0.2570	6.5278	0.0063	
G		0.2610	6.6294	0.0063	
–	17/64	0.2656	6.7469	0.0064	
H		0.2660	6.7564	0.0064	
I		0.2720	6.9088	0.0064	.002
J		0.2770	7.0358	0.0065	
K		0.2810	7.1374	0.0065	
–	9/32	0.2812	7.1437	0.0065	
L		0.2900	7.3660	0.0066	
M		0.2950	7.4930	0.0066	
–	19/64	0.2969	7.5406	0.0066	

Table 16 (Continued)

DRILLED HOLE TOLERANCE (UNDER NORMAL SHOP CONDITIONS)

STANDARD DRILL SIZE				TOLERANCE IN DECIMALS	
DRILL SIZE				PLUS	MINUS
Letter	Fraction	Decimal	Metric (MM)		
N		0.3020	7.6708	0.0067	
—	5/16	0.3125	7.9375	0.0067	
O		0.3160	8.0264	0.0068	
P		0.3230	8.2042	0.0068	
—	21/64	0.3281	8.3344	0.0068	
Q		0.3320	8.4328	0.0069	
R		0.3390	8.6106	0.0069	
—	11/32	0.3437	8.7312	0.0070	
S		0.3480	8.8392	0.0070	
T		0.3580	9.0932	0.0071	
—	23/64	0.3594	9.1281	0.0071	
U		0.3680	9.3472	0.0071	
—	3/8	0.3750	9.5250	0.0072	
V		0.3770	9.5758	0.0072	
W		0.3860	9.8044	0.0072	
—	25/64	0.3906	9.9219	0.0073	
X		0.3970	10.0838	0.0073	
Y		0.4040	10.2616	0.0073	
—	13/32	0.4062	10.3187	0.0074	
Z		0.4130	10.4902	0.0074	.002
	27/64	0.4219	10.7156	0.0075	
	7/16	0.4375	10.1125	0.0075	
	29/64	0.4531	11.5094	0.0076	
	15/32	0.4687	11.9062	0.0077	
	31/64	0.4844	12.3031	0.0078	
	1/2	0.5000	12.7000	0.0079	
	33/64	0.5156	13.0968	0.0080	
	17/32	0.5312	13.4937	0.0081	
	35/64	0.5469	13.8906	0.0081	
	9/16	0.5625	14.2875	0.0082	
	37/64	0.5781	14.6844	0.0083	
	19/32	0.5927	15.0812	0.0084	
	39/64	0.6094	15.4781	0.0084	
	5/8	0.6250	15.8750	0.0085	
	41/64	0.6406	16.2719	0.0086	
	21/32	0.6562	16.6687	0.0086	
	43/64	0.6719	17.0656	0.0087	
	11/16	0.6875	17.4625	0.0088	
	45/64	0.7031	17.8594	0.0088	
	23/32	0.7187	18.2562	0.0089	
	47/64	0.7344	18.6532	0.0090	
	3/4	0.7500	19.0500	0.0090	
	49/64	0.7656	19.4469	0.0091	
	25/32	0.7812	19.8433	0.0092	
	51/64	0.7969	20.2402	0.0092	
	13/16	0.8125	20.6375	0.0093	
	53/64	0.8281	21.0344	0.0093	
	27/32	0.8437	21.4312	0.0094	
	55/64	0.8594	21.8281	0.0095	
	7/8	0.8750	22.2250	0.0095	.003
	57/64	0.8906	22.6219	0.0096	
	29/32	0.9062	23.0187	0.0096	
	59/64	0.9219	23.4156	0.0097	
	15/16	0.9375	23.8125	0.0097	
	61/64	0.9531	24.2094	0.0098	
	31/32	0.9687	24.6062	0.0098	
	63/64	0.9844	25.0031	0.0099	
	1	1.0000	25.4000	0.0100	

From Drafting for Trades and Industry—Mechanical and Electronic. Nelson. Delmar Publishers, Inc.

Table 16 (Concluded)

INCH – METRIC THREAD COMPARISON

INCH SERIES			METRIC			
Size	Dia.(In.)	TPI	Size	Dia. (In.)	Pitch (MM)	TPI (Approx)
			M1.4	.055	.3 / .2	85 / 127
#0	.060	80				
			M1.6	.063	.35 / .2	74 / 127
#1	.073	64 / 72				
			M2	.079	.4 / .25	64 / 101
#2	.086	56 / 64				
			M2.5	.098	.45 / .35	56 / 74
#3	.099	48 / 56				
#4	.112	40 / 48				
			M3	.118	.5 / .35	51 / 74
#5	.125	40 / 44				
#6	.138	32 / 40				
			M4	.157	.7 / .5	36 / 51
#8	.164	32 / 36				
#10	.190	24 / 32				
			M5	.196	.8 / .5	32 / 51
			M6	.236	1.0 / .75	25 / 34
¼	.250	20 / 28				
5/16	.312	18 / 24				
			M8	.315	1.25 / 1.0	20 / 25
⅜	.375	16 / 24				
			M10	.393	1.5 / 1.25	17 / 20
7/16	.437	14 / 20				
			M12	.472	1.75 / 1.25	14.5 / 20
½	.500	13 / 20				
			M14	.551	2 / 1.5	12.5 / 17
⅝	.625	11 / 18				
			M16	.630	2 / 1.5	12.5 / 17
			M18	.709	2.5 / 1.5	10 / 17
¾	.750	10 / 16				
			M20	.787	2.5 / 1.5	10 / 17
			M22	.866	2.5 / 1.5	10 / 17
⅞	.875	9 / 14				
			M24	.945	3 / 2	8.5 / 12.5
1"	1.000	8 / 12				
			M27	1.063	3 / 2	8.5 / 12.5

Table 17

I.S.O. BASIC METRIC THREAD INFORMATION

Basic Major DIA & Pitch	Tap Drill DIA	INTERNAL THREADS		EXTERNAL THREADS		Clearance Hole
		Minor DIA MAX	Minor DIA MIN	Major DIA MAX	Major DIA MIN	
M1.6 × 0.35	1.25	1.321	1.221	1.576	1.491	1.9
M2 × 0.4	1.60	1.679	1.567	1.976	1.881	2.4
M2.5 × 0.45	2.05	2.138	2.013	2.476	2.013	2.9
M3 × 0.5	2.50	2.599	2.459	2.976	2.870	3.4
M3.5 × 0.6	2.90	3.010	2.850	3.476	3.351	4.0
M4 × 0.7	3.30	3.422	3.242	3.976	3.836	4.5
M5 × 0.8	4.20	4.334	4.134	4.976	4.826	5.5
M6 × 1	5.00	5.153	4.917	5.974	5.794	6.6
M8 × 1.25	6.80	6.912	6.647	7.972	7.760	9.0
M10 × 1.5	8.50	8.676	8.376	9.968	9.732	11.0
M12 × 1.75	10.20	10.441	10.106	11.966	11.701	13.5
M14 × 2	12.00	12.210	11.835	13.962	13.682	15.5
M16 × 2	14.00	14.210	13.835	15.962	15.682	17.5
M20 × 2.5	17.50	17.744	17.294	19.958	19.623	22.0
M24 × 3	21.00	21.252	20.752	23.952	23.577	26.0
M30 × 3.5	26.50	26.771	26.211	29.947	29.522	33.0
M36 × 4	32.00	32.270	31.670	35.940	35.465	39.0
M42 × 4.5	37.50	37.799	37.129	41.937	41.437	45.0
M48 × 5	43.00	43.297	42.587	47.929	47.399	52.0
M56 × 5.5	50.50	50.796	50.046	55.925	55.365	62.0
M64 × 6	58.00	58.305	57.505	63.920	63.320	70.0
M72 × 6	66.00	66.305	65.505	71.920	71.320	78.0
M80 × 6	74.00	74.305	73.505	79.920	79.320	86.0
M90 × 6	84.00	84.305	83.505	89.920	89.320	96.0
M100 × 6	94.00	94.305	93.505	99.920	99.320	107.0

Table 18

AMERICAN NATIONAL STANDARD
SQUARE AND HEX BOLTS AND SCREWS — INCH SERIES

ANSI B18.2.1—1981

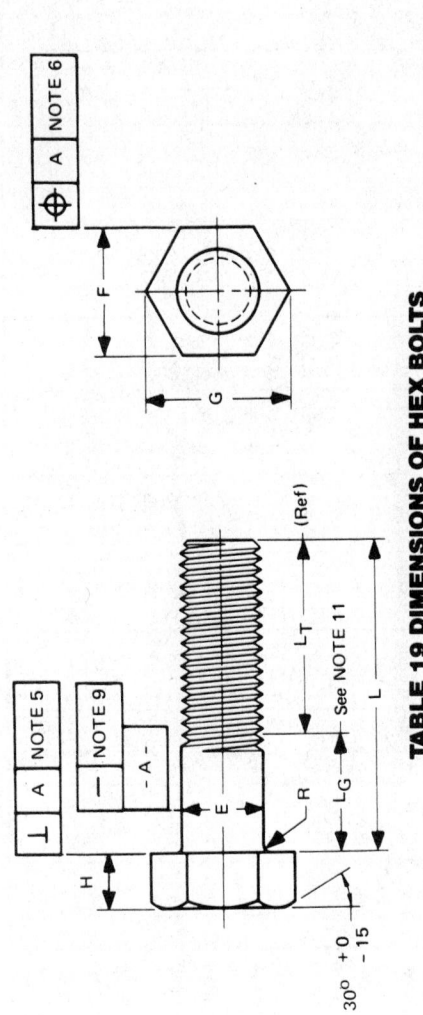

TABLE 19 DIMENSIONS OF HEX BOLTS

Nominal Size or Basic Product Dia (17)		E Body Dia (7) Max	F Width Across Flats (4) Basic	F Max	F Min	G Width Across Corners Max	G Min	H Height Basic	H Max	H Min	R Radius of Fillet Max	R Min	L_T Thread Length For Bolt Lengths (11) 6 in. and shorter Basic	L_T over 6 in. Basic
1/4	0.2500	0.260	7/16	0.438	0.425	0.505	0.484	11/64	0.188	0.150	0.03	0.01	0.750	1.000
5/16	0.3125	0.324	1/2	0.500	0.484	0.577	0.552	7/32	0.235	0.195	0.03	0.01	0.875	1.125
3/8	0.3750	0.388	9/16	0.562	0.544	0.650	0.620	1/4	0.268	0.226	0.03	0.01	1.000	1.250
7/16	0.4375	0.452	5/8	0.625	0.603	0.722	0.687	19/64	0.316	0.272	0.03	0.01	1.125	1.375
1/2	0.5000	0.515	3/4	0.750	0.725	0.866	0.826	11/32	0.364	0.302	0.03	0.01	1.250	1.500
5/8	0.6250	0.642	15/16	0.928	0.906	1.083	1.033	27/64	0.444	0.378	0.06	0.02	1.500	1.750
3/4	0.7500	0.768	1 1/8	1.125	1.088	1.299	1.240	1/2	0.524	0.455	0.06	0.02	1.750	2.000
7/8	0.8750	0.895	1 5/16	1.312	1.269	1.516	1.447	37/64	0.604	0.531	0.06	0.02	2.000	2.250
1	1.0000	1.022	1 1/2	1.500	1.450	1.732	1.653	43/64	0.700	0.591	0.09	0.03	2.250	2.500
1 1/8	1.1250	1.149	1 11/16	1.688	1.631	1.949	1.859	3/4	0.780	0.658	0.09	0.03	2.500	2.750
1 1/4	1.2500	1.277	1 7/8	1.875	1.812	2.165	2.066	27/32	0.876	0.749	0.09	0.03	2.750	3.000
1 3/8	1.3750	1.404	2 1/16	2.062	1.994	2.382	2.273	29/32	0.940	0.810	0.09	0.03	3.000	3.250
1 1/2	1.5000	1.531	2 1/4	2.250	2.175	2.598	2.480	1	1.036	0.902	0.09	0.03	3.250	3.500
1 3/4	1.7500	1.785	2 5/8	2.625	2.538	3.031	2.893	1 5/32	1.196	1.054	0.12	0.04	3.750	4.000
2	2.0000	2.039	3	3.000	2.900	3.464	3.306	1 11/32	1.388	1.175	0.12	0.04	4.250	4.500
2 1/4	2.2500	2.305	3 3/8	3.375	3.262	3.897	3.719	1 1/2	1.548	1.327	0.19	0.06	4.750	5.000
2 1/2	2.5000	2.559	3 3/4	3.750	3.625	4.330	4.133	1 21/32	1.708	1.479	0.19	0.06	5.250	5.500
2 3/4	2.7500	2.827	4 1/8	4.125	3.988	4.763	4.546	1 13/16	1.869	1.632	0.19	0.06	5.750	6.000
3	3.0000	3.081	4 1/2	4.500	4.350	5.196	4.959	2	2.060	1.815	0.19	0.06	6.250	6.500
3 1/4	3.2500	3.335	4 7/8	4.875	4.712	5.629	5.372	2 3/16	2.251	1.936	0.19	0.06	6.750	7.000
3 1/2	3.5000	3.589	5 1/4	5.250	5.075	6.062	5.786	2 5/16	2.380	2.057	0.19	0.06	7.250	7.500
3 3/4	3.7500	3.858	5 5/8	5.625	5.437	6.495	6.198	2 1/2	2.572	2.241	0.19	0.06	7.750	8.000
4	4.0000	4.111	6	6.000	5.800	6.928	6.612	2 11/16	2.764	2.424	0.19	0.06	8.250	8.500

From The American Society of Mechanical Engineers - ANSI B18.2.1 - 1981

AMERICAN NATIONAL STANDARD
SQUARE AND HEX BOLTS AND SCREWS – INCH SERIES

ANSI B18.2.1–1981

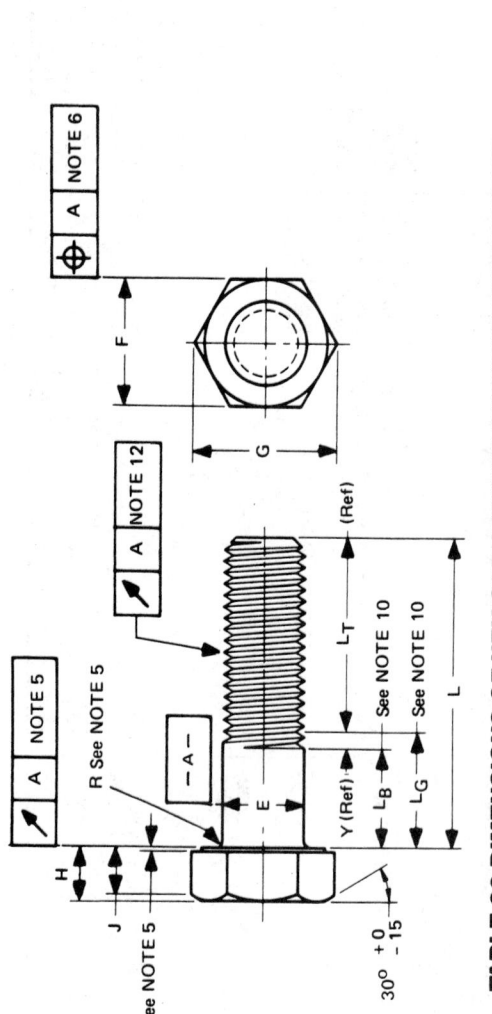

TABLE 20 DIMENSIONS OF HEX CAP SCREWS (FINISHED HEX BOLTS)

Nominal Size or Basic Product Dia (18)		E Body Dia (8) Max	E Body Dia (8) Min	F Width Across Flats Basic	F Width Across Flats Max	F Width Across Flats Min	G Width Across Corners (4) Max	G Width Across Corners (4) Min	H Height Basic	H Height Max	H Height Min	J Wrenching Height (4) Min	LT Thread Length For Screw Lengths (10) Basic 6 in. and Shorter	LT Thread Length For Screw Lengths (10) Basic Over 6 in.	Y Transition Thread Length (10) Max	Runout of Bearing Surface FIM (5) Max
1/4	0.2500	0.2500	0.2450	7/16	0.438	0.428	0.505	0.488	5/32	0.163	0.150	0.106	0.750	1.000	0.250	0.010
5/16	0.3125	0.3125	0.3065	1/2	0.500	0.489	0.577	0.557	13/64	0.211	0.195	0.140	0.875	1.125	0.278	0.011
3/8	0.3750	0.3750	0.3690	9/16	0.562	0.551	0.650	0.628	15/64	0.243	0.226	0.160	1.000	1.250	0.312	0.012
7/16	0.4375	0.4375	0.4305	5/8	0.625	0.612	0.722	0.698	9/32	0.291	0.272	0.195	1.125	1.375	0.357	0.013
1/2	0.5000	0.5000	0.4930	3/4	0.750	0.736	0.866	0.840	5/16	0.323	0.302	0.215	1.250	1.500	0.385	0.014
9/16	0.5625	0.5625	0.5545	13/16	0.812	0.798	0.938	0.910	23/64	0.371	0.348	0.250	1.375	1.625	0.417	0.015
5/8	0.6250	0.6250	0.6170	15/16	0.938	0.922	1.083	1.051	25/64	0.403	0.378	0.269	1.500	1.750	0.455	0.017
3/4	0.7500	0.7500	0.7410	1 1/8	1.125	1.100	1.299	1.254	15/32	0.483	0.455	0.324	1.750	2.000	0.500	0.020
7/8	0.8750	0.8750	0.8660	1 5/16	1.312	1.285	1.516	1.465	35/64	0.563	0.531	0.378	2.000	2.250	0.556	0.023
1	1.0000	1.0000	0.9900	1 1/2	1.500	1.469	1.732	1.675	39/64	0.627	0.591	0.416	2.250	2.500	0.625	0.026
1 1/8	1.1250	1.1250	1.1140	1 11/16	1.688	1.631	1.949	1.859	11/16	0.718	0.658	0.461	2.500	2.750	0.714	0.029
1 1/4	1.2500	1.2500	1.2390	1 7/8	1.875	1.812	2.165	2.066	25/32	0.813	0.749	0.530	2.750	3.000	0.714	0.033
1 3/8	1.3750	1.3750	1.3630	2 1/16	2.062	1.994	2.382	2.273	27/32	0.878	0.810	0.569	3.000	3.250	0.833	0.036
1 1/2	1.5000	1.5000	1.4880	2 1/4	2.230	2.175	2.598	2.480	1 5/16	0.974	0.902	0.640	3.250	3.500	0.833	0.039
1 3/4	1.7500	1.7500	1.7380	2 5/8	2.625	2.538	3.031	2.893	1 3/32	1.134	1.054	0.748	3.750	4.000	1.000	0.046
2	2.0000	2.0000	1.9880	3	3.000	2.900	3.464	3.306	1 7/32	1.263	1.175	0.825	4.250	4.500	1.111	0.052
2 1/4	2.2500	2.2500	2.2380	3 3/8	3.375	3.262	3.897	3.719	1 3/8	1.423	1.327	0.933	4.750	5.000	1.111	0.059
2 1/2	2.5000	2.5000	2.4880	3 3/4	3.750	3.625	4.330	4.133	1 17/32	1.583	1.479	1.042	5.250	5.500	1.250	0.065
2 3/4	2.7500	2.7500	2.7380	4 1/8	4.125	3.988	4.763	4.546	1 11/16	1.744	1.632	1.151	5.750	6.000	1.250	0.072
3	3.0000	3.0000	2.9880	4 1/2	4.500	4.350	5.196	4.959	1 7/8	1.935	1.815	1.290	6.250	6.500	1.250	0.079

From The American Society of Mechanical Engineers · ANSI B18.2.1 · 1981

AMERICAN NATIONAL STANDARD
SQUARE AND HEX BOLTS AND SCREWS – INCH SERIES

ANSI B18.2.1–1981

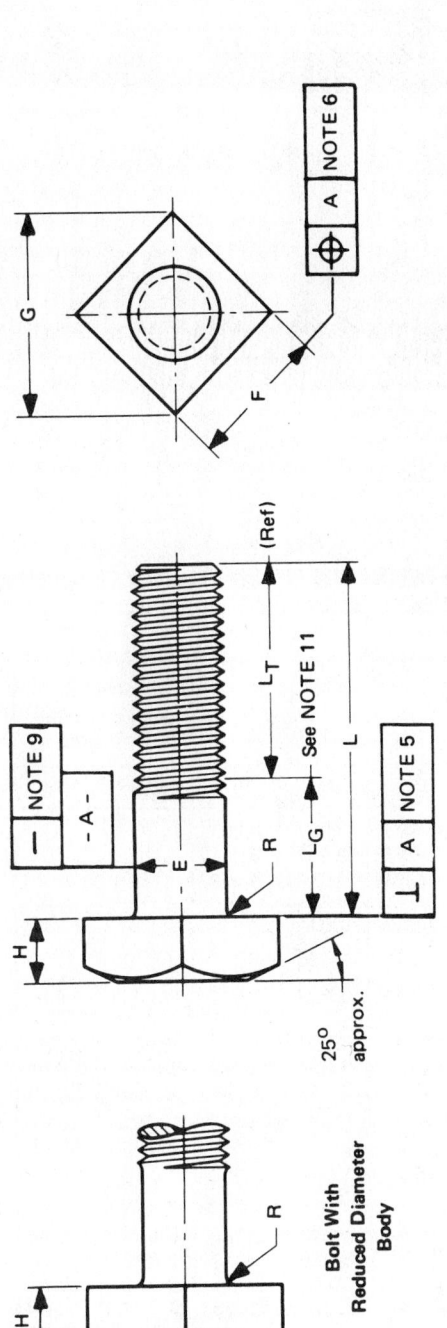

Bolt With Reduced Diameter Body

TABLE 21 DIMENSIONS OF SQUARE BOLTS

Nominal Size or Basic Product Dia (17)		E Body Dia (7),(14) Max	F Width Across Flats (4) Basic	F Max	F Min	G Width Across Corners Max	G Min	H Height Basic	H Max	H Min	R Radius of Fillet Max	R Min	L_T Thread Length For Bolt Lengths (11) 6 in. and shorter Basic	L_T over 6 in. Basic
1/4	0.2500	0.260	3/8	0.375	0.362	0.530	0.498	11/64	0.188	0.156	0.03	0.01	0.750	1.000
5/16	0.3125	0.324	1/2	0.500	0.484	0.707	0.665	13/64	0.220	0.186	0.03	0.01	0.875	1.125
3/8	0.3750	0.388	9/16	0.562	0.544	0.795	0.747	1/4	0.268	0.232	0.03	0.01	1.000	1.250
7/16	0.4375	0.452	5/8	0.625	0.603	0.884	0.828	19/64	0.316	0.278	0.03	0.01	1.125	1.375
1/2	0.5000	0.515	3/4	0.750	0.725	1.061	0.995	21/64	0.348	0.308	0.03	0.01	1.250	1.500
5/8	0.6250	0.642	15/16	0.938	0.906	1.326	1.244	27/64	0.444	0.400	0.06	0.02	1.500	1.750
3/4	0.7500	0.768	1 1/8	1.125	1.088	1.591	1.494	1/2	0.524	0.476	0.06	0.02	1.750	2.000
7/8	0.8750	0.895	1 5/16	1.312	1.269	1.856	1.742	19/32	0.620	0.568	0.06	0.02	2.000	2.250
1	1.0000	1.022	1 1/2	1.500	1.450	2.121	1.991	21/32	0.684	0.628	0.09	0.03	2.250	2.500
1 1/8	1.1250	1.149	1 11/16	1.688	1.631	2.386	2.239	3/4	0.780	0.720	0.09	0.03	2.500	2.750
1 1/4	1.2500	1.277	1 7/8	1.875	1.812	2.652	2.489	27/32	0.876	0.812	0.09	0.03	2.750	3.000
1 3/8	1.3750	1.404	2 1/16	2.062	1.994	2.917	2.738	29/32	0.940	0.872	0.09	0.03	3.000	3.250
1 1/2	1.5000	1.531	2 1/4	2.250	2.175	3.182	2.986	1	1.036	0.964	0.09	0.03	3.250	3.500

From The American Society of Mechanical Engineers - ANSI B18.2.1 - 1981

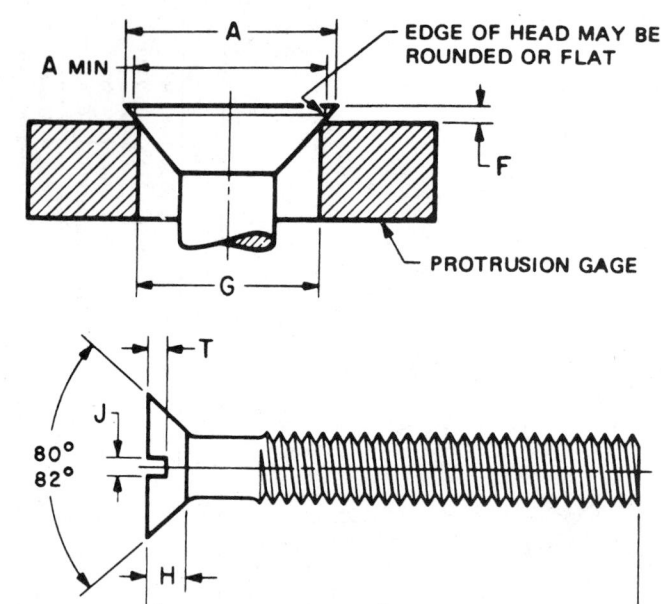

AMERICAN NATIONAL STANDARD
MACHINE SCREWS AND MACHINE SCREW NUTS

ANSI B18.6.3–1972–R1991

TABLE 22 DIMENSIONS OF SLOTTED FLAT COUNTERSUNK HEAD MACHINE SCREWS

Nominal Size[1] or Basic Screw Diameter		L[2] These Lengths or Shorter are Undercut.	A Head Diameter Max, Edge Sharp	A Head Diameter Min, Edge Rounded or Flat	H[3] Head Height Ref	J Slot Width Max	J Slot Width Min	T Slot Depth Max	T Slot Depth Min	F[4] Protrusion Above Gaging Diameter Max	F[4] Protrusion Above Gaging Diameter Min	G[4] Gaging Diameter
0000	0.0210	—	0.043	0.037	0.011	0.008	0.004	0.007	0.003	*	*	*
000	0.0340	—	0.064	0.058	0.016	0.011	0.007	0.009	0.005	*	*	*
00	0.0470	—	0.093	0.085	0.028	0.017	0.010	0.014	0.009	*	*	*
0	0.0600	1/8	0.119	0.099	0.035	0.023	0.016	0.015	0.010	0.026	0.016	0.078
1	0.0730	1/8	0.146	0.123	0.043	0.026	0.019	0.019	0.012	0.028	0.016	0.101
2	0.0860	1/8	0.172	0.147	0.051	0.031	0.023	0.023	0.015	0.029	0.017	0.124
3	0.0990	1/8	0.199	0.171	0.059	0.035	0.027	0.027	0.017	0.031	0.018	0.148
4	0.1120	3/16	0.225	0.195	0.067	0.039	0.031	0.030	0.020	0.032	0.019	0.172
5	0.1250	3/16	0.252	0.220	0.075	0.043	0.035	0.034	0.022	0.034	0.020	0.196
6	0.1380	3/16	0.279	0.244	0.083	0.048	0.039	0.038	0.024	0.036	0.021	0.220
8	0.1640	1/4	0.332	0.292	0.100	0.054	0.045	0.045	0.029	0.039	0.023	0.267
10	0.1900	5/16	0.385	0.340	0.116	0.060	0.050	0.053	0.034	0.042	0.025	0.313
12	0.2160	3/8	0.438	0.389	0.132	0.067	0.056	0.060	0.039	0.045	0.027	0.362
1/4	0.2500	7/16	0.507	0.452	0.153	0.075	0.064	0.070	0.046	0.050	0.029	0.424
5/16	0.3125	1/2	0.635	0.568	0.191	0.084	0.072	0.088	0.058	0.057	0.034	0.539
3/8	0.3750	9/16	0.762	0.685	0.230	0.094	0.081	0.106	0.070	0.065	0.039	0.653
7/16	0.4375	5/8	0.812	0.723	0.223	0.094	0.081	0.103	0.066	0.073	0.044	0.690
1/2	0.5000	3/4	0.875	0.775	0.223	0.106	0.091	0.103	0.065	0.081	0.049	0.739
9/16	0.5625	—	1.000	0.889	0.260	0.118	0.102	0.120	0.077	0.089	0.053	0.851
5/8	0.6250	—	1.125	1.002	0.298	0.133	0.116	0.137	0.088	0.097	0.058	0.962
3/4	0.7500	—	1.375	1.230	0.372	0.149	0.131	0.171	0.111	0.112	0.067	1.186

[1]Where specifying nominal size in decimals, zeros preceding decimal and in the fourth decimal place shall be omitted.
[2]Screws of these lengths and shorter shall have undercut heads as shown in Table 5.
[3]Tabulated values determined from formula for maximum H, Appendix V.
[4]No tolerance for gaging diameter is given. If the gaging diameter of the gage used differs from tabulated value, the protrusion will be affected accordingly and the proper protrusion values must be recalculated using the formulas shown in Appendix I.
*Not practical to gage.

From The American Society of Mechanical Engineers–ANSI B18.6.3–1972–R1991

AMERICAN NATIONAL STANDARD – SLOTTED HEAD CAP SCREWS,
SQUARE HEAD SET SCREWS, AND SLOTTED HEADLESS SET SCREWS

ANSI B18.6.2–1972–R1983

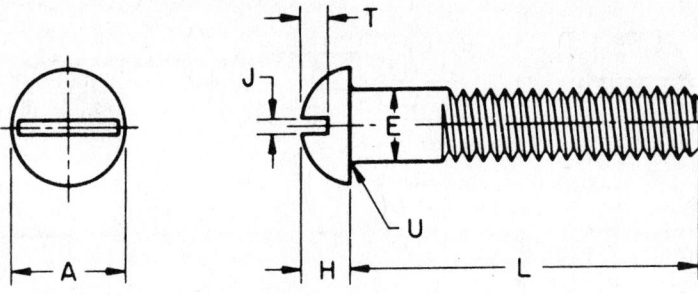

TABLE 23 DIMENSIONS OF SLOTTED ROUND HEAD CAP SCREWS

Nominal Size[1] or Basic Screw Diameter		E Body Diameter		A Head Diameter		H Head Height		J Slot Width		T Slot Depth		U Fillet Radius	
		Max	Min	Max	Min	Max	Min	Max	Min	Max	Min	Max	Min
1/4	0.2500	0.2500	0.2450	0.437	0.418	0.191	0.175	0.075	0.064	0.117	0.097	0.031	0.016
5/16	0.3125	0.3125	0.3070	0.562	0.540	0.245	0.226	0.084	0.072	0.151	0.126	0.031	0.016
3/8	0.3750	0.3750	0.3690	0.625	0.603	0.273	0.252	0.094	0.081	0.168	0.138	0.031	0.016
7/16	0.4375	0.4375	0.4310	0.750	0.725	0.328	0.302	0.094	0.081	0.202	0.167	0.047	0.016
1/2	0.5000	0.5000	0.4930	0.812	0.786	0.354	0.327	0.106	0.091	0.218	0.178	0.047	0.016
9/16	0.5625	0.5625	0.5550	0.937	0.909	0.409	0.378	0.118	0.102	0.252	0.207	0.047	0.016
5/8	0.6250	0.6250	0.6170	1.000	0.970	0.437	0.405	0.133	0.116	0.270	0.220	0.062	0.031
3/4	0.7500	0.7500	0.7420	1.250	1.215	0.546	0.507	0.149	0.131	0.338	0.278	0.062	0.031

[1]Where specifying nominal size in decimals, zeros preceding decimal and in the fourth decimal place shall be omitted.

AMERICAN NATIONAL STANDARD – SLOTTED HEAD CAP SCREWS,
SQUARE HEAD SET SCREWS, AND SLOTTED HEADLESS SET SCREWS

ANSI B18.6.2–1972–R1983

| CAP SCREWS |
| FILLISTER |

Type of Head

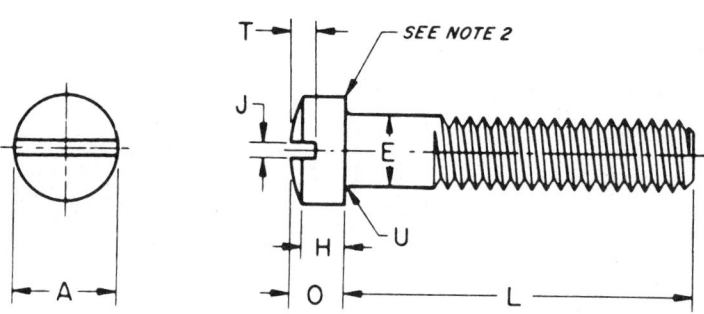

TABLE 24 DIMENSIONS OF SLOTTED FILLISTER HEAD CAP SCREWS

Nominal Size[1] or Basic Screw Diameter		E		A		H		O		J		T		U	
		Body Diameter		Head Diameter		Head Side Height		Total Head Height		Slot Width		Slot Depth		Fillet Radius	
		Max	Min	Max	Min	Max	Min	Max	Min	Max	Min	Max	Min	Max	Min
1/4	0.2500	0.2500	0.2450	0.375	0.363	0.172	0.157	0.216	0.194	0.075	0.064	0.097	0.077	0.031	0.016
5/16	0.3125	0.3125	0.3070	0.437	0.424	0.203	0.186	0.253	0.230	0.084	0.072	0.115	0.090	0.031	0.016
3/8	0.3750	0.3750	0.3690	0.562	0.547	0.250	0.229	0.314	0.284	0.094	0.081	0.142	0.112	0.031	0.016
7/16	0.4375	0.4375	0.4310	0.625	0.608	0.297	0.274	0.368	0.336	0.094	0.081	0.168	0.133	0.047	0.016
1/2	0.5000	0.5000	0.4930	0.750	0.731	0.328	0.301	0.413	0.376	0.106	0.091	0.193	0.153	0.047	0.016
9/16	0.5625	0.5625	0.5550	0.812	0.792	0.375	0.346	0.467	0.427	0.118	0.102	0.213	0.168	0.047	0.016
5/8	0.6250	0.6250	0.6170	0.875	0.853	0.422	0.391	0.521	0.478	0.133	0.116	0.239	0.189	0.062	0.031
3/4	0.7500	0.7500	0.7420	1.000	0.976	0.500	0.466	0.612	0.566	0.149	0.131	0.283	0.223	0.062	0.031
7/8	0.8750	0.8750	0.8660	1.125	1.098	0.594	0.556	0.720	0.668	0.167	0.147	0.334	0.264	0.062	0.031
1	1.0000	1.0000	0.9900	1.312	1.282	0.656	0.612	0.803	0.743	0.188	0.166	0.371	0.291	0.062	0.031

[1]Where specifying nominal size in decimals, zeros preceding decimal and in the fourth decimal place shall be omitted.
[2]A slight rounding of the edges at periphery of head shall be permissible provided the diameter of the bearing circle is equal to no less than 90 percent of the specified miminum head diameter.

From The American Society of Mechanical Engineers–ANSI B18.6.2–1972–R1983

AMERICAN NATIONAL STANDARD
MACHINE SCREWS AND MACHINE SCREW NUTS

ANSI B18.6.3–1972–R1991

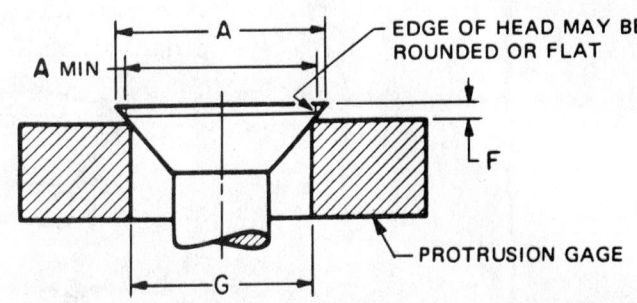

TYPE I RECESS

FLAT

Type of Head

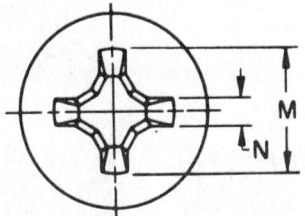

This type of recess has a large center opening, tapered wings, and blunt bottom, with all edges relieved or rounded.

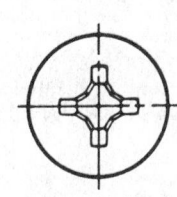

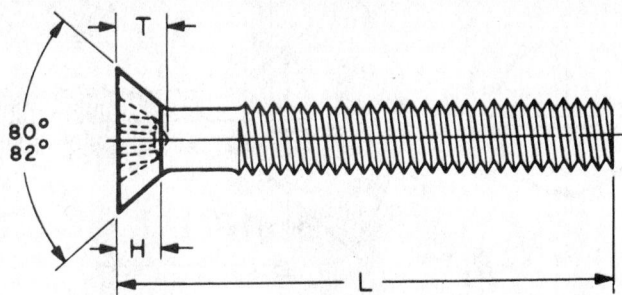

TABLE 25 DIMENSIONS OF TYPE I CROSS RECESSED FLAT COUNTERSUNK HEAD MACHINE SCREWS

Nominal Size[1] or Basic Screw Diameter		L^2 These Lengths or Shorter are Undercut	A Head Diameter		H^3 Head Height	M Recess Diameter		T Recess Depth		N Recess Width	Driver Size	Recess Penetration Gaging Depth		F^4 Protrusion Above Gaging Diameter		G^4 Gaging Diameter
			Max, Edge Sharp	Min, Edge Rounded or Flat	Ref	Max	Min	Max	Min	Min		Max	Min	Max	Min	
0	0.0600	1/8	0.119	0.099	0.035	0.069	0.056	0.043	0.027	0.014	0	0.036	0.020	0.026	0.016	0.078
1	0.0730	1/8	0.146	0.123	0.043	0.077	0.064	0.051	0.035	0.015	0	0.044	0.028	0.028	0.016	0.101
2	0.0860	1/8	0.172	0.147	0.051	0.102	0.089	0.063	0.047	0.017	1	0.056	0.040	0.029	0.017	0.124
3	0.0990	1/8	0.199	0.171	0.059	0.107	0.094	0.068	0.052	0.018	1	0.061	0.045	0.031	0.018	0.148
4	0.1120	3/16	0.225	0.195	0.067	0.128	0.115	0.089	0.073	0.018	1	0.082	0.066	0.032	0.019	0.172
5	0.1250	3/16	0.252	0.220	0.075	0.154	0.141	0.086	0.063	0.027	2	0.075	0.052	0.034	0.020	0.196
6	0.1380	3/16	0.279	0.244	0.083	0.174	0.161	0.106	0.083	0.029	2	0.095	0.072	0.036	0.021	0.220
8	0.1640	1/4	0.332	0.292	0.100	0.189	0.176	0.121	0.098	0.030	2	0.110	0.087	0.039	0.023	0.267
10	0.1900	5/16	0.385	0.340	0.116	0.204	0.191	0.136	0.113	0.032	2	0.125	0.102	0.042	0.025	0.313
12	0.2160	3/8	0.438	0.389	0.132	0.268	0.255	0.156	0.133	0.035	3	0.139	0.116	0.045	0.027	0.362
1/4	0.2500	7/16	0.507	0.452	0.153	0.283	0.270	0.171	0.148	0.036	3	0.154	0.131	0.050	0.029	0.424
5/16	0.3125	1/2	0.635	0.568	0.191	0.365	0.352	0.216	0.194	0.061	4	0.196	0.174	0.057	0.034	0.539
3/8	0.3750	9/16	0.762	0.685	0.230	0.393	0.380	0.245	0.223	0.065	4	0.225	0.203	0.065	0.039	0.653
7/16	0.4375	5/8	0.812	0.723	0.223	0.409	0.396	0.261	0.239	0.068	4	0.241	0.219	0.073	0.044	0.690
1/2	0.5000	3/4	0.875	0.775	0.223	0.424	0.411	0.276	0.254	0.069	4	0.256	0.234	0.081	0.049	0.739
9/16	0.5625	—	1.000	0.889	0.260	0.454	0.431	0.300	0.278	0.073	4	0.280	0.258	0.089	0.053	0.851
5/8	0.6250	—	1.125	1.002	0.298	0.576	0.553	0.342	0.316	0.079	5	0.309	0.283	0.097	0.058	0.962
3/4	0.7500	—	1.375	1.230	0.372	0.640	0.617	0.406	0.380	0.087	5	0.373	0.347	0.112	0.067	1.186

[1]Where specifying nominal size in decimals, zeros preceding decimal and in the fourth decimal place shall be omitted.
[2]Screws of these lengths and shorter shall have undercut heads.
[3]Tabulated values determined from formula for maximum H, Appendix V, ANSI B18.6.3-1972.
[4]No tolerance for gaging diameter is given. If the gaging diameter of the gage used differs from tabulated value, the protrusion will be affected accordingly and the proper protrusion values must be recalculated using the formulas shown in Appendix I, ANSI B18.6.3-1972.

From The American Society of Mechanical Engineers–ANSI B18.6.3–1972–R1991

AMERICAN NATIONAL STANDARD
MACHINE SCREWS AND MACHINE SCREW NUTS

ANSI B18.6.3–1972–R1991

SLOTTED
OVAL

Type of Head

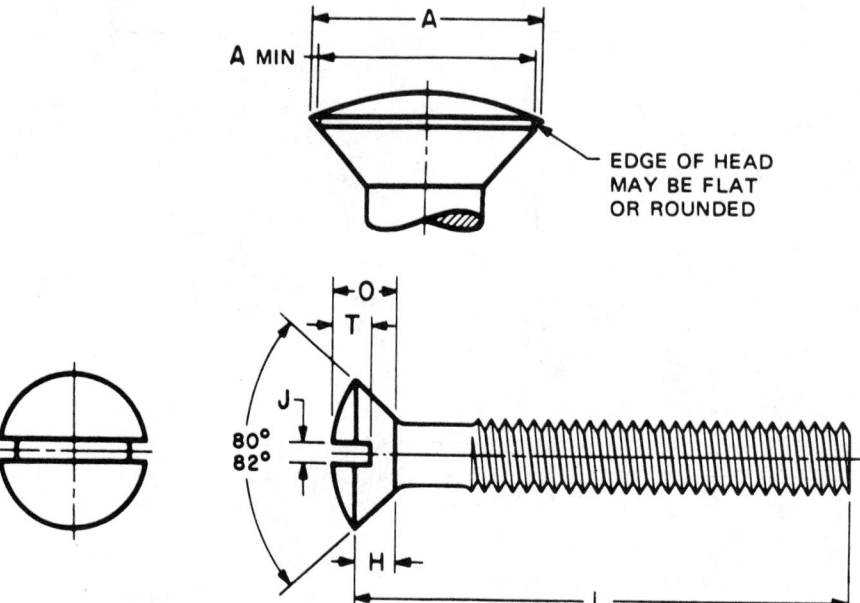

TABLE 26 DIMENSIONS OF SLOTTED OVAL COUNTERSUNK HEAD MACHINE SCREWS

Nominal Size[1] or Basic Screw Diameter		L[2] These Lengths or Shorter are Undercut	A Head Diameter Max, Edge Sharp	A Head Diameter Min, Edge Rounded or Flat	H[3] Head Side Height Ref	O Total Head Height Max	O Total Head Height Min	J Slot Width Max	J Slot Width Min	T Slot Depth Max	T Slot Depth Min
00	0.0470	—	0.093	0.085	0.028	0.042	0.034	0.017	0.010	0.023	0.016
0	0.0600	1/8	0.119	0.099	0.035	0.056	0.041	0.023	0.016	0.030	0.025
1	0.0730	1/8	0.146	0.123	0.043	0.068	0.052	0.026	0.019	0.038	0.031
2	0.0860	1/8	0.172	0.147	0.051	0.080	0.063	0.031	0.023	0.045	0.037
3	0.0990	1/8	0.199	0.171	0.059	0.092	0.073	0.035	0.027	0.052	0.043
4	0.1120	3/16	0.225	0.195	0.067	0.104	0.084	0.039	0.031	0.059	0.049
5	0.1250	3/16	0.252	0.220	0.075	0.116	0.095	0.043	0.035	0.067	0.055
6	0.1380	3/16	0.279	0.244	0.083	0.128	0.105	0.048	0.039	0.074	0.060
8	0.1640	1/4	0.332	0.292	0.100	0.152	0.126	0.054	0.045	0.088	0.072
10	0.1900	5/16	0.385	0.340	0.116	0.176	0.148	0.060	0.050	0.103	0.084
12	0.2160	3/8	0.438	0.389	0.132	0.200	0.169	0.067	0.056	0.117	0.096
1/4	0.2500	7/16	0.507	0.452	0.153	0.232	0.197	0.075	0.064	0.136	0.112
5/16	0.3125	1/2	0.635	0.568	0.191	0.290	0.249	0.084	0.072	0.171	0.141
3/8	0.3750	9/16	0.762	0.685	0.230	0.347	0.300	0.094	0.081	0.206	0.170
7/16	0.4375	5/8	0.812	0.723	0.223	0.345	0.295	0.094	0.081	0.210	0.174
1/2	0.5000	3/4	0.875	0.775	0.223	0.354	0.299	0.106	0.091	0.216	0.176
9/16	0.5625	—	1.000	0.889	0.260	0.410	0.350	0.118	0.102	0.250	0.207
5/8	0.6250	—	1.125	1.002	0.298	0.467	0.399	0.133	0.116	0.285	0.235
3/4	0.7500	—	1.375	1.230	0.372	0.578	0.497	0.149	0.131	0.353	0.293

[1]Where specifying nominal size in decimals, zeros preceding decimal and in the fourth decimal place shall be omitted.
[2]Screws of these lengths and shorter shall have undercut heads.
[3]Tabulated values determined from formula for maximum H, Appendix V, ANSI B18.6.3-1972.

From The American Society of Mechanical Engineers–ANSI B18.6.3–1972–R1991

SLOTTED
PAN

Type of Head

AMERICAN NATIONAL STANDARD
MACHINE SCREWS AND MACHINE SCREW NUTS

ANSI B18.6.3–1972–R1991

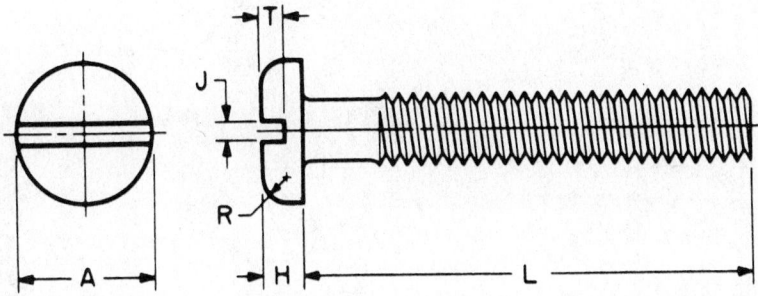

TABLE 27 DIMENSIONS OF SLOTTED PAN HEAD MACHINE SCREWS

Nominal Size[1] or Basic Screw Diameter		A Head Diameter		H Head Height		R Head Radius	J Slot Width		T Slot Depth	
		Max	Min	Max	Min	Max	Max	Min	Max	Min
0000	0.0210	0.042	0.036	0.016	0.010	0.007	0.008	0.004	0.008	0.004
000	0.0340	0.066	0.060	0.023	0.017	0.010	0.012	0.008	0.012	0.008
00	0.0470	0.090	0.082	0.032	0.025	0.015	0.017	0.010	0.016	0.010
0	0.0600	0.116	0.104	0.039	0.031	0.020	0.023	0.016	0.022	0.014
1	0.0730	0.142	0.130	0.046	0.038	0.025	0.026	0.019	0.027	0.018
2	0.0860	0.167	0.155	0.053	0.045	0.035	0.031	0.023	0.031	0.022
3	0.0990	0.193	0.180	0.060	0.051	0.037	0.035	0.027	0.036	0.026
4	0.1120	0.219	0.205	0.068	0.058	0.042	0.039	0.031	0.040	0.030
5	0.1250	0.245	0.231	0.075	0.065	0.044	0.043	0.035	0.045	0.034
6	0.1380	0.270	0.256	0.082	0.072	0.046	0.048	0.039	0.050	0.037
8	0.1640	0.322	0.306	0.096	0.085	0.052	0.054	0.045	0.058	0.045
10	0.1900	0.373	0.357	0.110	0.099	0.061	0.060	0.050	0.068	0.053
12	0.2160	0.425	0.407	0.125	0.112	0.078	0.067	0.056	0.077	0.061
1/4	0.2500	0.492	0.473	0.144	0.130	0.087	0.075	0.064	0.087	0.070
5/16	0.3125	0.615	0.594	0.178	0.162	0.099	0.084	0.072	0.106	0.085
3/8	0.3750	0.740	0.716	0.212	0.195	0.143	0.094	0.081	0.124	0.100
7/16	0.4375	0.863	0.837	0.247	0.228	0.153	0.094	0.081	0.142	0.116
1/2	0.5000	0.987	0.958	0.281	0.260	0.175	0.106	0.091	0.161	0.131
9/16	0.5625	1.041	1.000	0.315	0.293	0.197	0.118	0.102	0.179	0.146
5/8	0.6250	1.172	1.125	0.350	0.325	0.219	0.133	0.116	0.197	0.162
3/4	0.7500	1.435	1.375	0.419	0.390	0.263	0.149	0.131	0.234	0.192

[1]Where specifying nominal size in decimals, zeros preceding decimal and in the fourth decimal place shall be omitted.

From The American Society of Mechanical Engineers–ANSI B18.6.3–1972–R1991

SET SCREWS

SLOTTED
HEADLESS

AMERICAN NATIONAL STANDARD – SLOTTED HEAD CAP SCREWS,
SQUARE HEAD SET SCREWS, AND SLOTTED HEADLESS SET SCREWS   ANSI B18.6.2–1972–R1983

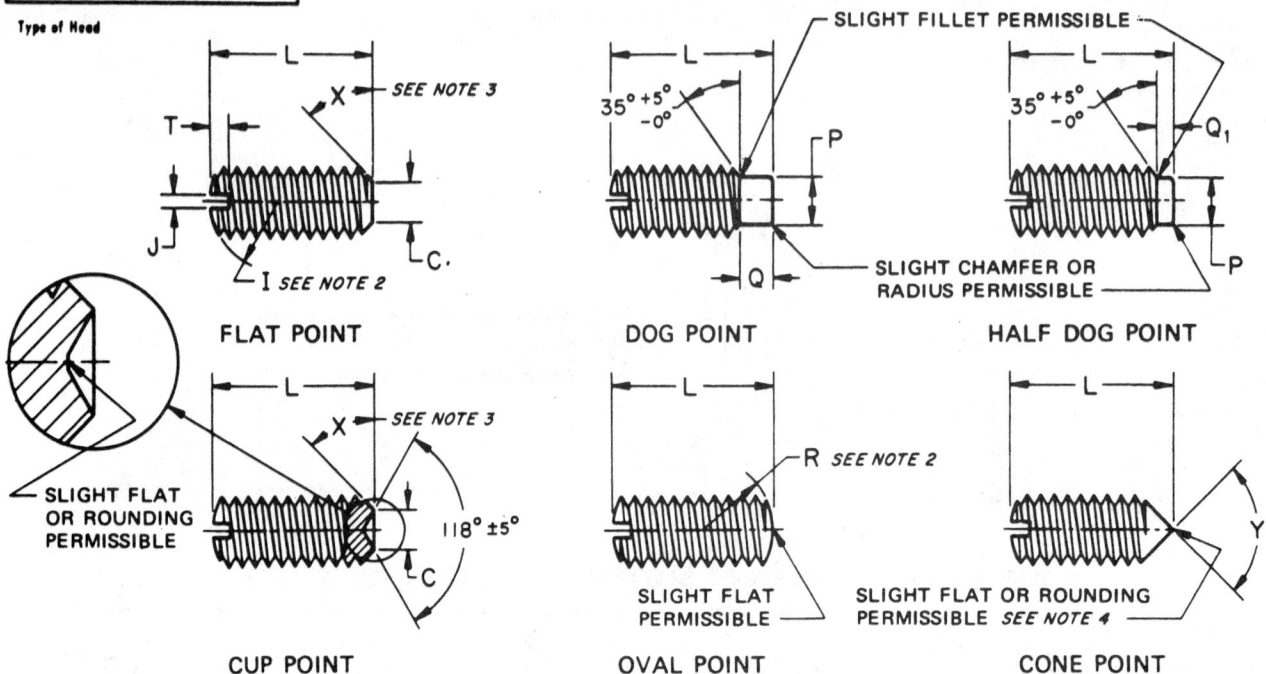

FLAT POINT   DOG POINT   HALF DOG POINT

CUP POINT   OVAL POINT   CONE POINT

TABLE 28 DIMENSIONS OF SLOTTED HEADLESS SET SCREWS

Nominal Size[1] or Basic Screw Diameter		I^2 Crown Radius	J Slot Width		T Slot Depth		C Cup and Flat Point Diameters		P Dog Point Diameters		Point Length Dog		Point Length Half Dog		R^2 Oval Point Radius	Y Cone Point Angle 90° ±2° For These Nominal Lengths or Longer; 118° ±2° For Shorter Screws
		Basic	Max	Min	Max	Min	Max	Min	Max	Min	Max	Min	Max	Min	Basic	
0	0.0600	0.060	0.014	0.010	0.020	0.016	0.033	0.027	0.040	0.037	0.032	0.028	0.017	0.013	0.045	5/64
1	0.0730	0.073	0.016	0.012	0.020	0.016	0.040	0.033	0.049	0.045	0.040	0.036	0.021	0.017	0.055	3/32
2	0.0860	0.086	0.018	0.014	0.025	0.019	0.047	0.039	0.057	0.053	0.046	0.042	0.024	0.020	0.064	7/64
3	0.0990	0.099	0.020	0.016	0.028	0.022	0.054	0.045	0.066	0.062	0.052	0.048	0.027	0.023	0.074	1/8
4	0.1120	0.112	0.024	0.018	0.031	0.025	0.061	0.051	0.075	0.070	0.058	0.054	0.030	0.026	0.084	5/32
5	0.1250	0.125	0.026	0.020	0.036	0.026	0.067	0.057	0.083	0.078	0.063	0.057	0.033	0.027	0.094	3/16
6	0.1380	0.138	0.028	0.022	0.040	0.030	0.074	0.064	0.092	0.087	0.073	0.067	0.038	0.032	0.104	3/16
8	0.1640	0.164	0.032	0.026	0.046	0.036	0.087	0.076	0.109	0.103	0.083	0.077	0.043	0.037	0.123	1/4
10	0.1900	0.190	0.035	0.029	0.053	0.043	0.102	0.088	0.127	0.120	0.095	0.085	0.050	0.040	0.142	1/4
12	0.2160	0.216	0.042	0.035	0.061	0.051	0.115	0.101	0.144	0.137	0.115	0.105	0.060	0.050	0.162	5/16
1/4	0.2500	0.250	0.049	0.041	0.068	0.058	0.132	0.118	0.156	0.149	0.130	0.120	0.068	0.058	0.188	5/16
5/16	0.3125	0.312	0.055	0.047	0.083	0.073	0.172	0.156	0.203	0.195	0.161	0.151	0.083	0.073	0.234	3/8
3/8	0.3750	0.375	0.068	0.060	0.099	0.089	0.212	0.194	0.250	0.241	0.193	0.183	0.099	0.089	0.281	7/16
7/16	0.4375	0.438	0.076	0.068	0.114	0.104	0.252	0.232	0.297	0.287	0.224	0.214	0.114	0.104	0.328	1/2
1/2	0.5000	0.500	0.086	0.076	0.130	0.120	0.291	0.270	0.344	0.334	0.255	0.245	0.130	0.120	0.375	9/16
9/16	0.5625	0.562	0.096	0.086	0.146	0.136	0.332	0.309	0.391	0.379	0.287	0.275	0.146	0.134	0.422	5/8
5/8	0.6250	0.625	0.107	0.097	0.161	0.151	0.371	0.347	0.469	0.456	0.321	0.305	0.164	0.148	0.469	3/4
3/4	0.7500	0.750	0.134	0.124	0.193	0.183	0.450	0.425	0.562	0.549	0.383	0.367	0.196	0.180	0.562	7/8

[1]Where specifying nominal size in decimals, zeros preceding decimal and in the fourth decimal place shall be omitted.
[2]Tolerance on radius for nominal sizes up to and including 5 (0.125 in.) shall be plus 0.015 in. and minus 0.000, and for larger sizes, plus 0.031 in. and minus 0.000. Slotted ends on screws may be flat at option of manufacturer.
[3]Point angle X shall be 45° plus 5°, minus 0°, for screws of nominal lengths equal to or longer than those listed in Column Y, and 30° minimum for screws of shorter nominal lengths.
[4]The extent of rounding or flat at apex of cone point shall not exceed an amount equivalent to 10 percent of the basic screw diameter.

From The American Society of Mechanical Engineers–ANSI B18.6.2–1972–R1983

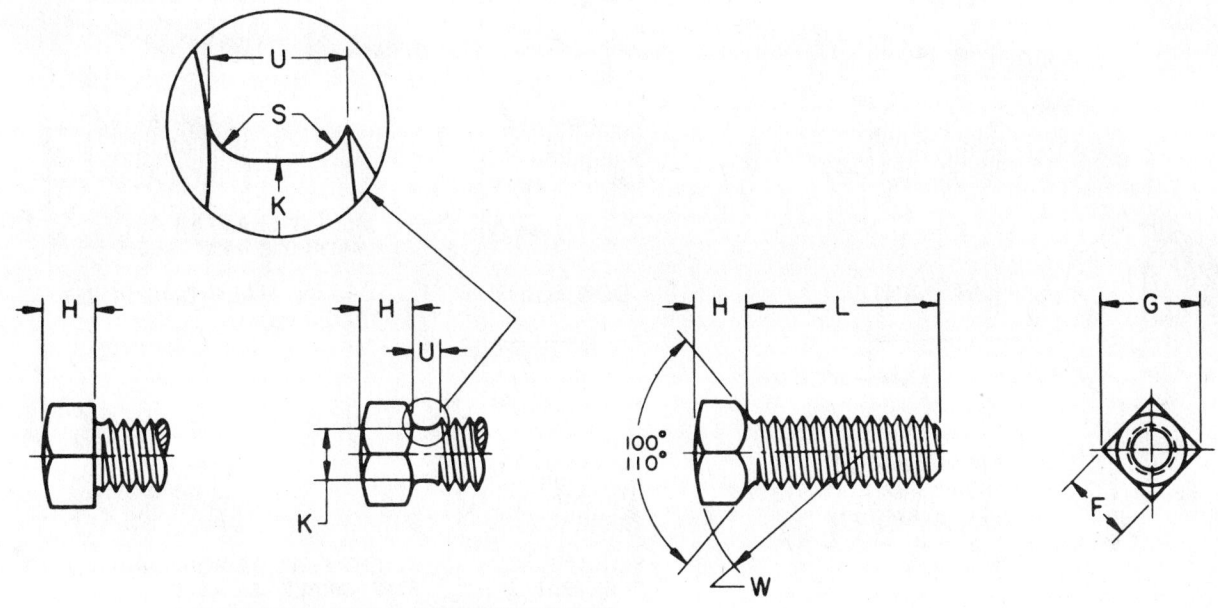

AMERICAN NATIONAL STANDARD – SLOTTED HEAD CAP SCREWS,
SQUARE HEAD SET SCREWS, AND SLOTTED HEADLESS SET SCREWS   ANSI B18.6.2–1972–R1983

SET SCREWS

SQUARE

Type of Head

OPTIONAL HEAD CONSTRUCTIONS

TABLE 29 DIMENSIONS OF SQUARE HEAD SET SCREWS

Nominal Size[1] or Basic Screw Diameter		F		G		H		K		S	U	W
		Width Across Flats		Width Across Corners		Head Height		Neck Relief Diameter		Neck Relief Fillet Radius	Neck Relief Width	Head Radius
		Max	Min	Max	Min	Max	Min	Max	Min	Max	Min	Min
10	0.1900	0.188	0.180	0.265	0.247	0.148	0.134	0.145	0.140	0.027	0.083	0.48
1/4	0.2500	0.250	0.241	0.354	0.331	0.196	0.178	0.185	0.170	0.032	0.100	0.62
5/16	0.3125	0.312	0.302	0.442	0.415	0.245	0.224	0.240	0.225	0.036	0.111	0.78
3/8	0.3750	0.375	0.362	0.530	0.497	0.293	0.270	0.294	0.279	0.041	0.125	0.94
7/16	0.4375	0.438	0.423	0.619	0.581	0.341	0.315	0.345	0.330	0.046	0.143	1.09
1/2	0.5000	0.500	0.484	0.707	0.665	0.389	0.361	0.400	0.385	0.050	0.154	1.25
9/16	0.5625	0.562	0.545	0.795	0.748	0.437	0.407	0.454	0.439	0.054	0.167	1.41
5/8	0.6250	0.625	0.606	0.884	0.833	0.485	0.452	0.507	0.492	0.059	0.182	1.56
3/4	0.7500	0.750	0.729	1.060	1.001	0.582	0.544	0.620	0.605	0.065	0.200	1.88
7/8	0.8750	0.875	0.852	1.237	1.170	0.678	0.635	0.731	0.716	0.072	0.222	2.19
1	1.0000	1.000	0.974	1.414	1.337	0.774	0.726	0.838	0.823	0.081	0.250	2.50
1 1/8	1.1250	1.125	1.096	1.591	1.505	0.870	0.817	0.939	0.914	0.092	0.283	2.81
1 1/4	1.2500	1.250	1.219	1.768	1.674	0.966	0.908	1.064	1.039	0.092	0.283	3.12
1 3/8	1.3750	1.375	1.342	1.945	1.843	1.063	1.000	1.159	1.134	0.109	0.333	3.44
1 1/2	1.5000	1.500	1.464	2.121	2.010	1.159	1.091	1.284	1.259	0.109	0.333	3.75

[1]Where specifying nominal size in decimals, zeros preceding decimal and in the fourth decimal place shall be omitted.

SET SCREWS

SQUARE

Type of Head

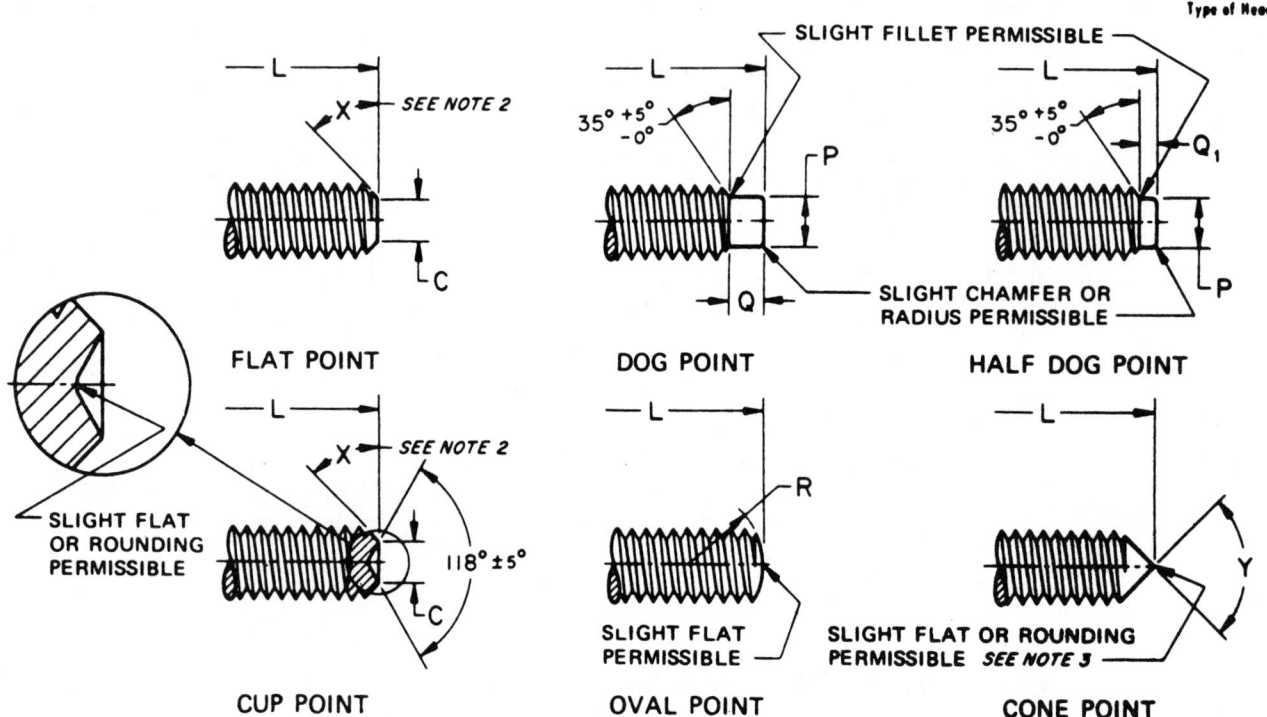

FLAT POINT   DOG POINT   HALF DOG POINT

SLIGHT FLAT OR ROUNDING PERMISSIBLE

CUP POINT   OVAL POINT   CONE POINT

TABLE 29 DIMENSIONS OF SQUARE HEAD SET SCREWS (CONTINUED)

Nominal Size[1] or Basic Screw Diameter		C — Cup and Flat Point Diameters		P — Dog and Half Dog Point Diameters		Q — Point Length Dog		Q₁ — Point Length Half Dog		R — Oval Point Radius +0.031 -0.000	Y — Cone Point Angle 90° ±2° For These Nominal Lengths or Longer; 118° ±2° For Shorter Screws
		Max	Min	Max	Min	Max	Min	Max	Min		
10	0.1900	0.102	0.088	0.127	0.120	0.095	0.085	0.050	0.040	0.142	1/4
1/4	0.2500	0.132	0.118	0.156	0.149	0.130	0.120	0.068	0.058	0.188	5/16
5/16	0.3125	0.172	0.156	0.203	0.195	0.161	0.151	0.083	0.073	0.234	3/8
3/8	0.3750	0.212	0.194	0.250	0.241	0.193	0.183	0.099	0.089	0.281	7/16
7/16	0.4375	0.252	0.232	0.297	0.287	0.224	0.214	0.114	0.104	0.328	1/2
1/2	0.5000	0.291	0.270	0.344	0.334	0.255	0.245	0.130	0.120	0.375	9/16
9/16	0.5625	0.332	0.309	0.391	0.379	0.287	0.275	0.146	0.134	0.422	5/8
5/8	0.6250	0.371	0.347	0.469	0.456	0.321	0.305	0.164	0.148	0.469	3/4
3/4	0.7500	0.450	0.425	0.562	0.549	0.383	0.367	0.196	0.180	0.562	7/8
7/8	0.8750	0.530	0.502	0.656	0.642	0.446	0.430	0.227	0.211	0.656	1
1	1.0000	0.609	0.579	0.750	0.734	0.510	0.490	0.260	0.240	0.750	1 1/8
1 1/8	1.1250	0.689	0.655	0.844	0.826	0.572	0.552	0.291	0.271	0.844	1 1/4
1 1/4	1.2500	0.767	0.733	0.938	0.920	0.635	0.615	0.323	0.303	0.938	1 1/2
1 3/8	1.3750	0.848	0.808	1.031	1.011	0.698	0.678	0.354	0.334	1.031	1 5/8
1 1/2	1.5000	0.926	0.886	1.125	1.105	0.760	0.740	0.385	0.365	1.125	1 3/4

[1]Where specifying nominal size in decimals, zeros preceding decimal and in the fourth decimal place shall be omitted.
[2]Point angle X shall be 45° plus 5°, minus 0°, for screws of nominal lengths equal to or longer than those listed in Column Y, and 30° minimum for screws of shorter nominal lengths.
[3]The extent of rounding or flat at apex of cone point shall not exceed an amount equivalent to 10 percent of the basic screw diameter.

From The American Society of Mechanical Engineers–ANSI B18.6.2–1972–R1983

AMERICAN STANDARD

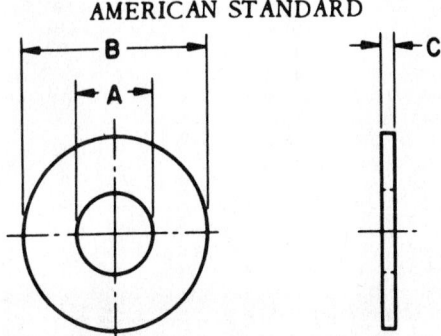

TABLE 30 DIMENSIONS OF PREFERRED SIZES OF TYPE A PLAIN WASHERS **

Nominal Washer Size***			Inside Diameter A			Outside Diameter B			Thickness C		
			Basic	Plus	Minus	Basic	Plus	Minus	Basic	Max	Min
				Tolerance			Tolerance				
–	–		0.078	0.000	0.005	0.188	0.000	0.005	0.020	0.025	0.016
–	–		0.094	0.000	0.005	0.250	0.000	0.005	0.020	0.025	0.016
–	–		0.125	0.008	0.005	0.312	0.008	0.005	0.032	0.040	0.025
No. 6	0.138		0.156	0.008	0.005	0.375	0.015	0.005	0.049	0.065	0.036
No. 8	0.164		0.188	0.008	0.005	0.438	0.015	0.005	0.049	0.065	0.036
No. 10	0.190		0.219	0.008	0.005	0.500	0.015	0.005	0.049	0.065	0.036
3/16	0.188		0.250	0.015	0.005	0.562	0.015	0.005	0.049	0.065	0.036
No. 12	0.216		0.250	0.015	0.005	0.562	0.015	0.005	0.065	0.080	0.051
1/4	0.250	N	0.281	0.015	0.005	0.625	0.015	0.005	0.065	0.080	0.051
1/4	0.250	W	0.312	0.015	0.005	0.734*	0.015	0.007	0.065	0.080	0.051
5/16	0.312	N	0.344	0.015	0.005	0.688	0.015	0.007	0.065	0.080	0.051
5/16	0.312	W	0.375	0.015	0.005	0.875	0.030	0.007	0.083	0.104	0.064
3/8	0.375	N	0.406	0.015	0.005	0.812	0.015	0.007	0.065	0.080	0.051
3/8	0.375	W	0.438	0.015	0.005	1.000	0.030	0.007	0.083	0.104	0.064
7/16	0.438	N	0.469	0.015	0.005	0.922	0.015	0.007	0.065	0.080	0.051
7/16	0.438	W	0.500	0.015	0.005	1.250	0.030	0.007	0.083	0.104	0.064
1/2	0.500	N	0.531	0.015	0.005	1.062	0.030	0.007	0.095	0.121	0.074
1/2	0.500	W	0.562	0.015	0.005	1.375	0.030	0.007	0.109	0.132	0.086
9/16	0.562	N	0.594	0.015	0.005	1.156*	0.030	0.007	0.095	0.121	0.074
9/16	0.562	W	0.625	0.015	0.005	1.469*	0.030	0.007	0.109	0.132	0.086
5/8	0.625	N	0.656	0.030	0.007	1.312	0.030	0.007	0.095	0.121	0.074
5/8	0.625	W	0.688	0.030	0.007	1.750	0.030	0.007	0.134	0.160	0.108
3/4	0.750	N	0.812	0.030	0.007	1.469	0.030	0.007	0.134	0.160	0.108
3/4	0.750	W	0.812	0.030	0.007	2.000	0.030	0.007	0.148	0.177	0.122
7/8	0.875	N	0.938	0.030	0.007	1.750	0.030	0.007	0.134	0.160	0.108
7/8	0.875	W	0.938	0.030	0.007	2.250	0.030	0.007	0.165	0.192	0.136
1	1.000	N	1.062	0.030	0.007	2.000	0.030	0.007	0.134	0.160	0.108
1	1.000	W	1.062	0.030	0.007	2.500	0.030	0.007	0.165	0.192	0.136
1 1/8	1.125	N	1.250	0.030	0.007	2.250	0.030	0.007	0.134	0.160	0.108
1 1/8	1.125	W	1.250	0.030	0.007	2.750	0.030	0.007	0.165	0.192	0.136
1 1/4	1.250	N	1.375	0.030	0.007	2.500	0.030	0.007	0.165	0.192	0.136
1 1/4	1.250	W	1.375	0.030	0.007	3.000	0.030	0.007	0.165	0.192	0.136
1 3/8	1.375	N	1.500	0.030	0.007	2.750	0.030	0.007	0.165	0.192	0.136
1 3/8	1.375	W	1.500	0.045	0.010	3.250	0.045	0.010	0.180	0.213	0.153
1 1/2	1.500	N	1.625	0.030	0.007	3.000	0.030	0.007	0.165	0.192	0.136
1 1/2	1.500	W	1.625	0.045	0.010	3.500	0.045	0.010	0.180	0.213	0.153
1 5/8	1.625		1.750	0.045	0.010	3.750	0.045	0.010	0.180	0.213	0.153
1 3/4	1.750		1.875	0.045	0.010	4.000	0.045	0.010	0.180	0.213	0.153
1 7/8	1.875		2.000	0.045	0.010	4.250	0.045	0.010	0.180	0.213	0.153
2	2.000		2.125	0.045	0.010	4.500	0.045	0.010	0.180	0.213	0.153
2 1/4	2.250		2.375	0.045	0.010	4.750	0.045	0.010	0.220	0.248	0.193
2 1/2	2.500		2.625	0.045	0.010	5.000	0.045	0.010	0.238	0.280	0.210
2 3/4	2.750		2.875	0.065	0.010	5.250	0.065	0.010	0.259	0.310	0.228
3	3.000		3.125	0.065	0.010	5.500	0.065	0.010	0.284	0.327	0.249

*The 0.734 in., 1.156 in., and 1.469 in. outside diameters avoid washers which could be used in coin operated devices.
**Preferred sizes are for the most part from series previously designated "Standard Plate" and "SAE." Where common sizes existed in the two series, the SAE size is designated "N" (narrow) and the Standard Plate "W" (wide). These sizes as well as all other sizes of Type A Plain Washers are to be ordered by ID, OD, and thickness dimensions.
***Nominal washer sizes are intended for use with comparable nominal screw or bolt sizes.

From The American Society of Mechanical Engineers–ANSI B18.22.1–1965–R1990

AMERICAN NATIONAL STANDARD
LOCK WASHERS

ANSI B18.21.1–1990

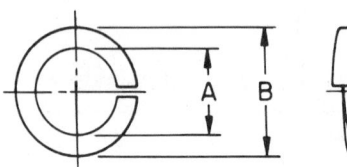

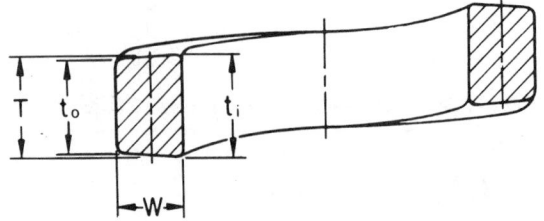

ENLARGED SECTION

TABLE 31 DIMENSIONS OF REGULAR HELICAL SPRING LOCK WASHERS[1]

Nominal Washer Size		A Inside Diameter		B Outside Diameter	T Mean Section Thickness $\left(\dfrac{t_i + t_o}{2}\right)$	W Section Width
		Max	Min	Max[2]	Min	Min
No. 4	0.112	0.120	0.114	0.173	0.022	0.022
No. 5	0.125	0.133	0.127	0.202	0.030	0.030
No. 6	0.138	0.148	0.141	0.216	0.030	0.030
No. 8	0.164	0.174	0.167	0.267	0.047	0.042
No. 10	0.190	0.200	0.193	0.294	0.047	0.042
¼	0.250	0.262	0.254	0.365	0.078	0.047
⁵⁄₁₆	0.312	0.326	0.317	0.460	0.093	0.062
³⁄₈	0.375	0.390	0.380	0.553	0.125	0.076
⁷⁄₁₆	0.438	0.455	0.443	0.647	0.140	0.090
½	0.500	0.518	0.506	0.737	0.172	0.103
⁵⁄₈	0.625	0.650	0.635	0.923	0.203	0.125
¾	0.750	0.775	0.760	1.111	0.218	0.154
⁷⁄₈	0.875	0.905	0.887	1.296	0.234	0.182
1	1.000	1.042	1.017	1.483	0.250	0.208
1⅛	1.125	1.172	1.144	1.669	0.313	0.236
1¼	1.250	1.302	1.271	1.799	0.313	0.236
1⅜	1.375	1.432	1.398	2.041	0.375	0.292
1½	1.500	1.561	1.525	2.170	0.375	0.292
1¾	1.750	1.811	1.775	2.602	0.469	0.383
2	2.000	2.061	2.025	2.852	0.469	0.383
2¼	2.250	2.311	2.275	3.352	0.508	0.508
2½	2.500	2.561	2.525	3.602	0.508	0.508
2¾	2.750	2.811	2.775	4.102	0.633	0.633
3	3.000	3.061	3.025	4.352	0.633	0.633

[1]For use with 1960 Series Socket Head Cap Screws specified in American National Standard, ANSI B18.3.
[2]The maximum outside diameters specified allow for the commercial tolerances on cold drawn wire.

From The American Society of Mechanical Engineers–ANSI B18.21.1–1990

AMERICAN NATIONAL STANDARD
LOCK WASHERS

ANSI B18.21.1–1990

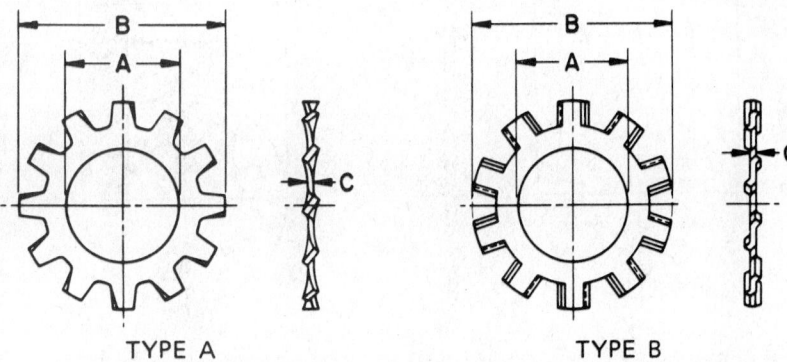

TYPE A   TYPE B

TABLE 32 DIMENSIONS OF EXTERNAL TOOTH LOCK WASHERS

Nominal Washer Size		A Inside Diameter		B Outside Diameter		C Thickness	
		Max	Min	Max	Min	Max	Min
No. 3	0.099	0.109	0.102	0.235	0.220	0.015	0.012
No. 4	0.112	0.123	0.115	0.260	0.245	0.019	0.015
No. 5	0.125	0.136	0.129	0.285	0.270	0.019	0.014
No. 6	0.138	0.150	0.141	0.320	0.305	0.022	0.016
No. 8	0.164	0.176	0.168	0.381	0.365	0.023	0.018
No. 10	0.190	0.204	0.195	0.410	0.395	0.025	0.020
No. 12	0.216	0.231	0.221	0.475	0.460	0.028	0.023
¼	0.250	0.267	0.256	0.510	0.494	0.028	0.023
5⁄16	0.312	0.332	0.320	0.610	0.588	0.034	0.028
3⁄8	0.375	0.398	0.384	0.694	0.670	0.040	0.032
7⁄16	0.438	0.464	0.448	0.760	0.740	0.040	0.032
½	0.500	0.530	0.513	0.900	0.880	0.045	0.037
9⁄16	0.562	0.596	0.576	0.985	0.960	0.045	0.037
5⁄8	0.625	0.663	0.641	1.070	1.045	0.050	0.042
11⁄16	0.688	0.728	0.704	1.155	1.130	0.050	0.042
¾	0.750	0.795	0.768	1.260	1.220	0.055	0.047
13⁄16	0.812	0.861	0.833	1.315	1.290	0.055	0.047
7⁄8	0.875	0.927	0.897	1.410	1.380	0.060	0.052
1	1.000	1.060	1.025	1.620	1.590	0.067	0.059

From The American Society of Mechanical Engineers—ANSI B18.21.1–1990

AMERICAN NATIONAL STANDARD
LOCK WASHERS

ANSI B18.21.1−1990

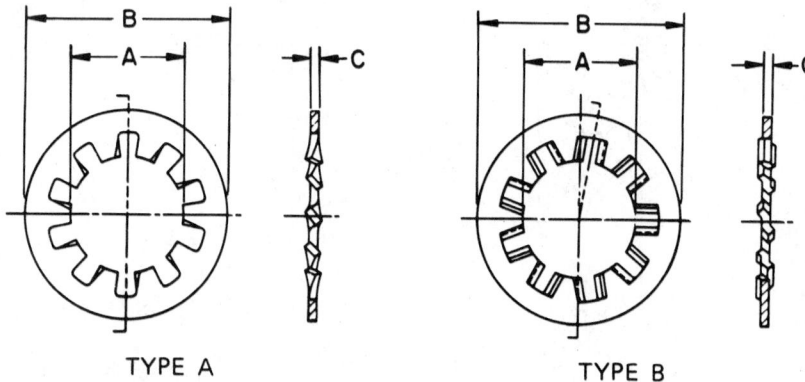

TYPE A   TYPE B

TABLE 33 DIMENSIONS OF INTERNAL TOOTH LOCK WASHERS

Nominal Washer Size		A Inside Diameter		B Outside Diameter		C Thickness	
		Max	Min	Max	Min	Max	Min
No. 2	0.086	0.095	0.089	0.200	0.175	0.015	0.010
No. 3	0.099	0.109	0.102	0.232	0.215	0.019	0.012
No. 4	0.112	0.123	0.115	0.270	0.255	0.019	0.015
No. 5	0.125	0.136	0.129	0.280	0.245	0.021	0.017
No. 6	0.138	0.150	0.141	0.295	0.275	0.021	0.017
No. 8	0.164	0.176	0.168	0.340	0.325	0.023	0.018
No. 10	0.190	0.204	0.195	0.381	0.365	0.025	0.020
No. 12	0.216	0.231	0.221	0.410	0.394	0.025	0.020
1/4	0.250	0.267	0.256	0.478	0.460	0.028	0.023
5/16	0.312	0.332	0.320	0.610	0.594	0.034	0.028
3/8	0.375	0.398	0.384	0.692	0.670	0.040	0.032
7/16	0.438	0.464	0.448	0.789	0.740	0.040	0.032
1/2	0.500	0.530	0.512	0.900	0.867	0.045	0.037
9/16	0.562	0.596	0.576	0.985	0.957	0.045	0.037
5/8	0.625	0.663	0.640	1.071	1.045	0.050	0.042
11/16	0.688	0.728	0.704	1.166	1.130	0.050	0.042
3/4	0.750	0.795	0.769	1.245	1.220	0.055	0.047
13/16	0.812	0.861	0.832	1.315	1.290	0.055	0.047
7/8	0.875	0.927	0.894	1.410	1.364	0.060	0.052
1	1.000	1.060	1.019	1.637	1.590	0.067	0.059
1 1/8	1.125	1.192	1.144	1.830	1.799	0.067	0.059
1 1/4	1.250	1.325	1.275	1.975	1.921	0.067	0.059

From The American Society of Mechanical Engineers—ANSI B18.21.1−1990

**AMERICAN NATIONAL STANDARD
SQUARE AND HEX NUTS**

ANSI B18.2.2–1987

TABLE 34 DIMENSIONS OF HEX NUTS AND HEX JAM NUTS

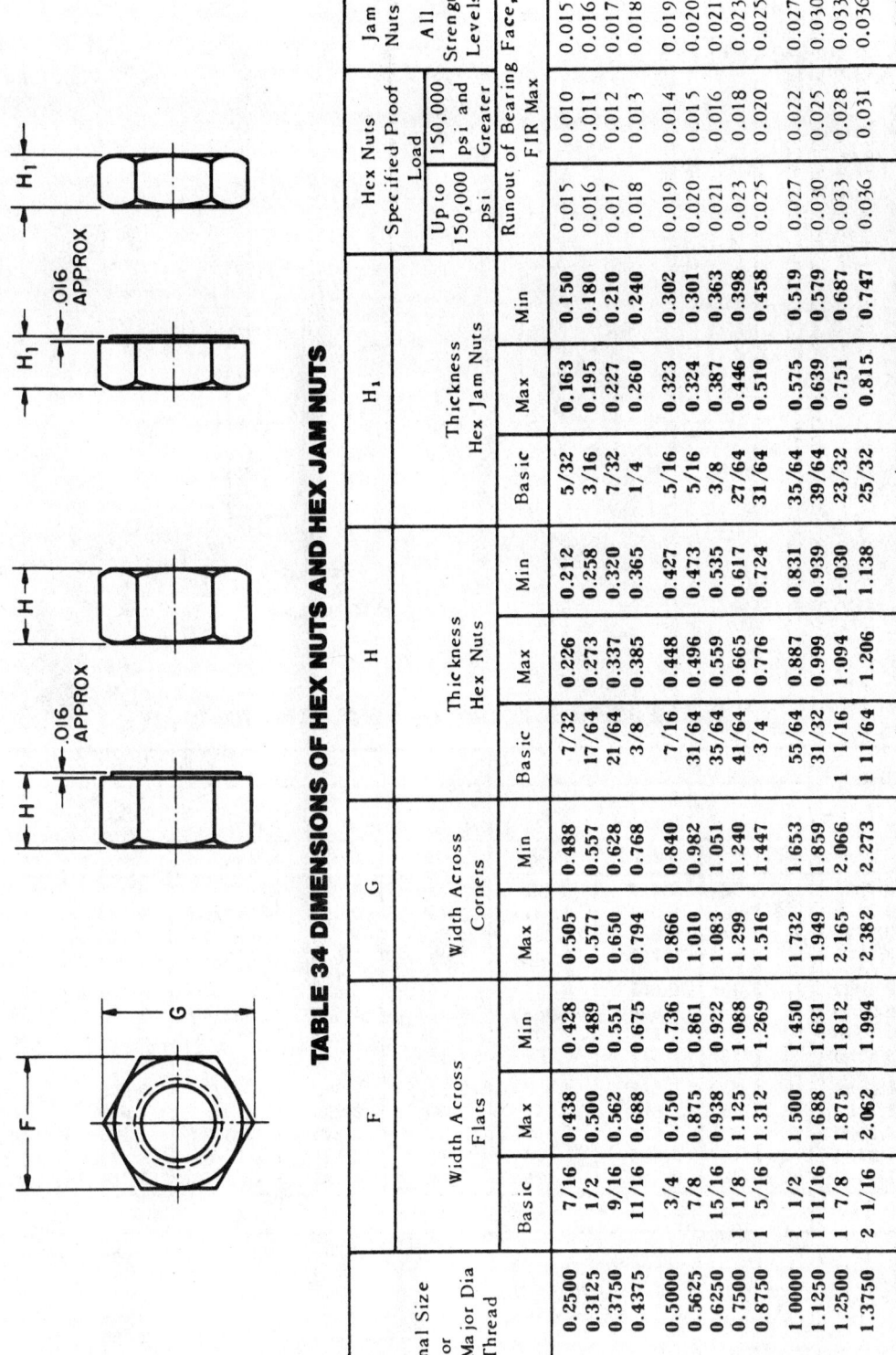

Nominal Size or Basic Major Dia of Thread		F Width Across Flats			G Width Across Corners		H Thickness Hex Nuts			H₁ Thickness Hex Jam Nuts			Runout of Bearing Face, FIR Max — Hex Nuts Specified Proof Load, Up to 150,000 psi	Runout of Bearing Face, FIR Max — Hex Nuts Specified Proof Load, 150,000 psi and Greater	Runout of Bearing Face, FIR Max — Jam Nuts, All Strength Levels
		Basic	Max	Min	Max	Min	Basic	Max	Min	Basic	Max	Min			
1/4	0.2500	7/16	0.438	0.428	0.505	0.488	7/32	0.226	0.212	5/32	0.163	0.150	0.015	0.010	0.015
5/16	0.3125	1/2	0.500	0.489	0.577	0.557	17/64	0.273	0.258	3/16	0.195	0.180	0.016	0.011	0.016
3/8	0.3750	9/16	0.562	0.551	0.650	0.628	21/64	0.337	0.320	7/32	0.227	0.210	0.017	0.012	0.017
7/16	0.4375	11/16	0.688	0.675	0.794	0.768	3/8	0.385	0.365	1/4	0.260	0.240	0.018	0.013	0.018
1/2	0.5000	3/4	0.750	0.736	0.866	0.840	7/16	0.448	0.427	5/16	0.323	0.302	0.019	0.014	0.019
9/16	0.5625	7/8	0.875	0.861	1.010	0.982	31/64	0.496	0.473	5/16	0.324	0.301	0.020	0.015	0.020
5/8	0.6250	15/16	0.938	0.922	1.083	1.051	35/64	0.559	0.535	3/8	0.387	0.363	0.021	0.016	0.021
3/4	0.7500	1 1/8	1.125	1.088	1.299	1.240	41/64	0.665	0.617	27/64	0.446	0.398	0.023	0.018	0.023
7/8	0.8750	1 5/16	1.312	1.269	1.516	1.447	3/4	0.776	0.724	31/64	0.510	0.458	0.025	0.020	0.025
1	1.0000	1 1/2	1.500	1.450	1.732	1.653	55/64	0.887	0.831	35/64	0.575	0.519	0.027	0.022	0.027
1 1/8	1.1250	1 11/16	1.688	1.631	1.949	1.859	31/32	0.999	0.939	39/64	0.639	0.579	0.030	0.025	0.030
1 1/4	1.2500	1 7/8	1.875	1.812	2.165	2.066	1 1/16	1.094	1.030	23/32	0.751	0.687	0.033	0.028	0.033
1 3/8	1.3750	2 1/16	2.062	1.994	2.382	2.273	1 11/64	1.206	1.138	25/32	0.815	0.747	0.036	0.031	0.036
1 1/2	1.5000	2 1/4	2.250	2.175	2.598	2.480	1 9/32	1.317	1.245	27/32	0.880	0.808	0.039	0.034	0.039
See Notes	9	3			4					2					

From The American Society of Mechanical Engineers—ANSI B18.2.2–1987

**AMERICAN NATIONAL STANDARD
SQUARE AND HEX NUTS**

ANSI B18.2.2—1987

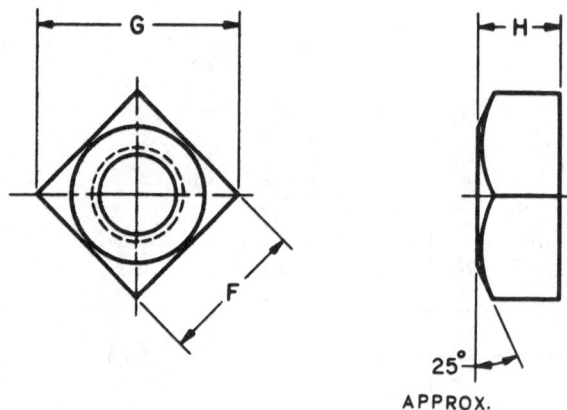

25°
APPROX.

TABLE 35 DIMENSIONS OF SQUARE NUTS

Nominal Size or Basic Major Dia of Thread		F Width Across Flats			G Width Across Corners		H Thickness		
		Basic	Max	Min	Max	Min	Basic	Max	Min
1/4	0.2500	7/16	0.438	0.425	0.619	0.584	7/32	0.235	0.203
5/16	0.3125	9/16	0.562	0.547	0.795	0.751	17/64	0.283	0.249
3/8	0.3750	5/8	0.625	0.606	0.884	0.832	21/64	0.346	0.310
7/16	0.4375	3/4	0.750	0.728	1.061	1.000	3/8	0.394	0.356
1/2	0.5000	13/16	0.812	0.788	1.149	1.082	7/16	0.458	0.418
5/8	0.6250	1	1.000	0.969	1.414	1.330	35/64	0.569	0.525
3/4	0.7500	1 1/8	1.125	1.088	1.591	1.494	21/32	0.680	0.632
7/8	0.8750	1 5/16	1.312	1.269	1.856	1.742	49/64	0.792	0.740
1	1.0000	1 1/2	1.500	1.450	2.121	1.991	7/8	0.903	0.847
1 1/8	1.1250	1 11/16	1.688	1.631	2.386	2.239	1	1.030	0.970
1 1/4	1.2500	1 7/8	1.875	1.812	2.652	2.489	1 3/32	1.126	1.062
1 3/8	1.3750	2 1/16	2.062	1.994	2.917	2.738	1 13/64	1.237	1.169
1 1/2	1.5000	2 1/4	2.250	2.175	3.182	2.986	1 5/16	1.348	1.276
See Notes	8	3							

From The American Society of Mechanical Engineers–ANSI B18.2.2–1987

TABLE 36 KEY & KEYWAY SIZES

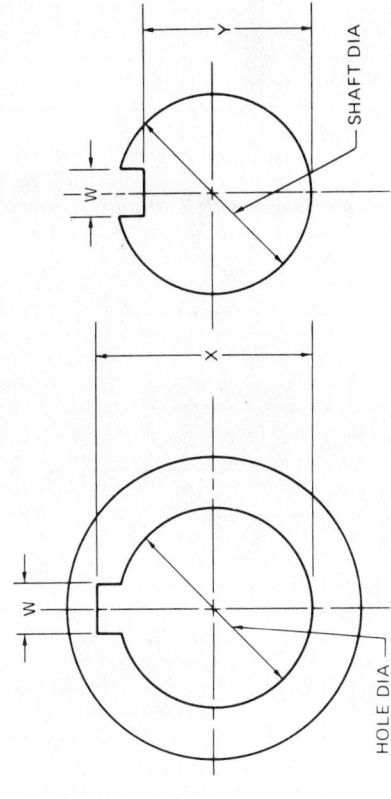

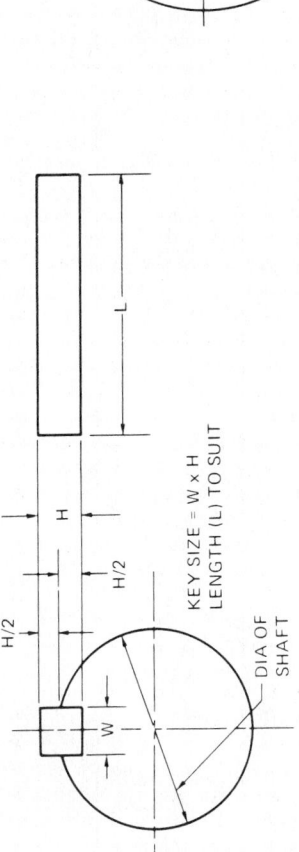

KEY SIZE = W x H
LENGTH (L) TO SUIT

Nom. Size (Inch)	– DIA. – (Shaft) Inch	mm	'X' (Collar) Inch	mm	'Y' (Shaft) Inch	mm
1/2	.500	12.700	.560	14.224	.430	10.922
9/16	.562	14.290	.623	15.824	.493	12.522
5/8	.625	15.875	.709	18.008	.517	13.132
11/16	.688	17.470	.773	18.618	.581	14.757
3/4	.750	19.050	.837	21.259	.644	16.357
13/16	.812	20.640	.900	22.860	.708	17.983
7/8	.875	22.225	.964	24.485	.771	19.583
15/16	.938	23.820	1.051	26.695	.791	20.091
1	1.000	25.400	1.114	28.295	.859	21.818
1 1/16	1.062	26.985	1.178	29.921	.923	23.444
1 1/8	1.125	28.575	1.241	31.521	.986	25.044
1 3/16	1.188	30.165	1.304	33.121	1.049	26.644
1 1/4	1.250	31.750	1.367	34.722	1.112	28.244
1 5/16	1.312	33.340	1.455	36.957	1.137	28.879
1 3/8	1.375	34.923	1.518	38.557	1.201	30.505

From Drafting for Trades and Industry—Mechanical and Electronic. Nelson. Delmar Publishers Inc.

Shaft Nom. Size – DIA. – From	To & Incl.	Square (W = H)	Type	Square Key From	To & Incl.	Tolerance
5/16 (8)	7/16 (11)	3/32 (2.38)	Bar Stock	–	3/4 (19.05)	+.000 – .002 (+.0000 – .0254)
7/16 (11)	9/16 (14)	1/8 (3.175)	Bar Stock	–	1 1/2 (38.1)	+.000 – .003 (+.0000 – .0762)
9/16 (14)	7/8 (22)	3/16 (4.76)	Bar Stock	3/4 (19.05)	2 1/2 (63.5)	+.000 – .004 (+.0000 – .1016)
7/8 (22)	1 1/4 (32)	1/4 (6.35)	Bar Stock	1 1/2 (38.1)	3 1/2 (88.9)	+.000 – .006 (+.0000 – .1524)
1 1/4 (32)	1 3/8 (35)	5/16 (7.94)	Bar Stock	2 1/2 (63.5)		
1 3/8 (35)	1 3/4 (44)	3/8 (9.53)	Keystock	–	1 1/4 (31.75)	+.001 – .000 (+.0254 – .0000)
1 3/4 (44)	2 1/4 (57)	1/2 (12.7)	Keystock	–		
2 1/4 (57)	2 3/4 (70)	5/8 (15.88)	Keystock	1 1/4 (31.75)	3 (76.2)	+.002 – .000 (+.0508 – .0000)
2 3/4 (70)	3 1/4 (82)	3/4 (19.05)	Keystock	3 (76.2)	3 1/2 (88.9)	+.003 – .000 (+.0762 – .0000)
3 1/4 (82)	3 3/4 (95)	7/8 (22.23)				

(Figures in parenthesis = mm)

USA STANDARD

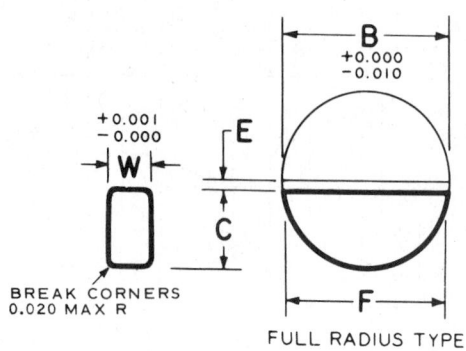

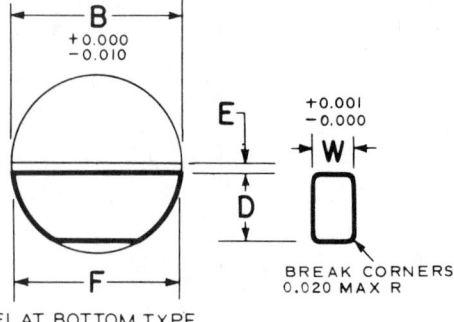

FULL RADIUS TYPE

FLAT BOTTOM TYPE

TABLE 37 WOODRUFF KEYS

Key No.	Nominal Key Size W × B	Actual Length F +0.000-0.010	Height of Key				Distance Below Center E
			C		D		
			Max	Min	Max	Min	
202	1/16 × 1/4	0.248	0.109	0.104	0.109	0.104	1/64
202.5	1/16 × 5/16	0.311	0.140	0.135	0.140	0.135	1/64
302.5	3/32 × 5/16	0.311	0.140	0.135	0.140	0.135	1/64
203	1/16 × 3/8	0.374	0.172	0.167	0.172	0.167	1/64
303	3/32 × 3/8	0.374	0.172	0.167	0.172	0.167	1/64
403	1/8 × 3/8	0.374	0.172	0.167	0.172	0.167	1/64
204	1/16 × 1/2	0.491	0.203	0.198	0.194	0.188	3/64
304	3/32 × 1/2	0.491	0.203	0.198	0.194	0.188	3/64
404	1/8 × 1/2	0.491	0.203	0.198	0.194	0.188	3/64
305	3/32 × 5/8	0.612	0.250	0.245	0.240	0.234	1/16
405	1/8 × 5/8	0.612	0.250	0.245	0.240	0.234	1/16
505	5/32 × 5/8	0.612	0.250	0.245	0.240	0.234	1/16
605	3/16 × 5/8	0.612	0.250	0.245	0.240	0.234	1/16
406	1/8 × 3/4	0.740	0.313	0.308	0.303	0.297	1/16
506	5/32 × 3/4	0.740	0.313	0.308	0.303	0.297	1/16
606	3/16 × 3/4	0.740	0.313	0.308	0.303	0.297	1/16
806	1/4 × 3/4	0.740	0.313	0.308	0.303	0.297	1/16
507	5/32 × 7/8	0.866	0.375	0.370	0.365	0.359	1/16
607	3/16 × 7/8	0.866	0.375	0.370	0.365	0.359	1/16
707	7/32 × 7/8	0.866	0.375	0.370	0.365	0.359	1/16
807	1/4 × 7/8	0.866	0.375	0.370	0.365	0.359	1/16
608	3/16 × 1	0.992	0.438	0.433	0.428	0.422	1/16
708	7/32 × 1	0.992	0.438	0.433	0.428	0.422	1/16
808	1/4 × 1	0.992	0.438	0.433	0.428	0.422	1/16
1008	5/16 × 1	0.992	0.438	0.433	0.428	0.422	1/16
1208	3/8 × 1	0.992	0.438	0.433	0.428	0.422	1/16
609	3/16 × 1 1/8	1.114	0.484	0.479	0.475	0.469	5/64
709	7/32 × 1 1/8	1.114	0.484	0.479	0.475	0.469	5/64
809	1/4 × 1 1/8	1.114	0.484	0.479	0.475	0.469	5/64
1009	5/16 × 1 1/8	1.114	0.484	0.479	0.475	0.469	5/64

WOODRUFF KEYS AND KEYSEATS

TABLE 37 WOODRUFF KEYS (Concluded)

Key No.	Nominal Key Size W × B	Actual Length F +0.000-0.010	Height of Key				Distance Below Center E
			C		D		
			Max	Min	Max	Min	
610	$\frac{3}{16} \times 1\frac{1}{4}$	1.240	0.547	0.542	0.537	0.531	$\frac{5}{64}$
710	$\frac{7}{32} \times 1\frac{1}{4}$	1.240	0.547	0.542	0.537	0.531	$\frac{5}{64}$
810	$\frac{1}{4} \times 1\frac{1}{4}$	1.240	0.547	0.542	0.537	0.531	$\frac{5}{64}$
1010	$\frac{5}{16} \times 1\frac{1}{4}$	1.240	0.547	0.542	0.537	0.531	$\frac{5}{64}$
1210	$\frac{3}{8} \times 1\frac{1}{4}$	1.240	0.547	0.542	0.537	0.531	$\frac{5}{64}$
811	$\frac{1}{4} \times 1\frac{3}{8}$	1.362	0.594	0.589	0.584	0.578	$\frac{3}{32}$
1011	$\frac{5}{16} \times 1\frac{3}{8}$	1.362	0.594	0.589	0.584	0.578	$\frac{3}{32}$
1211	$\frac{3}{8} \times 1\frac{3}{8}$	1.362	0.594	0.589	0.584	0.578	$\frac{3}{32}$
812	$\frac{1}{4} \times 1\frac{1}{2}$	1.484	0.641	0.636	0.631	0.625	$\frac{7}{64}$
1012	$\frac{5}{16} \times 1\frac{1}{2}$	1.484	0.641	0.636	0.631	0.625	$\frac{7}{64}$
1212	$\frac{3}{8} \times 1\frac{1}{2}$	1.484	0.641	0.636	0.631	0.625	$\frac{7}{64}$

All dimensions given are in inches.

The key numbers indicate nominal key dimensions. The last two digits give the nominal diameter B in eighths of an inch and the digits preceding the last two give the nominal width W in thirty-seconds of an inch.

Example:

No. 204 indicates a key $\frac{2}{32} \times \frac{4}{8}$ or $\frac{1}{16} \times \frac{1}{2}$.
No. 808 indicates a key $\frac{8}{32} \times \frac{8}{8}$ or $\frac{1}{4} \times 1$.
No. 1212 indicates a key $\frac{12}{32} \times \frac{12}{8}$ or $\frac{3}{8} \times 1\frac{1}{2}$.

From The American Society of Mechanical Engineers—ANSI B17.2—1967—R1990

WOODRUFF KEYS AND KEYSEATS

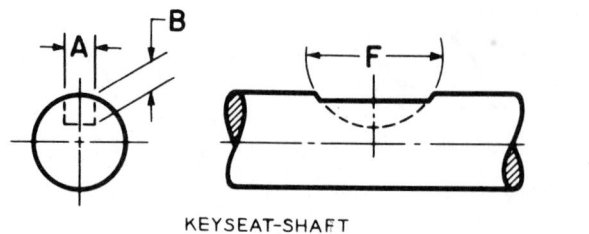

KEYSEAT-SHAFT

KEY ABOVE
SHAFT

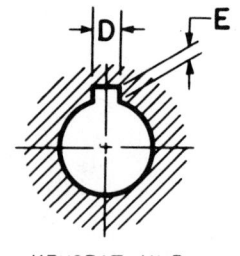

KEYSEAT-HUB

Key Number	Nominal Size Key	Keyseat – Shaft					Key Above Shaft	Keyseat – Hub	
		Width A*		Depth B	Diameter F		Height C	Width D	Depth E
		Min	Max	+0.005 -0.000	Min	Max	+0.005 -0.005	+0.002 -0.000	+0.005 -0.000
202	1/16 × 1/4	0.0615	0.0630	0.0728	0.250	0.268	0.0312	0.0635	0.0372
202.5	1/16 × 5/16	0.0615	0.0630	0.1038	0.312	0.330	0.0312	0.0635	0.0372
302.5	3/32 × 5/16	0.0928	0.0943	0.0882	0.312	0.330	0.0469	0.0948	0.0529
203	1/16 × 3/8	0.0615	0.0630	0.1358	0.375	0.393	0.0312	0.0635	0.0372
303	3/32 × 3/8	0.0928	0.0943	0.1202	0.375	0.393	0.0469	0.0948	0.0529
403	1/8 × 3/8	0.1240	0.1255	0.1045	0.375	0.393	0.0625	0.1260	0.0685
204	1/16 × 1/2	0.0615	0.0630	0.1668	0.500	0.518	0.0312	0.0635	0.0372
304	3/32 × 1/2	0.0928	0.0943	0.1511	0.500	0.518	0.0469	0.0948	0.0529
404	1/8 × 1/2	0.1240	0.1255	0.1355	0.500	0.518	0.0625	0.1260	0.0685
305	3/32 × 5/8	0.0928	0.0943	0.1981	0.625	0.643	0.0469	0.0948	0.0529
405	1/8 × 5/8	0.1240	0.1255	0.1825	0.625	0.643	0.0625	0.1260	0.0685
505	5/32 × 5/8	0.1553	0.1568	0.1669	0.625	0.643	0.0781	0.1573	0.0841
605	3/16 × 5/8	0.1863	0.1880	0.1513	0.625	0.643	0.0937	0.1885	0.0997
406	1/8 × 3/4	0.1240	0.1255	0.2455	0.750	0.768	0.0625	0.1260	0.0685
506	5/32 × 3/4	0.1553	0.1568	0.2299	0.750	0.768	0.0781	0.1573	0.0841
606	3/16 × 3/4	0.1863	0.1880	0.2143	0.750	0.768	0.0937	0.1885	0.0997
806	1/4 × 3/4	0.2487	0.2505	0.1830	0.750	0.768	0.1250	0.2510	0.1310
507	5/32 × 7/8	0.1553	0.1568	0.2919	0.875	0.895	0.0781	0.1573	0.0841
607	3/16 × 7/8	0.1863	0.1880	0.2763	0.875	0.895	0.0937	0.1885	0.0997
707	7/32 × 7/8	0.2175	0.2193	0.2607	0.875	0.895	0.1093	0.2198	0.1153
807	1/4 × 7/8	0.2487	0.2505	0.2450	0.875	0.895	0.1250	0.2510	0.1310
608	3/16 × 1	0.1863	0.1880	0.3393	1.000	1.020	0.0937	0.1885	0.0997
708	7/32 × 1	0.2175	0.2193	0.3237	1.000	1.020	0.1093	0.2198	0.1153
808	1/4 × 1	0.2487	0.2505	0.3080	1.000	1.020	0.1250	0.2510	0.1310
1008	5/16 × 1	0.3111	0.3130	0.2768	1.000	1.020	0.1562	0.3135	0.1622
1208	3/8 × 1	0.3735	0.3755	0.2455	1.000	1.020	0.1875	0.3760	0.1935
609	3/16 × 1 1/8	0.1863	0.1880	0.3853	1.125	1.145	0.0937	0.1885	0.0997
709	7/32 × 1 1/8	0.2175	0.2193	0.3697	1.125	1.145	0.1093	0.2198	0.1153
809	1/4 × 1 1/8	0.2487	0.2505	0.3540	1.125	1.145	0.1250	0.2510	0.1310
1009	5/16 × 1 1/8	0.3111	0.3130	0.3228	1.125	1.145	0.1562	0.3135	0.1622

From The American Society of Mechanical Engineers—ANSI B17.2–1967–R1990

Table 38

WOODRUFF KEY SIZES FOR DIFFERENT SHAFT DIAMETERS

Shaft Diameter	5/16 to 3/8	7/16 to 1/2	9/16 to 3/4	13/16 to 15/16	1 to 13/16	1 1/4 to 1 7/16	1 1/2 to 1 3/4	1 13/16 to 2 1/8	2 3/16 to 2 1/2
Key Numbers	204	304 305	404 405 406	505 506 507	606 607 608 609	807 808 809	810 811 812	1011 1012	1211 1212

Table 39

KEYS AND KEYSEATS

4. KEY DIMENSIONS AND TOLERANCES

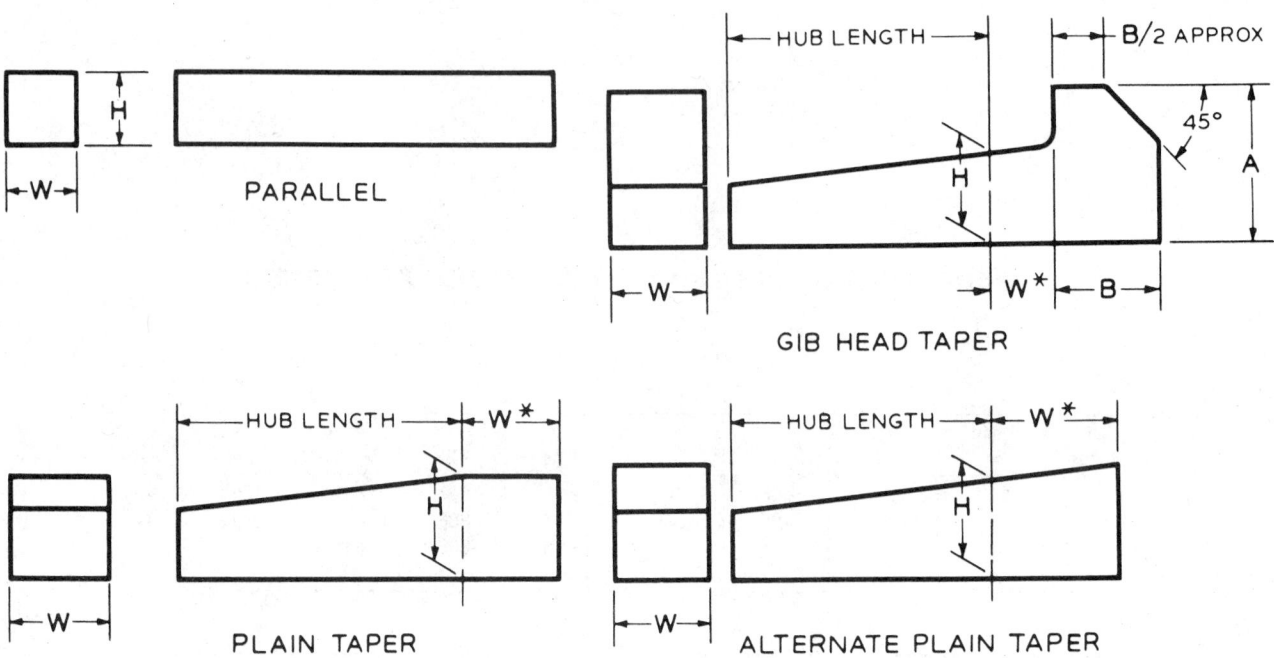

PARALLEL

GIB HEAD TAPER

PLAIN TAPER

ALTERNATE PLAIN TAPER

Plain and Gib Head Taper Keys Have a 1/8″ Taper in 12″

KEY			NOMINAL KEY SIZE		TOLERANCE			
			Width, W		Width, W		Height, H	
			Over	To (Incl)				
Parallel	Square	Bar Stock	— 3/4 1-1/2 2-1/2	3/4 1-1/2 2-1/2 3-1/2	+0.000 +0.000 +0.000 +0.000	−0.002 −0.003 −0.004 −0.006	+0.000 +0.000 +0.000 +0.000	−0.002 −0.003 −0.004 −0.006
		Keystock	— 1-1/4 3	1-1/4 3 3-1/2	+0.001 +0.002 +0.003	−0.000 −0.000 −0.000	+0.001 +0.002 +0.003	−0.000 −0.000 −0.000
	Rectangular	Bar Stock	— 3/4 1-1/2 3 4 6	3/4 1-1/2 3 4 6 7	+0.000 +0.000 +0.000 +0.000 +0.000 +0.000	−0.003 −0.004 −0.005 −0.006 −0.008 −0.013	+0.000 +0.000 +0.000 +0.000 +0.000 +0.000	−0.003 −0.004 −0.005 −0.006 −0.008 −0.013
		Keystock	— 1-1/4 3	1-1/4 3 7	+0.001 +0.002 +0.003	−0.000 −0.000 −0.000	+0.005 +0.005 +0.005	−0.005 −0.005 −0.005
Taper	Plain or Gib Head Square or Rectangular		— 1-1/4 3	1-1/4 3 7	+0.001 +0.002 +0.003	−0.000 −0.000 −0.000	+0.005 +0.005 +0.005	−0.000 −0.000 −0.000

*For locating position of dimension H. Tolerance does not apply.
See Table 41 for dimensions on gib heads.
All dimensions given in inches.

From The American Society of Mechanical Engineers—ANSI B17.1—1967—R1989

TABLE 40 KEY DIMENSIONS AND TOLERANCES

USA STANDARD

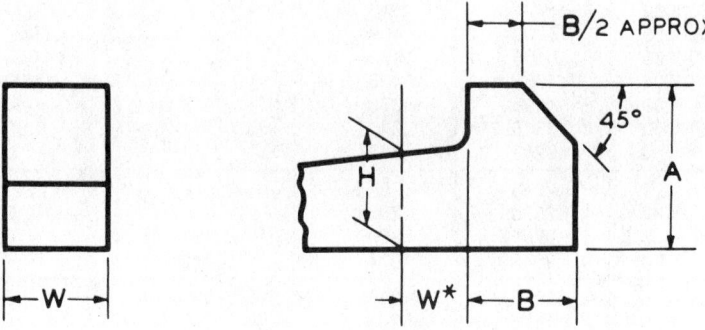

TABLE 41 GIB HEAD NOMINAL DIMENSIONS

Nominal Key Size Width, W	SQUARE			RECTANGULAR		
	H	A	B	H	A	B
1/8	1/8	1/4	1/4	3/32	3/16	1/8
3/16	3/16	5/16	5/16	1/8	1/4	1/4
1/4	1/4	7/16	3/8	3/16	5/16	5/16
5/16	5/16	1/2	7/16	1/4	7/16	3/8
3/8	3/8	5/8	1/2	1/4	7/16	3/8
1/2	1/2	7/8	5/8	3/8	5/8	1/2
5/8	5/8	1	3/4	7/16	3/4	9/16
3/4	3/4	1-1/4	7/8	1/2	7/8	5/8
7/8	7/8	1-3/8	1	5/8	1	3/4
1	1	1-5/8	1-1/8	3/4	1-1/4	7/8
1-1/4	1-1/4	2	1-7/16	7/8	1-3/8	1
1-1/2	1-1/2	2-3/8	1-3/4	1	1-5/8	1-1/8
1-3/4	1-3/4	2-3/4	2	1-1/2	2-3/8	1-3/4
2	2	3-1/2	2-1/4	1-1/2	2-3/8	1-3/4
2-1/2	2-1/2	4	3	1-3/4	2-3/4	2
3	3	5	3-1/2	2	3-1/2	2-1/4
3-1/2	3-1/2	6	4	2-1/2	4	3

*For locating position of dimension H.

For larger sizes the following relationships are suggested as guides for establishing A and B.

$$A = 1.8 H \qquad B = 1.2 H$$

All dimensions given in inches.

From The American Society of Mechanical Engineers—ANSI B17.1–1967–R1989

SHEET METAL AND WIRE GAGE DESIGNATION

GAGE NO.	AMERICAN OR BROWN & SHARPE'S A.W.G. OR B. & S.	UNITED STATES STANDARD	MANU-FACTURERS' STANDARD FOR SHEET STEEL	GAGE NO.
0000000	. . . .	.500	. . . .	0000000
000000	.5800	.469	. . . .	000000
00000	.5165	.438	. . . .	00000
0000	.4600	.406	. . . .	0000
000	.4096	.375	. . . .	000
00	.3648	.344	. . . .	00
0	.3249	.312	. . . .	0
1	.2893	.281	. . . .	1
2	.2576	.266	. . . .	2
3	.2294	.250	.2391	3
4	.2043	.234	.2242	4
5	.1819	.219	.2092	5
6	.1620	.203	.1943	6
7	.1443	.188	.1793	7
8	.1285	.172	.1644	8
9	.1144	.156	.1495	9
10	.1019	.141	.1345	10
11	.0907	.125	.1196	11
12	.0808	.109	.1046	12
13	.0720	.0938	.0897	13
14	.0642	.0781	.0747	14
15	.0571	.0703	.0673	15
16	.0508	.0625	.0598	16
17	.0453	.0562	.0538	17
18	.0403	.0500	.0478	18
19	.0359	.0438	.0418	19
20	.0320	.0375	.0359	20
21	.0285	.0344	.0329	21
22	.0253	.0312	.0299	22
23	.0226	.0281	.0269	23
24	.0201	.0250	.0239	24
25	.0179	.0219	.0209	25
26	.0159	.0188	.0179	26
27	.0142	.0172	.0164	27
28	.0126	.0156	.0149	28
29	.0113	.0141	.0135	29
30	.0100	.0125	.0120	30
31	.0089	.0109	.0105	31
32	.0080	.0102	.0097	32
33	.0071	.00938	.0090	33
34	.0063	.00859	.0082	34
35	.0056	.00781	.0075	35
36	.0050	.00703	.0067	36

Table 42

BEND ALLOWANCE FOR 90° BENDS (INCH)

Radii Thickness	.031	.063	.094	.125	.156	.188	.219	.250	.281	.313	.344	.375	.438	.500	.531	.625
.013	.058	.108	.157	.205	.254	.304	.353	.402	.450	.501	.549	.598	.697	.794	.843	.991
.016	.060	.110	.159	.208	.256	.307	.355	.404	.453	.503	.552	.600	.699	.796	.845	.993
.020	.062	.113	.161	.210	.259	.309	.358	.406	.455	.505	.554	.603	.702	.799	.848	.995
.022	.064	.114	.163	.212	.260	.311	.359	.408	.457	.507	.556	.604	.703	.801	.849	.997
.025	.066	.116	.165	.214	.263	.313	.362	.410	.459	.509	.558	.607	.705	.803	.851	.999
.028	.068	.119	.167	.216	.265	.315	.364	.412	.461	.511	.560	.609	.708	.805	.854	1.001
.032	.071	.121	.170	.218	.267	.317	.366	.415	.463	.514	.562	.611	.710	.807	.856	1.004
.038	.075	.126	.174	.223	.272	.322	.371	.419	.468	.518	.567	.616	.715	.812	.861	1.008
.040	.077	.127	.176	.224	.273	.323	.372	.421	.469	.520	.568	.617	.716	.813	.862	1.010
.050		.134	.183	.232	.280	.331	.379	.428	.477	.527	.576	.624	.723	.821	.869	1.017
.064		.144	.192	.241	.290	.340	.389	.437	.486	.536	.585	.634	.732	.830	.878	1.026
.072			.198	.247	.296	.346	.394	.443	.492	.542	.591	.639	.738	.836	.885	1.032
.078			.202	.251	.300	.350	.399	.447	.496	.546	.595	.644	.743	.840	.889	1.036
.081			.204	.253	.302	.352	.401	.449	.498	.548	.598	.646	.745	.842	.891	1.038
.091			.212	.260	.309	.359	.408	.456	.505	.555	.604	.653	.752	.849	.898	1.045
.094			.214	.262	.311	.361	.410	.459	.507	.558	.606	.655	.754	.851	.900	1.048
.102				.268	.317	.367	.416	.464	.513	.563	.612	.661	.760	857	.906	1.053
.109				.273	.321	.372	.420	.469	.518	.568	.617	.665	.764	.862	.910	1.058
.125				.284	.333	.383	.432	.480	.529	.579	.628	.677	.776	.873	.922	1.069
.156					.355	.405	.453	.502	.551	.601	.650	.698	.797	.895	.943	1.091
.188						.427	.476	.525	.573	.624	.672	.721	.820	.917	.966	1.114
.203								.535	.584	.634	.683	.731	.830	.928	.976	1.124
.218								.546	.594	.645	.693	.742	.841	.938	.987	1.135
.234								.557	.606	.656	.705	.753	.852	.950	.998	1.146
.250								.568	.617	.667	.716	.764	.863	.961	1.009	1.157

EXAMPLE: MATERIAL THICKNESS = 1/8", INSIDE RADII = 1/4" R. WHERE THESE CROSS = .284" BEND ALLOWANCE (B/A) PLUS TOTAL OF STRAIGHT LENGTHS = DEVELOPED LENGTH.

BEND ALLOWANCE FOR 90° BENDS (MILLIMETRE)

Radii Thickness	0.80	1.58	2.38	3.18	3.96	4.76	5.56	6.35	7.15	7.94	8.74	9.52	11.12	12.70	13.50	15.85
.330	1.46	2.74	3.98	5.22	6.45	7.73	8.96	10.20	11.44	12.71	13.95	15.19	17.70	20.18	21.42	25.16
.406	1.52	2.80	4.04	5.27	6.51	7.78	9.02	10.26	11.50	12.77	14.01	15.24	17.76	20.23	21.47	25.22
.508	1.59	2.87	4.11	5.34	6.58	7.86	9.09	10.33	11.57	12.84	14.08	15.32	17.83	20.30	21.54	25.29
.559	1.63	2.91	4.14	5.38	6.62	7.89	9.13	10.37	11.60	12.88	14.12	15.35	17.86	20.34	21.57	25.32
.635	1.68	2.96	4.20	5.43	6.67	7.95	9.18	10.42	11.66	12.93	14.17	15.40	17.92	20.39	21.63	25.38
.711	1.74	3.02	4.25	5.49	6.72	8.00	9.24	10.47	11.71	12.99	14.22	15.46	17.98	20.44	21.68	25.43
.813	1.81	3.08	4.32	5.56	6.79	8.07	9.30	10.54	11.78	13.05	14.29	15.53	18.04	20.52	21.75	25.50
.965	1.91	3.19	4.42	5.66	6.90	8.18	9.41	10.65	11.89	13.16	14.40	15.64	18.15	20.62	21.86	25.61
1.016	1.95	2.23	4.46	5.70	6.94	8.21	9.45	10.69	11.92	13.20	14.44	15.67	18.19	20.66	21.90	25.64
1.270		3.40	4.64	5.88	7.11	8.39	9.63	10.86	12.10	13.38	14.61	15.85	18.36	20.84	22.07	25.82
1.625		3.65	4.89	6.13	7.36	8.64	9.88	11.11	12.35	13.63	14.86	16.10	18.61	21.08	22.32	26.07
1.829			5.03	6.27	7.51	8.78	10.02	11.26	12.49	13.77	15.01	16.24	18.76	21.23	22.47	26.22
1.981			5.14	6.38	7.61	8.89	10.13	11.36	12.60	13.88	15.11	16.35	18.86	21.34	22.57	26.32
2.058			5.19	6.43	7.67	8.94	10.18	11.42	12.65	13.93	15.17	16.40	18.92	21.39	22.63	26.38
2.311			5.37	6.60	7.85	9.12	10.36	11.60	12.83	14.11	15.35	16.58	19.09	21.57	22.80	26.55
2.388			5.43	6.66	7.90	9.18	10.41	11.65	12.89	14.16	15.40	16.64	19.15	21.62	22.86	26.61
2.591				6.81	8.04	9.31	10.55	11.79	13.03	14.30	15.54	16.78	19.29	21.76	23.00	26.75
2.769				6.93	8.17	9.44	10.68	11.92	13.15	14.43	15.67	16.90	19.42	21.89	23.13	26.88
3.175				7.22	8.45	9.73	10.96	12.20	13.44	14.71	15.95	17.19	19.70	22.17	23.41	27.16
3.962					9.00	10.28	11.52	12.75	13.99	15.27	16.74	17.74	20.25	22.72	23.96	27.71
4.775					9.58	10.85	12.09	13.32	14.56	15.84	17.07	18.31	20.82	23.30	24.53	28.28
5.156						11.12	12.36	13.59	14.83	16.11	17.34	18.58	21.09	23.57	24.80	28.55
5.537								13.85	15.10	16.37	17.61	18.85	21.36	23.83	25.07	28.82
5.944								14.15	15.38	16.66	17.89	19.13	21.64	24.12	25.35	29.10
6.350								14.43	15.67	16.94	18.18	19.42	21.93	24.40	25.64	29.39

EXAMPLE: MATERIAL THICKNESS = 3.18 MM, INSIDE RADII = 6.35 MM R. WHERE THESE CROSS = 12.20 MM BEND ALLOWANCE (B/A) PLUS TOTAL OF STRAIGHT LENGTHS = DEVELOPED LENGTH.

From Drafting for Trades and Industry—Basic Skills. Nelson. Delmar Publishers Inc.

Table 43

BEND ALLOWANCE FOR EACH 1° OF BEND (INCH)

Radii / Thickness	.031	.063	.094	.125	.156	.188	.219	.250	.281	.313	.344	.375	.438	.500	.531	.625
.013	.00064	.00120	.00174	.00228	.00282	.00338	.00392	.00446	.00500	.00556	.00610	.00664	.00774	.00883	.00937	.01101
.016	.00067	.00122	.00176	.00231	.00285	.00342	.00395	.00449	.00503	.00559	.00613	.00667	.00777	.00885	.00939	.01103
.020	.00069	.00125	.00179	.00233	.00287	.00343	.00397	.00452	.00506	.00561	.00616	.00670	.00780	.00888	.00942	.01106
.022	.00071	.00127	.00181	.00235	.00289	.00345	.00399	.00453	.00508	.00563	.00617	.00672	.00782	.00890	.00944	.01108
.025	.00074	.00129	.00184	.00238	.00292	.00348	.00402	.00456	.00510	.00566	.00620	.00674	.00784	.00892	.00946	.01110
.028	.00076	.00132	.00186	.00240	.00294	.00350	.00404	.00458	.00512	.00568	.00622	.00676	.00786	.00894	.00948	.01112
.032	.00079	.00134	.00189	.00243	.00297	.00353	.00407	.00461	.00515	.00571	.00625	.00679	.00789	.00897	.00951	.01115
.038	.00084	.00140	.00194	.00248	.00302	.00358	.00412	.00466	.00520	.00576	.00630	.00684	.00794	.00902	.00956	.01120
.040	.00085	.00141	.00195	.00249	.00303	.00359	.00413	.00468	.00522	.00577	.00632	.00686	.00796	.00904	.00958	.01122
.050		.00149	.00203	.00258	.00312	.00368	.00422	.00476	.00530	.00586	.00640	.00694	.00804	.00912	.00966	.01130
.064		.00160	.00214	.00268	.00322	.00378	.00432	.00486	.00540	.00596	.00650	.00704	.00814	.00922	.00976	.01140
.072			.00220	.00274	.00328	.00384	.00438	.00492	.00546	.00602	.00656	.00710	.00820	.00929	.00983	.01147
.078			.00225	.00279	.00333	.00389	.00443	.00497	.00551	.00607	.00661	.00715	.00825	.00933	.00987	.01152
.081			.00227	.00281	.00335	.00391	.00445	.00499	.00554	.00609	.00664	.00718	.00828	.00936	.00990	.01154
.091			.00235	.00289	.00343	.00399	.00453	.00507	.00561	.00617	.00671	.00725	.00835	.00944	.00998	.01162
.094			.00237	.00291	.00346	.00401	.00456	.00510	.00564	.00620	.00674	.00728	.00838	.00946	.00999	.01164
.102				.00298	.00352	.00408	.00462	.00516	.00570	.00626	.00680	.00734	.00844	.00952	.01006	.01170
.109				.00303	.00357	.00413	.00467	.00521	.00575	.00631	.00685	.00739	.00849	.00958	.01012	.01176
.125				.00316	.00370	.00426	.00480	.00534	.00588	.00644	.00698	.00752	.00862	.00970	.01024	.01188
.156					.00394	.00450	.00504	.00558	.00612	.00668	.00722	.00776	.00886	.00994	.01048	.01212
.188						.00475	.00529	.00583	.00637	.00693	.00747	.00802	.00911	.01019	.01073	.01237
.203								.00595	.00649	.00704	.00759	.00813	.00923	.01031	.01085	.01249
.218								.00606	.00660	.00716	.00770	.00824	.00934	.01042	.01097	.01261
.234								.00619	.00673	.00729	.00783	.00837	.00947	.01055	.01109	.01273
.250								.00631	.00685	.00741	.00795	.00849	.00959	.01068	.01122	.01286

EXAMPLE: MATERIAL THICKNESS = 1/8", INSIDE RADII = 1/4" R. WHERE THESE CROSS = .00534" IF THE INSIDE OF YOUR BEND IS 20° THE BEND ALLOWANCE (B/A) = .1068 (.00534 X 20°) BEND ALLOWANCE (B/A) PLUS TOTAL OF STRAIGHT LENGTHS = DEVELOPED LENGTH.

BEND ALLOWANCE FOR EACH 1° OF BEND (MILLIMETRE)

Radii / Thickness	0.80	1.58	2.38	3.18	3.96	4.76	5.56	6.35	7.15	7.94	8.74	9.52	11.12	12.70	13.50	15.85
.330	.0161	.0305	.0442	.0580	.0717	.0859	.0996	.1134	.1271	.1413	.1550	.1668	.1967	.2242	.2379	.2796
.406	.0169	.0311	.0448	.0586	.0723	.0865	.1002	.1140	.1277	.1419	.1556	.1694	.1973	.2248	.2385	.2802
.508	.0177	.0319	.0456	.0594	.0731	.0873	.1010	.1148	.1285	.1427	.1564	.1702	.1981	.2256	.2393	.2810
.559	.0181	.0323	.0460	.0598	.0735	.0877	.1014	.1152	.1289	.1431	.1568	.1706	.1985	.2260	.2397	.2814
.635	.0187	.0329	.0466	.0604	.0741	.0883	.1020	.1158	.1295	.1437	.1574	.1712	.1991	.2266	.2403	.2820
.711	.0193	.0335	.0472	.0610	.0747	.0889	.1026	.1164	.1301	.1443	.1580	.1718	.1997	.2272	.2409	.2826
.813	.0201	.0343	.0480	.0617	.0755	.0897	.1034	.1174	.1309	.1451	.1588	.1726	.2005	.2280	.2417	.2834
.965	.0213	.0355	.0492	.0629	.0767	.0909	.1046	.1183	.1321	.1463	.1600	.1737	.2017	.2291	.2429	.2845
1.016	.0217	.0358	.0496	.0633	.0771	.0913	.1050	.1187	.1325	.1467	.1604	.1741	.2021	.2295	.2433	.2849
1.270		.0378	.0516	.0653	.0790	.0932	.1070	.1207	.1345	.1486	.1624	.1761	.2040	.2315	.2453	.2869
1.625		.0406	.0543	.0681	.0818	.0960	.1097	.1235	.1372	.1514	.1652	.1789	.2066	.2343	.2480	.2897
1.829			.0559	.0697	.0834	.0976	.1113	.1251	.1388	.1530	.1667	.1805	.2084	.2359	.2496	.2913
1.981			.0571	.0709	.0846	.0988	.1125	.1263	.1400	.1542	.1679	.1817	.2096	.2371	.2508	.2925
2.058			.0577	.0715	.0852	.0994	.1131	.1269	.1406	.1548	.1685	.1823	.2102	.2377	.2514	.2931
2.311			.0597	.0734	.0872	.1014	.1151	.1288	.1426	.1568	.1705	.1842	.2122	.2396	.2534	.2950
2.388			.0603	.0740	.0878	.1020	.1157	.1294	.1432	.1574	.1711	.1848	.2128	.2402	.2540	.2956
2.591				.0756	.0894	.1035	.1173	.1310	.1448	.1589	.1727	.1864	.2143	.2418	.2556	.2972
2.769				.0770	.0907	.1049	.1187	.1324	.1461	.1603	.1741	.1878	.2157	.2432	.2570	.2986
3.175				.0802	.0939	.1081	.1218	.1356	.1493	.1635	.1772	.1910	.2189	.2464	.2601	.3018
3.962					.1001	.1142	.1280	.1417	.1555	.1696	.1834	.1971	.2250	.2525	.2663	.3079
4.775					.1064	.1206	.1343	.1481	.1618	.1760	.1897	.2035	.2314	.2589	.2726	.3143
5.156						.1235	.1373	.1510	.1648	.1789	.1927	.2064	.2344	.2618	.2756	.3172
5.537								.1540	.1677	.1819	.1957	.2094	.2373	.2648	.2785	.3202
5.944								.1572	.1709	.1851	.1988	.2126	.2405	.2680	.2817	.3234
6.350								.1603	.1741	.1883	.2020	.2157	.2437	.2711	.2849	.3265

EXAMPLE: MATERIAL THICKNESS = 3.175 MM, INSIDE RADII = 6.35 MM R. WHERE THESE CROSS = .1356 MM IF THE INSIDE OF YOUR BEND IS 20° THE BEND ALLOWANCE (B/A) = 2.712 MM (.1356 X 20°) BEND ALLOWANCE (B/A) PLUS TOTAL OF STRAIGHT LENGTHS = DEVELOPED LENGTH

From Drafting for Trades and Industry—Basic Skills, Nelson. Delmar Publishers Inc.

Table 44

SPUR/PINION GEAR TOOTH PARTS

20 Degree Pressure Angles

Diametral Pitch	Circular Pitch	Circular Thickness	Addend.	Dedend.	Standard Fillet Radius
D.P.	C.P.	C.T.	A	D	
1	3.1416	1.5708	1.0000	1.2500	0.3000
2	1.5708	0.7854	0.5000	0.6250	0.1500
2.5	1.2566	0.6283	0.4000	0.5000	0.1200
3	1.0472	0.5236	0.3333	0.4167	0.1000
3.5	0.8976	0.4488	0.2857	0.3571	0.0857
4	0.7854	0.3927	0.2500	0.3125	0.0750
4.5	0.6981	0.3491	0.2222	0.2778	0.0667
5	0.6283	0.3142	0.2000	0.2500	0.0600
5.5	0.5712	0.2856	0.1818	0.2273	0.0545
6	0.5236	0.2618	0.1667	0.2083	0.0500
6.5	0.4833	0.2417	0.1538	0.1923	0.0462
7	0.4488	0.2244	0.1429	0.1786	0.0429
7.5	0.4189	0.2094	0.1333	0.1667	0.0400
8	0.3927	0.1963	0.1250	0.1563	0.0375
8.5	0.3696	0.1848	0.1176	0.1471	0.0353
9	0.3491	0.1745	0.1111	0.1389	0.0333
9.5	0.3307	0.1653	0.1053	0.1316	0.0316
10	0.3142	0.1571	0.1000	0.1250	0.0300
11	0.2856	0.1428	0.0909	0.1136	0.0273
12	0.2618	0.1309	0.0833	0.1042	0.0250
13	0.2417	0.1208	0.0769	0.0962	0.0231
14	0.2244	0.1122	0.0714	0.0893	0.0214
15	0.2094	0.1047	0.0667	0.0833	0.0200

Table 45

SURFACE TEXTURE ROUGHNESS

Surface Roughness Average Obtainable by Common Production Methods
Roughness Height Rating in N Series of Roughness Grade, Microinches, μ in. AA

Process	N12 2000	N11 1000	N10 500	N9 250	N8 125	N7 63	N6 32	N5 16	N4 8	N3 4	N2 2	N1 1	.5
Flame Cutting	▨	█	▨										
Snagging	▨	█	█	▨									
Sawing	▨	█	█	█	█	▨							
Planing, Shaping			▨	█	█	▨							
Drilling				▨	█	▨							
Chemical Milling				▨	█	▨							
Elect. Discharge Machining				▨	█	▨							
Milling			▨	█	█	█	▨	▨					
Broaching					▨	█	▨						
Reaming					▨	█	▨						
Electron Beam					█	█	▨	▨					
Laser					█	█	▨	▨					
Electro-Chemical				▨	█	█	█	█		▨	▨		
Boring, Turning			▨	█	█	█	█	█	▨	▨	▨		
Barrel Finishing					▨	█	█	█	▨				
Electrolytic Grinding								█	▨				
Roller Burnishing								█	▨				
Grinding					▨	█	█	█	█		▨	▨	
Honing							▨	█	█		▨		
Electro-Polishing						▨	█	█	█	▨	▨	▨	▨
Polishing								█	█	▨	▨	▨	
Lapping								▨	█	█	▨	▨	
Superfinishing								▨	█	█	▨	▨	
Sand Casting	▨	█	▨										
Hot Rolling	▨	█	▨										
Forging			▨	█	▨								
Perm. Mold Casting				▨	█	▨							
Investment Casting				▨	█	▨							
Extruding			▨	█	█	▨							
Cold Rolling, Drawing				▨	█	▨							
Die Casting					▨	█	▨						

TYPICAL APPLICATION

- (1000) Very rough surface. Equiv. to sand casting.
- (500) Rough surface. Rarely used.
- (250) Coarse finish. Equiv. to rolled surfaces & forgings.
- (125) Medium finish. Commonly used. Reasonable appear.
- (63) Good for close fits. Unsuitable for fast rotating members.
- (32) Used on shafts & bearings with light loads & moderate speeds.
- (16) Used on high speed shafts & bearings.
- (8) Used on precision gauge & instrument work. Costly.
- (4) Refined finish. Costly to produce.
- (2) Super-finish. Costly. Seldom used.

The ranges shown above are typical of the processes listed.

Higher or lower values may be obtained under special conditions.

KEY
- █ Average Application
- ▨ Less Frequent Application

From Interpreting Engineering Drawings. Jensen. Delmar Publishers Inc.

Table 46

PROPERTIES, GRADE NUMBERS, AND USAGE OF STEEL ALLOYS

Class of Steel	*Grade Number	Properties	Uses
Carbon - Low 0.3% carbon	10xx	Tough - Less Strength	Rivets - Hooks - Chains - Shafts - Pressed Steel Products
Carbon - Medium 0.3% to 0.6% carbon	10xx	Tough & Strong	Gears - Shafts - Studs - Various Machine Parts
Carbon - High 1.6% to 1.7%	10xx	Less Tough - Much Harder	Drills - Knives - Saws
Nickel	20xx	Tough & Strong	Axles - Connecting Rods - Crank Shafts
Nickel Chromium	30xx	Tough & Strong	Rings Gears - Shafts - Piston Pins - Bolts - Studs - Screws
Molybdenum	40xx	Very Strong	Forgings - Shafts - Gears - Cams
Chromium	50xx	Hard W/Strength & Toughness	Ball Bearings - Roller Bearing - Springs - Gears - Shafts
Chromium Vanadium	60xx	Hard & Strong	Shafts - Axles - Gears - Dies - Punches - Drills
Chromium Nickel Stainless	60xx	Rust Resistance	Food Containers - Medical/Dental Surgical Instruments
Silicon - Manganese	90xx	Springiness	Large Springs

*The first two numbers indicate type of steel, the last two numbers indicate the approx. average carbon content — 1010 steel indicates, carbon steel w/approx. 0.10% carbon.

From Drafting for Trades and Industry—Mechanical and Electronic. Nelson. Delmar Publishers, Inc.

Table 47

AMERICAN NATIONAL STANDARDS OF INTEREST TO DESIGNERS, ARCHITECTS, AND DRAFTERS

TITLE OF STANDARD

Abbreviations Y1.1-1989
American National Standard Drafting Practices
   Size and Format...Y14.1-1980 R1987
   Line Conventions and Lettering...Y14.2M-1979 R1987
   Multi and Sectional View Drawings...Y14.3-1975 R1987
   Pictorial Drawing Y14.4-1957 Y14.4M 1989
   Dimensioning and Tolerancing Y14.5M-1982 R1988
   Screw Threads...Y14.6-1978 R1987
   Screw Threads (Metric Supplement)...Y14.6aM-1981 R1987
   Gears and Splines
      Spur, Helical, and Racks...Y14.7.1-1971 R1988
      Bevel and Hyphoid Y14.7.2-1978 R1984
   Castings and Forgings..Y14.8M-1989
   Springs..Y14.13M-1981 R1987
   Electrical and Electronic Diagram Y14.15-1966 R1988
   Interconnection Diagrams Y14.15a-1971 R1988
   Information Sheet Y14.15b-1973 R1988
   Fluid Power Diagrams Y14.17-1966 R1980
   Digital Representation for Communication of Product Definition Data Y14.26M-1989
   Computer-Aided Preparation of Product Definition Data Dictionary of Terms..........Y14.26.3-1975
   Digital Representation of Physical Object Shapes ....Y14 Report
   Guideline–User Instructions ....Y14 Report No. 2
   Guideline–Design Requirements...Y14 Report No. 3
   Ground Vehicle Drawing Practices..In Preparation
   Chasis Frames...Y14.32.1-1974
   Parts Lists, Data Lists, and Index Lists...Y14.34M-1989
   Surface Texture Symbols..Y14.36-1978 R1987
Illustrations for Publication and Projection...Y15.1M-1979 R1986
Time Series Charts Y15.2M-1979 R1986
Process Charts Y15.3M-1979 R1986
Graphic Symbols for:
   Electrical and Electronics Diagrams ....with 315A-1986 (a supplement to Y32.2-1975)
   Plumbing...Y32.4-1977 R1987
   Use on Railroad Maps and Profiles Y32.7-1972 R1987
   Fluid Power Diagrams Y32.10-1967 R1987
   Process Flow Diagrams in Petroleum and Chemical Industries.............Y32.11-1961 R1985
   Mechanical and Acoustical Element as Used in Schematic Diagrams Y32.18-1972 R1985
   Pipe Fittings, Valves, and Piping Y32.2.3-1949 R1988
   Heating, Ventilating, and Air Conditioning Y32.2.4-1949 R1984
   Heat Power Apparatus..Y32.2.6-1950 R1984
Letter Symbols for:
   Glossary of Terms Concerning Letter Symbols Y10.1-1972 R1988
   Hydraulics...Y10.2-1958
   Quantities Used in Mechanics for Solid Bodies Y10.3M-1984
   Heat and Thermodynamics Y10.4 1988
   Quantities Used in Electrical Science and Electrical Engineering Y10.5-1968
   Aeronautical Sciences..Y10.7-1954
   Structural Analysis..Y10.8-1962
   Meteorology Y10.10-1953 R1973
   Acoustics Y10.11-1984
   Chemical Engineering ....Y10.12 -1955 R1988
   Rocket Propulsion Y10.14-1959
   Petroleum Reservoir Engineering and Electric Logging Y10.15-1958 R1973
   Shell Theory Y10.16-1964 R1973
   Guide for Selecting Greek Letters Used as Symbols for Engineering Mathematics.....Y10.17-1961 R1988
   Illuminating Engineering Y10.18-1967 R1987
   Mathematical Signs and Symbols for Use in Physical Sciences and Technology..........Y10.20-1990

Table 48

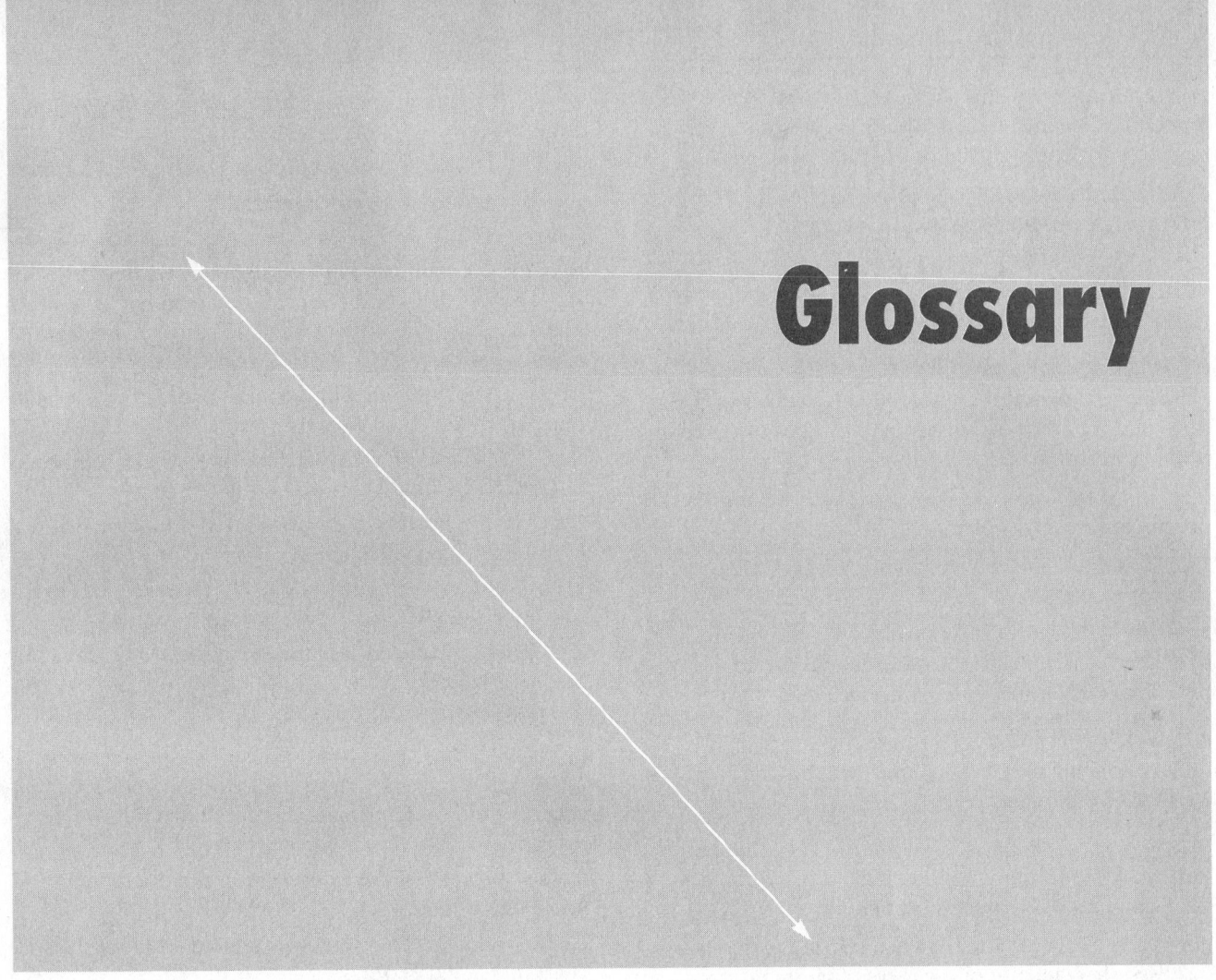

Glossary

Absolute System A type of numerical control system in which all coordinates are located from a fixed or absolute zero point

Acme Screw thread form

Actual Size Actual measured size of a feature

Acute Angle An angle less than 90°

Addendum Radial distance from pitch circle to top of gear tooth

Allen Screw Special set screw or cap screw with hexagon socket in head

Allowance Minimum clearance between mating parts

Alloy Two or more metals in combination, usually a fine metal with a baser metal

Alloy Steel Any steel that contains one or more alloying elements added to produce characteristics not available in unalloyed carbon steels

Alphanumeric Refers to the totality of characters that are either alphabetic or numeric

Aluminum A lightweight but relatively strong metal; often alloyed with copper to increase hardness and strength

Ammonia A colorless gas used in the development process of diazo and sepia prints

Analysis The study and evaluation of an object's characteristics and actions

Angle Iron A structural shape whose section is a right angle

Angularity Tolerancing of a feature at a specified angle from another feature

Anneal To soften metals by heating to remove internal stresses caused by rolling and forging

Anodizing The process of protecting aluminum by oxidizing in an acid bath using a direct current

Application A definable set of drafting tasks to be accomplished in a given drafting area; may be accomplished partly through manual procedures and partly through computerized procedures

Arc A part of a circle

Arc-weld To weld by electric arc; the work is usually the positive terminal

Artificial Intelligence The ability of a machine to improve its own operation or the ability of a machine to perform functions that are normally associated with human intelligence such as learning, adapting, reasoning, self-correction, and automatic improvement

Assembled Isometric Assembly Drawings Show how a single product looks when assembled

Assembly Drawing A drawing showing the working relationship of the various parts of a machine or structure as they fit together

Attributes The characteristics of a part

Automated Assembly Assembly by means of operations performed automatically by machines that are controlled and monitored by computers

Automated Process Planning Using a computer in developing a process plan

Automation The conversion of a process or system to automatic operation

Auxiliary View An additional view of an object, usually of a surface inclined to the principal surfaces of the object to provide a true size and shape view

Axes Plural of axis

Axis An imaginary line around which parts rotate or are regularly arranged

Babbitt A soft alloy for bearings, mostly of tin with small amounts of copper and antimony

Backlash Lost motion between moving parts, such as threaded shaft and nut or the teeth of meshing gears

Baseline Dimensioning A system of dimensioning where as many features of a part as are functionally practical are located from a common set of datums

Basic Dimension A theoretically "perfect" dimension similar to a reference or nominal dimension. It is used to identify the exact location, size, or shape of a feature.

Basic Size The size from which the limits of size are derived by the application of allowances and tolerances

Bearing A supporting member for a rotating shaft

Bend Allowance The amount of sheet metal required to make a bend over a specific radius

Bevel An inclined edge; not a right angle to joining surface

Bilateral Tolerances Tolerances that are applied to a nominal dimension in the positive and negative directions

Bisect To divide into two equal parts

Blanking A stamping operation in which a press uses a die to cut blanks from flat sheets or strips of metal

Blind Hole A plain hole that is machined to a specific depth

Blueprint A copy of a drawing

Bolt Circle A circular center line on a drawing, containing the centers of holes about a common center

Bore To enlarge a hole with a boring bar or tool in a lathe, drill press or boring mill

Boss A cylindrical projection on a casting or a forging

Brass An alloy of copper and zinc

Braze To join with hard solder of brass or zinc

Broach A tool for removing metal by pulling or pushing it across the work; the most common use is producing irregular hole shapes such as squares, hexagons, ovals or splines

Bronze An alloy of eight or nine parts of copper and one part of tin

Buff To finish or polish on a buffing wheel composed of fabric with abrasive powders

Burnish To finish or polish by pressure upon a smooth rolling or sliding tool

Burr The ragged edge or ridge left on metal after a cutting operation

Bushing A metal lining which acts as a bearing between rotating parts such as a shaft and pulley; also used on jigs to guide cutting tool

Cabinet Drawings Have the same range of receding axes, but the receding distances are drawn to half size

CAD (Computer-aided drafting) The use of computers and peripheral devices to aid in the documentation for design projects

CAD System The hardware, software, and user in computer-aided drafting

CAE (Computer-aided engineering) Any use of a computer to assist in engineering tasks, analysis testing, calculating, design review, and experimentation

Callout A note on the drawing giving a dimension, specification or machine process

Calipers Instrument for measuring diameters

Cam A rotating shape for changing circular motion to reciprocating motion

Carburize To heat a low-carbon steel to approximately 2000°F in contact with material which adds carbon to the surface of the steel

Case Harden To harden the outer surface of a carburized steel by heating and then quenching

Cast Iron Metals in a family of cast ferrous metals

Cast Steel Steel produced in ingot form by melting all component elements together and pouring the molten metal into a mold

Cavalier Drawings Have a receding axis drawn between 30° and 60° to the horizon, and the receding distances are drawn full size

Center Drill A special drill to produce bearing holes in the ends of a workpiece to be mounted between centers

Central Processing Unit (CPU) A unit of a computer that includes circuits controlling the interpretation and execution of instructions

Ceramics Materials made by combining metallic and non-metallic elements such as oxides, carbides, and nitrates

Chain Dimensioning Successive dimensions that extend from one feature to another, rather than each originating at a datum

Chamfer A bevel on an external edge or corner, usually at 45°

Character A letter, digit or other symbol that is used as part of the organization, control, or representation of data

Chase To cut threads with an external cutting tool

Chemical Milling A nontraditional machining process that etches or shapes workpieces to close tolerances by chemically induced material removal

Chill To harden the outer surface of cast iron by quick cooling, as in a metal mold

Chuck A mechanism for holding a rotating tool or workpiece

Circularity A tolerance that controls elemental tolerance zones around a circular feature that is independent of other features

Circular Pitch The length of the arc along the pitch circle between the center of one gear tooth to the center of the next

Circular Runout A tolerance that identifies an infinite number of elemental tolerance zones on a feature when the feature is rotated 360° for each element

Clearance Fit A condition between mating parts in which there is always a clearance assembly

Clockwise Rotation in the same direction as hands of a clock

Coin To form a part in one stamping operation

Cold Rolled Steel Bessemer steel containing .12% to .20% carbon that has been rolled while cold to produce a smooth, quite accurate stock

Collar A round flange or ring fitted on a shaft to prevent sliding

Composites Materials created by combining two or more materials

Compressive Force A force that "goes into" the joint

Computational Fluid Dynamics (CFD) The study and analysis of air, liquid, and gas using computer technology

Computational Prototype A three-dimensional computer-generated model that possesses the characteristics of a comparable physical object

Concentric Having a common center as circles or diameters

Concentricity A tolerance in which the axis of a feature must be coaxial to a specified datum regardless of the datum's and the feature's size

Configuration A group of machines, devices, parts, etc. that makes up a system

Contour Interval The difference in elevation between neighboring contour lines

Contour Line A line on a plot or map that is at a constant elevation on the surface of the earth, above sea level

Conventional Break An effective way of drawing long objects of constant cross section on a relatively small sheet of drawing paper

Coordinate An ordered set of data values, either absolute or relative, that specifies a location

Coordinate Dimensioning A type of rectangular datum dimensioning in which all dimensions are measured from two or three mutually perpendicular datum planes; all dimensions originate at a datum and include regular extension and dimension lines and arrowheads

Cope The upper portion of a flask used in molding

Copper A heavy metal that is salmon pink in its pure form and abundant

Core To form a hollow portion in a casting by using a dry-sand core or a green-sand core in a mold

Counterbore The enlargement of the end of a hole to a specified diameter and depth

Counterbored Hole Usually a plain through hole with an upper portion enlarged cylindrically to a specified diameter and depth

Countersink The chamfered end of a hole to receive a flat head screw

Countersunk Hole Usually a plain through hole with an upper portion enlarged conically, most often to 82° to accommodate standard, flat head machine screws

CPU Central processing unit

Cursor (1) On CRT, a movable marker that is visible on the viewing surface and is used to indicate a position at which an action is to take place or the position on which the next device operation would normally be directed. (2) On digitizers, a movable reference, usually optical cross hairs used by an operator to indicate manually the position of a reference point or line where an action is to take place

Cuts and Fills Excavating the earth of some hills and using it to fill in the valleys

Cutting Plane Line Indicates that path an imaginary cutting plane follows to slice through an object

Cylindricity A tolerance that simultaneously controls the surface of revolution for straightness, parallelism, and circularity of a feature, and is independent of any other features on a part

DFA Design for Assembly

DFD Design for Disassembly

DFM Design for Manufacturability

DFR Design for Recyclability

Data A representation of facts, concepts or instructions in formalized manner suitable for communication, interpretation or processing by human or automatic means

Datum   Points, lines, planes, cylinders and the like, assumed to be exact for purposes of computation from which the location or geometric relationship (form) of features of a part may be established

Datum Feature   A feature that is used to establish a datum

Datum Identification Symbol   A special rectangular box that contains the datum reference letter

Dedendum   The radial distance between the pitch circle and the bottom of the tooth

Delineation   Pictorial representations; a chart, a diagram or a sketch

Design Size   The size of a feature after an allowance for clearance has been applied and tolerances have been assigned

Detail Drawing   A drawing of a single part that provides all the information necessary in the production of that part

Development   Drawing of the surface of an object unfolded on a plane

Deviation   The variance from a specified dimension or design requirement

Diagram   A figure or drawing which is marked out by lines; a chart or outline

Diameter   The length of a straight line passing through the center of a circle and terminating at the circumference on each end

Diazo   Material that is either a film or paper sensitized by means of azo dyes used for photocopying

Die   Hardened metal piece shaped to cut or form a required shape in a sheet or metal by pressing it against a mating die

Die Casting   Process of forcing molten metal under pressure into metal dies or molds, producing a very accurate and smooth casting

Die Stamping   Process of cutting or forming a piece of sheet metal with a die

Digitize   To use numeric characters to express or represent data

Digitizer   A device for converting positional information into digital signals; typically, a drawing or other graphic is placed on the measuring surface of the digitizer and traced by the operator using a cursor

Dimension   Measurements given on a drawing such as size and location

Dimetric Drawing   Only two angles are equal

Dip   The downward angle of the vein with respect to a horizontal plane at the site of the ore vein

Direct Numerical Control (DNC)   The use of a common host computer for distribution of part programs to remote machine tools

Display   A visual presentation of data on an output device

Dowel   A cylindrical pin, commonly used to prevent sliding between two contacting flat surfaces

Draft   The tapered shape of the parts of a pattern to permit it to be easily withdrawn from the sand or withdrawn from the dies

Drag   Lower portion of a flask used in molding

Draw   To temper steel

Drill   To cut a cylindrical hole with a drill; a blind hole does not go through the piece

Drill Press   A machine for drilling and other hole-forming operations

Drop Forge   To form a piece while hot between dies in a drop hammer or with great pressure

Eccentric   Not having the same center; off center

Elastomers   Linear polymers that are able to stretch and return to their original shape and size (i.e., rubber bands)

Electrical Discharge Grinding (EDG)   A nontraditional machining process that employs rapidly recurring electrical discharges between an electrode (a rotating wheel made of graphite or brass) and the workpiece in a dielectric fluid

Electrical Discharge Machining (EDM)   A nontraditional machining process that removes metal using a series of rapidly recurring electrical discharges between an electrically charged cutting tool and the workpiece submerged in a dielectric fluid

Electrical Discharge Wire Cutting (EDCW)   A nontraditional machining process that is similar to the bandsawing process

Electrical Plan   Shows the house's electrical system superimposed on a floor plan without dimensions

Electrochemical Discharge Grinding (ECDG)   A combination of two nontraditional machining processes: electrochemical grinding and electrical discharge grinding; in ECDG alternating current or pulsating direct current moves from a bonded graphite wheel to a positively charged workpiece

Electrochemical Grinding (ECG)   A nontraditional machining process that combines an electrochemical reaction that occurs on the surface of the metal with abrasion

Electrochemical Honing (ECH)   A nontraditional machining process that combines electrochemical oxidation of surface metal and abrasion

Electrochemical Machining (ECM)   A broad concept that applies to a number of process variations that all use an electrolytic action in removing metal

Electrochemical Turning (ECT)   A nontraditional machining process that is a variation of electrochemical machining

Electromechanical Machining (EMM)   A nontraditional machining process that improves the performance of traditional machining processes

Electron Beam Machining (EBM)   A thermal energy process that uses thermal energy to remove metal

Electronic Method   A method of approximating areas by employing a computer and a digitizer

Elevations   Orthographic views of the front, back, and sides of the structure

Engineering Materials   Metals and non-metals used in manufacturing

Equilibrant   The equal and opposite reaction to the resultant if the vectors being added or subtracted are force vectors

Exploded Isometric Drawings   Show all the components of an assembly as individual isometric drawings

Extrusion   Metal which has been shaped by forcing it in its hot or cold state through dies of the desired shape

Fabrication   A term used to distinguish production operations from assembly operations

Face   To finish a surface at right angles, or nearly so, to the center line of rotation on a lathe

FAO   Finish all over

Fastener   A mechanical device for holding two or more bodies in definite positions with respect to each other

Feature   A portion of a part, such as a diameter, hole, keyway or flat surface

Ferrous   Having iron as a metal's base material

File   To finish or smooth with a file

Fillet   An interior rounded intersection between two surfaces

Fin   A thin extrusion of metal at the intersection of dies or sand molds

Finish   General finish requirements such as paint, chemical or electroplating, rather than surface texture or roughness (see surface texture)

Finite Element Analysis (FEA)   A means of designing, modifying, and analyzing objects and products

Finite Element Model (FEM)   A 3-D model created with nodes throughout the object

Fit   The clearance or interference between two mating parts

Fixture   A special device for holding the work in a machine tool

Flange   A relatively thin rim around a piece

Flask   A box made of two or more parts for holding the sand in sand molding

Flatbed Plotter   A plotter that draws an image on a data medium such as paper or film mounted on a table

Flat Pattern   A layout showing true dimensions of a part before bending; may be actual size pattern on polyester film for shop use

Flatness   A tolerance that controls the amount of variation from the perfect plane on a feature independent of any other features on the part

Floor Plan   A plan view of the structure in section with the ceiling and roof removed

Flute   Groove, as on twist drills, reamers, and taps

Fold Line   The place where the imaginary perpendicular-orthographic viewing planes meet each other along the straight edge of intersection

Foreshortening   A phenomenon wherein the lengths of edges are shorter in the image than their actual lengths

Forge   To force metal while it is hot to take on a desired shape by hammering or pressing

Form Tolerancing   The permitted variation of a feature from the perfect form indicated on the drawing

Foundation Plan   A plan view in section that shows the understructure of a building

Front View   The orthographic view showing the front elevation

Fusion Weld   The intimate mixing of molten metals

Gage   The thickness of sheet metal by number

Galvanize   To cover a surface with a thin layer of molten alloy, composed mainly of zinc

Gasket   A thin piece of rubber, metal or some other material, placed between surfaces to make a tight joint

Gate   The opening in a sand mold at the bottom of the sprue through which the molten metal passes to enter the cavity or mold

General Oblique Drawing   Similar to cabinet drawing except that the lines drawn along the direction of the receding may be at any scale the drafter chooses

Geometric Dimensioning and Tolerancing   A means of dimensioning and tolerancing a drawing with respect to the actual function or relationship or part features which can be most economically produced; it includes positional and form dimensioning and tolerancing

Graphic Communication   Using visual material to relate ideas

Graphical Method   A method of approximating areas by dividing them into convenient shapes, e.g., trapezoids and triangles, scaling appropriate lengths, and calculating the areas

Grind   To remove metal by means of an abrasive wheel, often made of carborundum, used where accuracy is required

Group Technology Grouping of parts into families based on similarities in design or production processes

Gusset A small plate used in reinforcing assemblies

Hard Copy A copy of output in a visually readable form (e.g., printed reports, listings, documents, summaries, and drawings)

Harden To heat steel above a critical temperature and then quench in water, oil or air

Hardness Test Techniques used to measure the degree of hardness of heat-treated materials

Hardware The tools available to users of CAD for input, processing, and output

Heat Treat To change the properties of metals by heating and then cooling

Heating-Ventilation-Air Conditioning Plan (HVAC) Shows the duct work for the house's heating and cooling system superimposed on the floor plan without dimensions

Hexagon A polygon having six angles and six sides

Hidden Lines Feature outlines, for example, intersections of plane surfaces, real but invisible; used in each viewing plane where appropriate

High-performance Alloys A group of metals that have high strength and/or high corrosion resistance over a broad range of temperatures

Holography Produces images that simulate the effect of seeing the three dimensionality of an object from multiple viewpoints

Hone A method of finishing a hole or other surface to a precise tolerance by using a spring-loaded abrasive block and rotary motion

Horizontal Parallel to the horizon

Hydrodynamic Machining (HDM) A nontraditional machining process that removes material using a high-velocity, high-pressure stream of liquid solution

Icon A standardized graphical image designed to represent and communicate a computer application operation or function

Inclined A line or plane at an angle to a horizontal line or plane

Incremental System A system of numerically controlled machining that always refers to the preceding point when making the next movement; also known as continuous path or contouring method of NC machining

Input The transfer of information into a computer or machine control unit (see "output")

Intelligent Robot Robot that can make decisions based on data received from sensing devices

Interactive The capability of a computer application to permit an operator to change variables or test a design

while the application is running. Virtual reality is an extreme example of an operator being able to interact with a computer application.

Interchangeable Refers to a part made to limit dimensions so that is will fit any mating part similarly manufactured

Interface The means of interacting with a computer. Graphical user interfaces are becoming more prevalent and are designed to simplify applications by using graphical icons to guide the interaction with computer applications

Intermediate A translucent reproduction made on vellum, cloth or film from an original drawing to serve in place of the original for making other prints

Invention Agreement A form a worker might sign that gives the company the right to any new invention designed while working for the company

Involute A spiral curve generated by a point on a chord as it unwinds from a circle of a polygon

Isometric Assembly Drawings that appear in exploded or assembled formats

Isometric Drawing A pictorial drawing of an object so positioned that all three axes make equal angles with the picture plane, and measurements on all three axes are made to the same scale

Isometric Projection Causes dimensions along the three principal edges to be foreshortened to approximately 82 percent of their true length

Jig A device for guiding a tool in cutting a piece

Journal Portion of a rotating shaft supported by a bearing

Just-in-Time Manufacturing (JIT) An approach to manufacturing that seeks to completely eliminate waste

Kerf Groove or cut made by a saw

Key A small piece of metal sunk partly into both shaft and hub to prevent rotation of mating parts

Keyboard The part of the CAD computer used for entering commands, text, annotation, and dimensions

Keying The process used for entering text on drawings, certain system commands, dimensions not entered automatically, and for logging-on to the system

Keyseat A slot in a shaft to hold a key

Keyslot The slot machined in a shaft for Woodruff-type keys

Keyway A slot in a hub or portion surrounding a shaft to receive a key

Laser Beam Machining (LBM) A nontraditional machining process that uses lasers for performing operations that are traditionally performed with cutting tools

Lathe A machine used to shape materials by rotating against a tool

Layout Lines drawn on material as guides for subsequent operations

Lead A dark grey metal that has high density, low melting point, corrosion resistance, chemical stability, malleability, lubricity, and electrical properties

Least Material Condition (LMC) A condition of a feature in which it contains the least amount of material

Light Pen A hand-held data-entry device used only with refresh displays. It consists of an optical lens and photocell, with associative circuitry mounted in a wand. Most light pens have a switch allowing the pen to be sensitive to light from the screen.

Limit Dimensions A tolerancing method showing only the maximum and minimum dimensions that establish the limits

Limits The maximum and minimum allowable sizes of a feature

Location Tolerance A tolerance that specifies the allowable variation from the perfect location

Lug An irregular projection of metal, but not round as in the case of a boss; usually with a hole in it for a bolt or screw

Magnaflux A nondestructive inspection technique that makes use of a magnetic field and magnetic particles to locate internal flaws in ferrous metal parts

Magnesium A soft silvery white metal that lacks the strength needed for structural uses

Main Storage The general-purpose storage of a computer; usually, main storage can be accessed directly by the operating registers

Malleable Casting A casting that has been made less brittle and tougher by annealing

Manufacturing The enterprise through which raw materials are converted into finished products

Manufacturing Resource Planning (MRPII) A concept that allows companies to systematically plan for the resources needed in manufacturing (i.e., material, capacity of production systems, personnel, time, etc.)

Maraging Steel A special class of high-strength steels in which nickel is the primary alloy

Mass Production Company An organization that produces large quantities of a product

Mass Properties Analysis A process involving performing analysis computations concerning mass properties such as weight, volume, surface area, moment of inertia, and center of gravity

Material Handling The movement and handling of materials, either manually or through the use of powered equipment

Maximum Material Condition (MMC) When a feature contains the maximum amount of material; that is, minimum hole diameter and maximum shaft diameter

Mechanical Method A method of approximating areas by utilizing a planimeter

Menu A list of options which are displayed on the CRT of a plastic or paper overlay

Microcomputer A computer that is constructed using a microprocessor as a basic element of the CPU; all electronic components are arranged on one printed circuit board

Mill To remove material by means of a rotating cutter on a milling machine

Modifier The application of MMC or RFS to alter the normally implied interpretation of a tolerance specification

Mold The mass of sand or other material that forms the cavity into which molten metal is poured

Mouse The part of the CAD computer used to manipulate a cursor that appears on the graphics display

Neck To cut a groove called a neck around a cylindrical piece

Nodes Points of an object that contain specific information about the characteristics at that particular point

Nonferrous A description of metals not derived from an iron base or an iron alloy base; nonferrous metals include aluminum, magnesium, and copper, among others

Nonisometric Line A line is not parallel or perpendicular to any of the three principal edges in an isometric drawing

Nontraditional Machining Encompasses a number of advance machining processes developed in response to the demands of modern manufacturing

Normalize To heat steel above its critical temperature, and then cooling it in air

Normalizing A process in which ferrous alloys are heated and then cooled in still air to room temperatures to restore the uniform grain structure free of strains caused by cold working or welding

Numerical Control A system of controlling a machine or tool by means of numeric codes that direct commands to control devices attached or built into the machine or tool

Oblique Drawing A pictorial drawing of an object so drawn that one of its principal faces is parallel to the plane of projection, and is projected in its true size and shape; the third set of edges is oblique to the plane of projection at some convenient angle

Obtuse Angle An angle larger than 90°

Octagon A polygon having eight angles and eight sides

Offset Measurements Used to locate one feature in relationship to another

Opitz System One of the many classification and coding systems

Optimization The process of improving a process or system to its maximum functionality

Ordinate The Y coordinate of a point; i.e., its vertical distance from the X axis measured parallel to the Y axis; the vertical axis of a graph or chart

Orthographic Projection A projection on a picture plane formed by perpendicular projectors from the object to the picture plane; third-angle projection is used in the United States and first-angle projection is used in most countries outside the United States

Output Data that have been processed from an internal storage to an external storage; opposite of "input"

Parallel Having the same direction, such as two lines which, if extended, would never meet

Parallelism A tolerance controlling interdependent surfaces and axes that must be an equal distance from a datum plane or axis

Parallelogram Law Two vectors, not parallel, drawn to scale, having the same units, and acting at the same point can be added or subtracted

Part Individual component of an assembly

Part Family A group of parts similar in design or the manufacturing processes used to produce them

Pattern A model, usually of wood, used in forming a mold for a casting

Peen To hammer into shape with a ballpeen hammer

Pentagon A polygon having five angles and five sides

Perpendicular Lines or planes at a right angle to a given line or plane

Perpendicularity A tolerance controlling surfaces and axes which must be at right angles with a datum axis

Perspective Drawing A pictorial drawing in which receding lines converge at vanishing points on the horizon; the most natural of all pictorial drawings

Perspective Projection A distortion phenomenon that occurs because the visual rays used to create the image in the viewer's eye are not parallel

Photochemical Machining (or chemical blanking) A nontraditional machining process that involves producing parts by chemical action

Physical Prototype A real, physical object that may be handled, analyzed, and tested

Pickle To clean forgings or castings in dilute sulphuric acid

Pictorial Drawing A drawing that depicts the object in a three-dimensional format; these drawings are typically isometric or perspective but may also be dimetric, trimetric, or oblique

Pilot A piece that guides a tool or machine part

Pilot Hole A small hole used to guide a cutting tool for making a larger hole

Pinion The smaller of two mating gears

Pitch The distance from a point on one thread to a corresponding point on the next thread; the slope of a surface

Pitch Circle An imaginary circle corresponding to the circumference of the friction gear from which the spur gear is derived

Placing Drawings Structural concrete drawings that show the placement of reinforcing bars in the on-site form

Plane To remove material by means of a planer

Plasma Arc Machining (PAM) A nontraditional machining process that makes use of a plasma arc to remove material

Plastics Non-metal materials made from either natural or synthetic resins

Plate To coat a metal piece with another metal, such as chrome or nickel, by electrochemical methods

Plot Plan A plan view of the building site showing where the structure is located on the lot

Plotter An output device that makes a drawing by converting digital data into text and graphics

Polish To produce a highly finished or polished surface by friction, using a very fine abrasive

Polygon A plane geometric figure with three or more sides

Positional Tolerancing The permitted variation of a feature from the exact or true position indicated on the drawing

Position Tolerance A tolerance that controls the position of a feature as related to a datum or datums

Powdered Metals Metals produced by atomization, electrolysis, chemical reduction, or oxide reduction

Precision The quality or state of being precise or accurate; mechanical exactness

Principal Viewing Planes The top, front, and right-side planes; they are all perpendicular to each other

Printed Circuit Board A special board on which a predetermined pattern of printed circuits has been formed and to which electronic components have been attached

Printer A device that prints the output of a computer

Prism A solid whose bases or ends are any congruent and parallel polygons, and whose sides are parallelograms

Prismatic Pertaining to or like a prism

Process Synchronization All processes involved in

producing a product are in balance so that production flows smoothly and continuously

Processor   The computer in a CAD system

Profile of a Line   A tolerance that controls the allowable variation along an elemental tolerance zone with regard to a basic profile

Profile of a Surface   A tolerance that controls the allowable variation of a surface from a basic profile or configuration

Program   A set of step-by-step instructions telling the computer to solve a problem with the information input to it or contained in memory

Projected Tolerance Zone   A tolerance zone that applies to the location of an axis above the surface of the feature being controlled

Punch   To cut an opening of a desired shape with a rigid tool having the same shape

Quench   To immerse a heated piece of metal in water or oil in order to harden it

Rack   A flat bar with gear teeth in a straight line to engage with teeth in a gear

Raster   The coordinate grid dividing the display area of a graphics display

Real Time   Response times in the interaction between humans and computers that are, for all intents and purposes, equal to real-world response times

Reference Dimension   A non-toleranced dimension used for information purposes only, which does not govern production or inspection operations

Refractory Metals   Metals with the exceptional ability to withstand high temperatures

Refresh Display   A CRT device that requires the refreshing of its screen presentation at a high rate so that the image will not fade or flicker

Regardless of Feature Size (RFS)   The condition where tolerance of position or form must be met irrespective of where the feature lies within its size tolerance

Relief   An offset of surfaces to provide clearance

Rendering   Finishing a drawing to give it a realistic appearance; a representation

Resistance Welding   The process of welding metals by using the resistance of the metals to the flow of electricity to produce the heat for fusion of the metals

Rib   A relatively thin flat member acting as a brace or support

Right-side View   The orthographic view that is projected to the right of the front view; the right profile

Rivet   To connect with rivets or to clench over the end of a pin by hammering

Roof Plan   A plan view showing the location and spacing of rafters and/or trusses

Rotary Ultrasonic Machining (RUM)   A nontraditional machining process similar to ultrasonic machining except that the grinding motion is rotary, RUM uses diamond tools, and RUM uses the diamond of the tool to remove material

Rotate   To revolve through an angle relative to an origin

Round   An exterior rounded intersection of two surfaces

Runout   The composite variation from the desired form of a part surface of revolution during full rotation of the part on a datum axis

Sandblast   To blow sand at high velocity with compressed air against castings or forgings to clean them

Scalar   Magnitudes without directions, such as temperature and pressure

Scrape   To remove metal by scraping with a hand scraper, usually to fit a bearing

Section Lining   Shows where an object is sliced or cut by the cutting plane line

Sectional View   A view of an object obtained by the imaginary cutting away of the front portion of the object to show the interior detail

Shape   To remove metal from a piece with a shaper

Shear   To cut metal by means of shearing with two blades in sliding contact

Shim   A thin piece of metal or other material used as a spacer in adjusting two parts

Simulation   The computer-generated emulation of a system or process

Size Tolerance   A tolerance that specifies how far individual features may vary from the desired size

Software   A set of programs, procedures, rules, and possibly associated documentation concerned with the operation of a CAD system

Solder   To join with solder, usually composed of lead and tin

Specification   A concise statement of the requirements of a material, an item, or a service, and of the procedure to be followed to determine if the requirements have been met

Spin   To form a rotating piece of sheet metal into a desired shape by pressing it with a smooth tool against a rotating form

Spline   A keyway, usually one of a series cut around a shaft or hole

Spotface   To produce a round spot or bearing surface around a hole, usually with a spotfacer; similar to a counterbore

Spotfaced Hole   A plain through hole with a shallow counterbore deep enough to provide a flat surface for a head of bolt, machine screw, or a nut for a threaded member

Spot Weld   A resistance-type weld that joins pieces of metal by welding separate spots rather than a continuous weld

Sprue   A hole in the sand leading to the gate which leads to the mold, through which the metal enters

Stainless Steel   Steel that contains 10.5 percent or more chromium

Stereolithography   A process whereby solid models are created layer by layer by an electronic-laser process in a special liquid solution

Straightness   A tolerance that controls the allowable variation of a surface or an axis from a theoretically perfect plane or line

Stress Relieving   To heat a metal part to a suitable temperature and holding that temperature for a determined time, then gradually cooling it in air; this treatment reduces the internal stresses induced by casting, quenching, machining, cold working or welding

Stretchout   A flat pattern development for use in laying out, cutting and folding lines on flat stock, such as paper or sheet metal, to be formed into an object

Strike   A compass direction of a horizontal line lying in the plane of an ore vein

Structure   Anything constructed of parts

Stylus   A penlike device that provides input or output of coordinate data usually for the purpose of indicating where the next entered character will be displayed

Surface Texture   The roughness, waviness, lay, and flaws of a surface

Swage   To hammer metal into shape while it is held over a swage or die that fits in a hole in the swage block or anvil

Sweat   To fasten metal together by the use of solder between the pieces and by the application of heat and pressure

Symbol   A letter, character or schematic design representing a unit or component

Tabular Dimension   A type of rectangular datum dimensioning in which dimensions from mutually perpendicular datum planes are listed in a table on the drawing instead of on the pictorial portion

Tap   A tool used to produce internal threads by hand or machine

Taper Pin   A small tapered pin for fastening, usually to prevent a collar or hub from rotating on a shaft

Taper Reamer   Produces accurately tapered holes, as for a taper pin

Tapered Hole   A hole that must be drilled first .015" smaller in diameter than the smallest diameter of the tapered hole, and then reamed

Temper   To reheat hardened steel to bring it to a desired degree of hardness

Template   A guide or pattern used to mark out the work

Tensile Force   A force that "goes away from" the joint

Tensile Strength   The maximum load (pull) a piece supports without breaking or failing

Thermal Analysis   A means of evaluating the thermal characteristics or a part or product

Through Hole   A plain hole that goes completely through an object

Time Dependant Variables   Variable characteristics that change over time; time-dependant variables are frequently best viewed as animations

Tin   A silvery metal used in alloys and for coating other metals with tin plate

Tolerance   Total amount of variation permitted in limit dimension of a part

Tooling   Standard and/or special tools for producing a particular part, including jigs, fixtures, gauges, cutting tools, etc.

Top View   A viewing plane that is usually above the object and perpendicular to the other two; the plan view

Torque   The rotational or twisting force in a turning shaft

Total Runout   A tolerance that simultaneously controls a surface related to a datum in all directions

Trammel   An instrument consisting of a straightedge with two adjustable fixed points for drawing curves and ellipses

Transition Fit   A condition in which two parts are toleranced so that either a clearance or an interference can result when the parts are assembled

Translucent   A quality of material that passes light but diffuses it so that objects are not identifiable

Trimetric Drawing   Uses axes that are rotated so that each axis is drawn at a different angle, and each axis uses a different scale

True Position   The basic or theoretically exact position of a feature

Truncated   Having the apex, vertex or end cut off by a plane

Trusses   Frames that are usually two-dimensional and are made of structural shapes, such as angles, tubes, rounds, and channels

Turn   To produce on a lathe a cylindrical surface parallel to the center line

Two Force Members   Members that have force at

each end, equal in magnitude, in the opposite direction, and collinear

Typical This term, when associated with any dimension or feature, means the dimension or feature applies to the locations that appear to be identical in size and shape

Ultrasonic Machining (USM) A nontraditional machining process that removes material using an abrasive slurry driven by a special tool vibrating at a high-frequency along a longitudinal axis

Ultrasonically Assisted Machining (UAM) A nontraditional machining process that involves assisting such traditional machining processes as drilling and turning by coupling the tool with an external source of vibrating energy

Uniform Having the same form or character; unvarying

Unilateral Tolerance A tolerance that allows variations in only one direction

Upset To form a head or enlarged end on a bar by pressure or by hammering between dies

User A person who operates a CAD system

Vector A directed line segment which, in computer graphics, is always defined by its two end points

Vernier Scale A small movable scale, attached to a larger fixed scale, for obtaining fractional subdivisions of the fixed scale

Vertex The highest point of something; the top; the summit; plural: vertices

Vertical Perpendicular to the horizon

Virtual Condition A condition of a feature in which all tolerances of location, size and form are considered

Virtual Reality A technology designed to simulate either a physical or hypothetical environment

Visualization A pictorial representation of data for the purpose of more clearly and concisely communicating information

Vorticity The stage of a fluid in vortical (rotational) motion; the whirling of a fluid (like blood in the heart)

Water Jet Machining (WJM) A nontraditional machining process that is used with non-metal workpieces; it involves cutting plastics, non-metal composites, ceramics, glass, and wood by directing a thin, pressurized jet of water against the workpiece

Web A thin flat part joining larger parts; also known as a rib

Welding Uniting metal pieces by pressure or fusion-welding processes

Woodruff Key A semicircular flat key

Working Drawings A set of drawings which provides details for the production of each part, and information for the correct assembly of the finished product

Workpiece A part in any stage of manufacturing before it becomes a finished part

Workstation The assigned location where a worker performs his or her job (i.e., the keyboard and the system display)

Wrought Iron Iron of low carbon content; useful because of its toughness, ductility, and malleability

Yards Volume in cubic yards, in construction

Zinc A bluish-white metal widely used in manufacturing

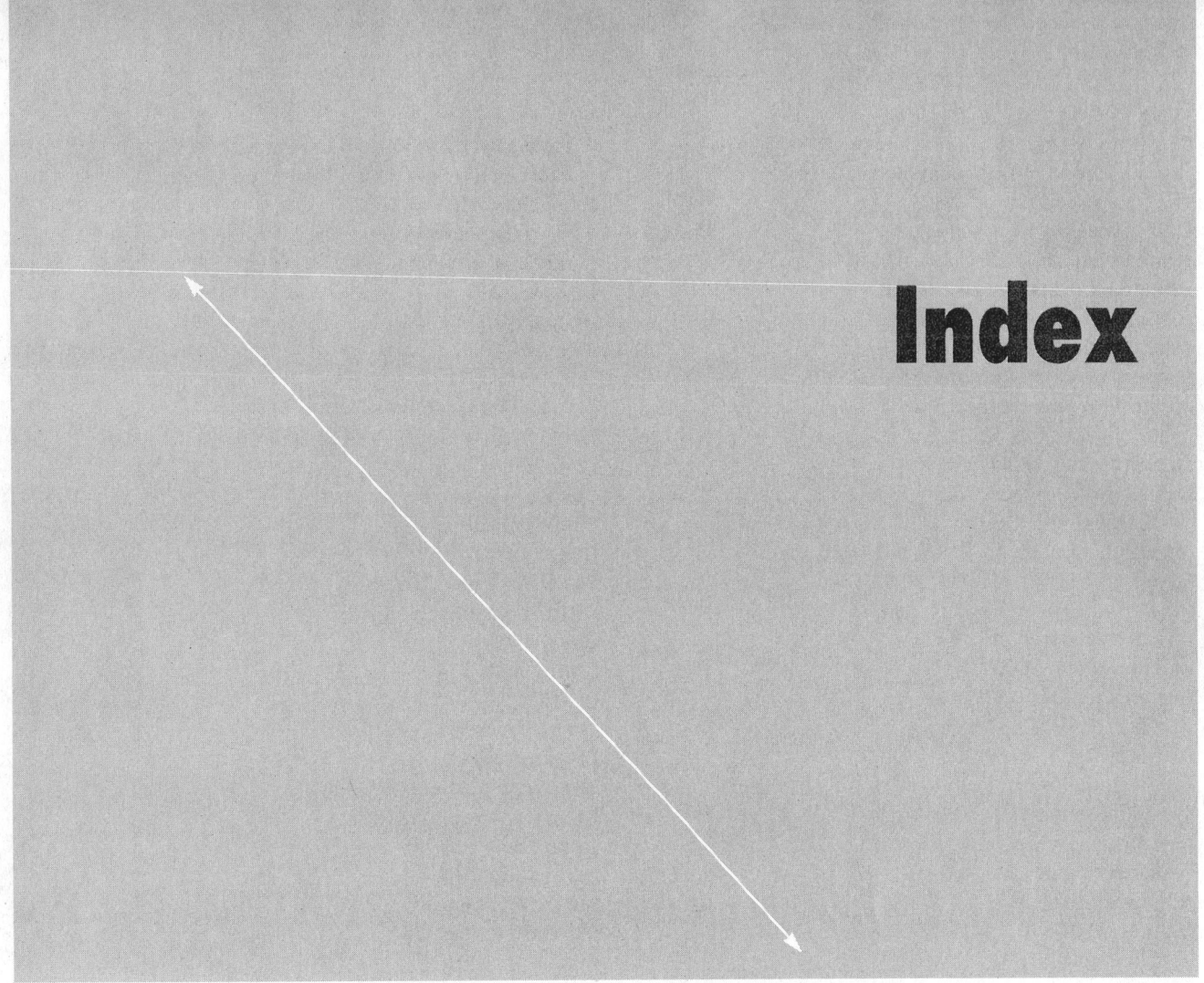

Index